A COMPLETE COURSE

Calculus

FOURTH EDITION

Instructor's Solutions Manual

ROBERT A. ADAMS

DEPARTMENT OF MATHEMATICS
UNIVERSITY OF BRITISH COLUMBIA

Addison-Wesley

An imprint of Addison Wesley Longman Ltd.

Don Mills, Ontario • Reading, Massachusetts • Harlow, England
Melbourne, Australia • Amsterdam, The Netherlands • Bonn, Germany

Publisher: Ron Doleman
Managing Editor: Linda Scott
Cover Design: Anthony Leung
Production Coordinator: Alexandra Odulak
Manufacturing Coordinator: Sharon Latta Paterson
Printing and Binding: Webcom

Copyright © 1999 Addison Wesley Longman Ltd.

All rights reserved. No part of this publication may be reproduced, stored in a retrieval system, or transmitted, in any form or by any means, electronic, mechanical, photocopying, recording, or otherwise, without the prior written permission of the publisher.

ISBN 0-201-47767-X

Printed and bound in Canada.

A B C D E - WC - 03 02 01 00 99

FOREWORD

These solutions are provided for the benefit of instructors using the textbook *Calculus: A Complete Course (Fourth Edition)* by R. A. Adams, published by Addison Wesley Longman Ltd. For the most part, the solutions are detailed, especially in exercises on core material and techniques. Occasionally some details are omitted — for example, in exercises on applications of integration, the evaluation of the integrals encountered is not always given with the same degree of detail as the evaluation of integrals found in those exercises dealing specifically with techniques of integration.

Instructors may wish to make these solutions available to their students. However, students should use such solutions with caution. It is always more beneficial for them to attempt exercises and problems on their own, before they look at solutions done by others. If they examine solutions as "study material" prior to attempting the exercises, they can lose much of the benefit that follows from diligent attempts to develop their own analytical powers. When they have tried unsuccessfully to solve a problem, then looking at a solution can give them a "hint" for a second attempt. A separate *Student Solutions Manual* is available for students. It contains the solutions to the even-numbered exercises only.

This manual also includes 33 miscellaneous student projects that appeared as "Explore" units in the 3rd edition of the text.

I am grateful to Joanna Kwan and Valerie Adams who assisted with the preparation and typesetting of some of the solutions for the second edition, and to Winnie Poon who helped with some of the new solutions in the third edition. Thanks also go to Ken MacKenzie of McGill University who did a very thorough checking of the solutions, thereby exposing for correction many errors that would otherwise have made it into this printed edition. Of course, the author accepts full responsibility for any errors that remain and would be grateful to readers who call them to his attention.

November 1998.

R. A. Adams
Department of Mathematics
The University of British Columbia
Vancouver, B.C., Canada. V6T 1Z2
adms@math.ubc.ca

CONTENTS

Solutions for Chapter	P	**1**
Solutions for Chapter	1	**19**
Solutions for Chapter	2	**36**
Solutions for Chapter	3	**77**
Solutions for Chapter	4	**102**
Solutions for Chapter	5	**170**
Solutions for Chapter	6	**204**
Solutions for Chapter	7	**258**
Solutions for Chapter	8	**303**
Solutions for Chapter	9	**335**
Solutions for Chapter	10	**377**
Solutions for Chapter	11	**402**
Solutions for Chapter	12	**428**
Solutions for Chapter	13	**468**
Solutions for Chapter	14	**507**
Solutions for Chapter	15	**548**
Solutions for Chapter	16	**578**
Solutions for Chapter	17	**604**
Solutions for Appendices		**628**
Miscellaneous Projects		**637**

CHAPTER P. LIMITS AND CONTINUITY

Section P.1 Real Numbers and the Real Line (page 11)

1. $\dfrac{2}{9} = 0.22222222\cdots = 0.\overline{2}$

2. $\dfrac{1}{11} = 0.09090909\cdots = 0.\overline{09}$

3. If $x = 0.121212\cdots$, then $100x = 12.121212\cdots = 12 + x$. Thus $99x = 12$ and $x = 12/99 = 4/33$.

4. If $x = 3.277777\cdots$, then $10x - 32 = 0.77777\cdots$ and $100x - 320 = 7 + (10x - 32)$, or $90x = 295$. Thus $x = 295/90 = 59/18$.

5. $1/7 = 0.142857142857\cdots = 0.\overline{142857}$
   $2/7 = 0.285714285714\cdots = 0.\overline{285714}$
   $3/7 = 0.428571428571\cdots = 0.\overline{428571}$
   $4/7 = 0.571428571428\cdots = 0.\overline{571428}$
   note the same cyclic order of the repeating digits
   $5/7 = 0.714285714285\cdots = 0.\overline{714285}$
   $6/7 = 0.857142857142\cdots = 0.\overline{857142}$

6. Two different decimal expansions can represent the same number. For instance, both $0.999999\cdots = 0.\overline{9}$ and $1.000000\cdots = 1.\overline{0}$ represent the number 1.

7. $x \geq 0$ and $x \leq 5$ define the interval $[0, 5]$.

8. $x < 2$ and $x \geq -3$ define the interval $[-3, 2)$.

9. $x > -5$ or $x < -6$ defines the union $(-\infty, -6) \cup (-5, \infty)$.

10. $x \leq -1$ defines the interval $(-\infty, -1]$.

11. $x > -2$ defines the interval $(-2, \infty)$.

12. $x < 4$ or $x \geq 2$ defines the interval $(-\infty, \infty)$, that is, the whole real line.

13. If $-2x > 4$, then $x < -2$. Solution: $(-\infty, -2)$

14. If $3x + 5 \leq 8$, then $3x \leq 8 - 5 - 3$ and $x \leq 1$. Solution: $(-\infty, 1]$

15. If $5x - 3 \leq 7 - 3x$, then $8x \leq 10$ and $x \leq 5/4$. Solution: $(-\infty, 5/4]$

16. If $\dfrac{6-x}{4} \geq \dfrac{3x-4}{2}$, then $6 - x \geq 6x - 8$. Thus $14 \geq 7x$ and $x \leq 2$. Solution: $(-\infty, 2]$

17. If $3(2 - x) < 2(3 + x)$, then $0 < 5x$ and $x > 0$. Solution: $(0, \infty)$

18. If $x^2 < 9$, then $|x| < 3$ and $-3 < x < 3$. Solution: $(-3, 3)$

19. Given: $1/(2 - x) < 3$.
    CASE I. If $x < 2$, then $1 < 3(2 - x) = 6 - 3x$, so $3x < 5$ and $x < 5/3$. This case has solutions $x < 5/3$.
    CASE II. If $x > 2$, then $1 > 3(2 - x) = 6 - 3x$, so $3x > 5$ and $x > 5/3$. This case has solutions $x > 2$.
    Solution: $(-\infty, 5/3) \cup (2, \infty)$.

20. Given: $(x + 1)/x \geq 2$.
    CASE I. If $x > 0$, then $x + 1 \geq 2x$, so $x \leq 1$.
    CASE II. If $x < 0$, then $x + 1 \leq 2x$, so $x \geq 1$. (not possible)
    Solution: $(0, 1]$.

21. Given: $x^2 - 2x \leq 0$. Then $x(x - 2) \leq 0$. This is only possible if $x \geq 0$ and $x \leq 2$. Solution: $[0, 2]$.

22. Given $6x^2 - 5x \leq -1$, then $(2x - 1)(3x - 1) \leq 0$, so either $x \leq 1/2$ and $x \geq 1/3$, or $x \leq 1/3$ and $x \geq 1/2$. The latter combination is not possible. The solution set is $[1/3, 1/2]$.

23. Given $x^3 > 4x$, we have $x(x^2 - 4) > 0$. This is possible if $x < 0$ and $x^2 < 4$, or if $x > 0$ and $x^2 > 4$. The possibilities are, therefore, $-2 < x < 0$ or $2 < x < \infty$. Solution: $(-2, 0) \cup (2, \infty)$.

24. Given $x^2 - x \leq 2$, then $x^2 - x - 2 \leq 0$ so $(x-2)(x+1) \leq 0$. This is possible if $x \leq 2$ and $x \geq -1$ or if $x \geq 2$ and $x \leq -1$. The latter situation is not possible. The solution set is $[-1, 2]$.

25. Given: $\dfrac{x}{2} \geq 1 + \dfrac{4}{x}$.
    CASE I. If $x > 0$, then $x^2 \geq 2x+8$, so that $x^2-2x-8 \geq 0$, or $(x - 4)(x + 2) \geq 0$. This is possible for $x > 0$ only if $x \geq 4$.
    CASE II. If $x < 0$, then we must have $(x - 4)(x + 2) \leq 0$, which is possible for $x < 0$ only if $x \geq -2$.
    Solution: $[-2, 0) \cup [4, \infty)$.

26. Given: $\dfrac{3}{x-1} < \dfrac{2}{x+1}$.
    CASE I. If $x > 1$ then $(x - 1)(x + 1) > 0$, so that $3(x + 1) < 2(x - 1)$. Thus $x < -5$. There are no solutions in this case.
    CASE II. If $-1 < x < 1$, then $(x - 1)(x + 1) < 0$, so $3(x + 1) > 2(x - 1)$. Thus $x > -5$. In this case all numbers in $(-1, 1)$ are solutions.
    CASE III. If $x < -1$, then $(x - 1)(x + 1) > 0$, so that $3(x + 1) < 2(x - 1)$. Thus $x < -5$. All numbers $x < -5$ are solutions.
    Solutions: $(-\infty, -5) \cup (-1, 1)$.

27. If $|x| = 3$ then $x = \pm 3$.

28. If $|x - 3| = 7$, then $x - 3 = \pm 7$, so $x = -4$ or $x = 10$.

29. If $|2t + 5| = 4$, then $2t + 5 = \pm 4$, so $t = -9/2$ or $t = -1/2$.

30. If $|1 - t| = 1$, then $1 - t = \pm 1$, so $t = 0$ or $t = 2$.

31. If $|8 - 3s| = 9$, then $8 - 3s = \pm 9$, so $3s = -1$ or 17, and $s = -1/3$ or $s = 17/3$.

1

32. If $\left|\dfrac{s}{2} - 1\right| = 1$, then $\dfrac{s}{2} - 1 = \pm 1$, so $s = 0$ or $s = 4$.

33. If $|x| < 2$, then x is in $(-2, 2)$.

34. If $|x| \leq 2$, then x is in $[-2, 2]$.

35. If $|s - 1| \leq 2$, then $1 - 2 \leq s \leq 1 + 2$, so s is in $[-1, 3]$.

36. If $|t + 2| < 1$, then $-2 - 1 < t < -2 + 1$, so t is in $(-3, -1)$.

37. If $|3x - 7| < 2$, then $7 - 2 < 3x < 7 + 2$, so x is in $(5/3, 3)$.

38. If $|2x + 5| < 1$, then $-5 - 1 < 2x < -5 + 1$, so x is in $(-3, -2)$.

39. If $\left|\dfrac{x}{2} - 1\right| \leq 1$, then $1 - 1 \leq \dfrac{x}{2} \leq 1 + 1$, so x is in $[0, 4]$.

40. If $\left|2 - \dfrac{x}{2}\right| < \dfrac{1}{2}$, then $x/2$ lies between $2 - (1/2)$ and $2 + (1/2)$. Thus x is in $(3, 5)$.

41. The inequality $|x + 1| > |x - 3|$ says that the distance from x to -1 is greater than the distance from x to 3, so x must be to the right of the point half-way between -1 and 3. Thus $x > 1$.

42. $|x - 3| < 2|x| \Leftrightarrow x^2 - 6x + 9 = (x - 3)^2 < 4x^2$
    $\Leftrightarrow 3x^2 + 6x - 9 > 0 \Leftrightarrow 3(x + 3)(x - 1) > 0$. This inequality holds if $x < -3$ or $x > 1$.

43. $|a| = a$ if and only if $a \geq 0$. It is false if $a < 0$.

44. The equation $|x - 1| = 1 - x$ holds if $|x - 1| = -(x - 1)$, that is, if $x - 1 < 0$, or, equivalently, if $x < 1$.

45. The triangle inequality $|x + y| \leq |x| + |y|$ implies that
    $$|x| \geq |x + y| - |y|.$$
    Apply this inequality with $x = a - b$ and $y = b$ to get
    $$|a - b| \geq |a| - |b|.$$
    Similarly, $|a - b| = |b - a| \geq |b| - |a|$. Since $\bigl||a| - |b|\bigr|$ is equal to either $|a| - |b|$ or $|b| - |a|$, depending on the sizes of a and b, we have
    $$|a - b| \geq \bigl||a| - |b|\bigr|.$$

Section P.2 Cartesian Coordinates in the Plane (page 18)

1. From $A(0, 3)$ to $B(4, 0)$, $\Delta x = 4 - 0 = 4$ and $\Delta y = 0 - 3 = -3$. $|AB| = \sqrt{4^2 + (-3)^2} = 5$.

2. From $A(-1, 2)$ to $B(4, -10)$, $\Delta x = 4 - (-1) = 5$ and $\Delta y = -10 - 2 = -12$. $|AB| = \sqrt{5^2 + (-12)^2} = 13$.

3. From $A(3, 2)$ to $B(-1, -2)$, $\Delta x = -1 - 3 = -4$ and $\Delta y = -2 - 2 = -4$. $|AB| = \sqrt{(-4)^2 + (-4)^2} = 4\sqrt{2}$.

4. From $A(0.5, 3)$ to $B(2, 3)$, $\Delta x = 2 - 0.5 = 1.5$ and $\Delta y = 3 - 3 = 0$. $|AB| = 1.5$.

5. Starting point: $(-2, 3)$. Increments $\Delta x = 4$, $\Delta y = -7$. New position is $(-2 + 4, 3 + (-7))$, that is, $(2, -4)$.

6. Arrival point: $(-2, -2)$. Increments $\Delta x = -5$, $\Delta y = 1$. Starting point was $(-2 - (-5), -2 - 1)$, that is, $(3, -3)$.

7. $x^2 + y^2 = 1$ represents a circle of radius 1 centred at the origin.

8. $x^2 + y^2 = 2$ represents a circle of radius $\sqrt{2}$ centred at the origin.

9. $x^2 + y^2 \leq 1$ represents points inside and on the circle of radius 1 centred at the origin.

10. $x^2 + y^2 = 0$ represents the origin.

11. $y \geq x^2$ represents all points lying on or above the parabola $y = x^2$.

12. $y < x^2$ represents all points lying below the parabola $y = x^2$.

13. The vertical line through $(-2, 5/3)$ is $x = -2$; the horizontal line through that point is $y = 5/3$.

14. The vertical line through $(\sqrt{2}, -1.3)$ is $x = \sqrt{2}$; the horizontal line through that point is $y = -1.3$.

15. Line through $(-1, 1)$ with slope $m = 1$ is $y = 1 + 1(x + 1)$, or $y = x + 2$.

16. Line through $(-2, 2)$ with slope $m = 1/2$ is $y = 2 + (1/2)(x + 2)$, or $x - 2y = -6$.

17. Line through $(0, b)$ with slope $m = 2$ is $y = b + 2x$.

18. Line through $(a, 0)$ with slope $m = -2$ is $y = 0 - 2(x - a)$, or $y = 2a - 2x$.

19. At $x = 2$, the height of the line $2x + 3y = 6$ is $y = (6 - 4)/3 = 2/3$. Thus $(2, 1)$ lies above the line.

20. At $x = 3$, the height of the line $x - 4y = 7$ is $y = (3 - 7)/4 = -1$. Thus $(3, -1)$ lies on the line.

21. The line through $(0, 0)$ and $(2, 3)$ has slope $m = (3 - 0)/(2 - 0) = 3/2$ and equation $y = (3/2)x$ or $3x - 2y = 0$.

22. The line through $(-2, 1)$ and $(2, -2)$ has slope $m = (-2 - 1)/(2 + 2) = -3/4$ and equation $y = 1 - (3/4)(x + 2)$ or $3x + 4y = -2$.

23. The line through $(4, 1)$ and $(-2, 3)$ has slope $m = (3 - 1)/(-2 - 4) = -1/3$ and equation $y = 1 - \dfrac{1}{3}(x - 4)$ or $x + 3y = 7$.

24. The line through $(-2, 0)$ and $(0, 2)$ has slope $m = (2 - 0)/(0 + 2) = 1$ and equation $y = 2 + x$.

25. If $m = -2$ and $b = \sqrt{2}$, then the line has equation $y = -2x + \sqrt{2}$.

26. If $m = -1/2$ and $b = -3$, then the line has equation $y = -(1/2)x - 3$, or $x + 2y = -6$.

27. $3x + 4y = 12$ has x-intercept $a = 12/3 = 4$ and y-intercept $b = 12/4 = 3$. Its slope is $-b/a = -3/4$.

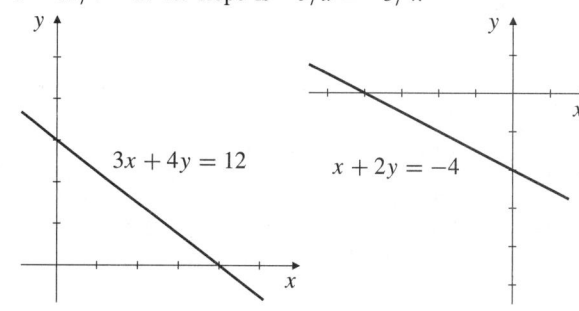

Fig. P.2.27   Fig. P.2.28

28. $x + 2y = -4$ has x-intercept $a = -4$ and y-intercept $b = -4/2 = -2$. Its slope is $-b/a = 2/(-4) = -1/2$.

29. $\sqrt{2}x - \sqrt{3}y = 2$ has x-intercept $a = 2/\sqrt{2} = \sqrt{2}$ and y-intercept $b = -2/\sqrt{3}$. Its slope is $-b/a = 2/\sqrt{6} = \sqrt{2/3}$.

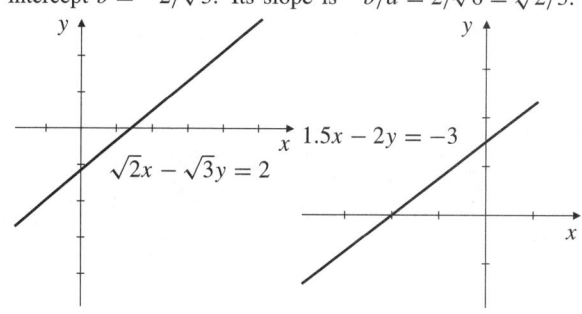

Fig. P.2.29   Fig. P.2.30

30. $1.5x - 2y = -3$ has x-intercept $a = -3/1.5 = -2$ and y-intercept $b = -3/(-2) = 3/2$. Its slope is $-b/a = 3/4$.

31. line through $(2, 1)$ parallel to $y = x + 2$ is $y = x - 1$; line perpendicular to $y = x + 2$ is $y = -x + 3$.

32. line through $(-2, 2)$ parallel to $2x + y = 4$ is $2x + y = -2$; line perpendicular to $2x + y = 4$ is $x - 2y = -6$.

33. We have
$$3x + 4y = -6 \implies 6x + 8y = -12$$
$$2x - 3y = 13 \qquad 6x - 9y = 39.$$

Subtracting these equations gives $17y = -51$, so $y = -3$ and $x = (13 - 9)/2 = 2$. The intersection point is $(2, -3)$.

34. We have
$$2x + y = 8 \implies 14x + 7y = 56$$
$$5x - 7y = 1 \qquad 5x - 7y = 1.$$

Adding these equations gives $19x = 57$, so $x = 3$ and $y = 8 - 2x = 2$. The intersection point is $(3, 2)$.

35. If $a \neq 0$ and $b \neq 0$, then $(x/a) + (y/b) = 1$ represents a straight line that is neither horizontal nor vertical, and does not pass through the origin. Putting $y = 0$ we get $x/a = 1$, so the x-intercept of this line is $x = a$; putting $x = 0$ gives $y/b = 1$, so the y-intercept is $y = b$.

36. The line $(x/2) - (y/3) = 1$ has x-intercept $a = 2$, and y-intercept $b = -3$.

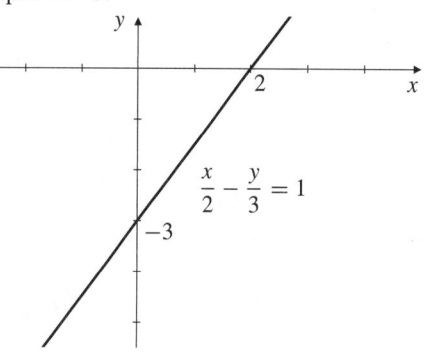

Fig. P.2.36

37. The line through $(2, 1)$ and $(3, -1)$ has slope $m = (-1 - 1)/(3 - 2) = -2$ and equation $y = 1 - 2(x - 2) = 5 - 2x$. Its y-intercept is 5.

38. The line through $(-2, 5)$ and $(k, 1)$ has x-intercept 3, so also passes through $(3, 0)$. Its slope m satisfies
$$\frac{1 - 0}{k - 3} = m = \frac{0 - 5}{3 + 2} = -1.$$

Thus $k - 3 = -1$, and so $k = 2$.

39. $C = Ax + B$. If $C = 5,000$ when $x = 10,000$ and $C = 6,000$ when $x = 15,000$, then
$$10,000A + B = 5,000$$
$$15,000A + B = 6,000$$

Subtracting these equations gives $5,000A = 1,000$, so $A = 1/5$. From the first equation, $2,000 + B = 5,000$, so $B = 3,000$. The cost of printing 100,000 pamphlets is $\$100,000/5 + 3,000 = \$23,000$.

40. $-40°$ and $-40°$ is the same temperature on both the Fahrenheit and Celsius scales.

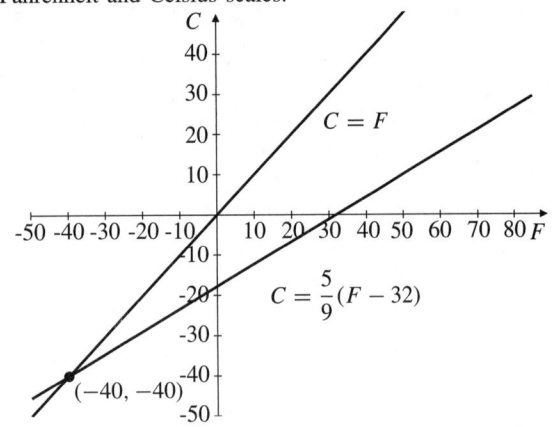

Fig. P.2.40

41. $A = (2,1)$, $B = (6,4)$, $C = (5,-3)$
$$|AB| = \sqrt{(6-2)^2 + (4-1)^2} = \sqrt{25} = 5$$
$$|AC| = \sqrt{(5-2)^2 + (-3-1)^2} = \sqrt{25} = 5$$
$$|BC| = \sqrt{(6-5)^2 + (4+3)^2} = \sqrt{50} = 5\sqrt{2}.$$
Since $|AB| = |AC|$, triangle ABC is isosceles.

42. $A = (0,0)$, $B = (1,\sqrt{3})$, $C = (2,0)$
$$|AB| = \sqrt{(1-0)^2 + (\sqrt{3}-0)^2} = \sqrt{4} = 2$$
$$|AC| = \sqrt{(2-0)^2 + (0-0)^2} = \sqrt{4} = 2$$
$$|BC| = \sqrt{(2-1)^2 + (0-\sqrt{3})^2} = \sqrt{4} = 2.$$
Since $|AB| = |AC| = |BC|$, triangle ABC is equilateral.

43. $A = (2,-1)$, $B = (1,3)$, $C = (-3,2)$
$$|AB| = \sqrt{(1-2)^2 + (3+1)^2} = \sqrt{17}$$
$$|AC| = \sqrt{(-3-2)^2 + (2+1)^2} = \sqrt{34} = \sqrt{2}\sqrt{17}$$
$$|BC| = \sqrt{(-3-1)^2 + (2-3)^2} = \sqrt{17}.$$
Since $|AB| = |BC|$ and $|AC| = \sqrt{2}|AB|$, triangle ABC is an isosceles right-angled triangle with right angle at B. Thus $ABCD$ is a square if D is displaced from C by the same amount A is from B, that is, by increments $\Delta x = 2 - 1 = 1$ and $\Delta y = -1 - 3 = -4$. Thus $D = (-3 + 1, 2 + (-4)) = (-2, -2)$.

44. If $M = (x_m, y_m)$ is the midpoint of $P_1 P_2$, then the displacement of M from P_1 equals the displacement of P_2 from M:
$$x_m - x_1 = x_2 - x_m, \quad y_m - y_1 = y_2 - y_m.$$
Thus $x_m = (x_1 + x_2)/2$ and $y_m = (y_1 + y_2)/2$.

45. If $Q = (x_q, y_q)$ is the point on $P_1 P_2$ that is two thirds of the way from P_1 to P_2, then the displacement of Q from P_1 equals twice the displacement of P_2 from Q:
$$x_q - x_1 = 2(x_2 - x_q), \quad y_q - y_1 = 2(y_2 - y_q).$$
Thus $x_q = (x_1 + 2x_2)/3$ and $y_q = (y_1 + 2y_2)/3$.

46. Let the coordinates of P be $(x, 0)$ and those of Q be $(X, -2X)$. If the midpoint of PQ is $(2, 1)$, then
$$(x + X)/2 = 2, \quad (0 - 2X)/2 = 1.$$
The second equation implies that $X = -1$, and the second then implies that $x = 5$. Thus P is $(5, 0)$.

47. $\sqrt{(x-2)^2 + y^2} = 4$ says that the distance of (x, y) from $(2, 0)$ is 4, so the equation represents a circle of radius 4 centred at $(2, 0)$.

48. $\sqrt{(x-2)^2 + y^2} = \sqrt{x^2 + (y-2)^2}$ says that (x, y) is equidistant from $(2, 0)$ and $(0, 2)$. Thus (x, y) must lie on the line that is the right bisector of the line from $(2, 0)$ to $(0, 2)$. A simpler equation for this line is $x = y$.

49. The line $2x + ky = 3$ has slope $m = -2/k$. This line is perpendicular to $4x + y = 1$, which has slope -4, provided $m = 1/4$, that is, provided $k = -8$. The line is parallel to $4x + y = 1$ if $m = -4$, that is, if $k = 1/2$.

50. For any value of k, the coordinates of the point of intersection of $x + 2y = 3$ and $2x - 3y = -1$ will also satisfy the equation
$$(x + 2y - 3) + k(2x - 3y + 1) = 0$$
because they cause both expressions in parentheses to be 0. The equation above is linear in x and y, and so represents a straight line for any choice of k. This line will pass through $(1, 2)$ provided $1 + 4 - 3 + k(2 - 6 + 1) = 0$, that is, if $k = 2/3$. Therefore, the line through the point of intersection of the two given lines and through the point $(1, 2)$ has equation
$$x + 2y - 3 + \frac{2}{3}(2x - 3y + 1) = 0,$$
or, on simplification, $x = 1$.

Section P.3 Graphs of Quadratic Equations (page 25)

1. $x^2 + y^2 = 16$
2. $x^2 + (y-2)^2 = 4$, or $x^2 + y^2 - 4y = 0$
3. $(x+2)^2 + y^2 = 9$, or $x^2 + y^2 + 4y = 5$
4. $(x-3)^2 + (y+4)^2 = 25$, or $x^2 + y^2 - 6x + 8y = 0$.
5. $x^2 + y^2 - 2x = 3$
   $x^2 - 2x + 1 + y^2 = 4$
   $(x-1)^2 + y^2 = 4$
   centre: $(1, 0)$; radius 2.

6. $x^2 + y^2 + 4y = 0$
   $x^2 + y^2 + 4y + 4 = 4$
   $x^2 + (y+2)^2 = 4$
   centre: $(0, -2)$; radius 2.

7. $x^2 + y^2 - 2x + 4y = 4$
   $x^2 - 2x + 1 + y^2 + 4y + 4 = 9$
   $(x-1)^2 + (y+2)^2 = 9$
   centre: $(1, -2)$; radius 3.

8. $x^2 + y^2 - 2x - y + 1 = 0$
   $x^2 - 2x + 1 + y^2 - y + \frac{1}{4} = \frac{1}{4}$
   $(x-1)^2 + \left(y - \frac{1}{2}\right)^2 = \frac{1}{4}$
   centre: $(1, 1/2)$; radius $1/2$.

9. $x^2 + y^2 > 1$ represents all points lying outside the circle of radius 1 centred at the origin.

10. $x^2 + y^2 < 4$ represents the open disk consisting of all points lying inside the circle of radius 2 centred at the origin.

11. $(x+1)^2 + y^2 \leq 4$ represents the closed disk consisting of all points lying inside or on the circle of radius 2 centred at the point $(-1, 0)$.

12. $x^2 + (y-2)^2 \leq 4$ represents the closed disk consisting of all points lying inside or on the circle of radius 2 centred at the point $(0, 2)$.

13. Together, $x^2 + y^2 > 1$ and $x^2 + y^2 < 4$ represent annulus (washer-shaped region) consisting of all points that are outside the circle of radius 1 centred at the origin and inside the circle of radius 2 centred at the origin.

14. Together, $x^2 + y^2 \leq 4$ and $(x+2)^2 + y^2 \leq 4$ represent the region consisting of all points that are inside or on both the circle of radius 2 centred at the origin and the circle of radius 2 centred at $(-2, 0)$.

15. Together, $x^2 + y^2 < 2x$ and $x^2 + y^2 < 2y$ (or, equivalently, $(x-1)^2 + y^2 < 1$ and $x^2 + (y-1)^2 < 1$) represent the region consisting of all points that are inside both the circle of radius 1 centred at $(1, 0)$ and the circle of radius 1 centred at $(0, 1)$.

16. $x^2 + y^2 - 4x + 2y > 4$ can be rewritten $(x-2)^2 + (y+1)^2 > 9$. This equation, taken together with $x + y > 1$, represents all points that lie both outside the circle of radius 3 centred at $(2, -1)$ and above the line $x + y = 1$.

17. The interior of the circle with centre $(-1, 2)$ and radius $\sqrt{6}$ is given by $(x+1)^2 + (y-2)^2 < 6$, or $x^2 + y^2 + 2x - 4y < 1$.

18. The exterior of the circle with centre $(2, -3)$ and radius 4 is given by $(x-2)^2 + (y+3)^2 > 16$, or $x^2 + y^2 - 4x + 6y > 3$.

19. $x^2 + y^2 < 2$, $x \geq 1$

20. $x^2 + y^2 > 4$, $(x-1)^2 + (y-3)^2 < 10$

21. The parabola with focus $(0, 4)$ and directrix $y = -4$ has equation $x^2 = 16y$.

22. The parabola with focus $(0, -1/2)$ and directrix $y = 1/2$ has equation $x^2 = -2y$.

23. The parabola with focus $(2, 0)$ and directrix $x = -2$ has equation $y^2 = 8x$.

24. The parabola with focus $(-1, 0)$ and directrix $x = 1$ has equation $y^2 = -4x$.

25. $y = x^2/2$ has focus $(0, 1/2)$ and directrix $y = -1/2$.

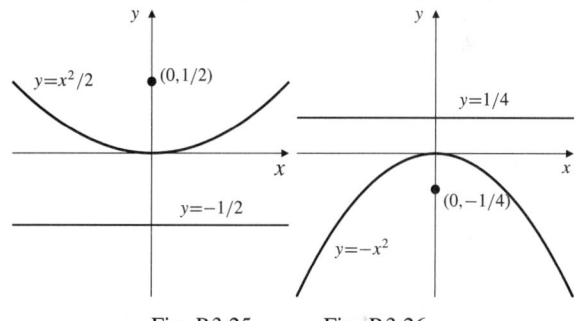

Fig. P.3.25   Fig. P.3.26

26. $y = -x^2$ has focus $(0, -1/4)$ and directrix $y = 1/4$.

27. $x = -y^2/4$ has focus $(-1, 0)$ and directrix $x = 1$.

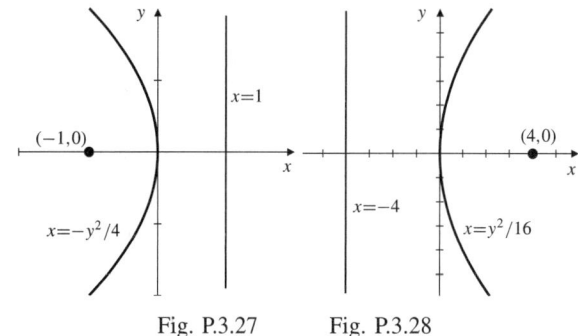

Fig. P.3.27   Fig. P.3.28

28. $x = y^2/16$ has focus $(4, 0)$ and directrix $x = -4$.

SECTION P.3 (PAGE 25)

29.

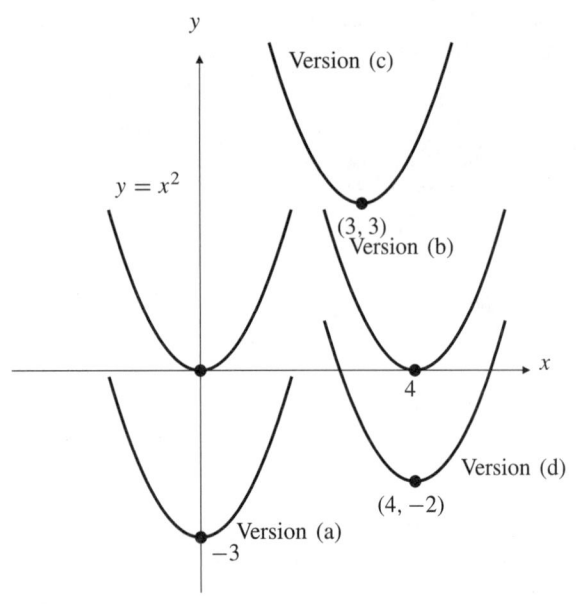

Fig. P.3.29

a) has equation $y = x^2 - 3$.

b) has equation $y = (x-4)^2$ or $y = x^2 - 8x + 16$.

c) has equation $y = (x-3)^2 + 3$ or $y = x^2 - 6x + 12$.

d) has equation $y = (x-4)^2 - 2$, or $y = x^2 - 8x + 14$.

30.
a) If $y = mx$ is shifted to the right by amount x_1, the equation $y = m(x - x_1)$ results. If (a, b) satisfies this equation, then $b = m(a - x_1)$, and so $x_1 = a - (b/m)$. Thus the shifted equation is
$y = m(x - a + (b/m)) = m(x-a) + b$.

b) If $y = mx$ is shifted vertically by amount y_1, the equation $y = mx + y_1$ results. If (a, b) satisfies this equation, then $b = ma + y_1$, and so $y_1 = b - ma$. Thus the shifted equation is $y = mx + b - ma = m(x-a) + b$, the same equation obtained in part (a).

31. $y = \sqrt{(x/3) + 1}$

32. $4y = \sqrt{x+1}$

33. $y = \sqrt{(3x/2) + 1}$

34. $(y/2) = \sqrt{4x+1}$

35. $y = 1 - x^2$ shifted down 1, left 1 gives $y = -(x+1)^2$.

36. $x^2 + y^2 = 5$ shifted up 2, left 4 gives $(x+4)^2 + (y-2)^2 = 5$.

37. $y = (x-1)^2 - 1$ shifted down 1, right 1 gives $y = (x-2)^2 - 2$.

38. $y = \sqrt{x}$ shifted down 2, left 4 gives $y = \sqrt{x+4} - 2$.

R. A. ADAMS: CALCULUS

39. $y = x^2 + 3$, $y = 3x + 1$. Subtracting these equations gives $x^2 - 3x + 2 = 0$, or $(x-1)(x-2) = 0$. Thus $x = 1$ or $x = 2$. The corresponding values of y are 4 and 7. The intersection points are $(1, 4)$ and $(2, 7)$.

40. $y = x^2 - 6$, $y = 4x - x^2$. Subtracting these equations gives
$2x^2 - 4x - 6 = 0$, or $2(x-3)(x+1) = 0$. Thus $x = 3$ or $x = -1$. The corresponding values of y are 3 and -5. The intersection points are $(3, 3)$ and $(-1, -5)$.

41. $x^2 + y^2 = 25$, $3x + 4y = 0$. The second equation says that $y = -3x/4$. Substituting this into the first equation gives $25x^2/16 = 25$, so $x = \pm 4$. If $x = 4$, then the second equation gives $y = -3$; if $x = -4$, then $y = 3$. The intersection points are $(4, -3)$ and $(-4, 3)$. Note that having found values for x, we substituted them into the linear equation rather than the quadratic equation to find the corresponding values of y. Had we substituted into the quadratic equation we would have got more solutions (four points in all), but two of them would have failed to satisfy $3x + 4y = 12$. When solving systems of nonlinear equations you should always verify that the solutions you find do satisfy the given equations.

42. $2x^2 + 2y^2 = 5$, $xy = 1$. The second equation says that $y = 1/x$. Substituting this into the first equation gives $2x^2 + (2/x^2) = 5$, or $2x^4 - 5x^2 + 2 = 0$. This equation factors to $(2x^2 - 1)(x^2 - 2) = 0$, so its solutions are $x = \pm 1/\sqrt{2}$ and $x = \pm\sqrt{2}$. The corresponding values of y are given by $y = 1/x$. Therefore, the intersection points are $(1/\sqrt{2}, \sqrt{2})$, $(-1/\sqrt{2}, -\sqrt{2})$, $(\sqrt{2}, 1/\sqrt{2})$, and $(-\sqrt{2}, -1/\sqrt{2})$.

43. $(x^2/4) + y^2 = 1$ is an ellipse with major axis between $(-2, 0)$ and $(2, 0)$ and minor axis between $(0, -1)$ and $(0, 1)$.

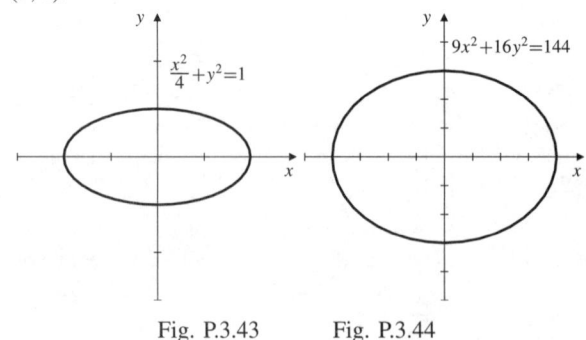

Fig. P.3.43   Fig. P.3.44

44. $9x^2 + 16y^2 = 144$ is an ellipse with major axis between $(-4, 0)$ and $(4, 0)$ and minor axis between $(0, -3)$ and $(0, 3)$.

45. $\dfrac{(x-3)^2}{9} + \dfrac{(y+2)^2}{4} = 1$ is an ellipse with centre at $(3, -2)$, major axis between $(0, -2)$ and $(6, -2)$ and minor axis between $(3, -4)$ and $(3, 0)$.

INSTRUCTOR'S SOLUTIONS MANUAL　　　　　　　　　　　　　　　　　SECTION P.4　(PAGE 34)

Fig. P.3.45　　Fig. P.3.46　　　　　　　　　　　　　　　　　　Fig. P.3.50

46. $(x-1)^2 + \dfrac{(y+1)^2}{4} = 4$ is an ellipse with centre at $(1, -1)$, major axis between $(1, -5)$ and $(1, 3)$ and minor axis between $(-1, -1)$ and $(3, -1)$.

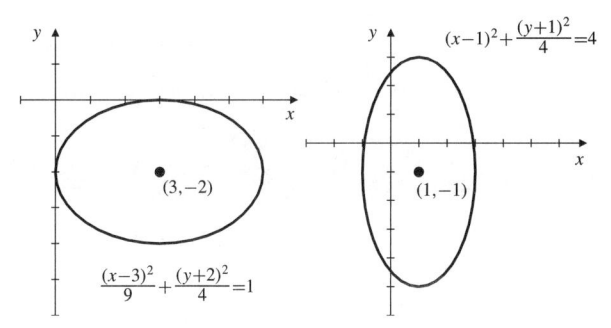

47. $(x^2/4) - y^2 = 1$ is a hyperbola with centre at the origin and passing through $(\pm 2, 0)$. Its asymptotes are $y = \pm x/2$.

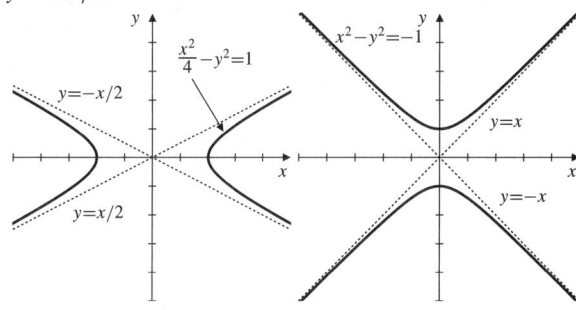

Fig. P.3.47　　Fig. P.3.48

48. $x^2 - y^2 = -1$ is a rectangular hyperbola with centre at the origin and passing through $(0, \pm 1)$. Its asymptotes are $y = \pm x$.

49. $xy = -4$ is a rectangular hyperbola with centre at the origin and passing through $(2, -2)$ and $(-2, 2)$. Its asymptotes are the coordinate axes.

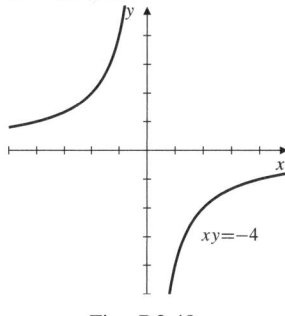

Fig. P.3.49

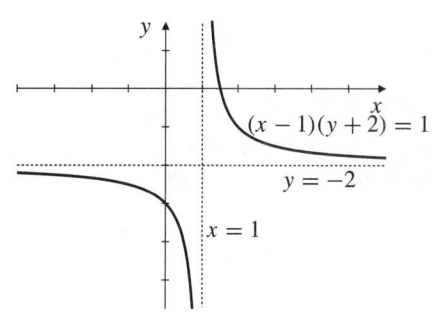

50. $(x-1)(y+2) = 1$ is a rectangular hyperbola with centre at $(1, -2)$ and passing through $(2, -1)$ and $(0, -3)$. Its asymptotes are $x = 1$ and $y = -2$.

51. a) Replacing x with $-x$ replaces a graph with its reflection across the y-axis.

    b) Replacing y with $-y$ replaces a graph with its reflection across the x-axis.

52. Replacing x with $-x$ and y with $-y$ reflects the graph in both axes. This is equivalent to rotating the graph $180°$ about the origin.

53. $|x| + |y| = 1$.
    In the first quadrant the equation is $x + y = 1$.
    In the second quadrant the equation is $-x + y = 1$.
    In the third quadrant the equation is $-x - y = 1$.
    In the fourth quadrant the equation is $x - y = 1$.

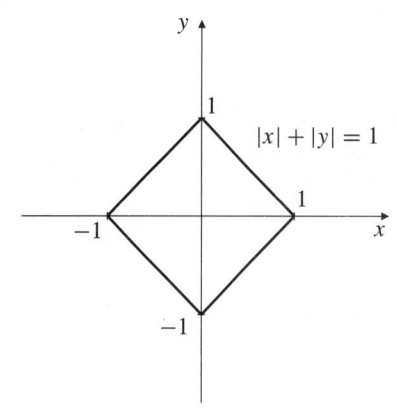

Fig. P.3.53

Section P.4　Functions and Their Graphs
(page 34)

1. $f(x) = 1 + x^2$; domain $\mathbb{R}$, range $[1, \infty)$

2. $f(x) = 1 - \sqrt{x}$; domain $[0, \infty)$, range $(-\infty, 1]$

3. $G(x) = \sqrt{8 - 2x}$; domain $(-\infty, 4]$, range $[0, \infty)$

4. $F(x) = 1/(x - 1)$; domain $(-\infty, 1) \cup (1, \infty)$, range $(-\infty, 0) \cup (0, \infty)$

7

5. $h(t) = \dfrac{t}{\sqrt{2-t}}$; domain $(-\infty, 2)$, range $\mathbb{R}$. (The equation $y = h(t)$ can be squared and rewritten as $t^2 + y^2 t - 2y^2 = 0$, a quadratic equation in t having real solutions for every real value of y. Thus the range of h contains all real numbers.)

6. $g(x) = \dfrac{1}{1 - \sqrt{x-2}}$; domain $(2, 3) \cup (3, \infty)$, range $(-\infty, 0) \cup (0, \infty)$. The equation $y = g(x)$ can be solved for $x = 2 - (1 - (1/y))^2$ so has a real solution provided $y \neq 0$.

7.

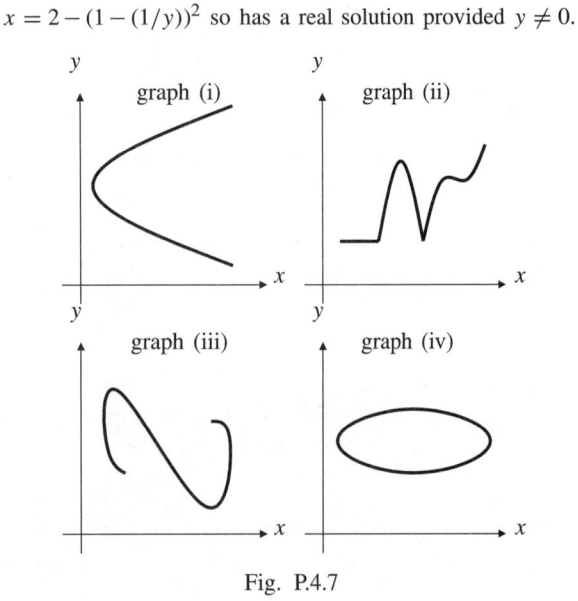

Fig. P.4.7

Graph (ii) is the graph of a function because vertical lines can meet the graph only once. Graphs (i), (iii), and (iv) do not have this property, so are not graphs of functions.

8.

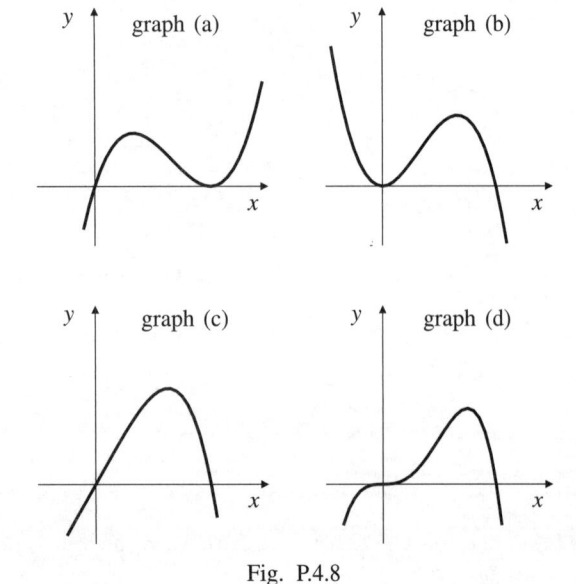

Fig. P.4.8

a) is the graph of $x(1-x)^2$, which is positive for $x > 0$.

b) is the graph of $x^2 - x^3 = x^2(1-x)$, which is positive if $x < 1$.

c) is the graph of $x - x^4$, which is positive if $0 < x < 1$ and behaves like x near 0.

d) is the graph of $x^3 - x^4$, which is positive if $0 < x < 1$ and behaves like x^3 near 0.

9.

x	$f(x) = x^4$
0	0
± 0.5	0.0625
± 1	1
± 1.5	5.0625
± 2	16

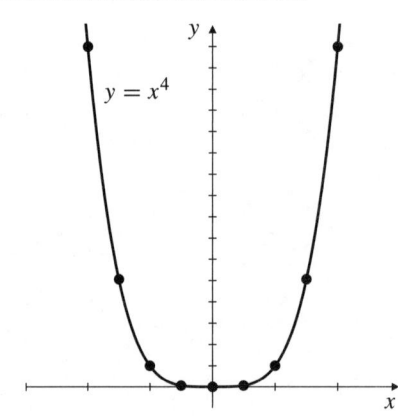

Fig. P.4.9

10.

x	$f(x) = x^{2/3}$
0	0
± 0.5	0.62996
± 1	1
± 1.5	1.3104
± 2	1.5874

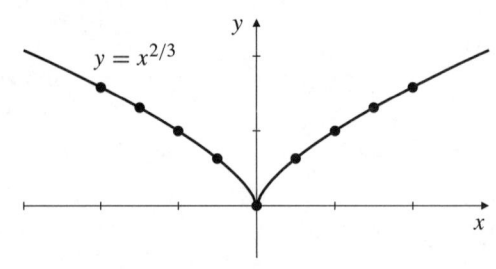

Fig. P.4.10

11. $f(x) = x^2 + 1$ is even: $f(-x) = f(x)$

12. $f(x) = x^3 + x$ is odd: $f(-x) = -f(x)$

13. $f(x) = \dfrac{x}{x^2 - 1}$ is odd: $f(-x) = -f(x)$

INSTRUCTOR'S SOLUTIONS MANUAL   SECTION P.4 (PAGE 34)

14. $f(x) = \dfrac{1}{x^2 - 1}$ is even: $f(-x) = f(x)$

15. $f(x) = \dfrac{1}{x - 2}$ is odd about $(2, 0)$: $f(2 - x) = -f(2 + x)$

16. $f(x) = \dfrac{1}{x + 4}$ is odd about $(-4, 0)$:
    $f(-4 - x) = -f(-4 + x)$

17. $f(x) = x^2 - 6x$ is even about $x = 3$: $f(3 - x) = f(3 + x)$

18. $f(x) = x^3 - 2$ is odd about $(0, -2)$:
    $f(-x) + 2 = -(f(x) + 2)$

19. $f(x) = |x|^3$ is even: $f(-x) = f(x)$

20. $f(x) = |x + 1|$ is even about $x = -1$:
    $f(-1 - x) = f(-1 + x)$

21. $f(x) = \sqrt{2x}$ has no symmetry.

22. $f(x) = \sqrt{(x - 1)^2}$ is even about $x = 1$:
    $f(1 - x) = f(1 + x)$

23. and 24.

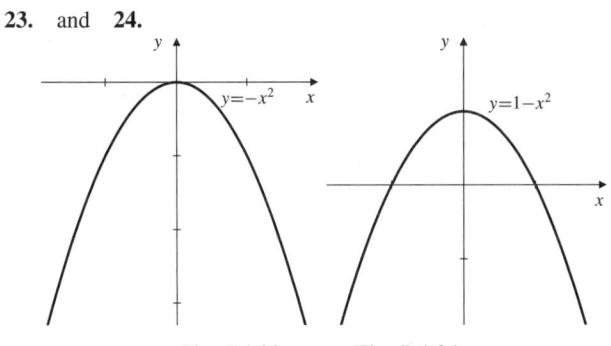

Fig. P.4.23   Fig. P.4.24

25. and 26.

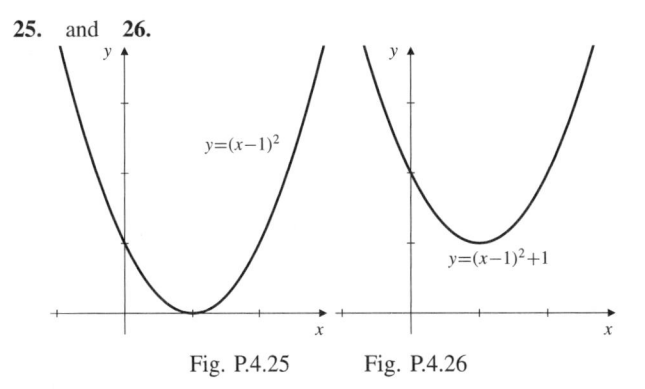

Fig. P.4.25   Fig. P.4.26

27. and 28.

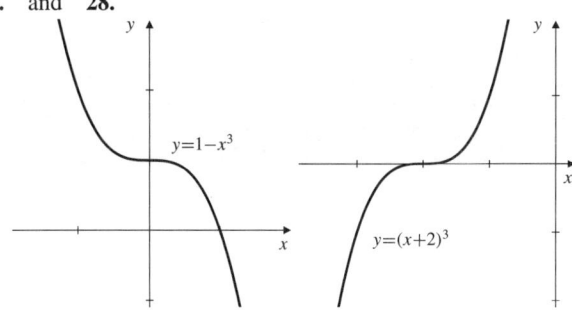

Fig. P.4.27   Fig. P.4.28

29. and 30.

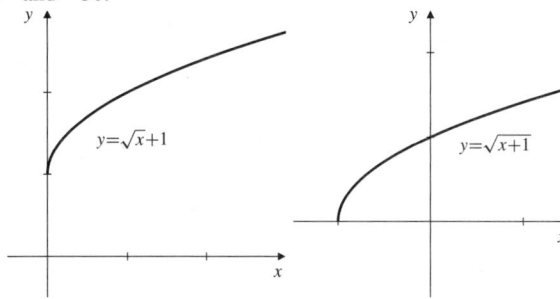

Fig. P.4.29   Fig. P.4.30

31. and 32.

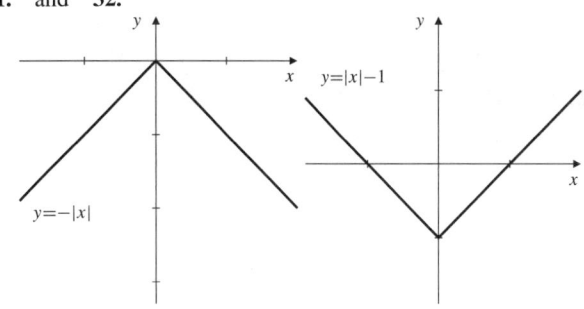

Fig. P.4.31   Fig. P.4.32

33. and 34.

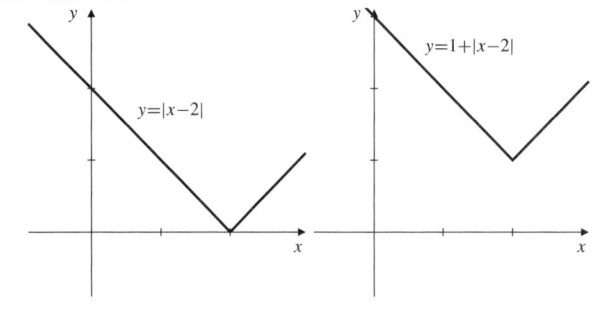

Fig. P.4.33   Fig. P.4.34

35. and **36.**

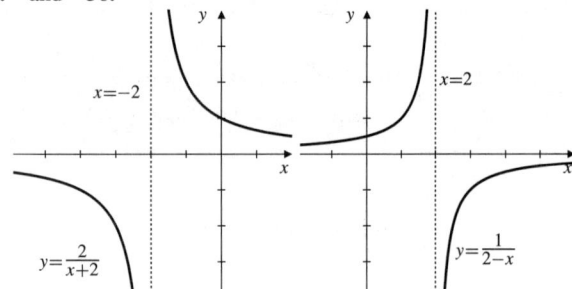

Fig. P.4.35   Fig. P.4.36

37. and **38.**

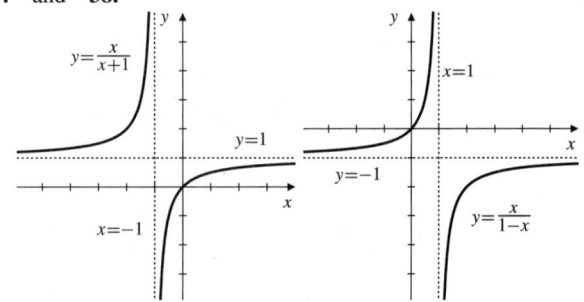

Fig. P.4.37   Fig. P.4.38

39(a). and **39(b).**

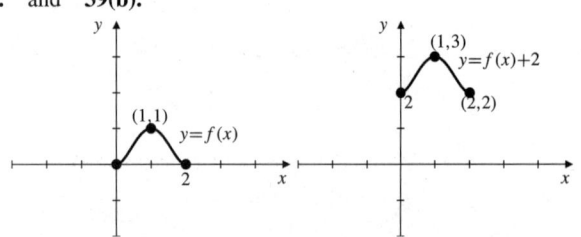

Fig. P.4.39(a)   Fig. P.4.39(b)

40.

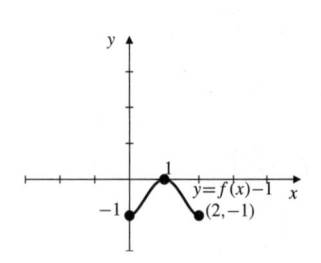

Fig. P.4.40

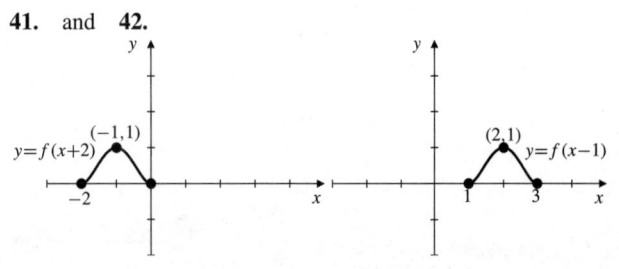

41. and **42.**

Fig. P.4.41   Fig. P.4.42

43. and **44.**

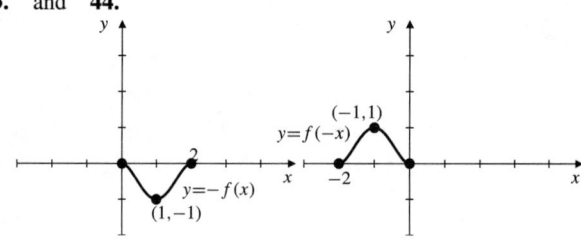

Fig. P.4.43   Fig. P.4.44

45. and **46.**

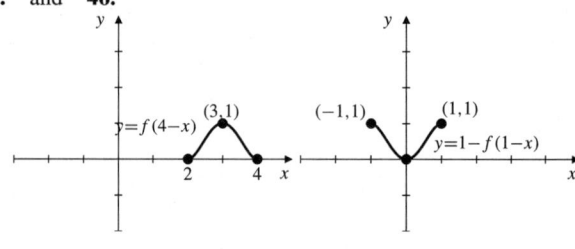

Fig. P.4.45   Fig. P.4.46

47. Range is approximately $[-0.18, 0.68]$.

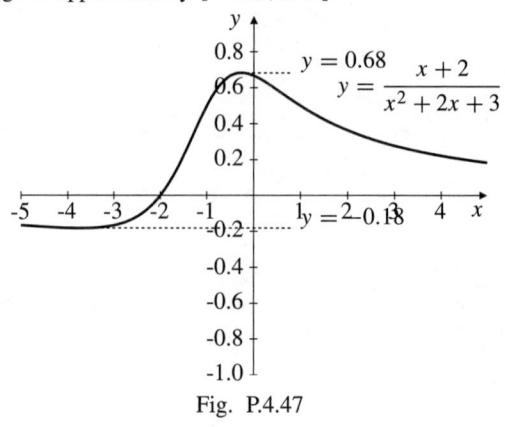

Fig. P.4.47

48. Range is approximately $(-\infty, 0.17]$.

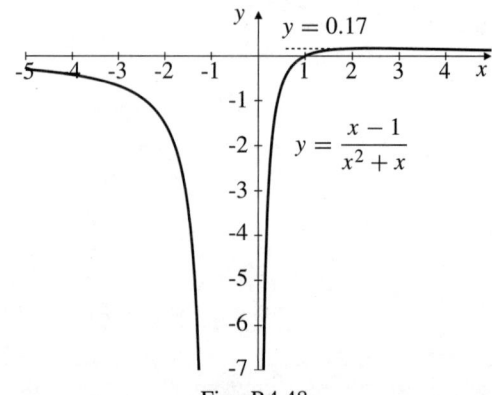

Fig. P.4.48

49.

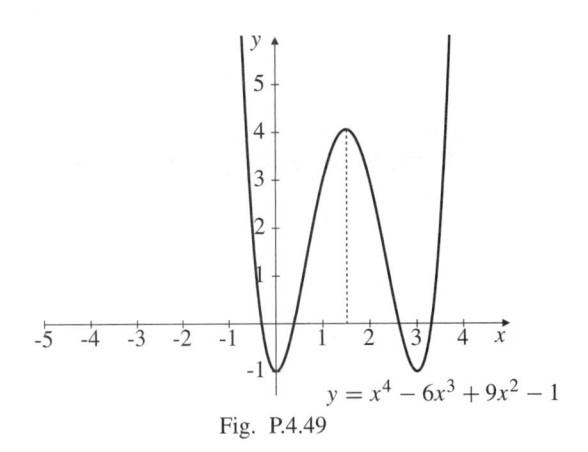

Fig. P.4.49 — $y = x^4 - 6x^3 + 9x^2 - 1$

Apparent symmetry about $x = 1.5$.
This can be confirmed by calculating $f(3-x)$, which turns out to be equal to $f(x)$.

50.

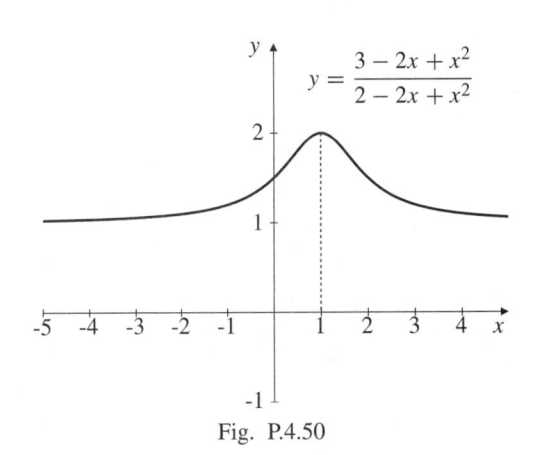

Fig. P.4.50 — $y = \dfrac{3 - 2x + x^2}{2 - 2x + x^2}$

Apparent symmetry about $x = 1$.
This can be confirmed by calculating $f(2-x)$, which turns out to be equal to $f(x)$.

51.

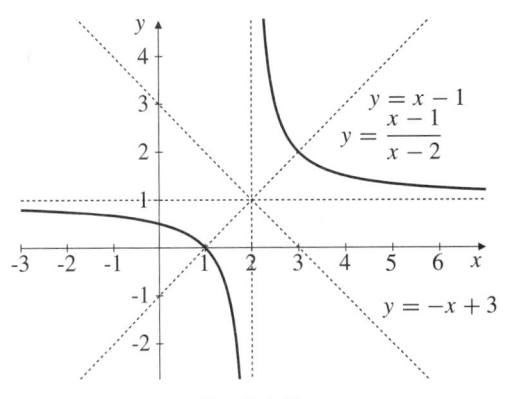

Fig. P.4.51 — $y = \dfrac{x-1}{x-2}$, $y = x-1$, $y = -x+3$

Apparent symmetry about $(2, 1)$, and about the lines $y = x - 1$ and $y = 3 - x$.
These can be confirmed by noting that $f(x) = 1 + \dfrac{1}{x - 2}$, so the graph is that of $1/x$ shifted right 2 units and up one.

52.

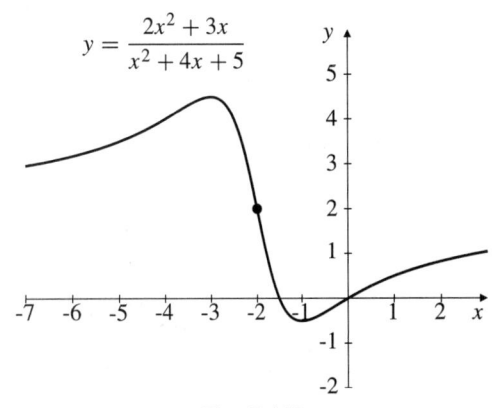

Fig. P.4.52 — $y = \dfrac{2x^2 + 3x}{x^2 + 4x + 5}$

Apparent symmetry about $(-2, 2)$.
This can be confirmed by calculating shifting the graph right by 2 (replace x with $x-2$) and then down 2 (subtract 2). The result is $-5x/(1 + x^2)$, which is odd.

53. If f is both even and odd the $f(x) = f(-x) = -f(x)$, so $f(x) = 0$ identically.

Section P.5 Combining Functions to Make New Functions (page 40)

1. $f(x) = x$, $g(x) = \sqrt{x - 1}$.
$Df = \mathbb{R}$, $D(g) = [1, \infty)$.
$D(f + g) = D(f - g) = D(fg) = D(g/f) = [1, \infty)$,
$D(f/g) = (1, \infty)$.
$(f + g)(x) = x + \sqrt{x - 1}$
$(f - g)(x) = x - \sqrt{x - 1}$
$(fg)(x) = x\sqrt{x - 1}$
$(f/g)(x) = x/\sqrt{x - 1}$
$(g/f)(x) = (\sqrt{1 - x})/x$

2. $f(x) = \sqrt{1 - x}$, $g(x) = \sqrt{1 + x}$.
$Df = (-\infty, 1]$, $D(g) = [-1, \infty)$.
$D(f + g) = D(f - g) = D(fg) = [-1, 1]$,
$D(f/g) = (-1, 1]$, $D(g/f) = [-1, 1)$.
$(f + g)(x) = \sqrt{1 - x} + \sqrt{1 + x}$
$(f - g)(x) = \sqrt{1 - x} - \sqrt{1 + x}$
$(fg)(x) = \sqrt{1 - x^2}$
$(f/g)(x) = \sqrt{(1 - x)/(1 + x)}$
$(g/f)(x) = \sqrt{(1 + x)/(1 - x)}$

3.

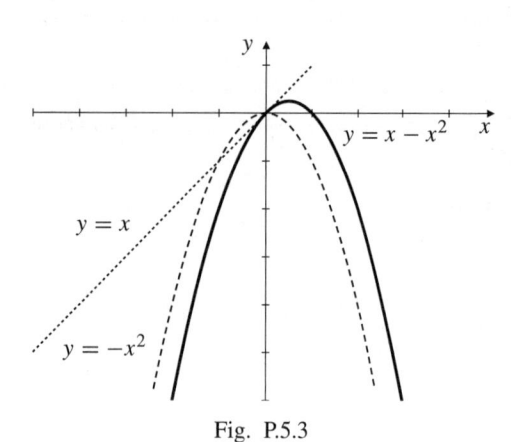

Fig. P.5.3

4.

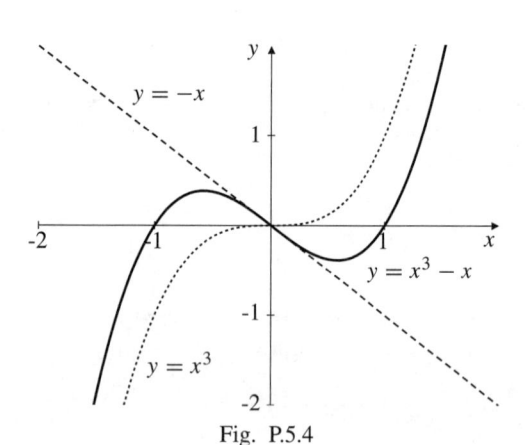

Fig. P.5.4

5.

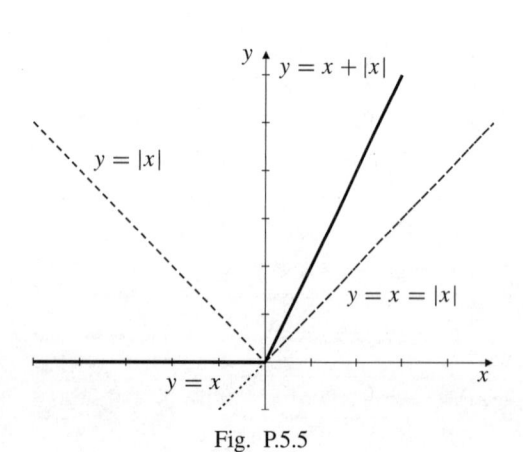

Fig. P.5.5

6.

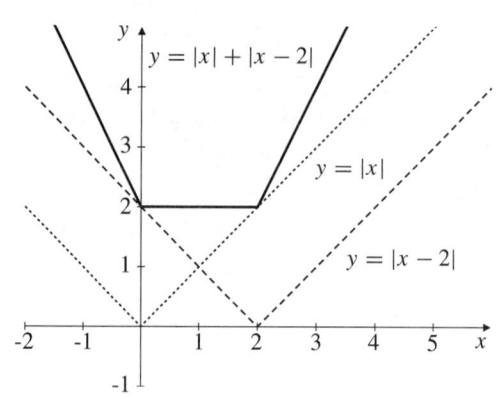

Fig. P.5.6

7. $f(x) = x + 5$, $g(x) = x^2 - 3$.
$f \circ g(0) = f(-3) = 2$, $\quad g(f(0)) = g(5) = 22$
$f(g(x)) = f(x^2 - 3) = x^2 + 2$
$g \circ f(x) = g(f(x)) = g(x + 5) = (x + 5)^2 - 3$
$f \circ f(-5) = f(0) = 5$, $\quad g(g(2)) = g(1) = -2$
$f(f(x)) = f(x + 5) = x + 10$
$g \circ g(x) = g(g(x)) = (x^2 - 3)^2 - 3$

8. $f(x) = 2/x$, $g(x) = x/(1 - x)$.
$f \circ f(x) = 2/(2/x) = x$; $\quad D(f \circ f) = \{x : x \neq 0\}$
$f \circ g(x) = 2/(x/(1 - x)) = 2(1 - x)/x$;
$\quad D(f \circ g) = \{x : x \neq 0, 1\}$
$g \circ f(x) = (2/x)/(1 - (2/x)) = 2/(x - 2)$;
$\quad D(g \circ f) = \{x : x \neq 0, 2\}$
$g \circ g(x) = (x/(1 - x))/(1 - (x/(1 - x))) = x/(1 - 2x)$;
$\quad D(g \circ g) = \{x : x \neq 1/2, 1\}$

9. $f(x) = 1/(1 - x)$, $g(x) = \sqrt{x - 1}$.
$f \circ f(x) = 1/(1 - (1/(1 - x))) = (x - 1)/x$;
$\quad D(f \circ f) = \{x : x \neq 0, 1\}$
$f \circ g(x) = 1/(1 - \sqrt{x - 1})$;
$\quad D(f \circ g) = \{x : x \geq 1, x \neq 2\}$
$g \circ f(x) = \sqrt{(1/(1 - x)) - 1} = \sqrt{x/(1 - x)}$;
$\quad D(g \circ f) = [0, 1)$
$g \circ g(x) = \sqrt{\sqrt{x - 1} - 1}$; $\quad D(g \circ g) = [2, \infty)$

10. $f(x) = (x + 1)/(x - 1) = 1 + 2/(x - 1)$, $g(x) = \text{sgn}(x)$.
$f \circ f(x) = 1 + 2/(1 + (2/(x - 1) - 1)) = x$;
$D(f \circ f) = \{x : x \neq 1\}$
$f \circ g(x) = \dfrac{\text{sgn}\, x + 1}{\text{sgn}\, x - 1} = 0$; $\quad D(f \circ g) = (-\infty, 0)$
$g \circ f(x) = \text{sgn}\left(\dfrac{x + 1}{x - 1}\right) = \begin{cases} 1 & \text{if } x < -1 \text{ or } x > 1 \\ -1 & \text{if } -1 < x < 1 \end{cases}$;
$D(g \circ f) = \{x : x \neq -1, 1\}$
$g \circ g(x) = \text{sgn}(\text{sgn}(x)) = \text{sgn}(x)$; $\quad D(g \circ g) = \{x : x \neq 0\}$

	$f(x)$	$g(x)$	$f \circ g(x)$
11.	x^2	$x+1$	$(x+1)^2$
12.	$x-4$	$x+4$	x
13.	$\sqrt{x}$	x^2	$|x|$
14.	$2x^3+3$	$x^{1/3}$	$2x+3$
15.	$(x+1)/x$	$1/(x-1)$	x
16.	$1/(x+1)^2$	$x-1$	$1/x^2$

17. $y = \sqrt{x}$.
$y = 2 + \sqrt{x}$: previous graph is raised 2 units.
$y = 2 + \sqrt{3+x}$: previous graph is shiftend left 3 units.
$y = 1/(2 + \sqrt{3+x})$: previous graph turned upside down and shrunk vertically.

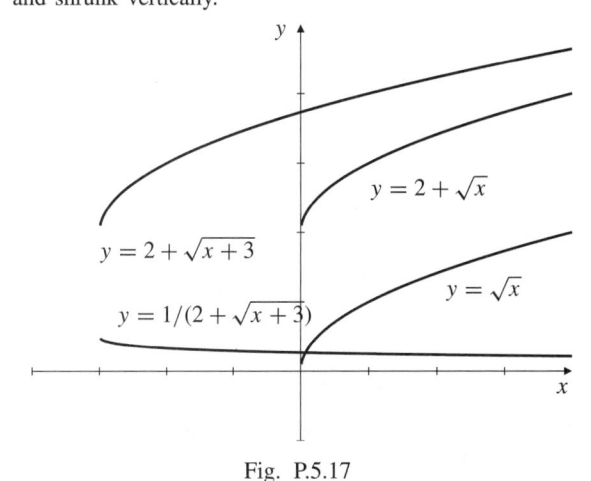

Fig. P.5.17

18.

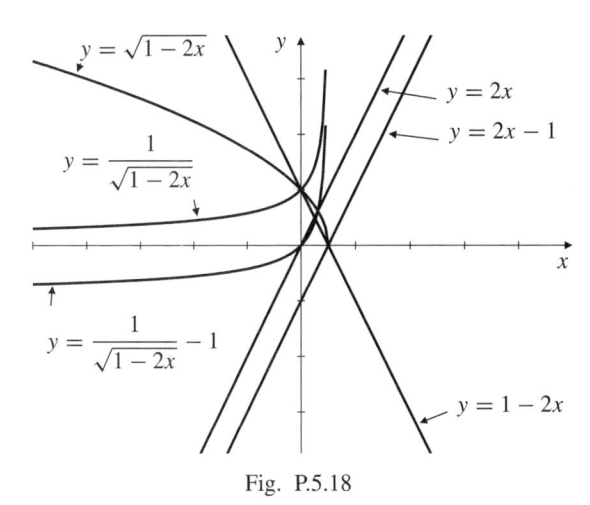

Fig. P.5.18

19. and 20.

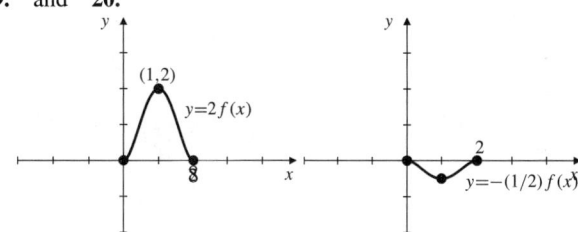

Fig. P.5.19   Fig. P.5.20

21. and 22.

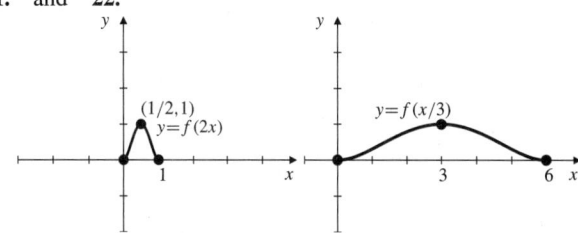

Fig. P.5.21   Fig. P.5.22

23. and 24.

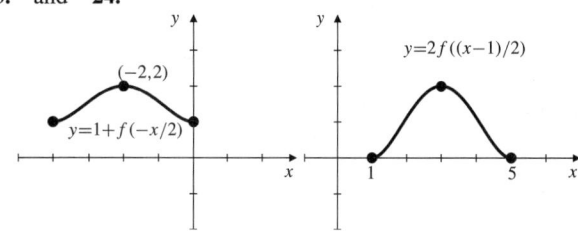

Fig. P.5.23   Fig. P.5.24

25. and 26.

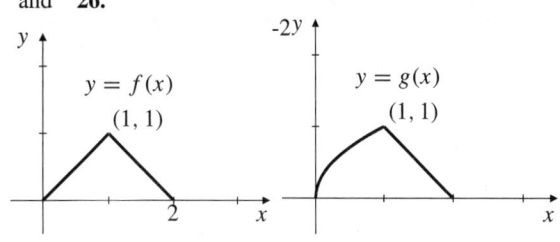

Fig. P.5.25   Fig. P.5.26

27. $F(x) = Ax + B$
(a) $F \circ F(x) = F(x)$
$\Rightarrow A(Ax + B) + B = Ax + B$
$\Rightarrow A[(A-1)x + B] = 0$
Thus, either $A = 0$ or $A = 1$ and $B = 0$.
(b) $F \circ F(x) = x$
$\Rightarrow A(Ax + B) + B = x$
$\Rightarrow (A^2 - 1)x + (A+1)B = 0$
Thus, either $A = -1$ or $A = 1$ and $B = 0$

28. $\lfloor x \rfloor = 0$ for $0 \le x < 1$; $\lceil x \rceil = 0$ for $-1 \le x < 0$.

29. $\lfloor x \rfloor = \lceil x \rceil$ for all integers x.

30. $\lceil -x \rceil = -\lfloor x \rfloor$ is true for all real x; if $x = n + y$ where n is an integer and $0 \le y < 1$, then $-x = -n - y$, so that $\lceil -x \rceil = -n$ and $\lfloor x \rfloor = n$.

31.

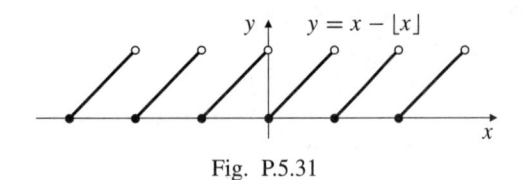

Fig. P.5.31

32. $f(x)$ is called the integer part of x because $|f(x)|$ is the largest integer that does not exceed x; i.e. $|x| = |f(x)| + y$, where $0 \le y < 1$.

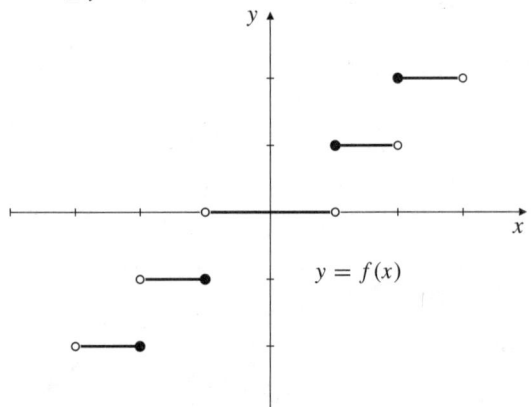

Fig. P.5.32

33. If f is even and g is odd, then: f^2, g^2, $f \circ g$, $g \circ f$, and $f \circ f$ are all even. fg, f/g, g/f, and $g \circ g$ are odd, and $f + g$ is neither even nor odd. Here are two typical verifications:

$$f \circ g(-x) = f(g(-x)) = f(-g(x)) = f(g(x)) = f \circ g(x)$$
$$(fg)(-x) = f(-x)g(-x) = f(x)[-g(x)]$$
$$= -f(x)g(x) = -(fg)(x).$$

The others are similar.

34. f even $\Leftrightarrow f(-x) = f(x)$
f odd $\Leftrightarrow f(-x) = -f(x)$
f even and odd $\Rightarrow f(x) = -f(x) \Rightarrow 2f(x) = 0$
$\Rightarrow f(x) = 0$

35. a) Let $E(x) = \frac{1}{2}[f(x) + f(-x)]$.
Then $E(-x) = \frac{1}{2}[f(-x) + f(x)] = E(x)$. Hence, $E(x)$ is even.
Let $O(x) = \frac{1}{2}[f(x) - f(-x)]$.
Then $O(-x) = \frac{1}{2}[f(-x) - f(x)] = -O(x)$ and $O(x)$ is odd.

$$E(x) + O(x)$$
$$= \frac{1}{2}[f(x) + f(-x)] + \frac{1}{2}[f(x) - f(-x)]$$
$$= f(x).$$

Hence, $f(x)$ is the sum of an even function and an odd function.

b) If $f(x) = E_1(x) + O_1(x)$ where E_1 is even and O_1 is odd, then

$$E_1(x) + O_1(x) = f(x) = E(x) + O(x).$$

Thus $E_1(x) - E(x) = O(x) - O_1(x)$. The left side of this equation is an even function and the right side is an odd function. Hence both sides are both even and odd, and are therefore identically 0 by Exercise 36. Hence $E_1 = E$ and $O_1 = O$. This shows that f can be written in only one way as the sum of an even function and an odd function.

Section P.6 The Trigonometric Functions (page 53)

1. $\cos\left(\dfrac{3\pi}{4}\right) = \cos\left(\pi - \dfrac{\pi}{4}\right) = -\cos\dfrac{\pi}{4} = -\dfrac{1}{\sqrt{2}}$

2. $\tan\dfrac{-3\pi}{4} = -\tan\dfrac{3\pi}{4} = -1$

3. $\sin\dfrac{2\pi}{3} = \sin\left(\pi - \dfrac{\pi}{3}\right) = \sin\dfrac{\pi}{3} = \dfrac{\sqrt{3}}{2}$

4. $\sin\left(\dfrac{7\pi}{12}\right) = \sin\left(\dfrac{\pi}{4} + \dfrac{\pi}{3}\right)$
$= \sin\dfrac{\pi}{4}\cos\dfrac{\pi}{3} + \cos\dfrac{\pi}{4}\sin\dfrac{\pi}{3}$
$= \dfrac{1}{\sqrt{2}}\dfrac{1}{2} + \dfrac{1}{\sqrt{2}}\dfrac{\sqrt{3}}{2} = \dfrac{1+\sqrt{3}}{2\sqrt{2}}$

5. $\cos\dfrac{5\pi}{12} = \cos\left(\dfrac{2\pi}{3} - \dfrac{\pi}{4}\right)$
$= \cos\dfrac{2\pi}{3}\cos\dfrac{\pi}{4} + \sin\dfrac{2\pi}{3}\sin\dfrac{\pi}{4}$
$= -\left(\dfrac{1}{2}\right)\left(\dfrac{1}{\sqrt{2}}\right) + \left(\dfrac{\sqrt{3}}{2}\right)\left(\dfrac{1}{\sqrt{2}}\right)$
$= \dfrac{\sqrt{3}-1}{2\sqrt{2}}$

6. $\sin\dfrac{11\pi}{12} = \sin\dfrac{\pi}{12}$
$= \sin\left(\dfrac{\pi}{3} - \dfrac{\pi}{4}\right)$
$= \sin\dfrac{\pi}{3}\cos\dfrac{\pi}{4} - \cos\dfrac{\pi}{3}\sin\dfrac{\pi}{4}$
$= \left(\dfrac{\sqrt{3}}{2}\right)\left(\dfrac{1}{\sqrt{2}}\right) - \left(\dfrac{1}{2}\right)\left(\dfrac{1}{\sqrt{2}}\right)$
$= \dfrac{\sqrt{3}-1}{2\sqrt{2}}$

7. $\cos(\pi + x) = \cos(2\pi - (\pi - x))$
$= \cos(-(\pi - x))$
$= \cos(\pi - x) = -\cos x$

8. $\sin(2\pi - x) = -\sin x$

9. $\sin\left(\dfrac{3\pi}{2} - x\right) = \sin\left(\pi - \left(x - \dfrac{\pi}{2}\right)\right)$
$= \sin\left(x - \dfrac{\pi}{2}\right)$
$= -\sin\left(\dfrac{\pi}{2} - x\right)$
$= -\cos x$

10. $\cos\left(\dfrac{3\pi}{2} + x\right) = \cos\dfrac{3\pi}{2}\cos x - \sin\dfrac{3\pi}{2}\sin x$
$= (-1)(-\sin x) = \sin x$

11. $\tan x + \cot x = \dfrac{\sin x}{\cos x} + \dfrac{\cos x}{\sin x}$
$= \dfrac{\sin^2 x + \cos^2 x}{\cos x \sin x}$
$= \dfrac{1}{\cos x \sin x}$

12. $\dfrac{\tan x - \cot x}{\tan x + \cot x} = \dfrac{\left(\dfrac{\sin x}{\cos x} - \dfrac{\cos x}{\sin x}\right)}{\left(\dfrac{\sin x}{\cos x} + \dfrac{\cos x}{\sin x}\right)}$
$= \dfrac{\left(\dfrac{\sin^2 x - \cos^2 x}{\cos x \sin x}\right)}{\left(\dfrac{\sin^2 x + \cos^2 x}{\cos x \sin x}\right)}$
$= \sin^2 x - \cos^2 x$

13. $\cos^4 x - \sin^4 x = (\cos^2 x - \sin^2 x)(\cos^2 x + \sin^2 x)$
$= \cos^2 x - \sin^2 x = \cos(2x)$

14. $(1 - \cos x)(1 + \cos x) = 1 - \cos^2 x = \sin^2 x$ implies $\dfrac{1 - \cos x}{\sin x} = \dfrac{\sin x}{1 + \cos x}$. Now
$\dfrac{1 - \cos x}{\sin x} = \dfrac{1 - \cos 2\left(\dfrac{x}{2}\right)}{\sin 2\left(\dfrac{x}{2}\right)}$
$= \dfrac{1 - \left(1 - 2\sin^2\left(\dfrac{x}{2}\right)\right)}{2\sin\dfrac{x}{2}\cos\dfrac{x}{2}}$
$= \dfrac{\sin\dfrac{x}{2}}{\cos\dfrac{x}{2}} = \tan\dfrac{x}{2}$

15. $\dfrac{1 - \cos x}{1 + \cos x} = \dfrac{2\sin^2\left(\dfrac{x}{2}\right)}{2\cos^2\left(\dfrac{x}{2}\right)} = \tan^2\left(\dfrac{x}{2}\right)$

16. $\dfrac{\cos x - \sin x}{\cos x + \sin x} = \dfrac{(\cos x - \sin x)^2}{(\cos x + \sin x)(\cos x - \sin x)}$
$= \dfrac{\cos^2 x - 2\sin x \cos x + \sin^2 x}{\cos^2 x - \sin^2 x}$
$= \dfrac{1 - \sin(2x)}{\cos(2x)}$
$= \sec(2x) - \tan(2x)$

17. $\sin 3x = \sin(2x + x)$
$= \sin 2x \cos x + \cos 2x \sin x$
$= 2\sin x \cos^2 x + \sin x(1 - 2\sin^2 x)$
$= 2\sin x(1 - \sin^2 x) + \sin x - 2\sin^3 x$
$= 3\sin x - 4\sin^3 x$

18. $\cos 3x = \cos(2x + x)$
$= \cos 2x \cos x - \sin 2x \sin x$
$= (2\cos^2 x - 1)\cos x - 2\sin^2 x \cos x$
$= 2\cos^3 x - \cos x - 2(1 - \cos^2 x)\cos x$
$= 4\cos^3 x - 3\cos x$

19. $\cos 2x$ has period π.

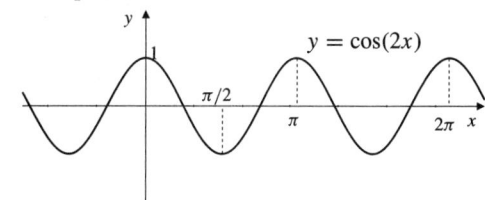

Fig. P.6.19

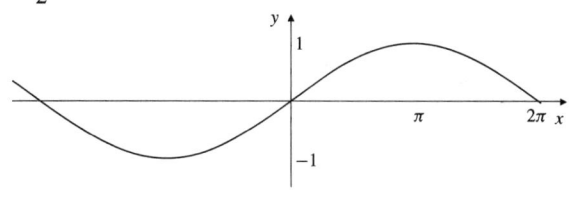

20. $\sin\dfrac{x}{2}$ has period 4π.

Fig. P.6.20

21. $\sin \pi x$ has period 2.

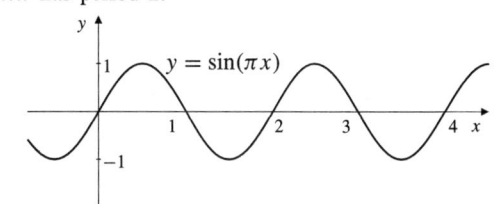

Fig. P.6.21

22. $\cos\dfrac{\pi x}{2}$ has period 4.

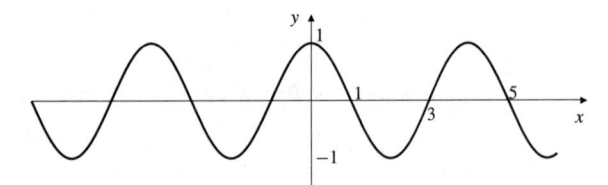

Fig. P.6.22

23.

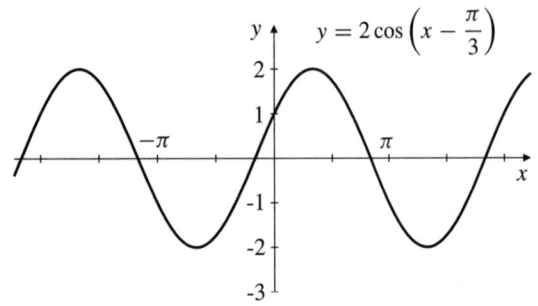

Fig. P.6.23 ($y = 2\cos\left(x - \frac{\pi}{3}\right)$)

24.

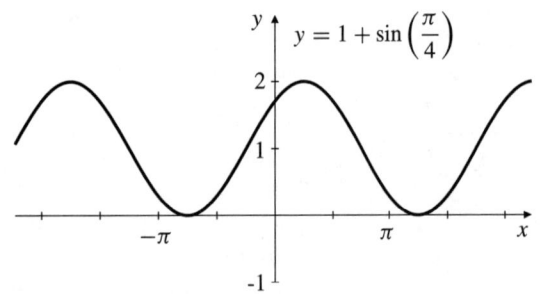

Fig. P.6.24 ($y = 1 + \sin\left(\frac{\pi}{4}\right)$)

25. $\sin x = \frac{3}{5}$, $\frac{\pi}{2} < x < \pi$

$\cos x = -\frac{4}{5}$, $\tan x = -\frac{3}{4}$

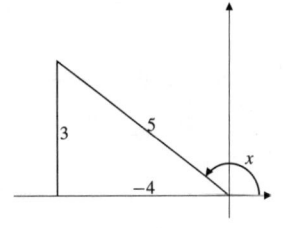

Fig. P.6.25

26. $\tan x = 2$ where x is in $[0, \frac{\pi}{2}]$. Then
$\sec^2 x = 1 + \tan^2 x = 1 + 4 = 5$. Hence,
$\sec x = \sqrt{5}$ and $\cos x = \frac{1}{\sec x} = \frac{1}{\sqrt{5}}$,
$\sin x = \tan x \cos x = \frac{2}{\sqrt{5}}$.

27. $\cos x = \frac{1}{3}$, $-\frac{\pi}{2} < x < 0$

$\sin x = -\frac{\sqrt{8}}{3} = -\frac{2}{3}\sqrt{2}$

$\tan x = -\frac{\sqrt{8}}{1} = -2\sqrt{2}$

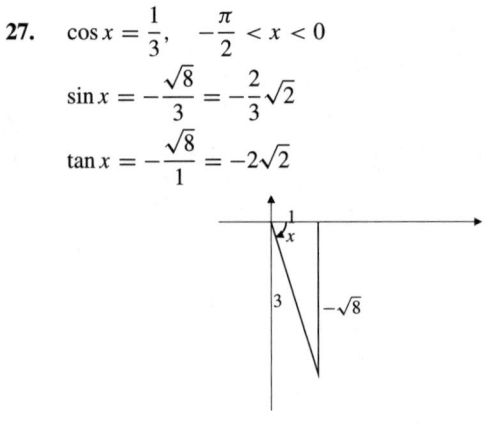

Fig. P.6.27

28. $\cos x = -\frac{5}{13}$ where x is in $\left[\frac{\pi}{2}, \pi\right]$. Hence,
$\sin x = \sqrt{1 - \cos^2 x} = \sqrt{1 - \frac{25}{169}} = \frac{12}{13}$,
$\tan x = -\frac{12}{5}$.

29. $\sin x = -\frac{1}{2}$, $\pi < x < \frac{3\pi}{2}$

$\cos x = -\frac{\sqrt{3}}{2}$

$\tan x = \frac{1}{\sqrt{3}}$

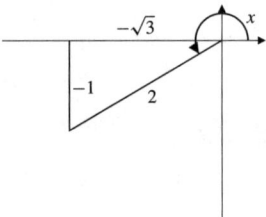

Fig. P.6.29

30. $\tan x = \frac{1}{2}$ where x is in $[\pi, \frac{3\pi}{2}]$. Then,
$\sec^2 x = 1 + \frac{1}{4} = \frac{5}{4}$. Hence,
$\sec x = -\frac{\sqrt{5}}{2}$, $\cos x = -\frac{2}{\sqrt{5}}$,
$\sin x = \tan x \cos x = -\frac{1}{\sqrt{5}}$.

31. $c = 2$, $B = \frac{\pi}{3}$

$a = c \cos B = 2 \times \frac{1}{2} = 1$

$b = c \sin B = 2 \times \frac{\sqrt{3}}{2} = \sqrt{3}$

INSTRUCTOR'S SOLUTIONS MANUAL SECTION P.6 (PAGE 53)

32. $b = 2$, $B = \dfrac{\pi}{3}$

$\dfrac{2}{a} = \tan B = \sqrt{3} \Rightarrow a = \dfrac{2}{\sqrt{3}}$

$\dfrac{2}{c} = \sin B = \dfrac{\sqrt{3}}{2} \Rightarrow c = \dfrac{4}{\sqrt{3}}$

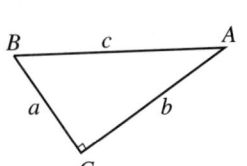

33. $a = 5$, $B = \dfrac{\pi}{6}$

$b = a \tan B = 5 \times \dfrac{1}{\sqrt{3}} = \dfrac{5}{\sqrt{3}}$

$c = \sqrt{a^2 + b^2} = \sqrt{25 + \dfrac{25}{3}} = \dfrac{10}{\sqrt{3}}$

34. $\sin A = \dfrac{a}{c} \Rightarrow a = c \sin A$

35. $\dfrac{a}{b} = \tan A \Rightarrow a = b \tan A$

36. $\cos B = \dfrac{a}{c} \Rightarrow a = c \cos B$

37. $\dfrac{b}{a} = \tan B \Rightarrow a = b \cot B$

38. $\sin A = \dfrac{a}{c} \Rightarrow c = \dfrac{a}{\sin A}$

39. $\dfrac{b}{c} = \cos A \Rightarrow c = b \sec A$

40. $\sin A = \dfrac{a}{c}$

41. $\sin A = \dfrac{a}{c} = \dfrac{\sqrt{c^2 - b^2}}{c}$

42. $\sin A = \dfrac{a}{c} = \dfrac{a}{\sqrt{a^2 + b^2}}$

43. $a = 4$, $b = 3$, $A = \dfrac{\pi}{4}$

$\sin B = b \dfrac{\sin A}{a} = \dfrac{3}{4} \dfrac{1}{\sqrt{2}} = \dfrac{3}{4\sqrt{2}}$

44. Given that $a = 2, b = 2, c = 3$.

Since $a^2 = b^2 + c^2 - 2bc \cos A$,

$\cos A = \dfrac{a^2 - b^2 - c^2}{-2bc}$

$= \dfrac{4 - 4 - 9}{-2(2)(3)} = \dfrac{3}{4}$.

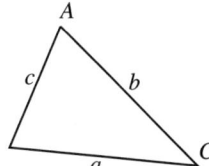

45. $a = 2$, $b = 3$, $c = 4$

$b^2 = a^2 + c^2 - 2ac \cos B$

Thus $\cos B = \dfrac{4 + 16 - 9}{2 \times 2 \times 4} = \dfrac{11}{16}$

$\sin B = \sqrt{1 - \dfrac{11^2}{16^2}} = \dfrac{\sqrt{256 - 121}}{16} = \dfrac{\sqrt{135}}{16}$

46. Given that $a = 2$, $b = 3$, $C = \dfrac{\pi}{4}$.

$c^2 = a^2 + b^2 - 2ab \cos C = 4 + 9 - 2(2)(3) \cos \dfrac{\pi}{4} = 13 - \dfrac{12}{\sqrt{2}}$.

Hence, $c = \sqrt{13 - \dfrac{12}{\sqrt{2}}} \approx 2.12479$.

47. $c = 3$, $A = \dfrac{\pi}{4}$, $B = \dfrac{\pi}{3}$ implies $C = \dfrac{5\pi}{12}$

$\dfrac{a}{\sin A} = \dfrac{c}{\sin C} \Rightarrow a = \dfrac{1}{\sqrt{2}} \dfrac{3}{\sin\left(\dfrac{5\pi}{12}\right)}$

$a = \dfrac{3}{\sqrt{2}} \dfrac{1}{\sin\left(\dfrac{7\pi}{12}\right)}$

$= \dfrac{3}{\sqrt{2}} \dfrac{2\sqrt{2}}{1 + \sqrt{3}}$ (by #5)

$= \dfrac{6}{1 + \sqrt{3}}$

48. Given that $a = 2$, $b = 3$, $C = 35°$. Then $c^2 = 4 + 9 - 2(2)(3) \cos 35°$, hence $c \approx 1.78050$.

49. $a = 4$, $B = 40°$, $C = 70°$

Thus $A = 70°$.

$\dfrac{b}{\sin 40°} = \dfrac{4}{\sin 70°}$ so $b = 4 \dfrac{\sin 40°}{\sin 70°} = 2.736$

50. If $a = 1, b = \sqrt{2}, A = 30°$, then $\dfrac{\sin B}{b} = \dfrac{\sin A}{a} = \dfrac{1}{2}$.

Thus $\sin B = \dfrac{\sqrt{2}}{2} = \dfrac{1}{\sqrt{2}}$, $B = \dfrac{\pi}{4}$ or $\dfrac{3\pi}{4}$, and

$C = \pi - \left(\dfrac{\pi}{4} + \dfrac{\pi}{6}\right) = \dfrac{7\pi}{12}$ or $C = \pi - \left(\dfrac{3\pi}{4} + \dfrac{\pi}{6}\right) = \dfrac{\pi}{12}$.

Thus, $\cos C = \cos \dfrac{7\pi}{12} = \cos\left(\dfrac{\pi}{4} + \dfrac{\pi}{3}\right) = \dfrac{1 - \sqrt{3}}{2\sqrt{2}}$ or

$\cos C = \cos \dfrac{\pi}{12} = \cos\left(\dfrac{\pi}{3} - \dfrac{\pi}{4}\right) = \dfrac{1 + \sqrt{3}}{2\sqrt{2}}$.

Hence,

$c^2 = a^2 + b^2 - 2ab \cos C$

$= 1 + 2 - 2\sqrt{2} \cos C$

$= 3 - (1 - \sqrt{3})$ or $3 - (1 + \sqrt{3})$

$= 2 + \sqrt{3}$ or $2 - \sqrt{3}$.

Hence, $c = \sqrt{2 + \sqrt{3}}$ or $\sqrt{2 - \sqrt{3}}$.

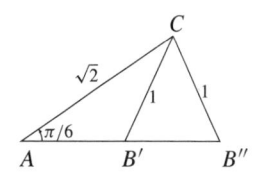

Fig. P.6.50

51. Let h be the height of the pole and x be the distance from C to the base of the pole.
Then $h = x \tan 50°$ and $h = (x + 10) \tan 35°$
Thus $x \tan 50° = x \tan 35° + 10 \tan 35°$ so

$$x = \frac{10 \tan 35°}{\tan 50° - \tan 35°}$$

$$h = \frac{10 \tan 50° \tan 35°}{\tan 50° - \tan 35°} \approx 16.98$$

The pole is about 16.98 metres high.

52. See the following diagram. Since $\tan 40° = h/a$, therefore $a = h/\tan 40°$. Similarly, $b = h/\tan 70°$.
Since $a + b = 2$ km, therefore,

$$\frac{h}{\tan 40°} + \frac{h}{\tan 70°} = 2$$

$$h = \frac{2(\tan 40° \tan 70°)}{\tan 70° + \tan 40°} \approx 1.286 \text{ km}.$$

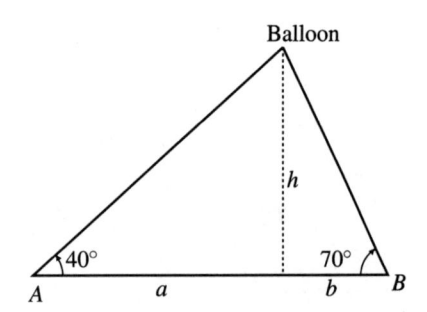

Fig. P.6.52

53. Area $\triangle ABC = \frac{1}{2}|BC|h = \frac{ah}{2} = \frac{ac \sin B}{2} = \frac{ab \sin C}{2}$

By symmetry, area $\triangle ABC$ also $= \frac{1}{2}bc \sin A$

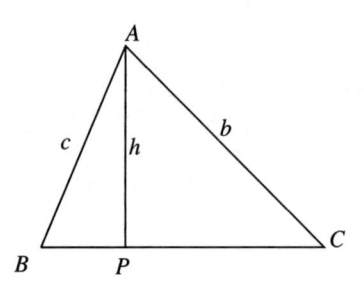

Fig. P.6.53

54. From Exercise 53, area $= \frac{1}{2}ac \sin B$. By Cosine Law, $\cos B = \frac{a^2 + c^2 - b^2}{2ac}$. Thus,

$$\sin B = \sqrt{1 - \left(\frac{a^2 + c^2 - b^2}{2ac}\right)^2}$$

$$= \frac{\sqrt{-a^4 - b^4 - c^4 + 2a^2b^2 + 2b^2c^2 + 2a^2c^2}}{2ac}.$$

Hence, Area $= \frac{\sqrt{-a^4 - b^4 - c^4 + 2a^2b^2 + 2b^2c^2 + 2a^2c^2}}{4}$

square units. Since,

$$s(s - a)(s - b)(s - c)$$
$$= \frac{b + c + a}{2} \cdot \frac{b + c - a}{2} \cdot \frac{a - b + c}{2} \cdot \frac{a + b - c}{2}$$
$$= \frac{1}{16}\left((b + c)^2 - a^2\right)\left(a^2 - (b - c)^2\right)$$
$$= \frac{1}{16}\left(a^2\left((b + c)^2 + (b - c)^2\right) - a^4 - (b^2 - c^2)^2\right)$$
$$= \frac{1}{16}\left(2a^2b^2 + 2a^2c^2 - a^4 - b^4 - c^4 + 2b^2c^2\right)$$

Thus $\sqrt{s(s - a)(s - b)(s - c)}$ = Area of triangle.

CHAPTER 1. LIMITS AND CONTINUITY

Section 1.1 Examples of Velocity, Growth Rate, and Area (page 59)

1. Average velocity $= \dfrac{\Delta x}{\Delta t} = \dfrac{(t+h)^2 - t^2}{h}$ ft/s.

2.

h	Avg. vel. over $[2, 2+h]$
1	5.0000
0.1	4.1000
0.01	4.0100
0.001	4.0010
0.0001	4.0001

3. Guess velocity is $v = 4$ ft/s at $t = 2$ s.

4. Average volocity on $[2, 2+h]$ is

$$\frac{(2+h)^2 - 4}{(2+h) - 2} = \frac{4 + 4h + h^2 - 4}{h} = \frac{4h + h^2}{h} = 4 + h.$$

As h approaches 0 this average velocity approaches 4 ft/s

5. $x = 3t^2 - 12t + 1$ m at time t s.
Average velocity over interval $[1, 2]$ is
$$\frac{(3 \times 2^2 - 12 \times 2 + 1) - (3 \times 1^2 - 12 \times 1 + 1)}{2 - 1} = -3 \text{ m/s}.$$
Average velocity over interval $[2, 3]$ is
$$\frac{(3 \times 3^2 - 12 \times 3 + 1) - (3 \times 2^2 - 12 \times 2 + 1)}{3 - 2} = 3 \text{ m/s}.$$
Average velocity over interval $[1, 3]$ is
$$\frac{(3 \times 3^2 - 12 \times 3 + 1) - (3 \times 1^2 - 12 \times 1 + 1)}{3 - 1} = 0 \text{ m/s}.$$

6. Average velocity over $[t, t+h]$ is

$$\frac{3(t+h)^2 - 12(t+h) + 1 - (3t^2 - 12t + 1)}{(t+h) - t}$$
$$= \frac{6th + 3h^2 - 12h}{h} = 6t + 3h - 12 \text{ m/s}.$$

This average velocity approaches $6t - 12$ m/s as h approaches 0.
At $t = 1$ the velocity is $6 \times 1 - 12 = -6$ m/s.
At $t = 2$ the velocity is $6 \times 2 - 12 = 0$ m/s.
At $t = 3$ the velocity is $6 \times 3 - 12 = 6$ m/s.

7. At $t = 1$ the velocity is $v = -6 < 0$ so the particle is moving to the left.
At $t = 2$ the velocity is $v = 0$ so the particle is stationary.
At $t = 3$ the velocity is $v = 6 > 0$ so the particle is moving to the right.

8. Average velocity over $[t - k, t + k]$ is

$$\frac{3(t+k)^2 - 12(t+k) + 1 - [3(t-k)^2 - 12(t-k) + 1]}{(t+k) - (t-k)}$$
$$= \frac{1}{2k}\Big(3t^2 + 6tk + 3k^2 - 12t - 12k + 1 - 3t^2 + 6tk - 3k^2$$
$$+ 12t - 12k + 1\Big)$$
$$= \frac{12tk - 24k}{2k} = 6t - 12 \text{ m/s},$$

which is the velocity at time t from Exercise 7.

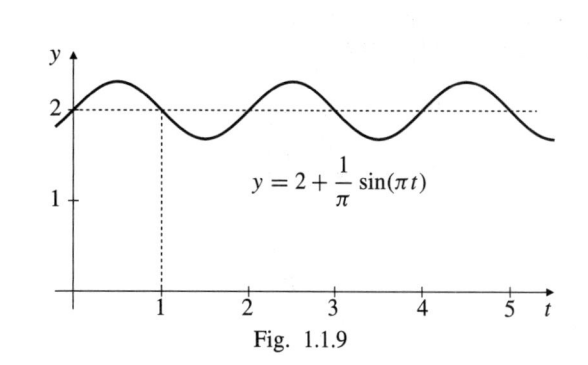

9.

Fig. 1.1.9

At $t = 1$ the height is $y = 2$ ft and the weight is moving downward.

10. Average velocity over $[1, 1+h]$ is

$$\frac{2 + \dfrac{1}{\pi}\sin\pi(1+h) - \left(2 + \dfrac{1}{\pi}\sin\pi\right)}{h}$$
$$= \frac{\sin(\pi + \pi h)}{\pi h} = \frac{\sin\pi\cos(\pi h) + \cos\pi\sin(\pi h)}{\pi h}$$
$$= -\frac{\sin(\pi h)}{\pi h}.$$

h	Avg. vel. on $[1, 1+h]$
1.0000	0
0.1000	-0.983631643
0.0100	-0.999835515
0.0010	-0.999998355

11. The velocity at $t = 1$ is about $v = -1$ ft/s. The "−" indicates that the weight is moving downward.

12. We sketched a tangent line to the graph on page 55 in the text at $t = 20$. The line appeared to pass through the points $(10, 0)$ and $(50, 1)$. On day 20 the biomass is growing at about $(1 - 0)/(50 - 10) = 0.025$ mm²/d.

13. The curve is steepest, and therefore the biomass is growing most rapidly, at about day 45.

14. a) profit

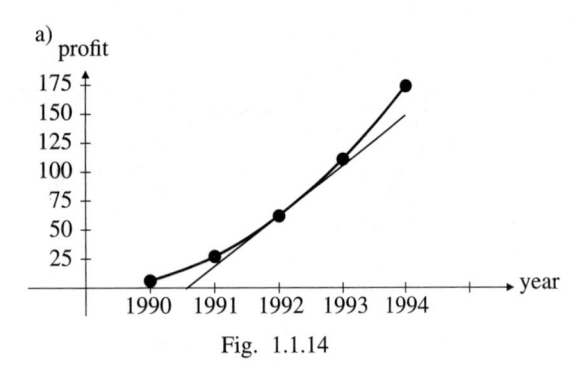

Fig. 1.1.14

b) Average rate of increase in profits between 1992 and 1994 is
$$\frac{174-62}{1994-1992} = \frac{112}{2} = 56 \text{ (thousand\$/yr)}.$$

c) Drawing a tangent line to the graph in (a) at $t = 1992$ and measuring its slope, we find that the rate of increase of profits in 1992 is about 43 thousand\$/year.

Section 1.2 Limits of Functions (page 68)

1. From inspecting the graph

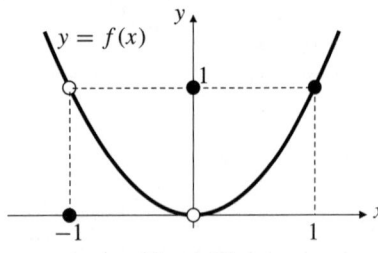

Fig. 1.2.1

we see that
$$\lim_{x \to -1} f(x) = 1, \quad \lim_{x \to 0} f(x) = 0, \quad \lim_{x \to 1} f(x) = 1.$$

2. From inspecting the graph

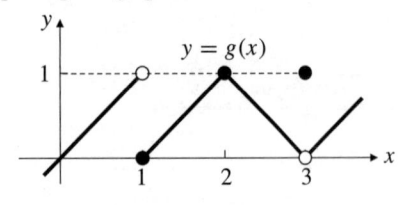

Fig. 1.2.2

we see that
$$\lim_{x \to 1} g(x) \text{ does not exist}$$
(left limit is 1, right limit is 0)
$$\lim_{x \to 2} g(x) = 1, \quad \lim_{x \to 3} g(x) = 0.$$

3. $\lim_{x \to 1-} g(x) = 1$

4. $\lim_{x \to 1+} g(x) = 0$

5. $\lim_{x \to 3+} g(x) = 0$

6. $\lim_{x \to 3-} g(x) = 0$

7. $\lim_{x \to 4}(x^2 - 4x + 1) = 4^2 - 4(4) + 1 = 1$

8. $\lim_{x \to 2} 3(1-x)(2-x) = 3(-1)(2-2) = 0$

9. $\lim_{x \to 3} \frac{x+3}{x+6} = \frac{3+3}{3+6} = \frac{2}{3}$

10. $\lim_{t \to -4} \frac{t^2}{4-t} = \frac{(-4)^2}{4+4} = 2$

11. $\lim_{x \to 1} \frac{x^2-1}{x+1} = \frac{1^2-1}{1+1} = \frac{0}{2} = 0$

12. $\lim_{x \to -1} \frac{x^2-1}{x+1} = \lim_{x \to -1}(x-1) = -2$

13. $\lim_{x \to 3} \frac{x^2-6x+9}{x^2-9} = \lim_{x \to 3} \frac{(x-3)^2}{(x-3)(x+3)}$
$= \lim_{x \to 3} \frac{x-3}{x+3} = \frac{0}{6} = 0$

14. $\lim_{x \to -2} \frac{x^2+2x}{x^2-4} = \lim_{x \to -2} \frac{x}{x-2} = \frac{-2}{-4} = \frac{1}{2}$

15. $\lim_{h \to 2} \frac{1}{4-h^2}$ does not exist; denominator approaches 0 but numerator does not approach 0.

16. $\lim_{h \to 0} \frac{3h+4h^2}{h^2-h^3} = \lim_{h \to 0} \frac{3+4h}{h-h^2}$ does not exist; denominator approaches 0 but numerator does not approach 0.

17. $\lim_{x \to 9} \frac{\sqrt{x}-3}{x-9} = \lim_{x \to 9} \frac{(\sqrt{x}-3)(\sqrt{x}+3)}{(x-9)(\sqrt{x}+3)}$
$= \lim_{x \to 9} \frac{x-9}{(x-9)(\sqrt{x}+3)} = \lim_{x \to 9} \frac{1}{\sqrt{x}+3} = \frac{1}{6}$

18. $\lim_{h \to 0} \frac{\sqrt{4+h}-2}{h}$
$= \lim_{h \to 0} \frac{4+h-4}{h(\sqrt{4+h}+2)}$
$= \lim_{h \to 0} \frac{1}{\sqrt{4+h}+2} = \frac{1}{4}$

19. $\lim_{x \to \pi} \frac{(x-\pi)^2}{\pi x} = \frac{0^2}{\pi^2} = 0$

20. $\lim_{x \to -2} |x-2| = |-4| = 4$

21. $\lim_{x \to 0} \frac{|x-2|}{x-2} = \frac{|-2|}{-2} = -1$

22. $\lim_{x\to 2} \frac{|x-2|}{x-2} = \lim_{x\to 2} \begin{cases} 1, & \text{if } x > 2 \\ -1, & \text{if } x < 2. \end{cases}$
Hence, $\lim_{x\to 2} \frac{|x-2|}{x-2}$ does not exist.

23. $\lim_{t\to 1} \frac{t^2-1}{t^2-2t+1}$
$\lim_{t\to 1} \frac{(t-1)(t+1)}{(t-1)^2} = \lim_{t\to 1} \frac{t+1}{t-1}$ does not exist
(denominator $\to 0$, numerator $\to 2$.)

24. $\lim_{x\to 2} \frac{\sqrt{4-4x+x^2}}{x-2}$
$= \lim_{x\to 2} \frac{|x-2|}{x-2}$ does not exist.

25. $\lim_{t\to 0} \frac{t}{\sqrt{4+t}-\sqrt{4-t}} = \lim_{t\to 0} \frac{t(\sqrt{4+t}+\sqrt{4-t})}{(4+t)-(4-t)}$
$= \lim_{t\to 0} \frac{\sqrt{4+t}+\sqrt{4-t}}{2} = 2$

26. $\lim_{x\to 1} \frac{x^2-1}{\sqrt{x+3}-2} = \lim_{x\to 1} \frac{(x-1)(x+1)(\sqrt{x+3}+2)}{(x+3)-4}$
$= \lim_{x\to 1}(x+1)(\sqrt{x+3}+2) = (2)(\sqrt{4}+2) = 8$

27. $\lim_{t\to 0} \frac{t^2+3t}{(t+2)^2-(t-2)^2}$
$= \lim_{t\to 0} \frac{t(t+3)}{t^2+4t+4-(t^2-4t+4)}$
$= \lim_{t\to 0} \frac{t+3}{8} = \frac{3}{8}$

28. $\lim_{s\to 0} \frac{(s+1)^2-(s-1)^2}{s} = \lim_{s\to 0} \frac{4s}{s} = 4$

29. $\lim_{y\to 1} \frac{y-4\sqrt{y}+3}{y^2-1}$
$= \lim_{y\to 1} \frac{(\sqrt{y}-1)(\sqrt{y}-3)}{(\sqrt{y}-1)(\sqrt{y}+1)(y+1)} = \frac{-2}{4} = \frac{-1}{2}$

30. $\lim_{x\to -1} \frac{x^3+1}{x+1}$
$= \lim_{x\to -1} \frac{(x+1)(x^2-x+1)}{x+1} = 3$

31. $\lim_{x\to 2} \frac{x^4-16}{x^3-8}$
$= \lim_{x\to 2} \frac{(x-2)(x+2)(x^2+4)}{(x-2)(x^2+2x+4)}$
$= \frac{(4)(8)}{4+4+4} = \frac{8}{3}$

32. $\lim_{x\to 8} \frac{x^{2/3}-4}{x^{1/3}-2}$
$= \lim_{x\to 8} \frac{(x^{1/3}-2)(x^{1/3}+2)}{(x^{1/3}-2)}$
$= \lim_{x\to 8}(x^{1/3}+2) = 4$

33. $\lim_{x\to 2}\left(\frac{1}{x-2}-\frac{4}{x^2-4}\right)$
$= \lim_{x\to 2} \frac{x+2-4}{(x-2)(x+2)} = \lim_{x\to 2} \frac{1}{x+2} = \frac{1}{4}$

34. $\lim_{x\to 2}\left(\frac{1}{x-2}-\frac{1}{x^2-4}\right)$
$= \lim_{x\to 2} \frac{x+2-1}{(x-2)(x+2)}$
$= \lim_{x\to 2} \frac{x+1}{(x-2)(x+2)}$ does not exist.

35. $\lim_{x\to 0} \frac{\sqrt{2+x^2}-\sqrt{2-x^2}}{x^2}$
$= \lim_{x\to 0} \frac{(2+x^2)-(2-x^2)}{x^2(\sqrt{2+x^2}+\sqrt{2-x^2})}$
$= \lim_{x\to 0} \frac{2x^2}{x^2(\sqrt{2+x^2}+\sqrt{2-x^2})}$
$= \frac{2}{\sqrt{2}+\sqrt{2}} = \frac{1}{\sqrt{2}}$

36. $\lim_{x\to 0} \frac{|3x-1|-|3x+1|}{x}$
$= \lim_{x\to 0} \frac{(3x-1)^2-(3x+1)^2}{x(|3x-1|+|3x+1|)}$
$= \lim_{x\to 0} \frac{-12x}{x(|3x-1|+|3x+1|)} = \frac{-12}{1+1} = -6$

37. $f(x) = x^2$
$\lim_{h\to 0} \frac{f(x+h)-f(x)}{h} = \lim_{h\to 0} \frac{(x+h)^2-x^2}{h}$
$= \lim_{h\to 0} \frac{2hx+h^2}{h} = \lim_{h\to 0} 2x+h = 2x$

38. $f(x) = x^3$
$\lim_{h\to 0} \frac{f(x+h)-f(x)}{h} = \lim_{h\to 0} \frac{(x+h)^3-x^3}{h}$
$= \lim_{h\to 0} \frac{3x^2h+3xh^2+h^3}{h}$
$= \lim_{h\to 0} 3x^2+3xh+h^2 = 3x^2$

39. $f(x) = 1/x$
$\lim_{h\to 0} \frac{f(x+h)-f(x)}{h} = \lim_{h\to 0} \frac{\frac{1}{x+h}-\frac{1}{x}}{h}$
$= \lim_{h\to 0} \frac{x-(x+h)}{h(x+h)x}$
$= \lim_{h\to 0} -\frac{1}{(x+h)x} = -\frac{1}{x^2}$

40. $f(x) = 1/x^2$

$$\lim_{h\to 0} \frac{f(x+h) - f(x)}{h} = \lim_{h\to 0} \frac{\frac{1}{(x+h)^2} - \frac{1}{x^2}}{h}$$
$$= \lim_{h\to 0} \frac{x^2 - (x^2 + 2xh + h^2)}{h(x+h)^2 x^2}$$
$$= \lim_{h\to 0} -\frac{2x+h}{(x+h)^2 x^2} = -\frac{2x}{x^4} = -\frac{2}{x^3}$$

41. $f(x) = \sqrt{x}$

$$\lim_{h\to 0} \frac{f(x+h) - f(x)}{h} = \lim_{h\to 0} \frac{\sqrt{x+h} - \sqrt{x}}{h}$$
$$= \lim_{h\to 0} \frac{x+h-x}{h(\sqrt{x+h} + \sqrt{x})}$$
$$= \lim_{h\to 0} \frac{1}{\sqrt{x+h} + \sqrt{x}} = \frac{1}{2\sqrt{x}}$$

42. $f(x) = 1/\sqrt{x}$

$$\lim_{h\to 0} \frac{f(x+h) - f(x)}{h} = \lim_{h\to 0} \frac{\frac{1}{\sqrt{x+h}} - \frac{1}{\sqrt{x}}}{h}$$
$$= \lim_{h\to 0} \frac{\sqrt{x} - \sqrt{x+h}}{h\sqrt{x}\sqrt{x+h}}$$
$$= \lim_{h\to 0} \frac{x - (x+h)}{h\sqrt{x}\sqrt{x+h}(\sqrt{x} + \sqrt{x+h})}$$
$$= \lim_{h\to 0} \frac{-1}{\sqrt{x}\sqrt{x+h}(\sqrt{x} + \sqrt{x+h})}$$
$$= \frac{-1}{2x^{3/2}}$$

43. $\lim_{x\to \pi/2} \sin x = \sin \pi/2 = 1$

44. $\lim_{x\to \pi/4} \cos x = \cos \pi/4 = 1/\sqrt{2}$

45. $\lim_{x\to \pi/3} \cos x = \cos \pi/3 = 1/2$

46. $\lim_{x\to 2\pi/3} \sin x = \sin 2\pi/3 = \sqrt{3}/2$

47.

x	$(\sin x)/x$
± 1.0	0.84147098
± 0.1	0.99833417
± 0.01	0.99998333
± 0.001	0.99999983
0.0001	1.00000000

It appears that $\lim_{x\to 0} \frac{\sin x}{x} = 1$.

48.

x	$(1 - \cos x)/x^2$
± 1.0	0.45969769
± 0.1	0.49958347
± 0.01	0.49999583
± 0.001	0.49999996
0.0001	0.50000000

It appears that $\lim_{x\to 0} \frac{1 - \cos x}{x^2} = \frac{1}{2}$.

49. $\lim_{x\to 2-} \sqrt{2-x} = 0$

50. $\lim_{x\to 2+} \sqrt{2-x}$ does not exist.

51. $\lim_{x\to -2-} \sqrt{2-x} = 2$

52. $\lim_{x\to -2+} \sqrt{2-x} = 2$

53. $\lim_{x\to 0} \sqrt{x^3 - x}$ does not exist.
$(x^3 - x < 0$ if $0 < x < 1)$

54. $\lim_{x\to 0-} \sqrt{x^3 - x} = 0$

55. $\lim_{x\to 0+} \sqrt{x^3 - x}$ does not exist. (See # 9.)

56. $\lim_{x\to 0+} \sqrt{x^2 - x^4} = 0$

57. $\lim_{x\to a-} \frac{|x-a|}{x^2 - a^2}$
$= \lim_{x\to a-} \frac{|x-a|}{(x-a)(x+a)} = -\frac{1}{2a} \quad (a \neq 0)$

58. $\lim_{x\to a+} \frac{|x-a|}{x^2 - a^2} = \lim_{x\to a+} \frac{x-a}{x^2 - a^2} = \frac{1}{2a}$

59. $\lim_{x\to 2-} \frac{x^2 - 4}{|x+2|} = \frac{0}{4} = 0$

60. $\lim_{x\to 2+} \frac{x^2 - 4}{|x+2|} = \frac{0}{4} = 0$

61. $f(x) = \begin{cases} x - 1 & \text{if } x \leq -1 \\ x^2 + 1 & \text{if } -1 < x \leq 0 \\ (x + \pi)^2 & \text{if } x > 0 \end{cases}$
$\lim_{x\to -1-} f(x) = \lim_{x\to -1-} x - 1 = -1 - 1 = -2$

62. $\lim_{x\to -1+} f(x) = \lim_{x\to -1+} x^2 + 1 = 1 + 1 = 2$

63. $\lim_{x\to 0+} f(x) = \lim_{x\to 0+} (x + \pi)^2 = \pi^2$

64. $\lim_{x\to 0-} f(x) = \lim_{x\to 0-} x^2 + 1 = 1$

65. If $\lim_{x\to 4} f(x) = 2$ and $\lim_{x\to 4} g(x) = -3$, then

a) $\lim_{x\to 4} (g(x) + 3) = -3 + 3 = 0$

INSTRUCTOR'S SOLUTIONS MANUAL SECTION 1.2 (PAGE 68)

b) $\lim_{x \to 4} xf(x) = 4 \times 2 = 8$

c) $\lim_{x \to 4} \bigl(g(x)\bigr)^2 = (-3)^2 = 9$

d) $\lim_{x \to 4} \dfrac{g(x)}{f(x) - 1} = \dfrac{-3}{2 - 1} = -3$

66. If $\lim_{x \to a} f(x) = 4$ and $\lim_{x \to a} g(x) = -2$, then

a) $\lim_{x \to a} \bigl(f(x) + g(x)\bigr) = 4 + (-2) = 2$

b) $\lim_{x \to a} f(x) \cdot g(x) = 4 \times (-2) = -8$

c) $\lim_{x \to a} 4g(x) = 4(-2) = -8$

d) $\lim_{x \to a} \dfrac{f(x)}{g(x)} = \dfrac{4}{-2} = -2$

67. If $\lim_{x \to 2} \dfrac{f(x) - 5}{x - 2} = 3$, then

$$\lim_{x \to 2}\bigl(f(x) - 5\bigr) = \lim_{x \to 2} \dfrac{f(x) - 5}{x - 2}(x - 2) = 3(2 - 2) = 0.$$

Thus $\lim_{x \to 2} f(x) = 5$.

68. If $\lim_{x \to 0} \dfrac{f(x)}{x^2} = -2$ then

$\lim_{x \to 0} f(x) = \lim_{x \to 0} x^2 \dfrac{f(x)}{x^2} = 0 \times (-2) = 0$,

and similarly, $\lim_{x \to 0} \dfrac{f(x)}{x} = \lim_{x \to 0} x \dfrac{f(x)}{x^2} = 0 \times (-2) = 0.$

69.

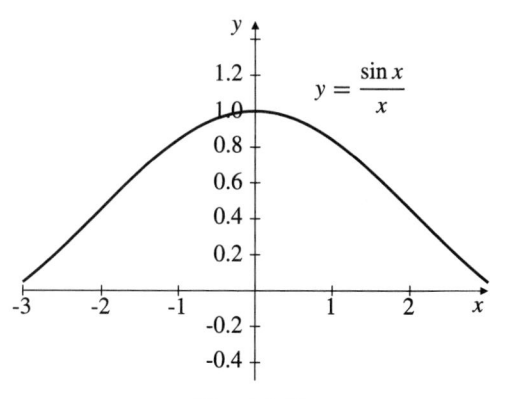

Fig. 1.2.69

$\lim_{x \to 0} \dfrac{\sin x}{x} = 1$

70.

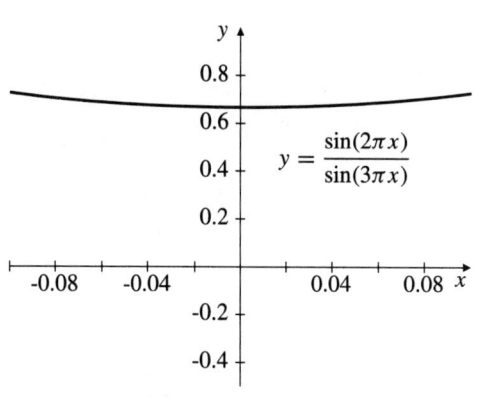

Fig. 1.2.70

$\lim_{x \to 0} \sin(2\pi x)/\sin(3\pi x) = 2/3$

71.

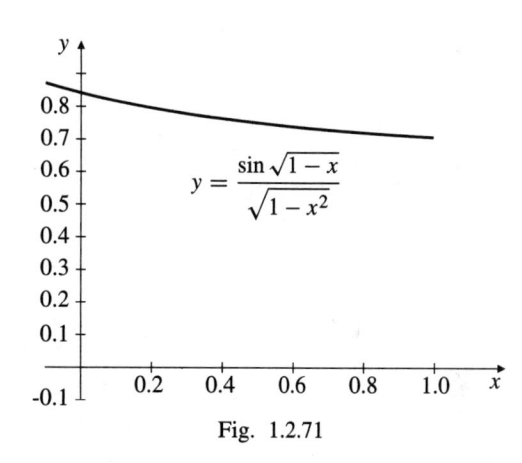

Fig. 1.2.71

$\lim_{x \to 1-} \dfrac{\sin\sqrt{1 - x}}{\sqrt{1 - x^2}} \approx 0.7071$

72.

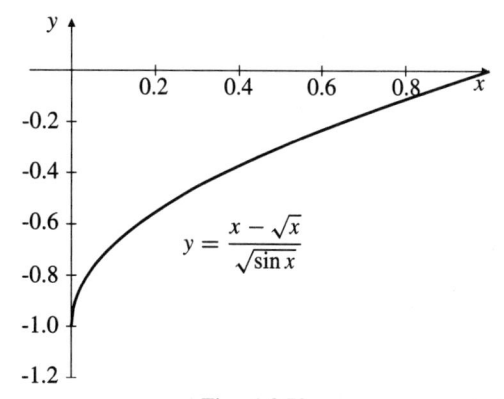

Fig. 1.2.72

$\lim_{x \to 0+} \dfrac{x - \sqrt{x}}{\sqrt{\sin x}} = -1$

23

SECTION 1.2 (PAGE 68)

73.

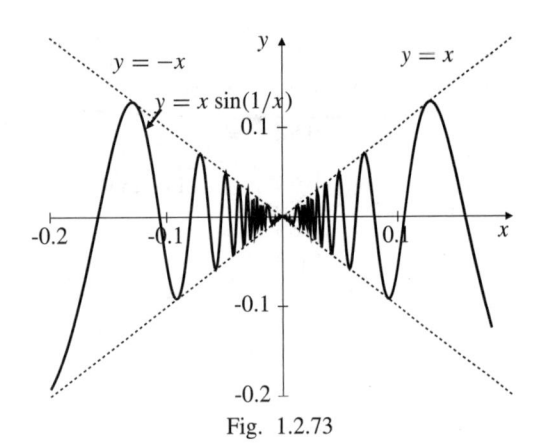

Fig. 1.2.73

$f(x) = x \sin(1/x)$ oscillates infinitely often as x approaches 0, but the amplitude of the oscillations decreases and, in fact, $\lim_{x \to 0} f(x) = 0$. This is predictable because $|x \sin(1/x)| \le |x|$. (See Exercise 95 below.)

74. Since $\sqrt{5 - 2x^2} \le f(x) \le \sqrt{5 - x^2}$ for $-1 \le x \le 1$, and $\lim_{x \to 0} \sqrt{5 - 2x^2} = \lim_{x \to 0} \sqrt{5 - x^2} = \sqrt{5}$, we have $\lim_{x \to 0} f(x) = \sqrt{5}$ by the squeeze theorem.

75. Since $2 - x^2 \le g(x) \le 2 \cos x$ for all x, and since $\lim_{x \to 0} (2 - x^2) = \lim_{x \to 0} 2 \cos x = 2$, we have $\lim_{x \to 0} g(x) = 2$ by the squeeze theorem.

76. a)

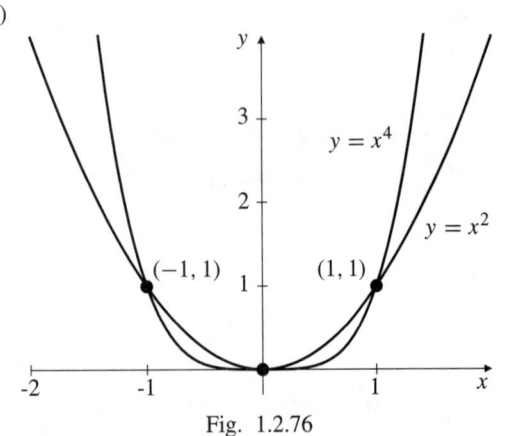

Fig. 1.2.76

b) Since the graph of f lies between those of x^2 and x^4, and since these latter graphs come together at $(\pm 1, 1)$ and at $(0, 0)$, we have $\lim_{x \to \pm 1} f(x) = 1$ and $\lim_{x \to 0} f(x) = 0$ by the squeeze theorem.

77. $x^{1/3} < x^3$ on $(-1, 0)$ and $(1, \infty)$. $x^{1/3} > x^3$ on $(-\infty, -1)$ and $(0, 1)$. The graphs of $x^{1/3}$ and x^3 intersect at $(-1, -1)$, $(0, 0)$, and $(1, 1)$. If the graph of $h(x)$ lies between those of $x^{1/3}$ and x^3, then we can determine $\lim_{x \to a} h(x)$ for $a = -1$, $a = 0$, and $a = 1$ by the squeeze theorem. In fact

$$\lim_{x \to -1} h(x) = -1, \quad \lim_{x \to 0} h(x) = 0, \quad \lim_{x \to 1} h(x) = 1.$$

78. $f(x) = s \sin \dfrac{1}{x}$ is defined for all $x \ne 0$; its domain is $(-\infty, 0) \cup (0, \infty)$. Since $|\sin t| \le 1$ for all t, we have $|f(x)| \le |x|$ and $-|x| \le f(x) \le |x|$ for all $x \ne 0$. Since $\lim_{x \to 0} = (-|x|) = 0 = \lim_{x \to 0} |x|$, we have $\lim_{x \to 0} f(x) = 0$ by the squeeze theorem.

79. $|f(x)| \le g(x) \Rightarrow -g(x) \le f(x) \le g(x)$
Since $\lim_{x \to a} g(x) = 0$, therefore $0 \le \lim_{x \to a} f(x) \le 0$.
Hence, $\lim_{x \to a} f(x) = 0$.
If $\lim_{x \to a} g(x) = 3$, then either $-3 \le \lim_{x \to a} f(x) \le 3$ or $\lim_{x \to a} f(x)$ does not exist.

Section 1.3 Limits at Infinity and Infinite Limits (page 74)

1. $\lim\limits_{x \to \infty} \dfrac{x}{2x - 3} = \lim\limits_{x \to \infty} \dfrac{1}{2 - (3/x)} = \dfrac{1}{2}$

2. $\lim\limits_{x \to \infty} \dfrac{x}{x^2 - 4} = \lim\limits_{x \to \infty} \dfrac{1/x}{1 - (4/x^2)} = \dfrac{0}{1} = 0$

3. $\lim\limits_{x \to \infty} \dfrac{3x^3 - 5x^2 + 7}{8 + 2x - 5x^3}$
$= \lim\limits_{x \to \infty} \dfrac{3 - \dfrac{5}{x} + \dfrac{7}{x^3}}{\dfrac{8}{x^3} + \dfrac{2}{x^2} - 5} = -\dfrac{3}{5}$

4. $\lim\limits_{x \to -\infty} \dfrac{x^2 - 2}{x - x^2}$
$= \lim\limits_{x \to -\infty} \dfrac{1 - \dfrac{2}{x^2}}{\dfrac{1}{x} - 1} = \dfrac{1}{-1} = -1$

5. $\lim\limits_{x \to -\infty} \dfrac{x^2 + 3}{x^3 + 2} = \lim\limits_{x \to -\infty} \dfrac{\dfrac{1}{x} + \dfrac{3}{x^3}}{1 + \dfrac{2}{x^3}} = 0$

6. $\lim\limits_{x \to \infty} \dfrac{x^2 + \sin x}{x^2 + \cos x} = \lim\limits_{x \to \infty} \dfrac{1 + \dfrac{\sin x}{x^2}}{1 + \dfrac{\cos x}{x^2}} = \dfrac{1}{1} = 1$

We have used the fact that $\lim_{x \to \infty} \dfrac{\sin x}{x^2} = 0$ (and similarly for cosine) because the numerator is bounded while the denominator grows large.

7. $\lim\limits_{x \to \infty} \dfrac{3x + 2\sqrt{x}}{1 - x}$
$= \lim\limits_{x \to \infty} \dfrac{3 + \dfrac{2}{\sqrt{x}}}{\dfrac{1}{x} - 1} = -3$

24

8. $\lim\limits_{x\to\infty} \dfrac{2x-1}{\sqrt{3x^2+x+1}}$

$= \lim\limits_{x\to\infty} \dfrac{x\left(2-\dfrac{1}{x}\right)}{|x|\sqrt{3+\dfrac{1}{x}+\dfrac{1}{x^2}}}$ (but $|x|=x$ as $x\to\infty$)

$= \lim\limits_{x\to\infty} \dfrac{2-\dfrac{1}{x}}{\sqrt{3+\dfrac{1}{x}+\dfrac{1}{x^2}}} = \dfrac{2}{\sqrt{3}}$

9. $\lim\limits_{x\to-\infty} \dfrac{2x-1}{\sqrt{3x^2+x+1}}$

$= \lim\limits_{x\to-\infty} \dfrac{2-\dfrac{1}{x}}{-\sqrt{3+\dfrac{1}{x}+\dfrac{1}{x^2}}} = -\dfrac{2}{\sqrt{3}}$,

because $x\to-\infty$ implies that $x<0$ and so $\sqrt{x^2}=-x$.

10. $\lim\limits_{x\to-\infty} \dfrac{2x-5}{|3x+2|} = \lim\limits_{x\to-\infty} \dfrac{2x-5}{-(3x+2)} = -\dfrac{2}{3}$

11. $\lim\limits_{x\to 3} \dfrac{1}{3-x}$ does not exist.

12. $\lim\limits_{x\to 3} \dfrac{1}{(3-x)^2} = \infty$

13. $\lim\limits_{x\to 3-} \dfrac{1}{3-x} = \infty$

14. $\lim\limits_{x\to 3+} \dfrac{1}{3-x} = -\infty$

15. $\lim\limits_{x\to -5/2} \dfrac{2x+5}{5x+2} = \dfrac{0}{-\dfrac{25}{2}+2} = 0$

16. $\lim\limits_{x\to -2/5} \dfrac{2x+5}{5x+2}$ does not exist.

17. $\lim\limits_{x\to -(2/5)-} \dfrac{2x+5}{5x+2} = -\infty$

18. $\lim\limits_{x\to -2/5+} \dfrac{2x+5}{5x+2} = \infty$

19. $\lim\limits_{x\to 1-} \dfrac{2x+3}{x^2+x-2} = \lim\limits_{x\to 1-} \dfrac{2x+3}{(x-1)(x+2)} = -\infty$

20. $\lim\limits_{x\to 1+} \dfrac{2x+3}{x^2+x-2} = \lim\limits_{x\to 1+} \dfrac{2x+3}{(x+2)(x-1)} = \infty$

21. $\lim\limits_{x\to 2+} \dfrac{x}{(2-x)^3} = -\infty$

22. $\lim\limits_{x\to 1-} \dfrac{x}{\sqrt{1-x^2}} = \infty$

23. $\lim\limits_{x\to 1+} \dfrac{1}{|x-1|} = \infty$

24. $\lim\limits_{x\to 1-} \dfrac{1}{|x-1|} = \infty$

25. $\lim\limits_{x\to 2} \dfrac{x-3}{x^2-4x+4} = \lim\limits_{x\to 2} \dfrac{x-3}{(x-2)^2} = -\infty$

26. $\lim\limits_{x\to 1+} \dfrac{\sqrt{x^2-x}}{x-x^2} = \lim\limits_{x\to 1+} \dfrac{-1}{\sqrt{x^2-x}} = -\infty$

27. $\lim\limits_{x\to\infty} \dfrac{x+x^3+x^5}{1+x^2+x^3}$

$= \lim\limits_{x\to\infty} \dfrac{\dfrac{1}{x^2}+1+x^2}{\dfrac{1}{x^3}+\dfrac{1}{x}+1} = \infty$

28. $\lim\limits_{x\to\infty} \dfrac{x^3+3}{x^2+2} = \lim\limits_{x\to\infty} \dfrac{x+\dfrac{3}{x^2}}{1+\dfrac{2}{x^2}} = \infty$

29. $\lim\limits_{x\to\infty} \dfrac{x\sqrt{x+1}\,(1-\sqrt{2x+3})}{7-6x+4x^2}$

$= \lim\limits_{x\to\infty} \dfrac{x^2\left(\sqrt{1+\dfrac{1}{x}}\right)\left(\dfrac{1}{\sqrt{x}}-\sqrt{2+\dfrac{3}{x}}\right)}{x^2\left(\dfrac{7}{x^2}-\dfrac{6}{x}+4\right)}$

$= \dfrac{1(-\sqrt{2})}{4} = -\dfrac{1}{4}\sqrt{2}$

30. $\lim\limits_{x\to\infty}\left(\dfrac{x^2}{x+1}-\dfrac{x^2}{x-1}\right) = \lim\limits_{x\to\infty}\dfrac{-2x^2}{x^2-1} = -2$

31. $\lim\limits_{x\to-\infty}\left(\sqrt{x^2+2x}-\sqrt{x^2-2x}\right)$

$= \lim\limits_{x\to-\infty}\dfrac{(x^2+2x)-(x^2-2x)}{\sqrt{x^2+2x}+\sqrt{x^2-2x}}$

$= \lim\limits_{x\to-\infty}\dfrac{4x}{(-x)\left(\sqrt{1+\dfrac{2}{x}}+\sqrt{1-\dfrac{2}{x}}\right)}$

$= -\dfrac{4}{1+1} = -2$

32. $\lim\limits_{x\to\infty}\left(\sqrt{x^2+2x}-\sqrt{x^2-2x}\right)$

$= \lim\limits_{x\to\infty}\dfrac{x^2+2x-x^2+2x}{\sqrt{x^2+2x}+\sqrt{x^2-2x}}$

$= \lim\limits_{x\to\infty}\dfrac{4x}{x\sqrt{1+\dfrac{2}{x}}+x\sqrt{1-\dfrac{2}{x}}}$

$= \lim\limits_{x\to\infty}\dfrac{4}{\sqrt{1+\dfrac{2}{x}}+\sqrt{1-\dfrac{2}{x}}} = \dfrac{4}{2} = 2$

33. $\displaystyle\lim_{x\to\infty}\frac{1}{\sqrt{x^2-2x}-x}$

$\displaystyle = \lim_{x\to\infty}\frac{\sqrt{x^2-2x}+x}{(\sqrt{x^2-2x}+x)(\sqrt{x^2-2x}-x)}$

$\displaystyle = \lim_{x\to\infty}\frac{\sqrt{x^2-2x}+x}{x^2-2x-x^2}$

$\displaystyle = \lim_{x\to\infty}\frac{x(\sqrt{1-(2/x)}+1)}{-2x} = \frac{2}{-2} = -1$

34. $\displaystyle\lim_{x\to-\infty}\frac{1}{\sqrt{x^2+2x}-x} = \lim_{x\to-\infty}\frac{1}{|x|(\sqrt{1+(2/x)}+1)} = 0$

35. By Exercise 35, $y=-1$ is a horizontal asymptote (at the right) of $y=\dfrac{1}{\sqrt{x^2-2x}-x}$. Since

$\displaystyle\lim_{x\to-\infty}\frac{1}{\sqrt{x^2-2x}-x} = \lim_{x\to-\infty}\frac{1}{|x|(\sqrt{1-(2/x)}+1)} = 0,$

$y=0$ is also a horizontal asymptote (at the left). Now $\sqrt{x^2-2x}-x=0$ if and only if $x^2-2x=x^2$, that is, if and only if $x=0$. The given function is undefined at $x=0$, and where $x^2-2x<0$, that is, on the interval $[0,2]$. Its only vertical asymptote is at $x=0$, where $\displaystyle\lim_{x\to 0-}\frac{1}{\sqrt{x^2-2x}-x} = \infty$.

36. Since $\displaystyle\lim_{x\to\infty}\frac{2x-5}{|3x+2|} = \frac{2}{3}$ and $\displaystyle\lim_{x\to-\infty}\frac{2x-5}{|3x+2|} = -\frac{2}{3}$, $y=\pm(2/3)$ are horizontal asymptotes of $y=(2x-5)/|3x+2|$. The only vertical asymptote is $x=-2/3$, which makes the denominator zero.

Exercises 37–48 refer to Figure 1.3.37.

37. $\displaystyle\lim_{x\to 0+} f(x) = 1$

38. $\displaystyle\lim_{x\to 1} f(x) = \infty$

Fig. 1.3.37

39. $\displaystyle\lim_{x\to 2+} f(x) = 1$

40. $\displaystyle\lim_{x\to 2-} f(x) = 2$

41. $\displaystyle\lim_{x\to 3-} f(x) = -\infty$

42. $\displaystyle\lim_{x\to 3+} f(x) = \infty$

43. $\displaystyle\lim_{x\to 4+} f(x) = 2$

44. $\displaystyle\lim_{x\to 4-} f(x) = 0$

45. $\displaystyle\lim_{x\to 5-} f(x) = -1$

46. $\displaystyle\lim_{x\to 5+} f(x) = 0$

47. $\displaystyle\lim_{x\to\infty} f(x) = 1$

48. horizontal: $y=1$; vertical: $x=1,\ x=3$.

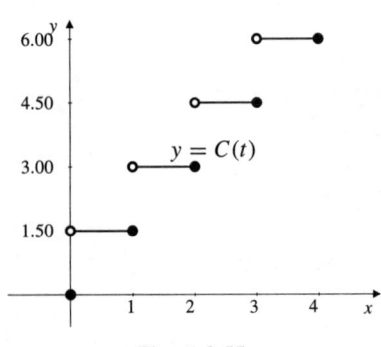

49. $\displaystyle\lim_{x\to 3+}\lfloor x\rfloor = 3$

50. $\displaystyle\lim_{x\to 3-}\lfloor x\rfloor = 2$

51. $\displaystyle\lim_{x\to 3}\lfloor x\rfloor$ does not exist

52. $\displaystyle\lim_{x\to 2.5}\lfloor x\rfloor = 2$

53. $\displaystyle\lim_{x\to 0+}\lfloor 2-x\rfloor = \lim_{x\to 2-}\lfloor x\rfloor = 1$

54. $\displaystyle\lim_{x\to -3-}\lfloor x\rfloor = -4$

55. $\displaystyle\lim_{t\to t_0} C(t) = C(t_0)$ except at integers t_0
$\displaystyle\lim_{t\to t_0-} C(t) = C(t_0)$ everywhere
$\displaystyle\lim_{t\to t_0+} C(t) = C(t_0)$ if $t_0\neq$ an integer
$\displaystyle\lim_{t\to t_0+} C(t) = C(t_0)+1.5$ if t_0 is an integer

Fig. 1.3.55

56. $\displaystyle\lim_{x\to 0+} f(x) = L$
(a) If f is even, then $f(-x)=f(x)$. Hence, $\displaystyle\lim_{x\to 0-} f(x) = L$.
(b) If f is odd, then $f(-x)=-f(x)$. Therefore, $\displaystyle\lim_{x\to 0-} f(x) = -L$.

57. $\lim_{x \to 0+} f(x) = A$, $\lim_{x \to 0-} f(x) = B$

   a) $\lim_{x \to 0+} f(x^3 - x) = B$ (since $x^3 - x < 0$ if $0 < x < 1$)

   b) $\lim_{x \to 0-} f(x^3 - x) = A$ (because $x^3 - x > 0$ if $-1 < x < 0$)

   c) $\lim_{x \to 0-} f(x^2 - x^4) = A$

   d) $\lim_{x \to 0+} f(x^2 - x^4) = A$ (since $x^2 - x^4 > 0$ for $0 < |x| < 1$)

Section 1.4 Continuity (page 85)

1. g is continuous at $x = -2$, discontinuous at $x = -1, 0, 1,$ and 2. It is left continuous at $x = 0$ and right continuous at $x = 1$.

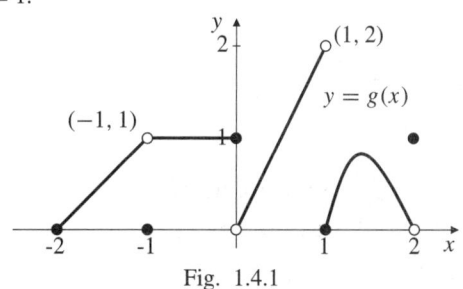

Fig. 1.4.1

2. g has removable discontinuities at $x = -1$ and $x = 2$. Redefine $g(-1) = 1$ and $g(2) = 0$ to make g continuous at those points.

3. g has no absolute maximum value on $[-2, 2]$. It takes on every positive real value less than 2, but does not take the value 2. It has absolute minimum value 0 on that interval, assuming this value at the three points $x = -2$, $x = -1$, and $x = 1$.

4. Function f is discontinuous at $x = 1, 2, 3, 4,$ and 5. f is left continuous at $x = 4$ and right continuous at $x = 2$ and $x = 5$.

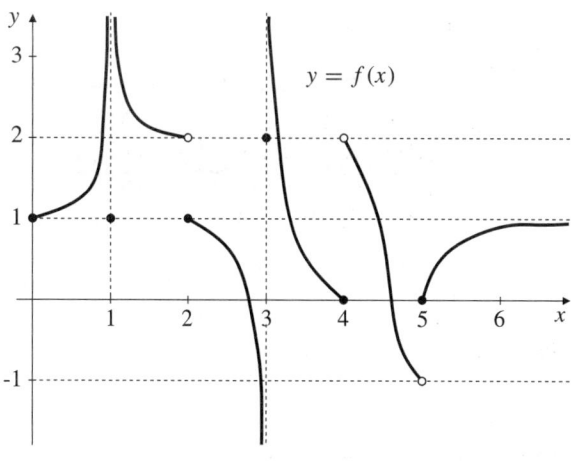

Fig. 1.4.4

5. f cannot be redefined at $x = 1$ to become continuous there because $\lim_{x \to 1} f(x) (= \infty)$ does not exist. (∞ is not a real number.)

6. $\text{sgn } x$ is not defined at $x = 0$, so cannot be either continuous or discontinuous there. (Functions can be continuous or discontinuous only at points in their domains!)

7. $f(x) = \begin{cases} x & \text{if } x < 0 \\ x^2 & \text{if } x \geq 0 \end{cases}$ is continuous everywhere on the real line, even at $x = 0$ where its left and right limits are both 0, which is $f(0)$.

8. $f(x) = \begin{cases} x & \text{if } x < -1 \\ x^2 & \text{if } x \geq -1 \end{cases}$ is continuous everywhere on the real line except at $x = -1$ where it is right continuous, but not left continuous.
$$\lim_{x \to -1-} f(x) = \lim_{x \to -1-} x = -1 \neq 1$$
$$= f(-1) = \lim_{x \to -1+} x^2 = \lim_{x \to -1+} f(x).$$

9. $f(x) = \begin{cases} 1/x^2 & \text{if } x \neq 0 \\ 0 & \text{if } x = 0 \end{cases}$ is continuous everywhere except at $x = 0$, where it is neither left nor right continuous since it does not have a real limit there.

10. $f(x) = \begin{cases} x^2 & \text{if } x \leq 1 \\ 0.987 & \text{if } x > 1 \end{cases}$ is continuous everywhere except at $x = 1$, where it is left continuous but not right continuous because $0.987 \neq 1$. Close, as they say, but no cigar.

11. The parking lot cost function $C(t)$ is continuous everywhere on $[0, \infty)$ except at the integers $t = 0, 1, \ldots$ where it is left continuous but not right continuous.

12. $C(t)$ is discontinuous only at the integers. It is continuous on the left at the integers, but not on the right.

13. Since $\dfrac{x^2 - 4}{x - 2} = x + 2$ for $x \neq 2$, we can define the function to be $2 + 2 = 4$ at $x = 2$ to make it continuous there. The continuous extension is $x + 2$.

14. Since $\dfrac{1+t^3}{1-t^2} = \dfrac{(1+t)(1-t+t^2)}{(1+t)(1-t)} = \dfrac{1-t+t^2}{1-t}$ for $t \neq -1$, we can define the function to be $3/2$ at $t = -1$ to make it continuous there. The continuous extension is $\dfrac{1-t+t^2}{1-t}$.

15. Since $\dfrac{t^2 - 5t + 6}{t^2 - t - 6} = \dfrac{(t-2)(t-3)}{(t+2)(t-3)} = \dfrac{t-2}{t+2}$ for $t \neq 3$, we can define the function to be $1/5$ at $t = 3$ to make it continuous there. The continuous extension is $\dfrac{t-2}{t+2}$.

16. Since
$$\dfrac{x^2 - 2}{x^4 - 4} = \dfrac{(x-\sqrt{2})(x+\sqrt{2})}{(x-\sqrt{2})(x+\sqrt{2})(x^2+2)} = \dfrac{x+\sqrt{2}}{(x+\sqrt{2})(x^2+2)}$$
for $x \neq \sqrt{2}$, we can define the function to be $1/4$ at $x = \sqrt{2}$ to make it continuous there. The continuous extension is $\dfrac{x+\sqrt{2}}{(x+\sqrt{2})(x^2+2)}$. (Note: cancelling the $x+\sqrt{2}$ factors provides a further continuous extension to $x = -\sqrt{2}$.

17. $\lim_{x\to 2+} f(x) = k - 4$ and $\lim_{x\to 2-} f(x) = 4 = f(2)$. Thus f will be continuous at $x = 2$ if $k - 4 = 4$, that is, if $k = 8$.

18. $\lim_{x\to 3-} g(x) = 3 - m$ and $\lim_{x\to 3+} g(x) = 1 - 3m = g(3)$. Thus g will be continuous at $x = 3$ if $3 - m = 1 - 3m$, that is, if $m = -1$.

19. x^2 has no maximum value on $-1 < x < 1$; it takes all positive real values less than 1, but it does not take the value 1. It does have a minimum value, namely 0 taken on at $x = 0$.

20. The Max-Min Theorem says that a continuous function defined on a closed, finite interval must have maximum and minimum values. It does not say that other functions cannot have such values. The Heaviside function is not continuous on $[-1, 1]$ (because it is discontinuous at $x = 0$), but it still has maximum and minimum values. Do not confuse a theorem with its converse.

21. Let the numbers be x and y, where $x \geq 0$, $y \geq 0$, and $x + y = 8$. If P is the product of the numbers, then
$$P = xy = x(8-x) = 8x - x^2 = 16 - (x-4)^2.$$
Therefore $P \leq 16$, so P is bounded. Clearly $P = 16$ if $x = y = 4$, so the largest value of P is 16.

22. Let the numbers be x and y, where $x \geq 0$, $y \geq 0$, and $x + y = 8$. If S is the sum of their squares then
$$S = x^2 + y^2 = x^2 + (8-x)^2$$
$$= 2x^2 - 16x + 64 = 2(x-4)^2 + 32.$$

Since $0 \leq x \leq 8$, the maximum value of S occurs at $x = 0$ or $x = 8$, and is 64. The minimum value occurs at $x = 4$ and is 32.

23. Since $T = 100 - 30x + 3x^2 = 3(x-5)^2 + 25$, T will be minimum when $x = 5$. Five programmers should be assigned, and the project will be completed in 25 days.

24. If x desks are shipped, the shipping cost per desk is
$$C = \dfrac{245x - 30x^2 + x^3}{x} = x^2 - 30x + 245$$
$$= (x-15)^2 + 20.$$
This cost is minimized if $x = 15$. The manufacturer should send 15 desks in each shipment, and the shipping cost will then be $20 per desk.

25. $f(x) = \dfrac{x^2 - 1}{x} = \dfrac{(x-1)(x+1)}{x}$
$f = 0$ at $x = \pm 1$. f is not defined at 0.
$f(x) > 0$ on $(-1, 0)$ and $(1, \infty)$.
$f(x) < 0$ on $(-\infty, -1)$ and $(0, 1)$.

26. $f(x) = x^2 + 4x + 3 = (x+1)(x+3)$
$f(x) > 0$ on $(-\infty, -3)$ and $(-1, \infty)$
$f(x) < 0$ on $(-3, -1)$.

27. $f(x) = \dfrac{x^2 - 1}{x^2 - 4} = \dfrac{(x-1)(x+1)}{(x-2)(x+2)}$
$f = 0$ at $x = \pm 1$.
f is not defined at $x = \pm 2$.
$f(x) > 0$ on $(-\infty, -2)$, $(-1, 1)$, and $(2, \infty)$.
$f(x) < 0$ on $(-2, -1)$ and $(1, 2)$.

28. $f(x) = \dfrac{x^2 + x - 2}{x^3} = \dfrac{(x+2)(x-1)}{x^3}$
$f(x) > 0$ on $(-2, 0)$ and $(1, \infty)$
$f(x) < 0$ on $(-\infty, -2)$ and $(0, 1)$.

29. $f(x) = x^3 + x - 1$, $f(0) = -1$, $f(1) = 1$.
Since f is continuous and changes sign between 0 and 1, it must be zero at some point between 0 and 1 by IVT.

30. $f(x) = x^3 - 15x + 1$ is continuous everywhere.
$f(-4) = -3$, $f(-3) = 19$, $f(1) = -13$, $f(4) = 5$.
Because of the sign changes f has a zero between -4 and -3, another zero between -3 and 1, and another between 1 and 4.

31. $F(x) = (x-a)^2(x-b)^2 + x$. Without loss of generality, we can assume that $a < b$. Being a polynomial, F is continuous on $[a, b]$. Also $F(a) = a$ and $F(b) = b$. Since $a < \tfrac{1}{2}(a+b) < b$, the Intermediate-Value Theorem guarantees that there is an x in (a, b) such that $F(x) = (a+b)/2$.

32. Let $g(x) = f(x) - x$. Since $0 \leq f(x) \leq 1$ if $0 \leq x \leq 1$, therefore, $g(0) \geq 0$ and $g(1) \leq 0$. If $g(0) = 0$ let $c = 0$, or if $g(1) = 0$ let $c = 1$. (In either case $f(c) = c$.) Otherwise, $g(0) > 0$ and $g(1) < 0$, and, by IVT, there exists c in $(0, 1)$ such that $g(c) = 0$, i.e., $f(c) = c$.

33. The domain of an even function is symmetric about the y-axis. Since f is continuous on the right at $x = 0$, therefore it must be defined on an interval $[0, h]$ for some $h > 0$. Being even, f must therefore be defined on $[-h, h]$. If $x = -y$, then
$$\lim_{x \to 0-} f(x) = \lim_{y \to 0+} f(-y) = \lim_{y \to 0+} f(y) = f(0).$$
Thus, f is continuous on the left at $x = 0$. Being continuous on both sides, it is therefore continuous.

34. f odd $\Leftrightarrow f(-x) = -f(x)$
f continuous on the right $\Leftrightarrow \lim_{x \to 0+} f(x) = f(0)$
Therefore, letting $t = -x$, we obtain
$$\lim_{x \to 0-} f(x) = \lim_{t \to 0+} f(-t) = \lim_{t \to 0+} -f(t)$$
$$= -f(0) = f(-0) = f(0).$$
Therefore f is continuous at 0 and $f(0) = 0$.

35. max 1.594 at $x = -0.832$; min -0.746 at $x = 0.629$

36. max 0.133 at $x = 1.437$; min -0.232 at $x = -1.805$

37. max 10.333 at $x = 3$; min 4.762 at $x = 1.260$

38. max 1.510 at $x = 0.465$; min 0 at $x = 0$ and $x = 1$

39. root $x = 0.682$

40. root $x = 0.739$

41. roots $x = -0.637$ and $x = 1.410$

42. roots $x = -0.725$ and $x = 1.221$

Section 1.5 The Formal Definition of Limit (page 91)

1. We require $39.9 \le L \le 40.1$. Thus
$$39.9 \le 39.6 + 0.025T \le 40.1$$
$$0.3 \le 0.025T \le 0.5$$
$$12 \le T \le 20.$$
The temperature should be kept between 12°C and 20°C.

2. Since 1.2% of 8,000 is 96, we require the edge length x of the cube to satisfy $7904 \le x^3 \le 8096$. It is sufficient that $19.920 \le x \le 20.079$. The edge of the cube must be within 0.079 cm of 20 cm.

3. $3 - 0.02 \le 2x - 1 \le 3 + 0.02$
$3.98 \le 2x \le 4.02$
$1.99 \le x \le 2.01$

4. $4 - 0.1 \le x^2 \le 4 + 0.1$
$1.9749 \le x \le 2.0024$

5. $1 - 0.1 \le \sqrt{x} \le 1.1$
$0.81 \le x \le 1.21$

6. $-2 - 0.01 \le \dfrac{1}{x} \le -2 + 0.01$
$-\dfrac{1}{2.01} \ge x \ge -\dfrac{1}{1.99}$
$-0.5025 \le x \le -0.4975$

7. We need $-0.03 \le (3x + 1) - 7 \le 0.03$, which is equivalent to $-0.01 \le x - 2 \le 0.01$ Thus $\delta = 0.01$ will do.

8. We need $-0.01 \le \sqrt{2x + 3} - 3 \le 0.01$. Thus
$$2.99 \le \sqrt{2x + 3} \le 3.01$$
$$8.9401 \le 2x + 3 \le 9.0601$$
$$2.97005 \le x \le 3.03005$$
$$3 - 0.02995 \le x - 3 \le 0.03005.$$
Here $\delta = 0.02995$ will do.

9. We need $8 - 0.2 \le x^3 \le 8.2$, or $1.9832 \le x \le 2.0165$. Thus, we need $-0.0168 \le x - 2 \le 0.0165$. Here $\delta = 0.0165$ will do.

10. We need $1 - 0.05 \le 1/(x + 1) \le 1 + 0.05$, or $1.0526 \ge x + 1 \ge 0.9524$. This will occur if $-0.0476 \le x \le 0.0526$. In this case we can take $\delta = 0.0476$.

11. To be proved: $\lim_{x \to 1}(3x + 1) = 4$.
Proof: Let $\epsilon > 0$ be given. Then $|(3x + 1) - 4| < \epsilon$ holds if $3|x - 1| < \epsilon$, and so if $|x - 1| < \delta = \epsilon/3$. This confirms the limit.

12. To be proved: $\lim_{x \to 2}(5 - 2x) = 1$.
Proof: Let $\epsilon > 0$ be given. Then $|(5 - 2x) - 1| < \epsilon$ holds if $|2x - 4| < \epsilon$, and so if $|x - 2| < \delta = \epsilon/2$. This confirms the limit.

13. To be proved: $\lim_{x \to 0} x^2 = 0$.
Let $\epsilon > 0$ be given. Then $|x^2 - 0| < \epsilon$ holds if $|x - 0| = |x| < \delta = \sqrt{\epsilon}$.

14. To be proved: $\lim_{x \to 2} \dfrac{x - 2}{1 + x^2} = 0$.
Proof: Let $\epsilon > 0$ be given. Then
$$\left| \dfrac{x - 2}{1 + x^2} - 0 \right| = \dfrac{|x - 2|}{1 + x^2} \le |x - 2| < \epsilon$$
provided $|x - 2| < \delta = \epsilon$.

15. To be proved: $\lim_{x \to 1/2} \dfrac{1 - 4x^2}{1 - 2x} = 2$.
Proof: Let $\epsilon > 0$ be given. Then if $x \ne 1/2$ we have
$$\left| \dfrac{1 - 4x^2}{1 - 2x} - 2 \right| = |(1 + 2x) - 2| = |2x - 1| = 2\left|x - \dfrac{1}{2}\right| < \epsilon$$
provided $|x - \tfrac{1}{2}| < \delta = \epsilon/2$.

16. To be proved: $\lim_{x\to -2} \dfrac{x^2+2x}{x+2} = -2$.
Proof: Let $\epsilon > 0$ be given. For $x \neq -2$ we have
$$\left|\dfrac{x^2+2x}{x+2} - (-2)\right| = |x+2| < \epsilon$$
provided $|x+2| < \delta = \epsilon$. This completes the proof.

17. To be proved: $\lim_{x\to 1} \dfrac{1}{x+1} = \dfrac{1}{2}$.
Proof: Let $\epsilon > 0$ be given. We have
$$\left|\dfrac{1}{x+1} - \dfrac{1}{2}\right| = \left|\dfrac{1-x}{2(x+1)}\right| = \dfrac{|x-1|}{2|x+1|}.$$
If $|x-1| < 1$, then $0 < x < 2$ and $1 < x+1 < 3$, so that $|x+1| > 1$. Let $\delta = \min(1, 2\epsilon)$. If $|x-1| < \delta$, then
$$\left|\dfrac{1}{x+1} - \dfrac{1}{2}\right| = \dfrac{|x-1|}{2|x+1|} < \dfrac{2\epsilon}{2} = \epsilon.$$
This establishes the required limit.

18. To be proved: $\lim_{x\to -1} \dfrac{x+1}{x^2-1} = -\dfrac{1}{2}$.
Proof: Let $\epsilon > 0$ be given. If $x \neq -1$, we have
$$\left|\dfrac{x+1}{x^2-1} - \dfrac{1}{2}\right| = \left|\dfrac{1}{x-1} - \left(-\dfrac{1}{2}\right)\right| = \dfrac{|x+1|}{2|x-1|}.$$
If $|x+1| < 1$, then $-2 < x < 0$, so $-3 < x-1 < -1$ and $|x-1| > 1$. Ler $\delta = \min(1, 2\epsilon)$. If $0 < |x-(-1)| < \delta$ then $|x-1| > 1$ and $|x+1| < 2\epsilon$. Thus
$$\left|\dfrac{x+1}{x^2-1} - \dfrac{1}{2}\right| = \dfrac{|x+1|}{2|x-1|} < \dfrac{2\epsilon}{2} = \epsilon.$$
This completes the required proof.

19. To be proved: $\lim_{x\to 1} \sqrt{x} = 1$.
Proof: Let $\epsilon > 0$ be given. We have
$$|\sqrt{x} - 1| = \left|\dfrac{x-1}{\sqrt{x}+1}\right| \leq |x-1| < \epsilon$$
provided $|x-1| < \delta = \epsilon$. This completes the proof.

20. To be proved: $\lim_{x\to 2} x^3 = 8$.
Proof: Let $\epsilon > 0$ be given. We have $|x^3-8| = |x-2||x^2+2x+4|$. If $|x-2| < 1$, then $1 < x < 3$ and $x^2 < 9$. Therefore $|x^2+2x+4| \leq 9+2\times 3+4 = 19$. If $|x-2| < \delta = \min(1, \epsilon/19)$, then
$$|x^3-8| = |x-2||x^2+2x+4| < \dfrac{\epsilon}{19} \times 19 = \epsilon.$$
This completes the proof.

21. We say that $\lim_{x\to a-} f(x) = L$ if the following condition holds: for every number $\epsilon > 0$ there exists a number $\delta > 0$, depending on ϵ, such that
$$a-\delta < x < a \quad \text{implies} \quad |f(x) - L| < \epsilon.$$

22. We say that $\lim_{x\to -\infty} f(x) = L$ if the following condition holds: for every number $\epsilon > 0$ there exists a number $R > 0$, depending on ϵ, such that
$$x < -R \quad \text{implies} \quad |f(x) - L| < \epsilon.$$

23. We say that $\lim_{x\to a} f(x) = -\infty$ if the following condition holds: for every number $B > 0$ there exists a number $\delta > 0$, depending on B, such that
$$0 < |x-a| < \delta \quad \text{implies} \quad f(x) < -B.$$

24. We say that $\lim_{x\to \infty} f(x) = \infty$ if the following condition holds: for every number $B > 0$ there exists a number $R > 0$, depending on B, such that
$$x > R \quad \text{implies} \quad f(x) > B.$$

25. We say that $\lim_{x\to a+} f(x) = -\infty$ if the following condition holds: for every number $B > 0$ there exists a number $\delta > 0$, depending on R, such that
$$a < x < a+\delta \quad \text{implies} \quad f(x) < -B.$$

26. We say that $\lim_{x\to a-} f(x) = \infty$ if the following condition holds: for every number $B > 0$ there exists a number $\delta > 0$, depending on B, such that
$$a-\delta < x < a \quad \text{implies} \quad f(x) > B.$$

27. To be proved: $\lim_{x\to 1+} \dfrac{1}{x-1} = \infty$. Proof: Let $B > 0$ be given. We have $\dfrac{1}{x-1} > B$ if $0 < x-1 < 1/B$, that is, if $1 < x < 1+\delta$, where $\delta = 1/B$. This completes the proof.

28. To be proved: $\lim_{x\to 1-} \dfrac{1}{x-1} = -\infty$. Proof: Let $B > 0$ be given. We have $\dfrac{1}{x-1} < -B$ if $0 > x-1 > -1/B$, that is, if $1-\delta < x < 1$, where $\delta = 1/B$.. This completes the proof.

29. To be proved: $\lim_{x\to \infty} \dfrac{1}{\sqrt{x^2+1}} = 0$. Proof: Let $\epsilon > 0$ be given. We have
$$\left|\dfrac{1}{\sqrt{x^2+1}}\right| = \dfrac{1}{\sqrt{x^2+1}} < \dfrac{1}{x} < \epsilon$$

provided $x > R$, where $R = 1/\epsilon$. This completes the proof.

30. To be proved: $\lim_{x \to \infty} \sqrt{x} = \infty$. Proof: Let $B > 0$ be given. We have $\sqrt{x} > B$ if $x > R$ where $R = B^2$. This completes the proof.

31. To be proved: if $\lim_{x \to a} f(x) = L$ and $\lim_{x \to a} f(x) = M$, then $L = M$.
Proof: Suppose $L \neq M$. Let $\epsilon = |L - M|/3$. Then $\epsilon > 0$. Since $\lim_{x \to a} f(x) = L$, there exists $\delta_1 > 0$ such that $|f(x) - L| < \epsilon$ if $|x - a| < \delta_1$. Since $\lim_{x \to a} f(x) = M$, there exists $\delta_2 > 0$ such that $|f(x) - M| < \epsilon$ if $|x - a| < \delta_2$. Let $\delta = \min(\delta_1, \delta_2)$. If $|x - a| < \delta$, then

$$3\epsilon = |L - M| = |(f(x) - M) + (L - f(x))|$$
$$\leq |f(x) - M| + |f(x) - L| < \epsilon + \epsilon = 2\epsilon.$$

This implies that $3 < 2$, a contradiction. Thus the original assumption that $L \neq M$ must be incorrect. Therefore $L = M$.

32. To be proved: if $\lim_{x \to a} g(x) = M$, then there exists $\delta > 0$ such that if $0 < |x - a| < \delta$, then $|g(x)| < 1 + |M|$.
Proof: Taking $\epsilon = 1$ in the definition of limit, we obtain a number $\delta > 0$ such that if $0 < |x - a| < \delta$, then $|g(x) - M| < 1$. It follows from this latter inequality that

$$|g(x)| = |(g(x) - M) + M| \leq |G(x) - M| + |M| < 1 + |M|.$$

33. To be proved: if $\lim_{x \to a} f(x) = L$ and $\lim_{x \to a} f(x) = M$, then $\lim_{x \to a} f(x)g(x) = LM$.
Proof: Let $\epsilon > 0$ be given. Since $\lim_{x \to a} f(x) = L$, there exists $\delta_1 > 0$ such that $|f(x) - L| < \epsilon/(2(1 + |M|))$ if $0 < |x - a| < \delta_1$. Since $\lim_{x \to a} g(x) = M$, there exists $\delta_2 > 0$ such that $|g(x) - M| < \epsilon/(2(1 + |L|))$ if $0 < |x - a| < \delta_2$. By Exercise 32, there exists $\delta_3 > 0$ such that $|g(x)| < 1 + |M|$ if $0 < |x - a| < \delta_3$. Let $\delta = \min(\delta_1, \delta_2, \delta_3)$. If $|x - a| < \delta$, then

$$|f(x)g(x) - LM| = |f(x)g(x) - Lg(x) + Lg(x) - LM|$$
$$= |(f(x) - L)g(x) + L(g(x) - M)|$$
$$\leq |(f(x) - L)g(x)| + |L(g(x) - M)|$$
$$= |f(x) - L||g(x)| + |L||g(x) - M|$$
$$< \frac{\epsilon}{2(1 + |M|)}(1 + |M|) + |L|\frac{\epsilon}{2(1 + |L|)}$$
$$\leq \frac{\epsilon}{2} + \frac{\epsilon}{2} = \epsilon.$$

Thus $\lim_{x \to a} f(x)g(x) = LM$.

34. To be proved: if $\lim_{x \to a} g(x) = M$ where $M \neq 0$, then there exists $\delta > 0$ such that if $0 < |x - a| < \delta$, then $|g(x)| > |M|/2$.
Proof: By the definition of limit, there exists $\delta > 0$ such that if $0 < |x - a| < \delta$, then $|g(x) - M| < |M|/2$ (since $|M|/2$ is a positive number). This latter inequality implies that

$$|M| = |g(x) + (M - g(x))| \leq |g(x)| + |g(x) - M| < |g(x)| + \frac{|M|}{2}.$$

It follows that $|g(x)| > |M| - (|M|/2) = |M|/2$, as required.

35. To be proved: if $\lim_{x \to a} g(x) = M$ where $M \neq 0$, then $\lim_{x \to a} \frac{1}{g(x)} = \frac{1}{M}$.
Proof: Let $\epsilon > 0$ be given. Since $\lim_{x \to a} g(x) = M \neq 0$, there exists $\delta_1 > 0$ such that $|g(x) - M| < \epsilon |M|^2/2$ if $0 < |x - a| < \delta_1$. By Exercise 34, there exists $\delta_2 > 0$ such that $|g(x)| > |M|/2$ if $0 < |x - a| < \delta_3$. Let $\delta = \min(\delta_1, \delta_2)$. If $0 < |x - a| < \delta$, then

$$\left| \frac{1}{g(x)} - \frac{1}{M} \right| = \frac{|M - g(x)|}{|M||g(x)|} < \frac{\epsilon |M|^2}{2} \cdot \frac{2}{|M|^2} = \epsilon.$$

This completes the proof.

36. To be proved: if $\lim_{x \to a} f(x) = L$ and $\lim_{x \to a} f(x) = M \neq 0$, then $\lim_{x \to a} \frac{f(x)}{g(x)} = \frac{L}{M}$.
Proof: By Exercises 33 and 35 we have

$$\lim_{x \to a} \frac{f(x)}{g(x)} = \lim_{x \to a} f(x) \times \frac{1}{g(x)} = L \times \frac{1}{M} = \frac{L}{M}.$$

37. To be proved: if f is continuous at L and $\lim_{x \to c} g(x) = L$, then $\lim_{x \to c} f(g(x)) = f(L)$.
Proof: Let $\epsilon > 0$ be given. Since f is continuous at L, there exists a number $\gamma > 0$ such that if $|y - L| < \gamma$, then $|f(y) - f(L)| < \epsilon$. Since $\lim_{x \to c} g(x) = L$, there exists $\delta > 0$ such that if $0 < |x - c| < \delta$, then $|g(x) - L| < \gamma$. Taking $y = g(x)$, it follows that if $0 < |x - c| < \delta$, then $|f(g(x)) - f(L)| < \epsilon$, so that $\lim_{x \to c} f(g(x)) = f(L)$.

38. To be proved: if $f(x) \le g(x) \le h(x)$ in an open interval containing $x = a$ (say, for $a - \delta_1 < x < a + \delta_1$, where $\delta_1 > 0$), and if $\lim_{x \to a} f(x) = \lim_{x \to a} h(x) = L$, then also $\lim_{x \to a} g(x) = L$.

Proof: Let $\epsilon > 0$ be given. Since $\lim_{x \to a} f(x) = L$, there exists $\delta_2 > 0$ such that if $0 < |x - a| < \delta_2$, then $|f(x) - L| < \epsilon/3$. Since $\lim_{x \to a} h(x) = L$, there exists $\delta_3 > 0$ such that if $0 < |x - a| < \delta_3$, then $|h(x) - L| < \epsilon/3$. Let $\delta = \min(\delta_1, \delta_2, \delta_3)$. If $0 < |x - a| < \delta$, then

$$|g(x) - L| = |g(x) - f(x) + f(x) - L|$$
$$\le |g(x) - f(x)| + |f(x) - L|$$
$$\le |h(x) - f(x)| + |f(x) - L|$$
$$= |h(x) - L + L - f(x)| + |f(x) - L|$$
$$\le |h(x) - L| + |f(x) - L| + |f(x) - L|$$
$$< \frac{\epsilon}{3} + \frac{\epsilon}{3} + \frac{\epsilon}{3} = \epsilon.$$

Thus $\lim_{x \to a} g(x) = L$.

Review Exercises 1 (page 93)

1. The average rate of change of x^3 over $[1, 3]$ is

$$\frac{3^3 - 1^3}{3 - 1} = \frac{26}{2} = 13.$$

2. The average rate of change of $1/x$ over $[-2, -1]$ is

$$\frac{(1/(-1)) - (1/(-2))}{-1 - (-2)} = \frac{-1/2}{1} = -\frac{1}{2}.$$

3. The rate of change of x^3 at $x = 2$ is

$$\lim_{h \to 0} \frac{(2+h)^3 - 2^3}{h} = \lim_{h \to 0} \frac{8 + 12h + 6h^2 + h^3 - 8}{h}$$
$$= \lim_{h \to 0} (12 + 6h + h^2) = 12.$$

4. The rate of change of $1/x$ at $x = -3/2$ is

$$\lim_{h \to 0} \frac{\dfrac{1}{-(3/2)+h} - \left(\dfrac{1}{-3/2}\right)}{h} = \lim_{h \to 0} \frac{\dfrac{2}{2h-3} + \dfrac{2}{3}}{h}$$
$$= \lim_{h \to 0} \frac{2(3 + 2h - 3)}{3(2h - 3)h}$$
$$= \lim_{h \to 0} \frac{4}{3(2h - 3)} = -\frac{4}{9}.$$

5. $\lim_{x \to 1} (x^2 - 4x + 7) = 1 - 4 + 7 = 4$

6. $\lim_{x \to 2} \dfrac{x^2}{1 - x^2} = \dfrac{2^2}{1 - 2^2} = -\dfrac{4}{3}$

7. $\lim_{x \to 1} \dfrac{x^2}{1 - x^2}$ does not exist. The denominator approaches 0 (from both sides) while the numerator does not.

8. $\lim_{x \to 2} \dfrac{x^2 - 4}{x^2 - 5x + 6} = \lim_{x \to 2} \dfrac{(x-2)(x+2)}{(x-2)(x-3)} = \lim_{x \to 2} \dfrac{x+2}{x-3} = -4$

9. $\lim_{x \to 2} \dfrac{x^2 - 4}{x^2 - 4x + 4} = \lim_{x \to 2} \dfrac{(x-2)(x+2)}{(x-2)^2} = \lim_{x \to 2} \dfrac{x+2}{x-2}$ does not exist. The denominator approaches 0 (from both sides) while the numerator does not.

10. $\lim_{x \to 2-} \dfrac{x^2 - 4}{x^2 - 4x + 4} = \lim_{x \to 2-} \dfrac{x+2}{x-2} = -\infty$

11. $\lim_{x \to -2+} \dfrac{x^2 - 4}{x^2 + 4x + 4} = \lim_{x \to -2+} \dfrac{x-2}{x+2} = -\infty$

12. $\lim_{x \to 4} \dfrac{2 - \sqrt{x}}{x - 4} = \lim_{x \to 4} \dfrac{4 - x}{(2 + \sqrt{x})(x - 4)} = -\dfrac{1}{4}$

13. $\lim_{x \to 3} \dfrac{x^2 - 9}{\sqrt{x} - \sqrt{3}} = \lim_{x \to 3} \dfrac{(x-3)(x+3)(\sqrt{x} + \sqrt{3})}{x - 3}$
$$= \lim_{x \to 3} (x+3)(\sqrt{x} + \sqrt{3}) = 12\sqrt{3}$$

14. $\lim_{h \to 0} \dfrac{h}{\sqrt{x+3h} - \sqrt{x}} = \lim_{h \to 0} \dfrac{h(\sqrt{x+3h} + \sqrt{x})}{(x+3h) - x}$
$$= \lim_{h \to 0} \dfrac{\sqrt{x+3h} + \sqrt{x}}{3} = \dfrac{2\sqrt{x}}{3}$$

15. $\lim_{x \to 0+} \sqrt{x - x^2} = 0$

16. $\lim_{x \to 0} \sqrt{x - x^2}$ does not exist because $\sqrt{x - x^2}$ is not defined for $x < 0$.

17. $\lim_{x \to 1} \sqrt{x - x^2}$ does not exist because $\sqrt{x - x^2}$ is not defined for $x > 1$.

18. $\lim_{x \to 1-} \sqrt{x - x^2} = 0$

19. $\lim_{x \to \infty} \dfrac{1 - x^2}{3x^2 - x - 1} = \lim_{x \to \infty} \dfrac{(1/x^2) - 1}{3 - (1/x) - (1/x^2)} = -\dfrac{1}{3}$

20. $\lim_{x \to -\infty} \dfrac{2x + 100}{x^2 + 3} = \lim_{x \to -\infty} \dfrac{(2/x) + (100/x^2)}{1 + (3/x^2)} = 0$

21. $\lim_{x \to -\infty} \dfrac{x^3 - 1}{x^2 + 4} = \lim_{x \to -\infty} \dfrac{x - (1/x^2)}{1 + (4/x^2)} = -\infty$

22. $\lim_{x \to \infty} \dfrac{x^4}{x^2 - 4} = \lim_{x \to \infty} \dfrac{x^2}{1 - (4/x^2)} = \infty$

23. $\lim_{x \to 0+} \dfrac{1}{\sqrt{x - x^2}} = \infty$

24. $\lim_{x \to 1/2} \dfrac{1}{\sqrt{x - x^2}} = \dfrac{1}{\sqrt{1/4}} = 2$

25. $\lim_{x\to\infty} \sin x$ does not exist; $\sin x$ takes the values -1 and 1 in any interval (R,∞), and limits, if they exist, must be unique.

26. $\lim_{x\to\infty} \dfrac{\cos x}{x} = 0$ by the squeeze theorem, since
$$-\dfrac{1}{x} \le \dfrac{\cos x}{x} \le \dfrac{1}{x} \quad \text{for all } x > 0$$
and $\lim_{x\to\infty}(-1/x) = \lim_{x\to\infty}(1/x) = 0$.

27. $\lim_{x\to 0} x \sin \dfrac{1}{x} = 0$ by the squeeze theorem, since
$$-|x| \le x \sin \dfrac{1}{x} \le |x| \quad \text{for all } x \ne 0$$
and $\lim_{x\to 0}(-|x|) = \lim_{x\to 0}|x| = 0$.

28. $\lim_{x\to 0} \sin \dfrac{1}{x^2}$ does not exist; $\sin(1/x^2)$ takes the values -1 and 1 in any interval $(-\delta,\delta)$, where $\delta > 0$, and limits, if they exist, must be unique.

29. $\lim_{x\to-\infty} [x + \sqrt{x^2 - 4x + 1}]$
$= \lim_{x\to-\infty} \dfrac{x^2 - (x^2 - 4x + 1)}{x - \sqrt{x^2 - 4x + 1}}$
$= \lim_{x\to-\infty} \dfrac{4x - 1}{x - |x|\sqrt{1 - (4/x) + (1/x^2)}}$
$= \lim_{x\to-\infty} \dfrac{x[4 - (1/x)]}{x + x\sqrt{1 - (4/x) + (1/x^2)}}$
$= \lim_{x\to-\infty} \dfrac{4 - (1/x)}{1 + \sqrt{1 - (4/x) + (1/x^2)}} = 2.$
Note how we have used $|x| = -x$ (in the second last line), because $x \to -\infty$.

30. $\lim_{x\to\infty} [x + \sqrt{x^2 - 4x + 1}] = \infty + \infty = \infty$

31. $f(x) = x^3 - 4x^2 + 1$ is continuous on the whole real line and so is discontinuous nowhere.

32. $f(x) = \dfrac{x}{x+1}$ is continuous everywhere on its domain, which consists of all real numbers except $x = -1$. It is discontinuous nowhere.

33. $f(x) = \begin{cases} x^2 & \text{if } x > 2 \\ x & \text{if } x \le 2 \end{cases}$ is defined everywhere and discontinuous at $x = 2$, where it is, however, left continuous since $\lim_{x\to 2-} f(x) = 2 = f(2)$.

34. $f(x) = \begin{cases} x^2 & \text{if } x > 1 \\ x & \text{if } x \le 1 \end{cases}$ is defined and continuous everywhere, and so discontinuous nowhere. Observe that $\lim_{x\to 1-} f(x) = 1 = \lim_{x\to 1+} f(x)$.

35. $f(x) = H(x-1) = \begin{cases} 1 & \text{if } x \ge 1 \\ 0 & \text{if } x < 1 \end{cases}$ is defined everywhere and discontinuous at $x = 1$ where it is, however, right continuous.

36. $f(x) = H(9 - x^2) = \begin{cases} 1 & \text{if } -3 \le x \le 3 \\ 0 & \text{if } x < -3 \text{ or } x > 3 \end{cases}$ is defined everywhere and discontinuous at $x = \pm 3$. It is right continuous at -3 and left continuous at 3.

37. $f(x) = |x| + |x+1|$ is defined and continuous everywhere. It is discontinuous nowhere.

38. $f(x) = \begin{cases} |x|/|x+1| & \text{if } x \ne -1 \\ 1 & \text{if } x = -1 \end{cases}$ is defined everywhere and discontinuous at $x = -1$ where it is neither left nor right continuous since $\lim_{x\to -1} f(x) = \infty$, while $f(-1) = 1$.

Challenging Problems 1 (page 93)

1. Let $0 < a < b$. The average rate of change of x^3 over $[a,b]$ is
$$\dfrac{b^3 - a^3}{b - a} = b^2 + ab + a^2.$$
The instantaneous rate of change of x^3 at $x = c$ is
$$\lim_{h\to 0} \dfrac{(c+h)^3 - c^3}{h} = \lim_{h\to 0} \dfrac{3c^2 h + 3ch^2 + h^3}{h} = 3c^2.$$
If $c = \sqrt{(a^2 + ab + b^2)/3}$, then $3c^2 = a^2 + ab + b^2$, so the average rate of change over $[a,b]$ is the instantaneous rate of change at $\sqrt{(a^2 + ab + b^2)/3}$.
Claim: $\sqrt{(a^2 + ab + b^2)/3} > (a+b)/2$.
Proof: Since $a^2 - 2ab + b^2 = (a-b)^2 > 0$, we have
$$4a^2 + 4ab + 4b^2 > 3a^2 + 6ab + 3b^2$$
$$\dfrac{a^2 + ab + b^2}{3} > \dfrac{a^2 + 2ab + b^2}{4} = \left(\dfrac{a+b}{2}\right)^2$$
$$\sqrt{\dfrac{a^2 + ab + b^2}{3}} > \dfrac{a+b}{2}.$$

2. For x near 0 we have $|x-1| = 1-x$ and $|x+1| = x+1$. Thus
$$\lim_{x\to 0} \dfrac{x}{|x-1| - |x+1|} = \lim_{x\to 0} \dfrac{x}{(1-x) - (x+1)} = -\dfrac{1}{2}.$$

3. For x near 3 we have $|5 - 2x| = 2x - 5$, $|x - 2| = x - 2$, $|x - 5| = 5 - x$, and $|3x - 7| = 3x - 7$. Thus
$$\lim_{x\to 3} \dfrac{|5 - 2x| - |x - 2|}{|x - 5| - |3x - 7|} = \lim_{x\to 3} \dfrac{2x - 5 - (x - 2)}{5 - x - (3x - 7)}$$
$$= \lim_{x\to 3} \dfrac{x - 3}{4(3 - x)} = -\dfrac{1}{4}.$$

4. Let $y = x^{1/6}$. Then we have
$$\lim_{x \to 64} \frac{x^{1/3} - 4}{x^{1/2} - 8} = \lim_{y \to 2} \frac{y^2 - 4}{y^3 - 8}$$
$$= \lim_{y \to 2} \frac{(y-2)(y+2)}{(y-2)(y^2 + 2y + 4)}$$
$$= \lim_{y \to 2} \frac{y+2}{y^2 + 2y + 4} = \frac{4}{12} = \frac{1}{3}.$$

5. Use $a - b = \dfrac{a^3 - b^3}{a^2 + ab + b^2}$ to handle the denominator. We have
$$\lim_{x \to 1} \frac{\sqrt{3+x} - 2}{\sqrt[3]{7+x} - 2}$$
$$= \lim_{x \to 1} \frac{3+x-4}{\sqrt{3+x} + 2} \times \frac{(7+x)^{2/3} + 2(7+x)^{1/3} + 4}{(7+x) - 8}$$
$$= \lim_{x \to 1} \frac{(7+x)^{2/3} + 2(7+x)^{1/3} + 4}{\sqrt{3+x} + 2} = \frac{4+4+4}{2+2} = 3.$$

6. $r_+(a) = \dfrac{-1 + \sqrt{1+a}}{a}$, $r_-(a) = \dfrac{-1 - \sqrt{1+a}}{a}$.

   a) $\lim_{a \to 0} r_-(a)$ does not exist. Observe that the right limit is $-\infty$ and the left limit is ∞.

   b) From the following table it appears that $\lim_{a \to 0} r_+(a) = 1/2$, the solution of the linear equation $2x - 1 = 0$ which results from setting $a = 0$ in the quadratic equation $ax^2 + 2x - 1 = 0$.

a	$r_+(a)$
1	0.41421
0.1	0.48810
-0.1	0.51317
0.01	0.49876
-0.01	0.50126
0.001	0.49988
-0.001	0.50013

   c) $\lim_{a \to 0} r_+(a) = \lim_{a \to 0} \dfrac{\sqrt{1+a} - 1}{a}$
   $$= \lim_{a \to 0} \frac{(1+a) - 1}{a(\sqrt{1+a} + 1)}$$
   $$= \lim_{a \to 0} \frac{1}{\sqrt{1+a} + 1} = \frac{1}{2}.$$

7. TRUE or FALSE

   a) If $\lim_{x \to a} f(x)$ exists and $\lim_{x \to a} g(x)$ does not exist, then $\lim_{x \to a}(f(x) + g(x))$ does not exist.
   TRUE, because if $\lim_{x \to a}(f(x) + g(x))$ were to exist then
   $$\lim_{x \to a} g(x) = \lim_{x \to a}\bigl(f(x) + g(x) - f(x)\bigr)$$
   $$= \lim_{x \to a}(f(x) + g(x)) - \lim_{x \to a} f(x)$$

   would also exist.

   b) If neither $\lim_{x \to a} f(x)$ nor $\lim_{x \to a} g(x)$ exists, then $\lim_{x \to a}(f(x) + g(x))$ does not exist.
   FALSE. Neither $\lim_{x \to 0} 1/x$ nor $\lim_{x \to 0}(-1/x)$ exist, but $\lim_{x \to 0}((1/x) + (-1/x)) = \lim_{x \to 0} 0 = 0$ exists.

   c) If f is continuous at a, then so is $|f|$.
   TRUE. For any two real numbers u and v we have
   $$\bigl||u| - |v|\bigr| \le |u - v|.$$
   This follows from
   $$|u| = |u - v + v| \le |u - v| + |v|, \quad \text{and}$$
   $$|v| = |v - u + u| \le |v - u| + |u| = |u - v| + |u|.$$
   Now we have
   $$\bigl||f(x)| - |f(a)|\bigr| \le |f(x) - f(a)|$$
   so the left side approaches zero whenever the right side does. This happens when $x \to a$ by the continuity of f at a.

   d) If $|f|$ is continuous at a, then so is f.
   FALSE. The function $f(x) = \begin{cases} -1 & \text{if } x < 0 \\ 1 & \text{if } x \ge 0 \end{cases}$ is discontinuous at $x = 0$, but $|f(x)| = 1$ everywhere, and so is continuous at $x = 0$.

   e) If $f(x) < g(x)$ in an interval around a and if $\lim_{x \to a} f(x) = L$ and $\lim_{x \to a} g(x) = M$ both exist, then $L < M$.
   FALSE. Let $g(x) = \begin{cases} x^2 & \text{if } x \ne 0 \\ 1 & \text{if } x = 0 \end{cases}$ and let $f(x) = -g(x)$. Then $f(x) < g(x)$ for all x, but $\lim_{x \to 0} f(x) = 0 = \lim_{x \to 0} g(x)$. (Note: under the given conditions, it is TRUE that $L \le M$, but not necessarily true that $L < M$.)

8. a) To be proved: if f is a continuous function defined on a closed interval $[a, b]$, then the range of f is a closed interval.
   Proof: By the Max-Min Theorem there exist numbers u and v in $[a, b]$ such that $f(u) \le f(x) \le f(v)$ for all x in $[a, b]$. By the Intermediate-Value Theorem, $f(x)$ takes on all values between $f(u)$ and $f(v)$ at values of x between u and v, and hence at points of $[a, b]$. Thus the range of f is $[f(u), f(v)]$, a closed interval.

   b) If the domain of the continuous function f is an open interval, the range of f can be any interval (open, closed, half open, finite, or infinite).

9. $f(x) = \dfrac{x^2-1}{|x^2-1|} = \begin{cases} -1 & \text{if } -1 < x < 1 \\ 1 & \text{if } x < -1 \text{ or } x > 1 \end{cases}$.

f is continuous wherever it is defined, that is at all points except $x = \pm 1$. f has left and right limits -1 and 1, respectively, at $x = 1$, and has left and right limits 1 and -1, respectively, at $x = -1$. It is not, however, discontinuous at any point, since -1 and 1 are not in its domain.

10. $f(x) = \dfrac{1}{x-x^2} = \dfrac{1}{\frac{1}{4} - \left(\frac{1}{4} - x + x^2\right)} = \dfrac{1}{\frac{1}{4} - \left(x - \frac{1}{2}\right)^2}$.

    Observe that $f(x) \geq f(1/2) = 4$ for all x in $(0, 1)$.

11. Suppose f is continuous on $[0, 1]$ and $f(0) = f(1)$.

    a) To be proved: $f(a) = f(a + \frac{1}{2})$ for some a in $[0, \frac{1}{2}]$.
       Proof: If $f(1/2) = f(0)$ we can take $a = 0$ and be done. If not, let
       $$g(x) = f(x + \tfrac{1}{2}) - f(x).$$
       Then $g(0) \neq 0$ and
       $$g(1/2) = f(1) - f(1/2) = f(0) - f(1/2) = -g(0).$$
       Since g is continuous and has opposite signs at $x = 0$ and $x = 1/2$, the Intermediate-Value Theorem assures us that there exists a between 0 and $1/2$ such that $g(a) = 0$, that is, $f(a) = f(a + \frac{1}{2})$.

    b) To be proved: if $n > 2$ is an integer, then $f(a) = f(a + \frac{1}{n})$ for some a in $[0, 1 - \frac{1}{n}]$.
       Proof: Let $g(x) = f(x + \frac{1}{n}) - f(x)$. Consider the numbers $x = 0, x = 1/n, x = 2/n, \ldots, x = (n-1)/n$. If $g(x) = 0$ for any of these numbers, then we can let a be that number. Otherwise, $g(x) \neq 0$ at any of these numbers. Suppose that the values of g at all these numbers has the same sign (say positive). Then we have
       $$f(1) > f(\tfrac{n-1}{n}) > \cdots > f(\tfrac{2}{n}) > \tfrac{1}{n} > f(0),$$
       which is a contradiction, since $f(0) = f(1)$. Therefore there exists j in the set $\{0, 1, 2, \ldots, n-1\}$ such that $g(j/n)$ and $g((j+1)/n)$ have opposite sign. By the Intermediate-Value Theorem, $g(a) = 0$ for some a between j/n and $(j+1)/n$, which is what we had to prove.

CHAPTER 2. DIFFERENTIATION

Section 2.1 Tangent Lines and Their Slopes (page 100)

1. Slope of $y = 3x - 1$ at $(1, 2)$ is

$$m = \lim_{h \to 0} \frac{3(1+h) - 1 - (3 \times 1 - 1)}{h} = \lim_{h \to 0} \frac{3h}{h} = 3.$$

The tangent line is $y - 2 = 3(x - 1)$, or $y = 3x - 1$. (The tangent to a straight line at any point on it is the same straight line.)

2. Since $y = x/2$ is a straight line, its tangent at any point $(a, a/2)$ on it is the same line $y = x/2$.

3. Slope of $y = 2x^2 - 5$ at $(2, 3)$ is

$$m = \lim_{h \to 0} \frac{2(2+h)^2 - 5 - (2(2^2) - 5)}{h}$$
$$= \lim_{h \to 0} \frac{8 + 8h + 2h^2 - 8}{h}$$
$$= \lim_{h \to 0} (8 + 2h) = 8$$

Tangent line is $y - 3 = 8(x - 2)$ or $y = 8x - 13$.

4. The slope of $y = 6 - x - x^2$ at $x = -2$ is

$$m = \lim_{h \to 0} \frac{6 - (-2+h) - (-2+h)^2 - 4}{h}$$
$$= \lim_{h \to 0} \frac{3h - h^2}{h} = \lim_{h \to 0} (3 - h) = 3.$$

The tangent line at $(-2, 4)$ is $y = 3x + 10$.

5. Slope of $y = x^3 + 8$ at $x = -2$ is

$$m = \lim_{h \to 0} \frac{(-2+h)^3 + 8 - (-8 + 8)}{h}$$
$$= \lim_{h \to 0} \frac{-8 + 12h - 6h^2 + h^3 + 8 - 0}{h}$$
$$= \lim_{h \to 0} \left(12 - 6h + h^2\right) = 12$$

Tangent line is $y - 0 = 12(x + 2)$ or $y = 12x + 24$.

6. The slope of $y = \dfrac{1}{x^2 + 1}$ at $(0, 1)$ is

$$m = \lim_{h \to 0} \frac{1}{h}\left(\frac{1}{h^2 + 1} - 1\right) = \lim_{h \to 0} \frac{-h}{h^2 + 1} = 0.$$

The tangent line at $(0, 1)$ is $y = 1$.

7. Slope of $y = \dfrac{1}{x^2}$ at $x = 3$ is

$$m = \lim_{h \to 0} \frac{\dfrac{1}{(3+h)^2} - \dfrac{1}{9}}{h}$$
$$= \lim_{h \to 0} \frac{1}{9h(3+h)^2}\left(9 - (9 + 6h + h^2)\right)$$
$$= \lim_{h \to 0} \frac{-6 - h}{9(3+h)^2} = -\frac{6}{81} = -\frac{2}{27}.$$

Tangent line is $y - \dfrac{1}{9} = -\dfrac{2}{27}(x - 3)$ or $2x + 27y = 9$.

8. The slope of $y = \dfrac{1}{x^2 + 1}$ at $x = -1$ is

$$m = \lim_{h \to 0} \frac{1}{h}\left[\frac{1}{(-1+h)^2 + 1} - \frac{1}{2}\right]$$
$$= \lim_{h \to 0} \frac{2 - [(-1+h)^2 + 1]}{2h[(-1+h)^2 + 1]}$$
$$= \lim_{h \to 0} \frac{2 - h}{2[(-1+h)^2 + 1]} = \frac{2}{4} = \frac{1}{2}.$$

The tangent line is $y = \frac{1}{2}x + 1$.

9. Slope of $y = \sqrt{x + 1}$ at $x = 3$ is

$$m = \lim_{h \to 0} \frac{\sqrt{4+h} - 2}{h} \cdot \frac{\sqrt{4+h} + 2}{\sqrt{4+h} + 2}$$
$$= \lim_{h \to 0} \frac{4 + h - 4}{h\left(\sqrt{h+h} + 2\right)}$$
$$= \lim_{h \to 0} \frac{1}{\sqrt{4+h} + 2} = \frac{1}{4}.$$

Tangent line is $y - 2 = \dfrac{1}{4}(x - 3)$, or $x - 4y = -5$.

10. The slope of $y = \dfrac{1}{\sqrt{x}}$ at $x = 9$ is

$$m = \lim_{h \to 0} \frac{1}{h}\left(\frac{1}{\sqrt{9+h}} - \frac{1}{3}\right)$$
$$= \lim_{h \to 0} \frac{3 - \sqrt{9+h}}{3h\sqrt{9+h}} \cdot \frac{3 + \sqrt{9+h}}{3 + \sqrt{9+h}}$$
$$= \lim_{h \to 0} \frac{9 - 9 - h}{3h\sqrt{9+h}(3 + \sqrt{9+h})}$$
$$= -\frac{1}{3(3)(6)} = -\frac{1}{54}.$$

The tangent line at $(9, \frac{1}{3})$ is $y = \frac{1}{3} - \frac{1}{54}(x - 9)$, or $y = \frac{1}{2} - \frac{1}{54}x$.

INSTRUCTOR'S SOLUTIONS MANUAL SECTION 2.1 (PAGE 100)

11. Slope of $y = \dfrac{2x}{x+2}$ at $x = 2$ is

$$m = \lim_{h \to 0} \dfrac{\dfrac{2(2+h)}{2+h+2} - 1}{h}$$
$$= \lim_{h \to 0} \dfrac{4 + 2h - 2 - h - 2}{h(2+h+2)}$$
$$= \lim_{h \to 0} \dfrac{h}{h(4+h)} = \dfrac{1}{4}.$$

Tangent line is $y - 1 = \dfrac{1}{4}(x - 2)$, or $x - 4y = -2$.

12. The slope of $y = \sqrt{5 - x^2}$ at $x = 1$ is

$$m = \lim_{h \to 0} \dfrac{\sqrt{5 - (1+h)^2} - 2}{h}$$
$$= \lim_{h \to 0} \dfrac{5 - (1+h)^2 - 4}{h\left(\sqrt{5 - (1+h)^2} + 2\right)}$$
$$= \lim_{h \to 0} \dfrac{-2 - h}{\sqrt{5 - (1+h)^2} + 2} = -\dfrac{1}{2}.$$

The tangent line at $(1, 2)$ is $y = 2 - \dfrac{1}{2}(x - 1)$, or $y = \dfrac{5}{2} - \dfrac{1}{2}x$.

13. Slope of $y = x^2$ at $x = x_0$ is

$$m = \lim_{h \to 0} \dfrac{(x_0 + h)^2 - x_0^2}{h} = \lim_{h \to 0} \dfrac{2x_0 h + h^2}{h} = 2x_0.$$

Tangent line is $y - x_0^2 = 2x_0(x - x_0)$, or $y = 2x_0 x - x_0^2$.

14. The slope of $y = \dfrac{1}{x}$ at $(a, \dfrac{1}{a})$ is

$$m = \lim_{h \to 0} \dfrac{1}{h}\left(\dfrac{1}{a+h} + \dfrac{1}{a}\right) = \lim_{h \to 0} \dfrac{a - a - h}{h(a+h)(a)} = -\dfrac{1}{a^2}.$$

The tangent line at $(a, \dfrac{1}{a})$ is $y = \dfrac{1}{a} - \dfrac{1}{a^2}(x - a)$, or $y = \dfrac{2}{a} - \dfrac{x}{a^2}$.

15. Since $\lim_{h \to 0} \dfrac{\sqrt{|0+h|} - 0}{h} = \lim_{h \to 0} \dfrac{1}{\sqrt{|h|} \operatorname{sgn}(h)}$ does not exist (and is not ∞ or $-\infty$), the graph of $f(x) = \sqrt{|x|}$ has no tangent at $x = 0$.

16. The slope of $f(x) = (x - 1)^{4/3}$ at $x = 1$ is

$$m = \lim_{h \to 0} \dfrac{(1 + h - 1)^{4/3} - 0}{h} = \lim_{h \to 0} h^{1/3} = 0.$$

The graph of f has a tangent line with slope 0 at $x = 1$. Since $f(1) = 0$, the tangent has equation $y = 0$.

17. The slope of $f(x) = (x + 2)^{3/5}$ at $x = -2$ is

$$m = \lim_{h \to 0} \dfrac{(-2 + h + 2)^{3/5} - 0}{h} = \lim_{h \to 0} h^{-2/5} = \infty.$$

The graph of f has vertical tangent $x = -2$ at $x = -2$.

18. The slope of $f(x) = |x^2 - 1|$ at $x = 1$ is

$$m = \lim_{h \to 0} \dfrac{|(1+h)^2 - 1| - |1 - 1|}{h} = \lim_{h \to 0} \dfrac{|2h + h^2|}{h},$$

which does not exist, and is not $-\infty$ or ∞. The graph of f has no tangent at $x = 1$.

19. If $f(x) = \begin{cases} \sqrt{x} & \text{if } x \geq 0 \\ -\sqrt{-x} & \text{if } x < 0 \end{cases}$, then

$$\lim_{h \to 0+} \dfrac{f(0+h) - f(0)}{h} = \lim_{h \to 0+} \dfrac{\sqrt{h}}{h} = \infty$$
$$\lim_{h \to 0-} \dfrac{f(0+h) - f(0)}{h} = \lim_{h \to 0-} \dfrac{-\sqrt{-h}}{h} = \infty$$

Thus the graph of f has a vertical tangent $x = 0$.

20. The slope of $y = x^2 - 1$ at $x = x_0$ is

$$m = \lim_{h \to 0} \dfrac{[(x_0 + h)^2 - 1] - (x_0^2 - 1)}{h}$$
$$= \lim_{h \to 0} \dfrac{2x_0 h + h^2}{h} = 2x_0.$$

If $m = -3$, then $x_0 = -\dfrac{3}{2}$. The tangent line with slope $m = -3$ at $(-\dfrac{3}{2}, \dfrac{5}{4})$ is $y = \dfrac{5}{4} - 3(x + \dfrac{3}{2})$, that is, $y = -3x - \dfrac{13}{4}$.

21. a) Slope of $y = x^3$ at $x = a$ is

$$m = \lim_{h \to 0} \dfrac{(a+h)^3 - a^3}{h}$$
$$= \lim_{h \to 0} \dfrac{a^3 + 3a^2 h + 3ah^2 + h^3 - a^3}{h}$$
$$= \lim_{h \to 0} (3a^2 + 3ah + h^2) = 3a^2$$

b) We have $m = 3$ if $3a^2 = 3$, i.e., if $a = \pm 1$. Lines of slope 3 tangent to $y = x^3$ are $y = 1 + 3(x - 1)$ and $y = -1 + 3(x + 1)$, or $y = 3x - 2$ and $y = 3x + 2$.

22. The slope of $y = x^3 - 3x$ at $x = a$ is

$$m = \lim_{h \to 0} \dfrac{1}{h}\left[(a+h)^3 - 3(a+h) - (a^3 - 3a)\right]$$
$$= \lim_{h \to 0} \dfrac{1}{h}\left[a^3 + 3a^2 h + 3ah^2 + h^3 - 3a - 3h - a^3 + 3a\right]$$
$$= \lim_{h \to 0} [3a^2 + 3ah + h^2 - 3] = 3a^2 - 3.$$

37

At points where the tangent line is parallel to the x-axis, the slope is zero, so such points must satisfy $3a^2 - 3 = 0$. Thus, $a = \pm 1$. Hence, the tangent line is parallel to the x-axis at the points $(1, -2)$ and $(-1, 2)$.

23. The slope of the curve $y = x^3 - x + 1$ at $x = a$ is

$$m = \lim_{h \to 0} \frac{(a+h)^3 - (a+h) + 1 - (a^3 - a + 1)}{h}$$
$$= \lim_{h \to 0} \frac{3a^2 h + 3ah^2 + a^3 - h}{h}$$
$$= \lim_{h \to 0} (3a^2 + 3ah + h^2 - 1) = 3a^2 - 1.$$

The tangent at $x = a$ is parallel to the line $y = 2x + 5$ if $3a^2 - 1 = 2$, that is, if $a = \pm 1$. The corresponding points on the curve are $(-1, 1)$ and $(1, 1)$.

24. The slope of the curve $y = 1/x$ at $x = a$ is

$$m = \lim_{h \to 0} \frac{\frac{1}{a+h} - \frac{1}{a}}{h} = \lim_{h \to 0} \frac{a - (a+h)}{ah(a+h)} = -\frac{1}{a^2}.$$

The tangent at $x = a$ is perpendicular to the line $y = 4x - 3$ if $-1/a^2 = -1/4$, that is, if $a = \pm 2$. The corresponding points on the curve are $(-2, -1/2)$ and $(2, 1/2)$.

25. The slope of the curve $y = x^2$ at $x = a$ is

$$m = \lim_{h \to 0} \frac{(a+h)^2 - a^2}{h} = \lim_{h \to 0} (2a + h) = 2a.$$

The normal at $x = a$ has slope $-1/(2a)$, and has equation

$$y - a^2 = -\frac{1}{2a}(x - a), \quad \text{or} \quad \frac{x}{2a} + y = \frac{1}{2} + a^2.$$

This is the line $x + y = k$ if $2a = 1$, and so $k = (1/2) + (1/2)^2 = 3/4$.

26. The curves $y = kx^2$ and $y = k(x - 2)^2$ intersect at $(1, k)$. The slope of $y = kx^2$ at $x = 1$ is

$$m_1 = \lim_{h \to 0} \frac{k(1+h)^2 - k}{h} = \lim_{h \to 0} (2 + h)k = 2k.$$

The slope of $y = k(x - 2)^2$ at $x = 1$ is

$$m_2 = \lim_{h \to 0} \frac{k(2 - (1+h))^2 - k}{h} = \lim_{h \to 0} (-2 + h)k = -2k.$$

The two curves intersect at right angles if $2k = -1/(-2k)$, that is, if $4k^2 = 1$, which is satisfied if $k = \pm 1/2$.

27. Horizontal tangents at $(0, 0)$, $(3, 108)$, and $(5, 0)$.

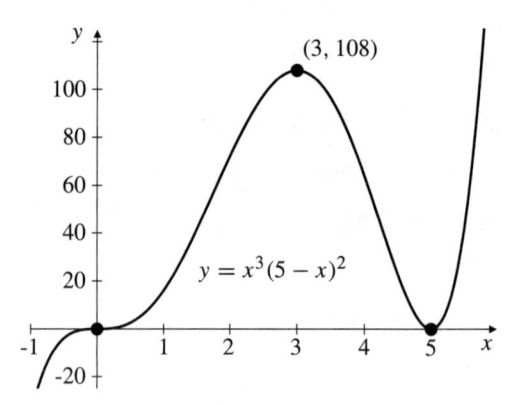

Fig. 2.1.27

28. Horizontal tangent at $(-1, 8)$ and $(2, -19)$.

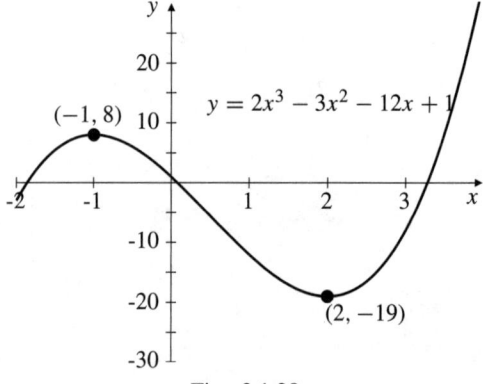

Fig. 2.1.28

29. Horizontal tangent at $(-1/2, 5/4)$. No tangents at $(-1, 1)$ and $(1, -1)$.

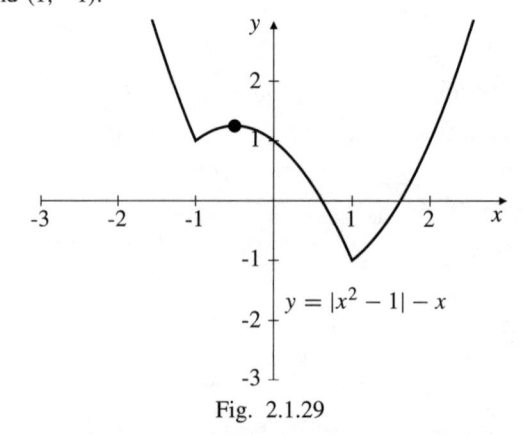

Fig. 2.1.29

30. Horizontal tangent at $(a, 2)$ and $(-a, -2)$ for all $a > 1$. No tangents at $(1, 2)$ and $(-1, -2)$.

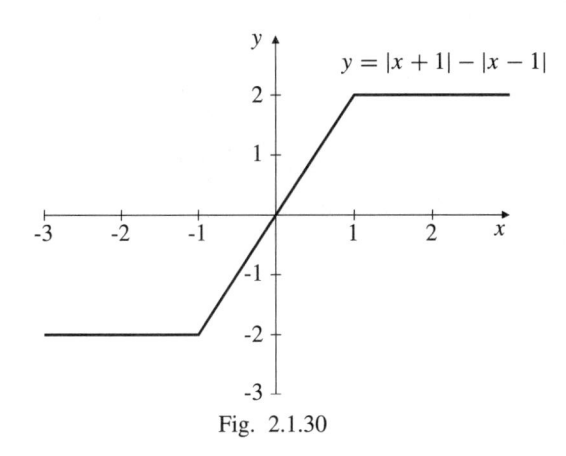

Fig. 2.1.30

31. Horizontal tangent at $(0, -1)$. The tangents at $(\pm 1, 0)$ are vertical.

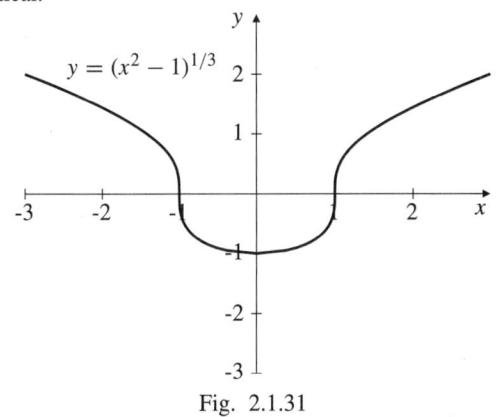

Fig. 2.1.31

32. Horizontal tangent at $(0, 1)$. No tangents at $(-1, 0)$ and $(1, 0)$.

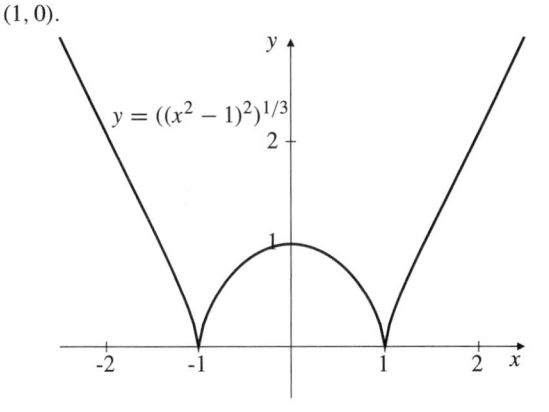

Fig. 2.1.32

33. The graph of the function $f(x) = x^{2/3}$ (see Figure 2.1.7 in the text) has a cusp at the origin O, so does not have a tangent line there. However, the angle between OP and the positive y-axis does $\to 0$ as P approaches 0 along the graph. Thus the answer is NO.

34. The slope of $P(x)$ at $x = a$ is

$$m = \lim_{h \to 0} \frac{P(a+h) - P(a)}{h}.$$

Since $P(a+h) = a_0 + a_1 h + a_2 h^2 + \cdots + a_n h^n$ and $P(a) = a_0$, the slope is

$$m = \lim_{h \to 0} \frac{a_0 + a_1 h + a_2 h^2 + \cdots + a_n h^n - a_0}{h}$$
$$= \lim_{h \to 0} a_1 + a_2 h + \cdots + a_n h^{n-1} = a_1.$$

Thus the line $y = \ell(x) = m(x - a) + b$ is tangent to $y = P(x)$ at $x = a$ if and only if $m = a_1$ and $b = a_0$, that is, if and only if

$$P(x) - \ell(x) = a_2(x-a)^2 + a_3(x-a)^3 + \cdots + a_n(x-a)^n$$
$$= (x-a)^2 \left[a_2 + a_3(x-a) + \cdots + a_n(x-a)^{n-2} \right]$$
$$= (x-a)^2 Q(x)$$

where Q is a polynomial.

Section 2.2   The Derivative   (page 108)

1.

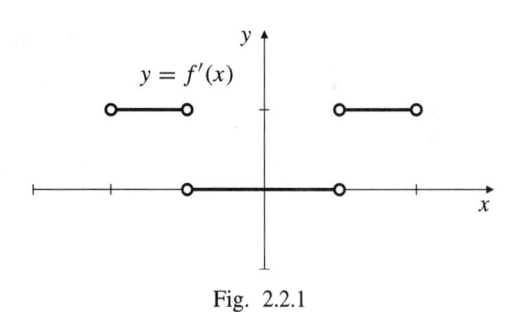

Fig. 2.2.1

2.

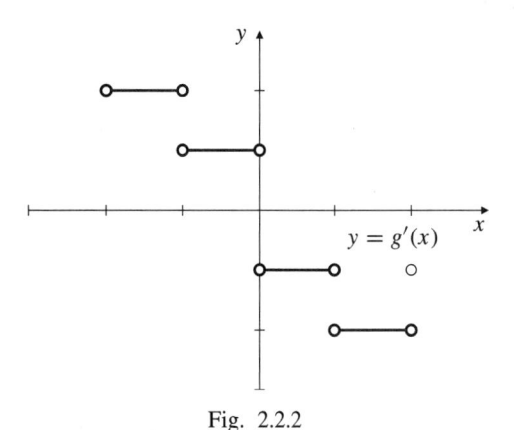

Fig. 2.2.2

3.

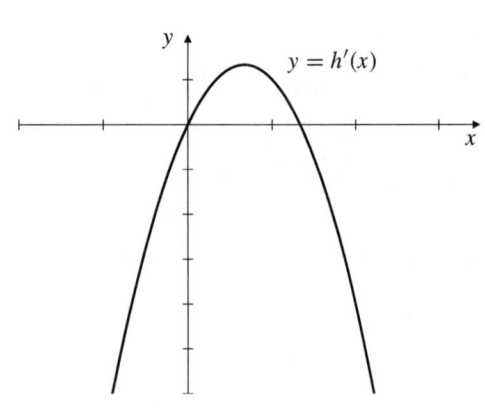

Fig. 2.2.3

4.

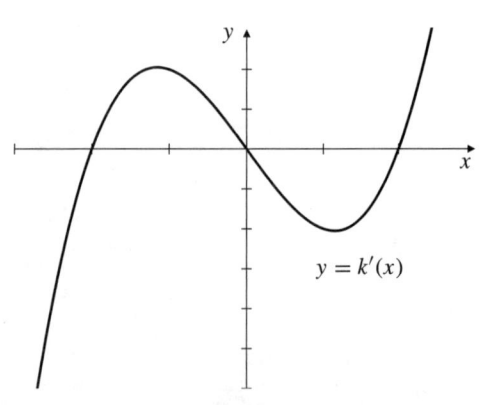

Fig. 2.2.4

5. Assuming the tick marks are spaced 1 unit apart, the function f is differentiable on the intervals $(-2, -1)$, $(-1, 1)$, and $(1, 2)$.

6. Assuming the tick marks are spaced 1 unit apart, the function g is differentiable on the intervals $(-2, -1)$, $(-1, 0)$, $(0, 1)$, and $(1, 2)$.

7. $y = f(x)$ has its minimum at $x = 3/2$ where $f'(x) = 0$

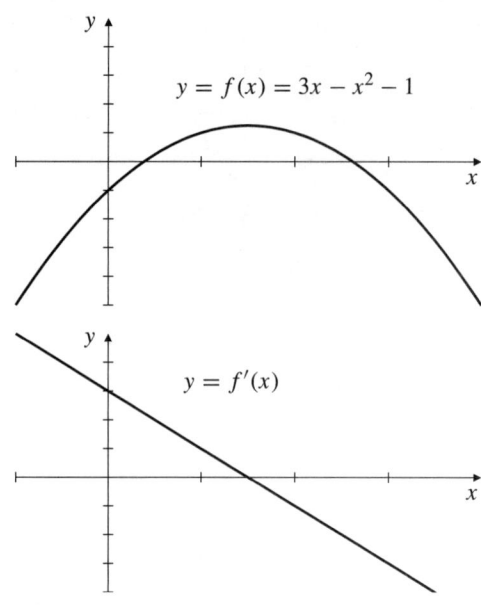

Fig. 2.2.7

8. $y = f(x)$ has horizontal tangents at the points near $1/2$ and $3/2$ where $f'(x) = 0$

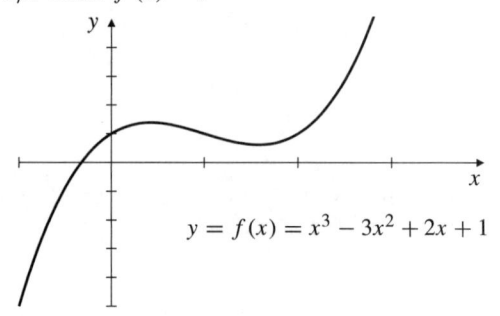

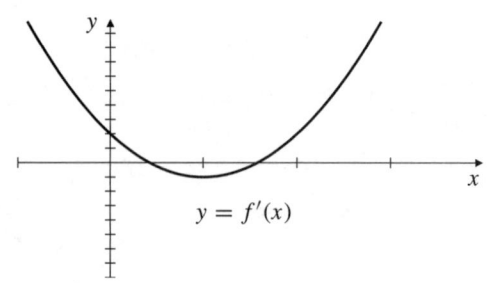

Fig. 2.2.8

9. $y = f(x)$ fails to be differentiable at $x = -1$, $x = 0$, and $x = 1$. It has horizontal tangents at two points, one between -1 and 0 and the other between 0 and 1.

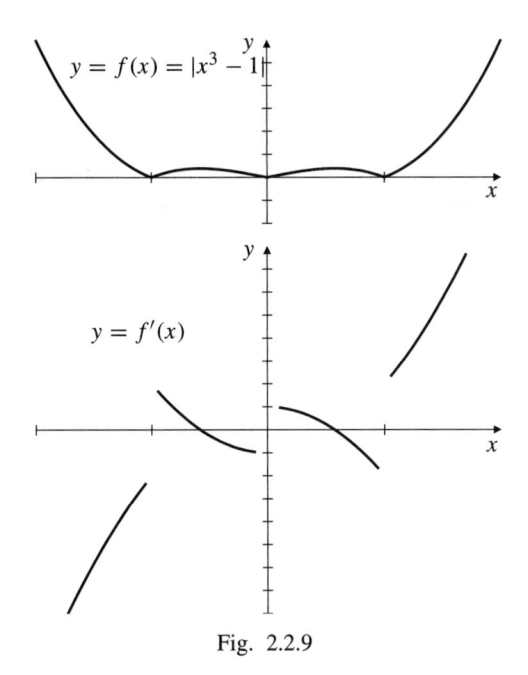

INSTRUCTOR'S SOLUTIONS MANUAL   SECTION 2.2 (PAGE 108)

$y = f(x) = |x^3 - 1|$

$y = f'(x)$

Fig. 2.2.9

10. $y = f(x)$ is constant on the intervals $(-\infty, -2)$, $(-1, 1)$, and $(2, \infty)$. It is not differentiable at $x = \pm 2$ and $x = \pm 1$.

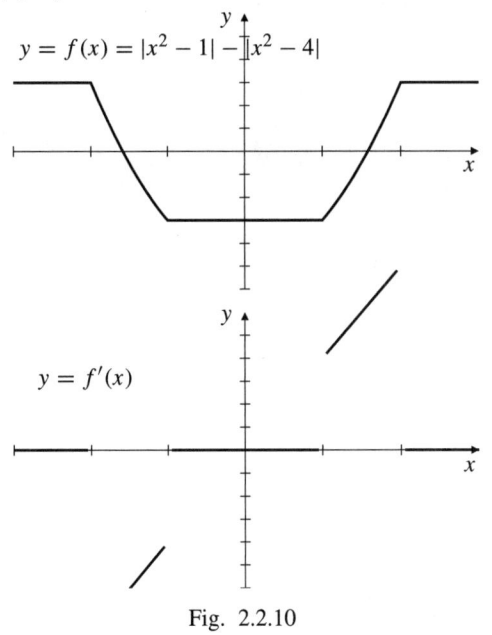

$y = f(x) = |x^2 - 1| - |x^2 - 4|$

$y = f'(x)$

Fig. 2.2.10

11. $y = x^2 - 3x$
$$y' = \lim_{h \to 0} \frac{(x+h)^2 - 3(x+h) - (x^2 - 3x)}{h}$$
$$= \lim_{h \to 0} \frac{2xh + h^2 - 3h}{h} = 2x - 3$$

12. $f(x) = 1 + 4x - 5x^2$
$$f'(x) = \lim_{h \to 0} \frac{1 + 4(x+h) - 5(x+h)^2 - (1 + 4x - 5x^2)}{h}$$
$$= \lim_{h \to 0} \frac{4h - 10xh - 5h^2}{h} = 4 - 10x$$

13. $f(x) = x^3$
$$f'(x) = \lim_{h \to 0} \frac{(x+h)^3 - x^3}{h}$$
$$= \lim_{h \to 0} \frac{3x^2 h + 3xh^2 + h^3}{h} = 3x^2$$

14. $y = \frac{1}{3}x^3 - x$
$$y' = \lim_{h \to 0} \frac{1}{h}\left[\frac{1}{3}(x+h)^3 - (x+h) - (\frac{1}{3}x^3 - x)\right]$$
$$= \lim_{h \to 0} \frac{1}{h}\left(x^2 h + xh^2 + \frac{1}{3}h^3 - h\right)$$
$$= \lim_{h \to 0}(x^2 + xh + \frac{1}{3}h^2 - 1) = x^2 - 1$$

15. $g(x) = \dfrac{2-x}{2+x}$
$$g'(x) = \lim_{h \to 0} \frac{\dfrac{2-(x+h)}{2+x+h} - \dfrac{2-x}{2+x}}{h}$$
$$= \lim_{h \to 0} \frac{(2-x-h)(2+x) - (2+x+h)(2-x)}{h(2+x+h)(2+x)}$$
$$= -\frac{4}{(2+x)^2}$$

16. $s = \dfrac{1}{3+4t}$
$$\frac{ds}{dt} = \lim_{h \to 0} \frac{1}{h}\left[\frac{1}{3+4(t+h)} - \frac{1}{3+4t}\right]$$
$$= \lim_{h \to 0} \frac{3+4t - 3 - 4t - 4h}{h(3+4t)[3+(4t+h)]} = -\frac{4}{(3+4t)^2}$$

17. $F(t) = \sqrt{2t+1}$
$$F'(t) = \lim_{h \to 0} \frac{\sqrt{2(t+h)+1} - \sqrt{2t+1}}{h}$$
$$= \lim_{h \to 0} \frac{2t + 2h + 1 - 2t - 1}{h\left(\sqrt{2(t+h)+1} + \sqrt{2t+1}\right)}$$
$$= \lim_{h \to 0} \frac{2}{\sqrt{2(t+h)+1} + \sqrt{2t+1}}$$
$$= \frac{1}{\sqrt{2t+1}}$$

18. $f(x) = \frac{3}{4}\sqrt{2-x}$
$$f'(x) = \lim_{h \to 0} \frac{\frac{3}{4}\sqrt{2-(x+h)} - \frac{3}{4}\sqrt{2-x}}{h}$$
$$= \lim_{h \to 0} \frac{3}{4}\left[\frac{2-x-h-2+x}{h(\sqrt{2-(x+h)} + \sqrt{2-x})}\right]$$
$$= -\frac{3}{8\sqrt{2-x}}$$

41

19. $y = x + \dfrac{1}{x}$

$$y' = \lim_{h \to 0} \dfrac{x + h + \dfrac{1}{x+h} - x - \dfrac{1}{x}}{h}$$

$$= \lim_{h \to 0} \left(1 + \dfrac{x - x - h}{h(x+h)x}\right)$$

$$= 1 + \lim_{h \to 0} \dfrac{-1}{(x+h)x} = 1 - \dfrac{1}{x^2}$$

20. $z = \dfrac{s}{1+s}$

$$\dfrac{dz}{ds} = \lim_{h \to 0} \dfrac{1}{h} \left[\dfrac{s+h}{1+s+h} - \dfrac{s}{1+s}\right]$$

$$= \lim_{h \to 0} \dfrac{(s+h)(1+s) - s(1+s+h)}{h(1+s)(1+s+h)} = \dfrac{1}{(1+s)^2}$$

21. $F(x) = \dfrac{1}{\sqrt{1+x^2}}$

$$F'(x) = \lim_{h \to 0} \dfrac{\dfrac{1}{\sqrt{1+(x+h)^2}} - \dfrac{1}{\sqrt{1+x^2}}}{h}$$

$$= \lim_{h \to 0} \dfrac{\sqrt{1+x^2} - \sqrt{1+(x+h)^2}}{h\sqrt{1+(x+h)^2}\sqrt{1+x^2}}$$

$$= \lim_{h \to 0} \dfrac{1 + x^2 - 1 - x^2 - 2hx - h^2}{h\sqrt{1+(x+h)^2}\sqrt{1+x^2}\left(\sqrt{1+x^2} + \sqrt{1+(x+h)^2}\right)}$$

$$= \dfrac{-2x}{2(1+x^2)^{3/2}} = -\dfrac{x}{(1+x^2)^{3/2}}$$

22. $y = \dfrac{1}{x^2}$

$$y' = \lim_{h \to 0} \dfrac{1}{h}\left[\dfrac{1}{(x+h)^2} - \dfrac{1}{x^2}\right]$$

$$= \lim_{h \to 0} \dfrac{x^2 - (x+h)^2}{hx^2(x+h)^2} = -\dfrac{2}{x^3}$$

23. $y = \dfrac{1}{\sqrt{1+x}}$

$$y'(x) = \lim_{h \to 0} \dfrac{\dfrac{1}{\sqrt{1+x+h}} - \dfrac{1}{\sqrt{1+x}}}{h}$$

$$= \lim_{h \to 0} \dfrac{\sqrt{1+x} - \sqrt{1+x+h}}{h\sqrt{1+x+h}\sqrt{1+x}}$$

$$= \lim_{h \to 0} \dfrac{1 + x - 1 - x - h}{h\sqrt{1+x+h}\sqrt{1+x}\left(\sqrt{1+x}+\sqrt{1+x+h}\right)}$$

$$= \lim_{h \to 0} -\dfrac{1}{\sqrt{1+x+h}\sqrt{1+x}\left(\sqrt{1+x}+\sqrt{1+x+h}\right)}$$

$$= -\dfrac{1}{2(1+x)^{3/2}}$$

24. $f(t) = \dfrac{t^2 - 3}{t^2 + 3}$

$$f'(t) = \lim_{h \to 0} \dfrac{1}{h}\left(\dfrac{(t+h)^2 - 3}{(t+h)^2 + 3} - \dfrac{t^2 - 3}{t^2 + 3}\right)$$

$$= \lim_{h \to 0} \dfrac{[(t+h)^2 - 3](t^2 + 3) - (t^2 - 3)[(t+h)^2 + 3]}{h(t^2 + 3)[(t+h)^2 + 3]}$$

$$= \lim_{h \to 0} \dfrac{12th + 6h^2}{h(t^2+3)[(t+h)^2 + 3]} = \dfrac{12t}{(t^2+3)^2}$$

25. Since $f(x) = x \operatorname{sgn} x = |x|$, for $x \ne 0$, f will become continuous at $x = 0$ if we define $f(0) = 0$. However, f will still not be differentiable at $x = 0$ since $|x|$ is not differentiable at $x = 0$.

26. Since $g(x) = x^2 \operatorname{sgn} x = x|x| = \begin{cases} x^2 & \text{if } x > 0 \\ -x^2 & \text{if } x < 0 \end{cases}$, g will become continuous and differentiable at $x = 0$ if we define $g(0) = 0$.

27. $h(x) = |x^2 + 3x + 2|$ fails to be differentiable where $x^2 + 3x + 2 = 0$, that is, at $x = -2$ and $x = -1$. Note: both of these are single zeros of $x^2 + 3x + 2$. If they were higher order zeros (i.e. if $(x+2)^n$ or $(x+1)^n$ were a factor of $x^2 + 3x + 2$ for some integer $n \ge 2$) then h would be differentiable at the corresponding point.

28. $y = x^3 - 2x$

x	$\dfrac{f(x) - f(1)}{x - 1}$	x	$\dfrac{f(x) - f(1)}{x - 1}$
0.9	0.71000	1.1	1.31000
0.99	0.97010	1.01	1.03010
0.999	0.99700	1.001	1.00300
0.9999	0.99970	1.0001	1.00030

$$\dfrac{d}{dx}(x^3 - 2x)\bigg|_{x=1} = \lim_{h \to 0} \dfrac{(1+h)^3 - 2(1+h) - (-1)}{h}$$

$$= \lim_{h \to 0} \dfrac{h + 3h^2 + h^3}{h}$$

$$= \lim_{h \to 0} 1 + 3h + h^2 = 1$$

29. $f(x) = 1/x$

x	$\dfrac{f(x) - f(2)}{x - 2}$	x	$\dfrac{f(x) - f(2)}{x - 2}$
1.9	−0.26316	2.1	−0.23810
1.99	−0.25126	2.01	−0.24876
1.999	−0.25013	2.001	−0.24988
1.9999	−0.25001	2.0001	−0.24999

$$f'(2) = \lim_{h \to 0} \dfrac{\dfrac{1}{2+h} - 2}{h} = \lim_{h \to 0} \dfrac{2 - (2+h)}{h(2+h)2}$$

$$= \lim_{h \to 0} -\dfrac{1}{(2+h)2} = -\dfrac{1}{4}$$

30. The slope of $y = 5 + 4x - x^2$ at $x = 2$ is

$$\left.\frac{dy}{dx}\right|_{x=2} = \lim_{h \to 0} \frac{5 + 4(2+h) - (2+h)^2 - 9}{h}$$
$$= \lim_{h \to 0} \frac{-h^2}{h} = 0.$$

Thus, the tangent line at $x = 2$ has the equation $y = 9$.

31. $y = \sqrt{x+6}$. Slope at $(3, 3)$ is

$$m = \lim_{h \to 0} \frac{\sqrt{9+h} - 3}{h} = \lim_{h \to 0} \frac{9+h-9}{h(\sqrt{9+h}+3)} = \frac{1}{6}.$$

Tangent line is $y - 3 = \frac{1}{6}(x - 3)$, or $x - 6y = -15$.

32. The slope of $y = \frac{t}{t^2 - 2}$ at $t = -2$ and $y = -1$ is

$$\left.\frac{dy}{dt}\right|_{t=-2} = \lim_{h \to 0} \frac{1}{h}\left[\frac{-2+h}{(-2+h)^2 - 2} - (-1)\right]$$
$$= \lim_{h \to 0} \frac{-2 + h + [(-2+h)^2 - 2]}{h[(-2+h)^2 - 2]} = -\frac{3}{2}.$$

Thus, the tangent line has the equation
$y = -1 - \frac{3}{2}(t + 2)$, that is, $y = -\frac{3}{2}t - 4$.

33. $y = \frac{2}{t^2 + t}$ Slope at $t = a$ is

$$m = \lim_{h \to 0} \frac{\frac{2}{(a+h)^2 + (a+h)} - \frac{2}{a^2 + a}}{h}$$
$$= \lim_{h \to 0} \frac{2(a^2 + a - a^2 - 2ah - h^2 - a - h)}{h[(a+h)^2 + a + h](a^2 + a)}$$
$$= \lim_{h \to 0} \frac{-4a - 2h - 2}{[(a+h)^2 + a + h](a^2 + a)}$$
$$= -\frac{4a + 2}{(a^2 + a)^2}$$

Tangent line is $y = \frac{2}{a^2 + a} - \frac{2(2a+1)}{(a^2+a)^2}(t - a)$

34. $f'(x) = -17x^{-18}$ for $x \neq 0$

35. $g'(t) = 22t^{21}$ for all t

36. $\frac{dy}{dx} = \frac{1}{3}x^{-2/3}$ for $x \neq 0$

37. $\frac{dy}{dx} = -\frac{1}{3}x^{-4/3}$ for $x \neq 0$

38. $\frac{d}{dt}t^{-2.25} = -2.25t^{-3.25}$ for $t > 0$

39. $\frac{d}{ds}s^{119/4} = \frac{119}{4}s^{115/4}$ for $s > 0$

40. $\left.\frac{d}{ds}\sqrt{s}\right|_{s=9} = \left.\frac{1}{2\sqrt{s}}\right|_{s=9} = \frac{1}{6}.$

41. $F(x) = \frac{1}{x}$, $F'(x) = -\frac{1}{x^2}$, $F'\left(\frac{1}{4}\right) = -16$

42. $f'(8) = \left.-\frac{2}{3}x^{-5/3}\right|_{x=8} = -\frac{1}{48}$

43. $\left.\frac{dy}{dt}\right|_{t=4} = \left.\frac{1}{4}t^{-3/4}\right|_{t=4} = \frac{1}{8\sqrt{2}}$

44. The slope of $y = \sqrt{x}$ at $x = x_0$ is

$$\left.\frac{dy}{dx}\right|_{x=x_0} = \frac{1}{2\sqrt{x_0}}.$$

Thus, the equation of the tangent line is
$y = \sqrt{x_0} + \frac{1}{2\sqrt{x_0}}(x - x_0)$, that is, $y = \frac{x + x_0}{2\sqrt{x_0}}$.

45. Slope of $y = \frac{1}{x}$ at $x = a$ is $\left.-\frac{1}{x^2}\right|_{x=a} = \frac{1}{a^2}$.

Normal has slope a^2, and equation $y - \frac{1}{a} = a^2(x - a)$,
or $y = a^2x - a^3 + \frac{1}{a}$

46. The intersection points of $y = x^2$ and $x + 4y = 18$ satisfy

$$4x^2 + x - 18 = 0$$
$$(4x + 9)(x - 2) = 0.$$

Therefore $x = -\frac{9}{4}$ or $x = 2$.
The slope of $y = x^2$ is $m_1 = \frac{dy}{dx} = 2x$.
At $x = -\frac{9}{4}$, $m_1 = -\frac{9}{2}$. At $x = 2$, $m_1 = 4$.
The slope of $x + 4y = 18$, i.e. $y = -\frac{1}{4}x + \frac{18}{4}$, is $m_2 = -\frac{1}{4}$.
Thus, at $x = 2$, the product of these slopes is
$(4)(-\frac{1}{4}) = -1$. So, the curve and line intersect at right angles at that point.

47. Let the point of tangency be (a, a^2). Slope of tangent is
$$\left.\frac{d}{dx}x^2\right|_{x=a} = 2a$$
This is the slope from (a, a^2) to $(1, -3)$, so
$$\frac{a^2 + 3}{a - 1} = 2a, \text{ and}$$

$$a^2 + 3 = 2a^2 - 2a$$
$$a^2 - 2a - 3 = 0$$
$$a = 3 \text{ or } -1$$

The two tangent lines are
(for $a = 3$): $y - 9 = 6(x - 3)$ or $6x - 9$
(for $a = -1$): $y - 1 = -2(x + 1)$ or $y = -2x - 1$

Fig. 2.2.47

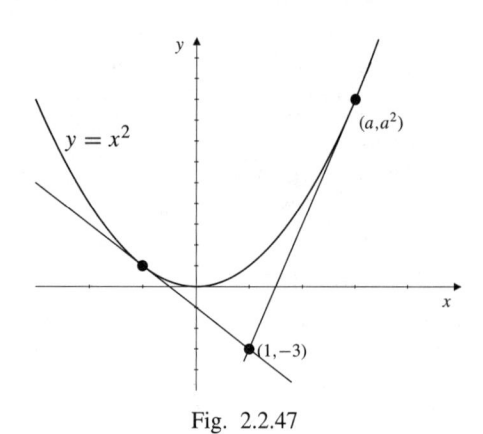

48. The slope of $y = \dfrac{1}{x}$ at $x = a$ is

$$\left.\dfrac{dy}{dx}\right|_{x=a} = -\dfrac{1}{a^2}.$$

If the slope is -2, then $-\dfrac{1}{a^2} = -2$, or $a = \pm\dfrac{1}{\sqrt{2}}$. Therefore, the equations of the two straight lines are

$$y = \sqrt{2} - 2\left(x - \dfrac{1}{\sqrt{2}}\right) \text{ and } y = -\sqrt{2} - 2\left(x + \dfrac{1}{\sqrt{2}}\right),$$

or $y = -2x \pm 2\sqrt{2}$.

49. Let the point of tangency be $(a, \sqrt{a})$

Slope of tangent is $\left.\dfrac{d}{dx}\sqrt{x}\right|_{x=a} = \dfrac{1}{2\sqrt{a}}$

Thus $\dfrac{1}{2\sqrt{a}} = \dfrac{\sqrt{a} - 0}{a + 2}$, so $a + 2 = 2a$, and $a = 2$.

The required slope is $\dfrac{1}{2\sqrt{2}}$.

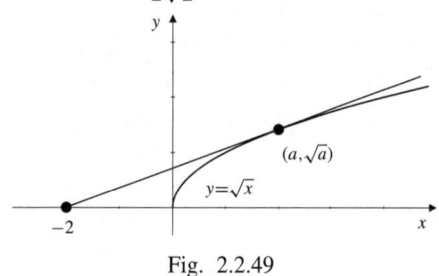

Fig. 2.2.49

50. If a line is tangent to $y = x^2$ at (t, t^2), then its slope is $\left.\dfrac{dy}{dx}\right|_{x=t} = 2t$. If this line also passes through (a, b), then its slope satisfies

$$\dfrac{t^2 - b}{t - a} = 2t, \quad \text{that is } t^2 - 2at + b = 0.$$

Hence $t = \dfrac{2a \pm \sqrt{4a^2 - 4b}}{2} = a \pm \sqrt{a^2 - b}$.

If $b < a^2$, i.e. $a^2 - b > 0$, then $t = a \pm \sqrt{a^2 - b}$ has two real solutions. Therefore, there will be two distinct tangent lines passing through (a, b) with equations $y = b + 2\left(a \pm \sqrt{a^2 - b}\right)(x - a)$. If $b = a^2$, then $t = a$. There will be only one tangent line with slope $2a$ and equation $y = b + 2a(x - a)$.
If $b > a^2$, then $a^2 - b < 0$. There will be no real solution for t. Thus, there will be no tangent line.

51. Suppose f is odd: $f(-x) = -f(x)$. Then

$$f'(-x) = \lim_{h \to 0} \dfrac{f(-x + h) - f(-x)}{h}$$

$$= \lim_{h \to 0} -\dfrac{f(x - h) - f(x)}{h}$$

(let $h = -k$)

$$= \lim_{k \to 0} \dfrac{f(x + k) - f(x)}{k} = f'(x)$$

Thus f' is even.
Now suppose f is even: $f(-x) = f(x)$. Then

$$f'(-x) = \lim_{h \to 0} \dfrac{f(-x + h) - f(-x)}{h}$$

$$= \lim_{h \to 0} \dfrac{f(x - h) - f(x)}{h}$$

$$= \lim_{k \to 0} \dfrac{f(x + k) - f(x)}{-k}$$

$$= -f'(x)$$

so f' is odd.

52. Let $f(x) = x^{-n}$. Then

$$f'(x) = \lim_{h \to 0} \dfrac{(x + h)^{-n} - x^{-n}}{h}$$

$$= \lim_{h \to 0} \dfrac{1}{h}\left(\dfrac{1}{(x + h)^n} - \dfrac{1}{x^n}\right)$$

$$= \lim_{h \to 0} \dfrac{x^n - (x + h)^n}{hx^n(x + h)^n}$$

$$= \lim_{h \to 0} \dfrac{x - (x + h)}{hx^n((x + h)^n} \times$$

$$\left(x^{n-1} + x^{n-2}(x + h) + \cdots + (x + h)^{n-1}\right)$$

$$= -\dfrac{1}{x^{2n}} \times nx^{n-1} = -nx^{-(n+1)}.$$

53. $f(x) = x^{1/3}$

$$f'(x) = \lim_{h \to 0} \frac{(x+h)^{1/3} - x^{1/3}}{h}$$
$$= \lim_{h \to 0} \frac{(x+h)^{1/3} - x^{1/3}}{h}$$
$$\times \frac{(x+h)^{2/3} + (x+h)^{1/3}x^{1/3} + x^{2/3}}{(x+h)^{2/3} + (x+h)^{1/3}x^{1/3} + x^{2/3}}$$
$$= \lim_{h \to 0} \frac{x+h-x}{h[(x+h)^{2/3} + (x+h)^{1/3}x^{1/3} + x^{2/3}]}$$
$$= \lim_{h \to 0} \frac{1}{(x+h)^{2/3} + (x+h)^{1/3}x^{1/3} + x^{2/3}}$$
$$= \frac{1}{3x^{2/3}} = \frac{1}{3}x^{-2/3}$$

54. Let $f(x) = x^{1/n}$. Then

$$f'(x) = \lim_{h \to 0} \frac{(x+h)^{1/n} - x^{1/n}}{h} \quad \text{(let } x+h = a^n, x = b^n\text{)}$$
$$= \lim_{a \to b} \frac{a-b}{a^n - b^n}$$
$$= \lim_{a \to b} \frac{1}{a^{n-1} + a^{n-2}b + a^{n-3}b^2 + \cdots + b^{n-1}}$$
$$= \frac{1}{nb^{n-1}} = \frac{1}{n}x^{(1/n)-1}.$$

55. $$\frac{d}{dx}x^n = \lim_{h \to 0} \frac{(x+h)^n - x^n}{h}$$
$$= \lim_{h \to 0} \frac{1}{h}\left[x^n + \frac{n}{1}x^{n-1}h + \frac{n(n-1)}{1 \times 2}x^{n-2}h^2 \right.$$
$$\left. + \frac{n(n-1)(n-2)}{1 \times 2 \times 3}x^{n-3}h^3 + \cdots + h^n - x^n\right]$$
$$= \lim_{h \to 0}\left(nx^{n-1} + h\left[\frac{n(n-1)}{1 \times 2}x^{n-2}h \right.\right.$$
$$\left.\left. + \frac{n(n-1)(n-2)}{1 \times 2 \times 3}x^{n-3}h^2 + \cdots + h^{n-1}\right]\right)$$
$$= nx^{n-1}$$

56. Let

$$f'(a+) = \lim_{h \to 0+} \frac{f(a+h) - f(a)}{h}$$
$$f'(a-) = \lim_{h \to 0-} \frac{f(a+h) - f(a)}{h}$$

If $f'(a+)$ is finite, call the half-line with equation $y = f(a) + f'(a+)(x - a)$, $(x \geq a)$, the *right tangent line* to the graph of f at $x = a$. Similarly, if $f'(a-)$ is finite, call the half-line $y = f(a) + f'(a-)(x-a)$, $(x \leq a)$, the *left tangent line*. If $f'(a+) = \infty$ (or $-\infty$), the right tangent line is the half-line $x = a$, $y \geq f(a)$ (or $x = a$, $y \leq f(a)$). If $f'(a-) = \infty$ (or $-\infty$), the right tangent line is the half-line $x = a$, $y \leq f(a)$ (or $x = a$, $y \geq f(a)$). The graph has a tangent line at $x = a$ if and only if $f'(a+) = f'(a-)$. (This includes the possibility that both quantities may be $+\infty$ or both may be $-\infty$.) In this case the right and left tangents are two opposite halves of the same straight line. For $f(x) = x^{2/3}$, $f'(x) = \frac{2}{3}x^{-1/3}$. At $(0, 0)$, we have $f'(0+) = +\infty$ and $f'(0-) = -\infty$. In this case both left and right tangents are the *positive y-axis*, and the curve does not have a tangent line at the origin. For $f(x) = |x|$, we have

$$f'(x) = \text{sgn}(x) = \begin{cases} 1 & \text{if } x > 0 \\ -1 & \text{if } x < 0. \end{cases}$$

At $(0, 0)$, $f'(0+) = 1$, and $f'(0-) = -1$. In this case the right tangent is $y = x$, $(x \geq 0)$, and the left tangent is $y = -x$, $(x \leq 0)$. There is no tangent line.

Section 2.3 Differentiation Rules (page 117)

1. $y = 3x^2 - 5x - 7$, $y' = 6x - 5$.
2. $y = x^8 - x^4$, $y' = 8x^7 - 4x^3$
3. $y = \frac{x^3}{3} - \frac{x^2}{2} + x$, $y' = x^2 - x + 1$
4. $y = 4x^{1/2} - \frac{5}{x}$, $y' = 2x^{-1/2} + 5x^{-2}$
5. $f(x) = Ax^2 + Bx + C$, $f'(x) = 2Ax + B$.
6. $f(x) = \frac{6}{x^3} + \frac{2}{x^2} - 2$, $f'(x) = -\frac{18}{x^4} - \frac{4}{x^3}$
7. $z = \frac{s^5 - s^3}{15}$, $\frac{dz}{dx} = \frac{1}{3}s^4 - \frac{1}{5}s^2$.
8. $y = x^{45} - x^{-45}$, $y' = 45x^{44} + 45x^{-46}$
9. $g(t) = t^{1/3} + 2t^{1/4} + 3t^{1/5}$

    $g'(t) = \frac{1}{3}t^{-2/3} + \frac{1}{2}t^{-3/4} + \frac{3}{5}t^{-4/5}$

10. $y = 3\sqrt[3]{t^2} - \frac{2}{\sqrt{t^3}} = 3t^{2/3} - 2t^{-3/2}$

    $\frac{dy}{dt} = 2t^{-1/3} + 3t^{-5/2}$

11. $u = \frac{3}{5}x^{5/3} - \frac{5}{3}x^{-3/5}$

    $\frac{du}{dx} = x^{2/3} + x^{-8/5}$

12. $F(x) = (3x-2)(1-5x)$
$F'(x) = 3(1-5x) + (3x-2)(-5) = 13 - 30x$

13. $y = \sqrt{x}\left(5 - x - \dfrac{x^2}{3}\right) = 5\sqrt{x} - x^{3/2} - \dfrac{1}{3}x^{5/2}$
$y' = \dfrac{5}{2\sqrt{x}} - \dfrac{3}{2}\sqrt{x} - \dfrac{5}{6}x^{3/2}$

14. $g(t) = \dfrac{1}{2t-3}, \quad g'(t) = -\dfrac{2}{(2t-3)^2}$

15. $y = \dfrac{1}{x^2 + 5x}$
$y' = -\dfrac{1}{(x^2+5x)^2}(2x+5) = -\dfrac{2x+5}{(x^2+5x)^2}$

16. $y = \dfrac{4}{3-x}, \quad y' = \dfrac{4}{(3-x)^2}$

17. $f(t) = \dfrac{\pi}{2 - \pi t}$
$f'(t) = -\dfrac{\pi}{(2-\pi t)^2}(-\pi) = \dfrac{\pi^2}{(2-\pi t)^2}$

18. $g(y) = \dfrac{2}{1-y^2}, \quad g'(y) = \dfrac{4y}{(1-y^2)^2}$

19. $f(x) = \dfrac{1 - 4x^2}{x^3} = x^{-3} - \dfrac{4}{x}$
$f'(x) = -3x^{-4} + 4x^{-2} = \dfrac{4x^2 - 3}{x^4}$

20. $g(u) = \dfrac{u\sqrt{u} - 3}{u^2} = u^{-1/2} - 3u^{-2}$
$g'(u) = -\dfrac{1}{2}u^{-3/2} + 6u^{-3} = \dfrac{12 - u\sqrt{u}}{2u^3}$

21. $y = \dfrac{2 + t + t^2}{\sqrt{t}} = 2t^{-1/2} + \sqrt{t} + t^{3/2}$
$\dfrac{dy}{dt} = -t^{-3/2} + \dfrac{1}{2\sqrt{t}} + \dfrac{3}{2}\sqrt{t} = \dfrac{3t^2 + t - 2}{2t\sqrt{t}}$

22. $z = \dfrac{x-1}{x^{2/3}} = x^{1/3} - x^{-2/3}$
$\dfrac{dz}{dx} = \dfrac{1}{3}x^{-2/3} + \dfrac{2}{3}x^{-5/3} = \dfrac{x+2}{3x^{5/3}}$

23. $y = \dfrac{3}{x + \sqrt{x}}$
$y' = \dfrac{-3}{(x+\sqrt{x})^2}\left(1 + \dfrac{1}{2\sqrt{x}}\right) = -\dfrac{3(2\sqrt{x}+1)}{2\sqrt{x}(x+\sqrt{x})^2}$

24. $y = \dfrac{x-2}{x+2}$
$y' = \dfrac{(x+2)(1) - (x-2)(1)}{(x+2)^2} = \dfrac{4}{(x+2)^2}$

25. $f(x) = \dfrac{3 - 4x}{3 + 4x}$
$f'(x) = \dfrac{(3+4x)(-4) - (3-4x)(4)}{(3+4x)^2}$
$= -\dfrac{24}{(3+4x)^2}$

26. $z = \dfrac{t^2 + 2t}{t^2 - 1}$
$z' = \dfrac{(t^2-1)(2t+2) - (t^2+2t)(2t)}{(t^2-1)^2}$
$= -\dfrac{2(t^2 + t + 1)}{(t^2-1)^2}$

27. $s = \dfrac{1 + \sqrt{t}}{1 - \sqrt{t}}$
$\dfrac{ds}{dt} = \dfrac{(1-\sqrt{t})\dfrac{1}{2\sqrt{t}} - (1+\sqrt{t})(-\dfrac{1}{2\sqrt{t}})}{(1-\sqrt{t})^2}$
$= \dfrac{1}{\sqrt{t}(1-\sqrt{t})^2}$

28. $f(x) = \dfrac{x^3 - 4}{x + 1}$
$f'(x) = \dfrac{(x+1)(3x^2) - (x^3-4)(1)}{(x+1)^2}$
$= \dfrac{2x^3 + 3x^2 + 4}{(x+1)^2}$

29. $f(x) = \dfrac{ax + b}{cx + d}$
$f'(x) = \dfrac{(cx+d)a - (ax+b)c}{(cx+d)^2}$
$= \dfrac{ad - bc}{(cx+d)^2}$

30. $F(t) = \dfrac{t^2 + 7t - 8}{t^2 - t + 1}$
$F'(t) = \dfrac{(t^2-t+1)(2t+7) - (t^2+7t-8)(2t-1)}{(t^2-t+1)^2}$
$= \dfrac{-8t^2 + 18t - 1}{(t^2-t+1)^2}$

31. $f(x) = (1+x)(1+2x)(1+3x)(1+4x)$
$f'(x) = (1+2x)(1+3x)(1+4x) + 2(1+x)(1+3x)(1+4x)$
$\quad + 3(1+x)(1+2x)(1+4x) + 4(1+x)(1+2x)(1+3x)$
OR
$f(x) = [(1+x)(1+4x)][(1+2x)(1+3x)]$
$= (1 + 5x + 4x^2)(1 + 5x + 6x^2)$
$= 1 + 10x + 25x^2 + 10x^2(1+5x) + 24x^4$
$= 1 + 10x + 35x^2 + 50x^3 + 24x^4$
$f'(x) = 10 + 70x + 150x^2 + 96x^3$

INSTRUCTOR'S SOLUTIONS MANUAL   SECTION 2.3 (PAGE 117)

32. $f(r) = (r^{-2} + r^{-3} - 4)(r^2 + r^3 + 1)$
$f'(r) = (-2r^{-3} - 3r^{-4})(r^2 + r^3 + 1)$
$\qquad + (r^{-2} + r^{-3} - 4)(2r + 3r^2)$
or
$f(r) = -2 + r^{-1} + r^{-2} + r^{-3} + r - 4r^2 - 4r^3$
$f'(r) = -r^{-2} - 2r^{-3} - 3r^{-4} + 1 - 8r - 12r^2$

33. $y = (x^2 + 4)(\sqrt{x} + 1)(5x^{2/3} - 2)$
$y' = 2x(\sqrt{x} + 1)(5x^{2/3} - 2)$
$\qquad + \frac{1}{2\sqrt{x}}(x^2 + 4)(5x^{2/3} - 2)$
$\qquad + \frac{10}{3}x^{-1/3}(x^2 + 4)(\sqrt{x} + 1)$

34. $y = \frac{(x^2+1)(x^3+2)}{(x^2+2)(x^3+1)}$
$= \frac{x^5 + x^3 + 2x^2 + 2}{x^5 + 2x^3 + x^2 + 2}$
$y' = \frac{(x^5+2x^3+x^2+2)(5x^4+3x^2+4x)}{(x^5+2x^3+x^2+2)^2}$
$\qquad - \frac{(x^5+x^3+2x^2+2)(5x^4+6x^2+2x)}{(x^5+2x^3+x^2+2)^2}$
$= \frac{2x^7 - 3x^6 - 3x^4 - 6x^2 + 4x}{(x^5+2x^3+x^2+2)^2}$
$= \frac{2x^7 - 3x^6 - 3x^4 - 6x^2 + 4x}{(x^2+2)^2(x^3+1)^2}$

35. $y = \dfrac{x}{2x + \dfrac{1}{3x+1}} = \dfrac{3x^2 + x}{6x^2 + 2x + 1}$
$y' = \frac{(6x^2+2x+1)(6x+1) - (3x^2+x)(12x+2)}{(6x^2+2x+1)^2}$
$= \frac{6x+1}{(6x^2+2x+1)^2}$

36. $f(x) = \dfrac{(\sqrt{x}-1)(2-x)(1-x^2)}{\sqrt{x}(3+2x)}$
$= \left(1 - \dfrac{1}{\sqrt{x}}\right) \cdot \dfrac{2 - x - 2x^2 + x^3}{3+2x}$
$f'(x) = \left(\dfrac{1}{2}x^{-3/2}\right)\dfrac{2-x-2x^2+x^3}{3+2x} + \left(1 - \dfrac{1}{\sqrt{x}}\right)$
$\qquad \times \dfrac{(3+2x)(-1-4x+3x^2) - (2-x-2x^2+x^3)(2)}{(3+2x)^2}$
$= \dfrac{(2-x)(1-x^2)}{2x^{3/2}(3+2x)}$
$\qquad + \left(1 - \dfrac{1}{\sqrt{x}}\right)\dfrac{4x^3 + 5x^2 - 12x - 7}{(3+2x)^2}$

37. $\dfrac{d}{dx}\left(\dfrac{x^2}{f(x)}\right)\bigg|_{x=2} = \dfrac{f(x)(2x) - x^2 f'(x)}{[f(x)]^2}\bigg|_{x=2}$
$= \dfrac{4f(2) - 4f'(2)}{[f(2)]^2} = -\dfrac{4}{4} = -1$

38. $\dfrac{d}{dx}\left(\dfrac{f(x)}{x^2}\right)\bigg|_{x=2} = \dfrac{x^2 f'(x) - 2x f(x)}{x^4}\bigg|_{x=2}$
$= \dfrac{4f'(2) - 4f(2)}{16} = \dfrac{4}{16} = \dfrac{1}{4}$

39. $\dfrac{d}{dx}(x^2 f(x))\bigg|_{x=2} = (2x f(x) + x^2 f'(x))\bigg|_{x=2}$
$= 4f(2) + 4f'(2) = 20$

40. $\dfrac{d}{dx}\left(\dfrac{f(x)}{x^2 + f(x)}\right)\bigg|_{x=2}$
$= \dfrac{(x^2+f(x))f'(x) - f(x)(2x+f'(x))}{(x^2+f(x))^2}\bigg|_{x=2}$
$= \dfrac{(4+f(2))f'(2) - f(2)(4+f'(2))}{(4+f(2))^2} = \dfrac{18-14}{6^2} = \dfrac{1}{9}$

41. $\dfrac{d}{dx}\left(\dfrac{x^2-4}{x^2+4}\right)\bigg|_{x=-2} = \dfrac{d}{dx}\left(1 - \dfrac{8}{x^2+4}\right)\bigg|_{x=-2}$
$= \dfrac{8}{(x^2+4)^2}(2x)\bigg|_{x=-2}$
$= -\dfrac{32}{64} = -\dfrac{1}{2}$

42. $\dfrac{d}{dt}\left[\dfrac{t(1+\sqrt{t})}{5-t}\right]\bigg|_{t=4}$
$= \dfrac{d}{dt}\left[\dfrac{t + t^{3/2}}{5-t}\right]\bigg|_{t=4}$
$= \dfrac{(5-t)(1+\frac{3}{2}t^{1/2}) - (t+t^{3/2})(-1)}{(5-t)^2}\bigg|_{t=4}$
$= \dfrac{(1)(4) - (12)(-1)}{(1)^2} = 16$

43. $f(x) = \dfrac{\sqrt{x}}{x+1}$
$f'(x) = \dfrac{(x+1)\dfrac{1}{2\sqrt{x}} - \sqrt{x}(1)}{(x+1)^2}$
$f'(2) = \dfrac{\dfrac{3}{2\sqrt{2}} - \sqrt{2}}{9} = -\dfrac{1}{18\sqrt{2}}$

44. $\dfrac{d}{dt}[(1+t)(1+2t)(1+3t)(1+4t)]\bigg|_{t=0}$
$= (1)(1+2t)(1+3t)(1+4t) + (1+t)(2)(1+3t)(1+4t) +$
$(1+t)(1+2t)(3)(1+4t) + (1+t)(1+2t)(1+3t)(4)\bigg|_{t=0}$
$= 1 + 2 + 3 + 4 = 10$

47

45. $y = \dfrac{2}{3 - 4\sqrt{x}}$, $y' = -\dfrac{2}{(3 - 4\sqrt{x})^2}\left(-\dfrac{4}{2\sqrt{x}}\right)$

Slope of tangent at $(1, -2)$ is $m = \dfrac{8}{(-1)^2 2} = 4$

Tangent line has the equation $y = -2 + 4(x - 1)$ or $y = 4x - 6$

46. For $y = \dfrac{x + 1}{x - 1}$ we calculate

$$y' = \dfrac{(x - 1)(1) - (x + 1)(1)}{(x - 1)^2} = -\dfrac{2}{(x - 1)^2}.$$

At $x = 2$ we have $y = 3$ and $y' = -2$. Thus, the equation of the tangent line is $y = 3 - 2(x - 2)$, or $y = -2x + 7$. The normal line is $y = 3 + \frac{1}{2}(x - 2)$, or $y = \frac{1}{2}x + 2$.

47. $y = x + \dfrac{1}{x}$, $y' = 1 - \dfrac{1}{x^2}$

For horizontal tangent: $0 = y' = 1 - \dfrac{1}{x^2}$ so $x^2 = 1$ and $x = \pm 1$

The tangent is horizontal at $(1, 2)$ and at $(-1, -2)$

48. If $y = x^2(4 - x^2)$, then

$$y' = 2x(4 - x^2) + x^2(-2x) = 8x - 4x^3 = 4x(2 - x^2).$$

The slope of a horizontal line must be zero, so $4x(2 - x^2) = 0$, which implies that $x = 0$ or $x = \pm\sqrt{2}$. At $x = 0$, $y = 0$ and at $x = \pm\sqrt{2}$, $y = 4$.
Hence, there are two horizontal lines that are tangent to the curve. Their equations are $y = 0$ and $y = 4$.

49. $y = \dfrac{1}{x^2 + x + 1}$, $y' = -\dfrac{2x + 1}{(x^2 + x + 1)^2}$

For horizontal tangent we want $0 = y' = -\dfrac{2x + 1}{(x^2 + x + 1)^2}$.

Thus $2x + 1 = 0$ and $x = -\dfrac{1}{2}$

The tangent is horizontal only at $\left(-\dfrac{1}{2}, \dfrac{4}{3}\right)$.

50. If $y = \dfrac{x + 1}{x + 2}$, then

$$y' = \dfrac{(x + 2)(1) - (x + 1)(1)}{(x + 2)^2} = \dfrac{1}{(x + 2)^2}.$$

In order to be parallel to $y = 4x$, the tangent line must have slope equal to 4, i.e.,

$$\dfrac{1}{(x + 2)^2} = 4, \quad \text{or } (x + 2)^2 = \tfrac{1}{4}.$$

Hence $x + 2 = \pm\frac{1}{2}$, and $x = -\frac{3}{2}$ or $-\frac{5}{2}$. At $x = -\frac{3}{2}$, $y = -1$, and at $x = -\frac{5}{2}$, $y = 3$.
Hence, the tangent is parallel to $y = 4x$ at the points $\left(-\frac{3}{2}, -1\right)$ and $\left(-\frac{5}{2}, 3\right)$.

51. Let the point of tangency be $(a, \frac{1}{a})$. The slope of the tangent is $-\dfrac{1}{a^2} = \dfrac{b - \frac{1}{a}}{0 - a}$. Thus $b - \frac{1}{a} = \frac{1}{a}$ and $a = \dfrac{2}{b}$.

Tangent has slope $-\dfrac{b^2}{4}$ so has equation $y = b - \dfrac{b^2}{4}x$.

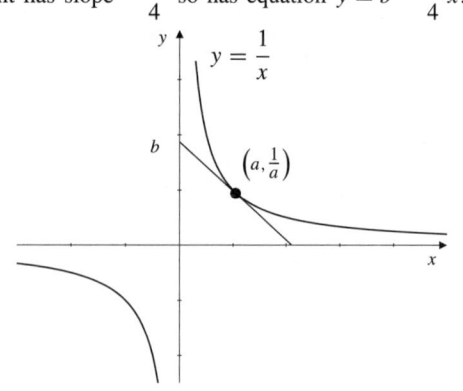

Fig. 2.3.51

52. Since $\dfrac{1}{\sqrt{x}} = y = x^2 \Rightarrow x^{5/2} = 1$, therefore $x = 1$ at the intersection point. The slope of $y = x^2$ at $x = 1$ is $2x\Big|_{x=1} = 2$. The slope of $y = \dfrac{1}{\sqrt{x}}$ at $x = 1$ is

$$\dfrac{dy}{dx}\bigg|_{x=1} = -\dfrac{1}{2}x^{-3/2}\bigg|_{x=1} = -\dfrac{1}{2}.$$

The product of the slopes is $(2)\left(-\frac{1}{2}\right) = -1$. Hence, the two curves intersect at right angles.

53. The tangent to $y = x^3$ at (a, a^3) has equation $y = a^3 + 3a^2(x - a)$, or $y = 3a^2 x - 2a^3$. This line passes through $(2, 8)$ if $8 = 6a^2 - 2a^3$ or, equivalently, if $a^3 - 3a^2 + 4 = 0$. Since $(2, 8)$ lies on $y = x^3$, $a = 2$ must be a solution of this equation. In fact it must be a double root; $(a - 2)^2$ must be a factor of $a^3 - 3a^2 + 4$. Dividing by this factor, we find that the other factor is $a + 1$, that is,

$$a^3 - 3a^2 + 4 = (a - 2)^2(a + 1).$$

The two tangent lines to $y = x^3$ passing through $(2, 8)$ correspond to $a = 2$ and $a = -1$, so their equations are $y = 12x - 16$ and $y = 3x + 2$.

54. The tangent to $y = x^2/(x - 1)$ at $(a, a^2/(a - 1))$ has slope

$$m = \dfrac{(x - 1)2x - x^2(1)}{(x - 1)^2}\bigg|_{x=a} = \dfrac{a^2 - 2a}{(a - 1)^2}.$$

The equation of the tangent is

$$y - \dfrac{a^2}{a - 1} = \dfrac{a^2 - 2a}{(a - 1)^2}(x - a).$$

48

This line passes through $(2, 0)$ provided

$$0 - \frac{a^2}{a-1} = \frac{a^2 - 2a}{(a-1)^2}(2-a),$$

or, upon simplification, $3a^2 - 4a = 0$. Thus we can have either $a = 0$ or $a = 4/3$. There are two tangents through $(2, 0)$. Their equations are $y = 0$ and $y = -8x + 16$.

55. $\frac{d}{dx}\sqrt{f(x)} = \lim_{h \to 0} \frac{\sqrt{f(x+h)} - \sqrt{f(x)}}{h}$

$= \lim_{h \to 0} \frac{f(x+h) - f(x)}{h} \cdot \frac{1}{\sqrt{f(x+h)} + \sqrt{f(x)}}$

$= \frac{f'(x)}{2\sqrt{f(x)}}$

$\frac{d}{dx}\sqrt{x^2+1} = \frac{2x}{2\sqrt{x^2+1}} = \frac{x}{\sqrt{x^2+1}}$

56. $f(x) = |x^3| = \begin{cases} x^3 & \text{if } x \geq 0 \\ -x^3 & \text{if } x < 0 \end{cases}$. Therefore f is differentiable everywhere except *possibly* at $x = 0$, However,

$\lim_{h \to 0+} \frac{f(0+h) - f(0)}{h} = \lim_{h \to 0+} h^2 = 0$

$\lim_{h \to 0-} \frac{f(0+h) - f(0)}{h} = \lim_{h \to 0-} (-h^2) = 0.$

Thus $f'(0)$ exists and equals 0. We have

$f'(x) = \begin{cases} 3x^2 & \text{if } x \geq 0 \\ -3x^2 & \text{if } x < 0. \end{cases}$

57. To be proved: $\frac{d}{dx}x^{n/2} = \frac{n}{2}x^{(n/2)-1}$ for $n = 1, 2, 3, \ldots$.
Proof: It is already known that the case $n = 1$ is true: the derivative of $x^{1/2}$ is $(1/2)x^{-1/2}$.
Assume that the formula is valid for $n = k$ for some positive integer k:

$\frac{d}{dx}x^{k/2} = \frac{k}{2}x^{(k/2)-1}.$

Then, by the Product Rule and this hypothesis,

$\frac{d}{dx}x^{(k+1)/2} = \frac{d}{dx}x^{1/2}x^{k/2}$

$= \frac{1}{2}x^{-1/2}x^{k/2} + \frac{k}{2}x^{1/2}x^{(k/2)-1} = \frac{k+1}{2}x^{(k+1)/2-1}.$

Thus the formula is also true for $n = k+1$. Therefore it is true for all positive integers n by induction.
For negative $n = -m$ (where $m > 0$) we have

$\frac{d}{dx}x^{n/2} = \frac{d}{dx}\frac{1}{x^{m/2}}$

$= \frac{-1}{x^m}\frac{m}{2}x^{(m/2)-1}$

$= -\frac{m}{2}x^{-(m/2)-1} = \frac{n}{2}x^{(n/2)-1}.$

58. To be proved:

$(f_1 f_2 \cdots f_n)'$
$= f_1' f_2 \cdots f_n + f_1 f_2' \cdots f_n + \cdots + f_1 f_2 \cdots f_n'$

Proof: The case $n = 2$ is just the Product Rule. Assume the formula holds for $n = k$ for some integer $k > 2$. Using the Product Rule and this hypothesis we calculate

$(f_1 f_2 \cdots f_k f_{k+1})'$
$= [(f_1 f_2 \cdots f_k) f_{k+1}]'$
$= (f_1 f_2 \cdots f_k)' f_{k+1} + (f_1 f_2 \cdots f_k) f'_{k+1}$
$= (f_1' f_2 \cdots f_k + f_1 f_2' \cdots f_k + \cdots + f_1 f_2 \cdots f_k') f_{k+1}$
$\quad + (f_1 f_2 \cdots f_k) f'_{k+1}$
$= f_1' f_2 \cdots f_k f_{k+1} + f_1 f_2' \cdots f_k f_{k+1} + \cdots$
$\quad + f_1 f_2 \cdots f_k' f_{k+1} + f_1 f_2 \cdots f_k f'_{k+1}$

so the formula is also true for $n = k+1$. The formula is therefore for all integers $n \geq 2$ by induction.

Section 2.4 The Chain Rule (page 123)

1. $y = (2x+3)^6$, $y' = 6(2x+3)^5 2 = 12(2x+3)^5$

2. $y = \left(1 - \frac{x}{3}\right)^{99}$

$y' = 99\left(1 - \frac{x}{3}\right)^{98}\left(-\frac{1}{3}\right) = -33\left(1 - \frac{x}{3}\right)^{98}$

3. $f(x) = (4 - x^2)^{10}$
$f'(x) = 10(4 - x^2)^9(-2x) = -20x(4 - x^2)^9$

4. $\frac{dy}{dx} = \frac{d}{dx}\sqrt{1 - 3x^2} = \frac{-6x}{2\sqrt{1 - 3x^2}} = -\frac{3x}{\sqrt{1 - 3x^2}}$

5. $F(t) = \left(2 + \frac{3}{t}\right)^{-10}$

$F'(t) = -10\left(2 + \frac{3}{t}\right)^{-11}\frac{-3}{t^2} = \frac{30}{t^2}\left(2 + \frac{3}{t}\right)^{-11}$

6. $z = (1 + x^{2/3})^{3/2}$
$z' = \frac{3}{2}(1 + x^{2/3})^{1/2}(\frac{2}{3}x^{-1/3}) = x^{-1/3}(1 + x^{2/3})^{1/2}$

7. $y = \frac{3}{5 - 4x}$

$y' = -\frac{3}{(5 - 4x)^2}(-4) = \frac{12}{(5 - 4x)^2}$

8. $y = (1 - 2x^2)^{-3/2}$
$y' = -\frac{3}{2}(1 - 2x^2)^{-5/2}(-4x) = 6x(1 - 2x^2)^{-5/2}$

SECTION 2.4 (PAGE 123)

9. $y = |1 - x^2|$, $y' = -2x\,\text{sgn}\,(1 - x^2) = \dfrac{2x^3 - 2x}{|1 - x^2|}$

10. $f(t) = |2 + t^3|$

   $f'(t) = [\text{sgn}\,(2 + t^3)](3t^2) = \dfrac{3t^2(2 + t^3)}{|2 + t^3|}$

11. $y = 4x + |4x - 1|$

   $y' = 4 + 4(\text{sgn}\,(4x - 1))$

   $= \begin{cases} 8 & \text{if } x > \frac{1}{4} \\ 0 & \text{if } x < \frac{1}{4} \end{cases}$

12. $y = (2 + |x|^3)^{1/3}$

   $y' = \frac{1}{3}(2 + |x|^3)^{-2/3}(3|x|^2)\text{sgn}\,(x)$

   $= |x|^2(2 + |x|^3)^{-2/3}\left(\dfrac{x}{|x|}\right) = x|x|(2 + |x|^3)^{-2/3}$

13. $y = \dfrac{1}{2 + \sqrt{3x + 4}}$

   $y' = -\dfrac{1}{(2 + \sqrt{3x + 4})^2}\left(\dfrac{3}{2\sqrt{3x + 4}}\right)$

   $= -\dfrac{3}{2\sqrt{3x + 4}(2 + \sqrt{3x + 4})^2}$

14. $f(x) = \left(1 + \sqrt{\dfrac{x - 2}{3}}\right)^4$

   $f'(x) = 4\left(1 + \sqrt{\dfrac{x - 2}{3}}\right)^3\left(\dfrac{1}{2}\sqrt{\dfrac{3}{x - 2}}\right)\left(\dfrac{1}{3}\right)$

   $= \dfrac{2}{3}\sqrt{\dfrac{3}{x - 2}}\left(1 + \sqrt{\dfrac{x - 2}{3}}\right)^3$

15. $z = \left(u + \dfrac{1}{u - 1}\right)^{-5/3}$

   $\dfrac{dz}{du} = -\dfrac{5}{3}\left(u + \dfrac{1}{u - 1}\right)^{-8/3}\left(1 - \dfrac{1}{(u - 1)^2}\right)$

   $= -\dfrac{5}{3}\left(1 - \dfrac{1}{(u - 1)^2}\right)\left(u + \dfrac{1}{u - 1}\right)^{-8/3}$

16. $y = \dfrac{x^5\sqrt{3 + x^6}}{(4 + x^2)^3}$

   $y' = \dfrac{1}{(4 + x^2)^6}\left((4 + x^2)^3\left[5x^4\sqrt{3 + x^6} + x^5\left(\dfrac{3x^5}{\sqrt{3 + x^6}}\right)\right]\right.$

   $\left. - x^5\sqrt{3 + x^6}[3(4 + x^2)^2(2x)]\right)$

   $= \dfrac{(4 + x^2)[5x^4(3 + x^6) + 3x^{10}] - x^5(3 + x^6)(6x)}{(4 + x^2)^4\sqrt{3 + x^6}}$

   $= \dfrac{60x^4 - 3x^6 + 32x^{10} + 2x^{12}}{(4 + x^2)^4\sqrt{3 + x^6}}$

17. and 18.

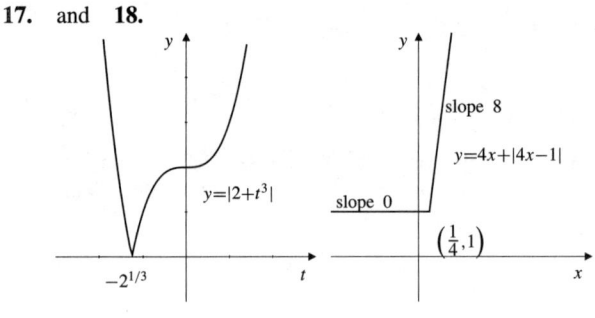

Fig. 2.4.17   Fig. 2.4.18

19. $\dfrac{d}{dx}x^{1/4} = \dfrac{d}{dx}\sqrt{\sqrt{x}} = \dfrac{1}{2\sqrt{\sqrt{x}}} \times \dfrac{1}{2\sqrt{x}} = \dfrac{1}{4}x^{-3/4}$

20. $\dfrac{d}{dx}x^{3/4} = \dfrac{d}{dx}\sqrt{x\sqrt{x}} = \dfrac{1}{2\sqrt{x\sqrt{x}}}\left(\sqrt{x} + \dfrac{x}{2\sqrt{x}}\right) = \dfrac{3}{4}x^{-1/4}$

21. $\dfrac{d}{dx}x^{3/2} = \dfrac{d}{dx}\sqrt{x^3} = \dfrac{1}{2\sqrt{x^3}}(3x^2) = \dfrac{3}{2}x^{1/2}$

22. $\dfrac{d}{dt}f(2t + 3) = 2f'(2t + 3)$

23. $\dfrac{d}{dx}f(5x - x^2) = (5 - 2x)f'(5x - x^2)$

24. $\dfrac{d}{dx}\left[f\left(\dfrac{2}{x}\right)\right]^3 = 3\left[f\left(\dfrac{2}{x}\right)\right]^2 f'\left(\dfrac{2}{x}\right)\left(\dfrac{-2}{x^2}\right)$

   $= -\dfrac{2}{x^2}f'\left(\dfrac{2}{x}\right)\left[f\left(\dfrac{2}{x}\right)\right]^2$

25. $\dfrac{d}{dx}\sqrt{3 + 2f(x)} = \dfrac{2f'(x)}{2\sqrt{3 + 2f(x)}} = \dfrac{f'(x)}{\sqrt{3 + 2f(x)}}$

26. $\dfrac{d}{dt}f(\sqrt{3 + 2t}) = f'(\sqrt{3 + 2t})\dfrac{2}{2\sqrt{3 + 2t}}$

   $= \dfrac{1}{\sqrt{3 + 2t}}f'(\sqrt{3 + 2t})$

27. $\dfrac{d}{dx}f(3 + 2\sqrt{x}) = \dfrac{1}{\sqrt{x}}f'(3 + 2\sqrt{x})$

29. $\dfrac{d}{dt}f\Big(2f\big(3f(x)\big)\Big)$

   $= f'\Big(2f\big(3f(x)\big)\Big) \cdot 2f'\big(3f(x)\big) \cdot 3f'(x)$

   $= 6f'(x)f'\big(3f(x)\big)f'\Big(2f\big(3f(x)\big)\Big)$

29. $\dfrac{d}{dx}f\big(2 - 3f(4 - 5t)\big)$

   $= f'\big(2 - 3f(4 - 5t)\big)\big(-3f'(4 - 5t)\big)(-5)$

   $= 15f'(4 - 5t)f'\big(2 - 3f(4 - 5t)\big)$

50

30. $\dfrac{d}{dx}\left(\dfrac{\sqrt{x^2-1}}{x^2+1}\right)\bigg|_{x=-2}$

$= \dfrac{(x^2+1)\dfrac{x}{\sqrt{x^2-1}} - \sqrt{x^2-1}(2x)}{(x^2+1)^2}\bigg|_{x=-2}$

$= \dfrac{(5)\left(-\dfrac{2}{\sqrt{3}}\right) - \sqrt{3}(-4)}{25} = \dfrac{2}{25\sqrt{3}}$

31. $\dfrac{d}{dt}\sqrt{3t-7}\bigg|_{t=3} = \dfrac{3}{2\sqrt{3t-7}}\bigg|_{t=3} = \dfrac{3}{2\sqrt{2}}$

32. $f(x) = \dfrac{1}{\sqrt{2x+1}}$

$f'(4) = -\dfrac{1}{(2x+1)^{3/2}}\bigg|_{x=4} = -\dfrac{1}{27}$

33. $y = (x^3+9)^{17/2}$

$y'\bigg|_{x=-2} = \dfrac{17}{2}(x^3+9)^{15/2}3x^2\bigg|_{x=-2} = \dfrac{17}{2}(12) = 102$

34. $F(x) = (1+x)(2+x)^2(3+x)^3(4+x)^4$

$F'(x) = (2+x)^2(3+x)^3(4+x)^4 +$
$\quad 2(1+x)(2+x)(3+x)^3(4+x)^4 +$
$\quad 3(1+x)(2+x)^2(3+x)^2(4+x)^4 +$
$\quad 4(1+x)(2+x)^2(3+x)^3(4+x)^3$

$F'(0) = (2^2)(3^3)(4^4) + 2(1)(2)(3^3)(4^4) +$
$\quad 3(1)(2^2)(3^2)(4^4) + 4(1)(2^2)(3^3)(4^3)$
$= 4(2^2 \cdot 3^3 \cdot 4^4) = 110{,}592$

35. $y = \left(x + \left((3x)^5 - 2\right)^{-1/2}\right)^{-6}$

$y' = -6\left(x + \left((3x)^5-2\right)^{-1/2}\right)^{-7}$
$\quad \times \left(1 - \dfrac{1}{2}\left((3x)^5-2\right)^{-3/2}\left(5(3x)^4 3\right)\right)$
$= -6\left(1 - \dfrac{15}{2}(3x)^4\left((3x)^5-2\right)^{-3/2}\right)$
$\quad \times \left(x + \left((3x)^5-2\right)^{-1/2}\right)^{-7}$

36. The slope of $y = \sqrt{1+2x^2}$ at $x=2$ is

$\dfrac{dy}{dx}\bigg|_{x=2} = \dfrac{4x}{2\sqrt{1+2x^2}}\bigg|_{x=2} = \dfrac{4}{3}.$

Thus, the equation of the tangent line at $(2,3)$ is $y = 3 + \tfrac{4}{3}(x-2)$, or $y = \tfrac{4}{3}x + \tfrac{1}{3}$.

37. Slope of $y = (1+x^{2/3})^{3/2}$ at $x=-1$ is

$\dfrac{3}{2}(1+x^{2/3})^{1/2}\left(\dfrac{2}{3}x^{-1/3}\right)\bigg|_{x=-1} = -\sqrt{2}$

The tangent line at $(-1, 2^{3/2})$ has equation
$y = 2^{3/2} - \sqrt{2}(x+1)$.

38. The slope of $y = (ax+b)^8$ at $x = \dfrac{b}{a}$ is

$\dfrac{dy}{dx}\bigg|_{x=b/a} = 8a(ax+b)^7\bigg|_{x=b/a} = 1024ab^7.$

The equation of the tangent line at $x = \dfrac{b}{a}$ and $y = (2b)^8 = 256b^8$ is
$y = 256b^8 + 1024ab^7\left(x - \dfrac{b}{a}\right)$, or $y = 2^{10}ab^7 x - 3 \times 2^8 b^8$.

39. Slope of $y = 1/(x^2 - x + 3)^{3/2}$ at $x=-2$ is

$-\dfrac{3}{2}(x^2-x+3)^{-5/2}(2x-1)\bigg|_{x=-2} = -\dfrac{3}{2}(9^{-5/2})(-5) = \dfrac{5}{162}$

The tangent line at $\left(-2, \dfrac{1}{27}\right)$ has equation
$y = \dfrac{1}{27} + \dfrac{5}{162}(x+2)$.

40. Given that $f(x) = (x-a)^m(x-b)^n$ then

$f'(x) = m(x-a)^{m-1}(x-b)^n + n(x-a)^m(x-b)^{n-1}$
$= (x-a)^{m-1}(x-b)^{n-1}(mx - mb + nx - na).$

If $x \neq a$ and $x \neq b$, then $f'(x) = 0$ if and only if

$mx - mb + nx - na = 0,$

which is equivalent to

$x = \dfrac{n}{m+n}a + \dfrac{m}{m+n}b.$

This point lies lies between a and b.

The solutions given for Exercises 41–44 are just the answers provided by Derive.

41. $x(x^4 + 2x^2 - 2)/(x^2+1)^{5/2}$

42. $4(7x^4 - 49x^2 + 54)/x^7$

43. $857{,}592$

44. $5/8$

45. The Chain Rule does *not* enable you to calculate the derivatives of $|x|^2$ and $|x^2|$ at $x=0$ directly as a composition of two functions, one of which is $|x|$, because $|x|$ is not differentiable at $x=0$. However, $|x|^2 = x^2$ and $|x^2| = x^2$, so both functions are differentiable at $x=0$ and have derivative 0 there.

46. It may happen that $k = g(x+h) - g(x) = 0$ for values of h arbitrarily close to 0 so that the division by k in the "proof" is not justified.

Section 2.5 Derivatives of Trigonometric Functions (page 129)

1. $\dfrac{d}{dx}\csc x = \dfrac{d}{dx}\dfrac{1}{\sin x} = -\dfrac{\cos x}{\sin^2 x} = -\csc x \cot x$

2. $\dfrac{d}{dx}\cot x = \dfrac{d}{dx}\dfrac{\cos x}{\sin x} = \dfrac{-\cos^2 x - \sin^2 x}{\sin^2 x} = -\csc^2 x$

3. $y = \cos 3x, \quad y' = -3\sin 3x$

4. $y = \sin 2x, \quad y' = 2\cos 2x.$

5. $y = \tan \pi x, \quad y' = \pi \sec^2 \pi x$

6. $y = \sin\dfrac{x}{5}, \quad y' = \dfrac{1}{5}\cos\dfrac{x}{5}.$

7. $y = \cos\dfrac{2x}{\pi}, \quad y' = -\dfrac{2}{\pi}\sin\dfrac{2x}{\pi}$

8. $y = \sec ax, \quad y' = a \sec ax \tan ax.$

9. $y = \cot(4 - 3x), \quad y' = 3\csc^2(4 - 3x)$

10. $\dfrac{d}{dx}\sin\dfrac{\pi - x}{3} = -\dfrac{1}{3}\cos\dfrac{\pi - x}{3}$

11. $f(x) = \cos(s - rx), \quad f'(x) = r\sin(s - rx)$

12. $y = \sin(Ax + B), \quad y' = A\cos(Ax + B)$

13. $\dfrac{d}{dx}\sin(\pi x^2) = 2\pi x \cos(\pi x^2)$

14. $\dfrac{d}{dx}\cos(\sqrt{x}) = -\dfrac{1}{2\sqrt{x}}\sin(\sqrt{x})$

15. $y = \sqrt{1 + \cos x}, \quad y' = \dfrac{-\sin x}{2\sqrt{1 + \cos x}}$

16. $\dfrac{d}{dx}\sin(2\cos x) = \cos(2\cos x)(-2\sin x)$
    $= -2\sin x \cos(2\cos x)$

17. $f(x) = \cos(x + \sin x)$
    $f'(x) = -(1 + \cos x)\sin(x + \sin x)$

18. $g(\theta) = \tan(\theta \sin \theta)$
    $g'(\theta) = (\sin\theta + \theta\cos\theta)\sec^2(\theta\sin\theta)$

19. $u = \sin^3(\pi x/2), \quad u' = \dfrac{3\pi}{2}\cos(\pi x/2)\sin^2(\pi x/2)$

20. $y = \sec(1/x), \quad y' = -(1/x^2)\sec(1/x)\tan(1/x)$

21. $F(t) = \sin at \cos at \quad (= \dfrac{1}{2}\sin 2at)$
    $F'(t) = a\cos at \cos at - a\sin at \sin at$
    $(= a\cos 2at)$

22. $G(\theta) = \dfrac{\sin a\theta}{\cos b\theta}$
    $G'(\theta) = \dfrac{a\cos b\theta \cos a\theta + b\sin a\theta \sin b\theta}{\cos^2 b\theta}.$

23. $\dfrac{d}{dx}\big(\sin(2x) - \cos(2x)\big) = 2\cos(2x) + 2\sin(2x)$

24. $\dfrac{d}{dx}(\cos^2 x - \sin^2 x) = \dfrac{d}{dx}\cos(2x)$
    $= -2\sin(2x) = -4\sin x \cos x$

25. $\dfrac{d}{dx}(\tan x + \cot x) = \sec^2 x - \csc^2 x$

26. $\dfrac{d}{dx}(\sec x - \csc x) = \sec x \tan x + \csc x \cot x$

27. $\dfrac{d}{dx}(\tan x - x) = \sec^2 x - 1 = \tan^2 x$

28. $\dfrac{d}{dx}\tan(3x)\cot(3x) = \dfrac{d}{dx}(1) = 0$

29. $\dfrac{d}{dt}(t\cos t - \sin t) = \cos t - t\sin t - \cos t = -t\sin t$

30. $\dfrac{d}{dt}(t\sin t + \cos t) = \sin t + t\cos t - \sin t = t\cos t$

31. $\dfrac{d}{dx}\dfrac{\sin x}{1 + \cos x} = \dfrac{(1 + \cos x)(\cos x) - \sin(x)(-\sin x)}{(1 + \cos x)^2}$
    $= \dfrac{\cos x + 1}{(1 + \cos x)^2} = \dfrac{1}{1 + \cos x}$

32. $\dfrac{d}{dx}\dfrac{\cos x}{1 + \sin x} = \dfrac{(1 + \sin x)(-\sin x) - \cos(x)(\cos x)}{(1 + \sin x)^2}$
    $= \dfrac{-\sin x - 1}{(1 + \sin x)^2} = \dfrac{-1}{1 + \sin x}$

33. $\dfrac{d}{dx}x^2\cos(3x) = 2x\cos(3x) - 3x^2\sin(3x)$

34. $g(t) = \sqrt{(\sin t)/t}$
    $g'(t) = \dfrac{1}{2\sqrt{(\sin t)/t}} \times \dfrac{t\cos t - \sin t}{t^2}$
    $= \dfrac{t\cos t - \sin t}{2t^{3/2}\sqrt{\sin t}}$

35. $v = \sec(x^2)\tan(x^2)$
    $v' = 2x\sec(x^2)\tan^2(x^2) + 2x\sec^3(x^2)$

36. $z = \dfrac{\sin\sqrt{x}}{1 + \cos\sqrt{x}}$
    $z' = \dfrac{(1 + \cos\sqrt{x})(\cos\sqrt{x}/2\sqrt{x}) - (\sin\sqrt{x})(-\sin\sqrt{x}/2\sqrt{x})}{(1 + \cos\sqrt{x})^2}$
    $= \dfrac{1 + \cos\sqrt{x}}{2\sqrt{x}(1 + \cos\sqrt{x})^2} = \dfrac{1}{2\sqrt{x}(1 + \cos\sqrt{x})}$

37. $\dfrac{d}{dt}\sin(\cos(\tan t)) = -(\sec^2 t)(\sin(\tan t))\cos(\cos(\tan t))$

38. $f(s) = \cos(s + \cos(s + \cos s))$
    $f'(s) = -[\sin(s + \cos(s + \cos s))]$
    $\times [1 - (\sin(s + \cos s))(1 - \sin s)]$

39. Differentiate both sides of $\sin(2x) = 2\sin x \cos x$ and divide by 2 to get $\cos(2x) = \cos^2 x - \sin^2 x$.

40. Differentiate both sides of $\cos(2x) = \cos^2 x - \sin^2 x$ and divide by -2 to get $\sin(2x) = 2\sin x \cos x$.

INSTRUCTOR'S SOLUTIONS MANUAL SECTION 2.5 (PAGE 129)

41. Slope of $y = \sin x$ at $(\pi, 0)$ is $\cos \pi = -1$. Therefore the tangent and normal lines to $y = \sin x$ at $(\pi, 0)$ have equations $y = -(x - \pi)$ and $y = x - \pi$, respectively.

42. The slope of $y = \tan(2x)$ at $(0, 0)$ is $2\sec^2(0) = 2$. Therefore the tangent and normal lines to $y = \tan(2x)$ at $(0, 0)$ have equations $y = 2x$ and $y = -x/2$, respectively.

43. The slope of $y = \sqrt{2}\cos(x/4)$ at $(\pi, 1)$ is $-(\sqrt{2}/4)\sin(\pi/4) = -1/4$. Therefore the tangent and normal lines to $y = \sqrt{2}\cos(x/4)$ at $(\pi, 1)$ have equations $y = 1 - (x - \pi)/4$ and $y = 1 + 4(x - \pi)$, respectively.

44. The slope of $y = \cos^2 x$ at $(\pi/3, 1/4)$ is $-\sin(2\pi/3) = -\sqrt{3}/2$. Therefore the tangent and normal lines to $y = \tan(2x)$ at $(0, 0)$ have equations
$y = (1/4) - (\sqrt{3}/2)(x - (\pi/3))$ and
$y = (1/4) + (2/\sqrt{3})(x - (\pi/3))$, respectively.

45. Slope of $y = \sin(x°) = \sin\left(\dfrac{\pi x}{180}\right)$ is $y' = \dfrac{\pi}{180}\cos\left(\dfrac{\pi x}{180}\right)$.
At $x = 45$ the tangent line has equation
$y = \dfrac{1}{\sqrt{2}} + \dfrac{\pi}{180\sqrt{2}}(x - 45)$.

46. For $y = \sec(x°) = \sec\left(\dfrac{x\pi}{180}\right)$ we have
$$\dfrac{dy}{dx} = \dfrac{\pi}{180}\sec\left(\dfrac{x\pi}{180}\right)\tan\left(\dfrac{x\pi}{180}\right).$$
At $x = 60$ the slope is $\dfrac{\pi}{180}(2\sqrt{3}) = \dfrac{\pi\sqrt{3}}{90}$.
Thus, the normal line has slope $-\dfrac{90}{\pi\sqrt{3}}$ and has equation
$y = 2 - \dfrac{90}{\pi\sqrt{3}}(x - 60)$.

47. The slope of $y = \tan x$ at $x = a$ is $\sec^2 a$. The tangent there is parallel to $y = 2x$ if $\sec^2 a = 2$, or $\cos a = \pm 1/\sqrt{2}$. The only solutions in $(-\pi/2, \pi/2)$ are $a = \pm \pi/4$. The corresponding points on the graph are $(\pi/4, 1)$ and $(-\pi/4, 1)$.

48. The slope of $y = \tan(2x)$ at $x = a$ is $2\sec^2(2a)$. The tangent there is normal to $y = -x/8$ if $2\sec^2(2a) = 8$, or $\cos(2a) = \pm 1/2$. The only solutions in $(-\pi/4, \pi/4)$ are $a = \pm \pi/6$. The corresponding points on the graph are $(\pi/6, \sqrt{3})$ and $(-\pi/6, -\sqrt{3})$.

49. $\dfrac{d}{dx}\sin x = \cos x = 0$ at odd multiples of $\pi/2$.
$\dfrac{d}{dx}\cos x = -\sin x = 0$ at multiples of π.
$\dfrac{d}{dx}\sec x = \sec x \tan x = 0$ at multiples of π.
$\dfrac{d}{dx}\csc x = -\csc x \cot x = 0$ at odd multiples of $\pi/2$.
Thus each of these functions has horizontal tangents at infinitely many points on its graph.

50. $\dfrac{d}{dx}\tan x = \sec^2 x = 0$ nowhere.
$\dfrac{d}{dx}\cot x = -\csc^2 x = 0$ nowhere.
Thus neither of these functions has a horizontal tangent.

51. $y = x + \sin x$ has a horizontal tangent at $x = \pi$ because $dy/dx = 1 + \cos x = 0$ there.

52. $y = 2x + \sin x$ has no horizontal tangents because $dy/dx = 2 + \cos x \geq 1$ everywhere.

53. $y = x + 2\sin x$ has horizontal tangents at $x = 2\pi/3$ and $x = 4\pi/3$ because $dy/dx = 1 + 2\cos x = 0$ at those points.

54. $y = x + 2\cos x$ has horizontal tangents at $x = \pi/6$ and $x = 5\pi/6$ because $dy/dx = 1 - 2\sin x = 0$ at those points.

55. $\lim\limits_{x\to 0}\dfrac{\tan(2x)}{x} = \lim\limits_{x\to 0}\dfrac{\sin(2x)}{2x}\dfrac{2}{\cos(2x)} = 1 \times 2 = 2$

56. $\lim\limits_{x\to\pi}\sec(1 + \cos x) = \sec(1 - 1) = \sec 0 = 1$

57. $\lim\limits_{x\to 0} x^2 \csc x \cot x = \lim\limits_{x\to 0}\left(\dfrac{x}{\sin x}\right)^2 \cos x = 1^2 \times 1 = 1$

58. $\lim\limits_{x\to 0}\cos\left(\dfrac{\pi - \pi\cos^2 x}{x^2}\right) = \lim\limits_{x\to 0}\cos\pi\left(\dfrac{\sin x}{x}\right)^2 = \cos\pi = -1$

59. $\lim\limits_{h\to 0}\dfrac{1 - \cos h}{h^2} = \lim\limits_{h\to 0}\dfrac{2\sin^2(h/2)}{h^2} = \lim\limits_{h\to 0}\dfrac{1}{2}\left(\dfrac{\sin(h/2)}{h/2}\right)^2 = \dfrac{1}{2}$

60. f will be differentiable at $x = 0$ if
$$2\sin 0 + 3\cos 0 = b, \quad \text{and}$$
$$\dfrac{d}{dx}(2\sin x + 3\cos x)\bigg|_{x=0} = a.$$
Thus we need $b = 3$ and $a = 2$.

61. There are infinitely many lines through the origin that are tangent to $y = \cos x$. The two with largest slope are shown in the figure.

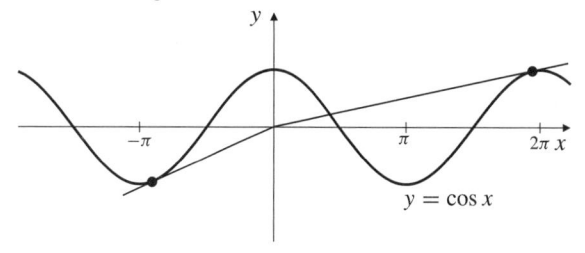

Fig. 2.5.61

The tangent to $y = \cos x$ at $x = a$ has equation $y = \cos a - (\sin a)(x - a)$. This line passes through the origin if $\cos a = -a\sin a$. We use a calculator with a "solve" function to find solutions of this equation near $a = -\pi$ and $a = 2\pi$ as suggested in the figure. The solutions are $a \approx -2.798386$ and $a \approx 6.121250$. The slopes of the corresponding tangents are given by $-\sin a$, so they are 0.336508 and 0.161228 to six decimal places.

53

62. 1

63. $-\sqrt{2\pi + 3}(2\pi^{3/2} - 4\pi + 3)/\pi$

64. a) As suggested by the figure in the problem, the square of the length of chord AP is $(1-\cos\theta)^2+(0-\sin\theta)^2$, and the square of the length of arc AP is θ^2. Hence

$$(1+\cos\theta)^2 + \sin^2\theta < \theta^2,$$

and, since squares cannot be negative, each term in the sum on the left is less than θ^2. Therefore

$$0 \le |1 - \cos\theta| < |\theta|, \quad 0 \le |\sin\theta| < |\theta|.$$

Since $\lim_{\theta\to 0} |\theta| = 0$, the squeeze theorem implies that

$$\lim_{\theta\to 0} 1 - \cos\theta = 0, \quad \lim_{\theta\to 0} \sin\theta = 0.$$

From the first of these, $\lim_{\theta\to 0} \cos\theta = 1$.

b) Using the result of (a) and the addition formulas for cosine and sine we obtain

$$\lim_{h\to 0} \cos(\theta_0 + h) = \lim_{h\to 0}(\cos\theta_0 \cos h - \sin\theta_0 \sin h) = \cos\theta_0$$
$$\lim_{h\to 0} \sin(\theta_0 + h) = \lim_{h\to 0}(\sin\theta_0 \cos h + \cos\theta_0 \sin h) = \sin\theta_0.$$

This says that cosine and sine are continuous at any point θ_0.

Section 2.6 The Mean-Value Theorem (page 137)

1. $f(x) = x^2, \quad f'(x) = 2x$

$$b + a = \frac{b^2 - a^2}{b-a} = \frac{f(b) - f(a)}{b-a}$$
$$= f'(c) = 2c \Rightarrow c = \frac{b+a}{2}$$

2. If $f(x) = \frac{1}{x}$, and $f'(x) = -\frac{1}{x^2}$ then

$$\frac{f(2) - f(1)}{2 - 1} = \frac{1}{2} - 1 = -\frac{1}{2} = -\frac{1}{c^2} = f'(c)$$

where $c = \sqrt{2}$ lies between 1 and 2.

3. $f(x) = x^3 - 3x + 1; \quad f'(x) = 3x^2 - 3, \quad a = -2, b = 2$

$$\frac{f(b) - f(a)}{b - a} = \frac{f(2) - f(-2)}{4}$$
$$= \frac{8 - 6 + 1 - (-8 + 6 + 1)}{4}$$
$$= \frac{4}{4} = 1$$

$f'(c) = 3c^2 - 3$

$3c^2 - 3 = 1 \Rightarrow 3c^2 = 4 \Rightarrow c = \pm\frac{2}{\sqrt{3}}$

(Both points will be in $(-2, 2)$.)

4. If $f(x) = \cos x + (x^2/2)$, then $f'(x) = x - \sin x > 0$ for $x > 0$. By the MVT, if $x > 0$, then $f(x) - f(0) = f'(c)(x - 0)$ for some $c > 0$, so $f(x) > f(0) = 1$. Thus $\cos x + (x^2/2) > 1$ and $\cos x > 1 - (x^2/2)$ for $x > 0$. Since both sides of the inequality are even functions, it must hold for $x < 0$ as well.

5. Let $f(x) = \tan x$. If $0 < x < \pi/2$, then by the MVT $f(x) - f(0) = f'(c)(x - 0)$ for some c in $(0, \pi/2)$. Thus $\tan x = x \sec^2 c > x$, since $\sec c > 1$.

6. Let $f(x) = (1+x)^r - 1 - rx$ where $r > 1$.
Then $f'(x) = r(1+x)^{r-1} - r$.
If $-1 \le x < 0$ then $f'(x) < 0$; if $x > 0$, then $f'(x) > 0$.
Thus $f(x) > f(0) = 0$ if $-1 \le x < 0$ or $x > 0$.
Thus $(1+x)^r > 1 + rx$ if $-1 \le x < 0$ or $x > 0$.

7. Let $f(x) = (1+x)^r$ where $0 < r < 1$. Thus, $f'(x) = r(1+x)^{r-1}$. By the Mean-Value Theorem, for $x \ge -1$, and $x \ne 0$,

$$\frac{f(x) - f(0)}{x - 0} = f'(c)$$
$$\Rightarrow \frac{(1-x)^r - 1}{x} = r(1+c)^{r-1}$$

for some c between 0 and x. Thus,
$(1 + x)^r = 1 + rx(1+c)^{r-1}$.
If $-1 \le x < 0$, then $c < 0$ and $0 < 1 + c < 1$. Hence

$$(1+c)^{r-1} > 1 \quad \text{(since } r - 1 < 0\text{)},$$
$$rx(1+c)^{r-1} < rx \quad \text{(since } x < 0\text{)}.$$

Hence, $(1 + x)^r < 1 + rx$.
If $x > 0$, then

$$c > 0$$
$$1 + c > 1$$
$$(1+c)^{r-1} < 1$$
$$rx(1+c)^{r-1} < rx.$$

Hence, $(1 + x)^r < 1 + rx$ in this case also.
Hence, $(1 + x)^r < 1 + rx$ for either $-1 \le x < 0$ or $x > 0$.

8. If $f(x) = x^2 + 2x + 2$ then $f'(x) = 2x + 2 = 2(x + 1)$.
Evidently, $f'(x) > 0$ if $x > -1$ and $f'(x) < 0$ if $x < -1$. Therefore, f is increasing on $(-1, \infty)$ and decreasing on $(-\infty, -1)$.

9. $f(x) = x^3 - 4x + 1$
$f'(x) = 3x^2 - 4$
$f'(x) > 0$ if $|x| > \frac{2}{\sqrt{3}}$
$f'(x) < 0$ if $|x| < \frac{2}{\sqrt{3}}$

f is increasing on $(-\infty, -\frac{2}{\sqrt{3}})$ and $(\frac{2}{\sqrt{3}}, \infty)$.

f is decreasing on $(-\frac{2}{\sqrt{3}}, \frac{2}{\sqrt{3}})$.

10. If $f(x) = x^3 + 4x + 1$, then $f'(x) = 3x^2 + 4$. Since $f'(x) > 0$ for all real x, hence $f(x)$ is increasing on the whole real line, i.e., on $(-\infty, \infty)$.

11. $f(x) = (x^2 - 4)^2$
    $f'(x) = 2x2(x^2 - 4) = 4x(x-2)(x+2)$
    $f'(x) > 0$ if $x > 2$ or $-2 < x < 0$
    $f'(x) < 0$ if $x < -2$ or $0 < x < 2$
    f is increasing on $(-2, 0)$ and $(2, \infty)$.
    f is decreasing on $(-\infty, -2)$ and $(0, 2)$.

12. If $f(x) = \dfrac{1}{x^2+1}$ then $f'(x) = \dfrac{-2x}{(x^2+1)^2}$. Evidently, $f'(x) > 0$ if $x < 0$ and $f'(x) < 0$ if $x > 0$. Therefore, f is increasing on $(-\infty, 0)$ and decreasing on $(0, \infty)$.

13. $f(x) = x^3(5-x)^2$
    $f'(x) = 3x^2(5-x)^2 + 2x^3(5-x)(-1)$
    $\quad = x^2(5-x)(15 - 5x)$
    $\quad = 5x^2(5-x)(3-x)$
    $f'(x) > 0$ if $x < 0$, $0 < x < 3$, or $x > 5$
    $f'(x) < 0$ if $3 < x < 5$
    f is increasing on $(-\infty, 3)$ and $(5, \infty)$.
    f is decreasing on $(3, 5)$.

14. If $f(x) = x - 2\sin x$, then $f'(x) = 1 - 2\cos x = 0$ at $x = \pm \pi/3 + 2n\pi$ for $n = 0, \pm 1, \pm 2, \ldots$.
    f is decreasing on $(-\pi/3 + 2n\pi, \pi + 2n\pi)$.
    f is increasing on $(\pi/3 + 2n\pi, -\pi/3 + 2(n+1)\pi)$ for integers n.

15. If $f(x) = x + \sin x$, then $f'(x) = 1 + \cos x \geq 0$
    $f'(x) = 0$ only at isolated points $x = \pm\pi, \pm 3\pi, \ldots$.
    Hence f is increasing everywhere.

16. If $x_1 < x_2 < \ldots < x_n$ belong to I, and $f(x_i) = 0$, $(1 \leq i \leq n)$, then there exists y_i in (x_i, x_{i+1}) such that $f'(y_i) = 0$, $(1 \leq i \leq n-1)$ by MVT.

17. Let a, b, and c be three points in I where f vanishes; that is, $f(a) = f(b) = f(c) = 0$. Suppose $a < b < c$. By the Mean-Value Theorem, there exist points r in (a, b) and s in (b, c) such that $f'(r) = f'(s) = 0$. By the Mean-Value Theorem applied to f' on $[r, s]$, there is some point t in (r, s) (and therefore in I) such that $f''(t) = 0$.

18. If $f^{(n)}$ exists on interval I and f vanishes at $n+1$ distinct points of I, then $f^{(n)}$ vanishes at at least one point of I.
    Proof: True for $n = 2$ by Exercise 8.
    Assume true for $n = k$. (Induction hypothesis)
    Suppose $n = k+1$, i.e., f vanishes at $k+2$ points of I and $f^{(k+1)}$ exists.
    By Exercise 7, f' vanishes at $k+1$ points of I.
    By the induction hypothesis, $f^{(k+1)} = (f')^{(k)}$ vanishes at a point of I so the statement is true for $n = k+1$.
    Therefore the statement is true for all $n \geq 2$ by induction. (case $n = 1$ is just MVT.)

19. Given that $f(0) = f(1) = 0$ and $f(2) = 1$:

    a) By MVT,
    $$f'(a) = \frac{f(2) - f(0)}{2 - 0} = \frac{1-0}{2-0} = \frac{1}{2}$$
    for some a in $(0, 2)$.

    b) By MVT, for some r in $(0, 1)$,
    $$f'(r) = \frac{f(1) - f(0)}{1 - 0} = \frac{0-0}{1-0} = 0.$$
    Also, for some s in $(1, 2)$,
    $$f'(s) = \frac{f(2) - f(1)}{2 - 1} = \frac{1-0}{2-1} = 1.$$
    Then, by MVT applied to f' on the interval $[r, s]$, for some b in (r, s),
    $$f''(b) = \frac{f'(s) - f'(r)}{s - r} = \frac{1-0}{s-r}$$
    $$= \frac{1}{s-r} > \frac{1}{2}$$
    since $s - r < 2$.

    c) Since $f''(x)$ exists on $[0, 2]$, therefore $f'(x)$ is continuous there. Since $f'(r) = 0$ and $f'(s) = 1$, and since $0 < \frac{1}{7} < 1$, the Intermediate-Value Theorem assures us that $f'(c) = \frac{1}{7}$ for some c between r and s.

20. There is no guarantee that the MVT applications for f and g yield the same c.

21. $f(x) = \begin{cases} x + 2x^2 \sin(1/x) & \text{if } x \neq 0 \\ 0 & \text{if } x = 0. \end{cases}$

    a) $f'(0) = \lim\limits_{h \to 0} \dfrac{f(0+h) - f(0)}{h}$
    $= \lim\limits_{h \to 0} \dfrac{h + 2h^2 \sin(1/h)}{h}$
    $= \lim\limits_{h \to 0} (1 + 2h \sin(1/h)) = 1$,
    because $|2h \sin(1/h)| \leq 2|h| \to 0$ as $h \to 0$.

    b) For $x \neq 0$, we have
    $$f'(x) = 1 + 4x \sin(1/x) - 2\cos(1/x).$$
    There are numbers x arbitrarily close to 0 where $f'(x) = -1$; namely, the numbers $x = \pm 1/(2n\pi)$, where $n = 1, 2, 3, \ldots$. Since $f'(x)$ is continuous at every $x \neq 0$, it is negative in a small interval about every such number. Thus f cannot be increasing on any interval containing $x = 0$.

Section 2.7 Using Derivatives (page 144)

1. If $y = x^2$, then $\Delta y \approx 2x \, \Delta x$. If $\Delta x = (2/100)x$, then $\Delta y \approx (4/100)x^2 = (4/100)y$, so y increases by about 4%.

2. If $y = 1/x$, then $\Delta y \approx (-1/x^2)\,\Delta x$. If $\Delta x = (2/100)x$, then $\Delta y \approx (-2/100)/x = (-2/100)y$, so y decreases by about 2%.

3. If $y = 1/x^2$, then $\Delta y \approx (-2/x^3)\,\Delta x$. If $\Delta x = (2/100)x$, then $\Delta y \approx (-4/100)/x^2 = (-4/100)y$, so y decreases by about 4%.

4. If $y = x^3$, then $\Delta y \approx 3x^2\,\Delta x$. If $\Delta x = (2/100)x$, then $\Delta y \approx (6/100)x^3 = (6/100)y$, so y increases by about 6%.

5. If $y = \sqrt{x}$, then $\Delta y \approx (1/2\sqrt{x})\,\Delta x$. If $\Delta x = (2/100)x$, then $\Delta y \approx (1/100)\sqrt{x} = (1/100)y$, so y increases by about 1%.

6. If $y = x^{-2/3}$, then $\Delta y \approx (-2/3)x^{-5/3}\,\Delta x$. If $\Delta x = (2/100)x$, then $\Delta y \approx (-4/300)x^{2/3} = (-4/300)y$, so y decreases by about 1.33%.

7. If $V = \frac{4}{3}\pi r^3$, then $\Delta V = 4\pi r^2\,\Delta r$. If r increases by 2%, then $\Delta r = 2r/100$ and $\Delta V \approx 8\pi r^3/100$. Therefore $\Delta V/V \approx 6/100$. The volume increases by about 6%.

8. If V is the volume and x is the edge length of the cube then $V = x^3$. Thus $\Delta V \approx 3x^2\,\Delta x$. $\Delta V = -(6/100)V$, then $-6x^3/100 = 3x^2\,\Delta x$, so $\Delta x \approx -(2/100)x$. The edge of the cube decreases by about 2%.

9. Rate change of Area A with respect to side s, where $A = s^2$, is $\dfrac{dA}{ds} = 2s$. When $s = 4$ ft, the area is changing at rate 8 ft^2/ft.

10. If $A = s^2$, then $s = \sqrt{A}$ and $ds/dA = 1/(2\sqrt{A})$. If $A = 16$ m^2, then the side is changing at rate $ds/dA = 1/8$ m/m^2.

11. The diameter D and area A of a circle are related by $D = 2\sqrt{A/\pi}$. The rate of change of diameter with respect to area is $dD/dA = \sqrt{1/(\pi A)}$ units per square unit.

12. Since $A = \pi D^2/4$, the rate of change of area with respect to diameter is $dA/dD = \pi D/2$ square units per unit.

13. Rate of change of $V = \dfrac{4}{3}\pi r^3$ with respect to radius r is $\dfrac{dV}{dr} = 4\pi r^2$. When $r = 2$ m, this rate of change is 16π m^3/m.

14. Let A be the area of a square, s be its side length and L be its diagonal. Then, $L^2 = s^2 + s^2 = 2s^2$ and $A = s^2 = \frac{1}{2}L^2$, so $\dfrac{dA}{dL} = L$. Thus, the rate of change of the area of a square with respect to its diagonal L is L.

15. If the radius of the circle is r then $C = 2\pi r$ and $A = \pi r^2$.
Thus $C = 2\pi\sqrt{\dfrac{A}{\pi}} = 2\sqrt{\pi}\sqrt{A}$.
Rate of change of C with respect to A is
$\dfrac{dC}{dA} = \dfrac{\sqrt{\pi}}{\sqrt{A}} = \dfrac{1}{r}$.

16. Let s be the side length and V be the volume of a cube. Then $V = s^3 \Rightarrow s = V^{1/3}$ and $\dfrac{ds}{dV} = \frac{1}{3}V^{-2/3}$. Hence, the rate of change of the side length of a cube with respect to its volume V is $\frac{1}{3}V^{-2/3}$.

17. If $f(x) = x^2 - 4$, then $f'(x) = 2x$. The critical point of f is $x = 0$. f is increasing on $(0, \infty)$ and decreasing on $(-\infty, 0)$.

18. If $f(x) = x^3 - 12x + 1$, then $f'(x) = 3(x^2 - 4)$. The critical points of f are $x = \pm 2$. f is increasing on $(-\infty, -2)$ and $(2, \infty)$ where $f'(x) > 0$, and is decreasing on $(-2, 2)$ where $f'(x) < 0$.

19. If $y = x^3 + 6x^2$, then $y' = 3x^2 + 12x = 3x(x+4)$. The critical points of y are $x = 0$ and $x = -4$. y is increasing on $(-\infty, -4)$ and $(0, \infty)$ where $y' > 0$, and is decreasing on $(-4, 0)$ where $y' < 0$.

20. If $y = 1 - x - x^5$, then $y' = -1 - 5x^4 < 0$ for all x. Thus y has no critical points and is decreasing on the whole real line.

21. $f(x) = x^3$ is increasing on $(-\infty, 0)$ and $(0, \infty)$ because $f'(x) = 3x^2 > 0$ there. But $f(x_1) < f(0) = 0 < f(x_2)$ whenever $x_1 < 0 < x_2$, so f is also increasing on intervals containing the origin.

22. If $f(x) = x + 2\sin x$, then $f'(x) = 1 + 2\cos x > 0$ if $\cos x > -1/2$. Thus f is increasing on the intervals $(-(4\pi/3) + 2n\pi, (4\pi/3) + 2n\pi)$ where n is any integer.

23. CPs $x = 0.535898$ and $x = 7.464102$

24. CPs $x = -1.366025$ and $x = 0.366025$

25. CPs $x = -0.518784$ and $x = 0$

26. CP $x = 0.521350$

27. Volume in tank is $V(t) = 350(20 - t)^2$ L at t min.

a) At $t = 5$, water volume is changing at rate
$$\left.\dfrac{dV}{dt}\right|_{t=5} = -700(20-t)\bigg|_{t=5} = -10{,}500.$$
Water is draining out at 10,500 L/min at that time. At $t = 15$, water volume is changing at rate
$$\left.\dfrac{dV}{dt}\right|_{t=15} = -700(20-t)\bigg|_{t=15} = -3{,}500.$$
Water is draining out at 3,500 L/min at that time.

b) Average rate of change between $t = 5$ and $t = 15$ is
$$\dfrac{V(15) - V(5)}{15 - 5} = \dfrac{350 \times (25 - 225)}{10} = -7{,}000.$$
The average rate of draining is 7,000 L/min over that interval.

28. Flow rate $F = kr^4$, so $\Delta F \approx 4kr^3 \Delta r$. If $\Delta F = F/10$, then
$$\Delta r \approx \frac{F}{40kr^3} = \frac{kr^4}{40kr^3} = 0.025r.$$
The flow rate will increase by 10% if the radius is increased by about 2.5%.

29. $F = k/r^2$ implies that $dF/dr = -2k/r^3$. Since $dF/dr = 1$ pound/mi when $r = 4,000$ mi, we have $2k = 4,000^3$. If $r = 8,000$, we have $dF/dr = -(4,000/8,000)^3 = -1/8$. At $r = 8,000$ mi F decreases with respect to r at a rate of 1/8 pounds/mi.

30. If price = $\$p$, then revenue is $\$R = 4,000p - 10p^2$.

    a) Sensitivity of R to p is $dR/dp = 4,000 - 20p$. If $p = 100, 200$, and 300, this sensitivity is $2,000$ \$/\$, 0 \$/\$, and $-2,000$ \$/\$ respectively.

    b) The distributor should charge \$200. This maximizes the revenue.

31. Cost is $\$C(x) = 8,000 + 400x - 0.5x^2$ if x units are manufactured.

    a) Marginal cost if $x = 100$ is
    $C'(100) = 400 - 100 = \$300$.

    b) $C(101) - C(100) = 43,299.50 - 43,000 = \$299.50$ which is approximately $C'(100)$.

32. Daily profit if production is x sheets per day is $\$P(x)$ where
$$P(x) = 8x - 0.005x^2 - 1,000.$$

    a) Marginal profit $P'(x) = 8 - 0.01x$. This is positive if $x < 800$ and negative if $x > 800$.

    b) To maximize daily profit, production should be 800 sheets/day.

33. $C = \dfrac{80,000}{n} + 4n + \dfrac{n^2}{100}$

    $\dfrac{dC}{dn} = -\dfrac{80,000}{n^2} + 4 + \dfrac{n}{50}.$

    (a) $n = 100$, $\dfrac{dC}{dn} = -2$. Thus, the marginal cost of production is $-\$2$.

    (b) $n = 300$, $\dfrac{dC}{dn} = \dfrac{82}{9} \approx 9.11$. Thus, the marginal cost of production is approximately \$9.11.

34. Daily profit $P = 13x - Cx = 13x - 10x - 20 - \dfrac{x^2}{1000}$
$$= 3x - 20 - \dfrac{x^2}{1000}$$
Graph of P is a parabola opening downward. P will be maximum where the slope is zero:
$$0 = \dfrac{dP}{dx} = 3 - \dfrac{2x}{1000} \text{ so } x = 1500$$

Should extract 1500 tonnes of ore per day to maximize profit.

35. One of the components comprising $C(x)$ is usually a fixed cost, \$S, for setting up the manufacturing operation. On a per item basis, this fixed cost \$S/x, decreases as the number x of items produced increases, especially when x is small. However, for large x other components of the total cost may increase on a per unit basis, for instance labour costs when overtime is required or maintenance costs for machinery when it is over used.

Let the average cost be $A(x) = \dfrac{C(x)}{x}$. The minimal average cost occurs at point where the graph of $A(x)$ has a horizontal tangent:
$$0 = \dfrac{dA}{dx} = \dfrac{xC'(x) - C(x)}{x^2}.$$
Hence, $xC'(x) - C(x) = 0 \Rightarrow C'(x) = \dfrac{C(x)}{x} = A(x).$
Thus the marginal cost $C'(x)$ equals the average cost at the minimizing value of x.

36. If $y = Cp^{-r}$, then the elasticity of y is
$$-\dfrac{p}{y}\dfrac{dy}{dp} = -\dfrac{p}{Cp^{-r}}(-r)Cp^{-r-1} = r.$$

Section 2.8 Higher-Order Derivatives (page 149)

1. $y = (3 - 2x)^7$
   $y' = -14(3 - 2x)^6$
   $y'' = 168(3 - 2x)^5$
   $y''' = -1680(3 - 2x)^4$

2. $y = x^2 - \dfrac{1}{x}$ $\quad y'' = 2 - \dfrac{2}{x^3}$
   $y' = 2x + \dfrac{1}{x^2}$ $\quad y''' = \dfrac{6}{x^4}$

3. $y = \dfrac{6}{(x-1)^2} = 6(x-1)^{-2}$
   $y' = -12(x-1)^{-3}$
   $y'' = 36(x-1)^{-4}$
   $y''' = -144(x-1)^{-5}$

4. $y = \sqrt{ax+b}$ $\quad y'' = -\dfrac{a^2}{4(ax+b)^{3/2}}$
   $y' = \dfrac{a}{2\sqrt{ax+b}}$ $\quad y''' = \dfrac{3a^3}{8(ax+b)^{5/2}}$

5. $y = x^{1/3} - x^{-1/3}$

$y' = \frac{1}{3}x^{-2/3} + \frac{1}{3}x^{-4/3}$

$y'' = -\frac{2}{9}x^{-5/3} - \frac{4}{9}x^{-7/3}$

$y''' = \frac{10}{27}x^{-8/3} + \frac{28}{27}x^{-10/3}$

6. $y = x^{10} + 2x^8 \qquad y'' = 90x^8 + 112x^6$

$y' = 10x^9 + 16x^7 \qquad y''' = 720x^7 + 672x^5$

7. $y = (x^2 + 3)\sqrt{x} = x^{5/2} + 3x^{1/2}$

$y' = \frac{5}{2}x^{3/2} + \frac{3}{2}x^{-1/2}$

$y'' = \frac{15}{4}x^{1/2} - \frac{3}{4}x^{-3/2}$

$y''' = \frac{15}{8}x^{-1/2} + \frac{9}{8}x^{-5/2}$

8. $y = \frac{x-1}{x+1} \qquad y'' = -\frac{4}{(x+1)^3}$

$y' = \frac{2}{(x+1)^2} \qquad y''' = \frac{12}{(x+1)^4}$

9. $y = \tan x \qquad y'' = 2\sec^2 x \tan x$

$y' = \sec^2 x \qquad y''' = 2\sec^4 x + 4\sec^2 x \tan^2 x$

10. $y = \sec x \qquad y'' = \sec x \tan^2 x + \sec^3 x$

$y' = \sec x \tan x \qquad y''' = \sec x \tan^3 x + 5\sec^3 x \tan x$

11. $y = \cos(x^2) \qquad y'' = -2\sin(x^2) - 4x^2 \cos(x^2)$

$y' = -2x \sin(x^2) \qquad y''' = -12x \cos(x^2) + 8x^3 \sin(x^2)$

12. $y = \frac{\sin x}{x}$

$y' = \frac{\cos x}{x} - \frac{\sin x}{x^2}$

$y'' = \frac{(2-x^2)\sin x}{x^3} - \frac{2\cos x}{x^2}$

$y''' = \frac{(6-x^2)\cos x}{x^3} + \frac{3(x^2-2)\sin x}{x^4}$

13. $f(x) = \frac{1}{x} = x^{-1}$

$f'(x) = -x^{-2}$

$f''(x) = 2x^{-3}$

$f'''(x) = -3!x^{-4}$

$f^{(4)}(x) = 4!x^{-5}$

Guess: $f^{(n)}(x) = (-1)^n n! x^{-(n+1)}$ (*)

Proof: (*) is valid for $n = 1$ (and 2, 3, 4).
Assume $f^{(k)}(x) = (-1)^k k! x^{-(k+1)}$ for some $k \geq 1$
Then $f^{(k+1)}(x) = (-1)^k k! \big(-(k+1)\big) x^{-(k+1)-1}$
$= (-1)^{k+1}(k+1)! x^{-((k+1)+1)}$ which is (*) for $n = k+1$.
Therefore, (*) holds for $n = 1, 2, 3, \ldots$ by induction.

14. $f(x) = \frac{1}{x^2} = x^{-2}$

$f'(x) = -2x^{-3}$

$f''(x) = -2(-3)x^{-4} = 3!x^{-4}$

$f^{(3)}(x) = -2(-3)(-4)x^{-5} = -4!x^{-5}$

Conjecture:

$f^{(n)}(x) = (-1)^n(n+1)! x^{-(n+2)} \qquad$ for $n = 1, 2, 3, \ldots$

Proof: Evidently, the above formula holds for $n = 1, 2$ and 3. Assume it holds for $n = k$,
i.e., $f^{(k)}(x) = (-1)^k(k+1)! x^{-(k+2)}$. Then

$f^{(k+1)}(x) = \frac{d}{dx} f^{(k)}(x)$

$= (-1)^k (k+1)! [(-1)(k+2)] x^{-(k+2)-1}$

$= (-1)^{k+1}(k+2)! x^{-[(k+1)+2]}$.

Thus, the formula is also true for $n = k + 1$. Hence it is true for $n = 1, 2, 3, \ldots$ by induction.

15. $f(x) = \frac{1}{2-x} = (2-x)^{-1}$

$f'(x) = +(2-x)^{-2}$

$f''(x) = 2(2-x)^{-3}$

$f'''(x) = +3!(2-x)^{-4}$

Guess: $f^{(n)}(x) = n!(2-x)^{-(n+1)}$ (*)

Proof: (*) holds for $n = 1, 2, 3$.
Assume $f^{(k)}(x) = k!(2-x)^{-(k+1)}$ (i.e., (*) holds for $n = k$)
Then $f^{(k+1)}(x) = k!\big(-(k+1)(2-x)^{-(k+1)-1}(-1)\big)$
$= (k+1)!(2-x)^{-((k+1)+1)}$.
Thus (*) holds for $n = k + 1$ if it holds for k.
Therefore, (*) holds for $n = 1, 2, 3, \ldots$ by induction.

16. $f(x) = \sqrt{x} = x^{1/2}$

$f'(x) = \frac{1}{2}x^{-1/2}$

$f''(x) = \frac{1}{2}(-\frac{1}{2})x^{-3/2}$

$f'''(x) = \frac{1}{2}(-\frac{1}{2})(-\frac{3}{2})x^{-5/2}$

$f^{(4)}(x) = \frac{1}{2}(-\frac{1}{2})(-\frac{3}{2})(-\frac{5}{2})x^{-7/2}$

Conjecture:

$f^{(n)}(x) = (-1)^{n-1} \frac{1 \cdot 3 \cdot 5 \cdots (2n-3)}{2^n} x^{-(2n-1)/2} \quad (n \geq 2)$.

Proof: Evidently, the above formula holds for $n = 2, 3$ and 4. Assume that it holds for $n = k$, i.e.

$f^{(k)}(x) = (-1)^{k-1} \frac{1 \cdot 3 \cdot 5 \cdots (2k-3)}{2^k} x^{-(2k-1)/2}$.

Then

$f^{(k+1)}(x) = \frac{d}{dx} f^{(k)}(x)$

$= (-1)^{k-1} \frac{1 \cdot 3 \cdot 5 \cdots (2k-3)}{2^k} \cdot \left[\frac{-(2k-1)}{2}\right] x^{-[(2k-1)/2]-1}$

$= (-1)^{(k+1)-1} \frac{1 \cdot 3 \cdot 5 \cdots (2k-3)[2(k+1)-3]}{2^{k+1}} x^{-[2(k+1)-1]/2}$.

Thus, the formula is also true for $n = k + 1$. Hence, it is true for $n \geq 2$ by induction.

17. $f(x) = \dfrac{1}{a + bx} = (a + bx)^{-1}$
$f'(x) = -b(a + bx)^{-2}$
$f''(x) = 2b^2(a + bx)^{-3}$
$f'''(x) = -3!b^3(a + bx)^{-4}$
Guess: $f^{(n)}(x) = (-1)^n n! b^n (a + bx)^{-(n+1)}$ (*)
Proof: (*) holds for $n = 1, 2, 3$
Assume (*) holds for $n = k$:
$f^{(k)}(x) = (-1)^k k! b^k (a + bx)^{-(k+1)}$
Then
$f^{(k+1)}(x) = (-1)^k k! b^k \big(-(k+1)\big)(a+bx)^{-(k+1)-1}(b)$
$= (-1)^{k+1}(k+1)! b^{k+1}(a+bx)^{((k+1)+1)}$
So (*) holds for $n = k+1$ if it holds for $n = k$.
Therefore, (*) holds for $n = 1, 2, 3, 4, \ldots$ by induction.

18. $f(x) = x^{2/3}$
$f'(x) = \tfrac{2}{3} x^{-1/3}$
$f''(x) = \tfrac{2}{3}(-\tfrac{1}{3}) x^{-4/3}$
$f'''(x) = \tfrac{2}{3}(-\tfrac{1}{3})(-\tfrac{4}{3}) x^{-7/3}$
Conjecture:
$f^{(n)}(x) = 2(-1)^{n-1} \dfrac{1 \cdot 4 \cdot 7 \cdots (3n-5)}{3^n} x^{-(3n-2)/3}$ for $n \geq 2$.
Proof: Evidently, the above formula holds for $n = 2$ and 3. Assume that it holds for $n = k$, i.e.
$f^{(k)}(x) = 2(-1)^{k-1} \dfrac{1 \cdot 4 \cdot 7 \cdots (3k-5)}{3^k} x^{-(3k-2)/3}.$

Then,
$f^{(k+1)}(x) = \dfrac{d}{dx} f^{(k)}(x)$
$= 2(-1)^{k-1} \dfrac{1 \cdot 4 \cdot 7 \cdots (3k-5)}{3^k} \cdot \left[\dfrac{-(3k-2)}{3}\right] x^{-[(3k-2)/3]-1}$
$= 2(-1)^{(k+1)-1} \dfrac{1 \cdot 4 \cdot 7 \cdots (3k-5)[3(k+1)-5]}{3^{(k+1)}} x^{-[3(k+1)-2]/3}.$

Thus, the formula is also true for $n = k + 1$. Hence, it is true for $n \geq 2$ by induction.

19. $f(x) = \cos(ax)$
$f'(x) = -a \sin(ax)$
$f''(x) = -a^2 \cos(ax)$
$f'''(x) = a^3 \sin(ax)$
$f^{(4)}(x) = a^4 \cos(ax) = a^4 f(x)$
It follows that $f^{(n)}(x) = a^4 f^{(n-4)}(x)$ for $n \geq 4$, and
$f^{(n)}(x) = \begin{cases} a^n \cos(ax) & \text{if } n = 4k \\ -a^n \sin(ax) & \text{if } n = 4k+1 \\ -a^n \cos(ax) & \text{if } n = 4k+2 \\ a^n \sin(ax) & \text{if } n = 4k+3 \end{cases}$ $(k = 0, 1, 2, \ldots)$

Differentiating any of these four formulas produces the one for the next higher value of n, so induction confirms the overall formula.

20. $f(x) = x \cos x$
$f'(x) = \cos x - x \sin x$
$f''(x) = -2 \sin x - x \cos x$
$f'''(x) = -3 \cos x + x \sin x$
$f^{(4)}(x) = 4 \sin x + x \cos x$
This suggests the formula (for $k = 0, 1, 2, \ldots$)
$f^{(n)}(x) = \begin{cases} n \sin x + x \cos x & \text{if } n = 4k \\ n \cos x - x \sin x & \text{if } n = 4k+1 \\ -n \sin x - x \cos x & \text{if } n = 4k+2 \\ -n \cos x + x \sin x & \text{if } n = 4k+3 \end{cases}$

Differentiating any of these four formulas produces the one for the next higher value of n, so induction confirms the overall formula.

21. $f(x) = x \sin(ax)$
$f'(x) = \sin(ax) + ax \cos(ax)$
$f''(x) = 2a \cos(ax) - a^2 x \sin(ax)$
$f'''(x) = -3a^2 \sin(ax) - a^3 x \cos(ax)$
$f^{4)}(x) = -4a^3 \cos(ax) + a^4 x \sin(ax)$
This suggests the formula
$f^{(n)}(x) = \begin{cases} -na^{n-1} \cos(ax) + a^n x \sin(ax) & \text{if } n = 4k \\ na^{n-1} \sin(ax) + a^n x \cos(ax) & \text{if } n = 4k+1 \\ na^{n-1} \cos(ax) - a^n x \sin(ax) & \text{if } n = 4k+2 \\ -na^{n-1} \sin(ax) - a^n x \cos(ax) & \text{if } n = 4k+3 \end{cases}$
for $k = 0, 1, 2, \ldots$. Differentiating any of these four formulas produces the one for the next higher value of n, so induction confirms the overall formula.

22. $f(x) = \dfrac{1}{|x|} = |x|^{-1}$. Recall that $\dfrac{d}{dx}|x| = \text{sgn}\,x$, so
$f'(x) = -|x|^{-2} \text{sgn}\,x.$
If $x \neq 0$ we have
$\dfrac{d}{dx} \text{sgn}\,x = 0 \quad \text{and} \quad (\text{sgn}\,x)^2 = 1.$

Thus we can calculate successive derivatives of f using the product rule where necessary, but will get only one nonzero term in each case:
$f''(x) = 2|x|^{-3}(\text{sgn}\,x)^2 = 2|x|^{-3}$
$f^{(3)}(x) = -3!|x|^{-4} \text{sgn}\,x$
$f^{(4)}(x) = 4!|x|^{-5}.$

The pattern suggests that
$f^{(n)}(x) = \begin{cases} -n!|x|^{-(n+1)} \text{sgn}\,x & \text{if } n \text{ is odd} \\ n!|x|^{-(n+1)} & \text{if } n \text{ is even} \end{cases}$

Differentiating this formula leads to the same formula with n replaced by $n+1$ so the formula is valid for all $n \geq 1$ by induction.

23. $f(x) = \sqrt{1-3x} = (1-3x)^{1/2}$

$f'(x) = \frac{1}{2}(-3)(1-3x)^{-1/2}$

$f''(x) = \frac{1}{2}\left(-\frac{1}{2}\right)(-3)^2(1-3x)^{-3/2}$

$f'''(x) = \frac{1}{2}\left(-\frac{1}{2}\right)\left(-\frac{3}{2}\right)(-3)^3(1-3x)^{-5/2}$

$f^{(4)}(x) = \frac{1}{2}\left(-\frac{1}{2}\right)\left(-\frac{3}{2}\right)\left(-\frac{5}{2}\right)(-3)^4(1-3x)^{-7/2}$

Guess: $f^{(n)}(x) = -\frac{1 \times 3 \times 5 \times \cdots \times (2n-3)}{2^n} 3^n (1-3x)^{-(2n-1)/2}$ (*)

Proof: (*) is valid for $n = 2, 3, 4$, (but not $n = 1$)
Assume (*) holds for $n = k$ for some integer $k \geq 2$
i.e., $f^{(k)}(x) = -\frac{1 \times 3 \times 5 \times \cdots \times (2k-3)}{2^k} 3^k (1-3x)^{-(2k-1)/2}$

Then $f^{(k+1)}(x) = -\frac{1 \times 3 \times 5 \times \cdots \times (2k-3)}{2^k} 3^k$
$\left(-\frac{2(k-1)}{2}\right)(1-3x)^{-(2k-1)/2-1}(-3)$

$= -\frac{1 \times 3 \times 5 \times \cdots (2(k+1)-1)}{2^{k+1}} 3^{k+1} (1-3x)^{-(2(k+1)-1)/2}$

Thus (*) holds for $n = k+1$ if it holds for $n = k$.
Therefore, (*) holds for $n = 2, 3, 4, \ldots$ by induction.

24. If $y = \tan(kx)$, then $y' = k\sec^2(kx)$ and

$y'' = 2k^2 \sec^2(kx) \tan(kx)$
$= 2k^2(1 + \tan^2(kx))\tan(kx) = 2k^2 y(1 + y^2)$.

25. If $y = \sec(kx)$, then $y' = k\sec(kx)\tan(kx)$ and

$y'' = k^2(\sec^2(kx)\tan^2(kx) + \sec^3(kx))$
$= k^2 y(2\sec^2(kx) - 1) = k^2 y(2y^2 - 1)$.

26. To be proved: if $f(x) = \sin(ax + b)$, then

$f^{(n)}(x) = \begin{cases} (-1)^k a^n \sin(ax+b) & \text{if } n = 2k \\ (-1)^k a^n \cos(ax+b) & \text{if } n = 2k+1 \end{cases}$

for $k = 0, 1, 2, \ldots$ Proof: The formula works for $k = 0$ ($n = 2 \times 0 = 0$ and $n = 2 \times 0 + 1 = 1$):

$\begin{cases} f^{(0)}(x) = f(x) = (-1)^0 a^0 \sin(ax+b) = \sin(ax+b) \\ f^{(1)}(x) = f'(x) = (-1)^0 a^1 \cos(ax+b) = a\cos(ax+b) \end{cases}$

Now assume the formula holds for some $k \geq 0$.
If $n = 2(k+1)$, then

$f^{(n)}(x) = \frac{d}{dx} f^{(n-1)}(x) = \frac{d}{dx} f^{(2k+1)}(x)$
$= \frac{d}{dx}\left((-1)^k a^{2k+1} \cos(ax+b)\right)$
$= (-1)^{k+1} a^{2k+2} \sin(ax+b)$

and if $n = 2(k+1) + 1 = 2k+3$, then

$f^{(n)}(x) = \frac{d}{dx}\left((-1)^{k+1} a^{2k+2} \sin(ax+b)\right)$
$= (-1)^{k+1} a^{2k+3} \cos(ax+b)$.

Thus the formula also holds for $k+1$. Therefore it holds for all positive integers k by induction.

27. If $y = \tan x$, then

$y' = \sec^2 x = 1 + \tan^2 x = 1 + y^2 = P_2(y)$,

where P_2 is a polynomial of degree 2. Assume that $y^{(n)} = P_{n+1}(y)$ where P_{n+1} is a polynomial of degree $n+1$. The derivative of any polynomial is a polynomial of one lower degree, so

$y^{(n+1)} = \frac{d}{dx} P_{n+1}(y) = P_n(y)\frac{dy}{dx} = P_n(y)(1+y^2) = P_{n+2}(y)$,

a polynomial of degree $n + 2$. By induction, $(d/dx)^n \tan x = P_{n+1}(\tan x)$, a polynomial of degree $n+1$ in $\tan x$.

28. $(fg)'' = (f'g + fg')' = f''g + f'g' + f'g' + fg''$
$= f''g + 2f'g' + fg''$

29. $(fg)^{(3)} = \frac{d}{dx}(fg)''$
$= \frac{d}{dx}[f''g + 2f'g' + fg'']$
$= f^{(3)}g + f''g' + 2f''g' + 2f'g'' + f'g'' + fg^{(3)}$
$= f^{(3)}g + 3f''g' + 3f'g'' + fg^{(3)}$.

$(fg)^{(4)} = \frac{d}{dx}(fg)^{(3)}$
$= \frac{d}{dx}[f^{(3)}g + 3f''g' + 3f'g'' + fg^{(3)}]$
$= f^{(4)}g + f^{(3)}g' + 3f^{(3)}g' + 3f''g'' + 3f''g''$
$+ 3f'g^{(3)} + f'g^{(3)} + fg^{(4)}$
$= f^{(4)}g + 4f^{(3)}g' + 6f''g'' + 4f'g^{(3)} + fg^{(4)}$.

$(fg)^{(n)} = f^{(n)}g + nf^{(n-1)}g' + \frac{n!}{2!(n-2)!}f^{(n-2)}g''$
$+ \frac{n!}{3!(n-3)!}f^{(n-3)}g^{(3)} + \cdots + nf'g^{(n-1)} + fg^{(n)}$
$= \sum_{k=0}^{n} \frac{n!}{k!(n-k)!} f^{(n-k)} g^{(k)}$.

Section 2.9 Implicit Differentiation (page 154)

1. $xy - x + 2y = 1$
   Differentiate with respect to x:
   $y + xy' - 1 + 2y' = 0$
   Thus $y' = \dfrac{1-y}{2+x}$

2. $x^3 + y^3 = 1$
   $3x^2 + 3y^2 y' = 0$, so $y' = -\dfrac{x^2}{y^2}$.

3. $x^2 + xy = y^3$
   Differentiate with respect to x:
   $2x + y + xy' = 3y^2 y'$
   $y' = \dfrac{2x+y}{3y^2 - x}$

4. $x^3 y + xy^5 = 2$
   $3x^2 y + x^3 y' + y^5 + 5xy^4 y' = 0$
   $y' = \dfrac{-3x^2 y - y^5}{x^3 + 5xy^4}$

5. $x^2 y^3 = 2x - y$
   $2xy^3 + 3x^2 y^2 y' = 2 - y'$
   $y' = \dfrac{2 - 2xy^3}{3x^2 y^2 + 1}$

6. $x^2 + 4(y-1)^2 = 4$
   $2x + 8(y-1)y' = 0$, so $y' = \dfrac{x}{4(1-y)}$

7. $\dfrac{x-y}{x+y} = \dfrac{x^2}{y} + 1 = \dfrac{x^2+y}{y}$
   Thus $xy - y^2 = x^3 + x^2 y + xy + y^2$, or $x^3 + x^2 y + 2y^2 = 0$
   Differentiate with respect to x:
   $3x^2 + 2xy + x^2 y' + 4yy' = 0$
   $y' = -\dfrac{3x^2 + 2xy}{x^2 + 4y}$

8. $x\sqrt{x+y} = 8 - xy$
   $\sqrt{x+y} + x\dfrac{1}{2\sqrt{x+y}}(1+y') = -y - xy'$
   $2(x+y) + x(1+y') = -2\sqrt{x+y}(y+xy')$
   $y' = -\dfrac{3x + 2y + 2y\sqrt{x+y}}{x + 2x\sqrt{x+y}}$

9. $2x^2 + 3y^2 = 5$
   $4x + 6yy' = 0$
   At $(1, 1)$: $4 + 6y' = 0$, $y' = -\dfrac{2}{3}$
   Tangent line: $y - 1 = -\dfrac{2}{3}(x-1)$ or $2x + 3y = 5$

10. $x^2 y^3 - x^3 y^2 = 12$
    $2xy^3 + 3x^2 y^2 y' - 3x^2 y^2 - 2x^3 yy' = 0$
    At $(-1, 2)$: $-16 + 12y' - 12 + 4y' = 0$, so the slope is
    $y' = \dfrac{12+16}{12+4} = \dfrac{28}{16} = \dfrac{7}{4}$.
    Thus, the equation of the tangent line is
    $y = 2 + \dfrac{7}{4}(x+1)$, or $7x - 4y + 15 = 0$.

11. $\dfrac{x}{y} + \left(\dfrac{y}{x}\right)^3 = 2$
    $x^4 + y^4 = 2x^3 y$
    $4x^3 + 4y^3 y' = 6x^2 y + 2x^3 y'$
    at $(-1, -1)$: $-4 - 4y' = -6 - 2y'$
    $2y' = 2$, $y' = 1$
    Tangent line: $y + 1 = 1(x+1)$ or $y = x$.

12. $x + 2y + 1 = \dfrac{y^2}{x-1}$
    $1 + 2y' = \dfrac{(x-1)2yy' - y^2(1)}{(x-1)^2}$
    At $(2, -1)$ we have $1 + 2y' = -2y' - 1$ so $y' = -\dfrac{1}{2}$.
    Thus, the equation of the tangent is
    $y = -1 - \dfrac{1}{2}(x-2)$, or $x + 2y = 0$.

13. $2x + y - \sqrt{2}\sin(xy) = \pi/2$
    $2 + y' - \sqrt{2}\cos(xy)(y + xy') = 0$
    At $(\pi/4, 1)$: $2 + y' - (1 + (\pi/4)y') = 0$, so $y' = -4/(4-\pi)$.
    The tangent has equation
    $$y = 1 - \dfrac{4}{4-\pi}\left(x - \dfrac{\pi}{4}\right).$$

14. $\tan(xy^2) = (2/\pi)xy$
    $(\sec^2(xy^2))(y^2 + 2xyy') = (2/\pi)(y + xy')$.
    At $(-\pi, 1/2)$: $2((1/4) - \pi y') = (1/\pi) - 2y'$, so
    $y' = (\pi - 2)/(4\pi(\pi - 1))$. The tangent has equation
    $$y = \dfrac{1}{2} + \dfrac{\pi - 2}{4\pi(\pi - 1)}(x + \pi).$$

15. $x\sin(xy - y^2) = x^2 - 1$
    $\sin(xy - y^2) + x(\cos(xy - y^2))(y + xy' - 2yy') = 2x$.
    At $(1, 1)$: $0 + (1)(1)(1 - y') = 2$, so $y' = -1$. The tangent has equation $y = 1 - (x - 1)$, or $y = 2 - x$.

16. $\cos\left(\dfrac{\pi y}{x}\right) = \dfrac{x^2}{y} - \dfrac{17}{2}$
    $\left[-\sin\left(\dfrac{\pi y}{x}\right)\right]\dfrac{\pi(xy' - y)}{x^2} = \dfrac{2xy - x^2 y'}{y^2}$.
    At $(3, 1)$: $-\dfrac{\sqrt{3}}{2}\dfrac{\pi(3y' - 1)}{9} = 6 - 9y'$,
    so $y' = (108 - \sqrt{3}\pi)/(162 - 3\sqrt{3}\pi)$. The tangent has equation
    $$y = 1 + \dfrac{108 - \sqrt{3}\pi}{162 - 3\sqrt{3}\pi}(x - 3).$$

17. $xy = x + y$

$y + xy' = 1 + y' \Rightarrow y' = \dfrac{y-1}{1-x}$

$y' + y' + xy'' = y''$

Therefore, $y'' = \dfrac{2y'}{1-x} = \dfrac{2(y-1)}{(1-x)^2}$

18. $x^2 + 4y^2 = 4$, $2x + 8yy' = 0$, $2 + 8(y')^2 + 8yy'' = 0$.

Thus, $y' = \dfrac{-x}{4y}$ and

$y'' = \dfrac{-2 - 8(y')^2}{8y} = -\dfrac{1}{4y} - \dfrac{x^2}{16y^3} = \dfrac{-4y^2 - x^2}{16y^3} = -\dfrac{1}{4y^3}.$

19. $x^3 - y^2 + y^3 = x$

$3x^2 - 2yy' + 3y^2 y' = 1 \Rightarrow y' = \dfrac{1 - 3x^2}{3y^2 - 2y}$

$6x - 2(y')^2 - 2yy'' + 6y(y')^2 + 3y^2 y'' = 0$

$y'' = \dfrac{(2-6y)(y')^2 - 6x}{3y^2 - 2y} = \dfrac{(2-6y)\dfrac{(1-3x^2)^2}{(3y^2-2y)^2} - 6x}{3y^2 - 2y}$

$= \dfrac{(2-6y)(1-3x^2)^2}{(3y^2-2y)^3} - \dfrac{6x}{3y^2-2y}$

20. $x^3 - 3xy + y^3 = 1$

$3x^2 - 3y - 3xy' + 3y^2 y' = 0$

$6x - 3y' - 3y' - 3xy'' + 6y(y')^2 + 3y^2 y'' = 0$

Thus

$y' = \dfrac{y - x^2}{y^2 - x}$

$y'' = \dfrac{-2x + 2y' - 2y(y')^2}{y^2 - x}$

$= \dfrac{2}{y^2-x}\left[-x + \left(\dfrac{y-x^2}{y^2-x}\right) - y\left(\dfrac{y-x^2}{y^2-x}\right)^2\right]$

$= \dfrac{2}{y^2-x}\left[\dfrac{-2xy}{(y^2-x)^2}\right] = \dfrac{4xy}{(x-y^2)^3}.$

21. $x^2 + y^2 = a^2$

$2x + 2yy' = 0$ so $x + yy' = 0$ and $y' = -\dfrac{x}{y}$

$1 + y'y' + yy'' = 0$ so

$y'' = -\dfrac{1 + (y')^2}{y} = -\dfrac{1 + \dfrac{x^2}{y^2}}{y}$

$= -\dfrac{y^2 + x^2}{y^3} = -\dfrac{a^2}{y^3}$

22. $Ax^2 + By^2 = C$

$2Ax + 2Byy' = 0 \Rightarrow y' = -\dfrac{Ax}{By}$

$2A + 2B(y')^2 + 2Byy'' = 0.$

Thus,

$y'' = \dfrac{-A - B(y')^2}{By} = \dfrac{-A - B\left(\dfrac{Ax}{By}\right)^2}{By}$

$= \dfrac{-A(By^2 + Ax^2)}{B^2 y^3} = -\dfrac{AC}{B^2 y^3}.$

23. Ellipse: $x^2 + 2y^2 = 2$

$2x + 4yy' = 0$

Slope of ellipse: $y'_E = -\dfrac{x}{2y}$

Hyperbola: $2x^2 - 2y^2 = 1$

$4x - 4yy' = 0$

Slope of hyperbola: $y'_H = \dfrac{x}{y}$

At intersection points $\begin{cases} x^2 + 2y^2 = 2 \\ 2x^2 - 2y^2 = 1 \end{cases}$

$3x^2 = 3$ so $x^2 = 1$, $y^2 = \dfrac{1}{2}$

Thus $y'_E y'_H = -\dfrac{x}{2y} \cdot \dfrac{x}{y} = -\dfrac{x^2}{2y^2} = -1$

Therefore the curves intersect at right angles.

24. The slope of the ellipse $\dfrac{x^2}{a^2} + \dfrac{y^2}{b^2} = 1$ is found from

$\dfrac{2x}{a^2} + \dfrac{2y}{b^2} y' = 0,$ i.e. $y' = -\dfrac{b^2 x}{a^2 y}.$

Similarly, the slope of the hyperbola $\dfrac{x^2}{A^2} - \dfrac{y^2}{B^2} = 1$ at (x, y) satisfies

$\dfrac{2x}{A^2} - \dfrac{2y}{B^2} y' = 0,$ or $y' = \dfrac{B^2 x}{A^2 y}.$

If the point (x, y) is an intersection of the two curves, then

$\dfrac{x^2}{a^2} + \dfrac{y^2}{b^2} = \dfrac{x^2}{A^2} - \dfrac{y^2}{B^2}$

$x^2 \left(\dfrac{1}{A^2} - \dfrac{1}{a^2}\right) = y^2 \left(\dfrac{1}{B^2} + \dfrac{1}{b^2}\right).$

Thus, $\dfrac{x^2}{y^2} = \dfrac{b^2 + B^2}{B^2 b^2} \cdot \dfrac{A^2 a^2}{a^2 - A^2}.$

Since $a^2 - b^2 = A^2 + B^2$, therefore $B^2 + b^2 = a^2 - A^2$, and $\dfrac{x^2}{y^2} = \dfrac{A^2 a^2}{B^2 b^2}$. Thus, the product of the slope of the two curves at (x, y) is

$-\dfrac{b^2 x}{a^2 y} \cdot \dfrac{B^2 x}{A^2 y} = -\dfrac{b^2 B^2}{a^2 A^2} \cdot \dfrac{A^2 a^2}{B^2 b^2} = -1.$

Therefore, the curves intersect at right angles.

25. If $z = \tan(x/2)$, then

$$1 = \sec^2(x/2)\frac{1}{2}\frac{dx}{dz} = \frac{1+\tan^2(x/2)}{2}\frac{dx}{dz} = \frac{1+z^2}{2}\frac{dx}{dz}.$$

Thus $dx/dz = 2/(1+z^2)$. Also

$$\cos x = 2\cos^2(x/2) - 1 = \frac{2}{\sec^2(x/2)} - 1$$
$$= \frac{2}{1+z^2} - 1 = \frac{1-z^2}{1+z^2}$$
$$\sin x = 2\sin(x/2)\cos(x/2) = \frac{2\tan(x/2)}{1+\tan^2(x/2)} = \frac{2z}{1+z^2}.$$

26. $\dfrac{x-y}{x+y} = \dfrac{x}{y} + 1 \Leftrightarrow xy - y^2 = x^2 + xy + xy + y^2$

$$\Leftrightarrow x^2 + 2y^2 + xy = 0$$

Differentiate with respect to x:

$$2x + 4yy' + y + xy' = 0 \quad \Rightarrow \quad y' = -\frac{2x+y}{4y+x}.$$

However, since $x^2 + 2y^2 + xy = 0$ can be written

$$x^2 + xy + \frac{1}{4}y^2 + \frac{7}{4}y^2 = 0, \text{ or } (x+\frac{y}{2})^2 + \frac{7}{4}y^2 = 0,$$

the only solution is $x = 0$, $y = 0$, and these values do not satisfy the original equation. There are no points on the given curve.

Section 2.10 Antiderivatives and Initial-Value Problems (page 160)

1. $\displaystyle\int 5\,dx = 5x + C$

2. $\displaystyle\int x^2\,dx = \frac{1}{3}x^3 + C$

3. $\displaystyle\int \sqrt{x}\,dx = \frac{2}{3}x^{3/2} + C$

4. $\displaystyle\int x^{12}\,dx = \frac{1}{13}x^{13} + C$

5. $\displaystyle\int x^3\,dx = \frac{1}{4}x^4 + C$

6. $\displaystyle\int (x + \cos x)\,dx = \frac{x^2}{2} + \sin x + C$

7. $\displaystyle\int \tan x \cos x\,dx = \int \sin x\,dx = -\cos x + C$

8. $\displaystyle\int \frac{1+\cos^3 x}{\cos^2 x}\,dx = \int(\sec^2 x + \cos x)\,dx = \tan x + \sin x + C$

9. $\displaystyle\int (a^2 - x^2)\,dx = a^2 x - \frac{1}{3}x^3 + C$

10. $\displaystyle\int (A + Bx + Cx^2)\,dx = Ax + \frac{B}{2}x^2 + \frac{C}{3}x^3 + K$

11. $\displaystyle\int (2x^{1/2} + 3x^{1/3})\,dx = \frac{4}{3}x^{3/2} + \frac{9}{4}x^{4/3} + C$

12. $\displaystyle\int \frac{6(x-1)}{x^{4/3}}\,dx = \int(6x^{-1/3} - 6x^{-4/3})\,dx$
$$= 9x^{2/3} + 18x^{-1/3} + C$$

13. $\displaystyle\int\left(\frac{x^3}{3} - \frac{x^2}{2} + x - 1\right)dx = \frac{1}{12}x^4 - \frac{1}{6}x^3 + \frac{1}{2}x^2 - x + C$

14. $105\displaystyle\int(1 + t^2 + t^4 + t^6)\,dt$
$$= 105(t + \tfrac{1}{3}t^3 + \tfrac{1}{5}t^5 + \tfrac{1}{7}t^7) + C$$
$$= 105t + 35t^3 + 21t^5 + 15t^7 + C$$

15. $\displaystyle\int \cos(2x)\,dx = \frac{1}{2}\sin(2x) + C$

16. $\displaystyle\int \sin\left(\frac{x}{2}\right)dx = -2\cos\left(\frac{x}{2}\right) + C$

17. $\displaystyle\int \frac{dx}{(1+x)^2} = -\frac{1}{1+x} + C$

18. $\displaystyle\int \sec(1-x)\tan(1-x)\,dx = -\sec(1-x) + C$

19. $\displaystyle\int \sqrt{2x+3}\,dx = \frac{1}{3}(2x+3)^{3/2} + C$

20. Since $\dfrac{d}{dx}\sqrt{x+1} = \dfrac{1}{2\sqrt{x+1}}$, therefore
$$\int \frac{4}{\sqrt{x+1}}\,dx = 8\sqrt{x+1} + C.$$

21. $\displaystyle\int 2x\sin(x^2)\,dx = -\cos(x^2) + C$

22. Since $\dfrac{d}{dx}\sqrt{x^2+1} = \dfrac{x}{\sqrt{x^2+1}}$, therefore
$$\int \frac{2x}{\sqrt{x^2+1}}\,dx = 2\sqrt{x^2+1} + C.$$

23. $\displaystyle\int \tan^2 x\,dx = \int(\sec^2 x - 1)\,dx = \tan x - x + C$

24. $\displaystyle\int \sin x \cos x\,dx = \int \frac{1}{2}\sin(2x)\,dx = -\frac{1}{4}\cos(2x) + C$

25. $\displaystyle\int \cos^2 x\,dx = \int \frac{1+\cos(2x)}{2}\,dx = \frac{x}{2} + \frac{\sin(2x)}{4} + C$

26. $\displaystyle\int \sin^2 x\,dx = \int \frac{1-\cos(2x)}{2}\,dx = \frac{x}{2} - \frac{\sin(2x)}{4} + C$

27. $\begin{cases} y' = x - 2 \\ y(0) = 3 \end{cases} \Rightarrow y = \frac{1}{2}x^2 - 2x + C \Rightarrow 3 = 0 + C$ therefore $C = 3$.
Thus $y = \frac{1}{2}x^2 - 2x + 3$ for all x.

28. Given that
$$\begin{cases} y' = x^{-2} - x^{-3} \\ y(-1) = 0, \end{cases}$$
then $y = \displaystyle\int (x^{-2} - x^{-3})\,dx = -x^{-1} + \frac{1}{2}x^{-2} + C$
and $0 = y(-1) = -(-1)^{-1} + \frac{1}{2}(-1)^{-2} + C$ so $C = -\frac{3}{2}$.
Hence, $y(x) = -\frac{1}{x} + \frac{1}{2x^2} - \frac{3}{2}$ which is valid on the interval $(-\infty, 0)$.

29. $\begin{cases} y' = 3\sqrt{x} \Rightarrow y = 2x^{3/2} + C \\ y(4) = 1 \Rightarrow 1 = 16 + C \text{ so } C = -15 \end{cases}$
Thus $y = 2x^{3/2} - 15$ for $x > 0$.

30. Given that
$$\begin{cases} y' = x^{1/3} \\ y(0) = 5, \end{cases}$$
then $y = \displaystyle\int x^{1/3}\,dx = \frac{3}{4}x^{4/3} + C$ and $5 = y(0) = C$.
Hence, $y(x) = \frac{3}{4}x^{4/3} + 5$ which is valid on the whole real line.

31. Since $y' = Ax^2 + Bx + C$ we have
$y = \frac{A}{3}x^3 + \frac{B}{2}x^2 + Cx + D$. Since $y(1) = 1$, therefore
$1 = y(1) = \frac{A}{3} + \frac{B}{2} + C + D$. Thus $D = 1 - \frac{A}{3} - \frac{B}{2} - C$,
and
$y = \frac{A}{3}(x^3 - 1) + \frac{B}{2}(x^2 - 1) + C(x - 1) + 1$ for all x

32. Given that
$$\begin{cases} y' = x^{-9/7} \\ y(1) = -4, \end{cases}$$
then $y = \displaystyle\int x^{-9/7}\,dx = -\frac{7}{2}x^{-2/7} + C$.
Also, $-4 = y(1) = -\frac{7}{2} + C$, so $C = -\frac{1}{2}$. Hence,
$y = -\frac{7}{2}x^{-2/7} - \frac{1}{2}$, which is valid in the interval $(0, \infty)$.

33. For $\begin{cases} y' = \cos x \\ y(\pi/6) = 2, \end{cases}$ we have
$y = \displaystyle\int \cos x\,dx = \sin x + C$
$2 = \sin\frac{\pi}{6} + C = \frac{1}{2} + C \Rightarrow C = \frac{3}{2}$
$y = \sin x + \frac{3}{2}$ (for all x).

34. For $\begin{cases} y' = \sin(2x) \\ y(\pi/2) = 1 \end{cases}$, we have
$y = \displaystyle\int \sin(2x)\,dx = -\frac{1}{2}\cos(2x) + C$
$1 = -\frac{1}{2}\cos\pi + C = \frac{1}{2} + C \Rightarrow C = \frac{1}{2}$
$y = \frac{1}{2}(1 - \cos(2x))$ (for all x).

35. For $\begin{cases} y' = \sec^2 x \\ y(0) = 1 \end{cases}$, we have
$y = \displaystyle\int \sec^2 x\,dx = \tan x + C$
$1 = \tan 0 + C = C \Rightarrow C = 1$
$y = \tan x + 1$ (for $-\pi/2 < x < \pi/2$).

36. For $\begin{cases} y' = \sec^2 x \\ y(\pi) = 1 \end{cases}$, we have
$y = \displaystyle\int \sec^2 x\,dx = \tan x + C$
$1 = \tan\pi + C = C \Rightarrow C = 1$
$y = \tan x + 1$ (for $\pi/2 < x < 3\pi/2$).

37. Since $y'' = 2$, therefore $y' = 2x + C_1$.
Since $y'(0) = 5$, therefore $5 = 0 + C_1$, and $y' = 2x + 5$.
Thus $y = x^2 + 5x + C_2$.
Since $y(0) = -3$, therefore $-3 = 0 + 0 + C_2$, and $C_2 = -3$.
Finally, $y = x^2 + 5x - 3$, for all x.

38. Given that
$$\begin{cases} y'' = x^{-4} \\ y'(1) = 2 \\ y(1) = 1, \end{cases}$$
then $y' = \displaystyle\int x^{-4}\,dx = -\frac{1}{3}x^{-3} + C$.
Since $2 = y'(1) = -\frac{1}{3} + C$, therefore $C = \frac{7}{3}$,
and $y' = -\frac{1}{3}x^{-3} + \frac{7}{3}$. Thus
$y = \displaystyle\int \left(-\frac{1}{3}x^{-3} + \frac{7}{3}\right)dx = \frac{1}{6}x^{-2} + \frac{7}{3}x + D$,
and $1 = y(1) = \frac{1}{6} + \frac{7}{3} + D$, so that $D = -\frac{3}{2}$. Hence,
$y(x) = \frac{1}{6}x^{-2} + \frac{7}{3}x - \frac{3}{2}$, which is valid in the interval $(0, \infty)$.

39. Since $y'' = x^3 - 1$, therefore $y' = \frac{1}{4}x^4 - x + C_1$.
Since $y'(0) = 0$, therefore $0 = 0 - 0 + C_1$, and $y' = \frac{1}{4}x^4 - x$.
Thus $y = \frac{1}{20}x^5 - \frac{1}{2}x^2 + C_2$.
Since $y(0) = 8$, we have $8 = 0 - 0 + C_2$.
Hence $y = \frac{1}{20}x^5 - \frac{1}{2}x^2 + 8$ for all x.

40. Given that
$$\begin{cases} y'' = 5x^2 - 3x^{-1/2} \\ y'(1) = 2 \\ y(1) = 0, \end{cases}$$

we have $y' = \int 5x^2 - 3x^{-1/2}\, dx = \frac{5}{3}x^3 - 6x^{1/2} + C$.

Also, $2 = y'(1) = \frac{5}{3} - 6 + C$ so that $C = \frac{19}{3}$. Thus, $y' = \frac{5}{3}x^3 - 6x^{1/2} + \frac{19}{3}$, and

$$y = \int \left(\frac{5}{3}x^3 - 6x^{1/2} + \frac{19}{3}\right) dx = \frac{5}{12}x^4 - 4x^{3/2} + \frac{19}{3}x + D.$$

Finally, $0 = y(1) = \frac{5}{12} - 4 + \frac{19}{3} + D$ so that $D = -\frac{11}{4}$. Hence, $y(x) = \frac{5}{12}x^4 - 4x^{3/2} + \frac{19}{3}x - \frac{11}{4}$.

41. For $\begin{cases} y'' = \cos x \\ y(0) = 0 \\ y'(0) = 1 \end{cases}$ we have

$$y' = \int \cos x\, dx = \sin x + C_1$$
$$1 = \sin 0 + C_1 \implies C_1 = 1$$
$$y = \int (\sin x + 1)\, dx = -\cos x + x + C_2$$
$$0 = -\cos 0 + 0 + C_2 \implies C_2 = 1$$
$$y = 1 + x - \cos x.$$

42. For $\begin{cases} y'' = x + \sin x \\ y(0) = 2 \\ y'(0) = 0 \end{cases}$ we have

$$y' = \int (x + \sin x)\, dx = \frac{x^2}{2} - \cos x + C_1$$
$$0 = 0 - \cos 0 + C_1 \implies C_1 = 1$$
$$y = \int \left(\frac{x^2}{2} - \cos x + 1\right) dx = \frac{x^3}{6} - \sin x + x + C_2$$
$$2 = 0 - \sin 0 + 0 + C_2 \implies C_2 = 2$$
$$y = \frac{x^3}{6} - \sin x + x + 2.$$

43. Let $y = Ax + \frac{B}{x}$. Then $y' = A - \frac{B}{x^2}$, and $y'' = \frac{2B}{x^3}$. Thus, for all $x \neq 0$,

$$x^2 y'' + xy' - y = \frac{2B}{x} + Ax - \frac{B}{x} - Ax - \frac{B}{x} = 0.$$

We will also have $y(1) = 2$ and $y'(1) = 4$ provided

$$A + B = 2, \quad \text{and} \quad A - B = 4.$$

These equations have solution $A = 3$, $B = -1$, so the initial value problem has solution $y = 3x - (1/x)$.

44. Let r_1 and r_2 be distinct rational roots of the equation $ar(r-1) + br + c = 0$

Let $y = Ax^{r_1} + Bx^{r_2}$ $(x > 0)$
Then $y' = Ar_1 x^{r_1-1} + Br_2 x^{r_2-1}$,
and $y'' = Ar_1(r_1-1)x^{r_1-2} + Br_2(r_2-1)x^{r_2-2}$. Thus
$ax^2 y'' + bxy' + cy$
$= ax^2(Ar_1(r_1-1)x^{r_1-2} + Br_2(r_2-1)x^{r_2-2}$
$\quad + bx(Ar_1 x^{r_1-1} + Br_2 x^{r_2-1}) + c(Ax^{r_1} + Bx^{r_2})$
$= A\big(ar_1(r_1-1) + br_1 + c\big)x^{r_1}$
$\quad + B\big(ar_2(r_2-1) + br_2 + c\big)x^{r_2}$
$= 0 x^{r_1} + 0 x^{r_2} \equiv 0 \quad (x > 0)$

45. $\begin{cases} 4x^2 y'' + 4xy' - y = 0 \quad (*) \\ y(4) = 2 \\ y'(4) = -2 \end{cases} \Rightarrow a = 4, b = 4, c = -1$

Auxilary Equation: $4r(r-1) + 4r - 1 = 0$
$$4r^2 - 1 = 0$$
$$r = \pm\frac{1}{2}$$

By #31, $y = Ax^{1/2} + Bx^{-1/2}$ solves $(*)$ for $x > 0$.

Now $y' = \frac{A}{2}x^{-1/2} - \frac{B}{2}x^{-3/2}$

Substitute the initial conditions:

$$2 = 2A + \frac{B}{2} \quad \Rightarrow 1 = A + \frac{B}{4}$$
$$-2 = \frac{A}{4} - \frac{B}{16} \quad \Rightarrow -8 = A - \frac{B}{4}.$$

Hence $9 = \frac{B}{2}$, so $B = 18$, $A = -\frac{7}{2}$.

Thus $y = -\frac{7}{2}x^{1/2} + 18x^{-1/2}$ (for $x > 0$).

46. Consider
$$\begin{cases} x^2 y'' - 6y = 0 \\ y(1) = 1 \\ y'(1) = 1. \end{cases}$$

Let $y = x^r$, $y' = rx^{r-1}$, $y'' = r(r-1)x^{r-2}$. Substituting these expressions into the differential equation we obtain
$$x^2[r(r-1)x^{r-2}] - 6x^r = 0$$
$$[r(r-1) - 6]x^r = 0.$$

Since this equation must hold for all $x > 0$, we must have
$$r(r-1) - 6 = 0$$
$$r^2 - r - 6 = 0$$
$$(r-3)(r+2) = 0.$$

There are two roots: $r_1 = -2$, and $r_2 = 3$. Thus the differential equation has solutions of the form $y = Ax^{-2} + Bx^3$. Then $y' = -2Ax^{-3} + 3Bx^2$. Since $1 = y(1) = A + B$ and $1 = y'(1) = -2A + 3B$, therefore $A = \frac{2}{5}$ and $B = \frac{3}{5}$. Hence, $y = \frac{2}{5}x^{-2} + \frac{3}{5}x^3$.

Section 2.11 Velocity and Acceleration (page 167)

1. $x = t^2 - 4t + 3$, $v = \dfrac{dx}{dt} = 2t - 4$, $a = \dfrac{dv}{dt} = 2$

   a) particle is moving: to the right for $t > 2$

   b) to the left for $t < 2$

   c) particle is always accelerating to the right

   d) never accelerating to the left

   e) particle is speeding up for $t > 2$

   f) slowing down for $t < 2$

   g) the acceleration is 2 at all times

   h) average velocity over $0 \le t \le 4$ is
   $$\frac{x(4) - x(0)}{4 - 0} = \frac{16 - 16 + 3 - 3}{4} = 0$$

2. $x = 4 + 5t - t^2$, $v = 5 - 2t$, $a = -2$.

   a) The point is moving to the right if $v > 0$, i.e., when $t < \frac{5}{2}$.

   b) The point is moving to the left if $v < 0$, i.e., when $t > \frac{5}{2}$.

   c) The point is accelerating to the right if $a > 0$, but $a = -2$ at all t; hence, the point never accelerates to the right.

   d) The point is accelerating to the left if $a < 0$, i.e., for all t.

   e) The particle is speeding up if v and a have the same sign, i.e., for $t > \frac{5}{2}$.

   f) The particle is slowing down if v and a have opposite sign, i.e., for $t < \frac{5}{2}$.

   g) Since $a = -2$ at all t, $a = -2$ at $t = \frac{5}{2}$ when $v = 0$.

   h) The average velocity over $[0, 4]$ is
   $$\frac{x(4) - x(0)}{4} = \frac{8 - 4}{4} = 1.$$

3. $x = t^3 - 4t + 1$, $v = \dfrac{dx}{dt} = 3t^2 - 4$, $a = \dfrac{dv}{dt} = 6t$

   a) particle moving: to the right for $t < -2/\sqrt{3}$ or $t > 2/\sqrt{3}$,

   b) to the left for $-2/\sqrt{3} < t < 2/\sqrt{3}$

   c) particle is accelerating: to the right for $t > 0$

   d) to the left for $t < 0$

   e) particle is speeding up for $t > 2/\sqrt{3}$ or for $-2/\sqrt{3} < t < 0$

   f) particle is slowing down for $t < -2/\sqrt{3}$ or for $0 < t < 2/\sqrt{3}$

   g) velocity is zero at $t = \pm 2/\sqrt{3}$. Acceleration at these times is $\pm 12/\sqrt{3}$.

   h) average velocity on $[0, 4]$ is $\dfrac{4^3 - 4 \times 4 + 1 - 1}{4 - 0} = 12$

4. $x = \dfrac{t}{t^2 + 1}$, $v = \dfrac{(t^2 + 1)(1) - (t)(2t)}{(t^2 + 1)^2} = \dfrac{1 - t^2}{(t^2 + 1)^2}$,

   $a = \dfrac{(t^2 + 1)^2(-2t) - (1 - t^2)(2)(t^2 + 1)(2t)}{(t^2 + 1)^4} = \dfrac{2t(t^2 - 3)}{(t^2 + 1)^3}.$

   a) The point is moving to the right if $v > 0$, i.e., when $1 - t^2 > 0$, or $-1 < t < 1$.

   b) The point is moving to the left if $v < 0$, i.e., when $t < -1$ or $t > 1$.

   c) The point is accelerating to the right if $a > 0$, i.e., when $2t(t^2 - 3) > 0$, that is, when $t > \sqrt{3}$ or $-\sqrt{3} < t < 0$.

   d) The point is accelerating to the left if $a < 0$, i.e., for $t < -\sqrt{3}$ or $0 < t < \sqrt{3}$.

   e) The particle is speeding up if v and a have the same sign, i.e., for $t < -\sqrt{3}$, or $-1 < t < 0$ or $1 < t < \sqrt{3}$.

   f) The particle is slowing down if v and a have opposite sign, i.e., for $-\sqrt{3} < t < -1$, or $0 < t < 1$ or $t > \sqrt{3}$.

   g) $v = 0$ at $t = \pm 1$. At $t = -1$, $a = \dfrac{-2(-2)}{(2)^3} = \dfrac{1}{2}$.

   At $t = 1$, $a = \dfrac{2(-2)}{(2)^3} = -\dfrac{1}{2}$.

   h) The average velocity over $[0, 4]$ is
   $$\frac{x(4) - x(0)}{4} = \frac{\frac{4}{17} - 0}{4} = \frac{1}{17}.$$

5. $y = 9.8t - 4.9t^2$ metres (t in seconds)

   velocity $v = \dfrac{dy}{dt} = 9.8 - 9.8t$

   acceleration $a = \dfrac{dv}{dt} = -9.8$

   The acceleration is 9.8 m/s^2 downward at all times.
   Ball is at maximum height when $v = 0$, i.e., at $t = 1$.
   Thus maximum height is $y\big|_{t=1} = 9.8 - 4.9 = 4.9$ metres.
   Ball strikes the ground when $y = 0$, $(t > 0)$, i.e.,
   $0 = t(9.8 - 4.9t)$ so $t = 2$.
   Velocity at $t = 2$ is $9.8 - 9.8(2) = -9.8$ m/s.
   Ball strikes the ground travelling at 9.8 m/s (downward).

6. Given that $y = 100 - 2t - 4.9t^2$, the time t at which the ball reaches the ground is the positive root of the equation $y = 0$, i.e., $100 - 2t - 4.9t^2 = 0$, namely,

$$t = \frac{-2 + \sqrt{4 + 4(4.9)(100)}}{9.8} \approx 4.318 \text{ s}.$$

The average velocity of the ball is $\dfrac{-100}{4.318} = -23.16$ m/s. Since $-23.159 = v = -2 - 9.8t$, then t 2.159 s.

7. $D = t^2$, D in metres, t in seconds
velocity $v = \dfrac{dD}{dt} = 2t$
Aircraft becomes airborne if
$v = 200$ km/h $= \dfrac{200,000}{3600} = \dfrac{500}{9}$ m/s.
Time for aircraft to become airborne is $t = \dfrac{250}{9}$ s, that is, about 27.8 s.
Distance travelled during takeoff run is $t^2 \approx 771.6$ metres.

8. Let $y(t)$ be the height of the projectile t seconds after it is fired upward from ground level with initial speed v_0. Then

$$y''(t) = -9.8, \quad y'(0) = v_0, \quad y(0) = 0.$$

Two antidifferentiations give

$$y = -4.9t^2 + v_0 t = t(v_0 - 4.9t).$$

Since the projectile returns to the ground at $t = 10$ s, we have $y(10) = 0$, so $v_0 = 49$ m/s. On Mars, the acceleration of gravity is 3.72 m/s^2 rather than 9.8 m/s^2, so the height of the projectile would be

$$y = -1.86t^2 + v_0 t = t(49 - 1.86t).$$

The time taken to fall back to ground level on Mars would be $t = 49/1.86 \approx 26.3$ s.

9. The height of the ball after t seconds is
$y(t) = -(g/2)t^2 + v_0 t$ m if its initial speed was v_0 m/s. Maximum height h occurs when $dy/dt = 0$, that is, at $t = v_0/g$. Hence

$$h = -\frac{g}{2} \cdot \frac{v_0^2}{g^2} + v_0 \cdot \frac{v_0}{g} = \frac{v_0^2}{2g}.$$

An initial speed of $2v_0$ means the maximum height will be $4v_0^2/2g = 4h$. To get a maximum height of $2h$ an initial speed of $\sqrt{2}v_0$ is required.

10. To get to $3h$ metres above Mars, the ball would have to be thrown upward with speed

$$v_M = \sqrt{6g_M h} = \sqrt{6g_M v_0^2/(2g)} = v_0\sqrt{3g_M/g}.$$

Since $g_M = 3.72$ and $g = 9.80$, we have $v_M \approx 1.067 v_0$ m/s.

11. If the cliff is h ft high, then the height of the rock t seconds after it falls is $y = h - 16t^2$ ft. The rock hits the ground ($y = 0$) at time $t = \sqrt{h/16}$ s. Its speed at that time is $v = -32t = -8\sqrt{h} = -160$ ft/s. Thus $\sqrt{h} = 20$, and the cliff is $h = 400$ ft high.

12. If the cliff is h ft high, then the height of the rock t seconds after it is thrown down is $y = h - 32t - 16t^2$ ft. The rock hits the ground ($y = 0$) at time

$$t = \frac{-32 + \sqrt{32^2 + 64h}}{32} = -1 + \frac{1}{4}\sqrt{16 + h} \text{ s}.$$

Its speed at that time is

$$v = -32 - 32t = -8\sqrt{16 + h} = -160 \text{ ft/s}.$$

Solving this equation for h gives the height of the cliff as 384 ft.

13. Let $x(t)$ be the distance travelled by the train in the t seconds after the brakes are applied. Since $d^2x/dt^2 = -1/6$ m/s^2 and since the initial speed is $v_0 = 60$ km/h $= 100/6$ m/s, we have

$$x(t) = -\frac{1}{12}t^2 + \frac{100}{6}t.$$

The speed of the train at time t is $v(t) = -(t/6) + (100/6)$ m/s, so it takes
the train 100 s to come to a stop. In that time it travels $x(100) = -100^2/12 + 100^2/6 = 100^2/12 \approx 833$ metres.

14. $x = At^2 + Bt + C$, $v = 2At + B$.
The average velocity over $[t_1, t_2]$ is
$\dfrac{x(t_2) - x(t_1)}{t_2 - t_1}$
$= \dfrac{At_2^2 + Bt_2 + C - At_1^2 - Bt_1 - C}{t_2 - t_1}$
$= \dfrac{A(t_2^2 - t_1^2) + B(t_2 - t_1)}{(t_2 - t_1)}$
$= \dfrac{A(t_2 + t_1)(t_2 - t_1) + B(t_2 - t_1)}{(t_2 - t_1)}$
$= A(t_2 + t_1) + B$.
The instantaneous velocity at the midpoint of $[t_1, t_2]$ is
$$v\left(\frac{t_2 + t_1}{2}\right) = 2A\left(\frac{t_2 + t_1}{2}\right) + B = A(t_2 + t_1) + B.$$
Hence, the average velocity over the interval is equal to the instantaneous velocity at the midpoint.

15. $s = \begin{cases} t^2 & 0 \le t \le 2 \\ 4t - 4 & 2 < t < 8 \\ -68 + 20t - t^2 & 8 \le t \le 10 \end{cases}$

Note: s is continuous at 2 and 8 since $2^2 = 4(2) - 4$ and $4(8) - 4 = -68 + 160 - 64$

velocity $v = \dfrac{ds}{dt} = \begin{cases} 2t & \text{if } 0 < t < 2 \\ 4 & \text{if } 2 < t < 8 \\ 20 - 2t & \text{if } 8 < t < 10 \end{cases}$

Since $2t \to 4$ as $t \to 2-$, therefore, v is continuous at 2 ($(v(2) = 4)$).
Since $20 - 2t \to 4$ as $t \to 8+$, therefore v is continuous at 8 ($v(8) = 4$). Hence the velocity is continuous for $0 < t < 10$

acceleration $a = \dfrac{dv}{dt} = \begin{cases} 2 & \text{if } 0 < t < 2 \\ 0 & \text{if } 2 < t < 8 \\ -2 & \text{if } 8 < t < 10 \end{cases}$

is discontinuous at $t = 2$ and $t = 8$
Maximum velocity is 4 and is attained on the interval $2 \le t \le 8$.

16. This exercise and the next three refer to the following figure depicting the velocity of a rocket fired from a tower as a function of time since firing.

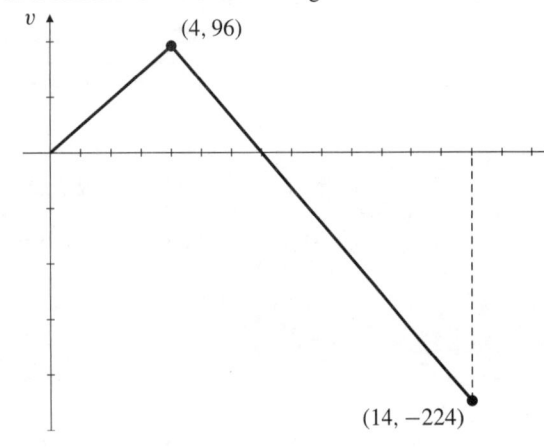

Fig. 2.11.16

The rocket's acceleration while its fuel lasted is the slope of the first part of the graph, namely $96/4 = 24$ ft/s.

17. The rocket was rising for the first 7 seconds.

18. As suggested in Example 1 on page 154 of the text, the distance travelled by the rocket while it was falling from its maximum height to the ground is the area between the velocity graph and the part of the t-axis where $v < 0$. The area of this triangle is $(1/2)(14 - 7)(224) = 784$ ft. This is the maximum height the rocket achieved.

19. The distance travelled upward by the rocket while it was rising is the area between the velocity graph and the part of the t-axis where $v > 0$, namely $(1/2)(7)(96) = 336$ ft. Thus the height of the tower from which the rocket was fired is $784 - 336 = 448$ ft.

20. Let $s(t)$ be the distance the car travels in the t seconds after the brakes are applied. Then $s''(t) = -t$ and the velocity at time t is given by

$$s'(t) = \int (-t)\, dt = -\frac{t^2}{2} + C_1,$$

where $C_1 = 20$ m/s (that is, 72km/h) as determined in Example 6. Thus

$$s(t) = \int \left(20 - \frac{t^2}{2}\right) dt = 20t - \frac{t^3}{6} + C_2,$$

where $C_2 = 0$ because $s(0) = 0$. The time taken to come to a stop is given by $s'(t) = 0$, so it is $t = \sqrt{40}$ s. The distance travelled is

$$s = 20\sqrt{40} - \frac{1}{6}40^{3/2} \approx 84.3 \text{ m.}$$

Review Exercises 2   (page 168)

1. $y = (3x + 1)^2$

$\dfrac{dy}{dx} = \lim_{h \to 0} \dfrac{(3x + 3h + 1)^2 - (3x + 1)^2}{h}$

$= \lim_{h \to 0} \dfrac{9x^2 + 18xh + 9h^2 + 6x + 6h + 1 - (9x^2 + 6x + 1)}{h}$

$= \lim_{h \to 0} (18x + 9h + 6) = 18x + 6$

2. $\dfrac{d}{dx}\sqrt{1 - x^2} = \lim_{h \to 0} \dfrac{\sqrt{1 - (x+h)^2} - \sqrt{1 - x^2}}{h}$

$= \lim_{h \to 0} \dfrac{1 - (x+h)^2 - (1 - x^2)}{h(\sqrt{1 - (x+h)^2} + \sqrt{1 - x^2})}$

$= \lim_{h \to 0} \dfrac{-2x - h}{\sqrt{1 - (x+h)^2} + \sqrt{1 - x^2}} = -\dfrac{x}{\sqrt{1 - x^2}}$

3. $f(x) = 4/x^2$

$f'(2) = \lim_{h \to 0} \dfrac{\dfrac{4}{(2+h)^2} - 1}{h}$

$= \lim_{h \to 0} \dfrac{4 - (4 + 4h + h^2)}{h(2+h)^2} = \lim_{h \to 0} \dfrac{-4 - h}{(2+h)^2} = -1$

4. $g(t) = \dfrac{t-5}{1+\sqrt{t}}$

$g'(9) = \lim\limits_{h \to 0} \dfrac{\dfrac{4+h}{1+\sqrt{9+h}} - 1}{h}$

$= \lim\limits_{h \to 0} \dfrac{(3+h-\sqrt{9+h})(3+h+\sqrt{9+h})}{h(1+\sqrt{9+h})(3+h+\sqrt{9+h})}$

$= \lim\limits_{h \to 0} \dfrac{9+6h+h^2-(9+h)}{h(1+\sqrt{9+h})(3+h+\sqrt{9+h})}$

$= \lim\limits_{h \to 0} \dfrac{5+h}{(1+\sqrt{9+h})(3+h+\sqrt{9+h})}$

$= \dfrac{5}{24}$

5. The tangent to $y = \cos(\pi x)$ at $x = 1/6$ has slope

$\left.\dfrac{dy}{dx}\right|_{x=1/6} = -\pi \sin \dfrac{\pi}{6} = -\dfrac{\pi}{2}.$

Its equation is

$y = \dfrac{\sqrt{3}}{2} - \dfrac{\pi}{2}\left(x - \dfrac{1}{6}\right).$

6. At $x = \pi$ the curve $y = \tan(x/4)$ has slope $(\sec^2(\pi/4))/4 = 1/2$. The normal to the curve there has equation $y = 1 - 2(x - \pi)$.

7. $\dfrac{d}{dx}\dfrac{1}{x - \sin x} = -\dfrac{1 - \cos x}{(x - \sin x)^2}$

8. $\dfrac{d}{dx}\dfrac{1 + x + x^2 + x^3}{x^4} = \dfrac{d}{dx}(x^{-4} + x^{-3} + x^{-2} + x^{-1})$

$= -4x^{-5} - 3x^{-4} - 2x^{-3} - x^{-2}$

$= -\dfrac{4 + 3x + 2x^2 + x^3}{x^5}$

9. $\dfrac{d}{dx}(4 - x^{2/5})^{-5/2} = -\dfrac{5}{2}(4 - x^{2/5})^{-7/2}\left(-\dfrac{2}{5}x^{-3/5}\right)$

$= x^{-3/5}(4 - x^{2/5})^{-7/2}$

10. $\dfrac{d}{dx}\sqrt{2 + \cos^2 x} = \dfrac{-2\cos x \sin x}{2\sqrt{2 + \cos^2 x}} = \dfrac{-\sin x \cos x}{\sqrt{2 + \cos^2 x}}$

11. $\dfrac{d}{d\theta}(\tan \theta - \theta \sec^2 \theta) = \sec^2 \theta - \sec^2 \theta - 2\theta \sec^2 \theta \tan \theta$

$= -2\theta \sec^2 \theta \tan \theta$

12. $\dfrac{d}{dt}\dfrac{\sqrt{1+t^2} - 1}{\sqrt{1+t^2} + 1}$

$= \dfrac{(\sqrt{1+t^2}+1)\dfrac{t}{\sqrt{1+t^2}} - (\sqrt{1+t^2}-1)\dfrac{t}{\sqrt{1+t^2}}}{(\sqrt{1+t^2}+1)^2}$

$= \dfrac{2t}{\sqrt{1+t^2}(\sqrt{1+t^2}+1)^2}$

13. $\lim\limits_{h \to 0} \dfrac{(x+h)^{20} - x^{20}}{h} = \dfrac{d}{dx}x^{20} = 20x^{19}$

14. $\lim\limits_{x \to 2} \dfrac{\sqrt{4x+1} - 3}{x - 2} = \lim\limits_{h \to 0} 4\dfrac{\sqrt{9+4h} - 3}{4h}$

$= \left.\dfrac{d}{dx}4\sqrt{x}\right|_{x=9} = \dfrac{4}{2\sqrt{9}} = \dfrac{2}{3}$

15. $\lim\limits_{x \to \pi/6} \dfrac{\cos(2x) - (1/2)}{x - \pi/6} = \lim\limits_{h \to 0} 2\dfrac{\cos((\pi/3)+2h) - \cos(\pi/3)}{2h}$

$= \left.2\dfrac{d}{dx}\cos x\right|_{x=\pi/3}$

$= -2\sin(\pi/3) = -\sqrt{3}$

16. $\lim\limits_{x \to -a} \dfrac{(1/x^2) - (1/a^2)}{x + a} = \lim\limits_{h \to 0} \dfrac{\dfrac{1}{(-a+h)^2} - \dfrac{1}{(-a)^2}}{h}$

$= \left.\dfrac{d}{dx}\dfrac{1}{x^2}\right|_{x=-a} = \dfrac{2}{a^3}$

17. $\dfrac{d}{dx}f(3 - x^2) = -2xf'(3 - x^2)$

18. $\dfrac{d}{dx}[f(\sqrt{x})]^2 = 2f(\sqrt{x})f'(\sqrt{x})\dfrac{1}{2\sqrt{x}} = \dfrac{f(\sqrt{x})f'(\sqrt{x})}{\sqrt{x}}$

19. $\dfrac{d}{dx}f(2x)\sqrt{g(x/2)} = 2f'(2x)\sqrt{g(x/2)} + \dfrac{f(2x)g'(x/2)}{4\sqrt{g(x/2)}}$

20. $\dfrac{d}{dx}\dfrac{f(x) - g(x)}{f(x) + g(x)}$

$= \dfrac{1}{(f(x)+g(x))^2}\Big[(f(x)+g(x))(f'(x)-g'(x))$

$\quad - (f(x)-g(x))(f'(x)+g'(x))\Big]$

$= \dfrac{2(f'(x)g(x) - f(x)g'(x))}{(f(x)+g(x))^2}$

21. $\dfrac{d}{dx}f(x + (g(x))^2) = (1 + 2g(x)g'(x))f'(x + (g(x))^2)$

22. $\dfrac{d}{dx}f\left(\dfrac{g(x^2)}{x}\right) = \dfrac{2x^2 g'(x^2) - g(x^2)}{x^2}f'\left(\dfrac{g(x^2)}{x}\right)$

23. $\dfrac{d}{dx}f(\sin x)g(\cos x)$

$= (\cos x)f'(\sin x)g(\cos x) - (\sin x)f(\sin x)g'(\cos x)$

24. $\dfrac{d}{dx}\sqrt{\dfrac{\cos f(x)}{\sin g(x)}}$

$= \dfrac{1}{2}\sqrt{\dfrac{\sin g(x)}{\cos f(x)}}$

$\times \dfrac{-f'(x)\sin f(x)\sin g(x) - g'(x)\cos f(x)\cos g(x)}{(\sin g(x))^2}$

25. If $x^3y + 2xy^3 = 12$, then $3x^2y + x^3y' + 2y^3 + 6xy^2y' = 0$.
At $(2, 1)$: $12 + 8y' + 2 + 12y' = 0$, so the slope there is $y' = -7/10$. The tangent line has equation
$y = 1 - \frac{7}{10}(x - 2)$ or $7x + 10y = 24$.

26. $3\sqrt{2}x\sin(\pi y) + 8y\cos(\pi x) = 2$
$3\sqrt{2}\sin(\pi y) + 3\pi\sqrt{2}x\cos(\pi y)y' + 8y'\cos(\pi x) - 8\pi y\sin(\pi x) = 0$
At $(1/3, 1/4)$: $3 + \pi y' + 4y' - \pi\sqrt{3} = 0$, so the slope there is $y' = \dfrac{\pi\sqrt{3} - 3}{\pi + 4}$.

27. $\displaystyle\int \frac{1+x^4}{x^2}\,dx = \int\left(\frac{1}{x^2} + x^2\right)dx = -\frac{1}{x} + \frac{x^3}{3} + C$

28. $\displaystyle\int \frac{1+x}{\sqrt{x}}\,dx = \int (x^{-1/2} + x^{1/2})\,dx = 2\sqrt{x} + \frac{2}{3}x^{3/2} + C$

29. $\displaystyle\int \frac{2 + 3\sin x}{\cos^2 x}\,dx = \int(2\sec^2 x + 3\sec x\tan x)\,dx$
$= 2\tan x + 3\sec x + C$

30. $\displaystyle\int (2x + 1)^4\,dx = \int (16x^4 + 32x^3 + 24x^2 + 8x + 1)\,dx$
$= \dfrac{16x^5}{5} + 8x^4 + 8x^3 + 4x^2 + x + C$
or, equivalently,
$\displaystyle\int (2x+1)^4\,dx = \frac{(2x+1)^5}{10} + C$

31. If $f'(x) = 12x^2 + 12x^3$, then $f(x) = 4x^3 + 3x^4 + C$.
If $f(1) = 0$, then $4 + 3 + C = 0$, so $C = -7$ and $f(x) = 4x^3 + 3x^4 - 7$.

32. If $g'(x) = \sin(x/3) + \cos(x/6)$, then
$$g(x) = -3\cos(x/3) + 6\sin(x/6) + C.$$
If $(\pi, 2)$ lies on $y = g(x)$, then $-(3/2) + 3 + C = 2$, so $C = 1/2$ and $g(x) = -3\cos(x/3) + 6\sin(x/6) + (1/2)$.

33. $\dfrac{d}{dx}(x\sin x + \cos x) = \sin x + x\cos x - \sin x = x\cos x$
$\dfrac{d}{dx}(x\cos x - \sin x) = \cos x - x\sin x - \cos x = -x\sin x$
$\displaystyle\int x\cos x\,dx = x\sin x + \cos x + C$
$\displaystyle\int x\sin x\,dx = -x\cos x + \sin x + C$

34. If $f'(x) = f(x)$ and $g(x) = xf(x)$, then
$$g'(x) = f(x) + xf'(x) = (1+x)f(x)$$
$$g''(x) = f(x) + (1+x)f'(x) = (2+x)f(x)$$
$$g'''(x) = f(x) + (2+x)f'(x) = (3+x)f(x)$$

Conjecture: $g^{(n)}(x) = (n + x)f(x)$ for $n = 1, 2, 3, \ldots$
Proof: The formula is true for $n = 1, 2$, and 3 as shown above. Suppose it is true for $n = k$; that is, suppose $g^{(k)}(x) = (k + x)f(x)$. Then
$$g^{(k+1)}(x) = \frac{d}{dx}\bigl((k+x)f(x)\bigr)$$
$$= f(x) + (k+x)f'(x) = ((k+1) + x)f(x).$$
Thus the formula is also true for $n = k + 1$. It is therefore true for all positive integers n by induction.

35. The tangent to $y = x^3 + 2$ at $x = a$ has equation $y = a^3 + 2 + 3a^2(x - a)$, or $y = 3a^2x - 2a^3 + 2$. This line passes through the origin if $0 = -2a^3 + 2$, that is, if $a = 1$. The line then has equation $y = 3x$.

36. The tangent to $y = \sqrt{2+x^2}$ at $x = a$ has slope $a/\sqrt{2+a^2}$ and equation
$$y = \sqrt{2+a^2} + \frac{a}{\sqrt{2+a^2}}(x - a).$$
This line passes through $(0, 1)$ provided
$$1 = \sqrt{2+a^2} - \frac{a^2}{\sqrt{2+a^2}}$$
$$\sqrt{2+a^2} = 2 + a^2 - a^2 = 2$$
$$2 + a^2 = 4$$
The possibilities are $a = \pm\sqrt{2}$, and the equations of the corrresponding tangent lines are $y = 1 \pm (x/\sqrt{2})$.

37. $\dfrac{d}{dx}\bigl(\sin^n x\sin(nx)\bigr)$
$= n\sin^{n-1} x\cos x\sin(nx) + n\sin^n x\cos(nx)$
$= n\sin^{n-1} x[\cos x\sin(nx) + \sin x\cos(nx)]$
$= n\sin^{n-1} x\sin((n+1)x)$
$y = \sin^n x\sin(nx)$ has a horizontal tangent at $x = m\pi/(n+1)$, for any integer m.

38. $\dfrac{d}{dx}\bigl(\sin^n x\cos(nx)\bigr)$
$= n\sin^{n-1} x\cos x\cos(nx) - n\sin^n x\sin(nx)$
$= n\sin^{n-1} x[\cos x\cos(nx) - \sin x\sin(nx)]$
$= n\sin^{n-1} x\cos((n+1)x)$

$\dfrac{d}{dx}\bigl(\cos^n x\sin(nx)\bigr)$
$= -n\cos^{n-1} x\sin x\sin(nx) + n\cos^n x\cos(nx)$
$= n\cos^{n-1} x[\cos x\cos(nx) - \sin x\sin(nx)]$
$= n\cos^{n-1} x\cos((n+1)x)$

$\dfrac{d}{dx}\bigl(\cos^n x\cos(nx)\bigr)$
$= -n\cos^{n-1} x\sin x\cos(nx) - n\cos^n x\sin(nx)$
$= -n\cos^{n-1} x[\sin x\cos(nx) + \cos x\sin(nx)]$
$= -n\cos^{n-1} x\sin((n+1)x)$

39. $Q = (0, 1)$. If $P = (a, a^2)$ on the curve $y = x^2$, then the slope of $y = x^2$ at P is $2a$, and the slope of PQ is $(a^2 - 1)/a$. PQ is normal to $y = x^2$ if $a = 0$ or $[(a^2 - 1)/a](2a) = -1$, that is, if $a = 0$ or $a^2 = 1/2$. The points P are $(0, 0)$ and $(\pm 1/\sqrt{2}, 1/2)$. The distances from these points to Q are 1 and $\sqrt{3}/2$, respectively. The distance from Q to the curve $y = x^2$ is the shortest of these distances, namely $\sqrt{3}/2$ units.

40. The average profit per tonne if x tonnes are exported is $P(x)/x$, that is the slope of the line joining $(x, P(x))$ to the origin. This slope is maximum if the line is tangent to the graph of $P(x)$. In this case the slope of the line is $P'(x)$, the marginal profit.

41. $F(r) = \begin{cases} \dfrac{mgR^2}{r^2} & \text{if } r \geq R \\ mkr & \text{if } 0 \leq r < R \end{cases}$

a) For continuity of $F(r)$ at $r = R$ we require $mg = mkR$, so $k = g/R$.

b) As r increases from R, F changes at rate
$$\frac{d}{dr}\frac{mgR^2}{r^2}\bigg|_{r=R} = -\frac{2mgR^2}{R^3} = -\frac{2mg}{R}.$$

As r decreases from R, F changes at rate
$$-\frac{d}{dr}(mkr)\bigg|_{r=R} = -mk = -\frac{mg}{R}.$$

Observe that this rate is half the rate at which F decreases when r increases from R.

42. $PV = kT$. Differentiate with respect to P holding T constant to get
$$V + P\frac{dV}{dP} = 0$$

Thus the isothermal compressibility of the gas is
$$\frac{1}{V}\frac{dV}{dP} = \frac{1}{V}\left(-\frac{V}{P}\right) = -\frac{1}{P}.$$

43. Let the building be h m high. The height of the first ball at time t during its motion is
$$y_1 = h + 10t - 4.9t^2.$$

It reaches maximum height when $dy_1/dt = 10 - 9.8t = 0$, that is, at $t = 10/9.8$ s. The maximum height of the first ball is
$$y_1 = h + \frac{100}{9.8} - \frac{4.9 \times 100}{(9.8)^2} = h + \frac{100}{19.6}.$$

The height of the second ball at time t during its motion is
$$y_2 = 20t - 4.9t^2.$$

It reaches maximum height when $dy_2/dt = 20 - 9.8t = 0$, that is, at $t = 20/9.8$ s. The maximum height of the second ball is
$$y_2 = \frac{400}{9.8} - \frac{4.9 \times 400}{(9.8)^2} = \frac{400}{19.6}.$$

These two maximum heights are equal, so
$$h + \frac{100}{19.6} = \frac{400}{19.6},$$

which gives $h = 300/19.6 \approx 15.3$ m as the height of the building.

44. The first ball has initial height 60 m and initial velocity 0, so its height at time t is
$$y_1 = 60 - 4.9t^2 \text{ m}.$$

The second ball has initial height 0 and initial velocity v_0, so its height at time t is
$$y_2 = v_0 t - 4.9t^2 \text{ m}.$$

The two balls collide at a height of 30 m (at time T, say). Thus
$$30 = 60 - 4.9T^2$$
$$30 = v_0 T - 4.9T^2.$$

Thus $v_0 T = 60$ and $T^2 = 30/4.9$. The initial upward speed of the second ball is
$$v_0 = \frac{60}{T} = 60\sqrt{\frac{4.9}{30}} \approx 24.25 \text{ m/s}.$$

At time T, the velocity of the first ball is
$$\frac{dy_1}{dt}\bigg|_{t=T} = -9.8T \approx -24.25 \text{ m/s}.$$

At time T, the velocity of the second ball is
$$\frac{dy_2}{dt}\bigg|_{t=T} = v_0 - 9.8T = 0 \text{ m/s}.$$

45. Let the car's initial speed be v_0. The car decelerates at 20 ft/s^2 starting at $t = 0$, and travels distance s in time t, where $d^2s/dt^2 = -20$. Thus
$$\frac{ds}{dt} = v_0 - 20t$$
$$x = v_0 t - 10t^2.$$

The car stops at time $t = v_0/20$. The stopping distance is $s = 160$ ft, so

$$160 = \frac{v_0^2}{20} - \frac{v_0^2}{40} = \frac{v_0^2}{40}.$$

The car's initial speed cannot exceed $v_0 = \sqrt{160 \times 40} = 80$ ft/s.

46. $P = 2\pi\sqrt{L/g} = 2\pi L^{1/2} g^{-1/2}$.

a) If L remains constant, then

$$\Delta P \approx \frac{dP}{dg} \Delta g = -\pi L^{1/2} g^{-3/2} \Delta g$$

$$\frac{\Delta P}{P} \approx \frac{-\pi L^{1/2} g^{-3/2}}{2\pi L^{1/2} g^{-1/2}} \Delta g = -\frac{1}{2}\frac{\Delta g}{g}.$$

If g increases by 1%, then $\Delta g/g = 1/100$, and $\Delta P/P = -1/200$. Thus P decreases by 0.5%.

b) If g remains constant, then

$$\Delta P \approx \frac{dP}{dL} \Delta L = \pi L^{-1/2} g^{-1/2} \Delta L$$

$$\frac{\Delta P}{P} \approx \frac{\pi L^{-1/2} g^{-1/2}}{2\pi L^{1/2} g^{-1/2}} \Delta L = \frac{1}{2}\frac{\Delta L}{L}.$$

If L increases by 2%, then $\Delta L/L = 2/100$, and $\Delta P/P = 1/100$. Thus P increases by 1%.

Challenging Problems 2   (page 170)

1. The line through (a, a^2) with slope m has equation $y = a^2 + m(x - a)$. It intersects $y = x^2$ at points x that satisfy

$$x^2 = a^2 + mx - ma, \quad \text{or}$$
$$x^2 - mx + ma - a^2 = 0$$

In order that this quadratic have only one solution $x = a$, the left side must be $(x-a)^2$, so that $m = 2a$. The tangent has slope $2a$.
This won't work for more general curves whose tangents can intersect them at more than one point.

2. $f'(x) = 1/x$, $f(2) = 9$.

a) $\displaystyle\lim_{x \to 2} \frac{f(x^2 + 5) - f(9)}{x - 2} = \lim_{h \to 0} \frac{f(9 + 4h + h^2) - f(9)}{h}$

$= \displaystyle\lim_{h \to 0} \frac{f(9 + 4h + h^2) - f(9)}{4h + h^2} \times \frac{4h + h^2}{h}$

$= \displaystyle\lim_{k \to 0} \frac{f(9 + k) - f(9)}{k} \times \lim_{h \to 0}(4 + h)$

$= f'(9) \times 4 = \dfrac{4}{9}.$

b) $\displaystyle\lim_{x \to 2} \frac{\sqrt{f(x)} - 3}{x - 2} = \lim_{h \to 0} \frac{\sqrt{f(2 + h)} - 3}{h}$

$= \displaystyle\lim_{h \to 0} \frac{f(2+h) - 9}{h} \times \frac{1}{\sqrt{f(2+h)} + 3}$

$= f'(2) \times \dfrac{1}{6} = \dfrac{1}{12}.$

3. $f'(4) = 3$, $g'(4) = 7$, $g(4) = 4$, $g(x) \neq 4$ if $x \neq 4$.

a) $\displaystyle\lim_{x \to 4}(f(x) - f(4)) = \lim_{x \to 4} \frac{f(x) - f(4)}{x - 4}(x - 4)$
$= f'(4)(4 - 4) = 0$

b) $\displaystyle\lim_{x \to 4} \frac{f(x) - f(4)}{x^2 - 16} = \lim_{x \to 4} \frac{f(x) - f(4)}{x - 4} \times \frac{1}{x + 4}$
$= f'(4) \times \dfrac{1}{8} = \dfrac{3}{8}$

c) $\displaystyle\lim_{x \to 4} \frac{f(x) - f(4)}{\sqrt{x} - 2} = \lim_{x \to 4} \frac{f(x) - f(4)}{x - 4} \times (\sqrt{x} + 2)$
$= f'(4) \times 4 = 12$

d) $\displaystyle\lim_{x \to 4} \frac{f(x) - f(4)}{\frac{1}{x} - \frac{1}{4}} = \lim_{x \to 4} \frac{f(x) - f(4)}{x - 4} \times \frac{x - 4}{(4 - x)/4x}$
$= f'(4) \times (-16) = -48$

e) $\displaystyle\lim_{x \to 4} \frac{f(x) - f(4)}{g(x) - 4} = \lim_{x \to 4} \frac{\frac{f(x) - f(4)}{x - 4}}{\frac{g(x) - g(4)}{x - 4}}$
$= \dfrac{f'(4)}{g'(4)} = \dfrac{3}{7}$

f) $\displaystyle\lim_{x \to 4} \frac{f(g(x)) - f(4)}{x - 4}$
$= \displaystyle\lim_{x \to 4} \frac{f(g(x)) - f(4)}{g(x) - 4} \times \frac{g(x) - g(4)}{x - 4}$
$= f'(g(4)) \times g'(4) = f'(4) \times g'(4) = 3 \times 7 = 21$

4. $f(x) = \begin{cases} x & \text{if } x = 1,\ 1/2,\ 1/3,\ \ldots \\ x^2 & \text{otherwise} \end{cases}.$

a) f is continuous except at $1/2, 1/3, 1/4, \ldots$. It is continuous at $x = 1$ and $x = 0$ (and everywhere else). Note that

$$\lim_{x \to 1} x^2 = 1 = f(1),$$
$$\lim_{x \to 0} x^2 = \lim_{x \to 0} x = 0 = f(0).$$

b) If $a = 1/2$ and $b = 1/3$, then

$$\frac{f(a) + f(b)}{2} = \frac{1}{2}\left(\frac{1}{2} + \frac{1}{3}\right) = \frac{5}{12}.$$

If $1/3 < x < 1/2$, then $f(x) = x^2 < 1/4 < 5/12$.
Thus the statement is FALSE.

c) By (a) f cannot be differentiable at $x = 1/2, 1/2$, ... It is not differentiable at $x = 0$ either, since

$$\lim_{h \to 0} h - 0h = 1 \neq 0 = \lim_{h \to 0} \frac{h^2 - 0}{h}.$$

f is differentiable elsewhere, including at $x = 1$ where its derivative is 2.

5. If $h \neq 0$, then

$$\left| \frac{f(h) - f(0)}{h} \right| = \frac{|f(h)|}{|h|} > \frac{\sqrt{|h|}}{|h|} \to \infty$$

as $h \to 0$. Therefore $f'(0)$ does not exist.

6. Given that $f'(0) = k$, $f(0) \neq 0$, and $f(x + y) = f(x)f(y)$, we have

$$f(0) = f(0+0) = f(0)f(0) \implies f(0) = 0 \text{ or } f(0) = 1.$$

Thus $f(0) = 1$.

$$f'(x) = \lim_{h \to 0} \frac{f(x+h) - f(x)}{h}$$
$$= \lim_{h \to 0} \frac{f(x)f(h) - f(x)}{h} = f(x)f'(0) = kf(x).$$

7. Given that $g'(0) = k$ and $g(x + y) = g(x) + g(y)$, then

   a) $g(0) = g(0+0) = g(0) + g(0)$. Thus $g(0) = 0$.

   b) $g'(x) = \lim_{h \to 0} \frac{g(x+h) - g(x)}{h}$
   $$= \lim_{h \to 0} \frac{g(x) + g(h) - g(x)}{h} = \lim_{h \to 0} \frac{g(h) - g(0)}{h}$$
   $$= g'(0) = k.$$

   c) If $h(x) = g(x) - kx$, then $h'(x) = g'(x) - k = 0$ for all x. Thus $h(x)$ is constant for all x. Since $h(0) = g(0) - 0 = 0$, we have $h(x) = 0$ for all x, and $g(x) = kx$.

8. a) $f'(x) = \lim_{k \to 0} \frac{f(x+k) - f(x)}{k}$ (let $k = -h$)
   $$= \lim_{h \to 0} \frac{f(x-h) - f(x)}{-h} = \lim_{h \to 0} \frac{f(x) - f(x-h)}{h}.$$
   $$f'(x) = \frac{1}{2}\bigl(f'(x) + f'(x)\bigr)$$
   $$= \frac{1}{2}\left(\lim_{h \to 0} \frac{f(x+h) - f(x)}{h} + \lim_{h \to 0} \frac{f(x) - f(x-h)}{h} \right)$$
   $$= \lim_{h \to 0} \frac{f(x+h) - f(x-h)}{2h}.$$

b) The change of variables used in the first part of (a) shows that

$$\lim_{h \to 0} \frac{f(x+h) - f(x)}{h} \text{ and } \lim_{h \to 0} \frac{f(x) - f(x-h)}{h}$$

are always equal if either exists.

c) If $f(x) = |x|$, then $f'(0)$ does not exist, but

$$\lim_{h \to 0} \frac{f(0+h) - f(0-h)}{2h} = \lim_{h \to 0} \frac{|h| - |h|}{h} = \lim_{h \to 0} \frac{0}{h} = 0.$$

9. The tangent to $y = x^3$ at $x = 3a/2$ has equation

$$y = \frac{27a^3}{8} + \frac{27}{4a^2}\left(x - \frac{3a}{2}\right).$$

This line passes through $(a, 0)$ because

$$\frac{27a^3}{8} + \frac{27}{4a^2}\left(a - \frac{3a}{2}\right) = 0.$$

If $a \neq 0$, the x-axis is another tangent to $y = x^3$ that passes through $(a, 0)$.

The number of tangents to $y = x^3$ that pass through (x_0, y_0) is

three, if $x_0 \neq 0$ and y_0 is between 0 and x_0^3;

two, if $x_0 \neq 0$ and either $y_0 = 0$ or $y_0 = x_0^3$;

one, otherwise.

This is the number of distinct real solutions b of the cubic equation $2b^3 - 3b^2 x_0 + y_0 = 0$, which states that the tangent to $y = x^3$ at (b, b^3) passes through (x_0, y_0).

10. By symmetry, any line tangent to both curves must pass through the origin.

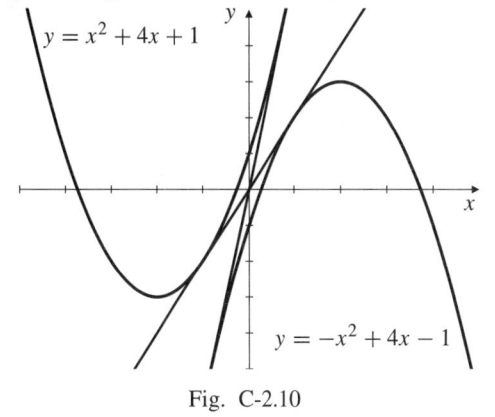

Fig. C-2.10

The tangent to $y = x^2 + 4x + 1$ at $x = a$ has equation
$$y = a^2 + 4a + 1 + (2a + 4)(x - a)$$
$$= (2a + 4)x - (a^2 - 1),$$

which passes through the origin if $a = \pm 1$. The two common tangents are $y = 6x$ and $y = 2x$.

11. The slope of $y = x^2$ at $x = a$ is $2a$.
    The slope of the line from $(0, b)$ to (a, a^2) is $(a^2 - b)/a$.
    This line is normal to $y = x^2$ if either $a = 0$ or $2a((a^2 - b)/a) = -1$, that is, if $a = 0$ or $2a^2 = 2b - 1$.
    There are three real solutions for a if $b > 1/2$ and only one ($a = 0$) if $b \le 1/2$.

12. The point $Q = (a, a^2)$ on $y = x^2$ that is closest to $P = (3, 0)$ is such that PQ is normal to $y = x^2$ at Q. Since PQ has slope $a^2/(a - 3)$ and $y = x^2$ has slope $2a$ at Q, we require
    $$\frac{a^2}{a-3} = -\frac{1}{2a},$$
    which simplifies to $2a^3 + a - 3 = 0$. Observe that $a = 1$ is a solution of this cubic equation. Since the slope of $y = 2x^3 + x - 3$ is $6x^2 + 1$, which is always positive, the cubic equation can have only one real solution. Thus $Q = (1, 1)$ is the point on $y = x^2$ that is closest to P. The distance from P to the curve is $|PQ| = \sqrt{5}$ units.

13. The curve $y = x^2$ has slope $m = 2a$ at (a, a^2). The tangent there has equation
    $$y = a^2 + m(x - a) = mx - \frac{m^2}{4}.$$
    The curve $y = Ax^2 + Bx + C$ has slope $m = 2Aa + B$ at $(a, Aa^2 + Ba + C)$. Thus $a = (m - B)/(2A)$, and the tangent has equation
    $$y = Aa^2 + Ba + C + m(x - a)$$
    $$= mx + \frac{(m-B)^2}{4A} + \frac{B(m-B)}{2A} + C - \frac{m(m-B)}{2A}$$
    $$= mx + C + \frac{(m-B)^2}{4A} - \frac{(m-B)^2}{2A}$$
    $$= mx + f(m),$$
    where $f(m) = C - (m - B)^2/(4A)$.

14. Parabola $y = x^2$ has tangent $y = 2ax - a^2$ at (a, a^2).
    Parabola $y = Ax^2 + Bx + C$ has tangent
    $$y = (2Ab + B)x - Ab^2 + C$$
    at $(b, Ab^2 + Bb + C)$. These two tangents coincide if
    $$2Ab + B = 2a \qquad (*)$$
    $$Ab^2 - C = a^2.$$
    The two curves have one (or more) common tangents if $(*)$ has real solutions for a and b. Eliminating a between the two equations leads to
    $$(2Ab + B)^2 = 4Ab^2 - 4C,$$
    or, on simplification,
    $$4A(A - 1)b^2 + 4ABb + (B^2 + 4C) = 0.$$
    This quadratic equation in b has discriminant
    $$D = 16A^2B^2 - 16A(A-1)(B^2+4C) = 16A(B^2-4(A-1)C).$$
    There are five cases to consider:

    CASE I. If $A = 1$, $B \ne 0$, then $(*)$ gives
    $$b = -\frac{B^2 + 4C}{4B}, \quad a = \frac{B^2 - 4C}{4B}.$$
    There is a single common tangent in this case.

    CASE II. If $A = 1$, $B = 0$, then $(*)$ forces $C = 0$, which is not allowed. There is no common tangent in this case.

    CASE III. If $A \ne 1$ but $B^2 = 4(A - 1)C$, then
    $$b = \frac{-B}{2(A-1)} = a.$$
    There is a single common tangent, and since the points of tangency on the two curves coincide, the two curves are tangent to each other.

    CASE IV. If $A \ne 1$ and $B^2 - 4(A - 1)C < 0$, there are no real solutions for b, so there can be no common tangents.

    CASE V. If $A \ne 1$ and $B^2 - 4(A - 1)C > 0$, there are two distinct real solutions for b, and hence two common tangent lines.

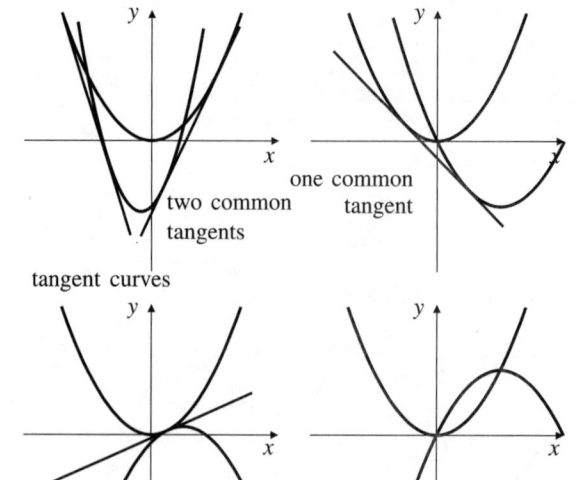

    Fig. C-2.14

15. a) The tangent to $y = x^3$ at (a, a^3) has equation
    $$y = 3a^2x - 2a^3.$$

For intersections of this line with $y = x^3$ we solve

$$x^3 - 3a^2 x + 2a^3 = 0$$
$$(x-a)^2(x+2a) = 0.$$

The tangent also intersects $y = x^3$ at (b, b^3), where $b = -2a$.

b) The slope of $y = x^3$ at $x = -2a$ is $3(-2a)^2 = 12a^2$, which is four times the slope at $x = a$.

c) If the tangent to $y = x^3$ at $x = a$ were also tangent at $x = b$, then the slope at b would be four times that at a and the slope at a would be four times that at b. This is clearly impossible.

d) No line can be tangent to the graph of a cubic polynomial $P(x)$ at two distinct points a and b, because if there was such a double tangent $y = L(x)$, then $(x-a)^2(x-b)^2$ would be a factor of the cubic polynomial $P(x) - L(x)$, and cubic polynomials do not have factors that are 4th degree polynomials.

16. a) $y = x^4 - 2x^2$ has horizontal tangents at points x satisfying $4x^3 - 4x = 0$, that is, at $x = 0$ and $x = \pm 1$. The horizontal tangents are $y = 0$ and $y = -1$. Note that $y = -1$ is a double tangent; it is tangent at the two points $(\pm 1, -1)$.

b) The tangent to $y = x^4 - 2x^2$ at $x = a$ has equation

$$y = a^4 - 2a^2 + (4a^3 - 4a)(x - a)$$
$$= 4a(a^2 - 1)x - 3a^4 + 2a^2.$$

Similarly, the tangent at $x = b$ has equation

$$y = 4b(b^2 - 1)x - 3b^4 + 2b^2.$$

These tangents are the same line (and hence a double tangent) if

$$4a(a^2 - 1) = 4b(b^2 - 1)$$
$$-3a^4 + 2a^2 = -3b^4 + 2b^2.$$

The second equation says that either $a^2 = b^2$ or $3(a^2 + b^2) = 2$; the first equation says that $a^3 - b^3 = a - b$, or, equivalently, $a^2 + ab + b^2 = 1$. If $a^2 = b^2$, then $a = -b$ ($a = b$ is not allowed). Thus $a^2 = b^2 = 1$ and the two points are $(\pm 1, -1)$ as discovered in part (a).
If $a^2 + b^2 = 2/3$, then $ab = 1/3$. This is not possible since it implies that

$$0 = a^2 + b^2 - 2ab = (a-b)^2 > 0.$$

Thus $y = -1$ is the only double tangent to $y = x^4 - 2x^2$.

c) If $y = Ax + B$ is a double tangent to $y = x^4 - 2x^2 + x$, then $y = (A-1)x + B$ is a double tangent to $y = x^4 - 2x^2$. By (b) we must have $A - 1 = 0$ and $B = -1$. Thus the only double tangent to $y = x^4 - 2x^2 + x$ is $y = x - 1$.

17. a) The tangent to

$$y = f(x) = ax^4 + bx^3 + cx^2 + dx + e$$

at $x = p$ has equation

$$y = (4ap^3 + 3bp^2 + 2cp + d)x - 3ap^4 - 2bp^3 - cp^2 + e.$$

This line meets $y = f(x)$ at $x = p$ (a double root), and

$$x = \frac{-2ap - b \pm \sqrt{b^2 - 4ac - 4abp - 8a^2p^2}}{2a}.$$

These two latter roots are equal (and hence correspond to a double tangent) if the expression under the square root is 0, that is, if

$$8a^2p^2 + 4abp + 4ac - b^2 = 0.$$

This quadratic has two real solutions for p provided its discriminant is positive, that is, provided

$$16a^2b^2 - 4(8a^2)(4ac - b^2) > 0.$$

This condition simplifies to

$$3b^2 > 8ac.$$

For example, for $y = x^4 - 2x^2 + x - 1$, we have $a = 1$, $b = 0$, and $c = -2$, so $3b^2 = 0 > -16 = 8ac$, and the curve has a double tangent.

b) From the discussion above, the second point of tangency is

$$q = \frac{-2ap - b}{2a} = -p - \frac{b}{2a}.$$

The slope of PQ is

$$\frac{f(q) - f(p)}{q - p} = \frac{b^3 - 4abc + 8a^2d}{8a^2}.$$

Calculating $f'((p+q)/2)$ leads to the same expression, so the double tangent PQ is parallel to the tangent at the point horizontally midway between P and Q.

c) The inflection points are the real zeros of

$$f''(x) = 2(6ax^2 + 3bx + c).$$

This equation has distinct real roots provided $9b^2 > 24ac$, that is, $3b^2 > 8ac$. The roots are

$$r = \frac{-3b - \sqrt{9b^2 - 24ac}}{12a}$$
$$s = \frac{-3b + \sqrt{9b^2 - 24ac}}{12a}.$$

The slope of the line joining these inflection points is

$$\frac{f(s) - f(r)}{s - r} = \frac{b^3 - 4abc + 8a^2d}{8a^2},$$

so this line is also parallel to the double tangent.

18.

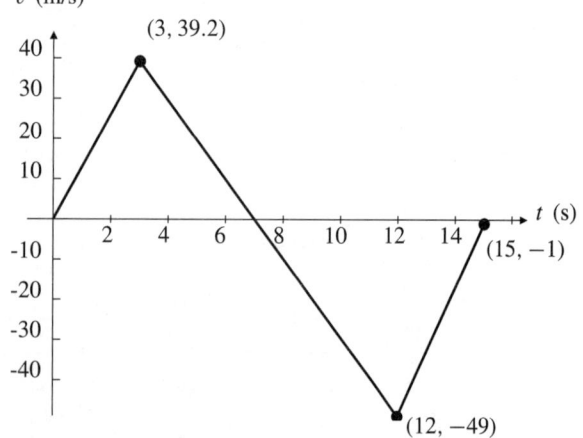

Fig. C-2.18

a) The fuel lasted for 3 seconds.

b) Maximum height was reached at $t = 7$ s.

c) The parachute was deployed at $t = 12$ s.

d) The upward acceleration in $[0, 3]$ was $39.2/3 \approx 13.07$ m/s^2.

e) The maximum height achieved by the rocket is the distance it fell from $t = 7$ to $t = 15$. This is the area under the t-axis and above the graph of v on that interval, that is,

$$\frac{12 - 7}{2}(49) + \frac{49 + 1}{2}(15 - 12) = 197.5 \text{ m}.$$

f) During the time interval $[0, 7]$, the rocket rose a distance equal to the area under the velocity graph and above the t-axis, that is,

$$\frac{1}{2}(7 - 0)(39.2) = 137.2 \text{ m}.$$

Therefore the height of the tower was $197.5 - 137.2 = 60.3$ m.

19. a) Claim: $\dfrac{d^n}{dx^n} \cos(ax) = a^n \cos\left(ax + \dfrac{n\pi}{2}\right)$.

Proof: For $n = 1$ we have

$$\frac{d}{dx} \cos(ax) = -a \sin(ax) = a \cos\left(ax + \frac{\pi}{2}\right),$$

so the formula above is true for $n = 1$. Assume it is true for $n = k$, where k is a positive integer. Then

$$\frac{d^{k+1}}{dx^{k+1}} \cos(ax) = \frac{d}{dx}\left[a^k \cos\left(ax + \frac{k\pi}{2}\right)\right]$$
$$= a^k \left[-a \sin\left(ax + \frac{k\pi}{2}\right)\right]$$
$$= a^{k+1} \cos\left(ax + \frac{(k+1)\pi}{2}\right).$$

Thus the formula holds for $n = 1, 2, 3, \ldots$ by induction.

b) Claim: $\dfrac{d^n}{dx^n} \sin(ax) = a^n \sin\left(ax + \dfrac{n\pi}{2}\right)$.

Proof: For $n = 1$ we have

$$\frac{d}{dx} \sin(ax) = a \cos(ax) = a \sin\left(ax + \frac{\pi}{2}\right),$$

so the formula above is true for $n = 1$. Assume it is true for $n = k$, where k is a positive integer. Then

$$\frac{d^{k+1}}{dx^{k+1}} \sin(ax) = \frac{d}{dx}\left[a^k \sin\left(ax + \frac{k\pi}{2}\right)\right]$$
$$= a^k \left[a \cos\left(ax + \frac{k\pi}{2}\right)\right]$$
$$= a^{k+1} \sin\left(ax + \frac{(k+1)\pi}{2}\right).$$

Thus the formula holds for $n = 1, 2, 3, \ldots$ by induction.

c) Note that

$$\frac{d}{dx}(\cos^4 x + \sin^4 x) = -4\cos^3 x \sin x + 4 \sin^3 x \cos x$$
$$= -4 \sin x \cos x (\cos^2 - \sin^2 x)$$
$$= -2 \sin(2x) \cos(2x)$$
$$= -\sin(4x) = \cos\left(4x + \frac{\pi}{2}\right).$$

It now follows from part (a) that

$$\frac{d^n}{dx^n}(\cos^4 x + \sin^4 x) = 4^{n-1} \cos\left(4x + \frac{n\pi}{2}\right).$$

CHAPTER 3. TRANSCENDENTAL FUNCTIONS

Section 3.1 Inverse Functions (page 178)

1. $f(x) = x - 1$
   $f(x_1) = f(x_2) \Rightarrow x_1 - 1 = x_2 - 1 \Rightarrow x_1 = x_2$.
   Thus f is one-to-one. Let $y = f^{-1}(x)$.
   Then $x = f(y) = y - 1$ and $y = x + 1$. Thus
   $f^{-1}(x) = x+1$. $D(f) = D(f^{-1}) = \mathbb{R} = R(f) = R(f^{-1})$.

2. $f(x) = 2x - 1$. If $f(x_1) = f(x_2)$, then $2x_1 - 1 = 2x_2 - 1$.
   Thus $2(x_1 - x_2) = 0$ and $x_1 = x_2$. Hence, f is one-to-one.
   Let $y = f^{-1}(x)$. Thus $x = f(y) = 2y - 1$, so
   $y = \frac{1}{2}(x + 1)$. Thus $f^{-1}(x) = \frac{1}{2}(x + 1)$.
   $D(f) = R(f^{-1}) = (-\infty, \infty)$.
   $R(f) = D(f^{-1}) = (-\infty, \infty)$.

3. $f(x) = \sqrt{x - 1}$
   $f(x_1) = f(x_2) \Leftrightarrow \sqrt{x_1 - 1} = \sqrt{x_2 - 1}, \quad (x_1, x_2 \geq 1)$
   $\Leftrightarrow x_1 - 1 = x_2 - 1 = 0$
   $\Leftrightarrow x_1 = x_2$
   Thus f is one-to-one. Let $y = f^{-1}(x)$.
   Then $x = f(y) = \sqrt{y - 1}$, and $y = 1 + x^2$. Thus
   $f^{-1}(x) = 1 + x^2$, $(x \geq 0)$.
   $D(f) = R(f^{-1}) = [1, \infty)$, $R(f) = D(f^{-1}) = [0, \infty)$.

4. $f(x) = -\sqrt{x - 1}$ for $x \geq 1$.
   If $f(x_1) = f(x_2)$, then $-\sqrt{x_1 - 1} = -\sqrt{x_2 - 1}$ and
   $x_1 - 1 = x_2 - 1$. Thus $x_1 = x_2$ and f is one-to-one.
   Let $y = f^{-1}(x)$. Then $x = f(y) = -\sqrt{y - 1}$ so
   $x^2 = y - 1$ and $y = x^2 + 1$. Thus, $f^{-1}(x) = x^2 + 1$.
   $D(f) = R(f^{-1}) = [1, \infty)$. $R(f) = D(f^{-1}) = (-\infty, 0]$.

5. $f(x) = x^3$
   $f(x_1) = f(x_2) \Leftrightarrow x_1^3 = x_2^3$
   $\Rightarrow (x_1 - x_2)(x_1^2 + x_1 x_2 + x_2^2) = 0$
   $\Rightarrow x_1 = x_2$
   Thus f is one-to-one. Let $y = f^{-1}(x)$.
   Then $x = f(y) = y^3$ so $y = x^{1/3}$.
   Thus $f^{-1}(x) = x^{1/3}$.
   $D(f) = D(f^{-1}) = \mathbb{R} = R(f) = R(f^{-1})$.

6. $f(x) = 1 + \sqrt[3]{x}$. If $f(x_1) = f(x_2)$, then $1 + \sqrt[3]{x_1} = 1 + \sqrt[3]{x_2}$
   so $x_1 = x_2$. Thus, f is one-to-one.
   Let $y = f^{-1}(x)$ so that $x = f(y) = 1 + \sqrt[3]{y}$. Thus
   $y = (x - 1)^3$ and $f^{-1}(x) = (x - 1)^3$.
   $D(f) = R(f^{-1}) = (-\infty, \infty)$.
   $R(f) = D(f^{-1}) = (-\infty, \infty)$.

7. $f(x) = x^2$, $(x \leq 0)$
   $f(x_1) = f(x_2) \Leftrightarrow x_1^2 = x_2^2$, $(x_1 \leq 0, \ x_2 \leq 0)$
   $\Leftrightarrow x_1 = x_2$
   Thus f is one-to-one. Let $y = f^{-1}(x)$.
   Then $x = f(y) = y^2$ $(y \leq 0)$.
   therefore $y = -\sqrt{x}$ and $f^{-1}(x) = -\sqrt{x}$.
   $D(f) = [-\infty, 0] = R(f^{-1})$, $D(f^{-1}) = [0, \infty) = R(f)$.

8. $f(x) = (1 - 2x)^3$. If $f(x_1) = f(x_2)$, then
   $(1 - 2x_1)^3 = (1 - 2x_2)^3$ and $x_1 = x_2$. Thus, f is one-to-one.
   Let $y = f^{-1}(x)$. Then $x = f(y) = (1 - 2y)^3$ so
   $y = \frac{1}{2}(1 - \sqrt[3]{x})$. Thus, $f^{-1}(x) = \frac{1}{2}(1 - \sqrt[3]{x})$.
   $D(f) = R(f^{-1}) = (-\infty, \infty)$.
   $R(f) = D(f^{-1}) = (-\infty, \infty)$.

9. $f(x) = \dfrac{1}{x + 1}$. $D(f) = \{x : x \neq -1\} = R(f^{-1})$.
   $f(x_1) = f(x_2) \Leftrightarrow \dfrac{1}{x_1 + 1} = \dfrac{1}{x_2 + 1}$
   $\Leftrightarrow x_2 + 1 = x_1 + 1$
   $\Leftrightarrow x_2 = x_1$
   Thus f is one-to-one; Let $y = f^{-1}(x)$.
   Then $x = f(y) = \dfrac{1}{y + 1}$
   so $y + 1 = \dfrac{1}{x}$ and $y = f^{-1}(x) = \dfrac{1}{x} - 1$.
   $D(f^{-1}) = \{x : x \neq 0\} = R(f)$.

10. $f(x) = \dfrac{x}{1 + x}$. If $f(x_1) = f(x_2)$, then $\dfrac{x_1}{1 + x_1} = \dfrac{x_2}{1 + x_2}$.
    Hence $x_1(1 + x_2) = x_2(1 + x_1)$ and, on simplification,
    $x_1 = x_2$. Thus, f is one-to-one.
    Let $y = f^{-1}(x)$. Then $x = f(y) = \dfrac{y}{1 + y}$ and
    $x(1 + y) = y$. Thus $y = \dfrac{x}{1 - x} = f^{-1}(x)$.
    $D(f) = R(f^{-1}) = (-\infty, -1) \cup (-1, \infty)$.
    $R(f) = D(f^{-1}) = (-\infty, 1) \cup (1, \infty)$.

11. $f(x) = \dfrac{1 - 2x}{1 + x}$. $D(f) = \{x : x \neq -1\} = R(f^{-1})$
    $f(x_1) = f(x_2) \Leftrightarrow \dfrac{1 - 2x_1}{1 + x_1} = \dfrac{1 - 2x_2}{1 + x_2}$
    $\Leftrightarrow 1 + x_2 - 2x_1 - 2x_1 x_2 = 1 + x_1 - 2x_2 - 2x_1 x_2$
    $\Leftrightarrow 3x_2 = 3x_1 \Leftrightarrow x_1 = x_2$
    Thus f is one-to-one. Let $y = f^{-1}(x)$.
    Then $x = f(y) = \dfrac{1 - 2y}{1 + y}$
    so $x + xy = 1 - 2y$
    and $f^{-1}(x) = y = \dfrac{1 - x}{2 + x}$.
    $D(f^{-1}) = \{x : x \neq -2\} = R(f)$.

12. $f(x) = \dfrac{x}{\sqrt{x^2+1}}$. If $f(x_1) = f(x_2)$, then
$$\dfrac{x_1}{\sqrt{x_1^2+1}} = \dfrac{x_2}{\sqrt{x_2^2+1}}. \quad (*)$$
Thus $x_1^2(x_2^2+1) = x_2^2(x_1^2+1)$ and $x_1^2 = x_2^2$.
From (*), x_1 and x_2 must have the same sign. Hence, $x_1 = x_2$ and f is one-to-one.
Let $y = f^{-1}(x)$. Then $x = f(y) = \dfrac{y}{\sqrt{y^2+1}}$, and
$x^2(y^2+1) = y^2$. Hence $y^2 = \dfrac{x^2}{1-x^2}$. Since $f(y)$ and y have the same sign, we must have $y = \dfrac{x}{\sqrt{1-x^2}}$, so
$f^{-1}(x) = \dfrac{x}{\sqrt{1-x^2}}$.
$D(f) = R(f^{-1}) = (-\infty, \infty)$.
$R(f) = D(f^{-1}) = (-1, 1)$.

13. $g(x) = f(x) - 2$
Let $y = g^{-1}(x)$. Then $x = g(y) = f(y) - 2$, so $f(y) = x + 2$ and $g^{-1}(x) = y = f^{-1}(x+2)$.

14. $h(x) = f(2x)$. Let $y = h^{-1}(x)$. Then $x = h(y) = f(2y)$ and $2y = f^{-1}(x)$. Thus $h^{-1}(x) = y = \tfrac{1}{2} f^{-1}(x)$.

15. $k(x) = -3f(x)$. Let $y = k^{-1}(x)$. Then $x = k(y) = -3f(y)$, so $f(y) = -\dfrac{x}{3}$ and $k^{-1}(x) = y = f^{-1}\left(-\dfrac{x}{3}\right)$.

16. $m(x) = f(x-2)$. Let $y = m^{-1}(x)$. Then $x = m(y) = f(y-2)$, and $y - 2 = f^{-1}(x)$.
Hence $m^{-1}(x) = y = f^{-1}(x) + 2$.

17. $p(x) = \dfrac{1}{1 + f(x)}$. Let $y = p^{-1}(x)$.
Then $x = p(y) = \dfrac{1}{1 + f(y)}$ so $f(y) = \dfrac{1}{x} - 1$,
and $p^{-1}(x) = y = f^{-1}\left(\dfrac{1}{x} - 1\right)$.

18. $q(x) = \dfrac{f(x) - 3}{2}$. Let $y = q^{-1}(x)$. Then
$x = q(y) = \dfrac{f(y) - 3}{2}$ and $f(y) = 2x + 3$. Hence
$q^{-1}(x) = y = f^{-1}(2x + 3)$.

19. $r(x) = 1 - 2f(3 - 4x)$
Let $y = r^{-1}(x)$. Then $x = r(y) = 1 - 2f(3 - 4y)$.
$$f(3 - 4y) = \dfrac{1-x}{2}$$
$$3 - 4y = f^{-1}\left(\dfrac{1-x}{2}\right)$$
and $r^{-1}(x) = y = \dfrac{1}{4}\left(3 - f^{-1}\left(\dfrac{1-x}{2}\right)\right)$.

20. $s(x) = \dfrac{1 + f(x)}{1 - f(x)}$. Let $y = s^{-1}(x)$.
Then $x = s(y) = \dfrac{1 + f(y)}{1 - f(y)}$. Solving for $f(y)$ we obtain
$f(y) = \dfrac{x-1}{x+1}$. Hence $s^{-1}(x) = y = f^{-1}\left(\dfrac{x-1}{x+1}\right)$.

21. $f(x) = x^2 + 1$ if $x \geq 0$, and $f(x) = x + 1$ if $x < 0$.
If $f(x_1) = f(x_2)$ then if $x_1 \geq 0$ and $x_2 \geq 0$ then
$x_1^2 + 1 = x_2^2 + 1$ so $x_1 = x_2$;
if $x_1 \geq 0$ and $x_2 < 0$ then $x_1^2 + 1 = x_2 + 1$ so $x_2 = x_1^2$ (not possible);
if $x_1 < 0$ and $x_2 \geq 0$ then $x_1 = x_2^2$ (not possible);
if $x_1 < 0$ and $x_2 < 0$ then $x_1 + 1 = x_2 + 1$ so $x_1 = x_2$.
Therefore f is one-to-one. Let $y = f^{-1}(x)$. Then
$$x = f(y) = \begin{cases} y^2 + 1 & \text{if } y \geq 0 \\ y + 1 & \text{if } y < 0. \end{cases}$$
Thus $f^{-1}(x) = y = \begin{cases} \sqrt{x-1} & \text{if } x \geq 1 \\ x - 1 & \text{if } x < 1. \end{cases}$

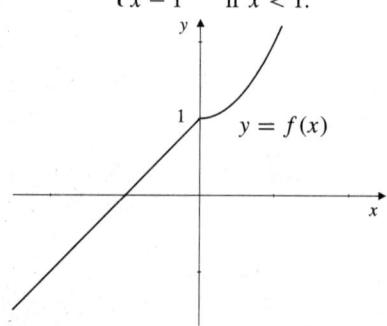

Fig. 3.1.21

22. $y = f^{-1}(x) \Leftrightarrow x = f(y) = y^3 + y$. To find $y = f^{-1}(2)$ we solve $y^3 + y = 2$ for y. Evidently $y = 1$ is the only solution, so $f^{-1}(2) = 1$.

23. If $y = f^{-1}(x)$ then $x = f(y)$.
Thus $1 = f'(y)\dfrac{dy}{dx}$ so $\dfrac{dy}{dx} = \dfrac{1}{f'(y)} = \dfrac{1}{\dfrac{1}{y}} = y$
(since $f'(x) = 1/x$).

We used a TI-85 solve routine to obtain the numerical values in the next two questions. Newton's Method could also be used on any scientific calculator.

24. If $f(x) = \dfrac{4x^3}{x^2 + 1}$, then
$$f'(x) = \dfrac{(x^2+1)(12x^2) - 4x^3(2x)}{(x^2+1)^2} = \dfrac{4x^2(x^2+3)}{(x^2+1)^2}.$$
Since $f'(x) > 0$ for all x, except $x = 0$, f must be one-to-one and so it has an inverse.
If $y = f^{-1}(x)$, then $x = f(y) = \dfrac{4y^3}{y^2+1}$, and
$$1 = f'(y) = \dfrac{(y^2+1)(12y^2 y') - 4y^3(2yy')}{(y^2+1)^2}.$$

INSTRUCTOR'S SOLUTIONS MANUAL   SECTION 3.2 (PAGE 182)

Thus $y' = \dfrac{(y^2+1)^2}{4y^4+12y^2}$. Since $f(1) = 2$, therefore $f^{-1}(2) = 1$ and

$$\left(f^{-1}\right)'(2) = \left.\dfrac{(y^2+1)^2}{4y^4+12y^2}\right|_{y=1} = \dfrac{1}{4}.$$

25. $f(x) = 1 + 2x^3$
    Let $y = f^{-1}(x)$.
    Thus $x = f(y) = 1 + 2y^3$.
    $1 = 6y^2 \dfrac{dy}{dx}$ so $(f^{-1})'(x) = \dfrac{dy}{dx} = \dfrac{1}{6y^2} = \dfrac{1}{6[f^{-1}(x)]^2}$

26. If $f(x) = x\sqrt{3+x^2}$ and $y = f^{-1}(x)$, then $x = f(y) = y\sqrt{3+y^2}$, so,

    $$1 = y'\sqrt{3+y^2} + y\dfrac{2yy'}{2\sqrt{3+y^2}} \Rightarrow y' = \dfrac{\sqrt{3+y^2}}{3+2y^2}.$$

    Since $f(-1) = -2$ implies that $f^{-1}(-2) = -1$, we have

    $$\left(f^{-1}\right)'(-2) = \left.\dfrac{\sqrt{3+y^2}}{3+2y^2}\right|_{y=-1} = \dfrac{2}{5}.$$

    Note: $f(x) = x\sqrt{3+x^2} = -2 \Rightarrow x^2(3+x^2) = 4$
    $\Rightarrow x^4 + 3x^2 - 4 = 0 \Rightarrow (x^2+4)(x^2-1) = 0$.
    Since $(x^2+4) = 0$ has no real solution, therefore $x^2 - 1 = 0$ and $x = 1$ or -1. Since it is given that $f(x) = -2$, therefore x must be -1.

We used a TI-85 solve routine to obtain the numerical values in the Exercises 27 and 28. Newton's Method could also be used on any scientific calculator.

27. $y = f^{-1}(2) \Leftrightarrow 2 = f(y) = y^2/(1+\sqrt{y})$. We must solve $2 + 2\sqrt{y} = y^2$ for y. There is a root between 2 and 3: $f^{-1}(2) \approx 2.23362$ to 5 decimal places.

28. $g(x) = 2x + \sin x \Rightarrow g'(x) = 2 + \cos x \geq 1$ for all x. Therefore g is increasing, and so one-to-one and invertible on the whole real line.
    $y = g^{-1}(x) \Leftrightarrow x = g(y) = 2y + \sin y$. For $y = g^{-1}(2)$, we need to solve $2y + \sin y - 2 = 0$. The root is between 0 and 1; to five decimal places $g^{-1}(2) = y \approx 0.68404$. Also

    $$1 = \dfrac{dx}{dx} = (2+\cos y)\dfrac{dy}{dx}$$
    $$(g^{-1})'(2) = \left.\dfrac{dy}{dx}\right|_{x=2} = \dfrac{1}{2+\cos y} \approx 0.36036.$$

29. If $f(x) = x \sec x$, then $f'(x) = \sec x + x \sec x \tan x \geq 1$ for x in $(-\pi/2, \pi/2)$. Thus f is increasing, and so one-to-one on that interval. Moreover, $\lim_{x \to -(\pi/2)+} f(x) = -\infty$ and $\lim_{x \to (\pi/2)+} f(x) = \infty$, so, being continuous, f has range $(-\infty, \infty)$, and so f^{-1} has domain $(-\infty, \infty)$.
    Since $f(0) = 0$, we have $f^{-1}(0) = 0$, and

    $$(f^{-1})'(0) = \dfrac{1}{f'(f^{-1}(0))} = \dfrac{1}{f'(0)} = 1.$$

30. If $y = (f \circ g)^{-1}(x)$, then $x = f \circ g(y) = f(g(y))$. Thus $g(y) = f^{-1}(x)$ and $y = g^{-1}(f^{-1}(x)) = g^{-1} \circ f^{-1}(x)$. That is, $(f \circ g)^{-1} = g^{-1} \circ f^{-1}$.

31. $f(x) = \dfrac{x-a}{bx-c}$
    Let $y = f^{-1}(x)$. Then $x = f(y) = \dfrac{y-a}{by-c}$ and $bxy - cx = y - a$ so $y = \dfrac{cx-a}{bx-1}$. We have $f^{-1}(x) = f(x)$ if $\dfrac{x-a}{bx-c} = \dfrac{cx-a}{bx-1}$. Evidently it is necessary and sufficient that $c = 1$. a and b may have any values.

32. Let $f(x)$ be an even function. Then $f(x) = f(-x)$. Hence, f is not one-to-one and it is not invertible. Therefore, it cannot be self-inverse.
    An odd function $g(x)$ may be self-inverse if its graph is symmetric about the line $x = y$. Examples are $g(x) = x$ and $g(x) = 1/x$.

Section 3.2 Exponential and Logarithmic Functions   (page 182)

1. $\dfrac{3^3}{\sqrt{3^5}} = 3^{3-5/2} = 3^{1/2} = \sqrt{3}$

2. $2^{1/2} 8^{1/2} = 2^{1/2} 2^{3/2} = 2^2 = 4$

3. $(x^{-3})^{-2} = x^6$

4. $(\tfrac{1}{2})^x 4^{x/2} = \dfrac{2^x}{2^x} = 1$

5. $\log_5 125 = \log_5 5^3 = 3$

6. If $\log_4(\tfrac{1}{8}) = y$ then $4^y = \tfrac{1}{8}$, or $2^{2y} = 2^{-3}$. Thus $2y = -3$ and $\log_4(\tfrac{1}{8}) = y = -\tfrac{3}{2}$.

7. $\log_{1/3} 3^{2x} = \log_{1/3}\left(\dfrac{1}{3}\right)^{-2x} = -2x$

8. $4^{3/2} = 8 \Rightarrow \log_4 8 = \tfrac{3}{2} \Rightarrow 2^{\log_4 8} = 2^{3/2} = 2\sqrt{2}$

9. $10^{-\log_{10}(1/x)} = \dfrac{1}{1/x} = x$

79

10. Since $\log_a \left(x^{1/(\log_a x)}\right) = \dfrac{1}{\log_a x} \log_a x = 1$, therefore $x^{1/(\log_a x)} = a^1 = a$.

11. $(\log_a b)(\log_b a) = \log_a a = 1$

12. $\log_x \left(x(\log_y y^2)\right) = \log_x(2x) = \log_x x + \log_x 2$
$$= 1 + \log_x 2 = 1 + \dfrac{1}{\log_2 x}$$

13. $(\log_4 16)(\log_4 2) = 2 \times \dfrac{1}{2} = 1$

14. $\log_{15} 75 + \log_{15} 3 = \log_{15} 225 = 2$
(since $15^2 = 225$)

15. $\log_6 9 + \log_6 4 = \log_6 36 = 2$

16. $2\log_3 12 - 4\log_3 6 = \log_3 \left(\dfrac{4^2 \cdot 3^2}{2^4 \cdot 3^4}\right)$
$$= \log_3(3^{-2}) = -2$$

17. $\log_a(x^4 + 3x^2 + 2) + \log_a(x^4 + 5x^2 + 6)$
$\quad - 4\log_a \sqrt{x^2 + 2}$
$= \log_a\left((x^2+2)(x^2+1)\right) + \log_a\left((x^2+2)(x^2+3)\right)$
$\quad - 2\log_a(x^2 + 2)$
$= \log_a(x^2 + 1) + \log_a(x^2 + 3)$
$= \log_a(x^4 + 4x^2 + 3)$

18. $\log_\pi(1 - \cos x) + \log_\pi(1 + \cos x) - 2\log_\pi \sin x$
$= \log_\pi \left[\dfrac{(1-\cos x)(1+\cos x)}{\sin^2 x}\right] = \log_\pi \dfrac{\sin^2 x}{\sin^2 x}$
$= \log_\pi 1 = 0$

19. $y = 3^{\sqrt{2}}$, $\log_{10} y = \sqrt{2} \log_{10} 3$,
$y = 10^{\sqrt{2} \log_{10} 3} \approx 4.72880$

20. $\log_3 5 = (\log_{10} 5)/(\log_{10} 3) \approx 1.46497$

21. $2^{2x} = 5^{x+1}$, $2x \log_{10} 2 = (x+1)\log_{10} 5$,
$x = (\log_{10} 5)/(2\log_{10} 2 - \log_{10} 5) \approx -7.21257$

22. $x^{\sqrt{2}} = 3$, $\sqrt{2} \log_{10} x = \log_{10} 3$,
$x = 10^{(\log_{10} 3)/\sqrt{2}} \approx 2.17458$

23. $\log_x 3 = 5$, $(\log_{10} 3)/(\log_{10} x) = 5$,
$\log_{10} x = (\log_{10} 3)/5$, $x = 10^{(\log_{10} 3)/5} \approx 1.24573$

24. $\log_3 x = 5$, $(\log_{10} x)/(\log_{10} 3) = 5$,
$\log_{10} x = 5\log_{10} 3$, $x = 10^{5\log_{10} 3} = 3^5 = 243$

25. Let $u = \log_a \left(\dfrac{1}{x}\right)$ then $a^u = \dfrac{1}{x} = x^{-1}$. Hence, $a^{-u} = x$ and $u = -\log_a x$.
Thus, $\log_a \left(\dfrac{1}{x}\right) = -\log_a x$.

26. Let $\log_a x = u$, $\log_a y = v$.
Then $x = a^u$, $y = a^v$.
Thus $\dfrac{x}{y} = \dfrac{a^u}{a^v} = a^{u-v}$
and $\log_a \left(\dfrac{x}{y}\right) = u - v = \log_a x - \log_a y$.

27. Let $u = \log_a(x^y)$, then $a^u = x^y$ and $a^{u/y} = x$.
Therefore $\dfrac{u}{y} = \log_a x$, or $u = y \log_a x$.
Thus, $\log_a(x^y) = y \log_a x$.

28. Let $\log_b x = u$, $\log_b a = v$.
Thus $b^u = x$ and $b^v = a$.
Therefore $x = b^u = b^{v(u/v)} = a^{u/v}$
and $\log_a x = \dfrac{u}{v} = \dfrac{\log_b x}{\log_b a}$.

29. $\log_4(x+4) - 2\log_4(x+1) = \dfrac{1}{2}$
$\log_4 \dfrac{x+4}{(x+1)^2} = \dfrac{1}{2}$
$\dfrac{x+4}{(x+1)^2} = 4^{1/2} = 2$
$2x^2 + 3x - 2 = 0$ so $x = 1/2$ or -2

30. First observe that $\log_9 x = \log_3 x/\log_3 9 = \dfrac{1}{2} \log_3 x$. Now
$2\log_3 x + \log_9 x = 10$
$\log_3 x^2 + \log_3 x^{1/2} = 10$
$\log_3 x^{5/2} = 10$
$x^{5/2} = 3^{10}$, so $x = (3^{10})^{2/5} = 3^4 = 81$

31. $f(x) = a^x$ and $f'(0) = \lim\limits_{h \to 0} \dfrac{a^h - 1}{h} = k$. Thus
$$f'(x) = \lim_{h \to 0} \dfrac{a^{x+h} - a^x}{h}$$
$$= \lim_{h \to 0} \dfrac{a^x a^h - a^x}{h}$$
$$= a^x \lim_{h \to 0} \dfrac{a^h - 1}{h} = a^x f'(0) = a^x k = kf(x).$$

32. $y = f^{-1}(x) \Rightarrow x = f(y) = a^y$
$\Rightarrow 1 = \dfrac{dx}{dx} = ka^y \dfrac{dy}{dx}$
$\Rightarrow \dfrac{dy}{dx} = \dfrac{1}{ka^y} = \dfrac{1}{kx}$.
Thus $(f^{-1})'(x) = 1/(kx)$.

Section 3.3 The Natural Logarithm and Exponential (page 192)

1. $\dfrac{e^3}{\sqrt{e^5}} = e^{3 - 5/2} = e^{1/2} = \sqrt{e}$

INSTRUCTOR'S SOLUTIONS MANUAL   SECTION 3.3 (PAGE 192)

2. $\ln(e^{1/2}e^{2/3}) = \frac{1}{2} + \frac{2}{3} = \frac{7}{6}$

3. $e^{5\ln x} = x^5$

4. $e^{(3\ln 9)/2} = 9^{3/2} = 27$

5. $\ln\frac{1}{e^{3x}} = \ln e^{-3x} = -3x$

6. $e^{2\ln\cos x} + \left(\ln e^{\sin x}\right)^2 = \cos^2 x + \sin^2 x = 1$

7. $3\ln 4 - 4\ln 3 = \ln\frac{4^3}{3^4} = \ln\frac{64}{81}$

8. $4\ln\sqrt{x} + 6\ln(x^{1/3}) = 2\ln x + 2\ln x = 4\ln x$

9. $2\ln x + 5\ln(x-2) = \ln\!\left(x^2(x-2)^5\right)$

10. $\ln(x^2 + 6x + 9) = \ln[(x+3)^2] = 2\ln(x+3)$

11. $2^{x+1} = 3^x$
    $(x+1)\ln 2 = x\ln 3$
    $x = \dfrac{\ln 2}{\ln 3 - \ln 2} = \dfrac{\ln 2}{\ln(3/2)}$

12. $3^x = 9^{1-x} \Rightarrow 3^x = 3^{2(1-x)}$
    $\Rightarrow x = 2(1-x) \Rightarrow x = \frac{2}{3}$

13. $\dfrac{1}{2^x} = \dfrac{5}{8^{x+3}}$
    $-x\ln 2 = \ln 5 - (x+3)\ln 8$
    $= \ln 5 - (3x+9)\ln 2$
    $2x\ln 2 = \ln 5 - 9\ln 2$
    $x = \dfrac{\ln 5 - 9\ln 2}{2\ln 2}$

14. $2^{x^2-3} = 4^x = 2^{2x} \Rightarrow x^2 - 3 = 2x$
    $x^2 - 2x - 3 = 0 \Rightarrow (x-3)(x+1) = 0$
    Hence, $x = -1$ or 3.

15. $\ln(x/(2-x))$ is defined if $x/(2-x) > 0$, that is, if $0 < x < 2$. The domain is the interval $(0, 2)$.

16. $\ln(x^2 - x - 2) = \ln[(x-2)(x+1)]$ is defined if $(x-2)(x+1) > 0$, that is, if $x < -1$ or $x > 2$. The domain is the union $(-\infty, -1) \cup (2, \infty)$.

17. $\ln(2x-5) > \ln(7-2x)$ holds if $2x-5 > 0$, $7-2x > 0$, and $2x-5 > 7-2x$, that is, if $x > 5/2$, $x < 7/2$, and $4x > 12$ (i.e., $x > 3$). The solution set is the interval $(3, 7/2)$.

18. $\ln(x^2 - 2) \leq \ln x$ holds if $x^2 > 2$, $x > 0$, and $x^2 - 2 \leq x$. Thus we need $x > \sqrt{2}$ and $x^2 - x - 2 \leq 0$. This latter inequality says that $(x-2)(x+1) \leq 0$, so it holds for $-1 \leq x \leq 2$. The solution set of the given inequality is $(\sqrt{2}, 2]$.

19. $y = e^{5x}$, $y' = 5e^{5x}$

20. $y = xe^x - x$, $y' = e^x + xe^x - 1$

21. $y = \dfrac{x}{e^{2x}} = xe^{-2x}$
    $y' = e^{-2x} - 2xe^{-2x}$
    $= (1 - 2x)e^{-2x}$

22. $y = x^2 e^{x/2}$, $y' = 2xe^{x/2} + \frac{1}{2}x^2 e^{x/2}$

23. $y = \ln(3x-2)$ $y' = \dfrac{3}{3x-2}$

24. $y = \ln|3x-2|$, $y' = \dfrac{3}{3x-2}$

25. $y = \ln(1 + e^x)$ $y' = \dfrac{e^x}{1 + e^x}$

26. $y = 2\ln\sqrt{x^2+2} = 2(\frac{1}{2})\ln(x^2+2) = \ln(x^2+2)$
    $y' = \dfrac{2x}{x^2+2}$

27. $y = \dfrac{e^x + e^{-x}}{2}$, $y' = \dfrac{e^x - e^{-x}}{2}$

28. $f(x) = e^{x^2}$, $f'(x) = (2x)e^{x^2}$

29. $y = e^{(e^x)}$, $y' = e^x e^{(e^x)} = e^{x+e^x}$

30. $x = e^{3t}\ln t$, $\dfrac{dx}{dt} = 3e^{3t}\ln t + \dfrac{1}{t}e^{3t}$

31. $y = \dfrac{e^x}{1+e^x} = 1 - \dfrac{1}{1+e^x}$, $y' = \dfrac{e^x}{(1+e^x)^2}$

32. $f(x) = \dfrac{e^x - e^{-x}}{e^x + e^{-x}}$
    $f'(x) = \dfrac{(e^x + e^{-x})(e^x + e^{-x}) - (e^x - e^{-x})(e^x - e^{-x})}{(e^x + e^{-x})^2}$
    $= \dfrac{4}{(e^x + e^{-x})^2}$

33. $y = e^x\sin x$, $y' = e^x(\sin x + \cos x)$

34. $y = e^{-x}\cos x$, $y' = -e^{-x}\cos x - e^{-x}\sin x$

35. $y = \ln\ln x$ $y' = \dfrac{1}{x\ln x}$

36. $y = x\ln x - x$
    $y' = \ln x + x\left(\dfrac{1}{x}\right) - 1 = \ln x$

37. $y = x^2\ln x - \dfrac{x^2}{2}$
    $y' = 2x\ln x + \dfrac{x^2}{x} - \dfrac{2x}{2} = 2x\ln x$

38. $y = \ln|\sin x|$, $y' = \dfrac{\cos x}{\sin x} = \cot x$

39. $y = 5^{2x+1}$
    $y' = 2(5^{2x+1})\ln 5 = (2\ln 5)5^{2x+1}$

40. $y = 2^{(x^2-3x+8)}$, $y' = (2x-3)(\ln 2)2^{(x^2-3x+8)}$

41. $g(x) = t^x x^t$, $g'(x) = t^x x^t \ln t + t^{x+1} x^{t-1}$

81

42. $h(t) = t^x - x^t$, $\quad h'(t) = xt^{x-1} - x^t \ln x$

43. $f(s) = \log_a(bs+c) = \dfrac{\ln(bs+c)}{\ln a}$

$f'(s) = \dfrac{b}{(bs+c)\ln a}$

44. $g(x) = \log_x(2x+3) = \dfrac{\ln(2x+3)}{\ln x}$

$g'(x) = \dfrac{\ln x \left(\dfrac{2}{2x+3}\right) - [\ln(2x+3)]\left(\dfrac{1}{x}\right)}{(\ln x)^2}$

$= \dfrac{2x \ln x - (2x+3)\ln(2x+3)}{x(2x+3)(\ln x)^2}$

45. $y = x^{\sqrt{x}} = e^{\sqrt{x}\ln x}$

$y' = e^{\sqrt{x}\ln x}\left(\dfrac{\ln x}{2\sqrt{x}} + \dfrac{\sqrt{x}}{x}\right)$

$= x^{\sqrt{x}}\left(\dfrac{1}{\sqrt{x}}\left(\dfrac{1}{2}\ln x + 1\right)\right)$

46. Given that $y = \left(\dfrac{1}{x}\right)^{\ln x}$, let $u = \ln x$. Then $x = e^u$ and

$y = \left(\dfrac{1}{e^u}\right)^u = (e^{-u})^u = e^{-u^2}$. Hence,

$\dfrac{dy}{dx} = \dfrac{dy}{du}\cdot\dfrac{du}{dx} = (-2ue^{-u^2})\left(\dfrac{1}{x}\right) = -\dfrac{2\ln x}{x}\left(\dfrac{1}{x}\right)^{\ln x}$.

47. $y = \ln|\sec x + \tan x|$

$y' = \dfrac{\sec x \tan x + \sec^2 x}{\sec x + \tan x}$

$= \sec x$

48. $y = \ln|x + \sqrt{x^2 - a^2}|$

$y' = \dfrac{1 + \dfrac{2x}{2\sqrt{x^2-a^2}}}{x + \sqrt{x^2-a^2}} = \dfrac{1}{\sqrt{x^2-a^2}}$

49. $y = \ln(\sqrt{x^2+a^2} - x)$

$y' = \dfrac{\dfrac{x}{\sqrt{x^2+a^2}} - 1}{\sqrt{x^2+a^2} - x}$

$= -\dfrac{1}{\sqrt{x^2+a^2}}$

50. $y = (\cos x)^x - x^{\cos x} = e^{x\ln\cos x} - e^{(\cos x)(\ln x)}$

$y' = e^{x\ln\cos x}\left[\ln\cos x + x\left(\dfrac{1}{\cos x}\right)(-\sin x)\right]$

$\quad - e^{(\cos x)(\ln x)}\left[-\sin x \ln x + \dfrac{1}{x}\cos x\right]$

$= (\cos x)^x(\ln\cos x - x\tan x)$

$\quad - x^{\cos x}\left(-\sin x \ln x + \dfrac{1}{x}\cos x\right)$

51. $f(x) = xe^{ax}$

$f'(x) = e^{ax}(1 + ax)$

$f''(x) = e^{ax}(2a + a^2x)$

$f'''(x) = e^{ax}(3a^2 + a^3x)$

$\vdots$

$f^{(n)}(x) = e^{ax}(na^{n-1} + a^n x)$

52. Since

$\dfrac{d}{dx}(ax^2 + bx + c)e^x = (2ax + b)e^x + (ax^2 + bx + c)e^x$

$= [ax^2 + (2a+b)x + (b+c)]e^x$

$= [Ax^2 + Bx + C]e^x$.

Thus, differentiating $(ax^2 + bx + c)e^x$ produces another function of the same type with different constants. Any number of differentiations will do likewise.

53. $y = e^{x^2}$

$y' = 2xe^{x^2}$

$y'' = 2e^{x^2} + 4x^2 e^{x^2} = 2(1 + 2x^2)e^{x^2}$

$y''' = 2(4x)e^{x^2} + 2(1+2x^2)2xe^{x^2} = 4(3x + 2x^3)e^{x^2}$

$y^{(4)} = 4(3 + 6x^2)e^{x^2} + 4(3x + 2x^3)2xe^{x^2}$

$= 4(3 + 12x^2 + 4x^4)e^{x^2}$

54. $f(x) = \ln(2x+1)$ $\qquad f'(x) = 2(2x+1)^{-1}$

$f''(x) = (-1)2^2(2x+1)^{-2}$ $\qquad f'''(x) = (2)2^3(2x+1)^{-3}$

$f^{(4)}(x) = -(3!)2^4(2x+1)^{-4}$

Thus, if $n = 1, 2, 3, \ldots$ we have

$f^{(n)}(x) = (-1)^{n-1}(n-1)!2^n(2x+1)^{-n}$.

55. a) $f(x) = (x^x)^x = x^{(x^2)}$
$\ln f(x) = x^2 \ln x$
$\dfrac{1}{f} f' = 2x \ln x + x$
$f' = x^{x^2+1}(2\ln x + 1)$

b) $g(x) = x^{x^x}$
$\ln g = x^x \ln x$
$\dfrac{1}{g} g' = x^x(1+\ln x)\ln x + \dfrac{x^x}{x}$
$g' = x^{x^x} x^x \left(\dfrac{1}{x} + \ln x + (\ln x)^2\right)$

Evidently g grows more rapidly than does f as x grows large.

56. Given that $x^{x^{x^{\cdot^{\cdot^{\cdot}}}}} = a$ where $a > 0$, then

$$\ln a = x^{x^{x^{\cdot^{\cdot^{\cdot}}}}} \ln x = a \ln x.$$

Thus $\ln x = \dfrac{1}{a} \ln a = \ln a^{1/a}$, so $x = a^{1/a}$.

57. $f(x) = (x-1)(x-2)(x-3)(x-4)$
$\ln f(x) = \ln(x-1) + \ln(x-2) + \ln(x-3) + \ln(x-4)$
$\dfrac{1}{f(x)} f'(x) = \dfrac{1}{x-1} + \dfrac{1}{x-2} + \dfrac{1}{x-3} + \dfrac{1}{x-4}$
$f'(x) = f(x)\left(\dfrac{1}{x-1} + \dfrac{1}{x-2} + \dfrac{1}{x-3} + \dfrac{1}{x-4}\right)$

58. $F(x) = \dfrac{\sqrt{1+x}\,(1-x)^{1/3}}{(1+5x)^{4/5}}$
$\ln F(x) = \tfrac{1}{2}\ln(1+x) + \tfrac{1}{3}\ln(1-x) - \tfrac{4}{5}\ln(1+5x)$
$\dfrac{F'(x)}{F(x)} = \dfrac{1}{2(1+x)} - \dfrac{1}{3(1-x)} - \dfrac{4}{(1+5x)}$
$F'(0) = F(0)\left[\dfrac{1}{2} - \dfrac{1}{3} - \dfrac{4}{1}\right] = (1)\left[\dfrac{1}{2} - \dfrac{1}{3} - 4\right] = -\dfrac{23}{6}$

59. $f(x) = \dfrac{(x^2-1)(x^2-2)(x^2-3)}{(x^2+1)(x^2+2)(x^2+3)}$
$f(2) = \dfrac{3 \times 2 \times 1}{5 \times 6 \times 7} = \dfrac{1}{35}, \qquad f(1) = 0$
$\ln f(x) = \ln(x^2-1) + \ln(x^2-2) + \ln(x^2-3)$
$\qquad - \ln(x^2+1) - \ln(x^2+2) - \ln(x^2+3)$
$\dfrac{1}{f(x)} f'(x) = \dfrac{2x}{x^2-1} + \dfrac{2x}{x^2-2} + \dfrac{2x}{x^2-3}$
$\qquad - \dfrac{2x}{x^2+1} - \dfrac{2x}{x^2+2} - \dfrac{2x}{x^2+3}$
$f'(x) = 2xf(x)\Big(\dfrac{1}{x^2-1} + \dfrac{1}{x^2-2} + \dfrac{1}{x^2-3}$
$\qquad - \dfrac{1}{x^2+1} - \dfrac{1}{x^2+2} - \dfrac{1}{x^2+3}\Big)$
$f'(2) = \dfrac{4}{35}\left(\dfrac{1}{3} + \dfrac{1}{2} + \dfrac{1}{1} - \dfrac{1}{5} - \dfrac{1}{6} - \dfrac{1}{7}\right)$
$= \dfrac{4}{35} \times \dfrac{139}{105} = \dfrac{556}{3675}$

Since $f(x) = (x^2-1)g(x)$ where $g(1) \neq 0$, then $f'(x) = 2xg(x) + (x^2-1)g'(x)$ and
$f'(1) = 2g(1) + 0 = 2 \times \dfrac{(-1)(-2)}{2 \times 3 \times 4} = \dfrac{1}{6}.$

60. Since $y = x^2 e^{-x^2}$, then
$$y' = 2xe^{-x^2} - 2x^3 e^{-x^2} = 2x(1-x)(1+x)e^{-x^2}.$$
The tangent is horizontal at $(0,0)$ and $\left(\pm 1, \dfrac{1}{e}\right)$.

61. $f(x) = xe^{-x}$
$f'(x) = e^{-x}(1-x), \quad$ C.P. $x = 1,\ f(1) = \dfrac{1}{e}$
$f'(x) > 0$ if $x < 1$ (f increasing)
$f'(x) < 0$ if $x > 1$ (f decreasing)

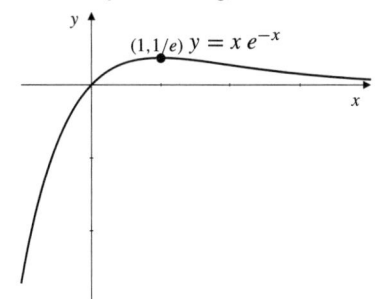

Fig. 3.3.61

62. Since $y = \ln x$ and $y' = \dfrac{1}{x} = 4$ then $x = \tfrac{1}{4}$ and $y = \ln \tfrac{1}{4} = -\ln 4$. The tangent line of slope 4 is $y = -\ln 4 + 4(x - \tfrac{1}{4})$, i.e., $y = 4x - 1 - \ln 4$.

63. Let the point of tangency be (a, e^a).
Tangent line has slope

$$\frac{e^a - 0}{a - 0} = \frac{d}{dx}e^x\Big|_{x=a} = e^a.$$

Therefore, $a = 1$ and line has slope e.
The line has equation $y = ex$.

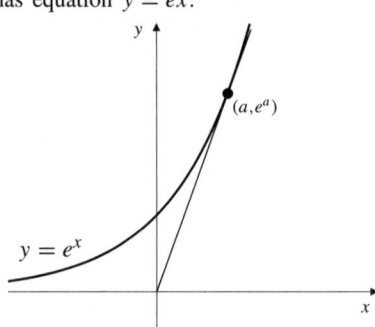

Fig. 3.3.63

64. The slope of $y = \ln x$ at $x = a$ is $y' = \frac{1}{x}\Big|_{x=a} = \frac{1}{a}$. The line from $(0, 0)$ to $(a, \ln a)$ is tangent to $y = \ln x$ if

$$\frac{\ln a - 0}{a - 0} = \frac{1}{a}$$

i.e., if $\ln a = 1$, or $a = e$. Thus, the line is $y = \frac{x}{e}$.

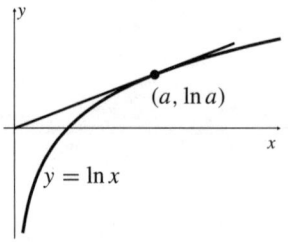

Fig. 3.3.64

65. Let the point of tangency be $(a, 2^a)$. Slope of the tangent is

$$\frac{2^a - 0}{a - 1} = \frac{d}{dx}2^x\Big|_{x=a} = 2^a \ln 2.$$

Thus $a - 1 = \frac{1}{\ln 2}$, $a = 1 + \frac{1}{\ln 2}$.
So the slope is $2^a \ln 2 = 2^{1+(1/\ln 2)} \ln 2 = 2e \ln 2$.
(Note: $\ln 2^{1/\ln 2} = \frac{1}{\ln 2} \ln 2 = 1 \Rightarrow 2^{1/\ln 2} = e$)
The tangent line has equation $y = 2e \ln 2(x - 1)$.

66. The tangent line to $y = a^x$ which passes through the origin is tangent at the point (b, a^b) where

$$\frac{a^b - 0}{b - 0} = \frac{d}{dx}a^x\Big|_{x=b} = a^b \ln a.$$

Thus $\frac{1}{b} = \ln a$, so $a^b = a^{1/\ln a} = e$. The line $y = x$ will intersect $y = a^x$ provided the slope of this tangent line does not exceed 1, i.e., provided $\frac{e}{b} \leq 1$, or $e \ln a \leq 1$.
Thus we need $a \leq e^{1/e}$.

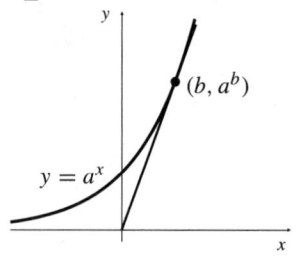

Fig. 3.3.66

67. $e^{xy} \ln \frac{x}{y} = x + \frac{1}{y}$

$$e^{xy}(y + xy') \ln \frac{x}{y} + e^{xy}\frac{y}{x}\left(\frac{y - xy'}{y^2}\right) = 1 - \frac{1}{y^2}y'$$

At $\left(e, \frac{1}{e}\right)$ we have

$$e\left(\frac{1}{e} + ey'\right)2 + e\frac{1}{e^2}(e - e^3 y') = 1 - e^2 y'$$
$$2 + 2e^2 y' + 1 - e^2 y' = 1 - e^2 y'.$$

Thus the slope is $y' = -\frac{1}{e^2}$.

68. $xe^y + y - 2x = \ln 2 \Rightarrow e^y + xe^y y' + y' - 2 = 0$.
At $(1, \ln 2)$, $2 + 2y' + y' - 2 = 0 \Rightarrow y' = 0$.
Therefore, the tangent line is $y = \ln 2$.

69. $f(x) = Ax \cos \ln x + Bx \sin \ln x$
$f'(x) = A \cos \ln x - A \sin \ln x + B \sin \ln x + B \cos \ln x$
$= (A + B) \cos \ln x + (B - A) \sin \ln x$
If $A = B = \frac{1}{2}$ then $f'(x) = \cos \ln x$.
Therefore $\int \cos \ln x \, dx = \frac{1}{2}x \cos \ln x + \frac{1}{2}x \sin \ln x + C$.
If $B = \frac{1}{2}$, $A = -\frac{1}{2}$ then $f'(x) = \sin \ln x$.
Therefore $\int \sin \ln x \, dx = \frac{1}{2}x \sin \ln x - \frac{1}{2}x \cos \ln x + C$.

70. $F_{A,B}(x) = Ae^x \cos x + Be^x \sin x$

$$\frac{d}{dx}F_{A,B}(x)$$
$$= Ae^x \cos x - Ae^x \sin x + Be^x \sin x + Be^x \cos x$$
$$= (A + B)e^x \cos x + (B - A)e^x \sin x = F_{A+B, B-A}(x)$$

71. Since $\frac{d}{dx}F_{A,B}(x) = F_{A+B, B-A}(x)$ we have

a) $\frac{d^2}{dx^2}F_{A,B}(x) = \frac{d}{dx}F_{A+B, B-A}(x) = F_{2B, -2A}(x)$

b) $\frac{d^3}{dx^3}e^x \cos x = \frac{d^3}{dx^3}F_{1,0}(x) = \frac{d}{dx}F_{0,-2}(x)$
$= F_{-2,-2}(x) = -2e^x \cos x - 2e^x \sin x$

72. $\dfrac{d}{dx}(Ae^{ax}\cos bx + Be^{ax}\sin bx)$
$= Aae^{ax}\cos bx - Abe^{ax}\sin bx + Bae^{ax}\sin bx$
$\quad + Bbe^{ax}\cos bx$
$= (Aa+Bb)e^{ax}\cos bx + (Ba-Ab)e^{ax}\sin bx.$

(a) If $Aa+Bb=1$ and $Ba-Ab=0$, then $A=\dfrac{a}{a^2+b^2}$
and $B=\dfrac{b}{a^2+b^2}$. Thus

$$\int e^{ax}\cos bx\, dx$$
$$= \dfrac{1}{a^2+b^2}\left(ae^{ax}\cos bx + be^{ax}\sin bx\right)+C.$$

(b) If $Aa+Bb=0$ and $Ba-Ab=1$, then $A=\dfrac{-b}{a^2+b^2}$
and $B=\dfrac{a}{a^2+b^2}$. Thus

$$\int e^{ax}\sin bx\, dx$$
$$= \dfrac{1}{a^2+b^2}\left(ae^{ax}\sin bx - be^{ax}\cos bx\right)+C.$$

73. $\dfrac{d}{dx}\left[\ln\dfrac{1}{x}+\ln x\right] = \dfrac{1}{1/x}\left(\dfrac{-1}{x^2}\right)+\dfrac{1}{x} = -\dfrac{1}{x}+\dfrac{1}{x}=0.$

Therefore $\ln\dfrac{1}{x}+\ln x = C$ (constant). Taking $x=1$, we get $C=\ln 1 + \ln 1 = 0$. Thus $\ln\dfrac{1}{x} = -\ln x.$

74. $\ln\dfrac{x}{y} = \ln\left(x\cdot\dfrac{1}{y}\right) = \ln x + \ln\dfrac{1}{y} = \ln x - \ln y.$

75. $\dfrac{d}{dx}[\ln(x^r) - r\ln x] = \dfrac{rx^{r-1}}{x^r} - \dfrac{r}{x} = \dfrac{r}{x} - \dfrac{r}{x} = 0.$

Therefore $\ln(x^r) - r\ln x = C$ (constant). Taking $x=1$, we get $C = \ln 1 - r\ln 1 = 0 - 0 = 0.$ Thus $\ln(x^r) = r\ln x.$

76. Let $x>0$, and $F(x)$ be the area bounded by $y=t^2$, the t-axis, $t=0$ and $t=x$. For $h>0$, $F(x+h)-F(x)$ is the shaded area in the following figure.

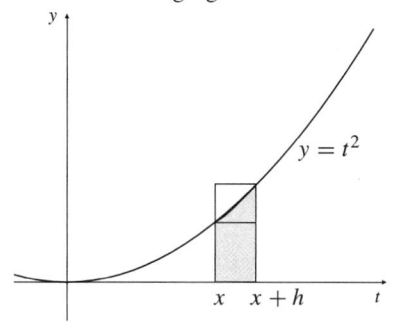

Fig. 3.3.76

Comparing this area with that of the two rectangles, we see that
$$hx^2 < F(x+h) - F(x) < h(x+h)^2.$$

Hence, the Newton quotient for $F(x)$ satisfies
$$x^2 < \dfrac{F(x+h)-F(x)}{h} < (x+h)^2.$$

Letting h approach 0 from the right (by the Squeeze Theorem applied to one-sided limits)
$$\lim_{h\to 0+}\dfrac{F(x+h)-F(x)}{h} = x^2.$$

If $h<0$ and $0<x+h<x$, then
$$(x+h)^2 < \dfrac{F(x+h)-F(x)}{h} < x^2,$$

so similarly,
$$\lim_{h\to 0-}\dfrac{F(x+h)-F(x)}{h} = x^2.$$

Combining these two limits, we obtain
$$\dfrac{d}{dx}F(x) = \lim_{h\to 0}\dfrac{F(x+h)-F(x)}{h} = x^2.$$

Therefore $F(x) = \int x^2\, dx = \tfrac{1}{3}x^3 + C.$ Since $F(0)=C=0$, therefore $F(x)=\tfrac{1}{3}x^3.$ For $x=2$, the area of the region is $F(2)=\tfrac{8}{3}$ square units.

77. a) The shaded area A in part (i) of the figure is less than the area of the rectangle (actually a square) with base from $t=1$ to $t=2$ and height $1/1 = 1$. Since $\ln 2 = A < 1$, we have $2 < e^1 = e$; i.e., $e > 2.$

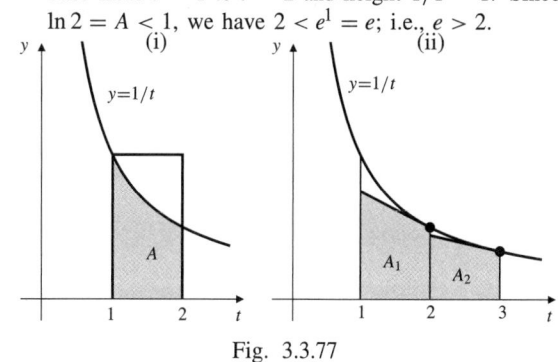

Fig. 3.3.77

b) If $f(t) = 1/t$, then $f'(t) = -1/t^2$ and $f''(t) = 2/t^3 > 0$ for $t>0$. Thus $f'(t)$ is an increasing function of t for $t>0$, and so the graph of $f(t)$ bends upward away from any of its tangent lines. (This kind of argument will be explored further in Chapter 5.)

c) The tangent to $y = 1/t$ at $t = 2$ has slope $-1/4$. Its equation is

$$y = \frac{1}{2} - \frac{1}{4}(x-2) \quad \text{or} \quad y = 1 - \frac{x}{4}.$$

The tangent to $y = 1/t$ at $t = 3$ has slope $-1/9$. Its equation is

$$y = \frac{1}{3} - \frac{1}{9}(x-3) \quad \text{or} \quad y = \frac{2}{3} - \frac{x}{9}.$$

d) The trapezoid bounded by $x = 1$, $x = 2$, $y = 0$, and $y = 1 - (x/4)$ has area

$$A_1 = \frac{1}{2}\left(\frac{3}{4} + \frac{1}{2}\right) = \frac{5}{8}.$$

The trapezoid bounded by $x = 2$, $x = 3$, $y = 0$, and $y = (2/3) - (x/9)$ has area

$$A_2 = \frac{1}{2}\left(\frac{4}{9} + \frac{1}{3}\right) = \frac{7}{18}.$$

e) $\ln 3 > A_1 + A_2 = \frac{5}{8} + \frac{7}{18} = \frac{73}{72} > 1$.

Thus $3 > e^1 = e$. Combining this with the result of (a) we conclude that $2 < e < 3$.

Section 3.4 Growth and Decay (page 200)

1. $\lim\limits_{x \to \infty} x^3 e^{-x} = \lim\limits_{x \to \infty} \dfrac{x^3}{e^x} = 0$ (exponential wins)

2. $\lim\limits_{x \to \infty} x^{-3} e^x = \lim\limits_{x \to \infty} \dfrac{e^x}{x^3} = \infty$

3. $\lim\limits_{x \to \infty} \dfrac{2e^x - 3}{e^x + 5} = \lim\limits_{x \to \infty} \dfrac{2 - 3e^{-x}}{1 + 5e^{-x}} = \dfrac{2-0}{1+0} = 2$

4. $\lim\limits_{x \to \infty} \dfrac{x - 2e^{-x}}{x + 3e^{-x}} = \lim\limits_{x \to \infty} \dfrac{1 - 2/(xe^x)}{1 + 3/(xe^x)} = \dfrac{1-0}{1+0} = 1$

5. $\lim\limits_{x \to 0+} x \ln x = 0$ (power wins)

6. $\lim\limits_{x \to 0+} \dfrac{\ln x}{x} = -\infty$

7. $\lim\limits_{x \to 0} x(\ln|x|)^2 = 0$

8. $\lim\limits_{x \to \infty} \dfrac{(\ln x)^3}{\sqrt{x}} = 0$ (power wins)

9. Let $N(t)$ be the number of bacteria present after t hours. Then $N(0) = 100$, $N(1) = 200$.
Since $\dfrac{dN}{dt} = kN$ we have $N(t) = N(0)e^{kt} = 100e^{kt}$.
Thus $200 = 100e^k$ and $k = \ln 2$.
Finally, $N\left(\dfrac{5}{2}\right) = 100e^{(5/2)\ln 2} \approx 565.685$.
There will be approximately 566 bacteria present after another $1\frac{1}{2}$ hours.

10. Let $y(t)$ be the number of kg undecomposed after t hours. Thus, $y(0) = 50$ and $y(5) = 20$. Since $y'(t) = ky(t)$, therefore $y(t) = y(0)e^{kt} = 50e^{kt}$. Then

$$20 = y(5) = 50e^{5k} \Rightarrow k = \tfrac{1}{5}\ln\tfrac{2}{5}.$$

If 90% of the sugar is decomposed at time T then $5 = y(T) = 50e^{kT}$, so

$$T = \frac{1}{k}\ln\frac{1}{10} = \frac{5\ln(0.1)}{\ln(0.4)} \approx 12.56.$$

Hence, 90% of the sugar will decomposed in about 12.56 hours.

11. Let $P(t)$ be the percentage undecayed after t years. Thus $P(0) = 100$, $P(15) = 70$.
Since $\dfrac{dP}{dt} = kP$, we have $P(t) = P(0)e^{kt} = 100e^{kt}$.
Thus $70 = P(15) = 100e^{15k}$ so $k = \dfrac{1}{15}\ln(0.7)$.
The half-life T satisfies if $50 = P(T) = 100e^{kT}$, so
$T = \dfrac{1}{k}\ln(0.5) = \dfrac{15\ln(0.5)}{\ln(0.7)} \approx 29.15$.
The half-life is about 29.15 years.

12. Let $P(t)$ be the percentage remaining after t years. Thus $P'(t) = kP(t)$ and $P(t) = P(0)e^{kt} = 100e^{kt}$. Then,

$$50 = P(1690) = 100e^{1690k} \Rightarrow k = \frac{1}{1690}\ln\frac{1}{2} \approx 0.0004101.$$

a) $P(100) = 100e^{100k} \approx 95.98$, i.e., about 95.98% remains after 100 years.

b) $P(1000) = 100e^{1000k} \approx 66.36$, i.e., about 66.36% remains after 1000 years.

13. Let $P(t)$ be the percentage of the initial amount remaining after t years.
Then $P(t) = 100e^{kt}$ and $99.57 = P(1) = 100e^k$.
Thus $k = \ln(0.9957)$.
The half-life T satisfies $50 = P(T) = 100e^{kT}$,
so $T = \dfrac{1}{k}\ln(0.5) = \dfrac{\ln(0.5)}{\ln(0.995)} \approx 160.85$.
The half-life is about 160.85 years.

14. Let $N(t)$ be the number of bacteria in the culture t days after the culture was set up. Thus $N(3) = 3N(0)$ and $N(7) = 10 \times 10^6$. Since $N(t) = N(0)e^{kt}$, we have

$$3N(0) = N(3) = N(0)e^{3k} \Rightarrow k = \tfrac{1}{3}\ln 3.$$
$$10^7 = N(7) = N(0)e^{7k} \Rightarrow N(0) = 10^7 e^{-(7/3)\ln 3} \approx 770400.$$

There were approximately 770,000 bacteria in the culture initially. (Note that we are approximating a discrete quantity (number of bacteria) by a continuous quantity $N(t)$ in this exercise.)

15. Let $W(t)$ be the weight t days after birth.
Thus $W(0) = 4000$ and $W(t) = 4000e^{kt}$.
Also $4400 = W(14) = 4000e^{14k}$, is $k = \frac{1}{14}\ln(1.1)$.
Five days after birth, the baby weighs
$W(5) = 4000e^{(5/14)\ln(1.1)} \approx 4138.50 \approx 4139$ grams.

16. Since
$$I'(t) = kI(t) \Rightarrow I(t) = I(0)e^{kt} = 40e^{kt},$$
$$15 = I(0.01) = 40e^{0.01k} \Rightarrow k = \frac{1}{0.01}\ln\frac{15}{40} = 100\ln\frac{3}{8},$$
thus,
$$I(t) = 40\exp\left(100t\ln\frac{3}{8}\right) = 40\left(\frac{3}{8}\right)^{100t}.$$

17. $\$P$ invested at 4% compounded continuously grows to $\$P(e^{0.04})^7 = \$Pe^{0.28}$ in 7 years. This will be $\$10,000$ if $\$P = \$10,000e^{-0.28} = \$7,557.84$.

18. Let $y(t)$ be the value of the investment after t years. Thus $y(0) = 1000$ and $y(5) = 1500$. Since $y(t) = 1000e^{kt}$ and $1500 = y(5) = 1000e^{5k}$, therefore, $k = \frac{1}{5}\ln\frac{3}{2}$.

   a) Let t be the time such that $y(t) = 2000$, i.e.,
   $$1000e^{kt} = 2000$$
   $$\Rightarrow t = \frac{1}{k}\ln 2 = \frac{5\ln 2}{\ln(\frac{3}{2})} = 8.55.$$
   Hence, the doubling time for the investment is about 8.55 years.

   b) Let $r\%$ be the effective annual rate of interest; then
   $$1000(1 + \frac{r}{100}) = y(1) = 1000e^{k}$$
   $$\Rightarrow r = 100(e^k - 1) = 100[\exp(\frac{1}{5}\ln\frac{3}{2}) - 1]$$
   $$= 8.447.$$
   The effective annual rate of interest is about 8.45%.

19. Let the purchasing power of the dollar be $P(t)$ cents after t years.
Then $P(0) = 100$ and $P(t) = 100e^{kt}$.
Now $91 = P(1) = 100e^k$ so $k = \ln(0.91)$.
If $25 = P(t) = 100e^{kt}$ then
$t = \frac{1}{k}\ln(0.25) = \frac{\ln(0.25)}{\ln(0.91)} \approx 14.7$.
The purchasing power will decrease to $\$0.25$ in about 14.7 years.

20. Let $i\%$ be the effective rate, then an original investment of $\$A$ will grow to $\$A\left(1 + \frac{i}{100}\right)$ in one year. Let $r\%$ be the nominal rate per annum compounded n times per year, then an original investment of $\$A$ will grow to

$$\$A\left(1 + \frac{r}{100n}\right)^n$$

in one year, if compounding is performed n times per year. For $i = 9.5$ and $n = 12$, we have

$$\$A\left(1 + \frac{9.5}{100}\right) = \$A\left(1 + \frac{r}{1200}\right)^{12}$$
$$\Rightarrow r = 1200\left(\sqrt[12]{1.095} - 1\right) = 9.1098.$$

The nominal rate of interest is about 9.1098%.

21. $f'(x) = a + bf(x)$.

   a) If $u(x) = a + bf(x)$, then $u'(x) = bf'(x) = b[a + bf(x)] = bu(x)$. This equation for u is the equation of exponential growth/decay. Thus
   $$u(x) = C_1 e^{bx},$$
   $$f(x) = \frac{1}{b}\left(C_1 e^{bx} - a\right) = Ce^{bx} - \frac{a}{b}.$$

   b) If $\frac{dy}{dx} = a + by$ and $y(0) = y_0$, then, from part (a),
   $$y = Ce^{bx} - \frac{a}{b}, \quad y_0 = Ce^0 - \frac{a}{b}.$$
   Thus $C = y_0 + (a/b)$, and
   $$y = \left(y_0 + \frac{a}{b}\right)e^{bx} - \frac{a}{b}.$$

22. a) The concentration $x(t)$ satisfies $\frac{dx}{dt} = a - bx(t)$. This says that $x(t)$ is increasing if it is less than a/b and decreasing if it is greater than a/b. Thus, the limiting concentration is a/b.

   b) The differential equation for $x(t)$ resembles that of Exercise 21(b), except that $y(x)$ is replaced by $x(t)$, and b is replaced by $-b$. Using the result of Exercise 21(b), we obtain, since $x(0) = 0$,
   $$x(t) = \left(x(0) - \frac{a}{b}\right)e^{-bt} + \frac{a}{b}$$
   $$= \frac{a}{b}\left(1 - e^{-bt}\right).$$

c) We will have $x(t) = \frac{1}{2}(a/b)$ if $1 - e^{-bt} = \frac{1}{2}$, that is, if $e^{-bt} = \frac{1}{2}$, or $-bt = \ln(1/2) = -\ln 2$. The time required to attain half the limiting concentration is $t = (\ln 2)/b$.

23. Let $T(t)$ be the reading t minutes after the Thermometer is moved outdoors. Thus $T(0) = 72$, $T(1) = 48$.
By Newton's law of cooling, $\dfrac{dT}{dt} = k(T - 20)$.
If $V(t) = T(t) - 20$, then $\dfrac{dV}{dt} = kV$, so
$V(t) = V(0)e^{kt} = 52e^{kt}$.
Also $28 = V(1) = 52e^k$, so $k = \ln(7/13)$.
Thus $V(5) = 52e^{5\ln(7/13)} \approx 2.354$. At $t = 5$ the thermometer reads about $T(5) = 20 + 2.354 = 22.35°C$.

24. Let $T(t)$ be the temperature of the object t minutes after its temperature was $45°$ C. Thus $T(0) = 45$ and $T(40) = 20$. Also $\dfrac{dT}{dt} = k(T+5)$. Let $u(t) = T(t) + 5$, so $u(0) = 50$, $u(40) = 25$, and $\dfrac{du}{dt} = \dfrac{dT}{dt} = k(T+5) = ku$.
Thus,
$$u(t) = 50e^{kt},$$
$$25 = u(40) = 50e^{40k},$$
$$\Rightarrow k = \frac{1}{40}\ln\frac{25}{50} = \frac{1}{40}\ln\frac{1}{2}.$$

We wish to know t such that $T(t) = 0$, i.e., $u(t) = 5$, hence
$$5 = u(t) = 50e^{kt}$$
$$t = \frac{40\ln\left(\dfrac{5}{50}\right)}{\ln\left(\dfrac{1}{2}\right)} = 132.88 \text{ min}.$$

Hence, it will take about $(132.88 - 40) = 92.88$ minutes more to cool to $0°$ C.

25. Let $T(t)$ be the temperature of the body t minutes after it was $5°$.
Thus $T(0) = 5$, $T(4) = 10$. Room temperature $= 20°$.
By Newton's law of cooling (warming) $\dfrac{dT}{dt} = k(T - 20)$.
If $V(t) = T(t) - 20$ then $\dfrac{dV}{dt} = kV$,
so $V(t) = V(0)e^{kt} = -15e^{kt}$.
Also $-10 = V(4) = -15e^{4k}$, so $k = \dfrac{1}{4}\ln\left(\dfrac{2}{3}\right)$.
If $T(t) = 15°$, then $-5 = V(t) = -15e^{kt}$
so $t = \dfrac{1}{k}\ln\left(\dfrac{1}{3}\right) = 4\dfrac{\ln\left(\dfrac{1}{3}\right)}{\ln\left(\dfrac{2}{3}\right)} \approx 10.838$.
It will take a further 6.84 minutes to warm to $15°$C.

26. By the solution given for the logistic equation, we have
$$y_1 = \frac{Ly_0}{y_0 + (L - y_0)e^{-k}}, \quad y_2 = \frac{Ly_0}{y_0 + (L - y_0)e^{-2k}}$$
Thus $y_1(L - y_0)e^{-k} = (L - y_1)y_0$, and
$y_2(L - y_0)e^{-2k} = (L - y_2)y_0$.
Square the first equation and thus eliminate e^{-k}:
$$\left(\frac{(L - y_1)y_0}{y_1(L - y_0)}\right)^2 = \frac{(L - y_2)y_0}{y_2(L - y_0)}$$
Now simplify: $y_0y_2(L - y_1)^2 = y_1^2(L - y_0)(L - y_2)$
$y_0y_2L^2 - 2y_1y_0y_2L + y_0y_1^2y_2 = y_1^2L^2 - y_1^2(y_0+y_2)L + y_0y_1^2y_2$
Assuming $L \neq 0$, $L = \dfrac{y_1^2(y_0 + y_2) - 2y_0y_1y_2}{y_1^2 - y_0y_2}$.
If $y_0 = 3$, $y_1 = 5$, $y_2 = 6$, then
$L = \dfrac{25(9) - 180}{25 - 18} = \dfrac{45}{7} \approx 6.429$.

27. The rate of growth of y in the logistic equation is
$$\frac{dy}{dt} = ky\left(1 - \frac{y}{L}\right).$$
Since
$$\frac{dy}{dt} = -\frac{k}{L}\left(y - \frac{L}{2}\right)^2 + \frac{kL}{4},$$
thus $\dfrac{dy}{dt}$ is greatest when $y = \dfrac{L}{2}$.

28. The solution $y = \dfrac{Ly_0}{y_0 + (L - y_0)e^{-kt}}$ is valid on the largest interval containing $t = 0$ on which the denominator does not vanish.
If $y_0 > L$ then $y_0 + (L - y_0)e^{-kt} = 0$ if
$t = t^* = -\dfrac{1}{k}\ln\dfrac{y_0}{y_0 - L}$.
Then the solution is valid on (t^*, ∞). $\lim_{t \to t^{*}+} y(t) = \infty$.

29. The solution
$$y = \frac{Ly_0}{y_0 + (L - y_0)e^{-kt}}$$
of the logistic equation is valid on any interval containing $t = 0$ and not containing any point where the denominator is zero. The denominator is zero if $y_0 = (y_0 - L)e^{-kt}$, that is, if
$$t = t^* = -\frac{1}{k}\ln\left(\frac{y_0}{y_0 - L}\right).$$
Assuming k and L are positive, but y_0 is negative, we have $t^* > 0$. The solution is therefore valid on $(-\infty, t^*)$. The solution approaches $-\infty$ as $t \to t^{*}-$.

30. $$y(t) = \frac{L}{1 + Me^{-kt}}$$
$$200 = y(0) = \frac{L}{1 + M}$$
$$1{,}000 = y(1) = \frac{L}{1 + Me^{-k}}$$
$$10{,}000 = \lim_{t \to \infty} y(t) = L$$

Thus $200(1 + M) = L = 10,000$, so $M = 49$. Also $1,000(1 + 49e^{-k}) = L = 10,000$, so $e^{-k} = 9/49$ and $k = \ln(49/9) \approx 1.695$.

31. $y(3) = \dfrac{L}{1 + Me^{-3k}} = \dfrac{10,000}{1 + 49(9/49)^3} \approx 7671$ cases

$y'(3) = \dfrac{LkMe^{-3k}}{(1 + Me^{-3k})^2} \approx 3,028$ cases/week.

Section 3.5 The Inverse Trigonometric Functions (page 209)

1. $\sin^{-1}\dfrac{\sqrt{3}}{2} = \dfrac{\pi}{3}$

2. $\cos^{-1}\left(-\dfrac{1}{2}\right) = \dfrac{2\pi}{3}$

3. $\tan^{-1}(-1) = -\dfrac{\pi}{4}$

4. $\sec^{-1}\sqrt{2} = \dfrac{\pi}{4}$

5. $\sin(\sin^{-1} 0.7) = 0.7$

6. $\cos(\sin^{-1} 0.7) = \sqrt{1 - \sin^2(\arcsin 0.7)}$
   $= \sqrt{1 - 0.49} = \sqrt{0.51}$

7. $\tan^{-1}\left(\tan\dfrac{2\pi}{3}\right) = \tan^{-1}(-\sqrt{3}) = -\dfrac{\pi}{3}$

8. $\sin^{-1}(\cos 40°) = 90° - \cos^{-1}(\cos 40°) = 50°$

9. $\cos^{-1}(\sin(-0.2)) = \dfrac{\pi}{2} - \sin^{-1}(\sin(-0.2))$
   $= \dfrac{\pi}{2} + 0.2$

10. $\sin(\cos^{-1}(-\tfrac{1}{3})) = \sqrt{1 - \cos^2(\arccos(-\tfrac{1}{3}))}$
    $= \sqrt{1 - \tfrac{1}{9}} = \dfrac{\sqrt{8}}{3} = \dfrac{2\sqrt{2}}{3}$

11. $\cos\left(\tan^{-1}\dfrac{1}{2}\right) = \dfrac{1}{\sec\left(\tan^{-1}\dfrac{1}{2}\right)}$
    $= \dfrac{1}{\sqrt{1 + \tan^2\left(\tan^{-1}\dfrac{1}{2}\right)}} = \dfrac{2}{\sqrt{5}}$

12. $\tan(\tan^{-1} 200) = 200$

13. $\sin(\cos^{-1} x) = \sqrt{1 - \cos^2(\cos^{-1} x)}$
    $= \sqrt{1 - x^2}$

14. $\cos(\sin^{-1} x) = \sqrt{1 - \sin^2(\sin^{-1} x)} = \sqrt{1 - x^2}$

15. $\cos(\tan^{-1} x) = \dfrac{1}{\sec(\tan^{-1} x)} = \dfrac{1}{\sqrt{1 + x^2}}$

16. $\tan(\arctan x) = x \Rightarrow \sec(\arctan x) = \sqrt{1 + x^2}$
    $\Rightarrow \cos(\arctan x) = \dfrac{1}{\sqrt{1 + x^2}}$
    $\Rightarrow \sin(\arctan x) = \dfrac{x}{\sqrt{1 + x^2}}$

17. $\tan(\cos^{-1} x) = \dfrac{\sin(\cos^{-1} x)}{\cos(\cos^{-1} x)}$
    $= \dfrac{\sqrt{1 - x^2}}{x}$ (by # 13)

18. $\cos(\sec^{-1} x) = \dfrac{1}{x} \Rightarrow \sin(\sec^{-1} x) = \sqrt{1 - \dfrac{1}{x^2}} = \dfrac{\sqrt{x^2 - 1}}{|x|}$
    $\Rightarrow \tan(\sec^{-1} x) = \sqrt{x^2 - 1}\,\text{sgn}\,x$
    $= \begin{cases} \sqrt{x^2 - 1} & \text{if } x \geq 1 \\ -\sqrt{x^2 - 1} & \text{if } x \leq -1 \end{cases}$

19. $y = \sin^{-1}\dfrac{2x - 1}{3}$
    $y' = \dfrac{1}{\sqrt{1 - \left(\dfrac{2x-1}{3}\right)^2}} \cdot \dfrac{2}{3}$
    $= \dfrac{2}{\sqrt{9 - (4x^2 - 4x + 1)}}$
    $= \dfrac{1}{\sqrt{2 + x - x^2}}$

20. $y = \tan^{-1}(ax + b)$, $\quad y' = \dfrac{a}{1 + (ax + b)^2}$.

21. $y = \cos^{-1}\dfrac{x - b}{a}$
    $y' = -\dfrac{1}{\sqrt{1 - \dfrac{(x-b)^2}{a^2}}} \cdot \dfrac{1}{a}$
    $= \dfrac{-1}{\sqrt{a^2 - (x - b)^2}}$ (assuming $a > 0$).

22. $f(x) = x \sin^{-1} x$
    $f'(x) = \sin^{-1} x + \dfrac{x}{\sqrt{1 - x^2}}$.

23. $f(t) = t \tan^{-1} t$
    $f'(t) = \tan^{-1} t + \dfrac{t}{1 + t^2}$

24. $u = z^2 \sec^{-1}(1 + z^2)$
    $\dfrac{du}{dz} = 2z \sec^{-1}(1 + z^2) + \dfrac{z^2(2z)}{(1 + z^2)\sqrt{(1 + z^2)^2 - 1}}$
    $= 2z \sec^{-1}(1 + z^2) + \dfrac{2z^2 \text{sgn}(z)}{(1 + z^2)\sqrt{z^2 + 2}}$

25. $F(x) = (1+x^2)\tan^{-1} x$
$F'(x) = 2x\tan^{-1} x + 1$

26. $y = \sin^{-1}\left(\dfrac{a}{x}\right)$ $(|x| > |a|)$
$y' = \dfrac{1}{\sqrt{1-\left(\dfrac{a}{x}\right)^2}}\left[-\dfrac{a}{x^2}\right] = -\dfrac{a}{|x|\sqrt{x^2-a^2}}$

27. $G(x) = \dfrac{\sin^{-1} x}{\sin^{-1}(2x)}$
$G'(x) = \dfrac{\sin^{-1}(2x)\dfrac{1}{\sqrt{1-x^2}} - \sin^{-1} x \dfrac{2}{\sqrt{1-4x^2}}}{\left(\sin^{-1}(2x)\right)^2}$
$= \dfrac{\sqrt{1-4x^2}\sin^{-1}(2x) - 2\sqrt{1-x^2}\sin^{-1} x}{\sqrt{1-x^2}\sqrt{1-4x^2}\left(\sin^{-1}(2x)\right)^2}$

28. $H(t) = \dfrac{\sin^{-1} t}{\sin t}$
$H'(t) = \dfrac{\sin t\left(\dfrac{1}{\sqrt{1-t^2}}\right) - \sin^{-1} t \cos t}{\sin^2 t}$
$= \dfrac{1}{(\sin t)\sqrt{1-t^2}} - \csc t \cot t \sin^{-1} t$

29. $f(x) = (\sin^{-1} x^2)^{1/2}$
$f'(x) = \dfrac{1}{2}(\sin^{-1} x^2)^{-1/2}\dfrac{2x}{\sqrt{1-x^4}}$
$= \dfrac{x}{\sqrt{1-x^4}\sqrt{\sin^{-1} x^2}}$

30. $y = \cos^{-1}\left(\dfrac{a}{\sqrt{a^2+x^2}}\right)$
$y' = -\left(1 - \dfrac{a^2}{a^2+x^2}\right)^{-1/2}\left[-\dfrac{a}{2}(a^2+x^2)^{-3/2}(2x)\right]$
$= \dfrac{a\,\text{sgn}(x)}{a^2+x^2}$

31. $y = \sqrt{a^2-x^2} + a\sin^{-1}\dfrac{x}{a}$
$y' = -\dfrac{x}{\sqrt{a^2-x^2}} + \dfrac{a}{\sqrt{1-\dfrac{x^2}{a^2}}}\dfrac{1}{a}$
$= \dfrac{a-x}{\sqrt{a^2-x^2}} = \sqrt{\dfrac{a-x}{a+x}}$ $(a>0)$

32. $y = a\cos^{-1}\left(1-\dfrac{x}{a}\right) - \sqrt{2ax-x^2}$ $(a>0)$
$y' = -a\left[1-\left(1-\dfrac{x}{a}\right)^2\right]^{-1/2}\left(-\dfrac{1}{a}\right) - \dfrac{2a-2x}{2\sqrt{2ax-x^2}}$
$= \dfrac{x}{\sqrt{2ax-x^2}}$

33. $\tan^{-1}\left(\dfrac{2x}{y}\right) = \dfrac{\pi x}{y^2}$
$\dfrac{1}{1+\dfrac{4x^2}{y^2}}\dfrac{2y-2xy'}{y^2} = \pi\dfrac{y^2-2xyy'}{y^4}$
At $(1,2)$ $\dfrac{1}{2}\dfrac{4-2y'}{4} = \pi\dfrac{4-4y'}{16}$
$8-4y' = 4\pi - 4\pi y' \Rightarrow y' = \dfrac{\pi-2}{\pi-1}$
At $(1,2)$ the slope is $\dfrac{\pi-2}{\pi-1}$

34. If $y = \sin^{-1} x$, then $y' = \dfrac{1}{\sqrt{1-x^2}}$. If the slope is 2 then
$\dfrac{1}{\sqrt{1-x^2}} = 2$ so that $x = \pm\dfrac{\sqrt{3}}{2}$. Thus the equations of the two tangent lines are
$y = \dfrac{\pi}{3} + 2\left(x - \dfrac{\sqrt{3}}{2}\right)$ and $y = -\dfrac{\pi}{3} + 2\left(x + \dfrac{\sqrt{3}}{2}\right)$.

35. $\dfrac{d}{dx}\sin^{-1} x = \dfrac{1}{\sqrt{1-x^2}} > 0$ on $(-1,1)$.
Therefore, $\sin^{-1}$ is increasing.
$\dfrac{d}{dx}\tan^{-1} x = \dfrac{1}{1+x^2} > 0$ on $(-\infty,\infty)$.
Therefore $\tan^{-1}$ is increasing.
$\dfrac{d}{dx}\cos^{-1} x = -\dfrac{1}{\sqrt{1-x^2}} < 0$ on $(-1,1)$.
Therefore $\cos^{-1}$ is decreasing.

36. Since the domain of $\sec^{-1}$ consists of two disjoint intervals $(-\infty,-1]$ and $[1,\infty)$, the fact that the derivative of $\sec^{-1}$ is positive wherever defined does not imply that $\sec^{-1}$ is increasing over its whole domain, only that it is increasing on each of those intervals taken independently. In fact, $\sec^{-1}(-1) = \pi > 0 = \sec^{-1}(1)$ even though $-1 < 1$.

37. $\dfrac{d}{dx}\csc^{-1} x = \dfrac{d}{dx}\sin^{-1}\dfrac{1}{x}$
$= \dfrac{1}{\sqrt{1-\dfrac{1}{x^2}}}\left(-\dfrac{1}{x^2}\right)$
$= -\dfrac{1}{|x|\sqrt{x^2-1}}$

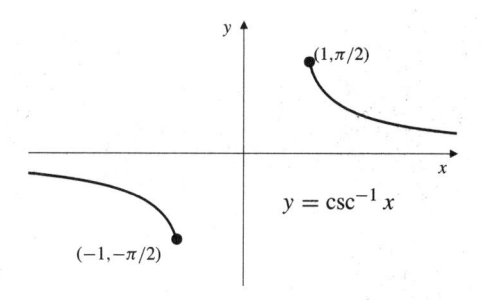

INSTRUCTOR'S SOLUTIONS MANUAL   SECTION 3.5 (PAGE 209)

Fig. 3.5.37

38. $\cot^{-1} x = \arctan(1/x)$;
$$\frac{d}{dx}\cot^{-1} x = \frac{1}{1+\frac{1}{x^2}}\cdot\frac{-1}{x^2} = -\frac{1}{1+x^2}$$

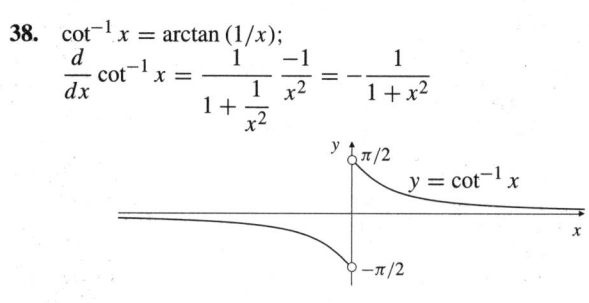

Fig. 3.5.38

Remark: the domain of $\cot^{-1}$ can be extended to include 0 by defining, say, $\cot^{-1} 0 = \pi/2$. This will make $\cot^{-1}$ right-continuous (but not continuous) at $x = 0$. It is also possible to define $\cot^{-1}$ in such a way that it is continuous on the whole real line, but we would then lose the identity $\cot^{-1} x = \tan^{-1}(1/x)$, which we prefer to maintain for calculation purposes.

39. $\dfrac{d}{dx}(\tan^{-1} x + \cot^{-1} x) = \dfrac{d}{dx}\left(\tan^{-1} x + \tan^{-1}\dfrac{1}{x}\right)$

$$= \frac{1}{1+x^2} + \frac{1}{1+\frac{1}{x^2}}\left(-\frac{1}{x^2}\right) = 0 \text{ if } x \neq 0$$

Thus $\tan^{-1} x + \cot^{-1} x = C_1$ (const. for $x > 0$)
At $x = 1$ we have $\dfrac{\pi}{4} + \dfrac{\pi}{4} = C_1$
Thus $\tan^{-1} x + \cot^{-1} x = \dfrac{\pi}{2}$ for $x > 0$.
Also $\tan^{-1} x + \cot^{-1} x = C_2$ for $(x < 0)$.
At $x = -1$, we get $-\dfrac{\pi}{4} - \dfrac{\pi}{4} = C_2$.
Thus $\tan^{-1} x + \cot^{-1} x = -\dfrac{\pi}{2}$ for $x < 0$.

40. If $g(x) = \tan(\tan^{-1} x)$ then
$$g'(x) = \frac{\sec^2(\tan^{-1} x)}{1+x^2}$$
$$= \frac{1+[\tan(\tan^{-1} x)]^2}{1+x^2} = \frac{1+x^2}{1+x^2} = 1.$$

If $h(x) = \tan^{-1}(\tan x)$ then h is periodic with period π, and
$$h'(x) = \frac{\sec^2 x}{1+\tan^2 x} = 1$$

provided that $x \neq (k+\tfrac{1}{2})\pi$ where k is an integer. $h(x)$ is not defined at odd multiples of $\dfrac{\pi}{2}$.

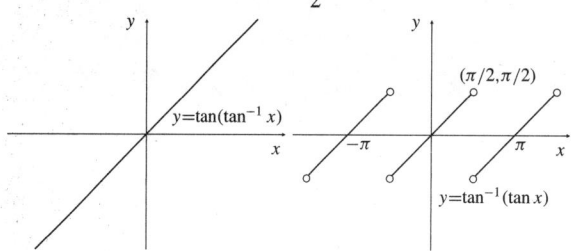

Fig. 4.5.40(a)   Fig. 4.5.40(b)

41. $\dfrac{d}{dx}\cos^{-1}(\cos x) = \dfrac{-1}{\sqrt{1-\cos^2 x}}(-\sin x)$

$$= \begin{cases} 1 & \text{if } \sin x > 0 \\ -1 & \text{if } \sin x < 0 \end{cases}$$

$\cos^{-1}(\cos x)$ is continuous everywhere and differentiable everywhere except at $x = n\pi$ for integers n.

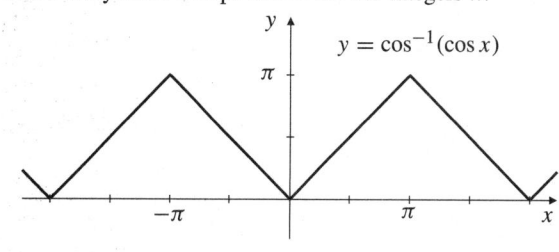

Fig. 3.5.41

42. $\dfrac{d}{dx}\sin^{-1}(\cos x) = \dfrac{1}{\sqrt{1-\cos^2 x}}(-\sin x)$

$$= \begin{cases} -1 & \text{if } \sin x > 0 \\ 1 & \text{if } \sin x < 0 \end{cases}$$

$\sin^{-1}(\cos x)$ is continuous everywhere and differentiable everywhere except at $x = n\pi$ for integers n.

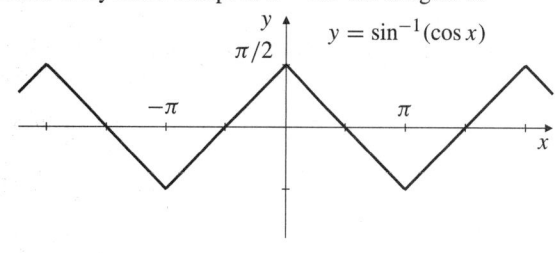

Fig. 3.5.42

43. $\dfrac{d}{dx}\tan^{-1}(\tan x) = \dfrac{1}{1+\tan^2 x}(\sec^2 x) = 1$ except at odd multiples of $\pi/2$.

$\tan^{-1}(\tan x)$ is continuous and differentiable everywhere except at $x = (2n+1)\pi/2$ for integers n. It is not defined at those points.

91

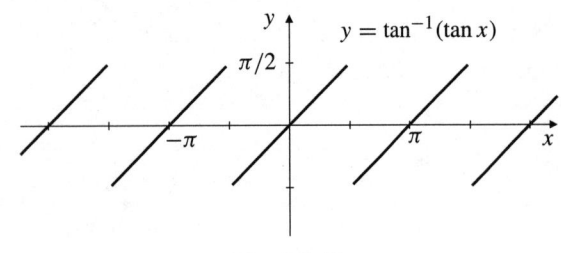

Fig. 3.5.43

44. $\frac{d}{dx}\tan^{-1}(\cot x) = \frac{1}{1+\cot^2 x}(-\csc^2 x) = -1$ except at integer multiples of π.

$\tan^{-1}(\cot x)$ is continuous and differentiable everywhere except at $x = n\pi$ for integers n. It is not defined at those points.

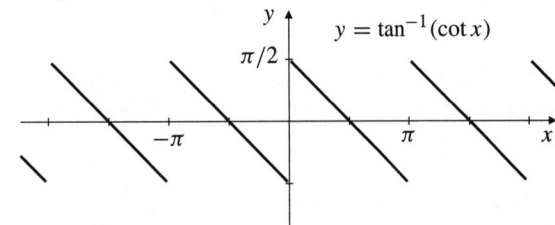

Fig. 3.5.44

45. If $|x| < 1$ and $y = \tan^{-1}\frac{x}{\sqrt{1-x^2}}$, then $y > 0 \Leftrightarrow x > 0$ and

$$\tan y = \frac{x}{\sqrt{1-x^2}}$$
$$\sec^2 y = 1 + \frac{x^2}{1-x^2} = \frac{1}{1-x^2}$$
$$\sin^2 y = 1 - \cos^2 y = 1 - (1-x^2) = x^2$$
$$\sin y = x.$$

Thus $y = \sin^{-1} x$ and $\sin^{-1} x = \tan^{-1}\frac{x}{\sqrt{1-x^2}}$.

An alternative method of proof involves showing that the derivative of the left side minus the right side is 0, and both sides are 0 at $x = 0$.

46. If $x \geq 1$ and $y = \tan^{-1}\sqrt{x^2-1}$, then $\tan y = \sqrt{x^2-1}$ and $\sec y = x$, so that $y = \sec^{-1} x$.
If $x \leq -1$ and $y = \pi - \tan^{-1}\sqrt{x^2-1}$, then $\frac{\pi}{2} < y < \frac{3\pi}{2}$, so $\sec y < 0$. Therefore

$$\tan y = \tan(\pi - \tan^{-1}\sqrt{x^2-1}) = -\sqrt{x^2-1}$$
$$\sec^2 y = 1 + (x^2-1) = x^2$$
$$\sec y = x,$$

because both x and $\sec y$ are negative. Thus $y = \sec^{-1} x$ in this case also.

47. If $y = \sin^{-1}\frac{x}{\sqrt{1+x^2}}$, then $y > 0 \Leftrightarrow x > 0$ and

$$\sin y = \frac{x}{\sqrt{1+x^2}}$$
$$\cos^2 y = 1 - \sin^2 y = 1 - \frac{x^2}{1+x^2} = \frac{1}{1+x^2}$$
$$\tan^2 y = \sec^2 y - 1 = 1 + x^2 - 1 = x^2$$
$$\tan y = x.$$

Thus $y = \tan^{-1} x$ and $\tan^{-1} x = \sin^{-1}\frac{x}{\sqrt{1+x^2}}$.

48. If $x \geq 1$ and $y = \sin^{-1}\frac{\sqrt{x^2-1}}{x}$, then $0 \leq y < \frac{\pi}{2}$ and

$$\sin y = \frac{\sqrt{x^2-1}}{x}$$
$$\cos^2 y = 1 - \frac{x^2-1}{x^2} = \frac{1}{x^2}$$
$$\sec^2 y = x^2.$$

Thus $\sec y = x$ and $y = \sec^{-1} x$.
If $x \leq -1$ and $y = \pi - \sin^{-1}\frac{\sqrt{x^2-1}}{x}$, then $\frac{\pi}{2} \leq y < \frac{3\pi}{2}$ and $\sec y < 0$. Therefore

$$\sin y = \sin\left(\pi - \sin^{-1}\frac{\sqrt{x^2-1}}{x}\right) = \frac{\sqrt{x^2-1}}{x}$$
$$\cos^2 y = 1 - \frac{x^2-1}{x^2} = \frac{1}{x^2}$$
$$\sec^2 y = x^2$$
$$\sec y = x,$$

because both x and $\sec y$ are negative. Thus $y = \sec^{-1} x$ in this case also.

49. $f'(x) \equiv 0$ on $(-\infty, -1)$
Thus $f(x) = \tan^{-1}\left(\frac{x-1}{x+1}\right) - \tan^{-1} x = C$ on $(-\infty, -1)$.

Evaluate the limit as $x \to -\infty$:

$$\lim_{x \to -\infty} f(x) = \tan^{-1} 1 - \left(-\frac{\pi}{2}\right) = \frac{3\pi}{4}$$

Thus $\tan^{-1}\left(\frac{x-1}{x+1}\right) - \tan^{-1} x = \frac{3\pi}{4}$ on $(-\infty, -1)$.

50. Since $f(x) = x - \tan^{-1}(\tan x)$ then

$$f'(x) = 1 - \frac{\sec^2 x}{1 + \tan^2 x} = 1 - 1 = 0$$

if $x \neq -(k + \frac{1}{2})\pi$ where k is an integer. Thus, f is constant on intervals not containing odd multiples of $\frac{\pi}{2}$. $f(0) = 0$ but $f(\pi) = \pi - 0 = \pi$. There is no contradiction here because $f'\left(\frac{\pi}{2}\right)$ is not defined, so f is not constant on the interval containing 0 and π.

51. $f(x) = x - \sin^{-1}(\sin x) \quad (-\pi \leq x \leq \pi)$

$f'(x) = 1 - \dfrac{1}{\sqrt{1 - \sin^2 x}} \cos x$

$= 1 - \dfrac{\cos x}{|\cos x|}$

$= \begin{cases} 0 & \text{if } -\dfrac{\pi}{2} < x < \dfrac{\pi}{2} \\ 2 & \text{if } -\pi < x < -\dfrac{\pi}{2} \text{ or } \dfrac{\pi}{2} < x < \pi \end{cases}$

Note: f is not differentiable at $\pm\dfrac{\pi}{2}$.

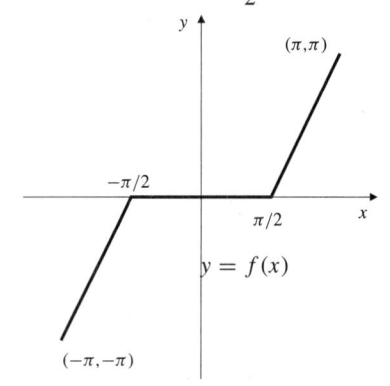

Fig. 3.5.51

52. $y' = \dfrac{1}{1 + x^2} \Rightarrow y = \tan^{-1} x + C$

$y(0) = C = 1$

Thus, $y = \tan^{-1} x + 1$.

53. $\begin{cases} y' = \dfrac{1}{9 + x^2} & \Rightarrow y = \dfrac{1}{3}\tan^{-1}\dfrac{x}{3} + C \\ y(3) = 2 & 2 = \dfrac{1}{3}\tan^{-1} 1 + C \quad C = 2 - \dfrac{\pi}{12} \end{cases}$

Thus $y = \dfrac{1}{3}\tan^{-1}\dfrac{x}{3} + 2 - \dfrac{\pi}{12}$.

54. $y' = \dfrac{1}{\sqrt{1 - x^2}} \Rightarrow y = \sin^{-1} x + C$

$y(\tfrac{1}{2}) = \sin^{-1}(\tfrac{1}{2}) + C = 1$

$\Rightarrow \dfrac{\pi}{6} + C = 1 \Rightarrow C = 1 - \dfrac{\pi}{6}$.

Thus, $y = \sin^{-1} x + 1 - \dfrac{\pi}{6}$.

55. $\begin{cases} y' = \dfrac{4}{\sqrt{25 - x^2}} & \Rightarrow y = 4\sin^{-1}\dfrac{x}{5} + C \\ y(0) = 0 & 0 = 0 + C \Rightarrow C = 0 \end{cases}$

Thus $y = 4\sin^{-1}\dfrac{x}{5}$.

Section 3.6 Hyperbolic Functions (page 215)

1. $\dfrac{d}{dx} \text{sech}\, x = \dfrac{d}{dx} \dfrac{1}{\cosh x}$

$= -\dfrac{1}{\cosh^2 x} \sinh x = -\text{sech}\, x \tanh x$

$\dfrac{d}{dx} \text{csch}\, x = \dfrac{d}{dx} \dfrac{1}{\sinh x}$

$= -\dfrac{1}{\sinh^2 x} \cosh x = -\text{csch}\, x \coth x$

$\dfrac{d}{dx} \coth x = \dfrac{d}{dx} \dfrac{\cosh}{\sinh x}$

$= \dfrac{\sinh^2 x - \cosh^2 x}{\sinh^2 x} = -\dfrac{1}{\sinh^2 x} = -\text{csch}^2 x$

2. $\cosh x \cosh y + \sinh x \sinh y$

$= \tfrac{1}{4}[(e^x + e^{-x})(e^y + e^{-y}) + (e^x - e^{-x})(e^y - e^{-y})]$

$= \tfrac{1}{4}(2e^{x+y} + 2e^{-x-y}) = \tfrac{1}{2}(e^{x+y} + e^{-(x+y)})$

$= \cosh(x + y)$.

$\sinh x \cosh y + \cosh x \sinh y$

$= \tfrac{1}{4}[(e^x - e^{-x})(e^y + e^{-y}) + (e^x + e^{-x})(e^y - e^{-y})]$

$= \tfrac{1}{2}(e^{x+y} - e^{-(x+y)}) = \sinh(x + y)$.

$\cosh(x - y) = \cosh[x + (-y)]$

$= \cosh x \cosh(-y) + \sinh x \sinh(-y)$

$= \cosh x \cosh y - \sinh x \sinh y$.

$\sinh(x - y) = \sinh[x + (-y)]$

$= \sinh x \cosh(-y) + \cosh x \sinh(-y)$

$= \sinh x \cosh y - \cosh x \sinh y$.

3. $\tanh(x \pm y) = \dfrac{\sinh(x \pm y)}{\cosh(x \pm y)}$

$= \dfrac{\sinh x \cosh y \pm \cosh x \sinh y}{\cosh x \cosh y \pm \sinh x \sinh y}$

$= \dfrac{\tanh x \pm \tanh y}{1 \pm \tanh x \tanh y}$

4. $y = \coth x = \dfrac{e^x + e^{-x}}{e^x - e^{-x}} \qquad y = \text{sech}\, x = \dfrac{2}{e^x + e^{-x}}$

Fig. 4.6.4(a) Fig. 4.6.4(b)

$y = \text{csch}\, x = \dfrac{2}{e^x - e^{-x}}$

$y = \operatorname{csch} x$

Fig. 3.6.4

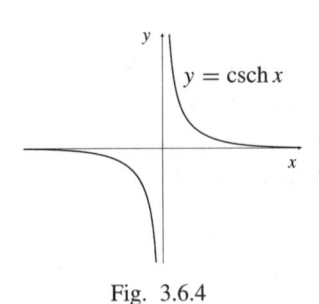

5. $\dfrac{d}{dx}\sinh^{-1} x = \dfrac{d}{dx}\ln(x+\sqrt{x^2+1}) = \dfrac{1+\dfrac{x}{\sqrt{x^2+1}}}{x+\sqrt{x^2+1}}$

$= \dfrac{1}{\sqrt{x^2+1}}$

$\dfrac{d}{dx}\cosh^{-1} x = \dfrac{d}{dx}\ln(x+\sqrt{x^2-1}) = \dfrac{1+\dfrac{x}{\sqrt{x^2-1}}}{x+\sqrt{x^2-1}}$

$= \dfrac{1}{\sqrt{x^2-1}}$

$\dfrac{d}{dx}\tanh^{-1} x = \dfrac{d}{dx}\dfrac{1}{2}\ln\left(\dfrac{1+x}{1-x}\right) = \dfrac{1}{2}\dfrac{1-x}{1+x}\dfrac{1-x-(1+x)(-1)}{(1-x)^2}$

$= \dfrac{1}{1-x^2}$

$\displaystyle\int \dfrac{dx}{\sqrt{x^2+1}} = \sinh^{-1} x + C$

$\displaystyle\int \dfrac{dx}{\sqrt{x^2-1}} = \cosh^{-1} x + C \quad (x > 1)$

$\displaystyle\int \dfrac{dx}{1-x^2} = \tanh^{-1} x + C \quad (-1 < x < 1)$

6. Let $y = \sinh^{-1}\left(\dfrac{x}{a}\right) \Leftrightarrow x = a\sinh y \Rightarrow 1 = a(\cosh y)\dfrac{dy}{dx}$.
Thus,

$\dfrac{d}{dx}\sinh^{-1}\left(\dfrac{x}{a}\right) = \dfrac{1}{a\cosh y}$

$= \dfrac{1}{a\sqrt{1+\sinh^2 y}} = \dfrac{1}{\sqrt{a^2+x^2}}$

$\displaystyle\int \dfrac{dx}{\sqrt{a^2+x^2}} = \sinh^{-1}\dfrac{x}{a} + C. \quad (a>0)$

Let $y = \cosh^{-1}\dfrac{x}{a} \Leftrightarrow x = a\operatorname{Cosh} y = a\cosh y$
for $y \geq 0$, $x \geq a$. We have $1 = a(\sinh y)\dfrac{dy}{dx}$. Thus,

$\dfrac{d}{dx}\cosh^{-1}\dfrac{x}{a} = \dfrac{1}{a\sinh y}$

$= \dfrac{1}{a\sqrt{\cosh^2 y - 1}} = \dfrac{1}{\sqrt{x^2-a^2}}$

$\displaystyle\int \dfrac{dx}{\sqrt{x^2-a^2}} = \cosh^{-1}\dfrac{x}{a} + C. \quad (a>0, x\geq a)$

Let $y = \tanh^{-1}\dfrac{x}{a} \Leftrightarrow x = a\tanh y \Rightarrow 1 = a(\operatorname{sech}^2 y)\dfrac{dy}{dx}$.
Thus,

$\dfrac{d}{dx}\tanh^{-1}\dfrac{x}{a} = \dfrac{1}{a\operatorname{sech}^2 y}$

$= \dfrac{a}{a^2 - a^2\tanh^2 x} = \dfrac{a}{a^2 - x^2}$

$\displaystyle\int \dfrac{dx}{a^2 - x^2} = \dfrac{1}{a}\tanh^{-1}\dfrac{x}{a} + C.$

7. a) $\sinh \ln x = \dfrac{1}{2}(e^{\ln x} - e^{-\ln x}) = \dfrac{1}{2}\left(x - \dfrac{1}{x}\right) = \dfrac{x^2 - 1}{2x}$

b) $\cosh \ln x = \dfrac{1}{2}(e^{\ln x} + e^{-\ln x}) = \dfrac{1}{2}\left(x + \dfrac{1}{x}\right) = \dfrac{x^2 + 1}{2x}$

c) $\tanh \ln x = \dfrac{\sinh \ln x}{\cosh \ln x} = \dfrac{x^2 - 1}{x^2 + 1}$

d) $\dfrac{\cosh \ln x + \sinh \ln x}{\cosh \ln x - \sinh \ln x} = \dfrac{x^2 + 1 + (x^2 - 1)}{(x^2 + 1) - (x^2 - 1)} = x^2$

8. $\operatorname{csch}^{-1} x = \sinh^{-1}(1/x) = \ln\left(\dfrac{1}{x} + \sqrt{\dfrac{1}{x^2} + 1}\right)$ has domain and range consisting of all real numbers x except $x = 0$. We have

$\dfrac{d}{dx}\operatorname{csch}^{-1} x = \dfrac{d}{dx}\sinh^{-1}\dfrac{1}{x}$

$= \dfrac{1}{\sqrt{1 + \left(\dfrac{1}{x}\right)^2}}\left(\dfrac{-1}{x^2}\right) = \dfrac{-1}{|x|\sqrt{x^2+1}}.$

$y = \operatorname{csch}^{-1} x$

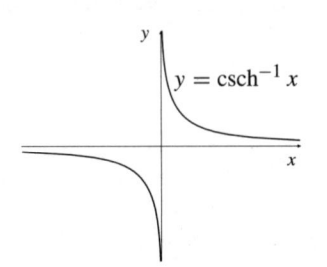

Fig. 3.6.8

9. $\operatorname{coth}^{-1} x = \tanh^{-1}\dfrac{1}{x} = \dfrac{1}{2}\ln\left(\dfrac{1+\dfrac{1}{x}}{1-\dfrac{1}{x}}\right) = \dfrac{1}{2}\ln\left(\dfrac{x+1}{x-1}\right)$,
for $|x| > 1$. Also

$\dfrac{d}{dx}\operatorname{coth}^{-1} x = \dfrac{d}{dx}\tanh^{-1}\dfrac{1}{x}$

$= \dfrac{1}{1 - (1/x)^2}\dfrac{-1}{x^2} = \dfrac{-1}{x^2 - 1}.$

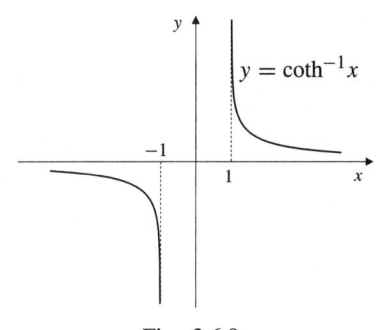

INSTRUCTOR'S SOLUTIONS MANUAL SECTION 3.7 (PAGE 223)

Fig. 3.6.9

10. Let $y = \text{Sech}^{-1} x$ where $\text{Sech}\, x = \text{sech}\, x$ for $x \geq 0$. Hence, for $y \geq 0$,

$$x = \text{sech}\, y \Leftrightarrow \frac{1}{x} = \cosh y$$
$$\Leftrightarrow \frac{1}{x} = \text{Cosh}\, y \Leftrightarrow y = \text{Cosh}^{-1} \frac{1}{x}.$$

Thus,
$$\text{Sech}^{-1} x = \text{Cosh}^{-1} \frac{1}{x}$$
$$D(\text{Sech}^{-1}) = R(\text{sech}) = (0, 1]$$
$$R(\text{Sech}^{-1}) = D(\text{sech}) = [0, \infty).$$

Also,
$$\frac{d}{dx} \text{Sech}^{-1} x = \frac{d}{dx} \text{Cosh}^{-1} \frac{1}{x}$$
$$= \frac{1}{\sqrt{\left(\frac{1}{x}\right)^2 - 1}} \left(\frac{-1}{x^2}\right) = \frac{-1}{x\sqrt{1-x^2}}.$$

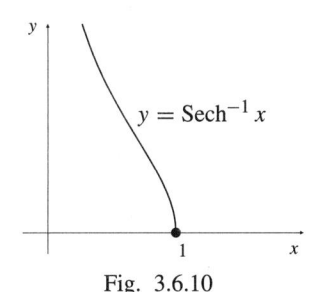

Fig. 3.6.10

11. $f_{A,B}(x) = Ae^{kx} + Be^{-kx}$
$f'_{A,B}(x) = kAe^{kx} - kBe^{-kx}$
$f''_{A,B}(x) = k^2 Ae^{kx} + k^2 Be^{-kx}$
Thus $f''_{A,B} - k^2 f_{A,B} = 0$

$g_{C,D}(x) = C \cosh kx + D \sinh kx$
$g'_{C,D}(x) = kC \cosh kx + kD \sinh kx$
$g''_{C,D}(x) = k^2 C \cosh kx + k^2 D \sinh kx$
Thus $g''_{C,D} - k^2 g_{C,D} = 0$
$\cosh kx + \sinh kx = e^{kx}$
$\cosh kx - \sinh kx = e^{-kx}$
Thus $f_{A,B}(x) = (A+B) \cosh kx + (A-B) \sinh kx$, that is, $f_{A,B}(x) = g_{A+B, A-B}(x)$, and
$$g_{C,D}(x) = \frac{C}{2}(e^{kx} + e^{-kx}) + \frac{D}{2}(e^{kx} - e^{-kx}),$$
that is $g_{C,D}(x) = f_{(C+D)/2,(C-D)/2}(x)$.

12. Since
$$h_{L,M}(x) = L \cosh k(x-a) + M \sinh k(x-a)$$
$$h''_{L,M}(x) = Lk^2 \cosh k(x-a) + Mk^2 \sinh k(x-a)$$
$$= k^2 h_{L,M}(x)$$

hence, $h_{L,M}(x)$ is a solution of $y'' - k^2 y = 0$ and
$$h_{L,M}(x)$$
$$= \frac{L}{2}\left(e^{kx-ka} + e^{-kx+ka}\right) + \frac{M}{2}\left(e^{kx-ka} - e^{-kx+ka}\right)$$
$$= \left(\frac{L}{2}e^{-ka} + \frac{M}{2}e^{-ka}\right)e^{kx} + \left(\frac{L}{2}e^{ka} - \frac{M}{2}e^{ka}\right)e^{-kx}$$
$$= Ae^{kx} + Be^{-kx} = f_{A,B}(x)$$
where $A = \frac{1}{2}e^{-ka}(L+M)$ and $B = \frac{1}{2}e^{ka}(L-M)$.

13. $y'' - k^2 y = 0 \Rightarrow y = h_{L,M}(x)$
$= L \cosh k(x-a) + M \sinh k(x-a)$
$y(a) = y_0 \Rightarrow y_0 = L + 0 \Rightarrow L = y_0$,
$y'(a) = v_0 \Rightarrow v_0 = 0 + Mk \Rightarrow M = \frac{v_0}{k}$
Therefore $y = h_{y_0, v_0/k}(x)$
$= y_0 \cosh k(x-a) + (v_0/k) \sinh k(x-a)$.

Section 3.7 Second Order Differential Equations with Constant Coefficients (page 223)

1. $y'' + 7y' + 10y = 0$
auxiliary eqn $r^2 + 7r + 10 = 0$
$(r+5)(r+2) = 0 \Rightarrow r = -5, -2$
$y = Ae^{-5t} + Be^{-2t}$

2. $y'' - 2y' - 3y = 0$
auxiliary eqn $r^2 - 2r - 3 = 0 \Rightarrow r = -1, r = 3$
$y = Ae^{-t} + Be^{3t}$

3. $y'' + 2y' = 0$
auxiliary eqn $r^2 + 2r = 0 \Rightarrow r = 0, -2$
$y = A + Be^{-2t}$

4. $4y'' - 4y' - 3y = 0$
$4r^2 - 4r - 3 = 0 \Rightarrow (2r+1)(2r-3) = 0$
Thus, $r_1 = -\frac{1}{2}$, $r_2 = \frac{3}{2}$, and $y = Ae^{-(1/2)t} + Be^{(3/2)t}$.

95

5. $$y'' + 8y' + 16y = 0$$
   auxiliary eqn $r^2 + 8r + 16 = 0 \Rightarrow r = -4, -4$
   $$y = Ae^{-4t} + Bte^{-4t}$$

6. $$y'' - 2y' + y = 0$$
   $r^2 - 2r + 1 = 0 \Rightarrow (r-1)^2 = 0$
   Thus, $r = 1, 1$, and $y = Ae^t + Bte^t$.

7. $$y'' - 6y' + 10y = 0$$
   auxiliary eqn $r^2 - 6r + 10 = 0 \Rightarrow r = 3 \pm i$
   $$y = Ae^{3t}\cos t + Be^{3t}\sin t$$

8. $9y'' + 6y' + y = 0$
   $9r^2 + 6r + 1 = 0 \Rightarrow (3r+1)^2 = 0$
   Thus, $r = -\frac{1}{3}, -\frac{1}{3}$, and $y = Ae^{-(1/3)t} + Bte^{-(1/3)t}$.

9. $$y'' + 2y' + 5y = 0$$
   auxiliary eqn $r^2 + 2r + 5 = 0 \Rightarrow r = -1 \pm 2i$
   $$y = Ae^{-t}\cos 2t + Be^{-t}\sin 2t$$

10. For $y'' - 4y' + 5y = 0$ the auxiliary equation is $r^2 - 4r + 5 = 0$, which has roots $r = 2 \pm i$. Thus, the general solution of the DE is $y = Ae^{2t}\cos t + Be^{2t}\sin t$.

11. For $y'' + 2y' + 3y = 0$ the auxiliary equation is $r^2 + 2r + 3 = 0$, which has solutions $r = -1 \pm \sqrt{2}i$. Thus the general solution of the given equation is $y = Ae^{-t}\cos(\sqrt{2}t) + Be^{-t}\sin(\sqrt{2}t)$.

12. Given that $y'' + y' + y = 0$, hence $r^2 + r + 1 = 0$. Since $a = 1, b = 1$ and $c = 1$, the discriminant is $D = b^2 - 4ac = -3 < 0$ and $-(b/2a) = -\frac{1}{2}$ and $\omega = \sqrt{3}/2$. Thus, the general solution is
$$y = Ae^{-(1/2)t}\cos\left(\frac{\sqrt{3}}{2}t\right) + Be^{-(1/2)t}\sin\left(\frac{\sqrt{3}}{2}t\right).$$

13. $\begin{cases} 2y'' + 5y' - 3y = 0 \\ y(0) = 1 \\ y'(0) = 0 \end{cases}$
    The DE has auxiliary equation $2r^2 + 5y - 3 = 0$, with roots $r = \frac{1}{2}$ and $r = -3$. Thus $y = Ae^{t/2} + Be^{-3t}$.
    Now $1 = y(0) = A + B$, and $0 = y'(0) = \frac{A}{2} - 3B$.
    Thus $B = 1/7$ and $A = 6/7$. The solution is
    $$y = \frac{6}{7}e^{t/2} + \frac{1}{7}e^{-3t}.$$

14. Given that $y'' + 10y' + 25y = 0$, hence $r^2 + 10r + 25 = 0 \Rightarrow (r+5)^2 = 0 \Rightarrow r = -5$. Thus,
    $$y = Ae^{-5t} + Bte^{-5t}$$
    $$y' = -5e^{-5t}(A + Bt) + Be^{-5t}.$$
    Since
    $$0 = y(1) = Ae^{-5} + Be^{-5}$$
    $$2 = y'(1) = -5e^{-5}(A + B) + Be^{-5},$$

we have $A = -2e^5$ and $B = 2e^5$.
Thus, $y = -2e^5 e^{-5t} + 2te^5 e^{-5t} = 2(t-1)e^{-5(t-1)}$.

15. $\begin{cases} y'' + 4y' + 5y = 0 \\ y(0) = 2 \\ y'(0) = 0 \end{cases}$
    The auxiliary equation for the DE is $r^2 + 4r + 5 = 0$, which has roots $r = -2 \pm i$. Thus
    $$y = Ae^{-2t}\cos t + Be^{-2t}\sin t$$
    $$y' = (-2Ae^{-2t} + Be^{-2t})\cos t - (Ae^{-2t} + 2Be^{-2t})\sin t.$$
    Now $2 = y(0) = A \Rightarrow A = 2$, and
    $2 = y'(0) = -2A + B \Rightarrow B = 6$.
    Therefore $y = e^{-2t}(2\cos t + 6\sin t)$.

16. Given that $a > 0$, $b > 0$ and $c > 0$:
    Case 1: If $D = b^2 - 4ac > 0$ then the two roots are
    $$r_{1,2} = \frac{-b \pm \sqrt{b^2 - 4ac}}{2a}.$$
    Since
    $$b^2 - 4ac < b^2$$
    $$\pm\sqrt{b^2 - 4ac} < b$$
    $$-b \pm \sqrt{b^2 - 4ac} < 0$$
    therefore r_1 and r_2 are negative. The general solution is
    $$y(t) = Ae^{r_1 t} + Be^{r_2 t}.$$
    If $t \to \infty$, then $e^{r_1 t} \to 0$ and $e^{r_2 t} \to 0$.
    Thus, $\lim_{t \to \infty} y(t) = 0$.
    Case 2: If $D = b^2 - 4ac = 0$ then the two equal roots $r_1 = r_2 = -b/(2a)$ are negative. The general solution is
    $$y(t) = Ae^{r_1 t} + Bte^{r_2 t}.$$
    If $t \to \infty$, then $e^{r_1 t} \to 0$ and $e^{r_2 t} \to 0$ at a faster rate than $Bt \to \infty$. Thus, $\lim_{t \to \infty} y(t) = 0$.
    Case 3: If $D = b^2 - 4ac < 0$ then the general solution is
    $$y = Ae^{-(b/2a)t}\cos(\omega t) + Be^{-(b/2a)t}\sin(\omega t)$$
    where $\omega = \dfrac{\sqrt{4ac - b^2}}{2a}$. If $t \to \infty$, then the amplitude of both terms $Ae^{-(b/2a)t} \to 0$ and $Be^{-(b/2a)t} \to 0$. Thus, $\lim_{t \to \infty} y(t) = 0$.

17. The auxiliary equation $ar^2 + br + c = 0$ has roots
    $$r_1 = \frac{-b - \sqrt{D}}{2a}, \quad r_2 = \frac{-b + \sqrt{D}}{2a},$$

where $D = b^2 - 4ac$. Note that $a(r_2 - r_1) = \sqrt{D} = -(2ar_1 + b)$. If $y = e^{r_1 t}u$, then $y' = e^{r_1 t}(u' + r_1 u)$, and $y'' = e^{r_1 t}(u'' + 2r_1 u' + r_1^2 u)$. Substituting these expressions into the DE $ay'' + by' + cy = 0$, and simplifying, we obtain

$$e^{r_1 t}(au'' + 2ar_1 u' + bu') = 0,$$

or, more simply, $u'' - (r_2 - r_1)u' = 0$. Putting $v = u'$ reduces this equation to first order:

$$v' = (r_2 - r_1)v,$$

which has general solution $v = Ce^{(r_2 - r_1)t}$. Hence

$$u = \int Ce^{(r_2 - r_1)t}\, dt = Be^{(r_2 - r_1)t} + A,$$

and $y = e^{r_1 t}u = Ae^{r_1 t} + Be^{r_2 t}$.

Exercises 18–21 constitute a derivation of the general solution of the differential equation

$$y'' + \omega^2 y = 0. \quad (\omega \neq 0) \quad (*)$$

18. If $y = A\cos\omega t + B\sin\omega t$ then

$$y'' + \omega^2 y = -A\omega^2 \cos\omega t - B\omega^2 \sin\omega t$$
$$+ \omega^2(A\cos\omega t + B\sin\omega t) = 0$$

for all t. So y is a solution of (*).

19. If $f(t)$ is any solution of (*) then $f''(t) = -\omega^2 f(t)$ for all t. Thus,

$$\frac{d}{dt}\left[\omega^2\bigl(f(t)\bigr)^2 + \bigl(f'(t)\bigr)^2\right]$$
$$= 2\omega^2 f(t)f'(t) + 2f'(t)f''(t)$$
$$= 2\omega^2 f(t)f'(t) - 2\omega^2 f(t)f'(t) = 0$$

for all t. Thus, $\omega^2\bigl(f(t)\bigr)^2 + \bigl(f'(t)\bigr)^2$ is constant. (This can be interpreted as a conservation of energy statement.)

20. If $g(t)$ satisfies (*) and also $g(0) = g'(0) = 0$, then by Exercise 23,

$$\omega^2\bigl(g(t)\bigr)^2 + \bigl(g'(t)\bigr)^2$$
$$= \omega^2\bigl(g(0)\bigr)^2 + \bigl(g'(0)\bigr)^2 = 0.$$

Since a sum of squares cannot vanish unless each term vanishes, $g(t) = 0$ for all t.

21. If $f(t)$ is any solution of (*), let $g(t) = f(t) - A\cos\omega t - B\sin\omega t$ where $A = f(0)$ and $B\omega = f'(0)$. By Exercise 21 and 22, g is a solution of (*). Also $g(0) = f(0) - A = 0$ and $g'(0) = f'(0) - B\omega = 0$. Thus, $g(t) = 0$ for all t by Exercise 24, and therefore $f(x) = A\cos\omega t + B\sin\omega t$. Thus, it is proved that every solution of (*) is of this form.

22. Because $y'' + 4y = 0$, therefore $y = A\cos 2t + B\sin 2t$. Now

$$y(0) = 2 \Rightarrow A = 2,$$
$$y'(0) = -5 \Rightarrow B = -\tfrac{5}{2}.$$

Thus, $y = 2\cos 2t - \tfrac{5}{2}\sin 2t$.

circular frequency $= \omega = 2$, frequency $= \dfrac{\omega}{2\pi} = \dfrac{1}{\pi} \approx 0.318$

period $= \dfrac{2\pi}{\omega} = \pi \approx 3.14$

amplitude $= \sqrt{(2)^2 + (-\tfrac{5}{2})^2}\ 3.20$

23. $\begin{cases} y'' + 100y = 0 \\ y(0) = 0 \\ y'(0) = 3 \end{cases}$
    $y = A\cos(10t) + B\sin(10t)$
    $A = y(0) = 0, \quad 10B = y'(0) = 3$
    $y = \dfrac{3}{10}\sin(10t)$

24. If $y'' + \dfrac{4\pi^2}{T^2} y = 0$, then $y = A\cos\dfrac{2\pi t}{T} + B\sin\dfrac{2\pi t}{T}$. Now

$$y(0) = R \Rightarrow A = R,$$
$$y'(0) = 0 \Rightarrow B = 0.$$

Thus, $y = R\cos\dfrac{2\pi t}{T}$.

circular frequency $= \omega = \dfrac{2\pi}{T}$, frequency $= \dfrac{1}{T}$

period $= T$, amplitude $= |R|$

25. $y = \mathcal{A}\cos\bigl(\omega(t - c)\bigr) + \mathcal{B}\sin\bigl(\omega(t - c)\bigr)$
    (easy to calculate $y'' + \omega^2 y = 0$)
    $y = \mathcal{A}\bigl(\cos(\omega t)\cos(\omega c) + \sin(\omega t)\sin(\omega c)\bigr)$
    $\quad + \mathcal{B}\bigl(\sin(\omega t)\cos(\omega c) - \cos(\omega t)\sin(\omega c)\bigr)$
    $= \bigl(\mathcal{A}\cos(\omega c) - \mathcal{B}\sin(\omega c)\bigr)\cos\omega t$
    $\quad + \bigl(\mathcal{A}\sin(\omega c) + \mathcal{B}\cos(\omega c)\bigr)\sin\omega t$
    $= A\cos\omega t + B\sin\omega t$
    where $A = \mathcal{A}\cos(\omega c) - \mathcal{B}\sin(\omega c)$ and $B = \mathcal{A}\sin(\omega c) + \mathcal{B}\cos(\omega c)$

26. For $y'' + y = 0$, we have $y = A\sin t + B\cos t$. Since,

$$y(2) = 3 = A\sin 2 + B\cos 2$$
$$y'(2) = -4 = A\cos 2 - B\sin 2,$$

therefore

$$A = 3\sin 2 - 4\cos 2$$
$$B = 4\sin 2 + 3\cos 2.$$

Thus,

$$y = (3\sin 2 - 4\cos 2)\sin t + (4\sin 2 + 3\cos 2)\cos t$$
$$= 3\cos(t-2) - 4\sin(t-2).$$

27. $\begin{cases} y'' + \omega^2 y = 0 \\ y(a) = A \\ y'(a) = B \end{cases}$

$$y = A\cos(\omega(t-a)) + \frac{B}{\omega}\sin(\omega(t-a))$$

28. From Example 9, the spring constant is $k = 9 \times 10^4$ gm/sec^2. For a frequency of 10 Hz (i.e., a circular frequency $\omega = 20\pi$ rad/sec.), a mass m satisfying $\sqrt{k/m} = 20\pi$ should be used. So,

$$m = \frac{k}{400\pi^2} = \frac{9 \times 10^4}{400\pi^2} = 22.8 \text{ gm.}$$

The motion is determined by

$$\begin{cases} y'' + 400\pi^2 y = 0 \\ y(0) = -1 \\ y'(0) = 2 \end{cases}$$

therefore, $y = A\cos 20\pi t + B\sin 20\pi t$ and

$$y(0) = -1 \Rightarrow A = -1$$
$$y'(0) = 2 \Rightarrow B = \frac{2}{20\pi} = \frac{1}{10\pi}.$$

Thus, $y = -\cos 20\pi t + \frac{1}{10\pi}\sin 20\pi t$, with y in cm and t in second, gives the displacement at time t. The amplitude is $\sqrt{(-1)^2 + (\frac{1}{10\pi})^2} \approx 1.0005$ cm.

29. Frequency $= \frac{\omega}{2\pi}$, $\omega^2 = \frac{k}{m}$ (k = spring const, m = mass)
Since the spring does not change, $\omega^2 m = k$ (constant)
For $m = 400$ gm, $\omega = 2\pi(24)$ (frequency = 24 Hz)
If $m = 900$ gm, then $\omega^2 = \frac{4\pi^2(24)^2(400)}{900}$
so $\omega = \frac{2\pi \times 24 \times 2}{3} = 32\pi.$
Thus frequency $= \frac{32\pi}{2\pi} = 16$ Hz
For $m = 100$ gm, $\omega = \frac{4\pi^2(24)^2 400}{100}$
so $\omega = 96\pi$ and frequency $= \frac{\omega}{2\pi} = 48$ Hz.

Review Exercises 3 (page 224)

1. $f(x) = 3x + x^3 \Rightarrow f'(x) = 3(1+x^2) > 0$ for all x, so f is increasing and therefore one-to-one and invertible. Since $f(0) = 0$, therefore $f^{-1}(0) = 0$, and

$$\frac{d}{dx}(f^{-1})(x)\Big|_{x=0} = \frac{1}{f'(f^{-1}(0))} = \frac{1}{f'(0)} = \frac{1}{3}.$$

2. $f(x) = \sec^2 x \tan x \Rightarrow f'(x) = 2\sec^2 x \tan^2 x + \sec^4 x > 0$ for x in $(-\pi/2, \pi/2)$, so f is increasing and therefore one-to-one and invertible there. The domain of f^{-1} is $(-\infty, \infty)$, the range of f. Since $f(\pi/4) = 2$, therefore $f^{-1}(2) = \pi/4$, and

$$(f^{-1})'(2) = \frac{1}{f'(f^{-1}(2))} = \frac{1}{f'(\pi/4)} = \frac{1}{8}.$$

3. $\lim_{x\to\pm\infty} f(x) = \lim_{x\to\pm\infty} \frac{x}{e^{x^2}} = 0.$

4. Observe $f'(x) = e^{-x^2}(1 - 2x^2)$ is positive if $x^2 < 1/2$ and is negative if $x^2 > 1/2$. Thus f is increasing on $(-1/\sqrt{2}, 1/\sqrt{2})$ and is decreasing on $(-\infty, -1/\sqrt{2})$ and on $(1/\sqrt{2}, \infty)$.

5. The max and min values of f are $1/\sqrt{2e}$ (at $x = 1/\sqrt{2}$) and $-1/\sqrt{2e}$ (at $x = -1/\sqrt{2}$).

6. $y = e^{-x}\sin x$, $(0 \le x \le 2\pi)$ has a horizontal tangent where

$$0 = \frac{dy}{dx} = e^{-x}(\cos x - \sin x).$$

This occurs if $\tan x = 1$, so $x = \pi/4$ or $x = 5\pi/4$. The points are $(\pi/4, e^{-\pi/4}/\sqrt{2})$ and $(5\pi/4, -e^{-5\pi/4}/\sqrt{2})$.

7. If $f'(x) = x$ for all x, then

$$\frac{d}{dx}\frac{f(x)}{e^{x^2/2}} = \frac{f'(x) - xf(x)}{e^{x^2/2}} = 0.$$

Thus $f(x)/e^{x^2/2} = C$ (constant) for all x. Since $f(2) = 3$, we have $C = 3/e^2$ and $f(x) = (3/e^2)e^{x^2/2} = 3e^{(x^2/2)-2}$.

8. Let the length, radius, and volume of the clay cylinder at time t be ℓ, r, and V, respectively. Then $V = \pi r^2 \ell$, and

$$\frac{dV}{dt} = 2\pi r \ell \frac{dr}{dt} + \pi r^2 \frac{d\ell}{dt}.$$

Since $dV/dt = 0$ and $d\ell/dt = k\ell$ for some constant $k > 0$, we have

$$2\pi r \ell \frac{dr}{dt} = -k\pi r^2 \ell, \quad \Rightarrow \quad \frac{dr}{dt} = -\frac{kr}{2}.$$

That is, r is decreasing at a rate proportional to itself.

INSTRUCTOR'S SOLUTIONS MANUAL   REVIEW EXERCISES 3 (PAGE 224)

9. a) An investment of $\$P$ at $r\%$ compounded continuously grows to $\$Pe^{rT/100}$ in T years. This will be $\$2P$ provided $e^{rT/100} = 2$, that is, $rT = 100\ln 2$. If $T = 5$, then $r = 20\ln 2 \approx 13.86\%$.

b) Since the doubling time is $T = 100\ln 2/r$, we have

$$\Delta T \approx \frac{dT}{dr}\Delta r = -\frac{100\ln 2}{r^2}\Delta r.$$

If $r = 13.863\%$ and $\Delta r = -0.5\%$, then

$$\Delta T \approx -\frac{100\ln 2}{13.863^2}(-0.5) \approx 0.1803 \text{ years.}$$

The doubling time will increase by about 66 days.

10. a) $\lim_{h\to 0}\dfrac{a^h - 1}{h} = \lim_{h\to 0}\dfrac{a^{0+h} - a^0}{h} = \dfrac{d}{dx}a^x\bigg|_{x=0} = \ln a.$

Putting $h = 1/n$, we get $\lim_{n\to\infty} n\left(a^{1/n} - 1\right) = \ln a.$

b) Using the technique described in the exercise, we calculate

$$2^{10}\left(2^{1/2^{10}} - 1\right) \approx 0.69338183$$

$$2^{11}\left(2^{1/2^{11}} - 1\right) \approx 0.69326449$$

Thus $\ln 2 \approx 0.693$.

11. $\dfrac{d}{dx}\left(f(x)\right)^2 = \left(f'(x)\right)^2$

$\Rightarrow 2f(x)f'(x) = \left(f'(x)\right)^2$

$\Rightarrow f'(x) = 0$ or $f'(x) = 2f(x)$.
Since $f(x)$ is given to be nonconstant, we have $f'(x) = 2f(x)$. Thus $f(x) = f(0)e^{2x} = e^{2x}$.

12. If $f(x) = (\ln x)/x$, then $f'(x) = (1-\ln x)/x^2$. Thus $f'(x) > 0$ if $\ln x < 1$ (i.e., $x < e$) and $f'(x) < 0$ if $\ln x > 1$ (i.e., $x > e$). Since f is increasing to the left of e and decreasing to the right, it has a maximum value $f(e) = 1/e$ at $x = e$. Thus, if $x > 0$ and $x \neq e$, then

$$\frac{\ln x}{x} < \frac{1}{e}.$$

Putting $x = \pi$ we obtain $(\ln\pi)/\pi < 1/e$. Thus

$$\ln(\pi^e) = e\ln\pi < \pi = \pi\ln e = \ln e^\pi,$$

and $\pi^e < e^\pi$ follows because $\ln$ is increasing.

13. $y = x^x = e^{x\ln x} \Rightarrow y' = x^x(1 + \ln x)$. The tangent to $y = x^x$ at $x = a$ has equation

$$y = a^a + a^a(1+\ln a)(x - a).$$

This line passes through the origin if $0 = a^a[1 - a(1+\ln a)]$, that is, if $(1+\ln a)a = 1$. Observe that $a = 1$ solves this equation. Therefore the slope of the line is $1^1(1+\ln 1) = 1$, and the line is $y = x$.

14. a) $\dfrac{\ln x}{x} = \dfrac{\ln 2}{2}$ is satisfied if $x = 2$ or $x = 4$ (because $\ln 4 = 2\ln 2$).

b) The line $y = mx$ through the origin intersects the curve $y = \ln x$ at $(b, \ln b)$ if $m = (\ln b)/b$. The same line intersects $y = \ln x$ at a different point $(x, \ln x)$ if $(\ln x)/x = m = (\ln b)/b$. This equation will have only one solution $x = b$ if the line $y = mx$ intersects the curve $y = \ln x$ only once, at $x = b$, that is, if the line is tangent to the curve at $x = b$. In this case m is the slope of $y = \ln x$ at $x = b$, so

$$\frac{1}{b} = m = \frac{\ln b}{b}.$$

Thus $\ln b = 1$, and $b = e$.

15. Let the rate be $r\%$. The interest paid by account A is $1,000(r/100) = 10r$.
The interest paid by account B is $1,000(e^{r/100} - 1)$. This is $\$10$ more than account A pays, so

$$1,000(e^{r/100} - 1) = 10r + 10.$$

A TI-85 solve routine gives $r \approx 13.8165\%$.

16. If $y = \cos^{-1} x$, then $x = \cos y$ and $0 \leq y \leq \pi$. Thus

$$\tan y = \text{sgn}\,x\sqrt{\sec^2 y - 1} = \text{sgn}\,x\sqrt{\frac{1}{x^2} - 1} = \frac{\sqrt{1-x^2}}{x}.$$

Thus $\cos^{-1} x = \tan^{-1}((\sqrt{1-x^2})/x)$.
Since $\cot x = 1/\tan x$, $\cot^{-1} x = \tan^{-1}(1/x)$.

$$\csc^{-1} x = \sin^{-1}\frac{1}{x} = \frac{\pi}{2} - \cos^{-1}\frac{1}{x}$$

$$= \frac{\pi}{2} - \tan^{-1}\frac{\sqrt{1-(1/x)^2}}{1/x}$$

$$= \frac{\pi}{2} - \text{sgn}\,x\tan^{-1}\sqrt{x^2 - 1}.$$

17. $\cos^{-1} x = \dfrac{\pi}{2} - \sin^{-1} x.$

If $y = \cot^{-1} x$, then $x = \cot y$ and $0 < y < \pi/2$. Thus

$$\csc y = \text{sgn}\,x\sqrt{1+\cot^2 y} = \text{sgn}\,x\sqrt{1+x^2}$$

$$\sin y = \frac{\text{sgn}\,x}{\sqrt{1+x^2}}.$$

Thus $\cot^{-1} x = \sin^{-1}\dfrac{\text{sgn}\,x}{\sqrt{1+x^2}} = \text{sgn}\,x\sin^{-1}\dfrac{1}{\sqrt{1+x^2}}.$

$$\csc^{-1} x = \sin^{-1}\frac{1}{x}.$$

99

18. Let $T(t)$ be the temperature of the milk t minutes after it is removed from the refrigerator. Let $U(t) = T(t) - 20$. By Newton's law,
$$U'(t) = kU(t) \Rightarrow U(t) = U(0)e^{kt}.$$

Now $T(0) = 5 \Rightarrow U(0) = -15$ and $T(12) = 12 \Rightarrow U(12) = -8$. Thus
$$-8 = U(12) = U(0)e^{12k} = -15e^{12k}$$
$$e^{12k} = 8/15, \quad k = \tfrac{1}{12}\ln(8/15).$$

If $T(s) = 18$, then $U(s) = -2$, so $-2 = -15e^{sk}$. Thus $sk = \ln(2/15)$, and
$$s = \frac{\ln(2/15)}{k} = 12\frac{\ln(2/15)}{\ln(8/15)} \approx 38.46.$$

It will take another $38.46 - 12 = 26.46$ min for the milk to warm up to $18°$.

19. Let R be the temperature of the room, Let $T(t)$ be the temperature of the water t minutes after it is brought into the room. Let $U(t) = T(t) - R$. Then
$$U'(t) = kU(t) \Rightarrow U(t) = U(0)e^{kt}.$$

We have
$T(0) = 96 \Rightarrow U(0) = 96 - R$
$T(10) = 60 \Rightarrow U(10) = 60 - R \Rightarrow 60 - R = (96 - R)e^{10k}$
$T(20) = 40 \Rightarrow U(20) = 40 - R \Rightarrow 40 - R = (96 - R)e^{20k}.$

Thus
$$\left(\frac{60 - R}{96 - R}\right)^2 = e^{20k} = \frac{40 - R}{96 - R}$$
$$(60 - R)^2 = (96 - R)(40 - R)$$
$$3600 - 120R + R^2 = 3840 - 136R + R^2$$
$$16R = 240 \quad R = 15.$$

Room temperature is $15°$.

20. Let $f(x) = e^x - 1 - x$. Then $f(0) = 0$ and by the MVT,
$$\frac{f(x)}{x} = \frac{f(x) - f(0)}{x - 0} = f'(c) = e^c - 1$$
for some c between 0 and x. If $x > 0$, then $c > 0$, and $f'(c) > 0$. If $x < 0$, then $c < 0$, and $f'(c) < 0$. In either case $f(x) = xf'(c) > 0$, which is what we were asked to show.

21. Suppose that for some positive integer k, the inequality
$$e^x > 1 + x + \frac{x^2}{2!} + \cdots + \frac{x^k}{k!}$$

holds for all $x > 0$. This is certainly true for $k = 1$, as shown in the previous exercise. Apply the MVT to
$$g(t) = e^t - 1 - t - \frac{t^2}{2!} - \cdots - \frac{t^{k+1}}{(k+1)!}$$
on the interval $(0, x)$ (where $x > 0$) to obtain
$$\frac{g(x)}{x} = \frac{g(x) - g(0)}{x - 0} = g'(c)$$
for some c in $(0, x)$. Since x and $g'(c)$ are both positive, so is $g(x)$. This completes the induction and shows the desired inequality holds for $x > 0$ for all positive integers k.

Challenging Problems 3   (page 224)

1. a) $(d/dx)x^x = x^x(1 + \ln x) > 0$ if $\ln x > -1$, that is, if $x > e^{-1}$. Thus x^x is increasing on $[e^{-1}, \infty)$.

   b) Being increasing on $[e^{-1}, \infty)$, $f(x) = x^x$ is invertible on that interval. Let $g = f^{-1}$. If $y = x^x$, then $x = g(y)$. Note that $y \to \infty$ if and only if $x \to \infty$. We have
$$\ln y = x \ln x$$
$$\ln(\ln y) = \ln x + \ln(\ln x)$$
$$\lim_{y \to \infty} \frac{g(y)\ln(\ln y)}{\ln y} = \lim_{x \to \infty} \frac{x(\ln x + \ln(\ln x))}{x \ln x}$$
$$= \lim_{x \to \infty} \left(1 + \frac{\ln(\ln x)}{\ln x}\right).$$

   Now $\ln x < \sqrt{x}$ for sufficiently large x, so $\ln(\ln x) < \sqrt{\ln x}$ for sufficiently large x. Therefore, $0 < \dfrac{\ln(\ln x)}{\ln x} < \dfrac{1}{\sqrt{\ln x}} \to 0$ as $x \to \infty$, and so
$$\lim_{y \to \infty} \frac{g(y)\ln(\ln y)}{\ln y} = 1 + 0 = 1.$$

2. $\dfrac{dv}{dt} = -g - kv$.

   a) Let $u(t) = -g - kv(t)$. Then $\dfrac{du}{dt} = -k\dfrac{dv}{dt} = -ku$, and
$$u(t) = u(0)e^{-kt} = -(g + kv_0)e^{-kt}$$
$$v(t) = -\frac{1}{k}\bigl(g + u(t)\bigr) = -\frac{1}{k}\bigl(g - (g + kv_0)e^{-kt}\bigr).$$

   b) $\lim_{t \to \infty} v(t) = -g/k$

INSTRUCTOR'S SOLUTIONS MANUAL — CHALLENGING PROBLEMS 3 (PAGE 224)

c) $\dfrac{dy}{dt} = v(t) = -\dfrac{g}{k} + \dfrac{g+kv_0}{k}e^{-kt}, \quad y(0) = y_0$

$y(t) = -\dfrac{gt}{k} - \dfrac{g+kv_0}{k^2}e^{-kt} + C$

$y_0 = -0 - \dfrac{g+kv_0}{k^2} + C \Rightarrow C = y_0 + \dfrac{g+kv_0}{k^2}$

$y(t) = y_0 - \dfrac{gt}{k} + \dfrac{g+kv_0}{k^2}\left(1 - e^{-kt}\right)$

3. $\dfrac{dv}{dt} = -g + kv^2 \quad (k > 0)$

a) Let $u = 2t\sqrt{gk}$. If $v(t) = \sqrt{\dfrac{g}{k}}\dfrac{1-e^u}{1+e^u}$, then

$\dfrac{dv}{dt} = \sqrt{\dfrac{g}{k}}\dfrac{(1+e^u)(-e^u) - (1-e^u)e^u}{(1+e^u)^2} 2\sqrt{gk}$

$= \dfrac{-4ge^u}{(1+e^u)^2}$

$kv^2 - g = g\left(\dfrac{(1-e^u)^2}{(1+e^u)^2} - 1\right)$

$= \dfrac{-4ge^u}{(1+e^u)^2} = \dfrac{dv}{dt}.$

Thus $v(t) = \sqrt{\dfrac{g}{k}}\dfrac{1-e^{2t\sqrt{gk}}}{1+e^{2t\sqrt{gk}}}$.

b) $\displaystyle\lim_{t\to\infty} v(t) = \lim_{t\to\infty} \sqrt{\dfrac{g}{k}}\dfrac{e^{-2t\sqrt{gk}}-1}{e^{-2t\sqrt{gk}}+1} = -\sqrt{\dfrac{g}{k}}$

c) If $y(t) = y_0 + \sqrt{\dfrac{g}{k}}t - \dfrac{1}{k}\ln\dfrac{1+e^{2t\sqrt{gk}}}{2}$, then $y(0) = y_0$ and

$\dfrac{dy}{dt} = \sqrt{\dfrac{g}{k}} - \dfrac{1}{k}\dfrac{2\sqrt{gk}e^{2t\sqrt{gk}}}{1+e^{2t\sqrt{gk}}}$

$= \sqrt{\dfrac{g}{k}}\dfrac{1-e^{2t\sqrt{gk}}}{1+e^{2t\sqrt{gk}}} = v(t).$

Thus $y(t)$ gives the height of the object at time t during its fall.

4. If $p = e^{-bt}y$, then $\dfrac{dp}{dt} = e^{-bt}\left(\dfrac{dy}{dt} - by\right)$.

The DE $\dfrac{dp}{dt} = kp\left(1 - \dfrac{p}{e^{-bt}M}\right)$ therefore transforms to

$\dfrac{dy}{dt} = by + kpe^{bt}\left(1 - \dfrac{p}{e^{-bt}M}\right)$

$= (b+k)y - \dfrac{ky^2}{M} = Ky\left(1 - \dfrac{y}{L}\right),$

where $K = b+k$ and $L = \dfrac{b+k}{k}M$. This is a standard Logistic equation with solution (as obtained in Section 3.4) given by

$y = \dfrac{Ly_0}{y_0 + (L-y_0)e^{-Kt}},$

where $y_0 = y(0) = p(0) = p_0$. Converting this solution back in terms of the function $p(t)$, we obtain

$p(t) = \dfrac{Lp_0 e^{-bt}}{p_0 + (L-p_0)e^{-(b+k)t}}$

$= \dfrac{(b+k)Mp_0}{p_0 ke^{bt} + \big((b+k)M - kp_0\big)e^{-kt}}.$

Since p represents a percentage, we must have $(b+k)M/k < 100$.

If $k = 10$, $b = 1$, $M = 90$, and $p_0 = 1$, then $\dfrac{b+k}{k}M = 99 < 100$. The numerator of the final expression for $p(t)$ given above is a constant. Therefore $p(t)$ will be largest when the derivative of the denominator,

$f(t) = p_0 ke^{bt} + \big((b+k)M - kp_0\big)e^{-kt} = 10e^t + 980e^{-10t}$

is zero. Since $f'(t) = 10e^t - 9,800e^{-10t}$, this will happen at $t = \ln(980)/11$. The value of p at this t is approximately 48.1. Thus the maximum percentage of potential clients who will adopt the technology is about 48.1%.

CHAPTER 4. SOME APPLICATIONS OF DERIVATIVES

Section 4.1 Related Rates (page 231)

1. If the side and area of the square at time t are x and A, respectively, then $A = x^2$, so
$$\frac{dA}{dt} = 2x\frac{dx}{dt}.$$
If $x = 8$ cm and $dx/dt = 2$ cm/min, then the area is increasing at rate $dA/dt = 32$ cm²/min.

2. As in Exercise 1, $dA/dt = 2x\,dx/dt$. If $dA/dt = -2$ ft²/s and $x = 8$ ft, then $dx/dt = -2/(16)$. The side length is decreasing at 1/8 ft/s.

3. Let the radius and area of the ripple t seconds after impact be r and A respectively. Then $A = \pi r^2$. We have
$$\frac{dA}{dt} = 2\pi r \frac{dr}{dt}.$$
If $r = 20$ cm and $\frac{dr}{dt} = 4$ cm/s, then
$\frac{dA}{dt} = 40\pi(4) = 160\pi$.
The area is increasing at 160π cm²/s.

4. Let A and r denote the area and radius of the circle. Then
$$A = \pi r^2 \Rightarrow r = \sqrt{\frac{A}{\pi}}$$
$$\Rightarrow \frac{dr}{dt} = \left(\frac{1}{2\sqrt{A\pi}}\right)\frac{dA}{dt}.$$
When $\frac{dA}{dt} = -2$, and $A = 100$, $\frac{dr}{dt} = -\frac{1}{10\sqrt{\pi}}$. The radius is decreasing at the rate $\frac{1}{10\sqrt{\pi}}$ cm/min when the area is 100 cm².

5. For $A = \pi r^2$, we have $dA/dt = 2\pi r\, dr/dt$. If $dA/dt = 1/3$ km²/h, then (a) $dr/dt = 1/(6\pi r)$ km/h, or (b) $dr/dt = 1/(6\pi\sqrt{A/\pi}) = 1/(6\sqrt{\pi A})$ km/h.

6. Let the length, width, and area be l, w, and A at time t. Thus $A = lw$.
$$\frac{dA}{dt} = l\frac{dw}{dt} + w\frac{dl}{dt}$$
When $l = 16$, $w = 12$, $\frac{dw}{dt} = 3$, $\frac{dA}{dt} = 0$, we have
$$0 = 16 \times 3 + 12\frac{dl}{dt} \Rightarrow \frac{dl}{dt} = -\frac{48}{12} = -4.$$
The length is decreasing at 4 m/s.

7. The volume V, surface area S, and edge length x of a cube are related by $V = x^3$ and $S = 6x^2$, so that
$$\frac{dV}{dt} = 3x^2\frac{dx}{dt}, \quad \frac{dS}{dt} = 12x\frac{dx}{dt}.$$
If $V = 64$ cm³ and $dV/dt = 2$ cm³/s, then $x = 4$ cm and $dx/dt = 2/(3 \times 16) = 1/24$ cm/s. Therefore, $dS/dt = 12(4)(1/24) = 2$. The surface area is increasing at 2 cm²/s.

8. Let V, r and h denote the volume, radius and height of the cylinder at time t. Thus, $V = \pi r^2 h$ and
$$\frac{dV}{dt} = 2\pi r h\frac{dr}{dt} + \pi r^2\frac{dh}{dt}.$$
If $V = 60$, $\frac{dV}{dt} = 2$, $r = 5$, $\frac{dr}{dt} = 1$, then
$$h = \frac{V}{\pi r^2} = \frac{60}{25\pi} = \frac{12}{5\pi}$$
$$\frac{dh}{dt} = \frac{1}{\pi r^2}\left(\frac{dV}{dt} - 2\pi r h\frac{dr}{dt}\right)$$
$$= \frac{1}{25\pi}\left(2 - 10\pi\frac{12}{5\pi}\right) = -\frac{22}{25\pi}.$$
The height is decreasing at the rate $\frac{22}{25\pi}$ cm/min.

9. Let the length, width, depth, and volume at time t be l, w, h and V respectively. Thus $V = lwh$, and
$$\frac{dV}{dt} = \frac{dl}{dt}wh + lh\frac{dw}{dt} + lw\frac{dh}{dt}.$$
If $l = 6$ cm, $w = 5$cm, $h = 4$cm, $\frac{dl}{dt} = \frac{dh}{dt} = 1$m/s, and $\frac{dw}{dt} = -2$cm/s, then
$$\frac{dV}{dt} = 20 - 48 + 30 = 2.$$
The volume is increasing at a rate of 2 cm³/s.

10. Let the length, width and area at time t be x, y and A respectively. Thus $A = xy$ and
$$\frac{dA}{dt} = x\frac{dy}{dt} + y\frac{dx}{dt}.$$
If $\frac{dA}{dt} = 5$, $\frac{dx}{dt} = 10$, $x = 20$, $y = 16$, then
$$5 = 20\frac{dy}{dt} + 16(10) \Rightarrow \frac{dy}{dt} = -\frac{31}{4}.$$
Thus, the width is decreasing at $\frac{31}{4}$ m/s.

11. $y = x^2$. Thus $\dfrac{dy}{dt} = 2x\dfrac{dx}{dt}$. If $x = -2$ and $\dfrac{dx}{dt} = -3$, then $\dfrac{dy}{dt} = -4(-3) = 12$. y is increasing at rate 12.

12. Since $x^2 y^3 = 72$, then
$$2xy^3 \frac{dx}{dt} + 3x^2 y^2 \frac{dy}{dt} = 0 \Rightarrow \frac{dy}{dt} = -\frac{2y}{3x}\frac{dx}{dt}.$$
If $x = 3$, $y = 2$, $\dfrac{dx}{dt} = 2$, then $\dfrac{dy}{dt} = -\dfrac{8}{9}$. Hence, the vertical velocity is $-\dfrac{8}{9}$ units/s.

13. We have
$$xy = t \Rightarrow x\frac{dy}{dt} + y\frac{dx}{dt} = 1$$
$$y = tx^2 \Rightarrow \frac{dy}{dt} = x^2 + 2xt\frac{dx}{dt}$$
At $t = 2$ we have $xy = 2$, $y = 2x^2 \Rightarrow 2x^3 = 2 \Rightarrow x = 1$, $y = 2$.
Thus $\dfrac{dy}{dt} + 2\dfrac{dx}{dt} = 1$, and $1 + 4\dfrac{dx}{dt} = \dfrac{dy}{dt}$.
So $1 + 6\dfrac{dx}{dt} = 1 \Rightarrow \dfrac{dx}{dt} = 0 \Rightarrow \dfrac{dy}{dt} = 1 \Rightarrow$.
Distance D from origin satisfies $D = \sqrt{x^2 + y^2}$. So
$$\frac{dD}{dt} = \frac{1}{2\sqrt{x^2+y^2}}\left(2x\frac{dx}{dt} + 2y\frac{dy}{dt}\right)$$
$$= \frac{1}{\sqrt{5}}\bigl(1(0) + 2(1)\bigr) = \frac{2}{\sqrt{5}}.$$
The distance from the origin is increasing at a rate of $2/\sqrt{5}$.

14. From the figure, $x^2 + k^2 = s^2$. Thus
$$x\frac{dx}{dt} = s\frac{ds}{dt}.$$
When angle $PCA = 45°$, $x = k$ and $s = \sqrt{2}k$. The radar gun indicates that $ds/dt = 100$ km/h. Thus $dx/dt = 100\sqrt{2}k/k \approx 141$. The car is travelling at about 141 km/h.

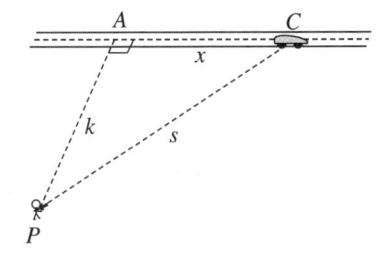

Fig. 4.1.14

15. We continue the notation of Exercise 16. If $dx/dt = 90$ km/h, and angle $PCA = 30°$, then $s = 2k$, $x = \sqrt{3}k$, and $ds/dt = (\sqrt{3}k/2k)(90) = 45\sqrt{3} = 77.94$. The radar gun will read about 78 km/h.

16. Let the distances x and y be as shown at time t. Thus $x^2 + y^2 = 25$ and $2x\dfrac{dx}{dt} + 2y\dfrac{dy}{dt} = 0$.
If $\dfrac{dx}{dt} = \dfrac{1}{3}$ and $y = 3$, then $x = 4$ and $\dfrac{4}{3} + 3\dfrac{dy}{dt} = 0$ so $\dfrac{dy}{dt} = -\dfrac{4}{9}$.
The top of the ladder is slipping down at a rate of $\dfrac{4}{9}$ m/s.

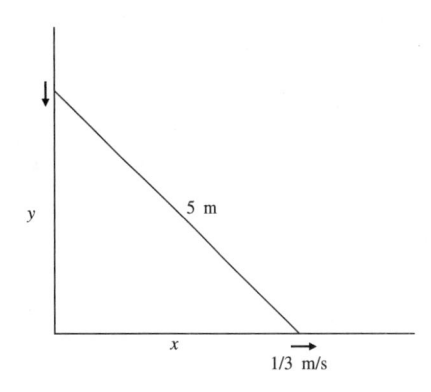

Fig. 4.1.16

17. Let x and y be the distances shown in the following figure. From similar triangles:
$$\frac{x}{2} = \frac{x+y}{5} \Rightarrow x = \frac{2y}{3} \Rightarrow \frac{dx}{dt} = \frac{2}{3}\frac{dy}{dt}.$$
Since $\dfrac{dy}{dt} = -\dfrac{1}{2}$, then
$$\frac{dx}{dt} = -\frac{1}{3} \quad \text{and} \quad \frac{d}{dt}(x+y) = -\frac{1}{2} - \frac{1}{3} = -\frac{5}{6}.$$
Hence, the man's shadow is decreasing at $\frac{1}{3}$ m/s and the shadow of his head is moving towards the lamppost at a rate of $\frac{5}{6}$ m/s.

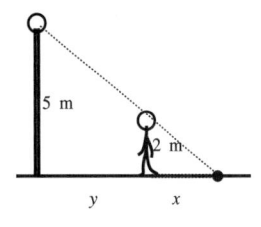

Fig. 4.1.17

SECTION 4.1 (PAGE 231)

18.

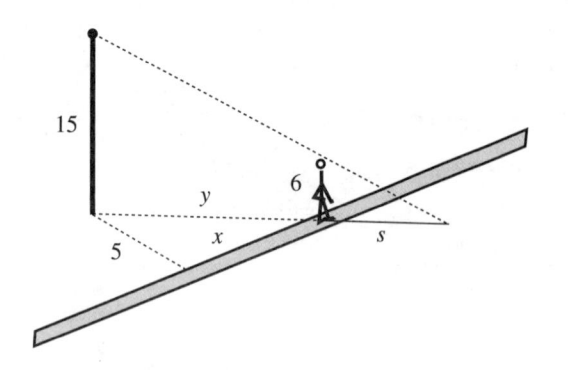

Fig. 4.1.18

Refer to the figure. s, y, and x are, respectively, the length of the woman's shadow, the distances from the woman to the lamppost, and the distances from the woman to the point on the path nearest the lamppost. From one of triangles in the figure we have

$$y^2 = x^2 + 25.$$

If $x = 12$, then $y = 13$. Moreover,

$$2y\frac{dy}{dt} = 2x\frac{dx}{dt}.$$

We are given that $dx/dt = 2$ ft/s, so $dy/dt = 24/13$ ft/s when $x = 12$ ft. Now the similar triangles in the figure show that

$$\frac{s}{6} = \frac{s+y}{15},$$

so that $s = 2y/3$. Hence $ds/dt = 48/39$. The woman's shadow is changing at rate $48/39$ ft/s when she is 12 ft from the point on the path nearest the lamppost.

19. $\quad C = 10,000 + 3x + \dfrac{x^2}{8,000}$

$\dfrac{dC}{dt} = \left(3 + \dfrac{x}{4,000}\right)\dfrac{dx}{dt}.$

If $dC/dt = 600$ when $x = 12,000$, then $dx/dt = 100$. The production is increasing at a rate of 100 tons per day.

20. Let x, y be distances travelled by A and B from their positions at 1:00 pm in t hours.
Thus $\dfrac{dx}{dt} = 16$ km/h, $\dfrac{dy}{dt} = 20$ km/h.
Let s be the distance between A and B at time t.
Thus $s^2 = x^2 + (25 + y)^2$

$$2s\frac{ds}{dt} = 2x\frac{dx}{dt} + 2(25+y)\frac{dy}{dt}$$

At 1:30 $\left(t = \tfrac{1}{2}\right)$ we have $x = 8$, $y = 10$,
$s = \sqrt{8^2 + 35^2} = \sqrt{1289}$ so

$$\sqrt{1289}\frac{ds}{dt} = 8 \times 16 + 35 \times 20 = 828$$

and $\dfrac{ds}{dt} = \dfrac{828}{\sqrt{1289}} \approx 23.06$. At 1:30, the ships are separating at about 23.06 km/h.

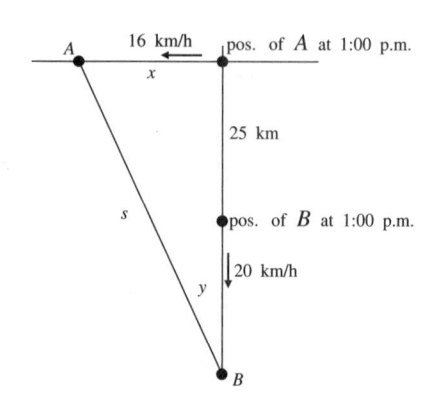

Fig. 4.1.20

21. Let θ and ω be the angles that the minute hand and hour hand made with the vertical t minutes after 3 o'clock. Then

$$\frac{d\theta}{dt} = \frac{\pi}{30}\text{ rad/min}$$
$$\frac{d\omega}{dt} = \frac{\pi}{360}\text{ rad/min}.$$

Since $\theta = 0$ and $\omega = \dfrac{\pi}{2}$ at $t = 0$, therefore

$$\theta = \frac{\pi}{30}t \quad \text{and} \quad \omega = \frac{\pi}{360}t + \frac{\pi}{2}.$$

At the first time after 3 o'clock when the hands of the clock are together, i.e., $\theta = \omega$,

$$\Rightarrow \frac{\pi}{30}t = \frac{\pi}{360}t + \frac{\pi}{2} \Rightarrow t = \frac{180}{11}.$$

Thus, the hands will be together at $16\tfrac{4}{11}$ minutes after 3 o'clock.

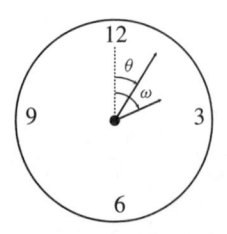

Fig. 4.1.21

INSTRUCTOR'S SOLUTIONS MANUAL SECTION 4.1 (PAGE 231)

22. Let y be the height of balloon t seconds after release.
Then $y = 5t$ m.
Let θ be angle of elevation at B of balloon at time t.
Then $\tan\theta = y/100$. Thus

$$\sec^2\theta \frac{d\theta}{dt} = \frac{1}{100}\frac{dy}{dt} = \frac{5}{100} = \frac{1}{20}$$

$$\left(1 + \tan^2\theta\right)\frac{d\theta}{dt} = \frac{1}{20}$$

$$\left[1 + \left(\frac{y}{100}\right)^2\right]\frac{d\theta}{dt} = \frac{1}{20}.$$

When $y = 200$ we have $5\dfrac{d\theta}{dt} = \dfrac{1}{20}$ so $\dfrac{d\theta}{dt} = \dfrac{1}{100}$.

The angle of elevation of balloon at B is increasing at a rate of $\dfrac{1}{100}$ rad/s.

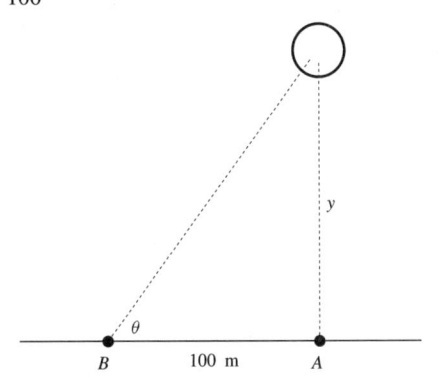

Fig. 4.1.22

23. Let V, r and h be the volume, radius and height of the cone. Since $h = r$, therefore

$$V = \tfrac{1}{3}\pi r^2 h = \tfrac{1}{3}\pi h^3$$

$$\frac{dV}{dt} = \pi h^2 \frac{dh}{dt} \Rightarrow \frac{dh}{dt} = \frac{1}{\pi h^2}\frac{dV}{dt}.$$

If $\dfrac{dV}{dt} = \dfrac{1}{2}$ and $h = 3$, then $\dfrac{dV}{dt} = \dfrac{1}{18\pi}$. Hence, the height of the pile is increasing at $\dfrac{1}{18\pi}$ m/min.

24. Let r, h, and V be the top radius, depth, and volume of the water in the tank at time t. Then $\dfrac{r}{h} = \dfrac{10}{8}$ and $V = \tfrac{1}{3}\pi r^2 h = \dfrac{\pi}{3}\dfrac{25}{16}h^3$. We have

$$\frac{1}{10} = \frac{\pi}{3}\frac{25}{16}3h^2\frac{dh}{dt} \Rightarrow \frac{dh}{dt} = \frac{16}{250\pi h^2}.$$

When $h = 4$ m, we have $\dfrac{dh}{dt} = \dfrac{1}{250\pi}$.

The water level is rising at a rate of $\dfrac{1}{250\pi}$ m/min when depth is 4 m.

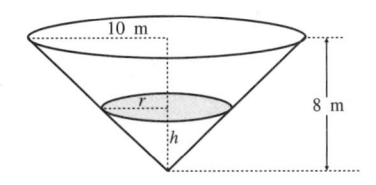

Fig. 4.1.24

25. Let r and h be the radius and height of the water in the tank at time t. By similar triangles,

$$\frac{r}{h} = \frac{10}{8} \Rightarrow r = \frac{5}{4}h.$$

The volume of water in the tank at time t is

$$V = \tfrac{1}{3}\pi r^2 h = \frac{25\pi}{48}h^3.$$

Thus,

$$\frac{dV}{dt} = \frac{25\pi}{16}h^2\frac{dh}{dt} \Rightarrow \frac{dh}{dt} = \frac{16}{25\pi h^2}\frac{dV}{dt}.$$

If $\dfrac{dV}{dt} = \dfrac{1}{10} - \dfrac{h^3}{1000}$ and $h = 4$, then

$$\frac{dh}{dt} = \frac{16}{(25\pi)(4)^2}\left(\frac{1}{10} - \frac{4^3}{1000}\right) = \frac{9}{6250\pi}.$$

Hence, the depth of water is increasing at $\dfrac{9}{6250\pi}$ m/min when the water is 4 m deep. The maximum depth occurs when $\dfrac{dh}{dt} = 0$, i.e.,

$$\frac{16}{25\pi h^2}\left(\frac{1}{10} - \frac{h^3}{1000}\right) = 0 \Rightarrow \frac{1}{10} - \frac{h^3}{1000} = 0$$
$$\Rightarrow h = \sqrt[3]{100}.$$

Thus, the maximum depth the water in the tank can get is $\sqrt[3]{100} \approx 4.64$ m.

26. Let r, h, and V be the top radius, depth, and volume of the water in the tank at time t. Then

$$\frac{r}{h} = \frac{3}{9} = \frac{1}{3}$$
$$V = \tfrac{1}{3}\pi r^2 h = \frac{\pi}{27}h^3$$
$$\frac{dV}{dt} = \frac{\pi}{9}h^2\frac{dh}{dt}.$$

105

If $\dfrac{dh}{dt} = 20$ cm/h $= \dfrac{2}{10}$ m/h when $h = 6$ m, then

$$\dfrac{dV}{dt} = \dfrac{\pi}{9} \times 36 \times \dfrac{2}{10} = \dfrac{4\pi}{5} \approx 2.51 \text{ m}^3/\text{h}.$$

Since water is coming in at a rate of 10 m³/h, it must be leaking out at a rate of $10 - 2.51 \approx 7.49$ m³/h.

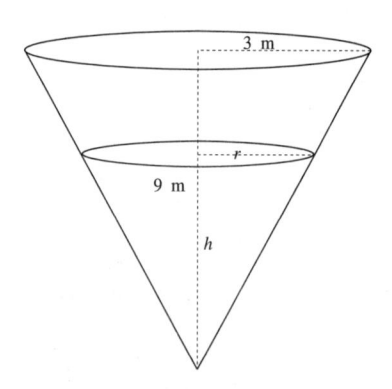

Fig. 4.1.26

27. Let x and s be the distance as shown. Then $s^2 = x^2 + 30^2$ and

$$2s\dfrac{ds}{dt} = 2x\dfrac{dx}{dt} \Rightarrow \dfrac{ds}{dt} = \dfrac{x}{s}\dfrac{dx}{dt}.$$

When $x = 40$, $\dfrac{dx}{dt} = 10$, $s = \sqrt{40^2 + 30^2} = 50$, then $\dfrac{ds}{dt} = \dfrac{40}{50}(10) = 8$. Hence, one must let out line at 8 m/min.

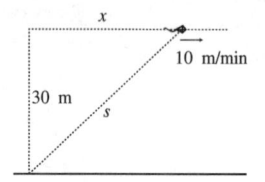

Fig. 4.1.27

28. Let P, x, and y be your position, height above centre, and horizontal distance from centre at time t. Let θ be the angle shown. Then $y = 10\sin\theta$, and $x = 10\cos\theta$. We have

$$\dfrac{dy}{dt} = 10\cos\theta \dfrac{d\theta}{dt}, \qquad \dfrac{d\theta}{dt} = 1 \text{ rpm} = 2\pi \text{ rad/min}.$$

When $x = 6$, then $\cos\theta = \dfrac{6}{10}$, so $\dfrac{dy}{dt} = 10 \times \dfrac{6}{10} \times 12\pi$. You are rising or falling at a rate of 12π m/min at the time in question.

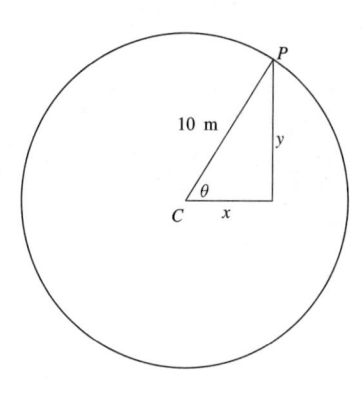

Fig. 4.1.28

29. Let x and y denote the distances of the two aircraft east and north of the airport respectively at time t as shown in the following diagram. Also let the distance between the two aircraft be s, then $s^2 = x^2 + y^2$. Thus,

$$2s\dfrac{ds}{dt} = 2x\dfrac{dx}{dt} + 2y\dfrac{dy}{dt}.$$

Since $\dfrac{dx}{dt} = -200$ and $\dfrac{dy}{dt} = 150$ when $x = 144$ and $y = 60$, we have $s = \sqrt{144^2 + 60^2} = 156$, and

$$\dfrac{ds}{dt} = \dfrac{1}{156}[144(-200) + 60(150)] \approx -126.9.$$

Thus, the distance between the aircraft is decreasing at about 126.9 km/h.

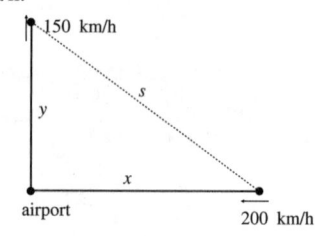

Fig. 4.1.29

30. $P = \dfrac{1}{3}x^{0.6}y^{0.4}$

$$\dfrac{dP}{dt} = \dfrac{0.6}{3}x^{-0.4}y^{0.4}\dfrac{dx}{dt} + \dfrac{0.4}{3}x^{0.6}y^{-0.6}\dfrac{dy}{dt}.$$

If $dP/dt = 0$, $x = 40$, $dx/dt = 1$, and $y = 10{,}000$, then

$$\dfrac{dy}{dt} = -\dfrac{6y^{0.4}}{x^{0.4}}\dfrac{y^{0.6}}{4x^{0.6}}\dfrac{dx}{dt} = -\dfrac{6y}{4x}\dfrac{dx}{dt} = -375.$$

The daily expenses are decreasing at \$375 per day.

31. Let the position of the ant be (x, y) and the position of its shadow be $(0, s)$. By similar triangles,

$$\dfrac{s - y}{x} = \dfrac{y}{3 - x} \Rightarrow s = \dfrac{3y}{3 - x}.$$

INSTRUCTOR'S SOLUTIONS MANUAL SECTION 4.1 (PAGE 231)

Then,
$$\frac{ds}{dt} = \frac{3(3-x)\frac{dy}{dt} + 3y\frac{dx}{dt}}{(3-x)^2}.$$

If the ant is at (1, 2) and $\frac{dx}{dt} = \frac{1}{3}$, $\frac{dy}{dt} = -\frac{1}{4}$, then

$$\frac{ds}{dt} = \frac{3(2)(-\frac{1}{4}) + 3(2)(\frac{1}{3})}{4} = \frac{1}{8}.$$

Hence, the ant's shadow is moving at $\frac{1}{8}$ units/s upwards along the y-axis.

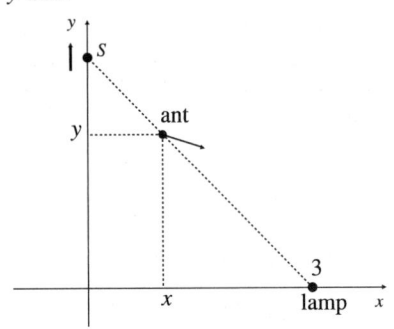

Fig. 4.1.31

32. Let x and y be the distances travelled from the intersection point by the boat and car respectively in t minutes. Then

$$\frac{dx}{dt} = 20 \times \frac{1000}{60} = \frac{1000}{3}\,\text{m/min}$$
$$\frac{dy}{dt} = 80 \times \frac{1000}{60} = \frac{4000}{3}\,\text{m/min}$$

The distance s between the boat and car satisfy

$$s^2 = x^2 + y^2 + 20^2, \quad s\frac{ds}{dt} = x\frac{dx}{dt} + y\frac{dy}{dt}.$$

After one minute, $x = \frac{1000}{3}$, $y = \frac{4000}{3}$ so $s \approx 1374.$ m. Thus

$$1374.5\frac{ds}{dt} = \frac{1000}{3}\frac{1000}{3} + \frac{4000}{3}\frac{4000}{3} \approx 1{,}888{,}889.$$

Hence $\frac{ds}{dt} \approx 1374.2$ m/min ≈ 82.45 km/h after 1 minute.

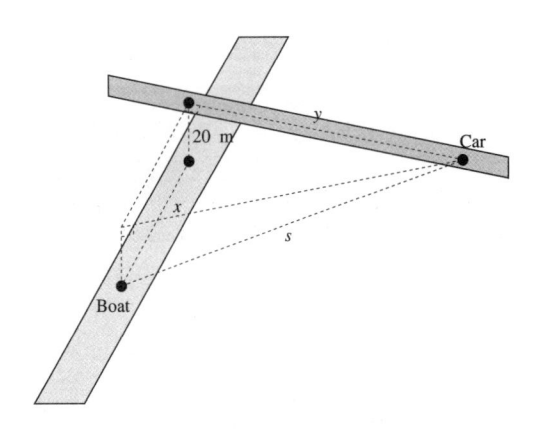

Fig. 4.1.32

33. Let h and b (measured in metres) be the depth and the surface width of the water in the tank at time t. We have

$$\frac{h}{(\frac{1}{2}b)} = \tan 60° = \sqrt{3} \Rightarrow b = \frac{2}{\sqrt{3}}h.$$

Thus, the volume of the water is

$$V = \left(\frac{1}{2}hb\right)(10) = \frac{10}{\sqrt{3}}h^2,$$

and

$$\frac{dV}{dt} = \frac{20}{\sqrt{3}}h\frac{dh}{dt} \Rightarrow \frac{dh}{dt} = \frac{\sqrt{3}}{20h}\frac{dV}{dt}.$$

If $\frac{dV}{dt} = \frac{1}{4}$ and $h = 0.2$ metres, then

$$\frac{dh}{dt} = \frac{\sqrt{3}}{20(0.2)}\left(\frac{1}{4}\right) = \frac{\sqrt{3}}{16}.$$

Hence, the water level is rising at $\frac{\sqrt{3}}{16}$ m/min.

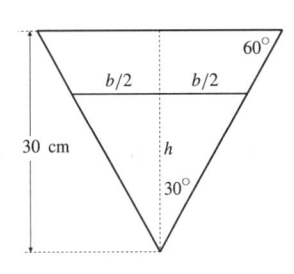

Fig. 4.1.33

34. Let V and h be the volume and depth of water in the tank at time t. If $h \leq 2$, then

$$\frac{x}{h} = \frac{20}{2} = 10, \quad \text{so } V = \frac{1}{2}xh8 = 40h^2.$$

107

SECTION 4.1 (PAGE 231)

If $2 \leq h \leq 3$, then $V = 160 + 160(h-2)$.

a) If $h = 2.5$m, then $-1 = \dfrac{dV}{dt} = 160\dfrac{dh}{dt}$.

So surface of water is dropping at a rate of $\dfrac{1}{160}$ m/min.

b) If $h = 1$m, then $-1 = \dfrac{dV}{dt} = 80h\dfrac{dh}{dt} = 80\dfrac{dh}{dt}$.

So surface of water is dropping at a rate of $\dfrac{1}{80}$ m/min.

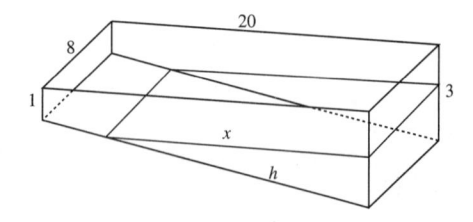

Fig. 4.1.34

35. Let the various distances be as shown in the figure.

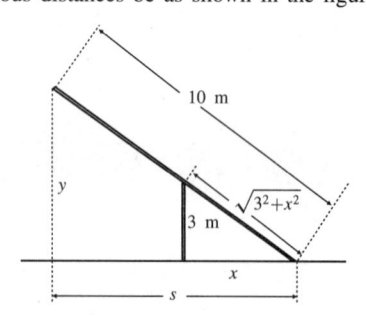

Fig. 4.1.35

a) By similar triangles,

$$\frac{y}{10} = \frac{3}{\sqrt{3^2+x^2}} \Rightarrow y = \frac{30}{\sqrt{9+x^2}}.$$

Thus,

$$\frac{dy}{dt} = \frac{dy}{dx}\frac{dx}{dt} = \frac{-30x}{(9+x^2)^{3/2}}\frac{dx}{dt}.$$

If $x = 4$ and $\dfrac{dx}{dt} = \dfrac{1}{5}$, then

$$\frac{dy}{dt} = \frac{-30(4)}{(9+16)^{3/2}}\left(\frac{1}{5}\right) = -\frac{24}{125}.$$

Hence, the free top end of the ladder is moving vertically downward at 24/125 m/s.

b) By similar triangles,

$$\frac{x}{\sqrt{3^2+x^2}} = \frac{s}{10} \Rightarrow s = \frac{10x}{\sqrt{9+x^2}}.$$

Then,

$$\frac{ds}{dt} = \frac{ds}{dx}\frac{dx}{dt}$$

$$= \frac{(\sqrt{9+x^2})(10) - (10x)\left(\dfrac{2x}{2\sqrt{9+x^2}}\right)}{(9+x^2)}\frac{dx}{dt}$$

$$= \frac{90}{(9+x^2)^{3/2}}\frac{dx}{dt}.$$

If $x = 4$ and $\dfrac{dx}{dt} = \dfrac{1}{5}$, then

$$\frac{ds}{dt} = \frac{90}{(9+16)^{3/2}}\left(\frac{1}{5}\right) = \frac{18}{125}.$$

This is the rate of change of the length of the horizontal projection of the ladder. The free top end of the ladder is moving horizontally to the right at rate

$$\frac{dx}{dt} - \frac{ds}{dt} = \frac{1}{5} - \frac{18}{125} = \frac{7}{125} \text{ m/s}.$$

36. Let x, y, and s be distances shown at time t. Then

$$s^2 = x^2 + 16, \qquad (15-s)^2 = y^2 + 16$$

$$s\frac{ds}{dt} = x\frac{dx}{dt}, \qquad -(15-s)\frac{ds}{dt} = y\frac{dy}{dt}.$$

When $x = 3$ and $\dfrac{dx}{dt} = \dfrac{1}{2}$, then $s = 5$ and $y = \sqrt{10^2 - 4^2} = \sqrt{84}$.

Also $\dfrac{ds}{dt} = \dfrac{3}{5}\left(\dfrac{1}{2}\right) = \dfrac{3}{10}$ so

$$\frac{dy}{dt} = -\frac{10}{\sqrt{84}}\frac{3}{10} = -\frac{3}{\sqrt{84}} \approx 0.327.$$

Crate B is moving toward Q at a rate of 0.327 m/s.

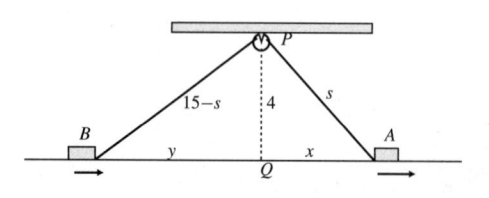

Fig. 4.1.36

INSTRUCTOR'S SOLUTIONS MANUAL SECTION 4.2 (PAGE 240)

37. Let θ be the angle of elevation, and x and y the horizontal and vertical distances from the launch site. We have

$$\tan\theta = \frac{y}{x} \quad \Rightarrow \quad \sec^2\theta \frac{d\theta}{dt} = \frac{x\frac{dy}{dt} - y\frac{dx}{dt}}{x^2}.$$

At the instant in question

$$\frac{dx}{dt} = 4\cos 30° = 2\sqrt{3}, \quad \frac{dy}{dt} = 4\sin 30° = 2,$$
$$x = 50\,\text{km}, \quad y = 100\,\text{km}.$$

Thus $\tan\theta = \frac{100}{50} = 2$, $\sec^2\theta = 1 + \tan^2\theta = 5$, and

$$\frac{d\theta}{dt} = \frac{1}{5} \frac{50(2) - 100(2\sqrt{3})}{(50)^2} = \frac{1 - 2\sqrt{3}}{125} \approx -0.0197.$$

Therefore, the angle of elevation is decreasing at about 0.0197 rad/s.

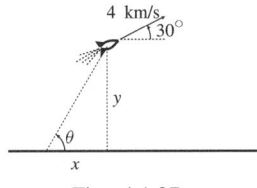

Fig. 4.1.37

38. Let y be height of ball t seconds after it drops. Thus $\frac{d^2y}{dt^2} = -9.8$, $\frac{dy}{dt}\big|_{t=0} = 0$, $y\big|_{t=0} = 20$, and

$$y = -4.9t^2 + 20, \quad \frac{dy}{dt} = -9.8t.$$

Let s be distance of shadow of ball from base of pole. By similar triangles, $\frac{s-10}{y} = \frac{s}{20}$.

$20s - 200 = sy$, $s = \frac{200}{20-y}$

$20\frac{ds}{dt} = y\frac{ds}{dt} + s\frac{dy}{dt}$.

a) At $t = 1$, we have $\frac{dy}{dt} = -9.8$, $y = 15.1$,

$4.9\frac{ds}{dt} = \frac{200}{4.9}(-9.8)$.

The shadow is moving at a rate of 81.63 m/s after one second.

b) As the ball hits the ground, $y = 0$, $s = 10$, $t = \sqrt{\frac{20}{4.9}}$,

and $\frac{dy}{dt} = -9.8\sqrt{\frac{20}{4.9}}$, so $20\frac{ds}{dt} = 0 + 10\frac{dy}{dt}$.

Now $y = 0$ implies that $t = \sqrt{\frac{20}{4.9}}$. Thus

$$\frac{ds}{dt} = -\frac{1}{2}(9.8)\sqrt{\frac{20}{4.9}} \approx -9.90.$$

The shadow is moving at about 9.90 m/s when the ball hits the ground.

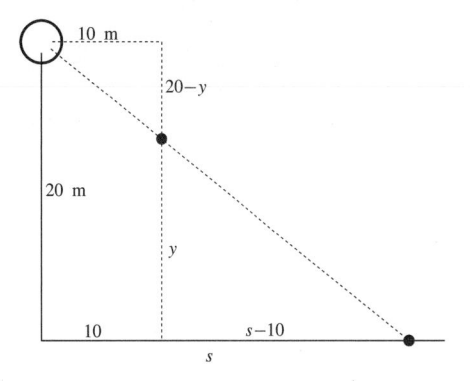

Fig. 4.1.38

39. Let $y(t)$ be the height of the rocket t seconds after it blasts off. We have

$$\frac{d^2y}{dt^2} = 10, \quad \frac{dy}{dt} = y = 0$$

at $t = 0$. Hence $y = 5t^2$, (y in metres, t in seconds). Now $\tan\theta = \frac{y}{2000}$, so $\sec^2\theta \frac{d\theta}{dt} = \frac{dy/dt}{2000}$, and

$$\left(1 + \left(\frac{y}{2000}\right)^2\right)\frac{d\theta}{dt} = \frac{10t}{2000} = \frac{t}{200}$$

$$\frac{d\theta}{dt} = \frac{t}{200} \cdot \frac{1}{1 + \frac{25t^4}{2000^2}}$$

$$= \frac{t}{200} \cdot \frac{1}{1 + \frac{t^4}{400^2}} = \frac{800t}{400^2 + t^4}.$$

At $t = 10$, we have $\frac{d\theta}{dt} = \frac{8000}{400^2 + 100^2} \approx 0.047$ rad/s.

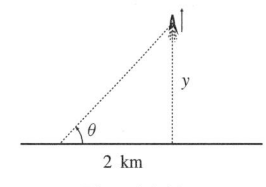

Fig. 4.1.39

Section 4.2 Extreme Values (page 240)

1. $f(x) = x + 2$ on $[-1, 1]$
$f'(x) = 1$ so f is increasing.
f has absolute minimum 1 at $x = -1$ and absolute maximum 3 at $x = 1$.

109

SECTION 4.2 (PAGE 240)

2. $f(x) = x + 2$ on $(-\infty, 0]$
   abs max 2 at $x = 0$, no min.

3. $f(x) = x + 2$ on $[-1, 1)$
   f has absolute minimum 1 at $x = -1$ and has no absolute maximum.

4. $f(x) = x^2 - 1$
   no max, abs min -1 at $x = 0$.

5. $f(x) = x^2 - 1$ on $[-2, 3]$
   f has abs min -1 at $x = 0$, abs max 8 at $x = 3$, and local max 3 at $x = -2$.

6. $f(x) = x^2 - 1$ on $(2, 3)$
   no max or min values.

7. $f(x) = x^3 + x - 4$ on $[a, b]$
   $f'(x) = 3x^2 + 1 > 0$ for all x.
   Therefore f has abs min $a^3 + a - 4$ at $x = a$ and abs max $b^3 + b - 4$ at $x = b$.

8. $f(x) = x^3 + x - 4$ on (a, b)
   Since $f'(x) = 3x^2 + 1 > 0$ for all x, therefore f is increasing. Since (a, b) is open, f has no max or min values.

9. $f(x) = x^5 + x^3 + 2x$ on $(a, b]$
   $f'(x) = 5x^4 + 3x^2 + 2 > 0$ for all x.
   f has no min value, but has abs max value $b^5 + b^3 + 2b$ at $x = b$.

10. $f(x) = \dfrac{1}{x-1}$. Since $f'(x) = \dfrac{-1}{(x-1)^2} < 0$ for all x in the domain of f, therefore f has no max or min values.

11. $f(x) = \dfrac{1}{x-1}$ on $(0, 1)$
    $f'(x) = -\dfrac{1}{(x-1)^2} < 0$ on $(0, 1)$
    f has no max or min values.

13. $f(x) = \dfrac{1}{x-1}$ on $[2, 3]$
    abs min $\frac{1}{2}$ at $x = 3$, abs max 1 at $x = 2$.

14. Let $f(x) = |x - 1|$ on $[-2, 2]$: $f(-2) = 3$, $f(2) = 1$.
    $f'(x) = \text{sgn}(x - 1)$. No CP; SP $x = 1$, $f(1) = 0$.
    Max value of f is 3 at $x = -2$; min value is 0 at $x = 1$.

15. $f(x) = \dfrac{1}{x^2 + 1}$, $f'(x) = -\dfrac{2x}{(x^2+1)^2}$
    f has abs max value 1 at $x = 0$; f has no min values.

16. $f(x) = (x + 2)^{2/3}$
    no max, abs min 0 at $x = -2$.

17. $f(x) = (x - 2)^{1/3}$, $f'(x) = \dfrac{1}{3}(x - 2)^{-2/3} > 0$
    f has no max or min values.

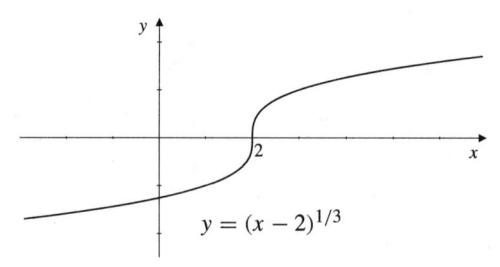

Fig. 4.2.17

18. $f(x) = x^2 + 2x$, $f'(x) = 2x + 2 = 2(x + 1)$
    Critical point: $x = -1$.
    $f(x) \to \infty$ as $x \to \pm\infty$.

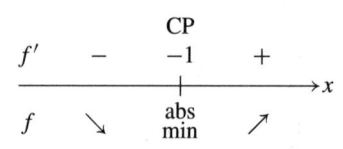

		CP	
f'	−	−1	+
f	↘	abs min	↗

    Hence, $f(x)$ has no max value, and the abs min is -1 at $x = -1$.

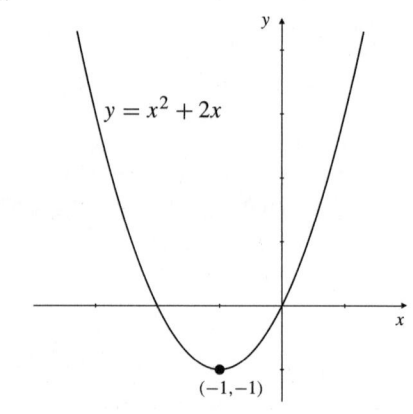

    Fig. 4.2.18

19. $f(x) = x^3 - 3x - 2$
    $f'(x) = 3x^2 - 3 = 3(x - 1)(x + 1)$

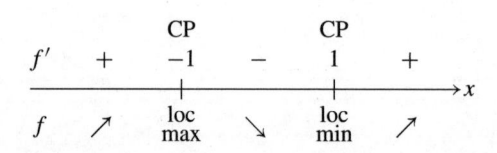

		CP		CP	
f'	+	−1	−	1	+
f	↗	loc max	↘	loc min	↗

    f has no absolute extrema.

110

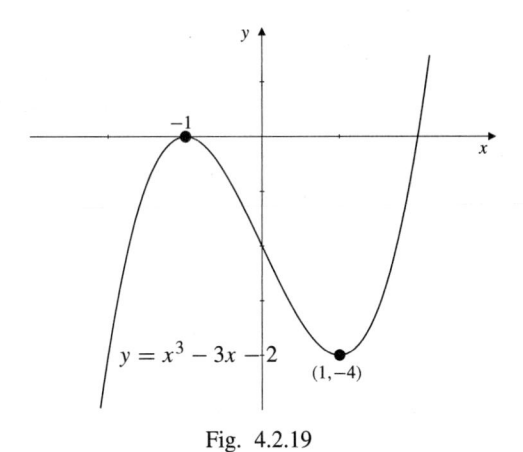

INSTRUCTOR'S SOLUTIONS MANUAL SECTION 4.2 (PAGE 240)

Fig. 4.2.19

Fig. 4.2.21

20. $f(x) = (x^2 - 4)^2$, $f'(x) = 4x(x^2 - 4) = 4x(x+2)(x-2)$
Critical points: $x = 0, \pm 2$.
$f(x) \to \infty$ as $x \to \pm\infty$.

```
              CP       CP      CP
   f'   -    -2   +    0   -   +2   +
   ─────────┼────────┼────────┼──────→ x
   f   ↘  abs  ↗  loc  ↘  abs  ↗
          min      max      min
```

Hence, $f(x)$ has abs min 0 at $x = \pm 2$ and loc max 16 at $x = 0$.

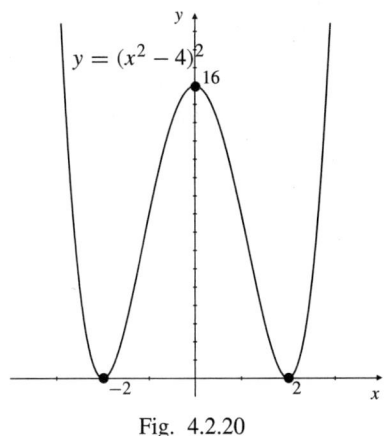

Fig. 4.2.20

21. $f(x) = x(x-1)^2$
$f'(x) = (x-1)^2 + 2x(x-1)$
$\quad\ = (x-1)(3x-1)$

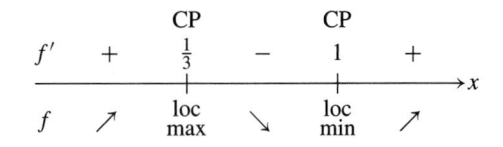

```
              CP      CP
   f'   +    1/3  -   1   +
   ─────────┼────────┼──────→ x
   f   ↗  loc  ↘  loc  ↗
          max      min
```

f has no absolute extrema.

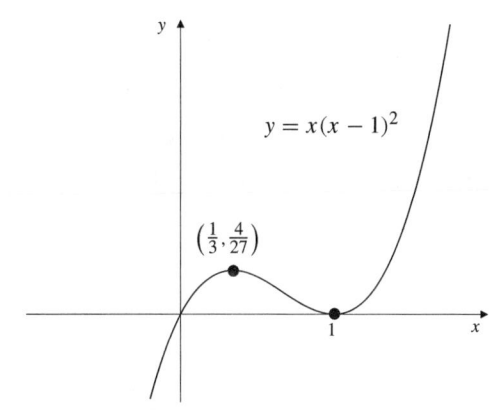

22. $f(x) = x^4 + 4x$, $f'(x) = 4x^3 + 4 = 4(x^3 + 1)$
Critical point: $x = -1$.
$f(x) \to \infty$ as $x \to \pm\infty$.

```
              CP
   f'   -    -1   +
   ─────────┼──────→ x
   f   ↘   abs   ↗
           min
```

Hence, $f(x)$ has no max value but has abs min -3 at $x = -1$.

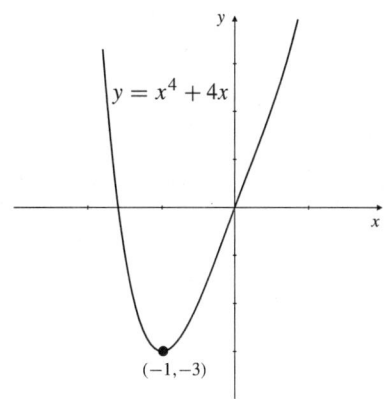

Fig. 4.2.22

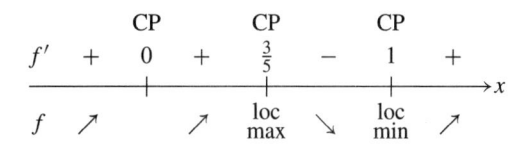

23. $f(x) = x^3(x-1)^2$
$f'(x) = 3x^2(x-1)^2 + 2x^3(x-1)$
$\quad\ = x^2(x-1)(5x-3)$
CP $x = 0, \dfrac{3}{5}, 1$

```
              CP      CP      CP
   f'   +    0   +   3/5  -   1   +
   ─────────┼────────┼────────┼──────→ x
   f   ↗       ↗  loc  ↘  loc  ↗
                  max      min
```

f has no absolute extrema.

111

SECTION 4.2 (PAGE 240) R. A. ADAMS: CALCULUS

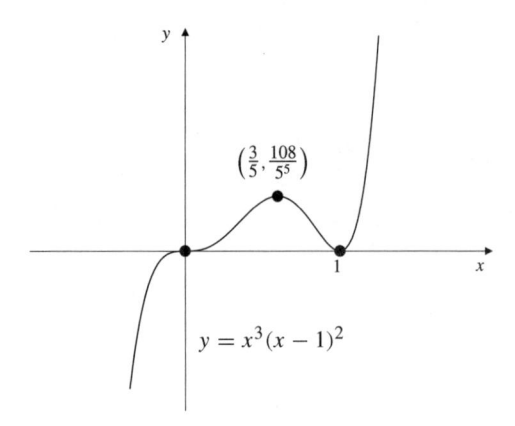

Fig. 4.2.23

24. $f(x) = x^2(x-1)^2$,
$f'(x) = 2x(x-1)^2 + 2x^2(x-1) = 2x(2x-1)(x-1)$
Critical points: $x = 0$, $\frac{1}{2}$ and 1.
$f(x) \to \infty$ as $x \to \pm\infty$.

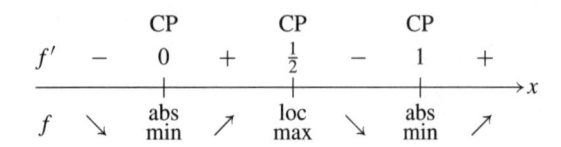

```
            CP        CP        CP
  f'  -    0    +    1/2   -    1   +
  ─────────┼─────────┼─────────┼────→ x
  f   ↘   abs   ↗   loc   ↘   abs   ↗
          min        max        min
```

Hence, $f(x)$ has loc max $\frac{1}{16}$ at $x = \frac{1}{2}$ and abs min 0 at $x = 0$ and $x = 1$.

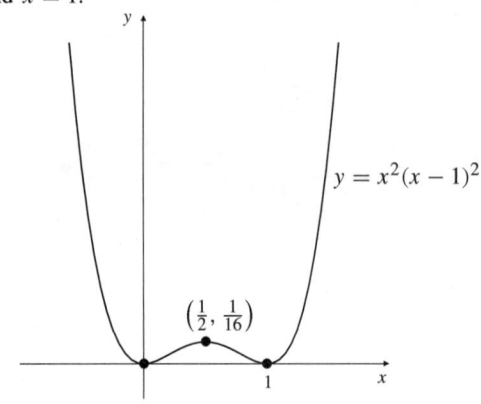

Fig. 4.2.24

25. $f(x) = x(x^2 - 1)^2$
$f'(x) = (x^2 - 1)^2 + 2x(x^2 - 1)2x$
$ = (x^2 - 1)(x^2 - 1 + 4x^2)$
$ = (x^2 - 1)(5x^2 - 1)$
$ = (x-1)(x+1)(\sqrt{5}x - 1)(\sqrt{5}x + 1)$

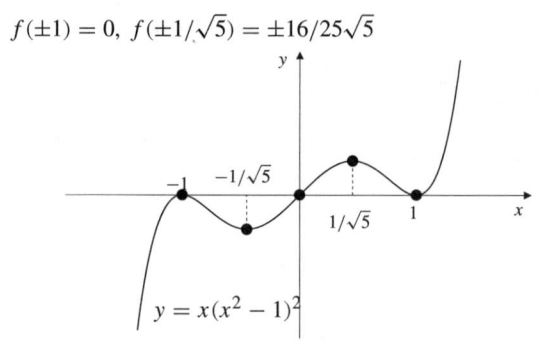

```
            CP        CP        CP        CP
  f'  +   -1    -   -1/√5  +   1/√5  -    1   +
  ────────┼─────────┼─────────┼─────────┼───→ x
  f   ↗  loc   ↘   loc   ↗   loc   ↘   loc   ↗
        max        min        max        min
```

$f(\pm 1) = 0$, $f(\pm 1/\sqrt{5}) = \pm 16/25\sqrt{5}$

Fig. 4.2.25

26. $f(x) = \dfrac{x}{x^2 + 1}$, $f'(x) = \dfrac{1 - x^2}{(x^2 + 1)^2}$
Critical point: $x = \pm 1$.
$f(x) \to 0$ as $x \to \pm\infty$.

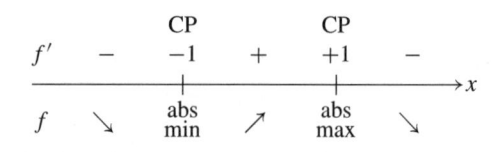

```
            CP        CP
  f'   -   -1    +   +1    -
  ─────────┼─────────┼────→ x
  f    ↘  abs   ↗   abs   ↘
         min        max
```

Hence, f has abs max $\frac{1}{2}$ at $x = 1$ and abs min $-\frac{1}{2}$ at $x = -1$.

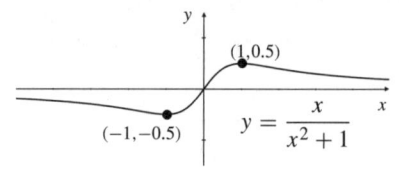

Fig. 4.2.26

27. $f(x) = \dfrac{x^2}{x^2 + 1} = 1 - \dfrac{1}{x^2 + 1} < 1$
$f'(x) = \dfrac{2x}{(x^2 + 1)^2}$

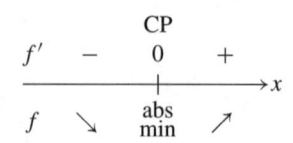

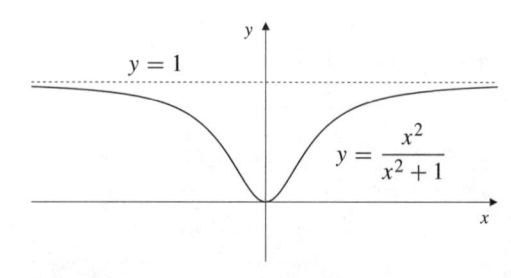

```
            CP
  f'   -    0    +
  ─────────┼────→ x
  f    ↘  abs   ↗
         min
```

Fig. 4.2.27

112

INSTRUCTOR'S SOLUTIONS MANUAL   SECTION 4.2 (PAGE 240)

28. $f(x) = \dfrac{x}{\sqrt{x^4+1}}$, $f'(x) = \dfrac{1-x^4}{(x^4+1)^{3/2}}$
    Critical points: $x = \pm 1$.
    $f(x) \to 0$ as $x \to \pm\infty$.

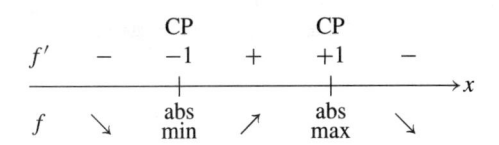

```
              CP        CP
   f'   -    -1   +    +1   -
   ─────┼─────────┼─────────→ x
   f    ↘   abs   ↗   abs   ↘
            min        max
```

Hence, f has abs max $\dfrac{1}{\sqrt{2}}$ at $x = 1$ and abs min $-\dfrac{1}{\sqrt{2}}$ at $x = -1$.

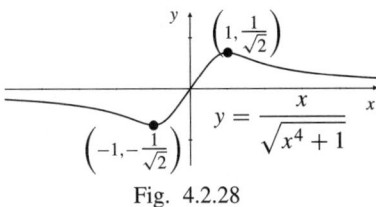

Fig. 4.2.28

29. $f(x) = x\sqrt{2-x^2}$ ($|x| \le \sqrt{2}$)
    $f'(x) = \sqrt{2-x^2} - \dfrac{x^2}{\sqrt{2-x^2}} = \dfrac{2(1-x^2)}{\sqrt{2-x^2}}$

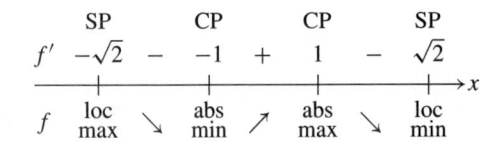

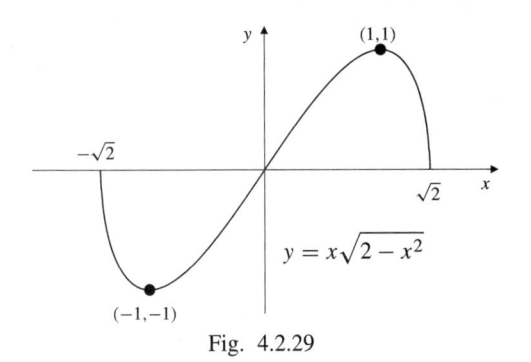

```
         SP       CP      CP      SP
   f'  -√2   -   -1   +   1   -   √2
   ────┼────────┼───────┼────────┼──→ x
   f   loc  ↘  abs  ↗  abs  ↘  loc
       max      min      max      min
```

Fig. 4.2.29

30. $f(x) = x + \sin x$, $f'(x) = 1 + \cos x \ge 0$
    $f'(x) = 0$ at $x = \pm\pi, \pm 3\pi, \ldots$
    $f(x) \to \pm\infty$ as $x \to \pm\infty$.
    Hence, f has no max or min values.

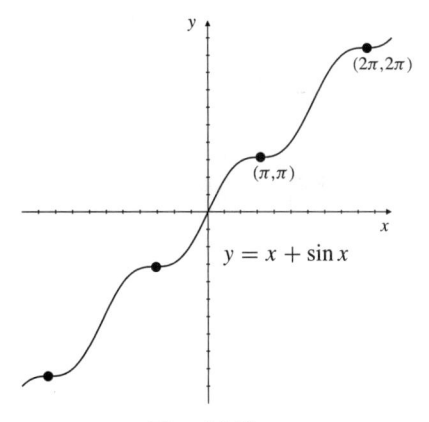

Fig. 4.2.30

31. $f(x) = x - 2\sin x$
    $f'(x) = 1 - 2\cos x$
    CP: $x = \pm\dfrac{\pi}{3} + 2n\pi$
    $n = 0, \pm 1, \pm 2, \cdots$
    alternating local maxima and minima

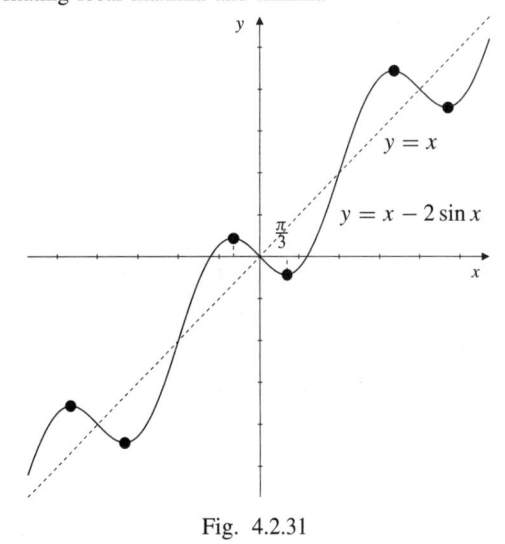

Fig. 4.2.31

32. $f(x) = x - 2\tan^{-1} x$, $f'(x) = 1 - \dfrac{2}{1+x^2} = \dfrac{x^2-1}{x^2+1}$
    Critical points: $x = \pm 1$.
    $f(x) \to \pm\infty$ as $x \to \pm\infty$.

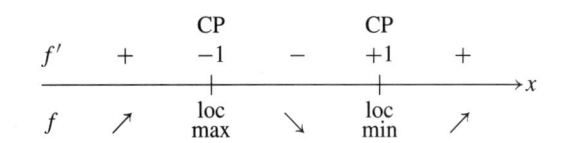

```
              CP        CP
   f'   +    -1   -    +1   +
   ─────┼─────────┼─────────→ x
   f    ↗   loc   ↘   loc   ↗
            max        min
```

Hence, f has loc max $-1 + \dfrac{\pi}{2}$ at $x = -1$ and loc min $1 - \dfrac{\pi}{2}$ at $x = 1$.

113

SECTION 4.2 (PAGE 240)

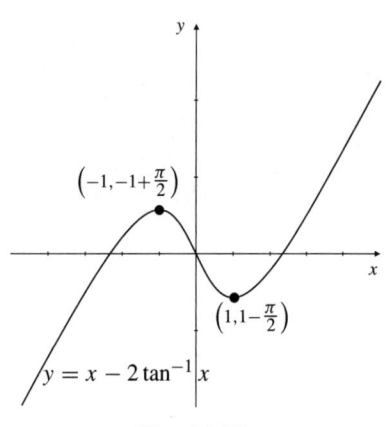

[Fig. 4.2.32: graph of $y = x - 2\tan^{-1} x$ with points $\left(-1, -1+\frac{\pi}{2}\right)$ and $\left(1, 1-\frac{\pi}{2}\right)$]

Fig. 4.2.32

33. $f(x) = 2x - \sin^{-1} x \qquad (-1 \le x \le 1)$

$f'(x) = 2 - \dfrac{1}{\sqrt{1-x^2}}$

$= \dfrac{2\sqrt{1-x^2} - 1}{\sqrt{1-x^2}}$

$= \dfrac{3 - 4x^2}{\sqrt{1-x^2}(2\sqrt{1-x^2}+1)}$

CP: $x = \pm\dfrac{\sqrt{3}}{2}$, SP: (EP:) $x = \pm 1$

$f\left(\pm\dfrac{\sqrt{3}}{2}\right) = \pm\left(\sqrt{3} - \dfrac{\pi}{3}\right)$

	SP		CP		CP		SP
f'	-1	$-$	$-\frac{\sqrt{3}}{2}$ $+$		$\frac{\sqrt{3}}{2}$ $-$		1
f	loc max	↘	abs min	↗	abs max	↘	loc min

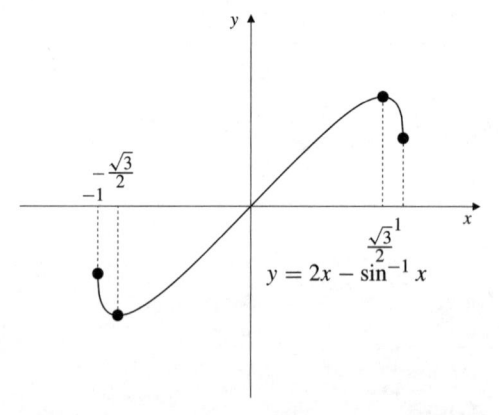

[Fig. 4.2.33: graph of $y = 2x - \sin^{-1} x$]

Fig. 4.2.33

34. $f(x) = e^{-x^2/2}$, $f'(x) = -xe^{-x^2/2}$
Critical point: $x = 0$.
$f(x) \to 0$ as $x \to \pm\infty$.

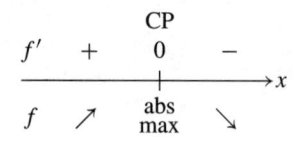

		CP	
f'	$+$	0	$-$
f	↗	abs max	↘

Hence, f has abs max 1 at $x = 0$ and no min value.

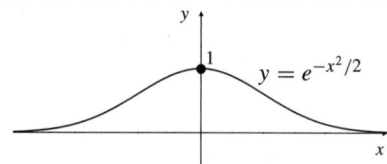

[Fig. 4.2.34: bell curve $y = e^{-x^2/2}$ with peak at $(0,1)$]

Fig. 4.2.34

35. $f(x) = x2^{-x}$
$f'(x) = 2^{-x} + x(-2^{-x}\ln 2)$
$\quad\; = 2^{-x}(1 - x\ln 2)$

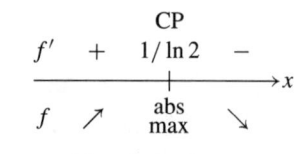

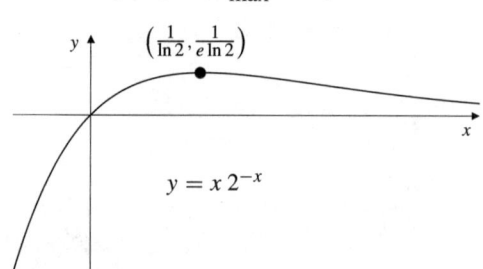

		CP	
f'	$+$	$1/\ln 2$	$-$
f	↗	abs max	↘

[Fig. 4.2.35: graph of $y = x2^{-x}$ with maximum at $\left(\frac{1}{\ln 2}, \frac{1}{e\ln 2}\right)$]

Fig. 4.2.35

36. $f(x) = x^2 e^{-x^2}$, $f'(x) = 2xe^{-x^2}(1 - x^2)$
Critical points: $x = 0, \pm 1$.
$f(x) \to 0$ as $x \to \pm\infty$.

		CP		CP		CP	
f'	$+$	-1	$-$	0	$+$	1	$-$
f	↗	abs max	↘	abs min	↗	abs max	↘

Hence, f has abs max $1/e$ at $x = \pm 1$ and abs min 0 at $x = 0$.

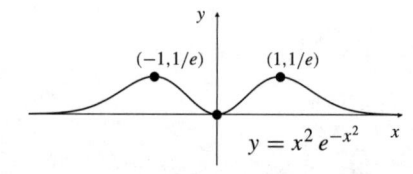

[Fig. 4.2.36: graph of $y = x^2 e^{-x^2}$ with peaks at $(-1, 1/e)$ and $(1, 1/e)$]

Fig. 4.2.36

37. $f(x) = \dfrac{\ln x}{x}$ $(x > 0)$

$f'(x) = \dfrac{\dfrac{x}{x} - \ln x}{x^2} = \dfrac{1 - \ln x}{x^2}$

$f(x) \to -\infty$ as $x \to 0+$ (vertical asymptote),
$f(x) \to 0$ as $x \to \infty$ (horizontal asymptote).

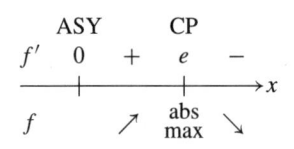

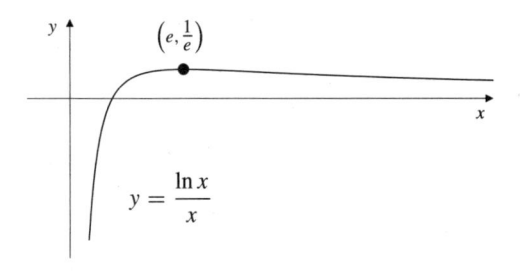

```
         ASY    CP
  f'      0  +  e   −
  ─────┼──────┼──────→ x
  f        ↗   abs   ↘
              max
```

$\left(e, \dfrac{1}{e}\right)$

$y = \dfrac{\ln x}{x}$

Fig. 4.2.37

38. Since $f(x) = |x + 1|$,

$$f'(x) = \operatorname{sgn}(x+1) = \begin{cases} 1, & \text{if } x > -1; \\ -1, & \text{if } x < -1. \end{cases}$$

-1 is a singular point; f has no max but has abs min 0 at $x = -1$.
$f(x) \to \infty$ as $x \to \pm\infty$.

$y = |x + 1|$

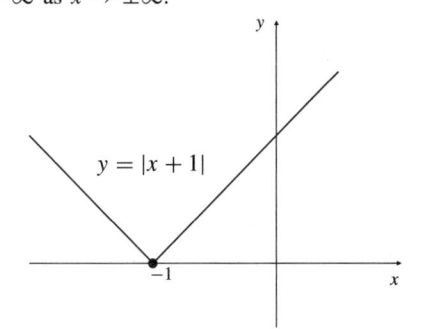

Fig. 4.2.38

39. $f(x) = |x^2 - 1|$

$f'(x) = 2x\operatorname{sgn}(x^2 - 1)$

CP: $x = 0$
SP: $x = \pm 1$

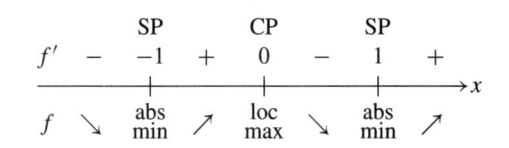

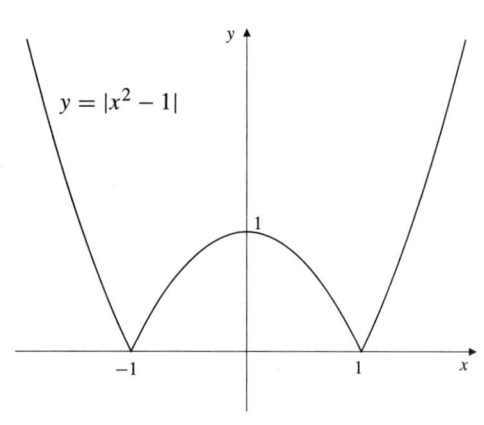

```
         SP     CP     SP
  f'  −  −1  +  0   −  1   +
  ─────┼──────┼──────┼──────→ x
  f  ↘ abs ↗ loc ↘ abs ↗
       min    max    min
```

$y = |x^2 - 1|$

Fig. 4.2.39

40. $f(x) = \sin|x|$

$f'(x) = \operatorname{sgn}(x)\cos|x| = 0$ at $x = \pm\dfrac{\pi}{2}, \pm\dfrac{3\pi}{2}, \pm\dfrac{5\pi}{2}, \ldots$

0 is a singular point. Since $f(x)$ is an even function, its graph is symmetric about the origin.

```
        CP      CP    SP    CP     CP
f'  −  -3π/2 + -π/2 − 0  +  π/2  − 3π/2  +
  ────┼──────┼──────┼──────┼──────┼────→ x
f  ↘ abs ↗ abs ↘ loc ↗ abs ↘ abs ↗
     min    max    min    max    min
```

Hence, f has abs max 1 at $x = \pm(4k+1)\dfrac{\pi}{2}$ and abs min -1 at $x = \pm(4k+3)\dfrac{\pi}{2}$ where $k = 0, 1, 2, \ldots$ and loc min 0 at $x = 0$.

$y = \sin|x|$

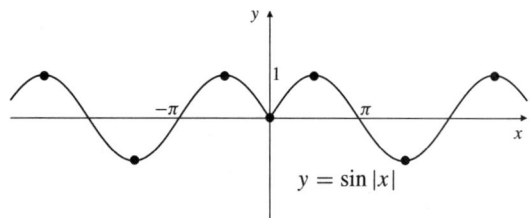

Fig. 4.2.40

41. $f(x) = |\sin x|$

CP: $x = \pm\dfrac{(2n+1)\pi}{2}$, SP $= \pm n\pi$

f has abs max 1 at all CP.
f has abs min 0 at all SP.

$y = |\sin x|$

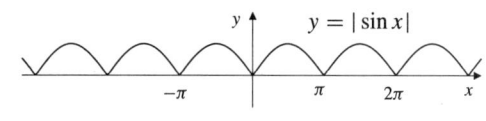

Fig. 4.2.41

42. $f(x) = (x-1)^{2/3} - (x+1)^{2/3}$
$f'(x) = \frac{2}{3}(x-1)^{-1/3} - \frac{2}{3}(x+1)^{-1/3}$
Singular point at $x = \pm 1$. For critical points:
$(x-1)^{-1/3} = (x+1)^{-1/3} \Rightarrow x-1 = x+1 \Rightarrow 2 = 0$, so there are no critical points.

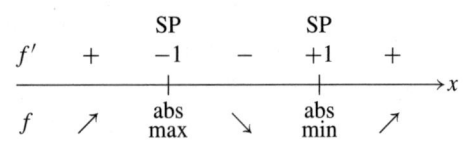

```
              SP        SP
  f'    +    -1    -    +1    +
  ─────────────┼─────────┼──────────→ x
  f    ↗    abs    ↘    abs    ↗
            max          min
```

Hence, f has abs max $2^{2/3}$ at $x = -1$ and abs min $-2^{2/3}$ at $x = 1$.

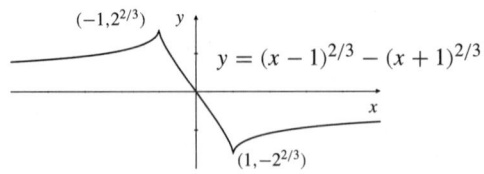

Fig. 4.2.42

43. $f(x) = (x-1)^{1/3} + (x+1)^{1/3}$
$f'(x) = \frac{1}{3}(x-1)^{-2/3} + \frac{1}{3}(x+1)^{-2/3}$
SP: $x = \pm 1$ $f(1) = 2^{1/3}$, $f(-1) = -2^{1/3}$
For CP: $(x-1)^{-2/3} = -(x+1)^{-2/3}$
$(x-1)^{-2} = -(x+1)^{-2}$ not possible (no CPs).

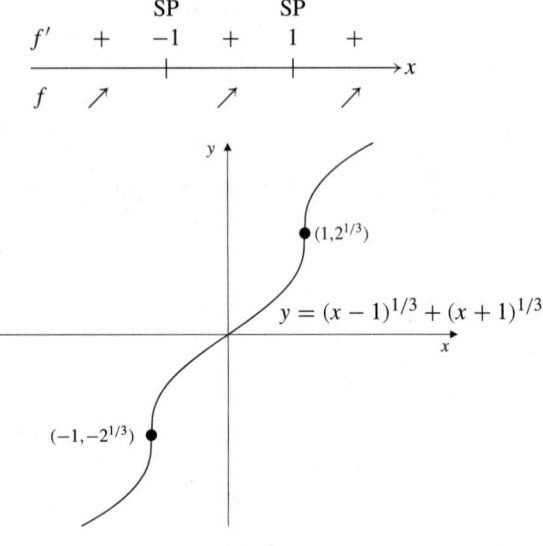

```
              SP        SP
  f'    +    -1    +    1    +
  ─────────────┼─────────┼──────────→ x
  f    ↗          ↗          ↗
```

Fig. 4.2.43

44. $f(x) = x - x^{1/3}$, $f'(x) = 1 - \frac{1}{3}x^{-2/3}$
Critical points: $x = \pm(\frac{1}{3})^{3/2}$
Singular point: $x = 0$.

```
           CP           SP           CP
  f'   +  -(1/3)^{3/2}  -    0    -  (1/3)^{3/2}  +
  ────────┼──────────────┼──────────────┼──────────→ x
  f   ↗     loc     ↘             ↘     loc    ↗
            max                         min
```

Hence, f has loc max $\frac{2}{(3)^{3/2}}$ at $x = -(\frac{1}{3})^{3/2}$ and loc min $-\frac{2}{(3)^{3/2}}$ at $x = (\frac{1}{3})^{3/2}$.

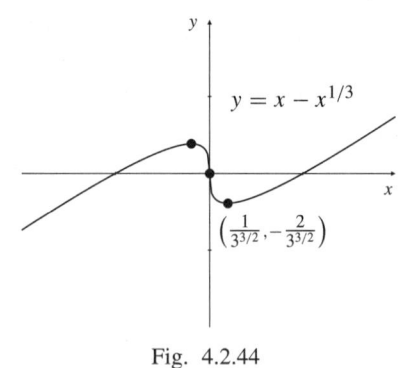

Fig. 4.2.44

45. $f(x) = x/\sqrt{x^2+1}$. Since
$$f'(x) = \frac{\sqrt{x^2+1} - x\frac{2x}{2\sqrt{x^2+1}}}{x^2+1} = \frac{1}{(x^2+1)^{3/2}} > 0,$$
for all x, f cannot have any maximum or minimum value.

46. $f(x) = x/\sqrt{x^4+1}$. f is continuous on $\mathbb{R}$, and $\lim_{x \to \pm\infty} f(x) = 0$. Since $f(1) > 0$ and $f(-1) < 0$, f must have both maximum and minimum values.
$$f'(x) = \frac{\sqrt{x^4+1} - x\frac{4x^3}{2\sqrt{x^4+1}}}{x^4+1} = \frac{1-x^4}{(x^4+1)^{3/2}}.$$

CP $x = \pm 1$. $f(\pm 1) = \pm 1/\sqrt{2}$. f has max value $1/\sqrt{2}$ and min value $-1/\sqrt{2}$.

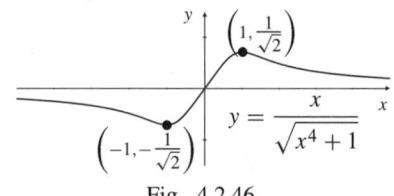

Fig. 4.2.46

47. $f(x) = x\sqrt{4-x^2}$ is continuous on $[-2, 2]$, and $f(\pm 2) = 0$.
$$f'(x) = \sqrt{4-x^2} + x\frac{-2x}{2\sqrt{4-x^2}} = \frac{2(2-x^2)}{\sqrt{4-x^2}}.$$

CP $x = \pm\sqrt{2}$. $f(\pm\sqrt{2}) = \pm 2$. f has maximum value 2 at $x = \sqrt{2}$ and min value -2 at $x = -\sqrt{2}$.

48. $f(x) = x^2/\sqrt{4-x^2}$ is continuous on $(-2, 2)$, and $\lim_{x\to -2+} f(x) = \lim_{x\to 2-} f(x) = \infty$. Thus f can have no maximum value, but will have a minimum value.

$$f'(x) = \frac{2x\sqrt{4-x^2} - x^2 \frac{-2x}{2\sqrt{4-x^2}}}{4-x^2} = \frac{8x - x^3}{(4-x^2)^{3/2}}.$$

CP $x = 0$, $x = \pm\sqrt{8}$. $f(0) = 0$, and $\pm\sqrt{8}$ is not in the domain of f. f has minimum value 0 at $x = 0$.

49. $f(x) = 1/[x \sin x]$ is continuous on $(0, \pi)$, and $\lim_{x\to 0+} f(x) = \infty = \lim_{x\to \pi-} f(x)$. Thus f can have no maximum value, but will have a minimum value. Since f is differentiable on $(0, \pi)$, the minimum value must occur at a CP in that interval.

50. $f(x) = (\sin x)/x$ is continuous and differentiable on $\mathbb{R}$ except at $x = 0$ where it is undefined.
Since $\lim_{x\to 0} f(x) = 1$, and $|f(x)| < 1$ for all $x \neq 0$ (because $|\sin x| < |x|$), f cannot have a maximum value. Since $\lim_{x\to \pm\infty} f(x) = 0$ and since $f(x) < 0$ at some points, f must have a minimum value occurring at a critical point. In fact, since $|f(x)| \leq 1/|x|$ for $x \neq 0$ and f is even, the minimum value will occur at the two critical points closest to $x = 0$. (See Figure 2.20 on page 124 of the text.)

51. An absolute max value is the max value of all the local max values. Hence, if a function has an absolute max value, it must have one or more local max values. On the other hand, if a function has a local max value, it may or may not have an absolute max value. Since a local max value, say $f(x_0)$ at the point x_0, is defined such that it is the max within some interval $|x - x_0| < h$ where $h > 0$, the function may have greater values, and may even approach ∞ outside this interval. There is no absolute max value in this latter case.

52. No. $f(x) = -x^2$ has abs max value 0, but $g(x) = |f(x)| = x^2$ has no abs max value.

53. $f(x) = \begin{cases} x \sin \dfrac{1}{x} & \text{if } x > 0 \\ 0 & \text{if } x < 0 \end{cases}$
$|f(x)| \leq |x|$ if $x > 0$ so $\lim_{x\to 0+} f(x) = 0 = f(0)$.
Therefore f is continuous at $x = 0$. Clearly $x \sin \dfrac{1}{x}$ is continuous at $x > 0$. Therefore f is continuous on $[0, \infty)$. Given any $h > 0$ there exists x_1 in $(0, h)$ and x_2 in $(0, h)$ such that $f(x_1) > 0 = f(0)$ and $f(x_2) < 0 = f(0)$. Therefore f cannot be a local max or min value at 0.
Specifically, let positive integer n satisfy $2n\pi > \dfrac{1}{h}$
and let $x_1 = \dfrac{1}{2n\pi + \dfrac{\pi}{2}}$, $x_2 = \dfrac{1}{2n\pi + \dfrac{3\pi}{2}}$.
Then $f(x_1) = x_1 > 0$ and $f(x_2) < 0$.

Section 4.3 Concavity and Inflections (page 245)

1. $f(x) = \sqrt{x}$, $f'(x) = \dfrac{1}{2\sqrt{x}}$, $f''(x) = -\dfrac{1}{4}x^{-3/2}$
$f''(x) < 0$ for all $x > 0$. f is concave down on $(0, \infty)$.

2. $f(x) = 2x - x^2$, $f'(x) = 2 - 2x$, $f''(x) = -2 < 0$.
Thus, f is concave down on $(-\infty, \infty)$.

3. $f(x) = x^2 + 2x + 3$, $f'(x) = 2x + 2$, $f''(x) = 2 > 0$.
f is concave up on $(-\infty, \infty)$.

4. $f(x) = x - x^3$, $f'(x) = 1 - 3x^2$,
$f''(x) = -6x$.

```
f''     +      0      -
              ---+---         →x
f       ⌣     infl    ⌢
```

5. $f(x) = 10x^3 - 3x^5$,
$f'(x) = 30x^2 - 15x^4$,
$f''(x) = 60(x - x^3) = 60x(1-x)(1+x)$.

```
f''   +   -1   -   0   +   1   -
          -+-      -+-     -+-       →x
f     ⌣  infl  ⌢  infl ⌣  infl  ⌢
```

6. $f(x) = 10x^3 + 3x^5$, $f'(x) = 30x^2 + 15x^4$,
$f''(x) = 60x + 60x^3 = 60x(1 + x^2)$.

```
f''     -      0      +
              ---+---         →x
f       ⌢     infl    ⌣
```

7. $f(x) = (3 - x^2)^2$,
$f'(x) = -4x(3 - x^2) = -12x + 4x^3$,
$f''(x) = -12 + 12x^2 = 12(x-1)(x+1)$.

```
f''   +   -1   -     1   +
          -+-        -+-           →x
f     ⌣  infl  ⌢    infl  ⌣
```

8. $f(x) = (2 + 2x - x^2)^2$, $f'(x) = 2(2 + 2x - x^2)(2 - 2x)$,
$f''(x) = 2(2 - 2x)^2 + 2(2 + 2x - x^2)(-2)$
$= 12x(x - 2)$.

```
f''   +   0   -    2   +
         -+-       -+-            →x
f     ⌣  infl  ⌢  infl  ⌣
```

SECTION 4.3 (PAGE 245)

9. $f(x) = (x^2 - 4)^3$,
$f'(x) = 6x(x^2 - 4)^2$,
$f''(x) = 6(x^2 - 4)^2 + 24x^2(x^2 - 4)$
$= 6(x^2 - 4)(5x^2 - 4)$.

```
f''   +   -2   -   -2/√5   +   2/√5   -   2   +
────────┼────────┼──────────┼────────┼────────→ x
f    ⌣     infl  ⌢   infl   ⌣  infl  ⌢  infl  ⌣
```

10. $f(x) = \dfrac{x}{x^2 + 3}$, $f'(x) = \dfrac{3 - x^2}{(x^2 + 3)^2}$,
$f''(x) = \dfrac{2x(x^2 - 9)}{(x^2 + 3)^3}$.

```
f''   -   -3   +   0   -   3   +
────────┼────────┼──────┼──────→ x
f    ⌢    infl  ⌣  infl  ⌢  infl  ⌣
```

11. $f(x) = \sin x$, $f'(x) = \cos x$, $f''(x) = -\sin x$.
f is concave down on intervals $(2n\pi, (2n+1)\pi)$ and concave up on intervals $((2n-1)\pi, 2n\pi)$, where n ranges over the integers. Points $x = n\pi$ are inflection points.

12. $f(x) = \cos 3x$, $f'(x) = -3\sin 3x$, $f''(x) = -9\cos 3x$.
Inflection points: $x = \left(n + \tfrac{1}{2}\right)\dfrac{\pi}{3}$ for $n = 0, \pm 1, \pm 2, \ldots$.
f is concave up on $\left(\dfrac{4n+1}{6}\pi, \dfrac{4n+3}{6}\pi\right)$ and concave down on $\left(\dfrac{4n+3}{6}\pi, \dfrac{4n+5}{6}\pi\right)$.

13. $f(x) = x + \sin 2x$,
$f'(x) = 1 + 2\cos 2x$,
$f''(x) = -4\sin 2x$.
f is concave up on intervals $\left(\dfrac{(2n-1)\pi}{2}, n\pi\right)$, and concave down on intervals $\left(n\pi, \dfrac{(2n+1)\pi}{2}\right)$. Points $\dfrac{n\pi}{2}$ are inflection points.

14. $f(x) = x - 2\sin x$, $f'(x) = 1 - 2\cos x$, $f''(x) = 2\sin x$.
Inflection points: $x = n\pi$ for $n = 0, \pm 1, \pm 2, \ldots$.
f is concave down on $((2n+1)\pi, (2n+2)\pi)$ and concave up on $((2n)\pi, (2n+1)\pi)$.

15. $f(x) = \tan^{-1} x$, $f'(x) = \dfrac{1}{1 + x^2}$,
$f''(x) = \dfrac{-2x}{(1 + x^2)^2}$.

```
f''   +   0   -
────────┼──────→ x
f    ⌣   infl  ⌢
```

16. $f(x) = xe^x$, $f'(x) = e^x(1 + x)$,
$f''(x) = e^x(2 + x)$.

```
f''   -   -2   +
────────┼──────→ x
f    ⌢   infl  ⌣
```

17. $f(x) = e^{-x^2}$, $f'(x) = -2xe^{-x^2}$,
$f''(x) = e^{-x^2}(4x^2 - 2)$.

```
f''   +   -1/√2   -   1/√2   +
────────┼────────┼──────────→ x
f    ⌣    infl   ⌢   infl   ⌣
```

18. $f(x) = \dfrac{\ln(x^2)}{x}$, $f'(x) = \dfrac{2 - \ln(x^2)}{x^2}$,
$f''(x) = \dfrac{-6 + 2\ln(x^2)}{x^3}$.
f has inflection point at $x = \pm e^{3/2}$ and f is undefined at $x = 0$. f is concave up on $(-e^{3/2}, 0)$ and $(e^{3/2}, \infty)$; and concave down on $(-\infty, -e^{3/2})$ and $(0, e^{3/2})$.

19. $f(x) = \ln(1 + x^2)$, $f'(x) = \dfrac{2x}{1 + x^2}$,
$f''(x) = \dfrac{(1 + x^2)(2) - 2x(2x)}{(1 + x^2)^2} = \dfrac{2(1 - x^2)}{(1 + x^2)^2}$.

```
f''   -   -1   +   1   -
────────┼──────┼──────→ x
f    ⌢   infl  ⌣  infl  ⌢
```

20. $f(x) = (\ln x)^2$, $f'(x) = \dfrac{2}{x}\ln x$,
$f''(x) = \dfrac{2(1 - \ln x)}{x^2}$ for all $x > 0$.

```
f''   0   +   e   -
────────┼──────→ x
f        ⌣   infl  ⌢
```

21. $f(x) = \dfrac{x^3}{3} - 4x^2 + 12x - \dfrac{25}{3}$,
$f'(x) = x^2 - 8x + 12$,
$f''(x) = 2x - 8 = 2(x - 4)$.

```
f''   -   4   +
────────┼──────→ x
f    ⌢   infl  ⌣
```

22. $f(x) = (x-1)^{1/3} + (x+1)^{1/3}$,
$f'(x) = \frac{1}{3}[(x-1)^{-2/3} + (x+1)^{-2/3}]$,
$f''(x) = -\frac{2}{9}[(x-1)^{-5/3} + (x+1)^{-5/3}]$.
$f(x) = 0 \Leftrightarrow x-1 = -(x+1) \Leftrightarrow x = 0$.
Thus, f has inflection point at $x = 0$. $f''(x)$ is undefined at $x = \pm 1$. f is defined at ± 1 and $x = \pm 1$ are also inflection points. f is concave up on $(-\infty, -1)$ and $(0, 1)$; and down on $(-1, 0)$ and $(1, \infty)$.

23. According to Definition 4.3.1 and the subsequent discussion, $f(x) = ax + b$ has no concavity and therefore no inflections.

24. $f(x) = 3x^3 - 36x - 3$, $f'(x) = 9(x^2 - 4)$, $f''(x) = 18x$.
The critical points are
$x = 2$, $f''(2) > 0 \Rightarrow$ local min;
$x = -2$, $f''(-2) < 0 \Rightarrow$ local max.

25. $f(x) = x(x-2)^2 + 1 = x^3 - 4x^2 + 4x + 1$
$f'(x) = 3x^2 - 8x + 4 = (x-2)(3x-2)$
CP: $x = 2$, $x = \frac{2}{3}$
$f''(x) = 6x - 8$, $f''(2) = 4 > 0$, $f''\left(\frac{2}{3}\right) = -4 < 0$.
Therefore, f has a loc min at $x = 2$ and a loc max at $x = \frac{2}{3}$.

26. $f(x) = x + \frac{4}{x}$, $f'(x) = 1 - \frac{4}{x^2}$, $f''(x) = 8x^{-3}$.
The critical points are
$x = 2$, $f''(2) > 0 \Rightarrow$ local min;
$x = -2$, $f''(-2) < 0 \Rightarrow$ local max.

27. $f(x) = x^3 + \frac{1}{x}$
$f'(x) = 3x^2 - \frac{1}{x^2} = \frac{3x^4 - 1}{x^2}$, CP: $x = \pm\frac{1}{\sqrt[4]{3}}$.
$f''(x) = 6x + \frac{2}{x^3}$.
$f''\left(\frac{1}{\sqrt[4]{3}}\right) > 0$, $f''\left(\frac{-1}{\sqrt[4]{3}}\right) < 0$.
Therefore f has a loc min at $\frac{1}{\sqrt[4]{3}}$ and a loc max at $\frac{-1}{\sqrt[4]{3}}$.

28. $f(x) = \frac{x}{2^x}$, $f'(x) = \frac{1 - x\ln 2}{2^x}$,
$f''(x) = \frac{\ln 2(x\ln 2 - 2)}{2^x}$.
The critical point is
$x = \frac{1}{\ln 2}$, $f''\left(\frac{1}{\ln 2}\right) < 0 \Rightarrow$ local max.

29. $f(x) = \frac{x}{1+x^2}$
$f'(x) = \frac{(1+x^2) - x2x}{(1+x^2)^2} = \frac{1-x^2}{(1+x^2)^2}$
CP: $x = \pm 1$
$f''(x) = \frac{(1+x^2)^2(-2x) - (1-x^2)2(1+x^2)2x}{(1+x^2)^4}$
$= \frac{-2x - 2x^3 - 4x + 4x^3}{(1+x^2)^3} = \frac{-6x + 2x^3}{(1+x^2)^3}$
$f''(1) = -\frac{1}{2}$, $f''(-1) = \frac{1}{2}$.
f has a loc max at 1 and a loc min at -1.

30. $f(x) = xe^x$, $f'(x) = e^x(1+x)$, $f''(x) = e^x(2+x)$.
The critical point is $x = -1$.
$f''(-1) > 0, \Rightarrow$ local min.

31. $f(x) = x \ln x$,
$f'(x) = 1 + \ln x$, CP: $x = \frac{1}{e}$
$f''(x) = \frac{1}{x}$, $f''\left(\frac{1}{e}\right) = e > 0$.
f has a loc min at $\frac{1}{e}$.

32. $f(x) = (x^2 - 4)^2$, $f'(x) = 4x^3 - 16x$, $f''(x) = 12x^2 - 16$.
The critical points are
$x = 0$, $f''(0) < 0 \Rightarrow$ local max;
$x = 2$, $f''(2) > 0 \Rightarrow$ local min;
$x = -2$, $f''(-2) > 0 \Rightarrow$ local min.

33. $f(x) = (x^2 - 4)^3$
$f'(x) = 6x(x^2 - 4)^2$
CP: $x = 0$, $x = \pm 2$
$f''(x) = 6(x^2 - 4)^2 + 24x^2(x^2 - 4)$
$= 6(x^2 - 4)(5x^2 - 4)$
$f''(0) > 0$, $f''(\pm 2) = 0$.
f has a loc min at $x = 0$. Second derivative test yields no direct information about ± 2. However, since f'' has opposite signs on opposite sides of the points 2 and -2, each of these points is an inflection point of f, and therefore f cannot have a local maximum or minimum value at either.

34. $f(x) = (x^2 - 3)e^x$,
$f'(x) = (x^2 + 2x - 3)e^x = (x+3)(x-1)e^x$,
$f''(x) = (x^2 + 4x - 1)e^x$.
The critical points are
$x = -3$, $f''(-3) < 0 \Rightarrow$ local max;
$x = 1$, $f''(1) > 0 \Rightarrow$ local min.

35. $f(x) = x^2 e^{-2x^2}$

$f'(x) = e^{-2x^2}(2x - 4x^3) = 2(x - 2x^3)e^{-2x^2}$

CP: $x = 0$, $x = \pm \dfrac{1}{\sqrt{2}}$

$f''(x) = e^{-2x^2}(2 - 20x^2 + 16x^4)$

$f''(0) > 0$, $f''\left(\pm \dfrac{1}{\sqrt{2}}\right) = -\dfrac{4}{e} < 0$.

Therefore, f has a loc (and abs) min value at 0, and loc (and abs) max values at $\pm \dfrac{1}{\sqrt{2}}$.

36. Since
$$f(x) = \begin{cases} x^2 & \text{if } x \geq 0 \\ -x^2 & \text{if } x < 0, \end{cases}$$
we have
$$f'(x) = \begin{cases} 2x & \text{if } x \geq 0 \\ -2x & \text{if } x < 0 \end{cases} = 2|x|$$
$$f''(x) = \begin{cases} 2 & \text{if } x > 0 \\ -2 & \text{if } x < 0 \end{cases} = 2\,\text{sgn}\,x.$$

$f'(x) = 0$ if $x = 0$. Thus, $x = 0$ is a critical point of f. It is also an inflection point since the conditions of Definition 4.3.3 are satisfied. $f''(0)$ does not exist. If a the graph of a function has a tangent line, vertical or not, at x_0, and has opposite concavity on opposite sides of x_0, the x_0 is an inflection point of f, whether or not $f''(x_0)$ even exists.

37. Suppose f is concave up (i.e., $f''(x) > 0$) on an open interval containing x_0.
Let $h(x) = f(x) - f(x_0) - f'(x_0)(x - x_0)$.
Since $h'(x) = f'(x) - f'(x_0) = 0$ at $x = x_0$, $x = x_0$ is a CP of h.
Now $h''(x) = f''(x)$. Since $h''(x_0) > 0$, therefore h has a min value at x_0, so $h(x) \geq h(x_0) = 0$ for x near x_0.
Since $h(x)$ measures the distance $y = f(x)$ lies above the tangent line $y = f(x_0) + f'(x_0)(x - x_0)$ at x, therefore $y = f(x)$ lies above that tangent line near x_0.
Note: we must have $h(x) > 0$ for x near x_0, $x \neq x_0$, for otherwise there would exist $x_1 \neq x_0$, x_1 near x_0, such that $h(x_1) = 0 = h(x_0)$. If $x_1 > x_0$, there would therefore exist x_2 such that $x_0 < x_2 < x_1$ and $f'(x_2) = f'(x_0)$. Therefore there would exist x_3 such that $x_0 < x_3 < x_2$ and $f'(x_3) = 0$, a contradiction.
The same contradiction can be obtained if $x_1 < x_0$.

38. Suppose that f has an inflection point at x_0. To be specific, suppose that $f''(x) < 0$ on (a, x_0) and $f''(x) > 0$ on (x_0, b) for some numbers a and b satisfying $a < x_0 < b$. If the graph of f has a non-vertical tangent line at x_0, then $f'(x_0)$ exists. Let
$$F(x) = f(x) - f(x_0) - f'(x_0)(x - x_0).$$

$F(x)$ represents the signed vertical distance between the graph of f and its tangent line at x_0. To show that the graph of f crosses its tangent line at x_0, it is sufficient to show that $F(x)$ has opposite signs on opposite sides of x_0. Observe that $F(x_0) = 0$, and $F'(x) = f'(x) - f'(x_0)$, so that $F'(x_0) = 0$ also. Since $F''(x) = f''(x)$, the assumptions above show that F' has a local minimum value at x_0 (by the First Derivative Test). Hence $F(x) > 0$ if $a < x < x_0$ or $x_0 < x < b$. It follows (by Theorem 6) that $F(x) < 0$ if $a < x < x_0$, and $F(x) > 0$ if $x_0 < x < b$. This completes the proof for the case of a nonvertical tangent.
If f has a vertical tangent at x_0, then its graph necessarily crosses the tangent (the line $x = x_0$) at x_0, since the graph of a function must cross any vertical line through a point of its domain that is not an endpoint.

39. $f(x) = x^n$
$g(x) = -x^n = -f(x)$, $n = 2, 3, 4, \ldots$
$f_n'(x) = nx^{n-1} = 0$ at $x = 0$
If n is even, f_n has a loc min, g_n has a loc max at $x = 0$.
If n is odd, f_n has an inflection at $x = 0$, and so does g_n.

40. Let there be a function f such that
$$f'(x_0) = f''(x_0) = \ldots = f^{(k-1)}(x_0) = 0,$$
$$f^{(k)}(x_0) \neq 0 \quad \text{for some } k \geq 2.$$

If k is even, then f has a local min value at $x = x_0$ when $f^{(k)}(x_0) > 0$, and f has a local max value at $x = x_0$ when $f^{(k)}(x_0) < 0$.
If k is odd, then f has an inflection point at $x = x_0$.

41. $f(x) = \begin{cases} e^{-1/x^2} & \text{if } x \neq 0 \\ 0 & \text{if } x = 0 \end{cases}$

a) $\lim\limits_{x \to 0+} x^{-n} f(x) = \lim\limits_{x \to 0+} \dfrac{e^{-1/x^2}}{x^n}$ (put $y = 1/x$)
$= \lim\limits_{y \to \infty} y^n e^{-y^2} = 0$ by Theorem 5 of Sec. 4.4
Similarly, $\lim_{x \to 0-} x^{-n} f(x) = 0$, and $\lim_{x \to 0} x^{-n} f(x) = 0$.

b) If $P(x) = \sum_{j=0}^{n} a_j x^j$ then by (a)
$$\lim_{x \to 0} P\left(\dfrac{1}{x}\right) f(x) = \sum_{j=0}^{n} a_j \lim_{x \to 0} x^{-j} f(x) = 0.$$

c) If $x \neq 0$ and $P_1(t) = 2t^3$, then
$$f'(x) = \dfrac{2}{x^3} e^{-1/x^2} = P_1\left(\dfrac{1}{x}\right) f(x).$$

Assume that $f^{(k)}(x) = P_k\left(\dfrac{1}{x}\right) f(x)$ for some $k \geq 1$, where P_k is a polynomial. Then

$$f^{(k+1)}(x) = -\dfrac{1}{x^2} P_k'\left(\dfrac{1}{x}\right) f(x) + P_k\left(\dfrac{1}{x}\right) P_1\left(\dfrac{1}{x}\right) f(x)$$
$$= P_{k+1}\left(\dfrac{1}{x}\right) f(x),$$

where $P_{k+1}(t) = t^2 P_k'(t) + P_1(t) P_k(t)$ is a polynomial. By induction, $f^{(n)} = P_n\left(\dfrac{1}{n}\right) f(x)$ for $n \neq 0$, where P_n is a polynomial.

d) $f'(0) = \lim_{h \to 0} \dfrac{f(h) - f(0)}{h} = \lim_{h \to 0} h^{-1} f(h) = 0$ by (a). Suppose that $f^{(k)}(0) = 0$ for some $k \geq 1$. Then

$$f^{(k+1)}(0) = \lim_{h \to 0} \dfrac{f^{(k)}(h) - f^{(k)}(0)}{h}$$
$$= \lim_{h \to 0} h^{-1} f^{(k)}(h)$$
$$= \lim_{h \to 0} h^{-1} P_k\left(\dfrac{1}{h}\right) f(h) = 0$$

by (b).
Thus $f^{(n)}(0) = 0$ for $n = 1, 2, \ldots$ by induction.

e) Since $f'(x) < 0$ if $x < 0$ and $f'(x) > 0$ if $x > 0$, therefore f has a local min value at 0 and $-f$ has a loc max value there.

f) If $g(x) = xf(x)$ then $g'(x) = f(x) + xf'(x)$, $g''(x) = 2f'(x) + xf''(x)$.
In general, $g^{(n)}(x) = nf^{(n-1)}(x) + xf^{(n)}(x)$ (by induction).
Then $g^{(n)}(0) = 0$ for all n (by (d)).
Since $g(x) < 0$ if $x < 0$ and $g(x) > 0$ if $x > 0$, g cannot have a max or min value at 0. It must have an inflection point there.

42. We are given that

$$f(x) = \begin{cases} x^2 \sin \dfrac{1}{x}, & \text{if } x \neq 0; \\ 0, & \text{if } x = 0. \end{cases}$$

If $x \neq 0$, then

$$f'(x) = 2x \sin \dfrac{1}{x} - \cos \dfrac{1}{x}$$
$$f''(x) = 2 \sin \dfrac{1}{x} - \dfrac{2}{x} \cos \dfrac{1}{x} - \dfrac{1}{x^2} \sin \dfrac{1}{x}.$$

If $x = 0$, then

$$f'(x) = \lim_{h \to 0} \dfrac{h^2 \sin \dfrac{1}{h} - 0}{h} = 0.$$

Thus 0 is a critical point of f. There are points x arbitrarily close to 0 where $f(x) > 0$, for example $x = \dfrac{2}{(4n+1)\pi}$, and other such points where $f(x) < 0$, for example $x = \dfrac{2}{(4n+3)\pi}$. Therefore f does not have a local max or min at $x = 0$. Also, there are points arbitrarily close to 0 where $f''(x) > 0$, for example $x = \dfrac{1}{(2n+1)\pi}$, and other such points where $f''(x) < 0$, for instance $x = \dfrac{1}{2n\pi}$. Therefore f does not have constant concavity on any interval $(0, a)$ where $a > 0$, so 0 is not an inflection point of f either.

Section 4.4 Sketching the Graph of a Function (page 255)

1. Function (d) appears to be the derivative of function (c), and function (b) appears to be the derivative of function (d). Thus graph (c) is the graph of f, (d) is the graph of f', (b) is the graph of f'', and (a) must be the graph of the other function g.

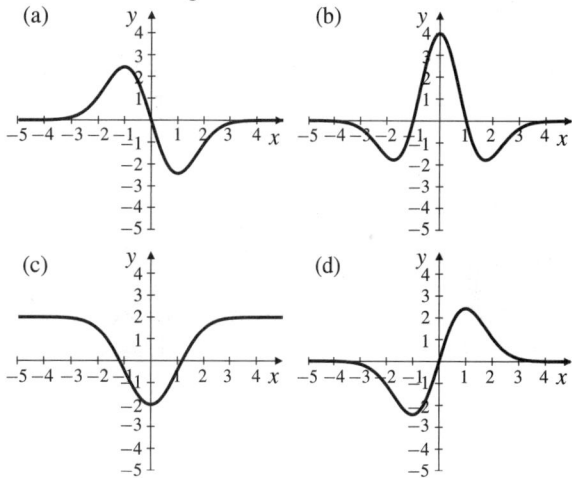

Fig. 4.4.1

SECTION 4.4 (PAGE 255)

2.

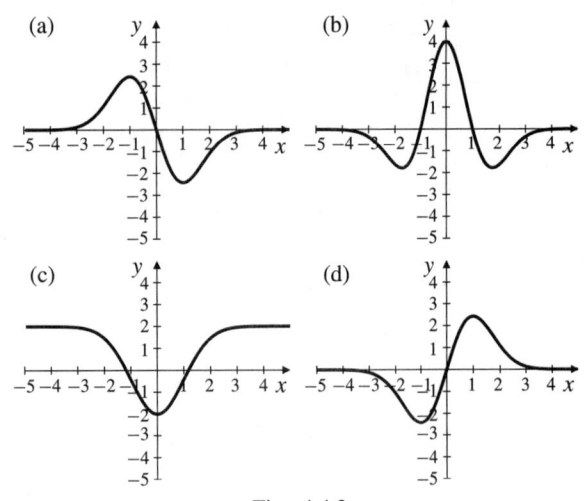

Fig. 4.4.2

The function graphed in Fig. 4.4.1(a):
is odd, is asymptotic to $y = 0$ at $\pm\infty$,
is increasing on $(-\infty, -1)$ and $(1, \infty)$,
is decreasing on $(-1, 1)$,
has CPs at $x = -1$ (max) and 1 (min),
is concave up on $(-\infty, -2)$ and $(0, 2)$ (approximately),
is concave down on $(-2, 0)$ and $(2, \infty)$ (approximately),
has inflections at $x = \pm 2$ (approximately).

The function graphed in Fig. 4.4.1(b):
is even, is asymptotic to $y = 0$ at $\pm\infty$,
is increasing on $(-1.7, 0)$ and $(1.7, \infty)$ (approximately),
is decreasing on $(-\infty, -1.7)$ and $(0, 1.7)$ (approximately),
has CPs at $x = 0$ (max) and ± 1.7 (min) (approximately),
is concave up on $(-2.5, -1)$ and $(1, 2.5)$ (approximately),
is concave down on $(-\infty, -2.5)$, $(-1, 1)$, and $(2.5, \infty)$ (approximately),
has inflections at ± 2.5 and ± 1 (approximately).

The function graphed in Fig. 4.4.1(c):
is even, is asymptotic to $y = 2$ at $\pm\infty$,
is increasing on $(0, \infty)$,
is decreasing on $(-\infty, 0)$,
has a CP at $x = 0$ (min),
is concave up on $(-1, 1)$ (approximately),
is concave down on $(-\infty, -1)$ and $(1, \infty)$ (approximately),
has inflections at $x = \pm 1$ (approximately).

The function graphed in Fig. 4.4.1(d):
is odd, is asymptotic to $y = 0$ at $\pm\infty$,
is increasing on $(-1, 1)$,
is decreasing on $(-\infty, -1)$ and $(1, \infty)$,
has CPs at $x = -1$ (min) and 1 (max),
is concave down on $(-\infty, -1.7)$ and $(0, 1.7)$ (approximately),
is concave up on $(-1.7, 0)$ and $(1.7, \infty)$ (approximately),
has inflections at 0 and ± 1.7 (approximately).

3. $f(x) = x/(1 - x^2)$ has slope 1 at the origin, so its graph must be (c).
$g(x) = x^3/(1 - x^4)$ has slope 0 at the origin, but has the same sign at all points as does $f(x)$, so its graph must be (b).
$h(x) = (x^3 - x)/\sqrt{1 + x^6}$ has no vertical asymptotes, so its graph must be (d).
$k(x) = x^3/\sqrt{|x^4 - 1|}$ is positive for all positive $x \neq 1$, so its graph must be (a).

4.

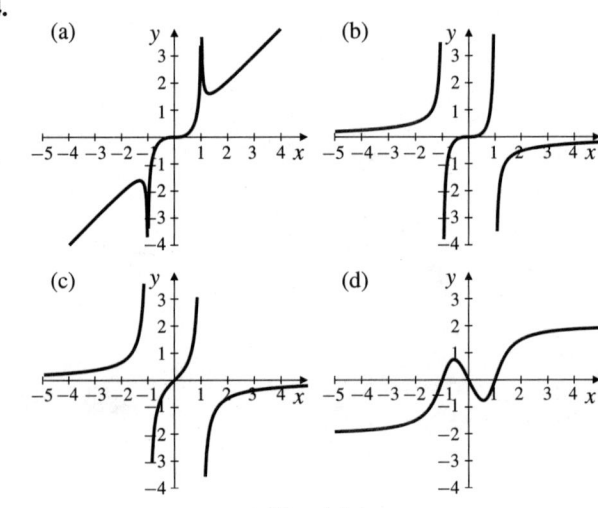

Fig. 4.4.4

The function graphed in Fig. 4.4.3(a):
is odd, is asymptotic to $x = \pm 1$ and $y = x$,
is increasing on $(-\infty, -1.5)$, $(-1, 1)$, and $(1.5, \infty)$ (approximately),
is decreasing on $(-1.5, -1)$ and $(1, 1.5)$ (approximately),
has CPs at $x = -1.5$, $x = 0$, and $x = 1.5$,
is concave up on $(0, 1)$ and $(1, \infty)$,
is concave down on $(-\infty, -1)$ and $(-1, 0)$,
has an inflection at $x = 0$.

The function graphed in Fig. 4.4.3(b):
is odd, is asymptotic to $x = \pm 1$ and $y = 0$,
is increasing on $(-\infty, -1)$, $(-1, 1)$, and $(1, \infty)$,
has a CP at $x = 0$,
is concave up on $(-\infty, -1)$ and $(0, 1)$,
is concave down on $(-1, 0)$ and $(1, \infty)$,
has an inflection at $x = 0$.

The function graphed in Fig. 4.4.3(c):
is odd, is asymptotic to $x = \pm 1$ and $y = 0$,
is increasing on $(-\infty, -1)$, $(-1, 1)$, and $(1, \infty)$,
has no CP,
is concave up on $(-\infty, -1)$ and $(0, 1)$,
is concave down on $(-1, 0)$ and $(1, \infty)$,
has an inflection at $x = 0$.

The function graphed in Fig. 4.4.3(d):
is odd, is asymptotic to $y = \pm 2$,
is increasing on $(-\infty, -0.7)$ and $(0.7, \infty)$ (approximately),
is decreasing on $(-0.7, 0.7)$ (approximately),
has CPs at $x = \pm 0.7$ (approximately),
is concave up on $(-\infty, -1)$ and $(0, 1)$ (approximately),
is concave down on $(-1, 0)$ and $(1, \infty)$ (approximately),
has an inflection at $x = 0$ and $x = \pm 1$ (approximately).

5. $f(0) = 1 \quad f(\pm 1) = 0 \quad f(2) = 1$
$\lim_{x \to \infty} f(x) = 2$, $\lim_{x \to -\infty} f(x) = -1$

```
           SP        CP
f'   +     0    -    1    +
     ─────┼────────┼─────→ x
           loc         loc
f    ↗    max   ↘    min   ↗
```

```
f''  +     0    +    2    -
     ─────┼────────┼─────→ x
f    ⌣          ⌣    infl  ⌢
```

0 must be a SP because $f'' > 0$ on both sides and it is a loc max. 1 must be a CP because f'' is defined there so f' must be too.

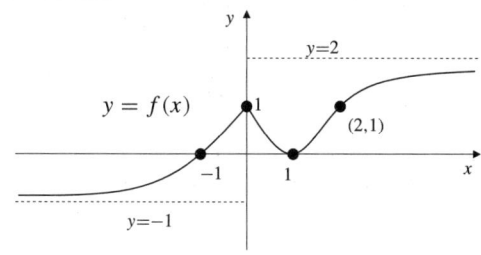

Fig. 4.4.5

6. According to the given properties:
Oblique asymptote: $y = x - 1$.
Critical points: $x = 0, 2$. Singular point: $x = -1$.
Local max 2 at $x = 0$; local min 0 at $x = 2$.

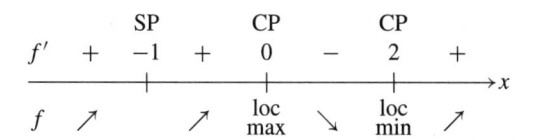

```
          SP        CP        CP
f'   +   -1    +    0    -    2    +
     ───┼────────┼────────┼─────→ x
                    loc        loc
f    ↗       ↗     max   ↘    min   ↗
```

Inflection points: $x = -1, 1, 3$.

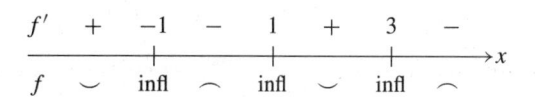

```
f'   +   -1   -    1    +    3    -
     ───┼────────┼────────┼─────→ x
f    ⌣    infl  ⌢   infl  ⌣   infl  ⌢
```

Since $\lim_{x \to \pm\infty} \big(f(x) + 1 - x\big) = 0$, the line $y = x - 1$ is an oblique asymptote.

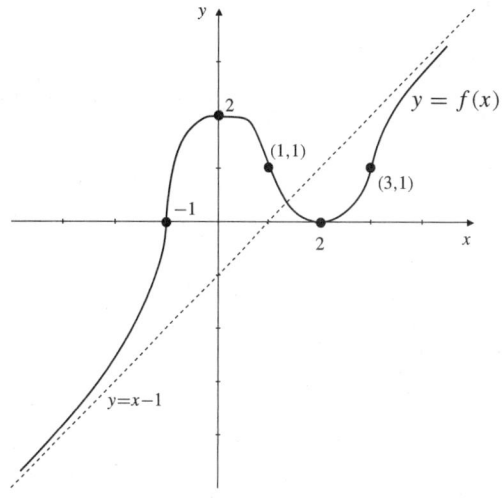

Fig. 4.4.6

7. $y = (x^2 - 1)^3$
$y' = 6x(x^2 - 1)^2$
$\quad = 6x(x - 1)^2(x + 1)^2$
$y'' = 6[(x^2 - 1)^2 + 4x^2(x^2 - 1)]$
$\quad = 6(x^2 - 1)(5x^2 - 1)$
$\quad = 6(x - 1)(x + 1)(\sqrt{5}x - 1)(\sqrt{5}x + 1)$

From y: Asymptotes: none. Symmetry: even. Intercepts: $x = \pm 1$.
From y': CP: $x = 0$, $x = \pm 1$. SP: none.

```
          CP       CP       CP
y'   -   -1   -    0    +   1    +
     ───┼────────┼────────┼─────→ x
                    abs
y    ↘        ↘    min   ↗       ↗
```

From y'': $y'' = 0$ at $x = \pm 1$, $x = \pm \dfrac{1}{\sqrt{5}}$.

```
y''  +  -1  -  -1/√5  +  1/√5  -  1  +
     ──┼──────┼────────┼───────┼──→ x
y    ⌣  infl ⌢  infl   ⌣  infl  ⌢ infl ⌣
```

SECTION 4.4 (PAGE 255)

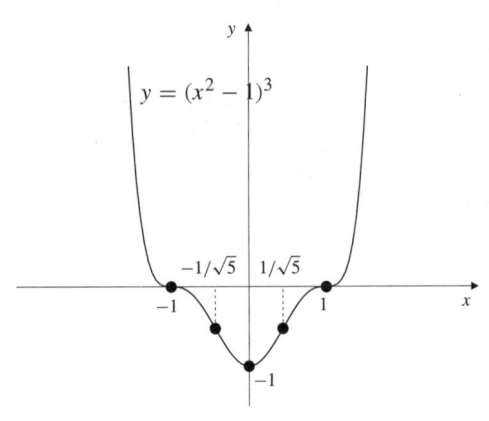

Fig. 4.4.7

8. $y = x(x^2-1)^2$, $y' = (x^2-1)(5x^2-1)$, $y'' = 4x(5x^2-3)$.
From y: Intercepts: $(0,0)$, $(1,0)$. Symmetry: odd (i.e., about the origin).

From y': Critical point: $x = \pm 1, \pm \dfrac{1}{\sqrt{5}}$.

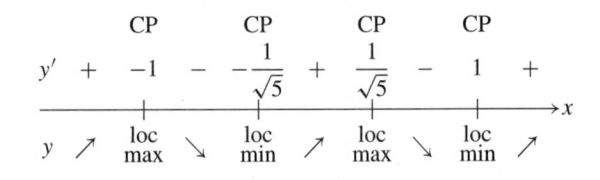

		CP		CP		CP		CP	
y'	$+$	-1	$-$	$-\dfrac{1}{\sqrt{5}}$	$+$	$\dfrac{1}{\sqrt{5}}$	$-$	1	$+$
y	↗	loc max	↘	loc min	↗	loc max	↘	loc min	↗

From y'': Inflection points at $x = 0$, $\pm\sqrt{\dfrac{3}{5}}$.

y''	$-$	$-\sqrt{\dfrac{3}{5}}$	$+$	0	$-$	$\sqrt{\dfrac{3}{5}}$	$+$
y	⌢	infl	⌣	infl	⌢	infl	⌣

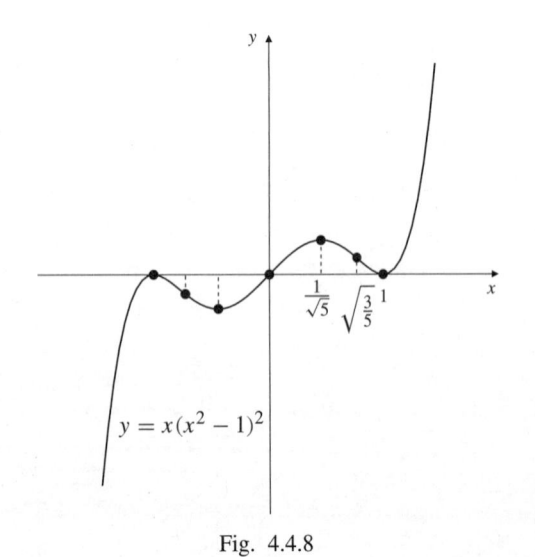

Fig. 4.4.8

9. $y = \dfrac{2-x}{x} = \dfrac{2}{x} - 1$, $y' = -\dfrac{2}{x^2}$, $y'' = \dfrac{4}{x^3}$.
From y: Asymptotes: $x = 0$, $y = -1$.
Symmetry: none obvious.
Intercept: $(2, 0)$. Points: $(-1, -3)$.
From y': CP: none. SP: none.

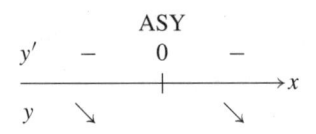

		ASY	
y'	$-$	0	$-$
y	↘		↘

From y'': $y'' = 0$ nowhere.

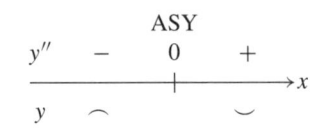

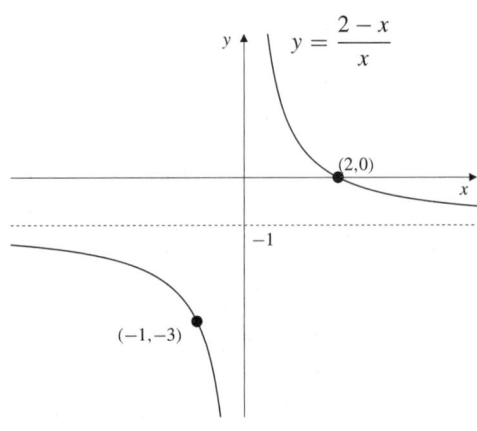

		ASY	
y''	$-$	0	$+$
y	⌢		⌣

Fig. 4.4.9

10. $y = \dfrac{x-1}{x+1} = 1 - \dfrac{2}{x+1}$, $y' = \dfrac{2}{(x+1)^2}$, $y'' = \dfrac{-4}{(x+1)^3}$.
From y: Intercepts: $(0, -1)$, $(1, 0)$. Asymptotes: $y = 1$ (horizontal), $x = -1$ (vertical). No obvious symmetry.
Other points: $(-2, 3)$.
From y': No critical point.

		ASY	
y'	$+$	-1	$+$
y	↗		↗

From y'': No inflection point.

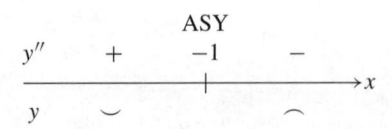

		ASY	
y''	$+$	-1	$-$
y	⌣		⌢

124

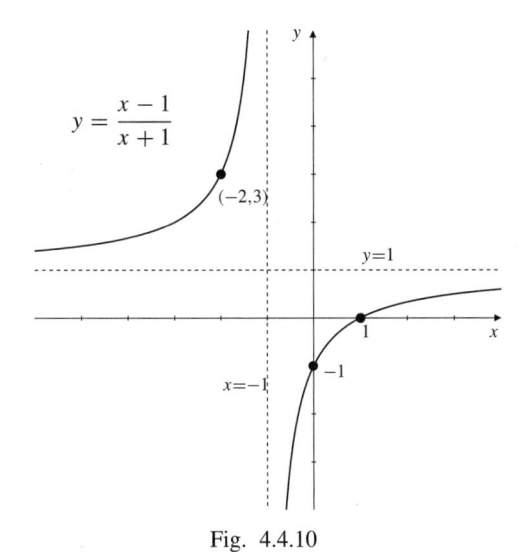

Fig. 4.4.10

11. $y = \dfrac{x^3}{1+x}$

$y' = \dfrac{(1+x)3x^2 - x^3}{(1+x)^2} = \dfrac{3x^2 + 2x^3}{(1+x)^2}$

$y'' = \dfrac{(1+x)^2(6x + 6x^2) - (3x^2 + 2x^3)2(1+x)}{(1+x)^4}$

$= \dfrac{6x(1+x)^2 - 6x^2 - 4x^3}{(1+x)^3} = \dfrac{6x + 6x^2 + 2x^3}{(1+x)^3}$

$= \dfrac{2x(3 + 3x + x^2)}{(1+x)^3}$

From y:
Asymptotes: $x = -1$. Symmetry: none.
Intercepts $(0, 0)$. Points $(-3/2, 27/4)$.
From y' CP: $x = 0$, $x = -\dfrac{3}{2}$.

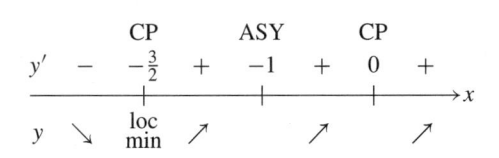

```
         CP        ASY       CP
y'  -   -3/2  +   -1    +    0    +
                                       → x
y   ↘   loc   ↗         ↗         ↗
         min
```

From y'': $y'' = 0$ only at $x = 0$.

```
              ASY
y''   +   -1   -   0   +
                           → x
y     ⌣        ⌢   infl ⌣
```

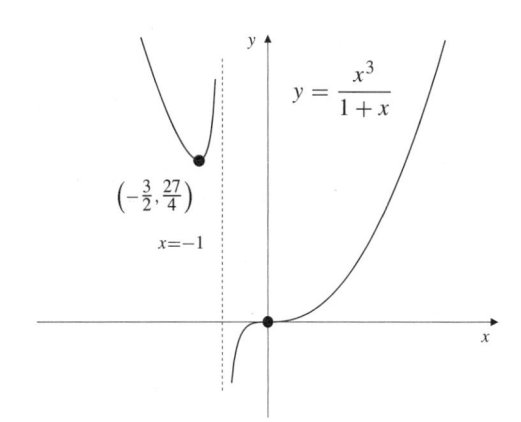

Fig. 4.4.11

12. $y = \dfrac{1}{4 + x^2}$, $y' = \dfrac{-2x}{(4 + x^2)^2}$, $y'' = \dfrac{6x^2 - 8}{(4 + x^2)^3}$.

From y: Intercept: $(0, \tfrac{1}{4})$. Asymptotes: $y = 0$ (horizontal). Symmetry: even (about y-axis).
From y': Critical point: $x = 0$.

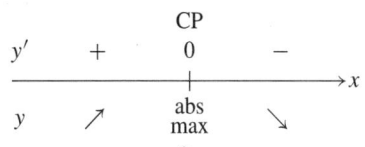

```
              CP
y'    +       0       -
                           → x
y     ↗      abs      ↘
             max
```

From y'': $y'' = 0$ at $x = \pm\dfrac{2}{\sqrt{3}}$.

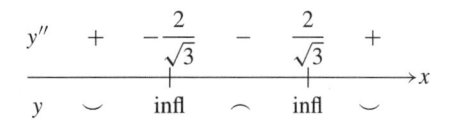

```
             -2         2
y''   +      ──   -    ──    +
             √3        √3
                                 → x
y     ⌣      infl ⌢   infl    ⌣
```

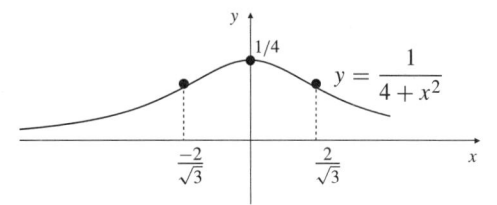

Fig. 4.4.12

13. $y = \dfrac{1}{2 - x^2}$, $y' = \dfrac{2x}{(2 - x^2)^2}$

$y'' = \dfrac{2}{(2 - x^2)^2} + \dfrac{8x^2}{(2 - x^2)^3} = \dfrac{4 + 6x^2}{(2 - x^2)^3}$

From y: Asymptotes: $y = 0$, $x = \pm\sqrt{2}$.
Symmetry: even.
Intercepts $(0, \tfrac{1}{2})$. Points $(\pm 2, -\tfrac{1}{2})$.
From y': CP $x = 0$.

```
           ASY       CP      ASY
y''   -   -√2   -    0   +   √2    +
                                       → x
y     ↘         ↘   loc  ↗         ↗
                    min
```

y'': $y'' = 0$ nowhere.

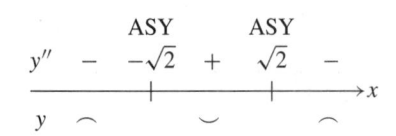

```
         ASY        ASY
y''   −  −√2   +    √2   −
─────────┼──────────┼────────→ x
y    ⌢        ⌣          ⌢
```

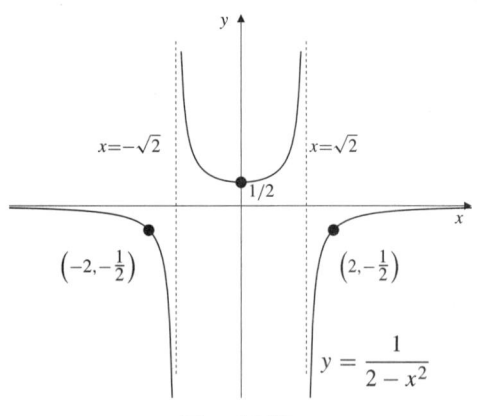

Fig. 4.4.13

14. $y = \dfrac{x}{x^2 - 1}$, $y' = -\dfrac{x^2 + 1}{(x^2 - 1)^2}$, $y'' = \dfrac{2x(x^2 + 3)}{(x^2 - 1)^3}$.
    From y: Intercept: $(0, 0)$. Asymptotes: $y = 0$ (horizontal), $x = \pm 1$ (vertical). Symmetry: odd. Other points: $(2, \tfrac{2}{3})$, $(-2, -\tfrac{2}{3})$.
    From y': No critical or singular points.

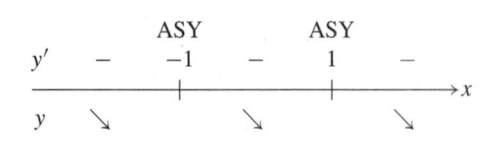

```
         ASY        ASY
y'   −   −1    −    1    −
─────────┼──────────┼────────→ x
y    ↘        ↘          ↘
```

From y'': $y'' = 0$ at $x = 0$.

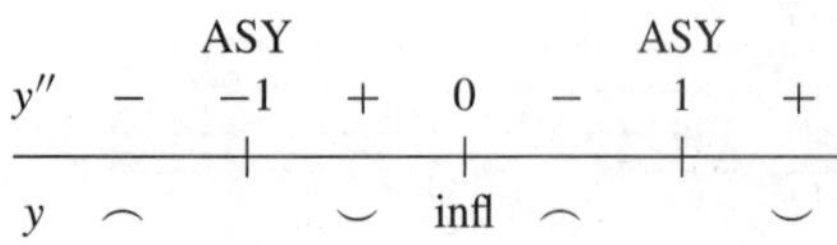

```
         ASY              ASY
y''  −   −1   +   0   −   1   +
─────────┼────────┼────────┼──────→ x
y    ⌢        ⌣   infl ⌢      ⌣
```

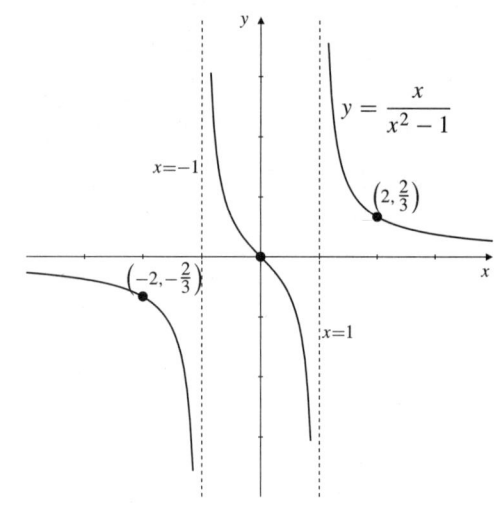

Fig. 4.4.14

15. $y = \dfrac{x^2}{x^2 - 1} = 1 + \dfrac{1}{x^2 - 1}$
    $y' = \dfrac{-2x}{(x^2 - 1)^2}$
    $y'' = -2\dfrac{(x^2 - 1)^2 - x \cdot 2(x^2 - 1)2x}{(x^2 - 1)^4} = \dfrac{2(3x^2 + 1)}{(x^2 - 1)^3}$
    From y: Asymptotes: $y = 1$, $x = \pm 1$. Symmetry: even.
    Intercepts $(0, 0)$. Points $\left(\pm 2, \tfrac{4}{3}\right)$.
    From y': CP $x = 0$.

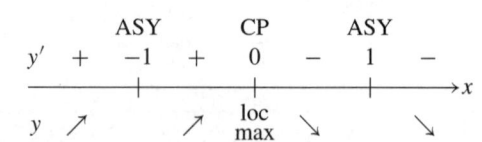

```
         ASY        CP       ASY
y'   +   −1    +    0   −    1    −
─────────┼──────────┼─────────┼──────→ x
y    ↗        ↗   loc  ↘          ↘
                  max
```

From y'': $y'' = 0$ nowhere.

```
         ASY        ASY
y''  +   −1    −    1    +
─────────┼──────────┼────────→ x
y    ⌣        ⌢          ⌣
```

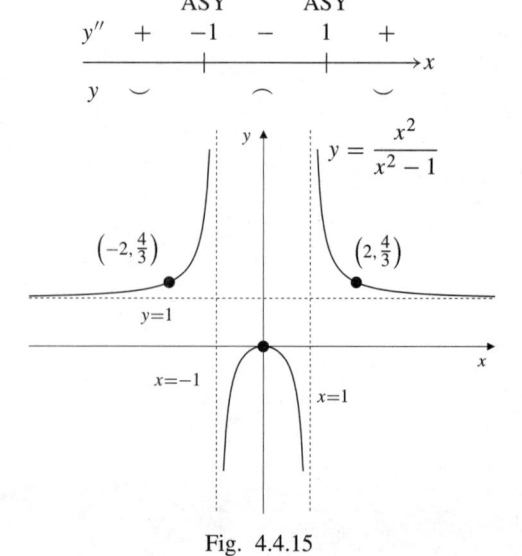

Fig. 4.4.15

16. $y = \dfrac{x^3}{x^2-1}$, $y' = \dfrac{x^2(x^2-3)}{(x^2-1)^2}$, $y'' = \dfrac{2x(x^2+3)}{(x^2-1)^3}$.
From y: Intercept: $(0,0)$. Asymptotes: $x = \pm 1$ (vertical), $y = x$ (oblique). Symmetry: odd. Other points: $\left(\pm\sqrt{3}, \pm\dfrac{3\sqrt{3}}{2}\right)$.
From y': Critical point: $x = 0, \pm\sqrt{3}$.

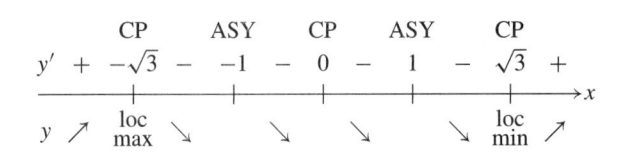

```
           CP      ASY    CP    ASY    CP
y'  +  -√3  -  -1  -  0  -  1  -  √3  +
                                          → x
y  ↗  loc  ↘     ↘     ↘     ↘  loc  ↗
       max                      min
```

From y'': $y'' = 0$ at $x = 0$.

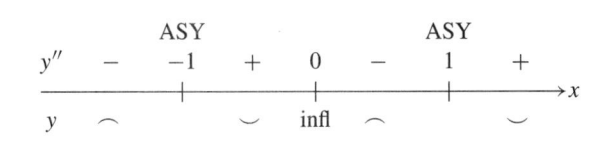

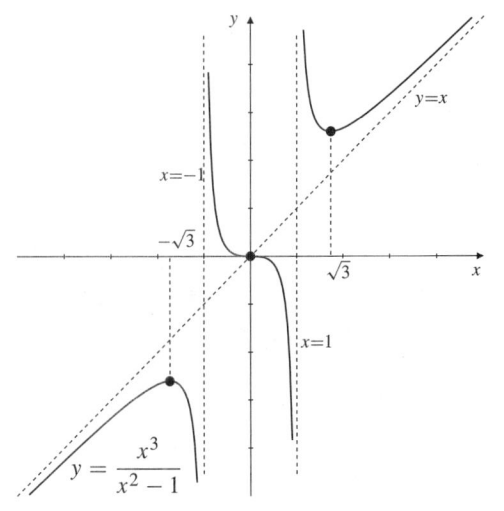

```
         ASY         ASY
y''  -  -1  +  0  -  1  +
                            → x
y   ⌢    ⌣  infl ⌢    ⌣
```

Fig. 4.4.16

17. $y = \dfrac{x^3}{x^2+1} = \dfrac{x^3+x-x}{x^2+1} = x - \dfrac{x}{x^2+1}$

$y' = \dfrac{(x^2+1)3x^2 - x^3 2x}{(x^2+1)^2} = \dfrac{x^4+3x^2}{(x^2+1)^2} = \dfrac{x^2(x^2+3)}{(x^2+1)^2}$

$y'' = \dfrac{(x^2+1)^2(4x^3+6x) - (x^4+3x^2)2(x^2+1)2x}{(x^2+1)^4}$

$= \dfrac{4x^5+10x^3+6x - 4x^5 - 12x^3}{(x^2+1)^3}$

$= \dfrac{2x(3-x^2)}{(x^2+1)^3}$

From y: Asymptotes: $y = x$ (oblique). Symmetry: odd. Intercepts $(0,0)$.
Points $(\pm\sqrt{3}, \pm\tfrac{3}{4}\sqrt{3})$.

From y': CP: $x = 0$.

```
              CP
y'    +       0       +
                        → x
y     ↗               ↗
```

From y'': $y'' = 0$ at $x = 0, x = \pm\sqrt{3}$.

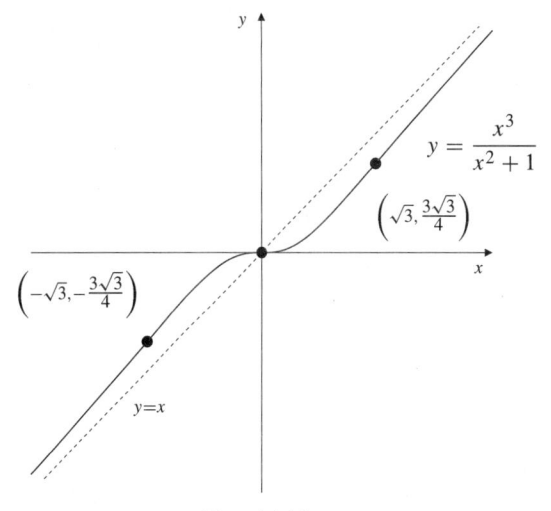

```
y''  +  -√3  -  0  +  √3  -
                              → x
y    ⌣  infl ⌢  infl ⌣  infl ⌢
```

Fig. 4.4.17

18. $y = \dfrac{x^2}{x^2+1}$, $y' = \dfrac{2x}{(x^2+1)^2}$, $y'' = \dfrac{2(1-3x^2)}{(x^2+1)^3}$.
From y: Intercept: $(0,0)$. Asymptotes: $y = 1$ (horizontal). Symmetry: even.
From y': Critical point: $x = 0$.

```
              CP
y'    -       0       +
                        → x
y     ↘      abs      ↗
             min
```

From y'': $y'' = 0$ at $x = \pm\dfrac{1}{\sqrt{3}}$.

```
y''   -   -1/√3   +   1/√3   -
                                → x
y     ⌢    infl   ⌣   infl   ⌢
```

127

SECTION 4.4 (PAGE 255)

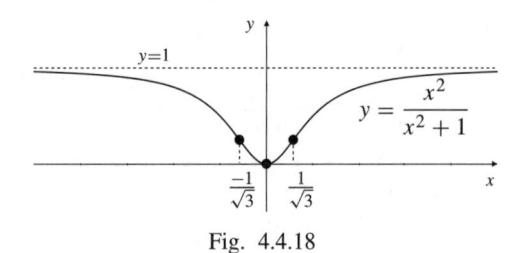

Fig. 4.4.18

19. $y = \dfrac{x^2 - 4}{x + 1} = x - 1 - \dfrac{3}{x+1}$

$y' = 1 + \dfrac{3}{(x+1)^2} = \dfrac{(x+1)^2 + 3}{(x+1)^2}$

$y'' = -\dfrac{6}{(x+1)^3}$

From y: Asymptotes: $y = x - 1$ (oblique), $x = -1$.
Symmetry: none.
Intercepts $(0, -4)$, $(\pm 2, 0)$.
From y': CP: none.

$$\begin{array}{c|ccc} & & \text{ASY} & \\ y' & + & -1 & + \\ \hline y & \nearrow & & \nearrow \end{array} \to x$$

From y'': $y'' = 0$ nowhere.

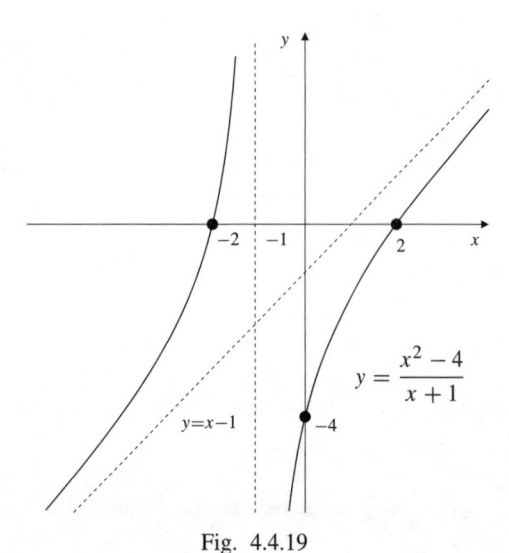

$$\begin{array}{c|ccc} & & \text{ASY} & \\ y'' & + & -1 & - \\ \hline y & \smile & & \frown \end{array} \to x$$

Fig. 4.4.19

20. $y = \dfrac{x^2 - 2}{x^2 - 1}$, $y' = \dfrac{2x}{(x^2 - 1)^2}$, $y'' = \dfrac{-2(3x^2 + 1)}{(x^2 - 1)^3}$.

From y: Intercept: $(0, 2)$, $(\pm\sqrt{2}, 0)$. Asymptotes: $y = 1$ (horizontal), $x = \pm 1$ (vertical). Symmetry: even.
From y': Critical point: $x = 0$.

$$\begin{array}{c|ccccc} & & \text{ASY} & & \text{CP} & & \text{ASY} & \\ f' & - & -1 & - & 0 & + & 1 & + \\ \hline f & \searrow & & \searrow & \text{loc min} & \nearrow & & \nearrow \end{array} \to x$$

From y'': $y'' = 0$ nowhere.

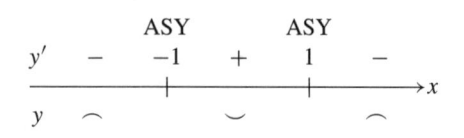

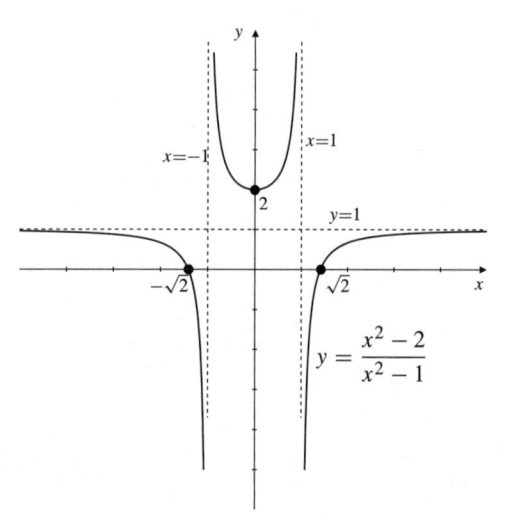

$$\begin{array}{c|ccccc} & & \text{ASY} & & \text{ASY} & \\ y' & - & -1 & + & 1 & - \\ \hline y & \frown & & \smile & & \frown \end{array} \to x$$

Fig. 4.4.20

21. $y = \dfrac{x^3 - 4x}{x^2 - 1} = \dfrac{x(x-2)(x+2)}{x^2 - 1}$

$y' = \dfrac{(x^2 - 1)(3x^2 - 4) - (x^3 - 4x)2x}{(x^2 - 1)^2}$

$= \dfrac{3x^4 - 7x^2 + 4 - 2x^4 + 8x^2}{(x^2 - 1)^2}$

$= \dfrac{x^4 + x^2 + 4}{(x^2 - 1)^2}$

$y'' = \dfrac{(x^2 - 1)^2(4x^3 + 2x) - (x^4 + x^2 + 4)2(x^2 - 1)2x}{(x^2 - 1)^4}$

$= \dfrac{4x^5 - 2x^3 - 2x - 4x^5 - 4x^3 - 16x}{(x^2 - 1)^3}$

$= \dfrac{-6x^3 - 18x}{(x^2 - 1)^3} = -6x\dfrac{x^2 + 3}{(x^2 - 1)^3}$

From y: Asymptotes: $y = x$ (oblique), $x = \pm 1$.
Symmetry: odd. Intercepts $(0, 0)$, $(\pm 2, 0)$.

INSTRUCTOR'S SOLUTIONS MANUAL   SECTION 4.4 (PAGE 255)

From y': CP: none.

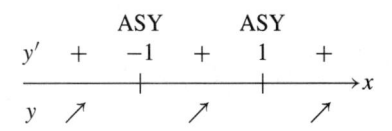

```
           ASY       ASY
y'  +  −1   +    1    +
   ────┼───────┼───────→ x
y   ↗      ↗       ↗
```

From y'': $y'' = 0$ at $x = 0$.

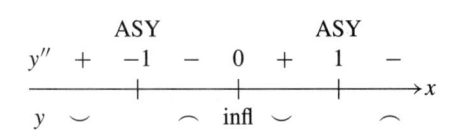

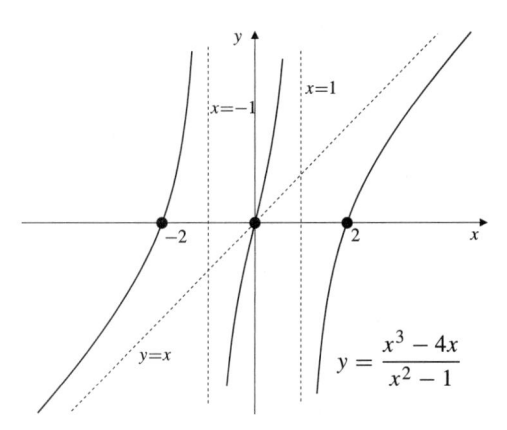

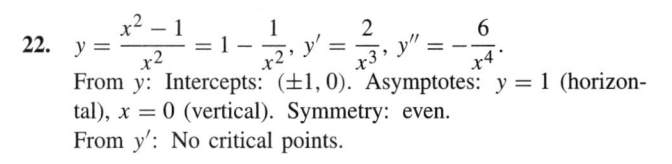

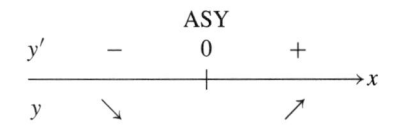

```
       ASY            ASY
y''  +  −1  −   0   +  1   −
   ────┼───────┼───────┼───→ x
y   ⌣       ⌢    infl ⌣    ⌢
```

Fig. 4.4.21 — $y = \dfrac{x^3 - 4x}{x^2 - 1}$, with asymptotes $x=-1$, $x=1$, $y=x$; intercepts at $-2, 0, 2$.

22. $y = \dfrac{x^2 - 1}{x^2} = 1 - \dfrac{1}{x^2}$, $y' = \dfrac{2}{x^3}$, $y'' = -\dfrac{6}{x^4}$.
From y: Intercepts: $(\pm 1, 0)$. Asymptotes: $y = 1$ (horizontal), $x = 0$ (vertical). Symmetry: even.
From y': No critical points.

```
        ASY
y'  −   0    +
   ────┼────→ x
y   ↘      ↗
```

From y'': y'' is negative for all x.

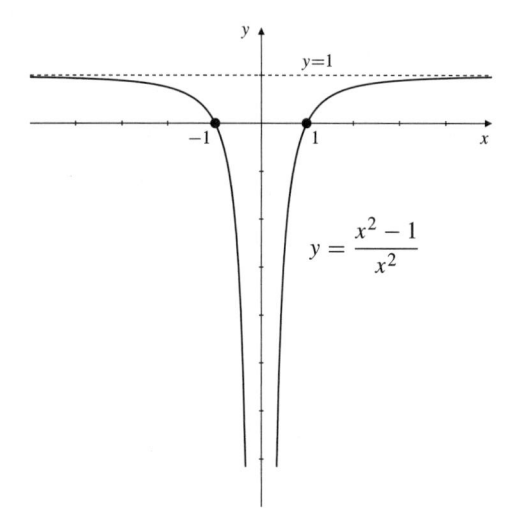

Fig. 4.4.22 — $y = \dfrac{x^2-1}{x^2}$, with horizontal asymptote $y=1$.

23. $y = \dfrac{x^5}{(x^2 - 1)^2} = x + \dfrac{2x^3 - x}{(x^2 - 1)^2}$

$y' = \dfrac{(x^2 - 1)^2 5x^4 - x^5 2(x^2 - 1)2x}{(x^2 - 1)^4}$

$= \dfrac{5x^6 - 5x^4 - 4x^6}{(x^2 - 1)^3} = \dfrac{x^4(x^2 - 5)}{(x^2 - 1)^3}$

$y'' = \dfrac{(x^2 - 1)^3(6x^5 - 20x^3) - (x^6 - 5x^4)3(x^2 - 1)^2 2x}{(x^2 - 1)^6}$

$= \dfrac{6x^7 - 26x^5 + 20x^3 - 6x^7 + 30x^5}{(x^2 - 1)^4}$

$= \dfrac{4x^3(x^2 + 5)}{(x^2 - 1)^4}$

From y: Asymptotes: $y = x$, $x = \pm 1$. Symmetry: odd. Intercepts $(0, 0)$. Points $\left(\pm\sqrt{5}, \pm\dfrac{25}{16}\sqrt{5}\right)$.
From y': CP $x = 0$, $x = \pm\sqrt{5}$.

```
      CP       ASY    CP    ASY      CP
y' +  −√5  −   −1  +  0  +  1   −   √5   +
  ────┼───────┼─────┼─────┼───────┼────→ x
y  ↗  loc  ↘     ↗    ↗    ↘    loc  ↗
      max                         min
```

From y'': $y'' = 0$ if $x = 0$.

```
         ASY            ASY
y''  −   −1   −    0   +    1   +
    ────┼────────┼─────────┼───→ x
y    ⌢       ⌢    infl ⌣       ⌣
```

129

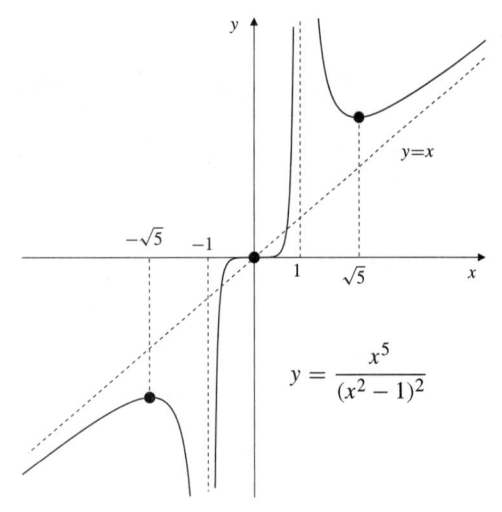

Fig. 4.4.23

24. $y = \dfrac{(2-x)^2}{x^3}$, $y' = -\dfrac{(x-2)(x-6)}{x^4}$,
$y'' = \dfrac{2(x^2 - 12x + 24)}{x^5} = \dfrac{2(x - 6 + 2\sqrt{3})(x - 6 - 2\sqrt{3})}{x^5}$.

From y: Intercept: $(2, 0)$. Asymptotes: $y = 0$ (horizontal), $x = 0$ (vertical). Symmetry: none obvious. Other points: $(-2, -2)$, $(-10, -0.144)$.

From y': Critical points: $x = 2, 6$.

```
         ASY         CP        CP
y'   −    0    −    2    +    6    −
                    |         |          →x
y    ↘         ↘   loc   ↗   loc   ↘
                    min       max
```

From y'': $y'' = 0$ at $x = 6 \pm 2\sqrt{3}$.

```
y''  −   0   +   6+2√3   −   6−2√3   +
                                              →x
y    ⌢       ⌣    infl    ⌢    infl    ⌣
```

Fig. 4.4.24

25. $y = \dfrac{1}{x^3 - 4x} = \dfrac{1}{x(x-2)(x+2)}$
$y' = -\dfrac{3x^2 - 4}{(x^3 - 4x)^2} = -\dfrac{3x^2 - 4}{x^2(x^2 - 4)^2}$
$y'' = -\dfrac{(x^3 - 4x)^2(6x) - (3x^2 - 4)2(x^3 - 4x)(3x^2 - 4)}{(x^3 - 4x)^4}$
$= -\dfrac{6x^4 - 24x^2 - 18x^4 + 48x^2 - 32}{(x^3 - 4x)^3}$
$= \dfrac{12(x^2 - 1)^2 + 20}{x^3(x^2 - 4)^3}$

From y: Asymptotes: $y = 0$, $x = 0, -2, 2$. Symmetry: odd. No intercepts.
Points: $\left(\pm \dfrac{2}{\sqrt{3}}, \pm \dfrac{16}{3\sqrt{3}}\right)$, $\left(\pm 3, \pm \dfrac{1}{15}\right)$

From y': CP: $x = \pm \dfrac{2}{\sqrt{3}}$.

```
      ASY       CP           CP      ASY
y'  −  −2  −  −2/√3  +   0   +  2/√3  −  2  −
                                                    →x
y   ↘      ↘   loc    ↗       ↗  loc   ↘      ↘
              min                   max
```

From y'': $y'' = 0$ nowhere.

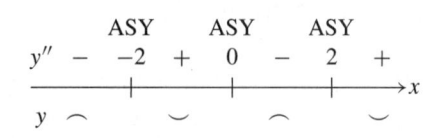

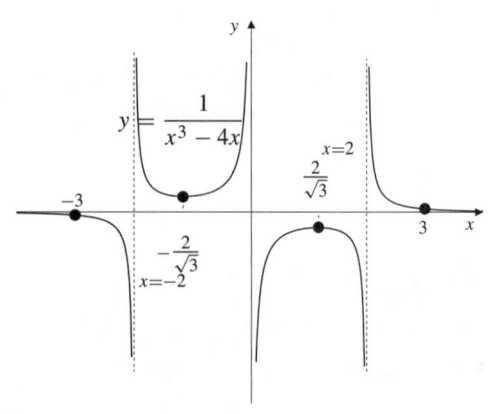

```
      ASY     ASY     ASY
y''  −   −2   +   0   −   2   +
                                  →x
y    ⌢        ⌣        ⌣        ⌣
```

Fig. 4.4.25

26. $y = \dfrac{x}{x^2 + x - 2} = \dfrac{x}{(2+x)(x-1)}$,
$y' = \dfrac{-(x^2 + 2)}{(x+2)^2(x-1)^2}$, $y'' = \dfrac{2(x^3 + 6x + 2)}{(x+2)^3(x-1)^3}$.

From y: Intercepts: $(0, 0)$. Asymptotes: $y = 0$ (horizontal), $x = 1$, $x = -2$ (vertical). Other points: $(-3, -\tfrac{3}{4})$, $(2, \tfrac{1}{2})$.

INSTRUCTOR'S SOLUTIONS MANUAL SECTION 4.4 (PAGE 255)

From y': No critical point.

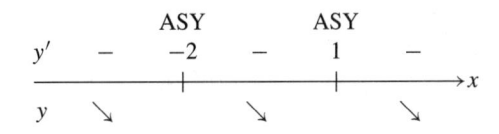

```
           ASY        ASY
  y'   -   -2    -    1     -
  ─────────┼──────────┼──────────→ x
  y    ↘          ↘         ↘
```

From y'': $y'' = 0$ if $f(x) = x^3 + 6x + 2 = 0$. Since $f'(x) = 3x^2 + 6 \geq 6$, f is increasing and can only have one root. Since $f(0) = 2$ and $f(-1) = -5$, that root must be between -1 and 0. Let the root be r.

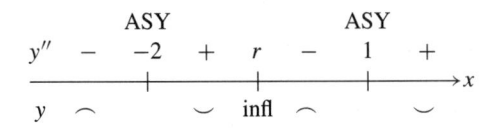

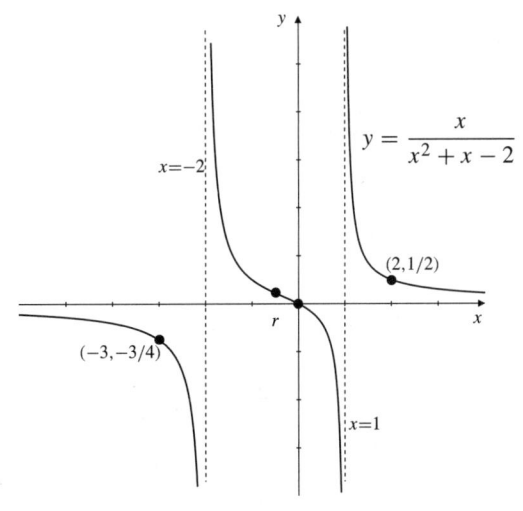

```
           ASY            ASY
  y''  -   -2    +    r    -    1    +
  ─────────┼──────────┼─────────┼──────→ x
  y    ⌢          ⌣    infl ⌢       ⌣
```

Fig. 4.4.26

27. $y = \dfrac{x^3 - 3x^2 + 1}{x^3} = 1 - \dfrac{3}{x} + \dfrac{1}{x^3}$

$y' = \dfrac{3}{x^2} - \dfrac{3}{x^4} = \dfrac{3(x^2 - 1)}{x^4}$

$y'' = -\dfrac{6}{x^3} + \dfrac{12}{x^5} = 6\dfrac{2 - x^2}{x^5}$

From y: Asymptotes: $y = 1$, $x = 0$. Symmetry: none. Intercepts: since $\lim_{x \to 0+} y = \infty$, and $\lim_{x \to 0-} y = -\infty$, there are intercepts between -1 and 0, between 0 and 1, and between 2 and 3.
Points: $(-1, 3)$, $(1, -1)$, $(2, -\frac{3}{8})$, $(3, \frac{1}{27})$.
From y': CP: $x = \pm 1$.

```
           CP        ASY        CP
  y'   +   -1    -    0    -    1    +
  ─────────┼──────────┼─────────┼──────→ x
  y    ↗   loc   ↘         ↘   loc   ↗
            max                min
```

From y'': $y'' = 0$ at $x = \pm\sqrt{2}$.

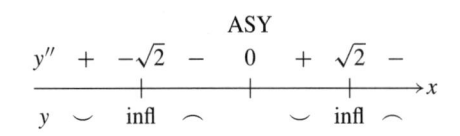

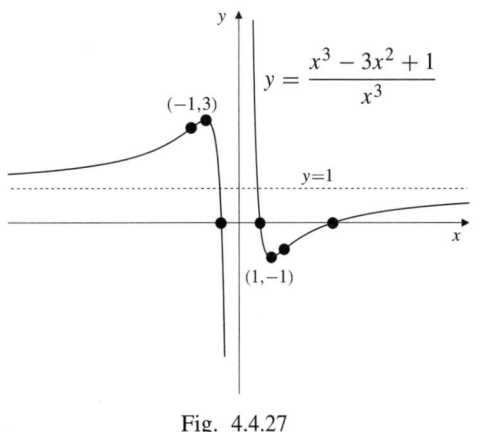

```
                     ASY
  y''  +   -√2   -    0    +    √2   -
  ─────────┼──────────┼─────────┼──────→ x
  y    ⌣   infl  ⌢         ⌣   infl  ⌢
```

Fig. 4.4.27

28. $y = x + \sin x$, $y' = 1 + \cos x$, $y'' = -\sin x$.
From y: Intercept: $(0, 0)$. Other points: $(k\pi, k\pi)$, where k is an integer. Symmetry: odd.
From y': Critical point: $x = (2k + 1)\pi$, where k is an integer.

```
           CP        CP        CP
  f'   +   -π    +    π    -    3π   +
  ─────────┼──────────┼─────────┼──────→ x
  f    ↗          ↗         ↗         ↗
```

From y'': $y'' = 0$ at $x = k\pi$, where k is an integer.

```
  y''  +  -2π  -  -π  +  0  -  π  +  2π  -
  ────────┼────────┼──────┼──────┼──────┼────→ x
  y    ⌣  infl ⌢ infl ⌣ infl ⌢ infl ⌣ infl ⌢
```

131

SECTION 4.4 (PAGE 255)　　　　　　　　　　　　　　R. A. ADAMS: CALCULUS

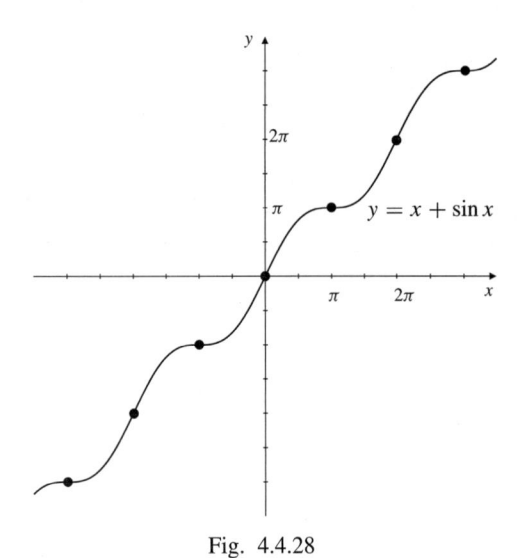

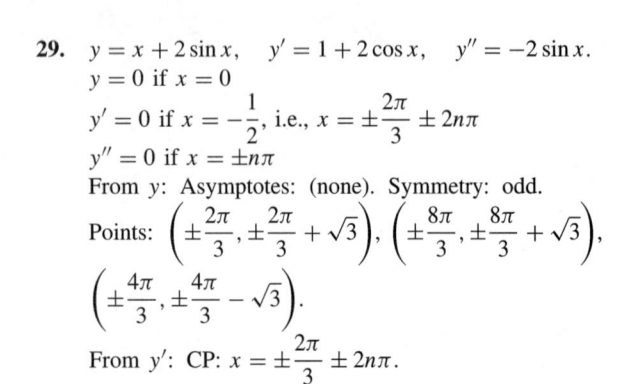

Fig. 4.4.28

29. $y = x + 2\sin x$, $\quad y' = 1 + 2\cos x$, $\quad y'' = -2\sin x$.
    $y = 0$ if $x = 0$
    $y' = 0$ if $x = -\dfrac{1}{2}$, i.e., $x = \pm\dfrac{2\pi}{3} \pm 2n\pi$
    $y'' = 0$ if $x = \pm n\pi$
    From y: Asymptotes: (none). Symmetry: odd.
    Points: $\left(\pm\dfrac{2\pi}{3}, \pm\dfrac{2\pi}{3} + \sqrt{3}\right)$, $\left(\pm\dfrac{8\pi}{3}, \pm\dfrac{8\pi}{3} + \sqrt{3}\right)$,
    $\left(\pm\dfrac{4\pi}{3}, \pm\dfrac{4\pi}{3} - \sqrt{3}\right)$.
    From y': CP: $x = \pm\dfrac{2\pi}{3} \pm 2n\pi$.

```
              CP        CP        CP        CP        CP
y'    -    -8π/3   +  -4π/3   +  -2π/3   +   2π/3   -   4π/3   +
                                                                   → x
y     ↘   loc    ↗   loc    ↘   loc    ↗   loc    ↘   loc    ↗
          min         max         min         max         min
```

From y'': $y'' = 0$ at $x = \pm n\pi$.

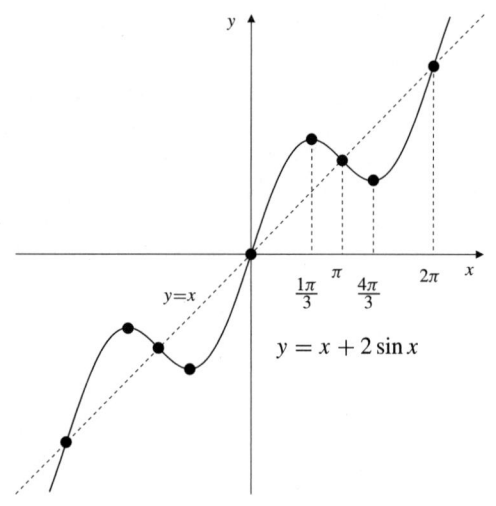

```
y''  +   -2π   -   -π   +   0   -   π   +   2π   -
                                                      → x
y    ⌣   infl  ⌢   infl  ⌣   infl  ⌢   infl  ⌣   infl  ⌢
```

Fig. 4.4.29

30. $y = e^{-x^2}$, $y' = -2xe^{-x^2}$, $y'' = (4x^2 - 2)e^{-x^2}$.
    From y: Intercept: $(0, 1)$. Asymptotes: $y = 0$ (horizontal). Symmetry: even.
    From y': Critical point: $x = 0$.

```
                    CP
y'      +           0           -
                                      → x
y       ↗         abs           ↘
                  max
```

From y'': $y'' = 0$ at $x = \pm\dfrac{1}{\sqrt{2}}$.

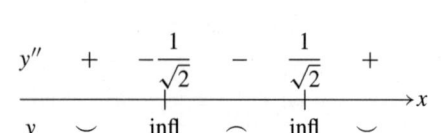

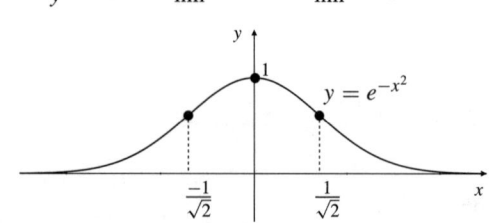

```
y''   +    -1/√2    -    1/√2    +
                                     → x
y     ⌣    infl    ⌢    infl    ⌣
```

Fig. 4.4.30

31. $y = xe^x$, $\quad y' = e^x(1 + x)$, $\quad y'' = e^x(2 + x)$.
    From y: Asymptotes: $y = 0$ (at $x = -\infty$).
    Symmetry: none. Intercept $(0, 0)$.
    Points: $\left(-1, -\dfrac{1}{e}\right)$, $\left(-2, -\dfrac{2}{e^2}\right)$.
    From y': CP: $x = -1$.

```
                 CP
y'       -       -1       +
                              → x
y        ↘      abs       ↗
                min
```

132

INSTRUCTOR'S SOLUTIONS MANUAL SECTION 4.4 (PAGE 255)

From y'': $y'' = 0$ at $x = -2$.

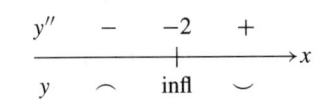

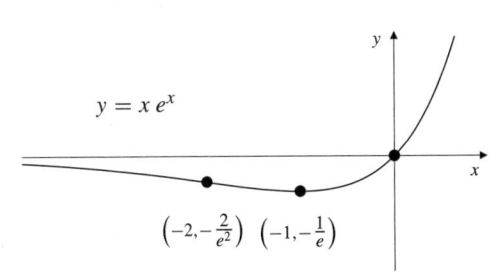

$$\begin{array}{c|ccc} y'' & - & -2 & + \\ \hline y & \frown & \text{infl} & \smile \end{array} \to x$$

Fig. 4.4.31

$y = xe^x$, with points $\left(-2, -\frac{2}{e^2}\right)$, $\left(-1, -\frac{1}{e}\right)$.

32. $y = e^{-x} \sin x$ $(x \geq 0)$,
$y' = e^{-x}(\cos x - \sin x)$, $y'' = -2e^{-x} \cos x$.
From y: Intercept: $(k\pi, 0)$, where k is an integer. Asymptotes: $y = 0$ as $x \to \infty$.
From y': Critical points: $x = \frac{\pi}{4} + k\pi$, where k is an integer.

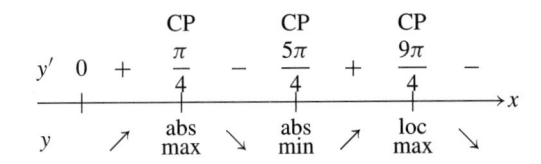

$$\begin{array}{c|ccccccc} & & \text{CP} & & \text{CP} & & \text{CP} & \\ y' & 0 & + & \frac{\pi}{4} & - & \frac{5\pi}{4} & + & \frac{9\pi}{4} & - \\ \hline y & & \nearrow & \text{abs max} & \searrow & \text{abs min} & \nearrow & \text{loc max} & \searrow \end{array} \to x$$

From y'': $y'' = 0$ at $x = (k + \frac{1}{2})\pi$, where k is an integer.

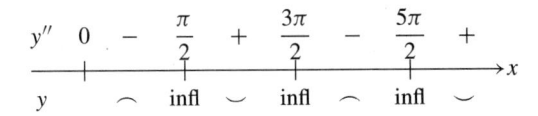

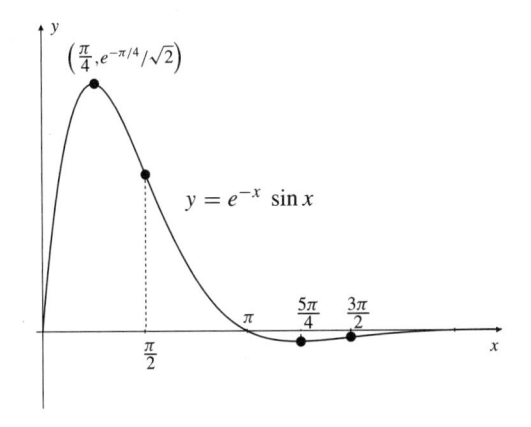

$$\begin{array}{c|ccccccc} y'' & 0 & - & \frac{\pi}{2} & + & \frac{3\pi}{2} & - & \frac{5\pi}{2} & + \\ \hline y & & \frown & \text{infl} & \smile & \text{infl} & \frown & \text{infl} & \smile \end{array} \to x$$

Graph: $y = e^{-x}\sin x$, with point $\left(\frac{\pi}{4}, e^{-\pi/4}/\sqrt{2}\right)$.

Fig. 4.4.32

33. $y = x^2 e^{-x^2}$
$y' = e^{-x^2}(2x - 2x^3) = 2x(1 - x^2)e^{-x^2}$
$y'' = e^{-x^2}(2 - 6x^2 - 2x(2x - 2x^3))$
$= (2 - 10x^2 + 4x^4)e^{-x^2}$
From y: Asymptotes: $y = 0$.
Intercept: $(0, 0)$. Symmetry: even.
Points $\left(\pm 1, \frac{1}{e}\right)$
From y': CP $x = 0$, $x = \pm 1$.

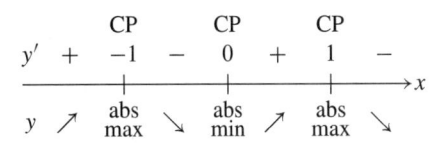

$$\begin{array}{c|ccccccc} & & \text{CP} & & \text{CP} & & \text{CP} & \\ y' & + & -1 & - & 0 & + & 1 & - \\ \hline y & \nearrow & \text{abs max} & \searrow & \text{abs min} & \nearrow & \text{abs max} & \searrow \end{array} \to x$$

From y'': $y'' = 0$ if
$2x^4 - 5x^2 + 1 = 0$
$x^2 = \frac{5 \pm \sqrt{25 - 8}}{4}$
$= \frac{5 \pm \sqrt{17}}{4}$.
so $x = \pm a = \pm\sqrt{\frac{5 + \sqrt{17}}{4}}$, $x = \pm b = \pm\sqrt{\frac{5 - \sqrt{17}}{4}}$.

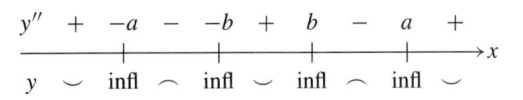

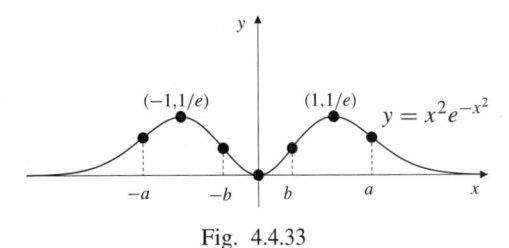

$$\begin{array}{c|ccccccccc} y'' & + & -a & - & -b & + & b & - & a & + \\ \hline y & \smile & \text{infl} & \frown & \text{infl} & \smile & \text{infl} & \frown & \text{infl} & \smile \end{array} \to x$$

Graph $y = x^2 e^{-x^2}$, with points $(-1, 1/e)$, $(1, 1/e)$, at $-a, -b, b, a$.

Fig. 4.4.33

34. $y = x^2 e^x$, $y' = (2x + x^2)e^x = x(2 + x)e^x$,
$y'' = (x^2 + 4x + 2)e^x = (x + 2 - \sqrt{2})(x + 2 + \sqrt{2})e^x$.
From y: Intercept: $(0, 0)$.
Asymptotes: $y = 0$ as $x \to -\infty$.
From y': Critical point: $x = 0$, $x = -2$.

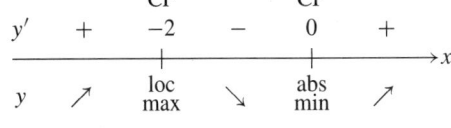

$$\begin{array}{c|ccccc} & & \text{CP} & & \text{CP} & \\ y' & + & -2 & - & 0 & + \\ \hline y & \nearrow & \text{loc max} & \searrow & \text{abs min} & \nearrow \end{array} \to x$$

From y'': $y'' = 0$ at $x = -2 \pm \sqrt{2}$.

$$\begin{array}{c|ccccc} y'' & + & -2-\sqrt{2} & - & -2+\sqrt{2} & + \\ \hline y & \smile & \text{infl} & \frown & \text{infl} & \smile \end{array} \to x$$

133

SECTION 4.4 (PAGE 255)

$y = x^2 e^x$

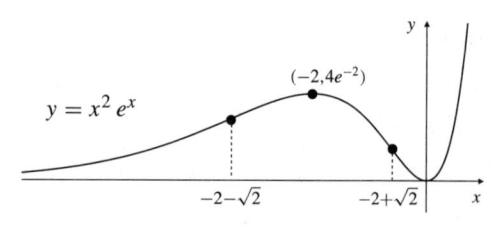

Fig. 4.4.34

35. $y = \dfrac{\ln x}{x}$, $y' = \dfrac{1 - \ln x}{x^2}$

$y'' = \dfrac{x^2\left(-\dfrac{1}{x}\right) - (1 - \ln x)2x}{x^4} = \dfrac{2\ln x - 3}{x^3}$

From y: Asymptotes: $x = 0$, $y = 0$.
Symmetry: none. Intercept: $(1, 0)$.
Points: $\left(e, \dfrac{1}{e}\right)$, $\left(e^{3/2}, \dfrac{3}{2e^{3/2}}\right)$.
From y': CP: $x = e$.

	ASY		CP	
y'	0	+	e	−
y		↗	abs max	↘

From y'': $y'' = 0$ at $x = e^{3/2}$.

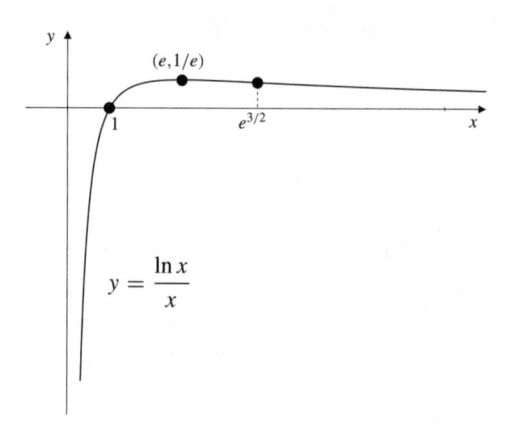

	ASY		$e^{3/2}$	
y''	0	−		+
y		⌢	infl	⌣

Fig. 4.4.35

36. $y = \dfrac{\ln x}{x^2}$ $(x > 0)$,
$y' = \dfrac{1 - 2\ln x}{x^3}$, $y'' = \dfrac{6\ln x - 5}{x^4}$.
From y: Intercepts: $(1, 0)$. Asymptotes: $y = 0$, since

$\displaystyle\lim_{x \to \infty} \dfrac{\ln x}{x^2} = 0$, and $x = 0$, since $\displaystyle\lim_{x \to 0+} \dfrac{\ln x}{x^2} = -\infty$.
From y': Critical point: $x = e^{1/2}$.

			CP	
y'	0	+	$\sqrt{e}$	−
y		↗	abs max	↘

From y'': $y'' = 0$ at $x = e^{5/6}$.

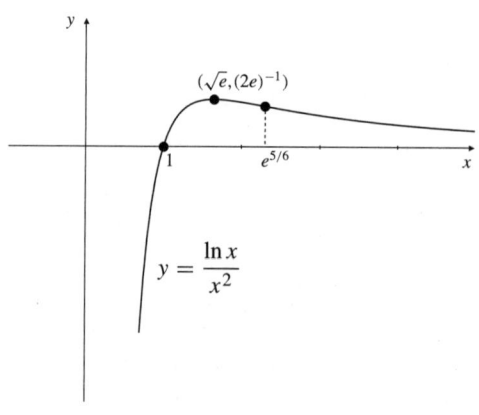

			$e^{5/6}$	
y''	0	−		+
y		⌢	infl	⌣

Fig. 4.4.36

37. $y = \dfrac{1}{\sqrt{4 - x^2}} = (4 - x^2)^{-1/2}$

$y' = -\dfrac{1}{2}(4 - x^2)^{-3/2}(-2x) = \dfrac{x}{(4 - x^2)^{3/2}}$

$y'' = \dfrac{(4-x^2)^{3/2} - x\dfrac{3}{2}(4-x^2)^{1/2}(-2x)}{(4-x^2)^3}$

$= \dfrac{4 + 2x^2}{(4 - x^2)^{5/2}}$

From y: Asymptotes: $x = \pm 2$. Domain $-2 < x < 2$.
Symmetry: even. Intercept: $(0, \tfrac{1}{2})$.
From y': CP: $x = 0$.

	ASY		CP		ASY
y'	−2	−	0	+	2
y		↘	abs min	↗	

From y'': $y'' = 0$ nowhere, $y'' > 0$ on $(-2, 2)$.
Therefore, y is concave up.

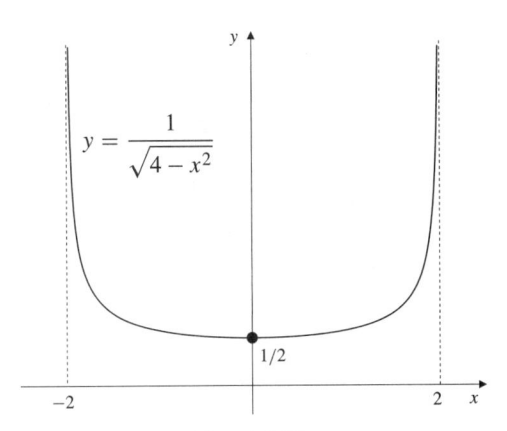

INSTRUCTOR'S SOLUTIONS MANUAL   SECTION 4.4 (PAGE 255)

Fig. 4.4.37

$y = \dfrac{1}{\sqrt{4-x^2}}$

38. $y = \dfrac{x}{\sqrt{x^2+1}}$, $y' = (x^2+1)^{-3/2}$, $y'' = -3x(x^2+1)^{-5/2}$.

From y: Intercept: $(0,0)$. Asymptotes: $y = 1$ as $x \to \infty$, and $y = -1$ as $x \to -\infty$. Symmetry: odd.
From y': No critical point. $y' > 0$ and y is increasing for all x.
From y'': $y'' = 0$ at $x = 0$.

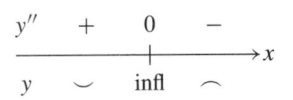

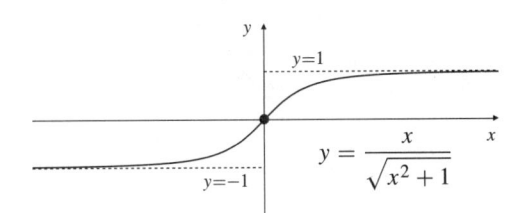

y''	+	0	−
y	⌣	infl	⌢

Fig. 4.4.38

39. $y = (x^2-1)^{1/3}$

$y' = \dfrac{2}{3}x(x^2-1)^{-2/3}$

$y'' = \dfrac{2}{3}[(x^2-1)^{-2/3} - \dfrac{2}{3}x(x^2-1)^{-5/3}2x]$

$= -\dfrac{2}{3}(x^2-1)^{-5/3}\left(1 + \dfrac{x^2}{3}\right)$

From y: Asymptotes: none.
Symmetry: even. Intercepts: $(\pm 1, 0)$, $(0, -1)$.
From y': CP: $x = 0$. SP: $x = \pm 1$.

	SP	CP	SP	
y'	− −1 −	0 +	1 +	
y	↘	↘ abs min	↗	↗

From y'': $y'' = 0$ nowhere.

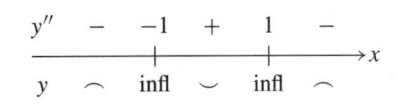

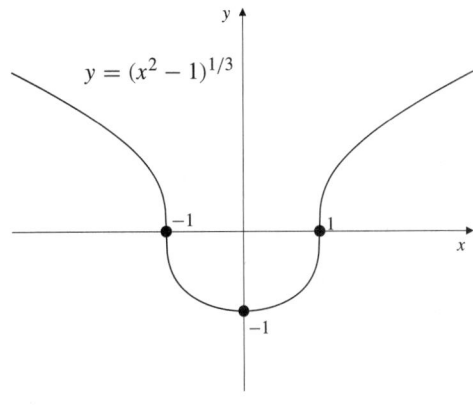

y''	− −1 +	1 −
y	⌢ infl ⌣	infl ⌢

Fig. 4.4.39

40. $y = x^{1/3} + (x-1)^{2/3}$,
$y' = \dfrac{1}{3}x^{-2/3} + \dfrac{2}{3}(x-1)^{-1/3}$,
$y'' = -\dfrac{2}{9}x^{-5/3} - \dfrac{2}{9}(x-1)^{-4/3}$.

From y: Intercept: $(0, 1)$. There are no other intercepts since $y = 0 \Leftrightarrow (x-1)^2 = -x \Leftrightarrow x^2 - x + 1 = 0$, which has no solutions. Symmetry: none. Asymptotes: none.
From y': $y' = 0$ if and only if

$$x^{-2/3} = -2(x-1)^{-1/3} \Rightarrow 8x^2 + x - 1 = 0$$

$$\Rightarrow x = \dfrac{-1 \pm \sqrt{1-4(8)(-1)}}{2(8)} \Rightarrow x = r_1 \text{ or } r_2$$

where $r_1 = \dfrac{-1-\sqrt{33}}{16}$ and $r_2 = \dfrac{-1+\sqrt{33}}{16}$, which are the critical points. Singular points: $x = 0$ and $x = 1$.

	CP	SP	CP	SP	
y'	− r_1 +	0 +	r_2 −	1 +	
y	↘ loc min ↗	↗	loc max ↘	loc min ↗	

From y'': $y'' = 0$ if and only if

$$x^{5/3} = -(x-1)^{4/3} \Leftrightarrow x^5 = -(x-1)^4.$$

If $f(x) = x^5 + (x-1)^4$ then $f(x)$ has only one real root, at a negative value of x. Since $f(-4) = -399$ and $f(-3) = 13$, this root lies between -4 and -3. Let this root be s. y'' is not defined at $x-0$ or $x = 1$.

y''	− s +	0 −	1 −
y	⌢ infl ⌣	infl ⌢	cusp ⌢

135

Fig. 4.4.40

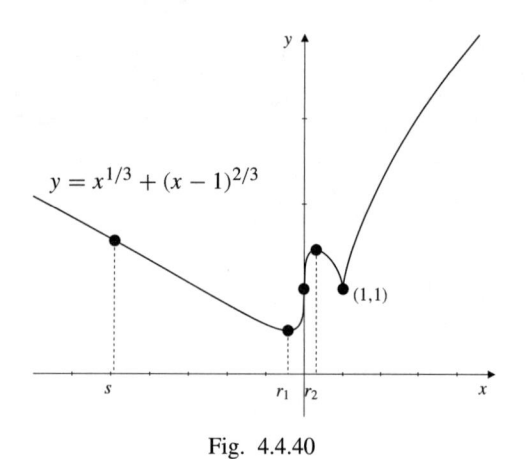

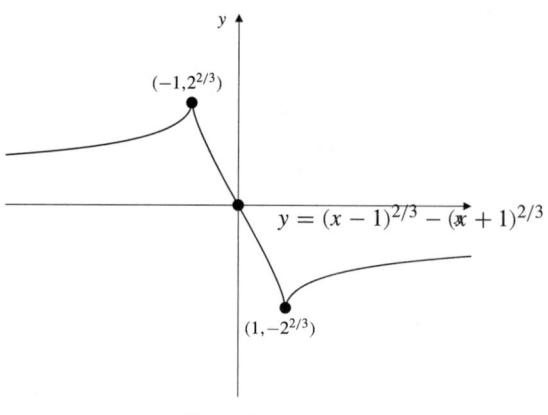

Fig. 4.4.41

41. $y = (x-1)^{2/3} - (x+1)^{2/3}$

$y' = \frac{2}{3}\big((x-1)^{-1/3} - (x+1)^{-1/3}\big)$

$y'' = -\frac{2}{9}\big((x-1)^{-4/3} - (x+1)^{-4/3}\big)$

$y = 0 \Leftrightarrow (x-1) = \pm(x+1)$

$ = \Leftrightarrow x - 1 = -x - 1 \Leftrightarrow x = 0$

$y' = 0 \Leftrightarrow x - 1 = x + 1 \quad \text{(nowhere)}$

$y'' = 0 \Leftrightarrow (x-1) = -(x+1)$

$ \Leftrightarrow x = 0$

From y: Asymptotes: $y = 0$ (as $x \to \pm\infty$). Symmetry: odd.
Intercept: $(0, 0)$. Points $(-1, 2^{2/3})$, $(1, -2^{2/3})$.
From y': CP: none. SP: $x = \pm 1$.

```
               SP        SP
    y'    +   -1   -    1   +
    ─────────┼────────┼────────→ x
    y    ↗   abs    ↘  abs   ↗
             max       min
```

From y'': $y'' = 0$ at $x = 0$.
y'' is undefined at $x = \pm 1$ (SP).

```
    y''  +   -1   +   0   -   1   -
    ─────────┼───────┼───────┼──────→ x
    y    ⌣  cusp  ⌣  infl  ⌢  cusp  ⌢
```

42. According to Theorem 5 of Section 4.4,

$$\lim_{x \to 0+} x \ln x = 0.$$

Thus,

$$\lim_{x \to 0} x \ln |x| = \lim_{x \to 0+} x \ln x = 0.$$

If $f(x) = x \ln |x|$ for $x \neq 0$, we may define $f(0)$ such that $f(0) = \lim_{x \to 0} x \ln |x| = 0$. Then f is continuous on the whole real line and

$$f'(x) = \ln |x| + 1, \qquad f''(x) = \frac{1}{|x|}\operatorname{sgn}(x).$$

From f: Intercept: $(0, 0)$, $(\pm 1, 0)$. Asymptotes: none. Symmetry: odd.
From f': CP: $x = \pm\dfrac{1}{e}$. SP: $x = 0$.

```
              CP           SP         CP
    f'   +   -1/e   -     0    -    1/e   +
    ─────────┼────────────┼────────────┼──────→ x
    f    ↗   loc    ↘           ↘    loc   ↗
             max                       min
```

From f'': f'' is undefined at $x = 0$.

```
    f''     -        0        +
    ────────────────┼──────────────→ x
    f       ⌢      infl       ⌣
```

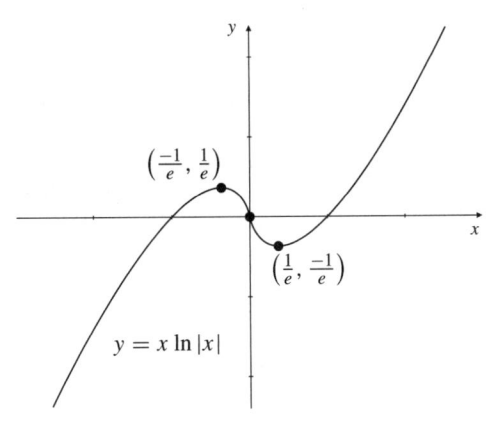

Fig. 4.4.42

43. $y = 0$ is an asymptote of $y = \dfrac{\sin x}{1+x^2}$.
Curve crosses asymptote at infinitely many points: $x = n\pi$ ($n = 0, \pm 1, \pm 2, \ldots$).

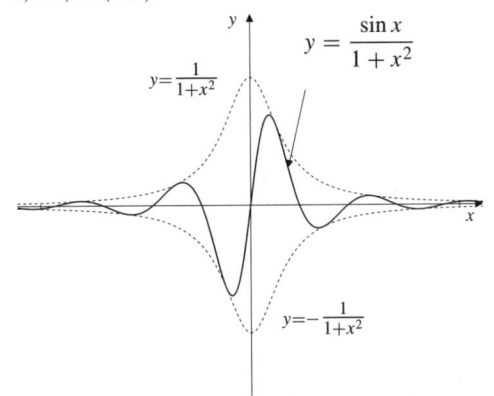

Fig. 4.4.43

Section 4.5 Extreme-Value Problems (page 263)

1. Let the numbers be x and $7-x$. Then $0 \le x \le 7$. The product is $P(x) = x(7-x) = 7x - x^2$.
$P(0) = P(7) = 0$ and $P(x) > 0$ if $0 < x < 7$. Thus maximum P occurs at a CP:
$$0 = \frac{dP}{dx} = 7 - 2x \Rightarrow x = \frac{7}{2}.$$
The maximum product is $P(7/2) = 49/4$.

2. Let the numbers be x and $\dfrac{8}{x}$ where $x > 0$. Their sum is $S = x + \dfrac{8}{x}$. Since $S \to \infty$ as $x \to \infty$ or $x \to 0+$, the minimum sum must occur at a critical point:
$$0 = \frac{dS}{dx} = 1 - \frac{8}{x^2} \Rightarrow x = 2\sqrt{2}.$$

Thus, the smallest possible sum is $2\sqrt{2} + \dfrac{8}{2\sqrt{2}} = 4\sqrt{2}$.

3. Let the numbers be x and $60-x$. Then $0 \le x \le 60$.
Let $P(x) = x^2(60-x) = 60x^2 - x^3$.
Clearly, $P(0) = P(60) = 0$ amd $P(x) > 0$ if $0 < x < 60$.
Thus maximum P occurs at a CP:
$$0 = \frac{dP}{dx} = 120x - 3x^2 = 3x(40-x).$$
Therefore, $x = 0$ or 40.
Max must correspond to $x = 40$. The numbers are 40 and 20.

4. Let the numbers be x and $16-x$. Let $P(x) = x^3(16-x)^5$.
Since $P(x) \to -\infty$ as $x \to \pm\infty$, so the maximum must occur at a critical point:
$$0 = P'(x) = 3x^2(16-x)^5 - 5x^3(16-x)^4$$
$$= x^2(16-x)^4(48-8x).$$
The critical points are 0, 6 and 16. Clearly, $P(0) = P(16) = 0$, and $P(6) = 216 \times 10^5$. Thus, $P(x)$ is maximum if the numbers are 6 and 10.

5. Let the numbers be x and $10-x$. We want to minimize
$$S(x) = x^3 + (10-x)^2, \quad 0 \le x \le 10.$$
$S(0) = 100$ and $S(10) = 1,000$. For CP:
$$0 = S'(x) = 3x^2 - 2(10-x) = 3x^2 + 2x - 20.$$
The only positive CP is $x = (-2 + \sqrt{4+240})/6 \approx 2.270$.
Since $S(2.270) \approx 71.450$, the minimum value of S is about 71.45.

6. If the numbers are x and $n-x$, then $0 \le x \le n$ and the sum of their squares is
$$S(x) = x^2 + (n-x)^2.$$
Observe that $S(0) = S(n) = n^2$. For critical points:
$$0 = S'(x) = 2x - 2(n-x) = 2(2x-n) \Rightarrow x = n/2.$$
Since $S(n/2) = n^2/2$, this is the smallest value of the sum of squares.

7. Let the dimensions of a rectangle be x and y. Then the area is $A = xy$ and the perimeter is $P = 2x + 2y$.
Given A we can express
$$P = P(x) = 2x + \frac{2A}{x}, \quad (0 < x < \infty).$$
Evidently, minimum P occurs at a CP. For CP:
$$0 = \frac{dP}{dx} = 2 - \frac{2A}{x^2} \Rightarrow x^2 = A = xy \Rightarrow x = y.$$

137

Thus min P occurs for $x = y$, i.e., for a square.

8. Let the width and the length of a rectangle of given perimeter $2P$ be x and $P - x$. Then the area of the rectangle is
$$A(x) = x(P - x) = Px - x^2.$$

Since $A(x) \to -\infty$ as $x \to \pm\infty$ the maximum must occur at a critical point:
$$0 = \frac{dA}{dx} = P - 2x \Rightarrow x = \frac{P}{2}$$

Hence, the width and the length are $\frac{P}{2}$ and $(P - \frac{P}{2}) = \frac{P}{2}$. Since the width equals the length, it is a square.

9. Let the dimensions of the isosceles triangle be as shown. Then $2x + 2y = P$ (given constant). The area is
$$A = xh = x\sqrt{y^2 - x^2} = x\sqrt{\left(\frac{P}{2} - x\right)^2 - x^2}.$$

Evidently, $y \geq x$ so $0 \leq x \leq P/4$. If $x = 0$ or $x = P/4$, then $A = 0$. Thus the maximum of A must occur at a CP. For max A:
$$0 = \frac{dA}{dx} = \sqrt{\frac{P^2}{4} - Px} - \frac{Px}{2\sqrt{\frac{P^2}{4} - Px}},$$

i.e., $\frac{P^2}{2} - 2Px - Px = 0$, or $x = \frac{P}{6}$. Thus $y = P/3$ and the triangle is equilateral since all three sides are $P/3$.

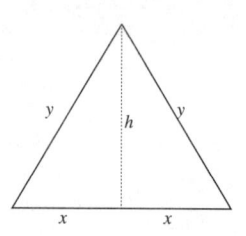

Fig. 4.5.9

10. Let the various dimensions be as shown in the figure. Since $h = 10\sin\theta$ and $b = 20\cos\theta$, the area of the triangle is
$$A(\theta) = \tfrac{1}{2}bh = 100\sin\theta\cos\theta$$
$$= 50\sin 2\theta \quad \text{for } 0 < \theta < \frac{\pi}{2}.$$

Since $A(\theta) \to 0$ as $\theta \to 0$ and $\theta \to \frac{\pi}{2}$, the maximum must be at a critial point:
$$0 = A'(\theta) = 100\cos 2\theta \Rightarrow 2\theta = \frac{\pi}{2} \Rightarrow \theta = \frac{\pi}{4}.$$

Hence, the largest possible area is
$$A(\pi/4) = 50\sin\left[2\left(\frac{\pi}{4}\right)\right] = 50\,\text{m}^2.$$

(Remark: alternatively, we may simply observe that the largest value of $\sin 2\theta$ is 1; therefore the largest possible area is $50(1) = 50\,\text{m}^2$.)

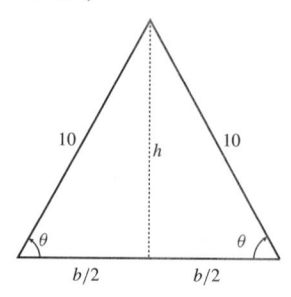

Fig. 4.5.10

11. Let the corners of the rectangle be as shown. The area of the rectangle is $A = 2xy = 2x\sqrt{R^2 - x^2}$ (for $0 \leq x \leq R$).
If $x = 0$ or $x = R$ then $A = 0$; otherwise $A > 0$. Thus maximum A must occur at a critical point:
$$0 = \frac{dA}{dx} = 2\left[\sqrt{R^2 - x^2} - \frac{x^2}{\sqrt{R^2 - x^2}}\right] \Rightarrow R^2 - 2x^2 = 0.$$
Thus $x = \frac{R}{\sqrt{2}}$ and the maximum area is
$$2\frac{R}{\sqrt{2}}\sqrt{R^2 - \frac{R^2}{2}} = R^2 \text{ square units.}$$

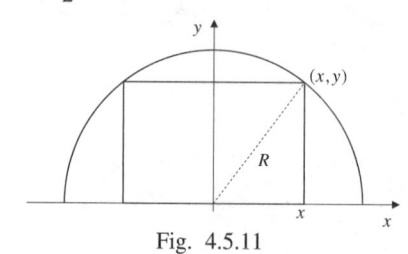

Fig. 4.5.11

12. Let x be as shown in the figure. The perimeter of the rectangle is
$$P(x) = 4x + 2\sqrt{R^2 - x^2} \qquad (0 \leq x \leq R).$$

For critical points:
$$0 = \frac{dP}{dx} = 4 + \frac{-2x}{\sqrt{R^2 - x^2}}$$
$$\Rightarrow 2\sqrt{R^2 - x^2} = x \Rightarrow x = \frac{2R}{\sqrt{5}}.$$

Since
$$\frac{d^2P}{dx^2} = \frac{-2R^2}{(R^2-x^2)^{3/2}} < 0$$
therefore $P(x)$ is concave down on $[0, R]$, so it must have an absolute maximum value at $x = \frac{2R}{\sqrt{5}}$. The largest perimeter is therefore
$$P\left(\frac{2R}{\sqrt{5}}\right) = 4\left(\frac{2R}{\sqrt{5}}\right) + \sqrt{R^2 - \frac{4R^2}{5}} = \frac{10R}{\sqrt{5}} \text{ units.}$$

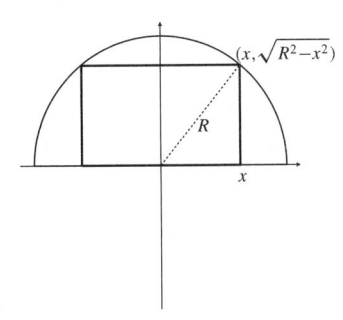

Fig. 4.5.12

13. Let the upper right corner be (x, y) as shown. Then $x \geq 0$ and $y = b\sqrt{1 - \frac{x^2}{a^2}}$, so $x \leq a$.
The area of the rectangle is
$$A(x) = 4xy = 4bx\sqrt{1 - \frac{x^2}{a^2}}, \quad (0 \leq x \leq a).$$
Clearly, $A = 0$ if $x = 0$ or $x = a$, so maximum A must occur at a critical point:
$$0 = \frac{dA}{dx} = 4b\left(\sqrt{1 - \frac{x^2}{a^2}} - \frac{\frac{2x^2}{a^2}}{2\sqrt{1 - \frac{x^2}{a^2}}}\right)$$
Thus $1 - \frac{x^2}{a^2} - \frac{x^2}{a^2} = 0$ and $x = \frac{a}{\sqrt{2}}$. Thus $y = \frac{b}{\sqrt{2}}$.
The largest area is $4\frac{a}{\sqrt{2}}\frac{b}{\sqrt{2}} = 2ab$ square units.

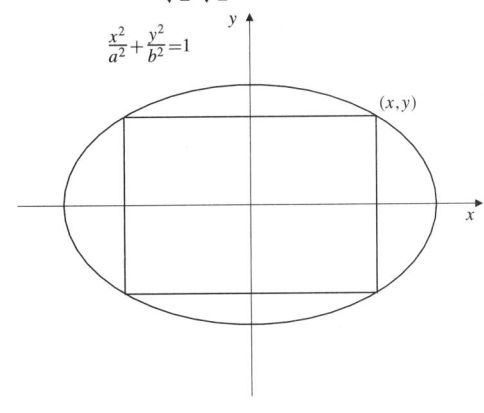

Fig. 4.5.13

14. See the diagrams below.

a) The area of the rectangle is $A = xy$. Since
$$\frac{y}{a-x} = \frac{b}{a} \Rightarrow y = \frac{b(a-x)}{a}.$$
Thus, the area is
$$A = A(x) = \frac{bx}{a}(a-x) \quad (0 \leq x \leq a).$$
For critical points:
$$0 = A'(x) = \frac{b}{a}(a - 2x) \Rightarrow x = \frac{a}{2}.$$
Since $A''(x) = -\frac{2b}{a} < 0$, A must have a maximum value at $x = \frac{a}{2}$. Thus, the largest area for the rectangle is
$$\frac{b}{a}\left(\frac{a}{2}\right)\left(a - \frac{a}{2}\right) = \frac{ab}{4} \text{ square units,}$$
that is, half the area of the triangle ABC.

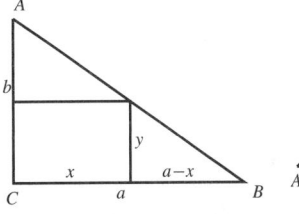

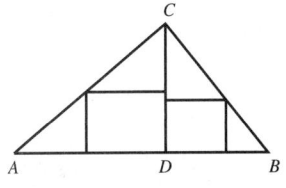

Fig. 3.4.16(a)   Fig. 3.4.16(b)

(b) This part has the same answer as part (a). To see this, let $CD \perp AB$, and solve separate problems for the largest rectangles in triangles ACD and BCD as shown. By part (a), both maximizing rectangles have the same height, namely half the length of CD. Thus, their union is a rectangle of area half of that of triangle ABC.

15. Let the width and the height of the billboard be w and h m respectively. The area of the board is $A = wh$. The printed area is $(w-8)(h-4) = 100$.
Thus $h = 4 + \frac{100}{w-8}$ and $A = 4w + \frac{100w}{w-8}$, $(w > 8)$.
Clearly, $A \to \infty$ if $w \to \infty$ or $w \to 8+$. Thus minimum A occurs at a critical point:
$$0 = \frac{dA}{dw} = 4 + \frac{100}{w-8} - \frac{100w}{(w-8)^2}$$
$$100w = 4(w^2 - 16w + 64) + 100w - 800$$
$$w^2 - 16w - 136 = 0$$
$$w = \frac{16 \pm \sqrt{800}}{2} = 8 \pm 10\sqrt{2}.$$

Since $w > 0$ we must have $w = 8 + 10\sqrt{2}$.
Thus $h = 4 + \dfrac{100}{10\sqrt{2}} = 4 + 5\sqrt{2}$.
The billboard should be $8 + 10\sqrt{2}$ m wide and $4 + 5\sqrt{2}$ m high.

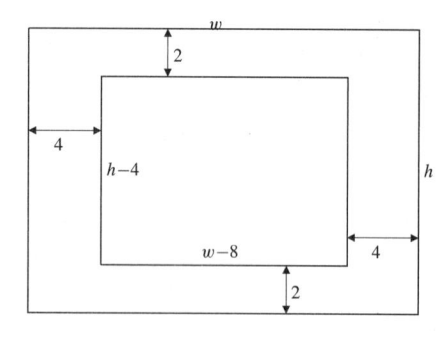

Fig. 4.5.15

16. Let x be the side of the cut-out squares. Then the volume of the box is

$$V(x) = x(70 - 2x)(150 - 2x) \qquad (0 \le x \le 35).$$

Since $V(0) = V(35) = 0$, the maximum value will occur at a critical point:

$$\begin{aligned} 0 = V'(x) &= 4(2625 - 220x + 3x^2) \\ &= 4(3x - 175)(x - 15) \\ &\Rightarrow x = 15 \text{ or } \frac{175}{3}. \end{aligned}$$

The only critical point in $[0, 35]$ is $x = 15$. Thus, the largest possible volume for the box is

$$V(15) = 15(70 - 30)(150 - 30) = 72{,}000 \text{ cm}^3.$$

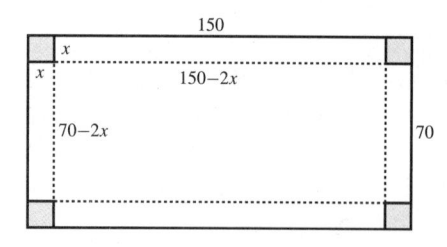

Fig. 4.5.16

17. Let the rebate be $\$x$. Then number of cars sold per month is

$$2000 + 200\left(\frac{x}{50}\right) = 2000 + 4x.$$

The profit per car is $1000 - x$, so the total monthly profit is

$$\begin{aligned} P &= (2000 + 4x)(1000 - x) = 4(500 + x)(1000 - x) \\ &= 4(500{,}000 + 500x - x^2). \end{aligned}$$

For maximum profit:

$$0 = \frac{dP}{dx} = 4(500 - 2x) \Rightarrow x = 250.$$

(Since $\dfrac{d^2P}{dx^2} = -8 < 0$ any critical point gives a local max.) The manufacturer should offer a rebate of $\$250$ to maximize profit.

18. If the manager charges $\$(40 + x)$ per room, then $(80 - 2x)$ rooms will be rented.
The total income will be $\$(80 - 2x)(40 + x)$ and the total cost will be $\$(80 - 2x)(10) + (2x)(2)$. Therefore, the profit is

$$\begin{aligned} P(x) &= (80 - 2x)(40 + x) - [(80 - 2x)(10) + (2x)(2)] \\ &= 2400 + 16x - 2x^2 \qquad \text{for } x > 0. \end{aligned}$$

If $P'(x) = 16 - 4x = 0$, then $x = 4$. Since $P''(x) = -4 < 0$, P must have a maximum value at $x = 4$. Therefore, the manager should charge $\$44$ per room.

19. Head for point C on road x km east of A. Travel time is

$$T = \frac{\sqrt{12^2 + x^2}}{15} + \frac{10 - x}{39}.$$

We have $T(0) = \dfrac{12}{15} + \dfrac{10}{39} = 1.0564$ hrs

$$T(10) = \frac{\sqrt{244}}{15} = 1.0414 \text{ hrs}$$

For critical points:

$$0 = \frac{dT}{dx} = \frac{1}{15}\frac{x}{\sqrt{12^2 + x^2}} - \frac{1}{39}$$

$$\Rightarrow 13x = 5\sqrt{12^2 + x^2}$$

$$\Rightarrow (13^2 - 5^2)x^2 = 5^2 \times 12^2 \Rightarrow x = 5$$

$T(5) = \dfrac{13}{15} + \dfrac{5}{39} = 0.9949 < \begin{cases} T(0) \\ T(10). \end{cases}$
(Or note that

$$\begin{aligned} \frac{d^2T}{dt^2} &= \frac{1}{15}\frac{\sqrt{12^2 + x^2} - \dfrac{x^2}{\sqrt{12^2 + x^2}}}{12^2 + x^2} \\ &= \frac{12^2}{15(12^2 + x^2)^{3/2}} > 0 \end{aligned}$$

so any critical point is a local minimum.)
To minimize travel time, head for point 5 km east of A.

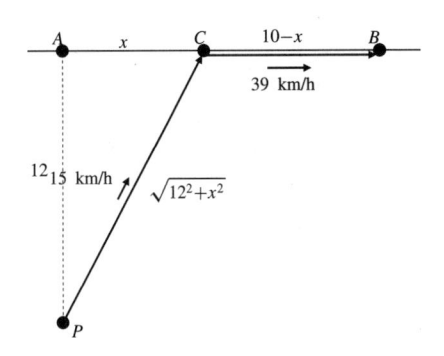

INSTRUCTOR'S SOLUTIONS MANUAL   SECTION 4.5 (PAGE 263)

Fig. 4.5.19

20. This problem is similar to the previous one except that the 10 in the numerator of the second fraction in the expression for T is replaced with a 4. This has no effect on the critical point of T, namely $x = 5$, which now lies outside the appropriate interval $0 \leq x \leq 4$. Minimum T must occur at an endpoint. Note that

$$T(0) = \frac{12}{15} + \frac{4}{39} = 0.9026$$
$$T(4) = \frac{1}{15}\sqrt{12^2 + 4^2} = 0.8433.$$

The minimum travel time corresponds to $x = 4$, that is, to driving in a straight line to B.

21. Use x m for the circle and $1 - x$ m for square. The sum of areas is

$$A = \pi r^2 + s^2 = \frac{\pi x^2}{4\pi^2} + \left(\frac{1-x}{4}\right)^2$$
$$= \frac{x^2}{4\pi} + \frac{(1-x)^2}{4^2} \qquad (0 \leq x \leq 1)$$

Now $A(0) = \frac{1}{16}$, $A(1) = \frac{1}{4\pi} > A(0)$. For CP:

$$0 = \frac{dA}{dx} = \frac{x}{2\pi} - \frac{1-x}{8} \Rightarrow x\left(\frac{1}{2\pi} + \frac{1}{8}\right) = \frac{1}{8} \Rightarrow x = \frac{\pi}{4+\pi}.$$

Since $\frac{d^2A}{dx^2} = \frac{1}{2\pi} + \frac{1}{8} > 0$, the CP gives local minimum for A.

a) For max total area use none of wire for the square, i.e., $x = 1$.

b) For minimum total area use $1 - \frac{\pi}{4+\pi} = \frac{4}{4+\pi}$ m for square.

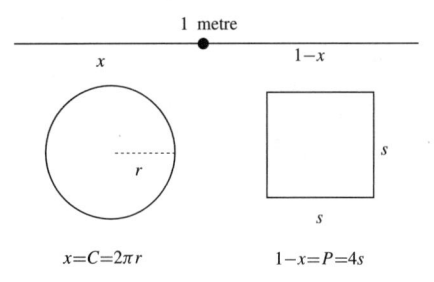

Fig. 4.5.21

22. Let the dimensions of the rectangle be as shown in the figure. Clearly,

$$x = a\sin\theta + b\cos\theta,$$
$$y = a\cos\theta + b\sin\theta.$$

Therefore, the area is

$$A(\theta) = xy$$
$$= (a\sin\theta + b\cos\theta)(a\cos\theta + b\sin\theta)$$
$$= ab + (a^2 + b^2)\sin\theta\cos\theta$$
$$= ab + \frac{1}{2}(a^2 + b^2)\sin 2\theta \qquad \text{for } 0 \leq \theta \leq \frac{\pi}{2}.$$

If $A'(\theta) = (a^2 + b^2)\cos 2\theta = 0$, then $\theta = \frac{\pi}{4}$. Since $A''(\theta) = -2(a^2 + b^2)\sin 2\theta < 0$ when $0 \leq \theta \leq \frac{\pi}{2}$, therefore $A(\theta)$ must have a maximum value at $\theta = \frac{\pi}{4}$. Hence, the area of the largest rectangle is

$$A\left(\frac{\pi}{4}\right) = ab + \frac{1}{2}(a^2 + b^2)\sin\left(\frac{\pi}{2}\right)$$
$$= ab + \frac{1}{2}(a^2 + b^2) = \frac{1}{2}(a+b)^2 \quad \text{sq. units.}$$

(Note: $x = y = \frac{a}{\sqrt{2}} + \frac{b}{\sqrt{2}}$ indicates that the rectangle containing the given rectangle with sides a and b, has largest area when it is a square.)

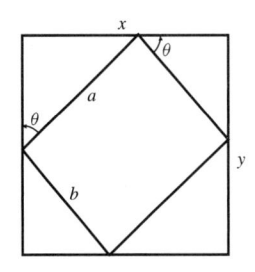

Fig. 4.5.22

23. Let the line have intercepts x, y as shown. Let θ be angle shown. The length of line is

$$L = \frac{9}{\cos\theta} + \frac{\sqrt{3}}{\sin\theta} \qquad (0 < \theta < \frac{\pi}{2}).$$

141

Clearly, $L \to \infty$ if $\theta \to 0+$ or $\theta \to \frac{\pi}{2}-$.
Thus the minimum length occurs at a critical point.
For CP:

$$0 = \frac{dL}{d\theta} = \frac{9\sin\theta}{\cos^2\theta} - \frac{\sqrt{3}\cos\theta}{\sin^2\theta} \Rightarrow \tan^3\theta = \left(\frac{1}{\sqrt{3}}\right)^3$$
$$\Rightarrow \theta = \frac{\pi}{6}$$

Shortest line segment has length

$$L = \frac{9}{\sqrt{3}/2} + \frac{\sqrt{3}}{1/2} = 8\sqrt{3} \text{ units.}$$

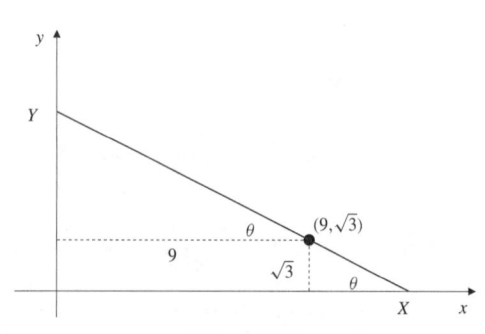

Fig. 4.5.23

24. The longest beam will have length equal to the minimum of $L = x + y$, where x and y are as shown in the figure below:

$$x = \frac{a}{\cos\theta}, \quad y = \frac{b}{\sin\theta}.$$

Thus,

$$L = L(\theta) = \frac{a}{\cos\theta} + \frac{b}{\sin\theta} \quad \left(0 < \theta < \frac{\pi}{2}\right).$$

If $L'(\theta) = 0$, then

$$\frac{a\sin\theta}{\cos^2\theta} - \frac{b\cos\theta}{\sin^2\theta} = 0$$
$$\Leftrightarrow \frac{a\sin^3\theta - b\cos^3\theta}{\cos^2\theta \sin^2\theta} = 0$$
$$\Leftrightarrow a\sin^3\theta - b\cos^3\theta = 0$$
$$\Leftrightarrow \tan^3\theta = \frac{b}{a}$$
$$\Leftrightarrow \tan\theta = \frac{b^{1/3}}{a^{1/3}}.$$

Clearly, $L(\theta) \to \infty$ as $\theta \to 0+$ or $\theta \to \frac{\pi}{2}-$. Thus, the minimum must occur at $\theta = \tan^{-1}\left(\frac{b^{1/3}}{a^{1/3}}\right)$. Using the triangle above for $\tan\theta = \frac{b^{1/3}}{a^{1/3}}$, it follows that

$$\cos\theta = \frac{a^{1/3}}{\sqrt{a^{2/3}+b^{2/3}}}, \quad \sin\theta = \frac{b^{1/3}}{\sqrt{a^{2/3}+b^{2/3}}}.$$

Hence, the minimum is

$$L(\theta) = \frac{a}{\left(\frac{a^{1/3}}{\sqrt{a^{2/3}+b^{2/3}}}\right)} + \frac{b}{\left(\frac{b^{1/3}}{\sqrt{a^{2/3}+b^{2/3}}}\right)}$$
$$= \left(a^{2/3}+b^{2/3}\right)^{3/2} \text{ units.}$$

25. If the largest beam that can be carried horizontally around the corner is l m long (by Exercise 26, $l = (a^{2/3}+b^{2/3})^{2/3}$ m), then at the point of maximum clearance, one end of the beam will be on the floor at the outer wall of one hall, and the other will be on the ceiling at the outer wall of the second hall. Thus the horizontal projection of the beam will be l. So the beam will have length

$$\sqrt{l^2+c^2} = [(a^{2/3}+b^{2/3})^3+c^2]^{1/2} \text{ units.}$$

26. Let θ be the angle of inclination of the ladder. The height of the fence is

$$h(\theta) = 6\sin\theta - 2\tan\theta \quad \left(0 < \theta < \frac{\pi}{2}\right).$$

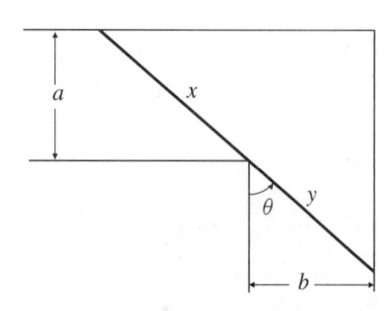

Fig. 4.5.24

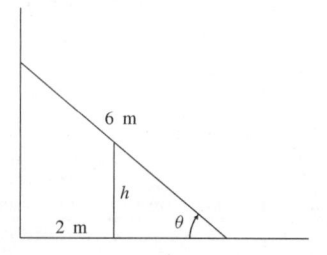

Fig. 4.5.26

For critical points:

$$0 = h'(\theta) = 6\cos\theta - 2\sec^2\theta$$
$$\Rightarrow 3\cos\theta = \sec^2\theta \Rightarrow 3\cos^3\theta = 1$$
$$\Rightarrow \cos\theta = \left(\tfrac{1}{3}\right)^{1/3}.$$

Since $h''(\theta) = -6\sin\theta - 4\sec^2\theta\tan\theta < 0$ for $0 < \theta < \dfrac{\pi}{2}$, therefore $h(\theta)$ must be maximum at $\theta = \cos^{-1}\left(\tfrac{1}{3}\right)^{1/3}$. Then

$$\sin\theta = \frac{\sqrt{3^{2/3} - 1}}{3^{1/3}}, \quad \tan\theta = \sqrt{3^{2/3} - 1}.$$

Thus, the maximum height of the fence is

$$h(\theta) = 6\left(\frac{\sqrt{3^{2/3}-1}}{3^{1/3}}\right) - 2\sqrt{3^{2/3}-1}$$
$$= 2(3^{2/3}-1)^{3/2} \approx 2.24\,\text{m}.$$

27. Let (x, y) be a point on $x^2y^4 = 1$. Then $x^2y^4 = 1$ and the square of distance from (x, y) to $(0, 0)$ is
$$S = x^2 + y^2 = \frac{1}{y^4} + y^2, \quad (y \neq 0)$$
Clearly, $S \to \infty$ as $y \to 0$ or $y \to \pm\infty$, so minimum S must occur at a critical point. For CP:

$$0 = \frac{dS}{dy} = \frac{-4}{y^5} + 2y \Rightarrow y^6 = 2 \Rightarrow y = \pm 2^{1/6}$$
$$\Rightarrow x = \pm \frac{1}{2^{1/3}}$$

Thus the shortest distance from origin to curve is

$$S = \sqrt{\frac{1}{2^{2/3}} + 2^{1/3}} = \sqrt{\frac{3}{2^{2/3}}} = \frac{3^{1/2}}{2^{1/3}}\,\text{units}.$$

28. The square of the distance from $(8, 1)$ to the curve $y = 1 + x^{3/2}$ is

$$S = (x-8)^2 + (y-1)^2$$
$$= (x-8)^2 + (1+x^{3/2}-1)^2$$
$$= x^3 + x^2 - 16x + 64.$$

Note that y, and therefore also S, is only defined for $x \geq 0$. If $x = 0$ then $S = 64$. Also, $S \to \infty$ if $x \to \infty$. For critical points:

$$0 = \frac{dS}{dx} = 3x^2 + 2x - 16 = (3x+8)(x-2)$$
$$\Rightarrow x = -\tfrac{8}{3} \text{ or } 2.$$

Only $x = 2$ is feasible. At $x = 2$ we have $S = 44 < 64$. Therefore the minimum distance is $\sqrt{44} = 2\sqrt{11}$ units.

29. Let the cylinder have radius r and height h. By symmetry, the centre of the cylinder is at the centre of the sphere. Thus
$$r^2 + \frac{h^2}{4} = R^2.$$

The volume of cylinder is

$$V = \pi r^2 h = \pi h\left(R^2 - \frac{h^2}{4}\right), \quad (0 \leq h \leq 2R).$$

Clearly, $V = 0$ if $h = 0$ or $h = 2R$, so maximum V occurs at a critical point. For CP:

$$0 = \frac{dV}{dh} = \pi\left[R^2 - \frac{h^2}{4} - \frac{2h^2}{4}\right]$$
$$\Rightarrow h^2 = \frac{4}{3}R^2 \quad \Rightarrow h = \frac{2R}{\sqrt{3}}$$
$$\Rightarrow r = \sqrt{\frac{2}{3}}R.$$

The largest cylinder has height $\dfrac{2R}{\sqrt{3}}$ units and radius $\sqrt{\dfrac{2}{3}}R$ units.

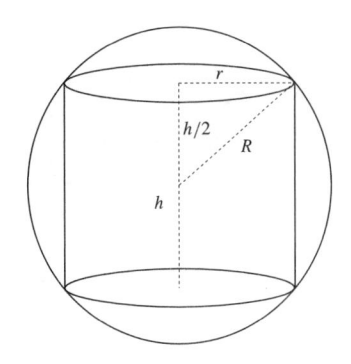

Fig. 4.5.29

30. Let the radius and the height of the circular cylinder be r and h. By similar triangles,

$$\frac{h}{R-r} = \frac{H}{R} \Rightarrow h = \frac{H(R-r)}{R}.$$

Hence, the volume of the circular cylinder is

$$V(r) = \pi r^2 h = \frac{\pi r^2 H(R-r)}{R}$$
$$= \pi H\left(r^2 - \frac{r^3}{R}\right) \quad \text{for } 0 \leq r \leq R.$$

Since $V(0) = V(R) = 0$, the maximum value of V must be at a critical point. If $\dfrac{dV}{dr} = \pi H\left(2r - \dfrac{3r^2}{R}\right) = 0$, then $r = \dfrac{2R}{3}$. Therefore the cylinder has maximum volume if its radius is $r = \dfrac{2R}{3}$ units, and its height is

$$h = \dfrac{H\left(R - \dfrac{2R}{3}\right)}{R} = \dfrac{H}{3} \text{ units.}$$

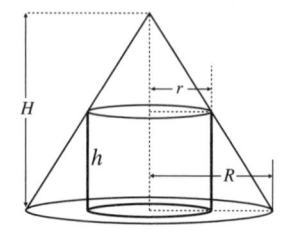

Fig. 4.5.30

31. Let the box have base dimensions x m and height y m. Then $x^2 y = $ volume $= 4$.
Most economical box has minimum surface area (bottom and sides). This area is

$$S = x^2 + 4xy = x^2 + 4x\left(\dfrac{4}{x^2}\right)$$
$$= x^2 + \dfrac{16}{x}, \quad (0 < x < \infty).$$

Clearly, $S \to \infty$ if $x \to \infty$ or $x \to 0+$. Thus minimum S occurs at a critical point. For CP:

$$0 = \dfrac{dS}{dx} = 2x - \dfrac{16}{x^2} \Rightarrow x^3 = 8 \Rightarrow x = 2 \Rightarrow y = 1.$$

Most economical box has base 2 m × 2 m and height 1 m.

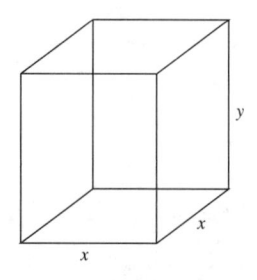

Fig. 4.5.31

32.

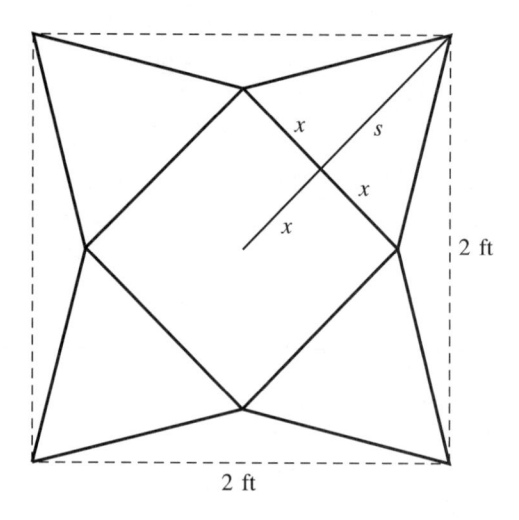

Fig. 4.5.32

From the figure, if the side of the square base of the pyramid is $2x$, then the slant height of triangular walls of the pyramid is $s = \sqrt{2} - x$. The vertical height of the pyramid is

$$h = \sqrt{s^2 - x^2} = \sqrt{2 - 2\sqrt{2}x + x^2 - x^2} = \sqrt{2}\sqrt{1 - \sqrt{2}x}.$$

Thus the volume of the pyramid is

$$V = \dfrac{4\sqrt{2}}{3} x^2 \sqrt{1 - \sqrt{2}x},$$

for $0 \leq x \leq 1/\sqrt{2}$. $V = 0$ at both endpoints, so the maximum will occur at an interior critical point. For CP:

$$0 = \dfrac{dV}{dx} = \dfrac{4\sqrt{2}}{3}\left[2x\sqrt{1 - \sqrt{2}x} - \dfrac{\sqrt{2}x^2}{2\sqrt{1 - \sqrt{2}x}}\right]$$
$$4x(1 - \sqrt{2}x) = \sqrt{2}x^2$$
$$4x = 5\sqrt{2}x^2 \quad, x = 4/(5\sqrt{2}).$$

$V(4/(5\sqrt{2})) = 32\sqrt{2}/(75\sqrt{5})$. The largest volume of such a pyramid is $32\sqrt{2}/(75\sqrt{5})$ ft^3.

33. Let the dimensions be as shown. The perimeter is $\pi\dfrac{x}{2} + x + 2y = 10$. Therefore,

$$\left(1 + \dfrac{\pi}{2}\right)x + 2y = 10, \quad \text{or} \quad (2 + \pi)x + 4y = 20.$$

The area of the window is

$$A = xy + \dfrac{1}{2}\pi\left(\dfrac{x}{2}\right)^2 = \pi\dfrac{x^2}{8} + x\left(5 - \dfrac{(2 + \pi)x}{4}\right).$$

To maximize light admitted, maximize the area A. For CP:

$$0 = \frac{dA}{dx} = \frac{\pi x}{4} + 5 - \frac{2+\pi}{4}x - \frac{2+\pi}{4}x \Rightarrow x = \frac{20}{4+\pi}$$
$$\Rightarrow y = \frac{10}{4+\pi}.$$

To admit greatest amount of light, let width $= \frac{20}{4+\pi}$ m and height (of the rectangular part) be $\frac{10}{4+\pi}$ m.

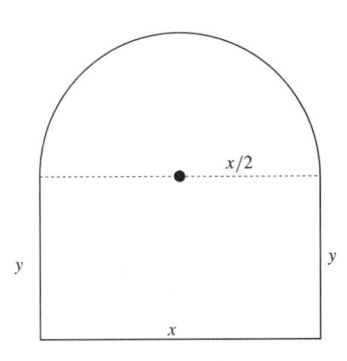

Fig. 4.5.33

34. Let h and r be the length and radius of the cylindrical part of the tank. The volume of the tank is

$$V = \pi r^2 h + \frac{4}{3}\pi r^3.$$

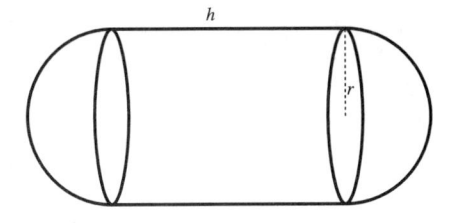

Fig. 4.5.34

If the cylindrical wall costs $\$k$ per unit area and the hemispherical wall $\$2k$ per unit area, then the total cost of the tank wall is

$$C = 2\pi rhk + 8\pi r^2 k$$
$$= 2\pi rk \frac{V - \frac{4}{3}\pi r^3}{\pi r^2} + 8\pi r^2 k$$
$$= \frac{2Vk}{r} + \frac{16}{3}\pi r^2 k \qquad (0 < r < \infty).$$

Since $C \to \infty$ as $r \to 0+$ or $r \to \infty$, the minimum cost must occur at a critical point. For critical points,

$$0 = \frac{dC}{dr} = -2Vkr^{-2} + \frac{32}{3}\pi rk \quad \Leftrightarrow \quad r = \left(\frac{3V}{16\pi}\right)^{1/3}.$$

Since $V = \pi r^2 h + \frac{4}{3}\pi r^3$,

$$r^3 = \frac{3}{16\pi}\left(\pi r^2 h + \frac{4}{3}\pi r^3\right) \Rightarrow r = \frac{1}{4}h$$
$$\Rightarrow h = 4r = 4\left(\frac{3V}{16\pi}\right)^{1/3}.$$

Hence, in order to minimize the cost, the radius and length of the cylindrical part of the tank should be $\left(\frac{3V}{16\pi}\right)^{1/3}$ and $4\left(\frac{3V}{16\pi}\right)^{1/3}$ units respectively.

35. Let D' be chosen so that mirror AB is the right bisector of DD'. Let CD' meet AB at X. Therefore, the travel time along CXD is

$$T_X = \frac{CX + XD}{\text{speed}} = \frac{CX + XD'}{\text{speed}} = \frac{CD'}{\text{speed}}.$$

If Y is any other point on AB, travel time along CYD is

$$T_Y = \frac{CY + YD}{\text{speed}} = \frac{CY + YD'}{\text{speed}} > \frac{CD'}{\text{speed}}.$$

(The sum of two sides of a triangle is greater than the third side.) Therefore, X minimizes travel time. Clearly, XN bisects $\angle CXD$.

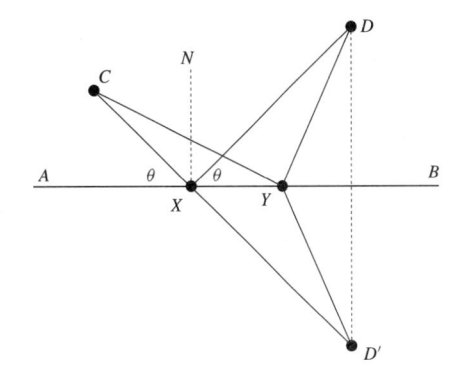

Fig. 4.5.35

36. If the path of the light ray is as shown in the figure then the time of travel from A to B is

$$T = T(x) = \frac{\sqrt{a^2 + x^2}}{v_1} + \frac{\sqrt{b^2 + (c-x)^2}}{v_2}.$$

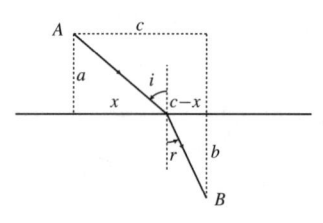

Fig. 4.5.36

To minimize T, we look for a critical point:

$$0 = \frac{dT}{dx} = \frac{1}{v_1}\frac{x}{\sqrt{a^2+x^2}} - \frac{1}{v_2}\frac{c-x}{\sqrt{b^2+(c-x)^2}}$$
$$= \frac{1}{v_1}\sin i - \frac{1}{v_2}\sin r.$$

Thus,
$$\frac{\sin i}{\sin r} = \frac{v_1}{v_2}.$$

37. Let the width be w, and the depth be h. Therefore

$$\left(\frac{h}{2}\right)^2 + \left(\frac{w}{2}\right)^2 = R^2.$$

The stiffness is $S = wh^3 = h^3\sqrt{4R^2-h^2}$ for $(0 \le h \le 2R)$. We have $S = 0$ if $h = 0$ or $h = 2R$. For maximum stiffness:

$$0 = \frac{dS}{dh} = 3h^2\sqrt{4R^2-h^2} - \frac{h^4}{\sqrt{4R^2-h^2}}.$$

Thus $3(4R^2 - h^2) = h^2$ so $h = \sqrt{3}R$, and $w = R$. The stiffest beam has width R and depth $\sqrt{3}R$.

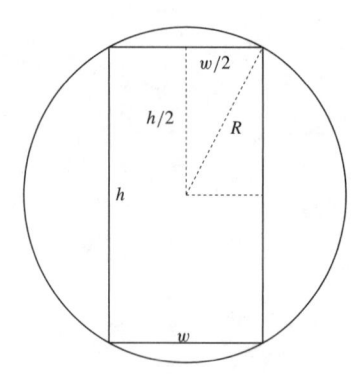

Fig. 4.5.37

38. The curve $y = 1 + 2x - x^3$ has slope $m = y' = 2 - 3x^2$. Evidently m is greatest for $x = 0$, in which case $y = 1$ and $m = 2$. Thus the tangent line with maximal slope has equation $y = 1 + 2x$.

39. $\dfrac{dQ}{dt} = kQ^3(L-Q)^5 \qquad (k, L > 0)$

Q grows at the greatest rate when $f(Q) = Q^3(L-Q)^5$ is maximum, i.e., when

$$0 = f'(Q) = 3Q^2(L-Q)^5 - 5Q^3(L-Q)^4$$
$$= Q^2(L-Q)^4(3L-8Q) \Rightarrow Q = 0, L, \frac{3L}{8}.$$

Since $f(0) = f(L) = 0$ and $f\left(\dfrac{3L}{8}\right) > 0$, Q is growing most rapidly when $Q = \dfrac{3L}{8}$.

40. Let h and r be the height and base radius of the cone and R be the radius of the sphere. From similar triangles,

$$\frac{r}{\sqrt{h^2+r^2}} = \frac{R}{h-R}$$
$$\Rightarrow h = \frac{2r^2 R}{r^2-R^2} \qquad (r > R).$$

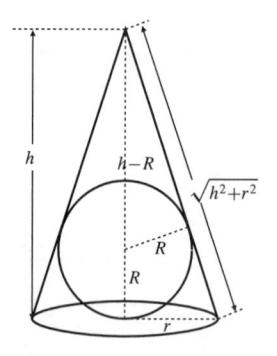

Fig. 4.5.40

Then the volume of the cone is

$$V = \frac{1}{3}\pi r^2 h = \frac{2}{3}\pi R \frac{r^4}{r^2-R^2} \qquad (R < r < \infty).$$

Clearly $V \to \infty$ if $r \to \infty$ or $r \to R+$. Therefore to minimize V, we look for a critical point:

$$0 = \frac{dV}{dr} = \frac{2}{3}\pi R\left[\frac{(r^2-R^2)(4r^3) - r^4(2r)}{(r^2-R^2)^2}\right]$$
$$\Leftrightarrow 4r^5 - 4r^3R^2 - 2r^5 = 0$$
$$\Leftrightarrow r = \sqrt{2}R.$$

Hence, the smallest possible volume of a right circular cone which can contain sphere of radius R is

$$V = \frac{2}{3}\pi R\left(\frac{4R^4}{2R^2-R^2}\right) = \frac{8}{3}\pi R^3 \text{ cubic units.}$$

41. Let distances and angles be as shown. Then $\tan\alpha = \dfrac{2}{x}$, $\tan(\theta+\alpha) = \dfrac{12}{x}$

$$\dfrac{12}{x} = \dfrac{\tan\theta + \tan\alpha}{1-\tan\theta\tan\alpha} = \dfrac{\tan\theta + \dfrac{2}{x}}{1 - \dfrac{2}{x}\tan\theta}$$

$$\dfrac{12}{x} - \dfrac{24}{x^2}\tan\theta = \tan\theta + \dfrac{2}{x}$$

$$\tan\theta\left(1 + \dfrac{24}{x^2}\right) = \dfrac{10}{x}, \quad \text{so } \tan\theta = \dfrac{10x}{x^2+24} = f(x).$$

To maximize θ (i.e., to get the best view of the mural), we can maximize $\tan\theta = f(x)$.
Since $f(0) = 0$ and $f(x) \to 0$ as $x \to \infty$, we look for a critical point.

$$0 = f'(x) = 10\left[\dfrac{x^2+24-2x^2}{(x^2+24)^2}\right] \Rightarrow x^2 = 24$$
$$\Rightarrow x = 2\sqrt{6}$$

Stand back $2\sqrt{6}$ ft (≈ 4.9 ft) to see the mural best.

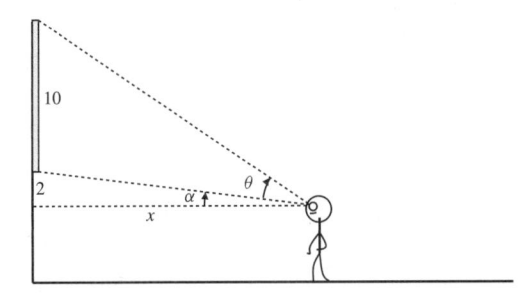

Fig. 4.5.41

42. Let r be the radius of the circular arc and θ be the angle shown in the left diagram below. Thus,

$$2r\theta = 100 \quad \Rightarrow \quad r = \dfrac{50}{\theta}.$$

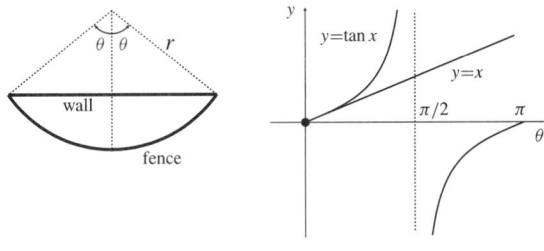

Fig. 3.4.44(a)   Fig. 3.4.44(b)

The area of the enclosure is

$$A = \dfrac{2\theta}{2\pi}\pi r^2 - (r\cos\theta)(r\sin\theta)$$
$$= \dfrac{50^2}{\theta} - \dfrac{50^2}{\theta^2}\dfrac{\sin 2\theta}{2}$$
$$= 50^2\left(\dfrac{1}{\theta} - \dfrac{\sin 2\theta}{2\theta^2}\right)$$

for $0 < \theta \le \pi$. Note that $A \to \infty$ as $\theta \to 0+$, and for $\theta = \pi$ we are surrounding the entire enclosure with fence (a circle) and not using the wall at all. Evidently this would not produce the greatest enclosure area, so the maximum area must correspond to a critical point of A:

$$0 = \dfrac{dA}{d\theta} = 50^2\left(-\dfrac{1}{\theta^2} - \dfrac{2\theta^2(2\cos 2\theta) - \sin 2\theta(4\theta)}{4\theta^4}\right)$$
$$\Leftrightarrow \dfrac{1}{\theta^2} + \dfrac{\cos 2\theta}{\theta^2} = \dfrac{\sin 2\theta}{\theta^3}$$
$$\Leftrightarrow 2\theta\cos^2\theta = 2\sin\theta\cos\theta$$
$$\Leftrightarrow \cos\theta = 0 \quad \text{or} \quad \tan\theta = \theta.$$

Observe that $\tan\theta = \theta$ has no solutions in $(0, \pi]$. (The graphs of $y = \tan\theta$ and $y = \theta$ cross at $\theta = 0$ but nowhere else between 0 and π.) Thus, the greatest enclosure area must correspond to $\cos\theta = 0$, that is, to $\theta = \dfrac{\pi}{2}$. The largest enclosure is thus semicircular, and has area $\dfrac{2}{\pi}(50)^2 = \dfrac{5000}{\pi}$ m^2.

43. Let the cone have radius r and height h.
Let sector of angle θ from disk be used.
Then $2\pi r = R\theta$ so $r = \dfrac{R}{2\pi}\theta$.
Also $h = \sqrt{R^2 - r^2} = \sqrt{R^2 - \dfrac{R^2\theta^2}{4\pi^2}} = \dfrac{R}{2\pi}\sqrt{4\pi^2 - \theta^2}$
The cone has volume

$$V = \dfrac{\pi r^2 h}{3} = \dfrac{\pi}{3}\dfrac{R^2}{4\pi^2}\theta^2 \dfrac{R}{2\pi}\sqrt{4\pi^2 - \theta}$$
$$= \dfrac{R^3}{24\pi^2}f(\theta) \quad \text{where } f(\theta) = \theta^2\sqrt{4\pi^2 - \theta^2} \quad (0 \le \theta \le 2\pi)$$

$V(0) = V(2\pi) = 0$ so maximum V must occur at a critical point. For CP:

$$0 = f'(\theta) = 2\theta\sqrt{4\pi^2 - \theta^2} - \dfrac{\theta^3}{\sqrt{4\pi^2 - \theta^2}}$$
$$\Rightarrow 2(4\pi^2 - \theta^2) = \theta^2 \quad \Rightarrow \theta^2 = \dfrac{8}{3}\pi^2.$$

The largest cone has volume $V\left(\pi\sqrt{\dfrac{8}{3}}\right) = \dfrac{2\pi R^3}{9\sqrt{3}}$ cu. units.

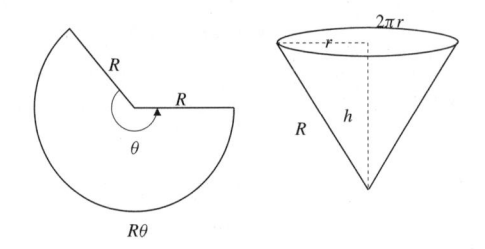

Fig. 4.5.43

44. Let the various distances be as labelled in the diagram.

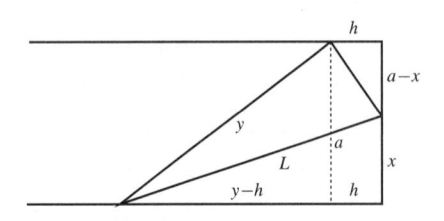

Fig. 4.5.44

From the geometry of the various triangles in the diagram we have

$$x^2 = h^2 + (a-x)^2 \Rightarrow h^2 = 2ax - a^2$$
$$y^2 = a^2 + (y-h)^2 \Rightarrow h^2 = 2hy - a^2$$

hence $hy = ax$. Then

$$L^2 = x^2 + y^2 = x^2 + \frac{a^2 x^2}{h^2}$$
$$= x^2 + \frac{a^2 x^2}{2ax - a^2} = \frac{2ax^3}{2ax - a^2}$$

for $\frac{a}{2} < x \leq a$. Clearly, $L \to \infty$ as $x \to \frac{a}{2}+$, and $L(a) = \sqrt{2}a$. For critical points of L^2:

$$0 = \frac{d(L^2)}{dx} = \frac{(2ax - a^2)(6ax^2) - (2ax^3)(2a)}{(2ax - a^2)^2}$$
$$= \frac{2a^2 x^2 (4x - 3a)}{(2ax - a^2)^2}.$$

The only critical point in $\left(\frac{a}{2}, a\right]$ is $x = \frac{3a}{4}$. Since $L\left(\frac{3a}{4}\right) = \frac{3\sqrt{3}a}{4} < L(a)$, therefore the least possible length for the fold is $\frac{3\sqrt{3}a}{4}$ cm.

Section 4.6 Finding Roots (page 272)

1. $f(x) = x^2 - 2$, $f'(x) = 2x$.
Newton's formula $x_{n+1} = g(x_n)$, where

$$g(x) = x - \frac{x^2 - 2}{2x} = \frac{x^2 + 2}{2x}.$$

Starting with $x_0 = 1.5$, get $x_3 = x_4 = 1.41421356237$.

2. $f(x) = x^2 - 3$, $f'(x) = 2x$.
Newton's formula $x_{n+1} = g(x_n)$, where

$$g(x) = x - \frac{x^2 - 3}{2x} = \frac{x^2 + 3}{2x}.$$

Starting with $x_0 = 1.5$, get $x_4 = x_5 = 1.73205080757$.

3. $f(x) = x^3 + 2x - 1$, $f'(x) = 3x^2 + 2$.
Newton's formula $x_{n+1} = g(x_n)$, where

$$g(x) = x - \frac{x^3 + 2x - 1}{3x^2 + 2} = \frac{2x^3 + 1}{3x^2 + 2}.$$

Starting with $x_0 = 0.5$, get $x_3 = x_4 = 0.45339765152$.

4. $f(x) = x^3 + 2x^2 - 2$, $f'(x) = 3x^2 + 4x$.
Newton's formula $x_{n+1} = g(x_n)$, where

$$g(x) = x - \frac{x^3 + 2x^2 - 2}{3x^2 + 4x} = \frac{2x^3 + 2x^2 + 2}{3x^2 + 4x}.$$

Starting with $x_0 = 1.5$, get $x_5 = x_6 = 0.839286755214$.

5. $f(x) = x^4 - 8x^2 - x + 16$, $f'(x) = 4x^3 - 16x - 1$.
Newton's formula $x_{n+1} = g(x_n)$, where

$$g(x) = x - \frac{x^4 - 8x^2 - x + 16}{4x^3 - 16x - 1} = \frac{3x^4 - 8x^2 - 16}{4x^3 - 16x - 1}.$$

Starting with $x_0 = 1.5$, get $x_4 = x_5 = 1.64809536561$.
Starting with $x_0 = 2.5$, get $x_5 = x_6 = 2.35239264766$.

6. $f(x) = x^3 + 3x^2 - 1$, $f'(x) = 3x^2 + 6x$.
Newton's formula $x_{n+1} = g(x_n)$, where

$$g(x) = x - \frac{x^3 + 3x^2 - 1}{3x^2 + 6x} = \frac{2x^3 + 3x^2 + 1}{3x^2 + 6x}.$$

Because $f(-3) = -1$, $f(-2) = 3$, $f(-1) = 1$, $f(0) = -1$, $f(1) = 3$, there are roots between -3 and -2, between -1 and 0, and between 0 and 1.
Starting with $x_0 = -2.5$, get $x_5 = x_6 = -2.87938524157$.
Starting with $x_0 = -0.5$, get $x_4 = x_5 = -0.652703644666$.
Starting with $x_0 = 0.5$, get $x_4 = x_5 = 0.532088886328$.

7. $f(x) = \sin x - 1 + x$, $f'(x) = \cos x + 1$.
   Newton's formula is $x_{n+1} = g(x_n)$, where
   $$g(x) = x - \frac{\sin x - 1 + x}{\cos x + 1}.$$
   The graphs of $\sin x$ and $1 - x$ suggest a root near $x = 0.5$. Starting with $x_0 = 0.5$, get
   $x_3 = x_4 = 0.510973429389$.

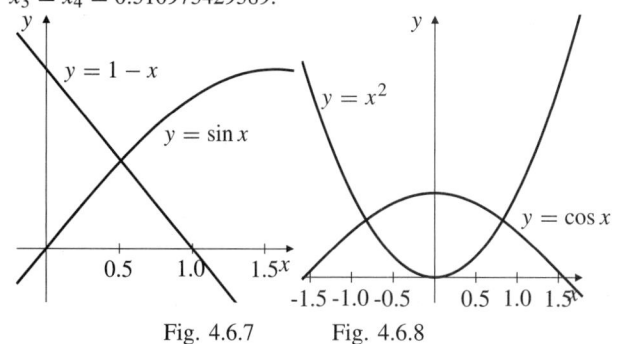

   Fig. 4.6.7   Fig. 4.6.8

8. $f(x) = x^2 - \cos x$, $f'(x) = 2x + \sin x$.
   Newton's formula is $x_{n+1} = g(x_n)$, where
   $$g(x) = x - \frac{x^2 - \cos x}{2x + \sin x}.$$
   The graphs of $\cos x$ and x^2, suggest a root near $x = \pm 0.8$. Starting with $x_0 = 0.8$, get
   $x_3 = x_4 = 0.824132312303$. The other root is the negative of this one, because $\cos x$ and x^2 are both even functions.

9. Since $\tan x$ takes all real values between any two consecutive odd multiples of $\pi/2$, its graph intersects $y = x$ infinitely often. Thus, $\tan x = x$ has infinitely many solutions. The one between $\pi/2$ and $3\pi/2$ is close to $3\pi/2$, so start with $x_0 = 4.5$. Newton's formula here is
   $$x_{n+1} = x_n - \frac{\tan x_n - x_n}{\sec^2 x_n - 1}.$$
   We get $x_3 = x_4 = 4.49340945791$.

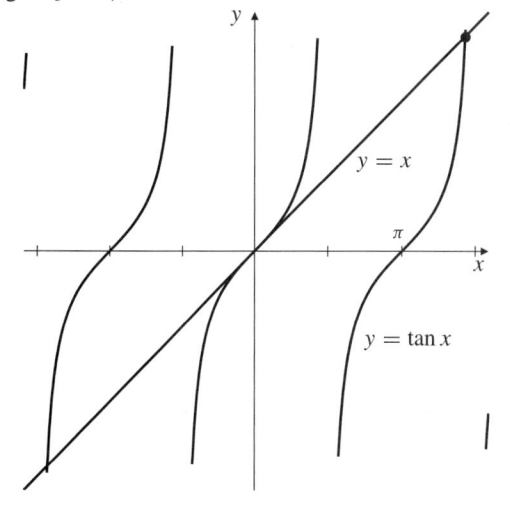

   Fig. 4.6.9

10. A graphing calculator shows that the equation
    $$(1 + x^2)\sqrt{x} - 1 = 0$$
    has a root near $x = 0.6$. Use of a solve routine or Newton's method gives $x = 0.56984029099806$.

12. Let $f(x) = \dfrac{\sin x}{1 + x^2}$. Since $|f(x)| \leq 1/(1 + x^2) \to 0$ as $x \to \pm\infty$ and $f(0) = 0$, the maximum and minimum values of f will occur at the two critical points of f that are closest to the origin on the right and left, respectively. For CP:
    $$0 = f'(x) = \frac{(1 + x^2)\cos x - 2x \sin x}{(1 + x^2)^2}$$
    $$0 = (1 + x^2)\cos x - 2x \sin x$$
    with $0 < x < \pi$ for the maximum and $-\pi < x < 0$ for the minimum. Solving this equation using a solve routine or Newton's Method starting, say, with $x_0 = 1.5$, we get $x = \pm 0.79801699184239$. The corresponding max and min values of f are ± 0.437414158279.

13. Let $f(x) = \dfrac{\cos x}{1 + x^2}$. Note that f is an even function, and that f has maximum value 1 at $x = 0$. (Clearly $f(0) = 1$ and $|f(x)| < 1$ if $x \neq 0$.) The minimum value will occur at the critical points closest to but not equal to 0. For CP:
    $$0 = f'(x) = \frac{(1 + x^2)(-\sin x) - 2x \cos x}{(1 + x^2)^2}$$
    $$0 = (1 + x^2)\sin x + 2x \cos x.$$
    The first CP to the right of zero is between $\pi/2$ and $3\pi/2$, so start with $x = 2.5$, say, and get $x = 2.5437321475261$. The minimum value is $f(x) = -0.110639672192$.

14. For $x^2 = 0$ we have $x_{n+1} = x_n - (x_n^2/(2x_n)) = x_n/2$.
    If $x_0 = 1$, then $x_1 = 1/2$, $x_2 = 1/4$, $x_3 = 1/8$.

    a) $x_n = 1/2^n$, by induction.

    b) x_n approximates the root $x = 0$ to within 0.0001 provided $2^n > 10,000$. We need $n \geq 14$ to ensure this.

    c) To ensure that x_n^2 is within 0.0001 of 0 we need $(1/2^n)^2 < 0.0001$, that is, $2^{2n} > 10,000$. We need $n \geq 7$.

    d) Convergence of Newton approximations to the root $x = 0$ of $x^2 = 0$ is slower than usual because the derivative $2x$ of x^2 is zero at the root.

15. $f(x) = \begin{cases} \sqrt{x} & \text{if } x \geq 0 \\ \sqrt{-x} & \text{if } x < 0 \end{cases}$,

$f'(x) = \begin{cases} 1/(2\sqrt{x}) & \text{if } x > 0 \\ -1/(2\sqrt{-x}) & \text{if } x < 0 \end{cases}$.

The Newton's Method formula says that

$$x_{n+1} = x_n - \frac{f(x_n)}{f'(x_n)} = x_n - 2x_n = -x_n.$$

If $x_0 = a$, then $x_1 = -a$, $x_2 = a$, and, in general, $x_n = (-1)^n a$. The approximations oscillate back and forth between two numbers.

If one observed that successive approximations were oscillating back and forth between two values a and b, one should try their average, $(a+b)/2$, as a new starting guess. It may even turn out to be the root!

16. Newton's Method formula for $f(x) = x^{1/3}$ is

$$x_{n+1} = x_n - \frac{x_n^{1/3}}{(1/3)x_n^{-2/3}} = x_n - 3x_n = -2x_n.$$

If $x_0 = 1$, then $x_1 = -2$, $x_2 = 4$, $x_3 = -8$, $x_4 = 16$, and, in general, $x_n = (-2)^n$. The successive "approximations" oscillate ever more widely, diverging from the root at $x = 0$.

17. Newton's Method formula for $f(x) = x^{2/3}$ is

$$x_{n+1} = x_n - \frac{x_n^{2/3}}{(2/3)x_n^{-1/3}} = x_n - \tfrac{3}{2}x_n = -\tfrac{1}{2}x_n.$$

If $x_0 = 1$, then $x_1 = -1/2$, $x_2 = 1/4$, $x_3 = -1/8$, $x_4 = 1/16$, and, in general, $x_n = (-1/2)^n$. The successive approximations oscillate around the root $x = 0$, but still converge to it (though more slowly than is usual for Newton's Method).

18. To solve $1 + \tfrac{1}{4}\sin x = x$, start with $x_0 = 1$ and iterate $x_{n+1} = 1 + \tfrac{1}{4}\sin x_n$. x_5 and x_6 round to 1.23613.

19. To solve $\cos(x/3) = x$, start with $x_0 = 0.9$ and iterate $x_{n+1} = \cos(x_n/3)$. x_4 and x_5 round to 0.95025.

20. To solve $(x+9)^{1/3} = x$, start with $x_0 = 2$ and iterate $x_{n+1} = (x_n + 9)^{1/3}$. x_4 and x_5 round to 2.24004.

21. To solve $1/(2+x^2) = x$, start with $x_0 = 0.5$ and iterate $x_{n+1} = 1/(2+x_n^2)$. x_6 and x_7 round to 0.45340.

22. To solve $x^3 + 10x - 10 = 0$, start with $x_0 = 1$ and iterate $x_{n+1} = 1 - \tfrac{1}{10}x_n^3$. x_7 and x_8 round to 0.92170.

23. r is a fixed point of $N(x)$

$$\begin{aligned}&\Longleftrightarrow\ r = N(r) = r - \frac{f(r)}{f'(r)}\\ &\Longleftrightarrow\ 0 = -f(r)/f'(r)\\ &\Longleftrightarrow\ f(r) = 0\end{aligned}$$

i.e., if and only if r is a root of $f(x) = 0$. In this case, $x_{n+1} = N(x_n)$ is the nth Newton's Method approximation to the root, starting from the initial guess x_0.

24. Let $g(x) = f(x) - x$ for $a \leq x \leq b$. g is continuous (because f is), and since $a \leq f(x) \leq b$ whenever $a \leq x \leq b$ (by condition (i)), we know that $g(a) \geq 0$ and $g(b) \leq 0$. By the Intermediate-Value Theorem there exists r in $[a, b]$ such that $g(r) = 0$, that is, such that $f(r) = r$.

25. We are given that there is a constant K satisfying $0 < K < 1$, such that

$$|f(u) - f(v)| \leq K|u - v|$$

holds whenever u and v are in $[a, b]$. Pick any x_0 in $[a, b]$, and let $x_1 = f(x_0)$, $x_2 = f(x_1)$, and, in general, $x_{n+1} = f(x_n)$. Let r be the fixed point of f in $[a, b]$ found in Exercise 24. Thus $f(r) = r$. We have

$$|x_1 - r| = |f(x_0) - f(r)| \leq K|x_0 - r|$$
$$|x_2 - r| = |f(x_1) - f(r)| \leq K|x_1 - r| \leq K^2|x_0 - r|,$$

and, in general, by induction

$$|x_n - r| \leq K^n|x_0 - r|.$$

Since $K < 1$, $\lim_{n \to \infty} K^n = 0$, so $\lim_{n \to \infty} x_n = r$. The iterates converge to the fixed point as claimed in Theorem 6.

Section 4.7 Linear Approximations (page 279)

1. $f(x) = x^2$, $f'(x) = 2x$, $f(3) = 9$, $f'(3) = 6$.
Linearization at $x = 3$: $L(x) = 9 + 6(x - 3)$.

2. $f(x) = x^{-3}$, $f'(x) = -3x^{-4}$, $f(2) = 1/8$, $f'(2) = -3/16$.
Linearization at $x = 2$: $L(x) = \tfrac{1}{8} - \tfrac{3}{16}(x-2)$.

3. $f(x) = \sqrt{4-x}$, $f'(x) = -1/(2\sqrt{4-x})$, $f(0) = 2$, $f'(0) = -1/4$.
Linearization at $x = 0$: $L(x) = 2 - \tfrac{1}{4}x$.

4. $f(x) = \sqrt{3+x^2}$, $f'(x) = x/\sqrt{3+x^2}$, $f(1) = 2$, $f'(1) = 1/2$.
Linearization at $x = 1$: $L(x) = 2 + \tfrac{1}{2}(x-1)$.

5. $f(x) = (1+x)^{-2}$, $f'(x) = -2(1+x)^{-3}$, $f(2) = 1/9$, $f'(2) = -2/27$.
Linearization at $x = 2$: $L(x) = \tfrac{1}{9} - \tfrac{2}{27}(x-2)$.

6. $f(x) = x^{-1/2}$, $f'(x) = (-1/2)x^{-3/2}$, $f(4) = 1/2$, $f'(4) = -1/16$.
Linearization at $x = 4$: $L(x) = \tfrac{1}{2} - \tfrac{1}{16}(x-4)$.

7. $f(x) = \sin x$, $f'(x) = \cos x$, $f(\pi) = 0$, $f'(\pi) = -1$.
Linearization at $x = \pi$: $L(x) = -(x - \pi)$.

INSTRUCTOR'S SOLUTIONS MANUAL   SECTION 4.7 (PAGE 279)

8. $f(x) = \cos(2x)$, $f'(x) = -2\sin(2x)$, $f(\pi/3) = -1/2$, $f'(\pi/3) = -\sqrt{3}$.
    Linearization at $x = \pi/3$: $L(x) = -\frac{1}{2} - \sqrt{3}\left(x - \frac{\pi}{3}\right)$.

9. $f(x) = \sin^2 x$, $f'(x) = 2\sin x \cos x$, $f(\pi/6) = 1/4$, $f'(\pi/6) = \sqrt{3}/2$.
    Linearization at $x = \pi/6$: $L(x) = \frac{1}{4} + (\sqrt{3}/2)\left(x - \frac{\pi}{6}\right)$.

10. $f(x) = \tan x$, $f'(x) = \sec^2 x$, $f(\pi/4) = 1$, $f'(\pi/4) = 2$.
    Linearization at $x = \pi/4$: $L(x) = 1 + 2\left(x - \frac{\pi}{4}\right)$.

11. If A and x are the area and side length of the square, then $A = x^2$. If $x = 10$ cm and $\Delta x = 0.4$ cm, then
    $$\Delta A \approx \frac{dA}{dx}\Delta x = 2x\,\Delta x = 20(0.4) = 8.$$
    The area increases by about 8 cm².

12. If V and x are the volume and side length of the cube, then $V = x^3$. If $x = 20$ cm and $\Delta V = -12$ cm³, then
    $$-12 = \Delta V \approx \frac{dV}{dx}\Delta x = 3x^2\,\Delta x = 1{,}200\,\Delta x,$$
    so that $\Delta x = -1/100$. The edge length must decrease by about 0.01 cm in to decrease the volume by 12 cm³.

13. The circumference C and radius r of the orbit are linked by $C = 2\pi r$. Thus $\Delta C = 2\pi\,\Delta r$. If $\Delta r = -10$ mi then $\Delta C \approx 2\pi\,\Delta r = 20\pi$. The circumference of the orbit will decrease by about $20\pi \approx 62.8$ mi if the radius decreases by 10 mi. Note that the answer does not depend on the actual radius of the orbit.

14. $a = g[R/(R+h)]^2$ implies that
    $$\Delta a \approx \frac{da}{dh}\Delta h = gR^2 \frac{-2}{(R+h)^3}\Delta h.$$
    If $h = 0$ and $\Delta h = 10$ mi, then
    $$\Delta a \approx -\frac{20g}{R} = -\frac{20 \times 32}{3960} \approx 0.16 \text{ ft/s}^2.$$

15. $f(x) = x^{1/2}$, $f'(x) = \frac{1}{2}x^{-1/2}$, $f''(x) = -\frac{1}{4}x^{-3/2}$
    $\sqrt{50} = f(50) \approx f(49) + f'(49)(50 - 49)$
    $= 7 + \frac{1}{14} = \frac{99}{14} \approx 7.071$.
    $f''(x) < 0$ on [49, 50], so error is negative: $\sqrt{50} < \frac{99}{14}$
    $|f''(x)| < \frac{1}{4 \times 49^{3/2}} = \frac{1}{4 \times 7^3} = \frac{1}{1372} \approx 0.00073 = k$ on (49, 50).
    Thus $|\text{error}| \leq \frac{k}{2}(50-49)^2 = \frac{1}{2744} = 0.00036$. We have
    $$\frac{99}{14} - \frac{1}{2744} \leq \sqrt{50} \leq \frac{99}{14},$$

    i.e., $7.071064 \leq \sqrt{50} \leq 7.071429$.

16. Let $f(x) = \sqrt{x}$, then $f'(x) = \frac{1}{2}x^{-1/2}$ and $f''(x) = -\frac{1}{4}x^{-3/2}$. Hence,
    $$\sqrt{47} = f(47) \approx f(49) + f'(49)(47 - 49)$$
    $$= 7 + \left(\frac{1}{14}\right)(-2) = \frac{48}{7} \approx 6.8571429.$$
    Clearly, if $x \geq 36$, then
    $$|f''(x)| \leq \frac{1}{4 \times 6^3} = \frac{1}{864} = K.$$
    Since $f''(x) < 0$, f is concave down. Therefore, the error $E = \sqrt{47} - \frac{48}{7} < 0$ and
    $$|E| < \frac{K}{2}(47-49)^2 = \frac{1}{432}.$$
    Thus,
    $$\frac{48}{7} - \frac{1}{432} < \sqrt{47} < \frac{48}{7}$$
    $$6.8548 < \sqrt{47} < 6.8572.$$

17. $f(x) = x^{1/4}$, $f'(x) = \frac{1}{4}x^{-3/4}$, $f''(x) = -\frac{3}{16}x^{-7/4}$
    $\sqrt[4]{85} = f(85) \approx f(81) + f'(81)(85-81)$
    $= 3 + \frac{4}{4 \times 27} = 3 + \frac{1}{27} = \frac{82}{27} \approx 3.037$.
    $f''(x) < 0$ on [81, 85] so error is negative: $\sqrt[4]{85} < \frac{82}{27}$.
    $|f''(x)| < \frac{3}{16 \times 3^7} = \frac{1}{11{,}664} = k$ on [81, 85].
    Thus $|\text{Error}| \leq \frac{k}{2}(85-81)^2 = 0.00069$.
    $$\frac{82}{27} - \frac{1}{1458} < \sqrt[4]{85} < \frac{82}{27},$$
    or $3.036351 \leq \sqrt[4]{85} \leq 3.037037$

18. Let $f(x) = \frac{1}{x}$, then $f'(x) = -\frac{1}{x^2}$ and $f''(x) = \frac{2}{x^3}$. Hence,
    $$\frac{1}{2.003} = f(2.003) \approx f(2) + f'(2)(0.003)$$
    $$= \frac{1}{2} + \left(-\frac{1}{4}\right)(0.003) = 0.49925.$$
    If $x \geq 2$, then $|f''(x)| \leq \frac{2}{8} = \frac{1}{4}$. Since $f''(x) > 0$ for $x > 0$, f is concave up. Therefore, the error
    $$E = \frac{1}{2.003} - 0.49925 > 0$$

151

and
$$|E| < \frac{1}{8}(0.003)^2 = 0.000001125.$$

Thus,
$$0.49925 < \frac{1}{2.003} < 0.49925 + 0.000001125$$
$$0.49925 < \frac{1}{2.003} < 0.499251125.$$

19. $f(x) = \cos x,\quad f'(x) = -\sin x,\quad f''(x) = -\cos x$
$$\cos 46° = \cos\left(\frac{\pi}{4} + \frac{\pi}{180}\right)$$
$$\approx \cos\frac{\pi}{4} - \sin\left(\frac{\pi}{4}\right)\left(\frac{\pi}{180}\right)$$
$$= \frac{1}{\sqrt{2}}\left(1 - \frac{\pi}{180}\right) \approx 0.694765.$$

$f''(0) < 0$ on $[45°, 46°]$ so
$$|\text{Error}| < \frac{1}{2\sqrt{2}}\left(\frac{\pi}{180}\right)^2 \approx 0.0001.$$

We have
$$\frac{1}{\sqrt{2}}\left(1 - \frac{\pi}{180} - \frac{\pi^2}{2 \times 180^2}\right) < \cos 46° < \frac{1}{\sqrt{2}}\left(1 - \frac{\pi}{180}\right)$$

i.e., $0.694658 \leq \cos 46° < 0.694765.$

20. Let $f(x) = \sin x$, then $f'(x) = \cos x$ and $f''(x) = -\sin x$. Hence,
$$\sin\left(\frac{\pi}{5}\right) = f\left(\frac{\pi}{6} + \frac{\pi}{30}\right) \approx f\left(\frac{\pi}{6}\right) + f'\left(\frac{\pi}{6}\right)\left(\frac{\pi}{30}\right)$$
$$= \frac{1}{2} + \frac{\sqrt{3}}{2}\left(\frac{\pi}{30}\right) \approx 0.5906900.$$

If $x \leq \frac{\pi}{4}$, then $|f''(x)| \leq \frac{1}{\sqrt{2}}$. Since $f''(x) < 0$ on $0 < x \leq 90°$, f is concave down. Therefore, the error E is negative and
$$|E| < \frac{1}{2\sqrt{2}}\left(\frac{\pi}{30}\right)^2 = 0.0038772.$$

Thus,
$$0.5906900 - 0.0038772 < \sin\left(\frac{\pi}{5}\right) < 0.5906900$$
$$0.5868128 < \sin\left(\frac{\pi}{5}\right) < 0.5906900.$$

21. Let $f(x) = \sin x$, then $f'(x) = \cos x$ and $f''(x) = -\sin x$. The linearization at $x = \pi$ gives:
$$\sin(3.14) \approx \sin\pi + \cos\pi(3.14-\pi) = \pi - 3.14 \approx 0.001592654.$$

Since $f''(x) < 0$ between 3.14 and π, the error E in the above approximation is negative: $\sin(3.14) < 0.001592654$. For $3.14 \leq t \leq \pi$, we have
$$|f''(t)| = \sin t \leq \sin(3.14) < 0.001592654.$$

Thus the error satisfies
$$|E| \leq \frac{0.001592654}{2}(3.14 - \pi)^2 < 0.000000002.$$

Therefore $0.001592652 < \sin(3.14) < 0.001592654.$

22. Let $f(x) = \sin x$, then $f'(x) = \cos x$ and $f''(x) = -\sin x$. The linearization at $x = 30° = \pi/6$ gives
$$\sin(33°) = \sin\left(\frac{\pi}{6} + \frac{\pi}{60}\right)$$
$$\approx \sin\frac{\pi}{6} + \cos\frac{\pi}{6}\left(\frac{\pi}{60}\right)$$
$$= \frac{1}{2} + \frac{\sqrt{3}}{2}\left(\frac{\pi}{60}\right) \approx 0.545345.$$

Since $f''(x) < 0$ between $30°$ and $33°$, the error E in the above approximation is negative: $\sin(33°) < 0.545345$. For $30° \leq t \leq 33°$, we have
$$|f''(t)| = \sin t \leq \sin(33°) < 0.545345.$$

Thus the error satisfies
$$|E| \leq \frac{0.545345}{2}\left(\frac{\pi}{60}\right)^2 < 0.000747.$$

Therefore
$$0.545345 - 0.000747 < \sin(33°) < 0.545345$$
$$0.544598 < \sin(33°) < 0.545345.$$

23. From the solution to Exercise 15, the linearization to $f(x) = x^{1/2}$ at $x = 49$ has value at $x = 50$ given by
$$L(50) = f(49) + f'(49)(50 - 49) \approx 7.071429.$$

Also, $7.071064 \leq \sqrt{50} \leq 7.071429$, and, since $f''(x) = -1/(4(\sqrt{x})^3)$,
$$\frac{-1}{4(7)^3} \leq f''(x) \leq \frac{-1}{4(\sqrt{50})^3} \leq \frac{-1}{4(7.071429)^3}$$

for $49 \leq x \leq 50$. Thus, on that interval, $M \leq f''(x) \leq N$, where $M = -0.000729$ and $N = -0.000707$. By Corollary C,
$$L(50) + \frac{M}{2}(50 - 49)^2 \leq f(50) \leq L(50) + \frac{N}{2}(50 - 49)^2$$
$$7.071064 \leq \sqrt{50} \leq 7.071075.$$

Using the midpoint of this interval as a new approximation for $\sqrt{50}$ ensures that the error is no greater than half the length of the interval:

$$\sqrt{50} \approx 7.071070, \quad |\text{error}| \leq 0.000006.$$

24. From the solution to Exercise 16, the linearization to $f(x) = x^{1/2}$ at $x = 49$ has value at $x = 47$ given by

$$L(47) = f(49) + f'(49)(47 - 49) \approx 6.8571429.$$

Also, $6.8548 \leq \sqrt{47} \leq 6.8572$, and, since $f''(x) = -1/(4(\sqrt{x})^3)$,

$$\frac{-1}{4(6.8548)^3} \leq \frac{-1}{4(\sqrt{47})^3} \leq f''(x) \leq \frac{-1}{4(7)^3}$$

for $47 \leq x \leq 49$. Thus, on that interval, $M \leq f''(x) \leq N$, where $M = -0.000776$ and $N = -0.000729$. By Corollary C,

$$L(47) + \frac{M}{2}(47 - 49)^2 \leq f(47) \leq L(47) + \frac{N}{2}(47 - 49)^2$$
$$6.855591 \leq \sqrt{47} \leq 6.855685.$$

Using the midpoint of this interval as a new approximation for $\sqrt{47}$ ensures that the error is no greater than half the length of the interval:

$$\sqrt{47} \approx 6.855638, \quad |\text{error}| \leq 0.000047.$$

25. From the solution to Exercise 17, the linearization to $f(x) = x^{1/4}$ at $x = 81$ has value at $x = 85$ given by

$$L(85) = f(81) + f'(81)(85 - 81) \approx 3.037037.$$

Also, $3.036351 \leq 85^{1/4} \leq 3.037037$, and, since $f''(x) = -3/(16(x^{1/4})^7)$,

$$\frac{-3}{16(3)^7} \leq f''(x) \leq \frac{-3}{16(85^{1/4})^7} \leq \frac{-3}{16(3.037037)^7}$$

for $81 \leq x \leq 85$. Thus, on that interval, $M \leq f''(x) \leq N$, where $M = -0.000086$ and $N = -0.000079$. By Corollary C,

$$L(85) + \frac{M}{2}(85 - 81)^2 \leq f(85) \leq L(85) + \frac{N}{2}(85 - 81)^2$$
$$3.036351 \leq 85^{1/4} \leq 3.036405.$$

Using the midpoint of this interval as a new approximation for $85^{1/4}$ ensures that the error is no greater than half the length of the interval:

$$85^{1/4} \approx 3.036378, \quad |\text{error}| \leq 0.000028.$$

26. From the solution to Exercise 22, the linearization to $f(x) = \sin x$ at $x = 30° = \pi/6$ has value at $x = 33° = \pi/6 + \pi/60$ given by

$$L(33°) = f(\pi/6) + f'(\pi/6)(\pi/60) \approx 0.545345.$$

Also, $0.544597 \leq \sin(33°) \leq 5.545345$, and, since $f''(x) = -\sin x$,

$$-\sin(33°) \leq f''(x) \leq -\sin(30°)$$

for $30° \leq x \leq 33°$. Thus, on that interval, $M \leq f''(x) \leq N$, where $M = -0.545345$ and $N = -0.5$. By Corollary C,

$$L(33°) + \frac{M}{2}(\pi/60)^2 \leq \sin(33°) \leq L(33°) + \frac{N}{2}(\pi/60)^2$$
$$0.544597 \leq \sin(33°) \leq 0.544660.$$

Using the midpoint of this interval as a new approximation for $\sin(33°)$ ensures that the error is no greater than half the length of the interval:

$$\sin(33°) \approx 0.544629, \quad |\text{error}| \leq 0.000031.$$

27. $f(2) = 4$, $f'(2) = -1$, $0 \leq f''(x) \leq \frac{1}{x}$ if $x > 0$.
   $f(3) \approx f(2) + f'(2)(3 - 2) = 4 - 1 = 3$.
   $f''(x) \geq 0 \Rightarrow \text{error} \geq 0 \Rightarrow f(3) \geq 3$.
   $|f''(x)| \leq \frac{1}{x} \leq \frac{1}{2}$ if $2 \leq x \leq 3$, so $|\text{Error}| \leq \frac{1}{4}(3-2)^2$.
   Thus $3 \leq f(3) \leq 3\frac{1}{4}$

28. The linearization of $f(x)$ about $x = 2$ is

$$L(x) = f(2) + f'(2)(x - 2) = 4 - (x - 2).$$

Thus $L(3) = 3$. Also, since $1/(2x) \leq f''(x) \leq 1/x$ for $x > 0$, we have for $2 \leq x \leq 3$, $(1/6) \leq f''(x) \leq (1/2)$. Thus

$$3 + \frac{1}{2}\left(\frac{1}{6}\right)(3-2)^2 \leq f(3) \leq 3 + \frac{1}{2}\left(\frac{1}{2}\right)(3-2)^2.$$

The best approximation for $f(3)$ is the midpoint of this interval: $f(3) \approx 3\frac{1}{6}$.

29. The linearization of $g(x)$ about $x = 2$ is

$$L(x) = g(2) + g'(2)(x - 2) = 1 + 2(x - 2).$$

Thus $L(1.8) = 0.6$.
If $|g''(x)| \leq 1 + (x - 2)^2$ for $x > 0$, then $|g''(x)| < 1 + (-0.2)^2 = 1.04$ for $1.8 \leq x \leq 2$. Hence $g(1.8) \approx 0.6$ with $|\text{error}| < \frac{1}{2}(1.04)(1.8 - 2)^2 = 0.0208$.

30. If $f(\theta) = \sin\theta$, then $f'(\theta) = \cos\theta$ and $f''(\theta) = -\sin\theta$. Since $f(0) = 0$ and $f'(0) = 1$, the linearization of f at $\theta = 0$ is $L(\theta) = 0 + 1(\theta - 0) = \theta$.
If $0 \le t \le \theta$, then $f''(t) \le 0$, so $0 \le \sin\theta \le \theta$.
If $0 \ge t \ge \theta$, then $f''(t) \ge 0$, so $0 \ge \sin\theta \ge \theta$.
In either case, $|\sin t| \le |\sin\theta| \le |\theta|$ if t is between 0 and θ. Thus the error $E(\theta)$ in the approximation $\sin\theta \approx \theta$ satisfies

$$|E(\theta)| \le \frac{|\theta|}{2}|\theta|^2 = \frac{|\theta|^3}{2}.$$

If $|\theta| \le 17° = 17\pi/180$, then

$$\frac{|E(\theta)|}{|\theta|} \le \frac{1}{2}\left(\frac{17\pi}{180}\right)^2 \approx 0.044.$$

Thus the percentage error is less than 5%.

31. $V = \frac{4}{3}\pi r^3 \Rightarrow \Delta V \approx 4\pi r^2 \Delta r$
If $r = 20.00$ and $\Delta r = 0.20$, then
$\Delta V \approx 4\pi(20.00)^2(0.20) \approx 1005$.
The volume has increased by about 1005 cm^2.

Section 4.8 Taylor Polynomials (page 287)

1. If $f(x) = e^{-x}$, then $f^{(k)}(x) = (-1)^k e^{-x}$, so $f^{(k)}(0) = (-1)^k$. Thus

$$P_4(x) = 1 - x + \frac{x^2}{2!} - \frac{x^3}{3!} + \frac{x^4}{4!}.$$

2. If $f(x) = \cos x$, then $f'(x) = -\sin x$, $f''(x) = -\cos x$, and $f'''(x) = \sin x$. In particular, $f(\pi/4) = f'''(\pi/4) = 1/\sqrt{2}$ and $f'(\pi/4) = f''(\pi/4) = -1/\sqrt{2}$. Thus

$$P_3(x) = \frac{1}{\sqrt{2}}\left[1 - \left(x - \frac{\pi}{4}\right) - \frac{1}{2}\left(x - \frac{\pi}{4}\right)^2 + \frac{1}{6}\left(x - \frac{\pi}{4}\right)^3\right].$$

3.
$f(x) = \ln x$ $f(e) = 1$
$f'(x) = \dfrac{1}{x}$ $f'(e) = \dfrac{1}{e}$
$f''(x) = \dfrac{-1}{x^2}$ $f''(e) = \dfrac{-1}{e^2}$
$f'''(x) = \dfrac{2}{x^3}$ $f'''(e) = \dfrac{2}{e^3}$
$f^{(4)}(x) = \dfrac{-6}{x^4}$ $f^{(4)}(e) = \dfrac{-6}{e^4}$

Thus

$$P_4(x) = 1 + \frac{1}{e}(x-e) - \frac{1}{2e^2}(x-e)^2 + \frac{1}{3e^3}(x-e)^3 - \frac{1}{4e^4}(x-e)^4.$$

4.
$f(x) = \sec x$ $f(0) = 1$
$f'(x) = \sec x \tan x$ $f'(0) = 0$
$f''(x) = 2\sec^3 x - \sec x$ $f''(0) = 1$
$f'''(x) = (6\sec^2 x - 1)\sec x \tan x$ $f'''(0) = 0$

Thus $P_3(x) = 1 + (x^2/2)$.

5.
$f(x) = x^{1/2}$ $f(4) = 2$
$f'(x) = \dfrac{1}{2}x^{-1/2}$ $f'(4) = \dfrac{1}{4}$
$f''(x) = \dfrac{-1}{4}x^{-3/2}$ $f''(4) = \dfrac{-1}{32}$
$f'''(x) = \dfrac{3}{8}x^{-5/2}$ $f'''(4) = \dfrac{3}{256}$

Thus

$$P_3(x) = 2 + \frac{1}{4}(x-4) - \frac{1}{64}(x-4)^2 + \frac{1}{512}(x-4)^3.$$

6.
$f(x) = \dfrac{1}{2+x}$ $f(1) = \dfrac{1}{3}$
$f'(x) = \dfrac{-1}{(2+x)^2}$ $f'(1) = \dfrac{-1}{9}$
$f''(x) = \dfrac{2!}{(2+x)^3}$ $f''(1) = \dfrac{2!}{27}$
$f'''(x) = \dfrac{-3!}{(2+x)^4}$ $f'''(1) = \dfrac{-3!}{3^4}$
$\vdots$ $\vdots$
$f^{(n)}(x) = \dfrac{(-1)^n n!}{(2+x)^{n+1}}$ $f^{(n)}(1) = \dfrac{(-1)^n n!}{3^{n+1}}$

Thus

$$P_n(x) = \frac{1}{3} - \frac{1}{9}(x-1) + \frac{1}{27}(x-1)^2 - \cdots + \frac{(-1)^n}{3^{n+1}}(x-1)^n.$$

7. $f(x) = x^{1/3}$, $f'(x) = \dfrac{1}{3}x^{-2/3}$,
$f''(x) = -\dfrac{2}{9}x^{-5/3}$, $f'''(x) = \dfrac{10}{27}x^{-8/3}$.

$a = 8$: $f(x) \approx f(8) + f'(8)(x-8) + \dfrac{f''(8)}{2}(x-8)^2$

$$= 2 + \frac{1}{12}(x-8) - \frac{1}{9 \times 32}(x-8)^2$$

$9^{1/2} \approx 2 + \dfrac{1}{12} - \dfrac{1}{288} \approx 2.07986$

Error $= \dfrac{f'''(X)}{3!}(9-8)^3 = \dfrac{10}{27 \times 6}\dfrac{1}{X^{8/3}}$ for some X in $[8, 9]$.
For $8 \le X \le 9$ we have $X^{8/3} \ge 8^{8/3} = 2^8 = 256$ so

$$0 < \text{Error} \le \frac{5}{81 \times 256} < 0.000241.$$

Thus $2.07986 < 9^{1/3} < 2.08010$.

8. Since $f(x) = \sqrt{x}$, then $f'(x) = \frac{1}{2}x^{-1/2}$, $f''(x) = -\frac{1}{4}x^{-3/2}$ and $f'''(x) = \frac{3}{8}x^{-5/2}$. Hence,

$$\sqrt{61} \approx f(64) + f'(64)(61-64) + \frac{1}{2}f''(64)(61-64)^2$$
$$= 8 + \frac{1}{16}(-3) - \frac{1}{2}\left(\frac{1}{2048}\right)(-3)^2 \approx 7.8103027.$$

The error is $R_2 = R_2(f; 64, 61) = \frac{f'''(X)}{3!}(61-64)^3$ for some X between 61 and 64. Clearly $R_2 < 0$. If $t \geq 49$, and in particular $61 \leq t \leq 64$, then

$$|f'''(t)| \leq \frac{3}{8}(49)^{-5/2} = 0.0000223 = K.$$

Hence,
$$|R_2| \leq \frac{K}{3!}|61-64|^3 = 0.0001004.$$

Since $R_2 < 0$, therefore,

$$7.8103027 - 0.0001004 < \sqrt{61} < 7.8103027$$
$$7.8102023 < \sqrt{61} < 7.8103027.$$

9. $f(x) = \frac{1}{x}$, $f'(x) = -\frac{1}{x^2}$, $f''(x) = \frac{2}{x^3}$, $f'''(x) = \frac{-6}{x^4}$.

$a = 1$: $f(x) \approx 1 - (x-1) + \frac{2}{2}(x-1)^2$

$\frac{1}{1.02} \approx 1 - (0.02) + (0.02)^2 = 0.9804.$

Error $= \frac{f'''(X)}{3!}(0.02)^3 = -\frac{1}{X^4}(0.02)^3$ where $1 \leq X \leq 1.02$.

Therefore, $-(0.02)^3 \leq \frac{1}{1.02} - 0.9804 < 0$,

i.e., $0.980392 \leq \frac{1}{1.02} < 0.980400.$

10. Since $f(x) = \tan^{-1} x$, then

$f'(x) = \frac{1}{1+x^2}$, $f''(x) = \frac{-2x}{(1+x^2)^2}$, $f'''(x) = \frac{-2+6x^2}{(1+x^2)^3}.$

Hence,
$\tan^{-1}(0.97) \approx f(1) + f'(1)(0.97-1) + \frac{1}{2}f''(1)(0.97-1)^2$
$= \frac{\pi}{4} + \frac{1}{2}(-0.03) + \left(-\frac{1}{4}\right)(-0.03)^2$
$= 0.7701731.$

The error is $R_2 = \frac{f'''(X)}{3!}(-0.03)^3$ for some X between 0.97 and 1. Note that $R_2 < 0$. If $0.97 \leq t \leq 1$, then

$$|f'''(t)| \leq f'''(1) = \frac{-2+6}{(1.97)^3} < 0.5232 = K.$$

Hence,
$$|R_2| \leq \frac{K}{3!}|0.97-1|^3 < 0.0000024.$$

Since $R_2 < 0$,

$$0.7701731 - 0.0000024 < \tan^{-1}(0.97) < 0.7701731$$
$$0.7701707 < \tan^{-1}(0.97) < 0.7701731.$$

11. $f(x) = e^x$, $f^{(k)}(x) = e^x$ for $k = 1, 2, 3 \ldots$

$a = 0$: $f(x) \approx 1 + x + \frac{x^2}{2}$

$e^{-0.5} \approx 1 - 0.5 + \frac{(0.5)^2}{2} = 0.625$

Error $= \frac{f'''(X)}{6}(0.5)^3 = \frac{e^X}{6}(-0.05)^3$ for some X between -0.5 and 0. Thus

$$|\text{Error}| < \frac{(0.5)^3}{6} < 0.020834,$$

and $-0.020833 < e^{-0.5} - 0.625 < 0$, or $0.604 < e^{-0.5} < 0.625.$

12. Since $f(x) = \sin x$, then $f'(x) = \cos x$, $f''(x) = -\sin x$ and $f'''(x) = -\cos x$. Hence,

$\sin(47°) = f\left(\frac{\pi}{4} + \frac{\pi}{90}\right)$
$\approx f\left(\frac{\pi}{4}\right) + f'\left(\frac{\pi}{4}\right)\left(\frac{\pi}{90}\right) + \frac{1}{2}f''\left(\frac{\pi}{4}\right)\left(\frac{\pi}{90}\right)^2$
$= \frac{1}{\sqrt{2}} + \frac{1}{\sqrt{2}}\left(\frac{\pi}{90}\right) - \frac{1}{2\sqrt{2}}\left(\frac{\pi}{90}\right)^2$
$\approx 0.7313587.$

The error is $R_2 = \frac{f'''(X)}{3!}\left(\frac{\pi}{90}\right)^3$ for some X between $45°$ and $47°$. Observe that $R_2 < 0$. If $45° \leq t \leq 47°$, then

$$|f'''(t)| \leq |-\cos 45°| = \frac{1}{\sqrt{2}} = K.$$

Hence,
$$|R_2| \leq \frac{K}{3!}\left(\frac{\pi}{90}\right)^3 < 0.0000051.$$

Since $R_2 < 0$, therefore

$$0.7313587 - 0.0000051 < \sin(47°) < 0.7313587$$
$$0.7313536 < \sin(47°) < 0.7313587.$$

13. $f(x) = \sin x$
$f'(x) = \cos x$
$f''(x) = -\sin x$
$f'''(x) = -\cos x$
$f^{(4)}(x) = \sin x$
$a = 0$; $n = 7$:

$$\sin x = 0 + x - 0 - \frac{x^3}{3!} + 0 + \frac{x^5}{5!} - 0 - \frac{x^7}{7!} + R_7,$$
$$= x - \frac{x^3}{3!} + \frac{x^5}{5!} - \frac{x^7}{7!} + R_7(x)$$

where $R_7(x) = \frac{\sin X}{8!} x^8$ for some X between 0 and x.

14. For $f(x) = \cos x$ we have

$$f'(x) = -\sin x \quad f''(x) = -\cos x \quad f'''(x) = \sin x$$
$$f^{(4)}(x) = \cos x \quad f^{(5)}(x) = -\sin x \quad f^{(6)}(x) = -\cos x.$$

The Taylor's Formula for f with $a = 0$ and $n = 6$ is

$$\cos x = 1 - \frac{x^2}{2!} + \frac{x^4}{4!} - \frac{x^6}{6!} + R_6(f; 0, x)$$

where the Lagrange remainder R_6 is given by

$$R_6 = R_6(f; 0, x) = \frac{f^{(7)}(X)}{7!} x^7 = \frac{\sin X}{7!} x^7,$$

for some X between 0 and x.

15. $f(x) = \sin x \quad a = \frac{\pi}{4}, \quad n = 4$

$$\sin x = \frac{1}{\sqrt{2}} + \frac{1}{\sqrt{2}} \left(x - \frac{\pi}{4} \right) - \frac{1}{\sqrt{2}} \frac{1}{2!} \left(x - \frac{\pi}{4} \right)^2$$
$$- \frac{1}{\sqrt{2}} \frac{1}{3!} \left(x - \frac{\pi}{4} \right)^3 + \frac{1}{\sqrt{2}} \frac{1}{4!} \left(x - \frac{\pi}{4} \right)^4 + R_4(x)$$

where $R_4(x) = \frac{1}{5!} (\cos X) \left(x - \frac{\pi}{4} \right)^5$

for some X between $\frac{\pi}{4}$ and x.

16. Given that $f(x) = \frac{1}{1-x}$, then

$$f'(x) = \frac{1}{(1-x)^2}, \quad f''(x) = \frac{2}{(1-x)^3}.$$

In general,

$$f^{(n)}(x) = \frac{n!}{(1-x)^{(n+1)}}.$$

Since $a = 0$, $f^{(n)}(0) = n!$. Hence, for $n = 6$, the Taylor's Formula is

$$\frac{1}{1-x} = f(0) + \sum_{n=1}^{6} \frac{f^{(n)}(0)}{n!} x^n + R_6(f; 0, x)$$
$$= 1 + x + x^2 + x^3 + x^4 + x^5 + x^6 + R_6(f; 0, x).$$

The Langrange remainder is

$$R_6(f; 0, x) = \frac{f^{(7)}(X)}{7!} x^7 = \frac{x^7}{(1-X)^8}$$

for some X between 0 and x.

17. $f(x) = \ln x$
$f'(x) = \frac{1}{x}$
$f''(x) = \frac{-1}{x^2}$
$f'''(x) = \frac{2!}{x^3}$
$f^{(4)}(x) = \frac{-3!}{x^4}$
$f^{(5)}(x) = \frac{4!}{x^5}$
$f^{(6)}(x) = \frac{-5!}{x^6}$
$f^{(7)} = \frac{6!}{x^7}$
$a = 1, \quad n = 6$

$$\ln x = 0 + 1(x-1) - \frac{1}{2!}(x-1)^2 + \frac{2!}{3!}(x-1)^3$$
$$- \frac{3!}{4!}(x-1)^4 + \frac{4!}{5!}(x-1)^5 - \frac{5!}{6!}(x-1)^6 + R_6(x)$$
$$= (x-1) - \frac{(x-1)^2}{2} + \frac{(x-1)^3}{3} - \frac{(x-1)^4}{4}$$
$$+ \frac{(x-1)^5}{5} - \frac{(x-1)^6}{6} + R_6(x)$$

where $R_6(x) = \frac{1}{7X^7}(x-1)^7$ for some X between 1 and x.

18. Given that $f(x) = \tan x$, then

$$f'(x) = \sec^2 x$$
$$f''(x) = 2 \sec^2 x \tan x$$
$$f^{(3)}(x) = 6 \sec^4 x - 4 \sec^2 x$$
$$f^{(4)}(x) = 8 \tan x (3 \sec^4 x - \sec^2 x).$$

Given that $a = 0$ and $n = 3$, the Taylor's Formula is

$$\tan x = f(0) + f'(0)x + \frac{f''(0)}{2!}x^2 + \frac{f'''(0)}{3!}x^3 + R_3(f; 0, x)$$
$$= x + \frac{2}{3!}x^3 + R_3(f; 0, x)$$
$$= x + \frac{1}{3}x^3 + \frac{2}{15}x^5.$$

The Lagrange remainder is

$$R_3(f; 0, x) = \frac{f^{(4)}(X)}{4!} x^4 = \frac{\tan X (3 \sec^4 X - \sec^2 C)}{3} x^4$$

for some X between 0 and x.

19. $e^{3x} = e^{3(x+1)}e^{-3}$

$P_3(x) = e^{-3}\left[1 + 3(x+1) + \frac{9}{2}(x+1)^2 + \frac{9}{2}(x+1)^3\right]$.

20. For e^u, $P_4(u) = 1 + u + \frac{u^2}{2!} + \frac{u^3}{3!} + \frac{u^4}{4!}$. Let $u = -x^2$. Then for e^{-x^2}:

$$P_8(x) = 1 - x^2 + \frac{x^4}{2!} - \frac{x^6}{3!} + \frac{x^8}{4!}.$$

21. For $\sin^2 x = \frac{1}{2}(1 - \cos(2x))$ about $x = 0$, we have

$$P_4(x) = \frac{1}{2}\left[1 - \left(1 - \frac{(2x)^2}{2!} + \frac{(2x)^4}{4!}\right)\right] = x^2 - \frac{x^4}{3}.$$

22. $\sin x = \sin(\pi + (x-\pi)) = -\sin(x-\pi)$

$$P_5(x) = -(x-\pi) + \frac{(x-\pi)^3}{3!} - \frac{(x-\pi)^5}{5!}$$

23. For $\frac{1}{1-u}$ about $u = 0$, $P_3(u) = 1 + u + u^2 + u^3$. Let $u = -2x^2$. Then for $\frac{1}{1+2x^2}$ about $x = 0$,

$$P_6(x) = 1 - 2x^2 + 4x^4 - 8x^6.$$

24. $\cos(3x - \pi) = -\cos(3x)$

$$P_8(x) = -1 + \frac{3^2 x^2}{2!} - \frac{3^4 x^4}{4!} + \frac{3^6 x^6}{6!} - \frac{3^8 x^8}{8!}.$$

25. $\sinh x = \frac{1}{2}(e^x - e^{-x})$

$$P_{2n+1}(x) = \frac{1}{2}\left(1 + x + \frac{x^2}{2!} + \cdots + \frac{x^{2n+1}}{(2n+1)!}\right)$$
$$- \frac{1}{2}\left(1 - x + \frac{x^2}{2!} + \cdots - \frac{x^{2n+1}}{(2n+1)!}\right)$$
$$= x + \frac{x^3}{3!} + \frac{x^5}{5!} + \cdots + \frac{x^{2n+1}}{(2n+1)!}.$$

26. For $\ln(1+x)$ about $x = 0$ we have

$$P_{2n+1}(x) = x - \frac{x^2}{2} + \frac{x^3}{3} - \cdots + \frac{x^{2n+1}}{2n+1}.$$

For $\ln(1-x)$ about $x = 0$ we have

$$P_{2n+1}(x) = -x - \frac{x^2}{2} - \frac{x^3}{3} - \cdots - \frac{x^{2n+1}}{2n+1}.$$

For $\tanh^{-1} x = \frac{1}{2}\ln(1+x) - \frac{1}{2}\ln(1-x)$,

$$P_{2n+1}(x) = x + \frac{x^3}{3} + \frac{x^5}{5} + \cdots + \frac{x^{2n+1}}{2n+1}.$$

27. $f(x) = e^{-x}$

$f^{(n)}(x) = \begin{cases} e^{-x} & \text{if } n \text{ is even} \\ -e^{-x} & \text{if } n \text{ is odd} \end{cases}$

$$e^{-x} = 1 - x + \frac{x^2}{2!} - \frac{x^3}{3!} + \cdots + (-1)^n \frac{x^5}{n!} + R_n(x)$$

where $R_n(x) = (-1)^{n+1}\frac{X^{n+1}}{(n+1)!}$ for some X between 0 and x.

For $x = 1$, we have
$$\frac{1}{e} = 1 - 1 + \frac{1}{2!} - \frac{1}{3!} + \cdots + (-1)^n \frac{1}{n!} + R_n(1)$$

where $R_n(1) = (-1)^{n+1}\frac{e^{-X}x^{n+1}}{(n+1)!}$ for some X between -1 and 0.

Therefore, $|R_n(1)| < \frac{1}{(n+1)!}$. We want $|R_n(1)| < 0.000005$ for 5 decimal places.
Choose n so that $\frac{1}{(n+1)!} < 0.000005$. $n = 8$ will do since $1/9! \approx 0.0000027$.
Thus $\frac{1}{e} \approx \frac{1}{2!} - \frac{1}{3!} + \frac{1}{4!} - \frac{1}{5!} + \frac{1}{6!} - \frac{1}{7!} + \frac{1}{8!}$
≈ 0.36788 (to 5 decimal places).

28. In Taylor's Formulas for $f(x) = \sin x$ with $a = 0$, only odd powers of x have nonzero coefficients. Accordingly we can take terms up to order x^{2n+1} and the remainder after the $0x^{2n+2}$. The formula is

$$\sin x = x - \frac{x^3}{3!} + \frac{x^5}{5!} - \cdots + (-1)^n \frac{x^{2n+1}}{(2n+1)!} + R_{2n+2},$$

where

$$R_{2n+2}(f;0,x) = (-1)^{n+1}\frac{\cos X}{(2n+3)!}x^{2n+3}$$

for some X between 0 and x.
In order to use the formula to approximate $\sin(1)$ correctly to 5 decimal places, we need $|R_{2n+2}(f;0,1)| < 0.000005$. Since $|\cos X| \leq 1$, it is sufficient to have $1/(2n+3)! < 0.000005$. $n = 3$ will do since $1/9! \approx 0.000003$. Thus

$$\sin(1) \approx 1 - \frac{1}{3!} + \frac{1}{5!} - \frac{1}{7!} \approx 0.84147$$

correct to five decimal places.

29. $f(x) = (x-1)^2$, $f'(x) = 2(x-1)$, $f''(x) = 2$.
$f(x) \approx 1 - 2x + \frac{2}{2}x^2 = 1 - 2x + x^2$
Error $= 0$
$g(x) = x^3 + 2x^2 + 3x + 4$
Quadratic approx.: $g(x) \approx 4 + 3x + 2x^2$
Error $= x^3$
Since $g'''(X) = 6 = 3!$, error $= \frac{g'''(X)}{3!}x^3$
so that constant $\frac{1}{3!}$ in the error formula for the quadratic approximation cannot be improved.

Section 4.9 Indeterminate Forms (page 294)

1. $\lim_{x\to 0} \dfrac{3x}{\tan 4x}\quad \left[\dfrac{0}{0}\right]$
   $= \lim_{x\to 0} \dfrac{3}{4\sec^2 4x} = \dfrac{3}{4}$

2. $\lim_{x\to 2} \dfrac{\ln(2x-3)}{x^2-4}\quad \left[\dfrac{0}{0}\right]$
   $= \dfrac{\left(\dfrac{2}{2x-3}\right)}{2x} = \dfrac{1}{2}.$

3. $\lim_{x\to 0} \dfrac{\sin ax}{\sin bx}\quad \left[\dfrac{0}{0}\right]$
   $= \lim_{x\to 0} \dfrac{a\cos ax}{b\cos bx} = \dfrac{a}{b}$

4. $\lim_{x\to 0} \dfrac{1-\cos ax}{1-\cos bx}\quad \left[\dfrac{0}{0}\right]$
   $= \lim_{x\to 0} \dfrac{a\sin ax}{b\sin bx}\quad \left[\dfrac{0}{0}\right]$
   $= \lim_{x\to 0} \dfrac{a^2\cos ax}{b^2\cos bx} = \dfrac{a^2}{b^2}.$

5. $\lim_{x\to 0} \dfrac{\sin^{-1} x}{\tan^{-1} x}\quad \left[\dfrac{0}{0}\right]$
   $= \lim_{x\to 0} \dfrac{1+x^2}{\sqrt{1-x^2}} = 1$

6. $\lim_{x\to 1} \dfrac{x^{1/3}-1}{x^{2/3}-1}\quad \left[\dfrac{0}{0}\right]$
   $= \lim_{x\to 1} \dfrac{(\frac{1}{3})x^{-2/3}}{(\frac{2}{3})x^{-1/3}} = \dfrac{1}{2}.$

7. $\lim_{x\to 0} x\cot x \quad [0\times\infty]$
   $= \lim_{x\to 0} \left(\dfrac{x}{\sin x}\right)\cos x$
   $= 1\times \lim_{x\to 0} \dfrac{x}{\sin x}\quad \left[\dfrac{0}{0}\right]$
   $= \lim_{x\to 0} \dfrac{1}{\cos x} = 1$

8. $\lim_{x\to 0} \dfrac{1-\cos x}{\ln(1+x^2)}\quad \left[\dfrac{0}{0}\right]$
   $= \lim_{x\to 0} \dfrac{\sin x}{\left(\dfrac{2x}{1+x^2}\right)}$
   $= \lim_{x\to 0}(1+x^2)\lim_{x\to 0}\dfrac{\sin x}{2x}$
   $= \lim_{x\to 0}\dfrac{\cos x}{2} = \dfrac{1}{2}.$

9. $\lim_{t\to \pi} \dfrac{\sin^2 t}{t-\pi}\quad \left[\dfrac{0}{0}\right]$
   $= \lim_{t\to \pi} \dfrac{2\sin t\cos t}{1} = 0$

10. $\lim_{x\to 0} \dfrac{10^x - e^x}{x}\quad \left[\dfrac{0}{0}\right]$
    $= \lim_{x\to 0} \dfrac{10^x\ln 10 - e^x}{1} = \ln 10 - 1.$

11. $\lim_{x\to \pi/2} \dfrac{\cos 3x}{\pi - 2x}\quad \left[\dfrac{0}{0}\right]$
    $= \lim_{x\to \pi/2} \dfrac{-3\sin 3x}{-2} = \dfrac{3}{2}(-1) = -\dfrac{3}{2}$

12. $\lim_{x\to 1} \dfrac{\ln(ex) - 1}{\sin \pi x}\quad \left[\dfrac{0}{0}\right]$
    $= \lim_{x\to 1} \dfrac{\dfrac{1}{x}}{\pi\cos(\pi x)} = -\dfrac{1}{\pi}.$

13. $\lim_{x\to \infty} x\sin\dfrac{1}{x}\quad [\infty\times 0]$
    $= \lim_{x\to \infty} \dfrac{\sin\dfrac{1}{x}}{\dfrac{1}{x}}\quad \left[\dfrac{0}{0}\right]$
    $= \lim_{x\to \infty} \dfrac{-\dfrac{1}{x^2}\cos\dfrac{1}{x}}{-\dfrac{1}{x^2}} = \lim_{x\to \infty} \cos\dfrac{1}{x} = 1.$

14. $\lim_{x\to 0} \dfrac{x-\sin x}{x^3}\quad \left[\dfrac{0}{0}\right]$
    $= \lim_{x\to 0} \dfrac{1-\cos x}{3x^2}\quad \left[\dfrac{0}{0}\right]$
    $= \lim_{x\to 0} \dfrac{\sin x}{6x}\quad \left[\dfrac{0}{0}\right]$
    $= \lim_{x\to 0} \dfrac{\cos x}{6} = \dfrac{1}{6}.$

15. $\lim_{x\to 0} \dfrac{x-\sin x}{x-\tan x}\quad \left[\dfrac{0}{0}\right]$
    $= \lim_{x\to 0} \dfrac{1-\cos x}{1-\sec^2 x}\quad \left[\dfrac{0}{0}\right]$
    $= \lim_{x\to 0}(\cos^2 x)\dfrac{1-\cos x}{\cos^2 x - 1}$
    $= -1\times \lim_{x\to 0} \dfrac{\cos x - 1}{(\cos x - 1)(\cos x + 1)}$
    $= -\dfrac{1}{2}$

16. $\lim_{x\to 0} \dfrac{2-x^2-2\cos x}{x^4}$ $\left[\dfrac{0}{0}\right]$

$= \lim_{x\to 0} \dfrac{-2x+2\sin x}{4x^3}$ $\left[\dfrac{0}{0}\right]$

$= -\dfrac{1}{2}\lim_{x\to 0} \dfrac{x-\sin x}{x^3}$

$= -\dfrac{1}{2}\left(\dfrac{1}{6}\right) = -\dfrac{1}{12}$ (by Exercise 14).

17. $\lim_{x\to 0+} \dfrac{\sin^2 x}{\tan x - x}$ $\left[\dfrac{0}{0}\right]$

$= \lim_{x\to 0+} \dfrac{2\sin x \cos x}{\sec^2 x - 1}$ $\left[\dfrac{0}{0}\right]$

$= 2 \times 1 \times \lim_{x\to 0+} \dfrac{\cos x}{2\sec^2 x \tan x} = \infty$

18. $\lim_{r\to \pi/2} \dfrac{\ln \sin r}{\cos r}$ $\left[\dfrac{0}{0}\right]$

$= \lim_{r\to \pi/2} \dfrac{\left(\dfrac{\cos r}{\sin r}\right)}{-\sin r} = 0.$

19. $\lim_{t\to \pi/2} \dfrac{\sin t}{t} = \dfrac{2}{\pi}$

20. $\lim_{x\to 1-} \dfrac{\cos^{-1} x}{x-1}$ $\left[\dfrac{0}{0}\right]$

$= \lim_{x\to 1-} \dfrac{-\left(\dfrac{1}{\sqrt{1-x^2}}\right)}{1} = -\infty.$

21. $\lim_{x\to \infty} x(2\tan^{-1} x - \pi)$ $[0\times\infty]$

$= \lim_{x\to \infty} \dfrac{2\tan^{-1} x - \pi}{\dfrac{1}{x}}$ $\left[\dfrac{0}{0}\right]$

$= \lim_{x\to \infty} \dfrac{2}{1+x^2} \Big/ -\dfrac{1}{x^2}$

$= \lim_{x\to \infty} -\dfrac{2x^2}{1+x^2} = -2$

22. $\lim_{t\to(\pi/2)-} (\sec t - \tan t)$ $[\infty - \infty]$

$= \lim_{t\to(\pi/2)-} \dfrac{1-\sin t}{\cos t}$ $\left[\dfrac{0}{0}\right]$

$= \lim_{t\to(\pi/2)-} \dfrac{-\cos t}{-\sin t} = 0.$

23. $\lim_{t\to 0} \left(\dfrac{1}{t} - \dfrac{1}{te^{at}}\right)$ $(\infty - \infty)$

$= \lim_{t\to 0} \dfrac{e^{at}-1}{te^{at}}$ $\left[\dfrac{0}{0}\right]$

$= \lim_{t\to 0} \dfrac{ae^{at}}{e^{at}+ate^{at}} = a$

24. Since $\lim_{x\to 0+} \sqrt{x}\ln x = \lim_{x\to 0+} \dfrac{\ln x}{x^{-1/2}}$ $\left[\dfrac{0}{0}\right]$

$= \lim_{x\to 0+} \dfrac{\left(\dfrac{1}{x}\right)}{\left(-\dfrac{1}{2}\right)x^{-3/2}} = 0,$

hence $\lim_{x\to 0+} x^{\sqrt{x}}$

$= \lim_{x\to 0+} e^{\sqrt{x}\ln x} = e^0 = 1.$

25. Let $y = (\csc x)^{\sin^2 x}$.
Then $\ln y = \sin^2 x \ln(\csc x)$
$\lim_{x\to 0+} \ln y = \lim_{x\to 0+} \dfrac{\ln(\csc x)}{\csc^2 x}$ $\left[\dfrac{\infty}{\infty}\right]$

$= \lim_{x\to 0+} \dfrac{-\dfrac{\csc x \cot x}{\csc x}}{-2\csc^2 x \cot x}$

$= \lim_{x\to 0+} \dfrac{1}{2\csc^2 x} = 0.$

Thus $\lim_{x\to 0+} (\csc x)^{\sin^2 x} = e^0 = 1.$

26. $\lim_{x\to 1+}\left(\dfrac{x}{x-1} - \dfrac{1}{\ln x}\right)$ $[\infty - \infty]$

$= \lim_{x\to 1+} \dfrac{x\ln x - x + 1}{(x-1)(\ln x)}$ $\left[\dfrac{0}{0}\right]$

$= \lim_{x\to 1+} \dfrac{\ln x}{\ln x + 1 - \dfrac{1}{x}}$ $\left[\dfrac{0}{0}\right]$

$= \lim_{x\to 1+} \dfrac{\dfrac{1}{x}}{\dfrac{1}{x} + \dfrac{1}{x^2}}$

$= \lim_{x\to 1+} \dfrac{x}{x+1} = \dfrac{1}{2}.$

27. $\lim_{t\to 0} \dfrac{3\sin t - \sin 3t}{3\tan t - \tan 3t}$ $\left[\dfrac{0}{0}\right]$

$= \lim_{t\to 0} \dfrac{3(\cos t - \cos 3t)}{3(\sec^2 t - \sec^2 3t)}$ $\left[\dfrac{0}{0}\right]$

$= \lim_{t\to 0} \dfrac{\cos t - \cos 3t}{\dfrac{\cos^2 3t - \cos^2 t}{\cos^2 t \cos^2 3t}}$

$= -\lim_{t\to 0} \dfrac{\cos 3t - \cos t}{\cos^2 3t - \cos^2 t}$

$= -\lim_{t\to 0} \dfrac{1}{\cos 3t + \cos t} = -\dfrac{1}{2}$

28. Let $y = \left(\dfrac{\sin x}{x}\right)^{1/x^2}$.

$$\lim_{x \to 0} \ln y = \lim_{x \to 0} \dfrac{\ln\left(\dfrac{\sin x}{x}\right)}{x^2} \quad \left[\dfrac{0}{0}\right]$$

$$= \lim_{x \to 0} \dfrac{\left(\dfrac{x}{\sin x}\right)\left(\dfrac{x \cos x - \sin x}{x^2}\right)}{2x}$$

$$= \lim_{x \to 0} \dfrac{x \cos x - \sin x}{2x^2 \sin x} \quad \left[\dfrac{0}{0}\right]$$

$$= \lim_{x \to 0} \dfrac{-x \sin x}{4x \sin x + 2x^2 \cos x}$$

$$= \lim_{x \to 0} \dfrac{-\sin x}{4 \sin x + 2x \cos x} \quad \left[\dfrac{0}{0}\right]$$

$$= \lim_{x \to 0} \dfrac{-\cos x}{6 \cos x - 2x \sin x} = -\dfrac{1}{6}.$$

Thus, $\lim_{x \to 0} \left(\dfrac{\sin x}{x}\right)^{1/x^2} = e^{-1/6}$.

29. Let $y = (\cos 2t)^{1/t^2}$.
Then $\ln y = \dfrac{\ln(\cos 2t)}{t^2}$. We have

$$\lim_{t \to 0} \ln y = \lim_{t \to 0} \dfrac{\ln(\cos 2t)}{t^2} \quad \left[\dfrac{0}{0}\right]$$

$$= \lim_{t \to 0} \dfrac{-2 \tan 2t}{2t} \quad \left[\dfrac{0}{0}\right]$$

$$= -\lim_{t \to 0} \dfrac{2 \sec^2 2t}{1} = -2.$$

Therefore $\lim_{t \to 0}(\cos 2t)^{1/t^2} = e^{-2}$.

30. $\lim_{x \to 0+} \dfrac{\csc x}{\ln x} \quad \left[-\dfrac{\infty}{\infty}\right]$

$$= \lim_{x \to 0+} \dfrac{-\csc x \cot x}{\dfrac{1}{x}} \quad \left[-\dfrac{\infty}{\infty}\right]$$

$$= \lim_{x \to 0+} \dfrac{-x \cos x}{\sin^2 x} \quad \left[\dfrac{0}{0}\right]$$

$$= -\left(\lim_{x \to 0+} \cos x\right) \lim_{x \to 0+} \dfrac{1}{2 \sin x \cos x}$$

$$= -\infty.$$

31. $\lim_{x \to 1-} \dfrac{\ln \sin \pi x}{\csc \pi x} \quad \left[\dfrac{\infty}{\infty}\right]$

$$= \lim_{x \to 1-} \dfrac{\dfrac{\pi \cos \pi x}{\sin \pi x}}{-\pi \csc \pi x \cot \pi x}$$

$$= \dfrac{-\pi}{\pi} \lim_{x \to 1-} \tan \pi x = 0.$$

32. Let $y = (1 + \tan x)^{1/x}$.

$$\lim_{x \to 0} \ln y = \lim_{x \to 0} \dfrac{\ln(1 + \tan x)}{x} \quad \left[\dfrac{0}{0}\right]$$

$$= \lim_{x \to 0} \dfrac{\sec^2 x}{1 + \tan x} = 1.$$

Thus, $\lim_{x \to 0}(1 + \tan x)^{1/x} = e$.

33. $\lim_{h \to 0} \dfrac{f(x+h) - 2f(x) + f(x-h)}{h^2} \quad \left[\dfrac{0}{0}\right]$

$$= \lim_{h \to 0} \dfrac{f'(x+h) - f'(x-h)}{2h} \quad \left[\dfrac{0}{0}\right]$$

$$= \lim_{h \to 0} \dfrac{f''(x+h) + f''(x-h)}{2}$$

$$= \dfrac{2f(x)}{2} = f''(x)$$

34. Suppose that f and g are continuous on $[a, b]$ and differentiable on (a, b) and $g(x) \neq 0$ there. Let $a < x < t < b$, and apply the Generalized Mean-Value Theorem; there exists c in (x, t) such that

$$\dfrac{f(x) - f(t)}{g(x) - g(t)} = \dfrac{f'(c)}{g'(c)}$$

$$\Rightarrow \left[\dfrac{f(x) - f(t)}{g(x)}\right]\left[\dfrac{g(x)}{g(x) - g(t)}\right] = \dfrac{f'(c)}{g'(c)}$$

$$\Rightarrow \dfrac{f(x)}{g(x)} - \dfrac{f(t)}{g(x)} = \dfrac{f'(c)}{g'(c)}\left[\dfrac{g(x) - g(t)}{g(x)}\right]$$

$$\Rightarrow \dfrac{f(x)}{g(x)} = \dfrac{f'(c)}{g'(c)} - \dfrac{g(t)}{g(x)}\dfrac{f'(c)}{g'(c)} + \dfrac{f(t)}{g(x)}$$

$$\Rightarrow \dfrac{f(x)}{g(x)} = \dfrac{f'(c)}{g'(c)} + \dfrac{1}{g(x)}\left[f(t) - g(t)\dfrac{f'(c)}{g'(c)}\right]$$

$$\Rightarrow \dfrac{f(x)}{g(x)} - L = \dfrac{f'(c)}{g'(c)} - L + \dfrac{1}{g(x)}\left[f(t) - g(t)\dfrac{f'(c)}{g'(c)}\right].$$

Since $|m + n| \leq |m| + |n|$, therefore,

$$\left|\dfrac{f(x)}{g(x)} - L\right| \leq \left|\dfrac{f'(c)}{g'(c)} - L\right| + \dfrac{1}{|g(x)|}\left[|f(t)| + |g(t)|\left|\dfrac{f'(c)}{g'(c)}\right|\right].$$

Now suppose that ϵ is an arbitrary small positive number. Since $\lim_{c \to a+} f'(c)/g'(c) = L$, and since $a < x < c < t$, we can choose t sufficiently close to a to ensure that

$$\left|\dfrac{f'(c)}{g'(c)} - L\right| < \dfrac{\epsilon}{2}.$$

In particular,

$$\left|\dfrac{f'(c)}{g'(c)}\right| < |L| + \dfrac{\epsilon}{2}.$$

Since $\lim_{x \to a+} |g(x)| = \infty$, we can choose x between a and t sufficiently close to a to ensure that

$$\dfrac{1}{|g(x)|}\left[|f(t)| + |g(t)|\left(|L| + \dfrac{\epsilon}{2}\right)\right] < \dfrac{\epsilon}{2}.$$

It follows that

$$\left|\frac{f(x)}{g(x)} - L\right| < \frac{\epsilon}{2} + \frac{\epsilon}{2} = \epsilon.$$

Thus $\lim_{x\to a+} \frac{f(x)}{g(x)} = L$.

Review Exercises 4 (page 295)

1. Let the building be h m high. The height of the first ball at time t during its motion is

$$y_1 = h + 10t - 4.9t^2.$$

It reaches maximum height when $dy_1/dt = 10 - 9.8t = 0$, that is, at $t = 10/9.8$ s. The maximum height of the first ball is

$$y_1 = h + \frac{100}{9.8} - \frac{4.9 \times 100}{(9.8)^2} = h + \frac{100}{19.6}.$$

The height of the second ball at time t during its motion is

$$y_2 = 20t - 4.9t^2.$$

It reaches maximum height when $dy_2/dt = 20 - 9.8t = 0$, that is, at $t = 20/9.8$ s. The maximum height of the second ball is

$$y_2 = \frac{400}{9.8} - \frac{4.9 \times 400}{(9.8)^2} = \frac{400}{19.6}.$$

These two maximum heights are equal, so

$$h + \frac{100}{19.6} = \frac{400}{19.6},$$

which gives $h = 300/19.6 \approx 15.3$ m as the height of the building.

2. The first ball has initial height 60 m and initial velocity 0, so its height at time t is

$$y_1 = 60 - 4.9t^2 \text{ m}.$$

The second ball has initial height 0 and initial velocity v_0, so its height at time t is

$$y_2 = v_0 t - 4.9t^2 \text{ m}.$$

The two balls collide at a height of 30 m (at time T, say). Thus

$$30 = 60 - 4.9T^2$$
$$30 = v_0 T - 4.9T^2.$$

Thus $v_0 T = 60$ and $T^2 = 30/4.9$. The initial upward speed of the second ball is

$$v_0 = \frac{60}{T} = 60\sqrt{\frac{4.9}{30}} \approx 24.25 \text{ m/s}.$$

At time T, the velocity of the first ball is

$$\left.\frac{dy_1}{dt}\right|_{t=T} = -9.8T \approx -24.25 \text{ m/s}.$$

At time T, the velocity of the second ball is

$$\left.\frac{dy_2}{dt}\right|_{t=T} = v_0 - 9.8T = 0 \text{ m/s}.$$

3. Let the car's initial speed be v_0. The car decelerates at 20 ft/s^2 starting at $t = 0$, and travels distance s in time t, where $d^2s/dt^2 = -20$. Thus

$$\frac{ds}{dt} = v_0 - 20t$$
$$x = v_0 t - 10t^2.$$

The car stops at time $t = v_0/20$. The stopping distance is $s = 160$ ft, so

$$160 = \frac{v_0^2}{20} - \frac{v_0^2}{40} = \frac{v_0^2}{40}.$$

The car's initial speed cannot exceed $v_0 = \sqrt{160 \times 40} = 80$ ft/s.

4. a) Since F must be continuous at $r = R$, we have

$$\frac{mgR^2}{R^2} = mkR, \quad \text{or} \quad k = \frac{g}{R}.$$

b) The rate of change of F as r decreases from R is

$$\left(-\frac{d}{dr}(mkr)\right)\bigg|_{r=R} = -mk = -\frac{mg}{R}.$$

The rate of change of F as r increases from R is

$$\left(-\frac{d}{dr}\frac{mgR^2}{r^2}\right)\bigg|_{r=R} = -\frac{2mgR^2}{R^3} = -2\frac{mg}{R}.$$

Thus F decreases as r increases from R at twice the rate at which it decreases as r decreases from R.

5. $1/R = 1/R_1 + 1/R_2$. If $R_1 = 250$ ohms and $R_2 = 1,000$ ohms, then $1/R = (1/250) + (1/1,000) = 1/200$, so $R = 200$ ohms. If $dR_1/dt = 100$ ohms/min, then

$$-\frac{1}{R^2}\frac{dR}{dt} = -\frac{1}{R_1^2}\frac{dR_1}{dt} - \frac{1}{R_2^2}\frac{dR_2}{dt}$$
$$\frac{1}{200^2}\frac{dR}{dt} = \frac{1}{250^2}(100) + \frac{1}{1,000^2}\frac{dR_2}{dt}.$$

a) If R remains constant, then $dR/dt = 0$, so

$$\frac{dR_2}{dt} = -\frac{1,000^2 \times 100}{250^2} = -1,600.$$

R_2 is decreasing at 1,600 ohms/min.

b) If R is increasing at 10 ohms/min, then then $dR/dt = 10$, and

$$\frac{dR_2}{dt} = 1,000^2 \left(\frac{10}{200^2} - \frac{100}{250^2}\right) = -1,350.$$

R_2 is decreasing at 1,350 ohms/min.

6. If $pV = 5.0T$, then

$$\frac{dp}{dt} V + p \frac{dV}{dt} = 5.0 \frac{dT}{dt}.$$

a) If $T = 400$ K, $dT/dt = 4$ K/min, and $V = 2.0$ m^3, then $dV/dt = 0$, so $dp/dt = 5.0(4)/2.0 = 10$. The pressure is increasing at 10 kPa/min.

b) If $T = 400$ K, $dT/dt = 0$, $V = 2$ m^3, and $dV/dt = 0.05$ m^3/min, then $p = 5.0(400)/2 = 1,000$ kPa, and $2\,dp/dt + 1,000(0.05) = 0$, so $dp/dt = -25$. The pressure is decreasing at 25 kPa/min.

7. If x copies of the book are printed, the cost of printing each book is

$$C = \frac{10,000}{x} + 8 + 6.25 \times 10^{-7} x^2.$$

Since $C \to \infty$ as $x \to 0+$ or $x \to \infty$, C will be minimum at a critical point. For CP:

$$0 = \frac{dC}{dx} = -\frac{10,000}{x^2} + 12.5 \times 10^{-7} x,$$

so $x^3 = 8 \times 10^9$ and $x = 2 \times 10^3$. 2,000 books should be printed.

8. If she charges $\$x$ per bicycle, her total profit is $\$P$, where

$$P = (x - 75)N(x) = 4.5 \times 10^6 \frac{x-75}{x^2}.$$

Evidently $P \leq 0$ if $x \leq 75$, and $P \to 0$ as $x \to \infty$. P will therefore have a maximum value at a critical point in $(75, \infty)$. For CP:

$$0 = \frac{dP}{dx} = 4.5 \times 10^6 \frac{x^2 - (x-75)2x}{x^4},$$

from which we obtain $x = 150$. She should charge $\$150$ per bicycle and order $N(150) = 200$ of them from the manufacturer.

9.

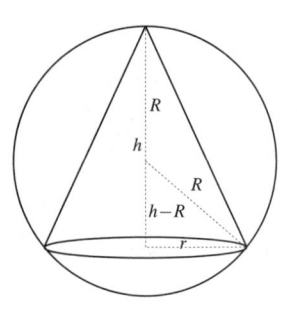

Fig. R-4.9

Let r, h and V denote the radius, height, and volume of the cone respectively. The volume of a cone is one-third the base area times the height, so

$$V = \frac{1}{3} \pi r^2 h.$$

From the small right-angled triangle in the figure,

$$(h - R)^2 + r^2 = R^2.$$

Thus $r^2 = R^2 - (h-R)^2$ and

$$V = V(h) = \frac{\pi}{3} h \left(R^2 - (h-R)^2\right) = \frac{\pi}{3}\left(2Rh^2 - h^3\right).$$

The height of any inscribed cone cannot exceed the diameter of the sphere, so $0 \leq h \leq 2R$. Being continuous, $V(h)$ must have a maximum value on this interval. Since $V = 0$ when $h = 0$ or $h = 2R$, and $V > 0$ if $0 < h < 2R$, the maximum value of V must occur at a critical point. (V has no singular points.) For a critical point,

$$0 = V'(h) = \frac{\pi}{3}(4Rh - 3h^2) = \frac{\pi}{3} h(4R - 3h),$$

$$h = 0 \quad \text{or} \quad h = \frac{4R}{3}.$$

$V'(h) > 0$ if $0 < h < 4R/3$ and $V'(h) < 0$ if $4R/3 < h < 2R$. Hence $h = 4R/3$ does indeed give the maximum value for V. The volume of the largest cone can be inscribed in a sphere of radius R is

$$V\left(\frac{4R}{3}\right) = \frac{\pi}{3}\left(2R\left(\frac{4R}{3}\right)^2 - \left(\frac{4R}{3}\right)^3\right)$$

$$= \frac{32}{81} \pi R^3 \text{ cubic units.}$$

10.

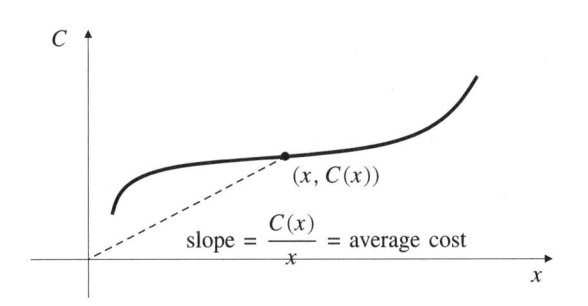

Fig. R-4.10

a) For minimum $C(x)/x$, we need
$$0 = \frac{d}{dx}\frac{C(x)}{x} = \frac{xC'(x) - C(x)}{x^2},$$
so $C'(x) = C(x)/x$; the marginal cost equals the average cost.

b) The line from $(0,0)$ to $(x, C(x))$ has smallest slope at a value of x which makes it tangent to the graph of $C(x)$. Thus $C'(x) = C(x)/x$, the slope of the line.

c) The line from $(0,0)$ to $(x, C(x))$ can be tangent to the graph of $C(x)$ at more than one point. Not all such points will provide a minimum value for the average cost. (In the figure, one such line will make the average cost maximum.)

11.

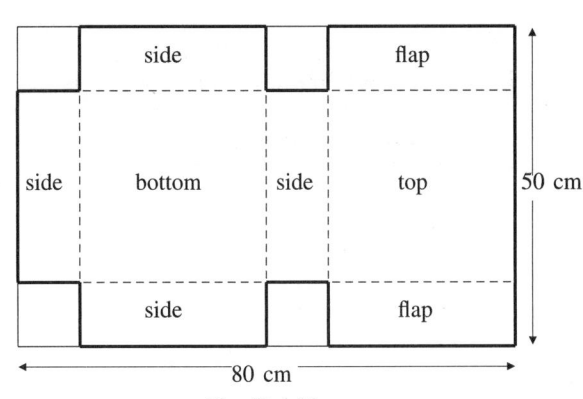

Fig. R-4.11

If the edge of the cutout squares is x cm, then the volume of the folded box is
$$V(x) = x(50 - 2x)(40 - x)$$
$$= 2x^3 - 130x^2 + 2{,}000x,$$
and is valid for $0 \le x \le 25$. Since $V(0) = V(25) = 0$, and $V(x) > 0$ if $0 < x < 25$, the maximum will occur at a CP:
$$0 = V'(x) = 6x^2 - 260x + 2{,}000$$
$$= 2(3x^2 - 130x + 1{,}000)$$
$$= 2(3x - 100)(x - 10).$$

Thus $x = 10$ or $x = 100/3$. The latter CP is not in the interval $[0, 25]$, so the maximum occurs at $x = 10$. The maximum volume of the box is $V(10) = 9{,}000$ cm^3.

12. If x more trees are planted, the yield of apples will be
$$Y = (60 + x)(800 - 10x)$$
$$= 10(60 + x)(80 - x)$$
$$= 10(4{,}800 + 20x - x^2).$$

This is a quadratic expression with graph opening downward; its maximum occurs at a CP:
$$0 = \frac{dY}{dx} = 10(20 - 2x) = 20(10 - x).$$

Thus 10 more trees should be planted to maximize the yield.

13.

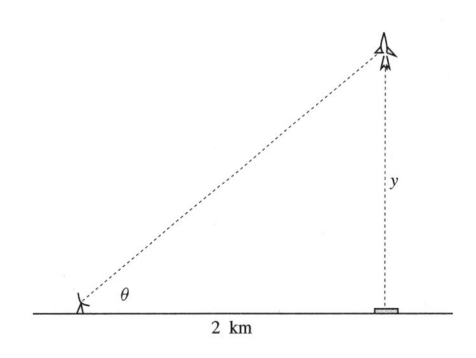

Fig. R-4.13

It was shown in the solution to Exercise 41 in Section 3.2 that at time t s after launch, the tracking antenna rotates upward at rate
$$\frac{d\theta}{dt} = \frac{800t}{400^2 + t^4} = f(t), \text{ say.}$$

Observe that $f(0) = 0$ and $f(t) \to 0$ as $t \to \infty$. For critical points,
$$0 = f'(t) = 800\left[\frac{(400^2 + t^4) - 4t^4}{(400^2 + t^4)^2}\right]$$
$$\Rightarrow 3t^4 = 400^2, \quad \text{or} \quad t \approx 15.197.$$

The maximum rate at which the antenna must turn is $f(15.197) \approx 0.057$ rad/s.

14. The narrowest hallway in which the table can be turned horizontally through $180°$ has width equal to twice the greatest distance from the origin (the centre of the table) to the curve $x^2 + y^4 = 1/8$ (the edge of the table). We maximize the square of this distance, which we express as a function of y:

$$S(y) = x^2 + y^2 = y^2 + \frac{1}{8} - y^4, \quad (0 \le y \le (1/8)^{1/4}).$$

Note that $S(0) = 1/8$ and $S((1/8)^{1/4}) = 1/\sqrt{8} > S(0)$. For CP:

$$0 = \frac{dS}{dy} = 2y - 4y^3 = 2y(1 - 2y^2).$$

The CPs are given by $y = 0$ (already considered), and $y^2 = 1/2$, where $S(y) = 3/8$. Since $3/8 > 1/\sqrt{8}$, this is the maximum value of S. The hallway must therefore be at least $2\sqrt{3/8} \approx 1.225$ m wide.

15. Let the ball have radius r cm. Its weight is proportional to the volume of metal it contains, so the condition of the problem states that

$$\frac{4\pi}{3}r^3 - \frac{4\pi}{3}(r-2)^3 = \frac{1}{2}\frac{4\pi}{3}r^3$$
$$r^3 - 12r^2 + 24r - 16 = 0.$$

Graphing the left side of this latter equation with a graphics calculator shows a root between 9 and 10. A "solve routine" or Newton's Method then refines an initial guess of, say, $r = 9.5$ to give $r = 9.69464420373$ cm for the radius of the ball.

16.

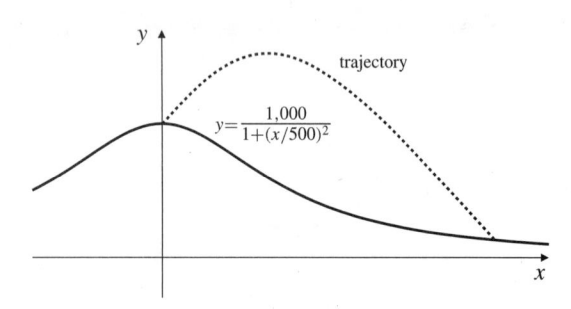

Fig. R-4.16

If the origin is at sea level under the launch point, and $x(t)$ and $y(t)$ are the horizontal and vertical coordinates of the cannon ball's position at time t s after it is fired, then

$$\frac{d^2x}{dt^2} = 0, \quad \frac{d^2y}{dt^2} = -32.$$

At $t = 0$, we have $dx/dt = dy/dt = 200/\sqrt{2}$, so

$$\frac{dx}{dt} = \frac{200}{\sqrt{2}}, \quad \frac{dy}{dt} = -32t + \frac{200}{\sqrt{2}}.$$

At $t = 0$, we have $x = 0$ and $y = 1,000$. Thus the position of the ball at time t is given by

$$x = \frac{200t}{\sqrt{2}}, \quad y = -16t^2 + \frac{200t}{\sqrt{2}} + 1,000.$$

We can obtain the Cartesian equation for the path of the cannon ball by solving the first equation for t and substituting into the second equation:

$$y = -16\frac{2x^2}{200^2} + x + 1,000.$$

The cannon ball strikes the ground when

$$-16\frac{2x^2}{200^2} + x + 1,000 = \frac{1,000}{1 + (x/500)^2}.$$

Graphing both sides of this equation suggests a solution near $x = 1,900$. Newton's Method or a solve routine then gives $x \approx 1,873$. The horizontal range is about 1,873 ft.

17. The percentage error in the approximation $-(g/L)\sin\theta \approx -(g.L)\theta$ is

$$100\left|\frac{\sin\theta - \theta}{\sin\theta}\right| = 100\left(\frac{\theta}{\sin\theta} - 1\right).$$

Since $\lim_{\theta \to 0} \theta/(\sin\theta) = 1$, the percentage error $\to 0$ as $\theta \to 0$. Also, $\theta/\sin\theta$ grows steadily larger as $|\theta|$ increases from 0 towards $\pi/2$. Thus the maximum percentage error for $|\theta| \le 20° = \pi/9$ will occur at $\theta = \pi/9$. This maximum percentage error is

$$100\left(\frac{\pi/9}{\sin(\pi/9)} - 1\right) \approx 2.06\%.$$

18. $\sin^2 x = \frac{1}{2}(1 - \cos(2x))$

$$= \frac{1}{2}\left[1 - \left(1 - \frac{2^2 x^2}{2!} + \frac{2^4 x^4}{4!} - \frac{2^6 x^6}{6!} + O(x^8)\right)\right]$$

$$= x^2 - \frac{x^4}{3} + \frac{2x^6}{45} + O(x^8)$$

$$\lim_{x \to 0} \frac{3\sin^2 x - 3x^2 + x^4}{x^6}$$

$$= \lim_{x \to 0} \frac{3x^2 - x^4 + \frac{2}{15}x^6 + O(x^8) - 3x^2 - x^4}{x^6}$$

$$= \lim_{x \to 0} \frac{2}{15} + O(x^2) = \frac{2}{15}.$$

19. $f(x) = \tan^{-1} x$, $f'(x) = \frac{1}{1+x^2}$, $f''(x) = \frac{-2x}{(1+x^2)^2}$,

$f'''(x) = \frac{6x^2 - 2}{(1+x^2)^3}.$

About $x = 1$, $P_2(x) = \dfrac{\pi}{4} + \dfrac{x-1}{2} - \dfrac{(x-1)^2}{4}$.

Thus $\tan^{-1}(1.1) \approx \dfrac{\pi}{4} + \dfrac{1}{20} - \dfrac{1}{400} \approx 0.832898$. On $[1, 1.1]$, we have

$$|f'''(x)| \le \dfrac{6(1.1)^2 - 2}{(1+1)^3} = 0.6575.$$

Thus the error does not exceed $\dfrac{0.6575}{3!}(1.1-1)^3 \approx .00011$ in absolute value.

20. The second approximation x_1 is the x-intercept of the tangent to $y = f(x)$ at $x = x_0 = 2$; it is the x-intercept of the line $2y = 10x - 19$. Thus $x_1 = 19/10 = 1.9$.

21.

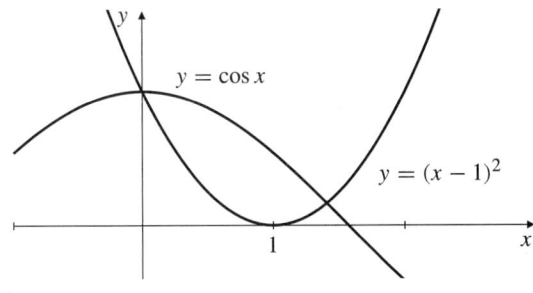

Fig. R-4.21

$y = \cos x$ and $y = (x-1)^2$ intersect at $x = 0$ and at a point x between $x = 1$ and $x = \pi/2 \approx 1.57$. Starting with an initial guess $x_0 = 1.3$, and iterating the Newton's Method formula

$$x_{n+1} = x_n - \dfrac{(x_n - 1)^2 - \cos x_n}{2(x_n - 1) + \sin x_n},$$

we get $x_4 = x_5 = 1.40556363276$. To 10 decimal places the two roots of the equation are $x = 0$ (exact), and $x = 1.4055636328$.

22. The square of the distance from $(2, 0)$ to $(x, \ln x)$ is $S(x) = (x-2)^2 + (\ln x)^2$, for $x > 0$. Since $S(x) \to \infty$ as $x \to \infty$ or $x \to 0+$, the minimum value of $S(x)$ will occur at a critical point. For CP:

$$0 = S'(x) = 2\left(x - 2 + \dfrac{\ln x}{x}\right).$$

We solve this equation using a TI-85 solve routine; $x \approx 1.6895797$. The minimum distance from the origin to $y = e^x$ is $\sqrt{S(x)} \approx 0.6094586$.

23. If the car is at (a, e^a), then its headlight beam lies along the tangent line to $y = e^x$ there, namely

$$y = e^a + e^a(x - a) = e^a(1 + x - a).$$

This line passes through $(1, 1)$ if $1 = e^a(2 - a)$. A solve routine gives $a \approx -1.1461932$. The corresponding value of e^a is about 0.3178444. The car is at (a, e^a).

24.
$$\lim_{h \to 0} \dfrac{f(x + 2h) - 2f(x + h) + 2f(x - h) - f(x - 2h)}{2h^3} \quad \left[\dfrac{0}{0}\right]$$

$$= \lim_{h \to 0} \dfrac{2f'(x + 2h) - 2f'(x + h) - 2f'(x - h) + 2f'(x - 2h)}{6h^2}$$

$$\left[\dfrac{0}{0}\right]$$

$$= \lim_{h \to 0} \dfrac{4f''(x + 2h) - 2f''(x + h) + 2f''(x - h) - 4f''(x - 2h)}{12h}$$

$$\left[\dfrac{0}{0}\right]$$

$$= \lim_{h \to 0} \dfrac{8f'''(x + 2h) - 2f'''(x + h) - 2f'''(x - h) + 8f'''(x - 2h)}{12}$$

$$= f'''(x).$$

Challenging Problems 4 (page 297)

1. $\dfrac{dV}{dt} = kx^2(V_0 - V)$.

a) If $V = x^3$, then $3x^2 \dfrac{dx}{dt} = \dfrac{dV}{dt} = kx^2(V_0 - x^3)$, so

$$\dfrac{dx}{dt} = \dfrac{k}{3}(V_0 - x^3).$$

b) The rate of growth of the edge is $(k/3)(V_0 - x^3)$, which is positive if $0 \le x < x_0 = V_0^{1/3}$. The time derivative of this rate is

$$-kx^2 \dfrac{dx}{dt} = -\dfrac{k^2}{3} x^2 (V_0 - x^3) < 0$$

for $0 < x < x_0$. Thus the edge length is increasing at a decreasing rate.

c) Initially, x grows at rate $kV_0/3$. The rate of growth of x will be half of this if

$$\dfrac{k}{3}(V_0 - x^3) = \dfrac{kV_0}{6},$$

that is, if $x = (V_0/2)^{1/3}$. Then $V = V_0/2$.

2. Let the speed of the tank be v where $v = \dfrac{dy}{dt} = ky$. Thus, $y = Ce^{kt}$. Given that at $t = 0$, $y = 4$, then $4 = y(0) = C$. Also given that at $t = 10$, $y = 2$, thus,

$$2 = y(10) = 4e^{10k} \Rightarrow k = -\tfrac{1}{10} \ln 2.$$

Hence, $y = 4e^{(-\frac{1}{10}\ln 2)t}$ and $v = \dfrac{dy}{dt} = (-\dfrac{1}{10}\ln 2)y$. The slope of the curve $xy = 1$ is $m = \dfrac{dy}{dx} = -\dfrac{1}{x^2}$. Thus, the equation of the tangent line at the point $\left(\dfrac{1}{y_0}, y_0\right)$ is

$$y = y_0 - \dfrac{1}{\left(\dfrac{1}{y_0}\right)^2}\left(x - \dfrac{1}{y_0}\right), \quad \text{i.e.,} \quad y = 2y_0 - xy_0^2.$$

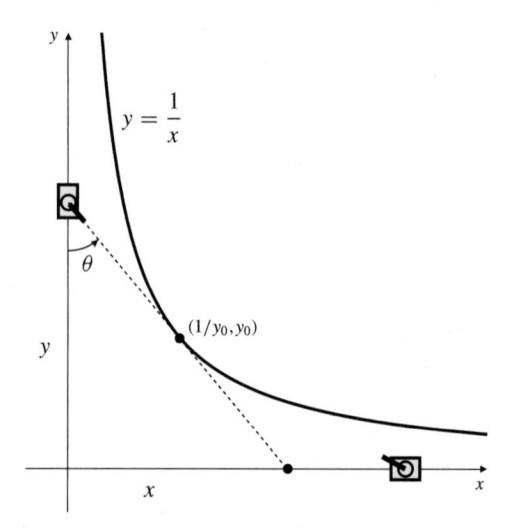

Fig. C-4.2

Hence, the x-intercept is $x = \dfrac{2}{y_0}$ and the y-intercept is $y = 2y_0$. Let θ be the angle between the gun and the y-axis. We have

$$\tan\theta = \dfrac{x}{y} = \dfrac{\left(\dfrac{2}{y_0}\right)}{2y_0} = \dfrac{1}{y_0^2} = \dfrac{4}{y^2}$$

$$\Rightarrow \sec^2\theta\,\dfrac{d\theta}{dt} = \dfrac{-8}{y^3}\dfrac{dy}{dt}.$$

Now

$$\sec^2\theta = 1 + \tan^2\theta = 1 + \dfrac{16}{y^4} = \dfrac{y^4 + 16}{y^4},$$

so

$$\dfrac{d\theta}{dt} = -\dfrac{8y}{y^4 + 16}\dfrac{dy}{dt} = -\dfrac{8ky^2}{y^4 + 16}.$$

The maximum value of $\dfrac{y^2}{y^4 + 16}$ occurs at a critical point:

$$0 = \dfrac{(y^4 + 16)2y - y^2(4y^3)}{(y^4 + 16)^2}$$

$$\Leftrightarrow 2y^5 = 32y,$$

or $y = 2$. Therefore the maximum rate of rotation of the gun turret must be

$$-8k\dfrac{2^2}{2^4 + 16} = -k = \dfrac{1}{10}\ln 2 \approx 0.0693 \text{ rad/m,}$$

and occurs when your tank is 2 km from the origin.

3. a) If $q = 0.99$, the number of tests required is $T = N((1/x) + 1 - 0.99^x)$. T is a decreasing function for small values of x because the term $1/x$ dominates. It is increasing for large x because -0.99^x dominates. Thus T will have a minimum value at a critical point, provided N is sufficiently large that the CP is in $(0, N)$. For CP:

$$0 = \dfrac{dT}{dx} = N\left(-\dfrac{1}{x^2} - 0.99^x \ln(0.99)\right)$$

$$x^2 = \dfrac{(0.99)^{-x}}{-\ln(0.99)}$$

$$x = \dfrac{(0.99)^{-x/2}}{\sqrt{-\ln(0.99)}} = f(x), \text{ say.}$$

b) Starting with $x_0 = 20$, we iterate $x_{n+1} = f(x_n)$. The first three iterations give

$$x_1 \approx 11.03, \quad x_2 \approx 10.54, \quad x_3 \approx 10.51.$$

This suggests the CP is near 10.5. Since x must be an integer, we test $x = 10$ and $x = 11$: $T(10) \approx 0.19562$ and $T(11) \approx 0.19557$. The minimum cost should arise by using groups of 11 individuals.

4. $P = 2\pi\sqrt{L/g} = 2\pi L^{1/2}g^{-1/2}$.

a) If L remains constant, then

$$\Delta P \approx \dfrac{dP}{dg}\Delta g = -\pi L^{1/2}g^{-3/2}\Delta g$$

$$\dfrac{\Delta P}{P} \approx \dfrac{-\pi L^{1/2}g^{-3/2}}{2\pi L^{1/2}g^{-1/2}}\Delta g = -\dfrac{1}{2}\dfrac{\Delta g}{g}.$$

If g increases by 1%, then $\Delta g/g = 1/100$, and $\Delta P/P = -1/200$. Thus P decreases by 0.5%.

b) If g remains constant, then

$$\Delta P \approx \dfrac{dP}{dL}\Delta L = \pi L^{-1/2}g^{-1/2}\Delta L$$

$$\dfrac{\Delta P}{P} \approx \dfrac{\pi L^{-1/2}g^{-1/2}}{2\pi L^{1/2}g^{-1/2}}\Delta L = \dfrac{1}{2}\dfrac{\Delta L}{L}.$$

If L increases by 2%, then $\Delta L/L = 2/100$, and $\Delta P/P = 1/100$. Thus P increases by 1%.

5. $\dfrac{dV}{dt} = -k\sqrt{y}, \quad V = Ay.$

INSTRUCTOR'S SOLUTIONS MANUAL   CHALLENGING PROBLEMS 4   (PAGE 297)

a) $A\dfrac{dy}{dt} = \dfrac{dV}{dt} = -k\sqrt{y}$, so $\dfrac{dy}{dt} = -\dfrac{k}{A}\sqrt{y}$.

b) If $y(t) = \left(\sqrt{y_0} - \dfrac{kt}{2A}\right)^2$, then $y(0) = y_0$, and

$$\dfrac{dy}{dt} = 2\left(\sqrt{y_0} - \dfrac{kt}{2A}\right)\left(-\dfrac{k}{2A}\right)$$
$$= -\dfrac{k}{A}\sqrt{y(t)}.$$

Thus the given expression does solve the initial-value problem for y.

c) If $y(T) = 0$, then $\dfrac{kT}{2A} = \sqrt{y_0}$, so $k = 2A\sqrt{y_0}/T$. Thus

$$y(t) = \left(\sqrt{y_0} - \dfrac{2A\sqrt{y_0}\,t}{2AT}\right)^2 = y_0\left(1 - \dfrac{t}{T}\right)^2.$$

d) Half the liquid drains out in time t_1, where

$$y_0\left(1 - \dfrac{t_1}{T}\right)^2 = \dfrac{y_0}{2}.$$

Thus $t_1 = T(1 - (1/\sqrt{2}))$.

6. If the depth of liquid in the tank at time t is $y(t)$, then the surface of the liquid has radius $r(t) = Ry(t)/H$, and the volume of liquid in the tank at that time is

$$V(t) = \dfrac{\pi}{3}\left(\dfrac{Ry(t)}{H}\right)^2 y(t) = \dfrac{\pi R^2}{3H^2}(y(t))^3.$$

By Torricelli's law, $dV/dt = -k\sqrt{y}$. Thus

$$\dfrac{\pi R^2}{3H^2} 3y^2 \dfrac{dy}{dt} = \dfrac{dV}{dt} = -k\sqrt{y},$$

or, $dy/dt = -k_1 y^{-3/2}$, where $k_1 = kH^2/(\pi R^2)$.
If $y(t) = y_0\left(1 - \dfrac{t}{T}\right)^{2/5}$, then $y(0) = y_0$, $y(T) = 0$, and

$$\dfrac{dy}{dt} = \dfrac{2}{5}y_0\left(1 - \dfrac{t}{T}\right)^{-3/5}\left(-\dfrac{1}{T}\right) = -k_1 y^{-3/2},$$

where $k_1 = 2y_0/(5T)$. Thus this function $y(t)$ satisfies the conditions of the problem.

7. If the triangle has legs x and y and hypotenuse $\sqrt{x^2+y^2}$, then

$$P = x + y + \sqrt{x^2+y^2}$$
$$(P - x - y)^2 = x^2 + y^2$$
$$P^2 + x^2 + y^2 + 2xy - 2Px - 2Py = x^2 + y^2$$
$$y(2P - 2x) = P^2 - 2Px$$
$$y = \dfrac{P(P - 2x)}{2(P - x)}.$$

The area of the triangle is

$$A = \dfrac{xy}{2} = \dfrac{P}{4}\cdot\dfrac{Px - 2x^2}{P - x}.$$

$A = 0$ if $x = 0$ or $x = P/2$ and $A > 0$ between these values of x. The maximum area will therefore occur at a critical point.

$$0 = \dfrac{dA}{dx} = \dfrac{P}{4}\cdot\dfrac{(P - x)(P - 4x) - x(P - 2x)(-1)}{(P - x)^2}$$
$$0 = P^2 - 5Px + 4x^2 + Px - 2x^2$$
$$2x^2 - 4Px + P^2 = 0.$$

This quadratic has two roots, but the only one in $[0, P/2]$ is

$$x = \dfrac{4P - \sqrt{16P^2 - 8P^2}}{4} = P\left(1 - \dfrac{1}{\sqrt{2}}\right).$$

This value of x gives $A(x) = \tfrac{1}{2}P^2\left(1 - \dfrac{1}{\sqrt{2}}\right)^2$ un² for the maximum area of the triangle. (Note that the maximal triangle is isosceles, as we might have guessed.)

8. The slope of $y = x^3 + ax^2 + bx + c$ is

$$y' = 3x^2 + 2ax + b,$$

which $\to \infty$ as $x \to \pm\infty$. The quadratic expression y' takes each of its values at two different points except its minimum value, which is achieved only at one point given by $y'' = 6x + 2a = 0$. Thus the tangent to the cubic at $x = -a/3$ is not parallel to any other tangent. This tangent has equation

$$y = -\dfrac{a^3}{27} + \dfrac{a^3}{9} - \dfrac{ab}{3} + c$$
$$\quad + \left(\dfrac{a^2}{3} - \dfrac{2a^2}{3} + b\right)\left(x + \dfrac{a}{3}\right)$$
$$= -\dfrac{a^3}{27} + c + \left(b - \dfrac{a^2}{3}\right)x.$$

9.

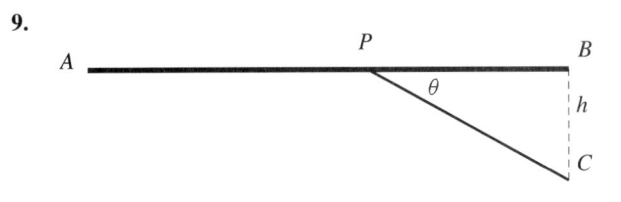

Fig. C-4.9

167

a) The total resistance of path APC is

$$R = \frac{k|AP|}{r_1^2} + \frac{k|PC|}{r_2^2}$$
$$= k\left(\frac{L - h\cot\theta}{r_1^2} + \frac{h\csc\theta}{r_2^2}\right).$$

We have

$$\frac{dR}{d\theta} = kh\left(\frac{\csc^2\theta}{r_1^2} - \frac{\csc\theta\cot\theta}{r_2^2}\right),$$

so the CP of R is given by $\dfrac{\csc\theta}{\cot\theta} = \dfrac{r_1^2}{r_2^2}$, that is, $\cos\theta = (r_2/r_1)^2$ or $\theta = \cos^{-1}((r_2/r_1)^2)$. This CP will give the minimum resistance if it is in the interval of possible values of θ, namely $[\tan^{-1}(h/L), \pi/2]$; otherwise the minimum will occur for $P = A$. Thus, for large L, P should be chosen to make $\cos\theta = (r_2/r_1)^2$.

b) This is the same problem as that in (a) except that r_1 and r_2 are replaced with r_1^2 and r_2^2, respectively. Thus the minimum resistance corresponds to choosing P so that $\cos\theta = (r_2/r_1)^4$. This puts P closer to B than it was in part (a), which is reasonable since the resistance ratio between the thin and thick pipes is greater than for the wires in part (a).

10.

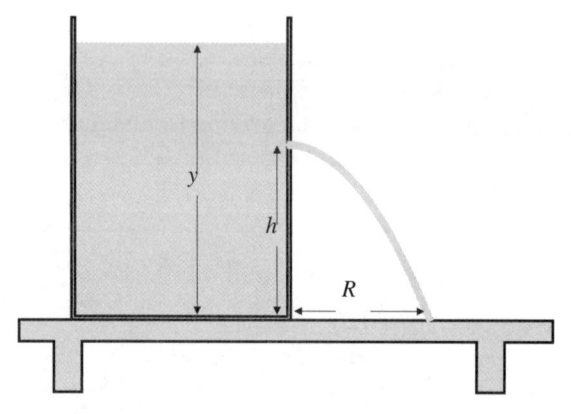

Fig. C-4.10

a) Let the origin be at the point on the table directly under the hole. If a water particle leaves the tank with horizontal velocity v, then its position $(X(t), Y(t))$, t seconds later, is given by

$$\frac{d^2X}{dt^2} = 0 \qquad \frac{d^2Y}{dt^2} = -g$$
$$\frac{dX}{dt} = v \qquad \frac{dY}{dt} = -gt$$
$$X = vt \qquad Y = -\frac{1}{2}gt^2 + h.$$

The range R of the particle (i.e., of the spurt) is the value of X when $Y = 0$, that is, at time $t = \sqrt{2h/g}$. Thus $R = v\sqrt{2h/g}$.

b) Since $v = k\sqrt{y - h}$, the range R is a function of y, the depth of water in the tank.

$$R = k\sqrt{\frac{2}{g}}\sqrt{h(y-h)}.$$

For a given depth y, R will be maximum if $h(y - h)$ is maximum. This occurs at the critical point $h = y/2$ of the quadratic $Q(h) = h(y - h)$.

c) By the result of part (c) of Problem 3 (with y replaced by $y - h$, the height of the surface of the water above the drain in the current problem), we have

$$y(t) - h = (y_0 - h)\left(1 - \frac{t}{T}\right)^2, \quad \text{for } 0 \le t \le T.$$

As shown above, the range of the spurt at time t is

$$R(t) = k\sqrt{\frac{2}{g}}\sqrt{h(y(t) - h)}.$$

Since $R = R_0$ when $y = y_0$, we have

$$k = \frac{R_0}{\sqrt{\dfrac{2}{g}}\sqrt{h(y_0 - h)}}.$$

Therefore $R(t) = R_0 \dfrac{\sqrt{h(y(t) - h)}}{\sqrt{h(y_0 - h)}} = R_0\left(1 - \dfrac{t}{T}\right)$.

11.

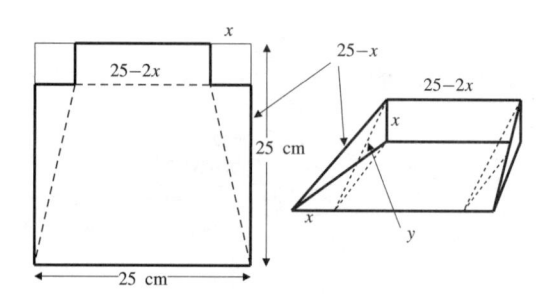

Fig. C-4.11

Note that the vertical back wall of the dustpan is perpendicular to the plane of the *top* of the pan, not the bottom. The volume of the pan is made up of three parts:

a triangular prism (the centre part) having height x, width $25 - 2x$, and depth y (all distances in cm), where $y^2 + x^2 = (25 - x)^2$, and so $y = \sqrt{625 - 50x} = 5\sqrt{25 - 2x}$, and two triangular pyramids (one on each side) each having height x and a right-triangular top with dimensions x and y.

The volume of the pan is, therefore,

$$V = \frac{1}{2}xy(25 - 2x) + 2\left(\frac{1}{3}\right)\left(\frac{1}{2}xy\right)x$$
$$= \frac{1}{2}xy\left(25 - 2x + \frac{2}{3}x\right)$$
$$= \frac{5}{6}x\sqrt{25 - 2x}(75 - 4x) = V(x).$$

The appropriate values for x are $0 \leq x \leq 25/2$. Note that $V(0) = V(25/2) = 0$ and $V(x) > 0$ in $(0, 25/2)$. The maximum volume will therefore occur at a critical point:

$$0 = \frac{dV}{dx} = -\frac{25}{6}\frac{4x^2 - 85x + 375}{\sqrt{25 - 2x}}$$

(after simplification). The quadratic in the numerator factors to $(x - 15)(4x - 25)$, so the CPs are $x = 15$ and $x = 25/4$. Only $x = 25/4$ is in the required interval. The maximum volume of the dustpan is $V(25/4) \approx 921$ cm^3.

CHAPTER 5. INTEGRATION

Section 5.1 Sums and Sigma Notation (page 303)

1. $\sum_{i=1}^{4} i^3 = 1^3 + 2^3 + 3^3 + 4^3$

2. $\sum_{j=1}^{100} \frac{j}{j+1} = \frac{1}{2} + \frac{2}{3} + \frac{3}{4} + \cdots + \frac{100}{101}$

3. $\sum_{i=1}^{n} 3^i = 3 + 3^2 + 3^3 + \cdots + 3^n$

4. $\sum_{i=0}^{n-1} \frac{(-1)^i}{i+1} = 1 - \frac{1}{2} + \frac{1}{3} - \cdots + \frac{(-1)^{n-1}}{n}$

5. $\sum_{j=3}^{n} \frac{(-2)^j}{(j-2)^2} = -\frac{2^3}{1^2} + \frac{2^4}{2^2} - \frac{2^5}{3^2} + \cdots + \frac{(-1)^n 2^n}{(n-2)^2}$

6. $\sum_{j=1}^{n} \frac{j^2}{n^3} = \frac{1}{n^3} + \frac{4}{n^3} + \frac{9}{n^3} + \cdots + \frac{n^2}{n^3}$

7. $\sum_{n=1}^{k} \sin \frac{n\pi}{3k} = \sin \frac{\pi}{3k} + \sin \frac{2\pi}{3k} + \cdots + \sin \frac{k\pi}{3k}$

8. $\sum_{k=2}^{n} \frac{e^{-k}}{n+k} = \frac{e^{-2}}{n+2} + \frac{e^{-3}}{n+3} + \cdots + \frac{e^{-n}}{n+n}$

9. $5 + 6 + 7 + 8 + 9 = \sum_{i=5}^{9} i$

10. $5^3 + 6^3 + 7^3 + 8^3 + 9^3 = \sum_{i=5}^{9} i^3$

11. $2 + 2 + 2 + \cdots + 2$ (200 terms) equals $\sum_{i=1}^{200} 2$

12. $\frac{1}{7} + \frac{1}{8} + \frac{1}{9} + \cdots + \frac{1}{99} = \sum_{i=7}^{99} \frac{1}{i}$

13. $2^2 - 3^2 + 4^2 - 5^2 + \cdots - 99^2 = \sum_{i=2}^{99} (-1)^i i^2$

14. $1 + 2x + 3x^2 + 4x^3 + \cdots + 100x^{99} = \sum_{i=1}^{100} ix^{i-1}$

15. $1 + x + x^2 + x^3 + \cdots + x^n = \sum_{i=0}^{n} x^i$

16. $1 - x + x^2 - x^3 + \cdots + x^{2n} = \sum_{i=0}^{2n} (-1)^i x^i$

17. $1 - \frac{1}{4} + \frac{1}{9} - \cdots + \frac{(-1)^{n-1}}{n^2} = \sum_{i=1}^{n} \frac{(-1)^{i-1}}{i^2}$

18. $\frac{1}{2} + \frac{2}{4} + \frac{3}{8} + \frac{4}{16} + \cdots + \frac{n}{2^n} = \sum_{i=1}^{n} \frac{i}{2^i}$

19. $\sum_{j=0}^{99} \sin j = \sum_{i=1}^{100} \sin(i-1)$

20. $\sum_{k=-5}^{m} \frac{1}{k^2+1} = \sum_{i=1}^{m+6} \frac{1}{((i-6)^2+1)}$

21. $\sum_{i=2}^{6} (i-1) = 1 + 2 + 3 + 4 + 5 = 15$

22. $\sum_{j=1}^{1,000} (2j+3) = \frac{2(1,000)(1,001)}{2} + 3,000 = 1,004,000$

23. $\sum_{i=1}^{n} (i^2 + 2i) = \frac{n(n+1)(2n+1)}{6} + 2\frac{n(n+1)}{2} = \frac{n(n+1)(2n+\ldots)}{6}$

24. $\sum_{k=1}^{n-1} (3k^2 - 4) = 3\frac{(n-1)n(2n-1)}{6} - 4(n-1) = n^3 - \frac{3}{2}n^2 - \frac{7}{2}n + 4$

25. $\sum_{i=m}^{n} (2i-1) = 2\frac{n(n+1)}{2} - n - 2\frac{(m-1)m}{2} + m - 1 = n^2 - (m-1)^2$

26. $\sum_{j=0}^{n-1} (4j + 4^j) = 4\frac{(n-1)n}{2} + \frac{4^n - 1}{3} = 2n^2 - 2n + \frac{4^n - 1}{3}$

27. $\sum_{k=1}^{n} (\pi^k - 3) = \frac{\pi(\pi^n - 1)}{\pi - 1} - 3n$

28. $\sum_{i=1}^{n} (2^i - i^2) = 2^{n+1} - 2 - \frac{1}{6}n(n+1)(2n+1)$

29. $\sum_{m=1}^{n} \ln m = \ln 1 + \ln 2 + \cdots + \ln n = \ln(n!)$

30. $\sum_{i=0}^{n} e^{i/n} = \frac{e^{(n+1)/n} - 1}{e^{1/n} - 1}$

31. $2 + 2 + \cdots + 2$ (200 terms) equals 400

32. $1 + x + x^2 + \cdots + x^n = \begin{cases} \frac{1 - x^{n+1}}{1 - x} & \text{if } x \neq 1 \\ n + 1 & \text{if } x = 1 \end{cases}$

33. $1 - x + x^2 - x^3 + \cdots + x^{2n} = \begin{cases} \frac{1 + x^{2n+1}}{1 + x} & \text{if } x \neq -1 \\ 2n + 1 & \text{if } x = -1 \end{cases}$

34. Let $f(x) = 1 + x + x^2 + \cdots + x^{100} = \dfrac{x^{101} - 1}{x - 1}$ if $x \neq 1$. Then

$$f'(x) = 1 + 2x + 3x^2 + \cdots + 100x^{99}$$
$$= \dfrac{d}{dx} \dfrac{x^{101} - 1}{x - 1} = \dfrac{100x^{101} - 101x^{100} + 1}{(x - 1)^2}.$$

35. $2^2 - 3^2 + 4^2 - 5^2 + \cdots + 98^2 - 99^2$

$$= \sum_{k=1}^{49} [(2k)^2 - (2k+1)^2] = \sum_{k=1}^{49} [4k^2 - 4k^2 - 4k - 1]$$

$$= -\sum_{k=1}^{49} [4k + 1] = -4\dfrac{49 \times 50}{2} - 49 = -4,949$$

36. Let $s = \dfrac{1}{2} + \dfrac{2}{4} + \dfrac{3}{8} + \cdots + \dfrac{n}{2^n}$. Then

$$\dfrac{s}{2} = \dfrac{1}{4} + \dfrac{2}{8} + \dfrac{3}{16} + \cdots + \dfrac{n}{2^{n+1}}.$$

Subtracting these two sums, we get

$$\dfrac{s}{2} = \dfrac{1}{2} + \dfrac{1}{4} + \dfrac{1}{8} + \cdots + \dfrac{1}{2^n} - \dfrac{n}{2^{n+1}}$$
$$= \dfrac{1}{2} \dfrac{1 - (1/2^n)}{1 - (1/2)} - \dfrac{n}{2^{n+1}}$$
$$= 1 - \dfrac{n+2}{2^{n+1}}.$$

Thus $s = 2 + (n+2)/2^n$.

37. $\displaystyle\sum_{i=m}^{n} (f(i+1) - f(i)) = \sum_{i=m}^{n} f(i+1) - \sum_{i=m}^{n} f(i)$
$$= \sum_{j=m+1}^{n+1} f(j) - \sum_{i=m}^{n} f(i)$$
$$= f(n+1) - f(m),$$

because each sum has only one term that is not cancelled by a term in the other sum. It is called "telescoping" because the sum "folds up" to a sum involving only part of the first and last terms.

38. $\displaystyle\sum_{n=1}^{10} (n^4 - (n-1)^4) = 10^4 - 0^4 = 10,000$

39. $\displaystyle\sum_{j=1}^{m} (2^j - 2^{j-1}) = 2^m - 2^0 = 2^m - 1$

40. $\displaystyle\sum_{i=m}^{2m} \left(\dfrac{1}{i} - \dfrac{1}{i+1} \right) = \dfrac{1}{m} - \dfrac{1}{2m+1} = \dfrac{m+1}{m(2m+1)}$

41. $\displaystyle\sum_{j=1}^{m} \dfrac{1}{j(j+1)} = \sum_{j=1}^{m} \left(\dfrac{1}{j} - \dfrac{1}{j+1} \right) = 1 - \dfrac{1}{n+1} = \dfrac{n}{n+1}$

42. The number of small shaded squares is $1 + 2 + \cdots + n$. Since each has area 1, the total area shaded is $\sum_{i=1}^{n} i$. But this area consists of a large right-angled triangle of area $n^2/2$ (below the diagonal), and n small triangles (above the diagonal) each of area $1/2$. Equating these areas, we get

$$\sum_{i=1}^{n} i = \dfrac{n^2}{2} + n\dfrac{1}{2} = \dfrac{n(n+1)}{2}.$$

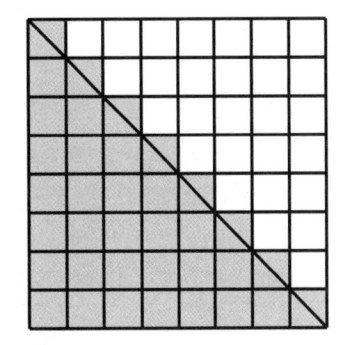

Fig. 5.1.42

43. To show that

$$\sum_{i=1}^{n} i = \dfrac{n(n+1)}{2},$$

we write n copies of the identity

$$(k+1)^2 - k^2 = 2k + 1,$$

one for each k from 1 to n:

$$2^2 - 1^2 = 2(1) + 1$$
$$3^2 - 2^2 = 2(2) + 1$$
$$4^2 - 3^2 = 2(3) + 1$$
$$\vdots$$
$$(n+1)^2 - n^2 = 2(n) + 1.$$

Adding the left and right sides of these formulas we get

$$(n+1)^2 - 1^2 = 2\sum_{i=1}^{n} i + n.$$

Hence, $\sum_{i=1}^{n} i = \dfrac{1}{2}(n^2 + 2n + 1 - 1 - n) = \dfrac{n(n+1)}{2}$.

44. The formula $\sum_{i=1}^{n} i = n(n+1)/2$ holds for $n = 1$, since it says $1 = 1$ in this case. Now assume that it holds for $n =$ some number $k \geq 1$; that is, $\sum_{i=1}^{k} i = k(k+1)/2$. Then for $n = k+1$, we have

$$\sum_{i=1}^{k+1} i = \sum_{i=1}^{k} i + (k+1) = \dfrac{k(k+1)}{2} + (k+1) = \dfrac{(k+1)(k+2)}{2}.$$

Thus the formula also holds for $n = k+1$. By induction, it holds for all positive integers n.

45. The formula $\sum_{i=1}^{n} i^2 = n(n+1)(2n+1)/6$ holds for $n = 1$, since it says $1 = 1$ in this case. Now assume that it holds for $n = $ some number $k \geq 1$; that is, $\sum_{i=1}^{k} i^2 = k(k+1)(2k+1)/6$. Then for $n = k+1$, we have

$$\sum_{i=1}^{k+1} i^2 = \sum_{i=1}^{k} i^2 + (k+1)^2$$
$$= \frac{k(k+1)(2k+1)}{6} + (k+1)^2$$
$$= \frac{k+1}{6}[2k^2 + k + 6k + 6]$$
$$= \frac{k+1}{6}(k+2)(2k+3)$$
$$= \frac{(k+1)((k+1)+1)(2(k+1)+1)}{6}.$$

Thus the formula also holds for $n = k+1$. By induction, it holds for all positive integers n.

46. The formula $\sum_{i=1}^{n} r^{i-1} = (r^n - 1)/(r-1)$ (for $r \neq 1$) holds for $n = 1$, since it says $1 = 1$ in this case. Now assume that it holds for $n = $ some number $k \geq 1$; that is, $\sum_{i=1}^{k} r^{i-1} = (r^k - 1)/(r-1)$. Then for $n = k+1$, we have

$$\sum_{i=1}^{k+1} r^{i-1} = \sum_{i=1}^{k} r^{i-1} + r^k = \frac{r^k - 1}{r-1} + r^k = \frac{r^{k+1} - 1}{r-1}.$$

Thus the formula also holds for $n = k+1$. By induction, it holds for all positive integers n.

47.

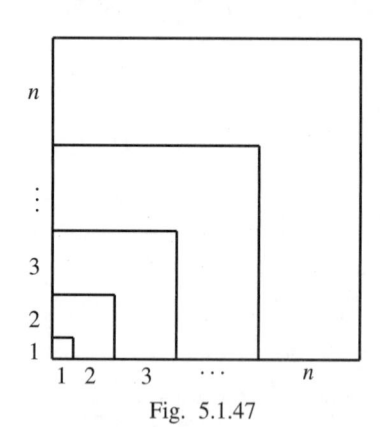

Fig. 5.1.47

The L-shaped region with short side i is a square of side $i(i+1)/2$ with a square of side $(i-1)i/2$ cut out. Since

$$\left(\frac{i(i+1)}{2}\right)^2 - \left(\frac{(i-1)i}{2}\right)^2$$
$$= \frac{i^4 + 2i^3 + i^2 - (i^4 - 2i^3 + i^2)}{4} = i^3,$$

that L-shaped region has area i^3. The sum of the areas of the n L-shaped regions is the area of the large square of side $n(n+1)/2$, so

$$\sum_{i=1}^{n} i^3 = \left(\frac{n(n+1)}{2}\right)^2.$$

48. To show that

$$\sum_{j=1}^{n} j^3 = 1^3 + 2^3 + 3^3 + \cdots + n^3 = \frac{n^2(n+1)^2}{4},$$

we write n copies of the identity

$$(k+1)^4 - k^4 = 4k^3 + 6k^2 + 4k + 1,$$

one for each k from 1 to n:

$$2^4 - 1^4 = 4(1)^3 + 6(1)^2 + 4(1) + 1$$
$$3^4 - 2^4 = 4(2)^3 + 6(2)^2 + 4(2) + 1$$
$$4^4 - 3^4 = 4(3)^3 + 6(3)^2 + 4(3) + 1$$
$$\vdots$$
$$(n+1)^4 - n^4 = 4(n)^3 + 6(n)^2 + 4(n) + 1.$$

Adding the left and right sides of these formulas we get

$$(n+1)^4 - 1^4 = 4\sum_{j=1}^{n} j^3 + 6\sum_{j=1}^{n} j^2 + 4\sum_{j=1}^{n} j + n$$
$$= 4\sum_{j=1}^{n} j^3 + \frac{6n(n+1)(2n+1)}{6} + \frac{4n(n+1)}{2} + n.$$

Hence,

$$4\sum_{j=1}^{n} j^3 = (n+1)^4 - 1 - n(n+1)(2n+1) - 2n(n+1) - n$$
$$= n^2(n+1)^2$$

so $\sum_{j=1}^{n} j^3 = \dfrac{n^2(n+1)^2}{4}$.

49. The formula $\sum_{i=1}^{n} i^3 = n^2(n+1)^2/4$ holds for $n = 1$, since it says $1 = 1$ in this case. Now assume that it holds for $n = $ some number $k \geq 1$; that is, $\sum_{i=1}^{k} i^3 = k^2(k+1)^2/4$. Then for $n = k+1$, we have

$$\sum_{i=1}^{k+1} i^3 = \sum_{i=1}^{k} i^3 + (k+1)^3$$
$$= \frac{k^2(k+1)^2}{4} + (k+1)^3 = \frac{(k+1)^2}{4}[k^2 + 4(k+1)]$$
$$= \frac{(k+1)^2}{4}(k+2)^2.$$

INSTRUCTOR'S SOLUTIONS MANUAL   SECTION 5.2 (PAGE 310)

Thus the formula also holds for $n = k + 1$. By induction, it holds for all positive integers n.

50. To find $\sum_{j=1}^{n} j^4 = 1^4 + 2^4 + 3^4 + \cdots + n^4$, we write n copies of the identity

$$(k+1)^5 - k^5 = 5k^4 + 10k^3 + 10k^2 + 5k + 1,$$

one for each k from 1 to n:

$$2^5 - 1^5 = 5(1)^4 + 10(1)^3 + 10(1)^2 + 5(1) + 1$$
$$3^5 - 2^5 = 5(2)^4 + 10(2)^3 + 10(2)^2 + 5(2) + 1$$
$$4^5 - 3^5 = 5(3)^4 + 10(3)^3 + 10(3)^2 + 5(3) + 1$$
$$\vdots$$
$$(n+1)^5 - n^5 = 5(n)^4 + 10(n)^3 + 10(n)^2 + 5(n) + 1.$$

Adding the left and right sides of these formulas we get

$$(n+1)^5 - 1^5 = 5\sum_{j=1}^{n} j^4 + 10\sum_{j=1}^{n} j^3 + 10\sum_{j=1}^{n} j^2 + 5\sum_{j=1}^{n} j + n.$$

Substituting the known formulas for all the sums except $\sum_{j=1}^{n} j^4$, and solving for this quantity, gives

$$\sum_{j=1}^{n} j^4 = \frac{n(n+1)(2n+1)(3n^2+3n-1)}{30}.$$

Of course we got Maple to do the donkey work!

51. $\sum_{i=1}^{n} i^5 = \frac{1}{6}n^6 + \frac{1}{2}n^5 + \frac{5}{12}n^4 - \frac{1}{12}n^2$

$\sum_{i=1}^{n} i^6 = \frac{1}{7}n^7 + \frac{1}{2}n^6 + \frac{1}{2}n^5 - \frac{1}{6}n^3 + \frac{1}{42}n$

$\sum_{i=1}^{n} i^7 = \frac{1}{8}n^8 + \frac{1}{2}n^7 + \frac{7}{12}n^6 - \frac{7}{24}n^4 + \frac{1}{12}n^2$

$\sum_{i=1}^{n} i^8 = \frac{1}{9}n^9 + \frac{1}{2}n^8 + \cdots$

We would guess (correctly) that

$$\sum_{i=1}^{n} i^{10} = \frac{1}{11}n^{11} + \frac{1}{2}n^{10} + \cdots.$$

Section 5.2 Areas as Limits of Sums (page 310)

1. The area is the limit of the sum of the areas of the rectangles shown in the figure. It is

$$A = \lim_{n\to\infty} \frac{1}{n}\left[\frac{3}{n} + \frac{3\times 2}{n} + \frac{3\times 3}{n} + \cdots + \frac{3n}{n}\right]$$
$$= \lim_{n\to\infty} \frac{3}{n^2}(1 + 2 + 3 + \cdots + n)$$
$$= \lim_{n\to\infty} \frac{3}{n^2} \cdot \frac{n(n+1)}{2} = \frac{3}{2} \text{ sq. units.}$$

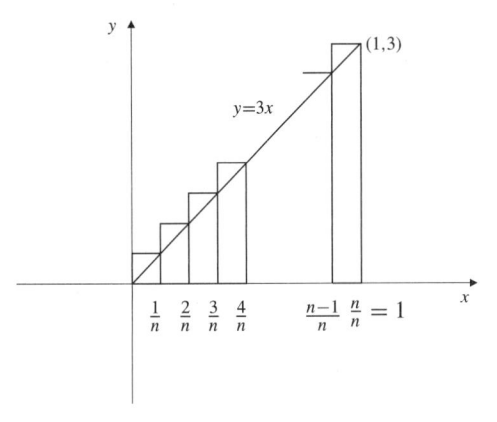

Fig. 5.2.1

2. This is similar to #1; the rectangles now have width $3/n$ and the ith has height $2(3i/n)+1$, the value of $2x+1$ at $x = 3i/n$. The area is

$$A = \lim_{n\to\infty} \sum_{i=1}^{n} \frac{3}{n}\left(2\frac{3i}{n} + 1\right)$$
$$= \lim_{n\to\infty} \frac{18}{n^2}\sum_{i=1}^{n} i + \frac{3}{n}n$$
$$= \lim_{n\to\infty} \frac{18}{n^2}\frac{n(n+1)}{2} + 3 = 9 + 3 = 12 \text{ sq. units.}$$

3. This is similar to #1; the rectangles have width $(3-1)/n = 2/n$ and the ith has height the value of $2x-1$ at $x = 1 + (2i/n)$. The area is

$$A = \lim_{n\to\infty} \sum_{i=1}^{n} \frac{2}{n}\left(2 + 2\frac{2i}{n} - 1\right)$$
$$= \lim_{n\to\infty} \frac{8}{n^2}\sum_{i=1}^{n} i + \frac{2}{n}n$$
$$= \lim_{n\to\infty} \frac{8}{n^2}\frac{n(n+1)}{2} + 2 = 4 + 2 = 6 \text{ sq. units.}$$

4. This is similar to #1; the rectangles have width $(2-(-1))/n = 3/n$ and the ith has height the value of $3x+4$ at $x = -1 + (3i/n)$. The area is

$$A = \lim_{n\to\infty} \sum_{i=1}^{n} \frac{3}{n}\left(-3 + 3\frac{3i}{n} + 4\right)$$

$$= \lim_{n\to\infty} \frac{27}{n^2}\sum_{i=1}^{n} i + \frac{3}{n}n$$

$$= \lim_{n\to\infty} \frac{27}{n^2}\frac{n(n+1)}{2} + 3 = \frac{27}{2} + 3 = \frac{33}{2} \text{ sq. units.}$$

5. The area is the limit of the sum of the areas of the rectangles shown in the figure. It is

$$A = \lim_{n\to\infty} \frac{2}{n}\left[\left(1+\frac{2}{n}\right)^2 + \left(1+\frac{4}{n}\right)^2 + \cdots + \left(1+\frac{2n}{n}\right)^2\right]$$

$$= \lim_{n\to\infty} \frac{2}{n}\left[1 + \frac{4}{n} + \frac{4}{n^2} + 1 + \frac{8}{n} + \frac{16}{n^2}\right.$$

$$\left.+ \cdots + 1 + \frac{4n}{n} + \frac{4n^2}{n^2}\right]$$

$$= \lim_{n\to\infty}\left(2 + \frac{8}{n^2}\cdot\frac{n(n+1)}{2} + \frac{8}{n^3}\cdot\frac{n(n+1)(2n+1)}{6}\right)$$

$$= 2 + 4 + \frac{8}{3} = \frac{26}{3} \text{ sq. units.}$$

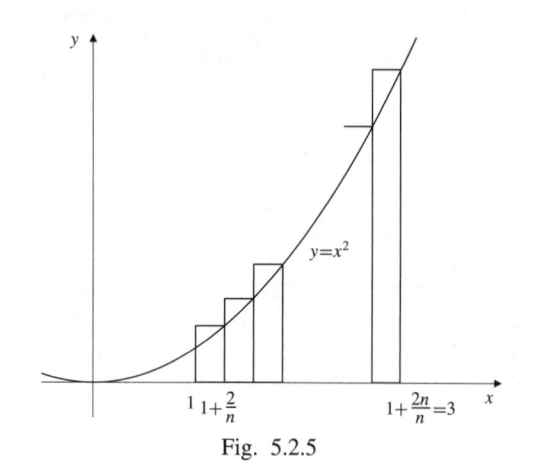

Fig. 5.2.5

6. Divide $[0, a]$ into n equal subintervals of length $\Delta x = \dfrac{a}{n}$ by points $x_i = \dfrac{ia}{n}$, $(0 \le i \le n)$. Then

$$S_n = \sum_{i=1}^{n}\left(\frac{a}{n}\right)\left[\left(\frac{ia}{n}\right)^2 + 1\right]$$

$$= \left(\frac{a}{n}\right)^3\sum_{i=1}^{n} i^2 + \frac{a}{n}\sum_{i=1}^{n}(1)$$

(Use Theorem 1(a) and 1(c).)

$$= \left(\frac{a}{n}\right)^3\frac{n(n+1)(2n+1)}{6} + \frac{a}{n}(n)$$

$$= \frac{a^3}{6}\frac{(n+1)(2n+1)}{n^2} + a.$$

Area $= \lim_{n\to\infty} S_n = \dfrac{a^3}{3} + a$ sq. units.

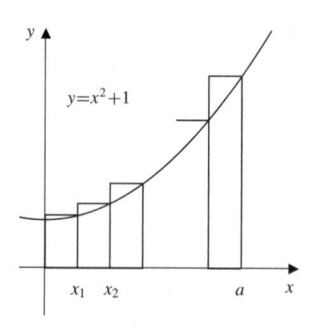

Fig. 5.2.6

7. The required area is (see the figure)

$$A = \lim_{n\to\infty}\frac{3}{n}\left[\left(-1+\frac{3}{n}\right)^2 + 2\left(-1+\frac{3}{n}\right) + 3\right.$$

$$+ \left(-1+\frac{6}{n}\right)^2 + 2\left(-1+\frac{6}{n}\right) + 3$$

$$\left.+ \cdots + \left(-1+\frac{3n}{n}\right)^2 + 2\left(-1+\frac{3n}{n}\right) + 3\right]$$

$$= \lim_{n\to\infty}\frac{3}{n}\left[\left(1 - \frac{6}{n} + \frac{3^2}{n^2} - 2 + \frac{6}{n} + 3\right)\right.$$

$$+ \left(1 - \frac{12}{n} + \frac{6^2}{n^2} - 2 + \frac{12}{n} + 3\right)$$

$$\left.+ \cdots + \left(1 - \frac{6n}{n} + \frac{9n^2}{n^2} - 2 + \frac{6n}{n} + 3\right)\right]$$

$$= \lim_{n\to\infty}\left(6 + \frac{27}{n^3}\cdot\frac{n(n+1)(2n+1)}{6}\right)$$

$$= 6 + 9 = 15 \text{ sq. units.}$$

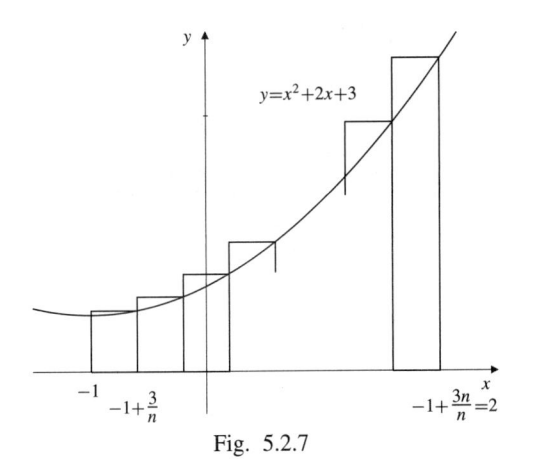

INSTRUCTOR'S SOLUTIONS MANUAL

SECTION 5.2 (PAGE 310)

Fig. 5.2.7

8.

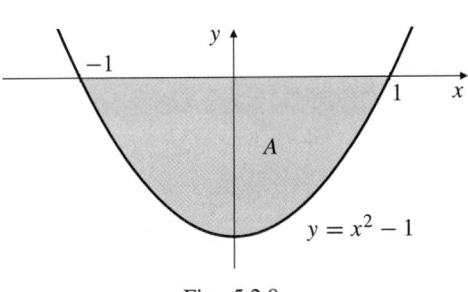

Fig. 5.2.8

The region in question lies between $x = -1$ and $x = 1$ and is symmetric about the y-axis. We can therefore double the area between $x = 0$ and $x = 1$. If we divide this interval into n equal subintervals of width $1/n$ and use the distance $0 - (x^2 - 1) = 1 - x^2$ between $y = 0$ and $y = x^2 - 1$ for the heights of rectangles, we find that the required area is

$$A = 2 \lim_{n \to \infty} \sum_{i=1}^{n} \frac{1}{n}\left(1 - \frac{i^2}{n^2}\right)$$
$$= 2 \lim_{n \to \infty} \sum_{i=1}^{n} \left(\frac{1}{n} - \frac{i^2}{n^3}\right)$$
$$= 2 \lim_{n \to \infty} \left(\frac{n}{n} - \frac{n(n+1)(2n+1)}{6n^3}\right) = 2 - \frac{4}{6} = \frac{4}{3} \text{ sq. units.}$$

9.

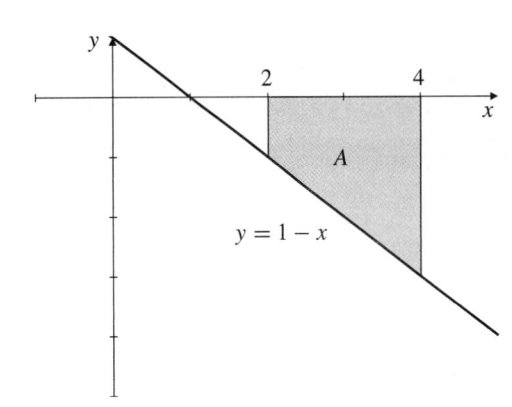

Fig. 5.2.9

The height of the region at position x is $0 - (1 - x) = x - 1$. The "base" is an interval of length 2, so we approximate using n rectangles of width $2/n$. The shaded area is

$$A = \lim_{n \to \infty} \sum_{i=1}^{n} \frac{2}{n}\left(2 + \frac{2i}{n} - 1\right)$$
$$= \lim_{n \to \infty} \sum_{i=1}^{n} \left(\frac{2}{n} + \frac{4i}{n^2}\right)$$
$$= \lim_{n \to \infty} \left(\frac{2n}{n} + 4\frac{n(n+1)}{2n^2}\right) = 2 + 2 = 4 \text{ sq. units.}$$

10.

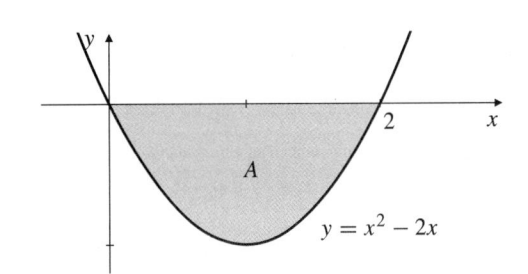

Fig. 5.2.10

The height of the region at position x is $0 - (x^2 - 2x) = 2x - x^2$. The "base" is an interval of length 2, so we approximate using n rectangles of width $2/n$. The shaded area is

$$A = \lim_{n \to \infty} \sum_{i=1}^{n} \frac{2}{n}\left(\frac{2i}{n} \cdot 2 - \frac{4i^2}{n^2}\right)$$
$$= \lim_{n \to \infty} \sum_{i=1}^{n} \left(\frac{8i}{n^2} - \frac{8i^2}{n^3}\right)$$
$$= \lim_{n \to \infty} \left(\frac{8}{n^2}\frac{n(n+1)}{2} - \frac{8}{n^3}\frac{n(n+1)(2n+1)}{6}\right)$$
$$= 4 - \frac{8}{3} = \frac{4}{3} \text{ sq. units.}$$

175

11.

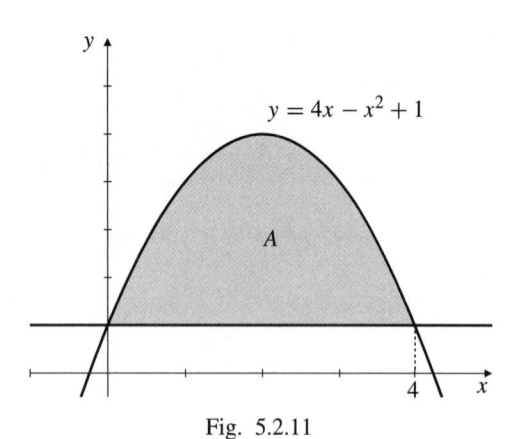

Fig. 5.2.11

The height of the region at position x is
$4x - x^2 + 1 - 1 = 4x - x^2$. The "base" is an interval of length 4, so we approximate using n rectangles of width $4/n$. The shaded area is

$$A = \lim_{n\to\infty} \sum_{i=1}^{n} \frac{4}{n}\left(4\frac{4i}{n} - \frac{16i^2}{n^2}\right)$$

$$= \lim_{n\to\infty} \sum_{i=1}^{n} \left(\frac{64i}{n^2} - \frac{64i^2}{n^3}\right)$$

$$= \lim_{n\to\infty} \left(\frac{64}{n^2}\frac{n(n+1)}{2} - \frac{64}{n^3}\frac{n(n+1)(2n+1)}{6}\right)$$

$$= 32 - \frac{64}{3} = \frac{32}{3} \text{ sq. units.}$$

12. Divide $[0, b]$ into n equal subintervals of length $\Delta x = \dfrac{b}{n}$ by points $x_i = \dfrac{ib}{n}$, $(0 \le i \le n)$. Then

$$S_n = \sum_{i=1}^{n} \frac{b}{n}\left(e^{(ib/n)}\right) = \frac{b}{n}\sum_{i=1}^{n}\left(e^{(b/n)}\right)^i$$

$$= \frac{b}{n}e^{(b/n)}\sum_{i=1}^{n}\left(e^{(b/n)}\right)^{i-1} \quad \text{(Use Thm. 6.1.2(d).)}$$

$$= \frac{b}{n}e^{(b/n)}\frac{e^{(b/n)n} - 1}{e^{(b/n)} - 1}$$

$$= \frac{b}{n}e^{(b/n)}\frac{e^b - 1}{e^{(b/n)} - 1}.$$

Let $r = \dfrac{b}{n}$.

$$\text{Area} = \lim_{n\to\infty} S_n = (e^b - 1)\lim_{r\to 0+} e^r \lim_{r\to 0+}\frac{r}{e^r - 1} \quad \left[\frac{0}{0}\right]$$

$$= (e^b - 1)(1)\lim_{r\to 0+}\frac{1}{e^r} = e^b - 1 \text{ sq. units.}$$

13. The required area is

$$A = \lim_{n\to\infty}\frac{2}{n}\left[2^{-1+(2/n)} + 2^{-1+(4/n)} + \cdots + 2^{-1+(2n/n)}\right]$$

$$= \lim_{n\to\infty}\frac{2^{2/n}}{n}\left[1 + (2^{2/n}) + (2^{2/n})^2 + \cdots + (2^{2/n})^{n-1}\right]$$

$$= \lim_{n\to\infty}\frac{2^{2/n}}{n}\cdot\frac{(2^{2/n})^n - 1}{2^{2/n} - 1}$$

$$= \lim_{n\to\infty} 2^{2/n} \times 3 \times \frac{1}{n(2^{2/n} - 1)}$$

$$= 3\lim_{n\to\infty}\frac{1}{n(2^{2/n} - 1)}.$$

Now we can use l'Hôpital's rule to evaluate

$$\lim_{n\to\infty} n(2^{2/n} - 1) = \lim_{n\to\infty}\frac{2^{2/n} - 1}{\frac{1}{n}} \quad \left[\frac{0}{0}\right]$$

$$= \lim_{n\to\infty}\frac{2^{2/n}\ln 2\left(\frac{-2}{n^2}\right)}{\frac{-1}{n^2}}$$

$$= \lim_{n\to\infty} 2^{(2/n)+1}\ln 2 = 2\ln 2.$$

Thus the area is $\dfrac{3}{2\ln 2}$ square units.

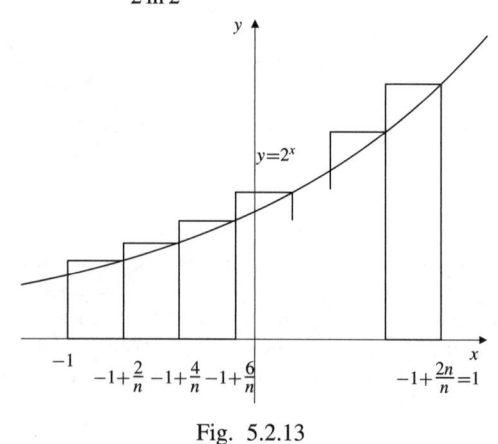

Fig. 5.2.13

14. $\text{Area} = \lim_{n\to\infty}\dfrac{b}{n}\left[\left(\dfrac{b}{n}\right)^3 + \left(\dfrac{2b}{n}\right)^3 + \cdots + \left(\dfrac{nb}{n}\right)^3\right]$

$$= \lim_{n\to\infty}\frac{b^4}{n^4}(1^3 + 2^3 + 3^3 + \cdots + n^3)$$

$$= \lim_{n\to\infty}\frac{b^4}{n^4}\cdot\frac{n^2(n+1)^2}{4} = \frac{b^4}{4} \text{ sq. units.}$$

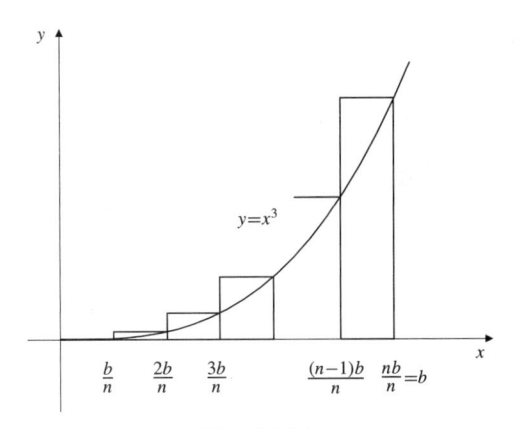

INSTRUCTOR'S SOLUTIONS MANUAL SECTION 5.2 (PAGE 310)

Fig. 5.2.14

15. Let $t = \left(\dfrac{b}{a}\right)^{1/n}$ and let

$$x_0 = a, \ x_1 = at, \ x_2 = at^2, \ \ldots, \ x_n = at^n = b.$$

The ith subinterval $[x_{i-1}, x_i]$ has length $\Delta x_i = at^{i-1}(t-1)$. Since $f(x_{i-1}) = \dfrac{1}{at^{i-1}}$, we form the sum

$$S_n = \sum_{i=1}^{n} at^{i-1}(t-1)\left(\dfrac{1}{at^{i-1}}\right)$$

$$= n(t-1) = n\left[\left(\dfrac{b}{a}\right)^{1/n} - 1\right].$$

Let $r = \dfrac{1}{n}$ and $c = \dfrac{b}{a}$. The area under the curve is

$$A = \lim_{n \to \infty} S_n = \lim_{r \to 0+} \dfrac{c^r - 1}{r} \quad \left[\dfrac{0}{0}\right]$$

$$= \lim_{r \to 0+} \dfrac{c^r \ln c}{1} = \ln c = \ln\left(\dfrac{b}{a}\right) \text{ square units.}$$

This is not surprising because it follows from the *definition* of ln.

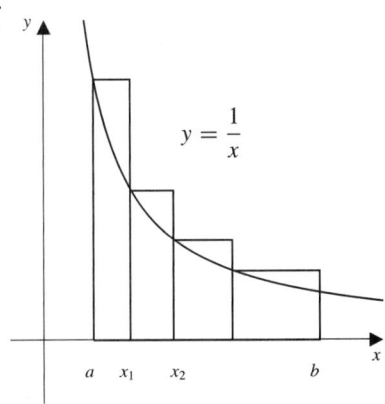

Fig. 5.2.15

16.

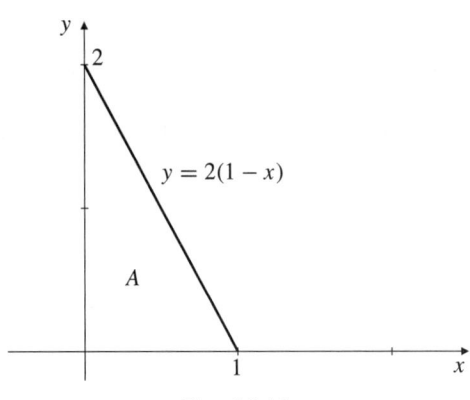

Fig. 5.2.16

$s_n = \sum_{i=1}^{n} \dfrac{2}{n}\left(1 - \dfrac{i}{n}\right)$ represents a sum of areas of n rectangles each of width $1/n$ and having heights equal to the height to the graph $y = 2(1-x)$ at the points $x = i/n$. Thus $\lim_{n \to \infty} S_n$ is the area A of the triangle in the figure above, and therefore has the value 1.

17.

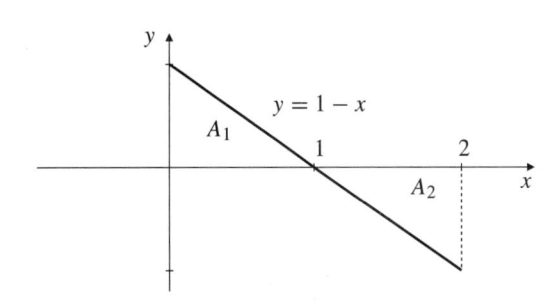

Fig. 5.2.17

$s_n = \sum_{i=1}^{n} \dfrac{2}{n}\left(1 - \dfrac{2i}{n}\right)$ represents a sum of areas of n rectangles each of width $2/n$ and having heights equal to the height to the graph $y = 1 - x$ at the points $x = 2i/n$. Half of these rectangles have negative height, and $\lim_{n \to \infty} S_n$ is the difference $A_1 - A_2$ of the areas of the two triangles in the figure above. It has the value 0 since the two triangles have the same area.

18.

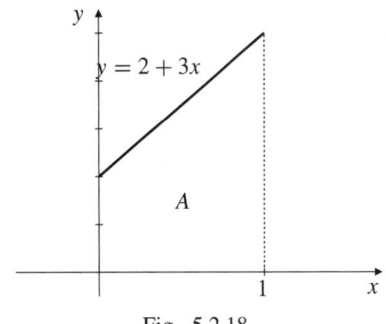

Fig. 5.2.18

177

$$s_n = \sum_{i=1}^{n} \frac{2n+3i}{n^2} = \sum_{i=1}^{n} \frac{1}{n}\left(2 + \frac{3i}{n}\right) \text{ represents a sum}$$

of areas of n rectangles each of width $1/n$ and having heights equal to the height to the graph $y = 2 + 3x$ at the points $x = i/n$. Thus $\lim_{n\to\infty} S_n$ is the area of the trapezoid in the figure above, and has the value $1(2+5)/2 = 7/2$.

19. $S_n = \sum_{j=1}^{n} \frac{1}{n}\sqrt{1-\left(\frac{j}{n}\right)^2}$

$=$ sum of areas of rectangles in the figure.

Thus the limit of S_n is the area of a quarter circle of unit radius:
$$\lim_{n\to\infty} S_n = \frac{\pi}{4}.$$

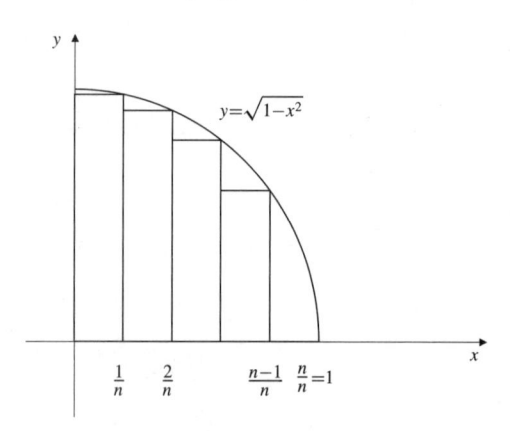

Fig. 5.2.19

Section 5.3 The Definite Integral
(page 316)

1. $f(x) = x$ on $[0, 2]$, $n = 8$.
$$P_8 = \left\{0, \frac{1}{4}, \frac{1}{2}, \frac{3}{4}, 1, \frac{5}{4}, \frac{3}{2}, \frac{7}{4}, 1\right\}$$
$$L(f, P_8) = \frac{2-0}{8}\left[0 + \frac{1}{4} + \frac{1}{2} + \frac{3}{4} + 1 + \frac{5}{4} + \frac{3}{2} + \frac{7}{4}\right] = \frac{7}{4}$$
$$U(f, P_8) = \frac{2-0}{8}\left[\frac{1}{4} + \frac{1}{2} + \frac{3}{4} + 1 + \frac{5}{4} + \frac{3}{2} + \frac{7}{4} + 2\right] = \frac{9}{4}$$

2. $f(x) = x^2$ on $[0, 4]$, $n = 4$.
$$L(f, P_4) = \left(\frac{4-0}{4}\right)[0 + (1)^2 + (2)^2 + (3)^2] = 14.$$
$$U(f, P_4) = \left(\frac{4-0}{4}\right)[(1)^2 + (2)^2 + (3)^2 + (4)^2] = 30.$$

3. $f(x) = e^x$ on $[-2, 2]$, $n = 4$.
$$L(f, P_4) = 1(e^{-2} + e^{-1} + e^0 + e^1) = \frac{e^4-1}{e^2(e-1)} \approx 4.22$$
$$U(f, P_4) = 1(e^{-1} + e^0 + e^1 + e^2) = \frac{e^4-1}{e(e-1)} \approx 11.48.$$

4. $f(x) = \ln x$ on $[1, 2]$, $n = 5$.
$$L(f, P_5) = \left(\frac{2-1}{5}\right)\left[\ln 1 + \ln\frac{6}{5} + \ln\frac{7}{5} + \ln\frac{8}{5} + \ln\frac{9}{5}\right]$$
$$\approx 0.3153168.$$
$$U(f, P_5) = \left(\frac{2-1}{5}\right)\left[\ln\frac{6}{5} + \ln\frac{7}{5} + \ln\frac{8}{5} + \ln\frac{9}{5} + \ln 2\right]$$
$$\approx 0.4539462.$$

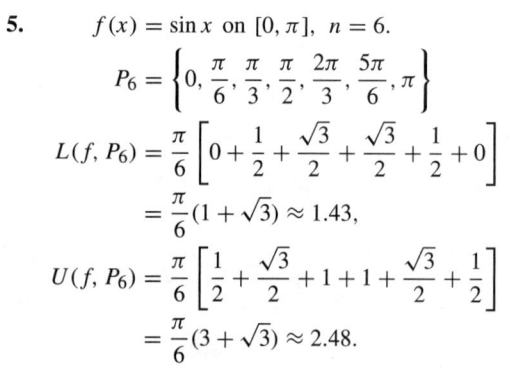

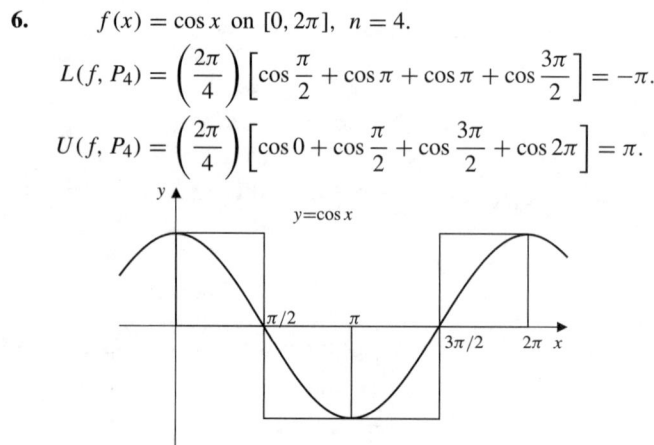

5. $f(x) = \sin x$ on $[0, \pi]$, $n = 6$.
$$P_6 = \left\{0, \frac{\pi}{6}, \frac{\pi}{3}, \frac{\pi}{2}, \frac{2\pi}{3}, \frac{5\pi}{6}, \pi\right\}$$
$$L(f, P_6) = \frac{\pi}{6}\left[0 + \frac{1}{2} + \frac{\sqrt{3}}{2} + \frac{\sqrt{3}}{2} + \frac{1}{2} + 0\right]$$
$$= \frac{\pi}{6}(1 + \sqrt{3}) \approx 1.43,$$
$$U(f, P_6) = \frac{\pi}{6}\left[\frac{1}{2} + \frac{\sqrt{3}}{2} + 1 + 1 + \frac{\sqrt{3}}{2} + \frac{1}{2}\right]$$
$$= \frac{\pi}{6}(3 + \sqrt{3}) \approx 2.48.$$

6. $f(x) = \cos x$ on $[0, 2\pi]$, $n = 4$.
$$L(f, P_4) = \left(\frac{2\pi}{4}\right)\left[\cos\frac{\pi}{2} + \cos\pi + \cos\pi + \cos\frac{3\pi}{2}\right] = -\pi.$$
$$U(f, P_4) = \left(\frac{2\pi}{4}\right)\left[\cos 0 + \cos\frac{\pi}{2} + \cos\frac{3\pi}{2} + \cos 2\pi\right] = \pi.$$

Fig. 5.3.6

7. $f(x) = x$ on $[0, 1]$. $P_n = \left\{0, \frac{1}{n}, \frac{2}{n}, \ldots, \frac{n-1}{n}, \frac{n}{n}\right\}$. We have
$$L(f, P_n) = \frac{1}{n}\left(0 + \frac{1}{n} + \frac{2}{n} + \cdots + \frac{n-1}{n}\right)$$
$$= \frac{1}{n^2} \cdot \frac{(n-1)n}{2} = \frac{n-1}{2n},$$
$$U(f, P_n) = \frac{1}{n}\left(\frac{1}{n} + \frac{2}{n} + \frac{3}{n} + \cdots + \frac{n}{n}\right)$$
$$= \frac{1}{n^2} \cdot \frac{n(n+1)n}{2} = \frac{n+1}{2n}.$$

Thus $\lim_{n\to\infty} L(f, P_n) = \lim_{n\to\infty} U(f, P_n) = 1/2$.
If P is any partition of $[0, 1]$, then

$$L(f, P) \leq U(f, P_n) = \frac{n+1}{2n}$$

for every n, so $L(f, P) \leq \lim_{n\to\infty} U(f, P_n) = 1/2$.
Similarly, $U(f, P) \geq 1/2$. If there exists any number I such that $L(f, P) \leq I \leq U(f, P)$ for all P, then I cannot be less than $1/2$ (or there would exist a P_n such that $L(f, P_n) > I$), and, similarly, I cannot be greater than $1/2$ (or there would exist a P_n such that $U(f, P_n) < I$).
Thus $I = 1/2$ and $\int_0^1 x\,dx = 1/2$.

8. $f(x) = 1 - x$ on $[0, 2]$. $P_n = \left\{0, \frac{2}{n}, \frac{4}{n}, \ldots, \frac{2n-2}{n}, \frac{2n}{n}\right\}$. We have

$$L(f, P_n) = \frac{2}{n}\left(\left(1-\frac{2}{n}\right) + \left(1-\frac{4}{n}\right) + \cdots + \left(1-\frac{2n}{n}\right)\right)$$

$$= \frac{2}{n}n - \frac{4}{n^2}\sum_{i=1}^{n} i$$

$$= 2 - \frac{4}{n^2}\frac{n(n+1)}{2} = -\frac{2}{n} \to 0 \text{ as } n \to \infty,$$

$$U(f, P_n) = \frac{2}{n}\left(\left(1-\frac{0}{n}\right) + \left(1-\frac{2}{n}\right) + \cdots + \left(1-\frac{2n-2}{n}\right)\right)$$

$$= \frac{2}{n}n - \frac{4}{n^2}\sum_{i=0}^{n-1} i$$

$$= 2 - \frac{4}{n^2}\frac{(n-1)n}{2} = \frac{2}{n} \to 0 \text{ as } n \to \infty.$$

Thus $\int_0^2 (1-x)\,dx = 0$.

9. $f(x) = x^3$ on $[0, 1]$. $P_n = \left\{0, \frac{1}{n}, \frac{2}{n}, \ldots, \frac{n-1}{n}, \frac{n}{n}\right\}$. We have (using the result of Exercise 51 (or 52) of Section 6.1)

$$L(f, P_n) = \frac{1}{n}\left(\left(\frac{0}{n}\right)^3 + \left(\frac{1}{n}\right)^3 + \cdots + \left(\frac{n-1}{n}\right)^3\right)$$

$$= \frac{1}{n^4}\sum_{i=0}^{n-1} i^3 = \frac{1}{n^4}\frac{(n-1)^2 n^2}{4}$$

$$= \frac{1}{4}\left(\frac{n-1}{n}\right)^2 \to \frac{1}{4} \text{ as } n \to \infty,$$

$$U(f, P_n) = \frac{1}{n}\left(\left(\frac{1}{n}\right)^3 + \left(\frac{2}{n}\right)^3 + \cdots + \left(\frac{n}{n}\right)^3\right)$$

$$= \frac{1}{n^4}\sum_{i=1}^{n} i^3 = \frac{1}{n^4}\frac{n^2(n+1)^2}{4}$$

$$= \frac{1}{4}\left(\frac{n+1}{n}\right)^2 \to \frac{1}{4} \text{ as } n \to \infty.$$

Thus $\int_0^1 x^3\,dx = \frac{1}{4}$.

10. $f(x) = e^x$ on $[0, 3]$. $P_n = \left\{0, \frac{3}{n}, \frac{6}{n}, \ldots, \frac{3n-3}{n}, \frac{3n}{n}\right\}$. We have (using the result of Exercise 51 (or 52) of Section 6.1)

$$L(f, P_n) = \frac{3}{n}\left(e^{0/n} + e^{3/n} + e^{6/n} + \cdots + e^{3(n-1)/n}\right)$$

$$= \frac{3}{n}\frac{e^{3n/n} - 1}{e^{3/n} - 1} = \frac{3(e^3 - 1)}{n(e^{3/n} - 1)},$$

$$U(f, P_n) = \frac{3}{n}\left(e^{3/n} + e^{6/n} + e^{9/n} + \cdots + e^{3n/n}\right) = e^{3/n} L(f, P_n).$$

By l'Hôpital's Rule,

$$\lim_{n\to\infty} n(e^{3/n} - 1) = \lim_{n\to\infty} \frac{e^{3/n} - 1}{1/n}$$

$$= \lim_{n\to\infty} \frac{e^{3/n}(-3/n^2)}{-1/n^2} = \lim_{n\to\infty} \frac{3e^{3/n}}{1} = 3.$$

Thus

$$\lim_{n\to\infty} L(f, P_n) = \lim_{n\to\infty} U(f, P_n) = e^3 - 1 = \int_0^3 e^x\,dx.$$

11. $\lim_{n\to\infty} \sum_{i=1}^{n} \frac{1}{n}\sqrt{\frac{i}{n}} = \int_0^1 \sqrt{x}\,dx$

12. $\lim_{n\to\infty} \sum_{i=1}^{n} \frac{1}{n}\sqrt{\frac{i-1}{n}} = \int_0^1 \sqrt{x}\,dx$

13. $\lim_{n\to\infty} \sum_{i=1}^{n} \frac{\pi}{n} \sin\frac{\pi i}{n} = \int_0^{\pi} \sin x\,dx$

14. $\lim_{n\to\infty} \sum_{i=1}^{n} \frac{2}{n} \ln\left(1 + \frac{2i}{n}\right) = \int_0^2 \ln(1+x)\,dx$

15. $\lim_{n\to\infty} \sum_{i=1}^{n} \frac{1}{n} \tan^{-1}\left(\frac{2i-1}{2n}\right) = \int_0^1 \tan^{-1}x\,dx$

Note that $\frac{2i-1}{2n}$ is the midpoint of $\left[\frac{i-1}{n}, \frac{i}{n}\right]$.

16. $\lim_{n\to\infty} \sum_{i=1}^{n} \frac{n}{n^2 + i^2} = \lim_{n\to\infty} \sum_{i=1}^{n} \frac{1}{n}\frac{1}{1+(i/n)^2} = \int_0^1 \frac{dx}{1+x^2}$

SECTION 5.3 (PAGE 316)

17. Let $\Delta x = \dfrac{b-a}{n}$ and $x_i = a + i\Delta x$ where $1 \le i \le n-1$.
Since f is continuous and nondecreasing,

$$L(f, P_n) = f(a)\Delta x + f(x_1)\Delta x + f(x_2)\Delta x + \cdots + f(x_{n-1})\Delta x$$
$$= \frac{b-a}{n}\left[f(a) + \sum_{i=1}^{n-1} f(x_i)\right],$$
$$U(f, P_n) = f(x_1)\Delta x + f(x_2)\Delta x + \cdots + f(x_{n-1})\Delta x + f(b)\Delta x$$
$$= \frac{b-a}{n}\left[\sum_{i=1}^{n-1} f(x_i) + f(b)\right].$$

Thus,

$$U(f, P_n) - L(f, P_n)$$
$$= \frac{b-a}{n}\left[\sum_{i=1}^{n-1} f(x_i) + f(b) - f(a) - \sum_{i=1}^{n-1} f(x_i)\right]$$
$$= \frac{(b-a)(f(b) - f(a))}{n}.$$

Since
$$\lim_{n\to\infty}[U(f, P_n) - L(f, P-n)] = 0,$$
therefore f must be integrable on $[a, b]$.

18. $P = \{x_0 < x_1 < \cdots < x_n\}$,
$P' = \{x_0 < x_1 < \cdots < x_{j-1} < x' < x_j < \cdots < x_n\}$.
Let m_i and M_i be, respectively, the minimum and maximum values of $f(x)$ on the interval $[x_{i-1}, x_i]$, for $1 \le i \le n$. Then

$$L(f, P) = \sum_{i=1}^n m_i(x_i - x_{i-1}),$$
$$U(f, P) = \sum_{i=1}^n M_i(x_i - x_{i-1}).$$

If m'_j and M'_j are the minimum and maximum values of $f(x)$ on $[x_{j-1}, x']$, and if m''_j and M''_j are the corresponding values for $[x', x_j]$, then

$$m'_j \ge m_j, \quad m''_j \ge m_j, \quad M'_j \le M_j, \quad M''_j \le M_j.$$

Therefore we have

$$m_j(x_j - x_{j-1}) \le m'_j(x' - x_{j-1}) + m''_j(x_j - x'),$$
$$M_j(x_j - x_{j-1}) \ge M'_j(x' - x_{j-1}) + M''_j(x_j - x').$$

Hence $L(f, P) \le L(f, P')$ and $U(f, P) \ge U(f, P')$.

If P'' is any refinement of P we can add the new points in P'' to those in P one at a time, and thus obtain

$$L(f, P) \le L(f, P''), \qquad U(f, P'') \le U(f, P).$$

Section 5.4 Properties of the Definite Integral (page 323)

1. $\displaystyle\int_a^b f(x)\,dx + \int_b^c f(x)\,dx + \int_c^a f(x)\,dx$
$$= \int_a^c f(x)\,dx - \int_a^c f(x)\,dx = 0$$

2. $\displaystyle\int_0^2 3f(x)\,dx + \int_1^3 3f(x)\,dx - \int_0^3 2f(x)\,dx - \int_1^2 3f(x)\,dx$
$$= \int_0^1 (3-2)f(x)\,dx + \int_1^2 (3+3-2-3)f(x)\,dx$$
$$+ \int_2^3 (3-2)f(x)\,dx$$
$$= \int_0^3 f(x)\,dx$$

3. $\displaystyle\int_{-2}^2 (x+2)\,dx = \frac{1}{2}(4)(4) = 8$

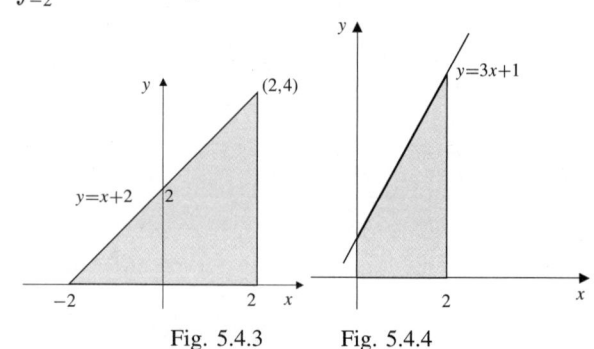

Fig. 5.4.3   Fig. 5.4.4

4. $\displaystyle\int_0^2 (3x+1)\,dx = \text{shaded area} = \frac{1}{2}(1+7)(2) = 8$

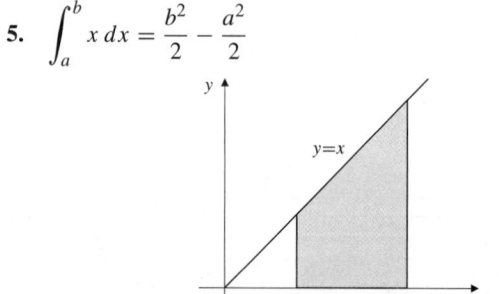

5. $\displaystyle\int_a^b x\,dx = \frac{b^2}{2} - \frac{a^2}{2}$

Fig. 5.4.5

INSTRUCTOR'S SOLUTIONS MANUAL — SECTION 5.4 (PAGE 323)

6. $\int_{-1}^{2}(1-2x)\,dx = A_1 - A_2 = 0$

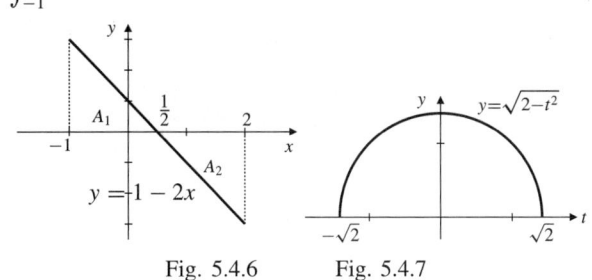

7. $\int_{-\sqrt{2}}^{\sqrt{2}}\sqrt{2-t^2}\,dt = \frac{1}{2}\pi(\sqrt{2})^2 = \pi$

8. $\int_{-\sqrt{2}}^{0}\sqrt{2-x^2}\,dx = \text{quarter disk} = \frac{1}{4}\pi(\sqrt{2})^2 = \frac{\pi}{2}$

9. $\int_{-\pi}^{\pi}\sin(x^3)\,dx = 0$. (The integrand is an odd function and the interval of integration is symmetric about $x = 0$.)

10. $\int_{-a}^{a}(a-|s|)\,ds = \text{shaded area} = 2(\frac{1}{2}a^2) = a^2$

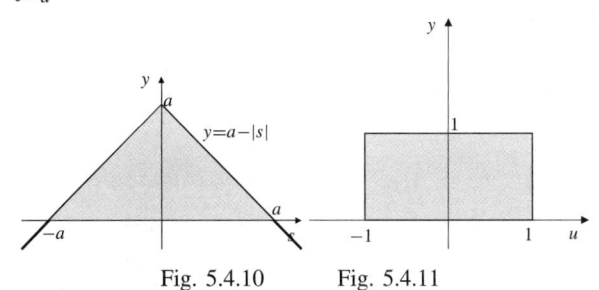

Fig. 5.4.10   Fig. 5.4.11

11. $\int_{-1}^{1}(u^5 - 3u^3 + \pi)\,du = \pi\int_{-1}^{1}du = 2\pi$

12. Let $y = \sqrt{2x - x^2} \Rightarrow y^2 + (x-1)^2 = 1$.

$\int_{0}^{2}\sqrt{2x-x^2}\,dx = \text{shaded area} = \frac{1}{2}\pi(1)^2 = \frac{\pi}{2}.$

Fig. 5.4.12

13. $\int_{-4}^{4}(e^x - e^{-x})\,dx = 0$ (odd function, symmetric interval)

14. $\int_{-3}^{3}(2+t)\sqrt{9-t^2}\,dt = 2\int_{-3}^{3}\sqrt{9-t^2}\,dt + \int_{-3}^{3}t\sqrt{9-t^2}\,dt$

$= 2\left(\frac{1}{2}\pi 3^2\right) + 0 = 9\pi$

15. $\int_{0}^{1}\sqrt{4-x^2}\,dx = \text{area } A_1 \text{ in figure below}$

$= \frac{1}{4}\text{area of circle} - \text{area } A_2$

(see #14 below)

$= \frac{1}{4}(\pi 2^2) - \left(\frac{2\pi}{3} - \frac{\sqrt{3}}{2}\right)$

$= \frac{\pi}{3} + \frac{\sqrt{3}}{2}$

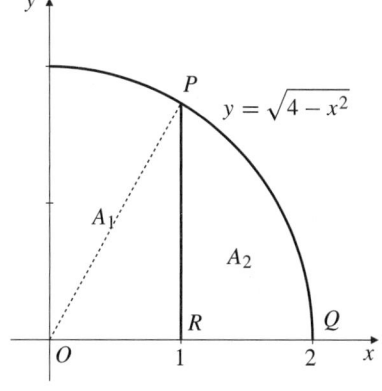

Fig. 5.4.15

16. $\int_{1}^{2}\sqrt{4-x^2}\,dx = \text{area } A_2 \text{ in figure above}$

$= \text{area sector POQ} - \text{area triangle POR}$

$= \frac{1}{6}(\pi 2^2) - \frac{1}{2}(1)\sqrt{3}$

$= \frac{2\pi}{3} - \frac{\sqrt{3}}{2}$

17. $\int_{0}^{2}6x^2\,dx = 6\int_{0}^{2}x^2\,dx = 6\frac{2^3}{3} = 16$

18. $\int_{1}^{2}6x^2\,dx = \int_{0}^{2}6x^2\,dx - \int_{0}^{1}6x^2\,dx$

$= 16 - 6\frac{1^3}{3} = 14$

19. $\int_{0}^{3}(1+x^2)\,dx = 3 + \frac{3^3}{3} = 12$

20. $\int_{2}^{3}(x^2 - 4)\,dx = \int_{0}^{3}x^2\,dx - \int_{0}^{2}x^2\,dx - 4(3-2)$

$= \frac{3^3}{3} - \frac{2^3}{3} - 4 = \frac{7}{3}$

181

21. $\int_{-2}^{2}(4-t^2)\,dt = 2\int_{0}^{2}(4-t^2)\,dt$
$= 2\left(2(4) - \dfrac{2^3}{3}\right) = \dfrac{32}{3}$

22. $\int_{0}^{2}(v^2 - v)\,dv = \dfrac{2^3}{3} - \dfrac{2^2}{2} = \dfrac{2}{3}$

23. $\int_{0}^{1}(x^2 + \sqrt{1-x^2})\,dx = \dfrac{1^3}{3} + \dfrac{1}{4}(\pi 1^2)$
$= \dfrac{1}{3} + \dfrac{\pi}{4}$

24. $\int_{-6}^{6} x^2(2 + \sin x)\,dx = \int_{-6}^{6} 2x^2\,dx + \int_{-6}^{6} x^2 \sin x\,dx$
$= 4\int_{0}^{6} x^2\,dx + 0 = \dfrac{4}{3}(6^3) = 288$

25. $\int_{1}^{2} \dfrac{1}{x}\,dx = \ln 2$

26. $\int_{2}^{4} \dfrac{1}{t}\,dt = \int_{1}^{4} \dfrac{1}{t}\,dt - \int_{1}^{2} \dfrac{1}{t}\,dt$
$= \ln 4 - \ln 2 = \ln(4/2) = \ln 2$

27. $\int_{1/3}^{1} \dfrac{1}{t}\,dt = -\int_{1}^{1/3} \dfrac{1}{t}\,dt = -\ln\dfrac{1}{3} = \ln 3$

28. $\int_{1/4}^{3} \dfrac{1}{s}\,ds = \int_{1}^{3} \dfrac{1}{s}\,ds - \int_{1}^{1/4} \dfrac{1}{s}\,ds$
$= \ln 3 - \ln\dfrac{1}{4} = \ln 3 + \ln 4 = \ln 12$

29. Average $= \dfrac{1}{4-0}\int_{0}^{4}(x+2)\,dx$
$= \dfrac{1}{4}\left[\dfrac{1}{2}(4^2) + 2(4)\right] = 4$

30. Average $= \dfrac{1}{b-a}\int_{a}^{b}(x+2)\,dx$
$= \dfrac{1}{b-a}\left[\dfrac{1}{2}(b^2 - a^2) + 2(b-a)\right]$
$= \dfrac{1}{2}(b+a) + 2 = \dfrac{4+a+b}{2}$

31. Average $= \dfrac{1}{\pi - (-\pi)}\int_{-\pi}^{\pi}(1 + \sin t)\,dt$
$= \dfrac{1}{2\pi}\left[\int_{-\pi}^{\pi} 1\,dt + \int_{-\pi}^{\pi} \sin t\,dt\right]$
$= \dfrac{1}{2\pi}[2\pi + 0] = 1$

32. Average $= \dfrac{1}{3-0}\int_{0}^{3} x^2\,dx = \dfrac{1}{3}\dfrac{3^3}{3} = 3$

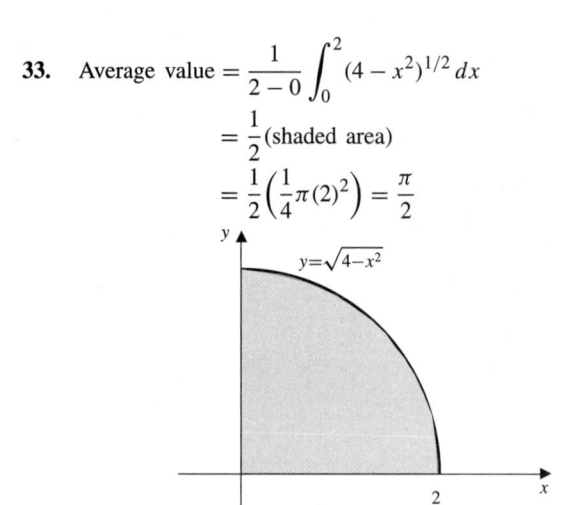

33. Average value $= \dfrac{1}{2-0}\int_{0}^{2}(4-x^2)^{1/2}\,dx$
$= \dfrac{1}{2}(\text{shaded area})$
$= \dfrac{1}{2}\left(\dfrac{1}{4}\pi(2)^2\right) = \dfrac{\pi}{2}$

Fig. 5.4.33

34. Average value $= \dfrac{1}{2 - (1/2)}\int_{1/2}^{2} \dfrac{1}{s}\,ds$
$= \dfrac{2}{3}\left(\ln 2 - \ln\dfrac{1}{2}\right) = \dfrac{4}{3}\ln 2$

35. $\int_{-1}^{2} \operatorname{sgn} x\,dx = 2 - 1 = 1$

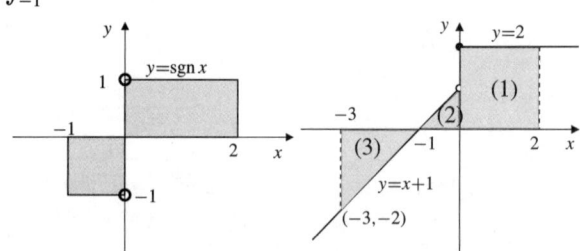

Fig. 6.4.37 Fig. 6.4.38

36. Let
$$f(x) = \begin{cases} 1+x & \text{if } x < 0 \\ 2 & \text{if } x \geq 0. \end{cases}$$
Then
$\int_{-3}^{2} f(x)\,dx = \text{area}(1) + \text{area}(2) - \text{area}(3)$
$= (2 \times 2) + \tfrac{1}{2}(1)(1) - \tfrac{1}{2}(2)(2) = 2\tfrac{1}{2}.$

37. $\int_{0}^{2} g(x)\,dx = \int_{0}^{1} x^2\,dx + \int_{1}^{2} x\,dx$
$= \dfrac{1^3}{3} + \dfrac{2^2 - 1^2}{2} = \dfrac{11}{6}$

38. $\int_{0}^{3}(2 - |x|)\,dx = \int_{0}^{3}(2-x)\,dx = 6 - \dfrac{3^2}{2} = \dfrac{3}{2}$

INSTRUCTOR'S SOLUTIONS MANUAL   SECTION 5.4 (PAGE 323)

39. $I = \int_0^2 \sqrt{4-x^2} \operatorname{sgn}(x-1)\, dx$
$= \text{area } A_1 - \text{area } A_2$.
Area $A_1 = \frac{1}{6}\pi 2^2 - \frac{1}{2}(1)(\sqrt{3}) = \frac{2}{3}\pi - \frac{1}{2}\sqrt{3}$.
Area $A_2 = \frac{1}{4}\pi 2^2 - \text{area } A_1 = \frac{1}{3}\pi + \frac{1}{2}\sqrt{3}$.
Therefore $I = (\pi/3) - \sqrt{3}$.

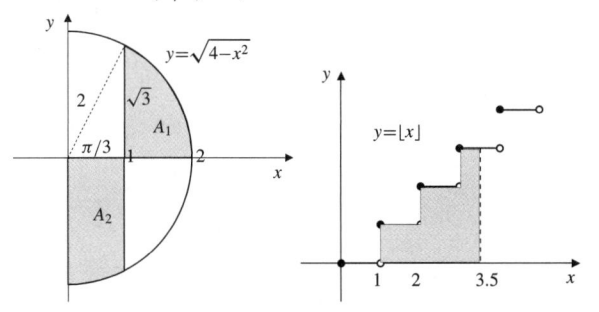

Fig. 5.4.39   Fig. 5.4.40

40. $\int_0^{3.5} \lfloor x \rfloor\, dx = \text{shaded area} = 1 + 2 + 1.5 = 4.5$.

41.

$y = |x+1| - |x-1| + |x+2|$

Fig. 5.4.41

$\int_{-3}^{4} (|x+1| - |x-1| + |x+2|)\, dx$
$= \text{area } A_1 - \text{area } A_2$
$= \frac{1}{2}\frac{5}{3}(5) + \frac{5+8}{2}(3) - \frac{1+2}{2}(1) - \frac{1+2}{2}(1) - \frac{1}{2}\frac{1}{3}(1) = \frac{41}{2}$

42.

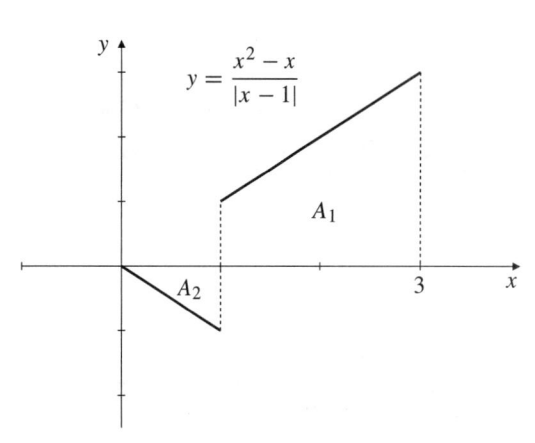

Fig. 5.4.42

$\int_0^3 \frac{x^2 - x}{|x-1|}\, dx$
$= \text{area } A_1 - \text{area } A_2$
$= \frac{1+3}{2}(2) - \frac{1}{2}(1)(1) = \frac{7}{2}$

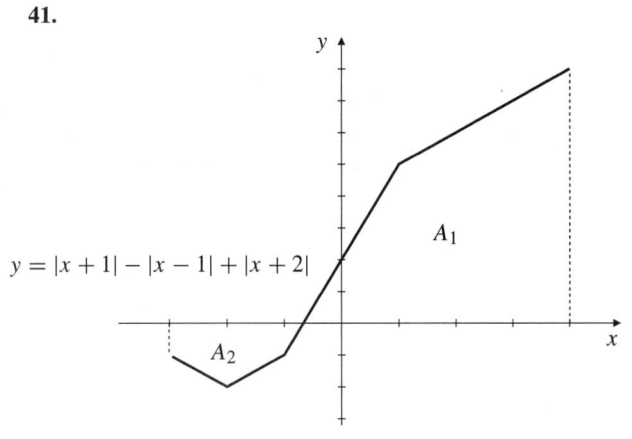

43. Average $= \frac{1}{4}\int_{-2}^{2} |x+1|\operatorname{sgn} x\, dx$
$= \frac{1}{4}\left(\int_0^2 (x+1)\, dx - \int_{-2}^0 |x+1|\, dx\right)$
$= \frac{1}{4}\left(\frac{1+3}{2} \times 2 - 2 \times \frac{1}{2} \times 1 \times 1\right)$
$= 1 - \frac{1}{4} = \frac{3}{4}$.

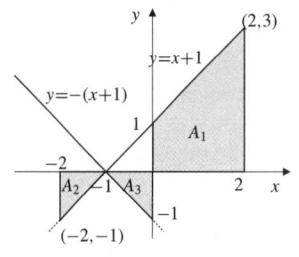

Fig. 5.4.43

44. $\int_a^b \left(f(x) - \overline{f}\right) dx = \int_a^b f(x)\, dx - \int_a^b \overline{f}\, dx$
$= (b-a)\overline{f} - \overline{f}\int_a^b dx$
$= (b-a)\overline{f} - (b-a)\overline{f} = 0$

183

45. $\int_a^b (f(x)-k)^2\, dx$

$= \int_a^b (f(x))^2\, dx - 2k\int_a^b f(x)\, dx + k^2\int_a^b dx$

$= \int_a^b (f(x))^2\, dx - 2k(b-a)\overline{f} + k^2(b-a)$

$= (b-a)(k-\overline{f})^2 + \int_a^b (f(x))^2\, dx - (b-a)\overline{f}^2$

This is minimum if $k = \overline{f}$.

Section 5.5 The Fundamental Theorem of Calculus (page 329)

1. $\int_0^2 x^3\, dx = \dfrac{x^4}{4}\Big|_0^2 = \dfrac{16-0}{4} = 4$

2. $\int_0^4 \sqrt{x}\, dx = \dfrac{2}{3}x^{3/2}\Big|_0^4 = \dfrac{16}{3}$

3. $\int_{1/2}^1 \dfrac{1}{x^2}\, dx = \dfrac{-1}{x}\Big|_{1/2}^1 = -1 - (-2) = 1$

4. $\int_4^9 s^{-1/2}\, ds = 2s^{1/2}\Big|_4^9 = 2\sqrt{9} - 2\sqrt{4} = 2$

5. $\int_0^1 (x^5 + x^3)\, dx = \left(\dfrac{x^6}{6} + \dfrac{x^4}{4}\right)\Big|_0^1$

$= \dfrac{1}{6} + \dfrac{1}{4} = \dfrac{5}{12}$

6. $\int_{-2}^{-1}\left(\dfrac{1}{x^2} - \dfrac{1}{x^3}\right) dx = \left(-\dfrac{1}{x} + \dfrac{1}{2x^2}\right)\Big|_{-2}^{-1}$

$= 1 + \dfrac{1}{2} - \left(\dfrac{1}{2} + \dfrac{1}{8}\right) = \dfrac{7}{8}$

7. $\int_{-1}^2 (3x^2 - 4x + 2)\, dx = (x^3 - 2x^2 + 2x)\Big|_{-1}^2 = 9$

8. $\int_1^2 \left(\dfrac{2}{x^3} - \dfrac{x^3}{2}\right) dx = \left(-\dfrac{1}{x^2} - \dfrac{x^4}{8}\right)\Big|_1^2 = -9/8$

9. $\int_{-2}^2 (x^2+3)^2\, dx = 2\int_0^2 (x^4 + 6x^2 + 9)\, dx$

$= 2\left(\dfrac{x^5}{5} + 2x^3 + 9x\right)\Big|_0^2$

$= 2\left(\dfrac{32}{5} + 16 + 18\right) = \dfrac{404}{5}$

10. $\int_4^9 \left(\sqrt{x} - \dfrac{1}{\sqrt{x}}\right) dx = \dfrac{2}{3}x^{3/2} - 2\sqrt{x}\,\Big|_4^9$

$= \left[\dfrac{2}{3}(9)^{3/2} - 2\sqrt{9}\right] - \left[\dfrac{2}{3}(4)^{3/2} - 2\sqrt{4}\right] = \dfrac{32}{3}$

11. $\int_{-\pi/4}^{-\pi/6} \cos x\, dx = \sin x\,\Big|_{-\pi/4}^{-\pi/6}$

$= -\dfrac{1}{2} + \dfrac{1}{\sqrt{2}} = \dfrac{2 - \sqrt{2}}{2\sqrt{2}}$

12. $\int_0^{\pi/3} \sec^2\theta\, d\theta = \tan\theta\,\Big|_0^{\pi/3} = \tan\dfrac{\pi}{3} = \sqrt{3}$

13. $\int_{\pi/4}^{\pi/3} \sin\theta\, d\theta = -\cos\theta\,\Big|_{\pi/4}^{\pi/3} = \dfrac{\sqrt{2} - 1}{2}$

14. $\int_0^{2\pi} (1 + \sin u)\, du = (u - \cos u)\Big|_0^{2\pi} = 2\pi$

15. $\int_{-\pi}^{\pi} e^x\, dx = e^x\Big|_{-\pi}^{\pi} = e^{\pi} - e^{-\pi}$

16. $\int_{-2}^2 (e^x - e^{-x})\, dx = 0$ (odd function, symmetric interval)

17. $\int_0^e a^x\, dx = \dfrac{a^x}{\ln a}\Big|_0^e = \dfrac{a^e - 1}{\ln a}$

18. $\int_{-1}^1 2^x\, dx = \dfrac{2^x}{\ln 2}\Big|_{-1}^1 = \dfrac{2}{\ln 2} - \dfrac{1}{2\ln 2} = \dfrac{3}{2\ln 2}$

19. $\int_{-1}^1 \dfrac{dx}{1+x^2} = \tan^{-1}x\,\Big|_{-1}^1 = \dfrac{\pi}{2}$

20. $\int_0^{1/2} \dfrac{dx}{\sqrt{1-x^2}} = \sin^{-1}x\,\Big|_0^{1/2} = \dfrac{\pi}{6}$

21. $\int_{-1}^1 \dfrac{dx}{\sqrt{4-x^2}} = \sin^{-1}\dfrac{x}{2}\Big|_{-1}^1$

$= \sin^{-1}\dfrac{1}{2} - \sin^{-1}\left(-\dfrac{1}{2}\right)$

$= \dfrac{\pi}{6} - \left(-\dfrac{\pi}{6}\right) = \dfrac{\pi}{3}$

22. $\int_{-2}^0 \dfrac{dx}{4+x^2} = \dfrac{1}{2}\tan^{-1}\dfrac{x}{2}\Big|_{-2}^0 = 0 - \dfrac{1}{2}\tan^{-1}(-1) = \dfrac{\pi}{8}$

23. Area $R = \int_0^1 x^4\, dx = \dfrac{x^5}{5}\Big|_0^1 = \dfrac{1}{5}$ sq. units.

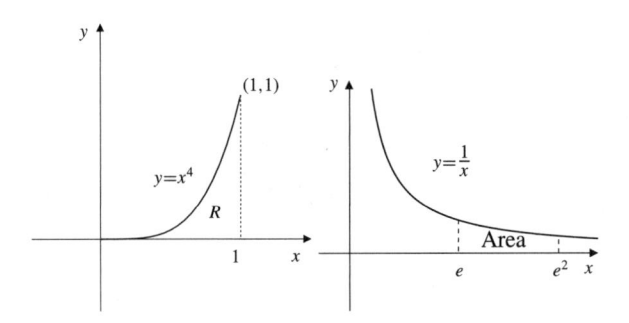

INSTRUCTOR'S SOLUTIONS MANUAL SECTION 5.5 (PAGE 329)

Fig. 5.5.23   Fig. 5.5.24

24. Area $= \int_e^{e^2} \frac{1}{x}\,dx = \ln x \Big|_e^{e^2}$
    $= \ln e^2 - \ln e = 2 - 1 = 1$ sq. units.

25. Area $R = -\int_0^4 (x^2 - 4x)\,dx$
    $= -\left(\frac{x^3}{3} - 2x^2\right)\Big|_0^4$
    $= -\left(\frac{64}{3} - 32\right) = \frac{32}{3}$ sq. units.

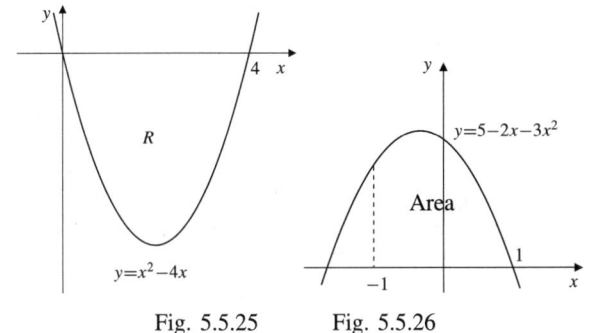

Fig. 5.5.25   Fig. 5.5.26

26. Since $y = 5 - 2x - 3x^2 = (5 + 3x)(1 - x)$, therefore $y = 0$ at $x = -\frac{5}{3}$ and 1, and $y > 0$ if $-\frac{5}{3} < x < 1$. Thus, the area is

$$\int_{-1}^1 (5 - 2x - 3x^2)\,dx = 2\int_0^1 (5 - 3x^2)\,dx$$
$$= 2(5x - x^3)\Big|_0^1$$
$$= 2(5 - 1) = 8 \text{ sq. units.}$$

27. For intersection of $y = x^2 - 3x + 3$ and $y = 1$, we have

$$x^2 - 3x + 3 = 1$$
$$x^2 - 3x + 2 = 0$$
$$(x - 2)(x - 1) = 0.$$

Thus $x = 1$ or $x = 2$. The indicated region has area

Area $R = 1 - \int_1^2 (x^2 - 3x + 3)\,dx$
$= 1 - \left(\frac{x^3}{3} - \frac{3x^2}{2} + 3x\right)\Big|_1^2$
$= 1 - \left(\frac{8}{3} - 6 + 6 - \left[\frac{1}{3} - \frac{3}{2} + 3\right]\right) = \frac{1}{6}$ sq. units.

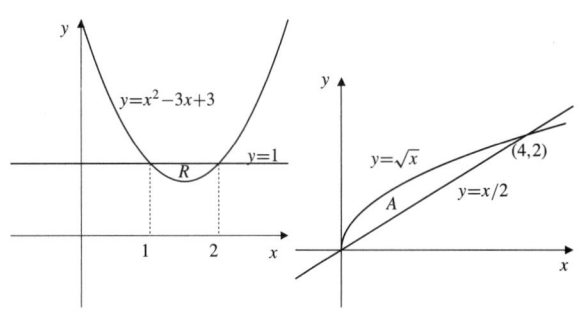

Fig. 5.5.27   Fig. 5.5.28

28. Since $y = \sqrt{x}$ and $y = \frac{x}{2}$ intersect where $\sqrt{x} = \frac{x}{2}$, that is, at $x = 0$ and $x = 4$, thus,

$$\text{Area} = \int_0^4 \sqrt{x}\,dx - \int_0^4 \frac{x}{2}\,dx$$
$$= \frac{2}{3}x^{3/2}\Big|_0^4 - \frac{x^2}{4}\Big|_0^4$$
$$= \frac{16}{3} - \frac{16}{4} = \frac{4}{3} \text{ sq. units.}$$

29. Area $R = 2 \times$ shaded area
$= 2\left(\frac{1}{2} - \int_0^1 x^2\,dx\right)$
$= 2\left(\frac{1}{2} - \frac{1}{3}\right) = \frac{1}{3}$ sq. units.

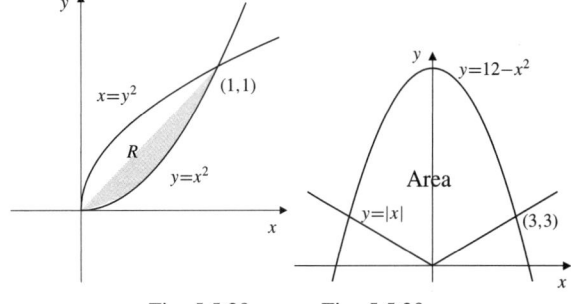

Fig. 5.5.29   Fig. 5.5.30

185

SECTION 5.5 (PAGE 329)

30. The two graphs intersect at $(\pm 3, 3)$, thus

$$\text{Area} = 2\int_0^3 (12 - x^2)\, dx - 2\int_0^3 x\, dx$$
$$= 2\left(12x - \frac{1}{3}x^3\right)\Big|_0^3 - 2\left(\frac{1}{2}x^2\right)\Big|_0^3$$
$$= 2(36 - 9) - 9 = 45 \text{ sq. units.}$$

31. Area $R = \int_0^1 x^{1/3}\, dx - \int_0^1 x^{1/2}\, dx$

$$= \frac{3}{4}x^{4/3}\Big|_0^1 - \frac{2}{3}x^{3/2}\Big|_0^1 = \frac{3}{4} - \frac{2}{3} = \frac{1}{12} \text{ sq. units.}$$

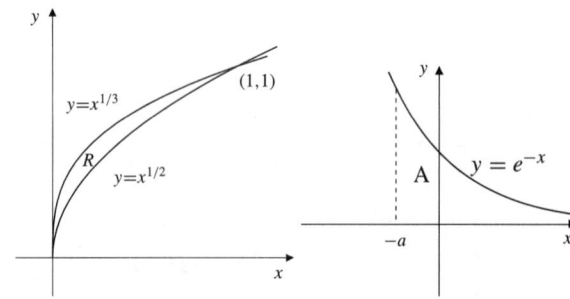

Fig. 5.5.31   Fig. 5.5.32

32. Area $= \int_{-a}^0 e^{-x}\, dx = -e^{-x}\Big|_{-a}^0 = e^a - 1$ sq. units.

33. Area $R = \int_0^{2\pi} (1 - \cos x)\, dx$

$$= (x - \sin x)\Big|_0^{2\pi} = 2\pi \text{ sq. units.}$$

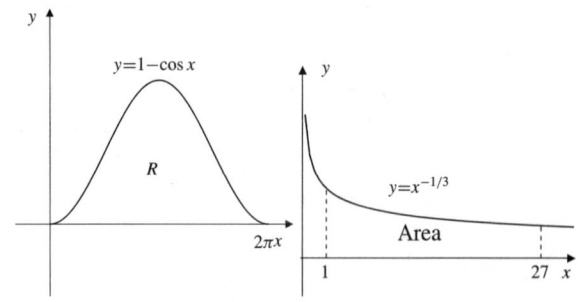

Fig. 5.5.33   Fig. 5.5.34

34. Area $= \int_1^{27} x^{-1/3}\, dx = \frac{3}{2}x^{2/3}\Big|_1^{27}$

$$= \frac{3}{2}(27)^{2/3} - \frac{3}{2} = 12 \text{ sq. units.}$$

35. $\int_0^{3\pi/2} |\cos x|\, dx = \int_0^{\pi/2} \cos x\, dx - \int_{\pi/2}^{3\pi/2} \cos x\, dx$

$$= \sin x\Big|_0^{\pi/2} - \sin x\Big|_{\pi/2}^{3\pi/2}$$
$$= 1 + 1 + 1 = 3$$

36. $\int_1^3 \frac{\text{sgn}(x-2)}{x^2}\, dx = -\int_1^2 \frac{dx}{x^2} + \int_2^3 \frac{dx}{x^2}$

$$= \frac{1}{x}\Big|_1^2 - \frac{1}{x}\Big|_2^3 = -\frac{1}{3}$$

37. Average value
$$= \frac{1}{2}\int_0^2 (1 + x + x^2 + x^3)\, dx$$
$$= \frac{1}{2}\left(x + \frac{x^2}{2} + \frac{x^3}{3} + \frac{x^4}{4}\right)\Big|_0^2$$
$$= \frac{1}{2}\left(2 + 2 + \frac{8}{3} + 4\right) = \frac{16}{3}.$$

38. Average value $= \dfrac{1}{2 - (-2)}\int_{-2}^2 e^{3x}\, dx$

$$= \frac{1}{4}\left(\frac{1}{3}e^{3x}\right)\Big|_{-2}^2$$
$$= \frac{1}{12}(e^6 - e^{-6}).$$

39. Avg. $= \dfrac{1}{1/\ln 2}\int_0^{1/\ln 2} 2^x\, dx = (\ln 2)\dfrac{2^x}{\ln 2}\Big|_0^{1/\ln 2} = e - 1$

40. Since
$$g(t) = \begin{cases} 0, & \text{if } 0 \leq t \leq 1, \\ 1, & \text{if } 1 < t \leq 3, \end{cases}$$

the average value of $g(t)$ over $[0,3]$ is

$$\frac{1}{3}\left[\int_0^1 (0)\, dt + \int_1^3 1\, dt\right] = \frac{1}{3}\left[0 + t\Big|_1^3\right]$$
$$= \frac{1}{3}(3 - 1) = \frac{2}{3}.$$

41. $\dfrac{d}{dx}\int_2^x \dfrac{\sin t}{t}\, dt = \dfrac{\sin x}{x}$

42. $\dfrac{d}{dt}\int_t^3 \dfrac{\sin x}{x}\, dx = \dfrac{d}{dt}\left[-\int_3^t \dfrac{\sin x}{x}\, dx\right] = -\dfrac{\sin t}{t}$

43. $\dfrac{d}{dx}\int_{x^2}^0 \dfrac{\sin t}{t}\, dt = -\dfrac{d}{dx}\int_0^{x^2} \dfrac{\sin t}{t}\, dt$

$$= -2x\frac{\sin x^2}{x^2} = -2\frac{\sin x^2}{x}$$

44. $\dfrac{d}{dx} x^2 \int_0^{x^2} \dfrac{\sin u}{u} \, du$

$= 2x \int_0^{x^2} \dfrac{\sin u}{u} \, du + x^2 \dfrac{d}{dx} \int_0^{x^2} \dfrac{\sin u}{u} \, du$

$= 2x \int_0^{x^2} \dfrac{\sin u}{u} \, du + x^2 \left[\dfrac{2x \sin x^2}{x^2} \right]$

$= 2x \int_0^{x^2} \dfrac{\sin u}{u} \, du + 2x \sin(x^2)$

45. $\dfrac{d}{dt} \int_{-\pi}^{t} \dfrac{\cos y}{1 + y^2} \, dy = \dfrac{\cos t}{1 + t^2}$

46. $\dfrac{d}{d\theta} \int_{\sin\theta}^{\cos\theta} \dfrac{1}{1 - x^2} \, dx$

$= \dfrac{d}{d\theta} \left[\int_a^{\cos\theta} \dfrac{1}{1 - x^2} \, dx - \int_a^{\sin\theta} \dfrac{1}{1 - x^2} \, dx \right]$

$= \dfrac{-\sin\theta}{1 - \cos^2\theta} - \dfrac{\cos\theta}{1 - \sin^2\theta}$

$= \dfrac{-1}{\sin\theta} - \dfrac{1}{\cos\theta} = -\csc\theta - \sec\theta$

47. $F(t) = \int_0^t \cos(x^2) \, dx$

$F(\sqrt{x}) = \int_0^{\sqrt{x}} \cos(u^2) \, du$

$\dfrac{d}{dx} F(\sqrt{x}) = \cos x \dfrac{1}{2\sqrt{x}} = \dfrac{\cos x}{2\sqrt{x}}$

48. $H(x) = 3x \int_4^{x^2} e^{-\sqrt{t}} \, dt$

$H'(x) = 3 \int_4^{x^2} e^{-\sqrt{t}} \, dt + 3x(2xe^{-|x|})$

$H'(2) = 3 \int_4^4 e^{-\sqrt{t}} \, dt + 3(2)(4e^{-2})$

$= 3(0) + 24 e^{-2} = \dfrac{24}{e^2}$

49. $f(x) = \pi + \pi \int_1^x f(t) \, dt$

$f'(x) = \pi f(x) \implies f(x) = Ce^{\pi x}$

$\pi = f(1) = Ce^\pi \implies C = \pi e^{-\pi}$

$f(x) = \pi e^{\pi(x-1)}$.

50. $f(x) = 1 - \int_0^x f(t) \, dt$

$f'(x) = -f(x) \implies f(x) = Ce^{-x}$

$1 = f(0) = C$

$f(x) = e^{-x}$.

51. The function $1/x^2$ is not *defined* (and therefore not continuous) at $x = 0$, so the Fundamental Theorem of Calculus cannot be applied to it on the interval $[-1, 1]$. Since $1/x^2 > 0$ wherever it is defined, we would expect $\int_{-1}^1 \dfrac{dx}{x^2}$ to be *positive* if it exists at all (which it doesn't).

52. If $F(x) = \int_{17}^x \dfrac{\sin t}{1 + t^2} \, dt$, then $F'(x) = \dfrac{\sin x}{1 + x^2}$ and $F(17) = 0$.

53. $F(x) = \int_0^{2x - x^2} \cos\left(\dfrac{1}{1 + t^2}\right) dt$.

Note that $0 < \dfrac{1}{1 + t^2} \leq 1$ for all t, and hence

$$0 < \cos(1) \leq \cos\left(\dfrac{1}{1 + t^2}\right) \leq 1.$$

The integrand is continuous for all t, so $F(x)$ is defined and differentiable for all x. Since $\lim_{x \to \pm\infty}(2x - x^2) = -\infty$, therefore $\lim_{x \to \pm\infty} F(x) = -\infty$. Now

$$F'(x) = (2 - 2x) \cos\left(\dfrac{1}{1 + (2x - x^2)^2}\right) = 0$$

only at $x = 1$. Therefore F must have a maximum value at $x = 1$, and no minimum value.

54. $\lim_{n \to \infty} \dfrac{1}{n} \left[\left(1 + \dfrac{1}{n}\right)^5 + \left(1 + \dfrac{2}{n}\right)^5 + \cdots + \left(1 + \dfrac{n}{n}\right)^5 \right]$

= area below $y = x^5$, above $y = 0$,
between $x = 1$ and $x = 2$

$= \int_1^2 x^5 \, dx = \dfrac{1}{6} x^6 \Big|_1^2 = \dfrac{1}{6}(2^6 - 1) = \dfrac{21}{2}$

55. $\lim_{n \to \infty} \dfrac{\pi}{n} \left(\sin \dfrac{\pi}{n} + \sin \dfrac{2\pi}{n} + \cdots + \sin \dfrac{n\pi}{n} \right)$

$= \lim_{n \to \infty}$ sum of areas of rectangles shown in figure

$= \int_0^\pi \sin x \, dx = -\cos x \Big|_0^\pi = 2$

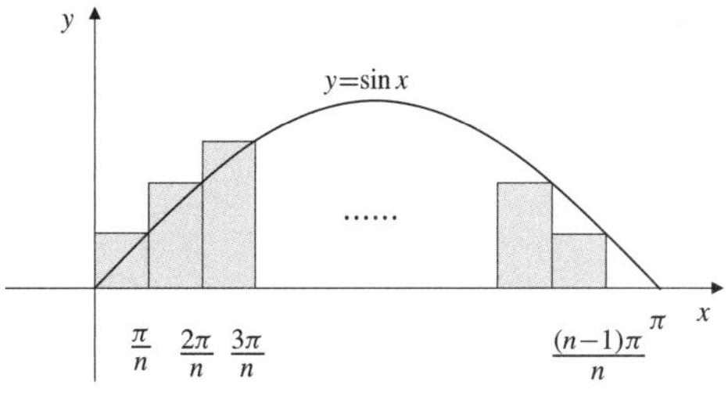

Fig. 5.5.55

56. $\lim_{n\to\infty}\left(\dfrac{n}{n^2+1}+\dfrac{n}{n^2+4}+\dfrac{n}{n^2+9}+\cdots+\dfrac{n}{2n^2}\right)$

$=\lim_{n\to\infty}\dfrac{1}{n}\left(\dfrac{n^2}{n^2+1}+\dfrac{n^2}{n^2+4}+\dfrac{n^2}{n^2+9}+\cdots+\dfrac{n^2}{2n^2}\right)$

$=\lim_{n\to\infty}\dfrac{1}{n}\left(\dfrac{1}{1+\left(\frac{1}{n}\right)^2}+\dfrac{1}{1+\left(\frac{2}{n}\right)^2}+\cdots+\dfrac{1}{1+\left(\frac{n}{n}\right)^2}\right)$

$=$ area below $y=\dfrac{1}{1+x^2}$, above $y=0$,

between $x=0$ and $x=1$

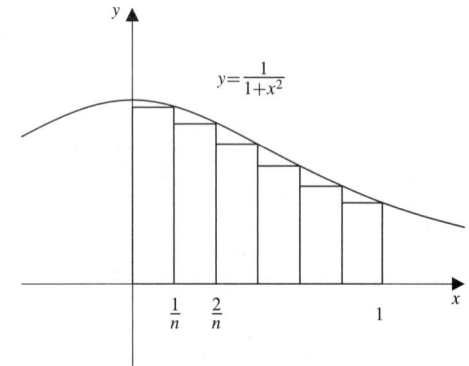

$=\displaystyle\int_0^1\dfrac{1}{1+x^2}\,dx=\tan^{-1}x\Big|_0^1=\dfrac{\pi}{4}$

Fig. 5.5.56

Section 5.6 The Method of Substitution (page 337)

1. $\displaystyle\int e^{5-2x}\,dx\quad$ Let $u=5-2x$
$\qquad\qquad\qquad\qquad du=-2\,dx$

$=-\dfrac{1}{2}\displaystyle\int e^u\,du=-\dfrac{1}{2}e^u+C=-\dfrac{1}{2}e^{5-2x}+C.$

2. $\displaystyle\int \cos(ax+b)\,dx\quad$ Let $u=ax+b$
$\qquad\qquad\qquad\qquad du=a\,dx$

$=\dfrac{1}{a}\displaystyle\int \cos u\,du=\dfrac{1}{a}\sin u+C$

$=\dfrac{1}{a}\sin(ax+b)+C.$

3. $\displaystyle\int \sqrt{3x+4}\,dx\quad$ Let $u=3x+4$
$\qquad\qquad\qquad\qquad du=3\,dx$

$=\dfrac{1}{3}\displaystyle\int u^{1/2}\,du=\dfrac{2}{9}u^{3/2}+C=\dfrac{2}{9}(3x+4)^{3/2}+C.$

4. $\displaystyle\int e^{2x}\sin(e^{2x})\,dx\quad$ Let $u=e^{2x}$
$\qquad\qquad\qquad\qquad du=2e^{2x}\,dx$

$=\dfrac{1}{2}\displaystyle\int \sin u\,du=-\dfrac{1}{2}\cos u+C$

$=-\dfrac{1}{2}\cos(e^{2x})+C.$

5. $\displaystyle\int \dfrac{x^2}{(x^3+2)^{5/2}}\,dx\quad$ Let $u=x^3+2$
$\qquad\qquad\qquad\qquad du=3x^2\,dx$

$=\dfrac{1}{3}\displaystyle\int u^{-5/2}\,du=-\dfrac{2}{9}u^{-3/2}+C=\dfrac{-2}{9(x^3+2)^{3/2}}+C.$

6. $\displaystyle\int (x+2)(x^2+4x+9)^{1/3}\,dx\quad$ Let $u=x^2+4x+9$
$\qquad\qquad\qquad\qquad du=2(x+2)\,dx$

$=\dfrac{1}{2}\displaystyle\int u^{1/3}\,du=\dfrac{1}{2}\dfrac{3}{4}u^{4/3}+C$

$=\dfrac{3}{8}(x^2+4x+9)^{4/3}+C.$

7. $\displaystyle\int \dfrac{x\,dx}{(4x^2+1)^5}\quad$ Let $u=4x^2+1$
$\qquad\qquad\qquad\qquad du=8x\,dx$

$=\dfrac{1}{8}\displaystyle\int u^{-5}\,du=-\dfrac{1}{32}u^{-4}+C=\dfrac{-1}{32(4x^2+1)^4}+C.$

8. $\displaystyle\int \dfrac{\sin\sqrt{x}}{\sqrt{x}}\,dx\quad$ Let $u=\sqrt{x}$
$\qquad\qquad\qquad\qquad du=\dfrac{dx}{2\sqrt{x}}$

$=2\displaystyle\int \sin u\,du=-2\cos u+C$

$=-2\cos\sqrt{x}+C.$

9. $\displaystyle\int xe^{x^2}\,dx\quad$ Let $u=x^2$
$\qquad\qquad\qquad\qquad du=2x\,dx$

$=\dfrac{1}{2}\displaystyle\int e^u\,du=\dfrac{1}{2}e^u+C=\dfrac{1}{2}e^{x^2}+C.$

10. $\displaystyle\int x^2 2^{x^3+1}\,dx\quad$ Let $u=x^3+1$
$\qquad\qquad\qquad\qquad du=3x^2\,dx$

$=\dfrac{1}{3}\displaystyle\int 2^u\,du=\dfrac{1}{3}\dfrac{2^u}{\ln 2}+C$

$=\dfrac{2^{x^3+1}}{3\ln 2}+C.$

11. $\displaystyle\int \dfrac{\cos x}{4+\sin^2 x}\,dx\quad$ Let $u=\sin x$
$\qquad\qquad\qquad\qquad du=\cos x\,dx$

$=\displaystyle\int \dfrac{du}{4+u^2}$

$=\dfrac{1}{2}\tan^{-1}\dfrac{u}{2}+C=\dfrac{1}{2}\tan^{-1}\left(\dfrac{1}{2}\sin x\right)+C.$

INSTRUCTOR'S SOLUTIONS MANUAL SECTION 5.6 (PAGE 337)

12. $\displaystyle\int \frac{\sec^2 x}{\sqrt{1-\tan^2 x}}\,dx$ Let $u = \tan x$
$\qquad du = \sec^2 x\,dx$
$= \displaystyle\int \frac{du}{\sqrt{1-u^2}}$
$= \sin^{-1} u + C$
$= \sin^{-1}(\tan x) + C.$

13. $\displaystyle\int \frac{e^x+1}{e^x-1}\,dx$
$= \displaystyle\int \frac{e^{x/2}+e^{-x/2}}{e^{x/2}-e^{x/2}}\,dx$ Let $u = e^{x/2} - e^{-x/2}$
$\qquad du = \tfrac{1}{2}\left(e^{x/2}+e^{-x/2}\right)dx$
$= 2\displaystyle\int \frac{du}{u} = 2\ln|u| + C$
$= 2\ln\left|e^{x/2}-e^{-x/2}\right| + C = \ln\left|e^x + e^{-x} - 2\right| + C.$

14. $\displaystyle\int \frac{\ln t}{t}\,dt$ Let $u = \ln t$
$\qquad du = \dfrac{dt}{t}$
$= \displaystyle\int u\,du = \tfrac{1}{2}u^2 + C = \tfrac{1}{2}(\ln t)^2 + C.$

15. $\displaystyle\int \frac{ds}{\sqrt{4-5s}}$ Let $u = 4 - 5s$
$\qquad du = -5\,ds$
$= -\tfrac{1}{5}\displaystyle\int \frac{du}{\sqrt{u}}$
$= -\tfrac{2}{5}u^{1/2} + C = -\tfrac{2}{5}\sqrt{4-5s} + C.$

16. $\displaystyle\int \frac{x+1}{\sqrt{x^2+2x+3}}\,dx$ Let $u = x^2 + 2x + 3$
$\qquad du = 2(x+1)\,dx$
$= \tfrac{1}{2}\displaystyle\int \frac{1}{\sqrt{u}}\,du = \sqrt{u} + C = \sqrt{x^2+2x+3} + C$

17. $\displaystyle\int \frac{t}{\sqrt{4-t^4}}\,dt$ Let $u = t^2$
$\qquad du = 2t\,dt$
$= \tfrac{1}{2}\displaystyle\int \frac{du}{\sqrt{4-u^2}}$
$= \tfrac{1}{2}\sin^{-1}\dfrac{u}{2} + C = \tfrac{1}{2}\sin^{-1}\left(\dfrac{t^2}{2}\right) + C.$

18. $\displaystyle\int \frac{x^2}{2+x^6}\,dx$ Let $u = x^3$
$\qquad du = 3x^2\,dx$
$= \tfrac{1}{3}\displaystyle\int \frac{du}{2+u^2} = \dfrac{1}{3\sqrt{2}}\tan^{-1}\left(\dfrac{u}{\sqrt{2}}\right) + C$
$= \dfrac{1}{3\sqrt{2}}\tan^{-1}\left(\dfrac{x^3}{\sqrt{2}}\right) + C.$

19. $\displaystyle\int \frac{dx}{e^x+1} = \displaystyle\int \frac{e^{-x}\,dx}{1+e^{-x}}$ Let $u = 1 + e^{-x}$
$\qquad du = -e^{-x}\,dx$
$= -\displaystyle\int \frac{du}{u} = -\ln|u| + C = -\ln(1+e^{-x}) + C.$

20. $\displaystyle\int \frac{dx}{e^x+e^{-x}} = \displaystyle\int \frac{e^x\,dx}{e^{2x}+1}$ Let $u = e^x$
$\qquad du = e^x\,dx$
$= \displaystyle\int \frac{du}{u^2+1} = \tan^{-1}u + C$
$= \tan^{-1}e^x + C.$

21. $\displaystyle\int \tan x \ln \cos x\,dx$ Let $u = \ln\cos x$
$\qquad du = -\tan x\,dx$
$= -\displaystyle\int u\,du = -\tfrac{1}{2}u^2 + C = -\tfrac{1}{2}(\ln\cos x)^2 + C.$

22. $\displaystyle\int \frac{x+1}{\sqrt{1-x^2}}\,dx$
$= \displaystyle\int \frac{x\,dx}{\sqrt{1-x^2}} + \displaystyle\int \frac{dx}{\sqrt{1-x^2}}$ Let $u = 1 - x^2$
$\qquad du = -2x\,dx$
in the first integral only
$= -\tfrac{1}{2}\displaystyle\int \frac{du}{\sqrt{u}} + \sin^{-1}x = -\sqrt{u} + \sin^{-1}x + C$
$= -\sqrt{1-x^2} + \sin^{-1}x + C.$

23. $\displaystyle\int \frac{2t+3}{t^2+9}\,dt = 3\displaystyle\int \frac{dt}{t^2+9} + \displaystyle\int \frac{2t\,dt}{t^2+9}$
$= \tfrac{3}{3}\tan^{-1}\dfrac{t}{3} + \displaystyle\int \frac{2t\,dt}{t^2+9}$ Let $u = t^2 + 9$
$\qquad du = 2t\,dt$
$= \tan^{-1}\dfrac{t}{3} + \displaystyle\int \frac{du}{u}$
$= \tan^{-1}\dfrac{t}{3} + \ln|u| + C = \tan^{-1}\dfrac{t}{3} + \ln(t^2+9) + C.$

24. $\displaystyle\int \frac{ax+b}{\sqrt{A^2-B^2x^2}}\,dx$
$= a\displaystyle\int \frac{x\,dx}{\sqrt{A^2-B^2x^2}} + b\displaystyle\int \frac{dx}{\sqrt{A^2-B^2x^2}}$
Let $u = A^2 - B^2 x^2$
$\qquad du = -2B^2 x\,dx$ first integral only
$= -\dfrac{a}{2B^2}\displaystyle\int \frac{du}{\sqrt{u}} + \dfrac{b}{|B|}\displaystyle\int \frac{dx}{\sqrt{(A/B)^2 - x^2}}$
$= -\dfrac{a}{B^2}\sqrt{u} + \dfrac{b}{|B|}\sin^{-1}\left(\left|\dfrac{B}{A}\right|x\right) + C$
$= -\dfrac{a}{B^2}\sqrt{A^2-B^2x^2} + \dfrac{b}{|B|}\sin^{-1}\left(\left|\dfrac{B}{A}\right|x\right) + C.$

189

25. $\displaystyle\int \frac{dx}{x^2+6x+13} = \int \frac{dx}{(x+3)^2+4}$ Let $u = x+3$
$du = dx$
$= \displaystyle\int \frac{du}{u^2+4} = \frac{1}{2}\tan^{-1}\frac{u}{2} + C$
$= \displaystyle\frac{1}{2}\tan^{-1}\frac{x+3}{2} + C.$

26. $\displaystyle\int \frac{dx}{\sqrt{4+2x-x^2}} = \int \frac{dx}{\sqrt{5-(1-x)^2}}$ Let $u = 1-x$
$du = -dx$
$= -\displaystyle\int \frac{du}{\sqrt{5-u^2}} = -\sin^{-1}\left(\frac{u}{\sqrt{5}}\right) + C$
$= -\sin^{-1}\left(\displaystyle\frac{1-x}{\sqrt{5}}\right) + C = \sin^{-1}\left(\displaystyle\frac{x-1}{\sqrt{5}}\right) + C.$

27. $\displaystyle\int \sin^3 x \cos^5 x\, dx$
$= \displaystyle\int \sin x (\cos^5 x - \cos^7 x)\, dx$ Let $u = \cos x$
$du = -\sin x\, dx$
$= \displaystyle\int (u^7 - u^5)\, du$
$= \displaystyle\frac{u^8}{8} - \frac{u^6}{6} + C = \frac{\cos^8 x}{8} - \frac{\cos^6 x}{6} + C.$

28. $\displaystyle\int \sin^4 t \cos^5 t\, dt$
$= \displaystyle\int \sin^4 t (1-\sin^2 t)^2 \cos t\, dt$ Let $u = \sin t$
$du = \cos t\, dt$
$= \displaystyle\int (u^4 - 2u^6 + u^8)\, du = \frac{u^5}{5} - \frac{2u^7}{7} + \frac{u^9}{9} + C$
$= \displaystyle\frac{1}{5}\sin^5 t - \frac{2}{7}\sin^7 t + \frac{1}{9}\sin^9 t + C.$

29. $\displaystyle\int \sin ax \cos^2 ax\, dx$ Let $u = \cos ax$
$du = -a \sin ax\, dx$
$= -\displaystyle\frac{1}{a}\int u^2\, du$
$= -\displaystyle\frac{u^3}{3a} + C = -\frac{1}{3a}\cos^3 ax + C.$

30. $\displaystyle\int \sin^2 x \cos^2 x\, dx = \int \left(\frac{\sin 2x}{2}\right)^2 dx$
$= \displaystyle\frac{1}{4}\int \frac{1-\cos 4x}{2}\, dx = \frac{x}{8} - \frac{\sin 4x}{32} + C.$

31. $\displaystyle\int \sin^6 x\, dx = \int \left(\frac{1-\cos 2x}{2}\right)^3 dx$
$= \displaystyle\frac{1}{8}\int (1 - 3\cos 2x + 3\cos^2 2x - \cos^3 2x)\, dx$
$= \displaystyle\frac{x}{8} - \frac{3\sin 2x}{16} + \frac{3}{16}\int(1+\cos 4x)\, dx$
$\quad - \displaystyle\frac{1}{8}\int \cos 2x(1-\sin^2 2x)\, dx$ Let $u = \sin 2x$
$du = 2\cos 2x\, dx$
$= \displaystyle\frac{5x}{16} - \frac{3\sin 2x}{16} + \frac{3\sin 4x}{64} - \frac{1}{16}\int(1-u^2)\, du$
$= \displaystyle\frac{5x}{16} - \frac{3\sin 2x}{16} + \frac{3\sin 4x}{64} - \frac{\sin 2x}{16} + \frac{\sin^3 2x}{48} + C$
$= \displaystyle\frac{5x}{16} - \frac{\sin 2x}{4} + \frac{3\sin 4x}{64} + \frac{\sin^3 2x}{48} + C.$

32. $\displaystyle\int \cos^4 x\, dx = \int \frac{[1+\cos(2x)]^2}{4}\, dx$
$= \displaystyle\frac{1}{4}\int[1 + 2\cos(2x) + \cos^2(2x)]\, dx$
$= \displaystyle\frac{x}{4} + \frac{\sin(2x)}{4} + \frac{1}{8}\int 1+\cos(4x)\, dx$
$= \displaystyle\frac{x}{4} + \frac{\sin(2x)}{4} + \frac{x}{8} + \frac{\sin(4x)}{32} + C$
$= \displaystyle\frac{3x}{8} + \frac{\sin(2x)}{4} + \frac{\sin(4x)}{32} + C.$

33. $\displaystyle\int \sec^5 x \tan x\, dx$ Let $u = \sec x$
$du = \sec x \tan x\, dx$
$= \displaystyle\int u^4\, du = \frac{u^5}{5} + C = \frac{\sec^5 x}{5} + C.$

34. $\displaystyle\int \sec^6 x \tan^2 x\, dx$
$= \displaystyle\int \sec^2 x \tan^2 x (1+\tan^2 x)^2\, dx$ Let $u = \tan x$
$du = \sec^2 x\, dx$
$= \displaystyle\int (u^2 + 2u^4 + u^6)\, du = \frac{1}{3}u^3 + \frac{2}{5}u^5 + \frac{1}{7}u^7 + C$
$= \displaystyle\frac{1}{3}\tan^3 x + \frac{2}{5}\tan^5 x + \frac{1}{7}\tan^7 x + C.$

35. $\displaystyle\int \sqrt{\tan x}\, \sec^4 x\, dx$
$= \displaystyle\int \sqrt{\tan x}(1+\tan^2 x)\sec^2 x\, dx$ Let $u = \tan x$
$du = \sec^2 x\, dx$
$= \displaystyle\int \left(u^{1/2} + u^{5/2}\right) du$
$= \displaystyle\frac{2u^{3/2}}{3} + \frac{2u^{7/2}}{7} + C$
$= \displaystyle\frac{2}{3}(\tan x)^{3/2} + \frac{2}{7}(\tan x)^{7/2} + C.$

36. $\displaystyle\int \sin^{-2/3} x \cos^3 x \, dx$ Let $u = \sin x$
$\qquad\qquad\qquad\qquad\qquad\qquad du = \cos x \, dx$

$= \displaystyle\int \frac{1-u^2}{u^{2/3}} \, du = 3u^{1/3} - \frac{3}{7} u^{7/3} + C$

$= 3 \sin^{1/3} x - \dfrac{3}{7} \sin^{7/3} x + C.$

37. $\displaystyle\int \cos x \sin^4(\sin x) \, dx$ Let $u = \sin x$
$\qquad\qquad\qquad\qquad\qquad du = \cos x \, dx$

$= \displaystyle\int \sin^4 u \, du = \int \left(\frac{1-\cos 2u}{2}\right)^2 du$

$= \dfrac{1}{4} \displaystyle\int \left(1 - 2\cos 2u + \dfrac{1 + \cos 4u}{2}\right) du$

$= \dfrac{3u}{8} - \dfrac{\sin 2u}{4} + \dfrac{\sin 4u}{32} + C$

$= \dfrac{3}{8} \sin x - \dfrac{1}{4} \sin(2 \sin x) + \dfrac{1}{32} \sin(4 \sin x) + C.$

38. $\displaystyle\int \frac{\sin^3(\ln x) \cos^3(\ln x)}{x} \, dx$ Let $u = \sin(\ln x)$
$\qquad\qquad\qquad\qquad\qquad\qquad du = \dfrac{\cos(\ln x)}{x} dx$

$= \displaystyle\int u^3(1-u^2) \, du = \dfrac{1}{4} u^4 - \dfrac{1}{6} u^6 + C$

$= \dfrac{1}{4} \sin^4(\ln x) - \dfrac{1}{6} \sin^6(\ln x) + C.$

39. $\displaystyle\int \frac{\sin^2 x}{\cos^4 x} \, dx$

$= \displaystyle\int \tan^2 x \sec^2 x \, dx$ Let $u = \tan x$
$\qquad\qquad\qquad\qquad du = \sec^2 x \, dx$

$= \displaystyle\int u^2 \, du = \dfrac{u^3}{3} + C = \dfrac{1}{3} \tan^3 x + C.$

40. $\displaystyle\int \frac{\sin^3 x}{\cos^4 x} \, dx = \int \tan^3 x \sec x \, dx$

$= \displaystyle\int (\sec^2 x - 1) \sec x \tan x \, dx$ Let $u = \sec x$
$\qquad\qquad\qquad\qquad\qquad\qquad du = \sec x \tan x \, dx$

$= \displaystyle\int (u^2 - 1) \, du = \tfrac{1}{3} u^3 - u + C$

$= \tfrac{1}{3} \sec^3 x - \sec x + C.$

41. $\displaystyle\int \csc^5 x \cot^5 x \, dx$

$= \displaystyle\int \csc x \cot x \csc^4 x (\csc^2 x - 1)^2 \, dx$

$\qquad\qquad$ Let $u = \csc x$
$\qquad\qquad\qquad du = -\csc x \cot x \, dx$

$= -\displaystyle\int (u^8 - 2u^6 + u^4) \, du$

$= -\dfrac{u^9}{9} + \dfrac{2u^7}{7} - \dfrac{u^5}{5} + C$

$= -\dfrac{1}{9} \csc^9 x + \dfrac{2}{7} \csc^7 x - \dfrac{1}{5} \csc^5 x + C.$

42. $\displaystyle\int \frac{\cos^4 x}{\sin^8 x} \, dx = \int \cot^4 x \csc^4 x \, dx$

$= \displaystyle\int \cot^4 x (1 + \cot^2 x) \csc^2 x \, dx$ Let $u = \cot x$
$\qquad\qquad\qquad\qquad\qquad\qquad\qquad du = -\csc^2 x \, dx$

$= -\displaystyle\int u^4(1+u^2) \, du = -\dfrac{u^5}{5} - \dfrac{u^7}{7} + C$

$= -\dfrac{1}{5} \cot^5 x - \dfrac{1}{7} \cot^7 x + C.$

43. $\displaystyle\int_0^4 x^3 (x^2+1)^{-1/2} \, dx$ Let $u = x^2 + 1$, $x^2 = u - 1$
$\qquad\qquad\qquad\qquad\qquad\qquad du = 2x \, dx$

$= \dfrac{1}{2} \displaystyle\int_1^{17} (u-1) u^{-1/2} \, du$

$= \dfrac{1}{2} \left(\dfrac{2}{3} u^{3/2} - 2u^{1/2}\right)\bigg|_1^{17}$

$= \dfrac{17\sqrt{17} - 1}{3} - (\sqrt{17} - 1) = \dfrac{14\sqrt{17}}{3} + \dfrac{2}{3}.$

44. $\displaystyle\int_1^{\sqrt{e}} \frac{\sin(\pi \ln x)}{x} \, dx$ Let $u = \pi \ln x$
$\qquad\qquad\qquad\qquad\qquad du = \dfrac{\pi}{x} dx$

$= \dfrac{1}{\pi} \displaystyle\int_0^{\pi/2} \sin u \, du = -\dfrac{1}{\pi} \cos u \bigg|_0^{\pi/2}$

$= -\dfrac{1}{\pi}(0 - 1) = \dfrac{1}{\pi}.$

45. $\displaystyle\int_0^{\pi/2} \sin^4 x \, dx = \int_0^{\pi/2} \left(\frac{1 - \cos 2x}{2}\right)^2 dx$

$= \dfrac{1}{4} \displaystyle\int_0^{\pi/2} \left(1 - 2\cos 2x + \dfrac{1 + \cos 4x}{2}\right) dx$

$= \dfrac{3x}{8}\bigg|_0^{\pi/2} - \dfrac{\sin 2x}{4}\bigg|_0^{\pi/2} + \dfrac{\sin 4x}{32}\bigg|_0^{\pi/2} = \dfrac{3\pi}{16}.$

46. $\int_{\pi/4}^{\pi} \sin^5 x \, dx$

$= \int_{\pi/4}^{\pi} (1 - \cos^2 x)^2 \sin x \, dx \quad$ Let $u = \cos x$
$\qquad du = -\sin x \, dx$

$= -\int_{1/\sqrt{2}}^{-1} (1 - 2u^2 + u^4) \, du = u - \frac{2}{3}u^3 + \frac{1}{5}u^5 \Big|_{-1}^{1/\sqrt{2}}$

$= \frac{1}{\sqrt{2}} - \frac{1}{3\sqrt{2}} + \frac{1}{20\sqrt{2}} - \left(-1 + \frac{2}{3} - \frac{1}{5}\right) = \frac{43}{60\sqrt{2}} + \frac{8}{15}.$

47. $\int_{e}^{e^2} \frac{dt}{t \ln t} \quad$ Let $u = \ln t$
$\qquad du = \frac{dt}{t}$

$= \int_{1}^{2} \frac{du}{u} = \ln u \Big|_{1}^{2} = \ln 2 - \ln 1 = \ln 2.$

48. $\int_{\pi^2/16}^{\pi^2/9} \frac{2^{\sin\sqrt{x}} \cos\sqrt{x}}{\sqrt{x}} \, dx \quad$ Let $u = \sin\sqrt{x}$
$\qquad du = \frac{\cos\sqrt{x}}{2\sqrt{x}} \, dx$

$= 2\int_{1/\sqrt{2}}^{\sqrt{3}/2} 2^u \, du = \frac{2(2^u)}{\ln 2} \Big|_{1/\sqrt{2}}^{\sqrt{3}/2}$

$= \frac{2}{\ln 2}(2^{\sqrt{3}/2} - 2^{1/\sqrt{2}}).$

49. $\int_{0}^{\pi/2} \sqrt{1 + \cos x} \, dx = \int_{0}^{\pi/2} \sqrt{2\cos^2 \frac{x}{2}} \, dx$

$= \sqrt{2} \int_{0}^{\pi/2} \cos\frac{x}{2} \, dx = 2\sqrt{2} \sin\frac{x}{2} \Big|_{0}^{\pi/2} = 2.$

$\int_{0}^{\pi/2} \sqrt{1 - \sin x} \, dx$

$= \int_{0}^{\pi/2} \sqrt{1 - \cos\left(\frac{\pi}{2} - x\right)} \, dx \quad$ Let $u = \frac{\pi}{2} - x$
$\qquad du = -dx$

$= -\int_{\pi/2}^{0} \sqrt{1 - \cos u} \, du$

$= \int_{0}^{\pi/2} \sqrt{2 \sin^2 \frac{u}{2}} \, du = \sqrt{2} \left(-2\cos\frac{u}{2}\right) \Big|_{0}^{\pi/2}$

$= -2 + 2\sqrt{2} = 2(\sqrt{2} - 1).$

50. Area $= \int_{0}^{2} \frac{x}{x^2 + 16} \, dx \quad$ Let $u = x^2 + 16$
$\qquad du = 2x \, dx$

$= \frac{1}{2} \int_{16}^{20} \frac{du}{u} = \frac{1}{2} \ln u \Big|_{16}^{20}$

$= \frac{1}{2}(\ln 20 - \ln 16) = \frac{1}{2} \ln\left(\frac{5}{4}\right)$ sq. units.

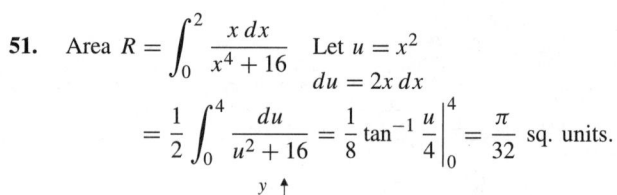

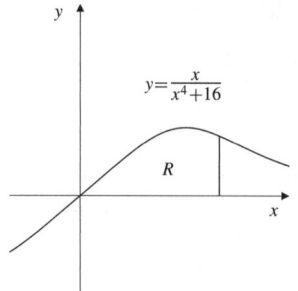

51. Area $R = \int_{0}^{2} \frac{x \, dx}{x^4 + 16} \quad$ Let $u = x^2$
$\qquad du = 2x \, dx$

$= \frac{1}{2} \int_{0}^{4} \frac{du}{u^2 + 16} = \frac{1}{8} \tan^{-1}\frac{u}{4} \Big|_{0}^{4} = \frac{\pi}{32}$ sq. units.

Fig. 5.6.51

52. The area bounded by the ellipse $(x^2/a^2) + (y^2/b^2) = 1$ is

$4 \int_{0}^{a} b\sqrt{1 - \frac{x^2}{a^2}} \, dx \quad$ Let $x = au$
$\qquad dx = a \, du$

$= 4ab \int_{0}^{1} \sqrt{1 - u^2} \, du.$

The integral is the area of a quarter circle of radius 1. Hence

Area $= 4ab \left(\frac{\pi(1)^2}{4}\right) = \pi ab$ sq. units.

53. We start with the addition formulas

$\cos(x + y) = \cos x \cos y - \sin x \sin y$
$\cos(x - y) = \cos x \cos y + \sin x \sin y$

and take half their sum and half their difference to obtain

$\cos x \cos y = \frac{1}{2}\big(\cos(x + y) + \cos(x - y)\big)$
$\sin x \sin y = \frac{1}{2}\big(\cos(x - y) - \cos(x + y)\big).$

Similarly, taking half the sum of the formulas

$\sin(x + y) = \sin x \cos y + \cos x \sin y$
$\sin(x - y) = \sin x \cos y - \cos x \sin y,$

we obtain

$\sin x \cos y = \frac{1}{2}\big(\sin(x + y) + \sin(x - y)\big).$

INSTRUCTOR'S SOLUTIONS MANUAL

SECTION 5.7 (PAGE 342)

54. We have

$$\int \cos ax \cos bx \, dx$$
$$= \frac{1}{2} \int [\cos(ax - bx) + \cos(ax + bx)] \, dx$$
$$= \frac{1}{2} \int \cos[(a - b)x] \, dx + \frac{1}{2} \int \cos[(a + b)x] \, dx$$

Let $u = (a - b)x$, $du = (a - b) \, dx$ in the first integral;
let $v = (a + b)x$, $dv = (a + b) \, dx$ in the second integral.

$$= \frac{1}{2(a - b)} \int \cos u \, du + \frac{1}{2(a + b)} \int \cos v \, dv$$
$$= \frac{1}{2} \left[\frac{\sin[(a - b)x]}{(a - b)} + \frac{\sin[(a + b)x]}{(a + b)} \right] + C.$$

$$\int \sin ax \sin bx \, dx$$
$$= \frac{1}{2} \int [\cos(ax - bx) - \cos(ax + bx)] \, dx$$
$$= \frac{1}{2} \left[\frac{\sin[(a - b)x]}{(a - b)} - \frac{\sin[(a + b)x]}{(a + b)} \right] + C.$$

$$\int \sin ax \cos bx \, dx$$
$$= \frac{1}{2} \int [\sin(ax + bx) + \sin(ax - bx)] \, dx$$
$$= \frac{1}{2} [\int \sin[(a + b)x] \, dx + \int \sin[(a - b)x] \, dx]$$
$$= -\frac{1}{2} \left[\frac{\cos[(a + b)x]}{(a + b)} + \frac{\cos[(a - b)x]}{(a - b)} \right] + C.$$

55. If m and n are integers, and $m \neq n$, then

$$\int_{-\pi}^{\pi} \left\{ \begin{array}{l} \cos mx \cos nx \\ \sin mx \sin nx \end{array} \right\} dx$$
$$= \frac{1}{2} \int_{-\pi}^{\pi} \left(\cos(m - n)x \pm \cos(m + n)x \right) dx$$
$$= \frac{1}{2} \left(\frac{\sin(m - n)x}{m - n} \pm \frac{\sin(m + n)x}{m + n} \right) \bigg|_{-\pi}^{\pi}$$
$$= 0 \pm 0 = 0.$$

$$\int_{-\pi}^{\pi} \sin mx \cos nx \, dx$$
$$= \frac{1}{2} \int_{-\pi}^{\pi} \left(\sin(m + n)x + \sin(m - n)x \right) dx$$
$$= -\frac{1}{2} \left(\frac{\cos(m + n)x}{m + n} + \frac{\cos(m - n)x}{m - n} \right) \bigg|_{-\pi}^{\pi}$$
$$= 0 \text{ (by periodicity)}.$$

If $m = n \neq 0$ then

$$\int_{-\pi}^{\pi} \sin mx \cos mx \, dx$$
$$= \frac{1}{2} \int_{-\pi}^{\pi} \sin 2mx \, dx$$
$$= -\frac{1}{4m} \cos 2mx \bigg|_{-\pi}^{\pi} = 0 \text{ (by periodicity)}.$$

Section 5.7 Areas of Plane Regions (page 342)

1. Area of $R = \int_0^1 (x - x^2) \, dx$

$$= \left(\frac{x^2}{2} - \frac{x^3}{3} \right) \bigg|_0^1 = \frac{1}{2} - \frac{1}{3} = \frac{1}{6} \text{ sq. units.}$$

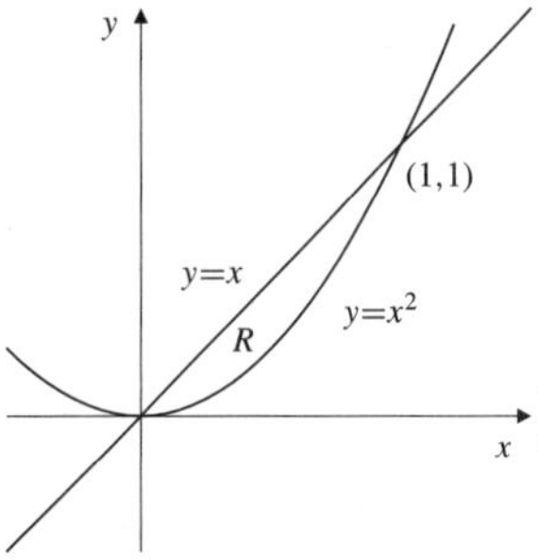

Fig. 5.7.1 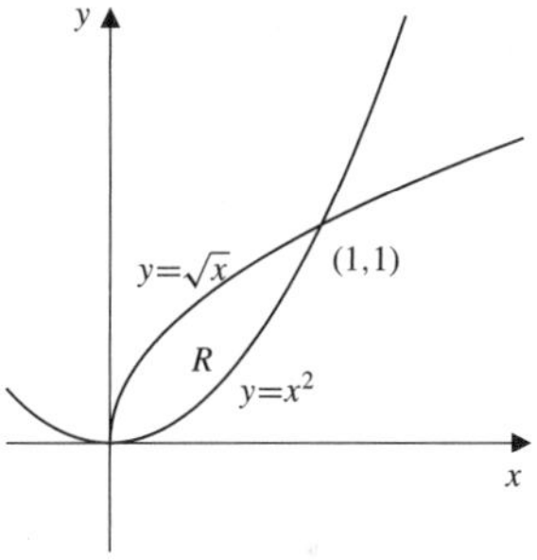   Fig. 5.7.2

2. Area of $R = \int_0^1 (\sqrt{x} - x^2) \, dx$

$$= \left(\frac{2}{3} x^{3/2} - \frac{1}{3} x^3 \right) \bigg|_0^1 = \frac{2}{3} - \frac{1}{3} = \frac{1}{3} \text{ sq. units.}$$

3. Area of $R = 2 \int_0^2 (8 - 2x^2) \, dx$

$$= \left(16x - \frac{4}{3} x^3 \right) \bigg|_0^2 = \frac{64}{3} \text{ sq. units.}$$

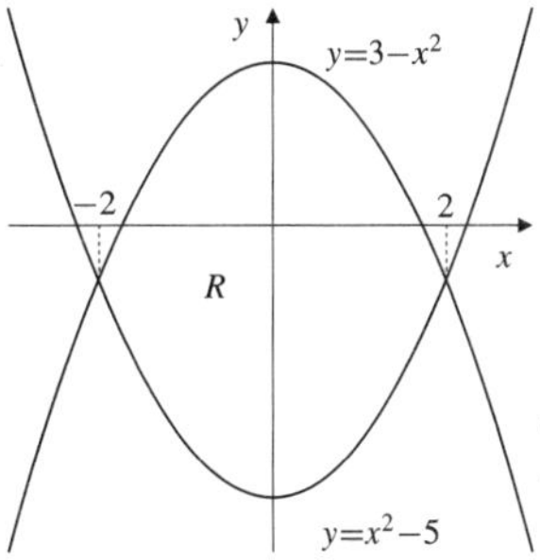

Fig. 5.7.3 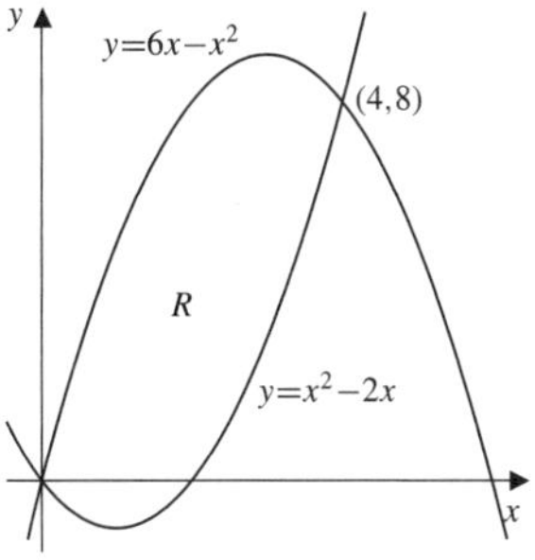   Fig. 5.7.4

193

SECTION 5.7 (PAGE 342)

4. For intersections:
$x^2 - 2x = 6x - x^2 \Rightarrow 2x^2 - 8x = 0$
i.e., $x = 0$ or 4.

$$\text{Area of } R = \int_0^4 [6x - x^2 - (x^2 - 2x)]\, dx$$

$$= \int_0^4 (8x - 2x^2)\, dx$$

$$= \left(4x^2 - \frac{2}{3}x^3\right)\bigg|_0^4 = \frac{64}{3} \text{ sq. units.}$$

5. For intersections:
$$\left.\begin{array}{l} 2y = 4x - x^2 \\ 2y + 3x = 6 \end{array}\right\} \Rightarrow \begin{array}{c} 4x - x^2 = 6 - 3x \\ x^2 - 7x + 6 = 0 \\ (x-1)(x-6) = 0 \end{array}$$

Thus intersections of the curves occur at $x = 1$ and $x = 6$. We have

$$\text{Area of } R = \int_1^6 \left(2x - \frac{x^2}{2} - 3 + \frac{3x}{2}\right) dx$$

$$= \left(\frac{7x^2}{4} - \frac{x^3}{6} - 3x\right)\bigg|_1^6$$

$$= \frac{245}{4} - 36 + \frac{1}{6} - 15 = \frac{125}{12} \text{ sq. units.}$$

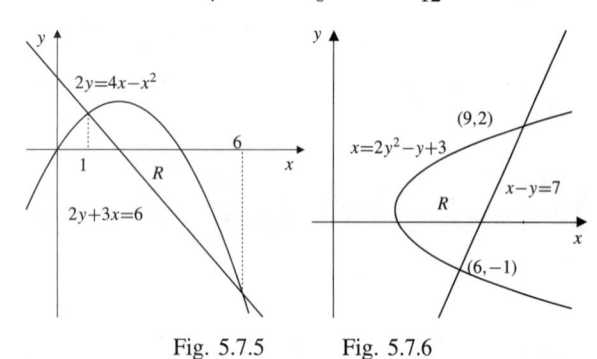

Fig. 5.7.5   Fig. 5.7.6

6. For intersections:
$7 + y = 2y^2 - y + 3 \Rightarrow 2y^2 - 2y - 4 = 0$
$2(y-2)(y+1) = 0 \Rightarrow$ i.e., $y = -1$ or 2.

$$\text{Area of } R = \int_{-1}^2 [(7+y) - (2y^2 - y + 3)]\, dy$$

$$= 2\int_{-1}^2 (2 + y - y^2)\, dy$$

$$= 2\left(2y + \frac{1}{2}y^2 - \frac{1}{3}y^3\right)\bigg|_{-1}^2 = 9 \text{ sq. units.}$$

7. $\text{Area of } R = 2\int_0^1 (x - x^3)\, dx$

$$= 2\left(\frac{x^2}{2} - \frac{x^4}{4}\right)\bigg|_0^1 = \frac{1}{2} \text{ sq. units.}$$

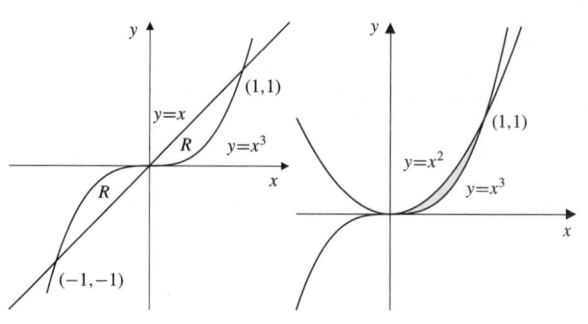

Fig. 5.7.7   Fig. 5.7.8

8. Shaded area $= \int_0^1 (x^2 - x^3)\, dx$

$$= \left(\frac{1}{3}x^3 - \frac{1}{4}x^4\right)\bigg|_0^1 = \frac{1}{12} \text{ sq. units.}$$

9. $\text{Area of } R = \int_0^1 (\sqrt{x} - x^3)\, dx$

$$= \left(\frac{2}{3}x^{3/2} - \frac{x^4}{4}\right)\bigg|_0^1 = \frac{5}{12} \text{ sq. units.}$$

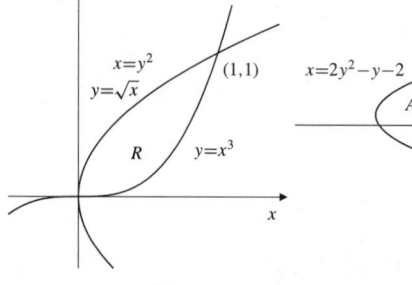

Fig. 5.7.9   Fig. 5.7.10

10. For intersections:
$y^2 = 2y^2 - y - 2 \Rightarrow y^2 - y - 2 = 0$
$(y-2)(y+1) = 0 \Rightarrow$ i.e., $y = -1$ or 2.

$$\text{Area of } R = \int_{-1}^2 [y^2 - (2y^2 - y - 2)]\, dy$$

$$= \int_{-1}^2 [2 + y - y^2]\, dy = \left(2y + \frac{1}{2}y^2 - \frac{1}{3}y^3\right)\bigg|_{-1}^2$$

$$= \frac{9}{2} \text{ sq. units.}$$

INSTRUCTOR'S SOLUTIONS MANUAL SECTION 5.7 (PAGE 342)

11. For intersections: $\dfrac{1}{x} = y = \dfrac{5-2x}{2}$.
Thus $2x^2 - 5x + 2 = 0$, i.e., $(2x-1)(x-2) = 0$. The graphs intersect at $x = 1/2$ and $x = 2$. Thus

$$\text{Area of } R = \int_{1/2}^{2} \left(\frac{5-2x}{2} - \frac{1}{x} \right) dx$$

$$= \left(\frac{5x}{2} - \frac{x^2}{2} - \ln x \right) \Big|_{1/2}^{2}$$

$$= \frac{15}{8} - 2\ln 2 \text{ sq. units.}$$

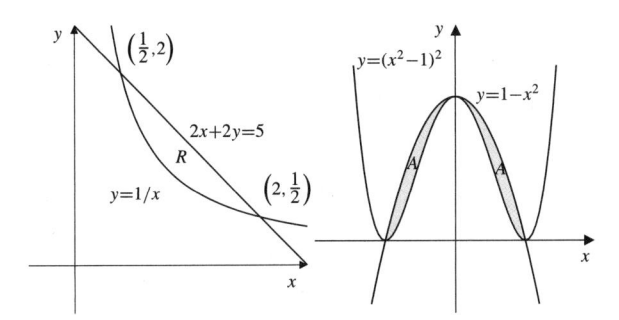

Fig. 5.7.11

Fig. 5.7.12

12. Area of shaded region $= 2 \int_0^1 [(1-x^2) - (x^2-1)^2] \, dx$

$= 2 \int_0^1 (x^2 - x^4) \, dx = 2\left(\dfrac{1}{3}x^3 - \dfrac{1}{5}x^5 \right) \Big|_0^1 = \dfrac{4}{15}$ sq. units.

13. The curves $y = \dfrac{x^2}{2}$ and $y = \dfrac{1}{1+x^2}$ intersect at $x = \pm 1$. Thus

$$\text{Area of } R = 2 \int_0^1 \left(\frac{1}{1+x^2} - \frac{x^2}{2} \right) dx$$

$$= 2 \left(\tan^{-1} x - \frac{x^3}{6} \right) \Big|_0^1 = \frac{\pi}{2} - \frac{1}{3} \text{ sq. units.}$$

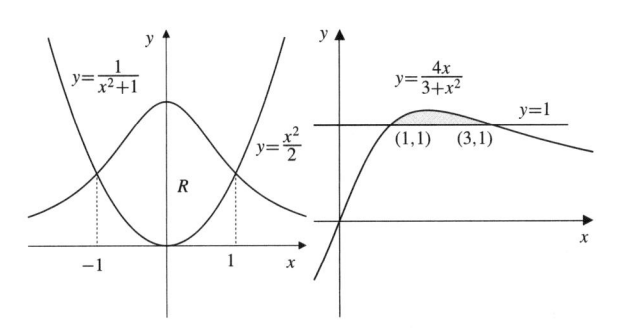

Fig. 5.7.13

Fig. 5.7.14

14. For intersections:
$\dfrac{4x}{3+x^2} = 1 \Rightarrow x^2 - 4x + 3 = 0$
i.e., $x = 1$ or 3.

$$\text{Shaded area} = \int_1^3 \left[\frac{4x}{3+x^2} - 1 \right] dx$$

$$= [2\ln(3+x^2) - x] \Big|_1^3 = 2\ln 3 - 2 \text{ sq. units.}$$

15. The curves $y = \dfrac{4}{x^2}$ and $y = 5 - x^2$ intersect where $x^4 - 5x^2 + 4 = 0$, i.e., where $(x^2-4)(x^2-1) = 0$. Thus the intersections are at $x = \pm 1$ and $x = \pm 2$. We have

$$\text{Area of } R = 2\int_1^2 \left(5 - x^2 - \frac{4}{x^2} \right) dx$$

$$= 2\left(5x - \frac{x^3}{3} + \frac{4}{x} \right) \Big|_1^2 = \frac{4}{3} \text{ sq. units.}$$

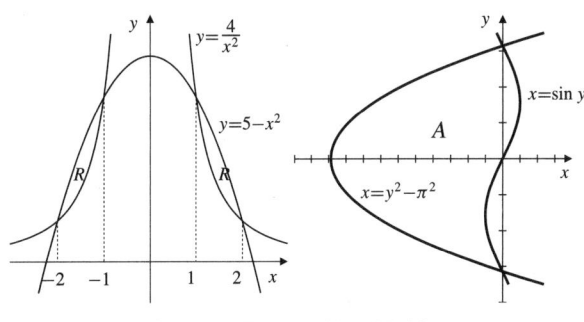

Fig. 5.7.15

Fig. 5.7.16

16. Area $A = \int_{-\pi}^{\pi} (\sin y - (y^2 - \pi^2)) \, dy$

$$= \left(-\cos y + \pi^2 y - \frac{y^3}{3} \right) \Big|_{-\pi}^{\pi} = \frac{4\pi^3}{3} \text{ sq. units.}$$

17. Area of $R = \int_{\pi/4}^{5\pi/4} (\sin x - \cos x) \, dx$

$$= -(\cos x + \sin x) \Big|_{\pi/4}^{5\pi/4}$$

$$= \sqrt{2} + \sqrt{2} = 2\sqrt{2} \text{ sq. units.}$$

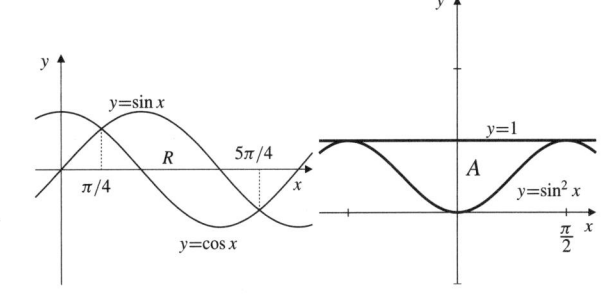

Fig. 5.7.17

Fig. 5.7.18

18. Area $= \int_{-\pi/2}^{\pi/2} (1-\sin^2 x)\,dx$

$= 2\int_0^{\pi/2} \dfrac{1+\cos(2x)}{2}\,dx$

$= \left(x + \dfrac{\sin(2x)}{2}\right)\bigg|_0^{\pi/2} = \dfrac{\pi}{2}$ sq. units.

19. Area $A = \int_0^{\pi/2} (\sin x - \sin^2 x)\,dx$

$= \left(-\cos x + \dfrac{\sin x \cos x - x}{2}\right)\bigg|_0^{\pi/2} = 1 - \dfrac{\pi}{4}$ sq. units.

Fig. 5.7.19

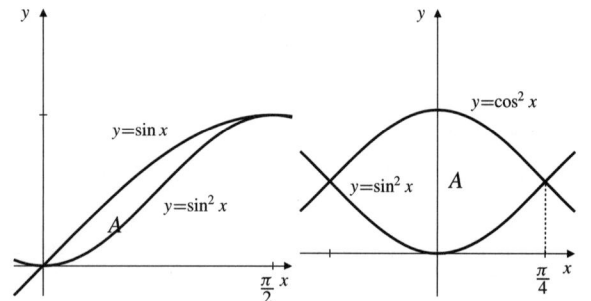

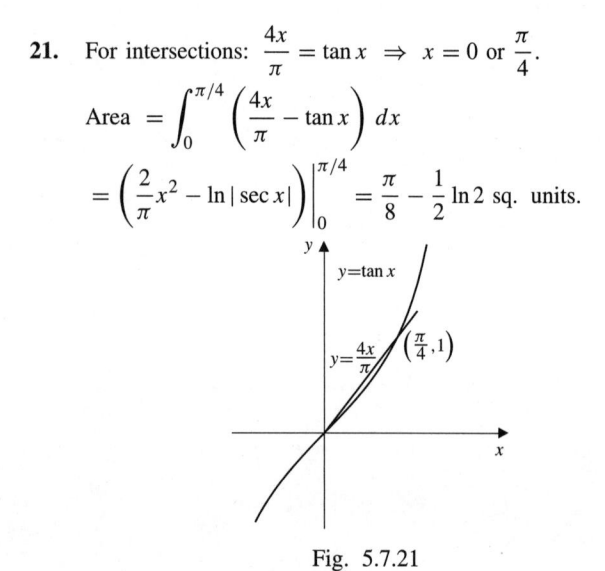

Fig. 5.7.20

20. Area $A = 2\int_0^{\pi/4} (\cos^2 x - \sin^2 x)\,dx$

$= 2\int_0^{\pi/4} \cos(2x)\,dx = \sin(2x)\bigg|_0^{\pi/4} = 1$ sq. units.

21. For intersections: $\dfrac{4x}{\pi} = \tan x \Rightarrow x = 0$ or $\dfrac{\pi}{4}$.

Area $= \int_0^{\pi/4} \left(\dfrac{4x}{\pi} - \tan x\right) dx$

$= \left(\dfrac{2}{\pi}x^2 - \ln|\sec x|\right)\bigg|_0^{\pi/4} = \dfrac{\pi}{8} - \dfrac{1}{2}\ln 2$ sq. units.

Fig. 5.7.21

22. For intersections: $x^{1/3} = \tan(\pi x/4)$. Thus $x = \pm 1$.

Area $A = 2\int_0^1 \left(x^{1/3} - \tan\dfrac{\pi x}{4}\right) dx$

$= 2\left(\dfrac{3}{4}x^{4/3} - \dfrac{4}{\pi}\ln\left|\sec\dfrac{\pi x}{4}\right|\right)\bigg|_0^1$

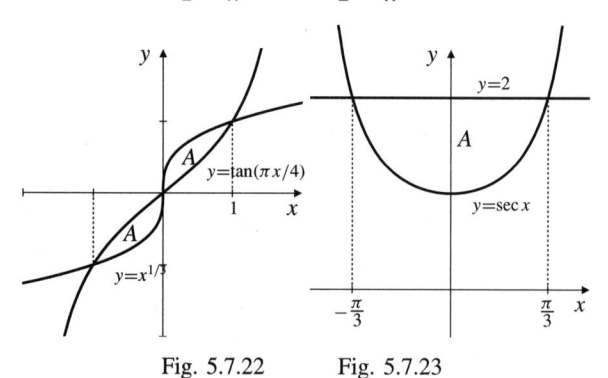

$= \dfrac{3}{2} - \dfrac{8}{\pi}\ln\sqrt{2} = \dfrac{3}{2} - \dfrac{4}{\pi}\ln 2$ sq. units.

Fig. 5.7.22

Fig. 5.7.23

23. For intersections: $\sec x = 2$. Thus $x = \pm \pi/3$.

Area $A = 2\int_0^{\pi/3} (2 - \sec x)\,dx$

$= (4x - 2\ln|\sec x + \tan x|)\bigg|_0^{\pi/3}$

$= \dfrac{4\pi}{3} - 2\ln(2+\sqrt{3})$ sq. units.

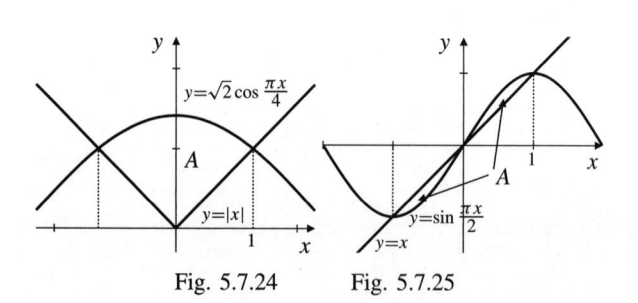

24. For intersections: $|x| = \sqrt{2}\cos(\pi x/4)$. Thus $x = \pm 1$.

Area $A = 2\int_0^1 \left(\sqrt{2}\cos\dfrac{\pi x}{4} - x\right) dx$

$= \left(\dfrac{8\sqrt{2}}{\pi}\sin\dfrac{\pi x}{4} - x^2\right)\bigg|_0^1$

$= \dfrac{8}{\pi} - 1$ sq. units.

Fig. 5.7.24

Fig. 5.7.25

INSTRUCTOR'S SOLUTIONS MANUAL SECTION 5.7 (PAGE 342)

25. For intersections: $x = \sin(\pi x/2)$. Thus $x = \pm 1$.

$$\text{Area } A = 2\int_0^1 \left(\sin\frac{\pi x}{2} - x\right) dx$$
$$= \left(-\frac{4}{\pi}\cos\frac{\pi x}{2} - x^2\right)\bigg|_0^1$$
$$= \frac{4}{\pi} - 1 \text{ sq. units.}$$

26. For intersections: $x^3 = \sin(\pi x/2)$. Thus $x = \pm 1$.

$$\text{Area } A = 2\int_0^1 \left(\sin\frac{\pi x}{2} - x^3\right) dx$$
$$= \left(-\frac{4}{\pi}\cos\frac{\pi x}{2} - \frac{x^4}{2}\right)\bigg|_0^1$$
$$= \frac{4}{\pi} - \frac{1}{2} \text{ sq. units.}$$

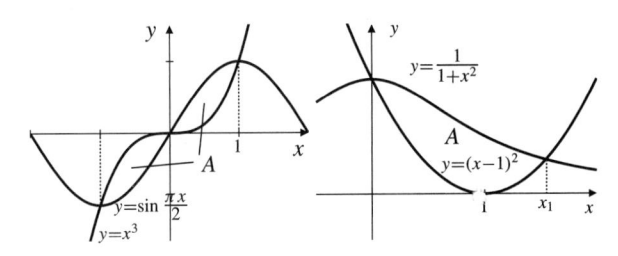

Fig. 5.7.26   Fig. 5.7.27

27. For intersections: $(x-1)^2 = 1/(1+x^2)$. One root is $x = 0$; the other root x_1 must be found numerically. We used a TI-85 solve routine to get $x_1 \approx 1.543689$. Thus

$$\text{Area } A = \int_0^{x_1} \left(\frac{1}{1+x^2} - (x-1)^2\right) dx$$
$$= \left(\tan^{-1}x - \frac{(x-1)^3}{3}\right)\bigg|_0^{x_1}$$
$$\approx 0.6090656 \text{ sq. units.}$$

28. For intersections: $e^x = x + 2$. There are two roots, both of which must be found numerically. We used a TI-85 solve routine to get $x_1 \approx -1.841406$ and $x_2 \approx 1.146193$. Thus

$$\text{Area } A = \int_{x_1}^{x_2} \left(x + 2 - e^x\right) dx$$
$$= \left(\frac{x^2}{2} + 2x - e^x\right)\bigg|_{x_1}^{x_2}$$
$$\approx 1.949091 \text{ sq. units.}$$

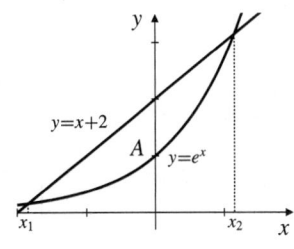

Fig. 5.7.28

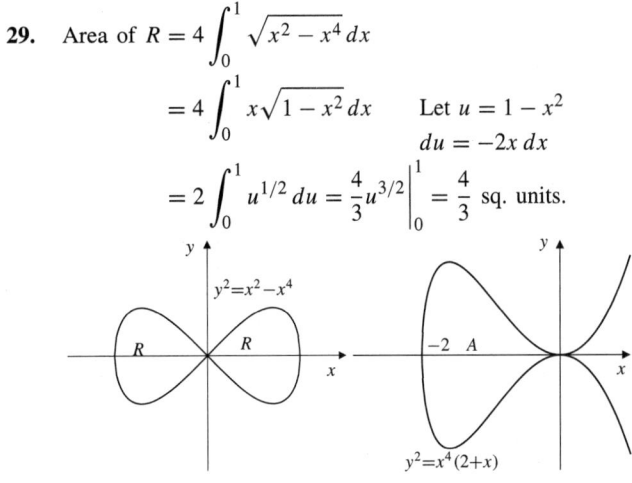

29. Area of $R = 4\int_0^1 \sqrt{x^2 - x^4}\, dx$

$$= 4\int_0^1 x\sqrt{1-x^2}\, dx \quad \text{Let } u = 1 - x^2$$
$$\quad\quad du = -2x\, dx$$
$$= 2\int_0^1 u^{1/2}\, du = \frac{4}{3}u^{3/2}\bigg|_0^1 = \frac{4}{3} \text{ sq. units.}$$

Fig. 5.7.29   Fig. 5.7.30

30. Loop area $= 2\int_{-2}^0 x^2\sqrt{2+x}\, dx \quad \text{Let } u^2 = 2 + x$
$$\quad\quad 2u\, du = dx$$
$$= 2\int_0^{\sqrt{2}} (u^2 - 2)^2 u(2u)\, du = 4\int_0^{\sqrt{2}} (u^6 - 4u^4 + 4u^2)\, du$$
$$= 4\left(\frac{1}{7}u^7 - \frac{4}{5}u^5 + \frac{4}{3}u^3\right)\bigg|_0^{\sqrt{2}} = \frac{256\sqrt{2}}{105} \text{ sq. units.}$$

31. The tangent line to $y = e^x$ at $x = 1$ is $y - e = e(x - 1)$, or $y = ex$. Thus

$$\text{Area of } R = \int_0^1 (e^x - ex)\, dx$$
$$= \left(e^x - \frac{ex^2}{2}\right)\bigg|_0^1 = \frac{e}{2} - 1 \text{ sq. units.}$$

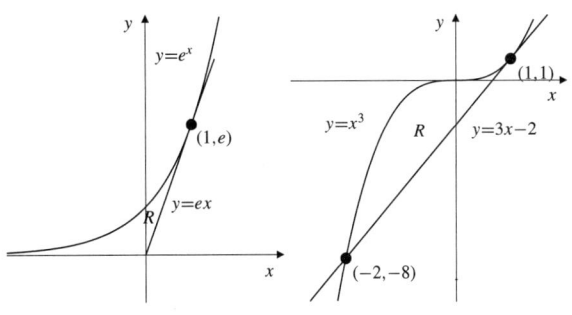

Fig. 5.7.31   Fig. 5.7.32

197

32. The tangent line to $y = x^3$ at $(1, 1)$ is $y - 1 = 3(x - 1)$, or $y = 3x - 2$. The intersections of $y = x^3$ and this tangent line occur where $x^3 - 3x + 2 = 0$. Of course $x = 1$ is a (double) root of this cubic equation, which therefore factors to $(x - 1)^2(x + 2) = 0$. The other intersection is at $x = -2$. Thus

$$\text{Area of } R = \int_{-2}^{1} (x^3 - 3x + 2) \, dx$$

$$= \left(\frac{x^4}{4} - \frac{3x^2}{2} + 2x \right) \Big|_{-2}^{1}$$

$$= -\frac{15}{4} - \frac{3}{2} + 6 + 2 + 4 = \frac{27}{4} \text{ sq. units.}$$

Review Exercises 5 (page 343)

1. $\dfrac{1}{j^2} - \dfrac{1}{(j+1)^2} = \dfrac{j^2 + 2j + 1 - j^2}{j^2(j+1)^2} = \dfrac{2j+1}{j^2(j+1)^2}$

$$\sum_{j=1}^{n} \frac{2j+1}{j^2(j+1)^2} = \sum_{j=1}^{n} \left(\frac{1}{j^2} - \frac{1}{(j+1)^2} \right)$$

$$= \frac{1}{1^2} - \frac{1}{(n+1)^2} = \frac{n^2 + 2n}{(n+1)^2}$$

2. The number of balls is

$$40 \times 30 + 39 \times 29 + \cdots + 12 \times 2 + 11 \times 1$$

$$= \sum_{i=1}^{30} i(i+10) = \frac{(30)(31)(61)}{6} + 10 \frac{(30)(31)}{2} = 14,105.$$

3. $x_i = 1 + (2i/n)$, $(i = 0, 1, 2, \ldots, n)$, $\Delta x_i = 2/n$.

$$\int_1^3 f(x) \, dx = \lim_{n \to \infty} \sum_{i=1}^{n} (x_i^2 - 2x_i + 3) \frac{2}{n}$$

$$= \lim_{n \to \infty} \frac{2}{n} \sum_{i=1}^{n} \left[\left(1 + \frac{4i}{n} + \frac{4i^2}{n^2} \right) - \left(2 + \frac{4i}{n} \right) + 3 \right]$$

$$= \lim_{n \to \infty} \frac{2}{n} \sum_{i=1}^{n} \left[2 + \frac{4}{n^2} i^2 \right]$$

$$= \lim_{n \to \infty} \left(\frac{4}{n} n + \frac{8}{n^3} \frac{n(n+1)(2n+1)}{6} \right)$$

$$= 4 + \frac{8}{3} = \frac{20}{3}$$

4. $R_n = \sum_{i=1}^{n} (1/n) \sqrt{1 + (i/n)}$ is a Riemann sum for $f(x) = \sqrt{1+x}$ on the interval $[0, 1]$. Thus

$$\lim_{n \to \infty} R_n = \int_0^1 \sqrt{1+x} \, dx$$

$$= \frac{2}{3}(1+x)^{3/2} \Big|_0^1 = \frac{4\sqrt{2} - 2}{3}.$$

5. $\displaystyle\int_{-\pi}^{\pi} (2 - \sin x) \, dx = 2(2\pi) - \int_{-\pi}^{\pi} \sin x \, dx = 4\pi - 0 = 4\pi$

6. $\displaystyle\int_0^{\sqrt{5}} \sqrt{5 - x^2} \, dx = 1/4$ of the area of a circle of radius $\sqrt{5}$

$$= \frac{1}{4} \pi (\sqrt{5})^2 = \frac{5\pi}{4}$$

7. $\displaystyle\int_1^3 \left(1 - \frac{x}{2} \right) dx = \text{area } A_1 - \text{area } A_2 = 0$

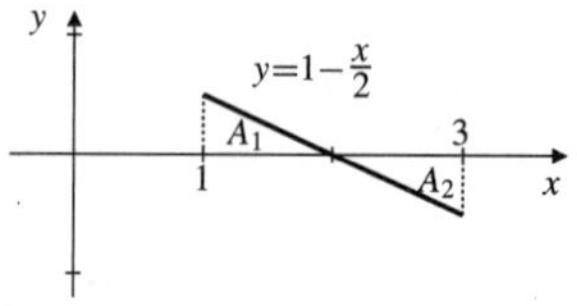

Fig. R-5.7 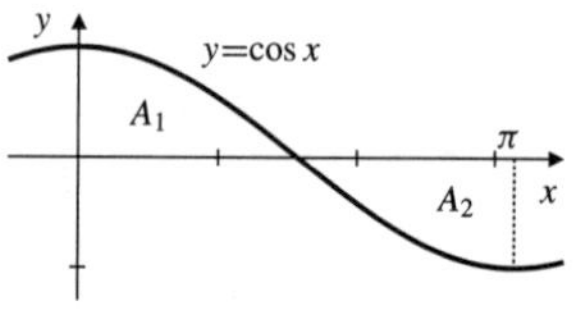   Fig. R-5.8

8. $\displaystyle\int_0^{\pi} \cos x \, dx = \text{area } A_1 - \text{area } A_2 = 0$

9. $\bar{f} = \dfrac{1}{2\pi} \displaystyle\int_{-\pi}^{\pi} (2 - \sin(x^3)) \, dx = \dfrac{1}{2\pi}[2(2\pi) - 0] = 2$

10. $\bar{h} = \dfrac{1}{3} \displaystyle\int_0^3 |x - 2| \, dx = \dfrac{1}{3} \dfrac{5}{2} = \dfrac{5}{6}$ (via #9)

11. $f(t) = \displaystyle\int_{13}^{t} \sin(x^2) \, dx$, $f'(t) = \sin(t^2)$

12. $f(x) = \displaystyle\int_{-13}^{\sin x} \sqrt{1 + t^2} \, dt$, $f'(x) = \sqrt{1 + \sin^2 x} (\cos x)$

13. $g(s) = \displaystyle\int_{4s}^{1} e^{\sin u} \, du$, $g'(s) = -4 e^{\sin(4s)}$

14. $g(\theta) = \displaystyle\int_{e^{\sin \theta}}^{e^{\cos \theta}} \ln x \, dx$

$g'(\theta) = (\ln(e^{\cos \theta})) e^{\cos \theta} (-\sin \theta) - (\ln(e^{\sin \theta})) e^{\sin \theta} \cos \theta$

$= -\sin \theta \cos \theta (e^{\cos \theta} + e^{\sin \theta})$

15. $2f(x) + 1 = 3 \int_x^1 f(t)\,dt$

$2f'(x) = -3f(x) \implies f(x) = Ce^{-3x/2}$

$2f(1) + 1 = 0$

$-\frac{1}{2} = f(1) = Ce^{-3/2} \implies C = -\frac{1}{2}e^{3/2}$

$f(x) = -\frac{1}{2}e^{(3/2)(1-x)}$.

16. $I = \int_0^\pi xf(\sin x)\,dx$ Let $x = \pi - u$

$dx = -du$

$= -\int_\pi^0 (\pi - u)f(\sin(\pi - u))\,du$ (but $\sin(\pi - u) = \sin u$)

$= \pi \int_0^\pi f(\sin u)\,du - \int_0^\pi uf(\sin u)\,du$

$= \pi \int_0^\pi f(\sin x)\,dx - I$.

Now, solving for I, we get

$\int_0^\pi xf(\sin x)\,dx = I = \frac{\pi}{2}\int_0^\pi f(\sin x)\,dx$.

17. $y = 2 + x - x^2$ and $y = 0$ intersect where $2 + x - x^2 = 0$, that is, where $(2-x)(1+x) = 0$, namely at $x = -1$ and $x = 2$. Since $2 + x - x^2 \geq 0$ on $[-1, 2]$, the required area is

$\int_{-1}^2 (2 + x - x^2)\,dx = \left(2x + \frac{x^2}{2} - \frac{x^3}{3}\right)\Big|_{-1}^2 = \frac{9}{2}$ sq. units..

18. The area bounded by $y = (x-1)^2$, $y = 0$, and $x = 0$ is

$\int_0^1 (x-1)^2\,dx = \frac{(x-1)^3}{3}\Big|_0^1 = \frac{1}{3}$ sq. units..

19. $x = y - y^4$ and $x = 0$ intersect where $y - y^4 = 0$, that is, at $y = 0$ and $y = 1$. Since $y - y^4 \geq 0$ on $[0, 1]$, the required area is

$\int_0^1 (y - y^4 - 0)\,dy = \left(\frac{y^2}{2} - \frac{y^5}{5}\right)\Big|_0^1 = \frac{3}{10}$ sq. units.

20. $y = 4x - x^2$ and $y = 3$ meet where $x^2 - 4x + 3 = 0$, that is, at $x = 1$ and $x = 3$. Since $4x - x^2 \geq 3$ on $[1, 3]$, the required area is

$\int_1^3 (4x - x^2 - 3)\,dx = \left(2x^2 - \frac{x^3}{3} - 3x\right)\Big|_1^3 = \frac{4}{3}$ sq. units.

21. $y = \sin x$ and $y = \cos(2x)$ intersect at $x = \pi/6$, but nowhere else in the interval $[0, \pi/6]$. The area between the curves in that interval is

$\int_0^{\pi/6} (\cos(2x) - \sin x)\,dx = \left(\frac{1}{2}\sin(2x) + \cos x\right)\Big|_0^{\pi/6}$

$= \frac{\sqrt{3}}{4} + \frac{\sqrt{3}}{2} - 1 = \frac{3\sqrt{3}}{4} - 1$ sq. units..

22. $y = 5 - x^2$ and $y = 4/x^2$ meet where $5 - x^2 = 4/x^2$, that is, where

$x^4 - 5x^2 + 4 = 0$

$(x^2 - 1)(x^2 - 4) = 0$.

There are four intersections: $x = \pm 1$ and $x = \pm 2$. By symmetry (see the figure) the total area bounded by the curves is

$2\int_1^2 \left(5 - x^2 - \frac{4}{x^2}\right)dx = 2\left(5x - \frac{x^3}{3} + \frac{4}{x}\right)\Big|_1^2 = \frac{4}{3}$ sq. unit

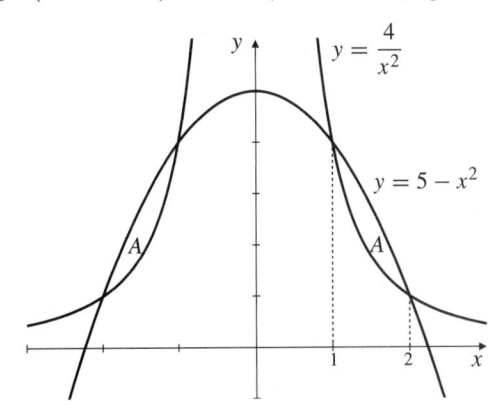

Fig. R-5.22

23. $\int x^2 \cos(2x^3 + 1)\,dx$ Let $u = 2x^3 + 1$

$du = 6x^2\,dx$

$= \frac{1}{6}\int \cos u\,du = \frac{\sin u}{6} + C = \frac{\sin(2x^3 + 1)}{6} + C$

24. $\int_1^e \frac{\ln x}{x}\,dx$ Let $u = \ln x$

$du = dx/x$

$= \int_0^1 u\,du = \frac{u^2}{2}\Big|_0^1 = \frac{1}{2}$

25. $\int_0^4 \sqrt{9t^2 + t^4}\,dt$

$= \int_0^4 t\sqrt{9 + t^2}\,dt$ Let $u = 9 + t^2$

$du = 2t\,dt$

$= \frac{1}{2}\int_9^{25} \sqrt{u}\,du = \frac{1}{3}u^{3/2}\Big|_9^{25} = \frac{98}{3}$

26. $\int \sin^3(\pi x)\,dx$

$= \int \sin(\pi x)\bigl(1 - \cos^2(\pi x)\bigr)\,dx$ Let $u = \cos(\pi x)$
$\qquad\qquad\qquad\qquad\qquad\qquad du = -\pi \sin(\pi x)\,dx$

$= -\dfrac{1}{\pi}\int (1 - u^2)\,du$

$= \dfrac{1}{\pi}\left(\dfrac{u^3}{3} - u\right) + C = \dfrac{1}{3\pi}\cos^3(\pi x) - \dfrac{1}{\pi}\cos(\pi x) + C$

27. $\displaystyle\int_0^{\ln 2} \dfrac{e^u}{4 + e^{2u}}\,du$ Let $v = e^u$
$\qquad\qquad\qquad\qquad\qquad dv = e^u\,du$

$= \displaystyle\int_1^2 \dfrac{dv}{4 + v^2}$

$= \dfrac{1}{2}\tan^{-1}\dfrac{v}{2}\bigg|_1^2 = \dfrac{\pi}{8} - \dfrac{1}{2}\tan^{-1}\dfrac{1}{2}$

28. $\displaystyle\int_1^{\sqrt[4]{e}} \dfrac{\tan^2(\pi \ln x)}{x}\,dx$ Let $u = \pi \ln x$
$\qquad\qquad\qquad\qquad\qquad\qquad du = (\pi/x)\,dx$

$= \dfrac{1}{\pi}\displaystyle\int_0^{\pi/4} \tan^2 u\,du = \dfrac{1}{\pi}\displaystyle\int_0^{\pi/4}(\sec^2 u - 1)\,du$

$= \dfrac{1}{\pi}(\tan u - u)\bigg|_0^{\pi/4} = \dfrac{1}{\pi} - \dfrac{1}{4}$

29. $\displaystyle\int \dfrac{\sin\sqrt{2s+1}}{\sqrt{2s+1}}\,ds$ Let $u = \sqrt{2s+1}$
$\qquad\qquad\qquad\qquad\qquad\qquad du = ds/\sqrt{2s+1}$

$= \displaystyle\int \sin u\,du = -\cos u + C = -\cos\sqrt{2s+1} + C$

30. $\displaystyle\int \cos^2\dfrac{t}{5}\sin^2\dfrac{t}{5}\,dt = \dfrac{1}{4}\int \sin^2\dfrac{2t}{5}\,dt$

$= \dfrac{1}{8}\displaystyle\int\left(1 - \cos\dfrac{4t}{5}\right)dt$

$= \dfrac{1}{8}\left(t - \dfrac{5}{4}\sin\dfrac{4t}{5}\right) + C$

31. $F(x) = \displaystyle\int_0^{x^2 - 2x} \dfrac{1}{1 + t^2}\,dt.$

Since $1/(1 + t^2) > 0$ for all t, $F(x)$ will be minimum when
$$x^2 - 2x = (x - 1)^2 - 1$$
is minimum, that is, when $x = 1$. The minimum value is

$$F(1) = \int_0^{-1} \dfrac{dt}{1 + t^2} = \tan^{-1} t\bigg|_0^{-1} = -\dfrac{\pi}{4}.$$

F has no maximum value; $F(x) < \pi/2$ for all x, but $F(x) \to \pi/2$ if $x^2 - 2x \to \infty$, which happens as $x \to \pm\infty$.

32. $f(x) = 4x - x^2 \geq 0$ if $0 \leq x \leq 4$, and $f(x) < 0$ otherwise. If $a < b$, then $\int_a^b f(x)\,dx$ will be maximum if $[a, b] = [0, 4]$; extending the interval to the left of 0 or to the right of 4 will introduce negative contributions to the integral. The maximum value is

$$\int_0^4 (4x - x^2)\,dx = \left(2x^2 - \dfrac{x^3}{3}\right)\bigg|_0^4 = \dfrac{32}{3}.$$

33. The average value of $v(t) = dx/dt$ over $[t_0, t_1]$ is

$$\dfrac{1}{t_1 - t_0}\int_{t_0}^{t_1} \dfrac{dx}{dt}\,dt = \dfrac{1}{t_1 - t_0}x(t)\bigg|_{t_0}^{t_1} = \dfrac{x(t_1) - x(t_0)}{t_1 - t_0} = v_{\text{av}}.$$

34. If $y(t)$ is the distance the object falls in t seconds from its release time, then

$$y''(t) = g, \quad y(0) = 0, \quad \text{and } y'(0) = 0.$$

Antidifferentiating twice and using the initial conditions leads to

$$y(t) = \dfrac{1}{2}gt^2.$$

The average height during the time interval $[0, T]$ is

$$\dfrac{1}{T}\int_0^T \dfrac{1}{2}gt^2\,dt = \dfrac{g}{2T}\dfrac{T^3}{3} = \dfrac{gT^2}{6} = y\left(\dfrac{T}{\sqrt{3}}\right).$$

Challenging Problems 5 (page 344)

1. $x_i = 2^{i/n}$, $0 \leq i \leq n$, $f(x) = 1/x$ on $[1, 2]$. Since f is decreasing, f is largest at the left endpoint and smallest at the right endpoint of any interval $[2^{(i-1)/n}, 2^{i/n}]$ of the partition. Thus

$$U(f, P_n) = \sum_{i=1}^n \dfrac{1}{2^{(i-1)/n}}(2^{i/n} - 2^{(i-1)/n})$$

$$= \sum_{i=1}^n (2^{1/n} - 1) = n(2^{1/n} - 1)$$

$$L(f, P_n) = \sum_{i=1}^n \dfrac{1}{2^{i/n}}(2^{i/n} - 2^{(i-1)/n})$$

$$= \sum_{i=1}^n (1 - 2^{-1/n}) = n(1 - 2^{-1/n}) = \dfrac{U(f, P_n)}{2^{1/n}}.$$

Now, by l'Hôpital's rule,

$$\lim_{n\to\infty} n(2^{1/n} - 1) = \lim_{x\to\infty} \dfrac{2^{1/x} - 1}{1/x} \quad \left[\dfrac{0}{0}\right]$$

$$= \lim_{x\to\infty} \dfrac{2^{1/x} \ln 2\,(-1/x^2)}{-1/x^2} = \ln 2.$$

INSTRUCTOR'S SOLUTIONS MANUAL CHALLENGING PROBLEMS 5 (PAGE 344)

Thus $\lim_{n\to\infty} U(f, P_n) = \lim_{n\to\infty} L(f, P_n) = \ln s$.

2. a) $\cos\big((j+\tfrac{1}{2})t\big) - \cos\big((j-\tfrac{1}{2})t\big)$
$= \cos(jt)\cos(\tfrac{1}{2}t) - \sin(jt)\sin(\tfrac{1}{2}t)$
$\quad - \cos(jt)\cos(\tfrac{1}{2}t) - \sin(jt)\sin(\tfrac{1}{2}t)$
$= -2\sin(jt)\sin(\tfrac{1}{2}t)$.
Therefore, we obtain a telescoping sum:

$\sum_{j=1}^{n} \sin(jt)$
$= -\dfrac{1}{2\sin(\tfrac{1}{2}t)} \sum_{j=1}^{n} \Big[\cos\big((j+\tfrac{1}{2})t\big) - \cos\big((j-\tfrac{1}{2})t\big)\Big]$
$= -\dfrac{1}{2\sin(\tfrac{1}{2}t)} \Big[\cos\big((n+\tfrac{1}{2})t\big) - \cos(\tfrac{1}{2}t)\Big]$
$= \dfrac{1}{2\sin(\tfrac{1}{2}t)} \Big[\cos(\tfrac{1}{2}t) - \cos\big((n+\tfrac{1}{2})t\big)\Big]$.

b) Let $P_n = \{0, \tfrac{\pi}{2n}, \tfrac{2\pi}{2n}, \tfrac{3\pi}{2n}, \ldots \tfrac{n\pi}{2n}\}$ be the partition of $[0, \pi/2]$ into n subintervals of equal length $\Delta x = \pi/2n$. Using $t = \pi/2n$ in the formula obtained in part (a), we get

$\int_0^{\pi/2} \sin x\, dx$
$= \lim_{n\to\infty} \sum_{j=1}^{n} \sin\left(\dfrac{j\pi}{2n}\right) \dfrac{\pi}{2n}$
$= \lim_{n\to\infty} \dfrac{\pi}{2n} \dfrac{1}{2\sin(\pi/(4n))} \left(\cos\dfrac{\pi}{4n} - \cos\dfrac{(2n+1)\pi}{4n}\right)$
$= \lim_{n\to\infty} \dfrac{\pi/(4n)}{\sin(\pi/(4n))} \lim_{n\to\infty} \left(\cos\dfrac{\pi}{4n} - \cos\dfrac{(2n+1)\pi}{4n}\right)$
$= 1 \times \left(\cos 0 - \cos\dfrac{\pi}{2}\right) = 1$.

3. a) $\sin\big((j+\tfrac{1}{2})t\big) - \sin\big((j-\tfrac{1}{2})t\big)$
$= \sin(jt)\cos(\tfrac{1}{2}t) + \cos(jt)\sin(\tfrac{1}{2}t)$
$\quad - \sin(jt)\cos(\tfrac{1}{2}t) + \cos(jt)\sin(\tfrac{1}{2}t)$
$= 2\cos(jt)\sin(\tfrac{1}{2}t)$.
Therefore, we obtain a telescoping sum:

$\sum_{j=1}^{n} \cos(jt)$
$= \dfrac{1}{2\sin(\tfrac{1}{2}t)} \sum_{j=1}^{n} \Big[\sin\big((j+\tfrac{1}{2})t\big) - \sin\big((j-\tfrac{1}{2})t\big)\Big]$
$= \dfrac{1}{2\sin(\tfrac{1}{2}t)} \Big[\sin\big((n+\tfrac{1}{2})t\big) - \sin(\tfrac{1}{2}t)\Big]$.

b) Let $P_n = \{0, \tfrac{\pi}{3n}, \tfrac{2\pi}{3n}, \tfrac{3\pi}{3n}, \ldots \tfrac{n\pi}{3n}\}$ be the partition of $[0, \pi/3]$ into n subintervals of equal length $\Delta x = \pi/3n$. Using $t = \pi/3n$ in the formula obtained in part (a), we get

$\int_0^{\pi/2} \cos x\, dx$
$= \lim_{n\to\infty} \sum_{j=1}^{n} \cos\left(\dfrac{j\pi}{3n}\right) \dfrac{\pi}{3n}$
$= \lim_{n\to\infty} \dfrac{\pi}{3n} \dfrac{1}{2\sin(\pi/(6n))} \left(\sin\dfrac{(2n+1)\pi}{6n} - \sin\dfrac{\pi}{6n}\right)$
$= \lim_{n\to\infty} \dfrac{\pi/(6n)}{\sin(\pi/(6n))} \lim_{n\to\infty} \left(\sin\dfrac{(2n+1)\pi}{6n} - \sin\dfrac{\pi}{6n}\right)$
$= 1 \times \left(\sin\dfrac{\pi}{3} - \sin 0\right) = \dfrac{\sqrt{3}}{2}$.

4. $f(x) = 1/x^2$, $1 = x_0 < x_1 < x_2 < \cdots < x_n = 2$. If $c_i = \sqrt{x_{i-1}x_i}$, then
$$x_{i-1}^2 < x_{i-1}x_i = c_i^2 < x_i^2,$$
so $x_{i-1} < c_i < x_i$. We have

$\sum_{i=1}^{n} f(c_i)\,\Delta x_i = \sum_{i=1}^{n} \dfrac{1}{x_{i-1}x_i}(x_i - x_{i-1})$
$= \sum_{i=1}^{n} \left(\dfrac{1}{x_{i-1}} - \dfrac{1}{x_i}\right)$ (telescoping)
$= \dfrac{1}{x_0} - \dfrac{1}{x_n} = 1 - \dfrac{1}{2} = \dfrac{1}{2}$.

Thus $\displaystyle\int_1^2 \dfrac{dx}{x^2} = \lim_{n\to\infty} \sum_{i=1}^{n} f(c_i)\,\Delta x_i = \dfrac{1}{2}$.

5. We want to prove that for each positive integer k,
$$\sum_{j=1}^{n} j^k = \dfrac{n^{k+1}}{k+1} + \dfrac{n^k}{2} + P_{k-1}(n),$$
where P_{k-1} is a polynomial of degree at most $k-1$. First check the case $k = 1$:
$$\sum_{j=1}^{n} j = \dfrac{n(n+1)}{2} = \dfrac{n^{1+1}}{1+1} + \dfrac{n}{2} + P_0(n),$$
where $P_0(n) = 0$ certainly has degree ≤ 0. Now assume that the formula above holds for $k = 1, 2, 3, \ldots, m$. We will show that it also holds for $k = m+1$. To this end, sum the the formula
$$(j+1)^{m+2} - j^{m+2} = (m+2)j^{m+1} + \dfrac{(m+2)(m+1)}{2} j^m + \cdots + 1$$

(obtained by the Binomial Theorem) for $j = 1, 2, \ldots, n$. The left side telescopes, and we get

$$(n+1)^{m+2} - 1^{m+2} = (m+2) \sum_{j=1}^{n} j^{m+1}$$
$$+ \frac{(m+2)(m+1)}{2} \sum_{j=1}^{n} j^m + \cdots + \sum_{j=1}^{n} 1.$$

Expanding the binomial power on the left and using the induction hypothesis on the other terms we get

$$n^{m+2} + (m+2)n^{m+1} + \cdots = (m+2) \sum_{j=1}^{n} j^{m+1}$$
$$+ \frac{(m+2)(m+1)}{2} \frac{n^{m+1}}{m+1} + \cdots,$$

where the $\cdots$ represent terms of degree m or lower in the variable n. Solving for the remaining sum, we get

$$\sum_{j=1}^{n} j^{m+1}$$
$$= \frac{1}{m+2} \left(n^{m+2} + (m+2)n^{m+1} + \cdots - \frac{m+2}{2} n^{m+1} - \cdots \right)$$
$$= \frac{n^{m+2}}{m+2} + \frac{n^{m+1}}{2} + \cdots$$

so that the formula is also correct for $k = m+1$. Hence it is true for all positive integers k by induction.

b) Using the technique of Example 2 in Section 6.2 and the result above,

$$\int_0^a x^k \, dx = \lim_{n \to \infty} \frac{a}{n} \sum_{j=1}^{n} \left(\frac{a}{n} \right)^j$$
$$= a^{k+1} \lim_{n \to \infty} \frac{1}{n^{k+1}} \sum_{j=1}^{n} j^k$$
$$= a^{k+1} \lim_{n \to \infty} \left(\frac{1}{k+1} + \frac{1}{2n} + \frac{P_{k-1}(n)}{n^{k+1}} \right)$$
$$= \frac{a^{k+1}}{k+1}.$$

6. Let $f(x) = ax^3 + bx^2 + cx + d$. We used Maple to calculate the following:

The tangent to $y = f(x)$ at $P = (p, f(p))$ has equation

$$y = g(x) = ap^3 + bp^2 + cp + d + (3ap^2 + 2bp + c)(x - p).$$

This line intersects $y = f(x)$ at $x = p$ (double root) and at $x = q$, where

$$q = -\frac{2ap + b}{a}.$$

Similarly, the tangent to $y = f(x)$ at $x = q$ has equation

$$y = h(x) = aq^3 + bq^2 + cq + d + (3aq^2 + 2bq + c)(x - q),$$

and intersects $y = f(x)$ at $x = q$ (double root) and $x = r$, where

$$r = -\frac{2aq + b}{a} = \frac{4ap + b}{a}.$$

The area between $y = f(x)$ and the tangent line at P is the absolute value of

$$\int_p^q (f(x) - g(x)) \, dx$$
$$= -\frac{1}{12} \left(\frac{81a^4 p^4 + 108a^3 b p^3 + 54a^2 b^2 p^2 + 12ab^3 p + b^4}{a^3} \right)$$

The area between $y = f(x)$ and the tangent line at $Q = (q, f(q))$ is the absolute value of

$$\int_q^r (f(x) - h(x)) \, dx$$
$$= -\frac{4}{3} \left(\frac{81a^4 p^4 + 108a^3 b p^3 + 54a^2 b^2 p^2 + 12ab^3 p + b^4}{a^3} \right),$$

which is 16 times the area between $y = f(x)$ and the tangent at P.

7. We continue with the calculations begun in the previous problem. P and Q are as they were in that problem, but $R = (r, f(r))$ is now the inflection point of $y = f(x)$, given by $f''(r) = 0$. Maple gives

$$r = -\frac{b}{3a}.$$

Since

$$p - r = \frac{b + 3ap}{a} \quad \text{and} \quad r - q = \frac{2(b + 3ap)}{a}$$

have the same sign, R must lie between Q and P on the curve $y = f(x)$. The line QR has a rather complicated equation $y = k(x)$, which we won't reproduce here, but the area between this line and the curve $y = f(x)$ is the absolute value of $\int_r^q (f(x) - k(x)) \, dx$, which Maple evaluates to be

$$-\frac{4}{81} \left(\frac{81a^4 p^4 + 108a^3 b p^3 + 54a^2 b^2 p^2 + 12ab^3 p + b^4}{a^3} \right),$$

which is 16/27 of the area between the curve and its tangent at P. This leaves 11/27 of that area to lie between the curve, QR, and the tangent, so QR divides the area between $y = f(x)$ and its tangent at P in the ratio 16/11.

INSTRUCTOR'S SOLUTIONS MANUAL — CHALLENGING PROBLEMS 5 (PAGE 344)

8. Let $f(x) = ax^4 + bx^3 + cx^2 + dx + e$. The tangent to $y = f(x)$ at $P = (p, f(p))$ has equation

$$y = g(x) = ap^4 + bp^3 + cp^2 + dp + e + (4ap^3 + 3bp^2 + 2cp + d)(x-p),$$

and intersects $y = f(x)$ at $x = p$ (double root) and at the two points

$$x = \frac{-2ap - b \pm \sqrt{b^2 - 4ac - 4abp - 8a^2p^2}}{2a}.$$

If these latter two points coincide, then the tangent is a "double tangent." This happens if

$$8a^2p^2 + 4abp + 4ac - b^2 = 0,$$

which has two solutions, which we take to be p and q:

$$p = \frac{-b + \sqrt{3b^2 - 8ac}}{4a}$$

$$q = \frac{-b - \sqrt{3b^2 - 8ac}}{4a} = -p - \frac{b}{2a}.$$

(Both roots exist and are distinct provided $3b^2 > 8ac$.)
The point T corresponds to $x = t = (p+q)/2 = -b/4a$.
The tangent to $y = f(x)$ at $x = t$ has equation

$$y = h(x) = -\frac{3b^4}{256a^3} + \frac{b^2c}{16a^2} - \frac{bd}{4a} + e + \left(\frac{b^3}{8a^2} - \frac{bc}{2a} + d\right)\left(x + \frac{b}{4a}\right)$$

and it intersects $y = f(x)$ at the points U and V with x-coordinates

$$u = \frac{-b - \sqrt{2}\sqrt{3b^2 - 8ac}}{4a},$$

$$v = \frac{-b + \sqrt{2}\sqrt{3b^2 - 8ac}}{4a}.$$

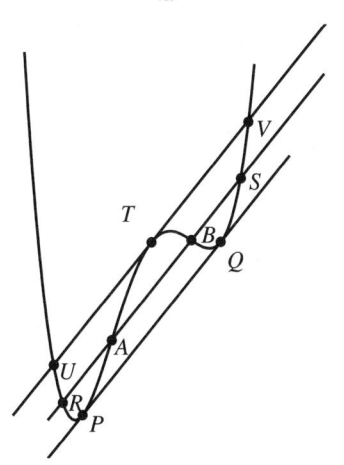

Fig. C-5.8

a) The areas between the curve $y = f(x)$ and the lines PQ and UV are, respectively, the absolute values of

$$A_1 = \int_p^q (f(x) - g(x))\, dx \quad \text{and} \quad A_2 = \int_u^v (h(x) - f(x))\, dx$$

Maple calculates these two integrals and simplifies the ratio A_1/A_2 to be $1/\sqrt{2}$.

b) The two inflection points A and B of f have x-coordinates shown by Maple to be

$$\alpha = \frac{-3b - \sqrt{3(3b^2 - 8ac)}}{12a} \quad \text{and}$$

$$\beta = \frac{-3b + \sqrt{3(3b^2 - 8ac)}}{12a}.$$

It then determines the four points of intersection of the line $y = k(x)$ through these inflection points and the curve. The other two points have x-coordinates

$$r = \frac{-3b - \sqrt{15(3b^2 - 8ac)}}{12a} \quad \text{and}$$

$$s = \frac{-3b + \sqrt{15(3b^2 - 8ac)}}{12a}.$$

The region bounded by RS and the curve $y = f(x)$ is divided into three parts by A and B. The areas of these three regions are the absolute values of

$$A_1 = \int_r^\alpha (k(x) - f(x))\, dx$$

$$A_2 = \int_\alpha^\beta (f(x) - k(x))\, dx$$

$$A_3 = \int_\beta^s (k(x) - f(x))\, dx.$$

The expressions calculated by Maple for $k(x)$ and for these three areas are very complicated, but Maple simplifies the rations A_3/A_1 and A_2/A_1 to 1 and 2 respectively, as was to be shown.

CHAPTER 6. TECHNIQUES OF INTEGRATION

Section 6.1 Integration by Parts (page 351)

1. $\int x \cos x \, dx$

$$U = x \quad dV = \cos x \, dx$$
$$dU = dx \quad V = \sin x$$
$$= x \sin x - \int \sin x \, dx$$
$$= x \sin x + \cos x + C.$$

2. $\int (x+3)e^{2x} \, dx$

$$U = x+3 \quad dV = e^{2x} \, dx$$
$$dU = dx \quad V = \tfrac{1}{2} e^{2x}$$
$$= \frac{1}{2}(x+3)e^{2x} - \frac{1}{2}\int e^{2x} \, dx$$
$$= \frac{1}{2}(x+3)e^{2x} - \frac{1}{4} e^{2x} + C.$$

3. $\int x^2 \cos \pi x \, dx$

$$U = x^2 \quad dV = \cos \pi x \, dx$$
$$dU = 2x \, dx \quad V = \frac{\sin \pi x}{\pi}$$
$$= \frac{x^2 \sin \pi x}{\pi} - \frac{2}{\pi} \int x \sin \pi x \, dx$$
$$U = x \quad dV = \sin \pi x \, dx$$
$$dU = dx \quad V = -\frac{\cos \pi x}{\pi}$$
$$= \frac{x^2 \sin \pi x}{\pi} - \frac{2}{\pi}\left(-\frac{x \cos \pi x}{\pi} + \frac{1}{\pi}\int \cos \pi x \, dx\right)$$
$$= \frac{1}{\pi} x^2 \sin \pi x + \frac{2}{\pi^2} x \cos \pi x - \frac{2}{\pi^3} \sin \pi x + C.$$

4. $\int (x^2 - 2x) e^{kx} \, dx$

$$U = x^2 - 2x \quad dV = e^{kx}$$
$$dU = (2x-2) \, dx \quad V = \frac{1}{k} e^{kx}$$
$$= \frac{1}{k}(x^2 - 2x)e^{kx} - \frac{1}{k}\int (2x-2) e^{kx} \, dx$$
$$U = x-1 \quad dV = e^{kx} \, dx$$
$$dU = dx \quad V = \frac{1}{k} e^{kx}$$
$$= \frac{1}{k}(x^2 - 2x)e^{kx} - \frac{2}{k}\left[\frac{1}{k}(x-1)e^{kx} - \frac{1}{k}\int e^{kx} \, dx\right]$$
$$= \frac{1}{k}(x^2 - 2x)e^{kx} - \frac{2}{k^2}(x-1)e^{kx} + \frac{2}{k^3} e^{kx} + C.$$

5. $\int x^3 \ln x \, dx$

$$U = \ln x \quad dV = x^3 \, dx$$
$$dU = \frac{dx}{x} \quad V = \frac{x^4}{4}$$
$$= \frac{1}{4} x^4 \ln x - \frac{1}{4} \int x^3 \, dx$$
$$= \frac{1}{4} x^4 \ln x - \frac{1}{16} x^4 + C.$$

6. $\int x(\ln x)^3 \, dx = I_3$ where

$$I_n = \int x(\ln x)^n \, dx$$
$$U = (\ln x)^n \quad dV = x \, dx$$
$$dU = \frac{n}{x}(\ln x)^{n-1} \, dx \quad V = \frac{1}{2} x^2$$
$$= \frac{1}{2} x^2 (\ln x)^n - \frac{n}{2} \int x (\ln x)^{n-1} \, dx$$
$$= \frac{1}{2} x^2 (\ln x)^n - \frac{n}{2} I_{n-1}$$
$$I_3 = \frac{1}{2} x^2 (\ln x)^3 - \frac{3}{2} I_2$$
$$= \frac{1}{2} x^2 (\ln x)^3 - \frac{3}{2}\left[\frac{1}{2} x^2 (\ln x)^2 - \frac{2}{2} I_1\right]$$
$$= \frac{1}{2} x^2 (\ln x)^3 - \frac{3}{4} x^2 (\ln x)^2 + \frac{3}{2}\left[\frac{1}{2} x^2 (\ln x) - \frac{1}{2} I_0\right]$$
$$= \frac{1}{2} x^2 (\ln x)^3 - \frac{3}{4} x^2 (\ln x)^2 + \frac{3}{4} x^2 (\ln x) - \frac{3}{4} \int x \, dx$$
$$= \frac{x^2}{2}\left[(\ln x)^3 - \frac{3}{2}(\ln x)^2 + \frac{3}{2}(\ln x) - \frac{3}{4}\right] + C.$$

7. $\int \tan^{-1} x \, dx$

$$U = \tan^{-1} x \quad dV = dx$$
$$dU = \frac{dx}{1+x^2} \quad V = x$$
$$= x \tan^{-1} x - \int \frac{x \, dx}{1+x^2}$$
$$= x \tan^{-1} x - \frac{1}{2} \ln(1+x^2) + C.$$

8. $\int x^2 \tan^{-1} x \, dx$

$$U = \tan^{-1} x \quad dV = x^2 \, dx$$
$$dU = \frac{dx}{1+x^2} \quad V = \frac{x^3}{3}$$
$$= \frac{x^3}{3} \tan^{-1} x - \frac{1}{3} \int \frac{x^3}{1+x^2} \, dx$$
$$= \frac{x^3}{3} \tan^{-1} x - \frac{1}{3} \int \left(x - \frac{x}{1+x^2}\right) dx$$
$$= \frac{x^3}{3} \tan^{-1} x - \frac{x^2}{6} + \frac{1}{6} \ln(1+x^2) + C.$$

INSTRUCTOR'S SOLUTIONS MANUAL SECTION 6.1 (PAGE 351)

9. $\int x \sin^{-1} x\, dx$

$$U = \sin^{-1} x \qquad dV = x\, dx$$
$$dU = \frac{dx}{\sqrt{1-x^2}} \qquad V = \frac{x^2}{2}$$

$$= \frac{1}{2}x^2 \sin^{-1} x - \frac{1}{2}\int \frac{x^2\, dx}{\sqrt{1-x^2}} \quad \text{Let } x = \sin\theta$$
$$\phantom{= \frac{1}{2}x^2 \sin^{-1} x - \frac{1}{2}\int \frac{x^2\, dx}{\sqrt{1-x^2}}\quad} dx = \cos\theta\, d\theta$$

$$= \frac{1}{2}x^2 \sin^{-1} x - \frac{1}{2}\int \sin^2\theta\, d\theta$$

$$= \frac{1}{2}x^2 \sin^{-1} x - \frac{1}{4}(\theta - \sin\theta\cos\theta) + C$$

$$= \left(\frac{1}{2}x^2 - \frac{1}{4}\right)\sin^{-1} x + \frac{1}{4}x\sqrt{1-x^2} + C.$$

10. $\int x^5 e^{-x^2}\, dx = I_2$ where

$$I_n = \int x^{(2n+1)} e^{-x^2}\, dx$$

$$U = x^{2n} \qquad dV = xe^{-x^2}\, dx$$
$$dU = 2nx^{(2n-1)}\, dx \qquad V = -\frac{1}{2}e^{-x^2}$$

$$= -\frac{1}{2}x^{2n}e^{-x^2} + n\int x^{(2n-1)}e^{-x^2}\, dx$$

$$= -\frac{1}{2}x^{2n}e^{-x^2} + nI_{n-1}$$

$$I_2 = -\frac{1}{2}x^4 e^{-x^2} + 2\left[-\frac{1}{2}x^2 e^{-x^2} + \int xe^{-x^2}\, dx\right]$$

$$= -\frac{1}{2}e^{-x^2}(x^4 + 2x^2 + 2) + C.$$

11. $I_n = \int_0^{\pi/4} \sec^n x\, dx$

$$U = \sec^{n-2} x \qquad dV = \sec^2 x\, dx$$
$$dU = (n-2)\sec^{n-2} x \tan x\, dx \qquad V = \tan x$$

$$= \tan x \sec^{n-2} x \Big|_0^{\pi/4} - (n-2)\int_0^{\pi/4} \sec^{n-2} x \tan^2 x\, dx$$

$$= (\sqrt{2})^{n-2} - (n-2)(I_n - I_{n-2}).$$
$$(n-1)I_n = (\sqrt{2})^{n-2} + (n-2)I_{n-2}.$$

Therefore

$$I_n = \frac{(\sqrt{2})^{n-2}}{n-1} + \frac{n-2}{n-1} I_{n-2}, \qquad (n \geq 2).$$

For $n = 5$ we have

$$\int_0^{\pi/4} \sec^5 x\, dx = I_5 = \frac{2\sqrt{2}}{4} + \frac{3}{4}I_3$$

$$= \frac{\sqrt{2}}{2} + \frac{3}{4}\left(\frac{\sqrt{2}}{2} + \frac{1}{2}I_1\right)$$

$$= \frac{7\sqrt{2}}{8} + \frac{3}{8}\ln|\sec x + \tan x|\Big|_0^{\pi/4}$$

$$= \frac{7\sqrt{2}}{8} + \frac{3}{8}\ln(1 + \sqrt{2}).$$

12. $I = \int \tan^2 x \sec x\, dx$

$$U = \tan x \qquad dV = \sec x \tan x\, dx$$
$$dU = \sec^2 x\, dx \qquad V = \sec x$$

$$= \sec x \tan x - \int \sec^3 x\, dx$$

$$= \sec x \tan x - \int (1 + \tan^2 x)\sec x\, dx$$

$$= \sec x \tan x - \ln|\sec x + \tan x| - I$$

Thus, $I = \frac{1}{2}\sec x \tan x - \frac{1}{2}\ln|\sec x + \tan x| + C$.

13. $I = \int e^{2x} \sin 3x\, dx$

$$U = e^{2x} \qquad dV = \sin 3x\, dx$$
$$dU = 2e^{2x}\, dx \qquad V = -\frac{1}{3}\cos 3x$$

$$= -\frac{1}{3}e^{2x}\cos 3x + \frac{2}{3}\int e^{2x}\cos 3x\, dx$$

$$U = e^{2x} \qquad dV = \cos 3x\, dx$$
$$dU = 2e^{2x}\, dx \qquad V = \frac{1}{3}\sin 3x$$

$$= -\frac{1}{3}e^{2x}\cos 3x + \frac{2}{3}\left(\frac{1}{3}e^{2x}\sin 3x - \frac{2}{3}I\right)$$

$$\frac{13}{9}I = -\frac{1}{3}e^{2x}\cos 3x + \frac{2}{9}e^{2x}\sin 3x + C_1$$

$$I = \frac{1}{13}e^{2x}(2\sin 3x - 3\cos 3x) + C.$$

14. $I = \int xe^{\sqrt{x}}\, dx$ Let $x = w^2$
$$dx = 2w\, dw$$

$$= 2\int w^3 e^w\, dw = 2I_3 \text{ where}$$

$$I_n = \int w^n e^w\, dw$$

$$U = w^n \qquad dV = e^w\, dw$$
$$dU = nw^{n-1}\, dw \qquad V = e^w$$

$$= w^n e^w - nI_{n-1}.$$

$$I = 2I_3 = 2w^3 e^w - 6[w^2 e^w - 2(we^w - I_0)]$$
$$= e^{\sqrt{x}}(2x\sqrt{x} - 6x + 12\sqrt{x} - 12) + C.$$

205

15. $\displaystyle\int_{1/2}^{1} \frac{\sin^{-1} x}{x^2}\, dx$

$$U = \sin^{-1} x \qquad dV = \frac{dx}{x^2}$$
$$dU = \frac{dx}{\sqrt{1-x^2}} \qquad V = -\frac{1}{x}$$

$$= -\frac{1}{x}\sin^{-1} x \Big|_{1/2}^{1} + \int_{1/2}^{1} \frac{dx}{x\sqrt{1-x^2}} \quad \text{Let } x = \sin\theta$$
$$\hspace{7cm} dx = \cos\theta\, d\theta$$

$$= -\frac{\pi}{2} + \frac{\pi}{3} + \int_{\pi/6}^{\pi/2} \csc\theta\, d\theta$$

$$= -\frac{\pi}{6} - \ln|\csc\theta + \cot\theta|\Big|_{\pi/6}^{\pi/2}$$

$$= -\frac{\pi}{6} - \ln 1 + \ln(2 + \sqrt{3}) = \ln(2 + \sqrt{3}) - \frac{\pi}{6}.$$

16. $\displaystyle\int_{0}^{1} \sqrt{x}\sin(\pi\sqrt{x})\, dx \quad \text{Let } x = w^2$
$$\hspace{5cm} dx = 2w\, dw$$

$$= 2\int_{0}^{1} w^2 \sin(\pi w)\, dw$$

$$U = w^2 \qquad dV = \sin(\pi w)\, dw$$
$$dU = 2w\, dw \qquad V = -\frac{\cos(\pi w)}{\pi}$$

$$= -\frac{2}{\pi} w^2 \cos(\pi w)\Big|_{0}^{1} + \frac{4}{\pi}\int_{0}^{1} w\cos(\pi w)\, dw$$

$$U = w \qquad dV = \cos(\pi w)\, dw$$
$$dU = dw \qquad V = \frac{\sin(\pi w)}{\pi}$$

$$= \frac{2}{\pi} + \frac{4}{\pi}\left[\frac{w}{\pi}\sin(\pi w)\right]_{0}^{1} - \frac{4}{\pi^2}\int_{0}^{1} \sin(\pi w)\, dw$$

$$= \frac{2}{\pi} + \frac{4}{\pi^3}\cos(\pi w)\Big|_{0}^{1} = \frac{2}{\pi} + \frac{4}{\pi^3}(-2) = \frac{2}{\pi} - \frac{8}{\pi^3}.$$

17. $\displaystyle\int x\sec^2 x\, dx$

$$U = x \qquad dV = \sec^2 x\, dx$$
$$dU = dx \qquad V = \tan x$$

$$= x\tan x - \int \tan x\, dx$$

$$= x\tan x - \ln|\sec x| + C.$$

18. $\displaystyle\int x\sin^2 x\, dx = \frac{1}{2}\int (x - x\cos 2x)\, dx$

$$= \frac{x^2}{4} - \frac{1}{2}\int x\cos 2x\, dx$$

$$U = x \qquad dV = \cos 2x\, dx$$
$$dU = dx \qquad V = \tfrac{1}{2}\sin 2x$$

$$= \frac{x^2}{4} - \frac{1}{2}\left[\frac{1}{2}x\sin 2x - \frac{1}{2}\int \sin 2x\, dx\right]$$

$$= \frac{x^2}{4} - \frac{x}{4}\sin 2x - \frac{1}{8}\cos 2x + C.$$

19. $I = \displaystyle\int \cos(\ln x)\, dx$

$$U = \cos(\ln x) \qquad dV = dx$$
$$dU = -\frac{\sin(\ln x)}{x}\, dx \qquad V = x$$

$$= x\cos(\ln x) + \int \sin(\ln x)\, dx$$

$$U = \sin(\ln x) \qquad dV = dx$$
$$dU = \frac{\cos(\ln x)}{x}\, dx \qquad V = x$$

$$= x\cos(\ln x) + x\sin(\ln x) - I$$

$$I = \frac{1}{2}\bigl(x\cos(\ln x) + x\sin(\ln x)\bigr) + C.$$

20. $I = \displaystyle\int_{1}^{e} \sin(\ln x)\, dx$

$$U = \sin(\ln x) \qquad dV = dx$$
$$dU = \frac{\cos(\ln x)}{x}\, dx \qquad V = x$$

$$= x\sin(\ln x)\Big|_{1}^{e} - \int_{1}^{e} \cos(\ln x)\, dx$$

$$U = \cos(\ln x) \qquad dV = dx$$
$$dU = -\frac{\sin(\ln x)}{x}\, dx \qquad V = x$$

$$= e\sin(1) - \left[x\cos(\ln x)\Big|_{1}^{e} + I\right]$$

Thus, $I = \dfrac{1}{2}[e\sin(1) - e\cos(1) + 1].$

21. $\displaystyle\int \frac{\ln(\ln x)}{x}\, dx \quad \text{Let } u = \ln x$
$$\hspace{4cm} du = \frac{dx}{x}$$

$$= \int \ln u\, du$$

$$U = \ln u \qquad dV = du$$
$$dU = \frac{du}{u} \qquad V = u$$

$$= u\ln u - \int du = u\ln u - u + C$$

$$= (\ln x)(\ln(\ln x)) - \ln x + C.$$

22. $\displaystyle\int_{0}^{4} \sqrt{x}\, e^{\sqrt{x}}\, dx \quad \text{Let } x = w^2$
$$\hspace{4cm} dx = 2w\, dw$$

$$= 2\int_{0}^{2} w^2 e^w\, dw = 2I_2$$

See solution #16 for the formula
$$I_n = \int w^n e^w\, dw = w^n e^w - nI_{n-1}.$$

$$= 2\left(w^2 e^w\Big|_{0}^{2} - 2I_1\right) = 8e^2 - 4\left(we^w\Big|_{0}^{2} - I_0\right)$$

$$= 8e^2 - 8e^2 + 4\int_{0}^{2} e^w\, dw = 4(e^2 - 1).$$

23. $\int \cos^{-1} x\, dx$

$U = \cos^{-1} x \qquad dV = dx$
$dU = -\dfrac{dx}{\sqrt{1-x^2}} \qquad V = x$

$= x\cos^{-1} x + \int \dfrac{x\, dx}{\sqrt{1-x^2}}$

$= x\cos^{-1} x - \sqrt{1-x^2} + C.$

24. $\int x\sec^{-1} x\, dx$

$U = \sec^{-1} x \qquad dV = x\, dx$
$dU = \dfrac{dx}{|x|\sqrt{x^2-1}} \qquad V = \dfrac{1}{2}x^2$

$= \dfrac{1}{2}x^2 \sec^{-1} x - \dfrac{1}{2}\int \dfrac{|x|}{\sqrt{x^2-1}}\, dx$

$= \dfrac{1}{2}x^2 \sec^{-1} x - \dfrac{1}{2}\operatorname{sgn}(x)\sqrt{x^2-1} + C.$

25. $\int_1^2 \sec^{-1} x\, dx$

$= \int_1^2 \cos^{-1}\dfrac{1}{x}\, dx$

$U = \cos^{-1}\dfrac{1}{x} \qquad dV = dx$
$dU = -\dfrac{1}{\sqrt{1-\dfrac{1}{x^2}}}\left(-\dfrac{1}{x^2}\right)dx \qquad V = x$

$= x\cos^{-1}\dfrac{1}{x}\Big|_1^2 - \int_1^2 \dfrac{dx}{\sqrt{x^2-1}} \quad \text{Let } x = \sec\theta$
$\qquad\qquad\qquad\qquad\qquad\qquad\qquad dx = \sec\theta\tan\theta\, d\theta$

$= \dfrac{2\pi}{3} - 0 - \int_0^{\pi/3} \sec\theta\, d\theta$

$= \dfrac{2\pi}{3} - \ln|\sec\theta + \tan\theta|\Big|_0^{\pi/3}$

$= \dfrac{2\pi}{3} - \ln(2+\sqrt{3}).$

26. $\int (\sin^{-1} x)^2\, dx \quad$ Let $x = \sin\theta$
$\qquad\qquad\qquad\qquad\quad dx = \cos\theta\, d\theta$

$= \int \theta^2 \cos\theta\, d\theta$

$U = \theta^2 \qquad dV = \cos\theta\, d\theta$
$dU = 2\theta\, d\theta \qquad V = \sin\theta$

$= \theta^2 \sin\theta - 2\int \theta\sin\theta\, d\theta$

$U = \theta \qquad dV = \sin\theta\, d\theta$
$dU = d\theta \qquad V = -\cos\theta$

$= \theta^2 \sin\theta - 2\left(-\theta\cos\theta + \int \cos\theta\, d\theta\right)$

$= \theta^2 \sin\theta + 2\theta\cos\theta - 2\sin\theta + C$

$= x(\sin^{-1} x)^2 + 2\sqrt{1-x^2}(\sin^{-1} x) - 2x + C.$

27. $\int x(\tan^{-1} x)^2\, dx$

$U = (\tan^{-1} x)^2 \qquad dV = x\, dx$
$dU = \dfrac{2\tan^{-1} x\, dx}{1+x^2} \qquad V = \dfrac{x^2}{2}$

$= \dfrac{x^2}{2}(\tan^{-1} x)^2 - \int \dfrac{x^2 \tan^{-1} x}{1+x^2}\, dx \quad \text{Let } u = \tan^{-1} x$
$\qquad\qquad\qquad\qquad\qquad\qquad\qquad\qquad du = \dfrac{dx}{1+x^2}$

$= \dfrac{x^2}{2}(\tan^{-1} x)^2 - \int u\tan^2 u\, du$

$= \dfrac{x^2}{2}(\tan^{-1} x)^2 + \int (u - u\sec^2 u)\, du$

$= \dfrac{x^2}{2}(\tan^{-1} x)^2 + \dfrac{u^2}{2} - \int u\sec^2 u\, du$

$U = u \qquad dV = \sec^2 u\, du$
$dU = du \qquad V = \tan u$

$= \dfrac{1}{2}(x^2+1)(\tan^{-1} x)^2 - u\tan u + \int \tan u\, du$

$= \dfrac{1}{2}(x^2+1)(\tan^{-1} x)^2 - x\tan^{-1} x + \ln|\sec u| + C$

$= \dfrac{1}{2}(x^2+1)(\tan^{-1} x)^2 - x\tan^{-1} x + \dfrac{1}{2}\ln(1+x^2) + C$

28. By the procedure used in Example 4 of Section 7.1,

$\int e^x \cos x\, dx = \dfrac{1}{2}e^x(\sin x + \cos x) + C;$

$\int e^x \sin x\, dx = \dfrac{1}{2}e^x(\sin x - \cos x) + C.$

Now

$\int xe^x \cos x\, dx$

$$U = x \qquad dV = e^x \cos x\, dx$$
$$dU = dx \qquad V = \tfrac{1}{2} e^x (\sin x + \cos x)$$
$$= \tfrac{1}{2} x e^x (\sin + \cos x) - \tfrac{1}{2} \int e^x (\sin x + \cos x)\, dx$$
$$= \tfrac{1}{2} x e^x (\sin + \cos x)$$
$$\quad - \tfrac{1}{4} e^x (\sin x - \cos x + \sin x + \cos x) + C$$
$$= \tfrac{1}{2} x e^x (\sin x + \cos x) - \tfrac{1}{2} e^x \sin x + C.$$

29. Area $= A = \displaystyle\int_0^\pi e^{-x} \sin x\, dx$

$$U = e^{-x} \qquad dV = \sin x\, dx$$
$$dU = -e^{-x} dx \qquad V = -\cos x$$
$$= -e^{-x} \cos x \Big|_0^\pi - \int_0^\pi e^{-x} \cos x\, dx$$
$$U = e^{-x} \qquad dV = \cos x\, dx$$
$$dU = -e^{-x} dx \qquad V = \sin x$$
$$= e^{-\pi} + 1 - \left(e^{-x} \sin x \Big|_0^\pi + A \right)$$

Thus Area $= A = \dfrac{1 + e^{-\pi}}{2}$ sq. units.

30. The tangent line to $y = \ln x$ at $x = 1$ is $y = x - 1$, Hence,

$$\text{Shaded area} = \tfrac{1}{2}(1)(1) + (1)(e - 2) - \int_1^e \ln x\, dx$$
$$= e - \tfrac{3}{2} - (x \ln x - x) \Big|_1^e$$
$$= e - \tfrac{3}{2} - e + e + 0 - 1 = e - \tfrac{5}{2} \text{ sq. units.}$$

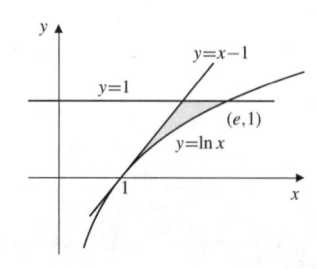

Fig. 6.1.30

31. $I_n = \displaystyle\int (\ln x)^n\, dx$

$$U = (\ln x)^n \qquad dV = dx$$
$$dU = n(\ln x)^{n-1} \dfrac{dx}{x} \qquad V = x$$
$$I_n = x(\ln x)^n - n I_{n-1}.$$
$$I_4 = x(\ln x)^4 - 4 I_3$$
$$= x(\ln x)^4 - 4\big(x(\ln x)^3 - 3 I_2\big)$$
$$= x(\ln x)^4 - 4x(\ln x)^3 + 12\big(x(\ln x)^2 - 2 I_1\big)$$
$$= x(\ln x)^4 - 4x(\ln x)^3 + 12 x(\ln x)^2$$
$$\quad - 24(x \ln x - x) + C$$
$$= x\big((\ln x)^4 - 4(\ln x)^3 + 12(\ln x)^2 - 24 \ln x + 24\big) + C.$$

32. $I_n = \displaystyle\int_0^{\pi/2} x^n \sin x\, dx$

$$U = x^n \qquad dV = \sin x\, dx$$
$$dU = n x^{n-1} dx \qquad V = -\cos x$$
$$= -x^n \cos x \Big|_0^{\pi/2} + n \int_0^{\pi/2} x^{n-1} \cos x\, dx$$
$$U = x^{n-1} \qquad dV = \cos x\, dx$$
$$dU = (n-1) x^{n-2} dx \qquad V = \sin x$$
$$= n \left[x^{n-1} \sin x \Big|_0^{\pi/2} - (n-1) \int_0^{\pi/2} x^{n-2} \sin x\, dx \right]$$
$$= n \left(\dfrac{\pi}{2} \right)^{n-1} - n(n-1) I_{n-2}, \quad (n \ge 2).$$
$$I_0 = \int_0^{\pi/2} \sin x\, dx = -\cos x \Big|_0^{\pi/2} = 1.$$
$$I_6 = 6 \left(\dfrac{\pi}{2} \right)^5 - 6(5) \left\{ 4 \left(\dfrac{\pi}{2} \right)^3 - 4(3) \left[2 \left(\dfrac{\pi}{2} \right) - 2(1) I_0 \right] \right\}$$
$$= \dfrac{3}{16} \pi^5 - 15 \pi^3 + 360 \pi - 720.$$

33. $I_n = \displaystyle\int \sin^n x\, dx \quad (n \ge 2)$

$$U = \sin^{n-1} x \qquad dV = \sin x\, dx$$
$$dU = (n-1) \sin^{n-2} x \cos x\, dx \qquad V = -\cos x$$
$$= -\sin^{n-1} x \cos x + (n-1) \int \sin^{n-2} x \cos^2 x\, dx$$
$$= -\sin^{n-1} x \cos x + (n-1)(I_{n-2} - I_n)$$
$$n I_n = -\sin^{n-1} x \cos x + (n-1) I_{n-2}$$
$$I_n = -\dfrac{1}{n} \sin^{n-1} x \cos x + \dfrac{n-1}{n} I_{n-2}.$$

INSTRUCTOR'S SOLUTIONS MANUAL SECTION 6.1 (PAGE 351)

Note: $I_0 = x + C$, $I_1 = -\cos x + C$. Hence

$$I_6 = -\frac{1}{6}\sin^5 x \cos x + \frac{5}{6}I_4$$
$$= -\frac{1}{6}\sin^5 x \cos x + \frac{5}{6}\left(-\frac{1}{4}\sin^3 x \cos x + \frac{3}{4}I_2\right)$$
$$= -\frac{1}{6}\sin^5 x \cos x - \frac{5}{24}\sin^3 x \cos x$$
$$\quad + \frac{5}{8}\left(-\frac{1}{2}\sin x \cos x + \frac{1}{2}I_0\right)$$
$$= -\frac{1}{6}\sin^5 x \cos x - \frac{5}{24}\sin^3 x \cos x - \frac{5}{16}\sin x \cos x$$
$$\quad + \frac{5}{16}x + C$$
$$= \frac{5x}{16} - \cos x \left(\frac{\sin^5 x}{6} + \frac{5\sin^3 x}{24} + \frac{5\sin x}{16}\right) + C.$$

$$I_7 = -\frac{1}{7}\sin^6 x \cos x + \frac{6}{7}I_5$$
$$= -\frac{1}{7}\sin^6 x \cos x + \frac{6}{7}\left(-\frac{1}{5}\sin^4 x \cos x + \frac{4}{5}I_3\right)$$
$$= -\frac{1}{7}\sin^6 x \cos x - \frac{6}{35}\sin^4 x \cos x$$
$$\quad + \frac{24}{35}\left(-\frac{1}{3}\sin^2 x \cos x + \frac{2}{3}I_1\right)$$
$$= -\frac{1}{7}\sin^6 x \cos x - \frac{6}{35}\sin^4 x \cos x - \frac{8}{35}\sin^2 x \cos x$$
$$\quad - \frac{16}{35}\cos x + C$$
$$= -\cos x \left(\frac{\sin^6 x}{7} + \frac{6\sin^4 x}{35} + \frac{8\sin^2 x}{35} + \frac{16}{35}\right) + C.$$

34. We have

$$I_n = \int \sec^n x \, dx \qquad (n \geq 3)$$

$$U = \sec^{n-2} x \qquad dV = \sec^2 x \, dx$$
$$dU = (n-2)\sec^{n-2} x \tan x \, dx \qquad V = \tan x$$

$$= \sec^{n-2} x \tan x - (n-2)\int \sec^{n-2} x \tan^2 x \, dx$$
$$= \sec^{n-2} x \tan x - (n-2)\int \sec^{n-2} x (\sec^2 x - 1) \, dx$$
$$= \sec^{n-2} x \tan x - (n-2)I_n + (n-2)I_{n-2} + C$$
$$I_n = \frac{1}{n-1}(\sec^{n-2} x \tan x) + \frac{n-2}{n-1}I_{n-2} + C.$$

$$I_1 = \int \sec x \, dx = \ln|\sec x + \tan x| + C;$$
$$I_2 = \int \sec^2 x \, dx = \tan x + C.$$
$$I_6 = \frac{1}{5}(\sec^4 x \tan x) + \frac{4}{5}\left(\frac{1}{3}\sec^2 x \tan x + \frac{2}{3}I_2\right) + C$$

$$= \frac{1}{5}\sec^4 x \tan x + \frac{4}{15}\sec^2 x \tan x + \frac{8}{15}\tan x + C.$$
$$I_7 = \frac{1}{6}(\sec^5 x \tan x) + \frac{5}{6}\bigg[\frac{1}{4}\sec^3 x \tan x +$$
$$\quad \frac{3}{4}\left(\frac{1}{2}\sec x \tan x + \frac{1}{2}I_1\right)\bigg] + C$$
$$= \frac{1}{6}\sec^5 x \tan x + \frac{5}{24}\sec^3 x \tan x + \frac{15}{48}\sec x \tan x +$$
$$\quad \frac{15}{48}\ln|\sec x + \tan x| + C.$$

35. $\displaystyle I_n = \int \frac{dx}{(x^2 + a^2)^n} = \frac{1}{a^2}\int \frac{x^2 + a^2 - x^2}{(x^2 + a^2)^n}\,dx$
$$= \frac{1}{a^2}\int \frac{dx}{(x^2+a^2)^{n-1}} - \frac{1}{a^2}\int \frac{x^2}{(x^2+a^2)^n}\,dx$$

$$U = x \qquad dV = \frac{x\,dx}{(x^2+a^2)^n}$$
$$dU = dx$$
$$\qquad V = \frac{-1}{2(n-1)(x^2+a^2)^{n-1}}$$

$$= \frac{1}{a^2}I_{n-1} - \frac{1}{a^2}\bigg(\frac{-x}{2(n-1)(x^2+a^2)^{n-1}}$$
$$\quad + \frac{1}{2(n-1)}\int \frac{dx}{(x^2+a^2)^{n-1}}\bigg)$$
$$I_n = \frac{x}{2(n-1)a^2(x^2+a^2)^{n-1}} + \frac{2n-3}{2(n-1)a^2}I_{n-1}.$$

Now $I_1 = \frac{1}{a}\tan^{-1}\frac{x}{a}$, so

$$I_3 = \frac{x}{4a^2(x^2+a^2)^2} + \frac{3}{4a^2}I_2$$
$$= \frac{x}{4a^2(x^2+a^2)^2} + \frac{3}{4a^2}\left(\frac{x}{2a^2(x^2+a^2)} + \frac{1}{2a^2}I_1\right)$$
$$= \frac{x}{4a^2(x^2+a^2)^2} + \frac{3x}{8a^4(x^2+a^2)} + \frac{3}{8a^5}\tan^{-1}\frac{x}{a} + C.$$

36. Given that $f(a) = f(b) = 0$.

$$\int_a^b (x-a)(b-x)f''(x)\,dx$$

$$U = (x-a)(b-x) \qquad dV = f''(x)\,dx$$
$$dU = (b+a-2x)\,dx \qquad V = f'(x)$$

$$= (x-a)(b-x)f'(x)\bigg|_a^b - \int_a^b (b+a-2x)f'(x)\,dx$$

$$U = b+a-2x \qquad dV = f'(x)\,dx$$
$$dU = -2\,dx \qquad V = f(x)$$

$$= 0 - \left[(b+a-2x)f(x)\bigg|_a^b + 2\int_a^b f(x)\,dx\right]$$
$$= -2\int_a^b f(x)\,dx.$$

209

37. Given: f'' and g'' are continuous on $[a, b]$, and $f(a) = g(a) = f(b) = g(b) = 0$. We have

$$\int_a^b f(x)g''(x)\,dx$$

$$\begin{aligned} U &= f(x) & dV &= g''(x)\,dx \\ dU &= f'(x)\,dx & V &= g'(x) \end{aligned}$$

$$= f(x)g'(x)\Big|_a^b - \int_a^b f'(x)g'(x)\,dx.$$

Similarly,

$$\int_a^b f''(x)g(x)\,dx = f'(x)g(x)\Big|_a^b - \int_a^b f'(x)g'(x)\,dx.$$

Thus we have

$$\int_a^b f(x)g''(x)\,dx - \int_a^b f''(x)g(x)\,dx$$

$$= \big(f(x)g'(x) - f'(x)g(x)\big)\Big|_a^b = 0$$

by the assumptions on f and g. Thus

$$\int_a^b f(x)g''(x)\,dx = \int_a^b f''(x)g(x)\,dx.$$

This equation is also valid for any (sufficiently smooth) functions f and g for which

$$f(b)g'(b) - f'(b)g(b) = f(a)g'(a) - f'(a)g(a).$$

Examples are functions which are periodic with period $b - a$, or if $f(a) = f(b) = f'(a) = f'(b) = 0$, or if instead g satisfies such conditions. Other combinations of conditions on f and g will also do.

38. $I_n = \int_0^{\pi/2} \cos^n x\,dx$.

a) For $0 \le x \le \pi/2$ we have $0 \le \cos x \le 1$, and so $0 \le \cos^{2n+2} x \le \cos^{2n+1} x \le \cos^{2n} x$. Therefore $0 \le I_{2n+2} \le I_{2n+1} \le I_{2n}$.

b) Since $I_n = \dfrac{n-1}{n} I_{n-2}$, we have $I_{2n+2} = \dfrac{2n+1}{2n+2} I_{2n}$. Combining this with part (a), we get

$$\frac{2n+1}{2n+2} = \frac{I_{2n+2}}{I_{2n}} \le \frac{I_{2n+1}}{I_{2n}} \le 1.$$

The left side approaches 1 as $n \to \infty$, so, by the Squeeze Theorem,

$$\lim_{n\to\infty} \frac{I_{2n+1}}{I_{2n}} = 1.$$

c) By Example 6 we have, since $2n+1$ is odd and $2n$ is even,

$$I_{2n+1} = \frac{2n}{2n+1} \cdot \frac{2n-2}{2n-1} \cdots \frac{4}{5} \cdot \frac{2}{3}$$

$$I_{2n} = \frac{2n-1}{2n} \cdot \frac{2n-3}{2n-2} \cdots \frac{3}{4} \cdot \frac{1}{2} \cdot \frac{\pi}{2}.$$

Multiplying the expression for I_{2n+1} by $\pi/2$ and dividing by the expression for I_{2n}, we obtain, by part (b),

$$\lim_{n\to\infty} \frac{\dfrac{2n}{2n+1} \cdot \dfrac{2n-2}{2n-1} \cdots \dfrac{4}{5} \cdot \dfrac{2}{3} \cdot \dfrac{\pi}{2}}{\dfrac{2n-1}{2n} \cdot \dfrac{2n-3}{2n-2} \cdots \dfrac{3}{4} \cdot \dfrac{1}{2} \cdot \dfrac{\pi}{2}} = \frac{\pi}{2} \times 1 = \frac{\pi}{2},$$

or, rearranging the factors on the left,

$$\lim_{n\to\infty} \frac{2}{1} \cdot \frac{2}{3} \cdot \frac{4}{3} \cdot \frac{4}{5} \cdots \frac{2n}{2n-1} \cdot \frac{2n}{2n+1} = \frac{\pi}{2}.$$

Section 6.2 Inverse Substitutions (page 359)

1. $\displaystyle\int \frac{dx}{\sqrt{1-4x^2}}$ Let $u = 2x$, $du = 2\,dx$

$$= \frac{1}{2}\int \frac{du}{\sqrt{1-u^2}} = \frac{1}{2}\sin^{-1} u + C = \frac{1}{2}\sin^{-1}(2x) + C.$$

2. $\displaystyle\int \frac{x^2\,dx}{\sqrt{1-4x^2}}$ Let $2x = \sin u$, $2\,dx = \cos u\,du$

$$= \frac{1}{8}\int \frac{\sin^2 u \cos u\,du}{\cos u}$$

$$= \frac{1}{16}\int (1 - \cos 2u)\,du = \frac{u}{16} - \frac{\sin 2u}{32} + C$$

$$= \frac{1}{16}\sin^{-1} 2x - \frac{1}{16}\sin u \cos u + C$$

$$= \frac{1}{16}\sin^{-1} 2x - \frac{1}{8}x\sqrt{1-4x^2} + C.$$

3. $\displaystyle\int \frac{x^2\,dx}{\sqrt{9-x^2}}$ Let $x = 3\sin\theta$, $dx = 3\cos\theta\,d\theta$

$$= \int \frac{9\sin^2\theta\, 3\cos\theta\,d\theta}{3\cos\theta}$$

$$= \frac{9}{2}(\theta - \sin\theta\cos\theta) + C$$

$$= \frac{9}{2}\sin^{-1}\frac{x}{3} - \frac{1}{2}x\sqrt{9-x^2} + C.$$

4. $\displaystyle\int \frac{dx}{x\sqrt{1-4x^2}}$ Let $x = \frac{1}{2}\sin\theta$
$dx = \frac{1}{2}\cos\theta\, d\theta$

$= \displaystyle\int \frac{\cos\theta\, d\theta}{\sin\theta\sqrt{1-\sin^2\theta}} = \int \csc\theta\, d\theta$

$= \ln|\csc\theta - \cot\theta| + C = \ln\left|\dfrac{1}{2x} - \dfrac{\sqrt{1-4x^2}}{2x}\right| + C$

$= \ln\left|\dfrac{1 - \sqrt{1-4x^2}}{x}\right| + C_1$.

5. $\displaystyle\int \frac{dx}{x^2\sqrt{9-x^2}}$ Let $x = 3\sin\theta$
$dx = 3\cos\theta\, d\theta$

$= \displaystyle\int \frac{3\cos\theta\, d\theta}{9\sin^2\theta\, 3\cos\theta}$

$= \dfrac{1}{9}\displaystyle\int \csc^2\theta\, d\theta$

$= -\dfrac{1}{9}\cot\theta + C = -\dfrac{1}{9}\dfrac{\sqrt{9-x^2}}{x} + C$.

6. $\displaystyle\int \frac{dx}{x\sqrt{9-x^2}}$ Let $x = 3\sin\theta$
$dx = 3\cos\theta\, d\theta$

$= \displaystyle\int \frac{3\cos\theta\, d\theta}{3\sin\theta\, 3\cos\theta} = \dfrac{1}{3}\int \csc\theta\, d\theta$

$= \dfrac{1}{3}\ln|\csc\theta - \cot\theta| + C = \dfrac{1}{3}\ln\left|\dfrac{3}{x} - \dfrac{\sqrt{9-x^2}}{x}\right| + C$

$= \dfrac{1}{3}\ln\left|\dfrac{3 - \sqrt{9-x^2}}{x}\right| + C$.

7. $\displaystyle\int \frac{x+1}{\sqrt{9-x^2}}\, dx = \int \frac{x\, dx}{\sqrt{9-x^2}} + \int \frac{dx}{\sqrt{9-x^2}}$

$= -\sqrt{9-x^2} + \sin^{-1}\dfrac{x}{3} + C$.

8. $\displaystyle\int \frac{dx}{\sqrt{9+x^2}}$ Let $x = 3\tan\theta$
$dx = 3\sec^2\theta\, d\theta$

$= \displaystyle\int \frac{3\sec^2\theta\, d\theta}{3\sec\theta} = \int \sec\theta\, d\theta$

$= \ln|\sec\theta + \tan\theta| + C = \ln(x + \sqrt{9+x^2}) + C_1$.

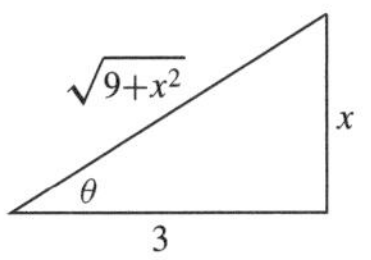

Fig. 6.2.8

9. $\displaystyle\int \frac{x^3\, dx}{\sqrt{9+x^2}}$ Let $u = 9 + x^2$
$du = 2x\, dx$

$= \dfrac{1}{2}\displaystyle\int \frac{(u-9)\, du}{\sqrt{u}} = \dfrac{1}{2}\int (u^{1/2} - 9u^{-1/2})\, du$

$= \dfrac{1}{3}u^{3/2} - 9u^{1/2} + C$

$= \dfrac{1}{3}(9+x^2)^{3/2} - 9\sqrt{9+x^2} + C$.

10. $\displaystyle\int \frac{\sqrt{9+x^2}}{x^4}\, dx$ Let $x = 3\tan\theta$
$dx = 3\sec^2\theta\, d\theta$

$= \displaystyle\int \frac{(3\sec\theta)(3\sec^2\theta)\, d\theta}{81\tan^4\theta}$

$= \dfrac{1}{9}\displaystyle\int \frac{\sec^3\theta}{\tan^4\theta}\, d\theta = \dfrac{1}{9}\int \frac{\cos\theta}{\sin^4\theta}\, d\theta$ Let $u = \sin\theta$
$du = \cos\theta\, d\theta$

$= \dfrac{1}{9}\displaystyle\int \frac{du}{u^4} = -\dfrac{1}{27u^3} + C = -\dfrac{1}{27\sin^3\theta} + C$

$= -\dfrac{(9+x^2)^{3/2}}{27x^3} + C$.

11. $\displaystyle\int \frac{dx}{(a^2-x^2)^{3/2}}$ Let $x = a\sin\theta$
$dx = a\cos\theta\, d\theta$

$= \displaystyle\int \frac{a\cos\theta\, d\theta}{a^3\cos^3\theta} = \dfrac{1}{a^2}\int \sec^2\theta\, d\theta$

$= \dfrac{1}{a^2}\tan\theta + C = \dfrac{1}{a^2}\dfrac{x}{\sqrt{a^2-x^2}} + C$.

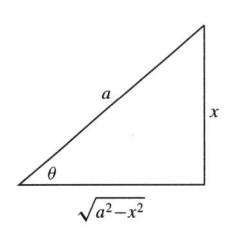

Fig. 6.2.11 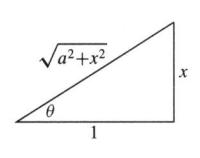   Fig. 6.2.12

12. $\displaystyle\int \frac{dx}{(a^2+x^2)^{3/2}}$ Let $x = a\tan\theta$
$dx = a\sec^2\theta\, d\theta$

$= \displaystyle\int \frac{a\sec^2\theta\, d\theta}{(a^2+a^2\tan^2\theta)^{3/2}} = \int \frac{a\sec^2\theta\, d\theta}{a^3\sec^3\theta}$

$= \dfrac{1}{a^2}\displaystyle\int \cos\theta\, d\theta = \dfrac{1}{a^2}\sin\theta + C = \dfrac{x}{a^2\sqrt{a^2+x^2}} + C$.

13. $\displaystyle\int \frac{x^2\,dx}{(a^2-x^2)^{3/2}}$ Let $x = a\sin\theta$
$$dx = a\cos\theta\,d\theta$$
$$= \int \frac{a^2\sin^2\theta\, a\cos\theta\,d\theta}{a^3\cos^3\theta}$$
$$= \int \tan^2\theta\,d\theta = \int (\sec^2\theta - 1)\,d\theta$$
$$= \tan\theta - \theta + C \quad \text{(see Fig. s6-5-17)}$$
$$= \frac{x}{\sqrt{a^2-x^2}} - \sin^{-1}\frac{x}{a} + C.$$

14. $\displaystyle\int \frac{dx}{(1+2x^2)^{5/2}}$ Let $x = \frac{1}{\sqrt{2}}\tan\theta$
$$dx = \frac{1}{\sqrt{2}}\sec^2\theta\,d\theta$$
$$= \frac{1}{\sqrt{2}}\int \frac{\sec^2\theta\,d\theta}{(1+\tan^2\theta)^{5/2}} = \frac{1}{\sqrt{2}}\int \cos^3\theta\,d\theta$$
$$= \frac{1}{\sqrt{2}}\int (1-\sin^2\theta)\cos\theta\,d\theta \quad \text{Let } u = \sin\theta$$
$$du = \cos\theta\,d\theta$$
$$= \frac{1}{\sqrt{2}}\int (1-u^2)\,du = \frac{1}{\sqrt{2}}\left(u - \frac{1}{3}u^3\right) + C$$
$$= \frac{1}{\sqrt{2}}\sin\theta - \frac{1}{3\sqrt{2}}\sin^3\theta + C$$
$$= \frac{\sqrt{2}x}{\sqrt{2}\sqrt{1+2x^2}} - \frac{1}{3\sqrt{2}}\left(\frac{\sqrt{2}x}{\sqrt{1+2x^2}}\right)^3 + C$$
$$= \frac{4x^3 + 3x}{3(1+2x^2)^{3/2}} + C.$$

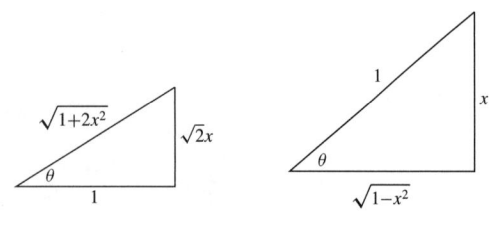

Fig. 6.2.14   Fig. 6.2.15

15. $\displaystyle\int \frac{dx}{x\sqrt{x^2-a^2}}$ Let $x = a\sec\theta$
$$dx = a\sec\theta\tan\theta\,d\theta$$
$$= \int \frac{a\sec\theta\tan\theta\,d\theta}{a\sec\theta\, a\tan\theta}$$
$$= \frac{1}{a}\int d\theta = \frac{\theta}{a} + C = \frac{1}{a}\sec^{-1}\frac{x}{a} + C.$$

16. $\displaystyle\int \frac{dx}{x^2\sqrt{x^2-a^2}}$ Let $x = a\sec\theta$ $(a>0)$
$$dx = a\sec\theta\tan\theta\,d\theta$$
$$= \int \frac{a\sec\theta\tan\theta\,d\theta}{a^2\sec^2\theta\, a\tan\theta}$$
$$= \frac{1}{a^2}\int \cos\theta\,d\theta = \frac{1}{a^2}\sin\theta + C$$
$$= \frac{1}{a^2}\frac{\sqrt{x^2-a^2}}{x} + C.$$

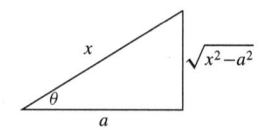

Fig. 6.2.16

17. $\displaystyle\int \frac{dx}{x^2+2x+10} = \int \frac{dx}{(x+1)^2+9} = \frac{1}{3}\tan^{-1}\frac{x+1}{3} + C.$

18. $\displaystyle\int \frac{dx}{x^2+x+1} = \int \frac{dx}{\left(x+\frac{1}{2}\right)^2 + \left(\frac{\sqrt{3}}{2}\right)^2}$ Let $u = x + \frac{1}{2}$
$$du = dx$$
$$= \int \frac{du}{u^2 + \left(\frac{\sqrt{3}}{2}\right)^2} = \frac{2}{\sqrt{3}}\tan^{-1}\left(\frac{2}{\sqrt{3}}u\right) + C$$
$$= \frac{2}{\sqrt{3}}\tan^{-1}\left(\frac{2x+1}{\sqrt{3}}\right) + C.$$

19. $\displaystyle\int \frac{dx}{(4x^2+4x+5)^2}$
$$= \int \frac{dx}{\left((2x+1)^2 + 4\right)^2} \quad \text{Let } 2x+1 = 2\tan\theta$$
$$2\,dx = 2\sec^2\theta\,d\theta$$
$$= \int \frac{\sec^2\theta\,d\theta}{16\sec^4\theta} = \frac{1}{16}\int \cos^2\theta\,d\theta$$
$$= \frac{1}{32}\left(\theta + \sin\theta\cos\theta\right)$$
$$= \frac{1}{32}\tan^{-1}\frac{2x+1}{2} + \frac{1}{16}\frac{2x+1}{4x^2+4x+5} + C.$$

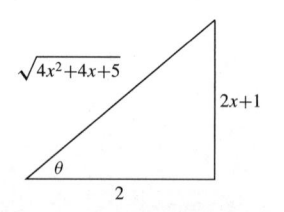

Fig. 6.2.19

INSTRUCTOR'S SOLUTIONS MANUAL SECTION 6.2 (PAGE 359)

20. $\int \dfrac{x\,dx}{x^2-2x+3} = \int \dfrac{(x-1)+1}{(x-1)^2+2}\,dx$ Let $u=x-1$
$du=dx$

$= \int \dfrac{u\,du}{u^2+2} + \int \dfrac{du}{u^2+2}$

$= \dfrac{1}{2}\ln(u^2+2) + \dfrac{1}{\sqrt{2}}\tan^{-1}\left(\dfrac{u}{\sqrt{2}}\right) + C$

$= \dfrac{1}{2}\ln(x^2-2x+3) + \dfrac{1}{\sqrt{2}}\tan^{-1}\left(\dfrac{x-1}{\sqrt{2}}\right) + C.$

21. $\int \dfrac{x\,dx}{\sqrt{2ax-x^2}}$

$= \int \dfrac{x\,dx}{\sqrt{a^2-(x-a)^2}}$ Let $x-a = a\sin\theta$
$dx = a\cos\theta\,d\theta$

$= \int \dfrac{(a+a\sin\theta)a\cos\theta\,d\theta}{a\cos\theta}$

$= a(\theta - \cos\theta) + C$

$= a\sin^{-1}\dfrac{x-a}{a} - \sqrt{2ax-x^2} + C.$

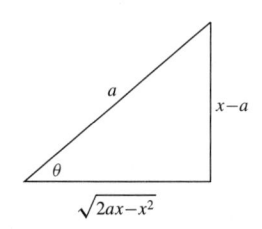

Fig. 6.2.21

22. $\int \dfrac{dx}{(4x-x^2)^{3/2}}$

$= \int \dfrac{dx}{[4-(2-x)^2]^{3/2}}$ Let $2-x = 2\sin u$
$-dx = 2\cos u\,du$

$= -\int \dfrac{2\cos u\,du}{8\cos^3 u} = -\dfrac{1}{4}\int \sec^2 u\,du$

$= -\dfrac{1}{4}\tan u + C = \dfrac{1}{4}\dfrac{x-2}{\sqrt{4x-x^2}} + C.$

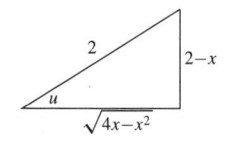

Fig. 6.2.22

23. $\int \dfrac{x\,dx}{(3-2x-x^2)^{3/2}}$

$= \int \dfrac{x\,dx}{\left(4-(x+1)^2\right)^{3/2}}$ Let $x+1 = 2\sin\theta$
$dx = 2\cos\theta\,d\theta$

$= \int \dfrac{(2\sin\theta-1)2\cos\theta\,d\theta}{8\cos^3\theta}$

$= \dfrac{1}{2}\int \sec\theta\tan\theta\,d\theta - \dfrac{1}{4}\int \sec^2\theta\,d\theta$

$= \dfrac{1}{2}\sec\theta - \dfrac{1}{4}\tan\theta + C$

$= \dfrac{1}{\sqrt{3-2x-x^2}} - \dfrac{1}{4}\dfrac{x+1}{\sqrt{3-2x-x^2}} + C$

$= \dfrac{1}{4}\cdot\dfrac{3-x}{\sqrt{3-2x-x^2}} + C.$

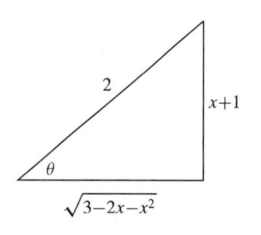

Fig. 6.2.23

24. $\int \dfrac{dx}{(x^2+2x+2)^2} = \int \dfrac{dx}{[(x+1)^2+1]^2}$ Let $x+1 = \tan u$
$dx = \sec^2 u\,du$

$= \int \dfrac{\sec^2 u\,du}{\sec^4 u} = \int \cos^2 u\,du$

$= \dfrac{1}{2}\int (1+\cos 2u)\,du = \dfrac{u}{2} + \dfrac{\sin 2u}{4} + C$

$= \dfrac{1}{2}\tan^{-1}(x+1) + \dfrac{1}{2}\sin u\cos u + C$

$= \dfrac{1}{2}\tan^{-1}(x+1) + \dfrac{1}{2}\dfrac{x+1}{x^2+2x+2} + C.$

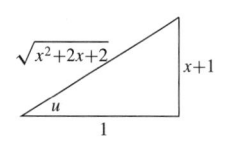

Fig. 6.2.24

25. $\displaystyle\int \frac{dx}{(1+x^2)^3}$ Let $x = \tan\theta$
$dx = \sec^2\theta\, d\theta$

$\displaystyle = \int \frac{\sec^2\theta}{\sec^6\theta}\, d\theta = \int \cos^4\theta\, d\theta$

$\displaystyle = \int \left(\frac{1+\cos 2\theta}{2}\right)^2 d\theta$

$\displaystyle = \frac{1}{4}\int \left(1 + 2\cos 2\theta + \frac{1+\cos 4\theta}{2}\right) d\theta$

$\displaystyle = \frac{3\theta}{8} + \frac{\sin 2\theta}{4} + \frac{\sin 4\theta}{32} + C$

$\displaystyle = \frac{3\theta}{8} + \frac{\sin\theta \cos\theta}{2} + \frac{\sin 2\theta \cos 2\theta}{16} + C$

$\displaystyle = \frac{3\theta}{8} + \frac{\sin\theta \cos\theta}{2} + \frac{1}{8}\sin\theta \cos\theta(2\cos^2\theta - 1) + C$

$\displaystyle = \frac{3}{8}\tan^{-1}x + \frac{1}{2}\cdot\frac{x}{1+x^2} + \frac{1}{8}\cdot\frac{x}{1+x^2}\left(\frac{2}{1+x^2} - 1\right) + C$

$\displaystyle = \frac{3}{8}\tan^{-1}x + \frac{3}{8}\cdot\frac{x}{1+x^2} + \frac{1}{4}\cdot\frac{x}{(1+x^2)^2} + C$

$\displaystyle = \frac{3}{8}\tan^{-1}x + \frac{3x^3 + 5x}{8(1+x^2)^2} + C.$

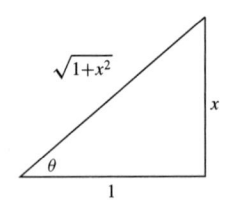

Fig. 6.2.25

26. $\displaystyle\int \frac{x^2\, dx}{(1+x^2)^2}$ Let $x = \tan u$
$dx = \sec^2 u\, du$

$\displaystyle = \int \frac{\tan^2 u\, \sec^2 u\, du}{\sec^4 u} = \int \frac{\tan^2 u\, du}{\sec^2 u}$

$\displaystyle = \int \sin^2 u\, du = \frac{1}{2}\int (1 - \cos 2u)\, du$

$\displaystyle = \frac{u}{2} - \frac{\sin u \cos u}{2} + C$

$\displaystyle = \frac{1}{2}\tan^{-1}x - \frac{1}{2}\frac{x}{1+x^2} + C.$

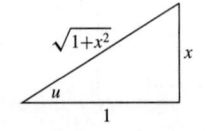

Fig. 6.2.26

27. $\displaystyle\int \frac{\sqrt{1-x^2}}{x^3}\, dx$ Let $x = \sin\theta$
$dx = \cos\theta\, d\theta$

$\displaystyle = \int \frac{\cos^2\theta}{\sin^3\theta}\, d\theta = I,$ where

$\displaystyle I = \int \cot^2\theta \csc\theta\, d\theta$

$\quad U = \cot\theta \qquad dV = \cot\theta \csc\theta\, d\theta$
$\quad dU = -\csc^2\theta\, d\theta \quad V = -\csc\theta$

$\displaystyle = -\csc\theta \cot\theta - \int \csc^3\theta\, d\theta$

$\displaystyle = -\csc\theta \cot\theta - \int \csc\theta\, d\theta - I.$

Therefore
$\displaystyle I = -\frac{1}{2}\csc\theta \cot\theta + \frac{1}{2}\ln|\csc\theta + \cot\theta| + C$

$\displaystyle = -\frac{1}{2}\frac{\sqrt{1-x^2}}{x^2} + \frac{1}{2}\ln\left|\frac{1}{x} + \frac{\sqrt{1-x^2}}{x}\right| + C$

$\displaystyle = \frac{1}{2}\ln(1 + \sqrt{1-x^2}) - \frac{1}{2}\ln|x| - \frac{1}{2}\frac{\sqrt{1-x^2}}{x^2} + C.$

28. $\displaystyle I = \int \sqrt{9+x^2}\, dx$ Let $x = 3\tan\theta$
$dx = 3\sec^2\theta\, d\theta$

$\displaystyle = \int 3\sec\theta\, 3\sec^2\theta\, d\theta$

$\displaystyle = 9\int \sec^3\theta\, d\theta$

$\quad U = \sec\theta \qquad dV = \sec^2\theta\, d\theta$
$\quad dU = \sec\theta \tan\theta\, d\theta \quad V = \tan\theta$

$\displaystyle = 9\sec\theta \tan\theta - 9\int \sec\theta \tan^2\theta\, d\theta$

$\displaystyle = 9\sec\theta \tan\theta - 9\int \sec\theta(\sec^2\theta - 1)\, d\theta$

$\displaystyle = 9\sec\theta \tan\theta + 9\int \sec\theta\, d\theta - 9\int \sec^3\theta\, d\theta$

$\displaystyle = 9\sec\theta \tan\theta + 9\ln|\sec\theta + \tan\theta| - I$

$\displaystyle I = \frac{9}{2}\left[\left(\frac{\sqrt{9+x^2}}{3}\right)\left(\frac{x}{3}\right)\right] + \frac{9}{2}\ln\left|\frac{\sqrt{9+x^2}}{3} + \frac{x}{3}\right| + C$

$\displaystyle = \frac{1}{2}x\sqrt{9+x^2} + \frac{9}{2}\ln(\sqrt{9+x^2} + x) + C_1.$
(where $C_1 = C - \frac{9}{2}\ln 3$)

29. $\displaystyle\int \frac{dx}{2+\sqrt{x}}$ Let $x = u^2$
$dx = 2u\, du$

$\displaystyle = \int \frac{2u\, du}{2+u} = 2\int \left(1 - \frac{2}{u+2}\right) du$

$\displaystyle = 2u - 4\ln|u+2| + C = 2\sqrt{x} - 4\ln(2 + \sqrt{x}) + C.$

INSTRUCTOR'S SOLUTIONS MANUAL SECTION 6.2 (PAGE 359)

30. $\displaystyle\int \frac{dx}{1+x^{1/3}}$ Let $x = u^3$
$\qquad dx = 3u^2\, du$

$\displaystyle= 3\int \frac{u^2\, du}{1+u}$ Let $v = 1 + u$
$\qquad dv = du$

$\displaystyle= 3\int \frac{v^2 - 2v + 1}{v}\, dv = 3\int\left(v - 2 + \frac{1}{v}\right) dv$

$\displaystyle= 3\left(\frac{v^2}{2} - 2v + \ln|v|\right) + C$

$\displaystyle= \frac{3}{2}(1 + x^{1/3})^2 - 6(1 + x^{1/3}) + 3\ln|1 + x^{1/3}| + C.$

31. $\displaystyle I = \int \frac{1 + x^{1/2}}{1 + x^{1/3}}\, dx$ Let $x = u^6$
$\qquad dx = 6u^5\, du$

$\displaystyle= \int \frac{1 + u^3}{1 + u^2} 6u^5\, du = 6\int \frac{u^8 + u^5}{1 + u^2}\, du.$

Division is required to render the last integrand as a polynomial with a remainder fraction of simpler form: observe that

$u^8 = u^8 + u^6 - u^6 - u^4 + u^4 + u^2 - u^2 - 1 + 1$
$\quad = (u^2 + 1)(u^6 - u^4 + u^2 - 1) + 1$
$u^5 = u^5 + u^3 - u^3 - u + u$
$\quad = (u^2 + 1)(u^3 - u) + u.$

Thus

$\displaystyle\frac{u^8 + u^5}{u^2 + 1} = u^6 - u^4 + u^3 + u^2 - u - 1 + \frac{u+1}{u^2+1}.$

Therefore

$\displaystyle I = 6\int\left(u^6 - u^4 + u^3 + u^2 - u - 1 + \frac{u+1}{u^2+1}\right) du$

$\displaystyle= 6\left(\frac{u^7}{7} - \frac{u^5}{5} + \frac{u^4}{4} + \frac{u^3}{3} - \frac{u^2}{2} - u\right.$

$\displaystyle\left. + \frac{1}{2}\ln(u^2 + 1) + \tan^{-1} u\right) + C$

$\displaystyle= \frac{6}{7}x^{7/6} - \frac{6}{5}x^{5/6} + \frac{3}{2}x^{2/3} + 2x^{1/2} - 3x^{1/3} - 6x^{1/6}$
$\displaystyle\quad + 3\ln(1 + x^{1/3}) + 6\tan^{-1} x^{1/6} + C.$

32. $\displaystyle\int \frac{x\sqrt{2 - x^2}}{\sqrt{x^2 + 1}}\, dx$ Let $u^2 = x^2 + 1$
$\qquad 2u\, du = 2x\, dx$

$\displaystyle= \int \frac{u\sqrt{3 - u^2}\, du}{u}$

$\displaystyle= \int \sqrt{3 - u^2}\, du$ Let $u = \sqrt{3}\sin v$
$\qquad du = \sqrt{3}\cos v\, dv$

$\displaystyle= \int (\sqrt{3}\cos v)\sqrt{3}\cos v\, dv = 3\int \cos^2 v\, dv$

$\displaystyle= \frac{3}{2}(v + \sin v \cos v) + C$

$\displaystyle= \frac{3}{2}\sin^{-1}\left(\frac{u}{\sqrt{3}}\right) + \frac{3}{2}\frac{u\sqrt{3 - u^2}}{3} + C$

$\displaystyle= \frac{3}{2}\sin^{-1}\left(\sqrt{\frac{x^2 + 1}{3}}\right) + \frac{1}{2}\sqrt{(x^2 + 1)(2 - x^2)} + C.$

33. $\displaystyle\int_{-\ln 2}^{0} e^x\sqrt{1 - e^{2x}}\, dx$ Let $e^x = \sin\theta$
$\qquad e^x\, dx = \cos\theta\, d\theta$

$\displaystyle= \int_{\pi/6}^{\pi/2} \cos^2\theta\, d\theta = \frac{1}{2}(\theta + \sin\theta\cos\theta)\Big|_{\pi/6}^{\pi/2}$

$\displaystyle= \frac{1}{2}\left(\frac{\pi}{3} - \frac{\sqrt{3}}{4}\right) = \frac{\pi}{6} - \frac{\sqrt{3}}{8}.$

34. $\displaystyle\int_0^{\pi/2} \frac{\cos x}{\sqrt{1 + \sin^2 x}}\, dx$ Let $u = \sin x$
$\qquad du = \cos x\, dx$

$\displaystyle= \int_0^1 \frac{du}{\sqrt{1 + u^2}}$ Let $u = \tan w$
$\qquad du = \sec^2 w\, dw$

$\displaystyle= \int_0^{\pi/4} \frac{\sec^2 w\, dw}{\sec w} = \int_0^{\pi/4} \sec w\, dw$

$\displaystyle= \ln|\sec w + \tan w|\Big|_0^{\pi/4}$

$\displaystyle= \ln|\sqrt{2} + 1| - \ln|1 + 0| = \ln(\sqrt{2} + 1).$

35. $\displaystyle\int_{-1}^{\sqrt{3}-1} \frac{dx}{x^2 + 2x + 2}$

$\displaystyle= \int_{-1}^{\sqrt{3}-1} \frac{dx}{(x+1)^2 + 1}$ Let $u = x + 1$
$\qquad du = dx$

$\displaystyle= \int_0^{\sqrt{3}} \frac{du}{u^2 + 1} = \tan^{-1} u\Big|_0^{\sqrt{3}} = \frac{\pi}{3}.$

215

36. $\displaystyle\int_1^2 \frac{dx}{x^2\sqrt{9-x^2}}$ Let $x = 3\sin u$,
$dx = 3\cos u\,du$
$= \displaystyle\int_{x=1}^{x=2} \frac{3\cos u\,du}{9\sin^2 u(3\cos u)} = \frac{1}{9}\int_{x=1}^{x=2} \csc^2 u\,du$
$= \dfrac{1}{9}(-\cot u)\Big|_{x=1}^{x=2} = -\dfrac{1}{9}\left(\dfrac{\sqrt{9-x^2}}{x}\right)\Big|_{x=1}^{x=2}$
$= -\dfrac{1}{9}\left(\dfrac{\sqrt{5}}{2} - \dfrac{\sqrt{8}}{1}\right) = \dfrac{2\sqrt{2}}{9} - \dfrac{\sqrt{5}}{18}.$

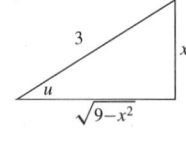

Fig. 6.2.36

37. $\displaystyle\int \frac{d\theta}{2+\sin\theta}$ Let $x = \tan(\theta/2)$,
$\sin\theta = \dfrac{2x}{1+x^2}, \quad d\theta = \dfrac{2\,dx}{1+x^2}$
$= \displaystyle\int \frac{\frac{2\,dx}{1+x^2}}{2+\frac{2x}{1+x^2}} = \int \frac{dx}{1+x+x^2}$
$= \displaystyle\int \frac{dx}{\left(x+\frac{1}{2}\right)^2 + \frac{3}{4}} = \frac{2}{\sqrt{3}}\tan^{-1}\frac{2x+1}{\sqrt{3}} + C$
$= \dfrac{2}{\sqrt{3}}\tan^{-1}\left(\dfrac{2\tan(\theta/2)+1}{\sqrt{3}}\right) + C.$

38. $\displaystyle\int_0^{\pi/2} \frac{d\theta}{1+\cos\theta+\sin\theta}$ Let $x = \tan\dfrac{\theta}{2}$, $d\theta = \dfrac{2}{1+x^2}dx$,
$\cos\theta = \dfrac{1-x^2}{1+x^2}, \quad \sin\theta = \dfrac{2x}{1+x^2}.$
$= \displaystyle\int_0^1 \frac{\left(\frac{2}{1+x^2}\right)dx}{1+\left(\frac{1-x^2}{1+x^2}\right)+\left(\frac{2x}{1+x^2}\right)}$
$= 2\displaystyle\int_0^1 \frac{dx}{2+2x} = \int_0^1 \frac{dx}{1+x}$
$= \ln|1+x|\Big|_0^1 = \ln 2.$

39. $\displaystyle\int \frac{d\theta}{3+2\cos\theta}$ Let $x = \tan(\theta/2)$,
$\cos\theta = \dfrac{1-x^2}{1+x^2}, \quad d\theta = \dfrac{2\,dx}{1+x^2}$
$= \displaystyle\int \frac{\frac{2\,dx}{1+x^2}}{3+\frac{2-2x^2}{1+x^2}} = \int \frac{2\,dx}{5+x^2}$
$= \dfrac{2}{\sqrt{5}}\tan^{-1}\dfrac{x}{\sqrt{5}} + C = \dfrac{2}{\sqrt{5}}\tan^{-1}\left(\dfrac{\tan(\theta/2)}{\sqrt{5}}\right) + C.$

40. Area $= \displaystyle\int_{1/2}^1 \frac{dx}{\sqrt{2x-x^2}} = \int_{1/2}^1 \frac{dx}{\sqrt{1-(x-1)^2}}$
Let $u = x - 1$
$du = dx$
$= \displaystyle\int_{-1/2}^0 \frac{du}{\sqrt{1-u^2}} = \sin^{-1}u\Big|_{-1/2}^0$
$= 0 - \left(-\dfrac{\pi}{6}\right) = \dfrac{\pi}{6}$ sq. units.

41. For intersection of $y = \dfrac{9}{x^4+4x^2+4}$ and $y = 1$ we have
$x^4 + 4x^2 + 4 = 9$
$x^4 + 4x^2 - 5 = 0$
$(x^2+5)(x^2-1) = 0,$

so the intersections are at $x = \pm 1$. The required area is

$A = 2\displaystyle\int_0^1 \left(\frac{9\,dx}{x^4+4x^2+4} - 1\right)dx$
$= 18\displaystyle\int_0^1 \frac{dx}{(x^2+2)^2} - 2$ Let $x = \sqrt{2}\tan\theta$
$dx = \sqrt{2}\sec^2\theta$
$= 18\displaystyle\int_{x=0}^{x=1} \frac{\sqrt{2}\sec^2\theta\,d\theta}{4\sec^4\theta} - 2$
$= \dfrac{9}{\sqrt{2}}\displaystyle\int_{x=0}^{x=1} \cos^2\theta\,d\theta - 2$
$= \dfrac{9}{2\sqrt{2}}(\theta + \sin\theta\cos\theta)\Big|_{x=0}^{x=1} - 2$
$= \dfrac{9}{2\sqrt{2}}\left(\tan^{-1}\dfrac{x}{\sqrt{2}} + \dfrac{\sqrt{2}x}{x^2+2}\right)\Big|_0^1 - 2$
$= \dfrac{9}{2\sqrt{2}}\left(\tan^{-1}\dfrac{1}{\sqrt{2}}\right) - \dfrac{1}{2}$ sq. units.

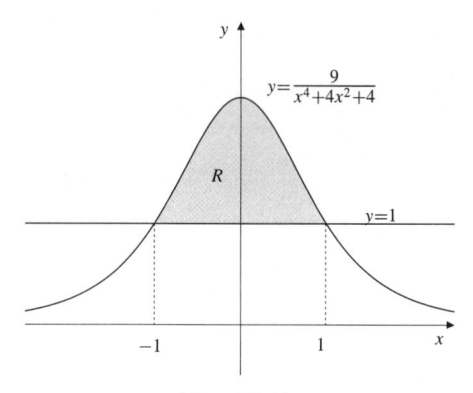

INSTRUCTOR'S SOLUTIONS MANUAL SECTION 6.2 (PAGE 359)

Fig. 6.2.41

Fig. 6.2.43   Fig. 6.2.44

42. Average value $= \dfrac{1}{4}\displaystyle\int_0^4 \dfrac{dx}{(x^2-4x+8)^{3/2}}$

$= \dfrac{1}{4}\displaystyle\int_0^4 \dfrac{dx}{[(x-2)^2+4]^{3/2}}$

Let $x-2 = 2\tan u$
$dx = 2\sec^2 u\, du$

$= \dfrac{1}{4}\displaystyle\int_{-\pi/4}^{\pi/4} \dfrac{2\sec^2 u\, du}{8\sec^3 u}$

$= \dfrac{1}{16}\displaystyle\int_{-\pi/4}^{\pi/4} \cos u\, du = \dfrac{1}{16}\sin u\Big|_{-\pi/4}^{\pi/4}$

$= \dfrac{1}{16}\left(\dfrac{1}{\sqrt{2}}+\dfrac{1}{\sqrt{2}}\right) = \dfrac{\sqrt{2}}{16}.$

43. Area of R

$= 2\displaystyle\int_0^{\sqrt{a^2-b^2}} \left(\sqrt{a^2-x^2}-b\right)dx$ Let $x = a\sin\theta$
$dx = a\cos\theta\, d\theta$

$= 2\displaystyle\int_{x=0}^{x=\sqrt{a^2-b^2}} a^2\cos^2\theta\, d\theta - 2b\sqrt{a^2-b^2}$

$= a^2(\theta+\sin\theta\cos\theta)\Big|_{x=0}^{x=\sqrt{a^2-b^2}} - 2b\sqrt{a^2-b^2}$

$= \left(a^2\sin^{-1}\dfrac{x}{a}+x\sqrt{a^2-x^2}\right)\Big|_0^{\sqrt{a^2-b^2}} - 2b\sqrt{a^2-b^2}$

$= a^2\sin^{-1}\sqrt{1-\dfrac{b^2}{a^2}}+b\sqrt{a^2-b^2}-2b\sqrt{a^2-b^2}$

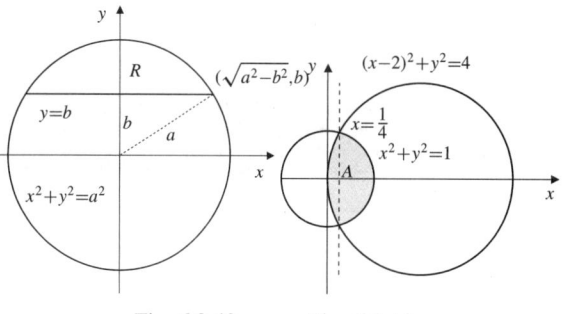

$= a^2\cos^{-1}\dfrac{b}{a}-b\sqrt{a^2-b^2}$ sq. units.

44. The circles intersect at $x = \tfrac{1}{4}$, so the common area is $A_1 + A_2$ where

$A_1 = 2\displaystyle\int_{1/4}^1 \sqrt{1-x^2}\,dx$ Let $x = \sin u$
$dx = \cos u\, du$

$= 2\displaystyle\int_{x=1/4}^{x=1} \cos^2 u\, du$

$= (u+\sin u\cos u)\Big|_{x=1/4}^{x=1}$

$= (\sin^{-1} x + x\sqrt{1-x^2})\Big|_{x=1/4}^{x=1}$

$= \dfrac{\pi}{2}-\sin^{-1}\dfrac{1}{4}-\dfrac{\sqrt{15}}{16}$ sq. units.

$A_2 = 2\displaystyle\int_0^{1/4}\sqrt{4-(x-2)^2}\,dx$ Let $x-2 = 2\sin v$
$dx = 2\cos v\, dv$

$= 8\displaystyle\int_{x=0}^{x=1/4}\cos^2 v\, dv$

$= 4(v+\sin v\cos v)\Big|_{x=0}^{x=1/4}$

$= 4\left[\sin^{-1}\left(\dfrac{x-2}{2}\right)+\left(\dfrac{x-2}{2}\right)\dfrac{\sqrt{4x-x^2}}{2}\right]\Big|_{x=0}^{x=1/4}$

$= 4\left[\sin^{-1}\left(-\dfrac{7}{8}\right)-\dfrac{7\sqrt{15}}{64}+\dfrac{\pi}{2}\right]$

$= -4\sin^{-1}\left(\dfrac{7}{8}\right)-\dfrac{7\sqrt{15}}{16}+2\pi$ sq. units.

Hence, the common area is

$A_1 + A_2 = \dfrac{5\pi}{2}-\dfrac{\sqrt{15}}{2}$
$-\sin^{-1}\left(\dfrac{1}{4}\right)-4\sin^{-1}\left(\dfrac{7}{8}\right)$ sq. units.

45. Required area $= \int_3^4 \left(\sqrt{25-x^2} - \frac{12}{x}\right) dx$

$= \int_3^4 \sqrt{25-x^2}\, dx - \int_3^4 \frac{12}{x}\, dx$

Let $x = 5\sin u$, $dx = 5\cos u\, du$ in the first integral.

$= \int_{x=3}^{x=4} 25\cos^2 u\, du - 12\ln x \Big|_3^4$

$= \frac{25}{2}(u + \sin u \cos u)\Big|_{x=3}^{x=4} - 12\ln\frac{4}{3}$

$= \frac{25}{2}\left(\sin^{-1}\frac{x}{5}\right) + \frac{1}{2}x\sqrt{25-x^2}\Big|_3^4 - 12\ln\frac{4}{3}$

$= \frac{25}{2}\left(\sin^{-1}\frac{4}{5} - \sin^{-1}\frac{3}{5}\right) - 12\ln\frac{4}{3}$ sq. units.

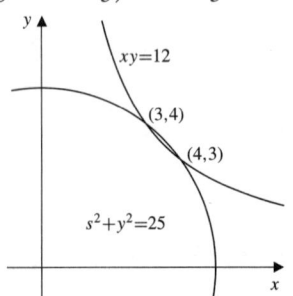

Fig. 6.2.45

46. Shaded area $= 2\int_c^a b\sqrt{1 - \left(\frac{x}{a}\right)^2}\, dx$ Let $x = a\sin u$
$dx = a\cos u\, du$

$= 2ab\int_{x=c}^{x=a} \cos^2 u\, du$

$= ab(u + \sin u \cos u)\Big|_{x=c}^{x=a}$

$= \left(ab\sin^{-1}\frac{x}{a} + \frac{b}{a}x\sqrt{a^2-x^2}\right)\Big|_c^a$

$= ab\left(\frac{\pi}{2} - \sin^{-1}\frac{c}{a}\right) - \frac{cb}{a}\sqrt{a^2-c^2}$ sq. units.

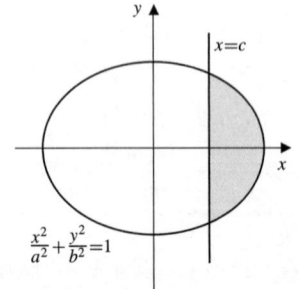

Fig. 6.2.46

47. Area of R

$= \frac{Y}{2}\sqrt{1+Y^2} - \int_1^{\sqrt{1+Y^2}} \sqrt{x^2-1}\, dx$

Let $x = \sec\theta$
$dx = \sec\theta\tan\theta\, d\theta$

$= \frac{Y}{2}\sqrt{1+Y^2} - \int_0^{\tan^{-1}Y} \sec\theta\tan^2\theta\, d\theta$

$= \frac{Y}{2}\sqrt{1+Y^2} - \int_0^{\tan^{-1}Y} \sec^3\theta\, d\theta$

$+ \int_0^{\tan^{-1}Y} \sec\theta\, d\theta$

$= \frac{Y}{2}\sqrt{1+Y^2} + \left(-\frac{1}{2}\sec\theta\tan\theta\right.$

$\left. -\frac{1}{2}\ln|\sec\theta + \tan\theta| + \ln|\sec\theta + \tan\theta|\right)\Big|_0^{\tan^{-1}Y}$

$= \frac{Y}{2}\sqrt{1+Y^2} - \frac{Y}{2}\sqrt{1+Y^2} + \frac{1}{2}\ln(Y + \sqrt{1+Y^2})$

$= \frac{1}{2}\ln(Y + \sqrt{1+Y^2})$ sq. units.

If $Y = \sinh t$, then we have

Area $= \frac{1}{2}\ln(\sinh t + \cosh t) = \frac{1}{2}\ln e^t = \frac{t}{2}$ sq. units.

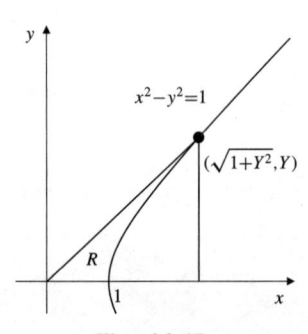

Fig. 6.2.47

48. $\displaystyle\int \frac{dx}{\sqrt{x^2-a^2}}$ Let $x = a\cosh u$
$dx = a\sinh u\,du$

$= \displaystyle\int \frac{a\sinh u\,du}{a\sinh u} = u + C$

$= \cosh^{-1}\dfrac{x}{a} + C = \ln(x + \sqrt{x^2-a^2}) + C, \quad (x \geq a).$

$\displaystyle\int \frac{dx}{x^2\sqrt{x^2-a^2}} = \int \frac{a\sinh u\,du}{a^2\cosh^2 u\, a\sinh u}$

$= \dfrac{1}{a^2}\displaystyle\int \operatorname{sech}^2 u\,du = \dfrac{1}{a^2}\tanh u + C$

$= \dfrac{1}{a^2}\tanh\left(\cosh^{-1}\dfrac{x}{a}\right) + C$

$= \dfrac{1}{a^2} \cdot \dfrac{\dfrac{x}{a} + \sqrt{\dfrac{x^2}{a^2}-1} - \dfrac{1}{\dfrac{x}{a} - \sqrt{\dfrac{x^2}{a^2}-1}}}{\dfrac{x}{a} + \sqrt{\dfrac{x^2}{a^2}-1} + \dfrac{1}{\dfrac{x}{a} - \sqrt{\dfrac{x^2}{a^2}-1}}} + C$

$= \dfrac{\sqrt{x^2-a^2}}{a^2 x} + C_1.$

Section 6.3 Integrals of Rational Functions (page 368)

1. $\displaystyle\int \frac{2\,dx}{2x-3} = \ln|2x-3| + C.$

2. $\displaystyle\int \frac{dx}{5-4x} = -\dfrac{1}{4}\ln|5-4x| + C.$

3. $\displaystyle\int \frac{x\,dx}{\pi x+2} = \dfrac{1}{\pi}\int \dfrac{\pi x+2-2}{\pi x+2}\,dx$

$= \dfrac{x}{\pi} - \dfrac{2}{\pi^2}\ln|\pi x+2| + C.$

4. $\displaystyle\int \frac{x^2}{x-4}\,dx = \int \left(x + 4 + \dfrac{16}{x-4}\right)dx$

$= \dfrac{x^2}{2} + 4x + 16\ln|x-4| + C.$

5. $\dfrac{1}{x^2-9} = \dfrac{A}{x-3} + \dfrac{B}{x+3}$

$= \dfrac{Ax + 3A + Bx - 3B}{x^2-9}$

$\Rightarrow \begin{cases} A+B = 0 \\ 3(A-B) = 1 \end{cases} \Rightarrow A = \dfrac{1}{6},\ B = -\dfrac{1}{6}.$

$\displaystyle\int \frac{dx}{x^2-9} = \dfrac{1}{6}\int \dfrac{dx}{x-3} - \dfrac{1}{6}\int \dfrac{dx}{x+3}$

$= \dfrac{1}{6}\left(\ln|x-3| - \ln|x+3|\right) + C$

$= \dfrac{1}{6}\ln\left|\dfrac{x-3}{x+3}\right| + C.$

6. $\dfrac{1}{5-x^2} = \dfrac{A}{\sqrt{5}-x} + \dfrac{B}{\sqrt{5}+x}$

$= \dfrac{(A+B)\sqrt{5} + (A-B)x}{5-x^2}$

$\Rightarrow \begin{cases} A+B = \dfrac{1}{\sqrt{5}} \\ A-B = 0 \end{cases} \Rightarrow A = B = \dfrac{1}{2\sqrt{5}}.$

$\displaystyle\int \dfrac{1}{5-x^2}\,dx = \dfrac{1}{2\sqrt{5}}\int \left(\dfrac{1}{\sqrt{5}-x} + \dfrac{1}{\sqrt{5}+x}\right)dx$

$= \dfrac{1}{2\sqrt{5}}\left(-\ln|\sqrt{5}-x| + \ln|\sqrt{5}+x|\right) + C$

$= \dfrac{1}{2\sqrt{5}}\ln\left|\dfrac{\sqrt{5}+x}{\sqrt{5}-x}\right| + C.$

7. $\dfrac{1}{a^2-x^2} = \dfrac{A}{a-x} + \dfrac{B}{a+x}$

$= \dfrac{Aa + Ax + Ba - Bx}{a^2-x^2}$

$\Rightarrow \begin{cases} Aa+Ba = 1 \\ A-B = 0 \end{cases} \Rightarrow A = B = \dfrac{1}{2a}.$

Thus

$\displaystyle\int \dfrac{dx}{a^2-x^2} = \dfrac{1}{2a}\int \dfrac{dx}{a-x} + \dfrac{1}{2a}\int \dfrac{dx}{a+x}$

$= \dfrac{1}{2a}\left(-\ln|a-x| + \ln|a+x|\right) + C$

$= \dfrac{1}{2a}\ln\left|\dfrac{a+x}{a-x}\right| + C.$

8. $\dfrac{1}{b^2-a^2 x^2} = \dfrac{A}{b-ax} + \dfrac{B}{b+ax}$

$= \dfrac{(A+B)b + (A-B)ax}{b^2-a^2 x^2}$

$\Rightarrow A = B = \dfrac{1}{2b}$

$\displaystyle\int \dfrac{dx}{b^2-a^2 x^2} = \dfrac{1}{2b}\int \left(\dfrac{1}{b-ax} + \dfrac{1}{b+ax}\right)dx$

$= \dfrac{1}{2b}\left(\dfrac{-\ln|b-ax|}{a} + \dfrac{\ln|b+ax|}{a}\right) + C$

$= \dfrac{1}{2ab}\ln\left|\dfrac{b+ax}{b-ax}\right| + C.$

9. $\displaystyle\int \dfrac{x^2\,dx}{x^2+x-2} = \int \left(1 - \dfrac{x-2}{x^2+x-2}\right)dx$

$= x - \displaystyle\int \dfrac{x-2}{x^2+x-2}\,dx.$

If $\dfrac{x-2}{x^2+x-2} = \dfrac{A}{x+2} + \dfrac{B}{x-1} = \dfrac{Ax - A + Bx + 2B}{x^2+x-2},$

then $A+B = 1$ and $-A + 2B = -2$, so that $A = 4/3$ and $B = -1/3.$ Thus

$\displaystyle\int \dfrac{x^2\,dx}{x^2+x-2} = x - \dfrac{4}{3}\int \dfrac{dx}{x+2} + \dfrac{1}{3}\int \dfrac{dx}{x-1}$

$= x - \dfrac{4}{3}\ln|x+2| + \dfrac{1}{3}\ln|x-1| + C.$

10. $\dfrac{x}{3x^2 + 8x - 3} = \dfrac{A}{3x-1} + \dfrac{B}{x+3}$
$= \dfrac{(A+3B)x + (3A-B)}{3x^2 + 8x - 3}$
$\Rightarrow \begin{cases} A + 3B = 1 \\ 3A - B = 0 \end{cases} \Rightarrow A = \dfrac{1}{10},\ B = \dfrac{3}{10}.$

$\displaystyle \int \dfrac{x\,dx}{3x^2 + 8x - 3} = \dfrac{1}{10} \int \left(\dfrac{1}{3x-1} + \dfrac{3}{x+3} \right) dx$
$= \dfrac{1}{30} \ln|3x-1| + \dfrac{3}{10} \ln|x+3| + C.$

11. $\dfrac{x-2}{x^2 + x} = \dfrac{A}{x} + \dfrac{B}{x+1} = \dfrac{Ax + A + Bx}{x^2 + x}$
$\Rightarrow \begin{cases} A + B = 1 \\ A = -2 \end{cases} \Rightarrow A = -2,\ B = 3.$

$\displaystyle \int \dfrac{x-2}{x^2 + x} dx = 3 \int \dfrac{dx}{x+1} - 2 \int \dfrac{dx}{x}$
$= 3 \ln|x+1| - 2 \ln|x| + C.$

12. $\dfrac{1}{x^3 + 9x} = \dfrac{A}{x} + \dfrac{Bx + C}{x^2 + 9}$
$= \dfrac{Ax^2 + 9A + Bx^2 + Cx}{x^3 + 9x}$
$\Rightarrow \begin{cases} A + B = 0 \\ C = 0 \\ 9A = 1 \end{cases} \Rightarrow A = \dfrac{1}{9},\ B = -\dfrac{1}{9},\ C = 0.$

$\displaystyle \int \dfrac{dx}{x^3 + 9x} = \dfrac{1}{9} \int \left(\dfrac{1}{x} - \dfrac{x}{x^2+9} \right) dx$
$= \dfrac{1}{9} \ln|x| - \dfrac{1}{18} \ln(x^2 + 9) + K.$

13. $\displaystyle \int \dfrac{dx}{1 - 6x + 9x^2} = \int \dfrac{dx}{(1-3x)^2} = \dfrac{1}{3(1-3x)} + C.$

14. $\displaystyle \int \dfrac{x}{2 + 6x + 9x^2} dx = \int \dfrac{x}{(3x+1)^2 + 1} dx$ Let $u = 3x + 1$, $du = 3\,dx$
$= \dfrac{1}{9} \int \dfrac{u-1}{u^2 + 1} du = \dfrac{1}{9} \int \dfrac{u}{u^2+1} du - \dfrac{1}{9} \int \dfrac{1}{u^2 + 1} du$
$= \dfrac{1}{18} \ln(u^2 + 1) - \dfrac{1}{9} \tan^{-1} u + C$
$= \dfrac{1}{18} \ln(2 + 6x + 9x^2) - \dfrac{1}{9} \tan^{-1}(3x+1) + C.$

15. $\displaystyle \int \dfrac{x^2 + 1}{6x - 9x^2} dx = \dfrac{1}{9} \int \dfrac{9x^2 - 6x + 6x + 9}{6x - 9x^2} dx$
$= -\dfrac{x}{9} + \dfrac{1}{9} \int \dfrac{2x + 3}{x(2-3x)} dx.$

Now

$\dfrac{2x+3}{x(2-3x)} = \dfrac{A}{x} + \dfrac{B}{2-3x} = \dfrac{2A - 3Ax + Bx}{x(2-3x)}$
$\Rightarrow 2A = 3,\quad -3A + B = 2$
$\Rightarrow A = \dfrac{3}{2},\quad B = \dfrac{13}{2}.$

Therefore we have

$\displaystyle \int \dfrac{x^2 + 1}{6x - 9x^2} dx$
$= -\dfrac{x}{9} + \dfrac{1}{6} \int \dfrac{dx}{x} + \dfrac{13}{18} \int \dfrac{dx}{2 - 3x}$
$= -\dfrac{x}{9} + \dfrac{1}{6} \ln|x| - \dfrac{13}{54} \ln|2 - 3x| + C.$

16. First divide to obtain

$\dfrac{x^3 + 1}{x^2 + 7x + 12} = x - 7 + \dfrac{37x + 85}{(x+4)(x+3)}$

$\dfrac{37x + 85}{(x+4)(x+3)} = \dfrac{A}{x+4} + \dfrac{B}{x+3}$
$= \dfrac{(A+B)x + 3A + 4B}{x^2 + 7x + 12}$
$\Rightarrow \begin{cases} A + B = 37 \\ 3A + 4B = 85 \end{cases} \Rightarrow A = 63,\ B = -26.$

Now we have

$\displaystyle \int \dfrac{x^3 + 1}{12 + 7x + x^2} dx = \int \left(x - 7 + \dfrac{63}{x+4} - \dfrac{26}{x+3} \right) dx$
$= \dfrac{x^2}{2} - 7x + 63 \ln|x+4| - 26 \ln|x+3| + C.$

17. $\dfrac{1}{x(x^2 - a^2)} = \dfrac{A}{x} + \dfrac{B}{x-a} + \dfrac{C}{x+a}$
$= \dfrac{Ax^2 - Aa^2 + Bx^2 + Bax + Cx^2 - Cax}{x(x^2 - a^2)}$
$\Rightarrow \begin{cases} A + B + C = 0 \\ B - C = 0 \\ -Aa^2 = 1 \end{cases} \Rightarrow \begin{cases} A = -1/a^2 \\ B = C = 1/(2a^2). \end{cases}$

Thus we have

$\displaystyle \int \dfrac{dx}{x(x^2 - a^2)}$
$= \dfrac{1}{2a^2} \left(-2 \int \dfrac{dx}{x} + \int \dfrac{dx}{x-a} + \int \dfrac{dx}{x+a} \right)$
$= \dfrac{1}{2a^2} \left(-2 \ln|x| + \ln|x-a| + \ln|x+a| \right) + K$
$= \dfrac{1}{2a^2} \ln \dfrac{|x^2 - a^2|}{x^2} + K.$

INSTRUCTOR'S SOLUTIONS MANUAL SECTION 6.3 (PAGE 368)

18. The partial fraction decomposition is

$$\frac{1}{x^4 - a^4} = \frac{A}{x-a} + \frac{B}{x+a} + \frac{Cx+D}{x^2+a^2}$$
$$= \frac{A(x^3 + ax^2 + a^2x + a^3) + B(x^3 - ax^2 + a^2x - a^3)}{x^4 - a^4}$$
$$+ \frac{C(x^3 - a^2x) + D(x^2 - a^2)}{x^4 - a^4}$$

$$\Rightarrow \begin{cases} A + B + C = 0 \\ aA - aB + D = 0 \\ a^2A + a^2B - a^2C = 0 \\ a^3A - a^3B - a^2D = 1 \end{cases}$$

$$\Rightarrow A = \frac{1}{4a^3},\ B = -\frac{1}{4a^3},\ C = 0,\ D = -\frac{1}{2a^2}.$$

$$\int \frac{dx}{x^4 - a^4} = \frac{1}{4a^3} \int \left(\frac{1}{x-a} - \frac{1}{x+a} - \frac{2a}{x^2+a^2} \right) dx$$
$$= \frac{1}{4a^3} \ln \left| \frac{x-a}{x+a} \right| - \frac{1}{2a^3} \tan^{-1}\left(\frac{x}{a}\right) + K.$$

19. $\dfrac{a^3}{x^3 - a^3} = \dfrac{A}{x-a} + \dfrac{Bx+C}{x^2 + ax + a^2}$

$$= \frac{Ax^2 + Aax + Aa^2 + Bx^2 - Bax + Cx - Ca}{x^3 - a^3}$$

$$\Rightarrow \begin{cases} A + B = 0 \\ Aa - Ba + C = 0 \\ Aa^2 - Ca = a^3 \end{cases} \Rightarrow \begin{cases} A = a/3 \\ B = -a/3 \\ C = -2a^2/3. \end{cases}$$

Therefore we have

$$\int \frac{x^3}{x^3 - a^3} = \int \left(1 + \frac{a^3}{x^3 - a^3} \right) dx$$
$$= x + \frac{a}{3} \int \frac{dx}{x - a} - \frac{a}{3} \int \frac{x + 2a}{x^2 + ax + a^2}$$
$$= x + \frac{a}{3} \ln|x - a| - \frac{a}{6} \int \frac{2x + a + 3a}{x^2 + ax + a^2}$$
$$= x + \frac{a}{3} \ln|x - a| - \frac{a}{6} \ln(x^2 + ax + a^2)$$
$$- \frac{a^2}{2} \int \frac{dx}{\left(x + \frac{a}{2}\right)^2 + \frac{3}{4}a^2}$$
$$= x + \frac{a}{3} \ln|x - a| - \frac{a}{6} \ln(x^2 + ax + a^2)$$
$$- \frac{a^2}{2} \cdot \frac{2}{\sqrt{3}a} \tan^{-1} \frac{x + (a/2)}{(\sqrt{3}a)/2} + K$$
$$= x + \frac{a}{3} \ln|x - a| - \frac{a}{6} \ln(x^2 + ax + a^2)$$
$$- \frac{a}{\sqrt{3}} \tan^{-1} \frac{2x + a}{\sqrt{3}a} + K.$$

20. Here the expansion is

$$\frac{1}{x^3 + 2x^2 + 2x} = \frac{A}{x} + \frac{Bx+C}{x^2 + 2x + 2}$$
$$= \frac{A(x^2 + 2x + 2) + Bx^2 + Cx}{x^3 + 2x^2 + 2}$$

$$\Rightarrow \begin{cases} A + B = 0 \\ 2A + C = 0 \\ 2A = 1 \end{cases} \Rightarrow A = -B = \frac{1}{2},\ C = -1,$$

so we have

$$\int \frac{dx}{x^3 + 2x^2 + 2x} = \frac{1}{2} \int \frac{dx}{x} - \frac{1}{2} \int \frac{x+2}{x^2+2x+2} dx$$

Let $u = x + 1$
$du = dx$

$$= \frac{1}{2} \ln|x| - \frac{1}{2} \int \frac{u+1}{u^2 + 1} du$$
$$= \frac{1}{2} \ln|x| - \frac{1}{4} \ln(u^2 + 1) - \frac{1}{2} \tan^{-1} u + K$$
$$= \frac{1}{2} \ln|x| - \frac{1}{4} \ln(x^2 + 2x + 2) - \frac{1}{2} \tan^{-1}(x+1) + K.$$

21. $\dfrac{1}{x^3 - 4x^2 + 3x} = \dfrac{A}{x} + \dfrac{B}{x-1} + \dfrac{C}{x-3}$

$$= \frac{A(x^2 - 4x + 3) + B(x^2 - 3x) + C(x^2 - x)}{x^3 - 4x^2 + 3x}$$

$$\Rightarrow \begin{cases} A + B + C = 0 \\ -4A - 3B - C = 0 \\ 3A = 1 \end{cases}$$

$$\Rightarrow A = \frac{1}{3},\ B = -\frac{1}{2},\ C = \frac{1}{6}.$$

Therefore we have

$$\int \frac{dx}{x^3 - 4x^2 + 3x}$$
$$= \frac{1}{3} \int \frac{dx}{x} - \frac{1}{2} \int \frac{dx}{x-1} + \frac{1}{6} \int \frac{dx}{x-3}$$
$$= \frac{1}{3} \ln|x| - \frac{1}{2} \ln|x-1| + \frac{1}{6} \ln|x-3| + K.$$

22. Here the expansion is

$$\frac{x^2 + 1}{x^3 + 8} = \frac{A}{x+2} + \frac{Bx+C}{x^2 - 2x + 4}$$
$$= \frac{A(x^2 - 2x + 4) + B(x^2 + 2x) + C(x+2)}{x^3 + 8}$$

$$\Rightarrow \begin{cases} A + B = 1 \\ -2A + 2B + C = 0 \Rightarrow A = \frac{5}{12},\ B = \frac{7}{12},\ C = -\frac{1}{3}, \\ 4A + 2C = 1 \end{cases}$$

221

so we have

$$\int \frac{x^2+1}{x^3+8}\,dx = \frac{5}{12}\int \frac{dx}{x+2} + \frac{1}{12}\int \frac{7x-4}{(x-1)^2+3}\,dx$$

Let $u = x-1$
$du = dx$

$$= \frac{5}{12}\ln|x+2| + \frac{1}{12}\int \frac{7u+3}{u^2+3}\,du$$

$$= \frac{5}{12}\ln|x+2| + \frac{7}{24}\ln(x^2-2x+4)$$
$$+ \frac{1}{4\sqrt{3}}\tan^{-1}\frac{x-1}{\sqrt{3}} + K.$$

23. $\dfrac{1}{(x^2-1)^2} = \dfrac{A}{x-1} + \dfrac{B}{(x-1)^2} + \dfrac{C}{x+1} + \dfrac{D}{(x+1)^2}$

$= \dfrac{1}{(x^2-1)^2}\Big(A(x-1)(x+1)^2 + B(x+1)^2$
$\quad + C(x+1)(x-1)^2 + D(x-1)^2\Big)$

$\Rightarrow \begin{cases} A+C=0 \\ A+B-C+D=0 \\ -A+2B-C-2D=0 \\ -A+B+C+D=1 \end{cases} \Rightarrow \begin{cases} A=-\dfrac{1}{4} \\ B=C=D=\dfrac{1}{4}. \end{cases}$

Thus

$$\int \frac{dx}{(x^2-1)^2}$$

$$= \frac{1}{4}\left(-\int \frac{dx}{x-1} + \int \frac{dx}{(x-1)^2}\right.$$
$$\left. + \int \frac{dx}{x+1} + \int \frac{dx}{(x+1)^2}\right)$$

$$= \frac{1}{4}\left(\ln|x+1| - \ln|x-1| - \frac{1}{x-1} - \frac{1}{x+1}\right) + K$$

$$= \frac{1}{4}\ln\left|\frac{x+1}{x-1}\right| - \frac{x}{2(x^2-1)} + K.$$

24. The expansion is

$$\frac{x^2}{(x^2-1)(x^2-4)} = \frac{A}{x-1} + \frac{B}{x+1} + \frac{C}{x-2} + \frac{D}{x+2}$$

$A = \lim\limits_{x\to 1}\dfrac{x^2}{(x+1)(x^2-4)} = \dfrac{1}{2(-3)} = -\dfrac{1}{6}$

$B = \lim\limits_{x\to -1}\dfrac{x^2}{(x-1)(x^2-4)} = \dfrac{1}{-2(-3)} = \dfrac{1}{6}$

$C = \lim\limits_{x\to 2}\dfrac{x^2}{(x^2-1)(x+2)} = \dfrac{4}{3(4)} = \dfrac{1}{3}$

$D = \lim\limits_{x\to -2}\dfrac{x^2}{(x^2-1)(x-2)} = \dfrac{4}{3(-4)} = -\dfrac{1}{3}.$

Therefore

$$\int \frac{x^2}{(x^2-1)(x^2-4)}\,dx = -\frac{1}{6}\ln|x-1| + \frac{1}{6}\ln|x+1| +$$
$$\frac{1}{3}\ln|x-2| - \frac{1}{3}\ln|x+2| + K.$$

25. $\dfrac{1}{x^4-3x^3} = \dfrac{1}{x^3(x-3)}$

$= \dfrac{A}{x} + \dfrac{B}{x^2} + \dfrac{C}{x^3} + \dfrac{D}{x-3}$

$= \dfrac{A(x^3-3x^2) + B(x^2-3x) + C(x-3) + Dx^3}{x^3(x-3)}$

$\Rightarrow \begin{cases} A+D=0 \\ -3A+B=0 \\ -3B+C=0 \\ -3C=1 \end{cases} \Rightarrow \begin{cases} A=-1/27 \\ B=-1/9 \\ C=-1/3 \\ D=1/27. \end{cases}$

Therefore

$$\int \frac{dx}{x^4-3x^3}$$

$$= -\frac{1}{27}\int \frac{dx}{x} - \frac{1}{9}\int \frac{dx}{x^2} - \frac{1}{3}\int \frac{dx}{x^3} + \frac{1}{27}\int \frac{dx}{x-3}$$

$$= \frac{1}{27}\ln\left|\frac{x-3}{x}\right| + \frac{1}{9x} + \frac{1}{6x^2} + K.$$

26. We have

$$\int \frac{x\,dx}{(x^2-x+1)^2} = \int \frac{x\,dx}{\left[(x-\frac{1}{2})^2+\frac{3}{4}\right]^2}$$

Let $u = x - \frac{1}{2}$
$du = dx$

$$= \int \frac{u\,du}{(u^2+\frac{3}{4})^2} + \frac{1}{2}\int \frac{du}{(u^2+\frac{3}{4})^2}$$

Let $u = \dfrac{\sqrt{3}}{2}\tan v$,
$du = \dfrac{\sqrt{3}}{2}\sec^2 v\,dv$ in the second integral.

$$= -\frac{1}{2}\left(\frac{1}{u^2+\frac{3}{4}}\right) + \frac{1}{2}\int \frac{\frac{\sqrt{3}}{2}\sec^2 v\,dv}{\frac{9}{16}\sec^4 v}$$

$$= \frac{-1}{2(x^2-x+1)} + \frac{4}{3\sqrt{3}}\int \cos^2 v\,dv$$

$$= \frac{-1}{2(x^2-x+1)} + \frac{2}{3\sqrt{3}}(v + \sin v\cos v) + C$$

$$= \frac{-1}{2(x^2-x+1)} + \frac{2}{3\sqrt{3}}\tan^{-1}\frac{2x-1}{\sqrt{3}} + \frac{2}{3\sqrt{3}}\cdot\frac{2(x-\frac{1}{2})\sqrt{3}}{(2\sqrt{x^2-x+1})^2} +$$

$$= \frac{2}{3\sqrt{3}}\tan^{-1}\frac{2x-1}{\sqrt{3}} + \frac{x-2}{3(x^2-x+1)} + C.$$

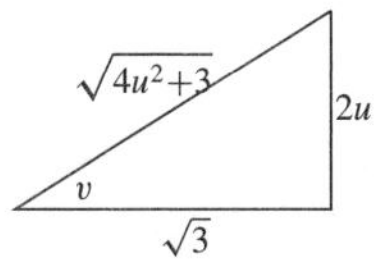

Fig. 6.3.26

27. $\dfrac{t}{(t+1)(t^2+1)^2} = \dfrac{A}{t+1} + \dfrac{Bt+C}{t^2+1} + \dfrac{Dt+E}{(t^2+1)^2}$

$= \dfrac{1}{(t+1)(t^2+1)^2}\Big(A(t^4+2t^2+1) + B(t^4+t^3+t^2+t)$
$\quad + C(t^3+t^2+t+1) + D(t^2+t) + E(t+1)\Big)$

$\Rightarrow \begin{cases} A+B=0 \\ B+C=0 \\ 2A+B+C+D=0 \\ B+C+D+E=1 \\ A+C+E=0 \end{cases} \Rightarrow \begin{cases} B=-A \\ C=A \\ D=-2A \\ 2D=1 \\ E=-2A. \end{cases}$

Thus $A=-1/4=C$, $B=1/4$, $D=E=1/2$. We have

$\displaystyle\int \dfrac{t\,dt}{(t+1)(t^2+1)^2}$

$= -\dfrac{1}{4}\int \dfrac{dt}{t+1} + \dfrac{1}{4}\int \dfrac{(t-1)\,dt}{t^2+1} + \dfrac{1}{2}\int \dfrac{(t+1)\,dt}{(t^2+1)^2}$

$= -\dfrac{1}{4}\ln|t+1| + \dfrac{1}{8}\ln(t^2+1) - \dfrac{1}{4}\tan^{-1}t$
$\quad - \dfrac{1}{4(t^2+1)} + \dfrac{1}{2}\int \dfrac{dt}{(t^2+1)^2}$ 　Let $t=\tan\theta$
$\hfill dt=\sec^2\theta\,d\theta$

$= -\dfrac{1}{4}\ln|t+1| + \dfrac{1}{8}\ln(t^2+1) - \dfrac{1}{4}\tan^{-1}t$
$\quad - \dfrac{1}{4(t^2+1)} + \dfrac{1}{2}\int \cos^2\theta\,d\theta$

$= -\dfrac{1}{4}\ln|t+1| + \dfrac{1}{8}\ln(t^2+1) - \dfrac{1}{4}\tan^{-1}t$
$\quad - \dfrac{1}{4(t^2+1)} + \dfrac{1}{4}(\theta + \sin\theta\cos\theta) + K$

$= -\dfrac{1}{4}\ln|t+1| + \dfrac{1}{8}\ln(t^2+1) - \dfrac{1}{4}\tan^{-1}t$
$\quad - \dfrac{1}{4(t^2+1)} + \dfrac{1}{4}\tan^{-1}t + \dfrac{1}{4}\dfrac{t}{t^2+1} + K$

$= \dfrac{1}{4}\dfrac{t-1}{t^2+1} - \dfrac{1}{4}\ln|t+1| + \dfrac{1}{8}\ln(t^2+1) + K.$

28. We have

$\displaystyle\int \dfrac{dt}{(t-1)(t^2-1)^2}$

$= \displaystyle\int \dfrac{dt}{(t-1)^3(t+1)^2}$ 　Let $u=t-1$
$\hfill du=dt$

$= \displaystyle\int \dfrac{du}{u^3(u+2)^2}$

$\dfrac{1}{u^3(u+2)^2} = \dfrac{A}{u} + \dfrac{B}{u^2} + \dfrac{C}{u^3} + \dfrac{D}{u+2} + \dfrac{E}{(u+2)^2}$

$= \dfrac{A(u^4+4u^3+4u^2) + B(u^3+4u^2+4u)}{u^3(u+2)^2}$
$\quad \dfrac{C(u^2+4u+4) + D(u^4+2u^3) + Eu^3}{u^3(u+2)^2}$

$\Rightarrow \begin{cases} A+D=0 \\ 4A+B+2D+E=0 \\ 4A+4B+C=0 \\ 4B+4C=0 \\ 4C=1 \end{cases}$

$\Rightarrow A=\dfrac{3}{16},\ B=-\dfrac{1}{4},\ C=\dfrac{1}{4},\ D=-\dfrac{3}{16},\ E=-\dfrac{1}{8}.$

$\displaystyle\int \dfrac{du}{u^3(u+2)^2}$

$= \dfrac{3}{16}\int \dfrac{du}{u} - \dfrac{1}{4}\int \dfrac{du}{u^2} + \dfrac{1}{4}\int \dfrac{du}{u^3}$
$\quad - \dfrac{3}{16}\int \dfrac{du}{u+2} - \dfrac{1}{8}\int \dfrac{du}{(u+2)^2}$

$= \dfrac{3}{16}\ln|t-1| + \dfrac{1}{4(t-1)} - \dfrac{1}{8(t-1)^2} -$
$\quad \dfrac{3}{16}\ln|t+1| + \dfrac{1}{8(t+1)} + K.$

29. $I \displaystyle\int \dfrac{dx}{x(3+x^2)\sqrt{1-x^2}}$ 　Let $1-x^2=u^2$
$\hfill -2x\,dx=2u\,du$

$= -\displaystyle\int \dfrac{du}{(1-u^2)(4-u^2)}.$

$\dfrac{1}{(1-u^2)(4-u^2)}$
$= \dfrac{A}{1-u} + \dfrac{B}{1+u} + \dfrac{C}{2-u} + \dfrac{D}{2+u}$

$A = \left.\dfrac{1}{(1+u)(4-u^2)}\right|_{u=1} = \dfrac{1}{6}$

$B = \left.\dfrac{1}{(1-u)(4-u^2)}\right|_{u=-1} = \dfrac{1}{6}$

$C = \left.\dfrac{1}{(1-u^2)(2+u)}\right|_{u=2} = -\dfrac{1}{12}$

$D = \left.\dfrac{1}{(1-u^2)(2-u)}\right|_{u=-2} = -\dfrac{1}{12}.$

Thus

$$I = -\left(\frac{1}{6}\int \frac{du}{1-u} + \frac{1}{6}\int \frac{du}{1+u}\right.$$
$$\left. - \frac{1}{12}\int \frac{du}{2-u} - \frac{1}{12}\int \frac{du}{2+u}\right)$$
$$= \frac{1}{6}\ln\left|\frac{1-u}{1+u}\right| + \frac{1}{12}\ln\left|\frac{2+u}{2-u}\right| + K$$
$$= \frac{1}{6}\ln\left|\frac{1-\sqrt{1-x^2}}{1+\sqrt{1-x^2}}\right| + \frac{1}{12}\ln\left|\frac{2+\sqrt{1-x^2}}{2-\sqrt{1-x^2}}\right| + K$$
$$= \frac{1}{6}\ln\frac{(1-\sqrt{1-x^2})^2}{x^2} + \frac{1}{12}\ln\frac{(2+\sqrt{1-x^2})^2}{3+x^2} + K.$$

30. $\displaystyle\int \frac{dx}{e^{2x} - 4e^x + 4} = \int \frac{dx}{(e^x-2)^2}$ Let $u = e^x$
$du = e^x\, dx$
$$= \int \frac{du}{u(u-2)^2}$$
$$\frac{1}{u(u-2)^2} = \frac{A}{u} + \frac{B}{u-2} + \frac{C}{(u-2)^2}$$
$$= \frac{A(u^2 - 4u + 4) + B(u^2 - 2u) + +Cu}{u(u-2)^2}$$
$$\Rightarrow \begin{cases} A + B = 0 \\ -4A - 2B + C = 0 \\ 4A = 1 \end{cases} \Rightarrow A = \frac{1}{4},\ B = -\frac{1}{4},\ C = \frac{1}{2}.$$
$$\int \frac{du}{u(u-2)^2} = \frac{1}{4}\int \frac{du}{u} - \frac{1}{4}\int \frac{du}{u-2} + \frac{1}{2}\int \frac{du}{(u-2)^2}$$
$$= \frac{1}{4}\ln|u| - \frac{1}{4}\ln|u-2| - \frac{1}{2}\frac{1}{(u-2)} + K$$
$$= \frac{x}{4} - \frac{1}{4}\ln|e^x - 2| - \frac{1}{2(e^x - 2)} + K.$$

31. $\displaystyle I = \int \frac{dx}{x(1+x^2)^{3/2}}$ Let $x = \tan\theta$
$dx = \sec^2\theta\, d\theta$
$$= \int \frac{\sec^2\theta\, d\theta}{\tan\theta \sec^3\theta} = \int \frac{\cos^2\theta\, d\theta}{\sin\theta}$$
$$= \int \frac{\cos^2\theta \sin\theta\, d\theta}{\sin^2\theta}\quad \text{Let } u = \cos\theta$$
$du = -\sin\theta\, d\theta$
$$= -\int \frac{u^2\, du}{1-u^2} = u + \int \frac{du}{u^2 - 1}.$$
We have
$$\frac{1}{u^2 - 1} = \frac{1}{2}\left(\frac{1}{u-1} - \frac{1}{u+1}\right).$$

Thus
$$I = u + \frac{1}{2}\ln\left|\frac{u-1}{u+1}\right| + C$$
$$= \cos\theta + \frac{1}{2}\ln\left|\frac{\cos\theta - 1}{\cos\theta + 1}\right| + C$$
$$= \frac{1}{\sqrt{1+x^2}} + \frac{1}{2}\ln\left|\frac{\frac{1}{\sqrt{1+x^2}} - 1}{\frac{1}{\sqrt{1+x^2}} + 1}\right| + C$$
$$= \frac{1}{\sqrt{1+x^2}} + \frac{1}{2}\ln\left(\frac{\sqrt{1+x^2} - 1}{\sqrt{1+x^2} + 1}\right) + C.$$

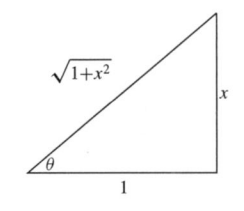

Fig. 6.3.31

32. We have
$$I = \int \frac{dx}{x(1-x^2)^{3/2}}\quad \text{Let } u^2 = 1 - x^2$$
$2u\, du = -2x\, dx$
$$= -\int \frac{u\, du}{(1-u^2)u^3} = -\int \frac{du}{(1-u^2)u^2}$$
$$\frac{1}{u^2(1-u^2)} = \frac{A}{u} + \frac{B}{u^2} + \frac{C}{1-u} + \frac{D}{1+u}$$
$$= \frac{A(u - u^3) + B(1 - u^2) + C(u^2 + u^3) + D(u^2 - u^3)}{u^2(1-u^2)}$$
$$\Rightarrow \begin{cases} -A + C - D = 0 \\ -B + C + D = 0 \\ A = 0 \\ B = 1 \end{cases}$$
$$\Rightarrow A = 0,\ B = 1,\ C = \frac{1}{2},\ D = \frac{1}{2}.$$
$$I = -\int \frac{du}{(1-u^2)u^2} = -\int \frac{du}{u^2} - \frac{1}{2}\int \frac{du}{1-u} - \frac{1}{2}\int \frac{du}{1+u}$$
$$= \frac{1}{u} + \frac{1}{2}\ln|1 - u| - \frac{1}{2}\ln|1 + u| + K$$
$$= \frac{1}{\sqrt{1-x^2}} + \frac{1}{2}\ln\left|\frac{1 - \sqrt{1-x^2}}{1 + \sqrt{1-x^2}}\right| + K$$
$$= \frac{1}{\sqrt{1-x^2}} + \ln\left(1 - \sqrt{1-x^2}\right) - \ln|x| + K.$$

33. $\int \dfrac{dx}{x^2(x^2-1)^{3/2}}$ Let $x = \sec\theta$
$\qquad dx = \sec\theta \tan\theta\, d\theta$

$= \int \dfrac{\sec\theta \tan\theta\, d\theta}{\sec^2\theta \tan^3\theta} = \int \dfrac{\cos^3\theta\, d\theta}{\sin^2\theta}$

$= \int \dfrac{1-\sin^2\theta}{\sin^2\theta}\cos\theta\, d\theta$ Let $u = \sin\theta$
$\qquad du = \cos\theta\, d\theta$

$= \int \dfrac{1-u^2}{u^2}\, du = -\dfrac{1}{u} - u + C$

$= -\left(\dfrac{1}{\sin\theta} + \sin\theta\right) + C$

$= -\left(\dfrac{x}{\sqrt{x^2-1}} + \dfrac{\sqrt{x^2-1}}{x}\right) + C.$

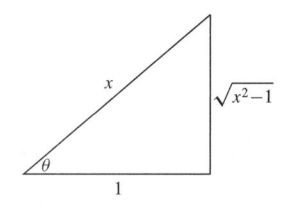

Fig. 6.3.33

34. $\int \dfrac{d\theta}{\cos\theta(1+\sin\theta)}$ Let $u = \sin\theta$
$\qquad du = \cos\theta\, d\theta$

$= \int \dfrac{du}{(1-u^2)(1+u)} = \int \dfrac{du}{(1-u)(1+u)^2}$

$\dfrac{1}{(1-u)(1+u)^2} = \dfrac{A}{1-u} + \dfrac{B}{1+u} + \dfrac{C}{(1+u)^2}$

$= \dfrac{A(1+2u+u^2) + B(1-u^2) + C(1-u)}{(1-u)(1+u)^2}$

$\Rightarrow \begin{cases} A - B = 0 \\ 2A - C = 0 \\ A + B + C = 1 \end{cases} \Rightarrow A = \dfrac{1}{4},\ B = \dfrac{1}{4},\ C = \dfrac{1}{2}.$

$\int \dfrac{du}{(1-u)(1+u)^2}$

$= \dfrac{1}{4}\int \dfrac{du}{1-u} + \dfrac{1}{4}\int \dfrac{du}{1+u} + \dfrac{1}{2}\int \dfrac{du}{(1+u)^2}$

$= \dfrac{1}{4}\ln\left|\dfrac{1+\sin\theta}{1-\sin\theta}\right| - \dfrac{1}{2(1+\sin\theta)} + C.$

35. Since $Q(x) = (x-a_1)(x-a_2)\cdots(x-a_n)$, we have

$\ln Q(x) = \ln(x-a_1) + \ln(x-a_2) + \cdots + \ln(x-a_n),$

and, differentiating both sides,

$\dfrac{Q'(x)}{Q(x)} = \dfrac{d}{dx}[\ln Q(x)] = \dfrac{1}{x-a_1} + \dfrac{1}{x-a_2} + \cdots + \dfrac{1}{x-a_n}$

$\dfrac{1}{Q(x)} = \dfrac{1}{Q'(x)}\left[\dfrac{1}{x-a_1} + \dfrac{1}{x-a_2} + \cdots + \dfrac{1}{x-a_n}\right].$

Since

$\dfrac{P(x)}{Q(x)} = \dfrac{A_1}{x-a_1} + \dfrac{A_2}{x-a_2} + \cdots + \dfrac{A_n}{x-a_n},$

we have

$\dfrac{P(x)}{Q'(x)}\left[\dfrac{1}{x-a_1} + \dfrac{1}{x-a_2} + \cdots + \dfrac{1}{x-a_n}\right]$

$= \dfrac{A_1}{x-a_1} + \dfrac{A_2}{x-a_2} + \cdots + \dfrac{A_n}{x-a_n}.$

Multiply both sides by $x - a_1$ and get

$\dfrac{P(x)}{Q'(x)}\left[1 + \dfrac{x-a_1}{x-a_2} + \cdots + \dfrac{x-a_1}{x-a_n}\right]$

$= A_1 + \dfrac{A_2(x-a_1)}{x-a_2} + \cdots + \dfrac{A_n(x-a_1)}{x-a_n}.$

Now let $x = a_1$ and obtain $\dfrac{P(a_1)}{Q'(a_1)} = A_1.$

Similarly, $A_j = \dfrac{P(a_j)}{Q'(a_j)}$ for $1 \le j \le n$.

Section 6.4 Integration Using Computer Algebra or Tables   (page 373)

2. According to Maple

$\int \dfrac{1+x+x^2}{(x^4-1)(x^4-16)^2}\, dx$

$= \dfrac{\ln(x-1)}{300} - \dfrac{\ln(x+1)}{900} - \dfrac{7}{15{,}360(x-2)}$

$\quad - \dfrac{613}{460{,}800}\ln(x-2) - \dfrac{1}{5{,}120(x+2)} + \dfrac{79}{153{,}600}\ln(x+2)$

$\quad - \dfrac{\ln(x^2+1)}{900} + \dfrac{47}{115{,}200}\ln(x^2+4)$

$\quad - \dfrac{23}{25{,}600}\tan^{-1}(x/2) - \dfrac{6x+8}{15{,}360(x^2+4)}$

One suspects it has forgotten to use absolute values in some of the logarithms.

3. Neither the author's version of Maple nor his version of Mathematics would do

$I = \int \dfrac{t^5}{\sqrt{3-2t^4}}\, dt$

as presented. Both did an integration by parts and left an unevaluated integral. Both managed to evaluate the integral after the substitution $u = t^2$ was made. (See Exercise 4.) However, Derive had no trouble doing the integral in its original form, giving as the answer

$\dfrac{3\sqrt{2}}{16}\sin^{-1}\dfrac{\sqrt{6}t^2}{3} - \dfrac{t^2\sqrt{3-2t^4}}{8}.$

SECTION 6.4 (PAGE 373)

All three programs readily gave

$$\int_0^1 \frac{1}{(x^2+1)^3}\,dx = \frac{3\pi}{32} + \frac{1}{4}.$$

5. Use the 6th integral in the list involving $\sqrt{x^2 \pm a^2}$.

$$\int \frac{x^2\,dx}{\sqrt{x^2-2}} = \frac{x}{2}\sqrt{x^2-2} + \ln|x+\sqrt{x^2-2}| + C$$

6. Use the last integral in the list involving $\sqrt{x^2 \pm a^2}$.

$$\int \sqrt{(x^2+4)^3}\,dx = \frac{x}{4}(x^2+10)\sqrt{x^2+4} + 6\ln|x+\sqrt{x^2+4}| + C$$

7. Use the 8th integral in the list involving $\sqrt{x^2 \pm a^2}$ after making the change of variable $x = \sqrt{3}t$.

$$\int \frac{dt}{t^2\sqrt{3t^2+5}} \quad \text{Let } x = \sqrt{3}t$$
$$\quad dx = \sqrt{3}\,dt$$
$$= \frac{3}{\sqrt{3}}\int \frac{dx}{x^2\sqrt{x^2+5}}$$
$$= -\sqrt{3}\,\frac{\sqrt{x^2+5}}{5x} + C = -\frac{\sqrt{3t^2+5}}{5t} + C$$

8. Use the 8th integral in the miscellaneous algebraic set.

$$\int \frac{dt}{t\sqrt{3t-5}} = \frac{2}{\sqrt{5}}\tan^{-1}\sqrt{\frac{3t-5}{5}} + C$$

9. The 5th and 4th integrals in the exponential/logarithmic set give

$$\int x^4(\ln x)^4\,dx = \frac{x^5(\ln x)^4}{5} - \frac{4}{5}\int x^4(\ln x)^3\,dx$$
$$= \frac{x^5(\ln x)^4}{5} - \frac{4}{5}\left(\frac{x^5(\ln x)^3}{5} - \frac{3}{5}\int x^4(\ln x)^2\,dx\right)$$
$$= x^5\left(\frac{(\ln x)^4}{5} - \frac{4(\ln x)^3}{25}\right)$$
$$\quad + \frac{12}{25}\left(\frac{x^5(\ln x)^2}{5} - \frac{2}{5}\int x^5 \ln x\,dx\right)$$
$$= x^5\left(\frac{(\ln x)^4}{5} - \frac{4(\ln x)^3}{25} + \frac{12(\ln x)^2}{125} - \frac{24\ln x}{625} + \frac{24}{3{,}125}\right) + C.$$

10. We make a change of variable and then use the first two integrals in the exponential/logarithmic set.

$$\int x^7 e^{x^2}\,dx \quad \text{Let } u = x^2$$
$$\quad du = 2x\,dx$$
$$= \frac{1}{2}\int u^3 e^u\,du$$
$$= \frac{1}{2}\left(u^3 e^u - 3\int u^2 e^u\,du\right)$$
$$= \frac{u^3 e^u}{2} - \frac{3}{2}\left(u^2 e^u - 2\int u e^u\,du\right)$$
$$= \left(\frac{u^3}{2} - \frac{3u^2}{2} + 3(u-1)\right)e^u + C$$
$$= \left(\frac{x^6}{2} - \frac{3x^4}{2} + 3x^2 - 3\right)e^{x^2} + C$$

11. Use integrals 14 and 12 in the miscellaneous algebraic set.

$$\int x\sqrt{2x-x^2}\,dx$$
$$= -\frac{(2x-x^2)^{3/2}}{3} + \frac{3}{3}\int \sqrt{2x-x^2}\,dx$$
$$= -\frac{(2x-x^2)^{3/2}}{3} + \frac{x-1}{2}\sqrt{2x-x^2} + \frac{1}{2}\sin^{-1}(x-1) + C$$

12. Use integrals 17 and 16 in the miscellaneous algebraic set.

$$\int \frac{\sqrt{2x-x^2}}{x^2}\,dx$$
$$= -\frac{(2x-x^2)^{3/2}}{x^2} - \frac{1}{1}\int \frac{\sqrt{2x-x^2}}{x}\,dx$$
$$= -\frac{(2x-x^2)^{3/2}}{x^2} - \sqrt{2x-x^2} - \sin^{-1}(x-1) + C$$

13. Use the last integral in the miscellaneous algebraic set.

$$\int \frac{dx}{(\sqrt{4x-x^2})^3} = \frac{x-2}{4}\frac{1}{\sqrt{4x-x^2}} + C$$

14. Use the last integral in the miscellaneous algebraic set. Then complete the square, change variables, and use the second last integral in the elementary list.

$$\int \frac{dx}{(\sqrt{4x-x^2})^4}$$
$$= \frac{x-2}{8}(\sqrt{4x-x^2})^{-2} + \frac{1}{8}\int \frac{dx}{4x-x^2}$$
$$= \frac{x-2}{8(4x-x^2)} + \frac{1}{8}\int \frac{dx}{4-(x-2)^2} \quad \text{Let } u = x-2$$
$$\qquad\qquad\qquad\qquad\qquad\qquad du = dx$$
$$= \frac{x-2}{8(4x-x^2)} + \frac{1}{8}\int \frac{du}{4-u^2}$$
$$= \frac{x-2}{8(4x-x^2)} + \frac{1}{32}\ln\left|\frac{u+2}{u-2}\right| + C$$
$$= \frac{x-2}{8(4x-x^2)} + \frac{1}{32}\ln\left|\frac{x}{x-4}\right| + C$$

Section 6.5 Improper Integrals (page 381)

1. $\int_2^\infty \frac{1}{(x-1)^3}\,dx$ Let $u = x-1$, $du = dx$
$$= \int_1^\infty \frac{du}{u^3} = \lim_{R\to\infty}\int_1^R \frac{du}{u^3}$$
$$= \lim_{R\to\infty}\frac{-1}{2u^2}\bigg|_1^R = \lim_{R\to\infty}\left(\frac{1}{2} - \frac{1}{2R^2}\right) = \frac{1}{2}$$

2. $\int_3^\infty \frac{1}{(2x-1)^{2/3}}\,dx$ Let $u = 2x-1$, $du = 2\,dx$
$$= \frac{1}{2}\int_5^\infty \frac{du}{u^{2/3}} = \frac{1}{2}\lim_{R\to\infty}\int_5^R u^{-2/3}\,du$$
$$= \frac{1}{2}\lim_{R\to\infty} 3u^{1/3}\bigg|_5^R = \infty \quad \text{(diverges)}$$

3. $\int_0^\infty e^{-2x}\,dx = \lim_{R\to\infty}\frac{e^{-2x}}{-2}\bigg|_0^R$
$$= \lim_{R\to\infty}\left(\frac{1}{2} - \frac{1}{2e^R}\right)$$
$$= \frac{1}{2}. \quad \text{This integral converges.}$$

4. $\int_{-\infty}^{-1} \frac{dx}{x^2+1} = \lim_{R\to-\infty}\int_R^{-1}\frac{dx}{x^2+1}$
$$= \lim_{R\to-\infty}\left[\tan^{-1}(-1) - \tan^{-1}(R)\right]$$
$$= -\frac{\pi}{4} - \left(-\frac{\pi}{2}\right) = \frac{\pi}{4}.$$
This integral converges.

5. $\int_{-1}^1 \frac{dx}{(x+1)^{2/3}} = \lim_{c\to -1+} 3(x+1)^{1/3}\bigg|_c^1$
$$= \lim_{c\to -1+} 3\left(2^{1/3} - (1+c)^{1/3}\right)$$
$$= 3\sqrt[3]{2}. \quad \text{This integral converges.}$$

6. $\int_0^a \frac{dx}{a^2-x^2} = \lim_{C\to a-}\int_0^C \frac{dx}{a^2-x^2}$
$$= \lim_{C\to a-}\frac{1}{2a}\ln\left|\frac{a+x}{a-x}\right|\bigg|_0^C$$
$$= \lim_{C\to a-}\frac{1}{2a}\ln\frac{a+C}{a-C} = \infty.$$
The integral diverges to infinity.

7. $\int_0^1 \frac{dx}{(1-x)^{1/3}}$ Let $u = 1-x$, $du = -dx$
$$= \int_0^1 \frac{du}{u^{1/3}} = \lim_{c\to 0+}\int_c^1 \frac{du}{u^{1/3}}$$
$$= \lim_{c\to 0+}\frac{3}{2}u^{2/3}\bigg|_c^1 = \frac{3}{2}$$

8. $\int_0^1 \frac{dx}{x\sqrt{1-x}}$ Let $u^2 = 1-x$, $2u\,du = -dx$
$$= \int_0^1 \frac{2u\,du}{(1-u^2)u} = 2\lim_{c\to 1-}\int_0^c \frac{du}{1-u^2}$$
$$= 2\lim_{c\to 1-}\frac{1}{2}\ln\left|\frac{u+1}{u-1}\right|\bigg|_0^c = \infty \quad \text{(diverges)}$$

9. $\int_0^{\pi/2} \frac{\cos x\,dx}{(1-\sin x)^{2/3}}$ Let $u = 1-\sin x$, $du = -\cos x\,dx$
$$= \int_0^1 u^{-2/3}\,du = \lim_{c\to 0+} 3u^{1/3}\bigg|_c^1 = 3.$$
The integral converges.

10. $\int_0^\infty xe^{-x}\,dx$
$$= \lim_{R\to\infty}\int_0^R xe^{-x}\,dx$$
$$\quad U = x \qquad dV = e^{-x}\,dx$$
$$\quad dU = dx \qquad V = -e^{-x}$$
$$= \lim_{R\to\infty}\left(-xe^{-x}\bigg|_0^R + \int_0^R e^{-x}\,dx\right)$$
$$= \lim_{R\to\infty}\left(-\frac{R}{e^R} - \frac{1}{e^R} + 1\right) = 1.$$
The integral converges.

11. $\displaystyle\int_0^1 \frac{dx}{\sqrt{x(1-x)}} = 2\int_0^{1/2} \frac{dx}{\sqrt{\frac{1}{4}-\left(x-\frac{1}{2}\right)^2}}$

$\displaystyle = 2\lim_{c\to 0+}\int_c^{1/2} \frac{dx}{\sqrt{\frac{1}{4}-\left(x-\frac{1}{2}\right)^2}}$

$\displaystyle = 2\lim_{c\to 0+} \sin^{-1}(2x-1)\Big|_c^{1/2} = \pi.$

The integral converges.

12. $\displaystyle\int_0^\infty \frac{x}{1+2x^2}\,dx = \lim_{R\to\infty}\int_0^R \frac{x}{1+2x^2}\,dx$

$\displaystyle = \lim_{R\to\infty} \frac{1}{4}\ln(1+2x^2)\Big|_0^R$

$\displaystyle = \lim_{R\to\infty}\left[\frac{1}{4}\ln(1+2R^2) - \frac{1}{4}\ln 1\right] = \infty.$

This integral diverges to infinity.

13. $\displaystyle\int_0^\infty \frac{x\,dx}{(1+2x^2)^{3/2}}$ Let $u = 1+2x^2$
$du = 4x\,dx$

$\displaystyle = \frac{1}{4}\int_1^\infty \frac{du}{u^{3/2}} = \frac{1}{4}\lim_{R\to\infty}\left(-\frac{2}{\sqrt{u}}\right)\Big|_1^R$

$\displaystyle = \frac{1}{2}.$ The integral converges.

14. $\displaystyle\int_0^{\pi/2} \sec x\,dx = \lim_{C\to(\pi/2)-}\ln|\sec x + \tan x|\Big|_0^C$

$\displaystyle = \lim_{C\to(\pi/2)-}\ln|\sec C + \tan C| = \infty.$

This integral diverges to infinity.

15. $\displaystyle\int_0^{\pi/2} \tan x\,dx = \lim_{c\to(\pi/2)-}\ln|\sec x|\Big|_0^c$

$\displaystyle = \lim_{c\to(\pi/2)-}\ln\sec c = \infty.$

This integral diverges to infinity.

16. $\displaystyle\int_e^\infty \frac{dx}{x(\ln x)}$ Let $u = \ln x$
$du = \dfrac{dx}{x}$

$\displaystyle = \lim_{R\to\infty}\int_1^{\ln R} \frac{du}{u} = \lim_{R\to\infty}\ln|u|\Big|_1^{\ln R}$

$\displaystyle = \lim_{R\to\infty}\ln(\ln R) - \ln 1 = \infty.$

This integral diverges to infinity.

17. $\displaystyle\int_1^e \frac{dx}{x\sqrt{\ln x}}$ Let $u = \ln x$
$du = dx/x$

$\displaystyle = \int_0^1 \frac{du}{\sqrt{u}} = \lim_{c\to 0+} 2\sqrt{u}\Big|_c^1 = 2.$

This integral converges.

18. $\displaystyle\int_e^\infty \frac{dx}{x(\ln x)^2}$ Let $u = \ln x$
$du = \dfrac{dx}{x}$

$\displaystyle = \lim_{R\to\infty}\int_1^{\ln R} \frac{du}{u^2} = \lim_{R\to\infty}\left(-\frac{1}{\ln R}+1\right) = 1.$

The integral converges.

19. $\displaystyle I = \int_{-\infty}^\infty \frac{x\,dx}{1+x^2} = \int_{-\infty}^0 + \int_0^\infty = I_1 + I_2$

$\displaystyle I_2 = \int_0^\infty \frac{x\,dx}{1+x^2}$ Let $u = 1+x^2$
$du = 2x\,dx$

$\displaystyle = \lim_{R\to\infty}\frac{1}{2}\int_1^R \frac{du}{u} = \infty$ (diverges)

20. $\displaystyle I = \int_{-\infty}^\infty \frac{x\,dx}{1+x^4} = \int_{-\infty}^0 + \int_0^\infty = I_1 + I_2$

$\displaystyle I_2 = \int_0^\infty \frac{x\,dx}{1+x^4}$ Let $u = x^2$
$du = 2x\,dx$

$\displaystyle = \frac{1}{2}\int_0^\infty \frac{du}{1+u^2} = \frac{1}{2}\lim_{R\to\infty}\tan^{-1}u\Big|_0^R = \frac{\pi}{4}$

Similarly, $I_1 = -\dfrac{\pi}{4}$. Therefore, $I = 0$.

21. $\displaystyle I = \int_{-\infty}^\infty xe^{-x^2}\,dx = \int_{-\infty}^0 + \int_0^\infty = I_1 + I_2$

$\displaystyle I_2 = \int_0^\infty xe^{-x^2}\,dx$ Let $u = x^2$
$du = 2x\,dx$

$\displaystyle = \frac{1}{2}\int_0^\infty e^{-u}\,du = \frac{1}{2}\lim_{R\to\infty}-e^{-u}\Big|_0^R = \frac{1}{2}$

Similarly, $I_1 = -\dfrac{1}{2}$. Therefore, $I = 0$.

22. $\displaystyle I = \int_{-\infty}^\infty e^{-|x|}\,dx = \int_{-\infty}^0 e^x\,dx + \int_0^\infty e^{-x}\,dx = I_1 + I_2$

$\displaystyle I_2 = \int_0^\infty e^{-x}\,dx = 1$

Similarly, $I_1 = 1$. Therefore, $I = 2$.

23. Area of $\displaystyle R = -\int_0^1 \ln x\,dx = -\lim_{c\to 0+}(x\ln x - x)\Big|_c^1$

$\displaystyle = -(0-1) + \lim_{c\to 0+}(c\ln c - c)$

$= 1 - 0 = 1$ sq. units.

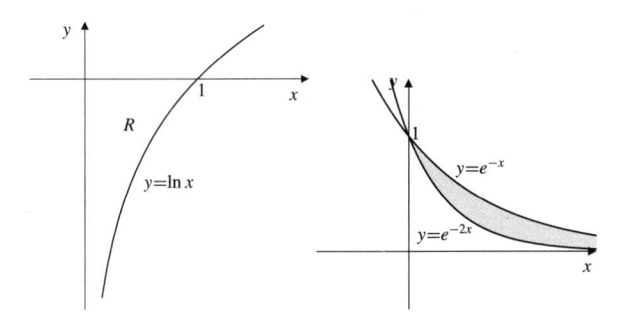

Fig. 6.5.23   Fig. 6.5.24

24. Area of shaded region $= \int_0^\infty (e^{-x} - e^{-2x})\,dx$

$$= \lim_{R\to\infty} \left(-e^{-x} + \frac{1}{2}e^{-2x}\right)\Big|_0^R$$

$$= \lim_{R\to\infty} \left(-e^{-R} + \frac{1}{2}e^{-2R} + 1 - \frac{1}{2}\right) = \frac{1}{2} \text{ sq. units.}$$

25. Area $= \int_1^\infty \left(\frac{4}{2x+1} - \frac{2}{x+2}\right) dx$

$$= \lim_{R\to\infty} 2\bigl(\ln(2x+1) - \ln(x+2)\bigr)\Big|_1^R$$

$$= \lim_{R\to\infty} 2\ln\left(\frac{2R+1}{R+2}\right) - 0 = 2\ln 2 \text{ sq. units.}$$

26. The required area is

$$\text{Area} = \int_0^\infty x^{-2} e^{-1/x}\,dx$$

$$= \int_0^1 x^{-2} e^{-1/x}\,dx + \int_1^\infty x^{-2} e^{-1/x}\,dx$$

$$= I_1 + I_2.$$

Then let $u = -\dfrac{1}{x}$ and $du = x^{-2}\,dx$ in both I_1 and I_2:

$$I_1 = \lim_{C\to 0+} \int_C^1 x^{-2} e^{-1/x}\,dx = \lim_{C\to 0+} \int_{-1/C}^{-1} e^u\,du$$

$$= \lim_{C\to 0+} (e^{-1} - e^{-1/C}) = \frac{1}{e}.$$

$$I_2 = \lim_{R\to\infty} \int_1^R x^{-2} e^{-1/x}\,dx = \lim_{R\to\infty} \int_{-1}^{-1/R} e^u\,du$$

$$= \lim_{R\to\infty} (e^{-1/R} - e^{-1}) = 1 - \frac{1}{e}.$$

Hence, the total area is $I_1 + I_2 = 1$ square unit.

27. First assume that $p \neq 1$. Since $a > 0$ we have

$$\int_a^\infty x^{-p}\,dx = \lim_{R\to\infty} \frac{x^{-p+1}}{-p+1}\Big|_a^R$$

$$= \frac{-a^{-p+1}}{1-p} + \lim_{R\to\infty} \frac{1}{(1-p)R^{p-1}}$$

$$= \begin{cases} \dfrac{1}{(p-1)a^{p-1}} & \text{if } p > 1 \\ \infty & \text{if } p < 1 \end{cases}$$

$$\int_0^a x^{-p}\,dx = \lim_{c\to 0+} \frac{x^{-p+1}}{-p+1}\Big|_c^a$$

$$= \frac{a^{-p+1}}{1-p} + \lim_{c\to 0+} \frac{c^{1-p}}{p-1}$$

$$= \begin{cases} \dfrac{a^{1-p}}{1-p} & \text{if } p < 1 \\ \infty & \text{if } p > 1. \end{cases}$$

If $p = 1$ both integrals diverge as shown in Examples 2 and 6(a).

28. $\displaystyle\int_{-1}^1 \frac{x\,\text{sgn}\,x}{x+2}\,dx = \int_{-1}^0 \frac{-x}{x+2}\,dx + \int_0^1 \frac{x}{x+2}\,dx$

$$= \int_{-1}^0 \left(-1 + \frac{2}{x+2}\right) dx + \int_0^1 \left(1 - \frac{2}{x+2}\right) dx$$

$$= (-x + 2\ln|x+2|)\Big|_{-1}^0 + (x - 2\ln|x+2|)\Big|_0^1 = \ln\frac{16}{9}.$$

29. $\displaystyle\int_0^2 x^2 \text{sgn}\,(x-1)\,dx$

$$= \int_0^1 -x^2\,dx + \int_1^2 x^2\,dx$$

$$= -\frac{x^3}{3}\Big|_0^1 + \frac{x^3}{3}\Big|_1^2 = -\frac{1}{3} + \frac{8}{3} - \frac{1}{3} = 2.$$

30. Since $\dfrac{x^2}{x^5+1} \leq \dfrac{1}{x^3}$ for all $x \geq 0$, therefore

$$I = \int_0^\infty \frac{x^2}{x^5+1}\,dx$$

$$= \int_0^1 \frac{x^2}{x^5+1}\,dx + \int_1^\infty \frac{x^2}{x^5+1}\,dx$$

$$\leq \int_0^1 \frac{x^2}{x^5+1}\,dx + \int_1^\infty \frac{dx}{x^3}$$

$$= I_1 + I_2.$$

Since I_1 is a proper integral (finite) and I_2 is a convergent improper integral, (see Theorem 2), therefore I converges.

31. $\frac{1}{1+\sqrt{x}} \geq \frac{1}{2\sqrt{x}}$ on $[1,\infty)$.

Since $\int_1^\infty \frac{dx}{\sqrt{x}}$ diverges to infinity, so must $\int_1^\infty \frac{dx}{1+\sqrt{x}}$.
Therefore $\int_0^\infty \frac{dx}{1+\sqrt{x}}$ also diverges to infinity.

32. Since $\frac{x\sqrt{x}}{x^2-1} \geq \frac{1}{\sqrt{x}}$ for all $x > 1$, therefore

$$I = \int_2^\infty \frac{x\sqrt{x}}{x^2-1}\,dx \geq \int_2^\infty \frac{dx}{\sqrt{x}} = I_1 = \infty.$$

Since I_1 is a divergent improper integral, I diverges.

33. $\int_0^\infty e^{-x^3}\,dx = \left(\int_0^1 + \int_1^\infty\right)e^{-x^3}\,dx.$

Now $\int_0^1 e^{-x^3}\,dx$ is a proper integral, and is therefore finite. Since $x^3 \geq x$ on $[1,\infty)$, we have

$$\int_1^\infty e^{-x^3}\,dx \leq \int_1^\infty e^{-x}\,dx = \frac{1}{e}.$$

Thus $\int_0^\infty e^{-x^3}\,dx$ converges.

34. On $[0,1]$, $\frac{1}{\sqrt{x}+x^2} \leq \frac{1}{\sqrt{x}}$. On $[1,\infty]$, $\frac{1}{\sqrt{x}+x^2} \leq \frac{1}{x^2}$.
Thus,

$$\int_0^1 \frac{dx}{\sqrt{x}+x^2} \leq \int_0^1 \frac{dx}{\sqrt{x}}$$
$$\int_1^\infty \frac{dx}{\sqrt{x}+x^2} \leq \int_1^\infty \frac{dx}{x^2}.$$

Since both of these integrals are convergent, therefore so is their sum $\int_0^\infty \frac{dx}{\sqrt{x}+x^2}$.

35. $\frac{e^x}{x+1} \geq \frac{e^{-1}}{x+1}$ on $[-1,1]$. Thus

$$\int_{-1}^1 \frac{e^x}{x+1}\,dx \geq \frac{1}{e}\int_{-1}^1 \frac{dx}{x+1} = \infty.$$

The given integral diverges to infinity.

36. Since $\sin x \leq x$ for all $x \geq 0$, thus $\frac{\sin x}{x} \leq 1$. Then

$$I = \int_0^\pi \frac{\sin x}{x}\,dx = \lim_{\epsilon \to 0+}\int_\epsilon^\pi \frac{\sin x}{x}\,dx \leq \int_0^\pi (1)\,dx = \pi.$$

Hence, I converges.

37. Since $\sin x \geq \frac{2x}{\pi}$ on $[0,\pi/2]$, we have

$$\int_0^\infty \frac{|\sin x|}{x^2}\,dx \geq \int_0^{\pi/2} \frac{\sin x}{x^2}\,dx$$
$$\geq \frac{2}{\pi}\int_0^{\pi/2} \frac{dx}{x} = \infty.$$

The given integral diverges to infinity.

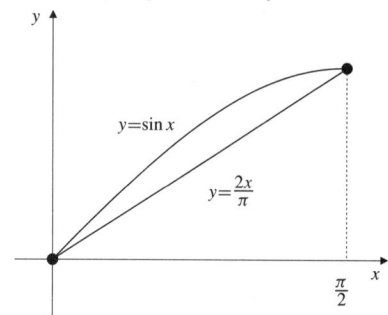

Fig. 6.5.37

38. Since $0 \leq 1 - \cos\sqrt{x} = 2\sin^2\left(\frac{\sqrt{x}}{2}\right) \leq 2\left(\frac{\sqrt{x}}{2}\right)^2 = \frac{x}{2}$,
for $x \geq 0$, therefore $\int_0^{\pi^2} \frac{dx}{1-\cos\sqrt{x}} \geq 2\int_0^{\pi^2} \frac{dx}{x}$, which diverges to infinity.

39. On $(0,\pi/2)$, $\sin x < x$, and so $\csc x \geq 1/x$. Thus

$$\int_0^{\pi/2} \csc x\,dx > \int_0^{\pi/2} \frac{dx}{x} = \infty.$$

Therefore $\int_{-\pi/2}^{\pi/2} \csc x\,dx$ must diverge. (It is of the form $\infty - \infty$.)

40. Since $\ln x$ grows more slowly than any positive power of x, therefore we have $\ln x \leq kx^{1/4}$ for some constant k and every $x \geq 2$. Thus, $\frac{1}{\sqrt{x}\ln x} \geq \frac{1}{kx^{3/4}}$ for $x \geq 2$ and $\int_2^\infty \frac{dx}{\sqrt{x}\ln x}$ diverges to infinity by comparison with $\frac{1}{k}\int_2^\infty \frac{dx}{x^{3/4}}$.

41. $\int_0^\infty \frac{dx}{xe^x} = \left(\int_0^1 + \int_1^\infty\right)\frac{dx}{xe^x}$. But

$$\int_0^1 \frac{dx}{xe^x} \geq \frac{1}{e}\int_0^1 \frac{dx}{x} = \infty.$$

Thus the given integral must diverge to infinity.

42. We are given that $\int_0^\infty e^{-x^2}\,dx = \frac{1}{2}\sqrt{\pi}$.

a) First we calculate

$$\int_0^\infty x^2 e^{-x^2}\,dx = \lim_{R\to\infty} \int_0^R x^2 e^{-x^2}\,dx$$

$$U = x \qquad dV = xe^{-x^2}\,dx$$
$$dU = dx \qquad V = -\tfrac{1}{2}e^{-x^2}$$

$$= \lim_{R\to\infty}\left[-\tfrac{1}{2}xe^{-x^2}\Big|_0^R + \tfrac{1}{2}\int_0^R e^{-x^2}\,dx\right]$$

$$= -\tfrac{1}{2}\lim_{R\to\infty} Re^{-R^2} + \tfrac{1}{2}\int_0^\infty e^{-x^2}\,dx$$

$$= 0 + \tfrac{1}{4}\sqrt{\pi} = \tfrac{1}{4}\sqrt{\pi}.$$

b) Similarly,

$$\int_0^\infty x^4 e^{-x^2}\,dx = \lim_{R\to\infty}\int_0^R x^4 e^{-x^2}\,dx$$

$$U = x^3 \qquad dV = xe^{-x^2}\,dx$$
$$dU = 3x^2\,dx \qquad V = -\tfrac{1}{2}e^{-x^2}$$

$$= \lim_{R\to\infty}\left[-\tfrac{1}{2}x^3 e^{-x^2}\Big|_0^R + \tfrac{3}{2}\int_0^R x^2 e^{-x^2}\,dx\right]$$

$$= -\tfrac{1}{2}\lim_{R\to\infty} R^3 e^{-R^2} + \tfrac{3}{2}\int_0^\infty x^2 e^{-x^2}\,dx$$

$$= 0 + \tfrac{3}{2}\left(\tfrac{1}{4}\sqrt{\pi}\right) = \tfrac{3}{8}\sqrt{\pi}.$$

43. Since f is continuous on $[a,b]$, there exists a positive constant K such that $|f(x)| \le K$ for $a \le x \le b$. If $a < c < b$, then

$$\left|\int_c^b f(x)\,dx - \int_a^b f(x)\,dx\right|$$
$$= \left|\int_c^a f(x)\,dx\right| \le K(c-a) \to 0 \text{ as } c \to a+.$$

Thus $\displaystyle\lim_{c\to a+}\int_c^b f(x)\,dx = \int_a^b f(x)\,dx.$
Similarly

$$\left|\int_a^c f(x)\,dx - \int_a^b f(x)\,dx\right|$$
$$= \left|\int_b^c f(x)\,dx\right| \le K(b-c) \to 0 \text{ as } c \to b-.$$

Thus $\displaystyle\lim_{c\to b-}\int_a^c f(x)\,dx = \int_a^b f(x)\,dx.$

44. $\Gamma(x) = \displaystyle\int_0^\infty t^{x-1}e^{-t}\,dt.$

a) Since $\lim_{t\to\infty} t^{x-1}e^{-t/2} = 0$, there exists $T > 0$ such that $t^{x-1}e^{-t/2} \le 1$ if $t \ge T$. Thus

$$0 \le \int_T^\infty t^{x-1}e^{-t}\,dt \le \int_T^\infty e^{-t}\,dt = 2e^{-T/2}$$

and $\displaystyle\int_T^\infty t^{x-1}e^{-t}\,dt$ converges by the comparison theorem.

If $x > 0$, then

$$0 \le \int_0^T t^{x-1}e^{-t}\,dt < \int_0^T t^{x-1}\,dt$$

converges by Theorem 2(b). Thus the integral defining $\Gamma(x)$ converges.

b) $\Gamma(x+1) = \displaystyle\int_0^\infty t^x e^{-t}\,dt$

$$= \lim_{\substack{c\to 0+\\ R\to\infty}} \int_c^R t^x e^{-t}\,dt$$

$$U = t^x \qquad dV = e^{-t}\,dt$$
$$dU = xt^{x-1}\,dx \qquad V = -e^{-t}$$

$$= \lim_{\substack{c\to 0+\\ R\to\infty}}\left(-t^x e^{-t}\Big|_c^R + x\int_c^R t^{x-1}e^{-t}\,dt\right)$$

$$= 0 + x\int_0^\infty t^{x-1}e^{-t}\,dt = x\Gamma(x).$$

c) $\Gamma(1) = \displaystyle\int_0^\infty e^{-t}\,dt = 1 = 0!.$

By (b), $\Gamma(2) = 1\Gamma(1) = 1 \times 1 = 1 = 1!.$
In general, if $\Gamma(k+1) = k!$ for some positive integer k, then
$\Gamma(k+2) = (k+1)\Gamma(k+1) = (k+1)k! = (k+1)!.$
Hence $\Gamma(n+1) = n!$ for all integers $n \ge 0$, by induction.

d) $\Gamma\left(\tfrac{1}{2}\right) = \displaystyle\int_0^\infty t^{-1/2}e^{-t}\,dt$ Let $t = x^2$
$$dt = 2x\,dx$$
$$= \int_0^\infty \tfrac{1}{x} e^{-x^2} 2x\,dx = 2\int_0^\infty e^{-x^2}\,dx = \sqrt{\pi}$$

$$\Gamma\left(\tfrac{3}{2}\right) = \tfrac{1}{2}\Gamma\left(\tfrac{1}{2}\right) = \tfrac{1}{2}\sqrt{\pi}.$$

Section 6.6 The Trapezoid and Midpoint Rules (page 389)

1. The exact value of I is

$$I = \int_0^2 (1+x^2)\,dx = \left(x + \frac{x^3}{3}\right)\Big|_0^2$$
$$= 2 + \frac{8}{3} \approx 4.6666667.$$

The approximations are

$$T_4 = \frac{1}{2}\left[\frac{1}{2} + \left(1 + \frac{1}{4}\right) + (1+1) + \left(1 + \frac{9}{4}\right) + \frac{5}{2}\right] = 4.75$$

$$M_4 = \frac{1}{2}\left[\left(1 + \frac{1}{16}\right) + \left(1 + \frac{9}{16}\right) + \left(1 + \frac{25}{16}\right) + \left(1 + \frac{49}{16}\right)\right]$$
$$= 4.625$$

$$T_8 = \frac{1}{2}(T_4 + M_4) = 4.6875$$

$$M_8 = \frac{1}{4}\left[\left(1 + \frac{1}{64}\right) + \left(1 + \frac{9}{64}\right) + \left(1 + \frac{25}{64}\right)\right.$$
$$+ \left(1 + \frac{49}{64}\right) + \left(1 + \frac{81}{64}\right) + \left(1 + \frac{121}{64}\right)$$
$$\left.+ \left(1 + \frac{169}{64}\right) + \left(1 + \frac{225}{64}\right)\right] = 4.65625$$

$$T_{16} = \frac{1}{2}(T_8 + M_8) = 4.671875.$$

The exact errors are

$$I - T_4 = -0.0833333; \quad I - M_4 = 0.0416667;$$
$$I - T_8 = -0.0208333; \quad I - M_8 = 0.0104167;$$
$$I - T_{16} = -0.0052083.$$

If $f(x) = 1 + x^2$, then $f''(x) = 2 = K$, and $\frac{K(2-0)}{12} = \frac{1}{3}$. Therefore, the error bounds are

$$\text{Trapezoid}: |I - T_4| \leq \frac{1}{3}\left(\frac{1}{2}\right)^2 \approx 0.0833333;$$

$$|I - T_8| \leq \frac{1}{3}\left(\frac{1}{4}\right)^2 \approx 0.0208333;$$

$$|I - T_{16}| \leq \frac{1}{3}\left(\frac{1}{8}\right)^2 \approx 0.0052083.$$

$$\text{Midpoint}: |I - M_4| \leq \frac{1}{6}\left(\frac{1}{2}\right)^2 \approx 0.0416667;$$

$$|I - M_8| \leq \frac{1}{6}\left(\frac{1}{4}\right)^2 \approx 0.0104167.$$

Note that the actual errors are equal to these estimates since f is a quadratic function.

2. The exact value of I is

$$I = \int_0^1 e^{-x}\,dx = -e^{-x}\Big|_0^1$$
$$= 1 - \frac{1}{e} \approx 0.6321206.$$

The approximations are

$$T_4 = \frac{1}{4}(\tfrac{1}{2}e^0 + e^{-1/4} + e^{-1/2} + e^{-3/4} + \tfrac{1}{2}e^{-1})$$
$$\approx 0.6354094$$

$$M_4 = \tfrac{1}{4}(e^{-1/8} + e^{-3/8} + e^{-5/8} + e^{-7/8})$$
$$\approx 0.6304774$$

$$T_8 = \tfrac{1}{2}(T_4 + M_4) \approx 0.6329434$$

$$M_8 = \tfrac{1}{8}(e^{-1/16} + e^{-3/16} + e^{-5/16} + e^{-7/16} +$$
$$e^{-9/16} + e^{-11/16} + e^{-13/16} + e^{-15/16})$$
$$\approx 0.6317092$$

$$T_{16} = \tfrac{1}{2}(T_8 + M_8) \approx 0.6323263.$$

The exact errors are

$$I - T_4 = -0.0032888; \quad I - M_4 = 0.0016432;$$
$$I - T_8 = -0.0008228; \quad I - M_8 = 0.0004114;$$
$$I - T_{16} = -0.0002057.$$

If $f(x) = e^{-x}$, then $f^{(2)}(x) = e^{-x}$. On $[0,1]$, $|f^{(2)}(x)| \leq 1$. Therefore, the error bounds are:

$$\text{Trapezoid}: |I - T_n| \leq \frac{1}{12}\left(\frac{1}{n}\right)^2$$

$$|I - T_4| \leq \frac{1}{12}\left(\frac{1}{16}\right) \approx 0.0052083;$$

$$|I - T_8| \leq \frac{1}{12}\left(\frac{1}{64}\right) \approx 0.001302;$$

$$|I - T_{16}| \leq \frac{1}{12}\left(\frac{1}{256}\right) \approx 0.0003255.$$

$$\text{Midpoint}: |I - M_n| \leq \frac{1}{24}\left(\frac{1}{n}\right)^2$$

$$|I - M_4| \leq \frac{1}{24}\left(\frac{1}{16}\right) \approx 0.0026041;$$

$$|I - M_8| \leq \frac{1}{24}\left(\frac{1}{64}\right) \approx 0.000651.$$

Note that the actual errors satisfy these bounds.

3. The exact value of I is

$$I = \int_0^{\pi/2} \sin x\,dx = 1.$$

The approximations are

$$T_4 = \frac{\pi}{8}\left(0 + \sin\frac{\pi}{8} + \sin\frac{\pi}{4} + \sin\frac{3\pi}{8} + \frac{1}{2}\right) \approx 0.9871158$$

$$M_4 = \frac{\pi}{8}\left(\sin\frac{\pi}{16} + \sin\frac{3\pi}{16} + \sin\frac{5\pi}{16} + \sin\frac{7\pi}{16}\right) \approx 1.0064545$$

$$T_8 = \frac{1}{2}(T_4 + M_4) \approx 0.9967852$$

$$M_8 = \frac{\pi}{16}\left(\sin\frac{\pi}{32} + \sin\frac{3\pi}{32} + \sin\frac{5\pi}{32}\right.$$
$$+ \sin\frac{7\pi}{32} + \sin\frac{9\pi}{32} + \sin\frac{11\pi}{32}$$
$$\left. + \sin\frac{13\pi}{32} + \sin\frac{15\pi}{32}\right) \approx 1.0016082$$

$$T_{16} = \frac{1}{2}(T_8 + M_8) \approx 0.9991967.$$

The actual errors are

$$I - T_4 \approx 0.0128842; \quad I - M_4 \approx -0.0064545;$$
$$I - T_8 \approx 0.0032148; \quad I - M_8 \approx -0.0016082;$$
$$I - T_{16} \approx 0.0008033.$$

If $f(x) = \sin x$, then $f''(x) = -\sin x$, and $|f''(x)| \leq 1 = K$. Therefore, the error bounds are:

$$\text{Trapezoid}: |I - T_4| \leq \frac{1}{12}\left(\frac{\pi}{2} - 0\right)\left(\frac{\pi}{8}\right)^2 \approx 0.020186;$$

$$|I - T_8| \leq \frac{1}{12}\left(\frac{\pi}{2} - 0\right)\left(\frac{\pi}{16}\right)^2 \approx 0.005047;$$

$$|I - T_{16}| \leq \frac{1}{12}\left(\frac{\pi}{2} - 0\right)\left(\frac{\pi}{32}\right)^2 \approx 0.001262.$$

$$\text{Midpoint}: |I - M_4| \leq \frac{1}{24}\left(\frac{\pi}{2} - 0\right)\left(\frac{\pi}{8}\right)^2 \approx 0.010093;$$

$$|I - M_8| \leq \frac{1}{24}\left(\frac{\pi}{2} - 0\right)\left(\frac{\pi}{16}\right)^2 \approx 0.002523.$$

Note that the actual errors satisfy these bounds.

4. The exact value of I is

$$I = \int_0^1 \frac{dx}{1 + x^2} = \tan^{-1} x \Big|_0^1 = \frac{\pi}{4} \approx 0.7853982.$$

The approximations are

$$T_4 = \frac{1}{4}\left[\frac{1}{2}(1) + \frac{16}{17} + \frac{4}{5} + \frac{16}{25} + \frac{1}{2}\left(\frac{1}{2}\right)\right]$$
$$\approx 0.7827941$$

$$M_4 = \frac{1}{4}\left[\frac{64}{65} + \frac{64}{73} + \frac{64}{89} + \frac{64}{113}\right]$$
$$\approx 0.7867001$$

$$T_8 = \frac{1}{2}(T_4 + M_4) \approx 0.7847471$$

$$M_8 = \frac{1}{8}\left[\frac{256}{257} + \frac{256}{265} + \frac{256}{281} + \frac{256}{305} + \right.$$
$$\left. \frac{256}{337} + \frac{256}{377} + \frac{256}{425} + \frac{256}{481}\right]$$
$$\approx 0.7857237$$

$$T_{16} = \frac{1}{2}(T_8 + M_8) \approx 0.7852354.$$

The exact errors are

$$I - T_4 = 0.0026041; \quad I - M_4 = -0.0013019;$$
$$I - T_8 = 0.0006511; \quad I - M_8 = -0.0003255;$$
$$I - T_{16} = 0.0001628.$$

Since $f(x) = \frac{1}{1+x^2}$, then $f'(x) = \frac{-2x}{(1+x^2)^2}$ and $f''(x) = \frac{6x^2 - 2}{(1+x^2)^3}$. On $[0,1]$, $|f''(x)| \leq 4$. Therefore, the error bounds are

$$\text{Trapezoid}: |I - T_n| \leq \frac{4}{12}\left(\frac{1}{n}\right)^2$$

$$|I - T_4| \leq \frac{4}{12}\left(\frac{1}{16}\right) \approx 0.0208333;$$

$$|I - T_8| \leq \frac{4}{12}\left(\frac{1}{64}\right) \approx 0.0052083;$$

$$|I - T_{16}| \leq \frac{4}{12}\left(\frac{1}{256}\right) \approx 0.001302.$$

$$\text{Midpoint}: |I - M_n| \leq \frac{4}{24}\left(\frac{1}{n}\right)^2$$

$$|I - M_4| \leq \frac{4}{24}\left(\frac{1}{16}\right) \approx 0.0104167;$$

$$|I - M_8| \leq \frac{4}{24}\left(\frac{1}{64}\right) \approx 0.0026042.$$

The exact errors are much smaller than these bounds. In part, this is due to very crude estimates made for $|f''(x)|$.

5. $T_4 = \frac{2}{2}[3 + 2(5 + 8 + 7) + 3] = 46$

$T_8 = \frac{1}{2}[3 + 2(3.8 + 5 + 6.7 + 8 + 8 + 7 + 5.2) + 3] = 46.7$

6. $M_4 = 2(3.8 + 6.7 + 8 + 5.2) = 47.4$

7. $T_4 = 100 \times \frac{2}{2}[0 + 2(5.5 + 5 + 4.5) + 0] = 3{,}000 \text{ km}^2$

$T_8 = 100 \times \frac{1}{2}[0 + 2(4 + 5.5 + 5.5 + 5 + 5.5 + 4.5 + 4) + 0]$
$= 3{,}400 \text{ km}^2$

8. $M_4 = 100 \times 2(4 + 5.5 + 5.5 + 4) = 3{,}800 \text{ km}^2$

9. We have

$T_4 = 0.4\left(\frac{1}{2}(1.4142) + 1.3860 + 1.3026 + 1.1772\right.$
$\left. + \frac{1}{2}(0.9853)\right) \approx 2.02622$
$M_4 = (0.4)(1.4071 + 1.3510 + 1.2411 + 1.0817) \approx 2.03236$
$T_8 = (T_4 + M_4)/2 \approx 2.02929$
$M_8 = (0.2)(1.4124 + 1.3983 + 1.3702 + 1.3285$
$+ 1.2734 + 1.2057 + 1.1258 + 1.0348) \approx 2.02982$
$T_{16} = (T_8 + M_8)/2 \approx 2.029555.$

10. The approximations for $I = \int_0^1 e^{-x^2}\,dx$ are

$M_8 = \frac{1}{8}\left(e^{-1/256} + e^{-9/256} + e^{-25/256} + e^{-49/256} + \right.$
$\left. e^{-81/256} + e^{-121/256} + e^{-169/256} + e^{-225/256}\right)$
≈ 0.7473

$T_{16} = \frac{1}{16}\left[\frac{1}{2}(1) + e^{-1/256} + e^{-1/64} + e^{-9/256} + e^{-1/16} + \right.$
$e^{-25/256} + e^{-9/64} + e^{-49/256} + e^{-1/4} + e^{-81/256} +$
$e^{-25/64} + e^{-121/256} + e^{-9/16} + e^{-169/256} + e^{-49/64} +$
$\left. e^{-225/256} + \frac{1}{2}e^{-1}\right]$
$\approx 0.74658.$

Since $f(x) = e^{-x^2}$, we have $f'(x) = -2xe^{-x^2}$, $f''(x) = 2(2x^2 - 1)e^{-x^2}$, and $f'''(x) = 4x(3 - 2x^2)e^{-x^2}$. Since $f'''(x) \neq 0$ on $(0,1)$, therefore the maximum value of $|f''(x)|$ on $[0, 1]$ must occur at an endpoint of that interval. We have $f''(0) = -2$ and $f''(1) = 2/e$, so $|f''(x)| \leq 2$ on $[0, 1]$. The error bounds are

$|I - M_n| \leq \frac{2}{24}\left(\frac{1}{n}\right)^2 \Rightarrow |I - M_8| \leq \frac{2}{24}\left(\frac{1}{64}\right)$
$\approx 0.00130.$

$|I - T_n| \leq \frac{2}{12}\left(\frac{1}{n}\right)^2 \Rightarrow |I - T_{16}| \leq \frac{2}{12}\left(\frac{1}{256}\right)$
$\approx 0.000651.$

According to the error bounds,

$$\int_0^1 e^{-x^2}\,dx = 0.747,$$

accurate to two decimal places, with error no greater than 1 in the third decimal place.

11. $I = \int_0^{\pi/2} \frac{\sin x}{x}\,dx.$ Note that $\lim_{x \to 0} \frac{\sin x}{x} = 1.$

$T_8 = \frac{\pi}{16}\left[\frac{1}{2} + \frac{16}{\pi}\sin\frac{\pi}{16} + \frac{8}{\pi}\sin\frac{\pi}{8} + \frac{16}{3\pi}\sin\frac{3\pi}{16} + \frac{4}{\pi}\sin\frac{\pi}{4}\right.$
$+ \frac{16}{5\pi}\sin\frac{5\pi}{16} + \frac{8}{3\pi}\sin\frac{3\pi}{8} + \frac{16}{7\pi}\sin\frac{7\pi}{16} + \left.\frac{1}{2}\left(\frac{2}{\pi}\right)\right]$
≈ 1.3694596

$M_8 = \frac{\pi}{16}\left[\frac{32}{\pi}\sin\frac{\pi}{32} + \frac{32}{3\pi}\sin\frac{3\pi}{32} + \frac{32}{5\pi}\sin\frac{5\pi}{32} + \frac{32}{7\pi}\sin\frac{7\pi}{32}\right.$
$+ \frac{32}{9\pi}\sin\frac{9\pi}{32} + \frac{32}{11\pi}\sin\frac{11\pi}{32} + \frac{32}{13\pi}\sin\frac{13\pi}{32}$
$+ \left.\frac{32}{15\pi}\sin\frac{15\pi}{32}\right] \approx 1.3714136$

$T_{16} = (T_8 + M_8)/2 \approx 1.3704366, \qquad I \approx 1.370.$

12. The exact value of I is

$$I = \int_0^1 x^2\,dx = \left.\frac{x^3}{3}\right|_0^1 = \frac{1}{3}.$$

The approximation is

$$T_1 = (1)\left[\frac{1}{2}(0)^2 + \frac{1}{2}(1)^2\right] = \frac{1}{2}.$$

The actual error is $I - T_1 = -\frac{1}{6}$. However, since $f(x) = x^2$, then $f''(x) = 2$ on $[0,1]$, so the error estimate here gives

$$|I - T_1| \leq \frac{2}{12}(1)^2 = \frac{1}{6}.$$

Since this is the actual size of the error in this case, the constant "12" in the error estimate cannot be improved (i.e., cannot be made larger).

13. $I = \int_0^1 x^2\,dx = \frac{1}{3}.$ $M_1 = \left(\frac{1}{2}\right)^2(1) = \frac{1}{4}.$ The actual error is $I - M_1 = \frac{1}{3} - \frac{1}{4} = \frac{1}{12}.$
Since the second derivative of x^2 is 2, the error estimate is

$$|I - M_1| \leq \frac{2}{24}(1 - 0)^2(1^2) = \frac{1}{12}.$$

Thus the constant in the error estimate for the Midpoint Rule cannot be improved; no smaller constant will work for $f(x) = x^2$.

14. Let $y = f(x)$. We are given that m_1 is the midpoint of $[x_0, x_1]$ where $x_1 - x_0 = h$. By tangent line approximate in the subinterval $[x_0, x_1]$,

$$f(x) \approx f(m_1) + f'(m_1)(x - m_1).$$

The error in this approximation is

$$E(x) = f(x) - f(m_1) - f'(m_1)(x - m_1).$$

If $f''(t)$ exists for all t in $[x_0, x_1]$ and $|f''(t)| \leq K$ for some constant K, then by Theorem 4 of Section 3.5,

$$|E(x)| \leq \frac{K}{2}(x - m_1)^2.$$

Hence,

$$|f(x) - f(m_1) - f'(m_1)(x - m_1)| \leq \frac{K}{2}(x - m_1)^2.$$

We integrate both sides of this inequality. Noting that $x_1 - m_1 = m_1 - x_0 = \frac{1}{2}h$, we obtain for the left side

$$\left| \int_{x_0}^{x_1} f(x)\,dx - \int_{x_0}^{x_1} f(m_1)\,dx \right.$$
$$\left. - \int_{x_0}^{x_1} f'(m_1)(x - m_1)\,dx \right|$$
$$= \left| \int_{x_0}^{x_1} f(x)\,dx - f(m_1)h - f'(m_1)\frac{(x - m_1)^2}{2} \Big|_{x_0}^{x_1} \right|$$
$$= \left| \int_{x_0}^{x_1} f(x)\,dx - f(m_1)h \right|.$$

Integrating the right-hand side, we get

$$\int_{x_0}^{x_1} \frac{K}{2}(x - m_1)^2\,dx = \frac{K}{2} \frac{(x - m_1)^3}{3} \Big|_{x_0}^{x_1}$$
$$= \frac{K}{6}\left(\frac{h^3}{8} + \frac{h^3}{8}\right) = \frac{K}{24}h^3.$$

Hence,

$$\left| \int_{x_0}^{x_1} f(x)\,dx - f(m_1)h \right|$$
$$= \left| \int_{x_0}^{x_1} [f(x) - f(m_1) - f'(m_1)(x - m_1)]\,dx \right|$$
$$\leq \frac{K}{24}h^3.$$

A similar estimate holds on each subinterval $[x_{j-1}, x_j]$ for $1 \leq j \leq n$. Therefore,

$$\left| \int_a^b f(x)\,dx - M_n \right| = \left| \sum_{j=1}^n \left(\int_{x_{j-1}}^{x_j} f(x)\,dx - f(m_j)h \right) \right|$$
$$\leq \sum_{j=1}^n \left| \int_{x_{j-1}}^{x_j} f(x)\,dx - f(m_j)h \right|$$
$$\leq \sum_{j=1}^n \frac{K}{24}h^3 = \frac{K}{24}nh^3 = \frac{K(b-a)}{24}h^2$$

because $nh = b - a$.

Section 6.7 Simpson's Rule   (page 394)

1. $I = \int_0^2 (1 + x^2)\,dx = \frac{14}{3} \approx 4.6666667$

$$S_4 = \frac{1}{6}\left[1 + 4\left(1 + \frac{1}{4}\right) + 2(1 + 1) + 4\left(1 + \frac{9}{4}\right)\right.$$
$$\left. + (1 + 4)\right] = \frac{14}{3}$$
$$S_8 = \frac{1}{12}\left[1 + 4\left(1 + \frac{1}{16}\right) + 2\left(1 + \frac{1}{4}\right) + 4\left(1 + \frac{9}{16}\right)\right.$$
$$+ 2(1 + 1) + 4\left(1 + \frac{25}{16}\right) + 2\left(1 + \frac{9}{4}\right)$$
$$\left. + 4\left(1 + \frac{49}{16}\right) + (1 + 4)\right] = \frac{14}{3}$$

The errors are zero because Simpson approximations are exact for polynomials of degree up to three.

2. The exact value of I is

$$I = \int_0^1 e^{-x}\,dx = -e^{-x}\Big|_0^1$$
$$= 1 - \frac{1}{e} \approx 0.6321206.$$

The approximations are

$$S_4 = \frac{1}{12}(e^0 + 4e^{-1/4} + 2e^{-1/2} + 4e^{-3/4} + e^{-1})$$
$$\approx 0.6321342$$
$$S_8 = \frac{1}{24}(e^0 + 4e^{-1/8} + 2e^{-1/4} + 4e^{-3/8} +$$
$$2e^{-1/2} + 4e^{-5/8} + 2e^{-3/4} + 4e^{-7/8} + e^{-1})$$
$$\approx 0.6321214.$$

The actual errors are

$$I - S_4 = -0.0000136; \quad I - S_8 = -0.0000008.$$

These errors are evidently much smaller than the corresponding errors for the corresponding Trapezoid Rule approximations.

3. $I = \int_0^{\pi/2} \sin x \, dx = 1.$

$$S_4 = \frac{\pi}{24}\left(0 + 4\sin\frac{\pi}{8} + 2\sin\frac{\pi}{4} + 4\sin\frac{3\pi}{8} + \sin\frac{\pi}{2}\right)$$
$$\approx 1.0001346$$

$$S_8 = \frac{\pi}{48}\left(0 + 4\sin\frac{\pi}{16} + 2\sin\frac{\pi}{8} + 4\sin\frac{3\pi}{16} + 2\sin\frac{\pi}{4}\right.$$
$$\left. + 4\sin\frac{5\pi}{16} + 2\sin\frac{3\pi}{8} + 4\sin\frac{7\pi}{16} + \sin\frac{\pi}{2}\right)$$
$$\approx 1.0000083.$$

Errors: $I - S_4 \approx -0.0001346$; $I - S_8 \approx -0.0000083$.

4. The exact value of I is

$$I = \int_0^1 \frac{dx}{1+x^2} = \tan^{-1} x \Big|_0^1 = \frac{\pi}{4} \approx 0.7853982.$$

The approximations are

$$S_4 = \frac{1}{12}\left[1 + 4\left(\frac{16}{17}\right) + 2\left(\frac{4}{5}\right) + 4\left(\frac{16}{25}\right) + \frac{1}{2}\right]$$
$$\approx 0.7853922$$

$$S_8 = \frac{1}{24}\left[1 + 4\left(\frac{64}{65}\right) + 2\left(\frac{16}{17}\right) + 4\left(\frac{64}{73}\right) + \right.$$
$$\left. 2\left(\frac{4}{5}\right) + 4\left(\frac{64}{89}\right) + 2\left(\frac{16}{25}\right) + 4\left(\frac{64}{113}\right) + \frac{1}{2}\right]$$
$$\approx 0.7853981.$$

The actual errors are

$$I - S_4 = 0.0000060; \quad I - S_8 = 0.0000001,$$

accurate to 7 decimal places. These errors are evidently much smaller than the corresponding errors for the corresponding Trapezoid Rule approximation.

5. $S_8 = \frac{1}{3}[3 + 4(3.8 + 6.7 + 8 + 5.2) + 2(5 + 8 + 7) + 3]$
≈ 46.93

6. $S_8 = 100 \times \frac{1}{3}[0 + 4(4 + 5.5 + 5.5 + 4) + 2(5.5 + 5 + 4.5) + 0]$
$\approx 3,533 \text{ km}^2$

7. If $f(x) = e^{-x}$, then $f^{(4)}(x) = e^{-x}$, and $|f^{(4)}(x)| \leq 1$ on $[0, 1]$. Thus

$$|I - S_4| \leq \frac{1(1-0)}{180}\left(\frac{1}{4}\right)^4 \approx 0.000022$$

$$|I - S_8| \leq \frac{1(1-0)}{180}\left(\frac{1}{8}\right)^4 \approx 0.0000014.$$

If $f(x) = \sin x$, then $f^{(4)}(x) = \sin x$, and $|f^{(4)}(x)| \leq 1$ on $[0, \pi/2]$. Thus

$$|I - S_4| \leq \frac{1((\pi/2)-0)}{180}\left(\frac{\pi}{8}\right)^4 \approx 0.00021$$

$$|I - S_8| \leq \frac{1((\pi/2)-0)}{180}\left(\frac{\pi}{16}\right)^4 \approx 0.000013.$$

8. Let $I = \int_a^b f(x)\, dx$, and the interval $[a, b]$ be subdivided into $2n$ subintervals of equal length $h = (b-a)/2n$. Let $y_j = f(x_j)$ and $x_j = a + jh$ for $0 \leq j \leq 2n$, then

$$S_{2n} = \frac{1}{3}\left(\frac{b-a}{2n}\right)\left[y_0 + 4y_1 + 2y_2 + \cdots \right.$$
$$\left. + 2y_{2n-2} + 4y_{2n-1} + y_{2n}\right]$$
$$= \frac{1}{3}\left(\frac{b-a}{2n}\right)\left[y_0 + 4\sum_{j=1}^{2n-1} y_j - 2\sum_{j=1}^{n-1} y_{2j} + y_{2n}\right]$$

and

$$T_{2n} = \frac{1}{2}\left(\frac{b-a}{2n}\right)\left(y_0 + 2\sum_{j=1}^{2n-1} y_j + y_{2n}\right)$$

$$T_n = \frac{1}{2}\left(\frac{b-a}{n}\right)\left(y_0 + 2\sum_{j=1}^{n-1} y_{2j} + y_{2n}\right).$$

Since $T_{2n} = \frac{1}{2}(T_n + M_n) \Rightarrow M_n = 2T_{2n} - T_n$, then

$$\frac{T_n + 2M_n}{3} = \frac{T_n + 2(2T_{2n} - T_n)}{3} = \frac{4T_{2n} - T_n}{3}$$
$$\frac{2T_{2n} + M_n}{3} = \frac{2T_{2n} + 2T_{2n} - T_n}{3} = \frac{4T_{2n} - T_n}{3}.$$

Hence,

$$\frac{T_n + 2M_n}{3} = \frac{2T_{2n} + M_n}{3} = \frac{4T_{2n} - T_n}{3}.$$

Using the formulas of T_{2n} and T_n obtained above,

$$\frac{4T_{2n} - T_n}{3}$$
$$= \frac{1}{3}\left[\frac{4}{2}\left(\frac{b-a}{2n}\right)\left(y_0 + 2\sum_{j=1}^{2n-1} y_j + y_{2n}\right)\right.$$
$$\left.- \frac{1}{2}\left(\frac{b-a}{n}\right)\left(y_0 + 2\sum_{j=1}^{n-1} y_{2j} + y_{2n}\right)\right]$$
$$= \frac{1}{3}\left(\frac{b-a}{2n}\right)\left[y_0 + 4\sum_{j=1}^{2n-1} y_j - 2\sum_{j=1}^{n-1} y_{2j} + y_{2n}\right]$$
$$= S_{2n}.$$

Hence,

$$S_{2n} = \frac{4T_{2n} - T_n}{3} = \frac{T_n + 2M_n}{3} = \frac{2T_{2n} + M_n}{3}.$$

9. We use the results of Exercise 9 of Section 7.6 and Exercise 8 of this section.

$$I = \int_0^{1.6} f(x)\,dx$$
$$S_4 = \frac{0.4}{3}(1.4142 + 4(1.3860) + 2(1.3026) + 4(1.1772)$$
$$+ 0.9853) \approx 2.0343333$$
$$S_8 = (T_4 + 2M_4)/3 \approx 2.0303133$$
$$S_{16} = (T_8 + 2M_8)/3 \approx 2.0296433.$$

10. The approximations for $I = \int_0^1 e^{-x^2}\,dx$ are

$$S_8 = \frac{1}{3}\left(\frac{1}{8}\right)\left[1 + 4\left(e^{-1/64} + e^{-9/64} + e^{-25/64} + e^{-49/64}\right) + 2\left(e^{-1/16} + e^{-1/4} + e^{-9/16}\right) + e^{-1}\right]$$
$$\approx 0.7468261$$

$$S_{16} = \frac{1}{3}\left(\frac{1}{16}\right)\left[1 + 4\left(e^{-1/256} + e^{-9/256} + e^{-25/256} + e^{-49/256} + e^{-81/256} + e^{-121/256} + e^{-169/256} + e^{-225/256}\right) + 2\left(e^{-1/64} + e^{-1/16} + e^{-9/64} + e^{-1/4} + e^{-25/64} + e^{-9/16} + e^{-49/64}\right) + e^{-1}\right]$$
$$\approx 0.7468243.$$

If $f(x) = e^{-x^2}$, then $f^{(4)}(x) = 4e^{-x^2}(4x^4 - 12x^2 + 3)$. On $[0,1]$, $|f^{(4)}(x)| \leq 12$, and the error bounds are

$$|I - S_n| \leq \frac{12(1)}{180}\left(\frac{1}{n}\right)^4$$

$$|I - S_8| \leq \frac{12}{180}\left(\frac{1}{8}\right)^4 \approx 0.0000163$$

$$|I - S_{16}| \leq \frac{12}{180}\left(\frac{1}{16}\right)^4 \approx 0.0000010.$$

Comparing the two approximations,

$$I = \int_0^1 e^{-x^2}\,dx = 0.7468,$$

accurate to 4 decimal places.

11. $I = \int_0^1 x^4\,dx = \frac{1}{5}$. $S_2 = \frac{1}{6}\left[0^4 + 4\left(\frac{1}{2}\right)^4 + 1^4\right] = \frac{5}{24}$.

If $f(x) = x^4$, then $f^{(4)}(x) = 24$.

Error estimate: $|I - S_2| \leq \frac{24(1-0)}{180}\left(\frac{1}{2}\right)^4 = \frac{1}{120}$.

Actual error: $|I - S_2| = \left|\frac{1}{5} - \frac{5}{24}\right| = \frac{1}{120}$.

Thus the error estimate cannot be improved.

12. The exact value of I is

$$I = \int_0^1 x^3\,dx = \frac{x^4}{4}\bigg|_0^1 = \frac{1}{4}.$$

The approximation is

$$S_2 = \frac{1}{3}\left(\frac{1}{2}\right)\left[0^3 + 4\left(\frac{1}{2}\right)^3 + 1^3\right] = \frac{1}{4}.$$

The actual error is zero. Hence, Simpson's Rule is exact for the cubic function $f(x) = x^3$. Since it is evidently exact for quadratic functions $f(x) = Bx^2 + Cx + D$, it must also be exact for arbitrary cubics $f(x) = Ax^3 + Bx^2 + Cx + D$.

Section 6.8 Other Aspects of Approximate Integration (page 401)

1. $\displaystyle\int_0^1 \frac{dx}{x^{1/3}(1+x)}$ Let $x = u^3$

$$= 3\int_0^1 \frac{u^2\,du}{u(1+u^3)} = 3\int_0^1 \frac{u\,du}{1+u^3}.$$

2. $\displaystyle\int_0^1 \frac{e^x}{\sqrt{1-x}}\,dx$ Let $t^2 = 1 - x$
$2t\,dt = -dx$

$$= -\int_1^0 \frac{e^{1-t^2}}{t}2t\,dt = 2\int_0^1 e^{1-t^2}\,dt.$$

3. One possibility: let $x = \sin\theta$ and get

$$I = \int_{-1}^{1} \frac{e^x\,dx}{\sqrt{1-x^2}} = \int_{-\pi/2}^{\pi/2} e^{\sin\theta}\,d\theta.$$

Another possibility:

$$I = \int_{-1}^{0} \frac{e^x\,dx}{\sqrt{1-x^2}} + \int_{0}^{1} \frac{e^x\,dx}{\sqrt{1-x^2}} = I_1 + I_2.$$

In I_1 put $1+x = u^2$; in I_2 put $1-x = u^2$:

$$I_1 = \int_0^1 \frac{2e^{u^2-1}u\,du}{u\sqrt{2-u^2}} = 2\int_0^1 \frac{e^{u^2-1}\,du}{\sqrt{2-u^2}}$$

$$I_2 = \int_0^1 \frac{2e^{1-u^2}u\,du}{u\sqrt{2-u^2}} = 2\int_0^1 \frac{e^{1-u^2}\,du}{\sqrt{2-u^2}}$$

so $I = 2\int_0^1 \dfrac{e^{u^2-1} + e^{1-u^2}}{\sqrt{2-u^2}}\,du.$

4. $\int_1^\infty \dfrac{dx}{x^2 + \sqrt{x} + 1}$ Let $x = \dfrac{1}{t^2}$

$$dx = -\frac{2\,dt}{t^3}$$

$$= \int_1^0 \frac{1}{\left(\frac{1}{t^2}\right)^2 + \sqrt{\frac{1}{t^2}} + 1}\left(-\frac{2\,dt}{t^3}\right)$$

$$= 2\int_0^1 \frac{t\,dt}{t^4 + t^3 + 1}.$$

5. $\int_0^{\pi/2} \dfrac{dx}{\sqrt{\sin x}}$ Let $\sin x = u^2$
$\quad 2u\,du = \cos x\,dx = \sqrt{1-u^4}\,dx$

$$= 2\int_0^1 \frac{u\,du}{u\sqrt{1-u^4}}$$

$$= 2\int_0^1 \frac{du}{\sqrt{(1-u)(1+u)(1+y^2)}} \quad \text{Let } 1-u = v^2$$
$$\qquad\qquad -du = 2v\,dv$$

$$= 4\int_0^1 \frac{v\,dv}{v\sqrt{(1+1-v^2)(1+(1-v^2)^2)}}$$

$$= 4\int_0^1 \frac{dv}{\sqrt{(2-v^2)(2-2v^2+v^4)}}.$$

6. Let

$$\int_0^\infty \frac{dx}{x^4+1} = \int_0^1 \frac{dx}{x^4+1} + \int_1^\infty \frac{dx}{x^4+1} = I_1 + I_2.$$

Let $x = \dfrac{1}{t}$ and $dx = -\dfrac{dt}{t^2}$ in I_2, then

$$I_2 = \int_1^0 \frac{1}{\left(\frac{1}{t}\right)^4 + 1}\left(-\frac{dt}{t^2}\right) = \int_0^1 \frac{t^2}{1+t^4}\,dt.$$

Hence,

$$\int_0^\infty \frac{dx}{x^4+1} = \int_0^1 \left(\frac{1}{x^4+1} + \frac{x^2}{1+x^4}\right)dx$$
$$= \int_0^1 \frac{x^2+1}{x^4+1}\,dx.$$

7. $I = \displaystyle\int_0^1 \sqrt{x}\,dx = \dfrac{2}{3} \approx 0.666667.$

$$T_2 = \frac{1}{2}\left(0 + \sqrt{\frac{1}{2}} + \frac{1}{2}\right) \approx 0.603553$$

$$T_4 = \frac{1}{4}\left(2T_2 + \sqrt{\frac{1}{4}} + \sqrt{\frac{3}{4}}\right) \approx 0.643283$$

$$T_8 = \frac{1}{8}\left(4T_4 + \sqrt{\frac{1}{8}} + \sqrt{\frac{3}{8}} + \sqrt{\frac{5}{8}} + \sqrt{\frac{7}{8}}\right) \approx 0.658130$$

$$T_{16} = \frac{1}{16}\left(8T_8 + \sqrt{\frac{1}{16}} + \sqrt{\frac{3}{16}} + \sqrt{\frac{5}{16}} + \sqrt{\frac{7}{16}}\right.$$
$$\left. + \sqrt{\frac{9}{16}} + \sqrt{\frac{11}{16}} + \sqrt{\frac{13}{16}} + \sqrt{\frac{15}{16}}\right) \approx 0.663581.$$

The errors are

$$I - T_2 \approx 0.0631$$
$$I - T_4 \approx 0.0234$$
$$I - T_8 \approx 0.0085$$
$$I - T_{16} \approx 0.0031.$$

Observe that, although these errors are decreasing, they are not decreasing like $1/n^2$; that is,

$$|I - T_{2n}| \gg \frac{1}{4}|I - T_n|.$$

This is because the second derivative of $f(x) = \sqrt{x}$ is $f''(x) = -1/(4x^{3/2})$, which is *not bounded* on $[0, 1]$.

8. Let

$$I = \int_1^\infty e^{-x^2}\,dx \quad \text{Let } x = \frac{1}{t}$$
$$dx = -\frac{dt}{t^2}$$

$$= \int_1^0 e^{-(1/t)^2}\left(-\frac{1}{t^2}\right)dt = \int_0^1 \frac{e^{-1/t^2}}{t^2}\,dt.$$

Observe that

$$\lim_{t\to 0+}\frac{e^{-1/t^2}}{t^2} = \lim_{t\to 0+}\frac{t^{-2}}{e^{1/t^2}} \quad \left[\frac{\infty}{\infty}\right]$$
$$= \lim_{t\to 0+}\frac{-2t^{-3}}{e^{1/t^2}(-2t^{-3})}$$
$$= \lim_{t\to 0+}\frac{1}{e^{1/t^2}} = 0.$$

Hence,

$$S_2 = \frac{1}{3}\left(\frac{1}{2}\right)\left[0 + 4(4e^{-4}) + e^{-1}\right]$$
$$\approx 0.1101549$$
$$S_4 = \frac{1}{3}\left(\frac{1}{4}\right)\left[0 + 4(16e^{-16}) + 2(4e^{-4})\right.$$
$$\left. + 4\left(\frac{16}{9}e^{-16/9}\right) + e^{-1}\right]$$
$$\approx 0.1430237$$
$$S_8 = \frac{1}{3}\left(\frac{1}{8}\right)\left[0 + 4\left(64e^{-64} + \frac{64}{9}e^{-64/9} + \frac{64}{25}e^{-64/25} + \right.\right.$$
$$\left.\left. \frac{64}{49}e^{-64/49}\right) + 2\left(16e^{-16} + 4e^{-4} + \frac{16}{9}e^{-16/9}\right) + e^{-1}\right]$$
$$\approx 0.1393877.$$

Hence, $I \approx 0.14$, accurate to 2 decimal places. These approximations do not converge very quickly, because the fourth derivative of e^{-1/t^2} has very large values for some values of t near 0. In fact, higher and higher derivatives behave more and more badly near 0, so higher order methods cannot be expected to work well either.

9. Referring to Example 5, we have

$$e^x = 1 + x + \frac{x^2}{2!} + \cdots + \frac{x^n}{n!} + R_n(f; 0, x),$$

where $R_n(f; 0, x) = \dfrac{e^X x^{n+1}}{(n+1)!}$, for some X between 0 and x. Now

$$|R_n(f; 0, -x^2)| \le \frac{x^{2n+2}}{(n+1)!}$$

if $0 \le x \le 1$ for any x, since $-x^2 \le X \le 0$. Therefore

$$\left|\int_0^1 R_n(f; 0, -x^2)\,dx\right| \le \frac{1}{(n+1)!}\int_0^1 x^{2n+2}\,dx$$
$$= \frac{1}{(2n+3)(n+1)!}.$$

This error will be less than 10^{-4} if $(2n+3)(n+1)! > 10,000$. Since $15 \times 7! > 10,000$, $n = 6$ will do. Thus we use seven terms of the series ($0 \le n \le 6$):

$$\int_0^1 e^{-x^2}\,dx$$
$$\approx \int_0^1 \left(1 - x^2 + \frac{x^4}{2!} - \frac{x^6}{3!} + \frac{x^8}{4!} - \frac{x^{10}}{5!} + \frac{x^{12}}{6!}\right)dx$$
$$= 1 - \frac{1}{3} + \frac{1}{5\times 2!} - \frac{1}{7\times 3!} + \frac{1}{9\times 4!} - \frac{1}{11\times 5!} + \frac{1}{13\times 6!}$$
$$\approx 0.74684 \quad \text{with error less than } 10^{-4}.$$

10. We are given that $\int_0^\infty e^{-x^2}\,dx = \frac{1}{2}\sqrt{\pi}$ and from the previous exercise $\int_0^1 e^{-x^2}\,dx = 0.74684$. Therefore,

$$\int_1^\infty e^{-x^2}\,dx = \int_0^\infty e^{-x^2}\,dx - \int_0^1 e^{-x^2}\,dx$$
$$= \frac{1}{2}\sqrt{\pi} - 0.74684$$
$$= 0.139 \quad \text{(to 3 decimal places)}.$$

11. If $f(x) = ax^3 + bx^2 + cx + d$, then, by symmetry,

$$\int_{-1}^1 f(x)\,dx = 2\int_0^1 (bx^2 + d)\,dx = 2\left(\frac{b}{3} + d\right)$$
$$Af(-u) + Af(u) = 2A(bu^2 + d).$$

These two expressions are identical provided $A = 1$ and $u^2 = 1/3$, so $u = 1/\sqrt{3}$.

12. For any function f we use the approximation

$$\int_{-1}^1 f(x)\,dx \approx f(-1/\sqrt{3}) + f(1/\sqrt{3}).$$

We have

$$\int_{-1}^1 x^4\,dx \approx \left(-\frac{1}{\sqrt{3}}\right)^4 + \left(\frac{1}{\sqrt{3}}\right)^4 = \frac{2}{9}$$
$$\text{Error} = \int_{-1}^1 x^4\,dx - \frac{2}{9} = \frac{2}{5} - \frac{2}{9} \approx 0.17778$$
$$\int_{-1}^1 \cos x\,dx \approx \cos\left(-\frac{1}{\sqrt{3}}\right) + \cos\left(\frac{1}{\sqrt{3}}\right) \approx 1.67582$$
$$\text{Error} = \int_{-1}^1 \cos x\,dx - 1.67582 \approx 0.00712$$
$$\int_{-1}^1 e^x\,dx \approx e^{-1/\sqrt{3}} + e^{1/\sqrt{3}} \approx 2.34270$$
$$\text{Error} = \int_{-1}^1 e^x\,dx - 2.34270 \approx 0.00771.$$

13. If $F(x) = ax^5 + bx^4 + cx^3 + dx^2 + ex + f$, then, by symmetry,

$$\int_{-1}^{1} F(x)\,dx = 2\int_0^1 (bx^4 + dx^2 + f)\,dx = 2\left(\frac{b}{5} + \frac{d}{3} + f\right)$$
$$AF(-u) + BF(0) + AF(u) = 2A(bu^4 + du^2 + f) + Bf.$$

These two expressions are identical provided

$$Au^4 = \frac{1}{5}, \quad Au^2 = \frac{1}{3}, \quad A + \frac{B}{2} = 1.$$

Dividing the first two equations gives $u^2 = 3/5$, so $u = \sqrt{3/5}$. Then $3A/5 = 1/3$, so $A = 5/9$, and finally, $B = 8/9$.

14. For any function f we use the approximation

$$\int_{-1}^{1} f(x)\,dx \approx \frac{5}{9}\left[f(-\sqrt{3/5}) + f(\sqrt{3/5})\right] + \frac{8}{9}f(0).$$

We have

$$\int_{-1}^{1} x^6\,dx \approx \frac{5}{9}\left[\left(-\sqrt{\frac{3}{5}}\right)^6 + \left(\sqrt{\frac{3}{5}}\right)^6\right] + 0 = 0.24000$$

$$\text{Error} = \int_{-1}^{1} x^6\,dx - 0.24000 \approx 0.04571$$

$$\int_{-1}^{1} \cos x\,dx \approx \frac{5}{9}\left[\cos\left(-\sqrt{\frac{3}{5}}\right) + \cos\left(\sqrt{\frac{3}{5}}\right)\right] + \frac{8}{9}$$
$$\approx 1.68300$$

$$\text{Error} = \int_{-1}^{1} \cos x\,dx - 1.68300 \approx 0.00006$$

$$\int_{-1}^{1} e^x\,dx \approx e^{-\sqrt{3/5}} + e^{\sqrt{3/5}} \approx 2.35034$$

$$\text{Error} = \int_{-1}^{1} e^x\,dx - 2.35034 \approx 0.00006.$$

15. $I = \int_0^1 e^{-x^2}\,dx$

$$T_0^0 = T_1 = R_0 = (1)\left(\frac{1}{2}e^0 + \frac{1}{2}e^{-1}\right) \approx 0.6839397$$

$$T_1^0 = T_2 = \frac{1}{2}\left(\frac{1}{2}e^0 + e^{-1/4} + \frac{1}{2}e^{-1}\right) \approx 0.7313703$$

$$T_2^0 = T_4 = \frac{1}{4}\left(2T_2 + e^{-1/16} + e^{-9/16}\right) \approx 0.7429841$$

$$T_3^0 = T_8 = \frac{1}{8}\left(4T_4 + e^{-1/64} + e^{-9/64} + e^{-25/64} + e^{-49/64}\right)$$
$$\approx 0.7458656$$

$$T_1^1 = S_2 = R_1 = \frac{4T_1^0 - T_0^0}{3} \approx 0.7471805$$

$$T_2^1 = S_4 = \frac{4T_2^0 - T_1^0}{3} \approx 0.7468554$$

$$T_3^1 = S_8 = \frac{4T_3^0 - T_2^0}{3} \approx 0.7468261$$

$$T_2^2 = R_2 = \frac{16T_2^1 - T_1^1}{15} \approx 0.7468337$$

$$T_3^2 = \frac{16T_3^1 - T_2^1}{15} \approx 0.7468242$$

$$T_3^3 = R_3 = \frac{64T_3^2 - T_2^2}{63} \approx 0.7468241$$

$I \approx 0.746824$ to 6 decimal places.

16. From Exercise 9 in Section 7.6, for $I = \int_0^{1.6} f(x)\,dx$,

$$T_0^0 = T_1 = 1.9196$$
$$T_1^0 = T_2 = 2.00188$$
$$T_2^0 = T_4 = 2.02622$$
$$T_3^0 = T_8 = 2.02929.$$

Hence,

$$R_1 = T_1^1 = \frac{4T_1^0 - T_0^0}{3} = 2.0346684$$

$$T_2^1 = \frac{4T_2^0 - T_1^0}{3} = 2.0343333 = S_4$$

$$R_2 = T_2^2 = \frac{16T_2^1 - T_1^1}{15} = 2.0346684$$

$$T_3^1 = \frac{4T_3^0 - T_2^0}{3} = 2.0303133 = S_8$$

$$T_3^2 = \frac{16T_3^1 - T_2^1}{15} = 2.0300453$$

$$R_3 = T_3^3 = \frac{64T_3^2 - T_2^2}{63} = 2.0299719.$$

17. $T_1^1 = S_2 = \dfrac{2h}{3}(y_0 + 4y_2 + y_4)$

$T_2^1 = S_4 = \dfrac{h}{3}(y_0 + 4y_1 + 2y_2 + 4y_3 + y_4)$

$R_2 = T_2^2 = \dfrac{16T_2^1 - T_1^1}{15}$

$= \dfrac{\frac{16h}{3}(y_0 + 4y_1 + 2y_2 + 4y_3 + y_4) - \frac{2h}{3}(y_0 + 4y_2 + y_4)}{15}$

$= \dfrac{h}{45}(14y_0 + 64y_1 + 24y_2 + 64y_3 + 14y_4)$

$= \dfrac{2h}{45}(7y_0 + 32y_1 + 12y_2 + 32y_3 + 7y_4)$

18. Let

$I = \displaystyle\int_\pi^\infty \dfrac{\sin x}{1+x^2}\,dx \quad \text{Let } x = \dfrac{1}{t}$

$\phantom{I = \int_\pi^\infty \dfrac{\sin x}{1+x^2}\,dx \quad } dx = -\dfrac{dt}{t^2}$

$= \displaystyle\int_{1/\pi}^0 \dfrac{\sin\left(\frac{1}{t}\right)}{1 + \left(\frac{1}{t^2}\right)}\left(-\dfrac{1}{t^2}\right)dt$

$= \displaystyle\int_0^{1/\pi} \dfrac{\sin\left(\frac{1}{t}\right)}{1+t^2}\,dt.$

The transformation is not suitable because the derivative of $\sin\left(\dfrac{1}{t}\right)$ is $-\dfrac{1}{t^2}\cos\left(\dfrac{1}{t}\right)$, which has very large values at some points close to 0.

In order to approximate the integral I to an desired degree of accuracy, say with error less than ϵ in absolute value, we have to divide the integral into two parts:

$I = \displaystyle\int_\pi^\infty \dfrac{\sin x}{1+x^2}\,dx$

$= \displaystyle\int_\pi^t \dfrac{\sin x}{1+x^2}\,dx + \int_t^\infty \dfrac{\sin x}{1+x^2}\,dx$

$= I_1 + I_2.$

If $t \geq \tan\dfrac{\pi - \epsilon}{2}$, then

$\displaystyle\int_t^\infty \dfrac{\sin x}{1+x^2}\,dx < \int_t^\infty \dfrac{dx}{1+x^2}$

$= \tan^{-1}(x)\Big|_t^\infty = \dfrac{\pi}{2} - \tan^{-1}(t) \leq \dfrac{\epsilon}{2}.$

Now let A be a numerical approximation to the proper integral $\displaystyle\int_\pi^t \dfrac{\sin x}{1+x^2}\,dx$, having error less than $\epsilon/2$ in absolute value. Then

$|I - A| = |I_1 + I_2 - A|$

$\leq |I_1 - A| + |I_2|$

$\leq \dfrac{\epsilon}{2} + \dfrac{\epsilon}{2} = \epsilon.$

Hence, A is an approximation to the integral I with the desired accuracy.

19. $f(x) = \dfrac{\sin x}{x}, \quad f'(x) = \dfrac{x\cos c - \sin x}{x^2},$

$f''(x) = \dfrac{x^2(\cos x - x\sin x - \cos x) - (x\cos x - \sin x)2x}{x^4}$

$= \dfrac{-x^2\sin x - 2x\cos x + 2\sin x}{x^3}.$

Now use l'Hôpital's Rule to get

$\displaystyle\lim_{x\to 0} f''(x)$

$= \displaystyle\lim_{x\to 0}\dfrac{-2x\sin x - x^2\cos x - 2\cos x + 2x\sin x + 2\cos x}{3x^2}$

$= \displaystyle\lim_{x\to 0}-\dfrac{\cos x}{3} = -\dfrac{1}{3}.$

Review Exercises on Techniques of Integration (page 403)

1. $\dfrac{x}{2x^2 + 5x + 2} = \dfrac{A}{2x+1} + \dfrac{B}{x+2}$

$= \dfrac{Ax + 2A + 2Bx + B}{2x^2 + 5x + 2}$

$\Rightarrow \begin{cases} A + 2B = 1 \\ 2A + B = 0 \end{cases}$

Thus $A = -1/3$ and $B = 2/3$. We have

$\displaystyle\int \dfrac{x\,dx}{2x^2 + 5x + 2} = -\dfrac{1}{3}\int\dfrac{dx}{2x+1} + \dfrac{2}{3}\int\dfrac{dx}{x+2}$

$= \dfrac{2}{3}\ln|x+2| - \dfrac{1}{6}\ln|2x+1| + C.$

2. $\displaystyle\int \dfrac{x}{(x-1)^3}\,dx \quad \text{Let } u = x - 1$

$\phantom{\int \dfrac{x}{(x-1)^3}\,dx \quad } du = dx$

$= \displaystyle\int \dfrac{u+1}{u^3}\,du = \int\left(\dfrac{1}{u^2} + \dfrac{1}{u^3}\right)du$

$= -\dfrac{1}{u} - \dfrac{1}{2u^2} + C = -\dfrac{1}{x-1} - \dfrac{1}{2(x-1)^2} + C.$

3. $\displaystyle\int \sin^3 x \cos^3 x\,dx$

$= \displaystyle\int \sin^3 x(1 - \sin^2 x)\cos x\,dx \quad \text{Let } u = \sin x$

$ du = \cos x\,dx$

$= \displaystyle\int (u^3 - u^5)\,du = \dfrac{u^4}{4} - \dfrac{u^6}{6} + C$

$= \dfrac{1}{4}\sin^4 x - \dfrac{1}{6}\sin^6 x + C.$

4. $\int \dfrac{(1+\sqrt{x})^{1/3}}{\sqrt{x}}\, dx$ Let $u = 1 + \sqrt{x}$

$$du = \dfrac{dx}{2\sqrt{x}}$$

$= 2\int u^{1/3}\, du = 2(\tfrac{3}{4})u^{4/3} + C$

$= \tfrac{3}{2}(1+\sqrt{x})^{4/3} + C.$

5. $\dfrac{3}{4x^2 - 1} = \dfrac{A}{2x-1} + \dfrac{B}{2x+1}$

$= \dfrac{2Ax + A + 2Bx - B}{4x^2 - 1}$

$\Rightarrow \begin{cases} 2A + 2B = 0 \\ A - B = 3 \end{cases} \Rightarrow A = -B = \dfrac{3}{2}$

$\int \dfrac{3\, dx}{4x^2 - 1} = \dfrac{3}{2}\left(\int \dfrac{dx}{2x-1} - \int \dfrac{dx}{2x+1}\right)$

$= \dfrac{3}{4} \ln\left|\dfrac{2x-1}{2x+1}\right| + C.$

6. $\int (x^2 + x - 2)\sin 3x\, dx$

$U = x^2 + x - 2 \quad\quad dV = \sin 3x$
$dU = (2x+1)\, dx \quad\quad V = -\tfrac{1}{3}\cos 3x$

$= -\tfrac{1}{3}(x^2 + x - 2)\cos 3x + \tfrac{1}{3}\int (2x+1)\cos 3x\, dx$

$U = 2x + 1 \quad\quad dV = \cos 3x\, dx$
$dU = 2\, dx \quad\quad V = \tfrac{1}{3}\sin 3x$

$= -\tfrac{1}{3}(x^2 + x - 2)\cos 3x + \tfrac{1}{9}(2x+1)\sin 3x$

$\quad\quad - \tfrac{2}{9}\int \sin 3x\, dx$

$= -\tfrac{1}{3}(x^2 + x - 2)\cos 3x + \tfrac{1}{9}(2x+1)\sin 3x$

$\quad\quad + \dfrac{2}{27}\cos 3x + C.$

7. $\int \dfrac{\sqrt{1-x^2}}{x^4}\, dx$ Let $x = \sin\theta$

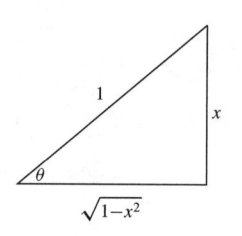

$$dx = \cos\theta\, d\theta$$

$= \int \dfrac{\cos^2\theta}{\sin^4\theta}\, d\theta$

$= \int \csc^2\theta \cot^2\theta\, d\theta$ Let $v = \cot\theta$

$$dv = -\csc^2\theta\, d\theta$$

$= -\int v^2\, dv = -\dfrac{v^3}{3} + C$

$= -\dfrac{\cot^3\theta}{3} + C = -\dfrac{1}{3}\left(\dfrac{\sqrt{1-x^2}}{x}\right)^3 + C.$

Fig. RT-6.7

8. $\int x^3 \cos(x^2)\, dx$ Let $w = x^2$

$$dw = 2x\, dx$$

$= \tfrac{1}{2}\int w \cos w\, dw$

$U = w \quad\quad dV = \cos w\, dw$
$dU = dw \quad\quad V = \sin w$

$= \tfrac{1}{2}w \sin w - \tfrac{1}{2}\int \sin w\, dw$

$= \tfrac{1}{2}x^2 \sin(x^2) + \tfrac{1}{2}\cos(x^2) + C.$

9. $\int \dfrac{x^2\, dx}{(5x^3 - 2)^{2/3}}$ Let $u = 5x^3 - 2$

$$du = 15x^2\, dx$$

$= \dfrac{1}{15}\int u^{-2/3}\, du = \dfrac{1}{5}u^{1/3} + C$

$= \dfrac{1}{5}(5x^3 - 2)^{1/3} + C.$

10. $\dfrac{1}{x^2 + 2x - 15} = \dfrac{A}{x-3} + \dfrac{B}{x+5} = \dfrac{(A+B)x + (5A - 3B)}{x^2 + 2x - 15}$

$\Rightarrow \begin{cases} A + B = 0 \\ 5A - 3B = 1 \end{cases} \Rightarrow A = \dfrac{1}{8},\ B = -\dfrac{1}{8}.$

$\int \dfrac{dx}{x^2 + 2x - 15} = \dfrac{1}{8}\int \dfrac{dx}{x-3} - \dfrac{1}{8}\int \dfrac{dx}{x+5}$

$= \dfrac{1}{8}\ln\left|\dfrac{x-3}{x+5}\right| + C.$

11. $\int \dfrac{dx}{(4+x^2)^2}$ Let $x = 2\tan\theta$

$$dx = 2\sec^2\theta\, d\theta$$

$= \int \dfrac{2\sec^2\theta\, d\theta}{16\sec^4\theta} = \dfrac{1}{8}\int \cos^2\theta\, d\theta$

$= \dfrac{1}{16}(\theta + \sin\theta \cos\theta) + C$

$= \dfrac{1}{16}\tan^{-1}\dfrac{x}{2} + \dfrac{1}{8}\left(\dfrac{x}{4+x^2}\right) + C.$

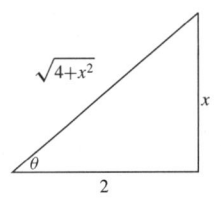

Fig. RT-6.11

12. $\int (\sin x + \cos x)^2 \, dx = \int (1 + \sin 2x) \, dx$
$= x - \tfrac{1}{2} \cos 2x + C.$

13. $\int 2^x \sqrt{1+4^x} \, dx \quad$ Let $2^x = \tan \theta$
$\qquad 2^x \ln 2 \, dx = \sec^2 \theta \, d\theta$
$= \dfrac{1}{\ln 2} \int \sec^3 \theta \, d\theta$
$= \dfrac{1}{2 \ln 2} \left(\sec \theta \tan \theta + \ln |\sec \theta + \tan \theta| \right) + C$
$= \dfrac{1}{2 \ln 2} \left(2^x \sqrt{1+4^x} + \ln(2^x + \sqrt{1+4^x}) \right) + C.$

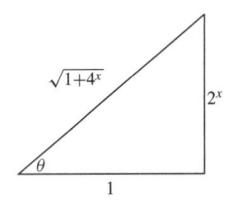

Fig. RT-6.13

14. $\int \dfrac{\cos x}{1 + \sin^2 x} \, dx \quad$ Let $u = \sin x$
$\qquad du = \cos x \, dx$
$= \int \dfrac{du}{1 + u^2} = \tan^{-1} u + C$
$= \tan^{-1}(\sin x) + C.$

15. $\int \dfrac{\sin^3 x}{\cos^7 x} \, dx = \int \tan^3 x \sec^4 x \, dx$
$= \int \tan^3 x (1 + \tan^2 x) \sec^2 x \, dx \quad$ Let $u = \tan x$
$\qquad du = \sec^2 x \, dx$
$= \int (u^3 + u^5) \, du = \dfrac{u^4}{4} + \dfrac{u^6}{6} + C$
$= \dfrac{1}{4} \tan^4 x + \dfrac{1}{6} \tan^6 x + C.$

16. We have

$\int \dfrac{x^2 \, dx}{(3 + 5x^2)^{3/2}} \quad$ Let $x = \sqrt{\tfrac{3}{5}} \tan u$
$\qquad dx = \sqrt{\tfrac{3}{5}} \sec^2 u \, du$
$= \int \dfrac{(\tfrac{3}{5} \tan^2 u)(\sqrt{\tfrac{3}{5}} \sec^2 u) \, du}{(3)^{3/2} \sec^3 u}$
$= \dfrac{1}{5\sqrt{5}} \int (\sec u - \cos u) \, du$
$= \dfrac{1}{5\sqrt{5}} (\ln |\sec u + \tan u| - \sin u) + C$
$= \dfrac{1}{5\sqrt{5}} \left(\ln \left| \dfrac{\sqrt{5x^2 + 3}}{\sqrt{3}} + \dfrac{\sqrt{5}x}{\sqrt{3}} \right| - \dfrac{\sqrt{5}x}{\sqrt{5x^2 + 3}} \right) + C$
$= \dfrac{1}{5\sqrt{5}} \ln(\sqrt{5x^2 + 3} + \sqrt{5}x) - \dfrac{x}{5\sqrt{5x^2 + 3}} + C_0,$

where $C_0 = C - \dfrac{1}{5\sqrt{5}} \ln \sqrt{3}.$

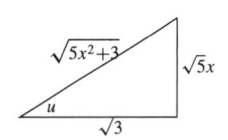

Fig. RT-6.16

17. $I = \int e^{-x} \sin 2x \, dx$
$\qquad U = e^{-x} \qquad\qquad dV = \sin 2x \, dx$
$\qquad dU = -e^{-x} \, dx \quad V = -\dfrac{1}{2} \cos 2x$
$= -\dfrac{1}{2} e^{-x} \cos 2x - \dfrac{1}{2} \int e^{-x} \cos 2x \, dx$

$\qquad U = e^{-x} \qquad\qquad dV = \cos 2x \, dx$
$\qquad dU = -e^{-x} \, dx \quad V = \dfrac{1}{2} \sin 2x$
$= -\dfrac{1}{2} e^{-x} \cos 2x - \dfrac{1}{2} \left(\dfrac{1}{2} e^{-x} \sin 2x + \dfrac{1}{2} I \right)$
$= -\dfrac{1}{2} e^{-x} \cos 2x - \dfrac{1}{4} e^{-x} \sin 2x - \dfrac{1}{4} I$
$I = -e^{-x} \left(\dfrac{2}{5} \cos 2x + \dfrac{1}{5} \sin 2x \right) + C.$

18. $I = \displaystyle\int \dfrac{2x^2 + 4x - 3}{x^2 + 5x}\,dx = \int \dfrac{2x^2 + 10x - 6x - 3}{x^2 + 5x}\,dx$

$= \displaystyle\int \left[2 - \dfrac{6x+3}{x(x+5)}\right] dx$

$\dfrac{6x+3}{x(x+5)} = \dfrac{A}{x} + \dfrac{B}{x+5} = \dfrac{(A+B)x + 5A}{x(x+5)}$

$\Rightarrow \begin{cases} A+B = 6 \\ 5A = 3 \end{cases} \Rightarrow A = \dfrac{3}{5},\ B = \dfrac{27}{5}.$

$I = \displaystyle\int 2\,dx - \dfrac{3}{5}\int \dfrac{dx}{x} - \dfrac{27}{5}\int \dfrac{dx}{x+5}$

$= 2x - \dfrac{3}{5}\ln|x| - \dfrac{27}{5}\ln|x+5| + C.$

19. $I = \displaystyle\int \cos(3\ln x)\,dx$

$U = \cos(3\ln x) \qquad dV = dx$
$dU = -\dfrac{3\sin(3\ln x)\,dx}{x} \qquad V = x$

$= x\cos(3\ln x) + 3\displaystyle\int \sin(3\ln x)\,dx$

$U = \sin(3\ln x) \qquad dV = dx$
$dU = \dfrac{3\cos(3\ln x)\,dx}{x} \qquad V = x$

$= x\cos(3\ln x) + 3\big(x\sin(3\ln x) - 3I\big)$

$I = \dfrac{1}{10}x\cos(3\ln x) + \dfrac{3}{10}x\sin(3\ln x) + C.$

20. $\dfrac{1}{4x^3 + x} = \dfrac{A}{x} + \dfrac{Bx + C}{4x^2 + 1}$

$= \dfrac{A(4x^2+1) + Bx^2 + Cx}{4x^3 + x}$

$\Rightarrow \begin{cases} 4A + B = 0 \\ C = 0,\ A = 1 \end{cases} \Rightarrow B = -4.$

$\displaystyle\int \dfrac{1}{4x^3+x}\,dx = \int \dfrac{dx}{x} - 4\int \dfrac{x\,dx}{4x^2+1}$

$= \ln|x| - \dfrac{1}{2}\ln(4x^2+1) + C.$

21. $\displaystyle\int \dfrac{x\ln(1+x^2)}{1+x^2}\,dx \quad$ Let $u = \ln(1+x^2)$

$du = \dfrac{2x\,dx}{1+x^2}$

$= \dfrac{1}{2}\displaystyle\int u\,du = \dfrac{u^2}{4} + C$

$= \dfrac{1}{4}\big(\ln(1+x^2)\big)^2 + C.$

22. $\displaystyle\int \sin^2 x \cos^4 x\,dx$

$= \displaystyle\int \tfrac{1}{2}(1 - \cos 2x)[\tfrac{1}{2}(1 + \cos 2x)]^2\,dx$

$= \dfrac{1}{8}\displaystyle\int (1 + \cos 2x - \cos^2 2x - \cos^3 2x)\,dx$

$= \dfrac{1}{8}x + \dfrac{1}{16}\sin 2x - \dfrac{1}{16}\displaystyle\int (1 + \cos 4x)\,dx$

$- \dfrac{1}{8}\displaystyle\int (1 - \sin^2 2x)\cos 2x\,dx$

$= \dfrac{x}{8} + \dfrac{1}{16}\sin 2x - \dfrac{x}{16} - \dfrac{1}{64}\sin 4x - \dfrac{1}{16}\sin 2x$

$+ \dfrac{1}{48}\sin^3 2x + C$

$= \dfrac{x}{16} - \dfrac{\sin 4x}{64} + \dfrac{\sin^3 2x}{48} + C.$

23. $\displaystyle\int \dfrac{x^2\,dx}{\sqrt{2-x^2}} \quad$ Let $x = \sqrt{2}\sin\theta$
$dx = \sqrt{2}\cos\theta\,d\theta$

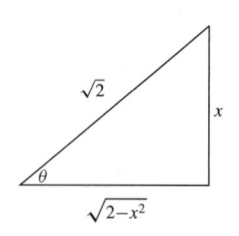

$= 2\displaystyle\int \sin^2\theta\,d\theta = \theta - \sin\theta\cos\theta + C$

$= \sin^{-1}\dfrac{x}{\sqrt{2}} - \dfrac{x\sqrt{2-x^2}}{2} + C.$

Fig. RT-6.23

24. We have

$I = \displaystyle\int \tan^4 x \sec x\,dx$

$U = \tan^3 x \qquad dV = \tan x \sec x\,dx$
$dU = 3\tan^2 x \sec^2 x\,dx \qquad V = \sec x$

$= \tan^3 x \sec x - 3\displaystyle\int \tan^2 x \sec^3 x\,dx$

$= \tan^3 x \sec x - 3\displaystyle\int \tan^2 x(\tan^2 x + 1)\sec x\,dx$

$= \tan^3 x \sec x - 3I - 3J \quad$ where

$J = \displaystyle\int \tan^2 x \sec x\,dx$

$U = \tan x \qquad dV = \tan x \sec x\,dx$
$dU = \sec^2 x\,dx \qquad V = \sec x$

$= \tan x \sec x - \displaystyle\int \sec^3 x\,dx$

$= \tan x \sec x - \int (\tan^2 x + 1) \sec x \, dx$
$= \tan x \sec x - J - \ln|\sec x + \tan x| + C$
$J = \frac{1}{2} \tan x \sec x - \frac{1}{2} \ln|\sec x + \tan x| + C.$
$I = \frac{1}{4} \tan^3 x \sec x - \frac{3}{8} \tan x \sec x$
$\quad + \frac{3}{8} \ln|\sec x + \tan x| + C.$

25. $\int \dfrac{x^2 \, dx}{(4x+1)^{10}}$ Let $u = 4x + 1$
$\qquad du = 4 \, dx$

$= \dfrac{1}{4} \int \left(\dfrac{u-1}{4} \right)^2 \dfrac{1}{u^{10}} \, du$

$= \dfrac{1}{64} \int (u^{-8} - 2u^{-9} + u^{-10}) \, du$

$= -\dfrac{1}{448} u^{-7} + \dfrac{1}{256} u^{-8} - \dfrac{1}{576} u^{-9} + C$

$= \dfrac{1}{64} \left(-\dfrac{1}{7(4x+1)^7} + \dfrac{1}{4(4x+1)^8} - \dfrac{1}{9(4x+1)^9} \right) + C.$

26. We have

$\int x \sin^{-1} \left(\dfrac{x}{2} \right) dx$

$U = \sin^{-1}\left(\dfrac{x}{2}\right) \qquad dV = x \, dx$
$dU = \dfrac{dx}{\sqrt{4-x^2}} \qquad V = \dfrac{x^2}{2}$

$= \dfrac{x^2}{2} \sin^{-1}\left(\dfrac{x}{2}\right) - \dfrac{1}{2} \int \dfrac{x^2 \, dx}{\sqrt{4-x^2}}$ Let $x = 2\sin u$
$\qquad dx = 2 \cos u \, du$

$= \dfrac{x^2}{2} \sin^{-1}\left(\dfrac{x}{2}\right) - 2 \int \sin^2 u \, du$

$= \dfrac{x^2}{2} \sin^{-1}\left(\dfrac{x}{2}\right) - \int (1 - \cos 2u) \, du$

$= \dfrac{x^2}{2} \sin^{-1}\left(\dfrac{x}{2}\right) - u + \sin u \cos u + C$

$= \left(\dfrac{x^2}{2} - 1 \right) \sin^{-1}\left(\dfrac{x}{2}\right) + \dfrac{1}{4} x \sqrt{4-x^2} + C.$

27. $\int \sin^5(4x) \, dx$

$= \int (1 - \cos^2 4x)^2 \sin 4x \, dx$ Let $u = \cos 4x$
$\qquad du = -4 \sin 4x \, dx$

$= -\dfrac{1}{4} \int (1 - 2u^2 + u^4) \, du$

$= -\dfrac{1}{4} \left(u - \dfrac{2}{3} u^3 + \dfrac{1}{5} u^5 \right) + C$

$= -\dfrac{1}{4} \cos 4x + \dfrac{1}{6} \cos^3 4x - \dfrac{1}{20} \cos^5 4x + C.$

28. We have

$I = \int \dfrac{dx}{x^5 - 2x^3 + x} = \int \dfrac{x \, dx}{x^6 - 2x^4 + x^2}$ Let $u = x^2$
$\qquad du = 2x \, dx$

$= \dfrac{1}{2} \int \dfrac{du}{u^3 - 2u^2 + u} = \dfrac{1}{2} \int \dfrac{du}{u(u-1)^2}$

$\dfrac{1}{u(u-1)^2} = \dfrac{A}{u} + \dfrac{B}{u-1} + \dfrac{C}{(u-1)^2}$

$= \dfrac{A(u^2 - 2u + 1) + B(u^2 - u) + Cu}{u^3 - 2u^2 + u}$

$\Rightarrow \begin{cases} A + B = 0 \\ -2A - B + C = 0 \\ A = 1 \end{cases} \Rightarrow A = 1, \ B = -1, \ C = 1.$

$\dfrac{1}{2} \int \dfrac{du}{u^3 - 2u^2 + u} = \dfrac{1}{2} \int \dfrac{du}{u} - \dfrac{1}{2} \int \dfrac{du}{u-1}$
$\qquad + \dfrac{1}{2} \int \dfrac{du}{(u-1)^2}$

$= \dfrac{1}{2} \ln|u| - \dfrac{1}{2} \ln|u-1| - \dfrac{1}{2} \dfrac{1}{u-1} + K$

$= \dfrac{1}{2} \ln \dfrac{x^2}{|x^2 - 1|} - \dfrac{1}{2(x^2 - 1)} + K.$

29. $\int \dfrac{dx}{2 + e^x}$

$= \int \dfrac{e^{-x} \, dx}{2e^{-x} + 1}$ Let $u = 2e^{-x} + 1$
$\qquad du = -2e^{-x} \, dx$

$= -\dfrac{1}{2} \int \dfrac{du}{u} = -\dfrac{1}{2} \ln(2e^{-x} + 1) + C.$

30. Let

$I_n = \int x^n 3^x \, dx$

$U = x^n \qquad dV = 3^x \, dx$
$dU = nx^{n-1} \, dx \qquad V = \dfrac{3^x}{\ln 3}$

$= \dfrac{x^n 3^x}{\ln 3} - \dfrac{n}{\ln 3} I_{n-1}.$

$I_0 = \int 3^x \, dx = \dfrac{3^x}{\ln 3} + C.$

Hence,

$I_3 = \int x^3 3^x \, dx$

$= \dfrac{x^3 3^x}{\ln 3} - \dfrac{3}{\ln 3} \left[\dfrac{x^2 3^x}{\ln 3} - \dfrac{2}{\ln 3} \left(\dfrac{x 3^x}{\ln 3} - \dfrac{1}{\ln 3} I_0 \right) \right] + C_1$

$= 3^x \left[\dfrac{x^3}{\ln 3} - \dfrac{3x^2}{(\ln 3)^2} + \dfrac{6x}{(\ln 3)^3} - \dfrac{6}{(\ln 3)^4} \right] + C_1.$

31. $\int \dfrac{\sin^2 x \cos x}{2 - \sin x}\, dx$ Let $u = \sin x$
$\qquad\qquad\qquad\qquad\quad du = \cos x\, dx$

$= \int \dfrac{u^2\, du}{2 - u}$ Let $2 - u = v$
$\qquad\qquad\qquad du = -dv$

$= -\int \dfrac{4 - 4v + v^2}{v}\, dv = \int \left(-\dfrac{4}{v} + 4 - v\right) dv$

$= -4\ln|v| + 4v - \dfrac{v^2}{2} + C$

$= -4\ln|2 - u| + 4(2 - u) - \dfrac{1}{2}(2 - u)^2 + C$

$= -4\ln(2 - \sin x) - 2\sin x - \dfrac{1}{2}\sin^2 x + C_1.$

32. We have

$\int \dfrac{x^2 + 1}{x^2 + 2x + 2}\, dx = \int \left(1 - \dfrac{2x + 1}{x^2 + 2x + 2}\right) dx$

$= x - \int \dfrac{2x + 1}{(x + 1)^2 + 1}\, dx$ Let $u = x + 1$
$\qquad\qquad\qquad\qquad\qquad\qquad du = dx$

$= x - \int \dfrac{2u - 1}{u^2 + 1}\, du$

$= x - \ln|u^2 + 1| + \tan^{-1} u + C$

$= x - \ln(x^2 + 2x + 2) + \tan^{-1}(x + 1) + C.$

33. $\int \dfrac{dx}{x^2\sqrt{1 - x^2}}$ Let $x = \sin\theta$
$\qquad\qquad\qquad\quad dx = \cos\theta\, d\theta$

$= \int \dfrac{\cos\theta\, d\theta}{\sin^2\theta \cos\theta} = \int \csc^2\theta\, d\theta$

$= -\cot\theta + C = -\dfrac{\sqrt{1 - x^2}}{x} + C.$

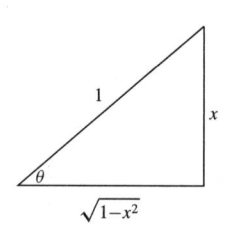

Fig. RT-6.33

34. We have

$\int x^3 (\ln x)^2\, dx$

$U = (\ln x)^2 \qquad dV = x^3\, dx$
$dU = \dfrac{2}{x}\ln x\, dx \qquad V = \dfrac{1}{4}x^4$

$= \dfrac{1}{4}x^4 (\ln x)^2 - \dfrac{1}{2}\int x^3 \ln x\, dx$

$\qquad\qquad U = \ln x \qquad dV = x^3\, dx$
$\qquad\qquad dU = \dfrac{1}{x}\, dx \qquad V = \dfrac{1}{4}x^4$

$= \dfrac{1}{4}x^4 (\ln x)^2 - \dfrac{1}{8}x^4 \ln x + \dfrac{1}{8}\int x^3\, dx$

$= \dfrac{x^4}{4}\left[(\ln x)^2 - \dfrac{1}{2}\ln x + \dfrac{1}{8}\right] + C.$

35. $\int \dfrac{x^3\, dx}{\sqrt{1 - 4x^2}}$ Let $2x = \sin\theta$
$\qquad\qquad\qquad\quad 2\, dx = \cos\theta\, d\theta$

$= \dfrac{1}{16} \int \dfrac{\sin^3\theta \cos\theta\, d\theta}{\cos\theta} = \dfrac{1}{16} \int (1 - \cos^2\theta) \sin\theta\, d\theta$

$= \dfrac{1}{16}\left(-\cos\theta + \dfrac{1}{3}\cos^3\theta\right) + C$

$= \dfrac{1}{48}(1 - 4x^2)^{3/2} - \dfrac{1}{16}\sqrt{1 - 4x^2} + C.$

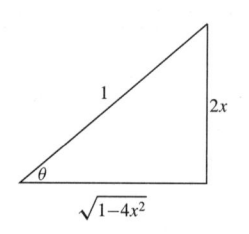

Fig. RT-6.35

36. $\int \dfrac{e^{1/x}}{x^2}\, dx$ Let $u = \dfrac{1}{x}$
$\qquad\qquad\qquad du = -\dfrac{1}{x^2}\, dx$

$= -\int e^u\, du = -e^u + C = -e^{1/x} + C.$

37. $\int \dfrac{x + 1}{\sqrt{x^2 + 1}}\, dx$

$= \sqrt{x^2 + 1} + \int \dfrac{dx}{\sqrt{x^2 + 1}}$ Let $x = \tan\theta$
$\qquad\qquad\qquad\qquad\qquad\qquad dx = \sec^2\theta\, d\theta$

$= \sqrt{x^2 + 1} + \int \sec\theta\, d\theta$

$= \sqrt{x^2 + 1} + \ln|\sec\theta + \tan\theta| + C$

$= \sqrt{x^2 + 1} + \ln(x + \sqrt{x^2 + 1}) + C.$

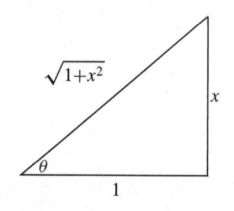

Fig. RT-6.37

INSTRUCTOR'S SOLUTIONS MANUAL REVIEW EXERCISES ON TECHNIQUES OF INTEGRATION (PAGE 403)

38. $\int e^{(x^{1/3})}$ Let $x = u^3$
    $dx = 3u^2\, du$

    $= 3\int u^2 e^u\, du = 3I_2$

    See solution to #16 of Section 6.6 for
    $I_n = \int u^n e^u\, dx = u^n e^u - nI_{n-1}$.

    $= 3[u^2 e^u - 2(ue^u - e^u)] + C$
    $= e^{(x^{1/3})}(3x^{2/3} - 6x^{1/3} + 6) + C$.

39. $I = \int \dfrac{x^3 - 3}{x^3 - 9x}\, dx = \int \left(1 + \dfrac{9x - 3}{x^3 - 9x}\right) dx$.

    $\dfrac{9x - 3}{x^3 - 9x} = \dfrac{A}{x} + \dfrac{B}{x - 3} + \dfrac{C}{x + 3}$
    $= \dfrac{Ax^2 - 9A + Bx^2 + 3Bx + Cx^2 - 3Cx}{x^3 - 9x}$

    $\Rightarrow \begin{cases} A + B + C = 0 \\ 3B - 3C = 9 \\ -9A = -3 \end{cases} \Rightarrow \begin{cases} A = 1/3 \\ B = 4/3 \\ C = -5/3 \end{cases}$

    Thus we have

    $I = x + \dfrac{1}{3}\int \dfrac{dx}{x} + \dfrac{4}{3}\int \dfrac{dx}{x - 3} - \dfrac{5}{3}\int \dfrac{dx}{x + 3}$
    $= x + \dfrac{1}{3}\ln|x| + \dfrac{4}{3}\ln|x - 3| - \dfrac{5}{3}\ln|x + 3| + K$.

40. $\int \dfrac{10^{\sqrt{x+2}}\, dx}{\sqrt{x + 2}}$ Let $u = \sqrt{x + 2}$
    $du = \dfrac{dx}{2\sqrt{x + 2}}$

    $= 2\int 10^u\, du = \dfrac{2}{\ln 10} 10^u + C = \dfrac{2}{\ln 10} 10^{\sqrt{x+2}} + C$.

41. $\int \sin^5 x \cos^9 x\, dx$

    $= \int (1 - \cos^2 x)^2 \cos^9 x \sin x\, dx$ Let $u = \cos x$
    $du = -\sin x\, dx$

    $= -\int (1 - 2u^2 + u^4)u^9\, du$

    $= -\dfrac{u^{10}}{10} + \dfrac{u^{12}}{6} - \dfrac{u^{14}}{14} + C$

    $= -\dfrac{\cos^{12} x}{6} - \dfrac{\cos^{10} x}{10} - \dfrac{\cos^{14} x}{14} + C$.

42. Assume that $x \geq 1$ and let $x = \sec u$ and $dx = \sec u \tan u\, du$. Then

    $\int \dfrac{x^2\, dx}{\sqrt{x^2 - 1}}$

    $= \int \dfrac{\sec^3 u \tan u\, du}{\tan u} = \int \sec^3 u\, du$

    $= \dfrac{1}{2}\sec u \tan u + \dfrac{1}{2}\ln|\sec u + \tan u| + C$

    $= \dfrac{1}{2}x\sqrt{x^2 - 1} + \dfrac{1}{2}\ln|x + \sqrt{x^2 - 1}| + C$.

    Differentiation shows that this solution is valid for $x \leq -1$ also.

43. $I = \int \dfrac{x\, dx}{x^2 + 2x - 1} = \int \dfrac{(x + 1 - 1)\, dx}{(x + 1)^2 - 2}$ Let $u = x + 1$
    $du = dx$

    $= \int \dfrac{u - 1}{u^2 - 2}\, du = \dfrac{1}{2}\ln|u^2 - 2| - \int \dfrac{du}{u^2 - 2}$.

    $\dfrac{1}{u^2 - 2} = \dfrac{A}{u - \sqrt{2}} + \dfrac{B}{u + \sqrt{2}}$
    $= \dfrac{Au + \sqrt{2}A + Bu - \sqrt{2}B}{u^2 - 2}$

    $\Rightarrow \begin{cases} A + B = 0 \\ \sqrt{2}(A - B) = 1 \end{cases}$
    $\Rightarrow A = -B = \dfrac{1}{2\sqrt{2}}$.

    Thus we have

    $I = \dfrac{1}{2}\ln|u^2 - 2| - \dfrac{1}{2\sqrt{2}}\ln\left|\dfrac{u - \sqrt{2}}{u + \sqrt{2}}\right| + K$

    $= \dfrac{1}{2}\ln|x^2 + 2x - 1| - \dfrac{1}{2\sqrt{2}}\ln\left|\dfrac{x + 1 - \sqrt{2}}{x + 1 + \sqrt{2}}\right| + K$.

44. $\int \dfrac{2x - 3}{\sqrt{4 - 3x + x^2}}\, dx$ Let $u = 4 - 3x + x^2$
    $du = (-3 + 2x)\, dx$

    $= \int \dfrac{du}{\sqrt{u}} = 2\sqrt{u} + C = 2\sqrt{4 - 3x + x^2} + C$.

247

45. $\int x^2 \sin^{-1} 2x \, dx$

$$U = \sin^{-1} 2x \qquad dV = x^2 \, dx$$
$$dU = \frac{2 \, dx}{\sqrt{1-4x^2}} \qquad V = \frac{x^3}{3}$$

$= \frac{x^3}{3} \sin^{-1} 2x - \frac{2}{3} \int \frac{x^3 \, dx}{\sqrt{1-4x^2}}$ Let $v = 1 - 4x^2$
$\qquad\qquad dv = -8x \, dx$

$= \frac{x^3}{3} \sin^{-1} 2x - \frac{2}{3} \int \frac{1-v}{4v^{1/2}} \left(-\frac{1}{8} dv\right)$

$= \frac{x^3}{3} \sin^{-1} 2x + \frac{1}{48} \int \left(v^{-1/2} - v^{1/2}\right) dv$

$= \frac{x^3}{3} \sin^{-1} 2x + \frac{1}{24} \sqrt{v} - \frac{1}{72} v^{3/2} + C$

$= \frac{x^3}{3} \sin^{-1} 2x + \frac{1}{24} \sqrt{1-4x^2} - \frac{1}{72}(1-4x^2)^{3/2} + C.$

46. Let $\sqrt{3} x = \sec u$ and $\sqrt{3} \, dx = \sec u \tan u \, du$. Then

$\int \frac{\sqrt{3x^2-1}}{x} dx$

$= \int \frac{\tan u \, \frac{1}{\sqrt{3}} \sec u \tan u \, du}{\frac{1}{\sqrt{3}} \sec u}$

$= \int \tan^2 u \, du = \int (\sec^2 u - 1) \, du$

$= \tan u - u + C = \sqrt{3x^2-1} - \sec^{-1}(\sqrt{3}x) + C$

$= \sqrt{3x^2-1} + \sin^{-1}\left(\frac{1}{\sqrt{3}x}\right) + C_1.$

47. $\int \cos^4 x \sin^4 x \, dx = \frac{1}{16} \int \sin^4 2x \, dx$

$= \frac{1}{64} \int (1 - \cos 4x)^2 \, dx$

$= \frac{1}{64} \int \left(1 - 2\cos 4x + \frac{1+\cos 8x}{2}\right) dx$

$= \frac{1}{64} \left(\frac{3x}{2} - \frac{\sin 4x}{2} + \frac{\sin 8x}{16}\right) + C$

$= \frac{1}{128}\left(3x - \sin 4x + \frac{\sin 8x}{8}\right) + C.$

48. $\int \sqrt{x - x^2} \, dx$

$= \int \sqrt{\frac{1}{4} - (x - \frac{1}{2})^2} \, dx$ Let $x - \frac{1}{2} = \frac{1}{2} \sin u$
$\qquad\qquad dx = \frac{1}{2} \cos u \, du$

$= \frac{1}{4} \int \cos^2 u \, du = \frac{1}{8} u + \frac{1}{8} \sin u \cos u + C$

$= \frac{1}{8} \sin^{-1}(2x-1) + \frac{1}{4}(2x-1)\sqrt{x-x^2} + C.$

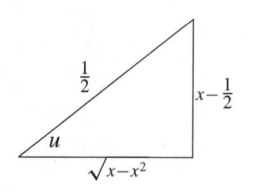

Fig. RT-6.48

49. $\int \frac{dx}{(4+x)\sqrt{x}}$ Let $x = u^2$
$\qquad\qquad dx = 2u \, du$

$= \int \frac{2u \, du}{(4+u^2)u} = 2 \int \frac{du}{4+u^2}$

$= \frac{2}{2} \tan^{-1} \frac{u}{2} + C = \tan^{-1} \frac{\sqrt{x}}{2} + C.$

50. $\int x \tan^{-1}\left(\frac{x}{3}\right) dx$

$$U = \tan^{-1}\left(\frac{x}{3}\right) \qquad dV = x \, dx$$
$$dU = \frac{3 \, dx}{9+x^2} \qquad V = \frac{x^2}{2}$$

$= \frac{x^2}{2} \tan^{-1}\left(\frac{x}{3}\right) - \frac{3}{2} \int \frac{x^2}{9+x^2} dx$

$= \frac{x^2}{2} \tan^{-1}\left(\frac{x}{3}\right) - \frac{3}{2} \int \left(1 - \frac{9}{9+x^2}\right) dx$

$= \frac{x^2}{2} \tan^{-1}\left(\frac{x}{3}\right) - \frac{3x}{2} + \frac{9}{2} \tan^{-1}\left(\frac{x}{3}\right) + C.$

51. $I = \int \frac{x^4 - 1}{x^3 + 2x^2} dx$

$= \int \frac{x^4 + 2x^3 - 2x^3 - 4x^2 + 4x^2 - 1}{x^3 + 2x^2} dx$

$= \int \left(x - 2 + \frac{4x^2 - 1}{x^3 + 2x^2}\right) dx.$

$\frac{4x^2 - 1}{x^3 + 2x^2} = \frac{A}{x} + \frac{B}{x^2} + \frac{C}{x+2}$

$= \frac{Ax^2 + 2Ax + Bx + 2B + Cx^2}{x^3 + 2x^2}$

$\Rightarrow \begin{cases} A + C = 4 \\ 2A + B = 0 \\ 2B = -1 \end{cases} \Rightarrow \begin{cases} A = 1/4 \\ B = -1/2 \\ C = 15/4. \end{cases}$

Thus

$I = \frac{x^2}{2} - 2x + \frac{1}{4} \int \frac{dx}{x} - \frac{1}{2} \int \frac{dx}{x^2} + \frac{15}{4} \int \frac{dx}{x+2}$

$= \frac{x^2}{2} - 2x + \frac{1}{4} \ln|x| + \frac{1}{2x} + \frac{15}{4} \ln|x+2| + K.$

52. Let $u = x^2$ and $du = 2x\,dx$; then we have

$$I = \int \frac{dx}{x(x^2+4)^2} = \int \frac{x\,dx}{x^2(x^2+4)^2} = \frac{1}{2}\int \frac{du}{u(u+4)^2}.$$

Since

$$\frac{1}{u(u+4)^2} = \frac{A}{u} + \frac{B}{u+4} + \frac{C}{(u+4)^2}$$

$$= \frac{A(u^2+8u+16) + B(u^2+4u) + Cu}{u(u+4)^2}$$

$$\Rightarrow \begin{cases} A + B = 0 \\ 8A + 4B + C = 0 \\ 16A = 1 \end{cases} \Rightarrow A = \frac{1}{16},\ B = -\frac{1}{16},\ C = -\frac{1}{4},$$

therefore

$$I = \frac{1}{32}\int \frac{du}{u} - \frac{1}{32}\int \frac{du}{u+4} - \frac{1}{8}\int \frac{du}{(u+4)^2}$$

$$= \frac{1}{32}\ln\left|\frac{u}{u+4}\right| + \frac{1}{8}\frac{1}{u+4} + C$$

$$= \frac{1}{32}\ln\left|\frac{x^2}{x^2+4}\right| + \frac{1}{8(x^2+4)} + C.$$

53. $\displaystyle\int \frac{\sin(2\ln x)}{x}\,dx$ Let $u = 2\ln x$

$$du = \frac{2}{x}\,dx$$

$$= \frac{1}{2}\int \sin u\,du = -\frac{1}{2}\cos u + C$$

$$= -\frac{1}{2}\cos(2\ln x) + C.$$

54. Since

$$I = \int \frac{\sin(\ln x)}{x^2}\,dx$$

$$\begin{array}{ll} U = \sin(\ln x) & dV = \dfrac{dx}{x^2} \\ dU = \dfrac{\cos(\ln x)}{x}dx & V = -\dfrac{1}{x} \end{array}$$

$$= -\frac{\sin(\ln x)}{x} + \int \frac{\cos(\ln x)}{x^2}\,dx$$

$$\begin{array}{ll} U = \cos(\ln x) & dV = \dfrac{dx}{x^2} \\ dU = -\dfrac{\sin(\ln x)}{x}dx & V = \dfrac{-1}{x} \end{array}$$

$$= -\frac{\sin(\ln x)}{x} - \frac{\cos(\ln x)}{x} - I,$$

therefore

$$I = -\frac{1}{2x}\Big[\sin(\ln x) + \cos(\ln x)\Big] + C.$$

55. $\displaystyle\int \frac{e^{2\tan^{-1}x}}{1+x^2}\,dx$ Let $u = 2\tan^{-1}x$

$$du = \frac{2\,dx}{1+x^2}$$

$$= \frac{1}{2}\int e^u\,du = \frac{1}{2}e^u + C = \frac{1}{2}e^{2\tan^{-1}x} + C.$$

56. We have

$$I = \int \frac{x^3 + x - 2}{x^2 - 7}\,dx = \int \frac{x^3 - 7x + 8x - 2}{x^2 - 7}\,dx$$

$$= \int\left(x + \frac{8x-2}{x^2-7}\right)dx.$$

Since

$$\frac{8x-2}{x^2-7} = \frac{A}{x+\sqrt{7}} + \frac{B}{x-\sqrt{7}} = \frac{(A+B)x + (B-A)\sqrt{7}}{x^2-7}$$

$$\Rightarrow \begin{cases} A+B = 8 \\ B - A = -\dfrac{2}{\sqrt{7}} \end{cases} \Rightarrow A = 4 + \frac{1}{\sqrt{7}},\ B = 4 - \frac{1}{\sqrt{7}},$$

therefore

$$I = \int\left(x + \frac{8x-2}{x^2-7}\right)dx$$

$$= \frac{x^2}{2} + \left(4 + \frac{1}{\sqrt{7}}\right)\int \frac{dx}{x+\sqrt{7}} + \left(4 - \frac{1}{\sqrt{7}}\right)\int \frac{dx}{x-\sqrt{7}}$$

$$= \frac{x^2}{2} + \left(4 + \frac{1}{\sqrt{7}}\right)\ln|x+\sqrt{7}| + \left(4 - \frac{1}{\sqrt{7}}\right)\ln|x-\sqrt{7}| + C.$$

57. $\displaystyle\int \frac{\ln(3+x^2)}{3+x^2}\,x\,dx$ Let $u = \ln(3+x^2)$

$$du = \frac{2x\,dx}{3+x^2}$$

$$= \frac{1}{2}\int u\,du = \frac{u^2}{4} + C = \frac{1}{4}\big(\ln(3+x^2)\big)^2 + C.$$

58. $\displaystyle\int \cos^7 x\,dx = \int (1-\sin^2 x)^3 \cos x\,dx$ Let $u = \sin x$

$$du = \cos x\,dx$$

$$= \int (1-u^2)^3\,du = \int (1 - 3u^2 + 3u^4 - u^6)\,du$$

$$= u - u^3 + \tfrac{3}{5}u^5 - \tfrac{1}{7}u^7 + C$$

$$= \sin x - \sin^3 x + \tfrac{3}{5}\sin^5 x - \tfrac{1}{7}\sin^7 x + C.$$

59. $\displaystyle\int \frac{\sin^{-1}(x/2)}{(4-x^2)^{1/2}}\,dx$ Let $u = \sin^{-1}(x/2)$

$$du = \frac{dx}{2\sqrt{1-(x^2/4)}} = \frac{dx}{\sqrt{4-x^2}}$$

$$= \int u\,du = \frac{u^2}{2} + C = \frac{1}{2}\big(\sin^{-1}(x/2)\big)^2 + C.$$

60. We have

$$\int \tan^4(\pi x)\,dx = \int \tan^2(\pi x)[\sec^2(\pi x) - 1]\,dx$$
$$= \int \tan^2(\pi x)\sec^2(\pi x)\,dx - \int [\sec^2(\pi x) - 1]\,dx$$
$$= \frac{1}{3\pi}\tan^3(\pi x) - \frac{1}{\pi}\tan(\pi x) + x + C.$$

61. $\displaystyle\int \frac{(x+1)\,dx}{\sqrt{x^2+6x+10}}$

$$= \int \frac{(x+3-2)\,dx}{\sqrt{(x+3)^2+1}} \quad \begin{array}{l} \text{Let } u = x+3 \\ du = dx \end{array}$$
$$= \int \frac{(u-2)\,du}{\sqrt{u^2+1}}$$
$$= \sqrt{u^2+1} - 2\int \frac{du}{\sqrt{u^2+1}} \quad \begin{array}{l} \text{Let } u = \tan\theta \\ du = \sec^2\theta\,d\theta \end{array}$$
$$= \sqrt{x^2+6x+10} - 2\int \sec\theta\,d\theta$$
$$= \sqrt{x^2+6x+10} - 2\ln|\sec\theta + \tan\theta| + C$$
$$= \sqrt{x^2+6x+10} - 2\ln\!\left(x+3+\sqrt{x^2+6x+10}\right) + C.$$

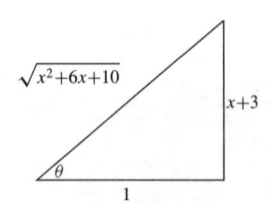

Fig. RT-6.61

62. $\displaystyle\int e^x(1-e^{2x})^{5/2}\,dx$ Let $e^x = \sin u$
$e^x\,dx = \cos u\,du$

$$= \int \cos^6 u\,du = \left(\frac{1}{2}\right)^3 \int (1+\cos 2u)^3\,du$$
$$= \frac{1}{8}\int (1 + 3\cos 2u + 3\cos^2 2u + \cos^3 2u)\,du$$
$$= \frac{u}{8} + \frac{3}{16}\sin 2u + \frac{3}{16}\int(1+\cos 4u)\,du +$$
$$\quad \frac{1}{8}\int(1-\sin^2 2u)\cos 2u\,du$$
$$= \frac{5u}{16} + \frac{3}{16}\sin 2u + \frac{3}{64}\sin 4u + \frac{\sin 2u}{16}$$
$$\quad - \frac{1}{48}\sin^3 2u + C$$
$$= \frac{5}{16}\sin^{-1}(e^x) + \frac{1}{4}\sin[2\sin^{-1}(e^x)] +$$
$$\quad \frac{3}{64}\sin[4\sin^{-1}(e^x)] - \frac{1}{48}\sin^3[2\sin^{-1}(e^x)] + C$$
$$= \frac{5}{16}\sin^{-1}(e^x) + \frac{1}{2}e^x\sqrt{1-e^{2x}}$$
$$\quad + \frac{3}{16}e^x\sqrt{1-e^{2x}}(1-2e^{2x})$$
$$\quad - \frac{1}{6}e^{3x}(1-e^{2x})^{3/2} + C.$$

63. $\displaystyle\int \frac{x^3\,dx}{(x^2+2)^{7/2}}$ Let $x = \sqrt{2}\tan\theta$
$dx = \sqrt{2}\sec^2\theta\,d\theta$

$$= \int \frac{2\sqrt{2}\tan^3\theta\,\sqrt{2}\sec^2\theta\,d\theta}{8\sqrt{2}\sec^7\theta}$$
$$= \frac{1}{2\sqrt{2}}\int \sin^3\theta\cos^2\theta\,d\theta$$
$$= \frac{1}{2\sqrt{2}}\int (1-\cos^2\theta)\cos^2\theta\sin\theta\,d\theta \quad \begin{array}{l}\text{Let } u = \cos\theta \\ du = -\sin\theta\,d\theta\end{array}$$
$$= \frac{1}{2\sqrt{2}}\int (u^4 - u^2)\,du = \frac{1}{2\sqrt{2}}\left(\frac{u^5}{5} - \frac{u^3}{3}\right) + C$$
$$= \frac{1}{2\sqrt{2}}\left(\frac{1}{5}\left(\frac{\sqrt{2}}{\sqrt{2+x^2}}\right)^5 - \frac{1}{3}\left(\frac{\sqrt{2}}{\sqrt{2+x^2}}\right)^3\right) + C$$
$$= \frac{2}{5(2+x^2)^{5/2}} - \frac{1}{3(2+x^2)^{3/2}} + C.$$

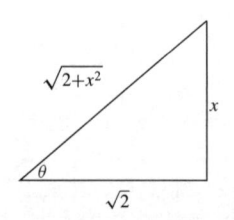

Fig. RT-6.63

64. $\displaystyle\int \frac{x^2}{2x^2-3}\,dx = \frac{1}{2}\int\left(1+\frac{3}{2x^2-3}\right)dx$

$\displaystyle = \frac{x}{2} + \frac{\sqrt{3}}{4}\int\left(\frac{1}{\sqrt{2}x-\sqrt{3}} - \frac{1}{\sqrt{2}x+\sqrt{3}}\right)dx$

$\displaystyle = \frac{x}{2} + \frac{\sqrt{3}}{4\sqrt{2}}\ln\left|\frac{\sqrt{2}x-\sqrt{3}}{\sqrt{2}x+\sqrt{3}}\right| + C.$

65. $\displaystyle\int \frac{x^{1/2}}{1+x^{1/3}}\,dx\quad$ Let $x = u^6$

$\qquad\qquad\qquad\qquad dx = 6u^5\,du$

$\displaystyle = 6\int \frac{u^8}{u^2+1}\,du$

$\displaystyle = 6\int \frac{u^8+u^6-u^6-u^4+u^4+u^2-u^2-1+1}{u^2+1}\,du$

$\displaystyle = 6\int\left(u^6-u^4+u^2-1+\frac{1}{u^2+1}\right)du$

$\displaystyle = 6\left(\frac{u^7}{7}-\frac{u^5}{5}+\frac{u^3}{3}-u+\tan^{-1}u\right) + C$

$\displaystyle = \frac{6}{7}x^{7/6} - \frac{6}{5}x^{5/6} + 2\sqrt{x} - 6x^{1/6} + 6\tan^{-1}x^{1/6} + C.$

66. We have

$\displaystyle\int \frac{dx}{x(x^2+x+1)^{1/2}}$

$\displaystyle = \int \frac{dx}{x[(x+\frac{1}{2})^2+\frac{3}{4}]^{1/2}}\quad$ Let $x+\frac{1}{2} = \frac{\sqrt{3}}{2}\tan\theta$

$\qquad\qquad\qquad\qquad\qquad dx = \frac{\sqrt{3}}{2}\sec^2\theta\,d\theta$

$\displaystyle = \int \frac{\frac{\sqrt{3}}{2}\sec^2\theta\,d\theta}{\left(\frac{\sqrt{3}}{2}\tan\theta - \frac{1}{2}\right)\left(\frac{\sqrt{3}}{2}\sec\theta\right)}$

$\displaystyle = \int \frac{2\sec\theta\,d\theta}{\sqrt{3}\tan\theta - 1} = 2\int \frac{d\theta}{\sqrt{3}\sin\theta - \cos\theta}$

$\displaystyle = 2\int \frac{\sqrt{3}\sin\theta + \cos\theta}{3\sin^2\theta - \cos^2\theta}\,d\theta$

$\displaystyle = 2\sqrt{3}\int \frac{\sin\theta\,d\theta}{3\sin^2\theta - \cos^2\theta} + 2\int \frac{\cos\theta\,d\theta}{3\sin^2\theta - \cos^2\theta}$

$\displaystyle = 2\sqrt{3}\int \frac{\sin\theta\,d\theta}{3 - 4\cos^2\theta} + 2\int \frac{\cos\theta\,d\theta}{4\sin^2\theta - 1}$

Let $u = \cos\theta$, $du = -\sin\theta\,d\theta$ in the first integral;
let $v = \sin\theta$, $dv = \cos\theta\,d\theta$ in the second integral.

$\displaystyle = -2\sqrt{3}\int \frac{du}{3-4u^2} + 2\int \frac{dv}{4v^2-1}$

$\displaystyle = -\frac{\sqrt{3}}{2}\int \frac{du}{\frac{3}{4}-u^2} - \frac{1}{2}\int \frac{du}{\frac{1}{4}-v^2}$

$\displaystyle = -\frac{\sqrt{3}}{2}\left(\frac{1}{2}\right)\left(\frac{2}{\sqrt{3}}\right)\ln\left|\frac{\cos\theta + \frac{\sqrt{3}}{2}}{\cos\theta - \frac{\sqrt{3}}{2}}\right|$

$\displaystyle\qquad -\frac{1}{2}\left(\frac{1}{2}\right)(2)\ln\left|\frac{\sin\theta + \frac{1}{2}}{\sin\theta - \frac{1}{2}}\right| + C$

$\displaystyle = \frac{1}{2}\ln\left|\frac{\left(\cos\theta - \frac{\sqrt{3}}{2}\right)\left(\sin\theta - \frac{1}{2}\right)}{\left(\cos\theta + \frac{\sqrt{3}}{2}\right)\left(\sin\theta + \frac{1}{2}\right)}\right| + C.$

Since $\displaystyle \sin\theta = \frac{2x+1}{2\sqrt{x^2+x+1}}$ and $\displaystyle \cos\theta = \frac{\sqrt{3}}{2\sqrt{x^2+x+1}}$, therefore

$\displaystyle\int \frac{dx}{x(x^2+x+1)^{1/2}} = \frac{1}{2}\ln\left|\frac{(x+2)-2\sqrt{x^2+x+1}}{(x+2)+2\sqrt{x^2+x+1}}\right| + C.$

67. $\displaystyle\int \frac{1+x}{1+\sqrt{x}}\,dx\quad$ Let $x = u^2$

$\qquad\qquad\qquad\qquad dx = 2u\,du$

$\displaystyle = 2\int \frac{u(1+u^2)}{1+u}\,du$

$\displaystyle = 2\int \frac{u^3+u^2-u^2-u+2u+2-2}{1+u}\,du$

$\displaystyle = 2\int \left(u^2 - u + 2 - \frac{2}{1+u}\right)du$

$\displaystyle = 2\left(\frac{u^3}{3} - \frac{u^2}{2} + 2u - 2\ln|1+u|\right) + C$

$\displaystyle = \frac{2}{3}x^{3/2} - x + 4\sqrt{x} - 4\ln(1+\sqrt{x}) + C.$

68. $\displaystyle\int \frac{x\,dx}{4x^4+4x^2+5}\quad$ Let $u = x^2$

$\qquad\qquad\qquad\qquad du = 2x\,dx$

$\displaystyle = \frac{1}{2}\int \frac{du}{4u^2+4u+5}$

$\displaystyle = \frac{1}{2}\int \frac{du}{(2u+1)^2+4}\quad$ Let $w = 2u+1$

$\qquad\qquad\qquad\qquad dw = 2\,du$

$\displaystyle = \frac{1}{4}\int \frac{dw}{w^2+4} = \frac{1}{8}\tan^{-1}\left(\frac{w}{2}\right) + C$

$\displaystyle = \frac{1}{8}\tan^{-1}\left(x^2 + \frac{1}{2}\right) + C.$

69. $\displaystyle\int \frac{x\,dx}{(x^2-4)^2}\quad$ Let $u = x^2 - 4$

$\qquad\qquad\qquad\qquad du = 2x\,dx$

$\displaystyle = \frac{1}{2}\int \frac{du}{u^2} = -\frac{1}{2u} + C$

$\displaystyle = -\frac{1}{2(x^2-4)} + C = -\frac{1}{2x^2-8} + C.$

70. Use the partial fraction decomposition

$$\frac{1}{x^3+x^2+x} = \frac{A}{x} + \frac{Bx+C}{x^2+x+1}$$

$$= \frac{A(x^2+x+1)+Bx^2+Cx}{x^3+x^2+x}$$

$$\Rightarrow \begin{cases} A+B=0 \\ A+C=0 \\ A=1 \end{cases} \Rightarrow A=1,\ B=-1,\ C=-1.$$

Therefore,

$$\int \frac{dx}{x^3+x^2+x}$$

$$= \int \frac{dx}{x} - \int \frac{x+1}{x^2+x+1}\,dx \quad \text{Let } u = x + \tfrac{1}{2}$$
$$\qquad\qquad\qquad\qquad\qquad\qquad du = dx$$

$$= \ln|x| - \int \frac{u+\tfrac{1}{2}}{u^2+\tfrac{3}{4}}\,du$$

$$= \ln|x| - \tfrac{1}{2}\ln(x^2+x+1) - \tfrac{1}{\sqrt{3}}\tan^{-1}\!\left(\frac{2x+1}{\sqrt{3}}\right) + C.$$

71. $\displaystyle\int x^2 \tan^{-1} x\,dx$

$$\begin{array}{ll} U = \tan^{-1} x & dV = x^2\,dx \\ dU = \dfrac{dx}{1+x^2} & V = \dfrac{x^3}{3} \end{array}$$

$$= \frac{x^3}{3}\tan^{-1} x - \frac{1}{3}\int \frac{x^3\,dx}{1+x^2}$$

$$= \frac{x^3}{3}\tan^{-1} x - \frac{1}{3}\int \frac{x^3+x-x}{x^2+1}\,dx$$

$$= \frac{x^3}{3}\tan^{-1} x - \frac{1}{6}x^2 + \frac{1}{6}\ln(1+x^2) + C.$$

72. $\displaystyle\int e^x \sec(e^x)\,dx \quad$ Let $u = e^x$
$$\qquad\qquad\qquad\qquad du = e^x\,dx$$

$$= \int \sec u\,du = \ln|\sec u + \tan u| + C$$

$$= \ln|\sec(e^x) + \tan(e^x)| + C.$$

73. $\displaystyle I = \int \frac{dx}{4\sin x - 3\cos x} \quad$ Let $z = \tan\dfrac{x}{2},\ dx = \dfrac{2\,dz}{1+z^2}$

$$\cos x = \frac{1-z^2}{1+z^2},\quad \sin x = \frac{2z}{1+z^2}$$

$$= \int \frac{\dfrac{2\,dz}{1+z^2}}{\dfrac{8z}{1+z^2} - \dfrac{3-3z^2}{1+z^2}}$$

$$= 2\int \frac{dz}{3z^2+8z-3} = 2\int \frac{dz}{(3z-1)(z+3)}.$$

$$\frac{1}{(3z-1)(z+3)} = \frac{A}{3z-1} + \frac{B}{z+3}$$

$$= \frac{Az+3A+3Bz-B}{(3z-1)(z+3)}$$

$$\Rightarrow \begin{cases} A+3B=0 \\ 3A-B=1 \end{cases} \Rightarrow \begin{cases} A=3/10 \\ B=-1/10 \end{cases}.$$

Thus

$$I = \frac{3}{5}\int \frac{dz}{3z-1} - \frac{1}{5}\int \frac{dz}{z+3}$$

$$= \frac{1}{5}\ln|3z-1| - \frac{1}{5}\ln|z+3| + C$$

$$= \frac{1}{5}\ln\left|\frac{3\tan^{-1}(x/2)-1}{\tan^{-1}(x/2)+3}\right| + C.$$

74. $\displaystyle\int \frac{dx}{x^{1/3}-1} \quad$ Let $x = (u+1)^3$
$$\qquad\qquad\qquad\qquad dx = 3(u+1)^2\,du$$

$$= 3\int \frac{(u+1)^2}{u}\,du = 3\int\!\left(u + 2 + \frac{1}{u}\right)du$$

$$= 3\left(\frac{u^2}{2} + 2u + \ln|u|\right) + C$$

$$= \frac{3}{2}(x^{1/3}-1)^2 + 6(x^{1/3}-1) + 3\ln|x^{1/3}-1| + C.$$

75. $\displaystyle\int \frac{dx}{\tan x + \sin x}$

$$= \int \frac{\cos x\,dx}{\sin x(1+\cos x)} \quad \text{Let } z = \tan(x/2),\ dx = \frac{2\,dz}{1+z^2}$$

$$\cos x = \frac{1-z^2}{1+z^2},\quad \sin x = \frac{2z}{1+z^2}$$

$$= \int \frac{\dfrac{1-z^2}{1+z^2}\cdot\dfrac{2\,dz}{1+z^2}}{\dfrac{2z}{1+z^2}\!\left(1+\dfrac{1-z^2}{1+z^2}\right)}$$

$$= \int \frac{(1-z^2)\,dz}{z(1+z^2+1-z^2)} = \frac{1}{2}\int \frac{1-z^2}{z}\,dz$$

$$= \frac{1}{2}\ln|z| - \frac{z^2}{4} + C$$

$$= \frac{1}{2}\ln\left|\tan\frac{x}{2}\right| - \frac{1}{4}\left(\tan\frac{x}{2}\right)^2 + C.$$

Remark: Since

$$\tan^2\frac{x}{2} = \frac{\sin^2\dfrac{x}{2}}{\cos^2\dfrac{x}{2}} = \frac{1-\cos x}{1+\cos x},$$

the answer can also be written as

$$\frac{1}{4}\ln\left|\frac{1-\cos x}{1+\cos x}\right| - \frac{1}{4}\cdot\frac{1-\cos x}{1+\cos x} + C.$$

252

76. $\displaystyle\int \frac{x\,dx}{\sqrt{3-4x-4x^2}} = \int \frac{x\,dx}{\sqrt{4-(2x+1)^2}}$ Let $u = 2x+1$
 $du = 2\,dx$

$= \frac{1}{4}\int \frac{u-1}{\sqrt{4-u^2}}\,du$

$= -\frac{1}{4}\sqrt{4-u^2} - \frac{1}{4}\sin^{-1}\left(\frac{u}{2}\right) + C$

$= -\frac{1}{4}\sqrt{3-4x-4x^2} - \frac{1}{4}\sin^{-1}\left(x+\frac{1}{2}\right) + C.$

77. $\displaystyle\int \frac{\sqrt{x}}{1+x}\,dx$ Let $x = u^2$
 $dx = 2u\,du$

$= 2\int \frac{u^2\,du}{1+u^2} = 2\int \left(1 - \frac{1}{1+u^2}\right) du$

$= 2\left(u - \tan^{-1} u\right) + C = 2\sqrt{x} - 2\tan^{-1}\sqrt{x} + C.$

78. $\displaystyle\int \sqrt{1+e^x}\,dx$ Let $u^2 = 1+e^x$
 $2u\,du = e^x\,dx$

$= \int \frac{2u^2\,du}{u^2-1} = \int \left(2 + \frac{2}{u^2-1}\right)du$

$= \int \left(2 + \frac{1}{u-1} - \frac{1}{u+1}\right)du$

$= 2u + \ln\left|\frac{u-1}{u+1}\right| + C$

$= 2\sqrt{1+e^x} + \ln\left|\frac{\sqrt{1+e^x}-1}{\sqrt{1+e^x}+1}\right| + C.$

79. $I = \displaystyle\int \frac{x^4\,dx}{x^3-8} = \int \left(x + \frac{8x}{x^3-8}\right)dx.$

$\dfrac{8x}{x^3-8} = \dfrac{A}{x-2} + \dfrac{Bx+C}{x^2+2x+4}$

$= \dfrac{Ax^2 + 2Ax + 4A + Bx^2 - 2Bx + Cx - 2C}{x^3-8}$

$\Rightarrow \begin{cases} A+B = 0 \\ 2A - 2B + C = 8 \\ 4A - 2C = 0 \end{cases} \Rightarrow \begin{cases} B = -A \\ C = 2A \\ 6A = 8 \end{cases}$

Thus $A = 4/3$, $B = -4/3$, $C = 8/3$. We have

$I = \dfrac{x^2}{2} + \dfrac{4}{3}\int \dfrac{dx}{x-2} - \dfrac{4}{3}\int \dfrac{x-2}{x^2+2x+4}\,dx$

$= \dfrac{x^2}{2} + \dfrac{4}{3}\ln|x-2| - \dfrac{4}{3}\int \dfrac{x+1-3}{(x+1)^2+3}\,dx$

$= \dfrac{x^2}{2} + \dfrac{4}{3}\ln|x-2| - \dfrac{2}{3}\ln(x^2+2x+4)$

$\quad + \dfrac{4}{\sqrt{3}}\tan^{-1}\dfrac{x+1}{\sqrt{3}} + K.$

80. By the procedure used in Example 4 of Section 7.1,

$\displaystyle\int e^x \cos x\,dx = \tfrac{1}{2}e^x(\sin x + \cos x) + C,$

$\displaystyle\int e^x \sin x\,dx = \tfrac{1}{2}e^x(\sin x - \cos x) + C.$

Now

$\displaystyle\int xe^x \cos x\,dx$

$U = x \qquad dV = e^x \cos x\,dx$
$dU = dx \qquad V = \tfrac{1}{2}e^x(\sin x + \cos x)$

$= \tfrac{1}{2}xe^x(\sin + \cos x) - \tfrac{1}{2}\int e^x(\sin x + \cos x)\,dx$

$= \tfrac{1}{2}xe^x(\sin + \cos x)$
$\quad - \tfrac{1}{4}e^x(\sin x - \cos x + \sin x + \cos x) + C$

$= \tfrac{1}{2}xe^x(\sin x + \cos x) - \tfrac{1}{2}e^x \sin x + C.$

Other Review Exercises   (page 404)

1. $\dfrac{d}{dx}e^x\left[(ax+b)\cos x + (cx+d)\sin x\right]$

$= e^x\big[(ax+b)\cos x + (cx+d)\sin x + a\cos x + c\sin x$
$\quad - (ax+b)\sin x + (cx+d)\cos x\big]$

$= e^x\big[\big((a+c)x + b + a + d\big)\cos x$
$\quad + \big((c-a)x + d + c - b\big)\sin x\big]$

If $a+c = 1$, $b+a+d = 0$, $c-a = 0$, and $d+c-b = 0$, then $a = c = -d = 1/2$ and $b = 0$. Thus

$I = \displaystyle\int xe^x \cos x\,dx = \dfrac{e^x}{2}\big[x\cos x + (x-1)\sin x\big] + C.$

If $a+c = 0$, $b+a+d = 0$, $c-a = 1$, and $d+c-b = 0$, then $b = c = -a = 1/2$ and $d = 0$. Thus

$J = \displaystyle\int xe^x \sin x\,dx = \dfrac{e^x}{2}\big[x\sin x - (x-1)\cos x\big] + C.$

2. $\displaystyle\int_0^\infty x^r e^{-x}\,dx$

$= \lim_{\substack{c\to 0+ \\ R\to\infty}} \int_c^R x^r e^{-x}\,dx$

$U = x^r \qquad dV = e^{-x}\,dx$
$dU = rx^{r-1}\,dr \qquad V = -e^{-x}$

$= \lim_{\substack{c\to 0+ \\ R\to\infty}} \left. -x^r e^{-x}\right|_c^R + r\int_0^\infty x^{r-1} e^{-x}\,dx$

$= \lim_{c\to 0+} c^r e^{-c} + r\int_0^\infty x^{r-1} e^{-x}\,dx$

because $\lim_{R\to\infty} R^r e^{-R} = 0$ for any r. In order to ensure that $\lim_{c\to 0+} c^r e^{-c} = 0$ we must have $\lim_{c\to 0+} c^r = 0$, so we need $r > 0$.

253

3. $\int_0^{\pi/2} \csc x\, dx = \lim_{c\to 0+} -\ln|\csc x + \cot x|\Big|_c^{\pi/2}$
$= \lim_{c\to 0+} \ln|\csc c + \cot c| = \infty$ (diverges)

4. $\int_1^\infty \dfrac{dx}{x+x^3} = \lim_{R\to\infty} \int_1^R \left(\dfrac{1}{x} - \dfrac{x}{1+x^2}\right) dx$
$= \lim_{R\to\infty} \left(\ln|x| - \dfrac{1}{2}\ln(1+x^2)\right)\Big|_1^R$
$= \lim_{R\to\infty} \dfrac{1}{2}\left(\ln \dfrac{R^2}{1+R^2} + \ln 2\right) = \dfrac{\ln 2}{2}$

5. $\int_0^1 \sqrt{x}\ln x\, dx$ Let $x = u^2$
$\qquad\qquad\qquad\qquad dx = 2u\, du$
$= \int_0^1 u(2\ln u)2u\, du$
$= 4\int_0^1 u^2 \ln u\, du$
$\quad U = \ln u \quad dV = u^2\, du$
$\quad dU = \dfrac{du}{u} \quad V = \dfrac{u^3}{3}$
$= 4 \lim_{c\to 0+}\left(\dfrac{u^3}{3}\ln u\Big|_c^1 - \dfrac{1}{3}\int_c^1 u^2\, du\right)$
$= -\dfrac{4}{3}\lim_{c\to 0+} c^3 \ln c - \dfrac{4}{9}(1-c^3) = -\dfrac{4}{9}$

6. $\int_0^1 \dfrac{dx}{x\sqrt{1-x^2}} > \int_0^1 \dfrac{dx}{x} = \infty$ (diverges)
Therefore $\int_{-1}^1 \dfrac{dx}{x\sqrt{1-x^2}}$ diverges.

7. $I = \int_0^\infty \dfrac{dx}{\sqrt{x}e^x} = \int_0^1 + \int_1^\infty = I_1 + I_2$
$I_1 = \int_0^1 \dfrac{dx}{\sqrt{x}e^x} < \int_0^1 \dfrac{dx}{\sqrt{x}} = 2$
$I_2 = \int_1^\infty \dfrac{dx}{\sqrt{x}e^x} < \int_1^\infty e^{-x}\, dx = \dfrac{1}{e}$
Thus I converges, and $I < 2 + (1/e)$.

8. Volume $= \int_0^{60} A(x)\, dx$. The approximation is
$T_6 = \dfrac{10}{2}\big[10,200 + 2(9,200 + 8,000 + 7,100$
$\qquad\qquad + 4,500 + 2,400) + 100\big]$
$\approx 364,000\text{ m}^3$.

9. $S_6 = \dfrac{10}{3}\big[10,200 + 4(9,200 + 7,100 + 2,400)$
$\qquad\qquad + 2(8,000 + 4,500) + 100\big]$
$\approx 367,000\text{ m}^3$

10. $I = \int_0^1 \sqrt{2+\sin(\pi x)}\, dx$
$T_4 = \dfrac{1}{8}\big[\sqrt{2} + 2(\sqrt{2+\sin(\pi/4)} + \sqrt{2+\sin(\pi/2)}$
$\qquad\qquad + \sqrt{2+\sin(3\pi/4)}) + \sqrt{2}\,\big]$
≈ 1.609230
$M_4 = \dfrac{1}{4}\big[\sqrt{2+\sin(\pi/8)} + \sqrt{2+\sin(3\pi/8)}$
$\qquad\qquad \sqrt{2+\sin(5\pi/8)} + \sqrt{2+\sin(7\pi/8)}\big]$
≈ 1.626765
$I \approx 1.6$

11. $T_8 = \dfrac{1}{2}(T_4 + M_4) \approx 1.617996$
$S_8 = \dfrac{1}{3}(T_4 + 2M_4) \approx 1.62092$
$I \approx 1.62$

12. $I = \int_{1/2}^\infty \dfrac{x^2}{x^5+x^3+1}\, dx$ Let $x = 1/t$
$\qquad\qquad\qquad\qquad\qquad dx = -(1/t^2)\, dt$
$= \int_0^2 \dfrac{(1/t^4)\, dt}{(1/t^5)+(1/t^3)+1} = \int_0^2 \dfrac{t\, dt}{t^5+t^2+1}$
$T_4 \approx 0.4444 \qquad M_4 \approx 0.4799$
$T_8 \approx 0.4622 \qquad M_8 \approx 0.4708$
$S_8 \approx 0.4681 \qquad S_{16} \approx 0.4680$
$I \approx 0.468$ to 3 decimal places

13. a) $T_4 = 1\left(\dfrac{0.730}{2} + 1.001 + 1.332 + 1.729 + \dfrac{2.198}{2}\right)$
$= 5.526$
$S_4 = \dfrac{1}{3}\big(0.730 + 2.198 + 4(1.001 + 1.729) + 2(1.332)\big)$
$= 5.504$.

b) If $T_8 = 5.5095$, then $S_8 = \dfrac{4T_8 - T_4}{3} = 5.504$.

c) Yes, $S_4 = S_8$ suggests that S_n may be independent of n, which is consistent with a polynomial of degree not exceeding 3.

Challenging Problems 6 (page 404)

1. a) Long division of $x^2 + 1$ into
$x^4(1-x)^4 = x^8 - 4x^7 + 6x^6 - 4x^5 + x^4$ yields
$\dfrac{x^4(1-x)^4}{x^2+1} = x^6 - 4x^5 + 5x^4 - 4x^2 + 4 - \dfrac{4}{x^2+1}$.

Integrating both sides over $[0,1]$ leads at once to
$\int_0^1 \dfrac{x^4(1-x)^4}{x^2+1}\, dx = \dfrac{22}{7} - 4\tan^{-1} 1 = \dfrac{22}{7} - \pi$.

Since $\dfrac{x^4(1-x)^4}{x^2+1} > 0$ on $(0,1)$, $\dfrac{22}{7} - \pi > 0$, and so $\pi < \dfrac{22}{7}$.

b) If $I = \displaystyle\int_0^1 x^4(1-x)^4\,dx$, then since $1 < x^2+1 < 2$ on $(0,1)$, we have

$$I > \int_0^1 \dfrac{x^4(1-x)^4}{x^2+1}\,dx > \dfrac{I}{2}.$$

Thus $I > (22/7) - \pi > I/2$, or

$$\dfrac{22}{7} - I < \pi < \dfrac{22}{7} - \dfrac{I}{2}.$$

c) $I = \displaystyle\int_0^1 (x^8 - 4x^7 + 6x^6 - 4x^5 + x^4)\,dx = \dfrac{1}{630}$. Thus

$$\dfrac{22}{7} - \dfrac{1}{630} < \pi < \dfrac{22}{7} - \dfrac{1}{1260}.$$

2. a) $I_n = \displaystyle\int (1-x^2)^n\,dx$

$$\begin{aligned} U &= (1-x^2)^n & dV &= dx \\ dU &= -2nx(1-x^2)^{n-1}\,dx & V &= x \end{aligned}$$

$$= x(1-x^2)^n + 2n\int x^2(1-x^2)^{n-1}\,dx$$

$$= x(1-x^2)^n - 2n\int (1-x^2-1)(1-x^2)^{n-1}\,dx$$

$$= x(1-x^2)^n - 2nI_n + 2nI_{n-1}, \quad \text{so}$$

$$I_n = \dfrac{1}{2n+1}x(1-x^2)^n + \dfrac{2n}{2n+1}I_{n-1}.$$

b) Let $J_n = \displaystyle\int_0^1 (1-x^2)^n\,dx$. Observe that $J_0 = 1$. By (a), if $n > 0$, then we have

$$J_n = \left.\dfrac{x(1-x^2)^n}{2n+1}\right|_0^1 + \dfrac{2n}{2n+1}J_{n-1} = \dfrac{2n}{2n+1}J_{n-1}.$$

Therefore,

$$\begin{aligned} J_n &= \dfrac{2n}{2n+1} \cdot \dfrac{2n-2}{2n-1} \cdots \dfrac{4}{5} \cdot \dfrac{2}{3} J_0 \\ &= \dfrac{[(2n)(2n-2)\cdots(4)(2)]^2}{(2n+1)!} = \dfrac{2^{2n}(n!)^2}{(2n+1)!}. \end{aligned}$$

c) From (a):

$$I_{n-1} = \dfrac{2n+1}{2n}I_n - \dfrac{1}{2n}x(1-x^2)^n.$$

Thus

$$\int (1-x^2)^{-3/2}\,dx = I_{-3/2}$$

$$= \dfrac{2(-1/2)+1}{-1}I_{-1/2} - \dfrac{1}{-1}x(1-x^2)^{-1/2}$$

$$= \dfrac{x}{\sqrt{1-x^2}}.$$

3. a) $x^4 + x^2 + 1 = (x^2+1)^2 - x^2 = (x^2-x+1)(x^2+x+1)$. Thus

$$\int \dfrac{x^2+1}{x^4+x^2+1} = \int \dfrac{x^2+1}{(x^2-x+1)(x^2+x+1)}\,dx$$

$$= \dfrac{1}{2}\int\left(\dfrac{1}{x^2-x+1} + \dfrac{1}{x^2+x+1}\right)dx$$

$$= \dfrac{1}{2}\int\left(\dfrac{1}{\left(x-\tfrac{1}{2}\right)^2 + \tfrac{3}{4}} + \dfrac{1}{\left(x+\tfrac{1}{2}\right)^2 + \tfrac{3}{4}}\right)dx$$

$$= \dfrac{1}{\sqrt{3}}\left(\tan^{-1}\dfrac{2x-1}{\sqrt{3}} + \tan^{-1}\dfrac{2x+1}{\sqrt{3}}\right) + C.$$

b) $x^4+1 = (x^2+1)^2 - 2x^2 = (x^2-\sqrt{2}x+1)(x^2+\sqrt{2}x+1)$. Thus

$$\int \dfrac{x^2+1}{x^4+1} = \int \dfrac{x^2+1}{(x^2-\sqrt{2}x+1)(x^2+\sqrt{2}x+1)}\,dx$$

$$= \dfrac{1}{2}\int\left(\dfrac{1}{x^2-\sqrt{2}x+1} + \dfrac{1}{x^2+\sqrt{2}x+1}\right)dx$$

$$= \dfrac{1}{2}\int\left(\dfrac{1}{\left(x-\tfrac{1}{\sqrt{2}}\right)^2 + \tfrac{3}{4}} + \dfrac{1}{\left(x+\tfrac{1}{\sqrt{2}}\right)^2 + \tfrac{3}{4}}\right)dx$$

$$= \dfrac{1}{\sqrt{3}}\left(\tan^{-1}\dfrac{2x-\sqrt{2}}{\sqrt{3}} + \tan^{-1}\dfrac{2x+\sqrt{2}}{\sqrt{3}}\right) + C.$$

4. $I_{m,n} = \displaystyle\int_0^1 x^m(\ln x)^n\,dx$ Let $x = e^{-t}$
$\qquad\qquad dx = -e^{-t}\,dt$

$$= \int_0^\infty e^{-mt}(-t)^n e^{-t}\,dt$$

$$= (-1)^n \int_0^\infty t^n e^{-(m+1)t}\,dt \quad \text{Let } u = (m+1)t$$
$\qquad\qquad\qquad du = (m+1)\,dt$

$$= \dfrac{(-1)^n}{(m+1)^n}\int_0^\infty u^n e^{-u}\,du$$

$$= \dfrac{(-1)^n}{(m+1)^n}\Gamma(n+1) \quad \text{(see \#50 in Section 7.5)}$$

$$= \dfrac{(-1)^n n!}{(m+1)^n}.$$

5. **a)** $0 < I_n = \int_0^1 x^n e^{-x}\,dx < \int_0^1 x^n\,dx = \dfrac{1}{n+1}$,
because $0 < e^{-x} < 1$ on $(0,1)$. Thus $\lim_{n\to\infty} I_n = 0$ by the Squeeze Theorem.

b) $I_0 = \int_0^1 e^{-x}\,dx = -e^{-x}\Big|_0^1 = 1 - \dfrac{1}{e}$

$I_n = \int_0^1 x^n e^{-x}\,dx$

$\quad U = x^n \qquad\qquad dV = e^{-x}\,dx$
$\quad dU = nx^{n-1}\,dx \quad V = -e^{-x}$

$\quad = -x^n e^{-x}\Big|_0^1 + n\int_0^1 x^{n-1} e^{-x}\,dx$

$\quad = nI_{n-1} - \dfrac{1}{e}\quad \text{if } n \geq 1$

c) The formula

$$I_n = n!\left(1 - \dfrac{1}{e}\sum_{j=0}^{n}\dfrac{1}{j!}\right)$$

holds for $n = 0$ by part (b). Assume that it holds for some integer $n = k \geq 0$. Then by (b),

$I_{k+1} = (k+1)I_k - \dfrac{1}{e} = (k+1)k!\left(1 - \dfrac{1}{e}\sum_{j=0}^{k}\dfrac{1}{j!}\right) - \dfrac{1}{e}$

$\quad = (k+1)!\left(1 - \dfrac{1}{e}\sum_{j=0}^{k}\dfrac{1}{j!} - \dfrac{1}{e(k+1)!}\right)$

$\quad = (k+1)!\left(1 - \dfrac{1}{e}\sum_{j=0}^{k+1}\dfrac{1}{j!}\right).$

Thus the formula holds for all $n \geq 0$, by induction.

d) Since $\lim_{n\to\infty} I_n = 0$, we must have

$$\lim_{n\to\infty}\left(1 - \dfrac{1}{e}\sum_{j=0}^{n}\dfrac{1}{j!}\right) = 0.$$

Thus $e = \lim_{n\to\infty}\sum_{j=0}^{n}\dfrac{1}{j!}$.

6. $I = \int_0^1 e^{-Kx}\,dx = \dfrac{e^{-Kx}}{-K}\Big|_0^1 = \dfrac{1}{K}\left(1 - \dfrac{1}{e^K}\right).$

For very large K, the value of I is very small ($I < 1/K$). However,

$T_{100} = \dfrac{1}{100}(1 + \cdots) > \dfrac{1}{100}$

$S_{100} = \dfrac{1}{300}(1 + \cdots) > \dfrac{1}{300}$

$M_{100} = \dfrac{1}{100}(e^{-K/200} + \cdots) < \dfrac{1}{100}.$

In each case the $\cdots$ represent terms much less than the first term (shown) in the sum. Evidently M_{100} is smallest if k is much greater than 100, and is therefore the best approximation. T_{100} appears to be the worst.

7. **a)** Let $f(x) = Ax^5 + Bx^4 + Cx^3 + Dx^2 + Ex + F$. Then

$$\int_{-h}^{h} f(x)\,dx = 2\left(\dfrac{Bh^5}{5} + \dfrac{Dh^3}{3} + Fh\right).$$

Also

$2h\big[af(-h) + bf(-h/2) + cf(0) + bf(h/2) + af(h)\big]$
$= 2\Big[a\left(2Bh^5 + 2Dh^3 + 2F\right)$
$\quad + b\left(\dfrac{2Bh^5}{16} + \dfrac{2Dh^3}{4} + 2F\right) + cFh\Big].$

These expressions will be identical if the coefficients of like powers of h on the two sides are identical. Thus

$2a + \dfrac{2b}{16} = \dfrac{1}{5},\quad 2a + \dfrac{2b}{4} = \dfrac{1}{3},\quad 2a + 2b + c = 1.$

Solving these equations, we get $a = 7/90$, $b = 16/45$, and $c = 2/15$. The approximation for the integral of any function f on $[m-h, m+h]$ is

$\int_{m-h}^{m+h} f(x)\,dx \approx 2h\Big[\dfrac{7}{90}f(m-h) + \dfrac{16}{45}f(m - \tfrac{1}{2}h)$
$\quad + \dfrac{2}{15}f(m) + \dfrac{16}{45}f(m + \tfrac{1}{2}h) + \dfrac{7}{90}f(m+h)\Big].$

b) If $m = h = 1/2$, we obtain

$\int_0^1 e^{-x}\,dx \approx 1\Big[\dfrac{7}{90}e^0 + \dfrac{16}{45}e^{-1/4} + \dfrac{2}{15}e^{-1/2}$
$\quad + \dfrac{16}{45}e^{-3/4} + \dfrac{7}{90}e^{-1}\Big]$

$\approx 0.63212087501.$

With two intervals having $h = 1/4$ and $m = 1/4$ and $m = 3/4$, we get

$\int_0^1 e^{-x}\,dx \approx \dfrac{1}{2}\Big[\dfrac{7}{90}e^0 + \dfrac{16}{45}e^{-1/8} + \dfrac{2}{15}e^{-1/4}$
$\quad + \dfrac{16}{45}e^{-3/8} + \dfrac{7}{45}e^{-1/2}$
$\quad + \dfrac{16}{45}e^{-5/8} + \dfrac{2}{15}e^{-3/4} + \dfrac{16}{45}e^{-7/8} + \dfrac{7}{90}e^{-1}\Big]$

$\approx 0.63212055883.$

8. a) $f'(x) < 0$ on $[1, \infty)$, and $\lim_{x\to\infty} f(x) = 0$. Therefore

$$\int_1^\infty |f'(x)|\, dx = -\int_1^\infty f'(x)\, dx$$
$$= -\lim_{R\to\infty} \int_1^R f'(x)\, dx$$
$$= \lim_{R\to\infty} (f(1) - f(R)) = f(1).$$

Thus

$$\left|\int_R^\infty f'(x) \cos x\, dx\right| \leq \int_R^\infty |f'(x)|\, dx \to 0 \text{ as } R \to \infty.$$

Thus $\lim_{R\to\infty} \int_1^R f'(x) \cos x\, dx$ exists.

b) $\int_1^\infty f(x) \sin x\, dx$

$U = f(x) \qquad dV = \sin x\, dx$
$dU = f'(x)\, dx \qquad V = -\cos x$

$$= \lim_{R\to\infty} f(x)\cos x \Big|_1^R + \int_1^\infty f'(x) \cos x\, dx$$
$$= -f(1)\cos(1) + \int_1^\infty f'(x) \cos x\, dx;$$
the integral converges.

c) $f(x) = 1/x$ satisfies the conditions of part (a), so

$$\int_1^\infty \frac{\sin x}{x}\, dx \quad \text{converges}$$

by part (b). Similarly, it can be shown that

$$\int_1^\infty \frac{\cos(2x)}{x}\, dx \quad \text{converges}.$$

But since $|\sin x| \geq \sin^2 x = \frac{1}{2}(1 - \cos(2x))$, we have

$$\int_1^\infty \frac{|\sin x|}{x}\, dx \geq \int_1^\infty \frac{1 - \cos(2x)}{2x}.$$

The latter integral diverges because $\int_1^\infty (1/x)\, dx$ diverges to infinity while $\int_1^\infty (\cos(2x))/(2x)\, dx$ converges. Therefore

$$\int_1^\infty \frac{|\sin x|}{x}\, dx \quad \text{diverges to infinity}.$$

CHAPTER 7. APPLICATIONS OF INTEGRATION

Section 7.1 Volumes of Solids of Revolution (page 415)

1. By slicing:
$$V = \pi \int_0^1 x^4\, dx = \frac{\pi}{5} \text{ cu. units.}$$

By shells:
$$V = 2\pi \int_0^1 y(1 - \sqrt{y})\, dy$$
$$= 2\pi \left(\frac{y^2}{2} - \frac{2y^{5/2}}{5} \right)\bigg|_0^1 = \frac{\pi}{5} \text{ cu. units.}$$

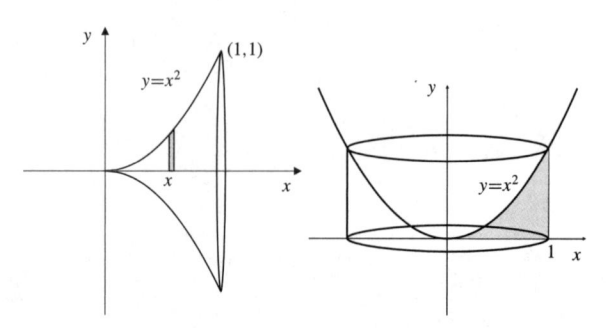

Fig. 7.1.1   Fig. 7.1.2

2. Slicing:
$$V = \pi \int_0^1 (1 - y)\, dy$$
$$= \pi \left(y - \frac{1}{2}y^2 \right)\bigg|_0^1 = \frac{\pi}{2} \text{ cu. units.}$$

Shells:
$$V = 2\pi \int_0^1 x^3\, dx$$
$$= 2\pi \left(\frac{x^4}{4} \right)\bigg|_0^1 = \frac{\pi}{2} \text{ cu. units.}$$

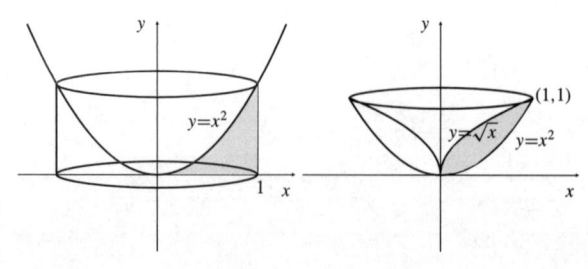

Fig. 7.1.2   Fig. 7.1.4

3. By slicing:
$$V = \pi \int_0^1 (x - x^4)\, dx$$
$$= \pi \left(\frac{x^2}{2} - \frac{x^5}{5} \right)\bigg|_0^1 = \frac{3\pi}{10} \text{ cu. units.}$$

By shells:
$$V = 2\pi \int_0^1 y(\sqrt{y} - y^2)\, dy$$
$$= 2\pi \left(\frac{2y^{5/2}}{5} - \frac{y^4}{4} \right)\bigg|_0^1 = \frac{3\pi}{10} \text{ cu. units.}$$

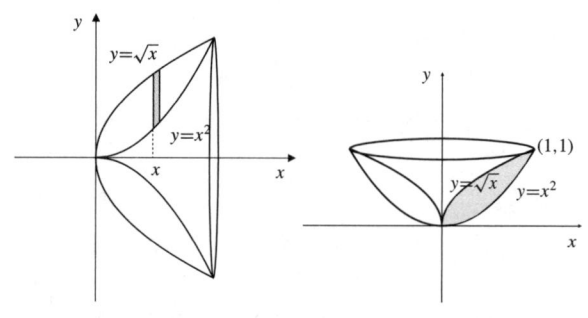

Fig. 7.1.3   Fig. 7.1.4

4. Slicing:
$$V = \pi \int_0^1 (y - y^4)\, dy$$
$$= \pi \left(\frac{1}{2}y^2 - \frac{1}{5}y^5 \right)\bigg|_0^1 = \frac{3\pi}{10} \text{ cu. units.}$$

Shells:
$$V = 2\pi \int_0^1 x(x^{1/2} - x^2)\, dx$$
$$= 2\pi \left(\frac{2}{5}x^{5/2} - \frac{1}{4}x^4 \right)\bigg|_0^1 = \frac{3\pi}{10} \text{ cu. units.}$$

5. a) About the x-axis:
$$V = \pi \int_0^2 x^2(2 - x)^2\, dx$$
$$= \pi \int_0^2 (4x^2 - 4x^3 + x^4)\, dx$$
$$= \pi \left(\frac{4x^3}{3} - x^4 + \frac{x^5}{5} \right)\bigg|_0^2 = \frac{16\pi}{15} \text{ cu. units.}$$

b) About the y-axis:
$$V = 2\pi \int_0^2 x^2(2-x)\,dy$$
$$= 2\pi \left(\frac{2x^3}{3} - \frac{x^4}{4}\right)\Big|_0^2 = \frac{8\pi}{3} \text{ cu. units.}$$

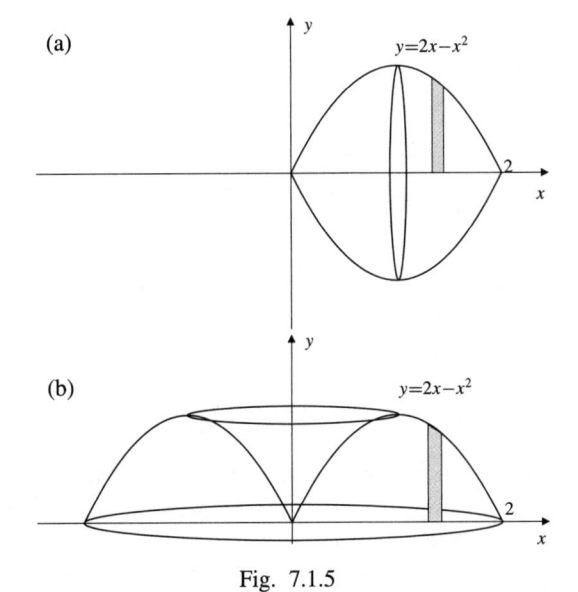

(a)

(b)

Fig. 7.1.5

6. Rotate about

a) the x-axis
$$V = \pi \int_0^1 (x^2 - x^4)\,dx$$
$$= \pi\left(\frac{1}{3}x^3 - \frac{1}{5}x^5\right)\Big|_0^1 = \frac{2\pi}{15} \text{ cu. units.}$$

b) the y-axis
$$V = 2\pi \int_0^1 x(x - x^2)\,dx$$
$$= 2\pi\left(\frac{1}{3}x^3 - \frac{1}{4}x^4\right)\Big|_0^1 = \frac{\pi}{6} \text{ cu. units.}$$

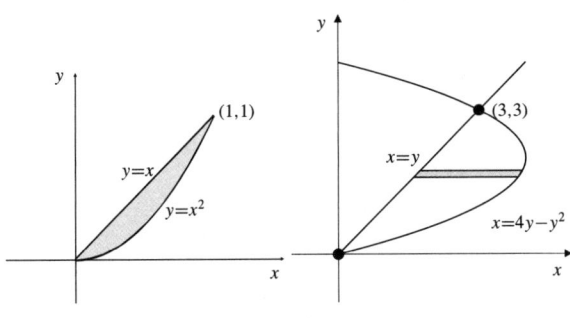

Fig. 7.1.6   Fig. 7.1.7

7. a) About the x-axis:
$$V = 2\pi \int_0^3 y(4y - y^2 - y)\,dy$$
$$= 2\pi\left(y^3 - \frac{y^4}{4}\right)\Big|_0^3 = \frac{27\pi}{2} \text{ cu. units.}$$

b) About the y-axis:
$$V = \pi \int_0^3 \left[(4y - y^2)^2 - y^2\right]dy$$
$$= \pi \int_0^3 (15y^2 - 8y^3 + y^4)\,dy$$
$$= \pi\left(5y^3 - 2y^4 + \frac{y^5}{5}\right)\Big|_0^3 = \frac{108\pi}{5} \text{ cu. units.}$$

8. Rotate about

a) the x-axis
$$V = \pi \int_0^\pi \left[(1 + \sin x)^2 - 1\right]dx$$
$$= \pi \int_0^\pi (2\sin x + \sin^2 x)\,dx$$
$$= \left(-2\pi\cos x + \frac{\pi}{2}x - \frac{\pi}{4}\sin 2x\right)\Big|_0^\pi$$
$$= 4\pi + \frac{1}{2}\pi^2 \text{ cu. units.}$$

b) the y-axis
$$V = 2\pi \int_0^\pi x\sin x\,dx$$
$$U = x \qquad dV = \sin x\,dx$$
$$dU = dx \qquad V = -\cos x$$
$$= 2\pi\left[-x\cos x\Big|_0^\pi + \int_0^\pi \cos x\,dx\right]$$
$$= 2\pi^2 \text{ cu. units.}$$

9. a) About the x-axis:
$$V = \pi \int_0^1 \left(4 - \frac{1}{(1+x^2)^2}\right)dx \quad \text{Let } x = \tan\theta$$
$$\qquad\qquad dx = \sec^2\theta\,d\theta$$
$$= 4\pi - \pi \int_0^{\pi/4} \frac{\sec^2\theta}{\sec^4\theta}\,d\theta$$
$$= 4\pi - \pi \int_0^{\pi/4} \cos^2\theta\,d\theta$$
$$= 4\pi - \frac{\pi}{2}(\theta + \sin\theta\cos\theta)\Big|_0^{\pi/4}$$
$$= 4\pi - \frac{\pi^2}{8} - \frac{\pi}{4} = \frac{15\pi}{4} - \frac{\pi^2}{8} \text{ cu. units.}$$

b) About the y-axis:

$$V = 2\pi \int_0^1 x\left(2 - \frac{1}{1+x^2}\right) dx$$
$$= 2\pi \left(x^2 - \frac{1}{2}\ln(1+x^2)\right)\Big|_0^1$$
$$= 2\pi\left(1 - \frac{1}{2}\ln 2\right) = 2\pi - \pi\ln 2 \text{ cu. units.}$$

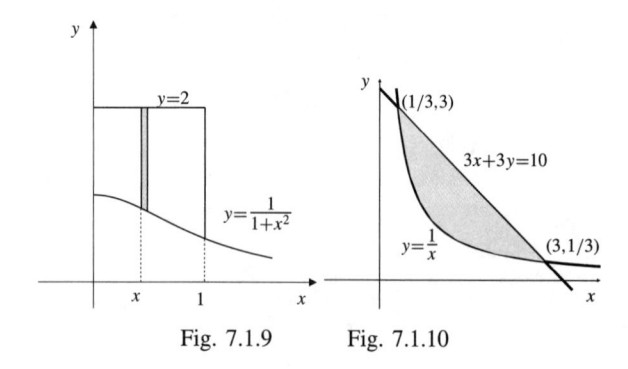

Fig. 7.1.9   Fig. 7.1.10

10. By symmetry, rotation about the x-axis gives the same volume as rotation about the y-axis, namely

$$V = 2\pi \int_{1/3}^3 x\left(\frac{10}{3} - x - \frac{1}{x}\right) dx$$
$$= 2\pi\left(\frac{5}{3}x^2 - \frac{1}{3}x^3 - x\right)\Big|_{1/3}^3$$
$$= \frac{512\pi}{81} \text{ cu. units.}$$

11. $V = 2 \times 2\pi \int_0^1 (2-x)(1-x)\, dx$
$$= 4\pi \int_0^1 (2 - 3x + x^2)\, dx$$
$$= 4\pi\left(2x - \frac{3x^2}{2} + \frac{x^3}{3}\right)\Big|_0^1 = \frac{10\pi}{3} \text{ cu. units.}$$

Fig. 7.1.11

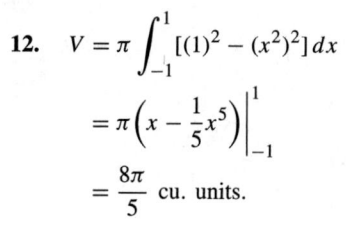

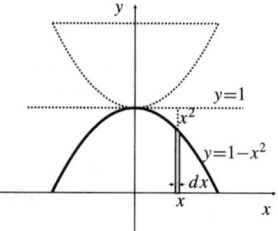

12. $V = \pi \int_{-1}^1 [(1)^2 - (x^2)^2]\, dx$
$$= \pi\left(x - \frac{1}{5}x^5\right)\Big|_{-1}^1$$
$$= \frac{8\pi}{5} \text{ cu. units.}$$

Fig. 7.1.12

13. The volume remaining is

$$V = 2 \times 2\pi \int_1^2 x\sqrt{4-x^2}\, dx \quad \text{Let } u = 4 - x^2$$
$$\hspace{5cm} du = -2x\, dx$$
$$= 2\pi \int_0^3 \sqrt{u}\, du = \frac{4\pi}{3} u^{3/2}\Big|_0^3 = 4\pi\sqrt{3} \text{ cu. units.}$$

Since the volume of the ball is $\frac{4}{3}\pi 2^3 = \frac{32\pi}{3}$ cu. units., therefore the volume removed is $\frac{32\pi}{3} - 4\pi\sqrt{3}$ cu. units. The percentage removed is

$$\frac{\frac{32\pi}{3} - 4\pi\sqrt{3}}{\frac{32\pi}{3}} \times 100 = 100\left(1 - \frac{3\sqrt{3}}{8}\right) \approx 35.$$

About 35% of the volume is removed.

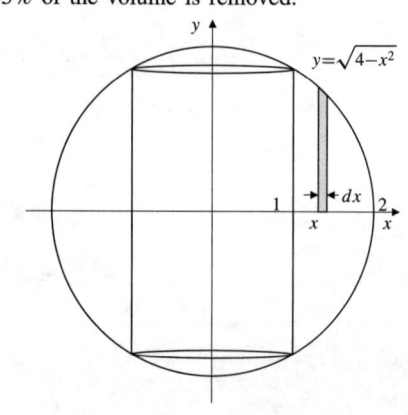

Fig. 7.1.13

14. The radius of the hole is $\sqrt{R^2 - \frac{1}{4}L^2}$. Thus, by slicing, the remaining volume is

$$V = \pi \int_{-L/2}^{L/2} \left[(R^2 - x^2) - \left(R^2 - \frac{L^2}{4}\right)\right] dx$$
$$= 2\pi \left(\frac{L^2}{4}x - \frac{1}{3}x^3\right)\bigg|_0^{L/2}$$
$$= \frac{\pi}{6}L^3 \text{ cu. units (independent of } R\text{)}.$$

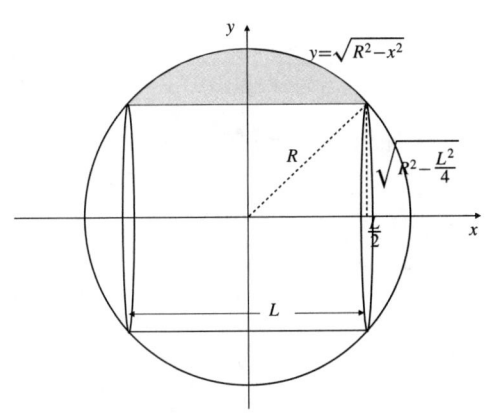

Fig. 7.1.14

15. The volume remaining is

$$V = 2\pi \int_a^b xh\left(1 - \frac{x}{b}\right) dx$$
$$= 2\pi h\left(\frac{x^2}{2} - \frac{x^3}{3b}\right)\bigg|_a^b$$
$$= \pi h(b^2 - a^2) - \frac{2}{3}\pi h\left(b^2 - \frac{a^3}{b}\right)$$
$$= \frac{1}{3}\pi h\left(b^2 - 3a^2 + \frac{2a^3}{b}\right) \text{ cu. units.}$$

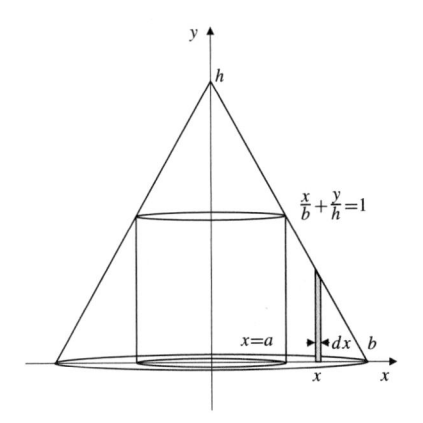

Fig. 7.1.15

16. Let a circular disk with radius a have centre at point $(a, 0)$. Then the disk is rotated about the y-axis which is one of its tangent lines. The volume is:

$$V = 2 \times 2\pi \int_0^{2a} x\sqrt{a^2 - (x-a)^2}\, dx \quad \text{Let } u = x - a$$
$$ \quad du = dx$$
$$= 4\pi \int_{-a}^a (u+a)\sqrt{a^2 - u^2}\, du$$
$$= 4\pi \int_{-a}^a u\sqrt{a^2 - u^2}\, du + 4\pi a \int_{-a}^a \sqrt{a^2 - u^2}\, du$$
$$= 0 + 4\pi a\left(\frac{1}{2}\pi a^2\right) = 2\pi^2 a^3 \text{ cu. units.}$$

(Note that the first integral is zero because the integrand is odd and the interval is symmetric about zero; the second integral is the area of a semicircle.)

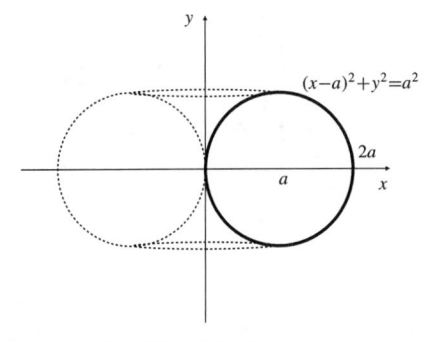

Fig. 7.1.16

17. Volume of the smaller piece:

$$V = \pi \int_b^a (a^2 - x^2)\, dx$$
$$= \pi \left(a^2 x - \frac{x^3}{3}\right)\bigg|_b^a$$
$$= \pi \left(a^2(a-b) - \frac{a^3 - b^3}{3}\right)$$
$$= \frac{\pi}{3}(a-b)[3a^2 - (a^2 + ab + b^2)]$$
$$= \frac{\pi}{3}(a-b)^2(2a+b) \text{ cu. units.}$$

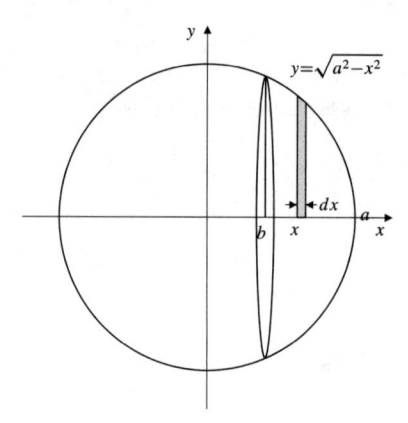

Fig. 7.1.17

18. Let the centre of the bowl be at (0, 30). Then the volume of the water in the bowl is

$$V = \pi \int_0^{20} \left[30^2 - (y - 30)^2\right] dy$$
$$= \pi \int_0^{20} 60y - y^2 \, dy$$
$$= \pi \left[30y^2 - \frac{1}{3}y^3\right]\Big|_0^{20}$$
$$\approx 29322 \text{ cm}^3.$$

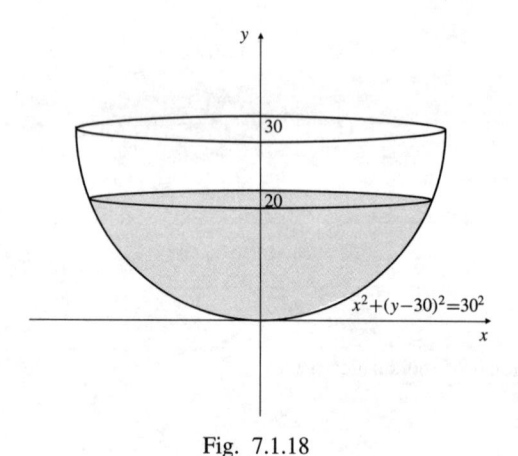

Fig. 7.1.18

19. The volume of the ellipsoid is

$$V = 2\pi \int_0^a b^2 \left(1 - \frac{x^2}{a^2}\right) dx$$
$$= 2\pi b^2 \left(x - \frac{x^3}{3a^2}\right)\Big|_0^a = \frac{4}{3}\pi ab^2 \text{ cu. units.}$$

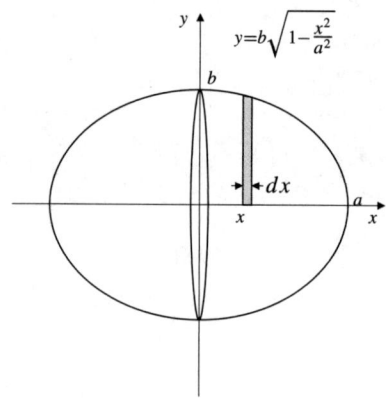

Fig. 7.1.19

20. The cross-section at height y is an annulus (ring) having inner radius $b - \sqrt{a^2 - y^2}$ and outer radius $b + \sqrt{a^2 - y^2}$. Thus the volume of the torus is

$$V = \pi \int_{-a}^{a} \left[(b + \sqrt{a^2 - y^2})^2 - (b - \sqrt{a^2 - y^2})^2\right] dy$$
$$= 2\pi \int_0^a 4b\sqrt{a^2 - y^2} \, dy$$
$$= 8\pi b \frac{\pi a^2}{4} = 2\pi^2 a^2 b \text{ cu. units..}$$

We used the area of a quarter-circle of radius a to evaluate the last integral.

21. a) Volume of revolution about the x-axis is

$$V = \pi \int_0^\infty e^{-2x} \, dx$$
$$= \pi \lim_{R \to \infty} \frac{e^{-2x}}{-2}\Big|_0^R = \frac{\pi}{2} \text{ cu. units.}$$

b) Volume of revolution about the y-axis is

$$V = 2\pi \int_0^\infty x e^{-x} \, dx$$
$$= 2\pi \lim_{R \to \infty} (-xe^{-x} - e^{-x})\Big|_0^R = 2\pi \text{ cu. units.}$$

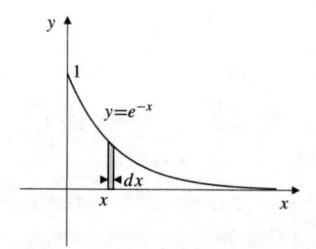

Fig. 7.1.21

INSTRUCTOR'S SOLUTIONS MANUAL   SECTION 7.1 (PAGE 415)

22. The volume is

$$V = \pi \int_1^\infty x^{-2k}\,dx = \pi \lim_{R\to\infty} \left.\frac{x^{1-2k}}{1-2k}\right|_1^R$$

$$= \pi \lim_{R\to\infty} \frac{R^{1-2k}}{1-2k} + \frac{\pi}{2k-1}.$$

In order for the solid to have finite volume we need

$$1 - 2k < 0, \quad \text{that is,} \quad k > \frac{1}{2}.$$

23. The volume is $V = 2\pi \int_1^\infty x^{1-k}\,dx$. This improper integral converges if $1-k < -1$, i.e., if $k > 2$. The solid has finite volume only if $k > 2$.

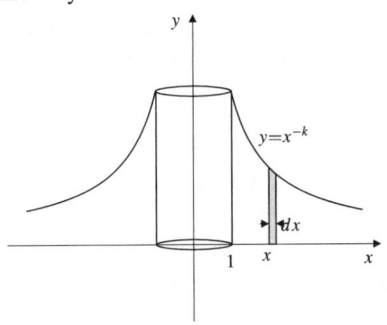

Fig. 7.1.23

24. Using heights $f(x)$ estimated from the given graph, we obtain

$$V = \pi \int_1^9 \bigl(f(x)\bigr)^2\,dx$$

$$\approx \frac{\pi}{3}\Bigl[3^2 + 4(3.8)^2 + 2(5)^2 + 4(6.7)^2 + 2(8)^2$$

$$+ 4(8)^2 + 2(7)^2 + 4(5.2)^2 + 3^2\Bigr] \approx 938 \text{ cu. units.}$$

25. Using heights $f(x)$ estimated from the given graph, we obtain

$$V = 2\pi \int_1^9 x f(x)\,dx$$

$$\approx \frac{2\pi}{3}\Bigl[1(3) + 4(2)(3.8) + 2(3)(5) + 4(4)(6.7) + 2(5)(8)$$

$$+ 4(6)(8) + 2(7)(7) + 4(8)(5.2) + 9(3)\Bigr] \approx 1537 \text{ cu. units.}$$

26. Using heights $f(x)$ estimated from the given graph, we obtain

$$V = 2\pi \int_1^9 (x+1) f(x)\,dx$$

$$\approx \frac{2\pi}{3}\Bigl[2(3) + 4(3)(3.8) + 2(4)(5) + 4(5)(6.7) + 2(6)(8)$$

$$+ 4(7)(8) + 2(8)(7) + 4(9)(5.2) + 10(3)\Bigr] \approx 1832 \text{ cu. units.}$$

27. The region is symmetric about $x = y$ so has the same volume of revolution about the two coordinate axes. The volume of revolution about the y-axis is

$$V = 2\pi \int_0^8 x(4 - x^{2/3})^{3/2}\,dx \quad \text{Let } x = 8\sin^3 u$$

$$\phantom{V = 2\pi \int_0^8 x(4 - x^{2/3})^{3/2}\,dx \quad} dx = 24\sin^2 u \cos u\,du$$

$$= 3072\pi \int_0^{\pi/2} \sin^5 u \cos^4 u\,du$$

$$= 3072\pi \int_0^{\pi/2} (1-\cos^2 u)^2 \cos^4 u \sin u\,du \quad \text{Let } v = \cos u$$

$$\phantom{= 3072\pi \int_0^{\pi/2} (1-\cos^2 u)^2 \cos^4 u \sin u\,du \quad} dv = -\sin u\,du$$

$$= 3072\pi \int_0^1 (1-v^2)^2 v^4\,dv$$

$$= 3072\pi \int_0^1 (v^4 - 2v^6 + v^8)\,dv$$

$$= 3072\pi \left(\frac{1}{5} - \frac{2}{7} + \frac{1}{9}\right) = \frac{8192\pi}{105} \text{ cu. units.}$$

28. The volume of the ball is $\frac{4}{3}\pi R^3$. Expressing this volume as the "sum" (i.e., integral) of volume elements that are concentric spherical shells of radius r and thickness dr, and therefore surface area kr^2 and volume $kr^2\,dr$, we obtain

$$\frac{4}{3}\pi R^3 = \int_0^R kr^2\,dr = \frac{k}{3}R^3.$$

Thus $k = 4\pi$.

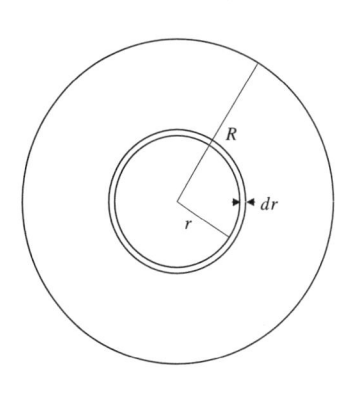

Fig. 7.1.28

29. Let the ball have radius R, and suppose its centre is x units above the top of the conical glass, as shown in the figure. (Clearly the ball which maximizes wine overflow from the glass must be tangent to the cone along some circle below the top of the cone — larger balls will have reduced displacement within the cone. Also, the ball will not be completely submerged.)

Note that $\dfrac{R}{x+h} = \sin\alpha$, so $R = (x+h)\sin\alpha$.

Using the result of Exercise #17, the volume of wine displaced by the ball is

$$V = \frac{\pi}{3}(R-x)^2(2R+x).$$

We would like to consider V as a function of x for $-2R \leq x \leq R$ since $V = 0$ at each end of this interval, and $V > 0$ inside the interval. However, the actual interval of values of x for which the above formulation makes physical sense is smaller: x must satisfy $-R \leq x \leq h\tan^2\alpha$. (The left inequality signifies nonsubmersion of the ball; the right inequality signifies that the ball is tangent to the glass somewhere below the rim.) We look for a critical point of V, considered as a function of x. (As noted above, R is a function of x.) We have

$$0 = \frac{dV}{dx} = \frac{\pi}{3}\left[2(R-x)\left(\frac{dR}{dx}-1\right)(2R+x) \right.$$
$$\left. + (R-x)^2\left(2\frac{dR}{dx}+1\right)\right]$$

$$\frac{dR}{dx}(4R+2x+2R-2x) = 4R+2x-(R-x).$$

Thus

$$6R\sin\alpha = 3(R+x) = 3\left(R+\frac{R}{\sin\alpha}-h\right)$$
$$2R\sin^2\alpha = R\sin\alpha + R - h\sin\alpha$$
$$R = \frac{h\sin\alpha}{1-2\sin^2\alpha+\sin\alpha} = \frac{h\sin\alpha}{\cos 2\alpha + \sin\alpha}.$$

This value of R yields a positive value of V, and corresponds to $x = R(2\sin\alpha - 1)$. Since $\sin\alpha \geq \sin^2\alpha$,

$$-R \leq x = \frac{h\sin\alpha(2\sin\alpha - 1)}{1+\sin\alpha - 2\sin^2\alpha} \leq \frac{h\sin^2\alpha}{\cos^2\alpha} = h\tan^2\alpha.$$

Therefore it gives the maximum volume of wine displaced.

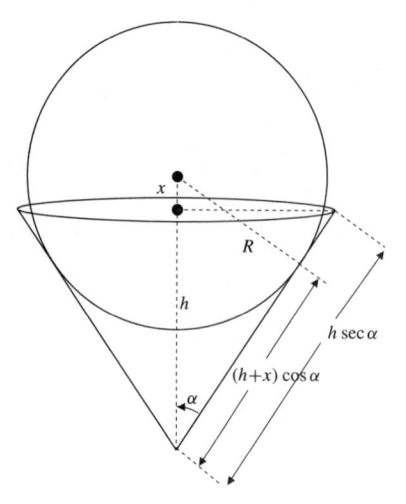

Fig. 7.1.29

30. Let P be the point $(t, \frac{5}{2}-t)$. The line through P perpendicular to AB has equation $y = x + \frac{5}{2} - 2t$, and meets the curve $xy = 1$ at point Q with x-coordinate s equal to the positive root of $s^2 + (\frac{5}{2}-2t)s = 1$. Thus,

$$s = \frac{1}{2}\left[2t - \frac{5}{2} + \sqrt{\left(\frac{5}{2}-2t\right)^2 + 4}\right].$$

The volume element at P has radius

$$PQ = \sqrt{2}(t-s)$$
$$= \sqrt{2}\left[\frac{5}{4} - \frac{1}{2}\sqrt{\left(\frac{5}{2}-2t\right)^2 + 4}\right]$$

and thickness $\sqrt{2}\,dt$. Hence, the volume of the solid is

$$V = \pi\int_{1/2}^{2}\left[\sqrt{2}\left(\frac{5}{4} - \frac{1}{2}\sqrt{\left(\frac{5}{2}-2t\right)^2+4}\right)\right]^2 \sqrt{2}\,dt$$
$$= 2\sqrt{2}\pi\int_{1/2}^{2}\left[\frac{25}{16} - \frac{5}{4}\left(\sqrt{\left(\frac{5}{2}-2t\right)^2+4}\right) + \right.$$
$$\left. \frac{1}{4}\left[\left(\frac{5}{2}-2t\right)^2+4\right]\right] dt \quad \text{Let } u = 2t - \frac{5}{2}$$
$$\qquad\qquad du = 2\,dt$$
$$= \sqrt{2}\pi\int_{-3/2}^{3/2}\left(\frac{41}{16} - \frac{5}{4}\sqrt{u^2+4} + \frac{u^2}{4}\right) du$$
$$= \sqrt{2}\pi\left(\frac{41}{16}u + \frac{1}{12}u^3\right)\bigg|_{-3/2}^{3/2} -$$
$$\quad \frac{5\sqrt{2}\pi}{4}\int_{-3/2}^{3/2}\sqrt{u^2+4}\,du \quad \text{Let } u = 2\tan v$$
$$\qquad\qquad du = 2\sec^2 v\,dv$$
$$= \frac{33\sqrt{2}\pi}{4} - 5\sqrt{2}\pi\int_{\tan^{-1}(-3/4)}^{\tan^{-1}(3/4)} \sec^3 v\,dv$$

INSTRUCTOR'S SOLUTIONS MANUAL   SECTION 7.2 (PAGE 419)

$$= \frac{33\sqrt{2}\pi}{4} - 10\sqrt{2}\pi \int_0^{\tan^{-1}(3/4)} \sec^3 v\, dv$$

$$= \frac{33\sqrt{2}\pi}{4} - 5\sqrt{2}\pi \left(\sec v \tan v + \ln|\sec v + \tan v|\right)\Big|_0^{\tan^{-1}(3/4)}$$

$$= \sqrt{2}\pi\left[\frac{33}{4} - 5\left(\frac{15}{16} + \ln 2 - 0 - \ln 1\right)\right]$$

$$= \sqrt{2}\pi\left(\frac{57}{16} - 5\ln 2\right) \text{ cu. units.}$$

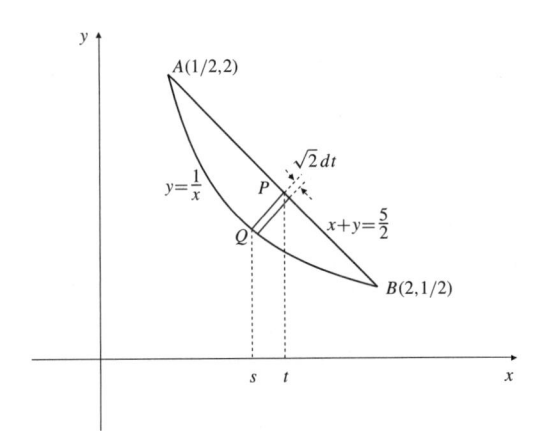

Fig. 7.1.30

Section 7.2 Other Volumes by Slicing (page 419)

1. $V = \int_0^2 3x\, dx = \frac{3}{2}x^2\Big|_0^2 = 6 \text{ m}^3$

2. $V = \int_1^3 x^2\, dx = \frac{x^3}{3}\Big|_1^3 = \frac{26}{3}$ cu. units

3. $V = \int_0^6 (2+z)(8-z)\, dz = \int_0^6 (16 + 6z - z^2)\, dz$
   $= \left(16z + 3z^2 - \frac{z^3}{3}\right)\Big|_0^6 = 132 \text{ ft}^3$

4. The area of an equilateral triangle of edge $\sqrt{x}$ is $A(x) = \frac{1}{2}\sqrt{x}\left(\frac{\sqrt{3}}{2}\sqrt{x}\right) = \frac{\sqrt{3}}{4}x$ sq. units. The volume of the solid is
   $$V = \int_1^4 \frac{\sqrt{3}}{4}x\, dx = \frac{\sqrt{3}}{8}x^2\Big|_1^4 = \frac{15\sqrt{3}}{8} \text{ cu. units.}$$

5. The area of cross-section at height y is
   $$A(y) = \frac{2\pi(1 - (y/h))}{2\pi}(\pi a^2) = \pi a^2\left(1 - \frac{y}{h}\right) \text{ sq. units.}$$
   The volume of the solid is
   $$V = \int_0^h \pi a^2\left(1 - \frac{y}{h}\right) dy = \frac{\pi a^2 h}{2} \text{ cu. units.}$$

6. Since $V = 4$, we have
   $$4 = \int_0^2 kx^3\, dx = k\left.\frac{x^4}{4}\right|_0^2 = 4k.$$
   Thus $k = 1$.

7. The volume between height 0 and height z is z^3. Thus
   $$z^3 = \int_0^z A(t)\, dt,$$
   where $A(t)$ is the cross-sectional area at height t. Differentiating the above equation with respect to z, we get $3z^2 = A(z)$. The cross-sectional area at height z is $3z^2$ sq. units.

8. This is similar to Exercise 7. We have $4z = \int_0^z A(t)\, dt$, so $A(z) = 4$. Thus the square cross-section at height z has side 2 units.

9. $V = 2\int_0^r \left(2\sqrt{r^2 - y^2}\right)^2 dy$
   $= 8\int_0^r (r^2 - y^2)\, dy = 8\left(r^2 y - \frac{y^3}{3}\right)\Big|_0^r = \frac{16r^3}{3}$ cu. units.

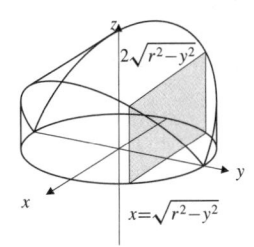

Fig. 7.2.9

10. The area of an equilateral triangle of base $2y$ is $\frac{1}{2}(2y)(\sqrt{3}y) = \sqrt{3}y^2$. Hence, the solid has volume
    $$V = 2\int_0^r \sqrt{3}(r^2 - x^2)\, dx$$
    $$= 2\sqrt{3}\left(r^2 x - \frac{1}{3}x^3\right)\Big|_0^r$$
    $$= \frac{4}{\sqrt{3}}r^3 \text{ cu. units.}$$

265

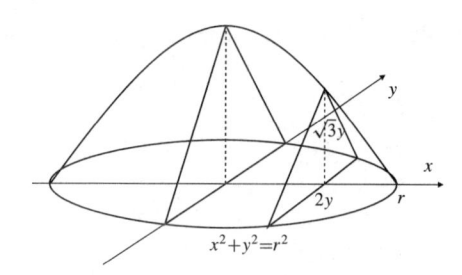

Fig. 7.2.10

11. The cross-section at distance y from the vertex of the partial cone is a semicircle of radius $y/2$ cm, and hence area $\pi y^2/8$ cm². The volume of the solid is

$$V = \int_0^{12} \frac{\pi}{8} y^2 \, dy = \frac{\pi 12^3}{24} = 72\pi \text{ cm}^3.$$

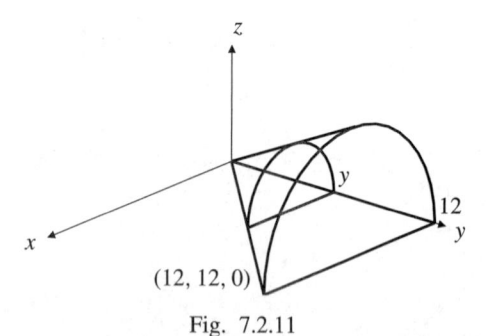

Fig. 7.2.11

12. The volume of a solid of given height h and given cross-sectional area $A(z)$ at height z above the base is given by

$$V = \int_0^h A(z) \, dz.$$

If two solids have the same height h and the same area function $A(z)$, then they must necessarily have the same volume.

13. Let the x-axis be along the diameter shown in the figure, with the origin at the centre of the base. The cross-section perpendicular to the x-axis at x is a rectangle having base $2\sqrt{r^2 - x^2}$ and height $h = \dfrac{a+b}{2} + \dfrac{a-b}{2r} x$. Thus the volume of the truncated cylinder is

$$V = \int_{-r}^{r} (2\sqrt{r^2 - x^2}) \left(\frac{a+b}{2} + \frac{a-b}{2r} x \right) dx$$
$$= \int_{-r}^{r} (a+b) \sqrt{r^2 - x^2} \, dx = \frac{\pi r^2 (a+b)}{2} \text{ cu. units.}$$

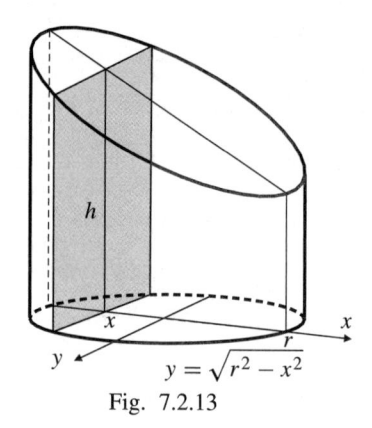

Fig. 7.2.13

14. The plane $z = k$ meets the ellipsoid in the ellipse

$$\left(\frac{x}{a}\right)^2 + \left(\frac{y}{b}\right)^2 = 1 - \left(\frac{k}{c}\right)^2$$

that is,
$$\frac{x^2}{a^2 \left[1 - \left(\frac{k}{c}\right)^2\right]} + \frac{y^2}{b^2 \left[1 - \left(\frac{k}{c}\right)^2\right]} = 1$$

which has area

$$A(k) = \pi ab \left[1 - \left(\frac{k}{c}\right)^2\right].$$

The volume of the ellipsoid is found by summing volume elements of thickness dk:

$$V = \int_{-c}^{c} \pi ab \left[1 - \left(\frac{k}{c}\right)^2\right] dk$$
$$= \pi ab \left[k - \frac{1}{3c^2} k^3\right]\Big|_{-c}^{c}$$
$$= \frac{4}{3} \pi abc \text{ cu. units.}$$

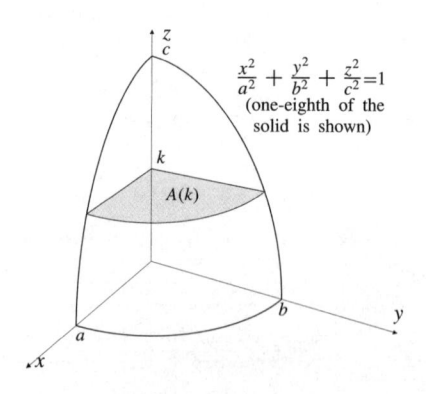

Fig. 7.2.14

INSTRUCTOR'S SOLUTIONS MANUAL   SECTION 7.3 (PAGE 427)

15. Cross-sections of the wedge removed perpendicular to the x-axis are isosceles, right triangles. The volume of the wedge removed from the log is

$$V = 2\int_0^{20} \frac{1}{2}(\sqrt{400-x^2})^2\,dx$$
$$= \left(400x - \frac{x^3}{3}\right)\bigg|_0^{20} = \frac{16,000}{3}\text{ cm}^3.$$

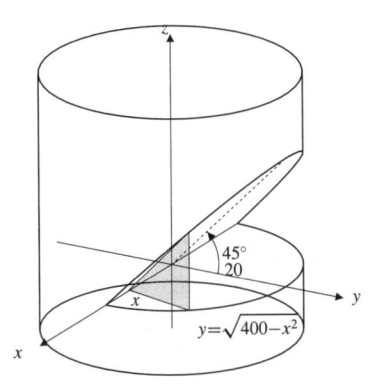

Fig. 7.2.15

16. The solution is similar to that of Exercise 15 except that the legs of the right-triangular cross-sections are $y-10$ instead of y, and x goes from $-10\sqrt{3}$ to $10\sqrt{3}$ instead of -20 to 20. The volume of the notch is

$$V = 2\int_0^{10\sqrt{3}} \frac{1}{2}(\sqrt{400-x^2}-10)^2\,dx$$
$$= \int_0^{10\sqrt{3}} \left(500 - x^2 - 20\sqrt{400-x^2}\right)dx$$
$$= 3,000\sqrt{3} - \frac{4,000\pi}{3} \approx 1,007\text{ cm}^3.$$

17. The hole has the shape of two copies of the truncated cylinder of Exercise 13, placed base to base, with $a+b = 3\sqrt{2}$ in and $r=2$ in. Thus the volume of wood removed (the volume of the hole) is $V = 2(\pi 2^2)(3\sqrt{2}/2) = 12\sqrt{2}\pi$ in^3.

18. One eighth of the region lying inside both cylinders is shown in the figure. If the region is sliced by a horizontal plane at height z, then the intersection is a rectangle with area
$$A(z) = \sqrt{b^2-z^2}\sqrt{a^2-z^2}.$$

The volume of the whole region is

$$V = 8\int_0^b \sqrt{b^2-z^2}\sqrt{a^2-z^2}\,dz.$$

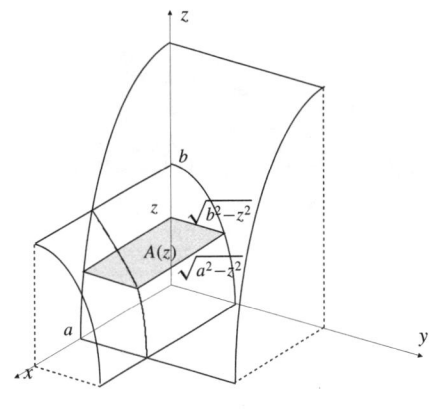

Fig. 7.2.18

19. By the result given in Exercise 18 with $a=4$ cm and $b=2$ cm, the volume of wood removed is

$$V = 8\int_0^2 \sqrt{4-z^2}\sqrt{16-z^2}\,dz \approx 97.28\text{ cm}^3.$$

(We used the numerical integration routine in Maple to evaluate the integral.)

Section 7.3 Arc Length and Surface Area (page 427)

1. $y = 2x-1$, $y'=2$, $ds = \sqrt{1+2^2}\,dx$
$$L = \int_1^3 \sqrt{5}\,dx = 2\sqrt{5}\text{ units.}$$

2. $y = ax+b$, $A \le x \le B$, $y'=a$. The length is
$$L = \int_A^B \sqrt{1+a^2}\,dx = \sqrt{1+a^2}(B-A)\text{ units.}$$

3. $y = \frac{2}{3}x^{3/2}$, $y' = \sqrt{x}$, $ds = \sqrt{1+x}\,dx$
$$L = \int_0^8 \sqrt{1+x}\,dx = \frac{2}{3}(1+x)^{3/2}\bigg|_0^8 = \frac{52}{3}\text{ units.}$$

4. $y^2 = (x-1)^3$, $y = (x-1)^{3/2}$, $y' = \frac{3}{2}\sqrt{x-1}$
$$L = \int_1^2 \sqrt{1+\frac{9}{4}(x-1)}\,dx = \frac{1}{2}\int_1^2 \sqrt{9x-5}\,dx$$
$$= \frac{1}{27}(9x-5)^{3/2}\bigg|_1^2 = \frac{13^{3/2}-8}{27}\text{ units.}$$

267

5. $y = x^{2/3}$, $y' = \frac{2}{3}x^{-1/3}$,

$$ds = \sqrt{1 + \frac{4}{9}x^{-2/3}}\, dx = \frac{\sqrt{9x^{2/3}+4}}{3|x|^{1/3}}\, dx$$

$$L = 2\int_0^1 \frac{\sqrt{9x^{2/3}+4}}{3x^{1/3}}\, dx \quad \text{Let } u = 9x^{2/3}+4$$
$$du = 6x^{-1/3}\, dx$$

$$= \frac{1}{9}\int_4^{13} \sqrt{u}\, du = \frac{2(13^{3/2}) - 16}{27} \text{ units.}$$

6. $2(x+1)^3 = 3(y-1)^2$, $y = 1 + \sqrt{\frac{2}{3}}(x+1)^{3/2}$

$y' = \sqrt{\frac{3}{2}}(x+1)^{1/2}$,

$$ds = \sqrt{1 + \frac{3x+3}{2}}\, dx = \sqrt{\frac{3x+5}{2}}\, dx$$

$$L = \frac{1}{\sqrt{2}}\int_{-1}^0 \sqrt{3x+5}\, dx = \left.\frac{\sqrt{2}}{9}(3x+5)^{3/2}\right|_{-1}^0$$

$$= \frac{\sqrt{2}}{9}\left(5^{3/2} - 2^{3/2}\right) \text{ units.}$$

7. $y = \frac{x^3}{12} + \frac{1}{x}$, $y' = \frac{x^2}{4} - \frac{1}{x^2}$

$$ds = \sqrt{1 + \left(\frac{x^2}{4} - \frac{1}{x^2}\right)^2}\, dx = \left(\frac{x^2}{4} + \frac{1}{x^2}\right) dx$$

$$L = \int_1^4 \left(\frac{x^2}{4} + \frac{1}{x^2}\right) dx = \left.\left(\frac{x^3}{12} - \frac{1}{x}\right)\right|_1^4 = 6 \text{ units.}$$

8. $y = \frac{x^3}{3} + \frac{1}{4x}$, $y' = x^2 - \frac{1}{4x^2}$

$$ds = \sqrt{1 + \left(x^2 - \frac{1}{4x^2}\right)^2}\, dx = \left(x^2 + \frac{1}{4x^2}\right) dx$$

$$L = \int_1^2 \left(x^2 + \frac{1}{4x^2}\right) dx = \left.\left(\frac{x^3}{3} - \frac{1}{4x}\right)\right|_1^2 = \frac{59}{24} \text{ units.}$$

9. $y = \frac{\ln x}{2} - \frac{x^2}{4}$, $y' = \frac{1}{2x} - \frac{x}{2}$

$$ds = \sqrt{1 + \left(\frac{1}{2x} - \frac{x}{2}\right)^2}\, dx = \left(\frac{1}{2x} + \frac{x}{2}\right) dx$$

$$L = \int_1^e \left(\frac{1}{2x} + \frac{x}{2}\right) dx = \left.\left(\frac{\ln x}{2} + \frac{x^2}{4}\right)\right|_1^e$$

$$= \frac{1}{2} + \frac{e^2 - 1}{4} = \frac{e^2 + 1}{4} \text{ units.}$$

10. If $y = x^2 - \frac{\ln x}{8}$ then $y' = 2x - \frac{1}{8x}$ and

$$1 + (y')^2 = \left(2x + \frac{1}{8x}\right)^2.$$

Thus the arc length is given by

$$s = \int_1^2 \sqrt{1 + \left(2x - \frac{1}{8x}\right)^2}\, dx$$

$$= \int_1^2 \left(2x + \frac{1}{8x}\right) dx$$

$$= \left.\left(x^2 + \frac{1}{8}\ln x\right)\right|_1^2 = 3 + \frac{1}{8}\ln 2 \text{ units.}$$

11. $y = \ln(1 - x^2)$, $-\frac{1}{2} \le x \le \frac{1}{2}$, $y' = \frac{-2x}{1 - x^2}$.

$$\text{length} = \int_{-1/2}^{1/2} \sqrt{1 + \frac{4x^2}{(1-x^2)^2}}\, dx$$

$$= \int_{-1/2}^{1/2} \frac{1+x^2}{1-x^2}\, dx$$

$$= \int_{-1/2}^{1/2} \left(-1 + \frac{2}{1-x^2}\right) dx$$

$$= \left.\left(-x + \ln\frac{1+x}{1-x}\right)\right|_{-1/2}^{1/2} = 2\ln 3 - 1 \text{ units.}$$

12. $s = \int_0^a \sqrt{1 + \sinh^2 x}\, dx = \int_0^a \cosh x\, dx$

$$= \left.\sinh x\right|_0^a = \sinh a = \frac{e^a - e^{-a}}{2} \text{ units.}$$

13. $s = \int_{\pi/6}^{\pi/4} \sqrt{1 + \tan^2 x}\, dx$

$$= \int_{\pi/6}^{\pi/4} \sec x\, dx = \left.\ln|\sec x + \tan x|\right|_{\pi/6}^{\pi/4}$$

$$= \ln(\sqrt{2}+1) - \ln\left(\frac{2}{\sqrt{3}} + \frac{1}{\sqrt{3}}\right)$$

$$= \ln\frac{\sqrt{2}+1}{\sqrt{3}} \text{ units.}$$

14. $y = x^2$, $0 \le x \le 2$, $y' = 2x$.

$$\text{length} = \int_0^2 \sqrt{1 + 4x^2}\, dx \quad \text{Let } 2x = \tan\theta$$
$$2\, dx = \sec^2\theta\, d\theta$$

$$= \frac{1}{2}\int_{x=0}^{x=2} \sec^3\theta$$

$$= \left.\frac{1}{4}\left(\sec\theta\tan\theta + \ln|\sec\theta + \tan\theta|\right)\right|_{x=0}^{x=2}$$

$$= \left.\frac{1}{4}\left(2x\sqrt{1+4x^2} + \ln(2x + \sqrt{1+4x^2})\right)\right|_0^2$$

$$= \frac{1}{4}\left(4\sqrt{17} + \ln(4 + \sqrt{17})\right)$$

$$= \sqrt{17} + \frac{1}{4}\ln(4 + \sqrt{17}) \text{ units.}$$

INSTRUCTOR'S SOLUTIONS MANUAL SECTION 7.3 (PAGE 427)

15. $y = \ln \dfrac{e^x - 1}{e^x + 1}$, $2 \le x \le 4$

$y' = \dfrac{e^x + 1}{e^x - 1} \cdot \dfrac{(e^x + 1)e^x - (e^x - 1)e^x}{(e^x + 1)^2}$

$= \dfrac{2e^x}{e^{2x} - 1}$.

The length of the curve is

$L = \displaystyle\int_2^4 \sqrt{1 + \dfrac{4e^{2x}}{(e^{2x} - 1)^2}}\, dx$

$= \displaystyle\int_2^4 \dfrac{e^{2x} + 1}{e^{2x} - 1}\, dx$

$= \displaystyle\int_2^4 \dfrac{e^x + e^{-x}}{e^x - e^{-x}}\, dx = \ln\left|e^x - e^{-x}\right|\Big|_2^4$

$= \ln\left(e^4 - \dfrac{1}{e^4}\right) - \ln\left(e^2 - \dfrac{1}{e^2}\right)$

$= \ln\left(\dfrac{e^8 - 1}{e^4} \cdot \dfrac{e^2}{e^4 - 1}\right) = \ln \dfrac{e^4 + 1}{e^2}$ units.

16. We have

$s = \displaystyle\int_1^e \sqrt{1 + \dfrac{1}{x^2}}\, dx$

$= \displaystyle\int_1^e \dfrac{\sqrt{x^2 + 1}}{x}\, dx$ Let $x = \tan u$

$dx = \sec^2 u\, du$

$= \displaystyle\int_{x=1}^{x=e} \dfrac{\sec^3 u}{\tan u}\, du = \displaystyle\int_{x=1}^{x=e} \dfrac{du}{\cos^2 u \sin u}$

$= \displaystyle\int_{x=1}^{x=e} \dfrac{\sin u\, du}{\cos^2 u \sin^2 u}$ Let $v = \cos u$

$dv = -\sin u\, du$

$= -\displaystyle\int_{x=1}^{x=e} \dfrac{dv}{v^2(1 - v^2)}$.

Since

$\dfrac{1}{v^2(1 - v^2)} = \dfrac{A}{v} + \dfrac{B}{v^2} + \dfrac{C}{1 - v} + \dfrac{D}{1 + v}$

$= \dfrac{A(v - v^3) + B(1 - v^2) + C(v^2 + v^3) + D(v^2 - v^3)}{v^2(1 - v^2)}$

$\Rightarrow \begin{cases} -A + C - D = 0 \\ -B + C + D = 0 \\ A = 0 \\ B = 1 \end{cases}$

$\Rightarrow A = 0,\ B = 1,\ C = D = \dfrac{1}{2}$,

therefore,

$s = -\displaystyle\int_{x=1}^{x=e}\left[\dfrac{1}{v^2} + \dfrac{1}{2}\left(\dfrac{1}{1 - v} + \dfrac{1}{1 + v}\right)\right] dv$

$= \dfrac{1}{v} - \dfrac{1}{2}\ln\left|\dfrac{1 + v}{1 - v}\right|\,\Big|_{x=1}^{x=e}$ (but $v = \cos u = \dfrac{1}{\sqrt{x^2 + 1}}$)

$= \left[\sqrt{x^2 + 1} - \dfrac{1}{2}\ln\dfrac{(1 + v)^2}{|1 - v^2|}\right]\Big|_{x=1}^{x=e}$

$= \left[\sqrt{x^2 + 1} - \dfrac{1}{2}\ln\dfrac{1 + \dfrac{2}{\sqrt{x^2+1}} + \dfrac{1}{x^2 + 1}}{1 - \dfrac{1}{x^2 + 1}}\right]\Big|_1^e$

$= \left[\sqrt{x^2 + 1} - \dfrac{1}{2}\ln\dfrac{2 + x^2 + 2\sqrt{1 + x^2}}{x^2}\right]\Big|_1^e$

$= \sqrt{e^2 + 1} - \dfrac{1}{2}\ln\dfrac{2 + e^2 + 2\sqrt{1 + e^2}}{e^2} - \sqrt{2} + \dfrac{1}{2}\ln(3 + 2\sqrt{2})$

$= \sqrt{e^2 + 1} - \sqrt{2} + \dfrac{1}{2}\ln\dfrac{(3 + 2\sqrt{2})e^2}{2 + e^2 + 2\sqrt{1 + e^2}}$ units.

17. $x^{2/3} + y^{2/3} = x^{2/3}$. By symmetry, the curve has congruent arcs in the four quadrants. For the first quadrant arc we have

$y = \left(a^{2/3} - x^{2/3}\right)^{3/2}$

$y' = \dfrac{3}{2}\left(a^{2/3} - x^{2/3}\right)^{1/2}\left(-\dfrac{2}{3}x^{-1/3}\right)$.

Thus the length of the whole curve is

$L = 4\displaystyle\int_0^a \sqrt{1 + \dfrac{a^{2/3} - x^{2/3}}{x^{2/3}}}\, dx$

$= 4a^{1/3}\displaystyle\int_0^a x^{-1/3}\, dx$

$= 4a^{1/3} \cdot \dfrac{3}{2}x^{2/3}\Big|_0^a = 6a$ units.

18. The required length is

$L = \displaystyle\int_0^1 \sqrt{1 + (4x^3)^2}\, dx = \displaystyle\int_0^1 \sqrt{1 + 16x^6}\, dx$.

Using a calculator we calculate some Simpson's Rule approximations as described in Section 7.2:

$S_2 \approx 1.59921$ $S_4 \approx 1.60110$

$S_8 \approx 1.60025$ $S_{16} \approx 1.60023$.

To four decimal places the length is 1.6002 units.

19. $y = x^{1/3}$, $1 \le x \le 2$, $y' = \dfrac{1}{3}x^{-2/3}$.

Length $= \int_1^2 f(x)\, dx$, where $f(x) = \sqrt{1 + \dfrac{1}{9x^{4/3}}}$. We have

$T_4 = 1.03406$ $M_4 = 1.03363$

$T_8 = 1.03385$ $M_8 = 1.03374$

$T_{16} = 1.03378$ $M_{16} = 1.00376$.

Thus the length is approximately 1.0338 units.

20. For the ellipse $3x^2 + y^2 = 3$, we have $6x + 2yy' = 0$, so $y' = -3x/y$. Thus

$$ds = \sqrt{1 + \frac{9x^2}{3 - 3x^2}}\, dx = \sqrt{\frac{3 + 6x^2}{3 - 3x^2}}\, dx.$$

The circumference of the ellipse is

$$4 \int_0^1 \sqrt{\frac{3 + 6x^2}{3 - 3x^2}}\, dx \approx 8.73775 \text{ units}$$

(with a little help from Maple's numerical integration routine.)

21. For the ellipse $x^2 + 2y^2 = 2$, we have $2x + 4yy' = 0$, so $y' = -x/(2y)$. Thus

$$ds = \sqrt{1 + \frac{x^2}{4 - 2x^2}}\, dx = \sqrt{\frac{4 - x^2}{4 - 2x^2}}\, dx$$

The length of the short arc from $(0, 1)$ to $(1, 1/\sqrt{2})$ is

$$\int_0^1 \sqrt{\frac{4 - x^2}{4 - 2x^2}}\, dx \approx 1.05810 \text{ units}$$

(with a little help from Maple's numerical integration routine).

22. $S = 2\pi \int_0^2 |x|\sqrt{1 + 4x^2}\, dx$ Let $u = 1 + 4x^2$
$\qquad\qquad\qquad\qquad\qquad\qquad du = 8x\, dx$
$= \dfrac{\pi}{4} \int_1^{17} \sqrt{u}\, du = \dfrac{\pi}{4}\left(\dfrac{2}{3}u^{3/2}\right)\Big|_1^{17}$
$= \dfrac{\pi}{6}(17\sqrt{17} - 1)$ sq. units.

23. $y = x^3$, $0 \leq x \leq 1$. $ds = \sqrt{1 + 9x^4}\, dx$.
The area of the surface of rotation about the x-axis is

$$S = 2\pi \int_0^1 x^3 \sqrt{1 + 9x^4}\, dx \quad \text{Let } u = 1 + 9x^4$$
$$\qquad\qquad\qquad\qquad\qquad\qquad du = 36x^3\, dx$$
$$= \frac{\pi}{18} \int_1^{10} \sqrt{u}\, du = \frac{\pi}{27}(10^{3/2} - 1) \text{ sq. units.}$$

24. $y = x^{3/2}$, $0 \leq x \leq 1$. $ds = \sqrt{1 + \frac{9}{4}x}\, dx$.
The area of the surface of rotation about the x-axis is

$$S = 2\pi \int_0^1 x^{3/2} \sqrt{1 + \frac{9x}{4}}\, dx \quad \text{Let } 9x = 4u^2$$
$$\qquad\qquad\qquad\qquad\qquad\qquad 9\, dx = 8u\, du$$
$$= \frac{128\pi}{243} \int_0^{3/2} u^4 \sqrt{1 + u^2}\, du \quad \text{Let } u = \tan v$$
$$\qquad\qquad\qquad\qquad\qquad\qquad du = \sec^2 v\, dv$$
$$= \frac{128\pi}{243} \int_0^{\tan^{-1}(3/2)} \tan^4 v \sec^3 v\, dv$$
$$= \frac{128\pi}{243} \int_0^{\tan^{-1}(3/2)} (\sec^7 v - 2\sec^5 v + \sec^3 v)\, dv.$$

At this stage it is convenient to use the reduction formula

$$\int \sec^n v\, dv = \frac{1}{n-1} \sec^{n-2} v \tan v + \frac{n-2}{n-1} \int \sec^{n-2} v\, dv$$

(see Exercise 36 of Section 7.1) to reduce the powers of secant down to 3, and then use

$$\int_0^a \sec^3 v\, dv = \frac{1}{2}(\sec a \tan a + \ln|\sec a + \tan a|).$$

We have

$$I = \int_0^a (\sec^7 v - 2\sec^5 v + \sec^3 v)\, dv$$
$$= \frac{\sec^5 v \tan v}{6}\bigg|_0^a + \left(\frac{5}{6} - 2\right) \int_0^a \sec^5 v\, dv + \int_0^a \sec^3 v\, dv$$
$$= \frac{\sec^5 a \tan a}{6} - \frac{7}{6}\left[\frac{\sec^3 v \tan v}{4}\bigg|_0^a + \frac{3}{4} \int_0^a \sec^3 v\, dv\right]$$
$$\qquad + \int_0^a \sec^3 v\, dv$$
$$= \frac{\sec^5 a \tan a}{6} - \frac{7 \sec^3 a \tan a}{24} + \frac{1}{8} \int_0^a \sec^3 v\, dv$$
$$= \frac{\sec^5 a \tan a}{6} - \frac{7 \sec^3 a \tan a}{24} + \frac{\sec a \tan a + \ln|\sec a + \tan a|}{16}$$

Substituting $a = \arctan(3/2)$ now gives the following value for the surface area:

$$S = \frac{28\sqrt{13}\pi}{81} + \frac{8\pi}{243} \ln\left(\frac{3 + \sqrt{13}}{2}\right) \text{ sq. units.}$$

25. If $y = x^{3/2}$, $0 \le x \le 1$, is rotated about the y-axis, the surface area generated is

$$S = 2\pi \int_0^1 x\sqrt{1 + \frac{9x}{4}}\, dx \qquad \text{Let } u = 1 + \frac{9x}{4}$$
$$du = \frac{9}{4} dx$$

$$= \frac{32\pi}{81} \int_1^{13/4} (u-1)\sqrt{u}\, du$$

$$= \frac{32\pi}{81} \left(\frac{2}{5}u^{5/2} - \frac{2}{3}u^{3/2}\right)\Big|_1^{13/4}$$

$$= \frac{64\pi}{81}\left(\frac{(13/4)^{5/2} - 1}{5} - \frac{(13/4)^{3/2} - 1}{3}\right) \text{ sq. units.}$$

26. We have

$$S = 2\pi \int_0^1 e^x \sqrt{1 + e^{2x}}\, dx \qquad \text{Let } e^x = \tan\theta$$
$$e^x\, dx = \sec^2\theta\, d\theta$$

$$= 2\pi \int_{x=0}^{x=1} \sqrt{1+\tan^2\theta}\, \sec^2\theta\, d\theta = 2\pi \int_{x=0}^{x=1} \sec^3\theta\, d\theta$$

$$= \pi\Big[\sec\theta\tan\theta + \ln|\sec\theta + \tan\theta|\Big]\Big|_{x=0}^{x=1}.$$

Since
$$x = 1 \Rightarrow \tan\theta = e,\ \sec\theta = \sqrt{1+e^2},$$
$$x = 0 \Rightarrow \tan\theta = 1,\ \sec\theta = \sqrt{2},$$

therefore

$$S = \pi\left[e\sqrt{1+e^2} + \ln|\sqrt{1+e^2} + e| - \sqrt{2} - \ln|\sqrt{2} + 1|\right]$$

$$= \pi\left[e\sqrt{1+e^2} - \sqrt{2} + \ln\frac{\sqrt{1+e^2} + e}{\sqrt{2}+1}\right] \text{ sq. units.}$$

27. If $y = \sin x$, $0 \le x \le \pi$, is rotated about the x-axis, the surface area generated is

$$S = 2\pi \int_0^\pi \sin x \sqrt{1 + \cos^2 x}\, dx \qquad \text{Let } u = \cos x$$
$$du = -\sin x\, dx$$

$$= 2\pi \int_{-1}^1 \sqrt{1+u^2}\, du \qquad \text{Let } u = \tan\theta$$
$$du = \sec^2\theta\, d\theta$$

$$= 2\pi \int_{-\pi/4}^{\pi/4} \sec^3\theta\, d\theta = 4\pi \int_0^{\pi/4} \sec^3\theta\, d\theta$$

$$= 2\pi \left(\sec\theta\tan\theta + \ln|\sec\theta + \tan\theta|\right)\Big|_0^{\pi/4}$$

$$= 2\pi\left(\sqrt{2} + \ln(1+\sqrt{2})\right) \text{ sq. units.}$$

28. $1 + (y')^2 = 1 + \left(\dfrac{x^2}{4} - \dfrac{1}{x^2}\right)^2 = \left(\dfrac{x^2}{4} + \dfrac{1}{x^2}\right)^2$

$$S = 2\pi \int_1^4 \left(\frac{x^3}{12} + \frac{1}{x}\right)\left(\frac{x^2}{4} + \frac{1}{x^2}\right) dx$$

$$= 2\pi \int_1^4 \left(\frac{x^5}{48} + \frac{x}{3} + \frac{1}{x^3}\right) dx$$

$$= 2\pi \left(\frac{x^6}{288} + \frac{x^2}{6} - \frac{1}{2x^2}\right)\Big|_1^4$$

$$= \frac{275}{8}\pi \text{ sq. units.}$$

29. For $y = \dfrac{x^3}{12} + \dfrac{1}{x}$, $1 \le x \le 4$, we have
$$ds = \left(\frac{x^2}{4} + \frac{1}{x^2}\right) dx.$$
The surface generated by rotating the curve about the y-axis has area

$$S = 2\pi \int_1^4 x\left(\frac{x^2}{4} + \frac{1}{x^2}\right) dx$$

$$= 2\pi \left(\frac{x^4}{16} + \ln|x|\right)\Big|_1^4$$

$$= 2\pi \left(\frac{255}{16} + \ln 4\right) \text{ sq. units.}$$

30. The area of the cone obtained by rotating the line $y = (h/r)x$, $0 \le x \le r$, about the y-axis is

$$S = 2\pi \int_0^r x\sqrt{1 + (h/r)^2}\, dx = 2\pi \frac{\sqrt{r^2+h^2}}{r}\frac{x^2}{2}\Big|_0^r$$
$$= \pi r\sqrt{r^2+h^2} \text{ sq. units.}$$

31. For the circle $(x-b)^2 + y^2 = a^2$ we have
$$2(x-b) + 2y\frac{dy}{dx} = 0 \quad \Rightarrow \quad \frac{dy}{dx} = -\frac{x-b}{y}.$$

Thus
$$ds = \sqrt{1 + \frac{(x-b)^2}{y^2}}\, dx = \frac{a}{y}\, dx = \frac{a}{\sqrt{a^2 - (x-b)^2}}\, dx \quad (\text{if } y > 0).$$

The surface area of the torus obtained by rotating the circle about the line $x = 0$ is

$$S = 2 \times 2\pi \int_{b-a}^{b+a} x \frac{a}{\sqrt{a^2 - (x-b)^2}}\, dx \qquad \text{Let } u = x - b$$
$$du = dx$$

$$= 4\pi a \int_{-a}^a \frac{u+b}{\sqrt{a^2-u^2}}\, du$$

$$= 8\pi ab \int_0^a \frac{du}{\sqrt{a^2-u^2}} \qquad \text{by symmetry}$$

$$= 8\pi ab \sin^{-1}\frac{u}{a}\Big|_0^a = 4\pi^2 ab \text{ sq. units.}$$

32. The top half of $x^2 + 4y^2 = 4$ is $y = \frac{1}{2}\sqrt{4 - x^2}$, so $\frac{dy}{dx} = \frac{-x}{2\sqrt{4 - x^2}}$, and

$$S = 2 \times 2\pi \int_0^2 \frac{\sqrt{4 - x^2}}{2} \sqrt{1 + \left(\frac{x}{2\sqrt{4 - x^2}}\right)^2} \, dx$$

$$= \pi \int_0^2 \sqrt{16 - 3x^2} \, dx \quad \text{Let } x = \sqrt{\frac{16}{3}} \sin\theta$$

$$dx = \sqrt{\frac{16}{3}} \cos\theta \, d\theta$$

$$= \pi \int_0^{\pi/3} (4\cos\theta) \frac{4}{\sqrt{3}} \cos\theta \, d\theta$$

$$= \frac{16\pi}{\sqrt{3}} \int_0^{\pi/3} \cos^2\theta \, d\theta$$

$$= \frac{8\pi}{\sqrt{3}} \left(\theta + \sin\theta \cos\theta\right)\Big|_0^{\pi/3}$$

$$= \frac{2\pi(4\pi + 3\sqrt{3})}{3\sqrt{3}} \text{ sq. units.}$$

33. For the ellipse $x^2 + 4y^2 = 4$ we have

$$2x\frac{dx}{dy} + 8y = 0 \quad \Rightarrow \quad \frac{dx}{dy} = -4\frac{y}{x}.$$

The arc length element on the ellipse is given by

$$ds = \sqrt{1 + \left(\frac{dx}{dy}\right)^2} \, dy$$

$$= \sqrt{1 + \frac{16y^2}{x^2}} \, dy = \frac{1}{x}\sqrt{4 + 12y^2} \, dy.$$

If the ellipse is rotated about the y-axis, the resulting surface has area

$$S = 2 \times 2\pi \int_0^1 x \frac{1}{x} \sqrt{4 + 12y^2} \, dy$$

$$= 8\pi \int_0^1 \sqrt{1 + 3y^2} \, dy \quad \text{Let } \sqrt{3}y = \tan\theta$$

$$\sqrt{3}\, dy = \sec^2\theta \, d\theta$$

$$= \frac{8\pi}{\sqrt{3}} \int_0^{\pi/3} \sec^3\theta \, d\theta$$

$$= \frac{8\pi}{2\sqrt{3}} \left(\sec\theta \tan\theta + \ln|\sec\theta + \tan\theta|\right)\Big|_0^{\pi/3}$$

$$= \frac{8\pi}{2\sqrt{3}} \left(2\sqrt{3} + \ln(2 + \sqrt{3})\right)$$

$$= 8\pi\left(1 + \frac{\ln(2 + \sqrt{3})}{2\sqrt{3}}\right) \text{ sq. units.}$$

34. As in Example 4, the arc length element for the ellipse is

$$ds = \sqrt{1 + \left(\frac{dy}{dx}\right)^2} \, dx = \sqrt{\frac{a^2 - \frac{a^2 - b^2}{a^2}x^2}{a^2 - x^2}} \, dx.$$

To get the area of the ellipsoid, we must rotate both the upper and lower semi-ellipses (see the figure for Exercise 20 of Section 8.1):

$$S = 2 \times 2\pi \int_0^a \left[\left(c - b\sqrt{1 - \left(\frac{x}{a}\right)^2}\right) + \left(c + b\sqrt{1 - \left(\frac{x}{a}\right)^2}\right)\right] ds$$

$$= 8\pi c \int_0^a \sqrt{\frac{a^2 - \frac{a^2 - b^2}{a^2}x^2}{a^2 - x^2}} \, dx$$

$$= 8\pi c\left[\frac{1}{4} \text{ of the circumference of the ellipse}\right]$$

$$= 8\pi c a E(\epsilon)$$

where $\epsilon = \frac{\sqrt{a^2 - b^2}}{a}$ and $E(\epsilon) = \int_0^{\pi/2} \sqrt{1 - \epsilon^2 \sin^2 t} \, dt$ as defined in Example 4.

35. From Example 3, the length is

$$s = \frac{10}{\pi} \int_0^{\pi/2} \sqrt{1 + \frac{\pi^2}{4} \cos^2 t} \, dt$$

$$= \frac{10}{\pi} \int_0^{\pi/2} \sqrt{1 + \frac{\pi^2}{4} - \frac{\pi^2}{4} \sin^2 t} \, dt$$

$$= \frac{5}{\pi}\sqrt{4 + \pi^2} \int_0^{\pi/2} \sqrt{1 - \frac{\pi^2}{4 + \pi^2} \sin^2 t} \, dt$$

$$= \frac{5}{\pi}\sqrt{4 + \pi^2}\, E\left(\frac{\pi}{\sqrt{4 + \pi^2}}\right).$$

36. Let the equation of the sphere be $x^2 + y^2 = R^2$. Then the surface area between planes $x = a$ and $x = b$ $(-R \leq a < b \leq R)$ is

$$S = 2\pi \int_a^b \sqrt{R^2 - x^2} \sqrt{1 + \left(\frac{dy}{dx}\right)^2} \, dx$$

$$= 2\pi \int_a^b \sqrt{R^2 - x^2} \frac{R}{\sqrt{R^2 - x^2}} \, dx$$

$$= 2\pi R \int_a^b dx = 2\pi R(b - a) \text{ sq. units.}$$

Thus, the surface area depends only on the radius R of the sphere, and the distance $(b-a)$ between the parallel planes.

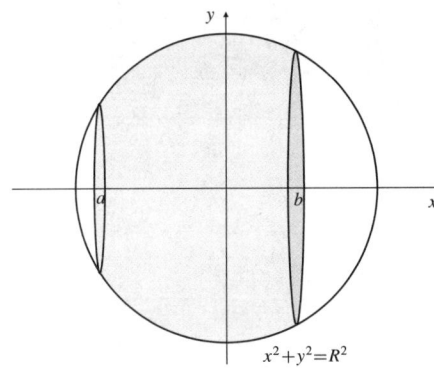

Fig. 7.3.36

37. If the curve $y = x^k$, $0 < x \leq 1$, is rotated about the y-axis, it generates a surface of area

$$S = 2\pi \int_0^1 x\sqrt{1+k^2 x^{2(k-1)}}\,dx$$
$$= 2\pi \int_0^1 \sqrt{x^2 + k^2 x^{2k}}\,dx.$$

If $k \leq -1$, we have $S \geq 2\pi k \int_0^1 x^k\,dx$, which is infinite.
If $k \geq 0$, the surface area S is finite, since x^k is bounded on $(0, 1]$ in that case.
Hence we need only consider the case $-1 < k < 0$. In this case $2 < 2 - 2k < 4$, and

$$S = 2\pi \int_0^1 x\sqrt{1+k^2 x^{2(k-1)}}\,dx$$
$$= 2\pi \int_0^1 \sqrt{x^{2-2k} + k^2}\,x^k\,dx$$
$$< 2\pi\sqrt{1+k^2} \int_0^1 x^k\,dx < \infty.$$

Thus the area is finite if and only if $k > -1$.

38. $S = 2\pi \int_0^1 |x|\sqrt{1+\dfrac{1}{x^2}}\,dx$

$= 2\pi \int_0^1 \sqrt{x^2+1}\,dx$ Let $x = \tan\theta$
$\qquad\qquad\qquad\qquad\qquad dx = \sec^2\theta\,d\theta$
$= 2\pi \int_0^{\pi/4} \sec^3\theta\,d\theta$
$= \pi\bigl(\sec\theta\tan\theta + \ln|\sec\theta + \tan\theta|\bigr)\Big|_0^{\pi/4}$
$= \pi[\sqrt{2} + \ln(\sqrt{2}+1)]$ sq. units.

39. a) Volume $V = \pi \int_1^\infty \dfrac{dx}{x^2} = \pi$ cu. units.

b) The surface area is

$$S = 2\pi \int_1^\infty \dfrac{1}{x}\sqrt{1+\dfrac{1}{x^4}}\,dx$$
$$> 2\pi \int_1^\infty \dfrac{dx}{x} = \infty.$$

c) Covering a surface with paint requires applying a layer of paint *of constant thickness* to the surface. Far to the right, the horn is thinner than any prescribed constant, so it can contain less paint than would be required to cover its surface.

Section 7.4 Mass, Moments, and Centres of Mass (page 435)

1. The mass of the wire is

$$m = \int_0^L \delta(s)\,ds = \int_0^L \sin\dfrac{\pi s}{L}\,ds$$
$$= -\dfrac{L}{\pi}\cos\dfrac{\pi s}{L}\bigg|_0^L = \dfrac{2L}{\pi}.$$

Since $\delta(s)$ is symmetric about $s = L/2$ (that is, $\delta((L/2)-s) = \delta((L/2)+s)$), the centre of mass is at the midpoint of the wire: $\bar{s} = L/2$.

2. A slice of the wire of width dx at x has volume $dV = \pi(a+bx)^2\,dx$. Therefore the mass of the whole wire is

$$m = \int_0^L \delta_0\pi(a+bx)^2\,dx$$
$$= \delta_0\pi \int_0^L (a^2 + 2abx + b^2x^2)\,dx$$
$$= \delta_0\pi\left(a^2 L + abL^2 + \dfrac{1}{3}b^2 L^3\right).$$

Its moment about $x = 0$ is

$$M_{x=0} = \int_0^L x\delta_0\pi(a+bx)^2\,dx$$
$$= \delta_0\pi \int_0^L (a^2 x + 2abx^2 + b^2 x^3)\,dx$$
$$= \delta_0\pi\left(\dfrac{1}{2}a^2 L^2 + \dfrac{2}{3}abL^3 + \dfrac{1}{4}b^2 L^4\right).$$

Thus, the centre of mass is

$$\bar{x} = \frac{\delta_0\pi\left(\frac{1}{2}a^2L^2 + \frac{2}{3}abL^3 + \frac{1}{4}b^2L^4\right)}{\delta_0\pi\left(a^2L + abL^2 + \frac{1}{3}b^2L^3\right)}$$

$$= \frac{L\left(\frac{1}{2}a^2 + \frac{2}{3}abL + \frac{1}{4}b^2L^2\right)}{a^2 + abL + \frac{1}{3}b^2L^2}.$$

3. The mass of the plate is $m = \delta_0 \times$ area $= \dfrac{\pi\delta_0 a^2}{4}$.
The moment about $x = 0$ is

$$M_{x=0} = \int_0^a x\delta_0\sqrt{a^2 - x^2}\,dx \quad \text{Let } u = a^2 - x^2$$
$$du = -2x\,dx$$

$$= \frac{\delta_0}{2}\int_0^{a^2}\sqrt{u}\,du$$

$$= \frac{\delta_0}{2}\cdot\frac{2}{3}u^{3/2}\bigg|_0^{a^2} = \frac{\delta_0 a^3}{3}.$$

Thus $\bar{x} = \dfrac{M_{x=0}}{m} = \dfrac{\delta_0 a^3}{3}\cdot\dfrac{4}{\pi\delta_0 a^2} = \dfrac{4a}{3\pi}$. By symmetry, $\bar{y} = \bar{x}$. Thus the centre of mass of the plate is $\left(\dfrac{4a}{3\pi}, \dfrac{4a}{3\pi}\right)$.

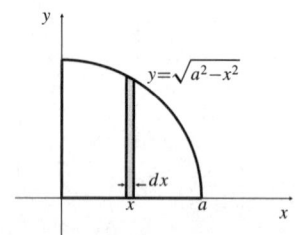

Fig. 7.4.3

4. A vertical strip has area $dA = \sqrt{a^2 - x^2}\,dx$. Therefore, the mass of the quarter-circular plate is

$$m = \int_0^a (\delta_0 x)\sqrt{a^2 - x^2}\,dx \quad \text{Let } u = a^2 - x^2$$
$$du = -2x\,dx$$

$$= \frac{1}{2}\delta_0\int_0^{a^2}\sqrt{u}\,du = \frac{1}{2}\delta_0\left(\frac{2}{3}u^{3/2}\right)\bigg|_0^{a^2} = \frac{1}{3}\delta_0 a^3.$$

The moment about $x = 0$ is

$$M_{x=0} = \int_0^a \delta_0 x^2\sqrt{a^2 - x^2}\,dx \quad \text{Let } x = a\sin\theta$$
$$dx = a\cos\theta\,d\theta$$

$$= \delta_0 a^4\int_0^{\pi/2}\sin^2\theta\cos^2\theta\,d\theta$$

$$= \frac{\delta_0 a^4}{4}\int_0^{\pi/2}\sin^2 2\theta\,d\theta$$

$$= \frac{\delta_0 a^4}{8}\int_0^{\pi/2}(1 - \cos 4\theta)\,d\theta = \frac{\pi\delta_0 a^4}{16}.$$

The moment about $y = 0$ is

$$M_{y=0} = \frac{1}{2}\delta_0\int_0^a x(a^2 - x^2)\,dx$$

$$= \frac{1}{2}\delta_0\left(\frac{a^2 x^2}{2} - \frac{x^4}{4}\right)\bigg|_0^a = \frac{1}{8}a^4\delta_0.$$

Thus, $\bar{x} = \dfrac{3}{16}\pi a$ and $\bar{y} = \dfrac{3}{8}a$. Hence, the centre of mass is located at $\left(\dfrac{3}{16}\pi a, \dfrac{3}{8}a\right)$.

5. The mass of the plate is

$$m = 2\int_0^4 ky\sqrt{4 - y}\,dy \quad \text{Let } u = 4 - y$$
$$du = -dy$$

$$= 2k\int_0^4 (4 - u)u^{1/2}\,du$$

$$= 2k\left(\frac{8}{3}u^{3/2} - \frac{2}{5}u^{5/2}\right)\bigg|_0^4 = \frac{256k}{15}.$$

By symmetry, $M_{x=0} = 0$, so $\bar{x} = 0$.

$$M_{y=0} = 2\int_0^4 ky^2\sqrt{4 - y}\,dy \quad \text{Let } u = 4 - y$$
$$du = -dy$$

$$= 2k\int_0^4 (16u^{1/2} - 8u^{3/2} + u^{5/2})\,du$$

$$= 2k\left(\frac{32}{3}u^{3/2} - \frac{16}{5}u^{5/2} + \frac{2}{7}u^{7/2}\right)\bigg|_0^4 = \frac{4096k}{105}.$$

Thus $\bar{y} = \dfrac{4096k}{105}\cdot\dfrac{15}{256k} = \dfrac{16}{7}$. The centre of mass of the plate is $(0, 16/7)$.

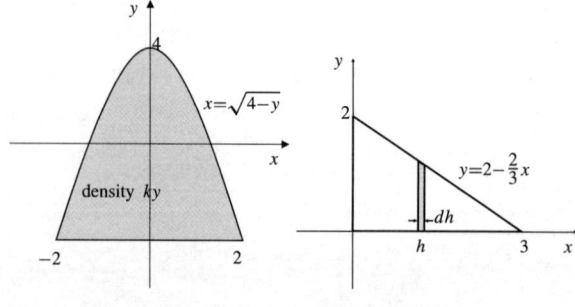

Fig. 7.4.5   Fig. 7.4.6

INSTRUCTOR'S SOLUTIONS MANUAL SECTION 7.4 (PAGE 435)

6. A vertical strip at h has area $dA = (2 - \frac{2}{3}h)\,dh$. Thus, the mass of the plate is

$$m = \int_0^3 (5h)\left(2 - \frac{2}{3}h\right) dh = 10\int_0^3 \left(h - \frac{h^2}{3}\right) dh$$

$$= 10\left(\frac{h^2}{2} - \frac{h^3}{9}\right)\Big|_0^3 = 15 \text{ kg}.$$

The moment about $x = 0$ is

$$M_{x=0} = 10\int_0^3 \left(h^2 - \frac{h^3}{3}\right) dh$$

$$= 10\left(\frac{h^3}{3} - \frac{h^4}{12}\right)\Big|_0^3 = \frac{45}{2} \text{ kg-m}.$$

The moment about $y = 0$ is

$$M_{y=0} = 10\int_0^3 \frac{1}{2}\left(2 - \frac{2}{3}h\right)\left(h - \frac{1}{3}h^2\right) dh$$

$$= 10\int_0^3 \left(h - \frac{2}{3}h^2 + \frac{1}{9}h^3\right) dh$$

$$= 10\left(\frac{h^2}{2} - \frac{2h^3}{9} + \frac{h^4}{36}\right)\Big|_0^3 = \frac{15}{2} \text{ kg-m}.$$

Thus, $\bar{x} = \dfrac{\left(\frac{45}{2}\right)}{15} = \dfrac{3}{2}$ and $\bar{y} = \dfrac{\left(\frac{15}{2}\right)}{15} = \dfrac{1}{2}$. The centre of mass is located at $(\frac{3}{2}, \frac{1}{2})$.

7. The mass of the plate is

$$m = \int_0^a kx\, a\, dx = \frac{ka^3}{2}.$$

By symmetry, $\bar{y} = a/2$.

$$M_{x=0} = \int_0^a kx^2 a\, dx = \frac{ka^4}{3}.$$

Thus $\bar{x} = \dfrac{ka^4}{3} \cdot \dfrac{2}{ka^3} = \dfrac{2a}{3}$. The centre of mass of the plate is $\left(\dfrac{2a}{3}, \dfrac{a}{2}\right)$.

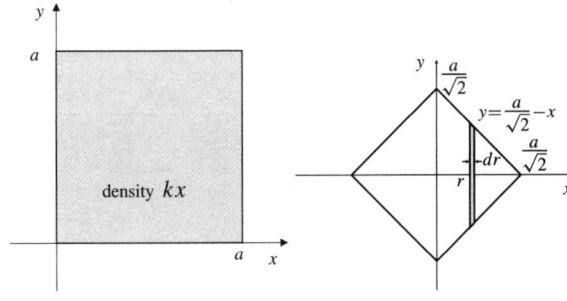

Fig. 7.4.7   Fig. 7.4.8

8. A vertical strip has area $dA = 2\left(\dfrac{a}{\sqrt{2}} - r\right) dr$. Thus, the mass is

$$m = 2\int_0^{a/\sqrt{2}} kr\left[2\left(\frac{a}{\sqrt{2}} - r\right)\right] dr$$

$$= 4k\int_0^{a/\sqrt{2}} \left(\frac{a}{\sqrt{2}}r - r^2\right) dr = \frac{k}{3\sqrt{2}}a^3 \text{ g}.$$

Since the mass is symmetric about the y-axis, and the plate is symmetric about both the x- and y-axis, therefore the centre of mass must be located at the centre of the square.

9.
$$m = \int_a^b \delta(x)\bigl(g(x) - f(x)\bigr) dx$$

$$M_{x=0} = \int_a^b x\delta(x)\bigl(g(x) - f(x)\bigr) dx$$

$$M_{y=0} = \frac{1}{2}\int_a^b x\delta(x)\bigl((g(x))^2 - (f(x))^2\bigr) dx$$

Centre of mass: $\left(\dfrac{M_{x=0}}{m}, \dfrac{M_{y=0}}{m}\right)$.

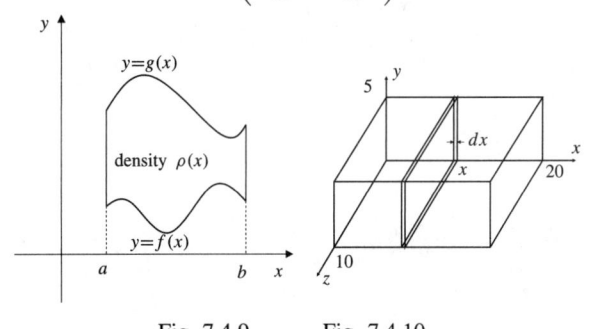

Fig. 7.4.9   Fig. 7.4.10

10. The slice of the brick shown in the figure has volume $dV = 50\,dx$. Thus, the mass of the brick is

$$m = \int_0^{20} kx 50\, dx = 25kx^2\Big|_0^{20} = 10000k \text{ g}.$$

The moment about $x = 0$, i.e., the yz-plane, is

$$M_{x=0} = 50k\int_0^{20} x^2\, dx = \frac{50}{3}kx^3\Big|_0^{20}$$

$$= \frac{50}{3}(8000)k \text{ g-cm}.$$

Thus, $\bar{x} = \dfrac{\frac{50}{3}(8000)k}{10000k} = \dfrac{40}{3}$. Since the density is independent of y and z, $\bar{y} = \dfrac{5}{2}$ and $\bar{z} = 5$. Hence, the centre of mass is located on the 20 cm long central axis of the brick, two-thirds of the way from the least dense 10×5 face to the most dense such face.

275

11. Choose axes through the centre of the ball as shown in the following figure. The mass of the ball is

$$m = \int_{-R}^{R} (y+2R)\pi(R^2-y^2)\,dy$$
$$= 4\pi R\left(R^2 y - \frac{y^3}{3}\right)\bigg|_0^R = \frac{8}{3}\pi R^4 \text{ kg.}$$

By symmetry, the centre of mass lies along the y-axis; we need only calculate $\bar{y}$.

$$M_{y=0} = \int_{-R}^{R} y(y+2R)\pi(R^2-y^2)\,dy$$
$$= 2\pi \int_0^R y^2(R^2-y^2)\,dy$$
$$= 2\pi\left(R^2\frac{y^3}{3} - \frac{y^5}{5}\right)\bigg|_0^R = \frac{4}{15}\pi R^5.$$

Thus $\bar{y} = \dfrac{4\pi R^5}{15}\cdot\dfrac{3}{8\pi R^4} = \dfrac{R}{10}$. The centre of mass is on the line through the centre of the ball perpendicular to the plane mentioned in the problem, at a distance $R/10$ from the centre of the ball on the side opposite to the plane.

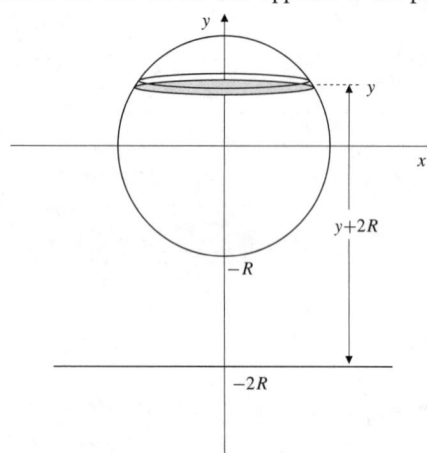

Fig. 7.4.11

12. A slice at height z has volume $dV = \pi y^2\,dz$ and density kz g/cm^3. Thus, the mass of the cone is

$$m = \int_0^b kz\pi y^2\,dz$$
$$= \pi ka^2 \int_0^b z\left(1-\frac{z}{b}\right)^2\,dz$$
$$= \pi ka^2\left(\frac{z^2}{2} - \frac{2z^3}{3b} + \frac{z^4}{4b^2}\right)\bigg|_0^b$$
$$= \frac{1}{12}\pi ka^2 b^2 \text{ g.}$$

The moment about $z=0$ is

$$M_{z=0} = \pi ka^2 \int_0^b z^2\left(1-\frac{z}{b}\right)^2\,dz = \frac{1}{30}\pi ka^2 b^3 \text{ g-cm.}$$

Thus, $\bar{z} = \dfrac{2b}{5}$. Hence, the centre of mass is on the axis of the cone at height $2b/5$ cm above the base.

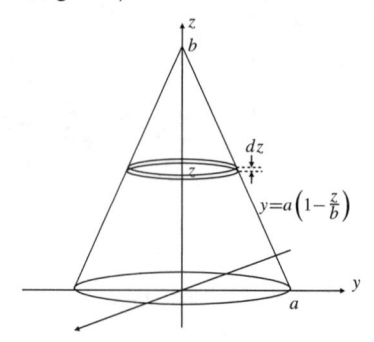

Fig. 7.4.12

13. By symmetry, $\bar{y} = 0$.
A horizontal slice of the solid at height z with thickness dz is a half-disk of radius $\sqrt{a^2-z^2}$ with centre of mass at $\bar{x} = \dfrac{4\sqrt{a^2-z^2}}{3\pi}$, by Exercise 3 above. Its mass is

$$dm = \delta_0 z\,dz\,\frac{\pi}{2}(a^2-z^2),$$

and its moment about $x=0$ is

$$dM_{x=0} = dm\,\bar{x} = \frac{\pi\delta_0}{2}z(a^2-z^2)\frac{4\sqrt{a^2-z^2}}{3\pi}$$
$$= \frac{2\delta_0}{3}z(a^2-z^2)^{3/2}.$$

Thus the mass of the solid is

$$m = \frac{\pi\delta_0}{2}\int_0^a (a^2 z - z^3)\,dz$$
$$= \frac{\pi\delta_0}{2}\left(\frac{a^2 z^2}{2} - \frac{z^4}{4}\right)\bigg|_0^a = \frac{\pi\delta_0 a^4}{8}.$$

Also,

$$M_{z=0} = \frac{\pi\delta_0}{2}\int_0^a (a^2 z^2 - z^4)\,dz$$
$$= \frac{\pi\delta_0}{2}\left(\frac{a^2 z^3}{3} - \frac{z^5}{5}\right)\bigg|_0^a = \frac{\pi\delta_0 a^5}{15},$$

INSTRUCTOR'S SOLUTIONS MANUAL   SECTION 7.4 (PAGE 435)

and $\bar{z} = \dfrac{\pi \delta_0 a^5}{15} \cdot \dfrac{8}{\pi \delta_0 a^4} = \dfrac{8a}{15}$.

Finally,

$$M_{x=0} = \dfrac{2\delta_0}{3} \int_0^a z(a^2 - z^2)^{3/2} dz \quad \text{Let } u = a^2 - z^2$$
$$\phantom{M_{x=0}} \qquad\qquad\qquad\qquad\qquad\qquad du = -2z\, dz$$

$$= \dfrac{\delta_0}{3} \int_0^{a^2} u^{3/2} du$$

$$= \dfrac{\delta_0}{3} \left(\dfrac{2}{5} u^{5/2}\right)\bigg|_0^{a^2} = \dfrac{2\delta_0 a^5}{15},$$

so $\bar{x} = \dfrac{2\delta_0 a^5}{15} \cdot \dfrac{8}{\pi \delta_0 a^4} = \dfrac{16a}{15}$.

The centre of mass is $\left(\dfrac{16a}{15}, 0, \dfrac{8a}{15}\right)$.

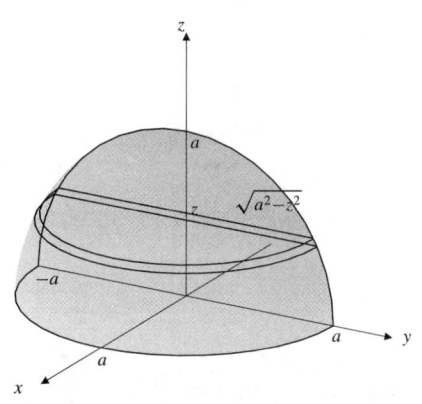

Fig. 7.4.13

14. Assume the cone has its base in the xy-plane and its vertex at height b on the z-axis. By symmetry, the centre of mass lies on the z-axis. A cylindrical shell of thickness dx and radius x about the z-axis has height $z = b(1 - (x/a))$. Since it's density is constant kx, its mass is

$$dm = 2\pi bkx^2 \left(1 - \dfrac{x}{a}\right) dx.$$

Also its centre of mass is at half its height,

$$\bar{y}_{\text{shell}} = \dfrac{b}{2}\left(1 - \dfrac{x}{a}\right).$$

Thus its moment about $z = 0$ is

$$dM_{z=0} = \bar{y}_{\text{shell}} dm = \pi bkx^2 \left(1 - \dfrac{x}{a}\right)^2 dx.$$

Hence

$$m = \int_0^a 2\pi bkx^2 \left(1 - \dfrac{x}{a}\right) dx = \dfrac{\pi k b a^3}{6}$$

$$M_{z=0} = \int_0^a \pi bkx^2 \left(1 - \dfrac{x}{a}\right)^2 dx = \dfrac{\pi k b^2 a^3}{30}$$

and $\bar{z} = M_{z=0}/m = b/5$. The centre of mass is on the axis of the cone at height $b/5$ cm above the base.

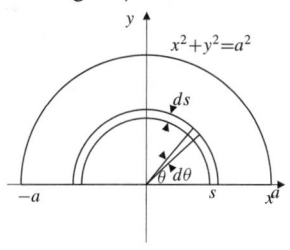

Fig. 7.4.14

15. Consider the area element which is the thin half-ring shown in the figure. We have

$$dm = ks\, \pi s\, ds = k\pi s^2\, ds.$$

Thus, $m = \dfrac{k\pi}{3} a^3$.

Regard this area element as itself composed of smaller elements at positions given by the angle θ as shown. Then

$$dM_{y=0} = \left(\int_0^\pi (s\sin\theta) s\, d\theta\right) ks\, ds$$
$$= 2ks^3\, ds,$$

$$M_{y=0} = 2k \int_0^a s^3\, ds = \dfrac{ka^4}{2}.$$

Therefore, $\bar{y} = \dfrac{ka^4}{2} \cdot \dfrac{3}{k\pi a^3} = \dfrac{3a}{2\pi}$. By symmetry, $\bar{x} = 0$.

Thus, the centre of mass of the plate is $\left(0, \dfrac{3a}{2\pi}\right)$.

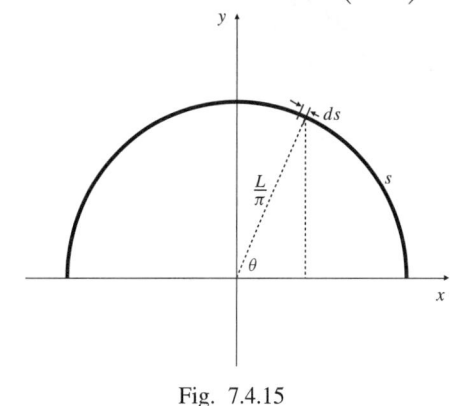

Fig. 7.4.15

16. The radius of the semicircle is $\dfrac{L}{\pi}$. Let s measure the distance along the wire from the point where it leaves the positive x-axis. Thus, the density at position s is $\delta\delta(s) = \sin\left(\dfrac{\pi s}{L}\right)$ g/cm. The mass of the wire is

$$m = \int_0^L \sin\dfrac{\pi s}{L}\, ds = -\dfrac{L}{\pi} \cos\dfrac{\pi s}{L}\bigg|_0^L = \dfrac{2L}{\pi}\ \text{g}.$$

277

Since an arc element ds at position s is at height $y = \frac{L}{\pi}\sin\theta = \frac{L}{\pi}\sin\frac{\pi s}{L}$, the moment of the wire about $y = 0$ is

$$M_{y=0} = \int_0^L \frac{L}{\pi}\sin^2\frac{\pi s}{L}\,ds \quad \text{Let } \theta = \pi s/L$$
$$d\theta = \pi\,ds/L$$
$$= \left(\frac{L}{\pi}\right)^2 \int_0^\pi \sin^2\theta\,d\theta$$
$$= \frac{L^2}{2\pi^2}(\theta - \sin\theta\cos\theta)\Big|_0^\pi = \frac{L^2}{2\pi}\text{ g-cm}.$$

Since the wire and the density function are both symmetric about the y-axis, we have $M_{x=0} = 0$.

Hence, the centre of mass is located at $\left(0, \frac{L}{4}\right)$.

17. $m = \int_0^\infty Ce^{-kr^2}(4\pi r^2)\,dr$

$$= 4\pi C\int_0^\infty r^2 e^{-kr^2}\,dr \quad \text{Let } u = \sqrt{k}\,r$$
$$du = \sqrt{k}\,dr$$
$$= \frac{4\pi C}{k^{3/2}}\int_0^\infty u^2 e^{-u^2}\,du$$

$$\begin{aligned} U &= u & dV &= ue^{-u^2}\,du \\ dU &= du & V &= -\tfrac{1}{2}e^{-u^2} \end{aligned}$$

$$= \frac{4\pi C}{k^{3/2}}\lim_{R\to\infty}\left(\frac{-ue^{-u^2}}{2}\Big|_0^R + \frac{1}{2}\int_0^R e^{-u^2}\,du\right)$$
$$= \frac{4\pi C}{k^{3/2}}\left(0 + \frac{1}{2}\int_0^\infty e^{-u^2}\,du\right)$$
$$= \frac{4\pi C}{k^{3/2}}\frac{\sqrt{\pi}}{4} = C\left(\frac{\pi}{k}\right)^{3/2} \approx \frac{5.57C}{k^{3/2}}.$$

18. $\bar{r} = \frac{1}{m}\int_0^\infty rCe^{-kr^2}(4\pi r^2)\,dr$

$$= \frac{4\pi C}{C\pi^{3/2}k^{-3/2}}\int_0^\infty r^3 e^{-kr^2}\,dr \quad \text{Let } u = kr^2$$
$$du = 2kr\,dr$$
$$= \frac{4k^{3/2}}{\sqrt{\pi}}\cdot\frac{1}{2k^2}\int_0^\infty ue^{-u}\,du$$

$$\begin{aligned} U &= u & dV &= e^{-u}\,du \\ dU &= du & V &= -e^{-u} \end{aligned}$$

$$= \frac{2}{\sqrt{\pi k}}\lim_{R\to\infty}\left(-ue^{-u}\Big|_0^R + \int_0^R e^{-u}\,du\right)$$
$$= \frac{2}{\sqrt{\pi k}}\left(0 + \lim_{R\to\infty}(e^0 - e^{-R})\right) = \frac{2}{\sqrt{\pi k}}.$$

Section 7.5 Centroids (page 441)

1. $A = \frac{\pi r^2}{4}$

$$M_{x=0} = \int_0^r x\sqrt{r^2 - x^2}\,dx \quad \text{Let } u = r^2 - x^2$$
$$du = -2x\,dx$$
$$= \frac{1}{2}\int_0^{r^2} u^{1/2}\,du = \frac{u^{3/2}}{3}\Big|_0^{r^2} = \frac{r^3}{3}$$
$$\bar{x} = \frac{r^3}{3}\cdot\frac{4}{\pi r^2} = \frac{4r}{3\pi} = \bar{y} \text{ by symmetry.}$$

The centroid is $\left(\frac{4r}{3\pi}, \frac{4r}{3\pi}\right)$.

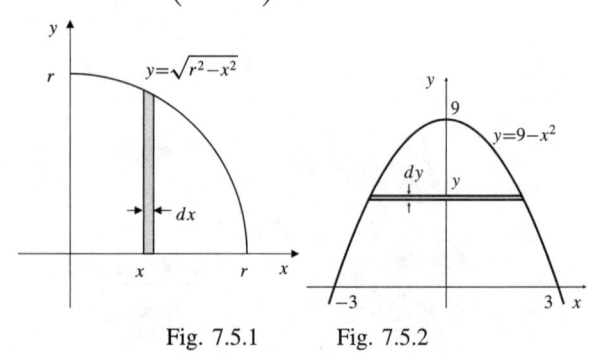

Fig. 7.5.1   Fig. 7.5.2

2. By symmetry, $\bar{x} = 0$. A horizontal strip at y has mass $dm = 2\sqrt{9-y}\,dy$ and moment $dM_{y=0} = 2y\sqrt{9-y}\,dy$ about $y = 0$. Thus,

$$m = 2\int_0^9 \sqrt{9-y}\,dy = -2\left(\frac{2}{3}\right)(9-y)^{3/2}\Big|_0^9 = 36$$

and

$$M_{y=0} = 2\int_0^9 y\sqrt{9-y}\,dy \quad \text{Let } u^2 = 9-y$$
$$2u\,du = -dy$$
$$= 4\int_0^3 (9u^2 - u^4)\,du = 4(3u^3 - \tfrac{1}{5}u^5)\Big|_0^3 = \frac{648}{5}.$$

Thus, $\bar{y} = \frac{648}{5\times 36} = \frac{18}{5}$. Hence, the centroid is at $\left(0, \frac{18}{5}\right)$.

3. The area and moments of the region are

$$A = \int_0^1 \frac{dx}{\sqrt{1+x^2}} \qquad \text{Let } x = \tan\theta$$
$$dx = \sec^2\theta\, d\theta$$
$$= \int_0^{\pi/4} \sec\theta\, d\theta$$
$$= \ln|\sec\theta + \tan\theta|\Big|_0^{\pi/4} = \ln(1+\sqrt{2})$$
$$M_{x=0} = \int_0^1 \frac{x\,dx}{\sqrt{1+x^2}} = \sqrt{1+x^2}\Big|_0^1 = \sqrt{2}-1$$
$$M_{y=0} = \frac{1}{2}\int_0^1 \frac{dx}{1+x^2} = \frac{1}{2}\tan^{-1}x\Big|_0^1 = \frac{\pi}{8}.$$

Thus $\bar{x} = \dfrac{\sqrt{2}-1}{\ln(1+\sqrt{2})}$, and $\bar{y} = \dfrac{\pi}{8\ln(1+\sqrt{2})}$. The centroid is $\left(\dfrac{\sqrt{2}-1}{\ln(1+\sqrt{2})}, \dfrac{\pi}{8\ln(1+\sqrt{2})}\right)$.

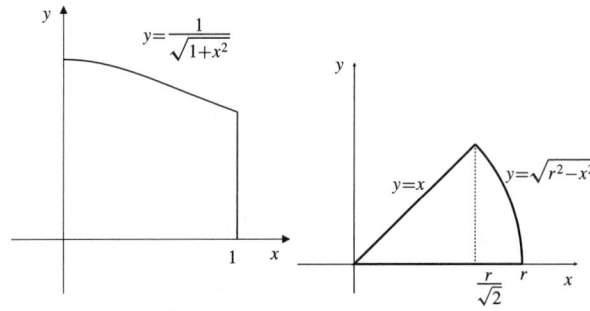

Fig. 7.5.3   Fig. 7.5.4

4. The area of the sector is $A = \frac{1}{8}\pi r^2$. Its moment about $x = 0$ is

$$M_{x=0} = \int_0^{r/\sqrt{2}} x^2\, dx + \int_{r/\sqrt{2}}^r x\sqrt{r^2-x^2}\, dx$$
$$= \frac{r^3}{6\sqrt{2}} - \frac{1}{3}(r^2-x^2)^{3/2}\Big|_{r/\sqrt{2}}^r = \frac{r^3}{3\sqrt{2}}.$$

Thus, $\bar{x} = \dfrac{r^3}{3\sqrt{2}} \times \dfrac{8}{\pi r^2} = \dfrac{8r}{3\sqrt{2}\pi}$. By symmetry, the centroid must lie on the line $y = x\left(\tan\dfrac{\pi}{8}\right) = x(\sqrt{2}-1)$.
Thus, $\bar{y} = \dfrac{8r(\sqrt{2}-1)}{3\sqrt{2}\pi}$.

5. By symmetry, $\bar{x} = 0$. We have

$$A = 2\int_0^{\sqrt{3}}\left(\sqrt{4-x^2}-1\right)dx \qquad \text{Let } x = 2\sin\theta$$
$$dx = 2\cos\theta\, d\theta$$
$$= 2\left(4\int_0^{\pi/3}\cos^2\theta\, d\theta - \sqrt{3}\right)$$
$$= 4(\theta + \sin\theta\cos\theta)\Big|_0^{\pi/3} - 2\sqrt{3}$$
$$= 4\left(\frac{\pi}{3} + \frac{\sqrt{3}}{4}\right) - 2\sqrt{3} = \frac{4\pi}{3} - \sqrt{3}$$
$$M_{y=0} = 2\times\frac{1}{2}\int_0^{\sqrt{3}}\left(\sqrt{4-x^2}-1\right)^2 dx$$
$$= \int_0^{\sqrt{3}}\left(5 - x^2 - 2\sqrt{4-x^2}\right)dx$$
$$= 5\sqrt{3} - \sqrt{3} - 2\int_0^{\sqrt{3}}\sqrt{4-x^2}\, dx$$
$$= 4\sqrt{3} - 4\left(\frac{\pi}{3} + \frac{\sqrt{3}}{4}\right) = 3\sqrt{3} - \frac{4\pi}{3}.$$

Thus $\bar{y} = \dfrac{9\sqrt{3}-4\pi}{3}\cdot\dfrac{3}{4\pi - 3\sqrt{3}} = \dfrac{9\sqrt{3}-4\pi}{4\pi - 3\sqrt{3}}$. The centroid is $\left(0, \dfrac{9\sqrt{3}-4\pi}{4\pi-3\sqrt{3}}\right)$.

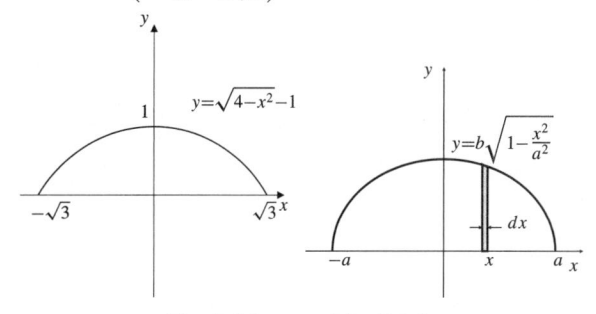

Fig. 7.5.5   Fig. 7.5.6

6. By symmetry, $\bar{x} = 0$. The area is $A = \frac{1}{2}\pi ab$. The moment about $y = 0$ is

$$M_{y=0} = \frac{1}{2}\int_{-a}^a b^2\left[1 - \left(\frac{x}{a}\right)^2\right]dx = b^2\int_0^a 1 - \frac{x^2}{a^2}\, dx$$
$$= b^2\left(x - \frac{x^3}{3a^2}\right)\Big|_0^a = \frac{2}{3}ab^2.$$

Thus, $\bar{y} = \dfrac{2ab^2}{3}\times\dfrac{2}{\pi ab} = \dfrac{4b}{3\pi}$.

7. The quadrilateral consists of two triangles, T_1 and T_2, as shown in the figure. The area and centroid of T_1 are given by

$$A_1 = \frac{4\times 1}{2} = 2,$$
$$\bar{x}_1 = \frac{0+3+4}{3} = \frac{7}{3}, \quad \bar{y}_1 = \frac{0+1+0}{3} = \frac{1}{3}.$$

279

SECTION 7.5 (PAGE 441)

The area and centroid of T_2 are given by

$$A_2 = \frac{4 \times 2}{2} = 4,$$
$$\bar{x}_2 = \frac{0+2+4}{3} = 2, \quad \bar{y}_2 = \frac{0-2+0}{3} = -\frac{2}{3}.$$

It follows that

$$M_{1,x=0} = \frac{7}{3} \times 2 = \frac{14}{3} \qquad M_{2,x=0} = 2 \times 4 = 8$$
$$M_{1,y=0} = \frac{1}{3} \times 2 = \frac{2}{3} \qquad M_{2,y=0} = -\frac{2}{3} \times 4 = -\frac{8}{3}.$$

Since areas and moments are additive, we have for the whole quadrilateral

$$A = 2 + 4 = 6,$$
$$M_{x=0} = \frac{14}{3} + 8 = \frac{38}{3}, \quad M_{y=0} = \frac{2}{3} - \frac{8}{3} = -2.$$

Thus $\bar{x} = \dfrac{38}{3 \times 6} = \dfrac{19}{9}$, and $\bar{y} = \dfrac{-2}{6} = -\dfrac{1}{3}$. The centroid of the quadrilateral is $\left(\dfrac{19}{9}, -\dfrac{1}{3}\right)$.

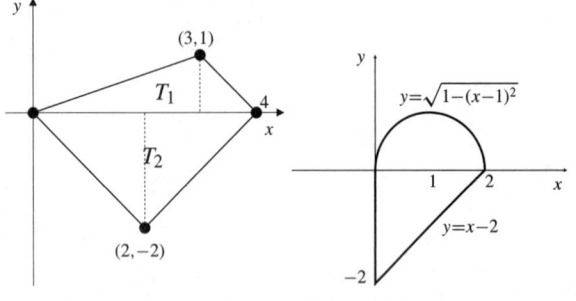

Fig. 7.5.7   Fig. 7.5.8

8. The region is the union of a half-disk and a triangle. The centroid of the half-disk is known to be at $\left(1, \dfrac{4}{3\pi}\right)$ and that of the triangle is at $\left(\dfrac{2}{3}, -\dfrac{2}{3}\right)$. The area of the semicircle is $\dfrac{\pi}{2}$ and the triangle is 2. Hence,

$$M_{x=0} = \left(\frac{\pi}{2}\right)(1) + (2)\left(\frac{2}{3}\right) = \frac{3\pi + 8}{6};$$
$$M_{y=0} = \left(\frac{\pi}{2}\right)\left(\frac{4}{3\pi}\right) + (2)\left(-\frac{2}{3}\right) = -\frac{2}{3}.$$

Since the area of the whole region is $\dfrac{\pi}{2} + 2$, then

$$\bar{x} = \frac{3\pi + 8}{3(\pi + 4)} \quad \text{and} \quad \bar{y} = -\frac{4}{3(\pi + 4)}.$$

9. A circular strip of the surface between heights y and $y + dy$ has area

$$dS = 2\pi x \frac{dy}{\cos\theta} = 2\pi x \frac{r}{x} dy = 2\pi r \, dy.$$

The total surface area is

$$S = 2\pi r \int_0^r dy = 2\pi r^2.$$

The moment about $y = 0$ is

$$M_{y=0} = 2\pi r \int_0^r y \, dy = \pi r (y^2)\Big|_0^r = \pi r^3.$$

Thus $\bar{y} = \dfrac{\pi r^3}{2\pi r^2} = \dfrac{r}{2}$. By symmetry, the centroid of the hemispherical surface is on the axis of symmetry of the hemisphere. It is halfway between the centre of the base circle and the vertex.

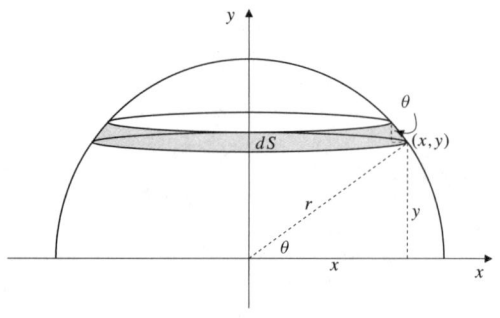

Fig. 7.5.9

10. By symmetry, $\bar{x} = \bar{y} = 0$. The volume is $V = \frac{2}{3}\pi r^3$. A thin slice of the solid at height z will have volume $dV = \pi y^2 \, dz = \pi(r^2 - z^2) \, dz$. Thus, the moment about $z = 0$ is

$$M_{z=0} = \int_0^r z\pi(r^2 - z^2) \, dz$$
$$= \pi \left(\frac{r^2 z^2}{2} - \frac{z^4}{4}\right)\Big|_0^r = \frac{\pi r^4}{4}.$$

Thus, $\bar{z} = \dfrac{\pi r^4}{4} \times \dfrac{3}{2\pi r^3} = \dfrac{3r}{8}$. Hence, the centroid is on the axis of the hemisphere at distance $3r/8$ from the base.

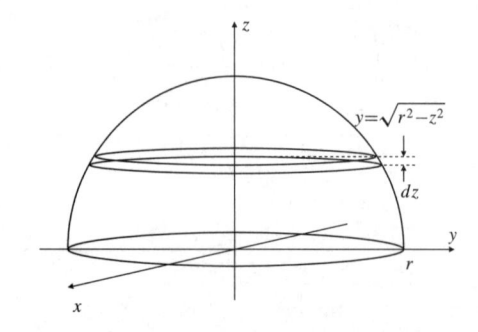

Fig. 7.5.10

11. The cone has volume $V = \frac{1}{3}\pi r^2 h$. (See the following figure.) The disk-shaped slice with vertical width dz has radius $y = r\left(1 - \frac{z}{h}\right)$, and therefore has volume

$$dV = \pi r^2 \left(1 - \frac{z}{h}\right)^2 dz = \pi \frac{r^2}{h^2}(h-z)^2 \, dz.$$

We have

$$M_{z=0} = \frac{\pi r^2}{h^2} \int_0^h z(h-z)^2 \, dz \quad \text{Let } u = h - z$$
$$du = -dz$$
$$= \frac{\pi r^2}{h^2} \int_0^h (h-u)u^2 \, du$$
$$= \frac{\pi r^2}{h^2} \left(\frac{hu^3}{3} - \frac{u^4}{4}\right)\bigg|_0^h = \frac{\pi r^2 h^2}{12}.$$

Therefore $\bar{z} = \frac{\pi r^2 h^2}{12} \cdot \frac{3}{\pi r^2 h} = \frac{h}{4}$. The centroid of the solid cone is on the axis of the cone, at a distance above the base equal to one quarter of the height of the cone.

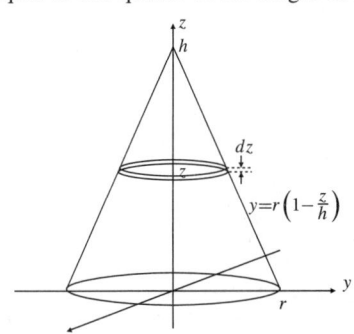

Fig. 7.5.11

12. A band at height z with vertical width dz has radius $y = r\left(1 - \frac{z}{h}\right)$, and has actual (slant) width

$$ds = \sqrt{1 + \left(\frac{dy}{dz}\right)^2}\, dz = \sqrt{1 + \frac{r^2}{h^2}}\, dz.$$

Its area is

$$dA = 2\pi r \left(1 - \frac{z}{h}\right)\sqrt{1 + \frac{r^2}{h^2}}\, dz.$$

Thus the area of the conical surface is

$$A = 2\pi r \sqrt{1 + \frac{r^2}{h^2}} \int_0^h \left(1 - \frac{z}{h}\right) dz = \pi r \sqrt{r^2 + h^2}.$$

The moment about $z = 0$ is

$$M_{z=0} = 2\pi r \sqrt{1 + \frac{r^2}{h^2}} \int_0^h z\left(1 - \frac{z}{h}\right) dz$$
$$= 2\pi r \sqrt{1 + \frac{r^2}{h^2}} \left(\frac{z^2}{2} - \frac{z^3}{3h}\right)\bigg|_0^h = \frac{1}{3}\pi rh\sqrt{r^2 + h^2}.$$

Thus, $\bar{z} = \frac{\pi rh\sqrt{r^2+h^2}}{3} \times \frac{1}{\pi r \sqrt{r^2+h^2}} = \frac{h}{3}$. By symmetry, $\bar{x} = \bar{y} = 0$. Hence, the centroid is on the axis of the conical surface, at distance $h/3$ from the base.

13. The region in figure (a) is the union of a rectangle of area 2 and centroid $(1, 3/2)$ and a triangle of area 1 and centroid $(2/3, 2/3)$. Therefore its area is 3 and its centroid is $(\bar{x}, \bar{y})$, where

$$3\bar{x} = 2(1) + 1\left(\frac{2}{3}\right) = \frac{8}{3}$$
$$3\bar{y} = 2\left(\frac{3}{2}\right) + 1\left(\frac{2}{3}\right) = \frac{11}{3}.$$

Therefore, the centroid is $(8/9, 11/9)$.

14. The region in figure (b) is the union of a square of area $(\sqrt{2})^2 = 2$ and centroid $(0, 0)$ and a triangle of area $1/2$ and centroid $(2/3, 2/3)$. Therefore its area is $5/2$ and its centroid is $(\bar{x}, \bar{y})$, where

$$\frac{5}{2}\bar{x} = 2(0) + \frac{1}{2}\left(\frac{2}{3}\right) = \frac{1}{3}.$$

Therefore, $\bar{x} = \bar{y} = 2/15$, and the centroid is $(2/15, 2/15)$.

15. The region in figure (c) is the union of a half-disk of area $\pi/2$ and centroid $(0, 4/(3\pi))$ (by Example 1) and a triangle of area 1 and centroid $(0, -1/3)$. Therefore its area is $(\pi/2) + 1$ and its centroid is $(\bar{x}, \bar{y})$, where $\bar{x} = 0$ and

$$\frac{\pi + 2}{2}\bar{y} = \frac{\pi}{2}\left(\frac{4}{3\pi}\right) + 1\left(\frac{-1}{3}\right) = \frac{1}{3}.$$

Therefore, the centroid is $(0, 2/[3(\pi + 2)])$.

16. The region in figure (d) is the union of three half-disks, one with area $\pi/2$ and centroid $(0, 4/(3\pi))$, and two with areas $\pi/8$ and centroids $(-1/2, -2/(3\pi))$ and $(1/2, -2/(3\pi))$. Therefore its area is $3\pi/4$ and its centroid is $(\bar{x}, \bar{y})$, where

$$\frac{3\pi}{4}(\bar{x}) = \frac{\pi}{2}(0) + \frac{\pi}{8}\left(\frac{-1}{2}\right) + \frac{\pi}{8}\left(\frac{1}{2}\right) = 0$$
$$\frac{3\pi}{4}(\bar{y}) = \frac{\pi}{2}\left(\frac{4}{3\pi}\right) + \frac{\pi}{8}\left(\frac{-2}{3\pi}\right) + \frac{\pi}{8}\left(\frac{-2}{3\pi}\right) = \frac{1}{2}.$$

Therefore, the centroid is $(0, 2/(3\pi))$.

17. By symmetry, $\bar{x} = \dfrac{\pi}{2}$. The area and y-moment of the region are given by

$$A = \int_0^\pi \sin x\, dx = 2$$

$$M_{y=0} = \dfrac{1}{2}\int_0^\pi \sin^2 x\, dx$$

$$= \dfrac{1}{4}(x - \sin x \cos x)\Big|_0^\pi = \dfrac{\pi}{4}.$$

Thus $\bar{y} = \dfrac{\pi}{8}$, and the centroid is $\left(\dfrac{\pi}{2}, \dfrac{\pi}{8}\right)$.

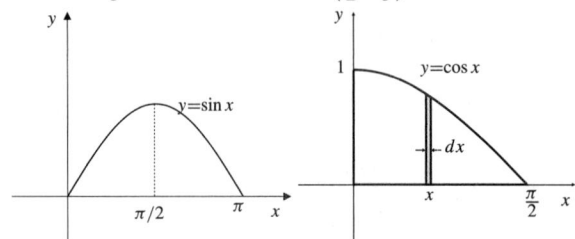

Fig. 7.5.17   Fig. 7.5.18

18. The area of the region is

$$A = \int_0^{\pi/2} \cos x\, dx = \sin x\Big|_0^{\pi/2} = 1.$$

The moment about $x = 0$ is

$$M_{x=0} = \int_0^{\pi/2} x \cos x\, dx$$

$$\begin{array}{ll} U = x & dV = \cos x\, dx \\ dU = dx & V = \sin x \end{array}$$

$$= x \sin x\Big|_0^{\pi/2} - \int_0^{\pi/2} \sin x\, dx = \dfrac{\pi}{2} - 1.$$

Thus, $\bar{x} = \dfrac{\pi}{2} - 1$. The moment about $y = 0$ is

$$M_{y=0} = \dfrac{1}{2}\int_0^{\pi/2} \cos^2 x\, dx$$

$$= \dfrac{1}{4}\left(x + \dfrac{1}{2}\sin 2x\right)\Big|_0^{\pi/2} = \dfrac{\pi}{8}.$$

Thus, $\bar{y} = \dfrac{\pi}{8}$. The centroid is $\left(\dfrac{\pi}{2} - 1, \dfrac{\pi}{8}\right)$.

19. The arc has length $L = \dfrac{\pi r}{2}$. By symmetry, $\bar{x} = \bar{y}$. An element of the arc between x and $x + dx$ has length

$$ds = \dfrac{dx}{\sin\theta} = \dfrac{r\, dx}{y} = \dfrac{r\, dx}{\sqrt{r^2 - x^2}}.$$

Thus

$$M_{x=0} = \int_0^r \dfrac{xr\, dx}{\sqrt{r^2 - x^2}} = -r\sqrt{r^2 - x^2}\Big|_0^r = r^2.$$

Hence $\bar{x} = r^2 \cdot \dfrac{2}{\pi r} = \dfrac{2r}{\pi}$, and the centroid is $\left(\dfrac{2r}{\pi}, \dfrac{2r}{\pi}\right)$.

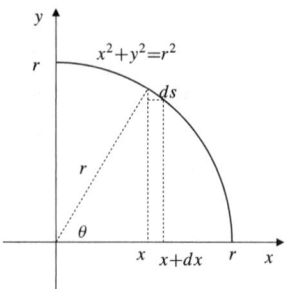

Fig. 7.5.19

20. The solid S in question consists of a solid cone C with vertex at the origin, height 1, and top a circular disk of radius 2, and a solid cylinder D of radius 2 and height 1 sitting on top of the cone. These solids have volumes $V_C = 4\pi/3$, $V_D = 4\pi$, and $V_S = V_C + V_D = 16\pi/3$.

By symmetry, the centroid of the solid lies on its vertical axis of symmetry; let us continue to call this the y-axis. We need only determine $\bar{y}_S$. Since D lies between $y = 1$ and $y = 2$, its centroid satisfies $\bar{y}_D = 3/2$. Also, by Exercise 11, the centroid of the solid cone satisfies $\bar{y}_C = 3/4$. Thus C and D have moments about $y = 0$:

$$M_{C,y=0} = \left(\dfrac{4\pi}{3}\right)\left(\dfrac{3}{4}\right) = \pi, \quad M_{D,y=0} = (4\pi)\left(\dfrac{3}{2}\right) = 6\pi.$$

Thus $M_{S,y=0} = \pi + 6\pi = 7\pi$, and $\bar{z}_S = 7\pi/(16\pi/3) = 21/16$. The centroid of the solid S is on its vertical axis of symmetry at height $21/16$ above the vertex of the conical part.

21. By symmetry the centroid is $(1, -2)$.

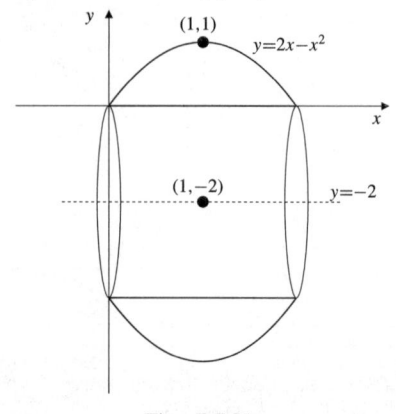

Fig. 7.5.21

22. The line segment from $(1, 0)$ to $(0, 1)$ has centroid $(\frac{1}{2}, \frac{1}{2})$ and length $\sqrt{2}$. By Pappus's Theorem, the surface area of revolution about $x = 2$ is

$$A = 2\pi \left(2 - \frac{1}{2}\right)\sqrt{2} = 3\pi\sqrt{2} \text{ sq. units.}$$

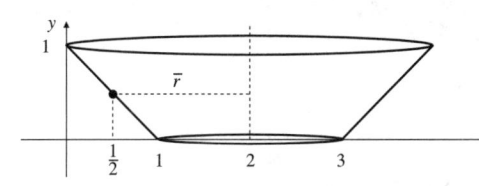

Fig. 7.5.22

23. The triangle T has centroid $(\frac{1}{3}, \frac{1}{3})$ and area $\frac{1}{2}$. By Pappus's Theorem the volume of revolution about $x = 2$ is

$$V = \frac{1}{2} \times 2\pi \left(2 - \frac{1}{3}\right) = \frac{5\pi}{3} \text{ cu. units.}$$

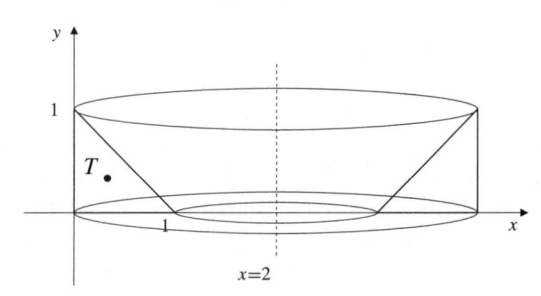

Fig. 7.5.23

24. The altitude h of the triangle is $\frac{s\sqrt{3}}{2}$. Its centroid is at height $\frac{h}{3} = \frac{s}{2\sqrt{3}}$ above the base side. Thus, by Pappus's Theorem, the volume of revolution is

$$V = 2\pi \left(\frac{s}{2\sqrt{3}}\right)\left(\frac{s}{2} \times \frac{\sqrt{3}s}{2}\right) = \frac{\pi s^3}{4} \text{ cu. units.}$$

The centroid of one side is $\frac{h}{2} = \frac{s\sqrt{3}}{4}$ above the base. Thus, the surface area of revolution is

$$S = 2 \times 2\pi \left(\frac{\sqrt{3}s}{4}\right)(s) = s^2\pi\sqrt{3} \text{ sq. units.}$$

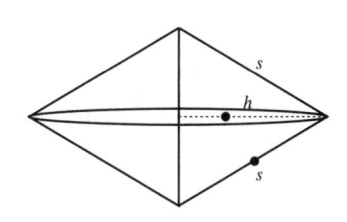

Fig. 7.5.24

25. For the purpose of evaluating the integrals in this problem and the next, the definite integral routine in the TI-85 calculator was used. For the region bounded by $y = 0$ and $y = \sqrt{x}\cos x$ between $x = 0$ and $x = \pi/2$, we have

$$A = \int_0^{\pi/2} \sqrt{x}\cos x \, dx \approx 0.704038$$

$$\bar{x} = \frac{1}{A}\int_0^{\pi/2} x^{3/2}\cos x \, dx \approx 0.71377$$

$$\bar{y} = \frac{1}{2A}\int_0^{\pi/2} x\cos^2 x \, dx \approx 0.26053.$$

26. The region bounded by $y = 0$ and $y = \ln(\sin x)$ between $x = 0$ and $x = \pi/2$ lies below the x-axis, so

$$A = -\int_0^{\pi/2} \ln(\sin x) \, dx \approx 1.088793$$

$$\bar{x} = \frac{-1}{A}\int_0^{\pi/2} x\ln(\sin x) \, dx \approx 0.30239$$

$$\bar{y} = \frac{-1}{2A}\int_0^{\pi/2} \left(\ln(\sin x)\right)^2 \, dx \approx -0.93986.$$

27. The area and moments of the region are

$$A = \int_0^\infty \frac{dx}{(1+x)^3} = \lim_{R\to\infty} \left.\frac{-1}{2(1+x)^2}\right|_0^R = \frac{1}{2}$$

$$M_{x=0} = \int_0^\infty \frac{x \, dx}{(1+x)^3} \quad \begin{array}{l} \text{Let } u = x+1 \\ du = dx \end{array}$$

$$= \int_1^\infty \frac{u-1}{u^3} \, du$$

$$= \lim_{R\to\infty} \left.\left(-\frac{1}{u} + \frac{1}{2u^2}\right)\right|_1^R = 1 - \frac{1}{2} = \frac{1}{2}$$

$$M_{y=0} = \frac{1}{2}\int_0^\infty \frac{dx}{(1+x)^6} = \lim_{R\to\infty} \left.\frac{-1}{10(1+x)^5}\right|_0^R = \frac{1}{10}.$$

The centroid is $\left(1, \frac{1}{5}\right)$.

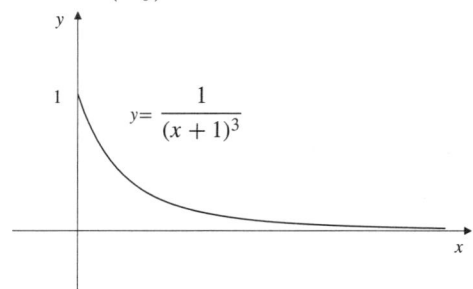

Fig. 7.5.27

28. The surface area is given by

$$S = 2\pi \int_{-\infty}^{\infty} e^{-x^2}\sqrt{1+4x^2 e^{-2x^2}}\, dx.$$ Since
$\lim_{x\to \pm\infty} 1 + 4x^2 e^{-2x^2} = 1$, this expression must be bounded for all x, that is, $1 \le 1 + 4x^2 e^{-2x^2} \le K^2$ for some constant K. Thus, $S \le 2\pi K \int_{-\infty}^{\infty} e^{-x^2}\, dx = 2K\pi\sqrt{\pi}$. The integral converges and the surface area is finite. Since the whole curve $y = e^{-x^2}$ lies above the x-axis, its centroid would have to satisfy $\overline{y} > 0$. However, Pappus's Theorem would then imply that the surface of revolution would have infinite area: $S = 2\pi \overline{y} \times$ (length of curve) $= \infty$. The curve cannot, therefore, have any centroid.

29. By analogy with the formulas for the region $a \le x \le b$, $f(x) \le y \le g(y)$, the region $c \le y \le d$, $f(y) \le x \le g(y)$ will have centroid $(M_{x=0}/A, M_{y=0}/A)$, where

$$A = \int_c^d \big(g(y) - f(y)\big)\, dy$$
$$M_{x=0} = \frac{1}{2}\int_c^d \left[\big(g(y)\big)^2 - \big(f(y)\big)^2\right] dy$$
$$M_{y=0} = \int_c^d y\big(g(y) - f(y)\big)\, dy.$$

30. Let us take L to be the y-axis and suppose that a plane curve C lies between $x = a$ and $x = b$ where $0 < a < b$. Thus, $\overline{r} = \overline{x}$, the x-coordinate of the centroid of C. Let ds denote an arc length element of C at position x. This arc length element generates, on rotation about L, a circular band of surface area $dS = 2\pi x\, ds$, so the surface area of the surface of revolution is

$$S = 2\pi \int_{x=a}^{x=b} x\, ds = 2\pi M_{x=0} = 2\pi \overline{r} s.$$

31.

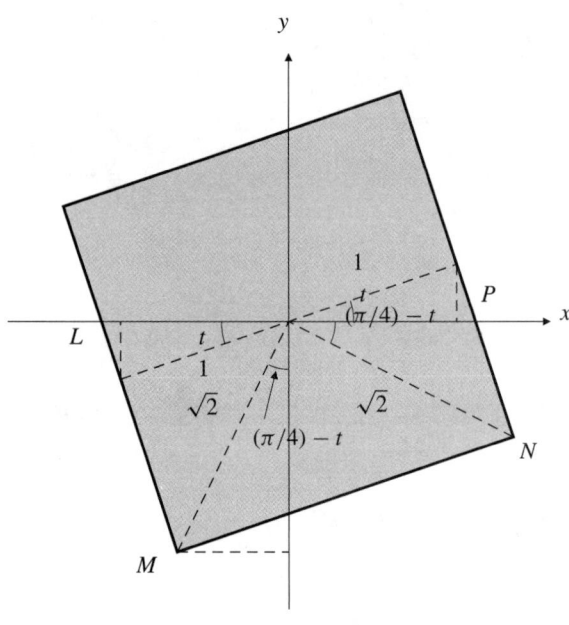

Fig. 7.5.31

We need to find the x-coordinate $\overline{x}_{LMNP}$ of the centre of buoyancy, that is, of the centroid of quadrilateral $LMNP$. From various triangles in the figure we can determine the x-coordinates of the four points:

$$x_L = -\sec t, \qquad x_P = \sec t,$$
$$x_M = -\sec t + (1 + \tan t)\sin t$$
$$x_N = \sec t + (1 - \tan t)\sin t$$

Triangle LMN has area $1 + \tan t$, and the x-coordinate of its centroid is

$$\overline{x}_{LMN}$$
$$= \frac{-\sec t - \sec t + (1 + \tan t)\sin t + \sec t + (1 - \tan t)\sin t}{3}$$
$$= \frac{2\sin t - \sec t}{3}.$$

Triangle LNP has area $1 - \tan t$, and the x-coordinate of its centroid is

$$\overline{x}_{LNP} = \frac{-\sec t + \sec t + \sec t + (1 - \tan t)\sin t}{3}$$
$$= \frac{\sec t + (1 - \tan t)\sin t}{3}.$$

Therefore,

$$\overline{x}_{LMNP} = \frac{1}{6}\big[(2\sin t - \sec t)(1 + \tan t)$$
$$\qquad + (\sec t + \sin t - \sin t\tan t)(1 - \tan t)\big]$$
$$= \frac{1}{6}\big[3\sin t - 2\sec t \tan t + \sin t \tan^2 t\big]$$
$$= \frac{\sin t}{6}\left[3 - \frac{2}{\cos^2 t} + \frac{\sin^2 t}{\cos^2 t}\right]$$
$$= \frac{\sin t}{6\cos^2 t}\big[3\cos^2 t + \sin^2 t - 2\big]$$
$$= \frac{\sin t}{6\cos^2 t}\big[2\cos^2 t - 1\big] = \frac{\sin t}{6\cos^2 t}\big[\cos(2t)\big].$$

INSTRUCTOR'S SOLUTIONS MANUAL   SECTION 7.6 (PAGE 449)

which is positive provided $0 < t < \pi/4$. Thus the beam will rotate counterclockwise until an edge is on top.

Section 7.6 Other Physical Applications (page 449)

1. a) The pressure at the bottom is $p = 9,800 \times 6$ N/m^2. The force on the bottom is $4 \times p = 235,200$ N.

   b) The pressure at depth h metres is $9,800h$ N/m^2. The force on a strip between depths h and $h + dh$ on one wall of the tank is

   $$dF = 9,800h \times 2\,dh = 19,600\,h\,dh \text{ N}.$$

   Thus, the total force on one wall is

   $$F = 19,600 \int_0^6 h\,dh = 19,600 \times 18 = 352,800 \text{ N}.$$

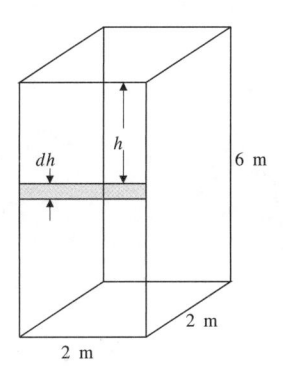

Fig. 7.6.1

2. A vertical slice of water at position y with thickness dy is in contact with the botttom over an area $8 \sec\theta\,dy = \frac{4}{5}\sqrt{101}\,dy$ m^2, which is at depth $x = \frac{1}{10}y + 1$ m. The force exerted on this area is then $dF = \rho g(\frac{1}{10}y + 1)\frac{4}{5}\sqrt{101}\,dy$. Hence, the total force exerted on the bottom is

$$F = \frac{4}{5}\sqrt{101}\,\rho g \int_0^{20} \left(\frac{1}{10}y + 1\right) dy$$
$$= \frac{4}{5}\sqrt{101}\,(1000)(9.8)\left(\frac{y^2}{20} + y\right)\Big|_0^{20}$$
$$\approx 3.1516 \times 10^6 \text{ N}.$$

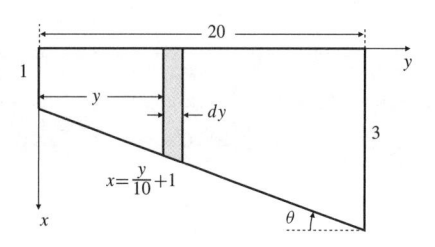

Fig. 7.6.2

3. A strip along the slant wall of the dam between depths h and $h + dh$ has area

$$dA = \frac{200\,dh}{\cos\theta} = 200 \times \frac{26}{24}\,dh.$$

The force on this strip is

$$dF = 9,800\,h\,dA \approx 2.12 \times 10^6\,h\,dh \text{ N}.$$

Thus the total force on the dam is

$$F = 2.12 \times 10^6 \int_0^{24} h\,dh \approx 6.12 \times 10^8 \text{ N}.$$

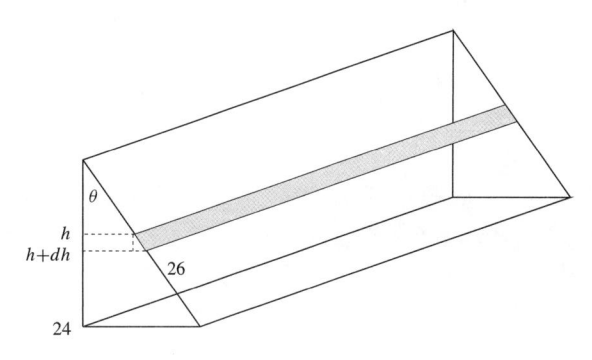

Fig. 7.6.3

4. The height of each triangular face is $2\sqrt{3}$ m and the height of the pyramid is $2\sqrt{2}$ m. Let the angle between the triangular face and the base be θ, then $\sin\theta = \sqrt{\frac{2}{3}}$ and $\cos\theta = \frac{1}{\sqrt{3}}$.

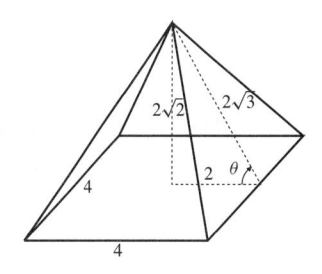

Fig. 7.6.4

285

SECTION 7.6 (PAGE 449)

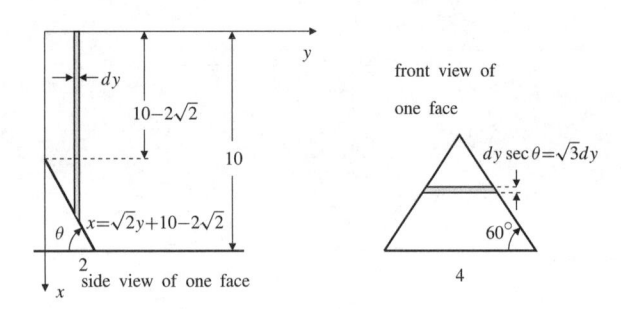

Fig. 7.6.4

A vertical slice of water with thickness dy at a distance y from the vertex of the pyramid exerts a force on the shaded strip shown in the front view, which has area $2\sqrt{3}y\,dy$ m^2 and which is at depth $\sqrt{2}y + 10 - 2\sqrt{2}$ m. Hence, the force exerted on the triangular face is

$$F = \rho g \int_0^2 (\sqrt{2}y + 10 - 2\sqrt{2})2\sqrt{3}y\,dy$$
$$= 2\sqrt{3}(9800)\left[\frac{\sqrt{2}}{3}y^3 + (5-\sqrt{2})y^2\right]\Big|_0^2$$
$$\approx 6.1495 \times 10^5 \text{ N}.$$

5. The unbalanced force is

$$F = 9,800 \times 5 \int_6^{20} h\,dh$$
$$= 9,800 \times 5 \left(\frac{h^2}{2}\right)\Big|_6^{20} \approx 8.92 \times 10^6 \text{ N}.$$

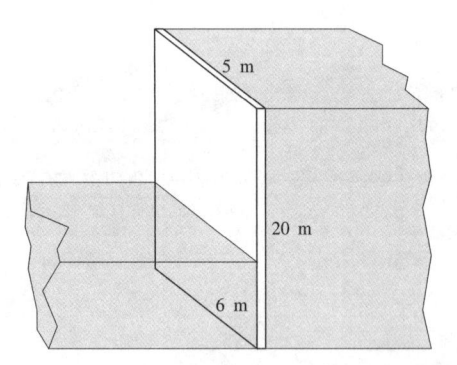

Fig. 7.6.5

6. The spring force is $F(x) = kx$, where x is the amount of compression. The work done to compress the spring 3 cm is

$$100 \text{ N·cm} = W = \int_0^3 kx\,dx = \frac{1}{2}kx^2\Big|_0^3 = \frac{9}{2}k.$$

Hence, $k = \dfrac{200}{9}$ N/cm. The work necessary to compress the spring a further 1 cm is

$$W = \int_3^4 kx\,dx = \left(\frac{200}{9}\right)\frac{1}{2}x^2\Big|_3^4 = \frac{700}{9} \text{ N·cm}.$$

7. A layer of water in the tank between depths h and $h + dh$ has weight $dF = \rho g\,dV = 4\rho g\,dh$. The work done to raise the water in this layer to the top of the tank is $dW = h\,dF = 4\rho gh\,dh$. Thus the total work done to pump all the water out over the top of the tank is

$$W = 4\rho g\int_0^6 h\,dh = 4 \times 9,800 \times 18 \approx 7.056 \times 10^5 \text{ N·m}.$$

8. The horizontal cross-sectional area of the pool at depth h is
$$A(h) = \begin{cases} 160, & \text{if } 0 \leq h \leq 1; \\ 240 - 80h, & \text{if } 1 < h \leq 3. \end{cases}$$

The work done to empty the pool is

$$W = \rho g \int_0^3 hA(h)\,dh$$
$$= \rho g\left[\int_0^1 160h\,dh + \int_1^3 240h - 80h^2\,dh\right]$$
$$= 9800\left[80h^2\Big|_0^1 + \left(120h^2 - \frac{80}{3}h^3\right)\Big|_1^3\right]$$
$$= 3.3973 \times 10^6 \text{ N·m}.$$

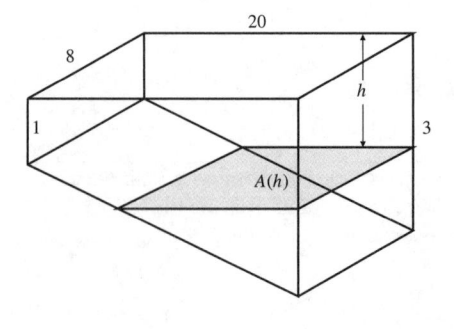

Fig. 7.6.8

9. A layer of water between depths y and $y + dy$ has volume $dV = \pi(a^2 - y^2)\,dy$ and weight $dF = 9,800\pi(a^2 - y^2)\,dy$ N. The work done to raise this water to height h m above the top of the bowl is

$$dW = (h+y)\,dF = 9,800\pi(h+y)(a^2-y^2)\,dy \text{ N·m}.$$

286

Thus the total work done to pump all the water in the bowl to that height is

$$W = 9,800\pi \int_0^a (ha^2 + a^2y - hy^2 - y^3)\,dy$$
$$= 9,800\pi \left[ha^2 y + \frac{a^2 y^2}{2} - \frac{hy^3}{3} - \frac{y^4}{4} \right]\Big|_0^a$$
$$= 9,800\pi \left[\frac{2a^3 h}{3} + \frac{a^4}{4} \right]$$
$$= 9,800\pi a^3 \frac{3a + 8h}{12} = 2450\pi a^3 \left(a + \frac{8h}{3} \right) \text{ N·m}.$$

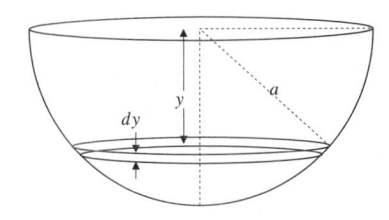

Fig. 7.6.9

10. Let the time required to raise the bucket to height h m be t minutes. Given that the velocity is 2 m/min, then $t = \frac{h}{2}$. The weight of the bucket at time t is
16 kg $-$ (1 kg/min)(t min) $= 16 - \frac{h}{2}$ kg. Therefore, the work done required to move the bucket to a height of 10 m is

$$W = g \int_0^{10} \left(16 - \frac{h}{2} \right) dh$$
$$= 9.8 \left(16h - \frac{h^2}{4} \right)\Big|_0^{10} = 1323 \text{ N·m}.$$

Section 7.7 Applications in Business, Finance, and Ecology (page 453)

1. Cost $= \$4,000 + \int_0^{1,000} \left(6 - \frac{2x}{10^3} + \frac{6x^2}{10^6} \right) dx$
   $= \$11,000.$

2. The number of chips sold in the first year was

$$1,000 \int_0^{52} te^{-t/10}\,dt = 100,000 - 620,000 e^{-26/5}$$

that is, about 96,580.

3. The monthly charge is

$$\int_0^x \frac{4}{1 + \sqrt{t}}\,dt \qquad \text{let } t = u^2$$
$$= 8 \int_0^{\sqrt{x}} \frac{u}{1+u}\,du = 8 \int_0^{\sqrt{x}} \left(1 - \frac{1}{1+u} \right) du$$
$$= \$8 \left(\sqrt{x} - \ln(1 + \sqrt{x}) \right).$$

4. The price per kg at time t (years) is $\$10 + 5t$. Thus the revenue per year at time t is $400(10+5t)/(1+0.1t)$ \$/year. The total revenue over the year is

$$\int_0^1 \frac{400(10 + 5t)}{1 + 0.1t}\,dt \approx \$4,750.37.$$

5. The present value of continuous payments of \$1,000 per year for 10 years at a discount rate of 2% is

$$V = \int_0^{10} 1,000 e^{-0.02t}\,dt = \frac{1,000}{-0.02} e^{-0.02t}\Big|_0^{10} = \$9,063.46.$$

6. The present value of continuous payments of \$1,000 per year for 10 years at a discount rate of 5% is

$$V = \int_0^{10} 1,000 e^{-0.05t}\,dt = \frac{1,000}{-0.05} e^{-0.05t}\Big|_0^{10} = \$7,869.39.$$

7. The present value of continuous payments of \$1,000 per year for 10 years beginning 2 years from now at a discount rate of 8% is

$$V = \int_2^{12} 1,000 e^{-0.08t}\,dt = \frac{1,000}{-0.08} e^{-0.08t}\Big|_2^{12} = \$5,865.64.$$

8. The present value of continuous payments of \$1,000 per year for 25 years beginning 10 years from now at a discount rate of 5% is

$$V = \int_{10}^{35} 1,000 e^{-0.05t}\,dt = \frac{1,000}{-0.05} e^{-0.05t}\Big|_{10}^{35} = \$8,655.13.$$

9. The present value of continuous payments of \$1,000 per year for all future time at a discount rate of 2% is

$$V = \int_0^{\infty} 1,000 e^{-0.02t}\,dt = \frac{1,000}{-0.02} = \$50,000.00.$$

10. The present value of continuous payments of $1,000 per year beginning 10 years from now and continuing for all future time at a discount rate of 5% is

$$V = \int_{10}^{\infty} 1{,}000 e^{-0.05t}\, dt = \frac{1{,}000}{-0.05} e^{-0.5} = \$12{,}130.61.$$

11. After t years, money is flowing at $\$(1{,}000 + 100t)$ per year. The present value of 10 years of payments discounted at 5% is

$$V = 100 \int_0^{10} (10+t) e^{-0.05t}\, dt$$

$$\begin{aligned} U &= 10+t & dV &= e^{-0.05t}\, dt \\ dU &= dt & V &= \frac{e^{-0.05t}}{-0.05} \end{aligned}$$

$$= 100(10+t)\frac{e^{-0.05t}}{-0.05}\bigg|_0^{10} + \frac{100}{0.05}\int_0^{10} e^{-0.05t}\, dt$$

$$= -4261.23 + \frac{100}{-(0.05)^2} e^{-0.05t}\bigg|_0^{10} = \$11{,}477.54.$$

12. After t years, money is flowing at $\$1{,}000(1.1)^t$ per year. The present value of 10 years of payments discounted at 5% is

$$V = 1{,}000 \int_0^{10} e^{t \ln(1.1)} e^{-0.05t}\, dt$$

$$= \frac{1{,}000}{\ln(1.1)-0.05} e^{t(\ln(1.1)-0.05)}\bigg|_0^{10} = \$12{,}650.23.$$

13. The amount after 10 years is

$$A = 5{,}000 \int_0^{10} e^{0.05t}\, dt = \frac{5{,}000}{0.05} e^{0.05t}\bigg|_0^{10} = \$64{,}872.13.$$

14. Let T be the time required for the account balance to reach $\$1{,}000{,}000$. The $\$5{,}000(1.1)^t\, dt$ deposited in the time interval $[t, t+dt]$ grows for $T-t$ years, so the balance after T years is

$$\int_0^T 5{,}000(1.1)^t (1.06)^{T-t}\, dt = 1{,}000{,}000$$

$$(1.06)^T \int_0^T \left(\frac{1.1}{1.06}\right)^t dt = \frac{1{,}000{,}000}{5{,}000} = 200$$

$$\frac{(1.06)^T}{\ln(1.1/1.06)} \left[\left(\frac{1.1}{1.06}\right)^T - 1\right] = 200$$

$$(1.1)^T - (1.06)^T = 200 \ln \frac{1.1}{1.06}.$$

This equation can be solved by Newton's method or using a calculator "solve" routine. The solution is $T \approx 26.05$ years.

15. Let $P(\tau)$ be the value at time $\tau < t$ that will grow to $\$P = P(t)$ at time t. If the discount rate at time τ is $\delta(\tau)$, then

$$\frac{d}{d\tau} P(\tau) = \delta(\tau) P(\tau),$$

or, equivalently,

$$\frac{dP(\tau)}{P(\tau)} = \delta(\tau)\, d\tau.$$

Integrating this from 0 to t, we get

$$\ln P(t) - \ln P(0) = \int_0^t \delta(\tau)\, d\tau = \lambda(t),$$

and, taking exponentials of both sides and solving for $P(0)$, we get

$$P(0) = P(t) e^{-\lambda(t)} = P e^{-\lambda(t)}.$$

The present value of a stream of payments due at a rate $P(t)$ at time t from $t=0$ to $t=T$ is

$$\int_0^T P(t) e^{-\lambda(t)}\, dt, \quad \text{where} \quad \lambda(t) = \int_0^t \delta(\tau)\, d\tau.$$

16. The analysis carried out in the text for the logistic growth model showed that the total present value of future harvests could be maximized by holding the population size x at a value that maximizes the quadratic expression

$$Q(x) = kx\left(1 - \frac{x}{L}\right) - \delta x.$$

If the logistic model $dx/dt = kx(1 - (x/L))$ is replaced with a more general growth model $dx/dt = F(x)$, exactly the same analysis leads us to maximize

$$Q(x) = F(x) - \delta x.$$

For realistic growth functions, the maximum will occur where $Q'(x) = 0$, that is, where $F'(x) = \delta$.

17. We are given $L = 80{,}000$, $k = 0.12$, and $\delta = 0.05$. According to the analysis in the text, the present value of future harvests will be maximized if the population level is maintained at

$$x = (k - \delta)\frac{L}{2k} = \frac{0.07}{0.24}(80{,}000) = 23{,}333.33$$

The annual revenue from harvesting to keep the population at this level (given a price of \$6 per fish) is

$$6(0.12)(23,333.33)\left(1 - \frac{23,333.33}{80,000}\right) = \$11,900.$$

18. We are given that $k = 0.02$, $L = 150,000$, $p = \$10,000$. The growth rate at population level x is

$$\frac{dx}{dt} = 0.02x\left(1 - \frac{x}{150,000}\right).$$

a) The maximum sustainable annual harvest is

$$\left.\frac{dx}{dt}\right|_{x=L/2} = 0.02(75,000)(0.5) = 750 \text{ whales}.$$

b) The resulting annual revenue is $\$750p = \$7,500,000$.

c) If the whole population of 75,000 is harvested and the proceeds invested at 2%, the annual interest will be

$$75,000(\$10,000)(0.02) = \$15,000,000.$$

d) At 5%, the interest would be $(5/2)(\$15,000) = \$37,500,000$.

e) The total present value of all future harvesting revenue if the population level is maintained at 75,000 and $\delta = 0.05$ is

$$\int_0^\infty e^{-0.05t} 7,500,000\, dt = \frac{7,500,000}{0.05} = \$150,000,000.$$

19. If we assume that the cost of harvesting 1 unit of population is $\$C(x)$ when the population size is x, then the effective income from 1 unit harvested is $\$(p-C(x))$. Using this expression in place of the constant p in the analysis given in the text, we are led to choose x to maximize

$$Q(x) = (p - C(x))\left[kx\left(1 - \frac{x}{L}\right) - \delta x\right].$$

A reasonable cost function $C(x)$ will increase as x decreases (the whales are harder to find), and will exceed p if $x \le x_0$, for some positive population level x_0. The value of x that maximizes $Q(x)$ must exceed x_0, so the model no longer predicts extinction, even for large discount rates δ. However, the optimizing population x may be so low that other factors not accounted for in the simple logistic growth model may still bring about extinction whether it is economically indicated or not.

Section 7.8 Probability (page 464)

1. We have $f(x) = Cx$ on $[0, 3]$.

a) C is given by

$$1 = \int_0^3 Cx\, dx = \frac{C}{2}x^2\bigg|_0^3 = \frac{9}{2}C.$$

Hence, $C = \frac{2}{9}$.

b) The mean is

$$\mu = E(X) = \frac{2}{9}\int_0^3 x^2\, dx = \frac{2}{27}x^3\bigg|_0^3 = 2.$$

Since $E(X^2) = \frac{2}{9}\int_0^3 x^3\, dx = \frac{2}{36}x^4\bigg|_0^3 = \frac{9}{2}$, the variance is

$$\sigma^2 = E(X^2) - \mu^2 = \frac{9}{2} - 4 = \frac{1}{2},$$

and the standard deviation is $\sigma = 1/\sqrt{2}$.

c) We have

$$\Pr(\mu - \sigma \le X \le \mu + \sigma) = \frac{2}{9}\int_{\mu-\sigma}^{\mu+\sigma} x\, dx$$

$$= \frac{(\mu+\sigma)^2 - (\mu-\sigma)^2}{9} = \frac{4\mu\sigma}{9} \approx 0.6285.$$

2. We have $f(x) = Cx$ on $[1, 2]$.

a) To find C, we have

$$1 = \int_1^2 Cx\, dx = \frac{C}{2}x^2\bigg|_1^2 = \frac{3}{2}C.$$

Hence, $C = \frac{2}{3}$.

b) The mean is

$$\mu = E(X) = \frac{2}{3}\int_1^2 x^2\, dx = \frac{2}{9}x^3\bigg|_1^2 = \frac{14}{9} \approx 1.556.$$

Since $E(X^2) = \frac{2}{3}\int_1^2 x^3\, dx = \frac{1}{6}x^4\bigg|_1^2 = \frac{5}{2}$, the variance is

$$\sigma^2 = E(X^2) - \mu^2 = \frac{5}{2} - \frac{196}{81} = \frac{13}{162}$$

and the standard deviation is

$$\sigma = \sqrt{\frac{13}{162}} \approx 0.283.$$

c) We have

$$\Pr(\mu - \sigma \leq X \leq \mu + \sigma) = \frac{2}{3}\int_{\mu-\sigma}^{\mu+\sigma} x\,dx$$
$$= \frac{(\mu+\sigma)^2 - (\mu-\sigma)^2}{3} = \frac{4\mu\sigma}{3} \approx 0.5875.$$

3. We have $f(x) = Cx^2$ on $[0, 1]$.

   a) C is given by

   $$1 = \int_0^1 Cx^2\,dx = \frac{C}{3}x^3\Big|_0^1 = \frac{C}{3}.$$

   Hence, $C = 3$.

   b) The mean, variance, and standard deviation are

   $$\mu = E(X) = 3\int_0^1 x^3\,dx = \frac{3}{4}$$
   $$\sigma^2 = E(X^2) - \mu^2 = 3\int_0^1 x^4\,dx - \frac{9}{16} = \frac{3}{5} - \frac{9}{16} = \frac{3}{80}$$
   $$\sigma = \sqrt{3/80}.$$

   c) We have

   $$\Pr(\mu - \sigma \leq X \leq \mu + \sigma) = 3\int_{\mu-\sigma}^{\mu+\sigma} x^2\,dx$$
   $$= (\mu+\sigma)^3 - (\mu-\sigma)^3$$
   $$= \left(\frac{3}{4} + \sqrt{\frac{3}{80}}\right)^3 - \left(\frac{3}{4} - \sqrt{\frac{3}{80}}\right)^3 \approx 0.668.$$

4. We have $f(x) = C\sin x$ on $[0, \pi]$.

   a) To find C, we calculate

   $$1 = \int_0^\pi C\sin x\,dx = -C\cos x\Big|_0^\pi = 2C.$$

   Hence, $C = \frac{1}{2}$.

   b) The mean is

   $$\mu = E(X) = \frac{1}{2}\int_0^\pi x\sin x\,dx$$
   $$U = x \qquad dV = \sin x\,dx$$
   $$dU = dx \qquad V = -\cos x$$
   $$= \frac{1}{2}\left[-x\cos x\Big|_0^\pi + \int_0^\pi \cos x\,dx\right]$$
   $$= \frac{\pi}{2} = 1.571.$$

Since

$$E(X^2) = \frac{1}{2}\int_0^\pi x^2 \sin x\,dx$$
$$U = x^2 \qquad dV = \sin x\,dx$$
$$dU = 2x\,dx \qquad V = -\cos x$$
$$= \frac{1}{2}\left[-x^2\cos x\Big|_0^\pi + 2\int_0^\pi x\cos x\,dx\right]$$
$$U = x \qquad dV = \cos x\,dx$$
$$dU = dx \qquad V = \sin x$$
$$= \frac{1}{2}\left[\pi^2 + 2\left(x\sin x\Big|_0^\pi - \int_0^\pi \sin x\,dx\right)\right]$$
$$= \frac{1}{2}(\pi^2 - 4).$$

Hence, the variance is

$$\sigma^2 = E(X^2) - \mu^2 = \frac{\pi^2 - 4}{2} - \frac{\pi^2}{4} = \frac{\pi^2 - 8}{4} \approx 0.467$$

and the standard deviation is

$$\sigma = \sqrt{\frac{\pi^2 - 8}{4}} \approx 0.684.$$

c) Then

$$\Pr(\mu - \sigma \leq X \leq \mu + \sigma) = \frac{1}{2}\int_{\mu-\sigma}^{\mu+\sigma} \sin x\,dx$$
$$= -\frac{1}{2}\left[\cos(\mu+\sigma) - \cos(\mu-\sigma)\right]$$
$$= \sin\mu\sin\sigma = \sin\sigma \approx 0.632.$$

5. We have $f(x) = C(x - x^2)$ on $[0, 1]$.

   a) C is given by

   $$1 = \int_0^1 C(x - x^2)\,dx = C\left(\frac{x^2}{2} - \frac{x^3}{3}\right)\Big|_0^1 = \frac{C}{6}.$$

   Hence, $C = 6$.

   b) The mean, variance, and standard deviation are

   $$\mu = E(X) = 6\int_0^1 (x^2 - x^3)\,dx = \frac{1}{2}$$
   $$\sigma^2 = E(X^2) - \mu^2 = 6\int_0^1 (x^3 - x^4)\,dx - \frac{1}{4}$$
   $$= \frac{3}{10} - \frac{1}{4} = \frac{1}{20}$$
   $$\sigma = \sqrt{1/20}.$$

c) We have

$$\Pr(\mu - \sigma \le X \le \mu + \sigma) = 6\int_{(1/2)-\sigma}^{(1/2)+\sigma} (x - x^2)\,dx$$

$$= 6\int_{(1/2)-\sigma}^{(1/2)+\sigma} \left[\frac{1}{4} - \left(x - \frac{1}{2}\right)^2\right]dx$$

Let $u = x - \tfrac{1}{2}$
$du = dx$

$$= 12\int_0^{\sigma}\left[\frac{1}{4} - u^2\right]du = 12\left[\frac{\sigma}{4} - \frac{\sigma^3}{3}\right]$$

$$= \frac{12}{\sqrt{20}}\left[\frac{1}{4} - \frac{1}{60}\right] \approx 0.626.$$

6. It was shown in Section 6.1 (p. 349) that

$$\int x^n e^{-x}\,dx = -x^n e^{-x} + n\int x^{n-1}e^{-x}\,dx.$$

If $I_n = \int_0^{\infty} x^n e^{-x}\,dx$, then

$$I_n = \lim_{R\to\infty} -R^n e^{-R} + nI_{n-1} = nI_{n-1} \quad \text{if } n \ge 1.$$

Since $I_0 = \int_0^{\infty} e^{-x}\,dx = 1$, therefore $I_n = n!$ for $n \ge 1$.
Let $u = kx$; then

$$\int_0^{\infty} x^n e^{-kx}\,dx = \frac{1}{k^{n+1}}\int_0^{\infty} u^n e^{-u}\,du = \frac{1}{k^{n+1}}I_n = \frac{n!}{k^{n+1}}.$$

Now let $f(x) = Cxe^{-kx}$ on $[0, \infty)$.

a) To find C, observe that

$$1 = C\int_0^{\infty} xe^{-kx}\,dx = \frac{C}{k^2}.$$

Hence, $C = k^2$.

b) The mean is

$$\mu = E(X) = k^2\int_0^{\infty} x^2 e^{-kx}\,dx = k^2\left(\frac{2}{k^3}\right) = \frac{2}{k}.$$

Since $E(X^2) = k^2\int_0^{\infty} x^3 e^{-kx}\,dx = k^2\left(\frac{6}{k^4}\right) = \frac{6}{k^2}$, then the variance is

$$\sigma^2 = E(X^2) - \mu^2 = \frac{6}{k^2} - \frac{4}{k^2} = \frac{2}{k^2}$$

and the standard deviation is $\sigma = \dfrac{\sqrt{2}}{k}$.

c) Finally,

$$\Pr(\mu - \sigma \le X \le \mu + \sigma)$$
$$= k^2 \int_{\mu-\sigma}^{\mu+\sigma} xe^{-kx}\,dx \quad \text{Let } u = kx$$
$$\qquad\qquad\qquad\qquad\qquad du = k\,dx$$
$$= \int_{k(\mu-\sigma)}^{k(\mu+\sigma)} ue^{-u}\,du$$
$$= -ue^{-u}\Big|_{k(\mu-\sigma)}^{k(\mu+\sigma)} + \int_{k(\mu-\sigma)}^{k(\mu+\sigma)} e^{-u}\,du$$
$$= -(2+\sqrt{2})e^{-(2+\sqrt{2})} + (2-\sqrt{2})e^{-(2-\sqrt{2})}$$
$$\quad - e^{-(2+\sqrt{2})} + e^{-(2-\sqrt{2})}$$
$$\approx 0.738.$$

7. a) We have

$$1 = C\int_0^{\infty} e^{-x^2}\,dx = \frac{C}{2}\int_{-\infty}^{\infty} e^{-x^2}\,dx = \frac{C\sqrt{\pi}}{2}.$$

Thus $C = 2/\sqrt{\pi}$.

b) The mean, variance, and standard deviation are

$$\mu = \frac{2}{\sqrt{\pi}}\int_0^{\infty} xe^{-x^2}\,dx = -\frac{e^{-x^2}}{\sqrt{\pi}}\bigg|_0^{\infty} = \frac{1}{\sqrt{\pi}}$$

$$\sigma^2 = -\frac{1}{\pi} + \frac{2}{\sqrt{\pi}}\int_0^{\infty} x^2 e^{-x^2}\,dx$$

$$U = x \qquad dV = xe^{-x^2}\,dx$$
$$dU = dx \qquad V = -\tfrac{1}{2}e^{-x^2}$$

$$= -\frac{1}{\pi} + \frac{2}{\sqrt{\pi}}\left(-\frac{x}{2}e^{-x^2}\bigg|_0^{\infty} + \frac{1}{2}\int_0^{\infty} e^{-x^2}\,dx\right)$$

$$= -\frac{1}{\pi} + \frac{2}{\sqrt{\pi}}\left(0 + \frac{1}{2}\cdot\frac{\sqrt{\pi}}{2}\right) = \frac{1}{2} - \frac{1}{\pi}$$

$$\sigma = \sqrt{\frac{1}{2} - \frac{1}{\pi}} \approx 0.426.$$

c) We have

$$\Pr(\mu - \sigma \le X \le \mu + \sigma) = \frac{2}{\sqrt{\pi}}\int_{\mu-\sigma}^{\mu+\sigma} e^{-x^2}\,dx$$

Let $x = z/\sqrt{2}$
$dx = dz/\sqrt{2}$

$$= \sqrt{\frac{2}{\pi}}\int_{\sqrt{2}(\mu-\sigma)}^{\sqrt{2}(\mu+\sigma)} e^{-z^2/2}\,dz.$$

But $\sqrt{2}(\mu-\sigma) \approx 0.195$ and $\sqrt{2}(\mu+\sigma) \approx 1.40$. Thus, if Z is a standard normal random variable, we obtain by interpolation in the table on page 386 in the text,

$$\Pr(\mu - \sigma \leq X \leq \mu + \sigma) = 2\Pr(0.195 \leq Z \leq 1.400)$$
$$\approx 2(0.919 - 0.577) \approx 0.68.$$

8. No. The identity $\int_{-\infty}^{\infty} C\, dx = 1$ is not satisfied for any constant C.

9. $f_{\mu,\sigma}(x) = \dfrac{1}{\sigma\sqrt{2\pi}} e^{-(x-\mu)^2/2\sigma^2}$

$\text{mean} = \dfrac{1}{\sigma\sqrt{2\pi}} \int_{-\infty}^{\infty} x e^{-(x-\mu)^2/2\sigma^2}\, dx \quad \text{Let } z = \dfrac{x-\mu}{\sigma}$
$$dz = \dfrac{1}{\sigma} dx$$

$$= \dfrac{1}{\sqrt{2\pi}} \int_{-\infty}^{\infty} (\mu + \sigma z) e^{-z^2/2}\, dz$$
$$= \dfrac{\mu}{\sqrt{2\pi}} \int_{-\infty}^{\infty} e^{-z^2/2}\, dz = \mu$$

$\text{variance} = E\big((x-\mu)^2\big)$
$$= \dfrac{1}{\sigma\sqrt{2\pi}} \int_{-\infty}^{\infty} (x-\mu)^2 e^{-(x-\mu)^2/2\sigma^2}\, dx$$
$$= \dfrac{1}{\sigma\sqrt{2\pi}} \int_{-\infty}^{\infty} \sigma^2 z^2 e^{-z^2/2}\, dz = \sigma \text{Var}(Z) = \sigma$$

10. Since $f(x) = \dfrac{2}{\pi(1+x^2)} > 0$ on $[0, \infty)$ and

$$\dfrac{2}{\pi}\int_0^{\infty} \dfrac{dx}{1+x^2} = \lim_{R\to\infty} \dfrac{2}{\pi} \tan^{-1}(R) = \dfrac{2}{\pi}\left(\dfrac{\pi}{2}\right) = 1,$$

therefore $f(x)$ is a probability density function on $[0, \infty)$. The expectation of X is

$$\mu = E(X) = \dfrac{2}{\pi}\int_0^{\infty} \dfrac{x\, dx}{1+x^2}$$
$$= \lim_{R\to\infty} \dfrac{1}{\pi}\ln(1+R^2) = \infty.$$

No matter what the cost per game, you should be willing to play (if you have an adequate bankroll). Your expected winnings per game in the long term is infinite.

11. a) The density function for the uniform distribution on $[a, b]$ is given by $f(x) = 1/(b-a)$, for $a \leq x \leq b$. By Example 5, the mean and standard deviation are given by

$$\mu = \dfrac{b+a}{2}, \qquad \sigma = \dfrac{b-a}{2\sqrt{3}}.$$

Since $\mu + 2\sigma = \dfrac{b+a}{2} + \dfrac{b-a}{\sqrt{3}} > b$, and similarly, $\mu - 2\sigma < a$, therefore $\Pr(|X - \mu| \geq 2\sigma) = 0$.

b) For $f(x) = ke^{-kx}$ on $[0, \infty)$, we know that $\mu = \sigma = \dfrac{1}{k}$ (Example 6). Thus $\mu - 2\sigma < 0$ and $\mu + 2\sigma = \dfrac{3}{k}$. We have

$$\Pr(|X-\mu| \geq 2\sigma) = \Pr\left(X \geq \dfrac{3}{k}\right)$$
$$= k\int_{3/k}^{\infty} e^{-kx}\, dx$$
$$= -e^{-kx}\Big|_{3/k}^{\infty} = e^{-3} \approx 0.050.$$

c) For $f_{\mu,\sigma}(x) = \dfrac{1}{\sigma\sqrt{2\pi}} e^{(x-\mu)^2/2\sigma^2}$, which has mean μ and standard deviation σ, we have

$$\Pr(|X - \mu| \geq 2\sigma) = 2\Pr(X \leq \mu - 2\sigma)$$
$$= 2\int_{-\infty}^{\mu - 2\sigma} \dfrac{1}{\sigma\sqrt{2\pi}} e^{-(x-\mu)^2/2\sigma^2}\, dx$$
$$\text{Let } z = \dfrac{x-\mu}{\sigma}$$
$$dz = \dfrac{1}{\sigma} dx$$
$$= \dfrac{2}{\sqrt{2\pi}} \int_{-\infty}^{-2} e^{-z^2}\, dz$$
$$= 2\Pr(Z \leq -2) \approx 2 \times 0.023 = 0.046$$

from the table on page 463.

12. The density function for T is $f(t) = ke^{-kt}$ on $[0, \infty)$, where $k = \dfrac{1}{\mu} = \dfrac{1}{20}$ (see Example 6). Then

$$\Pr(T \geq 12) = \dfrac{1}{20}\int_{12}^{\infty} e^{-t/20}\, dt = 1 - \dfrac{1}{20}\int_0^{12} e^{-t/20}\, dt$$
$$= 1 + e^{-t/20}\Big|_0^{12} = e^{-12/20} \approx 0.549.$$

The probability that the system will last at least 12 hours is about 0.549.

13. If X is distributed normally, with mean $\mu = 5{,}000$, and standard deviation $\sigma = 200$, then

$$\Pr(X \geq 5500)$$
$$= \frac{1}{200\sqrt{2\pi}} \int_{5500}^{\infty} e^{-(x-5000)^2/(2\times 200^2)}\, dx$$

Let $z = \dfrac{x-5000}{200}$

$dz = \dfrac{dx}{200}$

$$= \frac{1}{\sqrt{2\pi}} \int_{5/2}^{\infty} e^{-z^2/2}\, dz$$
$$= \Pr(Z \geq 5/2) = \Pr(Z \leq -5/2) \approx 0.006$$

from the table on page 463.

Section 7.9 First-Order Differential Equations   (page 472)

1. $\dfrac{dy}{dx} = \dfrac{y}{2x}$

   $2\dfrac{dy}{y} = \dfrac{dx}{x}$

   $2\ln y = \ln x + C_1 \quad \Rightarrow \quad y^2 = Cx$

2. $\dfrac{dy}{dx} = \dfrac{3y-1}{x}$

   $\displaystyle\int \dfrac{dy}{3y-1} = \int \dfrac{dx}{x}$

   $\dfrac{1}{3}\ln|3y-1| = \ln|x| + \dfrac{1}{3}\ln C$

   $\dfrac{3y-1}{x^3} = C$

   $\Rightarrow y = \dfrac{1}{3}(1+Cx^3)$.

3. $\dfrac{dy}{dx} = \dfrac{x^2}{y^2} \quad \Rightarrow \quad y^2\, dy = x^2\, dx$

   $\dfrac{y^3}{3} = \dfrac{x^3}{3} + C_1,\quad$ or $\quad x^3 - y^3 = C$

4. $\dfrac{dy}{dx} = x^2 y^2$

   $\displaystyle\int \dfrac{dy}{y^2} = \int x^2\, dx$

   $-\dfrac{1}{y} = \dfrac{1}{3}x^3 + \dfrac{1}{3}C$

   $\Rightarrow y = -\dfrac{3}{x^3+C}$.

5. $\dfrac{dY}{dt} = tY \quad \Rightarrow \quad \dfrac{dY}{Y} = t\, dt$

   $\ln Y = \dfrac{t^2}{2} + C_1,\quad$ or $\quad Y = Ce^{t^2/2}$

6. $\dfrac{dx}{dt} = e^x \sin t$

   $\displaystyle\int e^{-x}\, dx = \int \sin t\, dt$

   $-e^{-x} = -\cos t - C$

   $\Rightarrow x = -\ln(\cos t + C)$.

7. $\dfrac{dy}{dx} = 1 - y^2 \quad \Rightarrow \quad \dfrac{dy}{1-y^2} = dx$

   $\dfrac{1}{2}\left(\dfrac{1}{1+y} + \dfrac{1}{1-y}\right) dy = dx$

   $\dfrac{1}{2}\ln\left|\dfrac{1+y}{1-y}\right| = x + C_1$

   $\dfrac{1+y}{1-y} = Ce^{2x} \quad$ or $\quad y = \dfrac{Ce^{2x}-1}{Ce^{2x}+1}$

8. $\dfrac{dy}{dx} = 1 + y^2$

   $\displaystyle\int \dfrac{dy}{1+y^2} = \int dx$

   $\tan^{-1} y = x + C$

   $\Rightarrow y = \tan(x+C)$.

9. $\dfrac{dy}{dt} = 2 + e^y \quad \Rightarrow \quad \dfrac{dy}{2+e^y} = dt$

   $\displaystyle\int \dfrac{e^{-y}\, dy}{2e^{-y}+1} = \int dt$

   $-\dfrac{1}{2}\ln(2e^{-y}+1) = t + C_1$

   $2e^{-y} + 1 = C_2 e^{-2t},\quad$ or $\quad y = -\ln\left(Ce^{-2t} - \dfrac{1}{2}\right)$

10. We have

    $$\dfrac{dy}{dx} = y^2(1-y)$$
    $$\int \dfrac{dy}{y^2(1-y)} = \int dx = x + K.$$

    Expand the left side in partial fractions:

    $$\dfrac{1}{y^2(1-y)} = \dfrac{A}{y} + \dfrac{B}{y^2} + \dfrac{C}{1-y}$$
    $$= \dfrac{A(y-y^2) + B(1-y) + Cy^2}{y^2(1-y)}$$

    $\Rightarrow \begin{cases} -A + C = 0; \\ A - B = 0; \\ B = 1. \end{cases} \Rightarrow A = B = C = 1.$

    Hence,

    $$\int \dfrac{dy}{y^2(1-y)} = \int \left(\dfrac{1}{y} + \dfrac{1}{y^2} + \dfrac{1}{1-y}\right) dy$$
    $$= \ln|y| - \dfrac{1}{y} - \ln|1-y|.$$

Therefore,
$$\ln\left|\frac{y}{1-y}\right| - \frac{1}{y} = x + K.$$

11. $\dfrac{dy}{dx} - \dfrac{2}{x}y = x^2$ (linear)

$$\mu = \exp\left(\int -\frac{2}{x}\,dx\right) = \frac{1}{x^2}$$

$$\frac{1}{x^2}\frac{dy}{dx} - \frac{2}{x^3}y = 1$$

$$\frac{d}{dx}\frac{y}{x^2} = 1$$

$$\frac{y}{x^2} = x + C, \quad \text{so} \quad y = x^3 + Cx^2$$

12. We have $\dfrac{dy}{dx} + \dfrac{2y}{x} = \dfrac{1}{x^2}$. Let

$$\mu = \int \frac{2}{x}\,dx = 2\ln x = \ln x^2, \text{ then } e^\mu = x^2, \text{ and}$$

$$\frac{d}{dx}(x^2 y) = x^2 \frac{dy}{dx} + 2xy$$

$$= x^2\left(\frac{dy}{dx} + \frac{2y}{x}\right) = x^2\left(\frac{1}{x^2}\right) = 1$$

$$\Rightarrow x^2 y = \int dx = x + C$$

$$\Rightarrow y = \frac{1}{x} + \frac{C}{x^2}.$$

13. $\dfrac{dy}{dx} + 2y = 3 \qquad \mu = \exp\left(\int 2\,dx\right) = e^{2x}$

$$\frac{d}{dx}(e^{2x} y) = e^{2x}(y' + 2y) = 3e^{2x}$$

$$e^{2x} y = \frac{3}{2}e^{2x} + C \Rightarrow y = \frac{3}{2} + Ce^{-2x}$$

14. We have $\dfrac{dy}{dx} + y = e^x$. Let $\mu = \int dx = x$, then $e^\mu = e^x$, and

$$\frac{d}{dx}(e^x y) = e^x \frac{dy}{dx} + e^x y = e^x\left(\frac{dy}{dx} + y\right) = e^{2x}$$

$$\Rightarrow e^x y = \int e^{2x}\,dx = \frac{1}{2}e^{2x} + C.$$

Hence, $y = \dfrac{1}{2}e^x + Ce^{-x}$.

15. $\dfrac{dy}{dx} + y = x \qquad \mu = \exp\left(\int 1\,dx\right) = e^x$

$$\frac{d}{dx}(e^x y) = e^x(y' + y) = xe^x$$

$$e^x y = \int xe^x\,dx = xe^x - e^x + C$$

$$y = x - 1 + Ce^{-x}$$

16. We have $\dfrac{dy}{dx} + 2e^x y = e^x$. Let $\mu = \int 2e^x\,dx = 2e^x$, then

$$\frac{d}{dx}(e^{2e^x} y) = e^{2e^x}\frac{dy}{dx} + 2e^x e^{2e^x} y$$

$$= e^{2e^x}\left(\frac{dy}{dx} + 2e^x y\right) = e^{2e^x} e^x.$$

Therefore,

$$e^{2e^x} y = \int e^{2e^x} e^x\,dx \quad \text{Let } u = 2e^x$$

$$\qquad\qquad du = 2e^x\,dx$$

$$= \frac{1}{2}\int e^u\,du = \frac{1}{2}e^{2e^x} + C.$$

Hence, $y = \dfrac{1}{2} + Ce^{-2e^x}$.

17. $y(x) = 2 + \displaystyle\int_0^x \frac{t}{y(t)}\,dt \Longrightarrow y(0) = 2$

$$\frac{dy}{dx} = \frac{x}{y}, \quad \text{i.e. } y\,dy = x\,dx$$

$$y^2 = x^2 + C$$

$$2^2 = 0^2 + C \Longrightarrow C = 4$$

$$y = \sqrt{4 + x^2}.$$

18. $y(x) = 1 + \displaystyle\int_0^x \frac{(y(t))^2}{1+t^2}\,dt \Longrightarrow y(0) = 1$

$$\frac{dy}{dx} = \frac{y^2}{1+x^2}, \quad \text{i.e. } dy/y^2 = dx/(1+x^2)$$

$$-\frac{1}{y} = \tan^{-1} x + C$$

$$-1 = 0 + C \Longrightarrow C = -1$$

$$y = 1/(1 - \tan^{-1} x).$$

19. $y(x) = 1 + \displaystyle\int_1^x \frac{y(t)}{t(t+1)}\,dt \Longrightarrow y(1) = 1$

$$\frac{dy}{dx} = \frac{y}{x(x+1)}, \quad \text{for } x > 0$$

$$\frac{dy}{y} = \frac{dx}{x(x+1)} = \frac{dx}{x} - \frac{dx}{x+1}$$

$$\ln y = \ln\frac{x}{x+1} + \ln C$$

$$y = \frac{Cx}{x+1}, \quad \Longrightarrow 1 = C/2$$

$$y = \frac{2x}{x+1}.$$

20. $y(x) = 3 + \displaystyle\int_0^x e^{-y}\,dt \Longrightarrow y(0) = 3$

$$\frac{dy}{dx} = e^{-y}, \quad \text{i.e. } e^y\,dy = dx$$

$$e^y = x + C \Longrightarrow y = \ln(x + C)$$

$$3 = y(0) = \ln C \Longrightarrow C = e^3$$

$$y = \ln(x + e^3).$$

21. The solution given, namely

$$x = \frac{ab\left(e^{(b-a)kt} - 1\right)}{be^{(b-a)kt} - a},$$

is indeterminate (0/0) if $a = b$.
If $a = b$ the original differential equation becomes

$$\frac{dx}{dt} = k(a-x)^2,$$

which is separable and yields the solution

$$\frac{1}{a-x} = \int \frac{dx}{(a-x)^2} = k\int dt = kt + C.$$

Since $x(0) = 0$, we have $C = \frac{1}{a}$, so $\frac{1}{a-x} = kt + \frac{1}{a}$.
Solving for x, we obtain

$$x = \frac{a^2 kt}{1 + akt}.$$

This solution also results from evaluating the limit of solution obtained for the case $a \neq b$ as b approaches a (using l'Hôpital's Rule, say).

22. Given that $m\dfrac{dv}{dt} = mg - kv$, then

$$\int \frac{dv}{g - \frac{k}{m}v} = \int dt$$

$$-\frac{m}{k} \ln\left|g - \frac{k}{m}v\right| = t + C.$$

Since $v(0) = 0$, therefore $C = -\dfrac{m}{k}\ln g$. Also, $g - \dfrac{k}{m}v$ remains positive for all $t > 0$, so

$$\frac{m}{k}\ln \frac{g}{g - \frac{k}{m}v} = t$$

$$\frac{g - \frac{k}{m}v}{g} = e^{-kt/m}$$

$$\Rightarrow \quad v = v(t) = \frac{mg}{k}\left(1 - e^{-kt/m}\right).$$

Note that $\lim_{t\to\infty} v(t) = \dfrac{mg}{k}$. This limiting velocity can be obtained directly from the differential equation by setting $\dfrac{dv}{dt} = 0$.

23. We proceed by separation of variables:

$$m\frac{dv}{dt} = mg - kv^2$$

$$\frac{dv}{dt} = g - \frac{k}{m}v^2$$

$$\frac{dv}{g - \frac{k}{m}v^2} = dt$$

$$\int \frac{dv}{\frac{mg}{k} - v^2} = \frac{k}{m}\int dt = \frac{kt}{m} + C.$$

Let $a^2 = mg/k$, where $a > 0$. Thus, we have

$$\int \frac{dv}{a^2 - v^2} = \frac{kt}{m} + C$$

$$\frac{1}{2a}\ln\left|\frac{a+v}{a-v}\right| = \frac{kt}{m} + C$$

$$\ln\left|\frac{a+v}{a-v}\right| = \frac{2akt}{m} + C_1 = 2\sqrt{\frac{kg}{m}}\,t + C_1$$

$$\frac{a+v}{a-v} = C_2 e^{2t\sqrt{kg/m}}.$$

Assuming $v(0) = 0$, we get $C_2 = 1$. Thus

$$a + v = e^{2t\sqrt{kg/m}}(a - v)$$

$$v\left(1 + e^{2t\sqrt{kg/m}}\right) = a\left(e^{2t\sqrt{kg/m}} - 1\right)$$

$$= \sqrt{\frac{mg}{k}}\left(e^{2t\sqrt{kg/m}} - 1\right)$$

$$v = \sqrt{\frac{mg}{k}}\,\frac{e^{2t\sqrt{kg/m}} - 1}{e^{2t\sqrt{kg/m}} + 1}$$

Clearly $v \to \sqrt{\dfrac{mg}{k}}$ as $t \to \infty$. This also follows from setting $\dfrac{dv}{dt} = 0$ in the given differential equation.

24. The balance in the account after t years is $y(t)$ and $y(0) = 1000$. The balance must satisfy

$$\frac{dy}{dt} = 0.1y - \frac{y^2}{1,000,000}$$

$$\frac{dy}{dt} = \frac{10^5 y - y^2}{10^6}$$

$$\int \frac{dy}{10^5 y - y^2} = \int \frac{dt}{10^6}$$

$$\frac{1}{10^5}\int \left(\frac{1}{y} + \frac{1}{10^5 - y}\right) dy = \frac{t}{10^6} - \frac{C}{10^5}$$

$$\ln|y| - \ln|10^5 - y| = \frac{t}{10} - C$$

$$\frac{10^5 - y}{y} = e^{C - (t/10)}$$

$$y = \frac{10^5}{e^{C - (t/10)} + 1}.$$

Since $y(0) = 1000$, we have

$$1000 = y(0) = \frac{10^5}{e^C + 1} \quad \Rightarrow \quad C = \ln 99,$$

and

$$y = \frac{10^5}{99e^{-t/10} + 1}.$$

The balance after 1 year is

$$y = \frac{10^5}{99e^{-1/10} + 1} \approx \$1,104.01.$$

As $t \to \infty$, the balance can grow to

$$\lim_{t \to \infty} y(t) = \lim_{t \to \infty} \frac{10^5}{e^{(4.60 - 0.1t)} + 1} = \frac{10^5}{0 + 1} = \$100,000.$$

For the account to grow to \$50,000, t must satisfy

$$50,000 = y(t) = \frac{100,000}{99e^{-t/10} + 1}$$
$$\Rightarrow \quad 99e^{-t/10} + 1 = 2$$
$$\Rightarrow \quad t = 10\ln 99 \approx 46 \text{ years}.$$

25. The hyperbolas $xy = C$ satisfy the differential equation

$$y + x\frac{dy}{dx} = 0, \quad \text{or} \quad \frac{dy}{dx} = -\frac{y}{x}.$$

Curves that intersect these hyperbolas at right angles must therefore satisfy $\frac{dy}{dx} = \frac{x}{y}$, or $x\,dx = y\,dy$, a separated equation with solutions $x^2 - y^2 = C$, which is also a family of rectangular hyperbolas. (Both families are degenerate at the origin for $C = 0$.)

26. Let $x(t)$ be the number of kg of salt in the solution in the tank after t minutes. Thus, $x(0) = 50$. Salt is coming into the tank at a rate of 10 g/L × 12 L/min = 0.12 kg/min. Since the contents flow out at a rate of 10 L/min, the volume of the solution is increasing at 2 L/min and thus, at any time t, the volume of the solution is $1000 + 2t$ L. Therefore the concentration of salt is $\frac{x(t)}{1000 + 2t}$ L. Hence, salt is being removed at a rate

$$\frac{x(t)}{1000 + 2t} \text{ kg/L} \times 10 \text{ L/min} = \frac{5x(t)}{500 + t} \text{ kg/min}.$$

Therefore,

$$\frac{dx}{dt} = 0.12 - \frac{5x}{500 + t}$$
$$\frac{dx}{dt} + \frac{5}{500 + t}x = 0.12.$$

Let $\mu = \int \frac{5}{500 + t}\,dt = 5\ln|500 + t| = \ln(500 + t)^5$ for $t > 0$. Then $e^\mu = (500 + t)^5$, and

$$\frac{d}{dt}\left[(500+t)^5 x\right] = (500+t)^5 \frac{dx}{dy} + 5(500+t)^4 x$$
$$= (500+t)^5\left(\frac{dx}{dy} + \frac{5x}{500 + t}\right)$$
$$= 0.12(500 + t)^5.$$

Hence,

$$(500+t)^5 x = 0.12\int (500+t)^5\,dt = 0.02(500+t)^6 + C$$
$$\Rightarrow x = 0.02(500 + t) + C(500 + t)^{-5}.$$

Since $x(0) = 50$, we have $C = 1.25 \times 10^{15}$ and

$$x = 0.02(500 + t) + (1.25 \times 10^{15})(500 + t)^{-5}.$$

After 40 min, there will be

$$x = 0.02(540) + (1.25 \times 10^{15})(540)^{-5} = 38.023 \text{ kg}$$

of salt in the tank.

Review Exercises 7 (page 473)

1.

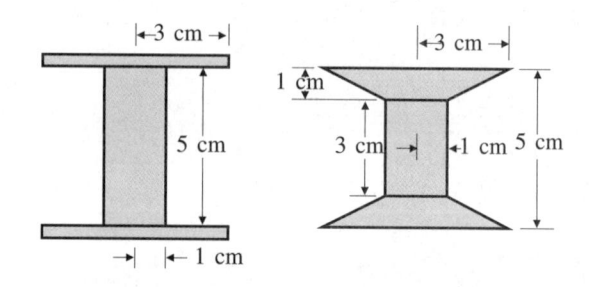

Fig. R-7.1

The volume of thread that can be wound on the left spool is $\pi(3^2 - 1^2)(5) = 40\pi$ cm^3.
The height of the winding region of the right spool at distance r from the central axis of the spool is of the form $h = A + Br$. Since $h = 3$ if $r = 1$, and $h = 5$ if $r = 3$, we have $A = 2$ and $B = 1$, so $h = 2 + r$. The volume of thread that can be wound on the right spool is

$$2\pi \int_1^3 r(2+r)\,dr = 2\pi\left(r^2 + \frac{r^3}{3}\right)\Big|_1^3 = \frac{100\pi}{3} \text{ cm}^3.$$

The right spool will hold $\frac{100}{3 \times 40}(1,000) = 833.33$ m of thread.

INSTRUCTOR'S SOLUTIONS MANUAL   REVIEW EXERCISES 7   (PAGE 473)

2. Let $A(y)$ be the cross-sectional area of the bowl at height y above the bottom. When the depth of water in the bowl is Y, then the volume of water in the bowl is

$$V(Y) = \int_0^Y A(y)\,dy.$$

The water evaporates at a rate proportional to exposed surface area. Thus

$$\frac{dV}{dt} = kA(Y)$$

$$\frac{dV}{dY}\frac{dY}{dt} = kA(Y)$$

$$A(Y)\frac{dY}{dt} = kA(Y).$$

Hence $dY/dt = k$; the depth decreases at a constant rate.

3. The barrel is generated by revolving $x = a - by^2$, $(-2 \le y \le 2)$, about the y-axis. Since the top and bottom disks have radius 1 ft, we have $a - 4b = 1$. The volume of the barrel is

$$V = 2\int_0^2 \pi(a - by^2)^2\,dy$$

$$= 2\pi\left(a^2 y - \frac{2aby^3}{3} + \frac{b^2 y^5}{5}\right)\bigg|_0^2$$

$$== 2\pi\left(2a^2 - \frac{16}{3}ab + \frac{32}{5}b^2\right).$$

Since $V = 16$ and $a = 1 + 4b$, we have

$$2\pi\left(2(1+4b)^2 - \frac{16}{3}b(1+4b) + \frac{32}{5}b^2\right) = 16$$

$$128b^2 + 80b + 15 - \frac{60}{\pi} = 0.$$

Solving this quadratic gives two solutions, $b \approx 0.0476$ and $b \approx -0.6426$. Since the second of these leads to an unacceptable negative value for a, we must have $b \approx 0.0476$, and so $a = 1 + 4b \approx 1.1904$.

4. A vertical slice parallel to the top ridge of the solid at distance x to the right of the centre is a rectangle of base $2\sqrt{100 - x^2}$ cm^and height $\sqrt{3}(10 - x)$ cm. Thus the solid has volume

$$V = 2\int_0^{10} \sqrt{3}(10 - x)2\sqrt{100 - x^2}\,dx$$

$$= 40\sqrt{3}\int_0^{10}\sqrt{100 - x^2}\,dx - 4\sqrt{3}\int_0^{10} x\sqrt{100 - x^2}\,dx$$

Let $u = 100 - x^2$
$du = -2x\,dx$

$$= 40\sqrt{3}\frac{100\pi}{4} - 2\sqrt{3}\int_0^{100}\sqrt{u}\,du$$

$$= 1{,}000\sqrt{3}\left(\pi - \frac{4}{3}\right)\text{ cm}^3.$$

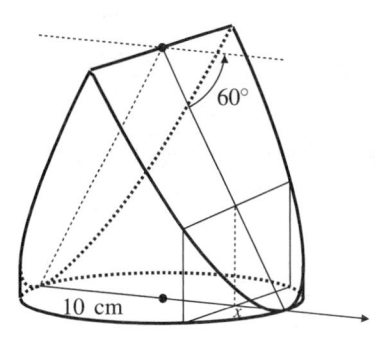

Fig. R-7.4

5. The arc length of $y = \dfrac{1}{a}\cosh(ax)$ from $x = 0$ to $x = 1$ is

$$s = \int_0^1 \sqrt{1 + \sinh^2(ax)}\,dx = \int_0^1 \cosh(ax)\,dx$$

$$= \frac{1}{a}\sinh(ax)\bigg|_0^1 = \frac{1}{a}\sinh a.$$

We want $\dfrac{1}{a}\sinh a = 2$, that is, $\sinh a = 2a$. Solving this by Newton's Method or a calculator solve function, we get $a \approx 2.1773$.

6. The area of revolution of $y = \sqrt{x}$, $(0 \le x \le 6)$, about the x-axis is

$$S = 2\pi\int_0^6 y\sqrt{1 + \left(\frac{dy}{dx}\right)^2}\,dx$$

$$= 2\pi\int_0^6 \sqrt{x}\sqrt{1 + \frac{1}{4x}}\,dx$$

$$= 2\pi\int_0^6 \sqrt{x + \frac{1}{4}}\,dx$$

$$= \frac{4\pi}{3}\left(x + \frac{1}{4}\right)^{3/2}\bigg|_0^6 = \frac{4\pi}{3}\left[\frac{125}{8} - \frac{1}{8}\right] = \frac{62\pi}{3}\text{ sq. units.}$$

7. The region is a quarter-elliptic disk with semi-axes $a = 2$ and $b = 1$. The area of the region is $A = \pi ab/4 = \pi/2$. The moments about the coordinate axes are

$$M_{x=0} = \int_0^2 x\sqrt{1 - \frac{x^2}{4}}\, dx \quad \text{Let } u = 1 - \frac{x^2}{4}$$
$$du = -\frac{x}{2}\, dx$$
$$= 2\int_0^1 \sqrt{u}\, du = \frac{4}{3}$$
$$M_{y=0} = \frac{1}{2}\int_0^2 \left(1 - \frac{x^2}{4}\right) dx$$
$$= \frac{1}{2}\left(x - \frac{x^3}{12}\right)\Big|_0^2 = \frac{2}{3}.$$

Thus $\bar{x} = M_{x=0}/A = 8/(3\pi)$ and $\bar{y} = M_{y=0}/A = 4/(3\pi)$. The centroid is $\big(8/(3\pi), 4/(3\pi)\big)$.

8.

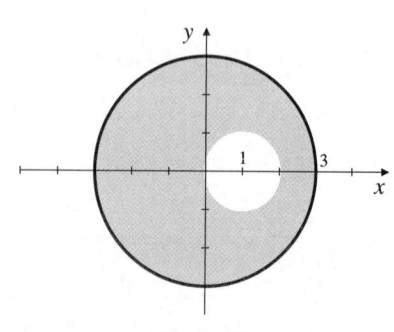

Fig. R-7.8

Let the disk have centre (and therefore centroid) at $(0, 0)$. Its area is 9π. Let the hole have centre (and therefore centroid) at $(1, 0)$. Its area is π. The remaining part has area 8π and centroid at $(\bar{x}, 0)$, where

$$(9\pi)(0) = (8\pi)\bar{x} + (\pi)(1).$$

Thus $\bar{x} = -1/8$. The centroid of the remaining part is 1/8 ft from the centre of the disk on the side opposite the hole.

9. Let the area of cross-section of the cylinder be A. When the piston is y cm above the base, the volume of gas in the cylinder is $V = Ay$, and its pressure $P(y)$ satisfies $P(y)V = k$ (constant). The force exerted by the piston is

$$F(y) = P(y)A = \frac{kA}{Ay} = \frac{k}{y}.$$

We are told that $F = 1,000$ N when $y = 20$ cm. Thus $k = 20,000$ N·cm. The work done by the piston as it descends to 5 cm is

$$W = \int_5^{20} \frac{20,000}{y}\, dy = 20,000 \ln \frac{20}{5} \approx 27,726 \text{ N·cm}.$$

10. We are told that for any $a > 0$,

$$\pi \int_0^a \left[\big(f(x)\big)^2 - \big(g(x)\big)^2\right] dx = 2\pi \int_0^a x\big[f(x) - g(x)\big] dx.$$

Differentiating both sides of this equation with respect to a, we get

$$\big(f(a)\big)^2 - \big(g(a)\big)^2 = 2a\big[f(a) - g(a)\big],$$

or, equivalently, $f(a) + g(a) = 2a$. Thus f and g must satisfy

$$f(x) + g(x) = 2x \quad \text{for every } x > 0.$$

11. $\dfrac{dy}{dx} = \dfrac{3y}{x-1} \Rightarrow \displaystyle\int \dfrac{dy}{y} = 3\int \dfrac{dx}{x-1}$
$\Rightarrow \ln|y| = \ln|x-1|^3 + \ln|C|$
$\Rightarrow y = C(x-1)^3.$
Since $y = 4$ when $x = 2$, we have $4 = C(2-1)^3 = C$, so the equation of the curve is $y = 4(x-1)^3$.

12. The ellipses $3x^2 + 4y^2 = C$ all satisfy the differential equation

$$6x + 8y \frac{dy}{dx} = 0, \quad \text{or} \quad \frac{dy}{dx} = -\frac{3x}{4y}.$$

A family of curves that intersect these ellipses at right angles must therefore have slopes given by $\dfrac{dy}{dx} = \dfrac{4y}{3x}$. Thus

$$3\int \frac{dy}{y} = 4\int \frac{dx}{x}$$
$$3\ln|y| = 4\ln|x| + \ln|C|.$$

The family is given by $y^3 = Cx^4$.

13. The original \$8,000 grows to \$8,000$e^{0.08}$ in two years. Between t and $t + dt$, an amount $\$10,000\sin(2\pi t)\, dt$ comes in, and this grows to $\$10,000\sin(2\pi t)e^{0.04(2-t)}\, dt$ by the end of two years. Thus the amount in the account after 2 years is

$$8,000 e^{0.08} + 10,000 \int_0^2 \sin(2\pi t)e^{0.04(2-t)}\, dt \approx \$8,798.85.$$

(We omit the details of evaluation of the integral, which is done by the method of Example 4 of Section 7.1.)

Challenging Problems 7 (page 473)

1. a) The nth bead extends from $x = (n-1)\pi$ to $x = n\pi$, and has volume

$$V_n = \pi \int_{(n-1)\pi}^{n\pi} e^{-2kx} \sin^2 x \, dx$$

$$= \frac{\pi}{2} \int_{(n-1)\pi}^{n\pi} e^{-2kx}(1 - \cos(2x)) \, dx$$

Let $x = u + (n-1)\pi$
$dx = du$

$$= \frac{\pi}{2} \int_0^{\pi} e^{-2ku} e^{-2k(n-1)\pi} \left[1 - \cos(2u + 2(n-1)\pi)\right] du$$

$$= \frac{\pi}{2} e^{-2k(n-1)\pi} \int_0^{\pi} e^{-2ku}(1 - \cos(2u)) \, du$$

$$= e^{-2k(n-1)\pi} V_1.$$

Thus $\dfrac{V_{n+1}}{V_n} = \dfrac{e^{-2kn\pi} V_1}{e^{-2k(n-1)\pi} V_1} = e^{-2k\pi}$, which depends on k but not n.

b) $V_{n+1}/V_n = 1/2$ if $-2k\pi = \ln(1/2) = -\ln 2$, that is, if $k = (\ln 2)/(2\pi)$.

c) Using the result of Example 4 in Section 7.1, we calculate the volume of the first bead:

$$V_1 = \frac{\pi}{2} \int_0^{\pi} e^{-2kx}(1 - \cos(2x)) \, dx$$

$$= \frac{\pi e^{-2kx}}{-4k}\bigg|_0^{\pi} - \frac{\pi}{2} \cdot \frac{e^{-2kx}(2\sin(2x) - 2k\cos(2x))}{4(1+k^2)}\bigg|_0^{\pi}$$

$$= \frac{\pi}{4k}(1 - e^{-2k\pi}) - \frac{\pi}{4(1+k^2)}(k - ke^{-2k\pi})$$

$$= \frac{\pi}{4k(1+k^2)}(1 - e^{-2k\pi}).$$

By part (a) and Theorem 1(d) of Section 6.1, the sum of the volumes of the first n beads is

$$S_n = \frac{\pi}{4k(1+k^2)}(1 - e^{-2k\pi})$$
$$\times \left[1 + e^{-2k\pi} + \left(e^{-2k\pi}\right)^2 + \cdots + \left(e^{-2k\pi}\right)^{n-1}\right]$$

$$= \frac{\pi}{4k(1+k^2)}(1 - e^{-2k\pi}) \frac{1 - e^{-2kn\pi}}{1 - e^{-2k\pi}}$$

$$= \frac{\pi}{4k(1+k^2)}(1 - e^{-2kn\pi}).$$

Thus the total volume of all the beads is

$$V = \lim_{n \to \infty} S_n = \frac{\pi}{4k(1+k^2)} \text{ cu. units..}$$

2.

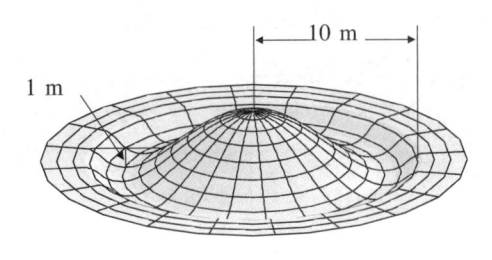

Fig. C-7.2

$h(r) = a(r^2 - 100)(r^2 - k^2)$, where $0 < k < 10$
$h'(r) = 2ar(r^2 - k^2) + 2ar(r^2 - 100) = 2ar(2r^2 - 100 - k^2)$

The deepest point occurs where $2r^2 = 100 + k^2$, i.e., $r^2 = 50 + (k^2/2)$. Since this depth must be 1 m, we require

$$a\left(\frac{k^2}{2} - 50\right)\left(50 - \frac{k^2}{2}\right) = -1,$$

or, equivalently, $a(100 - k^2)^2 = 4$. The volume of the pool is

$$V_P = 2\pi a \int_k^{10} r(100 - r^2)(r^2 - k^2) \, dr$$

$$= 2\pi a \left(\frac{250,000}{3} - 2,500k^2 + 25k^4 - \frac{1}{12}k^6\right).$$

The volume of the hill is

$$V_H = 2\pi a \int_0^k r(r^2 - 100)(r^2 - k^2) \, dr = 2\pi a \left(25k^4 - \frac{1}{12}k^6\right).$$

These two volumes must be equal, so $k^2 = 100/3$ and $k \approx 5.77$ m. Thus $a = 4/(100 - k^2)^2 = 0.0009$. The volume of earth to be moved is V_H with these values of a and k, namely

$$2\pi(0.0009)\left[25\left(\frac{100}{3}\right)^2 - \frac{1}{12}\left(\frac{100}{3}\right)^4\right] \approx 140 \text{ m}^3.$$

3.

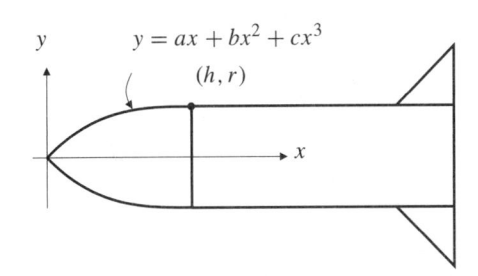

Fig. C-7.3

$f(x) = ax + bx^2 + cx^3$ must satisfy $f(h) = r$, $f'(h) = 0$, and $f'(x) > 0$ for $0 < x < h$. The first two conditions require that
$$ah + bh^2 + ch^3 = r$$
$$a + 2bh + 3ch^2 = 0,$$
from which we obtain by solving for b and c,
$$b = \frac{3r - 2ah}{h^2}, \quad c = \frac{ah - 2r}{h^3}.$$

The volume of the nose cone is then
$$V(a) = \pi \int_0^h \big(f(x)\big)^2 dx = \frac{\pi h}{210}(13ahr + 78r^2 + 2a^2h^2).$$

Solving $dV/da = 0$ gives only one critical point, $a = -13r/(4h)$. This is unacceptable, because the condition $f'(x) > 0$ on $(0, h)$ forces us to require $a \geq 0$. In fact
$$f'(x) = a + \frac{2(3r - 2ah)}{h^2}x + \frac{3(ah - 2r)}{h^3}x^2$$
is clearly positive for small x if $a > 0$. Its two roots are $x_1 = h$ and $x_2 = h^2 a/(3ah - 6r)$. a must be restricted so that x_2 is not in the interval $(0, h)$. If $a < 2r/h$, then $x_2 < 0$. If $2r/h < a < 3r/h$, then $x_2 > h$. If $a > 3r/h$, then $0 < x_2 < h$. Hence the interval of acceptable values of a is $0 \leq a \leq 3r/h$. We have
$$V(0) = \frac{13\pi r^2 h}{35}, \quad V\left(\frac{3r}{h}\right) = \frac{9\pi r^2 h}{14}.$$

The largest volume corresponds to $a = 3r/h$, which is the largest allowed value for a and so corresponds to the bluntest possible nose. The corresponding cubic $f(x)$ is
$$f(x) = \frac{r}{h^3}(3h^2 x - 3hx^2 + x^3).$$

4. a) If $f(x) = \begin{cases} a + bx + cx^2 & \text{for } 0 \leq x \leq 1 \\ p + qx + rx^2 & \text{for } 1 \leq x \leq 3 \end{cases}$, then $f'(x) = \begin{cases} b + 2cx & \text{for } 0 < x < 1 \\ q + 2rx & \text{for } 1 < x < 3 \end{cases}$. We require that

$$a = 1 \qquad p + 3q + 9r = 0$$
$$a + b + c = 2 \qquad p + q + r = 2$$
$$b + 2c = m \qquad q + 2r = m.$$

The solutions of these systems are $a = 1$, $b = 2 - m$, $c = m - 1$, $p = \frac{3}{2}(1 - m)$, $q = 2m + 1$, and $r = -\frac{1}{2}(1 + m)$. $f(x, m)$ is $f(x)$ with these values of the six constants.

b) The length of the spline is
$$L(m) = \int_0^1 \sqrt{1 + (b + 2cx)^2}\, dx + \int_1^3 \sqrt{1 + (q + 2rx)^2}\, dx$$
with the values of b, c, q, and r determined above. A plot of the graph of $L(m)$ reveals a minimum value in the neighbourhood of $m = -0.3$. The derivative of $L(m)$ is a horrible expression, but Mathematica determined its zero to be about $m = -0.281326$, and the corresponding minimum value of L is about 4.41748. The polygonal line ABC has length $3\sqrt{2} \approx 4.24264$, which is only slightly shorter.

5. Starting with $V_1(r) = 2r$, and using repeatedly the formula
$$V_n(r) = \int_{-r}^r V_{n-1}(\sqrt{r^2 - x^2})\, dx,$$
Maple gave the following results:

$$V_1(r) = 2r \qquad V_2(r) = \pi r^2$$
$$V_3(r) = \frac{4}{3}\pi r^3 \qquad V_4(r) = \frac{1}{2}\pi^2 r^4$$
$$V_5(r) = \frac{8}{15}\pi^2 r^5 \qquad V_6(r) = \frac{1}{6}\pi^3 r^6$$
$$V_7(r) = \frac{16}{105}\pi^3 r^7 \qquad V_8(r) = \frac{1}{24}\pi^4 r^8$$
$$V_9(r) = \frac{32}{945}\pi^4 r^9 \qquad V_{10}(r) = \frac{1}{120}\pi^5 r^{10}$$

It appears that
$$V_{2n}(r) = \frac{1}{n!}\pi^n r^{2n}, \quad \text{and}$$
$$V_{2n-1}(r) = \frac{2^n}{1 \cdot 3 \cdot 5 \cdots (2n-1)}\pi^{n-1} r^{2n-1}$$
$$= \frac{2^{2n-1}(n-1)!}{(2n-1)!}\pi^{n-1} r^{2n-1}.$$

These formulas predict that
$$V_{11}(r) = \frac{2^{11} 5!}{11!}\pi^5 r^{11} \quad \text{and} \quad V_{12}(r) = \frac{1}{6!}\pi^6 r^{12},$$
both of which Maple is happy to confirm.

6. With y and θ as defined in the statement of the problem, we have
$$0 \leq y \leq 10 \quad \text{and} \quad 0 \leq \theta < \pi.$$

The needle crosses a line if $y < 5\sin\theta$. The probability of this happening is the ratio of the area under the curve to the area of the rectangle in the figure, that is,
$$\Pr = \frac{1}{10\pi}\int_0^\pi 5\sin\theta\, d\theta = \frac{1}{\pi}.$$

INSTRUCTOR'S SOLUTIONS MANUAL   CHALLENGING PROBLEMS 7   (PAGE 473)

Fig. C-7.6

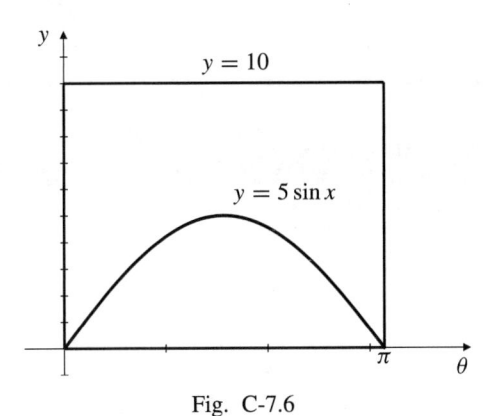

7.

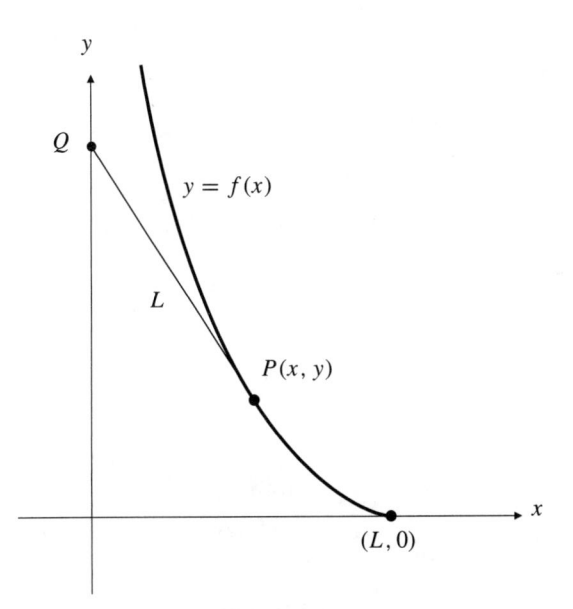

Fig. C-7.7

If $Q = (0, Y)$, then the slope of PQ is

$$\frac{y - Y}{x - 0} = f'(x) = \frac{dy}{dx}.$$

Since $|PQ| = L$, we have $(y - Y)^2 = L^2 - x^2$. Since the slope dy/dx is negative at P, $dy/dx = -\sqrt{L^2 - x^2}/x$. Thus

$$y = -\int \frac{\sqrt{L^2 - x^2}}{x}\,dx = L\ln\left(\frac{L + \sqrt{L^2 - x^2}}{x}\right) - \sqrt{L^2 - x^2} + C.$$

Since $y = 0$ when $x = L$, we have $C = 0$ and the equation of the tractrix is

$$y = L\ln\left(\frac{L + \sqrt{L^2 - x^2}}{x}\right) - \sqrt{L^2 - x^2}.$$

Note that the first term can be written in an alternate way:

$$y = L\ln\left(\frac{x}{L - \sqrt{L^2 - x^2}}\right) - \sqrt{L^2 - x^2}.$$

8.   a) $S(a, a, c)$ is the area of the surface obtained by rotating the ellipse $(x^2/a^2) + (y^2/c^2) = 1$ (where $a > c$) about the y-axis. Since $y' = -cx/(a\sqrt{a^2 - x^2})$, we have

$$S(a, a, c) = 2 \times 2\pi \int_0^a x\sqrt{1 + \frac{c^2 x^2}{a^2(a^2 - x^2)}}\,dx$$

$$= \frac{4\pi}{a}\int_0^a x\frac{\sqrt{a^4 - (a^2 - c^2)x^2}}{\sqrt{a^2 - x^2}}\,dx$$

Let $x = a\sin u$
$dx = a\cos u\,du$

$$= \frac{4\pi}{a}\int_0^{\pi/2} a\sin u\sqrt{a^4 - (a^2 - c^2)a^2\sin^2 u}\,du$$

$$= 4\pi a\int_0^{\pi/2} \sin u\sqrt{a^2 - (a^2 - c^2)(1 - \cos^2 u)}\,du$$

Let $v = \cos u$
$dv = -\sin u\,du$

$$= 4\pi a\int_0^1 \sqrt{c^2 + (a^2 - c^2)v^2}\,dv.$$

This integral can now be handled using tables or computer algebra. It evaluates to

$$S(a, a, c) = 2\pi a^2 + \frac{2\pi a c^2}{\sqrt{a^2 - c^2}}\ln\left(\frac{a + \sqrt{a^2 - c^2}}{c}\right).$$

b) $S(a, c, c)$ is the area of the surface obtained by rotating the ellipse of part (a) about the y-axis. Since $y' = -cx/(a\sqrt{a^2 - x^2})$, we have

$$S(a, c, c) = 2 \times 2\pi\int_0^a y\sqrt{1 + \frac{c^2 x^2}{a^2(a^2 - x^2)}}\,dx$$

$$= \frac{4\pi c}{a^2}\int_0^a \sqrt{a^2 - x^2}\frac{\sqrt{a^4 - (a^2 - c^2)x^2}}{\sqrt{a^2 - x^2}}\,dx$$

$$= \frac{4\pi c}{a^2}\int_0^a \sqrt{a^4 - (a^2 - c^2)x^2}\,dx$$

$$= 4\pi c\int_0^a \sqrt{1 - \frac{a^2 - c^2}{a^4}x^2}\,dx$$

$$= 2\pi c^2 + \frac{2\pi a^2 c}{\sqrt{a^2 - c^2}}\cos^{-1}\frac{c}{a}.$$

301

c) Since $b = \left(\dfrac{b-c}{a-c}\right)a + \left(\dfrac{a-b}{a-c}\right)c$, we use

$$S(a,b,c) \approx \left(\dfrac{b-c}{a-c}\right)S(a,a,c) + \left(\dfrac{a-b}{a-c}\right)S(a,c,c).$$

d) We cannot evaluate $S(3, 2, 1)$ even numerically at this stage. The double integral necessary to calculate it is not treated until a later chapter. (The value is approximately 48.882 sq. units.) However, using the formulas obtained above,

$$\begin{aligned}
S(3,2,1) &\approx \dfrac{S(3,3,1) + S(3,1,1)}{2} \\
&= \dfrac{1}{2}\left(18\pi + \dfrac{6\pi}{\sqrt{8}}\ln(3+\sqrt{8}) + 2\pi + \dfrac{18\pi}{\sqrt{8}}\cos^{-1}(1/3)\right) \\
&\approx 49.595 \text{ sq. units.}
\end{aligned}$$

INSTRUCTOR'S SOLUTIONS MANUAL   SECTION 8.1 (PAGE 487)

CHAPTER 8. CONICS, PARAMETRIC CURVES, AND POLAR CURVES

Section 8.1 Conics (page 487)

1. The ellipse with foci $(0, \pm 2)$ has major axis along the y-axis and $c = 2$. If $a = 3$, then $b^2 = 9 - 4 = 5$. The ellipse has equation
$$\frac{x^2}{5} + \frac{y^2}{9} = 1.$$

2. The ellipse with foci $(0, 1)$ and $(4, 1)$ has $c = 2$, centre $(2, 1)$, and major axis along $y = 1$. If $\epsilon = 1/2$, then $a = c/\epsilon = 4$ and $b^2 = 16 - 4 = 12$. The ellipse has equation
$$\frac{(x-2)^2}{16} + \frac{(y-1)^2}{12} = 1.$$

3. A parabola with focus $(2, 3)$ and vertex $(2, 4)$ has $a = -1$ and principal axis $x = 2$. Its equation is $(x - 2)^2 = -4(y - 4) = 16 - 4y$.

4. A parabola with focus at $(0, -1)$ and principal axis along $y = -1$ will have vertex at a point of the form $(v, -1)$. Its equation will then be of the form $(y + 1)^2 = \pm 4v(x - v)$. The origin lies on this curve if $1 = \pm 4(-v^2)$. Only the $-$ sign is possible, and in this case $v = \pm 1/2$. The possible equations for the parabola are $(y + 1)^2 = 1 \pm 2x$.

5. The hyperbola with semi-transverse axis $a = 1$ and foci $(0, \pm 2)$ has transverse axis along the y-axis, $c = 2$, and $b^2 = c^2 - a^2 = 3$. The equation is
$$y^2 - \frac{x^2}{3} = 1.$$

6. The hyperbola with foci at $(\pm 5, 1)$ and asymptotes $x = \pm(y - 1)$ is rectangular, has centre at $(0, 1)$ and has transverse axis along the line $y = 1$. Since $c = 5$ and $a = b$ (because the asymptotes are perpendicular to each other) we have $a^2 = b^2 = 25/2$. The equation of the hyperbola is
$$x^2 - (y-1)^2 = \frac{25}{2}.$$

7. If $x^2 + y^2 + 2x = -1$, then $(x + 1)^2 + y^2 = 0$. This represents the single point $(-1, 0)$.

8. If $x^2 + 4y^2 - 4y = 0$, then
$$x^2 + 4\left(y^2 - y + \frac{1}{4}\right) = 1, \quad \text{or} \quad \frac{x^2}{1} + \frac{(y - \frac{1}{2})^2}{\frac{1}{4}} = 1.$$

This represents an ellipse with centre at $\left(0, \frac{1}{2}\right)$, semi-major axis 1, semi-minor axis $\frac{1}{2}$, and foci at $\left(\pm\frac{\sqrt{3}}{2}, \frac{1}{2}\right)$.

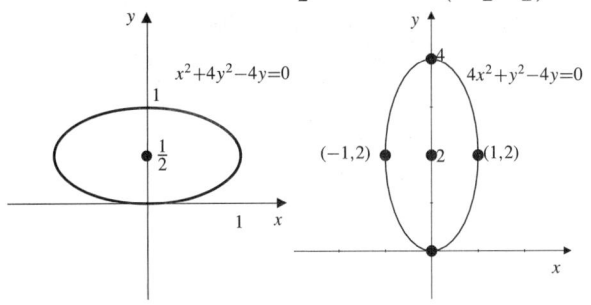

Fig. 8.1.8   Fig. 8.1.9

9. If $4x^2 + y^2 - 4y = 0$, then
$$4x^2 + y^2 - 4y + 4 = 4$$
$$4x^2 + (y - 2)^2 = 4$$
$$x^2 + \frac{(y-2)^2}{4} = 1$$

This is an ellipse with semi-axes 1 and 2, centred at $(0, 2)$.

10. If $4x^2 - y^2 - 4y = 0$, then
$$4x^2 - (y^2 + 4y + 4) = -4, \quad \text{or} \quad \frac{x^2}{1} - \frac{(y+2)^2}{4} = -1.$$

This represents a hyperbola with centre at $(0, -2)$, semi-transverse axis 2, semi-conjugate axis 1, and foci at $(0, -2 \pm \sqrt{5})$. The asymptotes are $y = \pm 2x - 2$.

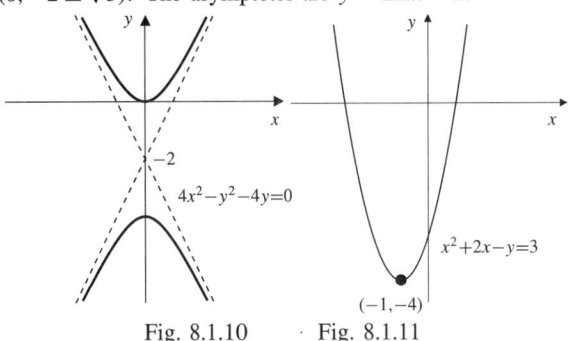

Fig. 8.1.10   Fig. 8.1.11

11. If $x^2 + 2x - y = 3$, then $(x + 1)^2 - y = 4$. Thus $y = (x + 1)^2 - 4$. This is a parabola with vertex $(-1, -4)$, opening upward.

12. If $x + 2y + 2y^2 = 1$, then
$$2\left(y^2 + y + \frac{1}{4}\right) = \frac{3}{2} - x$$
$$\Leftrightarrow x = \frac{3}{2} - 2\left(y + \frac{1}{2}\right)^2.$$

303

This represents a parabola with vertex at $(\frac{3}{2}, -\frac{1}{2})$, focus at $(\frac{11}{8}, -\frac{1}{2})$ and directrix $x = \frac{13}{8}$.

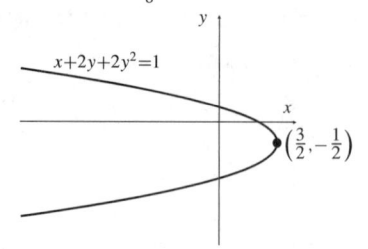

Fig. 8.1.12

13. If $x^2 - 2y^2 + 3x + 4y = 2$, then

$$\left(x + \frac{3}{2}\right)^2 - 2(y-1)^2 = \frac{9}{4}$$

$$\frac{\left(x + \frac{3}{2}\right)^2}{\frac{9}{4}} - \frac{(y-1)^2}{\frac{9}{8}} = 1$$

This is a hyperbola with centre $\left(-\frac{3}{2}, 1\right)$, and asymptotes the straight lines $2x + 3 = \pm 2\sqrt{2}(y-1)$.

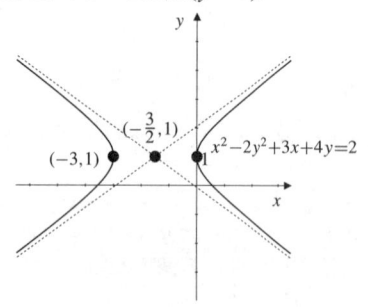

Fig. 8.1.13

14. If $9x^2 + 4y^2 - 18x + 8y = -13$, then

$$9(x^2 - 2x + 1) + 4(y^2 + 2y + 1) = 0$$
$$\Leftrightarrow 9(x-1)^2 + 4(y+1)^2 = 0.$$

This represents the single point $(1, -1)$.

15. If $9x^2 + 4y^2 - 18x + 8y = 23$, then

$$9(x^2 - 2x + 1) + 4(y^2 + 2y + 1) = 23 + 9 + 4 = 36$$
$$9(x-1)^2 + 4(y+1)^2 = 36$$
$$\frac{(x-1)^2}{4} + \frac{(y+1)^2}{9} = 1.$$

This is an ellipse with centre $(1, -1)$, and semi-axes 2 and 3.

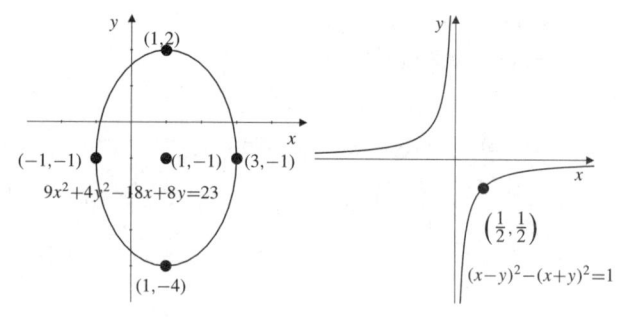

Fig. 8.1.15   Fig. 8.1.16

16. The equation $(x-y)^2 - (x+y)^2 = 1$ simplifies to $4xy = -1$ and hence represents a rectangular hyperbola with centre at the origin, asymptotes along the coordinate axes, transverse axis along $y = -x$, conjugate axis along $y = x$, vertices at $\left(\frac{1}{2}, -\frac{1}{2}\right)$ and $\left(-\frac{1}{2}, \frac{1}{2}\right)$, semi-transverse and semi-conjugate axes equal to $1/\sqrt{2}$, semi-focal separation equal to $\sqrt{\frac{1}{2} + \frac{1}{2}} = 1$, and hence foci at the points $\left(\frac{1}{\sqrt{2}}, -\frac{1}{\sqrt{2}}\right)$ and $\left(-\frac{1}{\sqrt{2}}, \frac{1}{\sqrt{2}}\right)$. The eccentricity is $\sqrt{2}$.

17. The parabola has focus at $(3, 4)$ and principal axis along $y = 4$. The vertex must be at a point of the form $(v, 4)$, in which case $a = \pm(3 - v)$ and the equation of the parabola must be of the form

$$(y - 4)^2 = \pm 4(3 - v)(x - v).$$

This curve passes through the origin if $16 = \pm 4(v^2 - 3v)$. We have two possible equations for v: $v^2 - 3v - 4 = 0$ and $v^2 - 3v + 4 = 0$. The first of these has solutions $v = -1$ or $v = 4$. The second has no real solutions. The two possible equations for the parabola are

$$(y - 4)^2 = 4(4)(x + 1) \quad \text{or} \quad y^2 - 8y = 16x$$
$$(y - 4)^2 = 4(-1)(x - 4) \quad \text{or} \quad y^2 - 8y = -4x$$

18. The foci of the ellipse are $(0, 0)$ and $(3, 0)$, so the centre is $(3/2, 0)$ and $c = 3/2$. The semi-axes a and b must satisfy $a^2 - b^2 = 9/4$. Thus the possible equations of the ellipse are

$$\frac{(x - (3/2))^2}{(9/4) + b^2} + \frac{y^2}{b^2} = 1.$$

19. For $xy + x - y = 2$ we have $A = C = 0$, $B = 1$. We therefore rotate the coordinate axes (see text pages 407–408) through angle $\theta = \pi/4$.
(Thus $\cot 2\theta = 0 = (A - C)/B$.) The transformation is

$$x = \frac{1}{\sqrt{2}}(u - v), \qquad y = \frac{1}{\sqrt{2}}(u + v).$$

INSTRUCTOR'S SOLUTIONS MANUAL SECTION 8.1 (PAGE 487)

The given equation becomes

$$\frac{1}{2}(u^2 - v^2) + \frac{1}{\sqrt{2}}(u - v) - \frac{1}{\sqrt{2}}(u + v) = 2$$
$$u^2 - v^2 - 2\sqrt{2}v = 4$$
$$u^2 - (v + \sqrt{2})^2 = 2$$
$$\frac{u^2}{2} - \frac{(v + \sqrt{2})^2}{2} = 1.$$

This is a rectangular hyperbola with centre $(0, -\sqrt{2})$, semi-axes $a = b = \sqrt{2}$, and eccentricity $\sqrt{2}$. The semi-focal separation is 2; the foci are at $(\pm 2, -\sqrt{2})$. The asymptotes are $u = \pm(v + \sqrt{2})$.
In terms of the original coordinates, the centre is $(1, -1)$, the foci are $(\pm\sqrt{2} + 1, \pm\sqrt{2} - 1)$, and the asymptotes are $x = 1$ and $y = -1$.

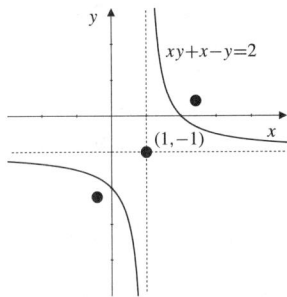

Fig. 8.1.19

20. We have $x^2 + 4xy + y^2 - 2\sqrt{2}(x - y) = 2$, then $A = 1$, $B = 4$, $C = 1$, $D = -2\sqrt{2}$, $E = 2\sqrt{2}$ and $F = -2$. We rotate the axes through angle θ satisfying $\tan 2\theta = B/(A - C) = \infty \Rightarrow \theta = \frac{\pi}{4}$. Then $A' = 3$, $B' = 0$, $C' = -1$, $D' = 0$, $E' = 4$ and the transformed equation is

$$3u^2 - v^2 + 4v - 2 = 0$$
$$3u^2 - (v^2 - 4v + 4) = 2 - 4$$
$$\frac{u^2}{\left(\frac{2}{3}\right)} - \frac{(v - 2)^2}{2} = -1,$$

which represents a hyperbola with centre $(u, v) = (0, 2)$, semi-transverse axis $a = \sqrt{2}$, semi-conjugate axis $b = \sqrt{\frac{2}{3}}$, semi-focal separation $c = \sqrt{2 + \frac{2}{3}} = \sqrt{\frac{8}{3}}$, eccentricity $\varepsilon = c/a = \frac{2}{\sqrt{3}}$, vertices at $(u, v) = (0, 2 \pm \sqrt{2})$, foci at $(u, v) = \left(0, 2 \pm \sqrt{\frac{8}{3}}\right)$, and asymptotes

$$\frac{u}{\sqrt{\frac{2}{3}}} \pm \frac{v - 2}{\sqrt{2}} = 0 \quad \Rightarrow \quad \sqrt{3}u \pm (v - 2) = 0.$$

Since the rotation of axes is specified by

$$x = \frac{1}{\sqrt{2}}(u - v) \text{ and } y = \frac{1}{\sqrt{2}}(u + v),$$

the centre of the hyperbola, in terms of the original coordinates (x, y), is $(-\sqrt{2}, \sqrt{2})$. The transverse and conjugate axes pass through this point, inclined at an angle $\frac{\pi}{4}$ to the x- and y-axes repectively. The vertices are at $(-\sqrt{2} - 1, \sqrt{2} + 1)$ and $(-\sqrt{2} + 1, \sqrt{2} - 1)$. The foci are at $\left(-\sqrt{2} - \frac{2}{\sqrt{3}}, \sqrt{2} + \frac{2}{\sqrt{3}}\right)$ and $\left(-\sqrt{2} + \frac{2}{\sqrt{3}}, \sqrt{2} - \frac{2}{\sqrt{3}}\right)$. The asymptotes are

$$\sqrt{3}\left[\frac{1}{\sqrt{2}}(x + y)\right] \pm \left[\frac{1}{\sqrt{2}}(x + y) - 2\right] = 0$$
$$\Rightarrow \begin{cases} (\sqrt{3} - 1)x + (\sqrt{3} + 1)y = 2\sqrt{2}; \\ (\sqrt{3} + 1)x + (\sqrt{3} - 1)y = -2\sqrt{2}. \end{cases}$$

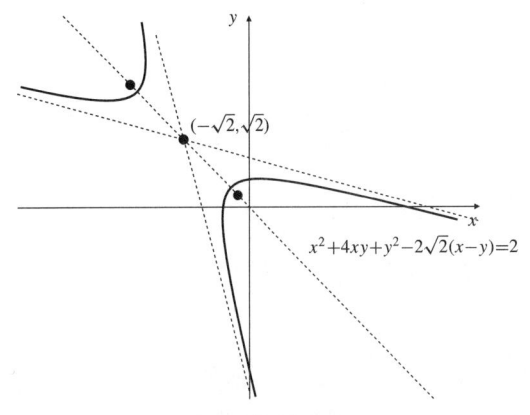

Fig. 8.1.20

21. For $x^2 + xy = \frac{1}{2\sqrt{2}}$, we have $A = B = 1$, $C = D = E = 0$, $F = -1/2\sqrt{2}$. We rotate the axes through angle θ where

$$\tan 2\theta = \frac{b}{A - C} = 1, \quad \text{i.e.,} \quad \theta = \frac{\pi}{8}.$$

Thus

$$x = \cos\frac{\pi}{8} u - \sin\frac{\pi}{8} v$$
$$y = \sin\frac{\pi}{8} u + \cos\frac{\pi}{8} v.$$

The transformed equation has coefficients (see text pages 485–486)

$$A' = \cos^2\frac{\pi}{8} + \cos\frac{\pi}{8}\sin\frac{\pi}{8} = \frac{\sqrt{2} + 2}{2\sqrt{2}}$$
$$B' = 0$$
$$C' = \sin^2\frac{\pi}{8} - \cos\frac{\pi}{8}\sin\frac{\pi}{8} = \frac{\sqrt{2} - 2}{2\sqrt{2}}$$
$$D' = E' = 0, \quad F' = F = -\frac{1}{2\sqrt{2}}.$$

305

Thus the transformed equation is

$$(2+\sqrt{2})u^2 - (2-\sqrt{2})v^2 = 1.$$

This is a hyperbola with centre $(0,0)$, $a = \dfrac{1}{\sqrt{2+\sqrt{2}}}$, and $b = \dfrac{1}{\sqrt{2-\sqrt{2}}}$. The semi-focal separation is c where

$$c^2 = a^2 + b^2 = \frac{1}{2+\sqrt{2}} + \frac{1}{2-\sqrt{2}} = \frac{4}{2} = 2,$$

so the foci are at $u = \pm 2$, $v = 0$. The asymptotes are

$$\sqrt{2+\sqrt{2}}\,u = \pm\sqrt{2-\sqrt{2}}\,v,$$

and have inclinations

$$\pm\tan^{-1}\sqrt{\frac{2+\sqrt{2}}{2-\sqrt{2}}} = \pm\tan^{-1}\frac{2+\sqrt{2}}{2\sqrt{2}} = \pm\frac{3\pi}{8}.$$

In terms of the original coordinates, the centre is $(0,0)$, the foci are $\left(\pm\cos\dfrac{\pi}{8}, \pm\sin\dfrac{\pi}{8}\right)$, and the asymptotes are lines through the origin with inclinations

$$\frac{\pi}{8} \pm \frac{3\pi}{8} = \frac{\pi}{2} \quad \text{or} \quad -\frac{\pi}{4}.$$

Thus they are the lines $x = 0$ and $y = -x$.

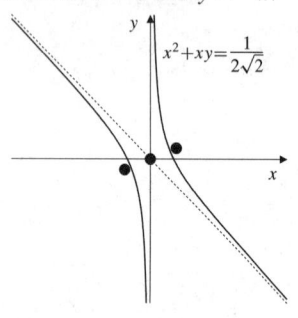

Fig. 8.1.21

22. We have $x^2 + 2xy + y^2 = 4x - 4y + 4$ and $A = 1$, $B = 2$, $C = 1$, $D = -4$, $E = 4$ and $F = -4$. We rotate the axes through angle θ satisfying $\tan 2\theta = B/(A-C) = \infty \Rightarrow \theta = \dfrac{\pi}{4}$. Then $A' = 2$, $B' = 0$, $C' = 0$, $D' = 0$, $E' = 4\sqrt{2}$ and the transformed equation is

$$2u^2 + 4\sqrt{2}v - 4 = 0 \quad\Rightarrow\quad u^2 = -2\sqrt{2}\left(v - \frac{1}{\sqrt{2}}\right)$$

which represents a parabola with vertex at $(u,v) = \left(0, \dfrac{1}{\sqrt{2}}\right)$ and principal axis along $u = 0$. The distance a from the focus to the vertex is given by $4a = 2\sqrt{2}$, so $a = 1/\sqrt{2}$ and the focus is at $(0,0)$. The directrix is $v = \sqrt{2}$.

Since $x = \dfrac{1}{\sqrt{2}}(u-v)$ and $y = \dfrac{1}{\sqrt{2}}(u+v)$, the vertex of the parabola in terms of xy-coordinates is $(-\tfrac{1}{2}, \tfrac{1}{2})$, and the focus is $(0,0)$. The directrix is $x - y = 2$. The principal axis is $y = -x$.

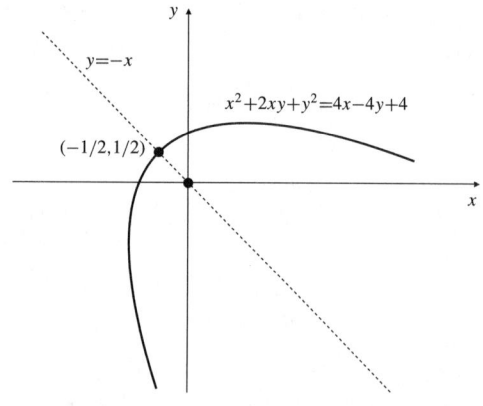

Fig. 8.1.22

23. For $8x^2 + 12xy + 17y^2 = 20$, we have $A = 8$, $B = 12$, $C = 17$, $F = -20$. Rotate the axes through angle θ where

$$\tan 2\theta = \frac{B}{A-C} = -\frac{12}{9} = -\frac{4}{3}.$$

Thus $\cos 2\theta = 3/5$, $\sin 2\theta = -4/5$, and

$$2\cos^2\theta - 1 = \cos 2\theta = \frac{3}{5} \quad\Rightarrow\quad \cos^2\theta = \frac{4}{5}.$$

We may therefore take $\cos\theta = \dfrac{2}{\sqrt{5}}$, and $\sin\theta = -\dfrac{1}{\sqrt{5}}$. The transformation is therefore

$$x = \frac{2}{\sqrt{5}}u + \frac{1}{\sqrt{5}}v \qquad u = \frac{2}{\sqrt{5}}x - \frac{1}{\sqrt{5}}y$$
$$y = -\frac{1}{\sqrt{5}}u + \frac{2}{\sqrt{5}}v \qquad v = \frac{1}{\sqrt{5}}x + \frac{2}{\sqrt{5}}y$$

The coefficients of the transformed equation are

$$A' = 8\left(\frac{4}{5}\right) + 12\left(-\frac{2}{5}\right) + 17\left(\frac{1}{5}\right) = 5$$
$$B' = 0$$
$$C' = 8\left(\frac{1}{5}\right) - 12\left(-\frac{2}{5}\right) + 17\left(\frac{4}{5}\right) = 20.$$

The transformed equation is

$$5u^2 + 20v^2 = 20, \quad \text{or} \quad \frac{u^2}{4} + v^2 = 1.$$

INSTRUCTOR'S SOLUTIONS MANUAL   SECTION 8.1 (PAGE 487)

This is an ellipse with centre $(0, 0)$, semi-axes $a = 2$ and $b = 1$, and foci at $u = \pm\sqrt{3}$, $v = 0$.
In terms of the original coordinates, the centre is $(0, 0)$, the foci are $\pm\left(\dfrac{2\sqrt{3}}{\sqrt{5}}, -\dfrac{\sqrt{3}}{\sqrt{5}}\right)$.

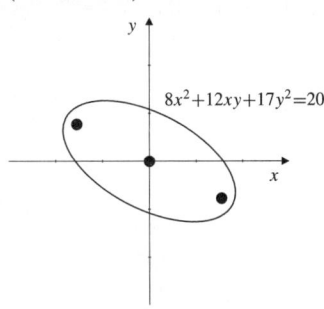

Fig. 8.1.23

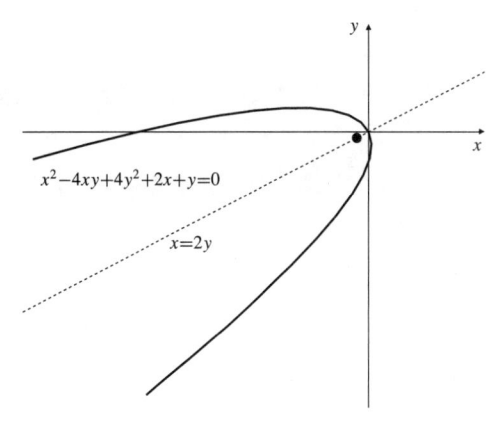

Fig. 8.1.24

24. We have $x^2 - 4xy + 4y^2 + 2x + y = 0$ and $A = 1$, $B = -4$, $C = 4$, $D = 2$, $E = 1$ and $F = 0$. We rotate the axes through angle θ satisfying $\tan 2\theta = B/(A - C) = \frac{4}{3}$. Then

$$\sec 2\theta = \sqrt{1 + \tan^2 2\theta} = \frac{5}{3} \quad \Rightarrow \quad \cos 2\theta = \frac{3}{5}$$

$$\Rightarrow \begin{cases} \cos\theta = \sqrt{\dfrac{1 + \cos 2\theta}{2}} = \sqrt{\dfrac{4}{5}} = \dfrac{2}{\sqrt{5}}; \\ \sin\theta = \sqrt{\dfrac{1 - \cos 2\theta}{2}} = \sqrt{\dfrac{1}{5}} = \dfrac{1}{\sqrt{5}}. \end{cases}$$

Then $A' = 0$, $B' = 0$, $C' = 5$, $D' = \sqrt{5}$, $E' = 0$ and the transformed equation is

$$5v^2 + \sqrt{5}u = 0 \quad \Rightarrow \quad v^2 = -\frac{1}{\sqrt{5}}u$$

which represents a parabola with vertex at $(u, v) = (0, 0)$, focus at $\left(-\dfrac{1}{4\sqrt{5}}, 0\right)$. The directrix is $u = \dfrac{1}{4\sqrt{5}}$ and the principal axis is $v = 0$. Since $x = \dfrac{2}{\sqrt{5}}u - \dfrac{1}{\sqrt{5}}v$ and $y = \dfrac{1}{\sqrt{5}}u + \dfrac{2}{\sqrt{5}}v$, in terms of the xy-coordinates, the vertex is at $(0, 0)$, the focus at $\left(-\dfrac{1}{10}, -\dfrac{1}{20}\right)$. The directrix is $2x + y = \frac{1}{4}$ and the principal axis is $2y - x = 0$.

25. The distance from P to F is $\sqrt{x^2 + y^2}$.
The distance from P to D is $x + p$. Thus

$$\frac{\sqrt{x^2 + y^2}}{x + p} = \epsilon$$

$$x^2 + y^2 = \epsilon^2(x^2 + 2px + p^2)$$

$$(1 - \epsilon^2)x^2 + y^2 - 2p\epsilon^2 x = \epsilon^2 p^2.$$

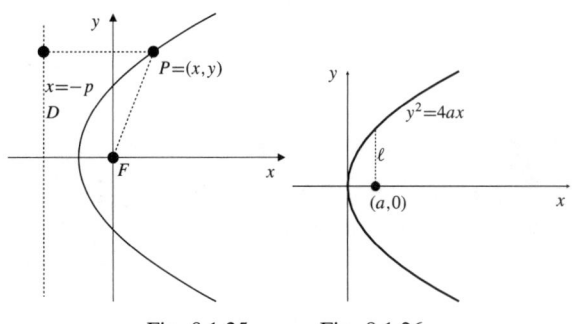

Fig. 8.1.25   Fig. 8.1.26

26. Let the equation of the parabola be $y^2 = 4ax$. The focus F is at $(a, 0)$ and vertex at $(0, 0)$. Then the distance from the vertex to the focus is a. At $x = a$, $y = \sqrt{4a(a)} = \pm 2a$. Hence, $\ell = 2a$, which is twice the distance from the vertex to the focus.

27. We have $\dfrac{c^2}{a^2} + \dfrac{\ell^2}{b^2} = 1$. Thus

$$\ell^2 = b^2\left(1 - \frac{c^2}{a^2}\right) \quad \text{but } c^2 = a^2 - b^2$$

$$= b^2\left(1 - \frac{a^2 - b^2}{a^2}\right) = b^2\frac{b^2}{a^2}.$$

Therefore $\ell = b^2/a$.

307

SECTION 8.1 (PAGE 487)

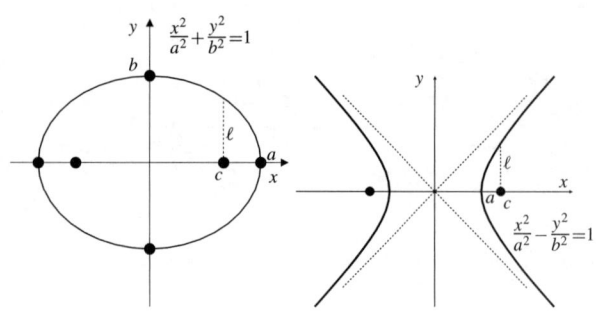

Fig. 8.1.27

Fig. 8.1.28

28. Suppose the hyperbola has equation $\dfrac{x^2}{a^2} - \dfrac{y^2}{b^2} = 1$. The vertices are at $(\pm a, 0)$ and the foci are at $(\pm c, 0)$ where $c = \sqrt{a^2+b^2}$. At $x = \sqrt{a^2+b^2}$,

$$\frac{a^2+b^2}{a^2} - \frac{y^2}{b^2} = 1$$
$$(a^2+b^2)b^2 - a^2 y^2 = a^2 b^2$$
$$y = \pm \frac{b^2}{a}.$$

Hence, $\ell = \dfrac{b^2}{a}$.

29.

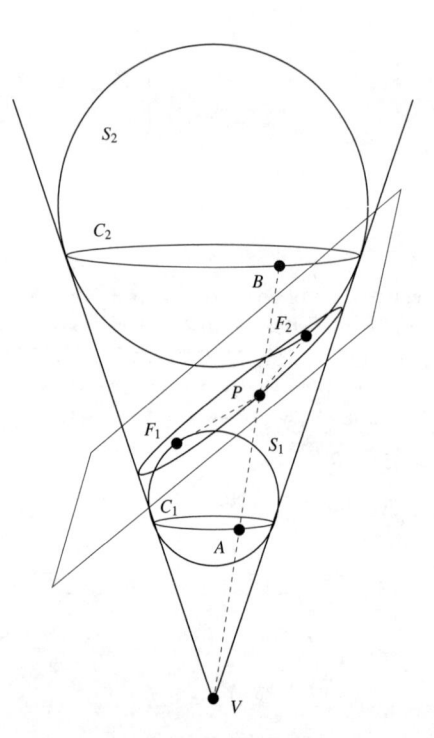

Fig. 8.1.29

Let the spheres S_1 and S_2 intersect the cone in the circles C_1 and C_2, and be tangent to the plane of the ellipse at the points F_1 and F_2, as shown in the figure.
Let P be any point on the ellipse, and let the straight line through P and the vertex of the cone meet C_1 and C_2 at A and B respectively. Then $PF_1 = PA$, since both segments are tangents to the sphere S_1 from P. Similarly, $PF_2 = PB$.
Thus $PF_1 + PF_2 = PA + PB = AB =$ constant (distance from C_1 to C_2 along all generators of the cone is the same.) Thus F_1 and F_2 are the foci of the ellipse.

30. Let F_1 and F_2 be the points where the plane is tangent to the spheres. Let P be an arbitrary point P on the hyperbola in which the plane intersects the cone. The spheres are tangent to the cone along two circles as shown in the figure. Let $PAVB$ be a generator of the cone (a straight line lying on the cone) intersecting these two circles at A and B as shown. (V is the vertex of the cone.) We have $PF_1 = PA$ because two tangents to a sphere from a point outside the sphere have equal lengths. Similarly, $PF_2 = PB$. Therefore

$$PF_2 - PF_1 = PB - PA = AB = \text{constant},$$

since the distance between the two circles in which the spheres intersect the cone, measured along the generators of the cone, is the same for all generators. Hence, F_1 and F_2 are the foci of the hyperbola.

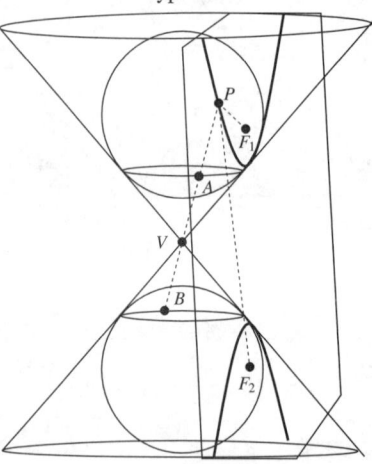

Fig. 8.1.30

31. Let the plane in which the sphere is tangent to the cone meet AV at X. Let the plane through F perpendicular to the axis of the cone meet AV at Y. Then $VF = VX$, and, if C is the centre of the sphere, $FC = XC$. Therefore VC is perpendicular to the axis of the cone. Hence YF is parallel to VC, and we have $YV = VX = VF$.
If P is on the parabola, $FP \perp VF$, and the line from P to the vertex A of the cone meets the circle of tangency of the sphere and the cone at Q, then

$$FP = PQ = YX = 2VX = 2VF.$$

308

INSTRUCTOR'S SOLUTIONS MANUAL SECTION 8.2 (PAGE 494)

Since $FP = 2VF$, FP is the semi-latus rectum of the parabola. (See Exercise 18.) Therefore F is the focus of the parabola.

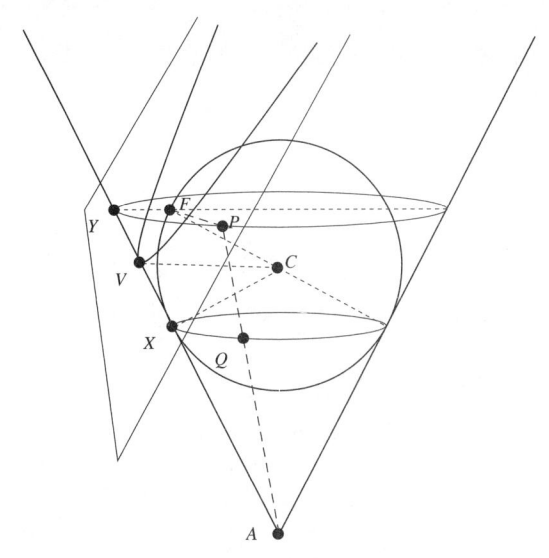

Fig. 8.1.31

Section 8.2 Parametric Curves (page 494)

1. If $x = t$, $y = 1 - t$, $(0 \leq t \leq 1)$ then $x + y = 1$. This is a straight line segment.

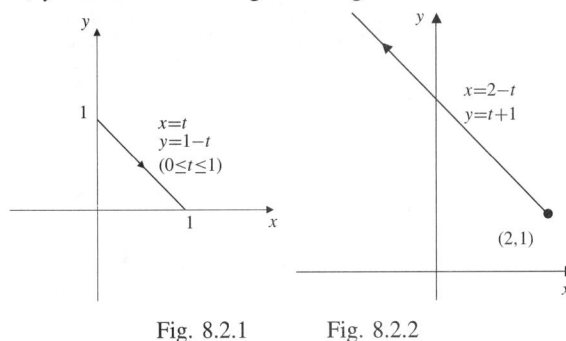

Fig. 8.2.1   Fig. 8.2.2

2. If $x = 2 - t$ and $y = t + 1$ for $0 \leq t < \infty$, then $y = 2 - x + 1 = 3 - x$ for $-\infty < x \leq 2$, which is a half line.

3. If $x = 1/t$, $y = t - 1$, $(0 < t < 4)$, then $y = \dfrac{1}{x} - 1$. This is part of a hyperbola.

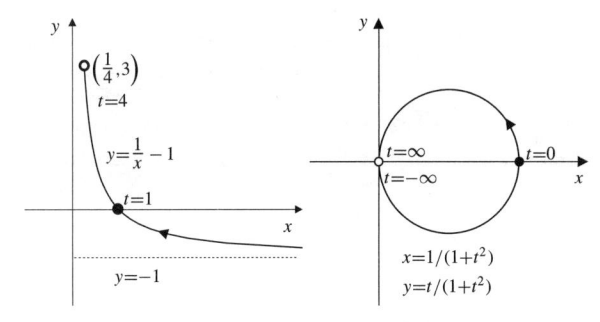

Fig. 8.2.3   Fig. 8.2.4

4. If $x = \dfrac{1}{1+t^2}$ and $y = \dfrac{t}{1+t^2}$ for $-\infty < t < \infty$, then

$$x^2 + y^2 = \dfrac{1+t^2}{(1+t^2)^2} = \dfrac{1}{1+t^2} = x$$

$$\Leftrightarrow \left(x - \dfrac{1}{2}\right)^2 + y^2 = \dfrac{1}{4}.$$

This curve consists of all points of the circle with centre at $(\tfrac{1}{2}, 0)$ and radius $\tfrac{1}{2}$ except the origin $(0, 0)$.

5. If $x = 3\sin 2t$, $y = 3\cos 2t$, $(0 \leq t \leq \pi/3)$, then $x^2 + y^2 = 9$. This is part of a circle.

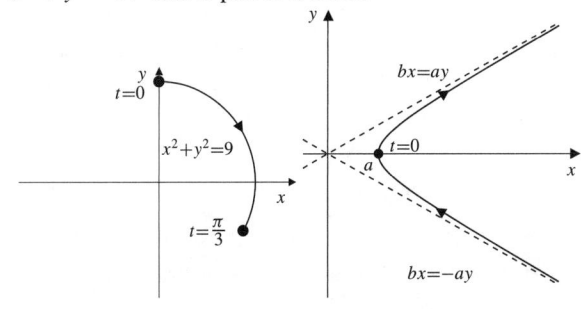

Fig. 8.2.5   Fig. 8.2.6

6. If $x = a\sec t$ and $y = b\tan t$ for $-\dfrac{\pi}{2} < t < \dfrac{\pi}{2}$, then

$$\dfrac{x^2}{a^2} - \dfrac{y^2}{b^2} = \sec^2 t - \tan^2 t = 1.$$

The curve is one arch of this hyperbola.

7. If $x = 3\sin \pi t$, $y = 4\cos \pi t$, $(-1 \leq t \leq 1)$, then $\dfrac{x^2}{9} + \dfrac{y^2}{16} = 1$. This is an ellipse.

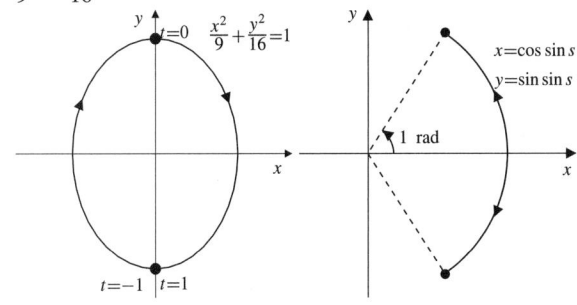

Fig. 8.2.7   Fig. 8.2.8

309

8. If $x = \cos \sin s$ and $y = \sin \sin s$ for $-\infty < s < \infty$, then $x^2 + y^2 = 1$. The curve consists of the arc of this circle extending from $(a, -b)$ through $(1, 0)$ to (a, b) where $a = \cos(1)$ and $b = \sin(1)$, traversed infinitely often back and forth.

9. If $x = \cos^3 t$, $y = \sin^3 t$, $(0 \leq t \leq 2\pi)$, then $x^{2/3} + y^{2/3} = 1$. This is an astroid.

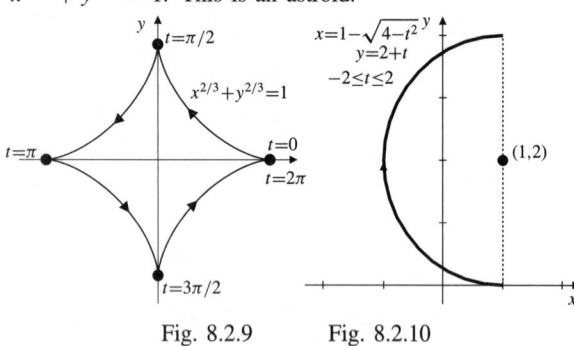

Fig. 8.2.9   Fig. 8.2.10

10. If $x = 1 - \sqrt{4 - t^2}$ and $y = 2 + t$ for $-2 \leq t \leq 2$ then
$$(x - 1)^2 = 4 - t^2 = 4 - (y - 2)^2.$$

The parametric curve is the left half of the circle of radius 4 centred at $(1, 2)$, and is traced in the direction of increasing y.

11. $x = \cosh t$, $y = \sinh t$ represents the right half (branch) of the rectangular hyperbola $x^2 - y^2 = 1$.

12. $x = 2 - 3 \cosh t$, $y = -1 + 2 \sinh t$ represents the left half (branch) of the hyperbola
$$\frac{(x - 2)^2}{9} - \frac{(y + 1)^2}{4} = 1.$$

13. $x = t \cos t$, $y = t \sin t$, $(0 \leq t \leq 4\pi)$ represents two revolutions of a spiral curve winding outwards from the origin in a counterclockwise direction. The point on the curve corresponding to parameter value t is t units distant from the origin in a direction making angle t with the positive x-axis.

14. (i) If $x = \cos^4 t$ and $y = \sin^4 t$, then
$$(x - y)^2 = (\cos^4 t - \sin^4 t)^2$$
$$= \left[(\cos^2 t + \sin^2 t)(\cos^2 t - \sin^2 t)\right]^2$$
$$= (\cos^2 t - \sin^2 t)^2$$
$$= \cos^4 t + \sin^4 t - 2\cos^2 t \sin^2 t$$

and
$$1 = (\cos^2 t + \sin^2 t)^2 = \cos^4 t + \sin^4 t + 2\cos^2 t \sin^2 t.$$

Hence,
$$1 + (x - y)^2 = 2(\cos^4 t + \sin^4 t) = 2(x + y).$$

(ii) If $x = \sec^4 t$ and $y = \tan^4 t$, then
$$(x - y)^2 = (\sec^4 t - \tan^4 t)^2$$
$$= (\sec^2 t + \tan^2 t)^2$$
$$= \sec^4 t + \tan^4 t + 2 \sec^2 t \tan^2 t$$

and
$$1 = (\sec^2 t - \tan^2 t)^2 = \sec^4 t + \tan^4 t - 2 \sec^2 t \tan^2 t.$$

Hence,
$$1 + (x - y)^2 = 2(\sec^4 t + \tan^4 t) = 2(x + y).$$

(iii) Similarly, if $x = \tan^4 t$ and $y = \sec^4 t$, then
$$1 + (x - y)^2 = 1 + (y - x)^2$$
$$= (\sec^2 t - \tan^2 t)^2 + (\sec^4 t - \tan^4 t)^2$$
$$= 2(\tan^4 t + \sec^4 t)$$
$$= 2(x + y).$$

These three parametric curves above correspond to different parts of the parabola $1 + (x - y)^2 = 2(x + y)$, as shown in the following diagram.

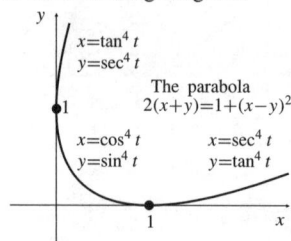

Fig. 8.2.14

15. The slope of $y = x^2$ at x is $m = 2x$. Hence the parabola can be parametrized $x = m/2$, $y = m^2/4$, $(-\infty < m < \infty)$.

16. If (x, y) is any point on the circle $x^2 + y^2 = R^2$ other than $(R, 0)$, then the line from (x, y) to $(R, 0)$ has slope $m = \dfrac{y}{x - R}$. Thus $y = m(x - R)$, and
$$x^2 + m^2(x - R)^2 = R^2$$
$$(m^2 + 1)x^2 - 2xRm^2 + (m^2 - 1)R^2 = 0$$
$$\left[(m^2 + 1)x - (m^2 - 1)R\right](x - R) = 0$$
$$\Rightarrow x = \frac{(m^2 - 1)R}{m^2 + 1} \text{ or } x = R.$$

INSTRUCTOR'S SOLUTIONS MANUAL SECTION 8.2 (PAGE 494)

The parametrization of the circle in terms of m is given by

$$x = \frac{(m^2-1)R}{m^2+1}$$
$$y = m\left[\frac{(m^2-1)R}{m^2+1} - R\right] = -\frac{2Rm}{m^2+1}$$

where $-\infty < m < \infty$. This parametrization gives every point on the circle except $(R, 0)$.

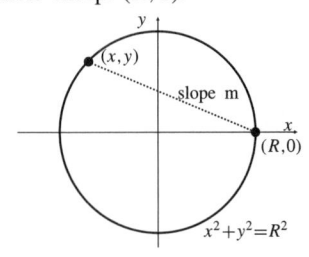

Fig. 8.2.16

17.

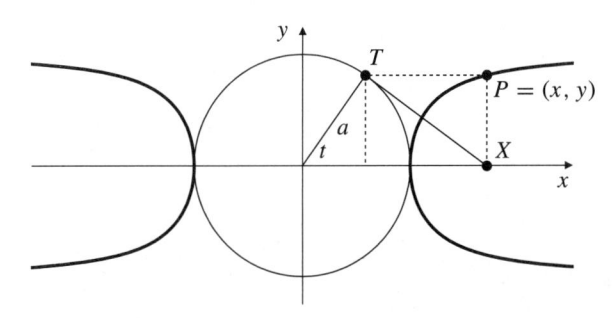

Fig. 8.2.17

Using triangles in the figure, we see that the coordinates of P satisfy

$$x = a\sec t, \quad y = a\sin t.$$

The Cartesian equation of the curve is

$$\frac{y^2}{a^2} + \frac{a^2}{x^2} = 1.$$

The curve has two branches extending to infinity to the left and right of the circle as shown in the figure.

18. The coordinates of P satisfy

$$x = a\sec t, \quad y = b\sin t.$$

The Cartesian equation is $\dfrac{y^2}{b^2} + \dfrac{a^2}{x^2} = 1$.

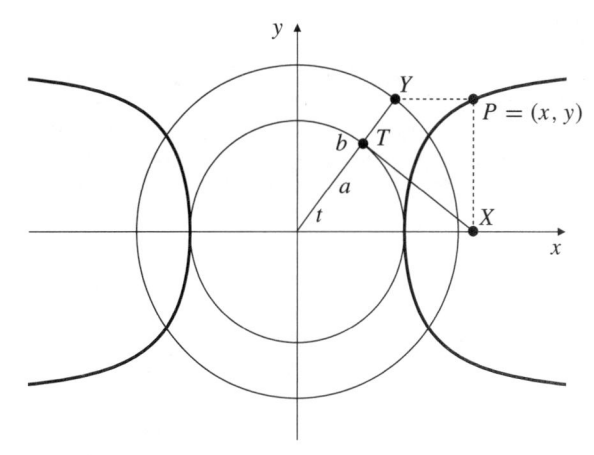

Fig. 8.2.18

19. If $x = \dfrac{3t}{1+t^3}$, $y = \dfrac{3t^2}{1+t^3}$, $(t \neq -1)$, then

$$x^3 + y^3 = \frac{27t^3}{(1+t^3)^3}(1+t^3) = \frac{27t^3}{(1+t^3)^2} = 3xy.$$

As $t \to -1$, we see that $|x| \to \infty$ and $|y| \to \infty$, but

$$x + y = \frac{3t(1+t)}{1+t^3} = \frac{3t}{1-t+t^2} \to -1.$$

Thus $x + y = -1$ is an asymptote of the curve.

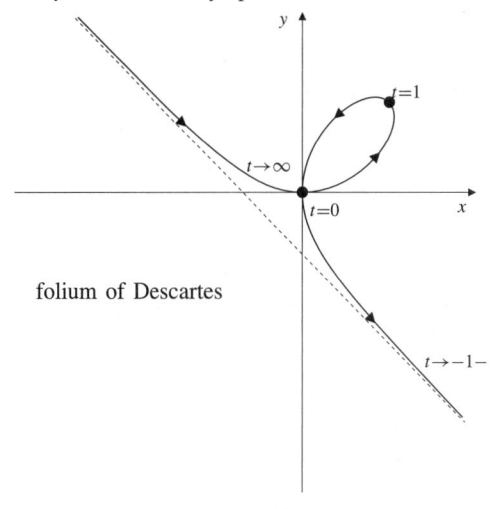

folium of Descartes

Fig. 8.2.19

20. Let C_0 and P_0 be the original positions of the centre of the wheel and a point at the bottom of the flange whose path is to be traced. The wheel is also shown in a subsequent position in which it makes contact with the rail at R. Since the wheel has been rotated by an angle θ,

$$OR = \text{arc } SR = a\theta.$$

311

SECTION 8.2 (PAGE 494)

Thus, the new position of the centre is $C = (a\theta, a)$. Let $P = (x, y)$ be the new position of the point; then

$$x = OR - PQ = a\theta - b\sin(\pi - \theta) = a\theta - b\sin\theta,$$
$$y = RC + CQ = a + b\cos(\pi - \theta) = a - b\cos\theta.$$

These are the parametric equations of the prolate cycloid.

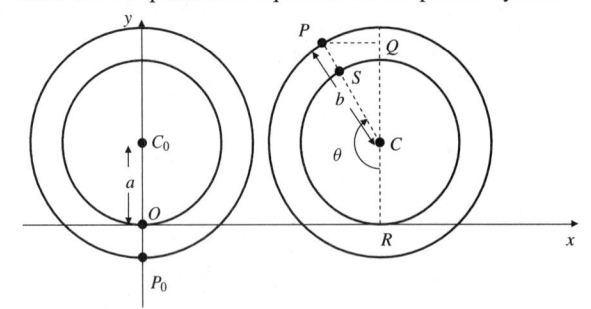

Fig. 8.2.20

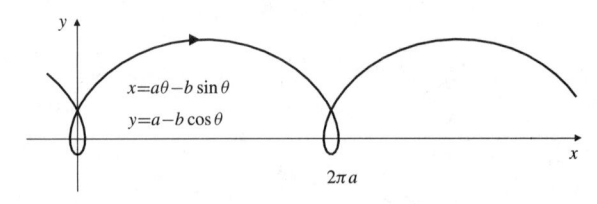

Fig. 8.2.20

21. Let t and θ_t be the angles shown in the figure below. Then arc AT_t = arc $T_t P_t$, that is, $at = b\theta_t$. The centre C_t of the rolling circle is $C_t = \big((a-b)\cos t, (a-b)\sin t\big)$. Thus

$$x - (a-b)\cos t = b\cos(\theta_t - t)$$
$$y - (a-b)\sin t = -b\sin(\theta_t - t).$$

Since $\theta_t - t = \dfrac{a}{b}t - t = \dfrac{a-b}{b}t$, therefore

$$x = (a-b)\cos t + b\cos\left(\frac{(a-b)t}{b}\right)$$
$$y = (a-b)\sin t - b\sin\left(\frac{(a-b)t}{b}\right).$$

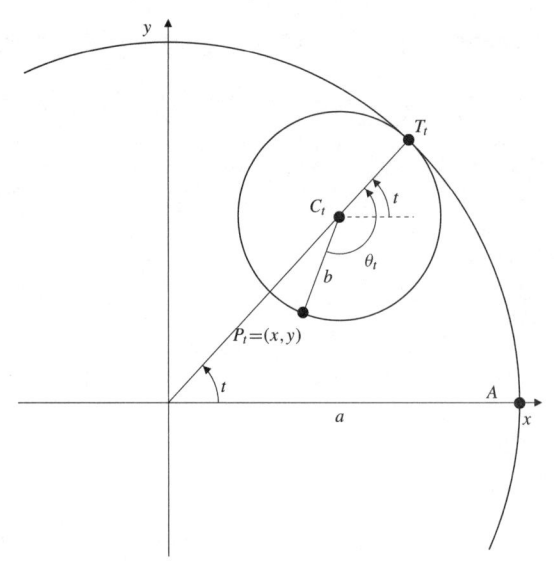

Fig. 8.2.21

If $a = 2$ and $b = 1$, then $x = 2\cos t$, $y = 0$. This is a straight line segment.
If $a = 4$ and $b = 1$, then

$$\begin{aligned} x &= 3\cos t + \cos 3t \\ &= 3\cos t + (\cos 2t \cos t - \sin 2t \sin t) \\ &= 3\cos t + \big((2\cos^2 t - 1)\cos t - 2\sin^2 t \cos t\big) \\ &= 2\cos t + 2\cos^3 t - 2\cos t(1 - \sin^2 t) = 4\cos^3 t \\ y &= 3\sin t + \sin 3t \\ &= 3\sin t - \sin 2t \cos t - (\cos 2t \sin t) \\ &= 3\sin t - 2\sin t \cos^2 t - \big((1 - 2\sin^2 t)\sin t\big) \\ &= 2\sin t - 2\sin t + 2\sin^3 t + 2\sin^3 t = 4\sin^3 t \end{aligned}$$

This is an astroid, similar to that of Exercise 11.

22. a) From triangles in the figure,
$$x = |TX| = |OT|\tan t = \tan t$$
$$y = |OY| = \sin\left(\frac{\pi}{2} - t\right) = |OY|\cos t$$
$$= |OT|\cos t \cos t = \cos^2 t.$$

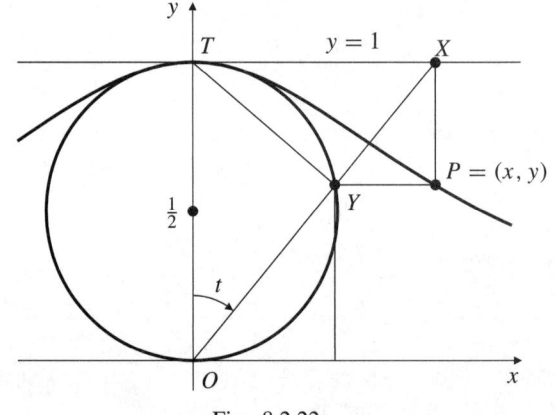

Fig. 8.2.22

b) $\frac{1}{y} = \sec^2 t = 1 + \tan^2 t = 1 + x^2$. Thus $y = \frac{1}{1+x^2}$.

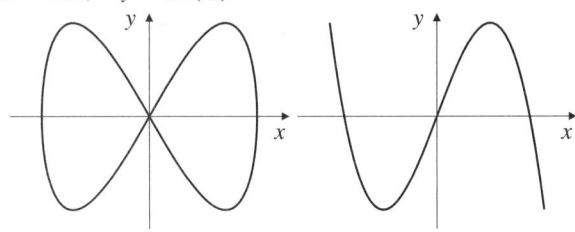

23. $x = \sin t$, $y = \sin(2t)$

Fig. 8.2.23   Fig. 8.2.24

24. $x = \sin t$, $y = \sin(3t)$

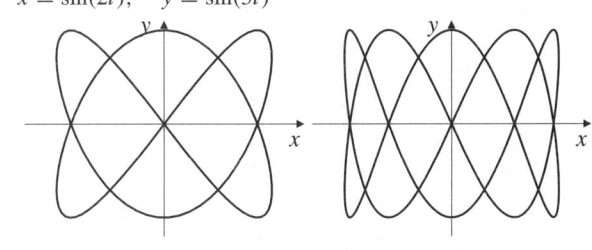

25. $x = \sin(2t)$, $y = \sin(3t)$

Fig. 8.2.25   Fig. 8.2.26

26. $x = \sin(2t)$, $y = \sin(5t)$

27. $x = \left(1 + \frac{1}{n}\right)\cos t - \frac{1}{n}\cos(nt)$

    $y = \left(1 + \frac{1}{n}\right)\sin t - \frac{1}{n}\sin(nt)$

    represents a cycloid-like curve that is wound around the circle $x^2 + y^2 = 1$ instead of extending along the x-axis. If $n \geq 2$ is an integer, the curve closes after one revolution and has $n-1$ cusps. The left figure below shows the curve for $n = 7$. If n is a rational number, the curve will wind around the circle more than once before it closes.

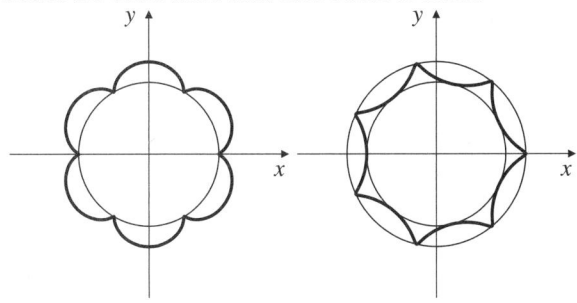

Fig. 8.2.27   Fig. 8.2.28

28. $x = \left(1 + \frac{1}{n}\right)\cos t + \frac{1}{n}\cos((n-1)t)$

    $y = \left(1 + \frac{1}{n}\right)\sin t - \frac{1}{n}\sin((n-1)t)$

    represents a cycloid-like curve that is wound around the inside circle $x^2 + y^2 = \left(1 + (2/n)\right)^2$ and is externally tangent to $x^2 + y^2 = 1$. If $n \geq 2$ is an integer, the curve closes after one revolution and has n cusps. The figure shows the curve for $n = 7$. If n is a rational number but not an integer, the curve will wind around the circle more than once before it closes.

Section 8.3 Smooth Parametric Curves and Their Slopes   (page 499)

1. $x = t^2 + 1 \qquad y = 2t - 4$
   $\frac{dx}{dt} = 2t \qquad \frac{dy}{dt} = 2$
   No horizontal tangents. Vertical tangent at $t = 0$, i.e., at $(1, -4)$.

2. $x = t^2 - 2t \qquad y = t^2 + 2t$
   $\frac{dx}{dt} = 2t - 2 \qquad \frac{dy}{dt} = 2t + 2$
   Horizontal tangent at $t = -1$, i.e., at $(3, -1)$.
   Vertical tangent at $t = 1$, i.e., at $(-1, 3)$.

3. $x = t^2 - 2t \qquad y = t^3 - 12t$
   $\frac{dx}{dt} = 2(t-1) \qquad \frac{dy}{dt} = 3(t^2 - 4)$
   Horizontal tangent at $t = \pm 2$, i.e., at $(0, -16)$ and $(8, 16)$.
   Vertical tangent at $t = 1$, i.e., at $(-1, -11)$.

4. $x = t^3 - 3t \qquad y = 2t^3 + 3t^2$
   $\frac{dx}{dt} = 3(t^2 - 1) \qquad \frac{dy}{dt} = 6t(t+1)$
   Horizontal tangent at $t = 0$, i.e., at $(0, 0)$.
   Vertical tangent at $t = 1$, i.e., at $(-2, 5)$.
   At $t = -1$ (i.e., at $(2, 1)$) both dx/dt and dy/dt change sign, so the curve is not smooth there. (It has a cusp.)

5. $x = te^{-t^2/2} \qquad\qquad y = e^{-t^2}$
   $\frac{dx}{dt} = (1 - t^2)e^{-t^2/2} \qquad \frac{dy}{dt} = -2te^{-t^2}$
   Horizontal tangent at $t = 0$, i.e., at $(0, 1)$.
   Vertical tangent at $t = \pm 1$, i.e. at $(\pm e^{-1/2}, e^{-1})$.

6. $x = \sin t \qquad y = \sin t - t \cos t$
   $\frac{dx}{dt} = \cos t \qquad \frac{dy}{dt} = t \sin t$
   Horizontal tangent at $t = n\pi$, i.e., at $(0, -(-1)^n n\pi)$ (for integers n).
   Vertical tangent at $t = (n + \frac{1}{2})\pi$, i.e. at $(1, 1)$ and $(-1, -1)$.

7. $x = \sin(2t) \qquad y = \sin t$
    $\dfrac{dx}{dt} = 2\cos(2t) \qquad \dfrac{dy}{dt} = \cos t$
    Horizontal tangent at $t = (n+\tfrac{1}{2})\pi$, i.e., at $(0, \pm 1)$.
    Vertical tangent at $t = \tfrac{1}{2}(n+\tfrac{1}{2})\pi$, i.e., at $(\pm 1, 1/\sqrt{2})$ and $(\pm 1, -1/\sqrt{2})$.

8. $x = \dfrac{3t}{1+t^3} \qquad y = \dfrac{3t^2}{1+t^3}$
    $\dfrac{dx}{dt} = \dfrac{3(1-2t^3)}{(1+t^3)^2} \qquad \dfrac{dy}{dt} = \dfrac{3t(2-t^3)}{(1+t^3)^2}$
    Horizontal tangent at $t=0$ and $t=2^{1/3}$, i.e., at $(0,0)$ and $(2^{1/3}, 2^{2/3})$.
    Vertical tangent at $t = 2^{-1/3}$, i.e., at $(2^{2/3}, 2^{1/3})$. The curve also approaches $(0,0)$ vertically as $t \to \pm\infty$.

9. $x = t^3 + t \qquad y = 1 - t^3$
    $\dfrac{dx}{dt} = 3t^2 + 1 \qquad \dfrac{dy}{dt} = -3t^2$
    At $t=1$; $\dfrac{dy}{dx} = \dfrac{-3(1)^2}{3(1)^2+1} = -\dfrac{3}{4}$.

10. $x = t^4 - t^2 \qquad y = t^3 + 2t$
    $\dfrac{dx}{dt} = 4t^3 - 2t \qquad \dfrac{dy}{dt} = 3t^2 + 2$
    At $t=-1$; $\dfrac{dy}{dx} = \dfrac{3(-1)^2+2}{4(-1)^3-2(-1)} = -\dfrac{5}{2}$.

11. $x = \cos(2t) \qquad y = \sin t$
    $\dfrac{dx}{dt} = -2\sin(2t) \qquad \dfrac{dy}{dt} = \cos t$
    At $t = \dfrac{\pi}{6}$; $\dfrac{dy}{dx} = \dfrac{\cos(\pi/6)}{-2\sin(\pi/3)} = -\dfrac{1}{2}$.

12. $x = e^{2t} \qquad y = te^{2t}$
    $\dfrac{dx}{dt} = 2e^{2t} \qquad \dfrac{dy}{dt} = e^{2t}(1+2t)$
    At $t=-2$; $\dfrac{dy}{dx} = \dfrac{e^{-4}(1-4)}{2e^{-4}} = -\dfrac{3}{2}$.

13. $x = t^3 - 2t = -1 \qquad y = t + t^3 = 2$ at $t=1$
    $\dfrac{dx}{dt} = 3t^2 - 2 = 1 \qquad \dfrac{dy}{dt} = 1 + 3t^2 = 4$ at $t=1$
    Tangent line: $x = -1 + t$, $y = 2 + 4t$. This line is at $(-1, 2)$ at $t=0$. If you want to be at that point at $t=1$ instead, use
    $$x = -1 + (t-1) = t - 2, \quad y = 2 + 4(t-1) = 4t - 2.$$

14. $x = t - \cos t = \dfrac{\pi}{4} - \dfrac{1}{\sqrt{2}}$
    $\dfrac{dx}{dt} = 1 + \sin t = 1 + \dfrac{1}{\sqrt{2}}$
    $y = 1 - \sin t = 1 - \dfrac{1}{\sqrt{2}}$ at $t = \dfrac{\pi}{4}$
    $\dfrac{dy}{dt} = -\cos t = -\dfrac{1}{\sqrt{2}}$ at $t = \dfrac{\pi}{4}$
    Tangent line: $x = \dfrac{\pi}{4} - \dfrac{1}{\sqrt{2}} + \left(1 + \dfrac{1}{\sqrt{2}}\right)t$,
    $y = 1 - \dfrac{1}{\sqrt{2}} - \dfrac{t}{\sqrt{2}}$.

15. $x = t^3 - t$, $y = t^2$ is at $(0,1)$ at $t=-1$ and $t=1$. Since
    $$\frac{dy}{dx} = \frac{2t}{3t^2-1} = \frac{\pm 2}{2} = \pm 1,$$
    the tangents at $(0,1)$ at $t = \pm 1$ have slopes ± 1.

16. $x = \sin t$, $y = \sin(2t)$ is at $(0,0)$ at $t=0$ and $t=\pi$. Since
    $$\frac{dy}{dx} = \frac{2\cos(2t)}{\cos t} = \begin{cases} 2 & \text{if } t = 0 \\ -2 & \text{if } t = \pi, \end{cases}$$
    the tangents at $(0,0)$ at $t=0$ and $t=\pi$ have slopes 2 and -2, respectively.

17. $x = t^3 \qquad y = t^2$
    $\dfrac{dx}{dt} = 3t^2 \qquad \dfrac{dy}{dt} = 2t$ both vanish at $t=0$.
    $\dfrac{dy}{dx} = \dfrac{2}{3t}$ has no limit as $t \to 0$. $\dfrac{dx}{dy} = \dfrac{3t}{2} \to 0$ as $t \to 0$, but dy/dt changes sign at $t=0$. Thus the curve is not smooth at $t=0$. (In this solution, and in the next five, we are using the Remark following Example 2 in the text.)

18. $x = (t-1)^4 \qquad y = (t-1)^3$
    $\dfrac{dx}{dt} = 4(t-1)^3 \qquad \dfrac{dy}{dt} = 3(t-1)^2$ both vanish at $t=1$.
    Since $\dfrac{dx}{dy} = \dfrac{4(t-1)}{3} \to 0$ as $t \to 1$, and dy/dt does not change sign at $t=1$, the curve is smooth at $t=1$ and therefore everywhere.

19. $x = t \sin t \qquad y = t^3$
    $\dfrac{dx}{dt} = \sin t + t \cos t \qquad \dfrac{dy}{dt} = 3t^2$ both vanish at $t=0$.
    $$\lim_{t \to 0} \frac{dy}{dx} = \lim_{t \to 0} \frac{3t^2}{\sin t + t \cos t} = \lim_{t \to 0} \frac{6t}{2\cos t - t \sin t} = 0,$$
    but dx/dt changes sign at $t=0$. dx/dy has no limit at $t=0$. Thus the curve is not smooth at $t=0$.

20. $x = t^3 \qquad y = t - \sin t$
    $\dfrac{dx}{dt} = 3t^2 \qquad \dfrac{dy}{dt} = 1 - \cos t$ both vanish at $t=0$.
    $$\lim_{t \to 0} \frac{dx}{dy} = \lim_{t \to 0} \frac{3t^2}{1 - \cos t} = \lim_{t \to 0} \frac{6t}{\sin t} = 6$$
    and dy/dt does not change sign at $t=0$. Thus the curve is smooth at $t=0$, and hence everywhere.

INSTRUCTOR'S SOLUTIONS MANUAL SECTION 8.3 (PAGE 499)

21. If $x = t^2 - 2t$ and $y = t^2 - 4t$, then

$$\frac{dx}{dt} = 2(t-1), \qquad \frac{dy}{dt} = 2(t-2)$$
$$\frac{d^2x}{dt^2} = \frac{d^2y}{dt^2} = 2$$
$$\frac{d^2y}{dx^2} = \frac{1}{dx/dt}\frac{d}{dt}\frac{dy}{dx}$$
$$= \frac{1}{2(t-1)}\frac{d}{dt}\frac{t-2}{t-1} = \frac{1}{2(t-1)^3}.$$

Directional information is as follows:

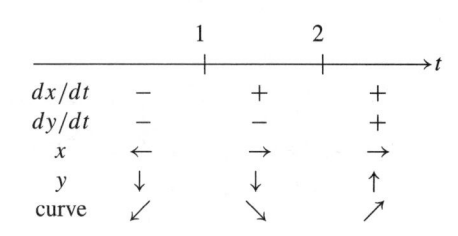

	1	2	
dx/dt	$-$	$+$	$+$
dy/dt	$-$	$-$	$+$
x	$\leftarrow$	$\rightarrow$	$\rightarrow$
y	$\downarrow$	$\downarrow$	$\uparrow$
curve	↙	↘	↗

The tangent is horizontal at $t = 2$, (i.e., $(0, -4)$), and is vertical at $t = 1$ (i.e., at $(-1, -3)$). Observe that $d^2y/dx^2 > 0$, and the curve is concave up, if $t > 1$. Similarly, $d^2y/dx^2 < 0$ and the curve is concave down if $t < 1$.

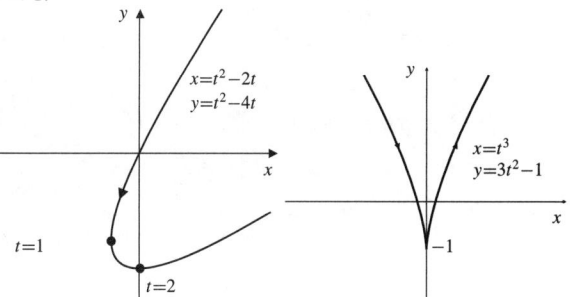

Fig. 8.3.21   Fig. 8.3.22

22. If $x = f(t) = t^3$ and $y = g(t) = 3t^2 - 1$, then

$$f'(t) = 3t^2, \ f''(t) = 6t;$$
$$g'(t) = 6t, \ g''(t) = 6.$$

Both $f'(t)$ and $g'(t)$ vanish at $t = 0$. Observe that

$$\frac{dy}{dx} = \frac{6t}{3t^2} = \frac{2}{t}.$$

Thus,

$$\lim_{t \to 0+}\frac{dy}{dx} = \infty, \qquad \lim_{t \to 0-}\frac{dy}{dx} = -\infty$$

and the curve has a cusp at $t = 0$, i.e., at $(0, -1)$. Since

$$\frac{d^2y}{dx^2} = \frac{(3t^2)(6) - (6t)(6t)}{(3t^2)^3} = -\frac{2}{3t^4} < 0$$

for all t, the curve is concave down everywhere.

23. $x = t^3 - 3t$, $y = 2/(1+t^2)$. Observe that $y \to 0$, $x \to \pm\infty$ as $t \to \pm\infty$.

$$\frac{dx}{dt} = 3(t^2 - 1), \qquad \frac{dy}{dt} = -\frac{4t}{(1+t^2)^2}$$
$$\frac{dy}{dx} = -\frac{4t}{3(t^2-1)(1+t^2)^2}$$
$$\frac{d^2x}{dt^2} = 6t, \qquad \frac{d^2y}{dt^2} = \frac{4(3t^2-1)}{(1+t^2)^3}$$
$$\frac{d^2y}{dx^2} = \frac{3(t^2-1)\dfrac{4(3t^2-1)}{(1+t^2)^3} - \dfrac{4t(6t)}{(1+t^2)^2}}{[3(t^2-1)]^3}$$
$$= \frac{60t^4 + 48t^2 + 12}{27(t^2-1)^3(1+t^2)^3}$$

Directional information:

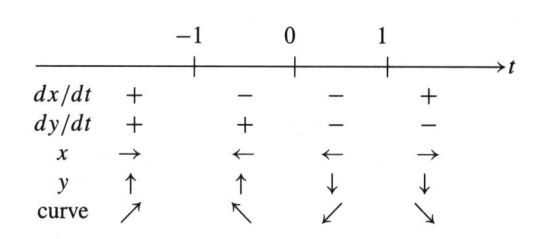

	-1	0	1	
dx/dt	$+$	$-$	$-$	$+$
dy/dt	$+$	$+$	$-$	$-$
x	$\rightarrow$	$\leftarrow$	$\leftarrow$	$\rightarrow$
y	$\uparrow$	$\uparrow$	$\downarrow$	$\downarrow$
curve	↗	↖	↙	↘

The tangent is horizontal at $t = 0$, i.e., $(0, 2)$, and vertical at $t = \pm 1$, i.e., $(\pm 2, 1)$.

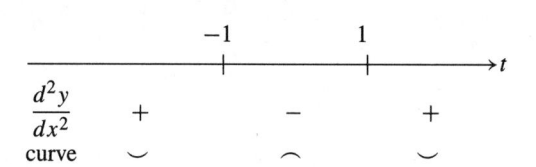

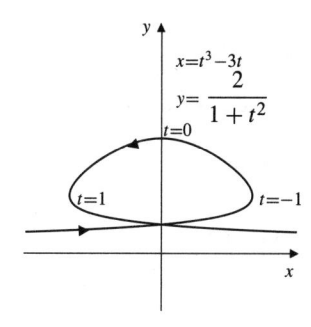

	-1	1	
$\dfrac{d^2y}{dx^2}$	$+$	$-$	$+$
curve	⌣	⌢	⌣

Fig. 8.3.23

24. If $x = f(t) = t^3 - 3t - 2$ and $y = g(t) = t^2 - t - 2$, then

$$f'(t) = 3t^2 - 3, \ f''(t) = 6t;$$
$$g'(t) = 2t - 1, \ g''(t) = 2.$$

315

The tangent is horizontal at $t = \frac{1}{2}$, i.e., at $\left(-\frac{27}{8}, -\frac{9}{4}\right)$.
The tangent is vertical at $t = \pm 1$, i.e., $(-4, -2)$ and $(0, 0)$.
Directional information is as follows:

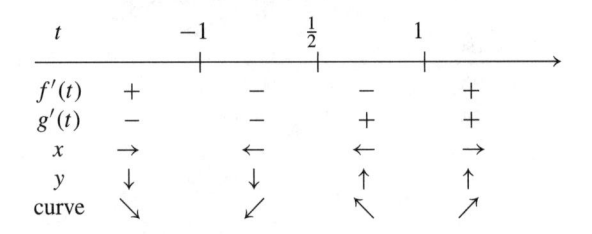

t		-1		$\frac{1}{2}$		1	
$f'(t)$	$+$		$-$		$-$		$+$
$g'(t)$	$-$		$-$		$+$		$+$
x	$\rightarrow$		$\leftarrow$		$\leftarrow$		$\rightarrow$
y	$\downarrow$		$\downarrow$		$\uparrow$		$\uparrow$
curve	↘		↙		↖		↗

For concavity,

$$\frac{d^2y}{dx^2} = \frac{3(t^2-1)(2) - (2t-1)(6t)}{[3(t^2-1)]^3} = -\frac{2(t^2-t+1)}{9(t^2-1)^3}$$

which is undefined at $t = \pm 1$, therefore

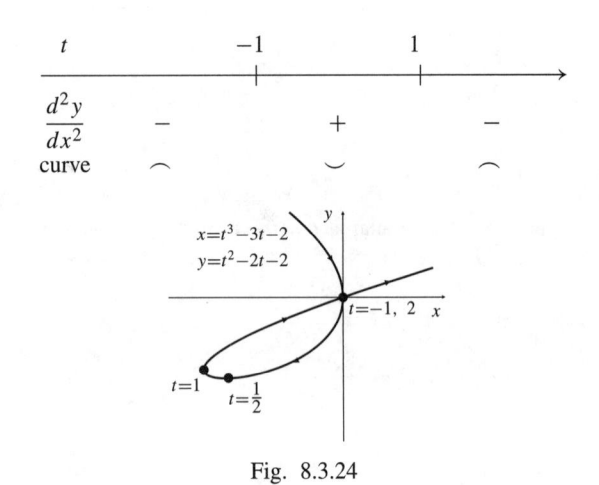

t		-1		1	
$\frac{d^2y}{dx^2}$	$-$		$+$		$-$
curve	⌢		⌣		⌢

Fig. 8.3.24

25. $x = \cos t + t \sin t$, $y = \sin t - t \cos t$, $(t \geq 0)$.

$$\frac{dx}{dt} = t \cos t, \quad \frac{dy}{dt} = t \sin t, \quad \frac{dy}{dx} = \tan t$$

$$\frac{d^2x}{dt^2} = \cos t - t \sin t$$

$$\frac{d^2y}{dt^2} = \sin t + t \cos t$$

$$\frac{d^2y}{dx^2} = \frac{\frac{dx}{dt}\frac{d^2y}{dt^2} - \frac{dy}{dt}\frac{d^2x}{dt^2}}{\left(\frac{dx}{dt}\right)^3}$$

$$= \frac{1}{t \cos^3 t}$$

Tangents are vertical at $t = \left(n + \frac{1}{2}\right)\pi$, and horizontal at $t = n\pi$ ($n = 0, 1, 2, \ldots$).

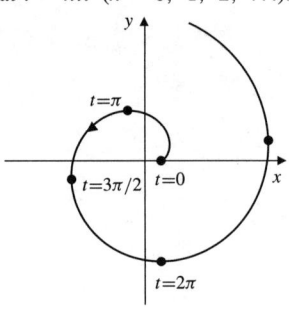

Fig. 8.3.25

Section 8.4 Arc Lengths and Areas for Parametric Curves (page 504)

1. $x = 3t^2 \quad y = 2t^3 \quad (0 \leq t \leq 1)$

$$\frac{dx}{dt} = 6t \quad \frac{dy}{dt} = 6t^2$$

$$\text{Length} = \int_0^1 \sqrt{(6t)^2 + (6t^2)^2}\, dt$$

$$= 6\int_0^1 t\sqrt{1+t^2}\, dt \quad \text{Let } u = 1+t^2$$
$$\qquad\qquad\qquad\qquad\qquad du = 2t\, dt$$

$$= 3\int_1^2 \sqrt{u}\, du = 2u^{3/2}\Big|_1^2 = 4\sqrt{2} - 2 \text{ units}$$

2. If $x = 1 + t^3$ and $y = 1 - t^2$ for $-1 \leq t \leq 2$, then the arc length is

$$s = \int_{-1}^{2} \sqrt{(3t^2)^2 + (-2t)^2}\, dt$$

$$= \int_{-1}^{2} |t|\sqrt{9t^2 + 4}\, dt$$

$$= \left(\int_0^1 + \int_0^2\right) t\sqrt{9t^2+4}\, dt \quad \text{Let } u = 9t^2+4$$
$$\qquad\qquad\qquad\qquad\qquad\qquad du = 18t\, dt$$

$$= \frac{1}{18}\left(\int_4^{13} + \int_4^{40}\right)\sqrt{u}\, du$$

$$= \frac{1}{27}\left(13\sqrt{13} + 40\sqrt{40} - 16\right) \text{ units.}$$

3. $x = a\cos^3 t$, $y = a\sin^3 t$, $(0 \leq t \leq 2\pi)$. The length is

$$\int_0^{2\pi} \sqrt{9a^2 \cos^4 t \sin^2 t + 9a^2 \sin^4 t \cos^2 t}\, dt$$

$$= 3a \int_0^{2\pi} |\sin t \cos t|\, dt$$

$$= 12a \int_0^{\pi/2} \frac{1}{2} \sin 2t\, dt$$

$$= 6a \left(-\frac{\cos 2t}{2} \right) \Big|_0^{\pi/2} = 6a \text{ units.}$$

4. If $x = \ln(1 + t^2)$ and $y = 2\tan^{-1} t$ for $0 \leq t \leq 1$, then

$$\frac{dx}{dt} = \frac{2t}{1 + t^2}; \qquad \frac{dy}{dt} = \frac{2}{1 + t^2}.$$

The arc length is

$$s = \int_0^1 \sqrt{\frac{4t^2 + 4}{(1 + t^2)^2}}\, dt$$

$$= 2 \int_0^1 \frac{dt}{\sqrt{1 + t^2}} \qquad \begin{aligned} &\text{Let } t = \tan\theta \\ &dt = \sec^2\theta\, d\theta \end{aligned}$$

$$= 2 \int_0^{\pi/4} \sec\theta\, d\theta$$

$$= 2 \ln|\sec\theta + \tan\theta| \Big|_0^{\pi/4} = 2\ln(1 + \sqrt{2}) \text{ units.}$$

5. $x = t^2 \sin t$, $y = t^2 \cos t$, $(0 \leq t \leq 2\pi)$.

$$\frac{dx}{dt} = 2t \sin t + t^2 \cos t$$

$$\frac{dy}{dt} = 2t \cos t - t^2 \sin t$$

$$\left(\frac{ds}{dt} \right)^2 = t^2 \Big[4\sin^2 t + 4t \sin t \cos t + t^2 \cos^2 t$$
$$\qquad\qquad + 4\cos^2 t - 4t \sin t \cos t + t^2 \sin^2 t \Big]$$

$$= t^2(4 + t^2).$$

The length of the curve is

$$\int_0^{2\pi} t\sqrt{4 + t^2}\, dt \qquad \begin{aligned} &\text{Let } u = 4 + t^2 \\ &du = 2t\, dt \end{aligned}$$

$$= \frac{1}{2} \int_4^{4 + 4\pi^2} u^{1/2}\, du = \frac{1}{3} u^{3/2} \Big|_4^{4 + 4\pi^2}$$

$$= \frac{8}{3} \left((1 + \pi^2)^{3/2} - 1 \right) \text{ units.}$$

6. $x = \cos t + t\sin t \qquad y = \sin t - t\cos t \quad (0 \leq t \leq 2\pi)$

$$\frac{dx}{dt} = t\cos t \qquad \frac{dy}{dt} = t\sin t$$

$$\text{Length} = \int_0^{2\pi} \sqrt{t^2 \cos^2 t + t^2 \sin^2 t}\, dt$$

$$= \int_0^{2\pi} t\, dt = \frac{t^2}{2} \Big|_0^{2\pi} = 2\pi^2 \text{ units.}$$

7. $x = t + \sin t \qquad y = \cos t \quad (0 \leq t \leq \pi)$

$$\frac{dx}{dt} = 1 + \cos t \qquad \frac{dy}{dt} = -\sin t$$

$$\text{Length} = \int_0^{\pi} \sqrt{1 + 2\cos t + \cos^2 t + \sin^2 t}\, dt$$

$$= \int_0^{\pi} \sqrt{4\cos^2(t/2)}\, dt = 2 \int_0^{\pi} \cos\frac{t}{2}\, dt$$

$$= 4 \sin\frac{t}{2} \Big|_0^{\pi} = 4 \text{ units.}$$

8. $x = \sin^2 t \qquad y = 2\cos t \quad (0 \leq t \leq \pi/2)$

$$\frac{dx}{dt} = 2\sin t \cos t \qquad \frac{dy}{dt} = -2\sin t$$

Length

$$= \int_0^{\pi/2} \sqrt{4\sin^2 t \cos^2 t + 4\sin^2 t}\, dt$$

$$= 2 \int_0^{\pi/2} \sin t \sqrt{1 + \cos^2 t}\, dt \quad \begin{aligned} &\text{Let } \cos t = \tan u \\ &-\sin t\, dt = \sec^2 u\, du \end{aligned}$$

$$= 2 \int_0^{\pi/4} \sec^3 u\, du$$

$$= \left(\sec u \tan u + \ln(\sec u + \tan u) \right) \Big|_0^{\pi/4}$$

$$= \sqrt{2} + \ln(1 + \sqrt{2}) \text{ units.}$$

9. $x = a(t - \sin t) \qquad y = a(1 - \cos t) \quad (0 \leq t \leq 2\pi)$

$$\frac{dx}{dt} = a(1 - \cos t) \qquad \frac{dy}{dt} = a\sin t$$

$$\text{Length} = \int_0^{2\pi} \sqrt{a^2(1 - 2\cos t + \cos^2 t + \sin^2 t)}\, dt$$

$$= a \int_0^{2\pi} \sqrt{2 - 2\cos t}\, dt = a \int_0^{2\pi} \sqrt{\sin^2 \frac{t}{2}}\, dt$$

$$= 2a \int_0^{\pi} \sin\frac{t}{2}\, dt = -4a \cos\frac{t}{2} \Big|_0^{\pi} = 4a \text{ units.}$$

10. If $x = at - a\sin t$ and $y = a - a\cos t$ for $0 \le t \le 2\pi$, then

$$\frac{dx}{dt} = a - a\cos t, \qquad \frac{dy}{dt} = a\sin t;$$

$$ds = \sqrt{(a - a\cos t)^2 + (a\sin t)^2}\, dt$$
$$= a\sqrt{2}\sqrt{1 - \cos t}\, dt = a\sqrt{2}\sqrt{2\sin^2\left(\frac{t}{2}\right)}\, dt$$
$$= 2a\sin\left(\frac{t}{2}\right)\, dt.$$

a) The surface area generated by rotating the arch about the x-axis is

$$S_x = 2\pi \int_0^{2\pi} |y|\, ds$$
$$= 4\pi \int_0^{\pi} (a - a\cos t) 2a\sin\left(\frac{t}{2}\right)\, dt$$
$$= 16\pi a^2 \int_0^{\pi} \sin^3\left(\frac{t}{2}\right)\, dt$$
$$= 16\pi a^2 \int_0^{\pi} \left[1 - \cos^2\left(\frac{t}{2}\right)\right]\sin\left(\frac{t}{2}\right)\, dt$$

Let $u = \cos\left(\frac{t}{2}\right)$

$du = -\frac{1}{2}\sin\left(\frac{t}{2}\right)\, dt$

$$= -32\pi a^2 \int_1^0 (1 - u^2)\, du$$
$$= 32\pi a^2 \left[u - \frac{1}{3}u^3\right]\Big|_0^1$$
$$= \frac{64}{3}\pi a^3 \text{ sq. units.}$$

b) The surface area generated by rotating the arch about the y-axis is

$$S_y = 2\pi \int_0^{2\pi} |x|\, ds$$
$$= 2\pi \int_0^{2\pi} (at - a\sin t) 2a\sin\left(\frac{t}{2}\right)\, dt$$
$$= 4\pi a^2 \int_0^{2\pi} \left[t - 2\sin\left(\frac{t}{2}\right)\cos\left(\frac{t}{2}\right)\right]\sin\left(\frac{t}{2}\right)\, dt$$
$$= 4\pi a^2 \int_0^{2\pi} t\sin\left(\frac{t}{2}\right)\, dt$$
$$\quad - 8\pi a^2 \int_0^{2\pi} \sin^2\left(\frac{t}{2}\right)\cos\left(\frac{t}{2}\right)\, dt$$
$$= 4\pi a^2\left[-2t\cos\left(\frac{t}{2}\right)\Big|_0^{2\pi} + 2\int_0^{2\pi}\cos\left(\frac{t}{2}\right)\, dt\right] - 0$$
$$= 4\pi a^2[4\pi + 0] = 16\pi^2 a^2 \text{ sq. units.}$$

11. $x = e^t \cos t \qquad y = e^t \sin t \quad (0 \le t \le \pi/2)$

$\frac{dx}{dt} = e^t(\cos t - \sin t) \qquad \frac{dy}{dt} = e^t(\sin t + \cos t)$

Arc length element:

$$ds = \sqrt{e^{2t}(\cos t - \sin t)^2 + e^{2t}(\sin t + \cos t)^2}\, dt$$
$$= \sqrt{2}e^t\, dt.$$

The area of revolution about the x-axis is

$$\int_{t=0}^{t=\pi/2} 2\pi y\, ds = 2\sqrt{2}\pi \int_0^{\pi/2} e^{2t}\sin t\, dt$$
$$= 2\sqrt{2}\pi \frac{e^{2t}}{5}(2\sin t - \cos t)\Big|_0^{\pi/2}$$
$$= \frac{2\sqrt{2}\pi}{5}(2e^\pi + 1) \text{ sq. units.}$$

12. The area of revolution of the curve in Exercise 11 about the y-axis is

$$\int_{t=0}^{t=\pi/2} 2\pi x\, ds = 2\sqrt{2}\pi \int_0^{\pi/2} e^{2t}\cos t\, dt$$
$$= 2\sqrt{2}\pi \frac{e^{2t}}{5}(2\cos t + \sin t)\Big|_0^{\pi/2}$$
$$= \frac{2\sqrt{2}\pi}{5}(e^\pi - 2) \text{ sq. units.}$$

13. $x = 3t^2 \qquad y = 2t^3 \quad (0 \le t \le 1)$

$\frac{dx}{dt} = 6t \qquad \frac{dy}{dt} = 6t^2$

Arc length element:
$ds = \sqrt{36(t^2 + t^4)}\, dt = 6t\sqrt{1 + t^2}\, dt.$

The area of revolution about the y-axis is

$$\int_{t=0}^{t=1} 2\pi x\, ds = 36\pi \int_0^1 t^3\sqrt{1 + t^2}\, dt \quad \text{Let } u = 1 + t^2$$
$$\qquad\qquad du = 2t\, dt$$
$$= 18\pi \int_1^2 (u - 1)\sqrt{u}\, du$$
$$= 18\pi \left(\frac{2}{5}u^{5/2} - \frac{2}{3}u^{3/2}\right)\Big|_1^2$$
$$= \frac{72\pi}{15}(1 + \sqrt{2}) \text{ sq. units.}$$

14. The area of revolution of the curve of Exercise 13 about the x-axis is

$$\int_{t=0}^{t=1} 2\pi y\, ds = 24\pi \int_0^1 t^4\sqrt{1+t^2}\, dt \quad \text{Let } t = \tan u$$
$$\hspace{6em} dt = \sec^2 u\, du$$
$$= 24\pi \int_0^{\pi/4} \tan^4 u \sec^3 u\, du$$
$$= 24\pi \int_0^{\pi/4} (\sec^7 u - 2\sec^5 u + \sec^3 u)\, du$$
$$= \frac{\pi}{2}\left(7\sqrt{2} + 3\ln(1+\sqrt{2})\right) \text{ sq. units.}$$

We have omitted the details of evaluation of the final integral. See Exercise 24 of Section 8.3 for a similar evaluation.

15. $x = t^3 - 4t$, $y = t^2$, $(-2 \le t \le 2)$.

$$\text{Area} = \int_{-2}^{2} t^2(3t^2 - 4)\, dt$$
$$= 2\int_0^2 (3t^4 - 4t^2)\, dt$$
$$= 2\left(\frac{3t^5}{5} - \frac{4t^3}{3}\right)\Big|_0^2 = \frac{256}{15} \text{ sq. units.}$$

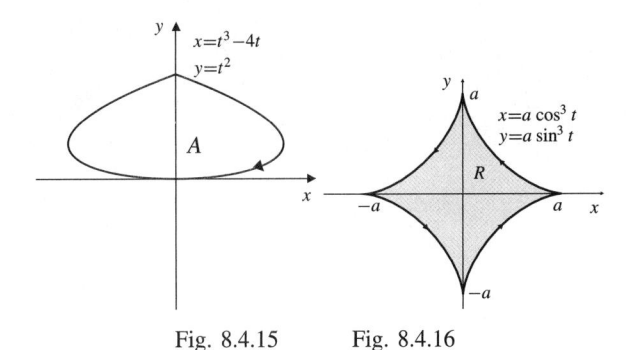

Fig. 8.4.15   Fig. 8.4.16

16. Area of $R = 4 \times \int_{\pi/2}^{0} (a\sin^3 t)(-3a\sin t \cos^2 t)\, dt$

$$= -12a^2 \int_{\pi/2}^{0} \sin^4 t \cos^2 t\, dt$$
$$= 12a^2 \left[\frac{t}{16} - \frac{\sin(4t)}{64} - \frac{\sin^3(2t)}{48}\right]\Big|_0^{\pi/2}$$
(See Exercise 34 of Section 6.4.)
$$= \frac{3}{8}\pi a^2 \text{ sq. units.}$$

17. $x = \sin^4 t$, $y = \cos^4 t$, $\left(0 \le t \le \frac{\pi}{2}\right)$.

$$\text{Area} = \int_0^{\pi/2} (\cos^4 t)(4\sin^3 t \cos t)\, dt$$
$$= 4\int_0^{\pi/2} \cos^5 t(1 - \cos^2 t)\sin t\, dt \quad \text{Let } u = \cos t$$
$$\hspace{10em} du = -\sin t\, dt$$
$$= 4\int_0^1 (u^5 - u^7)\, du = 6\left(\frac{1}{6} - \frac{1}{8}\right) = \frac{1}{6} \text{ sq. units.}$$

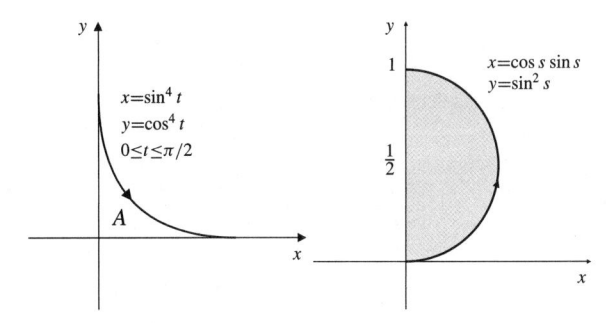

Fig. 8.4.17   Fig. 8.4.18

18. If $x = \cos s \sin s = \frac{1}{2}\sin 2s$ and $y = \sin^2 s = \frac{1}{2} - \frac{1}{2}\cos 2s$ for $0 \le s \le \frac{1}{2}\pi$, then

$$x^2 + \left(y - \frac{1}{2}\right)^2 = \frac{1}{4}\sin^2 2s + \frac{1}{4}\cos^2 2s = \frac{1}{4}$$

which is the right half of the circle with radius $\frac{1}{2}$ and centre at $(0, \frac{1}{2})$. Hence, the area of R is

$$\frac{1}{2}\left[\pi\left(\frac{1}{2}\right)^2\right] = \frac{\pi}{8} \text{ sq. units.}$$

19. $x = (2 + \sin t)\cos t$, $y = (2 + \sin t)\sin t$, $(0 \le t \le 2\pi)$. This is just the polar curve $r = 2 + \sin\theta$.

$$\text{Area} = -\int_0^{2\pi} (2 + \sin t)\sin t \frac{d}{dt}\big((2 + \sin t)\cos t\big)\, dt$$
$$= -\int_0^{2\pi} (2\sin t + \sin^2 t)(\cos^2 t - 2\sin t - \sin^2 t)\, dt$$
$$= \int_0^{2\pi} \big[4\sin^2 t + 4\sin^3 t + \sin^4 t$$
$$\hspace{4em} - 2\sin t \cos^2 t - \sin^2 t \cos^2 t\big]\, dt$$
$$= \int_0^{2\pi} \left[2(1 - \cos 2t) + \frac{1 - \cos 2t}{2}(-\cos 2t)\right]\, dt$$
$$\hspace{2em} + \int_0^{2\pi} \sin t\big[4 - 6\cos^2 t\big]\, dt$$
$$= 4\pi + \frac{\pi}{2} + 0 = \frac{9\pi}{2} \text{ sq. units.}$$

SECTION 8.4 (PAGE 504)

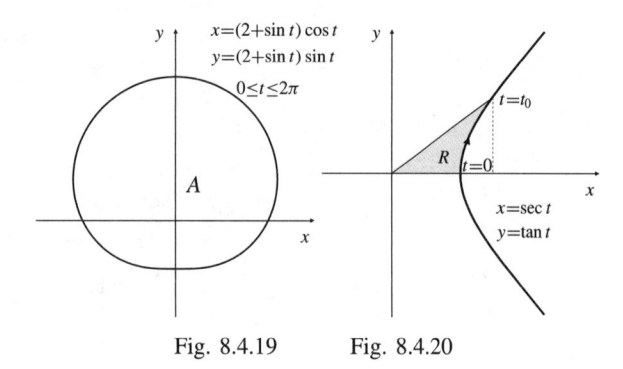

Fig. 8.4.19   Fig. 8.4.20

20. To find the shaded area we subtract the area under the upper half of the hyperbola from that of a right triangle:

Shaded area = Area $\triangle ABC$ − Area sector ABC

$= \dfrac{1}{2}\sec t_0 \tan t_0 - \displaystyle\int_0^{t_0} \tan t(\sec t \tan t)\,dt$

$= \dfrac{1}{2}\sec t_0 \tan t_0 - \displaystyle\int_0^{t_0} (\sec^3 t - \sec t)\,dt$

$= \dfrac{1}{2}\sec t_0 \tan t_0 - \left[\dfrac{1}{2}\sec t \tan t + \dfrac{1}{2}\ln|\sec t + \tan t| - \ln|\sec t + \tan t|\right]\Big|_0^{t_0}$

$= \dfrac{1}{2}\ln|\sec t_0 + \tan t_0|$ sq. units.

21. See the figure below. The area is the area of a triangle less the area under the hyperbola:

$A = \dfrac{1}{2}\cosh t_0 \sinh t_0 - \displaystyle\int_0^{t_0} \sinh t \sinh t\,dt$

$= \dfrac{1}{4}\sinh 2t_0 - \displaystyle\int_0^{t_0} \dfrac{\cosh 2t - 1}{2}\,dt$

$= \dfrac{1}{4}\sinh 2t_0 - \dfrac{1}{4}\sinh 2t_0 + \dfrac{1}{2}t_0$

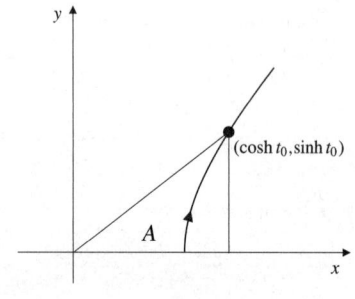

$= \dfrac{t_0}{2}$ sq. units.

Fig. 8.4.21

22. If $x = f(t) = at - a\sin t$ and $y = g(t) = a - a\cos t$, then the volume of the solid obtained by rotating about the x-axis is

$V = \displaystyle\int_{t=0}^{t=2\pi} \pi y^2\,dx = \pi \int_{t=0}^{t=2\pi} [g(t)]^2 f'(t)\,dt$

$= \pi \displaystyle\int_0^{2\pi} (a - a\cos t)^2 (a - a\cos t)\,dt$

$= \pi a^3 \displaystyle\int_0^{2\pi} (1 - \cos t)^3\,dt$

$= \pi a^3 \displaystyle\int_0^{2\pi} (1 - 3\cos t + 3\cos^2 t - \cos^3 t)\,dt$

$= \pi a^3 \left[2\pi - 0 + \dfrac{3}{2}\displaystyle\int_0^{2\pi}(1 + \cos 2t)\,dt - 0\right]$

$= \pi a^3 \left[2\pi + \dfrac{3}{2}(2\pi)\right] = 5\pi^2 a^3$ cu. units.

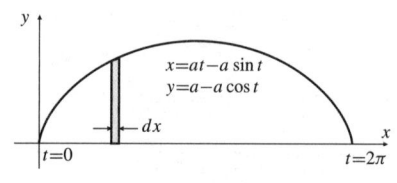

Fig. 8.4.22

23. Half of the volume corresponds to rotating $x = a\cos^3 t$, $y = a\sin^3 t$ ($0 \le t \le \pi/2$) about the x-axis. The whole volume is

$V = 2\displaystyle\int_0^{\pi/2} \pi y^2 (-dx)$

$= 2\pi \displaystyle\int_0^{\pi/2} a^2 \sin^6 t (3a\cos^2 t \sin t)\,dt$

$= 6\pi a^3 \displaystyle\int_0^{\pi/2} (1 - \cos^2 t)^3 \cos^2 t \sin t\,dt$ Let $u = \cos t$
$\qquad\qquad\qquad\qquad\qquad\qquad\qquad\qquad du = -\sin t\,dt$

$= 6\pi a^3 \displaystyle\int_0^1 (1 - 3u^2 + 3u^4 - u^6)u^2\,du$

$= 6\pi a^3 \left(\dfrac{1}{3} - \dfrac{3}{5} + \dfrac{3}{7} - \dfrac{1}{9}\right) = \dfrac{32\pi a^3}{105}$ cu. units.

Section 8.5 Polar Coordinates and Polar Curves (page 511)

1. $r = 3\sec\theta$
$r\cos\theta = 3$
$x = 3$ vertical straight line.

2. $r = -2\csc\theta \Rightarrow r\sin\theta = -2$
$\Leftrightarrow y = -2$ a horizontal line.

INSTRUCTOR'S SOLUTIONS MANUAL SECTION 8.5 (PAGE 511)

3. $r = 5/(3\sin\theta - 4\cos\theta)$
   $3r\sin\theta - 4r\cos\theta = 5$
   $3y - 4x = 5$ straight line.

4. $r = \sin\theta + \cos\theta$
   $r^2 = r\sin\theta + r\cos\theta$
   $x^2 + y^2 = y + x$
   $\left(x - \frac{1}{2}\right)^2 + \left(y - \frac{1}{2}\right)^2 = \frac{1}{2}$
   a circle with centre $\left(\frac{1}{2}, \frac{1}{2}\right)$ and radius $\frac{1}{\sqrt{2}}$.

5. $r^2 = \csc 2\theta$
   $r^2 \sin 2\theta = 1$
   $2r^2 \sin\theta \cos\theta = 1$
   $2xy = 1$ a rectangular hyperbola.

6. $r = \sec\theta \tan\theta \Rightarrow r\cos\theta = \dfrac{r\sin\theta}{r\cos\theta}$
   $x^2 = y$ a parabola.

7. $r = \sec\theta(1 + \tan\theta)$
   $r\cos\theta = 1 + \tan\theta$
   $x = 1 + \dfrac{y}{x}$
   $x^2 - x - y = 0$ a parabola.

8. $r = \dfrac{2}{\sqrt{\cos^2\theta + 4\sin^2\theta}}$
   $r^2\cos^2\theta + 4r^2\sin^2\theta = 4$
   $x^2 + 4y^2 = 4$ an ellipse.

9. $r = \dfrac{1}{1 - \cos\theta}$
   $r - x = 1$
   $r^2 = (1+x)^2$
   $x^2 + y^2 = 1 + 2x + x^2$
   $y^2 = 1 + 2x$ a parabola.

10. $r = \dfrac{2}{2 - \cos\theta}$
    $2r - r\cos\theta = 2$
    $4r^2 = (2+x)^2$
    $4x^2 + 4y^2 = 4 + 4x + x^2$
    $3x^2 + 4y^2 - 4x = 4$ an ellipse.

11. $r = \dfrac{2}{1 - 2\sin\theta}$
    $r - 2y = 2$
    $x^2 + y^2 = r^2 = 4(1+y)^2 = 4 + 8y + 4y^2$
    $x^2 - 3y^2 - 8y = 4$ a hyperbola.

12. $r = \dfrac{2}{1 + \sin\theta}$
    $r + r\sin\theta = 2$
    $r^2 = (2 - y)^2$
    $x^2 + y^2 = 4 - 4y + y^2$
    $x^2 = 4 - 4y$ a parabola.

13. $r = 1 + \sin\theta$ (cardioid)

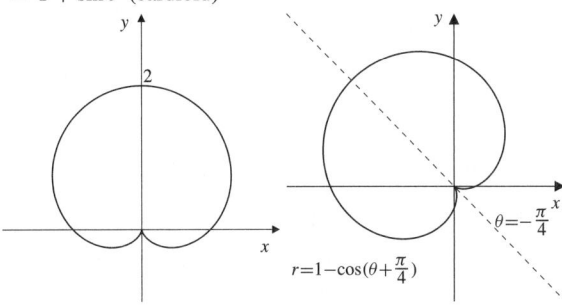

Fig. 8.5.13   Fig. 8.5.14

14. If $r = 1 - \cos\left(\theta + \dfrac{\pi}{4}\right)$, then $r = 0$ at $\theta = -\dfrac{\pi}{4}$ and $\dfrac{7\pi}{4}$. This is a cardioid.

15. $r = 1 + 2\cos\theta$
    $r = 0$ if $\theta = \pm 2\pi/3$.

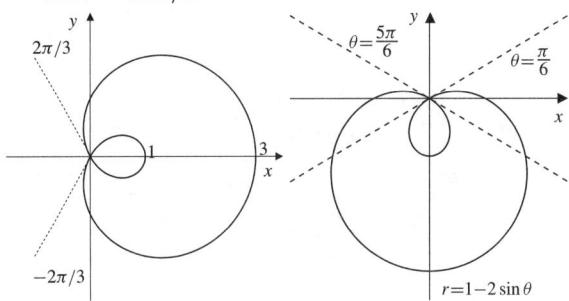

Fig. 8.5.15   Fig. 8.5.16

16. If $r = 1 - 2\sin\theta$, then $r = 0$ at $\theta = \dfrac{\pi}{6}$ and $\dfrac{5\pi}{6}$.

17. $r = 2 + \cos\theta$

Fig. 8.5.17   Fig. 8.5.18

18. If $r = 2\sin 2\theta$, then $r = 0$ at $\theta = 0$, $\pm\dfrac{\pi}{2}$ and π.

19. $r = \cos 3\theta$ (three leaf rosette)
$r = 0$ at $\theta = \pm\pi/6$, $\pm\pi/2$, $\pm 5\pi/6$.

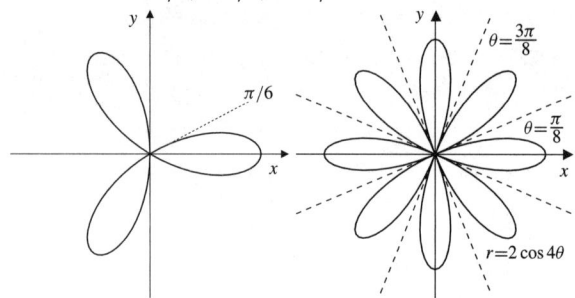

Fig. 8.5.19   Fig. 8.5.20

20. If $r = 2\cos 4\theta$, then $r = 0$ at $\theta = \pm\dfrac{\pi}{8}$, $\pm\dfrac{3\pi}{8}$, $\pm\dfrac{5\pi}{8}$ and $\pm\dfrac{7\pi}{8}$. (an eight leaf rosette)

21. $r^2 = 4\sin 2\theta$. Thus $r = \pm 2\sqrt{\sin 2\theta}$. This is a lemniscate. $r = 0$ at $\theta = 0$, $\theta = \pm\pi/2$, and $\theta = \pi$.

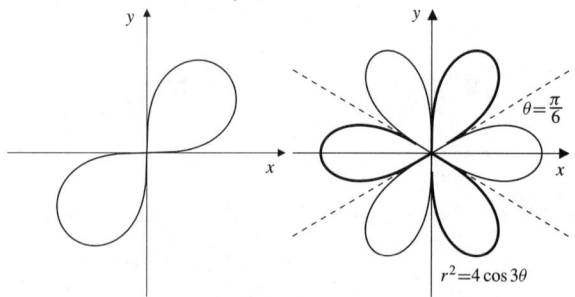

Fig. 8.5.21   Fig. 8.5.22

22. If $r^2 = 4\cos 3\theta$, then $r = 0$ at $\theta = \pm\dfrac{\pi}{6}$, $\pm\dfrac{\pi}{2}$ and $\pm\dfrac{5\pi}{6}$. This equation defines two functions of r, namely $r = \pm 2\sqrt{\cos 3\theta}$. Each contributes 3 leaves to the graph.

23. $r^2 = \sin 3\theta$. Thus $r = \pm\sqrt{\sin 3\theta}$. This is a lemniscate. $r = 0$ at $\theta = 0$, $\pm\pi/3$, $\pm 2\pi/3$, π.

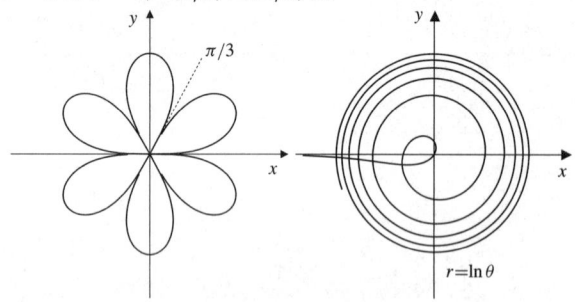

Fig. 8.5.23   Fig. 8.5.24

24. If $r = \ln\theta$, then $r = 0$ at $\theta = 1$. Note that
$$y = r\sin\theta = \ln\theta \sin\theta = (\theta\ln\theta)\left(\dfrac{\sin\theta}{\theta}\right) \to 0$$

as $\theta \to 0+$. Therefore, the (negative) x-axis is an asymptote of the curve.

25. $r = \sqrt{3}\cos\theta$, and $r = \sin\theta$ both pass through the origin, and so intersect there. Also
$\sin\theta = \sqrt{3}\cos\theta \;\Rightarrow\; \tan\theta = \sqrt{3} \;\Rightarrow\; \theta = \pi/3,\; 4\pi/3$.
Both of these give the same point $[\sqrt{3}/2, \pi/3]$.
Intersections: the origin and $[\sqrt{3}/2, \pi/3]$.

26. $r^2 = 2\cos(2\theta)$, $r = 1$.
$\cos(2\theta) = 1/2 \;\Rightarrow\; \theta = \pm\pi/6$ or $\theta = \pm 5\pi/6$.
Intersections: $[1, \pm\pi/6]$ and $[1, \pm 5\pi/6]$.

27. $r = 1 + \cos\theta$, $r = 3\cos\theta$. Both curves pass through the origin, so intersect there. Also
$3\cos\theta = 1 + \cos\theta \;\Rightarrow\; \cos\theta = 1/2 \;\Rightarrow\; \theta = \pm\pi/3$.
Intersections: the origin and $[3/2, \pm\pi/3]$.

28. Let $r_1(\theta) = \theta$ and $r_2(\theta) = \theta + \pi$. Although the equation $r_1(\theta) = r_2(\theta)$ has no solutions, the curves $r = r_1(\theta)$ and $r = r_2(\theta)$ can still intersect if $r_1(\theta_1) = -r_2(\theta_2)$ for two angles θ_1 and θ_2 having the opposite directions in the polar plane. Observe that $\theta_1 = -n\pi$ and $\theta_2 = (n-1)\pi$ are two such angles provided n is any integer. Since
$$r_1(\theta_1) = -n\pi = -r_2((n-1)\pi),$$
the curves intersect at any point of the form $[n\pi, 0]$ or $[n\pi, \pi]$.

29. If $r = 1/\theta$ for $\theta > 0$, then
$$\lim_{\theta\to 0+} y = \lim_{\theta\to 0+} \dfrac{\sin\theta}{\theta} = 1.$$
Thus $y = 1$ is a horizontal asymptote.

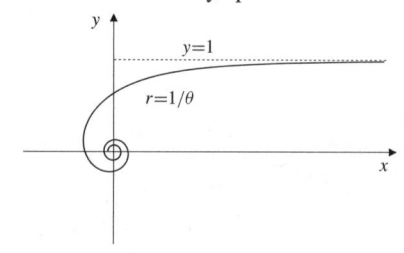

Fig. 8.5.29

30. The graph of $r = \cos n\theta$ has $2n$ leaves if n is an even integer and n leaves if n is an odd integer. The situation for $r^2 = \cos n\theta$ is reversed. The graph has $2n$ leaves if n is an odd integer (provided negative values of r are allowed), and it has n leaves if n is even.

31. If $r = f(\theta)$, then
$$x = r\cos\theta = f(\theta)\cos\theta$$
$$y = r\sin\theta = f(\theta)\sin\theta.$$

32. $r = \cos\theta\cos(m\theta)$
For odd m this flower has $2m$ petals, 2 large ones and 4 each of $(m-1)/2$ smaller sizes.
For even m the flower has $m+1$ petals, one large and 2 each of $m/2$ smaller sizes.

33. $r = 1 + \cos\theta\cos(m\theta)$
These are similar to the ones in Exercise 32, but the curve does not approach the origin except for $\theta = \pi$ in the case of even m. The petals are joined, and less distinct. The smaller ones cannot be distinguished.

34. $r = \sin(2\theta)\sin(m\theta)$
For odd m there are $m+1$ petals, 2 each of $(m+1)/2$ different sizes.
For even m there are always $2m$ petals. They are of n different sizes if $m = 4n - 2$ or $m = 4n$.

35. $r = 1 + \sin(2\theta)\sin(m\theta)$
These are similar to the ones in Exercise 34, but the petals are joined, and less distinct. The smaller ones cannot be distinguished. There appear to be $m+2$ petals in both the even and odd cases.

36. $r = C + \cos\theta\cos(2\theta)$
The curve always has 3 bulges, one larger than the other two. For $C = 0$ these are 3 distinct petals. For $0 < C < 1$ there is a fourth supplementary petal inside the large one. For $C = 1$ the curve has a cusp at the origin. For $C > 1$ the curve does not approach the origin, and the petals become less distinct as C increases.

37. $r = C + \cos\theta\sin(3\theta)$
For $C < 1$ there appear to be 6 petals of 3 different sizes. For $C \geq 1$ there are only 4 of 2 sizes, and these coalesce as C increases.

38.

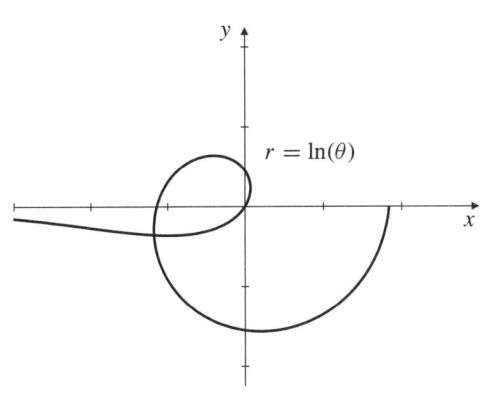

Fig. 8.5.38

We will have $[\ln\theta_1, \theta_1] = [\ln\theta_2, \theta_2]$ if

$$\theta_2 = \theta_1 + \pi \quad \text{and} \quad \ln\theta_1 = -\ln\theta_2,$$

that is, if $\ln\theta_1 + \ln(\theta_1 + \pi) = 0$. This equation has solution $\theta_1 \approx 0.29129956$. The corresponding intersection point has Cartesian coordinates $(\ln\theta_1\cos\theta_1, \ln\theta_1\sin\theta_1) \approx (-1.181442, -0.354230)$.

39.

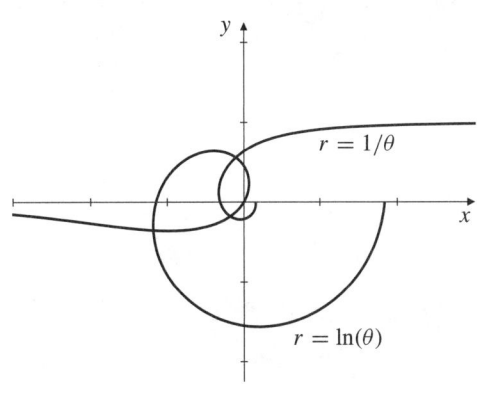

Fig. 8.5.39

The two intersections of $r = \ln\theta$ and $r = 1/\theta$ for $0 < \theta \leq 2\pi$ correspond to solutions θ_1 and θ_2 of

$$\ln\theta_1 = \frac{1}{\theta_1}, \quad \ln\theta_2 = -\frac{1}{\theta_2 + \pi}.$$

The first equation has solution $\theta_1 \approx 1.7632228$, giving the point $(-0.108461, 0.556676)$, and the second equation has solution $\theta_2 \approx 0.7746477$, giving the point $(-0.182488, -0.178606)$.

Section 8.6 Slopes, Areas, and Arc Lengths for Polar Curves (page 515)

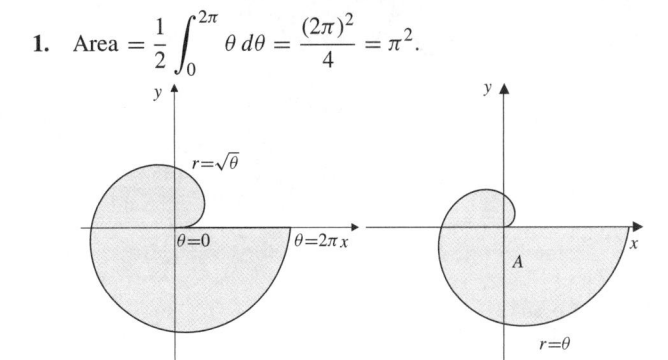

1. Area $= \dfrac{1}{2}\displaystyle\int_0^{2\pi}\theta\,d\theta = \dfrac{(2\pi)^2}{4} = \pi^2$.

Fig. 8.6.1   Fig. 8.6.2

2. Area $= \dfrac{1}{2}\displaystyle\int_0^{2\pi}\theta^2\,d\theta = \left.\dfrac{\theta^3}{6}\right|_0^{2\pi} = \dfrac{4}{3}\pi^3$ sq. units.

3. Area $= 4 \times \dfrac{1}{2}\displaystyle\int_0^{\pi/4} a^2\cos 2\theta\,d\theta$
$= \left.2a^2 \dfrac{\sin 2\theta}{2}\right|_0^{\pi/4} = a^2$ sq. units.

SECTION 8.6 (PAGE 515)

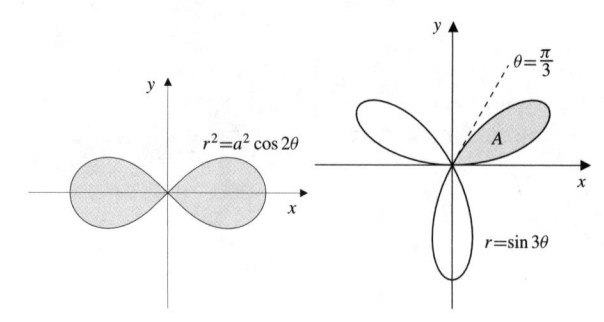

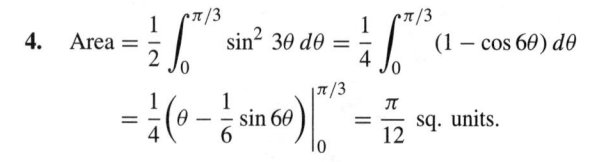

Fig. 8.6.3 Fig. 8.6.4 Fig. 8.6.7 Fig. 8.6.8

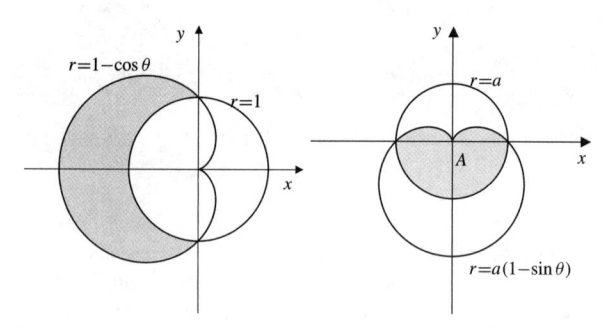

4. Area $= \dfrac{1}{2}\int_0^{\pi/3} \sin^2 3\theta\, d\theta = \dfrac{1}{4}\int_0^{\pi/3}(1-\cos 6\theta)\, d\theta$

$= \dfrac{1}{4}\left(\theta - \dfrac{1}{6}\sin 6\theta\right)\Big|_0^{\pi/3} = \dfrac{\pi}{12}$ sq. units.

5. Total area $= 16 \times \dfrac{1}{2}\int_0^{\pi/8} \cos^2 4\theta\, d\theta$

$= 4\int_0^{\pi/8}(1+\cos 8\theta)\, d\theta$

$= 4\left(\theta + \dfrac{\sin 8\theta}{8}\right)\Big|_0^{\pi/8} = \dfrac{\pi}{2}$ sq. units.

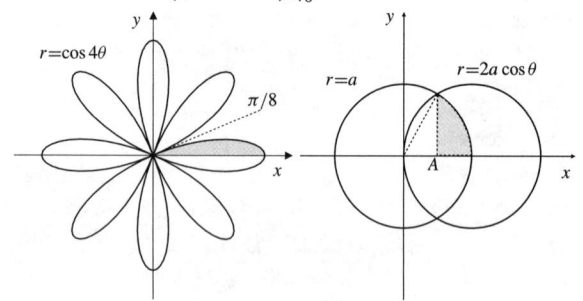

Fig. 8.6.5 Fig. 8.6.6

6. The circles $r = a$ and $r = 2a\cos\theta$ intersect at $\theta = \pm\pi/3$. By symmetry, the common area is $4 \times$ (area of sector $-$ area of right triangle) (see the figure), i.e.,

$4 \times \left[\left(\dfrac{1}{6}\pi a^2\right) - \left(\dfrac{1}{2}\dfrac{a}{2}\dfrac{\sqrt{3}a}{2}\right)\right] = \dfrac{4\pi - 3\sqrt{3}}{6}a^2$ sq. units.

7. Area $= 2 \times \dfrac{1}{2}\int_{\pi/2}^{\pi}(1-\cos\theta)^2\, d\theta - \dfrac{\pi}{2}$

$= \int_{\pi/2}^{\pi}\left(1 - 2\cos\theta + \dfrac{1+\cos 2\theta}{2}\right)d\theta - \dfrac{\pi}{2}$

$= \dfrac{3}{2}\left(\pi - \dfrac{\pi}{2}\right) - \left(2\sin\theta - \dfrac{\sin 2\theta}{4}\right)\Big|_{\pi/2}^{\pi} - \dfrac{\pi}{2}$

$= \dfrac{\pi}{4} + 2$ sq. units.

8. Area $= \dfrac{1}{2}\pi a^2 + 2 \times \dfrac{1}{2}\int_0^{\pi/2} a^2(1-\sin\theta)^2\, d\theta$

$= \dfrac{\pi a^2}{2} + a^2\int_0^{\pi/2}\left(1 - 2\sin\theta + \dfrac{1-\cos 2\theta}{2}\right)d\theta$

$= \dfrac{\pi a^2}{2} + a^2\left(\dfrac{3}{2}\theta + 2\cos\theta - \dfrac{1}{4}\sin 2\theta\right)\Big|_0^{\pi/2}$

$= \left(\dfrac{5\pi}{4} - 2\right)a^2$ sq. units.

9. For intersections: $1 + \cos\theta = 3\cos\theta$. Thus $2\cos\theta = 1$ and $\theta = \pm\pi/3$. The shaded area is given by

$2 \times \dfrac{1}{2}\left[\int_{\pi/3}^{\pi}(1+\cos\theta)^2\, d\theta - 9\int_{\pi/3}^{\pi/2}\cos^2\theta\, d\theta\right]$

$= \int_{\pi/3}^{\pi}\left(1 + 2\cos\theta + \dfrac{1+\cos 2\theta}{2}\right)d\theta$

$\quad - \dfrac{9}{2}\int_{\pi/3}^{\pi/2}(1+\cos 2\theta)\, d\theta$

$= \dfrac{3}{2}\left(\dfrac{2\pi}{3}\right) + \left(2\sin\theta + \dfrac{\sin 2\theta}{4}\right)\Big|_{\pi/3}^{\pi}$

$\quad - \dfrac{9}{2}\left(\theta + \dfrac{\sin 2\theta}{2}\right)\Big|_{\pi/3}^{\pi/2}$

$= \dfrac{\pi}{4} - \sqrt{3} - \dfrac{\sqrt{3}}{8} + \dfrac{9}{4}\left(\dfrac{\sqrt{3}}{2}\right) = \dfrac{\pi}{4}$ sq. units.

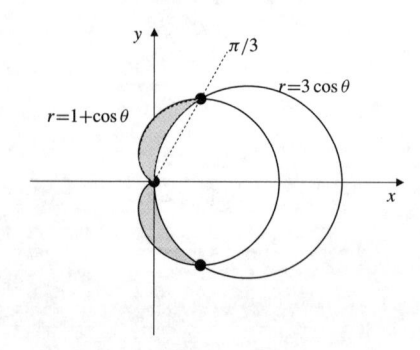

Fig. 8.6.9

10. Since $r^2 = 2\cos 2\theta$ meets $r = 1$ at $\theta = \pm\dfrac{\pi}{6}$ and $\pm\dfrac{5\pi}{6}$, the area inside the lemniscate and outside the circle is

$$4 \times \frac{1}{2}\int_0^{\pi/6} \left[2\cos 2\theta - 1^2\right] d\theta$$

$$= 2\sin 2\theta \Big|_0^{\pi/6} - \frac{\pi}{3} = \sqrt{3} - \frac{\pi}{3} \text{ sq. units.}$$

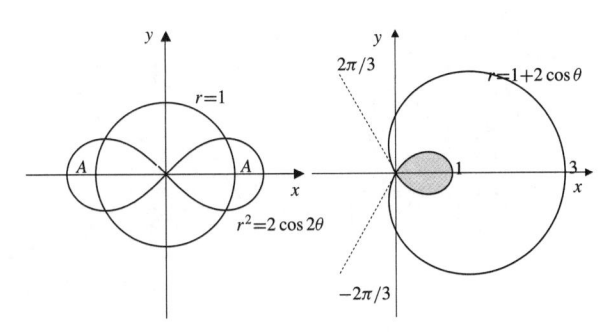

Fig. 8.6.10   Fig. 8.6.11

11. $r = 0$ at $\theta = \pm 2\pi/3$. The shaded area is

$$2 \times \frac{1}{2}\int_{2\pi/3}^{\pi} (1 + 2\cos\theta)^2 \, d\theta$$

$$= \int_{2\pi/3}^{\pi} \left(1 + 4\cos\theta + 2(1 + \cos 2\theta)\right) d\theta$$

$$= 3\left(\frac{\pi}{3}\right) + 4\sin\theta \Big|_{2\pi/3}^{\pi} + \sin 2\theta \Big|_{2\pi/3}^{\pi}$$

$$= \pi - 2\sqrt{3} + \frac{\sqrt{3}}{2} = \pi - \frac{3\sqrt{3}}{2} \text{ sq. units.}$$

12. $s = \int_0^{\pi} \sqrt{\left(\dfrac{dr}{d\theta}\right)^2 + r^2}\, d\theta = \int_0^{\pi} \sqrt{4\theta^2 + \theta^4}\, d\theta$

$$= \int_0^{\pi} \theta\sqrt{4 + \theta^2}\, d\theta \quad \text{Let } u = 4 + \theta^2$$
$$\qquad\qquad\qquad\qquad\qquad\quad du = 2\theta\, d\theta$$

$$= \frac{1}{2}\int_4^{4+\pi^2} \sqrt{u}\, du = \frac{1}{3}u^{3/2} \Big|_4^{4+\pi^2}$$

$$= \frac{1}{3}\left[(4 + \pi^2)^{3/2} - 8\right] \text{ units.}$$

13. $r = e^{a\theta}$, $(-\pi \le \theta \le \pi)$. $\dfrac{dr}{d\theta} = ae^{a\theta}$.
$ds = \sqrt{e^{2a\theta} + a^2 e^{2a\theta}}\, d\theta = \sqrt{1 + a^2}\, e^{a\theta}\, d\theta$. The length of the curve is

$$\int_{-\pi}^{\pi} \sqrt{1 + a^2}\, e^{a\theta}\, d\theta = \frac{\sqrt{1 + a^2}}{a}(e^{a\pi} - e^{-a\pi}) \text{ units.}$$

14. $s = \int_0^{2\pi} \sqrt{a^2 + a^2\theta^2}\, d\theta$

$$= a\int_0^{2\pi} \sqrt{1 + \theta^2}\, d\theta \quad \text{Let } \theta = \tan u$$
$$\qquad\qquad\qquad\qquad\qquad d\theta = \sec^2 u\, d\theta$$

$$= a\int_{\theta=0}^{\theta=2\pi} \sec^3 u\, du$$

$$= \frac{a}{2}\left(\sec u \tan u + \ln|\sec u + \tan u|\right)\Big|_{\theta=0}^{\theta=2\pi}$$

$$= \frac{a}{2}\left[\theta\sqrt{1 + \theta^2} + \ln|\sqrt{1 + \theta^2} + \theta|\right]\Big|_{\theta=0}^{\theta=2\pi}$$

$$= \frac{a}{2}\left[2\pi\sqrt{1 + 4\pi^2} + \ln(2\pi + \sqrt{1 + 4\pi^2})\right] \text{ units.}$$

15. $r^2 = \cos 2\theta$

$$2r\frac{dr}{d\theta} = -2\sin 2\theta \quad \Rightarrow \quad \frac{dr}{d\theta} = -\frac{\sin 2\theta}{r}$$

$$ds = \sqrt{\cos 2\theta + \frac{\sin^2 2\theta}{\cos 2\theta}}\, d\theta = \sqrt{\sec 2\theta}\, d\theta$$

$$\text{Length} = 4\int_0^{\pi/4} \sqrt{\sec 2\theta}\, d\theta.$$

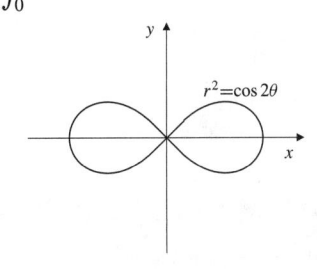

Fig. 8.6.15

16. If $r^2 = \cos 2\theta$, then

$$2r\frac{dr}{d\theta} = -2\sin 2\theta \Rightarrow \frac{dr}{d\theta} = -\frac{\sin 2\theta}{\sqrt{\cos 2\theta}}$$

and

$$ds = \sqrt{\cos 2\theta + \frac{\sin^2 2\theta}{\cos 2\theta}}\, d\theta = \frac{d\theta}{\sqrt{\cos 2\theta}}.$$

a) Area of the surface generated by rotation about the x-axis is

$$S_x = 2\pi \int_0^{\pi/4} r\sin\theta\, ds$$

$$= 2\pi \int_0^{\pi/4} \sqrt{\cos 2\theta}\, \sin\theta\, \frac{d\theta}{\sqrt{\cos 2\theta}}$$

$$= -2\pi \cos\theta \Big|_0^{\pi/4} = (2 - \sqrt{2})\pi \text{ sq. units.}$$

b) Area of the surface generated by rotation about the y-axis is

$$S_y = 2\pi \int_{-\pi/4}^{\pi/4} r\cos\theta \, ds$$
$$= 4\pi \int_0^{\pi/4} \sqrt{\cos 2\theta}\, \cos\theta \, \frac{d\theta}{\sqrt{\cos 2\theta}}$$
$$= 4\pi \sin\theta \Big|_0^{\pi/4} = 2\sqrt{2}\pi \text{ sq. units.}$$

17. For $r = 1 + \sin\theta$,

$$\tan\psi = \frac{r}{dr/d\theta} = \frac{1+\sin\theta}{\cos\theta}.$$

If $\theta = \pi/4$, then $\tan\psi = \sqrt{2}+1$ and $\psi = 3\pi/8$.
If $\theta = 5\pi/4$, then $\tan\psi = 1 - \sqrt{2}$ and $\psi = -\pi/8$.
The line $y = x$ meets the cardioid $r = 1 + \sin\theta$ at the origin at an angle of $45°$, and also at first and third quadrant points at angles of $67.5°$ and $-22.5°$ as shown in the figure.

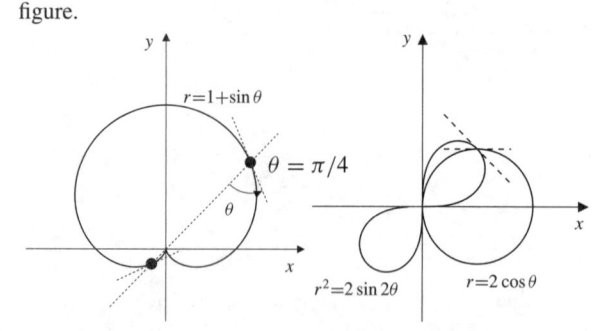

Fig. 8.6.17   Fig. 8.6.18

18. The two curves $r^2 = 2\sin 2\theta$ and $r = 2\cos\theta$ intersect where

$$2\sin 2\theta = 4\cos^2\theta$$
$$4\sin\theta\cos\theta = 4\cos^2\theta$$
$$(\sin\theta - \cos\theta)\cos\theta = 0$$
$$\Leftrightarrow \sin\theta = \cos\theta \text{ or } \cos\theta = 0,$$

i.e., at $P_1 = \left[\sqrt{2}, \frac{\pi}{4}\right]$ and $P_2 = (0,0)$.
For $r^2 = 2\sin 2\theta$ we have $2r\dfrac{dr}{d\theta} = 4\cos 2\theta$. At P_1 we have $r = \sqrt{2}$ and $dr/d\theta = 0$. Thus the angle ψ between the curve and the radial line $\theta = \pi/4$ is $\psi = \pi/2$.
For $r = 2\cos\theta$ we have $dr/d\theta = -2\sin\theta$, so the angle between this curve and the radial line $\theta = \pi/4$ satisfies

$$\tan\psi = \frac{r}{dr/d\theta}\bigg|_{\theta=\pi/4} = -1, \text{ and } \theta = 3\pi/4. \text{ The two}$$

curves intersect at P_1 at angle $\dfrac{3\pi}{4} - \dfrac{\pi}{2} = \dfrac{\pi}{4}$.
The Figure shows that at the origin, P_2, the circle meets the lemniscate twice, at angles 0 and $\pi/2$.

19. The curves $r = 1 - \cos\theta$ and $r = 1 - \sin\theta$ intersect on the rays $\theta = \pi/4$ and $\theta = 5\pi/4$, as well as at the origin. At the origin their cusps clearly intersect at right angles.
For $r = 1 - \cos\theta$, $\tan\psi_1 = (1 - \cos\theta)/\sin\theta$.
At $\theta = \pi/4$, $\tan\psi_1 = \sqrt{2} - 1$, so $\psi_1 = \pi/8$.
At $\theta = 5\pi/4$, $\tan\psi_1 = -(\sqrt{2}+1)$, so $\psi_1 = -3\pi/8$.
For $r = 1 - \sin\theta$, $\tan\psi_2 = (1 - \sin\theta)/(-\cos\theta)$.
At $\theta = \pi/4$, $\tan\psi_2 = 1 - \sqrt{2}$, so $\psi_2 = -\pi/8$.
At $\theta = 5\pi/4$, $\tan\psi_2 = \sqrt{2}+1$, so $\psi_2 = 3\pi/8$.
At $\pi/4$ the curves intersect at angle $\pi/8 - (-\pi/8) = \pi/4$.
At $5\pi/4$ the curves intersect at angle $3\pi/8 - (-3\pi/8)$
$= 3\pi/4$ (or $\pi/4$ if you use the supplementary angle).

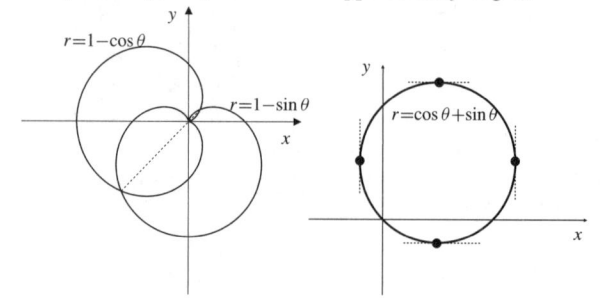

Fig. 8.6.19   Fig. 8.6.20

20. We have $r = \cos\theta + \sin\theta$. For horizontal tangents:

$$0 = \frac{dy}{d\theta} = \frac{d}{d\theta}\left(\cos\theta\sin\theta + \sin^2\theta\right)$$
$$= \cos^2\theta - \sin^2\theta + 2\sin\theta\cos\theta$$
$$\Leftrightarrow \cos 2\theta = -\sin 2\theta \Leftrightarrow \tan 2\theta = -1.$$

Thus $\theta = -\dfrac{\pi}{8}$ or $\dfrac{3\pi}{8}$. The tangents are horizontal at

$$\left[\cos\left(\frac{\pi}{8}\right) - \sin\left(\frac{\pi}{8}\right), -\frac{\pi}{8}\right] \text{ and}$$
$$\left[\cos\left(\frac{3\pi}{8}\right) + \sin\left(\frac{3\pi}{8}\right), \frac{3\pi}{8}\right].$$

For vertical tangent:

$$0 = \frac{dx}{d\theta} = \frac{d}{d\theta}\left(\cos^2\theta + \cos\theta\sin\theta\right)$$
$$= -2\cos\theta\sin\theta + \cos^2\theta - \sin^2\theta$$
$$\Leftrightarrow \sin 2\theta = \cos 2\theta \Leftrightarrow \tan 2\theta = 1.$$

Thus $\theta = \pi/8$ of $5\pi/8$. There are vertical tangents at

$$\left[\cos\left(\frac{\pi}{8}\right) + \sin\left(\frac{\pi}{8}\right), \frac{\pi}{8}\right] \text{ and}$$
$$\left[\cos\left(\frac{5\pi}{8}\right) + \sin\left(\frac{5\pi}{8}\right), \frac{5\pi}{8}\right].$$

21. $r = 2\cos\theta$. $\tan\psi = \dfrac{r}{dr/d\theta} = -\cot\theta$.
For horizontal tangents we want $\tan\psi = -\tan\theta$. Thus we want $-\tan\theta = -\cot\theta$, and so $\theta = \pm\pi/4$ or $\pm 3\pi/4$. The tangents are horizontal at $[\sqrt{2}, \pm\pi/4]$.
For vertical tangents we want $\tan\psi = \cot\theta$. Thus we want $-\cot\theta = \cot\theta$, and so $\theta = 0, \pm\pi/2$, or π. There are vertical tangents at the origin and at $[2, 0]$.

326

INSTRUCTOR'S SOLUTIONS MANUAL   SECTION 8.6 (PAGE 515)

Fig. 8.6.21   Fig. 8.6.22

22. We have $r^2 = \cos 2\theta$, and $2r\dfrac{dr}{d\theta} = -2\sin 2\theta$. For horizontal tangents:

$$0 = \frac{d}{d\theta} r\sin\theta = r\cos\theta + \sin\theta\left(-\frac{\sin 2\theta}{r}\right)$$

$\Leftrightarrow \quad \cos 2\theta \cos\theta = \sin 2\theta \sin\theta$

$\Leftrightarrow \quad (\cos^2\theta - \sin^2\theta)\cos\theta = 2\sin^2\theta\cos\theta$

$\Leftrightarrow \quad \cos\theta = 0 \quad \text{or} \quad \cos^2\theta = 3\sin^2\theta.$

There are no points on the curve where $\cos\theta = 0$. Therefore, horizontal tangents occur only where $\tan^2\theta = 1/3$. There are horizontal tangents at $\left[\dfrac{1}{\sqrt{2}}, \pm\dfrac{\pi}{6}\right]$ and $\left[\dfrac{1}{\sqrt{2}}, \pm\dfrac{5\pi}{6}\right]$.

For vertical tangents:

$$0 = \frac{d}{d\theta} r\cos\theta = -r\sin\theta + \cos\theta\left(-\frac{\sin 2\theta}{r}\right)$$

$\Leftrightarrow \quad \cos 2\theta \sin\theta = -\sin 2\theta \cos\theta$

$\Leftrightarrow \quad (\cos^2\theta - \sin^2\theta)\sin\theta = -2\sin\theta\cos^2\theta$

$\Leftrightarrow \quad \sin\theta = 0 \quad \text{or} \quad 3\cos^2\theta = \sin^2\theta.$

There are no points on the curve where $\tan^2\theta = 3$, so the only vertical tangents occur where $\sin\theta = 0$, that is, at the points with polar coordinates $[1, 0]$ and $[1, \pi]$.

23. $r = \sin 2\theta$. $\tan\psi = \dfrac{\sin 2\theta}{2\cos 2\theta} = \dfrac{1}{2}\tan 2\theta$.

For horizontal tangents:

$$\tan 2\theta = -2\tan\theta$$
$$\frac{2\tan\theta}{1 - \tan^2\theta} = -2\tan\theta$$
$$\tan\theta\left(1 + (1 - \tan^2\theta)\right) = 0$$
$$\tan\theta(2 - \tan^2\theta) = 0.$$

Thus $\theta = 0, \pi, \pm\tan^{-1}\sqrt{2}, \pi\pm\tan^{-1}\sqrt{2}$.

There are horizontal tangents at the origin and the points

$\left[\dfrac{2\sqrt{2}}{3}, \pm\tan^{-1}\sqrt{2}\right]$ and $\left[\dfrac{2\sqrt{2}}{3}, \pi\pm\tan^{-1}\sqrt{2}\right].$

Since the rosette $r = \sin 2\theta$ is symmetric about $x = y$, there must be vertical tangents at the origin and at the points

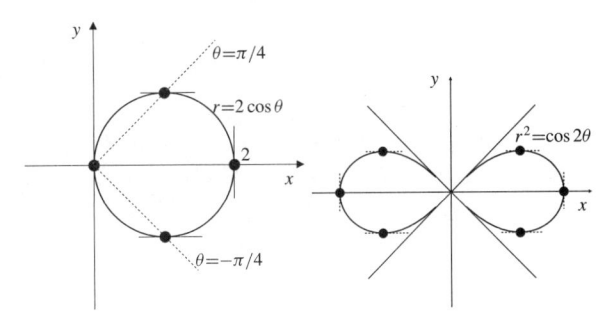

$\left[\dfrac{2\sqrt{2}}{3}, \pm\tan^{-1}\dfrac{1}{\sqrt{2}}\right]$ and $\left[\dfrac{2\sqrt{2}}{3}, \pi\pm\tan^{-1}\dfrac{1}{\sqrt{2}}\right].$

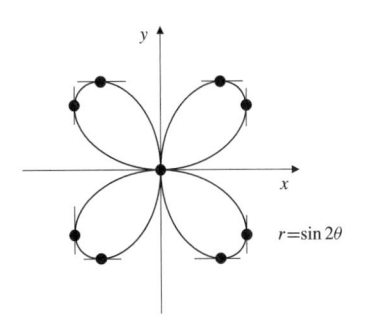

Fig. 8.6.23

24. We have $r = e^\theta$ and $\dfrac{dr}{d\theta} = e^\theta$. For horizontal tangents:

$$0 = \frac{d}{d\theta} r\sin\theta = e^\theta\cos\theta + e^\theta\sin\theta$$

$\Leftrightarrow \quad \tan\theta = -1 \quad \Leftrightarrow \quad \theta = -\dfrac{\pi}{4} + k\pi,$

where $k = 0, \pm 1, \pm 2, \ldots$. At the points $[e^{k\pi - \pi/4}, k\pi - \pi/4]$ the tangents are horizontal.

For vertical tangents:

$$0 = \frac{d}{d\theta} r\cos\theta = e^\theta\cos\theta - e^\theta\sin\theta$$

$\Leftrightarrow \quad \tan\theta = 1 \quad \Leftrightarrow \quad \theta = \dfrac{\pi}{4} + k\pi.$

At the points $[e^{k\pi + \pi/4}, k\pi + \pi/4]$ the tangents are vertical.

25. $r = 2(1 - \sin\theta)$, $\tan\psi = -\dfrac{1 - \sin\theta}{\cos\theta}$.

For horizontal tangents $\tan\psi = -\cot\theta$, so

$$-\frac{1 - \sin\theta}{\cos\theta} = -\frac{\sin\theta}{\cos\theta}$$
$$\cos\theta = 0, \quad \text{or} \quad 2\sin\theta = 1.$$

The solutions are $\theta = \pm\pi/2, \pm\pi/6$, and $\pm 5\pi/6$. $\theta = \pi/2$ corresponds to the origin where the cardioid has a cusp, and therefore no tangent. There are horizontal tangents at $[4, -\pi/2]$, $[1, \pi/6]$, and $[1, 5\pi/6]$.

For vertical tangents $\tan\psi = \cot\theta$, so

$$-\frac{1 - \sin\theta}{\cos\theta} = \frac{\cos\theta}{\sin\theta}$$
$$\sin^2\theta - \sin\theta = \cos^2\theta = 1 - \sin^2\theta$$
$$2\sin^2\theta - \sin\theta - 1 = 0$$
$$(\sin\theta - 1)(2\sin\theta + 1) = 0$$

327

The solutions here are $\theta = \pi/2$ (the origin again), $\theta = -\pi/6$ and $\theta = -5\pi/6$. There are vertical tangents at $[3, -\pi/6]$ and $[3, -5\pi/6]$.

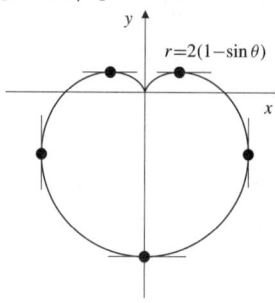

Fig. 8.6.25

26. $x = r\cos\theta = f(\theta)\cos\theta,\ y = r\sin\theta = f(\theta)\sin\theta.$

$$\frac{dx}{d\theta} = f'(\theta)\cos\theta - f(\theta)\sin\theta,\quad \frac{dy}{d\theta} = f'(\theta)\sin\theta + f(\theta)\cos\theta$$

$$ds = \sqrt{\big(f'(\theta)\cos\theta - f(\theta)\sin\theta\big)^2 + \big(f'(\theta)\sin\theta + f(\theta)\cos\theta\big)^2}\, d\theta$$

$$= \Big[\big(f'(\theta)\big)^2\cos^2\theta - 2f'(\theta)f(\theta)\cos\theta\sin\theta + \big(f(\theta)\big)^2\sin^2\theta$$

$$+ \big(f'(\theta)\big)^2\sin^2\theta + 2f'(\theta)f(\theta)\sin\theta\cos\theta + \big(f(\theta)\big)^2\cos^2\theta\Big]^{1/2}\, d\theta$$

$$= \sqrt{\big(f'(\theta)\big)^2 + \big(f(\theta)\big)^2}\, d\theta.$$

Review Exercises 8 (page 516)

1. $x^2 + 2y^2 = 2\ \Leftrightarrow\ \dfrac{x^2}{2} + y^2 = 1$
Ellipse, semi-major axis $a = \sqrt{2}$, along the x-axis. Semi-minor axis $b = 1$.
$c^2 = a^2 - b^2 = 1$. Foci: $(\pm 1, 0)$.

2. $9x^2 - 4y^2 = 36\ \Leftrightarrow\ \dfrac{x^2}{4} - \dfrac{y^2}{9} = 1$
Hyperbola, transverse axis along the x-axis.
Semi-transverse axis $a = 2$, semi-conjugate axis $b = 3$.
$c^2 = a^2 + b^2 = 13$. Foci: $(\pm\sqrt{13}, 0)$.
Asymptotes: $3x \pm 2y = 0$.

3. $x + y^2 = 2y + 3\ \Leftrightarrow\ (y-1)^2 = 4 - x$
Parabola, vertex $(4, 1)$, opening to the left, principal axis $y = 1$.
$a = -1/4$. Focus: $(15/4, 1)$.

4. $2x^2 + 8y^2 = 4x - 48y$
$2(x^2 - 2x + 1) + 8(y^2 + 6y + 9) = 74$

$$\frac{(x-1)^2}{37} + \frac{(y+3)^2}{37/4} = 1.$$

Ellipse, centre $(1, -3)$, major axis along $y = -3$.
$a = \sqrt{37},\ b = \sqrt{37}/2,\ c^2 = a^2 - b^2 = 111/4$.
Foci: $(1 \pm \sqrt{111}/2, -3)$.

5. $x = t,\ y = 2 - t,\ (0 \le t \le 2)$.
Straight line segment from $(0, 2)$ to $(2, 0)$.

6. $x = 2\sin(3t),\ y = 2\cos(3t),\ (0 \le t \le 2)$
Part of a circle of radius 2 centred at the origin from the point $(0, 2)$ clockwise to $(2\sin 6, 2\cos 6)$.

7. $x = \cosh t,\ y = \sinh^2 t$.
Parabola $x^2 - y = 1$, or $y = x^2 - 1$, traversed left to right.

8. $x = e^t,\ y = e^{-2t},\ (-1 \le t \le 1)$.
Part of the curve $x^2 y = 1$ from $(1/e, e^2)$ to $(e, 1/e^2)$.

9. $x = \cos(t/2),\ y = 4\sin(t/2),\ (0 \le t \le \pi)$.
The first quadrant part of the ellipse $16x^2 + y^2 = 16$, traversed counterclockwise.

10. $x = \cos t + \sin t,\ y = \cos t - \sin t,\ (0 \le t \le 2\pi)$
The circle $x^2 + y^2 = 2$, traversed clockwise, starting and ending at $(1, 1)$.

11. $x = \dfrac{4}{1+t^2}\qquad y = t^3 - 3t$

$\dfrac{dx}{dt} = -\dfrac{8t}{(1+t^2)^2}\qquad \dfrac{dy}{dt} = 3(t^2 - 1)$

Horizontal tangent at $t = \pm 1$, i.e., at $(2, \pm 2)$.
Vertical tangent at $t = 0$, i.e., at $(4, 0)$.
Self-intersection at $t = \pm\sqrt{3}$, i.e., at $(1, 0)$.

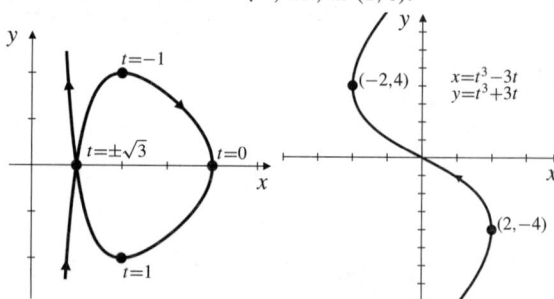

Fig. R-8.11   Fig. R-8.12

12. $x = t^3 - 3t\qquad y = t^3 + 3t$

$\dfrac{dx}{dt} = 3(t^2 - 1)\qquad \dfrac{dy}{dt} = 3(t^2 + 1)$

Horizontal tangent: none.
Vertical tangent at $t = \pm 1$, i.e., at $(2, -4)$ and $(-2, 4)$.

Slope $\dfrac{dy}{dx} = \dfrac{t^2 + 1}{t^2 - 1}\quad \begin{cases} > 0 & \text{if } |t| > 1 \\ < 0 & \text{if } |t| < 1 \end{cases}$
Slope $\to 1$ as $t \to \pm\infty$.

13. $x = t^3 - 3t\qquad y = t^3$

$\dfrac{dx}{dt} = 3(t^2 - 1)\qquad \dfrac{dy}{dt} = 3t^2$

Horizontal tangent at $t = 0$, i.e., at $(0, 0)$.
Vertical tangent at $t = \pm 1$, i.e., at $(2, -1)$ and $(-2, 1)$.

Slope $\dfrac{dy}{dx} = \dfrac{t^2}{t^2 - 1}\quad \begin{cases} > 0 & \text{if } |t| > 1 \\ < 0 & \text{if } |t| < 1 \end{cases}$
Slope $\to 1$ as $t \to \pm\infty$.

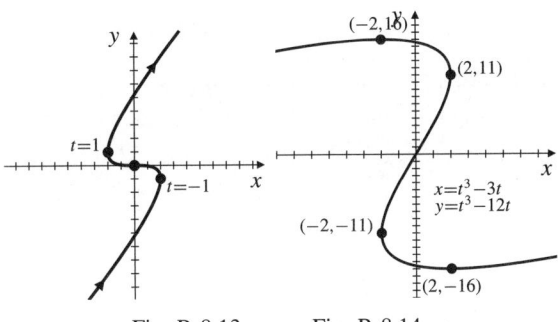

INSTRUCTOR'S SOLUTIONS MANUAL   REVIEW EXERCISES 8   (PAGE 516)

Fig. R-8.13   Fig. R-8.14

14. $x = t^3 - 3t$ $y = t^3 - 12t$
$\dfrac{dx}{dt} = 3(t^2 - 1)$ $\dfrac{dy}{dt} = 3(t^2 - 4)$
Horizontal tangent at $t = \pm 2$, i.e., at $(2, -16)$ and $(-2, 16)$.
Vertical tangent at $t = \pm 1$, i.e., at $(2, 11)$ and $(-2, -11)$.

Slope $\dfrac{dy}{dx} = \dfrac{t^2 - 4}{t^2 - 1}$ $\begin{cases} > 0 & \text{if } |t| > 2 \text{ or } |t| < 1 \\ < 0 & \text{if } 1 < |t| < 2 \end{cases}$
Slope $\to 1$ as $t \to \pm \infty$.

15. The curve $x = t^3 - t$, $y = |t^3|$ is symmetric about $x = 0$ since x is an odd function and y is an even function. Its self-intersection occurs at a nonzero value of t that makes $x = 0$, namely, $t = \pm 1$. The area of the loop is

$$A = 2 \int_{t=0}^{t=1} (-x)\, dy = -2 \int_0^1 (t^3 - t) 3t^2\, dt$$
$$= \left(-t^6 + \frac{3}{2} t^4\right) \Big|_0^1 = \frac{1}{2} \text{ sq. units.}$$

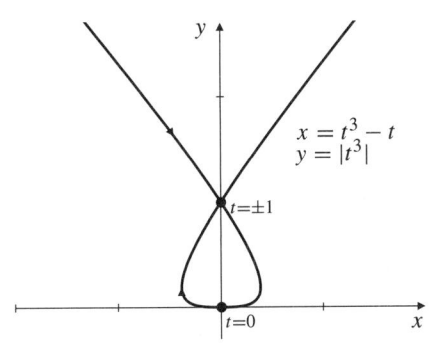

Fig. R-8.15

16. The volume of revolution about the y-axis is

$$V = \pi \int_{t=0}^{t=1} x^2\, dy$$
$$= \pi \int_0^1 (t^6 - 2t^4 + t^2) 3t^2\, dt$$
$$= 3\pi \int_0^1 (t^8 - 2t^6 + t^4)\, dt$$
$$= 3\pi \left(\frac{1}{9} - \frac{2}{7} + \frac{1}{5}\right) = \frac{8\pi}{105} \text{ cu. units.}$$

17. $x = e^t - t$, $y = 4e^{t/2}$, $(0 \le t \le 2)$. Length is

$$L = \int_0^2 \sqrt{(e^t - 1)^2 + 4e^t}\, dt$$
$$= \int_0^2 \sqrt{(e^t + 1)^2}\, dt = \int_0^2 (e^t + 1)\, dt$$
$$= (e^t + t)\Big|_0^2 = e^2 + 1 \text{ units.}$$

18. Area of revolution about the x-axis is

$$S = 2\pi \int 4e^{t/2}(e^t + 1)\, dt$$
$$= 8\pi \left(\frac{2}{3} e^{3t/2} + 2e^{t/2}\right)\Big|_0^2$$
$$= \frac{16\pi}{3}(e^3 + 3e - 4) \text{ sq. units.}$$

19. $r = \theta$, $\left(\dfrac{-3\pi}{2} \le \theta \le \dfrac{3\pi}{2}\right)$

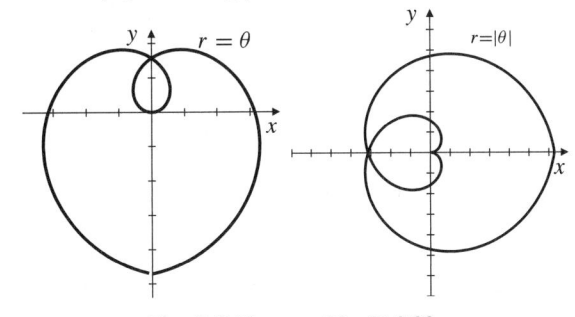

Fig. R-8.21   Fig. R-8.22

20. $r = |\theta|$, $(-2\pi \le \theta \le 2\pi)$

21. $r = 1 + \cos(2\theta)$

329

REVIEW EXERCISES 8 (PAGE 516)   R. A. ADAMS: CALCULUS

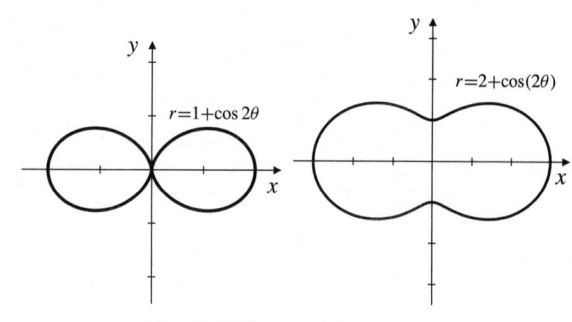

Fig. R-8.23

Fig. R-8.24

22. $r = 2 + \cos(2\theta)$

23. $r = 1 + 2\cos(2\theta)$

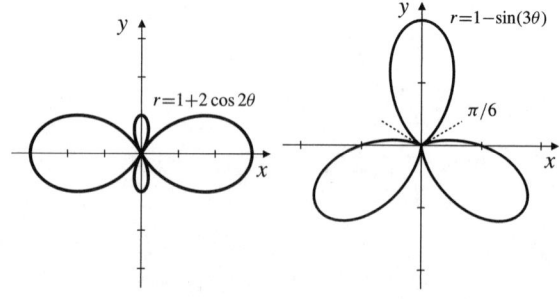

Fig. R-8.23

Fig. R-8.24

24. $r = 1 - \sin(3\theta)$

25. Area of a large loop:

$$A = 2 \times \frac{1}{2} \int_0^{\pi/3} (1 + 2\cos(2\theta))^2 \, d\theta$$

$$= \int_0^{\pi/3} [1 + 4\cos(2\theta) + 2(1 + \cos(4\theta))] \, d\theta$$

$$= \left(3\theta + 2\sin(2\theta) + \frac{1}{2}\sin(4\theta)\right)\bigg|_0^{\pi/3}$$

$$= \pi + \frac{3\sqrt{3}}{4} \text{ sq. units.}$$

26. Area of a small loop:

$$A = 2 \times \frac{1}{2} \int_{\pi/3}^{\pi/2} (1 + 2\cos(2\theta))^2 \, d\theta$$

$$= \int_{\pi/3}^{\pi/2} [1 + 4\cos(2\theta) + 2(1 + \cos(4\theta))] \, d\theta$$

$$= \left(3\theta + 2\sin(2\theta) + \frac{1}{2}\sin(4\theta)\right)\bigg|_{\pi/3}^{\pi/2}$$

$$= \frac{\pi}{2} - \frac{3\sqrt{3}}{4} \text{ sq. units.}$$

27. $r = 1 + \sqrt{2}\sin\theta$ approaches the origin in the directions for which $\sin\theta = -1/\sqrt{2}$, that is, $\theta = -3\pi/4$ and $\theta = -\pi/4$. The smaller loop corresponds to values of θ between these two values. By symmetry, the area of the loop is

$$A = 2 \times \frac{1}{2} \int_{-\pi/2}^{-\pi/4} (1 + 2\sqrt{2}\sin\theta + 2\sin^2\theta) \, d\theta$$

$$= \int_{-\pi/2}^{-\pi/4} (2 + 2\sqrt{2}\sin\theta - \cos(2\theta)) \, d\theta$$

$$= \left(2\theta - 2\sqrt{2}\cos\theta - \frac{1}{2}\sin(2\theta)\right)\bigg|_{-\pi/2}^{-\pi/4}$$

$$= \frac{\pi}{2} - 2 + \frac{1}{2} = \frac{\pi - 3}{2} \text{ sq. units.}$$

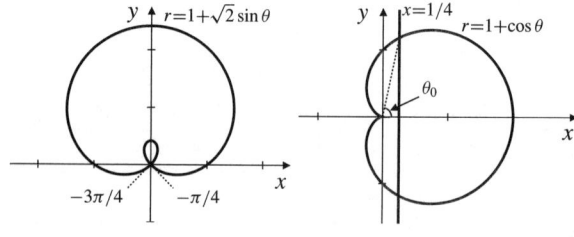

Fig. R-8.27

Fig. R-8.28

28. $r\cos\theta = x = 1/4$ and $r = 1 + \cos\theta$ intersect where

$$1 + \cos\theta = \frac{1}{4\cos\theta}$$

$$4\cos^2\theta + 4\cos\theta - 1 = 0$$

$$\cos\theta = \frac{-4 \pm \sqrt{16 + 16}}{8} = \frac{\pm\sqrt{2} - 1}{2}.$$

Only $(\sqrt{2} - 1)/2$ is between -1 and 1, so is a possible value of $\cos\theta$. Let $\theta_0 = \cos^{-1}\dfrac{\sqrt{2} - 1}{2}$. Then

$$\sin\theta_0 = \sqrt{1 - \left(\frac{\sqrt{2} - 1}{2}\right)^2} = \frac{\sqrt{1 + 2\sqrt{2}}}{2}.$$

By symmetry, the area inside $r = 1 + \cos\theta$ to the left of the line $x = 1/4$ is

$$A = 2 \times \frac{1}{2} \int_{\theta_0}^{\pi} \left(1 + 2\cos\theta + \frac{1 + \cos(2\theta)}{2}\right) d\theta + \cos\theta_0 \sin\theta_0$$

$$= \frac{3}{2}(\pi - \theta_0) + \left(2\sin\theta + \frac{1}{4}\sin(2\theta)\right)\bigg|_{\theta_0}^{\pi}$$

$$+ \frac{(\sqrt{2} - 1)\sqrt{1 + 2\sqrt{2}}}{4}$$

$$= \frac{3}{2}\left(\pi - \cos^{-1}\frac{\sqrt{2} - 1}{2}\right) + \sqrt{1 + 2\sqrt{2}}\left(\frac{\sqrt{2} - 9}{8}\right) \text{ sq. units.}$$

Challenging Problems 8 (page 516)

1. The surface of the water is elliptical (see Problem 2 below) whose semi-minor axis is 4 cm, the radius of the cylinder, and whose semi-major axis is $4\sec\theta$ cm because of the tilt of the glass. The surface area is that of the ellipse

$$x = 4\sec\theta\cos t, \quad y = 4\sin t, \quad (0 \le t \le 2\pi).$$

This area is

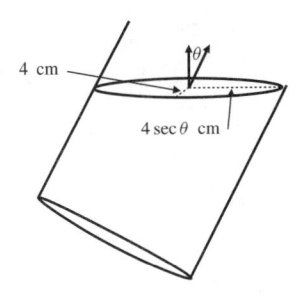

$$\begin{aligned}A &= 4\int_{t=0}^{t=\pi/2} x\,dy \\ &= 4\int_0^{\pi/2}(4\sec\theta\cos t)(4\cos t)\,dt \\ &= 32\sec\theta\int_0^{\pi/2}(1+\cos(2t))\,dt = 16\pi\sec\theta \text{ cm}^2.\end{aligned}$$

Fig. C-8.1

2. Let S_1 and S_2 be two spheres inscribed in the cylinder, one on each side of the plane that intersects the cylinder in the curve C that we are trying to show is an ellipse. Let the spheres be tangent to the cylinder around the circles C_1 and C_2, and suppose they are also tangent to the plane at the points F_1 and F_2, respectively, as shown in the figure.

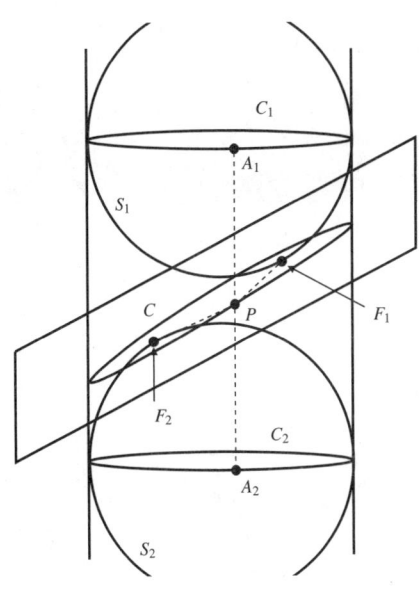

Fig. C-8.2

Let P be any point on C. Let A_1A_2 be the line through P that lies on the cylinder, with A_1 on C_1 and A_2 on C_2. Then $PF_1 = PA_1$ because both lengths are of tangents drawn to the sphere S_1 from the same exterior point P. Similarly, $PF_2 = PA_2$. Hence

$$PF_1 + PF_2 = PA_1 + PA_2 = A_1A_2,$$

which is constant, the distance between the centres of the two spheres. Thus C must be an ellipse, with foci at F_1 and F_2.

3. Given the foci F_1 and F_2, and the point P on the ellipse, construct N_1PN_2, the bisector of the angle F_1PF_2. Then construct T_1PT_2 perpendicular to N_1N_2 at P. By the reflection property of the ellipse, N_1N_2 is normal to the ellipse at P. Therefore T_1T_2 is tangent there.

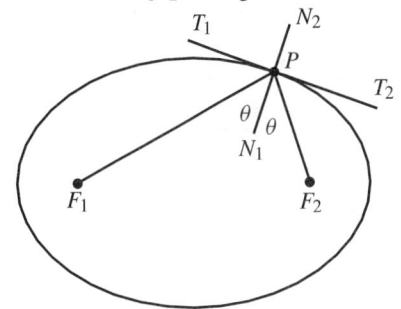

Fig. C-8.3

4. Without loss of generality, choose the axes and axis scales so that the parabola has equation $y = x^2$. If P is the point (x_0, x_0^2) on it, then the tangent to the parabola at P has equation

$$y = x_0^2 + 2x_0(x - x_0),$$

which intersects the principal axis $x = 0$ at $(0, -x_0^2)$. Thus $R = (0, -x_0^2)$ and $Q = (0, x_0^2)$. Evidently the vertex $V = (0, 0)$ bisects RQ.

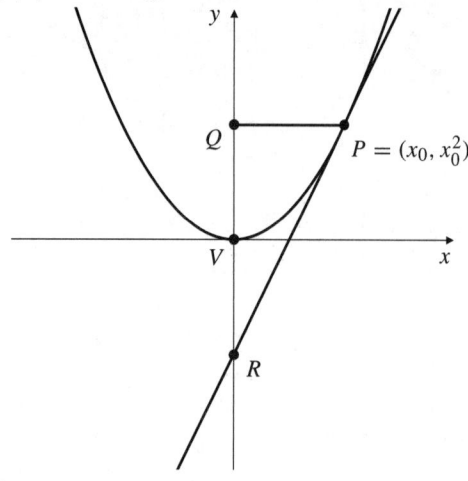

Fig. C-8.4

To construct the tangent at a given point P on a parabola with given vertex V and principal axis L, drop a perpendicular from P to L, meeting L at Q. Then find R on L on the side of V opposite Q and such that $QV = VR$. Then PR is the desired tangent.

5.

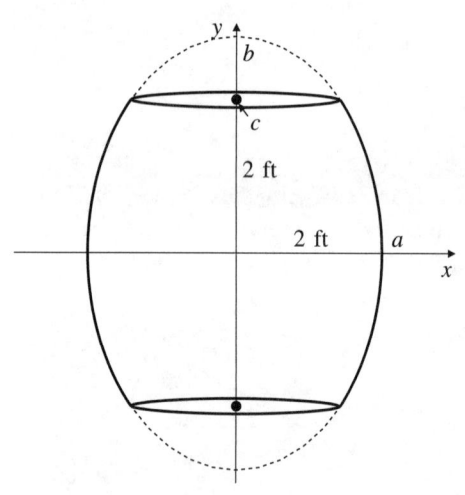

Fig. C-8.5

Let the ellipse be $\dfrac{x^2}{a^2} + \dfrac{y^2}{b^2} = 1$, with $a = 2$ and foci at $(0, \pm 2)$ so that $c = 2$ and $b^2 = a^2 + c^2 = 8$. The volume of the barrel is

$$V = 2\int_0^2 \pi x^2\, dy = 2\pi \int_0^2 4\left(1 - \frac{y^2}{8}\right) dy$$

$$= 8\pi \left(y - \frac{y^3}{24}\right)\Big|_0^2 = \frac{40\pi}{3} \text{ ft}^3.$$

6.

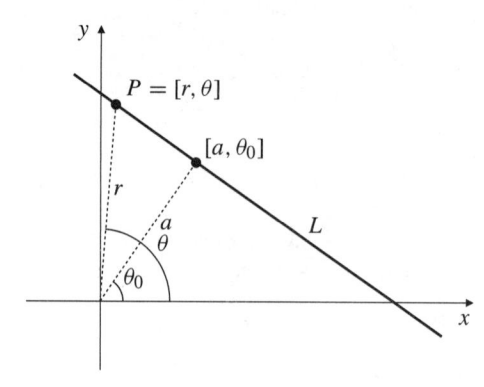

Fig. C-8.6

a) Let L be a line not passing through the origin, and let $[a, \theta_0]$ be the polar coordinates of the point on L that is closest to the origin. If $P = [r, \theta]$ is any point on the line, then, from the triangle in the figure,

$$\frac{a}{r} = \cos(\theta - \theta_0), \quad \text{or} \quad r = \frac{a}{\cos(\theta - \theta_0)}.$$

b) As shown in part (a), any line not passing through the origin has equation of the form

$$r = g(\theta) = \frac{a}{\cos(\theta - \theta_0)} = a\sec(\theta - \theta_0),$$

for some constants a and θ_0. We have

$$g'(\theta) = a\sec(\theta - \theta_0)\tan(\theta - \theta_0)$$
$$g''(\theta) = a\sec(\theta - \theta_0)\tan^2(\theta - \theta_0)$$
$$\qquad + a\sec^3(\theta - \theta_0)$$
$$\left(g(\theta)\right)^2 + 2\left(g'(\theta)\right)^2 - g(\theta)g''(\theta)$$
$$= a^2\sec^2(\theta - \theta_0) + 2a^2\sec^2(\theta - \theta_0)\tan^2(\theta - \theta_0)$$
$$\qquad - a^2\sec^2(\theta - \theta_0)\tan^2(\theta - \theta_0) - a^2\sec^4(\theta - \theta_0)$$
$$= a^2\left[\sec^2(\theta - \theta_0)\left(1 + \tan^2(\theta - \theta_0)\right) - \sec^4(\theta - \theta_0)\right]$$
$$= 0.$$

c) If $r = g(\theta)$ is the polar equation of the tangent to $r = f(\theta)$ at $\theta = \alpha$, then $g(\alpha) = f(\alpha)$ and $g'(\alpha) = f'(\alpha)$. Suppose that

$$\left(f(\alpha)\right)^2 + 2\left(f'(\alpha)\right)^2 - f(\alpha)f''(\alpha) > 0.$$

By part (b) we have

$$\left(g(\alpha)\right)^2 + 2\left(g'(\alpha)\right)^2 - g(\alpha)g''(\alpha) = 0.$$

INSTRUCTOR'S SOLUTIONS MANUAL

CHALLENGING PROBLEMS 8 (PAGE 516)

Subtracting, and using $g(\alpha) = f(\alpha)$ and $g'(\alpha) = f'(\alpha)$, we get $f''(\alpha) < g''(\alpha)$. It follows that $f(\theta) < g(\theta)$ for values of θ near α; that is, the graph of $r = f(\theta)$ is curving to the origin side of its tangent at α. Similarly, if

$$\bigl(f(\alpha)\bigr)^2 + 2\bigl(f'(\alpha)\bigr)^2 - f(\alpha)f''(\alpha) < 0,$$

then the graph is curving to the opposite side of the tangent, away from the origin.

7.

Fig. C-8.7

When the vehicle is at position x, as shown in the figure, the component of the gravitational force on it in the direction of the tunnel is

$$ma(r)\cos\theta = -\frac{mgr}{R}\cos\theta = -\frac{mg}{R}x.$$

By Newton's Law of Motion, this force produces an acceleration d^2x/dt^2 along the tunnel given by

$$m\frac{d^2x}{dt^2} = -\frac{mg}{R}x,$$

that is

$$\frac{d^2x}{dt^2} + \omega^2 x = 0, \quad \text{where} \quad \omega^2 = \frac{g}{R}.$$

This is the equation of simple harmonic motion, with period $T = 2\pi/\omega = 2\pi\sqrt{R/g}$.
For $R \approx 3960$ mi $\approx 2.09 \times 10^7$ ft, and $g \approx 32$ ft/s^2, we have $T \approx 5079$ s ≈ 84.6 minutes. This is a rather short time for a round trip between Atlanta and Baghdad, or any other two points on the surface of the earth.

8. Take the origin at station O as shown in the figure. Both of the lines L_1 and L_2 pass at distance $100\cos\epsilon$ from the origin. Therefore, by Problem 6(a), their equations are

$$L_1: \quad r = \frac{100\cos\epsilon}{\cos\left[\theta - \left(\frac{\pi}{2} - \epsilon\right)\right]} = \frac{100\cos\epsilon}{\sin(\theta + \epsilon)}$$

$$L_2: \quad r = \frac{100\cos\epsilon}{\cos\left[\theta - \left(\frac{\pi}{2} + \epsilon\right)\right]} = \frac{100\cos\epsilon}{\sin(\theta - \epsilon)}.$$

The search area $A(\epsilon)$ is, therefore,

$$A(\epsilon) = \frac{1}{2}\int_{\frac{\pi}{4}-\epsilon}^{\frac{\pi}{4}+\epsilon}\left(\frac{100^2\cos^2\epsilon}{\sin^2(\theta-\epsilon)} - \frac{100^2\cos^2\epsilon}{\sin^2(\theta+\epsilon)}\right)d\theta$$

$$= 5,000\cos^2\epsilon \int_{\frac{\pi}{4}-\epsilon}^{\frac{\pi}{4}+\epsilon}\left(\csc^2(\theta-\epsilon) - \csc^2(\theta+\epsilon)\right)d\theta$$

$$= 5,000\cos^2\epsilon \left[\cot\left(\tfrac{\pi}{4}+2\epsilon\right) - 2\cot\tfrac{\pi}{4} + \cot\left(\tfrac{\pi}{4}-2\epsilon\right)\right]$$

$$= 5,000\cos^2\epsilon \left[\frac{\cos\left(\tfrac{\pi}{4}+2\epsilon\right)}{\sin\left(\tfrac{\pi}{4}+2\epsilon\right)} + \frac{\sin\left(\tfrac{\pi}{4}+2\epsilon\right)}{\cos\left(\tfrac{\pi}{4}+2\epsilon\right)} - 2\right]$$

$$= 10,000\cos^2\epsilon \left[\csc\left(\tfrac{\pi}{2}+4\epsilon\right) - 1\right]$$

$$= 10,000\cos^2\epsilon\,(\sec(4\epsilon)-1) \text{ mi}^2.$$

For $\epsilon = 3° = \pi/60$, we have $A(\epsilon) \approx 222.8$ square miles. Also

$$A'(\epsilon) = -20,000\cos\epsilon\sin\epsilon(\sec(4\epsilon)-1)$$
$$\quad + 40,000\cos^2\epsilon\sec(4\epsilon)\tan(4\epsilon)$$

$$A'(\pi/60) \approx 8645.$$

When $\epsilon = 3°$, the search area increases at about $8645(\pi/180) \approx 151$ square miles per degree increase in ϵ.

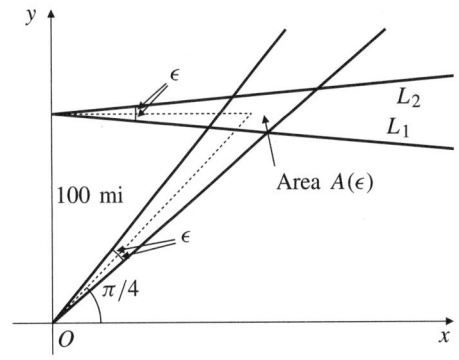

Fig. C-8.8

9. The easiest way to determine which curve is which is to calculate both their areas; the outer curve bounds the larger area.
The curve C_1 with parametric equations

$$x = \sin t, \quad y = \frac{1}{2}\sin(2t), \quad (0 \le t \le 2\pi)$$

333

has area

$$A_1 = 4\int_{t=0}^{t=\pi/2} y\,dx$$
$$= 4\int_0^{\pi/2} \frac{1}{2}\sin(2t)\cos t\,dt$$
$$= 4\int_0^{\pi/2} \sin t \cos^2 t\,dt$$

Let $u = \cos t$
$du = -\sin t\,dt$

$$= 4\int_0^1 u^2\,du = \frac{4}{3} \text{ sq. units.}$$

The curve C_2 with polar equation $r^2 = \cos(2\theta)$ has area

$$A_2 = \frac{4}{2}\int_0^{\pi/4} \cos(2\theta)\,d\theta = \sin(2\theta)\Big|_0^{\pi/4} = 1 \text{ sq. units.}$$

C_1 is the outer curve, and the area between the curves is 1/3 sq. units.

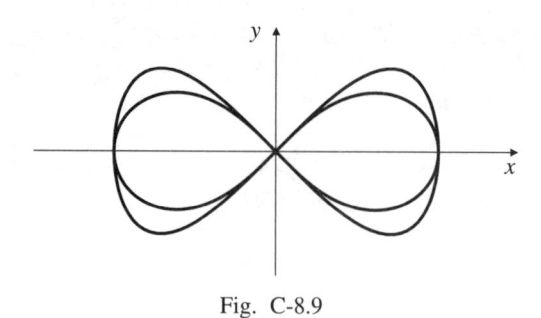

Fig. C-8.9

CHAPTER 9. SEQUENCES, SERIES, AND POWER SERIES

Section 9.1 Sequences and Convergence (page 526)

1. $\left\{\dfrac{2n^2}{n^2+1}\right\} = \left\{2 - \dfrac{2}{n^2+1}\right\} = \left\{1, \dfrac{8}{5}, \dfrac{9}{5}, \ldots\right\}$ is bounded, positive, increasing, and converges to 2.

2. $\left\{\dfrac{2n}{n^2+1}\right\} = \left\{1, \dfrac{4}{5}, \dfrac{3}{5}, \dfrac{8}{17}, \ldots\right\}$ is bounded, positive, decreasing, and converges to 0.

3. $\left\{4 - \dfrac{(-1)^n}{n}\right\} = \left\{5, \dfrac{7}{2}, \dfrac{13}{3}, \ldots\right\}$ is bounded, positive, and converges to 4.

4. $\left\{\sin\dfrac{1}{n}\right\} = \left\{\sin 1, \sin\left(\dfrac{1}{2}\right), \sin\left(\dfrac{1}{3}\right), \ldots\right\}$ is bounded, positive, decreasing, and converges to 0.

5. $\left\{\dfrac{n^2-1}{n}\right\} = \left\{n - \dfrac{1}{n}\right\} = \left\{0, \dfrac{3}{2}, \dfrac{8}{3}, \dfrac{15}{4}, \ldots\right\}$ is bounded below, positive, increasing, and diverges to infinity.

6. $\left\{\dfrac{e^n}{\pi^n}\right\} = \left\{\dfrac{e}{\pi}, \left(\dfrac{e}{\pi}\right)^2, \left(\dfrac{e}{\pi}\right)^3, \ldots\right\}$ is bounded, positive, decreasing, and converges to 0, since $e < \pi$.

7. $\left\{\dfrac{e^n}{\pi^{n/2}}\right\} = \left\{\left(\dfrac{e}{\sqrt{\pi}}\right)^n\right\}$. Since $e/\sqrt{\pi} > 1$, the sequence is bounded below, positive, increasing, and diverges to infinity.

8. $\left\{\dfrac{(-1)^n n}{e^n}\right\} = \left\{\dfrac{-1}{e}, \dfrac{2}{e^2}, \dfrac{-3}{e^3}, \ldots\right\}$ is bounded, alternating, and converges to 0.

9. $\{2^n/n^n\}$ is bounded, positive, decreasing, and converges to 0.

10. $\dfrac{(n!)^2}{(2n)!} = \dfrac{1}{n+1} \dfrac{2}{n+2} \dfrac{3}{n+3} \cdots \dfrac{n}{2n} \le \left(\dfrac{1}{2}\right)^n$.

    Also, $\dfrac{a_{n+1}}{a_n} = \dfrac{(n+1)^2}{(2n+2)(2n+1)} < \dfrac{1}{2}$. Thus the sequence $\left\{\dfrac{(n!)^2}{(2n)!}\right\}$ is positive, decreasing, bounded, and convergent to 0.

11. $\{n\cos(n\pi/2)\} = \{0, -2, 0, 4, 0, -6, \ldots\}$ is divergent.

12. $\left\{\dfrac{\sin n}{n}\right\} = \left\{\sin 1, \dfrac{\sin 2}{2}, \dfrac{\sin 3}{3}, \ldots\right\}$ is bounded and converges to 0.

13. $\{1, 1, -2, 3, 3, -4, 5, 5, -6, \ldots\}$ is divergent.

14. $\lim \dfrac{5-2n}{3n-7} = \lim \dfrac{\dfrac{5}{n}-2}{3-\dfrac{7}{n}} = -\dfrac{2}{3}$.

15. $\lim \dfrac{n^2-4}{n+5} = \lim \dfrac{n-\dfrac{4}{n}}{1+\dfrac{5}{n}} = \infty$.

16. $\lim \dfrac{n^2}{n^3+1} = \lim \dfrac{\dfrac{1}{n}}{1+\dfrac{1}{n^3}} = 0$.

17. $\lim(-1)^n \dfrac{n}{n^3+1} = 0$.

18. $\lim \dfrac{n^2 - 2\sqrt{n}+1}{1-n-3n^2} = \lim \dfrac{1-\dfrac{2}{n\sqrt{n}}+\dfrac{1}{n^2}}{\dfrac{1}{n^2}-\dfrac{1}{n}-3} = -\dfrac{1}{3}$.

19. $\lim \dfrac{e^n - e^{-n}}{e^n + e^{-n}} = \lim \dfrac{1-e^{-2n}}{1+e^{-2n}} = 1$.

20. $\lim n\sin\dfrac{1}{n} = \lim_{x\to 0+}\dfrac{\sin x}{x} = \lim_{x\to 0+}\dfrac{\cos x}{1} = 1$.

21. $\lim\left(\dfrac{n-3}{n}\right)^n = \lim\left(1+\dfrac{-3}{n}\right)^n = e^{-3}$ by l'Hôpital's Rule.

22. $\lim \dfrac{n}{\ln(n+1)} = \lim_{x\to\infty}\dfrac{x}{\ln(x+1)}$
    $= \lim_{x\to\infty}\dfrac{1}{\left(\dfrac{1}{x+1}\right)} = \lim_{x\to\infty} x+1 = \infty$.

23. $\lim(\sqrt{n+1}-\sqrt{n}) = \lim \dfrac{n+1-n}{\sqrt{n+1}+\sqrt{n}} = 0$.

24. $\lim(n - \sqrt{n^2-4n}) = \lim \dfrac{n^2 - (n^2-4n)}{n+\sqrt{n^2-4n}}$
    $= \lim \dfrac{4n}{n+\sqrt{n^2-4n}} = \lim \dfrac{4}{1+\sqrt{1-\dfrac{4}{n}}} = 2$.

25. $\lim(\sqrt{n^2+n} - \sqrt{n^2-1})$
    $= \lim \dfrac{n^2+n-(n^2-1)}{\sqrt{n^2+n}+\sqrt{n^2-1}}$
    $= \lim \dfrac{n+1}{n\left(\sqrt{1+\dfrac{1}{n}}+\sqrt{1-\dfrac{1}{n^2}}\right)}$
    $= \lim \dfrac{1+\dfrac{1}{n}}{\sqrt{1+\dfrac{1}{n}}+\sqrt{1-\dfrac{1}{n^2}}} = \dfrac{1}{2}$.

26. If $a_n = \left(\dfrac{n-1}{n+1}\right)^n$, then

$$\lim a_n = \lim \left(\dfrac{n-1}{n}\right)^n \left(\dfrac{n}{n+1}\right)^n$$
$$= \lim \left(1 - \dfrac{1}{n}\right)^n \Big/ \lim \left(1 + \dfrac{1}{n}\right)^n$$
$$= \dfrac{e^{-1}}{e} = e^{-2} \quad \text{(by Theorem 6 of Section 3.4)}.$$

27. $a_n = \dfrac{(n!)^2}{(2n)!} = \dfrac{(1 \cdot 2 \cdot 3 \cdots n)(1 \cdot 2 \cdot 3 \cdots n)}{1 \cdot 2 \cdot 3 \cdots n \cdot (n+1) \cdot (n+2) \cdots 2n}$
$$= \dfrac{1}{n+1} \cdot \dfrac{2}{n+2} \cdot \dfrac{3}{n+3} \cdots \dfrac{n}{n+n} \le \left(\dfrac{1}{2}\right)^n.$$
Thus $\lim a_n = 0$.

28. We have $\lim \dfrac{n^2}{2^n} = 0$ since 2^n grows much faster than n^2 and $\lim \dfrac{4^n}{n!} = 0$ by Theorem 3(b). Hence,
$$\lim \dfrac{n^2 2^n}{n!} = \lim \dfrac{n^2}{2^n} \cdot \dfrac{2^{2n}}{n!} = \left(\lim \dfrac{n^2}{2^n}\right)\left(\lim \dfrac{4^n}{n!}\right) = 0.$$

29. $a_n = \dfrac{\pi^n}{1 + 2^{2n}} \Rightarrow 0 < a_n < (\pi/4)^n$. Since $\pi/4 < 1$, therefore $(\pi/4)^n \to 0$ as $n \to \infty$. Thus $\lim a_n = 0$.

30. Let $a_1 = 1$ and $a_{n+1} = \sqrt{1 + 2a_n}$ for $n = 1, 2, 3, \ldots$. Then we have $a_2 = \sqrt{3} > a_1$. If $a_{k+1} > a_k$ for some k, then
$$a_{k+2} = \sqrt{1 + 2a_{k+1}} > \sqrt{1 + 2a_k} = a_{k+1}.$$
Thus, $\{a_n\}$ is increasing by induction. Observe that $a_1 < 3$ and $a_2 < 3$. If $a_k < 3$ then
$$a_{k+1} = \sqrt{1 + 2a_k} < \sqrt{1 + 2(3)} = \sqrt{7} < \sqrt{9} = 3.$$
Therefore, $a_n < 3$ for all n, by induction. Since $\{a_n\}$ is increasing and bounded above, it converges. Let $\lim a_n = a$. Then
$$a = \sqrt{1 + 2a} \Rightarrow a^2 - 2a - 1 = 0 \Rightarrow a = 1 \pm \sqrt{2}.$$
Since $a = 1 - \sqrt{2} < 0$, it is not appropriate. Hence, we must have $\lim a_n = 1 + \sqrt{2}$.

31. Let $a_1 = 3$ and $a_{n+1} = \sqrt{15 + 2a_n}$ for $n = 1, 2, 3, \ldots$. Then we have $a_2 = \sqrt{21} > 3 = a_1$. If $a_{k+1} > a_k$ for some k, then
$$a_{k+2} = \sqrt{15 + 2a_{k+1}} > \sqrt{15 + 2a_k} = a_{k+1}.$$

Thus, $\{a_n\}$ is increasing by induction. Observe that $a_1 < 5$ and $a_2 < 5$. If $a_k < 5$ then
$$a_{k+1} = \sqrt{15 + 2a_k} < \sqrt{15 + 2(5)} = \sqrt{25} = 5.$$
Therefore, $a_n < 5$ for all n, by induction. Since $\{a_n\}$ is increasing and bounded above, it converges. Let $\lim a_n = a$. Then
$$a = \sqrt{15 + 2a} \Rightarrow a^2 - 2a - 15 = 0 \Rightarrow a = -3, \text{ or } a = 5.$$
Since $a > a_1$, we must have $\lim a_n = 5$.

32. Let $a_n = \left(1 + \dfrac{1}{n}\right)^n$ so $\ln a_n = n \ln\left(1 + \dfrac{1}{n}\right)$.

a) If $f(x) = x \ln\left(1 + \dfrac{1}{x}\right) = x \ln(x+1) - x \ln x$, then
$$f'(x) = \ln(x+1) + \dfrac{x}{x+1} - \ln x - 1$$
$$= \ln\left(\dfrac{x+1}{x}\right) - \dfrac{1}{x+1}$$
$$= \int_x^{x+1} \dfrac{dt}{t} - \dfrac{1}{x+1}$$
$$> \dfrac{1}{x+1} \int_x^{x+1} dt - \dfrac{1}{x+1}$$
$$= \dfrac{1}{x+1} - \dfrac{1}{x+1} = 0.$$
Since $f'(x) > 0$, $f(x)$ must be an increasing function. Thus, $\{a_n\} = \{e^{f(x_n)}\}$ is increasing.

b) Since $\ln x \le x - 1$,
$$\ln a_k = k \ln\left(1 + \dfrac{1}{k}\right) \le k\left(1 + \dfrac{1}{k} - 1\right) = 1$$
which implies that $a_k \le e$ for all k. Since $\{a_n\}$ is increasing, e is an upper bound for $\{a_n\}$.

33. Suppose $\{a_n\}$ is ultimately increasing, say $a_{n+1} \ge a_n$ if $n \ge N$.
Case I. If there exists a real number K such that $a_n \le K$ for all n, then $\lim a_n = a$ exists by completeness.
Case II. Otherwise, for every integer K, there exists $n \ge N$ such that $a_n > K$, and hence $a_j > K$ for all $j \ge n$. Thus $\lim a_n = \infty$.

If $\{a_n\}$ is ultimately decreasing, then either it is bounded below, and therefore converges, or else it is unbounded below, and therefore diverges to negative infinity.

34. If $\{|a_n|\}$ is bounded then it is bounded above, and there exists a constant K such that $|a_n| \le K$ for all n. Therefore, $-K \le a_n \le K$ for all n, and so $\{a_n\}$ is bounded above and below, and is therefore bounded.

35. Suppose $\lim_{n\to\infty} |a_n| = 0$. Given any $\epsilon > 0$, there exists an integer $N = N(\epsilon)$ such that if $n > N$, then $||a_n|-0| < \epsilon$. In this case $|a_n - 0| = |a_n| = ||a_n|-0| < \epsilon$, so $\lim_{n\to\infty} a_n = 0$.

36. a) "If $\lim a_n = \infty$ and $\lim b_n = L > 0$, then $\lim a_n b_n = \infty$" is TRUE. Let R be an arbitrary, large positive number. Since $\lim a_n = \infty$, and $L > 0$, it must be true that $a_n \geq \dfrac{2R}{L}$ for n sufficiently large. Since $\lim b_n = L$, it must also be that $b_n \geq \dfrac{L}{2}$ for n sufficiently large. Therefore $a_n b_n \geq \dfrac{2R}{L}\dfrac{L}{2} = R$ for n sufficiently large. Since R is arbitrary, $\lim a_n b_n = \infty$.

b) "If $\lim a_n = \infty$ and $\lim b_n = -\infty$, then $\lim(a_n + b_n) = 0$" is FALSE. Let $a_n = 1 + n$ and $b_n = -n$; then $\lim a_n = \infty$ and $\lim b_n = -\infty$ but $\lim(a_n + b_n) = 1$.

c) "If $\lim a_n = \infty$ and $\lim b_n = -\infty$, then $\lim a_n b_n = -\infty$" is TRUE. Let R be an arbitrary, large positive number. Since $\lim a_n = \infty$ and $\lim b_n = -\infty$, we must have $a_n \geq \sqrt{R}$ and $b_n \leq -\sqrt{R}$, for all sufficiently large n. Thus $a_n b_n \leq -R$, and $\lim a_n b_n = -\infty$.

d) "If neither $\{a_n\}$ nor $\{b_n\}$ converges, then $\{a_n b_n\}$ does not converge" is FALSE. Let $a_n = b_n = (-1)^n$; then $\lim a_n$ and $\lim b_n$ both diverge. But $a_n b_n = (-1)^{2n} = 1$ and $\{a_n b_n\}$ does converge (to 1).

e) "If $\{|a_n|\}$ converges, then $\{a_n\}$ converges" is FALSE. Let $a_n = (-1)^n$. Then $\lim_{n\to\infty} |a_n| = \lim_{n\to\infty} 1 = 1$, but $\lim_{n\to\infty} a_n$ does not exist.

Section 9.2 Infinite Series (page 533)

1. $\dfrac{1}{3} + \dfrac{1}{9} + \dfrac{1}{27} + \cdots = \dfrac{1}{3}\left(1 + \dfrac{1}{3} + \left(\dfrac{1}{3}\right)^2 + \cdots\right)$
$= \dfrac{1}{3} \cdot \dfrac{1}{1-\dfrac{1}{3}} = \dfrac{1}{2}.$

2. $3 - \dfrac{3}{4} + \dfrac{3}{16} - \dfrac{3}{64} + \cdots = \sum_{n=1}^{\infty} 3\left(-\dfrac{1}{4}\right)^{n-1} = \dfrac{3}{1+\dfrac{1}{4}} = \dfrac{12}{5}.$

3. $\sum_{n=5}^{\infty} \dfrac{1}{(2+\pi)^{2n}}$
$= \dfrac{1}{(2+\pi)^{10}} + \dfrac{1}{(2+\pi)^{12}} + \dfrac{1}{(2+\pi)^{14}} + \cdots$
$= \dfrac{1}{(2+\pi)^{10}}\left[1 + \dfrac{1}{(2+\pi)^2} + \dfrac{1}{(2+\pi)^4} + \cdots\right]$
$= \dfrac{1}{(2+\pi)^{10}} \cdot \dfrac{1}{1-\dfrac{1}{(2+\pi)^2}} = \dfrac{1}{(2+\pi)^8\left[(2+\pi)^2 - 1\right]}.$

4. $\sum_{n=0}^{\infty} \dfrac{5}{10^{3n}} = 5\left[1 + \dfrac{1}{1000} + \left(\dfrac{1}{1000}\right)^2 + \cdots\right]$
$= \dfrac{5}{1-\dfrac{1}{1000}} = \dfrac{5000}{999}.$

5. $\sum_{n=2}^{\infty} \dfrac{(-5)^n}{8^{2n}} = \dfrac{(-5)^2}{8^4} + \dfrac{(-5)^3}{8^6} + \dfrac{(-5)^4}{8^8} + \cdots$
$= \dfrac{25}{8^4}\left[1 - \dfrac{5}{64} + \dfrac{5^2}{64^2} - \cdots\right]$
$= \dfrac{25}{8^4} \cdot \dfrac{1}{1+\dfrac{5}{64}} = \dfrac{25}{64 \times 69} = \dfrac{25}{4416}.$

6. $\sum_{n=0}^{\infty} \dfrac{1}{e^n} = 1 + \dfrac{1}{e} + \left(\dfrac{1}{e}\right)^2 + \cdots = \dfrac{1}{1-\dfrac{1}{e}} = \dfrac{e}{e-1}.$

7. $\sum_{k=0}^{\infty} \dfrac{2^{k+3}}{e^{k-3}} = 8e^3 \sum_{k=0}^{\infty} \left(\dfrac{2}{e}\right)^k = \dfrac{8e^3}{1-\dfrac{2}{e}} = \dfrac{8e^4}{e-2}.$

8. $\sum_{j=1}^{\infty} \pi^{j/2} \cos(j\pi) = \sum_{j=2}^{\infty} (-1)^j \pi^{j/2}$ diverges because $\lim_{j\to\infty} (-1)^j \pi^{j/2}$ does not exist.

9. $\sum_{n=1}^{\infty} \dfrac{3+2^n}{2^{n+2}}$ diverges to ∞ because
$\lim_{n\to\infty} \dfrac{3+2^n}{2^{n+2}} = \lim_{n\to\infty} \dfrac{\dfrac{3}{2^n}+1}{4} = \dfrac{1}{4} > 0.$

10. $\sum_{n=0}^{\infty} \dfrac{3+2^n}{3^{n+2}} = \dfrac{1}{3}\sum_{n=0}^{\infty} \left(\dfrac{1}{3}\right)^n + \dfrac{1}{9}\sum_{n=0}^{\infty} \left(\dfrac{2}{3}\right)^n$
$= \dfrac{1}{3} \cdot \dfrac{1}{1-\dfrac{1}{3}} + \dfrac{1}{9} \cdot \dfrac{1}{1-\dfrac{2}{3}} = \dfrac{1}{2} + \dfrac{1}{3} = \dfrac{5}{6}.$

11. Since $\dfrac{1}{n(n+2)} = \dfrac{1}{2}\left(\dfrac{1}{n} - \dfrac{1}{n+2}\right)$, therefore

$$s_n = \dfrac{1}{1\times 3} + \dfrac{1}{2\times 4} + \dfrac{1}{3\times 5} + \cdots + \dfrac{1}{n(n+2)}$$

$$= \dfrac{1}{2}\left[\dfrac{1}{1} - \dfrac{1}{3} + \dfrac{1}{2} - \dfrac{1}{4} + \dfrac{1}{3} - \dfrac{1}{5} + \dfrac{1}{4} - \dfrac{1}{6} + \cdots \right.$$

$$\left. + \dfrac{1}{n-2} - \dfrac{1}{n} + \dfrac{1}{n-1} - \dfrac{1}{n+1} + \dfrac{1}{n} - \dfrac{1}{n+2}\right]$$

$$= \dfrac{1}{2}\left[1 + \dfrac{1}{2} - \dfrac{1}{n+1} - \dfrac{1}{n+2}\right].$$

Thus $\lim s_n = \dfrac{3}{4}$, and $\displaystyle\sum_{n=1}^{\infty} \dfrac{1}{n(n+2)} = \dfrac{3}{4}$.

12. Let
$$\sum_{n=1}^{\infty} \dfrac{1}{(2n-1)(2n+1)} = \dfrac{1}{1\times 3} + \dfrac{1}{3\times 5} + \dfrac{1}{5\times 7} + \cdots.$$

Since $\dfrac{1}{(2n-1)(2n+1)} = \dfrac{1}{2}\left(\dfrac{1}{2n-1} - \dfrac{1}{2n+1}\right)$, the partial sum is

$$s_n = \dfrac{1}{2}\left(1 - \dfrac{1}{3}\right) + \dfrac{1}{2}\left(\dfrac{1}{3} - \dfrac{1}{5}\right) + \cdots$$
$$+ \dfrac{1}{2}\left(\dfrac{1}{2n-3} - \dfrac{1}{2n-1}\right) + \dfrac{1}{2}\left(\dfrac{1}{2n-1} - \dfrac{1}{2n+1}\right)$$
$$= \dfrac{1}{2}\left(1 - \dfrac{1}{2n+1}\right).$$

Hence,
$$\sum_{n=1}^{\infty} \dfrac{1}{(2n-1)(2n+1)} = \lim s_n = \dfrac{1}{2}.$$

13. Since $\dfrac{1}{(3n-2)(3n+1)} = \dfrac{1}{3}\left(\dfrac{1}{3n-2} - \dfrac{1}{3n+1}\right)$, therefore

$$s_n = \dfrac{1}{1\times 4} + \dfrac{1}{4\times 7} + \dfrac{1}{7\times 10} + \cdots + \dfrac{1}{(3n-2)(3n+1)}$$

$$= \dfrac{1}{3}\left[\dfrac{1}{1} - \dfrac{1}{4} + \dfrac{1}{4} - \dfrac{1}{7} + \dfrac{1}{7} - \dfrac{1}{10} + \cdots \right.$$

$$\left. + \dfrac{1}{3n-5} - \dfrac{1}{3n-2} + \dfrac{1}{3n-2} - \dfrac{1}{3n+1}\right]$$

$$= \dfrac{1}{3}\left(1 - \dfrac{1}{3n+1}\right) \to \dfrac{1}{3}.$$

Thus $\displaystyle\sum_{n=1}^{\infty} \dfrac{1}{(3n-2)(3n+1)} = \dfrac{1}{3}$.

14. Since
$$\dfrac{1}{n(n+1)(n+2)} = \dfrac{1}{2}\left[\dfrac{1}{n} - \dfrac{2}{n+1} + \dfrac{1}{n+2}\right],$$
the partial sum is

$$s_n = \dfrac{1}{2}\left(1 - \dfrac{2}{2} + \dfrac{1}{3}\right) + \dfrac{1}{2}\left(\dfrac{1}{2} - \dfrac{2}{3} + \dfrac{1}{4}\right) + \cdots$$
$$+ \dfrac{1}{2}\left(\dfrac{1}{n-1} - \dfrac{2}{n} + \dfrac{1}{n+1}\right) + \dfrac{1}{2}\left(\dfrac{1}{n} - \dfrac{2}{n+1} + \dfrac{1}{n+2}\right)$$
$$= \dfrac{1}{2}\left(\dfrac{1}{2} - \dfrac{1}{n+1} + \dfrac{1}{n+2}\right).$$

Hence,
$$\sum_{n=1}^{\infty} \dfrac{1}{n(n+1)(n+2)} = \lim s_n = \dfrac{1}{4}.$$

15. Since $\dfrac{1}{2n-1} > \dfrac{1}{2n} = \dfrac{1}{2}\cdot\dfrac{1}{n}$, therefore the partial sums of the given series exceed half those of the divergent harmonic series $\sum (1/2n)$. Hence the given series diverges to infinity.

16. $\displaystyle\sum_{n=1}^{\infty} \dfrac{n}{n+2}$ diverges to infinity since $\lim \dfrac{n}{n+2} = 1 > 0$.

17. Since $n^{-1/2} = \dfrac{1}{\sqrt{n}} \geq \dfrac{1}{n}$ for $n \geq 1$, we have

$$\sum_{k=1}^{n} k^{-1/2} \geq \sum_{k=1}^{n} \dfrac{1}{k} \to \infty,$$

as $n \to \infty$ (harmonic series). Thus $\sum n^{-1/2}$ diverges to infinity.

18. $\displaystyle\sum_{n=1}^{\infty} \dfrac{2}{n+1} = 2\left(\dfrac{1}{2} + \dfrac{1}{3} + \dfrac{1}{4} + \cdots\right)$ diverges to infinity since it is just twice the harmonic series with the first term omitted.

19. $s_n = -1 + 1 - 1 + \cdots + (-1)^n = \begin{cases} -1 & \text{if } n \text{ is odd} \\ 0 & \text{if } n \text{ is even} \end{cases}.$
Thus $\lim s_n$ does not exist, and $\sum (-1)^n$ diverges.

20. Since $1 + 2 + 3 + \cdots + n = \dfrac{n(n+1)}{2}$, the given series is $\displaystyle\sum_{n=1}^{\infty} \dfrac{2}{n(n+1)}$ which converges to 2 by the result of Example 3 of this section.

21. The total distance is

$$2 + 2\left[2\times\dfrac{3}{4} + 2\times\left(\dfrac{3}{4}\right)^2 + \cdots\right]$$

$$= 2 + 2\times\dfrac{3}{2}\left[1 + \dfrac{3}{4} + \left(\dfrac{3}{4}\right)^2 + \cdots\right]$$

$$= 2 + \dfrac{3}{1 - \dfrac{3}{4}} = 14 \text{ metres}.$$

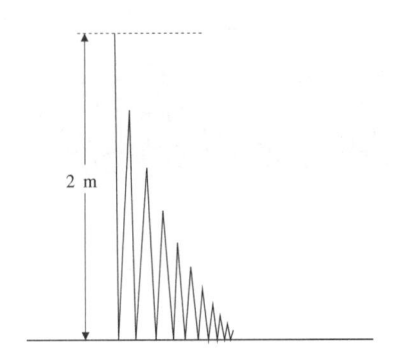

INSTRUCTOR'S SOLUTIONS MANUAL   SECTION 9.3 (PAGE 545)

2 m

Fig. 9.2.21

22. The balance at the end of 8 years is

$$s_n = 1000\big[(1.1)^8 + (1.1)^7 + \cdots + (1.1)^2 + (1.1)\big]$$
$$= 1000(1.1)\left(\frac{(1.1)^8 - 1}{1.1 - 1}\right) \approx \$12,579.48.$$

23. For $n > N$ let $s_n = \sum_{j=1}^n a_j$, and $S_n = \sum_{j=N}^n a_j$.
Then $s_n = S_n + C$, where $C = \sum_{j=1}^{N-1} a_j$. We have

$$\lim_{n\to\infty} s_n = \lim_{n\to\infty} S_n + C:$$

either both sides exist or neither does. Hence $\sum_{n=1}^\infty a_n$ and $\sum_{n=N}^\infty$ both converge or neither does.

24. If $\{a_n\}$ is ultimately positive, then the sequence $\{s_n\}$ of partial sums of the series must be ultimately increasing. By Theorem 2, if $\{s_n\}$ is ultimately increasing, then either it is bounded above, and therefore convergent, or else it is not bounded above and diverges to infinity. Since $\sum a_n = \lim s_n$, $\sum a_n$ must either converge when $\{s_n\}$ converges and $\lim s_n = s$ exists, or diverge to infinity when $\{s_n\}$ diverges to infinity.

25. If $\{a_n\}$ is ultimately negative, then the series $\sum a_n$ must either converge (if its partial sums are bounded below), or diverge to $-\infty$ (if its partial sums are not bounded below).

26. "If $a_n = 0$ for every n, then $\sum a_n$ converge" is TRUE because $s_n = \sum_{k=0}^n 0 = 0$, for every n, and so $\sum a_n = \lim s_n = 0$.

27. "If $\sum a_n$ converges, then $\sum 1/a_n$ diverges to infinity" is FALSE. A counterexample is $\sum (-1)^n/2^n$.

28. "If $\sum a_n$ and $\sum b_n$ both diverge, then so does $\sum (a_n + b_n)$" is FALSE. Let $a_n = \dfrac{1}{n}$ and $b_n = -\dfrac{1}{n}$, then $\sum a_n = \infty$ and $\sum b_n = -\infty$ but $\sum (a_n + b_n) = \sum (0) = 0$.

29. "If $a_n \geq c > 0$ for all n, then $\sum a_n$ diverges to infinity" is TRUE. We have

$$s_n = a_1 + a_2 + a_3 + \cdots + a_n \geq c + c + c + \cdots + c = nc,$$

and $nc \to \infty$ as $n \to \infty$.

30. "If $\sum a_n$ diverges and $\{b_n\}$ is bounded, then $\sum a_n b_n$ diverges" is FALSE. Let $a_n = \dfrac{1}{n}$ and $b_n = \dfrac{1}{n+1}$. Then $\sum a_n = \infty$ and $0 \leq b_n \leq 1/2$. But $\sum a_n b_n = \sum \dfrac{1}{n(n+1)}$ which converges by Example 3.

31. "If $a_n > 0$ and $\sum a_n$ converges, then $\sum a_n^2$ converges" is TRUE.
Since $\sum a_n$ converges, therefore $\lim a_n = 0$.
Thus there exists N such that $0 < a_n \leq 1$ for $n \geq N$.
Thus $0 < a_n^2 \leq a_n$ for $n \geq N$.
If $S_n = \sum_{k=N}^n a_k^2$ and $s_n = \sum_{k=N}^n a_k$, then $\{S_n\}$ is increasing and bounded above:

$$S_n \leq s_n \leq \sum_{k=1}^\infty a_k < \infty.$$

Thus $\sum_{k=N}^\infty a_k^2$ converges, and so $\sum_{k=1}^\infty a_k^2$ converges.

Section 9.3 Convergence Tests for Positive Series (page 545)

1. $\sum \dfrac{1}{n^2 + 1}$ converges by comparison with $\sum \dfrac{1}{n^2}$ since $0 < \dfrac{1}{n^2 + 1} < \dfrac{1}{n^2}$.

2. $\sum_{n=1}^\infty \dfrac{n}{n^4 - 2}$ converges by comparison with $\sum_{n=1}^\infty \dfrac{1}{n^3}$ since

$$\lim \frac{\left(\dfrac{n}{n^4 - 2}\right)}{\left(\dfrac{1}{n^3}\right)} = 1, \quad \text{and} \quad 0 < 1 < \infty.$$

3. $\sum \dfrac{n^2 + 1}{n^3 + 1}$ diverges to infinity by comparison with $\sum \dfrac{1}{n}$, since $\dfrac{n^2 + 1}{n^3 + 1} > \dfrac{1}{n}$.

339

4. $\sum_{n=1}^{\infty} \dfrac{\sqrt{n}}{n^2+n+1}$ converges by comparison with $\sum_{n=1}^{\infty} \dfrac{1}{n^{3/2}}$ since

$$\lim \dfrac{\left(\dfrac{\sqrt{n}}{n^2+n+1}\right)}{\left(\dfrac{1}{n^{3/2}}\right)} = 1, \quad \text{and} \quad 0 < 1 < \infty.$$

5. Since $\sin x \le x$ for $x \ge 0$, we have

$$\left|\sin \dfrac{1}{n^2}\right| = \sin \dfrac{1}{n^2} \le \dfrac{1}{n^2},$$

so $\sum \left|\sin \dfrac{1}{n^2}\right|$ converges by comparison with $\sum \dfrac{1}{n^2}$.

6. $\sum_{n=8}^{\infty} \dfrac{1}{\pi^n + 5}$ converges by comparison with the geometric series $\sum_{n=8}^{\infty} \left(\dfrac{1}{\pi}\right)^n$ since $0 < \dfrac{1}{\pi^n + 5} < \dfrac{1}{\pi^n}$.

7. Since $(\ln n)^3 < n$ for large n, $\sum \dfrac{1}{(\ln n)^3}$ diverges to infinity by comparison with $\sum \dfrac{1}{n}$.

8. $\sum_{n=1}^{\infty} \dfrac{1}{\ln(3n)}$ diverges to infinity by comparison with the harmonic series $\sum_{n=1}^{\infty} \dfrac{1}{3n}$ since $\dfrac{1}{\ln(3n)} > \dfrac{1}{3n}$ for $n \ge 1$.

9. Since $\lim_{n\to\infty} \dfrac{\pi^n}{\pi^n - n^\pi} = \lim \dfrac{1}{1 - \dfrac{n^\pi}{\pi^n}} = 1$, the series $\sum \dfrac{1}{\pi^n - n^\pi}$ converges by comparison with the geometric series $\sum \dfrac{1}{\pi^n}$.

10. $\sum_{n=0}^{\infty} \dfrac{1+n}{2+n}$ diverges to infinity since $\lim \dfrac{1+n}{2+n} = 1 > 0$.

11. $\sum \dfrac{1+n^{4/3}}{2+n^{5/3}}$ diverges to infinity by comparison with the divergent p-series $\sum \dfrac{1}{n^{1/3}}$, since

$$\lim_{n\to\infty} \dfrac{1+n^{4/3}}{2+n^{5/3}} \bigg/ \dfrac{1}{n^{1/3}} = \lim \dfrac{n^{1/3}+n^{5/3}}{2+n^{5/3}} = 1.$$

12. $\sum_{n=1}^{\infty} \dfrac{n^2}{1+n\sqrt{n}}$ diverges to infinity since $\lim \dfrac{n^2}{1+n\sqrt{n}} = \infty$.

13. $\sum_{n=3}^{\infty} \dfrac{1}{n \ln n \sqrt{\ln \ln n}}$ diverges to infinity by the integral test, since

$$\int_3^{\infty} \dfrac{dt}{t \ln t \sqrt{\ln \ln t}} = \int_{\ln \ln 3}^{\infty} \dfrac{du}{\sqrt{u}} = \infty.$$

14. $\sum_{n=2}^{\infty} \dfrac{1}{n \ln n (\ln \ln n)^2}$ converges by the integral test:

$$\int_a^{\infty} \dfrac{dt}{t \ln t (\ln \ln t)^2} = \int_{\ln \ln a}^{\infty} \dfrac{du}{u^2} < \infty \quad \text{if} \quad \ln \ln a > 0.$$

15. $\sum \dfrac{1-(-1)^n}{n^4}$ converges by comparison with $\sum \dfrac{1}{n^4}$, since $0 \le \dfrac{1-(-1)^n}{n^4} \le \dfrac{2}{n^4}$.

16. The series

$$\sum_{n=1}^{\infty} \dfrac{1+(-1)^n}{\sqrt{n}} = 0 + \dfrac{2}{\sqrt{2}} + 0 + \dfrac{2}{\sqrt{4}} + 0 + \dfrac{2}{\sqrt{6}} + \cdots$$

$$= 2 \sum_{k=1}^{\infty} \dfrac{1}{\sqrt{2k}} = \sqrt{2} \sum_{k=1}^{\infty} \dfrac{1}{\sqrt{k}}$$

diverges to infinity.

17. Since $\dfrac{1}{2^n(n+1)} < \dfrac{1}{2^n}$, the series $\sum \dfrac{1}{2^n(n+1)}$ converges by comparison with the geometric series $\sum \dfrac{1}{2^n}$.

18. $\sum_{n=1}^{\infty} \dfrac{n^4}{n!}$ converges by the ratio test since

$$\lim \dfrac{\dfrac{(n+1)^4}{(n+1)!}}{\dfrac{n^4}{n!}} = \lim \left(\dfrac{n+1}{n}\right)^4 \dfrac{1}{n+1} = 0.$$

19. $\sum \dfrac{n!}{n^2 e^n}$ diverges to infinity by the ratio test, since

$$\rho = \lim \dfrac{(n+1)!}{(n+1)^2 e^{n+1}} \cdot \dfrac{n^2 e^n}{n!} = \dfrac{1}{e} \lim \dfrac{n^2}{n+1} = \infty.$$

20. $\sum_{n=1}^{\infty} \dfrac{(2n)! 6^n}{(3n)!}$ converges by the ratio test since

$$\lim \dfrac{(2n+2)! 6^{n+1}}{(3n+3)!} \bigg/ \dfrac{(2n)! 6^n}{(3n)!}$$

$$= \lim \dfrac{(2n+2)(2n+1)6}{(3n+3)(3n+2)(3n+1)} = 0.$$

21. $\sum_{n=2}^{\infty} \dfrac{\sqrt{n}}{3^n \ln n}$ converges by the ratio test, since

$$\rho = \lim \dfrac{\sqrt{n+1}}{3^{n+1} \ln(n+1)} \cdot \dfrac{3^n \ln n}{\sqrt{n}}$$

$$= \dfrac{1}{3} \lim \sqrt{\dfrac{n+1}{n}} \cdot \lim \dfrac{\ln n}{\ln(n+1)} = \dfrac{1}{3} < 1.$$

22. $\sum_{n=0}^{\infty} \dfrac{n^{100} 2^n}{\sqrt{n!}}$ converges by the ratio test since

$$\lim \dfrac{(n+1)^{100} 2^{n+1}}{\sqrt{(n+1)!}} \bigg/ \dfrac{n^{100} 2^n}{\sqrt{n!}}$$

$$= \lim 2 \left(\dfrac{n+1}{n}\right)^{100} \dfrac{1}{\sqrt{n+1}} = 0.$$

23. $\sum \dfrac{(2n)!}{(n!)^3}$ converges by the ratio test, since

$$\rho = \lim \dfrac{(2n+2)!}{((n+1)!)^3} \cdot \dfrac{(n!)^3}{(2n)!} = \lim \dfrac{(2n+2)(2n+1)}{(n+1)^3} = 0 < 1.$$

24. $\sum_{n=1}^{\infty} \dfrac{1+n!}{(1+n)!}$ diverges by comparison with the harmonic series $\sum_{n=1}^{\infty} \dfrac{1}{n+1}$ since $\dfrac{1+n!}{(1+n)!} > \dfrac{n!}{(1+n)!} = \dfrac{1}{n+1}$.

25. $\sum \dfrac{2^n}{3^n - n^3}$ converges by the ratio test since

$$\rho = \lim \dfrac{2^{n+1}}{3^{n+1} - (n+1)^3} \cdot \dfrac{3^n - n^3}{2^n}$$

$$= \dfrac{2}{3} \lim \dfrac{3^n - n^3}{3^n - \dfrac{(n+1)^3}{3}} = \dfrac{2}{3} \lim \dfrac{1 - \dfrac{n^3}{3^n}}{1 - \dfrac{(n+1)^3}{3^{n+1}}} = \dfrac{2}{3} < 1.$$

26. $\sum_{n=1}^{\infty} \dfrac{n^n}{\pi^n n!}$ converges by the ratio test since

$$\lim \dfrac{(n+1)^{n+1}}{\pi^{(n+1)}(n+1)!} \bigg/ \dfrac{n^n}{\pi^n n!} = \dfrac{1}{\pi} \lim \left(1 + \dfrac{1}{n}\right)^n = \dfrac{e}{\pi} < 1.$$

27. $f(x) = 1/x^4$ is positive, continuous, and decreasing on $[1, \infty)$. Let

$$A_n = \int_n^{\infty} \dfrac{dx}{x^4} = \lim_{R \to \infty} \left(-\dfrac{1}{3x^3}\right)\bigg|_n^R = \dfrac{1}{3n^3}.$$

We use the approximation

$$s \approx s_n^* = s_n + \dfrac{1}{2}\left(\dfrac{1}{3(n+1)^3} + \dfrac{1}{3n^3}\right).$$

The error satisfies

$$|s - s_n^*| \leq \dfrac{1}{2}\left(\dfrac{1}{3n^3} - \dfrac{1}{3(n+1)^3}\right)$$

$$= \dfrac{1}{6} \dfrac{(n+1)^3 - n^3}{n^3(n+1)^3}$$

$$= \dfrac{1}{6} \dfrac{3n^2 + 3n + 1}{n^3(n+1)^3} < \dfrac{7}{6n^4}.$$

We have used $3n^2 + 3n + 1 \leq 7n^2$ and $n^3(n+1)^3 > n^6$ to obtain the last inequality. We will have $|s - s_n^*| < 0.001$ provided

$$\dfrac{7}{6n^4} < 0.001,$$

that is, if $n^4 > 7000/6$. Since $6^4 = 1296 > 7000/6$, $n = 6$ will do. Thus

$$\sum_{n=1}^{\infty} \dfrac{1}{n^4} \approx s_6^* = 1 + \dfrac{1}{2^4} + \dfrac{1}{3^4} + \dfrac{1}{4^4} + \dfrac{1}{5^4} + \dfrac{1}{6^4} + \dfrac{1}{6}\left(\dfrac{1}{7^3} + \dfrac{1}{6^3}\right)$$

$$\approx 1.082 \quad \text{with error less than 0.001 in absolute value.}$$

28. Since $f(x) = \dfrac{1}{x^3}$ is positive, continuous and decreasing on $[1, \infty)$, for any $n = 1, 2, 3, \ldots$, we have

$$s_n + A_{n+1} \leq s \leq s_n + A_n$$

where $s_n = \sum_{k=1}^{n} \dfrac{1}{k^3}$ and $A_n = \int_n^{\infty} \dfrac{dx}{x^3} = \dfrac{1}{2n^2}$. If $s_n^* = s_n + \dfrac{1}{2}(A_{n+1} + A_n)$, then

$$|s_n - s_n^*| \leq \dfrac{A_n - A_{n+1}}{2} = \dfrac{1}{4}\left[\dfrac{1}{n^2} - \dfrac{1}{(n+1)^2}\right]$$

$$= \dfrac{1}{4} \dfrac{2n+1}{n^2(n+1)^2} < 0.001$$

if $n = 8$. Thus, the error in the approximation $s \approx s_8^*$ is less than 0.001.

29. Since $f(x) = \dfrac{1}{x^{3/2}}$ is positive, continuous and decreasing on $[1, \infty)$, for any $n = 1, 2, 3, \ldots$, we have

$$s_n + A_{n+1} \leq s \leq s_n + A_n$$

where $s_n = \sum_{k=1}^{n} \frac{1}{k^{3/2}}$ and $A_n = \int_n^\infty \frac{dx}{x^{3/2}} = \frac{2}{\sqrt{n}}$. If $s_n^* = s_n + \frac{1}{2}(A_{n+1} + A_n) = s_n + \left(\frac{1}{\sqrt{n}} + \frac{1}{\sqrt{n+1}}\right)$, then

$$|s_n - s_n^*| \leq \frac{A_n - A_{n+1}}{2}$$
$$= \frac{1}{2}\left(\frac{2}{\sqrt{n}} - \frac{2}{\sqrt{n+1}}\right)$$
$$= \frac{\sqrt{n+1} - \sqrt{n}}{\sqrt{n}\sqrt{n+1}} = \frac{1}{\sqrt{n}\sqrt{n+1}(\sqrt{n} + \sqrt{n+1})}$$
$$< \frac{1}{2n^{3/2}} < 0.001$$

if $n \geq 63$. Thus, the error in the approximation $s \approx s_{63}^*$ is less than 0.001.

30. Again, we have $s_n + A_{n+1} \leq s \leq s_n + A_n$ where $s_n = \sum_{k=1}^n \frac{1}{k^2+4}$ and

$$A_n = \int_n^\infty \frac{dx}{x^2+4} = \frac{1}{2}\tan^{-1}\left(\frac{x}{2}\right)\bigg|_n^\infty = \frac{\pi}{4} - \frac{1}{2}\tan^{-1}\left(\frac{n}{2}\right).$$

If $s_n^* = s_n + \frac{1}{2}(A_{n+1} + A_n)$, then

$$|s_n - s_n^*| \leq \frac{A_n - A_{n+1}}{2}$$
$$= \frac{1}{2}\left[\frac{\pi}{4} - \frac{1}{2}\tan^{-1}\left(\frac{n}{2}\right) - \frac{\pi}{4} + \frac{1}{2}\tan^{-1}\left(\frac{n+1}{2}\right)\right]$$
$$= \frac{1}{4}\left[\tan^{-1}\left(\frac{n+1}{2}\right) - \tan^{-1}\left(\frac{n}{2}\right)\right] = \frac{1}{4}(a-b),$$

where $a = \tan^{-1}\left(\frac{n+1}{2}\right)$ and $b = \tan^{-1}\left(\frac{n}{2}\right)$. Now

$$\tan(a-b) = \frac{\tan a - \tan b}{1 + \tan a \tan b}$$
$$= \frac{\left(\frac{n+1}{2}\right) - \left(\frac{n}{2}\right)}{1 + \left(\frac{n+1}{2}\right)\left(\frac{n}{2}\right)}$$
$$= \frac{2}{n^2+n+4}$$
$$\Leftrightarrow a - b = \tan^{-1}\left(\frac{2}{n^2+n+4}\right).$$

We want error less than 0.001:

$$\frac{1}{4}(a-b) = \frac{1}{4}\tan^{-1}\left(\frac{2}{n^2+n+4}\right) < 0.001$$
$$\Leftrightarrow \frac{2}{n^2+n+4} < \tan 0.004$$
$$\Leftrightarrow n^2 + n > 2\cot(0.004) - 4 \approx 496.$$

$n = 22$ will do. The approximation $s \approx s_{22}^*$ has error less than 0.001.

31. We have $s = \sum_{k=1}^\infty \frac{1}{2^k k!}$ and

$$s_n = \sum_{k=1}^n \frac{1}{2^k k!} = \frac{1}{2} + \frac{1}{2^2 2!} + \frac{1}{2^3 3!} + \cdots + \frac{1}{2^n n!}.$$

Then

$$0 < s - s_n$$
$$= \frac{1}{2^{n+1}(n+1)!} + \frac{1}{2^{n+2}(n+2)!} + \frac{1}{2^{n+3}(n+3)!} + \cdots$$
$$= \frac{1}{2^{n+1}(n+1)!}\left[1 + \frac{1}{2(n+2)} + \frac{1}{2^2(n+2)(n+3)} + \cdots\right]$$
$$< \frac{1}{2^{n+1}(n+1)!}\left[1 + \frac{1}{2(n+2)} + \left(\frac{1}{2(n+2)}\right)^2 + \cdots\right]$$
$$= \frac{1}{2^{n+1}(n+1)!} \cdot \frac{1}{1 - \frac{1}{2(n+2)}}$$
$$= \frac{n+2}{2^n(n+1)!(2n+3)} < 0.001$$

if $n = 4$. Thus, $s \approx s_4 = \frac{1}{2} + \frac{1}{2^2 2!} + \frac{1}{2^3 3!} + \frac{1}{2^4 4!}$ with error less than 0.001.

32. We have $s = \sum_{k=1}^\infty \frac{1}{(2k-1)!}$ and

$$s_n = \sum_{k=1}^n \frac{1}{(2k-1)!} = \frac{1}{1!} + \frac{1}{3!} + \frac{1}{5!} + \cdots + \frac{1}{(2n-1)!}.$$

Then

$$0 < s - s_n = \frac{1}{(2n+1)!} + \frac{1}{(2n+3)!} + \frac{1}{(2n+5)!} + \cdots$$
$$= \frac{1}{(2n+1)!}\left[1 + \frac{1}{(2n+2)(2n+3)} + \frac{1}{(2n+2)(2n+3)(2n+4)(2n+5)} + \cdots\right]$$
$$< \frac{1}{(2n+1)!}\left[1 + \frac{1}{(2n+2)(2n+3)} + \frac{1}{[(2n+2)(2n+3)]^2} + \cdots\right]$$
$$= \frac{1}{(2n+1)!}\left[\frac{1}{1 - \frac{1}{(2n+2)(2n+3)}}\right]$$
$$= \frac{1}{(2n+1)!}\frac{4n^2+10n+6}{4n^2+10n+5} < 0.001$$

if $n = 3$. Thus, $s \approx s_3 = 1 + \frac{1}{3!} + \frac{1}{5!} = 1.175$ with error less than 0.001.

33. We have $s = \sum_{k=0}^{\infty} \frac{2^k}{(2k)!}$ and $s_n = \sum_{k=0}^{n-1} \frac{2^k}{(2k)!}$. Thus

$0 < s - s_n$
$= \frac{2^n}{(2n)!} + \frac{2^{n+1}}{(2n+2)!} + \frac{2^{n+2}}{(2n+4)!} + \cdots$
$= \frac{2^n}{(2n)!}\left[1 + \frac{2}{(2n+1)(2n+2)}\right.$
$\left. + \frac{2^2}{(2n+1)(2n+2)(2n+3)(2n+4)} + \cdots\right]$
$< \frac{2^n}{(2n)!}\left[1 + \frac{2}{(2n+1)(2n+2)} + \left(\frac{2}{(2n+1)(2n+2)}\right)^2 + \cdots\right]$
$= \frac{2^n}{(2n)!} \cdot \frac{1}{1 - \frac{2}{(2n+1)(2n+2)}}$
$= \frac{2^n}{(2n)!} \cdot \frac{4n^2 + 6n + 2}{4n^2 + 6n} < 0.001$

if $n = 4$. Thus, $s \approx s_4$ with error less than 0.001.

34. We have $s = \sum_{k=1}^{\infty} \frac{1}{k^k}$ and

$s_n = \sum_{k=1}^{n} \frac{1}{k^k} = \frac{1}{1} + \frac{1}{2^2} + \frac{1}{3^3} + \cdots + \frac{1}{n^n}.$

Then

$0 < s - s_n = \frac{1}{(n+1)^{n+1}} + \frac{1}{(n+2)^{n+2}} + \frac{1}{(n+3)^{n+3}} + \cdots$
$< \frac{1}{(n+1)^{n+1}}\left[1 + \frac{1}{n+1} + \frac{1}{(n+1)^2} + \cdots\right]$
$= \frac{1}{(n+1)^{n+1}}\left[\frac{1}{1 - \frac{1}{n+1}}\right]$
$= \frac{1}{n(n+1)^n} < 0.001$

if $n = 4$. Thus, $s \approx s_4 = 1 + \frac{1}{2^2} + \frac{1}{3^3} + \frac{1}{4^4} = 1.291$ with error less than 0.001.

35. Let $f(x) = \frac{1}{1+x^2}$. Then f is decreasing on $[1, \infty)$. Since $\sum_{n=1}^{\infty} f(n)$ is a right Riemann sum for

$\int_0^{\infty} f(x)\,dx = \lim_{R \to \infty} \tan^{-1} x \Big|_0^R = \frac{\pi}{2},$

$\sum_{n=1}^{\infty} \frac{1}{1+n^2} = \sum_{n=1}^{\infty} f(n)$ converges by the integral test, and its sum is less than $\pi/2$.

36. Let $u = \ln \ln t$, $du = \frac{dt}{t \ln t}$ and $\ln \ln a > 0$; then

$$\int_a^{\infty} \frac{dt}{t \ln t (\ln \ln t)^p} = \int_{\ln \ln a}^{\infty} \frac{du}{u^p}$$

will converge if and only if $p > 1$. Thus, $\sum_{n=3}^{\infty} \frac{1}{n \ln n (\ln \ln n)^p}$ will converge if and only if $p > 1$. Similarly,

$$\sum_{n=N}^{\infty} \frac{1}{n(\ln n)(\ln \ln n) \cdots (\ln_j n)(\ln_{j+1} n)^p}$$

converges if and only if $p > 1$, where N is large enough that $\ln_j N > 1$.

37. Let $a_n > 0$ for all n. (Let's forget the "ultimately" part.) Let $\sigma = \lim (a_n)^{1/n}$.

CASE I. Suppose $\sigma < 1$. Pick λ such that $\sigma < \lambda < 1$. Then there exists N such that $(a_n)^{1/n} \leq \lambda$ for all $n \geq N$. Therefore

$a_N \leq \lambda^N$, $a_{N+1} \leq \lambda^{N+1}$, $a_{N+2} \leq \lambda^{N+2}, \ldots.$

Thus $\sum_{n=N}^{\infty} a_n$ converges by comparison with the geometric series $\sum_{n=N}^{\infty} \lambda^n$, and $\sum_{n=1}^{\infty} a_n$ also converges.

CASE II. Suppose $\sigma > 1$. Then $(a_n)^{1/n} \geq 1$, and $a_n \geq 1$, for all sufficiently large values of n. Therefore $\lim a_n \neq 0$ and $\sum a_n$ must diverge. Since $a_n > 0$ it diverges to infinity.

CASE III. Let $a_n = \frac{1}{n}$ and $b_n = \frac{1}{n^2}$. Since $\lim n^{1/n} = 1$ (because $\lim \frac{\ln n}{n} = 0$), we have $\lim (a_n)^{1/n} = 1$ and $\lim (b_n)^{1/n} = 1$. That is, $\sigma = 1$ for both series. But $\sum a_n$ diverges to infinity, while $\sum b_n$ converges. Thus the case $\sigma = 1$ provides no information on the convergence or divergence of a series.

38. Let $a_n = 2^{n+1}/n^n$. Then

$$\lim_{n \to \infty} \sqrt[n]{a_n} = \lim_{n \to \infty} \frac{2 \times 2^{1/n}}{n} = 0.$$

Since this limit is less than 1, $\sum_{n=1}^{\infty} a_n$ converges by the root test.

39. $\sum_{n=1}^{\infty} \left(\dfrac{n}{n+1}\right)^{n^2}$ converges by the root test of Exercise 31 since

$$\sigma = \lim_{n\to\infty} \left[\left(\dfrac{n}{n+1}\right)^{n^2}\right]^{1/n} = \lim_{n\to\infty} \dfrac{1}{\left(1+\dfrac{1}{n}\right)^n} = \dfrac{1}{e} < 1.$$

40. Let $a_n = \dfrac{2^{n+1}}{n^n}$. Then

$$\dfrac{a_{n+1}}{a_n} = \dfrac{2^{n+2}}{(n+1)^{n+1}} \cdot \dfrac{n^n}{2^{n+1}}$$

$$= \dfrac{2}{(n+1)\left(\dfrac{n}{n+1}\right)^n} = \dfrac{2}{n+1} \cdot \dfrac{1}{\left(1+\dfrac{1}{n}\right)^n}$$

$$\to 0 \times \dfrac{1}{e} = 0 \text{ as } n \to \infty.$$

Thus $\sum_{n=1}^{\infty} a_n$ converges by the ratio test.
(Remark: the question contained a typo. It was intended to ask that #33 be repeated, using the ratio test. That is a little harder.)

41. Trying to apply the ratio test to $\sum \dfrac{2^{2n}(n!)^2}{(2n)!}$, we obtain

$$\rho = \lim \dfrac{2^{2n+2}((n+1)!)^2}{(2n+2)!} \cdot \dfrac{(2n)!}{2^{2n}(n!)^2} = \lim \dfrac{4(n+1)^2}{(2n+2)(2n+1)} = 1.$$

Thus the ratio test provides no information. However,

$$\dfrac{2^{2n}(n!)^2}{(2n)!} = \dfrac{[2n(2n-2)\cdots 6\cdot 4\cdot 2]^2}{2n(2n-1)(2n-2)\cdots 3\cdot 2\cdot 1}$$

$$= \dfrac{2n}{2n-1} \cdot \dfrac{2n-2}{2n-3} \cdots \dfrac{4}{3} \cdot \dfrac{2}{1} > 1.$$

Since the terms exceed 1, the series diverges to infinity.

42. We have

$$a_n = \dfrac{(2n)!}{2^{2n}(n!)^2} = \dfrac{1\times 2\times 3\times 4\times \cdots \times 2n}{(2\times 4\times 6\times 8\times \cdots \times 2n)^2}$$

$$= \dfrac{1\times 3\times 5\times \cdots \times (2n-1)}{2\times 4\times 6\times \cdots \times (2n-2)\times 2n}$$

$$= 1\times \dfrac{3}{2}\times \dfrac{5}{4}\times \dfrac{7}{6} \times \cdots \times \dfrac{2n-1}{2n-2} \times \dfrac{1}{2n} > \dfrac{1}{2n}.$$

Therefore $\sum_{n=1}^{\infty} \dfrac{(2n)!}{2^{2n}(n!)^2}$ diverges to infinity by comparison with the harmonic series $\sum_{n=1}^{\infty} \dfrac{1}{2n}$.

43. a) If n is a positive integer and $k > 0$, then $(1+k)^n \geq 1 + nk > nk$, so $n < \dfrac{1}{k}(1+k)^n$.

b) Let $s_N = \sum_{n=0}^{N} \dfrac{n}{2^n} < \dfrac{1}{k} \sum_{n=0}^{N} \left(\dfrac{1+k}{2}\right)^n$

$$= \dfrac{1}{k} \sum_{n=0}^{N} r^n = \dfrac{1}{k} \cdot \dfrac{1-r^{N+1}}{1-r},$$

where $r = (1+k)/n$. Thus

$$s_n < \dfrac{1}{k} \cdot \dfrac{1-\left(\dfrac{1+k}{2}\right)^{N+1}}{1-\dfrac{1+k}{2}}$$

$$= \dfrac{2}{k(1-k)}\left(1-\left(\dfrac{1+k}{2}\right)^{N+1}\right) \leq \dfrac{2}{k(1-k)}.$$

Therefore, $s = \sum_{n=0}^{\infty} \dfrac{n}{2^n} \leq \dfrac{2}{k(1-k)}$.

Since the maximum value of $k(1-k)$ is $1/4$ (at $k = 1/2$), the best upper bound we get for s by this method is $s \leq 8$.

c) $s - s_n = \sum_{j=n+1}^{\infty} \dfrac{j}{2^j} < \dfrac{1}{k} \sum_{j=n+1}^{\infty} \left(\dfrac{1+k}{2}\right)^j$

$$= \dfrac{1}{k}\left(\dfrac{1+k}{2}\right)^{n+1} \cdot \dfrac{1}{1-\dfrac{1+k}{2}}$$

$$= \dfrac{(1+k)^{n+1}}{k(1-k)2^n} = \dfrac{G(k)}{2^n},$$

where $G(k) = \dfrac{(1+k)^{n+1}}{k(1-k)}$. For minimum $G(k)$, look for a critical point:

$$\dfrac{k(1-k)(n+1)(1+k)^n - (1+k)^{n+1}(1-2k)}{k^2(1-k)^2} = 0$$

$$(k-k^2)(n+1) - (1+k)(1-2k) = 0$$

$$k^2(n+1) - k(n+1) + 1 - k - 2k^2 = 0$$

$$(n-1)k^2 - (n+2)k + 1 = 0$$

$$k = \dfrac{(n+2) \pm \sqrt{(n+2)^2 - 4(n-1)}}{2(n-1)}$$

$$= \dfrac{n+2 \pm \sqrt{n^2+8}}{2(n-1)}.$$

For given n, the upper bound is minimal if

$$k = \dfrac{n+2 - \sqrt{n^2+8}}{2(n-1)} \quad \text{(for } n \geq 2\text{)}.$$

44. If $s = \sum_{k=1}^{\infty} c_k = \sum_{k=1}^{\infty} \dfrac{1}{k^2(k+1)}$, then we have

$$s_n + A_{n+1} \leq s \leq s_n + A_n$$

where $s_n = \sum_{k=1}^{n} \dfrac{1}{k^2(k+1)}$ and

$$A_n = \int_n^{\infty} \dfrac{dx}{x^2(x+1)} = \int_n^{\infty} \left(\dfrac{-1}{x} + \dfrac{1}{x^2} + \dfrac{1}{x+1}\right) dx$$

$$= -\ln x - \dfrac{1}{x} + \ln(x+1) \Big|_n^{\infty}$$

$$= \ln\left(1+\dfrac{1}{x}\right) - \dfrac{1}{x} \Big|_n^{\infty}$$

$$= \dfrac{1}{n} - \ln\left(1+\dfrac{1}{n}\right).$$

If $s_n^* = s_n + \dfrac{1}{2}(A_{n+1} + A_n)$, then

$$|s_n - s_n^*| \le \dfrac{A_n - A_{n+1}}{2}$$

$$= \dfrac{1}{2}\left[\dfrac{1}{n} - \ln\left(1+\dfrac{1}{n}\right) - \dfrac{1}{n+1} + \ln\left(1+\dfrac{1}{n+1}\right)\right]$$

$$= \dfrac{1}{2}\left[\dfrac{1}{n(n+1)} + \ln\left(\dfrac{n^2+2n}{n^2+2n+1}\right)\right]$$

$$\le \dfrac{1}{2}\left[\dfrac{1}{n(n+1)} + \left(\dfrac{n^2+2n}{n^2+2n+1} - 1\right)\right]$$

$$= \dfrac{1}{2n(n+1)^2} < 0.001$$

if $n = 8$. Thus,

$$\sum_{n=1}^{\infty} \dfrac{1}{n^2} = 1 + s_8^* = 1 + s_8 + \dfrac{1}{2}(A_9 + A_8)$$

$$= 1 + \left[\dfrac{1}{2} + \dfrac{1}{2^2(3)} + \dfrac{1}{3^2(4)} + \cdots + \dfrac{1}{8^2(9)}\right] + \dfrac{1}{2}\left[\left(\dfrac{1}{9} - \ln\dfrac{10}{9}\right) + \left(\dfrac{1}{8} - \ln\dfrac{9}{8}\right)\right]$$

$$= 1.6450$$

with error less than 0.001.

Section 9.4 Absolute and Conditional Convergence (page 552)

1. $\sum \dfrac{(-1)^n}{\sqrt{n}}$ converges by the alternating series test (since the terms alternate in sign, decrease in size, and approach 0). However, the convergence is only conditional, since $\sum \dfrac{1}{\sqrt{n}}$ diverges to infinity.

2. $\sum_{n=1}^{\infty} \dfrac{(-1)^n}{n^2 + \ln n}$ converges absolutely since $\left|\dfrac{(-1)^n}{n^2 + \ln n}\right| \le \dfrac{1}{n^2}$ and $\sum_{n=1}^{\infty} \dfrac{1}{n^2}$ converges.

3. $\sum \dfrac{\cos(n\pi)}{(n+1)\ln(n+1)} = \sum \dfrac{(-1)^n}{(n+1)\ln(n+1)}$ converges by the alternating series test, but only conditionally since $\sum \dfrac{1}{(n+1)\ln(n+1)}$ diverges to infinity (by the integral test).

4. $\sum_{n=1}^{\infty} \dfrac{(-1)^{2n}}{2^n} = \sum_{n=1}^{\infty} \dfrac{1}{2^n}$ is a positive, convergent geometric series so must converge absolutely.

5. $\sum \dfrac{(-1)^n(n^2-1)}{n^2+1}$ diverges since its terms do not approach zero.

6. $\sum_{n=1}^{\infty} \dfrac{(-2)^n}{n!}$ converges absolutely by the ratio test since

$$\lim \left|\dfrac{(-2)^{n+1}}{(n+1)!} \cdot \dfrac{n!}{(-2)^n}\right| = 2\lim \dfrac{1}{n+1} = 0.$$

7. $\sum \dfrac{(-1)^n}{n\pi^n}$ converges absolutely, since, for $n \ge 1$,

$$\left|\dfrac{(-1)^n}{n\pi^n}\right| \le \dfrac{1}{\pi^n},$$

and $\sum \dfrac{1}{\pi^n}$ is a convergent geometric series.

8. $\sum_{n=0}^{\infty} \dfrac{-n}{n^2+1}$ diverges to $-\infty$ since all terms are negative and $\sum_{n=0}^{\infty} \dfrac{n}{n^2+1}$ diverges to infinity by comparison with $\sum_{n=0}^{\infty} \dfrac{1}{n}$.

9. $\sum (-1)^n \dfrac{20n^2 - n - 1}{n^3 + n^2 + 33}$ converges by the alternating series test (the terms are ultimately decreasing in size, and approach zero), but the convergence is only conditional since $\sum \dfrac{20n^2 - n - 1}{n^3 + n^2 + 33}$ diverges to infinity by comparison with $\sum \dfrac{1}{n}$.

10. $\sum_{n=1}^{\infty} \dfrac{100\cos(n\pi)}{2n+3} = \sum_{n=1}^{\infty} \dfrac{100(-1)^n}{2n+3}$ converges by the alternating series test but only conditionally since

$$\left|\dfrac{100(-1)^n}{2n+3}\right| = \dfrac{100}{2n+3}$$

and $\sum_{n=1}^{\infty} \dfrac{100}{2n+3}$ diverges to infinity.

11. $\sum \dfrac{n!}{(-100)^n}$ diverges since $\lim \dfrac{n!}{100^n} = \infty$.

12. $\sum_{n=10}^{\infty} \dfrac{\sin(n+\tfrac{1}{2})\pi}{\ln \ln n} = \sum_{n=10}^{\infty} \dfrac{(-1)^n}{\ln \ln n}$ converges by the alternating series test but only conditionally since $\sum_{n=10}^{\infty} \dfrac{1}{\ln \ln n}$ diverges to infinity by comparison with $\sum_{n=10}^{\infty} \dfrac{1}{n}$.
($\ln \ln n < n$ for $n \geq 10$.)

13. If $s = \sum_{k=1}^{\infty} (-1)^{k-1} \dfrac{k}{k^2+1}$, and $s_n = \sum_{k=1}^{n} (-1)^{k-1} \dfrac{k}{k^2+1}$, then
$$|s - s_n| < \dfrac{n+1}{(n+1)^2+1} < 0.001$$
if $n = 999$, because the series satisfies the conditions of the alternating series test.

14. Since the terms of the series $s = \sum_{n=0}^{\infty} \dfrac{(-1)^n}{(2n)!}$ are alternating in sign and decreasing in size, the size of the error in the approximation $s \approx s_n$ does not exceed that of the first omitted term:
$$|s - s_n| \leq \dfrac{1}{(2n+2)!} < 0.001$$
if $n = 3$. Hence $s \approx 1 - \dfrac{1}{2!} + \dfrac{1}{4!} - \dfrac{1}{6!}$; four terms will approximate s with error less than 0.001 in absolute value.

15. If $s = \sum_{k=1}^{\infty} (-1)^{k-1} \dfrac{k}{2^k}$, and $s_n = \sum_{k=1}^{n} (-1)^{k-1} \dfrac{k}{2^k}$, then
$$|s - s_n| < \dfrac{n+1}{2^{n+1}} < 0.001$$
if $n = 13$, because the series satisfies the conditions of the alternating series test from the second term on.

16. Since the terms of the series $s = \sum_{n=0}^{\infty} (-1)^n \dfrac{3^n}{n!}$ are alternating in sign and ultimately decreasing in size (they decrease after the third term), the size of the error in the approximation $s \approx s_n$ does not exceed that of the first omitted term (provided $n \geq 3$): $|s-s_n| \leq \dfrac{3^{n+1}}{(n+1)!} < 0.001$ if $n = 12$. Thus twelve terms will suffice to approximate s with error less than 0.001 in absolute value.

17. Applying the ratio test to $\sum \dfrac{x^n}{\sqrt{n+1}}$, we obtain
$$\rho = \lim \left| \dfrac{x^{n+1}}{\sqrt{n+2}} \cdot \dfrac{\sqrt{n+1}}{x^n} \right| = |x| \lim \sqrt{\dfrac{n+1}{n+2}} = |x|.$$
Hence the series converges absolutely if $|x| < 1$, that is, if $-1 < x < 1$. The series converges conditionally for $x = -1$, but diverges for all other values of x.

18. Let $a_n = \dfrac{(x-2)^n}{n^2 2^{2n}}$. Apply the ratio test
$$\rho = \lim \left| \dfrac{(x-2)^{n+1}}{(n+1)^2 2^{2n+2}} \times \dfrac{n^2 2^{2n}}{(x-2)^n} \right| = \dfrac{|x-2|}{4} < 1$$
if and only if $|x-2| < 4$, that is $-2 < x < 6$. If $x = -2$, then $\sum_{n=1}^{\infty} a_n = \sum_{n=1}^{\infty} \dfrac{(-1)^n}{n^2}$, which converges absolutely. If $x = 6$, then $\sum_{n=1}^{\infty} a_n = \sum_{n=1}^{\infty} \dfrac{1}{n^2}$, which also converges absolutely. Thus, the series converges absolutely if $-2 \leq x \leq 6$ and diverges elsewhere.

19. Apply the ratio test to $\sum (-1)^n \dfrac{(x-1)^n}{2n+3}$:
$$\rho = \lim \left| \dfrac{(x-1)^{n+1}}{2n+5} \cdot \dfrac{2n+3}{(x-1)^n} \right| = |x-1|.$$
The series converges absolutely if $|x-1| < 1$, that is, if $0 < x < 2$, and converges conditionally if $x = 2$. It diverges for all other values of x.

20. Let $a_n = \dfrac{1}{2n-1} \left(\dfrac{3x+2}{-5} \right)^n$. Apply the ratio test
$$\rho = \lim \left| \dfrac{1}{2n+1} \left(\dfrac{3x+2}{-5} \right)^{n+1} \times \dfrac{2n-1}{1} \left(\dfrac{3x+2}{-5} \right)^{-n} \right|$$
$$= \left| \dfrac{3x+2}{5} \right| < 1$$
if and only if $\left| x + \dfrac{2}{3} \right| < \dfrac{5}{3}$, that is $-\dfrac{7}{3} < x < 1$. If $x = -\dfrac{7}{3}$, then $\sum_{n=1}^{\infty} a_n = \sum_{n=1}^{\infty} \dfrac{1}{2n-1}$, which diverges. If $x = 1$, then $\sum_{n=1}^{\infty} a_n = \sum_{n=1}^{\infty} \dfrac{(-1)^n}{2n-1}$, which converges conditionally. Thus, the series converges absolutely if $-\dfrac{7}{3} < x < 1$, converges conditionally if $x = 1$ and diverges elsewhere.

21. Apply the ratio test to $\sum \dfrac{x^n}{2^n \ln n}$:
$$\rho = \lim \left| \dfrac{x^{n+1}}{2^{n+1} \ln(n+1)} \cdot \dfrac{2^n \ln n}{x^n} \right| = \dfrac{|x|}{2} \lim \dfrac{\ln n}{\ln(n+1)} = \dfrac{|x|}{2}.$$

(The last limit can be evaluated by l'Hôpital's Rule.) The given series converges absolutely if $|x| < 2$, that is, if $-2 < x < 2$. By the alternating series test, it converges conditionally if $x = -2$. It diverges for all other values of x.

22. Let $a_n = \dfrac{(4x+1)^n}{n^3}$. Apply the ratio test

$$\rho = \lim \left| \dfrac{(4x+1)^{n+1}}{(n+1)^3} \times \dfrac{n^3}{(4x+1)^n} \right| = |4x+1| < 1$$

if and only if $-\dfrac{1}{2} < x < 0$. If $x = -\dfrac{1}{2}$, then $\sum_{n=1}^{\infty} a_n = \sum_{n=1}^{\infty} \dfrac{(-1)^n}{n^3}$, which converges absolutely.

If $x = 0$, then $\sum_{n=1}^{\infty} a_n = \sum_{n=1}^{\infty} \dfrac{1}{n^3}$, which also converges absolutely. Thus, the series converges absolutely if $-\dfrac{1}{2} \le x \le 0$ and diverges elsewhere.

23. Apply the ratio test to $\sum \dfrac{(2x+3)^n}{n^{1/3} 4^n}$:

$$\rho = \lim \left| \dfrac{(2x+3)^{n+1}}{(n+1)^{1/3} 4^{n+1}} \cdot \dfrac{n^{1/3} 4^n}{(2x+3)^n} \right| = \dfrac{|2x+3|}{4} = \dfrac{|x+\tfrac{3}{2}|}{2}.$$

The series converges absolutely if $\left|x + \dfrac{3}{2}\right| < 2$, that is, if $-\dfrac{7}{2} < x < \dfrac{1}{2}$. By the alternating series test it converges conditionally at $x = -\dfrac{7}{2}$. It diverges elsewhere.

24. Let $a_n = \dfrac{1}{n}\left(1 + \dfrac{1}{x}\right)^n$. Apply the ratio test

$$\rho = \lim \left| \dfrac{1}{n+1}\left(1 + \dfrac{1}{x}\right)^{n+1} \times \dfrac{n}{1}\left(1 + \dfrac{1}{x}\right)^{-n} \right| = \left|1 + \dfrac{1}{x}\right| < 1$$

if and only if $|x+1| < |x|$, that is, $-2 < \dfrac{1}{x} < 0 \Rightarrow x < -\dfrac{1}{2}$. If $x = -\dfrac{1}{2}$, then $\sum_{n=1}^{\infty} a_n = \sum_{n=1}^{\infty} \dfrac{(-1)^n}{n}$, which converges conditionally.

Thus, the series converges absolutely if $x < -\dfrac{1}{2}$, converges conditionally if $x = -\dfrac{1}{2}$ and diverges elsewhere. It is undefined at $x = 0$.

25. $\sum_{n=1}^{\infty} \dfrac{\sin(n\pi/2)}{n} = 1 + 0 - \dfrac{1}{3} + 0 + \dfrac{1}{5} + 0 - \dfrac{1}{7} + 0 + \cdots$

The alternating series test does not apply directly, but does apply to the modified series with the zero terms deleted. Since this latter series converges conditionally, the given series also converges conditionally.

26. If
$$a_n = \begin{cases} \dfrac{10}{n^2}, & \text{if } n \text{ is even}; \\ \dfrac{-1}{10n^3}, & \text{if } n \text{ is odd}; \end{cases}$$

then $|a_n| \le \dfrac{10}{n^2}$ for every $n \ge 1$. Hence, $\sum_{n=1}^{\infty} a_n$ converges absolutely by comparison with $\sum_{n=1}^{\infty} \dfrac{10}{n^2}$.

27. a) "$\sum a_n$ converges implies $\sum (-1)^n a_n$ converges" is FALSE. $a_n = \dfrac{(-1)^n}{n}$ is a counterexample.

b) "$\sum a_n$ converges and $\sum (-1)^n a_n$ converges implies $\sum a_n$ converges absolutely" is FALSE. The series of Exercise 25 is a counterexample.

c) "$\sum a_n$ converges absolutely implies $\sum (-1)^n a_n$ converges absolutely" is TRUE, because $|(-1)^n a_n| = |a_n|$.

28. a) We have
$$\ln(n!) = \ln 1 + \ln 2 + \ln 3 + \cdots + \ln n$$
$$= \text{sum of area of the shaded rectangles}$$
$$> \int_1^n \ln t \, dt = (t \ln t - t)\Big|_1^n$$
$$= n \ln n - n + 1.$$

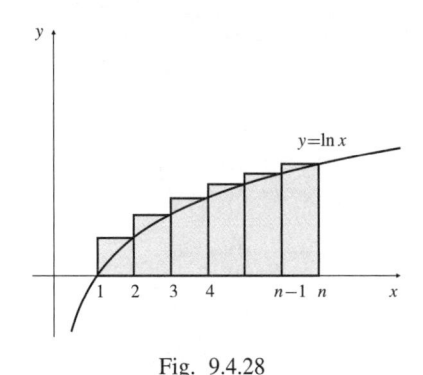

Fig. 9.4.28

b) Let $a_n = \dfrac{n! x^n}{n^n}$. Apply the ratio test

$$\rho = \lim \left| \dfrac{(n+1)! x^{n+1}}{(n+1)^{n+1}} \times \dfrac{n^n}{n! x^n} \right|$$
$$= \lim \dfrac{|x|}{\left(1 + \dfrac{1}{n}\right)^n} = \dfrac{|x|}{e} < 1$$

if and only if $-e < x < e$. If $x = \pm e$, then, by (a),

$$\ln\left|\frac{n!e^n}{n^n}\right| = \ln(n!) + \ln e^n - \ln n^n$$
$$> (n\ln n - n + 1) + n - n\ln n = 1.$$
$$\Rightarrow \left|\frac{n!e^n}{n^n}\right| > e.$$

Hence, $\sum_{n=1}^{\infty} a_n$ converges absolutely if $-e < x < e$ and diverges elsewhere.

29. Applying the ratio test to $\sum \frac{(2n)!x^n}{2^{2n}(n!)^2} = \sum a_n x^n$, we obtain
$$\rho = \lim |x|\frac{(2n+2)(2n+1)}{4(n+1)^2} = |x|.$$

Thus $\sum a_n x^n$ converges absolutely if $-1 < x < 1$, and diverges if $x > 1$ or $x < -1$. In Exercise 36 of Section 9.3 it was shown that $a_n \geq \frac{1}{2n}$, so the given series definitely diverges at $x = 1$ and may at most converge conditionally at $x = -1$. To see whether it does converge at -1, we write, as in Exercise 36 of Section 9.3,

$$a_n = \frac{(2n)!}{2^{2n}(n!)^2} = \frac{1 \times 2 \times 3 \times 4 \times \cdots \times 2n}{(2 \times 4 \times 6 \times 8 \times \cdots \times 2n)^2}$$
$$= \frac{1 \times 3 \times 5 \times \cdots \times (2n-1)}{2 \times 4 \times 6 \times \cdots \times (2n-2) \times 2n}$$
$$= \frac{1}{2} \times \frac{3}{4} \times \cdots \times \frac{2n-3}{2n-2} \times \frac{2n-1}{2n}$$
$$= \left(1 - \frac{1}{2}\right)\left(1 - \frac{1}{4}\right)\cdots\left(1 - \frac{1}{2n-2}\right)\left(1 - \frac{1}{2n}\right).$$

It is evident that a_n decreases as n increases. To see whether $\lim a_n = 0$, take logarithms and use the inequality $\ln(1+x) \leq x$:

$$\ln a_n = \ln\left(1 - \frac{1}{2}\right) + \ln\left(1 - \frac{1}{4}\right) + \cdots + \ln\left(1 - \frac{1}{2n}\right)$$
$$\leq -\frac{1}{2} - \frac{1}{4} - \cdots - \frac{1}{2n}$$
$$= -\frac{1}{2}\left(1 + \frac{1}{2} + \cdots + \frac{1}{n}\right) \to -\infty \text{ as } n \to \infty.$$

Thus $\lim a_n = 0$, and the given series converges conditionally at $x = -1$ by the alternating series test.

30. Let $p_n = \frac{1}{2n-1}$ and $q_n = -\frac{1}{2n}$. Then $\sum p_n$ diverges to ∞ and $\sum q_n$ diverges to $-\infty$. Also, the alternating harmonic series is the sum of all the p_ns and q_ns in a specific order:

$$\sum_{n=1}^{\infty} \frac{(-1)^{n-1}}{n} = \sum_{n=1}^{\infty}(p_n + q_n).$$

a) Rearrange the terms as follows: first add terms of $\sum p_n$ until the sum exceeds 2. Then add q_1. Then add more terms of $\sum p_n$ until the sum exceeds 3. Then add q_2. Continue in this way; at the nth stage, add new terms from $\sum p_n$ until the sum exceeds $n+1$, and then add q_n. All partial sums after the nth stage exceed n, so the rearranged series diverges to infinity.

b) Rearrange the terms of the original alternating harmonic series as follows: first add terms of $\sum q_n$ until the sum is less than -2. Then add p_1. The sum will now be greater than -2. (Why?) Then resume adding new terms from $\sum q_n$ until the sum is less than -2 again, and add p_2, which will raise the sum above -2 again. Continue in this way. After the nth stage, all succeeding partial sums will differ from -2 by less than $1/n$, so the rearranged series will converge to -2.

Section 9.5 Power Series (page 564)

1. For $\sum_{n=0}^{\infty} \frac{x^{2n}}{\sqrt{n+1}}$ we have $R = \lim\left|\frac{\sqrt{n+2}}{\sqrt{n+1}}\right| = 1$. The radius of convergence is 1; the centre of convergence is 0; the interval of convergence is $(-1, 1)$. (The series does not converge at $x = -1$ or $x = 1$.)

2. We have $\sum_{n=0}^{\infty} 3n(x+1)^n$. The centre of convergence is $x = -1$. The radius of convergence is
$$R = \lim \frac{3n}{3(n+1)} = 1.$$

The series converges absolutely on $(-2, 0)$ and diverges on $(-\infty, -2)$ and $(0, \infty)$. At $x = -2$, the series is $\sum_{n=0}^{\infty} 3n(-1)^n$, which diverges. At $x = 0$, the series is $\sum_{n=0}^{\infty} 3n$, which diverges to infinity. Hence, the interval of convergence is $(-2, 0)$.

3. For $\sum_{n=1}^{\infty} \frac{1}{n}\left(\frac{x+2}{2}\right)^n$ we have $R = \lim \frac{2^{n+1}(n+1)}{2^n n} = 2$. The radius of convergence is 2; the centre of convergence is -2. For $x = -4$ the series is an alternating harmonic series, so converges. For $x = 0$, the series is a divergent harmonic series. Therefore the interval of convergence is $[-4, 0)$.

4. We have $\sum_{n=1}^{\infty} \frac{(-1)^n}{n^4 2^{2n}} x^n$. The centre of convergence is $x = 0$. The radius of convergence is

$$R = \lim \left| \frac{(-1)^n}{n^4 2^{2n}} \cdot \frac{(n+1)^4 2^{2n+2}}{(-1)^{n+1}} \right|$$
$$= \lim \left| \left(\frac{n+1}{n}\right)^4 \cdot 4 \right| = 4.$$

At $x = 4$, the series is $\sum_{n=1}^{\infty} \frac{(-1)^n}{n^4}$, which converges.

At $x = -4$, the series is $\sum_{n=1}^{\infty} \frac{1}{n^4}$, which also converges.

Hence, the interval of convergence is $[-4, 4]$.

5. $\sum_{n=0}^{\infty} n^3 (2x-3)^n = \sum_{n=0}^{\infty} 2^n n^3 \left(x - \frac{3}{2}\right)^n$. Here
$R = \lim \frac{2^n n^3}{2^{n+1}(n+1)^3} = \frac{1}{2}$. The radius of convergence is $1/2$; the centre of convergence is $3/2$; the interval of convergence is $(1, 2)$.

6. We have $\sum_{n=1}^{\infty} \frac{e^n}{n^3}(4-x)^n$. The centre of convergence is $x = 4$. The radius of convergence is

$$R = \lim \frac{e^n}{n^3} \cdot \frac{(n+1)^3}{e^{n+1}} = \frac{1}{e}.$$

At $x = 4 + \frac{1}{e}$, the series is $\sum_{n=1}^{\infty} \frac{(-1)^n}{n^3}$, which converges.

At $x = 4 - \frac{1}{e}$, the series is $\sum_{n=1}^{\infty} \frac{1}{n^3}$, which also converges.

Hence, the interval of convergence is $\left[4 - \frac{1}{e}, 4 + \frac{1}{e}\right]$.

7. For $\sum_{n=0}^{\infty} \frac{1+5^n}{n!} x^n$
we have $R = \lim \frac{1+5^n}{n!} \cdot \frac{(n+1)!}{1+5^{n+1}} = \infty$. The radius of convergence is infinite; the centre of convergence is 0; the interval of convergence is the whole real line $(-\infty, \infty)$.

8. We have $\sum_{n=1}^{\infty} \frac{(4x-1)^n}{n^n} = \sum_{n=1}^{\infty} \left(\frac{4}{n}\right)^n \left(x - \frac{1}{4}\right)^n$. The centre of convergence is $x = \frac{1}{4}$. The radius of convergence is

$$R = \lim \frac{4^n}{n^n} \cdot \frac{(n+1)^{n+1}}{4^{n+1}}$$
$$= \frac{1}{4} \lim \left(\frac{n+1}{n}\right)^n (n+1) = \infty.$$

Hence, the interval of convergence is $(-\infty, \infty)$.

9. By Example 5(a),

$$\begin{array}{rcrcrcrcrcl}
  & 1 & + & 2x & + & 3x^2 & + & 4x^3 & + & \cdots & = \frac{1}{(1-x)^2} \\
\times & 1 & + & x & + & x^2 & + & x^3 & + & \cdots & = \frac{1}{1-x} \\
\hline
  & 1 & + & 2x & + & 3x^2 & + & 4x^3 & + & \cdots & \\
  &   &   & x & + & 2x^2 & + & 3x^3 & + & \cdots & \\
  &   &   &   &   & x^2 & + & 2x^3 & + & \cdots & \\
  &   &   &   &   &   &   & x^3 & + & \cdots & \\
  &   &   &   &   &   &   &   &   & \cdots & \\
\hline
  & 1 & + & 3x & + & 6x^2 & + & 10x^3 & + & \cdots & = \frac{1}{(1-x)^3}
\end{array}$$

Thus
$$\frac{1}{(1-x)^3} = \sum_{n=0}^{\infty} \frac{(n+1)(n+2)}{2} x^n,$$
for $-1 < x < 1$.

10. We have
$$1 + x + x^2 + x^3 + \cdots = \frac{1}{1-x} = \sum_{n=0}^{\infty} x^n$$
and
$$1 - x + x^2 - x^3 + \cdots = \frac{1}{1+x} = \sum_{n=0}^{\infty} (-1)^n x^n$$
holds for $-1 < x < 1$. Since $a_n = 1$ and $b_n = (-1)^n$ for $n = 0, 1, 2, \ldots$, we have
$$c_n = \sum_{j=0}^{n} (-1)^{n-j} = \begin{cases} 0, & \text{if } n \text{ is odd}; \\ 1, & \text{if } n \text{ is even.} \end{cases}$$

Then the Cauchy product is
$$1 + x^2 + x^4 + \cdots = \sum_{n=0}^{\infty} x^{2n} = \frac{1}{1-x} \cdot \frac{1}{1+x} = \frac{1}{1-x^2}$$
for $-1 < x < 1$.

11. By long division:

$$\begin{array}{r}
1 + 2x + 3x^2 + 4x^3 + \cdots \\
1 - 2x + x^2 \overline{\smash{\big)}\ 1 } \\
\underline{1 - 2x + x^2} \\
2x - x^2 \\
\underline{2x - 4x^2 + 2x^3 + \cdots} \\
3x^2 - 2x^3 + \cdots \\
\underline{3x^2 - 6x^3 + \cdots} \\
4x^3 + \cdots
\end{array}$$

Thus $\frac{1}{(1-x)^2} = \sum_{n=0}^{\infty} (n+1) x^n$, for $-1 < x < 1$.

12. $\dfrac{1}{2-x} = \dfrac{1}{2}\dfrac{1}{\left(1-\dfrac{x}{2}\right)} = \dfrac{1}{2}\sum_{n=0}^{\infty}\left(\dfrac{x}{2}\right)^n$

$= \dfrac{1}{2} + \dfrac{x}{2^2} + \dfrac{x^2}{2^3} + \dfrac{x^3}{2^4} + \cdots \quad (-2 < x < 2).$

13. $\dfrac{1}{2-x} = \dfrac{1}{2}\cdot\dfrac{1}{1-\dfrac{x}{2}}$

$= \dfrac{1}{2} + \dfrac{x}{2^2} + \dfrac{x^2}{2^3} + \dfrac{x^3}{2^4}\cdots$

for $-2 < x < 2$. Now differentiate to get

$\dfrac{1}{(2-x)^2} = \dfrac{1}{2^2} + \dfrac{2x}{2^3} + \dfrac{3x^2}{2^4} + \cdots$

$= \sum_{n=0}^{\infty}\dfrac{(n+1)x^n}{2^{n+2}}, \quad (-2 < x < 2).$

14. $\dfrac{1}{1+2x} = \sum_{n=0}^{\infty}(-2x)^n$

$= 1 - 2x + 2^2x^2 - 2^3x^3 + \cdots \quad (-\tfrac{1}{2} < x < \tfrac{1}{2}).$

15. $\displaystyle\int_0^x \dfrac{dt}{2-t} = \int_0^x \sum_{n=0}^{\infty}\dfrac{t^n}{2^{n+1}}\,dt$

$-\ln(2-t)\Big|_0^x = \sum_{n=0}^{\infty}\dfrac{t^{n+1}}{2^{n+1}(n+1)}\Big|_0^x$

$-\ln(2-x) + \ln 2 = \sum_{n=0}^{\infty}\dfrac{x^{n+1}}{2^{n+1}(n+1)}$

$\ln(2-x) = \ln 2 - \sum_{n=1}^{\infty}\dfrac{x^n}{2^n n}. \quad (-2 \le x < 2).$

16. Let $y = x - 1$. Then $x = 1 + y$ and

$\dfrac{1}{x} = \dfrac{1}{1+y} = \sum_{n=0}^{\infty}(-y)^n \quad (-1 < y < 1)$

$= \sum_{n=0}^{\infty}[-(x-1)]^n$

$= 1 - (x-1) + (x-1)^2 - (x-1)^3 + (x-1)^4 - \cdots$
(for $0 < x < 2$).

17. Let $x + 2 = t$, so $x = t - 2$. Then

$\dfrac{1}{x^2} = \dfrac{1}{(2-t)^2} = \sum_{n=0}^{\infty}\dfrac{(n+1)t^n}{2^{n+2}}$

$= \sum_{n=0}^{\infty}\dfrac{(n+1)(x+2)^n}{2^{n+2}}, \quad (-4 < x < 0).$

18. $\dfrac{1-x}{1+x} = \dfrac{2}{1+x} - 1$

$= 2(1 - x + x^2 - x^3 + \cdots) - 1$

$= 1 + 2\sum_{n=1}^{\infty}(-x)^n \quad (-1 < x < 1).$

19. We have

$\dfrac{x^3}{1-2x^2} = x^3\left(\sum_{n=0}^{\infty}(2x^2)^n\right)$

$= \sum_{n=0}^{\infty}2^n x^{2n+3}, \quad \left(-\dfrac{1}{\sqrt{2}} < x < \dfrac{1}{\sqrt{2}}\right).$

20. Let $y = x - 4$. Then $x = 4 + y$ and

$\dfrac{1}{x} = \dfrac{1}{4+y} = \dfrac{1}{4}\dfrac{1}{\left(1+\dfrac{y}{4}\right)} = \dfrac{1}{4}\sum_{n=0}^{\infty}\left(-\dfrac{y}{4}\right)^n$

$= \dfrac{1}{4}\sum_{n=0}^{\infty}\left[-\dfrac{(x-4)}{4}\right]^n$

$= \dfrac{1}{4} - \dfrac{(x-4)}{4^2} + \dfrac{(x-4)^2}{4^3} - \dfrac{(x-4)^3}{4^4} + \cdots$

for $0 < x < 8$. Therefore,

$\ln x = \displaystyle\int_1^x \dfrac{dt}{t} = \int_1^4 \dfrac{dt}{t} + \int_4^x \dfrac{dt}{t}$

$= \ln 4 + \displaystyle\int_4^x\left[\dfrac{1}{4} - \dfrac{(t-4)}{4^2} + \dfrac{(t-4)^2}{4^3} - \dfrac{(t-4)^3}{4^4} + \cdots\right]dt$

$= \ln 4 + \dfrac{x-4}{4} - \dfrac{(x-4)^2}{2\cdot 4^2} + \dfrac{(x-4)^3}{3\cdot 4^3} - \dfrac{(x-4)^4}{4\cdot 4^4} + \cdots$
(for $0 < x \le 8$).

21. $1 = 4x + 16x^2 - 64x^3 + \cdots$

$= 1 + (-4x) + (-4x)^2 - (-4x)^3 + \cdots$

$= \dfrac{1}{1-(-4x)} = \dfrac{1}{1+4x}, \quad \left(-\tfrac{1}{4} < x < \tfrac{1}{4}\right).$

22. We differentiate the series

$\sum_{n=0}^{\infty}x^n = 1 + x + x^2 + x^3 + \cdots = \dfrac{1}{1-x}$

and multiply by x to get

$\sum_{n=0}^{\infty}nx^n = x + 2x^2 + 3x^3 + \cdots = \dfrac{x}{(1-x)^2}$

INSTRUCTOR'S SOLUTIONS MANUAL   SECTION 9.5 (PAGE 564)

for $-1 < x < 1$. Therefore,

$$\sum_{n=0}^{\infty}(n+3)x^n = \sum_{n=0}^{\infty} nx^n + 3\sum_{n=0}^{\infty} x^n$$
$$= \frac{x}{(1-x)^2} + \frac{3}{1-x}$$
$$= \frac{3-2x}{(1-x)^2} \quad (-1 < x < 1).$$

23. $\frac{1}{3} + \frac{x}{4} + \frac{x^2}{5} + \frac{x^3}{6} + \cdots$
$= \frac{1}{x^3}\left(\frac{x^3}{3} + \frac{x^4}{4} + \frac{x^5}{5} + \cdots\right)$
$= \frac{1}{x^3}\left(x + \frac{x^2}{2} + \frac{x^3}{3} + \frac{x^4}{4} + + \cdots - x - \frac{x^2}{2}\right)$
$= \frac{1}{x^3}\left[-\ln(1-x) - x - \frac{x^2}{2}\right]$
$= -\frac{1}{x^3}\ln(1-x) - \frac{1}{x^2} - \frac{1}{2x}. \quad (-1 \le x < 1, \ x \ne 0).$

24. We start with
$$1 - x + x^2 - x^3 + x^4 - \cdots = \frac{1}{1+x}$$

and differentiate to get
$$-1 + 2x - 3x^3 + 4x^3 - \cdots = -\frac{1}{(1+x)^2}.$$

Now we multiply by $-x^3$:
$$x^3 - 2x^4 + 3x^5 - 4x^6 + \cdots = \frac{x^3}{(1+x)^2}.$$

Differentiating again we get
$$3x^2 - 2 \times 4x^3 + 3 \times 5x^4 - 4 \times 6x^5 + \cdots = \frac{x^3 + 3x^2}{(1+x)^3}.$$

Finally, we remove the factor x^2:
$$3 - 2 \times 4x + 3 \times 5x^2 - 4 \times 6x^3 + \cdots = \frac{x+3}{(1+x)^3}.$$

All steps are valid for $-1 < x < 1$.

25. Since $1 + x^2 + x^4 + x^6 + \cdots = \frac{1}{1-x^2}$, for $-1 < x < 1$, we obtain by differentiation
$$2x + 4x^3 + 6x^5 + 8x^7 + \cdots = \frac{2x}{(1-x^2)^2},$$

or, on division by x,
$$2 + 4x^2 + 6x^4 + 8x^6 + \cdots = \frac{2}{(1-x^2)^2},$$

for $-1 < x < 1$.

26. Since $x - \frac{x^2}{2} + \frac{x^3}{3} - \frac{x^4}{4} + \cdots = \ln(1+x)$ for $-1 < x \le 1$, therefore
$$x^2 - \frac{x^4}{2} + \frac{x^6}{3} - \frac{x^8}{4} + \cdots = \ln(1+x^2)$$

for $-1 \le x \le 1$, and, dividing by x^2,
$$1 - \frac{x^2}{2} + \frac{x^4}{3} - \frac{x^6}{4} + \cdots = \begin{cases} \frac{\ln(1+x^2)}{x^2} & \text{if } -1 \le x \le 1, x \ne 0 \\ 1 & \text{if } x = 0. \end{cases}$$

27. From Example 5(a),
$$\sum_{n=1}^{\infty} nx^{n-1} = \frac{1}{(1-x)^2}, \quad (-1 < x < 1).$$

Putting $x = 1/3$, we get
$$\sum_{n=1}^{\infty} \frac{n}{3^{n-1}} = \frac{1}{\left(1-\frac{1}{3}\right)^2} = \frac{9}{4}.$$

Thus $\sum_{n=1}^{\infty} \frac{n}{3^n} = \frac{1}{3} \cdot \frac{9}{4} = \frac{3}{4}.$

28. From Example 5(a) with $x = 1/2$,
$$\sum_{n=0}^{\infty} \frac{n+1}{2^n} = \sum_{k=1}^{\infty} k\left(\frac{1}{2}\right)^{k-1} = \frac{1}{\left(1-\frac{1}{2}\right)^2} = 4.$$

29. From Example 7, $\sum_{n=1}^{\infty} n^2 x^{n-1} = \frac{1+x}{(1-x)^3}$ for $-1 < x < 1$.
Putting $x = 1/\pi$, we get
$$\sum_{n=0}^{\infty} \frac{(n+1)^2}{\pi^n} = \sum_{k=1}^{\infty} \frac{k^2}{\pi^{k-1}} = \frac{1+\frac{1}{\pi}}{(1-\frac{1}{\pi})^3} = \frac{\pi^2(\pi+1)}{(\pi-1)^3}.$$

30. From Example 5(a),
$$\sum_{n=1}^{\infty} nx^{n-1} = \frac{1}{(1-x)^2}, \quad (-1 < x < 1).$$

351

Differentiate with respect to x and then replace n by $n+1$:

$$\sum_{n=2}^{\infty} n(n-1)x^{n-2} = \frac{2}{(1-x)^3}, \quad (-1 < x < 1)$$

$$\sum_{n=1}^{\infty} (n+1)nx^{n-1} = \frac{2}{(1-x)^3}, \quad (-1 < x < 1).$$

Now let $x = -1/2$:

$$\sum_{n=1}^{\infty} (-1)^{n-1} \frac{n(n+1)}{2^{n-1}} = \frac{16}{27}.$$

Finally, multiply by $-1/2$:

$$\sum_{n=1}^{\infty} (-1)^n \frac{n(n+1)}{2^n} = -\frac{8}{27}.$$

31. Since $\displaystyle\sum_{n=1}^{\infty} (-1)^{n-1} \frac{x^n}{n} = \ln(1+x)$ for $-1 < x \leq 1$, therefore

$$\sum_{n=1}^{\infty} \frac{(-1)^{n-1}}{n2^n} = \ln\left(1 + \frac{1}{2}\right) = \ln\frac{3}{2}.$$

32. In the series for $\ln(1+x)$ in Example 5(c), put $x = -1/2$ to get

$$\sum_{n=1}^{\infty} (-1) \frac{1}{n2^n} = \sum_{k=0}^{\infty} \frac{(-1)^k}{k+1}\left(-\frac{1}{2}\right)^{k+1} = \ln\left(1 - \frac{1}{2}\right) = -\ln 2.$$

Therefore

$$\sum_{n=1}^{\infty} \frac{1}{n2^n} = \ln 2$$

$$\sum_{n=3}^{\infty} \frac{1}{n2^n} = \ln 2 - \frac{1}{2} - \frac{1}{8} = \ln 2 - \frac{5}{8}.$$

Section 9.6 Taylor and Maclaurin Series (page 573)

1. $\displaystyle e^{3x+1} = e \cdot e^{3x} = e\left(\sum_{n=0}^{\infty} \frac{(3x)^n}{n!}\right)$

$$= \sum_{n=0}^{\infty} \frac{e3^n x^n}{n!} \quad \text{(for all } x\text{)}.$$

2. $\displaystyle \cos(2x^3) = 1 - \frac{(2x^3)^2}{2!} + \frac{(2x^3)^4}{4!} - \frac{(2x^3)^6}{6!} + \cdots$

$$= 1 - \frac{2^2 x^6}{2!} + \frac{2^4 x^{12}}{4!} - \frac{2^6 x^{18}}{6!} + \cdots$$

$$= \sum_{n=0}^{\infty} \frac{(-1)^n 4^n}{(2n)!} x^{6n} \quad \text{(for all } x\text{)}.$$

3. $\displaystyle \sin\left(x - \frac{\pi}{4}\right) = \sin x \cos\frac{\pi}{4} - \cos x \sin\frac{\pi}{4}$

$$= \frac{1}{\sqrt{2}} \sum_{n=0}^{\infty} (-1)^n \frac{x^{2n+1}}{(2n+1)!} - \frac{1}{\sqrt{2}} \sum_{n=0}^{\infty} (-1)^n \frac{x^{2n}}{(2n)!}$$

$$= \frac{1}{\sqrt{2}} \sum_{n=0}^{\infty} (-1)^n \left[-\frac{x^{2n}}{(2n)!} + \frac{x^{2n+1}}{(2n+1)!}\right] \quad \text{(for all } x\text{)}.$$

4. $\cos(2x - \pi) = -\cos(2x)$

$$= -1 + \frac{2^2 x^2}{2!} - \frac{2^4 x^4}{4!} + \frac{2^6 x^6}{6!} - \cdots$$

$$= -\sum_{n=0}^{\infty} \frac{(-1)^n}{(2n)!} (2x)^{2n}$$

$$= \sum_{n=0}^{\infty} \frac{(-1)^{n+1}}{(2n)!} 4^n (x)^{2n} \quad \text{(for all } x\text{)}.$$

5. $\displaystyle x^2 \sin\frac{x}{3} = \sum_{n=0}^{\infty} \frac{(-1)^n x^{2n+3}}{3^{2n+1}(2n+1)!} \quad \text{(for all } x\text{)}.$

6. $\displaystyle \cos^2\left(\frac{x}{2}\right) = \frac{1}{2}(1 + \cos x)$

$$= \frac{1}{2}\left(1 + 1 - \frac{x^2}{2!} + \frac{x^4}{4!} - \frac{x^6}{6!} + \cdots\right)$$

$$= 1 + \frac{1}{2} \sum_{n=1}^{\infty} \frac{(-1)^n}{(2n)!} x^{2n} \quad \text{(for all } x\text{)}.$$

7. $\displaystyle \sin x \cos x = \frac{1}{2} \sin(2x)$

$$= \sum_{n=0}^{\infty} (-1)^n \frac{2^{2n} x^{2n+1}}{(2n+1)!} \quad \text{(for all } x\text{)}.$$

8. $\displaystyle \tan^{-1}(5x^2) = (5x^2) - \frac{(5x^2)^3}{3} + \frac{(5x^2)^5}{5} - \frac{(5x^2)^7}{7} + \cdots$

$$= \sum_{n=0}^{\infty} \frac{(-1)^n}{(2n+1)} (5x^2)^{2n+1}$$

$$= \sum_{n=0}^{\infty} \frac{(-1)^n 5^{2n+1}}{(2n+1)} x^{4n+2}$$

$$\left(\text{for } -\frac{1}{\sqrt{5}} \leq x \leq \frac{1}{\sqrt{5}}\right).$$

9. $\dfrac{1+x^3}{1+x^2} = (1+x^3)\left(1 - x^2 + x^4 - x^6 + \cdots\right)$
$= 1 - x^2 + x^3 + x^4 - x^5 - x^6 + x^7 + x^8 - \cdots$
$= 1 - x^2 + \sum_{n=2}^{\infty}(-1)^n\left(x^{2n-1} + x^{2n}\right) \qquad (|x| < 1).$

10. $\ln(2 + x^2) = \ln 2\left(1 + \dfrac{x^2}{2}\right)$
$= \ln 2 + \ln\left(1 + \dfrac{x^2}{2}\right)$
$= \ln 2 + \left[\dfrac{x^2}{2} - \dfrac{1}{2}\left(\dfrac{x^2}{2}\right)^2 + \dfrac{1}{3}\left(\dfrac{x^2}{2}\right)^3 - \cdots\right]$
$= \ln 2 + \sum_{n=1}^{\infty} \dfrac{(-1)^{n-1}}{n} \cdot \dfrac{x^{2n}}{2^n}$
(for $-\sqrt{2} \le x \le \sqrt{2}$).

11. $\ln\dfrac{1-x}{1+x} = \ln(1-x) - \ln(1+x)$
$= -\sum_{n=1}^{\infty} \dfrac{x^n}{n} - \sum_{n=1}^{\infty}(-1)^{n-1}\dfrac{x^n}{n}$
$= -2\sum_{n=1}^{\infty} \dfrac{x^{2n-1}}{2n-1} \qquad (-1 < x < 1).$

12. $\dfrac{e^{2x^2} - 1}{x^2} = \dfrac{1}{x^2}\left(e^{2x^2} - 1\right)$
$= \dfrac{1}{x^2}\left(1 + 2x^2 + \dfrac{(2x^2)^2}{2!} + \dfrac{(2x^2)^3}{3!} + \cdots - 1\right)$
$= 2 + \dfrac{2^2 x^2}{2!} + \dfrac{2^3 x^4}{3!} + \dfrac{2^4 x^6}{4!} + \cdots$
$= \sum_{n=0}^{\infty} \dfrac{2^{n+1}}{(n+1)!} x^{2n} \qquad$ (for all $x \ne 0$).

13. $\cosh x - \cos x = \sum_{n=0}^{\infty}\left[1 - (-1)^n\right]\dfrac{x^{2n}}{(2n)!}$
$= 2\left(\dfrac{x^2}{2!} + \dfrac{x^6}{6!} + \dfrac{x^{10}}{10!} + \cdots\right)$
$= 2\sum_{n=0}^{\infty} \dfrac{x^{4n+2}}{(4n+2)!} \qquad$ (for all x).

14. $\sinh x - \sin x = \sum_{n=0}^{\infty}\left[1 - (-1)^n\right]\dfrac{x^{2n+1}}{(2n+1)!}$
$= 2\left(\dfrac{x^2}{2!} + \dfrac{x^6}{6!} + \dfrac{x^{10}}{10!} + \cdots\right)$
$= 2\sum_{n=0}^{\infty} \dfrac{x^{4n+3}}{(4n+3)!} \qquad$ (for all x).

15. Let $t = x + 1$, so $x = t - 1$. We have
$f(x) = e^{-2x} = e^{-2(t-1)}$
$= e^2 \sum_{n=0}^{\infty} \dfrac{(-2)^n t^n}{n!}$
$= e^2 \sum_{n=0}^{\infty} \dfrac{(-1)^n 2^n (x+1)^n}{n!} \qquad$ (for all x).

16. Let $y = x - \dfrac{\pi}{2}$; then $x = y + \dfrac{\pi}{2}$. Hence,
$\sin x = \sin\left(y + \dfrac{\pi}{2}\right) = \cos y$
$= 1 - \dfrac{y^2}{2!} + \dfrac{y^4}{4!} - \cdots \qquad$ (for all y)
$= 1 - \dfrac{1}{2!}\left(x - \dfrac{\pi}{2}\right)^2 + \dfrac{1}{4!}\left(x - \dfrac{\pi}{2}\right)^4 - \cdots$
$= \sum_{n=0}^{\infty} \dfrac{(-1)^n}{(2n)!}\left(x - \dfrac{\pi}{2}\right)^{2n} \qquad$ (for all x).

17. Let $t = x - \pi$, so $x = t + \pi$. Then
$f(x) = \cos x = \cos(t + \pi) = -\cos t = -\sum_{n=0}^{\infty}(-1)^n \dfrac{t^{2n}}{(2n)!}$
$= \sum_{n=0}^{\infty} \dfrac{(-1)^{n+1}}{(2n)!}(x - \pi)^{2n} \qquad$ (for all x).

18. Let $y = x - 3$; then $x = y + 3$. Hence,
$\ln x = \ln(y + 3) = \ln 3 + \ln\left(1 + \dfrac{y}{3}\right)$
$= \ln 3 + \dfrac{y}{3} - \dfrac{1}{2}\left(\dfrac{y}{3}\right)^2 + \dfrac{1}{3}\left(\dfrac{y}{3}\right)^3 - \dfrac{1}{4}\left(\dfrac{y}{3}\right)^4 + \cdots$
$= \ln 3 + \dfrac{(x-3)}{3} - \dfrac{(x-3)^2}{2 \cdot 3^2} + \dfrac{(x-3)^3}{3 \cdot 3^3} - \dfrac{(x-3)^4}{4 \cdot 3^4} + \cdots$
$= \ln 3 + \sum_{n=1}^{\infty} \dfrac{(-1)^{n-1}}{n \cdot 3^n}(x-3)^n \qquad (0 < x \le 6).$

19. $\ln(2+x) = \ln[4+(x-2)] = \ln\left[4\left(1 + \dfrac{x-2}{4}\right)\right]$
$= \ln 4 + \ln\left(1 + \dfrac{x-2}{4}\right)$
$= \ln 4 + \sum_{n=1}^{\infty}(-1)^{n-1}\dfrac{(x-2)^n}{n 4^n} \qquad (-2 < x \le 6).$

20. Let $t = x + 1$. Then $x = t - 1$, and
$e^{2x+3} = e^{2t+1} = e \, e^{2t}$
$= e\sum_{n=0}^{\infty} \dfrac{2^n t^n}{n!} \qquad$ (for all t)
$= \sum_{n=0}^{\infty} \dfrac{e 2^n (x+1)^n}{n!} \qquad$ (for all x).

21. Let $t = x - (\pi/4)$, so $x = t + (\pi/4)$. Then
$$f(x) = \sin x - \cos x$$
$$= \sin\left(t + \frac{\pi}{4}\right) - \cos\left(t + \frac{\pi}{4}\right)$$
$$= \frac{1}{\sqrt{2}}\left[(\sin t + \cos t) - (\cos t - \sin t)\right]$$
$$= \sqrt{2}\sin t = \sqrt{2}\sum_{n=0}^{\infty}(-1)^n \frac{t^{2n+1}}{(2n+1)!}$$
$$= \sqrt{2}\sum_{n=0}^{\infty}\frac{(-1)^n}{(2n+1)!}\left(x - \frac{\pi}{4}\right)^{2n+1} \quad \text{(for all } x\text{).}$$

22. Let $y = x - \frac{\pi}{8}$; then $x = y + \frac{\pi}{8}$. Thus,

$$\cos^2 x = \cos^2\left(y + \frac{\pi}{8}\right)$$
$$= \frac{1}{2}\left[1 + \cos\left(2y + \frac{\pi}{4}\right)\right]$$
$$= \frac{1}{2}\left[1 + \frac{1}{\sqrt{2}}\cos(2y) - \frac{1}{\sqrt{2}}\sin(2y)\right]$$
$$= \frac{1}{2} + \frac{1}{2\sqrt{2}}\left[1 - \frac{(2y)^2}{2!} + \frac{(2y)^4}{4!} - \cdots\right]$$
$$\quad - \frac{1}{2\sqrt{2}}\left[2y - \frac{(2y)^3}{3!} + \frac{(2y)^5}{5!} - \cdots\right]$$
$$= \frac{1}{2} + \frac{1}{2\sqrt{2}}\left[1 - 2y - \frac{(2y)^2}{2!} + \frac{(2y)^3}{3!}\right.$$
$$\quad \left. + \frac{(2y)^4}{4!} - \frac{(2y)^5}{5!} - \cdots\right]$$
$$= \frac{1}{2} + \frac{1}{2\sqrt{2}}\left[1 - 2\left(x - \frac{\pi}{8}\right) - \frac{2^2}{2!}\left(x - \frac{\pi}{8}\right)^2\right.$$
$$\quad \left. + \frac{2^3}{3!}\left(x - \frac{\pi}{8}\right)^3 + \frac{2^4}{4!}\left(x - \frac{\pi}{8}\right)^4 - \frac{2^5}{5!}\left(x - \frac{\pi}{8}\right)^5 - \cdots\right]$$
$$= \frac{1}{2} + \frac{1}{2\sqrt{2}} + \frac{1}{2\sqrt{2}}\sum_{n=1}^{\infty}(-1)^n\left[\frac{2^{2n-1}}{(2n-1)!}\left(x - \frac{\pi}{8}\right)^{2n-1}\right.$$
$$\quad \left. + \frac{2^{2n}}{(2n)!}\left(x - \frac{\pi}{8}\right)^{2n}\right] \quad \text{(for all } x\text{).}$$

23. Let $t = x + 2$, so $x = t - 2$. We have
$$f(x) = \frac{1}{x^2} = \frac{1}{(t-2)^2} = \frac{1}{4\left(1 - \frac{t}{2}\right)^2}$$
$$= \frac{1}{4}\sum_{n=1}^{\infty} n\frac{t^{n-1}}{2^{n-1}} \quad (-2 \le t < 2)$$
$$= \frac{1}{4}\sum_{n=1}^{\infty}\frac{n(x+2)^{n-1}}{2^{n-1}}$$
$$= \frac{1}{4}\sum_{n=0}^{\infty}\frac{(n+1)(x+2)^n}{2^n} \quad (-4 < x < 0).$$

24. Let $y = x - 1$; then $x = y + 1$. Thus,
$$\frac{x}{1+x} = \frac{1+y}{2+y} = 1 - \frac{1}{2\left(1 + \frac{y}{2}\right)}$$
$$= 1 - \frac{1}{2}\left[1 - \frac{y}{2} + \left(\frac{y}{2}\right)^2 - \left(\frac{y}{2}\right)^3 + \cdots\right]$$
$$= \frac{1}{2}\left[1 + \frac{y}{2} - \frac{y^2}{2^2} + \frac{y^3}{2^3} - \frac{y^4}{2^4} + \cdots\right] \quad (-1 < y < 1)$$
$$= \frac{1}{2} + \frac{1}{2^2}(x-1) - \frac{1}{2^3}(x-1)^2 + \frac{1}{2^4}(x-1)^3 - \cdots$$
$$= \frac{1}{2} + \sum_{n=1}^{\infty}\frac{(-1)^{n-1}}{2^{n+1}}(x-1)^n \quad \text{(for } 0 < x < 2\text{).}$$

25. Let $u = x - 1$. Then $x = 1 + u$, and
$$x\ln x = (1+u)\ln(1+u)$$
$$= (1+u)\sum_{n=1}^{\infty}(-1)^{n-1}\frac{u^n}{n} \quad (-1 < u \le 1)$$
$$= \sum_{n=1}^{\infty}(-1)^{n-1}\frac{u^n}{n} + \sum_{n=1}^{\infty}(-1)^{n-1}\frac{u^{n+1}}{n}.$$

Replace n by $n-1$ in the last sum.

$$x\ln x = \sum_{n=1}^{\infty}(-1)^{n-1}\frac{u^n}{n} + \sum_{n=2}^{\infty}(-1)^{n-2}\frac{u^n}{n-1}$$
$$= u + \sum_{n=2}^{\infty}(-1)^{n-1}\left(\frac{1}{n} - \frac{1}{n-1}\right)u^n$$
$$= (x-1) + \sum_{n=2}^{\infty}\frac{(-1)^n}{n(n-1)}(x-1)^n \quad (0 \le x \le 2).$$

26. Let $u = x + 2$. Then $x = u - 2$, and
$$xe^x = (u-2)e^{u-2}$$
$$= (u-2)e^{-2}\sum_{n=0}^{\infty}\frac{u^n}{n!} \quad \text{(for all } u\text{)}$$
$$= \sum_{n=0}^{\infty}\frac{e^{-2}u^{n+1}}{n!} - \sum_{n=0}^{\infty}\frac{2e^{-2}u^n}{n!}.$$

In the first sum replace n by $n-1$.

$$xe^x = \sum_{n=1}^{\infty}\frac{e^{-2}u^n}{(n-1)!} - \sum_{n=0}^{\infty}\frac{2e^{-2}u^n}{n!}$$
$$= -\frac{2}{e^2} + \sum_{n=1}^{\infty}\frac{1}{e^2}\left(\frac{1}{(n-1)!} - \frac{2}{n!}\right)u^n$$
$$= -\frac{2}{e^2} + \sum_{n=1}^{\infty}\frac{1}{e^2}\left(\frac{1}{(n-1)!} - \frac{2}{n!}\right)(x+2)^n \quad \text{(for all } x\text{).}$$

INSTRUCTOR'S SOLUTIONS MANUAL   SECTION 9.6 (PAGE 573)

27. $\cos x = 1 - \dfrac{x^2}{2} + \dfrac{x^4}{24} - \cdots$.

$$
\begin{array}{r}
1 + \dfrac{x^2}{2} + \dfrac{5x^4}{24} + \cdots \\
1 - \dfrac{x^2}{2} + \dfrac{x^4}{24} - \cdots \overline{\big)\, 1} \\
\underline{1 - \dfrac{x^2}{2} + \dfrac{x^4}{24} - \cdots} \\
\dfrac{x^2}{2} - \dfrac{x^4}{24} + \cdots \\
\underline{\dfrac{x^2}{2} - \dfrac{x^4}{4} + \cdots} \\
\dfrac{5x^4}{24} - \cdots
\end{array}
$$

Thus $\sec x = 1 + \dfrac{x^2}{2} + \dfrac{5x^4}{24} + \cdots$.

28. If we divide the first four terms of the series

$$\cos x = 1 - \dfrac{x^2}{2} + \dfrac{x^4}{24} - \dfrac{x^6}{720} + \cdots$$

into 1 we obtain

$$\sec x = 1 + \dfrac{x^2}{2} + \dfrac{5x^4}{24} + \dfrac{61x^6}{720} + \cdots.$$

Now we can differentiate and obtain

$$\sec x \tan x = x + \dfrac{5x^3}{6} + \dfrac{61x^5}{120} + \cdots.$$

(Note: the same result can be obtained by multiplying the first three nonzero terms of the series for $\sec x$ (from Exercise 25) and $\tan x$ (from Example 6(b)).)

29.
$$e^x - 1 = x + \dfrac{x^2}{2} + \dfrac{x^3}{6} + \cdots$$
$$\tan^{-1}(e^x - 1) = (e^x - 1) - \dfrac{(e^x - 1)^3}{3}$$
$$+ \dfrac{(e^x - 1)^5}{5} - \cdots$$
$$= x + \dfrac{x^2}{2} + \dfrac{x^3}{6} + \cdots$$
$$- \dfrac{1}{3}\left(x + \dfrac{x^2}{2} + \dfrac{x^3}{6} + \cdots\right)^3$$
$$+ \dfrac{1}{5}\left(x + \dfrac{x^2}{2} + \dfrac{x^3}{6} + \cdots\right)^5 + \cdots$$
$$= x + \dfrac{x^2}{2} + \dfrac{x^3}{6} - \dfrac{1}{3}\left(x^3 + \cdots\right) + \cdots$$
$$= x + \dfrac{x^2}{2} - \dfrac{x^3}{6} + \cdots$$

30. We have
$$e^{\tan^{-1} x} - 1 = \exp\left[x - \dfrac{x^3}{3} + \dfrac{x^5}{5} - \dfrac{x^7}{7} + \cdots\right] - 1$$
$$= 1 + \left(x - \dfrac{x^3}{3} + \dfrac{x^5}{5} - \cdots\right) + \dfrac{1}{2!}\left(x - \dfrac{x^3}{3} + \cdots\right)^2$$
$$+ \dfrac{1}{3!}(x - \cdots)^3 + \cdots - 1$$
$$= x - \dfrac{x^3}{3} + \dfrac{x^2}{2} + \dfrac{x^3}{6} + \text{higher degree terms}$$
$$= x + \dfrac{x^2}{2} - \dfrac{x^3}{6} + \cdots.$$

31. Let $\sqrt{1+x} = 1 + ax + bx^2 + \cdots$.
Then $1 + x = 1 + 2ax + (a^2 + 2b)x^2 + \cdots$, so $2a = 1$, and $a^2 + 2b = 0$. Thus $a = 1/2$ and $b = -1/8$.
Therefore $\sqrt{1+x} = 1 + (x/2) - (x^2/8) + \cdots$.

32. $\csc x$ does not have a Maclaurin series because $\lim_{x \to 0} \csc x$ does not exist.
Let $y = x - \dfrac{\pi}{2}$. Then $x = y + \dfrac{\pi}{2}$ and $\sin x = \cos y$.
Therefore, using the result of Exercise 25,
$$\csc x = \sec y = 1 + \dfrac{y^2}{2} + \dfrac{5y^4}{24} + \cdots$$
$$= 1 + \dfrac{1}{2}\left(x - \dfrac{\pi}{2}\right)^2 + \dfrac{5}{24}\left(x - \dfrac{\pi}{2}\right)^4 + \cdots.$$

33. $1 + x^2 + \dfrac{x^4}{2!} + \dfrac{x^6}{3!} + \cdots = e^{x^2}$ (for all x).

34. $x^3 - \dfrac{x^9}{3! \times 4} + \dfrac{x^{15}}{5! \times 16} - \dfrac{x^{21}}{7! \times 64} + \dfrac{x^{27}}{9! \times 256} - \cdots$
$$= 2\left[\dfrac{x^3}{2} - \dfrac{1}{3!}\left(\dfrac{x^3}{2}\right)^3 + \dfrac{1}{5!}\left(\dfrac{x^3}{2}\right)^5 - \cdots\right]$$
$$= 2\sin\left(\dfrac{x^3}{2}\right) \quad \text{(for all } x\text{)}.$$

35. $1 + \dfrac{x^2}{3!} + \dfrac{x^4}{5!} + \dfrac{x^6}{7!} + \cdots$
$$= \dfrac{1}{x}\sinh x = \dfrac{e^x - e^{-x}}{2x}$$
if $x \neq 0$. The sum is 1 if $x = 0$.

36. $1 + \dfrac{1}{2 \times 2!} + \dfrac{1}{4 \times 3!} + \dfrac{1}{8 \times 4!} + \cdots$
$$= 2\left[\dfrac{1}{2} + \dfrac{1}{2!}\left(\dfrac{1}{2}\right)^2 + \dfrac{1}{3!}\left(\dfrac{1}{2}\right)^3 + \cdots\right]$$
$$= 2\left(e^{1/2} - 1\right).$$

37. $P(x) = 1 + x + x^2$.

a) The Maclaurin series for $P(x)$ is $1 + x + x^2$ (for all x).

355

b) Let $t = x - 1$, so $x = t + 1$. Then

$$P(x) = P(t+1) = 1 + t + 1 + (t+1)^2 = 3 + 3t + t^2.$$

The Taylor series for $P(x)$ about $x = 1$ is $3 + 3(x - 1) + (x - 1)^2$.

38. If $a \neq 0$ and $|x - a| < |a|$, then

$$\frac{1}{x} = \frac{1}{a + (x-a)} = \frac{1}{a} \frac{1}{1 + \frac{x-a}{a}}$$

$$= \frac{1}{a}\left[1 - \frac{x-a}{a} + \frac{(x-a)^2}{a^2} - \frac{(x-a)^3}{a^3} + \cdots\right].$$

The radius of convergence of this series is $|a|$, and the series converges to $1/x$ throughout its interval of convergence. Hence, $1/x$ is analytic at $x = a$.

39. If $a > 0$ and $t = x - a$, then $x = t + a$ and

$$\ln x = \ln(a + t) = \ln a + \ln\left(1 + \frac{t}{a}\right)$$

$$= \ln a + \sum_{n=1}^{\infty}(-1)^{n-1}\frac{t^n}{a^n} \qquad (-a < t \leq a)$$

$$= \ln a + \sum_{n=1}^{\infty}(-1)^{n-1}\frac{(x-a)^n}{a^n} \qquad (0 < x < 2a).$$

Since the series converges to $\ln x$ on an interval of positive radius (a), centred at a, $\ln$ is analytic at $x = a$.

40. If

$$f(x) = \begin{cases} e^{-1/x^2}, & \text{if } x \neq 0; \\ 0, & \text{if } x = 0; \end{cases}$$

then the Maclaurin series for $f(x)$ is the identically zero series $0 + 0x + 0x^2 + \cdots$ since $f^{(k)}(0) = 0$ for every k. The series converges for every x, but converges to $f(x)$ only at $x = 0$, since $f(x) \neq 0$ if $x \neq 0$. Hence, f cannot be analytic at $x = 0$.

41. $e^x e^y = \left(\sum_{n=0}^{\infty}\frac{x^n}{n!}\right)\left(\sum_{m=0}^{\infty}\frac{y^m}{m!}\right)$

$$e^{x+y} = \sum_{k=0}^{\infty}\frac{(x+y)^k}{k!} = \sum_{k=0}^{\infty}\frac{1}{k!}\sum_{j=0}^{k}\frac{k!}{j!(k-j)!}x^j y^{k-j}$$

$$= \sum_{j=0}^{\infty}\frac{x^j}{j!}\sum_{k=j}^{\infty}\frac{y^{k-j}}{(k-j)!} \qquad (\text{let } k - j = m)$$

$$= \sum_{j=0}^{\infty}\frac{x^j}{j!}\sum_{m=0}^{\infty}\frac{y^m}{m!} = e^x e^y.$$

Section 9.7 Applications of Taylor and Maclaurin Series (page 577)

1. $e^{0.2} \approx 1 + 0.2 + \frac{(0.2)^2}{2!} + \cdots + \frac{(0.2)^n}{n!} = s_n$

Error estimate:

$$0 < e^{0.2} - s_n = \frac{(0.2)^{n+1}}{(n+1)!} + \frac{(0.2)^{n+2}}{(n+2)!} + \cdots$$

$$\leq \frac{(0.2)^{n+1}}{(n+1)!}\left[1 + \frac{0.2}{n+2} + \frac{(0.2)^2}{(n+2)^2} + \cdots\right]$$

$$= \frac{(0.2)^{n+1}}{(n+1)!} \cdot \frac{10n + 20}{10n + 18} < 5 \times 10^{-5} \text{ if } n = 4.$$

$$e^{0.2} \approx 1 + 0.2 + \frac{(0.2)^2}{2!} + \frac{(0.2)^3}{3!} + \frac{(0.2)^4}{4!}$$

$$\approx 1.221400$$

2. We have

$$\frac{1}{e} = e^{-1} = 1 - \frac{1}{1!} + \frac{1}{2!} - \frac{1}{3!} + \frac{1}{4!} - \cdots$$

which satisfies the conditions for the alternating series test, and the error incurred in using a partial sum to approximate e^{-1} is less than the first omitted term in absolute value. Now $\frac{1}{(n+1)!} < 5 \times 10^{-5}$ if $n = 7$, so

$$\frac{1}{e} \approx \frac{1}{2} - \frac{1}{6} + \frac{1}{24} - \frac{1}{120} + \frac{1}{720} - \frac{1}{5040} \approx 0.36786$$

with error less than 5×10^{-5} in absolute value.

3. $e^{1.2} = ee^{0.2}$. From Exercise 1: $e^{0.2} \approx 1.221400$, with error less than $\frac{(0.2)^5}{5!} \cdot \frac{60}{58} \approx 0.000003$. Since $e = 2.718281828\cdots$, it follows that $e^{1.2} \approx 3.3201094\cdots$, with error less than $3 \times 0.000003 = 0.000009 < \frac{1}{20,000}$. Thus $e^{1.2} \approx 3.32011$ with error less than $1/20,000$.

4. We have

$$\sin(0.1) = 0.1 - \frac{(0.1)^3}{3!} + \frac{(0.1)^5}{5!} - \frac{(0.1)^7}{7!} + \cdots.$$

Since $\frac{(0.1)^5}{5!} = 8.33 \times 10^{-8} < 5 \times 10^{-5}$, therefore

$$\sin(0.1) = 0.1 - \frac{(0.1)^3}{3!} \approx 0.09983$$

with error less than 5×10^{-5} in absolute value.

5. $\cos 5° = \cos \dfrac{5\pi}{180} = \cos \dfrac{\pi}{36}$

$\approx 1 - \dfrac{1}{2!}\left(\dfrac{\pi}{36}\right)^2 + \dfrac{1}{4!}\left(\dfrac{\pi}{36}\right)^4 - \cdots + \dfrac{(-1)^n}{(2n)!}\left(\dfrac{\pi}{36}\right)^{2n}$

$|\text{Error}| < \dfrac{1}{(2n+2)!}\left(\dfrac{\pi}{36}\right)^{2n+2}$

$< \dfrac{1}{(2n+2)! \, 9^{2n+2}} < 0.00005$ if $n = 1$.

$\cos 5° \approx 1 - \dfrac{1}{2!}\left(\dfrac{\pi}{36}\right)^2 \approx 0.996192$

with error less than 0.00005.

6. We have

$\ln\left(\dfrac{6}{5}\right) = \ln\left(1 + \dfrac{1}{5}\right)$

$= \dfrac{1}{5} - \dfrac{1}{2}\left(\dfrac{1}{5}\right)^2 + \dfrac{1}{3}\left(\dfrac{1}{5}\right)^3 - \dfrac{1}{4}\left(\dfrac{1}{5}\right)^4 + \cdots$.

Since $\dfrac{1}{n}\left(\dfrac{1}{5}\right)^n < 5 \times 10^{-5}$ if $n = 6$, therefore

$\ln\left(\dfrac{6}{5}\right) \approx \dfrac{1}{5} - \dfrac{1}{2}\left(\dfrac{1}{5}\right)^2 + \dfrac{1}{3}\left(\dfrac{1}{5}\right)^3 - \dfrac{1}{4}\left(\dfrac{1}{5}\right)^4 + \dfrac{1}{5}\left(\dfrac{1}{5}\right)^5$

≈ 0.18233

with error less than 5×10^{-5} in absolute value.

7. $\ln(0.9) = \ln(1 - 0.1)$

$\approx -0.1 - \dfrac{(0.1)^2}{2} - \dfrac{(0.1)^3}{3} - \cdots - \dfrac{(0.1)^n}{n}$

$|\text{Error}| < \dfrac{(0.1)^{n+1}}{n+1} + \dfrac{(0.1)^{n+2}}{n+2} + \cdots$

$< \dfrac{(0.1)^{n+1}}{n+1}\left[1 + 0.1 + (0.1)^2 + \cdots\right]$

$= \dfrac{(0.1)^{n+1}}{n+1} \cdot \dfrac{10}{9} < 0.00005$ if $n = 3$.

$\ln(0.9) \approx -0.1 - \dfrac{(0.1)^2}{2} - \dfrac{(0.1)^3}{3} \approx -0.10533$

with error less than 0.00005.

8. We have

$\sin 80° = \cos 10° = \cos\left(\dfrac{\pi}{18}\right)$

$= 1 - \dfrac{1}{2!}\left(\dfrac{\pi}{18}\right)^2 + \dfrac{1}{4!}\left(\dfrac{\pi}{18}\right)^4 - \cdots$.

Since $\dfrac{1}{4!}\left(\dfrac{\pi}{18}\right)^4 < 5 \times 10^{-5}$, therefore

$\sin 80° \approx 1 - \dfrac{1}{2!}\left(\dfrac{\pi}{18}\right)^2 \approx 0.98477$

with error less than 5×10^{-5} in absolute value.

9. $\cos 65° = \cos\left(\dfrac{\pi}{3} + \dfrac{5\pi}{180}\right)$

$= \dfrac{1}{2}\cos\dfrac{5\pi}{180} - \dfrac{\sqrt{3}}{2}\sin\dfrac{5\pi}{180}$

From Exercise 5, $\cos(5\pi/180) \approx 0.996192$ with error less than 0.000003. Also

$\sin\dfrac{5\pi}{180} = \dfrac{5\pi}{180} - \dfrac{1}{3!}\left(\dfrac{5\pi}{180}\right)^3 \approx 0.0871557$

with error less than $\dfrac{5^5 \pi^5}{5! \, 180^5} < 0.00000005$. Thus

$\cos 65° \approx \dfrac{0.996192}{2} - \dfrac{\sqrt{3}(0.0871557)}{2} \approx 0.42262$

with error less than 0.00005.

10. We have

$\tan^{-1}(0.2) = 0.2 - \dfrac{(0.2)^3}{3} + \dfrac{(0.2)^5}{5} - \dfrac{(0.2)^7}{7} + \cdots$.

Since $\dfrac{(0.2)^7}{7} < 5 \times 10^{-5}$, therefore

$\tan^{-1}(0.2) \approx 0.2 - \dfrac{(0.2)^3}{3} + \dfrac{(0.2)^5}{5} \approx 0.19740$

with error less than 5×10^{-5} in absolute value.

11. $\cosh 1 \approx 1 + \dfrac{1}{2!} + \dfrac{1}{4!} + \cdots + \dfrac{1}{(2n)!}$ with error less than

$\dfrac{1}{(2n+2)!}\left[1 + \dfrac{1}{(2n+3)^2} + \dfrac{1}{(2n+3)^4} + \cdots\right]$

$= \dfrac{1}{(2n+2)!} \cdot \dfrac{1}{1 - \dfrac{1}{(2n+3)^2}} < 0.00005$ if $n = 3$.

Thus $\cosh 1 \approx 1 + \dfrac{1}{2} + \dfrac{1}{24} + \dfrac{1}{720} \approx 1.54306$ with error less than 0.00005.

12. We have

$\ln\left(\dfrac{3}{2}\right) = \ln\left(1 + \dfrac{1}{2}\right)$

$= \dfrac{1}{2} - \dfrac{1}{2}\left(\dfrac{1}{2}\right)^2 + \dfrac{1}{3}\left(\dfrac{1}{2}\right)^3 - \dfrac{1}{4}\left(\dfrac{1}{2}\right)^4 + \cdots$.

Since $\dfrac{1}{n}\left(\dfrac{1}{2}\right)^n < \dfrac{1}{20000}$ if $n = 11$, therefore

$\ln\left(\dfrac{3}{2}\right) \approx \dfrac{1}{2} - \dfrac{1}{2}\left(\dfrac{1}{2}\right)^2 + \dfrac{1}{3}\left(\dfrac{1}{2}\right)^3 - \cdots - \dfrac{1}{10}\left(\dfrac{1}{2}\right)^{10}$

≈ 0.40543

with error less than 5×10^{-5} in absolute value.

13. $I(x) = \displaystyle\int_0^x \frac{\sin t}{t} dt$

    $= \displaystyle\int_0^x \left[1 - \frac{t^2}{3!} + \frac{t^4}{5!} - \frac{t^6}{7!} + \cdots\right] dt$

    $= x - \dfrac{x^3}{3 \times 3!} + \dfrac{x^5}{5 \times 5!} - \cdots$

    $= \displaystyle\sum_{n=0}^{\infty} (-1)^n \frac{x^{2n+1}}{(2n+1)(2n+1)!}$ for all x.

14. $J(x) = \displaystyle\int_0^x \frac{e^t - 1}{t} dt$

    $= \displaystyle\int_0^x \left(1 + \frac{t}{2!} + \frac{t^2}{3!} + \frac{t^3}{4!} + \cdots\right) dt$

    $= x + \dfrac{x^2}{2! \cdot 2} + \dfrac{x^3}{3! \cdot 3} + \dfrac{x^4}{4! \cdot 4} + \cdots$

    $= \displaystyle\sum_{n=1}^{\infty} \frac{x^n}{n! \cdot n}$.

15. $K(x) = \displaystyle\int_1^{1+x} \frac{\ln t}{t-1} dt$ let $u = t - 1$

    $= \displaystyle\int_0^x \frac{\ln(1+u)}{u} du$

    $= \displaystyle\int_0^x \left[1 - \frac{u}{2} + \frac{u^2}{3} - \frac{u^3}{4} + \cdots\right] du$

    $= x - \dfrac{x^2}{2^2} + \dfrac{x^3}{3^2} - \dfrac{x^4}{4^2} + \cdots$

    $= \displaystyle\sum_{n=0}^{\infty} (-1)^n \frac{x^{n+1}}{(n+1)^2}$ ($-1 \leq x \leq 1$)

16. $L(x) = \displaystyle\int_0^x \cos(t^2) dt$

    $= \displaystyle\int_0^x \left(1 - \frac{t^4}{2!} + \frac{t^8}{4!} - \frac{t^{12}}{6!} + \cdots\right) dt$

    $= x - \dfrac{x^5}{2! \cdot 5} + \dfrac{x^9}{4! \cdot 9} - \dfrac{x^{13}}{6! \cdot 13} + \cdots$

    $= \displaystyle\sum_{n=0}^{\infty} (-1)^n \frac{x^{4n+1}}{(2n)! \cdot (4n+1)}$.

17. $M(x) = \displaystyle\int_0^x \frac{\tan^{-1}(t^2)}{t^2} dt$

    $= \displaystyle\int_0^x \left[1 - \frac{t^4}{3} + \frac{t^8}{5} - \frac{t^{12}}{7} + \cdots\right] dt$

    $= x - \dfrac{x^5}{3 \times 5} + \dfrac{x^9}{5 \times 9} - \dfrac{x^{13}}{7 \times 13} + \cdots$

    $= \displaystyle\sum_{n=0}^{\infty} (-1)^n \frac{x^{4n+1}}{(2n+1)(4n+1)}$ ($-1 \leq x \leq 1$)

18. We have

    $L(0.5) = 0.5 - \dfrac{(0.5)^5}{2! \cdot 5} + \dfrac{(0.5)^9}{4! \cdot 9} - \dfrac{(0.5)^{13}}{6! \cdot 13} + \cdots$.

    Since $\dfrac{(0.5)^{4n+1}}{(2n)! \cdot (4n+1)} < 5 \times 10^{-4}$ if $n = 2$, therefore

    $L(0.5) \approx 0.5 - \dfrac{(0.5)^5}{2! \cdot 5} \approx 0.497$

    rounded to three decimal places.

19. From Exercise 13:

    $I(x) = x - \dfrac{x^3}{3!3} + \dfrac{x^5}{5!5} - \cdots$

    $I(1) \approx 1 - \dfrac{1}{3!3} + \dfrac{1}{5!5} - \cdots + (-1)^n \dfrac{1}{(2n+1)!(2n+1)}$

    $|\text{Error}| \leq \dfrac{1}{(2n+3)!(2n+3)} < 0.0005$ if $n = 2$.

    Thus $I(1) \approx 1 - \dfrac{1}{3!3} + \dfrac{1}{5!5} \approx 0.946$ correct to three decimal places.

20. $\displaystyle\lim_{x \to 0} \frac{\sin(x^2)}{\sinh x} = \lim_{x \to 0} \frac{x^2 - \dfrac{x^6}{3!} + \dfrac{x^{10}}{5!} - \cdots}{x + \dfrac{x^3}{3!} + \dfrac{x^5}{5!} + \cdots}$

    $= \displaystyle\lim_{x \to 0} \frac{x - \dfrac{x^5}{3!} + \dfrac{x^9}{5!} - \cdots}{1 + \dfrac{x^2}{3!} + \dfrac{x^4}{5!} + \cdots} = 0$.

21. $\displaystyle\lim_{x \to 0} \frac{1 - \cos(x^2)}{(1 - \cos x)^2} = \lim_{x \to 0} \frac{1 - 1 + \dfrac{x^4}{2!} - \dfrac{x^8}{4!} + \cdots}{\left(1 - 1 + \dfrac{x^2}{2!} - \dfrac{x^4}{4!} + \cdots\right)^2}$

    $= \displaystyle\lim_{x \to 0} \frac{\dfrac{1}{2!} + \mathcal{O}(x^2)}{\dfrac{1}{4} + \mathcal{O}(x^2)} = 2$.

22. We have

    $\displaystyle\lim_{x \to 0} \frac{(e^x - 1 - x)^2}{x^2 - \ln(1 + x^2)} = \lim_{x \to 0} \frac{\left(\dfrac{x^2}{2!} + \dfrac{x^3}{3!} + \dfrac{x^4}{4!} + \cdots\right)^2}{\dfrac{x^4}{2} - \dfrac{x^6}{3} + \dfrac{x^8}{4} - \cdots}$

    $= \displaystyle\lim_{x \to 0} \frac{\dfrac{x^4}{4}\left(1 + \dfrac{x}{3} + \dfrac{x^2}{12} + \cdots\right)^2}{\dfrac{x^4}{2} - \dfrac{x^6}{3} + \dfrac{x^8}{4} - \cdots} = \dfrac{\left(\dfrac{1}{4}\right)}{\left(\dfrac{1}{2}\right)} = \dfrac{1}{2}$.

23. $\displaystyle\lim_{x\to 0}\frac{2\sin 3x - 3\sin 2x}{5x - \tan^{-1} 5x}$

$\displaystyle = \lim_{x\to 0}\frac{2\left(3x - \frac{3^3 x^3}{3!} + \cdots\right) - 3\left(2x - \frac{2^3 x^3}{3!} + \cdots\right)}{5x - \left(5x - \frac{5^3 x^3}{3} + \cdots\right)}$

$\displaystyle = \lim_{x\to 0}\frac{-9 + 4 + \mathcal{O}(x^2)}{\frac{125}{3} + \mathcal{O}(x^2)} = -\frac{5\times 3}{125} = -\frac{3}{25}.$

24. We have

$\displaystyle\lim_{x\to 0}\frac{\sin(\sin x) - x}{x[\cos(\sin x) - 1]}$

$\displaystyle = \lim_{x\to 0}\frac{\left(\sin x - \frac{1}{3!}\sin^3 x + \frac{1}{5!}\sin^5 x - \cdots\right) - x}{x\left[1 - \frac{1}{2!}\sin^2 x + \frac{1}{4!}\sin^4 x - \cdots - 1\right]}$

$\displaystyle = \lim_{x\to 0}\frac{\left(x - \frac{x^3}{3!} + \cdots\right) - \frac{1}{3!}\left(x - \frac{x^3}{3!} + \cdots\right)^3 + \frac{1}{5!}\left(x - \cdots\right)^5 - \cdots - x}{x\left[-\frac{1}{2!}\left(x - \frac{x^3}{3!} + \cdots\right)^2 + \frac{1}{4!}\left(x - \cdots\right)^4 - \cdots\right]}$

$\displaystyle = \lim_{x\to 0}\frac{-\frac{2}{3!}x^3 + \text{higher degree terms}}{-\frac{1}{2!}x^3 + \text{higher degree terms}} = \frac{\frac{2}{3!}}{\frac{1}{2!}} = \frac{2}{3}.$

25. $\displaystyle\lim_{x\to 0}\frac{\sinh x - \sin x}{\cosh x - \cos x}$

$\displaystyle = \lim_{x\to 0}\frac{\left(x + \frac{x^3}{3!} + \cdots\right) - \left(x - \frac{x^3}{3!} + \cdots\right)}{\left(1 + \frac{x^2}{2!} + \cdots\right) - \left(1 - \frac{x^2}{2!} + \cdots\right)}$

$\displaystyle = \lim_{x\to 0}\frac{\frac{x^3}{3} + \mathcal{O}(x^5)}{x^2 + \mathcal{O}(x^4)} = 0.$

26. If $y = \displaystyle\sum_{n=0}^{\infty} a_n x^n$, then $y' = \sum_{n=1}^{\infty} n a_n x^{n-1}$ and

$\displaystyle y'' = \sum_{n=2}^{\infty} n(n-1)a_n x^{n-2} = \sum_{n=0}^{\infty}(n+2)(n+1)a_{n+2}x^n.$

Thus,

$0 = y'' + xy' + y$

$\displaystyle = \sum_{n=0}^{\infty}(n+2)(n+1)a_{n+2}x^n + x\sum_{n=1}^{\infty} n a_n x^{n-1} + \sum_{n=0}^{\infty} a_n x^n$

$\displaystyle = 2a_2 + a_0 + \sum_{n=1}^{\infty}\left[(n+2)(n+1)a_{n+2} + (n+1)a_n\right]x^n.$

Since coefficients of all powers of x must vanish, therefore $2a_2 + a_0 = 0$ and, for $n \geq 1$,

$(n+2)(n+1)a_{n+2} + (n+1)a_n = 0,$

that is, $a_{n+2} = \dfrac{-a_n}{n+2}.$

If $y(0) = 1$, then $a_0 = 1$, $a_2 = \dfrac{-1}{2}$, $a_4 = \dfrac{1}{2^2\cdot 2!}$, $a_6 = \dfrac{-1}{2^3\cdot 3!}$, $a_8 = \dfrac{1}{2^4\cdot 4!},\ldots$. If $y'(0) = 0$, then $a_1 = a_3 = a_5 = \cdots = 0$. Hence,

$\displaystyle y = 1 - \frac{1}{2}x^2 + \frac{1}{8}x^4 - \frac{1}{48}x^6 + \cdots = \sum_{n=0}^{\infty}\frac{(-1)^n}{2^n\cdot n!}x^{2n}.$

27. $\begin{cases} y'' + xy' + 2y = 0 \\ y(0) = 1 \\ y'(0) = 2 \end{cases}$

Let

$\displaystyle y = \sum_{n=0}^{\infty} a_n x^n \qquad y' = \sum_{n=1}^{\infty} n a_n x^{n-1}$

$\displaystyle y'' = \sum_{n=2}^{\infty} n(n-1)a_n x^{n-2} = \sum_{n=0}^{\infty}(n+2)(n+1)a_{n+2}x^n.$

Substituting these expressions into the differential equation, we get

$\displaystyle \sum_{n=0}^{\infty}(n+2)(n+1)a_{n+2}x^n + \sum_{n=1}^{\infty} n a_n x^n$

$\displaystyle + 2\sum_{n=0}^{\infty} a_n x^n = 0,$ so

$\displaystyle 2a_2 + 2 + \sum_{n=1}^{\infty}[(n+2)(n+1)a_{n+2} + (n+2)a_n]x^n = 0.$

It follows that

$a_2 = -1, \qquad a_{n+2} = -\dfrac{a_n}{n+1}, \quad n = 1, 2, 3, \ldots.$

Since $a_0 = y(0) = 1$, and $a_1 = y'(0) = 2$, we have

$a_0 = 1 \qquad\qquad a_1 = 2$
$a_2 = -1 \qquad\quad\ a_3 = -\dfrac{2}{2}$
$a_4 = \dfrac{1}{3} \qquad\qquad a_5 = \dfrac{2}{2\times 4}$
$a_6 = -\dfrac{1}{3\times 5} \quad\ a_7 = -\dfrac{2}{2\times 4\times 6}$
$a_8 = \dfrac{1}{3\times 5\times 7} \quad a_9 = \dfrac{2}{2\times 4\times 6\times 8}.$

The patterns here are obvious:

$$a_{2n} = \frac{(-1)^n}{3 \times 5 \times \cdots \times (2n-1)} \qquad a_{2n+1} = \frac{(-1)^n 2}{2^n n!}$$

$$= \frac{(-1)^n 2^n n!}{(2n)!}$$

Thus $y = \sum_{n=0}^{\infty} (-1)^n \left[\frac{2^n n! x^{2n}}{(2n)!} + \frac{x^{2n+1}}{2^{n-1} n!} \right]$.

Section 9.8 Taylor's Formula Revisited (page 581)

1. $f(x) = \sin x$, $\quad P_5(x) = x - \frac{x^3}{6} + \frac{x^5}{120}$

   $|f(x) - P_5(x)| = \left| \frac{f^{(6)}(X)}{6!} x^6 \right|$

   for some X between 0 and x

   $|f(0.2) - P_5(0.2)| \leq \frac{\sin(0.2)}{720}(0.2)^6 < \frac{(0.2)^7}{720} < 2 \times 10^{-8}$.

2. Let $f(x) = \cos x$, then

   $f'(x) = -\sin x \quad f''(x) = -\cos x \quad f'''(x) = \sin x$
   $f^{(4)}(x) = \cos x \quad f^{(5)}(x) = -\sin x \quad f^{(6)}(x) = -\cos x$
   $f^{(7)}(x) = \sin x$.

   Since $P_6(1)$ is used to approximate $f(1)$, the error is

   $$R_6(1) = \frac{f^{(7)}(X)}{7!}(1)^7 = \frac{\sin(X)}{7!}$$

   for some X between 0 and 1. Since $1 < \frac{\pi}{3}$, we have $\sin X < \sin \frac{\pi}{3} = \frac{\sqrt{3}}{2}$. Thus,

   $$|R_6(1)| < \frac{\sqrt{3}}{2 \cdot 7!} \approx 0.000172.$$

3. $f(x) = e^{-x}$, $\quad P_4(x) = 1 - x + \frac{x^2}{2} - \frac{x^3}{6} + \frac{x^4}{24}$

   $|f(x) - P_4(x)| \leq \frac{e^{-X}}{5!}|x|^5$

   for some X between 0 and x

   $|f(-0.5) - P_4(-0.5)| \leq \frac{1}{120}(0.5)^5 < 2.61 \times 10^{-4}$.

4. Let $f(x) = \sec x$, then

   $f'(x) = \sec x \tan x$,
   $f''(x) = \sec x \tan^2 x + \sec^3 x$,
   $f'''(x) = \sec x \tan^3 x + 5 \sec^3 x \tan x$.

   If $P_2(0.2)$ is used to approximate $f(0.2)$, then

   $$R_2(0.2) = \frac{f'''(X)}{3!}(0.2)^3$$

   for some X between 0 and 0.2. Since $0.2 < \pi/6$,

   $0 \leq \sec X \leq \sec(0.2) < \sec \frac{\pi}{6} = \frac{2}{\sqrt{3}}$;

   $0 \leq \tan X \leq \tan(0.2) < \tan \frac{\pi}{6} = \frac{1}{\sqrt{3}}$.

   Thus,

   $|f'''(X)| = \sec X \tan^3 X + 5 \sec^3 X \tan X$

   $< \left(\frac{2}{\sqrt{3}}\right)\left(\frac{1}{\sqrt{3}}\right)^3 + 5\left(\frac{2}{\sqrt{3}}\right)^3\left(\frac{1}{\sqrt{3}}\right) = \frac{42}{9}$

   and

   $$|R_2(0.2)| < \frac{\left(\frac{42}{9}\right)}{3!}(0.2)^3 \approx 0.0062.$$

5. If $f(x) = \ln(\cos x)$, then $f'(x) = -\tan x$, $f''(x) = -\sec^2 x$, $f'''(x) = -2\sec^2 x \tan x$, and $f^{(4)}(x) = -4\sec^2 x \tan^2 x - 2\sec^4 x$. Hence

   $|f(0.1) - P_3(0.1)| \leq \frac{1}{4!}|f^{(4)}(X)|(0.1)^4$

   for some X between 0 and 0.1
   $|f^{(4)}(X)| \leq |f^{(4)}(0.1)| < 2.082$
   $|\text{error}| = |f(0.1) - P_3(0.1)| \leq \frac{2.082}{24}(0.1)^4 = 8.675 \times 10^{-6}$.

6. If $f(x) = \tan^{-1}(x)$, then $f^{(4)}(x) = \frac{24(x - x^3)}{(x^2+1)^4}$. If $P_3(0.99)$ is used to approximate $f(0.99)$, the error will be

   $$R_3(0.99) = \frac{f^{(4)}(X)}{4!}(0.99 - 1)^4$$

   for some X between 0.99 and 1. For such X,

   $$|F^{(4)}(x)| \leq \frac{24[1-(0.99)^3]}{[(0.99)^2+1]^4} \approx 0.0464.$$

Hence,

$$|R_3(0.99)| \leq \frac{0.0464}{4!}(0.99-1)^4 = 1.93 \times 10^{-11}.$$

7. If $f(x) = \ln x$, then $f'(x) = 1/x$, $f''(x) = -1/x^2$, $f'''(x) = 2/x^3$, $f^{(4)}(x) = -6/x^4$, and $f^{(5)}(x) = 24/x^5$. If $P_4(x)$ is the Taylor polynomial for f about $x = 2$, then for some X between 1.95 and 2 we have

$$|f(1.95) - P_4(1.95)| = \frac{24}{X^5} \cdot \frac{(0.05)^5}{5!}$$
$$\leq \frac{24(0.05)^5}{(1.95)^5 120} \approx 2.22 \times 10^{-9}.$$

8. If $f(x) = e^{-x}$, then $f^{(k)}(x) = (-1)^k e^{-x}$ and $f^{(k)}(0) = (-1)^k$ for $k = 0, 1, 2, \ldots$. The Lagrange remainder in Taylor's Formula is

$$R_n(x) = \frac{f^{(n+1)}(X)}{(n+1)!} x^{(n+1)} = (-1)^{n+1} e^{-X} \frac{x^{(n+1)}}{(n+1)!}$$

for some X between 0 and x. Clearly,

$$|R_n(x)| \leq e^{|x|} \frac{|x|^{(n+1)}}{(n+1)!} \to 0$$

as $n \to \infty$ for all real x. Thus,

$$e^{-x} = 1 - x + \frac{x^2}{2!} - \frac{x^3}{3!} + \cdots = \sum_{k=0}^{\infty} \frac{(-1)^k x^k}{k!}.$$

9. If $f(x) = 2^x$, then $f'(x) = 2^x \ln 2$, $f''(x) = 2^x (\ln 2)^2, \ldots$, $f^{(k)}(x) = 2^x (\ln 2)^k$. Here Taylor's Formula is

$$2^x = 1 + (\ln 2)x + \frac{(\ln 2)^2}{2!} x^2 + \cdots + \frac{(\ln 2)^n}{n!} x^n + R_n,$$

where $R_n = \dfrac{(\ln 2)^{n+1} 2^X x^{n+1}}{(n+1)!}$ for some X between 0 and x. We have

$$|R_n| \leq 2^{|x|} \frac{((\ln 2)|x|)^{n+1}}{(n+1)!} \to 0 \quad \text{as } n \to \infty$$

for all x. Hence

$$2^x = \sum_{n=0}^{\infty} \frac{(x \ln 2)^n}{n!}$$

for all x.

10. If $f(x) = \cos x$, then for $k = 0, 1, 2, \ldots$,

$$f^{(n)}(0) = \begin{cases} 1, & \text{if } n = 4k; \\ 0, & \text{if } n = 4k+1 \text{ or } 4k+3; \\ -1, & \text{if } n = 4k+2. \end{cases}$$

The Taylor Formula with Lagrange remainder is

$$f(x) = 1 - \frac{x^2}{2!} + \frac{x^4}{4!} - \cdots + (-1)^n \frac{x^{2n}}{2n!} + R_{2n}(x)$$

where

$$R_{2n}(x) = \frac{f^{(2n+1)}(X)}{(2n+1)!} x^{2n+1} = (-1)^{n+1} \sin X \frac{x^{2n+1}}{(2n+1)!}$$

for some X between 0 and x. Since

$$|R_{2n}(x)| < \frac{|x|^{2n+1}}{(2n+1)!} \to 0 \text{ as } n \to \infty,$$

therefore

$$\cos x = 1 - \frac{x^2}{2!} + \frac{x^4}{4!} - \frac{x^6}{6!} + \cdots = \sum_{n=0}^{\infty} \frac{(-1)^n x^{2n}}{(2n)!}$$

for all real x.

11. For $f(x) = \sin x$ Taylor's Formula gives

$$\sin x = x - \frac{x^3}{3!} + \frac{x^5}{5!} - \cdots + (-1)^n \frac{x^{2n+1}}{(2n+1)!} + R_n,$$

where $R_n = \dfrac{f^{(2n+2)}(X) x^{2n+2}}{(2n+2)!}$, for some X between 0 and x. Since $f^{(2n+2)}(x) = \pm \sin x$, we have

$$|R_n| \leq \frac{|x|^{2n+2}}{(2n+2)!} \to 0 \quad \text{as } n \to \infty$$

for all x. Thus, for all x,

$$\sin x = \sum_{n=0}^{\infty} \frac{x^{2n+1}}{(2n+1)!}.$$

12. If $f(x) = \sin^2 x = \dfrac{1 - \cos 2x}{2}$, then, for $k = 0, 1, 2, \ldots$,

$$f^{(n)}(x) = \begin{cases} 2^{(n-1)} \sin 2x, & \text{if } n = 4k+1, \\ 2^{(n-1)} \cos 2x, & \text{if } n = 4k+2, \\ -2^{(n-1)} \sin 2x, & \text{if } n = 4k+3, \\ -2^{(n-1)} \cos 2x, & \text{if } n = 4k+4; \end{cases}$$

and

$$f^{(n)}(0) = \begin{cases} 2^{(n-1)}, & \text{if } n = 4k+2, \\ 0, & \text{if } n = 4k+1 \text{ or } 4k+3, \\ -2^{(n-1)}, & \text{if } n = 4k+4. \end{cases}$$

The Taylor Formula is

$$f(x) = \frac{2}{2!}x^2 - \frac{2^3}{4!}x^4 + \frac{2^5}{6!}x^6 - \frac{2^7}{8!}x^8 + \cdots + (-1)^{n-1}\frac{2^{2n-1}}{(2n)!}x^{2n} + R_{2n}$$

where

$$R_{2n}(x) = \frac{f^{(2n+1)}(X)}{(2n+1)!}x^{2n+1} = (-1)^n \sin 2X \frac{x^{2n+1}}{(2n+1)!}$$

for some X between 0 and x. Since

$$|R_{2n}(x)| < \frac{|x|^{2n+1}}{(2n+1)!} \to 0 \text{ as } n \to \infty,$$

therefore

$$f(x) = \frac{2}{2!}x^2 - \frac{2^3}{4!}x^4 + \frac{2^5}{6!}x^6 - \cdots$$
$$= \sum_{n=1}^{\infty} \frac{(-1)^{n-1} 2^{2n-1}}{(2n)!} x^{2n}.$$

13. If $f(x) = \frac{1}{1-x}$, then $f'(x) = f(x) = \frac{1}{(1-x)^2}$,

$f''(x) = f(x) = \frac{2!}{(1-x)^3}, \ldots,$

$f^{(k)}(x) = f(x) = \frac{3!}{(1-x)^{k+1}}$. Thus Taylor's Formula gives

$$\frac{1}{1-x} = 1 + x + x^2 + \cdots + x^n + R_n.$$

In this case we already know that the geometric sum

$$1 + x + x^2 + \cdots + x^n = \frac{1-x^{n+1}}{1-x},$$

so the remainder R_n must be given by $R_n = \frac{x^{n+1}}{1-x}$. We have

$$|R_n| \leq \frac{|x|^{n+1}}{|1-x|} \to 0 \text{ as } x \to \infty$$

provided $-1 < x < 1$. Hence (as we already knew),

$$\frac{1}{1-x} = 1 + x + x^2 + x^3 + \cdots$$

for $-1 < x < 1$.

14. If $f(x) = \ln(1+x)$, then

$f'(x) = \frac{1}{1+x}, \ f''(x) = \frac{-1}{(1+x)^2}, \ f'''(x) = \frac{2}{(1+x)^3},$

$f^{(4)}(x) = \frac{-3!}{(1+x)^4}, \ldots, \ f^{(n)} = \frac{(-1)^{n-1}(n-1)!}{(1+x)^n}$

and

$$f(0) = 0, \ f'(0) = 1, \ f''(0) = -1, \ f'''(0) = 2,$$
$$f^{(4)}(0) = -3!, \ldots, \ f^{(n)}(0) = (-1)^{n-1}(n-1)!.$$

Therefore, the Taylor Formula is

$$f(x) = x + \frac{-1}{2!}x^2 + \frac{2}{3!}x^3 + \frac{-3!}{4!}x^4 + \cdots + \frac{(-1)^{n-1}(n-1)!}{n!}x^n + R_n(x)$$

where

$$R_n(x) = \frac{1}{n!}\int_0^x (x-t)^n f^{(n+1)}(t)\,dt$$
$$= \frac{1}{n!}\int_0^x (x-t)^n \frac{(-1)^n n!}{(1+t)^{n+1}}\,dt$$
$$= (-1)^n \int_0^x \frac{(x-t)^n}{(1+t)^{n+1}}\,dt.$$

If $0 \leq t \leq x \leq 1$, then $1+t \geq 1$ and

$$|R_n(x)| \leq \int_0^x (x-t)^n\,dt = \frac{x^{n+1}}{n+1} \leq \frac{1}{n+1} \to 0$$

as $n \to \infty$.
If $-1 < x \leq t \leq 0$, then

$$\left|\frac{x-t}{1+t}\right| = \frac{t-x}{1+t} \leq |x|,$$

because $\frac{t-x}{1+t}$ increases from 0 to $-x = |x|$ as t increases from x to 0. Thus,

$$|R_n(x)| < \frac{1}{1+x}\int_0^{|x|} |x|^n\,dt = \frac{|x|^{n+1}}{1+x} \to 0$$

as $n \to \infty$ since $|x| < 1$. Therefore,

$$f(x) = x - \frac{x^2}{2} + \frac{x^3}{3} - \frac{x^4}{4} + \cdots = \sum_{n=1}^{\infty}(-1)^{n-1}\frac{x^n}{n},$$

for $-1 < x \leq 1$.

15. We obtain the desired result from Exercise 27 as follows:

$$\frac{x}{2+3x} = \frac{1}{3}\cdot\frac{3x}{2+3x} = \frac{1}{3}\left(\frac{2+3x-2}{2+3x}\right)$$
$$= \frac{1}{3} - \frac{2}{3}\cdot\frac{1}{2+3x}$$
$$= \frac{1}{3}\left[1 - \frac{1}{1+\frac{3x}{2}}\right]$$
$$= \frac{1}{3}\left[1 - \left(1 - \frac{3x}{2} + \left(\frac{3x}{2}\right)^2 - \cdots\right)\right]$$
$$= \frac{x}{2} - \frac{3x^2}{2^2} + \frac{3^2 x^3}{2^3} - \cdots$$
$$= \sum_{n=1}^{\infty}(-1)^{n-1} 3^{n-1}\left(\frac{x}{2}\right)^n.$$

INSTRUCTOR'S SOLUTIONS MANUAL

The series is valid for $-\frac{2}{3} < x < \frac{2}{3}$.

16. If $f(x) = e^x$, then $f^{(k)}(x) = e^x$ and $f^{(k)}(a) = e^a$ for all k. The Lagrange remainder is

$$R_n(x) = \frac{f^{(n+1)}(X)}{(n+1)!}(x-a)^{n+1} = e^X \frac{(x-a)^{n+1}}{(n+1)!}$$

for some X between a and x. Since $e^{|X|} < e^{|x|}$, clearly,

$$|R_n(x)| < e^{|x|} \frac{|x-a|^{n+1}}{(n+1)!} \to 0 \text{ as } n \to \infty.$$

Therefore,

$$f(x) = e^a + e^a(x-a) + \frac{e^a}{2!}(x-a)^2 + \cdots = e^a \sum_{n=0}^{\infty} \frac{(x-a)^n}{n!}.$$

17. For $f(x) = \sin x$ we have

$$f^{(4n)}(\pi/6) = 1/2 \qquad f^{(4n+1)}(\pi/6) = \sqrt{3}/2$$
$$f^{(4n+2)}(\pi/6) = -1/2 \qquad f^{(4n+3)}(\pi/6) = -\sqrt{3}/2$$

Thus Taylor's Formula gives

$$\sin x = \frac{1}{2} + \frac{\sqrt{3}}{2}\left(x - \frac{\pi}{6}\right) - \frac{1}{2}\frac{1}{2!}\left(1 - \frac{\pi}{6}\right)^2$$
$$- \frac{\sqrt{3}}{2}\left(1 - \frac{\pi}{6}\right)^3 + \cdots + \frac{f^{(n)}(\pi/6)}{n!}\left(1 - \frac{\pi}{6}\right)^n + R_n.$$

We have

$$|R_n| = \frac{|f^{(n+1)}(X)|}{(n+1)!}\left|x - \frac{\pi}{6}\right|^{n+1}$$
$$\leq \frac{1}{(n+1)!}\left|x - \frac{\pi}{6}\right|^{n+1} \to 0 \quad \text{as } n \to \infty$$

for all x. Thus

$$\sin x = \frac{1}{2}\left[1 + \sqrt{3}\left(x - \frac{\pi}{6}\right) - \frac{1}{2!}\left(x - \frac{\pi}{6}\right)^2 \right.$$
$$\left. - \frac{\sqrt{3}}{3!}\left(x - \frac{\pi}{6}\right)^3 + \cdots \right]$$

for all x.

18. If $f(x) = \cos x$, then for $k = 0, 1, 2, \ldots,$

$$f^{(n)}(x) = \begin{cases} \cos x, & \text{if } n = 4k; \\ -\sin x, & \text{if } n = 4k+1; \\ -\cos x, & \text{if } n = 4k+2; \\ \sin x, & \text{if } n = 4k+3. \end{cases}$$

SECTION 9.8 (PAGE 581)

Therefore,

$$f^{(n)}\left(\frac{\pi}{4}\right) = \begin{cases} \frac{1}{\sqrt{2}}, & \text{if } n = 4k \text{ or } 4k+3; \\ -\frac{1}{\sqrt{2}}, & \text{if } n = 4k+1 \text{ or } 4k+2. \end{cases}$$

The Lagrange remainder is

$$|R_n(x)| = \left|\frac{f^{(n+1)}(X)}{(n+1)!}\left(x - \frac{\pi}{4}\right)^{n+1}\right| \leq \frac{\left(x - \frac{\pi}{4}\right)^{n+1}}{(n+1)!},$$

since $|f^{(n+1)}(X)| \leq 1$ for any X. Clearly, $|R_n(x)| \to 0$ as $n \to \infty$ for all real x. Hence,

$$f(x) = \frac{1}{\sqrt{2}} - \frac{1}{\sqrt{2}}\left(x - \frac{\pi}{4}\right) - \frac{1}{\sqrt{2}}\left(\frac{1}{2!}\right)\left(x - \frac{\pi}{4}\right)^2$$
$$+ \frac{1}{\sqrt{2}}\left(\frac{1}{3!}\right)\left(x - \frac{\pi}{4}\right)^3 + \cdots$$
$$= \frac{1}{\sqrt{2}}\sum_{n=0}^{\infty}(-1)^n\left[\frac{1}{(2n)!}\left(x - \frac{\pi}{4}\right)^{2n}\right.$$
$$\left. - \frac{1}{(2n+1)!}\left(x - \frac{\pi}{4}\right)^{2n+1}\right].$$

19. For $f(x) = \ln x$ we have $f^{(k)}(x) = (-1)^{k-1}\frac{(k-1)!}{x^k}$, as suggested by Exercise 21. Hence $f^{(k)}(1) = (-1)^{k-1}(k-1)!$. Taylor's Formula gives

$$\ln x = (x-1) - \frac{(x-1)^2}{2} + \frac{(x-1)^3}{3} - \frac{(x-1)^4}{4}$$
$$+ \cdots + (-1)^{n-1}\frac{(x-1)^n}{n} + R_n,$$

where the remainder, R_n, is given by the integral formula

$$R_n = \frac{1}{n!}\int_1^x (x-t)^n \frac{(-1)^n n!}{t^{n+1}}\, dt.$$

If $1 \leq x \leq 2$, then

$$|R_n| \leq \int_1^x (x-t)^n\, dt = \frac{(x-1)^{n+1}}{n+1},$$

so $\lim R_n = 0$ in this case.
If $0 < x < 1$, then

$$|R_n| = \int_x^1 \frac{(t-x)^n}{t^{n+1}}\, dt$$
$$= \int_x^1 \left(1 - \frac{x}{t}\right)^n \frac{dt}{t} \qquad \begin{aligned} \text{Let } u &= 1 - (x/t) \\ du &= (x/t^2)\, dt \end{aligned}$$
$$= \int_0^{1-x} \frac{u^n}{1-u}\, du \leq \frac{1}{x}\frac{(1-x)^{n+1}}{n+1}$$

363

which also approaches zero as $n \to \infty$. Thus

$$\ln x = \sum_{n=1}^{\infty} (-1)^{n-1} \frac{(x-1)^n}{n},$$

for $0 < x \le 2$.

20. Let $f(x) = \ln x$, then $f^{(k)}(x) = (-1)^{k-1}(k-1)!x^{-k}$. The Taylor Formula in powers of $x - 2$ is

$$f(x) = \ln 2 + \frac{x-2}{2} - \frac{1}{2}\left(\frac{x-2}{2}\right)^2 + \frac{1}{3}\left(\frac{x-2}{2}\right)^3 + \cdots +$$
$$\frac{(-1)^{n-1}}{n}\left(\frac{x-2}{2}\right)^n + R_n(x)$$

where

$$R_n(x) = \frac{1}{n!}\int_2^x (x-t)^n (-1)^n (n!) t^{-(n+1)}\, dt$$
$$= (-1)^n \int_2^x \frac{(x-t)^n}{t^{n+1}}\, dt.$$

If $2 \le t \le x \le 4$ then $\left|\dfrac{x-t}{t}\right|^n \le \left(\dfrac{4-t}{2}\right)^n$, so

$$|R_n| \le \frac{1}{2^{n+1}}\int_2^4 (4-t)^n\, dt = \frac{1}{n+1} \to 0$$

as $n \to \infty$.
If $0 < x \le t \le 2$ then

$$\left|\frac{x-t}{t}\right| = \frac{t-x}{t} \le \frac{2-x}{2},$$

since $\dfrac{t-x}{t}$ increases from 0 to $\dfrac{2-x}{2}$ as t increases from x to 2. Therefore

$$|R_n| \le \frac{1}{x}\left(\frac{2-x}{2}\right)\int_x^2 dt = \frac{(2-x)^{n+1}}{2^n x} \to 0$$

as $n \to \infty$ (since $2 - x < 2$). Therefore

$$\ln x = \ln 2 + \frac{x-2}{2} - \frac{1}{2}\left(\frac{x-2}{2}\right)^2 + \frac{1}{3}\left(\frac{x-2}{2}\right)^3 - \cdots$$
$$= \ln 2 + \sum_{n=1}^{\infty} \frac{(-1)^{n-1}}{n}\left(\frac{x-2}{2}\right)^n,$$

for $0 < x \le 4$.

21. We use the result of Exercise 27 as follows: let $x = t - 2$, so that $t = x + 2$. Then

$$\frac{1}{x} = \frac{1}{t-2} = -\frac{1}{2\left(1-\dfrac{t}{2}\right)}$$
$$= -\frac{1}{2}\left[1 + \frac{t}{2} + \frac{t^2}{2^2} + \frac{t^3}{2^3} + \cdots\right] = -\frac{1}{2}\sum_{n=0}^{\infty}\left(\frac{x+2}{2}\right)^n$$

for $-4 < x < 0$.

Section 9.9 The Binomial Theorem and Binomial Series (page 585)

1. $\sqrt{1+x} = (1+x)^{1/2}$

$$= 1 + \frac{x}{2} + \frac{1}{2}\left(-\frac{1}{2}\right)\frac{x^2}{2!} + \frac{1}{2}\left(-\frac{1}{2}\right)\left(-\frac{3}{2}\right)\frac{x^3}{3!} + \cdots$$
$$= 1 + \frac{x}{2} + \sum_{n=2}^{\infty}(-1)^{n-1}\frac{1 \cdot 3 \cdot 5 \cdots (2n-3)}{2^n n!}x^n$$
$$= 1 + \frac{x}{2} + \sum_{n=2}^{\infty}(-1)^{n-1}\frac{(2n-2)!}{2^{2n-1}(n-1)!n!}x^n \quad (-1 < x < 1).$$

2. $x\sqrt{1-x} = x(1-x)^{1/2}$

$$= x - \frac{x^2}{2} + \frac{1}{2}\left(-\frac{1}{2}\right)\frac{(-1)^2 x^3}{2!}$$
$$+ \frac{1}{2}\left(-\frac{1}{2}\right)\left(-\frac{3}{2}\right)\frac{(-1)^3 x^4}{3!} + \cdots$$
$$= x - \frac{x^2}{2} - \sum_{n=2}^{\infty}\frac{1 \cdot 3 \cdot 5 \cdots (2n-3)}{2^n n!}x^{n+1}$$
$$= x - \frac{x^2}{2} - \sum_{n=2}^{\infty}(-1)^{n-1}\frac{(2n-2)!}{2^{2n-1}(n-1)!n!}x^{n+1} \quad (-1 < x \le 1)$$

3. $\sqrt{4+x} = 2\sqrt{1 + \dfrac{x}{4}}$

$$= 2\left[1 + \frac{1}{2} \cdot \frac{x}{4} + \frac{\dfrac{1}{2}\left(-\dfrac{1}{2}\right)}{2!}\left(\frac{x}{4}\right)^2\right.$$
$$\left. + \frac{\dfrac{1}{2}\left(-\dfrac{1}{2}\right)\left(-\dfrac{3}{2}\right)}{3!}\left(\frac{x}{4}\right)^3 + \cdots\right]$$
$$= 2 + \frac{x}{4} + 2\sum_{n=2}^{\infty}(-1)^{n-1}\frac{1 \cdot 3 \cdot 5 \cdots (2n-3)}{2^{3n}n!}x^n$$
$$= 2 + \frac{x}{4} + 2\sum_{n=2}^{\infty}(-1)^{n-1}\frac{(2n-1)!}{2^{4n-1}n!(n-1)!}x^n$$
$$(-4 < x < 4).$$

4. $\dfrac{1}{\sqrt{4+x^2}} = \dfrac{1}{2\sqrt{1+\left(\frac{x}{2}\right)^2}} = \dfrac{1}{2}\left[1+\left(\dfrac{x}{2}\right)^2\right]^{-1/2}$

$= \dfrac{1}{2}\left[1+\left(-\dfrac{1}{2}\right)\left(\dfrac{x}{2}\right)^2 + \dfrac{1}{2!}\left(-\dfrac{1}{2}\right)\left(-\dfrac{3}{2}\right)\left(\dfrac{x}{2}\right)^4 + \dfrac{1}{3!}\left(-\dfrac{1}{2}\right)\left(-\dfrac{3}{2}\right)\left(-\dfrac{5}{2}\right)\left(\dfrac{x}{2}\right)^6 + \cdots\right]$

$= \dfrac{1}{2} - \dfrac{1}{2^4}x^2 + \dfrac{3}{2^7 2!}x^4 - \dfrac{3\times 5}{2^{10} 3!}x^6 + \cdots$

$= \dfrac{1}{2} + \sum_{n=1}^{\infty}(-1)^n \dfrac{1\times 2\times 3\times \cdots \times (2n-1)}{2^{3n+1}n!} x^{2n}$

$(-2 \le x \le 2)$.

5. $(1-x)^{-2}$

$= 1 - 2(-x) + \dfrac{(-2)(-3)}{2!}(-x)^2 + \dfrac{(-2)(-3)(-4)}{3!}(-x)^3 + \cdots$

$= 1 + 2x + 3x^2 + 4x^3 + \cdots = \sum_{n=1}^{\infty} nx^{n-1} \quad (-1 < x < 1)$.

6. $(1+x)^{-3} = 1 - 3x + \dfrac{(-3)(-4)}{2!}x^2 + \dfrac{(-3)(-4)(-5)}{3!}x^3 + \cdots$

$= 1 - 3x + \dfrac{(3)(4)}{2}x^2 - \dfrac{(4)(5)}{2}x^3 + \cdots$

$= \sum_{n=0}^{\infty}(-1)^n \dfrac{(n+2)(n+1)}{2} x^n \quad (-1 < x < 1)$.

7. i) $\binom{n}{0} = \dfrac{n!}{0!n!} = 1, \quad \binom{n}{n} = \dfrac{n!}{n!0!} = 1$.

ii) If $0 \le k \le n$, then

$\binom{n}{k-1} + \binom{n}{k} = \dfrac{n!}{(k-1)!(n-k+1)!} + \dfrac{n!}{k!(n-k)!}$

$= \dfrac{n!}{k!(n-k+1)!}(k + (n-k+1))$

$= \dfrac{(n+1)!}{k!(n+1-k)!} = \binom{n+1}{k}$.

8. The formula $(a+b)^n = \sum_{k=0}^{n}\binom{n}{k}a^{n-k}b^k$

holds for $n = 1$; it says $a + b = a + b$ in this case. Suppose the formula holds for $n = m$, where m is some positive integer. Then

$(a+b)^{m+1} = (a+b)\sum_{k=0}^{m}\binom{m}{k}a^{m-k}b^k$

$= \sum_{k=0}^{m}\binom{m}{k}a^{m+1-k}b^k + \sum_{k=0}^{m}\binom{m}{k}a^{m-k}b^{k+1}$

(replace k by $k-1$ in the latter sum)

$= \sum_{k=0}^{m}\binom{m}{k}a^{m+1-k}b^k + \sum_{k=1}^{m+1}\binom{m}{k-1}a^{m+1-k}b^k$

$= a^{m+1} + \sum_{k=1}^{m}\left[\binom{m}{k} + \binom{m}{k-1}\right]a^{m+1-k}b^k + b^{m+1}$

(by #13(i))

$= a^{m+1} + \sum_{k=1}^{m}\binom{m+1}{k}a^{m+1-k}b^k + b^{m+1}$ (by #13(ii))

$= \sum_{k=0}^{m+1}\binom{m+1}{k}a^{m+1-k}b^k$ (by #13(i) again).

Thus the formula holds for $n = m+1$. By induction it holds for all positive integers n.

9. Consider the Leibniz Rule:

$(fg)^{(n)} = \sum_{k=0}^{n}\binom{n}{k}f^{(n-k)}g^{(k)}$.

This holds for $n = 1$; it says $(fg)' = f'g + fg'$ in this case. Suppose the formula holds for $n = m$, where m is some positive integer. Then

$(fg)^{(m+1)} = \dfrac{d}{dx}(fg)^{(m)}$

$= \dfrac{d}{dx}\sum_{k=0}^{m}\binom{m}{k}f^{(m-k)}g^{(k)}$

$= \sum_{k=0}^{m}\binom{m}{k}f^{(m+1-k)}g^{(k)} + \sum_{k=0}^{m}\binom{m}{k}f^{(m-k)}g^{(k+1)}$

(replace k by $k-1$ in the latter sum)

$= \sum_{k=0}^{m}\binom{m}{k}f^{(m+1-k)}g^{(k)} + \sum_{k=1}^{m+1}\binom{m}{k-1}f^{(m+1-k)}g^{(k)}$

$= f^{(m+1)}g^{(0)} + \sum_{k=1}^{m}\left[\binom{m}{k} + \binom{m}{k-1}\right] \times f^{(m+1-k)}g^{(k)} + f^{(0)}g^{(m+1)}$

(by Exercise 13(i))

$= f^{(m+1)}g^{(0)} + \sum_{k=1}^{m}\binom{m+1}{k}f^{(m+1-k)}g^{(k)} + f^{(0)}g^{(m+1)}$

(by Exercise 13(ii))

$= \sum_{k=0}^{m+1}\binom{m+1}{k}f^{(m+1-k)}g^{(k)}$ (by 13(i) again).

Thus the Rule holds for $n = m + 1$. By induction, it holds for all positive integers n.

Section 9.10 Fourier Series (page 592)

1. $f(t) = \sin(3t)$ has fundamental period $2\pi/3$ since $\sin t$ has fundamental period 2π:

$$f\left(t + \tfrac{2\pi}{3}\right) = \sin\left(3\left(t + \tfrac{2\pi}{3}\right)\right) = \sin(3t + 2\pi)$$
$$= \sin(3t) = f(t).$$

2. $g(t) = \cos(3 + \pi t)$ has fundamental period 2 since $\cos t$ has fundamental period 2π:

$$g(t + 2) = \cos\bigl(3 + \pi(t + 2)\bigr) = \cos(3 + \pi t + 2\pi)$$
$$= \cos(3 + \pi t) = g(t).$$

3. $h(t) = \cos^2 t = \tfrac{1}{2}(1 + \cos 2t)$ has fundamental period π:

$$h(t + \pi) = \frac{1 + \cos(2t + 2\pi)}{2} = \frac{1 + \cos 2t}{2} = h(t).$$

4. Since $\sin 2t$ has periods $\pi, 2\pi, 3\pi, \ldots$, and $\cos 3t$ has periods $\tfrac{2\pi}{3}, \tfrac{4\pi}{3}, \tfrac{6\pi}{3} = 2\pi, \tfrac{8\pi}{3}, \ldots$, the sum $k(t) = \sin(2t) + \cos(3t)$ has periods $2\pi, 4\pi, \ldots$. Its fundamental period is 2π.

5. Since $f(t) = t$ is odd on $(-\pi, \pi)$ and has period 2π, its cosine coefficients are 0 and its sine coefficients are given by

$$b_n = \frac{2}{2\pi}\int_{-\pi}^{\pi} t\sin(nt)\,dt = \frac{2}{\pi}\int_0^{\pi} t\sin(nt)\,dt.$$

This integral can be evaluated by a single integration by parts. Instead we used Maple to do the integral:

$$b_n = -\frac{2}{n}\cos(n\pi) = (-1)^{n+1}\frac{2}{n}.$$

The Fourier series of f is $\displaystyle\sum_{n=1}^{\infty}(-1)^{n+1}\frac{2}{n}\sin(nt)$.

6. $f(t) = \begin{cases} 0 & \text{if } 0 \le t < 1 \\ 1 & \text{if } 1 \le t < 2 \end{cases}$, f has period 2.

The Fourier coefficients of f are as follows:

$$\frac{a_0}{2} = \frac{1}{2}\int_0^2 f(t)\,dt = \frac{1}{2}\int_1^2 dt = \frac{1}{2}$$

$$a_n = \int_0^2 f(t)\cos(n\pi t)\,dt = \int_1^2 \cos(n\pi t)\,dt$$
$$= \frac{1}{n\pi}\sin(n\pi t)\Big|_1^2 = 0, \quad (n \ge 1)$$

$$b_n = \int_1^2 \sin(n\pi t)\,dt = -\frac{1}{n\pi}\cos(n\pi t)\Big|_1^2$$
$$= -\frac{1 - (-1)^n}{n\pi} = \begin{cases} -\dfrac{2}{n\pi} & \text{if } n \text{ is odd} \\ 0 & \text{if } n \text{ is even} \end{cases}$$

The Fourier series of f is

$$\frac{1}{2} - \sum_{n=1}^{\infty} \frac{2}{(2n-1)\pi}\sin\bigl((2n-1)\pi t\bigr).$$

7. $f(t) = \begin{cases} 0 & \text{if } -1 \le t < 0 \\ t & \text{if } 0 \le t < 1 \end{cases}$, f has period 2.

The Fourier coefficients of f are as follows:

$$\frac{a_0}{2} = \frac{-1}{1}\int_{-1}^1 f(t)\,dt = \frac{1}{2}\int_0^1 t\,dt = \frac{1}{4}$$

$$a_n = \int_{-1}^1 f(t)\cos(n\pi t)\,dt = \int_0^1 t\cos(n\pi t)\,dt$$
$$= \frac{(-1)^n - 1}{n^2\pi^2} = \begin{cases} -2/(n\pi)^2 & \text{if } n \text{ is odd} \\ 0 & \text{if } n \text{ is even} \end{cases}$$

$$b_n = \int_0^1 t\sin(n\pi t)\,dt$$
$$= -\frac{(-1)^n}{n\pi}.$$

The Fourier series of f is

$$\frac{1}{4} - \frac{2}{\pi^2}\sum_{n=1}^{\infty} \frac{1}{(2n-1)^2}\cos\bigl((2n-1)\pi t\bigr) - \frac{1}{\pi}\sum_{n=1}^{\infty}\frac{(-1)^n}{n}\sin(n\pi t).$$

8. $f(t) = \begin{cases} t & \text{if } 0 \le t < 1 \\ 1 & \text{if } 1 \le t < 2 \\ 3 - t & \text{if } 2 \le t < 3 \end{cases}$, f has period 3.

f is even, so its Fourier sine coefficients are all zero. Its

INSTRUCTOR'S SOLUTIONS MANUAL SECTION 9.10 (PAGE 592)

cosine coefficients are

$$\frac{a_0}{2} = \frac{1}{2} \cdot \frac{2}{3} \int_0^3 f(t)\,dt = \frac{2}{3}(2) = \frac{2}{3}$$

$$a_n = \frac{2}{3} \int_0^3 f(t) \cos \frac{2n\pi t}{3}\,dt$$

$$= \frac{2}{3}\left[\int_0^1 t \cos \frac{2n\pi t}{3}\,dt + \int_1^2 \cos \frac{2n\pi t}{3}\,dt \right.$$

$$\left. + \int_2^3 (3-t) \cos \frac{2n\pi t}{3}\,dt\right]$$

$$= \frac{3}{2n^2\pi^2}\left[\cos \frac{2n\pi}{3} - 1 - \cos(2n\pi) + \cos \frac{4n\pi}{3}\right].$$

The latter expression was obtained using Maple to evaluate the integrals. If $n = 3k$, where k is an integer, then $a_n = 0$. For other integers n we have $a_n = -9/(2\pi^2 n^2)$. Thus the Fourier series of f is

$$\frac{2}{3} - \frac{9}{2\pi^2} \sum_{n=1}^\infty \frac{1}{n^2} \cos \frac{2n\pi t}{3} + \frac{1}{2\pi^2} \sum_{n=1}^\infty \frac{1}{n^2} \cos(2n\pi t).$$

9. The even extension of $h(t) = 1$ on $[0, 1]$ to $[-1, 1]$ has the value 1 everywhere. Therefore all the coefficients a_n and b_n are zero except a_0, which is 2. The Fourier series is $a_0/2 = 1$.

10. The Fourier sine series of $g(t) = \pi - t$ on $[0, \pi]$ has coefficients

$$b_n = \frac{2}{\pi} \int_0^\pi (\pi - t) \sin nt\,dt = \frac{2}{n}.$$

The required Fourier sine series is

$$\sum_{n=1}^\infty \frac{2}{n} \sin nt.$$

11. The Fourier sine series of $f(t) = t$ on $[0, 1]$ has coefficients

$$b_n = 2 \int_0^1 t \sin(n\pi t)\,dt = -2\frac{(-1)^n}{n\pi}.$$

The required Fourier sine series is

$$\sum_{n=1}^\infty \frac{2(-1)^n}{n\pi} \sin(n\pi t).$$

12. The Fourier cosine series of $f(t) = t$ on $[0, 1]$ has coefficients

$$\frac{a_0}{2} = \int_0^1 t\,dt = \frac{1}{2}$$

$$a_n = 2 \int_0^1 t \cos(n\pi t)\,dt$$

$$= \frac{2(-1)^n - 2}{n^2\pi^2} = \begin{cases} 0 & \text{if } n \text{ is even} \\ \frac{-4}{n^2\pi^2} & \text{if } n \text{ is odd.} \end{cases}$$

The required Fourier cosine series is

$$\frac{1}{2} - \frac{4}{\pi^2} \sum_{n=1}^\infty \frac{\cos\bigl((2n-1)\pi t\bigr)}{(2n-1)^2}.$$

13. From Example 3,

$$\frac{\pi}{2} + \sum_{n=1}^\infty \frac{4}{\pi(2n-1)^2} \cos\bigl((2n-1)\pi t\bigr) = \pi - |t|$$

for $-\pi \le t \le \pi$. Putting $t = \pi$, we obtain

$$\frac{\pi}{2} + \sum_{n=1}^\infty \frac{4}{\pi(2n-1)^2}(-1) = 0.$$

Thus $\displaystyle\sum_{n=1}^\infty \frac{1}{(2n-1)^2} = \frac{\pi}{2} \cdot \frac{\pi}{4} = \frac{\pi^2}{8}.$

14. If f is even and has period T, then

$$b_n = \frac{2}{T} \int_{-T/2}^{T/2} f(t) \sin \frac{2n\pi t}{T}\,dt$$

$$= \frac{2}{T}\left[\int_{-T/2}^0 f(t) \sin \frac{2n\pi t}{T}\,dt + \int_0^{T/2} f(t) \sin \frac{2n\pi t}{T}\,dt\right].$$

In the first integral in the line above replace t with $-t$. Since $f(-t) = f(t)$ and sine is odd, we get

$$b_n = \frac{2}{T}\left[\int_{T/2}^0 f(t)\left(-\sin \frac{2n\pi t}{T}\right)(-dt) \right.$$

$$\left. + \int_0^{T/2} f(t) \sin \frac{2n\pi t}{T}\,dt\right]$$

$$= \frac{2}{T}\left[-\int_0^{T/2} f(t) \sin \frac{2n\pi t}{T}\,dt + \int_0^{T/2} f(t) \sin \frac{2n\pi t}{T}\,dt\right]$$

$$= 0.$$

Similarly,

$$a_n = \frac{2}{T}\left[\int_{-T/2}^0 f(t) \cos \frac{2n\pi t}{T}\,dt + \int_0^{T/2} f(t) \cos \frac{2n\pi t}{T}\,dt\right]$$

$$= \frac{2}{T}\left[\int_{T/2}^0 f(t) \cos \frac{2n\pi t}{T}(-dt) + \int_0^{T/2} f(t) \cos \frac{2n\pi t}{T}\,dt\right]$$

$$= \frac{4}{T} \int_0^{T/2} f(t) \cos \frac{2n\pi t}{T}\,dt.$$

The corresponding result for an odd function f states that $a_n = 0$ and

$$b_n = \frac{4}{T} \int_0^{T/2} f(t) \sin \frac{2n\pi t}{T}\, dt,$$

and is proved similarly.

Review Exercises 9 (page 593)

1. $\lim_{n\to\infty} \dfrac{(-1)^n e^n}{n!} = 0$. The sequence converges.

2. $\lim_{n\to\infty} \dfrac{n^{100} + 2^n \pi}{2^n} = \lim_{n\to\infty}\left(\pi + \dfrac{n^{100}}{2^n}\right) = \pi$.
   The sequence converges.

3. $\lim_{n\to\infty} \dfrac{\ln n}{\tan^{-1} n} \geq \lim_{n\to\infty} \dfrac{\ln n}{\pi/2} = \infty$.
   The sequence diverges to infinity.

4. $\lim_{n\to\infty} \dfrac{(-1)^n n^2}{\pi n(n-\pi)} = \lim_{n\to\infty} \dfrac{(-1)^n}{1-(\pi/n)}$ does not exist.
   The sequence diverges (oscillates).

5. Let $a_1 > \sqrt{2}$ and $a_{n+1} = \dfrac{a_n}{2} + \dfrac{1}{a_n}$.

   If $f(x) = \dfrac{x}{2} + \dfrac{1}{x}$, then $f'(x) = \dfrac{1}{2} - \dfrac{1}{x^2} > 0$ if $x > \sqrt{2}$.
   Since $f(\sqrt{2}) = \sqrt{2}$, we have $f(x) > \sqrt{2}$ if $x > \sqrt{2}$.
   Therefore, if $a_n > \sqrt{2}$, then $a_{n+1} = f(a_n) > \sqrt{2}$.
   Thus $a_n > \sqrt{2}$ for all $n \geq 1$, by induction.
   $a_n > \sqrt{2} \Rightarrow 2 < a_n^2 \Rightarrow a_n^2 + 2 < 2a_n^2$
   $\Rightarrow \dfrac{a_n^2 + 2}{2a_n} < a_n \Rightarrow a_{n+1} < a_n$.
   Thus $\{a_n\}$ is decreasing and $a_n > \sqrt{2}$ for all n.
   Being decreasing and bounded below by $\sqrt{2}$, $\{a_n\}$ must converge by the completeness axiom. Let $\lim_{n\to\infty} a_n = a$. Then $a \geq \sqrt{2}$, and

   $$\lim_{n\to\infty} a_{n+1} = \lim_{n\to\infty}\left(\frac{a_n}{2} + \frac{1}{a_n}\right)$$
   $$a = \frac{a}{2} + \frac{1}{a}.$$

   Thus $a/2 = 1/a$, so $a^2 = 2$, and $\lim_{n\to\infty} a_n = a = \sqrt{2}$.

6. By l'Hôpital's Rule,

   $$\lim_{x\to\infty} \frac{\ln(x+1)}{\ln x} = \lim_{x\to\infty} \frac{1/(x+1)}{1/x} = \lim_{x\to\infty} \frac{x}{x+1} = 1.$$

   Thus

   $$\lim_{n\to\infty}\left(\ln\ln(n+1) - \ln\ln n\right) = \lim_{n\to\infty} \ln \frac{\ln(n+1)}{\ln n} = \ln 1 = 0.$$

7. $\displaystyle\sum_{n=1}^{\infty} 2^{-(n-5)/2} = 2^2 \left(1 + \frac{1}{\sqrt{2}} + \frac{1}{2} + \cdots\right)$
   $= \dfrac{4}{1 - (1/\sqrt{2})} = \dfrac{4\sqrt{2}}{\sqrt{2} - 1}.$

8. $\displaystyle\sum_{n=0}^{\infty} \frac{4^{n-1}}{(\pi-1)^{2n}} = \frac{1}{4}\sum_{n=0}^{\infty}\left(\frac{4}{(\pi-1)^2}\right)^n$
   $= \dfrac{1}{4} \cdot \dfrac{1}{1 - \dfrac{4}{(\pi-1)^2}} = \dfrac{(\pi-1)^2}{4(\pi-1)^2 - 16},$
   since $(\pi-1)^2 > 4$.

9. $\displaystyle\sum_{n=1}^{\infty} \frac{1}{n^2 - \frac{1}{4}} = \sum_{n=1}^{\infty}\left(\frac{1}{n - \frac{1}{2}} - \frac{1}{n + \frac{1}{2}}\right)$ (telescoping)
   $= 2 - \lim_{N\to\infty} \dfrac{1}{N + \frac{1}{2}} = 2.$

10. $\displaystyle\sum_{n=1}^{\infty} \frac{1}{n^2 - \frac{9}{4}} = \sum_{n=1}^{\infty} \frac{1}{3}\left(\frac{1}{n - \frac{3}{2}} - \frac{1}{n + \frac{3}{2}}\right)$ (telescoping)
    $= \dfrac{1}{3}\left[\dfrac{1}{-1/2} - \dfrac{1}{5/2} + \dfrac{1}{1/2} - \dfrac{1}{7/2}\right.$
    $\left.\quad + \dfrac{1}{3/2} - \dfrac{1}{9/2} + \dfrac{1}{5/2} - \dfrac{1}{11/2} + \cdots\right]$
    $= \dfrac{1}{3}\left[-2 + 2 + \dfrac{2}{3}\right] = \dfrac{2}{9}.$

11. Since $0 \leq \dfrac{n-1}{n^3} \leq \dfrac{1}{n^2}$ for $n \geq 1$ and $\displaystyle\sum_{n=1}^{\infty}\dfrac{1}{n^2}$ converges, $\displaystyle\sum_{n=1}^{\infty} \frac{n-1}{n^3}$ must also converge.

12. $\displaystyle\sum_{n=1}^{\infty} \frac{n + 2^n}{1 + 3^n}$ converges by comparison with the convergent geometric series $\displaystyle\sum_{n=1}^{\infty}\left(\frac{2}{3}\right)^n$ because

    $$\lim_{n\to\infty} \frac{\frac{n + 2^n}{1 + 3^n}}{(2/3)^n} = \lim_{n\to\infty} \frac{(n/2^n) + 1}{(1/3^n) + 1} = 1.$$

13. $\displaystyle\sum_{n=1}^{\infty} \frac{n}{(1+n)(1 + n\sqrt{n})}$ converges by comparison with the convergent p-series $\displaystyle\sum_{n=1}^{\infty} \frac{1}{n^{3/2}}$ because

    $$\lim_{n\to\infty} \frac{\dfrac{n}{(1+n)(1+n\sqrt{n})}}{\dfrac{1}{n^{3/2}}} = \lim_{n\to\infty} \frac{1}{\left(\dfrac{1}{n}+1\right)\left(\dfrac{1}{n^{3/2}}+1\right)} = 1.$$

14. $\sum_{n=1}^{\infty} \dfrac{n^2}{(1+2^n)(1+n\sqrt{n})}$ converges by comparison with the convergent series $\sum_{n=1}^{\infty} \dfrac{\sqrt{n}}{2^n}$ (which converges by the ratio test) because

$$\lim_{n\to\infty} \dfrac{\dfrac{n^2}{(1+2^n)(1+n\sqrt{n})}}{\dfrac{\sqrt{n}}{2^n}} = \lim_{n\to\infty} \dfrac{1}{\left(\dfrac{1}{2^n}+1\right)\left(\dfrac{1}{n^{3/2}}+1\right)} = 1.$$

15. $\sum_{n=1}^{\infty} \dfrac{3^{2n+1}}{n!}$ converges by the ratio test, because

$$\lim_{n\to\infty} \dfrac{3^{2(n+1)+1}}{(n+1)!} \cdot \dfrac{n!}{3^{2n+1}} = \lim_{n\to\infty} \dfrac{9}{n+1} = 0 < 1.$$

16. $\sum_{n=1}^{\infty} \dfrac{n!}{(n+2)!+1}$ converges by comparison with the convergent p-series $\sum_{n=1}^{\infty} \dfrac{1}{n^2}$, because

$$0 \le \dfrac{n!}{(n+2)!+1} < \dfrac{n!}{(n+2)!} = \dfrac{1}{(n+2)(n+1)} < \dfrac{1}{n^2}.$$

17. $\sum_{n=1}^{\infty} \dfrac{(-1)^{n-1}}{1+n^3}$ converges absolutely by comparison with the convergent p-series $\sum_{n=1}^{\infty} \dfrac{1}{n^3}$, because

$$0 \le \left|\dfrac{(-1)^{n-1}}{1+n^3}\right| \le \dfrac{1}{n^3}.$$

18. $\sum_{n=1}^{\infty} \dfrac{(-1)^n}{2^n - n}$ converges absolutely by comparison with the convergent geometric series $\sum_{n=1}^{\infty} \dfrac{1}{2^n}$, because

$$\lim_{n\to\infty} \dfrac{\left|\dfrac{(-1)^n}{2^n-n}\right|}{\dfrac{1}{2^n}} = \lim_{n\to\infty} \dfrac{1}{1-\dfrac{n}{2^n}} = 1.$$

19. $\sum_{n=1}^{\infty} \dfrac{(-1)^{n-1}}{\ln\ln n}$ converges by the alternating series test, but the convergence is only conditional since $\sum_{n=1}^{\infty} \dfrac{1}{\ln\ln n}$ diverges to infinity by comparison with the divergent harmonic series $\sum_{n=1}^{\infty} \dfrac{1}{n}$. (Note that $\ln\ln n < n$ for all $n \ge 1$.)

20. $\sum_{n=1}^{\infty} \dfrac{n^2 \cos(n\pi)}{1+n^3}$ converges by the alternating series test (note that $\cos(n\pi) = (-1)^n$), but the convergence is only conditional because

$$\left|\dfrac{n^2\cos(n\pi)}{1+n^3}\right| = \dfrac{n^2}{1+n^3} \ge \dfrac{1}{2n}$$

for $n \ge 1$, and $\sum_{n=1}^{\infty} \dfrac{1}{2n}$ is a divergent harmonic series.

21. $\lim_{n\to\infty} \left|\dfrac{\dfrac{(x-2)^{n+1}}{3^{n+1}\sqrt{n+1}}}{\dfrac{(x-2)^n}{3^n\sqrt{n}}}\right| = \lim_{n\to\infty} \dfrac{|x-2|}{3}\sqrt{\dfrac{n}{n+1}} = \dfrac{|x-2|}{3}.$

$\sum_{n=1}^{\infty} \dfrac{(x-2)^n}{3^n\sqrt{n}}$ converges absolutely if $\dfrac{|x-2|}{3} < 1$, that is, if $-1 < x < 5$, and diverges if $x < -1$ or $x > 5$.

If $x = -1$ the series is $\sum \dfrac{(-1)^n}{\sqrt{n}}$, which converges conditionally.

If $x = 5$ the series is $\sum \dfrac{1}{\sqrt{n}}$, which diverges (to ∞).

22. $\lim_{n\to\infty} \left|\dfrac{\dfrac{(5-2x)^{n+1}}{n+1}}{\dfrac{(5-2x)^n}{n}}\right| = \lim_{n\to\infty} |5-2x|\dfrac{n}{n+1} = |5-2x|.$

$\sum_{n=1}^{\infty} \dfrac{(5-2x)^n}{n}$ converges absolutely if $|5-2x| < 1$, that is, if $2 < x < 3$, and diverges if $x < 2$ or $x > 3$.

If $x = 2$ the series is $\sum \dfrac{1}{n}$, which diverges.

If $x = 3$ the series is $\sum \dfrac{(-1)^n}{n}$, which converges conditionally.

23. Let $s = \sum_{k=1}^{\infty} \dfrac{1}{k^3}$ and $s_n = \sum_{k=1}^{n} \dfrac{1}{k^3}$. Then

$$\int_{n+1}^{\infty} \dfrac{dt}{t^3} < s - s_n < \int_n^{\infty} \dfrac{dt}{t^3}$$

$$s_n + \dfrac{1}{2(n+1)^2} < s < s_n + \dfrac{1}{2n^2}.$$

Let

$$s_n^* = \frac{1}{2}\left[s_n + \frac{1}{2(n+1)^2} + s_n + \frac{1}{2n^2}\right] = s_n + \frac{n^2+(n+1)^2}{4n^2(n+1)^2}.$$

Then $s \approx s_n^*$ with error satisfying

$$|s - s_n^*| < \frac{1}{2}\left[\frac{1}{2n^2} - \frac{1}{2(n+1)^2}\right] = \frac{2n+1}{4n^2(n+1)^2}.$$

This error is less than 0.001 if $n \geq 8$. Hence

$$s \approx \frac{1}{1^3} + \frac{1}{2^3} + \frac{1}{3^3} + \frac{1}{4^3} + \frac{1}{5^3} + \frac{1}{6^3} + \frac{1}{7^3} + \frac{1}{8^3}$$
$$+ \frac{64+81}{4(64)(81)} \approx 1.202$$

with error less than 0.001.

24. Let $s = \sum_{k=1}^{\infty} \frac{1}{4+k^2}$ and $s_n = \sum_{k=1}^{n} \frac{1}{4+k^2}$. Then

$$\int_{n+1}^{\infty} \frac{dt}{4+t^2} < s - s_n < \int_{n}^{\infty} \frac{dt}{4+t^2}$$

$$s_n + \frac{\pi}{4} - \frac{1}{2}\tan^{-1}\frac{n+1}{2} < s < s_n + \frac{\pi}{4} - \frac{1}{2}\tan^{-1}\frac{n}{2}.$$

Let

$$s_n^* = s_n + \frac{\pi}{4} - \frac{1}{4}\left[\tan^{-1}\frac{n+1}{2} + \tan^{-1}\frac{n}{2}\right].$$

Then $s \approx s_n^*$ with error satisfying

$$|s - s_n^*| < \frac{1}{4}\left[\tan^{-1}\frac{n+1}{2} - \tan^{-1}\frac{n}{2}\right].$$

This error is less than 0.001 if $n \geq 22$. Hence

$$s \approx \sum_{k=1}^{22}\frac{1}{4+k^2} + \frac{\pi}{4} - \frac{1}{4}\left[\tan^{-1}\frac{23}{2} + \tan^{-1}(11)\right] \approx 0.6605$$

with error less than 0.001.

25. $\dfrac{1}{3-x} = \dfrac{1}{3\left(1-\dfrac{x}{3}\right)}$

$$= \frac{1}{3}\sum_{n=0}^{\infty}\left(\frac{x}{3}\right)^n = \sum_{n=0}^{\infty}\frac{x^n}{3^{n+1}} \quad (-3 < x < 3).$$

26. Replace x with x^2 in Exercise 25 and multiply by x to get

$$\frac{x}{3-x^2} = \sum_{n=0}^{\infty}\frac{x^{2n+1}}{3^{n+1}} \quad (-\sqrt{3} < x < \sqrt{3}).$$

27. $\ln(e+x^2) = \ln e + \ln\left(1+\dfrac{x^2}{e}\right)$

$$= \ln e + \sum_{n=1}^{\infty}(-1)^{n-1}\frac{x^{2n}}{ne^n} \quad (-\sqrt{e} < x \leq \sqrt{e}).$$

28. $\dfrac{1-e^{-2x}}{x} = \dfrac{1}{x}\left(1 - 1 - \sum_{n=1}^{\infty}\dfrac{(-2x)^n}{n!}\right)$

$$= \sum_{n=1}^{\infty}(-1)^{n-1}\frac{2^n x^{n-1}}{n!} \quad \text{(for all } x \neq 0\text{)}.$$

29. $x\cos^2 x = \dfrac{x}{2}(1+\cos(2x))$

$$= \frac{x}{2}\left(1 + \sum_{n=0}^{\infty}(-1)^n\frac{(2x)^{2n}}{(2n)!}\right)$$

$$= x + \sum_{n=1}^{\infty}(-1)^n\frac{2^{2n-1}x^{2n+1}}{(2n)!} \quad \text{(for all } x\text{)}.$$

30. $\sin\left(x+\dfrac{\pi}{3}\right) = \sin x\cos\dfrac{\pi}{3} + \cos x\sin\dfrac{\pi}{3}$

$$= \frac{1}{2}\sum_{n=0}^{\infty}(-1)^n\frac{x^{2n+1}}{(2n+1)!} + \frac{\sqrt{3}}{2}\sum_{n=0}^{\infty}(-1)^n\frac{x^{2n}}{(2n)!}$$

$$= \sum_{n=0}^{\infty}\frac{(-1)^n}{2}\left(\frac{\sqrt{3}x^{2n}}{(2n)!} + \frac{x^{2n+1}}{(2n+1)!}\right) \quad \text{(for all } x\text{)}.$$

31. $(8+x)^{-1/3} = \dfrac{1}{2}\left(1+\dfrac{x}{8}\right)^{-1/3}$

$$= \frac{1}{2}\left[1 - \frac{1}{3}\left(\frac{x}{8}\right) + \frac{\left(-\frac{1}{3}\right)\left(-\frac{4}{3}\right)}{2!}\left(\frac{x}{8}\right)^2\right.$$
$$\left. + \frac{\left(-\frac{1}{3}\right)\left(-\frac{4}{3}\right)\left(-\frac{7}{3}\right)}{3!}\left(\frac{x}{8}\right)^3 + \cdots\right]$$

$$= \frac{1}{2} + \sum_{n=1}^{\infty}(-1)^n\frac{1\cdot 4\cdot 7\cdots(3n-2)}{2\cdot 3^n\cdot 8^n\cdot n!}x^n \quad (-8 < x < 8).$$

(Remark: Examining the ln of the absolute value of the nth term at $x=8$ shows that this term $\to 0$ as $n \to \infty$. Therefore the series also converges at $x=8$.)

32. $(1+x)^{1/3} = 1 + \dfrac{1}{3}x + \dfrac{\left(\dfrac{1}{3}\right)\left(-\dfrac{2}{3}\right)}{2!}x^2$

$$+ \frac{\left(\frac{1}{3}\right)\left(-\frac{2}{3}\right)\left(-\frac{5}{3}\right)}{3!}x^3 + \cdots$$

$$= 1 + \frac{x}{3} + \sum_{n=2}^{\infty}(-1)^{n-1}\frac{2\cdot 5\cdot 8\cdots(3n-4)}{3^n n!}x^n \quad (-1 < x < 1$$

(Remark: the series also converges at $x=1$.)

33. $\dfrac{1}{x} = \dfrac{1}{\pi+(x-\pi)} = \dfrac{1}{\pi}\cdot\dfrac{1}{1+\dfrac{x-\pi}{\pi}}$

$= \dfrac{1}{\pi}\sum_{n=0}^{\infty}(-1)^n\left(\dfrac{x-\pi}{\pi}\right)^n$

$= \sum_{n=0}^{\infty}(-1)^n\dfrac{(x-\pi)^n}{\pi^{n+1}} \quad (0 < x < 2\pi).$

34. Let $u = x - (\pi/4)$, so $x = u + (\pi/4)$. Then

$\sin x + \cos x = \sin\left(u+\dfrac{\pi}{4}\right) + \cos\left(u+\dfrac{\pi}{4}\right)$

$= \dfrac{1}{\sqrt{2}}\big((\sin u + \cos u) + (\cos u - \sin u)\big)$

$= \sqrt{2}\cos u = \sqrt{2}\sum_{n=0}^{\infty}(-1)^n\dfrac{u^{2n}}{(2n)!}$

$= \sqrt{2}\sum_{n=0}^{\infty}\dfrac{(-1)^n}{(2n)!}\left(x-\dfrac{\pi}{4}\right)^{2n} \quad \text{(for all } x\text{)}.$

35. $e^{x^2+2x} = e^{x^2}e^{2x}$

$= (1+x^2+\cdots)\left(1+2x+\dfrac{4x^2}{2!}+\dfrac{8x^3}{3!}+\cdots\right)$

$= 1 + 2x + 2x^2 + \dfrac{4}{3}x^3 + x^2 + 2x^3 + \cdots$

$P_3(x) = 1 + 2x + 3x^2 + \dfrac{10}{3}x^3.$

36. $\sin(1+x) = \sin(1)\cos x + \cos(1)\sin x$

$= \sin(1)\left(1-\dfrac{x^2}{2!}+\cdots\right) + \cos(1)\left(x-\dfrac{x^3}{3!}+\cdots\right)$

$P_3(x) = \sin(1) + \cos(1)x - \dfrac{\sin(1)}{2}x^2 - \dfrac{\cos(1)}{6}x^3.$

37. $\cos(\sin x) = 1 - \dfrac{\left(x-\dfrac{x^3}{3!}+\cdots\right)^2}{2!} + \dfrac{(x-\cdots)^4}{4!} - \cdots$

$= 1 - \dfrac{1}{2}\left(x^2 - \dfrac{x^4}{3}+\cdots\right) + \dfrac{x^4}{24}+\cdots$

$P_4(x) = 1 - \dfrac{1}{2}x^2 + \dfrac{5}{24}x^4.$

38. $\sqrt{1+\sin x} = 1 + \dfrac{1}{2}\sin x + \dfrac{\left(\dfrac{1}{2}\right)\left(-\dfrac{1}{2}\right)}{2!}(\sin x)^2$

$+ \dfrac{\left(\dfrac{1}{2}\right)\left(-\dfrac{1}{2}\right)\left(-\dfrac{3}{2}\right)}{3!}(\sin x)^3$

$+ \dfrac{\left(\dfrac{1}{2}\right)\left(-\dfrac{1}{2}\right)\left(-\dfrac{3}{2}\right)\left(-\dfrac{5}{2}\right)}{4!}(\sin x)^4 + \cdots$

$= 1 + \dfrac{1}{2}\left(x-\dfrac{x^3}{6}+\cdots\right) - \dfrac{1}{8}\left(x-\dfrac{x^3}{6}+\cdots\right)^2$

$+ \dfrac{1}{16}(x-\cdots)^3 - \dfrac{5}{128}(x-\cdots)^4 + \cdots$

$= 1 + \dfrac{x}{2} - \dfrac{x^3}{12} - \dfrac{x^2}{8} + \dfrac{x^4}{24} + \dfrac{x^3}{16} - \dfrac{5x^4}{128} + \cdots$

$P_4(x) = 1 + \dfrac{x}{2} - \dfrac{x^2}{8} - \dfrac{x^3}{48} + \dfrac{x^4}{384}.$

39. The series $\sum_{n=0}^{\infty}\dfrac{(-1)^n x^n}{(2n)!}$ is the Maclaurin series for $\cos x$ with x^2 replaced by x. For $x > 0$ the series therefore represents $\cos\sqrt{x}$. For $x < 0$, the series is $\sum_{n=0}^{\infty}\dfrac{|x|^n}{(2n)!}$, which is the Maclaurin series for $\cosh\sqrt{|x|}$. Thus the given series is the Maclaurin series for

$f(x) = \begin{cases}\cos\sqrt{x} & \text{if } x \geq 0 \\ \cosh\sqrt{|x|} & \text{if } x < 0.\end{cases}$

40. Since

$1 + \sum_{n=1}^{\infty}\dfrac{x^{2n}}{n^2} = \sum_{k=0}^{\infty}\dfrac{f^{(k)}(0)}{k!}x^k$

for x near 0, we have, for $n = 1, 2, 3, \ldots$

$f^{(2n)}(0) = \dfrac{(2n)!}{n^2}, \quad f^{(2n-1)}(0) = 0.$

41. $\sum_{n=0}^{\infty}x^n = \dfrac{1}{1-x}$

$\sum_{n=0}^{\infty}nx^{n-1} = \sum_{n=1}^{\infty}nx^{n-1} = \dfrac{1}{(1-x)^2}$

$\sum_{n=0}^{\infty}nx^n = \dfrac{x}{(1-x)^2}$

$\sum_{n=0}^{\infty}\dfrac{n+1}{\pi^n} = \dfrac{\dfrac{1}{\pi}}{\left(1-\dfrac{1}{\pi}\right)^2} + \dfrac{1}{1-\dfrac{1}{\pi}}$

$= \dfrac{\pi}{(\pi-1)^2} + \dfrac{\pi}{\pi-1} = \left(\dfrac{\pi}{\pi-1}\right)^2.$

42. $\sum_{n=0}^{\infty} nx^n = \dfrac{x}{(1-x)^2}$ as in Exercise 23

$\sum_{n=0}^{\infty} n^2 x^{n-1} = \dfrac{d}{dx}\dfrac{x}{(1-x)^2} = \dfrac{1+x}{(1-x)^3}$

$\sum_{n=0}^{\infty} n^2 x^n = \dfrac{x(1+x)}{(1-x)^3}$

$\sum_{n=0}^{\infty} \dfrac{n^2}{\pi^n} = \dfrac{\dfrac{1}{\pi}\left(1+\dfrac{1}{\pi}\right)}{\left(1-\dfrac{1}{\pi}\right)^3} = \dfrac{\pi(\pi+1)}{(\pi-1)^3}.$

43. $\sum_{n=1}^{\infty} \dfrac{x^n}{n} = -\ln(1-x)$

$\sum_{n=1}^{\infty} \dfrac{1}{ne^n} = -\ln\left(1-\dfrac{1}{e}\right) = 1 - \ln(e-1).$

44. $\sum_{n=1}^{\infty} \dfrac{(-1)^{n-1} x^{2n-1}}{(2n-1)!} = \sin x$

$\sum_{n=1}^{\infty} \dfrac{(-1)^n \pi^{2n-1}}{(2n-1)!} = -\sin\pi = 0$

$\sum_{n=2}^{\infty} \dfrac{(-1)^n \pi^{2n-4}}{(2n-1)!} = \dfrac{1}{\pi^3}\left(0 - \dfrac{(-1)\pi}{1!}\right) = \dfrac{1}{\pi^2}.$

45. $S(x) = \int_0^x \sin(t^2)\,dt$

$= \int_0^x \left(t^2 - \dfrac{t^6}{3!} + \cdots\right) dt$

$= \dfrac{x^3}{3} - \dfrac{x^7}{7 \cdot 3!} + \cdots$

$\lim_{x\to 0} \dfrac{x^3 - 3S(x)}{x^7} = \lim_{x\to 0} \dfrac{x^3 - x^3 + \dfrac{x^7}{14} - \cdots}{x^7} = \dfrac{1}{14}.$

46. $\lim_{x\to 0} \dfrac{(x-\tan^{-1}x)(e^{2x}-1)}{2x^2 - 1 + \cos(2x)}$

$= \lim_{x\to 0} \dfrac{\left(x - x + \dfrac{x^3}{3} - \dfrac{x^5}{5} + \cdots\right)\left(2x + \dfrac{4x^2}{2!} + \cdots\right)}{2x^2 - 1 + 1 - \dfrac{4x^2}{2!} + \dfrac{16x^4}{4!} - \cdots}$

$= \lim_{x\to 0} \dfrac{x^4\left(\dfrac{2}{3} + \cdots\right)}{x^4\left(\dfrac{2}{3} + \cdots\right)} = 1.$

47. $\int_0^{1/2} e^{-x^4}\,dx = \int_0^{1/2} \sum_{n=0}^{\infty} \dfrac{(-x^4)^n}{n!}\,dx$

$= \sum_{n=0}^{\infty} \dfrac{(-1)^n x^{4n+1}}{(4n+1)n!}\bigg|_0^{1/2}$

$= \sum_{n=0}^{\infty} \dfrac{(-1)^n}{2^{4n+1}(4n+1)n!}.$

The series satisfies the conditions of the alternating series test, so if we truncate after the term for $n = k - 1$, then the error will satisfy

$$|\text{error}| \le \dfrac{1}{2^{4k+1}(4k+1)k!}.$$

This is less than 0.000005 if $2^{4k+1}(4k+1)k! > 200{,}000$, which happens if $k \ge 3$. Thus, rounded to five decimal places,

$\int_0^{1/2} e^{-x^4}\,dx \approx \dfrac{1}{2\cdot 1\cdot 1} - \dfrac{1}{32\cdot 5\cdot 1} + \dfrac{1}{512\cdot 9\cdot 2} \approx 0.49386.$

48. If $f(x) = \ln(\sin x)$, then calculation of successive derivatives leads to

$$f^{(5)}(x) = 24\csc^4 x \cot x - 8\csc^2 \cot x.$$

Observe that $1.5 < \pi/2 \approx 1.5708$, that $\csc x \ge 1$ and $\cot x \ge 0$, and that both functions are decreasing on that interval. Thus

$$|f^{(5)}(x)| \le 24\csc^4(1.5)\cot(1.5) \le 2$$

for $1.5 \le x \le \pi/2$. Therefore, the error in the approximation

$$\ln(\sin 1.5) \approx P_4(x),$$

where P_4 is the 4th degree Taylor polynomial for $f(x)$ about $x = \pi/2$, satisfies

$$|\text{error}| \le \dfrac{2}{5!}\left|1.5 - \dfrac{\pi}{2}\right|^5 \le 3 \times 10^{-8}.$$

49. The Fourier sine series of $f(t) = \pi - t$ on $[0, \pi]$ has coefficients

$$b_n = \dfrac{2}{\pi}\int_0^{\pi}(\pi - t)\sin(nt)\,dt = \dfrac{2}{n}.$$

The series is $\sum_{n=1}^{\infty} \dfrac{2}{n}\sin(nt).$

50. $f(t) = \begin{cases} 1 & \text{if } -\pi < t \leq 0 \\ t & \text{if } 0 < t \leq \pi \end{cases}$ has period 2π. Its Fourier coefficients are

$$\frac{a_0}{2} = \frac{1}{2\pi} \int_{-\pi}^{\pi} f(t)\,dt$$

$$= \frac{1}{2\pi}\left[\int_{-\pi}^{0} dt + \int_{0}^{\pi} t\,dt\right] = \frac{1}{2} + \frac{\pi}{4}$$

$$a_n = \frac{1}{\pi}\left[\int_{-\pi}^{0} \cos(nt)\,dt + \int_{0}^{\pi} t\cos(nt)\,dt\right]$$

$$= \frac{1}{\pi}\int_{0}^{\pi}(1+t)\cos(nt)\,dt$$

$$= \frac{(-1)^n - 1}{\pi n^2} = \begin{cases} -2/(\pi n^2) & \text{if } n \text{ is odd} \\ 0 & \text{if } n \text{ is even} \end{cases}$$

$$b_n = \frac{1}{\pi}\left[\int_{-\pi}^{0}\sin(nt)\,dt + \int_{0}^{\pi} t\sin(nt)\,dt\right]$$

$$= \frac{1}{\pi}\int_{0}^{\pi}(t-1)\sin(nt)\,dt$$

$$= -\frac{1 + (-1)^n(\pi - 1)}{\pi n} = \begin{cases} (\pi - 2/(\pi n) & \text{if } n \text{ is odd} \\ -(1/n) & \text{if } n \text{ is even} \end{cases}$$

The required Fourier series is, therefore,

$$\frac{2+\pi}{4}$$
$$-\sum_{n=1}^{\infty}\left[\frac{2\cos((2n-1)t)}{\pi(2n-1)^2} + \frac{(2-\pi)\sin((2n-1)t)}{\pi(2n-1)} + \frac{\sin(2nt)}{2n}\right].$$

Challenging Problems 9 (page 594)

1. If $a_n > 0$ and $\dfrac{a_{n+1}}{a_n} > \dfrac{n}{n+1}$ for all n, then

$$\frac{a_2}{a_1} > \frac{1}{2} \;\Rightarrow\; a_2 > \frac{a_1}{2}$$
$$\frac{a_3}{a_2} > \frac{2}{3} \;\Rightarrow\; a_3 > \frac{2a_2}{3} > \frac{a_1}{3}$$
$$\vdots$$
$$\frac{a_n}{a_{n-1}} > \frac{n-1}{n} \;\Rightarrow\; a_n > \frac{a_1}{n}.$$

(This can be verified by induction.)
Therefore $\sum_{n=1}^{\infty} a_n$ diverges by comparison with the harmonic series $\sum_{n=1}^{\infty} \frac{1}{n}$.

2. a) If $s_n = \sum_{k=1}^{n} v_k$ for $n \geq 1$, and $s_0 = 0$, then $v_k = s_k - s_{k-1}$ for $k \geq 1$, and

$$\sum_{k=1}^{n} u_k v_k = \sum_{k=1}^{n} u_k s_k - \sum_{k=1}^{n} u_k s_{k-1}.$$

In the second sum on the right replace k with $k+1$:

$$\sum_{k=1}^{n} u_k v_k = \sum_{k=1}^{n} u_k s_k - \sum_{k=0}^{n-1} u_{k+1} s_k$$

$$= \sum_{k=1}^{n}(u_k - u_{k+1})s_k - u_1 s_0 + u_{n+1} s_n$$

$$= u_{n+1} s_n + \sum_{k=1}^{n}(u_k - u_{k+1})s_k.$$

b) If $\{u_n\}$ is positive and decreasing, and $\lim_{n\to\infty} u_n = 0$, then

$$\sum_{k=1}^{n}(u_k - u_{k+1}) = u_1 - u_2 + u_2 - u_3 + \cdots + u_n - u_{n+1}$$

$$= u_1 - u_{n+1} \to u_1 \text{ as } n \to \infty.$$

Thus $\sum_{k=1}^{n}(u_k - u_{k+1})$ is a convergent, positive, telescoping series.

If the partial sums s_n of $\{v_n\}$ are bounded, say $|s_n| \leq K$ for all n, then

$$|(u_n - u_{n+1})s_n| \leq K(u_n - u_{n+1}),$$

so $\sum_{n=1}^{\infty}(u_n - u_{n+1})s_n$ is absolutely convergent (and therefore convergent) by the comparison test. Therefore, by part (a),

$$\sum_{k=1}^{\infty} u_k v_k = \lim_{n\to\infty}\left(u_{n+1}s_n + \sum_{k=1}^{n}(u_k - u_{k+1})s_k\right)$$

$$= \sum_{k=1}^{\infty}(u_k - u_{k+1})s_k$$

converges.

3. If $x = m\pi$ for some integer m, then all the terms of the series $\sum_{n=1}^{\infty}(1/n)\sin(nx)$ are 0, so the series converges to 0.

If $x \neq m\pi$ for any integer m, then $\sin(x/2) \neq 0$. Using the addition formulas we obtain

$$\sin(nx)\sin(x/2) = \frac{1}{2}\left[\cos\left((n - \tfrac{1}{2})x\right) - \cos\left((n + \tfrac{1}{2})x\right)\right].$$

Therefore, using the telescoping property of these terms,

$$\sum_{n=1}^{N}\sin(nx) = \sum_{n=1}^{N}\frac{\left[\cos\left((n-\tfrac{1}{2})x\right) - \cos\left((n+\tfrac{1}{2})x\right)\right]}{2\sin(x/2)}$$

$$= \frac{\cos(x/2) - \cos\left((N+\tfrac{1}{2})x\right)}{2\sin(x/2)}.$$

Therefore, the partial sums of $\sum_{n=1}^{\infty} \sin(nx)$ are bounded. Since the sequence $\{1/n\}$ is positive, decreasing, and has limit 0, part (b) of Problem 2 shows that $\sum_{n=1}^{\infty} \sin(nx)/n$ converges in this case too. Therefore the series converges for all x.

4. Let a_n be the nth integer that has no zeros in its decimal representation. The number of such integers that have m digits is 9^m. (There are nine possible choices for each of the m digits.) Also, each such m-digit number is greater than 10^{m-1} (the smallest m-digit number). Therefore the sum of all the terms $1/a_n$ for which a_n has m digits is less than $9^m/(10^{m-1})$. Therefore,

$$\sum_{n=1}^{\infty} \frac{1}{a_n} < 9 \sum_{m=1}^{\infty} \left(\frac{9}{10}\right)^{m-1} = 90.$$

5. $\int_{k-1/2}^{k+1/2} f(x)\,dx - f(k) = \frac{f''(c)}{24}$, for some c in the interval $[k - \tfrac{1}{2}, k + \tfrac{1}{2}]$.

a) By the Mean-Value Theorem,

$$f'\left(k + \tfrac{3}{2}\right) - f'\left(k + \tfrac{1}{2}\right) = \left(\tfrac{3}{2} - \tfrac{1}{2}\right) f''(u) = f''(u)$$

for some u in $[k + \tfrac{1}{2}, k + \tfrac{3}{2}]$. Similarly,

$$f'\left(k - \tfrac{1}{2}\right) - f'\left(k - \tfrac{3}{2}\right) = \left(-\tfrac{1}{2} + \tfrac{3}{2}\right) f''(v) = f''(v)$$

for some v in $[k - \tfrac{3}{2}, k - \tfrac{1}{2}]$. Since f'' is decreasing and $v \le c \le u$, we have $f''(u) \le f''(c) \le f''(v)$, and so

$$f'\left(k + \tfrac{3}{2}\right) - f'\left(k + \tfrac{1}{2}\right) \le f''(c) \le f'\left(k - \tfrac{1}{2}\right) - f'\left(k - \tfrac{3}{2}\right).$$

b) If f'' is decreasing, $\int_{N+\tfrac{1}{2}}^{\infty} f(x)\,dx$ converges, and $f'(x) \to 0$ as $x \to \infty$, then

$$\sum_{n=N+1}^{\infty} f(n) - \int_{N+\tfrac{1}{2}}^{\infty} f(x)\,dx$$

$$= \sum_{n=N+1}^{\infty} \left(f(n) - \int_{n-\tfrac{1}{2}}^{n+\tfrac{1}{2}} f(x)\,dx\right)$$

$$= -\frac{1}{24} \sum_{n=N+1}^{\infty} f''(c_n),$$

for some numbers c_n in $[n - \tfrac{1}{2}, n + \tfrac{1}{2}]$. Using the result of part (a), we see that

$$\sum_{n=N+1}^{\infty} \left[f'(n + \tfrac{3}{2}) - f'(n + \tfrac{1}{2}) \right] \le \sum_{n=N+1}^{\infty} f''(c_n)$$

$$\le \sum_{n=N+1}^{\infty} \left[f'(n - \tfrac{1}{2}) - f'(n - \tfrac{3}{2}) \right]$$

$$-f'(N + \tfrac{3}{2}) \le \sum_{n=N+1}^{\infty} f''(c_n) \le -f'(N - \tfrac{1}{2})$$

$$\frac{f'(N - \tfrac{1}{2})}{24} \le \sum_{n=N+1}^{\infty} f(n) - \int_{N+\tfrac{1}{2}}^{\infty} f(x)\,dx \le \frac{f'(N + \tfrac{3}{2})}{24}$$

c) Let $f(x) = 1/x^2$. Then $f'(x) = -2/x^3 \to 0$ as $x \to \infty$, $f''(x) = 6/x^4$ is decreasing, and

$$\int_{N+\tfrac{1}{2}}^{\infty} f(x)\,dx = \int_{N+\tfrac{1}{2}}^{\infty} \frac{dx}{x^2} = \frac{1}{N + \tfrac{1}{2}}$$

converges. From part (b) we obtain

$$\left| \sum_{n=N+1}^{\infty} \frac{1}{n^2} - \frac{1}{N + \tfrac{1}{2}} \right| \le \frac{1}{12 \left(N - \tfrac{1}{2}\right)^3}.$$

The right side is less than 0.001 if $N = 5$. Therefore

$$\sum_{n=1}^{\infty} \frac{1}{n^2} = \sum_{n=1}^{5} \frac{1}{n^2} + \frac{1}{5.5} \approx 1.6454$$

correct to within 0.001.

6. a) Since $e = \sum_{j=0}^{\infty} \dfrac{1}{j!}$, we have

$$0 < e - \sum_{j=0}^{n} \frac{1}{j!} = \sum_{j=n+1}^{\infty} \frac{1}{j!}$$

$$= \frac{1}{(n+1)!} \left(1 + \frac{1}{n+2} + \frac{1}{(n+2)(n+3)} + \cdots \right)$$

$$\le \frac{1}{(n+1)!} \left(1 + \frac{1}{n+2} + \frac{1}{(n+2)^2} + \cdots \right)$$

$$= \frac{1}{(n+1)!} \cdot \frac{1}{1 - \dfrac{1}{n+2}} = \frac{n+2}{(n+1)!(n+1)} < \frac{1}{n!n}.$$

The last inequality follows from $\dfrac{n+2}{(n+1)^2} < \dfrac{1}{n}$, that is, $n^2 + 2n < n^2 + 2n + 1$.

b) Suppose e is rational, say $e = M/N$ where M and N are positive integers. Then $N!e$ is an integer and $N! \sum_{j=0}^{N} (1/j!)$ is an integer (since each $j!$ is a factor of $N!$). Therefore the number

$$Q = N! \left(e - \sum_{j=0}^{N} \frac{1}{j!} \right)$$

is a difference of two integers and so is an integer.

c) By part (a), $0 < Q < \dfrac{1}{N} \le 1$. By part (b), Q is an integer. This is not possible; there are no integers between 0 and 1. Therefore e cannot be rational.

7. Let $f(x) = \displaystyle\sum_{k=0}^{\infty} a_k x^{2k+1}$, where $a_k = \dfrac{2^{2k} k!}{(2k+1)!}$.

   a) Since
   $$\lim_{k\to\infty} \left|\dfrac{a_{k+1} x^{2k+3}}{a_k x^{2k+1}}\right|$$
   $$= |x|^2 \lim_{k\to\infty} \dfrac{2^{2k+2}}{2^{2k}} \cdot \dfrac{(k+1)!}{k!} \cdot \dfrac{(2k+1)!}{(2k+3)!}$$
   $$= |x|^2 \lim_{k\to\infty} \dfrac{4k+4}{(2k+3)(2k+2)} = 0$$
   for all x, the series for $f(x)$ converges for all x. Its radius of convergence is infinite.

   b) $f'(x) = \displaystyle\sum_{k=0}^{\infty} \dfrac{2^{2k} k!}{(2k+1)!}(2k+1) x^{2k} = 1 + \sum_{k=1}^{\infty} \dfrac{2^{2k} k!}{(2k)!} x^{2k}$
   $$1 + 2x f(x) = 1 + \sum_{k=0}^{\infty} \dfrac{2^{2k+1} k!}{(2k+1)!} x^{2k+2}$$
   (replace k with $k-1$)
   $$= 1 + \sum_{k=1}^{\infty} \dfrac{2^{2k-1}(k-1)!}{(2k-1)!} x^{2k}$$
   $$= 1 + \sum_{k=1}^{\infty} \dfrac{2^{2k} k!}{(2k)!} x^{2k} = f'(x).$$

   c) $\dfrac{d}{dx}\left(e^{-x^2} f(x)\right) = e^{-x^2}\left(f'(x) - 2x f(x)\right) = e^{-x^2}$.

   d) Since $f(0) = 0$, we have
   $$e^{-x^2} f(x) - f(0) = \int_0^x \dfrac{d}{dt}\left(e^{-t^2} f(t)\right) dt = \int_0^x e^{-t^2} dt$$
   $$f(x) = e^{x^2} \int_0^x e^{-t^2} dt.$$

8. Let f be a polynomial and let
   $$g(x) = \sum_{j=0}^{\infty} (-1)^j f^{(2j)}(x).$$
   This "series" is really just a polynomial since sufficiently high derivatives of f are all identically zero.

   a) By replacing j with $j-1$, observe that
   $$g''(x) = \sum_{j=0}^{\infty} (-1)^j f^{(2j+2)}(x)$$
   $$= \sum_{j=1}^{\infty} (-1)^{j-1} f^{(2j)}(x) = -\bigl(g(x) - f(x)\bigr).$$

Also
$$\dfrac{d}{dx}\bigl(g'(x) \sin x - g(x) \cos x\bigr)$$
$$= g''(x) \sin x + g'(x) \cos x - g'(x) \cos x + g(x) \sin x$$
$$= \bigl(g''(x) + g(x)\bigr) \sin x = f(x) \sin x.$$

Thus
$$\int_0^{\pi} f(x) \sin x \, dx = \bigl(g'(x) \sin x - g(x) \cos x\bigr)\Big|_0^{\pi} = g(\pi) + g(0).$$

   b) Suppose that $\pi = m/n$, where m and n are positive integers. Since $\lim_{k\to\infty} x^k/k! = 0$ for any x, there exists an integer k such that $(\pi m)^k/k! < 1/2$. Let
   $$f(x) = \dfrac{x^k (m - nx)^k}{k!} = \dfrac{1}{k!} \sum_{j=0}^{k} \binom{k}{j} m^{k-j} (-n)^j x^{j+k}.$$
   The sum is just the binomial expansion. For $0 < x < \pi = m/n$ we have
   $$0 < f(x) < \dfrac{\pi^k m^k}{k!} < \dfrac{1}{2}.$$
   Thus $0 < \int_0^{\pi} f(x) \sin x \, dx < \dfrac{1}{2} \int_0^{\pi} \sin x \, dx = 1$, and so $0 < g(\pi) + g(0) < 1$.

   c) $f^{(i)}(x) = \dfrac{1}{k!} \displaystyle\sum_{j=0}^{k} \binom{k}{j} m^{k-j} (-n)^j$
   $$\times (j+k)(j+k-1) \cdots (j+k-i+1) x^{j+k-i}$$
   $$= \dfrac{1}{k!} \sum_{j=0}^{k} \binom{k}{j} m^{k-j} (-n)^j \dfrac{(j+k)!}{(j+k-i)!} x^{j+k-i}.$$

   d) Evidently $f^{(i)}(0) = 0$ if $i < k$ or if $i > 2k$. If $k \le i \le 2k$, the only term in the sum for $f^{(i)}(0)$ that is not zero is the term for which $j = i - k$. This term is the constant
   $$\dfrac{1}{k!} \binom{k}{i-k} m^{k-j} (-n)^j \dfrac{i!}{0!}.$$
   This constant is an integer because the binomial coefficient $\binom{k}{i-k}$ is an integer and $i!/k!$ is an integer. (The other factors are also integers.) Hence $f^{(i)}(0)$ is an integer, and so $g(0)$ is an integer.

375

e) Observe that $f(\pi - x) = f((m/n) - x) = f(x)$ for all x. Therefore $f^{(i)}(\pi)$ is an integer (for each i), and so $g(\pi)$ is an integer. Thus $g(\pi) + g(0)$ is an integer, which contradicts the conclusion of part (b). (There is no integer between 0 and 1.) Therefore, π cannot be rational.

9. Let $x > 0$, and let

$$I_k = \int_0^x t^k e^{-1/t}\, dt$$

$$U = t^{k+2} \qquad dV = \frac{1}{t^2} e^{-1/t}\, dt$$
$$dU = (k+2) t^{k+1}\, dt \qquad V = e^{-1/t}$$

$$= t^{k+2} e^{-1/t} \Big|_0^x - (k+2) \int_0^x t^{k+1} e^{-1/t}\, dt$$

$$I_k = x^{k+2} e^{-1/x} - (k+2) I_{k+1}.$$

Therefore,

$$\int_0^x e^{-1/t}\, dt = I_0 = x^2 e^{-1/x} - 2 I_1$$
$$= x^2 e^{-1/x} - 2\big(x^3 e^{-1/x} - 3 I_2\big)$$
$$= e^{-1/x}[x^2 - 2! x^3] + 3!\big(x^4 e^{-1/x} - 4 I_3\big)$$
$$= e^{-1/x}[x^2 - 2! x^3 + 3! x^4] - 4!\big(x^5 e^{-1/x} - 5 I_4\big)$$
$$\vdots$$
$$= e^{-1/x} \sum_{n=2}^N (-1)^n (n-1)! x^n$$
$$\quad + (-1)^{N+1} N! \int_0^x t^{N-1} e^{-1/t}\, dt.$$

The Maclaurin series for $e^{-1/t}$ does not exist. The function is not defined at $t = 0$.
For $x = 0.1$ and $N = 5$, the approximation

$$I = \int_0^{0.1} e^{-1/t}\, dt \approx e^{-10} \sum_{n=2}^5 (-1)^n (n-1)! (0.1)^n$$
$$= e^{-10}\big((0.1)^2 - 2(0.1)^3 + 6(0.1)^4 - 24(0.1)^5\big)$$
$$\approx 0.00836 e^{-10}$$

has error E given by

$$E = (-1)^6 5! \int_0^{0.1} t^4 e^{-1/t}\, dt.$$

Since $e^{-1/t} \leq e^{-10}$ for $0 \leq t \leq 0.1$, we have

$$|E| \leq 120 e^{-10} \int_0^{0.1} t^4\, dt \approx 2.4 \times 10^{-4} e^{-10},$$

which is about 3% of the size of I.

For $N = 10$, the error estimate is

$$|E| \leq 10! e^{-10} \int_0^{0.1} t^9\, dt \approx 3.6 \times 10^{-5} e^{-10},$$

which is about 0.4% of the size of I.
For $N = 20$, the error estimate is

$$|E| \leq 20! e^{-10} \int_0^{0.1} t^{19}\, dt \approx 1.2 \times 10^{-3} e^{-10},$$

which is about 15% of the size of I.
Observe, therefore, that the sum for $N = 10$ does a better job of approximating I than those for $N = 5$ or $N = 20$.

CHAPTER 10. VECTORS AND COORDINATE GEOMETRY IN 3-SPACE

Section 10.1 Analytic Geometry in Three Dimensions (page 603)

1. The distance between $(0, 0, 0)$ and $(2, -1, -2)$ is
$$\sqrt{2^2 + (-1)^2 + (-2)^2} = 3 \text{ units.}$$

2. The distance between $(-1, -1, -1)$ and $(1, 1, 1)$ is
$$\sqrt{(1+1)^2 + (1+1)^2 + (1+1)^2} = 2\sqrt{3} \text{ units.}$$

3. The distance between $(1, 1, 0)$ and $(0, 2, -2)$ is
$$\sqrt{(0-1)^2 + (2-1)^2 + (-2-0)^2} = \sqrt{6} \text{ units.}$$

4. The distance between $(3, 8, -1)$ and $(-2, 3, -6)$ is
$$\sqrt{(-2-3)^2 + (3-8)^2 + (-6+1)^2} = 5\sqrt{3} \text{ units.}$$

5. a) The shortest distance from (x, y, z) to the xy-plane is $|z|$ units.

   b) The shortest distance from (x, y, z) to the x-axis is $\sqrt{y^2 + z^2}$ units.

6. If $A = (1, 2, 3)$, $B = (4, 0, 5)$, and $C = (3, 6, 4)$, then
$$|AB| = \sqrt{3^2 + (-2)^2 + 2^2} = \sqrt{17}$$
$$|AC| = \sqrt{2^2 + 4^2 + 1^2} = \sqrt{21}$$
$$|BC| = \sqrt{(-1)^2 + 6^2 + (-1)^2} = \sqrt{38}.$$

   Since $|AB|^2 + |AC|^2 = 17 + 21 = 38 = |BC|^2$, the triangle ABC has a right angle at A.

7. If $A = (2, -1, -1)$, $B = (0, 1, -2)$, and $C = (1, -3, 1)$, then
$$c = |AB| = \sqrt{(0-2)^2 + (1+1)^2 + (-2+1)^2} = 3$$
$$b = |AC| = \sqrt{(1-2)^2 + (-3+1)^2 + (1+1)^2} = 3$$
$$a = |BC| = \sqrt{(1-0)^2 + (-3-1)^2 + (1+2)^2} = \sqrt{26}.$$

   By the Cosine Law,
$$a^2 = b^2 + c^2 - 2bc \cos \angle A$$
$$26 = 9 + 9 - 18 \cos \angle A$$
$$\angle A = \cos^{-1} \frac{26 - 18}{-18} \approx 116.4°.$$

8. If $A = (1, 2, 3)$, $B = (1, 3, 4)$, and $C = (0, 3, 3)$, then
$$|AB| = \sqrt{(1-1)^2 + (3-2)^2 + (4-3)^2} = \sqrt{2}$$
$$|AC| = \sqrt{(0-1)^2 + (3-2)^2 + (3-3)^2} = \sqrt{2}$$
$$|BC| = \sqrt{(0-1)^2 + (3-3)^2 + (3-4)^2} = \sqrt{2}.$$

   All three sides being equal, the triangle is equilateral.

9. If $A = (1, 1, 0)$, $B = (1, 0, 1)$, and $C = (0, 1, 1)$, then
$$|AB| = |AC| = |BC| = \sqrt{2}.$$

   Thus the triangle ABC is equilateral with sides $\sqrt{2}$. Its area is, therefore,
$$\frac{1}{2} \times \sqrt{2} \times \sqrt{2 - \frac{1}{2}} = \frac{\sqrt{3}}{2} \text{ sq. units.}$$

10. The distance from the origin to $(1, 1, 1, \ldots, 1)$ in $\mathbb{R}^n$ is
$$\sqrt{1^2 + 1^2 + 1^2 + \cdots + 1} = \sqrt{n} \text{ units.}$$

11. The point on the x_1-axis closest to $(1, 1, 1, \ldots, 1)$ is $(1, 0, 0, \ldots, 0)$. The distance between these points is
$$\sqrt{0^2 + 1^2 + 1^2 + \cdots + 1^2} = \sqrt{n-1} \text{ units.}$$

12. $z = 2$ is a plane, perpendicular to the z-axis at $(0, 0, 2)$.

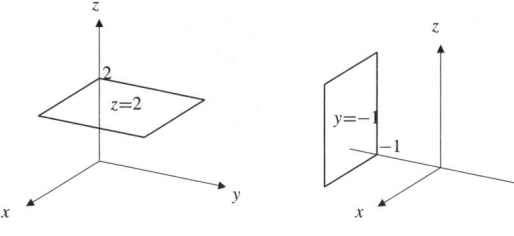

Fig. 10.1.12   Fig. 10.1.13

13. $y \geq -1$ is the half-space consisting of all points on the plane $y = -1$ (which is perpendicular to the y-axis at $(0, -1, 0)$) and all points on the same side of that plane as the origin.

14. $z = x$ is a plane containing the y-axis and making $45°$ angles with the positive directions of the x- and z-axes.

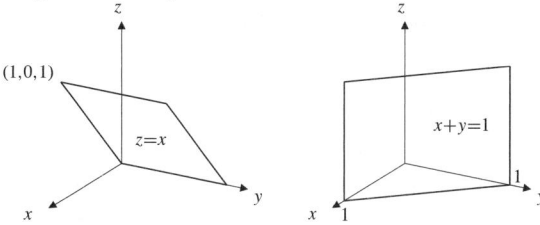

Fig. 10.1.14   Fig. 10.1.15

15. $x + y = 1$ is a vertical plane (parallel to the z-axis) passing through the points $(1, 0, 0)$ and $(0, 1, 0)$.

16. $x^2 + y^2 + z^2 = 4$ is a sphere centred at the origin and having radius 2 (i.e., all points at distance 2 from the origin).

17. $(x - 1)^2 + (y + 2)^2 + (z - 3)^2 = 4$ is a sphere of radius 2 with centre at the point $(1, -2, 3)$.

18. $x^2 + y^2 + z^2 = 2z$ can be rewritten
$$x^2 + y^2 + (z - 1)^2 = 1,$$
and so it represents a sphere with radius 1 and centre at $(0, 0, 1)$. It is tangent to the xy-plane at the origin.

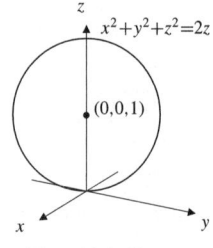

Fig. 10.1.18

19. $y^2 + z^2 \leq 4$ represents all points inside and on the circular cylinder of radius 2 with central axis along the x-axis (a solid cylinder).

20. $x^2 + z^2 = 4$ is a circular cylindrical surface of radius 2 with axis along the y-axis.

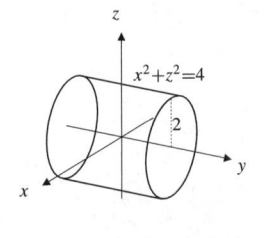

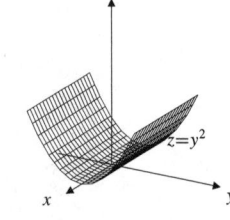

Fig. 10.1.20   Fig. 10.1.21

21. $z = y^2$ is a "parabolic cylinder" — a surface all of whose cross-sections in planes perpendicular to the x-axis are parabolas.

22. $z \geq \sqrt{x^2 + y^2}$ represents every point whose distance above the xy-plane is not less than its horizontal distance from the z-axis. It therefore consists of all points inside and on a circular cone with axis along the positive z-axis, vertex at the origin, and semi-vertical angle $45°$.

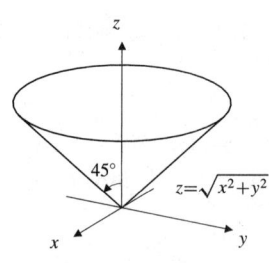

Fig. 10.1.22

23. $x + 2y + 3z = 6$ represents the plane that intersects the coordinate axes at the three points $(6, 0, 0)$, $(0, 3, 0)$, and $(0, 0, 2)$. Only the part of the plane in the first octant is shown in the figure.

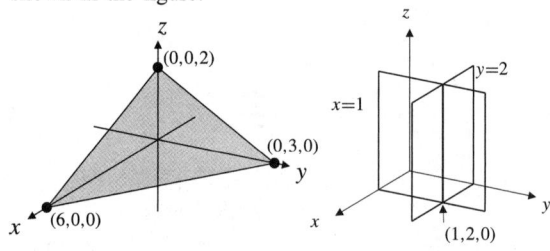

Fig. 10.1.23   Fig. 10.1.24

24. $\begin{cases} x = 1 \\ y = 2 \end{cases}$ represents the vertical straight line in which the plane $x = 1$ intersects the plane $y = 2$.

25. $\begin{cases} x = 1 \\ y = z \end{cases}$ is the straight line in which the plane $z = 1$ intersects the plane $y = z$. It passes through the points $(1, 0, 0)$ and $(1, 1, 1)$.

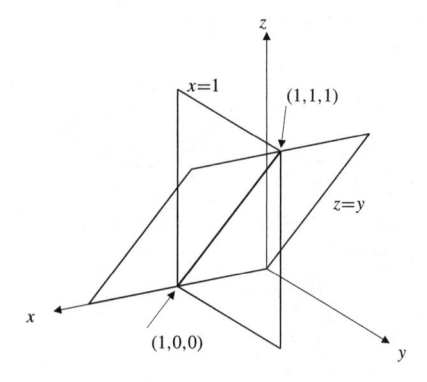

Fig. 10.1.25

26. $\begin{cases} x^2 + y^2 + z^2 = 4 \\ z = 1 \end{cases}$ is the circle in which the horizontal plane $z = 1$ intersects the sphere of radius 2 centred at the origin. The circle has centre $(0, 0, 1)$ and radius $\sqrt{4 - 1} = \sqrt{3}$.

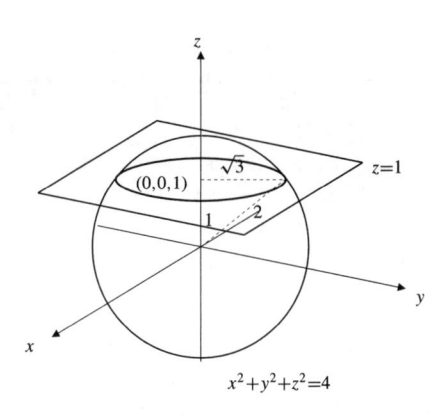

INSTRUCTOR'S SOLUTIONS MANUAL   SECTION 10.1 (PAGE 603)

Fig. 10.1.26

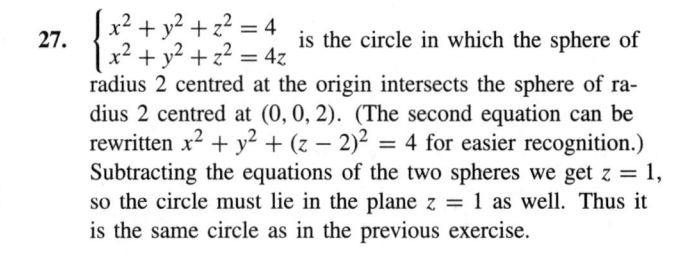

27. $\begin{cases} x^2 + y^2 + z^2 = 4 \\ x^2 + y^2 + z^2 = 4z \end{cases}$ is the circle in which the sphere of radius 2 centred at the origin intersects the sphere of radius 2 centred at $(0, 0, 2)$. (The second equation can be rewritten $x^2 + y^2 + (z-2)^2 = 4$ for easier recognition.) Subtracting the equations of the two spheres we get $z = 1$, so the circle must lie in the plane $z = 1$ as well. Thus it is the same circle as in the previous exercise.

28. $\begin{cases} x^2 + y^2 + z^2 = 4 \\ x^2 + z^2 = 1 \end{cases}$ represents the two circles in which the cylinder $x^2 + z^2 - 1$ intersects the sphere $x^2 + y^2 + z^2 = 4$. Subtracting the two equations, we get $y^2 = 3$. Thus, one circle lies in the plane $y = \sqrt{3}$ and has centre $(0, \sqrt{3}, 0)$ and the other lies in the plane $y = -\sqrt{3}$ and has centre $(0, -\sqrt{3}, 0)$. Both circles have radius 1.

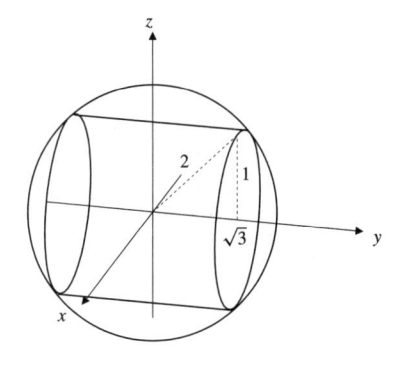

Fig. 10.1.28

29. $\begin{cases} x^2 + y^2 = 1 \\ z = x \end{cases}$ is the ellipse in which the slanted plane $z = x$ intersects the vertical cylinder $x^2 + y^2 = 1$.

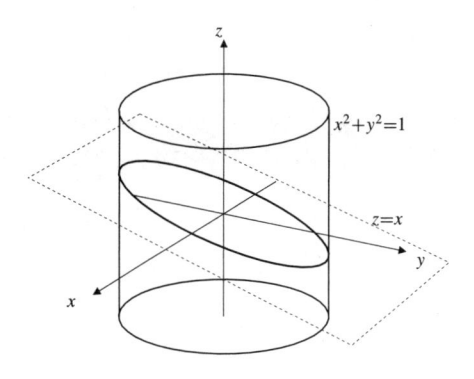

Fig. 10.1.29

30. $\begin{cases} y \geq x \\ z \leq y \end{cases}$ is the quarter-space consisting of all points lying on or on the same side of the planes $y = x$ and $z = y$ as does the point $(0, 1, 0)$.

31. $\begin{cases} x^2 + y^2 \leq 1 \\ z \geq y \end{cases}$ represents all points which are inside or on the vertical cylinder $x^2 + y^2 = 1$, and are also above or on the plane $z = y$.

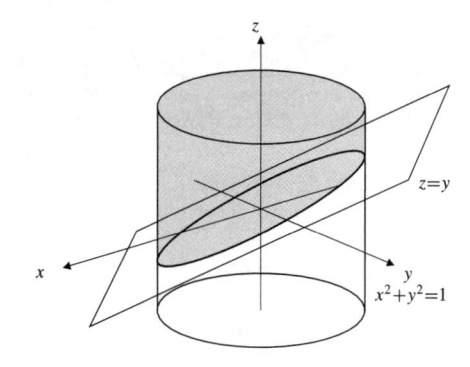

Fig. 10.1.31

32. $\begin{cases} x^2 + y^2 + z^2 \leq 1 \\ \sqrt{x^2 + y^2} \leq z \end{cases}$ represents all points which are inside or on the sphere of radius 1 centred at the origin and which are also inside or on the upper half of the circular cone with axis along the z-axis, vertex at the origin, and semi-vertical angle $45°$.

379

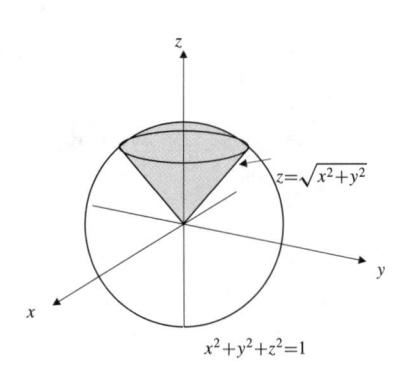

Fig. 10.1.32

33. $S = \{(x, y) : 0 < x^2 + y^2 < 1\}$
The boundary of S consists of the origin and all points on the circle $x^2 + y^2 = 1$. The interior of S is S, which is therefore open. S is bounded; all points in it are at distance less than 1 from the origin.

34. $S = \{(x, y) : x \geq 0, y < 0\}$
The boundary of S consists of points $(x, 0)$ where $x \geq 0$, and points $(0, y)$ where $y \leq 0$.
The interior of S consists of all points of S that are not on the y-axis, that is, all points (x, y) satisfying $x > 0$ and $y < 0$.
S is neither open nor closed; it contains some, but not all, of its boundary points.
S is not bounded; $(x, -1)$ belongs to S for $0 < x < \infty$.

35. $S = \{(x, y) : x + y = 1\}$
The boundary of S is S. The interior of S is the empty set. S is closed, but not bounded. There are points on the line $x + y = 1$ arbitrarily far away from the origin.

36. $S = \{(x, y) : |x| + |y| \leq 1\}$
The boundary of S consists of all points on the edges of the square with vertices $(\pm 1, 0)$ and $(0, \pm 1)$.
The interior of S consists of all points inside that square.
S is closed since it contains all its boundary points. It is bounded since all points in it are at distance not greater than 1 from the origin.

37. $S = \{(x, y, z) : 1 \leq x^2 + y^2 + z^2 \leq 4\}$
Boundary: the spheres of radii 1 and 2 centred at the origin.
Interior: the region between these spheres. S is closed.

38. $S = \{(x, y, z) : x \geq 0, y > 1, z < 2\}$
Boundary: the quarter planes $x = 0$, $(y \geq 1, z \leq 2)$, $y = 1$, $(x \geq 0, z \leq 2)$, and $z = 2$, $(x \geq 0, y \geq 1)$.
Interior: the set of points (x, y, z) such that $x > 0$, $y > 1$, $z < 2$.
S is neither open nor closed.

39. $S = \{(x, y, z) : (x - z)^2 + (y - z)^2 = 0\}$
The boundary of S is S, that is, the line $x = y = z$. The interior of S is empty. S is closed.

40. $S = \{(x, y, z) : x^2 + y^2 < 1, y + z > 2\}$
Boundary: the part of the cylinder $x^2 + y^2 = 1$ that lies on or above the plane $y + z = 2$ together with the part of that plane that lies inside the cylinder.
Interior: all points that are inside the cylinder $x^2 + y^2 = 1$ and above the plane $y + z = 2$. S is open.

Section 10.2 Vectors (page 613)

1. $A = (-1, 2)$, $B = (2, 0)$, $C = (1, -3)$, $D = (0, 4)$.

(a) $\overrightarrow{AB} = 3\mathbf{i} - 2\mathbf{j}$ (b) $\overrightarrow{BA} = -3\mathbf{i} + 2\mathbf{j}$

(c) $\overrightarrow{AC} = 2\mathbf{i} - 5\mathbf{j}$ (d) $\overrightarrow{BD} = -2\mathbf{i} + 4\mathbf{j}$

(e) $\overrightarrow{DA} = -\mathbf{i} - 2\mathbf{j}$ (f) $\overrightarrow{AB} - \overrightarrow{BC} = 4\mathbf{i} + \mathbf{j}$

(g) $\overrightarrow{AC} - 2\overrightarrow{AB} + 3\overrightarrow{CD} = -7\mathbf{i} + 20\mathbf{j}$

(h) $\dfrac{1}{3}(\overrightarrow{AB} + \overrightarrow{AC} + \overrightarrow{AD}) = 2\mathbf{i} - \dfrac{5}{3}\mathbf{j}$

2. $\mathbf{u} = \mathbf{i} - \mathbf{j}$
$\mathbf{v} = \mathbf{j} + 2\mathbf{k}$

a) $\mathbf{u} + \mathbf{v} = \mathbf{i} + 2\mathbf{k}$
$\mathbf{u} - \mathbf{v} = \mathbf{i} - 2\mathbf{j} - 2\mathbf{k}$
$2\mathbf{u} - 3\mathbf{v} = 2\mathbf{i} - 5\mathbf{j} - 6\mathbf{k}$

b) $|\mathbf{u}| = \sqrt{1+1} = \sqrt{2}$
$|\mathbf{v}| = \sqrt{1+4} = \sqrt{5}$

c) $\hat{\mathbf{u}} = \dfrac{1}{\sqrt{2}}(\mathbf{i} - \mathbf{j})$
$\hat{\mathbf{v}} = \dfrac{1}{\sqrt{5}}(\mathbf{j} + 2\mathbf{k})$

d) $\mathbf{u} \bullet \mathbf{v} = 0 - 1 + 0 = -1$

e) The angle between $\mathbf{u}$ and $\mathbf{v}$ is
$\cos^{-1} \dfrac{-1}{\sqrt{10}} \approx 108.4°$.

f) The scalar projection of $\mathbf{u}$ in the direction of $\mathbf{v}$ is
$\dfrac{\mathbf{u} \bullet \mathbf{v}}{|\mathbf{v}|} = \dfrac{-1}{\sqrt{5}}$.

g) The vector projection of $\mathbf{v}$ along $\mathbf{u}$ is
$\dfrac{(\mathbf{v} \bullet \mathbf{u})\mathbf{u}}{|\mathbf{u}|^2} = -\dfrac{1}{2}(\mathbf{i} - \mathbf{j})$.

3. $\mathbf{u} = 3\mathbf{i} + 4\mathbf{j} - 5\mathbf{k}$
$\mathbf{v} = 3\mathbf{i} - 4\mathbf{j} - 5\mathbf{k}$

a) $\mathbf{u} + \mathbf{v} = 6\mathbf{i} - 10\mathbf{k}$
$\mathbf{u} - \mathbf{v} = 8\mathbf{j}$
$2\mathbf{u} - 3\mathbf{v} = -3\mathbf{i} + 20\mathbf{j} + 5\mathbf{k}$

b) $|\mathbf{u}| = \sqrt{9 + 16 + 25} = 5\sqrt{2}$
$|\mathbf{v}| = \sqrt{9 + 16 + 25} = 5\sqrt{2}$

c) $\hat{\mathbf{u}} = \dfrac{1}{5\sqrt{2}}(3\mathbf{i} + 4\mathbf{j} - 5\mathbf{k})$
$\hat{\mathbf{v}} = \dfrac{1}{5\sqrt{2}}(3\mathbf{i} - 4\mathbf{j} - 5\mathbf{k})$

d) $\mathbf{u} \bullet \mathbf{v} = 9 - 16 + 25 = 18$

e) The angle between $\mathbf{u}$ and $\mathbf{v}$ is
$\cos^{-1}\dfrac{18}{50} \approx 68.9°$.

f) The scalar projection of $\mathbf{u}$ in the direction of $\mathbf{v}$ is
$\dfrac{\mathbf{u} \bullet \mathbf{v}}{|\mathbf{v}|} = \dfrac{18}{5\sqrt{2}}$.

g) The vector projection of $\mathbf{v}$ along $\mathbf{u}$ is
$\dfrac{(\mathbf{v} \bullet \mathbf{u})\mathbf{u}}{|\mathbf{u}|^2} = \dfrac{9}{25}(3\mathbf{i} + 4\mathbf{j} - 5\mathbf{k})$.

4. If $a = (-1, 1)$, $B = (2, 5)$ and $C = (10, -1)$, then $\overrightarrow{AB} = 3\mathbf{i} + 4\mathbf{j}$ and $\overrightarrow{BC} = 8\mathbf{i} - 6\mathbf{j}$. Since $\overrightarrow{AB} \bullet \overrightarrow{BC} = 0$, therefore, $\overrightarrow{AB} \perp \overrightarrow{BC}$. Hence, $\triangle ABC$ has a right angle at B.

5. Let the triangle be ABC. If M and N are the midpoints of AB and AC respectively, then $\overrightarrow{AM} = \frac{1}{2}\overrightarrow{AB}$, and $\overrightarrow{AN} = \frac{1}{2}\overrightarrow{AC}$. Thus

$$\overrightarrow{MN} = \overrightarrow{AN} - \overrightarrow{AM} = \dfrac{\overrightarrow{AC} - \overrightarrow{AB}}{2} = \dfrac{\overrightarrow{BC}}{2}.$$

Thus MN is parallel to and half as long as BC.

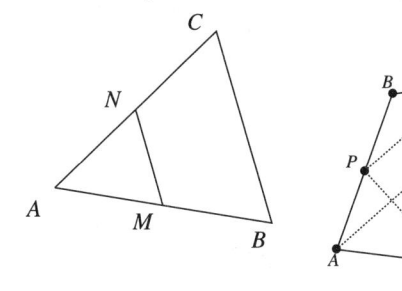

Fig. 10.2.5   Fig. 10.2.6

6. We have

$$\overrightarrow{PQ} = \overrightarrow{PB} + \overrightarrow{BQ} = \tfrac{1}{2}\overrightarrow{AB} + \tfrac{1}{2}\overrightarrow{BC} = \tfrac{1}{2}\overrightarrow{AC};$$
$$\overrightarrow{SR} = \overrightarrow{SD} + \overrightarrow{DR} = \tfrac{1}{2}\overrightarrow{AD} + \tfrac{1}{2}\overrightarrow{DC} = \tfrac{1}{2}\overrightarrow{AC}.$$

Therefore, $\overrightarrow{PQ} = \overrightarrow{SR}$. Similarly,

$$\overrightarrow{QR} = \overrightarrow{QC} + \overrightarrow{CR} = \tfrac{1}{2}\overrightarrow{BD};$$
$$\overrightarrow{PS} = \overrightarrow{PA} + \overrightarrow{AS} = \tfrac{1}{2}\overrightarrow{BD}.$$

Therefore, $\overrightarrow{QR} = \overrightarrow{PS}$. Hence, $PQRS$ is a parallelogram.

7. Let the parallelogram be $ABCO$. Take the origin at O. The position vector of the midpoint of OB is

$$\dfrac{\overrightarrow{OB}}{2} = \dfrac{\overrightarrow{OB} + \overrightarrow{CB}}{2} = \dfrac{\overrightarrow{OC} + \overrightarrow{OA}}{2}.$$

The position vector of the midpoint of CA is

$$\overrightarrow{OC} + \dfrac{\overrightarrow{CA}}{2} = \overrightarrow{OC} + \dfrac{\overrightarrow{OA} - \overrightarrow{OC}}{2}$$
$$= \dfrac{\overrightarrow{OC} + \overrightarrow{OA}}{2}.$$

Thus the midpoints of the two diagonals coincide, and the diagonals bisect each other.

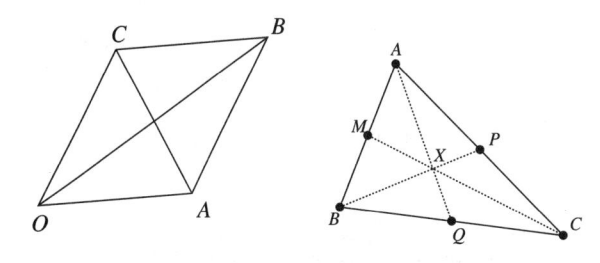

Fig. 10.2.7   Fig. 10.2.8

8. Let X be the point of intersection of the medians AQ and BP as shown. We must show that CX meets AB in the midpoint of AB. Note that $\overrightarrow{PX} = \alpha\overrightarrow{PB}$ and $\overrightarrow{QX} = \beta\overrightarrow{QA}$ for certain real numbers α and β. Then

$$\overrightarrow{CX} = \tfrac{1}{2}\overrightarrow{CB} + \beta\overrightarrow{QA} = \tfrac{1}{2}\overrightarrow{CB} + \beta\left(\tfrac{1}{2}\overrightarrow{CB} + \overrightarrow{BA}\right)$$
$$= \dfrac{1+\beta}{2}\overrightarrow{CB} + \beta\overrightarrow{BA};$$
$$\overrightarrow{CX} = \tfrac{1}{2}\overrightarrow{CA} + \alpha\overrightarrow{PB} = \tfrac{1}{2}\overrightarrow{CA} + \alpha\left(\tfrac{1}{2}\overrightarrow{CA} + \overrightarrow{AB}\right)$$
$$= \dfrac{1+\alpha}{2}\overrightarrow{CA} + \alpha\overrightarrow{AB}.$$

Thus,

$$\dfrac{1+\beta}{2}\overrightarrow{CB} + \beta\overrightarrow{BA} = \dfrac{1+\alpha}{2}\overrightarrow{CA} + \alpha\overrightarrow{AB}$$
$$(\beta + \alpha)\overrightarrow{BA} = \dfrac{1+\alpha}{2}\overrightarrow{CA} - \dfrac{1+\beta}{2}\overrightarrow{CB}$$
$$(\beta + \alpha)(\overrightarrow{CA} - \overrightarrow{CB}) = \dfrac{1+\alpha}{2}\overrightarrow{CA} - \dfrac{1+\beta}{2}\overrightarrow{CB}$$
$$\left(\beta + \alpha - \dfrac{1+\alpha}{2}\right)\overrightarrow{CA} = \left(\beta + \alpha - \dfrac{1+\beta}{2}\right)\overrightarrow{CB}.$$

381

Since $\overrightarrow{CA}$ is not parallel to $\overrightarrow{CB}$,

$$\beta + \alpha - \frac{1+\alpha}{2} = \beta + \alpha - \frac{1+\beta}{2} = 0$$
$$\Rightarrow \alpha = \beta = \frac{1}{3}.$$

Since $\alpha = \beta$, x divides AQ and BP in the same ratio. By symmetry, the third median CM must also divide the other two in this ratio, and so must pass through X and $MX = \frac{1}{3}MC$.

9. Let $\mathbf{i}$ point east and $\mathbf{j}$ point north. Let the wind velocity be

$$\mathbf{v}_{\text{wind}} = a\mathbf{i} + b\mathbf{j}.$$

Now $\mathbf{v}_{\text{wind}} = \mathbf{v}_{\text{wind rel car}} + \mathbf{v}_{\text{car}}$.
When $\mathbf{v}_{\text{car}} = 50\mathbf{j}$, the wind appears to come from the west, so $\mathbf{v}_{\text{wind rel car}} = \lambda\mathbf{i}$. Thus

$$a\mathbf{i} + b\mathbf{j} = \lambda\mathbf{i} + 50\mathbf{j},$$

so $a = \lambda$ and $b = 50$.
When $\mathbf{v}_{\text{car}} = 100\mathbf{j}$, the wind appears to come from the northwest, so $\mathbf{v}_{\text{wind rel car}} = \mu(\mathbf{i}-\mathbf{j})$. Thus

$$a\mathbf{i} + b\mathbf{j} = \mu(\mathbf{i}-\mathbf{j}) + 100\mathbf{j},$$

so $a = \mu$ and $b = 100 - \mu$.
Hence $50 = 100 - \mu$, so $\mu = 50$. Thus $a = b = 50$. The wind is from the southwest at $50\sqrt{2}$ km/h.

10. Let the x-axis point east and the y-axis north. The velocity of the water is

$$\mathbf{v}_{\text{water}} = 3\mathbf{i}.$$

If you row through the water with speed 5 in the direction making angle θ west of north, then your velocity relative to the water will be

$$\mathbf{v}_{\text{boat rel water}} = -5\sin\theta\mathbf{i} + 5\cos\theta\mathbf{j}.$$

Therefore, your velocity relative to the land will be

$$\mathbf{v}_{\text{boat rel land}} = \mathbf{v}_{\text{boat rel water}} + \mathbf{v}_{\text{water}}$$
$$= (3 - 5\sin\theta)\mathbf{i} + 5\cos\theta\mathbf{j}.$$

To make progress in the direction $\mathbf{j}$, choose θ so that $3 = 5\sin\theta$. Thus $\theta = \sin^{-1}(3/5) \approx 36.87°$. In this case, your actual speed relative to the land will be

$$5\cos\theta = \frac{4}{5} \times 5 = 4 \text{ km/h}.$$

To row from A to B, head in the direction $36.87°$ west of north. The $1/2$ km crossing will take $(1/2)/4 = 1/8$ of an hour, or about $7\frac{1}{2}$ minutes.

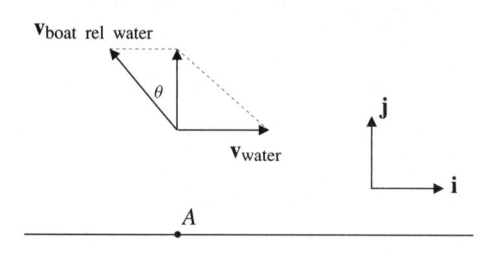

Fig. 10.2.10

11. We use the notations of the solution to Exercise 4. You now want to make progress in the direction $k\mathbf{i} + \frac{1}{2}\mathbf{j}$, that is, in the direction making angle

$$\phi = \tan^{-1}\frac{1}{2k}$$

with vector $\mathbf{i}$. Head at angle θ upstream of this direction. Since your rowing speed is 2, the triangle with angles θ and ϕ has sides 2 and 3 as shown in the figure. By the Sine Law, $\dfrac{3}{\sin\theta} = \dfrac{2}{\sin\phi}$, so

$$\sin\theta = \frac{3}{2}\sin\phi = \frac{3}{2}\frac{1}{2\sqrt{k^2 + \frac{1}{4}}} = \frac{3}{2\sqrt{4k^2 + 1}}.$$

This is only possible if $\dfrac{3}{2\sqrt{4k^2 + 1}} \leq 1$, that is, if

$$k \geq \frac{\sqrt{5}}{4}.$$

Head in the direction $\theta = \sin^{-1}\dfrac{3}{2\sqrt{4k^2 + 1}}$ upstream of the direction of AC, as shown in the figure. The trip is not possible if $k < \sqrt{5}/4$.

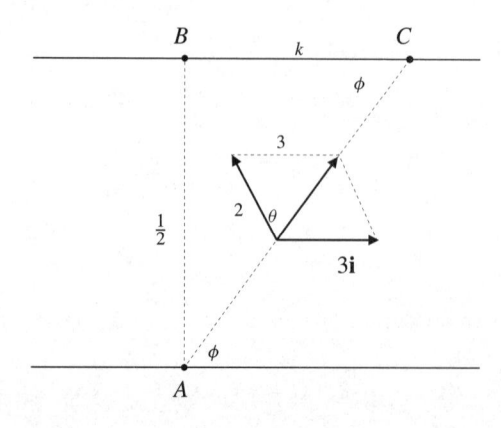

Fig. 10.2.11

12. Let **i** point east and **j** point north. If the aircraft heads in a direction θ north of east, then its velocity relative to the air is
$$750\cos\theta\,\mathbf{i} + 750\sin\theta\,\mathbf{j}.$$

The velocity of the air relative to the ground is
$$-\frac{100}{\sqrt{2}}\mathbf{i} + -\frac{100}{\sqrt{2}}\mathbf{j}.$$

Thus the velocity of the aircraft relative to the ground is
$$\left(750\cos\theta - \frac{100}{\sqrt{2}}\right)\mathbf{i} + \left(750\sin\theta - \frac{100}{\sqrt{2}}\right)\mathbf{j}.$$

If this velocity is true easterly, then
$$750\sin\theta = \frac{100}{\sqrt{2}},$$

so $\theta \approx 5.41°$. The speed relative to the ground is
$$750\cos\theta - \frac{100}{\sqrt{2}} \approx 675.9 \text{ km/h}.$$

The time for the 1500 km trip is $\dfrac{1500}{675.9} \approx 2.22$ hours.

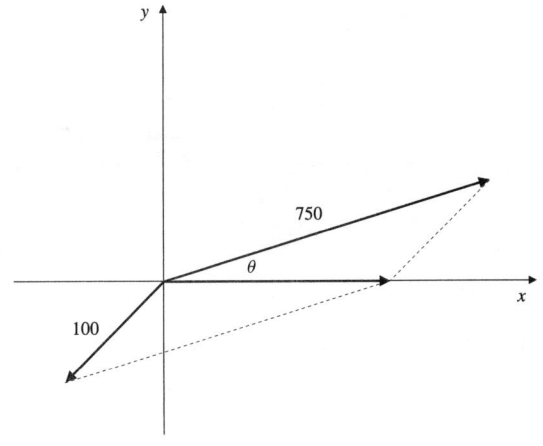

Fig. 10.2.12

13. The two vectors are perpendicular if their dot product is zero:
$$(2t\mathbf{i} + 4\mathbf{j} - (10+t)\mathbf{k}) \bullet (\mathbf{i} + t\mathbf{j} + \mathbf{k}) = 0$$
$$2t + 4t - 10 - t = 0 \Rightarrow t = 2.$$

The vectors are perpendicular if $t = 2$.

14. The cube with edges **i**, **j**, and **k** has diagonal $\mathbf{i}+\mathbf{j}+\mathbf{k}$. The angle between **i** and the diagonal is
$$\cos^{-1}\frac{\mathbf{i}\bullet(\mathbf{i}+\mathbf{j}+\mathbf{k})}{\sqrt{3}} = \cos^{-1}\frac{1}{\sqrt{3}} \approx 54.7°.$$

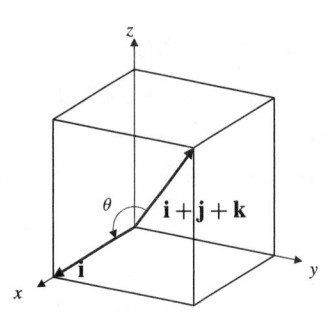

Fig. 10.2.14

15. The cube of Exercise 10 has six faces, each with 2 diagonals. The angle between $\mathbf{i}+\mathbf{j}+\mathbf{k}$ and the face diagonal $\mathbf{i}+\mathbf{j}$ is
$$\cos^{-1}\frac{(\mathbf{i}+\mathbf{j})\bullet(\mathbf{i}+\mathbf{j}+\mathbf{k})}{\sqrt{2}\sqrt{3}} = \cos^{-1}\frac{2}{\sqrt{6}} \approx 35.26°.$$

Six of the face diagonals make this angle with $\mathbf{i}+\mathbf{j}+\mathbf{k}$. The face diagonal $\mathbf{i}-\mathbf{j}$ (and five others) make angle
$$\cos^{-1}\frac{(\mathbf{i}-\mathbf{j})\bullet(\mathbf{i}+\mathbf{j}+\mathbf{k})}{\sqrt{2}\sqrt{3}} = \cos^{-1} 0 = 90°$$

with the cube diagonal $\mathbf{i}+\mathbf{j}+\mathbf{k}$.

16. If $\mathbf{u} = u_1\mathbf{i} + u_2\mathbf{j} + u_3\mathbf{k}$, then $\cos\alpha = \dfrac{\mathbf{u}\bullet\mathbf{i}}{|\mathbf{u}|} = \dfrac{u_1}{|\mathbf{u}|}$.
Similarly, $\cos\beta = \dfrac{u_2}{|\mathbf{u}|}$ and $\cos\gamma = \dfrac{u_3}{|\mathbf{u}|}$.
Thus, the unit vector in the direction of **u** is
$$\hat{\mathbf{u}} = \frac{\mathbf{u}}{|\mathbf{u}|} = \cos\alpha\,\mathbf{i} + \cos\beta\,\mathbf{j} + \cos\gamma\,\mathbf{k},$$

and so $\cos^2\alpha + \cos^2\beta + \cos^2\gamma = |\hat{\mathbf{u}}|^2 = 1$.

17. If $\hat{\mathbf{u}}$ makes equal angles $\alpha = \beta = \gamma$ with the coordinate axes, then $3\cos^2\alpha = 1$, and $\cos\alpha = 1/\sqrt{3}$. Thus
$$\hat{\mathbf{u}} = \frac{\mathbf{i}+\mathbf{j}+\mathbf{k}}{\sqrt{3}}.$$

18. If $A = (1, 0, 0)$, $B = (0, 2, 0)$, and $C = (0, 0, 3)$, then
$$\angle ABC = \cos^{-1}\frac{\overrightarrow{BA}\bullet\overrightarrow{BC}}{|BA||BC|} = \cos^{-1}\frac{4}{\sqrt{5}\sqrt{13}} \approx 60.26°$$
$$\angle BCA = \cos^{-1}\frac{\overrightarrow{CB}\bullet\overrightarrow{CA}}{|CB||CA|} = \cos^{-1}\frac{9}{\sqrt{10}\sqrt{13}} \approx 37.87°$$
$$\angle CAB = \cos^{-1}\frac{\overrightarrow{AC}\bullet\overrightarrow{AB}}{|AC||AB|} = \cos^{-1}\frac{1}{\sqrt{10}\sqrt{5}} \approx 81.87°.$$

19. Since $\mathbf{r} - \mathbf{r}_1 = \lambda \mathbf{r}_1 + (1-\lambda)\mathbf{r}_2 - \mathbf{r}_1 = (1-\lambda)(\mathbf{r}_1 - \mathbf{r}_2)$, therefore $\mathbf{r} - \mathbf{r}_1$ is parallel to $\mathbf{r}_1 - \mathbf{r}_2$, that is, parallel to the line $P_1 P_2$. Since P_1 is on that line, so must P be on it.

If $\lambda = \dfrac{1}{2}$, then $\mathbf{r} = \dfrac{1}{2}(\mathbf{r}_1 + \mathbf{r}_2)$, so P is midway between P_1 and P_2.

If $\lambda = \dfrac{2}{3}$, then $\mathbf{r} = \dfrac{2}{3}\mathbf{r}_1 + \dfrac{1}{3}\mathbf{r}_2$, so P is two-thirds of the way from P_2 towards P_1 along the line.

If $\lambda = -1$, the $\mathbf{r} = -\mathbf{r}_1 + 2\mathbf{r}_2 = \mathbf{r}_2 + (\mathbf{r}_2 - \mathbf{r}_1)$, so P is such that P_2 bisects the segment $P_1 P$.

If $\lambda = 2$, then $\mathbf{r} = 2\mathbf{r}_1 - \mathbf{r}_2 = \mathbf{r}_1 + (\mathbf{r}_1 - \mathbf{r}_2)$, so P is such that P_1 bisects the segment $P_2 P$.

20. If $\mathbf{a} \neq \mathbf{0}$, then $\mathbf{a} \bullet \mathbf{r} = 0$ implies that the position vector $\mathbf{r}$ is perpenducular to $\mathbf{a}$. Thus the equation is satisfied by all points on the plane through the origin that is normal (perpendicular) to $\mathbf{a}$.

21. If $\mathbf{r} \bullet \mathbf{a} = b$, then the vector projection of $\mathbf{r}$ along $\mathbf{a}$ is the constant vector

$$\frac{\mathbf{r} \bullet \mathbf{a}}{|\mathbf{a}|} \frac{\mathbf{a}}{|\mathbf{a}|} = \frac{b}{|\mathbf{a}|^2} \mathbf{a} = \mathbf{r}_0, \quad \text{say.}$$

Thus $\mathbf{r} \bullet \mathbf{a} = b$ is satisfied by all points on the plane through $\mathbf{r}_0$ that is normal to $\mathbf{a}$.

In Exercises 22–24, $\mathbf{u} = 2\mathbf{i} + \mathbf{j} - 2\mathbf{k}$, $\mathbf{v} = \mathbf{i} + 2\mathbf{j} - 2\mathbf{k}$, and $\mathbf{w} = 2\mathbf{i} - 2\mathbf{j} + \mathbf{k}$.

22. Vector $\mathbf{x} = x\mathbf{i} + y\mathbf{j} + z\mathbf{k}$ is perpendicular to both $\mathbf{u}$ and $\mathbf{v}$ if

$$\mathbf{u} \bullet \mathbf{x} = 0 \Leftrightarrow 2x + y - 2z = 0$$
$$\mathbf{v} \bullet \mathbf{x} = 0 \Leftrightarrow x + 2y - 2z = 0.$$

Subtracting these equations, we get $x - y = 0$, so $x = y$. The first equation now gives $3x = 2z$. Now $\mathbf{x}$ is a unit vector if $x^2 + y^2 + z^2 = 1$, that is, if $x^2 + x^2 + \frac{9}{4}x^2 = 1$, or $x = \pm 2/\sqrt{17}$. The two unit vectors are

$$\mathbf{x} = \pm \left(\frac{2}{\sqrt{17}}\mathbf{i} + \frac{2}{\sqrt{17}}\mathbf{j} + \frac{3}{\sqrt{17}}\mathbf{k} \right).$$

23. Let $\mathbf{x} = x\mathbf{i} + y\mathbf{j} + z\mathbf{k}$. Then

$$\mathbf{x} \bullet \mathbf{u} = 9 \Leftrightarrow 2x + y - 2z = 9$$
$$\mathbf{x} \bullet \mathbf{v} = 4 \Leftrightarrow x + 2y - 2z = 4$$
$$\mathbf{x} \bullet \mathbf{w} = 6 \Leftrightarrow 2x - 2y + z = 6.$$

This system of linear equations has solution $x = 2$, $y = -3$, $z = -4$. Thus $\mathbf{x} = 2\mathbf{i} - 3\mathbf{j} - 4\mathbf{k}$.

24. Since $\mathbf{u}$, $\mathbf{v}$, and $\mathbf{w}$ all have the same length (3), a vector $\mathbf{x} = x\mathbf{i} + y\mathbf{j} + z\mathbf{k}$ will make equal angles with all three if it has equal dot products with all three, that is, if

$$2x + y - 2z = x + 2y - 2z \Leftrightarrow x - y = 0$$
$$2x + y - 2z = 2x - 2y + z \Leftrightarrow 3y - 3z = 0.$$

Thus $x = y = z$. Two unit vectors satisfying this condition are

$$\mathbf{x} = \pm \left(\frac{1}{\sqrt{3}}\mathbf{i} + \frac{1}{\sqrt{3}}\mathbf{j} + \frac{1}{\sqrt{3}}\mathbf{k} \right).$$

25. Let $\hat{\mathbf{u}} = \mathbf{u}/|\mathbf{u}|$ and $\hat{\mathbf{v}} = \mathbf{v}/|\mathbf{v}|$.
Then $\hat{\mathbf{u}} + \hat{\mathbf{v}}$ bisects the angle between $\mathbf{u}$ and $\mathbf{v}$. A unit vector which bisects this angle is

$$\frac{\hat{\mathbf{u}} + \hat{\mathbf{v}}}{|\hat{\mathbf{u}} + \hat{\mathbf{v}}|} = \frac{\dfrac{\mathbf{u}}{|\mathbf{u}|} + \dfrac{\mathbf{v}}{|\mathbf{v}|}}{\left| \dfrac{\mathbf{u}}{|\mathbf{u}|} + \dfrac{\mathbf{v}}{|\mathbf{v}|} \right|}$$

$$= \frac{|\mathbf{v}|\mathbf{u} + |\mathbf{u}|\mathbf{v}}{\big| |\mathbf{v}|\mathbf{u} + |\mathbf{u}|\mathbf{v} \big|}.$$

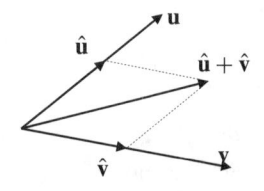

Fig. 10.2.25

26. If $\mathbf{u}$ and $\mathbf{v}$ are not parallel, then neither is the zero vector, and the origin and the two points with position vectors $\mathbf{u}$ and $\mathbf{v}$ lie on a unique plane. The equation $\mathbf{r} = \lambda \mathbf{u} + \mu \mathbf{v}$ (λ, μ real) gives the position vector of an arbitrary point on that plane.

27. a) $|\mathbf{u} + \mathbf{v}|^2 = (\mathbf{u} + \mathbf{v}) \bullet (\mathbf{u} + \mathbf{v})$
$\phantom{|\mathbf{u} + \mathbf{v}|^2} = \mathbf{u} \bullet \mathbf{u} + \mathbf{u} \bullet \mathbf{v} + \mathbf{v} \bullet \mathbf{u} + \mathbf{v} \bullet \mathbf{v}$
$\phantom{|\mathbf{u} + \mathbf{v}|^2} = |\mathbf{u}|^2 + 2\mathbf{u} \bullet \mathbf{v} + |\mathbf{v}|^2.$

b) If θ is the angle between $\mathbf{u}$ and $\mathbf{v}$, then $\cos\theta \leq 1$, so

$$\mathbf{u} \bullet \mathbf{v} = |\mathbf{u}||\mathbf{v}|\cos\theta \leq |\mathbf{u}||\mathbf{v}|.$$

c) $|\mathbf{u} + \mathbf{v}|^2 = |\mathbf{u}|^2 + 2\mathbf{u} \bullet \mathbf{v} + |\mathbf{v}|^2$
$\phantom{|\mathbf{u} + \mathbf{v}|^2} \leq |\mathbf{u}|^2 + 2|\mathbf{u}||\mathbf{v}| + |\mathbf{v}|^2$
$\phantom{|\mathbf{u} + \mathbf{v}|^2} = (|\mathbf{u}| + |\mathbf{v}|)^2.$
Thus $|\mathbf{u} + \mathbf{v}| \leq |\mathbf{u}| + |\mathbf{v}|$.

28. a) $\mathbf{u}$, $\mathbf{v}$, and $\mathbf{u} + \mathbf{v}$ are the sides of a triangle. The triangle inequality says that the length of one side cannot exceed the sum of the lengths of the other two sides.

b) If $\mathbf{u}$ and $\mathbf{v}$ are parallel and point in the *same direction*, (or if at least one of them is the zero vector), then $|\mathbf{u} + \mathbf{v}| = |\mathbf{u}| + |\mathbf{v}|$.

29. $\mathbf{u} = \frac{3}{5}\mathbf{i} + \frac{4}{5}\mathbf{j}$, $\mathbf{v} = \frac{4}{5}\mathbf{i} - \frac{3}{5}\mathbf{j}$, $\mathbf{w} = \mathbf{k}$.

a) $|\mathbf{u}| = \sqrt{\frac{9}{25} + \frac{16}{25}} = 1$, $|\mathbf{v}| = \sqrt{\frac{16}{25} + \frac{9}{25}} = 1$, $|\mathbf{w}| = 1$,
$\mathbf{u} \bullet \mathbf{v} = \frac{12}{25} - \frac{12}{25} = 0$, $\mathbf{u} \bullet \mathbf{w} = 0$, $\mathbf{v} \bullet \mathbf{w} = 0$.

b) If $\mathbf{r} = x\mathbf{i} + y\mathbf{j} + z\mathbf{k}$, then

$$(\mathbf{r} \bullet \mathbf{u})\mathbf{u} + (\mathbf{r} \bullet \mathbf{v})\mathbf{v} + (\mathbf{r} \bullet \mathbf{w})\mathbf{w}$$
$$= \left(\frac{3}{5}x + \frac{4}{5}y\right)\left(\frac{3}{5}\mathbf{i} + \frac{4}{5}\mathbf{j}\right)$$
$$+ \left(\frac{4}{5}x - \frac{3}{5}y\right)\left(\frac{4}{5}\mathbf{i} - \frac{3}{5}\mathbf{j}\right) + z\mathbf{k}$$
$$= \frac{9x + 16x}{25}\mathbf{i} + \frac{16y + 9y}{25}\mathbf{j} + z\mathbf{k}$$
$$= x\mathbf{i} + y\mathbf{j} + z\mathbf{k} = \mathbf{r}.$$

30. Suppose $|\mathbf{u}| = |\mathbf{v}| = |\mathbf{w}| = 1$, and $\mathbf{u} \bullet \mathbf{v} = \mathbf{u} \bullet \mathbf{w} = \mathbf{v} \bullet \mathbf{w} = 0$, and let $\mathbf{r} = a\mathbf{u} + b\mathbf{v} + w\mathbf{w}$. Then

$$\mathbf{r} \bullet \mathbf{u} = a\mathbf{u} \bullet \mathbf{u} + b\mathbf{v} \bullet \mathbf{u} + c\mathbf{w} \bullet \mathbf{u} = a|\mathbf{u}|^2 + 0 + 0 = a.$$

Similarly, $\mathbf{r} \bullet \mathbf{v} = b$ and $\mathbf{r} \bullet \mathbf{w} = c$.

31. Let $\mathbf{u} = \dfrac{\mathbf{w} \bullet \mathbf{a}}{|\mathbf{a}|^2}\mathbf{a}$, (the vector projection of $\mathbf{w}$ along $\mathbf{a}$). Let $\mathbf{v} = \mathbf{w} - \mathbf{u}$. Then $\mathbf{w} = \mathbf{u} + \mathbf{v}$. Clearly $\mathbf{u}$ is parallel to $\mathbf{a}$, and

$$\mathbf{v} \bullet \mathbf{a} = \mathbf{w} \bullet \mathbf{a} - \frac{\mathbf{w} \bullet \mathbf{a}}{|\mathbf{a}|^2}\mathbf{a} \bullet \mathbf{a} = \mathbf{w} \bullet \mathbf{a} - \mathbf{w} \bullet \mathbf{a} = 0,$$

so $\mathbf{v}$ is perpendicular to $\mathbf{a}$.

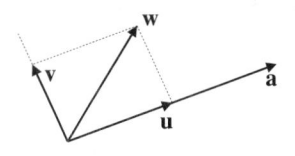

Fig. 10.2.31

32. Let $\hat{\mathbf{n}}$ be a unit vector that is perpendicular to $\mathbf{u}$ and lies in the plane containing the origin and the points U, V, and P. Then $\hat{\mathbf{u}} = \mathbf{u}/|\mathbf{u}|$ and $\hat{\mathbf{n}}$ constitute a standard basis in that plane, so each of the vectors $\mathbf{v}$ and $\mathbf{r}$ can be expressed in terms of them:

$$\mathbf{v} = s\hat{\mathbf{u}} + t\hat{\mathbf{n}}$$
$$\mathbf{r} = x\hat{\mathbf{u}} + y\hat{\mathbf{n}}.$$

Since $\mathbf{v}$ is not parallel to $\mathbf{u}$, we have $t \neq 0$. Thus $\hat{\mathbf{n}} = (1/t)(\mathbf{v} - s\hat{\mathbf{u}})$ and

$$\mathbf{r} = x\hat{\mathbf{u}} + \frac{y}{t}(\mathbf{v} - s\hat{\mathbf{u}}) = \lambda\mathbf{u} + \mu\mathbf{v},$$

where $\lambda = (tx - ys)/(t|\mathbf{u}|)$ and $\mu = y/t$.

33. Let $|\mathbf{a}|^2 - 4rst = K^2$, where $K > 0$. Now

$$|\mathbf{a}|^2 = \mathbf{a} \bullet \mathbf{a} = (r\mathbf{x} + s\mathbf{y}) \bullet (r\mathbf{x} + s\mathbf{y})$$
$$= r^2|\mathbf{x}|^2 + s^2|\mathbf{y}|^2 + 2rs\mathbf{x} \bullet \mathbf{y}$$
$$K^2 = |\mathbf{a}|^2 - 4rs\mathbf{x} \bullet \mathbf{y}$$
$$= |r\mathbf{x} - s\mathbf{y}|^2$$

(since $\mathbf{x} \bullet \mathbf{y} = t$).
Therefore $r\mathbf{x} - s\mathbf{y} = K\hat{\mathbf{u}}$, for some unit vector $\hat{\mathbf{u}}$.
Since $r\mathbf{x} + s\mathbf{y} = \mathbf{a}$, we have

$$2r\mathbf{x} = \mathbf{a} + K\hat{\mathbf{u}}$$
$$2s\mathbf{y} = \mathbf{a} - K\hat{\mathbf{u}}.$$

Thus
$$\mathbf{x} = \frac{\mathbf{a} + K\hat{\mathbf{u}}}{2r}, \quad \mathbf{y} = \frac{\mathbf{a} - K\hat{\mathbf{u}}}{2s},$$

where $K = \sqrt{|\mathbf{a}|^2 - 4rst}$, and $\hat{\mathbf{u}}$ is any unit vector. (The solution is not unique.)

34. The derivation of the equation of the hanging cable given in the text needs to be modified by replacing $\mathbf{W} = -\delta gs\mathbf{j}$ with $\mathbf{W} = -\delta gx\mathbf{j}$. Thus $T_v = \delta gx$, and the slope of the cable satisfies

$$\frac{dy}{dx} = \frac{\delta gx}{H} = ax$$

where $a = \delta g/H$. Thus

$$y = \frac{1}{2}ax^2 + C;$$

the cable hangs in a parabola.

35. If $y = \dfrac{1}{a}\cosh(ax)$, then $y' = \sinh(ax)$, so

$$s = \int_0^x \sqrt{1 + \sinh^2(au)}\,du = \int_0^x \cosh(au)\,du$$
$$= \left.\frac{\sinh(au)}{a}\right|_0^x = \frac{1}{a}\sinh(ax).$$

As shown in the text, the tension $\mathbf{T}$ at P has horizontal and vertical components that satisfy $T_h = H = \dfrac{\delta g}{a}$ and $T_v = \delta gs = \dfrac{\delta g}{a}\sinh(ax)$. Hence

$$|\mathbf{T}| = \sqrt{T_h^2 + T_v^2} = \frac{\delta g}{a}\cosh(ax) = \delta gy.$$

36. The cable hangs along the curve $y = \dfrac{1}{a}\cosh(ax)$, and its length from the lowest point at $x = 0$ to the support tower at $x = 45$ m is 50 m. Thus

$$50 = \int_0^{45} \sqrt{1 + \sinh^2(ax)}\,dx = \frac{1}{a}\sinh(45a).$$

The equation $\sinh(45a) = 50a$ has approximate solution $a \approx 0.0178541$. The vertical distance between the lowest point on the cable and the support point is

$$\frac{1}{a}\big(\cosh(45a) - 1\big) \approx 19.07 \text{ m}.$$

37. The equation of the cable is of the form $y = \frac{1}{a}\cosh(ax)$. At the point P where $x = 10$ m, the slope of the cable is $\sinh(10a) = \tan(55°)$. Thus

$$a = \frac{1}{10}\sinh^{-1}(\tan(55°)) \approx 0.115423.$$

The length of the cable between $x = 0$ and $x = 10$ m is

$$L = \int_0^{10} \sqrt{1 + \sinh^2(ax)}\, dx$$
$$= \int_0^{10} \cosh(ax)\, dx = \frac{1}{a}\sinh(ax)\Big|_0^{10}$$
$$= \frac{1}{a}\sinh(10a) \approx 12.371 \text{ m}.$$

Section 10.3 The Cross Product in 3-Space (page 622)

1. $(\mathbf{i} - 2\mathbf{j} + 3\mathbf{k}) \times (3\mathbf{i} + \mathbf{j} - 4\mathbf{k}) = 5\mathbf{i} + 13\mathbf{j} + 7\mathbf{k}$

2. $(\mathbf{j} + 2\mathbf{k}) \times (-\mathbf{i} - \mathbf{j} + \mathbf{k}) = 3\mathbf{i} - 2\mathbf{j} + \mathbf{k}$

3. If $A = (1, 2, 0)$, $B = (1, 0, 2)$, and $C = (0, 3, 1)$, then $\overrightarrow{AB} = -2\mathbf{j} + 2\mathbf{k}$, $\overrightarrow{AC} = -\mathbf{i} + \mathbf{j} + \mathbf{k}$, and the area of triangle ABC is

$$\frac{|\overrightarrow{AB} \times \overrightarrow{AC}|}{2} = \frac{|-4\mathbf{i} - 2\mathbf{j} - 2\mathbf{k}|}{2} = \sqrt{6} \text{ sq. units}.$$

4. A vector perpendicular to the plane containing the three given points is

$$(-a\mathbf{i} + b\mathbf{j}) \times (-a\mathbf{i} + c\mathbf{k}) = bc\mathbf{i} + ac\mathbf{j} + ab\mathbf{k}.$$

A unit vector in this direction is

$$\frac{bc\mathbf{i} + ac\mathbf{j} + ab\mathbf{k}}{\sqrt{b^2c^2 + a^2c^2 + a^2b^2}}.$$

The triangle has area $\frac{1}{2}\sqrt{b^2c^2 + a^2c^2 + a^2b^2}$.

5. A vector perpendicular to $\mathbf{i} + \mathbf{j}$ and $\mathbf{j} + 2\mathbf{k}$ is

$$\pm(\mathbf{i} + \mathbf{j}) \times (\mathbf{j} + 2\mathbf{k}) = \pm(2\mathbf{i} - 2\mathbf{j} + \mathbf{k}),$$

which has length 3. A unit vector in that direction is

$$\pm\left(\frac{2}{3}\mathbf{i} - \frac{2}{3}\mathbf{j} + \frac{1}{3}\mathbf{k}\right).$$

6. A vector perpendicular to $\mathbf{u} = 2\mathbf{i} - \mathbf{j} - 2\mathbf{k}$ and to $\mathbf{v} = 2\mathbf{i} - 3\mathbf{j} + \mathbf{k}$ is the cross product

$$\mathbf{u} \times \mathbf{v} = \begin{vmatrix} \mathbf{i} & \mathbf{j} & \mathbf{k} \\ 2 & -1 & -2 \\ 2 & -3 & 1 \end{vmatrix} = -7\mathbf{i} - 6\mathbf{j} - 4\mathbf{k},$$

which has length $\sqrt{101}$. A unit vector with positive $\mathbf{k}$ component that is perpenducular to $\mathbf{u}$ and $\mathbf{v}$ is

$$\frac{-1}{\sqrt{101}}\mathbf{u} \times \mathbf{v} = \frac{1}{\sqrt{101}}(7\mathbf{i} + 6\mathbf{j} + 4\mathbf{k}).$$

7. Since $\mathbf{u}$ makes zero angle with itself, $|\mathbf{u} \times \mathbf{u}| = 0$ and $\mathbf{u} \times \mathbf{u} = \mathbf{0}$.

8. $\mathbf{u} \times \mathbf{v} = \begin{vmatrix} \mathbf{i} & \mathbf{j} & \mathbf{k} \\ u_1 & u_2 & u_3 \\ v_1 & v_2 & v_3 \end{vmatrix}$
$= -\begin{vmatrix} \mathbf{i} & \mathbf{j} & \mathbf{k} \\ v_1 & v_2 & v_3 \\ u_1 & u_2 & u_3 \end{vmatrix} = -\mathbf{v} \times \mathbf{u}.$

9. $(\mathbf{u} + \mathbf{v}) \times \mathbf{w} = \begin{vmatrix} \mathbf{i} & \mathbf{j} & \mathbf{k} \\ u_1 + v_1 & u_2 + v_2 & u_3 + v_3 \\ w_1 & w_2 & w_3 \end{vmatrix}$
$= \begin{vmatrix} \mathbf{i} & \mathbf{j} & \mathbf{k} \\ u_1 & u_2 & u_3 \\ w_1 & w_2 & w_3 \end{vmatrix} + \begin{vmatrix} \mathbf{i} & \mathbf{j} & \mathbf{k} \\ v_1 & v_2 & v_3 \\ w_1 & w_2 & w_3 \end{vmatrix}$
$= \mathbf{u} \times \mathbf{w} + \mathbf{v} \times \mathbf{w}.$

10. $(t\mathbf{u}) \times \mathbf{v} = \begin{vmatrix} \mathbf{i} & \mathbf{j} & \mathbf{k} \\ tu_1 & tu_2 & tu_3 \\ v_1 & v_2 & v_3 \end{vmatrix}$
$= t\begin{vmatrix} \mathbf{i} & \mathbf{j} & \mathbf{k} \\ u_1 & u_2 & u_3 \\ v_1 & v_2 & v_3 \end{vmatrix} = t(\mathbf{u} \times \mathbf{v}),$
$\mathbf{u} \times (t\mathbf{v}) = -(t\mathbf{v}) \times \mathbf{u}$
$= -t(\mathbf{v} \times \mathbf{u}) = t(\mathbf{u} \times \mathbf{v}).$

11. $\mathbf{u} \bullet (\mathbf{u} \times \mathbf{v})$
$= u_1\begin{vmatrix} u_2 & u_3 \\ v_2 & v_3 \end{vmatrix} - u_2\begin{vmatrix} u_1 & u_3 \\ v_1 & v_3 \end{vmatrix} + u_3\begin{vmatrix} u_1 & u_2 \\ v_1 & v_2 \end{vmatrix}$
$= u_1u_2v_3 - u_1v_2u_3 - u_2u_1v_3$
$\quad + u_2v_1u_3 + u_3u_1v_2 - u_3v_1u_2 = 0,$
$\mathbf{v} \bullet (\mathbf{u} \times \mathbf{v}) = -\mathbf{v} \bullet (\mathbf{v} \times \mathbf{u}) = 0.$

12. Both $\mathbf{u} = \cos\beta\,\mathbf{i} + \sin\beta\,\mathbf{j}$ and $\mathbf{v} = \cos\alpha\,\mathbf{i} + \sin\alpha\,\mathbf{j}$ are unit vectors. They make angles β and α, respectively, with the positive x-axis, so the angle between them is $|\alpha - \beta| = \alpha - \beta$, since we are told that $0 \le \alpha - \beta \le \pi$. They span a parallelogram (actually a rhombus) having area

$$|\mathbf{u} \times \mathbf{v}| = |\mathbf{u}||\mathbf{v}|\sin(\alpha - \beta) = \sin(\alpha - \beta).$$

But

$$\mathbf{u} \times \mathbf{v} = \begin{vmatrix} \mathbf{i} & \mathbf{j} & \mathbf{k} \\ \cos\beta & \sin\beta & 0 \\ \cos\alpha & \sin\alpha & 0 \end{vmatrix} = (\sin\alpha\cos\beta - \cos\alpha\sin\beta)\mathbf{k}.$$

Because $\mathbf{v}$ is displaced counterclockwise from $\mathbf{u}$, the cross product above must be in the positive k direction. Therefore its length is the k component. Therefore

$$\sin(\alpha - \beta) = \sin\alpha\cos\beta - \cos\alpha\sin\beta.$$

13. Suppose that $\mathbf{u} + \mathbf{v} + \mathbf{w} = \mathbf{0}$. Then

$$\mathbf{u} \times \mathbf{v} + \mathbf{v} \times \mathbf{v} + \mathbf{w} \times \mathbf{v} = \mathbf{0} \times \mathbf{v} = \mathbf{0}.$$

Thus $\mathbf{u} \times \mathbf{v} + \mathbf{w} \times \mathbf{v} = \mathbf{0}$.
Thus $\mathbf{u} \times \mathbf{v} = -\mathbf{w} \times \mathbf{v} = \mathbf{v} \times \mathbf{w}$.
By symmetry, we also have $\mathbf{v} \times \mathbf{w} = \mathbf{w} \times \mathbf{u}$.

14. The base of the tetrahedron is a triangle spanned by $\mathbf{v}$ and $\mathbf{w}$, which has area

$$A = \frac{1}{2}|\mathbf{v} \times \mathbf{w}|.$$

The altitude h of the tetrahedron (measured perpendicular to the plane of the base) is equal to the length of the projection of $\mathbf{u}$ onto the vector $\mathbf{v} \times \mathbf{w}$ (which is perpendicular to the base). Thus

$$h = \frac{|\mathbf{u} \bullet (\mathbf{v} \times \mathbf{w})|}{|\mathbf{v} \times \mathbf{w}|}.$$

The volume of the tetrahedron is

$$V = \frac{1}{3}Ah = \frac{1}{6}|\mathbf{u} \bullet (\mathbf{v} \times \mathbf{w})|$$

$$= \frac{1}{6}\left| \begin{vmatrix} u_1 & u_2 & u_3 \\ v_1 & v_2 & v_3 \\ w_1 & w_2 & w_3 \end{vmatrix} \right|.$$

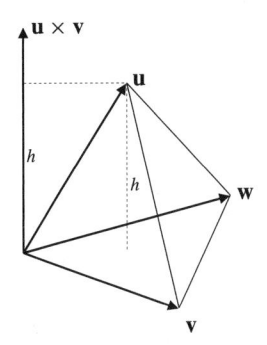

Fig. 10.3.14

15. The tetrahedron with vertices $(1, 0, 0)$, $(1, 2, 0)$, $(2, 2, 2)$, and $(0, 3, 2)$ is spanned by $\mathbf{u} = 2\mathbf{j}$, $\mathbf{v} = \mathbf{i} + 2\mathbf{j} + 2\mathbf{k}$, and $\mathbf{w} = -\mathbf{i} + 3\mathbf{j} + 2\mathbf{k}$. By Exercise 14, its volume is

$$V = \frac{1}{6}\left| \begin{vmatrix} 0 & 2 & 0 \\ 1 & 2 & 2 \\ -1 & 3 & 2 \end{vmatrix} \right| = \frac{4}{3} \text{ cu. units.}$$

16. Let the cube be as shown in the figure. The required parallelepiped is spanned by $a\mathbf{i} + a\mathbf{j}$, $a\mathbf{j} + a\mathbf{k}$, and $a\mathbf{i} + a\mathbf{k}$. Its volume is

$$V = \left| \begin{vmatrix} a & a & 0 \\ 0 & a & a \\ a & 0 & a \end{vmatrix} \right| = 2a^3 \text{ cu. units.}$$

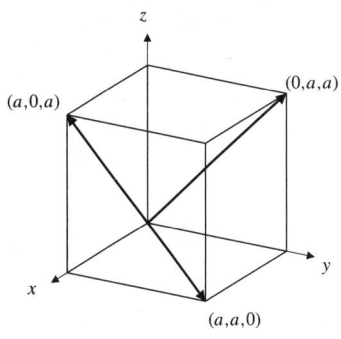

Fig. 10.3.16

17. The points $A = (1, 1, -1)$, $B = (0, 3, -2)$, $C = (-2, 1, 0)$, and $D = (k, 0, 2)$ are coplanar if $(\overrightarrow{AB} \times \overrightarrow{AC}) \bullet \overrightarrow{AD} = 0$. Now

$$\overrightarrow{AB} \times \overrightarrow{AC} = \begin{vmatrix} \mathbf{i} & \mathbf{j} & \mathbf{k} \\ -1 & 2 & -1 \\ -3 & 0 & 1 \end{vmatrix} = 2\mathbf{i} + 4\mathbf{j} + 6\mathbf{k}.$$

Thus the four points are coplanar if

$$2(k - 1) + 4(0 - 1) + 6(2 + 1) = 0,$$

that is, if $k = -6$.

18.
$$\mathbf{u} \bullet (\mathbf{v} \times \mathbf{w}) = \begin{vmatrix} u_1 & u_2 & u_3 \\ v_1 & v_2 & v_3 \\ w_1 & w_2 & w_3 \end{vmatrix}$$

$$= -\begin{vmatrix} v_1 & v_2 & v_3 \\ u_1 & u_2 & u_3 \\ w_1 & w_2 & w_3 \end{vmatrix}$$

$$= \begin{vmatrix} v_1 & v_2 & v_3 \\ w_1 & w_2 & w_3 \\ u_1 & u_2 & u_3 \end{vmatrix}$$

$$= \mathbf{v} \bullet (\mathbf{w} \times \mathbf{u})$$

$$= \mathbf{w} \bullet (\mathbf{u} \times \mathbf{v}) \quad \text{(by symmetry).}$$

19. If $\mathbf{u} \bullet (\mathbf{v} \times \mathbf{w}) \neq 0$, and $\mathbf{x} = \lambda\mathbf{u} + \mu\mathbf{v} + \nu\mathbf{w}$, then

$$\mathbf{x} \bullet (\mathbf{v} \times \mathbf{w})$$
$$= \lambda\mathbf{u} \bullet (\mathbf{v} \times \mathbf{w}) + \mu\mathbf{v} \bullet (\mathbf{v} \times \mathbf{w}) + \nu\mathbf{w} \bullet (\mathbf{v} \times \mathbf{w})$$
$$= \lambda\mathbf{u} \bullet (\mathbf{v} \times \mathbf{w}).$$

Thus
$$\lambda = \frac{\mathbf{x} \bullet (\mathbf{v} \times \mathbf{w})}{\mathbf{u} \bullet (\mathbf{v} \times \mathbf{w})}.$$

Since $\mathbf{u} \bullet (\mathbf{v} \times \mathbf{w}) = \mathbf{v} \bullet (\mathbf{w} \times \mathbf{u}) = \mathbf{w} \bullet (\mathbf{u} \times \mathbf{v})$, we have, by symmetry,

$$\mu = \frac{\mathbf{x} \bullet (\mathbf{w} \times \mathbf{u})}{\mathbf{u} \bullet (\mathbf{v} \times \mathbf{w})}, \quad \nu = \frac{\mathbf{x} \bullet (\mathbf{u} \times \mathbf{v})}{\mathbf{u} \bullet (\mathbf{v} \times \mathbf{w})}.$$

20. If $\mathbf{v} \times \mathbf{w} \neq \mathbf{0}$, then $(\mathbf{v} \times \mathbf{w}) \bullet (\mathbf{v} \times \mathbf{w}) \neq 0$. By the previous exercise, there exist constants λ, μ and ν such that

$$\mathbf{u} = \lambda\mathbf{v} + \mu\mathbf{w} + \nu(\mathbf{v} \times \mathbf{w}).$$

But $\mathbf{v} \times \mathbf{w}$ is perpendicular to both $\mathbf{v}$ and $\mathbf{w}$, so

$$\mathbf{u} \bullet (\mathbf{v} \times \mathbf{w}) = 0 + 0 + \nu(\mathbf{v} \times \mathbf{w}) \bullet (\mathbf{v} \times \mathbf{w}).$$

If $\mathbf{u} \bullet (\mathbf{v} \times \mathbf{w}) = 0$, then $\nu = 0$, and

$$\mathbf{u} = \lambda\mathbf{v} + \mu\mathbf{w}.$$

21. $\mathbf{u} = \mathbf{i} + 2\mathbf{j} + 3\mathbf{k}$
$\mathbf{v} = 2\mathbf{i} - 3\mathbf{j}$
$\mathbf{w} = \mathbf{j} - \mathbf{k}$
$\mathbf{u} \times (\mathbf{v} \times \mathbf{w}) = \mathbf{u} \times (3\mathbf{i} + 2\mathbf{j} + 2\mathbf{k}) = -2\mathbf{i} + 7\mathbf{j} - 4\mathbf{k}$
$(\mathbf{u} \times \mathbf{v}) \times \mathbf{w} = (9\mathbf{i} + 6\mathbf{j} - 7\mathbf{k}) \times \mathbf{w} = \mathbf{i} + 9\mathbf{j} + 9\mathbf{k}$.
$\mathbf{u} \times (\mathbf{v} \times \mathbf{w})$ lies in the plane of $\mathbf{v}$ and $\mathbf{w}$;
$(\mathbf{u} \times \mathbf{v}) \times \mathbf{w}$ lies in the plane of $\mathbf{u}$ and $\mathbf{v}$.

22. $\mathbf{u} \bullet \mathbf{v} \times \mathbf{w}$ makes sense in that it must mean $\mathbf{u} \bullet (\mathbf{v} \times \mathbf{w})$. $((\mathbf{u} \bullet \mathbf{v}) \times \mathbf{w}$ makes no sense since it is the cross product of a scalar and a vector.)

$\mathbf{u} \times \mathbf{v} \times \mathbf{w}$ makes no sense. It is ambiguous, since $(\mathbf{u} \times \mathbf{v}) \times \mathbf{w}$ and $\mathbf{u} \times (\mathbf{v} \times \mathbf{w})$ are not in general equal.

23. As suggested in the hint, let the x-axis lie in the direction of $\mathbf{v}$, and let the y-axis be such that $\mathbf{w}$ lies in the xy-plane. Thus

$$\mathbf{v} = v_1\mathbf{i}, \qquad \mathbf{w} = w_1\mathbf{i} + w_2\mathbf{j}.$$

Thus $\mathbf{v} \times \mathbf{w} = v_1 w_2 \mathbf{i} \times \mathbf{j} = v_1 w_2 \mathbf{k}$, and

$$\mathbf{u} \times (\mathbf{v} \times \mathbf{w}) = (u_1\mathbf{i} + u_2\mathbf{j} + u_3\mathbf{k}) \times (v_1 w_2 \mathbf{k})$$
$$= u_1 v_1 w_2 \mathbf{i} \times \mathbf{k} + u_2 v_1 w_2 \mathbf{j} \times \mathbf{k}$$
$$= -u_1 v_1 w_2 \mathbf{j} - u_1 v_1 w_2 \mathbf{j}.$$

But
$$(\mathbf{u} \bullet \mathbf{w})\mathbf{v} - (\mathbf{u} \bullet \mathbf{v})\mathbf{w}$$
$$= (u_1 w_1 + u_2 w_2)v_1\mathbf{i} - u_1 v_1 (w_1\mathbf{i} + w_2\mathbf{j})$$
$$= u_2 v_1 w_2 \mathbf{i} - u_1 v_1 w_2 \mathbf{j}.$$

Thus $\mathbf{u} \times (\mathbf{v} \times \mathbf{w}) = (\mathbf{u} \bullet \mathbf{w})\mathbf{v} - (\mathbf{u} \bullet \mathbf{v})\mathbf{w}$.

24. If $\mathbf{u}$, $\mathbf{v}$, and $\mathbf{w}$ are mutually perpendicular, then $\mathbf{v} \times \mathbf{w}$ is parallel to $\mathbf{u}$, so $\mathbf{u} \times (\mathbf{v} \times \mathbf{w}) = \mathbf{0}$. In this case, $\mathbf{u} \bullet (\mathbf{v} \times \mathbf{w}) = \pm|\mathbf{u}||\mathbf{v}||\mathbf{w}|$; the sign depends on whether $\mathbf{u}$ and $\mathbf{v} \times \mathbf{w}$ are in the same or opposite directions.

25. Applying the result of Exercise 23 three times, we obtain

$$\mathbf{u} \times (\mathbf{v} \times \mathbf{w}) + \mathbf{v} \times (\mathbf{w} \times \mathbf{u}) + \mathbf{w} \times (\mathbf{u} \times \mathbf{v})$$
$$= (\mathbf{u} \bullet \mathbf{w})\mathbf{v} - (\mathbf{u} \bullet \mathbf{v})\mathbf{w} + (\mathbf{v} \bullet \mathbf{u})\mathbf{w} - (\mathbf{v} \bullet \mathbf{w})\mathbf{u}$$
$$\quad + (\mathbf{w} \bullet \mathbf{v})\mathbf{u} - (\mathbf{w} \bullet \mathbf{u})\mathbf{v}$$
$$= \mathbf{0}.$$

26. If $\mathbf{a} = -\mathbf{i} + 2\mathbf{j} + 3\mathbf{k}$ and $\mathbf{x} = x\mathbf{i} + y\mathbf{j} + z\mathbf{k}$, then

$$\mathbf{a} \times \mathbf{x} = \begin{vmatrix} \mathbf{i} & \mathbf{j} & \mathbf{k} \\ -1 & 2 & 3 \\ x & y & z \end{vmatrix}$$
$$= (2z - 3y)\mathbf{i} + (3x + z)\mathbf{y} - (y + 2x)\mathbf{k}$$
$$= \mathbf{i} + 5\mathbf{j} - 3\mathbf{k},$$

provided $2z - 3y = 1$, $3x + z = 5$, and $-y - 2x = -3$. This system is satisfied by $x = t$, $y = 3 - 2t$, $z = 5 - 3t$, for any real number t. Thus

$$\mathbf{x} = t\mathbf{i} + (3 - 2t)\mathbf{j} + (5 - 3t)\mathbf{k}$$

gives a solution of $\mathbf{a} \times \mathbf{x} = \mathbf{i} + 5\mathbf{j} - 3\mathbf{k}$ for any t. These solutions constitute a line parallel to $\mathbf{a}$.

27. Let $\mathbf{a} = -\mathbf{i} + 2\mathbf{j} + 3\mathbf{k}$ and $\mathbf{b} = \mathbf{i} + 5\mathbf{j}$. If $\mathbf{x}$ is a solution of $\mathbf{a} \times \mathbf{x} = \mathbf{b}$, then

$$\mathbf{a} \bullet \mathbf{b} = \mathbf{a} \bullet (\mathbf{a} \times \mathbf{x}) = 0.$$

However, $\mathbf{a} \bullet \mathbf{b} \neq 0$, so there can be no such solution $\mathbf{x}$.

28. The equation $\mathbf{a} \times \mathbf{x} = \mathbf{b}$ can be solved for $\mathbf{x}$ if and only if $\mathbf{a} \bullet \mathbf{b} = 0$. The "only if" part is demonstrated in the previous solution. For the "if" part, observe that if $\mathbf{a} \bullet \mathbf{b} = 0$ and $\mathbf{x}_0 = (\mathbf{b} \times \mathbf{a})/|\mathbf{a}|^2$, then by Exercise 23,

$$\mathbf{a} \times \mathbf{x}_0 = \frac{1}{|a|^2}\mathbf{a} \times (\mathbf{b} \times \mathbf{a}) = \frac{(\mathbf{a} \bullet \mathbf{a})\mathbf{b} - (\mathbf{a} \bullet \mathbf{b})\mathbf{a}}{|a|^2} = \mathbf{b}.$$

The solution $\mathbf{x}_0$ is not unique; as suggested by the example in Exercise 26, any multiple of $\mathbf{a}$ can be added to it and the result will still be a solution. If $\mathbf{x} = \mathbf{x}_0 + t\mathbf{a}$, then

$$\mathbf{a} \times \mathbf{x} = \mathbf{a} \times \mathbf{x}_0 + t\mathbf{a} \times \mathbf{a} = \mathbf{b} + \mathbf{0} = \mathbf{b}.$$

Section 10.4 Planes and Lines (page 631)

1. a) $x^2 + y^2 + z^2 = z^2$ represents a line in 3-space, namely the z-axis.

   b) $x + y + z = x + y + z$ is satisfied by every point in 3-space.

   c) $x^2 + y^2 + z^2 = -1$ is satisfied by no points in (real) 3-space.

2. The plane through $(0, 2, -3)$ normal to $4\mathbf{i} - \mathbf{j} - 2\mathbf{k}$ has equation
$$4(x - 0) - (y - 2) - 2(z + 3) = 0,$$
or $4x - y - 2z = 4$.

3. The plane through the origin having normal $\mathbf{i} - \mathbf{j} + 2\mathbf{k}$ has equation $x - y + 2z = 0$.

4. The plane passing through $(1, 2, 3)$, parallel to the plane $3x + y - 2z = 15$, has equation $3z + y - 2z = 3 + 2 - 6$, or $3x + y - 2z = -1$.

5. The plane through $(1, 1, 0)$, $(2, 0, 2)$, and $(0, 3, 3)$ has normal
$$(\mathbf{i} - \mathbf{j} + 2\mathbf{k}) \times (\mathbf{i} - 2\mathbf{j} - 3\mathbf{k}) = 7\mathbf{i} + 5\mathbf{j} - \mathbf{k}.$$
It therefore has equation
$$7(x - 1) + 5(y - 1) - (z - 0) = 0,$$
or $7x + 5y - z = 12$.

6. The plane passing through $(-2, 0, 0)$, $(0, 3, 0)$, and $(0, 0, 4)$ has equation
$$\frac{x}{-2} + \frac{y}{3} + \frac{z}{4} = 1,$$
or $6x - 4y - 3z = -12$.

7. The normal $\mathbf{n}$ to a plane through $(1, 1, 1)$ and $(2, 0, 3)$ must be perpendicular to the vector $\mathbf{i} - \mathbf{j} + 2\mathbf{k}$ joining these points. If the plane is perpendicular to the plane $x + 2y - 3z = 0$, then $\mathbf{n}$ must also be perpendicular to $\mathbf{i} + 2\mathbf{j} - 3\mathbf{k}$, the normal to this latter plane. Hence we can use
$$\mathbf{n} = (\mathbf{i} - \mathbf{j} + 2\mathbf{k}) \times (\mathbf{i} + 2\mathbf{j} - 3\mathbf{k}) = -\mathbf{i} + 5\mathbf{j} + 3\mathbf{k}.$$
The plane has equation
$$1(x - 1) + 5(y - 1) + 3(z - 1) = 0,$$
or $x - 5y - 3z = -7$.

8. Since $(-2, 0, -1)$ does not lie on $x - 4y + 2z = -5$, the required plane will have an equation of the form
$$2x + 3y - z + \lambda(x - 4y + 2z + 5) = 0$$
for some λ. Thus
$$-4 + 1 + \lambda(-2 - 2 + 5) = 0,$$
so $\lambda = 3$. The required plane is $5x - 9y + 5z = -15$.

9. A plane through the line $x + y = 2$, $y - z = 3$ has equation of the form
$$x + y - 2 + \lambda(y - z - 3) = 0.$$
This plane will be perpendicular to $2x + 3y + 4z = 5$ if
$$(2)(1) + (1 + \lambda)(3) - (\lambda)(4) = 0,$$
that is, if $\lambda = 5$. The equation of the required plane is
$$x + 6y - 5z = 17.$$

10. Three distinct points will not determine a unique plane through them if they all lie on a straight line. If the points have position vectors $\mathbf{r}_1$, $\mathbf{r}_2$, and $\mathbf{r}_3$, then they will all lie on a straight line if
$$(\mathbf{r}_2 - \mathbf{r}_1) \times (\mathbf{r}_3 - \mathbf{r}_1) = \mathbf{0}.$$

11. If the four points have position vectors $\mathbf{r}_i$, $(1 \leq i \leq 4)$, then they are coplanar if, for example,
$$(\mathbf{r}_2 - \mathbf{r}_1) \bullet \left[(\mathbf{r}_3 - \mathbf{r}_1) \times (\mathbf{r}_4 - \mathbf{r}_1)\right] = 0$$
(or if they satisfy any similar such condition that asserts that the tetrahedron whose vertices they are has zero volume).

12. $x + y + z = \lambda$ is the family of all (parallel) planes normal to the vector $\mathbf{i} + \mathbf{j} + \mathbf{k}$.

13. $x + \lambda y + \lambda z = \lambda$ is the family of all planes containing the line of intersection of the planes $x = 0$ and $y + z = 1$, except the plane $y + z = 1$ itself. All these planes pass through the points $(0, 1, 0)$ and $(0, 0, 1)$.

14. The distance from the planes
$$\lambda x + \sqrt{1 - \lambda^2}\, y = 1$$
to the origin is $1/\sqrt{\lambda^2 + 1 - \lambda^2} = 1$. Hence the equation represents the family of all vertical planes at distance 1 from the origin. All such planes are tangent to the cylinder $x^2 + y^2 = 1$.

15. The line through $(1, 2, 3)$ parallel to $2\mathbf{i} - 3\mathbf{j} - 4\mathbf{k}$ has equations given in vector parametric form by
$$\mathbf{r} = (1 + 2t)\mathbf{i} + (2 - 3t)\mathbf{j} + (3 - 4t)\mathbf{k},$$

or in scalar parametric form by

$$x = 1 + 2t, \quad y = 2 - 3t, \quad z = 3 - 4t,$$

or in standard form by

$$\frac{x-1}{2} = \frac{y-2}{-3} = \frac{z-3}{-4}.$$

16. The line through $(-1, 0, 1)$ perpendicular to the plane $2x - y + 7z = 12$ is parallel to the normal vector $2\mathbf{i} - \mathbf{j} + 7\mathbf{k}$ to that plane. The equations of the line are, in vector parametric form,

$$\mathbf{r} = (-1 + 2t)\mathbf{i} - t\mathbf{j} + (1 + 7t)\mathbf{k},$$

or in scalar parametric form,

$$x = -1 + 2t, \quad y = -t, \quad z = 1 + 7t,$$

or in standard form

$$\frac{x+1}{2} = \frac{y}{-1} = \frac{z-1}{7}.$$

17. A line parallel to the line with equations

$$x + 2y - z = 2, \quad 2x - y + 4z = 5$$

is parallel to the vector

$$(\mathbf{i} + 2\mathbf{j} - \mathbf{k}) \times (2\mathbf{i} - \mathbf{j} + 4\mathbf{k}) = 7\mathbf{i} - 6\mathbf{j} - 5\mathbf{k}.$$

Since the line passes through the origin, it has equations

$$\mathbf{r} = 7t\mathbf{i} - 6t\mathbf{j} - 5t\mathbf{k} \quad \text{(vector parametric)}$$
$$x = 7t, \quad y = -6t, \quad z = -5t \quad \text{(scalar parametric)}$$
$$\frac{x}{7} = \frac{y}{-6} = \frac{z}{-5} \quad \text{(standard form)}.$$

18. A line parallel to $x + y = 0$ and to $x - y + 2z = 0$ is parallel to the cross product of the normal vectors to these two planes, that is, to the vector

$$(\mathbf{i} + \mathbf{j}) \times (\mathbf{i} - \mathbf{j} + 2\mathbf{k}) = 2(\mathbf{i} - \mathbf{j} - \mathbf{k}).$$

Since the line passes through $(2, -1, -1)$, its equations are, in vector parametric form

$$\mathbf{r} = (2 + t)\mathbf{i} - (1 + t)\mathbf{j} - (1 + t)\mathbf{k},$$

or in scalar parametric form

$$x = 2 + t, \quad y = -(1 + t), \quad z = -(1 + t),$$

or in standard form

$$x - 2 = -(y + 1) = -(z + 1).$$

19. A line making equal angles with the positive directions of the coordinate axes is parallel to the vector $\mathbf{i} + \mathbf{j} + \mathbf{k}$. If the line passes through the point $(1, 2, -1)$, then it has equations

$$\mathbf{r} = (1 + t)\mathbf{i} + (2 + t)\mathbf{j} + (-1 + t)\mathbf{k} \quad \text{(vector parametric)}$$
$$x = 1 + t, \quad y = 2 + t, \quad z = -1 + t \quad \text{(scalar parametric)}$$
$$x - 1 = y - 2 = z + 1 \quad \text{(standard form)}.$$

20. The line $\mathbf{r} = (1 - 2t)\mathbf{i} + (4 + 3t)\mathbf{j} + (9 - 4t)\mathbf{k}$ has standard form

$$\frac{x-1}{-2} = \frac{y-4}{3} = \frac{z-9}{-4}.$$

21. The line $\begin{cases} x = 4 - 5t \\ y = 3t \\ z = 7 \end{cases}$ has standard form

$$\frac{x-4}{-5} = \frac{y}{3}, \quad z = 7.$$

22. The line $\begin{cases} x - 2y + 3z = 0 \\ 2x + 3y - 4z = 4 \end{cases}$ is parallel to the vector

$$(\mathbf{i} - 2\mathbf{j} + 3\mathbf{k}) \times (2\mathbf{i} + 3\mathbf{j} - 4\mathbf{k}) = -\mathbf{i} + 10\mathbf{j} + 7\mathbf{k}.$$

We need a point on this line. Putting $z = 0$, we get

$$x - 2y = 0, \quad 2x + 3y = 4.$$

The solution of this system is $y = 4/7$, $x = 8/7$. A possible standard form for the given line is

$$\frac{x - \frac{8}{7}}{-1} = \frac{y - \frac{4}{7}}{10} = \frac{z}{7},$$

though, of course, this answer is not unique as the coordinates of any point on the line could have been used.

23. Let $\mathbf{r}_i$ be the position vector of P_i ($1 \le i \le 4$). The line $P_1 P_2$ intersects the line $P_3 P_4$ in a unique point if the four points are coplanar, and $P_1 P_2$ is not parallel to $P_3 P_4$. It is therefore sufficient that

$$(\mathbf{r}_2 - \mathbf{r}_1) \times (\mathbf{r}_4 - \mathbf{r}_3) \ne \mathbf{0}, \text{ and}$$
$$(\mathbf{r}_3 - \mathbf{r}_1) \bullet \big[(\mathbf{r}_2 - \mathbf{r}_1) \times (\mathbf{r}_4 - \mathbf{r}_3)\big] = 0.$$

(Other similar answers are possible.)

24. The distance from $(0, 0, 0)$ to $x + 2y + 3z = 4$ is

$$\frac{4}{\sqrt{1^2 + 2^2 + 3^2}} = \frac{4}{\sqrt{14}} \text{ units.}$$

25. The distance from $(1, 2, 0)$ to $3x - 4y - 5z = 2$ is

$$\frac{|3 - 8 - 0 - 2|}{\sqrt{3^2 + 4^2 + 5^2}} = \frac{7}{5\sqrt{2}} \text{ units.}$$

26. A vector parallel to the line $x + y + z = 0$, $2x - y - 5z = 1$ is

$$\mathbf{a} = (\mathbf{i} + \mathbf{j} + \mathbf{k}) \times (2\mathbf{i} - \mathbf{j} - 5\mathbf{k}) = -4\mathbf{i} + 7\mathbf{j} - 3\mathbf{k}.$$

We need a point on this line: if $z = 0$ then $x + y = 0$ and $2x - y = 1$, so $x = 1/3$ and $y = -1/3$. The position vector of this point is

$$\mathbf{r}_1 = \frac{1}{3}\mathbf{i} - \frac{1}{3}\mathbf{j}.$$

The distance from the origin to the line is

$$s = \frac{|\mathbf{r}_1 \times \mathbf{a}|}{|\mathbf{a}|} = \frac{|\mathbf{i} + \mathbf{j} + \mathbf{k}|}{\sqrt{74}} = \sqrt{\frac{3}{74}} \text{ units.}$$

27. The line $\begin{cases} x + 2y = 3 \\ y + 2z = 3 \end{cases}$ contains the points $(1, 1, 1)$ and $(3, 0, 3/2)$, so is parallel to the vector $2\mathbf{i} - \mathbf{j} + \frac{1}{2}\mathbf{k}$, or to $4\mathbf{i} - 2\mathbf{j} + \mathbf{k}$.

The line $\begin{cases} x + y + z = 6 \\ x - 2z = -5 \end{cases}$ contains the points $(-5, 11, 0)$ and $(-1, 5, 2)$, and so is parallel to the vector $4\mathbf{i} - 6\mathbf{j} + 2\mathbf{k}$, or to $2\mathbf{i} - 3\mathbf{j} + \mathbf{k}$.

Using the values

$$\mathbf{r}_1 = \mathbf{i} + \mathbf{j} + \mathbf{k} \qquad \mathbf{a}_1 = 4\mathbf{i} - 2\mathbf{j} + \mathbf{k}$$
$$\mathbf{r}_2 = -\mathbf{i} + 5\mathbf{j} + 2\mathbf{k} \qquad \mathbf{a}_2 = 2\mathbf{i} - 3\mathbf{j} + \mathbf{k},$$

we calculate the distance between the two lines by the formula in Section 10.4 as

$$s = \frac{|(\mathbf{r}_1 - \mathbf{r}_2) \bullet (\mathbf{a}_1 \times \mathbf{a}_2)|}{|\mathbf{a}_1 \times \mathbf{a}_2|}$$
$$= \frac{|(2\mathbf{i} - 4\mathbf{j} - \mathbf{k}) \bullet (\mathbf{i} - 2\mathbf{j} - 8\mathbf{k})|}{|\mathbf{i} - 2\mathbf{j} - 8\mathbf{k}|}$$
$$= \frac{18}{\sqrt{69}} \text{ units.}$$

28. The line $x - 2 = \frac{y + 3}{2} = \frac{z - 1}{4}$ passes through the point $(2, -3, 1)$, and is parallel to $\mathbf{a} = \mathbf{i} + 2\mathbf{j} + 4\mathbf{k}$.
The plane $2y - z = 1$ has normal $\mathbf{n} = 2\mathbf{j} - \mathbf{k}$.
Since $\mathbf{a} \bullet \mathbf{n} = 0$, the line is parallel to the plane.
The distance from the line to the plane is equal to the distance from $(2, -3, 1)$ to the plane $2y - z = 1$, so is

$$D = \frac{|-6 - 1 - 1|}{\sqrt{4 + 1}} = \frac{8}{\sqrt{5}} \text{ units.}$$

29. $(1 - \lambda)(x - x_0) = \lambda(y - y_0)$ represents any line in the xy-plane passing through (x_0, y_0). Therefore, in 3-space the pair of equations

$$(1 - \lambda)(x - x_0) = \lambda(y - y_0), \qquad z = z_0$$

represents all straight lines in the plane $z = z_0$ which pass through the point (x_0, y_0, z_0).

30. $\frac{x - x_0}{\sqrt{1 - \lambda^2}} = \frac{y - y_0}{\lambda} = z - z_0$ represents all lines through (x_0, y_0, z_0) parallel to the vectors

$$\mathbf{a} = \sqrt{1 - \lambda^2}\mathbf{i} + \lambda\mathbf{j} + \mathbf{k}.$$

All such lines are generators of the circular cone

$$(z - z_0)^2 = (x - x_0)^2 + (y - y_0)^2,$$

so the given equations specify all straight lines lying on that cone.

31. The equation

$$(A_1 x + B_1 y + C_1 z + D_1)(A_2 x + B_2 y + C_2 z + D_2) = 0$$

is satisfied if either $A_1 x + B_1 y + C_1 z + D_1 = 0$ or $A_2 x + B_2 y + C_2 z + D_2 = 0$, that is, if (a, y, z) lies on either of these planes. It is not necessary that the point lie on both planes, so the given equation represents all the points on each of the planes, not just those on the line of intersection of the planes.

Section 10.5 Quadric Surfaces (page 635)

1. $x^2 + 4y^2 + 9z^2 = 36$
$$\frac{x^2}{6^2} + \frac{y^2}{3^2} + \frac{z^2}{2^2} = 1$$
This is an ellipsoid with centre at the origin and semi-axes 6, 3, and 2.

2. $x^2 + y^2 + 4z^2 = 4$ represents an oblate spheroid, that is, an ellipsoid with its two longer semi-axes equal. In this case the longer semi-axes have length 2, and the shorter one (in the z direction) has length 1. Cross-sections in planes perpendicular to the z-axis between $z = -1$ and $z = 1$ are circles.

3. $2x^2 + 2y^2 + 2z^2 - 4x + 8y - 12z + 27 = 0$
$2(x^2 - 2x + 1) + 2(y^2 + 4y + 4) + 2(z^2 - 6z + 9)$
$= -27 + 2 + 8 + 18$
$$(x - 1)^2 + (y + 2)^2 + (z - 3)^2 = \frac{1}{2}$$
This is a sphere with radius $1/\sqrt{2}$ and centre $(1, -2, 3)$.

391

4. $x^2 + 4y^2 + 9z^2 + 4x - 8y = 8$
$(x+2)^2 + 4(y-1)^2 + 9z^2 = 8 + 8 = 16$
$$\frac{(x+2)^2}{4^2} + \frac{(y-1)^2}{2^2} + \frac{z^2}{(4/3)^2} = 1$$
This is an ellipsoid with centre $(-2, 1, 0)$ and semi-axes 4, 2, and 4/3.

5. $z = x^2 + 2y^2$ represents an elliptic paraboloid with vertex at the origin and axis along the positive z-axis. Cross-sections in planes $z = k > 0$ are ellipses with semi-axes $\sqrt{k}$ and $\sqrt{k/2}$.

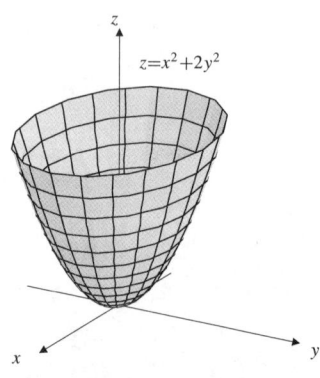

Fig. 10.5.5

6. $z = x^2 - 2y^2$ represents a hyperbolic paraboloid.

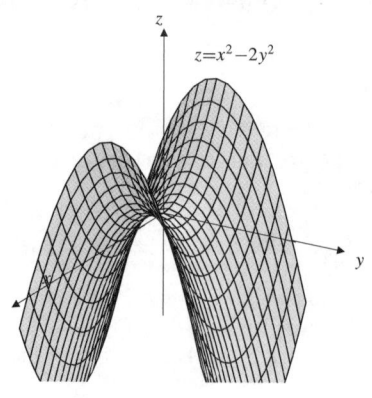

Fig. 10.5.6

7. $x^2 - y^2 - z^2 = 4$ represents a hyperboloid of two sheets with vertices at $(\pm 2, 0, 0)$ and circular cross-sections in planes $x = k$, where $|k| > 2$.

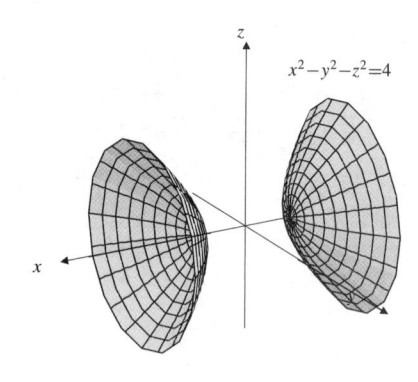

Fig. 10.5.7

8. $-x^2 + y^2 + z^2 = 4$ represents a hyperboloid of one sheet, with circular cross-sections in all planes perpendicular to the x-axis.

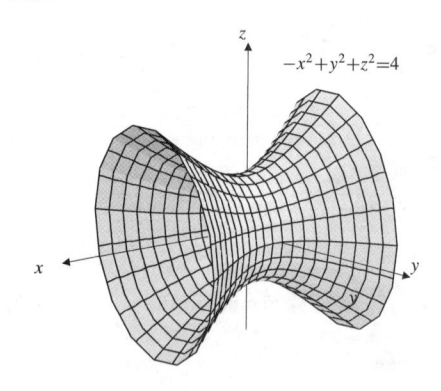

Fig. 10.5.8

9. $z = xy$ represents a hyperbolic paraboloid containing the x- and y-axes.

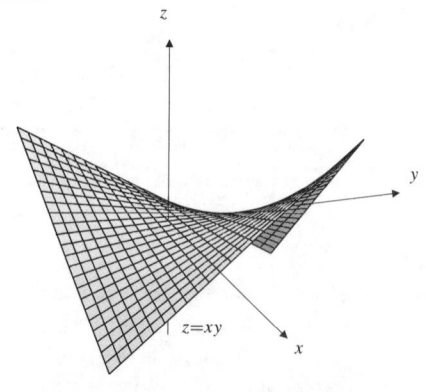

Fig. 10.5.9

10. $x^2 + 4z^2 = 4$ represents an elliptic cylinder with axis along the y-axis.

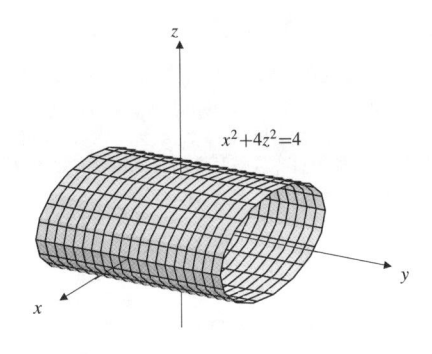

INSTRUCTOR'S SOLUTIONS MANUAL   SECTION 10.5 (PAGE 635)

Fig. 10.5.10

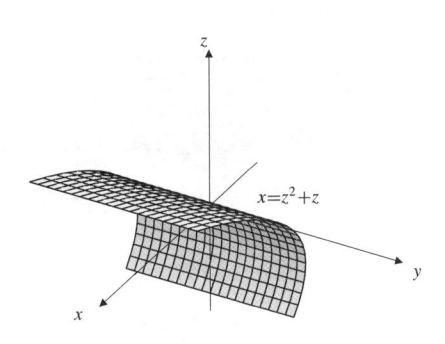

Fig. 10.5.13

11. $x^2 - 4z^2 = 4$ represents a hyperbolic cylinder with axis along the y-axis.

14. $x^2 = y^2 + 2z^2$ represents an elliptic cone with vertex at the origin and axis along the x-axis.

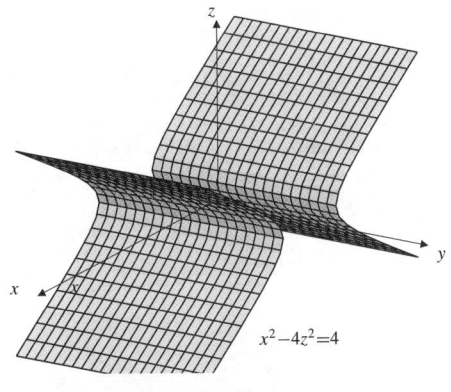

Fig. 10.5.11

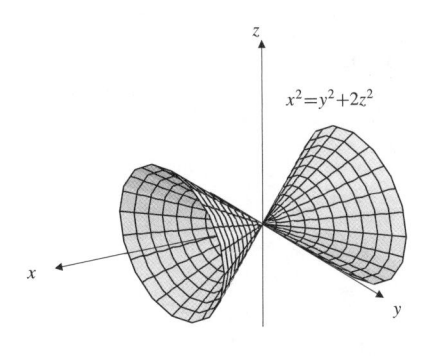

Fig. 10.5.14

12. $y = z^2$ represents a parabolic cylinder with vertex line along the x-axis.

15. $(z-1)^2 = (x-2)^2 + (y-3)^2$ represents a circular cone with axis along the line $x = 2$, $y = 3$, and vertex at $(2, 3, 1)$

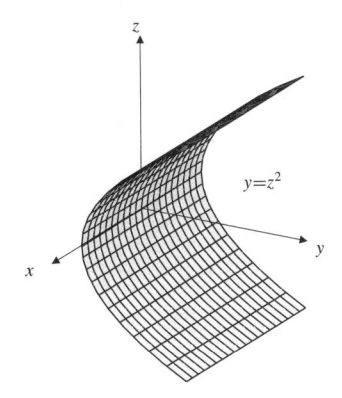

Fig. 10.5.12

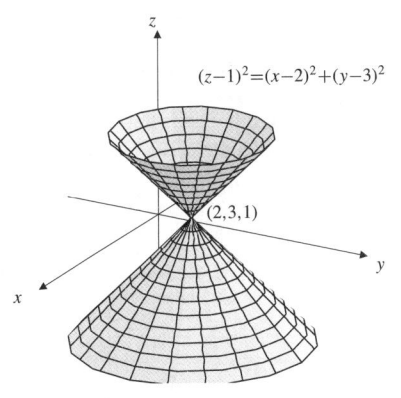

Fig. 10.5.15

13. $x = z^2 + z = \left(z + \dfrac{1}{2}\right)^2 - \dfrac{1}{4}$ represents a parabolic cylinder with vertex line along the line $z = -1/2$, $x = -1/4$.

16. $(z-1)^2 = (x-2)^2 + (y-3)^2 + 4$ represents a hyperboloid of two sheets with centre at $(2, 3, 1)$, axis along the line $x = 2$, $y = 3$, and vertices at $(2, 3, -1)$ and $(2, 3, 3)$.

393

SECTION 10.5 (PAGE 635)

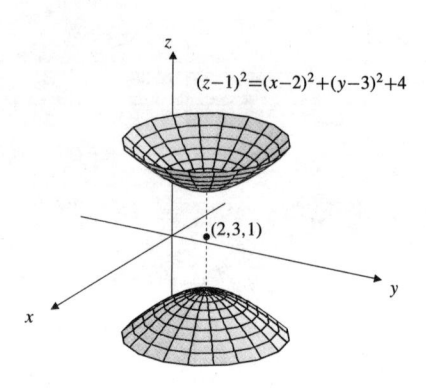

Fig. 10.5.16

17. $\begin{cases} x^2 + y^2 + z^2 = 4 \\ x + y + z = 1 \end{cases}$ represents the circle of intersection of a sphere and a plane. The circle lies in the plane $x + y + z = 1$, and has centre $(1/3, 1/3, 1/3)$ and radius $\sqrt{4 - (3/9)} = \sqrt{11/3}$.

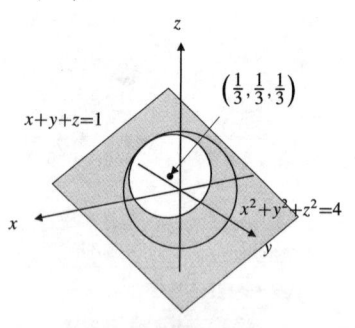

Fig. 10.5.17

18. $\begin{cases} x^2 + y^2 = 1 \\ z = x + y \end{cases}$ is the ellipse of intersection of the plane $z = x + y$ and the circular cylinder $x^2 + y^2 = 1$. The centre of the ellipse is at the origin, and the ends of the major axis are $\pm(1/\sqrt{2}, 1/\sqrt{2}, \sqrt{2})$.

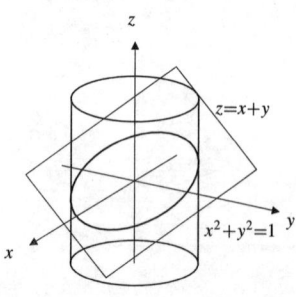

Fig. 10.5.18

19. $\begin{cases} z^2 = x^2 + y^2 \\ z = 1 + x \end{cases}$ is the parabola in which the plane $z = 1 + x$ intersects the circular cone $z^2 = x^2 + y^2$. (It is a parabola because the plane is parallel to a generator of the cone, namely the line $z = x$, $y = 0$.) The vertex of the parabola is $(-1/2, 0, 1/2)$, and its axis is along the line $y = 0$, $z = 1 + x$.

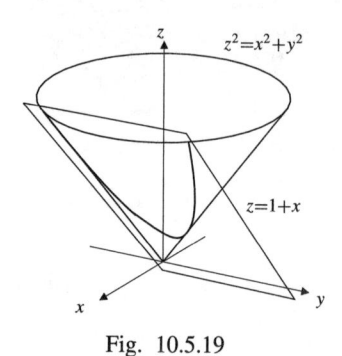

Fig. 10.5.19

20. $\begin{cases} x^2 + 2y^2 + 3z^2 = 6 \\ y = 1 \end{cases}$ is an ellipse in the plane $y = 1$. Its projection onto the xz-plane is the ellipse $x^2 + 3z^2 = 4$. One quarter of the ellipse is shown in the figure.

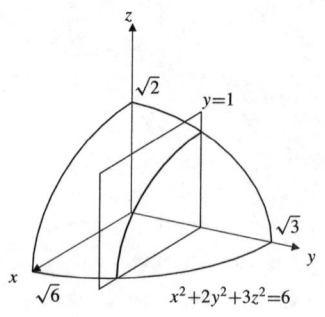

Fig. 10.5.20

21. $\dfrac{x^2}{a^2} + \dfrac{y^2}{b^2} - \dfrac{z^2}{c^2} = 1$

$\dfrac{x^2}{a^2} - \dfrac{z^2}{c^2} = 1 - \dfrac{y^2}{b^2}$

$\left(\dfrac{x}{a} + \dfrac{z}{c}\right)\left(\dfrac{x}{a} - \dfrac{z}{c}\right) = \left(1 + \dfrac{y}{b}\right)\left(1 - \dfrac{y}{b}\right)$

Family 1: $\begin{cases} \dfrac{x}{a} + \dfrac{z}{c} = \lambda\left(1 + \dfrac{y}{b}\right) \\ \lambda\left(\dfrac{x}{a} - \dfrac{z}{c}\right) = 1 - \dfrac{y}{b}. \end{cases}$

Family 2: $\begin{cases} \dfrac{x}{a} + \dfrac{z}{c} = \mu\left(1 - \dfrac{y}{b}\right) \\ \mu\left(\dfrac{x}{a} - \dfrac{z}{c}\right) = 1 + \dfrac{y}{b}. \end{cases}$

22. $z = xy$

Family 1: $\begin{cases} z = \lambda x \\ \lambda = y. \end{cases}$

394

Family 2: $\begin{cases} z = \mu y \\ \mu = x. \end{cases}$

23. The cylinder $2x^2 + y^2 = 1$ intersects horizontal planes in ellipses with semi-axes 1 in the y direction and $1/\sqrt{2}$ in the x direction. Tilting the plane in the x direction will cause the shorter semi-axis to increase in length. The plane $z = cx$ intersects the cylinder in an ellipse with principal axes through the points $(0, \pm 1, 0)$ and $(\pm 1/\sqrt{2}, 0, \pm c/\sqrt{2})$. The semi-axes will be equal (and the ellipse will be a circle) if $(1/2) + (c^2/2) = 1$, that is, if $c = \pm 1$. Thus cross-sections of the cylinder perpendicular to the vectors $\mathbf{a} = \mathbf{i} \pm \mathbf{k}$ are circular.

24. The plane $z = cx + k$ intersects the elliptic cone $z^2 = 2x^2 + y^2$ on the cylinder

$$c^2 x^2 + 2ckx + k^2 = 2x^2 + y^2$$
$$(2 - c^2) x^2 - 2ckx + y^2 = k^2$$
$$(2 - c^2)\left(x - \frac{ck}{2 - c^2}\right)^2 + y^2 = k^2 + \frac{c^2 k^2}{2 - c^2} = \frac{2k^2}{2 - c^2}$$
$$\frac{(x - x_0)^2}{a^2} + \frac{y^2}{b^2} = 1,$$

where $x_0 = \frac{ck}{2 - c^2}$, $a^2 = \frac{2k^2}{(2 - c^2)^2}$, and $b^2 = \frac{2k^2}{2 - c^2}$.

As in the previous exercise, $z = cx + k$ intersects the cylinder (and hence the cone) in an ellipse with principal axes joining the points

$(x_0 - a, 0, c(x_0 - a) + k)$ to $(x_0 + a, 0, c(x_0 + a) + k)$,
and $(x_0, -b, cx_0 + k)$ to $(x_0, b, cx_0 + k)$.

The centre of this ellipse is $(x_0, 0, cx_0 + k)$. The ellipse is a circle if its two semi-axes have equal lengths, that is, if

$$a^2 + c^2 a^2 = b^2,$$

that is,

$$(1 + c^2) \frac{2k^2}{(2 - c^2)^2} = \frac{2k^2}{2 - c^2},$$

or $1 + c^2 = 2 - c^2$. Thus $c = \pm 1/\sqrt{2}$. A vector normal to the plane $z = \pm(x/\sqrt{2}) + k$ is $\mathbf{a} = \mathbf{i} \pm \sqrt{2}\mathbf{k}$.

Section 10.6 A Little Matrix Algebra (page 643)

1. $\begin{pmatrix} 3 & 0 & -2 \\ 1 & 1 & 2 \\ -1 & 1 & -1 \end{pmatrix} \begin{pmatrix} 2 & 1 \\ 3 & 0 \\ 0 & -2 \end{pmatrix} = \begin{pmatrix} 6 & 7 \\ 5 & -3 \\ 1 & 1 \end{pmatrix}$

2. $\begin{pmatrix} 1 & 1 & 1 \\ 0 & 1 & 1 \\ 0 & 0 & 1 \end{pmatrix} \begin{pmatrix} 1 & 1 & 1 \\ 0 & 1 & 1 \\ 0 & 0 & 1 \end{pmatrix} = \begin{pmatrix} 1 & 2 & 3 \\ 0 & 1 & 2 \\ 0 & 0 & 1 \end{pmatrix}$

3. $\begin{pmatrix} a & b \\ c & d \end{pmatrix} \begin{pmatrix} w & x \\ y & z \end{pmatrix} = \begin{pmatrix} aw + by & ax + bz \\ cw + dy & cx + dz \end{pmatrix}$

4. $\begin{pmatrix} w & x \\ y & z \end{pmatrix} \begin{pmatrix} a & b \\ c & d \end{pmatrix} = \begin{pmatrix} aw + cx & bw + dx \\ ay + cz & by + dz \end{pmatrix}$

5. $\mathcal{A}\mathcal{A}^T = \begin{pmatrix} 1 & 1 & 1 & 1 \\ 0 & 1 & 1 & 1 \\ 0 & 0 & 1 & 1 \\ 0 & 0 & 0 & 1 \end{pmatrix} \begin{pmatrix} 1 & 0 & 0 & 0 \\ 1 & 1 & 0 & 0 \\ 1 & 1 & 1 & 0 \\ 1 & 1 & 1 & 1 \end{pmatrix}$

$= \begin{pmatrix} 4 & 3 & 2 & 1 \\ 3 & 3 & 2 & 1 \\ 2 & 2 & 2 & 1 \\ 1 & 1 & 1 & 1 \end{pmatrix}$

$\mathcal{A}^2 = \begin{pmatrix} 1 & 1 & 1 & 1 \\ 0 & 1 & 1 & 1 \\ 0 & 0 & 1 & 1 \\ 0 & 0 & 0 & 1 \end{pmatrix} \begin{pmatrix} 1 & 1 & 1 & 1 \\ 0 & 1 & 1 & 1 \\ 0 & 0 & 1 & 1 \\ 0 & 0 & 0 & 1 \end{pmatrix}$

$= \begin{pmatrix} 1 & 2 & 3 & 4 \\ 0 & 1 & 2 & 3 \\ 0 & 0 & 1 & 2 \\ 0 & 0 & 0 & 1 \end{pmatrix}$

6. $\mathbf{x} = \begin{pmatrix} x \\ y \\ z \end{pmatrix}, \quad \mathcal{A} = \begin{pmatrix} a & p & q \\ p & b & r \\ q & r & c \end{pmatrix}$

$\mathbf{x}\mathbf{x}^T = \begin{pmatrix} x \\ y \\ z \end{pmatrix} (x, y, z) = \begin{pmatrix} x^2 & xy & xz \\ xy & y^2 & yz \\ xz & yz & z^2 \end{pmatrix}$

$\mathbf{x}^T \mathbf{x} = (x, y, z) \begin{pmatrix} x \\ y \\ z \end{pmatrix} = (x^2 + y^2 + z^2)$

$\mathbf{x}^T \mathcal{A} \mathbf{x} = (x, y, z) \begin{pmatrix} a & p & q \\ p & b & r \\ q & r & c \end{pmatrix} \begin{pmatrix} x \\ y \\ z \end{pmatrix}$

$= (x, y, z) \begin{pmatrix} ax + py + qz \\ px + by + rz \\ qx + ry + cz \end{pmatrix}$

$= ax^2 + by^2 + cz^2 + 2pxy + 2qxz + 2ryz$

7. $\begin{vmatrix} 2 & 3 & -1 & 0 \\ 4 & 0 & 2 & 1 \\ 1 & 0 & -1 & 1 \\ -2 & 0 & 0 & 1 \end{vmatrix} = -3 \begin{vmatrix} 4 & 2 & 1 \\ 1 & -1 & 1 \\ -2 & 0 & 1 \end{vmatrix}$

$= -3 \left(-2 \begin{vmatrix} 1 & 1 \\ -2 & 1 \end{vmatrix} - 1 \begin{vmatrix} 4 & 1 \\ -2 & 1 \end{vmatrix} \right)$

$= 6(3) + 3(6) = 36$

8. $\begin{vmatrix} 1 & 1 & 1 & 1 \\ 1 & 2 & 3 & 4 \\ -2 & 0 & 2 & 4 \\ 3 & -3 & 2 & -2 \end{vmatrix}$

$= -2 \begin{vmatrix} 1 & 1 & 1 \\ 2 & 3 & 4 \\ -3 & 2 & -2 \end{vmatrix} + 2 \begin{vmatrix} 1 & 1 & 1 \\ 1 & 2 & 4 \\ 3 & -3 & -2 \end{vmatrix}$

$\quad - 4 \begin{vmatrix} 1 & 1 & 1 \\ 1 & 2 & 3 \\ 3 & -3 & 2 \end{vmatrix}$

$= -2 \begin{vmatrix} 1 & 1 & 1 \\ 0 & 1 & 2 \\ 0 & 5 & 1 \end{vmatrix} + 2 \begin{vmatrix} 1 & 1 & 1 \\ 0 & 1 & 3 \\ 0 & -6 & -5 \end{vmatrix}$

$\quad - 4 \begin{vmatrix} 1 & 1 & 1 \\ 0 & 1 & 2 \\ 0 & -6 & -1 \end{vmatrix}$

$= -2 \begin{vmatrix} 1 & 2 \\ 5 & 1 \end{vmatrix} + 2 \begin{vmatrix} 1 & 3 \\ -6 & -5 \end{vmatrix} - 4 \begin{vmatrix} 1 & 2 \\ -6 & -1 \end{vmatrix}$

$= -2(-9) + 2(13) - 4(11) = 0$

9. $\begin{vmatrix} a_{11} & a_{12} & a_{13} & \cdots & a_{1n} \\ 0 & a_{22} & a_{23} & \cdots & a_{2n} \\ 0 & 0 & a_{33} & \cdots & a_{3n} \\ \vdots & \vdots & \vdots & \ddots & \vdots \\ 0 & 0 & 0 & \cdots & a_{nn} \end{vmatrix}$

$= a_{11} \begin{vmatrix} a_{22} & a_{23} & \cdots & a_{2n} \\ 0 & a_{33} & \cdots & a_{3n} \\ \vdots & \vdots & \ddots & \vdots \\ 0 & 0 & \cdots & a_{nn} \end{vmatrix}$

$= a_{11} a_{22} \begin{vmatrix} a_{33} & \cdots & a_{3n} \\ \vdots & \ddots & \vdots \\ 0 & \cdots & a_{nn} \end{vmatrix}$

$= a_{11} a_{22} a_{33} \cdots a_{nn}$

(or use induction on n)

10. $\begin{vmatrix} 1 & 1 \\ x & y \end{vmatrix} = y - x$. If

$$f(x, y, z) = \begin{vmatrix} 1 & 1 & 1 \\ x & y & z \\ x^2 & y^2 & z^2 \end{vmatrix},$$

then f is a polynomial of degree 2 in z.
Since $f(x, y, x) = 0$ and $f(x, y, y) = 0$, we must have $f(x, y, z) = A(z - x)(z - y)$ for some A independent of z. But

$$Axy = f(x, y, 0) = \begin{vmatrix} 1 & 1 & 1 \\ x & y & 0 \\ x^2 & y^2 & 0 \end{vmatrix} = xy(y - x),$$

so $A = y - x$ and

$$f(x, y, z) = (y - x)(z - x)(z - y).$$

Generalization:

$$\begin{vmatrix} 1 & 1 & 1 & \cdots & 1 \\ x_1 & x_2 & x_3 & \cdots & x_n \\ x_1^2 & x_2^2 & x_3^2 & \cdots & x_n^2 \\ \vdots & \vdots & \vdots & \ddots & \vdots \\ x_1^{n-1} & x_2^{n-1} & x_3^{n-1} & \cdots & x_n^{n-1} \end{vmatrix} = \prod_{1 \le i < j \le n} (x_j - x_i).$$

11. Let $\mathcal{A} = \begin{pmatrix} a & b \\ c & d \end{pmatrix}$, $\mathcal{B} = \begin{pmatrix} \ell & m \\ n & p \end{pmatrix}$, $\mathcal{C} = \begin{pmatrix} w & x \\ y & z \end{pmatrix}$. Then

$(\mathcal{AB})\mathcal{C} = \begin{pmatrix} a\ell + bn & am + bp \\ c\ell + dn & cm + dp \end{pmatrix} \begin{pmatrix} w & x \\ y & z \end{pmatrix}$

$= \begin{pmatrix} a\ell w + bnw + amy + bpy & a\ell x + bnx + amz + bpz \\ c\ell w + dnw + cmy + dpy & c\ell x + dnx + cmz + dpz \end{pmatrix}$

$\mathcal{A}(\mathcal{BC}) = \begin{pmatrix} a & b \\ c & d \end{pmatrix} \begin{pmatrix} \ell w + my & \ell x + mz \\ nw + py & nx + pz \end{pmatrix}$

$= \begin{pmatrix} a\ell w + amy + bnw + bpy & a\ell x + amz + bnx + bpz \\ c\ell w + cmy + dnw + dpy & c\ell x + cmz + dnx + dpz \end{pmatrix}$

Thus $(\mathcal{AB})\mathcal{C} = \mathcal{A}(\mathcal{BC})$.

12. If $\mathcal{A} = \begin{pmatrix} a & b \\ c & d \end{pmatrix}$, then $\mathcal{A}^T = \begin{pmatrix} a & c \\ b & d \end{pmatrix}$, and

$$\det(\mathcal{A}) = ad - bc = \det(\mathcal{A}^T).$$

We generalize this by induction.
Suppose $\det(\mathcal{B}^T) = \det(\mathcal{B})$ for any $(n-1) \times (n-1)$ matrix, where $n \ge 3$. Let

$$\mathcal{A} = \begin{pmatrix} a_{11} & a_{12} & \cdots & a_{1n} \\ a_{21} & a_{22} & \cdots & a_{2n} \\ \vdots & \vdots & \ddots & \vdots \\ a_{n1} & a_{n2} & \cdots & a_{nn} \end{pmatrix}$$

be an $n \times n$ matrix. If $\det(\mathcal{A})$ is expanded in minors about the first row, and $\det(\mathcal{A}^T)$ is expanded in minors about the first column, the corresponding terms in these expansions are equal by the induction hypothesis. (The $(n-1)\times(n-1)$ matrices whose determinants appear in one expansion are the transposes of those in the other expansion.) Therefore $\det(\mathcal{A}^T) = \det(\mathcal{A})$ for any square matrix $\mathcal{A}$.

13. Let $\mathcal{A} = \begin{pmatrix} a & b \\ c & d \end{pmatrix}$ and $\mathcal{B} = \begin{pmatrix} w & x \\ y & z \end{pmatrix}$. Then

$$\mathcal{AB} = \begin{pmatrix} aw + by & ax + bz \\ cw + dy & cx + dz \end{pmatrix}.$$

Therefore,

$\det(\mathcal{A})\det(\mathcal{B}) = (ad - bc)(wz - xy)$
$= adwz - adxy - bcwz + bcxy$
$\det(\mathcal{AB}) = (aw + by)(cx + dz) - (ax + bz)(cw + dy)$
$= awcx + awdz + bycx + bydz$
$\quad - axwc - axdy - bzcw - bzdy$
$= adwz - adxy - bcwz + bcxy$
$= \det(\mathcal{A})\det(\mathcal{B}).$

14. If $\mathcal{A}_\theta = \begin{pmatrix} \cos\theta & \sin\theta \\ -\sin\theta & \cos\theta \end{pmatrix}$, then

$$\mathcal{A}_{-\theta} = \begin{pmatrix} \cos(-\theta) & \sin(-\theta) \\ -\sin(-\theta) & \cos(-\theta) \end{pmatrix} = \begin{pmatrix} \cos\theta & -\sin\theta \\ \sin\theta & \cos\theta \end{pmatrix},$$

and

$$\mathcal{A}_\theta \mathcal{A}_{-\theta} = \begin{pmatrix} 1 & 0 \\ 0 & 1 \end{pmatrix} = I.$$

Thus $\mathcal{A}_{-\theta} = (\mathcal{A}_\theta)^{-1}$.

15. Let $\mathcal{A} = \begin{pmatrix} 1 & 1 & 1 \\ 0 & 1 & 1 \\ 0 & 0 & 1 \end{pmatrix}$, $\mathcal{A}^{-1} = \begin{pmatrix} a & b & c \\ d & e & f \\ g & h & i \end{pmatrix}$. Since $\mathcal{A}\mathcal{A}^{-1} = I$ we must have

$$\begin{array}{lll} a+d+g=1 & b+e+h=0 & c+f+i=0 \\ d+g=0 & e+h=1 & f+i=0 \\ g=0 & h=0 & i=1. \end{array}$$

Thus $a=1$, $d=g=0$, $h=0$, $e=1$, $b=-1$, $i=1$, $f=-1$, $c=0$, and so

$$\mathcal{A}^{-1} = \begin{pmatrix} 1 & -1 & 0 \\ 0 & 1 & -1 \\ 0 & 0 & 1 \end{pmatrix}.$$

16. Let $\mathcal{A} = \begin{pmatrix} 1 & 0 & -1 \\ -1 & 1 & 0 \\ 2 & 1 & 3 \end{pmatrix}$, $\mathcal{A}^{-1} = \begin{pmatrix} a & b & c \\ d & e & f \\ g & h & i \end{pmatrix}$. Since $\mathcal{A}\mathcal{A}^{-1} = I$ we must have

$$\begin{array}{lll} a-g=1 & b-h=0 & c-i=0 \\ -a+d=0 & -b+e=1 & -c+f=0 \\ 2a+d+3g=0 & 2b+e+3h=0 & 2c+f+3i=1. \end{array}$$

Solving these three systems of equations, we get

$$\mathcal{A}^{-1} = \begin{pmatrix} \tfrac{1}{2} & -\tfrac{1}{6} & \tfrac{1}{6} \\ \tfrac{1}{2} & \tfrac{5}{6} & \tfrac{1}{6} \\ -\tfrac{1}{2} & -\tfrac{1}{6} & \tfrac{1}{6} \end{pmatrix}.$$

17. The given system of equations is

$$\mathcal{A} \begin{pmatrix} x \\ y \\ z \end{pmatrix} = \begin{pmatrix} -2 \\ 1 \\ 13 \end{pmatrix}.$$

Thus

$$\begin{pmatrix} x \\ y \\ z \end{pmatrix} = \mathcal{A}^{-1} \begin{pmatrix} -2 \\ 1 \\ 13 \end{pmatrix} = \begin{pmatrix} 1 \\ 2 \\ 3 \end{pmatrix},$$

so $x=1$, $y=2$, and $z=3$.

18. If $\mathcal{A}$ is the matrix of Exercises 16 and 17 then $\det(\mathcal{A}) = 6$. By Cramer's Rule,

$$x = \frac{1}{6} \begin{vmatrix} -2 & 0 & -1 \\ 1 & 1 & 0 \\ 13 & 1 & 3 \end{vmatrix} = \frac{6}{6} = 1$$

$$y = \frac{1}{6} \begin{vmatrix} 1 & -2 & -1 \\ -1 & 1 & 0 \\ 2 & 13 & 3 \end{vmatrix} = \frac{12}{6} = 2$$

$$z = \frac{1}{6} \begin{vmatrix} 1 & 0 & -2 \\ -1 & 1 & 1 \\ 2 & 1 & 13 \end{vmatrix} = \frac{18}{6} = 3.$$

19. $\mathcal{A} = \begin{pmatrix} 1 & 1 & 1 & 1 \\ 1 & 1 & 1 & -1 \\ 1 & 1 & -1 & -1 \\ 1 & -1 & -1 & -1 \end{pmatrix}$

$$\det(\mathcal{A}) = \begin{vmatrix} 0 & 0 & 0 & 2 \\ 0 & 0 & 2 & 0 \\ 0 & 2 & 0 & 0 \\ 1 & -1 & -1 & -1 \end{vmatrix}$$

$$= -2 \begin{vmatrix} 0 & 0 & 2 \\ 0 & 2 & 0 \\ 1 & -1 & -1 \end{vmatrix} = -4 \begin{vmatrix} 0 & 2 \\ 1 & -1 \end{vmatrix} = 8$$

$$x_1 = \frac{1}{8} \begin{vmatrix} 0 & 1 & 1 & 1 \\ 4 & 1 & 1 & -1 \\ 6 & 1 & -1 & -1 \\ 2 & -1 & -1 & -1 \end{vmatrix}$$

$$= \frac{1}{8} \begin{vmatrix} 0 & 1 & 1 & 1 \\ 4 & 0 & 0 & -2 \\ 6 & 2 & 0 & 0 \\ 2 & 0 & 0 & 0 \end{vmatrix}$$

$$= -\frac{2}{8} \begin{vmatrix} 1 & 1 & 1 \\ 0 & 0 & -2 \\ 2 & 0 & 0 \end{vmatrix} = -\frac{4}{8} \begin{vmatrix} 1 & 1 \\ 0 & -2 \end{vmatrix} = 1$$

$$x_2 = \frac{1}{8} \begin{vmatrix} 1 & 0 & 1 & 1 \\ 1 & 4 & 1 & -1 \\ 1 & 6 & -1 & -1 \\ 1 & 2 & -1 & -1 \end{vmatrix}$$

$$= \frac{1}{8} \begin{vmatrix} 2 & 0 & 0 & 1 \\ 0 & 4 & 2 & -1 \\ 0 & 6 & 0 & -1 \\ 0 & 2 & 0 & -1 \end{vmatrix}$$

$$= \frac{2}{8} \begin{vmatrix} 4 & 2 & -1 \\ 6 & 0 & -1 \\ 2 & 0 & -1 \end{vmatrix} = \frac{-4}{8} \begin{vmatrix} 6 & -1 \\ 2 & -1 \end{vmatrix} = 2$$

$$x_3 = \frac{1}{8} \begin{vmatrix} 1 & 1 & 0 & 1 \\ 1 & 1 & 4 & -1 \\ 1 & 1 & 6 & -1 \\ 1 & -1 & 2 & -1 \end{vmatrix}$$

$$= \frac{1}{8}\begin{vmatrix} 0 & 2 & 0 & 1 \\ 0 & 0 & 4 & -1 \\ 0 & 0 & 6 & -1 \\ 2 & -2 & 2 & -1 \end{vmatrix}$$

$$= -\frac{2}{8}\begin{vmatrix} 2 & 0 & 1 \\ 0 & 4 & -1 \\ 0 & 6 & -1 \end{vmatrix} = -\frac{4}{8}\begin{vmatrix} 4 & -1 \\ 6 & -1 \end{vmatrix} = -1$$

$x_4 = -(x_1 + x_2 + x_3) = -2.$

20. Let $F(x_1, x_2) = \mathcal{F}\begin{pmatrix} x_1 \\ x_2 \end{pmatrix}$, where $\mathcal{F} = \begin{pmatrix} a & b \\ c & d \end{pmatrix}$.
Let $G(y_1, y_2) = \mathcal{G}\begin{pmatrix} y_1 \\ y_2 \end{pmatrix}$, where $\mathcal{G} = \begin{pmatrix} p & q \\ r & s \end{pmatrix}$.
If $y_1 = ax_1 + bx_2$ and $y_2 = cx_1 + dx_2$, then

$$G \circ F(x_1, x_2) = G(y_1, y_2)$$
$$= \begin{pmatrix} p & q \\ r & s \end{pmatrix}\begin{pmatrix} ax_1 + bx_2 \\ cx_1 + dx_2 \end{pmatrix}$$
$$= \begin{pmatrix} pax_1 + pbx_2 + qcx_1 + qdx_2 \\ rax_1 + rbx_2 + scx_1 + sdx_2 \end{pmatrix}$$
$$= \begin{pmatrix} pa + qc & pb + qd \\ ra + sc & rb + sd \end{pmatrix}\begin{pmatrix} x_1 \\ x_2 \end{pmatrix}$$
$$= \begin{pmatrix} p & q \\ r & s \end{pmatrix}\begin{pmatrix} a & b \\ c & d \end{pmatrix}\begin{pmatrix} x_1 \\ x_2 \end{pmatrix}$$
$$= \mathcal{GF}\begin{pmatrix} x_1 \\ x_2 \end{pmatrix}.$$

Thus, $G \circ F$ is represented by the matrix $\mathcal{GF}$.

Review Exercises 10   (page 644)

1. $x + 3z = 3$ represents a plane parallel to the y-axis and passing through the points $(3, 0, 0)$ and $(0, 0, 1)$.

2. $y - z \geq 1$ represents all points on or below the plane parallel to the x-axis that passes through the points $(0, 1, 0)$ and $(0, 0, -1)$.

3. $x + y + z \geq 0$ represents all points on or above the plane through the origin having normal vector $\mathbf{i} + \mathbf{j} + \mathbf{k}$.

4. $x - 2y - 4z = 8$ represents all points on the plane passing through the three points $(8, 0, 0)$, $(0, -4, 0)$, and $(0, 0, -2)$.

5. $y = 1 + x^2 + z^2$ represents the circular paraboloid obtained by rotating about the y-axis the parabola in the xy-plane having equation $y = 1 + x^2$.

6. $y = z^2$ represents the parabolic cylinder parallel to the x-axis containing the curve $y = z^2$ in the yz-plane.

7. $x = y^2 - z^2$ represents the hyperbolic paraboloid whose intersections with the xy- and xz-planes are the parabolas $x = y^2$ and $x = -z^2$, respectively.

8. $z = xy$ is the hyperbolic paraboloid containing the x- and y-axes that results from rotating the hyperbolic paraboloid $z = (x^2 - y^2)/2$ through $45°$ about the z-axis.

9. $x^2 + y^2 + 4z^2 < 4$ represents the interior of the circular ellipsoid (oblate spheroid) centred at the origin with semi-axes 2, 2, and 1 in the x, y, and z directions, respectively.

10. $x^2 + y^2 - 4z^2 = 4$ represents a hyperboloid of one sheet with circular cross-sections in planes perpendicular to the z-axis, and asymptotic to the cone obtained by rotating the line $x = 2z$ about the z-axis.

11. $x^2 - y^2 - 4z^2 = 0$ represents an elliptic cone with axis along the x-axis whose cross-sections in planes $x = k$ are ellipses with semi-axes $|k|$ and $|k|/2$ in the y and z directions, respectively.

12. $x^2 - y^2 - 4z^2 = 4$ represents a hyperboloid of two sheets asymptotic to the cone of the previous exercise.

13. $(x-z)^2 + y^2 = 1$ represents an elliptic cylinder with oblique axis along the line $z = x$ in the xz-plane, having circular cross-sections of radius 1 in horizontal planes $z = k$.

14. $(x - z)^2 + y^2 = z^2$ represents an elliptic cone with oblique axis along the line $z = x$ in the xz-plane, having circular cross-sections of radius $|k|$ in horizontal planes $z = k$. The z-axis lies on the cone.

15. $x + 2y = 0$, $z = 3$ together represent the horizontal straight line through the point $(0, 0, 3)$ parallel to the vector $2\mathbf{i} - \mathbf{j}$.

16. $x + y + 2z = 1$, $x + y + z = 0$ together represent the straight line through the points $(-1, 0, 1)$ and $(0, -1, 1)$.

17. $x^2 + y^2 + z^2 = 4$, $x + y + z = 3$ together represent the circle in which the sphere of radius 2 centred at the origin intersects the plane through $(1, 1, 1)$ with normal $\mathbf{i} + \mathbf{j} + \mathbf{k}$. Since this plane lies at distance $\sqrt{3}$ from the origin, the circle has radius $\sqrt{4 - 3} = 1$.

18. $x^2 + z^2 \leq 1$, $x - y \geq 0$ together represent all points that lie inside or on the circular cylinder of radius 1 and axis along the y-axis and also either on the vertical plane $x - y = 0$ or on the side of that plane containing the positive x-axis.

19. The given line is parallel to the vector $\mathbf{a} = 2\mathbf{i} - \mathbf{j} + 3\mathbf{k}$. The plane through the origin perpendicular to $\mathbf{a}$ has equation $2x - y + 3z = 0$.

20. A plane through $(2, -1, 1)$ and $(1, 0, -1)$ is parallel to $\mathbf{b} = (2-1)\mathbf{i} + (-1-0)\mathbf{j} + (1-(-1))\mathbf{k} = \mathbf{i} - \mathbf{j} + 2\mathbf{k}$. If it is also parallel to the vector $\mathbf{a}$ in the previous solution, then it is normal to

$$\mathbf{a} \times \mathbf{b} = \begin{vmatrix} \mathbf{i} & \mathbf{j} & \mathbf{k} \\ 2 & -1 & 3 \\ 1 & -1 & 2 \end{vmatrix} = \mathbf{i} - \mathbf{j} - \mathbf{k}.$$

The plane has equation $(x - 1) - (y - 0) - (z + 1) = 0$, or $x - y - z = 2$.

21. A plane perpendicular to $x - y + z = 0$ and $2x + y - 3z = 2$ has normal given by the cross product of the normals of these two planes, that is, by

$$\begin{vmatrix} \mathbf{i} & \mathbf{j} & \mathbf{k} \\ 1 & -1 & 1 \\ 2 & 1 & -3 \end{vmatrix} = 2\mathbf{i} + 5\mathbf{j} + 3\mathbf{k}.$$

If the plane also passes through $(2, -1, 1)$, then its equation is
$$2(x - 2) + 5(y + 1) + 3(z - 1) = 0,$$
or $2x + 5y + 3z = 2$.

22. The plane through $A = (-1, 1, 0)$, $B = (0, 4, -1)$ and $C = (2, 0, 0)$ has normal

$$\overrightarrow{AC} \times \overrightarrow{AB} = \begin{vmatrix} \mathbf{i} & \mathbf{j} & \mathbf{k} \\ 3 & -1 & 0 \\ 1 & 3 & -1 \end{vmatrix} = \mathbf{i} + 3\mathbf{j} + 10\mathbf{k}.$$

Its equation is $(x - 2) + 3y + 10z = 0$, or $x + 3y + 10z = 2$.

23. A plane containing the line of intersection of the planes $x + y + z = 0$ and $2x + y - 3z = 2$ has equation
$$2x + y - 3z - 2 + \lambda(x + y + z - 0) = 0.$$
This plane passes through $(2, 0, 1)$ if $-1 + 3\lambda = 0$. In this case, the equation is $7x + 4y - 8z = 6$.

24. A plane containing the line of intersection of the planes $x + y + z = 0$ and $2x + y - 3z = 2$ has equation
$$2x + y - 3z - 2 + \lambda(x + y + z - 0) = 0.$$
This plane is perpendicular to $x - 2y - 5z = 17$ if their normals are perpendicular, that is, if
$$1(2 + \lambda) - 2(1 + \lambda) - 5(-3 + \lambda) = 0,$$
or $9x + 7y - z = 4$.

25. The line through $(2, 1, -1)$ and $(-1, 0, 1)$ is parallel to the vector $3\mathbf{i} + \mathbf{j} - 2\mathbf{k}$, and has vector parametric equation
$$\mathbf{r} = (2 + 3t)\mathbf{i} + (1 + t)\mathbf{j} - (1 + 2t)\mathbf{k}.$$

26. A vector parallel to the planes $x - y = 3$ and $x + 2y + z = 1$ is $(\mathbf{i} - \mathbf{j}) \times (\mathbf{i} + 2\mathbf{j} + \mathbf{k}) = -\mathbf{i} - \mathbf{j} + 3\mathbf{k}$. A line through $(1, 0, -1)$ parallel to this vector is
$$\frac{x - 1}{-1} = \frac{y}{-1} = \frac{z + 1}{3}.$$

27. The line through the origin perpendicular to the plane $3x - 2y + 4z = 5$ has equations $x = 3t$, $y = -2t$, $z = 4t$.

28. The vector
$$\mathbf{a} = (1 + t)\mathbf{i} - t\mathbf{j} - (2 + 2t)\mathbf{k} - \bigl(2s\mathbf{i} + (s - 2)\mathbf{j} - (1 + 3s)\mathbf{k}\bigr)$$
$$= (1 + t - 2s)\mathbf{i} - (t + s - 2)\mathbf{j} - (1 + 2t - 3s)\mathbf{k}$$

joins points on the two lines and is perpendicular to both lines if $\mathbf{a} \bullet (\mathbf{i} - \mathbf{j} - 2\mathbf{k}) = 0$ and $\mathbf{a} \bullet (2\mathbf{i} + \mathbf{j} - 3\mathbf{k}) = 0$, that is, if

$$1 + t - 2s + t + s - 2 + 2 + 4t - 6s = 0$$
$$2 + 2t - 4s - t - s + 2 + 3 + 6t - 9s = 0,$$

or, on simplification,

$$6t - 7s = -1$$
$$7t - 14s = -7.$$

This system has solution $t = 1$, $s = 1$. We would expect to use $\mathbf{a}$ as a vector perpendicular to both lines, but, as it happens, $\mathbf{a} = \mathbf{0}$ if $t = s = 1$, because the two given lines intersect at $(2, -1, -4)$. A nonzero vector perpendicular to both lines is

$$\begin{vmatrix} \mathbf{i} & \mathbf{j} & \mathbf{k} \\ 1 & -1 & -2 \\ 2 & 1 & -3 \end{vmatrix} = 5\mathbf{i} - \mathbf{j} + 3\mathbf{k}.$$

Thus the required line is parallel to this vector and passes through $(2, -1, -4)$, so its equation is

$$\mathbf{r} = (2 + 5t)\mathbf{i} - (1 + t)\mathbf{j} + (-4 + 3t)\mathbf{k}.$$

29. The points with position vectors $\mathbf{r}_1$, $\mathbf{r}_2$, and $\mathbf{r}_3$ are collinear if the triangle having these points as vertices has zero area, that is, if
$$(\mathbf{r}_2 - \mathbf{r}_1) \times (\mathbf{r}_3 - \mathbf{r}_1) = \mathbf{0}.$$
(Any permutation of the subscripts 1, 2, and 3 in the above equation will do as well.)

30. The points with position vectors $\mathbf{r}_1$, $\mathbf{r}_2$, $\mathbf{r}_3$, and $\mathbf{r}_4$ are coplanar if the tetrahedron having these points as vertices has zero volume, that is, if
$$\bigl[(\mathbf{r}_2 - \mathbf{r}_1) \times (\mathbf{r}_3 - \mathbf{r}_1)\bigr] \bullet (\mathbf{r}_4 - \mathbf{r}_1) = 0.$$
(Any permutation of the subscripts 1, 2, 3, and 4 in the above equation will do as well.)

31. The triangle with vertices $A = (1, 2, 1)$, $B = (4, -1, 1)$, and $C = (3, 4, -2)$ has area

$$\frac{1}{2}|\overrightarrow{AB} \times \overrightarrow{AC}| = \frac{1}{2}\begin{vmatrix} \mathbf{i} & \mathbf{j} & \mathbf{k} \\ 3 & -3 & 0 \\ 2 & 2 & -3 \end{vmatrix}$$
$$= \frac{1}{2}|9\mathbf{i} + 9\mathbf{j} + 12\mathbf{k}| = \frac{3\sqrt{34}}{2} \text{ sq. units.}$$

32. The tetrahedron with vertices $A = (1, 2, 1)$, $B = (4, -1, 1)$, $C = (3, 4, -2)$, and $D = (2, 2, 2)$ has volume

$$\frac{1}{6}|(\overrightarrow{AB} \times \overrightarrow{AC}) \bullet \overrightarrow{AD}| = \frac{1}{6}|(9\mathbf{i} + 9\mathbf{j} + 12\mathbf{k}) \bullet (\mathbf{i} + \mathbf{k})|$$
$$= \frac{9 + 12}{6} = \frac{7}{2} \text{ cu. units.}$$

33. The inverse of $\mathcal{A}$ satisfies

$$\begin{pmatrix} 1 & 0 & 0 & 0 \\ 2 & 1 & 0 & 0 \\ 3 & 2 & 1 & 0 \\ 4 & 3 & 2 & 1 \end{pmatrix} \begin{pmatrix} a & b & c & d \\ e & f & g & h \\ i & j & k & l \\ m & n & o & p \end{pmatrix} = \begin{pmatrix} 1 & 0 & 0 & 0 \\ 0 & 1 & 0 & 0 \\ 0 & 0 & 1 & 0 \\ 0 & 0 & 0 & 1 \end{pmatrix}.$$

Expanding the product on the left we get four systems of equations:

$a = 1, \quad 2a + e = 0, \quad 3a + 2e + i = 0, \quad 4a + 3e + 2i + m = 0.$
$b = 0, \quad 2b + f = 1, \quad 3b + 2f + j = 0, \quad 4b + 3f + 2j + n = 0.$
$c = 0, \quad 2c + g = 0, \quad 3c + 2g + k = 1, \quad 4c + 3g + 2k + o = 0.$
$d = 0, \quad 2d + h = 0, \quad 3d + 2h + l = 0, \quad 4d + 3h + 2l + p = 1.$

These systems have solutions

$a = 1, \quad e = -2, \quad i = 1, \quad m = 0,$
$b = 0, \quad f = 1, \quad j = -2, \quad n = 1,$
$c = 0, \quad g = 0, \quad k = 1, \quad o = -2,$
$d = 0, \quad h = 0, \quad l = 0, \quad p = 1.$

Thus

$$\mathcal{A}^{-1} = \begin{pmatrix} 1 & 0 & 0 & 0 \\ -2 & 1 & 0 & 0 \\ 1 & -2 & 1 & 0 \\ 0 & 1 & -2 & 1 \end{pmatrix}.$$

34. Let $\mathcal{A} = \begin{pmatrix} 1 & 1 & 1 \\ 2 & 1 & 0 \\ 1 & 0 & -1 \end{pmatrix}$, $\mathbf{x} = \begin{pmatrix} x_1 \\ x_2 \\ x_3 \end{pmatrix}$, and $\mathbf{b} = \begin{pmatrix} b_1 \\ b_2 \\ b_3 \end{pmatrix}$.
Then

$$\mathcal{A}\mathbf{x} = \mathbf{b} \Leftrightarrow \begin{aligned} x_1 + x_2 + x_3 &= b_1 \\ 2x_1 + x_2 \quad\quad &= b_2 \\ x_1 \quad\quad - x_3 &= b_3. \end{aligned}$$

The sum of the first and third equations is $2x_1 + x_2 = b_1 + b_3$, which is incompatible with the second equation unless $b_2 = b_1 + b_3$, that is, unless

$$\mathbf{b} \bullet (\mathbf{i} - \mathbf{j} + \mathbf{k}) = 0.$$

If $\mathbf{b}$ satisfies this condition then there will be a line of solutions; if $x_1 = t$, then $x_2 = b_2 - 2t$, and $x_3 = t - b_3$, so

$$\mathbf{x} = \begin{pmatrix} t \\ b_2 - 2t \\ t - b_3 \end{pmatrix}$$

is a solution for any t.

Challenging Problems 10 (page 645)

1. If d is the distance from P to the line AB, then d is the altitude of the triangle APB measured perpendicular to the base AB. Thus the area of the triangle is

$$(1/2)d|\overrightarrow{BA}| = (1/2)d|\mathbf{r}_A - \mathbf{r}_B|.$$

On the other hand, the area is also given by

$$(1/2)|\overrightarrow{PA} \times \overrightarrow{PB}| = (1/2)|(\mathbf{r}_A - \mathbf{r}_P) \times (\mathbf{r}_B - \mathbf{r}_P)|.$$

Equating these two expressions for the area of the triangle and solving for d we get

$$d = \frac{|(\mathbf{r}_A - \mathbf{r}_P) \times (\mathbf{r}_B - \mathbf{r}_P)|}{|\mathbf{r}_A - \mathbf{r}_B|}.$$

2. By the formula for the vector triple product given in Exercise 23 of Section 10.3,

$$(\mathbf{u} \times \mathbf{v}) \times (\mathbf{w} \times \mathbf{x}) = [(\mathbf{u} \times \mathbf{v}) \bullet \mathbf{x}]\mathbf{w} - [(\mathbf{u} \times \mathbf{v}) \bullet \mathbf{w}]\mathbf{x}$$
$$(\mathbf{u} \times \mathbf{v}) \times (\mathbf{w} \times \mathbf{x}) = -(\mathbf{w} \times \mathbf{x}) \times (\mathbf{u} \times \mathbf{v})$$
$$= -[(\mathbf{w} \times \mathbf{x}) \bullet \mathbf{v}]\mathbf{u} + [(\mathbf{w} \times \mathbf{x}) \bullet \mathbf{u}]\mathbf{v}.$$

In particular, if $\mathbf{w} = \mathbf{u}$, then, since $(\mathbf{u} \times \mathbf{v}) \bullet \mathbf{u} = 0$, we have

$$(\mathbf{u} \times \mathbf{v}) \times (\mathbf{u} \times \mathbf{x}) = [(\mathbf{u} \times \mathbf{v}) \bullet \mathbf{x}]\mathbf{u},$$

or, replacing x with w,

$$(\mathbf{u} \times \mathbf{v}) \times (\mathbf{u} \times \mathbf{w}) = [(\mathbf{u} \times \mathbf{v}) \bullet \mathbf{w}]\mathbf{u}.$$

3. The triangle with vertices $(x_1, y_1, 0)$, $(x_2, y_2, 0)$, and $(x_3, y_3, 0)$, has two sides corresponding to the vectors $(x_2 - x_1)\mathbf{i} + (y_2 - y_1)\mathbf{j}$ and $(x_3 - x_1)\mathbf{i} + (y_3 - y_1)\mathbf{j}$. Thus the triangle has area given by

$$A = \frac{1}{2}\begin{vmatrix} \mathbf{i} & \mathbf{j} & \mathbf{k} \\ x_2 - x_1 & y_2 - y_1 & 0 \\ x_3 - x_1 & y_3 - y_1 & 0 \end{vmatrix}$$
$$= \frac{1}{2}|[(x_2 - x_1)(y_3 - y_1) - (x_3 - x_1)(y_2 - y_1)]\mathbf{k}|$$

$$= \frac{1}{2}|x_2 y_3 - x_2 y_1 - x_1 y_3 - x_3 y_2 + x_3 y_1 + x_1 y_2|$$
$$= \frac{1}{2}\begin{vmatrix} x_1 & y_1 & 1 \\ x_2 & y_2 & 1 \\ x_3 & y_3 & 1 \end{vmatrix}.$$

4. a) Let Q_1 and Q_2 be the points on lines L_1 and L_2, respectively, that are closest together. As observed in Example 9 of Section 10.4, $\overrightarrow{Q_1Q_2}$ is perpendicular to both lines.
Therefore, the plane P_1 through Q_1 having normal $\overrightarrow{Q_1Q_2}$ contains the line L_1. Similarly, the plane P_2 through Q_2 having normal $\overrightarrow{Q_1Q_2}$ contains the line L_2. These planes are parallel since they have the same normal. They are different planes because $Q_1 \neq Q_2$ (because the lines are skew).

b) Line L_1 through $(1, 1, 0)$ and $(2, 0, 1)$ is parallel to $\mathbf{i} - \mathbf{j} + \mathbf{k}$, and has parametric equation

$$\mathbf{r}_1 = (1+t)\mathbf{i} + (1-t)\mathbf{j} + t\mathbf{k}.$$

Line L_2 through $(0, 1, 1)$ and $(1, 2, 2)$ is parallel to $\mathbf{i} + \mathbf{j} + \mathbf{k}$, and has parametric equation

$$\mathbf{r}_2 = s\mathbf{i} + (1+s)\mathbf{j} + (1+s)\mathbf{k}.$$

Now $\mathbf{r}_2 - \mathbf{r}_1 = (s - t - 1)\mathbf{i} + (s + t)\mathbf{j} + (1 + s - t)\mathbf{k}$.

To find the points Q_1 on L_1 and Q_2 on L_2 for which $\overrightarrow{Q_1Q_2}$ is perpendicular to both lines, we solve

$$(s - t - 1) - (s + t) + (1 + s - t) = 0$$
$$(s - t - 1) + (s + t) + (1 + s - t) = 0.$$

Subtracting these equations gives $s + t = 0$, so $t = -s$. Then substituting into either equation gives $2s - 1 + 1 + 2s = 0$, so $s = -t = 0$. Thus $Q_1 = (1, 1, 0)$ and $Q_2 = (0, 1, 1)$, and $\overrightarrow{Q_1Q_2} = -\mathbf{i} + \mathbf{k}$. The required planes are $x - z = 1$ (containing L_1) and $x - z = -1$ (containing L_2).

5. This problem is similar to Exercise 28 of Section 10.3. The equation $\mathbf{a} \times \mathbf{x} = \mathbf{b}$ has no solution $\mathbf{x}$ unless $\mathbf{a} \bullet \mathbf{b} = 0$. If this condition is satisfied, then $\mathbf{x} = \mathbf{x}_0 + t\mathbf{a}$ is a solution for any scalar t, where $\mathbf{x}_0 = (\mathbf{b} \times \mathbf{a})/|\mathbf{a}|^2$.

CHAPTER 11. VECTOR FUNCTIONS AND CURVES

Section 11.1 Vector Functions of One Variable (page 652)

1. Position: $\mathbf{r} = \mathbf{i} + t\mathbf{j}$
   Velocity: $\mathbf{v} = \mathbf{j}$
   Speed: $v = 1$
   Acceleration: $\mathbf{a} = \mathbf{0}$
   Path: the line $x = 1$ in the xy-plane.

2. Position: $\mathbf{r} = t^2\mathbf{i} + \mathbf{k}$
   Velocity: $\mathbf{v} = 2t\mathbf{i}$
   Speed: $v = 2|t|$
   Acceleration: $\mathbf{a} = 2\mathbf{i}$
   Path: the line $z = 1$, $y = 0$.

3. Position: $\mathbf{r} = t^2\mathbf{j} + t\mathbf{k}$
   Velocity: $\mathbf{v} = 2t\mathbf{j} + \mathbf{k}$
   Speed: $v = \sqrt{4t^2 + 1}$
   Acceleration: $\mathbf{a} = 2\mathbf{j}$
   Path: the parabola $y = z^2$ in the plane $x = 0$.

4. Position: $\mathbf{r} = \mathbf{i} + t\mathbf{j} + t\mathbf{k}$
   Velocity: $\mathbf{v} = \mathbf{j} + \mathbf{k}$
   Speed: $v = \sqrt{2}$
   Acceleration: $\mathbf{a} = \mathbf{0}$
   Path: the straight line $x = 1$, $y = z$.

5. Position: $\mathbf{r} = t^2\mathbf{i} - t^2\mathbf{j} + \mathbf{k}$
   Velocity: $\mathbf{v} = 2t\mathbf{i} - 2t\mathbf{j}$
   Speed: $v = 2\sqrt{2}t$
   Acceleration: $\mathbf{a} = 2\mathbf{i} - 2\mathbf{j}$
   Path: the half-line $x = -y \geq 0$, $z = 1$.

6. Position: $\mathbf{r} = t\mathbf{i} + t^2\mathbf{j} + t^2\mathbf{k}$
   Velocity: $\mathbf{v} = \mathbf{i} + 2t\mathbf{j} + 2t\mathbf{k}$
   Speed: $v = \sqrt{1 + 8t^2}$
   Acceleration: $\mathbf{a} = 2\mathbf{j} + 2\mathbf{k}$
   Path: the parabola $y = z = x^2$.

7. Position: $\mathbf{r} = a\cos t\mathbf{i} + a\sin t\mathbf{j} + ct\mathbf{k}$
   Velocity: $\mathbf{v} = -a\sin t\mathbf{i} + a\cos t\mathbf{j} + c\mathbf{k}$
   Speed: $v = \sqrt{a^2 + c^2}$
   Acceleration: $\mathbf{a} = -a\cos t\mathbf{i} - a\sin t\mathbf{j}$
   Path: a circular helix.

8. Position: $\mathbf{r} = a\cos\omega t\mathbf{i} + b\mathbf{j} + a\sin\omega t\mathbf{k}$
   Velocity: $\mathbf{v} = -a\omega\sin\omega t\mathbf{i} + a\omega\cos\omega t\mathbf{k}$
   Speed: $v = |a\omega|$
   Acceleration: $\mathbf{a} = -a\omega^2\cos\omega t\mathbf{i} - a\omega^2\sin\omega t\mathbf{k}$
   Path: the circle $x^2 + z^2 = a^2$, $y = b$.

9. Position: $\mathbf{r} = 3\cos t\mathbf{i} + 4\cos t\mathbf{j} + 5\sin t\mathbf{k}$
   Velocity: $\mathbf{v} = -3\sin t\mathbf{i} - 4\sin t\mathbf{j} + 5\cos t\mathbf{k}$
   Speed: $v = \sqrt{9\sin^2 t + 16\sin^2 t + 25\cos^2 t} = 5$
   Acceleration: $\mathbf{a} = -3\cos t\mathbf{i} - 4\cos t\mathbf{j} - 5\sin t\mathbf{k} = -\mathbf{r}$
   Path: the circle of intersection of the sphere $x^2 + y^2 + z^2 = 25$ and the plane $4x = 3y$.

10. Position: $\mathbf{r} = 3\cos t\mathbf{i} + 4\sin t\mathbf{j} + t\mathbf{k}$
    Velocity: $\mathbf{v} = -3\sin t\mathbf{i} + 4\cos t\mathbf{j} + \mathbf{k}$
    Speed: $v = \sqrt{9\sin^2 t + 16\cos^2 t + 1} = \sqrt{10 + 7\cos^2 t}$
    Acceleration: $\mathbf{a} = -3\cos t\mathbf{i} - 4\sin t\mathbf{j} = t\mathbf{k} - \mathbf{r}$
    Path: a helix (spiral) wound around the elliptic cylinder $(x^2/9) + (y^2/16) = 1$.

11. Position: $\mathbf{r} = ae^t\mathbf{i} + be^t\mathbf{j} + ce^t\mathbf{k}$
    Velocity and acceleration: $\mathbf{v} = \mathbf{a} = \mathbf{r}$
    Speed: $v = e^t\sqrt{a^2 + b^2 + c^2}$
    Path: the half-line $\dfrac{x}{a} = \dfrac{y}{b} = \dfrac{z}{c} > 0$.

12. Position: $\mathbf{r} = at\cos\omega t\mathbf{i} + at\sin\omega t\mathbf{j} + b\ln t\mathbf{k}$
    Velocity: $\mathbf{v} = a(\cos\omega t - \omega t\sin\omega t)\mathbf{i}$
    $\qquad + a(\sin\omega t + \omega t\cos\omega t)\mathbf{j} + (b/t)\mathbf{k}$
    Speed: $v = \sqrt{a^2(1 + \omega^2 t^2) + (b^2/t^2)}$
    Acceleration: $\mathbf{a} = -a\omega(2\sin\omega t + \omega\cos\omega t)\mathbf{i}$
    $\qquad + a\omega(2\cos\omega t - \omega\sin\omega t)\mathbf{j} - (b/t^2)\mathbf{k}$
    Path: a spiral on the surface $x^2 + y^2 = a^2 e^{z/b}$.

13. Position: $\mathbf{r} = e^{-t}\cos(e^t)\mathbf{i} + e^{-t}\sin(e^t)\mathbf{j} - e^t\mathbf{k}$
    Velocity: $\mathbf{v} = -\bigl(e^{-t}\cos(e^t) + \sin(e^t)\bigr)\mathbf{i}$
    $\qquad - \bigl(e^{-t}\sin(e^t) - \cos(e^t)\bigr)\mathbf{j} - e^t\mathbf{k}$
    Speed: $v = \sqrt{1 + e^{-2t} + e^{2t}}$
    Acceleration: $\mathbf{a} = \bigl((e^{-t} - e^t)\cos(e^t) + \sin(e^t)\bigr)\mathbf{i}$
    $\qquad + \bigl((e^{-t} - e^t)\sin(e^t) - \cos(e^t)\bigr)\mathbf{j} - e^t\mathbf{k}$
    Path: a spiral on the surface $z\sqrt{x^2 + y^2} = -1$.

14. Position: $\mathbf{r} = a\cos t\sin t\mathbf{i} + a\sin^2 t\mathbf{j} + a\cos t\mathbf{k}$
    $\qquad = \dfrac{a}{2}\sin 2t\mathbf{i} + \dfrac{a}{2}(1 - \cos 2t)\mathbf{j} + a\cos t\mathbf{k}$
    Velocity: $\mathbf{v} = a\cos 2t\mathbf{i} + a\sin 2t\mathbf{j} - a\sin t\mathbf{k}$
    Speed: $v = a\sqrt{1 + \sin^2 t}$
    Acceleration: $\mathbf{a} = -2a\sin 2t\mathbf{i} + 2a\cos 2t\mathbf{j} - a\cos t\mathbf{k}$
    Path: the path lies on the sphere $x^2 + y^2 + z^2 = a^2$, on the surface defined in terms of spherical polar coordinates by $\phi = \theta$, on the circular cylinder $x^2 + y^2 = ay$, and on the parabolic cylinder $ay + z^2 = a^2$. Any two of these surfaces serve to pin down the shape of the path.

15. The position of the particle is given by
    $$\mathbf{r} = 5\cos(\omega t)\mathbf{i} + 5\sin(\omega t)\mathbf{j},$$
    where $\omega = \pi$ to ensure that $\mathbf{r}$ has period $2\pi/\omega = 2$ s. Thus
    $$\mathbf{a} = \frac{d^2\mathbf{r}}{dt^2} = -\omega^2\mathbf{r} = -\pi^2\mathbf{r}.$$
    At $(3, 4)$, the acceleration is $-3\pi^2\mathbf{i} - 4\pi^2\mathbf{j}$.

INSTRUCTOR'S SOLUTIONS MANUAL
SECTION 11.1 (PAGE 652)

16. When its x-coordinate is x, the particle is at position $\mathbf{r} = x\mathbf{i} + (3/x)\mathbf{j}$, and its velocity and speed are

$$\mathbf{v} = \frac{d\mathbf{r}}{dt} = \frac{dx}{dt}\mathbf{i} - \frac{3}{x^2}\frac{dx}{dt}\mathbf{j}$$

$$v = \left|\frac{dx}{dt}\right|\sqrt{1 + \frac{9}{x^4}}.$$

We know that $dx/dt > 0$ since the particle is moving to the right. When $x = 2$, we have $10 = v = (dx/dt)\sqrt{1 + (9/16)} = (5/4)(dx/dt)$. Thus $dx/dt = 8$. The velocity at that time is $\mathbf{v} = 8\mathbf{i} - 6\mathbf{j}$.

17. The particle moves along the curve $z = x^2$, $x + y = 2$, in the direction of increasing y. Thus its position at time t is

$$\mathbf{r} = (2 - y)\mathbf{i} + y\mathbf{j} + (2 - y)^2\mathbf{k},$$

where y is an increasing function of time t. Thus

$$\mathbf{v} = \frac{dy}{dt}\left[-\mathbf{i} + \mathbf{j} - 2(2 - y)\mathbf{k}\right]$$

$$v = \frac{dy}{dt}\sqrt{1 + 1 + 4(2 - y)^2} = 3$$

since the speed is 3. When $y = 1$, we have $dy/dt = 3/\sqrt{6} = \sqrt{3/2}$. Thus

$$\mathbf{v} = \sqrt{\frac{3}{2}}(-\mathbf{i} + \mathbf{j} - 2\mathbf{k}).$$

18. The position of the object when its x-coordinate is x is

$$\mathbf{r} = x\mathbf{i} + x^2\mathbf{j} + x^3\mathbf{k},$$

so its velocity is $\mathbf{v} = \dfrac{dx}{dt}\left[\mathbf{i} + 2x\mathbf{j} + 3x^2\mathbf{k}\right]$. Since $dz/dt = 3x^2\,dx/dt = 3$, when $x = 2$ we have $12\,dx/dt = 3$, so $dx/dt = 1/4$. Thus

$$\mathbf{v} = \frac{1}{4}\mathbf{i} + \mathbf{j} + 3\mathbf{k}.$$

19. $\mathbf{r} = 3u\mathbf{i} + 3u^2\mathbf{j} + 2u^3\mathbf{k}$

$$\mathbf{v} = \frac{du}{dt}(3\mathbf{i} + 6u\mathbf{j} + 6u^2\mathbf{k})$$

$$\mathbf{a} = \frac{d^2u}{dt^2}(3\mathbf{i} + 6u\mathbf{j} + 6u^2\mathbf{k}) + \left(\frac{du}{dt}\right)^2(6\mathbf{j} + 12u\mathbf{k}).$$

Since u is increasing and the speed of the particle is 6,

$$6 = |\mathbf{v}| = 3\frac{du}{dt}\sqrt{1 + 4u^2 + 4u^4} = 3(1 + 2u^2)\frac{du}{dt}.$$

Thus $\dfrac{du}{dt} = \dfrac{2}{1 + 2u^2}$, and

$$\frac{d^2u}{dt^2} = \frac{-2}{(1 + 2u^2)^2}\,4u\,\frac{du}{dt} = \frac{-16u}{(1 + 2u^2)^3}.$$

The particle is at $(3, 3, 2)$ when $u = 1$. At this point $du/dt = 2/3$ and $d^2u/dt^2 = -16/27$, and so

$$\mathbf{v} = \frac{2}{3}(3\mathbf{i} + 6u\mathbf{j} + 6u^2\mathbf{k}) = 2\mathbf{i} + 4\mathbf{j} + 4\mathbf{k}$$

$$\mathbf{a} = \frac{-16}{27}(3\mathbf{i} + 6\mathbf{j} + 6\mathbf{k}) + \left(\frac{2}{3}\right)^2(6\mathbf{j} + 12\mathbf{k})$$

$$= \frac{8}{9}(-2\mathbf{i} - \mathbf{j} + 2\mathbf{k}).$$

20. $\mathbf{r} = x\mathbf{i} - x^2\mathbf{j} + +x^2\mathbf{k}$

$$\mathbf{v} = \frac{dx}{dt}(\mathbf{i} - 2x\mathbf{j} + 2x\mathbf{k})$$

$$\mathbf{a} = \frac{d^2x}{dt^2}(\mathbf{i} - 2x\mathbf{j} + 2x\mathbf{k}) + \left(\frac{dx}{dt}\right)^2(-2\mathbf{j} + 2\mathbf{k}).$$

Thus $|\mathbf{v}| = \left|\dfrac{dx}{dt}\right|\sqrt{1 + 4x^4 + 4x^4} = \sqrt{1 + 8x^4}\,\dfrac{dx}{dt}$, since x is increasing. At $(1, -1, 1)$, $x = 1$ and $|\mathbf{v}| = 9$, so $dx/dt = 3$, and the velocity at that point is $\mathbf{v} = 3\mathbf{i} - 6\mathbf{j} + 6\mathbf{k}$. Now

$$\frac{d}{dt}|\mathbf{v}| = \sqrt{1 + 8x^4}\,\frac{d^2x}{dt^2} + \frac{16x^3}{\sqrt{1 + 8x^4}}\left(\frac{dx}{dt}\right)^2.$$

The left side is 3 when $x = 1$, so $3(d^2x/dt^2) + 48 = 3$, and $d^2x/dt^2 = -15$ at that point, and the acceleration there is

$$\mathbf{a} = -15(\mathbf{i} - 2\mathbf{j} + 2\mathbf{k}) + 9(-2\mathbf{j} + 2\mathbf{k}) = -15\mathbf{i} + 12\mathbf{j} - 12\mathbf{k}.$$

21. $\dfrac{d}{dt}|\mathbf{v}|^2 = \dfrac{d}{dt}\mathbf{v}\bullet\mathbf{v} = 2\mathbf{v}\bullet\mathbf{a}$.
If $\mathbf{v}\bullet\mathbf{a} > 0$ then the speed $v = |\mathbf{v}|$ is increasing.
If $\mathbf{v}\bullet\mathbf{a} < 0$ then the speed is decreasing.

22. If $\mathbf{u}(t) = u_1(t)\mathbf{i} + u_2(t)\mathbf{j} + u_3(t)\mathbf{k}$
$\mathbf{v}(t) = v_1(t)\mathbf{i} + v_2(t)\mathbf{j} + v_3(t)\mathbf{k}$
then $\mathbf{u}\bullet\mathbf{v} = u_1v_1 + u_2v_2 + u_3v_3$, so

$$\frac{d}{dt}\mathbf{u}\bullet\mathbf{v} = \frac{du_1}{dt}v_1 + u_1\frac{dv_1}{dt} + \frac{du_2}{dt}v_2 + u_2\frac{dv_2}{dt}$$
$$+ \frac{du_3}{dt}v_3 + u_3\frac{dv_3}{dt}$$
$$= \frac{d\mathbf{u}}{dt}\bullet\mathbf{v} + \mathbf{u}\bullet\frac{d\mathbf{v}}{dt}.$$

403

23. $\dfrac{d}{dt}\begin{vmatrix} a_{11} & a_{12} & a_{13} \\ a_{21} & a_{22} & a_{23} \\ a_{31} & a_{32} & a_{33} \end{vmatrix}$

$= \dfrac{d}{dt}\big[a_{11}a_{22}a_{33} + a_{12}a_{23}a_{31} + a_{13}a_{21}a_{32}$
$\quad - a_{11}a_{23}a_{32} - a_{12}a_{21}a_{33} - a_{13}a_{22}a_{31}\big]$

$= a'_{11}a_{22}a_{33} + a_{11}a'_{22}a_{33} + a_{11}a_{22}a'_{33}$
$+ a'_{12}a_{23}a_{31} + a_{12}a'_{23}a_{31} + a_{12}a_{23}a'_{31}$
$+ a'_{13}a_{21}a_{32} + a_{13}a'_{21}a_{32} + a_{13}a_{21}a'_{32}$
$- a'_{11}a_{23}a_{32} - a_{11}a'_{23}a_{32} - a_{11}a_{23}a'_{32}$
$- a'_{12}a_{21}a_{33} - a_{12}a'_{21}a_{33} - a_{12}a_{21}a'_{33}$
$- a'_{13}a_{22}a_{31} - a_{13}a'_{22}a_{31} - a_{13}a_{22}a'_{31}$

$= \begin{vmatrix} a'_{11} & a'_{12} & a'_{13} \\ a_{21} & a_{22} & a_{23} \\ a_{31} & a_{32} & a_{33} \end{vmatrix} + \begin{vmatrix} a_{11} & a_{12} & a_{13} \\ a'_{21} & a'_{22} & a'_{23} \\ a_{31} & a_{32} & a_{33} \end{vmatrix}$
$+ \begin{vmatrix} a_{11} & a_{12} & a_{13} \\ a_{21} & a_{22} & a_{23} \\ a'_{31} & a'_{32} & a'_{33} \end{vmatrix}$

24. $\dfrac{d}{dt}|\mathbf{r}|^2 = \dfrac{d}{dt}\mathbf{r}\bullet\mathbf{r} = 2\mathbf{r}\bullet\mathbf{v} = 0$ implies that $|\mathbf{r}|$ is constant.
Thus $\mathbf{r}(t)$ lies on a sphere centred at the origin.

25. $\dfrac{d}{dt}|\mathbf{r} - \mathbf{r}_0|^2 = \dfrac{d}{dt}(\mathbf{r} - \mathbf{r}_0)\bullet(\mathbf{r} - \mathbf{r}_0)$
$= 2(\mathbf{r} - \mathbf{r}_0)\bullet\dfrac{d\mathbf{r}}{dt} = 0$

implies that $|\mathbf{r} - \mathbf{r}_0|$ is constant. Thus $\mathbf{r}(t)$ lies on a sphere centred at the point P_0 with position vector $\mathbf{r}_0$.

26. If $\mathbf{r}\bullet\mathbf{v} > 0$ then $|\mathbf{r}|$ is increasing. (See Exercise 16 above.) Thus $\mathbf{r}$ is moving farther away from the origin. If $\mathbf{r}\bullet\mathbf{v} < 0$ then $\mathbf{r}$ is moving closer to the origin.

27. $\dfrac{d}{dt}\left(\dfrac{d\mathbf{u}}{dt}\times\dfrac{d^2\mathbf{u}}{dt^2}\right) = \dfrac{d^2\mathbf{u}}{dt^2}\times\dfrac{d^2\mathbf{u}}{dt^2} + \dfrac{d\mathbf{u}}{dt}\times\dfrac{d^3\mathbf{u}}{dt^3}$
$= \dfrac{d\mathbf{u}}{dt}\times\dfrac{d^3\mathbf{u}}{dt^3}.$

28. $\dfrac{d}{dt}\big(\mathbf{u}\bullet(\mathbf{v}\times\mathbf{w})\big)$
$= \mathbf{u}'\bullet(\mathbf{v}\times\mathbf{w}) + \mathbf{u}\bullet(\mathbf{v}'\times\mathbf{w}) + \mathbf{u}\bullet(\mathbf{v}\times\mathbf{w}').$

29. $\dfrac{d}{dt}\big(\mathbf{u}\times(\mathbf{v}\times\mathbf{w})\big)$
$= \mathbf{u}'\times(\mathbf{v}\times\mathbf{w}) + \mathbf{u}\times(\mathbf{v}'\times\mathbf{w}) + \mathbf{u}\times(\mathbf{v}\times\mathbf{w}').$

30. $\dfrac{d}{dt}\left(\mathbf{u}\times\left(\dfrac{d\mathbf{u}}{dt}\times\dfrac{d^2\mathbf{u}}{dt^2}\right)\right)$
$= \dfrac{d\mathbf{u}}{dt}\times\left(\dfrac{d\mathbf{u}}{dt}\times\dfrac{d^2\mathbf{u}}{dt^2}\right) + \mathbf{u}\times\left(\dfrac{d^2\mathbf{u}}{dt^2}\times\dfrac{d^2\mathbf{u}}{dt^2}\right)$
$\quad + \mathbf{u}\times\left(\dfrac{d\mathbf{u}}{dt}\times\dfrac{d^3\mathbf{u}}{dt^3}\right)$
$= \dfrac{d\mathbf{u}}{dt}\times\left(\dfrac{d\mathbf{u}}{dt}\times\dfrac{d^2\mathbf{u}}{dt^2}\right) + \mathbf{u}\times\left(\dfrac{d\mathbf{u}}{dt}\times\dfrac{d^3\mathbf{u}}{dt^3}\right).$

31. $\dfrac{d}{dt}\big[(\mathbf{u}+\mathbf{u}'')\bullet(\mathbf{u}\times\mathbf{u}')\big]$
$= (\mathbf{u}'+\mathbf{u}''')\bullet(\mathbf{u}\times\mathbf{u}') + (\mathbf{u}+\mathbf{u}'')\bullet(\mathbf{u}'\times\mathbf{u}')$
$\quad + (\mathbf{u}+\mathbf{u}'')\bullet(\mathbf{u}\times\mathbf{u}'')$
$= \mathbf{u}'''\bullet(\mathbf{u}\times\mathbf{u}').$

32. $\dfrac{d}{dt}\big[(\mathbf{u}\times\mathbf{u}')\bullet(\mathbf{u}'\times\mathbf{u}'')\big]$
$= (\mathbf{u}'\times\mathbf{u}')\bullet(\mathbf{u}'\times\mathbf{u}'') + (\mathbf{u}\times\mathbf{u}'')\bullet(\mathbf{u}'\times\mathbf{u}'')$
$\quad + (\mathbf{u}\times\mathbf{u}')\bullet(\mathbf{u}''\times\mathbf{u}'') + (\mathbf{u}\times\mathbf{u}')\bullet(\mathbf{u}'\times\mathbf{u}''')$
$= (\mathbf{u}\times\mathbf{u}'')\bullet(\mathbf{u}'\times\mathbf{u}'') + (\mathbf{u}\times\mathbf{u}')\bullet(\mathbf{u}'\times\mathbf{u}''').$

33. Since $\dfrac{d\mathbf{r}}{dt} = \mathbf{v}(t) = 2\mathbf{r}(t)$ and $\mathbf{r}(0) = \mathbf{r}_0$, we have

$$\mathbf{r}(t) = \mathbf{r}(0)e^{2t} = \mathbf{r}_0 e^{2t},$$
$$\mathbf{a}(t) = \dfrac{d\mathbf{v}}{dt} = 2\dfrac{d\mathbf{r}}{dt} = 4\mathbf{r}_0 e^{2t}.$$

The path is the half-line from the origin in the direction of $\mathbf{r}_0$.

34. $\mathbf{r} = \mathbf{r}_0\cos\omega t + \left(\dfrac{\mathbf{v}_0}{\omega}\right)\sin\omega t$

$\dfrac{d\mathbf{r}}{dt} = -\omega\mathbf{r}_0\sin\omega t + \mathbf{v}_0\cos\omega t$

$\dfrac{d^2\mathbf{r}}{dt^2} = -\omega^2\mathbf{r}_0\cos\omega t - \omega\mathbf{v}_0\sin\omega t = -\omega^2\mathbf{r}$

$\mathbf{r}(0) = \mathbf{r}_0, \quad \dfrac{d\mathbf{r}}{dt}\bigg|_{t=0} = \mathbf{v}_0.$

Observe that $\mathbf{r}\bullet(\mathbf{r}_0\times\mathbf{v}_0) = 0$ for all t. Therefore the path lies in a plane through the origin having normal $\mathbf{N} = \mathbf{r}_0\times\mathbf{v}_0$.
Let us choose our coordinate system so that $\mathbf{r}_0 = a\mathbf{i}$ $(a > 0)$ and $\mathbf{v}_0 = \omega b\mathbf{i} + \omega c\mathbf{j}$ $(c > 0)$. Therefore, $\mathbf{N}$ is in the direction of $\mathbf{k}$. The path has parametric equations

$$x = a\cos\omega t + b\sin\omega t$$
$$y = c\sin\omega t.$$

The curve is a conic section since it has a quadratic equation:

$$\dfrac{1}{a^2}\left(x - \dfrac{by}{c}\right)^2 + \dfrac{y^2}{c^2} = 1.$$

Since the path is bounded ($|\mathbf{r}(t)| \leq |\mathbf{r}_0| + (|\mathbf{v}_0|/\omega)$), it must be an ellipse.

If $\mathbf{r}_0$ is perpendicular to $\mathbf{v}_0$, then $b = 0$ and the path is the ellipse $(x/a)^2 + (y/c)^2 = 1$ having semi-axes $a = |\mathbf{r}_0|$ and $c = |\mathbf{v}_0|/\omega$.

35. $\dfrac{d^2\mathbf{r}}{dt^2} = -g\mathbf{k} - c\dfrac{d\mathbf{r}}{dt}$

$\mathbf{r}(0) = \mathbf{r}_0, \quad \dfrac{d\mathbf{r}}{dt}\bigg|_{t=0} = \mathbf{v}_0.$

Let $\mathbf{w} = e^{ct}\dfrac{d\mathbf{r}}{dt}$. Then

$$\dfrac{d\mathbf{w}}{dt} = ce^{ct}\dfrac{d\mathbf{r}}{dt} + e^{ct}\dfrac{d^2\mathbf{r}}{dt^2}$$
$$= ce^{ct}\dfrac{d\mathbf{r}}{dt} - e^{ct}g\mathbf{k} - ce^{ct}\dfrac{d\mathbf{r}}{dt}$$
$$= -e^{ct}g\mathbf{k}$$
$$\mathbf{w}(t) = -\int e^{ct}g\mathbf{k}\,dt = -\dfrac{e^{ct}}{c}g\mathbf{k} + \mathbf{C}.$$

Put $t = 0$ and get $\mathbf{v}_0 = -\dfrac{g}{c}\mathbf{k} + \mathbf{C}$, so

$$e^{ct}\dfrac{d\mathbf{r}}{dt} = \mathbf{w} = \mathbf{v}_0 + \dfrac{g}{c}(1 - e^{ct})\mathbf{k}$$
$$\dfrac{d\mathbf{r}}{dt} = e^{-ct}\mathbf{v}_0 - \dfrac{g}{c}(1 - e^{-ct})\mathbf{k}$$
$$\mathbf{r} = -\dfrac{e^{-ct}}{c}\mathbf{v}_0 - \dfrac{g}{c}\left(t + \dfrac{e^{-ct}}{c}\right)\mathbf{k} + \mathbf{D}$$
$$\mathbf{r}_0 = \mathbf{r}(0) = -\dfrac{1}{c}\mathbf{v}_0 - \dfrac{g}{c^2}\mathbf{k} + \mathbf{D}.$$

Thus we have

$$\mathbf{r} = \mathbf{r}_0 + \dfrac{1 - e^{-ct}}{c}\mathbf{v}_0 - \dfrac{g}{c^2}(ct + e^{-ct} - 1)\mathbf{k}.$$

The limit of this solution, as $c \to 0$, is calculated via l'Hôpital's Rule:

$$\lim_{c\to 0}\mathbf{r}(t) = \mathbf{r}_0 + \mathbf{v}_0\lim_{c\to 0}\dfrac{te^{-ct}}{1} - g\mathbf{k}\lim_{c\to 0}\dfrac{t - te^{-ct}}{2c}$$
$$= \mathbf{r}_0 + \mathbf{v}_0 t - g\mathbf{k}\lim_{c\to 0}\dfrac{t^2 e^{-ct}}{2}$$
$$= \mathbf{r}_0 + \mathbf{v}_0 t - \dfrac{1}{2}gt^2\mathbf{k},$$

which is the solution obtained in Example 4.

Section 11.2 Some Applications of Vector Differentiation (page 659)

1. It was shown in the text that

$$\mathbf{v}(T) - \mathbf{v}(0) = -\ln\left(\dfrac{m(0)}{m(T)}\right)\mathbf{v}_e.$$

If $\mathbf{v}(0) = \mathbf{0}$ and $\mathbf{v}(T) = -\mathbf{v}_e$ then $\ln(m(0)/m(T)) = 1$ and $m(T) = (1/e)m(0)$. The rocket must therefore burn fraction $\dfrac{e-1}{e}$ of its initial mass to accelerate to the speed of its exhaust gases.

Similarly, if $\mathbf{v}(T) = -2\mathbf{v}_e$, then $m(T) = (1/e^2)m(0)$, so the rocket must burn fraction $\dfrac{e^2-1}{e^2}$ of its initial mass to accelerate to twice the speed of its exhaust gases.

2. Let $v(t)$ be the speed of the tank car at time t seconds. The mass of the car at time t is $m(t) = M - kt$ kg. At full power, the force applied to the car is $F = Ma$ (since the motor can accelerate the full car at a m/s^2). By Newton's Law, this force is the rate of change of the momentum of the car. Thus

$$\dfrac{d}{dt}\big[(M - kt)v\big] = Ma$$
$$(M - kt)\dfrac{dv}{dt} - kv = Ma$$
$$\dfrac{dv}{Ma + kv} = \dfrac{dt}{M - kt}$$
$$\dfrac{1}{k}\ln(Ma + kv) = -\dfrac{1}{k}\ln(M - kt) + \dfrac{1}{k}\ln C$$
$$Ma + kv = \dfrac{C}{M - kt}.$$

At $t = 0$ we have $v = 0$, so $Ma = C/M$. Thus $C = M^2 a$ and

$$kv = \dfrac{M^2 a}{M - kt} - Ma = \dfrac{Makt}{M - kt}.$$

The speed of the tank car at time t (before it is empty) is

$$v(t) = \dfrac{Mat}{M - kt} \text{ m/s}.$$

3. Given: $\dfrac{d\mathbf{r}}{dt} = \mathbf{k} \times \mathbf{r}$, $\mathbf{r}(0) = \mathbf{i} + \mathbf{k}$.
Let $\mathbf{r}(t) = x(t)\mathbf{i} + y(t)\mathbf{j} + z(t)\mathbf{k}$. Then $x(0) = z(0) = 1$ and $y(0) = 0$.
Since $\mathbf{k} \bullet (d\mathbf{r}/dt) = \mathbf{k} \bullet (\mathbf{k} \times \mathbf{r}) = 0$, the velocity is always perpendicular to $\mathbf{k}$, so $z(t)$ is constant: $z(t) = z(0) = 1$ for all t. Thus

$$\dfrac{dx}{dt}\mathbf{i} + \dfrac{dy}{dt}\mathbf{j} = \dfrac{d\mathbf{r}}{dt} = \mathbf{k} \times \mathbf{r} = x\mathbf{j} - y\mathbf{i}.$$

Separating this equation into components,

$$\dfrac{dx}{dt} = -y, \qquad \dfrac{dy}{dt} = x.$$

Therefore,

$$\dfrac{d^2 x}{dt^2} = -\dfrac{dy}{dt} = -x,$$

and $x = A\cos t + B\sin t$. Since $x(0) = 1$ and $y(0) = 0$, we have $A = 1$ and $B = 0$. Thus $x(t) = \cos t$ and $y(t) = \sin t$. The path has equation

$$\mathbf{r} = \cos t\,\mathbf{i} + \sin t\,\mathbf{j} + \mathbf{k}.$$

Remark: This result also follows from comparing the given differential equation with that obtained for circular motion in the text. This shows that the motion is a rotation with angular velocity $\mathbf{k}$, that is, rotation about the z-axis with angular speed 1. The initial value given for $\mathbf{r}$ then forces
$$\mathbf{r} = \cos t\,\mathbf{i} + \sin t\,\mathbf{j} + \mathbf{k}.$$

4. First observe that
$$\frac{d}{dt}|\mathbf{r}-\mathbf{b}|^2 = 2(\mathbf{r}-\mathbf{b})\cdot\frac{d\mathbf{r}}{dt} = 2(\mathbf{r}-\mathbf{b})\cdot\big(\mathbf{a}\times(\mathbf{r}-\mathbf{b})\big) = 0,$$
so $|\mathbf{r}-\mathbf{b}|$ is constant; for all t the object lies on the sphere centred at the point with position vector $\mathbf{b}$ having radius $\mathbf{r}_0 - \mathbf{b}$.
Next, observe that
$$\frac{d}{dt}(\mathbf{r}-\mathbf{r}_0)\cdot\mathbf{a} = \big(\mathbf{a}\times(\mathbf{r}-\mathbf{b})\big)\cdot\mathbf{a} = 0,$$
so $\mathbf{r}-\mathbf{r}_0 \perp \mathbf{a}$; for all t the object lies on the plane through $\mathbf{r}_0$ having normal $\mathbf{a}$. Hence the path of the object lies on the circle in which this plane intersects the sphere described above. The angle between $\mathbf{r}-\mathbf{b}$ and $\mathbf{a}$ must therefore also be constant, and so the object's speed $|d\mathbf{r}/dt|$ is constant. Hence the path must be the whole circle.

5. Use a coordinate system with origin at the observer, $\mathbf{i}$ pointing east, and $\mathbf{j}$ pointing north. The angular velocity of the earth is $2\pi/24$ radians per hour northward:
$$\mathbf{\Omega} = \frac{\pi}{12}\mathbf{j}.$$

Because the earth is rotating west to east, the true north to south velocity of the satellite will appear to the observer to be shifted to the west by $\pi R/12$ km/h, where R is the radius of the earth in kilometres. Since the satellite circles the earth at a rate of π radians/h, its velocity, as observed at the moving origin, is
$$\mathbf{v}_\mathcal{R} = -\pi R\mathbf{j} - \frac{\pi R}{12}\mathbf{i}.$$

$\mathbf{v}_\mathcal{R}$ makes angle $\tan^{-1}\left(\dfrac{\pi R/12}{\pi R}\right) = \tan^{-1}(1/12) \approx 4.76°$ with the southward direction. Thus the satellite appears to the observer to be moving in a direction $4.76°$ west of south.

The apparent Coriolis force is
$$-2\mathbf{\Omega}\times\mathbf{v}_\mathcal{R} = -\frac{2\pi}{12}\mathbf{j}\times\left(-\pi R\mathbf{j} - \frac{\pi R}{12}\mathbf{i}\right) = -\frac{\pi^2 R}{72}\mathbf{k},$$

which is pointing towards the ground.

6. The angular velocity of the earth is $\mathbf{\Omega}$, pointing due north. For a particle moving with horizontal velocity $\mathbf{v}$, the tangential and normal components of the Coriolis force $\mathbf{C}$, and of $\mathbf{\Omega}$, are related by
$$\mathbf{C}_T = -2\mathbf{\Omega}_N\times\mathbf{v},\qquad \mathbf{C}_N = -2\mathbf{\Omega}_T\times\mathbf{v}.$$

At the north or south pole, $\mathbf{\Omega}_T = \mathbf{0}$ and $\mathbf{\Omega}_N = \mathbf{\Omega}$. Thus $\mathbf{C}_N = \mathbf{0}$ and $\mathbf{C}_T = -2\mathbf{\Omega}\times\mathbf{v}$. The Coriolis force is horizontal. It is 90° east of $\mathbf{v}$ at the north pole and 90° west of $\mathbf{v}$ at the south pole.

At the equator, $\mathbf{\Omega}_N = \mathbf{0}$ and $\mathbf{\Omega}_T = \mathbf{\Omega}$. Thus $\mathbf{C}_T = \mathbf{0}$ and $\mathbf{C}_N = -2\mathbf{\Omega}\times\mathbf{v}$. The Coriolis force is vertical.

Section 11.3 Curves and Parametrizations (page 666)

1. On the first quadrant part of the circle $x^2 + y^2 = a^2$ we have $x = \sqrt{a^2 - y^2}$, $0 \le y \le a$. The required parametrization is
$$\mathbf{r} = \mathbf{r}(y) = \sqrt{a^2 - y^2}\,\mathbf{i} + y\mathbf{j},\quad (0 \le y \le a).$$

2. On the first quadrant part of the circle $x^2 + y^2 = a^2$ we have $y = \sqrt{a^2 - x^2}$, $0 \le x \le a$. The required parametrization is
$$\mathbf{r} = \mathbf{r}(x) = x\mathbf{i} + \sqrt{a^2 - x^2}\,\mathbf{j},\quad (0 \le x \le a).$$

3. From the figure we see that
$$\phi = \theta + \frac{\pi}{2},\qquad 0 \le \theta \le \frac{\pi}{2}$$
$$x = a\cos\theta = a\cos\left(\phi - \frac{\pi}{2}\right) = a\sin\phi$$
$$y = a\sin\theta = a\sin\left(\phi - \frac{\pi}{2}\right) = -a\cos\phi.$$

The required parametrization is
$$\mathbf{r} = a\sin\phi\,\mathbf{i} - a\cos\phi\,\mathbf{j},\qquad \left(\frac{\pi}{2} \le \phi \le \pi\right).$$

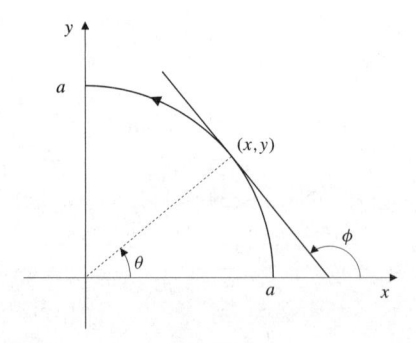

Fig. 11.3.3

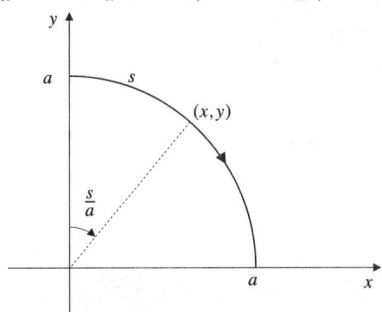

4. $x = a\sin\dfrac{s}{a}$, $y = a\cos\dfrac{s}{a}$, $0 \le \dfrac{s}{a} \le \dfrac{\pi}{2}$
   $\mathbf{r} = a\sin\dfrac{s}{a}\mathbf{i} + a\cos\dfrac{s}{a}\mathbf{j}$, $\left(0 \le s \le \dfrac{a\pi}{2}\right)$.

Fig. 11.3.4

5. $z = x^2$, $z = 4y^2$. If $t = y$, then $z = 4t^2$, so $x = \pm 2t$. The curve passes through $(2, -1, 4)$ when $t = -1$, so $x = -2t$. The parametrization is $\mathbf{r} = -2t\mathbf{i} + t\mathbf{j} + 4t^2\mathbf{k}$.

6. $z = x^2$, $x + y + z = 1$. If $t = x$, then $z = t^2$ and $y = 1 - t - t^2$. The parametrization is $\mathbf{r} = t\mathbf{i} + (1 - t - t^2)\mathbf{j} + t^2\mathbf{k}$.

7. $z = x + y$, $x^2 + y^2 = 9$. One possible parametrization is $\mathbf{r} = 3\cos t\,\mathbf{i} + 3\sin t\,\mathbf{j} + 3(\cos t + \sin t)\mathbf{k}$.

8. $x + y = 1$, $z = \sqrt{1 - x^2 - y^2}$. If $x = t$, then $y = 1 - t$ and $z = \sqrt{1 - t^2 - (1 - t)^2} = \sqrt{2(t - t^2)}$. One possible parametrization is
$$\mathbf{r} = t\mathbf{i} + (1 - t)\mathbf{j} + \sqrt{2(t - t^2)}\,\mathbf{k}.$$

9. $z = x^2 + y^2$, $2x - 4y - z - 1 = 0$. These surfaces intersect on the vertical cylinder
$$x^2 + y^2 = 2x - 4y - 1, \quad \text{that is}$$
$$(x - 1)^2 + (y + 2)^2 = 4.$$

One possible parametrization is
$x = 1 + 2\cos t$
$y = -2 + 2\sin t$
$z = -1 + 2(1 + 2\cos t) - 4(-2 + 2\sin t) = 9 + 4\cos t - 8\sin t$
$\mathbf{r} = (1 + 2\cos t)\mathbf{i} - 2(1 - \sin t)\mathbf{j} + (9 + 4\cos t - 8\sin t)\mathbf{k}$

10. $yz + x = 1$, $xz - x = 1$. One possible parametrization is $x = t$, $z = (1 + t)/t$, and $y = (1 - t)/z = (1 - t)t/(1 + t)$, that is,
$$\mathbf{r} = t\mathbf{i} + \dfrac{t - t^2}{1 + t}\mathbf{j} + \dfrac{1 + t}{t}\mathbf{k}.$$

11. $z^2 = x^2 + y^2$, $z = 1 + x$.

    a) If $t = x$, then $z = 1 + t$, so $1 + 2t + t^2 = t^2 + y^2$, and $y = \pm\sqrt{1 + 2t}$. Two parametrizations are needed to get the whole parabola, one for $y \le 0$ and one for $y \ge 0$.

    b) If $t = y$, then $x^2 + t^2 = z^2 = 1 + 2x + x^2$, so $2x + 1 = t^2$, and $x = (t^2 - 1)/2$. Thus $z = 1 + x = (t^2 + 1)/2$. The whole parabola is parametrized by
    $$\mathbf{r} = \dfrac{t^2 - 1}{2}\mathbf{i} + t\mathbf{j} + \dfrac{t^2 + 1}{2}\mathbf{k}.$$

    c) If $t = z$, then $x = t - 1$ and $t^2 = t^2 - 2t + 1 + y^2$, so $y = \pm\sqrt{2t - 1}$. Again two parametrizations are needed to get the whole parabola.

12. By symmetry, the centre of the circle $\mathcal{C}$ of intersection of the plane $x + y + z = 1$ and the sphere $x^2 + y^2 + z^2 = 1$ must lie on the plane and must have its three coordinates equal. Thus the centre has position vector
$$\mathbf{r}_0 = \dfrac{1}{3}(\mathbf{i} + \mathbf{j} + \mathbf{k}).$$

Since $\mathcal{C}$ passes through the point $(0, 0, 1)$, its radius is
$$\sqrt{\left(0 - \dfrac{1}{3}\right)^2 + \left(0 - \dfrac{1}{3}\right)^2 + \left(1 - \dfrac{1}{3}\right)^2} = \sqrt{\dfrac{2}{3}}.$$

Any vector $\mathbf{v}$ that satisfies $\mathbf{v} \bullet (\mathbf{i} + \mathbf{j} + \mathbf{k}) = 0$ is parallel to the plane $x + y + z = 1$ containing $\mathcal{C}$. One such vector is $\mathbf{v}_1 = \mathbf{i} - \mathbf{j}$. A second one, perpendicular to $\mathbf{v}_1$, is
$$\mathbf{v}_2 = (\mathbf{i} + \mathbf{j} + \mathbf{k}) \times (\mathbf{i} - \mathbf{j}) = \mathbf{i} + \mathbf{j} - 2\mathbf{k}.$$

Two perpendicular unit vectors that are parallel to the plane of $\mathcal{C}$ are
$$\hat{\mathbf{v}}_1 = \dfrac{\mathbf{i} - \mathbf{j}}{\sqrt{2}}, \quad \hat{\mathbf{v}}_2 = \dfrac{\mathbf{i} + \mathbf{j} - 2\mathbf{k}}{\sqrt{6}}.$$

Thus one possible parametrization of $\mathcal{C}$ is
$$\mathbf{r} = \mathbf{r}_0 + \sqrt{\dfrac{2}{3}}(\cos t\,\hat{\mathbf{v}}_1 + \sin t\,\hat{\mathbf{v}}_2)$$
$$= \dfrac{\mathbf{i} + \mathbf{j} + \mathbf{k}}{3} + \dfrac{\cos t}{\sqrt{3}}(\mathbf{i} - \mathbf{j}) + \dfrac{\sin t}{3}(\mathbf{i} + \mathbf{j} - 2\mathbf{k}).$$

13. $\mathbf{r} = t^2\mathbf{i} + t^2\mathbf{j} + t^3\mathbf{k}$, $(0 \le t \le 1)$
$v = \sqrt{(2t)^2 + (2t)^2 + (3t^2)^2} = t\sqrt{8 + 9t^2}$
Length $= \displaystyle\int_0^1 t\sqrt{8 + 9t^2}\,dt$ Let $u = 8 + 9t^2$
                                   $du = 18t\,dt$
$= \dfrac{1}{18}\cdot\dfrac{2}{3}u^{3/2}\Big|_8^{17} = \dfrac{17\sqrt{17} - 16\sqrt{2}}{27}$ units.

14. $\mathbf{r} = t\mathbf{i} + \lambda t^2 \mathbf{j} + t^3 \mathbf{k}$, $(0 \le t \le T)$
$v = \sqrt{1 + (2\lambda t)^2 + 9t^4} = \sqrt{(1+3t^2)^2}$
if $4\lambda^2 = 6$, that is, if $\lambda = \pm\sqrt{3/2}$. In this case, the length of the curve is

$$s(T) = \int_0^T (1 + 3t^2)\, dt = T + T^3.$$

15. Length $= \int_1^T \left|\dfrac{d\mathbf{r}}{dt}\right| dt$

$= \int_1^T \sqrt{4a^2 t^2 + b^2 + \dfrac{c^2}{t^2}}\, dt$ units.

If $b^2 = 4ac$ then
Length $= \int_1^T \sqrt{\left(2at + \dfrac{c}{t}\right)^2}\, dt$
$= \int_1^T \left(2at + \dfrac{c}{t}\right) dt$
$= a(T^2 - 1) + c \ln T$ units.

16. $x = a \cos t \sin t = \dfrac{a}{2} \sin 2t$,
$y = a \sin^2 t = \dfrac{a}{2}(1 - \cos 2t)$,
$z = bt$.
The curve is a circular helix lying on the cylinder

$$x^2 + \left(y - \dfrac{a}{2}\right)^2 = \dfrac{a^2}{4}.$$

Its length, from $t = 0$ to $t = T$, is

$$L = \int_0^T \sqrt{a^2 \cos^2 2t + a^2 \sin^2 2t + b^2}\, dt$$
$= T\sqrt{a^2 + b^2}$ units.

17. $\mathbf{r} = t \cos t\,\mathbf{i} + t \sin t\,\mathbf{j} + t\mathbf{k}$, $0 \le t \le 2\pi$
$\mathbf{v} = (\cos t - t \sin t)\mathbf{i} + (\sin t + t \cos t)\mathbf{j} + \mathbf{k}$
$v = |\mathbf{v}| = \sqrt{(1 + t^2) + 1} = \sqrt{2 + t^2}$.
The length of the curve is

$L = \int_0^{2\pi} \sqrt{2 + t^2}\, dt$ Let $t = \sqrt{2} \tan\theta$
  $dt = \sqrt{2} \sec^2\theta\, d\theta$
$= 2 \int_{t=0}^{t=2\pi} \sec^3\theta\, d\theta$
$= \left(\sec\theta \tan\theta + \ln|\sec\theta + \tan\theta|\right)\Big|_{t=0}^{t=2\pi}$
$= \dfrac{t\sqrt{2+t^2}}{2} + \ln\left(\dfrac{\sqrt{2+t^2}}{\sqrt{2}} + \dfrac{t}{\sqrt{2}}\right)\Big|_0^{2\pi}$
$= \pi\sqrt{2 + 4\pi^2} + \ln(\sqrt{1 + 2\pi^2} + \sqrt{2}\pi)$ units.

The curve is called a conical helix because it is a spiral lying on the cone $x^2 + y^2 = z^2$.

18. One-eighth of the curve $\mathcal{C}$ lies in the first octant. That part can be parametrized

$$x = \cos t, \quad z = \dfrac{1}{\sqrt{2}} \sin t, \quad (0 \le t \le \pi/2)$$
$$y = \sqrt{1 - \cos^2 t - \dfrac{1}{2}\sin^2 t} = \dfrac{1}{\sqrt{2}} \sin t.$$

Since the first octant part of $\mathcal{C}$ lies in the plane $y = z$, it must be a quarter of a circle of radius 1. Thus the length of all of $\mathcal{C}$ is $8 \times (\pi/2) = 4\pi$ units.
If you wish to use an integral, the length is

$$8 \int_0^{\pi/2} \sqrt{\sin^2 t + \dfrac{1}{2}\cos^2 t + \dfrac{1}{2}\cos^2 t}\, dt$$
$= 8 \int_0^{\pi/2} dt = 4\pi$ units.

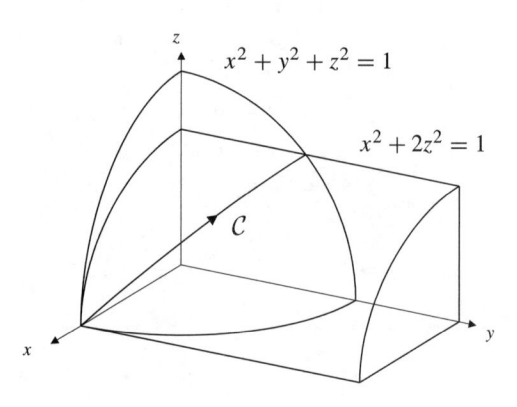

Fig. 11.3.18

19. If $\mathcal{C}$ is the curve

$$x = e^t \cos t, \quad y = e^t \sin t, \quad z = t, \quad (0 \le t \le 2\pi),$$

then the length of $\mathcal{C}$ is

$L = \int_0^{2\pi} \sqrt{\left(\dfrac{dx}{dt}\right)^2 + \left(\dfrac{dy}{dt}\right)^2 + \left(\dfrac{dz}{dt}\right)^2}\, dt$
$= \int_0^{2\pi} \sqrt{e^{2t}(\cos t - \sin t)^2 + e^{2t}(\sin t + \cos t)^2 + 1}\, dt$
$= \int_0^{2\pi} \sqrt{2e^{2t} + 1}\, dt$ Let $2e^{2t} + 1 = v^2$
  $2e^{2t}\, dt = v\, dv$
$= \int_{t=0}^{t=2\pi} \dfrac{v^2\, dv}{v^2 - 1} = \int_{t=0}^{t=2\pi} \left(1 + \dfrac{1}{v^2 - 1}\right) dv$

$$= \left(v + \frac{1}{2}\ln\left|\frac{v-1}{v+1}\right|\right)\Big|_{t=0}^{t=2\pi}$$

$$= \sqrt{2e^{4\pi}+1} - \sqrt{3} + \frac{1}{2}\ln\frac{\sqrt{2e^{2t}+1}-1}{\sqrt{2e^{2t}+1}+1}\Big|_0^{2\pi}$$

$$= \sqrt{2e^{4\pi}+1} - \sqrt{3} + \ln\frac{\sqrt{2e^{2t}+1}-1}{\sqrt{2}e^t}\Big|_0^{2\pi}$$

$$= \sqrt{2e^{4\pi}+1} - \sqrt{3} + \ln\left(\sqrt{2e^{4\pi}+1}-1\right)$$
$$\quad - 2\pi - \ln(\sqrt{3}-1) \text{ units.}$$

Remark: This answer appears somewhat different from that given in the answers section of the text. The two are, however, equal. Somewhat different simplifications were used in the two.

20. $\mathbf{r} = t^3\mathbf{i} + t^2\mathbf{j}$

$\mathbf{v} = 3t^2\mathbf{i} + 2t\mathbf{j}$

$v = |\mathbf{v}| = \sqrt{9t^4+4t^2} = |t|\sqrt{9t^2+4}$

The length L between $t=-1$ and $t=2$ is

$$L = \int_{-1}^{0}(-t)\sqrt{9t^2+4}\,dt + \int_0^2 t\sqrt{9t^2+4}\,dt.$$

Making the substitution $u = 9t^2+4$ in each integral, we obtain

$$L = \frac{1}{18}\left[\int_4^{13} u^{1/2}\,du + \int_4^{40} u^{1/2}\,du\right]$$
$$= \frac{1}{27}\left(13^{3/2} + 40^{3/2} - 16\right) \text{ units.}$$

21. $\mathbf{r}_1 = t\mathbf{i} + t\mathbf{j}$, $(0 \le t \le 1)$ represents the straight line segment from the origin to $(1,1)$ in the xy-plane.

$\mathbf{r}_2 = (1-t)\mathbf{i} + (1+t)\mathbf{j}$, $(0 \le t \le 1)$ represents the straight line segment from $(1,1)$ to $(0,2)$.

Thus $\mathcal{C} = \mathcal{C}_1 + \mathcal{C}_2$ is the 2-segment polygonal line from the origin to $(1,1)$ and then to $(0,2)$.

22. (Solution due to Roland Urbanek, a student at Okanagan College.) Suppose the spool is vertical and the cable windings make angle θ with the horizontal at each point.

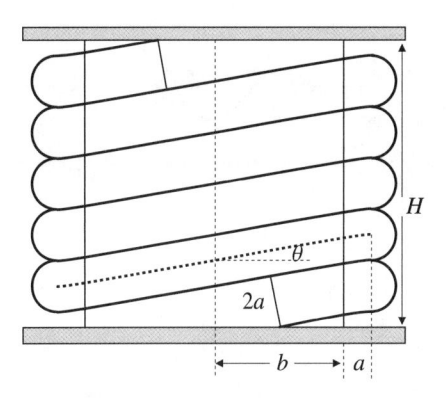

Fig. 11.3.22

The centreline of the cable is wound around a cylinder of radius $a+b$ and must rise a vertical distance $\dfrac{2a}{\cos\theta}$ in one revolution. The figure below shows the cable unwound from the spool and inclined at angle θ. The total length of spool required is the total height H of the cable as shown in that figure.

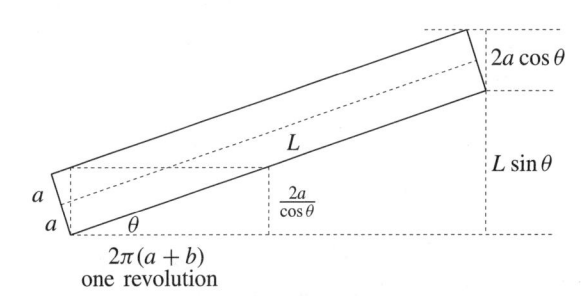

Fig. 11.3.22

Observe that $\tan\theta = \dfrac{2a}{\cos\theta} \times \dfrac{1}{2\pi(a+b)}$. Therefore

$$\sin\theta = \frac{a}{\pi(a+b)}$$

$$\cos\theta = \sqrt{1 - \frac{a^2}{\pi^2(a+b)^2}} = \frac{\sqrt{\pi^2(a+b)^2 - a^2}}{\pi(a+b)}.$$

The total length of spool required is

$$H = L\sin\theta + 2a\cos\theta$$
$$= \frac{a}{\pi(a+b)}\left(L + 2\sqrt{\pi^2(a+b)^2-a^2}\right) \text{ units.}$$

23. $\mathbf{r} = At\mathbf{i} + Bt\mathbf{j} + Ct\mathbf{k}$.

The arc length from the point where $t=0$ to the point corresponding to arbitrary t is

$$s = s(t) = \int_0^t \sqrt{A^2+B^2+C^2}\,du = \sqrt{A^2+B^2+C^2}\,t.$$

Thus $t = s/\sqrt{A^2 + B^2 + C^2}$. The required parametrization is
$$\mathbf{r} = \frac{As\mathbf{i} + Bs\mathbf{j} + Cs\mathbf{k}}{\sqrt{A^2 + B^2 + C^2}}.$$

24. $\mathbf{r} = e^t\mathbf{i} + \sqrt{2}t\mathbf{j} - e^{-t}\mathbf{k}$
$\mathbf{v} = e^t\mathbf{i} + \sqrt{2}\mathbf{j} + e^{-t}\mathbf{k}$
$v = |\mathbf{v}| = \sqrt{e^{2t} + 2 + e^{-2t}} = e^t + e^{-t}$.
The arc length from the point where $t = 0$ to the point corresponding to arbitrary t is
$$s = s(t) = \int_0^t (e^u + e^{-u})\, du = e^t - e^{-t} = 2\sinh t.$$
Thus $t = \sinh^{-1}(s/2) = \ln\left(\dfrac{s + \sqrt{s^2+4}}{2}\right)$,
and $e^t = \dfrac{s + \sqrt{s^2+4}}{2}$. The required parametrization is
$$\mathbf{r} = \frac{s + \sqrt{s^2+4}}{2}\mathbf{i} + \sqrt{2}\ln\left(\frac{s+\sqrt{s^2+4}}{2}\right)\mathbf{j} - \frac{2\mathbf{k}}{s+\sqrt{s^2+4}}.$$

25. $\mathbf{r} = a\cos^3 t\,\mathbf{i} + a\sin^3 t\,\mathbf{j} + b\cos 2t\,\mathbf{k}, \quad 0 \le t \le \dfrac{\pi}{2}$
$\mathbf{v} = -3a\cos^2 t\sin t\,\mathbf{i} + 3a\sin^2 t\cos t\,\mathbf{j} - 4b\sin t\cos t\,\mathbf{k}$
$v = \sqrt{9a^2 + 16b^2}\sin t\cos t$
$s = \displaystyle\int_0^t \sqrt{9a^2+16b^2}\sin u\cos u\, du$
$= \dfrac{1}{2}\sqrt{9a^2+16b^2}\sin^2 t = K\sin^2 t$
where $K = \dfrac{1}{2}\sqrt{9a^2+16b^2}$.
Therefore $\sin t = \sqrt{\dfrac{s}{K}}$, $\cos t = \sqrt{1 - \dfrac{s}{K}}$,
$\cos 2t = 1 - 2\sin^2 t = 1 - \dfrac{2s}{K}$.
The required parametrization is
$$\mathbf{r} = a\left(1 - \frac{s}{K}\right)^{3/2}\mathbf{i} + a\left(\frac{s}{K}\right)^{3/2}\mathbf{j} + b\left(1 - \frac{2s}{K}\right)\mathbf{k}$$
for $0 \le s \le K$, where $K = \dfrac{1}{2}\sqrt{9a^2+16b^2}$.

26. $\mathbf{r} = 3t\cos t\,\mathbf{i} + 3t\sin t\,\mathbf{j} + 2\sqrt{2}t^{3/2}\mathbf{k}, \quad (t \ge 0)$
$\mathbf{v} = 3(\cos t - t\sin t)\mathbf{i} + 3(\sin t + t\cos t)\mathbf{j} + 3\sqrt{2}\sqrt{t}\,\mathbf{k}$
$v = |\mathbf{v}| = 3\sqrt{1 + t^2 + 2t} = 3(1+t)$
$s = \displaystyle\int_0^t 3(1+u)\, du = 3\left(t + \dfrac{t^2}{2}\right)$

Thus $t^2 + 2t = \dfrac{2s}{3}$, so $t = -1 + \sqrt{1 + \dfrac{2s}{3}}$ since $t \ge 0$. The required parametrization is the given one with t replaced by $-1 + \sqrt{1 + (2s)/3}$.

27. As claimed in the statement of the problem,
$\mathbf{r}_1(t) = \mathbf{r}_2(u(t))$, where u is a function from $[a, b]$ to $[c, d]$, having $u(a) = c$ and $u(b) = d$. We assume u is differentiable. Since u is one-to-one and orientation-preserving, $du/dt \ge 0$ on $[a, b]$. By the Chain Rule:
$$\frac{d}{dt}\mathbf{r}_1(t) = \frac{d}{du}\mathbf{r}_2(u)\,\frac{du}{dt},$$
and so
$$\int_a^b \left|\frac{d}{dt}\mathbf{r}_1(t)\right| dt = \int_a^b \left|\frac{d}{du}\mathbf{r}_2(u(t))\right|\frac{du}{dt}\, dt = \int_c^d \left|\frac{d}{du}\mathbf{r}_2(u)\right| du.$$

28. If $\mathbf{r} = \mathbf{r}(t)$ has nonvanishing velocity $\mathbf{v} = d\mathbf{r}/dt$ on $[a, b]$, then for any t_0 in $[a, b]$, the function
$$s = g(t) = \int_{t_0}^t |\mathbf{v}(u)|\, du,$$
which gives the (signed) arc length s measured from $\mathbf{r}(t_0)$ along the curve, is an increasing function:
$$\frac{ds}{dt} = g'(t) = |\mathbf{v}(t)| > 0$$
on $[a, b]$, by the Fundamental Theorem of Calculus. Hence g is invertible, and defines t as a function of arc length s:
$$t = g^{-1}(s) \Leftrightarrow s = g(t).$$
Then
$$\mathbf{r} = \mathbf{r}_2(s) = \mathbf{r}(g^{-1}(s))$$
is a parametrization of the curve $\mathbf{r} = \mathbf{r}(t)$ in terms of arc length.

Section 11.4 Curvature, Torsion, and the Frenet Frame (page 675)

1. $\mathbf{r} = t\mathbf{i} - 2t^2\mathbf{j} + 3t^3\mathbf{k}$
$\mathbf{v} = \mathbf{i} - 4t\mathbf{j} + 9t^2\mathbf{k}$
$v = \sqrt{1 + 16t^2 + 81t^4}$
$\hat{\mathbf{T}} = \dfrac{\mathbf{v}}{v} = \dfrac{\mathbf{i} - 4t\mathbf{j} + 9t^2\mathbf{k}}{\sqrt{1 + 16t^2 + 81t^4}}$.

2. $\mathbf{r} = a\sin\omega t\,\mathbf{i} + a\cos\omega t\,\mathbf{k}$
$\mathbf{v} = a\omega\cos\omega t\,\mathbf{i} - a\omega\sin\omega t\,\mathbf{k}, \quad v = |a\omega|$
$\hat{\mathbf{T}} = \text{sgn}(a\omega)[\cos\omega t\,\mathbf{i} - \sin\omega t\,\mathbf{k}]$.

3. $\mathbf{r} = \cos t \sin t \mathbf{i} + \sin^2 t + \cos t \mathbf{k}$
$= \dfrac{1}{2}\sin 2t\mathbf{i} + \dfrac{1}{2}(1-\cos 2t)\mathbf{j} + \cos t\mathbf{k}$
$\mathbf{v} = \cos 2t\mathbf{i} + \sin 2t\mathbf{j} - \sin t\mathbf{k}$
$v = |\mathbf{v}| = \sqrt{1 + \sin^2 t}$
$\hat{\mathbf{T}} = \dfrac{1}{\sqrt{1+\sin^2 t}}(\cos 2t\mathbf{i} + \sin 2t\mathbf{j} - \sin t\mathbf{k}).$

4. $\mathbf{r} = a\cos t\mathbf{i} + b\sin t\mathbf{j} + t\mathbf{k}$
$\mathbf{v} = -a\sin t\mathbf{i} + b\cos t\mathbf{j} + \mathbf{k}$
$v = \sqrt{a^2\sin^2 t + b^2\cos^2 t + 1}$
$\hat{\mathbf{T}} = \dfrac{\mathbf{v}}{v} = \dfrac{-a\sin t\mathbf{i} + b\cos t\mathbf{j} + \mathbf{k}}{\sqrt{a^2\sin^2 t + b^2\cos^2 t + 1}}.$

5. If $\kappa(s) = 0$ for all s, then $\dfrac{d\hat{\mathbf{T}}}{ds} = \kappa\hat{\mathbf{N}} = 0$, so $\hat{\mathbf{T}}(s) = \hat{\mathbf{T}}(0)$ is constant. This says that $\dfrac{d\mathbf{r}}{ds} = \hat{\mathbf{T}}(0)$, so $\mathbf{r} = \hat{\mathbf{T}}(0)s + \mathbf{r}(0)$, which is the vector parametric equation of a straight line.

6. If $\tau(s) = 0$ for all s, then
$\dfrac{d\hat{\mathbf{B}}}{ds} = -\tau\hat{\mathbf{N}} = 0$, so $\hat{\mathbf{B}}(s) = \hat{\mathbf{B}}(0)$ is constant. Therefore,
$\dfrac{d}{ds}\big(\mathbf{r}(s) - \mathbf{r}(0)\big)\bullet\hat{\mathbf{B}}(s) = \dfrac{d\mathbf{r}}{ds}\bullet\hat{\mathbf{B}}(s) = \hat{\mathbf{T}}(s)\bullet\hat{\mathbf{B}}(s) = 0.$

It follows that
$\big(\mathbf{r}(s) - \mathbf{r}(0)\big)\bullet\hat{\mathbf{B}}(0) = \big(\mathbf{r}(s) - \mathbf{r}(0)\big)\bullet\hat{\mathbf{B}}(s) = 0$

for all s. This says that $\mathbf{r}(s)$ lies in the plane through $\mathbf{r}(0)$ having normal $\hat{\mathbf{B}}(0)$.

7. The circle $\mathcal{C}_1$ given by
$$\mathbf{r} = \dfrac{1}{C}\cos Cs\mathbf{i} + \dfrac{1}{C}\sin Cs\mathbf{j}$$
is parametrized in terms of arc length, and has curvature C and torsion 0. (See Examples 2 and 3.)
If curve $\mathcal{C}$ has constant curvature $\kappa(s) = C$ and constant torsion $\tau(s) = 0$, then $\mathcal{C}$ is congruent to $\mathcal{C}_1$ by Theorem 3. Thus $\mathcal{C}$ must itself be a circle (with radius $1/C$).

8. The circular helix $\mathcal{C}_1$ given by
$$\mathbf{r} = a\cos t\mathbf{i} + a\sin t\mathbf{j} + bt\mathbf{k}$$
has curvature and torsion given by
$$\kappa(s) = \dfrac{a}{a^2+b^2}, \qquad \tau(s) = \dfrac{b}{a^2+b^2},$$

by Example 3.
if a curve $\mathcal{C}$ has constant curvature $\kappa(s) = C > 0$, and constant torsion $\tau(s) = T \neq 0$, then we can choose a and b so that
$$\dfrac{a}{a^2+b^2} = C, \qquad \dfrac{b}{a^2+b^2} = T.$$
(Specifically, $a = \dfrac{C}{C^2+T^2}$, and $b = \dfrac{T}{C^2+T^2}$.) By Theorem 3, $\mathcal{C}$ is itself a circular helix, congruent to $\mathcal{C}_1$.

Section 11.5 Curvature and Torsion for General Parametrizations (page 681)

1. For $y = x^2$ we have
$$\kappa(x) = \dfrac{|d^2y/dx^2|}{(1+(dy/dx)^2)^{3/2}} = \dfrac{2}{(1+4x^2)^{3/2}}.$$
Hence $\kappa(0) = 2$ and $\kappa(\sqrt{2}) = 2/27$. The radii of curvature at $x = 0$ and $x = \sqrt{2}$ are $1/2$ and $27/2$, respectively.

2. For $y = \cos$ we have
$$\kappa(x) = \dfrac{|d^2y/dx^2|}{(1+(dy/dx)^2)^{3/2}} = \dfrac{|\cos x|}{(1+\sin^2 x)^{3/2}}.$$
Hence $\kappa(0) = 1$ and $\kappa(\pi/2) = 0$. The radius of curvature at $x = 0$ is 1. The radius of curvature at $x = \pi/2$ is infinite.

3. $\mathbf{r} = 2t\mathbf{i} + (1/t)\mathbf{j} - 2t\mathbf{k}$
$\mathbf{v} = 2\mathbf{i} - (1/t^2)\mathbf{j} - 2\mathbf{k}$
$\mathbf{a} = (2/t^3)\mathbf{j}$
$\mathbf{v} \times \mathbf{a} = (4/t^3)\mathbf{i} + (4/t^3)\mathbf{k}$
At $(2,1,-2)$, that is, at $t = 1$, we have
$$\kappa = \kappa(1) = \dfrac{|\mathbf{v}\times\mathbf{a}|}{v^3} = \dfrac{4\sqrt{2}}{27}.$$
Thus the radius of curvature is $27/(4\sqrt{2})$.

4. $\mathbf{r} = t^3\mathbf{i} + t^2\mathbf{j} + t\mathbf{k}$
$\mathbf{v} = 3t^2\mathbf{i} + 2t\mathbf{j} + \mathbf{k}$
$\mathbf{a} = 6t\mathbf{i} + 2\mathbf{j}$
$\mathbf{v}(1) = 3\mathbf{i} + 2\mathbf{j} + \mathbf{k}, \quad \mathbf{a}(1) = 6\mathbf{i} + 2\mathbf{j}$
$\mathbf{v}(1) \times \mathbf{a}(1) = -2\mathbf{i} + 6\mathbf{j} - 6\mathbf{k}$
$$\kappa(1) = \dfrac{\sqrt{4+36+36}}{(9+4+1)^{3/2}} = \dfrac{2\sqrt{19}}{14^{3/2}}$$
At $t = 1$ the radius of curvature is $14^{3/2}/(2\sqrt{19})$.

5. $\mathbf{r} = t\mathbf{i} + t^2\mathbf{j} + 2\mathbf{k}$
$\mathbf{v} = \mathbf{i} + 2t\mathbf{j}$
$\mathbf{a} = 2\mathbf{j}$
$\mathbf{v} \times \mathbf{a} = 2\mathbf{k}$
At $(1,1,2)$, where $t = 1$, we have
$\hat{\mathbf{T}} = \mathbf{v}/|\mathbf{v}| = (\mathbf{i}+2\mathbf{j})/\sqrt{5}$
$\hat{\mathbf{B}} = (\mathbf{v}\times\mathbf{a})/|\mathbf{v}\times\mathbf{a}| = \mathbf{k}$
$\hat{\mathbf{N}} = \hat{\mathbf{B}}\times\hat{\mathbf{T}} = (-2\mathbf{i}+\mathbf{j})/\sqrt{5}.$

6. $\mathbf{r} = t\mathbf{i} + t^2\mathbf{j} + t\mathbf{k}$
$\mathbf{v} = \mathbf{i} + 2t\mathbf{j} + \mathbf{k}$
$\mathbf{a} = 2\mathbf{j}$
$\mathbf{v} \times \mathbf{a} = -2\mathbf{i} + 2\mathbf{k}$
At $(1, 1, 1)$, where $t = 1$, we have
$\hat{\mathbf{T}} = \mathbf{v}/|\mathbf{v}| = (\mathbf{i} + 2\mathbf{j} + \mathbf{k})/\sqrt{6}$
$\hat{\mathbf{B}} = (\mathbf{v} \times \mathbf{a})/|\mathbf{v} \times \mathbf{a}| = -(\mathbf{i} - \mathbf{k})/\sqrt{2}$
$\hat{\mathbf{N}} = \hat{\mathbf{B}} \times \hat{\mathbf{T}} = -(\mathbf{i} - \mathbf{j} + \mathbf{k})/\sqrt{3}$.

7. $\mathbf{r} = t\mathbf{i} + \dfrac{t^2}{2}\mathbf{j} + \dfrac{t^3}{3}\mathbf{k}$
$\mathbf{v} = \mathbf{i} + t\mathbf{j} + t^2\mathbf{k}$
$\mathbf{a} = \mathbf{j} + 2t\mathbf{k}, \qquad \dfrac{d\mathbf{a}}{dt} = 2\mathbf{k}$
$\mathbf{v} \times \mathbf{a} = t^2\mathbf{i} - 2t\mathbf{j} + \mathbf{k}$
$v = |\mathbf{v}| = \sqrt{1 + t^2 + t^4}, \qquad |\mathbf{v} \times \mathbf{a}| = \sqrt{1 + 4t^2 + t^4}$
$(\mathbf{v} \times \mathbf{a}) \bullet \dfrac{d\mathbf{a}}{dt} = 2$

$\hat{\mathbf{T}} = \dfrac{\mathbf{v}}{v} = \dfrac{\mathbf{i} + t\mathbf{j} + t^2\mathbf{k}}{\sqrt{1 + t^2 + t^4}}$
$\hat{\mathbf{B}} = \dfrac{\mathbf{v} \times \mathbf{a}}{|\mathbf{v} \times \mathbf{a}|} = \dfrac{t^2\mathbf{i} - 2t\mathbf{j} + \mathbf{k}}{\sqrt{1 + 4t^2 + t^4}}$
$\hat{\mathbf{N}} = \hat{\mathbf{B}} \times \hat{\mathbf{T}} = \dfrac{-(2t^3 + t)\mathbf{i} + (1 - t^4)\mathbf{j} + (t^3 + 2t)\mathbf{k}}{\sqrt{(1 + t^2 + t^4)(1 + 4t^2 + t^4)}}$
$\kappa = \dfrac{|\mathbf{v} \times \mathbf{a}|}{v^3} = \dfrac{\sqrt{1 + 4t^2 + t^4}}{(1 + t^2 + t^4)^{3/2}}$
$\tau = \dfrac{(\mathbf{v} \times \mathbf{a}) \bullet \dfrac{d\mathbf{a}}{dt}}{|\mathbf{v} \times \mathbf{a}|^2} = \dfrac{2}{1 + 4t^2 + t^4}$.

8. $\mathbf{r} = e^t \cos t\,\mathbf{i} + e^t \sin t\,\mathbf{j} + e^t\mathbf{k}$
$\mathbf{v} = e^t(\cos t - \sin t)\mathbf{i} + e^t(\sin t + \cos t)\mathbf{j} + e^t\mathbf{k}$
$\mathbf{a} = -2e^t \sin t\,\mathbf{i} + 2e^t \cos t\,\mathbf{j} + e^t\mathbf{k}$
$\dfrac{d\mathbf{a}}{dt} = -2e^t(\cos t + \sin t)\mathbf{i} + 2e^t(\cos t - \sin t)\mathbf{j} + e^t\mathbf{k}$
$\mathbf{v} \times \mathbf{a} = e^{2t}(\sin t - \cos t)\mathbf{i} - e^{2t}(\cos t + \sin t)\mathbf{j} + 2e^{2t}\mathbf{k}$
$v = |\mathbf{v}| = \sqrt{3}e^t, \qquad |\mathbf{v} \times \mathbf{a}| = \sqrt{6}e^{2t}$
$(\mathbf{v} \times \mathbf{a}) \bullet \dfrac{d\mathbf{a}}{dt} = 2e^{3t}$
$\hat{\mathbf{T}} = \dfrac{\mathbf{v}}{v} = \dfrac{(\cos t - \sin t)\mathbf{i} + (\cos t + \sin t)\mathbf{j} + \mathbf{k}}{\sqrt{3}}$
$\hat{\mathbf{B}} = \dfrac{\mathbf{v} \times \mathbf{a}}{|\mathbf{v} \times \mathbf{a}|} = \dfrac{(\sin t - \cos t)\mathbf{i} - (\cos t + \sin t)\mathbf{j} + 2\mathbf{k}}{\sqrt{6}}$
$\hat{\mathbf{N}} = \hat{\mathbf{B}} \times \hat{\mathbf{T}} = -\dfrac{(\cos t + \sin t)\mathbf{i} - (\cos t - \sin t)\mathbf{j}}{\sqrt{2}}$
$\kappa = \dfrac{|\mathbf{v} \times \mathbf{a}|}{v^3} = \dfrac{\sqrt{2}}{3e^t}$
$\tau = \dfrac{(\mathbf{v} \times \mathbf{a}) \bullet \dfrac{d\mathbf{a}}{dt}}{|\mathbf{v} \times \mathbf{a}|^2} = \dfrac{1}{3e^t}$.

9. $\mathbf{r} = (2 + \sqrt{2}\cos t)\mathbf{i} + (1 - \sin t)\mathbf{j} + (3 + \sin t)\mathbf{k}$
$\mathbf{v} = -\sqrt{2}\sin t\,\mathbf{i} - \cos t\,\mathbf{j} + \cos t\,\mathbf{k}$
$v = \sqrt{2\sin^2 t + \cos^2 t + \cos^2 t} = \sqrt{2}$
$\mathbf{a} = -\sqrt{2}\cos t\,\mathbf{i} + \sin t\,\mathbf{j} - \sin t\,\mathbf{k}$
$\dfrac{d\mathbf{a}}{dt} = \sqrt{2}\sin t\,\mathbf{i} + \cos t\,\mathbf{j} - \cos t\,\mathbf{k}$
$\mathbf{v} \times \mathbf{a} = -\sqrt{2}\mathbf{j} - \sqrt{2}\mathbf{k}$
$\kappa = \dfrac{|\mathbf{v} \times \mathbf{a}|}{v^3} = \dfrac{2}{2\sqrt{2}} = \dfrac{1}{\sqrt{2}}$
$(\mathbf{v} \times \mathbf{a}) \bullet \dfrac{d\mathbf{a}}{dt} = -\sqrt{2}\cos t + \sqrt{2}\cos t = 0$
$\tau = 0$.
Since $\kappa = 1/\sqrt{2}$ is constant, and $\tau = 0$, the curve is a circle. Its centre is $(2, 1, 3)$ and its radius is $\sqrt{2}$. It lies in a plane with normal $\mathbf{j} + \mathbf{k}(= -\sqrt{2}\hat{\mathbf{B}})$.

10. $\mathbf{r} = x\mathbf{i} + \sin x\,\mathbf{j}$
$\mathbf{v} = \dfrac{dx}{dt}\mathbf{i} + \cos x \dfrac{dx}{dt}\mathbf{j} = k(\mathbf{i} + \cos x\,\mathbf{j})$
$v = k\sqrt{1 + \cos^2 x}$
$\mathbf{a} = -k \sin x \dfrac{dx}{dt}\mathbf{j} = -k^2 \sin x\,\mathbf{j}$
$\mathbf{v} \times \mathbf{a} = -k^3 \sin x\,\mathbf{k}$
$\kappa = \dfrac{|\mathbf{v} \times \mathbf{a}|}{v^3} = \dfrac{|\sin x|}{(1 + \cos^2 x)^{3/2}}$.

The tangential and normal components of acceleration are

$$\frac{dv}{dt} = \frac{k}{2\sqrt{1+\cos^2 x}} 2\cos x)(-\sin x)\frac{dx}{dt} = -\frac{k^2 \cos x \sin x}{\sqrt{1+\cos^2 x}}$$

$$v^2 \kappa = \frac{k^2 |\sin x|}{\sqrt{1+\cos^2 x}}.$$

11. $\mathbf{r} = \sin t \cos t \mathbf{i} + \sin^2 t \mathbf{j} + \cos t \mathbf{k}$
$\mathbf{v} = \cos 2t \mathbf{i} + \sin 2t \mathbf{j} - \sin t \mathbf{k}$
$\mathbf{a} = -2\sin 2t \mathbf{i} + 2\cos 2t \mathbf{j} - \cos t \mathbf{k}$
$\frac{d\mathbf{a}}{dt} = -4\cos 2t \mathbf{i} - 4\sin 2t \mathbf{j} + \sin t \mathbf{k}.$

At $t = 0$ we have $\mathbf{v} = \mathbf{i}$, $\mathbf{a} = 2\mathbf{j} - \mathbf{k}$, $\frac{d\mathbf{a}}{dt} = -4\mathbf{i}$,
$\mathbf{v} \times \mathbf{a} = \mathbf{j} + 2\mathbf{k}$, $(\mathbf{v} \times \mathbf{a}) \bullet \frac{d\mathbf{a}}{dt} = 0.$
Thus $\hat{\mathbf{T}} = \mathbf{i}$, $\hat{\mathbf{B}} = (\mathbf{j} + 2\mathbf{k})/\sqrt{5}$, $\hat{\mathbf{N}} = (2\mathbf{j} - \mathbf{k})/\sqrt{5}$,
$\kappa = \sqrt{5}$, and $\tau = 0$.

At $t = \pi/4$ we have $\mathbf{v} = \mathbf{j} - \frac{1}{\sqrt{2}}\mathbf{k}$, $\mathbf{a} = -2\mathbf{i} - \frac{1}{\sqrt{2}}\mathbf{k}$,
$\frac{d\mathbf{a}}{dt} = -4\mathbf{j} + \frac{1}{\sqrt{2}}\mathbf{k}$, $\mathbf{v} \times \mathbf{a} = -\frac{1}{\sqrt{2}}\mathbf{i} + \sqrt{2}\mathbf{j} + 2\mathbf{k}$,
$(\mathbf{v} \times \mathbf{a}) \bullet \frac{d\mathbf{a}}{dt} = -3\sqrt{2}.$
Thus

$$\hat{\mathbf{T}} = \frac{1}{\sqrt{3}}(\sqrt{2}\mathbf{j} - \mathbf{k})$$

$$\hat{\mathbf{B}} = \frac{1}{\sqrt{13}}(-\mathbf{i} + 2\mathbf{j} + 2\sqrt{2}\mathbf{k})$$

$$\hat{\mathbf{N}} = -\frac{1}{\sqrt{39}}(6\mathbf{i} + \mathbf{j} + \sqrt{2}\mathbf{k})$$

$$\kappa = \frac{2\sqrt{39}}{9}, \qquad \tau = -\frac{6\sqrt{2}}{13}.$$

12. $\mathbf{r} = a\cos t \mathbf{i} + b\sin t \mathbf{j}$
$\mathbf{v} = -a\sin t \mathbf{i} + b\cos t \mathbf{j}$
$\mathbf{a} = -a\cos t \mathbf{i} - b\sin t \mathbf{j}$
$\mathbf{v} \times \mathbf{a} = ab\mathbf{k}$
$v = \sqrt{a^2 \sin^2 t + b^2 \cos^2 t}.$

The tangential component of acceleration is

$$\frac{dv}{dt} = \frac{(a^2 - b^2)\sin t \cos t}{\sqrt{a^2 \sin^2 t + b^2 \cos^2 t}},$$

which is zero if t is an integer multiple of $\pi/2$, that is, at the ends of the major and minor axes of the ellipse. The normal component of acceleration is

$$v^2 \kappa = v^2 \frac{|\mathbf{v} \times \mathbf{a}|}{v^3} = \frac{ab}{\sqrt{a^2 \sin^2 t + b^2 \cos^2 t}}.$$

13. The ellipse is the same one considered in Exercise 16, so its curvature is

$$\kappa = \frac{ab}{(a^2 \sin^2 t + b^2 \cos^2 t)^{3/2}}$$

$$= \frac{ab}{\left((a^2 - b^2)\sin^2 t + b^2\right)^{3/2}}.$$

If $a > b > 0$, then the maximum curvature occurs when $\sin t = 0$, and is a/b^2. The minimum curvature occurs when $\sin t = \pm 1$, and is b/a^2.

14. The path of the road, $y = x^2$, has curvature

$$\kappa = \frac{\left|\frac{d^2 y}{dx^2}\right|}{\left(1 + \left(\frac{dy}{dx}\right)^2\right)^{3/2}} = \frac{2}{(1 + 4x^2)^{3/2}}.$$

The normal component of the acceleration of a vehicle travelling at speed v_0 along the road is

$$a_N = v_0^2 \kappa = \frac{2v_0^2}{(1 + 4x^2)^{3/2}}.$$

If the road is banked at angle θ (see the figure), then the resultant of the centrifugal force $-ma_N \hat{\mathbf{N}}$ and the gravitational force $-mg\mathbf{k}$ is normal to the roadway provided

$$\tan \theta = \frac{ma_N}{mg}, \quad \text{i.e.,} \quad \theta = \tan^{-1} \frac{2v_0^2}{g(1 + 4x^2)^{3/2}}.$$

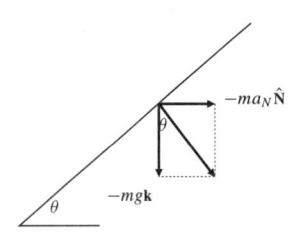

Fig. 11.5.14

15. Curve: $\mathbf{r} = x\mathbf{i} + e^x \mathbf{j}$.
Velocity: $\mathbf{v} = \mathbf{i} + e^x \mathbf{j}$. Speed: $v = \sqrt{1 + e^{2x}}$.
Acceleration: $\mathbf{a} = e^x \mathbf{j}$. We have

$$\mathbf{v} \times \mathbf{a} = e^x \mathbf{k}, \qquad |\mathbf{v} \times \mathbf{a}| = e^x.$$

The curvature is $\kappa = \dfrac{e^x}{(1 + e^{2x})^{3/2}}$. Therefore, the radius of curvature is $\rho = \dfrac{(1 + e^{2x})^{3/2}}{e^x}$.

The unit normal is

$$\hat{\mathbf{N}} = \hat{\mathbf{B}} \times \hat{\mathbf{T}} = \frac{(\mathbf{v} \times \mathbf{a}) \times \mathbf{v}}{|(\mathbf{v} \times \mathbf{a}) \times \mathbf{v}|} = \frac{-e^x \mathbf{i} + \mathbf{j}}{\sqrt{1 + e^{2x}}}.$$

The centre of curvature is

$$\mathbf{r}_c = \mathbf{r} + \rho \hat{\mathbf{N}}$$
$$= x\mathbf{i} + e^x \mathbf{j} + (1 + e^{2x})\left(-\mathbf{i} + \frac{1}{e^x}\mathbf{j}\right)$$
$$= (x - 1 - e^{2x})\mathbf{i} + (2e^x + e^{-x})\mathbf{j}.$$

This is the equation of the evolute.

16. The curve with polar equation $r = f(\theta)$ is given parametrically by

$$\mathbf{r} = f(\theta) \cos\theta \mathbf{i} + f(\theta) \sin\theta \mathbf{j}.$$

Thus we have

$$\mathbf{v} = \bigl(f'(\theta) \cos\theta - f(\theta) \sin\theta\bigr)\mathbf{i}$$
$$+ \bigl(f'(\theta) \sin\theta + f(\theta) \cos\theta\bigr)\mathbf{j}$$
$$\mathbf{a} = \bigl(f''(\theta) \cos\theta - 2f'(\theta) \sin\theta - f(\theta) \cos\theta\bigr)\mathbf{i}$$
$$+ \bigl(f''(\theta) \sin\theta + 2f'(\theta) \cos\theta - f(\theta) \sin\theta\bigr)\mathbf{j}$$
$$v = |\mathbf{v}| = \sqrt{\bigl(f'(\theta)\bigr)^2 + \bigl(f(\theta)\bigr)^2}$$
$$\mathbf{v} \times \mathbf{a} = \bigl[2\bigl(f'(\theta)\bigr)^2 + \bigl(f(\theta)\bigr)^2 - f(\theta)f''(\theta)\bigr]\mathbf{k}.$$

The curvature is, therefore,

$$\frac{|2\bigl(f'(\theta)\bigr)^2 + \bigl(f(\theta)\bigr)^2 - f(\theta)f''(\theta)|}{\bigl[\bigl(f'(\theta)\bigr)^2 + \bigl(f(\theta)\bigr)^2\bigr]^{3/2}}.$$

17. If $r = a(1 - \cos\theta)$, then $r' = a \sin\theta$, and $r'' = a \cos\theta$. By the result of Exercise 20, the curvature of this cardioid is

$$\kappa = \frac{1}{\bigl(a^2 \sin^2\theta + a^2(1 - \cos\theta)^2\bigr)^{3/2}} \times \bigl|2a^2 \sin^2\theta$$
$$+ a^2(1 - \cos\theta)^2 - a^2(\cos\theta - \cos^2\theta)\bigr|$$
$$= \frac{3a^2(1 - \cos\theta)}{\bigl(2a^2(1 - \cos\theta)\bigr)^{3/2}} = \frac{3}{2\sqrt{2ar}}.$$

18. By Exercise 8 of Section 11.4, the required curve must be a circular helix with parameters $a = 1/2$ (radius), and $b = 1/2$. Its equation will be

$$\mathbf{r} = \frac{1}{2} \cos t \mathbf{i}_1 + \frac{1}{2} \sin t \mathbf{j}_1 + \frac{1}{2} t \mathbf{k}_1 + \mathbf{r}_0$$

for some right-handed basis $\{\mathbf{i}_1, \mathbf{j}_1, \mathbf{k}_1\}$, and some constant vector $\mathbf{r}_0$. Example 3 of Section 11.4 provides values for $\hat{\mathbf{T}}(0)$, $\hat{\mathbf{N}}(0)$, and $\hat{\mathbf{B}}(0)$, which we can equate to the given values of these vectors:

$$\mathbf{i} = \hat{\mathbf{T}}(0) = \frac{1}{\sqrt{2}}\mathbf{j}_1 + \frac{1}{\sqrt{2}}\mathbf{k}_1$$
$$\mathbf{j} = \hat{\mathbf{N}}(0) = -\mathbf{i}_1$$
$$\mathbf{k} = \hat{\mathbf{B}}(0) = -\frac{1}{\sqrt{2}}\mathbf{j}_1 + \frac{1}{\sqrt{2}}\mathbf{k}_1.$$

Solving these equations for $\mathbf{i}_1$, $\mathbf{j}_1$, and $\mathbf{k}_1$ in terms of the given basis vectors, we obtain

$$\mathbf{i}_1 = -\mathbf{j}$$
$$\mathbf{j}_1 = \frac{1}{\sqrt{2}}\mathbf{i} - \frac{1}{\sqrt{2}}\mathbf{k}$$
$$\mathbf{k}_1 = \frac{1}{\sqrt{2}}\mathbf{i} + \frac{1}{\sqrt{2}}\mathbf{k}.$$

Therefore

$$\mathbf{r}(t) = \frac{t + \sin t}{2\sqrt{2}}\mathbf{i} - \frac{\cos t}{2}\mathbf{j} + \frac{t - \sin t}{2\sqrt{2}}\mathbf{k} + \mathbf{r}_0.$$

We also require that $\mathbf{r}(0) = \mathbf{i}$, so $\mathbf{r}_0 = \mathbf{i} + \frac{1}{2}\mathbf{j}$. The required equation is, therefore,

$$\mathbf{r}(t) = \left(\frac{t + \sin t}{2\sqrt{2}} + 1\right)\mathbf{i} + \frac{1 - \cos t}{2}\mathbf{j} + \frac{t - \sin t}{2\sqrt{2}}\mathbf{k}.$$

19. Given that $\dfrac{d\mathbf{r}}{dt} = \mathbf{c} \times \mathbf{r}(t)$, we have

$$\frac{d}{dt}|\mathbf{r}|^2 = \frac{d}{dt}\mathbf{r} \bullet \mathbf{r} = 2\mathbf{r} \bullet (\mathbf{c} \times \mathbf{r}) = 0$$
$$\frac{d}{dt}\bigl(\mathbf{r}(t) - \mathbf{r}(0)\bigr) \bullet \mathbf{c} = \frac{d\mathbf{r}}{dt} \bullet \mathbf{c} = (\mathbf{c} \times \mathbf{r}) \bullet \mathbf{c} = 0.$$

Thus $|\mathbf{r}(t)| = |\mathbf{r}(0)|$ is constant, and $\bigl(\mathbf{r}(t) - \mathbf{r}(0)\bigr) \bullet \mathbf{c} = 0$ is constant. Thus $\mathbf{r}(t)$ lies on the sphere centred at the origin with radius $|\mathbf{r}(0)|$, and also on the plane through $\mathbf{r}(0)$ with normal $\mathbf{c}$. The curve is the circle of intersection of this sphere and this plane.

20. For $\mathbf{r} = a\cos t\mathbf{i} + a\sin t\mathbf{j} + bt\mathbf{k}$, we have, by Example 3 of Section 11.4,

$$\hat{\mathbf{N}} = -\cos t\mathbf{i} - \sin t\mathbf{j}, \qquad \kappa = \frac{a}{a^2 + b^2}.$$

The centre of curvature $\mathbf{r}_c$ is given by

$$\mathbf{r}_c = \mathbf{r} + \rho\hat{\mathbf{N}} = \mathbf{r} + \frac{1}{\kappa}\hat{\mathbf{N}}.$$

Thus the evolute has equation

$$\mathbf{r} = a\cos t\mathbf{i} + a\sin t\mathbf{j} + bt\mathbf{k}$$
$$- \frac{a^2+b^2}{a}(\cos t\mathbf{i} + \sin t\mathbf{j})$$
$$= -\frac{b^2}{a}\cos t\mathbf{i} - \frac{b^2}{a}\sin t\mathbf{j} + bt\mathbf{k}.$$

The evolute is also a circular helix.

21. The parabola $y = x^2$ has curvature

$$\kappa = \frac{2}{(1+4x^2)^{3/2}},$$

by Exercise 18. The normal at (x, x^2) is perpendicular to the tangent, so has slope $-1/(2x)$. Since the unit normal points upward (the concave side of the parabola), we have

$$\hat{\mathbf{N}} = \frac{-2x\mathbf{i} + \mathbf{j}}{\sqrt{1+4x^2}}.$$

Thus the evolute of the parabola has equation

$$\mathbf{r} = x\mathbf{i} + x^2\mathbf{j} + \frac{(1+4x^2)^{3/2}}{2}\left(\frac{-2x\mathbf{i}+\mathbf{j}}{\sqrt{1+4x^2}}\right)$$
$$= x\mathbf{i} + x^2\mathbf{j} - (1+4x^2)x\mathbf{i} + \frac{1+4x^2}{2}\mathbf{j}$$
$$= -4x^3\mathbf{i} + \left(3x^2 + \frac{1}{2}\right)\mathbf{j}.$$

22. For the ellipse $\mathbf{r} = 2\cos t\mathbf{i} + \sin t\mathbf{j}$, we have

$$\mathbf{v} = -2\sin t\mathbf{i} + \cos t\mathbf{j}$$
$$\mathbf{a} = -2\cos t\mathbf{i} - \sin t\mathbf{j}$$
$$\mathbf{v} \times \mathbf{a} = 2\mathbf{k}$$
$$v = \sqrt{4\sin^2 t + \cos^2 t} = \sqrt{3\sin^2 t + 1}.$$

The curvature is $\kappa = \dfrac{2}{(3\sin^2 t+1)^{3/2}}$, so the radius of curvature is $\rho = \dfrac{(3\sin^2 t+1)^{3/2}}{2}$. We have

$$\hat{\mathbf{T}} = \frac{-2\sin t\mathbf{i} + \cos t\mathbf{j}}{\sqrt{3\sin^2 t+1}}, \quad \hat{\mathbf{B}} = \mathbf{k}$$
$$\hat{\mathbf{N}} = -\frac{\cos t\mathbf{i} + 2\sin t\mathbf{j}}{\sqrt{3\sin^2 t+1}}.$$

Therefore the evolute has equation

$$\mathbf{r} = 2\cos t\mathbf{i} + \sin t\mathbf{j} - \frac{3\sin^2 t+1}{2}(\cos t\mathbf{i} + 2\sin t\mathbf{j})$$
$$= \frac{3}{2}\cos^3 t\mathbf{i} - 3\sin^3 t\mathbf{j}.$$

23. We require that

$$f(1) = 1, \quad f'(1) = 0, \quad f''(1) = 0,$$
$$f(-1) = -1, \quad f'(-1) = 0, \quad f''(-1) = 0.$$

As in Example 5, we try a polynomial of degree 5. However, here it is clear that an odd function will do, and we need only impose the conditions at $x = 1$. Thus we try

$$f(x) = Ax + Bx^3 + Cx^5$$
$$f'(x) = A + 3Bx^2 + 5Cx^4$$
$$f''(x) = 6Bx + 20Cx^3.$$

The conditions at $x = 1$ become

$$\begin{array}{rcrcrcl} A & + & B & + & C & = & 1 \\ A & + & 3B & + & 5C & = & 0 \\ & & 6B & + & 20C & = & 0. \end{array}$$

This system has solution $A = 15/8$, $B = -5/4$, and $C = 3/8$. Thus

$$f(x) = \frac{15}{8}x - \frac{5}{4}x^3 + \frac{3}{8}x^5$$

is one possible solution.

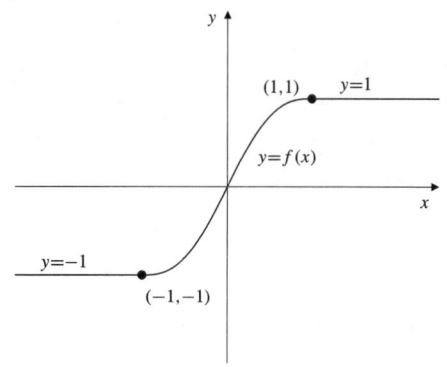

Fig. 11.5.23

24. We require

$$f(0) = 1, \quad f'(0) = 0, \quad f''(0) = -1,$$
$$f(-1) = 1, \quad f'(-1) = 0, \quad f''(-1) = 0.$$

The condition $f''(0) = -1$ follows from the fact that

$$\left.\frac{d^2}{dx^2}\sqrt{1-x^2}\right|_{x=0} = -1.$$

As in Example 5, we try

$$f(x) = A + Bx + Cx^2 + Dx^3 + Ex^4 + Fx^5$$
$$f'(x) = B + 2Cx + 3Dx^2 + 4Ex^3 + 5Fx^4$$
$$f'' = 2C + 6Dx + 12Ex^2 + 20Fx^3.$$

The required conditions force the coefficients to satisfy the system of equations

$$A - B + C - D + E - F = 1$$
$$B - 2C + 3D - 4E + 5F = 0$$
$$2C - 6D + 12E - 20F = 0$$
$$A = 1$$
$$B = 0$$
$$2C = -1$$

which has solution $A = 1$, $B = 0$, $C = -1/2$, $D = -3/2$, $E = -3/2$, $F = -1/2$. Thus we can use a track section in the shape of the graph of

$$f(x) = 1 - \frac{1}{2}x^2 - \frac{3}{2}x^3 - \frac{3}{2}x^4 - \frac{1}{2}x^5 = 1 - \frac{1}{2}x^2(1+x)^3.$$

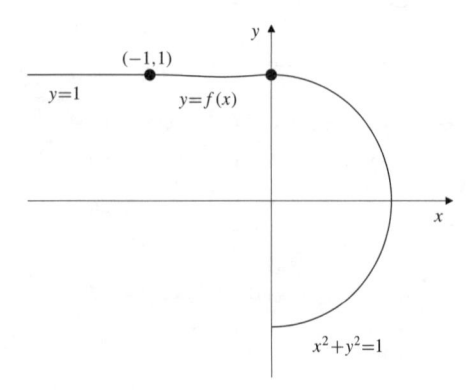

Fig. 11.5.24

25. Given: $\mathbf{a}(t) = \lambda(t)\mathbf{r}(t) + \mu(t)\mathbf{v}(t)$, $\mathbf{v} \times \mathbf{a} \neq \mathbf{0}$. We have

$$\mathbf{v} \times \mathbf{a} = \lambda \mathbf{v} \times \mathbf{r} + \mu \mathbf{v} \times \mathbf{v} = \lambda \mathbf{v} \times \mathbf{r}$$
$$\frac{d\mathbf{a}}{dt} = \lambda'\mathbf{r} + \lambda\mathbf{v} + \mu'\mathbf{v} + \mu\mathbf{a}$$
$$= \lambda'\mathbf{r} + (\lambda + \mu')\mathbf{v} + \mu(\lambda\mathbf{r} + \mu\mathbf{v})$$
$$= (\lambda' + \mu\lambda)\mathbf{r} + (\lambda + \mu' + \mu^2)\mathbf{v}.$$

Since $\mathbf{v} \times \mathbf{r}$ is perpendicular to both $\mathbf{v}$ and $\mathbf{r}$, we have

$$(\mathbf{v} \times \mathbf{a}) \bullet \frac{d\mathbf{a}}{dt} = 0.$$

Thus the torsion $\tau(t)$ of the curve is identically zero. It remains zero when expressed in terms of arc length: $\tau(s) = 0$. By Exercise 6 of Section 11.4, $\mathbf{r}(t)$ must be a plane curve.

Section 11.6 Kepler's Laws of Planetary Motion (page 691)

1. $r = \dfrac{\ell}{1 + \epsilon \cos\theta} \implies r + \epsilon x = \ell$

   $r = \ell - \epsilon x$
   $x^2 + y^2 = r^2 = \ell^2 - 2\ell\epsilon x + \epsilon^2 x^2$
   $(1 - \epsilon^2)x^2 + 2\ell\epsilon x + y^2 = \ell^2$
   $(1 - \epsilon^2)\left(x + \dfrac{\ell\epsilon}{1-\epsilon^2}\right)^2 + y^2 = \ell^2 + \dfrac{\ell^2\epsilon^2}{1-\epsilon^2} = \dfrac{\ell^2}{1-\epsilon^2}$

   $$\dfrac{\left(x + \dfrac{\ell\epsilon}{1-\epsilon^2}\right)^2}{\left(\dfrac{\ell}{1-\epsilon^2}\right)^2} + \dfrac{y^2}{\left(\dfrac{\ell}{\sqrt{1-\epsilon^2}}\right)^2} = 1.$$

2. Position: $\mathbf{r} = r\hat{\mathbf{r}} = k\hat{\mathbf{r}}$.
   Velocity: $\mathbf{v} = k\dot{\hat{\mathbf{r}}} = k\dot{\theta}\hat{\boldsymbol{\theta}}$; speed: $v = k\dot{\theta}$.
   Acceleration: $k\ddot{\theta}\hat{\boldsymbol{\theta}} + k\dot{\theta}\dot{\hat{\boldsymbol{\theta}}} = -k\dot{\theta}^2\hat{\mathbf{r}} + k\ddot{\theta}\hat{\boldsymbol{\theta}}$.
   Radial component of acceleration: $-k\dot{\theta}^2$.
   Transverse component of acceleration: $k\ddot{\theta} = \dot{v}$ (the rate of change of the speed).

3. Position: on the curve $r = e^\theta$.
   Radial velocity: $\dot{r} = e^\theta\dot{\theta}$.
   Transverse velocity: $r\dot{\theta} = e^\theta\dot{\theta}$.
   Speed $v = \sqrt{2}e^\theta\dot{\theta} = 1 \implies \dot{\theta} = (1/\sqrt{2})e^{-\theta}$.
   Thus $\ddot{\theta} = -(1/\sqrt{2})e^{-\theta}\dot{\theta} = -e^{-2\theta}/2$.
   Radial velocity = transverse velocity = $1/\sqrt{2}$.
   Radial acceleration:
   $\ddot{r} - r\dot{\theta}^2 = e^\theta\dot{\theta}^2 + e^\theta\ddot{\theta} - e^\theta\dot{\theta}^2 = e^\theta\ddot{\theta} = -e^{-\theta}/2.$
   Transverse acceleration:
   $r\ddot{\theta} + 2\dot{r}\dot{\theta} = -(e^{-\theta})/2 + e^{-\theta} = e^{-\theta}/2.$

4. Path: $r = \theta$. Thus $\dot{r} = \dot{\theta}$, $\ddot{r} = \ddot{\theta}$.
   Speed: $v = \sqrt{(\dot{r})^2 + (r\dot{\theta})^2} = \dot{\theta}\sqrt{1 + r^2}$.
   Transverse acceleration = 0 (central force). Thus $r\ddot{\theta} + 2\dot{r}\dot{\theta} = 0$, or $\ddot{\theta} = -2\dot{\theta}^2/r$.
   Radial acceleration:

   $$\ddot{r} - r\dot{\theta}^2 = \ddot{\theta} - r\dot{\theta}^2$$
   $$= -\left(\dfrac{2}{r} + r\right)\dot{\theta}^2 = -\dfrac{(2+r^2)v^2}{r(1+r^2)}.$$

   The magnitude of the acceleration is, therefore,
   $\dfrac{(2+r^2)v^2}{r(1+r^2)}.$

5. The distance r from $(0,0)$ to (x, y) on the curve $2y = x^2 - 1$ satisfies

   $$r^2 = x^2 + \left(\dfrac{x^2-1}{2}\right)^2 = \left(\dfrac{x^2+1}{2}\right)^2.$$

Thus $r = (x^2 + 1)/2$. Since $\dot{y} = x\dot{x}$, the speed is given by
$$v = \sqrt{\dot{x}^2 + \dot{y}^2} = \sqrt{1+x^2}|\dot{x}| = \sqrt{2r}|\dot{x}|.$$

Since the acceleration is directed towards the origin, its components are proportional to those of the position:
$$\frac{\ddot{x}}{x} = \frac{\ddot{y}}{y} = \frac{2(x\ddot{x} + \dot{x}^2)}{x^2 - 1}$$
$$x^2\ddot{x} - \ddot{x} = 2x^2\ddot{x} + 2x\dot{x}^2$$
$$(1+x^2)\ddot{x} + 2x\dot{x}^2 = 0.$$

Thus $\dfrac{d}{dt}\left[(1+x^2)\dot{x}\right] = 0$, and
$$(1+x^2)\dot{x} = C_1 \quad \text{(a constant)}.$$

Therefore, $v = \dfrac{\sqrt{2r}|C_1|}{2r} = \dfrac{C_2}{\sqrt{r}}$; the speed is proportional to $1/\sqrt{r}$.

The magnitude of the acceleration $\mathbf{a}$ is given by
$$|\mathbf{a}|^2 = \ddot{x}^2 + \ddot{y}^2 = \ddot{x}^2 + (x\ddot{x} + \dot{x}^2)^2$$
$$= (1+x^2)\ddot{x}^2 + 2x\ddot{x}\dot{x}^2 + \dot{x}^4$$
$$= \frac{4x^2\dot{x}^4}{1+x^2} - \frac{4x^2\dot{x}^4}{1+x^2} + \dot{x}^4$$
$$= \dot{x}^4 = \left(\frac{C_1}{1+x^2}\right)^4 = \frac{C_1^4}{r^4}.$$

Thus the magnitude of the acceleration is proportional to $1/r^2$.

6. Let the period and the semi-major axis of the orbit of Halley's comet be $T_H = 76$ years and a_H km respectively. Similar parameters for the earth's orbit are $T_E = 1$ year and $a_E = 150 \times 10^6$ km. By Kepler's third law
$$\frac{T_H^2}{a_H^3} = \frac{T_E^2}{a_E^3}.$$

Thus
$$a_H = 150 \times 10^6 \times 76^{2/3} \approx 2.69 \times 10^9.$$

The major axis of Halley's comet's orbit is $2a_H \approx 5.38 \times 10^9$ km.

7. The period and semi-major axis of the moon's orbit around the earth are
$$T_M \approx 27 \text{ days}, \qquad a_M \approx 385,000 \text{ km}.$$

The satellite has a circular orbit of radius a_S and period $T_S = 1$ day. (If the orbit is in the plane of the equator, the satellite will remain above the same point on the earth.) By Kepler's third law,
$$\frac{T_S^2}{a_S^3} = \frac{T_M^2}{a_M^3}.$$

Thus $a_S = 385,000 \times (1/27)^{2/3} \approx 42,788$. The satellite's orbit should have radius about 42,788 km, and should lie in the equatorial plane.

8. The period T (in years) and radius R (in km) of the asteroid's orbit satisfies
$$\frac{T^2}{R^3} = \frac{T_{\text{earth}}^2}{R_{\text{earth}}^3} = \frac{1^2}{(150 \times 10^6)^3}.$$

Thus the radius of the asteroid's orbit is
$R \approx 150 \times 10^6 T^{2/3}$ km.

9. If R is the radius and T is the period of the asteroid's circular orbit, then *almost* stopping the asteroid causes it to drop into a very eccentric elliptical orbit with major axis approximately R. (Thus, $a = R/2$.) The period T_e of the new elliptical orbit satisfies
$$\frac{T_e^2}{T^2} = \frac{(R/2)^3}{R^3} = \frac{1}{8}.$$

Thus $T_e = T/(2\sqrt{2})$. The time the asteroid will take to fall into the sun is half of T_e. Thus it is $T/(4\sqrt{2})$.

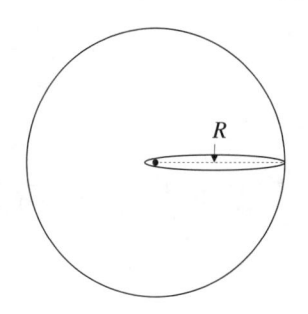

Fig. 11.6.9

10. At perihelion, $r = a - c = (1-\epsilon)a$.
At aphelion $r = a + c = (1+\epsilon)a$.
Since $\dot{r} = 0$ at perihelion and aphelion, the speed is $v = r\dot{\theta}$ at each point. Since $r^2\dot{\theta} = h$ is constant over the orbit, $v = h/r$. Therefore
$$v_{\text{perihelion}} = \frac{h}{a(1-\epsilon)}, \qquad v_{\text{aphelion}} = \frac{h}{a(1+\epsilon)}.$$

If $v_{\text{perihelion}} = 2v_{\text{aphelion}}$ then
$$\frac{h}{a(1-\epsilon)} = \frac{2h}{a(1+\epsilon)}.$$

Hence $1+\epsilon = 2(1-\epsilon)$, and $\epsilon = 1/3$. The eccentricity of the orbit is 1/3.

11. The orbital speed v of a planet satisfies (by conservation of energy)

$$\frac{v^2}{2} - \frac{k}{r} = K \quad \text{(total energy)}.$$

If v is constant so must be r, and the orbit will therefore be circular.

12. Since $r^2\dot{\theta} = h = $ constant for the planet's orbit, and since the speed is $v = r\dot{\theta}$ at perihelion and at aphelion (the radial velocity is zero at these points), we have

$$r_p v_p = r_a v_a,$$

where the subscripts p and a refer to perihelion and aphelion, respectively. Since $r_p/r_a = 8/10$, we must have $v_p/v_a = 10/8 = 1.25$. Also,

$$r_p = \frac{\ell}{1+\epsilon\cos 0} = \frac{\ell}{1+\epsilon}, \quad r_a = \frac{\ell}{1+\epsilon\cos\pi} = \frac{\ell}{1-\epsilon}.$$

Thus $\ell/(1+\epsilon) = (8/10)\,\ell/(1-\epsilon)$, and so $10-10\epsilon = 8+8\epsilon$. Hence $2 = 18\epsilon$. The eccentricity of the orbit is $\epsilon = 1/9$.

13. Let the radius of the circular orbit be R, and let the parameters of the new elliptical orbit be a and c, as shown in the figure. Then $R = a + c$. At the moment of the collision, r does not change ($r = R$), but the speed $r\dot{\theta}$ is cut in half. Therefore $\dot{\theta}$ is cut in half, and so $h = r^2\dot{\theta}$ is cut in half. Let H be the value of $r^2\dot{\theta}$ for the circular orbit, and let h be the value for the new elliptical orbit. Thus $h = H/2$. We have

$$R = \frac{H^2}{k}, \quad a = \frac{h^2}{k(1-\epsilon^2)} = \frac{H^2}{4k(1-\epsilon^2)} = \frac{R}{4(1-\epsilon^2)}.$$

Similarly, $c = \epsilon a = \dfrac{\epsilon R}{4(1-\epsilon^2)}$, so

$$R = c + a = \frac{(1+\epsilon)R}{4(1-\epsilon^2)} = \frac{R}{4(1-\epsilon)}.$$

It follows that $1 = 4 - 4\epsilon$, so $\epsilon = 3/4$. The new elliptical orbit has eccentricity $\epsilon = 3/4$.

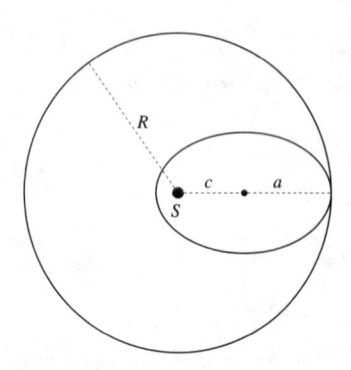

Fig. 11.6.13

14. As in Exercise 12, $r_P v_P = r_A v_A$, where $r_A = \ell/(1-\epsilon)$ and $r_P = \ell/(1+\epsilon)$, ϵ being the eccentricity of the orbit. Thus

$$\frac{v_P}{v_A} = \frac{r_A}{r_P} = \frac{1+\epsilon}{1-\epsilon}.$$

Solving this equation for ϵ in terms of v_P and v_A, we get

$$\epsilon = \frac{v_P - v_A}{v_P + v_A}.$$

By conservation of energy the speed v at the ends of the minor axis of the orbit (where $r = a$) satisfies

$$\frac{v^2}{2} - \frac{k}{a} = \frac{v_P^2}{2} - \frac{k}{r_P} = \frac{v_A^2}{2} - \frac{k}{r_A}.$$

The latter equality shows that

$$v_P^2 - v_A^2 = 2k\left(\frac{1}{r_P} - \frac{1}{r_A}\right) = \frac{4k\epsilon}{\ell}.$$

Using this result and the parameters of the orbit given in the text, we obtain

$$\begin{aligned}v^2 &= v_P^2 + 2k\left(\frac{1}{a} - \frac{1}{r_P}\right)\\&= v_P^2 + \frac{2k}{\ell}\left(1 - \epsilon^2 - (1+\epsilon)\right)\\&= v_P^2 - \frac{2k\epsilon}{\ell}(1+\epsilon)\\&= v_P^2 - \frac{v_P^2 - v_A^2}{2}\left(1 + \frac{v_P - v_A}{v_P + v_A}\right)\\&= v_P^2 - \frac{v_P - v_A}{2}(2v_P) = v_P v_A.\end{aligned}$$

Thus $v = \sqrt{v_P v_A}$.

15. Since the radial line from the sun to the planet sweeps out equal areas in equal times, the fraction of the planet's period spend on the same side of the minor axis as the sun is equal to the shaded area in the figure to the total area of the ellipse, that is,

$$\frac{\frac{1}{2}\pi ab - \frac{1}{2}(2bc)}{\pi ab} = \frac{\frac{1}{2}\pi ab - \epsilon ab}{\pi ab} = \frac{1}{2} - \frac{\epsilon}{\pi},$$

where $\epsilon = c/a$ is the eccentricity of the orbit.

Fig. 11.6.15

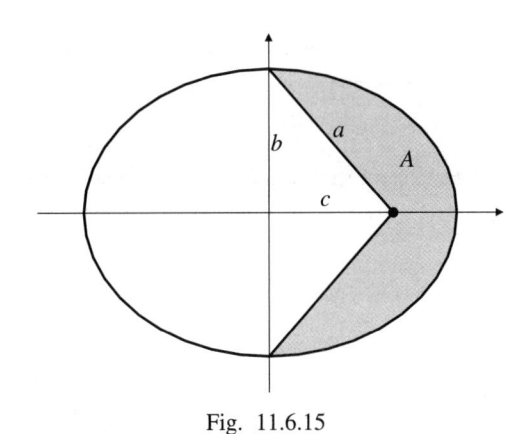

16. By conservation of energy, we have

$$\frac{k}{r} - \frac{1}{2}\left(\dot{r}^2 + \frac{h^2}{r^2}\right) = -K$$

where K is a constant for the orbit (the total energy). The term in the parentheses is v^2, the square of the speed. Thus

$$\frac{k}{r} - \frac{1}{2}v^2 = -K = \frac{k}{r_0} - \frac{1}{2}v_0^2,$$

where r_0 and v_0 are the given distance and speed. We evaluate $-K$ at perihelion.
The parameters of the orbit are

$$\ell = \frac{h^2}{k}, \quad a = \frac{h^2}{k(1-\epsilon^2)}, \quad b = \frac{h^2}{k\sqrt{1-\epsilon^2}}, \quad c = \epsilon a.$$

At perihelion P we have

$$r = a - c = (1-\epsilon)a = \frac{h^2}{k(1+\epsilon)}.$$

Since $\dot{r} = 0$ at perihelion, the speed there is $v = r\dot\theta$. By Kepler's second law, $r^2\dot\theta = h$, so $v = h/r = k(1+\epsilon)/h$. Thus

$$-K = \frac{k}{r} - \frac{v^2}{2}$$
$$= \frac{k^2}{h^2}(1+\epsilon) - \frac{1}{2}\frac{k^2}{h^2}(1+\epsilon)^2$$
$$= \frac{k^2}{2h^2}(1+\epsilon)\bigl[2 - (1+\epsilon)\bigr]$$
$$= \frac{k^2}{2h^2}(1-\epsilon^2) = \frac{k}{2a}.$$

Thus $a = \dfrac{k}{-2K}$. By Kepler's third law,

$$T^2 = \frac{4\pi^2}{k}a^3 = \frac{4\pi^2}{k}\left(\frac{k}{-2K}\right)^3.$$

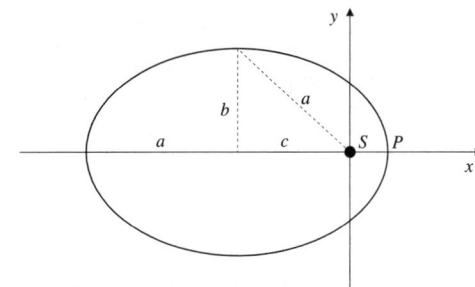

Thus $T = \dfrac{2\pi}{\sqrt{k}}\left(\dfrac{2}{r_0} - \dfrac{v_0^2}{k}\right)^{-3/2}$.

Fig. 11.6.16

17. Let $r_1(s)$ and $r_2(s)$ be the distances from the point $P = \mathbf{r}(s)$ on the ellipse $\mathcal{E}$ to the two foci. (Here s denotes arc length on $\mathcal{E}$, measured from any convenient point.) By symmetry

$$\int_{\mathcal{E}} r_1(s)\,ds = \int_{\mathcal{E}} r_2(s)\,ds.$$

But $r_1(s) + r_2(s) = 2a$ for any s. Therefore,

$$\int_{\mathcal{E}} r_1(s)\,ds + \int_{\mathcal{E}} r_2(s)\,ds = \int_{\mathcal{E}} 2a\,ds = 2ac(\mathcal{E}).$$

Hence $\int_{\mathcal{E}} r_1(s)\,ds = ac(\mathcal{E})$, and

$$\frac{1}{c(\mathcal{E})}\int_{\mathcal{E}} r_1(s)\,ds = a.$$

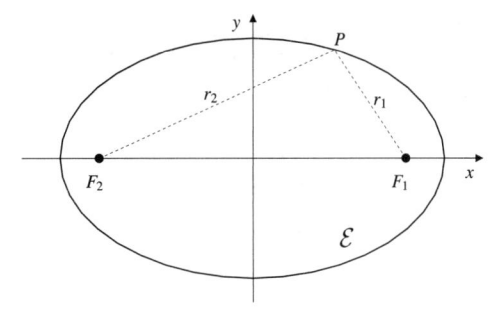

Fig. 11.6.17

18. Start with

$$\ddot{r} - \frac{h^2}{r^3} = -\frac{k}{r^2}.$$

Let $r(t) = \dfrac{1}{u(\theta)}$, where $\theta = \theta(t)$. Since $r^2\dot\theta = h$ (constant), we have

$$\dot{r} = -\frac{1}{u^2}\frac{du}{d\theta}\dot\theta = -r^2\frac{du}{d\theta}\frac{h}{r^2} = -h\frac{du}{d\theta}$$
$$\ddot{r} = -h\frac{d^2u}{d\theta^2}\dot\theta = -\frac{h^2}{r^2}\frac{d^2u}{d\theta^2} = -h^2u^2\frac{d^2u}{d\theta^2}.$$

419

Thus $-h^2 u^2 \dfrac{d^2 u}{d\theta^2} - h^2 u^3 = -ku^2$, or

$$\frac{d^2 u}{d\theta^2} + u = \frac{k}{h^2}.$$

This is the DE for simple harmonic motion with a constant forcing term (nonhomogeneous term) on the right-hand side. It is easily verified that

$$u = \frac{k}{h^2}\bigl(1 + \epsilon \cos(\theta - \theta_0)\bigr)$$

is a solution for any choice of the constants ϵ and θ_0. Expressing the solution in terms of r, we have

$$r = \frac{h^2/k}{1 + \epsilon \cos(\theta - \theta_0)},$$

which is an ellipse if $|\epsilon| < 1$.

19. For inverse cube attraction, the equation of motion is

$$\ddot r - \frac{h^2}{r^3} = -\frac{k}{r^3},$$

where $r^2 \dot\theta = h$ is constant, since the force is central. Making the same change of variables used in Exercise 18, we obtain

$$-h^2 u^2 \frac{d^2 u}{d\theta^2} - h^2 u^3 = -ku^3,$$

or

$$\frac{d^2 u}{d\theta^2} - \frac{k - h^2}{h^2} u = 0.$$

There are three cases to consider.

CASE I. If $k < h^2$ the DE is $\dfrac{d^2 u}{d\theta^2} + \omega^2 u = 0$, where $\omega^2 = (h^2 - k)/h^2$. This has solution $u = A\cos\omega(\theta - \theta_0)$. Thus

$$r = \frac{1}{A\cos\omega(\theta - \theta_0)}.$$

Note that $r \to \infty$ as $\theta \to \theta_0 + \dfrac{\pi}{2\omega}$. There are no bounded orbits in this case.

CASE II. If $k > h^2$ the DE is $\dfrac{d^2 u}{d\theta^2} - \omega^2 u = 0$, where $\omega^2 = (k - h^2)/h^2$. This has solution $u = Ae^{\omega\theta} + Be^{-\omega\theta}$. Since $u \to 0$ or ∞ as $\theta \to \infty$, the corresponding solution $r = 1/u$ cannot be both bounded and bounded away from zero. (Note that $\dot\theta = h/r^2 \geq K > 0$ for any orbit which is bounded away from zero, so we can be sure $\theta \to \infty$ on such an orbit.)

CASE III. If $k = h^2$ the DE is $\dfrac{d^2 u}{d\theta^2} = 0$, which has solutions $u = A\theta + B$, corresponding to

$$r = \frac{1}{A\theta + B}.$$

Such orbits are bounded away from zero and infinity only if $A = 0$, in which case they are circular.

Thus, the only possible orbits which are bounded away from zero and infinity (i.e., which do not escape to infinity or plunge into the sun) in a universe with an inverse cube gravitational attraction are some circular orbits for which $h^2 = k$. Such orbits cannot be considered "stable" since even slight loss of energy would result in decreased h and the condition $h^2 = k$ would no longer be satisfied. Now aren't you glad you live in an inverse square universe?

20. Since $\dfrac{k}{r} = \dfrac{1}{2}v^2 - K$ by conservation of energy, if $K < 0$, then

$$\frac{k}{r} \geq -K > 0,$$

so $r \leq -\dfrac{k}{K}$. The orbit is, therefore, bounded.

21. $r = \dfrac{\ell}{1 + \epsilon \cos\theta}$, $(\epsilon > 1)$.

See the following figure.
Vertices: At V_1, $\theta = 0$ and $r = \ell/(1 + \epsilon)$.
At V_2, $\theta = \pi$ and $r = \ell/(1 - \epsilon) = -\ell/(\epsilon - 1)$.
Semi-focal separation:

$$c = \frac{1}{2}\left(\frac{\ell}{1+\epsilon} + \frac{\ell}{1-\epsilon}\right) = \frac{\ell\epsilon}{\epsilon^2 - 1}.$$

The centre is $(c, 0)$.
Semi-transverse axis:

$$a = \frac{\ell\epsilon}{\epsilon^2 - 1} - \frac{\ell}{\epsilon + 1} = \frac{\ell}{\epsilon^2 - 1}.$$

Semi-conjugate axis:

$$b = \sqrt{c^2 - a^2} = \frac{\ell}{\sqrt{\epsilon^2 - 1}}.$$

Direction of asymptotes (see figure):

$$\theta = \tan^{-1}\frac{b}{a} = \cos^{-1}\frac{a}{c} = \cos^{-1}\frac{1}{\epsilon}.$$

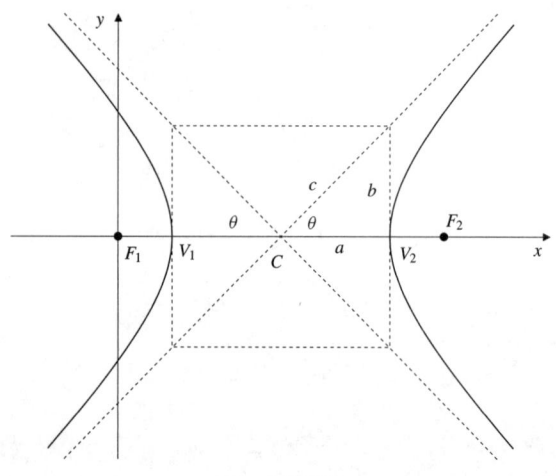

Fig. 11.6.21

22. By Exercise 17, the asymptotes make angle $\theta = \cos^{-1}(1/\epsilon)$ with the transverse axis, as shown in the figure. The angle of deviation δ satisfies $2\theta + \delta = \pi$, so $\theta = \dfrac{\pi}{2} - \dfrac{\delta}{2}$, and

$$\cos\theta = \sin\frac{\delta}{2}, \qquad \sin\theta = \cos\frac{\delta}{2}.$$

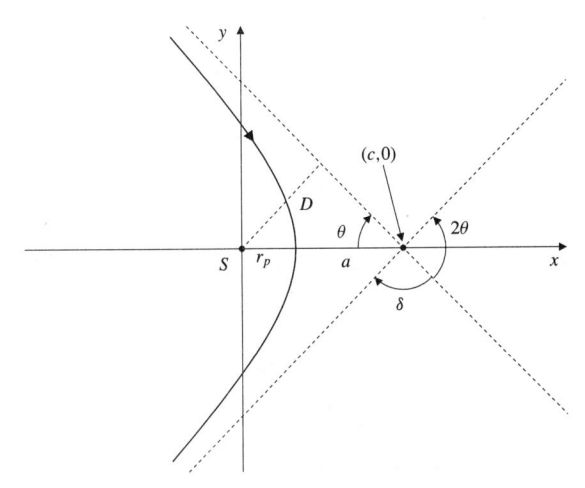

Fig. 11.6.22

By conservation of energy,

$$\frac{v^2}{2} - \frac{k}{r} = \text{constant} = \frac{v_\infty^2}{2}$$

for all points on the orbit. At perihelion,

$$r = r_p = c - a = (\epsilon - 1)a = \frac{\ell}{\epsilon+1},$$
$$v = v_p = r_p\dot\theta = \frac{h}{r_p} = \frac{h(\epsilon+1)}{\ell}.$$

Since $h^2 = k\ell$, we have

$$v_\infty^2 = v_p^2 - \frac{2k}{r_p}$$
$$= \frac{h^2}{\ell^2}(\epsilon+1)^2 - \frac{2k}{\ell}(\epsilon+1)$$
$$= \frac{k}{\ell}\big[(\epsilon+1)^2 - 2(\epsilon+1)\big]$$
$$= \frac{k}{\ell}(\epsilon^2 - 1) = \frac{k}{a}.$$

Thus $av_\infty^2 = k$.
If D is the perpendicular distance from the sun S to an asymptote of the orbit (see the figure) then

$$D = c\sin\theta = \epsilon a\sin\theta = a\frac{\sin\theta}{\cos\theta}$$
$$= a\frac{\cos(\delta/2)}{\sin(\delta/2)} = a\cot\frac{\delta}{2}.$$

Therefore

$$\frac{Dv_\infty^2}{k} = \frac{v_\infty^2 a}{k}\cot\frac{\delta}{2} = \cot\frac{\delta}{2}.$$

Review Exercises 11 (page 692)

1. Given that $\mathbf{a}\bullet\mathbf{r} = 0$ and $\mathbf{a}\bullet\mathbf{v} = 0$, we have

$$\frac{d}{dt}|\mathbf{r}(t) - t\mathbf{v}(t)|^2$$
$$= 2\big(\mathbf{r}(t) - t\mathbf{v}(t)\big)\bullet\big(\mathbf{v}(t) - \mathbf{v}(t) - t\mathbf{a}(t)\big)$$
$$= 2\big(\mathbf{r}(t) - t\mathbf{v}(t)\big)\bullet\mathbf{a}(t) = 0 - 0 = 0.$$

2. $\mathbf{r} = t\cos t\,\mathbf{i} + t\sin t\,\mathbf{j} + (2\pi - t)\mathbf{k}$, $(0 \le t \le 2\pi)$ is a conical helix wound around the cone $z = 2\pi - \sqrt{x^2 + y^2}$ starting at the vertex $(0, 0, 2\pi)$, and completing one revolution to end up at $(2\pi, 0, 0)$. Since

$$\mathbf{v} = (\cos t - t\sin t)\mathbf{i} + (\sin t + t\cos t)\mathbf{j} - \mathbf{k},$$

the length of the curve is

$$L = \int_0^{2\pi}\sqrt{2 + t^2}\,dt = \pi\sqrt{2 + 4\pi^2} + \ln\left(\frac{2\pi + \sqrt{2 + 4\pi^2}}{\sqrt{2}}\right)$$

units.

3. The position of the particle at time t is

$$\mathbf{r} = x\mathbf{i} + x^2\mathbf{j} + \tfrac{2}{3}x^3\mathbf{k},$$

where x is an increasing function of t. The velocity is

$$\mathbf{v} = \frac{dx}{dt}\big(\mathbf{i} + 2x\mathbf{j} + 2x^2\mathbf{k}\big).$$

Since the speed is 6, we have

$$6 = \frac{dx}{dt}\sqrt{1 + 4x^2 + 4x^4} = (2x^2 + 1)\frac{dx}{dt},$$

so that $dx/dt = 6/(2x^2 + 1)$. The particle is at $(1, 1, \tfrac{2}{3})$ when $x = 1$. At this time its velocity is

$$\mathbf{v}(1) = 2(\mathbf{i} + 2\mathbf{j} + 2\mathbf{k}).$$

Also

$$\frac{d^2x}{dt^2} = -\frac{6}{(2x^2 + 1)^2}(4x)\frac{dx}{dt} = -\frac{144x}{(2x^2 + 1)^3}$$
$$\mathbf{a} = \frac{d^2x}{dt^2}(\mathbf{i} + 2x\mathbf{j} + 2x^2\mathbf{k})$$
$$+ \frac{dx}{dt}\left(2\frac{dx}{dt}\mathbf{j} + 4x\frac{dx}{dt}\mathbf{k}\right).$$

At $x = 1$, we have

$$\mathbf{a}(1) = -\frac{16}{3}(\mathbf{i} + 2\mathbf{j} + 2\mathbf{k}) + 2(4\mathbf{j} + 8\mathbf{k})$$
$$= \frac{8}{3}(-2\mathbf{i} - \mathbf{j} + 2\mathbf{k}).$$

4. The position, velocity, speed, and acceleration of the particle are given by

$$\mathbf{r} = x\mathbf{i} + x^2\mathbf{j}$$
$$\mathbf{v} = \frac{dx}{dt}(\mathbf{i} + 2x\mathbf{j}), \quad v = \left|\frac{dx}{dt}\right|\sqrt{1 + 4x^2}$$
$$\mathbf{a} = \frac{d^2x}{dt^2}(\mathbf{i} + 2x\mathbf{j}) + 2\left(\frac{dx}{dt}\right)^2\mathbf{j}.$$

Let us assume that the particle is moving to the right, so that $dx/dt > 0$. Since the speed is t, we have

$$\frac{dx}{dt} = \frac{t}{\sqrt{1 + 4x^2}}$$
$$\frac{d^2x}{dt^2} = \frac{\sqrt{1 + 4x^2} - \frac{4tx}{\sqrt{1 + 4x^2}}\frac{dx}{dt}}{1 + 4x^2}.$$

If the particle is at $(\sqrt{2}, 2)$ at $t = 3$, then $dx/dt = 1$ at that time, and

$$\frac{d^2x}{dt^2} = \frac{3 - 4\sqrt{2}}{9}.$$

Hence the acceleration is

$$\mathbf{a} = \frac{3 - 4\sqrt{2}}{9}(\mathbf{i} + 2\sqrt{2}\mathbf{j}) + 2\mathbf{j}.$$

If the particle is moving to the left, so that $dx/dt < 0$, a similar calculation shows that at $t = 3$ its acceleration is

$$\mathbf{a} = -\frac{3 + 4\sqrt{2}}{9}(\mathbf{i} + 2\sqrt{2}\mathbf{j}) + 2\mathbf{j}.$$

5.
$$\mathbf{r} = e^t\mathbf{i} + \sqrt{2}t\mathbf{j} + e^{-t}\mathbf{k}$$
$$\mathbf{v} = e^t\mathbf{i} + \sqrt{2}\mathbf{j} - e^{-t}\mathbf{k}$$
$$\mathbf{a} = e^t\mathbf{i} + e^{-t}\mathbf{k}$$
$$\frac{d\mathbf{a}}{dt} = e^t\mathbf{i} - e^{-t}\mathbf{k}$$
$$\mathbf{v} \times \mathbf{a} = \sqrt{2}e^{-t}\mathbf{i} - 2\mathbf{j} - \sqrt{2}e^t\mathbf{k}$$
$$v = \sqrt{e^{2t} + 2 + e^{-2t}} = e^t + e^{-t}$$
$$|\mathbf{v} \times \mathbf{a}| = \sqrt{2}(e^t + e^{-t})$$
$$\kappa = \frac{|\mathbf{v} \times \mathbf{a}|}{v^3} = \frac{\sqrt{2}}{(e^t + e^{-t})^2}$$
$$\tau = \frac{(\mathbf{v} \times \mathbf{a}) \bullet \frac{d\mathbf{a}}{dt}}{|\mathbf{v} \times \mathbf{a}|^2} = \frac{\sqrt{2}}{(e^t + e^{-t})^2} = \kappa.$$

6. Tangential acceleration: $dv/dt = e^t - e^{-t}$.
Normal acceleration: $v^2\kappa = \sqrt{2}$.
Since $v = 2\cosh t$, the minimum speed is 2 at time $t = 0$.

7. For $x(s) = \int_0^s \cos\frac{kt^2}{2}\,dt$, $y(s) = \int_0^s \sin\frac{kt^2}{2}\,dt$, we have

$$\frac{dx}{ds} = \cos\frac{ks^2}{2}, \quad \frac{dy}{ds} = \sin\frac{ks^2}{2},$$

so that the speed is unity:

$$v = \sqrt{\left(\frac{dx}{ds}\right)^2 + \left(\frac{dy}{ds}\right)^2} = 1.$$

Since $x(0) = y(0) = 0$, the arc length along the curve, measured from the origin, is s. Also,

$$\mathbf{v} = \cos\frac{ks^2}{2}\mathbf{i} + \sin\frac{ks^2}{2}\mathbf{j}$$
$$\mathbf{a} = -ks\sin\frac{ks^2}{2}\mathbf{i} + ks\cos\frac{ks^2}{2}\mathbf{j}$$
$$\mathbf{v} \times \mathbf{a} = ks\mathbf{k}.$$

Therefore the curvature at position s is
$\kappa = |\mathbf{v} \times \mathbf{a}|/v^3 = ks$.

8. If $r = e^{-\theta}$, and $\dot{\theta} = k$, then $\dot{r} = -e^{-\theta}\dot{\theta} = -kr$, and $\ddot{r} = k^2r$. Since $\mathbf{r} = r\hat{\mathbf{r}}$, we have

$$\mathbf{v} = \dot{r}\hat{\mathbf{r}} + r\dot{\theta}\hat{\boldsymbol{\theta}} = -kr\hat{\mathbf{r}} + kr\hat{\boldsymbol{\theta}}$$
$$\mathbf{a} = (\ddot{r} - r\dot{\theta}^2)\hat{\mathbf{r}} + (r\ddot{\theta} + 2\dot{r}\dot{\theta})\hat{\boldsymbol{\theta}}$$
$$= (k^2r - k^2r)\hat{\mathbf{r}} + (0 - 2k^2r)\hat{\boldsymbol{\theta}} = -2k^2r\hat{\boldsymbol{\theta}}.$$

9. $\mathbf{r} = a(t - \sin t)\mathbf{i} + a(1 - \cos t)\mathbf{j}$
$\mathbf{v} = a(1 - \cos t)\mathbf{i} + a\sin t\mathbf{j}$
$v = a\sqrt{1 - 2\cos t + \cos^2 t + \sin^2 t}$
$= a\sqrt{2}\sqrt{1 - \cos t} = 2a\sin\frac{t}{2}$ if $0 \le t \le 2\pi$.
The length of the cycloid from $t = 0$ to $t = T \le 2\pi$ is

$$s(T) = \int_0^T 2a\sin\frac{t}{2}\,dt = 4a\left(1 - \cos\frac{T}{2}\right) \text{ units.}$$

10. $s = 4a\left(1 - \cos\frac{t}{2}\right) \Rightarrow t = 2\cos^{-1}\left(1 - \frac{s}{4a}\right) = t(s)$.

The required arc length parametrization of the cycloid is

$$\mathbf{r} = a\big(t(s) - \sin t(s)\big)\mathbf{i} + a\big(1 - \cos t(s)\big)\mathbf{j}.$$

11. From Exercise 9 we have

$$\hat{\mathbf{T}}(t) = \frac{\mathbf{v}}{v} = \frac{(1-\cos t)\mathbf{i} + \sin t\,\mathbf{j}}{2\sin(t/2)}$$

$$= \sin\frac{t}{2}\mathbf{i} + \cos\frac{t}{2}\mathbf{j}$$

$$\frac{d\hat{\mathbf{T}}}{ds} = \frac{1}{v}\frac{d\hat{\mathbf{T}}}{dt} = \frac{\frac{1}{2}\cos\frac{t}{2}\mathbf{i} - \frac{1}{2}\sin\frac{t}{2}\mathbf{j}}{2a\sin\frac{t}{2}}$$

$$= \frac{1}{4a}\left(\cot\frac{t}{2}\mathbf{i} - \mathbf{j}\right)$$

$$\kappa(t) = \left|\frac{d\hat{\mathbf{T}}}{ds}\right| = \frac{1}{4a\sin(t/2)}$$

$$\mathbf{r}_C(t) = \mathbf{r}(t) + \rho(t)\hat{\mathbf{N}}(t) = \mathbf{r}(t) + \frac{1}{(\kappa(t))^2}\frac{d\hat{\mathbf{T}}}{ds}$$

$$= \mathbf{r}(t) + \frac{16a^2\sin^2(t/2)}{4a}\left(\cot\frac{t}{2}\mathbf{i} - \mathbf{j}\right)$$

$$= \mathbf{r}(t) + 4a\cos\frac{t}{2}\sin\frac{t}{2}\mathbf{i} - 4a\sin^2\frac{t}{2}\mathbf{j}$$

$$= a(t-\sin t)\mathbf{i} + a(1-\cos t)\mathbf{j}$$
$$\quad + 2a\sin t\,\mathbf{i} - 2a(1-\cos t)\mathbf{j}$$

$$= a(t+\sin t)\mathbf{i} - a(1-\cos t)\mathbf{j} \quad (\text{let } t = u - \pi)$$

$$= a(u-\sin u - \pi)\mathbf{i} + a(1-\cos u - 2)\mathbf{j}.$$

This is the same cycloid as given by $\mathbf{r}(t)$ but translated πa units to the right and $2a$ units downward.

12. Let P be the point with position vector $\mathbf{r}(t)$ on the cycloid. By Exercise 9, the arc OP has length $4a - 4a\cos(t/2)$, and so PQ has length $4a$ - arc $OP = 4a\cos(t/2)$ units. Thus

$$\overrightarrow{PQ} = 4a\cos\frac{t}{2}\hat{\mathbf{T}}(t)$$

$$= 4a\cos\frac{t}{2}\left(\sin\frac{t}{2}\mathbf{i} + \cos\frac{t}{2}\mathbf{j}\right)$$

$$= 2a\sin t\,\mathbf{i} + 2a(1+\cos t)\mathbf{j}.$$

It follows that Q has position vector

$$\mathbf{r}_Q = \mathbf{r} + \overrightarrow{PQ}$$
$$= a(t-\sin t)\mathbf{i} + a(1-\cos t)\mathbf{j} + 2a\sin t\,\mathbf{i} + 2a(1+\cos t)\mathbf{j}$$
$$= a(t+\sin t)\mathbf{i} + a(1+\cos t + 2)\mathbf{j} \quad (\text{let } t = u + \pi)$$
$$= a(u-\sin u + \pi)\mathbf{i} + a(1-\cos u + 2)\mathbf{j}.$$

Thus $\mathbf{r}_Q(t)$ represents the same cycloid as $\mathbf{r}(t)$, but translated πa units to the left and $2a$ units upward. From Exercise 11, the given cycloid is the evolute of its involute.

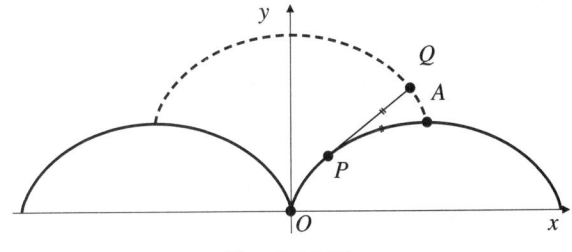

Fig. R-11.12

13. The position vector of P is given by

$$\mathbf{r} = \rho\sin\phi\cos\theta\mathbf{i} + \rho\sin\phi\sin\theta\mathbf{j} + \rho\cos\phi\mathbf{k}.$$

Mutually perpendicular unit vectors in the directions of increasing ρ, ϕ and θ can be found by differentiating $\mathbf{r}$ with respect to each of these coordinates and dividing the resulting vectors by their lengths. They are

$$\hat{\boldsymbol{\rho}} = \frac{d\mathbf{r}}{d\rho} = \sin\phi\cos\theta\mathbf{i} + \sin\phi\sin\theta\mathbf{j} + \cos\phi\mathbf{k}$$

$$\hat{\boldsymbol{\phi}} = \frac{1}{\rho}\frac{d\mathbf{r}}{d\phi} = \cos\phi\cos\theta\mathbf{i} + \cos\phi\sin\theta\mathbf{j} - \sin\phi\mathbf{k}$$

$$\hat{\boldsymbol{\theta}} = \frac{1}{\rho\sin\phi}\frac{d\mathbf{r}}{d\theta} = -\sin\theta\mathbf{i} + \cos\theta\mathbf{j}.$$

The triad $\{\hat{\boldsymbol{\rho}}, \hat{\boldsymbol{\phi}}, \hat{\boldsymbol{\theta}}\}$ is right-handed. This is the reason for ordering the spherical polar coordinates (ρ, ϕ, θ) rather than (ρ, θ, ϕ).

14. By Kepler's Second Law the position vector $\mathbf{r}$ from the origin (the sun) to the planet sweeps out area at a constant rate, say $h/2$:

$$\frac{dA}{dt} = \frac{h}{2}.$$

As observed in the text, $dA/dt = r^2\dot{\theta}/2$, so $r^2\dot{\theta} = h$, and

$$\mathbf{r}\times\mathbf{v} = (r\hat{\mathbf{r}})\times(\dot{r}\hat{\mathbf{r}} + r\dot{\theta}\hat{\boldsymbol{\theta}}) = r^2\dot{\theta}\hat{\mathbf{r}}\times\hat{\boldsymbol{\theta}} = h\mathbf{k} = \mathbf{h}$$

is a constant vector.

15. By Exercise 14, $\mathbf{r}\times\dot{\mathbf{r}} = \mathbf{r}\times\mathbf{v} = \mathbf{h}$ is constant, so, by Newton's second law of motion,

$$\mathbf{r}\times\mathbf{F}(\mathbf{r}) = m\mathbf{r}\times\ddot{\mathbf{r}} = m\frac{d}{dt}(\mathbf{r}\times\dot{\mathbf{r}}) = \mathbf{0}.$$

Thus $\mathbf{F}(\mathbf{r})$ is parallel to $\mathbf{r}$, and therefore has zero transverse component:

$$\mathbf{F}(\mathbf{r}) = -f(\mathbf{r})\hat{\mathbf{r}}$$

for some scalar function $f(\mathbf{r})$.

16. By Exercise 15, $\mathbf{F}(\mathbf{r}) = m(\ddot{r} - r\dot{\theta}^2)\hat{\mathbf{r}} = -f(\mathbf{r})\hat{\mathbf{r}}$. We are given that $r = \ell/(1+\epsilon\cos\theta)$. Thus

$$\dot{r} = -\frac{\ell}{(1+\epsilon\cos\theta)^2}(-\epsilon\sin\theta)\dot{\theta}$$

$$= \frac{\epsilon\ell\sin\theta}{(1+\epsilon\cos\theta)^2}\dot{\theta}$$

$$= \frac{\epsilon\sin\theta}{\ell}r^2\dot{\theta} = \frac{h\epsilon}{\ell}\sin\theta$$

$$\ddot{r} = \frac{h\epsilon}{\ell}(\cos\theta)\dot{\theta} = \frac{h^2\epsilon\cos\theta}{\ell r^2}.$$

It follows that

$$\ddot{r} - r\dot{\theta}^2 = \frac{h^2 \epsilon \cos\theta}{\ell r^2} - \frac{h^2}{r^3}$$
$$= \frac{h^2}{\ell r^2}\left(\epsilon \cos\theta - \frac{\ell}{r}\right) = -\frac{h^2}{\ell r^2},$$

(because $(\ell/r) = 1 + \epsilon \cos\theta$). Hence

$$f(\mathbf{r}) = \frac{mh^2}{\ell r^2}.$$

This says that the magnitude of the force on the planet is inversely proportional to the square of its distance from the sun. Thus Newton's law of gravitation follows from Kepler's laws and the second law of motion.

Challenging Problems 11 (page 693)

1. a) The angular velocity Ω of the earth points northward in the direction of the earth's axis; in terms of the basis vectors defined at a point P at 45° north latitude, it points in the direction of $\mathbf{j} + \mathbf{k}$:

$$\boldsymbol{\Omega} = \Omega \frac{\mathbf{j}+\mathbf{k}}{\sqrt{2}}, \quad \Omega = \frac{2\pi}{24 \times 3{,}600} \text{ rad/s}.$$

b) If $\mathbf{v} = -v\mathbf{k}$, then

$$\mathbf{a}_C = 2\boldsymbol{\Omega}\times\mathbf{v} = -\frac{2\Omega v}{\sqrt{2}}(\mathbf{j}+\mathbf{k})\times\mathbf{k} = -\sqrt{2}\Omega v\mathbf{i}.$$

c) If $\mathbf{r}(t) = x(t)\mathbf{i} + y(t)\mathbf{j} + z(t)\mathbf{k}$ is the position of the falling object at time t, then $\mathbf{r}(t)$ satisfies the DE

$$\frac{d^2\mathbf{r}}{dt^2} = -g\mathbf{k} + 2\boldsymbol{\Omega}\times\frac{d\mathbf{r}}{dt}$$

and the initial conditions $\mathbf{r}(0) = 100\mathbf{k}$, $\mathbf{r}'(0) = \mathbf{0}$. If we use the approximation

$$\frac{d\mathbf{r}}{dt} \approx \frac{dz}{dt}\mathbf{k},$$

which is appropriate since Ω is much smaller than g, then

$$2\boldsymbol{\Omega}\times\frac{d\mathbf{r}}{dt} \approx \sqrt{2}\Omega\frac{dz}{dt}\mathbf{i}.$$

Breaking the DE into its components, we get

$$\frac{d^2x}{dt^2} = \sqrt{2}\Omega\frac{dz}{dt}, \quad \frac{d^2y}{dt^2} = 0, \quad \frac{d^2z}{dt^2} = -g.$$

Solving these equations (beginning with the last one), using the initial conditions, we get

$$z(t) = 100 - \frac{gt^2}{2}, \quad y(t) = 0, \quad x(t) = -\frac{\Omega g t^3}{3\sqrt{2}}.$$

Since $g \approx 9.8$ m/s^2, the time of fall is

$$t = \sqrt{\frac{200}{g}} \approx 4.52,$$

at which time we have

$$x \approx -\frac{2\pi}{24\times 3{,}600}\,\frac{9.8}{3\sqrt{2}}(4.52)^3 \approx -0.0155 \text{ m}.$$

The object strikes the ground about 15.5 cm west of P.

2. $\begin{cases} \dfrac{d\mathbf{v}}{dt} = \mathbf{k}\times\mathbf{v} - 32\mathbf{k} \\ \mathbf{v}(0) = 70\mathbf{i} \end{cases}$

a) If $\mathbf{v} = v_1\mathbf{i} + v_2\mathbf{j} + v_3\mathbf{k}$, then $\mathbf{k}\times\mathbf{v} = v_1\mathbf{j} - v_2\mathbf{i}$. Thus the initial-value problem breaks down into component equations as

$$\begin{cases}\dfrac{dv_1}{dt}=-v_2\\ v_1(0)=70\end{cases}\quad \begin{cases}\dfrac{dv_2}{dt}=v_1\\ v_2(0)=0\end{cases}\quad \begin{cases}\dfrac{dv_3}{dt}=-32\\ v_3(0)=0.\end{cases}$$

b) If $\mathbf{r} = x\mathbf{i} + y\mathbf{j} + z\mathbf{k}$ denotes the position of the baseball t s after it is thrown, then $x(0) = y(0) = z(0) = 0$ and we have

$$\frac{dz}{dt} = v_3 = -32t \Rightarrow z = -16t^2.$$

Also, $\dfrac{d^2v_1}{dt^2} = -\dfrac{dv_2}{dt} = -v_1$ (the equation of simple harmonic motion), so

$$v_1(t) = A\cos t + B\sin t, \quad v_2(t) = A\sin t - B\cos t.$$

Since $v_1(0) = 70$, $v_2(0) = 0$, $x(0) = 0$, and $y(0) = 0$, we have

$$\frac{dx}{dt} = v_1 = 70\cos t \quad \frac{dy}{dt} = v_2 = 70\sin t$$
$$x(t) = 70\sin t \quad y(t) = 70(1-\cos t).$$

At time t seconds after it is thrown, the ball is at position

$$\mathbf{r} = 70\sin t\,\mathbf{i} + 70(1-\cos t)\mathbf{j} - 16t^2\mathbf{k}.$$

c) At $t = 1/5$ s, the ball is at about $(13.9, 1.40, -0.64)$. If it had been thrown without the vertical spin, its position at time t would have been

$$\mathbf{r} = 70t\mathbf{i} - 16t^2\mathbf{k},$$

so its position at $t = 1/5$ s would have been $(14, 0, -0.64)$. Thus the spin has deflected the ball approximately 1.4 ft to the left (as seen from above) of what would have been its parabolic path had it not been given the spin.

3. $\begin{cases} \dfrac{d\mathbf{v}}{dt} = \omega \mathbf{v} \times \mathbf{k}, \quad \omega = \dfrac{qB}{m} \\ \mathbf{v}(0) = \mathbf{v}_0 \end{cases}$

a) $\dfrac{d}{dt}(\mathbf{v} \bullet \mathbf{k}) = \dfrac{d\mathbf{v}}{dt} \bullet \mathbf{k} = \omega(\mathbf{v} \times \mathbf{k}) \bullet \mathbf{k} = 0$.
Thus $\mathbf{v} \bullet \mathbf{k} = $ constant $= \mathbf{v}_0 \bullet \mathbf{k}$.

Also, $\dfrac{d}{dt}|v|^2 = 2\dfrac{d\mathbf{v}}{dt} \bullet \mathbf{v} = 2\omega(\mathbf{v} \times \mathbf{k}) \bullet \mathbf{v} = 0$,
so $|\mathbf{v}| = $ constant $= |\mathbf{v}_0|$ for all t.

b) If $\mathbf{w}(t) = \mathbf{v}(t) - (\mathbf{v}_0 \bullet \mathbf{k})\mathbf{k}$, then $\mathbf{w} \bullet \mathbf{k} = 0$ by part (a). Also, using the result of Exercise 23 of Section 10.3, we have

$$\dfrac{d^2\mathbf{w}}{dt^2} = \dfrac{d^2\mathbf{v}}{dt^2} = \omega \dfrac{d\mathbf{v}}{dt} \times \mathbf{k} = \omega^2 (\mathbf{v} \times \mathbf{k}) \times \mathbf{k}$$
$$= -\omega^2 \big[(\mathbf{k} \bullet \mathbf{k})\mathbf{v} - (\mathbf{k} \bullet \mathbf{v})\mathbf{k}\big]$$
$$= -\omega^2 \big[\mathbf{v} - (\mathbf{v}_0 \bullet \mathbf{k})\mathbf{k}\big] = -\omega^2 \mathbf{w},$$

the equation of simple harmonic motion. Also,

$$\mathbf{w}(0) = \mathbf{v}_0 - (\mathbf{v}_0 \bullet \mathbf{k})\mathbf{k}$$
$$\mathbf{w}'(0) = \omega \mathbf{v}_0 \times \mathbf{k}.$$

c) Solving the above initial-value problem for $\mathbf{w}$, we get

$$\mathbf{w} = \mathbf{A}\cos(\omega t) + \mathbf{B}\sin(\omega t), \quad \text{where}$$
$$\mathbf{A} = \mathbf{w}(0) = \mathbf{v}_0 - (\mathbf{v}_0 \bullet \mathbf{k})\mathbf{k}, \quad \text{and}$$
$$\omega \mathbf{B} = \mathbf{w}'(0) = \omega \times \mathbf{k}.$$

Therefore,

$$\mathbf{v}(t) = \mathbf{w}(t) + (\mathbf{v}_0 \bullet \mathbf{k})\mathbf{k}$$
$$= \big[\mathbf{v}_0 - (\mathbf{v}_0 \bullet \mathbf{k})\mathbf{k}\big]\cos(\omega t) + (\mathbf{v}_0 \times \mathbf{k})\sin(\omega t)$$
$$+ (\mathbf{v}_0 \bullet \mathbf{k})\mathbf{k}.$$

d) If $d\mathbf{r}/dt = \mathbf{v}$ and $\mathbf{r}(0) = \mathbf{0}$, then

$$\mathbf{r}(t) = \dfrac{\mathbf{v}_0 - (\mathbf{v}_0 \bullet \mathbf{k})\mathbf{k}}{\omega} \sin(\omega t)$$
$$+ \dfrac{\mathbf{v}_0 \times \mathbf{k}}{\omega}\big(1 - \cos(\omega t)\big) + (\mathbf{v}_0 \bullet \mathbf{k})t\mathbf{k}.$$

Since the three constant vectors

$$\dfrac{\mathbf{v}_0 - (\mathbf{v}_0 \bullet \mathbf{k})\mathbf{k}}{\omega}, \quad \dfrac{\mathbf{v}_0 \times \mathbf{k}}{\omega}, \quad \text{and } (\mathbf{v}_0 \bullet \mathbf{k})\mathbf{k}$$

are mutually perpendicular, and the first two have the same length because

$$|\mathbf{v}_0 - (\mathbf{v}_0 \bullet \mathbf{k})\mathbf{k}| = |\mathbf{v}_0|\sin\theta = |\mathbf{v}_0 \times \mathbf{k}|,$$

where θ is the angle between $\mathbf{v}_0$ and $\mathbf{k}$, the curve $\mathbf{r}(t)$ is generally a circular helix with axis in the z direction. However, it will be a circle if $\mathbf{v}_0 \bullet \mathbf{k} = 0$, that is, if $\mathbf{v}_0$ is horizontal, and it will be a straight line if $\mathbf{v}_0 \times \mathbf{k} = \mathbf{0}$, that is, if $\mathbf{v}_0$ is vertical.

4. The arc length element on $x = a(\theta - \sin\theta)$, $y = a(\cos\theta - 1)$ is (for $\theta \leq \pi$)

$$ds = a\sqrt{(1 - \cos\theta)^2 + \sin^2\theta}\, d\theta$$
$$= a\sqrt{2(1 - \cos\theta)}\, d\theta = 2a\sin(\theta/2)\, d\theta.$$

If the bead slides downward from rest at height $y(\theta_0)$ to height $y(\theta)$, its gravitational potential energy has decreased by

$$mg\big[y(\theta_0) - y(\theta)\big] = mga(\cos\theta_0 - \cos\theta).$$

Since there is no friction, all this potential energy is converted to kinetic energy, so its speed v at height $y(\theta)$ is given by

$$\tfrac{1}{2}mv^2 = mga(\cos\theta_0 - \cos\theta),$$

and so $v = \sqrt{2ga(\cos\theta_0 - \cos\theta)}$. The time required for the bead to travel distance ds at speed v is $dt = ds/v$, so the time T required for the bead to slide from its starting position at $\theta = \theta_0$ to the lowest point on the wire, $\theta = \pi$, is

$$T = \int_{\theta=\theta_0}^{\theta=\pi} \dfrac{ds}{v} = \int_{\theta_0}^{\pi} \dfrac{1}{v}\dfrac{ds}{d\theta}\, d\theta$$
$$= \sqrt{\dfrac{2a}{g}} \int_{\theta_0}^{\pi} \dfrac{\sin(\theta/2)}{\sqrt{\cos\theta_0 - \cos\theta}}\, d\theta$$
$$= \sqrt{\dfrac{2a}{g}} \int_{\theta_0}^{\pi} \dfrac{\sin(\theta/2)}{\sqrt{2\cos^2(\theta_0/2) - 2\cos^2(\theta/2)}}\, d\theta$$

Let $u = \cos(\theta/2)$
$du = -\tfrac{1}{2}\sin(\theta/2)\, d\theta$

$$= 2\sqrt{\dfrac{a}{g}} \int_0^{\cos(\theta_0/2)} \dfrac{du}{\sqrt{\cos^2(\theta_0/2) - u^2}}$$
$$= 2\sqrt{\dfrac{a}{g}} \sin^{-1}\left(\dfrac{u}{\cos(\theta_0/2)}\right)\bigg|_0^{\cos(\theta_0/2)}$$
$$= \pi\sqrt{a g}$$

which is independent of θ_0.

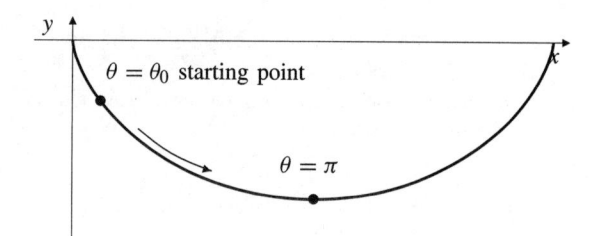

Fig. C-11.4

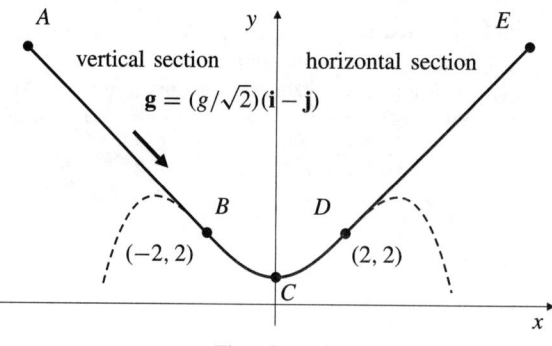

Fig. C-11.5

5. a) The curve BCD is the graph of an even function; a fourth degree polynomial with terms of even degree only will enable us to match the height, slope, and curvature at D, and therefore also at C. We have

$$f(x) = ax^4 + bx^2 + c$$
$$f'(x) = 4ax^3 + 2bx$$
$$f''(x) = 12ax^2 + 2b.$$

At D we have $x = 2$, so we need

$$2 = f(2) = 16a + 4b + c$$
$$1 = f'(2) = 32a + 4b$$
$$0 = f''(2) = 48a + 2b.$$

These equations yield $a = -1/64$, $b = 3/8$, $c = 3/4$, so the curved track BCD is the graph of

$$y = f(x) = \frac{1}{64}(-x^4 + 24x^2 + 48).$$

b) Since we are ignoring friction, the speed v of the car during its drop is given by $v = \sqrt{2gs}$, where s is the vertical distance dropped. (See the previous solution.) At B the car has dropped about 7.2 m, so its speed there is $v \approx \sqrt{2(9.8)(7.2)} \approx 11.9$ m/s. At C the car has dropped $10 - (c/\sqrt{2}) \approx 9.47$ m, so its speed there is $v = 13.6$ m/s. At D the car has dropped 10 m, so its speed is $v = 14.0$ m/s.

c) At C we have $x = 0$, $f'(0) = 0$, and $f''(0) = 2b = 3/4$. Thus the curvature of the track at C is

$$\kappa = \frac{|f''(0)|}{(1 + (f'(0))^2)^{3/2}} = \frac{3}{4}.$$

The normal acceleration is $v^2\kappa \approx 138.7$ m/s² (or about $14g$). Since $v = \sqrt{2gs}$, we have

$$\frac{dv}{dt} = \frac{\sqrt{2g}}{2\sqrt{s}} \frac{ds}{dt} = \frac{\sqrt{2g}}{2\sqrt{s}} v \approx \frac{\sqrt{19.6}}{2\sqrt{9.47}}(13.6) \approx 9.78 \text{ m/s}^2,$$

so the total acceleration has magnitude approximately

$$\sqrt{(138.7)^2 + (9.78)^2} \approx 139 \text{ m/s}^2,$$

which is again about $14g$.

6. a) At time t, the hare is at $P = (0, vt)$ and the fox is at $Q = (x(t), y(t))$, where x and y are such that the slope dy/dx of the fox's path is the slope of the line PQ:

$$\frac{dy}{dx} = \frac{y - vt}{x}.$$

b) Since $\dfrac{d}{dt}\dfrac{dy}{dx} = \dfrac{d^2y}{dx^2}\dfrac{dx}{dt}$, we have

$$\frac{dx}{dt}\frac{d^2y}{dx^2} = \frac{d}{dt}\left(\frac{y - vt}{x}\right)$$

$$= \frac{x\left(\dfrac{dy}{dt} - v\right) - (y - vt)\dfrac{dx}{dt}}{x^2}$$

$$= \frac{1}{x}\left(\frac{dy}{dx}\frac{dx}{dt} - v\right) - \frac{1}{x^2}(y - vt)\frac{dx}{dt}$$

$$= \frac{1}{x^2}(y - vt)\frac{dx}{dt} - \frac{v}{x} - \frac{1}{x^2}(y - vt)\frac{dx}{dt}$$

$$= -\frac{v}{x}.$$

Thus $x\dfrac{d^2y}{dx^2} = -\dfrac{v}{dx/dt}$.

Since the fox's speed is also v, we have

$$\left(\frac{dx}{dt}\right)^2 + \left(\frac{dy}{dt}\right)^2 = v^2.$$

Also, the fox is always running to the left (towards the y-axis from points where $x > 0$), so $dx/dt < 0$. Hence

$$\frac{v}{-\left(\dfrac{dx}{dt}\right)} = \sqrt{1 + \frac{(dy/dt)^2}{(dx/dt)^2}} = \sqrt{1 + \left(\frac{dy}{dx}\right)^2},$$

and so the fox's path $y = y(x)$ satisfies the DE

$$x\frac{d^2y}{dx^2} = \sqrt{1 + \left(\frac{dy}{dx}\right)^2}.$$

c) If $u = dy/dx$, then $u = 0$ and $y = 0$ when $x = a$, and

$$x \frac{du}{dx} = \sqrt{1+u^2}$$

$$\int \frac{du}{\sqrt{1+u^2}} = \int \frac{dx}{x} \quad \begin{array}{l} \text{Let } u = \tan\theta \\ du = \sec^2\theta \, d\theta \end{array}$$

$$\int \sec\theta \, d\theta = \ln x + \ln C$$

$$\ln(\tan\theta + \sec\theta) = \ln(Cx)$$

$$u + \sqrt{1+u^2} = Cx.$$

Since $u = 0$ when $x = a$, we have $C = 1/a$.

$$\sqrt{1+u^2} = \frac{x}{a} - u$$

$$1 + u^2 = \frac{x^2}{a^2} - \frac{2xu}{a} + u^2$$

$$\frac{2xu}{a} = \frac{x^2}{a^2} - 1$$

$$\frac{dy}{dx} = u = \frac{x}{2a} - \frac{a}{2x}$$

$$y = \frac{x^2}{4a} - \frac{a}{2} \ln x + C_1.$$

Since $y = 0$ when $x = a$, we have $C_1 = -\frac{a}{4} + \frac{a}{2}\ln a$, so

$$y = \frac{x^2 - a^2}{4} - \frac{a}{2} \ln \frac{x}{a}$$

is the path of the fox.

7. a) Since you are always travelling northeast at speed v, you are always moving north at rate $v/\sqrt{2}$. Therefore you will reach the north pole in finite time

$$T = \frac{\pi a/2}{v/\sqrt{2}} = \frac{\pi a}{\sqrt{2}v}.$$

b) Since your velocity at any point has a northward component $v/\sqrt{2}$, and progress northward is measured along a circle of radius a (a meridian), your colatitude $\phi(t)$ satisfies

$$a \frac{d\phi}{dt} = -\frac{v}{\sqrt{2}}.$$

Since $\phi(0) = \pi/2$, it follows that

$$\phi(t) = \frac{\pi}{2} - \frac{vt}{a\sqrt{2}}.$$

Since your velocity also has an eastward component $v/\sqrt{2}$ measured along a parallel of latitude that is a circle of radius $a \sin\phi$, your longitude coordinate θ satisfies

$$(a \sin\phi) \frac{d\theta}{dt} = \frac{v}{\sqrt{2}}$$

$$\left(\cos \frac{vt}{a\sqrt{2}}\right) \frac{d\theta}{dt} = \frac{v}{a\sqrt{2}}$$

$$\theta = \frac{v}{a\sqrt{2}} \int \sec\left(\frac{vt}{a\sqrt{2}}\right) dt$$

$$= \ln\left(\sec \frac{vt}{a\sqrt{2}} + \tan \frac{vt}{a\sqrt{2}}\right) + C.$$

As $\theta = 0$ at $t = 0$, we have $C = 0$, and so

$$\theta(t) = \ln\left(\sec \frac{vt}{a\sqrt{2}} + \tan \frac{vt}{a\sqrt{2}}\right).$$

c) As $t \to T = \pi a/(\sqrt{2}v)$, the expression for $\theta(t) \to \infty$, so your path spirals around the north pole, crossing any meridian infinitely often.

CHAPTER 12. PARTIAL DIFFERENTIATION

Section 12.1 Functions of Several Variables (page 701)

1. $f(x, y) = \dfrac{x+y}{x-y}$.
   The domain consists of all points in the xy-plane not on the line $x = y$.

2. $f(x, y) = \sqrt{xy}$.
   Domain is the set of points (x, y) for which $xy \geq 0$, that is, points on the coordinate axes and in the first and third quadrants.

3. $f(x, y) = \dfrac{x}{x^2 + y^2}$.
   The domain is the set of all points in the xy-plane except the origin.

4. $f(x, y) = \dfrac{xy}{x^2 - y^2}$.
   The domain consists of all points not on the lines $x = \pm y$.

5. $f(x, y) = \sqrt{4x^2 + 9y^2 - 36}$.
   The domain consists of all points (x, y) lying on or outside the ellipse $4x^2 + 9y^2 = 36$.

6. $f(x, y) = 1/\sqrt{x^2 - y^2}$.
   The domain consists of all points in the part of the plane where $|x| > |y|$.

7. $f(x, y) = \ln(1 + xy)$.
   The domain consists of all points satisfying $xy > -1$, that is, points lying between the two branches of the hyperbola $xy = -1$.

8. $f(x, y) = \sin^{-1}(x + y)$.
   The domain consists of all points in the strip $-1 \leq x + y \leq 1$.

9. $f(x, y, z) = \dfrac{xyz}{x^2 + y^2 + z^2}$.
   The domain consists of all points in 3-dimensional space except the origin.

10. $f(x, y, z) = \dfrac{e^{xyz}}{\sqrt{xyz}}$.
    The domain consists of all points (x, y, z) where $xyz > 0$, that is, all points in the four octants $x > 0, y > 0, z > 0$; $x > 0, y < 0, z < 0$; $x < 0, y > 0, z < 0$; and $x < 0, y < 0, z > 0$.

11. $z = f(x, y) = x$

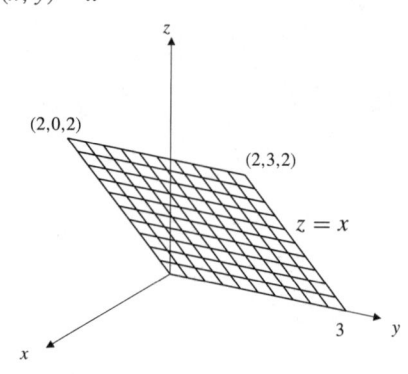

Fig. 12.1.11

12. $f(x, y) = \sin x$, $0 \leq x \leq 2\pi$, $0 \leq y \leq 1$

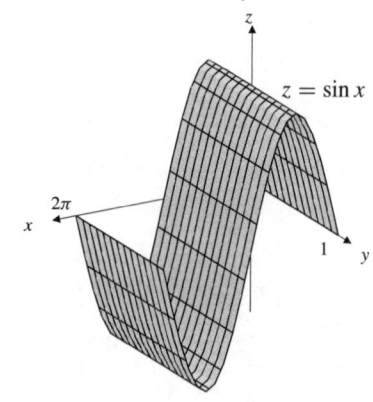

Fig. 12.1.12

13. $z = f(x, y) = y^2$

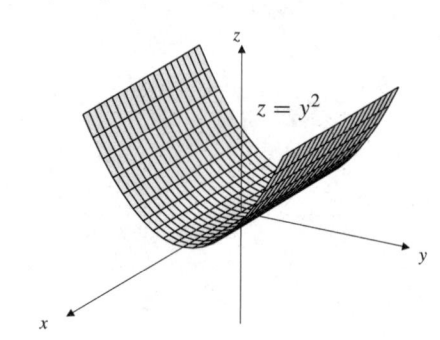

Fig. 12.1.13

14. $f(x,y) = 4 - x^2 - y^2$, $(x^2 + y^2 \leq 4, x \geq 0, y \geq 0)$

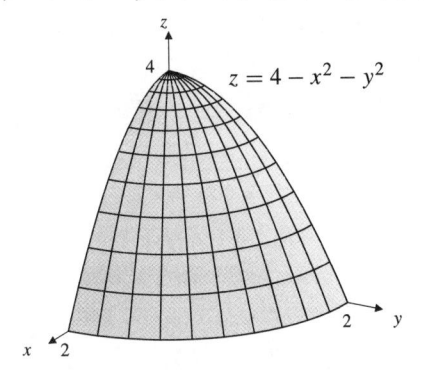

Fig. 12.1.14

15. $z = f(x,y) = \sqrt{x^2 + y^2}$

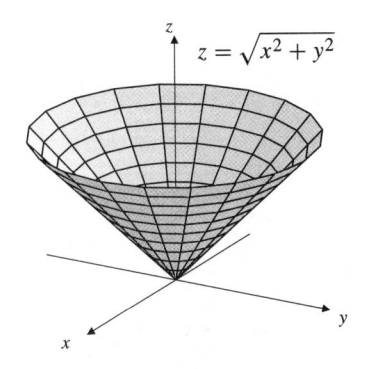

Fig. 12.1.15

16. $f(x,y) = 4 - x^2$

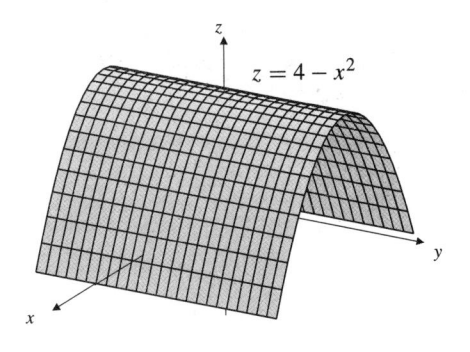

Fig. 12.1.16

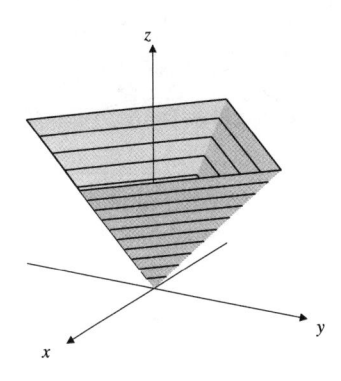

17. $z = f(x,y) = |x| + |y|$

Fig. 12.1.17

18. $f(x,y) = 6 - x - 2y$

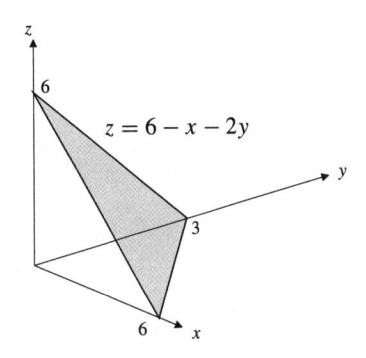

Fig. 12.1.18

19. $f(x,y) = x - y = C$, a family of straight lines of slope 1.

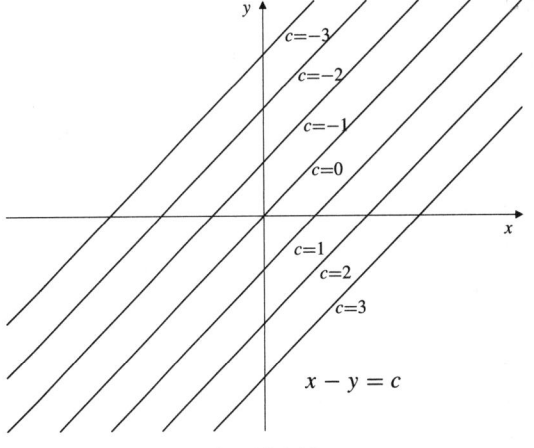

Fig. 12.1.19

20. $f(x,y) = x^2 + 2y^2 = C$, a family of similar ellipses centred at the origin.

Fig. 12.1.20

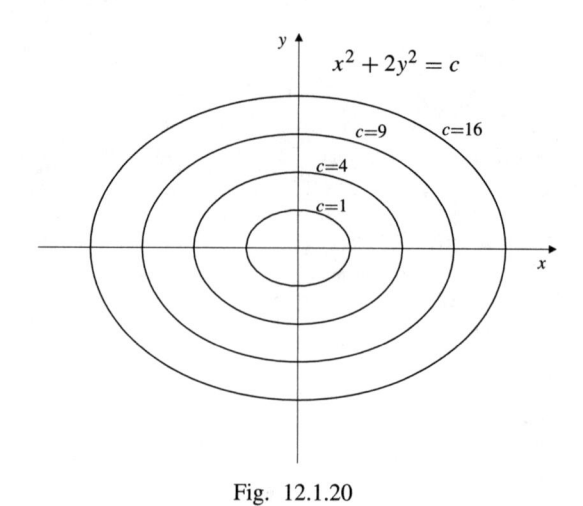

21. $f(x, y) = xy = C$, a family of rectangular hyperbolas with the coordinate axes as asymptotes.

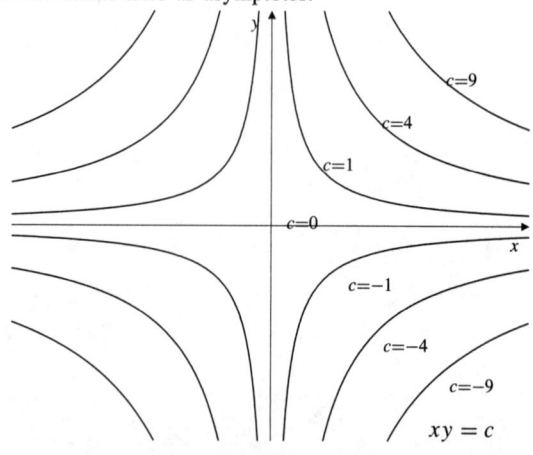

Fig. 12.1.21

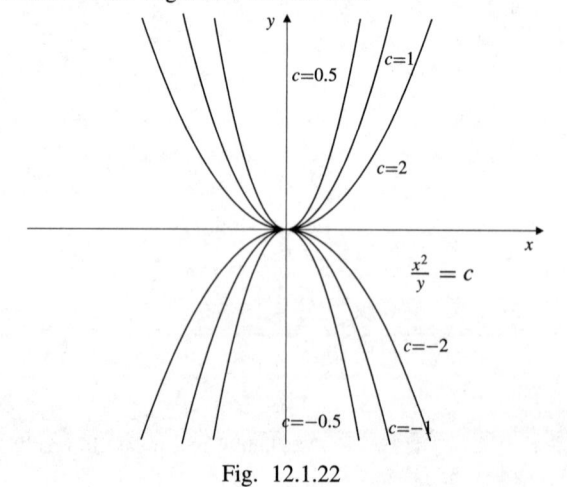

22. $f(x, y) = \dfrac{x^2}{y} = C$, a family of parabolas, $y = x^2/C$, with vertices at the origin and vertical axes.

Fig. 12.1.22

23. $f(x, y) = \dfrac{x - y}{x + y} = C$, a family of straight lines through the origin, but not including the origin.

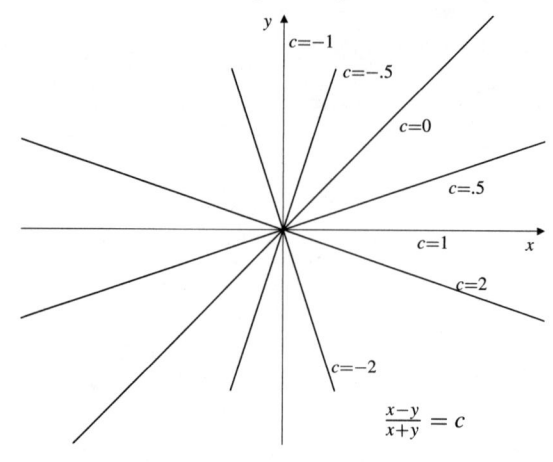

Fig. 12.1.23

24. $f(x, y) = \dfrac{y}{x^2 + y^2} = C$.

This is the family $x^2 + \left(y - \dfrac{1}{2C}\right)^2 = \dfrac{1}{4C^2}$ of circles passing through the origin and having centres on the y-axis. The origin itself is, however, not on any of the level curves.

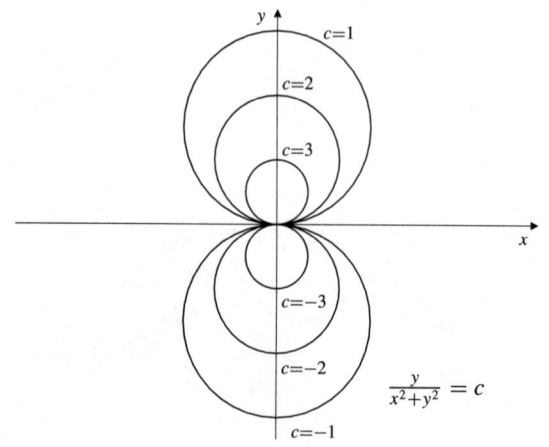

Fig. 12.1.24

25. $f(x, y) = xe^{-y} = C$.
This is the family of curves $y = \ln \dfrac{x}{C}$.

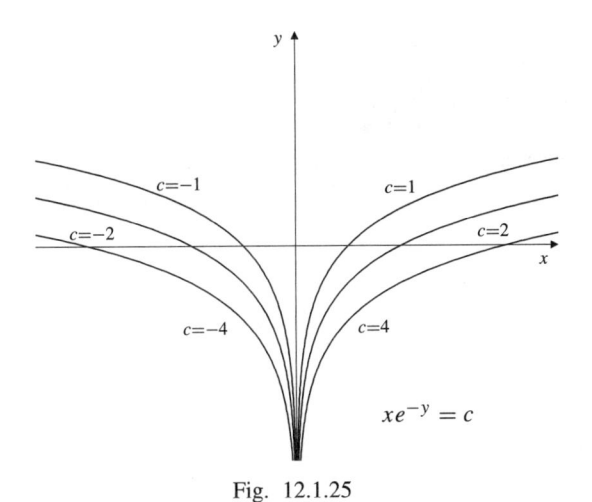

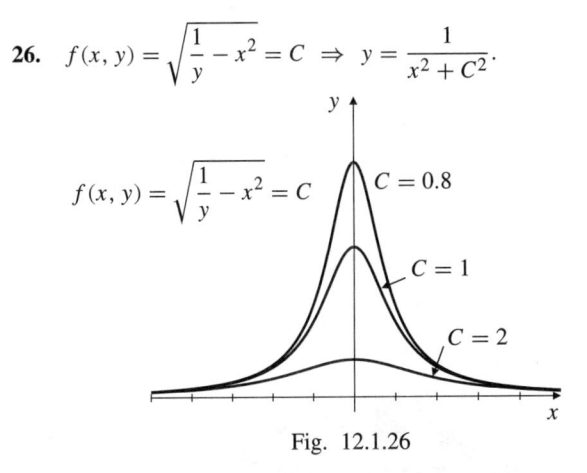

INSTRUCTOR'S SOLUTIONS MANUAL SECTION 12.1 (PAGE 701)

Fig. 12.1.25

$xe^{-y} = c$

26. $f(x, y) = \sqrt{\dfrac{1}{y} - x^2} = C \Rightarrow y = \dfrac{1}{x^2 + C^2}$.

Fig. 12.1.26

27. The landscape is steepest at B where the level curves are closest together.

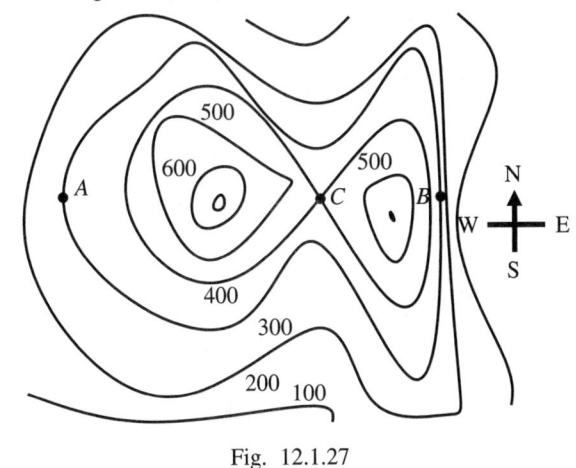

Fig. 12.1.27

28. C is a "pass" between two peaks to the east and west. The land is level at C and rises as you move to the east or west, but falls as you move to the north or south.

29. The graph of the function whose level curves are as shown in part (a) of Figure 12.1.29 is a plane containing the y-axis and sloping uphill to the right. It is consistent with, say, a function of the form $f(x, y) = y$.

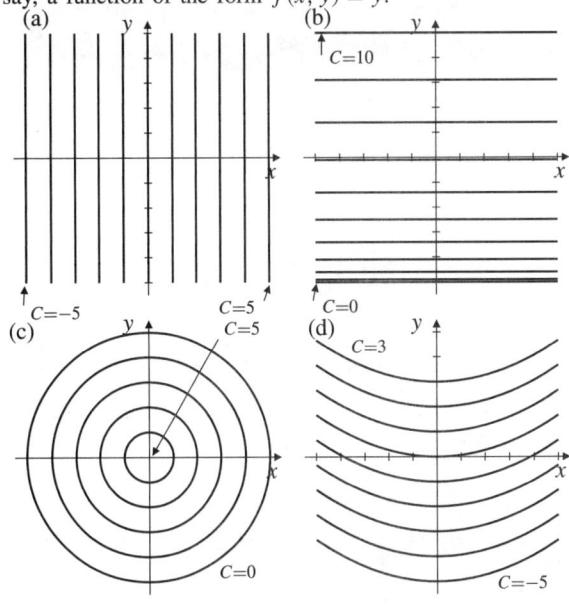

Fig. 12.1.29

30. The graph of the function whose level curves are as shown in part (b) of Figure 12.1.29 is a cylinder parallel to the x-axis, rising from height zero first steeply and then more and more slowly as y increases. It is consistent with, say, a function of the form $f(x, y) = \sqrt{y + 5}$.

31. The graph of the function whose level curves are as shown in part (c) of Figure 12.1.29 is an upside down circular cone with vertex at height 5 on the z-axis and base circle in the xy-plane. It is consistent with, say, a function of the form $f(x, y) = 5 - \sqrt{x^2 + y^2}$.

32. The graph of the function whose level curves are as shown in part (d) of Figure 12.1.29 is a cylinder (possibly parabolic) with axis in the yz-plane, sloping upwards in the direction of increasing y. It is consistent with, say, a function of the form $f(x, y) = y - x^2$.

33. The curves $y = (x - C)^2$ are all horizontally shifted versions of the parabola $y = x^2$, and they all lie in the half-plane $y \geq 0$. Since each of these curves intersects all of the others, they cannot be level curves of a function $f(x, y)$ defined in $y \geq 0$. To be a family of level curves of a function $f(x, y)$ in a region, the various curves in the family cannot intersect one another in that region.

34. $4z^2 = (x - z)^2 + (y - z)^2$.
If $z = c > 0$, we have $(x - c)^2 + (y - c)^2 = 4c^2$, which is a circle in the plane $z = c$, with centre (c, c, c) and radius $2c$.

431

$(x-c)^2 + (y-c)^2 = 4c^2$

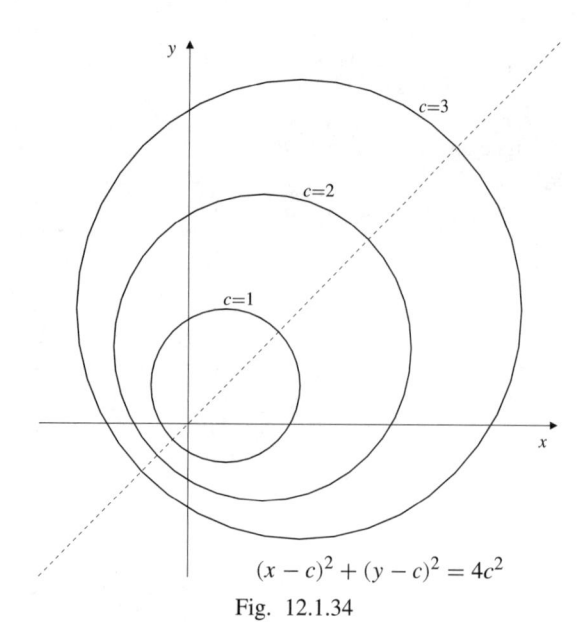

Fig. 12.1.34

The graph of the function $z = z(x, y) \geq 0$ defined by the given equation is (the upper half of) an elliptic cone with axis along the line $x = y = z$, and circular cross-sections in horizontal planes.

35. a) $f(x, y) = C$ is $x^2 + y^2 = C^2$ implies that $f(x, y) = \sqrt{x^2 + y^2}$.

   b) $f(x, y) = C$ is $x^2 + y^2 = C^4$ implies that $f(x, y) = (x^2 + y^2)^{1/4}$.

   c) $f(x, y) = C$ is $x^2 + y^2 = C$ implies that $f(x, y) = x^2 + y^2$.

   d) $f(x, y) = C$ is $x^2 + y^2 = (\ln C)^2$ implies that $f(x, y) = e^{\sqrt{x^2+y^2}}$.

36. If the level surface $f(x, y, z) = C$ is the plane
$$\frac{x}{C^3} + \frac{y}{2C^3} + \frac{z}{3C^3} = 1,$$
that is, $x + \frac{y}{2} + \frac{z}{3} = C^3$, then
$$f(x, y, z) = \left(x + \frac{y}{2} + \frac{z}{3}\right)^{1/3}.$$

37. $f(x, y, z) = x^2 + y^2 + z^2$.
The level surface $f(x, y, z) = c > 0$ is a sphere of radius $\sqrt{c}$ centred at the origin.

38. $f(x, y, z) = x + 2y + 3z$.
The level surfaces are parallel planes having common normal vector $\mathbf{i} + 2\mathbf{j} + 3\mathbf{k}$.

39. $f(x, y, z) = x^2 + y^2$.
The level surface $f(x, y, z) = c > 0$ is a circular cylinder of radius $\sqrt{c}$ with axis along the z-axis.

40. $f(x, y, z) = \dfrac{x^2 + y^2}{z^2}$.
The equation $f(x, y, z) = c$ can be rewritten $x^2 + y^2 = C^2 z^2$. The level surfaces are circular cones with vertices at the origin and axes along the z-axis.

41. $f(x, y, z) = |x| + |y| + |z|$.
The level surface $f(x, y, z) = c > 0$ is the surface of the octahedron with vertices $(\pm c, 0, 0)$, $(0, \pm c, 0)$, and $(0, 0, \pm c)$. (An octahedron is a solid with eight planar faces.)

42. $f(x, y, z, t) = x^2 + y^2 + z^2 + t^2$.
The "level hypersurface" $f(x, y, z, t) = c > 0$ is the "4-sphere" of radius $\sqrt{c}$ centred at the origin in $\mathbb{R}^4$. That is, it consists of all points in $\mathbb{R}^4$ at distance $\sqrt{c}$ from the origin.

43.

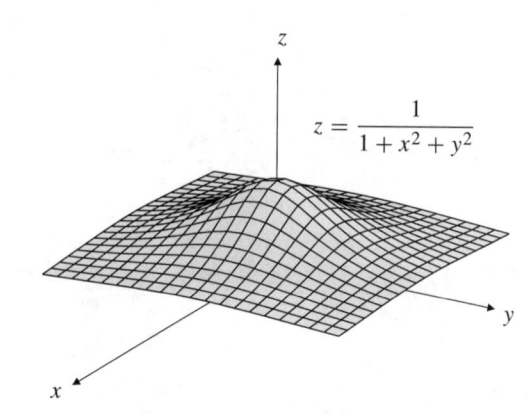

$z = \dfrac{1}{1 + x^2 + y^2}$

Fig. 12.1.43

44.

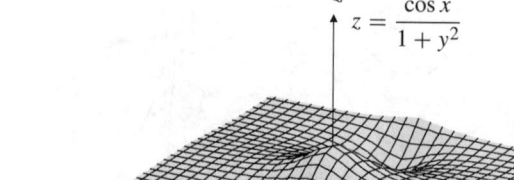

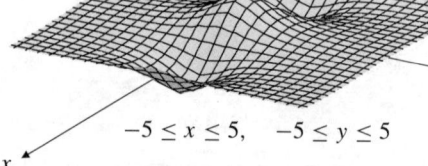

$z = \dfrac{\cos x}{1 + y^2}$

$-5 \leq x \leq 5$, $-5 \leq y \leq 5$

Fig. 12.1.44

INSTRUCTOR'S SOLUTIONS MANUAL

SECTION 12.2 (PAGE 706)

45.

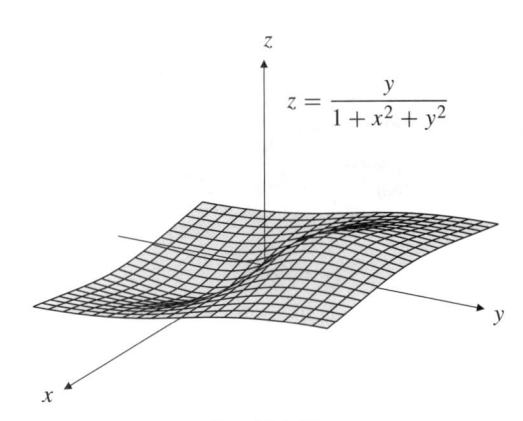

$z = \dfrac{y}{1+x^2+y^2}$

Fig. 12.1.45

46.

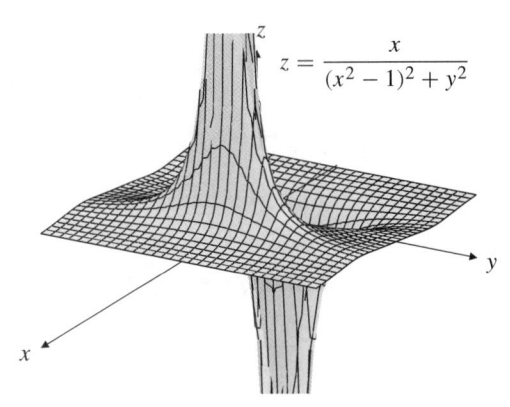

$z = \dfrac{x}{(x^2-1)^2+y^2}$

Fig. 12.1.46

47.

$z = xy$

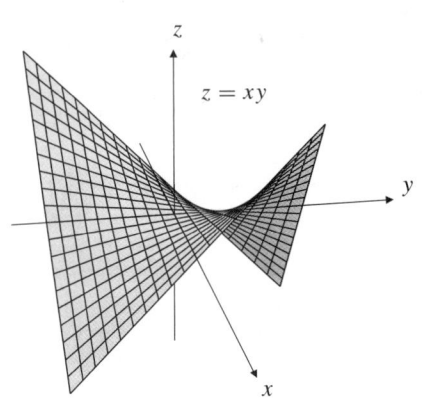

Fig. 12.1.47

48. The graph is asymptotic to the coordinate planes.

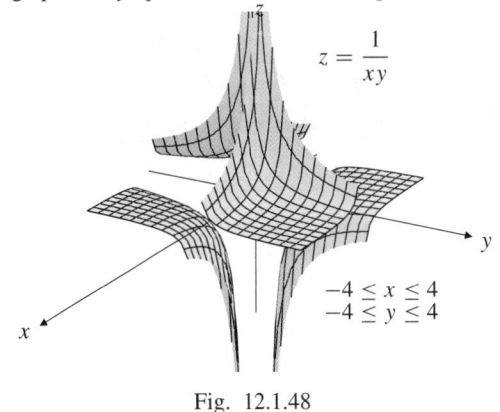

$z = \dfrac{1}{xy}$

$-4 \le x \le 4$
$-4 \le y \le 4$

Fig. 12.1.48

Section 12.2 Limits and Continuity (page 706)

1. $\lim\limits_{(x,y)\to(2,-1)} xy + x^2 = 2(-1) + 2^2 = 2$

2. $\lim\limits_{(x,y)\to(0,0)} \sqrt{x^2+y^2} = 0$

3. $\lim\limits_{(x,y)\to(0,0)} \dfrac{x^2+y^2}{y}$ does not exist.

If $(x, y) \to (0,0)$ along $x = 0$, then $\dfrac{x^2+y^2}{y} = y \to 0$.

If $(x, y) \to (0, 0)$ along $y = x^2$, then $\dfrac{x^2+y^2}{y} = 1 + x^2 \to 1$.

4. Let $f(x, y) = \dfrac{x}{x^2+y^2}$.
Then $|f(x, 0)| = |1/x| \to \infty$ as $x \to 0$.
But $|f(0, y)| = 0 \to 0$ as $y \to 0$.
Thus $\lim\limits_{(x,y)\to(0,0)} f(x, y)$ does not exist.

5. $\lim\limits_{(x,y)\to(1,\pi)} \dfrac{\cos(xy)}{1-x-\cos y} = \dfrac{\cos\pi}{1-1-\cos\pi} = -1$

6. $\lim\limits_{(x,y)\to(0,1)} \dfrac{x^2(y-1)^2}{x^2+(y-1)^2} = 0$, because

$$0 \le \left|\dfrac{x^2(y-1)^2}{x^2+(y-1)^2}\right| \le x^2$$

and $x^2 \to 0$ as $(x, y) \to (0, 1)$.

7. $\left|\dfrac{y^3}{x^2+y^2}\right| \le \dfrac{y^2}{x^2+y^2}|y| \le |y| \to 0$

as $(x, y) \to (0, 0)$. Thus $\lim\limits_{(x,y)\to(0,0)} \dfrac{y^3}{x^2+y^2} = 0$.

8. $\lim\limits_{(x,y)\to(0,0)} \dfrac{\sin(x-y)}{\cos(x+y)} = \dfrac{\sin 0}{\cos 0} = 0$.

433

9. Let $f(x, y) = \dfrac{\sin(xy)}{x^2 + y^2}$.
   Now $f(0, y) = 0/x^2 = 0 \to 0$ as $x \to 0$.
   However, $f(x, x) = \dfrac{\sin x^2}{2x^2} \to \dfrac{1}{2}$ as $x \to 0$.
   Therefore $\lim\limits_{(x,y)\to(0,0)} f(x, y)$ does not exist.

10. The fraction is not defined at points of the line $y = 2x$ and so cannot have a limit at $(1, 2)$ by Definition 4. However, if we use the extended Definition 6, then, cancelling the common factor $2x - y$, we get
$$\lim_{(x,y)\to(1,2)} \frac{2x^2 - xy}{4x^2 - y^2} = \lim_{(x,y)\to(1,2)} \frac{x}{2x + y} = \frac{1}{4}.$$

11. $x^2 \le x^2 + y^4$. Thus $\dfrac{x^2 y^2}{x^2 + y^4} \le y^2 \to 0$ as $y \to 0$. Thus
$$\lim_{(x,y)\to(0,0)} \frac{x^2 y^2}{x^2 + y^4} = 0.$$

12. If $x = 0$ and $y \ne 0$, then $\dfrac{x^2 y^2}{2x^4 + y^4} = 0$.
    If $x = y \ne 0$, then $\dfrac{x^2 y^2}{2x^4 + y^4} = \dfrac{x^4}{2x^4 + x^4} = \dfrac{1}{3}$.
    Therefore $\lim\limits_{(x,y)\to(0,0)} \dfrac{x^2 y^2}{2x^4 + y^4}$ does not exist.

13. $f(x, y) = \dfrac{x^2 + y^2 - x^3 y^3}{x^2 + y^2} = 1 - \dfrac{x^3 y^3}{x^2 + y^2}$. But
$$\left| \frac{x^3 y^3}{x^2 + y^2} \right| = \left| \frac{x^2}{x^2 + y^2} \right| |xy^3| \le |xy^3| \to 0$$
as $(x, y) \to (0, 0)$. Thus $\lim\limits_{(x,y)\to(0,0)} f(x, y) = 1 - 0 = 1$. Define $f(0, 0) = 1$.

14. For $x \ne y$, we have
$$f(x, y) = \frac{x^3 - y^3}{x - y} = x^2 + xy + y^2.$$

The latter expression has the value $3x^2$ at points of the line $x = y$. Therefore, if we extend the definition of $f(x, y)$ so that $f(x, x) = 3x^2$, then the resulting function will be equal to $x^2 + xy + y^2$ everywhere, and so continuous everywhere.

15. $f(x, y) = \dfrac{x - y}{x^2 - y^2} = \dfrac{x - y}{(x - y)(x + y)}$.
    Since f is not defined at points of the line $x = y$, it is not defined at all points sufficiently near $(1, 1)$. Therefore $\lim_{(x,y)\to(1,1)} f(x, y)$ does not exist according to Definition 4.
    However, if $x \ne \pm y$, then $f(x, y) = \dfrac{1}{x + y}$.
    Thus $\lim_{(x,y)\to(1,1)} f(x, y) = 1/2$, according to Definition 6. If we define $f(x, x) = \dfrac{1}{2x}$, for $x \ne 0$, then f becomes continuous at all points except those on the line $y = -x$. There is no way to define $f(x, -x)$ so that f becomes continuous on $y = -x$, since $|f(x, y)| = 1/|x + y| \to \infty$ as $y \to -x$.

16. Let f be the function of Example 3 of Section 12.2:
$$f(x, y) = \begin{cases} \dfrac{2xy}{x^2 + y^2} & \text{if } (x, y) \ne (0, 0) \\ 0 & \text{if } (x, y) = (0, 0). \end{cases}$$

Let $a = b = 0$. If $g(x) = f(x, 0)$ and $h(y) = f(0, y)$, then $g(x) = 0$ for all x, and $h(y) = 0$ for all y, so g and h are continuous at 0. But, as shown in Example 3 of Section 12.2, f is not continuous at $(0, 0)$.

If $f(x, y)$ is continuous at (a, b), then $g(x) = f(x, b)$ is continuous at $x = a$ because
$$\lim_{x \to a} g(x) = \lim_{\substack{x \to a \\ y = b}} f(x, y) = f(a, b).$$

Similarly, $h(y) = f(a, y)$ is continuous at $y = b$.

17. $f_{\mathbf{u}}(t) = f(a + tu, b + tv)$, where $\mathbf{u} = u\mathbf{i} + v\mathbf{j}$ is a unit vector.

$f(x, y)$ may not be continuous at (a, b) even if $f_{\mathbf{u}}(t)$ is continuous at $t = 0$ for every unit vector $\mathbf{u}$. A counterexample is the function f of Example 4 in this section. Here $a = b = 0$. The condition that each $f_{\mathbf{u}}$ should be continuous is the condition that f should be continuous on each straight line through $(0, 0)$, which it is if we extend the domain of f to include $(0, 0)$ by defining $f(0, 0) = 0$. (We showed that $f(x, y) \to 0$ as $(x, y) \to (0, 0)$ along every straight line.) However, we also showed that $\lim_{(x,y)\to(0,0)} f(x, y)$ did not exist.

On the other hand, if $f(x, y)$ is continuous at (a, b), then $f(x, y) \to f(a, b)$ if (x, y) approaches (a, b) in any way, in particular, along the line through (a, b) parallel to $\mathbf{u}$. Thus all such functions $f_{\mathbf{u}}(t)$ must be continuous at $t = 0$.

18. Since $|x| \le \sqrt{x^2 + y^2}$ and $|y| \le \sqrt{x^2 + y^2}$, we have
$$\left| \frac{x^m y^n}{(x^2 + y^2)^p} \right| \le \frac{(x^2 + y^2)^{(m+n)/2}}{(x^2 + y^2)^p} = (x^2 + y^2)^{-p + (m+n)/2}.$$

The expression on the right $\to 0$ as $(x, y) \to (0, 0)$, provided $m + n > 2p$. In this case

$$\lim_{(x,y)\to(0,0)} \frac{x^m y^n}{(x^2 + y^2)^p} = 0.$$

19. Suppose $(x, y) \to (0, 0)$ along the ray $y = kx$. Then

$$f(x, y) = \frac{xy}{ax^2 + bxy + cy^2} = \frac{k}{a + bk + ck^2}.$$

Thus $f(x, y)$ has different constant values along different rays from the origin unless $a = c = 0$ and $b \neq 0$. If this condition is not satisfied, $\lim_{(x,y)\to(0,0)} f(x, y)$ does not exist. If the condition is satisfied, then $\lim_{(x,y)\to(0,0)} f(x, y) = 1/b$ does exist.

20. $f(x, y) = \dfrac{\sin x \sin^3 y}{1 - \cos(x^2 + y^2)}$ cannot be defined at $(0, 0)$ so as to become continuous there, because $f(x, y)$ has no limit as $(x, y) \to (0, 0)$. To see this, observe that $f(x, 0) = 0$, so the limit must be 0 if it exists at all. However,

$$f(x, x) = \frac{\sin^4 x}{1 - \cos(2x^2)} = \frac{\sin^4 x}{2 \sin^2(x^2)}$$

which approaches 1/2 as $x \to 0$ by l'Hôpital's Rule or by using Maclaurin series.

21.

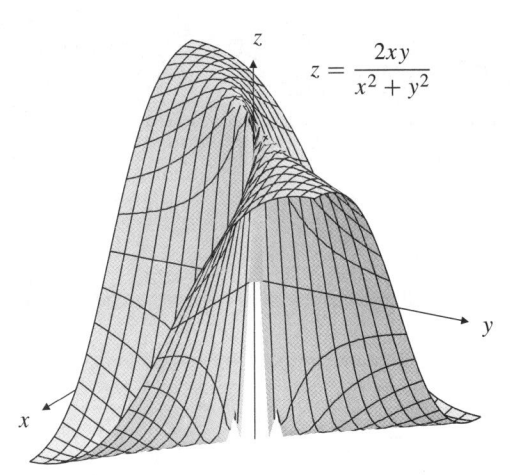

Fig. 12.2.21

The graphing software is unable to deal effectively with the discontinuity at $(x, y) = (0, 0)$ so it leaves some gaps and rough edges near the z-axis. The surface lies between a ridge of height 1 along $y = x$ and a ridge of height -1 along $y = -x$. It appears to be creased along the z-axis. The level curves are straight lines through the origin.

22. The graphing software is unable to deal effectively with the discontinuity at $(x, y) = (0, 0)$ so it leaves some gaps and rough edges near the z-axis. The surface lies between a ridge along $y = x^2$, $z = 1$, and a ridge along $y = -x^2$, $z = -1$. It appears to be creased along the z-axis. The level curves are parabolas $y = kx^2$ through the origin. One of the families of rulings on the surface is the family of contours corresponding to level curves.

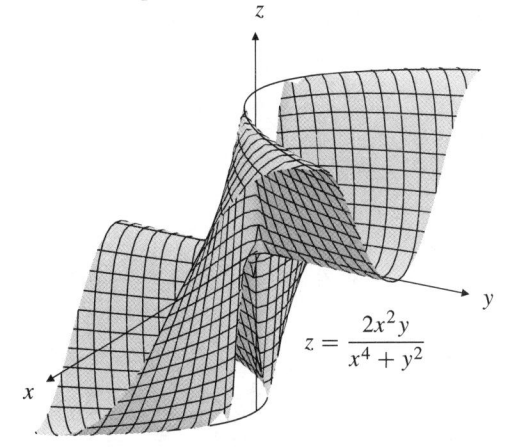

Fig. 12.2.22

23. The graph of a function $f(x, y)$ that is continuous on region R in the xy-plane is a surface with no breaks or tears in it that intersects each line parallel to the z-axis through a point (x, y) of R at exactly one point.

Section 12.3 Partial Derivatives (page 714)

1. $f(x, y) = x - y + 2$,
$f_1(x, y) = 1 = f_1(3, 2)$, $f_2(x, y) = -1 = f_2(3, 2)$.

2. $f(x, y) = xy + x^2$,
$f_1(x, y) = y + 2x$, $f_2(x, y) = x$,
$f_1(2, 0) = 4$, $f_2(2, 0) = 2$.

3. $f(x, y, z) = x^3 y^4 z^5$,
$f_1(x, y, z) = 3x^2 y^4 z^5$, $f_1(0, -1, -1) = 0$,
$f_2(x, y, z) = 4x^3 y^3 z^5$, $f_2(0, -1, -1) = 0$,
$f_3(x, y, z) = 5x^3 y^4 z^4$, $f_3(0, -1, -1) = 0$.

4. $g(x, y, z) = \dfrac{xz}{y + z}$,
$g_1(x, y, z) = \dfrac{z}{y + z}$, $g_1(1, 1, 1) = \dfrac{1}{2}$,
$g_2(x, y, z) = \dfrac{-xz}{(y + z)^2}$, $g_2(1, 1, 1) = -\dfrac{1}{4}$,
$g_3(x, y, z) = \dfrac{xy}{(y + z)^2}$, $g_3(1, 1, 1) = \dfrac{1}{4}$.

5. $z = \tan^{-1}\left(\dfrac{y}{x}\right)$

$\dfrac{\partial z}{\partial x} = \dfrac{1}{1+\dfrac{y^2}{x^2}}\left(-\dfrac{y}{x^2}\right) = -\dfrac{y}{x^2+y^2}$

$\dfrac{\partial z}{\partial y} = \dfrac{1}{1+\dfrac{y^2}{x^2}}\left(\dfrac{1}{x}\right) = \dfrac{x}{x^2+y^2}$

$\left.\dfrac{\partial z}{\partial x}\right|_{(-1,1)} = -\dfrac{1}{2}, \quad \left.\dfrac{\partial z}{\partial y}\right|_{(-1,1)} = -\dfrac{1}{2}.$

6. $w = \ln(1 + e^{xyz})$, $\dfrac{\partial w}{\partial x} = \dfrac{yze^{xyz}}{1+e^{xyz}}$,

$\dfrac{\partial w}{\partial y} = \dfrac{xze^{xyz}}{1+e^{xyz}}$, $\dfrac{\partial w}{\partial z} = \dfrac{xye^{xyz}}{1+e^{xyz}}$,

At $(2, 0, -1)$: $\dfrac{\partial w}{\partial x} = 0$, $\dfrac{\partial w}{\partial y} = -1$, $\dfrac{\partial w}{\partial z} = 0$.

7. $f(x, y) = \sin(x\sqrt{y})$,

$f_1(x, y) = \sqrt{y}\cos(x\sqrt{y})$, $f_1\left(\dfrac{\pi}{3}, 4\right) = -1$,

$f_2(x, y) = \dfrac{x}{2\sqrt{y}}\cos(x\sqrt{y})$, $f_2\left(\dfrac{\pi}{3}, 4\right) = -\dfrac{\pi}{24}.$

8. $f(x, y) = \dfrac{1}{\sqrt{x^2+y^2}}$,

$f_1(x, y) = -\dfrac{1}{2}(x^2+y^2)^{-3/2}(2x) = -\dfrac{x}{(x^2+y^2)^{3/2}}$,

By symmetry, $f_2(x, y) = -\dfrac{y}{(x^2+y^2)^{3/2}}$,

$f_1(-3, 4) = \dfrac{3}{125}$, $f_2(-3, 4) = -\dfrac{4}{125}.$

9. $w = x^{y \ln z}$,

$\dfrac{\partial w}{\partial x} = y\ln z\, x^{y\ln z-1}$, $\left.\dfrac{\partial w}{\partial x}\right|_{(e,2,e)} = 2e$,

$\dfrac{\partial w}{\partial y} = \ln x\ln z\, x^{y\ln z}$, $\left.\dfrac{\partial w}{\partial y}\right|_{(e,2,e)} = e^2$,

$\dfrac{\partial w}{\partial z} = \dfrac{y}{z}\ln x\, x^{y\ln z}$, $\left.\dfrac{\partial w}{\partial z}\right|_{(e,2,e)} = 2e.$

10. If $g(x_1, x_2, x_3, x_4) = \dfrac{x_1 - x_2^2}{x_3 + x_4^2}$, then

$g_1(x_1, x_2, x_3, x_4) = \dfrac{1}{x_3+x_4^2}$ $g_1(3, 1, -1, -2) = \dfrac{1}{3}$

$g_2(x_1, x_2, x_3, x_4) = \dfrac{-2x_2}{x_3+x_4^2}$ $g_2(3, 1, -1, -2) = -\dfrac{2}{3}$

$g_3(x_1, x_2, x_3, x_4) = \dfrac{x_2^2 - x_1}{(x_3+x_4^2)^2}$ $g_3(3, 1, -1, -2) = -\dfrac{2}{9}$

$g_4(x_1, x_2, x_3, x_4) = \dfrac{(x_2^2 - x_1)2x_4}{(x_3+x_4^2)^2}$ $g_4(3, 1, -1, -2) = \dfrac{8}{9}.$

11. $f(x, y) = \begin{cases} \dfrac{2x^3 - y^3}{x^2 + 3y^2} & \text{if } (x, y) \neq (0, 0) \\ 0 & \text{if } (x, y) = (0, 0) \end{cases}$

$f_1(0, 0) = \lim_{h \to 0} \dfrac{2h^3 - 0}{h(h^2 + 0)} = 2$

$f_2(0, 0) = \lim_{k \to 0} \dfrac{-k^3 - 0}{k(0+3k^2)} = -\dfrac{1}{3}.$

12. $f(x, y) = \begin{cases} \dfrac{x^2 - 2y^2}{x - y} & \text{if } x \neq y \\ 0 & \text{if } x = y \end{cases}$

$f_1(0, 0) = \lim_{h \to 0} \dfrac{f(h, 0) - f(0, 0)}{h} = \lim_{h \to 0} \dfrac{h - 0}{h} = 1$,

$f_2(0, 0) = \lim_{k \to 0} \dfrac{f(0, k) - f(0, 0)}{k} = \lim_{k \to 0} \dfrac{2k}{k} = 2.$

13. $f(x, y) = x^2 - y^2$ $f(-2, 1) = 3$
$f_1(x, y) = 2x$ $f_1(-2, 1) = -4$
$f_2(x, y) = -2y$ $f_2(-2, 1) = -2$
Tangent plane: $z = 3 - 4(x+2) - 2(y-1)$, or
$4x + 2y + z = -3.$
Normal line: $\dfrac{x+2}{-4} = \dfrac{y-1}{-2} = \dfrac{z-3}{-1}.$

14. $f(x, y) = \dfrac{x-y}{x+y}$, $f(1, 1) = 0$,

$f_1(x, y) = \dfrac{(x+y) - (x-y)}{(x+y)^2}$, $f_1(1, 1) = \dfrac{1}{2}$

$f_2(x, y) = \dfrac{(x+y)(-1) - (x-y)}{(x+y)^2}$, $f_2(1, 1) = -\dfrac{1}{2}.$

Tangent plane to $z = f(x, y)$ at $(1, 1)$ has equation
$z = \dfrac{x-1}{2} - \dfrac{y-1}{2}$, or $2z = x - y.$
Normal line: $2(x-1) = -2(y-1) = -z.$

15. $f(x, y) = \cos\dfrac{x}{y}$ $f(\pi, 4) = \dfrac{1}{\sqrt{2}}$

$f_1(x, y) = -\dfrac{1}{y}\sin\dfrac{x}{y}$ $f_1(\pi, 4) = -\dfrac{1}{4\sqrt{2}}$

$f_2(x, y) = \dfrac{x}{y^2}\sin\dfrac{x}{y}$ $f_2(\pi, 4) = \dfrac{\pi}{16\sqrt{2}}$

The tangent plane at $x = \pi$, $y = 4$ is

$z = \dfrac{1}{\sqrt{2}}\left(1 - \dfrac{1}{4}(x-\pi) + \dfrac{\pi}{16}(y-4)\right),$

or $4x - \pi y + 16\sqrt{2}z = 16.$
Normal line:
$-4\sqrt{2}(x-\pi) = \dfrac{16\sqrt{2}}{\pi}(y-4) = -\left(z - (1/\sqrt{2})\right).$

16. $f(x, y) = e^{xy}$, $f_1(x, y) = ye^{xy}$, $f_2(x, y) = xe^{xy}$,
$f(2, 0) = 1$, $f_1(2, 0) = 0$, $f_2(2, 0) = 2.$
Tangent plane to $z = e^{xy}$ at $(2, 0)$ has equation $z = 1 + 2y.$
Normal line: $x = 2$, $y = 2 - 2z.$

17. $f(x, y) = \dfrac{x}{x^2 + y^2}$

$f_1(x, y) = \dfrac{(x^2 + y^2)(1) - x(2x)}{(x^2 + y^2)^2} = \dfrac{y^2 - x^2}{(x^2 + y^2)^2}$

$f_2(x, y) = -\dfrac{2xy}{(x^2 + y^2)^2}$

$f(1, 2) = \dfrac{1}{5}, \quad f_1(1, 2) = \dfrac{3}{25}, \quad f_2(1, 2) = -\dfrac{4}{25}.$

The tangent plane at $x = 1$, $y = 2$ is
$$z = \dfrac{1}{5} + \dfrac{3}{25}(x - 1) - \dfrac{4}{25}(y - 2),$$
or $3x - 4y - 25z = -10$.

Normal line: $\dfrac{x - 1}{3} = \dfrac{y - 2}{-4} = \dfrac{5z - 1}{-125}$.

18. $f(x, y) = ye^{-x^2}$, $f_1 = -2xye^{-x^2}$, $f_2 = e^{-x^2}$,
$f(0, 1) = 1$, $f_1(0, 1) = 0$, $f_2(0, 1) = 1$.
Tangent plane to $z = f(x, y)$ at $(0, 1)$ has equation
$z = 1 + 1(y - 1)$, or $z = y$.
Normal line: $x = 0$, $y + z = 2$.

19. $f(x, y) = \ln(x^2 + y^2) \quad f(1, -2) = \ln 5$

$f_1(x, y) = \dfrac{2x}{x^2 + y^2} \quad f_1(1, -2) = \dfrac{2}{5}$

$f_2(x, y) = \dfrac{2y}{x^2 + y^2} \quad f_2(1, -2) = -\dfrac{4}{5}$

The tangent plane at $(1, -2, \ln 5)$ is
$$z = \ln 5 + \dfrac{2}{5}(x - 1) - \dfrac{4}{5}(y + 2),$$
or $2x - 4y - 5z = 10 - 5\ln 5$.

Normal line: $\dfrac{x - 1}{2/5} = \dfrac{y + 2}{-4/5} = \dfrac{z - \ln 5}{-1}$.

20. $f(x, y) = \dfrac{2xy}{x^2 + y^2}, \quad f(0, 2) = 0$

$f_1(x, y) = \dfrac{(x^2 + y^2)2y - 2xy(2x)}{(x^2 + y^2)^2} = \dfrac{2y(y^2 - x^2)}{(x^2 + y^2)^2}$

$f_2(x, y) = \dfrac{2x(x^2 - y^2)}{(x^2 + y^2)^2}$ (by symmetry)

$f_1(0, 2) = 1, \quad f_2(0, 2) = 0$.
Tangent plane at $(0, 2)$: $z = x$.
Normal line: $z + x = 0$, $y = 2$.

21. $f(x, y) = \tan^{-1}\left(\dfrac{y}{x}\right)$, $f(1, -1) = -\dfrac{\pi}{4}$,

$f_1(x, y) = \dfrac{1}{1 + \dfrac{y^2}{x^2}}\left(-\dfrac{y}{x^2}\right) = -\dfrac{y}{x^2 + y^2}$,

$f_2(x, y) = \dfrac{1}{1 + \dfrac{y^2}{x^2}}\left(\dfrac{1}{x}\right) = \dfrac{x}{x^2 + y^2}$,

$f_1(1, -1) = f_2(1, -1) = \dfrac{1}{2}$. The tangent plane is
$z = -\dfrac{\pi}{4} + \dfrac{1}{2}(x - 1) + \dfrac{1}{2}(y + 1)$, or $z = -\dfrac{\pi}{4} + \dfrac{1}{2}(x + y)$.
Normal line: $2(x - 1) = 2(y + 1) = -z - \dfrac{\pi}{4}$.

22. $f(x, y) = \sqrt{1 + x^3 y^2} \qquad f(2, 1) = 3$

$f_1(x, y) = \dfrac{3x^2 y^2}{2\sqrt{1 + x^3 y^2}} \qquad f_1(2, 1) = 2$

$f_2(x, y) = \dfrac{2x^3 y}{2\sqrt{1 + x^3 y^2}} \qquad f_2(2, 1) = \dfrac{8}{3}$

Tangent plane: $z = 3 + 2(x - 2) + \dfrac{8}{3}(y - 1)$, or $6x + 8y - 3z = 11$.

Normal line: $\dfrac{x - 2}{2} = \dfrac{y - 1}{8/3} = \dfrac{z - 3}{-1}$.

23. $z = x^4 - 4xy^3 + 6y^2 - 2$

$\dfrac{\partial z}{\partial x} = 4x^3 - 4y^3 = 4(x - y)(x^2 + xy + y^2)$

$\dfrac{\partial z}{\partial y} = -12xy^2 + 12y = 12y(1 - xy)$.

The tangent plane will be horizontal at points where both first partials are zero. Thus we require $x = y$ and either $y = 0$ or $xy = 1$.
If $x = y$ and $y = 0$, then $x = 0$.
If $x = y$ and $xy = 1$, then $x^2 = 1$, so $x = y = \pm 1$.
The tangent plane is horizontal at the points $(0, 0)$, $(1, 1)$, and $(-1, -1)$.

24. $z = xye^{-(x^2+y^2)/2}$

$\dfrac{\partial z}{\partial x} = ye^{-(x^2+y^2)/2} - x^2 ye^{-(x^2+y^2)/2} = y(1 - x^2)e^{-(x^2+y^2)/2}$

$\dfrac{\partial z}{\partial y} = x(1 - y^2)e^{-(x^2+y^2)/2}$ (by symmetry)

The tangent planes are horizontal at points where both of these first partials are zero, that is, points satisfying
$$y(1 - x^2) = 0 \quad \text{and} \quad x(1 - y^2) = 0.$$

These points are $(0, 0)$, $(1, 1)$, $(-1, -1)$, $(1, -1)$ and $(-1, 1)$.
At $(0,0)$ the tangent plane is $z = 0$.
At $(1, 1)$ and $(-1, -1)$ the tangent plane is $z = 1/e$.
At $(1, -1)$ and $(-1, 1)$ the tangent plane is $z = -1/e$.

25. If $z = xe^y$, then $\dfrac{\partial z}{\partial x} = e^y$ and $\dfrac{\partial z}{\partial y} = xe^y$.

Thus $x\dfrac{\partial z}{\partial x} = xe^y = \dfrac{\partial z}{\partial y}$.

26. $z = \dfrac{x + y}{x - y}$,

$\dfrac{\partial z}{\partial x} = \dfrac{(x - y)(1) - (x + y)(1)}{(x - y)^2} = \dfrac{-2y}{(x - y)^2}$,

$\dfrac{\partial z}{\partial y} = \dfrac{(x - y)(1) - (x + y)(-1)}{(x - y)^2} = \dfrac{2x}{(x - y)^2}$.

Therefore
$x\dfrac{\partial z}{\partial x} + y\dfrac{\partial z}{\partial y} = -\dfrac{2xy}{(x - y)^2} + \dfrac{2xy}{(x - y)^2} = 0$.

27. If $z = \sqrt{x^2 + y^2}$, then $\dfrac{\partial z}{\partial x} = \dfrac{x}{\sqrt{x^2 + y^2}}$, and $\dfrac{\partial z}{\partial y} = \dfrac{y}{\sqrt{x^2 + y^2}}$. Thus

$$x\frac{\partial z}{\partial x} + y\frac{\partial z}{\partial y} = \frac{x^2 + y^2}{\sqrt{x^2 + y^2}} = z.$$

28. $w = x^2 + yz$, $\dfrac{\partial w}{\partial x} = 2x$, $\dfrac{\partial w}{\partial y} = z$, $\dfrac{\partial w}{\partial z} = y$.
Therefore

$$x\frac{\partial w}{\partial x} + y\frac{\partial w}{\partial y} + z\frac{\partial w}{\partial z}$$
$$= 2x^2 + yz + yz$$
$$= 2(x^2 + yz) = 2w.$$

29. If $w = \dfrac{1}{x^2 + y^2 + z^2}$, then $\dfrac{\partial w}{\partial x} = -\dfrac{2x}{(x^2 + y^2 + z^2)^2}$, $\dfrac{\partial w}{\partial y} = -\dfrac{2y}{(x^2 + y^2 + z^2)^2}$, and $\dfrac{\partial w}{\partial z} = -\dfrac{2z}{(x^2 + y^2 + z^2)^2}$. Thus

$$x\frac{\partial w}{\partial x} + y\frac{\partial w}{\partial y} + z\frac{\partial w}{\partial z} = -2\frac{x^2 + y^2 + z^2}{(x^2 + y^2 + z^2)^2} = -2w.$$

30. $z = f(x^2 + y^2)$,
$\dfrac{\partial z}{\partial x} = f'(x^2 + y^2)(2x)$, $\dfrac{\partial z}{\partial y} = f'(x^2 + y^2)(2y)$.
Thus $y\dfrac{\partial z}{\partial x} - x\dfrac{\partial z}{\partial y} = 2xyf'(x^2 + y^2) - 2xyf'(x^2 + y^2) = 0$.

31. $z = f(x^2 - y^2)$,
$\dfrac{\partial z}{\partial x} = f'(x^2 - y^2)(2x)$, $\dfrac{\partial z}{\partial y} = f'(x^2 + y^2)(-2y)$.
Thus $y\dfrac{\partial z}{\partial x} + x\dfrac{\partial z}{\partial y} = (2xy - 2xy)f'(x^2 + y^2) = 0$.

32. $f_1(x, y, z) = \lim_{h \to 0} \dfrac{f(x+h, y, z) - f(x, y, z)}{h}$
$f_2(x, y, z) = \lim_{k \to 0} \dfrac{f(x, y+k, z) - f(x, y, z)}{k}$
$f_3(x, y, z) = \lim_{\ell \to 0} \dfrac{f(x, y, z+\ell) - f(x, y, z)}{\ell}$

33. At $(a, b, c, f(a, b, c))$ the graph of $w = f(x, y, z)$ has tangent hyperplane

$$w = f(a, b, c) + f_1(a, b, c)(x - a) + f_2(a, b, c)(y - b) + f_3(a, b, c)(z - c).$$

34. If $Q = (X, Y, Z)$ is the point on the surface $z = x^2 + y^2$ that is closest to $P = (1, 1, 0)$, then

$$\overrightarrow{PQ} = (X - 1)\mathbf{i} + (Y - 1)\mathbf{j} + Z\mathbf{k}$$

must be normal to the surface at Q, and hence must be parallel to $\mathbf{n} = 2X\mathbf{i} + 2Y\mathbf{j} - \mathbf{k}$. Hence $\overrightarrow{PQ} = t\mathbf{n}$ for some real number t, so

$$X - 1 = 2tX, \qquad Y - 1 = 2tY, \qquad Z = -t.$$

Thus $X = Y = \dfrac{1}{1 - 2t}$, and, since $Z = X^2 + Y^2$, we must have

$$-t = \frac{2}{(1 - 2t)^2}.$$

Evidently this equation is satisfied by $t = -\dfrac{1}{2}$. Since the left and right sides of the equation have graphs similar to those in Figure 12.18(b) (in the text), the equation has only this one real solution. Hence $X = Y = \dfrac{1}{2}$, and so $Z = \dfrac{1}{2}$.
The distance from $(1, 1, 0)$ to $z = x^2$ is the distance from $(1, 1, 0)$ to $\left(\dfrac{1}{2}, \dfrac{1}{2}, \dfrac{1}{2}\right)$, which is $\sqrt{3}/2$ units.

35. If $Q = (X, Y, Z)$ is the point on the surface $z = x^2 + 2y^2$ that is closest to $P = (0, 0, 1)$, then

$$\overrightarrow{PQ} = X\mathbf{i} + Y\mathbf{j} + (Z - 1)\mathbf{k}$$

must be normal to the surface at Q, and hence must be parallel to $\mathbf{n} = 2X\mathbf{i} + 4Y\mathbf{j} - \mathbf{k}$. Hence $\overrightarrow{PQ} = t\mathbf{n}$ for some real number t, so

$$X = 2tX, \qquad Y = 4tY, \qquad Z - 1 = -t.$$

If $X \neq 0$, then $t = 1/2$, so $Y = 0$, $Z = 1/2$, and $X = \sqrt{Z} = 1/\sqrt{2}$. The distance from $(1/\sqrt{2}, 0, 1/2)$ to $(0, 0, 1)$ is $\sqrt{3}/2$ units.
If $Y \neq 0$, then $t = 1/4$, so $X = 0$, $Z = 3/4$, and $Y = \sqrt{Z/2} = \sqrt{3/8}$. The distance from $(0, \sqrt{3/8}, 3/4)$ to $(0, 0, 1)$ is $\sqrt{7}/4$ units.
If $X = Y = 0$, then $Z = 0$ (and $t = 1$). The distance from $(0, 0, 0)$ to $(0, 0, 1)$ is 1 unit.
Since

$$\frac{\sqrt{7}}{4} < \frac{\sqrt{3}}{2} < 1,$$

the closest point to $(0, 0, 1)$ on $z = x^2 + 2y^2$ is $(0, \sqrt{3/8}, 3/4)$, and the distance from $(0, 0, 1)$ to that surface is $\sqrt{7}/4$ units.

36. $f(x, y) = \dfrac{2xy}{x^2 + y^2}$ if $(x, y) \neq (0, 0)$, $f(0, 0) = 0$
$f_1(0, 0) = \lim_{h \to 0} \dfrac{f(h, 0) - f(0, 0)}{h} = \lim_{h \to 0} \dfrac{0 - 0}{h} = 0$
$f_2(0, 0) = \lim_{k \to 0} \dfrac{f(0, k) - f(0, 0)}{k} = \lim_{k \to 0} \dfrac{0 - 0}{k} = 0$
Thus $f_1(0, 0)$ and $f_2(0, 0)$ both exist even though f is not continuous at $(0, 0)$ (as shown in Example 2 of Section 12.2).

37. $f(x, y) = \begin{cases} (x^3 + y) \sin \dfrac{1}{x^2 + y^2} & \text{if } (x, y) \neq (0, 0) \\ 0 & \text{if } (x, y) = (0, 0) \end{cases}$

$f_1(0, 0) = \lim_{h \to 0} \dfrac{1}{h} \left(h^3 \sin \dfrac{1}{h^2} \right)$

$= \lim_{h \to 0} h^2 \sin \dfrac{1}{h^2} = 0$

$f_2(0, 0) = \lim_{k \to 0} \dfrac{1}{k} \left(k \sin \dfrac{1}{k^2} \right)$

$= \lim_{k \to 0} \sin \dfrac{1}{k^2}$ does not exist.

38. If $(x, y) \neq (0, 0)$, then

$f_1(x, y) = 3x^2 \sin \dfrac{1}{x^2 + y^2} - \dfrac{(x^3 + y)2x}{(x^2 + y^2)^2} \cos \dfrac{1}{x^2 + y^2}.$

The first term on the right $\to 0$ as $(x, y) \to (0, 0)$, but the second term has no limit at $(0, 0)$. (It is 0 along $x = 0$, but along $x = y$ it is

$-\dfrac{2x^4 + 2x^2}{4x^4} \cos \dfrac{1}{2x^2} = -\dfrac{1}{2}\left(1 + \dfrac{1}{x^2}\right) \cos \dfrac{1}{2x^2},$

which has no limit as $x \to 0$.) Thus $f_1(x, y)$ has no limit at $(0, 0)$ and is not continuous there.

39. $f(x, y) = \begin{cases} \dfrac{x^3 - y^3}{x^2 + y^2} & \text{if } (x, y) \neq (0, 0) \\ 0 & \text{if } (x, y) = (0, 0). \end{cases}$

If $(x, y) \neq (0, 0)$, then

$f_1(x, y) = \dfrac{(x^2 + y^2)3x^2 - (x^3 - y^3)2x}{(x^2 + y^2)^2}$

$= \dfrac{x^4 + 3x^2y^2 + 2xy^3}{(x^2 + y^2)^2}$

$f_2(x, y) = \dfrac{(x^2 + y^2)(-3y^2) - (x^3 - y^3)2y}{(x^2 + y^2)^2}$

$= -\dfrac{y^4 + 3x^2y^2 + 2x^3y}{(x^2 + y^2)^2}.$

Also, at $(0, 0)$,

$f_1(0, 0) = \lim_{h \to 0} \dfrac{h^3}{h \cdot h^2} = 1, \quad f_2(0, 0) = \lim_{k \to 0} \dfrac{-k^3}{k \cdot k^2} = -1.$

Neither f_1 nor f_2 has a limit at $(0, 0)$ (the limits along $x = 0$ and $y = 0$ are different in each case), so neither function is continuous at $(0, 0)$. However, f is continuous at $(0, 0)$ because

$|f(x, y)| \leq \left| \dfrac{x^3}{x^2 + y^2} \right| + \left| \dfrac{y^3}{x^2 + y^2} \right| \leq |x| + |y|,$

which $\to 0$ as $(x, y) \to (0, 0)$.

40. $f(x, y, z) = \begin{cases} \dfrac{xy^2z}{x^4 + y^4 + z^4} & \text{if } (x, y, z) \neq (0, 0, 0) \\ 0 & \text{if } (x, y, z) = (0, 0, 0). \end{cases}$

By symmetry we have

$f_3(0, 0, 0) = f_1(0, 0, 0) = \lim_{h \to 0} \dfrac{0}{h^5} = 0.$

Also,

$f_2(0, 0, 0) = \lim_{k \to 0} \dfrac{0}{k^5} = 0.$

f is not continuous at $(0, 0, 0)$; it has different limits as $(x, y, z) \to (0, 0, 0)$ along $x = 0$ and along $x = y = z$. None of f_1, f_2, and f_3 is continuous at $(0, 0, 0)$ either. For example,

$f_1(x, y, z) = \dfrac{(y^4 + z^4 - 3x^4)y^2z}{(x^4 + y^4 + z^4)^2},$

which has no limit as $(x, y, z) \to (0, 0, 0)$ along the line $x = y = z$.

Section 12.4 Higher-Order Derivatives (page 719)

1. $z = x^2(1 + y^2)$

$\dfrac{\partial z}{\partial x} = 2x(1 + y^2), \quad \dfrac{\partial z}{\partial y} = 2x^2y,$

$\dfrac{\partial^2 z}{\partial x^2} = 2(1 + y^2), \quad \dfrac{\partial^2 z}{\partial y^2} = 2x^2,$

$\dfrac{\partial^2 z}{\partial y \partial x} = 4xy = \dfrac{\partial^2 z}{\partial x \partial y}.$

2. $f(x, y) = x^2 + y^2, \quad f_1(x, y) = 2x, \quad f_2(x, y) = 2y,$
$f_{11}(x, y) = f_{22}(x, y) = 2, \quad f_{12}(x, y) = f_{21}(x, y) = 0.$

3. $w = x^3y^3z^3,$

$\dfrac{\partial w}{\partial x} = 3x^2y^3z^3, \quad \dfrac{\partial w}{\partial y} = 3x^3y^2z^3, \quad \dfrac{\partial w}{\partial z} = 3x^3y^3z^2,$

$\dfrac{\partial^2 w}{\partial x^2} = 6xy^3z^3, \quad \dfrac{\partial^2 w}{\partial y^2} = 6x^3yz^3, \quad \dfrac{\partial^2 w}{\partial z^2} = 6x^3y^3z,$

$\dfrac{\partial^2 w}{\partial x \partial y} = 9x^2y^2z^3 = \dfrac{\partial^2 w}{\partial y \partial x},$

$\dfrac{\partial^2 w}{\partial x \partial z} = 9x^2y^3z^2 = \dfrac{\partial^2 w}{\partial z \partial x},$

$\dfrac{\partial^2 w}{\partial y \partial z} = 9x^3y^2z^2 = \dfrac{\partial^2 w}{\partial z \partial y}.$

4. $z = \sqrt{3x^2 + y^2}$,
$\dfrac{\partial z}{\partial x} = \dfrac{3x}{\sqrt{3x^2 + y^2}}$, $\dfrac{\partial z}{\partial y} = \dfrac{y}{\sqrt{3x^2 + y^2}}$,

$\dfrac{\partial^2 z}{\partial x^2} = \dfrac{\sqrt{3x^2 + y^2}(3) - 3x \dfrac{3x}{\sqrt{3x^2 + y^2}}}{3x^2 + y^2} = \dfrac{3y^2}{(3x^2 + y^2)^{3/2}}$,

$\dfrac{\partial^2 z}{\partial y^2} = \dfrac{\sqrt{3x^2 + y^2} - y \dfrac{y}{\sqrt{3x^2 + y^2}}}{3x^2 + y^2} = \dfrac{3x^2}{(3x^2 + y^2)^{3/2}}$,

$\dfrac{\partial^2 z}{\partial x \partial y} = \dfrac{\partial^2 z}{\partial y \partial x} = -\dfrac{3xy}{(3x^2 + y^2)^{3/2}}$.

5. $z = xe^y - ye^x$,
$\dfrac{\partial z}{\partial x} = e^y - ye^x$, $\dfrac{\partial z}{\partial y} = xe^y - e^x$,
$\dfrac{\partial^2 z}{\partial x^2} = -ye^x$, $\dfrac{\partial^2 z}{\partial y^2} = xe^y$,
$\dfrac{\partial^2 z}{\partial y \partial x} = e^y - e^x = \dfrac{\partial^2 z}{\partial x \partial y}$.

6. $f(x, y) = \ln(1 + \sin(xy))$
$f_1(x, y) = \dfrac{y \cos(xy)}{1 + \sin(xy)}$, $f_2(x, y) = \dfrac{x \cos(xy)}{1 + \sin(xy)}$
$f_{11}(x, y)$
$= \dfrac{(1 + \sin(xy))(-y^2 \sin(xy)) - (y \cos(xy))(y \cos(xy))}{(1 + \sin(xy))^2}$
$= -\dfrac{y^2}{1 + \sin(xy)}$
$f_{22}(x, y) = -\dfrac{x^2}{1 + \sin(xy)}$ (by symmetry)
$f_{12}(x, y) =$
$\dfrac{(1 + \sin(xy))(\cos(xy) - xy \sin(xy)) - (y \cos(xy))(x \cos(xy))}{(1 + \sin(xy))^2}$
$= \dfrac{\cos(xy) - xy}{1 + \sin(xy)} = f_{21}(x, y)$.

7. A function $f(x, y, z)$ of three variables can have $3^3 = 27$ partial derivatives of order 3. Of these, ten can have different values, namely $f_{111}, f_{222}, f_{333}, f_{112}, f_{122}, f_{223}, f_{233}, f_{113}, f_{133},$ and f_{123}.
For $f(x, y, z) = xe^{xy} \cos(xz)$, we have

$f_{133} = f_{313} = f_{331} = \dfrac{\partial}{\partial x}\left(-x^3 e^{xy} \cos(xz)\right)$
$= -(3x^2 + x^3 y)e^{xy} \cos(xz) + x^3 z e^{xy} \sin(xz)$.

8. $f(x, y) = A(x^2 - y^2) + Bxy$, $f_1 = 2Ax + By$,
$f_2 = -2Ay + Bx$,
$f_{11} = 2A$, $f_{22} = -2A$,
Thus $f_{11} + f_{22} = 0$, and f is harmonic.

9. $f(x, y) = 3x^2 y - y^3$,
$f_1(x, y) = 6xy$, $f_{11}(x, y) = 6y$,
$f_2(x, y) = 3x^2 - 3y^2$, $f_{22}(x, y) = -6y$.
Thus $f_{11} + f_{22} = 0$ and f is harmonic.
Also $g(x, y) = x^3 - 3xy^2$ is harmonic.

10. $f(x, y) = \dfrac{x}{x^2 + y^2}$
$f_1(x, y) = \dfrac{x^2 + y^2 - 2x^2}{(x^2 + y^2)^2} = \dfrac{y^2 - x^2}{(x^2 + y^2)^2}$
$f_2(x, y) = -\dfrac{2xy}{(x^2 + y^2)^2}$
$f_{11}(x, y) = \dfrac{(x^2 + y^2)^2(-2x) - (y^2 - x^2)2(x^2 + y^2)(2x)}{(x^2 + y^2)^4}$
$= \dfrac{2x^3 - 6xy^2}{(x^2 + y^2)^3}$
$f_{22}(x, y) = -\dfrac{(x^2 + y^2)^2(2x) - 2xy2(x^2 + y^2)(2y)}{(x^2 + y^2)^4}$
$= \dfrac{-2x^3 + 6xy^2}{(x^2 + y^2)^3}$.
Evidently $f_{11}(x, y) + f_{22}(x, y) = 0$ for $(x, y) \neq (0, 0)$.
Hence f is harmonic except at the origin.

11. $f(x, y) = \ln(x^2 + y^2)$, $f_1 = \dfrac{2x}{x^2 + y^2}$, $f_2 = \dfrac{2y}{x^2 + y^2}$
$f_{11} = \dfrac{(x^2 + y^2)(2) - 2x(2x)}{(x^2 + y^2)^2} = \dfrac{2(y^2 - x^2)}{(x^2 + y^2)^2}$
$f_{22} = \dfrac{2(x^2 - y^2)}{(x^2 + y^2)^2}$ (by symmetry)
Thus $f_{11} + f_{22} = 0$ (everywhere except at the origin), and f is harmonic.

12. $f(x, y) = \tan^{-1}\left(\dfrac{y}{x}\right)$, $(x \neq 0)$.
$f_1(x, y) = \dfrac{1}{1 + \dfrac{y^2}{x^2}}\left(-\dfrac{y}{x^2}\right) = -\dfrac{y}{x^2 + y^2}$,
$f_2(x, y) = \dfrac{1}{1 + \dfrac{y^2}{x^2}}\left(\dfrac{1}{x}\right) = \dfrac{x}{x^2 + y^2}$,
$f_{11} = \dfrac{2xy}{(x^2 + y^2)^2}$, $f_{22} = -\dfrac{2xy}{(x^2 + y^2)^2}$.
Thus $f_{11} + f_{22} = 0$ and f is harmonic.

13. $w = e^{3x+4y} \sin(5z)$,
$w_1 = 3w$, $w_2 = 4w$, $w_{11} = 9w$, $w_{22} = 16w$,
$w_3 = 5e^{3x+4y} \cos(5z)$, $w_{33} = -25w$.
Thus $w_{11} + w_{22} + w_{33} = (9 + 16 - 25)w = 0$, and w is harmonic in 3-space.

14. Let $g(x, y, z) = zf(x, y)$. Then
$g_1(x, y, z) = zf_1(x, y)$, $g_{11}(x, y, z) = zf_{11}(x, y)$
$g_2(x, y, z) = zf_2(x, y)$, $g_{22}(x, y, z) = zf_{22}(x, y)$
$g_3(x, y, z) = f(x, y)$, $g_{33}(x, y, z) = 0$.

Thus $g_{11} + g_{22} + g_{33} = z(f_{11} + f_{22}) = 0$ and g is harmonic because f is harmonic. This proves (a). The proofs of (b) and (c) are similar.

If $h(x, y, z) = f(ax + by, cz)$, then $h_{11} = a^2 f_{11}$, $h_{22} = b^2 f_{11}$ and $h_{33} = c^2 f_{22}$. If $a^2 + b^2 = c^2$ and f is harmonic then

$$h_{11} + h_{22} + h_{33} = c^2(f_{11} + f_{22}) = 0,$$

so h is harmonic.

15. Since $\dfrac{\partial u}{\partial x} = \dfrac{\partial v}{\partial y}$, $\dfrac{\partial u}{\partial y} = -\dfrac{\partial v}{\partial x}$, and the second partials of u are continuous, we have

$$\frac{\partial^2 u}{\partial x^2} = \frac{\partial}{\partial x}\frac{\partial v}{\partial y} = \frac{\partial}{\partial y}\frac{\partial v}{\partial x} = -\frac{\partial^2 u}{\partial y^2}.$$

Thus $\dfrac{\partial^2 u}{\partial x^2} + \dfrac{\partial^2 u}{\partial y^2} = 0$, and u is harmonic. The proof that v is harmonic is similar.

16. Let
$$f(x, y) = \begin{cases} \dfrac{2xy}{x^2 + y^2} & \text{if } (x, y) \neq (0, 0) \\ 0 & \text{if } (x, y) = (0, 0). \end{cases}$$

For $(x, y) \neq (0, 0)$, we have

$$f_1(x, y) = \frac{(x^2 + y^2)2y - 2xy(2x)}{(x^2 + y^2)^2} = \frac{2y(y^2 - x^2)}{(x^2 + y^2)^2}$$

$$f_2(x, y) = \frac{2x(x^2 - y^2)}{(x^2 + y^2)^2} \quad \text{(by symmetry)}.$$

Let $F(x, y) = (x^2 - y^2)f(x, y)$. Then we calculate

$$F_1(x, y) = 2xf(x, y) + (x^2 - y^2)f_1(x, y)$$
$$= 2xf(x, y) - \frac{2y(y^2 - x^2)^2}{(x^2 + y^2)^2}$$
$$F_2(x, y) = -2yf(x, y) + (x^2 - y^2)f_2(x, y)$$
$$= -2yf(x, y) + \frac{2x(x^2 - y^2)^2}{(x^2 + y^2)^2}$$
$$F_{12}(x, y) = \frac{2(x^6 + 9x^4y^2 - 9x^2y^4 - y^6)}{(x^2 + y^2)^3} = F_{21}(x, y).$$

For the values at $(0, 0)$ we revert to the definition of derivative to calculate the partials:

$$F_1(0, 0) = \lim_{h \to 0} \frac{F(h, 0) - F(0, 0)}{h} = 0 = F_2(0, 0)$$

$$F_{12}(0, 0) = \lim_{k \to 0} \frac{F_1(0, k) - F_1(0, 0)}{k} = \lim_{k \to 0} \frac{-2k(k^4)}{k(k^4)} = -2$$

$$F_{21}(0, 0) = \lim_{h \to 0} \frac{F_2(h, 0) - F_2(0, 0)}{h} = \lim_{h \to 0} \frac{2h(h^4)}{h(h^4)} = 2$$

This does not contradict Theorem 1 since the partials F_{12} and F_{21} are not continuous at $(0, 0)$. (Observe, for instance, that $F_{12}(x, x) = 0$, while $F_{12}(x, 0) = 2$ for $x \neq 0$.)

17. $u(x, t) = t^{-1/2} e^{-x^2/4t}$
$$\frac{\partial u}{\partial t} = \left(-\frac{1}{2}t^{-3/2} + \frac{1}{4}t^{-5/2}x^2\right)e^{-x^2/4t}$$
$$\frac{\partial u}{\partial x} = -\frac{1}{2}xt^{-3/2}e^{-x^2/4t}$$
$$\frac{\partial^2 u}{\partial x^2} = \left(-\frac{1}{2}t^{-3/2} + \frac{1}{4}t^{-5/2}x^2\right)e^{-x^2/4t}$$
$$= \frac{\partial u}{\partial t}.$$

18. $u(x, y, t) = t^{-1} e^{-(x^2+y^2)/4t}$
$$\frac{\partial u}{\partial t} = -\frac{1}{t^2}e^{-(x^2+y^2)/4t} + \frac{x^2+y^2}{4t^3}e^{-(x^2+y^2)/4t}$$
$$\frac{\partial u}{\partial x} = -\frac{x}{2t^2}e^{-(x^2+y^2)/4t}$$
$$\frac{\partial^2 u}{\partial x^2} = -\frac{1}{2t^2}e^{-(x^2+y^2)/4t} + \frac{x^2}{4t^3}e^{-(x^2+y^2)/4t}$$
$$\frac{\partial^2 u}{\partial y^2} = -\frac{1}{2t^2}e^{-(x^2+y^2)/4t} + \frac{y^2}{4t^3}e^{-(x^2+y^2)/4t}$$
Thus $\dfrac{\partial u}{\partial t} = \dfrac{\partial^2 u}{\partial x^2} + \dfrac{\partial^2 u}{\partial y^2}$.

19. For $\dfrac{\partial u}{\partial t} = \dfrac{\partial^2 u}{\partial x^2} + \dfrac{\partial^2 u}{\partial y^2} + \dfrac{\partial^2 u}{\partial z^2}$ the solution is

$$u(x, y, z, t) = t^{-3/2} e^{-(x^2+y^2+z^2)/4t},$$

which is verified similarly to the previous Exercise.

20. $u(x, y)$ is biharmonic $\Leftrightarrow \dfrac{\partial^2 u}{\partial x^2} + \dfrac{\partial^2 u}{\partial y^2}$ is harmonic

$$\Leftrightarrow \left(\frac{\partial^2}{\partial x^2} + \frac{\partial^2}{\partial y^2}\right)\left(\frac{\partial^2 u}{\partial x^2} + \frac{\partial^2 u}{\partial y^2}\right) = 0$$

$$\Leftrightarrow \frac{\partial^4 u}{\partial x^4} + 2\frac{\partial^4 u}{\partial x^2 \partial y^2} + \frac{\partial^4 u}{\partial y^4} = 0$$

by the equality of mixed partials.

21. If $u(x, y) = x^4 - 3x^2y^2$, then

$$\frac{\partial^2 u}{\partial x^2} = \frac{\partial}{\partial x}(4x^3 - 6xy^2) = 12x^2 - 6y^2$$
$$\frac{\partial^2 u}{\partial y^2} = \frac{\partial}{\partial y}(-6x^2y) = -6x^2$$
$$\frac{\partial^4 u}{\partial x^4} = \frac{\partial}{\partial x}(24x) = 24$$
$$\frac{\partial^4 u}{\partial x^2 \partial y^2} = \frac{\partial}{\partial x}(-12x) = -12$$
$$\frac{\partial^4 u}{\partial y^4} = 0$$
$$\frac{\partial^4 u}{\partial x^4} + 2\frac{\partial^4 u}{\partial x^2 \partial y^2} + \frac{\partial^4 u}{\partial y^4} = 24 - 24 = 0.$$

Thus u is biharmonic.

22. If u is harmonic, then $\frac{\partial^2 u}{\partial x^2} + \frac{\partial^2 u}{\partial y^2} = 0$. If $v(x, y) = xu(x, y)$, then

$$\frac{\partial^2 v}{\partial x^2} = \frac{\partial}{\partial x}\left(u + x\frac{\partial u}{\partial x}\right) = 2\frac{\partial u}{\partial x} + x\frac{\partial^2 u}{\partial x^2}$$
$$\frac{\partial^2 v}{\partial y^2} = \frac{\partial}{\partial y}\left(x\frac{\partial u}{\partial y}\right) = x\frac{\partial^2 u}{\partial y^2}$$
$$\frac{\partial^2 v}{\partial x^2} + \frac{\partial^2 v}{\partial y^2} = 2\frac{\partial u}{\partial x} + x\left(\frac{\partial^2 u}{\partial x^2} + \frac{\partial^2 u}{\partial y^2}\right) = 2\frac{\partial u}{\partial x}.$$

Since u is harmonic, so is $\partial u/\partial x$:

$$\left(\frac{\partial^2}{\partial x^2} + \frac{\partial^2}{\partial y^2}\right)\frac{\partial u}{\partial x} = \frac{\partial}{\partial x}\left(\frac{\partial^2 u}{\partial x^2} + \frac{\partial^2 u}{\partial y^2}\right) = \frac{\partial}{\partial x}(0) = 0.$$

Thus $\frac{\partial^2 v}{\partial x^2} + \frac{\partial^2 v}{\partial y^2}$ is harmonic, and so v is biharmonic. The proof that $w(x, y) = yu(x, y)$ is biharmonic is similar.

23. By Example 3, $e^x \sin y$ is harmonic. Therefore $xe^x \sin y$ is biharmonic by Exercise 22.

24. By Exercise 11, $\ln(x^2 + y^2)$ is harmonic (except at the origin). Therefore $y \ln(x^2 + y^2)$ is biharmonic by Exercise 22.

25. By Exercise 10, $\dfrac{x}{x^2 + y^2}$ is harmonic (except at the origin). Therefore $\dfrac{xy}{x^2 + y^2}$ is biharmonic by Exercise 22.

26. $u(x, y, z)$ is biharmonic $\Leftrightarrow \dfrac{\partial^2 u}{\partial x^2} + \dfrac{\partial^2 u}{\partial y^2} + \dfrac{\partial^2 u}{\partial z^2}$ is harmonic

$$\Leftrightarrow \left(\frac{\partial^2}{\partial x^2} + \frac{\partial^2}{\partial y^2} + \frac{\partial^2}{\partial z^2}\right)\left(\frac{\partial^2 u}{\partial x^2} + \frac{\partial^2 u}{\partial y^2} + \frac{\partial^2 u}{\partial z^2}\right) = 0$$
$$\Leftrightarrow \frac{\partial^4 u}{\partial x^4} + \frac{\partial^4 u}{\partial y^4} + \frac{\partial^4 u}{\partial z^4} + 2\left(\frac{\partial^4 u}{\partial x^2 \partial y^2} + \frac{\partial^4 u}{\partial x^2 \partial z^2} + \frac{\partial^4 u}{\partial y^2 \partial z^2}\right) = 0$$

by the equality of mixed partials.

If $u(x, y, z)$ is harmonic then the functions $xu(x, y, z)$, $yu(x, y, z)$, and $zu(x, y, z)$ are all biharmonic. The proof is almost identical to that given in Exercise 22.

Section 12.5 The Chain Rule (page 730)

1. If $w = f(x, y, z)$ where $x = g(s, t)$, $y = h(s, t)$, and $z = k(s, t)$, then

$$\frac{\partial w}{\partial t} = f_1(x, y, z)g_2(s, t) + f_2(x, y, z)h_2(s, t)$$
$$+ f_3(x, y, z)k_2(s, t).$$

2. If $w = f(x, y, z)$ where $x = g(s)$, $y = h(s, t)$ and $z = k(t)$, then

$$\frac{\partial w}{\partial t} = f_2(x, y, z)h_2(s, t) + f_3(x, y, z)k'(t).$$

3. If $z = g(x, y)$ where $y = f(x)$ and $x = h(u, v)$, then

$$\frac{\partial z}{\partial u} = g_1(x, y)h_1(u, v) + g_2(x, y)f'(x)h_1(u, v).$$

4. If $w = f(x, y)$ where $x = g(r, s)$, $y = h(r, t)$, $r = k(s, t)$ and $s = m(t)$, then

$$\frac{dw}{dt} = f_1(x, y)\big[g_1(r, s)\big(k_1(s, t)m'(t)$$
$$+ k_2(s, t)\big) + g_2(r, s)m'(t)\big]$$
$$+ f_2(x, y)\big[h_1(r, t)\big(k_1(s, t)m'(t)$$
$$+ k_2(s, t)\big) + h_2(r, t)\big].$$

5. If $w = f(x, y, z)$ where $x = g(y, z)$ and $y = h(z)$, then

$$\frac{dw}{dz} = f_1(x, y, z)\big[g_1(y, z)h'(z) + g_2(y, z)\big]$$
$$+ f_2(x, y, z)h'(z) + f_3(x, y, z)$$
$$\frac{\partial w}{\partial z}\bigg|_x = f_2(x, y, z)h'(z) + f_3(x, y, z)$$
$$\frac{\partial w}{\partial z}\bigg|_{x,y} = f_3(x, y, z).$$

6. If $u = \sqrt{x^2 + y^2}$, where $x = e^{st}$ and $y = 1 + s^2 \cos t$, then

Method I.

$$\frac{\partial u}{\partial t} = \frac{x}{\sqrt{x^2 + y^2}}se^{st} + \frac{y}{\sqrt{x^2 + y^2}}(-s^2 \sin t)$$
$$= \frac{xse^{st} - ys^2 \sin t}{\sqrt{x^2 + y^2}}.$$

Method II.

$$u = \sqrt{e^{2st} + (1 + s^2 \cos t)^2}$$
$$\frac{\partial u}{\partial t} = \frac{2se^{2st} - 2s^2 \sin t(1 + s^2 \cos t)}{2\sqrt{e^{2st} + (1 + s^2 \cos t)^2}}$$
$$= \frac{x^2 s - y s^2 \sin t}{\sqrt{x^2 + y^2}}.$$

7. If $z = \tan^{-1}\frac{u}{v}$, where $u = 2x + y$ and $v = 3x - y$, then

Method I.

$$\frac{\partial z}{\partial x} = \frac{\partial z}{\partial u}\frac{\partial u}{\partial x} + \frac{\partial z}{\partial v}\frac{\partial v}{\partial x}$$
$$= \frac{1}{1 + \frac{u^2}{v^2}}\left(\frac{1}{v}\right)(2) + \frac{1}{1 + \frac{u^2}{v^2}}\left(\frac{-u}{v^2}\right)(3)$$
$$= \frac{2v - 3u}{u^2 + v^2} = -\frac{5y}{13x^2 - 2xy + 2y^2}.$$

Method II.

$$z = \tan^{-1}\frac{2x + y}{3x - y}$$
$$\frac{\partial z}{\partial x} = \frac{1}{1 + \frac{(2x+y)^2}{(3x-y)^2}} \cdot \frac{(3x-y)(2) - (2x+y)(3)}{(3x-y)^2}$$
$$= \frac{-5y}{(3x-y)^2 + (2x+y)^2} = \frac{-5y}{13x^2 - 2xy + 2y^2}.$$

8. If $z = txy^2$, where $x = t + \ln(y + t^2)$ and $y = e^t$, then

Method I.

$$\frac{dz}{dt} = \frac{\partial z}{\partial t} + \frac{\partial z}{\partial x}\left(\frac{\partial x}{\partial t} + \frac{\partial x}{\partial y}\frac{\partial y}{\partial t}\right)$$
$$\quad + \frac{\partial z}{\partial y}\frac{\partial y}{\partial t}$$
$$= xy^2 + ty^2\left(1 + \frac{y + 2t}{y + t^2}\right) + 2txy^2.$$

Method II.

$$z = t\bigl(t + \ln(e^t + t^2)\bigr)e^{2t}$$
$$\frac{\partial z}{\partial t} = \bigl(t + \ln(e^t + t^2)\bigr)e^{2t} + te^{2t}\left(1 + \frac{e^t + 2t}{e^t + t^2}\right)$$
$$\quad + 2te^{2t}\bigl(t + \ln(e^t + t^2)\bigr)$$
$$= xy^2 + ty^2\left(1 + \frac{y + 2t}{y + t^2}\right) + 2txy^2.$$

9. $\dfrac{\partial}{\partial x}f(2x, 3y) = 2f_1(2x, 3y).$

10. $\dfrac{\partial}{\partial x}f(2y, 3x) = 3f_2(2y, 3x).$

11. $\dfrac{\partial}{\partial x}f(y^2, x^2) = 2xf_2(y^2, x^2).$

12. $\dfrac{\partial}{\partial y}f\bigl(yf(x,t), f(y,t)\bigr)$
$$= f(x,t)f_1\bigl(yf(x,t), f(y,t)\bigr)$$
$$\quad + f_1(y,t)f_2\bigl(yf(x,t), f(y,t)\bigr).$$

13. $T = e^{-t}z$, where $z = f(t)$.
$$\frac{dT}{dt} = \frac{\partial T}{\partial t} + \frac{\partial T}{\partial z}\frac{dz}{dt} = -e^{-t}f(t) + e^{-t}f'(t).$$

If $f(t) = e^t$, then $f'(t) = e^t$ and $\dfrac{dT}{dt} = 0$. The temperature is rising with respect to depth at the same rate at which it is falling with respect to time.

14. If $S = S(x, y, z, t)$, where $x = \sin t$, $y = \cos t$ and $z = t$, then the rate of change of S is

$$\frac{dS}{dt} = \frac{\partial S}{\partial x}\cos t - \frac{\partial S}{\partial y}\sin t + \frac{\partial S}{\partial z} + \frac{\partial S}{\partial t}.$$

15. $z = f(x, y)$, where $x = 2s + 3t$ and $y = 3s - 2t$.

a) $\dfrac{\partial^2 z}{\partial s^2} = \dfrac{\partial}{\partial s}\bigl(2f_1(x,y) + 3f_2(x,y)\bigr)$
$$= 2(2f_{11} + 3f_{12}) + 3(2f_{21} + 3f_{22})$$
$$= 4f_{11} + 12f_{12} + 9f_{22}$$

b) $\dfrac{\partial^2 z}{\partial s\partial t} = \dfrac{\partial^2 z}{\partial t\partial s} = \dfrac{\partial}{\partial t}(2f_1 + 3f_2)$
$$= 2(3f_{11} - 2f_{12}) + 3(3f_{21} - 2f_{22})$$
$$= 6f_{11} + 5f_{12} - 6f_{22}$$

c) $\dfrac{\partial^2 z}{\partial t^2} = \dfrac{\partial}{\partial t}(3f_1 - 2f_2)$
$$= 3(3f_{11} - 2f_{12}) - 2(3f_{21} - 2f_{22})$$
$$= 9f_{11} - 12f_{12} + 4f_{22}$$

16. Let $u = \dfrac{x}{x^2 + y^2}$, $v = -\dfrac{y}{x^2 + y^2}$. Then

$$\frac{\partial u}{\partial x} = \frac{y^2 - x^2}{(x^2 + y^2)^2} \qquad \frac{\partial v}{\partial x} = \frac{2xy}{(x^2 + y^2)^2}$$
$$\frac{\partial u}{\partial y} = -\frac{2xy}{(x^2 + y^2)^2} \qquad \frac{\partial v}{\partial y} = \frac{y^2 - x^2}{(x^2 + y^2)^2}.$$

We have

$$\frac{\partial}{\partial x}f(u,v) = f_1(u,v)\frac{\partial u}{\partial x} + f_2(u,v)\frac{\partial v}{\partial x}$$

$$\frac{\partial}{\partial y}f(u,v) = f_1(u,v)\frac{\partial u}{\partial y} + f_2(u,v)\frac{\partial v}{\partial y}$$

$$\frac{\partial^2}{\partial x^2}f(u,v) = f_{11}\left(\frac{\partial u}{\partial x}\right)^2 + f_{12}\frac{\partial u}{\partial x}\frac{\partial v}{\partial x} + f_1\frac{\partial^2 u}{\partial x^2}$$
$$+ f_{21}\frac{\partial u}{\partial x}\frac{\partial v}{\partial x} + f_{22}\left(\frac{\partial v}{\partial x}\right)^2 + f_2\frac{\partial^2 v}{\partial x^2}$$

$$\frac{\partial^2}{\partial y^2}f(u,v) = f_{11}\left(\frac{\partial u}{\partial y}\right)^2 + f_{12}\frac{\partial u}{\partial y}\frac{\partial v}{\partial y} + f_1\frac{\partial^2 u}{\partial y^2}$$
$$+ f_{21}\frac{\partial u}{\partial y}\frac{\partial v}{\partial y} + f_{22}\left(\frac{\partial v}{\partial y}\right)^2 + f_2\frac{\partial^2 v}{\partial y^2}.$$

Noting that

$$\left(\frac{\partial u}{\partial x}\right)^2 + \left(\frac{\partial u}{\partial y}\right)^2 = \frac{1}{(x^2+y^2)^2} = \left(\frac{\partial v}{\partial x}\right)^2 + \left(\frac{\partial v}{\partial y}\right)^2$$

$$\frac{\partial u}{\partial x}\frac{\partial v}{\partial x} + \frac{\partial u}{\partial y}\frac{\partial v}{\partial y} = 0,$$

we have

$$\frac{\partial^2}{\partial x^2}f(u,v) + \frac{\partial^2}{\partial y^2}f(u,v)$$

$$= f_{11}\left[\left(\frac{\partial u}{\partial x}\right)^2 + \left(\frac{\partial u}{\partial y}\right)^2\right]$$
$$+ f_{22}\left[\left(\frac{\partial v}{\partial x}\right)^2 + \left(\frac{\partial v}{\partial y}\right)^2\right]$$
$$+ 2f_{12}\left[\frac{\partial u}{\partial x}\frac{\partial v}{\partial x} + \frac{\partial u}{\partial y}\frac{\partial v}{\partial y}\right]$$
$$+ f_1\left[\frac{\partial^2 u}{\partial x^2} + \frac{\partial^2 u}{\partial y^2}\right] + f_2\left[\frac{\partial^2 v}{\partial x^2} + \frac{\partial^2 v}{\partial y^2}\right]$$
$$= f_1\left[\frac{\partial^2 u}{\partial x^2} + \frac{\partial^2 u}{\partial y^2}\right] + f_2\left[\frac{\partial^2 v}{\partial x^2} + \frac{\partial^2 v}{\partial y^2}\right],$$

because we are given that f is harmonic, that is, $f_{11}(u,v) + f_{22}(u,v) = 0$.

Finally, u is harmonic by Exercise 10 of Section 12.4, and, by symmetry, so is v. Thus

$$\frac{\partial^2}{\partial x^2}f(u,v) + \frac{\partial^2}{\partial y^2}f(u,v) = 0$$

and $f\left(\dfrac{x}{x^2+y^2}, -\dfrac{y}{x^2+y^2}\right)$ is harmonic for $(x,y) \neq (0,0)$.

17. If $x = t\sin s$ and $y = t\cos s$, then

$$\frac{\partial^2}{\partial s \partial t}f(x,y) = \frac{\partial}{\partial s}\left(\sin s f_1(x,y) + \cos s f_2(x,y)\right)$$
$$= \cos s f_1 + t\sin s\cos s f_{11} - t\sin^2 s f_{12}$$
$$- \sin s f_2 + t\cos^2 s f_{12} - t\sin s\cos s f_{22}$$
$$= \cos s f_1 - \sin s f_2 + t\cos s\sin s(f_{11} - f_{22})$$
$$+ t(\cos^2 s - \sin^2 s)f_{12},$$

where all partials of f are evaluated at $(t\sin s, t\cos s)$.

18. $\dfrac{\partial^3}{\partial x \partial y^2}f(2x+3y, xy) = \dfrac{\partial^2}{\partial x \partial y}(3f_1 + xf_2)$

$$= \frac{\partial}{\partial x}(9f_{11} + 3xf_{12} + 3xf_{21} + x^2 f_{22})$$
$$= \frac{\partial}{\partial x}(9f_{11} + 6xf_{12} + x^2 f_{22})$$
$$= 18f_{111} + 9yf_{112} + 6f_{12} + 12xf_{121} + 6xyf_{122}$$
$$+ 2xf_{22} + 2x^2 f_{221} + x^2 y f_{222}$$
$$= 18f_{111} + (12x+9y)f_{112} + (6xy+2x^2)f_{122} + x^2 y f_{222}$$
$$+ 6f_{12} + 2xf_{22},$$

where all partials are evaluated at $(2x+3y, xy)$.

19. $\dfrac{\partial^2}{\partial y \partial x}f(y^2, xy, -x^2) = \dfrac{\partial}{\partial y}(yf_2 - 2xf_3)$

$$= f_2 + 2y^2 f_{21} + xyf_{22} - 4xyf_{31} - 2x^2 f_{32},$$

where all partials are evaluated at $(y^2, xy, -x^2)$.

20. $\dfrac{\partial^3}{\partial t^2 \partial s}f(s^2 - t, s + t^2) = \dfrac{\partial^2}{\partial t^2}(2sf_1 + f_2)$

$$= \frac{\partial}{\partial t}(-2sf_{11} + 4stf_{12} - f_{21} + 2tf_{22})$$
$$= \frac{\partial}{\partial t}(-2sf_{11} + (4st-1)f_{12} + 2tf_{22})$$
$$= 2sf_{111} - 4stf_{112} + 4sf_{12} - (4st-1)f_{121}$$
$$+ 2t(4st-1)f_{122} + 2f_{22} - 2tf_{221} + 4t^2 f_{222}$$
$$= 2sf_{111} + (1-8st)f_{112} + 4t(2st-1)f_{122} + 4t^2 f_{222}$$
$$+ 4sf_{12} + 2f_{22},$$

where all partials are evaluated at $(s^2 - t, s + t^2)$.

21. Let $g(x, y) = f(u, v)$, where $u = u(x, y)$, $v = v(x, y)$. Then

$$g_1(x, y) = f_1(u, v)u_1(x, y) + f_2(u, v)v_1(x, y)$$
$$g_2(x, y) = f_1(u, v)u_2(x, y) + f_2(u, v)v_2(x, y)$$
$$g_{11}(x, y) = f_1(u, v)u_{11}(x, y) + f_{11}(u, v)(u_1(x, y))^2$$
$$+ f_{12}(u, v)u_1(x, y)v_1(x, y) + f_2(u, v)v_{11}(x, y)$$
$$+ f_{21}(u, v)u_1(x, y)v_1(x, y) + f_{22}(u, v)(v_1(x, y))^2$$
$$g_{22}(x, y) = f_1(u, v)u_{22}(x, y) + f_{11}(u, v)(u_2(x, y))^2$$
$$+ f_{12}(u, v)u_2(x, y)v_2(x, y) + f_2(u, v)v_{22}(x, y)$$
$$+ f_{21}(u, v)u_2(x, y)v_2(x, y) + f_{22}(u, v)(v_2(x, y))^2$$
$$g_{11}(x, y) + g_{22}(x, y)$$
$$= f_1(u, v)[u_{11}(x, y) + u_{22}(x, y)]$$
$$+ f_2(u, v)[v_{11}(x, y) + v_{22}(x, y)]$$
$$+ [(u_1(x, y))^2 + (u_2(x, y))^2]f_{11}(u, v)$$
$$+ [(v_1(x, y))^2 + (v_2(x, y))^2]f_{22}(u, v)$$
$$+ 2[u_1(x, y)v_1(x, y) + u_2(x, y)v_2(x, y)]f_{12}(u, v).$$

The first two terms on the right are zero because u and v are harmonic. The next two terms simplify to $[(v_1)^2 + (v_2)^2][f_{11} + f_{22}] = 0$ because u and v satisfy the Cauchy-Riemann equations and f is harmonic. The last term is zero because u and v satisfy the Cauchy-Riemann equations. Thus g is harmonic.

22. If $r^2 = x^2 + y^2 + z^2$, then $2r\dfrac{\partial r}{\partial x} = 2x$, so $\dfrac{\partial r}{\partial x} = \dfrac{x}{r}$. Similarly, $\dfrac{\partial r}{\partial y} = \dfrac{y}{r}$ and $\dfrac{\partial r}{\partial z} = \dfrac{z}{r}$. If $u = \dfrac{1}{r}$, then

$$\frac{\partial u}{\partial x} = -\frac{1}{r^2}\frac{\partial r}{\partial x} = -\frac{x}{r^3}$$
$$\frac{\partial^2 u}{\partial x^2} = -\frac{1}{r^3} + \frac{3x}{r^4}\frac{x}{r} = \frac{3x^2 - r^2}{r^5}.$$

Similarly,

$$\frac{\partial^2 u}{\partial y^2} = \frac{3y^2 - r^2}{r^5}, \quad \frac{\partial^2 u}{\partial z^2} = \frac{3z^2 - r^2}{r^5}.$$

Adding these three expressions, we get

$$\frac{\partial^2 u}{\partial x^2} + \frac{\partial^2 u}{\partial y^2} + \frac{\partial^2 u}{\partial z^2} = 0,$$

so u is harmonic except at $r = 0$.

23. If $x = e^s \cos t$ and $y = e^s \sin t$, then

$$\frac{\partial x}{\partial s} = e^s \cos t \qquad \frac{\partial y}{\partial s} = e^s \sin t$$
$$\frac{\partial x}{\partial t} = -e^x \sin t \qquad \frac{\partial y}{\partial t} = e^s \cos t.$$

Therefore we have

$$\frac{\partial z}{\partial s} = e^s \cos t \frac{\partial z}{\partial x} + e^s \sin t \frac{\partial z}{\partial y}$$
$$\frac{\partial z}{\partial t} = -e^s \sin t \frac{\partial z}{\partial x} + e^s \cos t \frac{\partial z}{\partial y}$$
$$\frac{\partial^2 z}{\partial s^2} = e^s \cos t \frac{\partial z}{\partial x} + e^s \sin t \frac{\partial z}{\partial y}$$
$$+ e^s \cos t \left(e^s \cos t \frac{\partial^2 z}{\partial x^2} + e^s \sin t \frac{\partial^2 z}{\partial y \partial x}\right)$$
$$+ e^s \sin t \left(e^s \cos t \frac{\partial^2 z}{\partial x \partial y} + e^s \sin t \frac{\partial^2 z}{\partial y^2}\right)$$
$$\frac{\partial^2 z}{\partial t^2} = -e^s \cos t \frac{\partial z}{\partial x} - e^s \sin t \frac{\partial z}{\partial y}$$
$$- e^s \sin t \left(-e^s \sin t \frac{\partial^2 z}{\partial x^2} + e^s \cos t \frac{\partial^2 z}{\partial y \partial x}\right)$$
$$+ e^s \cos t \left(-e^s \sin t \frac{\partial^2 z}{\partial x \partial y} + e^s \cos t \frac{\partial^2 z}{\partial y^2}\right).$$

It follows that

$$\frac{\partial^2 z}{\partial s^2} + \frac{\partial^2 z}{\partial t^2} = e^{2s}(\cos^2 t + \sin^2 t)\left(\frac{\partial^2 z}{\partial x^2} + \frac{\partial^2 z}{\partial y^2}\right)$$
$$= (x^2 + y^2)\left(\frac{\partial^2 z}{\partial x^2} + \frac{\partial^2 z}{\partial y^2}\right).$$

24. If $x = r\cos\theta$ and $y = r\sin\theta$, then $r^2 = x^2 + y^2$ and $\tan\theta = y/x$. Thus $2r\dfrac{\partial r}{\partial x} = 2x$, so $\dfrac{\partial r}{\partial x} = \dfrac{x}{r} = \cos\theta$, and similarly, $\dfrac{\partial r}{\partial y} = \dfrac{y}{r} = \sin\theta$. Also

$$\sec^2\theta \frac{\partial \theta}{\partial x} = -\frac{y}{x^2} \qquad \sec^2\theta \frac{\partial \theta}{\partial y} = \frac{1}{x}$$
$$\frac{\partial \theta}{\partial x} = -\frac{y}{x^2 + y^2} \qquad \frac{\partial \theta}{\partial x} = \frac{x}{x^2 + y^2}$$
$$= -\frac{\sin\theta}{r} \qquad\qquad = \frac{\cos\theta}{r}.$$

Now

$$\frac{\partial u}{\partial x} = \frac{\partial u}{\partial r}\frac{\partial r}{\partial x} + \frac{\partial u}{\partial \theta}\frac{\partial \theta}{\partial x} = \cos\theta\frac{\partial u}{\partial r} - \frac{\sin\theta}{r}\frac{\partial u}{\partial \theta}$$

$$\frac{\partial u}{\partial y} = \frac{\partial u}{\partial r}\frac{\partial r}{\partial y} + \frac{\partial u}{\partial \theta}\frac{\partial \theta}{\partial y} = \sin\theta\frac{\partial u}{\partial r} + \frac{\cos\theta}{r}\frac{\partial u}{\partial \theta}$$

$$\frac{\partial^2 u}{\partial x^2} = \left(\frac{\partial}{\partial x}\cos\theta\right)\frac{\partial u}{\partial r} + \cos\theta\left(\cos\theta\frac{\partial^2 u}{\partial r^2} - \frac{\sin\theta}{r}\frac{\partial^2 u}{\partial \theta \partial r}\right)$$
$$- \left(\frac{\partial}{\partial x}\frac{\sin\theta}{r}\right)\frac{\partial u}{\partial \theta} - \frac{\sin\theta}{r}\left(\cos\theta\frac{\partial^2 u}{\partial r \partial \theta} - \frac{\sin\theta}{r}\frac{\partial^2 u}{\partial \theta^2}\right)$$
$$= \frac{\sin^2\theta}{r}\frac{\partial u}{\partial r} + \frac{2\sin\theta\cos\theta}{r^2}\frac{\partial u}{\partial \theta} + \cos^2\theta\frac{\partial^2 u}{\partial r^2}$$
$$- \frac{2\sin\theta\cos\theta}{r}\frac{\partial^2 u}{\partial r \partial \theta} + \frac{\sin^2\theta}{r^2}\frac{\partial^2 u}{\partial \theta^2}$$

$$\frac{\partial^2 u}{\partial y^2} = \left(\frac{\partial}{\partial y}\sin\theta\right)\frac{\partial u}{\partial r} + \sin\theta\left(\sin\theta\frac{\partial^2 u}{\partial r^2} + \frac{\cos\theta}{r}\frac{\partial^2 u}{\partial \theta \partial r}\right)$$
$$+ \left(\frac{\partial}{\partial y}\frac{\cos\theta}{r}\right)\frac{\partial u}{\partial \theta} + \frac{\cos\theta}{r}\left(\sin\theta\frac{\partial^2 u}{\partial r \partial \theta} + \frac{\cos\theta}{r}\frac{\partial^2 u}{\partial \theta^2}\right)$$
$$= \frac{\cos^2\theta}{r}\frac{\partial u}{\partial r} - \frac{2\sin\theta\cos\theta}{r^2}\frac{\partial u}{\partial \theta} + \sin^2\theta\frac{\partial^2 u}{\partial r^2}$$
$$+ \frac{2\sin\theta\cos\theta}{r}\frac{\partial^2 u}{\partial r \partial \theta} + \frac{\cos^2\theta}{r^2}\frac{\partial^2 u}{\partial \theta^2}.$$

Therefore

$$\frac{\partial^2 u}{\partial x^2} + \frac{\partial^2 u}{\partial y^2} = \frac{\partial^2 u}{\partial r^2} + \frac{1}{r}\frac{\partial u}{\partial r} + \frac{1}{r^2}\frac{\partial^2 u}{\partial \theta^2},$$

as was to be shown.

25. If $u = r^2 \ln r$, where $r^2 = x^2 + y^2$, then, since $\partial r/\partial x = x/r$ and $\partial r/\partial y = y/r$, we have

$$\frac{\partial u}{\partial x} = (2r \ln r + r)\frac{x}{r} = x(1 + 2\ln r)$$

$$\frac{\partial^2 u}{\partial x^2} = 1 + 2\ln r + \frac{2x^2}{r^2}$$

$$\frac{\partial^2 u}{\partial y^2} = 1 + 2\ln r + \frac{2y^2}{r^2} \text{ (similarly)}$$

$$\frac{\partial^2 u}{\partial x^2} + \frac{\partial^2 u}{\partial y^2} = 2 + 4\ln r + \frac{2(x^2+y^2)}{r^2} = 4 + 4\ln r.$$

The constant 4 is harmonic, and so is $4\ln r$ by Exercise 11 of Section 12.4. Therefore $\frac{\partial^2 u}{\partial x^2} + \frac{\partial^2 u}{\partial y^2}$ is harmonic, and so u is biharmonic.

26. $f(tx, ty) = t^k f(x, y)$
$xf_1(tx, ty) + yf_2(tx, ty) = kt^{k-1}f(x, y)$
$x\bigl(xf_{11}(tx, ty) + yf_{12}(tx, ty)\bigr)$
$\quad + y\bigl(xf_{21}(tx, ty) + yf_{22}(tx, ty)\bigr)$
$\quad\quad = k(k-1)t^{k-2}f(x, y)$
Put $t = 1$ and get

$$x^2 f_{11}(x, y) + 2xy f_{12}(x, y) + y^2 f_{22}(x, y) = k(k-1)f(x, y).$$

27. If $f(x_1, \cdots, x_n)$ is positively homogeneous of degree k and has continuous partial derivatives of second order, then

$$\sum_{i,j=1}^n x_1 x_j f_{ij}(x_1, \cdots, x_n) = k(k-1)f(x_1, \cdots, x_n).$$

Proof: Differentiate $f(tx_1, \cdots, tx_n) = t^k f(x_1, \cdots, x_n)$ twice with respect to t:

$$\sum_{i=1}^n x_i f_i(tx_1, \cdots, tx_n) = kt^{k-1}f/xn$$

$$\sum_{i,j=1}^n x_i x_j f_{ij}(tx_1, \cdots, tx_n) = k(k-1)t^{k-2}f(x_1, \cdots, x_n),$$

and then put $t = 1$.

28. If $f(x_1, \cdots, x_n)$ is positively homogeneous of degree k and has continuous partial derivatives of mth order, then

$$\sum_{i_1,\ldots,i_m=1}^n x_{i_1}\cdots x_{i_m} f_{i_1\ldots i_m}(x_1, \cdots, x_n)$$
$$= k(k-1)\cdots(k-m+1)f(x_1, \cdots, x_n).$$

The proof is identical to those of Exercises 26 or 27, except that you differentiate m times before putting $t = 1$.

29. $F(x, y) = \begin{cases} \dfrac{2xy(x^2-y^2)}{x^2+y^2} & \text{if } (x, y) \neq (0, 0) \\ 0 & \text{if } (x, y) = (0, 0) \end{cases}$

a) For $(x, y) \neq (0, 0)$,

$$F(x, y) = \frac{2xy(x^2-y^2)}{x^2+y^2} = -\frac{2xy(y^2-x^2)}{x^2+y^2} = -F(y, x).$$

Since $0 = -0$, this holds for $(x, y) = (0, 0)$ also.

b) For $(x, y) \neq (0, 0)$,

$$F_1(x, y) = \frac{\partial}{\partial x}F(x, y) = -\frac{\partial}{\partial x}F(y, x) = -F_2(y, x)$$

$$F_{12}(x, y) = \frac{\partial}{\partial y}F_1(x, y) = -\frac{\partial}{\partial y}F_2(y, x) = -F_{21}(y, x).$$

c) If $(x, y) \neq (0, 0)$,
then $F_1(x, y) = \dfrac{2y(x^2-y^2)}{x^2+y^2} + 2xy\dfrac{\partial}{\partial x}\dfrac{x^2-y^2}{x^2+y^2}$.
Thus $F_1(0, y) = -2y + 0 = -2y$ for $y \neq 0$. This result holds for $y = 0$ also, since $F_1(0, 0) = \lim_{h\to 0}(0-0)/h = 0$.

d) By (b) and (c), $F_2(x, 0) = -F_1(0, x) = 2x$, and $F_{21}(0, 0) = 2$.

30. a) Since $F_{12}(x, y) = -F_{21}(y, x)$ for $(x, y) \neq (0, 0)$, we have $F_{12}(x, x) = -F_{21}(x, x)$ for $x \neq 0$. However, all partial derivatives of the rational function F are continuous except possibly at the origin. Thus $F_{12}(x, x) = F_{21}(x, x)$ for $x \neq 0$. Therefore, $F_{12}(x, x) = 0$ for $x \neq 0$.

b) F_{12} cannot be continuous at $(0, 0)$ because its value there (which is -2) differs from the value of $F_{21}(0, 0)$ (which is 2). Alternatively, $F_{12}(0, 0)$ is not the limit of $F_{12}(x, x)$ as $x \to 0$.

31. If $\xi = x + ct$, $\eta = x$, and $v(\xi, \eta) = v(x + ct, x) = u(x, t)$, then
$$\frac{\partial u}{\partial t} = \frac{\partial v}{\partial \xi}\frac{\partial \xi}{\partial t} = c\frac{\partial v}{\partial \xi}$$
$$\frac{\partial u}{\partial x} = \frac{\partial v}{\partial \xi}\frac{\partial \xi}{\partial x} + \frac{\partial v}{\partial \eta}\frac{\partial \eta}{\partial x} = \frac{\partial v}{\partial \xi} + \frac{\partial v}{\partial \eta}.$$

If u satisfies $\dfrac{\partial u}{\partial t} = c\dfrac{\partial u}{\partial x}$, then v satisfies

$$c\frac{\partial v}{\partial \xi} = c\frac{\partial v}{\partial \xi} + c\frac{\partial v}{\partial \eta}, \quad \text{that is,} \quad \frac{\partial v}{\partial \eta} = 0.$$

Thus v is independent of η, so $v(\xi, \eta) = f(\xi)$ for an arbitrary differentiable function f of one variable. The original differential equation has solution

$$u(x, t) = f(x + ct).$$

32. If $w(r) = f(r) + g(s)$, where f and g are arbitrary twice differentiable functions, then

$$\frac{\partial^2 w}{\partial r \partial s} = \frac{\partial}{\partial r}g'(s) = 0.$$

33. If $r = x + ct$, $s = x - ct$, and $w(r, s) = w(x + ct, x - ct) = u(x, t)$, then
$$\frac{\partial u}{\partial t} = c\frac{\partial w}{\partial r} - c\frac{\partial w}{\partial s}$$
$$\frac{\partial^2 w}{\partial t^2} = c^2\frac{\partial^2 w}{\partial r^2} - 2c^2\frac{\partial^2 w}{\partial r \partial s} + c^2\frac{\partial^2 w}{\partial s^2}$$
$$\frac{\partial u}{\partial x} = \frac{\partial w}{\partial r} + \frac{\partial w}{\partial s}$$
$$\frac{\partial^2 w}{\partial x^2} = \frac{\partial^2 w}{\partial r^2} + 2\frac{\partial^2 w}{\partial r \partial s} + \frac{\partial^2 w}{\partial s^2}.$$

If u satisfies $\dfrac{\partial^2 u}{\partial t^2} = c^2\dfrac{\partial^2 u}{\partial x^2}$, then w satisfies

$$c^2\left(\frac{\partial^2 w}{\partial r^2} - 2\frac{\partial^2 w}{\partial r \partial s} + \frac{\partial^2 w}{\partial s^2}\right) = c^2\left(\frac{\partial^2 w}{\partial r^2} + 2\frac{\partial^2 w}{\partial r \partial s} + \frac{\partial^2 w}{\partial s^2}\right)$$

and hence

$$\frac{\partial^2 w}{\partial r \partial s} = 0.$$

By Exercise 38, $w(r, s) = f(r) + g(s)$, where f and g are arbitrary twice differentiable functions. Hence the original differential equation has solution

$$u(x, t) = f(x + ct) + g(x - ct).$$

34. By Exercise 39, the DE $u_t = c^2 u_{xx}$ has solution

$$u(x, t) = f(x + ct) + g(x - ct),$$

for arbitrary sufficiently smooth functions f and g. The initial conditions imply that

$$p(x) = u(x, 0) = f(x) + g(x)$$
$$q(x) = u_t(x, 0) = cf'(x) - cg'(x).$$

Integrating the second of these equations, we get

$$f(x) - g(x) = \frac{1}{c}\int_a^x q(s)\,ds,$$

where a is a constant. Solving the two equations for f and g we obtain

$$f(x) = \frac{1}{2}p(x) + \frac{1}{2c}\int_a^x q(s)\,ds$$
$$g(x) = \frac{1}{2}p(x) - \frac{1}{2c}\int_a^x q(s)\,ds.$$

Thus the solution to the initial-value problem is

$$u(x, t) = \frac{p(x + ct) + p(x - ct)}{2} + \frac{1}{2c}\int_{x-ct}^{x+ct} q(s)\,ds.$$

Section 12.6 Linear Approximation, Differentiability, and Differentials (page 738)

1. $f(x, y) = x^2 y^3$ $f(3, 1) = 9$
$f_1(x, y) = 2xy^3$ $f_1(3, 1) = 6$
$f_2(x, y) = 3x^2 y^2$ $f_2(3, 1) = 27$
$f(3.1, 0.9) = f(3 + 0.1, 1 - 0.1)$
$\approx f(3, 1) + 0.1 f_1(3, 1) - 0.1 f_2(3, 1)$
$= 9 + 0.6 - 2.7 = 6.9$

2. $f(x, y) = \tan^{-1}\dfrac{y}{x}$ $f(3, 3) = \dfrac{\pi}{4}$

$f_1(x, y) = -\dfrac{y}{x^2 + y^2}$ $f_1(3, 3) = -\dfrac{1}{6}$

$f_2(x, y) = \dfrac{x}{x^2 + y^2}$ $f_2(3, 3) = \dfrac{1}{6}$

$f(3.01, 2.99) = f(3 + 0.01, 3 - 0.01)$
$\approx f(3, 3) + 0.01 f_1(3, 3) - 0.01 f_2(3, 3)$
$= \dfrac{\pi}{4} - \dfrac{0.01}{6} - \dfrac{0.01}{6} = \dfrac{\pi}{4} - \dfrac{0.01}{3}$
≈ 0.7820648

3. $f(x, y) = \sin(\pi xy + \ln y), \ f(0, 1) = 0$

$f_1(x, y) = \pi y \cos(\pi xy + \ln y), \ f_1(0, 1) = \pi$

$f_2(x, y) = \left(\pi x + \dfrac{1}{y}\right)\cos(\pi xy + \ln y), \ f_2(0, 1) = 1$

$f(0.01, 1.05) \approx f(0, 1) + 0.01 f_1(0, 1) + 0.05 f_2(0, 1)$
$= 0 + 0.01\pi + 0.05 \approx 0.081416$

4. $f(x, y) = \dfrac{24}{x^2 + xy + y^2}$

$f_1(x, y) = \dfrac{-24(2x + y)}{(x^2 + xy + y^2)^2}, \quad f_2(x, y) = \dfrac{-24(x + 2y)}{(x^2 + xy + y^2)^2}$

$f(2, 2) = 2, \quad f_1(2, 2) = -1, \quad f_2(2, 2) = -1$

$f(2.1, 1.8) \approx f(2, 2) + 0.1 f_1(2, 2) - 0.2 f_2(2, 2)$
$= 2 - 0.1 + 0.2 = 2.1$

5. $f(x, y, z) = \sqrt{x + 2y + 3z}, \ f(2, 2, 1) = 3$

$f_1(x, y, z) = \dfrac{1}{2\sqrt{x + 2y + 3z}}, \quad f_2(x, y, z) = \dfrac{1}{\sqrt{x + 2y + 3z}}$

$f_3(x, y, z) = \dfrac{3}{2\sqrt{x + 2y + 3z}}$

$f(1.9, 1.8, 1.1) \approx f(2, 2, 1)$
$\quad = 0.1 f_1(2, 2, 1) - 0.2 f_2(2, 2, 1) + 0.1 f_3(2, 2, 1)$
$= 3 - \dfrac{0.1}{6} - \dfrac{0.2}{3} + \dfrac{0.1}{2} \approx 2.967$

6. $f(x, y) = xe^{y+x^2}$ $f(2, -4) = 2$

$f_1(x, y) = e^{y+x^2}(1 + 2x^2)$ $f_1(2, -4) = 9$

$f_2(x, y) = xe^{y+x^2}$ $f_2(2, -4) = 2$

$f(2.05, -3.92) \approx f(2, -4) + 0.05 f_1(2, -4) + 0.08 f_2(2, -4)$
$= 2 + 0.45 + 0.16 = 2.61$

7. If the edges are x, y, and z, and $\dfrac{dx}{x} = \dfrac{dy}{y} = \dfrac{dz}{z} = \dfrac{1}{100}$, then

a) $V = xyz \Rightarrow dV = yz\,dx + xz\,dy + xy\,dz$

$\Rightarrow \dfrac{dV}{V} = \dfrac{dx}{x} + \dfrac{dy}{y} + \dfrac{dz}{z} = \dfrac{3}{100}$.

The volume can be in error by about 3%.

b) $A = xy \Rightarrow dA = y\,dx + x\,dy$

$\Rightarrow \dfrac{dA}{A} = \dfrac{dx}{x} + \dfrac{dy}{y} = \dfrac{2}{100}$.

The area of a face can be in error by about 2%.

c) $D^2 = x^2 + y^2 + z^2 \Rightarrow 2D\,dD = 2x\,dx + 2y\,dy + 2z\,dz$

$\Rightarrow \dfrac{dD}{D} = \dfrac{x^2}{D^2}\dfrac{dx}{x} + \dfrac{y^2}{D^2}\dfrac{dy}{y} + \dfrac{z^2}{D^2}\dfrac{dz}{z} = \dfrac{1}{100}$.

The diagonal can be in error by about 1%.

8. $V = \tfrac{1}{3}\pi r^2 h \Rightarrow dV = \tfrac{2}{3}\pi rh\,dr + \tfrac{1}{3}\pi r^2\,dh$. If $r = 25$ ft, $h = 21$ ft, and $dr = dh = 0.5/12$ ft, then

$dV = \dfrac{\pi}{3}(2 \times 25 \times 21 + 25^2)\dfrac{0.5}{12} \approx 73.08$.

The calculated volume can be in error by about 73 cubic feet.

9. $S = \pi r\sqrt{r^2 + h^2}$, so

$dS = \left(\pi\sqrt{r^2 + h^2} + \dfrac{\pi r^2}{\sqrt{r^2 + h^2}}\right)dr + \dfrac{\pi rh}{\sqrt{r^2 + h^2}}\,dh$

$= \pi\left(\sqrt{25^2 + 21^2} + \dfrac{25^2 + 25 \times 21}{\sqrt{25^2 + 21^2}}\right)\dfrac{0.5}{12} \approx 8.88$.

The surface area can be in error by about 9 square feet.

10. If the sides and contained angle of the triangle are x and y m and θ radians, then its area A satisfies

$A = \dfrac{1}{2}xy\sin\theta$

$dA = \dfrac{1}{2}y\sin\theta\,dx + \dfrac{1}{2}x\sin\theta\,dy + \dfrac{1}{2}xy\cos\theta\,d\theta$

$\dfrac{dA}{A} = \dfrac{dx}{x} + \dfrac{dy}{y} + \cot\theta\,d\theta$.

For $x = 224$, $y = 158$, $\theta = 64° = 64\pi/180$, $dx = dy = 0.4$, and $d\theta = 2° = 2\pi/180$, we have

$\dfrac{dA}{A} = \dfrac{0.4}{224} + \dfrac{0.4}{158} + (\cot 64°)\dfrac{2\pi}{180} \approx 0.0213$.

The calculated area of the plot can be in error by a little over 2%.

11. From the figure we have

$h = s\tan\theta$

$h = (s + x)\tan\phi = \left(\dfrac{h}{\tan\theta} + x\right)\tan\phi$.

Solving the latter equation for h, we obtain

$h = \dfrac{x\tan\phi\tan\theta}{\tan\theta - \tan\phi}$.

We calculate the values of h and its first partials at $x = 100$, $\theta = 50°$, $\phi = 35°$:

$$h \approx 170$$
$$\frac{\partial h}{\partial x} = \frac{\tan\phi \tan\theta}{\tan\theta - \tan\phi} \approx 1.70$$
$$\frac{\partial h}{\partial \theta} = x\tan\phi \,\frac{(\tan\theta - \tan\phi)\sec^2\theta - \tan\theta \sec^2\theta}{(\tan\theta - \tan\phi)^2}$$
$$= -\frac{x\tan^2\phi \sec^2\theta}{(\tan\theta - \tan\phi)^2} \approx -491.12$$
$$\frac{\partial h}{\partial \phi} = \frac{x\tan^2\theta \sec^2\phi}{(\tan\theta - \tan\phi)^2} \approx 876.02.$$

Thus $dh \approx 1.70\,dx - 491\,d\theta + 876\,d\phi$. For $dx = 0.1$ m and $|d\theta| = |d\phi| = 1° = \pi/180$, the largest value of dh will come from taking $d\theta$ negative and $d\phi$ positive:

$$dh \approx (1.70)(0.1) + (491 + 876)\frac{\pi}{180} \approx 24.03.$$

The calculated height of the tower is 170 m and can be in error by as much as 24 m. The calculation of the height is most sensitive to the accuracy of the measurement of ϕ.

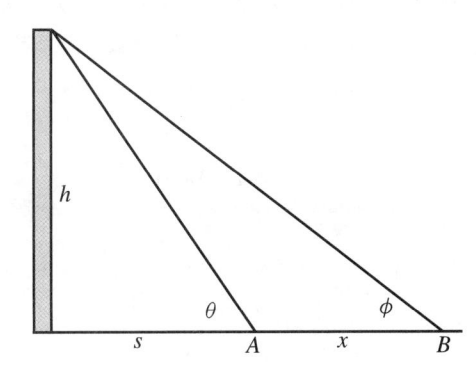

Fig. 12.6.11

12. $\quad w = \dfrac{x^2 y^3}{z^4} \qquad \dfrac{\partial w}{\partial x} = \dfrac{2xy^3}{z^4} = \dfrac{2w}{x}$

$\dfrac{\partial w}{\partial y} = \dfrac{3x^2 y^2}{z^4} = \dfrac{3w}{y} \qquad \dfrac{\partial w}{\partial z} = -\dfrac{4x^2 y^3}{z^5} = -\dfrac{4w}{x}.$

$$dw = \frac{\partial w}{\partial x}dx + \frac{\partial w}{\partial y}dy + \frac{\partial w}{\partial z}dz$$
$$\frac{dw}{w} = 2\frac{dx}{x} + 3\frac{dy}{y} - 4\frac{dz}{z}.$$

Since x increases by 1%, then $\dfrac{dx}{x} = \dfrac{1}{100}$. Similarly, $\dfrac{dy}{y} = \dfrac{2}{100}$ and $\dfrac{dz}{z} = \dfrac{3}{100}$. Therefore

$$\frac{\Delta w}{w} \approx \frac{dw}{w} = \frac{2+6-12}{100} = -\frac{4}{100},$$

and w decreases by about 4%.

13. $\mathbf{f}(r, \theta) = (r\cos\theta, r\sin\theta)$
$$D\mathbf{f}(r,\theta) = \begin{pmatrix} \cos\theta & -r\sin\theta \\ \sin\theta & r\cos\theta \end{pmatrix}$$

14. $\mathbf{f}(\rho, \phi, \theta) = (\rho\sin\phi\cos\theta, \rho\sin\phi\sin\theta, \rho\cos\phi)$
$$D\mathbf{f}(\rho,\phi,\theta) = \begin{pmatrix} \sin\phi\cos\theta & \rho\cos\phi\cos\theta & -\rho\sin\phi\sin\theta \\ \sin\phi\sin\theta & \rho\cos\phi\sin\theta & \rho\sin\phi\cos\theta \\ \cos\phi & -\rho\sin\phi & 0 \end{pmatrix}$$

15. $\mathbf{f}(x, y, z) = \begin{pmatrix} x^2 + yz \\ y^2 - x\ln z \end{pmatrix}$
$$D\mathbf{f}(x,y,z) = \begin{pmatrix} 2x & z & y \\ -\ln z & 2y & -x/z \end{pmatrix}$$
$$D\mathbf{f}(2,2,1) = \begin{pmatrix} 4 & 1 & 2 \\ 0 & 4 & -2 \end{pmatrix}$$
$$\mathbf{f}(1.98, 2.01, 1.03) \approx \mathbf{f}(2,2,1) + D\mathbf{f}(2,2,1)\begin{pmatrix} -0.02 \\ 0.01 \\ 0.03 \end{pmatrix}$$
$$= \begin{pmatrix} 6 \\ 4 \end{pmatrix} + \begin{pmatrix} -0.01 \\ -0.02 \end{pmatrix} = \begin{pmatrix} 5.99 \\ 3.98 \end{pmatrix}$$

16. $\mathbf{g}(r, s, t) = \begin{pmatrix} r^2 s \\ r^2 t \\ s^2 - t^2 \end{pmatrix}$
$$D\mathbf{g}(r,s,t) = \begin{pmatrix} 2rs & r^2 & 0 \\ 2rt & 0 & r^2 \\ 0 & 2s & -2t \end{pmatrix}$$
$$D\mathbf{g}(1,3,3) = \begin{pmatrix} 6 & 1 & 0 \\ 6 & 0 & 1 \\ 0 & 6 & -6 \end{pmatrix}$$
$$\mathbf{g}(0.99, 3.02, 2.97) \approx \mathbf{g}(1,3,3) + D\mathbf{g}(1,3,3)\begin{pmatrix} -0.01 \\ 0.02 \\ -0.03 \end{pmatrix}$$
$$= \begin{pmatrix} 3 \\ 3 \\ 0 \end{pmatrix} + \begin{pmatrix} -0.04 \\ -0.09 \\ 0.30 \end{pmatrix} = \begin{pmatrix} 2.96 \\ 2.91 \\ 0.30 \end{pmatrix}$$

17. If f is differentiable at (a, b), then

$$\frac{f(a+h, b+k) - f(a,b) - hf_1(a,b) - kf_2(a,b)}{\sqrt{h^2 + k^2}}$$

approaches 0 as $(h, k) \to (0, 0)$. Since the denominator of this fraction approaches zero, the numerator must also approach 0 or the fraction would not have a limit. Since the terms $hf_1(a, b)$ and $kf_2(a, b)$ both approach 0, we must have

$$\lim_{(h,k) \to (0,0)} [f(a+h, b+k) - f(a, b)] = 0.$$

Thus f is continuous at (a, b).

18. Let $g(t) = f(a + th, b + tk)$. Then

$$g'(t) = hf_1(a+th, b+tk) + kf_2(a+th, b+tk).$$

If h and k are small enough that $(a+h, b+k)$ belongs to the disk referred to in the statement of the problem, then we can apply the (one-variable) Mean-Value Theorem to $g(t)$ on $[0, 1]$ and obtain

$$g(1) = g(0) + g'(\theta),$$

for some θ satisfying $0 < \theta < 1$, i.e.,

$$f(a+h, b+k) = f(a,b) + hf_1(a+\theta h, b+\theta k) + kf_2(a+\theta h, b+\theta k).$$

19. Apply Taylor's Formula:

$$g(1) = g(0) + g'(0) + \frac{g''(\theta)}{2!}$$

for some θ between 0 and 1 to $g(t) = f(a+th, b+tk)$. We have

$$g'(t) = hf_1(a+th, b+tk) + kf_2(a+th, b+tk)$$
$$g'(0) = hf_1(a,b) + kf_2(a,b)$$
$$g''(t) = h^2 f_{11}(a+th, b+tk) + 2hk f_{12}(a+th, b+tk) + k^2 f_{22}(a+th, b+tk).$$

Thus

$$f(a+h, b+k) = f(a,b) + hf_1(a,b) + kf_2(a,b)$$
$$+ \frac{1}{2}\Big(h^2 f_{11}(a+\theta h, b+\theta k) + 2hk f_{12}(a+\theta h, b+\theta k)$$
$$+ k^2 f_{22}(a+\theta h, b+\theta k)\Big)$$
$$\Delta f = f(a+h, b+k) - f(a,b)$$
$$df = hf_1(a,b) + kf_2(a,b)$$
$$|\Delta f - df|$$
$$\leq \frac{1}{2}\Big|h^2 f_{11}(a+\theta h, b+\theta k) + 2hk f_{12}(a+\theta h, b+\theta k)$$
$$+ k^2 f_{22}(a+\theta h, b+\theta k)\Big|$$
$$\leq K(h^2 + k^2) \quad \text{(since } 2hk \leq h^2 + k^2\text{)},$$

for some K depending on f, and valid in some disk $h^2 + k^2 \leq R^2$ of positive radius R.

Section 12.7 Gradients and Directional Derivatives (page 749)

1. $f(x, y) = x^2 - y^2$, $f(2, -1) = 3$.
$\nabla f(x, y) = 2x\mathbf{i} - 2y\mathbf{j}$, $\nabla f(2, -1) = 4\mathbf{i} + 2\mathbf{j}$.
Tangent plane to $z = f(x, y)$ at $(2, -1, 3)$ has equation $4(x-2) + 2(y+1) = z - 3$, or $4x + 2y - z = 3$.
Tangent line to $f(x, y) = 3$ at $(2, -1)$ has equation $4(x-2) + 2(y+1) = 0$, or $2x + y = 3$.

2. $f(x, y) = \dfrac{x-y}{x+y}$, $f(1, 1) = 0$.
$\nabla f = \dfrac{2y\mathbf{i} - 2x\mathbf{j}}{(x+y)^2}$,
$\nabla f(1, 1) = \dfrac{1}{2}(\mathbf{i} - \mathbf{j})$. Tangent plane to $z = f(x, y)$ at $(1, 1, 0)$ has equation $\tfrac{1}{2}(x-1) - \tfrac{1}{2}(y-1) = z$, or $x - y - 2z = 0$.
Tangent line to $f(x, y) = 0$ at $(1, 1)$ has equation $\tfrac{1}{2}(x-1) - \tfrac{1}{2}(y-1)$, or $x = y$.

3. $f(x, y) = \dfrac{x}{x^2 + y^2}$,
$f_1(x, y) = \dfrac{(x^2+y^2)(1) - x(2x)}{(x^2+y^2)^2} = \dfrac{y^2 - x^2}{(x^2+y^2)^2}$,
$f_2(x, y) = -\dfrac{2xy}{(x^2+y^2)^2}$.
$\nabla f(x, y) = \dfrac{1}{(x^2+y^2)^2}\big((y^2 - x^2)\mathbf{i} - 2xy\mathbf{j}\big)$,
$\nabla f(1, 2) = \tfrac{3}{25}\mathbf{i} - \tfrac{4}{25}\mathbf{j}$.
Tangent plane to $z = f(x, y)$ at $(1, 2, \tfrac{1}{5})$ has equation
$\tfrac{3}{25}(x-1) - \tfrac{4}{25}(y-2) = z - \tfrac{1}{5}$, or $3x - 4y - 25z = -10$.
Tangent line to $f(x, y) = 1/5$ at $(1, 2)$ has equation
$\tfrac{3}{25}(x-1) - \tfrac{4}{25}(y-2) = 0$, or $3x - 4y = -5$.

4. $f(x, y) = e^{xy}$, $\nabla f = ye^{xy}\mathbf{i} + xe^{xy}\mathbf{j}$.
$\nabla f(2, 0) = 2\mathbf{j}$. Tangent plane to $z = f(x, y)$ at $(2, 0, 1)$ has equation $2y = z - 1$, or $2y - z = -1$.
Tangent line to $f(x, y) = 1$ at $(2, 0)$ has equation $y = 0$.

5. $f(x, y) = \ln(x^2 + y^2)$, $\nabla f(x, y) = \dfrac{2x\mathbf{i} + 2y\mathbf{j}}{x^2 + y^2}$,
$\nabla f(1, -2) = \tfrac{2}{5}\mathbf{i} - \tfrac{4}{5}\mathbf{j}$. Tangent plane to $z = f(x, y)$ at $(1, -2, \ln 5)$ has equation $\tfrac{2}{5}(x-1) - \tfrac{4}{5}(y+2) = z - \ln 5$, or $2x - 4y - 5z = 10 - 5\ln 5$.
Tangent line to $f(x, y) = \ln 5$ at $(1, -2)$ has equation $\tfrac{2}{5}(x-1) - \tfrac{4}{5}(y+2) = 0$, or $x - 2y = 5$.

6. $f(x, y) = \sqrt{1 + xy^2}$, $f(2, -2) = 3$.
$\nabla f(x, y) = \dfrac{y^2\mathbf{i} + 2xy\mathbf{j}}{2\sqrt{1 + xy^2}}$,
$\nabla f(2, -2) = \tfrac{2}{3}\mathbf{i} - \tfrac{4}{3}\mathbf{j}$.
Tangent plane to $z = f(x, y)$ at $(2, -2, 3)$ has equation $\tfrac{2}{3}(x-2) - \tfrac{4}{3}(y+2) = z - 3$, or $2x - 4y - 3z = 3$.
Tangent line to $f(x, y) = 3$ at $(2, -2)$ has equation $\tfrac{2}{3}(x-2) - \tfrac{4}{3}(y+2) = 0$, or $x - 2y = 6$.

INSTRUCTOR'S SOLUTIONS MANUAL SECTION 12.7 (PAGE 749)

7. $f(x, y, z) = x^2y + y^2z + z^2x$, $\quad f(1, -1, 1) = 1$.
$\nabla f(x, y, z) = (2xy + z^2)\mathbf{i} + (x^2 + 2yz)\mathbf{j} + (y^2 + 2zx)\mathbf{k}$,
$\nabla f(1, -1, 1) = -\mathbf{i} - \mathbf{j} + 3\mathbf{k}$.
Tangent plane to $f(x, y, z) = 1$ at $(1, -1, 1)$ has equation
$-(x - 1) - (y + 1) + 3(z - 1) = 0$, or $x + y - 3z = -3$.

8. $f(x, y, z) = \cos(x + 2y + 3z)$,
$f\left(\dfrac{\pi}{2}, \pi, \pi\right) = \cos\dfrac{11\pi}{2} = 0$.
$\nabla f(x, y, z) = -\sin(x + 2y + 3z)(\mathbf{i} + 2\mathbf{j} + 3\mathbf{k})$,
$\nabla f\left(\dfrac{\pi}{2}, \pi, \pi\right) = -\sin\dfrac{11\pi}{2}(\mathbf{i} + 2\mathbf{j} + 3\mathbf{k}) = \mathbf{i} + 2\mathbf{j} + 3\mathbf{k}$.
Tangent plane to $f(x, y, z) = 0$ at $\left(\dfrac{\pi}{2}, \pi, \pi\right)$ has equation
$$x - \dfrac{\pi}{2} + 2(y - \pi) + 3(z - \pi) = 0,$$
or $x + 2y + 3z = \dfrac{11\pi}{2}$.

9. $f(x, y, z) = ye^{-x^2}\sin z$, $\quad f(0, 1, \pi/3) = \sqrt{3}/2$.
$\nabla f(x, y, z) = -2xye^{-x^2}\sin z\mathbf{i} + e^{-x^2}\sin z\mathbf{j} + ye^{-x^2}\cos z\mathbf{k}$,
$\nabla f(0, 1, \pi/3) = \dfrac{\sqrt{3}}{2}\mathbf{j} + \dfrac{1}{2}\mathbf{k}$.
The tangent plane to $f(x, y, z) = \dfrac{\sqrt{3}}{2}$ at $0, 1, \pi/3$ has equation
$$\dfrac{\sqrt{3}}{2}(y - 1) + \dfrac{1}{2}\left(z - \dfrac{\pi}{3}\right) = 0,$$
or $\sqrt{3}y + z = \sqrt{3} + \dfrac{\pi}{3}$.

10. $f(x, y) = 3x - 4y$, $\quad \nabla f(0, 2) = \nabla f(x, y) = 3\mathbf{i} - 4\mathbf{j}$,
$D_{-\mathbf{i}}f(0, 2) = -\mathbf{i} \bullet (3\mathbf{i} - 4\mathbf{j}) = -3$.

11. $f(x, y) = x^2y$, $\quad \nabla f = 2xy\mathbf{i} + x^2\mathbf{j}$,
$\nabla f(-1, -1) = 2\mathbf{i} + \mathbf{j}$.
Rate of change of f at $(-1, -1)$ in the direction of $\mathbf{i} + 2\mathbf{j}$ is
$$\dfrac{\mathbf{i} + 2\mathbf{j}}{\sqrt{5}} \bullet (2\mathbf{i} + \mathbf{j}) = \dfrac{4}{\sqrt{5}}.$$

12. $f(x, y) = \dfrac{x}{1 + y}$, $\quad \nabla f(x, y) = \dfrac{1}{1 + y}\mathbf{i} - \dfrac{x}{(1 + y)^2}\mathbf{j}$,
$\nabla f(0, 0) = \mathbf{i}$, $\quad \mathbf{u} = \dfrac{\mathbf{i} - \mathbf{j}}{\sqrt{2}}$,
$D_\mathbf{u}f(0, 0) = \mathbf{i} \bullet \left(\dfrac{\mathbf{i} - \mathbf{j}}{\sqrt{2}}\right) = \dfrac{1}{\sqrt{2}}$.

13. $f(x, y) = x^2 + y^2$, $\quad \nabla f = 2x\mathbf{i} + 2y\mathbf{j}$,
$\nabla f(1, -2) = 2\mathbf{i} - 4\mathbf{j}$.
A unit vector in the direction making a $60°$ angle with the positive x-axis is $\mathbf{u} = \frac{1}{2}\mathbf{i} + \frac{\sqrt{3}}{2}\mathbf{j}$.
The rate of change of f at $(1, -2)$ in the direction of $\mathbf{u}$ is
$\mathbf{u} \bullet \nabla f(1, -2) = 1 - 2\sqrt{3}$.

14. $f(x, y) = \ln |\mathbf{r}|$, where $\mathbf{r} = x\mathbf{i} + y\mathbf{j}$. Since $|\mathbf{r}| = \sqrt{x^2 + y^2}$, we have
$$\nabla f(x, y) = \dfrac{1}{|\mathbf{r}|}\left(\dfrac{x}{|\mathbf{r}|}\mathbf{i} + \dfrac{y}{|\mathbf{r}|}\mathbf{j}\right) = \dfrac{\mathbf{r}}{|\mathbf{r}|^2}.$$

15. $f(x, y, z) = |\mathbf{r}|^{-n}$, where $\mathbf{r} = x\mathbf{i} + y\mathbf{j} + z\mathbf{k}$. Since $|\mathbf{r}| = \sqrt{x^2 + y^2 + z^2}$, we have
$$\nabla f(x, y, z) = -n|\mathbf{r}|^{-n-1}\left(\dfrac{x}{|\mathbf{r}|}\mathbf{i} + \dfrac{y}{|\mathbf{r}|}\mathbf{j} + \dfrac{z}{|\mathbf{r}|}\mathbf{k}\right)$$
$$= -\dfrac{n\mathbf{r}}{|\mathbf{r}|^{n+2}}.$$

16. Since $x = r\cos\theta$ and $y = r\sin\theta$, we have
$$\dfrac{\partial f}{\partial r} = \cos\theta\dfrac{\partial f}{\partial x} + \sin\theta\dfrac{\partial f}{\partial y}$$
$$\dfrac{\partial f}{\partial \theta} = -r\sin\theta\dfrac{\partial f}{\partial x} + r\cos\theta\dfrac{\partial f}{\partial y}.$$
Also,
$$\hat{\mathbf{r}} = \dfrac{x\mathbf{i} + y\mathbf{j}}{r} = (\cos\theta)\mathbf{i} + (\sin\theta)\mathbf{j}$$
$$\hat{\boldsymbol{\theta}} = \dfrac{-y\mathbf{i} + x\mathbf{j}}{r} = -(\sin\theta)\mathbf{i} + (\cos\theta)\mathbf{j}.$$
Therefore,
$$\dfrac{\partial f}{\partial r}\hat{\mathbf{r}} + \dfrac{1}{r}\dfrac{\partial f}{\partial \theta}\hat{\boldsymbol{\theta}}$$
$$= \left(\cos^2\theta\dfrac{\partial f}{\partial x} + \sin\theta\cos\theta\dfrac{\partial f}{\partial y}\right)\mathbf{i}$$
$$+ \left(\cos\theta\sin\theta\dfrac{\partial f}{\partial x} + \sin^2\theta\dfrac{\partial f}{\partial y}\right)\mathbf{j}$$
$$+ \left(\sin^2\theta\dfrac{\partial f}{\partial x} - \sin\theta\cos\theta\dfrac{\partial f}{\partial y}\right)\mathbf{i}$$
$$+ \left(-\cos\theta\sin\theta\dfrac{\partial f}{\partial x} + \cos^2\theta\dfrac{\partial f}{\partial y}\right)\mathbf{j}$$
$$= \dfrac{\partial f}{\partial x}\mathbf{i} + \dfrac{\partial f}{\partial y}\mathbf{j} = \nabla f.$$

17. $f(x, y) = xy$, $\quad \nabla f(x, y) = y\mathbf{i} + x\mathbf{j}$, $\quad \nabla f(2, 0) = 2\mathbf{j}$.
Let $\mathbf{u} = u_1\mathbf{i} + u_2\mathbf{j}$ be a unit vector. Thus $u_1^2 + u_2^2 = 1$. We have
$$-1 = D_\mathbf{u}f(2, 0)\mathbf{u} \bullet \nabla f(2, 0) = 2u_2$$
if $u_2 = -\dfrac{1}{2}$, and therefore $u_1 = \pm\dfrac{\sqrt{3}}{2}$. At $(2, 0)$, f has rate of change -1 in the directions $\pm\dfrac{\sqrt{3}}{2}\mathbf{i} - \dfrac{1}{2}\mathbf{j}$.

If $-3 = D_\mathbf{u}f(2, 0) = 2u_2$, then $u_2 = -\dfrac{3}{2}$. This is not possible for a unit vector $\mathbf{u}$, so there is no direction at $(2, 0)$ in which f changes at rate -3.

If $-2 = D_\mathbf{u}f(2, 0) = 2u_2$, then $u_2 = -1$ and $u_1 = 0$. At $(2, 0)$, f has rate of change -2 in the direction $-\mathbf{j}$.

451

18. $f(x, y, z) = x^2 + y^2 - z^2$.
$\nabla f(a, b, c) = 2a\mathbf{i} + 2b\mathbf{j} - 2c\mathbf{k}$. The maximum rate of change of f at (a, b, c) is in the direction of $\nabla f(a, b, c)$, and is equal to $|\nabla f(a, b, c)|$.
Let $\mathbf{u}$ be a unit vector making an angle θ with $\nabla f(a, b, c)$. The rate of change of f at (a, b, c) in the direction of $\mathbf{u}$ will be half of the maximum rate of change of f at that point provided

$$\frac{1}{2}|\nabla f(a, b, c)| = \mathbf{u} \bullet \nabla f(a, b, c) = |\nabla f(a, b, c)| \cos\theta,$$

that is, if $\cos\theta = \frac{1}{2}$, which means $\theta = 60°$. At (a, b, c), f increases at half its maximal rate in all directions making $60°$ angles with the direction $a\mathbf{i} + b\mathbf{j} - c\mathbf{k}$.

19. Let $\nabla f(a, b) = u\mathbf{i} + v\mathbf{j}$. Then

$$3\sqrt{2} = D_{(\mathbf{i}+\mathbf{j})/\sqrt{2}} f(a, b) = \frac{\mathbf{i}+\mathbf{j}}{\sqrt{2}} \bullet (u\mathbf{i}+v\mathbf{j}) = \frac{u+v}{\sqrt{2}}$$

$$5 = D_{(3\mathbf{i}-4\mathbf{j})/5} f(a, b) = \frac{3\mathbf{i}-4\mathbf{j}}{5} \bullet (u\mathbf{i}+v\mathbf{j}) = \frac{3u-4v}{5}.$$

Thus $u + v = 6$ and $3u - 4v = 25$. This system has solution $u = 7$, $v = -1$. Thus $\nabla f(a, b) = 7\mathbf{i} - \mathbf{j}$.

20. Given the values $D_{\phi_1} f(a, b)$ and $D_{\phi_2} f(a, b)$, we can solve the equations

$$f_1(a, b) \cos\phi_1 + f_2(a, b) \sin\phi_1 = D_{\phi_1} f(a, b)$$
$$f_1(a, b) \cos\phi_2 + f_2(a, b) \sin\phi_2 = D_{\phi_2} f(a, b)$$

for unique values of $f_1(a, b)$ and $f_2(a, b)$ (and hence determine $\nabla f(a, b)$ uniquely), provided the coefficients satisfy

$$0 \neq \begin{vmatrix} \cos\phi_1 & \sin\phi_1 \\ \cos\phi_2 & \sin\phi_2 \end{vmatrix} = \sin(\phi_2 - \phi_1).$$

Thus ϕ_1 and ϕ_2 must not differ by an integer multiple of π.

21. a) $T(x, y) = x^2 - 2y^2$.

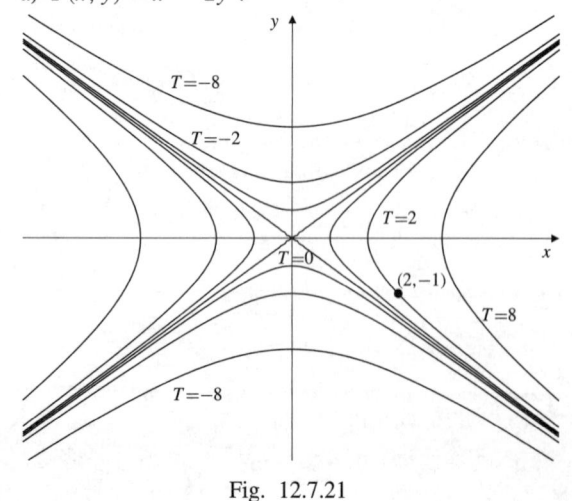

Fig. 12.7.21

b) $\nabla T = 2x\mathbf{i} - 4y\mathbf{j}$, $\nabla T(2, -1) = 4\mathbf{i} + 4\mathbf{j}$.
An ant at $(2, -1)$ should move in the direction of $-\nabla T(2, -1)$, that is, in the direction $-\mathbf{i} - \mathbf{j}$, in order to cool off as rapidly as possible.

c) If the ant moves at speed k in the direction $-\mathbf{i} - \mathbf{j}$, it will experience temperature decreasing at rate $|\nabla T(2, -1)|k = 4\sqrt{2}k$ degrees per unit time.

d) If the ant moves at speed k in the direction $-\mathbf{i} - 2\mathbf{j}$, it experiences temperature changing at rate

$$\frac{-\mathbf{i} - 2\mathbf{j}}{\sqrt{5}} \bullet (4\mathbf{i} + 4\mathbf{j})k = -\frac{12k}{\sqrt{5}},$$

that is, decreasing at rate $12k/\sqrt{5}$ degrees per unit time.

e) To continue to experience maximum rate of cooling, the ant should crawl along the curve $x = x(t)$, $y = y(t)$, which is everywhere tangent to $\nabla T(x, y)$. Thus we want

$$\frac{dx}{dt}\mathbf{i} + \frac{dy}{dt}\mathbf{j} = \lambda(2x\mathbf{i} - 4y\mathbf{j}).$$

Thus $\frac{1}{y}\frac{dy}{dt} = -\frac{2}{x}\frac{dx}{dt}$, from which we obtain, on integration,

$$\ln|y(t)| = -2\ln|x(t)| + \ln|C|,$$

or $yx^2 = C$. Since the curve passes through $(2, -1)$, we have $yx^2 = -4$. Thus, the ant should crawl along the path $y = -4/x^2$.

22. Let the curve be $y = g(x)$. At (x, y) this curve has normal $\nabla(g(x) - y) = g'(x)\mathbf{i} - \mathbf{j}$.
A curve of the family $x^4 + y^2 = C$ has normal $\nabla(x^4 + y^2) = 4x^3\mathbf{i} + 2y\mathbf{j}$.
These curves will intersect at right angles if their normals are perpendicular. Thus we require that

$$0 = 4x^3 g'(x) - 2y = 4x^3 g'(x) - 2g(x),$$

or, equivalently,

$$\frac{g'(x)}{g(x)} = \frac{1}{2x^3}.$$

Integration gives $\ln|g(x)| = -\frac{1}{4x^2} + \ln|C|$,

or $g(x) = Ce^{-(1/4x^2)}$.
Since the curve passes through $(1, 1)$, we must have $1 = g(1) = Ce^{-1/4}$, so $C = e^{1/4}$.
The required curve is $y = e^{(1/4)-(1/4x^2)}$.

23. Let the curve be $y = f(x)$. At (x, y) it has normal $\frac{dy}{dx}\mathbf{i} - \mathbf{j}$.
The curve $x^2y^3 = K$ has normal $2xy^3\mathbf{i} + 3x^2y^2\mathbf{j}$.
These curves will intersect at right angles if their normals are perpendicular, that is, if

$$2xy^3 \frac{dy}{dx} - 3x^2y^2 = 0$$
$$\frac{dy}{dx} = \frac{3x}{2y}$$
$$2y\, dy = 3x\, dx$$
$$y^2 = \frac{3}{2}x^2 + C.$$

Since the curve must pass through $(2, -1)$, we have $1 = 6 + C$, so $C = -5$.
The required curve is $3x^2 - 2y^2 = 10$.

24. Let $f(x, y) = e^{-(x^2+y^2)}$. Then

$$\nabla f(x, y) = -2e^{-(x^2+y^2)}(x\mathbf{i} + y\mathbf{j}).$$

The vector $\mathbf{u} = \dfrac{a\mathbf{i} + b\mathbf{j}}{\sqrt{a^2 + b^2}}$ is a unit vector in the direction directly away from the origin at (a, b).
The first directional derivative of f at (x, y) in the direction of $\mathbf{u}$ is

$$\mathbf{u} \bullet \nabla f(x, y) = -\frac{2}{\sqrt{a^2 + b^2}}(ax + by)e^{-(x^2+y^2)}.$$

The second directional derivative is

$$\mathbf{u} \bullet \nabla \left(-\frac{2}{\sqrt{a^2+b^2}}(ax+by)e^{-(x^2+y^2)} \right)$$
$$= -\frac{2}{a^2+b^2}(a\mathbf{i}+b\mathbf{j}) \bullet e^{-(x^2+y^2)}$$
$$\left[\big(a - 2x(ax+by)\big)\mathbf{i} + \big(b - 2y(ax+by)\big)\mathbf{j} \right].$$

At (a, b) this second directional derivative is

$$-\frac{2e^{-(a^2+b^2)}}{a^2+b^2}\big(a^2 - 2a^4 - 2a^2b^2 + b^2 - 2a^2b^2 - 2b^4\big)$$
$$= \frac{2}{a^2+b^2}\big(2(a^2+b^2)^2 - a^2 - b^2\big)e^{-(a^2+b^2)}$$
$$= 2\big(2(a^2+b^2) - 1\big)e^{-(a^2+b^2)}.$$

Remark: Since $f(x, y) = e^{-r^2}$ (expressed in terms of polar coordinates), the second directional derivative of f at (a, b) in the direction directly away from the origin (i.e., the direction of increasing r) can be more easily calculated as

$$\left. \frac{d^2}{dr^2} e^{-r^2} \right|_{r^2 = a^2+b^2}.$$

25. $f(x, y, z) = xyz$, $\nabla f(x, y, z) = yz\mathbf{i} + xz\mathbf{j} + xy\mathbf{k}$.
The first directional derivative of f in the direction $\mathbf{i} - \mathbf{j} - \mathbf{k}$ is

$$\frac{\mathbf{i} - \mathbf{j} - \mathbf{k}}{\sqrt{3}} \bullet \nabla f(x, y, z) = \frac{1}{\sqrt{3}}(yz - xz - xy).$$

The second directional derivative in that direction is

$$\frac{\mathbf{i} - \mathbf{j} - \mathbf{k}}{\sqrt{3}} \bullet \frac{1}{\sqrt{3}} \nabla(yz - xz - xy)$$
$$= \frac{\mathbf{i} - \mathbf{j} - \mathbf{k}}{3} \bullet \big[-(y+z)\mathbf{i} + (z-x)\mathbf{j} + (y-x)\mathbf{k} \big]$$
$$= \frac{1}{3}\big[-(y+z) - (z-x) - (y-x) \big] = \frac{2x - 2y - 2z}{3}.$$

At $(2, 3, 1)$ this second directional derivative has value $-4/3$.

26. At $(1, -1, 1)$ the surface $x^2 + y^2 = 2$ has normal

$$\mathbf{n}_1 = \nabla(x^2 + y^2)\Big|_{(1,-1,1)} = 2\mathbf{i} - 2\mathbf{j},$$

and $y^2 + z^2 = 2$ has normal

$$\mathbf{n}_2 = \nabla(y^2 + z^2)\Big|_{(1,-1,1)} = -2\mathbf{j} + 2\mathbf{k}.$$

A vector tangent to the curve of intersection of the two surfaces at $(1, -1, 1)$ must be perpendicular to both these normals. Since

$$(\mathbf{i} - \mathbf{j}) \times (-\mathbf{j} + \mathbf{k}) = -(\mathbf{i} + \mathbf{j} + \mathbf{k}),$$

the vector $\mathbf{i} + \mathbf{j} + \mathbf{k}$, or any scalar multiple of this vector, is tangent to the curve at the given point.

27. The vector $\mathbf{n}_1 = \mathbf{i} + \mathbf{j} + \mathbf{k}$ is normal to the plane $x + y + z = 6$ at $(1, 2, 3)$. A normal to the sphere $x^2 + y^2 + z^2 = 14$ at that point is

$$\mathbf{n}_2 = \nabla(x^2 + y^2 + z^2)\Big|_{(1,2,3)} = 2\mathbf{i} + 4\mathbf{j} + 6\mathbf{k}.$$

A vector tangent to the circle of intersection of the two surfaces at $(1, 2, 3)$ is

$$\mathbf{n}_1 \times \mathbf{n}_2 = \begin{vmatrix} \mathbf{i} & \mathbf{j} & \mathbf{k} \\ 1 & 1 & 1 \\ 2 & 4 & 6 \end{vmatrix} = 2\mathbf{i} - 4\mathbf{j} + 2\mathbf{k}.$$

Any vector parallel to $\mathbf{i} - 2\mathbf{j} + \mathbf{k}$ is tangent to the circle at $(1, 2, 3)$.

28. A vector tangent to the path of the fly at $(1, 1, 2)$ is given by

$$\mathbf{v} = \nabla(3x^2 - y^2 - z) \times \nabla(2x^2 + 2y^2 - z^2)\Big|_{(1,1,2)}$$
$$= (6x\mathbf{i} - 2y\mathbf{j} - \mathbf{k}) \times (4x\mathbf{i} + 4y\mathbf{j} - 2z\mathbf{k})\Big|_{(1,1,2)}$$
$$= (6\mathbf{i} - 2\mathbf{j} - \mathbf{k}) \times (4\mathbf{i} + 4\mathbf{j} - 4\mathbf{k})$$
$$= 4\begin{vmatrix} \mathbf{i} & \mathbf{j} & \mathbf{k} \\ 6 & -2 & -1 \\ 1 & 1 & -1 \end{vmatrix} = 4(3\mathbf{i} + 5\mathbf{j} + 8\mathbf{k}).$$

The temperature $T = x^2 - y^2 + z^2 + xz^2$ has gradient at $(1, 1, 2)$ given by

$$\nabla T(1, 1, 2) = (2x + z^2)\mathbf{i} - 2y\mathbf{j} + 2z(1 + x)\mathbf{k}\Big|_{(1,1,2)}$$
$$= 6\mathbf{i} - 2\mathbf{j} + 8\mathbf{k}.$$

Thus the fly, passing through $(1, 1, 2)$ with speed 7, experiences temperature changing at rate

$$7 \times \frac{\mathbf{v}}{|\mathbf{v}|} \bullet \nabla T(1, 1, 2) = 7\frac{3\mathbf{i} + 5\mathbf{j} + 8\mathbf{k}}{\sqrt{98}} \bullet (6\mathbf{i} - 2\mathbf{j} + 8\mathbf{k})$$
$$= \frac{1}{\sqrt{2}}(18 - 10 + 64) = \frac{72}{\sqrt{2}}.$$

We don't know which direction the fly is moving along the curve, so all we can say is that it experiences temperature changing at rate $36\sqrt{2}$ degrees per unit time.

29. If $f(x, y, z)$ is differentiable at the point (a, b, c) and $\nabla f(a, b, c) \neq \mathbf{0}$, then $\nabla f(a, b, c)$ is normal to the level surface of f which passes through (a, b, c).

The proof is very similar to that of Theorem 6 of Section 12.7, modified to include the extra variable. The angle θ between $\nabla f(a, b, c)$ and the secant vector from (a, b, c) to a neighbouring point $(a + h, b + k, c + \ell)$ on the level surface of f passing through (a, b, c) satisfies

$$\cos\theta = \frac{\nabla f(a, b, c) \bullet (h\mathbf{i} + k\mathbf{j} + \ell\mathbf{k})}{|\nabla f(a, b, c)|\sqrt{h^2 + k^2 + \ell^2}}$$
$$= \frac{hf_1(a, b, c) + kf_2(a, b, c) + \ell f_3(a, b, c)}{|\nabla f(a, b, c)|\sqrt{h^2 + k^2 + \ell^2}}$$
$$= \frac{-1}{|\nabla f(a, b, c)|\sqrt{h^2 + k^2 + \ell^2}}\Big[f(a + h, b + k, c + \ell)$$
$$\qquad - f(a, b, c) - hf_1(a, b, c) - kf_2(a, b, c) - \ell f_3(a, b, c)\Big]$$
$$\to 0 \quad \text{as } (h, k, \ell) \to (0, 0, 0)$$

because f is differentiable at (a, b, c). Thus $\theta \to \frac{\pi}{2}$, and $\nabla f(a, b, c)$ is normal to the level surface of f through (a, b, c).

30. The level surface of $f(x, y, z) = \cos(x + 2y + 3z)$ through (π, π, π) has equation $\cos(x + 2y + 3z) = \cos(6\pi) = 1$, which simplifies to $x + 2y + 3z = 6\pi$. This level surface is a plane, and is therefore its own tangent plane. We cannot determine this plane by the method used to find the tangent plane to the level surface of f through $(\pi/2, \pi, \pi)$ in Exercise 10, because $\nabla f(\pi, \pi, \pi) = \mathbf{0}$, so the gradient does not provide a usable normal vector to define the tangent plane.

31. By the version of the Mean-Value Theorem in Exercise 18 of Section 12.6,

$$f(x, y) = f(0, 0) + xf_1(\theta x, \theta y) + yf_2(\theta x, \theta y)$$

for some θ between 0 and 1. Since ∇f is assumed to vanish throughout the disk $x^2 + y^2 < r^2$, this implies that $f(x, y) = f(0, 0)$ throughout the disk, that is, f is constant there. (Note that Theorem 3 of Section 12.6 can be used instead of Exercise 18 of Section 12.6 in this argument.)

32. Let $f(x, y) = x^3 - y^2$. Then $\nabla f(x, y) = 3x^2\mathbf{i} - 2y\mathbf{j}$ exists everywhere, but equals $\mathbf{0}$ at $(0, 0)$. The level curve of f passing through $(0, 0)$ is $y^2 = x^3$, which has a cusp at $(0, 0)$, so is not smooth there.

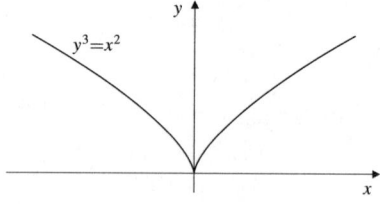

Fig. 12.7.32

33. Let $\mathbf{v} = v_1\mathbf{i} + v_2\mathbf{j} + v_3\mathbf{k}$. Thus

$$D_\mathbf{v} f = v_1\frac{\partial f}{\partial x} + v_2\frac{\partial f}{\partial y} + v_3\frac{\partial f}{\partial z}$$
$$\nabla(D_\mathbf{v} f) = \left(v_1\frac{\partial^2 f}{\partial x^2} + v_2\frac{\partial^2 f}{\partial x\partial y} + v_3\frac{\partial^2 f}{\partial x\partial z}\right)\mathbf{i}$$
$$\qquad + \left(v_1\frac{\partial^2 f}{\partial y\partial x} + v_2\frac{\partial^2 f}{\partial y^2} + v_3\frac{\partial^2 f}{\partial y\partial z}\right)\mathbf{j}$$
$$\qquad + \left(v_1\frac{\partial^2 f}{\partial x\partial z} + v_2\frac{\partial^2 f}{\partial y\partial z} + v_3\frac{\partial^2 f}{\partial z^2}\right)\mathbf{k}$$
$$D_\mathbf{v}(D_\mathbf{v} f) = \mathbf{v} \bullet \nabla(D_\mathbf{v} f)$$
$$= v_1^2\frac{\partial^2 f}{\partial x^2} + 2v_1 v_2\frac{\partial^2 f}{\partial x\partial y} + 2v_1 v_3\frac{\partial^2 f}{\partial x\partial z}$$
$$\qquad + v_2^2\frac{\partial^2 f}{\partial y^2} + 2v_2 v_3\frac{\partial^2 f}{\partial y\partial z} + v_3^2\frac{\partial^2 f}{\partial z^2}$$

(assuming all second partials are continuous).

$D_\mathbf{v}(D_\mathbf{v} f)$ gives the second time derivative of the quantity f as measured by an observer moving with constant velocity $\mathbf{v}$.

34. $T = T(x, y, z)$. As measured by the observer,

$$\frac{dT}{dt} = D_{\mathbf{v}(t)}T = \mathbf{v}(t) \bullet \nabla T$$

$$\frac{d^2T}{dt^2} = \mathbf{a}(t) \bullet \nabla T + \mathbf{v}(t) \bullet \frac{d}{dt}\nabla T$$

$$= D_{\mathbf{a}(t)}T + \left(v_1(t)\frac{d}{dt}\frac{\partial T}{\partial x} + \cdots\right)$$

$$= D_{\mathbf{a}(t)}T + \left(v_1(t)\mathbf{v}(t) \bullet \nabla \frac{\partial T}{\partial x} + \cdots\right)$$

$$= D_{\mathbf{a}(t)}T + \left(\left(v_1(t)\right)^2\frac{\partial^2 T}{\partial x^2} + v_1(t)v_2(t)\frac{\partial^2 T}{\partial y \partial x} + \cdots\right)$$

$$= D_{\mathbf{a}(t)}T + D_{\mathbf{v}(t)}(D_{\mathbf{v}(t)}T)$$

(as in Exercise 37 above).

35. $T = T(x, y, z, t)$. The calculation is similar to that of Exercise 38, but produces a few more terms because of the dependence of T explicitly on time t. We continue to use ∇ to denote the gradient with respect to the spatial variables only. Using the result of Exercise 38, we have

$$\frac{dT}{dt} = \frac{\partial T}{\partial t} + \mathbf{v}(t) \bullet \nabla T$$

$$\frac{d^2T}{dt^2} = \frac{d}{dt}\frac{\partial T}{\partial t} + \frac{d}{dt}\mathbf{v}(t) \bullet \nabla T$$

$$= \frac{\partial^2 T}{\partial t^2} + \mathbf{v}(t) \bullet \frac{\partial T}{\partial t}$$

$$+ \mathbf{v}(t) \bullet \frac{\partial}{\partial t}\nabla T + D_{\mathbf{a}(t)}T + D_{\mathbf{v}(t)}(D_{\mathbf{v}(t)}T)$$

$$= \frac{\partial^2 T}{\partial t^2} + 2D_{\mathbf{v}(t)}\left(\frac{\partial T}{\partial t}\right) + D_{\mathbf{a}(t)}T + D_{\mathbf{v}(t)}(D_{\mathbf{v}(t)}T).$$

36. $f(x, y) = \begin{cases} \dfrac{\sin(xy)}{\sqrt{x^2+y^2}} & \text{if } (x, y) \neq (0, 0) \\ 0 & \text{if } (x, y) = (0, 0) \end{cases}$

a) $f_1(0, 0) = \lim_{h \to 0} \dfrac{0-0}{h} = 0 = f_2(0, 0)$. Thus $\nabla f(0, 0) = \mathbf{0}$.

b) If $\mathbf{u} = (\mathbf{i}+\mathbf{j})/\sqrt{2}$, then

$$D_{\mathbf{u}}f(0, 0) = \lim_{h \to 0+} \frac{1}{h}\frac{\sin(h^2/2)}{\sqrt{h^2}} = \frac{1}{2}.$$

c) f cannot be differentiable at $(0, 0)$; if it were, then the directional derivative obtained in part (b) would have been $\mathbf{u} \bullet \nabla f(0, 0) = 0$.

37. $f(x, y) = \begin{cases} \dfrac{2x^2y}{x^4+y^2} & \text{if } (x, y) \neq (0, 0) \\ 0 & \text{if } (x, y) = (0, 0) \end{cases}$.

Let $\mathbf{u} = u\mathbf{i} + v\mathbf{j}$ be a unit vector. If $v \neq 0$, then

$$D_{\mathbf{u}}f(0, 0) = \lim_{h \to 0+} \frac{1}{h}\frac{2(h^2u^2)(hv)}{h^4u^4+h^2v^2}$$

$$= \lim_{h \to 0+} \frac{2u^2v}{h^2u^4+v^2} = \frac{2u^2}{v}.$$

If $v = 0$, then $u = \pm 1$ and

$$D_{\mathbf{u}}f(0, 0) = \lim_{h \to 0+} \frac{1}{h}\frac{0}{h^2} = 0.$$

Thus f has a directional derivative in every direction at the origin even though it is not continuous there.

Section 12.8 Implicit Functions (page 760)

1. $xy^3 + x^4y = 2$ defines x as a function of y.

$$y^3\frac{dx}{dy} + 3xy^2 + 4x^3y\frac{dx}{dy} + x^4 = 0$$

$$\frac{dx}{dy} = -\frac{x^4+3xy^2}{y^3+4x^3y}.$$

The given equation has a solution $x = x(y)$ with this derivative near any point where $y^3 + 4x^3y \neq 0$, i.e., $y \neq 0$ and $y^2 + 4x^3 \neq 0$.

2. $xy^3 = y - z$: $\quad x = x(y, z)$

$$y^3\frac{\partial x}{\partial y} + 3xy^2 = 1$$

$$\frac{\partial x}{\partial y} = \frac{1-3xy^2}{y^3}.$$

The given equation has a solution $x = x(y, z)$ with this partial derivative near any point where $y \neq 0$.

3. $z^2 + xy^3 = \dfrac{xz}{y}$: $\quad z = z(x, y)$

$$2x\frac{\partial z}{\partial y} + 3xy^2 = \frac{x}{y}\frac{\partial z}{\partial y} - \frac{xz}{y^2}$$

$$\frac{\partial z}{\partial y} = \frac{\dfrac{xz}{y^2}+3xy^2}{\dfrac{x}{y}-2z} = \frac{xz+3xy^4}{xy-2y^2z}.$$

The given equation has a solution $z = z(x, y)$ with this derivative near any point where $y \neq 0$ and $x \neq 2yz$.

4. $e^{yz} - x^2 \ln y = \pi$: $\quad y = y(x, z)$

$$e^{yz}\left(z\frac{\partial y}{\partial z} + y\right) - x^2 \ln y - \frac{x^2z}{y}\frac{\partial y}{\partial z} = 0$$

$$\frac{\partial y}{\partial z} = \frac{x^2 \ln y - ye^{yz}}{ze^{yz}-\dfrac{x^2z}{y}} = \frac{x^2y \ln y - y^2e^{yz}}{yze^{yz}-x^2z}.$$

The given equation has a solution $y = y(x, z)$ with this derivative near any point where $y > 0$, $z \neq 0$, and $ye^{yz} \neq x^2$.

5. $x^2y^2 + y^2z^2 + z^2t^2 + t^2w^2 - xw = 0$: $x = x(y, z, t, w)$

$$2xy^2\frac{\partial x}{\partial w} + 2t^2w - w\frac{\partial x}{\partial w} - x = 0$$

$$\frac{\partial x}{\partial w} = \frac{x - 2t^2w}{2xy^2 - w}.$$

The given equation has a solution with this derivative wherever $w \neq 2xy^2$.

6. $F(x, y, x^2 - y^2) = 0$: $y = y(x)$

$$F_1 + F_2\frac{dy}{dx} + F_3\left(2x - 2y\frac{dy}{dx}\right) = 0$$

$$\frac{dy}{dx} = \frac{F_1(x, y, x^2 - y^2) + 2xF_3(x, y, x^2 - y^2)}{2yF_3(x, y, x^2 - y^2) - F_2(x, y, x^2 - y^2)}.$$

The given equation has a solution with this derivative near any point where $F_2(x, y, x^2 - y^2) \neq 2yF_3(x, y, x^2 - y^2)$.

7. $G(x, y, z, u, v) = 0$: $u = u(x, y, z, v)$

$$G_1 + G_4\frac{\partial u}{\partial x}$$

$$\frac{\partial u}{\partial x} = -\frac{G_1(x, y, z, u, v)}{G_4(x, y, z, u, v)}.$$

The given equation has a solution with this derivative near any point where $G_4(x, y, z, u, v) \neq 0$.

8. $F(x^2 - z^2, y^2 + xz) = 0$: $z = z(x, y)$

$$F_1\left(2x - 2z\frac{\partial z}{\partial x}\right) + F_2\left(x\frac{\partial z}{\partial x} + z\right) = 0$$

$$\frac{\partial z}{\partial x} = \frac{2xF_1(x^2 - z^2, y^2 + xz) + zF_2(x^2 - z^2, y^2 + xz)}{2zF_1(x^2 - z^2, y^2 + xz) - xF_2(x^2 - z^2, y^2 + xz)}.$$

The given equation has a solution with this derivative near any point where $xF_2(x^2 - z^2, y^2 + xz) \neq 2zF_1(x^2 - z^2, y^2 + xz)$.

9. $H(u^2w, v^2t, wt) = 0$: $w = w(u, v, t)$

$$H_1u^2\frac{\partial w}{\partial t} + H_2v^2 + H_3\left(t\frac{\partial w}{\partial t} + w\right) = 0$$

$$\frac{\partial w}{\partial t} = -\frac{H_2(u^2w, v^2t, wt)v^2 + H_3(u^2w, v^2t, wt)w}{H_1(u^2w, v^2t, wt)u^2 + H_3(u^2w, v^2t, wt)t}.$$

The given equation has a solution with this derivative near any point where $tH_3(u^2w, v^2t, wt) \neq -u^2H_1(u^2w, v^2t, wt)$.

10. $\begin{cases} xyuv = 1 \\ x + y + u + v = 0 \end{cases} \Rightarrow \begin{cases} y = y(x, u) \\ v = v(x, u) \end{cases}$

Differentiate the given equations with respect to x:

$$yuv + xuv\frac{\partial y}{\partial x} + xyu\frac{\partial v}{\partial x} = 0$$

$$1 + \frac{\partial y}{\partial x} + \frac{\partial v}{\partial x} = 0$$

Multiply the last equation by xyu and subtract the two equations:

$$yuv - xyu + (xuv - xyu)\frac{\partial y}{\partial x} = 0$$

$$\left(\frac{\partial y}{\partial x}\right)_u = \frac{y(x - v)}{x(v - y)}.$$

The given equations have a solution of the indicated form with this derivative near any point where $u \neq 0$, $x \neq 0$ and $y \neq v$.

11. $\begin{cases} x^2 + y^2 + z^2 + w^2 = 1 \\ x + 2y + 3z + 4w = 2 \end{cases} \Rightarrow \begin{cases} x = x(y, z) \\ w = w(y, z) \end{cases}$

$$2x\frac{\partial x}{\partial y} + 2y + 2w\frac{\partial w}{\partial y} = 0 \quad \times 2$$

$$\frac{\partial x}{\partial y} + 2 + 4\frac{\partial w}{\partial y} = 0 \quad \times w$$

$$(4x - w)\frac{\partial x}{\partial y} + 4y - 2w = 0$$

$$\left(\frac{\partial x}{\partial y}\right)_z = \frac{2w - 4y}{4x - w}.$$

The given equations have a solution of the indicated form with this derivative near any point where $w \neq 4x$.

12. $\begin{cases} x^2y + y^2u - u^3 = 0 \\ x^2 + yu = 1 \end{cases} \Rightarrow \begin{cases} u = u(x) \\ y = y(x) \end{cases}$

$$2xy + (x^2 + 2yu)\frac{dy}{dx} + (y^2 - 3u^2)\frac{du}{dx} = 0$$

$$2x + u\frac{dy}{dx} + y\frac{du}{dx} = 0$$

Multiply the first equation by u and the second by $x^2 + 2yu$ and subtract:

$$2x(x^2 + yu) + (x^2y + y^2u + 3u^3)\frac{du}{dx} = 0$$

$$\frac{du}{dx} = -\frac{2x(x^2 + yu)}{3u^3 + x^2y + y^2u} = -\frac{x}{2u^3}.$$

The given equations have a solution with the indicated derivative near any point where $u \neq 0$.

13. $\begin{cases} x = u^3 + v^3 \\ y = uv - v^2 \end{cases} \Rightarrow \begin{cases} u = u(x, y) \\ v = v(x, y) \end{cases}$

Take partials with respect to x:

$$1 = 3u^2\frac{\partial u}{\partial x} + 3v^2\frac{\partial v}{\partial x}$$

$$0 = v\frac{\partial u}{\partial x} + (u - 2v)\frac{\partial v}{\partial x}.$$

At $u = v = 1$ we have

$$1 = 3\frac{\partial u}{\partial x} + 3\frac{\partial v}{\partial x}$$

$$0 = \frac{\partial u}{\partial x} - \frac{\partial v}{\partial x}.$$

INSTRUCTOR'S SOLUTIONS MANUAL SECTION 12.8 (PAGE 760)

Thus $\dfrac{\partial u}{\partial x} = \dfrac{\partial v}{\partial x} = \dfrac{1}{6}$.
Similarly, differentiating the given equations with respect to y and putting $u = v = 1$, we get

$$0 = 3\dfrac{\partial u}{\partial y} + 3\dfrac{\partial v}{\partial y}$$
$$1 = \dfrac{\partial u}{\partial y} - \dfrac{\partial v}{\partial y}.$$

Thus $\dfrac{\partial u}{\partial y} = -\dfrac{\partial v}{\partial y} = \dfrac{1}{2}$.
Finally,

$$\dfrac{\partial(u, v)}{\partial(x, y)} = \begin{vmatrix} \tfrac{1}{6} & \tfrac{1}{6} \\ \tfrac{1}{2} & -\tfrac{1}{2} \end{vmatrix} = -\dfrac{1}{6}.$$

14. $\begin{cases} x = r^2 + 2s \\ y = s^2 - 2r \end{cases}$

$$\dfrac{\partial(x, y)}{\partial(r, s)} = \begin{vmatrix} 2r & 2 \\ -2 & 2s \end{vmatrix} = 4(rs + 1).$$

The given system can be solved for r and s as functions of x and y near any point (r, s) where $rs \neq -1$.
We have
$$1 = 2r\dfrac{\partial r}{\partial x} + 2\dfrac{\partial s}{\partial x}$$
$$0 = -2\dfrac{\partial r}{\partial x} + 2s\dfrac{\partial s}{\partial x}$$
$$0 = 2r\dfrac{\partial r}{\partial y} + 2\dfrac{\partial s}{\partial y}$$
$$1 = -2\dfrac{\partial r}{\partial y} + 2s\dfrac{\partial s}{\partial y}.$$

Thus
$$\dfrac{\partial r}{\partial x} = \dfrac{s}{2(rs+1)} \qquad \dfrac{\partial r}{\partial y} = -\dfrac{1}{2(rs+1)}$$
$$\dfrac{\partial s}{\partial x} = \dfrac{1}{2(rs+1)} \qquad \dfrac{\partial s}{\partial y} = \dfrac{r}{2(rs+1)}.$$

15. $x = r\cos\theta, \qquad y = r\sin\theta$

$$\dfrac{\partial(x, y)}{\partial(r, \theta)} = \begin{vmatrix} \cos\theta & -r\sin\theta \\ \sin\theta & r\cos\theta \end{vmatrix} = r.$$

The transformation is one-to-one (and hence invertible) near any point where $r \neq 0$, that is, near any point except the origin.

16. $x = \rho\sin\phi\cos\theta$, $y = \rho\sin\phi\sin\theta$, $z = \rho\cos\phi$.

$$\dfrac{\partial(x, y, z)}{\partial(\rho, \phi, \theta)} = \begin{vmatrix} \sin\phi\cos\theta & \rho\cos\phi\cos\theta & -\rho\sin\phi\sin\theta \\ \sin\phi\sin\theta & \rho\cos\phi\sin\theta & \rho\sin\phi\cos\theta \\ \cos\phi & -\rho\sin\phi & 0 \end{vmatrix}$$
$$= \cos\phi \begin{vmatrix} \rho\cos\phi\cos\theta & -\rho\sin\phi\sin\theta \\ \rho\cos\phi\sin\theta & \rho\sin\phi\cos\theta \end{vmatrix}$$
$$+ \rho\sin\phi \begin{vmatrix} \sin\phi\cos\theta & -\rho\sin\phi\sin\theta \\ \sin\phi\sin\theta & \rho\sin\phi\cos\theta \end{vmatrix}$$
$$= \rho^2\cos\phi\big[\cos\phi\sin\phi\cos^2\theta + \sin\phi\cos\phi\sin^2\theta\big]$$
$$+ \rho^2\sin\phi\big[\sin^2\phi\cos^2\theta + \sin^2\phi\sin^2\theta\big]$$
$$= \rho^2\cos^2\phi\sin\phi + \rho^2\sin^3\phi = \rho^2\sin\phi.$$

The transformation is one-to-one (and invertible) near any point where $\rho^2\sin\phi \neq 0$, that is, near any point not on the z-axis.

17. Let $F(x, y, z, u, v) = xy^2 + zu + v^2 - 3$
$G(x, y, z, u, v) = x^3z + 2y - uv - 2$
$H(x, y, z, u, v) = xu + yv - xyz - 1$.
Then
$$\dfrac{\partial(F, G, H)}{\partial(x, y, z)} = \begin{vmatrix} y^2 & 2xy & u \\ 3x^2z & 2 & x^3 \\ u - yz & v - xz & -xy \end{vmatrix}.$$

At point P_0 where $x = y = z = u = v = 1$, we have
$$\dfrac{\partial(F, G, H)}{\partial(x, y, z)} = \begin{vmatrix} 1 & 2 & 1 \\ 3 & 2 & 1 \\ 0 & 0 & -1 \end{vmatrix} = 4.$$

Since this Jacobian is not zero, the equations $F = G = H = 0$ can be solved for x, y, and z as functions of u and v near P_0. Also,

$$\left(\dfrac{\partial y}{\partial u}\right)_v\bigg|_{(1,1)} = -\dfrac{1}{4}\dfrac{\partial(F, G, H)}{\partial(x, u, z)}\bigg|_{P_0}$$
$$= -\dfrac{1}{4}\begin{vmatrix} y^2 & z & u \\ 3x^2z & -v & x^3 \\ u - yz & x & -xy \end{vmatrix}\bigg|_{P_0}$$
$$= -\dfrac{1}{4}\begin{vmatrix} 1 & 1 & 1 \\ 3 & -1 & 1 \\ 0 & 1 & -1 \end{vmatrix} = -\dfrac{3}{2}.$$

18. Let $F(x, y, z, u, v) = xe^y + uz - \cos v - 2$
$G(x, y, z, u, v) = u\cos y + x^2v - yz^2 - 1$.
If P_0 is the point where $(x, y, z) = (2, 0, 1)$ and $(u, v) = (1, 0)$, then

$$\dfrac{\partial(F, G)}{\partial(u, v)}\bigg|_{P_0} = \begin{vmatrix} z & \sin v \\ \cos y & x^2 \end{vmatrix}\bigg|_{P_0}$$
$$= \begin{vmatrix} 1 & 0 \\ 1 & 4 \end{vmatrix} = 4.$$

Since this Jacobian is not zero, the equations $F = G = 0$ can be solved for u, and v in terms of x, y and z near P_0. Also,

$$\left(\dfrac{\partial u}{\partial z}\right)_{x,y}\bigg|_{(2,0,1)} = -\dfrac{1}{4}\dfrac{\partial(F, G)}{\partial(z, v)}\bigg|_{P_0}$$
$$= -\dfrac{1}{4}\begin{vmatrix} u & \sin v \\ -2yz & x^2 \end{vmatrix}\bigg|_{P_0}$$
$$= -\dfrac{1}{4}\begin{vmatrix} 1 & 0 \\ 0 & 4 \end{vmatrix} = -1.$$

457

19. $\begin{cases} F(x,y,z,w) = 0 \\ G(x,y,z,w) = 0 \\ H(x,y,x,w) = 0 \end{cases} \Rightarrow \begin{cases} x = x(y) \\ z = z(y) \\ w = w(y) \end{cases}$

$$F_1 \frac{dx}{dy} + F_2 + F_3 \frac{dz}{dy} + F_4 \frac{dw}{dy} = 0$$
$$G_1 \frac{dx}{dy} + G_2 + G_3 \frac{dz}{dy} + G_4 \frac{dw}{dy} = 0$$
$$H_1 \frac{dx}{dy} + H_2 + H_3 \frac{dz}{dy} + H_4 \frac{dw}{dy} = 0$$

By Cramer's Rule,

$$\frac{dx}{dy} = -\frac{\frac{\partial(F,G,H)}{\partial(y,z,w)}}{\frac{\partial(F,G,H)}{\partial(x,z,w)}}.$$

20. $F(x,y,z,u,v) = 0$
$G(x,y,z,u,v) = 0$
$H(x,y,z,u,v) = 0$

To calculate $\dfrac{\partial x}{\partial y}$ we require that x be one of three dependent variables, and y be one of two independent variables. The other independent variable can be z or u or v. The possible interpretations for this partial, and their values, are

$$\left(\frac{\partial x}{\partial y}\right)_z = -\frac{\frac{\partial(F,G,H)}{\partial(y,u,v)}}{\frac{\partial(F,G,H)}{\partial(x,u,v)}}$$

$$\left(\frac{\partial x}{\partial y}\right)_u = -\frac{\frac{\partial(F,G,H)}{\partial(y,z,v)}}{\frac{\partial(F,G,H)}{\partial(x,z,v)}}$$

$$\left(\frac{\partial x}{\partial y}\right)_v = -\frac{\frac{\partial(F,G,H)}{\partial(y,z,u)}}{\frac{\partial(F,G,H)}{\partial(x,z,u)}}.$$

21. $F(x_1, x_2, \ldots, x_8) = 0$
$G(x_1, x_2, \ldots, x_8) = 0$
$H(x_1, x_2, \ldots, x_8) = 0$

To find $\dfrac{\partial x_1}{\partial x_2}$ we require that x_1 be one of three dependent variables, and that x_2 be one of five independent variables. The other four independent variables must be chosen from among the six remaining variables. This can be done in

$$\binom{6}{4} = \frac{6!}{4!2!} = 15 \text{ ways.}$$

There are 15 possible interpretations for $\dfrac{\partial x_1}{\partial x_2}$. We have

$$\left(\frac{\partial x_1}{\partial x_2}\right)_{x_4 x_6 x_7 x_8} = -\frac{\frac{\partial(F,G,H)}{\partial(x_2, x_3, x_5)}}{\frac{\partial(F,G,H)}{\partial(x_1, x_3, x_5)}}.$$

22. If $F(x,y,z) = 0 \Rightarrow z = z(x,y)$, then

$$F_1 + F_3 \frac{\partial z}{\partial x} = 0, \qquad F_2 + F_3 \frac{\partial z}{\partial y} = 0$$

$$F_{11} + F_{13} \frac{\partial z}{\partial x} + F_{31} \frac{\partial z}{\partial x} + F_{33}\left(\frac{\partial z}{\partial x}\right)^2 + F_3 \frac{\partial^2 z}{\partial x^2} = 0.$$

Thus

$$\frac{\partial^2 z}{\partial x^2} = -\frac{1}{F_3}\left[F_{11} + 2F_{13}\left(-\frac{F_1}{F_3}\right) + F_{33}\left(-\frac{F_1}{F_3}\right)^2\right]$$
$$= -\frac{1}{F_3^3}\left[F_{11}F_3^2 - 2F_1 F_3 F_{13} + F_1^2 F_{33}\right].$$

Similarly,

$$\frac{\partial^2 z}{\partial y^2} = -\frac{1}{F_3^3}\left[F_{22}F_3^2 - 2F_2 F_3 F_{23} + F_2^2 F_{33}\right].$$

Also,

$$F_{12} + F_{13}\frac{\partial z}{\partial y} + \left(F_{32} + F_{33}\frac{\partial z}{\partial y}\right)\frac{\partial z}{\partial x} + F_3 \frac{\partial^2 z}{\partial y \partial x}.$$

Therefore

$$\frac{\partial^2 z}{\partial x \partial y} = -\frac{1}{F_3}\left[F_{12} + F_{13}\left(-\frac{F_2}{F_3}\right) + F_{23}\left(-\frac{F_1}{F_3}\right) \right.$$
$$\left. + F_{33}\left(\frac{F_1 F_2}{F_3^2}\right)\right]$$
$$= -\frac{1}{F_3^3}\left[F_3^2 F_{12} - F_2 F_3 F_{13} - F_1 F_3 F_{23} + F_1 F_2 F_{33}\right].$$

23. $x = u + v$, $y = uv$, $z = u^2 + v^2$.
The first two equations define u and v as functions of x and y, and therefore derivatives of z with respect to x and y can be determined by the Chain Rule.
Differentiate the first two equations with respect to x:

$$1 = \frac{\partial u}{\partial x} + \frac{\partial v}{\partial x}$$
$$0 = v\frac{\partial u}{\partial x} + u\frac{\partial v}{\partial x}.$$

Thus $\dfrac{\partial u}{\partial x} = \dfrac{u}{u-v}$ and $\dfrac{\partial v}{\partial x} = \dfrac{v}{v-u}$, and

$$\dfrac{\partial z}{\partial x} = \dfrac{\partial z}{\partial u}\dfrac{\partial u}{\partial x} + \dfrac{\partial z}{\partial v}\dfrac{\partial v}{\partial x}$$
$$= 2u\dfrac{u}{u-v} + 2v\dfrac{v}{v-u} = \dfrac{2(u^2-v^2)}{u-v} = 2(u+v) = 2x.$$

Similarly, differentiating the first two of the given equations with respect to y, we get

$$0 = \dfrac{\partial u}{\partial y} + \dfrac{\partial v}{\partial y}$$
$$1 = v\dfrac{\partial u}{\partial y} + u\dfrac{\partial v}{\partial y}.$$

Thus $\dfrac{\partial u}{\partial y} = \dfrac{1}{v-u}$ and $\dfrac{\partial v}{\partial y} = \dfrac{1}{u-v}$, and

$$\dfrac{\partial z}{\partial y} = \dfrac{2u}{v-u} + \dfrac{2v}{u-v} = \dfrac{2(u-v)}{v-u} = -2$$
$$\dfrac{\partial^2 z}{\partial x \partial y} = 0.$$

24. $pV = T - \dfrac{4p}{T^2}$, $T = T(p, V)$

a) $V = \dfrac{\partial T}{\partial p} - \dfrac{4}{T^2} + \dfrac{8p}{T^3}\dfrac{\partial T}{\partial p}$

$p = \dfrac{\partial T}{\partial V} + \dfrac{8p}{T^3}\dfrac{\partial T}{\partial V}.$

Putting $p = V = 1$ and $T = 2$, we obtain

$$2\dfrac{\partial T}{\partial p} = 2, \qquad 2\dfrac{\partial T}{\partial V} = 1,$$

so $\dfrac{\partial T}{\partial p} = 1$ and $\dfrac{\partial T}{\partial V} = \dfrac{1}{2}$.

b) $dT = \dfrac{\partial T}{\partial p}dp + \dfrac{\partial T}{\partial V}dV.$
If $p = 1$, $|dp| \leq 0.001$, $V = 1$, and $|dV| \leq 0.002$, then $T = 2$ and

$$|dT| \leq (1)(0.001) + \dfrac{1}{2}(0.002) = 0.002.$$

The approximate maximum error in T is 0.002.

25. $F(x, y, z) = 0$

$$F_1\left(\dfrac{\partial x}{\partial y}\right)_z + F_2 = 0, \quad \Rightarrow \quad \left(\dfrac{\partial x}{\partial y}\right)_z = -\dfrac{F_2}{F_1}.$$

Similarly, $\left(\dfrac{\partial y}{\partial z}\right)_x = -\dfrac{F_3}{F_2}$, and $\left(\dfrac{\partial z}{\partial x}\right)_y = -\dfrac{F_1}{F_3}$. Hence

$$\left(\dfrac{\partial x}{\partial y}\right)_z \left(\dfrac{\partial y}{\partial z}\right)_x \left(\dfrac{\partial z}{\partial x}\right)_y = (-1)^3 = -1.$$

For $F(x, y, z, u) = 0$ we have, similarly,

$$\left(\dfrac{\partial x}{\partial y}\right)_{z,u}\left(\dfrac{\partial y}{\partial z}\right)_{u,x}\left(\dfrac{\partial z}{\partial u}\right)_{x,y}\left(\dfrac{\partial u}{\partial x}\right)_{y,z} = (-1)^4 = 1.$$

For $F(x, y, z, u, v) = 0$ we have, similarly,

$$\left(\dfrac{\partial x}{\partial y}\right)_{z,u,v}\left(\dfrac{\partial y}{\partial z}\right)_{u,v,x}\left(\dfrac{\partial z}{\partial u}\right)_{v,x,y}\left(\dfrac{\partial u}{\partial v}\right)_{x,y,z}\left(\dfrac{\partial v}{\partial x}\right)_{y,z,u}$$
$$= (-1)^5 = -1.$$

In general, if $F(x_1, x_2, \ldots, x_n) = 0$, then

$$\left(\dfrac{\partial x_1}{\partial x_2}\right)_{x_3,\ldots,x_n}\left(\dfrac{\partial x_2}{\partial x_3}\right)_{x_4,\ldots,x_n,x_1}\cdots\left(\dfrac{\partial x_n}{\partial x_1}\right)_{x_2,\ldots,x_{n-1}}$$
$$= (-1)^n.$$

26. Given $F(x, y, u, v) = 0$, $G(x, y, u, v) = 0$, let

$$\Delta = \dfrac{\partial(F, G)}{\partial(x, y)} = \dfrac{\partial F}{\partial x}\dfrac{\partial G}{\partial y} - \dfrac{\partial F}{\partial y}\dfrac{\partial G}{\partial x}.$$

Then, regarding the given equations as defining x and y as functions of u and v, we have

$$\dfrac{\partial x}{\partial u} = -\dfrac{1}{\Delta}\dfrac{\partial(F, G)}{\partial(u, y)} \qquad \dfrac{\partial y}{\partial u} = -\dfrac{1}{\Delta}\dfrac{\partial(F, G)}{\partial(x, u)}$$
$$\dfrac{\partial x}{\partial v} = -\dfrac{1}{\Delta}\dfrac{\partial(F, G)}{\partial(v, y)} \qquad \dfrac{\partial y}{\partial v} = -\dfrac{1}{\Delta}\dfrac{\partial(F, G)}{\partial(x, v)}.$$

Therefore,

$$\dfrac{\partial(x, y)}{\partial(u, v)} = \dfrac{1}{\Delta^2}\left(\dfrac{\partial(F, G)}{\partial(u, y)}\dfrac{\partial(F, G)}{\partial(x, v)} - \dfrac{\partial(F, G)}{\partial(v, y)}\dfrac{\partial(F, G)}{\partial(x, u)}\right)$$
$$= \dfrac{1}{\Delta^2}\left[\left(\dfrac{\partial F}{\partial u}\dfrac{\partial G}{\partial y} - \dfrac{\partial F}{\partial y}\dfrac{\partial G}{\partial u}\right)\left(\dfrac{\partial F}{\partial x}\dfrac{\partial G}{\partial v} - \dfrac{\partial F}{\partial v}\dfrac{\partial G}{\partial x}\right)\right.$$
$$\left. -\left(\dfrac{\partial F}{\partial v}\dfrac{\partial G}{\partial y} - \dfrac{\partial F}{\partial y}\dfrac{\partial G}{\partial v}\right)\left(\dfrac{\partial F}{\partial x}\dfrac{\partial G}{\partial u} - \dfrac{\partial F}{\partial u}\dfrac{\partial G}{\partial x}\right)\right]$$
$$= \dfrac{1}{\Delta^2}\left[\dfrac{\partial F}{\partial u}\dfrac{\partial G}{\partial y}\dfrac{\partial F}{\partial x}\dfrac{\partial G}{\partial v} - \dfrac{\partial F}{\partial y}\dfrac{\partial G}{\partial u}\dfrac{\partial F}{\partial x}\dfrac{\partial G}{\partial v}\right.$$
$$-\dfrac{\partial F}{\partial u}\dfrac{\partial G}{\partial y}\dfrac{\partial F}{\partial v}\dfrac{\partial G}{\partial x} + \dfrac{\partial F}{\partial y}\dfrac{\partial G}{\partial u}\dfrac{\partial F}{\partial v}\dfrac{\partial G}{\partial x}$$
$$-\dfrac{\partial F}{\partial v}\dfrac{\partial G}{\partial y}\dfrac{\partial F}{\partial x}\dfrac{\partial G}{\partial u} + \dfrac{\partial F}{\partial v}\dfrac{\partial G}{\partial y}\dfrac{\partial F}{\partial u}\dfrac{\partial G}{\partial x}$$
$$\left.+\dfrac{\partial F}{\partial y}\dfrac{\partial G}{\partial v}\dfrac{\partial F}{\partial x}\dfrac{\partial G}{\partial u} - \dfrac{\partial F}{\partial y}\dfrac{\partial G}{\partial v}\dfrac{\partial F}{\partial u}\dfrac{\partial G}{\partial x}\right]$$
$$= \dfrac{1}{\Delta^2}\left[\dfrac{\partial F}{\partial u}\dfrac{\partial G}{\partial y}\dfrac{\partial F}{\partial x}\dfrac{\partial G}{\partial v} + \dfrac{\partial F}{\partial y}\dfrac{\partial G}{\partial u}\dfrac{\partial F}{\partial v}\dfrac{\partial G}{\partial x}\right.$$
$$\left.-\dfrac{\partial F}{\partial v}\dfrac{\partial G}{\partial y}\dfrac{\partial F}{\partial x}\dfrac{\partial G}{\partial u} - \dfrac{\partial F}{\partial y}\dfrac{\partial G}{\partial v}\dfrac{\partial F}{\partial u}\dfrac{\partial G}{\partial x}\right]$$
$$= \dfrac{1}{\Delta^2}\left(\dfrac{\partial F}{\partial x}\dfrac{\partial G}{\partial y} - \dfrac{\partial F}{\partial y}\dfrac{\partial G}{\partial x}\right)\left(\dfrac{\partial F}{\partial u}\dfrac{\partial G}{\partial v} - \dfrac{\partial F}{\partial v}\dfrac{\partial G}{\partial u}\right)$$
$$= \dfrac{1}{\Delta^2}\dfrac{\partial(F, G)}{\partial(x, y)}\dfrac{\partial(F, G)}{\partial(u, v)}$$
$$= \dfrac{1}{\Delta}\dfrac{\partial(F, G)}{\partial(u, v)} = \dfrac{\partial(F, G)}{\partial(u, v)}\bigg/\dfrac{\partial(F, G)}{\partial(x, y)}.$$

27. By Exercise 26, with the roles of (x, y) and (u, v) reversed, we have

$$\frac{\partial(u, v)}{\partial(x, y)} = \frac{\partial(F, G)}{\partial(x, y)} \bigg/ \frac{\partial(F, G)}{\partial(u, v)}.$$

Apply this with

$$F(x, y, u, v) = f(u, v) - x = 0$$
$$G(x, y, u, v) = g(u, v) - y = 0$$

so that

$$\frac{\partial(F, G)}{\partial(x, y)} = \begin{vmatrix} -1 & 0 \\ 0 & -1 \end{vmatrix} = 1$$

and

$$\frac{\partial(F, G)}{\partial(u, v)} = \frac{\partial(f, g)}{\partial(u, v)} = \frac{\partial(x, y)}{\partial(u, v)}$$

and we obtain

$$\frac{\partial(u, v)}{\partial(x, y)} = 1 \bigg/ \frac{\partial(x, y)}{\partial(u, v)}.$$

28. By the Chain Rule,

$$\begin{pmatrix} \frac{\partial x}{\partial r} & \frac{\partial x}{\partial s} \\ \frac{\partial y}{\partial r} & \frac{\partial y}{\partial s} \end{pmatrix}$$
$$= \begin{pmatrix} \frac{\partial x}{\partial u}\frac{\partial u}{\partial r} + \frac{\partial x}{\partial v}\frac{\partial v}{\partial r} & \frac{\partial x}{\partial u}\frac{\partial u}{\partial s} + \frac{\partial x}{\partial v}\frac{\partial v}{\partial s} \\ \frac{\partial y}{\partial u}\frac{\partial u}{\partial r} + \frac{\partial y}{\partial v}\frac{\partial v}{\partial r} & \frac{\partial y}{\partial u}\frac{\partial u}{\partial s} + \frac{\partial y}{\partial v}\frac{\partial v}{\partial s} \end{pmatrix}$$
$$= \begin{pmatrix} \frac{\partial x}{\partial u} & \frac{\partial x}{\partial v} \\ \frac{\partial y}{\partial u} & \frac{\partial y}{\partial v} \end{pmatrix} \begin{pmatrix} \frac{\partial u}{\partial r} & \frac{\partial u}{\partial s} \\ \frac{\partial v}{\partial r} & \frac{\partial v}{\partial s} \end{pmatrix}.$$

Since the determinant of a product of matrices is the product of their determinants, we have

$$\frac{\partial(x, y)}{\partial(r, s)} = \frac{\partial(x, y)}{\partial(u, v)} \frac{\partial(u, v)}{\partial(r, s)}.$$

29. If $f(x, y) = k\big(g(x, y)\big)$, then

$$\frac{\partial f}{\partial x} = k'\big(g(x, y)\big) \frac{\partial g}{\partial x}, \quad \frac{\partial f}{\partial y} = k'\big(g(x, y)\big) \frac{\partial g}{\partial y}.$$

Therefore,

$$\frac{\partial(f, g)}{\partial(x, y)} = k'\big(g(x, y)\big) \frac{\partial(g, g)}{\partial(r, s)} = 0.$$

30. Let $u = f(x, y)$ and $v = g(x, y)$, and suppose that

$$\frac{\partial(u, v)}{\partial(x, y)} = \frac{\partial(f, g)}{\partial(x, y)} = 0$$

for all (x, y). Thus

$$\frac{\partial f}{\partial x} \frac{\partial g}{\partial y} - \frac{\partial f}{\partial y} \frac{\partial g}{\partial x} = 0.$$

Now consider the equations $u = f(x, y)$ and $v = g(x, y)$ as defining u and y as functions of x and v. Holding v constant and differentiating with respect to x, we get

$$\frac{\partial g}{\partial x} + \frac{\partial g}{\partial y} \frac{\partial y}{\partial x} = 0,$$

and

$$\left(\frac{\partial u}{\partial x}\right)_v = \frac{\partial f}{\partial x} + \frac{\partial f}{\partial y} \frac{\partial y}{\partial x}$$
$$= \frac{1}{\frac{\partial g}{\partial y}} \left(\frac{\partial f}{\partial x} \frac{\partial g}{\partial y} - \frac{\partial f}{\partial y} \frac{\partial g}{\partial x} \right) = 0.$$

This says that $u = u(x, v)$ is independent of x, and so depends only on v: $u = k(v)$ for some function k of one variable. Thus $f(x, y) = k\big(g(x, y)\big)$, so f and g are functionally dependent.

Section 12.9 Taylor Series and Approximations (page 766)

1. Since the Maclaurin series for $\dfrac{1}{1+t}$ is

$$1 - t + t^2 - \cdots = \sum_{n=0}^{\infty} (-1)^n t^n,$$

the Taylor series for

$$f(x, y) = \frac{1}{2 + xy^2} = \frac{1}{2} \frac{1}{1 + \frac{xy^2}{2}}$$

about $(0, 0)$ is $\displaystyle\sum_{n=0}^{\infty} (-1)^n \frac{x^n y^{2n}}{2^{n+1}}$.

2. Since $f(x, y) = \ln(1 + x + y + xy)$
$$= \ln\big((1+x)(1+y)\big)$$
$$= \ln(1+x) + \ln(1+y),$$
the Taylor series for f about $(0, 0)$ is

$$\sum_{n=1}^{\infty} (-1)^{n-1} \frac{x^n + y^n}{n}.$$

3. Since $f(x,y) = \tan^{-1}(x+xy) = \tan^{-1}(ux)$, where $u = y+1$, the Taylor series for f about $(0,-1)$ is

$$\sum_{n=0}^{\infty}(-1)^n \frac{(ux)^{2n+1}}{2n+1} = \sum_{n=0}^{\infty}(-1)^n \frac{x^{2n+1}(1+y)^{2n+1}}{2n+1}.$$

4. Let $u = x-1$, $v = y+1$. Thus

$$\begin{aligned}f(x,y) &= x^2 + xy + y^3 \\ &= (u+1)^2 + (u+1)(v-1) + (v-1)^3 \\ &= 1+2u+u^2 - 1 + v - u + uv + v^3 - 3v^2 + 3v - 1 \\ &= -1 + u + 4v + u^2 + uv - 3v^2 + v^3 \\ &= -1 + (x-1) + 4(y+1) + (x-1)^2 \\ &\quad + (x-1)(y+1) - 3(y+1)^2 + (y+1)^3.\end{aligned}$$

This is the Taylor series for f about $(1,-1)$.

5. $f(x,y) = e^{x^2+y^2}$

$$= \sum_{n=0}^{\infty} \frac{(x^2+y^2)^n}{n!}$$

$$= \sum_{n=0}^{\infty} \frac{1}{n!} \sum_{j=0}^{n} \frac{n!}{j!(n-j)!} x^{2j} y^{2n-2j}$$

$$= \sum_{n=0}^{\infty} \sum_{j=0}^{n} \frac{x^{2j} y^{2n-2j}}{j!(n-j)!}.$$

This is the Taylor series for f about $(0,0)$.

6. $f(x,y) = \sin(2x+3y) = \sum_{n=0}^{\infty}(-1)^n \frac{(2x+3y)^{2n+1}}{(2n+1)!}$

$$= \sum_{n=0}^{\infty} \frac{(-1)^n}{(2n+1)!} \sum_{j=0}^{2n+1} \frac{(2n+1)!}{j!(2n+1-j)!}(2x)^j (3y)^{2n+1-j}$$

$$= \sum_{n=0}^{\infty} \sum_{j=0}^{2n+1} \frac{(-1)^n 2^j 3^{2n+1-j}}{j!(2n+1-j)!} x^j y^{2n+1-j}.$$

This is the Taylor series for f about $(0,0)$.

7. Let $u = x-2$, $v = y-1$. Then

$$f(x,y) = \frac{1}{2+x-2y} = \frac{1}{2+(2+u)-2(v+1)}$$

$$= \frac{1}{2+u-2v} = \frac{1}{2\left(1 + \frac{u-2v}{2}\right)}$$

$$= \frac{1}{2}\left[1 - \frac{u-2v}{2} + \left(\frac{u-2v}{2}\right)^2 - \left(\frac{u-2v}{2}\right)^3 + \cdots\right]$$

$$= \frac{1}{2} - \frac{u}{4} + \frac{v}{2} + \frac{u^2}{8} - \frac{uv}{2}$$

$$\quad + \frac{v^2}{2} - \frac{u^3}{16} + \frac{3u^2 v}{8} - \frac{3uv^2}{4} + \frac{v^3}{2} + \cdots.$$

The Taylor polynomial of degree 3 for f about $(2,1)$ is

$$\frac{1}{2} - \frac{x-2}{4} + \frac{y-1}{2} + \frac{(x-1)^2}{8}$$
$$- \frac{(x-2)(y-1)}{2} + \frac{(y-1)^2}{2} - \frac{(x-2)^3}{16}$$
$$+ \frac{3(x-2)^2(y-1)}{8} - \frac{3(x-2)(y-1)^2}{4} + \frac{(y-1)^3}{2}.$$

8. Let $u = x-1$. Then

$$\begin{aligned}f(x,y) &= \ln(x^2+y^2) = \ln(1+2u+u^2+y^2) \\ &= (2u+u^2+y^2) - \frac{(2u+u^2+y^2)^2}{2} \\ &\quad + \frac{(2u+u^2+y^2)^3}{3} - \cdots \\ &= 2u + u^2 + y^2 - 2u^2 - 2u^3 - 2uy^2 + \frac{8u^3}{3} + \cdots.\end{aligned}$$

The Taylor polynomial of degree 3 for f near $(1,0)$ is

$$2(x-1) - (x-1)^2 + y^2 - 2(x-1)^3$$
$$- 2(x-1)y^2 + \frac{8}{3}(x-1)^3.$$

9. $f(x,y) = \int_0^{x+y^2} e^{-t^2}\,dt$

$$= \int_0^{x+y^2}\left(1 - t^2 + \cdots\right)dt$$

$$= \left(t - \frac{t^3}{3} + \cdots\right)\Big|_0^{x+y^2}$$

$$= x + y^2 - \frac{1}{3}(x+y^2)^3 + \cdots$$

$$= x + y^2 - \frac{x^3}{3} + \cdots.$$

The Taylor polynomial of degree 3 for f near $(0,0)$ is

$$x + y^2 - \frac{x^3}{3}.$$

10. $f(x,y) = \cos(x + \sin y)$

$$= 1 - \frac{(x+\sin y)^2}{2!} + \frac{(x+\sin y)^4}{4!} - \cdots$$

$$= 1 - \frac{\left(x+y-\frac{y^3}{6}+\cdots\right)^2}{2} + \frac{(x+y-\cdots)^4}{4} - \cdots$$

$$= 1 - \frac{1}{2}\left(x^2+y^2+2xy - \frac{xy^3}{3} - \frac{y^4}{3} + \cdots\right)$$

$$\quad + \frac{1}{4}(x^4 + 4x^3 y + 6x^2 y^2 + 4xy^3 + y^4 + \cdots).$$

The Taylor polynomial of degree 4 for f near $(0,0)$ is

$$1 - \frac{x^2}{2} - xy - \frac{y^2}{2} + \frac{x^4}{4} + x^3 y$$
$$+ \frac{3x^2 y^2}{2} + \frac{7xy^3}{6} + \frac{5y^4}{12}.$$

11. Let $u = x - \frac{\pi}{2}$, $v = y - 1$. Then

$$f(x, y) = \frac{\sin x}{y} = \frac{\sin(u + \pi/2)}{1 + v} = \frac{\cos u}{1 + v}$$
$$= \left(1 - \frac{u^2}{2} + \cdots\right)(1 - v + v^2 - \cdots)$$
$$= 1 - v - \frac{u^2}{2} + v^2 + \cdots.$$

The Taylor polynomial of degree 2 for f near $(\pi/2, 1)$ is

$$1 - (y - 1) - \frac{1}{2}\left(x - \frac{\pi}{2}\right)^2 + (y - 1)^2.$$

12. $f(x, y) = \dfrac{1 + x}{1 + x^2 + y^4}$
$$= (1 + x)\left(1 - (x^2 + y^4) + \cdots\right)$$
$$= 1 + x - x^2 - \cdots.$$

The Taylor polynomial of degree 2 for f near $(0, 0)$ is

$$1 + x - x^2.$$

13. The equation $x \sin y = y + \sin x$ can be written $F(x, y) = 0$ where $F(x, y) = x \sin y - y - \sin x$. Since $F(0, 0) = 0$, and $F_2(0, 0) = -1 \neq 0$, the given equation has a solution of the form $y = f(x)$ where $f(0) = 0$.
Try $y = a_1 x + a_2 x^2 + a_3 x^3 + a_4 x^4 + \cdots$. Then

$$\sin y = y - \frac{1}{6} y^3 + \cdots$$
$$= a_1 x + a_2 x^2 + a_3 x^3 + a_4 x^4 + \cdots - \frac{1}{6}(a_1 x + \cdots)^3 + \cdots.$$

Substituting into the given equation we obtain

$$a_1 x^2 + a_2 x^3 + \left(a_3 - \frac{1}{6} a_1^3\right) x^4 + \cdots$$
$$= a_1 x + a_2 x^2 + a_3 x^3 + a_4 x^4 + \cdots + x - \frac{1}{6} x^3 + \cdots.$$

Comparing coefficients of various powers of x on both sides, we get

$$a_1 + 1 = 0, \quad a_2 = a_1, \quad a_3 - \frac{1}{6} = a_2.$$

Thus $a_1 = -1$, $a_2 = -1$, and $a_3 = -5/6$. The required solution is

$$y = -x - x^2 - \frac{5}{6} x^3 + \cdots.$$

14. The equation $\sqrt{1 + xy} = 1 + x + \ln(1 + y)$ can be rewritten $F(x, y) = 0$, where $F(x, y) = \sqrt{1 + xy} - 1 - x - \ln(1 + y)$. Since $F(0, 0) = 0$ and $F_2(0, 0) = -1 \neq 0$, the given equation has a solution of the form $y = f(x)$ where $f(0) = 0$.

Try $y = a_1 x + a_2 x^2 + a_3 x^3 + a_4 x^4 + \cdots$. We have

$$\sqrt{1 + xy}$$
$$= \sqrt{1 + a_1 x^2 + a_2 x^3 + a_3 x^4 + \cdots}$$
$$= 1 + \frac{1}{2}(a_1 x^2 + a_2 x^3 + a_3 x^4 + \cdots)$$
$$\quad - \frac{1}{8}(a_1 x^2 + \cdots)^2 + \cdots$$

$1 + x + \ln(1 + y)$
$$= 1 + x + (a_1 x + a_2 x^2 + a_3 x^3 + a_4 x^4 + \cdots)$$
$$\quad - \frac{1}{2}(a_1 x + a_2 x^2 + a_3 x^3 + \cdots)^2 + \frac{1}{3}(a_1 x + a_2 x^2 \cdots)^3 - \cdots$$

Thus we must have

$$0 = 1 + a_1$$
$$\frac{1}{2} a_1 = a_2 - \frac{1}{2} a_1^2$$
$$\frac{1}{2} a_2 = a_3 - a_1 a_2 + \frac{1}{3} a_1^3$$
$$\frac{1}{2} a_3 - \frac{1}{8} a_1^2 = a_4 - \frac{1}{2} a_2^2 - a_1 a_3 + a_1^2 a_2,$$

and $a_1 = -1$, $a_2 = 0$, $a_3 = \dfrac{1}{3}$, $a_4 = -\dfrac{7}{24}$. The required solution is

$$y = -x + \frac{1}{3} x^3 - \frac{7}{24} x^4 + \cdots.$$

15. The equation $x + 2y + z + e^{2z} = 1$ can be written $F(x, y, z) = 0$, where $F(x, y, z) = x + 2y + z + e^{2z} - 1$. Since $F(0, 0, 0) = 0$ and $F_3(0, 0, 0) = 3 \neq 0$, the given equation has a solution of the form $z = f(x, y)$, where $f(0, 0) = 0$.

Try $z = Ax + By + Cx^2 + Dxy + Ey^2 + \cdots$. Then

$$x + 2y + Ax + By + Cx^2 + Dxy + Ey^2 + \cdots$$
$$\quad + 1 + 2(Ax + By + Cx^2 + Dxy + Ey^2 + \cdots)$$
$$\quad + 2(Ax + By + \cdots)^2 + \cdots = 1.$$

Thus

$$1 + A + 2A = 0 \quad \Rightarrow \quad A = -1/3$$
$$2 + B + 2B = 0 \quad \Rightarrow \quad B = -2/3$$
$$C + 2C + 2A^2 = 0 \quad \Rightarrow \quad C = -2/27$$
$$D + 2D + 4AB = 0 \quad \Rightarrow \quad D = -8/27$$
$$E + 2E + 2B^2 = 0 \quad \Rightarrow \quad E = -8/27.$$

The Taylor polynomial of degree 2 for z is

$$-\frac{1}{3} x - \frac{2}{3} y - \frac{2}{27} x^2 - \frac{8}{27} xy - \frac{8}{27} y^2.$$

INSTRUCTOR'S SOLUTIONS MANUAL   REVIEW EXERCISES 12   (PAGE 767)

16. The coefficient of $x^2 y$ in the Taylor series for $f(x, y) = \tan^{-1}(x+y)$ about $(0, 0)$ is

$$\frac{1}{2!1!} f_{112}(0, 0) = \frac{1}{2} f_{112}(0, 0).$$

But

$$\tan^{-1}(x+y) = x + y - \frac{1}{3}(x+y)^3 + \cdots$$
$$= x + y - \frac{1}{3}(x^3 + 3x^2 y + 3xy^2 + y^3) + \cdots$$

so the coefficient of $x^2 y$ is -1. Hence $f_{112}(0, 0) = -2$.

17. Let $f(x, y) = \dfrac{1}{1 + x^2 + y^2}$.

The coefficient of $x^{2n} y^{2n}$ in the Taylor series for $f(x, y)$ about $(0, 0)$ is

$$\frac{1}{(2n)!(2n)!} \left.\frac{\partial^{4n}}{\partial x^{2n} \partial y^{2n}} f(x, y)\right|_{(0,0)}.$$

However,

$$f(x, y) = \sum_{j=0}^{\infty} (-1)^j (x^2 + y^2)^j$$
$$= \sum_{j=0}^{\infty} (-1)^j \sum_{k=0}^{j} \frac{j!}{k!(j-k)!} x^{2k} y^{2j-2k}.$$

The coefficient of $x^{2n} y^{2n}$ is

$$(-1)^{2n} \frac{(2n)!}{n! n!} = \frac{(2n)!}{(n!)^2}.$$

Thus $\left.\dfrac{\partial^{4n}}{\partial x^{2n} \partial y^{2n}} f(x, y)\right|_{(0,0)} = \dfrac{[(2n)!]^3}{(n!)^2}$.

Review Exercises 12   (page 767)

1. $x + \dfrac{4y^2}{x} = C$
   $x^2 + 4y^2 = Cx$
   $\dfrac{(x - (C/2))^2}{(C/2)^2} + \dfrac{y^2}{(C/4)^2} = 1$
   Ellipse: centre $((C/2), 0)$, semi-axes: $C/2$, $C/4$, with the origin deleted.

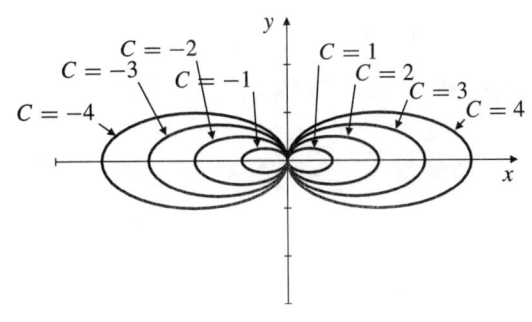

Fig. R-12.1

2. $T = \dfrac{140 + 30x^2 - 60x + 120y^2}{8 + x^2 - 2x + 4y^2}$
   $= 30 - \dfrac{100}{(x-1)^2 + 4y^2 + 7}$
   Ellipses: centre $(1, 0)$, values of T between $30 - (100/7)$ (minimum) at $(1, 0)$ and 30 (at infinite distance from $(1, 0)$).

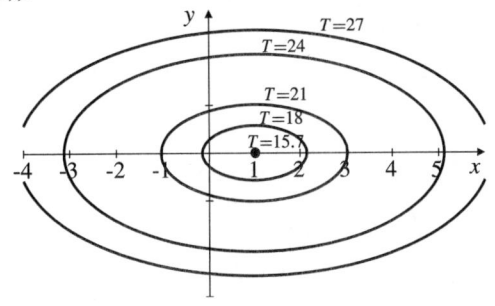

Fig. R-12.2

3. The graph is a saddle-like surface with downward slopes for legs and a tail, thus *monkey saddle*.

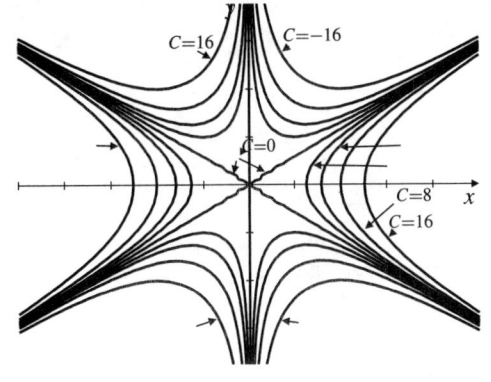

Fig. R-12.3

4. $f(x, y) = \begin{cases} x^3/(x^2 + y^2) & \text{if } (x, y) \neq (0, 0) \\ 0 & \text{if } (x, y) = (0, 0) \end{cases}$.

$f_1(0, 0) = \lim_{h \to 0} \dfrac{(h^3 - 0)/h^2}{h} = 1$

$f_2(0, 0) = \lim_{k \to 0} \dfrac{0 - 0}{k} = 0$.

463

For $(x, y) \neq (0, 0)$, we have

$$f_1(x, y) = \frac{x^4 + 3x^2y^2}{(x^2 + y^2)^2}$$

$$f_2(x, y) = -\frac{2x^3y}{(x^2 + y^2)^2}$$

$$f_{12}(0, 0) = \lim_{k \to 0} \frac{f_1(0, k) - f_1(0, 0)}{k} = \lim_{k \to 0} \frac{0 - 1}{k} \text{ does not exist}$$

$$f_{21}(0, 0) = \lim_{h \to 0} \frac{f_2(h, 0) - f_2(0, 0)}{h} = \lim_{h \to 0} \frac{0 - 0}{h} = 0.$$

5. $f(x, y) = \dfrac{x^3 - y^3}{x^2 - y^2} = \dfrac{(x - y)(x^2 + xy + y^2)}{(x - y)(x + y)}.$

f is continuous except on the lines $x = y$ and $x = -y$ where it is not defined. It has a continuous extension, namely $\dfrac{x^2 + xy + y^2}{x + y}$, to all points of $x = y$ except the origin. It cannot be extended so as to be continuous at the origin. For example, if $(x, y) \to (0, 0)$ along the curve $y = -x + x^4$, then

$$f(x, y) = \frac{x^2 - x^2 + x^5 + (x^4 - x)^2}{x^4} = \frac{x^6 - x^3 + 1}{x^2},$$

which $\to \infty$ as $x \to 0$.

If we define $f(0, 0) = 0$, then

$$f_1(0, 0) = \lim_{h \to 0} \frac{f(h, 0) - f(0, 0)}{h} = \lim_{h \to 0} \frac{h}{h} = 1$$

$$f_2(0, 0) = \lim_{k \to 0} \frac{f(0, k) - f(0, 0)}{k} = \lim_{k \to 0} \frac{k}{k} = 1.$$

6. $f(x, y) = e^{x^2 - 2x - 4y^2 + 5}$ $\qquad f(1, -1) = 1$
   $f_1(x, y) = 2(x - 1)e^{x^2 - 2x - 4y^2 + 5}$ $\qquad f_1(1, -1) = 0$
   $f_2(x, y) = -8ye^{x^2 - 2x - 4y^2 + 5}$ $\qquad f_2(1, -1) = 8.$

   a) The tangent plane to $z = f(x, y)$ at $(1, -1, 1)$ has equation $z = 1 + 8(y + 1)$, or $z = 8y + 9$.

   b) $f(x, y) = C \Rightarrow (x - 1)^2 - 4y^2 + 4 = \ln C$
   $\Rightarrow (x - 1)^2 - 4y^2 = \ln C - 4.$
   These are hyperbolas with centre $(1, 0)$ and asymptotes $x = 1 \pm 2y$.

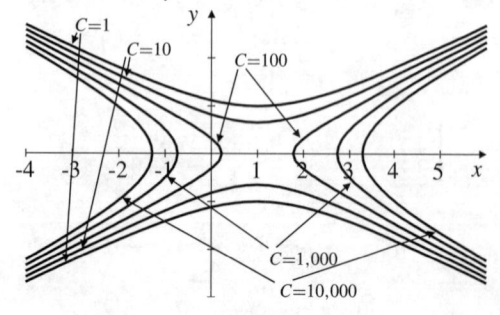

Fig. R-12.6

7. Let $f(x, y, z) = x^2 + y^2 + 4z^2$. Then S has equation $f(x, y, z) = 16$.

   a) $\nabla f(a, b, c) = 2a\mathbf{i} + 2b\mathbf{j} + 8c\mathbf{k}$. The tangent plane to S at (a, b, c) has equation
   $$2a(x - a) + 2b(y - b) + 4c(z - c) = 0, \quad \text{or}$$
   $$ax + by + 4cz = a^2 + b^2 + 4c^2 = 16.$$

   b) The tangent plane $ax + by + 4cz = 16$ passes through $(0, 0, 4)$ if $16c = 16$, that is, if $c = 1$. In this case $a^2 + b^2 = 16 - 4c^2 = 12$. These points (a, b, c) lie on a horizontal circle of radius $\sqrt{12}$ centred at $(0, 0, 1)$ in the plane $z = 1$.

   c) The tangent plane of part (a) is parallel to the plane $x + y + 2\sqrt{2}z = 97$ if
   $$a\mathbf{i} + b\mathbf{j} + 4c\mathbf{k} = t(\mathbf{i} + \mathbf{j} + 2\sqrt{2}\mathbf{k}),$$
   that is, $a = t$, $b = t$, $c = t/\sqrt{2}$. Then $16 = a^2 + b^2 + 4c^2 = 4t^2$, so $t = \pm 2$. The two points on S where the tangent plane is parallel to $x + y + 2\sqrt{2}z = 97$ are $(2, 2, \sqrt{2})$ and $(-2, -2, -\sqrt{2})$.

8. $\dfrac{1}{R} = \dfrac{1}{R_1} + \dfrac{1}{R_2}$

   $-\dfrac{1}{R^2} dR = -\dfrac{1}{R_1^2} dR_1 - \dfrac{1}{R_2^2} dR_2$

   If $R_1 = 100$ and $R_2 = 25$, so that $R = 20$, and if $|dR_1/R_1| = 5/100$ and $|dR_2/R_2| = 2/100$, then

   $$\frac{1}{20} \left| \frac{dR}{R} \right| \leq \frac{1}{100} \cdot \frac{5}{100} + \frac{1}{25} \cdot \frac{2}{100} = \frac{13}{100^2}.$$

   Thus $|dR/R| \leq 13/500$; R can be in error by about 2.6%.

9. The measured sides of the field are $x = 150$ m and $y = 200$ m with $|dx| = 1$ and $|dy| = 1$, and the contained angle between them is $\theta = 30°$ with $|d\theta| = 2° = \pi/90$ rad. The area A of the field satisfies

   $$A = \frac{1}{2} xy \sin \theta \approx 7,500$$
   $$dA = \frac{y}{2} \sin \theta \, dx + \frac{x}{2} \sin \theta \, dy + \frac{xy}{2} \cos \theta \, d\theta$$
   $$= \frac{175}{2} + 15,000 \frac{\sqrt{3}}{2} \cdot \frac{\pi}{90} \approx 541.$$

   The area is 7,500 m^2, accurate to within about 540 m^2 for a percentage error of about 7.2%.

10. $T = x^3 y + y^3 z + z^3 x$.

INSTRUCTOR'S SOLUTIONS MANUAL   REVIEW EXERCISES 12 (PAGE 767)

a) $\nabla T = (3x^2y + z^3)\mathbf{i} + (3y^2z + x^3)\mathbf{j} + (3z^2x + y^3)\mathbf{k}$
$\nabla T(2, -1, 0) = -12\mathbf{i} + 8\mathbf{j} - \mathbf{k}$.
A unit vector in the direction from $(2, -1, 0)$ towards $(1, 1, 2)$ is $\mathbf{u} = (-\mathbf{i} + 2\mathbf{j} + 2\mathbf{k})/3$. The directional derivative of T at $(2, -1, 0)$ in the direction of $\mathbf{u}$ is

$$\mathbf{u} \bullet \nabla T(2, -1, 0) = \frac{12 + 16 - 2}{3} = \frac{26}{3}.$$

b) Since $\nabla(2x^2 + 3y^2 + z^2) = 4x\mathbf{i} + 6y\mathbf{j} + 2z\mathbf{k}$, at $t = 0$ the fly is at $(2, -1, 0)$ and is moving in the direction $\pm(8\mathbf{i} - 6\mathbf{j})$, so its velocity is

$$\pm 5 \frac{8\mathbf{i} - 6\mathbf{j}}{10} = \pm(4\mathbf{i} - 3\mathbf{j}).$$

Since the fly is moving in the direction of increasing T, the rate at which it experiences T increasing is

$$\frac{dT}{dt} = |(4\mathbf{i} - 3\mathbf{j}) \bullet (-12\mathbf{i} + 8\mathbf{j} - \mathbf{k})| = 48 + 24 = 72.$$

11. $f(x, y, z) = x^2y + yz + z^2$.

a) $\nabla f(x, y, z) = 2xy\mathbf{i} + (x^2 + z)\mathbf{j} + (y + 2z)\mathbf{k}$
$\nabla f(1, -1.1) = -2\mathbf{i} + 2\mathbf{j} + \mathbf{k}$.
The directional derivative of f in the direction $\mathbf{i} + \mathbf{k}$ at $(1, -1, 1)$ is

$$\frac{\mathbf{i} + \mathbf{k}}{\sqrt{2}} \bullet (-2\mathbf{i} + 2\mathbf{j} + \mathbf{k}) = -\frac{1}{\sqrt{2}}.$$

b) The plane $x + y + z = 1$ intersects the level surface of f through $(1, -1, 1)$ in a curve whose tangent vector at $(1, -1, 1)$ is perpendicular to both $\nabla f(1, -1, 1)$ and the normal vector $\mathbf{i} + \mathbf{j} + \mathbf{k}$ to the plane. Thus the ant is crawling in the direction of the cross product of these vectors:

$$\pm \begin{vmatrix} \mathbf{i} & \mathbf{j} & \mathbf{k} \\ -2 & 2 & 1 \\ 1 & 1 & 1 \end{vmatrix} = \pm(\mathbf{i} + 3\mathbf{j} - 4\mathbf{k}).$$

c) The second ant is crawling in the direction of the vector projection of $\nabla f(1, -1, 1)$ onto the plane $x + y + z = 1$, which is $\nabla f(1, -1, 1)$ minus its vector projection onto the normal to that plane:

$$\nabla f(1, -1, 1) - \frac{\nabla f(1, -1, 1) \bullet (\mathbf{i} + \mathbf{j} + \mathbf{k})}{|\mathbf{i} + \mathbf{j} + \mathbf{k}|^2} (\mathbf{i} + \mathbf{j} + \mathbf{k})$$
$$= -2\mathbf{i} + 2\mathbf{j} + \mathbf{k} - \frac{1}{3}(\mathbf{i} + \mathbf{j} + \mathbf{k}) = \frac{-7\mathbf{i} + 5\mathbf{j} + 2\mathbf{k}}{3},$$

that is, in the direction $-7\mathbf{i} + 5\mathbf{j} + 2\mathbf{k}$.

12. $f(x, y, z) = (x^2 + z^2)\sin\frac{\pi xy}{2} + yz^2$, $P_0 = (1, 1, -1)$.

a) $\nabla f = \left(2x \sin\frac{\pi xy}{2} + \frac{\pi y}{2}(x^2 + z^2)\cos\frac{\pi xy}{2}\right)\mathbf{i}$
$+ \left(\frac{\pi x}{2}(x^2 + z^2)\cos\frac{\pi xy}{2} + z^2\right)\mathbf{j}$
$+ 2z\left(\sin\frac{\pi xy}{2} + y\right)\mathbf{k}$
$\nabla f(P_0) = 2\mathbf{i} + \mathbf{j} - 4\mathbf{k}$.

b) Since $f(P_0) = 2 + 1 = 3$, the linearization of f at P_0 is

$$L(x, y, z) = 3 + 2(x - 1) + (y - 1) - 4(z + 1).$$

c) The tangent plane at P_0 to the level surface of f through P_0 has equation

$\nabla f(P_0) \bullet ((x - 1)\mathbf{i} + (y - 1)\mathbf{j} + (z + 1)\mathbf{k}) = 0$
$2(x - 1) + (y - 1) - 4(z + 1) = 0$
$2x + y - 4z = 7.$

d) The bird is flying in direction

$$(2 - 1)\mathbf{i} + (-1 - 1)\mathbf{j} + (1 + 1)\mathbf{k} = \mathbf{i} - 2\mathbf{j} + 2\mathbf{k},$$

a vector of length 3. Since the bird's speed is 5, its velocity is

$$\mathbf{v} = \frac{5}{3}(\mathbf{i} - 2\mathbf{j} + 2\mathbf{k}).$$

The rate of change of f as experienced by the bird is

$$\frac{df}{dt} = \mathbf{v} \bullet \nabla f(P_0) = \frac{5}{3}(2 - 2 - 8) = -\frac{40}{3}.$$

e) To experience the greatest rate of increase of f while flying through P_0 at speed 5, the bird should fly in the direction of $\nabla f(P_0)$, that is, $2\mathbf{i} + \mathbf{j} - 4\mathbf{k}$.

13. $u = k\left(\ln \cos\frac{x}{k} - \ln \cos\frac{y}{k}\right)$

$u_x = k\left(-\frac{1}{k}\tan\frac{x}{k}\right) = -\tan\frac{x}{k}$

$u_y = k\left(\frac{1}{k}\tan\frac{y}{k}\right) = \tan\frac{y}{k}$

$u_{xx} = -\frac{1}{k}\sec^2\frac{x}{k}$

$u_{yy} = \frac{1}{k}\sec^2\frac{y}{k}$

$u_{xy} = 0$

$(1 + u_x^2)u_{yy} - uu_xu_yu_{xy} + (1 + u_y^2)u_{xx}$
$= \frac{1}{k}\sec^2\frac{x}{k}\sec^2\frac{y}{k} - 0 - \frac{1}{k}\sec^2\frac{y}{k}\sec^2\frac{x}{k} = 0.$

465

14. If $F(x, y, z) = 0$, $G(x, y, z) = 0$ are solved for $x = x(y)$, $z = z(y)$, then

$$F_1 \frac{dx}{dy} + F_2 + F_3 \frac{dz}{dy} = 0$$
$$G_1 \frac{dx}{dy} + G_2 + G_3 \frac{dz}{dy} = 0.$$

Eliminating dz/dy from these equations, we obtain

$$\frac{dx}{dy} = -\frac{F_2 G_3 - F_3 G_2}{F_1 G_3 - F_3 G_1}.$$

Similarly, if the equations are solved for $x = x(z)$, $y = y(z)$, then

$$\frac{dy}{dz} = -\frac{F_3 G_1 - F_1 G_3}{F_2 G_1 - F_1 G_2},$$

and if the equations are solved for $y = y(x)$, $z = z(x)$, then

$$\frac{dz}{dx} = -\frac{F_1 G_2 - F_2 G_1}{F_3 G_2 - F_2 G_3}.$$

Hence

$$\frac{dx}{dy} \cdot \frac{dy}{dz} \cdot \frac{dz}{dx}$$
$$= -\frac{F_2 G_3 - F_3 G_2}{F_1 G_3 - F_3 G_1} \cdot \frac{F_3 G_1 - F_1 G_3}{F_2 G_1 - F_1 G_2} \cdot \frac{F_1 G_2 - F_2 G_1}{F_3 G_2 - F_2 G_3} = 1.$$

15. $x = u^3 - uv$
$y = 3uv + 2v^2$
Assume these equations define $u = u(x, y)$ and $v = v(x, y)$ near the point P where $(u, v, x, y) = (-1, 2, 1, 2)$.

a) Differentiating both equations with respect to x, we get

$$1 = 3u^2 \frac{\partial u}{\partial x} - v \frac{\partial u}{\partial x} - u \frac{\partial v}{\partial x}$$
$$0 = 3v \frac{\partial u}{\partial x} + 3u \frac{\partial v}{\partial x} + 4v \frac{\partial v}{\partial x}.$$

At P, these equations become

$$1 = \frac{\partial u}{\partial x} + \frac{\partial v}{\partial x}, \quad 0 = 6 \frac{\partial u}{\partial x} + 5 \frac{\partial v}{\partial x},$$

from which we obtain $\partial u / \partial x \big|_P = -5$.
Similarly, differentiating the given equations with respect to y leads to

$$0 = \frac{\partial u}{\partial y} + \frac{\partial v}{\partial y}, \quad 1 = 6 \frac{\partial u}{\partial y} + 5 \frac{\partial v}{\partial y},$$

from which we obtain $\partial u / \partial y \big|_P = 1$.

b) Since $u(1, 2) = -1$, we have

$$u(1.02, 1.97) \approx -1 + \frac{\partial u}{\partial x}\bigg|_P (0.02) + \frac{\partial u}{\partial y}\bigg|_P (-0.03)$$
$$= -1 - 5(0.02) + 1(-0.03) = -1.13.$$

16. $u = x^2 + y^2$
$v = x^2 - 2xy^2$
Assume these equations define $x = x(u, v)$ and $y = y(u, v)$ near the point $(u, v) = (5, -7)$, with $x = 1$ and $y = 2$ at that point.

a) Differentiate the given equations with respect to u to obtain

$$1 = 2x \frac{\partial x}{\partial u} + 2y \frac{\partial y}{\partial u}$$
$$0 = 2(x - y^2) \frac{\partial x}{\partial u} - 4xy \frac{\partial y}{\partial u}.$$

At $x = 1$, $y = 2$,

$$2 \frac{\partial x}{\partial u} + 4 \frac{\partial y}{\partial u} = 1$$
$$-6 \frac{\partial x}{\partial u} - 8 \frac{\partial y}{\partial u} = 0,$$

from which we obtain $\partial x / \partial u = -1$ and $\partial y / \partial u = 3/4$ at $(5, -7)$.

b) If $z = \ln(y^2 - x^2)$, then

$$\frac{\partial z}{\partial u} = \frac{1}{y^2 - x^2} \left[-2x \frac{\partial x}{\partial u} + 2y \frac{\partial y}{\partial u} \right].$$

At $(u, v) = (5, -7)$, we have $(x, y) = (1, 2)$, and so

$$\frac{\partial z}{\partial u} = \frac{1}{3} \left[-2(-1) + 4 \left(\frac{3}{4} \right) \right] = \frac{5}{3}.$$

Challenging Problems 12   (page 768)

1. a) If f is differentiable at (a, b), then its graph has a nonvertical tangent plane at $(a, b, f(a, b))$. Any line through that point, part of which lies on the surface $z = f(x, y)$ near (a, b), must be tangent to that surface at (a, b), so must lie in the tangent plane.

b) The surface S with equation $z = y g(x/y)$ has the property that if $P = (x_0, y_0, z_0)$ is any point on it, then all points other than the origin on the line joining P_0 to the origin also lie on S. Specifically, if $t \neq 0$, then (tx_0, ty_0, tz_0) lies on S, because

$$tz_0 = ty_0 \, g\left(\frac{tx_0}{ty_0} \right) \Leftrightarrow z_0 = y_0 \, g\left(\frac{x_0}{y_0} \right).$$

Thus S consists entirely of lines through the origin; it is some kind of "cone" with vertex at the origin. By part (a), all tangent planes to S contain the lines on S through the points of contact, so all tangent planes must pass through the origin.

2. Let the position vector of the particle at time t be $\mathbf{r} = x(t)\mathbf{i} + y(t)\mathbf{j} + z(t)\mathbf{k}$. Then the velocity of the particle is
$$\mathbf{v} = \frac{dx}{dt}\mathbf{i} + \frac{dy}{dt}\mathbf{j} + \frac{dz}{dt}\mathbf{k}.$$
This velocity must be parallel to
$$\nabla f(x, y, z) = -2x\mathbf{i} - 4y\mathbf{j} + 6z\mathbf{k}$$
at every point of the path, that is,
$$\frac{dx}{dt} = -2tx, \quad \frac{dy}{dt} = -4ty, \quad \frac{dz}{dt} = 6tz,$$
so that $\dfrac{dx}{-2x} = \dfrac{dy}{-4y} = \dfrac{dz}{6z}$. Integrating these equations, we get
$$\ln|y| = 2\ln|x| + C_1, \quad \ln|z| = -3\ln|x| + C_2.$$
Since the path passes through $(1, 1, 8)$, C_1 and C_2 are determined by
$$\ln 1 = 2\ln 1 + C_1, \quad \ln 8 = -3\ln 1 + C_2.$$
Thus $C_1 = 0$ and $C_2 = \ln 8$. The path therefore has equations $y = x^2$, $z = 8/x^3$. Evidently $(2, 4, 1)$ lies on the path, and $(3, 7, 0)$ does not.

3. We used Maple V to verify the stated identity. Using r, p, and t to represent ρ, ϕ, and θ, respectively, we defined

```
v:=(r,p,t)->
u(r*sin(p)*cos(t), r*sin(p)*sin(t),
r*cos(p));
```

and then asked Maple to calculate the left side of the identity:

```
D[1](D[1](v))(r,p,t)
+(2/r)*D[1](v)(r,p,t)
+(cot(p)/r^2)*D[2](v)(r,p,t)
+(1/r^2)*D[2](D[2](v))(r,p,t)
+(1/(r*sin(p))^2)*D[3](D[3](v))(r,p,t);
```

Maple responded with a very complicated expression about two screenfulls in length. However, when asked to simplify the result:

```
simplify(");
```

Maple obliged with the much simpler form

```
D[3, 3](u)(r sin(p) cos(t), r sin(p)
sin(t), r cos(p))
+ D[2, 2](u)(r sin(p) cos(t), r sin(p)
sin(t), r cos(p))
+ D[1, 1](u)(r sin(p) cos(t), r sin(p)
sin(t), r cos(p))
```

thus confirming that the identity is true.

4. If $u(x, y, z, t) = v(\rho, t) = \dfrac{f(\rho - ct)}{\rho}$ is independent of θ and ϕ, then
$$\frac{\partial^2 u}{\partial x^2} + \frac{\partial^2 u}{\partial y^2} + \frac{\partial^2 u}{\partial z^2} = \frac{\partial^2 v}{\partial \rho^2} + \frac{2}{\rho}\frac{\partial v}{\partial \rho}$$
by Problem 3. We have
$$\frac{\partial v}{\partial \rho} = \frac{f'(\rho - ct)}{\rho} - \frac{f(\rho - ct)}{\rho^2}$$
$$\frac{\partial^2 v}{\partial \rho^2} = \frac{f''(\rho - ct)}{\rho} - \frac{2f'(\rho - ct)}{\rho^2} + \frac{2f(\rho - ct)}{\rho^3}$$
$$\frac{\partial v}{\partial t} = -\frac{cf'(\rho - ct)}{\rho}$$
$$\frac{\partial^2 v}{\partial t^2} = \frac{c^2 f''(\rho - ct)}{\rho}$$
$$\frac{\partial^2 v}{\partial \rho^2} + \frac{2}{\rho}\frac{\partial v}{\partial \rho}$$
$$= \frac{f''(\rho - ct)}{\rho} - \frac{2f'(\rho - ct)}{\rho^2} + \frac{2f(\rho - ct)}{\rho^3}$$
$$+ \frac{2f'(\rho - ct)}{\rho^2} - \frac{2f(\rho - ct)}{\rho^3}$$
$$= \frac{f''(\rho - ct)}{\rho}$$
$$= \frac{1}{c^2}\frac{\partial^2 v}{\partial t^2} = \frac{1}{c^2}\frac{\partial^2 u}{\partial t^2}.$$

The function $f(\rho - ct)/\rho$ represents the shape of a symmetrical wave travelling uniformly away from the origin at speed c. Its amplitude at distance ρ from the origin decreases as ρ increases; it is proportional to the reciprocal of ρ.

CHAPTER 13. APPLICATIONS OF PARTIAL DERIVATIVES

Section 13.1 Extreme Values (page 778)

1. $f(x, y) = x^2 + 2y^2 - 4x + 4y$
   $f_1(x, y) = 2x - 4 = 0$ if $x = 2$
   $f_2(x, y) = 4y + 4 = 0$ if $y = -1$.
   Critical point is $(2, -1)$. Since $f(x, y) \to \infty$ as $x^2 + y^2 \to \infty$, f has a local (and absolute) minimum value at that critical point.

2. $f(x, y) = xy - x + y$, $f_1 = y - 1$, $f_2 = x + 1$
   $A = f_{11} = 0$, $B = f_{12} = 1$, $C = f_{22} = 0$.
   Critical point $(-1, 1)$ is a saddle point since $B^2 - AC > 0$.

3. $f(x, y) = x^3 + y^3 - 3xy$
   $f_1(x, y) = 3(x^2 - y)$, $f_2(x, y) = 3(y^2 - x)$.
   For critical points: $x^2 = y$ and $y^2 = x$. Thus $x^4 - x = 0$, that is, $x(x - 1)(x^2 + x + 1) = 0$. Thus $x = 0$ or $x = 1$. The critical points are $(0, 0)$ and $(1, 1)$. We have
   $$A = f_{11}(x, y) = 6x, \quad B = f_{12}(x, y) = -3,$$
   $$C = f_{22}(x, y) = 6y.$$
   At $(0, 0)$: $A = C = 0$, $B = -3$. Thus $AC < B^2$, and $(0, 0)$ is a saddle point of f.
   At $(1, 1)$: $A = C = 6$, $B = -3$, so $AC > B^2$. Thus f has a local minimum value at $(1, 1)$.

4. $f(x, y) = x^4 + y^4 - 4xy$, $f_1 = 4(x^3 - y)$, $f_2 = 4(y^3 - x)$
   $A = f_{11} = 12x^2$, $B = f_{12} = -4$, $C = f_{22} = 12y^2$.
   For critical points: $x^3 = y$ and $y^3 = x$. Thus $x^9 = x$, or $x(x^8 - 1) = 0$, and $x = 0, 1$, or -1. The critical points are $(0, 0)$, $(1, 1)$ and $(-1, -1)$.
   At $(0, 0)$, $B^2 - AC = 16 - 0 > 0$, so $(0, 0)$ is a saddle point.
   At $(1, 1)$ and $(-1, -1)$, $B^2 - AC = 16 - 144 < 0$, $A > 0$, so f has local minima at these points.

5. $f(x, y) = \dfrac{x}{y} + \dfrac{8}{x} - y$
   $f_1(x, y) = \dfrac{1}{y} - \dfrac{8}{x^2} = 0$ if $8y = x^2$
   $f_2(x, y) = -\dfrac{x}{y^2} - 1 = 0$ if $x = -y^2$.
   For critical points: $8y = x^2 = y^4$, so $y = 0$ or $y = 2$. $f(x, y)$ is not defined when $y = 0$, so the only critical point is $(-4, 2)$. At $(-4, 2)$ we have
   $$A = f_{11} = \dfrac{16}{x^3} = -\dfrac{1}{4}, \quad B = f_{12} = -\dfrac{1}{y^2} = -\dfrac{1}{4},$$
   $$C = f_{22} = \dfrac{2x}{y^3} = -1.$$

Thus $B^2 - AC = \dfrac{1}{16} - \dfrac{1}{4} < 0$, and $(-4, 2)$ is a local maximum.

6. $f(x, y) = \cos(x + y)$, $f_1 = -\sin(x + y) = f_2$.
   All points on the lines $x + y = n\pi$ (n is an integer) are critical points. If n is even, $f = 1$ at such points; if n is odd, $f = -1$ there. Since $-1 \leq f(x, y) \leq 1$ at all points in $\mathbb{R}^2$, f must have local and absolute maximum values at points $x + y = n\pi$ with n even, and local and absolute minimum values at such points with n odd.

7. $f(x, y) = x \sin y$. For critical points we have
   $$f_1 = \sin y = 0, \quad f_2 = x \cos y = 0.$$
   Since $\sin y$ and $\cos y$ cannot vanish at the same point, the only critical points correspond to $x = 0$ and $\sin y = 0$. They are $(0, n\pi)$, for all integers n. All are saddle points.

8. $f(x, y) = \cos x + \cos y$, $f_1 = -\sin x$, $f_2 = -\sin y$
   $A = f_{11} = -\cos x$, $B = f_{12} = 0$, $C = f_{22} = -\cos y$.
   The critical points are points $(m\pi, n\pi)$, where m and n are integers.
   Here $B^2 - AC = -\cos(m\pi)\cos(n\pi) = (-1)^{m+n+1}$ which is negative if $m + n$ is even, and positive if $m + n$ is odd. If $m + n$ is odd then f has a saddle point at $(m\pi, n\pi)$. If $m + n$ is even and m is odd then f has a local (and absolute) minimum value, -2, at $(m\pi, n\pi)$. If $m + n$ is even and m is even then f has a local (and absolute) maximum value, 2, at $(m\pi, n\pi)$.

9. $f(x, y) = x^2 y e^{-(x^2 + y^2)}$
   $f_1(x, y) = 2xy(1 - x^2)e^{-(x^2 + y^2)}$
   $f_2(x, y) = x^2(1 - 2y^2)e^{-(x^2 + y^2)}$
   $A = f_{11}(x, y) = 2y(1 - 5x^2 + 2x^4)e^{-(x^2 + y^2)}$
   $B = f_{12}(x, y) = 2x(1 - x^2)(1 - 2y^2)e^{-(x^2 + y^2)}$
   $C = f_{22}(x, y) = 2x^2 y(2y^2 - 3)e^{-(x^2 + y^2)}$.

   For critical points:
   $$xy(1 - x^2) = 0$$
   $$x^2(1 - 2y^2) = 0.$$

   The critical points are $(0, y)$ for all y, $(\pm 1, 1/\sqrt{2})$, and $(\pm 1, -1/\sqrt{2})$.
   Evidently, $f(0, y) = 0$. Also $f(x, y) > 0$ if $y > 0$ and $x \neq 0$, and $f(x, y) < 0$ if $y < 0$ and $x \neq 0$. Thus f has a local minimum at $(0, y)$ if $y > 0$, and a local maximum if $y < 0$. The origin is a saddle point.
   At $(\pm 1, 1/\sqrt{2})$: $A = C = -2\sqrt{2}e^{-3/2}$, $B = 0$, and so $AC > B^2$. Thus f has local maximum values at these two points.
   At $(\pm 1, -1/\sqrt{2})$: $A = C = 2\sqrt{2}e^{-3/2}$, $B = 0$, and so $AC > B^2$. Thus f has local minimum values at these two points.

Since $f(x, y) \to 0$ as $x^2 + y^2 \to \infty$, the value $f(\pm 1, 1/\sqrt{2}) = e^{-3/2}/\sqrt{2}$ is the absolute maximum value for f, and the value $f(\pm 1, -1/\sqrt{2}) = -e^{-3/2}/\sqrt{2}$ is the absolute minimum value.

10. $f(x, y) = \dfrac{xy}{2 + x^4 + y^4}$

$f_1 = \dfrac{(2 + x^4 + y^4)y - xy4x^3}{(2 + x^4 + y^4)^2} = \dfrac{y(2 + y^4 - 3x^4)}{(2 + x^4 + y^4)^2}$

$f_2 = \dfrac{x(2 + x^4 - 3y^4)}{(2 + x^4 + y^4)^2}.$

For critical points, $y(2 + y^4 - 3x^4) = 0$ and $x(2 + x^4 - 3y^4) = 0$.
One critical point is $(0, 0)$. Since $f(0, 0) = 0$ but $f(x, y) > 0$ in the first quadrant and $f(x, y) < 0$ in the second quadrant, $(0, 0)$ must be a saddle point of f. Any other critical points must satisfy $2 + y^4 - 3x^4 = 0$ and $2 + x^4 - 3y^4 = 0$, that is, $y^4 = x^4$, or $y = \pm x$. Thus $2 - 2x^4 = 0$ and $x = \pm 1$. Therefore there are four other critical points: $(1, 1)$, $(-1, -1)$, $(1, -1)$ and $(-1, 1)$. f is positive at the first two of these, and negative at the other two. Since $f(x, y) \to 0$ as $x^2 + y^2 \to \infty$, f must have maximum values at $(1, 1)$ and $(-1, -1)$, and minimum values at $(1, -1)$ and $(-1, 1)$.

11. $f(x, y) = xe^{-x^3 + y^3}$

$f_1(x, y) = (1 - 3x^3)e^{-x^3 + y^3}$

$f_2(x, y) = 3xy^2 e^{-x^3 + y^3}$

$A = f_{11}(x, y) = 3x^2(3x^3 - 4)e^{-x^3 + y^3}$

$B = f_{12}(x, y) = -3y^2(3x^3 - 1)e^{-x^3 + y^3}$

$C = f_{22}(x, y) = 3xy(3y^3 + 2)e^{-x^3 + y^3}$

For critical points: $3x^3 = 1$ and $3xy^2 = 0$. The only critical point is $(3^{-1/3}, 0)$. At that point we have $B = C = 0$ so the second derivative test is inconclusive.
However, note that $f(x, y) = f(x, 0)e^{y^3}$, and e^{y^3} has an inflection point at $y = 0$. Therefore $f(x, y)$ has neither a maximum nor a minimum value at $(3^{-1/3}, 0)$, so has a saddle point there.

12. $f(x, y) = \dfrac{1}{1 - x + y + x^2 + y^2}$

$= \dfrac{1}{\left(x - \dfrac{1}{2}\right)^2 + \left(y + \dfrac{1}{2}\right)^2 + \dfrac{1}{2}}.$

Evidently f has absolute maximum value 2 at $\left(\dfrac{1}{2}, -\dfrac{1}{2}\right)$.
Since

$f_1(x, y) = \dfrac{1 - 2x}{(1 - x + y + x^2 + y^2)^2}$

$f_2(x, y) = -\dfrac{1 + 2y}{(1 - x + y + x^2 + y^2)^2},$

$\left(\dfrac{1}{2}, -\dfrac{1}{2}\right)$ is the only critical point of f.

13. $f(x, y) = \left(1 + \dfrac{1}{x}\right)\left(1 + \dfrac{1}{y}\right)\left(\dfrac{1}{x} + \dfrac{1}{y}\right)$

$= \dfrac{(x + 1)(y + 1)(x + y)}{x^2 y^2}$

$f_1(x, y) = -\dfrac{(y + 1)(xy + x + 2y)}{x^3 y^2}$

$f_2(x, y) = -\dfrac{(x + 1)(xy + y + 2x)}{x^2 y^3}$

$A = f_{11}(x, y) = \dfrac{2(y + 1)(xy + x + 3y)}{x^4 y^2}$

$B = f_{12}(x, y) = \dfrac{2(xy + x + y)}{x^3 y^3}$

$C = f_{22}(x, y) = \dfrac{2(x + 1)(xy + y + 3x)}{x^2 y^4}.$

For critical points:

$y = -1$ or $xy + x + 2y = 0,$
and $x = -1$ or $xy + y + 2x = 0.$

If $y = -1$, then $x = -1$ or $x - 1 = 0$.
If $x = -1$, then $y = -1$ or $y - 1 = 0$.
If $x \ne -1$ and $y \ne -1$, then $x - y = 0$, so $x^2 + 3x = 0$. Thus $x = 0$ or $x = -3$. However, the definition of f excludes $x = 0$. Thus, the only critical points are

$(1, -1), \quad (-1, 1), \quad (-1, -1), \quad \text{and} \quad (-3, -3).$

At $(1, -1)$, $(-1, 1)$, and $(-1, -1)$ we have $AC = 0$ and $B \ne 0$. Therefore these three points are saddle points of f.
At $(-3, -3)$, $A = C = 4/243$ and $B = 2/243$, so $AC > B^2$. Therefore f has a local minimum value at $(-3, -3)$.

14. $f(x, y, z) = xyz - x^2 - y^2 - z^2$. For critical points we have

$0 = f_1 = yz - 2x, \quad 0 = f_2 = xz - 2y, \quad 0 = f_3 = xy - 2z.$

Thus $xyz = 2x^2 = 2y^2 = 2z^2$, so $x^2 = y^2 = z^2$. Hence $x^3 = \pm 2x^2$, and $x = \pm 2$ or 0. Similarly for y and z.
The only critical points are $(0, 0, 0)$, $(2, 2, 2)$, $(-2, -2, 2)$, $(-2, 2, -2)$, and $(2, -2, -2)$.

Let $\mathbf{u} = u\mathbf{i} + v\mathbf{j} + w\mathbf{k}$, where $u^2 + v^2 + w^2 = 1$. Then

$D_{\mathbf{u}} f(x, y, z) = (yz - 2x)u + (xz - 2y)v + (xy - 2z)w$

$D_{\mathbf{u}}\big(D_{\mathbf{u}} f(x, y, z)\big) = (-2u + zv + yw)u$
$\qquad + (zu - 2v + xw)v + (yu + xv - 2w)w.$

At $(0, 0, 0)$, $D_{\mathbf{u}}\big(D_{\mathbf{u}} f(0, 0, 0)\big) = -2u^2 - 2v^2 - 2w^2 < 0$ for $\mathbf{u} \ne \mathbf{0}$, so f has a local maximum value at $(0, 0, 0)$.

At $(2, 2, 2)$, we have

$$D_{\mathbf{u}}(D_{\mathbf{u}}f(2,2,2)) = (-2u + 2v + 2w)u + (2u - 2v + 2w)v$$
$$+ (2u + 2v - 2w)w$$
$$= -2(u^2 + v^2 + w^2) + 4(uv + vw + wu)$$
$$= -2[(u - v - w)^2 - 4vw]$$
$$\begin{cases} < 0 & \text{if } v = w = 0, u \neq 0 \\ > 0 & \text{if } v = w \neq 0, u - v - w = 0. \end{cases}$$

Thus $(2, 2, 2)$ is a saddle point.

At $(2, -2, -2)$, we have

$$D_{\mathbf{u}}(D_{\mathbf{u}}f) = -2(u^2 + v^2 + w^2 + 2uv + 2uw - 2vw)$$
$$= -2[(u + v + w)^2 - 4vw]$$
$$\begin{cases} < 0 & \text{if } v = w = 0, u \neq 0 \\ > 0 & \text{if } v = w \neq 0, u + v + w = 0. \end{cases}$$

Thus $(2, -2, -2)$ is a saddle point. By symmetry, so are the remaining two critical points.

15. $f(x, y, z) = xy + x^2 z - x^2 - y - z^2$
$f_1(x, y, z) = y + 2x(z - 1)$
$f_2(x, y, z) = x - 1$
$f_3(x, y, z) = x^2 - 2z.$

The only critical point is $(1, 1, \tfrac{1}{2})$. We have

$$D = f\left(1 + h, 1 + k, \tfrac{1}{2} + m\right) - f\left(1, 1, \tfrac{1}{2}\right)$$
$$= 1 + h + k + hk + \frac{1 + 2h + h^2}{2} + (1 + 2h + h^2)m$$
$$- 1 - 2h - h^2 - 1 - k - \frac{1}{4} - m - m^2 - \left(-\frac{3}{4}\right)$$
$$= \frac{h^2(2m - 1) + 2h(k + 2m) - 2m^2}{2}.$$

If $m = h$ and $k = 0$, then $D = \dfrac{h^2(1 + 2h)}{2} > 0$ for small $|h|$.
If $h = k = 0$, then $D = -m^2 < 0$ for $m \neq 0$.
Thus f has a saddle point at $\left(1, 1, \tfrac{1}{2}\right)$.

16. $f(x, y, z) = 4xyz - x^4 - y^4 - z^4$
$D = f(1 + h, 1 + k, 1 + m) - f(1, 1, 1)$
$= 4(1 + h)(1 + k)(1 + m) - (1 + h)^4 - (1 + k)^4$
$\qquad - (1 + m)^4 - 1$
$= 4(1 + h + k + m + hk + hm + km + hkm)$
$\qquad - (1 + 4h + 6h^2 + 4h^3 + h^4)$
$\qquad - (1 + 4k + 6k^2 + 4k^3 + k^4)$
$\qquad - (1 + 4m + 6m^2 + 4m^3 + m^4) - 1$
$= 4(hk + hm + km) - 6(h^2 + k^2 + m^2) + \cdots,$

where $\cdots$ stands for terms of degree 3 and 4 in the variables h, k, and m. Completing some squares among the quadratic terms we obtain

$$D = -2\left[(h-k)^2 + (k-m)^2 + (h-m)^2 + h^2 + k^2 + m^2\right] + \cdots$$

which is negative if $|h|$, $|k|$ and $|m|$ are small and not all 0. (This is because the terms of degree 3 and 4 are smaller in size than the quadratic terms for small values of the variables.)
Hence f has a local maximum value at $(1, 1, 1)$.

17. $f(x, y) = xye^{-(x^2 + y^4)}$
$f_1(x, y) = y(1 - 2x^2)e^{-(x^2 + y^4)}$
$f_2(x, y) = x(1 - 4y^4)e^{-(x^2 + y^4)}$

For critical points $y(1 - 2x^2) = 0$ and $x(1 - 4y^4) = 0$. The critical points are

$$(0, 0), \quad \left(\pm\frac{1}{\sqrt{2}}, \frac{1}{\sqrt{2}}\right), \quad \left(\pm\frac{1}{\sqrt{2}}, -\frac{1}{\sqrt{2}}\right).$$

We have

$f(0, 0) = 0$

$f\left(\frac{1}{\sqrt{2}}, \frac{1}{\sqrt{2}}\right) = f\left(-\frac{1}{\sqrt{2}}, -\frac{1}{\sqrt{2}}\right) = \frac{1}{2}e^{-3/4} > 0$

$f\left(-\frac{1}{\sqrt{2}}, \frac{1}{\sqrt{2}}\right) = f\left(\frac{1}{\sqrt{2}}, -\frac{1}{\sqrt{2}}\right) = -\frac{1}{2}e^{-3/4} < 0$

Since $f(x, y) \to 0$ as $x^2 + y^2 \to \infty$, the maximum and minimum values of f are $\tfrac{1}{2}e^{-3/4}$ and $-\tfrac{1}{2}e^{-3/4}$ respectively.

18. $f(x, y) = \dfrac{x}{1 + x^2 + y^2}$
$f_1(x, y) = \dfrac{1 + y^2 - x^2}{(1 + x^2 + y^2)^2}$
$f_2(x, y) = \dfrac{-2xy}{(1 + x^2 + y^2)^2}.$

For critical points, $x^2 - y^2 = 1$, and $xy = 0$. The critical points are $(\pm 1, 0)$. $f(\pm 1, 0) = \pm\tfrac{1}{2}$.
Since $f(x, y) \to 0$ as $x^2 + y^2 \to \infty$, the maximum and minimum values of f are $1/2$ and $-1/2$ respectively.

19. $f(x, y, z) = xyze^{-(x^2 + y^2 + z^2)}$
$f_1(x, y, z) = yz(1 - 2x^2)e^{-(x^2 + y^2 + z^2)}$
$f_2(x, y, z) = xz(1 - 2y^2)e^{-(x^2 + y^2 + z^2)}$
$f_3(x, y, z) = xy(1 - 2z^2)e^{-(x^2 + y^2 + z^2)}.$

Any critical point must satisfy

$yz(1 - 2x^2) = 0 \quad \text{i.e., } y = 0 \text{ or } z = 0 \text{ or } x = \pm\dfrac{1}{\sqrt{2}}$

$xz(1 - 2y^2) = 0 \quad \text{i.e., } x = 0 \text{ or } z = 0 \text{ or } y = \pm\dfrac{1}{\sqrt{2}}$

$xy(1 - 2z^2) = 0 \quad \text{i.e., } x = 0 \text{ or } y = 0 \text{ or } z = \pm\dfrac{1}{\sqrt{2}}.$

Since $f(x, y, z)$ is positive at some points, negative at others, and approaches 0 as (x, y, z) recedes to infinity, f must have maximum and minimum values at critical points. Since $f(x, y, z) = 0$ if $x = 0$ or $y = 0$ or $z = 0$, the maximum and minimum values must occur among the eight critical points where $x = \pm 1/\sqrt{2}$, $y = \pm 1/\sqrt{2}$, and $z = \pm 1/\sqrt{2}$. At four of these points, f has the value $\frac{1}{2\sqrt{2}} e^{-3/2}$, the maximum value. At the other four f has the value $-\frac{1}{2\sqrt{2}} e^{-3/2}$, the minimum value.

20. $f(x, y) = x + 8y + \frac{1}{xy}$, $(x > 0, \ y > 0)$

$f_1(x, y) = 1 - \frac{1}{x^2 y} = 0 \ \Rightarrow \ x^2 y = 1$

$f_2(x, y) = 8 - \frac{1}{xy^2} = 0 \ \Rightarrow \ 8xy^2 = 1$.

The critical points must satisfy

$$\frac{x}{y} = \frac{x^2 y}{xy^2} = 8,$$

that is, $x = 8y$. Also, $x^2 y = 1$, so $64 y^3 = 1$. Thus $y = 1/4$, and $x = 2$; the critical point is $(2, \frac{1}{4})$. Since $f(x, y) \to \infty$ if $x \to 0+$, $y \to 0+$, or $x^2 + y^2 \to \infty$, the critical point must give a minimum value for f. The minimum value is $f(2, \frac{1}{4}) = 2 + 2 + 2 = 6$.

21. Let the length, width, and height of the box be x, y, and z, respectively. Then $V = xyz$. The total surface area of the bottom and sides is

$$S = xy + 2xz + 2yz = xy + 2(x+y)\frac{V}{xy}$$
$$= xy + \frac{2V}{x} + \frac{2V}{y},$$

where $x > 0$ and $y > 0$. Since $S \to \infty$ as $x \to 0+$ or $y \to 0+$ or $x^2 + y^2 \to \infty$, S must have a minimum value at a critical point in the first quadrant. For CP:

$$0 = \frac{\partial S}{\partial x} = y - \frac{2V}{x^2}$$
$$0 = \frac{\partial S}{\partial y} = x - \frac{2V}{y^2}.$$

Thus $x^2 y = 2V = xy^2$, so that $x = y = (2V)^{1/3}$ and $z = V/(2V)^{2/3} = 2^{-2/3} V^{1/3}$.

22. Let the length, width, and height of the box be x, y, and z, respectively. Then $V = xyz$. If the top and side walls cost \$k per unit area, then the total cost of materials for the box is

$$C = 2kxy + kxy + 2kxz + 2kyz$$
$$= k\left[3xy + 2(x+y)\frac{V}{xy}\right] = k\left[3xy + \frac{2V}{x} + \frac{2V}{y}\right],$$

where $x > 0$ and $y > 0$. Since $C \to \infty$ as $x \to 0+$ or $y \to 0+$ or $x^2 + y^2 \to \infty$, C must have a minimum value at a critical point in the first quadrant. For CP:

$$0 = \frac{\partial C}{\partial x} = k\left(3y - \frac{2V}{x^2}\right)$$
$$0 = \frac{\partial C}{\partial y} = k\left(3x - \frac{2V}{y^2}\right).$$

Thus $3x^2 y = 2V = 3xy^2$, so that $x = y = (2V/3)^{1/3}$ and $z = V/(2V/3)^{2/3} = (9V/4)^{1/3}$.

23. Let (x, y, z) be the coordinates of the corner of the box that is in the first octant of space. Thus $x, y, z \geq 0$, and

$$\frac{x^2}{a^2} + \frac{y^2}{b^2} + \frac{z^2}{c^2} = 1.$$

The volume of the box is

$$V = (2x)(2y)(2z) = 8cxy\sqrt{1 - \frac{x^2}{a^2} - \frac{y^2}{b^2}}$$

for $x \geq 0$, $y \geq 0$, and $(x^2/a^2) + (y^2/b^2) \leq 1$. For analysis it is easier to deal with V^2 than with V:

$$V^2 = 64c^2 \left(x^2 y^2 - \frac{x^4 y^2}{a^2} - \frac{x^2 y^4}{b^2}\right).$$

Since $V = 0$ if $x = 0$ or $y = 0$ or $(x^2/a^2) + (y^2/b^2) = 1$, the maximum value of V^2, and hence of V, will occur at a critical point of V^2 where $x > 0$ and $y > 0$. For CP:

$$0 = \frac{\partial V^2}{\partial x} = 64c^2\left(2xy^2 - \frac{4x^3 y^2}{a^2} - \frac{2xy^4}{b^2}\right)$$
$$= 128c^2 xy^2 \left(1 - \frac{2x^2}{a^2} - \frac{y^2}{b^2}\right)$$
$$0 = \frac{\partial V^2}{\partial y} = 128c^2 x^2 y \left(1 - \frac{x^2}{a^2} - \frac{2y^2}{b^2}\right).$$

Hence we must have

$$\frac{2x^2}{a^2} + \frac{y^2}{b^2} = 1 = \frac{x^2}{a^2} + \frac{2y^2}{b^2},$$

so that $x^2/a^2 = y^2/b^2 = 1/3$, and $x = a/\sqrt{3}$, $y = b/\sqrt{3}$. The largest box has volume

$$V = \frac{8abc}{3}\sqrt{1 - \frac{1}{3} - \frac{1}{3}} = \frac{8abc}{3\sqrt{3}} \ \text{cubic units}.$$

24. Given that $a > 0$, $b > 0$, $c > 0$, and $a + b + c = 30$, we want to maximize

$$P = ab^2c^3 = (30 - b - c)b^2c^3 = 30b^2c^3 - b^3c^3 - b^2c^4.$$

Since $P = 0$ if $b = 0$ or $c = 0$ or $b + c = 30$ (i.e., $a = 0$), the maximum value of P will occur at a critical point (b, c) satisfying $b > 0$, $c > 0$, and $b + c < 30$. For CP:

$$0 = \frac{\partial P}{\partial b} = 60bc^3 - 3b^2c^3 - 2bc^4 = bc^3(60 - 3b - 2c)$$

$$0 = \frac{\partial P}{\partial c} = 90b^2c^2 - 3b^3c^2 - 4b^2c^3 = b^2c^2(90 - 3b - 4c).$$

Hence $9b + 6c = 180 = 6b + 8c$, from which we obtain $3b = 2c = 30$. The three numbers are $b = 10$, $c = 15$, and $a = 30 - 10 - 15 = 5$.

25. Differentiate the given equation

$$e^{2zx-x^2} - 3e^{2zy+y^2} = 2$$

with respect to x and y, regarding z as a function of x and y:

$$e^{2zx-x^2}\left(2x\frac{\partial z}{\partial x} + 2z - 2x\right) - 3e^{2zy+y^2}\left(2y\frac{\partial z}{\partial x}\right) = 0 \quad (*)$$

$$e^{2zx-x^2}\left(2x\frac{\partial z}{\partial y}\right) - 3e^{2zy+y^2}\left(2y\frac{\partial z}{\partial y} + 2z + 2y\right) = 0 \quad (**)$$

For a critical point we have $\frac{\partial z}{\partial x} = 0$ and $\frac{\partial z}{\partial y} = 0$, and it follows from the equations above that $z = x$ and $z = -y$. Substituting these into the given equation, we get

$$e^{z^2} - 3e^{-z^2} = 2$$
$$(e^{z^2})^2 - 2e^{z^2} - 3 = 0$$
$$(e^{z^2} - 3)(e^{z^2} + 1) = 0.$$

Thus $e^{z^2} = 3$ or $e^{z^2} = -1$. Since $e^{z^2} = -1$ is not possible, we have $e^{z^2} = 3$, so $z = \pm\sqrt{\ln 3}$.
The critical points are $(\sqrt{\ln 3}, -\sqrt{\ln 3})$, and $(-\sqrt{\ln 3}, \sqrt{\ln 3})$.

26. We will use the second derivative test to classify the two critical points calculated in Exercise 25. To calculate the second partials

$$A = \frac{\partial^2 z}{\partial x^2}, \qquad B = \frac{\partial^2 z}{\partial x \partial y}, \qquad C = \frac{\partial^2 z}{\partial y^2},$$

we differentiate the expressions $(*)$, and $(**)$ obtained in Exercise 25.
Differentiating $(*)$ with respect to x, we obtain

$$e^{2zx-x^2}\left[\left(2x\frac{\partial z}{\partial x} + 2z - 2x\right)^2 + 4\frac{\partial z}{\partial x} + 2x\frac{\partial^2 z}{\partial x^2} - 2\right]$$
$$- 3e^{2zy+y^2}\left[\left(2y\frac{\partial z}{\partial x}\right)^2 + 2y\frac{\partial^2 z}{\partial x^2}\right] = 0.$$

At a critical point, $\frac{\partial z}{\partial x} = 0$, $z = x$, $z = -y$, and $z^2 = \ln 3$, so

$$3\left(2x\frac{\partial^2 z}{\partial x^2} - 2\right) - \frac{3}{3}\left(2y\frac{\partial^2 z}{\partial x^2}\right) = 0,$$

$$A = \frac{\partial^2 z}{\partial x^2} = \frac{6}{6x - 2y}.$$

Differentiating $(**)$ with respect to y gives

$$e^{2zx-x^2}\left[\left(2x\frac{\partial z}{\partial y}\right)^2 + 2x\frac{\partial^2 z}{\partial y^2}\right]$$
$$- 3e^{2zy+y^2}\left[\left(2y\frac{\partial z}{\partial y} + 2z + 2y\right)^2 + 4\frac{\partial z}{\partial y} + 2y\frac{\partial^2 z}{\partial y^2} + 2\right] = 0,$$

and evaluation at a critical point gives

$$3\left(2x\frac{\partial^2 z}{\partial y^2}\right) - \frac{3}{3}\left(2y\frac{\partial^2 z}{\partial y^2} + 2\right) = 0,$$

$$C = \frac{\partial^2 z}{\partial y^2} = \frac{2}{6x - 2y}.$$

Finally, differentiating $(*)$ with respect to y gives

$$e^{2zx-x^2}\left[\left(2x\frac{\partial z}{\partial x} + 2z - 2x\right)\left(2x\frac{\partial z}{\partial y}\right) + 2x\frac{\partial^2 z}{\partial x \partial y} + 2\frac{\partial z}{\partial y}\right]$$
$$- 3e^{2zy+y^2}\left[\left(2y\frac{\partial z}{\partial y} + 2z + 2y\right)\left(2y\frac{\partial z}{\partial x}\right) + 2\frac{\partial z}{\partial x} + 2y\frac{\partial^2 z}{\partial x \partial y}\right] = 0,$$

and, evaluating at a critical point,

$$(6x - 2y)\frac{\partial^2 z}{\partial x \partial y} = 0,$$

so that

$$B = \frac{\partial^2 z}{\partial x \partial y} = 0.$$

At the critical point $(\sqrt{\ln 3}, -\sqrt{\ln 3})$ we have

$$A = \frac{6}{8\ln 3}, \qquad B = 0, \qquad C = \frac{2}{8\ln 3},$$

so $B^2 - AC < 0$, and f has a local minimum at that critical point.
At the critical point $(-\sqrt{\ln 3}, \sqrt{\ln 3})$ we have

$$A = -\frac{6}{8\ln 3}, \quad B = 0, \quad C = -\frac{2}{8\ln 3},$$

so $B^2 - AC < 0$, and f has a local maximum at that critical point.

27. $f(x, y) = (y - x^2)(y - 3x^2) = y^2 - 4x^2 y + 3x^4$
$f_1(x, y) = -8xy + 12x^3 = 4x(3x^2 - 2y)$
$f_2(x, y) = 2y - 4x^2$.

Since $f_1(0, 0) = f_2(0, 0) = 0$, therefore $(0, 0)$ is a critical point of f.
Let $g(x) = f(x, kx) = k^2 x^2 - 4kx^3 + 3x^4$. Then

$$g'(x) = 2k^2 x - 12kx^2 + 12x^3$$
$$g''(x) = 2k^2 - 24kx + 36x^2.$$

Since $g'(0) = 0$ and $g''(0) = 2k^2 > 0$ for $k \neq 0$, g has a local minimum value at $x = 0$. Thus $f(x, kx)$ has a local minimum at $x = 0$ if $k \neq 0$. Since $f(x, 0) = 3x^4$ and $f(0, y) = y^2$ both have local minimum values at $(0, 0)$, f has a local minimum at $(0, 0)$ when restricted to any straight line through the origin.
However, on the curve $y = 2x^2$ we have

$$f(x, 2x^2) = x^2(-x^2) = -x^4,$$

which has a local maximum value at the origin. Therefore f does *not* have an (unrestricted) local minimum value at $(0, 0)$.

Note that $A = f_{11}(0, 0) = (-8y + 36x^2)\big|_{(0,0)} = 0$
$B = f_{12}(0, 0) = -8x\big|_{(0,0)} = 0$.
Thus $AC = B^2$, and the second derivative test is indeterminate at the origin.

28. Given that

$$Q(u, v, w) = Au^2 + Bv^2 + Cw^2 + 2Duv + 2Euw + 2Fvw$$

and that

$$A > 0, \quad \begin{vmatrix} A & D \\ D & B \end{vmatrix} = AB - D^2 > 0$$

$$\begin{vmatrix} A & D & E \\ D & B & F \\ E & F & C \end{vmatrix} = ABC + 2DEF - BE^2 - CD^2 - AF^2 > 0,$$

we want to show that $Q(u, v, w) > 0$ if $(u, v, w) \neq (0, 0, 0)$. We have

$$Q(u, v, w)$$
$$= A\left[u^2 + 2u\frac{Dv + Ew}{A} + \left(\frac{Dv + Ew}{A}\right)^2\right]$$
$$+ \frac{AB - D^2}{A}v^2 + \frac{AC - E^2}{A}w^2 + \frac{2(AF - DE)}{A}vw$$
$$= A\left(u + \frac{Dv + Ew}{A}\right)^2$$
$$+ \frac{AB - D^2}{A}\left(v^2 + \frac{2(AF - DE)}{AB - D^2}vw + \left(\frac{AF - DE}{AB - D^2}\right)^2 w^2\right)$$
$$+ \left[\frac{AC - E^2}{A} - \frac{(AF - DE)^2}{A(AB - D^2)}\right]w^2$$
$$= A\left(u + \frac{Dv + Ew}{A}\right)^2 + \frac{AB - D^2}{A}\left(v + \frac{AF - DE}{AB - D^2}w\right)^2$$
$$+ \frac{A(ABC - BE^2 - AF^2 - CD^2 + 2DEF)}{A(AB - D^2)}w^2.$$

Under the given conditions, this expression is a sum of squares which is ≥ 0, and $\neq 0$ unless $w = 0$, $v = 0$ and $u = 0$.

29. $Q(w, x, y, z) = \mathbf{u}^T Q \mathbf{u}$, where

$$\mathbf{u} = \begin{pmatrix} w \\ x \\ y \\ z \end{pmatrix}, \quad Q = \begin{pmatrix} A & E & F & G \\ E & B & H & I \\ F & H & C & J \\ G & I & J & D \end{pmatrix}.$$

Q is positive definite if

$$A > 0, \quad \begin{vmatrix} A & E \\ E & B \end{vmatrix} > 0, \quad \begin{vmatrix} A & E & F \\ E & B & H \\ F & H & C \end{vmatrix} > 0,$$

$$\begin{vmatrix} A & E & F & G \\ E & B & H & I \\ F & H & C & J \\ G & I & J & D \end{vmatrix} > 0.$$

Section 13.2 Extreme Values of Functions Defined on Restricted Domains (page 785)

1. $f(x, y) = x - x^2 + y^2$ on
$R = \{(x, y) : 0 \leq x \leq 2, \ 0 \leq y \leq 1\}$.
For critical points:

$$0 = f_1(x, y) = 1 - 2x, \quad 0 = f_2(x, y) = 2y.$$

The only CP is $(1/2, 0)$, which lies on the boundary of R. The boundary consists of four segments; we investigate each.
On $x = 0$ we have $f(x, y) = f(0, y) = y^2$ for $0 \le y \le 1$, which has minimum value 0 and maximum value 1.
On $y = 0$ we have $f(x, y) = f(x, 0) = x - x^2 = g(x)$ for $0 \le x \le 2$. Since $g'(x) = 1 - 2x = 0$ at $x = 1/2$, $g(1/2) = 1/4$, $g(0) = 0$, and $g(2) = -2$, the maximum and minimum values of f on the boundary segment $y = 0$ are $1/4$ and -2 respectively.
On $x = 2$ we have $f(x, y) = f(2, y) = -2 + y^2$ for $0 \le y \le 1$, which has minimum value -2 and maximum value -1.
On $y = 1$, $f(x, y) = f(x, 1) = x - x^2 + 1 = g(x) + 1$ for $0 \le x \le 2$. Thus the maximum and minimum values of f on the boundary segment $y = 1$ are $5/4$ and -1 respectively.
Overall, f has maximum value $5/4$ and minimum value -2 on the rectangle R.

2. $f(x, y) = xy - 2x$ on
$R = \{(x, y) : -1 \le x \le 1, \; 0 \le y \le 1\}$.
For critical points:
$$0 = f_1(x, y) = y - 2, \quad 0 = f_2(x, y) = x.$$

The only CP is $(0, 2)$, which lies outside R. Therefore the maximum and minimum values of f on R lie on one of the four boundary segments of R.
On $x = -1$ we have $f(-1, y) = 2 - y$ for $0 \le y \le 1$, which has maximum value 2 and minimum value 1.
On $x = 1$ we have $f(1, y) = y - 2$ for $0 \le y \le 1$, which has maximum value -1 and minimum value -2.
On $y = 0$ we have $f(x, 0) = -2x$ for $-1 \le x \le 1$, which has maximum value 2 and minimum value -2.
On $y = 1$ we have $f(x, 1) = -x$ for $-1 \le x \le 1$, which has maximum value 1 and minimum value -1.
Thus the maximum and minimum values of f on the rectangle R are 2 and -2 respectively.

3. $f(x, y) = xy - y^2$ on $D = \{(x, y) : x^2 + y^2 \le 1\}$.
For critical points:
$$0 = f_1(x, y) = y, \quad 0 = f_2(x, y) = x - 2y.$$

The only CP is $(0, 0)$, which lies inside D. We have $f(0, 0) = 0$.
The boundary of D is the circle $x = \cos t$, $y = \sin t$, $-\pi \le t \le \pi$. On this circle we have
$$g(t) = f(\cos t, \sin t) = \cos t \sin t - \sin^2 t$$
$$= \frac{1}{2}\big[\sin 2t + \cos 2t - 1\big], \quad (-\pi \le t \le \pi).$$
$$g(0) = g(2\pi) = 0$$
$$g'(t) = \cos 2t - \sin 2t.$$

The critical points of g satisfy $\cos 2t = \sin 2t$, that is, $\tan 2t = 1$, so $2t = \pm\dfrac{\pi}{4}$ or $\pm\dfrac{5\pi}{4}$, and $t = \pm\dfrac{\pi}{8}$ or $\pm\dfrac{5\pi}{8}$.
We have
$$g\left(\frac{\pi}{8}\right) = \frac{1}{2\sqrt{2}} - \frac{1}{2} + \frac{1}{2\sqrt{2}} = \frac{1}{\sqrt{2}} - \frac{1}{2} > 0$$
$$g\left(-\frac{\pi}{8}\right) = -\frac{1}{2\sqrt{2}} - \frac{1}{2} + \frac{1}{2\sqrt{2}} = -\frac{1}{2}$$
$$g\left(\frac{5\pi}{8}\right) = -\frac{1}{2\sqrt{2}} - \frac{1}{2} - \frac{1}{2\sqrt{2}} = -\frac{1}{\sqrt{2}} - \frac{1}{2}$$
$$g\left(-\frac{5\pi}{8}\right) = \frac{1}{2\sqrt{2}} - \frac{1}{2} - \frac{1}{2\sqrt{2}} = -\frac{1}{2}.$$

Thus the maximum and minimum values of f on the disk D are $\dfrac{1}{\sqrt{2}} - \dfrac{1}{2}$ and $-\dfrac{1}{\sqrt{2}} - \dfrac{1}{2}$ respectively.

4. $f(x, y) = x + 2y$ on the closed disk $x^2 + y^2 \le 1$. Since $f_1 = 1$ and $f_2 = 2$, f has no critical points, and the maximum and minimum values of f, which must exist because f is continuous on a closed, bounded set in the plane, must occur at boundary points of the domain, that is, points of the circle $x^2 + y^2 = 1$. This circle can be parametrized $x = \cos t$, $y = \sin t$, so that
$$f(x, y) = f(\cos t, \sin t) = \cos t + 2\sin t = g(t), \text{ say.}$$
For critical points of g: $0 = g'(t) = -\sin t + 2\cos t$. Thus $\tan t = 2$, and $x = \pm 1/\sqrt{5}$, $y = \pm 2/\sqrt{5}$. The critical points are $(-1/\sqrt{5}, -2/\sqrt{5})$, where f has value $-\sqrt{5}$, and $(1/\sqrt{5}, 2/\sqrt{5})$, where f has value $\sqrt{5}$. Thus the maximum and minimum values of $f(x, y)$ on the disk are $\sqrt{5}$ and $-\sqrt{5}$ respectively.

5. $f(x, y) = xy - x^3 y^2$ on the square S: $0 \le x \le 1$, $0 \le y \le 1$.
$f_1 = y - 3x^2 y^2 = y(1 - 3x^2 y)$,
$f_2 = x - 2x^3 y = x(1 - 2x^2 y)$.
$(0, 0)$ is a critical point. Any other critical points must satisfy $3x^2 y = 1$ and $2x^2 y = 1$, that is, $x^2 y = 0$. Therefore $(0, 0)$ is the only critical point, and it is on the boundary of S. We need therefore only consider the values of f on the boundary of S.
On the sides $x = 0$ and $y = 0$ of S, $f(x, y) = 0$.
On the side $x = 1$ we have $f(1, y) = y - y^2 = g(y)$, $(0 \le y \le 1)$. g has maximum value $1/4$ at its critical point $y = 1/2$.
On the side $y = 1$ we have $f(x, 1) = x - x^3 = h(x)$, $(0 \le x \le 1)$. h has critical point given by $1 - 3x^2 = 0$; only $x = 1/\sqrt{3}$ is on the side of S.
$$h\left(\frac{1}{\sqrt{3}}\right) = \frac{2}{3\sqrt{3}} > \frac{1}{4}.$$
On the square S, $f(x, y)$ has minimum value 0 (on the sides $x = 0$ and $y = 0$ and at the corner $(1, 1)$ of the square), and maximum value $2/(3\sqrt{3})$ at the point $(1/\sqrt{3}, 1)$. There is a smaller local maximum value at $(1, 1/2)$.

6. $f(x, y) = xy(1 - x - y)$ on the triangle T shown in the figure. Evidently $f(x, y) = 0$ on all three boundary segments of T, and $f(x, y) > 0$ inside T. Thus the minimum value of f on T is 0, and the maximum value must occur at an interior critical point. For critical points:
$$0 = f_1(x, y) = y(1-2x-y), \qquad 0 = f_2(x, y) = x(1-x-2y).$$

The only critical points are $(0, 0)$, $(1, 0)$ and $(0, 1)$, which are on the boundary of T, and $(1/3, 1/3)$, which is inside T. The maximum value of f over T is $f(1/3, 1/3) = 1/27$.

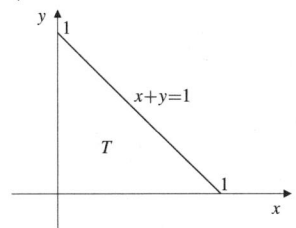

Fig. 13.2.6

7. Since $-1 \leq f(x, y) = \sin x \cos y \leq 1$ everywhere, and since $f(\pi/2, 0) = 1$, $f(3\pi/2, 0) = -1$, and both $(\pi/2, 0)$ and $(3\pi/2, 0)$ belong to the triangle bounded by $x = 0$, $y = 0$ and $x + y = 2\pi$, therefore the maximum and minimum values of f over that triangle must be 1 and -1 respectively.

8. $f(x, y) = \sin x \sin y \sin(x + y)$ on the triangle T shown in the figure. Evidently $f(x, y) = 0$ on the boundary of T, and $f(x, y) > 0$ at all points inside T. Thus the minimum value of f on T is zero, and the maximum value must occur at an interior critical point. For critical points inside T we must have
$$0 = f_1(x, y) = \cos x \sin y \sin(x + y) + \sin x \sin y \cos(x + y)$$
$$0 = f_2(x, y) = \sin x \cos y \sin(x + y) + \sin x \sin y \cos(x + y).$$
Therefore $\cos x \sin y = \cos y \sin x$, which implies $x = y$ for points inside T, and
$$\cos x \sin x \sin 2x + \sin^2 x \cos 2x = 0$$
$$2 \sin^2 x \cos^2 x + 2 \sin^2 x \cos^2 x - \sin^2 x = 0$$
$$4 \cos^2 x = 1.$$

Thus $\cos x = \pm 1/2$, and $x = \pm \pi/3$. The interior critical point is $(\pi/3, \pi/3)$, where f has the value $3\sqrt{3}/8$. This is the maximum value of f on T.

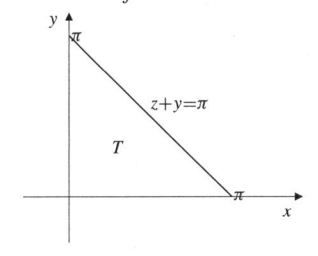

Fig. 13.2.8

9. $T = (x + y)e^{-x^2 - y^2}$ on $D = \{(x, y) : x^2 + y^2 \leq 1\}$. For critical points:
$$0 = \frac{\partial T}{\partial x} = \left(1 - 2x(x + y)\right)e^{-x^2 - y^2}$$
$$0 = \frac{\partial T}{\partial y} = \left(1 - 2y(x + y)\right)e^{-x^2 - y^2}.$$

The critical points are given by $2x(x+y) = 1 = 2y(x+y)$, which forces $x = y$ and $4x^2 = 1$, so $x = y = \pm\frac{1}{2}$.
The two critical points are $\left(\frac{1}{2}, \frac{1}{2}\right)$ and $\left(-\frac{1}{2}, -\frac{1}{2}\right)$, both of which lie inside D. T takes the values $\pm e^{-1/2}$ at these points.

On the boundary of D, $x = \cos t$, $y = \sin t$, $0 \leq t \leq 2\pi$, so that
$$T = (\cos t + \sin t)e^{-1} = g(t), \qquad (0 \leq t \leq 2\pi).$$

We have $g(0) = g(2\pi) = e^{-1}$. For critical points of g:
$$0 = g'(t) = (\cos t - \sin t)e^{-1},$$

so $\tan t = 1$ and $t = \pi/4$ or $t = 5\pi/4$. Observe that $g(\pi/4) = \sqrt{2}e^{-1}$, and $g(5\pi/4) = -\sqrt{2}e^{-1}$.
Since $e^{-1/2} > \sqrt{2}e^{-1}$ (because $e > 2$), the maximum and minimum values of T on the disk are $\pm e^{-1/2}$, the values at the interior critical points.

10. $f(x, y) = \dfrac{x - y}{1 + x^2 + y^2}$ on the half-plane $y \geq 0$. For critical points:
$$0 = f_1(x, y) = \frac{1 - x^2 + y^2 + 2xy}{(1 + x^2 + y^2)^2}$$
$$0 = f_2(x, y) = \frac{-1 - x^2 + y^2 - 2xy}{(1 + x^2 + y^2)^2}.$$

Any critical points must satisfy $1 - x^2 + y^2 + 2xy = 0$ and $-1 - x^2 + y^2 - 2xy = 0$, and hence $x^2 = y^2$ and $2xy = -1$. Therefore $y = -x = \pm 1/\sqrt{2}$. The only critical point in the region $y \geq 0$ is $(-1/\sqrt{2}, 1/\sqrt{2})$, where f has the value $-1/\sqrt{2}$.

On the boundary $y = 0$ we have
$$f(x, 0) = \frac{x}{1 + x^2} = g(x), \qquad (-\infty < x < \infty).$$

Evidently, $g(x) \to 0$ as $x \to \pm\infty$.
Since $g'(x) = \dfrac{1-x^2}{(1+x^2)^2}$, the critical points of g are $x = \pm 1$. We have $g(\pm 1) = \pm\dfrac{1}{2}$.
The maximum and minimum values of f on the upper half-plane $y \geq 0$ are $1/2$ and $-1/\sqrt{2}$ respectively.

11. Let $f(x, y, z) = xy^2 + yz^2$ on the ball B: $x^2 + y^2 + z^2 \leq 1$. First look for interior critical points:
$$0 = f_1 = y^2, \quad 0 = f_2 = 2xy + z^2, \quad 0 = f_3 = 2yz.$$

All points on the x-axis are CPs, and $f = 0$ at all such points.

Now consider the boundary sphere $z^2 = 1 - x^2 - y^2$. On it
$$f(x, y, z) = xy^2 + y(1-x^2-y^2) = xy^2 + y - x^2 y - y^3 = g(x, y),$$
where g is defined for $x^2 + y^2 \leq 1$. Look for interior CPs of g:
$$0 = g_1 = y^2 - 2xy = y(y - 2x)$$
$$0 = g_2 = 2xy + 1 - x^2 - 3y^2.$$
Case I: $y = 0$. Then $g = 0$ and $f = 0$.
Case II: $y = 2x$. Then $4x^2 + 1 - x^2 - 12x^2 = 0$, so $9x^2 = 1$ and $x = \pm 1/3$. This case produces critical points
$$\left(\frac{1}{3}, \frac{2}{3}, \pm\frac{2}{3}\right), \quad \text{where } f = \frac{12}{27}, \quad \text{and}$$
$$\left(-\frac{1}{3}, -\frac{2}{3}, \pm\frac{2}{3}\right), \quad \text{where } f = -\frac{12}{27}.$$

Now we must consider the boundary $x^2 + y^2 = 1$ of the domain of g. Here
$$g(x, y) = xy^2 = x(1 - x^2) = x - x^3 = h(x)$$
for $-1 \leq x \leq 1$. At the endpoints $x = \pm 1$, $h = 0$, so $g = 0$ and $f = 0$. For CPs of h:
$$0 = h'(x) = 1 - 3x^2,$$
so $x = \pm 1/\sqrt{3}$ and $y = \pm\sqrt{2/3}$. The value of h at such points is $\pm 2/(3\sqrt{3})$. However $2/(3\sqrt{3}) < 12/27$, so the maximum value of f is $12/27$, and the minimum value is $-12/27$.

12. Let $f(x, y, z) = xz + yz$ on the ball $x^2 + y^2 + z^2 \leq 1$. First look for interior critical points:
$$0 = f_1 = z, \quad 0 = f_2 = z, \quad 0 = f_3 = x + y.$$

All points on the line $z = 0$, $x + y = 0$ are CPs, and $f = 0$ at all such points.

Now consider the boundary sphere $x^2 + y^2 + z^2 = 1$. On it
$$f(x, y, z) = (x+y)z = \pm(x+y)\sqrt{1-x^2-y^2} = g(x, y),$$
where g has domain $x^2 + y^2 \leq 1$. On the boundary of its domain, g is identically 0, although g takes both positive and negative values at some points inside its domain. Therefore, we need consider only critical points of g in $x^2 + y^2 < 1$. For such CPs:
$$0 = g_1 = \sqrt{1-x^2-y^2} + \frac{(x+y)(-2x)}{2\sqrt{1-x^2-y^2}}$$
$$= \frac{1-x^2-y^2-x^2-xy}{\sqrt{1-x^2-y^2}}$$
$$0 = g_2 = \frac{1-x^2-y^2-xy-y^2}{\sqrt{1-x^2-y^2}}.$$

Therefore $2x^2 + y^2 + xy = 1 = x^2 + 2y^2 + xy$, from which $x^2 = y^2$.
Case I: $x = -y$. Then $g = 0$, so $f = 0$.
Case II: $x = y$. Then $2x^2 + x^2 + x^2 = 1$, so $x^2 = 1/4$ and $x = \pm 1/2$. g (which is really two functions depending on our choice of the "+" or "−" sign) has four CPs, two corresponding to $x = y = 1/2$ and two to $x = y = -1/2$. The values of g at these four points are $\pm 1/\sqrt{2}$.

Since we have considered all points where f can have extreme values, we conclude that the maximum value of f on the ball is $1/\sqrt{2}$ (which occurs at the boundary points $\pm(\frac{1}{2}, \frac{1}{2}, \frac{1}{\sqrt{2}}))$ and minimum value $-1/\sqrt{2}$ (which occurs at the boundary points $\pm(\frac{1}{2}, \frac{1}{2}, -\frac{1}{\sqrt{2}}))$.

13. $f(x, y) = xye^{-xy}$ on $Q = \{(x, y) : x \geq 0, y \geq 0\}$.
Since $f(x, kx) = kx^2 e^{-kx^2} \to 0$ as $x \to \infty$ if $k > 0$, and $f(x, 0) = f(0, y) = 0$, we have $f(x, y) \to 0$ as (x, y) recedes to infinity along any straight line from the origin lying in the first quadrant Q.
However, $f\left(x, \dfrac{1}{x}\right) = 1$ and $f(x, 0) = 0$ for all $x > 0$, even though the points $\left(x, \dfrac{1}{x}\right)$ and $(x, 0)$ become arbitrarily close together as x increases. Thus f does not have a limit as $x^2 + y^2 \to \infty$.
Observe that $f(x, y) = re^{-r} = g(r)$ on the hyperbola $xy = r > 0$. Since $g(r) \to 0$ as r approaches 0 or ∞, and
$$g'(r) = (1-r)e^{-r} = 0 \quad \Rightarrow \quad r = 1,$$
$f(x, y)$ is everywhere on Q less than $g(1) = 1/e$. Thus f does have a maximum value on Q.

14. $f(x, y) = xy^2 e^{-xy}$ on $Q = \{(x, y) : x \geq 0, y \geq 0\}$.
As in Exercise 13, $f(x, 0) = f(0, y) = 0$ and $\lim_{x \to \infty} f(x, kx) = k^2 x^3 e^{-x^2} = 0$.
Also, $f(0, y) = 0$ while $f\left(\dfrac{1}{y}, y\right) = \dfrac{y}{e} \to \infty$ as $y \to \infty$, so that f has no limit as $x^2 + y^2 \to \infty$ in Q, and f has no maximum value on Q.

15. If brewery A produces x litres per month and brewery B produces y litres per month, then the monthly profits of the two breweries are given by

$$P = 2x - \frac{2x^2 + y^2}{10^6}, \qquad Q = 2y - \frac{4y^2 + x^2}{2 \times 10^6}.$$

STRATEGY I. Each brewery selects its production level to maximize its own profit, and assumes its competitor does the same.
Then A chooses x to satisfy

$$0 = \frac{\partial P}{\partial x} = 2 - \frac{4x}{10^6} \quad \Rightarrow \quad x = 5 \times 10^5.$$

B chooses y to satisfy

$$0 = \frac{\partial Q}{\partial y} = 2 - \frac{8y}{2 \times 10^6} \quad \Rightarrow \quad y = 5 \times 10^5.$$

The total profit of the two breweries under this strategy is

$$P + Q = 10^6 - \frac{3 \times 25 \times 10^{10}}{10^6} + 10^6 - \frac{5 \times 25 \times 10^{10}}{2 \times 10^6}$$
$$= \$625,000.$$

STRATEGY II. The two breweries cooperate to maximize the total profit

$$T = P + Q = 2x + 2y - \frac{5x^2 + 6y^2}{2 \times 10^6}$$

by choosing x and y to satisfy

$$0 = \frac{\partial T}{\partial x} = 2 - \frac{10x}{2 \times 10^6},$$
$$0 = \frac{\partial T}{\partial y} = 2 - \frac{12y}{2 \times 10^6}.$$

Thus $x = 4 \times 10^5$ and $y = \frac{1}{3} \times 10^6$.
In this case the total monthly profit is

$$P + Q = 8 \times 10^5 + \frac{2}{3} \times 10^6 - \frac{80 \times 10^{10} + \frac{2}{3} \times 10^{12}}{2 \times 10^6}$$
$$\approx \$733,333.$$

Observe that the total profit is larger if the two breweries cooperate and fix prices to maximize it.

16. Let the dimensions be as shown in the figure. Then $2x + y = 100$, the length of the fence. For maximum area A of the enclosure we will have $x > 0$ and $0 < \theta < \pi/2$. Since $h = x \cos \theta$, the area A is

$$A = xy \cos \theta + 2 \times \frac{1}{2}(x \sin \theta)(x \cos \theta)$$
$$= x(100 - 2x) \cos \theta + x^2 \sin \theta \cos \theta$$
$$= (100x - 2x^2) \cos \theta + \frac{1}{2}x^2 \sin 2\theta.$$

We look for a critical point of A satisfying $x > 0$ and $0 < \theta < \pi/2$.

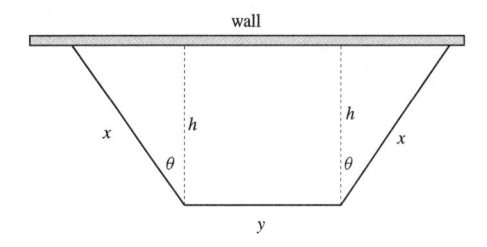

Fig. 13.2.16

$$0 = \frac{\partial A}{\partial x} = (100 - 4x) \cos \theta + x \sin 2\theta$$
$$\Rightarrow \cos \theta(100 - 4x + 2x \sin \theta) = 0$$
$$\Rightarrow 4x - 2x \sin \theta = 100 \Rightarrow x = \frac{50}{2 - \sin \theta}$$
$$0 = \frac{\partial A}{\partial \theta} = -(100x - 2x^2) \sin \theta + x^2 \cos 2\theta$$
$$\Rightarrow x(1 - 2\sin^2 \theta) + 2x \sin \theta - 100 \sin \theta = 0.$$

Substituting the first equation into the second we obtain

$$\frac{50}{2 - \sin \theta}(1 - 2\sin^2 \theta + 2\sin \theta) - 100 \sin \theta = 0$$
$$50(1 - 2\sin^2 \theta + 2\sin \theta) = 100(2\sin \theta - \sin^2 \theta)$$
$$50 = 100 \sin \theta.$$

Thus $\sin \theta = 1/2$, and $\theta = \pi/6$.
Therefore $x = \frac{50}{2 - (1/2)} = \frac{100}{3}$, and $y = 100 - 2x = \frac{100}{3}$.
The maximum area for the enclosure is

$$A = \left(\frac{100}{3}\right)^2 \frac{\sqrt{3}}{2} + \left(\frac{100}{3}\right)^2 \frac{1}{2}\frac{\sqrt{3}}{2} = \frac{2500}{\sqrt{3}}$$

square units. All three segments of the fence will be the same length, and the bend angles will be $120°$.

17. To maximize $Q(x, y) = 2x + 3y$ subject to

$$x \geq 0, \quad y \geq 0, \quad y \leq 5, \quad x + 2y \leq 12, \quad 4x + y \leq 12.$$

The constraint region is shown in the figure.

SECTION 13.2 (PAGE 785)

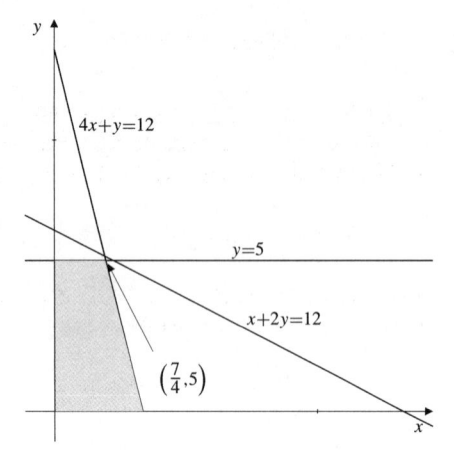

Fig. 13.2.17

Observe that any point satisfying $y \le 5$ and $4x + y \le 12$ automatically satisfies $x + 2y \le 12$. Since $y = 5$ and $4x + y = 12$ intersect at $\left(\frac{7}{4}, 5\right)$, the maximum value of $Q(x, y)$ subject to the given constraints is

$$Q\left(\frac{7}{4}, 5\right) = \frac{7}{2} + 15 = \frac{37}{2}.$$

18. Minimize $F(x, y, z) = 2x + 3y + 4z$ subject to

$$x \ge 0, \qquad y \ge 0, \qquad z \ge 0,$$
$$x + y \ge 2, \qquad y + z \ge 2, \qquad x + z \ge 2.$$

Here the constraint region has vertices $(1, 1, 1)$, $(2, 2, 0)$, $(2, 0, 2)$, and $(0, 2, 2)$. Since $F(1, 1, 1) = 9$, $F(2, 2, 0) = 10$, $F(2, 0, 2) = 12$, and $F(0, 2, 2) = 14$, the minimum value of F subject to the constraints is 9.

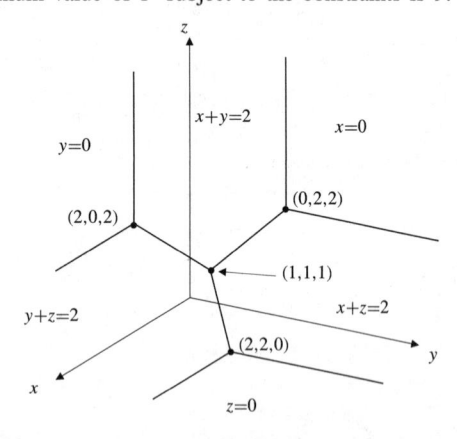

Fig. 13.2.18

19. Suppose that x kg of deluxe fabric and y kg of standard fabric are produced. Then the total revenue is

$$R = 3x + 2y.$$

The constraints imposed by raw material availability are

$$\frac{20}{100}x + \frac{10}{100}y \le 2,000, \Leftrightarrow 2x + y \le 20,000$$
$$\frac{50}{100}x + \frac{40}{100}y \le 6,000, \Leftrightarrow 5x + 4y \le 60,000$$
$$\frac{30}{100}x + \frac{50}{100}y \le 6,000, \Leftrightarrow 3x + 5y \le 60,000.$$

The lines $2x + y = 20,000$ and $5x + 4y = 60,000$ intersect at the point $\left(\frac{20,000}{3}, \frac{20,000}{3}\right)$, which satisfies $3x + 5y \le 60,000$, so lies in the constraint region. We have

$$f\left(\frac{20,000}{3}, \frac{20,000}{3}\right) \approx 33,333.$$

The lines $2x + y = 20,000$ and $3x + 5y = 60,000$ intersect at the point $\left(\frac{40,000}{7}, \frac{60,000}{7}\right)$, which does not satisfy $5x + 4y \le 60,000$ and so does not lie in the constraint region.

The lines $5x + 4y = 60,000$ and $3x + 5y = 60,000$ intersect at the point $\left(\frac{60,000}{13}, \frac{120,000}{13}\right)$, which satisfies $2x + y \le 20,000$ and so lies in the constraint region. We have

$$f\left(\frac{60,000}{13}, \frac{120,000}{13}\right) \approx 32,307.$$

To produce the maximum revenue, the manufacturer should produce $20,000/3 \approx 6,667$ kg of each grade of fabric.

20. If the developer builds x houses, y duplex units, and z apartments, his profit will be

$$P = 40,000x + 20,000y + 16,000z.$$

The legal constraints imposed require that

$$\frac{x}{6} + \frac{y}{8} + \frac{z}{12} \le 10, \quad \text{that is } 4x + 3y + 2z \le 240,$$

and also

$$z \ge x + y.$$

Evidently we must also have $x \ge 0$, $y \ge 0$, and $z \ge 0$. The planes $4x + 3y + 2z = 240$ and $z = x + y$ intersect where $6x + 5y = 240$. Thus the constraint region has vertices $(0, 0, 0)$, $(40, 0, 40)$, $(0, 48, 48)$, and $(0, 0, 120)$, which yield revenues of \$0, \$2,240,000, \$1,728,000, and \$1,920,000 respectively.

For maximum profit, the developer should build 40 houses, no duplex units, and 40 apartments.

Section 13.3 Lagrange Multipliers
(page 793)

1. First we observe that $f(x, y) = x^3 y^5$ must have a maximum value on the line $x + y = 8$ because if $x \to -\infty$ then $y \to \infty$ and if $x \to \infty$ then $y \to -\infty$. In either case $f(x, y) \to -\infty$.
Let $L = x^3 y^5 + \lambda(x + y - 8)$. For CPs of L:

$$0 = \frac{\partial L}{\partial x} = 3x^2 y^5 + \lambda$$

$$0 = \frac{\partial L}{\partial y} = 5x^3 y^4 + \lambda$$

$$0 = \frac{\partial L}{\partial \lambda} = x + y - 8.$$

The first two equations give $3x^2 y^5 = 5x^3 y^4$, so that either $x = 0$ or $y = 0$ or $3y = 5x$. If $x = 0$ or $y = 0$ then $f(x, y) = 0$. If $3y = 5x$, then $x + \frac{5}{3}x = 8$, so $8x = 24$ and $x = 3$. Then $y = 5$, and $f(x, y) = 3^3 5^5 = 84,375$. This is the maximum value of f on the line.

2. a) Let D be the distance from $(3, 0)$ to the point (x, y) on the curve $y = x^2$. Then

$$D^2 = (x - 3)^2 + y^2 = (x - 3)^2 + x^4.$$

For a minimum, $0 = \dfrac{dD^2}{dx} = 2(x - 3) + 4x^3$. Thus $2x^3 + x - 3 = 0$. Clearly $x = 1$ is a root of this cubic equation. Since

$$\frac{2x^3 + x - 3}{x - 1} = 2x^2 + 2x + 3,$$

and $2x^2 + 2x + 3$ has negative discriminant, $x = 1$ is the only critical point. Thus the minimum distance from $(3, 0)$ to $y = x^2$ is $D = \sqrt{(-2)^2 + 1^4} = \sqrt{5}$ units.

b) We want to minimize $D^2 = (x-3)^2 + y^2$ subject to the constraint $y = x^2$. Let $L = (x - 3)^2 + y^2 + \lambda(x^2 - y)$. For critical points of L we want

$$0 = \frac{\partial L}{\partial x} = 2(x - 3) + 2\lambda x$$
$$\Rightarrow (1 + \lambda)x - 3 = 0 \qquad (A)$$

$$0 = \frac{\partial L}{\partial y} = 2y - \lambda \qquad (B)$$

$$0 = \frac{\partial L}{\partial \lambda} = x^2 - y. \qquad (C)$$

Eliminating λ from (A) and (B), we get
$x + 2xy - 3 = 0$.
Substituting (C) then leads to $2x^3 + x - 3 = 0$, or $(x - 1)(2x^2 + 2x + 3) = 0$. The only real solution is $x = 1$, so the point on $y = x^2$ closest to $(3, 0)$ is $(1, 1)$.
Thus the minimum distance from $(3, 0)$ to $y = x^2$ is $D = \sqrt{(1 - 3)^2 + 1^2} = \sqrt{5}$ units.

3. Let (X, Y, Z) be the point on the plane $x + 2y + 2z = 3$ closest to $(0, 0, 0)$.

a) The vector $\nabla(x+2y+2z) = \mathbf{i}+2\mathbf{j}+2\mathbf{k}$ is perpendicular to the plane, so must be parallel to the vector $X\mathbf{i} + Y\mathbf{j} + Z\mathbf{k}$ from the origin to (X, Y, Z). Thus

$$X\mathbf{i} + Y\mathbf{j} + Z\mathbf{k} = t(\mathbf{i} + 2\mathbf{j} + 2\mathbf{k}),$$

for some scalar t. Thus $X = t$, $Y = 2t$, $Z = 2t$, and, since (X, Y, Z) lies on the plane,

$$3 = X + 2Y + 2Z = t + 4t + 4t = 9t.$$

Thus $t = \frac{1}{3}$, and we have $X = \frac{1}{3}$ and $Y = Z = \frac{2}{3}$.
The minimum distance from the origin to the plane is therefore $\frac{1}{3}\sqrt{1 + 4 + 4} = 1$ unit.

b) (X, Y, Z) must minimize the square of the distance from the origin to (x, y, z) on the plane. Thus it is a critical point of $S = x^2 + y^2 + z^2$. Since $x + 2y + 2z = 3$, we have $x = 3 - 2(y + z)$, and

$$S = S(y, z) = \big(3 - 2(y + z)\big)^2 + y^2 + z^2.$$

The critical points of this function are given by

$$0 = \frac{\partial S}{\partial y} = -4\big(3 - 2(y + z)\big) + 2y = -12 + 10y + 8z$$

$$0 = \frac{\partial S}{\partial z} = -4\big(3 - 2(y + z)\big) + 2z = -12 + 8y + 10z.$$

Therefore $Y = Z = \frac{2}{3}$ and $X = \frac{1}{3}$, and the distance is 1 unit as in part (a).

c) The point (X, Y, Z) must be a critical point of the Lagrangian function

$$L = x^2 + y^2 + z^2 + \lambda(x + 2y + 2z - 3).$$

To find these critical points we have

$$0 = \frac{\partial L}{\partial x} = 2x + \lambda$$

$$0 = \frac{\partial L}{\partial y} = 2y + 2\lambda$$

$$0 = \frac{\partial L}{\partial z} = 2z + 2\lambda$$

$$0 = \frac{\partial L}{\partial \lambda} = x + 2y + 2z - 3.$$

The first three equations yield $y = z = -\lambda$, $x = -\lambda/2$. Substituting these into the fourth equation we get $\lambda = -\frac{2}{3}$, so that the critical point is once again $\left(\frac{1}{3}, \frac{2}{3}, \frac{2}{3}\right)$, whose distance from the origin is 1 unit.

4. Let $f(x, y, z) = x + y - z$, and define the Lagrangian

$$L = x + y - z + \lambda(x^2 + y^2 + z^2 - 1).$$

Solutions to the constrained problem will be found among the critical points of L. To find these we have

$$0 = \frac{\partial L}{\partial x} = 1 + 2\lambda x,$$
$$0 = \frac{\partial L}{\partial y} = 1 + 2\lambda y,$$
$$0 = \frac{\partial L}{\partial z} = -1 + 2\lambda z,$$
$$0 = \frac{\partial L}{\partial \lambda} = x^2 + y^2 + z^2 - 1.$$

Therefore $2\lambda x = 2\lambda y = -2\lambda z$. Either $\lambda = 0$ or $x = y = -z$. $\lambda = 0$ is not possible. (It implies $0 = 1$ from the first equation.) From $x = y = -z$ we obtain $1 = x^2 + y^2 + z^2 = 3x^2$, so $x = \pm\frac{1}{\sqrt{3}}$. L has critical points at $\left(\frac{1}{\sqrt{3}}, \frac{1}{\sqrt{3}}, -\frac{1}{\sqrt{3}}\right)$ and $\left(\frac{1}{-\sqrt{3}}, -\frac{1}{\sqrt{3}}, \frac{1}{\sqrt{3}}\right)$. At the first $f = \sqrt{3}$, which is the maximum value of f on the sphere; at the second $f = -\sqrt{3}$, which is the minimum value.

5. The distance D from $(2, 1, -2)$ to (x, y, z) is given by

$$D^2 = (x - 2)^2 + (y - 1)^2 + (z + 2)^2.$$

We can extremize D by extremizing D^2. If (x, y, z) lies on the sphere $x^2 + y^2 + z^2 = 1$, we should look for critical points of the Lagrangian

$$L = (x - 2)^2 + (y - 1)^2 + (z + 2)^2 + \lambda(x^2 + y^2 + z^2 - 1).$$

Thus

$$0 = \frac{\partial L}{\partial x} = 2(x - 2) + 2\lambda x \quad \Leftrightarrow \quad x = \frac{2}{1 + \lambda}$$
$$0 = \frac{\partial L}{\partial y} = 2(y - 1) + 2\lambda y \quad \Leftrightarrow \quad y = \frac{1}{1 + \lambda}$$
$$0 = \frac{\partial L}{\partial z} = 2(z + 2) + 2\lambda z \quad \Leftrightarrow \quad z = \frac{-2}{1 + \lambda}$$
$$0 = \frac{\partial L}{\partial \lambda} = x^2 + y^2 + z^2 - 1.$$

Substituting the solutions of the first three equations into the fourth, we obtain

$$\frac{1}{(1 + \lambda)^2}(4 + 1 + 4) = 1$$
$$(1 + \lambda)^2 = 9$$
$$1 + \lambda = \pm 3.$$

Thus we must consider the two points $P = \left(\frac{2}{3}, \frac{1}{3}, -\frac{2}{3}\right)$, and $Q = \left(-\frac{2}{3}, -\frac{1}{3}, \frac{2}{3}\right)$ for giving extreme values for D. At P, $D = 2$. At Q, $D = 4$. Thus the greatest and least distances from $(2, 1, -2)$ to the sphere $x^2 + y^2 + z^2 = 1$ are 4 units and 2 units respectively.

6. Let $L = x^2 + y^2 + z^2 + \lambda(xyz^2 - 2)$. For critical points:

$$0 = \frac{\partial L}{\partial x} = 2x + \lambda yz^2 \quad \Leftrightarrow \quad -\lambda xyz^2 = 2x^2$$
$$0 = \frac{\partial L}{\partial y} = 2y + \lambda xz^2 \quad \Leftrightarrow \quad -\lambda xyz^2 = 2y^2$$
$$0 = \frac{\partial L}{\partial z} = 2z + 2\lambda xyz \quad \Leftrightarrow \quad -\lambda xyz^2 = z^2$$
$$0 = \frac{\partial L}{\partial \lambda} = xyz^2 - 2.$$

From the first three equations, $x^2 = y^2$ and $z^2 = 2x^2$. The fourth equation then gives $x^2 y^2 4z^4 = 4$, or $x^8 = 1$. Thus $x^2 = y^2 = 1$ and $z^2 = 2$.
The shortest distance from the origin to the surface $xyz^2 = 2$ is

$$\sqrt{1 + 1 + 2} = 2 \text{ units.}$$

7. We want to minimize $V = \frac{4\pi abc}{3}$ subject to the constraint $\frac{1}{a^2} + \frac{4}{b^2} + \frac{1}{c^2} = 1$. Note that abc cannot be zero. Let

$$L = \frac{4\pi abc}{3} + \lambda\left(\frac{1}{a^2} + \frac{4}{b^2} + \frac{1}{c^2} - 1\right).$$

For critical points of L:

$$0 = \frac{\partial L}{\partial a} = \frac{4\pi bc}{3} - \frac{2\lambda}{a^3} \quad \Leftrightarrow \quad \frac{2\pi abc}{3} = \frac{\lambda}{a^2}$$
$$0 = \frac{\partial L}{\partial b} = \frac{4\pi ac}{3} - \frac{8\lambda}{b^3} \quad \Leftrightarrow \quad \frac{2\pi abc}{3} = \frac{4\lambda}{b^2}$$
$$0 = \frac{\partial L}{\partial c} = \frac{4\pi ab}{3} - \frac{2\lambda}{c^3} \quad \Leftrightarrow \quad \frac{2\pi abc}{3} = \frac{\lambda}{c^2}$$
$$0 = \frac{\partial L}{\partial \lambda} = \frac{1}{a^2} + \frac{4}{b^2} + \frac{1}{c^2} - 1.$$

$abc \neq 0$ implies $\lambda \neq 0$, and so we must have

$$\frac{1}{a^2} = \frac{4}{b^2} = \frac{1}{c^2} = \frac{1}{3},$$

so $a = \pm\sqrt{3}$, $b = \pm 2\sqrt{3}$, and $c = \pm\sqrt{3}$.

8. Let $L = x^2 + y^2 + \lambda(3x^2 + 2xy + 3y^2 - 16)$. We have

$$0 = \frac{\partial L}{\partial x} = 2x + 6\lambda x + 2\lambda y \qquad (A)$$
$$0 = \frac{\partial L}{\partial y} = 2y + 6\lambda y + 2\lambda x. \qquad (B)$$

Multiplying (A) by y and (B) by x and subtracting we get

$$2\lambda(y^2 - x^2) = 0.$$

Thus, either $\lambda = 0$, or $y = x$, or $y = -x$.
$\lambda = 0$ is not possible, since it implies $x = 0$ and $y = 0$, and the point $(0, 0)$ does not lie on the given ellipse.
If $y = x$, then $8x^2 = 16$, so $x = y = \pm\sqrt{2}$.
If $y = -x$, then $4x^2 = 16$, so $x = -y = \pm 2$.
The points on the ellipse nearest the origin are $(\sqrt{2}, \sqrt{2})$ and $(-\sqrt{2}, -\sqrt{2})$. The points farthest from the origin are $(2, -2)$ and $(-2, 2)$. The major axis of the ellipse lies along $y = -x$ and has length $4\sqrt{2}$. The minor axis lies along $y = x$ and has length 4.

9. Let $L = xyz + \lambda(x^2 + y^2 + z^2 - 12)$. For CPs of L:

$$0 = \frac{\partial L}{\partial x} = yz + 2\lambda x \qquad (A)$$

$$0 = \frac{\partial L}{\partial y} = xz + 2\lambda y \qquad (B)$$

$$0 = \frac{\partial L}{\partial z} = xy + 2\lambda z \qquad (C)$$

$$0 = \frac{\partial L}{\partial \lambda} = x^2 + y^2 + z^2 - 12. \qquad (D)$$

Multiplying equations (A), (B), and (C) by x, y, and z, respectively, and subtracting in pairs, we conclude that $\lambda x^2 = \lambda y^2 = \lambda z^2$, so that either $\lambda = 0$ or $x^2 = y^2 = z^2$. If $\lambda = 0$, then (A) implies that $yz = 0$, so $xyz = 0$. If $x^2 = y^2 = z^2$, then (D) gives $3x^2 = 12$, so $x^2 = 4$. We obtain eight points (x, y, z) where each coordinate is either 2 or -2. At four of these points $xyz = 8$, which is the maximum value of xyz on the sphere. At the other four $xyz = -8$, which is the minimum value.

10. Let $L = x + 2y - 3z + \lambda(x^2 + 4y^2 + 9z^2 - 108)$. For CPs of L:

$$0 = \frac{\partial L}{\partial x} = 1 + 2\lambda x \qquad (A)$$

$$0 = \frac{\partial L}{\partial y} = 2 + 8\lambda y \qquad (B)$$

$$0 = \frac{\partial L}{\partial z} = -3 + 18\lambda z \qquad (C)$$

$$0 = \frac{\partial L}{\partial \lambda} = x^2 + 4y^2 + 9z^2 - 108. \qquad (D)$$

From (A), (B), and (C),

$$\lambda = -\frac{1}{2x} = -\frac{2}{8y} = \frac{3}{18z},$$

so $x = 2y = -3z$. From (D):

$$x^2 + 4\left(\frac{x^2}{4}\right) + 9\left(\frac{x^2}{9}\right) = 108,$$

so $x^2 = 36$, and $x = \pm 6$. There are two CPs: $(6, 3, -2)$ and $(-6, -3, 2)$. At the first, $x + 2y - 3z = 18$, the maximum value, and at the second, $x + 2y - 3z = -18$, the minimum value.

11. Let $L = x + \lambda(x + y - z) + \mu(x^2 + 2y^2 + 2z^2 - 8)$. For critical points of L:

$$0 = \frac{\partial L}{\partial x} = 1 + \lambda + 2\mu x \qquad (A)$$

$$0 = \frac{\partial L}{\partial y} = \lambda + 4\mu y \qquad (B)$$

$$0 = \frac{\partial L}{\partial z} = -\lambda + 4\mu z \qquad (C)$$

$$0 = \frac{\partial L}{\partial \lambda} = x + y - z \qquad (D)$$

$$0 = \frac{\partial L}{\partial \mu} = x^2 + 2y^2 + 2z^2 - 8. \qquad (E)$$

From (B) and (C) we have $\mu(y + z) = 0$. Thus $\mu = 0$ or $y + z = 0$.

CASE I. $\mu = 0$. Then $\lambda = 0$ by (B), and $1 = 0$ by (A), so this case is not possible.

CASE II. $y + z = 0$. Then $z = -y$ and, by (D), $x = -2y$. Therefore, by (E), $4y^2 + 2y^2 + 2y^2 = 8$, and so $y = \pm 1$. From this case we obtain two points: $(2, -1, 1)$ and $(-2, 1, -1)$.
The function $f(x, y, z) = x$ has maximum value 2 and minimum value -2 when restricted to the curve $x + y = z$, $x^2 + 2y^2 + 2z^2 = 8$.

12. Let $L = x^2 + y^2 + z^2 + \lambda(x^2 + y^2 - z^2) + \mu(x - 2z - 3)$. For critical points of L:

$$0 = \frac{\partial L}{\partial x} = 2x(1 + \lambda) + \mu \qquad (A)$$

$$0 = \frac{\partial L}{\partial y} = 2y(1 + \lambda) \qquad (B)$$

$$0 = \frac{\partial L}{\partial z} = 2z(1 - \lambda) - 2\mu \qquad (C)$$

$$0 = \frac{\partial L}{\partial \lambda} = x^2 + y^2 - z^2 \qquad (D)$$

$$0 = \frac{\partial L}{\partial \mu} = x - 2z - 3. \qquad (E)$$

From (B), either $y = 0$ or $\lambda = -1$.

CASE I. $y = 0$. Then (D) implies $x = \pm z$.
If $x = z$ then (E) implies $z = -3$, so we get the point $(-3, 0, -3)$.
If $x = -z$ then (E) implies $z = -1$, so we get the point $(1, 0, -1)$.

CASE II. $\lambda = -1$. Then (A) implies $\mu = 0$ and (C) implies $z = 0$. By (D), $x = y = 0$, and this contradicts (E), so this case is not possible.

If $f(x, y, z) = x^2 + y^2 + z^2$, then $f(-3, 0, -3) = 18$ is the maximum value of f on the ellipse $x^2 + y^2 = z^2$, $x - 2z = 3$, and $f(1, 0, -1) = 2$ is the minimum value.

13. Let $L = 4 - z + \lambda(x^2 + y^2 - 8) + \mu(x + y + z - 1)$. For critical points of L:

$$0 = \frac{\partial L}{\partial x} = 2\lambda x + \mu \quad (A)$$

$$0 = \frac{\partial L}{\partial y} = 2\lambda y + \mu \quad (B)$$

$$0 = \frac{\partial L}{\partial z} = -1 + \mu \quad (C)$$

$$0 = \frac{\partial L}{\partial \lambda} = x^2 + y^2 - 8 \quad (D)$$

$$0 = \frac{\partial L}{\partial \mu} = x + y + z - 1. \quad (E)$$

From (C), $\mu = 1$. From (A) and (B), $\lambda(x - y) = 0$, so either $\lambda = 0$ or $x = y$.

CASE I. $\lambda = 0$. Then $\mu = 0$ by (A), and this contradicts (C), so this case is not possible.

CASE II. $x = y$. Then $x = y = \pm 2$ by (D).
If $x = y = 2$, then $z = -3$ by (E).
If $x = y = -2$, then $z = 5$ by (E).
Thus we have two points, $(2, 2, -3)$ and $(-2, -2, 5)$, where $f(x, y, z) = 4 - z$ takes the values 7 (maximum), and -1 (minimum) respectively.

14. The max and min values of $f(x, y, z) = x + y^2 z$ subject to the constraints $y^2 + z^2 = 2$ and $z = x$ will be found among the critical points of

$$L = x + y^2 z + \lambda(y^2 + z^2 - 2) + \mu(z - x).$$

Thus

$$0 = \frac{\partial L}{\partial x} = 1 - \mu = 0,$$

$$0 = \frac{\partial L}{\partial y} = 2yz + 2\lambda y = 0,$$

$$0 = \frac{\partial L}{\partial z} = y^2 + 2\lambda z + \mu = 0,$$

$$0 = \frac{\partial L}{\partial \lambda} = y^2 + z^2 - 2,$$

$$0 = \frac{\partial L}{\partial \mu} = z - x.$$

From the first equation $\mu = 1$. From the second, either $y = 0$ or $z = -\lambda$.
If $y = 0$ then $z^2 = 2$, $z = x$, so critical points are $(\sqrt{2}, 0, \sqrt{2})$ and $(-\sqrt{2}, 0, -\sqrt{2})$. f has the values $\pm\sqrt{2}$ at these points. If $z = -\lambda$ then $y^2 - 2z^2 + 1 = 0$. Thus $2z^2 - 1 = 2 - z^2$, or $z^2 = 1$, $z = \pm 1$. This leads to critical points $(1, \pm 1, 1)$ and $(-1, \pm 1, -1)$ where f has values ± 2. The maximum value of f subject to the constraints is 2; the minimum value is -2.

15. Let

$$L = (x - a)^2 + (y - b)^2 + (z - c)^2 + \lambda(x - y) + \mu(y - z) + \sigma(a + b) + \tau(c - 2).$$

For critical points of L, we have

$$0 = \frac{\partial L}{\partial x} = 2(x - a) + \lambda \quad (A)$$

$$0 = \frac{\partial L}{\partial y} = 2(y - b) - \lambda + \mu \quad (B)$$

$$0 = \frac{\partial L}{\partial z} = 2(z - c) - \mu \quad (C)$$

$$0 = \frac{\partial L}{\partial a} = -2(x - a) + \sigma \quad (D)$$

$$0 = \frac{\partial L}{\partial b} = -2(y - b) + \sigma \quad (E)$$

$$0 = \frac{\partial L}{\partial c} = -2(z - c) + \tau \quad (F)$$

$$0 = \frac{\partial L}{\partial \lambda} = x - y \quad (G)$$

$$0 = \frac{\partial L}{\partial \mu} = y - z \quad (H)$$

$$0 = \frac{\partial L}{\partial \sigma} = a + b \quad (I)$$

$$0 = \frac{\partial L}{\partial \tau} = c - 2. \quad (J)$$

Subtracting (D) and (E) we get $x - y = a - b$. From (G), $x = y$, and therefore $a = b$. From (I), $a = b = 0$, and from (J), $c = 2$.
Adding (A), (B) and (C), we get $x + y + z = a + b + c = 2$.
From (G) and (H), $x = y = z = 2/3$.
The minimum distance between the two lines is

$$\sqrt{\left(\frac{2}{3} - 0\right)^2 + \left(\frac{2}{3} - 0\right)^2 + \left(\frac{2}{3} - 2\right)^2} = \sqrt{\frac{24}{9}} = \frac{2\sqrt{6}}{3} \text{ units}$$

16. Let $L = x_1 + x_2 + \cdots + x_n + \lambda(x_1^2 + x_2^2 + \cdots + x_n^2 - 1)$.
For critical points of L we have

$$0 = \frac{\partial L}{\partial x_1} = 1 + 2\lambda x_1, \quad \ldots \quad 0 = \frac{\partial L}{\partial x_n} = 1 + 2\lambda x_n$$

$$0 = \frac{\partial L}{\partial \lambda} = x_1^2 + x_2^2 + \cdots + x_n^2 - 1.$$

The first n equations give

$$x_1 = x_2 = \cdots = x_n = -\frac{1}{2\lambda},$$

and the final equation gives

$$\frac{1}{4\lambda^2} + \frac{1}{4\lambda^2} + \cdots + \frac{1}{4\lambda^2} = 1,$$

so that $4\lambda^2 = n$, and $\lambda = \pm\sqrt{n}/2$.
The maximum and minimum values of $x_1 + x_2 + \cdots + x_n$ subject to $x_1^2 + \cdots + x_n^2 = 1$ are $\pm\frac{n}{2\lambda}$, that is, $\sqrt{n}$ and $-\sqrt{n}$ respectively.

INSTRUCTOR'S SOLUTIONS MANUAL SECTION 13.3 (PAGE 793)

17. Let $L = x_1 + 2x_2 + \cdots + nx_n + \lambda(x_1^2 + x_2^2 + \cdots + x_n^2 - 1)$.
For critical points of L we have

$$0 = \frac{\partial L}{\partial x_1} = 1 + 2\lambda x_1 \quad \Leftrightarrow \quad x_1 = -\frac{1}{2\lambda}$$

$$0 = \frac{\partial L}{\partial x_2} = 2 + 2\lambda x_2 \quad \Leftrightarrow \quad x_2 = -\frac{2}{2\lambda}$$

$$0 = \frac{\partial L}{\partial x_3} = 3 + 2\lambda x_3 \quad \Leftrightarrow \quad x_3 = -\frac{3}{2\lambda}$$

$$\vdots$$

$$0 = \frac{\partial L}{\partial x_n} = n + 2\lambda x_n \quad \Leftrightarrow \quad x_n = -\frac{n}{2\lambda}$$

$$0 = \frac{\partial L}{\partial \lambda} = x_1^2 + x_2^2 + \cdots + x_n^2 - 1.$$

Thus

$$\frac{1}{4\lambda^2} + \frac{4}{4\lambda^2} + \frac{9}{4\lambda^2} + \cdots + \frac{n^2}{4\lambda^2} = 1$$

$$4\lambda^2 = 1 + 4 + 9 + \cdots + n^2 = \frac{n(n+1)(2n+1)}{6}$$

$$\lambda = \pm \frac{1}{2}\sqrt{\frac{n(n+1)(2n+1)}{6}}.$$

Thus the maximum and minimum values of $x_1 + 2x_2 + \cdots + nx_n$ over the hypersphere $x_1^2 + x_2^2 + \cdots + x_n^2 = 1$ are

$$\pm \sqrt{\frac{6}{n(n+1)(2n+1)}}(1^2 + 2^2 + 3^2 + \cdots + n^2)$$

$$= \pm \sqrt{\frac{n(n+1)(2n+1)}{6}}.$$

18. Let the width, depth, and height of the box be x, y and z respectively. We want to minimize the surface area

$$S = xy + 2xz + 2yz$$

subject to the constraint that $xyz = V$, where V is a given positive volume. Let

$$L = xy + 2xz + 2yz + \lambda(xyz - V).$$

For critical points of L,

$$0 = \frac{\partial L}{\partial x} = y + 2z + \lambda yz \quad \Leftrightarrow \quad -\lambda xyz = xy + 2xz$$

$$0 = \frac{\partial L}{\partial y} = x + 2z + \lambda xz \quad \Leftrightarrow \quad -\lambda xyz = xy + 2yz$$

$$0 = \frac{\partial L}{\partial z} = 2x + 2y + \lambda xy \quad \Leftrightarrow \quad -\lambda xyz = 2xz + 2yz$$

$$0 = \frac{\partial L}{\partial \lambda} = xyz - V.$$

From the first three equations, $xy = 2xz = 2yz$. Since x, y, and z are all necessarily positive, we must therefore have $x = y = 2z$. Thus the most economical box with no top has width and depth equal to twice the height.

19. We want to maximize $V = xyz$ subject to $4x + 2y + z = 2$. Let

$$L = xyz + \lambda(4x + 2y + z - 2).$$

For critical points of L,

$$0 = \frac{\partial L}{\partial x} = yz + 4\lambda \quad \Leftrightarrow \quad xyz + 4\lambda x = 0$$

$$0 = \frac{\partial L}{\partial y} = xz + 2\lambda \quad \Leftrightarrow \quad xyz + 2\lambda y = 0$$

$$0 = \frac{\partial L}{\partial z} = xy + \lambda \quad \Leftrightarrow \quad xyz + \lambda z = 0$$

$$0 = \frac{\partial L}{\partial \lambda} = 4x + 2y + z - 2 = 0.$$

The first three equations imply that $z = 2y = 4x$ (since we cannot have $\lambda = 0$ if V is positive). The fourth equation then implies that $12x = 2$. Hence $x = 1/6$, $y = 1/3$, and $z = 2/3$.
The largest box has volume

$$V = \frac{1}{6} \times \frac{1}{3} \times \frac{2}{3} = \frac{1}{27} \text{ cubic units.}$$

20. We want to maximize xyz subject to $xy + 2yz + 3xz = 18$. Let

$$L = xyz + \lambda(xy + 2yz + 3xz - 18).$$

For critical points of L,

$$0 = \frac{\partial L}{\partial x} = yz + \lambda(y + 3z) \quad \Leftrightarrow \quad -xyz = \lambda(xy + 3xz)$$

$$0 = \frac{\partial L}{\partial y} = xz + \lambda(x + 2z) \quad \Leftrightarrow \quad -xyz = \lambda(xy + 2yz)$$

$$0 = \frac{\partial L}{\partial z} = xy + \lambda(2y + 3x) \quad \Leftrightarrow \quad -xyz = \lambda(2yz + 3xz)$$

$$0 = \frac{\partial L}{\partial \lambda} = xy + 2yz + 3xz - 18.$$

From the first three equations $xy = 2yz = 3xz$. From the fourth equation, the sum of these expressions is 18. Thus

$$xy = 2yz = 3xz = 6.$$

Thus the maximum volume of the box is

$$V = xyz = \sqrt{(xy)(yz)(xz)} = \sqrt{6 \times 3 \times 2} = 6 \text{ cubic units.}$$

21. Let the width, depth, and height of the box be x, y, and z as shown in the figure. Let the cost per unit area of the back and sides be k. Then the cost per unit area of the front and bottom is $5k$. We want to minimize

$$C = 5k(xz + xy) + k(2yz + xz)$$

483

subject to the constraint $xyz = V$ (constant). Let

$$L = k(5xy + 6xz + 2yz) + \lambda(xyz - V).$$

For critical points of L,

$$0 = \frac{\partial L}{\partial x} = 5ky + 6kz + \lambda yz \Leftrightarrow -\lambda xyz = 5kxy + 6kxz$$
$$0 = \frac{\partial L}{\partial y} = 5kx + 2kz + \lambda xz \Leftrightarrow -\lambda xyz = 5kxy + 2kyz$$
$$0 = \frac{\partial L}{\partial z} = 6kx + 2ky + \lambda xy \Leftrightarrow -\lambda xyz = 6kxz + 2kyz$$
$$0 = \frac{\partial L}{\partial \lambda} = xyz - V.$$

From the first three of these equations we obtain $5xy = 6xz = 2yz$. Thus $y = 3x$ and $z = \frac{5x}{2}$. From the fourth equation, $V = xyz = \frac{15}{2}x^3$.

The largest box has width $\left(\frac{2V}{15}\right)^{1/3}$, depth $3\left(\frac{2V}{15}\right)^{1/3}$, and height $\frac{5}{2}\left(\frac{2V}{15}\right)^{1/3}$.

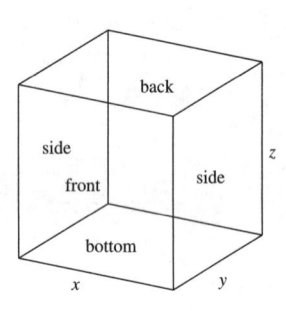

Fig. 13.3.21

22. $f(x, y, z) = xy + z^2$ on $B = \{(x, y, z) : x^2 + y^2 + z^2 \leq 1\}$. For critical points of f,

$$0 = f_1(x, y, z) = y, \qquad 0 = f_2(x, y, z) = x,$$
$$0 = f_3(x, y, z) = 2z.$$

Thus the only critical point is the interior point $(0, 0, 0)$, where f has the value 0, evidently neither a maximum nor a minimum. The maximum and minimum must therefore occur on the boundary of B, that is, on the sphere $x^2 + y^2 + z^2 = 1$. Let

$$L = xy + z^2 + \lambda(x^2 + y^2 + z^2 - 1).$$

For critical points of L,

$$0 = \frac{\partial L}{\partial x} = y + 2\lambda x \qquad (A)$$
$$0 = \frac{\partial L}{\partial y} = x + 2\lambda y \qquad (B)$$
$$0 = \frac{\partial L}{\partial z} = 2z(1 + \lambda) \qquad (C)$$
$$0 = \frac{\partial L}{\partial \lambda} = x^2 + y^2 + z^2 - 1. \qquad (D)$$

From (C) either $z = 0$ or $\lambda = -1$.

CASE I. $z = 0$. (A) and (B) imply that $y^2 = x^2$ and (D) then implies that $x^2 = y^2 = 1/2$. At the four points

$$\left(\frac{1}{\sqrt{2}}, \pm\frac{1}{\sqrt{2}}, 0\right) \quad \text{and} \quad \left(-\frac{1}{\sqrt{2}}, \pm\frac{1}{\sqrt{2}}, 0\right)$$

f takes the values $\frac{1}{2}$ and $-\frac{1}{2}$.

CASE II. $\lambda = -1$. (A) and (B) imply that $x = y = 0$, and so by (D), $z = \pm 1$. f has the value 1 at the points $(0, 0, \pm 1)$.

Thus the maximum and minimum values of f on B are 1 and $-1/2$ respectively.

23. In this problem we do the boundary analysis for Exercise 22 using the suggested parametrization of the sphere $x^2 + y^2 + z^2 = 1$. We have

$$f(x, y, z) = xy + z^2$$
$$= \sin^2\phi \sin\theta \cos\theta + \cos^2\phi$$
$$= \frac{1}{2}\sin^2\phi \sin 2\theta + \cos^2\phi$$
$$= g(\phi, \theta)$$

for $0 \leq \phi \leq \pi$ and $0 \leq \theta \leq 2\pi$. For critical points of g,

$$0 = g_1(\phi, \theta) = \sin\phi \cos\phi \sin 2\theta - 2\sin\phi \cos\phi$$
$$= \sin\phi \cos\phi (\sin 2\theta - 2)$$
$$0 = g_2(\phi, \theta) = \sin^2\phi \cos 2\theta.$$

The first of these equations implies that either $\sin\phi = 0$ or $\cos\phi = 0$.

If $\sin\phi = 0$, then both equations are satisfied. Since $\cos\phi = \pm 1$ in this case, we have $g(\phi, \theta) = 1$.

If $\cos\phi = 0$, then $\sin\phi = \pm 1$, and the second equation requires $\cos 2\theta = 0$. Thus $\theta = \pm\frac{\pi}{4}$ or $\pm\frac{3\pi}{4}$. In this case $g(\phi, \theta) = \pm\frac{1}{2}$.

Again we find that $f(x, y, z) = xy + z^2$ has maximum value 1 and minimum value $-\frac{1}{2}$ when restricted to the surface of the ball B. These are the maximum and minimum values for the whole ball as noted in Exercise 22.

24. Let $L = \sin\frac{x}{2}\sin\frac{y}{2}\sin\frac{z}{2} + \lambda(x+y+z-\pi)$. Then

$$0 = \frac{\partial L}{\partial x} = \frac{1}{2}\cos\frac{x}{2}\sin\frac{y}{2}\sin\frac{z}{2} + \lambda \quad (A)$$
$$0 = \frac{\partial L}{\partial y} = \frac{1}{2}\sin\frac{x}{2}\cos\frac{y}{2}\sin\frac{z}{2} + \lambda \quad (B)$$
$$0 = \frac{\partial L}{\partial z} = \frac{1}{2}\sin\frac{x}{2}\sin\frac{y}{2}\cos\frac{z}{2} + \lambda. \quad (C)$$

For any triangle we must have $0 \le x \le \pi$, $0 \le y \le \pi$ and $0 \le z \le \pi$. Also

$$P = \sin\frac{x}{2}\sin\frac{y}{2}\sin\frac{z}{2}$$

is 0 if any of x, y or z is 0 or π. Subtracting equations (A) and (B) gives

$$\frac{1}{2}\sin\frac{z}{2}\sin\frac{x-y}{2} = 0.$$

It follows that we must have $x = y$; all other possibilities lead to a zero value for P. Similarly, $y = z$. Thus the triangle for which P is maximum must be equilateral: $x = y = z = \pi/3$. Since $\sin(\pi/3) = 1/2$, the maximum value of P is $1/8$.

25. We are given that $g_2(a,b) \neq 0$, and therefore that the equation $g(x,y) = C$ has a solution of the form $y = h(x)$ valid near (a,b). Since $g(x, h(x)) = C$ holds identically for x near a, we must have

$$0 = \left(\frac{d}{dx}g(x, h(x))\right)\bigg|_{x=a} = g_1(a,b) + g_2(a,b)h'(a).$$

If $f(x,y)$, subject to the constraint $g(x,y) = C$, has an extreme value at (a,b), then $F(x) = f(x, h(x))$ has an extreme value at $x = a$, so

$$0 = F'(a) = f_1(a,b) + f_2(a,b)h'(a).$$

Together these equations imply that $g_1(a,b)f_2(a,b) = g_2(a,b)f_1(a,b)$, and therefore that

$$\frac{f_1(a,b)}{g_1(a,b)} = \frac{f_2(a,b)}{g_2(a,b)} = -\lambda \quad \text{(say)}.$$

(Since $g_2(a,b) \neq 0$, therefore, if $g_1(a,b) = 0$, then $f_1(a,b) = 0$ also.) It follows that

$$0 = f_1(a,b) + \lambda g_1(a,b), \quad 0 = f_2(a,b) + \lambda g_2(a,b),$$

so (a,b) is a critical point of $L = f(x,y) + \lambda g(x,y)$.

26. As can be seen in the figure, the minimum distance from $(0,-1)$ to points of the semicircle $y = \sqrt{1-x^2}$ is $\sqrt{2}$, the closest points to $(0,-1)$ on the semicircle being $(\pm 1, 0)$. These points will not be found by the method of Lagrange multipliers because the level curve $f(x,y) = 2$ of the function f giving the square of the distance from (x,y) to $(0,-1)$ is not tangent to the semicircle at $(\pm 1, 0)$. This could only have happened because $(\pm 1, 0)$ are *endpoints* of the semicircle.

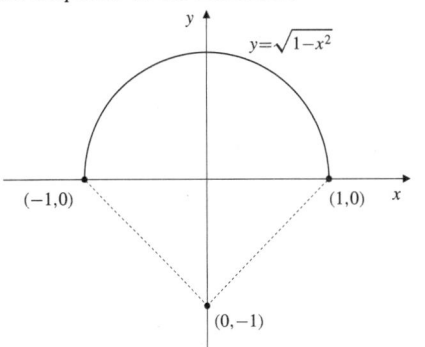

Fig. 13.3.26

27. If $f(x,y)$ has an extreme value on $g(x,y) = 0$ at a point (x_0, y_0) where $\nabla g \neq \mathbf{0}$, and if ∇f exists at that point, then $\nabla f(x_0, y_0)$ must be parallel to $\nabla g(x_0, y_0)$;

$$\nabla f(x_0, y_0) + \lambda \nabla g(x_0, y_0) = \mathbf{0}$$

as shown in the text. The argument given there holds whether or not $\nabla f(x_0, y_0)$ is $\mathbf{0}$. However, if

$$\nabla f(x_0, y_0) = \mathbf{0}$$

then we will have $\lambda = 0$.

Section 13.4 The Method of Least Squares (page 800)

1. If the power plant is located at (x,y), then x and y should minimize (and hence be a critical point of)

$$S = \sum_{i=1}^{n}\left[(x-x_i)^2 + (y-y_2)^2\right].$$

Thus we must have

$$0 = \frac{\partial S}{\partial x} = 2\sum_{i=1}^{n}(x-x_i) = 2\left(nx - \sum_{i=1}^{n}x_i\right)$$
$$0 = \frac{\partial S}{\partial y} = 2\sum_{i=1}^{n}(y-y_i) = 2\left(ny - \sum_{i=1}^{n}y_i\right).$$

Thus $x = \frac{1}{n}\sum_{i=1}^{n}x_i = \overline{x}$, and $y = \frac{1}{n}\sum_{i=1}^{n}y_i = \overline{y}$.

Place the power plant at the position whose coordinates are the averages of the coordinates of the machines.

2. We want to minimize $S = \sum_{i=1}^{n}(ax_i^2 - y_i)^2$. Thus

$$0 = \frac{dS}{da} = \sum_{i=1}^{n} 2(ax_i^2 - y_i)x_i^2$$
$$= 2\sum_{i=1}^{n}(ax_i^4 - x_i^2 y_i),$$

and $a = \left(\sum_{i=1}^{n} x_i^2 y_i\right) \Big/ \left(\sum_{i=1}^{n} x_i^4\right)$.

3. We minimize $S = \sum_{i=1}^{n}(ae^{x_i} - y_i)^2$. Thus

$$0 = \frac{dS}{da} = 2\sum_{i=1}^{n}(ae^{x_i} - y_i)e^{x_i},$$

and $a = \left(\sum_{i=1}^{n} y_i e^{x_i}\right) \Big/ \left(\sum_{i=1}^{n} e^{2x_i}\right)$.

4. We choose a, b, and c to minimize

$$S = \sum_{i=1}^{n}(ax_i + by_i + c - z_i)^2.$$

Thus

$$0 = \frac{\partial S}{\partial a} = 2\sum_{i=1}^{n}(ax_i + by_i + c - z_i)x_i$$
$$0 = \frac{\partial S}{\partial b} = 2\sum_{i=1}^{n}(ax_i + by_i + c - z_i)y_i$$
$$0 = \frac{\partial S}{\partial c} = 2\sum_{i=1}^{n}(ax_i + by_i + c - z_i).$$

Let $A = \sum x_i^2$, $B = \sum x_i y_i$, $C = \sum x_i$, $D = \sum y_i^2$, $E = \sum y_i$, $F = \sum x_i z_i$, $G = \sum y_i z_i$, and $H = \sum z_i$. In terms of these quantities the above equations become

$$\begin{array}{ccccccc} Aa & + & Bb & + & Cc & = & F \\ Ba & + & Db & + & Ec & = & G \\ Ca & + & Eb & + & nc & = & H \end{array}$$

By Cramer's Rule (Theorem 5 of Section 10.6) the solution is

$$a = \frac{1}{\Delta}\begin{vmatrix} F & B & C \\ G & D & E \\ H & E & n \end{vmatrix}, \quad b = \frac{1}{\Delta}\begin{vmatrix} A & F & C \\ B & G & E \\ C & H & n \end{vmatrix},$$

$$c = \frac{1}{\Delta}\begin{vmatrix} A & B & F \\ B & D & G \\ C & E & H \end{vmatrix}, \quad \text{where } \Delta = \begin{vmatrix} A & B & C \\ B & D & E \\ C & E & n \end{vmatrix}.$$

5. If $\mathbf{x} = (x_1, \ldots, x_n)$, $\mathbf{y} = (y_1, \ldots, y_n)$, $\mathbf{z} = (z_1, \ldots, z_n)$, $\mathbf{w} = (1, \ldots, 1)$, and $\mathbf{p} = a\mathbf{x} + b\mathbf{y} + c\mathbf{w}$, we want to choose a, b, and c so that $\mathbf{p}$ is the vector projection of $\mathbf{z}$ onto the subspace of $\mathbb{R}^3$ spanned by $\mathbf{x}$, $\mathbf{y}$ and $\mathbf{w}$. Thus $\mathbf{p} - \mathbf{z}$ must be perpendicular to each of $\mathbf{x}$, $\mathbf{y}$, and $\mathbf{w}$:

$$(\mathbf{p} - \mathbf{z}) \bullet \mathbf{x} = 0, \quad (\mathbf{p} - \mathbf{z}) \bullet \mathbf{y} = 0, \quad (\mathbf{p} - \mathbf{z}) \bullet \mathbf{w} = 0.$$

When written in terms of the components of the vectors involved, these three equations are the same as the equations for a, b, and c encountered in Exercise 4, and so they have the same solution as given for that exercise.

6. The relationship $y = p + qx^2$ is linear in p and q, so we choose p and q to minimize

$$S = \sum_{i=1}^{n}(p + qx_i^2 - y_i)^2.$$

Thus

$$0 = \frac{\partial S}{\partial p} = 2\sum_{i=1}^{n}(p + qx_i^2 - y_i)$$
$$0 = \frac{\partial S}{\partial q} = 2\sum_{i=1}^{n}(p + qx_i^2 - y_i)x_i^2,$$

that is,

$$\begin{array}{ccccc} np & + & \left(\sum x_i^2\right) q & = & \sum y_i \\ \left(\sum x_i^2\right) p & + & \left(\sum x_i^4\right) q & = & \sum x_i^2 y_i, \end{array}$$

so

$$p = \frac{\left(\sum y_i\right)\left(\sum x_i^4\right) - \left(\sum x_i^2 y_i\right)\left(\sum x_i^2\right)}{n\left(\sum x_i^4\right) - \left(\sum x_i^2\right)^2}$$

$$q = \frac{n\left(\sum x_i^2 y_i\right) - \left(\sum y_i\right)\left(\sum x_i^2\right)}{n\left(\sum x_i^4\right) - \left(\sum x_i^2\right)^2}.$$

This is the result obtained by direct linear regression. (No transformation of variables was necessary.)

7. We transform $y = pe^{qx}$ into the form $\ln y = \ln p + qx$, which is linear in $\ln p$ and q. We let $\eta_i = \ln y_i$ and use the regression line $\eta = a + bx$ obtained from the data (x_i, η_i), with $b = q$ and $a = \ln p$.
Using the formulas for a and b obtained in the text, we have

$$\ln p = a = \frac{n\left(\sum x_i \ln y_i\right) - \left(\sum x_i\right)\left(\sum \ln y_i\right)}{n\left(\sum x_i^2\right) - \left(\sum x_i\right)^2}$$

$$q = b = \frac{\left(\sum x_i^2\right)\left(\sum \ln y_i\right) - \left(\sum x_i\right)\left(\sum x_i \ln y_i\right)}{n\left(\sum x_i^2\right) - \left(\sum x_i\right)^2}$$

$$p = e^a.$$

These values of p and q are not the same values that minimize the expression

$$S = \sum_{i=1}^{n}(y_i - pe^{qx_i})^2.$$

8. We transform $y = \ln(p + qx)$ into the form $e^y = p + qx$, which is linear in p and q. We let $\eta_i = e^{y_i}$ and use the regression line $\eta = ax + b$ obtained from the data (x_i, η_i), with $a = q$ and $b = p$.
Using the formulas for a and b obtained in the text, we have

$$q = a = \frac{n\left(\sum x_i e^{y_i}\right) - \left(\sum x_i\right)\left(\sum e^{y_i}\right)}{n\left(\sum x_i^2\right) - \left(\sum x_i\right)^2}$$

$$p = b = \frac{\left(\sum x_i^2\right)\left(\sum e^{y_i}\right) - \left(\sum x_i\right)\left(\sum x_i e^{y_i}\right)}{n\left(\sum x_i^2\right) - \left(\sum x_i\right)^2}.$$

These values of p and q are not the same values that minimize the expression

$$S = \sum_{i=1}^{n}\left(\ln(p + qx_i) - y_i\right)^2.$$

9. The relationship $y = px + qx^2$ is linear in p and q, so we choose p and q to minimize

$$S = \sum_{i=1}^{n}(px_i + qx_i^2 - y_i)^2.$$

Thus

$$0 = \frac{\partial S}{\partial p} = 2\sum_{i=1}^{n}(px_i + qx_i^2 - y_i)x_i$$

$$0 = \frac{\partial S}{\partial q} = 2\sum_{i=1}^{n}(px_i + qx_i^2 - y_i)x_i^2,$$

that is,

$$\begin{aligned}\left(\sum x_i^2\right)p + \left(\sum x_i^3\right)q &= \sum x_i y_i \\ \left(\sum x_i^3\right)p + \left(\sum x_i^4\right)q &= \sum x_i^2 y_i,\end{aligned}$$

so

$$p = \frac{\left(\sum x_i y_i\right)\left(\sum x_i^4\right) - \left(\sum x_i^2 y_i\right)\left(\sum x_i^3\right)}{\left(\sum x_i^2\right)\left(\sum x_i^4\right) - \left(\sum x_i^3\right)^2}$$

$$q = \frac{\left(\sum x_i^2\right)\left(\sum x_i^2 y_i\right) - \left(\sum x_i y_i\right)\left(\sum x_i^3\right)}{\left(\sum x_i^2\right)\left(\sum x_i^4\right) - \left(\sum x_i^3\right)^2}.$$

This is the result obtained by direct linear regression. (No transformation of variables was necessary.)

10. We transform $y = \sqrt{(px + q)}$ into the form $y^2 = px + q$, which is linear in p and q. We let $\eta_i = y_i^2$ and use the regression line $\eta = ax + b$ obtained from the data (x_i, η_i), with $a = p$ and $b = q$.
Using the formulas for a and b obtained in the text, we have

$$p = a = \frac{n\left(\sum x_i y_i^2\right) - \left(\sum x_i\right)\left(\sum y_i^2\right)}{n\left(\sum x_i^2\right) - \left(\sum x_i\right)^2}$$

$$q = b = \frac{\left(\sum x_i^2\right)\left(\sum y_i^2\right) - \left(\sum x_i\right)\left(\sum x_i y_i^2\right)}{n\left(\sum x_i^2\right) - \left(\sum x_i\right)^2}.$$

These values of p and q are not the same values that minimize the expression

$$S = \sum_{i=1}^{n}\left(\sqrt{px_i + q} - y_i\right)^2.$$

11. The relationship $y = pe^x + qe^{-x}$ is linear in p and q, so we choose p and q to minimize

$$S = \sum_{i=1}^{n}\left(pe^{x_i} + qe^{-x_i} - y_i\right)^2.$$

Thus

$$0 = \frac{\partial S}{\partial p} = 2\sum_{i=1}^{n}\left(pe^{x_i} + qe^{-x_i} - y_i\right)e^{x_i}$$

$$0 = \frac{\partial S}{\partial q} = 2\sum_{i=1}^{n}\left(pe^{x_i} + qe^{-x_i} - y_i\right)e^{-x_i}.$$

that is,

$$\begin{aligned}\left(\sum e^{2x_i}\right)p + nq &= \sum e^{x_i} y_i \\ np + \left(\sum e^{-2x_i}\right)q &= \sum e^{-x_i} y_i,\end{aligned}$$

so

$$p = \frac{\left(\sum e^{-2x_i}\right)\left(\sum e^{x_i} y_i\right) - n\left(\sum e^{-x_i} y_i\right)}{\left(\sum e^{2x_i}\right)\left(\sum e^{-2x_i}\right) - n^2}$$

$$q = \frac{\left(\sum e^{2x_i}\right)\left(\sum e^{-x_i} y_i\right) - n\left(\sum e^{x_i} y_i\right)}{\left(\sum e^{2x_i}\right)\left(\sum e^{-2x_i}\right) - n^2}.$$

This is the result obtained by direct linear regression. (No transformation of variables was necessary.)

12. We use the result of Exercise 6. We have $n = 6$ and

$$\sum x_i^2 = 115, \qquad \sum x_i^4 = 4051,$$
$$\sum y_i = 55.18, \qquad \sum x_i^2 y_i = 1984.50.$$

Therefore

$$p = \frac{\left(\sum y_i\right)\left(\sum x_i^4\right) - \left(\sum x_i^2 y_i\right)\left(\sum x_i^2\right)}{n\left(\sum x_i^4\right) - \left(\sum x_i^2\right)^2}$$

$$= \frac{55.18 \times 4051 - 1984.50 \times 115}{6 \times 4051 - 115^2} \approx -0.42$$

$$q = \frac{n\left(\sum x_i^2 y_i\right) - \left(\sum y_i\right)\left(\sum x_i^2\right)}{n\left(\sum x_i^4\right) - \left(\sum x_i^2\right)^2}$$

$$= \frac{6 \times 1984.50 - 55.18 \times 115}{6 \times 4051 - 115^2} \approx 0.50.$$

We have (approximately) $y = -0.42 + 0.50x^2$. The predicted value of y at $x = 5$ is $-0.42 + 0.50 \times 25 \approx 12.1$.

13. Choose a, b, and c to minimize

$$S = \sum_{i=1}^{n} (ax_i^2 + bx_i + c - y_i)^2.$$

Thus

$$0 = \frac{\partial S}{\partial a} = 2\sum_{i=1}^{n} (ax_i^2 + bx_i + c - y_i)x_i^2$$

$$0 = \frac{\partial S}{\partial b} = 2\sum_{i=1}^{n} (ax_i^2 + bx_i + c - y_i)x_i$$

$$0 = \frac{\partial S}{\partial c} = 2\sum_{i=1}^{n} (ax_i^2 + bx_i + c - y_i).$$

Let $A = \sum x_i^4$, $B = \sum x_i^3$, $C = \sum x_i^2$, $D = \sum x_i$, $H = \sum x_i^2 y_i$, $I = \sum x_i y_i$, and $J = \sum y_i$. In terms of these quantities the above equations become

$$\begin{aligned} Aa + Bb + Cc &= H \\ Ba + Cb + Dc &= I \\ Ca + Db + nc &= J. \end{aligned}$$

By Cramer's Rule (Theorem 5 of Section 10.6) the solution is

$$a = \frac{1}{\Delta}\begin{vmatrix} H & B & C \\ I & C & D \\ J & D & n \end{vmatrix}, \qquad b = \frac{1}{\Delta}\begin{vmatrix} A & H & C \\ B & I & D \\ C & J & n \end{vmatrix},$$

$$c = \frac{1}{\Delta}\begin{vmatrix} A & B & H \\ B & C & I \\ C & D & J \end{vmatrix}, \quad \text{where } \Delta = \begin{vmatrix} A & B & C \\ B & C & D \\ C & D & n \end{vmatrix}.$$

14. Since $y = pe^x + q + re^{-x}$ is equivalent to

$$e^x y = p(e^x)^2 + qe^x + r,$$

we let $\xi_i = e^{x_i}$ and $\eta_i = e^{x_i} y_i$ for $i = 1, 2, \ldots, n$. We then have $p = a$, $q = b$, and $r = c$, where a, b, and c are the values calculated by the formulas in Exercise 13, but for the data (ξ_i, η_i) instead of (x_i, y_i).

15. To minimize $I = \int_0^1 (ax^2 - x^3)^2 dx$, we choose a so that

$$0 = \frac{dI}{da} = \int_0^1 2(ax^2 - x^3)x^2 dx$$

$$= \left(2a\frac{x^5}{5} - \frac{2x^6}{6}\right)\bigg|_0^1 = \frac{2a}{5} - \frac{1}{3}.$$

Thus $a = 5/6$, and the minimum value of I is

$$\int_0^1 \left(\frac{25x^4}{36} - \frac{5x^5}{3} + x^6\right) dx$$

$$= \frac{5}{36} - \frac{5}{18} + \frac{1}{7} = \frac{1}{252}.$$

16. To maximize $I = \int_0^\pi \bigl(ax(\pi - x) - \sin x\bigr)^2 dx$, we choose a so that

$$0 = \frac{dI}{da} = \int_0^\pi 2\bigl(ax(\pi - x) - \sin x\bigr)x(\pi - x)\, dx$$

$$= 2a\int_0^\pi x^2(\pi - x)^2 dx - 2\int_0^\pi x(\pi - x)\sin x\, dx$$

$$= \frac{\pi^5 a}{15} - 8.$$

(We have omitted the details of evaluation of these integrals.) Hence $a = 120/\pi^5$. The minimum value of I is

$$\int_0^\pi \left(\frac{120}{\pi^5}x(\pi - x) - \sin x\right)^2 dx = \frac{\pi}{2} - \frac{480}{\pi^5} \approx 0.00227.$$

17. To minimize $I = \int_0^1 (ax^2 + b - x^3)^2 dx$, we choose a and b so that

$$0 = \frac{\partial I}{\partial a} = \int_0^1 2(ax^2 + b - x^3)x^2 dx = \frac{2a}{5} + \frac{2b}{3} - \frac{1}{3}$$

$$0 = \frac{\partial I}{\partial b} = \int_0^1 2(ax^2 + b - x^3) dx = \frac{2a}{3} + 2b - \frac{1}{2}.$$

Solving these two equations, we get $a = 15/16$ and $b = -1/16$. The minimum value of I is

$$\int_0^1 \left(\frac{15x^2}{16} - \frac{1}{16} - x^3\right)^2 dx = \frac{1}{448}.$$

18. To minimize $\int_0^1 (x^3 - ax^2 - bx - c)^2 dx$, choose a, b and c so that

$$0 = 2\int_0^1 (x^3 - ax^2 - bx - c)(-x^2)\, dx$$

$$0 = 2\int_0^1 (x^3 - ax^2 - bx - c)(-x)\, dx$$

$$0 = 2\int_0^1 (x^3 - ax^2 - bx - c)(-1)\, dx,$$

INSTRUCTOR'S SOLUTIONS MANUAL   SECTION 13.4 (PAGE 800)

that is,
$$\frac{a}{5} + \frac{b}{4} + \frac{c}{3} = \frac{1}{6}$$
$$\frac{a}{4} + \frac{b}{3} + \frac{c}{2} = \frac{1}{5}$$
$$\frac{a}{3} + \frac{b}{2} + c = \frac{1}{4}$$

for which the solution is $a = \frac{3}{2}$, $b = -\frac{3}{5}$, and $c = \frac{1}{20}$.

19. To minimize $\int_0^\pi (\sin x - ax^2 - bx)^2\, dx$ we choose a and b so that

$$0 = 2\int_0^\pi (\sin x - ax^2 - bx)(-x^2)\, dx$$
$$0 = 2\int_0^\pi (\sin x - ax^2 - bx)(-x)\, dx.$$

We omit the details of the evaluation of the integrals. The result of the evaluation is that a and b satisfy

$$\frac{\pi^5}{5}a + \frac{\pi^4}{4}b = \pi^2 - 4$$
$$\frac{\pi^4}{4}a + \frac{\pi^3}{3}b = \pi,$$

for which the solution is

$$a = \frac{20}{\pi^5}(\pi^2 - 16)$$
$$b = \frac{12}{\pi^4}(20 - \pi^2).$$

20. $J = \int_{-1}^{1} (x - a\sin\pi x - b\sin 2\pi x - c\sin 3\pi x)^2\, dx.$

To minimize J, choose a, b, and c to satisfy

$$0 = \frac{\partial J}{\partial a}$$
$$= -2\int_{-1}^{1} (x - a\sin\pi x - b\sin 2\pi x - c\sin 3\pi x)\sin\pi x\, dx$$
$$= \frac{2}{\pi}(\pi a - 2)$$
$$0 = \frac{\partial J}{\partial b}$$
$$= -2\int_{-1}^{1} (x - a\sin\pi x - b\sin 2\pi x - c\sin 3\pi x)\sin 2\pi x\, dx$$
$$= \frac{2}{\pi}(\pi b + 1)$$
$$0 = \frac{\partial J}{\partial c}$$
$$= -2\int_{-1}^{1} (x - a\sin\pi x - b\sin 2\pi x - c\sin 3\pi x)\sin 3\pi x\, dx$$
$$= \frac{2}{3\pi}(3\pi c - 2).$$

We have omitted the details of evaluation of these integrals, but note that

$$\int_{-1}^{1} \sin m\pi x \sin n\pi x\, dx = 0$$

if m and n are different integers.

The equations above imply that $a = 2/\pi$, $b = -1/\pi$, and $c = 2/(3\pi)$. These are the values that minimize J.

21. To minimize

$$I = \int_0^\pi \left(f(x) - \frac{a_0}{2} - \sum_{k=1}^{n} a_k \cos kx\right)^2 dx$$

we require

$$0 = \frac{\partial I}{\partial a_0} = 2\int_0^\pi \left(f(x) - \frac{a_0}{2} - \sum_{k=1}^{n} a_k \cos kx\right)\left(-\frac{1}{2}\right) dx,$$

and

$$0 = \frac{\partial I}{\partial a_n} = 2\int_0^\pi \left(f(x) - \frac{a_0}{2} - \sum_{k=1}^{n} a_k \cos kx\right)(-\cos nx)\, dx$$

for $n = 1, 2, \ldots$. Thus

$$a_0 = \frac{2}{\pi}\int_0^\pi f(x)\, dx,$$

and, since

$$\int_0^\pi \cos kx \cos nx\, dx = \begin{cases} 0 & \text{if } k \neq n \\ \frac{\pi}{2} & \text{if } k = n = 1, 2, \ldots \end{cases}$$

we also have

$$a_n = \frac{2}{\pi}\int_0^\pi f(x) \cos nx\, dx \qquad (n = 1, 2, \ldots).$$

22. The Fourier sine series coefficients for $f(x) = x$ on $(0, \pi)$ are

$$b_n = \frac{2}{\pi}\int_0^\pi x \sin(nx)\, dx = (-1)^{n-1}\frac{2}{n}$$

for $n = 1, 2, \ldots$. Thus the series is

$$\sum_{n=0}^{\infty} (-1)^{n-1}\frac{2}{n} \sin nx.$$

Since x and the functions $\sin nx$ are all odd functions, we would also expect the series to converge to x on $(-\pi, 0)$.

489

23. The Fourier cosine series coefficients for $f(x) = x$ on $(0, \pi)$ are

$$a_0 = \frac{2}{\pi} \int_0^\pi x \, dx = \pi$$

$$a_n = \frac{2}{\pi} \int_0^\pi x \cos(nx) \, dx = -\frac{2(1-(-1)^n)}{n^2 \pi}$$

$$= \begin{cases} 0 & \text{if } n \geq 2 \text{ is even} \\ -\dfrac{4}{n^2 \pi} & \text{if } n \geq 1 \text{ is odd.} \end{cases}$$

Thus the Fourier cosine series is

$$\pi - \frac{4}{\pi} \sum_{n=0}^\infty \frac{\cos((2n+1)x)}{(2n+1)^2}.$$

Since the terms of this series are all even functions, and the series converges to x if $0 < x < \pi$, it will converge to $-x = |x|$ if $-\pi < x < 0$.

Remark: since $|x|$ is continuous at $x = 0$, the series also converges at $x = 0$ to 0. It follows that

$$1 + \frac{1}{3^2} + \frac{1}{5^2} + \cdots = \sum_{n=0}^\infty \frac{1}{(2n+1)^2} = \frac{\pi^2}{8}.$$

24. We are given that $x_1 \leq x_2 \leq x_3 \leq \ldots \leq x_n$. To motivate the method, look at a special case, $n = 5$ say.

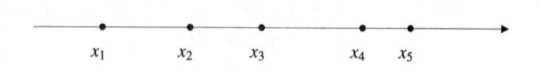

Fig. 13.4.24

If $x = x_3$, then

$$\sum_{i=1}^5 |x - x_i|$$
$$= (x_3 - x_1) + (x_3 - x_2) + 0 + (x_4 - x_3) + (x_5 - x_3)$$
$$= (x_5 - x_1) + (x_4 - x_2).$$

If x moves away from x_3 in either direction, then

$$\sum_{i=1}^5 |x - x_i| = (x_5 - x_1) + (x_4 - x_2) + |x - x_3|.$$

Thus the minimum sum occurs if $x = x_3$.
In general, if n is odd, then $\sum_{i=1}^n |x - x_i|$ is minimum if $x = x_{(n+1)/2}$, the middle point of the set of points $\{x_1, x_2, \ldots, x_n\}$. The value of x is unique in this case. If n is even and x satisfies $x_{n/2} \leq x \leq x_{(n/2)+1}$, then

$$\sum_{i=1}^n |x - x_i| = \sum_{i=1}^{n/2} |x_{n+1-i} - x_i|,$$

and the sum will increase if x is outside that interval. In this case the value of x which minimizes the sum is not unique unless it happens that $x_{n/2} = x_{(n/2)+1}$.

Section 13.5 Parametric Problems (page 810)

1. $F(x) = \displaystyle\int_0^1 t^x \, dt = \dfrac{1}{x+1} \quad (x > -1)$

$F'(x) = \displaystyle\int_0^1 t^x \ln t \, dt = -\dfrac{1}{(x+1)^2}$

$F''(x) = \displaystyle\int_0^1 t^x (\ln t)^2 \, dt = \dfrac{2}{(x+1)^3}$

$\vdots$

$F^{(n)}(x) = \displaystyle\int_0^1 t^x (\ln t)^n \, dt = \dfrac{(-1)^n n!}{(x+1)^{n+1}}.$

2. $\displaystyle\int_{-\infty}^\infty e^{-u^2} \, du = \sqrt{\pi}$ Let $u = xt$
$\phantom{\displaystyle\int_{-\infty}^\infty e^{-u^2} \, du = \sqrt{\pi}}$ $du = x \, dt$

$\displaystyle\int_{-\infty}^\infty e^{-x^2 t^2} \, dt = \dfrac{\sqrt{\pi}}{x}.$

Differentiate with respect to x:

$\displaystyle\int_{-\infty}^\infty -2xt^2 e^{-t^2 x^2} \, dt = -\dfrac{\sqrt{\pi}}{x^2}$

$\displaystyle\int_{-\infty}^\infty t^2 e^{-x^2 t^2} \, dt = \dfrac{\sqrt{\pi}}{2x^3}.$ (∗)

If $x = 1$ we get $\displaystyle\int_{-\infty}^\infty t^2 e^{-t^2} \, dt = \dfrac{\sqrt{\pi}}{2}$.

Differentiate (∗) with respect to x again:

$\displaystyle\int_{-\infty}^\infty -2xt^4 e^{-x^2 t^2} \, dt = -\dfrac{3\sqrt{\pi}}{2x^4}.$

Divide by -2 and let $x = 1$:

$\displaystyle\int_{-\infty}^\infty t^4 e^{-t^2} \, dt = \dfrac{3\sqrt{\pi}}{4}.$

3. Let $I(x, y) = \displaystyle\int_{-\infty}^\infty \dfrac{e^{-xt^2} - e^{-yt^2}}{t^2} \, dt$, where $x > 0$ and $y > 0$. Then

$\dfrac{\partial I}{\partial x} = -\displaystyle\int_{-\infty}^\infty e^{-xt^2} \, dt$ Let $\sqrt{x} t = s$
$\phantom{\dfrac{\partial I}{\partial x}}$ $\sqrt{x} \, dt = ds$

$= -\dfrac{1}{\sqrt{x}} \displaystyle\int_{-\infty}^\infty e^{-s^2} \, ds = -\dfrac{\sqrt{\pi}}{\sqrt{x}}.$

Similarly, $\dfrac{\partial I}{\partial y} = \dfrac{\sqrt{\pi}}{\sqrt{y}}$. Now

$$I(x,y) = -\sqrt{\pi}\int \dfrac{dx}{\sqrt{x}} = -2\sqrt{\pi x} + C_1(y)$$

$$\dfrac{\sqrt{\pi}}{\sqrt{y}} = \dfrac{\partial I}{\partial y} = \dfrac{\partial C_1}{\partial y} \Rightarrow C_1(y) = 2\sqrt{\pi y} + C_2$$

$$I(x,y) = 2\sqrt{\pi}\left(\sqrt{y} - \sqrt{x}\right) + C_2.$$

But $I(x,x) = 0$. Therefore $C_2 = 0$, and

$$I(x,y) = \int_{-\infty}^{\infty} \dfrac{e^{-xt^2} - e^{-yt^2}}{t^2}\,dt = 2\sqrt{\pi}\left(\sqrt{y}-\sqrt{x}\right).$$

4. Let $I(x,y) = \displaystyle\int_0^1 \dfrac{t^x - t^y}{\ln t}\,dt$, where $x > -1$ and $y > -1$.
Then

$$\dfrac{\partial I}{\partial x} = \int_0^1 t^x\,dt = \dfrac{1}{x+1}$$

$$\dfrac{\partial I}{\partial y} = -\dfrac{1}{y+1}.$$

Thus

$$I(x,y) = \int \dfrac{dx}{x+1} = \ln(x+1) + C_1(y)$$

$$\dfrac{-1}{y+1} = \dfrac{\partial I}{\partial y} = \dfrac{\partial C_1}{\partial y} \Rightarrow C_1(y) = -\ln(y+1) + C_2$$

$$I(x,y) = \ln\left(\dfrac{x+1}{y+1}\right) + C_2.$$

But $I(x,x) = 0$, so $C_2 = 0$. Thus

$$I(x,y) = \int_0^1 \dfrac{t^x - t^y}{\ln t}\,dt = \ln\left(\dfrac{x+1}{y+1}\right)$$

for $x > -1$ and $y > -1$.

5. $\displaystyle\int_0^\infty e^{-xt}\sin t\,dt = \dfrac{1}{1+x^2}$ if $x > 0$.
Multiply by -1 and differentiate with respect to x twice:

$$\int_0^\infty t e^{-xt}\sin t\,dt = \dfrac{2x}{(1+x^2)^2}$$

$$\int_0^\infty t^2 e^{-xt}\sin t\,dt = \dfrac{2(3x^2-1)}{(1+x^2)^3}.$$

6. $F(x) = \displaystyle\int_0^\infty e^{-xt}\dfrac{\sin t}{t}\,dt$

$F'(x) = \displaystyle\int_0^\infty -e^{-xt}\sin t\,dt = -\dfrac{1}{1+x^2}$ $(x > 0)$.

Therefore $F(x) = -\displaystyle\int \dfrac{dx}{1+x^2} = -\tan^{-1}x + C$.
Now, make the change of variable $xt = s$ in the integral defining $F(x)$, and obtain

$$F(x) = \int_0^\infty e^{-s}\dfrac{\sin(s/x)}{s/x}\dfrac{ds}{x} = \int_0^\infty \dfrac{e^{-s}}{s}\sin\dfrac{s}{x}\,ds.$$

Since $|\sin(s/x)| \le s/x$ if $s > 0$, $x > 0$, we have

$$|F(x)| \le \dfrac{1}{|x|}\int_0^\infty e^{-s}\,ds = \dfrac{1}{|x|} \to 0 \quad \text{as } x \to \infty.$$

Hence $-\dfrac{\pi}{2} + C = 0$, and $C = \dfrac{\pi}{2}$. Therefore

$$F(x) = \int_0^\infty e^{-xt}\dfrac{\sin t}{t}\,dt = \dfrac{\pi}{2} - \tan^{-1} x.$$

In particular, $\displaystyle\int_0^\infty \dfrac{\sin t}{t}\,dt = \lim_{x\to 0} F(x) = \dfrac{\pi}{2}$.

7. $\displaystyle\int_0^\infty \dfrac{dt}{x^2+t^2} = \dfrac{1}{x}\tan^{-1}\dfrac{t}{x}\bigg|_0^\infty = \dfrac{\pi}{2x}$ for $x > 0$.
Differentiate with respect to x:

$$\int_0^\infty \dfrac{-2x\,dt}{(x^2+t^2)^2} = -\dfrac{\pi}{2x^2}$$

$$\int_0^\infty \dfrac{dt}{(x^2+t^2)^2} = \dfrac{\pi}{4x^3}.$$

Differentiate with respect to x again:

$$\int_0^\infty \dfrac{-4x\,dt}{(x^2+t^2)^3} = -\dfrac{3\pi}{4x^4}$$

$$\int_0^\infty \dfrac{dt}{(x^2+t^2)^3} = \dfrac{3\pi}{16x^5}.$$

8. $\displaystyle\int_0^x \dfrac{dt}{x^2+t^2} = \dfrac{1}{x}\tan^{-1}\dfrac{t}{x}\bigg|_0^x = \dfrac{\pi}{4x}$ for $x > 0$.
Differentiate with respect to x:

$$\dfrac{1}{2x^2} + \int_0^x \dfrac{-2x\,dt}{(x^2+t^2)^2} = -\dfrac{\pi}{4x^2}$$

$$\int_0^x \dfrac{dt}{(x^2+t^2)^2} = -\dfrac{1}{2x}\left[-\dfrac{\pi}{4x^2} - \dfrac{1}{2x^2}\right]$$

$$= \dfrac{\pi}{8x^3} + \dfrac{1}{4x^3}.$$

Differentiate with respect to x again:

$$\dfrac{1}{4x^4} + \int_0^x \dfrac{-4x\,dt}{(x^2+t^2)^3} = -\dfrac{3}{x^4}\left[\dfrac{\pi}{8} + \dfrac{1}{4}\right]$$

$$\int_0^x \dfrac{dt}{(x^2+t^2)^3} = -\dfrac{1}{4x}\left[-\dfrac{3\pi}{8x^4} - \dfrac{3}{4x^4} - \dfrac{1}{4x^4}\right]$$

$$= \dfrac{3\pi}{32x^5} + \dfrac{1}{4x^5}.$$

SECTION 13.5 (PAGE 810)

9. $f(x) = 1 + \int_a^x (x-t)^n f(t)\,dt \Rightarrow f(a) = 1$
    $f'(x) = n \int_a^x (x-t)^{n-1} f(t)\,dt$
    $f''(x) = n(n-1) \int_a^x (x-t)^{n-2} f(t)\,dt$
    $\vdots$
    $f^{(n)}(x) = n! \int_a^x f(t)\,dt$
    $f^{(n+1)}(x) = n! f(x) \Rightarrow f^{(n+1)}(a) = n! f(a) = n!.$

10. $f(x) = Cx + D + \int_0^x (x-t) f(t)\,dt \Rightarrow f(0) = D$
    $f'(x) = C + \int_0^x f(t)\,dt \Rightarrow f'(0) = C$
    $f''(x) = f(x) \Rightarrow f(x) = A\cosh x + B\sinh x$
    $D = f(0) = A, \quad C = f'(0) = B$
    $\Rightarrow f(x) = D\cosh x + C\sinh x.$

11. $f(x) = x + \int_0^x (x-2t) f(t)\,dt \Rightarrow f(0) = 0$
    $f'(x) = 1 - xf(x) + \int_0^x f(t)\,dt \Rightarrow f'(0) = 1$
    $f''(x) = -f(x) - xf'(x) + f(x) = -xf'(x).$
    If $u = f'(x)$, then $\dfrac{du}{u} = -x\,dx$, so $\ln u = -\dfrac{x^2}{2} + \ln C_1$.
    Therefore
    $$f'(x) = u = C_1 e^{-x^2/2}.$$
    We have $1 = f'(0) = C_1$, so $f'(x) = e^{-x^2/2}$ and
    $$f(x) = \int_0^x e^{-t^2/2}\,dt + C_2.$$
    But $0 = f(0) = C_2$, and so
    $$f(x) = \int_0^x e^{-t^2/2}\,dt.$$

12. $f(x) = 1 + \int_0^1 (x+t) f(t)\,dt$
    $f'(x) = \int_0^1 f(t)\,dt = C$, say,
    since the integral giving $f'(x)$ does not depend on x.
    Thus $f(x) = A + Cx$, where $A = f(0)$. Substituting this expression into the given equation, we obtain
    $$A + Cx = 1 + \int_0^1 (x+t)(A + Ct)\,dt$$
    $$= 1 + Ax + \dfrac{A}{2} + \dfrac{Cx}{2} + \dfrac{C}{3}.$$

Therefore
$$\dfrac{A}{2} - 1 - \dfrac{C}{3} + x\left(\dfrac{C}{2} - A\right) = 0.$$
This can hold for all x only if
$$\dfrac{A}{2} - 1 - \dfrac{C}{3} = 0 \quad \text{and} \quad \dfrac{C}{2} - A = 0.$$
Thus $C = 2A$ and $\dfrac{A}{2} - \dfrac{2A}{3} = 1$, so that $A = -6$ and $C = -12$. Therefore $f(x) = -6 - 12x$.

13. We eliminate c from the pair of equations
    $$f(x, y, c) = 2cx - c^2 - y = 0$$
    $$\dfrac{\partial}{\partial c} f(x, y, c) = 2x - 2c = 0.$$
    Thus $c = x$ and $2x^2 - x^2 - y = 0$. The envelope is $y = x^2$.

14. We eliminate c from the pair of equations
    $$f(x, y, c) = y - (x - c)\cos c - \sin c = 0$$
    $$\dfrac{\partial}{\partial c} f(x, y, c) = \cos c + (x - c)\sin c - \cos c = 0.$$
    Thus $c = x$ and $y - 0 - \sin x = 0$.
    The envelope is $y = \sin x$.

15. We eliminate c from the pair of equations
    $$f(x, y, c) = x\cos c + y\sin c - 1 = 0$$
    $$\dfrac{\partial}{\partial c} f(x, y, c) = -x\sin c + y\cos c = 0.$$
    Squaring and adding these equations yields $x^2 + y^2 = 1$, which is the equation of the envelope.

16. We eliminate c from the pair of equations
    $$f(x, y, c) = \dfrac{x}{\cos c} + \dfrac{y}{\sin c} - 1 = 0$$
    $$\dfrac{\partial}{\partial c} f(x, y, c) = \dfrac{x\sin c}{\cos^2 c} - \dfrac{y\cos c}{\sin^2 c} = 0.$$
    From the second equation, $y = x\tan^3 c$. Thus
    $$\dfrac{x}{\cos c}(1 + \tan^2 c) = 1$$
    which implies that $x = \cos^3 c$, and hence $y = \sin^3 c$. The envelope is the astroid $x^{2/3} + y^{2/3} = 1$.

17. We eliminate c from the pair of equations
    $$f(x, y, c) = c + (x - c)^2 - y = 0$$
    $$\dfrac{\partial}{\partial c} f(x, y, c) = 1 + 2(c - x) = 0.$$

INSTRUCTOR'S SOLUTIONS MANUAL SECTION 13.5 (PAGE 810)

Thus $c = x - \dfrac{1}{2}$. The envelope is the line $y = x - \dfrac{1}{4}$.

18. We eliminate c from the pair of equations

$$f(x, y, c) = (x - c)^2 + (y - c)^2 - 1 = 0$$
$$\frac{\partial}{\partial c} f(x, y, c) = 2(c - x) + 2(c - y) = 0.$$

Thus $c = (x + y)/2$, and

$$\left(\frac{x - y}{2}\right)^2 + \left(\frac{y - x}{2}\right)^2 = 1$$

or $x - y = \pm\sqrt{2}$. These two parallel lines constitute the envelope of the given family which consists of circles of radius 1 with centres along the line $y = x$.

19. Not every one-parameter family of curves in the plane has an envelope. The family of parabolas $y = x^2 + c$ evidently does not. (See the figure.) If we try to calculate the envelope by eliminating c from the equations

$$f(x, y, c) = y - x^2 - c = 0$$
$$\frac{\partial}{\partial c} f(x, y, c) = -1 = 0,$$

we fail because the second equation is contradictory.

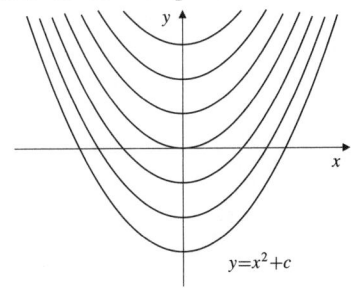

Fig. 13.5.19

20. The curve $x^2 + (y - c)^2 = kc^2$ is a circle with centre $(0, c)$ and radius $\sqrt{k}c$, provided $k > 0$. Consider the system:

$$f(x, y, c) = x^2 + (y - c)^2 - kc^2 = 0$$
$$\frac{\partial}{\partial c} f(x, y, c) = -2(y - c) - 2kc = 0.$$

The second equation implies that $y - c = -kc$, and the first equation then says that $x^2 = k(1 - k)c^2$. This is only possible if $0 \le k \le 1$.
The cases $k = 0$ and $k = 1$ are degenerate. If $k = 0$ the "curves" are just points on the y-axis. If $k = 1$ the curves are circles, all of which are tangent to the x-axis at the origin. There is no reasonable envelope in either case. If $0 < k < 1$, the envelope is the pair of lines given by $x^2 = \dfrac{k}{1-k} y^2$, that is, the lines $\sqrt{1-k}\,x = \pm\sqrt{k}\,y$. These lines make angle $\sin^{-1}\sqrt{k}$ with the y-axis.

circles
$x^2 + (y-c)^2 = kc^2$

envelope
$(1-k)x^2 = ky^2$

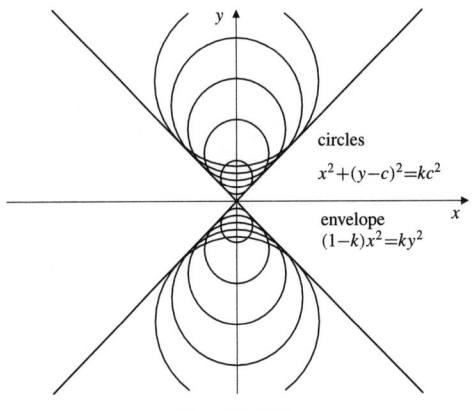

Fig. 13.5.20

21. We eliminate c from the equations

$$f(x, y, c) = y^3 - (x + c)^2 = 0$$
$$\frac{\partial}{\partial c} f(x, y, c) = -2(x + c) = 0.$$

Thus $x = -c$, and we obtain the equation $y = 0$ for the envelope. However, this is not really an envelope at all. The curves $y^3 = (x + c)^2$ all have cusps along the x-axis; none of them is tangent to the axis.

$f(x,y,c) = y^3 - (x+c)^2 = 0$

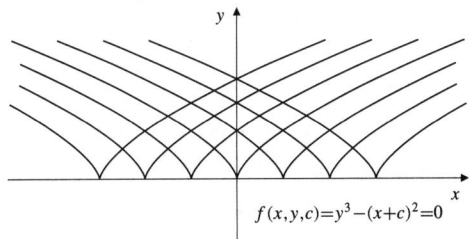

Fig. 13.5.21

22. If the family of surfaces $f(x, y, z, \lambda, \mu) = 0$ has an envelope, that envelope will have parametric equations

$$x = x(\lambda, \mu), \qquad y = y(\lambda, \mu), \qquad z = z(\lambda, \mu),$$

giving the point on the envelope where the envelope is tangent to the particular surface in the family having parameter values λ and μ. Thus

$$f\big(x(\lambda, \mu), y(\lambda, \mu), z(\lambda, \mu), \lambda, \mu\big) = 0.$$

Differentiating with respect to λ, we obtain

$$f_1 \frac{\partial x}{\partial \lambda} + f_2 \frac{\partial y}{\partial \lambda} + f_3 \frac{\partial z}{\partial \lambda} + f_4 = 0.$$

However, since for fixed μ, the parametric curve

$$x = x(t, \mu), \qquad y = y(t, \mu), \qquad z = z(t, \mu)$$

493

is tangent to the surface $f(x, y, z, \lambda, \mu) = 0$ at $t = \lambda$, its tangent vector there,

$$\mathbf{T} = \frac{\partial x}{\partial \lambda}\mathbf{i} + \frac{\partial y}{\partial \lambda}\mathbf{j} + \frac{\partial z}{\partial \lambda}\mathbf{k},$$

is perpendicular to the normal

$$\mathbf{N} = \nabla f = f_1\mathbf{i} + f_2\mathbf{j} + f_3\mathbf{k},$$

so

$$f_1\frac{\partial x}{\partial \lambda} + f_2\frac{\partial y}{\partial \lambda} + f_3\frac{\partial z}{\partial \lambda} = 0.$$

Hence we must also have $\dfrac{\partial f}{\partial \lambda} = f_4(x, y, z, \lambda, \mu) = 0$.
Similarly, $\dfrac{\partial f}{\partial \mu} = 0$.
The parametric equations of the envelope must therefore satisfy the three equations

$$f(x, y, z, \lambda, \mu) = 0$$
$$\frac{\partial}{\partial \lambda} f(x, y, z, \lambda, \mu) = 0$$
$$\frac{\partial}{\partial \mu} f(x, y, z, \lambda, \mu) = 0.$$

The envelope can be found by eliminating λ and μ from these three equations.

23. To find the envelope we eliminate λ and μ from the equations

$$x \sin\lambda \cos\mu + y \sin\lambda \sin\mu + z \cos\lambda = 1 \quad (1)$$
$$x \cos\lambda \cos\mu + y \cos\lambda \sin\mu - z \sin\lambda = 0 \quad (2)$$
$$-x \sin\lambda \sin\mu + y \sin\lambda \cos\mu = 0. \quad (3)$$

Multiplying (1) by $\cos\lambda$ and (2) by $\sin\lambda$ and subtracting the two gives

$$z = \cos\lambda.$$

Therefore (2) and (3) can be rewritten

$$x \cos\mu + y \sin\mu = \sin\lambda$$
$$x \sin\mu - y \cos\mu = 0.$$

Squaring and adding these equations gives

$$x^2 + y^2 = \sin^2\lambda.$$

Therefore

$$x^2 + y^2 + z^2 = \sin^2\lambda + \cos^2\lambda = 1;$$

the envelope is the sphere of radius 1 centred at the origin.

24. $(x - \lambda)^2 + (y - \mu)^2 + z^2 = \dfrac{\lambda^2 + \mu^2}{2}$.
Differentiate with respect to λ and μ:

$$-2(x - \lambda) = \lambda, \qquad -2(y - \mu) = \mu.$$

Thus $\lambda = 2x$, $\mu = 2y$, and

$$x^2 + y^2 + z^2 = 2x^2 + 2y^2.$$

The envelope is the cone $z^2 = x^2 + y^2$.

25. $y + \epsilon \sin(\pi y) = x \;\Rightarrow\; y = y(\epsilon, x)$

$$\frac{\partial y}{\partial \epsilon} + \sin(\pi y) + \pi\epsilon \cos(\pi y)\frac{\partial y}{\partial \epsilon} = 0$$

$$\frac{\partial^2 y}{\partial \epsilon^2} + 2\pi \cos(\pi y)\frac{\partial y}{\partial \epsilon} - \pi^2\epsilon \sin(\pi y)\left(\frac{\partial y}{\partial \epsilon}\right)^2$$
$$+ \pi\epsilon \cos(\pi y)\frac{\partial^2 y}{\partial \epsilon^2} = 0.$$

If $\epsilon = 0$ then $y = x$, so $y(x, 0) = x$. Also, at $\epsilon = 0$,

$$y_\epsilon(x, 0)(1 + 0) = -\sin(\pi y(x, 0)) = -\sin(\pi x),$$

that is, $y_\epsilon(x, 0) = -\sin(\pi x)$. Also,

$$y_{\epsilon\epsilon}(x, 0)(1 + 0) = -2\pi \cos(\pi x) y_\epsilon(x, 0) + 0$$
$$= 2\pi \cos(\pi x)\sin(\pi x) = \pi \sin(2\pi x).$$

Thus

$$y = y(x, \epsilon) = y(x, 0) + \epsilon y_\epsilon(x, 0) + \frac{\epsilon^2}{2!} y_{\epsilon\epsilon}(x, 0) + \cdots$$
$$= x - \epsilon \sin(\pi x) + \frac{\epsilon^2}{2}\pi \sin(2\pi x) + \cdots$$

26. $y^2 + \epsilon e^{-y^2} = 1 + x^2$
$2yy_\epsilon + e^{-y^2} - 2y\epsilon e^{-y^2} y_\epsilon = 0$
$2y\left(1 - \epsilon e^{-y^2}\right) y_\epsilon + e^{-y^2} = 0$
$2y_\epsilon\left(1 - \epsilon e^{-y^2}\right) y_\epsilon - 2ye^{-y^2} y_\epsilon + 2y\left(2y\epsilon e^{-y^2} y_\epsilon\right) y_\epsilon$
$\qquad + 2y\left(1 - \epsilon e^{-y^2}\right) y_{\epsilon\epsilon} - 2ye^{-y^2} y_\epsilon = 0.$

At $\epsilon = 0$ we have $y(x, 0) = \sqrt{1 + x^2}$, and

$$2\sqrt{1 + x^2}\, y_\epsilon(x, 0) + e^{-(1+x^2)} = 0$$
$$y_\epsilon(x, 0) = -\frac{1}{2\sqrt{1 + x^2}} e^{-(1+x^2)}$$
$$2y_\epsilon^2 - 4ye^{-y^2} y_\epsilon + 2yy_{\epsilon\epsilon} = 0$$
$$yy_{\epsilon\epsilon} = 2yy_\epsilon e^{-y^2} - y_\epsilon^2$$
$$y_{\epsilon\epsilon}(x, 0) = -\left(\frac{1}{\sqrt{1 + x^2}} + \frac{1}{4(1 + x^2)^{3/2}}\right) e^{-2(1+x^2)}.$$

Thus

$$y = y(x, \epsilon) = y(x, 0) + \epsilon y_\epsilon(x, 0) + \frac{\epsilon^2}{2!} y_{\epsilon\epsilon}(x, 0) + \cdots$$
$$= \sqrt{1 + x^2} - \frac{\epsilon}{2\sqrt{1 + x^2}} e^{-(1+x^2)}$$
$$- \frac{\epsilon^2}{2} \left(\frac{1}{\sqrt{1 + x^2}} + \frac{1}{4(1 + x^2)^{3/2}} \right) e^{-2(1+x^2)} + \cdots.$$

27. $2y + \dfrac{\epsilon x}{1 + y^2} = 1$

$2y_\epsilon + \dfrac{x}{1 + y^2} - \dfrac{2\epsilon x y y_\epsilon}{(1 + y^2)^2} = 0$

$2y_{\epsilon\epsilon} - \dfrac{4xyy_\epsilon}{(1 + y^2)^2} - \epsilon \dfrac{\partial}{\partial \epsilon}\left(\dfrac{2xyy_\epsilon}{(1+y^2)^2}\right) = 0.$

At $\epsilon = 0$ we have $y(x, 0) = \dfrac{1}{2}$, and

$$y_\epsilon(x, 0) = -\frac{1}{2} \frac{x}{1 + \frac{1}{4}} = -\frac{2x}{5}$$

$$y_{\epsilon\epsilon} = \frac{1}{2} \frac{4x\left(\frac{1}{2}\right)\left(-\frac{2x}{5}\right)}{\left(1 + \frac{1}{4}\right)^2} = -\frac{32x^2}{125}.$$

Thus

$$y = y(x, \epsilon) = y(x, 0) + \epsilon y_\epsilon(x, 0) + \frac{\epsilon^2}{2!} y_{\epsilon\epsilon}(x, 0) + \cdots$$
$$= \frac{1}{2} - \frac{2\epsilon x}{5} - \frac{16\epsilon^2 x^2}{125} + \cdots.$$

28. Let $y(x, \epsilon)$ be the solution of $y + \epsilon y^5 = \dfrac{1}{2}$. Then we have

$$y_\epsilon\left(1 + 5\epsilon y^4\right) + y^5 = 0$$
$$y_{\epsilon\epsilon}\left(1 + 5\epsilon y^4\right) + 20\epsilon y^3 y_\epsilon^2 + 10 y^4 y_\epsilon = 0$$
$$y_{\epsilon\epsilon\epsilon}\left(1 + 5\epsilon y^4\right) + y_{\epsilon\epsilon}\left(60\epsilon y^3 y_\epsilon + 15 y^4\right)$$
$$+ 60\epsilon y_\epsilon^3 y^2 + 60 y^3 y_\epsilon^2 = 0.$$

At $\epsilon = 0$ we have

$$y(x, 0) = \frac{1}{2}$$
$$y_\epsilon(x, 0) = -\frac{1}{32}$$
$$y_{\epsilon\epsilon}(x, 0) = -\frac{10}{16}\left(-\frac{1}{32}\right) = \frac{5}{16^2}$$
$$y_{\epsilon\epsilon\epsilon}(x, 0) = -\frac{5}{16^2}\left(\frac{15}{16}\right) - \frac{60}{8}\left(-\frac{1}{32}\right)^2 = -\frac{105}{4096}.$$

For $\epsilon = \dfrac{1}{100}$ we have

$$y = \frac{1}{2} - \frac{1}{32} \times \frac{1}{100} + \frac{5}{256} \times \frac{1}{2 \times 100^2}$$
$$- \frac{105}{4096} \times \frac{1}{6 \times 100^3} + \cdots$$
$$\approx 0.49968847$$

with error less than 10^{-8} in magnitude.

29. Let $x(\epsilon)$ and $y(\epsilon)$ be the solution of

$$\begin{array}{rcl} x + 2y + \epsilon e^{-x} &=& 3 \\ x - y + \epsilon e^{-y} &=& 0. \end{array}$$

Thus

$$\begin{array}{rcl} x' + 2y' + e^{-x} - \epsilon e^{-x} x' &=& 0 \\ x' - y' + e^{-y} - \epsilon e^{-y} y' &=& 0 \\ x'' + 2y'' - 2e^{-x} x' + \epsilon e^{-x}(x')^2 - \epsilon e^{-x} x'' &=& 0 \\ x'' - y'' - 2e^{-y} y' + \epsilon e^{-y}(y')^2 - \epsilon e^{-y} y'' &=& 0. \end{array}$$

At $\epsilon = 0$ we have

$$\left.\begin{array}{rcl} x + 2y &=& 3 \\ x - y &=& 0 \end{array}\right\} \Rightarrow x = y = 1$$

$$\left.\begin{array}{rcl} x' + 2y' &=& -\frac{1}{e} \\ x' - y' &=& -\frac{1}{e} \end{array}\right\} \Rightarrow \begin{array}{rcl} x' &=& -\frac{1}{e} \\ y' &=& 0 \end{array}$$

$$\left.\begin{array}{rcl} x'' + 2y'' &=& -\frac{2}{e^2} \\ x'' - y'' &=& 0 \end{array}\right\} \Rightarrow x'' = y'' = \frac{-2}{3e^2}.$$

Thus

$$x = 1 - \frac{\epsilon}{e} - \frac{\epsilon^2}{3e^2} + \cdots, \quad y = 1 - \frac{\epsilon^2}{3e^2} + \cdots.$$

For $\epsilon = \dfrac{1}{100}$ we have

$$x = 1 - \frac{1}{100e} + \frac{1}{30,000e^2} + \cdots$$
$$y = 1 - \frac{1}{30,000e^2} + \cdots.$$

Section 13.6 Newton's Method (page 815)

For each of Exercises 1–6, and 9, we sketch the graphs of the two given equations, $f(x, y) = 0$ and $g(x, y) = 0$, and use their intersections to make initial guesses x_0 and y_0 for the solutions. These guesses are then refined using the formulas

$$x_{n+1} = x_n - \left.\frac{fg_2 - gf_2}{f_1g_2 - g_1f_2}\right|_{(x_n, y_n)}, \quad y_{n+1} = y_n - \left.\frac{f_1g - g_1f}{f_1g_2 - g_1f_2}\right|_{(x_n, y_n)}.$$

NOTE: The numerical values in the tables below were obtained by programming a microcomputer to calculate the iterations of the above formulas. In most cases the computer was using more significant digits than appear in the tables, and did not truncate the values obtained at one step before using them to calculate the next step. If you use a calculator, and use the numbers as quoted on one line of a table to calculate the numbers on the next line, your results may differ slightly (in the last one or two decimal places).

1. $f(x, y) = y - e^x$ $f_1(x, y) = -e^x$ $g_1(x, y) = 1$
   $g(x, y) = x - \sin y$ $f_2(x, y) = 1$ $g_2(x, y) = -\cos y$

   We start with $x_0 = 0.9$, $y_0 = 2.0$.

n	x_n	y_n	$f(x_n, y_n)$	$g(x_n, y_n)$
0	0.9000000	2.0000000	-0.4596031	-0.0092974
1	0.8100766	2.2384273	-0.0096529	0.0247861
2	0.7972153	2.2191669	-0.0001851	0.0001464
3	0.7971049	2.2191071	0.0000000	0.0000000
4	0.7971049	2.2191071	0.0000000	0.0000000

   Thus $x = 0.7971049$, $y = 2.2191071$.

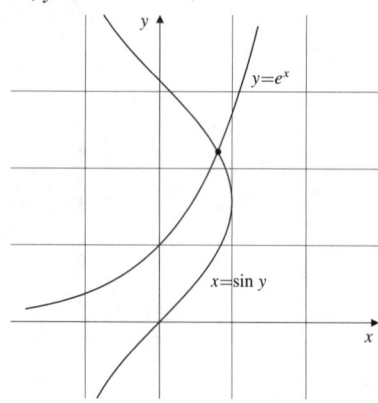

   Fig. 13.6.1 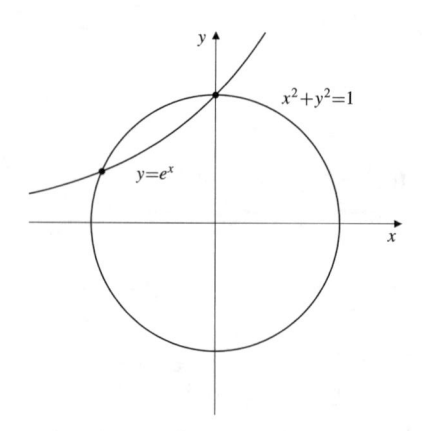   Fig. 13.6.2

2. $f(x, y) = x^2 + y^2 - 1$ $f_1(x, y) = 2x$ $g_1(x, y) = -e^x$.
   $g(x, y) = y - e^x$ $f_2(x, y) = 2y$ $g_2(x, y) = 1$

   Evidently one solution is $x = 0$, $y = 1$. The second solution is near $(-1, 0)$. We try $x_0 = -0.9$, $y_0 = 0.2$.

n	x_n	y_n	$f(x_n, y_n)$	$g(x_n, y_n)$
0	-0.9000000	0.2000000	-0.1500000	-0.2065697
1	-0.9411465	0.3898407	0.0377325	-0.0003395
2	-0.9170683	0.3995751	0.0006745	-0.0001140
3	-0.9165628	0.3998911	0.0000004	-0.0000001
4	-0.9165626	0.3998913	0.0000000	0.0000000

   The second solution is $x = -0.9165626$, $y = 0.3998913$.

INSTRUCTOR'S SOLUTIONS MANUAL SECTION 13.6 (PAGE 815)

3. $f(x, y) = x^4 + y^2 - 16$ $f_1(x, y) = 4x^3$ $g_1(x, y) = y$.
   $g(x, y) = xy - 1$ $f_2(x, y) = 2y$ $g_2(x, y) = x$

   There are four solutions as shown in the figure. We will find the two in the first quadrant; the other two are the negatives of these by symmetry.
   The first quadrant solutions appear to be near $(1.9, 0.5)$ and $(0.25, 3.9)$.

n	x_n	y_n	$f(x_n, y_n)$	$g(x_n, y_n)$
0	1.9000000	0.5000000	-2.7179000	-0.0500000
1	1.9990542	0.5002489	0.2200049	0.0000247
2	1.9921153	0.5019730	0.0011548	-0.0000120
3	1.9920783	0.5019883	0.0000000	0.0000000
4	1.9920783	0.5019883	0.0000000	0.0000000

n	x_n	y_n	$f(x_n, y_n)$	$g(x_n, y_n)$
0	0.2500000	3.9000000	-0.7860937	-0.0250000
1	0.2499499	4.0007817	0.0101569	-0.0000050
2	0.2500305	3.9995117	0.0000016	-0.0000001
3	0.2500305	3.9995115	0.0000000	0.0000000

   The four solutions are $x = \pm 1.9920783$, $\pm y = 0.5019883$, and $x = \pm 0.2500305$, $y = \pm 3.9995115$.

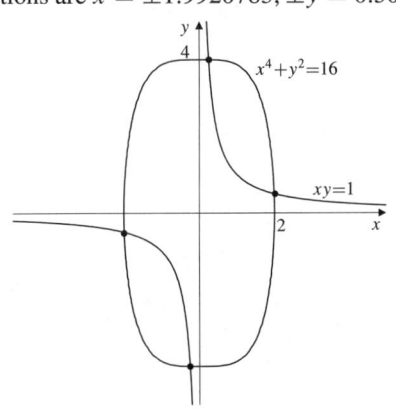

   Fig. 13.6.3

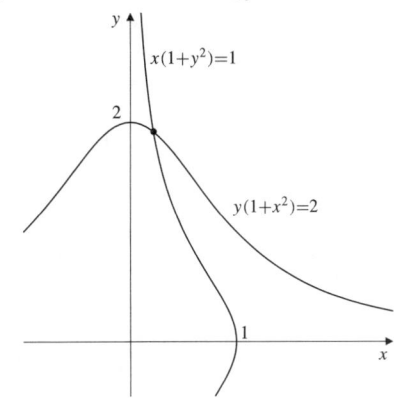

   Fig. 13.6.4

4. $f(x, y) = x(1 + y^2) - 1$ $f_1(x, y) = 1 + y^2$ $g_1(x, y) = 2xy$.
   $g(x, y) = y(1 + x^2) - 2$ $f_2(x, y) = 2xy$ $g_2(x, y) = 1 + x^2$

   The solution appears to be near $x = 0.2$, $y = 1.8$.

n	x_n	y_n	$f(x_n, y_n)$	$g(x_n, y_n)$
0	0.2000000	1.8000000	-0.1520000	-0.1280000
1	0.2169408	1.9113487	0.0094806	0.0013031
2	0.2148268	1.9117785	-0.0000034	0.0000081
3	0.2148292	1.9117688	0.0000000	0.0000000

   The solution is $x = 0.2148292$, $y = 1.9117688$.

5. $f(x, y) = y - \sin x$ $f_1(x, y) = -\cos x$ $g_1(x, y) = 2x$.
   $g(x, y) = x^2 + (y + 1)^2 - 2$ $f_2(x, y) = 1$ $g_2(x, y) = 2(y + 1)$

   Solutions appear to be near $(0.5, 0.3)$ and $(-1.5, -1)$.

497

n	x_n	y_n	$f(x_n, y_n)$	$g(x_n, y_n)$
0	0.5000000	0.3000000	−0.1794255	−0.0600000
1	0.3761299	0.3707193	0.0033956	0.0203450
2	0.3727877	0.3642151	0.0000020	0.0000535
3	0.3727731	0.3641995	0.0000000	0.0000000
4	0.3727731	0.3641995	0.0000000	0.0000000

n	x_n	y_n	$f(x_n, y_n)$	$g(x_n, y_n)$
0	−1.5000000	−1.0000000	−0.0025050	0.2500000
1	−1.4166667	−0.9916002	−0.0034547	0.0070150
2	−1.4141680	−0.9877619	−0.0000031	0.0000210
3	−1.4141606	−0.9877577	0.0000000	0.0000000
4	−1.4141606	−0.9877577	0.0000000	0.0000000

The solutions are $x = 0.3727731$, $y = 0.3641995$, and $x = -1.4141606$, $y = -0.9877577$.

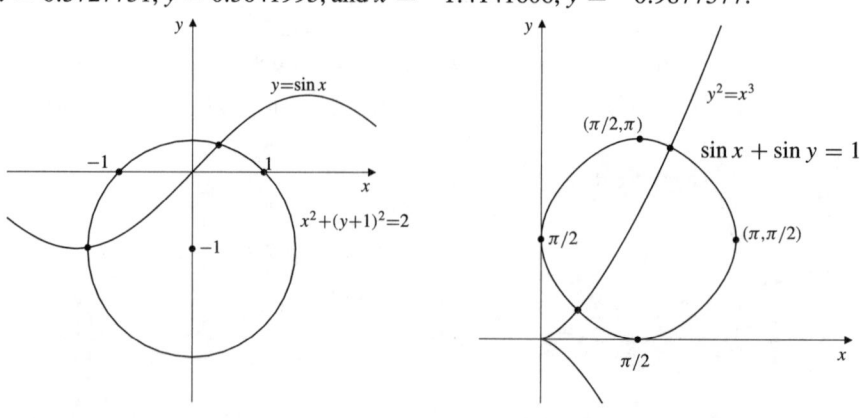

Fig. 13.6.5

Fig. 13.6.6

6. $f(x, y) = \sin x + \sin y - 1$ $f_1(x, y) = \cos x$ $g_1(x, y) = -3x^2$.
$g(x, y) = y^2 - x^3$ $f_2(x, y) = \cos y$ $g_2(x, y) = 2y$

There are infinitely many solutions for the given pair of equations, since the level curve of $f(x, y) = 0$ is repeated periodically throughout the plane. We will find the two solutions closest to the origin in the first quadrant. From the figure, it appears that these solutions are near $(0.6, 0.4)$ and $(2, 3)$.

n	x_n	y_n	$f(x_n, y_n)$	$g(x_n, y_n)$
0	0.6000000	0.4000000	−0.0459392	−0.0560000
1	0.5910405	0.4579047	−0.0007050	0.0032092
2	0.5931130	0.4567721	−0.0000015	−0.0000063
3	0.5931105	0.4567761	0.0000000	0.0000000
4	0.5931105	0.4567761	0.0000000	0.0000000

n	x_n	y_n	$f(x_n, y_n)$	$g(x_n, y_n)$
0	2.0000000	3.0000000	0.0504174	1.0000000
1	2.0899016	3.0131366	−0.0036336	−0.0490479
2	2.0854887	3.0116804	−0.0000086	−0.0001199
3	2.0854779	3.0116770	0.0000000	0.0000000
4	2.0854779	3.0116770	0.0000000	0.0000000

The solutions are $x = 0.5931105$, $y = 0.4567761$, and $x = 2.0854779$, $y = 3.0116770$.

INSTRUCTOR'S SOLUTIONS MANUAL SECTION 13.6 (PAGE 815)

7. By analogy with the two-dimensional case, the Newton's Method iteration formulas are

$$x_{n+1} = x_n - \frac{1}{\Delta}\begin{vmatrix} f & f_2 & f_3 \\ g & g_2 & g_3 \\ h & h_2 & h_3 \end{vmatrix}_{(x_n,y_n,z_n)} \qquad y_{n+1} = y_n - \frac{1}{\Delta}\begin{vmatrix} f_1 & f & f_3 \\ g_1 & g & g_3 \\ h_1 & h & h_3 \end{vmatrix}_{(x_n,y_n,z_n)}$$

$$z_{n+1} = z_n - \frac{1}{\Delta}\begin{vmatrix} f_1 & f_2 & f \\ g_1 & g_2 & g \\ h_1 & h_2 & h \end{vmatrix}_{(x_n,y_n,z_n)} \qquad \text{where } \Delta = \begin{vmatrix} f_1 & f_2 & f_3 \\ g_1 & g_2 & g_3 \\ h_1 & h_2 & h_3 \end{vmatrix}_{(x_n,y_n,z_n)}$$

8. $f(x,y,z) = y^2 + z^2 - 3$ $\quad g(x,y,z) = x^2 + z^2 - 2$ $\quad h(x,y,z) = x^2 - z$
   $f_1(x,y,z) = 0$ $\quad g_1(x,y,z) = 2x$ $\quad h_1(x,y,z) = 2x$
   $f_2(x,y,z) = 2y$ $\quad g_2(x,y,z) = 0$ $\quad h_2(x,y,z) = 0$
   $f_3(x,y,z) = 2z$ $\quad g_3(x,y,z) = 2z$ $\quad h_3(x,y,z) = -1$

   It is easily seen that the system

   $$f(x,y,z) = 0, \qquad g(x,y,z) = 0, \qquad h(x,y,z) = 0$$

   has first-quadrant solution $x = z = 1$, $y = \sqrt{2}$. Let us start at the "guess" $x_0 = y_0 = z_0 = 2$.

n	x_n	y_n	z_n	$f(x_n,y_n,z_n)$	$g(x_n,y_n,z_n)$	$h(x_n,y_n,z_n)$
0	2.0000000	2.0000000	2.0000000	5.0000000	6.0000000	2.0000000
1	1.3000000	1.5500000	1.2000000	0.8425000	1.1300000	0.4900000
2	1.0391403	1.4239564	1.0117647	0.0513195	0.1034803	0.0680478
3	1.0007592	1.4142630	1.0000458	0.0002313	0.0016104	0.0014731
4	1.0000003	1.4142136	1.0000000	0.0000000	0.0000006	0.0000006
5	1.0000000	1.4142136	1.0000000	0.0000000	0.0000000	0.0000000

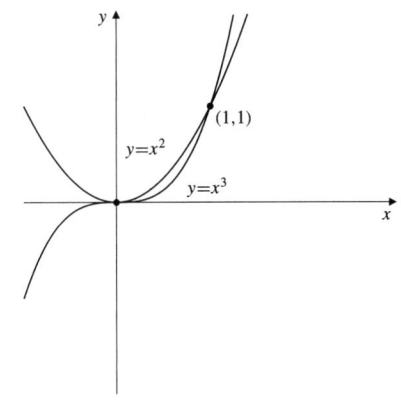

9. $f(x,y) = y - x^2$ $\quad f_1(x,y) = -2x$ $\quad g_1(x,y) = -3x^2$
   $g(x,y) = y - x^3$ $\quad f_2(x,y) = 1$ $\quad g_2(x,y) = 1$

Fig. 13.6.9

n	x_n	y_n
0	0.1000000	0.1000000
1	0.0470588	−0.0005882
2	0.0229337	−0.0000561
3	0.0113307	−0.0000062
4	0.0056327	−0.0000007
5	0.0028083	−0.0000001
⋮		
15	0.0000027	0.0000000
16	0.0000014	0.0000000
17	0.0000007	0.0000000
18	0.0000003	0.0000000

n	x_n	y_n
0	0.9000000	0.9000000
1	1.0285714	1.0414286
2	1.0015038	1.0022771
3	1.0000045	1.0000068
4	1.0000000	1.0000000

Eighteen iterations were needed to obtain the solution $x = y = 0$ correct to six decimal places, starting from $x = y = 0.1$. This slow convergence is due to the fact that the curves $y = x^2$ and $y = x^3$ are tangent at $(0, 0)$. Only four iterations were needed to obtain the solution $x = y = 1$ starting from $x = y = 0$, because, although the angle between the curves is small at $(1, 1)$, it is not 0. The curves are not tangent there.

Review Exercises 13 (page 816)

1. $f(x, y) = xye^{-x+y}$
$f_1(x, y) = (y - xy)e^{-x+y} = y(1 - x)e^{-x+y}$
$f_2(x, y) = (x + xy)e^{-x+y} = x(1 + y)e^{-x+y}$
$A = f_{11} = (-2y + xy)e^{-x+y}$
$B = f_{12} = (1 - x + y - xy)e^{-x+y}$
$C = f_{22} = (2x + xy)e^{-x+y}$.
For CP: either $y = 0$ or $x = 1$, and either $x = 0$ or $y = -1$. The CPs are $(0, 0)$ and $(1, -1)$.

CP	A	B	C	$AC - B^2$	class
$(0, 0)$	0	1	0	−1	saddle
$(1, -1)$	e^{-2}	0	e^{-2}	e^{-4}	loc. min

2. $f(x, y) = x^2 y - 2xy^2 + 2xy$
$f_1(x, y) = 2xy - 2y^2 + 2y = 2y(x - y + 1)$
$f_2(x, y) = x^2 - 4xy + 2x = x(x - 4y + 2)$
$A = f_{11} = 2y$
$B = f_{12} = 2x - 4y + 2$
$C = f_{22} = -4x$.
For CP: either $y = 0$ or $x - y + 1 = 0$, and either $x = 0$ or $x - 4y + 2 = 0$. The CPs are $(0, 0)$, $(0, 1)$, $(-2, 0)$, and $(-2/3, 1/3)$.

CP	A	B	C	$AC - B^2$	class
$(0, 0)$	0	2	0	−4	saddle
$(0, 1)$	2	−2	0	−4	saddle
$(-2, 0)$	0	−2	8	−4	saddle
$(-\frac{2}{3}, \frac{1}{3})$	$\frac{2}{3}$	$-\frac{2}{3}$	$\frac{8}{3}$	$\frac{4}{3}$	loc. min

3. $f(x, y) = \dfrac{1}{x} + \dfrac{4}{y} + \dfrac{9}{4 - x - y}$
$f_1(x, y) = -\dfrac{1}{x^2} + \dfrac{9}{(4 - x - y)^2}$
$f_2(x, y) = -\dfrac{4}{y^2} + \dfrac{9}{(4 - x - y)^2}$
$A = f_{11} = \dfrac{2}{x^3} + \dfrac{18}{(4 - x - y)^3}$
$B = f_{12} = \dfrac{18}{(4 - x - y)^3}$
$C = f_{22} = \dfrac{8}{y^3} + \dfrac{18}{(4 - x - y)^3}$.
For CP: $y^2 = 4x^2$ so that $y = \pm 2x$. If $y = 2x$, then $9x^2 = (4 - 3x)^2$, from which $x = 2/3$, $y = 4/3$. If $y = -2x$, then $9x^2 = (4+x)^2$, from which $x = -1$ or $x = 2$. The CPs are $(2/3, 4/3)$, $(-1, 2)$, and $(2, -4)$.

CP	A	B	C	$AC - B^2$	class
$(-1, 2)$	$-\frac{4}{3}$	$\frac{2}{3}$	$\frac{5}{3}$	$-\frac{8}{3}$	saddle
$(2, -4)$	$\frac{1}{3}$	$\frac{1}{12}$	$-\frac{1}{24}$	$-\frac{1}{48}$	saddle
$(\frac{2}{3}, \frac{4}{3})$	9	$\frac{9}{4}$	$\frac{45}{8}$	$\frac{729}{16}$	loc. min

4. $f(x, y) = x^2 y(2 - x - y) = 2x^2 y - x^3 y - x^2 y^2$
$f_1(x, y) = 4xy - 3x^2 y - 2xy^2 = xy(4 - 3x - 2y)$
$f_2(x, y) = 2x^2 - x^3 - 2x^2 y = x^2(2 - x - 2y)$
$A = f_{11} = 4y - 6xy - 2y^2$
$B = f_{12} = 4x - 3x^2 - 4xy$
$C = f_{22} = -2x^2$.
$(0, y)$ is a CP for any y. If $x \neq 0$ but $y = 0$, then $x = 2$ from the second equation. Thus $(2, 0)$ is a CP.

500

INSTRUCTOR'S SOLUTIONS MANUAL   REVIEW EXERCISES 13 (PAGE 816)

If neither x nor y is 0, then $x + 2y = 2$ and $3x + 2y = 4$, so that $x = 1$ and $y = 1/2$. The third CP is $(1, 1/2)$.

CP	A	B	C	$AC - B^2$	class
$(0, y)$	$4y - 2y^2$	0	0	0	?
$(2, 0)$	0	-4	-8	-16	saddle
$(1, \frac{1}{2})$	$-\frac{3}{2}$	-1	-2	2	loc. max

The second derivative test is unable to classify the line of critical points along the y-axis. However, direct inspection of $f(x, y)$ shows that these are local minima if $y(2 - y) > 0$ (that is, if $0 < y < 2$) and local maxima if $y(2 - y) < 0$ (that is, if $y < 0$ or $y > 2$). The points $(0, 0)$ and $(0, 2)$ are neither maxima nor minima, so they are saddle points.

5. $f(x, y, z) = g(s) = s + (1/s)$, where $s = x^2 + y^2 + z^2$. Since $g(s) \to \infty$ as $s \to \infty$ or $s \to 0+$, g must have a minimum value at a critical point in $(0, \infty)$. For CP: $0 = g'(s) = 1 - (1/s^2)$, that is, $s = 1$. $g(1) = 2$. The minimum value of f is 2, and is assumed at every point of the sphere $x^2 + y^2 + z^2 = 1$.

6. $x^2 + y^2 + z^2 - xy - xz - yz$
$= \frac{1}{2}\left[(x^2 - 2xy + y^2) + (x^2 - 2xz + z^2) + (y^2 - 2yz + z^2)\right]$
$= \frac{1}{2}\left[(x - y)^2 + (x - z)^2 + (y - z)^2\right] \geq 0$.
The minimum value, 0, is assumed at the origin and at all points of the line $x = y = z$.

7. $f(x, y) = xye^{-x^2-4y^2}$ satisfies $\lim_{x^2+y^2 \to \infty} f(x, y) = 0$. Since $f(1, 1) > 0$ and $f(-1, 1) < 0$, f must have maximum and minimum values and these must occur at critical points. For CP:

$0 = f_1 = e^{-x^2-4y^2}(y - 2x^2y) = e^{-x^2-4y^2}y(1 - 2x^2)$
$0 = f_2 = e^{-x^2-4y^2}(x - 8xy^2) = e^{-x^2-4y^2}x(1 - 8y^2)$.

The CPs are $(0, 0)$ (where $f = 0$), $\pm(\frac{1}{\sqrt{2}}, \frac{1}{2\sqrt{2}})$ (where $f = 1/4e$), and $\pm(\frac{1}{\sqrt{2}}, -\frac{1}{2\sqrt{2}})$ (where $f = -1/4e$). Thus f has maximum value $1/4e$ and minimum value $-1/4e$.

8. $f(x, y) = (4x^2 - y^2)e^{-x^2+y^2}$
$f_1(x, y) = e^{-x^2+y^2}2x(4 - 4x^2 + y^2)$
$f_2(x, y) = e^{-x^2+y^2}(-2y)(1 - 4x^2 + y^2)$.
f has CPs $(0, 0)$, $(\pm 1, 0)$. $f(0, 0) = 0$.
$f(\pm 1, 0) = 4/e$.

a) Since $f(0, y) = -y^2 e^{y^2} \to -\infty$ as $y \to \pm\infty$, and since $f(x, x) = 3x^2 e^0 = 3x^2 \to \infty$ as $x \to \pm\infty$, f does not have a minimum or a maximum value on the xy-plane.

b) On $y = 3x$, $f(x, 3x) = -5x^2 e^{8x^2} \to -\infty$ as $x \to \infty$. Thus f can have no minimum value on the wedge $0 \leq y \leq 3x$. However, as noted in (a), $f(x, x) \to \infty$ as $x \to \infty$. Since (x, x) is in the wedge for $x > 0$, f cannot have a maximum value on the wedge either.

9. Let the three pieces of wire have lengths x, y, and $L - x - y$ cm, respectively. The sum of areas of the squares is

$$S = \frac{1}{16}(x^2 + y^2 + (L - x - y)^2),$$

for which we must find extreme values over the triangle $x \geq 0$, $y \geq 0$, $x + y \leq L$. For critical points:

$$0 = \frac{\partial S}{\partial x} = \frac{1}{8}(x - (L - x - y))$$
$$0 = \frac{\partial S}{\partial y} = \frac{1}{8}(y - (L - x - y)),$$

from which we obtain $x = y = L/3$. This CP is inside the triangle, and $S = L^2/48$ at it.

On the boundary segment $x = 0$, we have

$$S = \frac{1}{16}(y^2 + (L - y)^2), \quad (0 \leq y \leq L).$$

At $y = 0$ or $y = L$, we have $S = L^2/16$. For critical points

$$0 = \frac{dS}{dy} = \frac{1}{8}(y - (L - y)),$$

so $y = L/2$ and $S = L^2/32$. By symmetry the extreme values of S on the other two boundary segments are the same.
Thus the minimum value of S is $L^2/48$, and corresponds to three equal squares. The maximum value of S is $L^2/16$, and corresponds to using the whole wire for one square.

10. Let the length, width, and height of the box be x, y, and z in, respectively. Then the girth is $g = 2x + 2y$. We require $g + z \leq 120$ in. The volume $V = xyz$ of the box will be maximized under the constraint $2x + 2y + z = 120$, so we look for CPs of

$$L = xyz + \lambda(2x + 2y + z - 120).$$

501

For CPs:

$$0 = \frac{\partial L}{\partial x} = yz + 2\lambda \quad \text{(A)}$$

$$0 = \frac{\partial L}{\partial y} = xz + 2\lambda \quad \text{(B)}$$

$$0 = \frac{\partial L}{\partial z} = xy + \lambda \quad \text{(C)}$$

$$0 = \frac{\partial L}{\partial \lambda} = 2x + 2y + z - 120. \quad \text{(D)}$$

Comparing (A), (B), and (C), we see that $x = y = z/2$. Then (D) implies that $3z = 120$, so $z = 40$ and $x = y = 20$ in. The largest box has volume

$$V = (20)(20)(40) = 16,000 \text{ in}^3,$$

or, about 9.26 cubic feet.

11. The ellipse $(x/a)^2 + (y/b)^2 = 1$ contains the rectangle $-1 \le x \le 1$, $-2 \le y \le 2$, if $(1/a^2) + (4/b^2) = 1$. The area of the ellipse is $A = \pi ab$. We minimize A by looking for critical points of

$$L = \pi ab + \lambda\left(\frac{1}{a^2} + \frac{4}{b^2} - 1\right).$$

For CPs:

$$0 = \frac{\partial L}{\partial a} = \pi b - \frac{2\lambda}{a^3} \quad \text{(A)}$$

$$0 = \frac{\partial L}{\partial b} = \pi a - \frac{8\lambda}{b^3} \quad \text{(B)}$$

$$0 = \frac{\partial L}{\partial \lambda} = \frac{1}{a^2} + \frac{4}{b^2} - 1. \quad \text{(C)}$$

Multiplying (A) by a and (B) by b, we obtain $2\lambda/a^2 = 8\lambda/b^2$, so that either $\lambda = 0$ or $b = 2a$. Now $\lambda = 0$ implies $b = 0$, which is inconsistent with (C). If $b = 2a$, then (C) implies that $2/a^2 = 1$, so $a = \sqrt{2}$. The smallest area of the ellipse is $V = 4\pi$ cubic units.

12. The ellipsoid $(x/a)^2 + (y/b)^2 + (z/c)^2 = 1$ contains the rectangle $-1 \le x \le 1$, $-2 \le y \le 2$, $-3 \le z \le 3$, provided $(1/a^2) + (4/b^2) + (9/c^2) = 1$. The volume of the ellipsoid is $V = 4\pi abc/3$. We minimize V by looking for critical points of

$$L = \frac{4\pi}{3}abc + \lambda\left(\frac{1}{a^2} + \frac{4}{b^2} + \frac{9}{c^2} - 1\right).$$

For CPs:

$$0 = \frac{\partial L}{\partial a} = \frac{4\pi}{3}bc - \frac{2\lambda}{a^3} \quad \text{(A)}$$

$$0 = \frac{\partial L}{\partial b} = \frac{4\pi}{3}ac - \frac{8\lambda}{b^3} \quad \text{(B)}$$

$$0 = \frac{\partial L}{\partial c} = \frac{4\pi}{3}ab - \frac{18\lambda}{c^3} \quad \text{(C)}$$

$$0 = \frac{\partial L}{\partial \lambda} = \frac{1}{a^2} + \frac{4}{b^2} + \frac{9}{c^2} - 1. \quad \text{(D)}$$

Multiplying (A) by a, (B) by b, and (C) by c, we obtain $2\lambda/a^2 = 8\lambda/b^2 = 18\lambda/c^2$, so that either $\lambda = 0$ or $b = 2a$, $c = 3a$. Now $\lambda = 0$ implies $bc = 0$, which is inconsistent with (D). If $b = 2a$ and $c = 3a$, then (D) implies that $3/a^2 = 1$, so $a = \sqrt{3}$. The smallest volume of the ellipsoid is

$$V = \frac{4\pi}{3}(\sqrt{3})(2\sqrt{3})(3\sqrt{3}) = 24\sqrt{3}\pi \text{ cubic units.}$$

13. The box $-1 \le x \le 1$, $-2 \le y \le 2$, $0 \le z \le 2$ is contained in the region

$$0 \le z \le a\left(1 - \frac{x^2}{b^2} - \frac{y^2}{c^2}\right)$$

provided that $(2/a) + (1/b^2) + (4/c^2) = 1$. The volume of the region would normally be calculated via a "double integral" which we have not yet encountered. (See Chapter 14.) It can also be done directly by slicing. A horizontal plane at height z (where $0 \le z \le a$) intersects the region in an elliptic disk bounded by the ellipse

$$\frac{x^2}{b^2} + \frac{y^2}{c^2} = 1 - \frac{z}{a}.$$

The area of this disk is

$$A(z) = \pi\left(b\sqrt{1 - \frac{z}{a}}\right)\left(c\sqrt{1 - \frac{z}{a}}\right) = \pi bc\left(1 - \frac{z}{a}\right).$$

Thus the region has volume

$$V = \pi bc \int_0^a \left(1 - \frac{z}{a}\right) dz = \frac{\pi abc}{2}.$$

Thus we look for critical points of

$$L = \frac{\pi abc}{2} + \lambda\left(\frac{2}{a} + \frac{1}{b^2} + \frac{4}{c^2} - 1\right).$$

INSTRUCTOR'S SOLUTIONS MANUAL   REVIEW EXERCISES 13 (PAGE 816)

For critical points:

$$0 = \frac{\partial L}{\partial a} = \frac{\pi}{2}bc - \frac{2\lambda}{a^2} \qquad \text{(A)}$$

$$0 = \frac{\partial L}{\partial b} = \frac{\pi}{2}ac - \frac{2\lambda}{b^3} \qquad \text{(B)}$$

$$0 = \frac{\partial L}{\partial c} = \frac{\pi}{2}ab - \frac{8\lambda}{c^3} \qquad \text{(C)}$$

$$0 = \frac{\partial L}{\partial \lambda} = \frac{2}{a} + \frac{1}{b^2} + \frac{4}{c^2} - 1. \qquad \text{(D)}$$

Multiplying (A) by a, (B) by b, and (C) by c, we obtain $2\lambda/a = 2\lambda/b^2 = 8\lambda/c^2$, so that either $\lambda = 0$ or $b^2 = a$, $c^2 = 4a$. Now $\lambda = 0$ implies $bc = 0$, which is inconsistent with (D). If $b^2 = a$ and $c^2 = 4a$, then (D) implies that $4/a = 1$, so $a = 4$. The smallest volume of the region is $V = \pi(4)(2)(4)/2 = 16\pi$ cubic units.

14. The area of the window is

$$A = xy + \frac{x}{2}\sqrt{z^2 - \frac{x^2}{4}},$$

or, since $x + 2y + 2z = L$,

$$A = \frac{x}{2}\left(L - x - 2z + \sqrt{z^2 - \frac{x^2}{4}}\right).$$

For maximum A, we look for critical points:

$$0 = \frac{\partial A}{\partial x} = \frac{1}{2}\left(L - x - 2z + \sqrt{z^2 - \frac{x^2}{4}}\right)$$

$$+ \frac{x}{2}\left(-1 - \frac{x}{4\sqrt{z^2 - \frac{x^2}{4}}}\right)$$

$$= \frac{L}{2} - x - z + \frac{2z^2 - x^2}{4\sqrt{z^2 - \frac{x^2}{4}}} \qquad \text{(A)}$$

$$0 = \frac{\partial A}{\partial z} = -x + \frac{xz}{2\sqrt{z^2 - \frac{x^2}{4}}}. \qquad \text{(B)}$$

Now (B) implies that either $x = 0$ or $z = 2\sqrt{z^2 - (x^2/4)}$. But $x = 0$ gives zero area rather than maximum area, so the second alternative must hold, and it implies that $z = x/\sqrt{3}$. Then (A) gives

$$\frac{L}{2} = \left(1 + \frac{1}{\sqrt{3}}\right)x + \frac{x}{2\sqrt{3}},$$

from which we obtain $x = L/(2 + \sqrt{3})$. The maximum area of the window is, therefore,

$$A\bigg|_{x=\frac{L}{2+\sqrt{3}}, \ z=\frac{L/\sqrt{3}}{2+\sqrt{3}}} = \frac{1}{4}\frac{L^2}{2 + \sqrt{3}}$$

$$\approx 0.0670L^2 \text{ sq. units.}$$

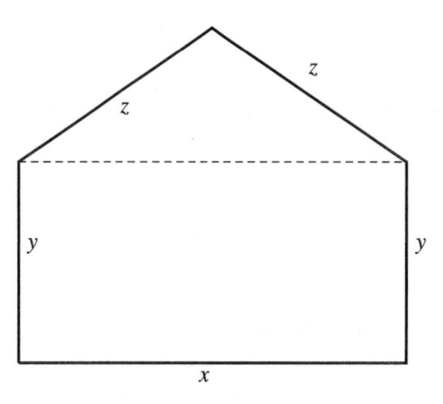

Fig. R-13.14

15. If $1,000x$ widgets per month are manufactured and sold for y per widget, then the monthly profit is $1,000P$, where

$$P = xy - \frac{x^2y^3}{27} - x.$$

We are required to maximize P over the rectangular region R satisfying $0 \le x \le 3$ and $0 \le y \le 2$. First look for critical points:

$$0 = \frac{\partial P}{\partial x} = y - \frac{2xy^3}{27} - 1 \qquad \text{(A)}$$

$$0 = \frac{\partial P}{\partial y} = x - \frac{x^2y^2}{9}. \qquad \text{(B)}$$

(B) implies that $x = 0$, which yields zero profit, or $xy^2 = 9$, which, when substituted into (A), gives $y = 3$ and $x = 1$. Unfortunately, the critical point $(1, 3)$ lies outside of R. Therefore the maximum P must occur on the boundary of R.
We consider all four boundary segments of R.
On segment $x = 0$, we have $P = 0$.
On segment $y = 0$, we have $P = -x \le 0$.
On segment $x = 3$, $0 \le y \le 2$, we have
$P = 3y - (y^3/3) - 3$, which has values $P = -3$ at $y = 0$ and $P = 1/3$ at $y = 2$. It also has a critical point given by

$$0 = \frac{dP}{dy} = 3 - y^2,$$

503

so $y = \sqrt{3}$ and $P = 2\sqrt{3} - 3 \approx 0.4641$.
On segment $y = 2$, $0 \leq x \leq 3$, we have
$P = x - (8x^2/27)$, which has values $P = 0$ at $x = 0$
and $P = 1/3$ at $x = 3$. It also has a critical point
given by
$$0 = \frac{dP}{dx} = 1 - \frac{16x}{27},$$
so $x = 27/16$ and $P = 27/32 \approx 0.84375$.
It appears that the greatest monthly profit corresponds
to manufacturing $27,000/16 \approx 1,688$ widgets/month
and selling them for $2 each.

16. The envelope of $y = (x - c)^3 + 3c$ is found by
eliminating c from that equation and
$$0 = \frac{\partial}{\partial c}[(x - c)^3 + 3c] = -3(x - c)^2 + 3.$$
This later equation implies that $(x - c)^2 = 1$, so
$x - c = \pm 1$.
The envelope is $y = (\pm 1)^3 + 3(x \mp 1)$, or $y = 3x \pm 2$.

17. Look for a solution of $y + \epsilon x e^y = -2x$ in the form
of a Maclaurin series
$$y = y(x, \epsilon) = y(x, 0) + \epsilon y_\epsilon(x, 0) + \frac{\epsilon^2}{2!} y_{\epsilon\epsilon}(x, 0) + \cdots.$$
Putting $\epsilon = 0$ in the given equation, we get
$y(x, 0) = -2x$. Now differentiate the given equation
with respect to ϵ twice:
$$y_\epsilon + xe^y + \epsilon x e^y y_\epsilon = 0$$
$$y_{\epsilon\epsilon} + 2xe^y y_\epsilon + \epsilon x e^y y_\epsilon^2 + \epsilon x e^y y_{\epsilon\epsilon} = 0.$$
The first of these equations gives
$$y_\epsilon(x, 0) = -xe^{y(x,0)} = -xe^{-2x}.$$
The second gives
$$y_{\epsilon\epsilon}(x, 0) = -2xe^{y(x,0)} y_\epsilon(x, 0) = 2x^2 e^{-4x}.$$
Thus $y = -2x - 2\epsilon x e^{-2x} + \epsilon^2 x^2 e^{-4x} + \cdots$.

Challenging Problems 13   (page 817)

1. To minimize
$$I_n = \int_{-\pi}^{\pi} \left[f(x) - \frac{a_0}{2} - \sum_{k=1}^{n}(a_k \cos kx + b_k \sin kx) \right]^2 dx$$

we choose a_k and b_k to satisfy
$$0 = \frac{\partial I_n}{\partial a_0}$$
$$= -\int_{-\pi}^{\pi} \left[f(x) - \frac{a_0}{2} - \sum_{k=1}^{n}(a_k \cos kx + b_k \sin kx) \right] dx$$
$$= \left[\pi a_0 - \int_{-\pi}^{\pi} f(x)\,dx \right]$$
$$0 = \frac{\partial I_n}{\partial a_m}$$
$$= -2 \int_{-\pi}^{\pi} \left[f(x) - \frac{a_0}{2} - \sum_{k=1}^{n}(a_k \cos kx + b_k \sin kx) \right] \cos mx\,dx$$
$$= 2a_m \int_{-\pi}^{\pi} \cos^2 mx\,dx - \int_{-\pi}^{\pi} f(x) \cos mx\,dx$$
$$0 = \frac{\partial I_n}{\partial b_m}$$
$$= -2 \int_{-\pi}^{\pi} \left[f(x) - \frac{a_0}{2} - \sum_{k=1}^{n}(a_k \cos kx + b_k \sin kx) \right] \sin mx\,dx$$
$$= 2b_m \int_{-\pi}^{\pi} \sin^2 mx\,dx - \int_{-\pi}^{\pi} f(x) \sin mx\,dx.$$

The simplifications in the integrals above resulted
from the facts that for any integers k and m,
$$\int_{-\pi}^{\pi} \cos kx \cos mx\,dx = 0 \text{ unless } k = m$$
$$\int_{-\pi}^{\pi} \sin kx \sin mx\,dx = 0 \text{ unless } k = m, \text{ and}$$
$$\int_{-\pi}^{\pi} \cos kx \sin mx\,dx = 0.$$

(See Exercise 59 in Section 5.6.) Since
$$\int_{-\pi}^{\pi} \cos^2 mx\,dx = \int_{-\pi}^{\pi} \sin^2 mx\,dx = \pi,$$
I_n is minimized when
$$a_m = \frac{1}{\pi} \int_{-\pi}^{\pi} f(x) \cos mx\,dx \text{ for } 0 \leq m \leq n, \text{ and}$$
$$b_m = \frac{1}{\pi} \int_{-\pi}^{\pi} f(x) \sin mx\,dx \text{ for } 1 \leq m \leq n.$$

CHALLENGING PROBLEMS 13 (PAGE 817)

2. If $f(x) = \begin{cases} 0 & \text{for } -\pi \leq x < 0 \\ x & \text{for } 0 \leq x \leq \pi \end{cases}$, then

$$a_0 = \frac{1}{\pi}\int_0^\pi x\,dx = \frac{\pi}{2}$$

$$a_k = \frac{1}{\pi}\int_0^\pi x\cos kx\,dx$$

$$\begin{aligned} U &= x & dV &= \cos kx\,dx \\ dU &= dx & V &= \frac{1}{k}\sin kx \end{aligned}$$

$$= \frac{1}{\pi k}\left(x\sin kx\Big|_0^\pi - \int_0^\pi \sin kx\,dx\right)$$

$$= \frac{\cos k\pi - 1}{\pi k^2} = \begin{cases} 0 & \text{if } k \text{ is even} \\ -\dfrac{2}{\pi k^2} & \text{if } k \text{ is odd} \end{cases}$$

$$b_k = \frac{1}{\pi}\int_0^\pi x\sin kx\,dx$$

$$\begin{aligned} U &= x & dV &= \sin kx\,dx \\ dU &= dx & V &= -\frac{1}{k}\cos kx \end{aligned}$$

$$= -\frac{1}{\pi k}\left(x\cos kx\Big|_0^\pi - \int_0^\pi \cos kx\,dx\right)$$

$$= \frac{(-1)^{k+1}}{k}.$$

Because of the properties of trigonometric integrals listed in the solution to Problem 1,

$$\int_{-\pi}^\pi \left(\frac{a_0}{2} + \sum_{k=1}^n (a_k\cos kx + b_k\sin kx)\right)^2 dx$$

$$= \frac{\pi a_0^2}{2} + \pi\sum_{k=0}^n (a_k^2 + b_k^2)$$

$$\int_{-\pi}^\pi f(x)\left(\frac{a_0}{2} + \sum_{k=1}^n (a_k\cos kx + b_k\sin kx)\right) dx$$

$$= \frac{\pi a_0^2}{2} + \pi\sum_{k=0}^n (a_k^2 + b_k^2).$$

Therefore

$$I_n = \int_{-\pi}^\pi \left[f(x) - \left(\frac{a_0}{2} + \sum_{k=1}^n (a_k\cos kx + b_k\sin kx)\right)\right]^2 dx$$

$$= \int_{-\pi}^\pi (f(x))^2\,dx - 2\left(\frac{\pi a_0^2}{2} + \pi\sum_{k=0}^n (a_k^2 + b_k^2)\right)$$

$$+ \frac{\pi a_0^2}{2} + \pi\sum_{k=0}^n (a_k^2 + b_k^2)$$

$$= \int_{-\pi}^\pi (f(x))^2\,dx - \left(\frac{\pi a_0^2}{2} + \pi\sum_{k=0}^n (a_k^2 + b_k^2)\right).$$

In fact, it can be shown that $I_n \to 0$ as $n \to \infty$.

3. Let $I(x) = \displaystyle\int_0^x \frac{\ln(1+tx)}{1+t^2}\,dt$. Then

$$I'(x) = \frac{\ln(1+x^2)}{1+x^2} + \int_0^x \frac{t}{(1+t^2)(1+tx)}\,dt.$$

If we expand the latter integrand in partial fractions with respect to t, we obtain

$$\frac{t}{(1+t^2)(1+tx)} = \frac{x+t}{(1+x^2)(1+t^2)} - \frac{x}{(1+x^2)(1+tx)}.$$

Now we have

$$\int_0^x \frac{(x+t)\,dt}{(1+x^2)(1+t^2)} = \frac{2x\tan^{-1}t + \ln(1+t^2)}{2(1+x^2)}\bigg|_0^x$$

$$= \frac{2x\tan^{-1}x + \ln(1+x^2)}{2(1+x^2)}$$

$$= \frac{1}{2}\frac{d}{dx}\tan^{-1}x\,\ln(1+x^2)$$

$$\int_0^x \frac{x\,dt}{(1+x^2)(1+tx)} = \frac{x}{1+x^2}\int_0^x \frac{dt}{1+tx}$$

Let $u = 1 + tx$
$du = x\,dt$

$$= \frac{1}{1+x^2}\int_1^{1+x^2} \frac{du}{u} = \frac{\ln(1+x^2)}{1+x^2}.$$

Thus

$$I'(x) = \frac{\ln(1+x^2)}{1+x^2} + \frac{1}{2}\frac{d}{dx}\tan^{-1}x\,\ln(1+x^2) - \frac{\ln(1+x)}{1+x^2}$$

$$= \frac{1}{2}\frac{d}{dx}\tan^{-1}x\,\ln(1+x^2).$$

Therefore, $I(x) = \dfrac{1}{2}\tan^{-1}x\,\ln(1+x^2) + C$. Since $I(0) = 0$, we have $C = 0$, and

$$\int_0^x \frac{\ln(1+tx)}{1+t^2}\,dx = \frac{1}{2}\tan^{-1}x\,\ln(1+x^2).$$

4.

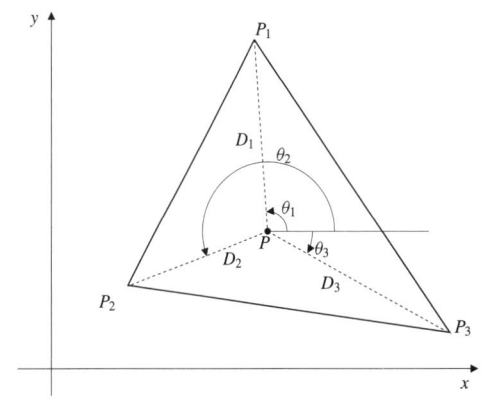

Fig. C-13.4

CHALLENGING PROBLEMS 13 (PAGE 817)

If $D_i = |PP_i|$ for $i = 1, 2, 3$, then

$$D_i^2 = (x - x_i)^2 + (y - y_i)^2$$
$$2D_i \frac{\partial D_i}{\partial x} = 2(x - x_i)$$
$$\frac{\partial D_i}{\partial x} = \frac{x - x_i}{D_i} = \cos \theta_i$$

where θ_i is the angle between $\overrightarrow{PP_i}$ and $\mathbf{i}$. Similarly $\partial D_i/\partial y = \sin \theta_i$. To minimize $S = D_1 + D_2 + D_3$ we look for critical points:

$$0 = \frac{\partial S}{\partial x} = \cos \theta_1 + \cos \theta_2 + \cos \theta_3$$
$$0 = \frac{\partial S}{\partial y} = \sin \theta_1 + \sin \theta_2 + \sin \theta_3.$$

Thus $\cos \theta_1 + \cos \theta_2 = -\cos \theta_3$ and $\sin \theta_1 + \sin \theta_2 = -\sin \theta_3$. Squaring and adding these two equations we get

$$2 + 2(\cos \theta_1 \cos \theta_2 + \sin \theta_1 \sin \theta_2) = 1,$$

or $\cos(\theta_1 - \theta_2) = -1/2$. Thus $\theta_1 - \theta_2 = \pm 2\pi/3$. Similarly $\theta_1 - \theta_3 = \theta_2 - \theta_3 = \pm 2\pi/3$. Thus P should be chosen so that $\overrightarrow{PP_1}$, $\overrightarrow{PP_2}$, and $\overrightarrow{PP_3}$ make $120°$ angles with each other. This is possible only if all three angles of the triangle are less than $120°$. If the triangle has an angle of $120°$ or more (say at P_1), then P should be that point on the side P_2P_3 such that $PP_1 \perp P_2P_3$.

INSTRUCTOR'S SOLUTIONS MANUAL SECTION 14.1 (PAGE 823)

CHAPTER 14. MULTIPLE INTEGRATION

Section 14.1 Double Integrals (page 823)

1. $f(x, y) = 5 - x - y$
   $R = 1 \times \big[f(0, 1) + f(0, 2) + f(1, 1) + f(1, 2)$
   $\quad + f(2, 1) + f(2, 2)\big]$
   $= 4 + 3 + 3 + 2 + 2 + 1 = 15$

2. $R = 1 \times \big[f(1, 1) + f(1, 2) + f(2, 1) + f(2, 2)$
   $\quad + f(3, 1) + f(3, 2)\big]$
   $= 3 + 2 + 2 + 1 + 1 + 0 = 9$

3. $R = 1 \times \big[f(0, 0) + f(0, 1) + f(1, 0) + f(1, 1)$
   $\quad + f(2, 0) + f(2, 1)\big]$
   $= 5 + 4 + 4 + 3 + 3 + 2 = 21$

4. $R = 1 \times \big[f(1, 0) + f(1, 1) + f(2, 0) + f(2, 1)$
   $\quad + f(3, 0) + f(3, 1)\big]$
   $= 4 + 3 + 3 + 2 + 2 + 1 = 15$

5. $R = 1 \times \big[f(\tfrac{1}{2}, \tfrac{1}{2}) + f(\tfrac{1}{2}, \tfrac{3}{2}) + f(\tfrac{3}{2}, \tfrac{1}{2}) + f(\tfrac{3}{2}, \tfrac{3}{2})$
   $\quad + f(\tfrac{5}{2}, \tfrac{1}{2}) + f(\tfrac{5}{2}, \tfrac{3}{2})\big]$
   $= 4 + 3 + 3 + 2 + 2 + 1 = 15$

6. $I = \iint_D (5 - x - y)\, dA$ is the volume of the solid in the figure.

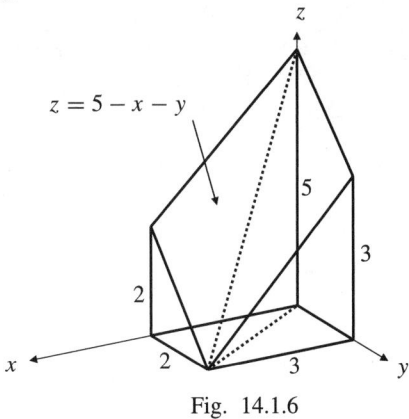

   $z = 5 - x - y$

   Fig. 14.1.6

   The solid is split by the vertical plane through the z-axis and the point $(3, 2, 0)$ into two pyramids, each with a trapezoidal base; one pyramid's base is in the plane $y = 0$ and the other's is in the plane $z = 0$. I is the sum of the volumes of these pyramids:

   $I = \dfrac{1}{3}\left(\dfrac{5+2}{2}(3)(2)\right) + \dfrac{1}{3}\left(\dfrac{5+3}{2}(2)(3)\right) = 15.$

7. $J = \iint_D 1\, dA$
   $R = 4 \times 1 \times [5 + 5 + 5 + 5 + 4] = 96$

8. $R = 4 \times 1 \times [4 + 4 + 4 + 3 + 0] = 60$

9. $R = 4 \times 1 \times [5 + 5 + 4 + 4 + 2] = 80$

10. $J = $ area of disk $= \pi(5^2) \approx 78.54$

11. $R = 1 \times (e^{1/2} + e^{1/2} + e^{3/2} + e^{3/2} + e^{5/2} + e^{5/2})$
    ≈ 32.63

12. $f(x, y) = x^2 + y^2$
    $R = 4 \times 1 \times \big[f(\tfrac{1}{2}, \tfrac{1}{2}) + f(\tfrac{3}{2}, \tfrac{1}{2}) + f(\tfrac{5}{2}, \tfrac{1}{2}) + f(\tfrac{7}{2}, \tfrac{1}{2})$
    $\quad + f(\tfrac{9}{2}, \tfrac{1}{2}) + f(\tfrac{1}{2}, \tfrac{3}{2}) + f(\tfrac{3}{2}, \tfrac{3}{2}) + f(\tfrac{5}{2}, \tfrac{3}{2})$
    $\quad + f(\tfrac{7}{2}, \tfrac{3}{2}) + f(\tfrac{9}{2}, \tfrac{3}{2})$
    $\quad + f(\tfrac{1}{2}, \tfrac{5}{2}) + f(\tfrac{3}{2}, \tfrac{5}{2}) + f(\tfrac{5}{2}, \tfrac{5}{2}) + f(\tfrac{7}{2}, \tfrac{5}{2})$
    $\quad + f(\tfrac{1}{2}, \tfrac{7}{2}) + f(\tfrac{3}{2}, \tfrac{7}{2}) + f(\tfrac{5}{2}, \tfrac{7}{2}) + f(\tfrac{1}{2}, \tfrac{9}{2}) + f(\tfrac{3}{2}, \tfrac{9}{2})\big]$
    $= 918$

13. $\iint_R dA = $ area of $R = 4 \times 5 = 20.$

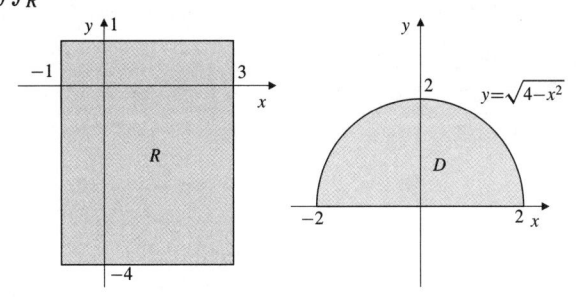

    Fig. 14.1.13   Fig. 14.1.14

14. $\iint_D (x + 3)\, dA = \iint_D x\, dA + 3 \iint_D dA$
    $= 0 + 3(\text{area of } D)$
    $= 3 \times \dfrac{\pi 2^2}{2} = 6\pi.$

    The integral of x over D is zero because D is symmetrical about $x = 0$.

15. T is symmetric about the line $x + y = 0$. Therefore,
    $\iint_T (x + y)\, dA = 0.$

SECTION 14.1 (PAGE 823)

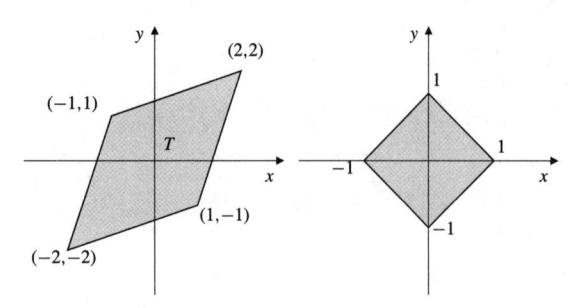

Fig. 14.1.15   Fig. 14.1.16

16. $\iint_{|x|+|y|\le 1}\left(x^3\cos(y^2)+3\sin y-\pi\right)dA$
$= 0+0-\pi\left(\text{area bounded by }|x|+|y|=1\right)$
$= -\pi\times 4\times\dfrac{1}{2}(1)(1) = -2\pi.$
(Each of the first two terms in the integrand is an odd function of one of the variables, and the square is symmetrical about each coordinate axis.)

17. $\iint_{x^2+y^2\le 1}(4x^2y^3-x+5)\,dA$
$= 0-0+5(\text{area of disk})\quad\text{(by symmetry)}$
$= 5\pi.$

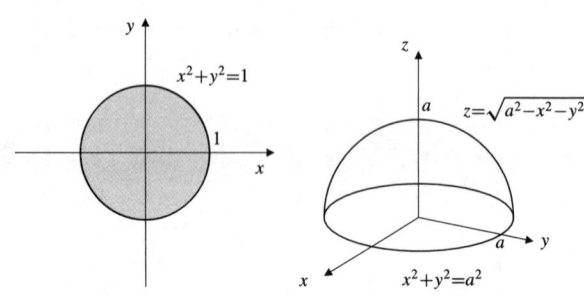

Fig. 14.1.17   Fig. 14.1.18

18. $\iint_{x^2+y^2\le a^2}\sqrt{a^2-x^2-y^2}\,dA$
$= \text{volume of hemisphere shown in the figure}$
$= \dfrac{1}{2}\left(\dfrac{4}{3}\pi a^3\right) = \dfrac{2}{3}\pi a^3.$

19. $\iint_{x^2+y^2\le a^2}\left(a-\sqrt{x^2+y^2}\right)dA$
$= \text{volume of cone shown in the figure}$
$= \dfrac{1}{3}\pi a^3.$

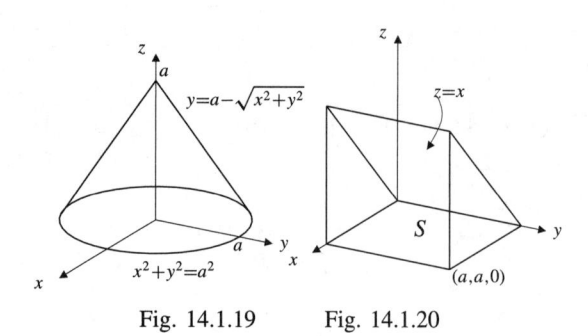

Fig. 14.1.19   Fig. 14.1.20

20. By the symmetry of S with respect to x and y we have
$\iint_S (x+y)\,dA = 2\iint_S x\,dA$
$= 2\times(\text{volume of wedge shown in the figure})$
$= 2\times\dfrac{1}{2}(a^2)a = a^3.$

21. $\iint_T (1-x-y)\,dA$
$= \text{volume of the tetrahedron shown in the figure}$
$= \dfrac{1}{3}\left(\dfrac{1}{2}(1)(1)\right)(1) = \dfrac{1}{6}.$

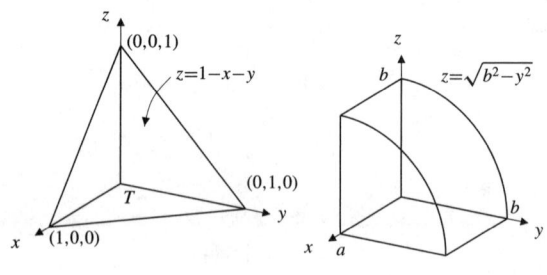

Fig. 14.1.21   Fig. 14.1.22

22. $\iint_R \sqrt{b^2-y^2}\,dA$
$= \text{volume of the quarter cylinder shown in the figure}$
$= \dfrac{1}{4}(\pi b^2)a = \dfrac{1}{4}\pi ab^2.$

Section 14.2 Iteration of Double Integrals in Cartesian Coordinates (page 830)

1. $\displaystyle\int_0^1 dx\int_0^x (xy+y^2)\,dy$
$= \displaystyle\int_0^1 dx\left(\dfrac{xy^2}{2}+\dfrac{y^3}{3}\right)\bigg|_{y=0}^{y=x}$
$= \dfrac{5}{6}\int_0^1 x^3\,dx = \dfrac{5}{24}.$

2. $\displaystyle\int_0^1\int_0^y (xy+y^2)\,dx\,dy$

$\displaystyle = \int_0^1 \left(\frac{x^2 y}{2} + xy^2\right)\bigg|_{x=0}^{x=y} dy$

$\displaystyle = \frac{3}{2}\int_0^1 y^3\,dy = \frac{3}{8}.$

3. $\displaystyle\int_0^\pi \int_{-x}^x \cos y\,dy\,dx$

$\displaystyle = \int_0^\pi \sin y\bigg|_{y=-x}^{y=x} dx$

$\displaystyle = 2\int_0^\pi \sin x\,dx = -2\cos x\bigg|_0^\pi = 4.$

4. $\displaystyle\int_0^2 dy \int_0^y y^2 e^{xy}\,dx$

$\displaystyle = \int_0^2 y^2\,dy \left(\frac{1}{y}e^{xy}\bigg|_{x=0}^{x=y}\right)$

$\displaystyle = \int_0^2 y(e^{y^2} - 1)\,dy = \frac{e^{y^2}-y^2}{2}\bigg|_0^2 = \frac{e^4 - 5}{2}.$

5. $\displaystyle\iint_R (x^2+y^2)\,dA = \int_0^a dx \int_0^b (x^2+y^2)\,dy$

$\displaystyle = \int_0^a dx \left(x^2 y + \frac{y^3}{3}\right)\bigg|_{y=0}^{y=b}$

$\displaystyle = \int_0^a \left(bx^2 + \frac{1}{3}b^3\right) dx$

$\displaystyle = \frac{1}{3}(bx^3 + b^3 x)\bigg|_0^a = \frac{1}{3}(a^3 b + ab^3).$

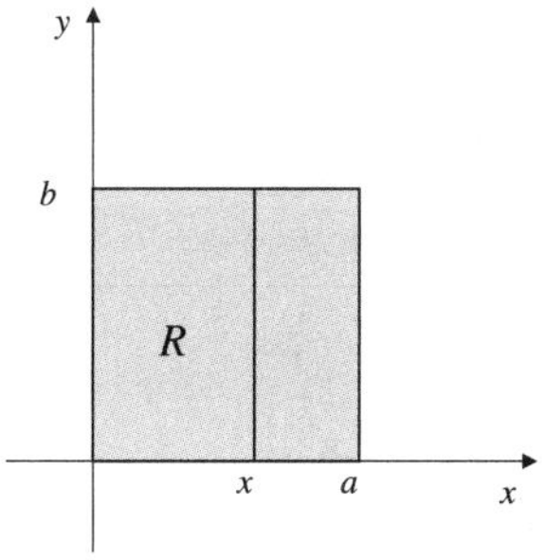

Fig. 14.2.5

6. $\displaystyle\iint_R x^2 y^2\,dA = \int_0^a x^2\,dx \int_0^b y^2\,dy$

$\displaystyle = \frac{a^3}{3}\cdot\frac{b^3}{3} = \frac{a^3 b^3}{9}.$

7. $\displaystyle\iint_S (\sin x + \cos y)\,dA$

$\displaystyle = \int_0^{\pi/2} dx \int_0^{\pi/2} (\sin x + \cos y)\,dy$

$\displaystyle = \int_0^{\pi/2} dx\,(y\sin x + \sin y)\bigg|_{y=0}^{y=\pi/2}$

$\displaystyle = \int_0^{\pi/2}\left(\frac{\pi}{2}\sin x + 1\right) dx$

$\displaystyle = \left(-\frac{\pi}{2}\cos x + x\right)\bigg|_0^{\pi/2} = \frac{\pi}{2} + \frac{\pi}{2} = \pi.$

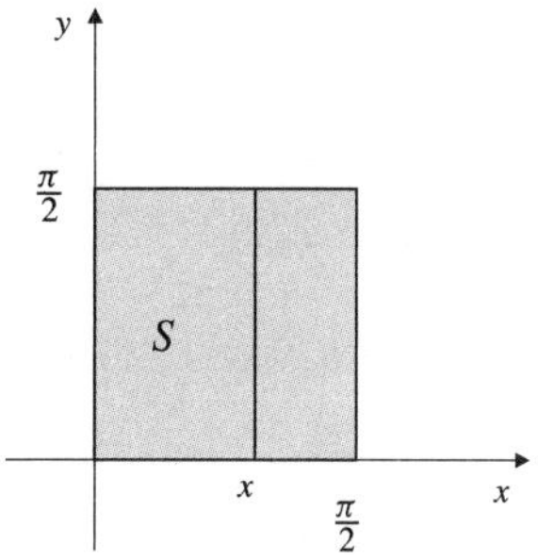

Fig. 14.2.7 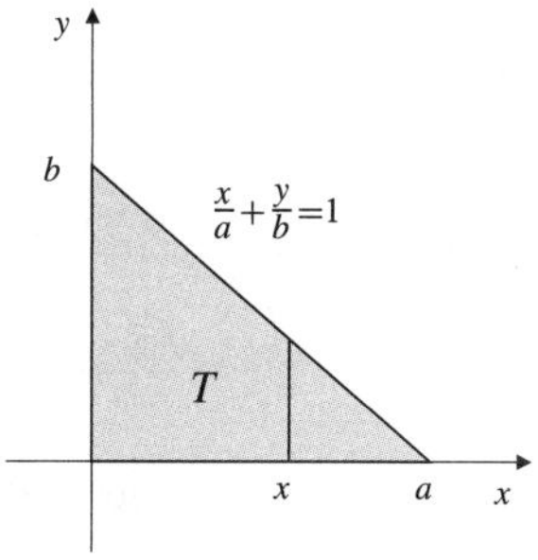   Fig. 14.2.8

8. $\displaystyle\iint_T (x - 3y)\,dA = \int_0^a dx \int_0^{b(1-(x/a))} (x-3y)\,dy$

$\displaystyle = \int_0^a dx\left(xy - \frac{3}{2}y^2\right)\bigg|_{y=0}^{y=b(1-(x/a))}$

$\displaystyle = \int_0^a \left[b\left(x - \frac{x^2}{a}\right) - \frac{3}{2}b^2\left(1 - \frac{2x}{a} + \frac{x^2}{a^2}\right)\right] dx$

$\displaystyle = \left(b\frac{x^2}{2} - \frac{b}{a}\frac{x^3}{3} - \frac{3}{2}b^2 x + \frac{3}{2}\frac{b^2 x^2}{a} - \frac{1}{2}\frac{b^2 x^3}{a^2}\right)\bigg|_0^a$

$\displaystyle = \frac{a^2 b}{6} - \frac{ab^2}{2}.$

9. $\displaystyle\iint_R xy^2\,dA = \int_0^1 x\,dx \int_{x^2}^{\sqrt{x}} y^2\,dy$

$\displaystyle = \int_0^1 x\,dx \left(\frac{1}{3}y^3\right)\bigg|_{y=x^2}^{y=\sqrt{x}}$

$\displaystyle = \frac{1}{3}\int_0^1 (x^{5/2} - x^7)\,dx$

$\displaystyle = \frac{1}{3}\left(\frac{2}{7}x^{7/2} - \frac{x^8}{8}\right)\bigg|_0^1$

$\displaystyle = \frac{1}{3}\left(\frac{2}{7} - \frac{1}{8}\right) = \frac{3}{56}.$

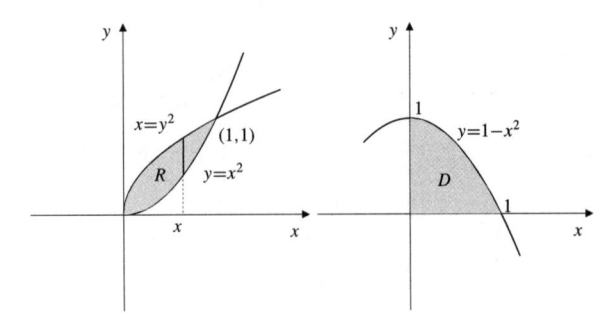

SECTION 14.2 (PAGE 830)

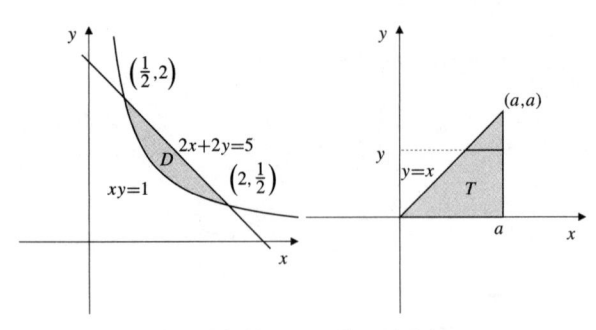

Fig. 14.2.9   Fig. 14.2.10   Fig. 14.2.11   Fig. 14.2.12

10. $\iint_D x \cos y \, dA$

$= \int_0^1 x \, dx \int_0^{1-x^2} \cos y \, dy$

$= \int_0^1 x \, dx (\sin y)\Big|_{y=0}^{y=1-x^2}$

$= \int_0^1 x \sin(1-x^2) \, dx \quad$ Let $u = 1 - x^2$

$\qquad\qquad\qquad\qquad du = -2x \, dx$

$= -\frac{1}{2}\int_1^0 \sin u \, du = \frac{1}{2}\cos u\Big|_1^0 = \frac{1-\cos(1)}{2}.$

11. For intersection: $xy = 1$, $2x + 2y = 5$.
Thus $2x^2 - 5x + 2 = 0$, or $(2x-1)(x-2) = 0$. The intersections are at $x = 1/2$ and $x = 2$. We have

$\iint_D \ln x \, dA = \int_{1/2}^2 \ln x \, dx \int_{1/x}^{(5/2)-x} dy$

$= \int_{1/2}^2 \ln x \left(\frac{5}{2} - x - \frac{1}{x}\right) dx$

$= \int_{1/2}^2 \ln x \left(\frac{5}{2} - x\right) dx - \frac{1}{2}(\ln x)^2\Big|_{1/2}^2$

$\qquad U = \ln x \quad dV = \left(\frac{5}{2} - x\right) dx$

$\qquad dU = \frac{dx}{x} \quad V = \frac{5}{2}x - \frac{x^2}{2}$

$= -\frac{1}{2}\left((\ln 2)^2 - (\ln \frac{1}{2})^2\right) + \left(\frac{5}{2}x - \frac{x^2}{2}\right)\ln x\Big|_{1/2}^2$

$\quad - \int_{1/2}^2 \left(\frac{5}{2} - \frac{x}{2}\right) dx$

$= (5-2)\ln 2 - \left(\frac{5}{4} - \frac{1}{8}\right)\ln \frac{1}{2} - \frac{15}{4} + \frac{15}{16}$

$= \frac{33}{8} \ln 2 - \frac{45}{16}.$

12. $\iint_T \sqrt{a^2 - y^2} \, dA = \int_0^a \sqrt{a^2 - y^2} \, dy \int_y^a dx$

$= \int_0^a (a-y)\sqrt{a^2 - y^2} \, dy$

$= a\int_0^a \sqrt{a^2 - y^2} \, dy - \int_0^a y\sqrt{a^2 - y^2} \, dy$

$\qquad\qquad\qquad$ Let $u = a^2 - y^2$

$\qquad\qquad\qquad du = -2y \, dy$

$= a\frac{\pi a^2}{4} + \frac{1}{2}\int_{a^2}^0 u^{1/2} \, du$

$= \frac{\pi a^3}{4} - \frac{1}{3}u^{3/2}\Big|_0^{a^2} = \left(\frac{\pi}{4} - \frac{1}{3}\right)a^3.$

13. $\iint_R \frac{x}{y}e^y \, dA = \int_0^1 \frac{e^y}{y} \, dy \int_y^{\sqrt{y}} x \, dx$

$= \frac{1}{2}\int_0^1 (1-y)e^y \, dy$

$\qquad U = 1-y \quad dV = e^y \, dy$

$\qquad dU = -dy \quad V = e^y$

$= \frac{1}{2}\left[(1-y)e^y\Big|_0^1 + \int_0^1 e^y \, dy\right]$

$= -\frac{1}{2} + \frac{1}{2}(e-1) = \frac{e}{2} - 1.$

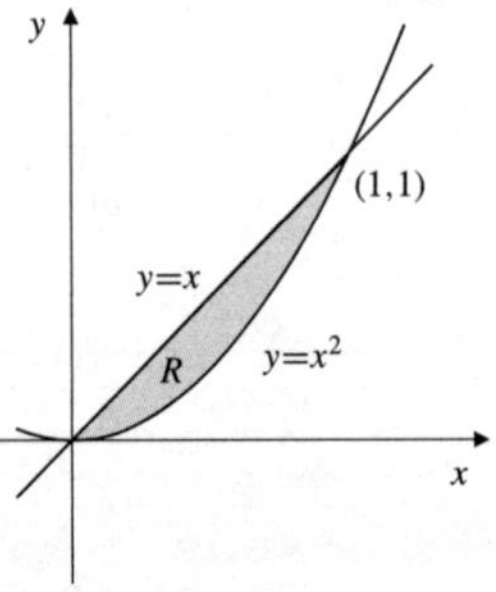

Fig. 14.2.13 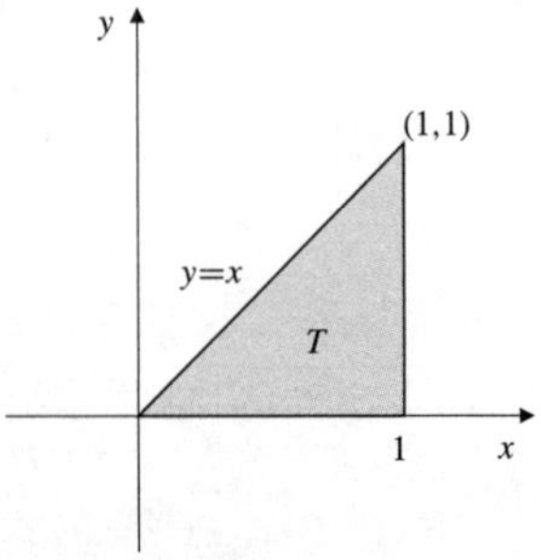   Fig. 14.2.14

INSTRUCTOR'S SOLUTIONS MANUAL

SECTION 14.2 (PAGE 830)

14. $\iint_T \dfrac{xy}{1+x^4} \, dA = \int_0^1 \dfrac{x}{1+x^4} \, dx \int_0^x y \, dy$
$= \dfrac{1}{2} \int_0^1 \dfrac{x^3}{1+x^4} \, dx$
$= \dfrac{1}{8} \ln(1+x^4) \Big|_0^1 = \dfrac{\ln 2}{8}.$

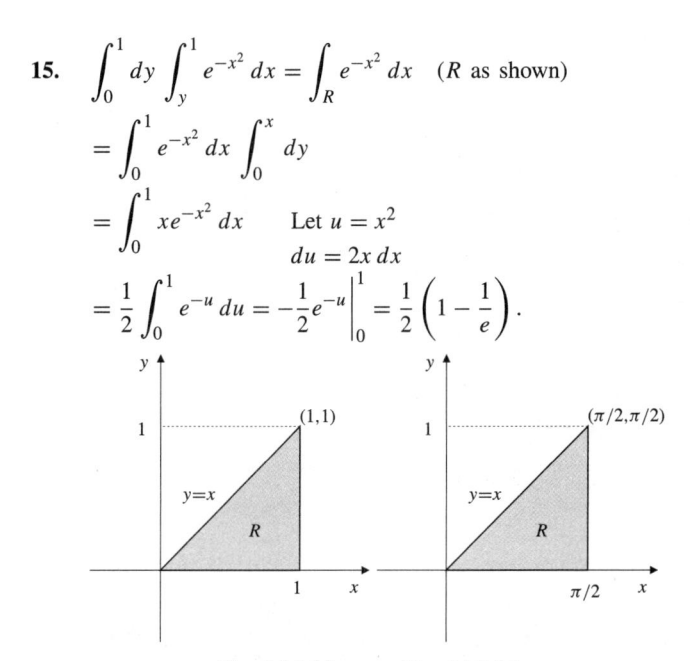

15. $\int_0^1 dy \int_y^1 e^{-x^2} \, dx = \int_R e^{-x^2} \, dx$ (R as shown)
$= \int_0^1 e^{-x^2} \, dx \int_0^x dy$
$= \int_0^1 x e^{-x^2} \, dx$ 　Let $u = x^2$
　　　　　　　　　　　　$du = 2x \, dx$
$= \dfrac{1}{2} \int_0^1 e^{-u} \, du = -\dfrac{1}{2} e^{-u} \Big|_0^1 = \dfrac{1}{2}\left(1 - \dfrac{1}{e}\right).$

Fig. 14.2.15　　Fig. 14.2.16

16. $\int_0^{\pi/2} dy \int_y^{\pi/2} \dfrac{\sin x}{x} \, dx = \iint_R \dfrac{\sin x}{x} \, dA$ (R as shown)
$= \int_0^{\pi/2} \dfrac{\sin x}{x} \, dx \int_0^x dy = \int_0^{\pi/2} \sin x \, dx = 1.$

17. $\int_0^1 dx \int_x^1 \dfrac{y^\lambda}{x^2+y^2} \, dy$ 　$(\lambda > 0)$
$= \iint_R \dfrac{y^\lambda}{x^2+y^2} \, dA$ (R as shown)
$= \int_0^1 y^\lambda \, dy \int_0^y \dfrac{dx}{x^2+y^2}$
$= \int_0^1 y^\lambda \, dy \, \dfrac{1}{y}\left(\tan^{-1} \dfrac{x}{y}\right)\Big|_{x=0}^{x=y}$
$= \dfrac{\pi}{4} \int_0^1 y^{\lambda-1} \, dy = \dfrac{\pi y^\lambda}{4\lambda}\Big|_0^1 = \dfrac{\pi}{4\lambda}.$

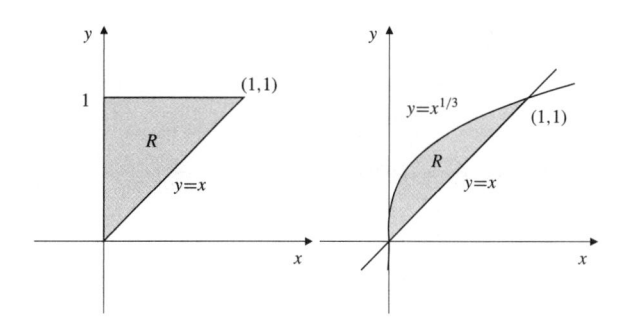

Fig. 14.2.17　　Fig. 14.2.18

18. $\int_0^1 dx \int_x^{x^{1/3}} \sqrt{1-y^4} \, dy$
$= \iint_R \sqrt{1-y^4} \, dA$ (R as shown)
$= \int_0^1 y\sqrt{1-y^4} \, dy - \int_0^1 y^3 \sqrt{1-y^4} \, dy$
Let $u = y^2$　　　Let $v = 1-y^4$
　$du = 2y \, dy$　　$dv = -4y^3 \, dy$
$= \dfrac{1}{2} \int_0^1 \sqrt{1-u^2} \, du + \dfrac{1}{4} \int_1^0 v^{1/2} \, dv$
$= \dfrac{1}{2}\left(\dfrac{\pi}{4} \times 1^2\right) + \dfrac{1}{6} v^{3/2}\Big|_1^0 = \dfrac{\pi}{8} - \dfrac{1}{6}.$

19. $V = \int_0^1 dx \int_0^x (1-x^2) \, dy$
$= \int_0^1 (1-x^2) x \, dx = \dfrac{1}{2} - \dfrac{1}{4} = \dfrac{1}{4}$ cu. units.

20. $V = \int_0^1 dy \int_0^y (1-x^2) \, dx$
$= \int_0^1 \left(y - \dfrac{y^3}{3}\right) dy = \dfrac{1}{2} - \dfrac{1}{12} = \dfrac{5}{12}$ cu. units.

21. $V = \int_0^1 dx \int_0^{1-x} (1-x^2-y^2) \, dy$
$= \int_0^1 \left((1-x^2)y - \dfrac{y^3}{3}\right)\Big|_{y=0}^{y=1-x} dx$
$= \int_0^1 \left((1-x^2)(1-x) - \dfrac{(1-x)^3}{3}\right) dx$
$= \int_0^1 \left(\dfrac{2}{3} - 2x^2 + \dfrac{4x^3}{3}\right) dx = \dfrac{2}{3} - \dfrac{2}{3} + \dfrac{1}{3} = \dfrac{1}{3}$ cu. units.

511

22. $z = 1-y^2$ and $z = x^2$ intersect on the cylinder $x^2+y^2 = 1$. The volume lying below $z = 1 - y^2$ and above $z = x^2$ is

$$V = \iint_{x^2+y^2 \le 1} (1 - y^2 - x^2) \, dA$$

$$= 4 \int_0^1 dx \int_0^{\sqrt{1-x^2}} (1 - x^2 - y^2) \, dy$$

$$= 4 \int_0^1 dx \left((1-x^2)y - \frac{y^3}{3} \right) \bigg|_{y=0}^{y=\sqrt{1-x^2}}$$

$$= \frac{8}{3} \int_0^1 (1-x^2)^{3/2} \, dx \quad \text{Let } x = \sin u$$
$$\qquad dx = \cos u \, du$$

$$= \frac{8}{3} \int_0^{\pi/2} \cos^4 u \, du = \frac{2}{3} \int_0^{\pi/2} (1 + \cos 2u)^2 \, du$$

$$= \frac{2}{3} \int_0^{\pi/2} \left(1 + 2\cos 2u + \frac{1 + \cos 4u}{2} \right) du$$

$$= \frac{2}{3} \cdot \frac{3}{2} \cdot \frac{\pi}{2} = \frac{\pi}{2} \text{ cu. units.}$$

23. $V = \int_1^2 dx \int_0^x \frac{1}{x+y} \, dy$

$$= \int_1^2 dx \left(\ln(x+y) \bigg|_{y=0}^{y=x} \right)$$

$$= \int_1^2 (\ln 2x - \ln x) \, dx = \ln 2 \int_1^2 dx = \ln 2 \text{ cu. units.}$$

24. $V = \int_0^{\pi^{1/4}} dy \int_0^y x^2 \sin(y^4) \, dx$

$$= \frac{1}{3} \int_0^{\pi^{1/4}} y^3 \sin(y^4) \, dy \quad \text{Let } u = y^4$$
$$\qquad du = 4y^3 \, dy$$

$$= \frac{1}{12} \int_0^\pi \sin u \, du = \frac{1}{6} \text{ cu. units.}$$

25. $\text{Vol} = \iint_E (1 - x^2 - 2y^2) \, dA$

$$= 4 \int_0^1 dx \int_0^{\sqrt{(1-x^2)/2}} (1 - x^2 - 2y^2) \, dy$$

$$= 4 \int_0^1 \left(\frac{1}{\sqrt{2}}(1-x^2)^{3/2} - \frac{2}{3} \frac{(1-x^2)^{3/2}}{2\sqrt{2}} \right) dx$$

$$= \frac{4\sqrt{2}}{3} \int_0^1 (1-x^2)^{3/2} \, dx \quad \text{Let } x = \sin\theta$$
$$\qquad dx = \cos\theta \, d\theta$$

$$= \frac{4\sqrt{2}}{3} \int_0^{\pi/2} \cos^4\theta \, d\theta = \frac{4\sqrt{2}}{3} \int_0^{\pi/2} \left(\frac{1 + \cos 2\theta}{2} \right)^2 d\theta$$

$$= \frac{\sqrt{2}}{3} \int_0^{\pi/2} \left(1 + 2\cos 2\theta + \frac{1 + \cos 4\theta}{2} \right) d\theta$$

$$= \frac{\sqrt{2}}{3} \left[\frac{3\theta}{2} + \sin 2\theta + \frac{1}{8} \sin 4\theta \right]_0^{\pi/2} = \frac{\pi}{2\sqrt{2}} \text{ cu. units.}$$

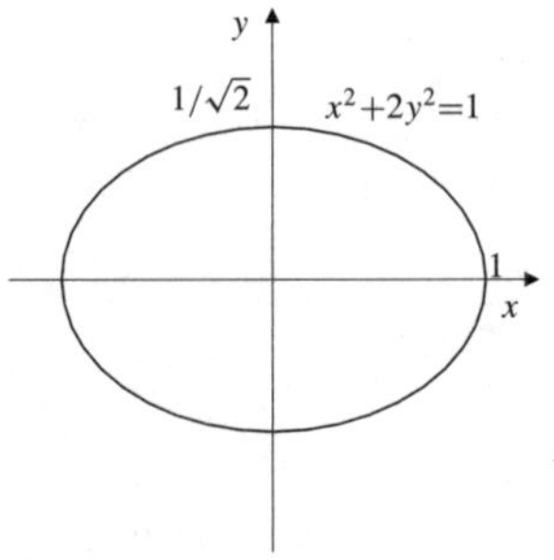

Fig. 14.2.25 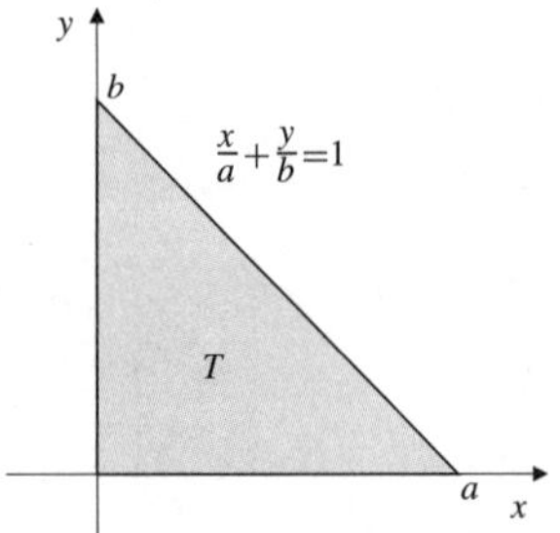   Fig. 14.2.26

26. $\text{Vol} = \iint_T \left(2 - \frac{x}{a} - \frac{y}{b} \right) dA$

$$= \int_0^a dx \int_0^{b(1-(x/a))} \left(2 - \frac{x}{a} - \frac{y}{b} \right) dy$$

$$= \int_0^a \left[\left(2 - \frac{x}{a} \right) b \left(1 - \frac{x}{a} \right) - \frac{1}{2b} b^2 \left(1 - \frac{x}{a} \right)^2 \right] dx$$

$$= \frac{b}{2} \int_0^a \left(3 - \frac{4x}{a} + \frac{x^2}{a^2} \right) dx$$

$$= \frac{b}{2} \left(3x - \frac{2x^2}{a} + \frac{x^3}{3a^2} \right) \bigg|_0^a = \frac{2}{3} ab \text{ cu. units.}$$

27. $\text{Vol} = 8 \times \text{part in the first octant}$

$$= 8 \int_0^a dx \int_0^{\sqrt{a^2-x^2}} \sqrt{a^2 - x^2} \, dy$$

$$= 8 \int_0^a (a^2 - x^2) \, dx$$

$$= 8 \left(a^2 x - \frac{x^3}{3} \right) \bigg|_0^a = \frac{16}{3} a^3 \text{ cu. units.}$$

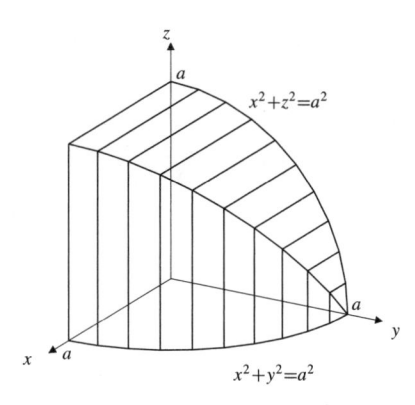

INSTRUCTOR'S SOLUTIONS MANUAL   SECTION 14.3 (PAGE 836)

Fig. 14.2.27

28. The part of the plane $z = 8 - x$ lying inside the elliptic cylinder $x^2 = 2y^2 = 8$ lies above $z = 0$. The part of the plane $z = y - 4$ inside the cylinder lies below $z = 0$. Thus the required volume is

$$\text{Vol} = \iint_{x^2+2y^2 \le 8} \left(8 - x - (y-4)\right) dA$$
$$= \iint_{x^2+2y^2 \le 8} 12 \, dA \quad \text{(by symmetry)}$$
$$= 12 \times \text{area of ellipse } \frac{x^2}{8} + \frac{y^2}{4} = 1$$
$$= 12 \times \pi(2\sqrt{2})(2) = 48\sqrt{2}\pi \text{ cu. units.}$$

29. With $g(x)$ and $G(x)$ defined as in the statement of the problem, we have

$$\int_a^x G(u) \, du = \int_a^x du \int_c^d f_1(u,t) \, dt$$
$$= \int_c^d dt \int_a^x f_1(u,t) \, du$$
$$= \int_c^d \left(f(x,t) - f(a,t)\right) dt = g(x) - C,$$

where $C = \int_c^d f(a,t) \, dt$ is independent of x. Applying the Fundamental Theorem of Calculus we obtain

$$g'(x) = \frac{d}{dx} \int_a^x G(u) \, du = G(x).$$

30. Since $F'(x) = f(x)$ and $G'(x) = g(x)$ on $a \le x \le b$, we have

$$\iint_T f(x)g(x) \, dA = \int_a^b f(x) \, dx \int_a^x G'(y) \, dy$$
$$= \int_a^b f(x)\left(G(x) - G(a)\right) dx$$
$$= \int_a^b f(x)G(x) \, dx - G(a)F(b) + G(a)F(a)$$
$$\iint_T f(x)g(x) \, dA = \int_a^b g(y) \, dy \int_y^b F'(x) \, dx$$
$$= \int_a^b g(y)\left(F(b) - F(y)\right) dy$$
$$= F(b)G(b) - F(b)G(a) - \int_a^b F(y)g(y) \, dx.$$

Thus

$$\int_a^b f(x)G(x) \, dx = F(b)G(b) - F(a)G(a) - \int_a^b g(y)F(y) \, dy.$$

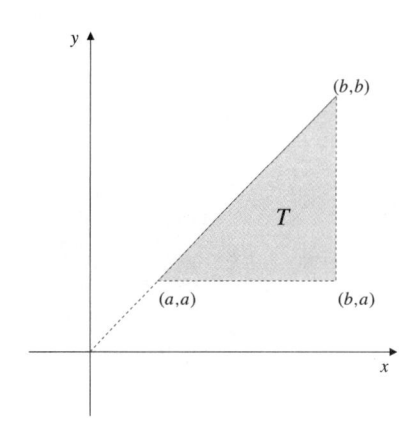

Fig. 14.2.30

Section 14.3 Improper Integrals and a Mean-Value Theorem (page 836)

1. $\iint_Q e^{-x-y} \, dA = \int_0^\infty e^{-x} \, dx \int_0^\infty e^{-y} \, dy$
$$= \left(\lim_{R \to \infty} (-e^{-x})\Big|_0^R\right)^2 = 1 \text{ (converges)}$$

2. $\iint_Q \frac{dA}{(1+x^2)(1+y^2)} = \int_0^\infty \frac{dx}{1+x^2} \int_0^\infty \frac{dy}{1+y^2}$
$$= \left(\lim_{R \to \infty} (\tan^{-1} x)\Big|_0^R\right)^2 = \frac{\pi^2}{4}$$
(converges)

513

3. $\iint_S \frac{y}{1+x^2} dA = \int_0^1 y \, dy \int_{-\infty}^{\infty} \frac{dx}{1+x^2}$
$= \frac{1}{2} \left(\lim_{\substack{S \to -\infty \\ R \to \infty}} \tan^{-1} x \right) \Big|_S^R = \frac{\pi}{2}$ (converges)

4. $\iint_T \frac{1}{x\sqrt{y}} dA = \int_0^1 \frac{dx}{x} \int_x^{2x} \frac{dy}{\sqrt{y}}$
$= \int_0^1 \frac{2(\sqrt{2x} - \sqrt{x})}{x} dx$
$= 2(\sqrt{2} - 1) \int_0^1 \frac{dx}{\sqrt{x}} = 4(\sqrt{2} - 1)$ (converges)

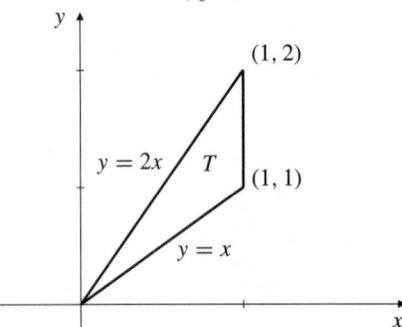

Fig. 14.3.4

5. $\iint_Q \frac{x^2 + y^2}{(1+x^2)(1+y^2)} dA$
$= 2 \iint_Q \frac{x^2 \, dA}{(1+x^2)(1+y^2)}$ (by symmetry)
$= 2 \int_0^{\infty} \frac{x^2 \, dx}{1+x^2} \int_0^{\infty} \frac{dy}{1+y^2} = \pi \int_0^{\infty} \frac{x^2 \, dx}{1+x^2},$
which diverges to infinity, since $x^2/(1+x^2) \geq 1/2$ on $[1, \infty)$.

6. $\iint_H \frac{dA}{1+x+y} = \int_0^{\infty} dx \int_0^1 \frac{1}{1+x+y} dy$
$= \int_0^{\infty} \left(\ln(1+x+y) \Big|_{y=0}^{y=1} \right) dx$
$= \int_0^{\infty} \ln \left(\frac{2+x}{1+x} \right) dx = \int_0^{\infty} \ln \left(1 + \frac{1}{1+x} \right) dx.$

Since $\lim_{u \to 0+} \frac{\ln(1+u)}{u} = 1$, we have $\ln(1+u) \geq u/2$ on some interval $(0, u_0)$. Therefore

$\ln \left(1 + \frac{1}{1+x} \right) \geq \frac{1}{2(1+x)}$

on some interval (x_0, ∞), and

$\int_0^{\infty} \ln \left(1 + \frac{1}{1+x} \right) dx \geq \int_{x_0}^{\infty} \frac{1}{2(1+x)} dx,$

which diverges to infinity. Thus the given double integral diverges to infinity by comparison.

7. $\iint_{\mathbb{R}^2} e^{-(|x|+|y|)} dA = 4 \iint_{\substack{x \geq 0 \\ y \geq 0}} e^{-(x+y)} dA$
$= 4 \int_0^{\infty} e^{-x} dx \int_0^{\infty} e^{-y} dy$
$= 4 \left(\lim_{R \to \infty} -e^{-x} \Big|_0^R \right)^2 = 4$
(The integral converges.)

8. On the strip S between the parallel lines $x + y = 0$ and $x + y = 1$ we have $e^{-|x+y|} = e^{-(x+y)} \geq 1/e$. Since S has infinite area,

$\iint_S e^{-|x+y|} dA = \infty.$

Since $e^{-|x+y|} > 0$ for all (x, y) in $\mathbb{R}^2$, we have

$\iint_{\mathbb{R}^2} e^{-|x+y|} dA > \iint_S e^{-|x+y|} dA,$

and the given integral diverges to infinity.

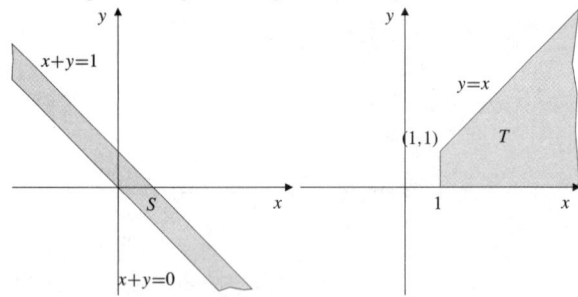

Fig. 14.3.8   Fig. 14.3.9

9. $\iint_T \frac{1}{x^3} e^{-y/x} dA = \int_1^{\infty} \frac{dx}{x^3} \int_0^x e^{-y/x} dy$
$= \int_1^{\infty} \frac{dx}{x^3} \left(-xe^{-y/x} \Big|_{y=0}^{y=x} \right)$
$= \left(1 - \frac{1}{e} \right) \int_1^{\infty} \frac{dx}{x^2}$
$= \left(1 - \frac{1}{e} \right) \lim_{R \to \infty} \left(-\frac{1}{x} \Big|_1^R \right) = 1 - \frac{1}{e}$
(The integral converges.)

10. $\iint_T \frac{dA}{x^2 + y^2} = \int_1^{\infty} dx \int_0^x \frac{dy}{x^2 + y^2}$
$= \int_1^{\infty} dx \left(\frac{1}{x} \tan^{-1} \frac{y}{x} \Big|_{y=0}^{y=x} \right)$
$= \frac{\pi}{4} \int_1^{\infty} \frac{dx}{x} = \infty$
(The integral diverges to infinity.)

INSTRUCTOR'S SOLUTIONS MANUAL   SECTION 14.3 (PAGE 836)

11. Since $e^{-xy} > 0$ on Q we have

$$\iint_Q e^{-xy}\,dA > \iint_R e^{-xy}\,dA,$$

where R satisfies $1 \le x < \infty$, $0 \le y \le 1/x$. Thus

$$\iint_Q e^{-xy}\,dA > \int_1^\infty dx \int_0^{1/x} e^{-xy}\,dy > \frac{1}{e}\int_1^\infty \frac{dx}{x} = \infty.$$

The given integral diverges to infinity.

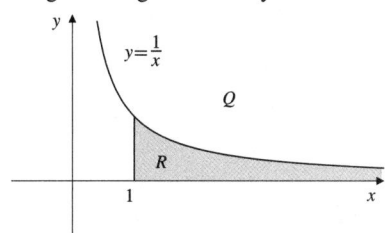

Fig. 14.3.11

12. $\displaystyle\iint_R \frac{1}{x}\sin\frac{1}{x}\,dA = \int_{2/\pi}^\infty \frac{1}{x}\sin\frac{1}{x}\,dx \int_0^{1/x} dy$

$= \displaystyle\int_{2/\pi}^\infty \frac{1}{x^2}\sin\frac{1}{x}\,dx$ Let $u = 1/x$
$\qquad\qquad\qquad\qquad\qquad du = -1/x^2\,dx$

$= -\displaystyle\int_{\pi/2}^0 \sin u\,du = \cos u\Big|_{\pi/2}^0 = 1$

(The integral converges.)

13. a) $I = \displaystyle\iint_S \frac{dA}{x+y} = \int_0^1 dx \int_0^1 \frac{dy}{x+y}$

$= \displaystyle\int_0^1 dx\left(\ln(x+y)\Big|_{y=0}^{y=1}\right)$

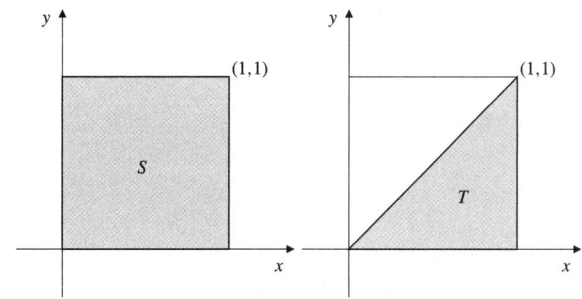

$= \displaystyle\lim_{c\to 0+}\left[(x+1)\ln(x+1) - x\ln x\right]\Big|_c^1$

$= \displaystyle\lim_{c\to 0+} 2\ln 2 - 0 - (c+1)\ln(c+1) + c\ln c = 2\ln 2.$

Fig. 14.3.13(a)   Fig. 14.3.13(b)

b) $I = 2\displaystyle\iint_T \frac{dA}{x+y} = 2\lim c \to 0+ \int_c^1 dx \int_0^x \frac{dy}{x+y}$

$= 2\lim c \to 0+ \displaystyle\int_c^1 dx\left(\ln(x+y)\Big|_{y=0}^{y=x}\right)$

$= 2\displaystyle\lim_{c\to 0+}\int_c^1 (\ln 2x - \ln x)\,dx = 2\ln 2\int_0^1 dx = 2\ln 2.$

14. Vol $= \displaystyle\iint_S \frac{2xy}{x^2+y^2}\,dA$

$= 4\displaystyle\iint_T \frac{2xy}{x^2+y^2}\,dA$ (T as in #9(b))

$= 4\displaystyle\int_0^1 x\,dx \int_0^x \frac{y\,dy}{x^2+y^2}$ Let $u = x^2 + y^2$
$\qquad\qquad\qquad\qquad\qquad\qquad du = 2y\,dy$

$= 2\displaystyle\int_0^1 x\,dx \int_{x^2}^{2x^2} \frac{du}{u}$

$= 2\ln 2 \displaystyle\int_0^1 x\,dx = \ln 2$ cu. units.

15. $\displaystyle\iint_{D_k} \frac{dA}{x^a} = \int_0^1 \frac{dx}{x^a}\int_0^{x^k} dy = \int_0^1 x^{k-a}\,dx$, which converges if $k - a > -1$, that is, if $k > a - 1$.

16. $\displaystyle\iint_{D_k} y^b\,dA = \int_0^1 dx \int_0^{x^k} y^b\,dy = \int_0^1 \frac{x^{k(b+1)}}{b+1}\,dx$ if $b > -1$. This latter integral converges if $k(b+1) > -1$. Thus, the given integral converges if $b > -1$ and $k > -1/(b+1)$.

17. $\displaystyle\iint_{R_k} x^a\,dA = \int_1^\infty x^a\,dx \int_0^{x^k} dy = \int_1^\infty x^{k+a}\,dx$, which converges if $k + a < -1$, that is, if $k < -(a+1)$.

18. $\displaystyle\iint_{R_k} \frac{dA}{y^b} = \int_1^\infty dx \int_0^{x^k}\frac{dy}{y^b} = \int_1^\infty \frac{x^{k(1-b)}}{1-b}\,dx$ if $b < 1$. This latter integral converges if $k(1-b) < -1$. Thus, the given integral converges if $b < 1$ and $k < -1/(1-b)$.

19. $\displaystyle\iint_{D_k} x^a y^b\,dA = \int_0^1 x^a\,dx \int_0^{x^k} y^b\,dy = \int_0^1 \frac{x^{a+(b+1)k}}{b+1}\,dx$, if $b > -1$. This latter integral converges if $a + (b+1)k > -1$. Thus, the given integral converges if $b > -1$ and $k > -(a+1)/(b+1)$.

20. $\displaystyle\iint_{R_k} x^a y^b\,dA = \int_1^\infty x^a\,dx \int_0^{x^k} y^b\,dy = \int_1^\infty \frac{x^{a+(b+1)k}}{b+1}\,dx$, if $b > -1$. This latter integral converges if $a + (b+1)k < -1$. Thus, the given integral converges if $b > -1$ and $k < -(a+1)/(b+1)$.

21. One iteration:

$$\iint_S \frac{x-y}{(x+y)^3}\,dA = \int_0^1 dx \int_0^1 \frac{x-y}{(x+y)^3}\,dy \quad \text{Let } u = x+y$$
$$du = dy$$
$$= \int_0^1 dx \int_x^{x+1} \frac{2x-u}{u^3}\,du$$
$$= \int_0^1 dx \left(\frac{1}{u} - \frac{x}{u^2}\right)\bigg|_{u=x}^{u=x+1}$$
$$= \int_0^1 \left(\frac{1}{x+1} - \frac{x}{(x+1)^2} - \frac{1}{x} + \frac{1}{x}\right)dx$$
$$= \int_0^1 \frac{dx}{(x+1)^2} = -\frac{1}{x+1}\bigg|_0^1 = \frac{1}{2}.$$

Other iteration:

$$\iint_S \frac{x-y}{(x+y)^3}\,dA = \int_0^1 dy \int_0^1 \frac{x-y}{(x+y)^3}\,dx \quad \text{Let } u = x+y$$
$$du = dx$$
$$= \int_0^1 dy \int_y^{y+1} \frac{u-2y}{u^3}\,du$$
$$= \int_0^1 dy \left(\frac{y}{u^2} - \frac{1}{u}\right)\bigg|_{u=y}^{u=y+1}$$
$$= \int_0^1 \left(\frac{y}{(y+1)^2} - \frac{1}{y+1} - \frac{1}{y} + \frac{1}{y}\right)dy$$
$$= -\int_0^1 \frac{dx}{(y+1)^2} = \frac{1}{y+1}\bigg|_0^1 = -\frac{1}{2}.$$

These seemingly contradictory results are explained by the fact that the given double integral is improper and does not, in fact, exist, that is, it does not converge. To see this, we calculate the integral over a certain subset of the square S, namely the triangle T defined by $0 < x < 1$, $0 < y < x$.

$$\iint_T \frac{x-y}{(x+y)^3}\,dA = \int_0^1 dx \int_0^x \frac{x-y}{(x+y)^3}\,dy$$
$$\text{Let } u = x+y$$
$$du = dy$$
$$= \int_0^1 dx \int_x^{2x} \frac{2x-u}{u^3}\,du$$
$$= \int_0^1 dx \left(\frac{1}{u} - \frac{x}{u^2}\right)\bigg|_{u=x}^{u=2x}$$
$$= \frac{1}{4}\int_0^1 \frac{dx}{x}$$

which diverges to infinity.

22. The average value of x^2 over the rectangle R is

$$\frac{1}{(b-a)(d-c)}\iint_R x^2\,dA$$
$$= \frac{1}{(b-a)(d-c)}\int_a^b x^2\,dx \int_c^d dy$$
$$= \frac{1}{b-a}\cdot\frac{b^3-a^3}{3} = \frac{a^2+ab+b^2}{3}.$$

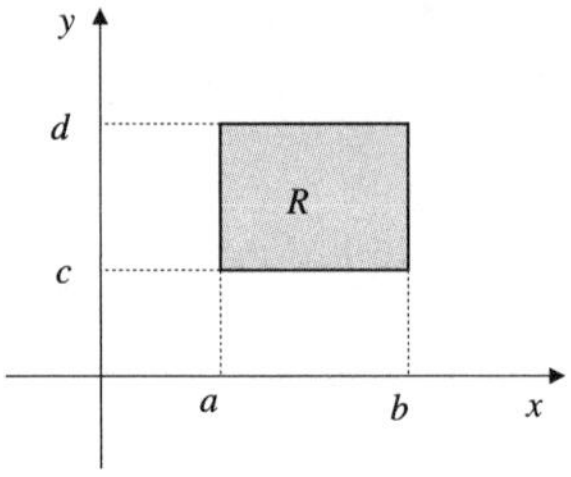

Fig. 14.3.22 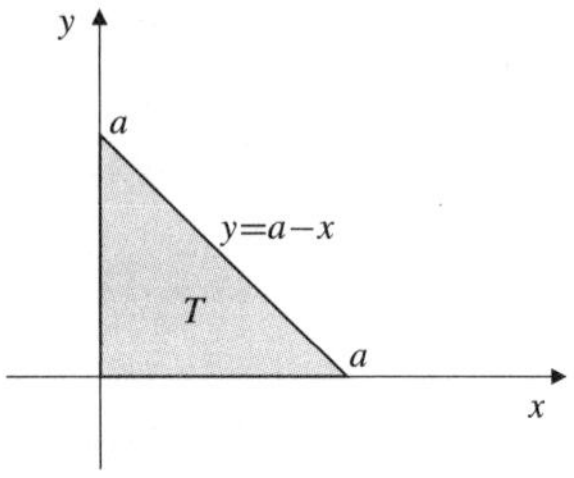   Fig. 14.3.23

23. The average value of $x^2 + y^2$ over the triangle T is

$$\frac{2}{a^2}\iint_T (x^2+y^2)\,dA$$
$$= \frac{2}{a^2}\int_0^a dx \int_0^{a-x}(x^2+y^2)\,dy$$
$$= \frac{2}{a^2}\int_0^a dx \left(x^2 y + \frac{y^3}{3}\right)\bigg|_{y=0}^{y=a-x}$$
$$= \frac{2}{3a^2}\int_0^a [3x^2(a-x) + (a-x)^3]\,dx$$
$$= \frac{2}{3a^2}\int_0^a [a^3 - 3a^2 x + 6ax^2 - 4x^3]\,dx = \frac{a^2}{3}.$$

24. The area of region R is

$$\int_0^1 (\sqrt{x} - x^2)\,dx = \frac{1}{3} \text{ sq. units.}$$

The average value of $1/x$ over R is

$$3\iint_R \frac{dA}{x} = 3\int_0^1 \frac{dx}{x}\int_{x^2}^{\sqrt{x}}dy$$
$$= 3\int_0^1 (x^{-1/2} - x)\,dx = \frac{9}{2}.$$

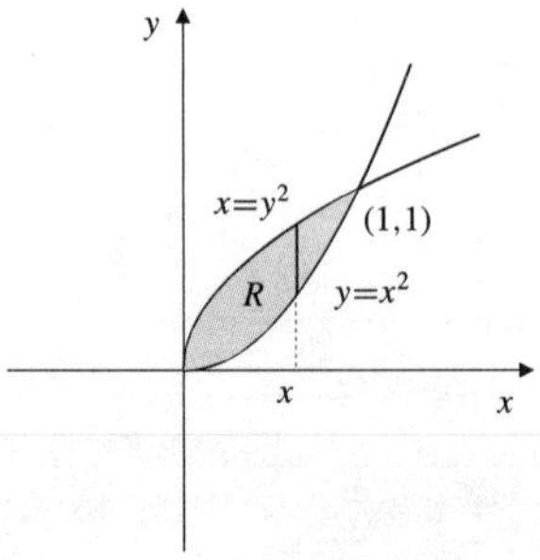

Fig. 14.3.24 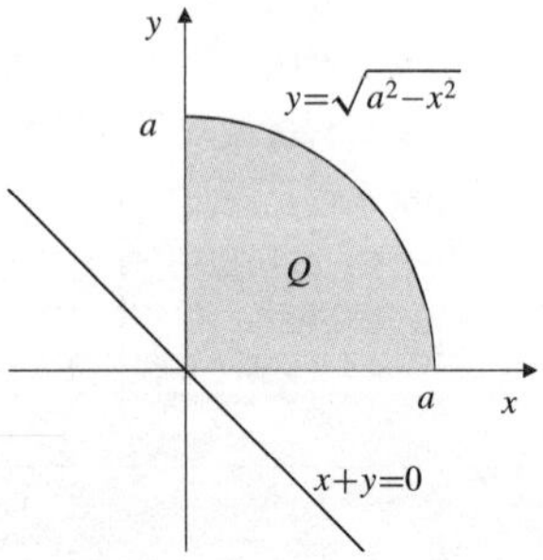   Fig. 14.3.25

25. The distance from (x, y) to the line $x + y = 0$ is $(x + y)/\sqrt{2}$. The average value of this distance over the quarter-disk Q is

$$\frac{4}{\pi a^2} \iint_Q \frac{x+y}{\sqrt{2}} \, dA = \frac{4\sqrt{2}}{\pi a^2} \iint_Q x \, dA$$

$$= \frac{4\sqrt{2}}{\pi a^2} \int_0^a x \, dx \int_0^{\sqrt{a^2-x^2}} dy$$

$$= \frac{4\sqrt{2}}{\pi a^2} \int_0^a x\sqrt{a^2 - x^2} \, dx \quad \text{Let } u = a^2 - x^2$$
$$\qquad\qquad\qquad\qquad\qquad\qquad du = -2x \, dx$$

$$= \frac{2\sqrt{2}}{\pi a^2} \int_0^{a^2} u^{1/2} \, du = \frac{4\sqrt{2}a}{3\pi}.$$

26. Let R be the region $0 \le x < \infty$, $0 \le y \le 1/(1 + x^2)$. If $f(x, y) = x$, then

$$\int_R f(x, y) \, dA = \int_0^\infty x \, dx \int_0^{1/(1+x^2)} dy = \int_0^\infty \frac{x \, dx}{1 + x^2}$$

which diverges to infinity. Thus f has no average value on R.

27. If $f(x, y) = xy$ on the region R of the previous exercise, then

$$\iint_R f(x, y) \, dA = \int_0^\infty x \, dx \int_0^{1/(1+x^2)} y \, dy$$

$$= \frac{1}{2} \int_0^\infty \frac{x \, dx}{(1 + x^2)^2} \quad \text{Let } u = 1 + x^2$$
$$\qquad\qquad\qquad\qquad\qquad du = 2x \, dx$$

$$= \frac{1}{4} \int_1^\infty \frac{du}{u^2} = \frac{1}{4}$$

$$\text{Area} = \int_0^\infty \frac{dx}{1 + x^2} = \frac{\pi}{2}.$$

Thus $f(x, y)$ has average value $\dfrac{2}{\pi} \times \dfrac{1}{4} = \dfrac{1}{2\pi}$ on R.

28. The integral in Example 2 reduced to

$$\int_1^\infty \ln\left(1 + \frac{1}{x^2}\right) dx$$

$$U = \ln\left(1 + \frac{1}{x^2}\right) \quad dV = dx$$
$$\qquad\qquad\qquad\qquad\quad V = x$$
$$dU = -\frac{2 \, dx}{x(x^2 + 1)}$$

$$= \lim_{R \to \infty} \left[x \ln\left(1 + \frac{1}{x^2}\right)\Big|_1^R + 2 \int_1^R \frac{dx}{1 + x^2} \right]$$

$$= 2\left(\frac{\pi}{2} - \frac{\pi}{4}\right) - \ln 2 + \lim_{R \to \infty} \frac{\ln(1 + (1/R^2))}{1/R}$$

$$= \frac{\pi}{2} - \ln 2 + \lim_{R \to \infty} \frac{-(2/R^3)}{(1 + (1/R^2))(-1/R^2)}$$

$$= \frac{\pi}{2} - \ln 2.$$

29. By the Mean-Value Theorem (Theorem 3),

$$\iint_{R_{hk}} f(x, y) \, dA = f(x_0, y_0) h k$$

for some point (x_0, y_0) in R_{hk}. Since $(x_0, y_0) \to (a, b)$ as $(h, k) \to (0, 0)$, and since f is continuous at (a, b), we have

$$\lim_{(h,k) \to (0,0)} \frac{1}{hk} \iint_{R_{hk}} f(x, y) \, dA$$
$$= \lim_{(h,k) \to (0,0)} f(x_0, y_0) = f(a, b).$$

30. If $R = \{(x, y) : a \le x \le a + h, \ b \le y \le b + k\}$, then

$$\iint_R f_{12}(x, y) \, dA = \int_a^{a+h} dx \int_b^{b+k} f_{12}(x, y) \, dy$$

$$= \int_a^{a+h} \left[f_1(x, b + k) - f_1(x, b) \right] dx$$

$$= f(a + h, b + k) - f(a, b + k) - f(a + h, b) + f(a, b)$$

$$\iint_R f_{21}(x, y) \, dA = \int_b^{b+k} dy \int_a^{a+h} f_{21}(x, y) \, dx$$

$$= \int_b^{b+k} \left[f_2(a + h, y) - f_2(a, y) \right] dy$$

$$= f(a + h, b + k) - f(a + h, b) - f(a, b + k) + f(a, b).$$

Thus

$$\iint_R f_{12}(x, y) \, dA = \iint_R f_{21}(x, y) \, dA.$$

Divide both sides of this identity by hk and let $(h, k) \to (0, 0)$ to obtain, using the result of Exercise 31,

$$f_{12}(a, b) = f_{21}(a, b).$$

Section 14.4 Double Integrals in Polar Coordinates (page 846)

1. $$\iint_D (x^2 + y^2) \, dA = \int_0^{2\pi} d\theta \int_0^a r^2 r \, dr$$
$$= 2\pi \frac{a^4}{4} = \frac{\pi a^4}{2}$$

2. $$\iint_D \sqrt{x^2 + y^2} \, dA = \int_0^{2\pi} d\theta \int_0^a r \, r \, dr = \frac{2\pi a^3}{3}$$

3. $$\iint_D \frac{dA}{\sqrt{x^2 + y^2}} = \int_0^{2\pi} d\theta \int_0^a \frac{r \, dr}{r} = 2\pi a$$

4. $\iint_D |x|\, dA = 4 \int_0^{\pi/2} d\theta \int_0^a r\cos\theta\, r\, dr$

$= 4\sin\theta \Big|_0^{\pi/2} \dfrac{a^3}{3} = \dfrac{4a^3}{3}$

5. $\iint_D x^2\, dA = \dfrac{\pi a^4}{4}$; by symmetry the value of this integral is half of that in Exercise 1.

6. $\iint_D x^2 y^2\, dA = 4\int_0^{\pi/2} d\theta \int_0^a r^4 \cos^2\theta \sin^2\theta\, r\, dr$

$= \dfrac{a^6}{6} \int_0^{\pi/2} \sin^2(2\theta)\, d\theta$

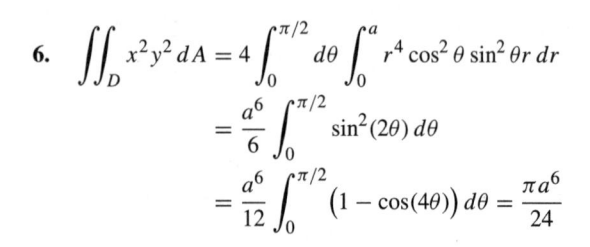

$= \dfrac{a^6}{12} \int_0^{\pi/2} \big(1 - \cos(4\theta)\big)\, d\theta = \dfrac{\pi a^6}{24}$

7. $\iint_Q y\, dA = \int_0^{\pi/2} d\theta \int_0^a r\sin\theta\, r\, dr$

$= (-\cos\theta)\Big|_0^{\pi/2} \dfrac{a^3}{3} = \dfrac{a^3}{3}$

8. $\iint_Q (x+y)\, dA = \dfrac{2a^3}{3}$; by symmetry, the value is twice that obtained in the previous exercise.

9. $\iint_Q e^{x^2+y^2}\, dA = \int_0^{\pi/2} d\theta \int_0^a e^{r^2} r\, dr$

$= \dfrac{\pi}{2}\left(\dfrac{1}{2} e^{r^2}\right)\bigg|_0^a = \dfrac{\pi(e^{a^2} - 1)}{4}$

10. $\iint_Q \dfrac{2xy}{x^2+y^2}\, dA = \int_0^{\pi/2} d\theta \int_0^a \dfrac{2r^2 \sin\theta \cos\theta}{r^2} r\, dr$

$= \dfrac{a^2}{2} \int_0^{\pi/2} \sin(2\theta)\, d\theta = -\dfrac{a^2 \cos(2\theta)}{4}\bigg|_0^{\pi/2} = \dfrac{a^2}{2}$

11. $\iint_S (x+y)\, dA = \int_0^{\pi/3} d\theta \int_0^a (r\cos\theta + r\sin\theta) r\, dr$

$= \int_0^{\pi/3} (\cos\theta + \sin\theta)\, d\theta \int_0^a r^2\, dr$

$= \dfrac{a^3}{3}(\sin\theta - \cos\theta)\Big|_0^{\pi/3}$

$= \left[\left(\dfrac{\sqrt{3}}{2} - \dfrac{1}{2}\right) - (-1)\right] \dfrac{a^3}{3} = \dfrac{(\sqrt{3}+1)a^3}{6}$

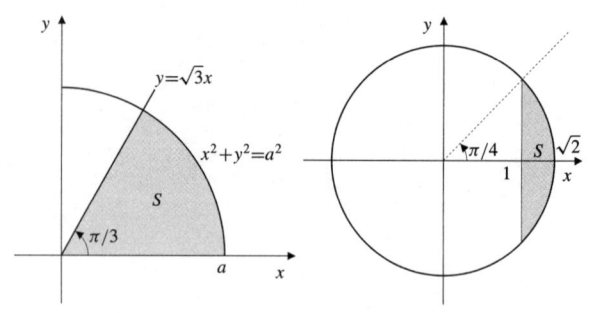

Fig. 14.4.11

Fig. 14.4.12

12. $\iint_S x\, dA = 2\int_0^{\pi/4} d\theta \int_{\sec\theta}^{\sqrt{2}} r\cos\theta\, r\, dr$

$= \dfrac{2}{3} \int_0^{\pi/4} \cos\theta \big(2\sqrt{2} - \sec^3\theta\big)\, d\theta$

$= \dfrac{4\sqrt{2}}{3} \sin\theta\bigg|_0^{\pi/4} - \dfrac{2}{3}\tan\theta\bigg|_0^{\pi/4}$

$= \dfrac{4}{3} - \dfrac{2}{3} = \dfrac{2}{3}$

13. $\iint_T (x^2+y^2)\, dA = \int_0^{\pi/4} d\theta \int_0^{\sec\theta} r^3\, dr$

$= \dfrac{1}{4} \int_0^{\pi/4} \sec^4\theta\, d\theta$

$= \dfrac{1}{4} \int_0^{\pi/4} (1+\tan^2\theta)\sec^2\theta\, d\theta$ Let $u = \tan\theta$
$\qquad\qquad du = \sec^2\theta\, d\theta$

$= \dfrac{1}{4} \int_0^1 (1+u^2)\, du$

$= \dfrac{1}{4}\left(u + \dfrac{u^3}{3}\right)\bigg|_0^1 = \dfrac{1}{3}$

Fig. 14.4.13

INSTRUCTOR'S SOLUTIONS MANUAL SECTION 14.4 (PAGE 846)

14. $\displaystyle\iint_{x^2+y^2\le 1} \ln(x^2+y^2)\,dA = \int_0^{2\pi} d\theta \int_0^1 (\ln r^2)r\,dr$

$= 4\pi \int_0^1 r\ln r\,dr$

$\quad U=\ln r \quad dV = r\,dr$
$\quad dU = \dfrac{dr}{r} \quad V = \dfrac{r^2}{2}$

$= 4\pi\left[\dfrac{r^2}{2}\ln r\Big|_0^1 - \dfrac{1}{2}\int_0^1 r\,dr\right]$

$= 4\pi\left[0 - 0 - \dfrac{1}{4}\right] = -\pi$

(Note that the integral is improper, but converges since $\lim_{r\to 0+} r^2\ln r = 0$.)

15. The average distance from the origin to points in the disk $D: x^2+y^2\le a^2$ is

$\dfrac{1}{\pi a^2}\iint_D \sqrt{x^2+y^2}\,dA = \dfrac{1}{\pi a^2}\int_0^{2\pi}d\theta\int_0^a r^2\,dr = \dfrac{2a}{3}.$

16. The annular region R: $0 < a \le \sqrt{x^2+y^2} \le b$ has area $\pi(b^2-a^2)$. The average value of $e^{-(x^2+y^2)}$ over the region is

$\dfrac{1}{\pi(b^2-a^2)}\iint_R e^{-(x^2+y^2)}\,dA$

$= \dfrac{1}{\pi(b^2-a^2)}\int_0^{2\pi}d\theta\int_a^b e^{-r^2}r\,dr\quad$ Let $u=r^2$
$\hfill du = 2r\,dr$

$= \dfrac{1}{\pi(b^2-a^2)}(2\pi)\dfrac{1}{2}\int_{a^2}^{b^2} e^{-u}\,du$

$= \dfrac{1}{b^2-a^2}\left(e^{-a^2}-e^{-b^2}\right).$

17. If D is the disk $x^2+y^2\le 1$, then

$\displaystyle\iint_D \dfrac{dA}{(x^2+y^2)^k} = \int_0^{2\pi}d\theta\int_0^1 r^{-2k}r\,dr = 2\pi\int_0^1 r^{1-2k}\,dr$

which converges if $1-2k > -1$, that is, if $k < 1$. In this case the value of the integral is

$2\pi\dfrac{r^{2-2k}}{2-2k}\Big|_0^1 = \dfrac{\pi}{1-k}.$

18. $\displaystyle\iint_{\mathbb{R}^2} \dfrac{dA}{(1+x^2+y^2)^k}$

$= \int_0^{2\pi}d\theta\int_0^\infty \dfrac{r\,dr}{(1+r^2)^k}\quad$ Let $u=1+r^2$
$\hfill du = 2r\,dr$

$= \pi\int_1^\infty u^{-k}\,du = \dfrac{-\pi}{1-k}$ if $k>1$.

The integral converges to $\dfrac{\pi}{k-1}$ if $k>1$.

19. $\displaystyle\iint_D xy\,dA = \int_0^{\pi/4}d\theta\int_0^a r\cos\theta\, r\sin\theta\, r\,dr$

$= \dfrac{1}{2}\int_0^{\pi/4}\sin 2\theta\,d\theta\int_0^a r^3\,dr$

$= \dfrac{a^4}{8}\left(-\dfrac{\cos 2\theta}{2}\right)\Big|_0^{\pi/4} = \dfrac{a^4}{16}.$

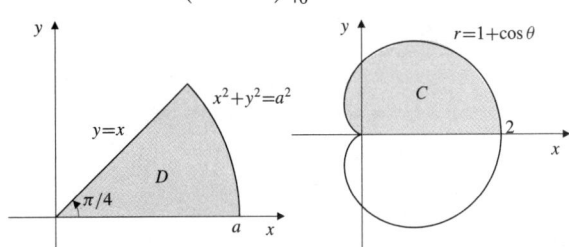

Fig. 14.4.19   Fig. 14.4.20

20. $\displaystyle\iint_C y\,dA = \int_0^\pi d\theta\int_0^{1+\cos\theta} r\sin\theta\, r\,dr$

$= \dfrac{1}{3}\int_0^\pi \sin\theta(1+\cos\theta)^3\,d\theta\quad$ Let $u=1+\cos\theta$
$\hfill du = -\sin\theta\,d\theta$

$= \dfrac{1}{3}\int_0^2 u^3\,du = \dfrac{u^4}{12}\Big|_0^2 = \dfrac{4}{3}$

21. The paraboloids $z=x^2+y^2$ and $3z=4-x^2-y^2$ intersect where $3(x^2+y^2)=4-(x^2+y^2)$, i.e., on the cylinder $x^2+y^2=1$. The volume they bound is given by

$V = \displaystyle\iint_{x^2+y^2\le 1}\left[\dfrac{4-x^2-y^2}{3}-(x^2+y^2)\right]dA$

$= \int_0^{2\pi}d\theta\int_0^1\left[\dfrac{4-r^2}{3}-r^2\right]r\,dr$

$= \dfrac{8\pi}{3}\int_0^1 (r-r^3)\,dr$

$= \dfrac{8\pi}{3}\left(\dfrac{r^2}{2}-\dfrac{r^4}{4}\right)\Big|_0^1 = \dfrac{2\pi}{3}$ cu. units.

519

22. One quarter of the required volume lies in the first octant. (See the figure.) In polar coordinates the cylinder $x^2 + y^2 = ax$ becomes $r = a\cos\theta$. Thus, the required volume is

$$V = 4\iint_D \sqrt{a^2 - x^2 - y^2}\, dA$$
$$= 4\int_0^{\pi/2} d\theta \int_0^{a\cos\theta} \sqrt{a^2 - r^2}\, r\, dr \quad \text{Let } u = a^2 - r^2$$
$$\hspace{6cm} du = -2r\, dr$$
$$= 2\int_0^{\pi/2} d\theta \int_{a^2\sin^2\theta}^{a^2} u^{1/2}\, du$$
$$= \frac{4}{3}\int_0^{\pi/2} d\theta \left(u^{3/2} \Big|_{a^2\sin^2\theta}^{a^2} \right)$$
$$= \frac{4}{3}a^3 \int_0^{\pi/2} (1 - \sin^3\theta)\, d\theta$$
$$= \frac{4}{3}a^3 \left(\frac{\pi}{2} - \int_0^{\pi/2} \sin\theta(1 - \cos^2\theta)\, d\theta \right)$$
$$\text{Let } v = \cos\theta$$
$$dv = -\sin\theta\, d\theta$$
$$= \frac{2\pi a^3}{3} - \frac{4a^3}{3}\int_0^1 (1 - v^2)\, dv$$
$$= \frac{2\pi a^3}{3} - \frac{4a^3}{3}\left(v - \frac{v^3}{3} \right)\Big|_0^1$$
$$= \frac{2\pi a^3}{3} - \frac{8a^3}{9} = \frac{2}{9}a^3(3\pi - 4) \text{ cu. units.}$$

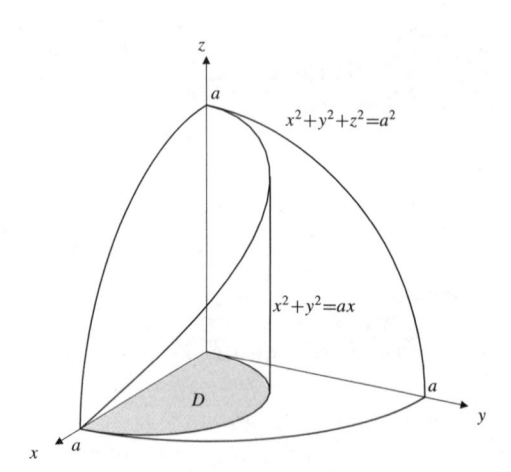

Fig. 14.4.22

23. The volume inside the sphere $x^2 + y^2 + z^2 = 2a^2$ and the cylinder $x^2 + y^2 = a^2$ is

$$V = 8\int_0^{\pi/2} d\theta \int_0^a \sqrt{2a^2 - r^2}\, r\, dr \quad \text{Let } u = 2a^2 - r^2$$
$$\hspace{6cm} du = -2r\, dr$$
$$= 2\pi \int_{a^2}^{2a^2} u^{1/2}\, du = \frac{4\pi a^3}{3}(2\sqrt{2} - 1) \text{ cu. units.}$$

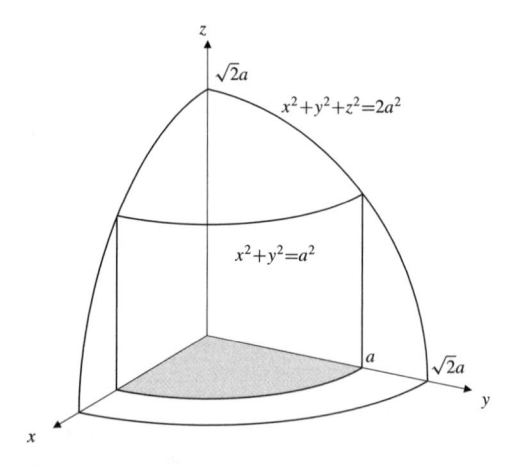

Fig. 14.4.23

24. Volume $= \int_0^{2\pi} d\theta \int_0^2 (r\cos\theta + r\sin\theta + 4)r\, dr$
$$= \int_0^{2\pi} (\cos\theta + \sin\theta)\, d\theta \int_0^2 r^2\, dr + 8\pi \int_0^2 r\, dr$$
$$= 0 + 4\pi(2^2) = 16\pi \text{ cu. units.}$$

25. One eighth of the required volume lies in the first octant. This eighth is divided into two equal parts by the plane $x = y$. One of these parts lies above the circular sector D in the xy-plane specified in polar coordinate by $0 \le r \le a$, $0 \le \theta \le \pi/4$, and beneath the cylinder $z = \sqrt{a^2 - x^2}$. Thus, the total volume lying inside all three cylinders is

$$V = 16\iint_D \sqrt{a^2 - x^2}\, dA$$
$$= 16\int_0^{\pi/4} d\theta \int_0^a \sqrt{a^2 - r^2\cos^2\theta}\, r\, dr$$
$$\text{Let } u = a^2 - r^2\cos^2\theta$$
$$du = -2r\cos^2\theta\, dr$$
$$= 8\int_0^{\pi/4} \frac{d\theta}{\cos^2\theta} \int_{a^2\sin^2\theta}^{a^2} u^{1/2}\, du$$
$$= \frac{16a^3}{3}\int_0^{\pi/4} \frac{1 - \sin^3\theta}{\cos^2\theta}\, d\theta$$
$$= \frac{16a^3}{3}\int_0^{\pi/4} \left(\sec^2\theta - \frac{1 - \cos^2\theta}{\cos^2\theta}\sin\theta \right) d\theta$$

INSTRUCTOR'S SOLUTIONS MANUAL SECTION 14.4 (PAGE 846)

$$= \frac{16a^3}{3}\left(\tan\theta - \frac{1}{\cos\theta} - \cos\theta\right)\Big|_0^{\pi/4}$$

$$= \frac{16a^3}{3}\left(1 - 0 - \sqrt{2} + 1 - \frac{1}{\sqrt{2}} + 1\right)$$

$$= 16\left(1 - \frac{1}{\sqrt{2}}\right)a^3 \text{ cu. units.}$$

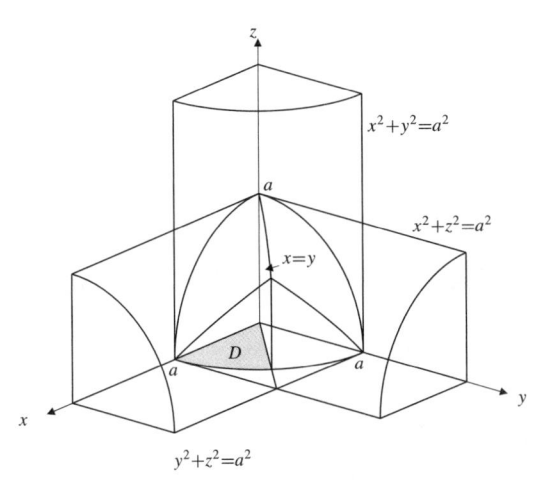

Fig. 14.4.25

26. One quarter of the required volume V is shown in the figure. We have

$$V = 4\iint_D \sqrt{y}\,dA$$

$$= 4\int_0^{\pi/2} d\theta \int_0^{2\sin\theta} \sqrt{r\sin\theta}\,r\,dr$$

$$= 4\int_0^{\pi/2} \sqrt{\sin\theta}\,d\theta \left(\frac{2}{5}r^{5/2}\Big|_0^{2\sin\theta}\right)$$

$$= \frac{32\sqrt{2}}{5}\int_0^{\pi/2} \sin^3\theta\,d\theta = \frac{64\sqrt{2}}{15} \text{ cu. units.}$$

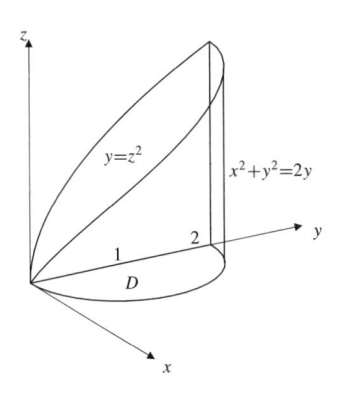

Fig. 14.4.26

27. By symmetry, we need only calculate the average distance from points in the sector S: $0 \leq \theta \leq \pi/4$, $0 \leq r \leq 1$ to the line $x = 1$. This average value is

$$\frac{8}{\pi}\iint_S (1-x)\,dA = \frac{8}{\pi}\int_0^{\pi/4} d\theta \int_0^1 (1 - r\cos\theta)r\,dr$$

$$= \frac{8}{\pi}\left[\frac{\pi}{8} - \int_0^{\pi/4} \cos\theta\,d\theta \int_0^1 r^2\,dr\right]$$

$$= 1 - \frac{8}{3\sqrt{2}\pi} = 1 - \frac{4\sqrt{2}}{3\pi} \text{ units.}$$

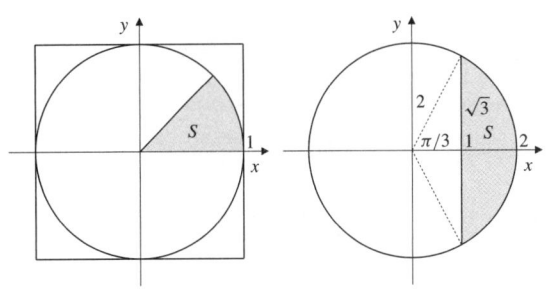

Fig. 14.4.27   Fig. 14.4.28

28. The area of S is $(4\pi - 3\sqrt{3})/3$ sq. units. Thus

$$\bar{x} = \frac{3}{4\pi - 3\sqrt{3}}\iint_S x\,dA$$

$$= \frac{6}{4\pi - 3\sqrt{3}}\int_0^{\pi/3} d\theta \int_{\sec\theta}^2 r\cos\theta\,r\,dr$$

$$= \frac{2}{4\pi - 3\sqrt{3}}\int_0^{\pi/3} \cos\theta(8 - \sec^3\theta)\,d\theta$$

$$= \frac{2}{4\pi - 3\sqrt{3}}\left(4\sqrt{3} - \tan\theta\Big|_0^{\pi/3}\right) = \frac{6\sqrt{3}}{4\pi - 3\sqrt{3}}.$$

The segment has centroid $\left(\dfrac{6\sqrt{3}}{4\pi - 3\sqrt{3}}, 0\right)$.

29. Let E be the region in the first quadrant of the xy-plane bounded by the coordinate axes and the ellipse $\dfrac{x^2}{a^2} + \dfrac{y^2}{b^2} = 1$. The volume of the ellipsoid is

$$V = 8c\iint_E \sqrt{1 - \frac{x^2}{a^2} - \frac{y^2}{b^2}}\,dx\,dy.$$

Let $x = au$, $y = bv$. Then

$$dx\,dy = \left|\frac{\partial(x,y)}{\partial(u,v)}\right|du\,dv = ab\,du\,dv.$$

521

SECTION 14.4 (PAGE 846)

The region E corresponds to the quarter disk Q: $u^2 + v^2 \leq 1$, $u, v \geq 0$ in the uv-plane. Thus

$$V = 8abc \iint_Q \sqrt{1 - u^2 - v^2}\, du\, dv$$

$$= 8abc \times \left(\frac{1}{8} \times \text{volume of ball of radius } 1\right)$$

$$= \frac{4}{3}\pi abc \text{ cu. units.}$$

30. We use the same regions and change of variables as in the previous exercise. The required volume is

$$V = \iint_E \left(1 - \frac{x^2}{a^2} - \frac{y^2}{b^2}\right) dx\, dy$$

$$= ab \iint_Q (1 - u^2 - v^2)\, du\, dv.$$

Now transform to polar coordinates in the uv-plane: $u = r\cos\theta$, $v = r\sin\theta$.

$$V = ab \int_0^{\pi/2} d\theta \int_0^1 (1 - r^2) r\, dr$$

$$= \frac{\pi ab}{2}\left(\frac{r^2}{2} - \frac{r^4}{4}\right)\bigg|_0^1 = \frac{\pi ab}{8} \text{ cu. units.}$$

31. Let $x = \dfrac{u+v}{2}$, $y = \dfrac{u-v}{2}$, so that $x + y = u$ and $x - y = v$. We have

$$dx\, dy = \left|\begin{matrix} \frac{1}{2} & \frac{1}{2} \\ \frac{1}{2} & -\frac{1}{2} \end{matrix}\right| du\, dv = \frac{1}{2} du\, dv.$$

Under the above transformation the square $|x| + |y| \leq a$ corresponds to the square S: $-a \leq u \leq a$, $-a \leq v \leq a$. Thus

$$\iint_{|x|+|y|\leq a} e^{x+y}\, dA = \frac{1}{2}\iint_S e^u\, du\, dv$$

$$= \frac{1}{2}\int_{-a}^{a} e^u\, du \int_{-a}^{a} dv$$

$$= a(e^a - e^{-a}) = 2a \sinh a.$$

32. The parallelogram P bounded by $x + y = 1$, $x + y = 2$, $3x + 4y = 5$, and $3x + 4y = 6$ corresponds to the square S bounded by $u = 1$, $u = 2$, $v = 5$, and $v = 6$ under the transformation

$$u = x + y, \quad v = 3x + 4y,$$

or, equivalently,

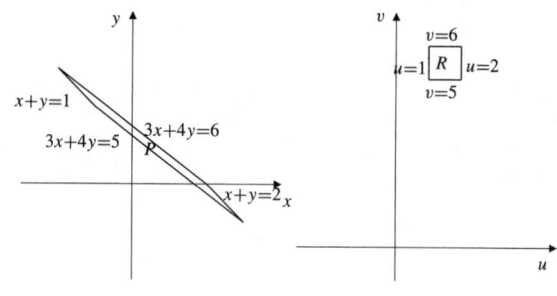

$$x = 4u - v, \quad y = v - 3u.$$

Fig. 14.4.32(a) Fig. 14.4.32(b)

We have

$$\frac{\partial(x, y)}{\partial(u, v)} = \left|\begin{matrix} 4 & -1 \\ -3 & 1 \end{matrix}\right| = 1,$$

so $dx\, dy = du\, dv$. Also

$$x^2 + y^2 = (4u - v)^2 + (v - 3u)^2 = 25u^2 - 14uv + 2v^2.$$

Thus we have

$$\iint_P (x^2 + y^2)\, dx\, dy = \iint_S (25u^2 - 14uv + 2v^2)\, du\, dv$$

$$= \int_1^2 du \int_5^6 (25u^2 - 14uv + 2v^2)\, dv = \frac{7}{2}.$$

33. Let $u = xy$, $v = y/x$. Then

$$\frac{\partial(u, v)}{\partial(x, y)} = \left|\begin{matrix} y & x \\ -y/x^2 & 1/x \end{matrix}\right| = 2\frac{y}{x} = 2v,$$

so that $\dfrac{\partial(x, y)}{\partial(u, v)} = \dfrac{1}{2v}$. The region D in the first quadrant of the xy-plane bounded by $xy = 1$, $xy = 4$, $y = x$, and $y = 2x$ corresponds to the rectangle R in the uv-plane bounded by $u = 1$, $u = 4$, $v = 1$, and $v = 2$. Thus the area of D is given by

$$\iint_D dx\, dy = \iint_R \frac{1}{2v}\, du\, dv$$

$$= \frac{1}{2}\int_1^4 du \int_1^2 \frac{dv}{v} = \frac{3}{2}\ln 2 \text{ sq. units.}$$

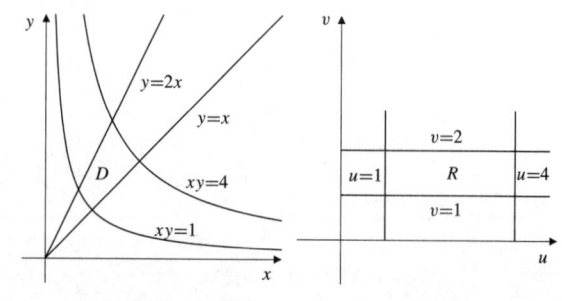

Fig. 14.4.33(a) Fig. 14.4.33(b)

34. Under the transformation $u = x^2 - y^2$, $v = xy$, the region R in the first quadrant of the xy-plane bounded by $y = 0$, $y = x$, $xy = 1$, and $x^2 - y^2 = 1$ corresponds to the square S in the uv-plane bounded by $u = 0$, $u = 1$, $v = 0$, and $v = 1$. Since

$$\frac{\partial(u,v)}{\partial(x,y)} = \begin{vmatrix} 2x & -2y \\ y & x \end{vmatrix} = 2(x^2 + y^2),$$

we therefore have

$$(x^2 + y^2)\,dx\,dy = \frac{1}{2}\,du\,dv.$$

Hence,

$$\iint_R (x^2 + y^2)\,dx\,dy = \iint_S \frac{1}{2}\,du\,dv = \frac{1}{2}.$$

35. $I = \iint_T e^{(y-x)/(y+x)}\,dA.$

a) $I = \int_0^{\pi/2} d\theta \int_0^{1/(\cos\theta+\sin\theta)} e^{\frac{\cos\theta-\sin\theta}{\sin\theta+\cos\theta}}\, r\,dr$

$= \frac{1}{2}\int_0^{\pi/2} e^{\frac{\cos\theta-\sin\theta}{\sin\theta+\cos\theta}} \frac{d\theta}{(\cos\theta+\sin\theta)^2}$

Let $u = \dfrac{\cos\theta - \sin\theta}{\sin\theta + \cos\theta}$

$du = -\dfrac{2\,d\theta}{(\sin\theta+\cos\theta)^2}$

$= \frac{1}{4}\int_{-1}^{1} e^u\,du = \frac{e - e^{-1}}{4}.$

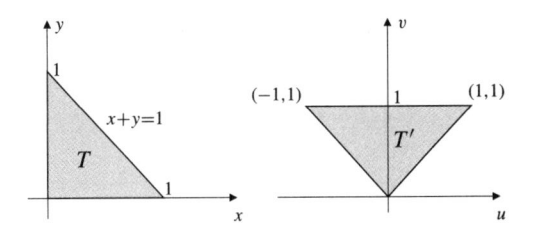

Fig. 14.4.35

b) If $u = y - x$, $v = y + x$ then

$$\frac{\partial(u,v)}{\partial(x,y)} = \begin{vmatrix} -1 & 1 \\ 1 & 1 \end{vmatrix} = -2,$$

so that $dA = dx\,dy = \dfrac{1}{2}\,du\,dv$. Also, T corresponds to the triangle T' bounded by $u = -v$, $u = v$, and $v = 1$. Thus

$$I = \frac{1}{2}\iint_{T'} e^{u/v}\,du\,dv$$

$= \dfrac{1}{2}\int_0^1 dv \int_{-v}^{v} e^{u/v}\,du$

$= \dfrac{1}{2}\int_0^1 dv\, \left(ve^{u/v}\right)\Big|_{-v}^{v}$

$= \dfrac{1}{2}(e - e^{-1})\int_0^1 v\,dv = \dfrac{e-e^{-1}}{4}.$

36. The region R whose area we must find is shown in part (a) of the figure. The change of variables $x = 3u$, $y = 2v$ maps the ellipse $4x^2 + 9y^2 = 36$ to the circle $u^2 + v^2 = 1$, and the line $2x + 3y = 1$ to the line $u + v = 1$. Thus it maps R to the region S in part (b) of the figure. Since

$$dx\,dy = \left|\begin{matrix} 3 & 0 \\ 0 & 2 \end{matrix}\right|\,du\,dv = 6\,du\,dv,$$

the area of R is

$$A = \iint_R dx\,dy = 6\iint_S du\,dv.$$

But the area of S is $(\pi/4) - (1/2)$, so $A = (3\pi/2) - 3$ square units.

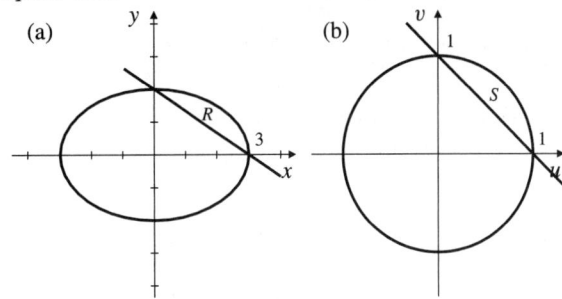

Fig. 14.4.36

37. $\text{Erf}(x) = \dfrac{2}{\sqrt{\pi}}\int_0^x e^{-t^2}\,dt = \dfrac{2}{\sqrt{\pi}}\int_0^x e^{-s^2}\,ds.$ Thus

$$\left(\text{Erf}(x)\right)^2 = \frac{4}{\pi}\iint_S e^{-(s^2+t^2)}\,ds\,dt,$$

where S is the square $0 \le s \le x$, $0 \le t \le x$. By symmetry,

$$\left(\text{Erf}(x)\right)^2 = \frac{8}{\pi}\iint_T e^{-(s^2+t^2)}\,ds\,dt,$$

where T is the triangle $0 \le s \le x$, $0 \le t \le s$.

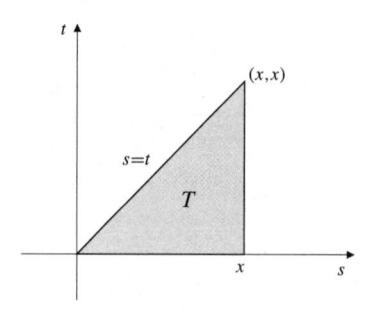

Fig. 14.4.37

Now transform to polar coordinates in the st-plane. We have

$$\left(\text{Erf}(x)\right)^2 = \frac{8}{\pi}\int_0^{\pi/4} d\theta \int_0^{x\sec\theta} e^{-r^2} r\, dr$$

$$= \frac{4}{\pi}\int_0^{\pi/4} d\theta \left(-e^{-r^2}\right)\Big|_0^{x\sec\theta}$$

$$= \frac{4}{\pi}\int_0^{\pi/4} \left(1 - e^{-x^2/\cos^2\theta}\right) d\theta.$$

Since $\cos^2\theta \le 1$, we have $e^{-x^2/\cos^2\theta} \le e^{-x^2}$, so

$$\left(\text{Erf}(x)\right)^2 \ge 1 - e^{-x^2}$$
$$\text{Erf}(x) \ge \sqrt{1 - e^{-x^2}}.$$

38. a) $\Gamma(x) = \displaystyle\int_0^\infty t^{x-1} e^{-t} dt \quad$ Let $t = s^2$
$$dt = 2s\, ds$$
$$= 2\int_0^\infty s^{2x-1} e^{-s^2} ds.$$

b) $\Gamma\left(\frac{1}{2}\right) = 2\displaystyle\int_0^\infty e^{-s^2} ds = 2\frac{\sqrt{\pi}}{2} = \sqrt{\pi}$

$\Gamma\left(\frac{3}{2}\right) = \frac{1}{2}\Gamma\left(\frac{1}{2}\right) = \frac{1}{2}\sqrt{\pi}.$

c) $B(x, y) = \displaystyle\int_0^1 t^{x-1}(1-t)^{y-1} dt \quad (x > 0,\ y > 0)$

let $t = \cos^2\theta,\ dt = -2\sin\theta\cos\theta\, d\theta$

$$= 2\int_0^{\pi/2} \cos^{2x-1}\theta \sin^{2y-1}\theta\, d\theta.$$

d) If Q is the first quadrant of the st-plane,

$$\Gamma(x)\Gamma(y) = \left(2\int_0^\infty s^{2x-1} e^{-s^2} ds\right)\left(2\int_0^\infty t^{2y-1} e^{-t^2} dt\right)$$

$$= 4\iint_Q s^{2x-1} t^{2y-1} e^{-(s^2+t^2)} ds\, dt$$

(change to polar coordinates)

$$= 4\int_0^{\pi/2} d\theta \int_0^\infty r^{2x-1}\cos^{2x-1}\theta\, r^{2y-1}\sin^{2y-1}\theta\, e^{-r^2} r\, dr$$

$$= \left(2\int_0^{\pi/2} \cos^{2x-1}\theta \sin^{2y-1}\theta\, d\theta\right)$$

$$\times \left(2\int_0^\infty r^{2(x+y)-1} e^{-r^2} dr\right)$$

$$= B(x,y)\Gamma(x+y) \qquad \text{by (a) and (c).}$$

Thus $B(x, y) = \dfrac{\Gamma(x)\Gamma(y)}{\Gamma(x+y)}.$

Section 14.5 Triple Integrals (page 854)

1. R is symmetric about the coordinate planes and has volume $8abc$. Thus

$$\iiint_R (1 + 2x - 3y)\, dV = \text{volume of } R + 0 - 0 = 8abc.$$

2. $\displaystyle\iiint_B xyz\, dV = \int_0^1 x\, dx \int_{-2}^0 y\, dy \int_1^4 z\, dz$

$$= \frac{1}{2}\left(-\frac{4}{2}\right)\left(\frac{16-1}{2}\right) = -\frac{15}{2}.$$

3. The hemispherical dome $x^2 + y^2 + z^2 \le 4,\ z \ge 0$, is symmetric about the planes $x = 0$ and $y = 0$. Therefore

$$\iiint_D (3 + 2xy)\, dV = 3\iiint_D dV + 2\iiint_D xy\, dV$$

$$= 3 \times \frac{2}{3}\pi(2^3) + 0 = 16\pi.$$

4. $\displaystyle\iiint_R x\, dV = \int_0^a x\, dx \int_0^{b\left(1-\frac{x}{a}\right)} dy \int_0^{c\left(1-\frac{x}{a}-\frac{y}{b}\right)} dz$

$$= c\int_0^a x\, dx \int_0^{b\left(1-\frac{x}{a}\right)} \left(1 - \frac{x}{a} - \frac{y}{b}\right) dy$$

$$= c\int_0^a x\left[b\left(1-\frac{x}{a}\right)^2 - \frac{b^2}{2b}\left(1-\frac{x}{a}\right)^2\right] dx$$

$$= \frac{bc}{2}\int_0^a \left(1-\frac{x}{a}\right)^2 x\, dx \quad \text{Let } u = 1 - (x/a)$$
$$\phantom{= \frac{bc}{2}\int_0^a \left(1-\frac{x}{a}\right)^2 x\, dx \quad} du = -(1/a)\, dx$$

$$= \frac{a^2 bc}{2}\int_0^1 u^2(1-u)\, du = \frac{a^2 bc}{24}.$$

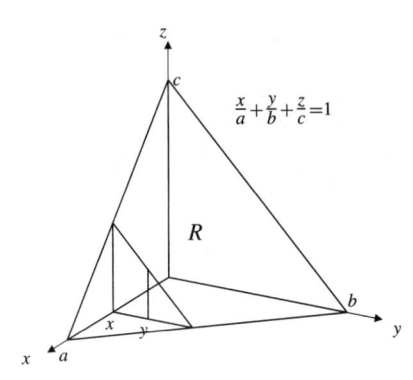

INSTRUCTOR'S SOLUTIONS MANUAL SECTION 14.5 (PAGE 854)

Fig. 14.5.4

5. R is the cube $0 \leq x, y, z \leq 1$. By symmetry,

$$\iiint_R (x^2 + y^2)\, dV = 2 \iiint_R x^2\, dV$$
$$= 2 \int_0^1 x^2\, dx \int_0^1 dy \int_0^1 dz = \frac{2}{3}.$$

6. As in Exercise 5,

$$\iiint_R (x^2 + y^2 + z^2)\, dV = 3 \iiint_R x^2\, dV = \frac{3}{3} = 1.$$

7. The set R: $0 \leq z \leq 1 - |x| - |y|$ is a pyramid, one quarter of which lies in the first octant and is bounded by the coordinate planes and the plane $x + y + z = 1$. (See the figure.) By symmetry, the integral of xy over R is 0. Therefore,

$$\iiint_R (xy + z^2)\, dV = \iiint_R z^2\, dV$$
$$= 4 \int_0^1 z^2\, dz \int_0^{1-z} dy \int_0^{1-z-y} dx$$
$$= 4 \int_0^1 z^2\, dz \int_0^{1-z} (1 - z - y)\, dy$$
$$= 4 \int_0^1 z^2 \left[(1-z)^2 - \frac{1}{2}(1-z)^2 \right] dz$$
$$= 2 \int_0^1 (z^2 - 2z^3 + z^4)\, dz = \frac{1}{15}.$$

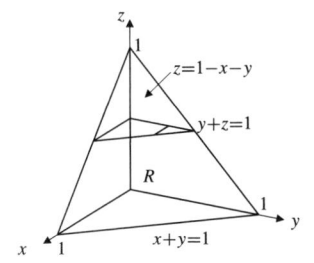

Fig. 14.5.7

8. R is the cube $0 \leq x, y, z \leq 1$. We have

$$\iiint_R yz^2 e^{-xyz}\, dV$$
$$= \int_0^1 z\, dz \int_0^1 dy \left(-e^{-xyz} \right) \Big|_{x=0}^{x=1}$$
$$= \int_0^1 z\, dz \int_0^1 (1 - e^{-yz})\, dy$$
$$= \int_0^1 z \left(1 + \frac{1}{z} e^{-yz} \Big|_{y=0}^{y=1} \right) dz$$
$$= \frac{1}{2} + \int_0^1 (e^{-z} - 1)\, dz$$
$$= \frac{1}{2} - 1 - e^{-z} \Big|_0^1 = \frac{1}{2} - \frac{1}{e}.$$

9. $\iiint_R \sin(\pi y^3)\, dV = \int_0^1 \sin(\pi y^3)\, dy \int_0^y dz \int_0^y dx$
$$= \int_0^1 y^2 \sin(\pi y^3)\, dy = -\frac{\cos(\pi y^3)}{3\pi} \Big|_0^1$$
$$= \frac{2}{3\pi}.$$

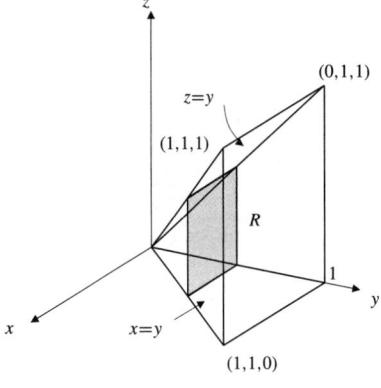

Fig. 14.5.9

10. $\iiint_R y\, dV = \int_0^1 y\, dy \int_{1-y}^1 dz \int_0^{2-y-z} dx$
$$= \int_0^1 y\, dy \int_{1-y}^1 (2 - y - z)\, dz$$
$$= \int_0^1 y\, dy \left((2-y)z - \frac{z^2}{2} \right) \Big|_{z=1-y}^{z=1}$$
$$= \int_0^1 y \left((2-y)y - \frac{1}{2}(1 - (1-y)^2) \right) dy$$
$$= \int_0^1 \frac{1}{2}(2y^2 - y^3)\, dy = \frac{5}{24}.$$

525

SECTION 14.5 (PAGE 854)

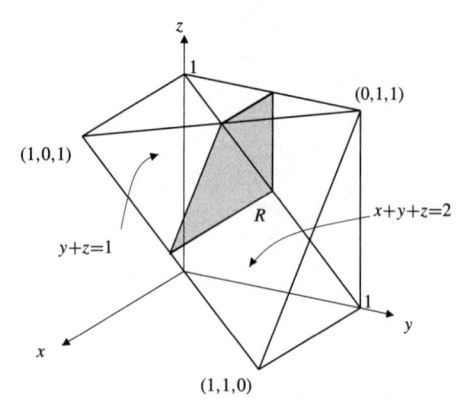

Fig. 14.5.10

11. R is bounded by $z = 1$, $z = 2$, $y = 0$, $y = z$, $x = 0$, and $x = y + z$. These bounds provide an iteration of the triple integral without our having to draw a diagram.

$$\iiint_R \frac{dV}{(x+y+z)^3}$$
$$= \int_1^2 dz \int_0^z dy \int_0^{y+z} \frac{dx}{(x+y+z)^3}$$
$$= \int_1^2 dz \int_0^z dy \left(\frac{-1}{2(x+y+z)^2} \right) \bigg|_{x=0}^{x=y+z}$$
$$= \frac{3}{8} \int_1^2 dz \int_0^z \frac{dy}{(y+z)^2}$$
$$= \frac{3}{8} \int_1^2 \left(\frac{-1}{y+z} \right) \bigg|_{y=0}^{y=z} dz$$
$$= \frac{3}{16} \int_1^2 \frac{dz}{z} = \frac{3}{16} \ln 2.$$

12. We have

$$\iiint_R \cos x \cos y \cos z \, dV$$
$$= \int_0^\pi \cos x \, dx \int_0^{\pi-x} \cos y \, dy \int_0^{\pi-x-y} \cos z \, dz$$
$$= \int_0^\pi \cos x \, dx \int_0^{\pi-x} \cos y \, dy \, (\sin z) \bigg|_{z=0}^{z=\pi-x-y}$$
$$= \int_0^\pi \cos x \, dx \int_0^{\pi-x} \cos y \sin(x+y) \, dy$$

recall that $\sin a \cos b = \frac{1}{2}(\sin(a+b) + \sin(a-b))$

$$= \int_0^\pi \cos x \, dx \int_0^{\pi-x} \frac{1}{2}\left[\sin(x+2y) + \sin x\right] dy$$
$$= \frac{1}{2} \int_0^\pi \cos x \, dx \left[-\frac{\cos(x+2y)}{2} + y \sin x\right]_{y=0}^{y=\pi-x}$$

R. A. ADAMS: CALCULUS

$$= \frac{1}{2} \int_0^\pi \left(-\frac{\cos x \cos(2\pi - x)}{2} + \frac{\cos^2 x}{2} + (\pi - x) \cos x \sin x \right) dx$$
$$= \frac{1}{2} \int_0^\pi \frac{\pi - x}{2} \sin 2x \, dx$$

$U = \pi - x \quad dV = \sin 2x \, dx$
$dU = -dx \quad V = -\frac{\cos 2x}{2}$

$$= \frac{1}{4} \left[-\frac{\pi - x}{2} \cos 2x \bigg|_0^\pi - \frac{1}{2} \int_0^\pi \cos 2x \, dx \right]$$
$$= \frac{1}{8} \left[\pi - \frac{\sin 2x}{2} \bigg|_0^\pi \right] = \frac{\pi}{8}.$$

13. By Example 4 of Section 14.4, $\int_{-\infty}^\infty e^{-u^2} du = \sqrt{\pi}$. If $k > 0$, let $u = \sqrt{k}t$, so that $du = \sqrt{k} \, dt$. Thus

$$\int_{-\infty}^\infty e^{-kt^2} dt = \sqrt{\frac{\pi}{k}}.$$

Thus

$$\iiint_{\mathbb{R}^3} e^{-x^2 - 2y^2 - 3z^2} dV$$
$$= \int_{-\infty}^\infty e^{-x^2} dx \int_{-\infty}^\infty e^{-2y^2} dy \int_{-\infty}^\infty e^{-3z^2} dz$$
$$= \sqrt{\pi} \sqrt{\frac{\pi}{2}} \sqrt{\frac{\pi}{3}} = \frac{\pi^{3/2}}{\sqrt{6}}.$$

14. Let E be the elliptic disk bounded by $x^2 + 4y^2 = 4$. Then E has area $\pi(2)(1) = 2\pi$ square units. The volume of the region of 3-space lying above E and beneath the plane $z = 2 + x$ is

$$V = \iint_E (2 + x) \, dA = 2 \iint_E dA = 4\pi \text{ cu. units,}$$

since $\iint_E x \, dA = 0$ by symmetry.

15. $$\iiint_T x \, dV = \int_0^1 x \, dx \int_{1-x}^1 dy \int_{2-x-y}^1 dz$$
$$= \int_0^1 x \, dx \int_{1-x}^1 (x + y - 1) \, dy$$
$$= \int_0^1 x \left[\frac{(x-1)^2}{2} + x - \frac{1}{2} \right] dx$$
$$= \int_0^1 \frac{x^3}{2} dx = \frac{1}{8}.$$

526

INSTRUCTOR'S SOLUTIONS MANUAL

SECTION 14.5 (PAGE 854)

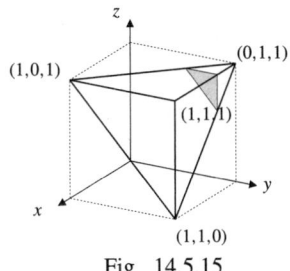

Fig. 14.5.15

16.

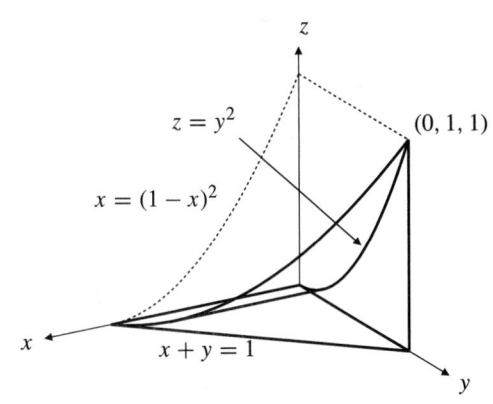

Fig. 14.5.16

$$\iiint_R f(x,y,z)\,dV = \int_0^1 dx \int_0^{1-x} dy \int_0^{y^2} f(x,y,z)\,dz$$
$$= \int_0^1 dy \int_0^{1-y} dx \int_0^{y^2} f(x,y,z)\,dz$$
$$= \int_0^1 dy \int_0^{y^2} dz \int_0^{1-y} f(x,y,z)\,dx$$
$$= \int_0^1 dz \int_{\sqrt{z}}^1 dy \int_0^{1-y} f(x,y,z)\,dx$$
$$= \int_0^1 dx \int_0^{(1-x)^2} dz \int_{\sqrt{z}}^{1-x} f(x,y,z)\,dy$$
$$= \int_0^1 dz \int_0^{1-\sqrt{z}} dx \int_{\sqrt{z}}^{1-x} f(x,y,z)\,dy.$$

17.
$$\int_0^1 dz \int_0^{1-z} dy \int_0^1 f(x,y,z)\,dx$$
$$= \iiint_R f(x,y,z)\,dV \qquad (R \text{ is the prism in the figure})$$
$$= \int_0^1 dx \int_0^1 dy \int_0^{1-y} f(x,y,z)\,dz.$$

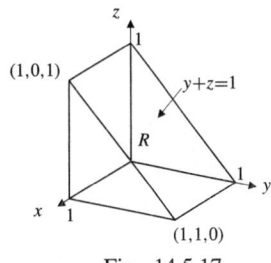

Fig. 14.5.17

18.
$$\int_0^1 dz \int_z^1 dy \int_0^y f(x,y,z)\,dx$$
$$= \iiint_R f(x,y,z)\,dV \qquad (R \text{ is the pyramid in the figure})$$
$$= \int_0^1 dx \int_x^1 dy \int_0^y f(x,y,z)\,dz.$$

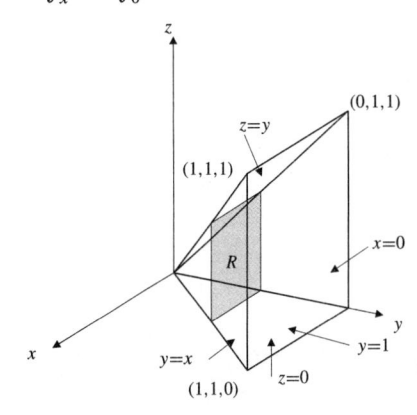

Fig. 14.5.18

19.
$$\int_0^1 dz \int_z^1 dx \int_0^{x-z} f(x,y,z)\,dy$$
$$= \iiint_R f(x,y,z)\,dV \qquad (R \text{ is the tetrahedron in the figure})$$
$$= \int_0^1 dx \int_0^x dy \int_0^{x-y} f(x,y,z)\,dz.$$

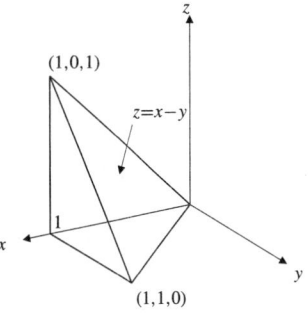

Fig. 14.5.19

527

SECTION 14.5 (PAGE 854)

20. $\int_0^1 dy \int_0^{\sqrt{1-y^2}} dz \int_{y^2+z^2}^1 f(x,y,z)\,dx$

    $= \iiint_R f(x,y,z)\,dV$ (R is the paraboloid in the figure)

    $= \int_0^1 dx \int_0^{\sqrt{x}} dy \int_0^{\sqrt{x-y^2}} f(x,y,z)\,dz.$

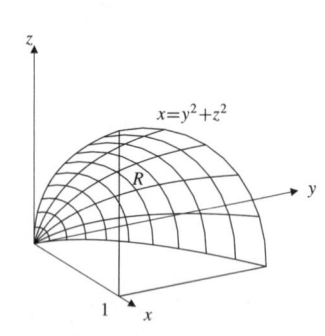

Fig. 14.5.20

21. $I = \int_0^1 dz \int_0^{1-z} dy \int_0^1 f(x,y,z)\,dx.$
    The given iteration corresponds to

    $0 \le z \le 1, \quad 0 \le y \le 1-z, \quad 0 \le x \le 1.$

    Thus $0 \le x \le 1$, $0 \le y \le 1-0 = 1$, $0 \le z \le 1-y$, and

    $I = \int_0^1 dx \int_0^1 dy \int_0^{1-y} f(x,y,z)\,dz.$

22. $I = \int_0^1 dz \int_z^1 dy \int_0^y f(x,y,z)\,dx.$
    The given iteration corresponds to

    $0 \le z \le 1, \quad z \le y \le 1, \quad 0 \le x \le y.$

    Thus $0 \le x \le 1$, $x \le y \le 1$, $0 \le z \le y$, and

    $I = \int_0^1 dx \int_x^1 dy \int_0^y f(x,y,z)\,dz.$

23. $I = \int_0^1 dz \int_z^1 dx \int_0^{x-z} f(x,y,z)\,dy.$
    The given iteration corresponds to

    $0 \le z \le 1, \quad z \le x \le 1, \quad 0 \le y \le x-z.$

    Thus $0 \le x \le 1$, $0 \le y \le x$, $0 \le z \le x-y$, and

    $I = \int_0^1 dx \int_0^x dy \int_0^{x-y} f(x,y,z)\,dz.$

24. $I = \int_0^1 dy \int_0^{\sqrt{1-y^2}} dz \int_{y^2+z^2}^1 f(x,y,z)\,dx.$
    The given iteration corresponds to

    $0 \le y \le 1, \quad 0 \le z \le \sqrt{1-y^2}, \quad y^2+z^2 \le x \le 1.$

    Thus $0 \le x \le 1$, $0 \le y \le \sqrt{x}$, $0 \le z \le \sqrt{x-y^2}$, and

    $I = \int_0^1 dx \int_0^{\sqrt{x}} dy \int_0^{\sqrt{x-y^2}} f(x,y,z)\,dz.$

25. $I = \int_0^1 dy \int_y^1 dz \int_0^z f(x,y,z)\,dx.$
    The given iteration corresponds to

    $0 \le y \le 1, \quad y \le z \le 1, \quad 0 \le x \le z.$

    Thus $0 \le x \le 1$, $x \le z \le 1$, $0 \le y \le z$, and

    $I = \int_0^1 dx \int_x^1 dz \int_0^z f(x,y,z)\,dy.$

26.

Fig. 14.5.26

$I = \int_0^1 dx \int_x^1 dy \int_x^y f(x,y,z)\,dz = \iiint_P f(x,y,z)\,dV,$

where P is the triangular pyramid (see the figure) with vertices at $(0,0,0)$, $(0,1,0)$, $(0,1,1)$, and $(1,1,1)$. If we we reiterate I to correspond to the horizontal slice shown then

$\int_0^1 dz \int_z^1 dy \int_0^z f(x,y,z)\,dx.$

27. $\int_0^1 dz \int_z^1 dx \int_0^x e^{x^3} dy$

$= \iiint_R e^{x^3} dV$ (R is the pyramid in the figure)

$= \int_0^1 e^{x^3} dx \int_0^x dy \int_0^x dz$

$= \int_0^1 x^2 e^{x^3} dx = \dfrac{e-1}{3}.$

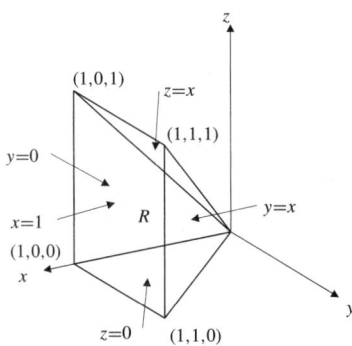

Fig. 14.5.27

28. $\int_0^1 dx \int_0^{1-x} dy \int_y^1 \dfrac{\sin(\pi z)}{z(2-z)} dz$

$= \iiint_R \dfrac{\sin(\pi z)}{z(2-z)} dV$ (R is the pyramid in the figure)

$= \int_0^1 \dfrac{\sin(\pi z)}{z(2-z)} dz \int_0^z dy \int_0^{1-y} dx$

$= \int_0^1 \dfrac{\sin(\pi z)}{z(2-z)} dz \int_0^z (1-y) dy$

$= \int_0^1 \dfrac{\sin(\pi z)}{z(2-z)} \left(z - \dfrac{z^2}{2}\right) dz$

$= \dfrac{1}{2} \int_0^1 \sin(\pi z) dz = \dfrac{1}{\pi}.$

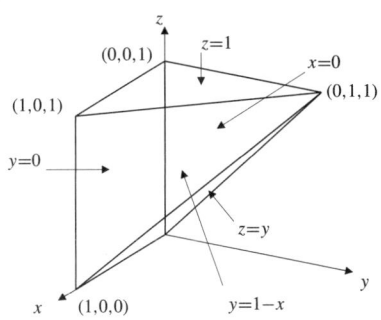

Fig. 14.5.28

29. The average value of $f(x, y, z)$ over R is

$$\overline{f} = \dfrac{1}{\text{volume of } R} \iiint_R f(x, y, z) dV.$$

If $f(x, y, z) = x^2 + y^2 + z^2$ and R is the cube $0 \le x, y, z \le 1$, then, by Exercise 6,

$$\overline{f} = \dfrac{1}{1} \iiint_R (x^2 + y^2 + z^2) dV = 1.$$

30. If the function $f(x, y, z)$ is continuous on a closed, bounded, connected set D in 3-space, then there exists a point (x_0, y_0, z_0) in D such that

$$\iiint_D f(x, y, z) dV = f(x_0, y_0, z_0) \times (\text{volume of } D).$$

Apply this with $D = B_\epsilon(a, b, c)$, which has volume $\dfrac{4}{3}\pi\epsilon^3$, to get

$$\iiint_{B_\epsilon(a,b,c)} f(x, y, z) dV = f(x_0, y_0, z_0) \dfrac{4}{3}\pi\epsilon^3$$

for some (x_0, y_0, z_0) in $B_\epsilon(a, b, c)$. Thus

$$\lim_{\epsilon \to 0} \dfrac{3}{4\pi\epsilon^3} \iiint_{B_\epsilon(a,b,c)} f(x, y, z) dV$$

$$= \lim_{\epsilon \to 0} f(x_0, y_0, z_0) = f(a, b, c)$$

since f is continuous at (a, b, c).

Section 14.6 Change of Variables in Triple Integrals (page 863)

1. Spherical: $[4, \pi/3, 2\pi/3]$;
   Cartesian: $(-\sqrt{3}, 3, -2)$; Cylindrical: $[2\sqrt{3}, 2\pi/3, 2]$.

2. Cartesian: $(2, -2, 1)$;
   Cylindrical: $[2\sqrt{2}, -\pi/4, 1]$;
   Spherical: $[3, \cos^{-1}(1/3), -\pi/4]$.

3. Cylindrical: $[2, \pi/6, -2]$;
   Cartesian: $(\sqrt{3}, 1, -2]$; Spherical: $[2\sqrt{2}, 3\pi/4, \pi/6]$.

4. Spherical: $[1, \phi, \theta]$; Cylindrical: $[r, \pi/4, r]$.

   $x = \sin\phi \cos\theta = r \cos\pi/4 = r/\sqrt{2}$
   $y = \sin\phi \sin\theta = r \sin\pi/4 = r/\sqrt{2}$
   $z = \cos\phi = r.$

   Thus $x = y$, $\theta = \pi/4$, and $r = \sin\phi = \cos\phi$. Hence $\phi = \pi/4$, so $r = 1/\sqrt{2}$. Finally: $x = y = 1/2$, $z = 1/\sqrt{2}$. Cartesian: $(1/2, 1/2, 1/\sqrt{2})$.

5. $\theta = \pi/2$ represents the half-plane $x = 0$, $y > 0$.

6. $\phi = 2\pi/3$ represents the lower half of the right-circular cone with vertex at the origin, axis along the z-axis, and semi-vertical angle $\pi/3$. Its Cartesian equation is $z = -\sqrt{(x^2 + y^2)/3}$.

7. $\phi = \pi/2$ represents the xy-plane.

8. $\rho = 4$ represents the sphere of radius 4 centred at the origin.

9. $r = 4$ represents the circular cylinder of radius 4 with axis along the z-axis.

10. $\rho = z$ represents the positive half of the z-axis.

11. $\rho = r$ represents the xy-plane.

12. $\rho = 2x$ represents the half-cone with vertex at the origin, axis along the positive x-axis, and semi-vertical angle $\pi/3$. Its Cartesian equation is $x = \sqrt{(y^2 + z^2)/3}$.

13. If $\rho = 2\cos\phi$, then $\rho^2 = 2\rho\cos\phi$, so
$$x^2 + y^2 + z^2 = 2z$$
$$x^2 + y^2 + z^2 - 2z + 1 = 1$$
$$x^2 + y^2 + (z-1)^2 = 1.$$

Thus $\rho = 2\cos\phi$ represents the sphere of radius 1 centred at $(0, 0, 1)$.

14. $r = 2\cos\theta \Rightarrow x^2 + y^2 = r^2 = 2r\cos\theta = 2x$, or $(x-1)^2 + y^2 = 1$. Thus the given equation represents the circular cylinder of radius 1 with axis along the vertical line $x = 1$, $y = 0$.

15. $V = \int_0^{2\pi} d\theta \int_0^{\pi/4} \sin\phi\, d\phi \int_0^a R^2\, dR$
$= \dfrac{2\pi a^3}{3}\left(1 - \dfrac{1}{\sqrt{2}}\right)$ cu. units.

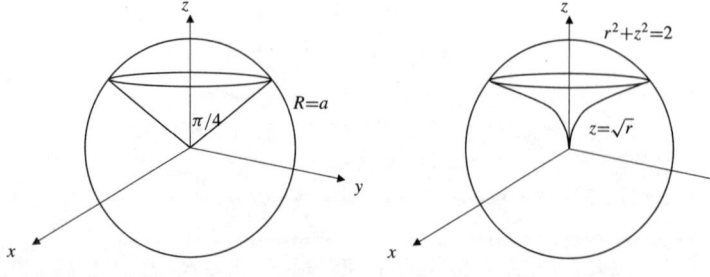

Fig. 14.6.15   Fig. 14.6.16

16. The surface $z = \sqrt{r}$ intersects the sphere $r^2 + z^2 = 2$ where $r^2 + r - 2 = 0$. This equation has positive root $r = 1$. The required volume is

$$V = \int_0^{2\pi} d\theta \int_0^1 r\, dr \int_{\sqrt{r}}^{\sqrt{2-r^2}} dz$$
$$= \int_0^{2\pi} d\theta \int_0^1 (\sqrt{2-r^2} - \sqrt{r})r\, dr$$
$$= 2\pi \left(\int_0^1 r\sqrt{2-r^2}\, dr - \dfrac{2}{5}\right) \quad \text{Let } u = 2 - r^2$$
$$\hspace{6cm} du = -2r\, dr$$
$$= \pi \int_1^2 u^{1/2}\, du - \dfrac{4\pi}{5}$$
$$= \dfrac{2\pi}{3}(2\sqrt{2} - 1) - \dfrac{4\pi}{5} = \dfrac{4\sqrt{2}\pi}{3} - \dfrac{22\pi}{15} \text{ cu. units.}$$

17. The paraboloids $z = 10 - r^2$ and $z = 2(r^2 - 1)$ intersect where $r^2 = 4$, that is, where $r = 2$. The volume lying between these surfaces is

$$V = \int_0^{2\pi} d\theta \int_0^2 [10 - r^2 - 2(r^2 - 1)]r\, dr$$
$$= 2\pi \int_0^2 (12r - 3r^3)\, dr = 24\pi \text{ cu. units.}$$

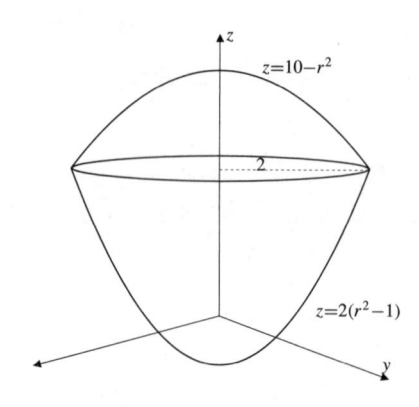

Fig. 14.6.17

18. The paraboloid $z = r^2$ intersects the sphere $r^2 + z^2 = 12$ where $r^4 + r^2 - 12 = 0$, that is, where $r = \sqrt{3}$. The required volume is

$$V = \int_0^{2\pi} d\theta \int_0^{\sqrt{3}} (\sqrt{12 - r^2} - r^2)r\, dr$$
$$= 2\pi \int_0^{\sqrt{3}} r\sqrt{12 - r^2}\, dr - \dfrac{9\pi}{2} \quad \text{Let } u = 12 - r^2$$
$$\hspace{6cm} du = -2r\, dr$$
$$= \pi \int_9^{12} u^{1/2}\, du - \dfrac{9\pi}{2}$$
$$= \dfrac{2\pi}{3}(12^{3/2} - 27) - \dfrac{9\pi}{2} = 16\sqrt{3}\pi - \dfrac{45\pi}{2} \text{ cu. units.}$$

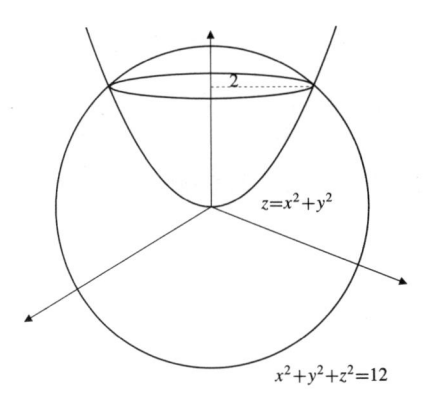

INSTRUCTOR'S SOLUTIONS MANUAL SECTION 14.6 (PAGE 863)

Fig. 14.6.18

19. One half of the required volume V lies in the first octant, inside the cylinder with polar equation $r = 2a\sin\theta$. Thus

$$V = 2\int_0^{\pi/2} d\theta \int_0^{2a\sin\theta} (2a-r)r\,dr$$

$$= 2a\int_0^{\pi/2} 4a^2\sin^2\theta\,d\theta - \frac{2}{3}\int_0^{\pi/2} 8a^3\sin^3\theta\,d\theta$$

$$= 4a^3\int_0^{\pi/2}(1-\cos 2\theta)\,d\theta - \frac{16a^3}{3}\int_0^{\pi/2}\sin^3\theta\,d\theta$$

$$= 2\pi a^3 - \frac{32a^3}{9} \text{ cu. units.}$$

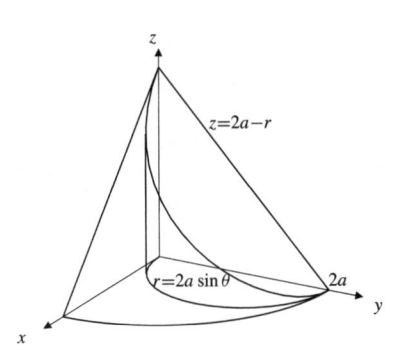

Fig. 14.6.19

20. The required volume V lies above $z=0$, below $z = 1-r^2$, and between $\theta = -\pi/4$ and $\theta = \pi/3$. Thus

$$V = \int_{-\pi/4}^{\pi/3} d\theta \int_0^1 (1-r^2)r\,dr$$

$$= \frac{7\pi}{12}\left(\frac{1}{2}-\frac{1}{4}\right) = \frac{7\pi}{48} \text{ cu. units.}$$

21. Let R be the region in the first octant, inside the ellipsoid

$$\frac{x^2}{a^2} + \frac{y^2}{b^2} + \frac{z^2}{c^2} = 1,$$

and between the planes $y=0$ and $y=x$. Under the transformation

$$x = au, \qquad y = bv, \qquad z = cw,$$

R corresponds to the region S in the first octant of uvw-space, inside the sphere

$$u^2 + v^2 + w^2 = 1,$$

and between the planes $v=0$ and $bv = au$. Therefore, the volume of R is

$$V = \iiint_R dx\,dy\,dz = abc\iiint_S du\,dv\,dw.$$

Using spherical coordinates in uvw-space, S corresponds to

$$0 \le R \le 1, \qquad 0 \le \phi \le \frac{\pi}{2}, \qquad 0 \le \theta \le \tan^{-1}\frac{a}{b}.$$

Thus

$$V = abc\int_0^{\tan^{-1}(a/b)} d\theta \int_0^{\pi/2}\sin\phi\,d\phi \int_0^1 R^2\,dR$$

$$= \frac{1}{3}abc\tan^{-1}\frac{a}{b} \text{ cu. units.}$$

22. One eighth of the required volume V lies in the first octant. Call this region R. Under the transformation

$$x = au, \qquad y = bv, \qquad z = cw,$$

R corresponds to the region S in the first octant of uvw-space bounded by $w = 0$, $w = 1$, and $u^2 + v^2 - w^2 = 1$. Thus

$$V = 8abc \times \text{(volume of } S\text{)}.$$

The volume of S can be determined by using horizontal slices:

$$V = 8abc\int_0^1 \frac{\pi}{4}(1+w^2)\,dw = \frac{8}{3}\pi abc \text{ cu. units.}$$

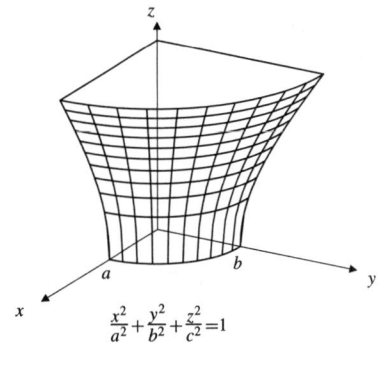

Fig. 14.6.22

531

23. Let $x = au$, $y = bv$, $z = w$. The indicated region R corresponds to the region S above the uv-plane and below the surface $w = 1 - u^2 - v^2$. We use polar coordinates in the uv-plane to calculate the volume V of R:

$$V = \iiint_R dV = ab \iiint_S du\, dv\, dw$$
$$= ab \int_0^{2\pi} d\theta \int_0^1 (1-r^2) r\, dr = \frac{\pi ab}{2} \text{ cu. units.}$$

24. $\iiint_R (x^2 + y^2 + z^2)\, dV$
$$= \int_0^{2\pi} d\theta \int_0^a r\, dr \int_0^h (r^2 + z^2)\, dz$$
$$= 2\pi \int_0^a \left(r^3 h + \frac{1}{3} r h^3 \right) dr$$
$$= 2\pi \left(\frac{a^4 h}{4} + \frac{a^2 h^3}{6} \right) = \frac{\pi a^4 h}{2} + \frac{\pi a^2 h^3}{3}.$$

25. $\iiint_B (x^2 + y^2)\, dV$
$$= \int_0^{2\pi} d\theta \int_0^{\pi} \sin\phi\, d\phi \int_0^a R^2 \sin^2\phi\, R^2\, dR$$
$$= 2\pi \int_0^{\pi} \sin^3\phi\, d\phi \int_0^a R^4\, dR$$
$$= 2\pi \left(\frac{4}{3} \right) \frac{a^5}{5} = \frac{8\pi a^5}{15}.$$

26. $\iiint_B (x^2 + y^2 + z^2)\, dV$
$$= \int_0^{2\pi} d\theta \int_0^{\pi} \sin\phi\, d\phi \int_0^a R^4\, dR = \frac{4\pi a^5}{5}.$$

27. $\iiint_R (x^2 + y^2 + z^2)\, dV$
$$= \int_0^{2\pi} d\theta \int_0^{\tan^{-1}(1/c)} \sin\phi\, d\phi \int_0^a R^4\, dR$$
$$= \frac{2\pi a^5}{5} \left[1 - \cos\left(\tan^{-1} \frac{1}{c} \right) \right] = \frac{2\pi a^5}{5} \left(1 - \frac{c}{\sqrt{c^2+1}} \right).$$

28. $\iiint_R (x^2 + y^2)\, dV$
$$= \int_0^{2\pi} d\theta \int_0^{\tan^{-1}(1/c)} \sin^3\phi\, d\phi \int_0^a R^4\, dR$$
$$= \frac{2\pi a^5}{5} \int_0^{\tan^{-1}(1/c)} \sin\phi (1 - \cos^2\phi)\, d\phi \quad \text{Let } u = \cos\phi$$
$$\qquad\qquad\qquad\qquad\qquad\qquad\qquad\qquad du = -\sin\phi\, d\phi$$
$$= \frac{2\pi a^5}{5} \int_{c/\sqrt{c^2+1}}^1 (1 - u^2)\, du$$
$$= \frac{2\pi a^5}{5} \left(u - \frac{u^3}{3} \right) \Big|_{c/\sqrt{c^2+1}}^1$$
$$= \frac{2\pi a^5}{5} \left(\frac{2}{3} - \frac{c}{\sqrt{c^2+1}} + \frac{c^3}{3(c^2+1)^{3/2}} \right).$$

29. $z = r^2$ and $z = \sqrt{2 - r^2}$ intersect where $r^4 + r^2 - 2 = 0$, that is, on the cylinder $r = 1$. Thus

$$\iiint_R z\, dV = \int_0^{2\pi} d\theta \int_0^1 r\, dr \int_{r^2}^{\sqrt{2-r^2}} z\, dz$$
$$= \pi \int_0^1 (2 - r^2 - r^4) r\, dr = \frac{7\pi}{12}.$$

30. By symmetry, both integrals have the same value:

$$\iiint_R x\, dV = \iiint_R z\, dV$$
$$= \int_0^{\pi/2} d\theta \int_0^{\pi/2} \cos\phi \sin\phi\, d\phi \int_0^a R^3\, dR$$
$$= \frac{\pi}{2} \left(\frac{1}{2} \right) \frac{a^4}{4} = \frac{\pi a^4}{16}.$$

31. $\iiint_R x\, dV = \int_0^{\pi/2} d\theta \int_0^a r\, dr \int_0^{h(1-(r/a))} r\cos\theta\, dz$
$$= h \int_0^{\pi/2} \cos\theta\, d\theta \int_0^a r^2 \left(1 - \frac{r}{a} \right) dr = \frac{ha^3}{12},$$
$$\iiint_R z\, dV = \int_0^{\pi/2} d\theta \int_0^a r\, dr \int_0^{h(1-(r/a))} z\, dz$$
$$= \frac{\pi h^2}{4} \int_0^a \left(1 - \frac{r}{a} \right)^2 r\, dr$$
$$= \frac{\pi h^2}{4} \left(\frac{r^2}{2} - \frac{2r^3}{3a} + \frac{r^4}{4a^2} \right) \Big|_0^a = \frac{\pi a^2 h^2}{48}.$$

32. If
$$x = au, \qquad y = bv, \qquad z = cw,$$
then the volume of a region R in xyz-space is abc times the volume of the corresponding region S in uvw-space.

If R is the region inside the ellipsoid
$$\frac{x^2}{a^2}+\frac{y^2}{b^2}+\frac{z^2}{c^2}=1$$
and above the plane $y+z=b$, then the corresponding region S lies inside the sphere
$$u^2+v^2+w^2=1$$
and above the plane $bv+cw=b$. The distance from the origin to this plane is
$$D=\frac{b}{\sqrt{b^2+c^2}} \quad \text{(assuming } b>0\text{)}$$
by Example 7 of Section 10.4. By symmetry, the volume of S is equal to the volume lying inside the sphere $u^2+v^2+w^2=1$ and above the plane $w=D$. We calculate this latter volume by slicing; it is
$$\pi\int_D^1 (1-w^2)\,dw = \pi\left(w-\frac{w^3}{3}\right)\bigg|_D^1$$
$$=\pi\left(\frac{2}{3}-D+\frac{D^3}{3}\right).$$
Hence, the volume of R is
$$\pi abc\left(\frac{2}{3}-\frac{b}{\sqrt{b^2+c^2}}+\frac{b^3}{3(b^2+c^2)^{3/2}}\right) \text{ cu. units.}$$

33. By Example 10 of Section 12.5, we know that
$$\frac{\partial^2 u}{\partial x^2}+\frac{\partial^2 u}{\partial y^2}=\frac{\partial^2 u}{\partial r^2}+\frac{1}{r}\frac{\partial u}{\partial r}+\frac{1}{r^2}\frac{\partial^2 u}{\partial \theta^2}.$$
The required result follows if we add $\frac{\partial^2 u}{\partial z^2}$ to both sides.

34. Cylindrical and spherical coordinates are related by
$$z=\rho\cos\phi, \qquad r=\rho\sin\phi.$$
(The θ coordinates are identical in the two systems.) Observe that z, r, ρ, and ϕ play, respectively, the same roles that x, y, r, and θ play in the transformation from Cartesian to polar coordinates in the plane. We can exploit this correspondence to avoid repeating the calculations of partial derivatives of a function u, since the results correspond to calculations made (for a function z) in Example 10 of Section 12.5. Comparing with the calculations in that Example, we have
$$\frac{\partial u}{\partial \rho}=\cos\phi\frac{\partial u}{\partial z}+\sin\phi\frac{\partial u}{\partial r}$$
$$\frac{\partial u}{\partial \phi}=-\rho\sin\phi\frac{\partial u}{\partial z}+\rho\cos\phi\frac{\partial u}{\partial r}$$
$$\frac{\partial^2 u}{\partial \rho^2}=\cos^2\phi\frac{\partial^2 u}{\partial z^2}+2\cos\phi\sin\phi\frac{\partial^2 u}{\partial z\partial r}+\sin^2\phi\frac{\partial^2 u}{\partial r^2}$$
$$\frac{\partial^2 u}{\partial \phi^2}=-\rho\frac{\partial u}{\partial \rho}+\rho^2\bigg(\sin^2\phi\frac{\partial^2 u}{\partial z^2}$$
$$-2\cos\phi\sin\phi\frac{\partial^2 u}{\partial z\partial r}+\cos^2\phi\frac{\partial^2 u}{\partial r^2}\bigg).$$

Substituting these expressions into the expression for Δu given in the statement of this exercise in terms of spherical coordinates, we obtain the expression in terms of cylindrical coordinates established in the previous exercise:
$$\frac{\partial^2 u}{\partial \rho^2}+\frac{2}{\rho}\frac{\partial u}{\partial \rho}+\frac{\cot\phi}{\rho^2}\frac{\partial u}{\partial \phi}+\frac{1}{\rho^2}\frac{\partial^2 u}{\partial \phi^2}+\frac{1}{\rho^2\sin^2\phi}\frac{\partial^2 u}{\partial \theta^2}$$
$$=\frac{\partial^2 u}{\partial r^2}+\frac{1}{r}\frac{\partial u}{\partial r}+\frac{1}{r^2}\frac{\partial^2 u}{\partial \theta^2}+\frac{\partial^2 u}{\partial z^2}$$
$$=\frac{\partial^2 u}{\partial x^2}+\frac{\partial^2 u}{\partial y^2}=\Delta u$$
by Exercise 33.

35. Consider the transformation
$$x=x(u,v,w),\quad y=y(u,v,w),\quad z=z(u,v,w),$$
and let P be the point in xyz-space corresponding to $u=a$, $v=b$, $w=c$. Fixing $v=b$, $w=c$, results in a parametric curve (with parameter u) through P. The vector
$$\overrightarrow{PQ}=\frac{\partial x}{\partial u}\mathbf{i}+\frac{\partial y}{\partial u}\mathbf{j}+\frac{\partial z}{\partial u}\mathbf{k}$$
and corresponding vectors
$$\overrightarrow{PR}=\frac{\partial x}{\partial v}\mathbf{i}+\frac{\partial y}{\partial v}\mathbf{j}+\frac{\partial z}{\partial v}\mathbf{k}$$
$$\overrightarrow{PS}=\frac{\partial x}{\partial w}\mathbf{i}+\frac{\partial y}{\partial w}\mathbf{j}+\frac{\partial z}{\partial w}\mathbf{k}$$
span a parallelepiped in xyz-space corresponding to a rectangular box with volume $du\,dv\,dw$ in uvw-space. The parallelepiped has volume
$$|(\overrightarrow{PQ}\times\overrightarrow{PR})\bullet\overrightarrow{PS}|=\left|\frac{\partial(x,y,z)}{\partial(u,v,w)}\right|du\,dv\,dw.$$
Thus
$$dV=dx\,dy\,dz=\left|\frac{\partial(x,y,z)}{\partial(u,v,w)}\right|du\,dv\,dw.$$

Section 14.7 Applications of Multiple Integrals (page 872)

1. $z = 2x + 2y$, $\dfrac{\partial z}{\partial x} = 2 = \dfrac{\partial z}{\partial y}$

$dS = \sqrt{1 + 2^2 + 2^2}\, dA = 3\, dA$

$S = \iint_{x^2+y^2\leq 1} 3\, dA = 3\pi(1^2) = 3\pi$ sq. units.

2. $z = (3x - 4y)/5$, $\dfrac{\partial z}{\partial x} = \dfrac{3}{5}$, $\dfrac{\partial z}{\partial y} = \dfrac{4}{5}$

$dS = \sqrt{1 + \dfrac{3^2 + 4^2}{5^2}}\, dA = \sqrt{2}\, dA$

$S = \iint_{(x/2)^2 + y^2 \leq 1} \sqrt{2}\, dA = \sqrt{2}\pi(2)(1) = 2\sqrt{2}\pi$ sq. units.

3. $z = \sqrt{a^2 - x^2 - y^2}$

$\dfrac{\partial z}{\partial x} = -\dfrac{x}{\sqrt{a^2 - x^2 - y^2}}$, $\dfrac{\partial z}{\partial y} = -\dfrac{y}{\sqrt{a^2 - x^2 - y^2}}$

$dS = \sqrt{1 + \dfrac{x^2 + y^2}{a^2 - x^2 - y^2}}\, dA = \dfrac{a}{\sqrt{a^2 - x^2 - y^2}}\, dA$

$S = \iint_{x^2 + y^2 \leq a^2} \dfrac{a\, dA}{\sqrt{a^2 - x^2 - y^2}}$ (use polars)

$= a\int_0^{2\pi} d\theta \int_0^a \dfrac{r\, dr}{\sqrt{a^2 - r^2}}$ Let $u = a^2 - r^2$, $du = -2r\, dr$

$= \pi a \int_0^{a^2} u^{-1/2}\, du = 2\pi a^2$ sq. units.

4. $z = 2\sqrt{1 - x^2 - y^2}$

$\dfrac{\partial z}{\partial x} = -\dfrac{2x}{\sqrt{1 - x^2 - y^2}}$, $\dfrac{\partial z}{\partial y} = -\dfrac{2y}{\sqrt{1 - x^2 - y^2}}$

$dS = \sqrt{1 + \dfrac{4(x^2 + y^2)}{1 - x^2 - y^2}}\, dA = \sqrt{\dfrac{1 + 3(x^2 + y^2)}{1 - x^2 - y^2}}\, dA$

$S = \iint_{x^2 + y^2 \leq 1} dS$

$= \int_0^{2\pi} d\theta \int_0^1 \sqrt{\dfrac{1 + 3r^2}{1 - r^2}}\, r\, dr$ Let $u^2 = 1 - r^2$, $u\, du = -r\, dr$

$= 2\pi \int_0^1 \sqrt{4 - 3u^2}\, du$ Let $\sqrt{3}u = 2\sin v$, $\sqrt{3}\, du = 2\cos v\, dv$

$= 2\pi \int_0^{\pi/3} (2\cos^2 v) \dfrac{2\, dv}{\sqrt{3}}$

$= \dfrac{4\pi}{\sqrt{3}} \int_0^{\pi/3} (1 + \cos 2v)\, dv$

$= \dfrac{4\pi}{\sqrt{3}}\left(v + \dfrac{\sin 2v}{2}\right)\Big|_0^{\pi/3} = \dfrac{4\pi^2}{3\sqrt{3}} + \pi$ sq. units.

5. $3z^2 = x^2 + y^2$, $6z\dfrac{\partial z}{\partial x} = 2x$, $\dfrac{\partial z}{\partial x} = \dfrac{x}{3z}$, $\dfrac{\partial z}{\partial y} = \dfrac{y}{3z}$

$dS = \sqrt{1 + \dfrac{x^2 + y^2}{9z^2}}\, dA = \sqrt{\dfrac{9z^2 + 3z^2}{9z^2}}\, dA = \dfrac{2}{\sqrt{3}}\, dA$

$S = \iint_{x^2 + y^2 \leq 12} \dfrac{2}{\sqrt{3}}\, dA = \dfrac{2}{\sqrt{3}}\pi(12) = \dfrac{24\pi}{\sqrt{3}}$ sq. units.

6. $z = 1 - x^2 - y^2$, $\dfrac{\partial z}{\partial x} = -2x$, $\dfrac{\partial z}{\partial y} = -2y$

$dS = \sqrt{1 + 4x^2 + 4y^2}\, dA$

$S = \iint_{x^2 + y^2 \leq 1,\ x \geq 0,\ y \geq 0} \sqrt{1 + 4(x^2 + y^2)}\, dA$

$= \int_0^{\pi/2} d\theta \int_0^1 \sqrt{1 + 4r^2}\, r\, dr$ Let $u = 1 + 4r^2$, $du = 8r\, dr$

$= \dfrac{\pi}{16} \int_1^5 u^{1/2}\, du$

$= \dfrac{\pi}{16}\left(\dfrac{2}{3} u^{3/2}\right)\Big|_1^5 = \dfrac{\pi(5\sqrt{5} - 1)}{24}$ sq. units.

7. The triangle is defined by $0 \leq y \leq 1$, $0 \leq x \leq y$.

$z = y^2$, $\dfrac{\partial z}{\partial y} = 2y$, $dS = \sqrt{1 + 4y^2}\, dA$

$S = \int_0^1 dy \int_0^y \sqrt{1 + 4y^2}\, dx$

$= \int_0^1 y\sqrt{1 + 4y^2}\, dy$ Let $u = 1 + 4y^2$, $du = 8y\, dy$

$= \dfrac{1}{8} \int_1^5 u^{1/2}\, du = \dfrac{1}{8}\left(\dfrac{2}{3}u^{3/2}\right)\Big|_1^5 = \dfrac{5\sqrt{5} - 1}{12}$ sq. units.

8. $z = \sqrt{x}$, $\dfrac{\partial z}{\partial x} = \dfrac{1}{2\sqrt{x}}$, $dS = \sqrt{1 + \dfrac{1}{4x}}\, dA$

$S = \int_0^1 dx \int_0^{\sqrt{x}} \sqrt{1 + \dfrac{1}{4x}}\, dy = \int_0^1 \sqrt{\dfrac{4x + 1}{4x}}\, \sqrt{x}\, dx$

$= \dfrac{1}{2} \int_0^1 \sqrt{4x + 1}\, dx$ Let $u = 4x + 1$, $du = 4\, dx$

$= \dfrac{1}{8} \int_1^5 u^{1/2}\, du = \dfrac{1}{8}\left(\dfrac{2}{3}u^{3/2}\right)\Big|_1^5 = \dfrac{5\sqrt{5} - 1}{12}$ sq. units.

9. $z^2 = 4 - x^2$, $2z\dfrac{\partial z}{\partial x} = -2x$, $\dfrac{\partial z}{\partial x} = -\dfrac{x}{z}$

$dS = \sqrt{1 + \dfrac{x^2}{z^2}}\, dA = \dfrac{2}{z}\, dA = \dfrac{2}{\sqrt{4-x^2}}\, dA$

(since $z \geq 0$ on the part of the surface whose area we want to find)

$S = \displaystyle\int_0^2 dx \int_0^x \dfrac{2}{\sqrt{4-x^2}}\, dy$

$= \displaystyle\int_0^2 \dfrac{2x}{\sqrt{4-x^2}}\, dx \quad$ Let $u = 4 - x^2$, $du = -2x\, dx$

$= \displaystyle\int_0^4 u^{-1/2}\, du = 2\sqrt{u}\,\Big|_0^4 = 4$ sq. units.

10. The area elements on $z = 2xy$ and $z = x^2 + y^2$, respectively, are

$dS_1 = \sqrt{1 + (2y)^2 + (2x)^2}\, dA = \sqrt{1 + 4x^2 + 4y^2}\, dx\, dy$,

$dS_2 = \sqrt{1 + (2x)^2 + (2y)^2}\, dA = \sqrt{1 + 4x^2 + 4y^2}\, dx\, dy$.

Since these elements are equal, the area of the parts of both surfaces defined over any region of the xy-plane will be equal.

11. If $z = \tfrac{1}{2}(x^2 + y^2)$, then $dS = \sqrt{1 + x^2 + y^2}\, dA$. One-eighth of the part of the surface above $-1 \leq x \leq 1$, $-1 \leq y \leq 1$, lies above the triangle T: given by $0 \leq x \leq 1$, $0 \leq y \leq x$, or, in polar coordinates, by $0 \leq \theta \leq \pi/4$, $0 \leq r \leq 1/\cos\theta = \sec\theta$. Thus

$S = 8\displaystyle\iint_T \sqrt{1 + x^2 + y^2}\, dA$

$= 8\displaystyle\int_0^{\pi/4} d\theta \int_0^{\sec\theta} \sqrt{1 + r^2}\, r\, dr \quad$ Let $u = 1 + r^2$, $du = 2r\, dr$

$= 4\displaystyle\int_0^{\pi/4} d\theta \int_0^{1+\sec^2\theta} \sqrt{u}\, du$

$= \dfrac{8}{3}\displaystyle\int_0^{\pi/4} \left[(1 + \sec^2\theta)^{3/2} - 1\right] d\theta$

$= \dfrac{8}{3}\displaystyle\int_0^{\pi/4} (1 + \sec^2\theta)^{3/2}\, d\theta - \dfrac{2\pi}{3}$.

Using a TI-85 numerical integration routine, we obtain the numerical value $S \approx 5.123$ sq. units.

12. As the figure suggests, the area of the canopy is the area of a hemisphere of radius $\sqrt{2}$ minus four times the area of half of a spherical cap cut off from the sphere $x^2 + y^2 + z^2 = 2$ by a plane at distance 1 from the origin, say the plane $z = 1$. Such a spherical cap, $z = \sqrt{2 - x^2 - y^2}$, lies above the disk $x^2 + y^2 \leq 2 - 1 = 1$. Since $\dfrac{\partial z}{\partial x} = -x/z$ and $\dfrac{\partial z}{\partial y} = -y/z$ on it, the area of the spherical cap is

$\displaystyle\iint_{x^2+y^2\leq 1} \sqrt{1 + \dfrac{x^2 + y^2}{z^2}}\, dA$

$= 2\sqrt{2}\pi \displaystyle\int_0^1 \dfrac{r\, dr}{\sqrt{2 - r^2}} \quad$ Let $u = 2 - r^2$, $du = -2r\, dr$

$= \sqrt{2}\pi \displaystyle\int_1^2 u^{-1/2}\, du = 2\sqrt{2}(\sqrt{2} - 1) = 4 - 2\sqrt{2}$.

Thus the area of the canopy is

$S = 2\pi(\sqrt{2})^2 - 4 \times \dfrac{1}{2} \times (4 - 2\sqrt{2}) = 4(\pi + \sqrt{2}) - 8$ sq. units.

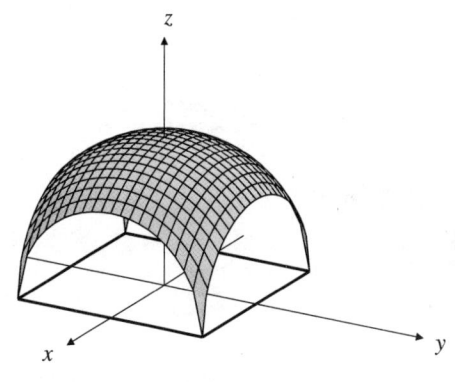

Fig. 14.7.12

13. Mass $= \displaystyle\int_0^{2\pi} d\theta \int_0^{\pi} \sin\phi\, d\phi \int_0^a \dfrac{A\rho^2\, d\rho}{B + \rho^2}$

$= 4\pi A \displaystyle\int_0^a \left(1 - \dfrac{B}{\rho^2 + B}\right) d\rho$

$= 4\pi A \left(a - \sqrt{B}\tan^{-1}\dfrac{a}{\sqrt{B}}\right)$ units.

14. A slice of the ball at height z, having thickness dz, is a circular disk of radius $\sqrt{a^2 - z^2}$ and areal density $\delta\, dz$. As calculated in the text, this disk attracts mass m at $(0, 0, b)$ with vertical force

$dF = 2\pi km\delta dz\left(1 - \dfrac{b - z}{\sqrt{a^2 - z^2 + (b - z)^2}}\right)$.

SECTION 14.7 (PAGE 872)

Thus the ball attracts m with vertical force

$$F = 2\pi km\delta \int_{-a}^{a} \left(1 - \frac{b-z}{\sqrt{a^2 + b^2 - 2bz}}\right) dz$$

$$\text{let } v = a^2 + b^2 - 2bz, \quad dv = -2b\, dz$$

$$\text{then } b - z = b - \frac{a^2 + b^2 - v}{2b} = \frac{b^2 - a^2 + v}{2b}$$

$$= 2\pi km\delta \left[2a - \frac{1}{4b^2} \int_{(b-a)^2}^{(b+a)^2} \frac{b^2 - a^2 + v}{\sqrt{v}} dv\right]$$

$$= 2\pi km\delta \left[2a - \frac{b^2 - a^2}{2b^2}(b + a - (b - a))\right.$$

$$\left. - \frac{1}{6b^2}\left((b+a)^3 - (b-a)^3\right)\right]$$

$$= \frac{4\pi km\delta a^3}{3b^2} = \frac{kmM}{b^2},$$

where $M = (4/3)\pi a^3 \delta$ is the mass of the ball. Thus the ball attracts the external mass m as though the ball were a point mass M located at its centre.

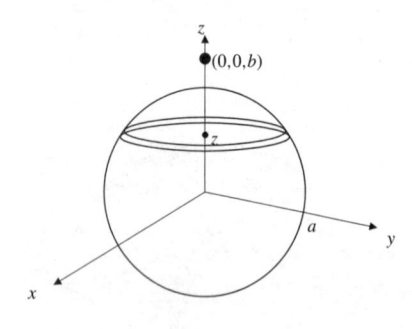

Fig. 14.7.14

15. The force is

$$F = 2\pi km\delta \int_{0}^{h} \left(1 - \frac{b-z}{\sqrt{a^2 + (b-z)^2}}\right) dz$$

$$\text{Let } u = a^2 + (b-z)^2$$

$$du = -2(b-z)\, dz$$

$$= 2\pi km\delta \left(h - \frac{1}{2}\int_{a^2+(b-h)^2}^{a^2+b^2} \frac{du}{\sqrt{u}}\right)$$

$$= 2\pi km\delta\left(h - \sqrt{a^2 + b^2} + \sqrt{a^2 + (b-h)^2}\right).$$

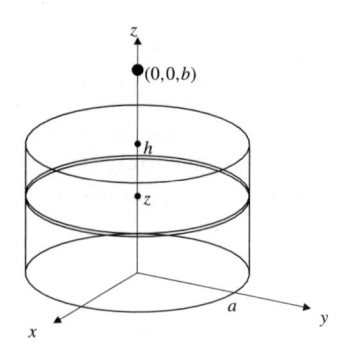

Fig. 14.7.15

16. The force is

$$F = 2\pi km\delta \int_{0}^{b} \left(1 - \frac{b-z}{\sqrt{a^2(b-z)^2 + (b-z)^2}}\right) dz$$

$$= 2\pi km\delta \int_{0}^{b} \left(1 - \frac{1}{\sqrt{a^2 + 1}}\right) dz$$

$$= 2\pi km\delta b \left(1 - \frac{1}{\sqrt{a^2 + 1}}\right).$$

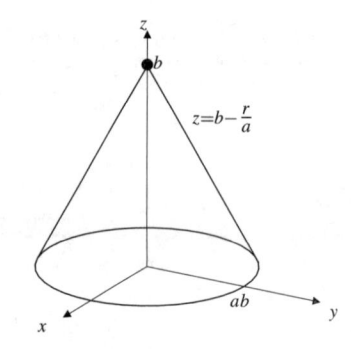

Fig. 14.7.16

17. The force is

$$F = 2\pi km\delta \int_{0}^{a} \left(1 - \frac{b-z}{\sqrt{a^2 + b^2 - 2bz}}\right) dz$$

use the same substitution as in Exercise 2)

$$= 2\pi km\delta \left(a - \frac{1}{4b^2}\int_{(b-a)^2}^{a^2+b^2} \frac{b^2 - a^2 + v}{\sqrt{v}} dv\right)$$

$$= 2\pi km\delta \left(a - \frac{b^2 - a^2}{2b^2}\left(\sqrt{a^2 + b^2} - (b-a)\right)\right.$$

$$\left. - \frac{1}{6b^2}\left((a^2 + b^2)^{3/2} - (b-a)^3\right)\right)$$

$$= \frac{2\pi km\delta}{3b^2}\left(2b^3 + a^3 - (2b^2 - a^2)\sqrt{a^2 + b^2}\right).$$

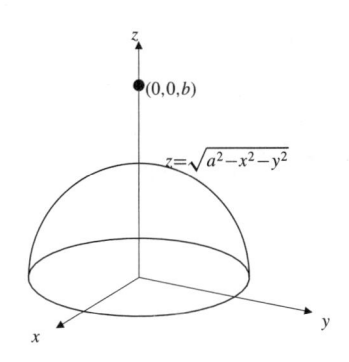

INSTRUCTOR'S SOLUTIONS MANUAL   SECTION 14.7 (PAGE 872)

Fig. 14.7.17

18. $m = \int_0^a dx \int_0^a dy \int_0^a (x^2 + y^2 + z^2)\, dz$
$= 3\int_0^a x^2 dx \int_0^a dy \int_0^a dz = a^5$
$M_{x=0} = \int_0^a x\, dx \int_0^a dy \int_0^a (x^2 + y^2 + z^2)\, dz$
$= \int_0^a x\, dx \int_0^a \left(a(x^2 + y^2) + \frac{a^3}{3}\right) dy$
$= \int_0^a \left(\frac{2a^4}{3} + a^2 x^2\right) x\, dx = \frac{7a^6}{12}.$

Thus $\bar{x} = M_{x=0}/m = \frac{7a}{12}$.

By symmetry, the centre of mass is $\left(\frac{7a}{12}, \frac{7a}{12}, \frac{7a}{12}\right)$.

19. Since the base triangle has centroid $\left(\frac{1}{3}, \frac{1}{3}, 0\right)$, the centroid of the prism is $\left(\frac{1}{3}, \frac{1}{3}, \frac{1}{2}\right)$.

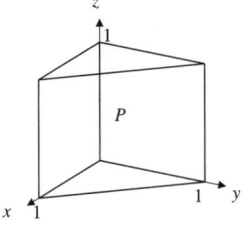

Fig. 14.7.19

20. Volume of region $= \int_0^{2\pi} d\theta \int_0^\infty e^{-r^2} r\, dr = \pi$. By symmetry, the moments about $x = 0$ and $y = 0$ are both zero. We have

$M_{z=0} = \int_0^{2\pi} d\theta \int_0^\infty r\, dr \int_0^{e^{-r^2}} z\, dz$
$= \pi \int_0^\infty re^{-2r^2}\, dr = \frac{\pi}{4}.$

The centroid is $(0, 0, 1/4)$.

21. The volume is $\frac{1}{8}\left(\frac{4}{3}\pi a^3\right) = \frac{\pi a^3}{6}$. By symmetry, the moments about all three coordinate planes are equal. We have

$M_{z=0} = \int_0^{\pi/2} d\theta \int_0^{\pi/2} \sin\phi\, d\phi \int_0^a \rho\cos\phi\, \rho^2\, d\rho$
$= \frac{\pi a^4}{8} \int_0^{\pi/2} \sin\phi\cos\phi\, d\phi = \frac{\pi a^4}{16}.$

Thus $\bar{z} = M_{z=0}/\text{volume} = 3a/8$.
The centroid is $\left(\frac{3a}{8}, \frac{3a}{8}, \frac{3a}{8}\right)$.

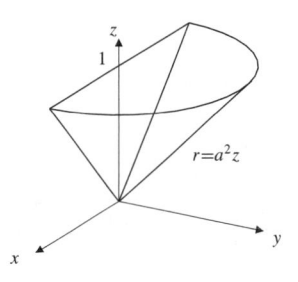

Fig. 14.7.21

22. The cube has centroid $(1/2, 1/2, 1/2)$. The tetrahedron lying above the plane $x + y + x = 2$ has centroid $(3/4, 3/4, 3/4)$ and volume $1/6$. Therefore the part of the cube lying below the plane has centroid (c, c, c) and volume $5/6$, where

$$\frac{5}{6}c + \frac{3}{4} \times \frac{1}{6} = \frac{1}{2} \times 1.$$

Thus $c = 9/20$; the centroid is $\left(\frac{9}{20}, \frac{9}{20}, \frac{9}{20}\right)$.

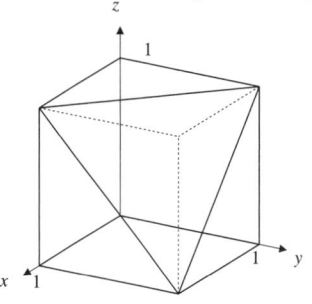

Fig. 14.7.22

23. The model still involves angular acceleration to spin the ball — it doesn't just fall. Part of the gravitational potential energy goes to producing this spin as the ball falls, even in the limiting case where the fall is vertical.

24. $I = \delta \int_0^{2\pi} d\theta \int_0^a r^3 \, dr \int_0^h dz$

$= 2\pi\delta h \left(\dfrac{a^4}{4}\right) = \dfrac{\pi\delta h a^4}{2}.$

$m = \pi\delta a^2 h, \qquad \overline{D} = \sqrt{I/m} = \dfrac{a}{\sqrt{2}}.$

25. $I = \delta \int_0^{2\pi} d\theta \int_0^a r \, dr \int_0^h (x^2 + z^2) \, dz$

$= \delta \int_0^{2\pi} d\theta \int_0^a \left(hr^2 \cos^2\theta + \dfrac{h^3}{3}\right) r \, dr$

$= \delta \int_0^{2\pi} \left(\dfrac{ha^4}{4} \cos^2\theta + \dfrac{h^3 a^2}{6}\right) d\theta$

$= \delta \left(\dfrac{\pi h a^4}{4} + \dfrac{\pi h^3 a^2}{3}\right) = \pi\delta a^2 h \left(\dfrac{a^2}{4} + \dfrac{h^2}{3}\right)$

$m = \pi\delta a^2 h, \qquad \overline{D} = \sqrt{I/m} = \sqrt{\dfrac{a^2}{4} + \dfrac{h^2}{3}}.$

26. $I = \delta \int_0^{2\pi} d\theta \int_0^a r^3 \, dr \int_0^{h(1-(r/a))} dz$

$= 2\pi\delta h \int_0^a r^3 \left(1 - \dfrac{r}{a}\right) dr = \dfrac{\pi\delta a^4 h}{10},$

$m = \dfrac{\pi\delta a^2 h}{3}, \qquad \overline{D} = \sqrt{I/m} = \sqrt{\dfrac{3}{10}} a.$

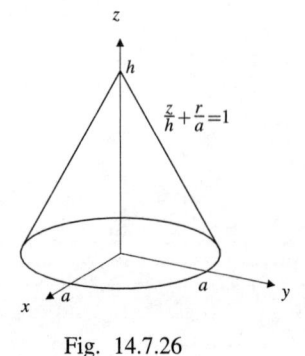

Fig. 14.7.26

27. $I = \delta \int_0^{2\pi} d\theta \int_0^a r \, dr \int_0^{h(1-(r/a))} (x^2 + z^2) \, dz$

$= \delta \int_0^{2\pi} d\theta \int_0^a \left[h\left(1 - \dfrac{r}{a}\right) r^2 \cos^2\theta \right.$

$\left. + \dfrac{h^3}{3}\left(1 - \dfrac{r}{a}\right)^3\right] r \, dr$

$= \pi\delta h \int_0^a \left(r^3 - \dfrac{r^4}{a}\right) dr + \dfrac{2\pi\delta h^3}{3} \int_0^a r \left(1 - \dfrac{r}{h}\right)^3 dr$

in the second integral put $u = 1 - (r/a)$

$= \dfrac{\pi\delta a^4 h}{20} + \dfrac{2\pi\delta a^2 h^3}{3} \int_0^1 (1-u) u^3 \, du$

$= \dfrac{\pi\delta a^4 h}{20} + \dfrac{2\pi\delta a^2 h^3}{60} = \dfrac{\pi\delta a^2 h}{60}(3a^2 + 2h^2),$

$m = \dfrac{\pi\delta a^2 h}{3}, \qquad \overline{D} = \sqrt{I/m} = \sqrt{\dfrac{3a^2 + 2h^2}{20}}.$

28. $I = \delta \iiint_Q (x^2 + y^2) \, dV$

$= 2\delta \int_0^a x^2 \, dx \int_0^a dy \int_0^a dz = \dfrac{2\delta a^5}{3},$

$m = \delta a^3, \qquad \overline{D} = \sqrt{I/m} = \sqrt{\dfrac{2}{3}} a.$

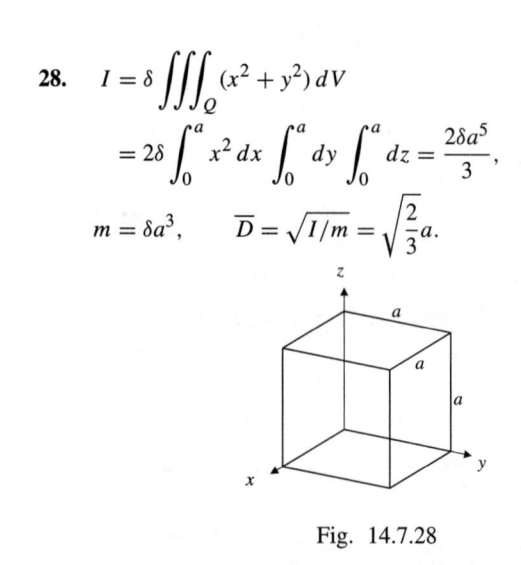

Fig. 14.7.28

29. The distance s from (x, y, z) to the line $x = y$, $z = 0$ satisfies $s^2 = u^2 + z^2$, where u is the distance from $(x, y, 0)$ to the line $x = y$ in the xy-plane. By Example 7 of Section 10.4 $u = |x - y|/\sqrt{2}$, so

$$s^2 = \dfrac{(x-y)^2}{2} + z^2.$$

INSTRUCTOR'S SOLUTIONS MANUAL SECTION 14.7 (PAGE 872)

The moment of inertia of the cube about this line is

$$I = \delta \int_0^a dx \int_0^a dy \int_0^a \left(\frac{(x-y)^2}{2} + z^2\right) dz$$

$$= \delta \int_0^a dx \int_0^a \left(\frac{a}{2}(x-y)^2 + \frac{a^3}{3}\right) dy \quad \text{Let } u = x - y$$
$$\quad\quad du = -dy$$

$$= \frac{\delta a^5}{3} + \frac{\delta a}{2} \int_0^a dx \int_{x-a}^x u^2 \, du$$

$$= \frac{\delta a^5}{3} + \frac{\delta a}{6} \int_0^a (3ax^2 - 3a^2 x + a^3) \, dx$$

$$= \frac{\delta a^5}{3} + \frac{\delta a}{6}\left(a^4 - \frac{3a^4}{2} + a^4\right) = \frac{5\delta a^5}{12},$$

$$m = \delta a^3, \quad \overline{D} = \sqrt{I/m} = \sqrt{\frac{5}{12}} a.$$

30. The line L through the origin parallel to the vector $\mathbf{v} = \mathbf{i} + \mathbf{j} + \mathbf{k}$ is a diagonal of the cube Q. By Example 8 of Section 10.4, the distance from the point with position vector $\mathbf{r} = x\mathbf{i} + y\mathbf{j} + z\mathbf{k}$ to L is $s = |\mathbf{v} \times \mathbf{r}|/|\mathbf{v}|$. Thus, the square of the distance from (x, y, z) to L is

$$s^2 = \frac{(x-y)^2 + (y-z)^2 + (z-x)^2}{3}$$

$$= \frac{2}{3}(x^2 + y^2 + z^2 - xy - xz - yz).$$

We have

$$\iiint_Q x^2 \, dV = \iiint_Q y^2 \, dV = \iiint_Q z^2 \, dV = \frac{a^5}{3}$$

$$\iiint_Q xy \, dV = \iiint_Q yz \, dV = \iiint_Q xz \, dV = \frac{a^5}{4}.$$

Therefore, the moment of inertia of Q about L is

$$I = \frac{2\delta}{3}\left(3 \times \frac{a^5}{3} - 3 \times \frac{a^5}{4}\right) = \frac{\delta a^5}{6}.$$

The mass of Q is $m = \delta a^3$, so the radius of gyration is

$$\overline{D} = \sqrt{I/m} = \frac{a}{\sqrt{6}}.$$

31. $I = \delta \int_{-a}^a dx \int_{-b}^b dy \int_{-c}^c (x^2 + y^2) \, dz$

$$= 2\delta c \int_{-a}^a \left(2bx^2 + \frac{2b^3}{3}\right) dx$$

$$= \frac{8\delta abc}{3}(a^2 + b^2),$$

$$m = 8\delta abc, \quad \overline{D} = \sqrt{I/m} = \sqrt{\frac{a^2 + b^2}{3}}.$$

32. $I = \delta \int_0^{2\pi} d\theta \int_0^c dz \int_a^b r^3 \, dr = \frac{\pi \delta c(b^4 - a^4)}{2},$

$$m = \pi \delta c(b^2 - a^2), \quad \overline{D} = \sqrt{\frac{b^2 + a^2}{2}}.$$

Fig. 14.7.32

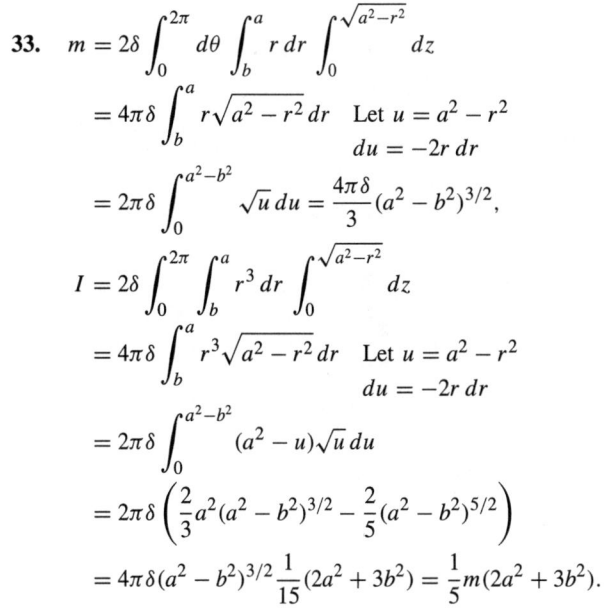

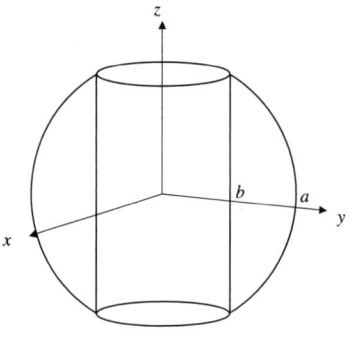

33. $m = 2\delta \int_0^{2\pi} d\theta \int_b^a r \, dr \int_0^{\sqrt{a^2-r^2}} dz$

$$= 4\pi \delta \int_b^a r\sqrt{a^2 - r^2} \, dr \quad \text{Let } u = a^2 - r^2$$
$$\quad\quad du = -2r \, dr$$

$$= 2\pi \delta \int_0^{a^2-b^2} \sqrt{u} \, du = \frac{4\pi \delta}{3}(a^2 - b^2)^{3/2},$$

$$I = 2\delta \int_0^{2\pi} \int_b^a r^3 \, dr \int_0^{\sqrt{a^2-r^2}} dz$$

$$= 4\pi \delta \int_b^a r^3 \sqrt{a^2 - r^2} \, dr \quad \text{Let } u = a^2 - r^2$$
$$\quad\quad du = -2r \, dr$$

$$= 2\pi \delta \int_0^{a^2-b^2} (a^2 - u)\sqrt{u} \, du$$

$$= 2\pi \delta \left(\frac{2}{3}a^2(a^2 - b^2)^{3/2} - \frac{2}{5}(a^2 - b^2)^{5/2}\right)$$

$$= 4\pi \delta (a^2 - b^2)^{3/2} \frac{1}{15}(2a^2 + 3b^2) = \frac{1}{5}m(2a^2 + 3b^2).$$

Fig. 14.7.33

34. By Exercise 26, the cylinder has moment of inertia

$$I = \frac{\pi \delta a^4 h}{2} = \frac{ma^2}{2},$$

where m is its mass. Following the method of Example 4(b), the kinetic energy of the cylinder rolling down the inclined plane with speed v is

$$KE = \frac{1}{2}mv^2 + \frac{1}{2}I\Omega^2$$
$$= \frac{1}{2}mv^2 + \frac{1}{4}ma^2 \frac{v^2}{a^2} = \frac{3}{4}mv^2.$$

The potential energy of the cylinder when it is at height h is mgh, so, by conservation of energy,

$$\frac{3}{4}mv^2 + mgh = \text{constant}.$$

Differentiating this equation with respect to time t, we obtain

$$0 = \frac{3}{2}mv\frac{dv}{dt} + mg\frac{dh}{dt}$$
$$= \frac{3}{2}mv\frac{dv}{dt} + mgv\sin\alpha.$$

Thus the cylinder rolls down the plane with acceleration

$$-\frac{dv}{dt} = \frac{2}{3}g\sin\alpha.$$

35. By Exercise 35, the ball with hole has moment of inertia

$$I = \frac{m}{5}(2a^2 + 3b^2)$$

about the axis of the hole. The kinetic energy of the rolling ball is

$$KE = \frac{1}{2}mv^2 + \frac{m}{10}(2a^2 + 3b^2)\frac{v^2}{a^2}$$
$$= mv^2\left(\frac{1}{2} + \frac{2a^2 + 3b^2}{10a^2}\right) = mv^2 \frac{7a^2 + 3b^2}{10a^2}.$$

By conservation of energy,

$$mv^2 \frac{7a^2 + 3b^2}{10a^2} + mgh = \text{constant}.$$

Differentiating with respect to time, we obtain

$$\frac{7a^2 + 3b^2}{5a^2} mv \frac{dv}{dt} + mgv\sin\alpha = 0.$$

Thus the ball rolls down the plane (with its hole remaining horizontal) with acceleration

$$-\frac{dv}{dt} = \frac{5a^2}{7a^2 + 3b^2} g\sin\alpha.$$

36. The kinetic energy of the oscillating pendulum is

$$KE = \frac{1}{2}I\left(\frac{d\theta}{dt}\right)^2.$$

The potential energy is mgh, where h is the distance of C above A. In this case, $h = -a\cos\theta$. By conservation of energy,

$$\frac{1}{2}I\left(\frac{d\theta}{dt}\right)^2 - mga\cos\theta = \text{constant}.$$

Differentiating with respect to time t, we obtain

$$I\left(\frac{d\theta}{dt}\right)\frac{d^2\theta}{dt^2} + mga\sin\theta\left(\frac{d\theta}{dt}\right) = 0,$$

or

$$\frac{d^2\theta}{dt^2} + \frac{mga}{I}\sin\theta = 0.$$

For small oscillations we have $\sin\theta \approx \theta$, and the above equation is approximated by

$$\frac{d^2\theta}{dt^2} + \omega^2\theta = 0,$$

where $\omega^2 = mga/I$. The period of oscillation is

$$T = \frac{2\pi}{\omega} = 2\pi\sqrt{\frac{I}{mga}}.$$

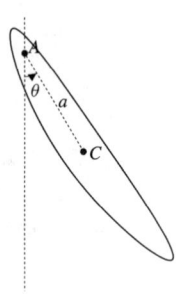

Fig. 14.7.36

37. If the centre of mass of B is at the origin, then

$$M_{x=0} = \iiint_B x\delta\, dV = 0.$$

If line L_0 is the z-axis, and L_k is the line $x = k$, $y = 0$, then the moment of inertia I_k of B about L_k is

$$I_k = \iiint_B \left((x-k)^2 + y^2\right) \delta \, dV$$
$$= \iiint_B (x^2 + y^2 + k^2 - 2kx) \, \delta \, dV$$
$$= I_0 + k^2 m - 2k M_{x=0} = I_0 + k^2 m,$$

where m is the mass of B and I_0 is the moment about L_0.

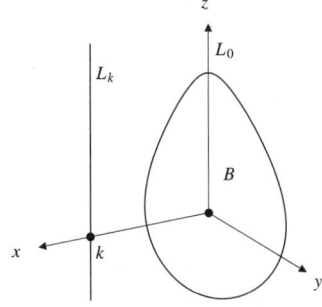

Fig. 14.7.37

38. The moment of inertia of the ball about the point where it contacts the plane is, by Example 4(b) and Exercise 39,

$$I = \frac{8}{15}\pi\delta a^5 + \left(\frac{4}{3}\pi\delta a^3\right) a^2$$
$$= \left(\frac{2}{5} + 1\right) ma^2 = \frac{7}{5} ma^2.$$

The kinetic energy of the ball, regarded as rotating about the point of contact with the plane, is therefore

$$KE = \frac{1}{2} I \Omega^2 = \frac{7}{10} ma^2 \frac{v^2}{a^2} = \frac{7}{10} mv^2.$$

39. By Example 7 of Section 10.4, the distance from the point with position vector $\mathbf{r} = x\mathbf{i} + y\mathbf{j} + z\mathbf{k}$ to the straight line L through the origin parallel to the vector $\mathbf{a} = A\mathbf{i} + B\mathbf{j} + C\mathbf{k}$ is

$$s = \frac{|\mathbf{a} \times \mathbf{r}|}{|\mathbf{a}|}.$$

The moment of inertia of the body occupying region R about L is, therefore,

$$I = \frac{1}{|\mathbf{a}|^2} \iiint_R |\mathbf{a} \times \mathbf{r}|^2 \delta \, dV$$
$$= \frac{1}{A^2 + B^2 + C^2} \iiint_R \left[(Bz - Cy)^2 + (Cx - Az)^2 + (Ay - Bx)^2\right] \delta \, dV$$
$$= \frac{1}{A^2 + B^2 + C^2} \left[(B^2 + C^2) P_{xx} + (A^2 + C^2) P_{yy} + (A^2 + B^2) P_{zz} - 2AB P_{xy} - 2AC P_{xz} - 2BC P_{yz}\right].$$

Review Exercises 14 (page 874)

1. By symmetry,

$$\iint_R (x+y) \, dA = 2 \iint_R x \, dA = 2 \int_0^1 x \, dx \int_{x^2}^{\sqrt{x}} dy$$
$$= 2 \int_0^1 (x^{3/2} - x^3) \, dx$$
$$= 2 \left(\frac{2}{5} x^{5/2} - \frac{x^4}{4}\right) \Bigg|_0^1 = 2 \left(\frac{2}{5} - \frac{1}{4}\right) = \frac{3}{10}$$

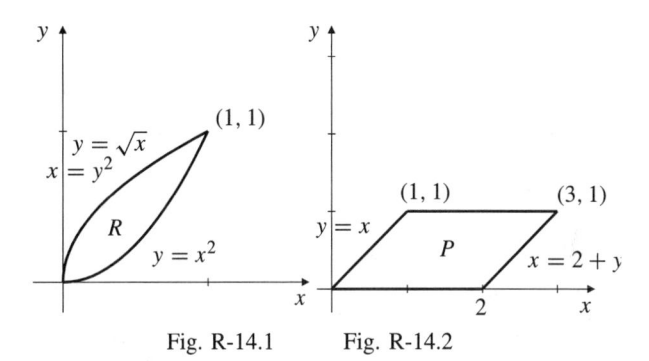

Fig. R-14.1

Fig. R-14.2

2. $$\iint_P (x^2 + y^2) \, dA = \int_0^1 dy \int_y^{2+y} (x^2 + y^2) \, dx$$
$$= \int_0^1 \left(\frac{x^3}{3} + xy^2\right) \Bigg|_{x=y}^{x=2+y} dy$$
$$= \int_0^1 \left(\frac{(2+y)^3}{3} + y^2(2+y) - \frac{y^3}{3} - y^3\right) dy$$
$$= \int_0^1 \left(\frac{8}{3} + 4y + 4y^2\right) dy = \frac{8}{3} + 2 + \frac{4}{3} = 6$$

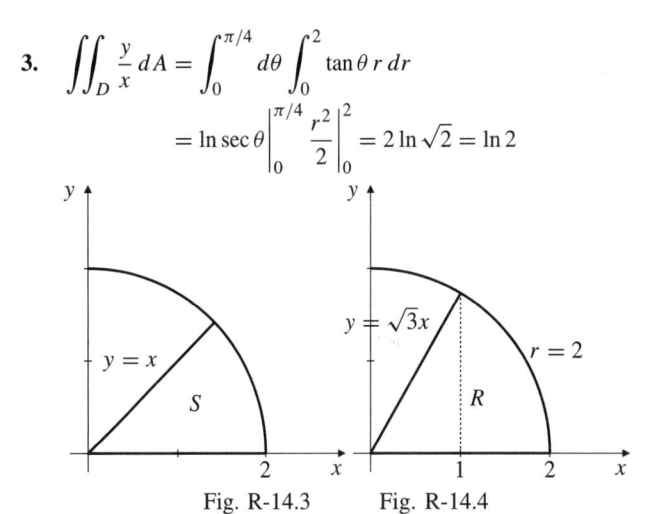

3. $$\iint_D \frac{y}{x} \, dA = \int_0^{\pi/4} d\theta \int_0^2 \tan\theta \, r \, dr$$
$$= \ln \sec\theta \Bigg|_0^{\pi/4} \frac{r^2}{2} \Bigg|_0^2 = 2\ln\sqrt{2} = \ln 2$$

Fig. R-14.3

Fig. R-14.4

541

REVIEW EXERCISES 14 (PAGE 874)

4. a) $I = \int_0^{\sqrt{3}} dy \int_{y/\sqrt{3}}^{\sqrt{4-y^2}} e^{-x^2-y^2} dx$
$= \iint_R e^{-x^2-y^2} dA$
where R is as shown in the figure.

b) $I = \int_0^1 dx \int_0^{\sqrt{3}x} e^{-x^2-y^2} dy$
$+ \int_1^2 dx \int_0^{\sqrt{4-x^2}} e^{-x^2-y^2} dy$

c) $I = \int_0^{\pi/3} d\theta \int_0^2 e^{-r^2} r\, dr$

d) $I = \frac{\pi}{3}\left(-\frac{e^{-r^2}}{2}\right)\Big|_0^2 = \frac{\pi(1-e^{-4})}{6}$

5. The cone $z = k\sqrt{x^2+y^2}$ has semi-vertical angle $\phi_0 = \tan^{-1}(1/k)$. Thus the volume inside the cone and inside the sphere $x^2+y^2+z^2 = a^2$ is

$V = \int_0^{2\pi} d\theta \int_0^{\phi_0} \sin\phi\, d\phi \int_0^a \rho^2 d\rho$
$= \frac{2\pi a^3}{3}(1-\cos\phi_0) = \frac{2\pi a^3}{3}\left(1 - \frac{k}{\sqrt{k^2+1}}\right).$

To have
$$V = \frac{1}{4}\left(\frac{4}{3}\pi a^3\right) = \frac{\pi a^3}{3},$$
we need to ensure that
$$2\left(1 - \frac{k}{\sqrt{k^2+1}}\right) = 1.$$
Thus $k^2+1 = (2k)^2$, and so $3k^2 = 1$, and $k = 1/\sqrt{3}$.

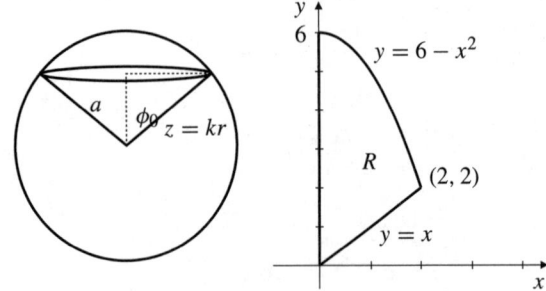

Fig. R-14.5   Fig. R-14.6

6. $I = \int_0^2 dy \int_0^y f(x,y)\, dx + \int_2^6 dy \int_0^{\sqrt{6-y}} f(x,y)\, dx$
$= \iint_R f(x,y)\, dA,$
where R is as shown in the figure. Thus
$I = \int_0^2 dx \int_x^{6-x^2} f(x,y)\, dy.$

7. $J = \int_0^1 dz \int_0^z dy \int_0^y f(x,y,z)\, dx$
corresponds to the region
$$0 \le z \le 1, \quad 0 \le y \le z, \quad 0 \le x \le y,$$
which can also be expressed in the form
$$0 \le x \le 1, \quad x \le y \le 1, \quad y \le z \le 1.$$
Thus $J = \int_0^1 dx \int_x^1 dy \int_y^1 f(x,y,z)\, dz.$

8. A horizontal slice of the object at height z above the base, and having thickness dz, is a disk of radius $r = \frac{1}{2}(10-z)$ m. Its volume is
$$dV = \pi\frac{(10-z)^2}{4} dz\ \text{m}^3.$$
The density of the slice is $\delta = kz^2$ kg/m^3. Since $\delta = 3,000$ when $z = 10$, we have $k = 30$.

a) The mass of the object is
$m = \int_0^{10} 30z^2 \frac{\pi}{4}(10-z)^2 dz$
$= \frac{15\pi}{2}\int_0^{10}(100z^2 - 20z^3 + z^4)\, dz$
$= \frac{15\pi}{2}\left(\frac{100,000}{3} - 50,000 + 20,000\right) \approx 78,540$ kg

b) The moment of inertia (about its central axis) of the disk-shaped slice at height z is
$dI = 30z^2\, dz \int_0^{2\pi} d\theta \int_0^{(10-z)/2} r^3\, dr.$

Thus the moment of inertia about the whole solid cone is
$I = \int_0^{10} 30z^2\, dz \int_0^{2\pi} d\theta \int_0^{(10-z)/2} r^3\, dr.$

9. $f(t) = \int_t^a e^{-x^2} dx$
$\bar{f} = \frac{1}{a}\int_0^a f(t)\, dt = \frac{1}{a}\int_0^a dt \int_t^a e^{-x^2} dx$
$= \frac{1}{a}\int_0^a e^{-x^2} dx \int_0^x dt = \frac{1}{a}\int_0^a xe^{-x^2} dx$
$= \frac{1}{a}\left(-\frac{e^{-x^2}}{2}\right)\Big|_0^a = \frac{1-e^{-a^2}}{2a}$

INSTRUCTOR'S SOLUTIONS MANUAL REVIEW EXERCISES 14 (PAGE 874)

10. If $f(x, y) = \lfloor x + y \rfloor$, then $f = 0$, 1, or 2, in parts of the quarter disk Q, as shown in the figure.

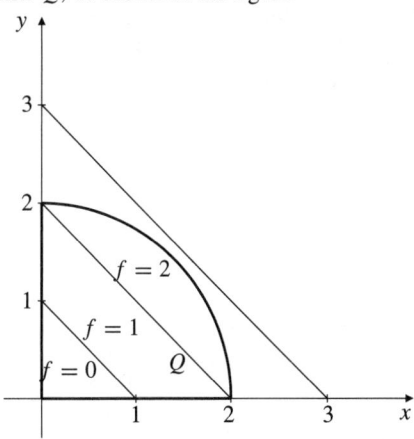

Fig. R-14.10

Thus

$$\iint_Q f(x, y)\, dA = 0\left(\frac{1}{2}\right) + 1\left(\frac{3}{2}\right) + 2(\pi - 2) = 2\pi - \frac{5}{2},$$

and $\overline{f} = \dfrac{1}{\pi}\left(2\pi - \dfrac{5}{2}\right) = 2 - \dfrac{5}{2\pi}$.

11. The sphere $x^2 + y^2 + z^2 = 6a^2$ and the paraboloid $z = (x^2 + y^2)/a$ intersect where $z^2 + az - 6a^2 = 0$, that is, where $(z + 3a)(z - 2a) = 0$. Only $z = 2a$ is possible; the plane $z = -3a$ does not intersect the sphere. If $z = 2a$, then $x^2 + y^2 = r^2 = 6a^2 - 4a^2 = 2a^2$, so the intersection is on the vertical cylinder of radius $\sqrt{2}a$ with axis on the z-axis. We have,

$$\iiint_D (x^2 + y^2)\, dV$$
$$= \int_0^{2\pi} d\theta \int_0^{\sqrt{2}a} r^3\, dr \int_{r^2/a}^{\sqrt{6a^2 - r^2}} dz$$
$$= 2\pi \int_0^{\sqrt{2}a} \left[r^3\sqrt{6a^2 - r^2} - \frac{r^5}{a} \right] dr$$

Let $u = 6a^2 - r^2$
$du = -2r\, dr$

$$= \pi \int_{4a^2}^{6a^2} (6a^2 - u)\sqrt{u}\, du - \frac{\pi}{3a}(\sqrt{2}a)^6$$
$$= \pi \left(4a^2 u^{3/2} - \frac{2}{5}u^{5/2} \right)\bigg|_{4a^2}^{6a^2} - \frac{8}{3}\pi a^5$$
$$= \frac{8\pi}{15}(18\sqrt{6} - 41)a^5$$

12. The solid S lies above the region in the xy-plane bounded by the circle $x^2 + y^2 = 2ay$, which has polar equation $r = 2a\sin\theta$, ($0 \le \theta \le \pi$). It lies below the cone $z = \sqrt{x^2 + y^2} = r$. The moment of inertia of S about the z-axis is

$$I = \iiint_S (x^2 + y^2)\, dV = \int_0^\pi d\theta \int_0^{2a\sin\theta} r^3\, dr \int_0^r dz$$
$$= \int_0^\pi d\theta \int_0^{2a\sin\theta} r^4\, dr = \frac{32a^5}{5}\int_0^\pi \sin^5\theta\, d\theta$$
$$= \frac{32a^5}{5}\int_0^\pi (1 - \cos^2\theta)^2 \sin\theta\, d\theta \quad \text{Let } u = \cos\theta$$
$$\hspace{6.5cm} du = -\sin\theta\, d\theta$$
$$= \frac{32a^5}{5}\int_{-1}^1 (1 - 2u^2 + u^4)\, du$$
$$= \frac{64a^5}{5}\left(1 - \frac{2}{3} + \frac{1}{5}\right) = \frac{512a^5}{75}.$$

13. A horizontal slice of D at height z is a right triangle with legs $(2 - z)/2$ and $2 - z$. Thus the volume of D is

$$V = \frac{1}{4}\int_0^1 (2 - z)^2\, dz = \frac{7}{12}.$$

Its moment about $z = 0$ is

$$M_{z=0} = \frac{1}{4}\int_0^1 z(2 - z)^2\, dz$$
$$= \frac{1}{4}\int_0^1 (4z - 4z^2 + z^3)\, dz = \frac{11}{48}.$$

The z-coordinate of the centroid of D is

$$\overline{z} = \frac{11}{48} \bigg/ \frac{7}{12} = \frac{11}{28}.$$

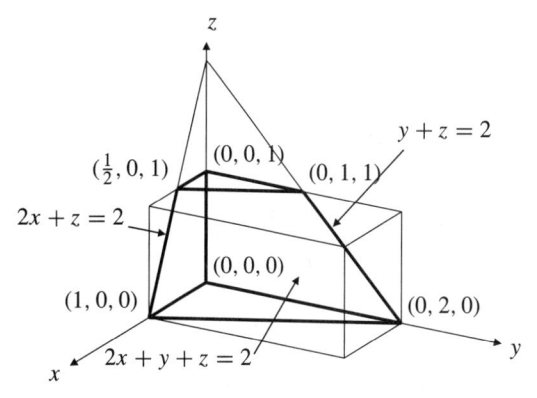

Fig. R-14.13

543

14. $V = \iiint_S dV = \int_0^1 dy \int_0^{1-y} dz \int_0^{2-y-2z} dx$

$= \int_0^1 dy \int_0^{1-y} (2-y-2z)\,dz$

$= \int_0^1 [(2-y)(1-y) - (1-y)^2]\,dy$

$= \int_0^1 (1-y)\,dy = \frac{1}{2}$

$M_{x=0} = \iiint_S x\,dV = \int_0^1 dy \int_0^{1-y} dz \int_0^{2-y-2z} x\,dx$

$= \frac{1}{2}\int_0^1 dy \int_0^{1-y} [(2-y)^2 - 4(2-y)z + 4z^2]\,dz$

$= \frac{1}{2}\int_0^1 \Big[(2-y)^2(1-y) - 2(2-y)(1-y)^2$

$\quad + \frac{4}{3}(1-y)^3\Big]\,dy \quad$ Let $u = 1-y$
$\qquad\qquad\qquad\qquad\qquad\qquad\quad du = -dy$

$= \frac{1}{2}\int_0^1 \Big[(u+1)^2 u - 2(u+1)u^2 + \frac{4}{3}u^3\Big]\,du$

$= \frac{1}{2}\int_0^1 \Big[\frac{1}{3}u^3 + u\Big]\,du = \frac{7}{24}$

$\bar{x} = \frac{7}{24}\Big/\frac{1}{2} = \frac{7}{12}$

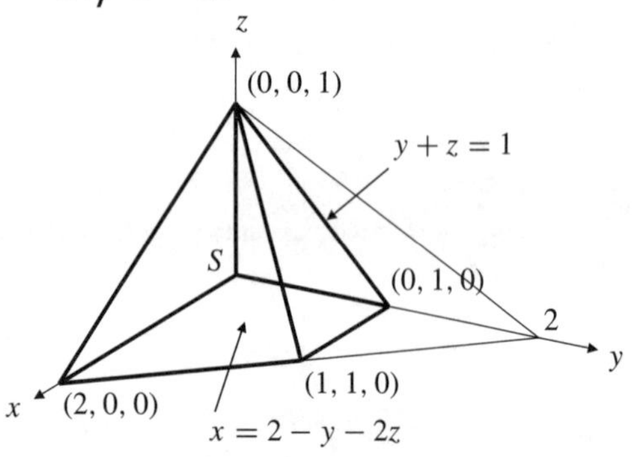

Fig. R-14.14

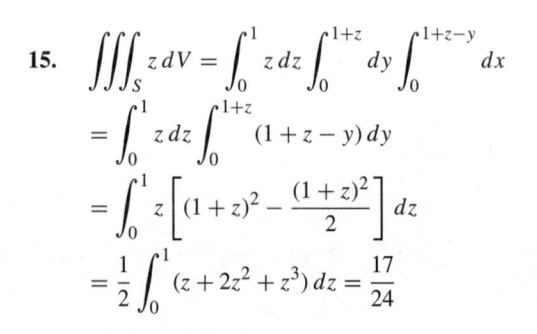

15. $\iiint_S z\,dV = \int_0^1 z\,dz \int_0^{1+z} dy \int_0^{1+z-y} dx$

$= \int_0^1 z\,dz \int_0^{1+z} (1+z-y)\,dy$

$= \int_0^1 z\Big[(1+z)^2 - \frac{(1+z)^2}{2}\Big]\,dz$

$= \frac{1}{2}\int_0^1 (z + 2z^2 + z^3)\,dz = \frac{17}{24}$

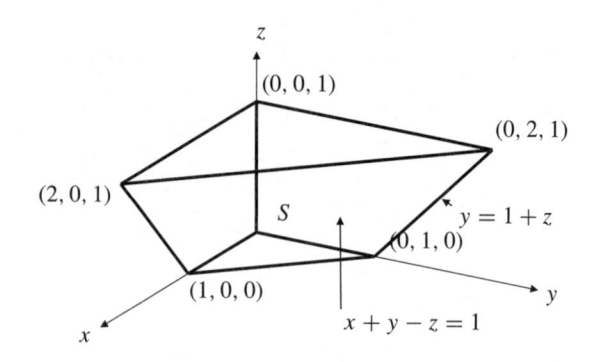

Fig. R-14.15

16. The plane $z = 2x$ intersects the paraboloid $z = x^2 + y^2$ on the circular cylinder $x^2 + y^2 = 2x$, (that is, $(x-1)^2 + y^2 = 1$), which has radius 1. Since $dS = \sqrt{1+2^2}\,dA = \sqrt{5}\,dA$ on the plane, the area of the part of the plane inside the paraboloid (and therefore inside the cylinder) is $\sqrt{5}$ times the area of a circle of radius 1, that is, $\sqrt{5}\pi$ square units.

17. As noted in the previous exercise, the part of the paraboloid $z = x^2 + y^2$ that lies below the plane $z = 2x$ is inside the vertical cylinder $x^2 + y^2 = 2x$, which has polar equation $r = 2\cos\theta$ ($-\pi/2 \le \theta \le \pi/2$). On the paraboloid:

$$dS = \sqrt{1 + (2x)^2 + (2y)^2}\,dA = \sqrt{1+4r^2}\,r\,dr\,d\theta.$$

The area of that part of the paraboloid is

$S = \int_{-\pi/2}^{\pi/2} d\theta \int_0^{2\cos\theta} \sqrt{1+4r^2}\,r\,dr \quad$ Let $u = 1+4r^2$
$\qquad\qquad\qquad\qquad\qquad\qquad\qquad\qquad\quad du = 8r\,dr$

$= \frac{1}{8}\int_{-\pi/2}^{\pi/2} d\theta \int_1^{1+16\cos^2\theta} u^{1/2}\,du$

$= \frac{1}{4}\int_0^{\pi/2} \frac{2}{3}[(1+16\cos^2\theta)^{3/2} - 1]\,d\theta$

$= \frac{1}{6}\int_0^{\pi/2} [(1+16\cos^2\theta)^{3/2} - 1]\,d\theta$

≈ 7.904 sq. units.

(using a TI-85 numerical integration function).

18. The region R inside the ellipsoid $\dfrac{x^2}{36} + \dfrac{y^2}{9} + \dfrac{z^2}{4} = 1$ and above the plane $x+y+z = 1$ is transformed by the change of variables

$$x = 6u, \quad y = 3v, \quad z = 2w$$

to the region S inside the sphere $u^2 + v^2 + w^2 = 1$ and above the plane $6u + 3v + 2w = 1$. The distance from the origin to this plane is

$$D = \frac{1}{\sqrt{6^2 + 3^2 + 2^2}} = \frac{1}{7},$$

so, by symmetry, the volume of S is equal to the volume inside the sphere and above the plane $w = 1/7$, that is,

$$\int_{1/7}^{1} \pi(1-w^2)\,dw = \pi\left(w - \frac{w^3}{3}\right)\bigg|_{1/7}^{1} = \frac{180\pi}{343} \text{ units}^3.$$

Since $|\partial(x,y,z)/\partial(u,v,w)| = 6 \cdot 3 \cdot 2 = 18$, the volume of R is $18 \times (180\pi/343) = 3240\pi/343 \approx 29.68$ cu. units.

Challenging Problems 14 (page 875)

1. This problem is similar to Review Exercise 18 above. The region R inside the ellipsoid $\frac{x^2}{a^2} + \frac{y^2}{b^2} + \frac{z^2}{c^2} = 1$ and above the plane $\frac{x}{a} + \frac{y}{b} + \frac{z}{c} = 1$ is transformed by the change of variables

$$x = au, \quad y = bv, \quad z = cw$$

to the region S inside the sphere $u^2 + v^2 + w^2 = 1$ and above the plane $u + v + w = 1$. The distance from the origin to this plane is $\frac{1}{\sqrt{3}}$, so, by symmetry, the volume of S is equal to the volume inside the sphere and above the plane $w = 1/\sqrt{3}$, that is,

$$\int_{1/\sqrt{3}}^{1} \pi(1-w^2)\,dw = \pi\left(w - \frac{w^3}{3}\right)\bigg|_{1/\sqrt{3}}^{1}$$
$$= \frac{2\pi(9 - 4\sqrt{3})}{27} \text{ cu. units.}$$

Since $|\partial(x,y,z)/\partial(u,v,w)| = abc$, the volume of R is $\frac{2\pi(9 - 4\sqrt{3})}{27}abc$ cu. units.

2. The plane $(x/a) + (y/b) + (z/c) = 1$ intersects the ellipsoid $(x/a)^2 + (y/b)^2 + (z/c)^2 = 1$ above the region R in the xy-plane bounded by the ellipse

$$\frac{x^2}{a^2} + \frac{y^2}{b^2} + \left(1 - \frac{x}{a} - \frac{y}{b}\right)^2 = 1,$$

or, equivalently,

$$\frac{x^2}{a^2} + \frac{y^2}{b^2} + \frac{xy}{ab} - \frac{x}{a} - \frac{y}{b} = 0.$$

Thus the area of the part of the plane lying inside the ellipsoid is

$$S = \iint_R \sqrt{1 + \frac{c^2}{a^2} + \frac{c^2}{b^2}}\,dx\,dy$$
$$= \frac{\sqrt{a^2b^2 + a^2c^2 + b^2c^2}}{ab} \text{ (area of } R\text{)}.$$

Under the transformation $x = a(u+v)$, $y = b(u-v)$, R corresponds to the ellipse in the uv-plane bounded by

$$(u+v)^2 + (u-v)^2 + (u^2 - v^2) - (u+v) - (u-v) = 0$$
$$3u^2 + v^2 - 2u = 0$$
$$3\left(u^2 - \frac{2}{3}u + \frac{1}{9}\right) + v^2 = \frac{1}{3}$$
$$\frac{(u-1/3)^2}{1/9} + \frac{v^2}{1/3} = 1,$$

an ellipse with area $\pi(1/3)(1/\sqrt{3}) = \pi/(3\sqrt{3})$ sq. units. Since

$$dx\,dy = \left|\begin{matrix} a & a \\ b & -b \end{matrix}\right| du\,dv = 2ab\,du\,dv,$$

we have

$$S = \frac{2\pi}{3\sqrt{3}}\sqrt{a^2b^2 + a^2c^2 + b^2c^2} \text{ sq. units.}$$

3. a)
$$\frac{1}{1-xy} = 1 + xy + (xy)^2 + \cdots = \sum_{n=1}^{\infty}(xy)^{n-1}$$

$$\int_0^1\int_0^1 \frac{dx\,dy}{1-xy} = \sum_{n=1}^{\infty}\int_0^1 x^{n-1}\,dx \int_0^1 y^{n-1}\,dy$$
$$= \sum_{n=1}^{\infty}\frac{1}{n^2}.$$

Remark: The series for $1/(1-xy)$ converges for $|xy| < 1$. Therefore the outer integral is improper (i.e., $\lim_{c \to 1-}\int_0^c dx$). We cannot do a detailed analysis of the convergence here, but the convergence of $\sum 1/n^2$ shows that the iterated double integral must converge.

b) Similarly,

$$\frac{1}{1+xy} = 1 - xy + (xy)^2 - \cdots = \sum_{n=1}^{\infty}(-xy)^{n-1}$$

$$\int_0^1\int_0^1 \frac{dx\,dy}{1+xy}$$
$$= \sum_{n=1}^{\infty}(-1)^{n-1}\int_0^1 x^{n-1}\,dx \int_0^1 y^{n-1}\,dy$$
$$= \sum_{n=1}^{\infty}\frac{(-1)^{n-1}}{n^2}$$

$$\int_0^1\int_0^1\int_0^1 \frac{dx\,dy}{1-xyz}$$
$$= \sum_{n=1}^{\infty}\int_0^1 x^{n-1}\,dx \int_0^1 y^{n-1}\,dy \int_0^1 z^{n-1}\,dz$$

$$= \sum_{n=1}^{\infty} \frac{1}{n^3}$$

$$\int_0^1 \int_0^1 \int_0^1 \frac{dx\,dy}{1+xyz}$$

$$= \sum_{n=1}^{\infty}(-1)^{n-1}\int_0^1 x^{n-1}\,dx \int_0^1 y^{n-1}\,dy \int_0^1 z^{n-1}\,dz$$

$$= \sum_{n=1}^{\infty} \frac{(-1)^{n-1}}{n^3}.$$

4. a) $G(y) = \int_0^{\infty} \frac{\tan^{-1}(xy)}{x}\,dx$

$G'(y) = \int_0^{\infty} \frac{1}{x}\frac{x}{1+x^2y^2}\,dx$ Let $u = xy$
   $du = y\,dx$

$= \frac{1}{y}\int_0^{\infty} \frac{du}{1+u^2} = \frac{\pi}{2y}$ for $y > 0$.

b) $\int_0^{\infty} \frac{\tan^{-1}(\pi x) - \tan^{-1}x}{x}\,dx$

$= G(\pi) - G(1) = \int_1^{\pi} G'(y)\,dy = \frac{\pi}{2}\int_1^{\pi} \frac{dy}{y} = \frac{\pi \ln \pi}{2}.$

5. Under the transformation $u = \mathbf{a} \bullet \mathbf{r}$, $v = \mathbf{b} \bullet \mathbf{r}$, $w = \mathbf{c} \bullet \mathbf{r}$, where $\mathbf{r} = x\mathbf{i} + y\mathbf{j} + z\mathbf{k}$, the parallelepiped P corresponds to the rectangle R specified by $0 \le u \le d_1$, $0 \le v \le d_2$, $0 \le w \le d_3$. If $\mathbf{a} = a_1\mathbf{i} + a_2\mathbf{j} + a_3\mathbf{k}$ and similar expressions hold for $\mathbf{b}$ and $\mathbf{c}$, then

$$\frac{\partial(u,v,w)}{\partial(x,y,z)} = \begin{vmatrix} a_1 & a_2 & a_3 \\ b_1 & b_2 & b_3 \\ c_1 & c_2 & c_3 \end{vmatrix} = \mathbf{a} \bullet (\mathbf{b} \times \mathbf{c}).$$

Therefore

$$dx\,dy\,dz = \left|\frac{\partial(x,y,z)}{\partial(u,v,w)}\right| du\,dv\,dw = \frac{du\,dv\,dw}{|\mathbf{a}\bullet(\mathbf{b}\times\mathbf{c})|},$$

and we have

$$\iiint_P (\mathbf{a}\bullet\mathbf{r})(\mathbf{b}\bullet\mathbf{r})(\mathbf{c}\bullet\mathbf{r})\,dx\,dy\,dz$$

$$= \iiint_R \frac{uvw}{|\mathbf{a}\bullet(\mathbf{b}\times\mathbf{c})|}\,du\,dv\,dw$$

$$= \frac{1}{|\mathbf{a}\bullet(\mathbf{b}\times\mathbf{c})|}\int_0^{d_1} u\,du \int_0^{d_2} v\,dv \int_0^{d_3} w\,dw$$

$$= \frac{d_1^2 d_2^2 d_3^2}{8|\mathbf{a}\bullet(\mathbf{b}\times\mathbf{c})|}.$$

6. The volume V_0 removed from the ball is eight times the part in the first octant, which is itself split into two equal parts by the plane $x = y$:

$$V_0 = 16\int_0^1 dx \int_0^x \sqrt{4 - x^2 - y^2}\,dy$$

$$= 16\int_0^{\pi/4} d\theta \int_0^{\sec\theta} \sqrt{4-r^2}\,r\,dr \quad \text{Let } u = 4 - r^2$$
$$\qquad du = -2r\,dr$$

$$= 8\int_0^{\pi/4} d\theta \int_{4-\sec^2\theta}^4 u^{1/2}\,du$$

$$= \frac{16}{3}\int_0^{\pi/4}\left[8 - (4 - \sec^2\theta)^{3/2}\right]d\theta$$

$$= \frac{32\pi}{3} - \frac{16}{3}\int_0^{\pi/4} \frac{(4\cos^2\theta - 1)^{3/2}}{\cos^3\theta}\,d\theta.$$

Now the volume of the whole ball is $(4\pi/3)2^3 = 32\pi/3$, so the volume remaining after the hole is cut is

$$V = \frac{32\pi}{3} - V_0$$

$$= \frac{16}{3}\int_0^{\pi/4} \frac{(3 - 4\sin^2\theta)^{3/2}}{(1-\sin^2\theta)^2}\cos\theta\,d\theta \quad \text{Let } v = \sin\theta$$
$$\qquad dv = \cos\theta\,d\theta$$

$$= \frac{16}{3}\int_0^{1/\sqrt{2}} \frac{(3-4v^2)^{3/2}}{(1-v^2)^2}\,dv.$$

We submitted this last integral to Mathematica to obtain

$$V = \frac{4}{3}\left(32\sin^{-1}\sqrt{\frac{2}{3}} - 2^{3/2} + 11\tan^{-1}(3 - 2^{3/2})\right.$$
$$\left. - 11\tan^{-1}(3 + 2^{3/2})\right) \approx 18.9349.$$

7. Under the transformation $x = u^3$, $y = v^3$, $z = w^3$, the region R bounded by the surface $x^{2/3} + y^{2/3} + z^{2/3} = a^{2/3}$ gets mapped to the ball B bounded by $u^2 + v^2 + w^2 = a^{2/3}$. Assume that $a > 0$. Since

$$\frac{\partial(x,y,z)}{\partial(u,v,w)} = 27u^2v^2w^2,$$

the volume of R is

$$V = 27\iiint_B u^2v^2w^2\,du\,dv\,dw.$$

Now switch to polar coordinates $[\rho, \phi, \theta]$ in uvw-space. Since

$$uvw = (\rho\sin\phi\cos\theta)(\rho\sin\phi\sin\theta)(\rho\cos\phi),$$

we have

$$V = 27 \int_0^{2\pi} \cos^2\theta \sin^2\theta \, d\theta \int_0^{\pi} \sin^5\phi \cos^2\phi \, d\phi \int_0^{a^{1/3}} \rho^8 \, d\rho$$

$$= 3a^3 \int_0^{2\pi} \frac{\sin^2(2\theta)}{4} d\theta \int_0^{\pi} (1-\cos^2\phi)^2 \cos^2\phi \sin\phi \, d\phi$$

Let $t = \cos\phi$, $dt = -\sin\phi \, d\phi$

$$= 3a^3 \int_0^{2\pi} \frac{1-\cos(4\theta)}{8} d\theta \int_{-1}^{1} (1-t^2)^2 t^2 \, dt$$

$$= \frac{3a^3}{8}(2\pi)2 \int_0^1 (t^2 - 2t^4 + t^6)\, dt = \frac{4\pi a^3}{35} \text{ cu. units.}$$

8. One-eighth of the required volume lies in the first octant. Under the transformation $x = u^6$, $y = v^6$, $z = w^6$, the region first-octant R bounded by the surface $x^{1/3} + y^{1/3} + z^{1/3} = a^{1/3}$ and the coordinate planes gets mapped to the first octant part B of the ball bounded by $u^2 + v^2 + w^2 \le a^{1/3}$. Assume that $a > 0$. Since

$$\frac{\partial(x,y,z)}{\partial(u,v,w)} = 6^3 u^5 v^5 w^5,$$

the required volume is

$$V = 8(6^3) \iiint_B u^5 v^5 w^5 \, du \, dv \, dw.$$

Now switch to polar coordinates $[\rho, \phi, \theta]$ in uvw-space. Since

$$uvw = (\rho \sin\phi \cos\theta)(\rho \sin\phi \sin\theta)(\rho \cos\phi),$$

we have

$$V = 1,728 \int_0^{\pi/2} (\cos\theta \sin\theta)^5 \, d\theta \int_0^{\pi/2} (\sin^2\phi \cos\phi)^5 \sin\phi \, d\phi$$

$$\times \int_0^{a^{1/6}} \rho^{17} \, d\rho$$

$$= 96a^3 \int_0^{\pi/2} \frac{\sin^5(2\theta)}{32} d\theta \int_0^{\pi/2} \sin^{11}\phi(1-\sin^2\phi)^2 \cos\phi \, d\phi$$

Let $s = \sin\phi$, $ds = \cos\phi \, d\phi$

$$= 3a^3 \int_0^{\pi/2} (1-\cos^2(2\theta))^2 \sin(2\theta) \, d\theta \int_0^1 s^{11}(1-s^2)^2 \, ds$$

Let $t = \cos(2\theta)$, $dt = -2\sin(2\theta) \, d\theta$

$$= \frac{3a^3}{2} \int_{-1}^1 (1-2t^2+t^4)\, dt \int_0^1 (s^{11} - 2s^{13} + s^{15})\, ds$$

$$= 3a^3 \left(1 - \frac{2}{3} + \frac{1}{5}\right)\left(\frac{1}{12} - \frac{1}{7} + \frac{1}{16}\right) = \frac{a^3}{210} \text{ cu. units.}$$

CHAPTER 15. VECTOR FIELDS

Section 15.1 Vector and Scalar Fields (page 881)

1. $\mathbf{F} = x\mathbf{i} + x\mathbf{j}$.
   The field lines satisfy $\dfrac{dx}{x} = \dfrac{dy}{x}$, i.e., $dy = dx$. The field lines are $y = x + C$, straight lines parallel to $y = x$.

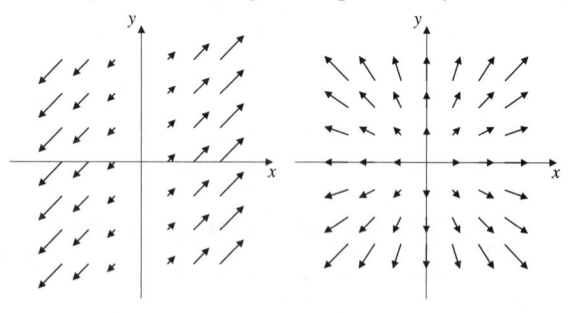

   Fig. 15.1.1   Fig. 15.1.2

2. $\mathbf{F} = x\mathbf{i} + y\mathbf{j}$.
   The field lines satisfy $\dfrac{dx}{x} = \dfrac{dy}{y}$.
   Thus $\ln y = \ln x + \ln C$, or $y = Cx$. The field lines are straight half-lines emanating from the origin.

3. $\mathbf{F} = y\mathbf{i} + x\mathbf{j}$.
   The field lines satisfy $\dfrac{dx}{y} = \dfrac{dy}{x}$.
   Thus $x\,dx = y\,dy$. The field lines are the rectangular hyperbolas (and their asymptotes) given by $x^2 - y^2 = C$.

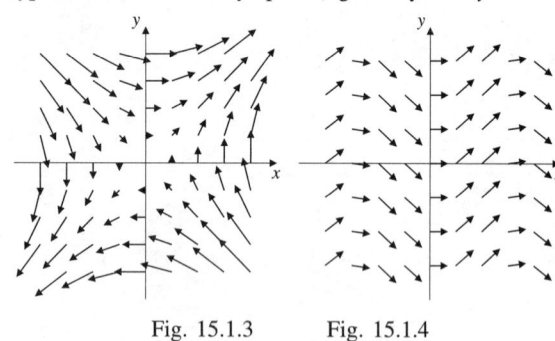

   Fig. 15.1.3   Fig. 15.1.4

4. $\mathbf{F} = \mathbf{i} + \sin x\,\mathbf{j}$.
   The field lines satisfy $dx = \dfrac{dy}{\sin x}$.
   Thus $\dfrac{dy}{dx} = \sin x$. The field lines are the curves $y = -\cos x + C$.

5. $\mathbf{F} = e^x\mathbf{i} + e^{-x}\mathbf{j}$.
   The field lines satisfy $\dfrac{dx}{e^x} = \dfrac{dy}{e^{-x}}$.
   Thus $\dfrac{dy}{dx} = e^{-2x}$. The field lines are the curves $y = -\dfrac{1}{2}e^{-2x} + C$.

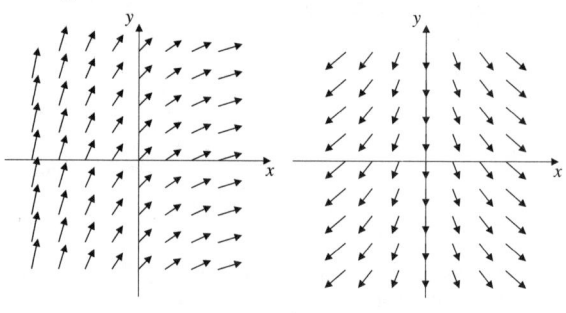

   Fig. 15.1.5   Fig. 15.1.6

6. $\mathbf{F} = \nabla(x^2 - y) = 2x\mathbf{i} - \mathbf{j}$.
   The field lines satisfy $\dfrac{dx}{2x} = \dfrac{dy}{-1}$. They are the curves $y = -\dfrac{1}{2}\ln x + C$.

7. $\mathbf{F} = \nabla \ln(x^2 + y^2) = \dfrac{2x\mathbf{i} + 2y\mathbf{j}}{x^2 + y^2}$.
   The field lines satisfy $\dfrac{dx}{x} = \dfrac{dy}{y}$. Thus they are radial lines $y = Cx$ (and $x = 0$)

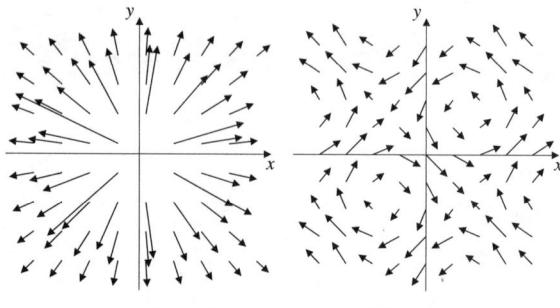

   Fig. 15.1.7   Fig. 15.1.8

8. $\mathbf{F} = \cos y\,\mathbf{i} - \cos x\,\mathbf{j}$.
   The field lines satisfy $\dfrac{dx}{\cos y} = -\dfrac{dy}{\cos x}$, that is, $\cos x\,dx + \cos y\,dy = 0$. Thus they are the curves $\sin x + \sin y = C$.

9. $\mathbf{v}(x, y, z) = y\mathbf{i} - y\mathbf{j} - y\mathbf{k}$.
   The streamlines satisfy $dx = -dy = -dz$. Thus $y + x = C_1$, $z + x = C_2$. The streamlines are straight lines parallel to $\mathbf{i} - \mathbf{j} - \mathbf{k}$.

10. $\mathbf{v}(x,y,z) = x\mathbf{i} + y\mathbf{j} - x\mathbf{k}$.
The streamlines satisfy $\dfrac{dx}{x} = \dfrac{dy}{y} = -\dfrac{dz}{x}$. Thus $z+x = C_1$, $y = C_2 x$. The streamlines are straight half-lines emanating from the z-axis and perpendicular to the vector $\mathbf{i} + \mathbf{k}$.

11. $\mathbf{v}(x,y,z) = y\mathbf{i} - x\mathbf{j} + \mathbf{k}$.
The streamlines satisfy $\dfrac{dx}{y} = -\dfrac{dy}{x} = dz$. Thus $x\,dx + y\,dy = 0$, so $x^2 + y^2 = C_1^2$. Therefore,
$$\frac{dz}{dx} = \frac{1}{y} = \frac{1}{\sqrt{C_1^2 - x^2}}.$$
This implies that $z = \sin^{-1}\dfrac{x}{C_1} + C_2$. The streamlines are the spirals in which the surfaces $x = C_1 \sin(z - C_2)$ intersect the cylinders $x^2 + y^2 = C_1^2$.

12. $\mathbf{v} = \dfrac{x\mathbf{i} + y\mathbf{j}}{(1+z^2)(x^2+y^2)}$.
The streamlines satisfy $dz = 0$ and $\dfrac{dx}{x} = \dfrac{dy}{y}$. Thus $z = C_1$ and $y = C_2 x$. The streamlines are horizontal half-lines emanating from the z-axis.

13. $\mathbf{v} = xz\mathbf{i} + yz\mathbf{j} + x\mathbf{k}$. The field lines satisfy
$$\frac{dx}{xz} = \frac{dy}{yz} = \frac{dz}{x},$$
or, equivalently, $dx/x = dy/y$ and $dx = z\,dz$. Thus the field lines have equations $y = C_1 x$, $2x = z^2 + C_2$, and are therefore parabolas.

14. $\mathbf{v} = e^{xyz}(x\mathbf{i} + y^2\mathbf{j} + z\mathbf{k})$. The field lines satisfy
$$\frac{dx}{x} = \frac{dy}{y^2} = \frac{dz}{z},$$
so they are given by $z = C_1 x$, $\ln|x| = \ln|C_2| - (1/y)$ (or, equivalently, $x = C_2 e^{-1/y}$).

15. $\mathbf{v}(x,y) = x^2\mathbf{i} - y\mathbf{j}$. The field lines satisfy $dx/x^2 = -dy/y$, so they are given by $\ln|y| = (1/x) + \ln|C|$, or $y = Ce^{1/x}$.

16. $\mathbf{v}(x,y) = x\mathbf{i} + (x+y)\mathbf{j}$. The field lines satisfy
$$\frac{dx}{x} = \frac{dy}{x+y}$$
$$\frac{dy}{dx} = \frac{x+y}{x} \quad \text{Let } y = xv(x)$$
$$\frac{dy}{dx} = v + x\frac{dv}{dx}$$
$$v + x\frac{dv}{dx} = \frac{x(1+v)}{x} = 1 + v.$$
Thus $dv/dx = 1/x$, and so $v(x) = \ln|x| + C$. The field lines have equations $y = x\ln|x| + Cx$.

17. $\mathbf{F} = \hat{\mathbf{r}} + r\hat{\boldsymbol{\theta}}$. The field lines satisfy $dr = d\theta$, so they are the spirals $r = \theta + C$.

18. $\mathbf{F} = \hat{\mathbf{r}} + \theta\hat{\boldsymbol{\theta}}$. The field lines satisfy $dr = r\,d\theta/\theta$, or $dr/r = d\theta/\theta$, so they are the spirals $r = C\theta$.

19. $\mathbf{F} = 2\hat{\mathbf{r}} + \theta\hat{\boldsymbol{\theta}}$. The field lines satisfy $dr/2 = r\,d\theta/\theta$, or $dr/r = 2d\theta/\theta$, so they are the spirals $r = C\theta^2$.

20. $\mathbf{F} = r\hat{\mathbf{r}} - \hat{\boldsymbol{\theta}}$. The field lines satisfy $dr/r = -r\,d\theta$, or $-dr/r^2 = d\theta$, so they are the spirals $1/r = \theta + C$, or $r = 1/(\theta + C)$.

Section 15.2 Conservative Fields (page 890)

1. $\mathbf{F} = x\mathbf{i} - 2y\mathbf{j} + 3z\mathbf{k}$, $F_1 = x$, $F_2 = -2y$, $F_3 = 3z$. We have
$$\frac{\partial F_1}{\partial y} = 0 = \frac{\partial F_2}{\partial x},$$
$$\frac{\partial F_1}{\partial z} = 0 = \frac{\partial F_3}{\partial x},$$
$$\frac{\partial F_2}{\partial z} = 0 = \frac{\partial F_3}{\partial y}.$$
Therefore, $\mathbf{F}$ may be conservative. If $\mathbf{F} = \nabla\phi$, then
$$\frac{\partial \phi}{\partial x} = x, \quad \frac{\partial \phi}{\partial y} = -2y, \quad \frac{\partial \phi}{\partial z} = 3z.$$
Evidently $\phi(x,y,z) = \dfrac{x^2}{2} - y^2 + \dfrac{3z^2}{2}$ is a potential for $\mathbf{F}$. Thus $\mathbf{F}$ is conservative on $\mathbb{R}^3$.

2. $\mathbf{F} = y\mathbf{i} + x\mathbf{j} + z^2\mathbf{k}$, $F_1 = y$, $F_2 = x$, $F_3 = z^2$. We have
$$\frac{\partial F_1}{\partial y} = 1 = \frac{\partial F_2}{\partial x},$$
$$\frac{\partial F_1}{\partial z} = 0 = \frac{\partial F_3}{\partial x},$$
$$\frac{\partial F_2}{\partial z} = 0 = \frac{\partial F_3}{\partial y}.$$
Therefore, $\mathbf{F}$ may be conservative. If $\mathbf{F} = \nabla\phi$, then
$$\frac{\partial \phi}{\partial x} = y, \quad \frac{\partial \phi}{\partial y} = x, \quad \frac{\partial \phi}{\partial z} = z^2.$$
Therefore,
$$\phi(x,y,z) = \int y\,dx = xy + C_1(y,z)$$
$$x = \frac{\partial \phi}{\partial y} = x + \frac{\partial C_1}{\partial y} \Rightarrow \frac{\partial C_1}{\partial y} = 0$$
$$C_1(y,z) = C_2(z), \quad \phi(x,y,z) = xy + C_2(z)$$
$$z^2 = \frac{\partial \phi}{\partial z} = C_2'(z) \Rightarrow C_2(z) = \frac{z^3}{3}.$$

Thus $\phi(x,y,z) = xy + \dfrac{z^3}{3}$ is a potential for **F**, and **F** is conservative on $\mathbb{R}^3$.

3. $\mathbf{F} = \dfrac{x\mathbf{i} - y\mathbf{j}}{x^2 + y^2}$, $F_1 = \dfrac{x}{x^2+y^2}$, $F_2 = -\dfrac{y}{x^2+y^2}$. We have

$$\dfrac{\partial F_1}{\partial y} = -\dfrac{2xy}{(x^2+y^2)^2}, \quad \dfrac{\partial F_2}{\partial x} = \dfrac{2xy}{(x^2+y^2)^2}.$$

Thus **F** cannot be conservative.

4. $\mathbf{F} = \dfrac{x\mathbf{i} + y\mathbf{j}}{x^2 + y^2}$, $F_1 = \dfrac{x}{x^2+y^2}$, $F_2 = \dfrac{y}{x^2+y^2}$. We have

$$\dfrac{\partial F_1}{\partial y} = -\dfrac{2xy}{(x^2+y^2)^2} = \dfrac{\partial F_2}{\partial x}.$$

Therefore, **F** may be conservative. If $\mathbf{F} = \nabla\phi$, then

$$\dfrac{\partial \phi}{\partial x} = \dfrac{x}{x^2+y^2}, \quad \dfrac{\partial \phi}{\partial y} = \dfrac{y}{x^2+y^2}.$$

Therefore,

$$\phi(x,y) = \int \dfrac{x}{x^2+y^2}\,dx = \dfrac{\ln(x^2+y^2)}{2} + C_1(y)$$

$$\dfrac{y}{x^2+y^2} = \dfrac{\partial\phi}{\partial y} = \dfrac{y}{x^2+y^2} + c_1'(y) \Rightarrow c_1'(y) = 0.$$

Thus we can choose $C_1(y) = 0$, and

$$\phi(x,y) = \dfrac{1}{2}\ln(x^2+y^2)$$

is a scalar potential for **F**, and **F** is conservative everywhere on $\mathbb{R}^2$ except at the origin.

5. $\mathbf{F} = (2xy - z^2)\mathbf{i} + (2yz + x^2)\mathbf{j} - (2zx - y^2)\mathbf{k}$,
$F_1 = 2xy - z^2$, $F_2 = 2yz + x^2$, $F_3 = y^2 - 2zx$. We have

$$\dfrac{\partial F_1}{\partial y} = 2x = \dfrac{\partial F_2}{\partial x},$$
$$\dfrac{\partial F_1}{\partial z} = -2z = \dfrac{\partial F_3}{\partial x},$$
$$\dfrac{\partial F_2}{\partial z} = 2y = \dfrac{\partial F_3}{\partial y}.$$

Therefore, **F** may be conservative. If $\mathbf{F} = \nabla\phi$, then

$$\dfrac{\partial\phi}{\partial x} = 2xy - z^2, \quad \dfrac{\partial\phi}{\partial y} = 2yz + x^2,$$
$$\dfrac{\partial\phi}{\partial z} = y^2 - 2zx.$$

Therefore,

$$\phi(x,y,z) = \int (2xy - z^2)\,dx = x^2y - xz^2 + C_1(y,z)$$
$$2yz + x^2 = \dfrac{\partial\phi}{\partial y} = x^2 + \dfrac{\partial C_1}{\partial y}$$
$$\Rightarrow \dfrac{\partial C_1}{\partial y} = 2yz \Rightarrow C_1(y,z) = y^2z + C_2(z)$$
$$\phi(x,y,z) = x^2y - xz^2 + y^2z + C_2(z)$$
$$y^2 - 2zx = \dfrac{\partial\phi}{\partial z} = -2xz + y^2 + C_2'(z)$$
$$\Rightarrow C_2'(z) = 0.$$

Thus $\phi(x,y,z) = x^2y - xz^2 + y^2z$ is a scalar potential for **F**, and **F** is conservative on $\mathbb{R}^3$.

6. $\mathbf{F} = e^{x^2+y^2+z^2}(xz\mathbf{i} + yz\mathbf{j} + xy\mathbf{k})$.
$F_1 = xze^{x^2+y^2+z^2}$, $F_2 = yze^{x^2+y^2+z^2}$,
$F_3 = xye^{x^2+y^2+z^2}$. We have

$$\dfrac{\partial F_1}{\partial y} = 2xyze^{x^2+y^2+z^2} = \dfrac{\partial F_2}{\partial x},$$
$$\dfrac{\partial F_1}{\partial z} = (x + 2xz^2)e^{x^2+y^2+z^2},$$
$$\dfrac{\partial F_3}{\partial x} = (y + 2x^2y)e^{x^2+y^2+z^2} \neq \dfrac{\partial F_1}{\partial z}.$$

Thus **F** cannot be conservative.

7. $\phi(\mathbf{r}) = \dfrac{1}{|\mathbf{r} - \mathbf{r}_0|^2}$
$$\dfrac{\partial\phi}{\partial x} = -\dfrac{2}{|\mathbf{r} - \mathbf{r}_0|^3}\dfrac{\partial}{\partial x}|\mathbf{r} - \mathbf{r}_0|$$
$$= -\dfrac{2}{|\mathbf{r} - \mathbf{r}_0|^3}\dfrac{(\mathbf{r} - \mathbf{r}_0)\bullet\dfrac{\partial\mathbf{r}}{\partial x}}{|\mathbf{r} - \mathbf{r}_0|}$$
$$= -\dfrac{2(x - x_0)}{|\mathbf{r} - \mathbf{r}_0|^4}.$$

Since similar formulas hold for the other first partials of ϕ, we have

$$\mathbf{F} = \nabla\phi$$
$$= -\dfrac{2}{|\mathbf{r}-\mathbf{r}_0|^4}\left[(x-x_0)\mathbf{i} + (y-y_0)\mathbf{j} + (z-z_0)\mathbf{k}\right]$$
$$= -2\dfrac{\mathbf{r} - \mathbf{r}_0}{|\mathbf{r} - \mathbf{r}_0|^4}.$$

This is the vector field whose scalar potential is ϕ.

8. $\dfrac{\partial}{\partial x}\ln|\mathbf{r}| = \dfrac{1}{|\mathbf{r}|}\dfrac{\mathbf{r}\bullet\dfrac{\partial\mathbf{r}}{\partial x}}{|\mathbf{r}|} = \dfrac{x}{|\mathbf{r}|^2}$

$\nabla\ln|\mathbf{r}| = \dfrac{x\mathbf{i} + y\mathbf{j} + z\mathbf{k}}{|\mathbf{r}|^2} = \dfrac{\mathbf{r}}{|\mathbf{r}|^2}.$

9. $\mathbf{F} = \dfrac{2x}{z}\mathbf{i} + \dfrac{2y}{z}\mathbf{j} - \dfrac{x^2+y^2}{z^2}\mathbf{k}$,

$F_1 = \dfrac{2x}{z}$, $F_2 = \dfrac{2y}{z}$, $F_3 = -\dfrac{x^2+y^2}{z^2}$. We have

$$\dfrac{\partial F_1}{\partial y} = 0 = \dfrac{\partial F_2}{\partial x},$$
$$\dfrac{\partial F_1}{\partial z} = -\dfrac{2x}{z^2} = \dfrac{\partial F_3}{\partial x},$$
$$\dfrac{\partial F_2}{\partial z} = -\dfrac{2y}{z^2} = \dfrac{\partial F_3}{\partial y}.$$

Therefore, $\mathbf{F}$ may be conservative in $\mathbb{R}^3$ except on the plane $z=0$ where it is not defined. If $\mathbf{F} = \nabla\phi$, then

$$\dfrac{\partial \phi}{\partial x} = \dfrac{2x}{z}, \quad \dfrac{\partial \phi}{\partial y} = \dfrac{2y}{z}, \quad \dfrac{\partial \phi}{\partial z} = -\dfrac{x^2+y^2}{z^2}.$$

Therefore,

$$\phi(x,y,z) = \int \dfrac{2x}{z}\,dx = \dfrac{x^2}{z} + C_1(y,z)$$
$$\dfrac{2y}{z} = \dfrac{\partial \phi}{\partial y} = \dfrac{\partial C_1}{\partial y} \Rightarrow C_1(y,z) = \dfrac{y^2}{z} + C_2(z)$$
$$\phi(x,y,z) = \dfrac{x^2+y^2}{z} + C_2(z)$$
$$-\dfrac{x^2+y^2}{z^2} = \dfrac{\partial \phi}{\partial z} = -\dfrac{x^2+y^2}{z^2} + C_2'(z)$$
$$\Rightarrow C_2(z) = 0.$$

Thus $\phi(x,y,z) = \dfrac{x^2+y^2}{z}$ is a potential for $\mathbf{F}$, and $\mathbf{F}$ is conservative on $\mathbb{R}^3$ except on the plane $z=0$.

The equipotential surfaces have equations

$$\dfrac{x^2+y^2}{z} = C, \quad \text{or} \quad Cz = x^2+y^2.$$

Thus the equipotential surfaces are circular paraboloids.

The field lines of $\mathbf{F}$ satisfy

$$\dfrac{dx}{\frac{2x}{z}} = \dfrac{dy}{\frac{2y}{z}} = \dfrac{dz}{-\frac{x^2+y^2}{z^2}}.$$

From the first equation, $\dfrac{dx}{x} = \dfrac{dy}{y}$, so $y = Ax$ for an arbitrary constant A. Therefore

$$\dfrac{dx}{2x} = \dfrac{z\,dz}{-(x^2+y^2)} = \dfrac{z\,dz}{-x^2(1+A^2)},$$

so $-(1+A^2)x\,dx = 2z\,dz$. Hence

$$\dfrac{1+A^2}{2}x^2 + z^2 = \dfrac{B}{2},$$

or $x^2 + y^2 + 2z^2 = B$, where B is a second arbitrary constant. The field lines of $\mathbf{F}$ are the ellipses in which the vertical planes containing the z-axis intersect the ellipsoids $x^2 + y^2 + 2z^2 = B$. These ellipses are orthogonal to all the equipotential surfaces of $\mathbf{F}$.

10. $\mathbf{F} = \dfrac{2x}{z}\mathbf{i} + \dfrac{2y}{z}\mathbf{j} - \dfrac{x^2+y^2}{z^2}\mathbf{k} = \mathbf{G} + \mathbf{k}$,

where $\mathbf{G}$ is the vector field $\mathbf{F}$ of Exercise 9. Since $\mathbf{G}$ is conservative (except on the plane $z=0$), so is $\mathbf{F}$, which has scalar potential

$$\phi(x,y,z) = \dfrac{x^2+y^2}{z} + z = \dfrac{x^2+y^2+z^2}{z},$$

since $\dfrac{x^2+y^2}{z}$ is a potential for $\mathbf{G}$ and z is a potential for the vector $\mathbf{k}$.

The equipotential surfaces of $\mathbf{F}$ are $\phi(x,y,z) = C$, or

$$x^2 + y^2 + z^2 = Cz$$

which are spheres tangent to the xy-plane having centres on the z-axis.

The field lines of $\mathbf{F}$ satisfy

$$\dfrac{dx}{\frac{2x}{z}} = \dfrac{dy}{\frac{2y}{z}} = \dfrac{dz}{1 - \frac{x^2+y^2}{z^2}}.$$

As in Exercise 9, the first equation has solutions $y = Ax$, representing vertical planes containing the z-axis. The remaining equations can then be written in the form

$$\dfrac{dz}{dx} = \dfrac{z^2 - x^2 - y^2}{2xz} = \dfrac{z^2 - (1+A^2)x^2}{2zx}.$$

This first order DE is of homogeneous type (see Section 17.2), and can be solved by a change of dependent variable: $z = xv(x)$. We have

$$v + x\dfrac{dv}{dx} = \dfrac{dz}{dx} = \dfrac{x^2v^2 - (1+A^2)x^2}{2x^2v}$$
$$x\dfrac{dv}{dx} = \dfrac{v^2 - (1+A^2)}{2v} - v = -\dfrac{v^2 + (1+A^2)}{2v}$$
$$\dfrac{2v\,dv}{v^2 + (1+A^2)} = -\dfrac{dx}{x}$$
$$\ln\bigl(v^2 + (1+A^2)\bigr) = -\ln x + \ln B$$
$$v^2 + 1 + A^2 = \dfrac{B}{x}$$
$$\dfrac{z^2}{x^2} + 1 + A^2 = \dfrac{B}{x}$$
$$z^2 + x^2 + y^2 = Bx.$$

These are spheres centred on the x-axis and passing through the origin. The field lines are the intersections of the planes $y = Ax$ with these spheres, so they are vertical circles passing through the origin and having centres in the xy-plane. (The technique used to find these circles excludes those circles with centres on the y-axis, but they are also field lines of **F**.)

Note: In two dimensions, circles passing through the origin and having centres on the x-axis intersect perpendicularly circles passing through the origin and having centres on the y-axis. Thus the nature of the field lines of **F** can be determined geometrically from the nature of the equipotential surfaces.

11. The scalar potential for the two-source system is

$$\phi(x, y, z) = \phi(\mathbf{r}) = -\frac{m}{|\mathbf{r} - \ell\mathbf{k}|} - \frac{m}{|\mathbf{r} + \ell\mathbf{k}|}.$$

Hence the velocity field is given by

$$\mathbf{v}(\mathbf{r}) = \nabla\phi(\mathbf{r})$$
$$= \frac{m(\mathbf{r} - \ell\mathbf{k})}{|\mathbf{r} - v\mathbf{k}|^3} + \frac{m(\mathbf{r} + \ell\mathbf{k})}{|\mathbf{r} + \ell\mathbf{k}|^3}$$
$$= \frac{m(x\mathbf{i} + y\mathbf{j} + (z - \ell)\mathbf{k})}{[x^2 + y^2 + (z - \ell)^2]^{3/2}} + \frac{m(x\mathbf{i} + y\mathbf{j} + (z + \ell)\mathbf{k})}{[x^2 + y^2 + (z - \ell)^2]^{3/2}}.$$

Observe that $v_1 = 0$ if and only if $x = 0$, and $v_2 = 0$ if and only if $y = 0$. Also

$$\mathbf{v}(0, 0, z) = m\left(\frac{z - \ell}{|z - \ell|^3} + \frac{z + \ell}{|z + \ell|^3}\right)\mathbf{k},$$

which is **0** if and only if $z = 0$. Thus $\mathbf{v} = \mathbf{0}$ only at the origin.

At points in the xy-plane we have

$$\mathbf{v}(x, y, 0) = \frac{2m(x\mathbf{i} + y\mathbf{j})}{(x^2 + y^2 + \ell^2)^{3/2}}.$$

The velocity is radially away from the origin in the xy-plane, as is appropriate by symmetry. The speed at $(x, y, 0)$ is

$$v(x, y, 0) = \frac{2m\sqrt{x^2 + y^2}}{(x^2 + y^2 + \ell^2)^{3/2}} = \frac{2ms}{(s^2 + \ell^2)^{3/2}} = g(s),$$

where $s = \sqrt{x^2 + y^2}$. For maximum $g(s)$ we set

$$0 = g'(s) = 2m\frac{(s^2 + \ell^2)^{3/2} - \frac{3}{2}s(s^2 + \ell^2)^{1/2}2s}{(s^2 + \ell^2)^3}$$
$$= \frac{2m(\ell^2 - 2s^2)}{(s^2 + \ell^2)^{5/2}}.$$

Thus, the speed in the xy-plane is greatest at points of the circle $x^2 + y^2 = \ell^2/2$.

12. The scalar potential for the source-sink system is

$$\phi(x, y, z) = \phi(\mathbf{r}) = -\frac{2}{|\mathbf{r}|} + \frac{1}{|\mathbf{r} - \mathbf{k}|}.$$

Thus, the velocity field is

$$\mathbf{v} = \nabla\phi = \frac{2\mathbf{r}}{|\mathbf{r}|^3} - \frac{\mathbf{r} - \mathbf{k}}{|\mathbf{r} - \mathbf{k}|^3}$$
$$= \frac{2(x\mathbf{i} + y\mathbf{j} + z\mathbf{k})}{(x^2 + y^2 + z^2)^{3/2}} - \frac{x\mathbf{i} + y\mathbf{j} + (z - 1)\mathbf{k}}{(x^2 + y^2 + (z - 1)^2)^{3/2}}.$$

For vertical velocity we require

$$\frac{2x}{(x^2 + y^2 + z^2)^{3/2}} = \frac{x}{(x^2 + y^2 + (z - 1)^2)^{3/2}},$$

and a similar equation for y. Both equations will be satisfied at all points of the z-axis, and also wherever

$$2\left(x^2 + y^2 + (z - 1)^2\right)^{3/2} = \left(x^2 + y^2 + z^2\right)^{3/2}$$
$$2^{2/3}\left(x^2 + y^2 + (z - 1)^2\right) = x^2 + y^2 + z^2$$
$$x^2 + y^2 + (z - K)^2 = K^2 - K,$$

where $K = 2^{2/3}/(2^{2/3} - 1)$. This latter equation represents a sphere, S, since $K^2 - K > 0$. The velocity is vertical at all points on S, as well as at all points on the z-axis.

Since the source at the origin is twice as strong as the sink at $(0, 0, 1)$, only half the fluid it emits will be sucked into the sink. By symmetry, this half will the half emitted into the half-space $z > 0$. The rest of the fluid emitted at the origin will flow outward to infinity. There is one point where $\mathbf{v} = \mathbf{0}$. This point (which is easily calculated to be $(0, 0, 2 + \sqrt{2})$) lies inside S. Streamlines emerging from the origin parallel to the xy-plane lead to this point. Streamlines emerging into $z > 0$ cross S and approach the sink. Streamlines emerging into $z < 0$ flow to infinity. Some of these cross S twice, some others are tangent to S, some do not intersect S anywhere.

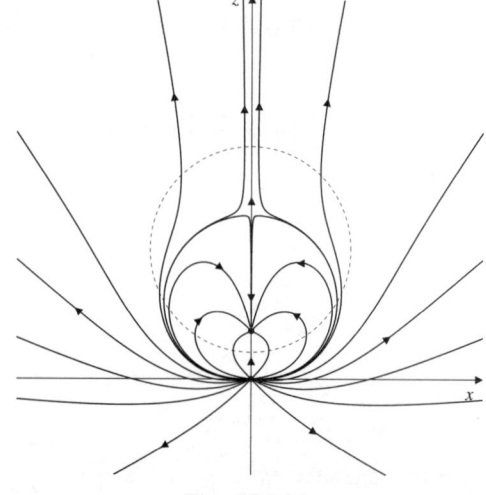

Fig. 15.2.12

13. Fluid emitted by interval Δz in time interval $[0, t]$ occupies, at time t, a cylinder of radius r, where
$$\pi r^2 \Delta Z = \text{vol. of cylinder} = 2\pi m t \Delta z.$$
Thus $r^2 = 2mt$, and $r\dfrac{dr}{dt} = m$. The surface of this cylinder is moving away from the z-axis at rate
$$\frac{dr}{dt} = \frac{m}{r} = \frac{m}{\sqrt{x^2+y^2}},$$
so the velocity at any point (x, y, z) is
$$\mathbf{v} = \frac{m}{\sqrt{x^2+y^2}} \times \text{unit vector in direction } x\mathbf{i}+y\mathbf{j}$$
$$= \frac{m(x\mathbf{i}+y\mathbf{j})}{x^2+y^2}.$$

14. For $\mathbf{v}(x, y) = \dfrac{m(x\mathbf{i}+y\mathbf{j})}{x^2+y^2}$, we have
$$\frac{\partial v_1}{\partial y} = -\frac{2mxy}{(x^2+y^2)^2} = \frac{\partial v_2}{\partial x},$$
so $\mathbf{v}$ may be conservative, except at $(0,0)$. We have
$$\phi(x, y) = m \int \frac{x\,dx}{x^2+y^2} = \frac{m}{2}\ln(x^2+y^2) + C_1(y)$$
$$\frac{my}{x^2+y^2} = \frac{\partial \phi}{\partial y} = \frac{my}{x^2+y^2} + \frac{dC_1}{dy}.$$
Thus we may take $C_1(y) = 0$, and obtain
$$\phi(x, y) = \frac{m}{2}\ln(x^2+y^2) = m\ln|\mathbf{r}|,$$
as a scalar potential for the velocity field $\mathbf{v}$ of a line source of strength of m.

15. The two-dimensional dipole of strength μ has potential
$$\phi(x, y)$$
$$= \lim_{\substack{\ell \to 0 \\ m\ell = \mu}} \frac{m}{2}\left[\ln\left(x^2 + \left(y-\frac{\ell}{2}\right)^2\right) - \ln\left(x^2 + \left(y+\frac{\ell}{2}\right)^2\right)\right]$$
$$= \frac{\mu}{2}\lim_{\ell \to 0} \frac{\ln\left(x^2 + \left(y-\frac{\ell}{2}\right)^2\right) - \ln\left(x^2 + \left(y+\frac{\ell}{2}\right)^2\right)}{\ell}$$
(apply l'Hôpital's Rule)
$$= \frac{\mu}{2}\lim_{\ell \to 0} \frac{-\left(y-\frac{\ell}{2}\right)}{x^2 + \left(y-\frac{\ell}{2}\right)^2} - \frac{\left(y+\frac{\ell}{2}\right)}{x^2 + \left(y+\frac{\ell}{2}\right)^2}$$
$$= -\frac{\mu y}{x^2+y^2} = -\frac{\mu y}{r^2}.$$

Now
$$\frac{\partial \phi}{\partial x} = \frac{2\mu y}{r^3}\frac{\partial r}{\partial x} = \frac{2\mu xy}{r^4}$$
$$\frac{\partial \phi}{\partial y} = -\mu\frac{r^2 - 2yr\frac{y}{r}}{r^4} = \frac{\mu(y^2-x^2)}{r^4}.$$
Thus
$$\mathbf{F} = \nabla\phi = \frac{\mu}{(x^2+y^2)^2}\left(2xy\mathbf{i} + (y^2-x^2)\mathbf{j}\right).$$

16. The equipotential curves for the two-dimensional dipole have equations $y = 0$ or
$$-\frac{\mu y}{x^2+y^2} = \frac{1}{C}$$
$$x^2 + y^2 + \mu C y = 0$$
$$x^2 + \left(y + \frac{\mu C}{2}\right)^2 = \frac{\mu^2 C^2}{4}.$$
These equipotentials are circles tangent to the x-axis at the origin.

17. All circles tangent to the y-axis at the origin intersect all circles tangent to the x-axis at the origin at right angles, so they must be the streamlines of the two-dimensional dipole.

As an alternative derivation of this fact, the streamlines must satisfy
$$\frac{dx}{2xy} = \frac{dy}{y^2-x^2},$$
or, equivalently,
$$\frac{dy}{dx} = \frac{y^2-x^2}{2xy}.$$
This homogeneous DE can be solved (as was that in Exercise 10) by a change in dependent variable. Let $y = xv(x)$. Then
$$v + x\frac{dv}{dx} = \frac{dy}{dx} = \frac{v^2 x^2 - x^2}{2vx^2}$$
$$x\frac{dv}{dx} = \frac{v^2-1}{2v} - v = -\frac{v^2+1}{2v}$$
$$\frac{2v\,dv}{v^2+1} = -\frac{dx}{x}$$
$$\ln(v^2+1) = -\ln x + \ln C$$
$$v^2 + 1 = \frac{C}{x} \quad \Rightarrow \quad \frac{y^2}{x^2} + 1 = \frac{C}{x}$$
$$x^2 + y^2 = Cx$$
$$(x - C)^2 + y^2 = C^2.$$
These streamlines are circles tangent to the y-axis at the origin.

18. The velocity field for a point source of strength $m\,dt$ at $(0,0,t)$ is
$$\mathbf{v}_t(x,y,z) = \frac{m(x\mathbf{i} + y\mathbf{j} + (z-t)\mathbf{k})}{(x^2 + y^2 + (z-t)^2)^{3/2}}.$$

Hence we have
$$\int_{-\infty}^{\infty} \mathbf{v}_t(x,y,z)\,dt$$
$$= m\int_{-\infty}^{\infty} \frac{x\mathbf{i} + y\mathbf{j} + (z-t)\mathbf{k}}{(x^2 + y^2 + (z-t)^2)^{3/2}}\,dt$$
$$= m(x\mathbf{i} + y\mathbf{j})\int_{-\infty}^{\infty} \frac{dt}{(x^2 + y^2 + (z-t)^2)^{3/2}}$$

Let $z - t = \sqrt{x^2 + y^2}\tan\theta$
$-dt = \sqrt{x^2 + y^2}\sec^2\theta\,d\theta$

$$= \frac{m(x\mathbf{i} + y\mathbf{j})}{x^2 + y^2}\int_{-\pi/2}^{\pi/2} \cos\theta\,d\theta$$
$$= \frac{2m(x\mathbf{i} + y\mathbf{j})}{x^2 + y^2},$$

which is the velocity field of a line source of strength $2m$ along the z-axis.

The definition of strength of a point source in 3-space was made to ensure that the velocity field of a source of strength 1 had speed 1 at distance 1 from the source. This corresponds to fluid being emitted from the source at a volume rate of 4π. Similarly, the definition of strength of a line source guaranteed that a source of strength 1 gives rise to fluid speed of 1 at unit distance 1 from the line source. This corresponds to a fluid emission at a volume rate 2π per unit length along the line. Thus, the integral of a 3-dimensional source gives twice the volume rate of a 2-dimensional source, per unit length along the line.

The potential of a point source $m\,dt$ at $(0,0,t)$ is
$$\phi(x,y,z) = -\frac{m}{\sqrt{x^2 + y^2 + (x-t)^2}}.$$

This potential cannot be integrated to give the potential for a line source along the z-axis because the integral
$$-m\int_{-\infty}^{\infty} \frac{dt}{\sqrt{x^2 + y^2 + (z-t)^2}}$$
does not converge, in the usual sense in which convergence of improper integrals was defined.

19. Since $x = r\cos\theta$ and $y = r\sin\theta$, we have
$$\frac{\partial\phi}{\partial r} = \cos\theta\,\frac{\partial\phi}{\partial x} + \sin\theta\,\frac{\partial\phi}{\partial y}$$
$$\frac{\partial\phi}{\partial\theta} = -r\sin\theta\,\frac{\partial\phi}{\partial x} + r\cos\theta\,\frac{\partial\phi}{\partial y}.$$

Also,
$$\hat{\mathbf{r}} = \frac{x\mathbf{i} + y\mathbf{j}}{r} = (\cos\theta)\mathbf{i} + (\sin\theta)\mathbf{j}$$
$$\hat{\boldsymbol{\theta}} = \frac{-y\mathbf{i} + x\mathbf{j}}{r} = -(\sin\theta)\mathbf{i} + (\cos\theta)\mathbf{j}.$$

Therefore,
$$\frac{\partial\phi}{\partial r}\hat{\mathbf{r}} + \frac{1}{r}\frac{\partial\phi}{\partial\theta}\hat{\boldsymbol{\theta}}$$
$$= \left(\cos^2\theta\,\frac{\partial\phi}{\partial x} + \sin\theta\cos\theta\,\frac{\partial\phi}{\partial y}\right)\mathbf{i}$$
$$+ \left(\cos\theta\sin\theta\,\frac{\partial\phi}{\partial x} + \sin^2\theta\,\frac{\partial\phi}{\partial y}\right)\mathbf{j}$$
$$+ \left(\sin^2\theta\,\frac{\partial\phi}{\partial x} - \sin\theta\cos\theta\,\frac{\partial\phi}{\partial y}\right)\mathbf{i}$$
$$+ \left(-\cos\theta\sin\theta\,\frac{\partial\phi}{\partial x} + \cos^2\theta\,\frac{\partial\phi}{\partial y}\right)\mathbf{j}$$
$$= \frac{\partial\phi}{\partial x}\mathbf{i} + \frac{\partial\phi}{\partial y}\mathbf{j} = \nabla\phi.$$

20. If $\mathbf{F} = F_r(r,\theta)\hat{\mathbf{r}} + F_\theta(r,\theta)\hat{\boldsymbol{\theta}}$ is conservative, then $\mathbf{F} = \nabla\phi$ for some scalar field $\phi(r,\theta)$, and by Exercise 19,
$$\frac{\partial\phi}{\partial r} = F_r, \quad \frac{1}{r}\frac{\partial\phi}{\partial\theta} = F_\theta.$$

For the equality of the mixed second partial derivatives of ϕ, we require that
$$\frac{\partial F_r}{\partial\theta} = \frac{\partial}{\partial r}(rF_\theta) = F_\theta + r\frac{\partial F_\theta}{\partial r},$$
that is, $\dfrac{\partial F_r}{\partial\theta} - r\dfrac{\partial F_\theta}{\partial r} = F_\theta$.

21. If $\mathbf{F} = r\sin(2\theta)\hat{\mathbf{r}} + r\cos(2\theta)\hat{\boldsymbol{\theta}} = \nabla\phi(r,\theta)$, then we must have
$$\frac{\partial\phi}{\partial r} = r\sin(2\theta), \quad \frac{1}{r}\frac{\partial\phi}{\partial\theta} = r\cos(2\theta).$$
Both of these equations are satisfied by
$$\phi(r\theta) = \frac{1}{2}r^2\sin(2\theta) + C,$$
so $\mathbf{F}$ is conservative and this ϕ is a potential for it.

22. If $\mathbf{F} = r^2\cos\theta\,\hat{\mathbf{r}} + \alpha r^\beta\sin\theta\,\hat{\boldsymbol{\theta}} = \nabla\phi(r,\theta)$, then we must have
$$\frac{\partial\phi}{\partial r} = r^2\cos\theta, \quad \frac{1}{r}\frac{\partial\phi}{\partial\theta} = \alpha r^\beta\sin\theta.$$

From the first equation
$$\phi(r,\theta) = \frac{r^3}{3}\cos\theta + C(\theta).$$

INSTRUCTOR'S SOLUTIONS MANUAL SECTION 15.3 (PAGE 896)

The second equation then gives

$$C'(\theta) - \frac{r^3}{3}\sin\theta = \frac{\partial \phi}{\partial \theta} = \alpha r^{\beta+1}\sin\theta.$$

This equation can be solved for a function $C(\theta)$ independent of r only if $\alpha = -1/3$ and $\beta = 2$. In this case, $C(\theta) = C$ (a constant). **F** is conservative if α and β have these values, and a potential for it is $\phi = \frac{1}{3}r^3\cos\theta + C$.

Section 15.3   Line Integrals   (page 896)

1. C: $\mathbf{r} = a\cos t\sin t\mathbf{i} + a\sin^2 t\mathbf{j} + a\cos t\mathbf{k}$, $0 \le t \le \pi/2$.
Since

$$|\mathbf{r}|^2 = a^2(\cos^2 t\sin^2 t + \sin^4 t + \cos^2 t) = a^2$$

for all t, C must lie on the sphere of radius a centred at the origin. We have

$$ds = a\sqrt{(\cos^2 t - \sin^2 t)^2 + 4\sin^2 t\cos^2 t + \sin^2 t}\,dt$$
$$= a\sqrt{\cos^2 2t + \sin^2 2t + \sin^2 t}\,dt$$
$$= a\sqrt{1 + \sin^2 t}\,dt.$$

Thus

$$\int_C z\,ds = \int_0^{\pi/2} a\cos t\,a\sqrt{1 + \sin^2 t}\,dt \quad \begin{aligned}\text{Let } u &= \sin t\\ du &= \cos t\,dt\end{aligned}$$
$$= a^2\int_0^1\sqrt{1 + u^2}\,du \quad \begin{aligned}\text{Let } u &= \tan\phi\\ du &= \sec^2\phi\,d\phi\end{aligned}$$
$$= a^2\int_0^{\pi/4}\sec^3\phi\,d\phi$$
$$= \frac{a^2}{2}\Big[\sec\phi\tan\phi + \ln|\sec\phi + \tan\phi|\Big]\Big|_0^{\pi/4}$$
$$= \frac{a^2}{2}\left(\sqrt{2} + \ln(1+\sqrt{2})\right).$$

2. C: $x = t\cos t$, $y = t\sin t$, $z = t$, $(0 \le t \le 2\pi)$. We have

$$ds = \sqrt{(\cos t - t\sin t)^2 + (\sin t + t\cos t)^2 + 1}\,dt$$
$$= \sqrt{2 + t^2}\,dt.$$

Thus

$$\int_C z\,ds = \int_0^{2\pi} t\sqrt{2 + t^2}\,dt \quad \begin{aligned}\text{Let } u &= 2 + t^2\\ du &= 2t\,dt\end{aligned}$$
$$= \frac{1}{2}\int_2^{2+4\pi^2} u^{1/2}\,du$$
$$= \frac{1}{3}u^{3/2}\Big|_2^{2+4\pi^2} = \frac{(2+4\pi^2)^{3/2} - 2^{3/2}}{3}.$$

3. Wire: $\mathbf{r} = 3t\mathbf{i} + 3t^2\mathbf{j} + 2t^3\mathbf{k}$, $(0 \le t \le 1)$
$\mathbf{v} = 3\mathbf{i} + 6t\mathbf{j} + 6t^2\mathbf{k}$
$v = 3\sqrt{1 + 4t^2 + 4t^4} = 3(1 + 2t^2)$.
If the wire has density $\delta(t) = 1 + t$ g/unit length, then its mass is

$$m = 3\int_0^1 (1 + 2t^2)(1 + t)\,dt$$
$$= 3\left(t + \frac{t^2}{2} + \frac{2t^3}{3} + \frac{t^4}{2}\right)\Big|_0^1 = 8\text{ g}.$$

4. The wire of Example 3 lies in the first octant on the surfaces $z = x^2$ and $z = 2 - x^2 - 2y^2$, and, therefore, also on the surface $x^2 = 2 - x^2 - 2y^2$, or $x^2 + y^2 = 1$, a circular cylinder. Since it goes from $(1, 0, 1)$ to $(0, 1, 0)$ it can be parametrized

$$\mathbf{r} = \cos t\mathbf{i} + \sin t\mathbf{j} + \cos^2 t\mathbf{k}, \quad (0 \le t \le \pi/2)$$
$$\mathbf{v} = -\sin t\mathbf{i} + \cos t\mathbf{j} - 2\cos t\sin t\mathbf{k}$$
$$v = \sqrt{1 + \sin^2(2t)} = \sqrt{2 - \cos^2(2t)}.$$

Since the wire has density $\delta = xy = \sin t\cos t = \frac{1}{2}\sin(2t)$, its mass is

$$m = \frac{1}{2}\int_0^{\pi/2}\sqrt{2 - \cos^2(2t)}\sin(2t)\,dt \quad \begin{aligned}\text{Let } v &= \cos(2t)\\ dv &= -2\sin(2t)\,dt\end{aligned}$$
$$= \frac{1}{4}\int_{-1}^1\sqrt{2 - v^2}\,dv = \frac{1}{2}\int_0^1\sqrt{2 - v^2}\,dv,$$

which is the same integral obtained in Example 3, and has value $(\pi + 2)/8$.

5. C: $\mathbf{r} = e^t\cos t\mathbf{i} + e^t\sin t\mathbf{j} + t\mathbf{k}$, $0 \le t \le 2\pi$).

$$ds = \sqrt{e^{2t}(\cos t - \sin t)^2 + e^{2t}(\sin t + \cos t)^2 + 1}\,dt$$
$$= \sqrt{1 + 2e^{2t}}\,dt.$$

The moment of inertia of C about the z-axis is

$$I = \delta\int_C (x^2 + y^2)\,ds$$
$$= \delta\int_0^{2\pi} e^{2t}\sqrt{1 + 2e^{2t}}\,dt \quad \begin{aligned}\text{Let } u &= 1 + 2e^{2t}\\ du &= 4e^{2t}\,dt\end{aligned}$$
$$= \frac{\delta}{4}\int_3^{1+2e^{4\pi}}\sqrt{u}\,du$$
$$= \frac{\delta}{6}u^{3/2}\Big|_3^{1+2e^{4\pi}} = \frac{\delta}{6}\left[(1 + 2e^{4\pi})^{3/2} - 3^{3/2}\right].$$

555

6. $\mathcal{C}$ is the same curve as in Exercise 5. We have

$$\int_\mathcal{C} e^z\, ds = \int_0^{2\pi} e^t\sqrt{1+2e^{2t}}\, dt \quad \text{Let } \sqrt{2}e^t = \tan\theta$$
$$\sqrt{2}e^t\, dt = \sec^2\theta\, d\theta$$
$$= \frac{1}{\sqrt{2}} \int_{t=0}^{t=2\pi} \sec^3\theta\, d\theta$$
$$= \frac{1}{2\sqrt{2}}\Big[\sec\theta\tan\theta + \ln|\sec\theta+\tan\theta|\Big]\Big|_{t=0}^{t=2\pi}$$
$$= \frac{\sqrt{2}e^t\sqrt{1+2e^{2t}} + \ln(\sqrt{2}e^t + \sqrt{1+2e^{2t}})}{2\sqrt{2}}\Big|_0^{2\pi}$$
$$= \frac{e^{2\pi}\sqrt{1+2e^{4\pi}} - \sqrt{3}}{2}$$
$$\quad + \frac{1}{2\sqrt{2}} \ln\frac{\sqrt{2}e^{2\pi} + \sqrt{1+2e^{4\pi}}}{\sqrt{2}+\sqrt{3}}.$$

7. The line of intersection of the planes $x - y + z = 0$ and $x + y + 2z = 0$ from $(0,0,0)$ to $(3,1,-2)$ can be parametrized

$$\mathbf{r} = 3t\mathbf{i} + t\mathbf{j} - 2t\mathbf{k}, \quad (0 \le t \le 1).$$

Thus $ds = \sqrt{14}\, dt$ and

$$\int_\mathcal{C} x^2\, ds = \sqrt{14} \int_0^1 9t^2\, dt = 3\sqrt{14}.$$

8. The curve $\mathcal{C}$ of intersection of $x^2+z^2=1$ and $y=x^2$ can be parametrized

$$\mathbf{r} = \cos t\mathbf{i} + \cos^2 t\mathbf{j} + \sin t\mathbf{k}, \quad (0 \le t \le 2\pi).$$

Thus

$$ds = \sqrt{\sin^2 t + 4\sin^2 t\cos^2 t + \cos^2 t}\, dt = \sqrt{1+\sin^2 2t}\, dt.$$

We have

$$\int_\mathcal{C} \sqrt{1+4x^2z^2}\, ds$$
$$= \int_0^{2\pi} \sqrt{1+4\cos^2 t\sin^2 t}\sqrt{1+\sin^2 2t}\, dt$$
$$= \int_0^{2\pi} (1+\sin^2 2t)\, dt$$
$$= \int_0^{2\pi} \left(1 + \frac{1-\cos 4t}{2}\right) dt$$
$$= \frac{3}{2}(2\pi) = 3\pi.$$

9. $\mathbf{r} = \cos t\mathbf{i} + \sin t\mathbf{j} + t\mathbf{k}, \quad (0 \le t \le 2\pi)$
$\mathbf{v} = -\sin t\mathbf{i} + \cos t\mathbf{j} + \mathbf{k}, \quad v = \sqrt{2}.$
If the density is $\delta = z = t$, then

$$m = \sqrt{2} \int_0^{2\pi} t\, dt = 2\pi^2\sqrt{2}$$
$$M_{x=0} = \sqrt{2} \int_0^{2\pi} t\cos t\, dt = 0$$
$$M_{y=0} = \sqrt{2} \int_0^{2\pi} t\sin t\, dt = -2\pi\sqrt{2}$$
$$M_{z=0} = \sqrt{2} \int_0^{2\pi} t^2\, dt = \frac{8\pi^3\sqrt{2}}{3}.$$

(We have omitted the details of the evaluation of these integrals.) The centre of mass is $\left(0, -\frac{1}{\pi}, \frac{4\pi}{3}\right)$.

10. Here the wire of Exercise 9 extends only from $t=0$ to $t=\pi$:

$$m = \sqrt{2} \int_0^\pi t\, dt = \frac{\pi^2\sqrt{2}}{2}$$
$$M_{x=0} = \sqrt{2} \int_0^\pi t\cos t\, dt = -2\sqrt{2}$$
$$M_{y=0} = \sqrt{2} \int_0^\pi t\sin t\, dt = \pi\sqrt{2}$$
$$M_{z=0} = \sqrt{2} \int_0^\pi t^2\, dt = \frac{\pi^3\sqrt{2}}{3}.$$

The centre of mass is $\left(-\frac{4}{\pi^2}, \frac{2}{\pi}, \frac{2\pi}{3}\right)$.

11. $\mathbf{r} = e^t\mathbf{i} + \sqrt{2}t\mathbf{j} + e^{-t}\mathbf{k}, \quad (0 \le t \le 1)$
$\mathbf{v} = e^t\mathbf{i} + \sqrt{2}\mathbf{j} - e^{-t}\mathbf{k}$
$v = \sqrt{e^{2t} + 2 + e^{-2t}} = e^t + e^{-t}$

$$\int_\mathcal{C} (x^2+z^2)\, ds = \int_0^1 (e^{2t}+e^{-2t})(e^t+e^{-t})\, dt$$
$$= \int_0^1 (e^{3t}+e^t+e^{-t}+e^{-3t})\, dt$$
$$= \frac{e^3}{3} + e - \frac{1}{e} - \frac{1}{3e^3}.$$

12. $$m = \int_0^1 (e^t+e^{-t})\, dt = \frac{e^2-1}{e}$$
$$M_{x=0} = \int_0^1 e^t(e^t+e^{-t})\, dt = \frac{e^2+1}{2}$$
$$M_{y=0} = \int_0^1 \sqrt{2}t(e^t+e^{-t})\, dt = \frac{2\sqrt{2}(e-1)}{e}$$
$$M_{z=0} = \int_0^1 e^{-t}(e^t+e^{-t})\, dt = \frac{3e^2-1}{2e^2}$$

The centroid is $\left(\dfrac{e^3+e}{2e^2-2}, \dfrac{2\sqrt{2}}{e+1}, \dfrac{3e^2-1}{2e^3-2e}\right)$.

13. The first octant part C of the curve $x^2 + y^2 = a^2$, $z = x$, can be parametrized

$$\mathbf{r} = a\cos t\,\mathbf{i} + a\sin t\,\mathbf{j} + a\cos t\,\mathbf{k}, \quad (0 \le t \le \pi/2).$$

We have $ds = a\sqrt{1 + \sin^2 t}\,dt$, so

$$\int_C x\,ds = a^2 \int_0^{\pi/2} \cos t\sqrt{1 + \sin^2 t}\,dt \quad \text{Let } \sin t = \tan\theta$$
$$\cos t\,dt = \sec^2\theta\,d\theta$$
$$= a^2 \int_{t=0}^{t=\pi/2} \sec^3\theta\,d\theta$$
$$= \frac{a^2}{2}\left[\sec\theta\tan\theta + \ln|\sec\theta + \tan\theta|\right]\Big|_{t=0}^{t=\pi/2}$$
$$= \frac{a^2}{2}\left[\sin t\sqrt{1+\sin^2 t} + \ln|\sin t + \sqrt{1+\sin^2 t}|\right]\Big|_0^{\pi/2}$$
$$= \frac{a^2}{2}\left[\sqrt{2} + \ln(1+\sqrt{2})\right].$$

14. On C, we have

$$z = \sqrt{1 - x^2 - y^2} = \sqrt{1 - x^2 - (1-x)^2} = \sqrt{2(x - x^2)}.$$

Thus C can be parametrized

$$\mathbf{r} = t\mathbf{i} + (1-t)\mathbf{j} + \sqrt{2(t - t^2)}\mathbf{k}, \quad (0 \le t \le 1).$$

Hence

$$ds = \sqrt{1 + 1 + \frac{(1-2t)^2}{2(t-t^2)}}\,dt = \frac{dt}{\sqrt{2(t-t^2)}}.$$

We have

$$\int_C z\,ds = \int_0^1 \sqrt{2(t-t^2)}\,\frac{dt}{\sqrt{2(t-t^2)}} = 1.$$

15. The parabola $z^2 = x^2 + y^2$, $x + z = 1$, can be parametrized in terms of $y = t$ since

$$(1-x)^2 = z^2 = x^2 + y^2 = x^2 + t^2$$
$$\Rightarrow \quad 1 - 2x = t^2 \quad \Rightarrow \quad x = \frac{1-t^2}{2}$$
$$\Rightarrow \quad z = 1 - x = \frac{1+t^2}{2}.$$

Thus $ds = \sqrt{t^2 + 1 + t^2}\,dt = \sqrt{1 + 2t^2}\,dt$, and

$$\int_C \frac{ds}{(2y^2+1)^{3/2}} = \int_{-\infty}^{\infty} \frac{\sqrt{1+2t^2}}{(2t^2+1)^{3/2}}\,dt$$
$$= 2\int_0^{\infty} \frac{dt}{1+2t^2}$$
$$= \sqrt{2}\tan^{-1}(\sqrt{2}t)\Big|_0^{\infty} = \sqrt{2}\frac{\pi}{2} = \frac{\pi}{\sqrt{2}}.$$

16. C: $y = x^2$, $z = y^2$, from $(0,0,0)$ to $(2,4,16)$. Parametrize C by

$$\mathbf{r} = t\mathbf{i} + t^2\mathbf{j} + t^4\mathbf{k}, \quad (0 \le t \le 2).$$

Since $ds = \sqrt{1 + 4t^2 + 16t^6}\,dt$, we have

$$\int_C xyz\,ds = \int_0^2 t^7\sqrt{1 + 4t^2 + 16t^6}\,dt.$$

17. Helix: $x = a\cos t$, $y = b\sin t$, $z = ct$ $(0 < a < b)$.

$$ds = \sqrt{a^2\sin^2 t + b^2\cos^2 t + c^2}\,dt$$
$$= \sqrt{c^2 + b^2 - (b^2 - a^2)\sin^2 t}\,dt$$
$$= \sqrt{b^2 + c^2}\sqrt{1 - k^2\sin^2 t}\,dt \quad \left(k^2 = \frac{b^2 - a^2}{b^2 + c^2}\right).$$

One complete revolution of the helix corresponds to $0 \le t \le 2\pi$, and has length

$$L = \sqrt{b^2 + c^2}\int_0^{2\pi} \sqrt{1 - k^2\sin^2 t}\,dt$$
$$= 4\sqrt{b^2 + c^2}\int_0^{\pi/2} \sqrt{1 - k^2\sin^2 t}\,dt$$
$$= 4\sqrt{b^2 + c^2}\,E(k) = 4\sqrt{b^2 + c^2}\,E\left(\sqrt{\frac{b^2 - a^2}{b^2 + c^2}}\right) \text{ units.}$$

The length of the part of the helix from $t = 0$ to $t = T < \pi/2$ is

$$L = \sqrt{b^2 + c^2}\int_0^T \sqrt{1 - k^2\sin^2 t}\,dt$$
$$= \sqrt{b^2 + c^2}\,E(k, T) = \sqrt{b^2 + c^2}\,E\left(\sqrt{\frac{b^2 - a^2}{b^2 + c^2}}, T\right) \text{ units.}$$

18. The straight line $\mathcal{L}$ with equation $Ax + By = C$, $(C \ne 0)$, lies at distance $D = \sqrt{|C|}/\sqrt{A^2 + B^2}$ from the origin. So does the line $\mathcal{L}_1$ with equation $y = D$. Since $x^2 + y^2$ depends only on distance from the origin, we have, by symmetry,

$$\int_{\mathcal{L}} \frac{ds}{x^2 + y^2} = \int_{\mathcal{L}_1} \frac{ds}{x^2 + y^2}$$
$$= \int_{-\infty}^{\infty} \frac{dx}{x^2 + D^2}$$
$$= \frac{2}{D}\tan^{-1}\frac{x}{D}\Big|_0^{\infty} = \frac{2}{D}\left(\frac{\pi}{2} - 0\right)$$
$$= \frac{\pi}{D} = \frac{\pi\sqrt{A^2 + B^2}}{|C|}.$$

Section 15.4 Line Integrals of Vector Fields (page 904)

1. $\mathbf{F} = xy\mathbf{i} - x^2\mathbf{j}$.
   $\mathcal{C}: \mathbf{r} = t\mathbf{i} + t^2\mathbf{j}, \quad (0 \le t \le 1)$.
   $$\int_{\mathcal{C}} \mathbf{F} \cdot d\mathbf{r} = \int_0^1 [t^3 - t^2(2t)]\,dt = -\int_0^1 t^3\,dt = -\frac{1}{4}.$$

2. $\mathbf{F} = \cos x\,\mathbf{i} - y\mathbf{j} = \nabla\left(\sin x - \dfrac{y^2}{2}\right)$.
   $\mathcal{C}: y = \sin x$ from $(0,0)$ to $(\pi, 0)$.
   $$\int_{\mathcal{C}} \mathbf{F} \cdot d\mathbf{r} = \left(\sin x - \frac{y^2}{2}\right)\bigg|_{(0,0)}^{(\pi,0)} = 0.$$

3. $\mathbf{F} = y\mathbf{i} + z\mathbf{j} - x\mathbf{k}$.
   $\mathcal{C}: \mathbf{r} = t\mathbf{i} + t\mathbf{j} + t\mathbf{k}, \quad (0 \le t \le 1)$.
   $$\int_{\mathcal{C}} \mathbf{F} \cdot d\mathbf{r} = \int_0^1 (t + t - t)\,dt = \frac{t^2}{2}\bigg|_0^1 = \frac{1}{2}.$$

4. $\mathbf{F} = z\mathbf{i} - y\mathbf{j} + 2x\mathbf{k}$.
   $\mathcal{C}: \mathbf{r} = t\mathbf{i} + t^2\mathbf{j} + t^3\mathbf{k}, \ (0 \le t \le 1)$.
   $$\int_{\mathcal{C}} \mathbf{F} \cdot d\mathbf{r} = \int_0^1 [t^3 - t^2(2t) + 2t(3t^2)]\,dt$$
   $$= \int_0^1 5t^3\,dt = \frac{5t^4}{4}\bigg|_0^1 = \frac{5}{4}.$$

5. $\mathbf{F} = yz\mathbf{i} + xz\mathbf{j} + xy\mathbf{k} = \nabla(xyz)$.
   $\mathcal{C}$: a curve from $(-1, 0, 0)$ to $(1, 0, 0)$. (Since $\mathbf{F}$ is conservative, it doesn't matter what curve.)
   $$\int_{\mathcal{C}} \mathbf{F} \cdot d\mathbf{r} = xyz\bigg|_{(-1,0,0)}^{(1,0,0)} = 0 - 0 = 0.$$

6. $\mathbf{F} = (x - z)\mathbf{i} + (y - z)\mathbf{j} - (x + y)\mathbf{k}$
   $= \nabla\left(\dfrac{x^2 + y^2}{2} - (x + y)z\right)$.
   $\mathcal{C}$ is a given polygonal path from $(0,0,0)$ to $(1,1,1)$ (but any other piecewise smooth path from the first point to the second would do as well).
   $$\int_{\mathcal{C}} \mathbf{F} \cdot d\mathbf{r} = \left(\frac{x^2 + y^2}{2} - (x + y)z\right)\bigg|_{(0,0,0)}^{(1,1,1)} = 1 - 2 = -1.$$

7. $\mathbf{F} = (x + y)\mathbf{i} + (x - z)\mathbf{j} + (z - y)\mathbf{k}$
   $= \nabla\left(\dfrac{x^2 + z^2}{2} + y(x - z)\right)$.
   The work done by $\mathbf{F}$ in moving an object from $(1, 0, -1)$ to $(0, -2, 3)$ is
   $$W = \int_{\mathcal{C}} \mathbf{F} \cdot d\mathbf{r} = \left(\frac{x^2 + z^2}{2} + y(x - z)\right)\bigg|_{(1,0,-1)}^{(0,-2,3)}$$
   $$= \frac{9}{2} - 2(-3) - (1 + 0) = \frac{19}{2} \text{ units.}$$

8. $\mathcal{C}$ is made up of four segments as shown in the figure.
   On $\mathcal{C}_1$, $y = 0$, $dy = 0$, and x goes from 0 to 1.
   On $\mathcal{C}_2$, $x = 1$, $dx = 0$, and y goes from 0 to 1.
   On $\mathcal{C}_3$, $y = 1$, $dy = 0$, and x goes from 1 to 0.
   On $\mathcal{C}_4$, $x = 0$, $dx = 0$, and y goes from 1 to 0.
   Thus
   $$\int_{\mathcal{C}_1} x^2 y^2\,dx + x^3 y\,dy = 0$$
   $$\int_{\mathcal{C}_2} x^2 y^2\,dx + x^3 y\,dy = \int_0^1 y\,dy = \frac{1}{2}$$
   $$\int_{\mathcal{C}_3} x^2 y^2\,dx + x^3 y\,dy = \int_1^0 x^2\,dx = -\frac{1}{3}$$
   $$\int_{\mathcal{C}_4} x^2 y^2\,dx + x^3 y\,dy = 0.$$
   Finally, therefore,
   $$\int_{\mathcal{C}} x^2 y^2\,dx + x^3 y\,dy = 0 + \frac{1}{2} - \frac{1}{3} + 0 = \frac{1}{6}.$$

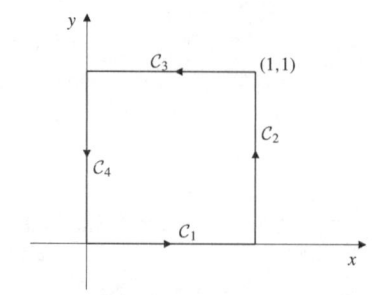

Fig. 15.4.8

9. Observe that if $\phi = e^{x+y}\sin(y + z)$, then
   $$\nabla\phi = e^{x+y}\sin(y + z)\mathbf{i} + e^{x+y}\big(\sin(y + z) + \cos(y + z)\big)\mathbf{j}$$
   $$+ e^{x+y}\cos(y + z)\mathbf{k}.$$
   Thus, for any piecewise smooth path from $(0, 0, 0)$ to $\left(1, \frac{\pi}{4}, \frac{\pi}{4}\right)$, we have
   $$\int_{\mathcal{C}} e^{x+y}\sin(y + z)\,dx + e^{x+y}\big(\sin(y + z) + \cos(y + z)\big)\,dy$$
   $$+ e^{x+y}\cos(y + z)\,dz$$
   $$= \int_{\mathcal{C}} \nabla\phi \cdot d\mathbf{r} = \phi(x, y, z)\bigg|_{(0,0,0)}^{(1,\pi/4,\pi/4)} = e^{1+(\pi/4)}.$$

10. $\mathbf{F} = (axy+z)\mathbf{i} + x^2\mathbf{j} + (bx+2z)\mathbf{k}$ is conservative if

$$\frac{\partial F_1}{\partial y} = \frac{\partial F_2}{\partial x} \Leftrightarrow a = 2$$
$$\frac{\partial F_1}{\partial z} = \frac{\partial F_3}{\partial x} \Leftrightarrow b = 1$$
$$\frac{\partial F_2}{\partial z} = \frac{\partial F_3}{\partial y} \Leftrightarrow 0 = 0.$$

If $a = 2$ and $b = 1$, then $\mathbf{F} = \nabla\phi$ where

$$\phi = \int (2xy+z)\,dx = x^2y + xz + C_2(y,z)$$
$$\frac{\partial C_1}{\partial y} + x^2 = F_2 = x^2 \Rightarrow C_1(y,z) = C_2(z)$$
$$\frac{dC_2}{dz} + x = F_3 = x + 2z \Rightarrow C_2(z) = z^2 + C.$$

Thus $\phi = x^2y + xz + z^2 + C$ is a potential for $\mathbf{F}$.

11. $\mathbf{F} = Ax\ln z\mathbf{i} + By^2z\mathbf{j} + \left(\dfrac{x^2}{z} + y^3\right)\mathbf{k}$ is conservative if

$$\frac{\partial F_1}{\partial y} = \frac{\partial F_2}{\partial x} \Leftrightarrow 0 = 0$$
$$\frac{\partial F_1}{\partial z} = \frac{\partial F_3}{\partial x} \Leftrightarrow A = 2$$
$$\frac{\partial F_2}{\partial z} = \frac{\partial F_3}{\partial y} \Leftrightarrow B = 3.$$

If $A = 2$ and $B = 3$, then $\mathbf{F} = \nabla\phi$ where
$\phi = x^2\ln z + y^3z$. If $\mathcal{C}$ is the straight line $x = t+1$, $y = 1$, $z = t + 1$, $(0 \le t \le 1)$, from $(1, 1, 1)$ to $(2, 1, 2)$, then

$$\int_{\mathcal{C}} 2x\ln z\,dx + 2y^2z\,dy + y^3\,dz$$
$$= \int_{\mathcal{C}} \nabla\phi \bullet d\mathbf{r} - \int_{\mathcal{C}} y^2z\,dy + \frac{x^2}{z}dz$$
$$= (x^2\ln z + y^3z)\Big|_{(1,1,1)}^{(2,1,2)} - \int_0^1 [(t+1)(0) + (t+1)]\,dt$$
$$= 4\ln 2 + 2 - 1 - \left(\frac{t^2}{2} + t\right)\Big|_0^1 = 4\ln 2 - \frac{1}{2}.$$

12. $\mathbf{F} = (y^2\cos x + z^3)\mathbf{i} + (2y\sin x - 4)\mathbf{j} + (3xz^2 + 2)\mathbf{k}$
$= \nabla(y^2\sin x + xz^3 - 4y + 2z).$
The curve $\mathcal{C}$: $x = \sin^{-1}t$, $y = 1 - 2t$, $z = 3t - 1$, $(0 \le t \le 1)$, goes from $(0, 1, -1)$ to $(\pi/2, -1, 2)$. The work done by $\mathbf{F}$ in moving a particle along $\mathcal{C}$ is

$$W = \int_{\mathcal{C}} \mathbf{F} \bullet d\mathbf{r}$$
$$= (y^2\sin x + xz^3 - 4y + 2z)\Big|_{(0,1,-1)}^{(\pi/2,-1,2)}$$
$$= 1 + 4\pi + 4 + 4 - 0 - 0 + 4 + 2 = 15 + 4\pi.$$

13. For $z = \ln(1+x)$, $y = x$, from $x = 0$ to $x = 1$, we have

$$\int_{\mathcal{C}} \left[(2x\sin(\pi y) - e^z)\,dx \right.$$
$$\left. + (\pi x^2\cos(\pi y) - 3e^z)\,dy - xe^z\,dz\right]$$
$$= \int_{\mathcal{C}} \nabla(x^2\sin(\pi y) - xe^z) \bullet d\mathbf{r} - 3\int_{\mathcal{C}} e^z\,dy$$
$$= (x^2\sin(\pi y) - xe^z)\Big|_{(0,0,0)}^{(1,1,\ln 2)} - 3\int_0^1 (1+x)\,dx$$
$$= -2 - 3\left(x + \frac{x^2}{2}\right)\Big|_0^1 = -2 - \frac{9}{2} = -\frac{13}{2}.$$

14. a) $S = \{(x, y) : x > 0, y \ge 0\}$ is a simply connected domain.

b) $S = \{(x, y) : x = 0, y \ge 0\}$ is not a domain. (It has empty interior.)

c) $S = \{(x, y) : x \ne 0, y > 0\}$ is a domain but is not connected. There is no path in S from $(-1, 1)$ to $(1, 1)$.

d) $S = \{(x, y, z) : x^2 > 1\}$ is a domain but is not connected. There is no path in S from $(-2, 0, 0)$ to $(2, 0, 0)$.

e) $S = \{(x, y, z) : x^2 + y^2 > 1\}$ is a connected domain but is not simply connected. The circle $x^2 + y^2 = 2$, $z = 0$ lies in S, but cannot be shrunk through S to a point since it surrounds the cylinder $x^2 + y^2 \le 1$ which is outside S.

f) $S = \{(x, y, z) : x^2 + y^2 + z^2 > 1\}$ is a simply connected domain even though it has a ball-shaped "hole" in it.

15. $\mathcal{C}$ is the curve $\mathbf{r} = a\cos t\mathbf{i} + a\sin t\mathbf{j}$, $(0 \le t \le 2\pi)$.

$$\oint_{\mathcal{C}} x\,dy = \int_0^{2\pi} a\cos t\, a\cos t\,dt = \pi a^2$$
$$\oint_{\mathcal{C}} y\,dx = \int_0^{2\pi} a\sin t(-a\sin t)\,dt = -\pi a^2.$$

16. $\mathcal{C}$ is the curve $\mathbf{r} = a\cos t\mathbf{i} + b\sin t\mathbf{j}$, $(0 \le t \le 2\pi)$.

$$\oint_{\mathcal{C}} x\,dy = \int_0^{2\pi} a\cos t\, b\cos t\,dt = \pi ab$$
$$\oint_{\mathcal{C}} y\,dx = \int_0^{2\pi} b\sin t(-a\sin t)\,dt = -\pi ab.$$

17. $\mathcal{C}$ consists of two parts:
On $\mathcal{C}_1$, $y = 0$, $dy = 0$, and x goes from $-a$ to a.
On $\mathcal{C}_2$, $x = a\cos t$, $y = a\sin t$, t goes from 0 to π.

$$\oint_\mathcal{C} x\,dy = \int_{\mathcal{C}_1} x\,dy + \int_{\mathcal{C}_2} x\,dy$$
$$= 0 + \int_0^\pi a^2\cos^2 t\,dt = \frac{\pi a^2}{2},$$
$$\oint_\mathcal{C} y\,dx = \int_{\mathcal{C}_1} y\,dx + \int_{\mathcal{C}_2} y\,dx$$
$$= 0 + \int_0^\pi (-a^2\cos^2 t)\,dt = -\frac{\pi a^2}{2}.$$

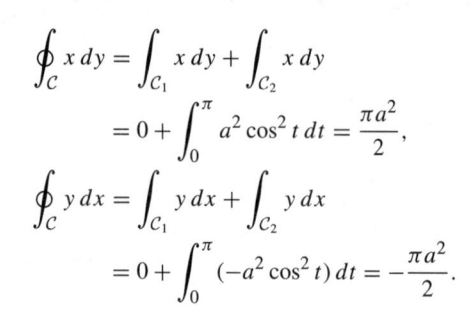

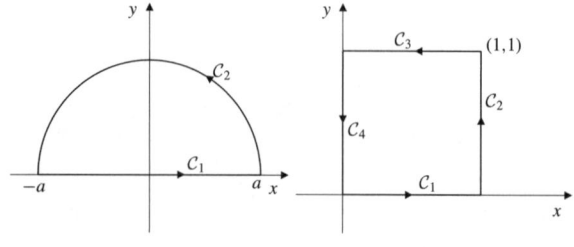

Fig. 15.4.17   Fig. 15.4.18

18. $\mathcal{C}$ is made up of four segments as shown in the figure.
On $\mathcal{C}_1$, $y = 0$, $dy = 0$, and x goes from 0 to 1.
On $\mathcal{C}_2$, $x = 1$, $dx = 0$, and y goes from 0 to 1.
On $\mathcal{C}_3$, $y = 1$, $dy = 0$, and x goes from 1 to 0.
On $\mathcal{C}_4$, $x = 0$, $dx = 0$, and y goes from 1 to 0.

$$\oint_\mathcal{C} x\,dy = \int_{\mathcal{C}_1} + \int_{\mathcal{C}_2} + \int_{\mathcal{C}_3} + \int_{\mathcal{C}_4}$$
$$= 0 + \int_0^1 dy + 0 + 0 = 1$$
$$\oint_\mathcal{C} y\,dx = \int_{\mathcal{C}_1} + \int_{\mathcal{C}_2} + \int_{\mathcal{C}_3} + \int_{\mathcal{C}_4}$$
$$= 0 + 0 + \int_1^0 dx + 0 = -1.$$

19. $\mathcal{C}$ is made up of three segments as shown in the figure.
On $\mathcal{C}_1$, $y = 0$, $dy = 0$, and x goes from 0 to a.
On $\mathcal{C}_2$, $y = bt$, $x = a(1-t)$, and t goes from 0 to 1.
On $\mathcal{C}_3$, $x = 0$, $dx = 0$, and y goes from b to 0.

$$\oint_\mathcal{C} x\,dy = \int_{\mathcal{C}_1} + \int_{\mathcal{C}_2} + \int_{\mathcal{C}_3}$$
$$= 0 + \int_0^1 a(1-t)b\,dt + 0 = \frac{ab}{2}$$
$$\oint_\mathcal{C} y\,dx = \int_{\mathcal{C}_1} + \int_{\mathcal{C}_2} + \int_{\mathcal{C}_3}$$
$$= 0 + \int_0^1 bt(-a\,dt) + 0 = -\frac{ab}{2}.$$

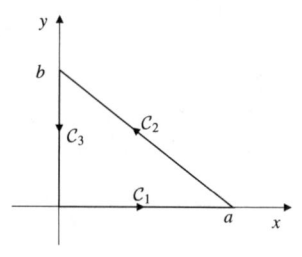

Fig. 15.4.19

20. Conjecture: If D is a domain in $\mathbb{R}^2$ whose boundary is a closed, non-self-intersecting curve $\mathcal{C}$, oriented counter-clockwise, then

$$\oint_\mathcal{C} x\,dy = \text{area of } D,$$
$$\oint_\mathcal{C} y\,dx = -\text{ area of } D.$$

Proof for a domain D that is x-simple and y-simple:
Since D is x-simple, it can be specified by the inequalities
$$c \leq y \leq d, \qquad f(y) \leq x \leq g(y).$$

Let $\mathcal{C}$ consist of the four parts shown in the figure. On $\mathcal{C}_1$ and $\mathcal{C}_3$, $dy = 0$.
On $\mathcal{C}_2$, $x = g(y)$, where y goes from c to d.
On $\mathcal{C}_2$, $x = f(y)$, where y goes from d to c. Thus

$$\oint_\mathcal{C} x\,dy = \int_{\mathcal{C}_1} + \int_{\mathcal{C}_2} + \int_{\mathcal{C}_3} + \int_{\mathcal{C}_4}$$
$$= 0 + \int_c^d g(y)\,dy + 0 + \int_d^c f(y)\,dy$$
$$= \big(g(y) - f(y)\big)\,dy = \text{area of } D.$$

The proof that $\oint_\mathcal{C} y\,dx = -(\text{area of } D)$ is similar, and uses the fact that D is y-simple.

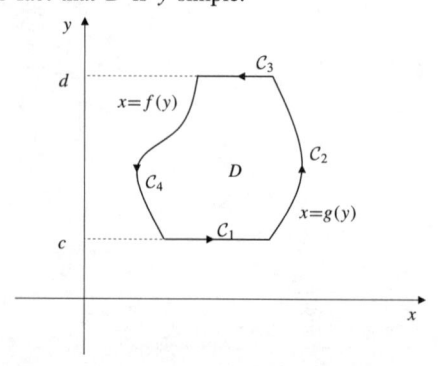

Fig. 15.4.20

21. $\nabla(fg) = +\left(f\dfrac{\partial g}{\partial x} + \dfrac{\partial f}{\partial x}g\right)\mathbf{i} + \left(f\dfrac{\partial g}{\partial y} + \dfrac{\partial f}{\partial y}g\right)\mathbf{j}$
$\qquad + \left(f\dfrac{\partial g}{\partial z} + \dfrac{\partial f}{\partial z}g\right)\mathbf{k}$
$= g\nabla f + f\nabla g.$
Thus, since C goes from P to Q,

$$\int_C f\nabla g \bullet d\mathbf{r} + \int_C g\nabla f \bullet d\mathbf{r}$$
$$= \int_C \nabla(fg) \bullet d\mathbf{r} = (fg)\Big|_P^Q$$
$$= f(Q)g(Q) - f(P)g(P).$$

22. a) $\mathcal{C}$: $x = a\cos t$, $x = a\sin t$, $0 \le t \le 2\pi$.

$$\dfrac{1}{2\pi}\oint_C \dfrac{x\,dy - y\,dx}{x^2 + y^2}$$
$$= \dfrac{1}{2\pi}\int_0^{2\pi} \dfrac{a^2\cos^2 t + a^2\sin^2 t}{a^2\cos^2 t + a^2\sin^2 t}\,dt = 1.$$

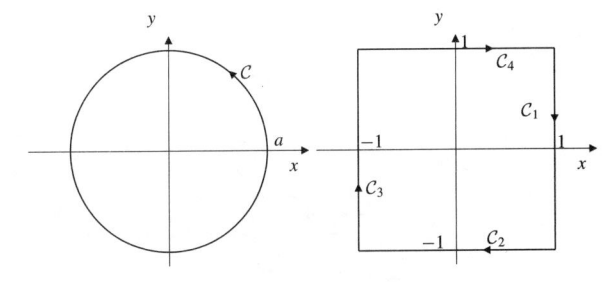

Fig. 15.4.22(a)   Fig. 15.4.22(b)

b) See the figure. C has four parts.
On C_1, $x = 1$, $dx = 0$, y goes from 1 to -1.
On C_2, $y = -1$, $dy = 0$, x goes from 1 to -1.
On C_3, $x = -1$, $dx = 0$, y goes from -1 to 1.
On C_4, $x = 1$, $dx = 0$, y goes from 1 to -1.

$$\dfrac{1}{2\pi}\oint_C \dfrac{x\,dy - y\,dx}{x^2 + y^2}$$
$$= \dfrac{1}{2\pi}\left[\int_1^{-1}\dfrac{dy}{1+y^2} + \int_1^{-1}\dfrac{dx}{x^2+1}\right.$$
$$\left.\int_{-1}^1 \dfrac{-dy}{1+y^2} + \int_{-1}^1 \dfrac{-dx}{x^2+1}\right]$$
$$= -\dfrac{2}{\pi}\int_{-1}^1 \dfrac{dt}{1+t^2}$$
$$= -\dfrac{2}{\pi}\tan^{-1} t\Big|_{-1}^1 = -\dfrac{2}{\pi}\left(\dfrac{\pi}{4} + \dfrac{\pi}{4}\right) = -1.$$

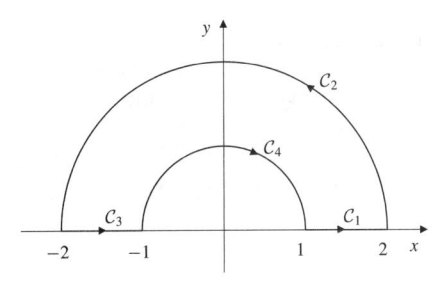

Fig. 15.4.22

c) See the figure. C has four parts.
On C_1, $y = 0$, $dy = 0$, x goes from 1 to 2.
On C_2, $x = 2\cos t$, $y = 2\sin t$, t goes from 0 to π.
On C_3, $y = 0$, $dy = 0$, x goes from -2 to -1.
On C_4, $x = \cos t$, $y = \sin t$, t goes from π to 0.

$$\dfrac{1}{2\pi}\oint_C \dfrac{x\,dy - y\,dx}{x^2+y^2}$$
$$= \dfrac{1}{2\pi}\left[0 + \int_0^\pi \dfrac{4\cos^2 t + 4\sin^2 t}{4\cos^2 t + 4\sin^2 t}\,dt\right.$$
$$\left.+ 0 + \int_\pi^0 \dfrac{\cos^2 t + \sin^2 t}{\cos^2 t + \sin^2 t}\,dt\right]$$
$$= \dfrac{1}{2\pi}(\pi - \pi) = 0.$$

23. Although
$$\dfrac{\partial}{\partial y}\left(\dfrac{-y}{x^2+y^2}\right) = \dfrac{\partial}{\partial x}\left(\dfrac{x}{x^2+y^2}\right)$$

for all $(x, y) \neq (0, 0)$, Theorem 1 does not imply that $\oint_C \dfrac{x\,dy - y\,dx}{x^2+y^2}$ is zero for all closed curves C in $\mathbb{R}^2$. The set consisting of points in $\mathbb{R}$ except the origin is *not simply connected*, and the vector field

$$\mathbf{F} = \dfrac{-y\mathbf{i} + x\mathbf{j}}{x^2 + y^2}$$

is not conservative on any domain in $\mathbb{R}^2$ that contains the origin in its interior. (See Example 5.) However, the integral will be 0 for any closed curve that does not contain the origin in its interior. (An example is the curve in Exercise 22(c).)

24. If C is a closed, piecewise smooth curve in $\mathbb{R}^2$ having equation $\mathbf{r} = \mathbf{r}(t)$, $a \le t \le b$, and if C does not pass through the origin, then the polar angle function $\theta = \theta(x(t), y(t)) = \theta(t)$ can be defined so as to vary continuously on C. Therefore,

$$\theta(x,y)\Big|_{t=a}^{t=b} = 2\pi \times w(C),$$

where $w(\mathcal{C})$ is the number of times $\mathcal{C}$ winds around the origin in a counterclockwise direction. For example, $w(\mathcal{C})$ equals 1, -1 and 0 respectively, for the curves $\mathcal{C}$ in parts (a), (b) and (c) of Exercise 22. Since

$$\nabla\theta = \frac{\partial\theta}{\partial x}\mathbf{i} + \frac{\partial\theta}{\partial y}\mathbf{j}$$
$$= \frac{-y\mathbf{i} + x\mathbf{j}}{x^2 + y^2},$$

we have

$$\frac{1}{2\pi}\oint_\mathcal{C} \frac{x\,dy - y\,dx}{x^2 + y^2} = \frac{1}{2\pi}\oint_\mathcal{C} \nabla\theta \cdot d\mathbf{r}$$
$$= \frac{1}{2\pi}\theta(x,y)\bigg|_{t=a}^{t=b} = w(\mathcal{C}).$$

Section 15.5 Surfaces and Surface Integrals (page 917)

1. The polar curve $r = g(\theta)$ is parametrized by

$$x = g(\theta)\cos\theta, \qquad y = g(\theta)\sin\theta.$$

Hence its arc length element is

$$ds = \sqrt{\left(\frac{dx}{d\theta}\right)^2 + \left(\frac{dy}{d\theta}\right)^2}\,d\theta$$
$$= \sqrt{\left(g'(\theta)\cos\theta - g(\theta)\sin\theta\right)^2 + \left(g'(\theta)\sin\theta + g(\theta)\cos\theta\right)^2}\,d\theta$$
$$= \sqrt{\left(g(\theta)\right)^2 + \left(g'(\theta)\right)^2}\,d\theta.$$

The area element on the vertical cylinder $r = g(\theta)$ is

$$dS = ds\,dz = \sqrt{\left(g(\theta)\right)^2 + \left(g'(\theta)\right)^2}\,d\theta\,dz.$$

2. The area element dS is bounded by the curves in which the coordinate planes at θ and $\theta + d\theta$ and the coordinate cones at ϕ and $\phi + d\phi$ intersect the sphere $R = a$. (See the figure.) The element is rectangular with sides $a\,d\phi$ and $a\sin\phi\,d\theta$. Thus

$$dS = a^2\sin\phi\,d\phi\,d\theta.$$

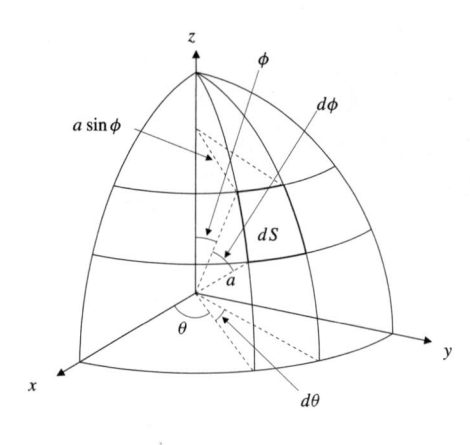

Fig. 15.5.2

3. The plane $Ax + By + Cz = D$ has normal $\mathbf{n} = A\mathbf{i} + B\mathbf{j} + C\mathbf{k}$, and so an area element on it is given by

$$dS = \frac{|\mathbf{n}|}{|\mathbf{n}\cdot\mathbf{k}|}\,dx\,dy = \frac{\sqrt{A^2 + B^2 + C^2}}{|C|}\,dx\,dy.$$

Hence the area S of that part of the plane lying inside the elliptic cylinder

$$\frac{x^2}{a^2} + \frac{y^2}{b^2} = 1$$

is given by

$$S = \iint_{\frac{x^2}{a^2} + \frac{y^2}{b^2} \le 1} \frac{\sqrt{A^2 + B^2 + C^2}}{|C|}\,dx\,dy$$
$$= \frac{\pi ab\sqrt{A^2 + B^2 + C^2}}{|C|} \text{ sq. units.}$$

4. One-quarter of the required area is shown in the figure. It lies above the semicircular disk R bounded by $x^2 + y^2 = 2ay$, or, in terms of polar coordinates, $r = 2a\sin\theta$. On the sphere $x^2 + y^2 + z^2 = 4a^2$, we have

$$2z\frac{\partial z}{\partial x} = -2x, \quad \text{or} \quad \frac{\partial z}{\partial x} = -\frac{x}{z}.$$

Similarly, $\dfrac{\partial z}{\partial y} = -\dfrac{y}{z}$, so the surface area element on the sphere can be written

$$dS = \sqrt{1 + \frac{x^2 + y^2}{z^2}}\,dx\,dy = \frac{2a\,dx\,dy}{\sqrt{4a^2 - x^2 - y^2}}.$$

The required area is

$$S = 4\iint_R \frac{2a}{\sqrt{4a^2 - x^2 - y^2}}\, dx\, dy$$

$$= 8a \int_0^{\pi/2} d\theta \int_0^{2a\sin\theta} \frac{r\, dr}{\sqrt{4a^2 - r^2}} \quad \text{Let } u = 4a^2 - r^2$$
$$\hspace{7cm} du = -2r\, dr$$

$$= 4a \int_0^{\pi/2} d\theta \int_{4a^2\cos^2\theta}^{4a^2} u^{-1/2}\, du$$

$$= 8a \int_0^{\pi/2} (2a - 2a\cos\theta)\, d\theta$$

$$= 16a^2 (\theta - \sin\theta)\Big|_0^{\pi/2} = 8a^2(\pi - 2) \text{ sq. units.}$$

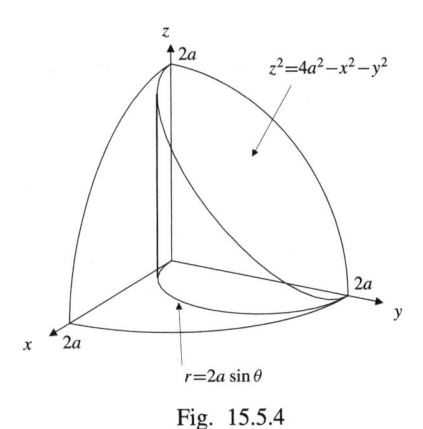

Fig. 15.5.4

5. $dS = \left|\dfrac{\nabla F(x,y,z)}{F_2(x,y,z)}\right| dx\, dz$

$dS = \left|\dfrac{\nabla F(x,y,z)}{F_1(x,y,z)}\right| dy\, dz$

6. The cylinder $x^2 + y^2 = 2ay$ intersects the sphere $x^2 + y^2 + z^2 = 4a^2$ on the parabolic cylinder $2ay + z^2 = 4a^2$. By Exercise 5, the area element on $x^2 + y^2 - 2ay = 0$ is

$$dS = \left|\frac{2x\mathbf{i} + (2y - 2a)\mathbf{j}}{2x}\right| dy\, dz$$

$$= \sqrt{1 + \frac{(y-a)^2}{2ay - y^2}}\, dy\, dz$$

$$= \sqrt{\frac{2ay - y^2 + y^2 - 2ay + a^2}{2ay - y^2}}\, dy\, dz = \frac{a}{\sqrt{2ay - y^2}}\, dy\, dz.$$

The area of the part of the cylinder inside the sphere is 4 times the part shown in Figure 15.23 in the text, that is, 4 times the double integral of dS over the region $0 \le y \le 2a$, $0 \le z \le \sqrt{4a^2 - 2ay}$, or

$$S = 4\int_0^{2a} \frac{a\, dy}{\sqrt{2ay - y^2}} \int_0^{\sqrt{4a^2 - 2ay}} dz$$

$$= 4a\int_0^{2a} \frac{\sqrt{2a(2a - y)}}{\sqrt{y(2a - y)}}\, dy = 4\sqrt{2}a^{3/2}\int_0^{2a} \frac{dy}{\sqrt{y}}$$

$$= 4\sqrt{2}a^{3/2}(2\sqrt{y})\Big|_0^{2a} = 16a^2 \text{ sq. units.}$$

7. On the surface $\mathcal{S}$ with equation $z = x^2/2$ we have $\partial z/\partial x = x$ and $\partial z/\partial y = 0$. Thus

$$dS = \sqrt{1 + x^2}\, dx\, dy.$$

If R is the first quadrant part of the disk $x^2 + y^2 \le 1$, then the required surface integral is

$$\iint_S x\, dS = \iint_R x\sqrt{1 + x^2}\, dx\, dy$$

$$= \int_0^1 x\sqrt{1 + x^2}\, dx \int_0^{\sqrt{1-x^2}} dy$$

$$= \int_0^1 x\sqrt{1 - x^4}\, dx \quad \text{Let } u = x^2$$
$$\hspace{4cm} du = 2x\, dx$$

$$= \frac{1}{2}\int_0^1 \sqrt{1 - u^2}\, du = \frac{1}{2}\frac{\pi}{4} = \frac{\pi}{8}.$$

8. The normal to the cone $z^2 = x^2 + y^2$ makes a 45° angle with the vertical, so $dS = \sqrt{2}\, dx\, dy$ is a surface area element for the cone. Both *nappes* (halves) of the cone pass through the interior of the cylinder $x^2 + y^2 = 2ay$, so the area of that part of the cone inside the cylinder is $2\sqrt{2}\pi a^2$ square units, since the cylinder has a circular cross-section of radius a.

9. One-quarter of the required area lies in the first octant. (See the figure.) In polar coordinates, the Cartesian equation $x^2 + y^2 = 2ay$ becomes $r = 2a\sin\theta$. The arc length element on this curve is

$$ds = \sqrt{r^2 + \left(\frac{dr}{d\theta}\right)^2}\, d\theta = 2a\, d\theta.$$

Thus $dS = \sqrt{x^2 + y^2}\, ds = 2ar\, d\theta = 4a^2 \sin\theta\, d\theta$ on the cylinder. The area of that part of the cylinder lying between the nappes of the cone is

$$4\int_0^{\pi/2} 4a^2 \sin\theta\, d\theta = 16a^2 \text{ sq. units..}$$

SECTION 15.5 (PAGE 917)

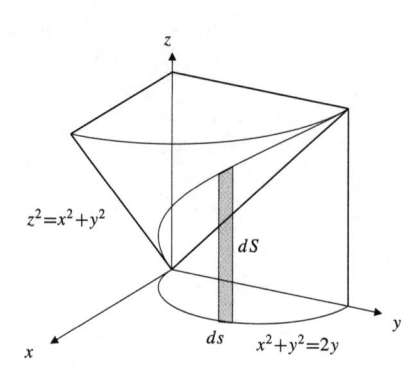

Fig. 15.5.9

10. One-eighth of the required area lies in the first octant, above the triangle T with vertices $(0, 0, 0)$, $(a, 0, 0)$ and $(a, a, 0)$. (See the figure.)
The surface $x^2 + z^2 = a^2$ has normal $\mathbf{n} = x\mathbf{i} + z\mathbf{k}$, so an area element on it can be written

$$dS = \frac{|\mathbf{n}|}{|\mathbf{n} \bullet \mathbf{k}|} \, dx \, dy = \frac{a}{z} \, dx \, dy = \frac{a \, dx \, dy}{\sqrt{a^2 - x^2}}.$$

The area of the part of that cylinder lying inside the cylinder $y^2 + z^2 = a^2$ is

$$S = 8 \iint_T \frac{a \, dx \, dy}{\sqrt{a^2 - x^2}} = 8a \int_0^a \frac{dx}{\sqrt{a^2 - x^2}} \int_0^x dy$$

$$= 8a \int_0^a \frac{x \, dx}{\sqrt{a^2 - x^2}}$$

$$= -8a\sqrt{a^2 - x^2} \Big|_0^a = 8a^2 \text{ sq. units.}$$

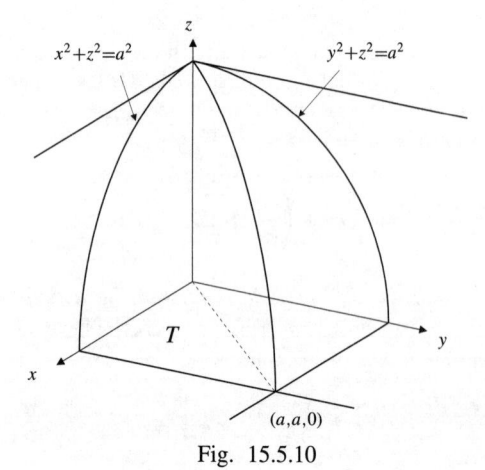

Fig. 15.5.10

11. Let the sphere be $x^2 + y^2 + z^2 = R^2$, and the cylinder be $x^2 + y^2 = R^2$. Let $\mathcal{S}_1$ and $\mathcal{S}_2$ be the parts of the sphere and the cylinder, respectively, lying between the planes $z = a$ and $z = b$, where $-R \leq a \leq b \leq R$.
Evidently, the area of $\mathcal{S}_2$ is $S_2 = 2\pi R(b - a)$ square units. An area element on the sphere is given in terms of spherical coordinates by

$$dS = R^2 \sin \phi \, d\phi \, d\theta.$$

On $\mathcal{S}_1$ we have $z = R \cos \phi$, so $\mathcal{S}_1$ lies between $\phi = \cos^{-1}(b/R)$ and $\phi = \cos^{-1}(a/R)$. Thus the area of $\mathcal{S}_1$ is

$$S_1 = R^2 \int_0^{2\pi} d\theta \int_{\cos^{-1}(b/R)}^{\cos^{-1}(a/R)} \sin \phi \, d\phi$$

$$= 2\pi R^2 (-\cos \phi) \Big|_{\cos^{-1}(b/R)}^{\cos^{-1}(a/R)} = 2\pi R(b - a) \text{ sq. units.}$$

Observe that $\mathcal{S}_1$ and $\mathcal{S}_2$ have the same area.

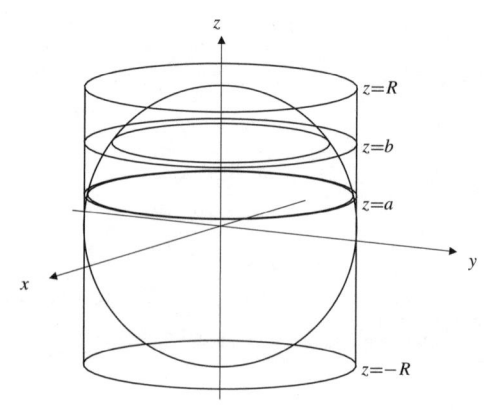

Fig. 15.5.11

12. We want to find A_1, the area of that part of the cylinder $x^2 + z^2 = a^2$ inside the cylinder $y^2 + z^2 = b^2$, and A_2, the area of that part of $y^2 + z^2 = b^2$ inside $x^2 + z^2 = a^2$. We have

$$A_1 = 8 \times (\text{area of } \mathcal{S}_1),$$
$$A_2 = 8 \times (\text{area of } \mathcal{S}_2),$$

where $\mathcal{S}_1$ and $\mathcal{S}_2$ are the parts of these surfaces lying in the first octant, as shown in the figure.
A normal to $\mathcal{S}_1$ is $\mathbf{n}_1 = x\mathbf{i} + z\mathbf{k}$, and the area element on $\mathcal{S}_1$ is

$$dS_1 = \frac{|\mathbf{n}_1|}{|\mathbf{n}_1 \bullet \mathbf{i}|} \, dy \, dz = \frac{a \, dy \, dz}{\sqrt{a^2 - z^2}}.$$

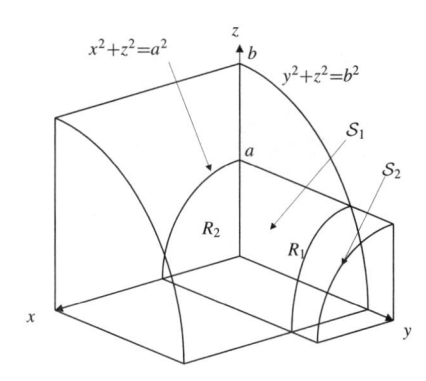

INSTRUCTOR'S SOLUTIONS MANUAL   SECTION 15.5 (PAGE 917)

Fig. 15.5.12

A normal to $\mathcal{S}_2$ is $\mathbf{n}_2 = x\mathbf{j} + z\mathbf{k}$, and the area element on $\mathcal{S}_2$ is

$$dS_2 = \frac{|\mathbf{n}_2|}{|\mathbf{n}_2 \bullet \mathbf{j}|} dx\, dz = \frac{b\, dx\, dz}{\sqrt{b^2 - z^2}}.$$

Let R_1 be the region of the first quadrant of the yz-plane bounded by $y^2 + z^2 = b^2$, $y = 0$, $z = 0$, and $z = a$.
Let R_2 be the quarter-disk in the first quadrant of the xz-plane bounded by $x^2 + z^2 = a^2$, $x = 0$, and $z = 0$. Then

$$A_1 = 8\iint_{R_1} dS_1 = 8a \int_0^a \frac{dz}{\sqrt{a^2 - z^2}} \int_0^{\sqrt{b^2-z^2}} dy$$

$$= 8a \int_0^a \frac{\sqrt{b^2-z^2}}{\sqrt{a^2-z^2}}\, dz \quad \text{Let } z = a\sin t$$
$$\quad dz = a\cos t\, dt$$

$$= 8a \int_0^{\pi/2} \sqrt{b^2 - a^2\sin^2 t}\, dt$$

$$= 8ab \int_0^{\pi/2} \sqrt{1 - \frac{a^2}{b^2}\sin^2 t}\, dt$$

$$= 8ab E\left(\frac{a}{b}\right) \text{ sq. units.}$$

$$A_2 = 8\iint_{R_2} dS_2 = 8b \int_0^a \frac{dz}{\sqrt{b^2-z^2}} \int_0^{\sqrt{a^2-z^2}} dx$$

$$= 8b \int_0^a \frac{\sqrt{a^2-z^2}}{\sqrt{b^2-z^2}}\, dz \quad \text{Let } z = b\sin t$$
$$\quad dz = b\cos t\, dt$$

$$= 8b \int_0^{\sin^{-1}(a/b)} \sqrt{a^2 - b^2 \sin^2 t}\, dt$$

$$= 8ab \int_0^{\sin^{-1}(a/b)} \sqrt{1 - \frac{b^2}{a^2}\sin^2 t}\, dt$$

$$= 8ab E\left(\frac{b}{a}, \sin^{-1}\frac{a}{b}\right) \text{ sq. units.}$$

13. The intersection of the plane $z = 1 + y$ and the cone $z = \sqrt{2(x^2 + y^2)}$ has projection onto the xy-plane the elliptic disk E bounded by

$$(1+y)^2 = 2(x^2 + y^2)$$
$$1 + 2y + y^2 = 2x^2 + 2y^2$$
$$2x^2 + y^2 - 2y + 1 = 2$$
$$x^2 + \frac{(y-1)^2}{2} = 1.$$

Note that E has area $A = \pi(1)(\sqrt{2})$ and centroid $(0, 1)$. If $\mathcal{S}$ is the part of the plane lying inside the cone, then the area element on $\mathcal{S}$ is

$$dS = \sqrt{1 + \left(\frac{\partial z}{\partial y}\right)^2}\, dx\, dy = \sqrt{2}\, dx\, dy.$$

Thus

$$\iint_{\mathcal{S}} y\, dS = \sqrt{2} \iint_E y\, dx\, dy = \sqrt{2} A\bar{y} = 2\pi.$$

14. Continuing the above solution, the cone $z = \sqrt{2(x^2 + y^2)}$ has area element

$$dS = \sqrt{1 + \left(\frac{\partial z}{\partial x}\right)^2 + \left(\frac{\partial z}{\partial y}\right)^2}\, dx\, dy$$

$$= \sqrt{1 + \frac{4(x^2 + y^2)}{z^2}}\, dx\, dy = \sqrt{3}\, dx\, dy.$$

If $\mathcal{S}$ is the part of the cone lying below the plane $z = 1 + y$, then

$$\iint_{\mathcal{S}} y\, dS = \sqrt{3} \iint_E y\, dx\, dy = \sqrt{3} A\bar{y} = \sqrt{6}\pi.$$

15. If $\mathcal{S}$ is the part of $z = x^2$ in the first octant and inside (that is, below) $z = 1 - 3x^2 - y^2$, then $\mathcal{S}$ has projection E onto the xy-plane bounded by $x^2 = 1 - 3x^2 - y^2$, or $4x^2 + y^2 = 1$, an ellipse. Since $z = x^2$ has area element $dS = \sqrt{1 + 4x^2}\, dx\, dy$, we have

$$\iint_{\mathcal{S}} xz\, dS = \iint_E x^3 \sqrt{1 + 4x^2}\, dx\, dy$$

$$= \int_0^{1/2} x^3 \sqrt{1 + 4x^2}\, dx \int_0^{\sqrt{1-4x^2}} dy$$

$$= \int_0^{1/2} x^3 \sqrt{1 - 16x^4}\, dx \quad \text{Let } u = 1 - 16x^4$$
$$\quad du = -64x^3\, dx$$

$$= \frac{1}{64} \int_0^1 u^{1/2}\, du = \frac{1}{96}.$$

565

16. The surface $z = \sqrt{2xy}$ has area element

$$dS = \sqrt{1 + \frac{y}{2x} + \frac{x}{2y}}\, dx\, dy$$

$$= \sqrt{\frac{2xy + y^2 + x^2}{2xy}}\, dx\, dy = \frac{|x+y|}{\sqrt{2xy}}\, dx\, dy.$$

If its density is kz, the mass of the specified part of the surface is

$$m = \int_0^5 dx \int_0^2 k\sqrt{2xy}\, \frac{x+y}{\sqrt{2xy}}\, dy$$

$$= k\int_0^5 dx \int_0^2 (x+y)\, dy$$

$$= k\int_0^5 (2x+2)\, dx = 35k \text{ units}.$$

17. The surface $\mathcal{S}$ is given by $x = e^u \cos v$, $y = e^u \sin v$, $z = u$, for $0 \le u \le 1$, $0 \le v \le \pi$. Since

$$\frac{\partial(y,z)}{\partial(u,v)} = \begin{vmatrix} e^u \sin v & e^u \cos v \\ 1 & 0 \end{vmatrix} = -e^u \cos v$$

$$\frac{\partial(z,x)}{\partial(u,v)} = \begin{vmatrix} 1 & 0 \\ e^u \cos v & -e^u \sin v \end{vmatrix} = -e^u \sin v$$

$$\frac{\partial(x,y)}{\partial(u,v)} = \begin{vmatrix} e^u \cos v & -e^u \sin v \\ e^u \sin v & e^u \cos v \end{vmatrix} = e^{2u}$$

the area element on $\mathcal{S}$ is

$$dS = \sqrt{e^{2u}\cos^2 v + e^{2u}\sin^2 v + e^{4u}}\, du\, dv = e^u \sqrt{1 + e^{2u}}\, du\, dv.$$

If the charge density on $\mathcal{S}$ is $\sqrt{1+e^{2u}}$, then the total charge is

$$\iint_S \sqrt{1+e^{2u}}\, dS = \int_0^1 e^u(1+e^{2u})\, du \int_0^\pi dv$$

$$= \pi \left(e^u + \frac{e^{3u}}{3}\right)\Bigg|_0^1 = \frac{\pi}{3}(3e + e^3 - 4).$$

18. The upper half of the spheroid $\frac{x^2}{a^2} + \frac{y^2}{a^2} + \frac{z^2}{c^2} = 1$ has a circular disk of radius a as projection onto the xy-plane. Since

$$\frac{2x}{a^2} + \frac{2z}{c^2}\frac{\partial z}{\partial x} = 0 \quad \Rightarrow \quad \frac{\partial z}{\partial x} = -\frac{c^2 x}{a^2 z},$$

and, similarly, $\frac{\partial z}{\partial y} = -\frac{c^2 y}{a^2 z}$, the area element on the spheroid is

$$dS = \sqrt{1 + \frac{c^4}{a^4}\frac{x^2+y^2}{z^2}}\, dx\, dy$$

$$= \sqrt{1 + \frac{c^2}{a^2}\frac{x^2+y^2}{a^2 - x^2 - y^2}}\, dx\, dy$$

$$= \sqrt{\frac{a^4 + (c^2-a^2)r^2}{a^2(a^2-r^2)}}\, r\, dr\, d\theta$$

in polar coordinates. Thus the area of the spheroid is

$$S = \frac{2}{a}\int_0^{2\pi} d\theta \int_0^a \sqrt{\frac{a^4 + (c^2-a^2)r^2}{a^2-r^2}}\, r\, dr$$

Let $u^2 = a^2 - r^2$

$u\, du = -r\, dr$

$$= \frac{4\pi}{a}\int_0^a \sqrt{a^4 + (c^2-a^2)(a^2-u^2)}\, du$$

$$= \frac{4\pi}{a}\int_0^a \sqrt{a^2 c^2 - (c^2-a^2)u^2}\, du$$

$$= 4\pi c \int_0^a \sqrt{1 - \frac{c^2-a^2}{a^2 c^2}u^2}\, du.$$

For the case of a prolate spheroid $0 < a < c$, let $k^2 = \frac{c^2 - a^2}{a^2 c^2}$. Then

$$S = 4\pi c \int_0^a \sqrt{1-k^2 u^2}\, du \quad \text{Let } ku = \sin v$$

$k\, du = \cos v\, dv$

$$= \frac{4\pi c}{k}\int_0^{\sin^{-1}(ka)} \cos^2 v\, dv$$

$$= \frac{2\pi c}{k}(v + \sin v \cos v)\Bigg|_0^{\sin^{-1}(ka)}$$

$$= \frac{2\pi a c^2}{\sqrt{c^2-a^2}}\sin^{-1}\frac{\sqrt{c^2-a^2}}{c} + 2\pi a^2 \text{ sq. units}.$$

19. We continue from the formula for the surface area of a spheroid developed part way through the solution above. For the case of an oblate spheroid $0 < c < a$, let

$k^2 = \dfrac{a^2 - c^2}{a^2 c^2}$. Then

$$S = 4\pi c \int_0^a \sqrt{1 + k^2 u^2}\, du \quad \text{Let } ku = \tan v$$
$$\phantom{S = 4\pi c \int_0^a \sqrt{1 + k^2 u^2}\, du \quad} k\, du = \sec^2 v\, dv$$
$$= \frac{4\pi c}{k} \int_0^{\tan^{-1}(ka)} \sec^3 v\, dv$$
$$= \frac{2\pi c}{k} \left(\sec v \tan v + \ln(\sec v + \tan v)\right)\Big|_0^{\tan^{-1}(ka)}$$
$$= \frac{2\pi a c^2}{\sqrt{a^2 - c^2}} \left[\frac{a\sqrt{a^2 - c^2}}{c^2} + \ln\left(\frac{a}{c} + \frac{\sqrt{a^2 - c^2}}{c}\right)\right]$$
$$= 2\pi a^2 + \frac{2\pi a c^2}{\sqrt{a^2 - c^2}} \ln\left(\frac{a + \sqrt{a^2 - c^2}}{c}\right) \text{ sq. units.}$$

20. $x = au \cos v$, $y = au \sin v$, $z = bv$, ($0 \le u \le 1$, $0 \le v \le 2\pi$). This surface is a spiral (helical) ramp of radius a and height $2\pi b$, wound around the z-axis. (It's like a circular staircase with a ramp instead of stairs.) We have

$$\frac{\partial(x, y)}{\partial(u, v)} = \begin{vmatrix} a\cos v & -au \sin v \\ a \sin v & au \cos v \end{vmatrix} = a^2 u$$
$$\frac{\partial(y, z)}{\partial(u, v)} = \begin{vmatrix} a \sin v & au \cos v \\ 0 & b \end{vmatrix} = ab \sin v$$
$$\frac{\partial(z, x)}{\partial(u, v)} = \begin{vmatrix} 0 & b \\ a \cos v & -au \sin v \end{vmatrix} = -ab \cos v$$
$$dS = \sqrt{a^4 u^2 + a^2 b^2 \sin^2 v + a^2 b^2 \cos^2 v}\, du\, dv$$
$$= a\sqrt{a^2 u^2 + b^2}\, du\, dv.$$

The area of the ramp is

$$A = a \int_0^1 \sqrt{a^2 u^2 + b^2}\, du \int_0^{2\pi} dv$$
$$= 2\pi a \int_0^1 \sqrt{a^2 u^2 + b^2}\, du \quad \text{Let } au = b\tan\theta$$
$$\phantom{= 2\pi a \int_0^1 \sqrt{a^2 u^2 + b^2}\, du \quad} a\, du = b\sec^2 \theta\, d\theta$$
$$= 2\pi b^2 \int_{u=0}^{u=1} \sec^3 \theta\, d\theta$$
$$= \pi b^2 \left(\sec\theta \tan\theta + \ln|\sec\theta + \tan\theta|\right)\Big|_{u=0}^{u=1}$$
$$= \pi b^2 \left(\frac{au\sqrt{a^2 u^2 + b^2}}{b^2} + \ln\left|\frac{au + \sqrt{a^2 u^2 + b^2}}{b}\right|\right)\Big|_0^1$$
$$= \pi a \sqrt{a^2 + b^2} + \pi b^2 \ln\left(\frac{a + \sqrt{a^2 + b^2}}{b}\right) \text{ sq. units.}$$

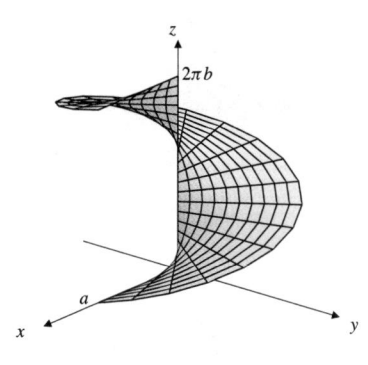

Fig. 15.5.20

21. The distance from the origin to the plane $\mathcal{P}$ with equation $Ax + By + Cz = D$, ($D \ne 0$) is

$$\delta = \frac{|D|}{\sqrt{A^2 + B^2 + C^2}}.$$

If $\mathcal{P}_1$ is the plane $z = \delta$, then, since the integrand depends only on distance from the origin, we have

$$\iint_{\mathcal{P}} \frac{dS}{(x^2 + y^2 + z^2)^{3/2}}$$
$$= \iint_{\mathcal{P}_1} \frac{dS}{(x^2 + y^2 + z^2)^{3/2}}$$
$$= \int_0^{2\pi} d\theta \int_0^\infty \frac{r\, dr}{(r^2 + \delta^2)^{3/2}} \quad \text{Let } u = r^2 + \delta^2$$
$$\phantom{= \int_0^{2\pi} d\theta \int_0^\infty \frac{r\, dr}{(r^2 + \delta^2)^{3/2}} \quad} du = 2r\, dr$$
$$= 2\pi \times \frac{1}{2} \int_{\delta^2}^\infty \frac{du}{u^{3/2}}$$
$$= \pi \left(-\frac{2}{\sqrt{u}}\right)\Big|_{\delta^2}^\infty$$
$$= \frac{2\pi}{\delta} = \frac{2\pi \sqrt{A^2 + B^2 + C^2}}{|D|}.$$

22. Use spherical coordinates. The area of the eighth-sphere $\mathcal{S}$ is

$$A = \frac{1}{8}(4\pi a^2) = \frac{\pi a^2}{2} \text{ sq. units.}$$

The moment about $z = 0$ is

$$M_{z=0} = \iint_{\mathcal{S}} z\, dS$$
$$= \int_0^{\pi/2} d\theta \int_0^{\pi/2} a\cos\phi\, a^2 \sin\phi\, d\phi$$
$$= \frac{\pi a^3}{2} \int_0^{\pi/2} \frac{\sin 2\phi}{2}\, d\phi = \frac{\pi a^3}{4}.$$

Thus $\bar{z} = \dfrac{M_{z=0}}{A} = \dfrac{a}{2}$. By symmetry, $\bar{x} = \bar{y} = \bar{z}$, so the centroid of that part of the surface of the sphere $x^2 + y^2 + z^2 = a^2$ lying in the first octant is $\left(\dfrac{a}{2}, \dfrac{a}{2}, \dfrac{a}{2}\right)$.

23. The cone $z = h\left(1 - \dfrac{\sqrt{x^2+y^2}}{a}\right)$ has normal

$$\mathbf{n} = -\dfrac{\partial z}{\partial x}\mathbf{i} - \dfrac{\partial z}{\partial y}\mathbf{j} + \mathbf{k}$$
$$= -\dfrac{h}{a}\left(\dfrac{x\mathbf{i}+y\mathbf{j}}{\sqrt{x^2+y^2}}\right) + \mathbf{k},$$

so its surface area element is

$$dS = \sqrt{\dfrac{h^2}{a^2}+1}\,dx\,dy = \dfrac{\sqrt{a^2+h^2}}{a}\,dx\,dy.$$

The mass of the conical shell is

$$m = \sigma\iint_{x^2+y^2\leq a^2} dS = \dfrac{\sigma\sqrt{a^2+h^2}}{a}(\pi a^2) = \pi\sigma a\sqrt{a^2+h^2}.$$

The moment about $z=0$ is

$$M_{z=0} = \sigma\iint_{x^2+y^2\leq a^2} h\left(1-\dfrac{\sqrt{x^2+y^2}}{a}\right)\dfrac{\sqrt{a^2+h^2}}{a}\,dx\,dy$$
$$= \dfrac{2\pi\sigma h\sqrt{a^2+h^2}}{a}\int_0^a \left(1-\dfrac{r}{a}\right)r\,dr$$
$$= \dfrac{\pi\sigma ha\sqrt{a^2+h^2}}{3}.$$

Thus $\bar{z} = \dfrac{h}{3}$. By symmetry, $\bar{x} = \bar{y} = 0$. The centre of mass is on the axis of the cone, one-third of the way from the base towards the vertex.

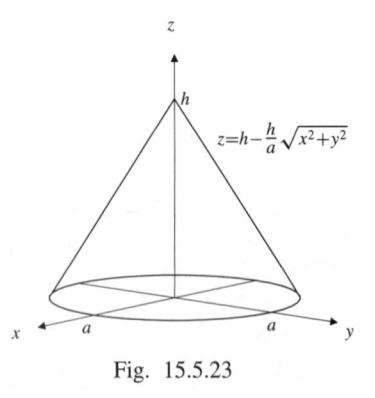

Fig. 15.5.23

24. By symmetry, the force of attraction of the hemisphere shown in the figure on the mass m at the origin is vertical. The vertical component of the force exerted by area element $dS = a^2\sin\phi\,d\phi\,d\theta$ at the position with spherical coordinates (a,ϕ,θ) is

$$dF = \dfrac{km\sigma\,dS}{a^2}\cos\phi = km\sigma\sin\phi\cos\phi\,d\phi\,d\theta.$$

Thus, the total force on m is

$$F = km\sigma\int_0^{2\pi} d\theta\int_0^{\pi/2}\sin\phi\cos\phi\,d\phi = \pi km\sigma \text{ units.}$$

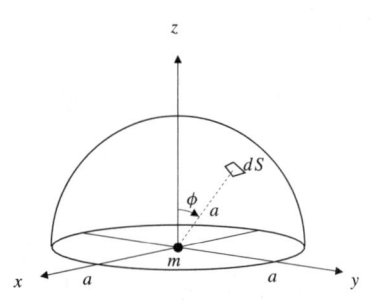

Fig. 15.5.24

25. The surface element $dS = a\,d\theta\,dz$ at the point with cylindrical coordinates (a,θ,z) attracts mass m at point $(0,0,b)$ with a force whose vertical component (see the figure) is

$$dF = \dfrac{km\sigma\,dS}{D^2}\cos\psi = \dfrac{km\sigma a(b-z)\,d\theta\,dz}{D^3}$$
$$= \dfrac{km\sigma a(b-z)\,d\theta\,dz}{\left(a^2+(b-z)^2\right)^{3/2}}.$$

The total force exerted by the cylindrical surface on the mass m is

$$F = -\int_0^{2\pi} d\theta\int_0^h \dfrac{km\sigma a(b-z)\,dz}{\left(a^2+(b-z)^2\right)^{3/2}} \quad \begin{array}{l}\text{Let } b-z = a\tan t\\ -dz = a\sec^2 t\,dt\end{array}$$
$$= 2\pi km\sigma a\int_{z=0}^{z=h}\dfrac{a\tan t\,a\sec^2 t\,dt}{a^3\sec^3 t}$$
$$= 2\pi km\sigma\int_{z=0}^{z=h}\sin t\,dt$$
$$= 2\pi km\sigma(-\cos t)\Big|_{z=0}^{z=h}$$
$$= 2\pi km\sigma\dfrac{a}{\sqrt{a^2+(b-z)^2}}\Big|_0^h$$
$$= 2\pi km\sigma a\left(\dfrac{1}{\sqrt{a^2+(b-h)^2}} - \dfrac{1}{\sqrt{a^2+b^2}}\right).$$

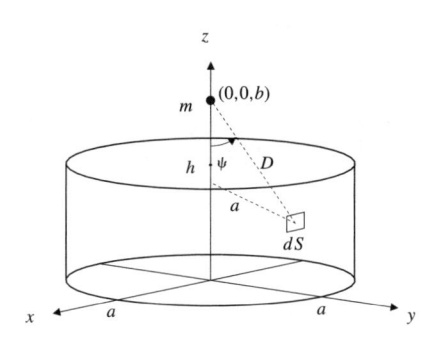

Fig. 15.5.25

26. S is the cylindrical surface $x^2 + y^2 = a^2$, $0 \le z \le h$, with areal density σ. Its mass is $m = 2\pi a h \sigma$. Since all surface elements are at distance a from the z-axis, the radius of gyration of the cylindrical surface about the z-axis is $\overline{D} = a$. Therefore the moment of inertia about that axis is
$$I = m\overline{D}^2 = ma^2 = 2\pi\sigma a^3 h.$$

27. S is the spherical shell, $x^2 + y^2 + z^2 = a^2$, with areal density σ. Its mass is $4\pi\sigma a^2$. Its moment of inertia about the z-axis is
$$\begin{aligned}I &= \sigma \iint_S (x^2 + y^2)\,dS \\ &= \sigma \int_0^{2\pi} d\theta \int_0^{\pi} a^2 \sin^2\phi\, a^2 \sin\phi\, d\phi \\ &= 2\pi\sigma a^4 \int_0^{\pi} \sin\phi(1 - \cos^2\phi)\,d\phi \quad \text{Let } u = \cos\phi \\ &\qquad\qquad\qquad\qquad\qquad\qquad\qquad du = -\sin\phi\,d\phi \\ &= 2\pi\sigma a^4 \int_{-1}^{1} (1 - u^2)\,du = \frac{8\pi\sigma a^4}{3}.\end{aligned}$$

The radius of gyration is $\overline{D} = \sqrt{I/m} = \sqrt{\frac{2}{3}}a$.

28. The surface area element for a conical surface S,
$$z = h\left(1 - \frac{\sqrt{x^2+y^2}}{a}\right),$$
having base radius a and height h, was determined in the solution to Exercise 23 to be
$$dS = \frac{\sqrt{a^2 + h^2}}{a}\,dx\,dy.$$

The mass of S, which has areal density σ, was also determined in that exercise: $m = \pi\sigma a\sqrt{a^2 + h^2}$. The moment of inertia of S about the z-axis is
$$\begin{aligned}I &= \sigma \iint_S (x^2 + y^2)\,dS \\ &= \frac{\sigma\sqrt{a^2+h^2}}{a} \int_0^{2\pi} d\theta \int_0^a r^2 r\,dr \\ &= \frac{2\pi\sigma\sqrt{a^2+h^2}}{a} \cdot \frac{a^4}{4} = \frac{\pi\sigma a^3 \sqrt{a^2+h^2}}{2}.\end{aligned}$$

The radius of gyration is $\overline{D} = \sqrt{I/m} = \frac{a}{\sqrt{2}}$.

29. By Exercise 27, the moment of inertia of a spherical shell of radius a about its diameter is $I = \frac{2}{3}ma^2$. Following the argument given in Example 4(b) of Section 14.7, the kinetic energy of the sphere, rolling with speed v down a plane inclined at angle α above the horizontal (and therefore rotating with angular speed $\Omega = v/a$) is
$$\begin{aligned}K.E. &= \frac{1}{2}mv^2 + \frac{1}{2}I\Omega^2 \\ &= \frac{1}{2}mv^2 + \frac{1}{2}\cdot\frac{2}{3}ma^2\cdot\frac{v^2}{a^2} \\ &= \frac{5}{6}mv^2.\end{aligned}$$

The potential energy is $P.E. = mgh$, so, by conservation of total energy,
$$\frac{5}{6}mv^2 + mgh = \text{constant}.$$

Differentiating with respect to time t, we get
$$0 = \frac{5}{6}m\,2v\frac{dv}{dt} + mg\frac{dh}{dt} = \frac{5}{3}mv\frac{dv}{dt} + mgv\sin\alpha.$$

Thus the sphere rolls with acceleration
$$\frac{dv}{dt} = \frac{3}{5}g\sin\alpha.$$

Section 15.6 Oriented Surfaces and Flux Integrals (page 924)

1. $\mathbf{F} = x\mathbf{i} + z\mathbf{j}$.
   The surface $\mathcal{S}$ of the tetrahedron has four faces:
   On $\mathcal{S}_1$, $x = 0$, $\hat{\mathbf{N}} = -\mathbf{i}$, $\mathbf{F} \bullet \hat{\mathbf{N}} = 0$.
   On $\mathcal{S}_2$, $y = 0$, $\hat{\mathbf{N}} = -\mathbf{j}$, $\mathbf{F} \bullet \hat{\mathbf{N}} = -z$, $dS = dx\,dz$.
   On $\mathcal{S}_3$, $z = 0$, $\hat{\mathbf{N}} = -\mathbf{k}$, $\mathbf{F} \bullet \hat{\mathbf{N}} = 0$.
   On $\mathcal{S}_4$, $x + 2y + 3z = 6$, $\hat{\mathbf{N}} = \dfrac{\mathbf{i} + 2\mathbf{j} + 3\mathbf{k}}{\sqrt{14}}$, $\mathbf{F} \bullet \hat{\mathbf{N}} = \dfrac{x + 2z}{\sqrt{14}}$,
   $dS = \dfrac{dx\,dy}{|\hat{\mathbf{N}} \bullet \mathbf{j}|} = \dfrac{\sqrt{14}}{2} dx\,dz$.
   We have

   $$\iint_{\mathcal{S}_1} \mathbf{F} \bullet \hat{\mathbf{N}}\,dS = \iint_{\mathcal{S}_3} \mathbf{F} \bullet \hat{\mathbf{N}}\,dS = 0$$

   $$\iint_{\mathcal{S}_2} \mathbf{F} \bullet \hat{\mathbf{N}}\,dS = -\int_0^2 z\,dz \int_0^{6-3z} dx$$

   $$= -\int_0^2 (6z - 3z^2)\,dz = -4$$

   $$\iint_{\mathcal{S}_4} \mathbf{F} \bullet \hat{\mathbf{N}}\,dS = \dfrac{\sqrt{14}}{2} \dfrac{1}{\sqrt{14}} \int_0^2 dz \int_0^{6-3z} (x + 2z)\,dx$$

   $$= \dfrac{1}{2} \int_0^2 \left(\dfrac{(6-3z)^2}{2} + 2z(6-3z) \right) dz$$

   $$= \dfrac{1}{4} \int_0^2 (6 - 3z)(6 + z)\,dz$$

   $$= \dfrac{1}{4}(36z - 6z^2 - z^3)\Big|_0^2 = 10.$$

   The flux of $\mathbf{F}$ out of the tetrahedron is

   $$\iint_{\mathcal{S}} \mathbf{F} \bullet \hat{\mathbf{N}}\,dS = 0 - 4 + 0 + 10 = 6.$$

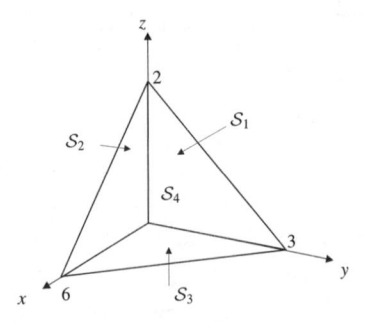

   Fig. 15.6.1

2. On the sphere $\mathcal{S}$ with equation $x^2 + y^2 + z^2 = a^2$ we have

   $$\hat{\mathbf{N}} = \dfrac{x\mathbf{i} + y\mathbf{j} + z\mathbf{k}}{a}.$$

   If $\mathbf{F} = x\mathbf{i} + y\mathbf{j} + z\mathbf{k}$, then $\mathbf{F} \bullet \hat{\mathbf{N}} = a$ on $\mathcal{S}$. Thus the flux of $\mathbf{F}$ out of $\mathcal{S}$ is

   $$\iint_{\mathcal{S}} \mathbf{F} \bullet \hat{\mathbf{N}}\,dS = a \times 4\pi a^2 = 4\pi a^3.$$

3. $\mathbf{F} = x\mathbf{i} + y\mathbf{j} + z\mathbf{k}$.
   The box has six faces. $\mathbf{F} \bullet \hat{\mathbf{N}} = 0$ on the three faces $x = 0$, $y = 0$, and $z = 0$. On the face $x = a$, we have $\hat{\mathbf{N}} = \mathbf{i}$, so $\mathbf{F} \bullet \hat{\mathbf{N}} = a$. Thus the flux of $\mathbf{F}$ out of that face is

   $$a \times (\text{area of the face}) = abc.$$

   By symmetry, the flux of $\mathbf{F}$ out of the faces $y = b$ and $z = c$ are also each abc. Thus the total flux of $\mathbf{F}$ out of the box is $3abc$.

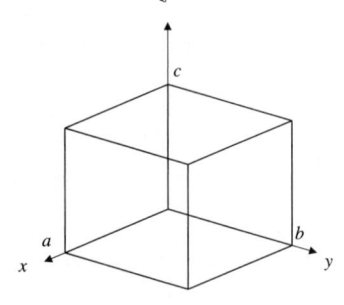

   Fig. 15.6.3

4. $\mathbf{F} = y\mathbf{i} + z\mathbf{k}$. Let $\mathcal{S}_1$ be the conical surface and $\mathcal{S}_2$ be the base disk. The flux of $\mathbf{F}$ outward through the surface of the cone is

   $$\iint_{\mathcal{S}} \mathbf{F} \bullet \hat{\mathbf{N}} = \iint_{\mathcal{S}_1} + \iint_{\mathcal{S}_2}.$$

   On $\mathcal{S}_1$: $\hat{\mathbf{N}} = \dfrac{1}{\sqrt{2}}\left(\dfrac{x\mathbf{i} + y\mathbf{j}}{\sqrt{x^2 + y^2}} + \mathbf{k} \right)$, $dS = \sqrt{2}\,dx\,dy$.
   Thus

   $$\iint_{\mathcal{S}_1} \mathbf{F} \bullet \hat{\mathbf{N}}\,dS$$

   $$= \iint_{x^2 + y^2 \leq 1} \left(\dfrac{xy}{\sqrt{x^2 + y^2}} + 1 - \sqrt{x^2 + y^2} \right) dx\,dy$$

   $$= 0 + \pi \times 1^2 - \int_0^{2\pi} d\theta \int_0^1 r^2\,dr$$

   $$= \pi - \dfrac{2\pi}{3} = \dfrac{\pi}{3}.$$

   On $\mathcal{S}_2$: $\hat{\mathbf{N}} = -\mathbf{k}$ and $z = 0$, so $\mathbf{F} \bullet \hat{\mathbf{N}} = 0$. Thus, the total flux of $\mathbf{F}$ out of the cone is $\pi/3$.

Fig. 15.6.4

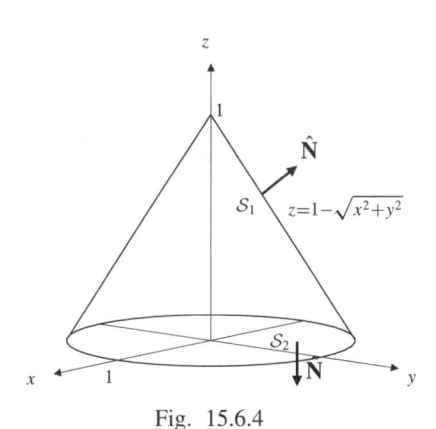

5. The part $\mathcal{S}$ of $z = a - x^2 - y^2$ lying above $z = b < a$ lies inside the vertical cylinder $x^2 + y^2 = a - b$. For $z = a - x^2 - y^2$, the upward vector surface element is

$$\hat{\mathbf{N}} dS = \frac{2x\mathbf{i} + 2y\mathbf{j} + \mathbf{k}}{1} dx\, dy.$$

Thus the flux of $\mathbf{F} = x\mathbf{i} + y\mathbf{j} + z\mathbf{k}$ upward through $\mathcal{S}$ is

$$\iint_{\mathcal{S}} \mathbf{F} \cdot \hat{\mathbf{N}} dS$$
$$= \iint_{x^2+y^2 \leq a-b} [2(x^2+y^2) + a - x^2 - y^2]\, dx\, dy$$
$$= \int_0^{2\pi} d\theta \int_0^{\sqrt{a-b}} (r^2 + a) r\, dr$$
$$= 2\pi \left(\frac{(a-b)^2}{4} + \frac{a(a-b)}{2} \right) = \frac{\pi}{2}(a-b)(3a-b).$$

6. For $z = x^2 - y^2$ the upward surface element is

$$\hat{\mathbf{N}} dS = \frac{-2x\mathbf{i} + 2y\mathbf{j} + \mathbf{k}}{1} dx\, dy.$$

The flux of $\mathbf{F} = x\mathbf{i} + x\mathbf{j} + \mathbf{k}$ upward through $\mathcal{S}$, the part of $z = x^2 - y^2$ inside $x^2 + y^2 = a^2$ is

$$\iint_{\mathcal{S}} \mathbf{F} \cdot \hat{\mathbf{N}} dS = \iint_{x^2+y^2 \leq a^2} (-2x^2 + 2xy + 1)\, dx\, dy$$
$$= -2 \int_0^{2\pi} \cos^2\theta\, d\theta \int_0^a r^3\, dr + 0 + \pi a^2$$
$$= \pi a^2 - 2(\pi)\frac{a^4}{4} = \frac{\pi}{2}a^2(2 - a^2).$$

7. The part $\mathcal{S}$ of $z = 4 - x^2 - y^2$ lying above $z = 2x + 1$ has projection onto the xy-plane the disk D bounded by

$$2x + 1 = 4 - x^2 - y^2, \quad \text{or } (x+1)^2 + y^2 = 4.$$

Note that D has area 4π and centroid $(-1, 0)$. For $z = 4 - x^2 - y^2$, the downward vector surface element is

$$\hat{\mathbf{N}} dS = \frac{-2x\mathbf{i} - 2y\mathbf{j} - \mathbf{k}}{1} dx\, dy.$$

Thus the flux of $\mathbf{F} = y^3\mathbf{i} + z^2\mathbf{j} + x\mathbf{k}$ downward through $\mathcal{S}$ is

$$\iint_{\mathcal{S}} \mathbf{F} \cdot \hat{\mathbf{N}} dS = -\iint_D \left(2xy^3 + 2y(4 - x^2 - y^2)^2 + x \right) dx\, dy$$

(use the symmetry of D about the x-axis)

$$= -\iint_D x\, dA = -(4\pi)(-1) = 4\pi.$$

8. The upward vector surface element on the top half of $x^2 + y^2 + z^2 = a^2$ is

$$\hat{\mathbf{N}} dS = \frac{2x\mathbf{i} + 2y\mathbf{j} + 2z\mathbf{k}}{2z}\, dx\, dy = \left(\frac{x\mathbf{i} + y\mathbf{j}}{z} + \mathbf{k} \right) dx\, dy.$$

The flux of $\mathbf{F} = z^2\mathbf{k}$ upward through the first octant part $\mathcal{S}$ of the sphere is

$$\iint_{\mathcal{S}} \mathbf{F} \cdot \hat{\mathbf{N}} dS = \int_0^{\pi/2} d\theta \int_0^a (a^2 - r^2) r\, dr = \frac{\pi a^4}{8}.$$

9. The upward vector surface element on $z = 2 - x^2 - 2y^2$ is

$$\hat{\mathbf{N}} dS = \frac{2x\mathbf{i} + 4y\mathbf{j} + \mathbf{k}}{1} dx\, dy.$$

If E is the elliptic disk bounded by $\frac{x^2}{2} + y^2 = 1$, then the flux of $\mathbf{F} = x\mathbf{i} + y\mathbf{j}$ through the required surface $\mathcal{S}$ is

$$\iint_{\mathcal{S}} \mathbf{F} \cdot \hat{\mathbf{N}} dS$$
$$= \iint_E (2x^2 + 4y^2)\, dx\, dy \quad \text{Let } x = \sqrt{2}u,\ y = v$$
$$\hspace{5cm} dx\, dy = \sqrt{2}\, du\, dv$$
$$= 4\sqrt{2} \iint_{u^2+v^2 \leq 1} (u^2 + v^2)\, du\, dv \quad \text{(now use polars)}$$
$$= 4\sqrt{2} \int_0^{2\pi} d\theta \int_0^1 r^3\, dr = 2\sqrt{2}\pi.$$

10. $\mathcal{S}: \mathbf{r} = u^2 v\mathbf{i} + uv^2\mathbf{j} + v^3\mathbf{k}$, $(0 \leq u \leq 1,\ 0 \leq v \leq 1)$, has upward surface element

$$\hat{\mathbf{N}} dS = \frac{\partial \mathbf{r}}{\partial u} \times \frac{\partial \mathbf{r}}{\partial v}\, du\, dv$$
$$= (2uv\mathbf{i} + v^2\mathbf{j}) \times (u^2\mathbf{i} + 2uv\mathbf{j} + 3v^2\mathbf{k})\, du\, dv$$
$$= (3v^4\mathbf{i} - 6uv^3\mathbf{j} + 3u^2v^2\mathbf{k})\, du\, dv.$$

SECTION 15.6 (PAGE 924)

The flux of $\mathbf{F} = 2x\mathbf{i} + y\mathbf{j} + z\mathbf{k}$ upward through S is

$$\iint_S \mathbf{F} \cdot \hat{\mathbf{N}}\, dS$$
$$= \int_0^1 du \int_0^1 (6u^2 v^5 - 6u^2 v^5 + 3u^2 v^5)\, dv$$
$$= \frac{1}{2}\int_0^1 u^2\, du = \frac{1}{6}.$$

11. S: $\mathbf{r} = u\cos v\,\mathbf{i} + u\sin v\,\mathbf{j} + u\mathbf{k}$, $(0 \le u \le 2,\ 0 \le v \le \pi)$, has upward surface element

$$\hat{\mathbf{N}}\, dS = \frac{\partial \mathbf{r}}{\partial u} \times \frac{\partial \mathbf{r}}{\partial v}\, du\, dv$$
$$= (-u\cos v\,\mathbf{i} - u\sin v\,\mathbf{j} + u\mathbf{k})\, du\, dv.$$

The flux of $\mathbf{F} = x\mathbf{i} + y\mathbf{j} + z^2\mathbf{k}$ upward through S is

$$\iint_S \mathbf{F} \cdot \hat{\mathbf{N}}\, dS$$
$$= \int_0^2 du \int_0^\pi (-u^2\cos^2 v - u^2\sin^2 v + u^3)\, dv$$
$$= \int_0^2 (u^3 - u^2)\, du \int_0^\pi dv = \frac{4\pi}{3}.$$

12. S: $\mathbf{r} = e^u\cos v\,\mathbf{i} + e^u\sin v\,\mathbf{j} + u\mathbf{k}$, $(0 \le u \le 1,\ 0 \le v \le \pi)$, has upward surface element

$$\hat{\mathbf{N}}\, dS = \frac{\partial \mathbf{r}}{\partial u} \times \frac{\partial \mathbf{r}}{\partial v}\, du\, dv$$
$$= (-e^u \cos v\,\mathbf{i} - e^u \sin v\,\mathbf{j} + e^{2u}\mathbf{k})\, du\, dv.$$

The flux of $\mathbf{F} = yz\mathbf{i} - xz\mathbf{j} + (x^2 + y^2)\mathbf{k}$ upward through S is

$$\iint_S \mathbf{F} \cdot \hat{\mathbf{N}}\, dS$$
$$= \int_0^1 du \int_0^\pi (-ue^{2u}\sin v\cos v + ue^{2u}\sin v\cos v + e^{4u})\, dv$$
$$= \int_0^1 e^{4u}\, du \int_0^\pi dv = \frac{\pi e^4}{4}.$$

13. $\mathbf{F} = \dfrac{m\mathbf{r}}{|\mathbf{r}|^3} = \dfrac{m(x\mathbf{i} + y\mathbf{j} + z\mathbf{k})}{(x^2 + y^2 + z^2)^{3/2}}.$

By symmetry, the flux of $\mathbf{F}$ out of the cube $-a \le x, y, z \le a$ is 6 times the flux out of the top face, $z = a$, where $\hat{\mathbf{N}} = \mathbf{k}$ and $dS = dx\, dy$. The total flux is

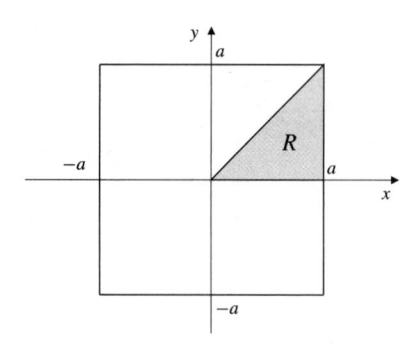

Fig. 15.6.13

$$6ma \int_{\substack{-a \le x \le a \\ -a \le y \le a}} \frac{dx\, dy}{(x^2 + y^2 + a^2)^{3/2}}$$
$$= 48ma \iint_R \frac{r\, dr\, d\theta}{(r^2 + a^2)^{3/2}}$$
(R as shown in the figure)
$$= 48ma \int_0^{\pi/4} d\theta \int_0^{a\sec\theta} \frac{r\, dr}{(r^2 + a^2)^{3/2}}$$
Let $u = r^2 + a^2$
$du = 2r\, dr$
$$= 24ma \int_0^{\pi/4} d\theta \int_{a^2}^{a^2(1+\sec^2\theta)} \frac{du}{u^{3/2}}$$
$$= 48ma \int_0^{\pi/4} \left(\frac{1}{a} - \frac{1}{a\sqrt{1 + \sec^2\theta}}\right) d\theta$$
$$= 48m\left(\frac{\pi}{4} - \int_0^{\pi/4} \frac{\cos\theta\, d\theta}{\sqrt{\cos^2\theta + 1}}\right)$$
$$= 48m\left(\frac{\pi}{4} - \int_0^{\pi/4} \frac{\cos\theta\, d\theta}{\sqrt{2 - \sin^2\theta}}\right)$$
Let $\sqrt{2}\sin v = \sin\theta$
$\sqrt{2}\cos v\, dv = \cos\theta\, d\theta$
$$= 48m\left(\frac{\pi}{4} - \int_0^{\pi/6} \frac{\sqrt{2}\cos v\, dv}{\sqrt{2}\cos v}\right)$$
$$= 48m\left(\frac{\pi}{4} - \frac{\pi}{6}\right) = 4\pi m.$$

14. The flux of $\mathbf{F} = \dfrac{m\mathbf{r}}{|\mathbf{r}|^3}$ out of the cube $1 \le x, y, z \le 2$ is equal to three times the total flux out of the pair of opposite faces $z = 1$ and $z = 2$, which have outward normals $-\mathbf{k}$ and $\mathbf{k}$ respectively. This latter flux is

$2mI_2 - mI_1$, where

$$I_k = \int_1^2 dx \int_1^2 \frac{dy}{(x^2+y^2+k^2)^{3/2}}$$

Let $y = \sqrt{x^2+k^2}\tan u$
$dy = \sqrt{x^2+k^2}\sec^2 u\, du$

$$= \int_1^2 \frac{dx}{x^2+k^2} \int_{y=1}^{y=2} \cos u\, du$$

$$= \int_1^2 \frac{dx}{x^2+k^2} (\sin u)\Big|_{y=1}^{y=2}$$

$$= \int_1^2 \frac{dx}{x^2+k^2} \left(\frac{y}{\sqrt{x^2+y^2+k^2}}\Big|_1^2\right) = J_{k2} - J_{k1},$$

where

$$J_{kn} = n\int_1^2 \frac{dx}{(x^2+k^2)\sqrt{x^2+n^2+k^2}}$$

Let $x = \sqrt{n^2+k^2}\tan v$
$dx = \sqrt{n^2+k^2}\sec^2 v\, dv$

$$= n\int_{x=1}^{x=2} \frac{\sec^2 v\, dv}{[(n^2+k^2)\tan^2 v + k^2]\sec v}$$

$$= n\int_{x=1}^{x=2} \frac{\cos v\, dv}{(n^2+k^2)\sin^2 v + k^2\cos^2 v}$$

$$= n\int_{x=1}^{x=2} \frac{\cos v\, dv}{k^2+n^2\sin^2 v}$$ Let $w = n\sin v$
$dw = n\cos v\, dv$

$$= \int_{x=1}^{x=2} \frac{dw}{k^2+w^2} = \frac{1}{k}\tan^{-1}\frac{w}{k}\Big|_{x=1}^{x=2}$$

$$= \frac{1}{k}\tan^{-1}\frac{n\sin v}{k}\Big|_{x=1}^{x=2}$$

$$= \frac{1}{k}\tan^{-1}\frac{nx}{k\sqrt{x^2+n^2+k^2}}\Big|_1^2$$

$$= \frac{1}{k}\left(\tan^{-1}\frac{2n}{k\sqrt{4+n^2+k^2}} - \tan^{-1}\frac{n}{k\sqrt{1+n^2+k^2}}\right).$$

Thus

$$I_k = \frac{1}{k}\left[\tan^{-1}\frac{4}{k\sqrt{8+k^2}} - 2\tan^{-1}\frac{2}{k\sqrt{5+k^2}}\right.$$

$$\left. + \tan^{-1}\frac{1}{k\sqrt{2+k^2}}\right].$$

The contribution to the total flux from the pair of surfaces $z=1$ and $z=2$ of the cube is

$$2mI_2 - mI_1$$

$$= m\left[\tan^{-1}\frac{1}{\sqrt{3}} - 2\tan^{-1}\frac{1}{3} + \tan^{-1}\frac{1}{2\sqrt{6}}\right.$$

$$\left. - \tan^{-1}\frac{4}{3} + 2\tan^{-1}\frac{2}{\sqrt{6}} - \tan^{-1}\frac{1}{\sqrt{3}}\right].$$

Using the identities

$$2\tan^{-1}a = \tan^{-1}\frac{2a}{1-a^2}, \text{ and}$$
$$\tan^{-1}a = \frac{\pi}{2} - \tan^{-1}\frac{1}{a},$$

we calculate

$$-2\tan^{-1}\frac{1}{3} = -\tan^{-1}\frac{3}{4} = -\frac{\pi}{2} + \tan^{-1}\frac{4}{3}$$

$$2\tan^{-1}\frac{2}{\sqrt{6}} = \tan^{-1}\frac{12}{\sqrt{6}} = \frac{\pi}{2} - \tan^{-1}\frac{1}{2\sqrt{6}}.$$

Thus the net flux out of the pair of opposite faces is 0. By symmetry this holds for each pair, and the total flux out of the cube is 0. (You were warned this would be a difficult calculation!)

15. The flux of the plane vector field **F** across the piecewise smooth curve $\mathcal{C}$, in the direction of the unit normal $\hat{\mathbf{N}}$ to the curve, is

$$\int_{\mathcal{C}} \mathbf{F}\bullet\mathbf{n}\, ds.$$

The flux of $\mathbf{F} = x\mathbf{i} + y\mathbf{j}$ outward across

a) the circle $x^2 + y^2 = a^2$ is

$$\oint_{\mathcal{C}} \mathbf{F}\bullet\left(\frac{x\mathbf{i}+y\mathbf{j}}{a}\right) ds = \frac{a^2}{a}\times 2\pi a = 2\pi a^2.$$

b) the boundary of the square $-1 \leq x, y \leq 1$ is

$$4\int_{-1}^{1}(\mathbf{i}+y\mathbf{j})\bullet\mathbf{i}\, dy = 4\int_{-1}^{1} dy = 8.$$

16. $\mathbf{F} = -\dfrac{x\mathbf{i}+y\mathbf{j}}{x^2+y^2}.$

a) The flux of **F** inward across the circle of Exercise 7(a) is

$$-\oint_{\mathcal{C}}\left(-\frac{x\mathbf{i}+y\mathbf{j}}{a^2}\right)\bullet\frac{x\mathbf{i}+y\mathbf{j}}{a} ds$$

$$= \oint_{\mathcal{C}} \frac{a^2}{a^3} ds = \frac{1}{a}\times 2\pi a = 2\pi.$$

b) The flux of **F** inward across the boundary of the square of Exercise 7(b) is four times the flux inward across the edge $x=1$, $-1 \le y \le 1$. Thus it is

$$-4\int_{-1}^{1}\left(-\frac{\mathbf{i}+y\mathbf{j}}{1+y^2}\right)\bullet\mathbf{i}\,dy = 4\int_{-1}^{1}\frac{dy}{1+y^2}$$
$$= 4\tan^{-1}y\bigg|_{-1}^{1} = 2\pi.$$

17. The flux of $\hat{\mathbf{N}}$ across $\mathcal{S}$ is

$$\iint_{\mathcal{S}}\hat{\mathbf{N}}\bullet\hat{\mathbf{N}}\,dS = \iint_{\mathcal{S}}dS = \text{area of } \mathcal{S}.$$

18. Let $\mathbf{F} = F_1\mathbf{i} + F_2\mathbf{j} + F_3\mathbf{k}$ be a constant vector field.

a) If R is a rectangular box, we can choose the origin and coordinate axes in such a way that the box is $0 \le x \le a$, $0 \le y \le b$, $0 \le z \le c$. On the faces $x=0$ and $x=a$ we have $\hat{\mathbf{N}} = -\mathbf{i}$ and $\hat{\mathbf{N}} = \mathbf{i}$ respectively. Since F_1 is constant, the total flux out of the box through these two faces is

$$\iint_{\substack{0\le y\le b\\ 0\le z\le c}}(F_1 - F_1)\,dy\,dz = 0.$$

The flux out of the other two pairs of opposite faces is also 0. Thus the total flux of **F** out of the box is 0.

b) If $\mathcal{S}$ is a sphere of radius a we can choose the origin so that $\mathcal{S}$ has equation $x^2+y^2+z^2 = a^2$, and so its outward normal is

$$\hat{\mathbf{N}} = \frac{x\mathbf{i}+y\mathbf{j}+z\mathbf{k}}{a}.$$

Thus the flux out of $\mathcal{S}$ is

$$\frac{1}{a}\iint_{\mathcal{S}}(F_1 x + F_2 y + F_3 z)\,ds = 0,$$

since the sphere $\mathcal{S}$ is symmetric about the origin.

Review Exercises 15 (page 925)

1. $\mathcal{C}:\ x=t$, $y=2e^t$, $z=e^{2t}$, $(-1 \le t \le 1)$
$$v = \sqrt{1+4e^{2t}+4e^{4t}} = 1+2e^{2t}$$
$$\int_{\mathcal{C}}\frac{ds}{y} = \int_{-1}^{1}\frac{1+2e^{2t}}{2e^t}\,dt$$
$$= \left(-\frac{e^{-t}}{2}+e^t\right)\bigg|_{-1}^{1} = \frac{3(e^2-1)}{2e}.$$

2. $\mathcal{C}$ can be parametrized $x=t$, $y=2t$, $z=t+4t^2$, $(0 \le t \le 2)$. Thus

$$\int_{\mathcal{C}}2y\,dx + x\,dy + 2\,dz$$
$$= \int_0^2 [4t(1)+t(2)+2(1+8t)]\,dt$$
$$= \int_0^2 (22t+2)\,dt = 48.$$

3. The cone $z=\sqrt{x^2+y^2}$ has area element

$$dS = \sqrt{1+\frac{x^2+y^2}{z^2}}\,dx\,dy = \sqrt{2}\,dx\,dy.$$

If $\mathcal{S}$ is the part of the cone in the region $0 \le x \le 1-y^2$ (which itself lies between $y=-1$ and $y=1$), then

$$\iint_{\mathcal{S}}x\,dS = \sqrt{2}\int_{-1}^{1}dy\int_0^{1-y^2}x\,dx$$
$$= 2\sqrt{2}\int_0^1 \frac{1-2y^2+y^4}{2}\,dy = \frac{8\sqrt{2}}{15}.$$

4. The plane $x+y+z=1$ has area element $dS = \sqrt{3}\,dx\,dy$. If $\mathcal{S}$ is the part of the plane in the first octant, then the projection of $\mathcal{S}$ on the xy-plane is the triangle $0 \le x \le 1$, $0 \le y \le 1-x$. Thus

$$\iint_{\mathcal{S}}xyz\,dS = \sqrt{3}\int_0^1 x\,dx\int_0^{1-x}y(1-x-y)\,dy$$
$$= \sqrt{3}\int_0^1 \frac{x(1-x)^3}{6}\,dx \quad \text{Let } u=1-x$$
$$\phantom{= \sqrt{3}\int_0^1 \frac{x(1-x)^3}{6}\,dx \quad} du=-dx$$
$$= \frac{\sqrt{3}}{6}\int_0^1 u^3(1-u)\,du = \frac{\sqrt{3}}{6}\left(\frac{1}{4}-\frac{1}{5}\right) = \frac{\sqrt{3}}{120}.$$

5. For $z=xy$, the upward vector surface element is

$$\hat{\mathbf{N}}\,dS = \frac{-y\mathbf{i}-x\mathbf{j}+\mathbf{k}}{1}\,dx\,dy.$$

The flux of $\mathbf{F} = x^2 y\mathbf{i} - 10xy^2\mathbf{j}$ upward through $\mathcal{S}$, the part of $z=xy$ satisfying $0 \le x \le 1$ and $0 \le y \le 1$ is

$$\iint_{\mathcal{S}}\mathbf{F}\bullet\hat{\mathbf{N}}\,dS = \int_0^1 dx\int_0^1 (-x^2 y^2 + 10x^2 y^2)\,dy$$
$$= \int_0^1 3x^2\,dx\int_0^1 3y^2\,dy = 1.$$

6. The plane $x+2y+3z=6$ has downward vector surface element

$$\hat{\mathbf{N}}\,dS = \frac{-\mathbf{i}-2\mathbf{j}-3\mathbf{k}}{3}\,dx\,dy.$$

INSTRUCTOR'S SOLUTIONS MANUAL   REVIEW EXERCISES 15 (PAGE 925)

If $\mathcal{S}$ is the part of the plane in the first octant, then the projection of $\mathcal{S}$ on the xy-plane is the triangle $0 \le y \le 3$, $0 \le x \le 6 - 2y$. Thus

$$\iint_{\mathcal{S}} (x\mathbf{i} + y\mathbf{j} + z\mathbf{k}) \bullet \hat{\mathbf{N}} \, dS$$
$$= -\frac{1}{3} \int_0^3 dy \int_0^{6-2y} (x + 2y + 6 - x - 2y) \, dx$$
$$= -2 \int_0^3 (6 - 2y) = -36 + 18 = -18.$$

7. $\mathbf{r} = a\sin t\mathbf{i} + a\cos t\mathbf{j} + bt\mathbf{k}$, $(0 \le t \le 6\pi)$
   $\mathbf{r}(0) = a\mathbf{j}$, $\mathbf{r}(6\pi) = a\mathbf{j} + 6\pi b\mathbf{k}$.

   a) The force $\mathbf{F} = -mg\mathbf{k} = -\nabla(mgz)$ is conservative, so the work done by $\mathbf{F}$ as the bead moves from $\mathbf{r}(6\pi)$ to $\mathbf{r}(0)$ is

   $$W = \int_{t=6\pi}^{t=0} \mathbf{F} \bullet d\mathbf{r} = -mgz \Big|_{z=6\pi b}^{z=0} = 6\pi mgb.$$

   b) $\mathbf{v} = a\cos t\mathbf{i} - a\sin t\mathbf{j} + b\mathbf{k}$, $|\mathbf{v}| = \sqrt{a^2 + b^2}$. A force of constant magnitude R opposing the motion of the bead is in the direction of $-\mathbf{v}$, so it is

   $$\mathbf{F} = -R\frac{\mathbf{v}}{|\mathbf{v}|} = -\frac{R}{\sqrt{a^2 + b^2}} \mathbf{v}.$$

   Since $d\mathbf{r} = \mathbf{v} \, dt$, the work done against the resistive force is

   $$W = \int_0^{6\pi} \frac{R}{\sqrt{a^2 + b^2}} |\mathbf{v}|^2 \, dt = 6\pi R\sqrt{a^2 + b^2}.$$

8. $\int_{\mathcal{C}} \mathbf{F} \bullet d\mathbf{r}$ can be determined using only the endpoints of $\mathcal{C}$, provided

   $$\mathbf{F} = (axy + 3yz)\mathbf{i} + (x^2 + 3xz + by^2z)\mathbf{j} + (bxy + cy^3)\mathbf{k}$$

   is conservative, that is, if

   $$ax + 3z = \frac{\partial F_1}{\partial y} = \frac{\partial F_2}{\partial x} = 2x + 3z$$
   $$3y = \frac{\partial F_1}{\partial z} = \frac{\partial F_3}{\partial x} = by$$
   $$3x + by^2 = \frac{\partial F_2}{\partial z} = \frac{\partial F_3}{\partial y} = bx + 3cy^2.$$

   Thus we need $a = 2$, $b = 3$, and $c = 1$.
   With these values, $\mathbf{F} = \nabla(x^2y + 3xyz + y^3z)$. Thus

   $$\int_{\mathcal{C}} \mathbf{F} \bullet d\mathbf{r} = (x^2y + 3xyz + y^3z)\Big|_{(0,1,-1)}^{(2,1,1)} = 11 - (-1) = 12.$$

9. $\mathbf{F} = (x^2/y)\mathbf{i} + y\mathbf{j} + \mathbf{k}$.
   The field lines satisfy $\dfrac{y\,dx}{x^2} = \dfrac{dy}{y} = dz$. Thus $dx/x^2 = dy/y^2$ and the field lines are given by

   $$\frac{1}{x} = \frac{1}{y} + C_1, \quad \ln y = z + C_2.$$

   The field line passes through $(1, 1, 0)$ provided $C_1 = 0$ and $C_2 = 0$. In this case the field line also passes through $(e, e, 1)$, and the segment from $(1, 1, 0)$ to $(e, e, 1)$ can be parametrized $\mathbf{r}(t) = e^t\mathbf{i} + e^t\mathbf{j} + t\mathbf{k}$, $(0 \le t \le 1)$. Then

   $$\int_{\mathcal{C}} \mathbf{F} \bullet d\mathbf{r} = \int_0^1 (e^{2t} + e^{2t} + 1) \, dt$$
   $$= (e^{2t} + t)\Big|_0^1 = e^2.$$

10. a) $\mathbf{F} = (1 + x)e^{x+y}\mathbf{i} + (xe^{x+y} + 2y)\mathbf{j} - 2z\mathbf{k}$
    $= \nabla(xe^{x+y} + y^2 - z^2)$.
    Thus $\mathbf{F}$ is conservative.

    b) $\mathbf{G} = (1 + x)e^{x+y}\mathbf{i} + (xe^{x+y} + 2z)\mathbf{j} - 2y\mathbf{k}$
    $= \mathbf{F} + 2(z - y)(\mathbf{j} + \mathbf{k})$.
    $\mathcal{C}: \mathbf{r} = (1 - t)e^t\mathbf{i} + t\mathbf{j} + 2t\mathbf{k}$, $(0 \le t \le 1)$.
    $\mathbf{r}(0) = (1, 0, 0)$, $\mathbf{r}(1) = (0, 1, 2)$. Thus

    $$\int_{\mathcal{C}} \mathbf{G} \bullet d\mathbf{r} = \int_{\mathcal{C}} \mathbf{F} \bullet d\mathbf{r} + \int_{\mathcal{C}} 2(z-y)(\mathbf{j} + \mathbf{k}) \bullet d\mathbf{r}$$
    $$= (xe^{x+y} + y^2 - z^2)\Big|_{(1,0,0)}^{(0,1,2)}$$
    $$+ 2\int_0^1 (2t - t)(1 + 2) \, dt$$
    $$= -3 - e + 3t^2 \Big|_0^1 = -e.$$

11. Since the field lines of $\mathbf{F}$ are $xy = C$, and so satisfy

    $$y\,dx + x\,dy = 0, \quad \text{or } \frac{dx}{x} = -\frac{dy}{y},$$

    thus $\mathbf{F} = \lambda(x, y)(x\mathbf{i} - y\mathbf{j})$. Since $|\mathbf{F}(x,y)| = 1$ if $(x, y) \neq (0, 0)$, $\lambda(x, y) = \pm 1/\sqrt{x^2 + y^2}$, and

    $$\mathbf{F}(x, y) = \pm \frac{x\mathbf{i} - y\mathbf{j}}{\sqrt{x^2 + y^2}}.$$

    Since $\mathbf{F}(1, 1) = (\mathbf{i} - \mathbf{j})/\sqrt{2}$, we need the plus sign. Thus

    $$\mathbf{F}(x, y) = \frac{x\mathbf{i} - y\mathbf{j}}{\sqrt{x^2 + y^2}},$$

REVIEW EXERCISES 15 (PAGE 925)

which is continuous everywhere except at $(0, 0)$.

12. The first octant part of the cylinder $y^2 + z^2 = 16$ has outward vector surface element

$$\hat{\mathbf{N}}\,dS = \frac{2y\mathbf{j} + 2z\mathbf{k}}{2z}\,dx\,dy = \left(\frac{y}{\sqrt{16-y^2}}\mathbf{j} + \mathbf{k}\right)dx\,dy.$$

The flux of $3z^2 x\mathbf{i} - x\mathbf{j} - y\mathbf{k}$ outward through the specified surface $\mathcal{S}$ is

$$\mathbf{F}\bullet\hat{\mathbf{N}}\,dS = \int_0^5 dx \int_0^4 \left(0 - \frac{xy}{\sqrt{16-y^2}} - y\right)dy$$

$$= \int_0^5 \left(x\sqrt{16-y^2} - \frac{y^2}{2}\right)\bigg|_{y=0}^{y=4}dx$$

$$= -\int_0^5 (4x+8)\,dx = -90.$$

Challenging Problems 15 (page 926)

1. Given: $x = (2+\cos v)\cos u$, $y = (2+\cos v)\sin u$, $z = \sin v$ for $0 \le u \le 2\pi$, $0 \le v \le \pi$.
The cylindrical coordinate r satisfies

$$r^2 = x^2 + y^2 = (2+\cos v)^2$$
$$r = 2 + \cos v$$
$$(r-2)^2 + z^2 = 1.$$

This equation represents the surface of a torus, obtained by rotating about the z-axis the circle of radius 1 in the xz-plane centred at $(2, 0, 0)$. Since $0 \le v \le \pi$ implies that $z \ge 0$, the given surface is only the top half of the toroidal surface.
By symmetry, $\bar{x} = 0$ and $\bar{y} = 0$.
A ring-shaped strip on the surface at angular position v with width dv has radius $2 + \cos v$, and so its surface area is $dS = 2\pi(2 + \cos v)\,dv$. The area of the whole given surface is

$$S = \int_0^\pi 2\pi(2 + \cos v)\,dv = 4\pi^2.$$

The strip has moment $z\,dS = 2\pi(2 + \cos v)\sin v\,dv$ about $z = 0$, so the moment of the whole surface about $z = 0$ is

$$M_{z=0} = 2\pi \int_0^\pi (2 + \cos v)\sin v\,dv$$

$$= 2\pi\left(-2\cos v - \frac{1}{4}\cos(2v)\right)\bigg|_0^\pi = 8\pi.$$

Thus $\bar{z} = \dfrac{8\pi}{4\pi^2} = \dfrac{2}{\pi}$. The centroid is $(0, 0, 2/\pi)$.

2. This is a trick question. Observe that the given parametrization $\mathbf{r}(u, v)$ satisfies

$$\mathbf{r}(u + \pi, v) = \mathbf{r}(u, -v).$$

Therefore the surface $\mathcal{S}$ is traced out twice as u goes from 0 to 2π. (It is a Möbius band. See Figure 15.28 in the text.) If $\mathcal{S}_1$ is the part of the surface corresponding to $0 \le u \le \pi$, and $\mathcal{S}_2$ is the part corresponding to $\pi \le u \le 2\pi$, then $\mathcal{S}_1$ and $\mathcal{S}_2$ coincide as point sets, but their normals are oppositely oriented: $\hat{\mathbf{N}}_2 = -\hat{\mathbf{N}}_1$ at corresponding points on the two surfaces. Hence

$$\iint_{\mathcal{S}_1} \mathbf{F}\bullet\hat{\mathbf{N}}_1\,dS = -\iint_{\mathcal{S}_2} \mathbf{F}\bullet\hat{\mathbf{N}}_2\,dS,$$

for any smooth vector field, and

$$\iint_{\mathcal{S}} \mathbf{F}\bullet\hat{\mathbf{N}}\,dS = \iint_{\mathcal{S}_1} \mathbf{F}\bullet\hat{\mathbf{N}}_1\,dS + \iint_{\mathcal{S}_2} \mathbf{F}\bullet\hat{\mathbf{N}}_2\,dS = 0.$$

3.

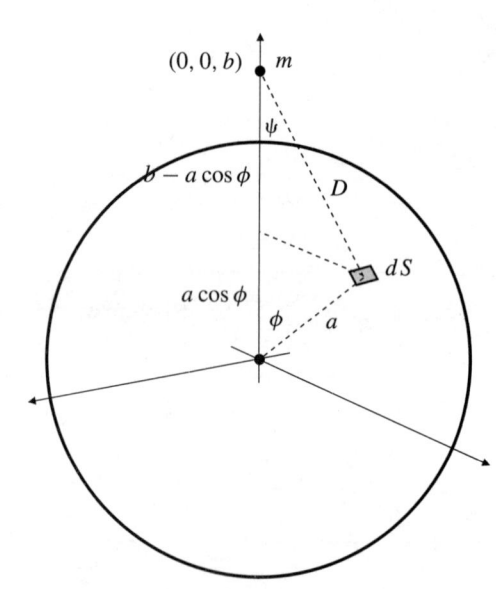

Fig. C-15.3

The mass element $\sigma\,dS$ at position $[a, \phi, \theta]$ on the sphere is at distance $D = \sqrt{a^2 + b^2 - 2ab\cos\phi}$ from the mass m located at $(0, 0, b)$, and thus it attracts m with a force of magnitude $dF = km\sigma\,dS/D^2$. By symmetry, the horizontal components of dF coresponding to mass elements on opposite sides of the sphere (i.e., at $[a, \phi, \theta]$ and $[a, \phi, \theta + \pi]$) cancel, but the vertical components

$$dF\cos\psi = \frac{km\sigma\,dS}{D^2}\frac{b - a\cos\phi}{D}$$

INSTRUCTOR'S SOLUTIONS MANUAL   **CHALLENGING PROBLEMS 15 (PAGE 926)**

reinforce. The total force on the mass m is the sum of all such vertical components. Since $dS = a^2 \sin\phi\, d\phi\, d\theta$, it is

$$F = km\sigma a^2 \int_0^{2\pi} d\theta \int_0^{\pi} \frac{(b - a\cos\phi)\sin\phi\, d\phi}{(a^2 + b^2 - 2ab\cos\phi)^{3/2}}$$

$$= 2\pi km\sigma a^2 \int_{-1}^{1} \frac{(b - at)\, dt}{(a^2 - 2abt + b^2)^{3/2}}.$$

We have made the change of variable $t = \cos\phi$ to get the last integral. This integral can be evaluated by using another substitution. Let $u = \sqrt{a^2 - 2abt + b^2}$. Thus

$$t = \frac{a^2 + b^2 - u^2}{2ab}, \quad dt = -\frac{u\, du}{ab}, \quad b - at = \frac{u^2 + b^2 - a^2}{2b}.$$

When $t = -1$ and $t = 1$ we have $u = a+b$ and $u = |a-b|$ respectively. Therefore

$$F = 2\pi km\sigma a^2 \int_{a+b}^{|a-b|} \frac{u^2 + b^2 - a^2}{2bu^3}\left(-\frac{u\, du}{ab}\right)$$

$$= \frac{\pi km\sigma a}{b^2} \int_{|a-b|}^{a+b} \left(1 + \frac{b^2 - a^2}{u^2}\right) du$$

$$= \frac{\pi km\sigma a}{b^2} \left(u - \frac{b^2 - a^2}{u}\right)\Bigg|_{|a-b|}^{a+b}.$$

There are now two cases to consider. If the mass m is *outside* the sphere, so that $b > a$ and $|a - b| = b - a$, then

$$F = \frac{\pi km\sigma a}{b^2}\Big((a+b) - (b-a) - (b-a) + (b+a)\Big) = 4\pi km\sigma \frac{a^2}{b^2}.$$

However, if m is *inside* the sphere, so that $b < a$ and $|a - b| = a - b$, then

$$F = \frac{\pi km\sigma a}{b^2}\Big((a+b) + (a-b) - (a-b) - (a+b)\Big) = 0.$$

CHAPTER 16. VECTOR CALCULUS

Section 16.1 Gradient, Divergence, and Curl (page 935)

1. $\mathbf{F} = x\mathbf{i} + y\mathbf{j}$

   $\operatorname{div} \mathbf{F} = \dfrac{\partial}{\partial x}(x) + \dfrac{\partial}{\partial y}(y) + \dfrac{\partial}{\partial z}(0) = 1 + 1 = 2$

   $\operatorname{curl} \mathbf{F} = \begin{vmatrix} \mathbf{i} & \mathbf{j} & \mathbf{k} \\ \dfrac{\partial}{\partial x} & \dfrac{\partial}{\partial y} & \dfrac{\partial}{\partial z} \\ x & y & 0 \end{vmatrix} = \mathbf{0}$

2. $\mathbf{F} = y\mathbf{i} + x\mathbf{j}$

   $\operatorname{div} \mathbf{F} = \dfrac{\partial}{\partial x}(y) + \dfrac{\partial}{\partial y}(x) + \dfrac{\partial}{\partial z}(0) = 0 + 0 = 0$

   $\operatorname{curl} \mathbf{F} = \begin{vmatrix} \mathbf{i} & \mathbf{j} & \mathbf{k} \\ \dfrac{\partial}{\partial x} & \dfrac{\partial}{\partial y} & \dfrac{\partial}{\partial z} \\ y & x & 0 \end{vmatrix} = (1-1)\mathbf{k} = \mathbf{0}$

3. $\mathbf{F} = y\mathbf{i} + z\mathbf{j} + x\mathbf{k}$

   $\operatorname{div} \mathbf{F} = \dfrac{\partial}{\partial x}(y) + \dfrac{\partial}{\partial y}(z) + \dfrac{\partial}{\partial z}(x) = 0$

   $\operatorname{curl} \mathbf{F} = \begin{vmatrix} \mathbf{i} & \mathbf{j} & \mathbf{k} \\ \dfrac{\partial}{\partial x} & \dfrac{\partial}{\partial y} & \dfrac{\partial}{\partial z} \\ y & z & x \end{vmatrix} = -\mathbf{i} - \mathbf{j} - \mathbf{k}$

4. $\mathbf{F} = yz\mathbf{i} + xz\mathbf{j} + xy\mathbf{k}$

   $\operatorname{div} \mathbf{F} = \dfrac{\partial}{\partial x}(yz) + \dfrac{\partial}{\partial y}(xz) + \dfrac{\partial}{\partial z}(xy) = 0$

   $\operatorname{curl} \mathbf{F} = \begin{vmatrix} \mathbf{i} & \mathbf{j} & \mathbf{k} \\ \dfrac{\partial}{\partial x} & \dfrac{\partial}{\partial y} & \dfrac{\partial}{\partial z} \\ yz & xz & xy \end{vmatrix}$

   $= (x-x)\mathbf{i} + (y-y)\mathbf{j} + (z-z)\mathbf{k} = \mathbf{0}$

5. $\mathbf{F} = x\mathbf{i} + x\mathbf{k}$

   $\operatorname{div} \mathbf{F} = \dfrac{\partial}{\partial x}(x) + \dfrac{\partial}{\partial y}(0) + \dfrac{\partial}{\partial z}(x) = 1$

   $\operatorname{curl} \mathbf{F} = \begin{vmatrix} \mathbf{i} & \mathbf{j} & \mathbf{k} \\ \dfrac{\partial}{\partial x} & \dfrac{\partial}{\partial y} & \dfrac{\partial}{\partial z} \\ x & 0 & x \end{vmatrix} = -\mathbf{j}$

6. $\mathbf{F} = xy^2\mathbf{i} - yz^2\mathbf{j} + zx^2\mathbf{k}$

   $\operatorname{div} \mathbf{F} = \dfrac{\partial}{\partial x}(xy^2) + \dfrac{\partial}{\partial y}(-yz^2) + \dfrac{\partial}{\partial z}(zx^2)$

   $= y^2 - z^2 + x^2$

   $\operatorname{curl} \mathbf{F} = \begin{vmatrix} \mathbf{i} & \mathbf{j} & \mathbf{k} \\ \dfrac{\partial}{\partial x} & \dfrac{\partial}{\partial y} & \dfrac{\partial}{\partial z} \\ xy^2 & -yz^2 & zx^2 \end{vmatrix}$

   $= 2yz\mathbf{i} - 2xz\mathbf{j} - 2xy\mathbf{k}$

7. $\mathbf{F} = f(x)\mathbf{i} + g(y)\mathbf{j} + h(z)\mathbf{k}$

   $\operatorname{div} \mathbf{F} = \dfrac{\partial}{\partial x}f(x) + \dfrac{\partial}{\partial y}g(y) + \dfrac{\partial}{\partial z}h(z)$

   $= f'(x) + g'(y) + h'(z)$

   $\operatorname{curl} \mathbf{F} = \begin{vmatrix} \mathbf{i} & \mathbf{j} & \mathbf{k} \\ \dfrac{\partial}{\partial x} & \dfrac{\partial}{\partial y} & \dfrac{\partial}{\partial z} \\ f(x) & g(y) & h(z) \end{vmatrix} = \mathbf{0}$

8. $\mathbf{F} = f(z)\mathbf{i} - f(z)\mathbf{j}$

   $\operatorname{div} \mathbf{F} = \dfrac{\partial}{\partial x}f(z) + \dfrac{\partial}{\partial y}\bigl(-f(z)\bigr) = 0$

   $\operatorname{curl} \mathbf{F} = \begin{vmatrix} \mathbf{i} & \mathbf{j} & \mathbf{k} \\ \dfrac{\partial}{\partial x} & \dfrac{\partial}{\partial y} & \dfrac{\partial}{\partial z} \\ f(z) & -f(z) & 0 \end{vmatrix} = f'(z)(\mathbf{i} + \mathbf{j})$

9. Since $x = r\cos\theta$, and $y = r\sin\theta$, we have $r^2 = x^2 + y^2$, and so

   $\dfrac{\partial r}{\partial x} = \dfrac{x}{r} = \cos\theta$

   $\dfrac{\partial r}{\partial y} = \dfrac{y}{r} = \sin\theta$

   $\dfrac{\partial}{\partial x}\sin\theta = \dfrac{\partial}{\partial x}\dfrac{y}{r} = \dfrac{-xy}{r^3} = -\dfrac{\cos\theta \sin\theta}{r}$

   $\dfrac{\partial}{\partial y}\sin\theta = \dfrac{\partial}{\partial y}\dfrac{y}{r} = \dfrac{1}{r} - \dfrac{y^2}{r^3}$

   $= \dfrac{x^2}{r^3} = \dfrac{\cos^2\theta}{r}$

   $\dfrac{\partial}{\partial x}\cos\theta = \dfrac{\partial}{\partial x}\dfrac{x}{r} = \dfrac{1}{r} - \dfrac{x^2}{r^3}$

   $= \dfrac{y^2}{r^3} = \dfrac{\sin^2\theta}{r}$

   $\dfrac{\partial}{\partial y}\cos\theta = \dfrac{\partial}{\partial y}\dfrac{x}{r} = \dfrac{-xy}{r^3} = -\dfrac{\cos\theta \sin\theta}{r}.$

   (The last two derivatives are not needed for this exercise, but will be useful for the next two exercises.) For

   $$\mathbf{F} = r\mathbf{i} + \sin\theta\, \mathbf{j},$$

   we have

   $\operatorname{div} \mathbf{F} = \dfrac{\partial r}{\partial x} + \dfrac{\partial}{\partial y}\sin\theta = \cos\theta + \dfrac{\cos^2\theta}{r}$

   $\operatorname{curl} \mathbf{F} = \begin{vmatrix} \mathbf{i} & \mathbf{j} & \mathbf{k} \\ \dfrac{\partial}{\partial x} & \dfrac{\partial}{\partial y} & \dfrac{\partial}{\partial z} \\ r & \sin\theta & 0 \end{vmatrix}$

   $= \left(-\dfrac{\sin\theta \cos\theta}{r} - \sin\theta\right)\mathbf{k}.$

10.
$$\mathbf{F} = \hat{\mathbf{r}} = \cos\theta\mathbf{i} + \sin\theta\mathbf{j}$$
$$\text{div}\,\mathbf{F} = \frac{\sin^2\theta}{r} + \frac{\cos^2\theta}{r} = \frac{1}{r} = \frac{1}{\sqrt{x^2+y^2}}$$
$$\text{curl}\,\mathbf{F} = \begin{vmatrix} \mathbf{i} & \mathbf{j} & \mathbf{k} \\ \frac{\partial}{\partial x} & \frac{\partial}{\partial y} & \frac{\partial}{\partial z} \\ \cos\theta & \sin\theta & 0 \end{vmatrix}$$
$$= -\left(\frac{\cos\theta\sin\theta}{r} - \frac{\cos\theta\sin\theta}{r}\right)\mathbf{k} = \mathbf{0}$$

11.
$$\mathbf{F} = \hat{\boldsymbol{\theta}} = -\sin\theta\mathbf{i} + \cos\theta\mathbf{j}$$
$$\text{div}\,\mathbf{F} = \frac{\cos\theta\sin\theta}{r} - \frac{\cos\theta\sin\theta}{r} = 0$$
$$\text{curl}\,\mathbf{F} = \begin{vmatrix} \mathbf{i} & \mathbf{j} & \mathbf{k} \\ \frac{\partial}{\partial x} & \frac{\partial}{\partial y} & \frac{\partial}{\partial z} \\ -\sin\theta & \cos\theta & 0 \end{vmatrix}$$
$$= \left(\frac{\sin^2\theta}{r} + \frac{\cos^2\theta}{r}\right)\mathbf{k} = \frac{1}{r}\mathbf{k} = \frac{1}{\sqrt{x^2+y^2}}\mathbf{k}$$

12. We use the Maclaurin expansion of $\mathbf{F}$, as presented in the proof of Theorem 1:
$$\mathbf{F} = \mathbf{F}_0 + \mathbf{F}_1 x + \mathbf{F}_2 y + \mathbf{F}_3 z + \cdots,$$
where
$$\mathbf{F}_0 = \mathbf{F}(0,0,0)$$
$$\mathbf{F}_1 = \left.\frac{\partial}{\partial x}\mathbf{F}(x,y,z)\right|_{(0,0,0)} = \left.\left(\frac{\partial F_1}{\partial x}\mathbf{i} + \frac{\partial F_2}{\partial x}\mathbf{j} + \frac{\partial F_3}{\partial x}\mathbf{k}\right)\right|_{(0,0,0)}$$
$$\mathbf{F}_2 = \left.\frac{\partial}{\partial y}\mathbf{F}(x,y,z)\right|_{(0,0,0)} = \left.\left(\frac{\partial F_1}{\partial y}\mathbf{i} + \frac{\partial F_2}{\partial y}\mathbf{j} + \frac{\partial F_3}{\partial y}\mathbf{k}\right)\right|_{(0,0,0)}$$
$$\mathbf{F}_3 = \left.\frac{\partial}{\partial z}\mathbf{F}(x,y,z)\right|_{(0,0,0)} = \left.\left(\frac{\partial F_1}{\partial z}\mathbf{i} + \frac{\partial F_2}{\partial z}\mathbf{j} + \frac{\partial F_3}{\partial z}\mathbf{k}\right)\right|_{(0,0,0)}$$

and where $\cdots$ represents terms of degree 2 and higher in x, y, and z.
On the top of the box $B_{a,b,c}$, we have $z = c$ and $\hat{\mathbf{N}} = \mathbf{k}$.
On the bottom of the box, we have $z = -c$ and $\hat{\mathbf{N}} = -\mathbf{k}$.
On both surfaces $dS = dx\,dy$. Thus

$$\left(\iint_{\text{top}} + \iint_{\text{bottom}}\right)\mathbf{F}\bullet\hat{\mathbf{N}}\,dS$$
$$= \int_{-a}^{a}dx\int_{-b}^{b}dy\left(c\mathbf{F}_3\bullet\mathbf{k} - c\mathbf{F}_3\bullet(-\mathbf{k})\right) + \cdots$$
$$= 8abc\mathbf{F}_3\bullet\mathbf{k} + \cdots = 8abc\left.\frac{\partial}{\partial z}F_3(x,y,z)\right|_{(0,0,0)} + \cdots,$$

where $\cdots$ represents terms of degree 4 and higher in a, b, and c.
Similar formulas obtain for the two other pairs of faces, and the three formulas combine into
$$\iint_{B_{a,b,c}} \mathbf{F}\bullet\hat{\mathbf{N}}\,dS = 8abc\,\text{div}\,\mathbf{F}(0,0,0) + \cdots.$$

It follows that
$$\lim_{a,b,c\to 0+}\frac{1}{8abc}\iint_{B_{a,b,c}}\mathbf{F}\bullet\hat{\mathbf{N}}\,dS = \text{div}\,\mathbf{F}(0,0,0).$$

13. This proof just mimics that of Theorem 1. $\mathbf{F}$ can be expanded in Maclaurin series
$$\mathbf{F} = \mathbf{F}_0 + \mathbf{F}_1 x + \mathbf{F}_2 y + \cdots,$$
where
$$\mathbf{F}_0 = \mathbf{F}(0,0)$$
$$\mathbf{F}_1 = \left.\frac{\partial}{\partial x}\mathbf{F}(x,y)\right|_{(0,0)} = \left.\left(\frac{\partial F_1}{\partial x}\mathbf{i} + \frac{\partial F_2}{\partial x}\mathbf{j}\right)\right|_{(0,0)}$$
$$\mathbf{F}_2 = \left.\frac{\partial}{\partial y}\mathbf{F}(x,y)\right|_{(0,0)} = \left.\left(\frac{\partial F_1}{\partial y}\mathbf{i} + \frac{\partial F_2}{\partial y}\mathbf{j}\right)\right|_{(0,0)}$$

and where $\cdots$ represents terms of degree 2 and higher in x and y.
On the curve $\mathcal{C}_\epsilon$ of radius ϵ centred at $(0,0)$, we have $\hat{\mathbf{N}} = \frac{1}{\epsilon}(x\mathbf{i} + y\mathbf{j})$. Therefore,

$$\mathbf{F}\bullet\hat{\mathbf{N}} = \frac{1}{\epsilon}\big(\mathbf{F}_0\bullet\mathbf{i}x + \mathbf{F}_0\bullet\mathbf{j}y + \mathbf{F}_1\bullet\mathbf{i}x^2$$
$$+ \mathbf{F}_1\bullet\mathbf{j}xy + \mathbf{F}_2\bullet\mathbf{i}xy + \mathbf{F}_2\bullet\mathbf{j}y^2 + \cdots\big)$$

where $\cdots$ represents terms of degree 3 or higher in x and y. Since

$$\oint_{\mathcal{C}_\epsilon} x\,ds = \oint_{\mathcal{C}_\epsilon} y\,ds = \oint_{\mathcal{C}_\epsilon} xy\,ds = 0$$
$$\oint_{\mathcal{C}_\epsilon} x^2\,ds = \oint_{\mathcal{C}_\epsilon} y^2\,ds = \int_0^{2\pi}\epsilon^2\cos^2\theta\,\epsilon\,d\theta = \pi\epsilon^3,$$

we have
$$\frac{1}{\pi\epsilon^2}\oint_{\mathcal{C}_\epsilon}\mathbf{F}\bullet\hat{\mathbf{N}}\,ds = \frac{1}{\pi\epsilon^2}\frac{\pi\epsilon^3}{\epsilon}(\mathbf{F}_1\bullet\mathbf{i} + \mathbf{F}_2\bullet\mathbf{j}) + \cdots$$
$$= \text{div}\,\mathbf{F}(0,0) + \cdots$$

where $\cdots$ represents terms of degree 1 or higher in ϵ. Therefore, taking the limit as $\epsilon \to 0$ we obtain

$$\lim_{\epsilon\to 0}\frac{1}{\pi\epsilon^2}\oint_{\mathcal{C}_\epsilon}\mathbf{F}\bullet\hat{\mathbf{N}}\,ds = \text{div}\,\mathbf{F}(0,0).$$

14. We use the same Maclaurin expansion for **F** as in Exercises 12 and 13. On $\mathcal{C}_\epsilon$ we have

$$\mathbf{r} = \epsilon\cos\theta\mathbf{i} + \epsilon\sin\theta\mathbf{j}, \quad (0 \le \theta \le 2\pi)$$
$$d\mathbf{r} = -\epsilon\sin\theta\mathbf{i} + \epsilon\cos\theta\mathbf{j}$$
$$\mathbf{F}\bullet d\mathbf{r} = \big(-\epsilon\sin\theta\mathbf{F}_0\bullet\mathbf{i} + \epsilon\cos\theta\mathbf{F}_0\bullet\mathbf{j}$$
$$-\epsilon^2\sin\theta\cos\theta\mathbf{F}_1\bullet\mathbf{i} + \epsilon^2\cos^2\theta\mathbf{F}_1\bullet\mathbf{j}$$
$$-\epsilon^2\sin^2\theta\mathbf{F}_2\bullet\mathbf{i} + \epsilon^2\sin\theta\cos\theta\mathbf{F}_2\bullet\mathbf{j} + \cdots\big)\,ds,$$

where $\cdots$ represents terms of degree 3 or higher in ϵ. Since

$$\int_0^{2\pi}\sin\theta\,d\theta = \int_0^{2\pi}\cos\theta\,d\theta = \int_0^{2\pi}\sin\theta\cos\theta\,d\theta = 0$$
$$\int_0^{2\pi}\cos^2\theta\,d\theta = \int_0^{2\pi}\sin^2\theta\,d\theta = \pi,$$

we have

$$\frac{1}{\pi\epsilon^2}\oint_{\mathcal{C}_\epsilon}\mathbf{F}\bullet d\mathbf{r} = \mathbf{F}_1\bullet\mathbf{j} - \mathbf{F}_2\bullet\mathbf{i} + \cdots,$$

where $\cdots$ represents terms of degree at least 1 in ϵ. Hence

$$\lim_{\epsilon\to 0+}\frac{1}{\pi\epsilon^2}\oint_{\mathcal{C}_\epsilon}\mathbf{F}\bullet d\mathbf{r} = \mathbf{F}_1\bullet\mathbf{j} - \mathbf{F}_2\bullet\mathbf{i}$$
$$= \frac{\partial F_2}{\partial x} - \frac{\partial F_1}{\partial y}$$
$$= \operatorname{curl}\mathbf{F}\bullet\mathbf{k} = \operatorname{curl}\mathbf{F}\bullet\hat{\mathbf{N}}.$$

Section 16.2 Some Identities Involving Grad, Div, and Curl (page 941)

1. Theorem 3(a):

$$\boldsymbol{\nabla}(\phi\psi) = \frac{\partial}{\partial x}(\phi\psi) + \frac{\partial}{\partial y}(\phi\psi) + \frac{\partial}{\partial z}(\phi\psi)$$
$$= \left(\phi\frac{\partial\psi}{\partial x} + \frac{\partial\phi}{\partial x}\psi\right)\mathbf{i} + \cdots + \left(\phi\frac{\partial\psi}{\partial z} + \frac{\partial\phi}{\partial z}\psi\right)\mathbf{k}$$
$$= \phi\boldsymbol{\nabla}\psi + \psi\boldsymbol{\nabla}\phi.$$

2. Theorem 3(b):

$$\boldsymbol{\nabla}\bullet(\phi\mathbf{F}) = \frac{\partial}{\partial x}(\phi F_1) + \frac{\partial}{\partial y}(\phi F_2) + \frac{\partial}{\partial z}(\phi F_3)$$
$$= \frac{\partial\phi}{\partial x}F_1 + \phi\frac{\partial F_1}{\partial x} + \cdots + \frac{\partial\phi}{\partial z}F_3 + \phi\frac{\partial F_3}{\partial z} + \cdots$$
$$= \boldsymbol{\nabla}\phi\bullet\mathbf{F} + \phi\boldsymbol{\nabla}\bullet\mathbf{F}.$$

3. Theorem 3(d):

$$\boldsymbol{\nabla}\bullet(\mathbf{F}\times\mathbf{G}) = \frac{\partial}{\partial x}(F_2G_3 - F_3G_2) + \cdots$$
$$= \frac{\partial F_2}{\partial x}G_3 + F_2\frac{\partial G_3}{\partial x} - \frac{\partial F_3}{\partial x}G_2 - F_3\frac{\partial G_2}{\partial x} + \cdots$$
$$= (\boldsymbol{\nabla}\times\mathbf{F})\bullet\mathbf{G} - \mathbf{F}\bullet(\boldsymbol{\nabla}\times\mathbf{G}).$$

4. Theorem 3(f). The first component of $\boldsymbol{\nabla}(\mathbf{F}\bullet\mathbf{G})$ is

$$\frac{\partial F_1}{\partial x}G_1 + F_1\frac{\partial G_1}{\partial x} + \frac{\partial F_2}{\partial x}G_2 + F_2\frac{\partial G_2}{\partial x} + \frac{\partial F_3}{\partial x}G_3 + F_3\frac{\partial G_3}{\partial x}.$$

We calculate the first components of the four terms on the right side of the identity to be proved.
The first component of $\mathbf{F}\times(\boldsymbol{\nabla}\times\mathbf{G})$ is

$$F_2\left(\frac{\partial G_2}{\partial x} - \frac{\partial G_1}{\partial y}\right) - F_3\left(\frac{\partial G_1}{\partial z} - \frac{\partial G_3}{\partial x}\right).$$

The first component of $\mathbf{G}\times(\boldsymbol{\nabla}\times\mathbf{F})$ is

$$G_2\left(\frac{\partial F_2}{\partial x} - \frac{\partial F_1}{\partial y}\right) - G_3\left(\frac{\partial F_1}{\partial z} - \frac{\partial F_3}{\partial x}\right).$$

The first component of $(\mathbf{F}\bullet\boldsymbol{\nabla})\mathbf{G}$ is

$$F_1\frac{\partial G_1}{\partial x} + F_2\frac{\partial G_1}{\partial y} + F_3\frac{\partial G_1}{\partial z}.$$

The first component of $(\mathbf{G}\bullet\boldsymbol{\nabla})\mathbf{F}$ is

$$G_1\frac{\partial F_1}{\partial x} + G_2\frac{\partial F_1}{\partial y} + G_3\frac{\partial F_1}{\partial z}.$$

When we add these four first components, eight of the fourteen terms cancel out and the six remaining terms are the six terms of the first component of $\boldsymbol{\nabla}(\mathbf{F}\bullet\mathbf{G})$, as calculated above. Similar calculations show that the second and third components of both sides of the identity agree. Thus

$$\boldsymbol{\nabla}(\mathbf{F}\bullet\mathbf{G}) = \mathbf{F}\times(\boldsymbol{\nabla}\times\mathbf{G}) + \mathbf{G}\times(\boldsymbol{\nabla}\times\mathbf{F}) + (\mathbf{F}\bullet\boldsymbol{\nabla})\mathbf{G} + (\mathbf{G}\bullet\boldsymbol{\nabla})\mathbf{F}.$$

5. Theorem 3(h). By equality of mixed partials,

$$\boldsymbol{\nabla}\times\boldsymbol{\nabla}\phi = \begin{vmatrix} \mathbf{i} & \mathbf{j} & \mathbf{k} \\ \dfrac{\partial}{\partial x} & \dfrac{\partial}{\partial y} & \dfrac{\partial}{\partial z} \\ \dfrac{\partial\phi}{\partial x} & \dfrac{\partial\phi}{\partial y} & \dfrac{\partial\phi}{\partial z} \end{vmatrix}$$
$$= \left(\frac{\partial}{\partial y}\frac{\partial\phi}{\partial z} - \frac{\partial}{\partial z}\frac{\partial\phi}{\partial y}\right)\mathbf{i} + \cdots = \mathbf{0}.$$

6. Theorem 3(i). We examine the first components of the terms on both sides of the identity

$$\boldsymbol{\nabla}\times(\boldsymbol{\nabla}\times\mathbf{F}) = \boldsymbol{\nabla}(\boldsymbol{\nabla}\bullet\mathbf{F}) - \boldsymbol{\nabla}^2\mathbf{F}.$$

The first component of $\nabla \times (\nabla \times \mathbf{F})$ is

$$\frac{\partial}{\partial y}\left(\frac{\partial F_2}{\partial x} - \frac{\partial F_1}{\partial y}\right) - \frac{\partial}{\partial z}\left(\frac{\partial F_1}{\partial z} - \frac{\partial F_3}{\partial x}\right)$$
$$= \frac{\partial^2 F_2}{\partial y \partial x} - \frac{\partial^2 F_1}{\partial y^2} - \frac{\partial^2 F_1}{\partial z^2} + \frac{\partial^2 F_3}{\partial z \partial x}.$$

The first component of $\nabla(\nabla \bullet \mathbf{F})$ is

$$\frac{\partial}{\partial x}\nabla \bullet \mathbf{F} = \frac{\partial^2 F_1}{\partial x^2} + \frac{\partial^2 F_2}{\partial x \partial y} + \frac{\partial^2 F_3}{\partial x \partial z}.$$

The first component of $-\nabla^2 \mathbf{F}$ is

$$-\nabla^2 F_1 = -\frac{\partial^2 F_1}{\partial x^2} - \frac{\partial^2 F_1}{\partial y^2} - \frac{\partial^2 F_1}{\partial z^2}.$$

Evidently the first components of both sides of the given identity agree. By symmetry, so do the other components.

7. If the field lines of $\mathbf{F}(x, y, z)$ are parallel straight lines, in the direction of the constant nonzero vector $\mathbf{a}$ say, then

$$\mathbf{F}(x, y, z) = \phi(x, y, z)\mathbf{a}$$

for some scalar field ϕ, which we assume to be smooth. By Theorem 3(b) and (c) we have

$$\text{div } \mathbf{F} = \text{div } (\phi \mathbf{a}) = \nabla \phi \bullet \mathbf{a}$$
$$\text{curl } \mathbf{F} = \text{curl } (\phi \mathbf{a}) = \nabla \phi \times \mathbf{a}.$$

Since $\nabla \phi$ is an arbitrary gradient, $\text{div } \mathbf{F}$ can have any value, but $\text{curl } \mathbf{F}$ is perpendicular to $\mathbf{a}$, and thereofore to $\mathbf{F}$.

8. If $\mathbf{r} = x\mathbf{i} + y\mathbf{j} + z\mathbf{k}$ and $r = |\mathbf{r}|$, then

$$\nabla \bullet \mathbf{r} = 3, \qquad \nabla \times \mathbf{r} = \mathbf{0}, \qquad \nabla r = \frac{\mathbf{r}}{r}.$$

If $\mathbf{c}$ is a constant vector, then its divergence and curl are both zero. By Theorem 3(d), (e), and (f) we have

$$\nabla \bullet (\mathbf{c} \times \mathbf{r}) = (\nabla \times \mathbf{c}) \bullet \mathbf{r} - \mathbf{c} \bullet (\nabla \times \mathbf{r}) = \mathbf{0}$$
$$\nabla \times (\mathbf{c} \times \mathbf{r}) = (\nabla \bullet \mathbf{r})\mathbf{c} + (\mathbf{r} \bullet \nabla)\mathbf{c} - (\nabla \bullet \mathbf{c})\mathbf{r} - (\mathbf{c} \bullet \nabla)\mathbf{r}$$
$$= 3\mathbf{c} + 0 - 0 - \mathbf{c} = 2\mathbf{c}$$

$$\nabla(\mathbf{c} \bullet \mathbf{r}) = \mathbf{c} \times (\nabla \times \mathbf{r}) + \mathbf{r} \times (\nabla \times \mathbf{c}) + (\mathbf{c} \bullet \nabla)\mathbf{r} + (\mathbf{r} \bullet \nabla)\mathbf{c}$$
$$= \mathbf{0} + \mathbf{0} + \mathbf{c} + \mathbf{0} = \mathbf{c}.$$

9. $\nabla \bullet \big(f(r)\mathbf{r}\big) = \big(\nabla f(r)\big) \bullet \mathbf{r} + f(r)(\nabla \bullet \mathbf{r})$
$$= f'(r)\frac{\mathbf{r} \bullet \mathbf{r}}{r} + 3f(r)$$
$$= rf'(r) + 3f(r).$$

If $f(r)\mathbf{r}$ is solenoidal then $\nabla \bullet \big(f(r)\mathbf{r}\big) = 0$, so that $u = f(r)$ satisfies

$$r\frac{du}{dr} + 3u = 0$$
$$\frac{du}{u} = -\frac{3\, dr}{r}$$
$$\ln|u| = -3\ln|r| + \ln|C|$$
$$u = Cr^{-3}.$$

Thus $f(r) = Cr^{-3}$, for some constant C.

10. Given that $\text{div } \mathbf{F} = 0$ and $\text{curl } \mathbf{F} = 0$, Theorem 3(i) implies that $\nabla^2 \mathbf{F} = 0$ too. Hence the components of $\mathbf{F}$ are harmonic functions.
If $\mathbf{F} = \nabla \phi$, then

$$\nabla^2 \phi = \nabla \bullet \nabla \phi = \nabla \bullet \mathbf{F} = 0,$$

so ϕ is also harmonic.

11. By Theorem 3(e) and 3(f),

$$\nabla \times (\mathbf{F} \times \mathbf{r}) = (\nabla \bullet \mathbf{r})\mathbf{F} + (\mathbf{r} \bullet \nabla)\mathbf{F} - (\nabla \bullet \mathbf{F})\mathbf{r} - (\mathbf{F} \bullet \nabla)\mathbf{r}$$
$$\nabla(\mathbf{F} \bullet \mathbf{r}) = \mathbf{F} \times (\nabla \times \mathbf{r}) + \mathbf{r} \times (\nabla \times \mathbf{F})$$
$$+ (\mathbf{F} \bullet \nabla)\mathbf{r} + (\mathbf{r} \bullet \nabla)\mathbf{F}.$$

If $\mathbf{r} = x\mathbf{i} + y\mathbf{j} + z\mathbf{k}$, then $\nabla \bullet \mathbf{r} = 3$ and $\nabla \times \mathbf{r} = \mathbf{0}$. Also,

$$(\mathbf{F} \bullet \nabla)\mathbf{r} = F_1\frac{\partial \mathbf{r}}{\partial x} + F_2\frac{\partial \mathbf{r}}{\partial y} + F_3\frac{\partial \mathbf{r}}{\partial z} = \mathbf{F}.$$

Combining all these results, we obtain

$$\nabla \times (\mathbf{F} \times \mathbf{r}) - \nabla(\mathbf{F} \bullet \mathbf{r}) = 3\mathbf{F} - 2(\mathbf{F} \bullet \nabla)\mathbf{r}$$
$$- (\nabla \bullet \mathbf{F})\mathbf{r} - \mathbf{r} \times (\nabla \times \mathbf{F})$$
$$= \mathbf{F} - (\nabla \bullet \mathbf{F})\mathbf{r} - \mathbf{r} \times (\nabla \times \mathbf{F}).$$

In particular, if $\nabla \bullet \mathbf{F} = 0$ and $\nabla \times \mathbf{F} = \mathbf{0}$, then

$$\nabla \times (\mathbf{F} \times \mathbf{r}) - \nabla(\mathbf{F} \bullet \mathbf{r}) = \mathbf{F}.$$

12. If $\nabla^2 \phi = 0$ and $\nabla^2 \psi = 0$, then

$$\nabla \bullet (\phi \nabla \psi - \psi \nabla \phi)$$
$$= \nabla \phi \bullet \nabla \psi + \phi \nabla^2 \psi - \nabla \psi \bullet \nabla \phi - \psi \nabla^2 \phi = 0,$$

so $\phi \nabla \psi - \psi \nabla \phi$ is solenoidal.

13. By Theorem 3(c) and (h),

$$\nabla \times (\phi \nabla \psi) = \nabla \phi \times \nabla \psi + \phi \nabla \times \nabla \psi = \nabla \phi \times \nabla \psi$$
$$-\nabla \times (\psi \nabla \phi) = -\nabla \psi \times \nabla \phi - \psi \nabla \times \nabla \phi = \nabla \phi \times \nabla \psi.$$

14. By Theorem 3(b), (d), and (h), we have

$$\nabla \bullet \left(f(\nabla g \times \nabla h)\right)$$
$$= \nabla f \bullet (\nabla g \times \nabla h) + f \nabla \bullet (\nabla g \times \nabla h)$$
$$= \nabla f \bullet (\nabla g \times \nabla h) + f\left((\nabla \times \nabla g) \bullet \nabla h - \nabla g \bullet (\nabla \times \nabla h)\right)$$
$$= \nabla f \bullet (\nabla g \times \nabla h) + 0 - 0 = \nabla f \bullet (\nabla g \times \nabla h).$$

15. If $\mathbf{F} = \nabla \phi$ and $\mathbf{G} = \nabla \psi$, then $\nabla \times \mathbf{F} = 0$ and $\nabla \times \mathbf{G} = 0$ by Theorem 3(h). Therefore, by Theorem 3(d) we have

$$\nabla \bullet (\mathbf{F} \times \mathbf{G}) = (\nabla \times \mathbf{F}) \bullet \mathbf{G} + \mathbf{F} \bullet (\nabla \times \mathbf{G}) = 0.$$

Thus $\mathbf{F} \times \mathbf{G}$ is solenoidal. By Exercise 13,

$$\nabla \times (\phi \nabla \psi) = \nabla \phi \times \nabla \psi = \mathbf{F} \times \mathbf{G},$$

so $\phi \nabla \psi$ is a vector potential for $\mathbf{F} \times \mathbf{G}$. (So is $-\psi \nabla \phi$.)

16. If $\nabla \times \mathbf{G} = \mathbf{F} = -y\mathbf{i} + x\mathbf{j}$, then

$$\frac{\partial G_3}{\partial y} - \frac{\partial G_2}{\partial z} = -y$$
$$\frac{\partial G_1}{\partial z} - \frac{\partial G_3}{\partial x} = x$$
$$\frac{\partial G_2}{\partial x} - \frac{\partial G_1}{\partial y} = 0.$$

As in Example 1, we try to find a solution with $G_2 = 0$. Then

$$G_3 = -\int y \, dy = -\frac{y^2}{2} + M(x, z).$$

Again we try $M(x, z) = 0$, so $G_3 = -\frac{y^2}{2}$. Thus $\frac{\partial G_3}{\partial x} = 0$ and

$$G_1 = \int x \, dz = xz + N(x, y).$$

Since $\frac{\partial G_1}{\partial y} = 0$ we may take $N(x, y) = 0$.

$\mathbf{G} = xz\mathbf{i} - \frac{1}{2}y^2\mathbf{k}$ is a vector potential for $\mathbf{F}$. (Of course, this answer is not unique.)

17. If $\mathbf{F} = xe^{2z}\mathbf{i} + ye^{2z}\mathbf{j} - e^{2z}\mathbf{k}$, then

$$\mathbf{div}\,\mathbf{F} = e^{2z} + e^{2z} - 2e^{2z} = 0,$$

so $\mathbf{F}$ is solenoidal.
If $\mathbf{F} = \nabla \times \mathbf{G}$, then

$$\frac{\partial G_3}{\partial y} - \frac{\partial G_2}{\partial z} = xe^{2z}$$
$$\frac{\partial G_1}{\partial z} - \frac{\partial G_3}{\partial x} = ye^{2z}$$
$$\frac{\partial G_2}{\partial x} - \frac{\partial G_1}{\partial y} = -e^{2z}.$$

Look for a solution with $G_2 = 0$. We have

$$G_3 = \int xe^{2z} \, dy = xye^{2z} + M(x, z).$$

Try $M(x, z) = 0$. Then $G_3 = xye^{2z}$, and

$$\frac{\partial G_1}{\partial z} = ye^{2z} + \frac{\partial G_3}{\partial x} = 2ye^{2z}.$$

Thus

$$G_1 = \int 2ye^{2z} \, dz = ye^{2z} + N(x, y).$$

Since

$$-e^{2z} = -\frac{\partial G_1}{\partial y} = -e^{2z} - \frac{\partial N}{\partial y},$$

we can take $N(x, y) = 0$.
Thus $\mathbf{G} = ye^{2z}\mathbf{i} + xye^{2z}\mathbf{k}$ is a vector potential for $\mathbf{F}$.

18. For (x, y, z) in D let $\mathbf{v} = x\mathbf{i} + y\mathbf{j} + z\mathbf{k}$. The line segment $\mathbf{r}(t) = t\mathbf{v}$, $(0 \le t \le 1)$, lies in D, so $\mathbf{div}\,\mathbf{F} = 0$ on the path. We have

$$\mathbf{G}(x, y, z) = \int_0^1 t\mathbf{F}(\mathbf{r}(t)) \times \mathbf{v} \, dt$$
$$= \int_0^1 t\mathbf{F}(\xi(t), \eta(t), \zeta(t)) \times \mathbf{v} \, dt$$

where $\xi = tx$, $\eta = ty$, $\zeta = tz$. The first component of $\mathbf{curl}\,\mathbf{G}$ is

$$(\mathbf{curl}\,\mathbf{G})_1$$
$$= \int_0^1 t\left(\mathbf{curl}\,(\mathbf{F} \times \mathbf{v})\right)_1 dt$$
$$= \int_0^1 t\left(\frac{\partial}{\partial y}(\mathbf{F} \times \mathbf{v})_3 - \frac{\partial}{\partial z}(\mathbf{F} \times \mathbf{v})_2\right) dt$$
$$= \int_0^1 t\left(\frac{\partial}{\partial y}(F_1 y - F_2 x) - \frac{\partial}{\partial z}(F_3 x - F_1 z)\right) dt$$
$$= \int_0^1 \left(tF_1 + t^2 y \frac{\partial F_1}{\partial \eta} - t^2 x \frac{\partial F_2}{\partial \eta} - t^2 x \frac{\partial F_3}{\partial \zeta}\right.$$
$$\left. + tF_1 + t^2 z \frac{\partial F_1}{\partial \zeta}\right) dt$$
$$= \int_0^1 \left(2tF_1 + t^2 x \frac{\partial F_1}{\partial \xi} + t^2 y \frac{\partial F_1}{\partial \eta} + t^2 z \frac{\partial F_1}{\partial \zeta}\right) dt.$$

To get the last line we used the fact that $div\,\mathbf{F} = 0$ to replace $-t^2 x \frac{\partial F_2}{\partial \eta} - t^2 x \frac{\partial F_3}{\partial \zeta}$ with $t^2 x \frac{\partial F_1}{\partial \xi}$. Continuing the calculation, we have

$$(\mathbf{curl}\,\mathbf{G})_1 = \int_0^1 \frac{d}{dt}\left(t^2 F_1(\xi, \eta, \zeta)\right) dt$$
$$= t^2 F_1(tx, ty, tz)\Big|_0^1 = F_1(x, y, z).$$

INSTRUCTOR'S SOLUTIONS MANUAL   SECTION 16.3 (PAGE 945)

Similarly, $(\text{curl}\,\mathbf{G})_2 = F_2$ and $(\text{curl}\,\mathbf{G})_3 = F_3$. Thus $\text{curl}\,\mathbf{G} = \mathbf{F}$, as required.

Section 16.3 Green's Theorem in the Plane (page 945)

1. $\oint_C (\sin x + 3y^2)\,dx + (2x - e^{-y^2})\,dy$
    $= \iint_R \left[\frac{\partial}{\partial x}(2x - e^{-y^2}) - \frac{\partial}{\partial y}(\sin x + 3y^2)\right]dA$
    $= \iint_R (2 - 6y)\,dA$
    $= \int_0^\pi d\theta \int_0^a (2 - 6r\sin\theta) r\,dr$
    $= \pi a^2 - 6 \int_0^\pi \sin\theta\,d\theta \int_0^a r^2\,dr$
    $= \pi a^2 - 4a^3.$

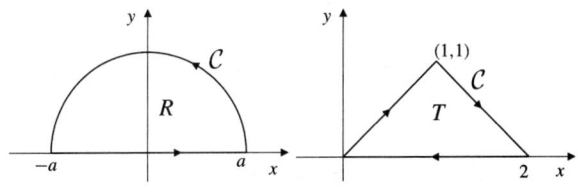

Fig. 16.4.1   Fig. 16.4.2

2. $\oint_C (x^2 - xy)\,dx + (xy - y^2)\,dy$
    $= -\iint_T \left[\frac{\partial}{\partial x}(xy - y^2) - \frac{\partial}{\partial y}(x^2 - xy)\right]dA$
    $= -\iint_T (y + x)\,dA$
    $= -(\bar{y} + \bar{x}) \times (\text{area of } T) = -\left(\frac{1}{3} + 1\right) \times 1 = -\frac{4}{3}.$

3. $\oint_C (x \sin y^2 - y^2)\,dx + (x^2 y \cos y^2 + 3x)\,dy$
    $= \iint_T [2xy \cos y^2 + 3 - (2xy \cos y^2 - 2y)]\,dA$
    $= \iint_T (3 + 2y)\,dA = 3\iint_T dA + 0 = 3 \times 3 = 9.$

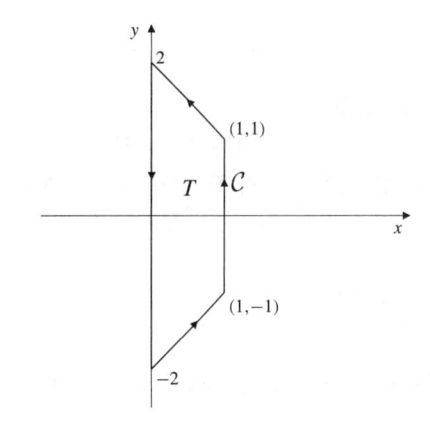

Fig. 16.3.3

4. Let D be the region $x^2 + y^2 \leq 9$, $y \geq 0$. Since C is the clockwise boundary of D,
    $\oint_C x^2 y\,dx - xy^2\,dy$
    $= -\iint_D \left[\frac{\partial}{\partial x}(-xy^2) - \frac{\partial}{\partial y}(x^2 y)\right]dx\,dy$
    $= \iint_D (y^2 + x^2)\,dA = \int_0^\pi d\theta \int_0^3 r^3\,dr = \frac{81\pi}{4}.$

5. By Example 1,
    $\text{Area} = \frac{1}{2} \oint_C x\,dy - y\,dx$
    $= \frac{1}{2} \int_0^{2\pi} \left[a \cos^3 t\, 3b \sin^2 t \cos t\right.$
    $\left. - b \sin^3 t(-3a \cos^2 t \sin t)\right] dt$
    $= \frac{3ab}{2} \int_0^{2\pi} \sin^2 t \cos^2 t\,dt$
    $= \frac{3ab}{2} \int_0^{2\pi} \frac{\sin^2(2t)}{4}\,dt = \frac{3\pi ab}{8}.$

6. Let R, C, and $\mathbf{F}$ be as in the statement of Green's Theorem. As noted in the proof of Theorem 7, the unit tangent $\hat{\mathbf{T}}$ to C and the unit exterior normal $\hat{\mathbf{N}}$ satisfy $\hat{\mathbf{N}} = \hat{\mathbf{T}} \times \mathbf{k}$. Let
    $$\mathbf{G} = F_2(x, y)\mathbf{i} - F_1(x, y)\mathbf{j}.$$
    Then $\mathbf{F} \bullet \hat{\mathbf{T}} = \mathbf{G} \bullet \hat{\mathbf{N}}$. Applying the 2-dimensional Divergence Theorem to $\mathbf{G}$, we obtain
    $\int_C F_1\,dx + F_2\,dy = \int_C \mathbf{F} \bullet \hat{\mathbf{T}}\,ds = \int_C \mathbf{G} \bullet \hat{\mathbf{N}}\,ds$
    $= \iint_R \text{div}\,\mathbf{G}\,dA$
    $= \iint_R \left(\frac{\partial F_2}{\partial x} - \frac{\partial F_1}{\partial y}\right) dA$
    as required

7. $\mathbf{r} = \sin t\,\mathbf{i} + \sin 2t\,\mathbf{j}$, $(0 \leq t \leq 2\pi)$

583

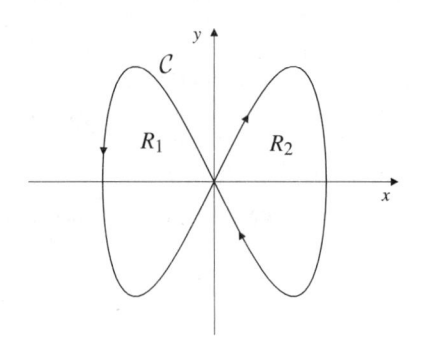

Fig. 16.3.7

$$\mathbf{F} = ye^{x^2}\mathbf{i} + x^3 e^y \mathbf{j}$$

$$\operatorname{curl}\mathbf{F} = \begin{vmatrix} \mathbf{i} & \mathbf{j} & \mathbf{k} \\ \frac{\partial}{\partial x} & \frac{\partial}{\partial y} & \frac{\partial}{\partial z} \\ ye^{x^2} & x^3 e^y & 0 \end{vmatrix} = (3x^2 e^y - e^{x^2})\mathbf{k}.$$

Observe that $\mathcal{C}$ bounds two congruent regions, R_1 and R_2, one counterclockwise and the other clockwise. For R_1, $\hat{\mathbf{N}} = \mathbf{k}$; for R_2, $\hat{\mathbf{N}} = -\mathbf{k}$. Since R_1 and R_2 are mirror images of each other in the y-axis, and since $\operatorname{curl}\mathbf{F}$ is an even function of x, we have

$$\iint_{R_1} \operatorname{curl}\mathbf{F} \bullet \hat{\mathbf{N}}\, dS = -\iint_{R_2} \operatorname{curl}\mathbf{F} \bullet \hat{\mathbf{N}}\, dS.$$

Thus

$$\oint_{\mathcal{C}} \mathbf{F} \bullet d\mathbf{r} = \left(\iint_{R_1} + \iint_{R_2}\right) \operatorname{curl}\mathbf{F} \bullet \hat{\mathbf{N}}\, dS = 0.$$

8. a) $\mathbf{F} = x^2 \mathbf{j}$

$$\oint_{\mathcal{C}} \mathbf{F} \bullet d\mathbf{r} = \oint_{\mathcal{C}} x^2\, dy = \iint_R 2x\, dA = 2A\bar{x}.$$

b) $\mathbf{F} = xy\mathbf{i}$

$$\oint_{\mathcal{C}} \mathbf{F} \bullet d\mathbf{r} = \oint_{\mathcal{C}} xy\, dx = -\iint_R x\, dA = -A\bar{x}.$$

c) $\mathbf{F} = y^2 \mathbf{i} + 3xy\mathbf{j}$

$$\oint_{\mathcal{C}} \mathbf{F} \bullet d\mathbf{r} = \oint_{\mathcal{C}} y^2\, dx + 3xy\, dy$$
$$= \iint_R (3y - 2y)\, dA = A\bar{y}.$$

9. The circle $\mathcal{C}_r$ of radius r and centre at $\mathbf{r}_0$ has parametrization

$$\mathbf{r} = \mathbf{r}_0 + r\cos t\, \mathbf{i} + r\sin t\, \mathbf{j}, \qquad (0 \le t \le 2\pi).$$

Note that $d\mathbf{r}/dt = \cos t\, \mathbf{i} + \sin t\, \mathbf{j} = \hat{\mathbf{N}}$, the unit normal to $\mathcal{C}_r$ exterior to the disk D_r of which $\mathcal{C}_r$ is the boundary. The average value of $u(x, y)$ on $\mathcal{C}_r$ is

$$\bar{u}_r = \frac{1}{2\pi} \int_0^{2\pi} u(x_0 + r\cos t, y_0 + r\sin t)\, dt,$$

and so

$$\frac{d\bar{u}_r}{dr} = \frac{1}{2\pi} \int_0^{2\pi} \left(\frac{\partial u}{\partial x}\cos t + \frac{\partial u}{\partial y}\sin t\right) dt$$
$$= \frac{1}{2\pi r} \oint_{\mathcal{C}_r} \nabla u \bullet \hat{\mathbf{N}}\, ds$$

since $ds = r\, dt$. By the (2-dimensional) divergence theorem, and since u is harmonic,

$$\frac{d\bar{u}_r}{dr} = \frac{1}{2\pi r} \iint_{D_r} \nabla \bullet \nabla u\, dx\, dy$$
$$= \frac{1}{2\pi r} \iint_{D_r} \left(\frac{\partial^2 u}{\partial x^2} + \frac{\partial^2 u}{\partial y^2}\right) dx\, dy = 0.$$

Thus $\bar{u}_r = \lim_{r \to 0} \bar{u}_r = u(x_0, y_0)$.

Section 16.4 The Divergence Theorem in 3-Space (page 951)

1. In this exercise, the sphere S bounds the ball B of radius a centred at the origin.
If $\mathbf{F} = x\mathbf{i} - 2y\mathbf{j} + 4z\mathbf{k}$, then $\operatorname{div}\mathbf{F} = 1 - 2 + 4 = 3$. Thus

$$\oiint_S \mathbf{F} \bullet \hat{\mathbf{N}}\, dS = \iiint_B 3\, dV = 4\pi a^3.$$

2. If $\mathbf{F} = ye^z\mathbf{i} + x^2 e^z \mathbf{j} + xy\mathbf{k}$, then $\operatorname{div}\mathbf{F} = 0$, and

$$\oiint_S \mathbf{F} \bullet \hat{\mathbf{N}}\, dS = \iiint_B 0\, dV = 0.$$

3. If $\mathbf{F} = (x^2 + y^2)\mathbf{i} + (y^2 - z^2)\mathbf{j} + z\mathbf{k}$, then $\operatorname{div}\mathbf{F} = 2x + 2y + 1$, and

$$\oiint_S \mathbf{F} \bullet \hat{\mathbf{N}}\, dS = \iiint_B (2x+2y+1)\, dV = \iiint_B 1\, dV = \frac{4}{3}\pi a^3.$$

4. If $\mathbf{F} = x^3 \mathbf{i} + 3yz^2 \mathbf{j} + (3y^2 z + x^2)\mathbf{k}$, then $\operatorname{div}\mathbf{F} = 3x^2 + 3z^2 + 3y^2$, and

$$\oiint_S \mathbf{F} \bullet \hat{\mathbf{N}}\, dS = 3\iiint_B (x^2 + y^2 + z^2)\, dV$$
$$= 3 \int_0^{2\pi} d\theta \int_0^{\pi} \sin\phi\, d\phi \int_0^a \rho^4\, d\rho$$
$$= \frac{12}{5}\pi a^5.$$

5. If $\mathbf{F} = x^2\mathbf{i} + y^2\mathbf{j} + z^2\mathbf{k}$, then $\text{div}\,\mathbf{F} = 2(x+y+z)$. Therefore the flux of $\mathbf{F}$ out of any solid region R is

$$\text{Flux} = \iiint_R \text{div}\,\mathbf{F}\,dV$$
$$= 2\iiint_R (x+y+z)\,dV = 2(\overline{x}+\overline{y}+\overline{z})V$$

where $(\overline{x}, \overline{y}, \overline{z})$ is the centroid of R and V is the volume of R. We use this result in this exercise and in the next three.

If R is the ball $(x-2)^2 + y^2 + (z-3)^2 \leq 9$, then $\overline{x} = 2$, $\overline{y} = 0$, $\overline{z} = 3$, and $V = (4\pi/3)3^3 = 36\pi$. The flux of $\mathbf{F}$ out of R is $2(2+0+3)(36\pi) = 360\pi$.

6. If R is the ellipsoid $x^2 + y^2 + 4(z-1)^2 \leq 4$, then $\overline{x} = 0$, $\overline{y} = 0$, $\overline{z} = 1$, and $V = (4\pi/3)(2)(2)(1) = 16\pi/3$. The flux of $\mathbf{F}$ out of R is $2(0+0+1)(16\pi/3) = 32\pi/3$.

7. If R is the tetrahedron with vertices $(3,0,0)$, $(0,3,0)$, $(0,0,3)$, and $(0,0,0)$, then $\overline{x} = \overline{y} = \overline{z} = 3/4$, and $V = (1/6)(3)(3)(3) = 9/2$. The flux of $\mathbf{F}$ out of R is $2((3/4) + (3/4) + (3/4))(9/2) = 81/4$.

8. If R is the cylinder $x^2 + y^2 \leq 2y$ (or, equivalently, $x^2 + (y-1)^2 \leq 1$), $0 \leq z \leq 4$, then $\overline{x} = 0$, $\overline{y} = 1$, $\overline{z} = 2$, and $V = (\pi 1^2)(4) = 4\pi$. The flux of $\mathbf{F}$ out of R is $2(0+1+2)(4\pi) = 24\pi$.

9. If $\mathbf{F} = x\mathbf{i} + y\mathbf{j} + z\mathbf{k}$, then $\text{div}\,\mathbf{F} = 3$. If C is any solid region having volume V, then

$$\iiint_C \text{div}\,\mathbf{F}\,dV = 3V.$$

The region C described in the statement of the problem is the part of a solid cone with vertex at the origin that lies inside a ball of radius R with centre at the origin. The surface $\mathcal{S}$ of C consists of two parts, the conical wall $\mathcal{S}_1$, and the region D on the spherical boundary of the ball. At any point P on $\mathcal{S}_1$, the outward normal field $\hat{\mathbf{N}}$ is perpendicular to the line OP, that is, to $\mathbf{F}$, so $\mathbf{F} \cdot \hat{\mathbf{N}} = 0$. At any point P on D, $\hat{\mathbf{N}}$ is parallel to $\mathbf{F}$, in fact $\hat{\mathbf{N}} = \mathbf{F}/|\mathbf{F}| = \mathbf{F}/R$. Thus

$$\oiint_{\mathcal{S}} \mathbf{F} \cdot \hat{\mathbf{N}}\,dS = \iint_{\mathcal{S}_1} \mathbf{F} \cdot \hat{\mathbf{N}}\,dS + \iint_D \mathbf{F} \cdot \hat{\mathbf{N}}\,dS$$
$$= 0 + \iint_D \frac{\mathbf{F} \cdot \mathbf{F}}{R}\,dS = \frac{R^2}{R}\iint_D dS = AR$$

where A is the area of D. By the Divergence Theorem, $3V = AR$, so $V = AR/3$.

10. The required surface integral,

$$I = \iint_{\mathcal{S}} \nabla\phi \cdot \hat{\mathbf{N}}\,dS,$$

can be calculated directly by the methods of Section 15.6. We will do it here by using the Divergence Theorem instead. $\mathcal{S}$ is one face of a tetrahedral domain D whose other faces are in the coordinate planes, as shown in the figure. Since $\phi = xy + z^2$, we have

$$\nabla\phi = y\mathbf{i} + x\mathbf{j} + 2z\mathbf{k}, \qquad \nabla \cdot \nabla\phi = \nabla^2\phi = 2.$$

Thus

$$\iiint_D \nabla \cdot \nabla\phi\,dV = 2 \times \frac{abc}{6} = \frac{abc}{3},$$

the volume of the tetrahedron D being $abc/6$ cubic units.

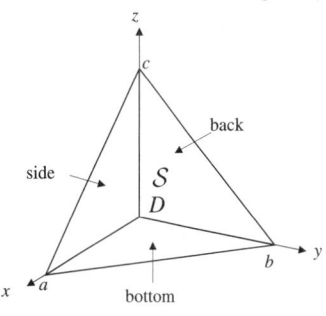

Fig. 16.4.10

The flux of $\nabla\phi$ out of D is the sum of its fluxes out of the four faces of the tetrahedron.

On the bottom, $\hat{\mathbf{N}} = -\mathbf{k}$ and $z = 0$, so $\nabla\phi \cdot \hat{\mathbf{N}} = 0$, and the flux out of the bottom face is 0.

On the side, $y = 0$ and $\hat{\mathbf{N}} = -\mathbf{j}$, so $\nabla\phi \cdot \hat{\mathbf{N}} = -x$. The flux out of the side face is

$$\iint_{\text{side}} \nabla\phi \cdot \hat{\mathbf{N}}\,dS = -\iint_{\text{side}} x\,dx\,dz = -\frac{ac}{2} \times \frac{a}{3} = -\frac{a^2c}{6}.$$

(We used the fact that $M_{x=0} = \text{area} \times \overline{x}$ and $\overline{x} = a/3$ for that face.)

On the back face, $x = 0$ and $\hat{\mathbf{N}} = -\mathbf{i}$, so the flux out of that face is

$$\iint_{\text{back}} \nabla\phi \cdot \hat{\mathbf{N}}\,dS = -\iint_{\text{back}} y\,dy\,dz = -\frac{bc}{2} \times \frac{b}{3} = -\frac{b^2c}{6}.$$

Therefore, by the Divergence Theorem

$$I - \frac{a^2c}{6} - \frac{b^2c}{6} + 0 = \frac{abc}{3},$$

so $\displaystyle\iint_{\mathcal{S}} \nabla\phi \cdot \hat{\mathbf{N}}\,dS = I = \frac{abc}{3} + \frac{c(a^2+b^2)}{6}$.

11. $\mathbf{F} = (x+y^2)\mathbf{i} + (3x^2y + y^3 - x^3)\mathbf{j} + (z+1)\mathbf{k}$
$\text{div}\,\mathbf{F} = 1 + 3(x^2+y^2) + 1 = 2 + 3(x^2+y^2)$.

Fig. 16.4.11

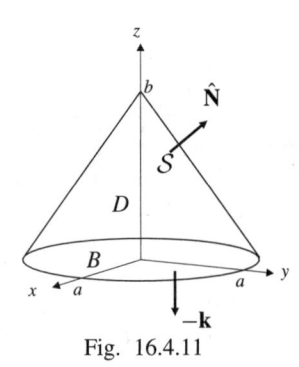

Let D be the conical domain, S its conical surface, and B its base disk, as shown in the figure. We have

$$\iiint_D \text{div}\,\mathbf{F}\,dV = \int_0^{2\pi} d\theta \int_0^a r\,dr \int_0^{b(1-(r/a))} (2+3r^2)\,dz$$

$$= 2\pi b \int_0^a r(2+3r^2)\left(1-\frac{r}{a}\right) dr$$

$$= 2\pi b \int_0^a \left(2r + 3r^3 - \frac{2r^2}{a} - \frac{3r^4}{a}\right) dr$$

$$= \frac{2\pi a^2 b}{3} + \frac{3\pi a^4 b}{10}.$$

On B we have $z=0$, $\hat{\mathbf{N}} = -\mathbf{k}$, $\mathbf{F} \bullet \hat{\mathbf{N}} = -1$, so

$$\iint_B \mathbf{F} \bullet \hat{\mathbf{N}}\,dS = -\text{area of } B = -\pi a^2.$$

By the Divergence Theorem,

$$\iint_S \mathbf{F} \bullet \hat{\mathbf{N}}\,dS + \iint_B \mathbf{F} \bullet \hat{\mathbf{N}}\,dS = \iiint_D \text{div}\,\mathbf{F}\,dV,$$

so the flux of $\mathbf{F}$ upward through the conical surface S is

$$\iint_S = \frac{2\pi a^2 b}{3} + \frac{3\pi a^4 b}{10} + \pi a^2.$$

12. $\mathbf{F} = (y+xz)\mathbf{i} + (y+yz)\mathbf{j} - (2x+z^2)\mathbf{k}$
    $\text{div}\,\mathbf{F} = z + (1+z) - 2z = 1$. Thus

$$\iiint_D \text{div}\,\mathbf{F}\,dV = \text{volume of } D = \frac{\pi a^3}{6},$$

where D is the region in the first octant bounded by the sphere and the coordinate planes. The boundary of D consists of the spherical part S and the four planar parts, called the bottom, side, and back in the figure.

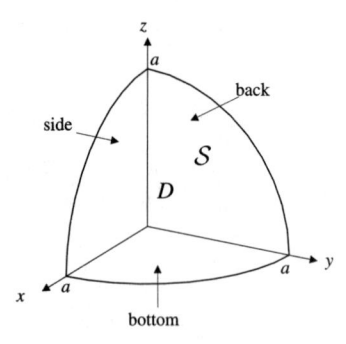

Fig. 16.4.12

On the side, $y=0$, $\hat{\mathbf{N}} = -\mathbf{j}$, $\mathbf{F} \bullet \hat{\mathbf{N}} = 0$, so

$$\iint_{\text{side}} \mathbf{F} \bullet \hat{\mathbf{N}}\,dS = 0.$$

On the back, $x=0$, $\hat{\mathbf{N}} = -\mathbf{i}$, $\mathbf{F} \bullet \hat{\mathbf{N}} = -y$, so

$$\iint_{\text{back}} \mathbf{F} \bullet \hat{\mathbf{N}}\,dS = -\int_0^{\pi/2} d\theta \int_0^a r\cos\theta\, r\,dr$$

$$= -\sin\theta\Big|_0^{\pi/2} \times \frac{a^3}{3} = -\frac{a^3}{3}.$$

On the bottom, $z=0$, $\hat{\mathbf{N}} = -\mathbf{k}$, $\mathbf{F} \bullet \hat{\mathbf{N}} = 2x$, so

$$\iint_{\text{bottom}} \mathbf{F} \bullet \hat{\mathbf{N}}\,dS = 2\int_0^{\pi/2} d\theta \int_0^a r\cos\theta\, r\,dr = \frac{2a^3}{3}.$$

By the Divergence Theorem

$$\iint_S \mathbf{F} \bullet \hat{\mathbf{N}}\,dS + 0 - \frac{a^3}{3} + \frac{2a^3}{3} = \frac{\pi a^3}{6}.$$

Hence the flux of $\mathbf{F}$ upward through S is

$$\iint_S \mathbf{F} \bullet \hat{\mathbf{N}}\,dS = \frac{\pi a^3}{6} - \frac{a^3}{3}.$$

13. $\mathbf{F} = (x+yz)\mathbf{i} + (y-xz)\mathbf{j} + (z-e^x \sin y)\mathbf{k}$
    $\text{div}\,\mathbf{F} = 1+1+1 = 3.$

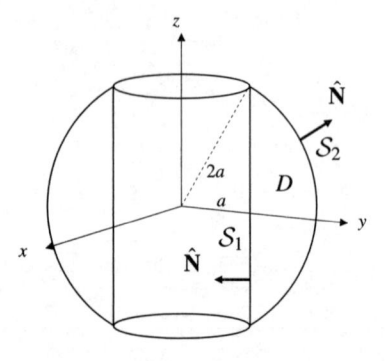

Fig. 16.4.13

a) The flux of **F** out of D through $\mathcal{S} = \mathcal{S}_1 \cup \mathcal{S}_2$ is

$$\oiint_{\mathcal{S}} \mathbf{F} \bullet \hat{\mathbf{N}} \, dS = \iiint_D \text{div } \mathbf{F} \, dV$$

$$= 3 \int_0^{2\pi} d\theta \int_a^{2a} r \, dr \int_0^{\sqrt{4a^2-r^2}} 2 \, dz$$

$$= 12\pi \int_a^{2a} r\sqrt{4a^2 - r^2} \, dr$$

$$\text{Let } u = 4a^2 - r^2$$
$$du = -2r \, dr$$

$$= 6\pi \int_0^{3a^2} u^{1/2} \, du = 12\sqrt{3}\pi a^3.$$

b) On $\mathcal{S}_1$, $\hat{\mathbf{N}} = -\dfrac{x\mathbf{i} + y\mathbf{j}}{a}$, $dS = a \, d\theta \, dz$. The flux of **F** out of D through $\mathcal{S}_1$ is

$$\iint_{\mathcal{S}_1} \mathbf{F} \bullet \hat{\mathbf{N}} \, dS = \iint_{\mathcal{S}_1} \frac{-x^2 - xyz - y^2 + xyz}{a} \, a \, d\theta \, dz$$

$$= -a^2 \int_0^{2\pi} d\theta \int_{-\sqrt{3}a}^{\sqrt{3}a} dz = -4\sqrt{3}\pi a^3.$$

c) The flux of **F** out of D through the spherical part $\mathcal{S}_2$ is

$$\iint_{\mathcal{S}_2} \mathbf{F} \bullet \hat{\mathbf{N}} \, dS = \oiint_{\mathcal{S}} \mathbf{F} \bullet \hat{\mathbf{N}} \, dS - \iint_{\mathcal{S}_1} \mathbf{F} \bullet \hat{\mathbf{N}} \, dS$$

$$= 12\sqrt{3}\pi a^3 + 4\sqrt{3}\pi a^3 = 16\sqrt{3}\pi a^3.$$

14. Let D be the domain bounded by $\mathcal{S}$, the coordinate planes, and the plane $x = 1$. If

$$\mathbf{F} = 3xz^2 \mathbf{i} - x\mathbf{j} - y\mathbf{k},$$

then $\text{div } \mathbf{F} = 3z^2$, so the total flux of **F** out of D is

$$\oiint_{\text{bdry of } D} \mathbf{F} \bullet \hat{\mathbf{N}} \, dS = \iiint_D 3z^2 \, dV$$

$$= 3 \int_0^1 dx \int_0^{\pi/2} d\theta \int_0^1 r^2 \cos^2\theta \, r \, dr$$

$$= 3 \times \frac{1}{4} \times \frac{\pi}{4} = \frac{3\pi}{16}.$$

The boundary of D consists of the cylindrical surface $\mathcal{S}$ and four planar surfaces, the side, bottom, back, and front.

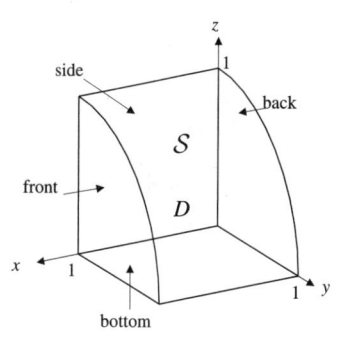

Fig. 16.4.14

On the side, $y = 0$, $\hat{\mathbf{N}} = -\mathbf{j}$, $\mathbf{F} \bullet \hat{\mathbf{N}} = x$, so

$$\iint_{\text{side}} \mathbf{F} \bullet \hat{\mathbf{N}} \, dS = \int_0^1 x \, dx \int_0^1 dz = \frac{1}{2}.$$

On the bottom, $z = 0$, $\hat{\mathbf{N}} = -\mathbf{k}$, $\mathbf{F} \bullet \hat{\mathbf{N}} = y$, so

$$\iint_{\text{bottom}} \mathbf{F} \bullet \hat{\mathbf{N}} \, dS = \int_0^1 y \, dy \int_0^1 dx = \frac{1}{2}.$$

On the back, $x = 0$, $\hat{\mathbf{N}} = -\mathbf{i}$, $\mathbf{F} \bullet \hat{\mathbf{N}} = 0$, so

$$\iint_{\text{back}} \mathbf{F} \bullet \hat{\mathbf{N}} \, dS = 0.$$

On the front, $x = 1$, $\hat{\mathbf{N}} = \mathbf{i}$, $\mathbf{F} \bullet \hat{\mathbf{N}} = 3z^2$, so

$$\iint_{\text{front}} \mathbf{F} \bullet \hat{\mathbf{N}} \, dS = 3 \int_0^{\pi/2} d\theta \int_0^1 r^2 \cos^2\theta \, r \, dr = \frac{3\pi}{16}.$$

Hence,

$$\iint_{\mathcal{S}} (3xz^2 \mathbf{i} - x\mathbf{j} - y\mathbf{k}) \bullet \hat{\mathbf{N}} \, dS = \frac{3\pi}{16} - \frac{1}{2} - \frac{1}{2} - 0 - \frac{3\pi}{16} = -1.$$

15. $\mathbf{F} = (x^2 - x - 2y)\mathbf{i} + (2y^2 + 3y - z)\mathbf{j} - (z^2 - 4z + xy)\mathbf{k}$
$\text{div } \mathbf{F} = 2x - 1 + 4y + 3 - 2z + 4 = 2x + 4y - 2z + 6.$

The flux of **F** out of R through its surface $\mathcal{S}$ is

$$\oiint_{\mathcal{S}} \mathbf{F} \bullet \hat{\mathbf{N}} \, dS = \iiint_R (2x + 4y - 2z + 6) \, dV.$$

Now $\iiint_R x \, dV = M_{x=0} = V\bar{x}$, where R has volume V and centroid $(\bar{x}, \bar{y}, \bar{z})$. Similar formulas obtain for the other variables, so the required flux is

$$\oiint_{\mathcal{S}} \mathbf{F} \bullet \hat{\mathbf{N}} \, dS = 2V\bar{x} + 4V\bar{y} - 2V\bar{z} + 6V.$$

16. $\mathbf{F} = x\mathbf{i} + y\mathbf{j} + z\mathbf{k}$ implies that $\text{div } \mathbf{F} = 3$. The total flux of **F** out of D is

$$\oiint_{\text{bdry of } D} \mathbf{F} \bullet \hat{\mathbf{N}} \, dS = 3 \iiint_D dV = 12,$$

since the volume of D is half that of a cube of side 2, that is, 4 square units.

D has three triangular faces, three pentagonal faces, and a hexagonal face. By symmetry, the flux of $\mathbf{F}$ out of each triangular face is equal to that out of the triangular face T in the plane $z = 1$. Since $\mathbf{F} \bullet \hat{\mathbf{N}} = \mathbf{k} \bullet \mathbf{k} = 1$ on that face, these fluxes are

$$\iint_T dx\, dy = \text{area of } T = \frac{1}{2}.$$

Similarly, the flux of $\mathbf{F}$ out of each pentagonal face is equal to the flux out of the pentagonal face P in the plane $z = -1$, where $\mathbf{F} \bullet \hat{\mathbf{N}} = -\mathbf{k} \bullet (-\mathbf{k}) = 1$; that flux is

$$\iint_P dx\, dy = \text{area of } P = 4 - \frac{1}{2} = \frac{7}{2}.$$

Thus the flux of $\mathbf{F}$ out of the remaining hexagonal face H is

$$12 - 3 \times \left(\frac{1}{2} + \frac{7}{2}\right) = 0.$$

(This can also be seen directly, since $\mathbf{F}$ radiates from the origin, so is everywhere tangent to the plane of the hexagonal face, the plane $x + y + z = 0$.)

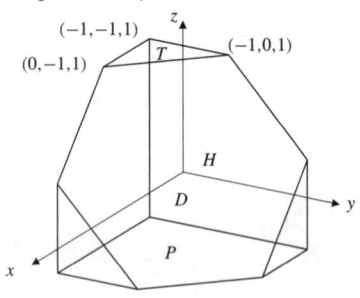

Fig. 16.4.16

17. The part of the sphere $\mathcal{S}$: $x^2 + y^2 + (z - a)^2 = 4a^2$ above $z = 0$ and the disk D: $x^2 + y^2 = 3a^2$ in the xy-plane form the boundary of a region R in 3-space. The outward normal from R on D is $-\mathbf{k}$. If

$$\mathbf{F} = (x^2 + y + 2 + z^2)\mathbf{i} + (e^{x^2} + y^2)\mathbf{j} + (3 + x)\mathbf{k},$$

then $div\mathbf{F} = 2x + 2y$. By the Divergence Theorem,

$$\iint_\mathcal{S} \mathbf{F} \bullet \hat{\mathbf{N}}\, dS + \iint_D \mathbf{F} \bullet (-\mathbf{k})\, dx\, dy = \iiint_R \text{div}\, \mathbf{F}\, dV = 0$$

because R is symmetric about $x = 0$ and $y = 0$. Thus the flux of $\mathbf{F}$ outward across $\mathcal{S}$ is

$$\iint_\mathcal{S} \mathbf{F} \bullet \hat{\mathbf{N}}\, dS = \iint_D (3 + x)\, dx\, dy = 3\pi(3a^2) = 9\pi a^2.$$

18. $\phi = x^2 - y^2 + z^2$, $\mathbf{G} = \frac{1}{3}(-y^3\mathbf{i} + x^3\mathbf{j} + z^3\mathbf{k})$.
$\mathbf{F} = \nabla\phi + \mu\,\text{curl}\,\mathbf{G}$.

Let R be the region of 3-space occupied by the sandpile. Then R is bounded by the upper surface $\mathcal{S}$ of the sandpile and by the disk D: $x^2 + y^2 \leq 1$ in the plane $z = 0$. The outward (from R) normal on D is $-\mathbf{k}$. The flux of $\mathbf{F}$ out of R is given by

$$\iint_\mathcal{S} \mathbf{F} \bullet \hat{\mathbf{N}}\, dS + \iint_D \mathbf{F} \bullet (-\mathbf{k})\, dA = \iiint_R \text{div}\, \mathbf{F}\, dV.$$

Now $\text{div}\,\text{curl}\,\mathbf{G} = 0$ by Theorem 3(g). Also $\text{div}\,\nabla\phi = \text{div}\,(2x\mathbf{i} - 2y\mathbf{j} + 2z\mathbf{k}) = 2 - 2 + 2 = 2$. Therefore

$$\iiint_R \text{div}\,\mathbf{F}\, dV = \iiint_R (2 + \mu \times 0)\, dV = 2(5\pi) = 10\pi.$$

In addition,

$$\text{curl}\,\mathbf{G} = \frac{1}{3}\begin{vmatrix} \mathbf{i} & \mathbf{j} & \mathbf{k} \\ \frac{\partial}{\partial x} & \frac{\partial}{\partial y} & \frac{\partial}{\partial z} \\ -y^3 & x^3 & z^3 \end{vmatrix} = 3(x^2 + y^2)\mathbf{k},$$

and $\nabla\phi \bullet \mathbf{k} = 2z = 0$ on D, so

$$\iint_D \mathbf{F} \bullet \mathbf{k}\, dA = 3\mu \int_0^{2\pi} d\theta \int_0^1 r^3\, dr = \frac{3\pi\mu}{2}.$$

The flux of $\mathbf{F}$ out of $\mathcal{S}$ is $10\pi + (3\pi\mu)/2$.

19. $\oiint_\mathcal{S} \text{curl}\,\mathbf{F} \bullet \hat{\mathbf{N}}\, dS = \iiint_D \text{div}\,\text{curl}\,\mathbf{F} = 0$, by Theorem 3(g).

20. If $\mathbf{r} = x\mathbf{i} + y\mathbf{j} + z\mathbf{k}$, then $\text{div}\,\mathbf{r} = 3$ and

$$\frac{1}{3}\oiint_\mathcal{S} \mathbf{r} \bullet \hat{\mathbf{N}}\, dS = \frac{1}{3}\iiint_D 3\, dV = V.$$

21. We use Theorem 7(b), the proof of which is given in Exercise 29. Taking $\phi(x, y, z) = x^2 + y^2 + z^2$, we have

$$\frac{1}{2V}\oiint_\mathcal{S} (x^2 + y^2 + z^2)\hat{\mathbf{N}}\, dS = \frac{1}{2V}\oiint_\mathcal{S} \phi\hat{\mathbf{N}}\, dS$$

$$= \frac{1}{2V}\iiint_D \text{grad}\,\phi\, dV$$

$$= \frac{1}{V}\iiint_D (x\mathbf{i} + y\mathbf{j} + z\mathbf{k})\, dV$$

$$= \bar{\mathbf{r}},$$

since $\iint x\, dV = M_{x=0} = V\bar{x}$.

22. Taking $\mathbf{F} = \nabla\phi$ in the first identity in Theorem 7(a), we have
$$\oiint_{\mathcal{S}} \nabla\phi \times \hat{\mathbf{N}}\, dS = -\iiint_{D} \operatorname{curl} \nabla\phi\, dV = 0,$$
since $\nabla \times \nabla\phi = 0$ by Theorem 3(h).

23. $\operatorname{div}(\phi\mathbf{F}) = \phi\operatorname{div}\mathbf{F} + \nabla\phi \bullet \mathbf{F}$ by Theorem 3(b). Thus
$$\iiint_{D} \phi\operatorname{div}\mathbf{F}\, dV + \iiint_{D} \nabla\phi \bullet \mathbf{F}\, dV = \iiint_{D} \operatorname{div}(\phi\mathbf{F})\, dV$$
$$= \oiint_{\mathcal{S}} \phi\mathbf{F} \bullet \hat{\mathbf{N}}\, dS$$
by the Divergence Theorem.

24. If $\mathbf{F} = \nabla\phi$ in the previous exercise, then $\operatorname{div}\mathbf{F} = \nabla^2\phi$ and
$$\iiint_{D} \phi\nabla^2\phi\, dV + \iiint_{D} |\nabla\phi|^2\, dV = \oiint_{\mathcal{S}} \phi\nabla\phi \bullet \hat{\mathbf{N}}\, dS.$$

If $\nabla^2\phi = 0$ in D and $\phi = 0$ on $\mathcal{S}$, then
$$\iiint_{D} |\nabla\phi|^2\, dV = 0.$$

Since ϕ is assumed to be smooth, $\nabla\phi = 0$ throughout D, and therefore ϕ is constant on each connected component of D. Since $\phi = 0$ on $\mathcal{S}$, these constants must all be 0, and $\phi = 0$ on D.

25. If u and v are two solutions of the given Dirichlet problem, and $\phi = u - v$, then
$$\nabla^2\phi = \nabla^2 u - \nabla^2 v = f - f = 0 \text{ on } D$$
$$\phi = u - v = g - g = 0 \text{ on } \mathcal{S}.$$

By the previous exercise, $\phi = 0$ on D, so $u = v$ on D. That is, solutions of the Dirichlet problem are unique.

26. Re-examine the solution to Exercise 24 above. If $\nabla^2\phi = 0$ in D and $\partial\phi/\partial n = \nabla\phi \bullet \hat{\mathbf{N}} = 0$ on $\mathcal{S}$, then we can again conclude that
$$\iiint_{D} |\nabla\phi|\, dV = 0$$
and $\nabla\phi = 0$ throughout D. Thus ϕ is constant on the connected components of D. (We can't conclude the constant is 0 because we don't know the value of ϕ on $\mathcal{S}$.) If u and v are solutions of the given Neumann problem, then $\phi = u - v$ satisfies
$$\nabla^2\phi = \nabla^2 u - \nabla^2 v = f - f = 0 \text{ on } D$$
$$\frac{\partial\phi}{\partial n} = \frac{\partial u}{\partial n} - \frac{\partial v}{\partial n} = g - g = 0 \text{ on } \mathcal{S},$$

so ϕ is constant on any connected component of $\mathcal{S}$, and u and v can only differ by a constant on $\mathcal{S}$.

27. Apply the Divergence Theorem to $\mathbf{F} = \nabla\phi$:
$$\iiint_{D} \nabla^2\phi\, dV = \iiint_{D} \nabla \bullet \nabla\phi\, dV$$
$$= \oiint_{\mathcal{S}} \nabla\phi \bullet \hat{\mathbf{N}}\, dS = \oiint_{\mathcal{S}} \frac{\partial\phi}{\partial n}\, dS.$$

28. By Theorem 3(b),
$$\operatorname{div}(\phi\nabla\psi - \psi\nabla\phi)$$
$$= \nabla\phi \bullet \nabla\psi + \phi\nabla^2\psi - \nabla\psi \bullet \nabla\phi - \psi\nabla^2\phi$$
$$= \phi\nabla^2\psi - \psi\nabla^2\phi.$$

Hence, by the Divergence Theorem,
$$\iiint_{D} (\phi\nabla^2\psi - \psi\nabla^2\phi)\, dV = \iiint_{D} \operatorname{div}(\phi\nabla\psi - \psi\nabla\phi)\, dV$$
$$= \oiint_{\mathcal{S}} (\phi\nabla\psi - \psi\nabla\phi) \bullet \hat{\mathbf{N}}\, dS$$
$$= \oiint_{\mathcal{S}} \left(\phi\frac{\partial\psi}{\partial n} - \psi\frac{\partial\phi}{\partial n}\right) dS.$$

29. If $\mathbf{F} = \phi\mathbf{c}$, where $\mathbf{c}$ is an arbitrary, constant vector, then $\operatorname{div}\mathbf{F} = \nabla\phi \bullet \mathbf{c}$, and by the Divergence Theorem,
$$\mathbf{c} \bullet \iiint_{D} \nabla\phi\, dV = \iiint_{D} \operatorname{div}\mathbf{F}\, dV$$
$$= \oiint_{\mathcal{S}} \mathbf{F} \bullet \hat{\mathbf{N}}\, dS$$
$$= \oiint_{\mathcal{S}} \phi\mathbf{c} \bullet \hat{\mathbf{N}}\, dS = \mathbf{c} \bullet \oiint_{\mathcal{S}} \phi\hat{\mathbf{N}}\, dS.$$

Thus
$$\mathbf{c} \bullet \left(\iiint_{D} \nabla\phi\, dV - \oiint_{\mathcal{S}} \phi\hat{\mathbf{N}}\, dS\right) = 0.$$

Since $\mathbf{c}$ is arbitrary, the vector in the large parentheses must be the zero vector. Hence
$$\iiint_{D} \nabla\phi\, dV = \oiint_{\mathcal{S}} \phi\hat{\mathbf{N}}\, dS.$$

30. $\dfrac{1}{\text{vol}(D_\epsilon)} \oiint_{\mathcal{S}_\epsilon} \mathbf{F} \bullet \hat{\mathbf{N}} \, dS = \dfrac{1}{\text{vol}(D_\epsilon)} \iiint_{D_\epsilon} \text{div}\,\mathbf{F}\, dV$

$= \dfrac{1}{\text{vol}(D_\epsilon)} \left[\iiint_{D_\epsilon} \text{div}\,\mathbf{F}(P_0)\, dV \right.$

$\left. + \iiint_{D_\epsilon} \big(\text{div}\,\mathbf{F} - \text{div}\,\mathbf{F}(P_0)\big)\, dV \right]$

$= \text{div}\,\mathbf{F}(P_0) + \dfrac{1}{\text{vol}(D_\epsilon)} \iiint_{D_\epsilon} \big(\text{div}\,\mathbf{F} - \text{div}\,\mathbf{F}(P_0)\big)\, dV.$

Thus

$\left| \dfrac{1}{\text{vol}(D_\epsilon)} \oiint_{\mathcal{S}_\epsilon} \mathbf{F} \bullet \hat{\mathbf{N}}\, dS - \text{div}\,\mathbf{F}(P_0) \right|$

$\le \dfrac{1}{\text{vol}(D_\epsilon)} \iiint_{D_\epsilon} |\text{div}\,\mathbf{F} - \text{div}\,\mathbf{F}(P_0)|\, dV$

$\le \max_{P \text{ in } D_\epsilon} |\text{div}\,\mathbf{F} - \text{div}\,\mathbf{F}(P_0)|$

$\to 0$ as $\epsilon \to 0+$ assuming $\text{div}\,\mathbf{F}$ is continuous.

$\displaystyle\lim_{\epsilon \to 0+} \dfrac{1}{\text{vol}(D_\epsilon)} \oiint_{\mathcal{S}_\epsilon} \mathbf{F} \bullet \hat{\mathbf{N}}\, dS = \text{div}\,\mathbf{F}(P_0).$

Section 16.5 Stokes's Theorem (page 957)

1. The triangle T lies in the plane $x + y + z = 1$. We use the downward normal

$$\hat{\mathbf{N}} = -\dfrac{\mathbf{i} + \mathbf{j} + \mathbf{k}}{\sqrt{3}}$$

on T, because of the given orientation of its boundary. If $\mathbf{F} = xy\mathbf{i} + yz\mathbf{j} + zx\mathbf{k}$, then

$$\text{curl}\,\mathbf{F} = \begin{vmatrix} \mathbf{i} & \mathbf{j} & \mathbf{k} \\ \dfrac{\partial}{\partial x} & \dfrac{\partial}{\partial y} & \dfrac{\partial}{\partial z} \\ xy & yz & zx \end{vmatrix} = -y\mathbf{i} - z\mathbf{j} - x\mathbf{k}.$$

Therefore

$\displaystyle\oint_\mathcal{C} xy\, dx + yz\, dy + zx\, dz = \oint_\mathcal{C} \mathbf{F} \bullet d\mathbf{r}$

$= \displaystyle\iint_T \text{curl}\,\mathbf{F} \bullet \hat{\mathbf{N}}\, dS = \iint_T \dfrac{y + z + x}{\sqrt{3}}\, dS$

$= \dfrac{1}{\sqrt{3}} \displaystyle\iint_T dS = \dfrac{1}{\sqrt{3}} \times (\text{area of } T)$

$= \dfrac{1}{\sqrt{3}} \times \left(\dfrac{1}{2} \times \sqrt{2} \times \dfrac{\sqrt{3}}{\sqrt{2}} \right) = \dfrac{1}{2}.$

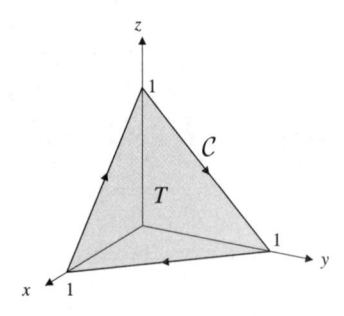

Fig. 16.5.1

2. Let $\mathcal{S}$ be the part of the surface $z = y^2$ lying inside the cylinder $x^2 + y^2 = 4$, and having upward normal $\hat{\mathbf{N}}$. Then $\mathcal{C}$ is the oriented boundary of $\mathcal{S}$. Let D be the disk $x^2 + y^2 \le 4$, $z = 0$, that is, the projection of $\mathcal{S}$ onto the xy-plane.

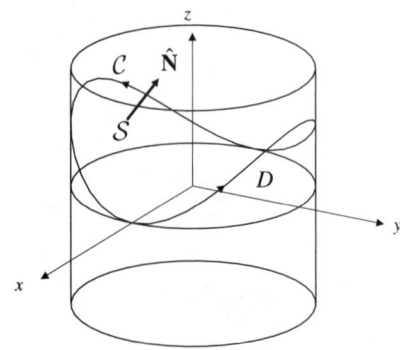

Fig. 16.5.2

If $\mathbf{F} = y\mathbf{i} - x\mathbf{j} + z^2\mathbf{k}$, then

$$\text{curl}\,\mathbf{F} = \begin{vmatrix} \mathbf{i} & \mathbf{j} & \mathbf{k} \\ \dfrac{\partial}{\partial x} & \dfrac{\partial}{\partial y} & \dfrac{\partial}{\partial z} \\ y & -x & z^2 \end{vmatrix} = -2\mathbf{k}.$$

Since $dS = \dfrac{dx\, dy}{\mathbf{k} \bullet \hat{\mathbf{N}}}$ on $\mathcal{S}$, we have

$\displaystyle\oint_\mathcal{C} y\, dx - x\, dy + z^2\, dz = \oint_\mathcal{C} \mathbf{F} \bullet d\mathbf{r} = \iint_\mathcal{S} \text{curl}\,\mathbf{F} \bullet \hat{\mathbf{N}}\, dS$

$= \displaystyle\iint_D -2\mathbf{k} \bullet \hat{\mathbf{N}}\, \dfrac{dx\, dy}{\mathbf{k} \bullet \hat{\mathbf{N}}} = -8\pi.$

3. Let $\mathcal{C}$ be the circle $x^2 + y^2 = a^2$, $z = 0$, oriented counterclockwise as seen from the positive z-axis. Let D be the disk bounded by $\mathcal{C}$, with normal $\mathbf{k}$. We have

$$\mathbf{F} = 3y\mathbf{i} - 2xz\mathbf{j} + (x^2 - y^2)\mathbf{k}$$

$$\text{curl}\,\mathbf{F} = \begin{vmatrix} \mathbf{i} & \mathbf{j} & \mathbf{k} \\ \dfrac{\partial}{\partial x} & \dfrac{\partial}{\partial y} & \dfrac{\partial}{\partial z} \\ 3y & -2xz & x^2 - y^2 \end{vmatrix}$$

$$= 2(x - y)\mathbf{i} - 2x\mathbf{j} - (2z + 3)\mathbf{k}.$$

Applying Stokes's Theorem (twice) we calculate

$$\iint_S = \oint_C \mathbf{F} \cdot d\mathbf{r} = \iint_D \operatorname{curl} \mathbf{F} \cdot \mathbf{k} \, dA$$
$$= -\iint_D 3 \, dA = -3\pi a^2.$$

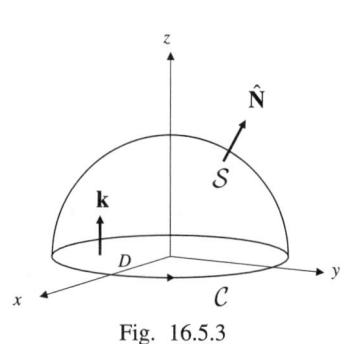

Fig. 16.5.3

4. The surface S with equation

$$x^2 + y^2 + 2(z-1)^2 = 6, \qquad z \geq 0,$$

with outward normal $\hat{\mathbf{N}}$, is that part of an ellipsoid of revolution about the z-axis, centred at $(0, 0, 1)$, and lying above the xy-plane. The boundary of S is the circle C: $x^2 + y^2 = 4$, $z = 0$, oriented counterclockwise as seen from the positive z-axis. C is also the oriented boundary of the disk $x^2 + y^2 \leq 4$, $z = 0$, with normal $\hat{\mathbf{N}} = \mathbf{k}$.
If $\mathbf{F} = (xz - y^3 \cos z)\mathbf{i} + x^3 e^z \mathbf{j} + xyze^{x^2+y^2+z^2}\mathbf{k}$, then, on $z = 0$, we have

$$\operatorname{curl} \mathbf{F} \cdot \mathbf{k} = \left(\frac{\partial}{\partial x} x^3 e^z - \frac{\partial}{\partial y}(xz - y^3 \cos z) \right) \bigg|_{z=0}$$
$$= (3x^2 e^z + 3y^2 \cos z) \bigg|_{z=0} = 3(x^2 + y^2).$$

Thus

$$\iint_S \operatorname{curl} \mathbf{F} \cdot \hat{\mathbf{N}} \, dS = \oint_C \mathbf{F} \cdot d\mathbf{r} = \iint_D \operatorname{curl} \mathbf{F} \cdot \mathbf{k} \, dA$$
$$= \int_0^{2\pi} d\theta \int_0^2 3r^2 \, r \, dr = 24\pi.$$

5. The circle C of intersection of $x^2 + y^2 + z^2 = a^2$ and $x + y + z = 0$ is the boundary of a circular disk of radius a in the plane $x + y + z = 0$.
If $\mathbf{F} = y\mathbf{i} + z\mathbf{j} + x\mathbf{k}$, then

$$\operatorname{curl} \mathbf{F} = \begin{vmatrix} \mathbf{i} & \mathbf{j} & \mathbf{k} \\ \frac{\partial}{\partial x} & \frac{\partial}{\partial y} & \frac{\partial}{\partial z} \\ y & z & x \end{vmatrix} = -(\mathbf{i} + \mathbf{j} + \mathbf{k}).$$

If C is oriented so that D has normal

$$\hat{\mathbf{N}} = -\frac{\mathbf{i} + \mathbf{j} + \mathbf{k}}{\sqrt{3}},$$

then $\operatorname{curl} \mathbf{F} \cdot \hat{\mathbf{N}} = \sqrt{3}$ on D, so

$$\oint_C y \, dx + z \, dy + x \, dz = \oint_C \mathbf{F} \cdot d\mathbf{r} = \iint_D \operatorname{curl} \mathbf{F} \cdot \hat{\mathbf{N}} \, dS$$
$$= \sqrt{3} \iint_D dS = \sqrt{3} \pi a^2,$$

since D has area πa^2.

6. The curve C:

$$\mathbf{r} = \cos t \mathbf{i} + \sin t \mathbf{j} + \sin 2t \mathbf{k}, \qquad 0 \leq t \leq 2\pi,$$

lies on the surface $z = 2xy$, since $\sin 2t = 2\cos t \sin t$. It also lies on the cylinder $x^2 + y^2 = 1$, so it is the boundary of that part of $z = 2xy$ lying inside that cylinder. Since C is oriented counterclockwise as seen from high on the z-axis, S should be oriented with upward normal,

$$\hat{\mathbf{N}} = \frac{-2y\mathbf{i} - 2x\mathbf{j} + \mathbf{k}}{\sqrt{1 + 4(x^2 + y^2)}},$$

and has area element

$$dS = \sqrt{1 + 4(x^2 + y^2)} \, dx \, dy.$$

If $\mathbf{F} = (e^x - y^3)\mathbf{i} + (e^y + x^3)\mathbf{j} + e^z \mathbf{k}$, then

$$\operatorname{curl} \mathbf{F} = \begin{vmatrix} \mathbf{i} & \mathbf{j} & \mathbf{k} \\ \frac{\partial}{\partial x} & \frac{\partial}{\partial y} & \frac{\partial}{\partial z} \\ e^x - y^3 & e^y + x^3 & e^z \end{vmatrix} = 3(x^2 + y^2)\mathbf{k}.$$

If D is the disk $x^2 + y^2 \leq 1$ in the xy-plane, then

$$\oint_C \mathbf{F} \cdot d\mathbf{r} = \iint_S \operatorname{curl} \mathbf{F} \cdot \hat{\mathbf{N}} \, dS = \iint_D 3(x^2 + y^2) \, dx \, dy$$
$$= 3 \int_0^{2\pi} d\theta \int_0^1 r^2 \, r \, dr = \frac{3\pi}{2}.$$

7. The part of the paraboloid $z = 9 - x^2 - y^2$ lying above the xy-plane having upward normal $\hat{\mathbf{N}}$ has boundary the circle C: $x^2 + y^2 = 9$, oriented counterclockwise as seen from above. C is also the oriented boundary of the plane disk $x^2 + y^2 \leq 9$, $z = 0$, oriented with normal field $\hat{\mathbf{N}} = \mathbf{k}$.
If $\mathbf{F} = -y\mathbf{i} + x^2\mathbf{j} + z\mathbf{k}$, then

$$\operatorname{curl} \mathbf{F} = \begin{vmatrix} \mathbf{i} & \mathbf{j} & \mathbf{k} \\ \frac{\partial}{\partial x} & \frac{\partial}{\partial y} & \frac{\partial}{\partial z} \\ -y & x^2 & z \end{vmatrix} = (2x + 1)\mathbf{k}.$$

By Stokes's Theorem, the circulation of $\mathbf{F}$ around $\mathcal{C}$ is

$$\oint_{\mathcal{C}} \mathbf{F} \bullet d\mathbf{r} = \iint_D (\operatorname{curl} \mathbf{F} \bullet \mathbf{k}) \, dA$$
$$= \iint_D (2x+1) \, dA = 0 + \pi(3^2) = 9\pi.$$

8. The closed curve

$$\mathbf{r} = (1+\cos t)\mathbf{i} + (1+\sin t)\mathbf{j} + (1-\cos t - \sin t)\mathbf{k},$$

$(0 \le t \le 2\pi)$, lies in the plane $x+y+z = 3$ and is oriented counterclockwise as seen from above. Therefore it is the boundary of a region $\mathcal{S}$ in that plane with normal field $\hat{\mathbf{N}} = (\mathbf{i}+\mathbf{j}+\mathbf{k})/\sqrt{3}$. The projection of $\mathcal{S}$ onto the xy-plane is the circular disk D of radius 1 with centre at $(1,1)$.

If $\mathbf{F} = ye^x\mathbf{i} + (x^2+e^x)\mathbf{j} + z^2 e^z \mathbf{k}$, then

$$\operatorname{curl} \mathbf{F} = \begin{vmatrix} \mathbf{i} & \mathbf{j} & \mathbf{k} \\ \frac{\partial}{\partial x} & \frac{\partial}{\partial y} & \frac{\partial}{\partial z} \\ ye^x & x^2+e^x & z^2+e^z \end{vmatrix} = 2x\mathbf{k}.$$

By Stokes's Theorem,

$$\oint_{\mathcal{C}} \mathbf{F} \bullet d\mathbf{r} = \iint_{\mathcal{S}} \operatorname{curl} \mathbf{F} \bullet \hat{\mathbf{N}} \, dS$$
$$= \iint_{\mathcal{S}} \frac{2x}{\sqrt{3}} \, dS = \iint_D \frac{2x}{\sqrt{3}}(\sqrt{3}) \, dx \, dy$$
$$= 2\bar{x}A = 2\pi,$$

where $\bar{x} = 1$ is the x-coordinate of the boundary of D, and $A = \pi 1^2 = \pi$ is the area of D.

9. If $\mathcal{S}_1$ and $\mathcal{S}_2$ are two surfaces joining $\mathcal{C}_1$ to $\mathcal{C}_2$, each having upward normal, then the closed surface $\mathcal{S}_3$ consisting of $\mathcal{S}_1$ and $-\mathcal{S}_2$ (that is, $\mathcal{S}_2$ with downward normal) bound a region R in 3-space. Then

$$\iint_{\mathcal{S}_1} \mathbf{F} \bullet \hat{\mathbf{N}} \, dS - \iint_{\mathcal{S}_2} \mathbf{F} \bullet \hat{\mathbf{N}} \, dS$$
$$= \iint_{\mathcal{S}_1} \mathbf{F} \bullet \hat{\mathbf{N}} \, dS + \iint_{-\mathcal{S}_2} \mathbf{F} \bullet \hat{\mathbf{N}} \, dS$$
$$= \oiint_{\mathcal{S}_3} \mathbf{F} \bullet \hat{\mathbf{N}} \, dS = \pm \iiint_R \operatorname{div} \mathbf{F} \, dV = 0,$$

provided that $\operatorname{div} \mathbf{F} = 0$ identically. Since

$$\mathbf{F} = (\alpha x^2 - z)\mathbf{i} + (xy + y^3 + z)\mathbf{j} + \beta y^2(z+1)\mathbf{k},$$

we have $\operatorname{div} \mathbf{F} = 2\alpha x + x + 3y^2 + \beta y^2 = 0$ if $\alpha = -1/2$ and $\beta = -3$. In this case we can evaluate $\iint_{\mathcal{S}} \mathbf{F} \bullet \hat{\mathbf{N}} \, dS$ for any such surface $\mathcal{S}$ by evaluating the special case where S is the half-disk H: $x^2+y^2 \le 1$, $z = 0$, $y \ge 0$, with upward normal $\hat{\mathbf{N}} = \mathbf{k}$. We have

$$\iint_{\mathcal{S}} \mathbf{F} \bullet \hat{\mathbf{N}} \, dS = -3 \iint_H y^2 \, dx \, dy$$
$$= -3 \int_0^\pi \sin^2 \theta \, d\theta \int_0^1 r^3 \, dr = -\frac{3\pi}{8}.$$

10. The curve $\mathcal{C}$: $(x-1)^2 + 4y^2 = 16$, $2x+y+z = 3$, oriented counterclockwise as seen from above, bounds an elliptic disk $\mathcal{S}$ on the plane $2x+y+z = 3$. $\mathcal{S}$ has normal $\hat{\mathbf{N}} = (2\mathbf{i}+\mathbf{j}+\mathbf{k})/\sqrt{6}$. Since its projection onto the xy-plane is an elliptic disk with centre at $(1,0,0)$ and area $\pi(4)(2) = 8\pi$, therefore $\mathcal{S}$ has area $8\sqrt{6}\pi$ and centroid $(1,0,1)$. If

$$\mathbf{F} = (z^2 + y^2 + \sin x^2)\mathbf{i} + (2xy+z)\mathbf{j} + (xz+2yz)\mathbf{k},$$

then

$$\operatorname{curl} \mathbf{F} = \begin{vmatrix} \mathbf{i} & \mathbf{j} & \mathbf{k} \\ \frac{\partial}{\partial x} & \frac{\partial}{\partial y} & \frac{\partial}{\partial z} \\ z^2+y^2+\sin x^2 & 2xy+z & xz+2yz \end{vmatrix}$$
$$= (2z-1)\mathbf{i} + z\mathbf{j}.$$

By Stokes's Theorem,

$$\oint_{\mathcal{C}} \mathbf{F} \bullet d\mathbf{r} = \iint_{\mathcal{S}} \operatorname{curl} \mathbf{F} \bullet \hat{\mathbf{N}} \, dS$$
$$= \frac{1}{\sqrt{6}} \iint_{\mathcal{S}} (2(2z-1)+z) \, dS$$
$$= \frac{5\bar{z}-2}{\sqrt{6}}(8\sqrt{6}\pi) = 24\pi.$$

11. As was shown in Exercise 13 of Section 16.2,

$$\nabla \times (\phi \nabla \psi) = -\nabla \times (\psi \times \phi) = \nabla \phi \times \nabla \psi.$$

Thus, by Stokes's Theorem,

$$\oint_{\mathcal{C}} \phi \nabla \psi = \iint_{\mathcal{S}} \nabla \times (\phi \nabla \psi) \bullet \hat{\mathbf{N}} \, dS$$
$$= \iint_{\mathcal{S}} (\nabla \phi \times \nabla \psi) \bullet \hat{\mathbf{N}} \, dS$$
$$-\oint_{\mathcal{C}} \psi \nabla \phi = \iint_{\mathcal{S}} -\nabla \times (\psi \nabla \phi) \bullet \hat{\mathbf{N}} \, dS$$
$$= \iint_{\mathcal{S}} (\nabla \phi \times \nabla \psi) \bullet \hat{\mathbf{N}} \, dS.$$

$\nabla \phi \times \nabla \psi$ is solenoidal, with potential $\phi \nabla \psi$, or $-\psi \nabla \phi$.

12. We are given that C bounds a region R in a plane P with unit normal $\hat{\mathbf{N}} = a\mathbf{i} + b\mathbf{j} + c\mathbf{k}$. Therefore, $a^2 + b^2 + c^2 = 1$. If $\mathbf{F} = (bz - cy)\mathbf{i} + (cx - az)\mathbf{j} + (ay - bx)\mathbf{k}$, then

$$\operatorname{curl} \mathbf{F} = \begin{vmatrix} \mathbf{i} & \mathbf{j} & \mathbf{k} \\ \frac{\partial}{\partial x} & \frac{\partial}{\partial y} & \frac{\partial}{\partial z} \\ bz - cy & cx - az & ay - bx \end{vmatrix}$$
$$= 2a\mathbf{i} + 2b\mathbf{j} + 2c\mathbf{k}.$$

Hence $\operatorname{curl} \mathbf{F} \bullet \hat{\mathbf{N}} = 2(a^2 + b^2 + c^2) = 2$. We have

$$\frac{1}{2} \oint_C (bz - cy)\, dx + (cx - az)\, dy + (ay - bx)\, dz$$
$$= \frac{1}{2} \oint_C \mathbf{F} \bullet d\mathbf{r} = \frac{1}{2} \iint_R \operatorname{curl} \mathbf{F} \bullet \hat{\mathbf{N}}\, dS$$
$$= \frac{1}{2} \iint_R 2\, dS = \text{area of } R.$$

13. The circle C_ϵ of radius ϵ centred at P is the oriented boundary of the disk S_ϵ of area $\pi\epsilon^2$ having constant normal field $\hat{\mathbf{N}}$. By Stokes's Theorem,

$$\oint_{C_\epsilon} \mathbf{F} \bullet d\mathbf{r} = \iint_{S_\epsilon} \operatorname{curl} \mathbf{F} \bullet \hat{\mathbf{N}}\, dS$$
$$= \iint_{S_\epsilon} \operatorname{curl} \mathbf{F}(P) \bullet \hat{\mathbf{N}}\, dS$$
$$\quad + \iint_{S_\epsilon} \bigl(\operatorname{curl} \mathbf{F} - \operatorname{curl} \mathbf{F}(P)\bigr) \bullet \hat{\mathbf{N}}\, dS$$
$$= \pi\epsilon^2 \operatorname{curl} \mathbf{F}(P) \bullet \hat{\mathbf{N}}$$
$$\quad + \iint_{S_\epsilon} \bigl(\operatorname{curl} \mathbf{F} - \operatorname{curl} \mathbf{F}(P)\bigr) \bullet \hat{\mathbf{N}}\, dS.$$

Since $\mathbf{F}$ is assumed smooth, its curl is continuous at P. Therefore

$$\left| \frac{1}{\pi\epsilon^2} \oint_{C_\epsilon} \mathbf{F} \bullet d\mathbf{r} - \operatorname{curl} \mathbf{F}(P) \bullet \hat{\mathbf{N}} \right|$$
$$\leq \frac{1}{\pi\epsilon^2} \iint_{S_\epsilon} \bigl| \bigl(\operatorname{curl} \mathbf{F} - \operatorname{curl} \mathbf{F}(P)\bigr) \bullet \hat{\mathbf{N}} \bigr|\, dS$$
$$\leq \max_{Q \text{ on } S_\epsilon} |\operatorname{curl} \mathbf{F}(Q) - \operatorname{curl} \mathbf{F}(P)|$$
$$\to 0 \text{ as } \epsilon \to 0+.$$

Thus $\lim_{\epsilon \to 0+} \oint_{C_\epsilon} \mathbf{F} \bullet d\mathbf{r} = \operatorname{curl} \mathbf{F}(P) \bullet \hat{\mathbf{N}}$.

Section 16.6 Some Physical Applications of Vector Calculus (page 965)

1. a) If we measure depth in the liquid by $-z$, so that the z-axis is vertical and $z = 0$ at the surface, then the pressure at depth $-z$ is $p = -\delta g z$, where δ is the density of the liquid. Thus

$$\nabla p = -\delta g \mathbf{k} = \delta \mathbf{g},$$

where $\mathbf{g} = -g\mathbf{k}$ is the constant downward vector acceleration of gravity.

The force of the liquid on surface element dS of the solid with outward (from the solid) normal $\hat{\mathbf{N}}$ is

$$d\mathbf{B} = -p\hat{\mathbf{N}}\, dS = -(-\delta g z)\hat{\mathbf{N}}\, dS = \delta g z \hat{\mathbf{N}}\, dS.$$

Thus, the total force of the liquid on the solid (the buoyant force) is

$$\mathbf{B} = \oiint_S \delta g z \hat{\mathbf{N}}\, dS$$
$$= \iiint_R \nabla(\delta g z)\, dV \quad \text{(see Theorem 7)}$$
$$= -\iiint_R \delta \mathbf{g}\, dV = -M\mathbf{g},$$

where $M = \iiint_R \delta\, dV$ is the mass of the liquid which would occupy the same space as the solid. Thus $\mathbf{B} = -\mathbf{F}$, where $\mathbf{F} = M\mathbf{g}$ is the weight of the liquid displaced by the solid.

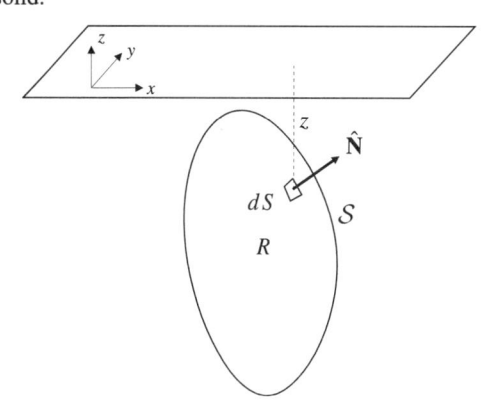

Fig. 16.6.1

b) The above argument extends to the case where the solid is only partly submerged. Let R^* be the part of the region occupied by the solid that is below the surface of the liquid. Let $\mathcal{S}^* = \mathcal{S}_1 \cup \mathcal{S}_2$ be the boundary of R^*, with $\mathcal{S}_1 \subset \mathcal{S}$ and $\mathcal{S}_2$ in the plane of the surface of the liquid. Since $p = -\delta gz = 0$ on $\mathcal{S}_2$, we have

$$\iint_{\mathcal{S}_2} \delta gz \hat{\mathbf{N}} \, dS = 0.$$

Therefore the buoyant force on the solid is

$$\begin{aligned}\mathbf{B} &= \iint_{\mathcal{S}_1} \delta gz \hat{\mathbf{N}} \, dS \\ &= \iint_{\mathcal{S}_1} \delta gz \hat{\mathbf{N}} \, dS + \iint_{\mathcal{S}_2} \delta gz \hat{\mathbf{N}} \, dS \\ &= \oiint_{\mathcal{S}^*} \delta gz \hat{\mathbf{N}} \, dS \\ &= -\iiint_{R^*} \delta \mathbf{g} \, dV = -M^* \mathbf{g},\end{aligned}$$

where $M^* = \iiint_{R^*} \delta \, dV$ is the mass of the liquid which would occupy R^*. Again we conclude that the buoyant force is the negative of the weight of the liquid displaced.

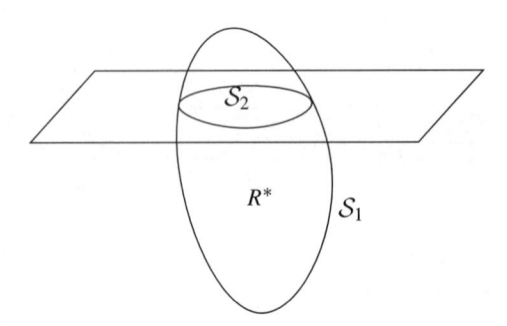

Fig. 16.6.1

2. The first component of $\mathbf{F}(\mathbf{G} \bullet \hat{\mathbf{N}})$ is $(F_1 \mathbf{G}) \bullet \hat{\mathbf{N}}$. Applying the Divergence Theorem and Theorem 3(b), we obtain

$$\begin{aligned}\oiint_{\mathcal{S}} (F_1 \mathbf{G}) \bullet \hat{\mathbf{N}} \, dS &= \iiint_D \operatorname{div}(F_1 \mathbf{G}) \, dV \\ &= \iiint_D (\boldsymbol{\nabla} F_1 \bullet \mathbf{G} + F_1 \boldsymbol{\nabla} \bullet \mathbf{G}) \, dS.\end{aligned}$$

But $\boldsymbol{\nabla} F_1 \bullet \mathbf{G}$ is the first component of $(\mathbf{G} \bullet \boldsymbol{\nabla})\mathbf{F}$, and $F_1 \boldsymbol{\nabla} \bullet \mathbf{G}$ is the first component of $\mathbf{F}\operatorname{div}\mathbf{G}$. Similar results obtain for the other components, so

$$\oiint_{\mathcal{S}} \mathbf{F}(\mathbf{G} \bullet \hat{\mathbf{N}}) \, dS = \iiint_D (\mathbf{F}\operatorname{div}\mathbf{G} + (\mathbf{G} \bullet \boldsymbol{\nabla})\mathbf{F}) \, dV.$$

3. Suppose the closed surface $\mathcal{S}$ bounds a region R in which charge is distributed with density ρ. Since the electric field $\mathbf{E}$ due to the charge satisfies $\operatorname{div}\mathbf{E} = k\rho$, the total flux of $\mathbf{E}$ out of R through $\mathcal{S}$ is, by the Divergence Theorem,

$$\oiint_{\mathcal{S}} \mathbf{E} \bullet \hat{\mathbf{N}} \, dS = \iiint_R \operatorname{div}\mathbf{E} \, dV = k\iiint_R \rho \, dV = kQ,$$

where $Q = \iiint_R \rho \, dV$ is the total charge in R.

4. If f is continuous and vanishes outside a bounded region (say the ball of radius R centred at $\mathbf{r}$), then $|f(\xi, \eta, \zeta)| \leq K$, and, if (ρ, ϕ, θ) denote spherical coordinates centred at $\mathbf{r}$, then

$$\iiint_{\mathbb{R}^3} \frac{|f(\mathbf{s})|}{|\mathbf{r} - \mathbf{s}|} \, dV_s \leq K \int_0^{2\pi} d\theta \int_0^{\pi} \sin\phi \, d\phi \int_0^R \frac{\rho^2}{\rho} \, d\rho$$
$$= 2\pi K R^2 \quad \text{a constant.}$$

5. This derivation is similar to that of the continuity equation for fluid motion given in the text. If $\mathcal{S}$ is an (imaginary) surface bounding an arbitrary region D, then the rate of change of total charge in D is

$$\frac{\partial}{\partial t} \iiint_D \rho \, dV = \iiint_D \frac{\partial \rho}{\partial t} \, dV,$$

where ρ is the charge density. By conservation of charge, this rate must be equal to the rate at which charge is crossing $\mathcal{S}$ into D, that is, to

$$\oiint_{\mathcal{S}} (-\mathbf{J}) \bullet \hat{\mathbf{N}} \, dS = -\iiint_D \operatorname{div}\mathbf{J} \, dV.$$

(The negative sign occurs because $\hat{\mathbf{N}}$ is the outward (from D) normal on $\mathcal{S}$.) Thus we have

$$\iiint_D \left(\frac{\partial \rho}{\partial t} + \operatorname{div}\mathbf{J}\right) dV = 0.$$

Since D is arbitrary and we are assuming the integrand is continuous, it must be 0 at every point:

$$\frac{\partial \rho}{\partial t} + \operatorname{div}\mathbf{J} = 0.$$

6. Since $\mathbf{r} = x\mathbf{i} + y\mathbf{j} + z\mathbf{k}$ and $\mathbf{b} = b_1\mathbf{i} + b_2\mathbf{j} + b_3\mathbf{k}$, we have

$$|\mathbf{r} - \mathbf{b}|^2 = (x - b_1)^2 + (y - b_2)^2 + (z - b_3)^2$$
$$2|\mathbf{r} - \mathbf{b}|\frac{\partial}{\partial x}|\mathbf{r} - \mathbf{b}| = 2(x - b_1)$$
$$\frac{\partial}{\partial x}|\mathbf{r} - \mathbf{b}| = \frac{x - b_1}{|\mathbf{r} - \mathbf{b}|}.$$

Similar formulas hold for the other first partials of $|\mathbf{r} - \mathbf{b}|$, so

$$\nabla\left(\frac{1}{|\mathbf{r}-\mathbf{b}|}\right)$$
$$= \frac{-1}{|\mathbf{r}-\mathbf{b}|^2}\left(\frac{\partial}{\partial x}|\mathbf{r}-\mathbf{b}|\mathbf{i} + \cdots + \frac{\partial}{\partial z}|\mathbf{r}-\mathbf{b}|\mathbf{k}\right)$$
$$= \frac{-1}{|\mathbf{r}-\mathbf{b}|^2}\frac{(x-b_1)\mathbf{i}+(y-b_2)\mathbf{j}+(z-b_3)\mathbf{k}}{|\mathbf{r}-\mathbf{b}|}$$
$$= -\frac{\mathbf{r}-\mathbf{b}}{|\mathbf{r}-\mathbf{b}|^3}.$$

7. Using the result of Exercise 4 and Theorem 3(d) and (h), we calculate, for constant $\mathbf{a}$,

$$\mathbf{div}\left(\mathbf{a} \times \frac{\mathbf{r}-\mathbf{b}}{|\mathbf{r}-\mathbf{b}|^3}\right)$$
$$= -\mathbf{div}\left(\mathbf{a} \times \nabla\frac{1}{|\mathbf{r}-\mathbf{b}|}\right)$$
$$= -(\nabla \times \mathbf{a}) \bullet \nabla\frac{1}{|\mathbf{r}-\mathbf{b}|} + \mathbf{a} \bullet \nabla \times \nabla\frac{1}{|\mathbf{r}-\mathbf{b}|} = 0 + 0 = 0.$$

8. For any element $d\mathbf{s}$ on the filament $\mathcal{F}$, we have

$$\mathbf{div}\left(d\mathbf{s} \times \frac{\mathbf{r}-\mathbf{s}}{|\mathbf{r}-\mathbf{s}|^3}\right) = 0$$

by Exercise 5, since the divergence is taken with respect to $\mathbf{r}$, and so $\mathbf{s}$ and $d\mathbf{s}$ can be regarded as constant. Hence

$$\mathbf{div}\oint_{\mathcal{F}} \frac{d\mathbf{s} \times (\mathbf{r}-\mathbf{s})}{|\mathbf{r}-\mathbf{s}|^3} = \oint_{\mathcal{F}} \mathbf{div}\left(d\mathbf{s} \times \frac{\mathbf{r}-\mathbf{s}}{|\mathbf{r}-\mathbf{s}|^3}\right) = 0.$$

9. By the result of Exercise 4 and Theorem 3(e), we calculate

$$\mathbf{curl}\left(\mathbf{a} \times \frac{\mathbf{r}-\mathbf{b}}{|\mathbf{r}-\mathbf{b}|^3}\right)$$
$$= -\mathbf{curl}\left(\mathbf{a} \times \nabla\frac{1}{|\mathbf{r}-\mathbf{b}|}\right)$$
$$= -\left(\nabla \bullet \nabla\frac{1}{|\mathbf{r}-\mathbf{b}|}\right)\mathbf{a} - \left(\nabla\frac{1}{|\mathbf{r}-\mathbf{b}|} \bullet \nabla\right)\mathbf{a}$$
$$\quad + (\nabla \bullet \mathbf{a})\nabla\frac{1}{|\mathbf{r}-\mathbf{b}|} + (\mathbf{a} \bullet \nabla)\nabla\frac{1}{|\mathbf{r}-\mathbf{b}|}.$$

Observe that $\nabla \bullet \nabla\frac{1}{|\mathbf{r}-\mathbf{b}|} = 0$ for $\mathbf{r} \neq \mathbf{b}$, either by direct calculation or by noting that $\nabla\frac{1}{|\mathbf{r}-\mathbf{b}|}$ is the field of a point source at $\mathbf{r} = \mathbf{b}$ and applying the result of Example 3 of Section 16.1.

Also $-\left(\nabla\frac{1}{|\mathbf{r}-\mathbf{b}|} \bullet \nabla\right)\mathbf{a} = \mathbf{0}$ and $\nabla \bullet \mathbf{a} = 0$, since $\mathbf{a}$ is constant. Therefore we have

$$\mathbf{curl}\left(\mathbf{a} \times \frac{\mathbf{r}-\mathbf{b}}{|\mathbf{r}-\mathbf{b}|^3}\right) = (\mathbf{a} \bullet \nabla)\nabla\frac{1}{|\mathbf{r}-\mathbf{b}|}$$
$$= -(\mathbf{a} \bullet \nabla)\frac{\mathbf{r}-\mathbf{b}}{|\mathbf{r}-\mathbf{b}|^3}.$$

10. The first component of $(d\mathbf{s} \bullet \nabla)\mathbf{F}(s)$ is $\nabla F_1(s) \bullet d\mathbf{s}$. Since $\mathcal{F}$ is closed and ∇F_1 is conservative,

$$\mathbf{i} \bullet \oint_{\mathcal{F}} (d\mathbf{s} \bullet \nabla)\mathbf{F}(s) = \oint_{\mathcal{F}} \nabla F_1(s) \bullet d\mathbf{s} = 0.$$

Similarly, the other components have zero line integrals, so

$$\oint_{\mathcal{F}} (d\mathbf{s} \bullet \nabla)\mathbf{F}(s) = \mathbf{0}.$$

11. Using the results of Exercises 7 and 8, we have

$$\mathbf{curl}\oint_{\mathcal{F}} \frac{d\mathbf{s} \times (\mathbf{r}-\mathbf{s})}{|\mathbf{r}-\mathbf{s}|^3} = \oint_{\mathcal{F}} \mathbf{curl}\left(d\mathbf{s} \times \frac{\mathbf{r}-\mathbf{s}}{|\mathbf{r}-\mathbf{s}|^3}\right) = \mathbf{0}$$

for $\mathbf{r}$ not on $\mathcal{F}$. (Again, this is because the curl is taken with respect to $\mathbf{r}$, so $\mathbf{s}$ and $d\mathbf{s}$ can be regarded as constant for the calculation of the curl.)

12. By analogy with the filament case, the current in volume element dV at position $\mathbf{s}$ is $\mathbf{J}(s)\,dV$, which gives rise at position $\mathbf{r}$ to a magnetic field

$$d\mathbf{H}(\mathbf{r}) = \frac{1}{4\pi}\frac{\mathbf{J}(s) \times (\mathbf{r}-\mathbf{s})}{|\mathbf{r}-\mathbf{s}|^3}\,dV.$$

If R is a region of 3-space outside which $\mathbf{J}$ is identically zero, then at any point $\mathbf{r}$ in 3-space, the total magnetic field is

$$\mathbf{H}(\mathbf{r}) = \frac{1}{4\pi}\iiint_R \frac{\mathbf{J}(s) \times (\mathbf{r}-\mathbf{s})}{|\mathbf{r}-\mathbf{s}|^3}\,dV.$$

Now $\mathbf{A}(\mathbf{r})$ was defined to be

$$\mathbf{A}(\mathbf{r}) = \frac{1}{4\pi}\iiint_R \frac{\mathbf{J}(s)}{|\mathbf{r}-\mathbf{s}|}\,dV.$$

We have

$$\mathbf{curl}\,\mathbf{A}(\mathbf{r}) = \frac{1}{4\pi}\iiint_R \nabla_{\mathbf{r}} \times \left(\frac{1}{|\mathbf{r}-\mathbf{s}|}\mathbf{J}(s)\right)dV$$
$$= \frac{1}{4\pi}\iiint_R \nabla_{\mathbf{r}}\frac{1}{|\mathbf{r}-\mathbf{s}|} \times \mathbf{J}(s)\,dV$$
$$\text{(by Theorem 3(c))}$$
$$= -\frac{1}{4\pi}\iiint_R \frac{(\mathbf{r}-\mathbf{s}) \times \mathbf{J}(s)}{|\mathbf{r}-\mathbf{s}|^3}\,dV$$
$$\text{(by Exercise 4)}$$
$$= \mathbf{H}(\mathbf{r}).$$

13. $$\mathbf{A}(\mathbf{r}) = \frac{I}{4\pi}\oint_{\mathcal{F}} \frac{d\mathbf{s}}{|\mathbf{r}-\mathbf{s}|}$$
$$\operatorname{div}\mathbf{A}(\mathbf{r}) = \frac{I}{4\pi}\oint_{\mathcal{F}} \operatorname{div}_{\mathbf{r}}\left(\frac{1}{|\mathbf{r}-\mathbf{s}|}\,d\mathbf{s}\right)$$
$$= \frac{I}{4\pi}\oint_{\mathcal{F}} \nabla\left(\frac{1}{|\mathbf{r}-\mathbf{s}|}\right)\bullet d\mathbf{s}$$
(by Theorem 3(b))
$$= 0 \quad \text{for } \mathbf{r} \text{ not on } \mathcal{F},$$
since $\nabla(1/|\mathbf{r}-\mathbf{s}|)$ is conservative.

14. $\mathbf{A}(\mathbf{r}) = \dfrac{1}{4\pi}\iiint_R \dfrac{\mathbf{J}(\mathbf{s})\,dV}{|\mathbf{r}-\mathbf{s}|}$, where R is a region of 3-space such that $\mathbf{J}(\mathbf{s}) = \mathbf{0}$ outside R. We assume that $\mathbf{J}(\mathbf{s})$ is continuous, so $\mathbf{J}(\mathbf{s}) = \mathbf{0}$ on the surface $\mathcal{S}$ of R.
In the following calculations we use subscripts $\mathbf{s}$ and $\mathbf{r}$ to denote the variables with respect to which derivatives are taken. By Theorem 3(b),
$$\operatorname{div}_{\mathbf{s}}\frac{\mathbf{J}(\mathbf{s})}{|\mathbf{r}-\mathbf{s}|} = \left(\nabla_{\mathbf{s}}\frac{1}{|\mathbf{r}-\mathbf{s}|}\right)\bullet\mathbf{J}(\mathbf{s}) + \frac{1}{|\mathbf{r}-\mathbf{s}|}\nabla_{\mathbf{s}}\bullet\mathbf{J}(\mathbf{s})$$
$$= -\nabla_{\mathbf{r}}\left(\frac{1}{|\mathbf{r}-\mathbf{s}|}\right)\bullet\mathbf{J}(\mathbf{s}) + 0$$
because $\nabla_{\mathbf{r}}|\mathbf{r}-\mathbf{s}| = -\nabla_{\mathbf{s}}|\mathbf{r}-\mathbf{s}|$, and because $\nabla\bullet\mathbf{J} = \nabla\bullet(\nabla\times\mathbf{H}) = 0$ by Theorem 3(g). Hence
$$\operatorname{div}\mathbf{A}(\mathbf{r}) = \frac{1}{4\pi}\iiint_R\left(\nabla_{\mathbf{r}}\frac{1}{|\mathbf{r}-\mathbf{s}|}\right)\bullet\mathbf{J}(\mathbf{s})\,dV$$
$$= -\frac{1}{4\pi}\iiint_R \nabla_{\mathbf{s}}\bullet\frac{\mathbf{J}(\mathbf{s})}{|\mathbf{r}-\mathbf{s}|}\,dV$$
$$= -\frac{1}{4\pi}\oiint_{\mathcal{S}}\frac{\mathbf{J}(\mathbf{s})}{|\mathbf{r}-\mathbf{s}|}\bullet\hat{\mathbf{N}}\,dS = 0$$
since $\mathbf{J}(\mathbf{s}) = \mathbf{0}$ on $\mathcal{S}$.
By Theorem 3(i),
$$\mathbf{J} = \nabla\times\mathbf{H} = \nabla\times(\nabla\times\mathbf{A}) = \nabla(\nabla\bullet\mathbf{A}) - \nabla^2\mathbf{A} = -\nabla^2\mathbf{A}.$$

15. By Maxwell's equations, since $\rho = 0$ and $\mathbf{J} = \mathbf{0}$,
$$\operatorname{div}\mathbf{E} = 0 \qquad \operatorname{div}\mathbf{H} = 0$$
$$\operatorname{curl}\mathbf{E} = -\mu_0\frac{\partial\mathbf{H}}{\partial t} \qquad \operatorname{curl}\mathbf{H} = \epsilon_0\frac{\partial\mathbf{E}}{\partial t}$$
Therefore,
$$\operatorname{curl}\operatorname{curl}\mathbf{E} = \operatorname{grad}\operatorname{div}\mathbf{E} - \nabla^2\mathbf{E} = -\nabla^2\mathbf{E}$$
$$\nabla^2\mathbf{E} = -\operatorname{curl}\operatorname{curl}\mathbf{E} = \mu_0\frac{\partial}{\partial t}\operatorname{curl}\mathbf{H} = \mu_0\epsilon_0\frac{\partial^2\mathbf{E}}{\partial t^2}.$$
Similarly,
$$\nabla^2\mathbf{H} = \mu_0\epsilon_0\frac{\partial^2\mathbf{H}}{\partial t^2}.$$

Thus $\mathbf{U} = \mathbf{E}$ and $\mathbf{U} = \mathbf{H}$ both satisfy the wave equation
$$\frac{\partial^2\mathbf{U}}{\partial t^2} = c^2\nabla^2\mathbf{U}, \quad \text{where} \quad c^2 = \frac{1}{\mu_0\epsilon_0}.$$

16. The heat content of an arbitrary region R (with surface $\mathcal{S}$) at time t is
$$H(t) = \delta c\iiint_R T(x,y,z,t)\,dV.$$
This heat content increases at (time) rate
$$\frac{dH}{dt} = \delta c\iiint_R \frac{\partial T}{\partial t}\,dV.$$
If heat is not "created" or "destroyed" (by chemical or other means) within R, then the increase in heat content must be due to heat flowing into R across $\mathcal{S}$.
The rate of flow of heat into R across surface element dS with outward normal $\hat{\mathbf{N}}$ is
$$-k\nabla T\bullet\hat{\mathbf{N}}\,dS.$$
Therefore, the rate at which heat enters R through $\mathcal{S}$ is
$$k\oiint_{\mathcal{S}}\nabla T\bullet\hat{\mathbf{N}}\,dS.$$
By conservation of energy and the Divergence Theorem we have
$$\delta c\iiint_R \frac{\partial T}{\partial t}\,dV = k\oiint_{\mathcal{S}}\nabla T\bullet\hat{\mathbf{N}}\,dS$$
$$= k\iiint_R \nabla\bullet\nabla T\,dV$$
$$= k\iiint_R \nabla^2 T\,dV.$$
Thus, $\displaystyle\iiint_R\left(\frac{\partial T}{\partial t} - \frac{k}{\delta c}\nabla^2 T\right)dV = 0.$

Since R is arbitrary, and the temperature T is assumed to be smooth, the integrand must vanish everywhere. Thus
$$\frac{\partial T}{\partial t} = \frac{k}{\delta c}\nabla^2 T = \frac{k}{\delta c}\left[\frac{\partial^2 T}{\partial x^2} + \frac{\partial^2 T}{\partial y^2} + \frac{\partial^2 T}{\partial z^2}\right].$$

Section 16.7 Orthogonal Curvilinear Coordinates (page 976)

1. $f(r, \theta, z) = r\theta z$ (cylindrical coordinates). By Example 9,
$$\nabla f = \frac{\partial f}{\partial r}\hat{\mathbf{r}} + \frac{1}{r}\frac{\partial f}{\partial \theta}\hat{\boldsymbol{\theta}} + \frac{\partial f}{\partial z}\mathbf{k}$$
$$= \theta z\,\hat{\mathbf{r}} + z\,\hat{\boldsymbol{\theta}} + r\theta\,\mathbf{k}.$$

2. $f(\rho, \phi, \theta) = \rho\phi\theta$ (spherical coordinates). By Example 10,
$$\nabla f = \frac{\partial f}{\partial \rho}\hat{\boldsymbol{\rho}} + \frac{1}{\rho}\frac{\partial f}{\partial \phi}\hat{\boldsymbol{\phi}} + \frac{1}{\rho\sin\phi}\frac{\partial f}{\partial \theta}\hat{\boldsymbol{\theta}}$$
$$= \phi\theta\,\hat{\boldsymbol{\rho}} + \theta\,\hat{\boldsymbol{\phi}} + \frac{\phi}{\sin\phi}\hat{\boldsymbol{\theta}}.$$

3. $\mathbf{F}(r, \theta, z) = r\hat{\mathbf{r}}$
$$\operatorname{div}\mathbf{F} = \frac{1}{r}\left[\frac{\partial}{\partial r}(r^2)\right] = 2$$
$$\operatorname{curl}\mathbf{F} = \frac{1}{r}\begin{vmatrix} \hat{\mathbf{r}} & r\hat{\boldsymbol{\theta}} & \mathbf{k} \\ \frac{\partial}{\partial r} & \frac{\partial}{\partial \theta} & \frac{\partial}{\partial z} \\ r & 0 & 0 \end{vmatrix} = \mathbf{0}.$$

4. $\mathbf{F}(r, \theta, z) = r\hat{\boldsymbol{\theta}}$
$$\operatorname{div}\mathbf{F} = \frac{1}{r}\left[\frac{\partial}{\partial \theta}(r)\right] = 0$$
$$\operatorname{curl}\mathbf{F} = \frac{1}{r}\begin{vmatrix} \hat{\mathbf{r}} & r\hat{\boldsymbol{\theta}} & \mathbf{k} \\ \frac{\partial}{\partial r} & \frac{\partial}{\partial \theta} & \frac{\partial}{\partial z} \\ 0 & r^2 & 0 \end{vmatrix} = 2\mathbf{k}.$$

5. $\mathbf{F}(\rho, \phi, \theta) = \sin\phi\,\hat{\boldsymbol{\rho}}$
$$\operatorname{div}\mathbf{F} = \frac{1}{\rho^2\sin\phi}\left[\frac{\partial}{\partial \rho}(\rho^2\sin^2\phi)\right] = \frac{2\sin\phi}{\rho}$$
$$\operatorname{curl}\mathbf{F} = \frac{1}{\rho^2\sin\phi}\begin{vmatrix} \hat{\boldsymbol{\rho}} & \rho\hat{\boldsymbol{\phi}} & \rho\sin\phi\,\hat{\boldsymbol{\theta}} \\ \frac{\partial}{\partial \rho} & \frac{\partial}{\partial \phi} & \frac{\partial}{\partial \theta} \\ \sin\phi & 0 & 0 \end{vmatrix}$$
$$= -\frac{\cos\phi}{\rho}\hat{\boldsymbol{\theta}}.$$

6. $\mathbf{F}(\rho, \phi, \theta) = \rho\,\hat{\boldsymbol{\phi}}$
$$\operatorname{div}\mathbf{F} = \frac{1}{\rho^2\sin\phi}\left[\frac{\partial}{\partial \phi}(\rho^2\sin\phi)\right] = \cot\phi$$
$$\operatorname{curl}\mathbf{F} = \frac{1}{\rho^2\sin\phi}\begin{vmatrix} \hat{\boldsymbol{\rho}} & \rho\hat{\boldsymbol{\phi}} & \rho\sin\phi\,\hat{\boldsymbol{\theta}} \\ \frac{\partial}{\partial \rho} & \frac{\partial}{\partial \phi} & \frac{\partial}{\partial \theta} \\ 0 & \rho^2 & 0 \end{vmatrix} = 2\hat{\boldsymbol{\theta}}.$$

7. $\mathbf{F}(\rho, \phi, \theta) = \rho\,\hat{\boldsymbol{\theta}}$
$$\operatorname{div}\mathbf{F} = \frac{1}{\rho^2\sin\phi}\left[\frac{\partial}{\partial \theta}(\rho^2)\right] = 0$$
$$\operatorname{curl}\mathbf{F} = \frac{1}{\rho^2\sin\phi}\begin{vmatrix} \hat{\boldsymbol{\rho}} & \rho\hat{\boldsymbol{\phi}} & \rho\sin\phi\,\hat{\boldsymbol{\theta}} \\ \frac{\partial}{\partial \rho} & \frac{\partial}{\partial \phi} & \frac{\partial}{\partial \theta} \\ 0 & 0 & \rho^2\sin\phi \end{vmatrix}$$
$$= \cot\phi\,\hat{\boldsymbol{\rho}} - 2\hat{\boldsymbol{\phi}}.$$

8. $\mathbf{F}(\rho, \phi, \theta) = \rho^2\,\hat{\boldsymbol{\rho}}$
$$\operatorname{div}\mathbf{F} = \frac{1}{\rho^2\sin\phi}\left[\frac{\partial}{\partial \rho}(\rho^4\sin\phi)\right] = 4\rho$$
$$\operatorname{curl}\mathbf{F} = \frac{1}{\rho^2\sin\phi}\begin{vmatrix} \hat{\boldsymbol{\rho}} & \rho\hat{\boldsymbol{\phi}} & \rho\sin\phi\,\hat{\boldsymbol{\theta}} \\ \frac{\partial}{\partial \rho} & \frac{\partial}{\partial \phi} & \frac{\partial}{\partial \theta} \\ \rho^2 & 0 & 0 \end{vmatrix} = \mathbf{0}.$$

9. Let $\mathbf{r} = x(u, v)\mathbf{i} + y(u, v)\mathbf{j}$. The scale factors are
$$h_u = \left|\frac{\partial \mathbf{r}}{\partial u}\right| \quad \text{and} \quad h_v = \left|\frac{\partial \mathbf{r}}{\partial v}\right|.$$
The local basis consists of the vectors
$$\hat{\mathbf{u}} = \frac{1}{h_u}\frac{\partial \mathbf{r}}{\partial u} \quad \text{and} \quad \hat{\mathbf{v}} = \frac{1}{h_v}\frac{\partial \mathbf{r}}{\partial v}.$$
The area element is $dA = h_u h_v\,du\,dv$.

10. Since (u, v, z) constitute orthogonal curvilinear coordinates in $\mathbb{R}^3$, with scale factors h_u, h_v and $h_z = 1$, we have, for a function $f(u, v)$ independent of z,
$$\nabla f(u, v) = \frac{1}{h_u}\frac{\partial f}{\partial u}\hat{\mathbf{u}} + \frac{1}{h_v}\frac{\partial f}{\partial v}\hat{\mathbf{v}} + \frac{1}{1}\frac{\partial f}{\partial z}\mathbf{k}$$
$$= \frac{1}{h_u}\frac{\partial f}{\partial u}\hat{\mathbf{u}} + \frac{1}{h_v}\frac{\partial f}{\partial v}\hat{\mathbf{v}}.$$
For $\mathbf{F}(u, v) = F_u(u, v)\hat{\mathbf{u}} + F_v(u, v)\hat{\mathbf{v}}$ (independent of z and having no $\mathbf{k}$ component), we have
$$\operatorname{div}\mathbf{F}(u, v) = \frac{1}{h_u h_v}\left[\frac{\partial}{\partial u}(h_u F_u) + \frac{\partial}{\partial v}(h_v F_v)\right]$$
$$\operatorname{curl}\mathbf{F}(u, v) = \frac{1}{h_u h_v}\begin{vmatrix} h_u\hat{\mathbf{u}} & h_v\hat{\mathbf{v}} & \mathbf{k} \\ \frac{\partial}{\partial u} & \frac{\partial}{\partial v} & \frac{\partial}{\partial z} \\ h_u F_u & h_v F_v & 0 \end{vmatrix}$$
$$= \frac{1}{h_u h_v}\left[\frac{\partial}{\partial u}(h_v F_v) - \frac{\partial}{\partial v}(h_u F_u)\right]\mathbf{k}.$$

11. We can use the expressions calculated in the text for cylindrical coordinates, applied to functions independent of z and having no $\mathbf{k}$ components:
$$\nabla f(r, \theta) = \frac{\partial f}{\partial r}\hat{\mathbf{r}} + \frac{1}{r}\frac{\partial f}{\partial \theta}\hat{\boldsymbol{\theta}}$$
$$\operatorname{div}\mathbf{F}(r, \theta) = \frac{\partial F_r}{\partial r} + \frac{F_r}{r} + \frac{1}{r}\frac{\partial F_\theta}{\partial \theta}$$
$$\operatorname{curl}\mathbf{F}(r, \theta) = \left[\frac{\partial F_\theta}{\partial r} + \frac{F_\theta}{r} - \frac{1}{r}\frac{\partial F_r}{\partial \theta}\right]\mathbf{k}.$$

12. $x = a \cosh u \cos v$, $y = a \sinh u \sin v$.

a) u-curves: If $A = a \cosh u$ and $B = a \sinh u$, then
$$\frac{x^2}{A^2} + \frac{y^2}{B^2} = \cos^2 v + \sin^2 v = 1.$$
Since $A^2 - B^2 = a^2(\cosh^2 u - \sinh^2 u) = a^2$, the u-curves are ellipses with foci at $(\pm a, 0)$.

b) v-curves: If $A = a \cos v$ and $B = a \sin v$, then
$$\frac{x^2}{A^2} - \frac{y^2}{B^2} = \cosh^2 u - \sinh^2 u = 1.$$
Since $A^2 + B^2 = a^2(\cos^2 v + \sin^2 v) = a^2$, the v-curves are hyperbolas with foci at $(\pm a, 0)$.

c) The u-curve $u = u_0$ has parametric equations
$$x = a \cosh u_0 \cos v, \qquad y = a \sinh u_0 \sin v,$$
and therefore has slope at (u_0, v_0) given by
$$m_u = \frac{dy}{dx} = \frac{dy}{dv} \bigg/ \frac{dx}{dv}\bigg|_{(u_0,v_0)} = \frac{a \sinh u_0 \cos v_0}{-a \cosh u_0 \sin v_0}.$$

The v-curve $v = v_0$ has parametric equations
$$x = a \cosh u \cos v_0, \qquad y = a \sinh u \sin v_0,$$
and therefore has slope at (u_0, v_0) given by
$$m_v = \frac{dy}{dx} = \frac{dy}{du} \bigg/ \frac{dx}{du}\bigg|_{(u_0,v_0)} = \frac{a \cosh u_0 \sin v_0}{a \sinh u_0 \cos v_0}.$$

Since the product of these slopes is $m_u m_v = -1$, the curves $u = u_0$ and $v = v_0$ intersect at right angles.

d) $\mathbf{r} = a \cosh u \cos v \, \mathbf{i} + a \sinh u \sin v \, \mathbf{j}$
$$\frac{\partial \mathbf{r}}{\partial u} = a \sinh u \cos v \, \mathbf{i} + a \cosh u \sin v \, \mathbf{j}$$
$$\frac{\partial \mathbf{r}}{\partial v} = -a \cosh u \sin v \, \mathbf{i} + a \sinh u \cos v \, \mathbf{j}.$$

The scale factors are
$$h_u = \left|\frac{\partial \mathbf{r}}{\partial u}\right| = a\sqrt{\sinh^2 u \cos^2 v + \cosh^2 u \sin^2 v}$$
$$h_v = \left|\frac{\partial \mathbf{r}}{\partial v}\right| = a\sqrt{\sinh^2 u \cos^2 v + \cosh^2 u \sin^2 v} = h_u.$$

The area element is
$$dA = h_u h_v \, du \, dv$$
$$= a^2 \left(\sinh^2 u \cos^2 v + \cosh^2 u \sin^2 v\right) du \, dv.$$

13. $x = a \cosh u \cos v$
$y = a \sinh u \sin v$
$z = z$.
Using the result of Exercise 12, we see that the coordinate surfaces are
$u = u_0$: vertical elliptic cylinders with focal axes $x = \pm a$, $y = 0$.
$v = v_0$: vertical hyperbolic cylinders with focal axes $x = \pm a$, $y = 0$.
$z = z_0$: horizontal planes.

The coordinate curves are
u-curves: the horizontal hyperbolas in which the $v = v_0$ cylinders intersect the $z = z_0$ planes.
v-curves: the horizontal ellipses in which the $u = u_0$ cylinders intersect the $z = z_0$ planes.
z-curves: sets of four vertical straight lines where the elliptic cylinders $u = u_0$ and hyperbolic cylinders $v = v_0$ intersect.

14. $\nabla f(r, \theta, z) = \frac{\partial f}{\partial r} \hat{\mathbf{r}} + \frac{1}{r} \frac{\partial f}{\partial \theta} \hat{\boldsymbol{\theta}} + \frac{\partial f}{\partial z} \mathbf{k}$
$\nabla^2 f(r, \theta, z) = \text{div}\left(\nabla f(r, \theta, z)\right)$
$$= \frac{1}{r}\left[\frac{\partial}{\partial r}\left(r\frac{\partial f}{\partial r}\right) + \frac{\partial}{\partial \theta}\left(\frac{1}{r}\frac{\partial f}{\partial \theta}\right) + \frac{\partial}{\partial z}\left(r\frac{\partial f}{\partial z}\right)\right]$$
$$= \frac{\partial^2 f}{\partial r^2} + \frac{1}{r}\frac{\partial f}{\partial r} + \frac{1}{r^2}\frac{\partial^2 f}{\partial \theta^2} + \frac{\partial^2 f}{\partial z^2}.$$

15. $\nabla f(\rho, \phi, \theta) = \frac{\partial f}{\partial \rho} \hat{\boldsymbol{\rho}} + \frac{1}{\rho}\frac{\partial f}{\partial \phi} \hat{\boldsymbol{\phi}} + \frac{1}{\rho \sin \phi}\frac{\partial f}{\partial \theta} \hat{\boldsymbol{\theta}}$
$\nabla^2 f(\rho, \phi, \theta) = \text{div}\left(f(\rho, \phi, \theta)\right)$
$$= \frac{1}{\rho^2 \sin \phi}\left[\frac{\partial}{\partial \rho}\left(\rho^2 \sin \phi \frac{\partial f}{\partial \rho}\right) + \frac{\partial}{\partial \phi}\left(\rho \sin \phi \frac{1}{\rho}\frac{\partial f}{\partial \phi}\right)\right.$$
$$\left. + \frac{\partial}{\partial \theta}\left(\frac{\rho}{\rho \sin \phi}\frac{\partial f}{\partial \theta}\right)\right]$$
$$= \frac{\partial^2 f}{\partial \rho^2} + \frac{2}{\rho}\frac{\partial f}{\partial \rho} + \frac{1}{\rho^2}\frac{\partial^2 f}{\partial \phi^2}$$
$$+ \frac{\cot \phi}{\rho^2}\frac{\partial f}{\partial \phi} + \frac{1}{\rho^2 \sin^2 \phi}\frac{\partial^2 f}{\partial \theta^2}.$$

16. $\nabla f(u, v, w) = \frac{1}{h_u} \frac{\partial f}{\partial u} \hat{\mathbf{u}} + \frac{1}{h_v} \frac{\partial f}{\partial v} \hat{\mathbf{v}} + \frac{1}{h_w} \frac{\partial f}{\partial w} \hat{\mathbf{w}}$

$\nabla^2 f(u, v, w) = \text{div}\left(\nabla f(u, v, w)\right)$

$= \frac{1}{h_u h_v h_w} \left[\frac{\partial}{\partial u}\left(\frac{h_v h_w}{h_u} \frac{\partial f}{\partial u}\right) + \frac{\partial}{\partial v}\left(\frac{h_u h_w}{h_v} \frac{\partial f}{\partial v}\right) \right.$

$\left. + \frac{\partial}{\partial w}\left(\frac{h_u h_v}{h_w} \frac{\partial f}{\partial w}\right) \right]$

$= \frac{1}{h_u^2}\left[\frac{\partial^2 f}{\partial u^2} + \left(\frac{1}{h_v}\frac{\partial h_v}{\partial u} + \frac{1}{h_w}\frac{\partial h_w}{\partial u} - \frac{1}{h_u}\frac{\partial h_u}{\partial u}\right)\frac{\partial f}{\partial u}\right]$

$+ \frac{1}{h_v^2}\left[\frac{\partial^2 f}{\partial v^2} + \left(\frac{1}{h_u}\frac{\partial h_u}{\partial v} + \frac{1}{h_w}\frac{\partial h_w}{\partial v} - \frac{1}{h_v}\frac{\partial h_v}{\partial v}\right)\frac{\partial f}{\partial v}\right]$

$+ \frac{1}{h_w^2}\left[\frac{\partial^2 f}{\partial w^2} + \left(\frac{1}{h_u}\frac{\partial h_u}{\partial w} + \frac{1}{h_v}\frac{\partial h_v}{\partial w} - \frac{1}{h_w}\frac{\partial h_w}{\partial w}\right)\frac{\partial f}{\partial w}\right]$.

Review Exercises 16 (page 977)

1. The semi-ellipsoid $\mathcal{S}$ with upward normal $\hat{\mathbf{N}}$ specified in the problem and the disk D given by $x^2 + y^2 \le 16$, $z = 0$, with downward normal $-\mathbf{k}$ together bound the solid region R: $0 \le z \le \frac{1}{2}\sqrt{16 - x^2 - y^2}$. By the Divergence Theorem:

$$\iint_{\mathcal{S}} \mathbf{F} \cdot \hat{\mathbf{N}} \, dS + \iint_D \mathbf{F} \cdot (-\mathbf{k}) \, dA = \iiint_R \text{div}\,\mathbf{F} \, dV.$$

For $\mathbf{F} = x^2 z \mathbf{i} + (y^2 z + 3y)\mathbf{j} + x^2 \mathbf{k}$ we have

$\iiint_R \text{div}\,\mathbf{F}\, dV = \iiint_R (2xz + 2yz + 3)\, dV$

$= 0 + 0 + 3 \iiint_R dV = 3 \times (\text{volume of } R)$

$= \frac{3}{2} \frac{4}{3} \pi 4^2 2 = 64\pi$.

The flux of $\mathbf{F}$ across $\mathcal{S}$ is

$\iint_{\mathcal{S}} \mathbf{F} \cdot \hat{\mathbf{N}} \, dS = 64\pi + \iint_D \mathbf{F} \cdot \mathbf{k}\, dA$

$= 64\pi + \iint_D x^2 \, dA$

$= 64\pi + \int_0^{2\pi} \cos^2\theta\, d\theta \int_0^4 r^3\, dr = 128\pi$.

2. Let R be the region inside the cylinder $\mathcal{S}$ and between the planes $z = 0$ and $z = b$. The oriented boundary of R consists of $\mathcal{S}$ and the disks D_1 with normal $\hat{\mathbf{N}}_1 = \mathbf{k}$ and D_2 with normal $\hat{\mathbf{N}}_2 = -\mathbf{k}$ as shown in the figure. For $\mathbf{F} = x\mathbf{i} + \cos(z^2)\mathbf{j} + e^z \mathbf{k}$ we have $\text{div}\,\mathbf{F} = 1 + e^z$ and

$\iiint_R \text{div}\,\mathbf{F}\, dV = \iint_{D_2} dx\, dy \int_0^b (1 + e^z)\, dz$

$= \iint_{D_2} [b + (e^b - 1)]\, dx\, dy$

$= \pi a^2 b + \pi a^2 (e^b - 1).$

Also $\iint_{D_2} \mathbf{F} \cdot (-\mathbf{k})\, dA = -\iint_{D_2} e^0\, dA = -\pi a^2$

$\iint_{D_1} \mathbf{F} \cdot \mathbf{k}\, dA = \iint_{D_1} e^b\, dA = \pi a^2 e^b.$

By the Divergence Theorem

$\iint_{\mathcal{S}} \mathbf{F} \cdot \hat{\mathbf{N}}\, dS + \iint_{D_1} \mathbf{F} \cdot \mathbf{k}\, dA + \iint_{D_2} \mathbf{F} \cdot (-\mathbf{k})\, dA$

$= \iiint_R \text{div}\,\mathbf{F}\, dV = \pi a^2 b + \pi a^2(e^b - 1).$

Therefore, $\iint_{\mathcal{S}} \mathbf{F} \cdot \hat{\mathbf{N}}\, dS = \pi a^2 b.$

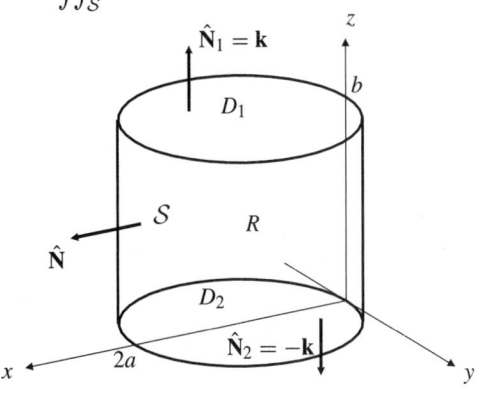

Fig. R-16.2

3. $\oint_{\mathcal{C}} (3y^2 + 2xe^{y^2})\, dx + (2x^2 y e^{y^2})\, dy$

$= \iint_P [4xye^{y^2} - (6y + 4xye^{y^2})]\, dA$

$= -6 \iint_P y\, dA = -6\bar{y}A = -6,$

since P has area $A = 2$ and its centroid has y-coordinate $\bar{y} = 1/2$.

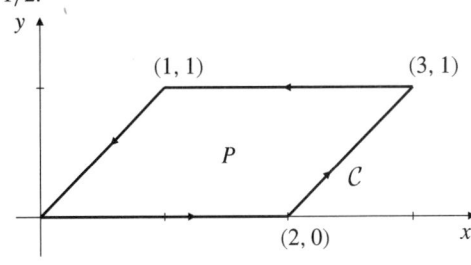

Fig. R-16.3

4. If $\mathbf{F} = -z\mathbf{i} + x\mathbf{j} + y\mathbf{k}$, then

$\text{curl}\,\mathbf{F} = \begin{vmatrix} \mathbf{i} & \mathbf{j} & \mathbf{k} \\ \frac{\partial}{\partial x} & \frac{\partial}{\partial y} & \frac{\partial}{\partial z} \\ -z & x & y \end{vmatrix} = \mathbf{i} - \mathbf{j} + \mathbf{k}.$

The unit normal $\hat{\mathbf{N}}$ to a region in the plane $2x+y+2z=7$ is
$$\hat{\mathbf{N}} = \pm \frac{2\mathbf{i}+\mathbf{j}+2\mathbf{k}}{3}.$$

If $\mathcal{C}$ is the boundary of a disk D of radius a in that plane, then
$$\oint_{\mathcal{C}} \mathbf{F} \bullet d\mathbf{r} = \iint_D \operatorname{curl} \mathbf{F} \bullet \hat{\mathbf{N}} \, dS$$
$$= \pm \iint_D \frac{2-1+2}{3} \, dS = \pm \pi a^2.$$

5. If $\mathcal{S}_a$ is the sphere of radius a centred at the origin, then
$$\operatorname{div} \mathbf{F}(0,0,0) = \lim_{a \to 0+} \frac{1}{\frac{4}{3}\pi a^3} \oiint_{\mathcal{S}_a} \mathbf{F} \bullet \hat{\mathbf{N}} \, dS$$
$$= \lim_{a \to 0+} \frac{3}{4\pi a^3}(\pi a^3 + 2a^4) = \frac{3}{4}.$$

6. If $\mathcal{S}$ is any surface with upward normal $\hat{\mathbf{N}}$ and boundary the curve $\mathcal{C}: x^2+y^2=1, z=2$, then $\mathcal{C}$ is oriented counterclockwise as seen from above, and it has parametrization
$$\mathbf{r} = \cos t\, \mathbf{i} + \sin t\, \mathbf{j} + 2\mathbf{k} \quad (0 \le 2 \le 2\pi).$$

Thus $d\mathbf{r} = (-\sin t\, \mathbf{i} + \cos t\, \mathbf{j})\, dt$, and if $\mathbf{F} = -y\mathbf{i} + x\cos(1-x^2-y^2)\mathbf{j} + yz\mathbf{k}$, then the flux of $\operatorname{curl} \mathbf{F}$ upward through $\mathcal{S}$ is
$$\iint_{\mathcal{S}} \operatorname{curl} \mathbf{F} \bullet \hat{\mathbf{N}} \, dS = \oint_{\mathcal{C}} \mathbf{F} \bullet d\mathbf{r}$$
$$= \int_0^{2\pi} (\sin^2 t + \cos^2 t + 0)\, dt = 2\pi.$$

7. $\mathbf{F}(\mathbf{r}) = r^\lambda \mathbf{r}$ where $\mathbf{r} = x\mathbf{i}+y\mathbf{j}+z\mathbf{k}$ and $r=|\mathbf{r}|$. Since $r^2 = x^2+y^2+z^2$, therefore $\partial r/\partial x = x/r$ and
$$\frac{\partial}{\partial x}(r^\lambda x) = \lambda r^{\lambda-1}\frac{x^2}{r} + r^\lambda = r^{\lambda-2}(\lambda x^2 + r^2).$$

Similar expressions hold for $(\partial/\partial y)(r^\lambda y)$ and $(\partial/\partial z)(r^\lambda z)$, so
$$\operatorname{div} \mathbf{F}(\mathbf{r}) = r^{\lambda-2}(\lambda r^2 + 3r^2) = (\lambda+3)r^\lambda.$$

$\mathbf{F}$ is solenoidal on any set in $\mathbb{R}^3$ that excludes the origin if an only if $\lambda = -3$. In this case $\mathbf{F}$ is not defined at $\mathbf{r}=\mathbf{0}$. There is no value of λ for which $\mathbf{F}$ is solenoidal on all of $\mathbb{R}^3$.

8. If $\operatorname{curl} \mathbf{F} = \mu \mathbf{F}$ on $\mathbb{R}^3$, where $\mu \ne 0$ is a constant, then
$$\operatorname{div} \mathbf{F} = \frac{1}{\mu} \operatorname{div} \operatorname{curl} \mathbf{F} = 0$$

by Theorem 3(g) of Section 16.2. By part (i) of the same theorem,
$$\nabla^2 \mathbf{F} = \nabla(\operatorname{div} \mathbf{F}) - \operatorname{curl} \operatorname{curl} \mathbf{F}$$
$$= 0 - \mu \operatorname{curl} \mathbf{F} = -\mu^2 \mathbf{F}.$$

Thus $\nabla^2 \mathbf{F} + \mu^2 \mathbf{F} = \mathbf{0}$.

9. Apply the variant of the Divergence Theorem given in Theorem 7(b) of Section 16.3, namely
$$\iiint_P \operatorname{\mathbf{grad}} \phi \, dV = \oiint_{\mathcal{S}} \phi \hat{\mathbf{N}} \, dS,$$
to the scalar field $\phi = 1$ over the polyhedron P. Here $\mathcal{S} = \bigcup_{i=1}^n F_i$ is the surface of P, oriented with outward normal field $\hat{\mathbf{N}}_i$ on the face F_i. If $\mathbf{N}_i = A_i \hat{\mathbf{N}}_i$, where A_i is the area of F_i, then, since $\operatorname{\mathbf{grad}} \phi = \mathbf{0}$, we have
$$\mathbf{0} = \oiint_{\mathcal{S}} \hat{\mathbf{N}} \, dS = \sum_{i=1}^n \iint_{F_i} \frac{\mathbf{N}_i}{A_i} \, dS = \sum_{i=1}^n \frac{\mathbf{N}_i}{A_i} A_i = \sum_{i=1}^n \mathbf{N}_i.$$

10. Let $\mathcal{C}$ be a simple, closed curve in the xy-plane bounding a region R. If
$$\mathbf{F} = (2y^3 - 3y + xy^2)\mathbf{i} + (x - x^3 + x^2 y)\mathbf{j},$$
then by Green's Theorem, the circulation of $\mathbf{F}$ around $\mathcal{C}$ is
$$\oint_{\mathcal{C}} \mathbf{F} \bullet d\mathbf{r}$$
$$= \iint_R \left[\frac{\partial}{\partial x}(x - x^3 + x^2 y) - \frac{\partial}{\partial y}(2y^3 - 3y + xy^2) \right] dA$$
$$= \iint_R (1 - 3x^2 + 2xy - 6y^2 + 3 - 2xy)\, dA$$
$$= \iint_R (4 - 3x^2 - 6y^2)\, dx\, dy.$$

The last integral has a maximum value when the region R is bounded by the ellipse $3x^2 + 6y^2 = 4$, oriented counterclockwise; this is the largest region in the xy-plane where the integrand is nonnegative.

11. Let $\mathcal{S}$ be a closed, oriented surface in $\mathbb{R}^3$ bounding a region R, and having outward normal field $\hat{\mathbf{N}}$. If
$$\mathbf{F} = (4x + 2x^3 z)\mathbf{i} - y(x^2 + z^2)\mathbf{j} - (3x^2 z^2 + 4y^2 z)\mathbf{k},$$
then by the Divergence Theorem, the flux of $\mathbf{F}$ through $\mathcal{S}$ is
$$\oiint_{\mathcal{S}} \mathbf{F} \bullet \hat{\mathbf{N}} \, dS = \iiint_R \operatorname{div} \mathbf{F} \, dV = \iiint_R (4 - x^2 - 4y^2 - z^2)\, dV.$$

The last integral has a maximum value when the region R is bounded by the ellipsoid $x^2 + 4y^2 + z^2 = 4$ with outward normal; this is the largest region in $\mathbb{R}^3$ where the integrand is nonnegative.

12. Let $\mathcal{C}$ be a simple, closed curve on the plane $x + y + z = 1$, oriented counterclockwise as seen from above, and bounding a plane region $\mathcal{S}$ on $x + y + z = 1$. Then $\mathcal{S}$ has normal $\hat{\mathbf{N}} = (\mathbf{i} + \mathbf{j} + \mathbf{k})/\sqrt{3}$. If $\mathbf{F} = xy^2\mathbf{i} + (3z - xy^2)\mathbf{j} + (4y - x^2 y)\mathbf{k}$, then

$$\operatorname{curl}\mathbf{F} = \begin{vmatrix} \mathbf{i} & \mathbf{j} & \mathbf{k} \\ \dfrac{\partial}{\partial x} & \dfrac{\partial}{\partial y} & \dfrac{\partial}{\partial z} \\ xy^2 & 3z - xy^2 & 4y - x^2 y \end{vmatrix}$$
$$= (1 - x^2)\mathbf{i} + 2xy\mathbf{j} - (y^2 + 2xy)\mathbf{k}.$$

By Stokes's Theorem we have

$$\oint_{\mathcal{C}} \mathbf{F} \cdot d\mathbf{r} = \iint_{\mathcal{S}} \operatorname{curl}\mathbf{F} \cdot \hat{\mathbf{N}} \, dS = \iint_{\mathcal{S}} \frac{1 - x^2 - y^2}{\sqrt{3}} \, dS.$$

The last integral will be maximum if the projection of $\mathcal{S}$ onto the xy-plane is the disk $x^2 + y^2 \le 1$. This maximum value is

$$\iint_{x^2 + y^2 \le 1} \frac{1 - x^2 - y^2}{\sqrt{3}} \sqrt{3} \, dx \, dy$$
$$= \int_0^{2\pi} d\theta \int_0^1 (1 - r^2) r \, dr = 2\pi \left(\frac{1}{2} - \frac{1}{4} \right) = \frac{\pi}{2}.$$

Challenging Problems 16 (page 978)

1. By Theorem 1 of Section 16.1, we have

$$\operatorname{div}\mathbf{v}(\mathbf{r}_1) = \lim_{\epsilon \to 0+} \frac{3}{4\pi \epsilon^3} \oiint_{\mathcal{S}_\epsilon} \mathbf{v}(\mathbf{r}) \cdot \hat{\mathbf{N}}(\mathbf{r}) \, dS.$$

Here $\mathcal{S}_\epsilon$ is the sphere of radius ϵ centred at the point (with position vector) $\mathbf{r}_1$ and having outward normal field $\hat{\mathbf{N}}(\mathbf{r})$. If $\mathbf{r}$ is (the position vector of) any point on $\mathcal{S}_\epsilon$, then $\mathbf{r} = \mathbf{r}_1 + \epsilon \hat{\mathbf{N}}(\mathbf{r})$, and

$$\oiint_{\mathcal{S}_\epsilon} \mathbf{v}(\mathbf{r}) \cdot \hat{\mathbf{N}}(\mathbf{r}) \, dS$$
$$= \oiint_{\mathcal{S}_\epsilon} \big[\mathbf{v}(\mathbf{r}_1) + \big(\mathbf{v}(\mathbf{r}) - \mathbf{v}(\mathbf{r}_1)\big)\big] \cdot \hat{\mathbf{N}}(\mathbf{r}) \, dS$$
$$= \mathbf{v}(\mathbf{r}_1) \cdot \oiint_{\mathcal{S}_\epsilon} \hat{\mathbf{N}}(\mathbf{r}) \, dS$$
$$+ \oiint_{\mathcal{S}_\epsilon} \big(\mathbf{v}(\mathbf{r}) - \mathbf{v}(\mathbf{r}_1)\big) \cdot \frac{\mathbf{r} - \mathbf{r}_1}{\epsilon} \, dS.$$

But $\oiint_{\mathcal{S}_\epsilon} \hat{\mathbf{N}}(\mathbf{r}) \, dS = \mathbf{0}$ by Theorem 7(b) of Section 16.3 with $\phi = 1$. Also, since $\mathbf{v}$ satisfies

$$\mathbf{v}(\mathbf{r}_2) - \mathbf{v}(\mathbf{r}_1) = C |\mathbf{r}_2 - \mathbf{r}_1|^2,$$

we have

$$\oiint_{\mathcal{S}_\epsilon} \big(\mathbf{v}(\mathbf{r}) - \mathbf{v}(\mathbf{r}_1)\big) \cdot \frac{\mathbf{r} - \mathbf{r}_1}{\epsilon} \, dS$$
$$= \oiint_{\mathcal{S}_\epsilon} \frac{C \epsilon^2}{\epsilon} \, dS = 4\pi C \epsilon^3.$$

Thus

$$\operatorname{div}\mathbf{v}(\mathbf{r}_1) = \lim_{\epsilon \to 0+} \frac{3}{4\pi \epsilon^3} (0 + 4\pi C \epsilon^3) = 3C.$$

The divergence of the large-scale velocity field of matter in the universe is three times Hubble's constant C.

2. a) The steradian measure of a half-cone of semi-vertical angle α is

$$\int_0^{2\pi} d\theta \int_0^\alpha \sin\phi \, d\phi = 2\pi(1 - \cos\alpha).$$

b) If $\mathcal{S}$ is the intersection of a smooth surface with the general half-cone K, and is oriented with normal field $\hat{\mathbf{N}}$ pointing away from the vertex P of K, and if $\mathcal{S}_a$ is the intersection with K of a sphere of radius a centred at P, with a chosen so that $\mathcal{S}$ and $\mathcal{S}_a$ do not intersect in K, then $\mathcal{S}$, $\mathcal{S}_a$, and the walls of K bound a solid region R that does not contain the origin. If $\mathbf{F} = \mathbf{r}/|\mathbf{r}|^3$, then $\operatorname{div}\mathbf{F} = 0$ in R (see Example 3 in Section 16.1), and $\mathbf{F} \cdot \hat{\mathbf{N}} = 0$ on the walls of K. It follows from the Divergence Theorem applied to $\mathbf{F}$ over R that

$$\iint_{\mathcal{S}} \mathbf{F} \cdot \hat{\mathbf{N}} \, dS = \iint_{\mathcal{S}_a} \mathbf{F} \cdot \frac{\mathbf{r}}{|\mathbf{r}|} \, dS$$
$$= \frac{a^2}{a^4} \iint_{\mathcal{S}_a} dS = \frac{1}{a^2} (\text{area of } \mathcal{S}_a)$$
$$= \text{area of } \mathcal{S}_1.$$

The area of $\mathcal{S}_1$ (the part of the sphere of radius 1 in K) is the measure (in steradians) of the solid angle subtended by K at its vertex P. Hence this measure is given by

$$\iint_{\mathcal{S}} \frac{\mathbf{r}}{|\mathbf{r}|^3} \cdot \hat{\mathbf{N}} \, dS.$$

3. a) Verification of the identity

$$\frac{\partial}{\partial t}\left(\mathbf{G} \cdot \frac{\partial \mathbf{r}}{\partial s}\right) - \frac{\partial}{\partial s}\left(\mathbf{G} \cdot \frac{\partial \mathbf{r}}{\partial t}\right)$$
$$= \frac{\partial \mathbf{F}}{\partial t} \cdot \frac{\partial \mathbf{r}}{\partial s} + \big((\nabla \times \mathbf{F}) \times \frac{\partial \mathbf{r}}{\partial t}\big) \cdot \frac{\partial \mathbf{r}}{\partial s}.$$

601

can be carried out using the following MapleV commands:

```
with(linalg):
F:=(x,y,z,t)->
[F1(x,y,z,t),F2(x,y,z,t),F3(x,y,z,t)];
r:=(s,t)->[x(s,t),y(s,t),z(s,t)];
G:=(s,t)->F(x(s,t),y(s,t),z(s,t),t);
g:=(s,t)->
dotprod(G(s,t),map(diff,r(s,t),s));
h:=(s,t)->
dotprod(G(s,t),map(diff,r(s,t),t));
LH1:=diff(g(s,t),t);
LH2:=diff(h(s,t),s);
LHS:=simplify(LH1-LH2);
RH1:=dotprod(subs(x=x(s,t),y=y(s,t),
z=z(s,t),diff(F(x,y,z,t),t)),
diff(r(s,t),s));
RH2:=dotprod(crossprod(subs(x=x(s,t),
y=y(s,t),z=z(s,t),
curl(F(x,y,z,t),[x,y,z])),
diff(r(s,t),t)),diff(r(s,t),s));
RHS:=RH1+RH2;
LHS-RHS;
simplify(");
```

We omit the output here; some of the commands produce screenfulls of output. The output of the final command is 0, indicating that the identity is valid.

b) As suggested by the hint,

$$\frac{d}{dt}\int_{C_t} \mathbf{F} \bullet d\mathbf{r} = \int_a^b \frac{\partial}{\partial t}\left(\mathbf{G} \bullet \frac{\partial \mathbf{r}}{\partial s}\right) ds$$

$$= \int_a^b \left[\frac{\partial}{\partial s}\left(\mathbf{G} \bullet \frac{\partial \mathbf{r}}{\partial t}\right)\right.$$

$$\left.+ \left(\frac{\partial}{\partial t}\left(\mathbf{G} \bullet \frac{\partial \mathbf{r}}{\partial s}\right) - \frac{\partial}{\partial s}\left(\mathbf{G} \bullet \frac{\partial \mathbf{r}}{\partial t}\right)\right)\right] ds$$

$$= \mathbf{G} \bullet \frac{\partial \mathbf{r}}{\partial t}\bigg|_{s=a}^{s=b}$$

$$+ \int_a^b \left[\frac{\partial \mathbf{F}}{\partial t} + \left((\nabla \times \mathbf{F}) \times \frac{\partial \mathbf{r}}{\partial t}\right)\right] \bullet \frac{\partial \mathbf{r}}{\partial s} ds$$

$$= \mathbf{F}(\mathbf{r}(b,t),t) \bullet \mathbf{v}_C(b,t) - \mathbf{F}(\mathbf{r}(a,t),t) \bullet \mathbf{v}_C(a,t)$$

$$+ \int_{C_t} \frac{\partial \mathbf{F}}{\partial t} \bullet d\mathbf{r} + \int_{C_t} ((\nabla \times \mathbf{F}) \times \mathbf{v}_C) \bullet d\mathbf{r}.$$

4. a) Verification of the identity

$$\frac{\partial}{\partial t}\left(\mathbf{G} \bullet \left[\frac{\partial \mathbf{r}}{\partial u} \times \frac{\partial \mathbf{r}}{\partial v}\right]\right) - \frac{\partial}{\partial u}\left(\mathbf{G} \bullet \left[\frac{\partial \mathbf{r}}{\partial t} \times \frac{\partial \mathbf{r}}{\partial v}\right]\right)$$
$$- \frac{\partial}{\partial v}\left(\mathbf{G} \bullet \left[\frac{\partial \mathbf{r}}{\partial u} \times \frac{\partial \mathbf{r}}{\partial t}\right]\right)$$
$$= \frac{\partial \mathbf{F}}{\partial t} \bullet \left[\frac{\partial \mathbf{r}}{\partial u} \times \frac{\partial \mathbf{r}}{\partial v}\right] + (\nabla \bullet \mathbf{F})\frac{\partial \mathbf{r}}{\partial t} \bullet \left[\frac{\partial \mathbf{r}}{\partial u} \times \frac{\partial \mathbf{r}}{\partial v}\right].$$

can be carried out using the following MapleV commands:

```
with(linalg):
F:=(x,y,z,t)->
[F1(x,y,z,t),F2(x,y,z,t),F3(x,y,z,t)];
r:=(u,v,t)
->[x(u,v,t),y(u,v,t),z(u,v,t)];
ru:=(u,v,t)->diff(r(u,v,t),u);
rv:=(u,v,t)->diff(r(u,v,t),v);
rt:=(u,v,t)->diff(r(u,v,t),t);
G:=(u,v,t)
->F(x(u,v,t),y(u,v,t),z(u,v,t),t);
ruxv:=(u,v,t)
->crossprod(ru(u,v,t),rv(u,v,t));
rtxv:=(u,v,t)
->crossprod(rt(u,v,t),rv(u,v,t));
ruxt:=(u,v,t)
->crossprod(ru(u,v,t),rt(u,v,t));
LH1:=
diff(dotprod(G(u,v,t),ruxv(u,v,t)),t);
LH2:=
diff(dotprod(G(u,v,t),rtxv(u,v,t)),u);
LH3:=
diff(dotprod(G(u,v,t),ruxt(u,v,t)),v);
LHS:=simplify(LH1-LH2-LH3);
RH1:=dotprod(subs(x=x(u,v,t),
y=y(u,v,t),z=z(u,v,t),
diff(F(x,y,z,t),t)),ruxv(u,v,t));
RH2:=(divf(u,v,t))*
(dotprod(rt(u,v,t),ruxv(u,v,t)));
RHS:=simplify(RH1+RH2);
```

```
simplify(LHS-RHS);
```

Again the final output is 0, indicating that the identity is valid.

b) If $\mathcal{C}_t$ is the oriented boundary of $\mathcal{S}_t$ and L_t is the corresponding counterclockwise boundary of the parameter region R in the uv-plane, then

$$\oint_{\mathcal{C}_t}\left(\mathbf{F}\times\frac{\partial \mathbf{r}}{\partial t}\right)\bullet d\mathbf{r}$$
$$=\oint_{L_t}\left(\mathbf{G}\times\frac{\partial \mathbf{r}}{\partial t}\right)\bullet\left(\frac{\partial \mathbf{r}}{\partial u}du+\frac{\partial \mathbf{r}}{\partial v}dv\right)$$
$$=\oint_{L_t}\left[-\mathbf{G}\bullet\left(\frac{\partial \mathbf{r}}{\partial u}\times\frac{\partial \mathbf{r}}{\partial t}\right)+\mathbf{G}\bullet\left(\frac{\partial \mathbf{r}}{\partial t}\times\frac{\partial \mathbf{r}}{\partial v}\right)\right]dt$$
$$=\iint_R\left[\frac{\partial}{\partial u}\left(\mathbf{G}\bullet\left(\frac{\partial \mathbf{r}}{\partial t}\times\frac{\partial \mathbf{r}}{\partial v}\right)\right)\right.$$
$$\left.+\frac{\partial}{\partial v}\left(\mathbf{G}\bullet\left(\frac{\partial \mathbf{r}}{\partial u}\times\frac{\partial \mathbf{r}}{\partial t}\right)\right)\right]du\,dv,$$

by Green's Theorem.

c) Using the results of (a) and (b), we calculate

$$\frac{d}{dt}\iint_{\mathcal{S}_t}\mathbf{F}\bullet\hat{\mathbf{N}}\,dS = \iint_R\frac{\partial}{\partial t}\left[\mathbf{G}\bullet\left(\frac{\partial \mathbf{r}}{\partial u}\times\frac{\partial \mathbf{r}}{\partial v}\right)\right]du\,dv$$
$$=\iint_R\frac{\partial \mathbf{F}}{\partial t}\bullet\left(\frac{\partial \mathbf{r}}{\partial u}\times\frac{\partial \mathbf{r}}{\partial v}\right)du\,dv$$
$$+\iint_R(\operatorname{div}\mathbf{F})\frac{\partial \mathbf{r}}{\partial t}\bullet\left(\frac{\partial \mathbf{r}}{\partial u}\times\frac{\partial \mathbf{r}}{\partial v}\right)du\,dv$$
$$+\iint_R\left[\frac{\partial}{\partial u}\left(\mathbf{G}\bullet\left(\frac{\partial \mathbf{r}}{\partial t}\times\frac{\partial \mathbf{r}}{\partial v}\right)\right)\right.$$
$$\left.+\frac{\partial}{\partial v}\left(\mathbf{G}\bullet\left(\frac{\partial \mathbf{r}}{\partial u}\times\frac{\partial \mathbf{r}}{\partial t}\right)\right)\right]du\,dv$$
$$=\iint_{\mathcal{S}_t}\frac{\partial \mathbf{F}}{\partial t}\bullet\hat{\mathbf{N}}\,dS+\iint_{\mathcal{S}_t}(\operatorname{div}\mathbf{F})\mathbf{v}_S\bullet\hat{\mathbf{N}}\,dS$$
$$+\oint_{\mathcal{C}_t}(\mathbf{F}\times\mathbf{v}_C)\bullet d\mathbf{r}.$$

5. We have

$$\frac{1}{\Delta t}\left[\iiint_{D_{t+\Delta t}}f(\mathbf{r},t+\Delta t)\,dV-\iiint_{D_t}f(\mathbf{r},t)\,dV\right]$$
$$=\iiint_{D_t}\frac{f(\mathbf{r},t+\Delta t)-f(\mathbf{r},t)}{\Delta t}\,dV$$
$$+\frac{1}{\Delta t}\iiint_{D_{t+\Delta t}-D_t}f(\mathbf{r},t+\Delta t)\,dV$$
$$-\frac{1}{\Delta t}\iiint_{D_t-D_{t+\Delta t}}f(\mathbf{r},t+\Delta t)\,dV$$
$$=I_1+I_2-I_3.$$

Evidently $I_1 \to \iiint_{D_t}\frac{\partial f}{\partial t}\,dV$ as $\Delta t \to 0$.

I_2 and I_3 are integrals over the parts of ΔD_t where the surface $\mathcal{S}_t$ is moving outwards and inwards, respectively, that is, where $\mathbf{v}_S\bullet\hat{\mathbf{N}}$ is, respectively, positive and negative. Since $dV=|\mathbf{v}_S\bullet\hat{\mathbf{N}}|\,dS\,\Delta T$, we have

$$I_2-I_3=\iint_{\mathcal{S}_t}f(\mathbf{r},t+\Delta t)\mathbf{v}_S\bullet\hat{\mathbf{N}}\,dS$$
$$=\iint_{\mathcal{S}_t}f(\mathbf{r},t)\mathbf{v}_S\bullet\hat{\mathbf{N}}\,dS$$
$$+\iint_{\mathcal{S}_t}\bigl(f(\mathbf{r},t+\Delta t)-f(\mathbf{r},t)\bigr)\mathbf{v}_S\bullet\hat{\mathbf{N}}\,dS.$$

The latter integral approaches 0 as $\Delta t \to 0$ because

$$\left|\iint_{\mathcal{S}_t}\bigl(f(\mathbf{r},t+\Delta t)-f(\mathbf{r},t)\bigr)\mathbf{v}_S\bullet\hat{\mathbf{N}}\,dS\right|$$
$$\le \max|\mathbf{v}_S|\left|\frac{\partial f}{\partial t}\right|(\text{area of }\mathcal{S}_t)\Delta t.$$

CHAPTER 17. ORDINARY DIFFERENTIAL EQUATIONS

Section 17.1 Classifying Differential Equations (page 983)

1. $\dfrac{dy}{dx} = 5y$: 1st order, linear, homogeneous.

2. $\dfrac{d^2y}{dx^2} + x = y$: 2nd order, linear, nonhomogeneous.

3. $y\dfrac{dy}{dx} = x$: 1st order, nonlinear.

4. $y''' + xy' = x \sin x$: 3rd order, linear, nonhomogeneous.

5. $y'' + x \sin x \, y' = y$: 2nd order, linear, homogeneous.

6. $y'' + 4y' - 3y = 2y^2$: 2nd order, nonlinear.

7. $\dfrac{d^3y}{dt^3} + t\dfrac{dy}{dt} + t^2 y = t^3$: 3rd order, linear, nonhomogeneous.

8. $\cos x \dfrac{dx}{dt} + x \sin t = 0$: 1st order, nonlinear, homogeneous.

9. $y^{(4)} + e^x y'' = x^3 y'$: 4th order, linear, homogeneous.

10. $x^2 y'' + e^x y' = \dfrac{1}{y}$: 2nd order, nonlinear.

11. If $y = \cos x$, then $y'' + y = -\cos x + \cos x = 0$.
    If $y = \sin x$, then $y'' + y = -\sin x + \sin x = 0$. Thus $y = \cos x$ and $y = \sin x$ are both solutions of $y'' + y = 0$. This DE is linear and homogeneous, so any function of the form
    $$y = A \cos x + B \sin x,$$
    where A and B are constants, is a solution also. Therefore $\sin x - \cos x$ is a solution ($A = -1$, $B = 1$), and
    $$\sin(x + 3) = \sin 3 \cos x + \cos 3 \sin x$$
    is a solution, but $\sin 2x$ is not since it cannot be represented in the form $A \cos x + B \sin x$.

12. If $y = e^x$, then $y'' - y = e^x - e^x = 0$; if $y = e^{-x}$, then $y'' - y = e^{-x} - e^{-x} = 0$. Thus e^x and e^{-x} are both solutions of $y'' - y = 0$. Since $y'' - y = 0$ is linear and homogeneous, any function of the form
    $$y = Ae^x + Be^{-x}$$
    is also a solution. Thus $\cosh x = \tfrac{1}{2}(e^x + e^{-x})$ is a solution, but neither $\cos x$ nor x^e is a solution.

13. Given that $y_1 = \cos(kx)$ is a solution of $y'' + k^2 y = 0$, we suspect that $y_2 = \sin(kx)$ is also a solution. This is easily verified since
    $$y_2'' + k^2 y_2 = -k^2 \sin(kx) + k^2 \sin(kx) = 0.$$

Since the DE is linear and homogeneous,
$$y = Ay_1 + By_2 = A\cos(kx) + B\sin(kx)$$
is a solution for any constants A and B. It will satisfy
$$3 = y(\pi/k) = A\cos(\pi) + B\sin(\pi) = -A$$
$$3 = y'(\pi/k) = -Ak\sin(\pi) + Bk\cos(\pi) = -Bk,$$
provided $A = -3$ and $B = -3/k$. The required solution is
$$y = -3\cos(kx) - \frac{3}{k}\sin(kx).$$

14. Given that $y_1 = e^{kx}$ is a solution of $y'' - k^2 y = 0$, we suspect that $y_2 = e^{-kx}$ is also a solution. This is easily verified since
    $$y_2'' - k^2 y_2 = k^2 e^{-kx} - k^2 e^{-kx} = 0.$$

Since the DE is linear and homogeneous,
$$y = Ay_1 + By_2 = Ae^{kx} + Be^{-kx}$$
is a solution for any constants A and B. It will satisfy
$$0 = y(1) = Ae^k + Be^{-k}$$
$$2 = y'(1) = Ake^k - Bke^{-k},$$
provided $A = e^{-k}/k$ and $B = -e^k/k$. The required solution is
$$y = \frac{1}{k}e^{k(x-1)} - \frac{1}{k}e^{-k(x-1)}.$$

15. By Exercise 11, $y = A\cos x + B\sin x$ is a solution of $y'' + y = 0$ for any choice of the constants A and B. This solution will satisfy
    $$0 = y(\pi/2) - 2y(0) = B - 2A,$$
    $$3 = y(\pi/4) = \frac{A}{\sqrt{2}} + \frac{B}{\sqrt{2}},$$
    provided $A = \sqrt{2}$ and $B = 2\sqrt{2}$. The required solution is
    $$y = \sqrt{2}\cos x + 2\sqrt{2}\sin x.$$

16. $y = e^{rx}$ is a solution of the equation $y'' - y' - 2y = 0$ if $r^2 e^{rx} - re^{rx} - 2e^{rx} = 0$, that is, if $r^2 - r - 2 = 0$. This quadratic has two roots, $r = 2$, and $r = -1$. Since the DE is linear and homogeneous, the function $y = Ae^{2x} + Be^{-x}$ is a solution for any constants A and B. This solution satisfies

$$1 = y(0) = A + B, \quad 2 = y'(0) = 2A - B,$$

provided $A = 1$ and $B = 0$. Thus, the required solution is $y = e^{2x}$.

17. If $y = y_1(x) = x$, then $y_1' = 1$ and $y_1'' = 0$. Thus $y_1'' + y_1 = 0 + x = x$. By Exercise 11 we know that $y_2 = A\cos x + B\sin x$ satisfies the homogeneous DE $y'' + y = 0$. Therefore, by Theorem 2,

$$y = y_1(x) + y_2(x) = x + A\cos x + B\sin x$$

is a solution of $y'' + y = x$. This solution satisfies

$$1 = y(\pi) = \pi - A, \quad 0 = y'(\pi) = 1 - B,$$

provided $A = \pi - 1$ and $B = 1$. Thus the required solution is $y = x + (\pi - 1)\cos x + \sin x$.

18. If $y = y_1(x) = -e$, then $y_1' = 0$ and $y_1'' = 0$. Thus $y_1'' - y_1 = 0 + e = e$. By Exercise 12 we know that $y_2 = Ae^x + Be^{-x}$ satisfies the homogeneous DE $y'' - y = 0$. Therefore, by Theorem 2,

$$y = y_1(x) + y_2(x) = -e + Ae^x + Be^{-x}$$

is a solution of $y'' - y = e$. This solution satisfies

$$0 = y(1) = Ae + \frac{B}{e} - e, \quad 1 = y'(1) = Ae - \frac{B}{e},$$

provided $A = (e+1)/(2e)$ and $B = e(e-1)/2$. Thus the required solution is $y = -e + \frac{1}{2}(e+1)e^{x-1} + \frac{1}{2}(e-1)e^{1-x}$.

Section 17.2 First-Order Separable and Homogeneous Equations (page 988)

1. $\dfrac{dy}{dx} = \dfrac{y}{2x}$
$2\dfrac{dy}{y} = \dfrac{dx}{x}$
$2\ln y = \ln x + C_1 \quad \Rightarrow \quad y^2 = Cx.$

2. $\dfrac{dy}{dx} = \dfrac{3y - 1}{x}$
$\displaystyle\int \dfrac{dy}{3y-1} = \int \dfrac{dx}{x}$
$\dfrac{1}{3}\ln|3y - 1| = \ln|x| + \dfrac{1}{3}\ln C$
$\dfrac{3y - 1}{x^3} = C$
$\Rightarrow \quad y = \dfrac{1}{3}(1 + Cx^3).$

3. $\dfrac{dy}{dx} = \dfrac{x^2}{y^2} \quad \Rightarrow \quad y^2\,dy = x^2\,dx$
$\dfrac{y^3}{3} = \dfrac{x^3}{3} + C_1, \quad \text{or} \quad x^3 - y^3 = C.$

4. $\dfrac{dy}{dx} = x^2 y^2$
$\displaystyle\int \dfrac{dy}{y^2} = \int x^2\,dx$
$-\dfrac{1}{y} = \dfrac{1}{3}x^3 + \dfrac{1}{3}C$
$\Rightarrow \quad y = -\dfrac{3}{x^3 + C}.$

5. $\dfrac{dy}{dx} = \dfrac{x^2}{y^3}$
$\displaystyle\int y^3\,dy = \int x^2\,dx$
$\dfrac{y^4}{4} = \dfrac{x^3}{3} + C_1$
$4x^3 - 3y^4 = C.$

6. $\dfrac{dy}{dx} = x^2 y^3$
$\displaystyle\int \dfrac{dy}{y^3} = \int x^2\,dx$
$-\dfrac{1}{2y^2} = \dfrac{x^3}{3} + C_1$
$y^2 = \dfrac{1}{C - (2/3)x^3}.$

7. $\dfrac{dY}{dt} = tY \quad \Rightarrow \quad \dfrac{dY}{Y} = t\,dt$
$\ln Y = \dfrac{t^2}{2} + C_1, \quad \text{or} \quad Y = Ce^{t^2/2}.$

8. $\dfrac{dx}{dt} = e^x \sin t$
$\displaystyle\int e^{-x}\,dx = \int \sin t\,dt$
$-e^{-x} = -\cos t - C$
$\Rightarrow \quad x = -\ln(\cos t + C).$

9. $\dfrac{dy}{dx} = 1 - y^2 \quad \Rightarrow \quad \dfrac{dy}{1 - y^2} = dx$
$\dfrac{1}{2}\left(\dfrac{1}{1+y} + \dfrac{1}{1-y}\right)dy = dx$
$\dfrac{1}{2}\ln\left|\dfrac{1+y}{1-y}\right| = x + C_1$
$\dfrac{1+y}{1-y} = Ce^{2x} \quad \text{or} \quad y = \dfrac{Ce^{2x} - 1}{Ce^{2x} + 1}.$

10. $\dfrac{dy}{dx} = 1 + y^2$

$\displaystyle\int \dfrac{dy}{1+y^2} = \int dx$

$\tan^{-1} y = x + C$

$\Rightarrow \quad y = \tan(x + C).$

11. $\dfrac{dy}{dt} = 2 + e^y \quad \Rightarrow \quad \dfrac{dy}{2 + e^y} = dt$

$\displaystyle\int \dfrac{e^{-y}\, dy}{2e^{-y} + 1} = \int dt$

$-\dfrac{1}{2}\ln(2e^{-y} + 1) = t + C_1$

$2e^{-y} + 1 = C_2 e^{-2t}, \quad\text{or}\quad y = -\ln\left(Ce^{-2t} - \dfrac{1}{2}\right).$

12. We have

$$\dfrac{dy}{dx} = y^2(1-y)$$

$$\int \dfrac{dy}{y^2(1-y)} = \int dx = x + K.$$

Expand the left side in partial fractions

$\dfrac{1}{y^2(1-y)} = \dfrac{A}{y} + \dfrac{B}{y^2} + \dfrac{C}{1-y}$

$= \dfrac{A(y - y^2) + B(1-y) + Cy^2}{y^2(1-y)}$

$\Rightarrow \begin{cases} -A + C = 0; \\ A - B = 0; \quad \Rightarrow A = B = C = 1. \\ B = 1. \end{cases}$

Hence,

$\displaystyle\int \dfrac{dy}{y^2(1-y)} = \int \left(\dfrac{1}{y} + \dfrac{1}{y^2} + \dfrac{1}{1-y}\right) dy$

$= \ln|y| - \dfrac{1}{y} - \ln|1-y|.$

Therefore,

$$\ln\left|\dfrac{y}{1-y}\right| - \dfrac{1}{y} = x + K.$$

13. $\dfrac{dy}{dx} = \sin x \cos^2 y$

$\displaystyle\int \sec^2 y\, dy = \int \sin x\, dx$

$\tan y = -\cos x + C$

$y = \tan^{-1}(C - \cos x) + n\pi.$

14. $x\dfrac{dy}{dx} = y \ln x$

$\displaystyle\int \dfrac{dy}{y} = \int \dfrac{\ln x}{x}\, dx$

$\ln|y| = \dfrac{1}{2}(\ln x)^2 + C_1$

$y = Ce^{(\ln x)^2/2}, \quad (x > 0).$

15. $\dfrac{dy}{dx} = \dfrac{x+y}{x-y} \qquad$ Let $y = vx$

$v + x\dfrac{dv}{dx} = \dfrac{x(1+v)}{x(1-v)}$

$x\dfrac{dv}{dx} = \dfrac{1+v}{1-v} - v = \dfrac{1+v^2}{1-v}$

$\displaystyle\int \dfrac{1-v}{1+v^2}\, dv = \int \dfrac{dx}{x}$

$\tan^{-1} v - \dfrac{1}{2}\ln(1+v^2) = \ln|x| + C_1$

$\tan^{-1}(y/x) - \dfrac{1}{2}\ln\dfrac{x^2+y^2}{x^2} = \ln|x| + C_1$

$2\tan^{-1}(y/x) - \ln(x^2 + y^2) = C.$

16. $\dfrac{dy}{dx} = \dfrac{xy}{x^2 + 2y^2} \qquad$ Let $y = vx$

$v + x\dfrac{dv}{dx} = \dfrac{vx^2}{(1+2v^2)x^2}$

$x\dfrac{dv}{dx} = \dfrac{v}{1+2v^2} - v = -\dfrac{2v^3}{1+2v^2}$

$\displaystyle\int \dfrac{1+2v^2}{v^3}\, dv = -2\int \dfrac{dx}{x}$

$-\dfrac{1}{2v^2} + 2\ln|v| = -2\ln|x| + C_1$

$-\dfrac{x^2}{2y^2} + 2\ln|y| = C_1$

$x^2 - 4y^2 \ln|y| = Cy^2.$

17. $\dfrac{dy}{dx} = \dfrac{x^2 + xy + y^2}{x^2} \qquad$ Let $y = vx$

$v + x\dfrac{dv}{dx} = \dfrac{x^2(1 + v + v^2)}{x^2}$

$\displaystyle\int \dfrac{dv}{1+v^2} = \int \dfrac{dx}{x}$

$\tan^{-1} v = \ln|x| + C$

$\dfrac{y}{x} = \tan(\ln|x| + C)$

$y = x\tan(\ln|x| + C).$

18. $\dfrac{dy}{dx} = \dfrac{x^3 + 3xy^2}{3x^2y + y^3}$ Let $y = vx$

$v + x\dfrac{dv}{dx} = \dfrac{x^3(1+3v^2)}{x^3(3v+v^3)}$

$x\dfrac{dv}{dx} = \dfrac{1+3v^2}{3v+v^3} - v = \dfrac{1-v^4}{v(3+v^2)}$

$\displaystyle\int \dfrac{(3+v^2)v\,dv}{1-v^4} = \int \dfrac{dx}{x}$ Let $u = v^2$
$du = 2v\,dv$

$\dfrac{1}{2}\displaystyle\int \dfrac{3+u}{1-u^2}\,du = \ln|x| + C_1$

$\dfrac{3}{4}\ln\left|\dfrac{u+1}{u-1}\right| - \dfrac{1}{4}\ln|1-u^2| = \ln|x| + C_1$

$3\ln\left|\dfrac{y^2+x^2}{y^2-x^2}\right| - \ln\left|\dfrac{x^4-y^4}{x^4}\right| = 4\ln|x| + C_2$

$\ln\left|\left(\dfrac{x^2+y^2}{x^2-y^2}\right)^3 \dfrac{1}{x^4-y^4}\right| = C_2$

$\ln\left|\dfrac{(x^2+y^2)^2}{(x^2-y^2)^4}\right| = C_2$

$x^2 + y^2 = C(x^2 - y^2)^2$.

19. $x\dfrac{dy}{dx} = y + x\cos^2\left(\dfrac{y}{x}\right)$ (let $y=vx$)

$xv + x^2\dfrac{dv}{dx} = vx + x\cos^2 v$

$x\dfrac{dv}{dx} = \cos^2 v$

$\sec^2 v\,dv = \dfrac{dx}{x}$

$\tan v = \ln|x| + \ln|C|$

$\tan\left(\dfrac{y}{x}\right) = \ln|Cx|$

$y = x\tan^{-1}(\ln|Cx|)$.

20. $\dfrac{dy}{dx} = \dfrac{y}{x} - e^{-y/x}$ (let $y=vx$)

$v + x\dfrac{dv}{dx} = v - e^{-v}$

$e^v\,dv = -\dfrac{dx}{x}$

$e^v = -\ln|x| + \ln|C|$

$e^{y/x} = \ln\left|\dfrac{C}{x}\right|$

$y = x\ln\ln\left|\dfrac{C}{x}\right|$.

21. We require $\dfrac{dy}{dx} = \dfrac{2x}{1+y^2}$. Thus

$\displaystyle\int (1+y^2)\,dy = \int 2x\,dx$

$y + \dfrac{1}{3}y^3 = x^2 + C$.

Since $(2,3)$ lies on the curve, $12 = 4 + C$. Thus $C = 8$ and $y + \dfrac{1}{3}y^3 - x^2 = 8$, or $3y + y^3 - 3x^2 = 24$.

22. $\dfrac{dy}{dx} = 1 + \dfrac{2y}{x}$ Let $y = vx$

$v + x\dfrac{dv}{dx} = 1 + 2v$

$x\dfrac{dv}{dx} = 1 + v$

$\displaystyle\int \dfrac{dv}{1+v} = \int \dfrac{dx}{x}$

$\ln|1+v| = \ln|x| + C_1$

$1 + \dfrac{y}{x} = Cx \;\Rightarrow\; x + y = Cx^2$.

Since $(1, 3)$ lies on the curve, $4 = C$. Thus the curve has equation $x + y = 4x^2$.

23. If $\xi = x - x_0$, $\eta = y - y_0$, and

$\dfrac{dy}{dx} = \dfrac{ax + by + c}{ex + fy + g}$,

then

$\dfrac{d\eta}{d\xi} = \dfrac{dy}{dx} = \dfrac{a(\xi+x_0) + b(\eta+y_0) + c}{e(\xi+x_0) + f(\eta+y_0) + g}$

$= \dfrac{a\xi + b\eta + (ax_0 + by_0 + c)}{e\xi + f\eta + (ex_0 + fy_0 + g)}$

$= \dfrac{a\xi + b\eta}{e\xi + f\eta}$

provided x_0 and y_0 are chosen such that

$ax_0 + by_0 + c = 0$,   and $ex_0 + fy_0 + g = 0$.

24. The system $x_0 + 2y_0 - 4 = 0$, $2x_0 - y_0 - 3 = 0$ has solution $x_0 = 2$, $y_0 = 1$. Thus, if $\xi = x - 2$ and $\eta = y - 1$, where

$\dfrac{dy}{dx} = \dfrac{x + 2y - 4}{2x - y - 3}$,

then

$\dfrac{d\eta}{d\xi} = \dfrac{\xi + 2\eta}{2\xi - \eta}$ Let $\eta = v\xi$

$v + \xi\dfrac{dv}{d\xi} = \dfrac{1+2v}{2-v}$

$\xi\dfrac{dv}{d\xi} = \dfrac{1+2v}{2-v} - v = \dfrac{1+v^2}{2-v}$

$\displaystyle\int \left(\dfrac{2-v}{1+v^2}\right) dv = \int \dfrac{d\xi}{\xi}$

$2\tan^{-1} v - \dfrac{1}{2}\ln(1+v^2) = \ln|\xi| + C_1$

$4\tan^{-1}\dfrac{\eta}{\xi} - \ln(\xi^2 + \eta^2) = C$.

Hence the solution of the original equation is

$$4\tan^{-1}\frac{y-1}{x-2} - \ln\bigl((x-2)^2 + (y-1)^2\bigr) = C.$$

Section 17.3 Exact Equations and Integrating Factors (page 991)

1. $(xy^2 + y)\,dx + (x^2 y + x)\,dy = 0$

$d\left(\frac{1}{2}x^2 y^2 + xy\right) = 0$

$x^2 y^2 + 2xy = C.$

2. $(e^x \sin y + 2x)\,dx + (e^x \cos y + 2y)\,dy = 0$

$d(e^x \sin y + x^2 + y^2) = 0$

$e^x \sin y + x^2 + y^2 = C.$

3. $e^{xy}(1 + xy)\,dx + x^2 e^{xy}\,dy = 0$

$d(xe^{xy}) = 0 \quad \Rightarrow \quad xe^{xy} = C.$

4. $\left(2x + 1 - \dfrac{y^2}{x^2}\right)dx + \dfrac{2y}{x}\,dy = 0$

$d\left(x^2 + x + \dfrac{y^2}{x}\right) = 0$

$x^2 + x + \dfrac{y^2}{x} = C.$

5. $(x^2 + 2y)\,dx - x\,dy = 0$

$M = x^2 + 2y, \qquad N = -x$

$\dfrac{1}{N}\left(\dfrac{\partial M}{\partial y} - \dfrac{\partial N}{\partial x}\right) = -\dfrac{3}{x}$ (indep. of y)

$\dfrac{d\mu}{\mu} = -\dfrac{3}{x}\,dx \quad \Rightarrow \quad \mu = \dfrac{1}{x^3}$

$\left(\dfrac{1}{x} + \dfrac{2y}{x^3}\right)dx - \dfrac{1}{x^2}\,dy = 0$

$d\left(\ln|x| - \dfrac{y}{x^2}\right) = 0$

$\ln|x| - \dfrac{y}{x^2} = C_1$

$y = x^2 \ln|x| + Cx^2.$

6. $(xe^x + x\ln y + y)\,dx + \left(\dfrac{x^2}{y} + x\ln x + x\sin y\right)dy = 0$

$M = xe^x + x\ln y + y, \qquad N = \dfrac{x^2}{y} + x\ln x + x\sin y$

$\dfrac{\partial M}{\partial y} = \dfrac{x}{y} + 1, \qquad \dfrac{\partial N}{\partial x} = \dfrac{2x}{y} + \ln x + 1 + \sin y$

$\dfrac{1}{N}\left(\dfrac{\partial M}{\partial y} - \dfrac{\partial N}{\partial x}\right) = \dfrac{1}{N}\left(-\dfrac{x}{y} - \ln x - \sin y\right) = -\dfrac{1}{x}$

$\dfrac{d\mu}{\mu} = -\dfrac{1}{x}\,dx \quad \Rightarrow \quad \mu = \dfrac{1}{x}$

$\left(e^x + \ln y + \dfrac{y}{x}\right)dx + \left(\dfrac{x}{y} + \ln x + \sin y\right)dy$

$d\bigl(e^x + x\ln y + y\ln x - \cos y\bigr) = 0$

$e^x + x\ln y + y\ln x - \cos y = C.$

7. If $\mu(y)M(x,y)\,dx + \mu(y)N(x,y)\,dy$ is exact, then

$$\dfrac{\partial}{\partial y}\bigl(\mu(y)M(x,y)\bigr) = \dfrac{\partial}{\partial x}\bigl(\mu(y)N(x,y)\bigr)$$

$\mu'(y)M + \mu\dfrac{\partial M}{\partial y} = \mu\dfrac{\partial N}{\partial x}$

$\dfrac{\mu'}{\mu} = \dfrac{1}{M}\left(\dfrac{\partial N}{\partial x} - \dfrac{\partial M}{\partial y}\right).$

Thus M and N must be such that

$$\dfrac{1}{M}\left(\dfrac{\partial N}{\partial x} - \dfrac{\partial M}{\partial y}\right)$$

depends only on y.

8. $2y^2(x + y^2)\,dx + xy(x + 6y^2)\,dy = 0$

$(2xy^2 + 2y^4)\mu(y)\,dx + (x^2 y + 6xy^3)\mu(y)\,dy = 0$

$\dfrac{\partial M}{\partial y} = (4xy + 8y^3)\mu(y) + (2xy^2 + 2y^4)\mu'(y)$

$\dfrac{\partial N}{\partial x} = (2xy + 6y^3)\mu(y).$

For exactness we require

$(2xy^2 + 2y^4)\mu'(y) = [(2xy + 6y^3) - (4xy + 8y^3)]\mu(y)$

$y(2xy + 2y^3)\mu'(y) = -(2xy + 2y^3)\mu(y)$

$y\mu'(y) = -\mu(y) \quad \Rightarrow \quad \mu(y) = \dfrac{1}{y}$

$(2xy + 2y^3)\,dx + (x^2 + 6xy^2)\,dy = 0$

$d(x^2 y + 2xy^3) = 0 \quad \Rightarrow \quad x^2 y + 2xy^3 = C.$

9. Consider $y\,dx - (2x + y^3 e^y)\,dy = 0$.
   Here $M = y$, $N = -2x - y^3 e^y$, $\dfrac{\partial M}{\partial y} = 1$, and $\dfrac{\partial N}{\partial x} = -2$.
   Thus
   $$\frac{\mu'}{\mu} = -\frac{3}{y} \quad\Rightarrow\quad \mu = \frac{1}{y^3}$$
   $$\frac{1}{y^2}\,dx - \left(\frac{2x}{y^3} + e^y\right)dy = 0$$
   $$d\left(\frac{x}{y^2} - e^y\right) = 0$$
   $$\frac{x}{y^2} - e^y = C, \quad\text{or}\quad x - y^2 e^y = Cy^2.$$

10. If $\mu(xy)$ is an integrating factor for $M\,dx + N\,dy = 0$, then
    $$\frac{\partial}{\partial y}(\mu M) = \frac{\partial}{\partial x}(\mu N), \quad\text{or}$$
    $$x\mu'(xy)M + \mu(xy)\frac{\partial M}{\partial y} = y\mu'(xy)N + \mu(xy)\frac{\partial N}{\partial x}.$$

    Thus M and N will have to be such that the right-hand side of the equation
    $$\frac{\mu'(xy)}{\mu(xy)} = \frac{1}{xM - yN}\left(\frac{\partial N}{\partial x} - \frac{\partial M}{\partial y}\right)$$
    depends only on the product xy.

11. For $\left(x\cos x + \dfrac{y^2}{x}\right)dx - \left(\dfrac{x\sin x}{y} + y\right)dy$ we have
    $$M = x\cos x + \frac{y^2}{x}, \quad N = -\frac{x\sin x}{y} - y$$
    $$\frac{\partial M}{\partial y} = \frac{2y}{x}, \quad \frac{\partial N}{\partial x} = -\frac{\sin x}{y} - \frac{x\cos x}{y}$$
    $$\frac{\partial N}{\partial x} - \frac{\partial M}{\partial y} = -\left(\frac{\sin x}{y} + \frac{x\cos x}{y} + \frac{2y}{x}\right)$$
    $$xM - yN = x^2\cos x + y^2 + x\sin x + y^2$$
    $$\frac{1}{xM - yN}\left(\frac{\partial N}{\partial x} - \frac{\partial M}{\partial y}\right) = -\frac{1}{xy}.$$

    Thus, an integrating factor is given by
    $$\frac{\mu'(t)}{\mu(t)} = -\frac{1}{t} \quad\Rightarrow\quad \mu(t) = \frac{1}{t}.$$

    We multiply the original equation by $1/(xy)$ to make it exact:
    $$\left(\frac{\cos x}{y} + \frac{y}{x^2}\right)dx - \left(\frac{\sin x}{y^2} + \frac{1}{x}\right)dy = 0$$
    $$d\left(\frac{\sin x}{y} - \frac{y}{x}\right) = 0$$
    $$\frac{\sin x}{y} - \frac{y}{x} = C.$$

The solution is $x\sin x - y^2 = Cxy$.

Section 17.4 First-Order Linear Equations (page 994)

1. $\dfrac{dy}{dx} - \dfrac{2}{x}y = x^2$ (linear)
   $$\mu = \exp\left(\int -\frac{2}{x}\,dx\right) = \frac{1}{x^2}$$
   $$\frac{1}{x^2}\frac{dy}{dx} - \frac{2}{x^3}y = 1$$
   $$\frac{d}{dx}\frac{y}{x^2} = 1$$
   $$\frac{y}{x^2} = x + C, \quad\text{so}\quad y = x^3 + Cx^2.$$

2. We have $\dfrac{dy}{dx} + \dfrac{2y}{x} = \dfrac{1}{x^2}$.
   Let $\mu = \displaystyle\int \frac{2}{x}\,dx = 2\ln x = \ln x^2$, then $e^\mu = x^2$, and
   $$\frac{d}{dx}(x^2 y) = x^2\frac{dy}{dx} + 2xy$$
   $$= x^2\left(\frac{dy}{dx} + \frac{2y}{x}\right) = x^2\left(\frac{1}{x^2}\right) = 1$$
   $$\Rightarrow\quad x^2 y = \int dx = x + C$$
   $$\Rightarrow\quad y = \frac{1}{x} + \frac{C}{x^2}.$$

3. $\dfrac{dy}{dx} + 2y = 3 \qquad \mu = \exp\left(\int 2\,dx\right) = e^{2x}$
   $$\frac{d}{dx}(e^{2x}y) = e^{2x}(y' + 2y) = 3e^{2x}$$
   $$e^{2x}y = \frac{3}{2}e^{2x} + C \quad\Rightarrow\quad y = \frac{3}{2} + Ce^{-2x}.$$

4. We have $\dfrac{dy}{dx} + y = e^x$. Let $\mu = \int dx = x$, then $e^\mu = e^x$, and
   $$\frac{d}{dx}(e^x y) = e^x\frac{dy}{dx} + e^x y = e^x\left(\frac{dy}{dx} + y\right) = e^{2x}$$
   $$\Rightarrow\quad e^x y = \int e^{2x}\,dx = \frac{1}{2}e^{2x} + C.$$

   Hence, $y = \dfrac{1}{2}e^x + Ce^{-x}$.

609

5. $\dfrac{dy}{dx}+y=x$, $\mu=\exp\left(\int 1\,dx\right)=e^x$

$\dfrac{d}{dx}(e^x y)=e^x(y'+y)=xe^x$

$e^x y=\int xe^x\,dx=xe^x-e^x+C$

$y=x-1+Ce^{-x}$.

6. We have $\dfrac{dy}{dx}+2e^x y=e^x$. Let $\mu=\int 2e^x\,dx=2e^x$; then

$$\dfrac{d}{dx}\left(e^{2e^x}y\right)=e^{2e^x}\dfrac{dy}{dx}+2e^x e^{2e^x}y$$
$$=e^{2e^x}\left(\dfrac{dy}{dx}+2e^x y\right)=e^{2e^x}e^x.$$

Therefore

$$e^{2e^x}y=\int e^{2e^x}e^x\,dx \quad \text{Let } u=2e^x$$
$$\phantom{e^{2e^x}y}\quad du=2e^x\,dx$$
$$=\dfrac{1}{2}\int e^u\,du=\dfrac{1}{2}e^{2e^x}+C.$$

Hence, $y=\dfrac{1}{2}+Ce^{-2e^x}$.

7. $\dfrac{dy}{dt}+10y=1$, $y\left(\tfrac{1}{10}\right)=\tfrac{2}{10}$

$\mu=\int 10\,dt=10t$

$\dfrac{d}{dt}(e^{10t}y)=e^{10t}\dfrac{dy}{dt}+10e^{10t}y=e^{10t}$

$e^{10t}y(t)=\dfrac{1}{10}e^{10t}+C$

$y\left(\tfrac{1}{10}\right)=\tfrac{2}{10} \Rightarrow \dfrac{2e}{10}=\dfrac{e}{10}+C \Rightarrow C=\dfrac{e}{10}$

$y=\dfrac{1}{10}+\dfrac{1}{10}e^{1-10t}$.

8. $\dfrac{dy}{dx}+3x^2 y=x^2$, $y(0)=1$

$\mu=\int 3x^2\,dx=x^3$

$\dfrac{d}{dx}(e^{x^3}y)=e^{x^3}\dfrac{dy}{dx}+3x^2 e^{x^3}y=x^2 e^{x^3}$

$e^{x^3}y=\int x^2 e^{x^3}\,dx=\dfrac{1}{3}e^{x^3}+C$

$y(0)=1 \Rightarrow 1=\dfrac{1}{3}+C \Rightarrow C=\dfrac{2}{3}$

$y=\dfrac{1}{3}+\dfrac{2}{3}e^{-x^3}$.

9. $y'+(\cos x)y=2xe^{-\sin x}$, $y(\pi)=0$

$\mu=\int \cos x\,dx=\sin x$

$\dfrac{d}{dx}(e^{\sin x}y)=e^{\sin x}(y'+(\cos x)y)=2x$

$e^{\sin x}y=\int 2x\,dx=x^2+C$

$y(\pi)=0 \Rightarrow 0=\pi^2+C \Rightarrow C=-\pi^2$

$y=(x^2-\pi^2)e^{-\sin x}$.

10. $x^2 y'+y=x^2 e^{1/x}$, $y(1)=3e$

$y'+\dfrac{1}{x^2}y=e^{1/x}$

$\mu=\int \dfrac{1}{x^2}\,dx=-\dfrac{1}{x}$

$\dfrac{d}{dx}\left(e^{-1/x}y\right)=e^{-1/x}\left(y'+\dfrac{1}{x^2}y\right)=1$

$e^{-1/x}y=\int 1\,dx=x+C$

$y(1)=3e \Rightarrow 3=1+C \Rightarrow C=2$

$y=(x+2)e^{1/x}$.

11. Given that $m\dfrac{dv}{dt}=mg-kv$, then

$$\int \dfrac{dv}{g-\dfrac{k}{m}v}=\int dt$$

$$-\dfrac{m}{k}\ln\left|g-\dfrac{k}{m}v\right|=t+C.$$

Since $v(0)=0$, therefore $C=-\dfrac{m}{k}\ln g$. Also, $g-\dfrac{k}{m}v$ remains positive for all $t>0$, so

$$\dfrac{m}{k}\ln \dfrac{g}{g-\dfrac{k}{m}v}=t$$

$$\dfrac{g-\dfrac{k}{m}v}{g}=e^{-kt/m}$$

$$\Rightarrow \quad v=v(t)=\dfrac{mg}{k}\left(1-e^{-kt/m}\right).$$

Note that $\lim\limits_{t\to\infty}v(t)=\dfrac{mg}{k}$. This limiting velocity can be obtained directly from the differential equation by setting $\dfrac{dv}{dt}=0$.

12. We proceed by separation of variables:

$$m\frac{dv}{dt} = mg - kv^2$$
$$\frac{dv}{dt} = g - \frac{k}{m}v^2$$
$$\frac{dv}{g - \frac{k}{m}v^2} = dt$$
$$\int \frac{dv}{\frac{mg}{k} - v^2} = \frac{k}{m}\int dt = \frac{kt}{m} + C.$$

Let $a^2 = mg/k$, where $a > 0$. Thus, we have

$$\int \frac{dv}{a^2 - v^2} = \frac{kt}{m} + C$$
$$\frac{1}{2a}\ln\left|\frac{a+v}{a-v}\right| = \frac{kt}{m} + C$$
$$\ln\left|\frac{a+v}{a-v}\right| = \frac{2akt}{m} + C_1 = 2\sqrt{\frac{kg}{m}}t + C_1$$
$$\frac{a+v}{a-v} = C_2 e^{2t\sqrt{kg/m}}$$

Assuming $v(0) = 0$, we get $C_2 = 1$.

$$a + v = e^{2t\sqrt{kg/m}}(a - v)$$
$$v\left(1 + e^{2t\sqrt{kg/m}}\right) = a\left(e^{2t\sqrt{kg/m}} - 1\right)$$
$$= \sqrt{\frac{mg}{k}}\left(e^{2t\sqrt{kg/m}} - 1\right)$$
$$v = \sqrt{\frac{mg}{k}}\frac{e^{2t\sqrt{kg/m}} - 1}{e^{2t\sqrt{kg/m}} + 1}.$$

Clearly $v \to \sqrt{\frac{mg}{k}}$ as $t \to \infty$. This also follows from setting $\frac{dv}{dt} = 0$ in the given differential equation.

13. The balance $\$y(t)$ in the account after t years satisfies

$$\frac{dy}{dt} = \frac{5}{100}y - 50 = \frac{y}{20} - 50, \quad y(0) = 900$$
$$\frac{d}{dt}(e^{-t/20}y) = e^{-t/20}\left(\frac{dy}{dt} - \frac{y}{20}\right) = -50e^{-t/20}$$
$$e^{-t/20}y = -\int 50e^{-t/20}\,dt = 1{,}000e^{-t/20} + C$$
$$900 = 1{,}000 + C \implies C = -100$$
$$y(t) = 1{,}000 - 100e^{t/20}.$$

After 5 years the balance is $y(5) = 1{,}000 - 100e^{5/20}$, or approximately \$871.60.

Section 17.5 Existence, Uniqueness, and Numerical Methods (page 1002)

A computer spreadsheet was used in Exercises 1–12. The intermediate results appearing in the spreadsheet are not shown in these solutions.

1. We start with $x_0 = 1$, $y_0 = 0$, and calculate

$$x_{n+1} = x_n + h, \quad y_{n+1} = y_n + h(x_n + y_n).$$

a) For $h = 0.2$ we get $x_5 = 2$, $y_5 = 1.97664$.

b) For $h = 0.1$ we get $x_{10} = 2$, $y_{10} = 2.187485$.

c) For $h = 0.05$ we get $x_{20} = 2$, $y_{20} = 2.306595$.

2. We start with $x_0 = 1$, $y_0 = 0$, and calculate

$$x_{n+1} = x_n + h, \quad u_{n+1} = y_n + h(x_n + y_n)$$
$$y_{n+1} = y_n + \frac{h}{2}(x_n + y_n + x_{n+1} + u_{n+1}).$$

a) For $h = 0.2$ we get $x_5 = 2$, $y_5 = 2.405416$.

b) For $h = 0.1$ we get $x_{10} = 2$, $y_{10} = 2.428162$.

c) For $h = 0.05$ we get $x_{20} = 2$, $y_{20} = 2.434382$.

3. We start with $x_0 = 1$, $y_0 = 0$, and calculate

$$x_{n+1} = x_n + h$$
$$p_n = x_n + y_n$$
$$q_n = x_n + \frac{h}{2} + y_n + \frac{h}{2}p_n$$
$$r_n = x_n + \frac{h}{2} + y_n + \frac{h}{2}q_n$$
$$q_n = x_n + h + y_n + hr_n$$
$$y_{n+1} = y_n + \frac{h}{6}(p_n + 2q_n + 2r_n + s_n).$$

a) For $h = 0.2$ we get $x_5 = 2$, $y_5 = 2.436502$.

b) For $h = 0.1$ we get $x_{10} = 2$, $y_{10} = 2.436559$.

c) For $h = 0.05$ we get $x_{20} = 2$, $y_{20} = 2.436563$.

4. We start with $x_0 = 0$, $y_0 = 0$, and calculate

$$x_{n+1} = x_n + h, \quad y_{n+1} = hx_n e^{-y_n}.$$

a) For $h = 0.2$ we get $x_{10} = 2$, $y_{10} = 1.074160$.

b) For $h = 0.1$ we get $x_{20} = 2$, $y_{20} = 1.086635$.

5. We start with $x_0 = 0$, $y_0 = 0$, and calculate

$$x_{n+1} = x_n + h, \quad u_{n+1} = y_n + hx_n e^{-y_n}$$
$$y_{n+1} = y_n + \frac{h}{2}(x_n e^{-y_n} + x_{n+1}e^{-u_{n+1}}).$$

a) For $h = 0.2$ we get $x_{10} = 2$, $y_{10} = 1.097897$.

SECTION 17.5 (PAGE 1002)

b) For $h = 0.1$ we get $x_{20} = 2$, $y_{20} = 1.098401$.

6. We start with $x_0 = 0$, $y_0 = 0$, and calculate

$$x_{n+1} = x_n + h$$
$$p_n = x_n e^{-y_n}$$
$$q_n = \left(x_n + \frac{h}{2}\right) e^{-(y_n + (h/2) p_n)}$$
$$r_n = \left(x_n + \frac{h}{2}\right) e^{-(y_n + (h/2) q_n)}$$
$$s_n = (x_n + h) e^{-(y_n + h r_n)}$$
$$y_{n+1} = y_n + \frac{h}{6}(p_n + 2q_n + 2r_n + s_n).$$

a) For $h = 0.2$ we get $x_{10} = 2$, $y_{10} = 1.098614$.

b) For $h = 0.1$ we get $x_{20} = 2$, $y_{20} = 1.098612$.

7. We start with $x_0 = 0$, $y_0 = 0$, and calculate

$$x_{n+1} = x_n + h, \qquad y_{n+1} = y_n + h \cos y_n.$$

a) For $h = 0.2$ we get $x_5 = 1$, $y_5 = 0.89441$.

b) For $h = 0.1$ we get $x_{10} = 1$, $y_{10} = 0.87996$.

c) For $h = 0.05$ we get $x_{20} = 1$, $y_{20} = 0.872831$.

8. We start with $x_0 = 0$, $y_0 = 0$, and calculate

$$x_{n+1} = x_n + h, \qquad u_{n+1} = y_n + h \cos y_n$$
$$y_{n+1} = y_n + \frac{h}{2}(\cos y_n + \cos u_{n+1}).$$

a) For $h = 0.2$ we get $x_5 = 1$, $y_5 = 0.862812$.

b) For $h = 0.1$ we get $x_{10} = 1$, $y_{10} = 0.865065$.

c) For $h = 0.05$ we get $x_{20} = 1$, $y_{20} = 0.865598$.

9. We start with $x_0 = 0$, $y_0 = 0$, and calculate

$$x_{n+1} = x_n + h$$
$$p_n = \cos y_n$$
$$q_n = \cos(y_n + (h/2) p_n)$$
$$r_n = \cos(y_n + (h/2) q_n)$$
$$q_n = \cos(y_n + h r_n)$$
$$y_{n+1} = y_n + \frac{h}{6}(p_n + 2q_n + 2r_n + s_n).$$

a) For $h = 0.2$ we get $x_5 = 1$, $y_5 = 0.865766$.

b) For $h = 0.1$ we get $x_{10} = 1$, $y_{10} = 0.865769$.

c) For $h = 0.05$ we get $x_{20} = 1$, $y_{20} = 0.865769$.

10. We start with $x_0 = 0$, $y_0 = 0$, and calculate

$$x_{n+1} = x_n + h, \qquad y_{n+1} = y_n + h \cos(x_n^2).$$

a) For $h = 0.2$ we get $x_5 = 1$, $y_5 = 0.944884$.

b) For $h = 0.1$ we get $x_{10} = 1$, $y_{10} = 0.926107$.

c) For $h = 0.05$ we get $x_{20} = 1$, $y_{20} = 0.915666$.

11. We start with $x_0 = 0$, $y_0 = 0$, and calculate

$$x_{n+1} = x_n + h, \qquad u_{n+1} = y_n + h \cos(x_n^2)$$
$$y_{n+1} = y_n + \frac{h}{2}(\cos(x_n^2) + \cos(x_{n+1}^2)).$$

a) For $h = 0.2$ we get $x_5 = 1$, $y_5 = 0.898914$.

b) For $h = 0.1$ we get $x_{10} = 1$, $y_{10} = 0.903122$.

c) For $h = 0.05$ we get $x_{20} = 1$, $y_{20} = 0.904174$.

12. We start with $x_0 = 0$, $y_0 = 0$, and calculate

$$x_{n+1} = x_n + h$$
$$p_n = \cos(x_n^2)$$
$$q_n = \cos((x_n + (h/2))^2)$$
$$r_n = \cos((x_n + (h/2))^2)$$
$$q_n = \cos((x_n + h)^2)$$
$$y_{n+1} = y_n + \frac{h}{6}(p_n + 2q_n + 2r_n + s_n).$$

a) For $h = 0.2$ we get $x_5 = 1$, $y_5 = 0.904524$.

b) For $h = 0.1$ we get $x_{10} = 1$, $y_{10} = 0.904524$.

c) For $h = 0.05$ we get $x_{20} = 1$, $y_{20} = 0.904524$.

13. $y(x) = 2 + \displaystyle\int_1^x \bigl(y(t)\bigr)^2 \, dt$

$$\frac{dy}{dx} = \bigl(y(x)\bigr)^2, \qquad y(1) = 2 + 0 = 2$$
$$\frac{dy}{y^2} = dx \quad\Rightarrow\quad -\frac{1}{y(x)} = x + C$$
$$-\frac{1}{2} = 1 + C \quad\Rightarrow\quad C = -\frac{3}{2}$$
$$y = -\frac{1}{x - (3/2)} = \frac{2}{3 - 2x}.$$

14. $u(x) = 1 + 3 \displaystyle\int_2^x t^2 u(t) \, dt$

$$\frac{du}{dx} = 3x^2 u(x), \qquad u(2) = 1 + 0 = 1$$
$$\frac{du}{u} = 3x^2 \, dx \quad\Rightarrow\quad \ln u = x^3 + C$$
$$0 = \ln 1 = \ln u(2) = 2^3 + C \quad\Rightarrow\quad C = -8$$
$$u = e^{x^3 - 8}.$$

15. For the problem $y' = f(x)$, $y(a) = 0$, the 1-step Runge-Kutta method with $h = b - a$ gives:

$$x_0 = a, \quad y_0 = 0, \quad x_1 = x_0 + h = b$$
$$p_0 = f(a), \quad q_0 = f\left(a + \frac{h}{2}\right) = f\left(\frac{a+b}{2}\right) = r_0$$
$$s_0 = f(a+h) = f(b)$$
$$y_1 = y_0 + \frac{h}{6}(p_0 + 2q_0 + 2r_0 + s_0)$$
$$= \frac{b-a}{6}\left(f(a) + 4f\left(\frac{a+b}{2}\right) + f(b)\right),$$

which is the Simpson's Rule approximation to $\int_a^b f(x)\,dx$ based on 2 subintervals of length $h/2$.

16. If $\phi(0) = A \geq 0$ and $\phi'(x) \geq k\phi(x)$ on an interval $[0, X]$, where $k > 0$ and $X > 0$, then

$$\frac{d}{dx}\left(\frac{\phi(x)}{e^{kx}}\right) = \frac{e^{kx}\phi'(x) - ke^{kx}\phi(x)}{e^{2kx}} \geq 0.$$

Thus $\phi(x)/e^{kx}$ is increasing on $[0, X]$. Since its value at $x = 0$ is $\phi(0) = A \geq 0$, therefore $\phi(x)/e^{kx} \geq A$ on $[0, X]$, and $\phi(x) \geq Ae^{kx}$ there.

17. a) Suppose $u' = u^2$, $y' = x + y^2$, and $v' = 1 + v^2$ on $[0, X]$, where $u(0) = y(0) = v(0) = 1$, and $X > 0$ is such that $v(x)$ is defined on $[0, X]$. (In part (b) below, we will show that $X < 1$, and we assume this fact now.) Since all three functions are increasing on $[0, X]$, we have $u(x) \geq 1$, $y(x) \geq 1$, and $v(x) \geq 1$ on $[0, X]$.

If $\phi(x) = y(x) - u(x)$, then $\phi(0) = 0$ and

$$\phi'(x) = x + y^2 - u^2 \geq y^2 - u^2$$
$$\geq (y+u)(y-u) \geq 2\phi$$

on $[0, X]$. By Exercise 16, $\phi(x) \geq 0$ on $[0, X]$, and so $u(x) \leq y(x)$ there.

Similarly, since $X < 1$, if $\phi(x) = v(x) - y(x)$, then $\phi(0) = 0$ and

$$\phi'(x) = 1 + v^2 - x - y^2 \geq v^2 - y^2$$
$$\geq (v+y)(v-y) \geq 2\phi$$

on $[0, X]$, so $y(x) \leq v(x)$ there.

b) The IVP $u' = u^2$, $u(0) = 1$ has solution
$$u(x) = \frac{1}{1-x},$$ obtained by separation of variables. This solution is valid for $x < 1$.

The IVP $v' = 1 + v^2$, $v(0) = 1$ has solution $v(x) = \tan\left(x + \frac{\pi}{4}\right)$, also obtained by separation of variables. It is valid only for $-3\pi/4 < x < \pi/4$. Observe that $\pi/4 < 1$, proving the assertion made about v in part (a). By the result of part (a), the solution of the IVP $y' = x + y^2$, $y(0) = 1$, increases on an interval $[0, X]$ and $\to \infty$ as $x \to X$ from the left, where X is some number in the interval $[\pi/4, 1]$.

c) Here are some approximations to $y(x)$ for values of x near 0.9 obtained by the Runge-Kutta method with $x_0 = 0$ and $y_0 = 1$:

For $h = 0.05$

$n = 17$	$x_n = 0.85$	$y_n = 12.37139$
$n = 18$	$x_n = 0.90$	$y_n = 31.777317$
$n = 19$	$x_n = 0.95$	$y_n = 4071.117315$.

For $h = 0.02$

$n = 43$	$x_n = 0.86$	$y_n = 14.149657$
$n = 44$	$x_n = 0.88$	$y_n = 19.756061$
$n = 45$	$x_n = 0.90$	$y_n = 32.651029$
$n = 46$	$x_n = 0.92$	$y_n = 90.770048$
$n = 47$	$x_n = 0.94$	$y_n = 34266.466629$.

For $h = 0.01$

$n = 86$	$x_n = 0.86$	$y_n = 14.150706$
$n = 87$	$x_n = 0.87$	$y_n = 16.493286$
$n = 88$	$x_n = 0.88$	$y_n = 19.761277$
$n = 89$	$x_n = 0.89$	$y_n = 24.638758$
$n = 90$	$x_n = 0.90$	$y_n = 32.703853$
$n = 91$	$x_n = 0.91$	$y_n = 48.591332$
$n = 92$	$x_n = 0.92$	$y_n = 94.087476$
$n = 93$	$x_n = 0.93$	$y_n = 636.786465$
$n = 94$	$x_n = 0.94$	$y_n = 2.8399 \times 10^{11}$.

The values are still in reasonable agreement at $x = 0.9$, but they start to diverge quickly thereafter. This suggests that X is slightly greater than 0.9.

Section 17.6 Differential Equations of Second Order (page 1006)

1. If $y_1 = e^x$, then $y_1'' - 3y_1' + 2y_1 = e^x(1 - 3 + 2) = 0$, so y_1 is a solution of the DE $y'' - 3y' + 2y = 0$. Let $y = e^x v$. Then

$$y' = e^x(v' + v), \quad y'' = e^x(v'' + 2v' + v)$$
$$y'' - 3y' + 2y = e^x(v'' + 2v' + v - 3v' - 3v + 2v)$$
$$= e^x(v'' - v').$$

SECTION 17.6 (PAGE 1006)

y satisfies $y'' - 3y' + 2y = 0$ provided $w = v'$ satisfies $w' - w = 0$. This equation has solution $v' = w = C_1 e^x$, so $v = C_1 e^x + C_2$. Thus the given DE has solution $y = e^x v = C_1 e^{2x} + C_2 e^x$.

2. If $y_1 = e^{-2x}$, then $y_1'' - y_1' - 6y_1 = e^{-2x}(4 + 2 - 6) = 0$, so y_1 is a solution of the DE $y'' - y' - 6y = 0$. Let $y = e^{-2x} v$. Then

$$y' = e^{-2x}(v' - 2v), \qquad y'' = e^{-2x}(v'' - 4v' + 4v)$$
$$y'' - y' - 6y = e^{-2x}(v'' - 4v' + 4v - v' + 2v - 6v)$$
$$= e^x(v'' - 5v').$$

y satisfies $y'' - y' - 6y = 0$ provided $w = v'$ satisfies $w' - 5w = 0$. This equation has solution $v' = w = (C_1/5)e^{5x}$, so $v = C_1 e^{5x} + C_2$. Thus the given DE has solution $y = e^{-2x} v = C_1 e^{3x} + C_2 e^{-2x}$.

3. If $y_1 = x$ on $(0, \infty)$, then

$$x^2 y_1'' + 2x y_1' - 2y_1 = 0 + 2x - 2x = 0,$$

so y_1 is a solution of the DE $x^2 y'' + 2xy' - 2y = 0$. Let $y = xv(x)$. Then

$$y' = xv' + v, \qquad y'' = xv'' + 2v'$$
$$x^2 y'' + 2xy' - 2y = x^3 v'' + 2x^2 v' + 2x^2 v' + 2xv - 2xv$$
$$= x^2(xv'' + 4v').$$

y satisfies $x^2 y'' + 2xy' - 2y = 0$ provided $w = v'$ satisfies $xw' + 4w = 0$.
This equation has solution $v' = w = -3C_1 x^{-4}$ (obtained by separation of variables), so $v = C_1 x^{-3} + C_2$. Thus the given DE has solution $y = xv = C_1 x^{-2} + C_2 x$.

4. If $y_1 = x^2$ on $(0, \infty)$, then

$$x^2 y_1'' - 3x y_1' + 4y_1 = 2x^2 - 6x^2 + 4x^2 = 0,$$

so y_1 is a solution of the DE $x^2 y'' - 3xy' + 4y = 0$. Let $y = x^2 v(x)$. Then

$$y' = x^2 v' + 2xv, \qquad y'' = x^2 v'' + 4xv' + 2v$$
$$x^2 y'' - 3xy' + 4y = x^4 v'' + 4x^3 v' + 2x^2 v$$
$$\qquad - 3x^3 v' - 6x^2 v + 4x^2 v$$
$$= x^3(xv'' + v').$$

y satisfies $x^2 y'' - 3xy' + 4y = 0$ provided $w = v'$ satisfies $xw' + w = 0$. This equation has solution $v' = w = C_1/x$ (obtained by separation of variables), so $v = C_1 \ln x + C_2$. Thus the given DE has solution $y = x^2 v = C_1 x^2 \ln x + C_2 x^2$.

5. If $y = x$, then $y' = 1$ and $y'' = 0$. Thus

$$x^2 y'' - x(x + 2)y' + (x + 2)y = 0.$$

R. A. ADAMS: CALCULUS

Now let $y = xv(x)$. Then

$$y' = v + xv', \qquad y'' = 2v' + xv''.$$

Substituting these expressions into the differential equation we get

$$2x^2 v' + x^3 v'' - x^2 v - 2xv - x^3 v'$$
$$\qquad - 2x^2 v' + x^2 v + 2xv = 0$$
$$x^3 v'' - x^3 v' = 0, \qquad \text{or } v'' - v' = 0,$$

which has solution $v = C_1 + C_2 e^x$. Hence the general solution of the given differential equation is

$$y = C_1 x + C_2 x e^x.$$

6. If $y = x^{-1/2} \cos x$, then

$$y' = -\frac{1}{2} x^{-3/2} \cos x - x^{-1/2} \sin x$$
$$y'' = \frac{3}{4} x^{-5/2} \cos x + x^{-3/2} \sin x - x^{-1/2} \cos x.$$

Thus

$$x^2 y'' + xy' + \left(x^2 - \frac{1}{4} \right) y$$
$$= \frac{3}{4} x^{-1/2} \cos x + x^{1/2} \sin x - x^{3/2} \cos x$$
$$\quad - \frac{1}{2} x^{-1/2} \cos x - x^{1/2} \sin x + x^{3/2} \cos x - \frac{1}{4} x^{-1/2} \cos x$$
$$= 0.$$

Therefore $y = x^{-1/2} \cos x$ is a solution of the Bessel equation

$$x^2 y'' + xy' + \left(x^2 - \frac{1}{4} \right) y = 0. \qquad (*)$$

Now let $y = x^{-1/2}(\cos x) v(x)$. Then

$$y' = -\frac{1}{2} x^{-3/2}(\cos x) v - x^{-1/2}(\sin x) v + x^{-1/2}(\cos x) v'$$
$$y'' = \frac{3}{4} x^{-5/2}(\cos x) v + x^{-3/2}(\sin x) v - x^{-3/2}(\cos x) v'$$
$$\quad - x^{-1/2}(\cos x) v - 2x^{-1/2}(\sin x) v' + x^{-1/2}(\cos x) v''.$$

If we substitute these expressions into the equation (∗), many terms cancel out and we are left with the equation

$$(\cos x) v'' - 2(\sin x) v' = 0.$$

614

INSTRUCTOR'S SOLUTIONS MANUAL SECTION 17.7 (PAGE 1015)

Substituting $u = v'$, we rewrite this equation in the form

$$(\cos x)\frac{du}{dx} = 2(\sin x)u$$

$$\int \frac{du}{u} = 2\int \tan x\, dx \Rightarrow \ln|u| = 2\ln|\sec x| + C_0.$$

Thus $v' = u = C_1 \sec^2 x$, from which we obtain

$$v = C_1 \tan x + C_2.$$

Thus the general solution of the Bessel equation (∗) is

$$y = x^{-1/2}(\cos x)v = C_1 x^{-1/2}\sin x + C_2 x^{-1/2}\cos x.$$

7. If $y_1 = y$ and $y_2 = y'$ where y satisfies

$$y'' + a_1(x)y' + a_0(x)y = f(x),$$

then $y_1' = y_2$ and $y_2' = -a_0 y_1 - a_1 y_2 + f$. Thus

$$\frac{d}{dx}\begin{pmatrix} y_1 \\ y_2 \end{pmatrix} = \begin{pmatrix} 0 & 1 \\ -a_0 & -a_1 \end{pmatrix}\begin{pmatrix} y_1 \\ y_2 \end{pmatrix} + \begin{pmatrix} 0 \\ f \end{pmatrix}.$$

8. If y satisfies

$$y^{(n)} + a_{n-1}(x)y^{(n-1)} + \cdots + a_1(x)y' + a_0(x)y = f(x),$$

then let

$$y_1 = y, \quad y_2 = y', \quad y_3 = y'', \quad \ldots \quad y_n = y^{(n-1)}.$$

Therefore

$$y_1' = y_2, \quad y_2' = y_3, \quad \ldots \quad y_{n-2}' = y_{n-1}, \quad \text{and}$$
$$y_n' = -a_0 y_1 - a_1 y_2 - a_2 y_3 - \cdots - a_{n-1} y_n + f,$$

and we have

$$\frac{d}{dx}\begin{pmatrix} y_1 \\ y_2 \\ \vdots \\ y_n \end{pmatrix} = \begin{pmatrix} 0 & 1 & 0 & \cdots & 0 \\ 0 & 0 & 1 & \cdots & 0 \\ \vdots & \vdots & \vdots & & \vdots \\ 0 & 0 & 0 & \cdots & 1 \\ -a_0 & -a_1 & -a_2 & \cdots & -a_n \end{pmatrix}\begin{pmatrix} y_1 \\ y_2 \\ \vdots \\ y_n \end{pmatrix}$$

$$+ \begin{pmatrix} 0 \\ 0 \\ \vdots \\ 0 \\ f \end{pmatrix}.$$

9. If $\mathbf{y} = C_1 e^{\lambda x}\mathbf{v}$, then

$$\mathbf{y}' = C_1 \lambda e^{\lambda x}\mathbf{v} = C_1 e^{\lambda x}\mathcal{A}\mathbf{v} = \mathcal{A}\mathbf{y}$$

provided λ and $\mathbf{v}$ satisfy $\mathcal{A}\mathbf{v} = \lambda\mathbf{v}$.

10. $\begin{vmatrix} 2-\lambda & 1 \\ 2 & 3-\lambda \end{vmatrix} = 6 - 5\lambda + \lambda^2 - 2$
$$= \lambda^2 - 5\lambda + 4$$
$$= (\lambda - 1)(\lambda - 4) = 0$$

if $\lambda = 1$ or $\lambda = 4$.

Let $\mathcal{A} = \begin{pmatrix} 2 & 1 \\ 2 & 3 \end{pmatrix}$.

If $\lambda = 1$ and $\mathcal{A}\mathbf{v} = \mathbf{v}$, then

$$\mathcal{A} = \begin{pmatrix} 2 & 1 \\ 2 & 3 \end{pmatrix}\begin{pmatrix} v_1 \\ v_2 \end{pmatrix} = \begin{pmatrix} v_1 \\ v_2 \end{pmatrix} \Leftrightarrow v_1 + v_2 = 0.$$

Thus we may take $\mathbf{v} = \mathbf{v}_1 = \begin{pmatrix} 1 \\ -1 \end{pmatrix}$.

If $\lambda = 4$ and $\mathcal{A}\mathbf{v} = 4\mathbf{v}$, then

$$\mathcal{A} = \begin{pmatrix} 2 & 1 \\ 2 & 3 \end{pmatrix}\begin{pmatrix} v_1 \\ v_2 \end{pmatrix} = 4\begin{pmatrix} v_1 \\ v_2 \end{pmatrix} \Leftrightarrow 2v_1 - v_2 = 0.$$

Thus we may take $\mathbf{v} = \mathbf{v}_2 = \begin{pmatrix} 1 \\ 2 \end{pmatrix}$.

By the result of Exercise 9, $\mathbf{y} = e^x \mathbf{v}_1$ and $\mathbf{y} = e^{4x}\mathbf{v}_2$ are solutions of the homogeneous linear system $\mathbf{y}' = \mathcal{A}\mathbf{y}$. Therefore the general solution of the system is

$$\mathbf{y} = C_1 e^x \mathbf{v}_1 + C_2 e^{4x}\mathbf{v}_2,$$

that is

$$\begin{pmatrix} y_1 \\ y_2 \end{pmatrix} = C_1 e^x \begin{pmatrix} 1 \\ -1 \end{pmatrix} + C_2 e^{4x}\begin{pmatrix} 1 \\ 2 \end{pmatrix}, \quad \text{or}$$
$$y_1 = C_1 e^x + C_2 e^{4x}$$
$$y_2 = -C_1 e^x + 2C_2 e^{4x}.$$

Section 17.7 Linear Differential Equations with Constant Coefficients (page 1015)

1. $\qquad y'' + 7y' + 10y = 0$
   auxiliary eqn $r^2 + 7r + 10 = 0$
   $(r+5)(r+2) = 0 \Rightarrow r = -5, -2$
   $y = Ae^{-5t} + Be^{-2t}$.

2. $\qquad y'' - 2y' - 3y = 0$
   auxiliary eqn $r^2 - 2r - 3 = 0 \Rightarrow r = -1, r = 3$
   $y = Ae^{-t} + Be^{3t}$.

3. $\qquad y'' + 2y' = 0$
   auxiliary eqn $r^2 + 2r = 0 \Rightarrow r = 0, -2$
   $y = A + Be^{-2t}$.

615

4. $4y'' - 4y' - 3y = 0$
   $4r^2 - 4r - 3 = 0 \Rightarrow (2r+1)(2r-3) = 0$
   Thus, $r_1 = -\frac{1}{2}$, $r_2 = \frac{3}{2}$, and $y = Ae^{-(1/2)t} + Be^{(3/2)t}$.

5. $\qquad y'' + 8y' + 16y = 0$
   auxiliary eqn $r^2 + 8r + 16 = 0 \Rightarrow r = -4, -4$
   $y = Ae^{-4t} + Bte^{-4t}$.

6. $y'' - 2y' + y = 0$
   $r^2 - 2r + 1 = 0 \Rightarrow (r-1)^2 = 0$
   Thus, $r = 1, 1$, and $y = Ae^t + Bte^t$.

7. $\qquad y'' - 6y' + 10y = 0$
   auxiliary eqn $r^2 - 6r + 10 = 0 \Rightarrow r = 3 \pm i$
   $y = Ae^{3t}\cos t + Be^{3t}\sin t$.

8. $9y'' + 6y' + y = 0$
   $9r^2 + 6r + 1 = 0 \Rightarrow (3r+1)^2 = 0$
   Thus, $r = -\frac{1}{3}, -\frac{1}{3}$, and $y = Ae^{-(1/3)t} + Bte^{-(1/3)t}$.

9. $\qquad y'' + 2y' + 5y = 0$
   auxiliary eqn $r^2 + 2r + 5 = 0 \Rightarrow r = -1 \pm 2i$
   $y = Ae^{-t}\cos 2t + Be^{-t}\sin 2t$.

10. For $y'' - 4y' + 5y = 0$ the auxiliary equation is $r^2 - 4r + 5 = 0$, which has roots $r = 2 \pm i$. Thus, the general solution of the DE is $y = Ae^{2t}\cos t + Be^{2t}\sin t$.

11. For $y'' + 2y' + 3y = 0$ the auxiliary equation is $r^2 + 2r + 3 = 0$, which has solutions $r = -1 \pm \sqrt{2}i$. Thus the general solution of the given equation is $y = Ae^{-t}\cos(\sqrt{2}t) + Be^{-t}\sin(\sqrt{2}t)$.

12. Given that $y'' + y' + y = 0$, hence $r^2 + r + 1 = 0$. Since $a = 1$, $b = 1$ and $c = 1$, the discriminant is $D = b^2 - 4ac = -3 < 0$ and $-(b/2a) = -\frac{1}{2}$ and $\omega = \sqrt{3}/2$. Thus, the general solution is
$$y = Ae^{-(1/2)t}\cos\left(\frac{\sqrt{3}}{2}t\right) + Be^{-(1/2)t}\sin\left(\frac{\sqrt{3}}{2}t\right).$$

13. $\begin{cases} 2y'' + 5y' - 3y = 0 \\ y(0) = 1 \\ y'(0) = 0 \end{cases}$
    The DE has auxiliary equation $2r^2 + 5y - 3 = 0$, with roots $r = \frac{1}{2}$ and $r = -3$. Thus $y = Ae^{t/2} + Be^{-3t}$.
    Now $1 = y(0) = A + B$, and $0 = y'(0) = \frac{A}{2} - 3B$.
    Thus $B = 1/7$ and $A = 6/7$. The solution is
    $y = \frac{6}{7}e^{t/2} + \frac{1}{7}e^{-3t}$.

14. Given that $y'' + 10y' + 25y = 0$, hence $r^2 + 10r + 25 = 0 \Rightarrow (r+5)^2 = 0 \Rightarrow r = -5$. Thus,
    $y = Ae^{-5t} + Bte^{-5t}$
    $y' = -5e^{-5t}(A + Bt) + Be^{-5t}$.

Since
$$0 = y(1) = Ae^{-5} + Be^{-5}$$
$$2 = y'(1) = -5e^{-5}(A+B) + Be^{-5},$$
we have $A = -2e^5$ and $B = 2e^5$.
Thus, $y = -2e^5 e^{-5t} + 2te^5 e^{-5t} = 2(t-1)e^{-5(t-1)}$.

15. $\begin{cases} y'' + 4y' + 5y = 0 \\ y(0) = 2 \\ y'(0) = 0 \end{cases}$
    The auxiliary equation for the DE is $r^2 + 4r + 5 = 0$, which has roots $r = -2 \pm i$. Thus
    $y = Ae^{-2t}\cos t + Be^{-2t}\sin t$
    $y' = (-2Ae^{-2t} + Be^{-2t})\cos t - (Ae^{-2t} + 2Be^{-2t})\sin t$.
    Now $2 = y(0) = A \Rightarrow A = 2$, and
    $2 = y'(0) = -2A + B \Rightarrow B = 6$.
    Therefore $y = e^{-2t}(2\cos t + 6\sin t)$.

16. $\begin{cases} y'' + y' + y = 0 \\ y(2\pi/\sqrt{3}) = 0 \\ y'(2\pi/\sqrt{3}) = 1 \end{cases}$
    The auxiliary equation for the DE is $r^2 + r + 1 = 0$, which has roots $r = \frac{1}{2}(-1 \pm i\sqrt{3})$. Thus
    $y = Ae^{-t/2}\cos\frac{\sqrt{3}t}{2} + Be^{-t/2}\sin\frac{\sqrt{3}t}{2}$
    $y' = \frac{-A + \sqrt{3}B}{2}e^{-t/2}\cos\frac{\sqrt{3}t}{2} - \frac{\sqrt{3}A + B}{2}e^{-t/2}\sin\frac{\sqrt{3}t}{2}$.
    Now $0 = y(2\pi/\sqrt{3}) = -e^{-\pi/\sqrt{3}}A \Rightarrow A = 0$, and
    $1 = y'(2\pi/\sqrt{3}) = \frac{A - \sqrt{3}B}{2}e^{-\pi/\sqrt{3}} \Rightarrow B = \frac{-2e^{\pi/\sqrt{3}}}{\sqrt{3}}$.
    Therefore $y = -\frac{2}{\sqrt{3}}e^{(\pi/\sqrt{3})-(t/2)}\sin\frac{\sqrt{3}t}{2}$.

17. Given that $a > 0$, $b > 0$ and $c > 0$:
    Case 1: If $D = b^2 - 4ac > 0$ then the two roots are
    $$r_{1,2} = \frac{-b \pm \sqrt{b^2 - 4ac}}{2a}.$$
    Since
    $$b^2 - 4ac < b^2$$
    $$\pm\sqrt{b^2 - 4ac} < b$$
    $$-b \pm \sqrt{b^2 - 4ac} < 0$$
    therefore r_1 and r_2 are negative. The general solution is
    $$y(t) = Ae^{r_1 t} + Be^{r_2 t}.$$
    If $t \to \infty$, then $e^{r_1 t} \to 0$ and $e^{r_2 t} \to 0$.
    Thus, $\lim_{t \to \infty} y(t) = 0$.
    Case 2: If $D = b^2 - 4ac = 0$ then the two equal roots $r_1 = r_2 = -b/(2a)$ are negative. The general solution is
    $$y(t) = Ae^{r_1 t} + Bte^{r_2 t}.$$

If $t \to \infty$, then $e^{r_1 t} \to 0$ and $e^{r_2 t} \to 0$ at a faster rate than $Bt \to \infty$. Thus, $\lim_{t \to \infty} y(t) = 0$.

Case 3: If $D = b^2 - 4ac < 0$ then the general solution is
$$y = Ae^{-(b/2a)t} \cos(\omega t) + Be^{-(b/2a)t} \sin(\omega t)$$
where $\omega = \dfrac{\sqrt{4ac - b^2}}{2a}$. If $t \to \infty$, then the amplitude of both terms $Ae^{-(b/2a)t} \to 0$ and $Be^{-(b/2a)t} \to 0$. Thus, $\lim_{t \to \infty} y(t) = 0$.

18. The auxiliary equation $ar^2 + br + c = 0$ has roots
$$r_1 = \frac{-b - \sqrt{D}}{2a}, \quad r_2 = \frac{-b + \sqrt{D}}{2a},$$
where $D = b^2 - 4ac$. Note that $a(r_2 - r_1) = \sqrt{D} = -(2ar_1 + b)$. If $y = e^{r_1 t} u$, then $y' = e^{r_1 t}(u' + r_1 u)$, and $y'' = e^{r_1 t}(u'' + 2r_1 u' + r_1^2 u)$. Substituting these expressions into the DE $ay'' + by' + cy = 0$, and simplifying, we obtain
$$e^{r_1 t}(au'' + 2ar_1 u' + bu') = 0,$$
or, more simply, $u'' - (r_2 - r_1)u' = 0$. Putting $v = u'$ reduces this equation to first order:
$$v' = (r_2 - r_1)v,$$
which has general solution $v = Ce^{(r_2 - r_1)t}$. Hence
$$u = \int Ce^{(r_2 - r_1)t}\, dt = Be^{(r_2 - r_1)t} + A,$$
and $y = e^{r_1 t} u = Ae^{r_1 t} + Be^{r_2 t}$.

19. $y''' - 4y'' + 3y' = 0$
Auxiliary: $r^3 - 4r^2 + 3r = 0$
$r(r-1)(r-3) = 0 \Rightarrow r = 0, 1, 3$
General solution: $y = C_1 + C_2 e^t + C_3 e^{3t}$.

20. $y^{(4)} - 2y'' + y = 0$
Auxiliary: $r^4 - 2r^2 + 1 = 0$
$(r^2 - 1)^2 = 0 \Rightarrow r = -1, -1, 1, 1$
General solution: $y = C_1 e^{-t} + C_2 t e^{-t} + C_3 e^t + C_4 t e^t$.

21. $y^{(4)} + 2y'' + y = 0$
Auxiliary: $r^4 + 2r^2 + 1 = 0$
$(r^2 + 1)^2 = 0 \Rightarrow r = -i, -i, i, i$
General solution:
$y = C_1 \cos t + C_2 \sin t + C_3 t \cos t + C_4 t \sin t$.

22. $y^{(4)} + 4y^{(3)} + 6y'' + 4y' + y = 0$
Auxiliary: $r^4 + 4r^3 + 6r^2 + 4r + 1 = 0$
$(r+1)^4 = 0 \Rightarrow r = -1, -1, -1, -1$
General solution: $y = e^{-t}(C_1 + C_2 t + C_3 t^2 + C_4 t^3)$.

23. If $y = e^{2t}$, then $y''' - 2y' - 4y = e^{2t}(8 - 4 - 4) = 0$. The auxiliary equation for the DE is $r^3 - 2r - 4 = 0$, for which we already know that $r = 2$ is a root. Dividing the left side by $r - 2$, we obtain the quotient $r^2 + 2r + 2$. Hence the other two auxiliary roots are $-1 \pm i$.
General solution: $y = C_1 e^{2t} + C_2 e^{-t} \cos t + C_3 e^{-t} \sin t$.

24. Aux. eqn: $(r^2 - r - 2)^2 (r^2 - 4)^2 = 0$
$(r+1)^2(r-2)^2(r-2)^2(r+2)^2 = 0$
$r = 2, 2, 2, 2, -1, -1, -2, -2$.
The general solution is
$y = e^{2t}(C_1 + C_2 t + C_3 t^2 + C_4 t^3) + e^{-t}(C_5 + C_6 t)$
$\quad + e^{-2t}(C_7 + C_8 t)$.

25. $\quad t^2 y'' - ty' + y = 0$
aux: $r(r-1) - r + 1 = 0$
$r^2 - 2r + 1 = 0$
$(r-1)^2 = 0, \quad r = 1, 1$.
Thus $y = At + Bt \ln t$.

26. $\quad t^2 y'' - ty' - 3y = 0$
$r(r-1) - r - 3 = 0 \Rightarrow r^2 - 2r - 3 = 0$
$\Rightarrow (r-3)(r+1) = 0 \Rightarrow r_1 = -1$ and $r_2 = 3$
Thus, $y = At^{-1} + Bt^3$.

27. $\quad t^2 y'' + ty' - y = 0$
aux: $r(r-1) + r - 1 = 0 \Rightarrow r = \pm 1$
$y = At + \dfrac{B}{t}$.

28. Consider $t^2 y'' - ty' + 5y = 0$. Since $a = 1$, $b = -1$, and $c = 5$, therefore $(b-a)^2 < 4ac$. Then $k = (a-b)/2a = 1$ and $\omega^2 = 4$. Thus, the general solution is
$y = At \cos(2 \ln t) + Bt \sin(2 \ln t)$.

29. $\quad t^2 y'' + ty' = 0$
aux: $r(r-1) + r = 0 \Rightarrow r = 0, 0$.
Thus $y = A + B \ln t$.

30. Given that $t^2 y'' + ty' + y = 0$. Since $a = 1$, $b = 1$, $c = 1$ therefore $(b-a)^2 < 4ac$. Then $k = (a-b)/2a = 0$ and $\omega^2 = 1$. Thus, the general solution is
$y = A \cos(\ln t) + B \sin(\ln t)$.

31. $x^3 y''' + xy' - y = 0$.
Trying $y = x^r$ leads to the auxiliary equation
$$r(r-1)(r-2) + r - 1 = 0$$
$$r^3 - 3r^2 + 3r - 1 = 0$$
$$(r-1)^3 = 0 \Rightarrow r = 1, 1, 1.$$

Thus $y = x$ is a solution. To find the general solution, try $y = xv(x)$. Then
$$y' = xv' + v, \quad y'' = xv'' + 2v', \quad y''' = xv''' + 3v''.$$

Now $x^3y''' + xy' - y = x^4v''' + 3x^3v'' + x^2v' + xv - xv$
$$= x^2(x^2v''' + 3xv'' + v'),$$
and y is a solution of the given equation if $v' = w$ is a solution of $x^2w'' + 3xw' + w = 0$. This equation has auxiliary equation $r(r-1) + 3r + 1 = 0$, that is $(r+1)^2 = 0$, so its solutions are

$$v' = w = \frac{C_2}{x} + \frac{2C_3 \ln x}{x}$$
$$v = C_1 + C_2 \ln x + C_3 (\ln x)^2.$$

The general solution of the given equation is, therefore,

$$y = C_1 x + C_2 x \ln x + C_3 x (\ln x)^2.$$

Section 17.8 Nonhomogeneous Linear Equations (page 1021)

1. $y'' + y' - 2y = 1$.
The auxiliary equation for $y'' + y' - 2y = 0$ is $r^2 + r - 2 = 0$, which has roots $r = -2$ and $r = 1$. Thus the complementary function is

$$y_h = C_1 e^{-2x} + C_2 e^x.$$

For a particular solution y_p of the given equation try $y = A$. This satisfies the given equation if $A = -1/2$. Thus the general solution of the given equation is

$$y = -\frac{1}{2} + C_1 e^{-2x} + C_2 e^x.$$

2. $y'' + y' - 2y = x$.
The complementary function is $y_h = C_1 e^{-2x} + C_2 e^x$, as shown in Exercise 1. For a particular solution try $y = Ax + B$. Then $y' = A$ and $y'' = 0$, so y satisfies the given equation if

$$x = A - 2(Ax + B) = A - 2B - 2Ax.$$

We require $A - 2B = 0$ and $-2A = 1$, so $A = -1/2$ and $B = -1/4$. The general solution of the given equation is

$$y = -\frac{2x+1}{4} + C_1 e^{-2x} + C_2 e^x.$$

3. $y'' + y' - 2y = e^{-x}$.
The complementary function is $y_h = C_1 e^{-2x} + C_2 e^x$, as shown in Exercise 1. For a particular solution try $y = Ae^{-x}$. Then $y' = -Ae^{-x}$ and $y'' = Ae^{-x}$, so y satisfies the given equation if

$$e^{-x} = e^{-x}(A - A - 2A) = -2Ae^{-x}.$$

We require $A = -1/2$. The general solution of the given equation is

$$y = -\frac{1}{2}e^{-x} + C_1 e^{-2x} + C_2 e^x.$$

4. $y'' + y' - 2y = e^x$.
The complementary function is $y_h = C_1 e^{-2x} + C_2 e^x$, as shown in Exercise 1. For a particular solution try $y = Axe^x$. Then

$$y' = Ae^x(1+x), \qquad y'' = Ae^x(2+x),$$

so y satisfies the given equation if

$$e^x = Ae^x(2 + x + 1 + x - 2x) = 3Ae^x.$$

We require $A = 1/3$. The general solution of the given equation is

$$y = \frac{1}{3}xe^x + C_1 e^{-2x} + C_3 e^x.$$

5. $y'' + 2y' + 5y = x^2$.
The homogeneous equation has auxiliary equation $r^2 + 2r + 5 = 0$ with roots $r = -1 \pm 2i$. Thus the complementary function is

$$y_h = C_1 e^{-x} \cos(2x) + C_2 e^{-x} \sin(2x).$$

For a particular solution, try $y = Ax^2 + Bx + C$. Then $y' = 2Ax + B$ and $y'' = 2A$. We have

$$x^2 = y'' + 2y' + 5y$$
$$= 2A + 4Ax + 2B + 5Ax^2 + 5Bx + 5C.$$

Thus we require $5A = 1$, $4A + 5B = 0$, and $2A + 2B + 5C = 0$. This gives $A = 1/5$, $B = -4/25$, and $C = -2/125$. The given equation has general solution

$$y = \frac{x^2}{5} - \frac{4x}{25} - \frac{2}{125} + e^{-x}(C_1 \cos(2x) + C_2 \sin(2x)).$$

6. $y'' + 4y = x^2$. The complementary function is $y = C_1 \cos(2x) + C_2 \sin(2x)$. For the given equation, try $y = Ax^2 + Bx + C$. Then

$$x^2 = y'' + 4y = 2A + 4Ax^2 + 4Bx + 4C$$

Thus $2A + 4C = 0$, $4A = 1$, $4B = 0$, and we have $A = \frac{1}{4}$, $B = 0$, and $C = -\frac{1}{8}$. The given equation has general solution

$$y = \frac{1}{4}x^2 - \frac{1}{8} + C_1 \cos(2x) + C_2 \sin(2x).$$

7. $y'' - y' - 6y = e^{-2x}$.
   The homogeneous equation has auxiliary equation $r^2 - r - 6 = 0$ with roots $r = -2$ and $r = 3$. Thus the complementary function is
   $$y_h = C_1 e^{-2x} + C_2 e^{3x}.$$
   For a particular solution, try $y = Axe^{-2x}$. Then $y' = e^{-2x}(A - 2Ax)$ and $y'' = e^{-2x}(-4A + 4Ax)$. We have
   $$e^{-2x} = y'' - y' - 6y$$
   $$= e^{-2x}(-4A + 4Ax - A + 2Ax - 6Ax) = -5Ae^{-2x}.$$
   Thus we require $A = -1/5$. The given equation has general solution
   $$y = -\frac{1}{5}xe^{-2x} + C_1 e^{-2x} + C_2 e^{3x}.$$

8. $y'' + 4y' + 4y = e^{-2x}$.
   The homogeneous equation has auxiliary equation $r^2 + 4r + 4 = 0$ with roots $r = -2, -2$. Thus the complementary function is
   $$y_h = C_1 e^{-2x} + C_2 xe^{-2x}.$$
   For a particular solution, try $y = Ax^2 e^{-2x}$. Then $y' = e^{-2x}(2Ax - 2Ax^2)$ and $y'' = e^{-2x}(2A - 8Ax + 4Ax^2)$. We have
   $$e^{-2x} = y'' + 4y' + 4y$$
   $$= e^{-2x}(2A - 8Ax + 4Ax^2 + 8Ax - 8Ax^2 + 4Ax^2)$$
   $$= 2Ae^{-2x}.$$
   Thus we require $A = 1/2$. The given equation has general solution
   $$y = e^{-2x}\left(\frac{x^2}{2} + C_1 + C_2 x\right).$$

9. $y'' + 2y' + 2y = e^x \sin x$.
   The homogeneous equation has auxiliary equation $r^2 + 2r + 2 = 0$ with roots $r = -1 \pm i$. Thus the complementary function is
   $$y_h = C_1 e^{-x} \cos x + C_2 e^{-x} \sin x.$$
   For a particular solution, try $y = Ae^x \cos x + Be^x \sin x$. Then
   $$y' = (A + B)e^x \cos x + (B - A)e^x \sin x$$
   $$y'' = 2Be^x \cos x - 2Ae^x \sin x.$$

   This satisfies the nonhomogeneous DE if
   $$e^x \sin x = y'' + 2y' + 2y$$
   $$= e^x \cos x(2B + 2(A + B) + 2A)$$
   $$+ e^x \sin x(-2A + 2(B - A) + 2B)$$
   $$= e^x \cos x(4A + 4B) + e^x \sin x(4B - 4A).$$
   Thus we require $A + B = 0$ and $4(B - A) = 1$, that is, $B = -A = 1/8$. The given equation has general solution
   $$y = \frac{e^x}{8}(\sin x - \cos x) + e^{-x}(C_1 \cos x + C_2 \sin x).$$

10. $y'' + 2y' + 2y = e^{-x} \sin x$.
    The complementary function is the same as in Exercise 9, but for a particular solution we try
    $$y = Axe^{-x} \cos x + Bxe^{-x} \sin x$$
    $$y' = e^{-x} \cos x(A - Ax + Bx) + e^{-x} \sin x(B - Bx - Ax)$$
    $$y'' = e^{-x} \cos x(2B - 2Bx - 2A)$$
    $$+ e^{-x} \sin x(2Ax - 2A - 2B).$$
    This satisfies the nonhomogeneous DE if
    $$e^{-x} \sin x = y'' + 2y' + 2y$$
    $$= 2Be^{-x} \cos x - 2Ae^{-x} \sin x.$$
    Thus we require $B = 0$ and $A = -1/2$. The given equation has general solution
    $$y = -\frac{1}{2}xe^{-x} \cos x + e^{-x}(C_1 \cos x + C_2 \sin x).$$

11. $y'' + y' = 4 + 2x + e^{-x}$.
    The homogeneous equation has auxiliary equation $r^2 + r = 0$ with roots $r = 0$ and $r = -1$. Thus the complementary function is $y_h = C_1 + C_2 e^{-x}$. For a particular solution, try $y = Ax + Bx^2 + Cxe^{-x}$. Then
    $$y' = A + 2Bx + e^{-x}(C - Cx)$$
    $$y'' = 2B + e^{-x}(-2C + Cx).$$
    This satisfies the nonhomogeneous DE if
    $$4 + 2x + e^{-x} = y'' + y'$$
    $$= A + 2B + 2Bx - Ce^{-x}.$$
    Thus we require $A + 2B = 4$, $2B = 2$, and $-C = 1$, that is, $A = 2$, $B = 1$, $C = -1$. The given equation has general solution
    $$y = 2x + x^2 - xe^{-x} + C_1 + C_2 e^{-x}.$$

12. $y'' + 2y' + y = xe^{-x}$.
The homogeneous equation has auxiliary equation $r^2 + 2r + 1 = 0$ with roots $r = -1$ and $r = -1$. Thus the complementary function is $y_h = C_1 e^{-x} + C_2 x e^{-x}$. For a particular solution, try $y = e^{-x}(Ax^2 + Bx^3)$. Then

$$y' = e^{-x}(2Ax + (3B - A)x^2 - Bx^3)$$
$$y'' = e^{-x}(2A + (6B - 4A)x - (6B - A)x^2 + Bx^3).$$

This satisfies the nonhomogeneous DE if

$$xe^{-x} = y'' + 2y' + y$$
$$= e^{-x}(2A + 6Bx).$$

Thus we require $A = 0$ and $B = 1/6$. The given equation has general solution

$$y = \frac{1}{6}x^3 e^{-x} + C_1 e^{-x} + C_2 x e^{-x}.$$

13. $y'' + y' - 2y = e^{-x}$.
The complementary function is $y_h = C_1 e^{-2x} + C_2 e^x$. For a particular solution use

$$y_p = e^{-2x} u_1(x) + e^x u_2(x),$$

where the coefficients u_1 and u_2 satisfy

$$-2e^{-2x} u_1' + e^x u_2' = e^{-x}$$
$$e^{-2x} u_1' + e^x u_2' = 0.$$

Thus

$$u_1' = -\frac{1}{3}e^x \qquad u_2' = \frac{1}{3}e^{-2x}$$
$$u_1 = -\frac{1}{3}e^x \qquad u_2 = -\frac{1}{6}e^{-2x}.$$

Thus $y_p = -\frac{1}{3}e^{-x} - \frac{1}{6}e^{-x} = -\frac{1}{2}e^{-x}$. The general solution of the given equation is

$$y = -\frac{1}{2}e^{-x} + C_1 e^{-2x} + C_2 e^x.$$

14. $y'' + y' - 2y = e^x$.
The complementary function is $y_h = C_1 e^{-2x} + C_2 e^x$. For a particular solution use

$$y_p = e^{-2x} u_1(x) + e^x u_2(x),$$

where the coefficients u_1 and u_2 satisfy

$$-2e^{-2x} u_1' + e^x u_2' = e^x$$
$$e^{-2x} u_1' + e^x u_2' = 0.$$

Thus

$$u_1' = -\frac{1}{3}e^{3x} \qquad u_2' = \frac{1}{3}$$
$$u_1 = -\frac{1}{9}e^{3x} \qquad u_2 = \frac{1}{3}x.$$

Thus $y_p = -\frac{1}{9}e^x + \frac{1}{3}xe^x$. The general solution of the given equation is

$$y = -\frac{1}{9}e^x + \frac{1}{3}xe^x + C_1 e^{-2x} + C_2 e^x$$
$$= \frac{1}{3}xe^x + C_1 e^{-2x} + C_3 e^x.$$

15. $x^2 y'' + xy' - y = x^2$.
If $y = Ax^2$, then $y' = 2Ax$ and $y'' = 2A$. Thus

$$x^2 = x^2 y'' + xy' - y$$
$$= 2Ax^2 + 2Ax^2 - Ax^2 = 3Ax^2,$$

so $A = 1/3$. A particular solution of the given equation is $y = x^2/3$. The auxiliary equation for the homogeneous equation $x^2 y'' + xy' - y = 0$ is $4r(r - 1) + r - 1 = 0$, or $r^2 - 1 = 0$, which has solutions $r = \pm 1$. Thus the general solution of the given equation is

$$y = \frac{1}{3}x^2 + C_1 x + \frac{C_2}{x}.$$

16. $x^2 y'' + xy' - y = x^r$ has a solution of the form $y = Ax^r$ provided $r \neq \pm 1$. If this is the case, then

$$x^r = Ax^r \bigl(r(r-1) + r - 1\bigr) = Ax^r (r^2 - 1).$$

Thus $A = 1/(r^2 - 1)$ and a particular solution of the DE is

$$y = \frac{1}{r^2 - 1} x^r.$$

17. $x^2 y'' + xy' - y = x$.
Try $y = Ax \ln x$. Then $y' = A(\ln x + 1)$ and $y'' = A/x$. We have

$$x = x^2 \frac{A}{x} + xA(\ln x + 1) - Ax \ln x = 2Ax.$$

Thus $A = 1/2$. The complementary function was obtained in Exercise 15. The given equation has general solution

$$y = \frac{1}{2}x \ln x + C_1 x + \frac{C_2}{x}.$$

18. $x^2 y'' + xy' - y = x$.
Try $y = xu_1(x) + \dfrac{1}{x}u_2(x)$, where u_1 and u_2 satisfy

$$xu_1' + \dfrac{u_2'}{x} = 0, \quad u_1' - \dfrac{u_2'}{x^2} = \dfrac{1}{x}.$$

Solving these equations for u_1' and u_2', we get

$$u_2' = -\dfrac{x}{2}, \quad u_1' = \dfrac{1}{2x}.$$

Thus $u_1 = \dfrac{1}{2}\ln x$ and $u_2 = -\dfrac{x^2}{4}$. A particular solution is

$$y = \dfrac{1}{2}x\ln x - \dfrac{x}{4}.$$

The term $-x/4$ can be absorbed into the term $C_1 x$ in the complementary function, so the general solution is

$$y = \dfrac{1}{2}x\ln x + C_1 x + \dfrac{C_2}{x}.$$

19. $x^2 y'' - (2x + x^2)y' + (2 + x)y = x^3$.
Since x and xe^x are independent solutions of the corresponding homogeneous equation, we can write a solution of the given equation in the form

$$y = xu_1(x) + xe^x u_2(x),$$

where u_1 and u_2 are chosen to satisfy

$$xu_1' + xe^x u_2' = 0, \quad u_1' + (1+x)e^x u_2' = x.$$

Solving these equations for u_1' and u_2', we get $u_1' = -1$ and $u_2' = e^{-x}$. Thus $u_1 = -x$ and $u_2 = -e^{-x}$. The particular solution is $y = -x^2 - x$. Since $-x$ is a solution of the homogeneous equation, we can absorb that term into the complementary function and write the general solution of the given DE as

$$y = -x^2 + C_1 x + C_2 x e^x.$$

20. $x^2 y'' + xy' + \left(x^2 - \dfrac{1}{4}\right)y = x^{3/2}$.
A particular solution can be obtained in the form

$$y = x^{-1/2}(\cos x)u_1(x) + x^{-1/2}(\sin x)u_2(x),$$

where u_1 and u_2 satisfy

$$x^{-1/2}(\cos x)u_1' + x^{-1/2}(\sin x)u_2' = 0$$
$$\left(-\dfrac{1}{2}x^{-3/2}\cos x - x^{-1/2}\sin x\right)u_1'$$
$$- \left(\dfrac{1}{2}x^{-3/2}\sin x - x^{-1/2}\cos x\right)u_2' = x^{-1/2}.$$

We can simplify these equations by dividing the first by $x^{-1/2}$, and adding the first to $2x$ times the second, then dividing the result by $2x^{1/2}$. The resulting equations are

$$(\cos x)u_1' + (\sin x)u_2' = 0$$
$$-(\sin x)u_1' + (\cos x)u_2' = 1,$$

which have solutions $u_1' = -\sin x$, $u_2' = \cos x$, so that $u_1 = \cos x$ and $u_2 = \sin x$. Thus a particular solution of the given equation is

$$y = x^{-1/2}\cos^2 x + x^{-1/2}\sin^2 x = x^{-1/2}.$$

The general solution is

$$y = x^{-1/2}\left(1 + C_2 \cos x + C_2 \sin x\right).$$

Section 17.9 Series Solutions (page 1026)

1. $y'' = (x-1)^2 y$. Try

$$y = \sum_{n=0}^{\infty} a_n (x-1)^n.$$

$$y'' = \sum_{n=2}^{\infty} n(n-1)a_n (x-1)^{n-2}$$

$$= \sum_{n=0}^{\infty} (n+2)(n+1)a_{n+2}(x-1)^n$$

$$0 = y'' - (x-1)^2 y$$

$$= \sum_{n=0}^{\infty} (n+2)(n+1)a_{n+2}(x-1)^n - \sum_{n=0}^{\infty} a_n (x-1)^{n+2}$$

$$= \sum_{n=0}^{\infty} (n+2)(n+1)a_{n+2}(x-1)^n - \sum_{n=2}^{\infty} a_{n-2}(x-1)^n$$

$$= 2a_2 + 6a_3(x-1)$$

$$+ \sum_{n=2}^{\infty} \left[(n+2)(n+1)a_{n+2} - a_{n-2}\right](x-1)^n.$$

621

Thus $a_2 = a_3 = 0$, and $a_{n+2} = \dfrac{a_{n-2}}{(n+1)(n+2)}$ for $n \geq 2$.
Given a_0 and a_1 we have

$$a_4 = \frac{a_0}{3 \times 4}$$
$$a_8 = \frac{a_4}{7 \times 8} = \frac{a_0}{3 \times 4 \times 7 \times 8}$$
$$\vdots$$
$$a_{4n} = \frac{a_0}{3 \times 4 \times 7 \times 8 \times \cdots \times (4n-1)(4n)}$$
$$= \frac{a_0}{4^n n! \times 3 \times 7 \times \cdots \times (4n-1)}$$
$$a_5 = \frac{a_1}{4 \times 5}$$
$$a_9 = \frac{a_5}{8 \times 9} = \frac{a_1}{4 \times 5 \times 8 \times 9}$$
$$\vdots$$
$$a_{4n+1} = \frac{a_1}{4 \times 5 \times 8 \times 9 \times \cdots \times (4n)(4n+1)}$$
$$= \frac{a_1}{4^n n! \times 5 \times 9 \times \cdots \times (4n+1)}$$
$$a_{4n+3} = a_{4n+2} = \cdots = a_3 = a_2 = 0.$$

The solution is

$$y = a_0 \left(1 + \sum_{n=1}^{\infty} \frac{(x-1)^{4n}}{4^n n! \times 3 \times 7 \times \cdots \times (4n-1)} \right)$$
$$+ a_1 \left(x - 1 + \sum_{n=1}^{\infty} \frac{(x-1)^{4n+1}}{4^n n! \times 5 \times 9 \times \cdots \times (4n+1)} \right).$$

2. $y'' = xy$. Try $\sum_{n=0}^{\infty} a_n x^n$. Then

$$y' = \sum_{n=0}^{\infty} n a_n x^{n-1} = \sum_{n=1}^{\infty} n a_n x^{n-1}$$
$$y'' = \sum_{n=2}^{\infty} n(n-1) a_n x^{n-2} = \sum_{n=0}^{\infty} (n+2)(n+1) a_{n+2} x^n.$$

Thus we have

$$0 = y'' - xy$$
$$= \sum_{n=0}^{\infty} (n+2)(n+1) a_{n+2} x^n - \sum_{n=0}^{\infty} a_n x^{n+1}$$
$$= \sum_{n=0}^{\infty} (n+2)(n+1) a_{n+2} x^n - \sum_{n=1}^{\infty} a_{n-1} x^n$$
$$= 2a_2 + \sum_{n=1}^{\infty} \left[(n+2)(n+1) a_{n+2} - a_{n-1} \right] x^n.$$

Thus $a_2 = 0$ and $a_{n+2} = \dfrac{a_{n-1}}{(n+2)(n+1)}$ for $n \geq 1$.
Given a_0 and a_1, we have

$$a_3 = \frac{a_0}{2 \times 3}$$
$$a_6 = \frac{a_3}{5 \times 6} = \frac{a_0}{2 \times 3 \times 5 \times 6} = \frac{1 \times 4 \times a_0}{6!}$$
$$a_9 = \frac{a_6}{8 \times 9} = \frac{1 \times 4 \times 7 \times a_0}{9!}$$
$$\vdots$$
$$a_{3n} = \frac{1 \times 4 \times \cdots \times (3n-2) a_0}{(3n)!}$$
$$a_4 = \frac{a_1}{3 \times 4} = \frac{2 \times a_1}{4!}$$
$$a_7 = \frac{a_4}{6 \times 7} = \frac{2 \times 5 \times a_1}{7!}$$
$$\vdots$$
$$a_{3n+1} = \frac{2 \times 5 \times \cdots \times (3n-1) a_1}{(3n+1)!}$$
$$0 = a_2 = a_5 = a_8 = \cdots = a_{3n+2}.$$

Thus the general solution of the given equation is

$$y = a_0 \left(1 + \sum_{n=1}^{\infty} \frac{1 \times 4 \times \cdots \times (3n-2)}{(3n)!} x^{3n} \right)$$
$$+ a_1 \sum_{n=1}^{\infty} \frac{2 \times 5 \times \cdots \times (3n-1)}{(3n+1)!} x^{3n+1}.$$

3. $y'' + (\sin x) y = 0$, $y(0) = 1$, $y'(0) = 0$. Try

$$y = a_0 + a_1 x + a_2 x^2 + a_3 x^3 + a_4 x^4 + a_5 x^5 + \cdots.$$

Then $a_0 = 1$ and $a_1 = 0$. We have

$$y'' = 2a_2 + 6a_3 x + 12 a_4 x^2 + 20 a_5 x^3 + \cdots$$
$$(\sin x) y = \left(x - \frac{x^3}{6} + \frac{x^5}{120} - \cdots \right)$$
$$\times (1 + a_2 x^2 + a_3 x^3 + a_4 x^4 + a_5 x^5 + \cdots)$$
$$= x + \left(a_2 - \frac{1}{6} \right) x^3 + a_3 x^4$$
$$+ \left(a_4 - \frac{1}{6} a_2 + \frac{1}{120} \right) x^5 + \cdots.$$

Hence we must have $2a_2 = 0$, $6a_3 + 1 = 0$, $12 a_4 = 0$, $20 a_5 + a_2 - \dfrac{1}{6} = 0$, ... That is, $a_2 = 0$, $a_4 = 0$, $a_3 = -\dfrac{1}{6}$, $a_5 = \dfrac{1}{120}$. The solution is

$$y = 1 - \frac{1}{6} x^3 + \frac{1}{120} x^5 + \cdots.$$

4. $(1-x^2)y'' - xy' + 9y = 0$, $y(0) = 0$, $y'(0) = 1$. Try

$$y = \sum_{n=0}^{\infty} a_n x^n.$$

Then $a_0 = 0$ and $a_1 = 1$. We have

$$y' = \sum_{n=1}^{\infty} n a_n x^{n-1}$$

$$y'' = \sum_{n=2}^{\infty} n(n-1) a_n x^{n-2}$$

$$0 = (1-x^2)y'' - xy' + 9y$$

$$= \sum_{n=0}^{\infty} (n+2)(n+1) a_{n+2} x^n - \sum_{n=2}^{\infty} n(n-1) a_n x^n$$

$$- \sum_{n=1}^{\infty} n a_n x^n + 9 \sum_{n=0}^{\infty} a_n x^n$$

$$= 2a_2 + 9a_0 + (6a_3 + 8a_1)x$$

$$+ \sum_{n=2}^{\infty} \left[(n+2)(n+1) a_{n+2} - (n^2 - 9) a_n\right] x^n.$$

Thus $2a_2 + 9a_0 = 0$, $6a_3 + 8a_1 = 0$, and

$$a_{n+2} = \frac{(n^2 - 9) a_n}{(n+1)(n+2)}.$$

Therefore we have

$$a_2 = a_4 = a_6 = \cdots = 0$$

$$a_3 = -\frac{4}{3}, \quad a_5 = 0 = a_7 = a_9 = \cdots.$$

The initial-value problem has solution

$$y = x - \frac{4}{3} x^3.$$

5. $3xy'' + 2y' + y = 0$.
Since $x = 0$ is a regular singular point of this equation, try

$$y = \sum_{n=0}^{\infty} a_n x^{n+\mu} \quad (a_0 = 1)$$

$$y' = \sum_{n=0}^{\infty} (n+\mu) a_n x^{n+\mu-1}$$

$$y'' = \sum_{n=0}^{\infty} (n+\mu)(n+\mu-1) a_n x^{n+\mu-2}.$$

Then we have

$$0 = 3xy'' + 2y' + y$$

$$= \sum_{n=0}^{\infty} \left[3(n+\mu)^2 - (n+\mu)\right] a_n x^{n+\mu-1} + \sum_{n=1}^{\infty} a_{n-1} x^{n+\mu-1}$$

$$= (3\mu^2 - \mu) x^{\mu-1}$$

$$+ \sum_{n=1}^{\infty} \left[\left(3(n+\mu)^2 - (n+\mu)\right) a_n + a_{n-1}\right] x^{n+\mu-1}.$$

Thus $3\mu^2 - \mu = 0$ and $a_n = -\dfrac{a_{n-1}}{3(n+\mu)^2 - (n+\mu)}$ for $n \geq 1$. There are two cases: $\mu = 0$ and $\mu = 1/3$.

CASE I. $\mu = 0$. Then $a_n = -\dfrac{a_{n-1}}{n(3n-1)}$. Since $a_0 = 1$ we have

$$a_1 = -\frac{1}{1 \times 2}, \quad a_2 = \frac{1}{1 \times 2 \times 2 \times 5}$$

$$a_3 = -\frac{1}{1 \times 2 \times 2 \times 5 \times 3 \times 8}$$

$$\vdots$$

$$a_n = \frac{(-1)^n}{n! \times 2 \times 5 \times \cdots \times (3n-1)}.$$

One series solution is

$$y = 1 + \sum_{n=1}^{\infty} \frac{(-1)^n x^n}{n! \times 2 \times 5 \times \cdots \times (3n-1)}.$$

CASE II. $\mu = \dfrac{1}{3}$. Then

$$a_n = \frac{-a_{n-1}}{3\left(n+\frac{1}{3}\right)^2 - \left(n+\frac{1}{3}\right)} = \frac{-a_{n-1}}{n(3n+1)}.$$

Since $a_0 = 1$ we have

$$a_1 = -\frac{1}{1 \times 4}, \quad a_2 = \frac{1}{1 \times 4 \times 2 \times 7}$$

$$a_3 = -\frac{1}{1 \times 4 \times 2 \times 7 \times 3 \times 10}$$

$$\vdots$$

$$a_n = \frac{(-1)^n}{n! \times 1 \times 4 \times 7 \times \cdots \times (3n+1)}.$$

A second series solution is

$$y = x^{1/3} \left(1 + \sum_{n=1}^{\infty} \frac{(-1)^n x^n}{n! \times 1 \times 4 \times 7 \times \cdots \times (3n+1)}\right).$$

6. $xy'' + y' + xy = 0$.
    Since $x = 0$ is a regular singular point of this equation, try

    $$y = \sum_{n=0}^{\infty} a_n x^{n+\mu} \quad (a_0 = 1)$$

    $$y' = \sum_{n=0}^{\infty} (n+\mu) a_n x^{n+\mu-1}$$

    $$y'' = \sum_{n=0}^{\infty} (n+\mu)(n+\mu-1) a_n x^{n+\mu-2}.$$

    Then we have

    $$0 = xy'' + y' + xy$$
    $$= \sum_{n=0}^{\infty} \left[(n+\mu)(n+\mu-1) + (n+\mu)\right] a_n x^{n+\mu-1}$$
    $$+ \sum_{n=0}^{\infty} a_n x^{n+\mu+1}$$
    $$= \sum_{n=0}^{\infty} (n+\mu)^2 a_n x^{n+\mu-1} + \sum_{n=2}^{\infty} a_{n-2} x^{n+\mu-1}$$
    $$= \mu^2 x^{\mu-1} + (1+\mu)^2 a_1 x^\mu$$
    $$+ \sum_{n=2}^{\infty} \left[(n+\mu)^2 a_n + a_{n-2}\right] x^{n+\mu-1}.$$

    Thus $\mu = 0$, $a_1 = 0$, and $a_n = -\dfrac{a_{n-2}}{n^2}$ for $n \geq 2$.
    It follows that $0 = a_1 = a_3 = a_5 = \cdots$, and, since $a_0 = 1$,

    $$a_2 = -\frac{1}{2^2}, \quad a_4 = \frac{1}{2^2 4^2}, \ldots$$
    $$a_{2n} = \frac{(-1)^n}{2^2 4^2 \cdots (2n)^2} = \frac{(-1)^n}{2^{2n}(n!)^2}.$$

    One series solution is

    $$y = 1 + \sum_{n=1}^{\infty} \frac{(-1)^n x^{2n}}{2^{2n}(n!)^2}.$$

Review Exercises 17 (page 1026)

1. $\dfrac{dy}{dx} = 2xy$

    $\dfrac{dy}{y} = 2x\,dx \Rightarrow \ln|y| = x^2 + C_1$

    $y = Ce^{x^2}$

2. $\dfrac{dy}{dx} = e^{-y} \sin x$

    $e^y\,dy = \sin x\,dx \Rightarrow e^y = -\cos x + C$

    $y = \ln(C - \cos x)$

3. $\dfrac{dy}{dx} = x + 2y \Rightarrow \dfrac{dy}{dx} - 2y = x$

    $\dfrac{d}{dx}(e^{-2x} y) = e^{-2x}\left(\dfrac{dy}{dx} - 2y\right) = xe^{-2x}$

    $e^{-2x} y = \int xe^{-2x}\,dx = -\dfrac{x}{2}e^{-2x} - \dfrac{1}{4}e^{-2x} + C$

    $y = -\dfrac{x}{2} - \dfrac{1}{4} + Ce^{2x}$

4. $\dfrac{dy}{dx} = \dfrac{x^2 + y^2}{2xy} \quad$ (let $y = xv(x)$)

    $v + x\dfrac{dv}{dx} = \dfrac{1+v^2}{2v}$

    $x\dfrac{dv}{dx} = \dfrac{1+v^2}{2v} - v = \dfrac{1-v^2}{2v}$

    $\dfrac{2v\,dv}{v^2 - 1} = -\dfrac{dx}{x}$

    $\ln(v^2 - 1) = \ln\dfrac{1}{x} + \ln C = \ln\dfrac{C}{x}$

    $\dfrac{y^2}{x^2} - 1 = \dfrac{C}{x} \Rightarrow y^2 - x^2 = Cx$

5. $\dfrac{dy}{dx} = \dfrac{x+y}{y-x}$

    $(x+y)\,dx + (x-y)\,dy = 0 \quad$ (exact)

    $d\left(\dfrac{x^2}{2} + xy - \dfrac{y^2}{2}\right) = 0$

    $x^2 + 2xy - y^2 = C$

6. $\dfrac{dy}{dx} = -\dfrac{y + e^x}{x + e^y}$

    $(y + e^x)\,dx + (x + e^y)\,dy = 0 \quad$ (exact)

    $d\left(xy + e^x + e^y\right) = 0$

    $xy + e^x + e^y = C$

7. $\dfrac{d^2 y}{dt^2} = \left(\dfrac{dy}{dt}\right)^2 \quad$ (let $p = dy/dt$)

    $\dfrac{dp}{dt} = p^2 \Rightarrow \dfrac{dp}{p^2} = dt$

    $\dfrac{1}{p} = C_1 - t$

    $\dfrac{dy}{dt} = p = \dfrac{1}{C_1 - t}$

    $y = \int \dfrac{dt}{C_1 - t} = -\ln|t - C_1| + C_2$

8. $2\dfrac{d^2 y}{dt^2} + 5\dfrac{dy}{dt} + 2y = 0$

    Aux: $2r^2 + 5r + 2 = 0 \Rightarrow r = -1/2, -2$

    $y = C_1 e^{-t/2} + C_2 e^{-2t}$

9. $4y'' - 4y' + 5y = 0$

   Aux: $4r^2 - 4r + 5 = 0$

   $(2r-1)^2 + 4 = 0 \Rightarrow r = \dfrac{1}{2} \pm i$

   $y = C_1 e^{x/2} \cos x + C_2 e^{x/2} \sin x$

10. $2x^2 y'' + y = 0$

    Aux: $2r(r-1) + 1 = 0$

    $2r^2 - 2r + 1 = 0 \Rightarrow r = \dfrac{1}{2}(1 \pm i)$

    $y = C_1 |x|^{1/2} \cos\left(\tfrac{1}{2} \ln |x|\right) + C_2 |x|^{1/2} \sin\left(\tfrac{1}{2} \ln |x|\right)$

11. $t^2 \dfrac{d^2 y}{dt^2} - t \dfrac{dy}{dt} + 5y = 0$

    Aux: $r(r-1) - r + 5 = 0$

    $(r-1)^2 + 4 = 0 \Rightarrow r = 1 \pm 2i$

    $y = C_1 t \cos(2 \ln |t|) + C_2 t \sin(2 \ln |t|)$

12. $\dfrac{d^3 y}{dt^3} + 8 \dfrac{d^2 y}{dt^2} + 16 \dfrac{dy}{dt} = 0$

    Aux: $r^3 + 8r^2 + 16r = 0$

    $r(r+4)^2 = 0 \Rightarrow r = 0, -4, -4$

    $y = C_1 + C_2 e^{-4t} + C_3 t e^{-4t}$

13. $\dfrac{d^2 y}{dx^2} - 5 \dfrac{dy}{dx} + 6y = e^x + e^{3x}$

    Aux: $r^2 - 5r + 6 = 0 \Rightarrow r = 2, 3$.

    Complementary function: $y = C_1 e^{2x} + C_2 e^{3x}$.

    Particular solution: $y = A e^x + B x e^{3x}$

    $y' = A e^x + B(1 + 3x) e^{3x}$

    $y'' = A e^x + B(6 + 9x) e^{3x}$

    $e^x + e^{3x} = A e^x (1 - 5 + 6)$

    $\qquad + B e^{3x}(6 + 9x - 5 - 15x + 6x)$

    $\qquad = 2A e^x + B e^{3x}$.

    Thus $A = 1/2$ and $B = 1$. The general solution is

    $$y = \dfrac{1}{2} e^x + x e^{3x} + C_1 e^{2x} + C_2 e^{3x}.$$

14. $\dfrac{d^2 y}{dx^2} - 5 \dfrac{dy}{dx} + 6y = x e^{2x}$

    Same complementary function as in Exercise 13: $C_1 e^{2x} + C_2 e^{3x}$. For a particular solution we try $y = (Ax^2 + Bx) e^{2x}$. Substituting this into the given DE leads to

    $$x e^{2x} = (2A - B) e^{2x} - 2A x e^{2x},$$

so that we need $A = -1/2$ and $B = 2A = -1$. The general solution is

$$y = -\left(\dfrac{1}{2} x^2 + x\right) e^{2x} + C_1 e^{2x} + C_2 e^{3x}.$$

15. $\dfrac{d^2 y}{dx^2} + 2 \dfrac{dy}{dx} + y = x^2$

    Aux: $r^2 + 2r + 1 = 0$ has solutions $r = -1, -1$.
    Complementary function: $y = C_1 e^{-x} + C_2 x e^{-x}$.
    Particular solution: try $y = Ax^2 + Bx + C$. Then

    $$x^2 = 2A + 2(2Ax + B) + Ax^2 + Bx + C.$$

    Thus $A = 1$, $B = -4$, $C = 6$. The general solution is

    $$y = x^2 - 4x + 6 + C_1 e^{-x} + C_2 x e^{-x}.$$

16. $x^2 \dfrac{d^2 y}{dx^2} - 2y = x^3$.

    The corresponding homogeneous equation has auxiliary equation $r(r-1) - 2 = 0$, with roots $r = 2$ and $r = -1$, so the complementary function is $y = C_1 x^2 + C_2 / x$. A particular solution of the nonhomogeneous equation can have the form $y = A x^3$. Substituting this into the DE gives

    $$6A x^3 - 2A x^3 = x^3,$$

    so that $A = 1/4$. The general solution is

    $$y = \dfrac{1}{4} x^3 + C_1 x^2 + \dfrac{C_2}{x}.$$

17. $\dfrac{dy}{dx} = \dfrac{x^2}{y^2}, \quad y(2) = 1$

    $y^2 \, dy = x^2 \, dx$

    $y^3 = x^3 + C$

    $1 = 8 + C \Rightarrow C = -7$

    $y^3 = x^3 - 7 \Rightarrow y = (x^3 - 7)^{1/3}$

18. $\dfrac{dy}{dx} = \dfrac{y^2}{x^2}, \quad y(2) = 1$

    $\dfrac{dy}{y^2} = \dfrac{dx}{x^2} \Rightarrow -\dfrac{1}{y} = -\dfrac{1}{x} - C$

    $1 = \dfrac{1}{2} + C \Rightarrow C = \dfrac{1}{2}$

    $y = \left(\dfrac{1}{x} + \dfrac{1}{2}\right)^{-1} = \dfrac{2x}{x + 2}$

625

REVIEW EXERCISES 17 (PAGE 1026)   R. A. ADAMS: CALCULUS

19. $\dfrac{dy}{dx} = \dfrac{xy}{x^2+y^2}$, $y(0) = 1$. Let $y = xv(x)$. Then

$$v + x\dfrac{dv}{dx} = \dfrac{v}{1+v^2}$$

$$x\dfrac{dv}{dx} = \dfrac{v}{1+v^2} - v = -\dfrac{v^3}{1+v^2}$$

$$-\dfrac{1+v^2}{v^3}\,dv = \dfrac{dx}{x}$$

$$\dfrac{1}{2v^2} - \ln|v| = \ln|x| + \ln C$$

$$\dfrac{x^2}{y^2} = \dfrac{1}{v^2} = \ln(Cvx)^2 = \ln(C^2y^2)$$

$$C^2 y^2 = e^{x^2/y^2}, \quad y(0) = 1 \Rightarrow C^2 = 1$$

$$y^2 = e^{x^2/y^2}, \quad \text{or } y = e^{x^2/(2y^2)}.$$

20. $\dfrac{dy}{dx} + (\cos x)y = 2\cos x$, $y(\pi) = 1$

$$\dfrac{d}{dx}\left(e^{\sin x}y\right) = e^{\sin x}\left(\dfrac{dy}{dx} + (\cos x)y\right) = 2\cos x\,e^{\sin x}$$

$$e^{\sin x}y = 2e^{\sin x} + C$$

$$y = 2 + Ce^{-\sin x}$$

$$1 = 2 + Ce^0 \Rightarrow C = -1$$

$$y = 2 - e^{-\sin x}$$

21. $y'' + 3y' + 2y = 0$, $y(0) = 1$, $y'(0) = 2$
Aux: $r^2 + 3r + 2 = 0 \Rightarrow r = -1, -2$.

$$y = Ae^{-x} + Be^{-2x} \Rightarrow 1 = A + B$$
$$y' = -Ae^{-x} - 2Be^{-2x} \Rightarrow 2 = -A - 2B.$$

Thus $B = -3$, $A = 4$. The solution is $y = 4e^{-x} - 3e^{-2x}$.

22. $y'' + 2y' + (1+\pi^2)y = 0$, $y(1) = 0$, $y'(1) = \pi$
Aux: $r^2 + 2r + 1 + \pi^2 = 0 \Rightarrow r = -1 \pm \pi i$.

$$y = Ae^{-x}\cos(\pi x) + Be^{-x}\sin(\pi x)$$
$$y' = e^{-x}\cos(\pi x)(-A + B\pi) + e^{-x}\sin(\pi x)(-B - A\pi).$$

Thus $-Ae^{-1} = 0$ and $(A - B\pi)e^{-1} = \pi$, so that $A = 0$ and $B = -e$. The solution is $y = -e^{1-x}\sin(\pi x)$.

23. $y'' + 10y' + 25y = 0$, $y(1) = e^{-5}$, $y'(1) = 0$
Aux: $r^2 + 10r + 25 = 0 \Rightarrow r = -5, -5$.

$$y = Ae^{-5x} + Bxe^{-5x}$$
$$y' = -5Ae^{-5x} + B(1-5x)e^{-5x}.$$

We require $e^{-5} = (A+B)e^{-5}$ and $0 = e^{-5}(-5A - 4B)$.
Thus $A + B = 1$ and $-5A = 4B$, so that $B = 5$ and $A = -4$. The solution is $y = -4e^{-5x} + 5xe^{-5x}$.

24. $x^2 y'' - 3xy' + 4y = 0$, $y(e) = e^2$, $y'(e) = 0$
Aux: $r(r-1) - 3r + 4 = 0$, or $(r-2)^2 = 0$, so that $r = 2, 2$.

$$y = Ax^2 + Bx^2 \ln x$$
$$y' = 2Ax + 2Bx \ln x + Bx.$$

We require $e^2 = Ae^2 + Be^2$ and $0 = 2Ae + 3Be$. Thus $A + B = 1$ and $2A = -3B$, so that $A = 3$ and $B = -2$. The solution is $y = 3x^2 - 2x^2 \ln x$, valid for $x > 0$.

25. $\dfrac{d^2 y}{dt^2} + 4y = 8e^{2t}$, $y(0) = 1$, $y'(0) = -2$
Complementary function: $y = C_1 \cos(2t) + C_2 \sin(2t)$.
Particular solution: $y = Ae^{2t}$, provided $4A + 4A = 8$, that is, $A = 1$. Thus

$$y = e^{2t} + C_1 \cos(2t) + C_2 \sin(2t)$$
$$y' = 2e^{2t} - 2C_1 \sin(2t) + 2C_2 \cos(2t).$$

We require $1 = y(0) = 1 + C_1$ and
$-2 = y'(0) = 2 + 2C_2$. Thus $C_1 = 0$ and $C_2 = -2$. The solution is $y = e^{2t} - 2\sin(2t)$.

26. $2\dfrac{d^2 y}{dx^2} + 5\dfrac{dy}{dx} - 3y = 6 + 7e^{x/2}$, $y(0) = 0$, $y'(0) = 1$
Aux: $2r^2 + 5r - 3 = 0 \Rightarrow r = 1/2, -3$.
Complementary function: $y = C_1 e^{x/2} + C_2 e^{-3x}$.
Particular solution: $y = A + Bxe^{x/2}$

$$y' = Be^{x/2}\left(1 + \dfrac{x}{2}\right)$$
$$y'' = Be^{x/2}\left(1 + \dfrac{x}{4}\right).$$

We need

$$Be^{x/2}\left(2 + \dfrac{x}{2} + 5 + \dfrac{5x}{2} - 3x\right) - 3A = 6 + 7e^{x/2}.$$

This is satisfied if $A = -2$ and $B = 1$. The general solution of the DE is

$$y = -2 + xe^{x/2} + C_1 e^{x/2} + C_2 e^{-3x}.$$

Now the initial conditions imply that

$$0 = y(0) = -2 + C_1 + C_2$$
$$1 = y'(0) = 1 + \dfrac{C_1}{2} - 3C_2,$$

which give $C_1 = 12/7$, $C_2 = 2/7$. Thus the IVP has solution

$$y = -2 + xe^{x/2} + \dfrac{1}{7}(12e^{x/2} + 2e^{-3x}).$$

27. $[(x+A)e^x \sin y + \cos y]\,dx + x[e^x \cos y + B \sin y]\,dy = 0$
is $M\,dx + N\,dy$. We have

$$\dfrac{\partial M}{\partial y} = (x+A)e^x \cos y - \sin y$$

$$\dfrac{\partial N}{\partial x} = e^x \cos y + B \sin y + xe^x \cos y.$$

626

These expressions are equal (and the DE is exact) if $A = 1$ and $B = -1$. If so, the left side of the DE is $d\phi(x, y)$, where

$$\phi(x, y) = xe^x \sin y + x \cos y.$$

The general solution is $xe^x \sin y + x \cos y = C$.

28. $(x^2 + 3y^2)\, dx + xy\, dy = 0$. Multiply by x^n:

$$x^n(x^2 + 3y^2)\, dx + x^{n+1} y\, dy = 0$$

is exact provided $6x^n y = (n + 1)x^n y$, that is, provided $n = 5$. In this case the left side is $d\phi$, where

$$\phi(x, y) = \frac{1}{2} x^6 y^2 + \frac{1}{8} x^8.$$

The general solution of the given DE is

$$4x^6 y^2 + x^8 = C.$$

29. $x^2 y'' - x(2 + x \cot x) y' + (2 + x \cot x) y = 0$
If $y = x$, then $y' = 1$ and $y'' = 0$, so the DE is clearly satisfied by y. To find a second, independent solution, try $y = xv(x)$. Then $y' = v + xv'$, and $y'' = 2v' + xv''$. Substituting these expressions into the given DE, we obtain

$$2x^2 v' + x^3 v'' - (xv + x^2 v')(2 + x \cot x)$$
$$+ xv(2 + x \cot x) = 0$$
$$x^3 v'' - x^3 v' \cot x = 0,$$

or, putting $w = v'$, $w' = (\cot x) w$, that is,

$$\frac{dw}{w} = \frac{\cos x\, dx}{\sin x}$$
$$\ln w = \ln \sin x + \ln C_2$$
$$v' = w = C_2 \sin x \;\Rightarrow\; v = C_1 - C_2 \cos x.$$

A second solution of the DE is $x \cos x$, and the general solution is

$$y = C_1 x + C_2 x \cos x.$$

30. $x^2 y'' - x(2 + x \cot x) y' + (2 + x \cot x) y = x^3 \sin x$
Look for a particular solution of the form
$y = xu_1(x) + x \cos x u_2(x)$, where

$$xu_1' + x \cos x u_2' = 0$$
$$u_1' + (\cos x - x \sin x) u_2' = x \sin x.$$

Divide the first equation by x and subtract from the second equation to get

$$-x \sin x u_2' = x \sin x.$$

Thus $u_2' = -1$ and $u_2 = -x$. The first equation now gives $u_1' = \cos x$, so that $u_1 = \sin x$. The general solution of the DE is

$$y = x \sin x - x^2 \cos x + C_1 x + C_2 x \cos x.$$

31. Suppose $y' = f(x, y)$ and $y(x_0) = y_0$, where $f(x, y)$ is continuous on the whole xy-plane and satisfies $|f(x, y)| \le K$ there. By the Fundamental Theorem of Calculus, we have

$$y(x) - y_0 = y(x) - y(x_0)$$
$$= \int_{x_0}^{x} y'(t)\, dt = \int_{x_0}^{x} f\big(t, y(t)\big)\, dt.$$

Therefore,

$$|y(x) - y_0| \le K |x - x_0|.$$

Thus $y(x)$ is bounded above and below by the lines $y = y_0 \pm K(x - x_0)$, and cannot have a vertical asymptote anywhere.

Remark: we don't seem to have needed the continuity of $\partial f / \partial y$, only the continuity of f (to enable the use of the Fundamental Theorem).

APPENDICES

Appendix I. Complex Numbers (page A-11)

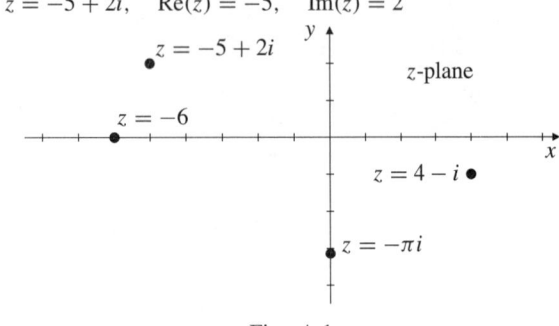

1. $z = -5 + 2i$, $\text{Re}(z) = -5$, $\text{Im}(z) = 2$

   Fig. A.1

2. $z = 4 - i$, $\text{Re}(z) = 4$, $\text{Im}(z) = -1$

3. $z = -\pi i$, $\text{Re}(z) = 0$, $\text{Im}(z) = -\pi$

4. $z = -6$, $\text{Re}(z) = -6$, $\text{Im}(z) = 0$

5. $z = -1 + i$, $|z| = \sqrt{2}$, $\text{Arg}(z) = 3\pi/4$
   $z = \sqrt{2}(\cos(3\pi/4) + i\sin(3\pi/4))$

6. $z = -2$, $|z| = 2$, $\text{Arg}(z) = \pi$
   $z = 2(\cos\pi + i\sin\pi)$

7. $z = 3i$, $|z| = 3$, $\text{Arg}(z) = \pi/2$
   $z = 3(\cos(\pi/2) + i\sin(\pi/2))$

8. $z = -5i$, $|z| = 5$, $\text{Arg}(z) = 3\pi/2$
   $z = 5(\cos(3\pi/2) + i\sin(3\pi/2))$

9. $z = 1 + 2i$, $|z| = \sqrt{5}$, $\theta = \text{Arg}(z) = \tan^{-1}2$
   $z = \sqrt{5}(\cos\theta + i\sin\theta)$

10. $z = -2 + i$, $|z| = \sqrt{5}$, $\theta = \text{Arg}(z) = \pi - \tan^{-1}(1/2)$
    $z = \sqrt{5}(\cos\theta + i\sin\theta)$

11. $z = -3 - 4i$, $|z| = 5$, $\theta = \text{Arg}(z) = \pi + \tan^{-1}(4/3)$
    $z = 5(\cos\theta + i\sin\theta)$

12. $z = 3 - 4i$, $|z| = 5$, $\theta = \text{Arg}(z) = 2\pi - \tan^{-1}(4/3)$
    $z = 5(\cos\theta + i\sin\theta)$

13. $z = \sqrt{3} - i$, $|z| = 2$, $\text{Arg}(z) = 11\pi/6$
    $z = 2(\cos(11\pi/6) + i\sin(11\pi/6))$

14. $z = -\sqrt{3} - 3i$, $|z| = 2\sqrt{3}$, $\text{Arg}(z) = 4\pi/3$
    $z = 2\sqrt{3}(\cos(4\pi/3) + i\sin(4\pi/3))$

15. $z = 3\cos\dfrac{4\pi}{5} + 3i\sin\dfrac{4\pi}{5}$
    $|z| = 3$, $\text{Arg}(z) = \dfrac{4\pi}{5}$

16. If $\text{Arg}(z) = \dfrac{7\pi}{4}$ and $\text{Arg}(w) = \dfrac{\pi}{2}$, then
    $\arg(zw) = \dfrac{7\pi}{4} + \dfrac{\pi}{2} = \dfrac{9\pi}{4}$, so $\text{Arg}(zw) = \dfrac{9\pi}{4} - 2\pi = \dfrac{\pi}{4}$.

17. If $\text{Arg}(z) = \dfrac{\pi}{6}$ and $\text{Arg}(w) = \dfrac{\pi}{4}$, then
    $\arg(z/w) = \dfrac{\pi}{6} - \dfrac{\pi}{4} = -\dfrac{\pi}{12}$, so
    $\text{Arg}(z/w) = -\dfrac{\pi}{12} + 2\pi = \dfrac{23\pi}{12}$.

18. $|z| = 2$, $\arg(z) = \pi \Rightarrow z = 2(\cos\pi + i\sin\pi) = -2$

19. $|z| = 5$, $\theta = \arg(z) = \pi \Rightarrow \sin\theta = 3/5$, $\cos\theta = 4/5$
    $z = 4 + 3i$

20. $|z| = 1$, $\arg(z) = \dfrac{3\pi}{4} \Rightarrow z = \left(\cos\dfrac{3\pi}{4} + i\sin\dfrac{3\pi}{4}\right)$
    $\Rightarrow z = -\dfrac{1}{\sqrt{2}} + \dfrac{1}{\sqrt{2}}i$

21. $|z| = \pi$, $\arg(z) = \dfrac{\pi}{6} \Rightarrow z = \pi\left(\cos\dfrac{\pi}{6} + i\sin\dfrac{\pi}{6}\right)$
    $\Rightarrow z = \dfrac{\pi\sqrt{3}}{2} + \dfrac{\pi}{2}i$

22. $|z| = 0 \Rightarrow z = 0$ for any value of $\arg(z)$

23. $|z| = \dfrac{1}{2}$, $\arg(z) = -\dfrac{\pi}{3} \Rightarrow z = \dfrac{1}{2}\left(\cos\dfrac{\pi}{3} - i\sin\dfrac{\pi}{3}\right)$
    $\Rightarrow z = \dfrac{1}{4} - \dfrac{\sqrt{3}}{4}i$

24. $\overline{5 + 3i} = 5 - 3i$

25. $\overline{-3 - 5i} = -3 + 5i$

26. $\overline{4i} = -4i$

27. $\overline{2 - i} = 2 + i$

28. $|z| = 2$ represents all points on the circle of radius 2 centred at the origin.

29. $|z| \leq 2$ represents all points in the closed disk of radius 2 centred at the origin.

30. $|z - 2i| \leq 3$ represents all points in the closed disk of radius 3 centred at the point $2i$.

31. $|z - 3 + 4i| \leq 5$ represents all points in the closed disk of radius 5 centred at the point $3 - 4i$.

32. $\arg(z) = \pi/3$ represents all points on the ray from the origin in the first quadrant, making angle 60° with the positive direction of the real axis.

33. $\pi \leq \arg(z) \leq 7\pi/4$ represents the closed wedge-shaped region in the third and fourth quadrants bounded by the ray from the origin to $-\infty$ on the real axis and the ray from the origin making angle $-45°$ with the positive direction of the real axis.

34. $(2 + 5i) + (3 - i) = 5 + 4i$

35. $i - (3 - 2i) + (7 - 3i) = -3 + 7 + i + 2i - 3i = 4$

36. $(4+i)(4-i) = 16 - i^2 = 17$

37. $(1+i)(2-3i) = 2 + 2i - 3i - 3i^2 = 5 - i$

38. $(a+bi)(\overline{2a-bi}) = (a+bi)(2a+bi) = 2a^2 - b^2 + 3abi$

39. $(2+i)^3 = 8 + 12i + 6i^2 + i^3 = 2 + 11i$

40. $\dfrac{2-i}{2+i} = \dfrac{(2-i)^2}{4-i^2} = \dfrac{3-4i}{5}$

41. $\dfrac{1+3i}{2-i} = \dfrac{(1+3i)(2+i)}{4-i^2} = \dfrac{-1+7i}{5}$

42. $\dfrac{1+i}{i(2+3i)} = \dfrac{1+i}{-3+2i} = \dfrac{(1+i)(-3-2i)}{9+4} = \dfrac{-1-5i}{13}$

43. $\dfrac{(1+2i)(2-3i)}{(2-i)(3+2i)} = \dfrac{8+i}{8+i} = 1$

44. If $z = x + yi$ and $w = u + vi$, where x, y, u, and v are real, then
$$\overline{z+w} = \overline{x+u+(y+v)i}$$
$$= x + u - (y+v)i = x - yi + u - vi = \bar{z} + \bar{w}.$$

45. Using the fact that $|zw| = |z||w|$, we have
$$\overline{\left(\dfrac{z}{w}\right)} = \overline{\left(\dfrac{z\bar{w}}{|w|^2}\right)} = \dfrac{\bar{z}\overline{\bar{w}}}{|w|^2} = \dfrac{\bar{z}w}{\bar{w}w} = \dfrac{\bar{z}}{\bar{w}}.$$

46. $z = 3 + i\sqrt{3} = 2\sqrt{3}\left(\cos\dfrac{\pi}{6} + i\sin\dfrac{\pi}{6}\right)$

$w = -1 + i\sqrt{3} = 2\left(\cos\dfrac{2\pi}{3} + i\sin\dfrac{2\pi}{3}\right)$

$zw = 4\sqrt{3}\left(\cos\dfrac{5\pi}{6} + i\sin\dfrac{5\pi}{6}\right)$

$\dfrac{z}{w} = \sqrt{3}\left(\cos\dfrac{-\pi}{2} + i\sin\dfrac{-\pi}{2}\right) = -i\sqrt{3}$

47. $z = -1 + i = \sqrt{2}\left(\cos\dfrac{3\pi}{4} + i\sin\dfrac{3\pi}{4}\right)$

$w = 3i = 3\left(\cos\dfrac{\pi}{2} + i\sin\dfrac{\pi}{2}\right)$

$zw = 3\sqrt{2}\left(\cos\dfrac{5\pi}{4} + i\sin\dfrac{5\pi}{4}\right) = -3 - 3i$

$\dfrac{z}{w} = \dfrac{\sqrt{2}}{3}\left(\cos\dfrac{\pi}{4} + i\sin\dfrac{\pi}{4}\right) = \dfrac{1}{3} + \dfrac{1}{3}i$

48. $\cos(3\theta) + i\sin(3\theta) = (\cos\theta + i\sin\theta)^3$
$= \cos^3\theta + 3i\cos^2\theta\sin\theta - 3\cos\theta\sin^2\theta - i\sin^3\theta$
Thus
$$\cos(3\theta) = \cos^3\theta - 3\cos\theta\sin^2\theta = 4\cos^3\theta - 3\cos\theta$$
$$\sin(3\theta) = 3\cos^2\theta\sin\theta - \sin^3\theta = 3\sin\theta - 4\sin^3\theta.$$

49. a) $\bar{z} = 2/z$ can be rewritten $|z|^2 = z\bar{z} = 2$, so is satisfied by all numbers z on the circle of radius $\sqrt{2}$ centred at the origin.

b) $\bar{z} = -2/z$ can be rewritten $|z|^2 = z\bar{z} = -2$, which has no solutions since the square of $|z|$ is nonnegative for all complex z.

50. If $z = w = -1$, then $zw = 1$, so $\sqrt{zw} = 1$. But if we use $\sqrt{z} = \sqrt{-1} = i$ and the same value for $\sqrt{w}$, then $\sqrt{z}\sqrt{w} = i^2 = -1 \neq \sqrt{zw}$.

51. The three cube roots of $-1 = \cos\pi + i\sin\pi$ are of the form $\cos\theta + i\sin\theta$ where $\theta = \pi/3$, $\theta = \pi$, and $\theta = 5\pi/3$. Thus they are
$$\dfrac{1}{2} + i\dfrac{\sqrt{3}}{2}, \quad -1, \quad \dfrac{1}{2} - i\dfrac{\sqrt{3}}{2}.$$

52. The three cube roots of $-8i = 8\left(\cos\dfrac{3\pi}{2} + i\sin\dfrac{3\pi}{2}\right)$ are of the form $2(\cos\theta + i\sin\theta)$ where $\theta = \pi/2$, $\theta = 7\pi/6$, and $\theta = 11\pi/6$. Thus they are
$$2i, \quad -\sqrt{3} - i, \quad \sqrt{3} - i.$$

53. The three cube roots of $-1 + i = \sqrt{2}\left(\cos\dfrac{3\pi}{4} + i\sin\dfrac{3\pi}{4}\right)$ are of the form $2^{1/6}(\cos\theta + i\sin\theta)$ where $\theta = \pi/4$, $\theta = 11\pi/12$, and $\theta = 19\pi/12$.

54. The four fourth roots of $4 = 4(\cos 0 + i\sin 0)$ are of the form $\sqrt{2}(\cos\theta + i\sin\theta)$ where $\theta = 0$, $\theta = \pi/2$, π, and $\theta = 3\pi/2$. Thus they are $\sqrt{2}, i\sqrt{2}, -\sqrt{2}$, and $-i\sqrt{2}$.

55. The equation $z^4 + 1 - i\sqrt{3} = 0$ has solutions that are the four fourth roots of $-1 + i\sqrt{3} = 2\left(\cos\dfrac{2\pi}{3} + i\sin\dfrac{2\pi}{3}\right)$. Thus they are of the form $2^{1/4}(\cos\theta + i\sin\theta)$, where $\theta = \pi/6$, $2\pi/3$, $7\pi/6$, and $5\pi/3$. They are the complex numbers
$$\pm 2^{1/4}\left(\dfrac{\sqrt{3}}{2} + \dfrac{1}{2}i\right), \quad \pm 2^{1/4}\left(\dfrac{1}{2} - \dfrac{\sqrt{3}}{2}i\right).$$

56. The equation $z^5 + a^5 = 0$ ($a > 0$) has solutions that are the five fifth roots of $-a^5 = a(\cos\pi + i\sin\pi)$; they are of the form $a(\cos\theta + i\sin\theta)$, where $\theta = \pi/5$, $3\pi/5$, π, $7\pi/5$, and $9\pi/5$.

57. The n nth roots of unity are

$$\omega_1 = 1$$
$$\omega_2 = \cos\frac{2\pi}{n} + i\sin\frac{2\pi}{n}$$
$$\omega_3 = \cos\frac{4\pi}{n} + i\sin\frac{4\pi}{n} = \omega_2^2$$
$$\omega_4 = \cos\frac{6\pi}{n} + i\sin\frac{6\pi}{n} = \omega_2^3$$
$$\vdots$$
$$\omega_n = \cos\frac{2(n-1)\pi}{n} + i\sin\frac{2(n-1)\pi}{n} = \omega_2^{n-1}.$$

Hence

$$\omega_1 + \omega_2 + \omega_3 + \cdots + \omega_n = 1 + \omega_2 + \omega_2^2 + \cdots + \omega_2^{n-1}$$
$$= \frac{1 - \omega_2^n}{1 - \omega_2} = \frac{0}{1 - \omega_2} = 0.$$

Appendix II. Complex Functions (page A-21)

In Solutions 1–12, $z = x + yi$ and $w = u + vi$, where x, y, u, and v are real.

1. The function $w = \bar{z}$ transforms the closed rectangle $0 \le x \le 1$, $0 \le y \le 2$ to the closed rectangle $0 \le u \le 1$, $-2 \le v \le 0$.

2. The function $w = \bar{z}$ transforms the line $x + y = 1$ to the line $u - v = 1$.

3. The function $w = z^2$ transforms the closed annular sector $1 \le |z| \le 2$, $\pi/2 \le \arg(z) \le 3\pi/4$ to the closed annular sector $1 \le |w| \le 4$, $\pi \le \arg(w) \le 3\pi/2$.

4. The function $w = z^3$ transforms the closed quarter-circular disk $0 \le |z| \le 2$, $0 \le \arg(z) \le \pi/2$ to the closed three-quarter disk $0 \le |w| \le 8$, $0 \le \arg(w) \le 3\pi/2$.

5. The function $w = 1/z = \bar{z}/|z|^2$ transforms the closed quarter-circular disk $0 \le |z| \le 2$, $0 \le \arg(z) \le \pi/2$ to the closed region lying on or outside the circle $|w| = 1/2$ and in the fourth quadrant, that is, having $-\pi/2 \le \arg(w) \le 0$.

6. The function $w = -iz$ rotates the z-plane $-90°$, so transforms the wedge $\pi/4 \le \arg(z) \le \pi/3$ to the wedge $-\pi/4 \le \arg(z) \le -\pi/6$.

7. The function $w = \sqrt{z}$ transforms the ray $\arg(z) = -\pi/3$ (that is, $\text{Arg}(z) = 5\pi/3$) to the ray $\arg(w) = 5\pi/6$.

8. The function $w = z^2 = x^2 - y^2 + 2xyi$ transforms the line $x = 1$ to $u = 1 - y^2$, $v = 2y$, which is the parabola $v^2 = 4 - 4u$ with vertex at $w = 1$, opening to the left.

9. The function $w = z^2 = x^2 - y^2 + 2xyi$ transforms the line $y = 1$ to $u = x^2 - 1$, $v = 2x$, which is the parabola $v^2 = 4u + 4$ with vertex at $w = -1$ and opening to the right.

10. The function $w = 1/z = (x - yi)/(x^2 + y^2)$ transforms the line $x = 1$ to the curve given parametrically by

$$u = \frac{1}{1+y^2}, \qquad v = \frac{-y}{1+y^2}.$$

This curve is, in fact, a circle,

$$u^2 + v^2 = \frac{1+y^2}{(1+y^2)^2} = u,$$

with centre $w = 1/2$ and radius $1/2$.

11. The function $w = e^z = e^x \cos y + ie^x \sin y$ transforms the horizontal strip $-\infty < x < \infty$, $\pi/4 \le y \le \pi/2$ to the wedge $\pi/4 \le \arg(w) \le \pi/2$, or, equivalently, $u \ge 0$, $v \ge u$.

12. The function $w = e^{iz} = e^{-y}(\cos x + i\sin x)$ transforms the vertical half-strip $0 < x < \pi/2$, $0 < y < \infty$ to the first-quadrant part of the unit open disk $|w| = e^{-y} < 1$, $0 < \arg(w) = x < \pi/2$, that is $u > 0$, $v > 0$, $u^2 + v^2 < 1$.

13. $f(z) = z^2 = (x + yi)^2 = x^2 - y^2 + 2xyi$
$u = x^2 - y^2, \qquad v = 2xy$
$$\frac{\partial u}{\partial x} = 2x = \frac{\partial v}{\partial y}, \qquad \frac{\partial u}{\partial y} = -2y = -\frac{\partial v}{\partial x}$$
$$f'(z) = \frac{\partial u}{\partial x} + i\frac{\partial v}{\partial x} = 2x + 2yi = 2z.$$

14. $f(z) = z^3 = (x + yi)^3 = x^3 - 3xy^2 + (3x^2y - y^3)i$
$u = x^3 - 3xy^2, \qquad v = 3x^2y - y^3$
$$\frac{\partial u}{\partial x} = 3(x^2 - y^2) = \frac{\partial v}{\partial y}, \qquad \frac{\partial u}{\partial y} = -6xy = -\frac{\partial v}{\partial x}$$
$$f'(z) = \frac{\partial u}{\partial x} + i\frac{\partial v}{\partial x} = 3(x^2 - y^2 + 2xyi) = 3z^2.$$

15. $f(z) = \dfrac{1}{z} = \dfrac{x - yi}{x^2 + y^2}$
$$u = \frac{x}{x^2 + y^2}, \qquad v = \frac{-y}{x^2 + y^2}$$
$$\frac{\partial u}{\partial x} = \frac{y^2 - x^2}{(x^2 + y^2)^2} = \frac{\partial v}{\partial y}, \qquad \frac{\partial u}{\partial y} = \frac{-2xy}{(x^2 + y^2)^2} = -\frac{\partial v}{\partial x}$$
$$f'(z) = \frac{\partial u}{\partial x} + i\frac{\partial v}{\partial x} = \frac{-(x^2 - y^2) + 2xyi}{(x^2 + y^2)^2} = \frac{-(\bar{z})^2}{(z\bar{z})^2} = -\frac{1}{z^2}.$$

INSTRUCTOR'S SOLUTIONS MANUAL   APPENDIX II. (PAGE A-21)

16. $f(z) = e^{z^2} = e^{x^2-y^2}(\cos(2xy) + i\sin(2xy))$
$u = e^{x^2-y^2}\cos(2xy)$, $v = e^{x^2-y^2}\sin(2xy)$
$\dfrac{\partial u}{\partial x} = e^{x^2-y^2}(2x\cos(2xy) - 2y\sin(2xy)) = \dfrac{\partial v}{\partial y}$
$\dfrac{\partial u}{\partial y} = -e^{x^2-y^2}(2y\cos(2xy) + 2x\sin(2xy)) = -\dfrac{\partial v}{\partial x}$
$f'(z) = \dfrac{\partial u}{\partial x} + i\dfrac{\partial v}{\partial x}$
$= e^{x^2-y^2}[2x\cos(2xy) - 2y\sin(2xy)$
$\quad + i(2y\cos(2xy) + 2x\sin(2xy))]$
$= (2x + 2yi)e^{x^2-y^2}(\cos(2xy) + i\sin(2xy)) = 2ze^{z^2}$.

17. $e^{yi} = \cos y + i\sin y$ (for real y). Replacing y by $-y$, we get $e^{-yi} = \cos y - i\sin y$ (since cos is even and sin is odd). Adding and subtracting these two formulas gives
$$e^{yi} + e^{-yi} = 2\cos y, \qquad e^{yi} - e^{-yi} = 2i\sin y.$$
Thus $\cos y = \dfrac{e^{yi} + e^{-yi}}{2}$ and $\sin y = \dfrac{e^{yi} - e^{-yi}}{2i}$.

18. $e^{z+2\pi i} = e^x(\cos(y + 2\pi) + i\sin(y + 2\pi))$
$= e^x(\cos y + i\sin y) = e^z$.
Thus e^z is periodic with period $2\pi i$. So is $e^{-z} = 1/e^z$. Since $e^{i(z+2\pi)} = e^{zi+2\pi i} = e^{zi}$, therefore e^{zi} and also e^{-zi} are periodic with period 2π. Hence
$$\cos z = \dfrac{e^{zi} + e^{-zi}}{2} \text{ and } \sin z = \dfrac{e^{zi} - e^{-zi}}{2i}$$
are periodic with period 2π, and
$$\cosh z = \dfrac{e^z + e^{-z}}{2} \text{ and } \sinh z = \dfrac{e^z - e^{-z}}{2}$$
are periodic with period $2\pi i$.

19. $\dfrac{d}{dz}\cos z = \dfrac{d}{dz}\dfrac{e^{zi} + e^{-zi}}{2} = \dfrac{ie^{zi} - e^{-zi}}{2} = -\sin z$
$\dfrac{d}{dz}\sin z = \dfrac{d}{dz}\dfrac{e^{zi} - e^{-zi}}{2i} = \dfrac{ie^{zi} + e^{-zi}}{2i} = \cos z$
$\dfrac{d}{dz}\cosh z = \dfrac{d}{dz}\dfrac{e^z + e^{-z}}{2} = \dfrac{e^z - e^{-z}}{2} = \sinh z$
$\dfrac{d}{dz}\sinh z = \dfrac{d}{dz}\dfrac{e^z - e^{-z}}{2} = \dfrac{e^z + e^{-z}}{2} = \cosh z$

20. $\cosh(iz) = \dfrac{e^{iz} + e^{-iz}}{2} = \cos z$
$-i\sinh(iz) = \dfrac{1}{i}\dfrac{e^{iz} - e^{-iz}}{2} = \sin z$
$\cos(iz) = \dfrac{e^{-z} + e^z}{2} = \cosh z$
$\sin(iz) = \dfrac{e^{-z} - e^z}{2i} = i\dfrac{-e^{-z} + e^z}{2} = i\sinh z$

21. $\cos z = 0 \Leftrightarrow e^{zi} = -e^{-zi} \Leftrightarrow e^{2zi} = -1$
$\Leftrightarrow e^{-2y}[\cos(2x) + i\sin(2x)] = -1$
$\Leftrightarrow \sin(2x) = 0, \quad e^{-2y}\cos(2x) = -1$
$\Leftrightarrow y = 0, \cos(2x) = -1$
$= \Leftrightarrow y = 0, \quad x = \pm\dfrac{\pi}{2}, \pm\dfrac{3\pi}{2}, \ldots$
Thus the only complex zeros of $\cos z$ are its real zeros at $z = (2n+1)\pi/2$ for integers n.

22. $\sin z = 0 \Leftrightarrow e^{zi} = e^{-zi} \Leftrightarrow e^{2zi} = 1$
$\Leftrightarrow e^{-2y}[\cos(2x) + i\sin(2x)] = 1$
$\Leftrightarrow \sin(2x) = 0, \quad e^{-2y}\cos(2x) = 1$
$\Leftrightarrow y = 0, \cos(2x) = 1$
$= \Leftrightarrow y = 0, \quad x = 0, \pm\pi, \pm 2\pi, \ldots$
Thus the only complex zeros of $\sin z$ are its real zeros at $z = n\pi$ for integers n.

23. By Exercises 20 and 21, $\cosh z = 0$ if and only if $\cos(iz) = 0$, that is, if and only if $z = (2n+1)\pi i/2$ for integer n.
Similarly, $\sinh z = 0$ if and only if $\sin(iz) = 0$, that is, if and only if $z = n\pi i$ for integer n.

24. $e^z = e^{x+yi} = e^x\cos y + ie^x\sin y$
$e^{-z} = e^{-x-yi} = e^{-x}\cos y - e^{-x}\sin y$
$\cosh z = \dfrac{e^z + e^{-z}}{2} = \dfrac{e^x + e^{-x}}{2}\cos y + i\dfrac{e^x - e^{-x}}{2}\sin y$
$= \cosh x\cos y + i\sinh x\sin y$
$\text{Re}(\cosh z) = \cosh x\cos y, \quad \text{Im}(\cosh z) = \sinh x\sin y$.

25. $\sinh z = \dfrac{e^z - e^{-z}}{2} = \dfrac{e^x - e^{-x}}{2}\cos y + i\dfrac{e^x + e^{-x}}{2}\sin y$
$= \sinh x\cos y + i\cosh x\sin y$
$\text{Re}(\sinh z) = \sinh x\cos y, \quad \text{Im}(\cosh z) = \cosh x\sin y$.

26. $e^{iz} = e^{-y+xi} = e^{-y}\cos x + ie^{-y}\sin x$
$e^{-iz} = e^{y-xi} = e^y\cos x - ie^y\sin x$
$\cos z = \dfrac{e^{iz} + e^{-iz}}{2} = \dfrac{e^{-y} + e^y}{2}\cos x + i\dfrac{e^{-y} - e^y}{2}\sin x$
$= \cos x\cosh y - i\sin x\sinh y$
$\text{Re}(\cos z) = \cos x\cosh y, \quad \text{Im}(\cos z) = -\sin x\sinh y$
$\sin z = \dfrac{e^{iz} - e^{-iz}}{2i} = \dfrac{e^{-y} - e^y}{2i}\cos x + i\dfrac{e^{-y} + e^y}{2i}\sin x$
$= \sin x\cosh y + i\cos x\sinh y$
$\text{Re}(\sin z) = \sin x\cosh y, \quad \text{Im}(\sin z) = \cos x\sinh y$.

27. $z^2 + 2iz = 0 \Rightarrow z = 0$ or $z = -2i$

28. $z^2 - 2z + i = 0 \Rightarrow (z-1)^2 = 1 - i$
$= \sqrt{2}\left(\cos\dfrac{7\pi}{4} + i\sin\dfrac{7\pi}{4}\right)$
$\Rightarrow z = 1 \pm 2^{1/4}\left(\cos\dfrac{7\pi}{8} + i\sin\dfrac{7\pi}{8}\right)$

29. $z^2 + 2z + 5 = 0 \Rightarrow (z+1)^2 = -4$
$\Rightarrow z = -1 \pm 2i$

30. $z^2 - 2iz - 1 = 0 \Rightarrow (z-i)^2 = 0$
$\Rightarrow z = i$ (double root)

31. $z^3 - 3iz^2 - 2z = z(z^2 - 3iz - 2) = 0$
$\Rightarrow z = 0$ or $z^2 - 3iz - 2 = 0$
$\Rightarrow z = 0$ or $\left(z - \frac{3}{2}i\right)^2 = -\frac{1}{4}$
$\Rightarrow z = 0$ or $z = \left(\frac{3}{2} \pm \frac{1}{2}\right)i$
$\Rightarrow z = 0$ or $z = i$ or $z = 2i$

32. $z^4 - 2z^2 + 4 = 0 \Rightarrow (z^2 - 1)^2 = -3$
$z^2 = 1 - i\sqrt{3}$ or $z^2 = 1 + i\sqrt{3}$
$z^2 = 2\left(\cos\frac{5\pi}{3} + i\sin\frac{5\pi}{3}\right)$, $z^2 = 2\left(\cos\frac{\pi}{3} + i\sin\frac{\pi}{3}\right)$
$z = \pm\sqrt{2}\left(\cos\frac{5\pi}{6} + i\sin\frac{5\pi}{6}\right)$, or
$z = \pm\sqrt{2}\left(\cos\frac{\pi}{6} + i\sin\frac{\pi}{6}\right)$
$z = \pm\left(\sqrt{\frac{3}{2}} - \frac{i}{\sqrt{2}}\right)$, $z = \pm\left(\sqrt{\frac{3}{2}} + \frac{i}{\sqrt{2}}\right)$

33. $z^4 + 1 = 0 \Rightarrow z^2 = i$ or $z^2 = -i$
$\Rightarrow z = \pm\frac{1+i}{\sqrt{2}}$, $z = \pm\frac{1-i}{\sqrt{2}}$
$z^4 + 1 = \left(z - \frac{1+i}{\sqrt{2}}\right)\left(z - \frac{1-i}{\sqrt{2}}\right)$
$\times \left(z + \frac{1+i}{\sqrt{2}}\right)\left(z + \frac{1-i}{\sqrt{2}}\right)$
$= \left(\left[z - \frac{1}{\sqrt{2}}\right]^2 + \frac{1}{2}\right)\left(\left[z + \frac{1}{\sqrt{2}}\right]^2 + \frac{1}{2}\right)$
$= (z^2 - \sqrt{2}z + 1)(z^2 + \sqrt{2}z + 1)$

34. Since $P(z) = z^4 - 4z^3 + 12z^2 - 16z + 16$ has real coefficients, if $z_1 = 1 - \sqrt{3}i$ is a zero of $P(z)$, then so is $\overline{z_1}$. Now

$(z - z_1)(z - \overline{z_1}) = (z-1)^2 + 3 = z^2 - 2z + 4.$

By long division (details omitted) we discover that

$\frac{z^4 - 4z^3 + 12z^2 - 16z + 16}{z^2 - 2z + 4} = z^2 - 2z + 4.$

Thus z_1 and $\overline{z_1}$ are both *double zeros* of $P(z)$. These are the only zeros.

35. Since $P(z) = z^5 + 3z^4 + 4z^3 + 4z^2 + 3z + 1$ has real coefficients, if $z_1 = i$ is a zero of $P(z)$, then so is $z_2 = -i$. Now

$(z - z_1)(z - z_2) = (z - i)(z + i) = z^2 + 1.$

By long division (details omitted) we discover that

$\frac{z^5 + 3z^4 + 4z^3 + 4z^2 + 3z + 1}{z^2 + 1} = z^3 + 3z^2 + 3z + 1$
$= (z+1)^3.$

Thus $P(z)$ has the five zeros: i, $-i$, -1, -1, and -1.

36. Since $P(z) = z^5 - 2z^4 - 8z^3 + 8z^2 + 31z - 30$ has real coefficients, if $z_1 = -2 + i$ is a zero of $P(z)$, then so is $z_2 = -2 - i$. Now

$(z - z_1)(z - z_2) = z^2 + 4z + 5.$

By long division (details omitted) we discover that

$\frac{z^5 - 2z^4 - 8z^3 + 8z^2 + 31z - 30}{z^2 + 4z + 5}$
$= z^3 - 6z^2 + 11z - 6.$

Observe that $z_3 = 1$ is a zero of $z^3 - 6z^2 + 11z - 6$. By long division again:

$\frac{z^3 - 6z^2 + 11z - 6}{z - 1} = z^2 - 5z + 6 = (z-2)(z-3).$

Hence $P(z)$ has the five zeros $-2 + i$, $-2 - i$, 1, 2, and 3.

37. If $w = z^4 + z^3 - 2iz - 3$ and $|z| = 2$, then $|z^4| = 16$ and

$|w - z^4| = |z^3 - 2iz - 3| \le 8 + 4 + 3 = 15 < 16.$

By the mapping principle described in the proof of Theorem 2, the image in the w-plane of the circle $|z| = 2$ is a closed curve that winds around the origin the same number of times that the image of z^4 does, namely 4 times.

Appendix III. Continuous Functions (page A-27)

1. To be proved: If $a < b < c$, $f(x) \le g(x)$ for $a \le x \le c$, $\lim_{x \to b} f(x) = L$, and $\lim_{x \to b} g(x) = M$, then $L \le M$.

   Proof: Suppose, to the contrary, that $L > M$. Let $\epsilon = (L-M)/3$, so $\epsilon > 0$. There exist numbers $\delta_1 > 0$ and $\delta_2 > 0$ such that if $a \le x \le b$, then
   $$|x-b| < \delta_1 \Rightarrow |f(x) - L| < \epsilon$$
   $$|x-b| < \delta_2 \Rightarrow |g(x) - M| < \epsilon.$$
   Thus if $|x-b| < \delta = \min\{\delta_1, \delta_2, b-a, c-b\}$, then
   $$f(x) - g(x) > L - \epsilon - M - \epsilon = L - M - 2\epsilon = \frac{L-M}{3} > 0.$$
   This contradicts the fact that $f(x) \le g(x)$ on $[a,b]$. Therefore $L \le M$.

2. To be proved: If $f(x) \le K$ on $[a,b)$ and $(b,c]$, and if $\lim_{x \to b} f(x) = L$, then $L \le K$.

   Proof: If $L > K$, then let $\epsilon = (L-K)/2$; thus $\epsilon > 0$. There exists $\delta > 0$ such that $\delta < b - a$ and $\delta < c - b$, and such that if $0 < |x-b| < \delta$, then $|f(x) - L| < \epsilon$. In this case
   $$f(x) > L - \epsilon = L - \frac{L-K}{2} > K,$$
   which contradicts the fact that $f(x) \le K$ on $[a,b)$ and $(b,c]$. Therefore $L \le K$.

3. Let $\epsilon > 0$ be given. Let $\delta = \epsilon^{1/r}$, $(r > 0)$. Then
   $$0 < x < \delta \quad \Rightarrow \quad 0 < x^r < \delta^r = \epsilon.$$
   Thus $\lim_{x \to 0+} x^r = 0$.

4. a) Let $f(x) = C$, $g(x) = x$. Let $\epsilon > 0$ be given and let $\delta = \epsilon$. For any real number x, if $|x-a| < \delta$, then
   $$|f(x) - f(a)| = |C - C| = 0 < \epsilon,$$
   $$|g(x) - g(a)| = |x - a| < \delta = \epsilon.$$
   Thus $\lim_{x \to a} f(x) = f(a)$ and $\lim_{x \to a} g(x) = g(a)$, and f and g are both continuous at every real number a.

5. A polynomial is constructed by adding and multiplying finite numbers of functions of the type of f and g in Exercise 4. By Theorem 1(a), such sums and products are continuous everywhere, since their components have been shown to be continuous everywhere.

6. If P and Q are polynomials, they are continuous everywhere by Exercise 5. If $Q(a) \ne 0$, then $\lim_{x \to a} \frac{P(x)}{Q(x)} = \frac{P(a)}{Q(a)}$ by Theorem 1(a). Hence P/Q is continuous everywhere except at the zeros of Q.

7. Suppose n is a positive integer and $a > 0$. Let $\epsilon > 0$ be given. Let $b = a^{1/n}$, and let $\delta = \min\{a(1 - 2^{-n}), b^{n-1}\epsilon\}$. If $|x-a| < \delta$, then $x > a/2^n$, and if $y = x^{1/n}$, then $y > b/2$. Thus
   $$\left|x^{1/n} - a^{1/n}\right| = |y - b|$$
   $$= \frac{|y^n - b^n|}{y^{n-1} + y^{n-2}b + \cdots + b^{n-1}}$$
   $$< \frac{|x-a|}{b^{n-1}} < \frac{b^{n-1}\epsilon}{b^{n-1}} = \epsilon.$$
   Thus $\lim_{x \to a} x^{1/n} = a^{1/n}$, and $x^{1/n}$ is continuous at $x = a$.

8. By Exercise 5, x^m is continuous everywhere. By Exercise 7, $x^{1/n}$ is continuous at each $a > 0$. Thus for $a > 0$ we have
   $$\lim_{x \to a} x^{m/n} = \lim_{x \to a} \left(x^{1/n}\right)^m = \left(\lim_{x \to a} x^{1/n}\right)^m$$
   $$= (a^{1/n})^m = a^{m/n},$$
   and $x^{m/n}$ is continuous at each positive number.

9. If m and n are integers and n is odd, then $(-x)^{m/n} = cx^{m/n}$, where $c = (-1)^{m/n}$ is either -1 or 1 depending on the parity of m. Since $x^{m/n}$ is continuous at each positive number a, so is $cx^{m/n}$. Thus $(-x)^{m/n}$ is continuous at each positive number, and $x^{m/n}$ is continuous at each negative number.

   If $r = m/n > 0$, then $\lim_{x \to 0+} x^r = 0$ by Exercise 3. Hence $\lim_{x \to 0-} x^r = (-1)^r \lim_{x \to 0+} x^r = 0$, also. Therefore $\lim_{x \to 0} x^r = 0$, and x^r is continuous at $x = 0$.

10. Let $\epsilon > 0$ be given. Let $\delta = \epsilon$. If a is any real number then
    $$\left||x| - |a|\right| \le |x - a| < \epsilon \quad \text{if} \quad |x - a| < \delta.$$
    Thus $\lim_{x \to a} |x| = |a|$, and the absolute value function is continuous at every real number.

11. By the definition of $\sin$, $P_t = (\cos t, \sin t)$, and $P_a = (\cos a, \sin a)$ are two points on the unit circle $x^2 + y^2 = 1$. Therefore
    $$|t - a| = \text{length of the arc from } P_t \text{ to } P_a$$
    $$> \text{length of the chord from } P_t \text{ to } P_a$$
    $$= \sqrt{(\cos t - \cos a)^2 + (\sin t - \sin a)^2}.$$
    If $\epsilon > 0$ is given, and $|t - a| < \delta = \epsilon$, then the above inequality implies that
    $$|\cos t - \cos a| \le |t - a| < \epsilon,$$
    $$|\sin t - \sin a| \le |t - a| < \epsilon.$$
    Thus $\sin$ is continuous everywhere.

12. The proof that cos is continuous everywhere is almost identical to that for sin in Exercise 11.

13. Let $a > 0$ and $\epsilon > 0$. Let $\delta = \min\left\{\dfrac{a}{2}, \dfrac{\epsilon a}{2}\right\}$.

    If $|x - a| < \delta$, then $x > \dfrac{a}{2}$, so $\dfrac{1}{t} < \dfrac{2}{a}$ whenever t is between a and x. Thus

    $|\ln x - \ln a|$
    $=$ area under $y = \dfrac{1}{t}$ between $t = a$ and $t = x$
    $< \dfrac{2}{a}|x - a| < \dfrac{2}{a}\dfrac{\epsilon a}{2} = \epsilon.$

    Thus $\lim_{x \to a} \ln x = \ln a$, and $\ln$ is continuous at each point a in its domain $(0, \infty)$.

14. Let a be any real number, and let $\epsilon > 0$ be given. Assume (making ϵ smaller if necessary) that $\epsilon < e^a$. Since

    $$\ln\left(1 - \frac{\epsilon}{e^a}\right) + \ln\left(1 + \frac{\epsilon}{e^a}\right) = \ln\left(1 - \frac{\epsilon^2}{e^{2a}}\right) < 0,$$

    we have $\ln\left(1 + \dfrac{\epsilon}{e^a}\right) < -\ln\left(1 - \dfrac{\epsilon}{e^a}\right)$.

    Let $\delta = \ln\left(1 + \dfrac{\epsilon}{e^a}\right)$. If $|x - a| < \delta$, then

    $$\ln\left(1 - \frac{\epsilon}{e^a}\right) < x - a < \ln\left(1 + \frac{\epsilon}{e^a}\right)$$
    $$1 - \frac{\epsilon}{e^a} < e^{x-a} < 1 + \frac{\epsilon}{e^a}$$
    $$\left|e^{x-a} - 1\right| < \frac{\epsilon}{e^a}$$
    $$|e^x - e^a| = e^a|e^{x-a} - 1| < \epsilon.$$

    Thus $\lim_{x \to a} e^x = e^a$ and e^x is continuous at every point a in its domain.

15. Suppose $a \le x_n \le b$ for each n, and $\lim x_n = L$. Then $a \le L \le b$ by Theorem 3. Let $\epsilon > 0$ be given. Since f is continuous on $[a, b]$, there exists $\delta > 0$ such that if $a \le x \le b$ and $|x - L| < \delta$ then $|f(x) - f(L)| < \epsilon$. Since $\lim x_n = L$, there exists an integer N such that if $n \ge N$ then $|x_n - L| < \delta$. Hence $|f(x_n) - f(L)| < \epsilon$ for such n. Therefore $\lim(f(x_n)) = f(L)$.

16. Let $g(t) = \dfrac{t}{1 + |t|}$. For $t \ne 0$ we have

    $$g'(t) = \frac{1 + |t| - t\,\text{sgn}\,t}{(1 + |t|)^2} = \frac{1 + |t| - |t|}{(1 + |t|)^2} = \frac{1}{(1 + |t|)^2} > 0.$$

    If $t = 0$, g is also differentiable, and has derivative 1:

    $$g'(0) = \lim_{h \to 0} \frac{g(h) - g(0)}{h} = \lim_{h \to 0} \frac{1}{1 + |h|} = 1.$$

Thus g is continuous and increasing on $\mathbb{R}$. If f is continuous on $[a, b]$, then

$$h(x) = g(f(x)) = \frac{f(x)}{1 + |f(x)|}$$

is also continuous there, being the composition of continuous functions. Also, $h(x)$ is bounded on $[a, b]$, since

$$|g(f(x))| \le \frac{|f(x)|}{1 + |f(x)|} \le 1.$$

By assumption in this problem, $h(x)$ must assume maximum and minimum values; there exist c and d in $[a, b]$ such that

$$g(f(c)) \le g(f(x)) \le g(f(d))$$

for all x in $[a, b]$. Since g is increasing, so is its inverse g^{-1}. Therefore

$$f(c) \le f(x) \le f(d)$$

for all x in $[a, b]$, and f is bounded on that interval.

Appendix IV. The Riemann Integral (page A-34)

1. $f(x) = \begin{cases} 1 & \text{if } 0 \le x \le 1 \\ 0 & \text{if } 1 < x \le 2 \end{cases}$
   Let $0 < \epsilon < 1$. Let $P = \{0, 1 - \frac{\epsilon}{3}, 1 + \frac{\epsilon}{3}, 2\}$. Then

   $$L(f, P) = 1\left(1 - \frac{\epsilon}{3}\right) + 0 + 0 = 1 - \frac{\epsilon}{3}$$
   $$U(f, P) = 1\left(1 - \frac{\epsilon}{3}\right) + 1\left(\frac{2\epsilon}{3}\right) + 0 = 1 + \frac{\epsilon}{3}.$$

   Since $U(f, P) - L(f, P) < \epsilon$, f is integrable on $[0, 2]$. Since $L(f, P) < 1 < U(f, P)$ for every ϵ, therefore

   $$\int_0^2 f(x)\,dx = 1.$$

2. $f(x) = \begin{cases} 1 & \text{if } x = 1/n \quad (n = 1, 2, 3, \ldots) \\ 0 & \text{otherwise} \end{cases}$
   If P is any partition of $[0, 1]$ then $L(f, P) = 0$. Let $0 < \epsilon \le 2$. Let N be an integer such that $N + 1 > \dfrac{2}{\epsilon} \ge N$. A partition P of $[0, 1]$ can be constructed so that the first two points of P are 0 and $\dfrac{\epsilon}{2}$, and such that each of the N points $\dfrac{1}{n}$ $(n = 1, 2, 3, \ldots, n)$ lies in a subinterval of P having length at most $\dfrac{\epsilon}{2N}$. Since every number $\dfrac{1}{n}$ with n a positive integer lies either in $\left[0, \dfrac{\epsilon}{2}\right]$ or one of these other N subintervals of P, and since $\max f(x) = 1$ for these subintervals and $\max f(x) = 0$ for all other subintervals of P, therefore $U(f, P) \le \dfrac{\epsilon}{2} + N\dfrac{\epsilon}{2N} = \epsilon$. By Theorem 3, f is integrable on $[0, 1]$. Evidently

   $$\int_0^1 f(x)\,dx = \text{least upper bound } L(f, P) = 0.$$

3. $f(x) = \begin{cases} 1/n & \text{if } x = m/n \text{ in lowest terms} \\ 0 & \text{otherwise} \end{cases}$

Clearly $L(f, P) = 0$ for every partition P of $[0, 1]$. Let $\epsilon > 0$ be given. To show that f is integrable we must exhibit a partition P for which $U(f, P) < \epsilon$. We can assume $\epsilon < 1$. Choose a positive integer N such that $2/N < \epsilon$. There are only finitely many integers n such that $1 \le n \le N$. For each such n, there are only finitely many integers m such that $0 \le m/n \le 1$. Therefore there are only finitely many points x in $[0, 1]$ where $f(x) > \epsilon/2$. Let P be a partition of $[0, 1]$ such that all these points are contained in subintervals of the partition having total length less than $\epsilon/2$. Since $f(x) \le 1$ on these subintervals, and $f(x) < \epsilon/2$ on all other subintervals P, therefore $U(f, P) \le 1 \times (\epsilon/2) + (\epsilon/2) \times 1 = \epsilon$, and f is integrable on $[0, 1]$. Evidently $\int_0^1 f(x)\, dx = 0$, since all lower sums are 0.

4. Suppose, to the contrary, that $I_* > I^*$. Let $\epsilon = \dfrac{I_* - I^*}{3}$, so $\epsilon > 0$. By the definition of I_* and I^*, there exist partitions P_1 and P_2 of $[a, b]$, such that $L(f, P_1) \ge I_* - \epsilon$ and $U(f, P_2) \le I^* + \epsilon$. By Theorem 2, $L(f, P_1) \le U(f, P_2)$, so

$$3\epsilon = I_* - I^* \le L(f, P_1) + \epsilon - U(f, P_2) + \epsilon \le 2\epsilon.$$

Since $\epsilon > 0$, it follows that $3 \le 2$. This contradiction shows that we must have $I_* \le I^*$.

5. Theorem 3 of Section 6.4: Proofs of parts (c)–(h).

   c) Multiplying a function by a constant multiplies all its Riemann sums by the same constant. If the constant is positive, upper and lower sums remain upper and lower; if the constant is negative upper sums become lower and vice versa. Therefore

   $$\int_a^b A f(x)\, dx = A \int_a^b f(x)\, dx.$$

   It therefore remains to be proved only that the integral of a sum of functions is the sum of the integrals. Suppose that

   $$\int_a^b f(x)\, dx = I, \quad \text{and} \quad \int_a^b g(x)\, dx = J.$$

   If $\epsilon > 0$, then there exist partitions P_1 and P_2 of $[a, b]$ such that

   $$U(f, P_1) - \frac{\epsilon}{2} \le I < L(f, P_1) + \frac{\epsilon}{2}$$
   $$U(g, P_2) - \frac{\epsilon}{2} \le J < L(g, P_2) + \frac{\epsilon}{2}.$$

   Let P be the common refinement of P_1 and P_2. Then the above inequalities hold with P replacing P_1 and P_2. If $m_1 \le f(x) \le M_1$ and $m_2 \le g(x) \le M_2$ on any interval, then $m_1 + m_2 \le f(x) + g(x) \le M_1 + M_2$ there. It follows that

   $$U(f + g, P) \le U(f, P) + U(g, P),$$
   $$L(f, P) + L(g, P) \le L(f + g, P).$$

   Therefore

   $$U(f + g, P) - \epsilon \le I + J \le L(f + g, P) + \epsilon.$$

   Hence $\int_a^b \bigl(f(x) + g(x)\bigr) dx = I + J$.

   d) Assume $a < b < c$; the other cases are similar. Let $\epsilon > 0$. If

   $$\int_a^b f(x)\, dx = I, \quad \text{and} \quad \int_b^c f(x)\, dx = J,$$

   then there exist partitions P_1 of $[a, b]$, and P_2 of $[b, c]$ such that

   $$L(f, P_1) \le I < L(f, P_1) + \frac{\epsilon}{2}$$
   $$L(f, P_2) \le J < L(f, P_2) + \frac{\epsilon}{2}$$

   (with similar inequalities for upper sums). Let P be the partition of $[a, c]$ formed by combining all the subdivision points of P_1 and P_2. Then

   $$L(f, P) = L(f, P_1) + L(f, P_2) \le I + J < L(f, P) + \epsilon.$$

   Similarly, $U(f, P) - \epsilon < I + J \le U(f, P)$. Therefore

   $$\int_a^c f(x)\, dx = I + J.$$

   e) Let

   $$\int_a^b f(x)\, dx = I, \quad \text{and} \quad \int_a^b g(x)\, dx = J,$$

   where $f(x) \le g(x)$ on $[a, b]$. We want to show that $I \le J$. Suppose, to the contrary, that $I > J$. Then there would exist a partition P of $[a, b]$ for which

   $$I < L(f, P) + \frac{I - J}{2}, \quad \text{and} \quad U(g, P) - \frac{I - J}{2} < J.$$

   Thus $L(f, P) > \dfrac{I + J}{2} > U(g, P) \ge L(g, P)$. However, $f(x) \le g(x)$ on $[a, b]$ implies that $L(f, P) \le L(g, P)$ for any partition. Thus we have a contradiction, and so $I \le J$.

f) Since $-|f(x)| \le f(x) \le |f(x)|$ for any x, we have by part (e), if $a \le b$,

$$-\int_a^b |f(x)|\,dx \le \int_a^b f(x)\,dx \le \int_a^b |f(x)|\,dx.$$

Therefore $\left|\int_a^b f(x)\,dx\right| \le \int_a^b |f(x)|\,dx.$

g) By parts (b), (c) and (d),

$$\int_{-a}^a f(x)\,dx = \int_{-a}^0 f(x)\,dx + \int_0^a f(x)\,dx$$
$$= \int_0^a f(-x)\,dx + \int_0^a f(x)\,dx$$
$$= \int_0^a [f(-x) + f(x)]\,dx.$$

If f is odd, the last integral is 0. If f is even, the last integral is $\int_0^a 2f(x)\,dx$. Thus both (g) and (h) are proved.

6. Let $\epsilon > 0$ be given. Let $\delta = \epsilon^2/2$. Let $0 \le x \le 1$ and $0 \le y \le 1$. If $x < \epsilon^2/4$ and $y < \epsilon^2/4$ then $|\sqrt{x} - \sqrt{y}| \le \sqrt{x} + \sqrt{y} < \epsilon$.
If $|x - y| < \delta$ and either $x \ge \epsilon^2/4$ or $y \ge \epsilon^2/4$ then

$$|\sqrt{x} - \sqrt{y}| = \frac{|x-y|}{\sqrt{x}+\sqrt{y}} < \frac{2}{\epsilon} \times \frac{\epsilon^2}{2} = \epsilon.$$

Thus $f(x) = \sqrt{x}$ is uniformly continuous on $[0, 1]$.

7. Suppose f is uniformly continuous on $[a, b]$. Taking $\epsilon = 1$ in the definition of uniform continuity, we can find a positive number δ such that $|f(x) - f(y)| < 1$ whenever x and y are in $[a, b]$ and $|x - y| < \delta$. Let N be a positive integer such that $h = (b - a)/N$ satisfies $h < \delta$.
If $x_k = a + kh$, $(0 \le k \le N)$, then each of the subintervals of the partition $P = \{x_0, x_1, \ldots, x_N\}$ has length less than δ. Thus

$$|f(x_k) - f(x_{k-1})| < 1 \qquad \text{for} \quad 1 \le k \le N.$$

By repeated applications of the triangle inequality,

$$|f(x_{k-1}) - f(a)| = |f(x_{k-1}) - f(x_0)| < k - 1.$$

If x is any point in $[a, b]$, then x belongs to one of the intervals $[x_{k-1}, x_k]$, so, by the triangle inequality again,

$$|f(x) - f(a)| \le |f(x) - f(x_{k-1})| + |f(x_{k-1}) - f(a)| < k \le N.$$

Thus $|f(x)| < |f(a)| + N$, and f is bounded on $[a, b]$.

8. Suppose that $|f(x)| \le K$ on $[a, b]$ (where $K > 0$), and that f is integrable on $[a, b]$. Let $\epsilon > 0$ be given, and let $\delta = \epsilon/K$. If x and y belong to $[a, b]$ and $|x - y| < \delta$, then

$$|F(x) - F(y)| = \left|\int_a^x f(t)\,dt - \int_a^y f(t)\,dt\right|$$
$$= \left|\int_y^x f(t)\,dt\right| \le K|x-y| < K\frac{\epsilon}{K} = \epsilon.$$

(See Theorem 3(f) of Section 6.4.) Thus F is uniformly continuous on $[a, b]$.

MISCELLANEOUS STUDENT PROJECTS

The third edition of *Calculus: a Complete Course* included within the text a number of suggested projects under the title "Explore." Several readers commented that these broke the continuity of the text. In the fourth edition some of these have been turned into examples or problems. Others are included here for the benefit of instructors who may want to assign them to their classes or to selected students. Many of them involve the use of calculators, computer graphing software, or computer algebra systems.

PROJECT #1 (Section P.3): Families of Curves

The equation $y = 1 + x + k(x^2 - 1)$ represents a family of curves, one for each value of the constant k, which is called a **parameter**. Since the equation involves only one parameter, we say that it represents a **one-parameter family of curves**. Use a graphing utility (graphing calculator or computer graphing software) to plot simultaneously several curves in the family, for example, those corresponding to $k = 0, \pm 1, \pm 2$. What do all these curves have in common? Describe how the curve changes as k varies. Write an equation of a one-parameter family of parabolas with vertical axes, all of which pass through the points (a, b) and (c, d).

PROJECT #2 (Section P.4): Graphs of Power Functions

(a) Use a graphing utility (a graphing calculator or computer graphing software) to plot the graphs of the functions
$$f_n(x) = x^n \quad \text{for} \quad n = 1, 2, 3, 4, 5, 6$$
on the interval $[-1.2, 1.2]$. Describe the behaviour of $f_n(x)$ as n increases for values of x near -1, 0, and 1.

(b) Do the same for $g_n(x) = x^{-n}$, and $h_n(x) = x^{1/n}$ on $[-2, 2]$.

PROJECT #3 (Section P.6): Adding Periodic Functions

The positive number p is a **period** of the function f if $f(x + p) = f(x)$ for all x. This means that if the x-axis is divided into consecutive intervals of length p, the graph of f behaves identically in each of these intervals. The smallest positive period p of a periodic function is called the **fundamental period** of that

PROJECTS

function. We know that $\cos x$ and $\sin x$ each have fundamental period 2π. Show that $\cos(2\pi x)$ and $\sin(2\pi x)$ have fundamental period 1. What are the fundamental periods of

$$\cos(\pi x)? \quad \cos(2\pi x/3)? \quad \cos(\pi x/2)?$$
$$\sin(\pi x)? \quad \sin(2\pi x/3)? \quad \sin(\pi x/2)?$$

Use a graphing utility to graph the following combinations of functions and determine their fundamental periods from the graphs:

$$\sin(2\pi x) + \sin(\pi x) \qquad \cos(2\pi x) + \sin(2\pi x/3)$$
$$\cos(2\pi x) + \sin(\pi x/2) \qquad \sin(2\pi x/3) + \sin(\pi x/2)$$
$$\sin(\pi x) + \cos(2\pi x/3) \qquad \cos(\pi x) + \cos(\pi x/2)$$

Try to guess a general formula for the fundamental period of a function of the form $f(px/m) + g(px/n)$, where each of f and g has fundamental period p, and where m and n are positive integers. Now try to prove the formula.

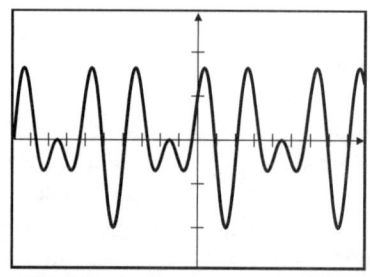

The graph of $\sin(\pi x) + \cos(2\pi x/3)$. The fundamental period is, for instance, the distance between the bottom points of two consecutive deep valleys. The tick marks on the axes are at 1 unit intervals

---

PROJECT #4 (Section 1.2): **Finding Limits Graphically**

(a) Use a graphing utility to plot the function

$$y = (1 + x^2)^{1/x^2}$$

in the window $-10 \leq x \leq 10$, $-4 \leq y \leq 4$. Trace the curve as $x \to 0$ from both sides, and estimate $L = \lim_{x \to 0} y$. Use the zoom feature, or redefine the range of plotting to get a more accurate result and determine L correct to five decimal places.

(b) Repeat (a) using $y = (1 + x^{-2})^{x^2}$.

(c) Plot the function $y = f(x) = \dfrac{2x^2 - 3x}{\sqrt{4x^2 - 12x + 9}}$ in the window $-3 \leq x \leq 3$, $-3 \leq y \leq 3$. Describe the behaviour of $f(x)$ near $x = 1.5$. Use graphical techniques to find $\lim_{x \to 1.5-} f(x)$ and $\lim_{x \to 1.5+} f(x)$. Now obtain the same results algebraically. Does $\lim_{x \to 1.5} f(x)$ exist?

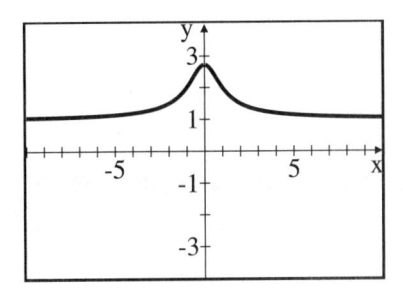

The graph of $(1+x^2)^{1/x^2}$

PROJECT #5   (Section 1.3):   **Finding Limits Involving Infinity Graphically**

(a) Plot the graph of the function
$$y = f(x) = \frac{3x^2 + 1{,}000x}{x^2 + 5{,}000}$$
in several windows, beginning with $-1{,}000 \le x \le 1{,}000$, $-10 \le y \le 10$, and then expanding the horizontal scale. Trace the curve to the left and right, and thus obtain graphical evidence for
$$\lim_{x \to -\infty} f(x) = \lim_{x \to \infty} f(x) = 3.$$

(b) Examine graphically the function
$$f(x) = \frac{x^2 - 1}{1 + \cos(\pi x)}$$
in the window $-2 \le x \le 2$, $-10 \le y \le 10$. Find
$$\lim_{x \to -1-} f(x) \qquad \lim_{x \to 1-} f(x)$$
$$\lim_{x \to -1+} f(x) \qquad \lim_{x \to 1+} f(x)$$

PROJECT #6   (Section 1.4):   **Finding Maxima and Minima Graphically**

Graphing utilities can be used to find maximum and minimum values of functions on intervals where they are continuous. Plot the graph of
$$y = f(x) = \frac{x+1}{x^2+1}$$
on the window $-5 \le x \le 5$, $-2 \le y \le 2$, and observe that f appears to have a maximum value near $x = 0.5$ and a minimum value near $x = -2.5$. By using the zooming and curve tracing features of your calculator or software, try to determine the maximum and minimum values of $f(x)$ on the interval $[-5, 5]$, and the points x where these extreme values occur, all to an accuracy of four decimal places. It is harder to get this accuracy for the values of x than for the values of $f(x)$. Why should this be the case? If you have a "zoom box" feature you can use it to expand the horizontal and vertical scales at different rates. Enclose a small part of the curve surrounding the high or low point in a zoom box (rectangle) whose height is much less than its width, and expand that box to fill the whole screen.

PROJECTS

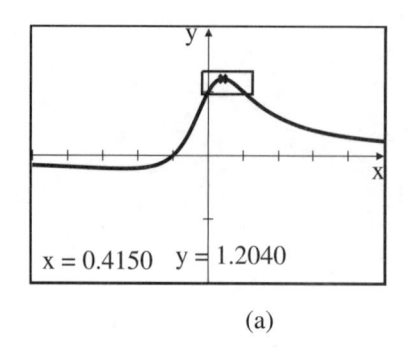

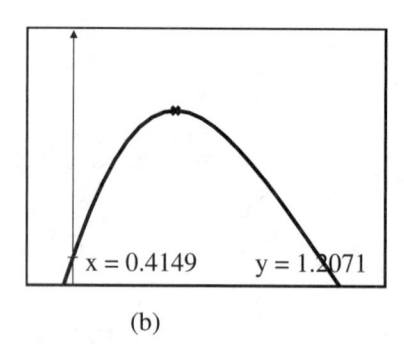

[Figure: Two graphing calculator screens labeled (a) and (b). Screen (a) shows a curve with a small zoom box near its maximum, with readout "x = 0.4150 y = 1.2040". Screen (b) shows the zoomed-in view of the curve near its maximum, with readout "x = 0.4149 y = 1.2071".]

(a)   (b)

Using a "zoom box" to zoom part of a curve (a) near a maximum value to fill the screen (b) without allowing the curve to become flattened

Continuing in this way, show that $f(x)$ has maximum value 1.2071 at $x = 0.4142$, each to four significant figures.

PROJECT #7   (Section 2.1):   Tangent Lines

Some graphing calculators will plot tangent lines to curves and tell you their slopes. If your calculator will do this, then:

(a) Plot the graph of $y = 5/(x^2 - 2x + 5)$ in the window defined by $-2 \leq x \leq 4$, $-0.5 \leq y \leq 2$, and plot the tangent lines to this curve at $x = 0, 1, 2,$ and 3, recording the slope of each line. At what value of x does the value of y seem to be greatest? What is the slope there?

(b) Plot the graph of $y = x^4 - 2x^2 + 2$ in the window $-2 \leq x \leq 2$, $-1 \leq y \leq 3$. Where does the curve have horizontal tangents? Plot these horizontal tangents. What are their equations?

PROJECT #8   (Section 2.2):   Graphs of Derivatives

Some graphing calculators and most computer algebra systems can plot the graph of a function and its derivative simultaneously, given only the formula for the function. Use such a device to plot the following functions and their derivatives:

(a) $y = x^2 + 3x - 1$. What is the graph of y' doing at the value of x where the graph of y is lowest?

(b) $y = (x^3 - x)/(x^2 + 1)$. What is the graph of y doing in the intervals where $y' > 0$? where $y' < 0$?

(c) $y = \sqrt{|x - 1|}$. Where does y' fail to be defined? How does the graphing device deal with this singular point?

(d) $y = |x^2 - x - 2|$. Where does y' fail to be defined? How does the graphing device deal with this singular point?

PROJECTS

PROJECT #9 (Section 2.4): **Finding Derivatives by Computer Algebra**

Computer algebra systems such as Derive, Maple, and Mathematica know the derivatives of elementary functions and can calculate the derivatives of combinations of these functions symbolically, using differentiation rules. For example, the commands for calculating the derivative of $\sqrt{1+2x^2}$ and then evaluating the result at $x = 2$ are:

```
for Derive:    dif(sqrt(1+2x^2),x) <Enter>
               S(implify)<Enter>
               M(anage)S(ubstitute)<Enter> 2<Enter>
               S(implify)<Enter>
for Maple V:   f := diff(sqrt(1+2*x^2),x); <Enter>
               simplify(subs(x=2,f));<Enter>
for Mathematica: D[Sqrt[1+2x^2],x] <Enter>
               %/.x->2 <Enter>
```

Use a computer algebra system to find and simplify the following derivatives:

1. dy/dx, if $y = (x^2 - x + \pi)^2/(x^2 + x + \pi)^2$;
2. $f'(2)$, if $f(t) = \sqrt{(t^2-3)(t^2+5)(t^2+12)}$;
3. $g'(\pi/4)$, if $g(x) = (\sin x)/\sqrt{1+\sin x} - (\cos x)/\sqrt{1+\cos x}$.

PROJECT #10 (Section 2.6): **Finding Horizontal Tangents**

Some of the most important applications of calculus involve finding the critical points of a differentiable function $f(x)$. This subject is developed further in Sections 3.3 and 3.4, but we can explore here the use of a calculator or computer to find such points. There are two ways to proceed.

(a) We can examine the graph of f to locate points where the graph is horizontal, zooming in on such points to obtain more accuracy in their coordinates. Unfortunately, zooming in causes the graph to straighten out and makes it difficult to increase the accuracy unless the vertical scale is expanded faster than the horizontal scale.

(b) We can examine the graph of f' to locate points where the graph crosses the x-axis, that is, where $f'(x) = 0$. This method is better for two reasons:

  (i) zooming the graph without changing the aspect ratio preserves the angle at which the graph of f' crosses the x-axis, and so each such zoom increases the accuracy with which the value of x can be determined.

  (ii) the calculator or computer software may have a "solve" routine for finding either exact or approximate solutions of an equation. Such solve routines usually require an initial approximation to, or an interval containing, the desired root; these can be determined from the graph. Some graphics calculators and all computer algebra systems are able to graph the derivative of a given function directly; they calculate derivatives symbolically (using the differentiation rules).

The best way to find critical points on the graph of a function is as follows:

(a) calculate the derivative $f'(x)$ (either by machine or manually), and plot its graph $y = f'(x)$;

PROJECTS

(b) use this graph to approximate the critical points of f, that is, the points x for which $f'(x) = 0$;

(c) use a "solve" routine or zoom in on the graph of f' to obtain more accuracy in these critical points.

Explore the use of these ideas to find the critical points on the graphs of the following functions:

1. $f(x) = x^3 - 3x^2 + x + 1$ 2. $f(x) = x^4 - 4x^3 + 2x^2 - 4x + 1$

PROJECT #11 (Section 3.1): **Values of Inverse Functions**

(a) Plot the graph of $f(x) = (x^3 + x + 1)/(x^2 + 1)$ in the window $-5 \leq x \leq 5$, $-5 \leq y \leq 5$. Does the graph suggest that f has an inverse?

(b) Show that $f'(x) > 0$ everywhere except at $x = 0$, and hence conclude that f does indeed have an inverse.

(c) By zooming in on various points of the graph in (a), or by using the equation solving feature on your calculator, find the values of $f^{-1}(-1)$, $f^{-1}(0)$, and $f^{-1}(2)$. What is the value of $f^{-1}(1)$? Do you need a calculator to find this value?

PROJECT #12 (Section 3.4): **Polynomial Approximations to the Exponential Function**

(a) Use a graphics utility to make six plots, each showing the graphs of $y = e^x$ and one of the curves (1)–(6) listed below. Make all plots in the window $-3 \leq x \leq 3$, $-1 \leq y \leq 12$.

(1) $y = 1$

(2) $y = 1 + x$

(3) $y = 1 + x + \dfrac{x^2}{2}$

(4) $y = 1 + x + \dfrac{x^2}{2} + \dfrac{x^3}{6}$

(5) $y = 1 + x + \dfrac{x^2}{2} + \dfrac{x^3}{6} + \dfrac{x^4}{24}$

(6) $y = 1 + x + \dfrac{x^2}{2} + \dfrac{x^3}{6} + \dfrac{x^4}{24} + \dfrac{x^5}{120}$

Describe the results. Over what interval does each curve (1)–(6) approximate $y = e^x$ well enough that they appear to overlap? Your answers will depend on the resolution of your graphics. A computer graphics program will give more reliable results visually than a graphing calculator. With a calculator you can use the trace facility to jump back and forth between the two curves and compare the y-values at any particular value of x. To what accuracy does each of the functions (1)–(6) approximate e when $x = 1$?

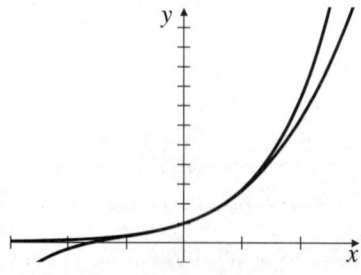

The graphs of: $y = e^x$ (upper curve), and $y = 1 + x + \frac{1}{2}x^2 + \frac{1}{6}x^3$ (lower curve)

(b) Continuing from (a), calculate the value of
$$y_n = 1 + 1 + \frac{1}{2} + \frac{1}{6} + \frac{1}{24} + \cdots + \frac{1}{n!}$$
for $n = 6, 7, 8, 9$, and 10. For each of these values of n, to how many decimal places is the approximation $e \approx y_n$ correct?

The polynomials used to approximate e^x in (a) are called Taylor polynomials or Maclaurin polynomials. They will be developed in Section 4.8 and further studied in Chapter 9.

PROJECT #13 (Section 4.2): **Finding Extreme Values Graphically**

A graphics utility can be used to find (approximately) the local and absolute extreme values of more complicated functions. Again it is usually easier to approximate critical points by looking at the graph of the derivative. Do the following functions have any local or absolute maximum and minimum values, and if so, where are they located?

1. $f(x) = x^3 - 15x^2 + 48x + 10$ on $[-1, 11]$
2. $f(x) = \dfrac{x^2 + x - 6}{x^4 + 1}$ on $\mathbb{R}$
3. $f(x) = x \sin x$ on $[\pi/4, \pi]$
4. $f(x) = \left(x^2(3 - x)\right)^{1/3}$ on $\mathbb{R}$
5. $x(1 - x^2)^{2/3} = x((1 - x^2)^2)^{1/3}$ on $\mathbb{R}$
6. $f(x) = |x + 3| - |x + 2| - 2|x - 1| + 2|x - 2|$ on $[-4, 4]$

PROJECT #14 (Section 4.7): **Behaviour of the Error in a Linearization**

(a) Calculate the linearization $L(x)$ for $f(x) = 2/(1 + x^2)$ at $x = 1$ and simultaneously plot the graphs of $f(x)$ and $L(x)$ on the interval $0 \le x \le 2$. Do the graphs continue to diverge as x moves further from 1 in either direction? Plot the graph of the error function $E(x) = f(x) - L(x)$ on $0 \le x \le 2$, and by tracing it or otherwise make a table of values of $E(x)$ for $x = 1.1, 1.2, 1.4$, and 1.8, and also for $x = 0.9, 0.8, 0.6$, and 0.2. How do the errors appear to behave as the distance from $x = 1$ increases?

(b) Repeat the same procedures for $f(x) = \sec(\pi x/4)$ at $x = 1$.

PROJECT #15 (Section 4.4): **Using a Graphing Utility**

The techniques for curve sketching developed above are useful only for graphs of functions that are simple enough to allow us to calculate and analyze their derivatives. In practice, if we want to study the behaviour of a function using its graph, we will likely want to use a graphing calculator or a computer graphing program to produce the graph quickly and painlessly. However, to make effective use of such a utility, we have to decide on a viewing window; what part of the graph do we want to see, and what horizontal and vertical scales should be used? An inappropriate choice of viewing window can cause us to miss significant features of the graph, as the following examples show.

PROJECTS

(a) Plot the graph of $y = (2/x^2) - (1/(50x^4))$ in a "standard" viewing window, say $-10 \le x \le 10$, $-10 \le y \le 10$. Describe the behaviour of the graph near $x = 0$. Does y appear to have a maximum value? What is $\lim_{x \to 0} y$?

Now repeat the plot, but using the window $-2 \le x \le 2$, $-60 \le y \le 60$. Reconsider your answers to the questions above.

(b) Plot the graph of $y = (x^3 + 1)/(x^2 + 1)$ in the "standard" window referred to in (a). Do you think that this function is one-to-one and has an inverse? Now replot it in the window $-1 \le x \le 2$, $0 \le y \le 2$, and reconsider your answer.

(c) Plot $y = 30x^4 - x^2 - 1$ in a "standard" window. How many critical points does this function have? Now replot it in the window $-0.3 \le x \le 0.3$, $-1.1 \le y \le -0.9$.

(d) Plot the graph of $y = x^4 - 16x^2 + \sin x$ in the window $-5 \le x \le 5$, $-100 \le y \le 225$. Where does this function have an absolute minimum value? Where does it have a local maximum? Now zoom in on the graph near these points and reconsider your answers.

PROJECT #16 (Section 4.8): Taylor Polynomials

(a) Use a graphing utility to plot simultaneously the graphs

$$y = \ln(1+x) \quad \text{and} \quad y = P_{10}(x) = \sum_{n=1}^{10} \frac{(-1)^{n-1} x^n}{n},$$

the 10th degree Taylor polynomial for $\ln(1 + x)$ about $x = 0$. Plot in the window $-2 \le x \le 2$, $-4 \le y \le 2$. On what interval does the graph suggest that $P_{10}(x)$ is a reasonably good approximation to $\ln(1+x)$?

(b) Repeat (a) for the function $y = e^x$ and its Taylor polynomials $P_3(x)$, $P_6(x)$, and $P_9(x)$ about $x = 0$. (Here $P_k(x) = \sum_{n=0}^{k} x^n/n!$.) Experiment with different window settings to best determine how far on each side of $x = 0$ each polynomial well approximates e^x.

PROJECT #17 (Section 4.9): Exploring a Limit Graphically

Consider the function $g(x) = (1+x^2)^{1/x^2}$, which was previously considered in Project #4. Use the techniques of Section 4.9 to show that $\lim_{x \to 0} g(x) = e$.

Now explore the behaviour of $g(x)$ near $x = 0$ by plotting the graph of g in, say, the window $-1 \le x \le 1$, $0 \le y \le 4$, and then successively zooming in on the point where the graph crosses the y-axis and tracing the curve near that point to try to get better and better approximations to e. What eventually happens to the graph? Why does it happen?

PROJECTS

PROJECT #18 (Section 5.2): Approximating Areas with a Calculator

(a) Write a simple program for a programmable calculator to calculate the sums $\sum_{i=1}^{n} \frac{1}{n} \left(\frac{i}{n}\right)^3$ for $n = 50$, 100, 200, 500, and 1,000. What do you think is the area under $y = x^3$, above $y = 0$, between $x = 0$ and $x = 1$?

(b) Calculate the sums $\sum_{i=1}^{n} \frac{\pi}{n} \sin\left(\frac{i\pi}{n}\right)$ for the same values of n as in (a). What do you think is the area under $y = \sin x$, above $y = 0$, between $x = 0$ and $x = \pi$?

PROJECT #19 (Section 5.3): Verifying Integrability Using Computer Algebra

In Section 5.1 the following closed-form summation formulas were verified:

$$\sum_{i=1}^{n} i = \frac{n(n+1)}{2}, \quad \sum_{i=1}^{n} i^2 = \frac{n(n+1)(2n+1)}{6}, \quad \sum_{i=1}^{n} i^3 = \frac{n^2(n+1)^2}{4}.$$

(a) Use the "sum" function of a computer algebra program to evaluate the following sums in closed form:

$$\sum_{i=1}^{n} i^4, \quad \sum_{i=1}^{n} i^5, \quad \sum_{i=1}^{n} i^6.$$

(b) Observe that for $1 \le k \le 6$, the closed form expression for $\sum_{i=1}^{n} i^k$ is a polynomial of degree $k + 1$ in the variable n. What is the coefficient of n^{k+1} in the polynomial?

(c) Use the technique of Examples 2 and 3 of Section 5.3 to show that $f(x) = x^k$ is integrable on $[0, a]$, where $a > 0$ and $k = 4, 5$, and 6, and find $\int_0^a x^k \, dx$.

(d) Find a closed-form expression for $\sum_{i=1}^{n} i \, r^i$, and use it to find $\int_0^a x \, e^x \, dx$. You will likely want to use the computer algebra system both to find the sum and to evaluate the limits involved.

PROJECT #20 (Section 5.7): Using Integration Utilities

Some graphics calculators (for instance the TI-85) and most computer algebra systems have built-in integration functions to evaluate definite integrals, either symbolically, or using numerical approximations. The use of such functions in applications of definite integrals is illustrated here.

(a) Simultaneously plot the graphs of the two curves $y = (\sin x)^2$ and $y = \sin(x^2)$ on the interval $-1 \le x \le 3$. Make a rough estimate of x_0, the first positive value of x where the two curves intersect. Which curve is higher between 0 and x_0?

(b) By zooming in or using a "solve" routine, find x_0 to 10 decimal places.

(c) Use the integration function to find the area between the two curves of part (a), between $x = 0$ and $x = x_0$.

PROJECTS

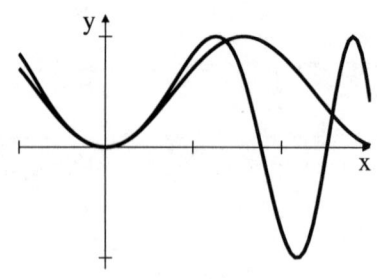

The graphs of $(\sin x)^2$ and $\sin(x^2)$

(d) In a similar way, find the total area of the region lying above the curve $y = x^2$ and below the curve $y = \cos(x^2)$.

PROJECT #21 (Section 6.1): **Using Computer Algebra to Iterate a Reduction Formula**

(a) A computer algebra system can be used to iterate a reduction formula a large number of times. For example, the reduction formula obtained above for $I_n = \int x^n e^{-x}\, dx$, namely

$$I_n = -x^n e^{-x} + n I_{n-1},$$

and its starting value $I_0 = -e^{-x} + C$ can be combined in a single function $f(n)$. For DERIVE, this takes the form

```
f(n) := if(n=0, -exp(-x)+C, -x^n*exp(-x)+n*f(n-1))
```

For MAPLE V, the form is

```
f := n -> if n=0 then -exp(-x)+C
              else -x^n*exp(-x)+n*f(n-1) fi;
```

Try one of these to obtain $f(4)$ and $f(10)$.

(b) Some computer algebra systems allow for specific integration by parts in an integral. Read the documentation for your system to see whether this is possible and how to carry it out if it is possible.

(c) Obtain a reduction formula for $I_n = \int \dfrac{1}{(1+x^2)^n}\, dx$ and find I_{10}. Check your answer by getting the system to integrate I_{10} directly.

PROJECTS

PROJECT #22 (Section 6.8): **Implementing the Romberg Method on a Spreadsheet**

Computer spreadsheet programs like Lotus 1-2-3 or Microsoft Excel have a feature that makes them very useful for implementing schemes, like the Romberg Method, which require repetetive calculations of the same formulas on different values. A formula in a cell of the spreadsheet that calculates a value in terms of values in other cells can be replicated (copied) to a row or column of cells and will then calculate the values for all those cells using values in cells having the same *relative position* to the cell being calculated as the cells used to calculate the original value have to the original cell. For instance, if cell C4 contains a formula calculating the value in that cell in terms of the values in cells A3 and B3, then copying C4 to the range of cells C5, C6, C7 will result in the use of the same formula to calculate:

$$C5 \text{ in terms of the values in A4 and B4,}$$
$$C6 \text{ in terms of the values in A5 and B5, and}$$
$$C7 \text{ in terms of the values in A6 and B6.}$$

Set up a spreadsheet (with column widths wide enough to accommodate 12-digit numbers) to approximate $\int_0^1 f(x)\,dx$, where $f(x) = e^{-x^2}$, as follows:

(a) Put the numbers 0, 1, 2, ..., 16 in cells A1, A2, ..., A17.

(b) In cells B1, B2, ..., B17, put the formulas to calculate the values $e^{-(i/16)^2}$, for $0 \le i \le 16$. Typically, this can be done by typing `@exp(-(a1/16)^2)` into cell B1, and replicating this formula down the column from B2 to B17.

(c) In cells C1, D1, E1, F1, and G1, put the labels `Trap1`, `Trap2`, `Trap4`, `Trap8`, and `Trap16`. In cells C2, D2, E2, F2, and G2 put formulas to calculate these approximations; in C2 put `(b1+b17)/2`, in D2 put `(c2+b9)/2`, in E2 put `(2d2+b5+b13)/4`, in F2 put `(4e2+b3+b7+b11+b15)/8`, and in G2 put `(8f2+b2+b4+b6+b8+b10+b12+b14+b16)/16`.

(d) In Cells D3, E3, F3, and G3, put the labels `T11`, `T12`, `T13`, and `T14`. In cell D4, put the formula `(4*d2-c2)/3`, and then replicate this formula to E4, F4, and G4.

(e) In cells E5, F5, and G5, put the labels `T22`, `T23`, and `T24`. In cell E6, put the formula `(16*e4-e3)/15`, and then replicate this formula to F6 and G6. Continue in this way to label the cells F7, G7, and G9, and to put appropriate formulas in F8, G8, and G10. The formula in F8 is `(64*F6-E6)/63`, and that in G10 is `(256*G8-f8)/255`. The values appearing on the "diagonal," that is, the values Trap1 = R0, T11 = R1, T22 = R2, T33 = R3, and T44 = R4 are the successive Romberg approximations. Observe how close they get to

$$\int_0^1 e^{-x^2}\,dx = 0.746824132782\ldots.$$

(d) You may wish to save the spreadsheet. With only minor modifications, it can be used to obtain Romberg approximations for other integrals. Evaluate $\int_1^3 \sqrt{1+x^2}\,dx$ by first making the change of variable $x = \sqrt{8y+1}$ so the interval of integration becomes [0, 1], and then modify the entries in column B only. How many decimal places do you think are correct in T44 = R4?

PROJECTS

PROJECT #23 (Section 7.2): **The Volume of a Ball in n Dimensions**

Euclidean n-dimensional space consists of *points* $(x_1, x_2, \ldots, x_n)$ with n real coordinates. By analogy with the 3-dimensional case, we call the set of such points that satisfy the inequality

$$x_1^2 + x_2^2 + \cdots + x_n^2 \leq r^2$$

the n-dimensional *ball* centred at the origin. For example:

The 1-dimensional ball is the interval $-r \leq x_1 \leq r$, which has *volume* (actually *length*), $V_1(r) = 2r$.

The 2-dimensional ball is the disk $x_1^2 + x_2^2 \leq r^2$, which has *volume* (actually *area*),

$$\begin{aligned} V_2(r) = \pi r^2 &= \int_{-r}^{r} 2\sqrt{r^2 - x^2}\, dx \\ &= \int_{-r}^{r} V_1\left(\sqrt{r^2 - x^2}\right) dx. \end{aligned}$$

The 3-dimensional ball $x_1^2 + x_2^2 + x_3^2 \leq r^2$ has volume

$$\begin{aligned} V_3(r) = \frac{4}{3}\pi r^3 &= \int_{-r}^{r} \pi \left(\sqrt{r^2 - x^2}\right)^2 dx \\ &= \int_{-r}^{r} V_2\left(\sqrt{r^2 - x^2}\right) dx. \end{aligned}$$

By analogy with these formulas, the volume $V_n(r)$ of the n-dimensional ball of radius r is the integral of the volume of the $(n - 1)$-dimensional ball of radius $\sqrt{r^2 - x^2}$ from $x = -r$ to $x = r$:

$$V_n(r) = \int_{-r}^{r} V_{n-1}\left(\sqrt{r^2 - x^2}\right) dx.$$

Calculate $V_4(r)$, $V_5(r)$, ..., $V_{10}(r)$, and guess formulas for $V_{2n}(r)$ (the even-dimensional balls) and $V_{2n+1}(r)$ (the odd-dimensional balls). If your computer algebra software is sufficiently powerful, you may be able to verify your guesses by induction. Otherwise, use them to predict $V_{11}(r)$ and $V_{12}(r)$, then check your predictions by starting from $V_{10}(r)$.

PROJECT #24 (Section 7.3): **Approximating the Circumference of an Ellipse**

Evaluating the complete elliptic integral function $E(\epsilon)$ for arbitrary values of ϵ between 0 and 1 requires either a table of values, or a calculator or computer program with a numerical integration function. Here we use the latter to discover an approximation $F(\epsilon)$ to $E(\epsilon)$ that will enable us to calculate the circumference of any ellipse to within about 5.6%. First, observe that

$$E(x) = \int_0^{\pi/2} \sqrt{1 - x^2 \sin^2 t}\, dt$$

satisfies $E(0) = \pi/2$ and $E(1) = 1$.

(a) Plot the graph of $y = E(x)$ in the window $0 \leq x \leq 1, 0 \leq y \leq 2$. This requires a definite integral function. For the TI-85, the function `fnInt`, located in the CALC menu, will do; you can define the function to be graphed by

```
y1=fnInt(√(1-(x*sin(t))^2),t,0,π/2)
```

Be prepared for the plot to take some time; a complete numerical integration must be performed for each point plotted.

(b) Observe the shape of the graph of $E(x)$. What appears to be the slope at $x = 0$? (Can you verify this from the formula for $E(x)$?) Describe the concavity of the graph. The shape suggests an approximation of the form
$$E(x) \approx F(x) = \frac{\pi}{2} - kx^2,$$
where the constant k is chosen to ensure that $F(1) = E(1) = 1$. Show that $k = (\pi - 1)/2$.

(c) Use graphing means to discover the maximum value of the function $100|E(x) - F(x)|/E(x)$ for $0 \leq x \leq 1$. This is the maximum percentage error in the approximation $E(x) \approx F(x)$.

(d) What are the values of a (the longer semiaxis), and ϵ (the eccentricity) for the ellipse
$$\frac{x^2}{9} + \frac{y^2}{4} = 1?$$
Find the circumference $C = 4aE(\epsilon)$ and the approximation $C_a = 4aF(\epsilon)$ for the ellipse. What is the percentage error?

PROJECT #25 (Section 8.2): Lissajous Figures

Some graphing calculators and most computer graphing software can plot parametric curves. Use such a utility to plot the following families of parametric curves. Plot each curve for parameter interval $0 \leq t \leq 2\pi$, and in the window $-1.2 \leq x \leq 1.2$, $-1.2 \leq y \leq 1.2$. For each family try to formulate a principle that governs the behaviour of the various curves in the family.

(a) $x = \sin t$, $y = \sin(t + C)$, for $C = 0, \frac{\pi}{6}, \frac{\pi}{3}, \frac{\pi}{2}, \frac{2\pi}{3}, \frac{5\pi}{6}, \pi$.

(b) $x = \sin t$, $y = \sin(mt)$, for $m = 1, 2, 3, 4, 5, 6$.

(c) $x = \sin t$, $y = \cos(mt)$, for $m = 1, 2, 3, 4, 5, 6$.

(d) $x = \sin(mt)$, $y = \sin(nt)$, for $(m, n) = (2, 3), (2, 4), (2, 5), (3, 4), (3, 5), (4, 5)$.

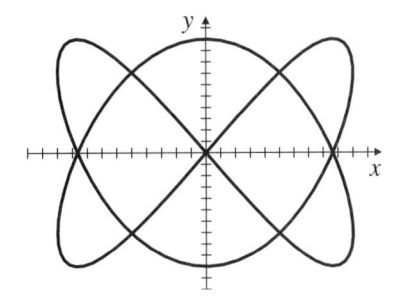

The Lissajous figure $x = \sin(2t)$, $y = \cos(3t)$

PROJECTS

Such graphs are called Lissajous figures. They arise, for example, in the analysis of electrical signals (alternating current voltages at various frequencies) using an oscilloscope. A signal of fixed but unknown frequency is applied to the vertical input, and a control signal is applied to the horizontal input. The horizontal frequency is varied until a stable Lissajous figure is observed. The (known) frequency of the control signal and the shape of the figure then determine the unknown frequency.

PROJECT #26 (Section 8.5): Polar Graphs

Some graphing calculators and most mathematical graphing software can plot polar curves with equations of the form $r = f(\theta)$. Use such a utility to carry out appropriate parts of the following investigation.

(a) In the window $-4 \leq x \leq 4$, $-4 \leq y \leq 4$ plot the curves $r = 1 + 2\cos(m\theta)$ for $m = 1, 2, 3,$ and 4. Predict the appearance of the graph for $m = 5$ and then check your prediction.

(b) In the window $-4 \leq x \leq 4$, $-4 \leq y \leq 4$ plot the curves $r = 2 + \cos(m\theta)$ for $m = 1, 2, 3,$ and 4. Predict the appearance of the graph for $m = 5$ and then check your prediction.

(c) Predict the appearance of the graphs of $r = 1 + \cos(m\theta)$ for $m = 2, 3,$ and 4, and check your predictions.

(d) In the window $-3 \leq x \leq 3$, $-3 \leq y \leq 3$, plot the polar curve $r = \ln \theta$ for $0 < \theta \leq 4\pi$. Use a step size $\Delta \theta = 0.01$. The curve has an asymptote. What is it? Prove your claim.

PROJECT #27 (Section 9.3): Using a Calculator to Find the Sum of a Series

Suppose you want to find the value of $S = \sum_{n=1}^{\infty} 1/n^2$ correct to three decimal places with the aid of a calculator. How many terms should you add?

Since $1/n^2 \leq 1/10{,}000 = 0.0001$ if $n \geq 100$, you might be *tempted* to use the first 100 terms. Use a programmable calculator or one with a built-in function for summing the terms of a sequence to calculate $\sum_{n=1}^{100} 1/n^2$. Round the answer to three decimal places.

Now, as a check, calculate $\sum_{n=1}^{500} 1/n^2$ and round the answer to three decimal places. Was your first answer correct to three decimal places? If not, why not?

The number of terms of a convergent series needed to estimate the sum to within a given error tolerance depends on the behaviour of the terms. It is usually *not* sufficient to add terms until the terms are smaller than the tolerance.

PROJECT #28 (Section 10.6): Linear Algebra Using Calculators or Computers

Computer algebra programs and some advanced calculators have built-in routines for handling matrices and solving linear systems. Among other features, such utilities can

(i) set up, transpose, and multiply rectangular matrices,

(ii) calculate determinants and inverses of square matrices, and

(iii) solve systems of linear equations.

Use such a utility to find $\det(\mathcal{A})$ and $\mathcal{A}^{-1}$, and to solve the system $\mathcal{A}\mathbf{x} = \mathbf{b}$, where

$$\mathcal{A} = \begin{pmatrix} 3 & 2 & 1 & -2 \\ 2 & 0 & 1 & 1 \\ 2 & 1 & 1 & 0 \\ -1 & -2 & 1 & 1 \end{pmatrix} \quad \text{and} \quad \mathbf{b} = \begin{pmatrix} 5 \\ 2 \\ 10 \\ -5 \end{pmatrix}.$$

PROJECT #29 (Section 11.1): **Graphs of Functions of Two Variables**

Some mathematical software packages will produce plots of three-dimensional graphs to help you get a feeling for how the corresponding functions behave.

Use a graphics program with such capabilities to explore the behaviour of the following functions. In each case you will have to make suitable choices for intervals of the variables and for other graphing parameters. Some experimentation will be necessary; don't expect to get a good graph showing all the important behaviour on your first try.

1. $y^3 - 3x^2 y$
2. $(x^2 - y^2)^2$
3. $\sin(x - y)$
4. $\sin x \sin y$
5. $0.1/\sqrt{x^2 + y^2} - 0.2x$
6. $0.1y/\sqrt{(x^2 + (y^2 - 1)^2}$

PROJECT #30 (Section 11.1): **Level Curves Using Graphics Software**

Computer programs that are able to plot graphs of functions of two variables can usually also plot level curves of such functions. Use such software to plot many level curves of the functions given in the previous Project.

PROJECT #31 (Section 14.2): **Evaluating Double Integrals Using Computer Algebra**

Iterated double integrals can be evaluated with the help of a computer algebra system such as Derive, Maple, or Mathematica. When the definite integrals involved can be evaluated symbolically, these programs will return a symbolic answer. For example, Maple can be given the task of evaluating the iterated integral in Example 4 of Section 14.2 by issuing the command

```
int(int(a+y,y=0..a-(x^2/a)),x=-a..a);
```

and it will respond with the value $\dfrac{28}{15}a^3$.

Even when the program cannot evaluate the integrals in symbolic form, you can ask for an approximate value. If charged to evaluate

$$\int_0^1 \int_0^1 e^{((x+y)/2)^4}\, dx\, dy,$$

PROJECTS

Maple will respond by returning the integral unevaluated, but the command

```
evalf(Int(Int(exp(((x+y)/2)^4),x=0..1),y=0..1),5);
```

will return the value of the integral to five significant figures, namely 1.1556. The calculation may take a considerable time, depending on your processor.

(a) Evaluate $\iint_T y \sin(x+y)\, dA$ over the triangle $0 \leq x \leq \pi/2, 0 \leq y \leq 1-x$.

(b) Evaluate $\int_0^1 \int_0^x \sqrt{1+e^{xy}}\, dy\, dx$ to five significant figures.

PROJECT #32 (Section 14.5): Computer Evaluation of Triple Integrals

Computer algebra systems can be used to evaluate iterated triple integrals just as they could be used to evaluate double integrals. Try to evaluate the following integrals using such software. If symbolic evaluation is not possible, find a numerical approximation.

(a) $\displaystyle\int_0^1 dx \int_1^{1+x} dy \int_2^{2+xy} (x^2+y^2+z^2)\, dz$

(b) $\displaystyle\int_0^\pi dz \int_0^{z^2} dy \int_0^{z^2} \sin z\, dx$

(c) $\displaystyle\int_0^1 dz \int_0^z dy \int_0^y \cos(xyz)\, dx$

PROJECT #33 (Section 15.5): Tubes Around Curves

If $\mathbf{r} = \mathbf{F}(t)$, $a \leq t \leq b$, is a parametric curve $\mathcal{C}$ in 3-space having unit normal $\hat{\mathbf{N}}(t)$ and binormal $\hat{\mathbf{B}}(t)$, then the parametric surface

$$\mathbf{r} = \mathbf{F}(u) + s\cos v\, \hat{\mathbf{N}}(u) + s\sin v\, \hat{\mathbf{B}}(u), \quad a \leq u \leq b, \quad 0 \leq v \leq 2\pi,$$

is a tube-shaped surface of radius s centred along the curve $\mathcal{C}$. (Why?) An example is given in Example 3 of Section 15.5. Use a three-dimensional graphing program with parametric surface capabilities to plot tubes of radius $s = 0.25$ about the following curves:

(a) the circle $\mathbf{r} = \cos t\, \mathbf{i} + \sin t\, \mathbf{j}$, $0 \leq t \leq 2\pi$,

(b) the helix $\mathbf{r} = \cos t\, \mathbf{i} + \sin t\, \mathbf{j} + (t/2\pi)\mathbf{k}$, $0 \leq t \leq 2\pi$.